Some Physical Constants

Quantity	Symbol	
Atomic mass unit	u	(83) $\times 10^{-27}$ kg
		931.494 028 (23) MeV/c^2
Avogadro's number	N_A	6.022 141 79 (30) $\times 10^{23}$ particles/mol
Bohr magneton	$\mu_B = \dfrac{e\hbar}{2m_e}$	9.274 009 15 (23) $\times 10^{-24}$ J/T
Bohr radius	$a_0 = \dfrac{\hbar^2}{m_e e^2 k_e}$	5.291 772 085 9 (36) $\times 10^{-11}$ m
Boltzmann's constant	$k_B = \dfrac{R}{N_A}$	1.380 650 4 (24) $\times 10^{-23}$ J/K
Compton wavelength	$\lambda_C = \dfrac{h}{m_e c}$	2.426 310 217 5 (33) $\times 10^{-12}$ m
Coulomb constant	$k_e = \dfrac{1}{4\pi\epsilon_0}$	8.987 551 788 ... $\times 10^9$ N·m^2/C^2 (exact)
Deuteron mass	m_d	3.343 583 20 (17) $\times 10^{-27}$ kg
		2.013 553 212 724 (78) u
Electron mass	m_e	9.109 382 15 (45) $\times 10^{-31}$ kg
		5.485 799 094 3 (23) $\times 10^{-4}$ u
		0.510 998 910 (13) MeV/c^2
Electron volt	eV	1.602 176 487 (40) $\times 10^{-19}$ J
Elementary charge	e	1.602 176 487 (40) $\times 10^{-19}$ C
Gas constant	R	8.314 472 (15) J/mol·K
Gravitational constant	G	6.674 28 (67) $\times 10^{-11}$ N·m^2/kg^2
Neutron mass	m_n	1.674 927 211 (84) $\times 10^{-27}$ kg
		1.008 664 915 97 (43) u
		939.565 346 (23) MeV/c^2
Nuclear magneton	$\mu_n = \dfrac{e\hbar}{2m_p}$	5.050 783 24 (13) $\times 10^{-27}$ J/T
Permeability of free space	μ_0	$4\pi \times 10^{-7}$ T·m/A (exact)
Permittivity of free space	$\epsilon_0 = \dfrac{1}{\mu_0 c^2}$	8.854 187 817 ... $\times 10^{-12}$ C^2/N·m^2 (exact)
Planck's constant	h	6.626 068 96 (33) $\times 10^{-34}$ J·s
	$\hbar = \dfrac{h}{2\pi}$	1.054 571 628 (53) $\times 10^{-34}$ J·s
Proton mass	m_p	1.672 621 637 (83) $\times 10^{-27}$ kg
		1.007 276 466 77 (10) u
		938.272 013 (23) MeV/c^2
Rydberg constant	R_H	1.097 373 156 852 7 (73) $\times 10^7$ m^{-1}
Speed of light in vacuum	c	2.997 924 58 $\times 10^8$ m/s (exact)

Note: These constants are the values recommended in 2006 by CODATA, based on a least-squares adjustment of data from different measurements. For a more complete list, see P. J. Mohr, B. N. Taylor, and D. B. Newell, "CODATA Recommended Values of the Fundamental Physical Constants: 2006." *Rev. Mod. Phys.* **80:**2, 633–730, 2008.

[a]The numbers in parentheses for the values represent the uncertainties of the last two digits.

Solar System Data

Body	Mass (kg)	Mean Radius (m)	Period (s)	Mean Distance from the Sun (m)
Mercury	3.30×10^{23}	2.44×10^6	7.60×10^6	5.79×10^{10}
Venus	4.87×10^{24}	6.05×10^6	1.94×10^7	1.08×10^{11}
Earth	5.97×10^{24}	6.37×10^6	3.156×10^7	1.496×10^{11}
Mars	6.42×10^{23}	3.39×10^6	5.94×10^7	2.28×10^{11}
Jupiter	1.90×10^{27}	6.99×10^7	3.74×10^8	7.78×10^{11}
Saturn	5.68×10^{26}	5.82×10^7	9.29×10^8	1.43×10^{12}
Uranus	8.68×10^{25}	2.54×10^7	2.65×10^9	2.87×10^{12}
Neptune	1.02×10^{26}	2.46×10^7	5.18×10^9	4.50×10^{12}
Pluto[a]	1.25×10^{22}	1.20×10^6	7.82×10^9	5.91×10^{12}
Moon	7.35×10^{22}	1.74×10^6	—	—
Sun	1.989×10^{30}	6.96×10^8	—	—

[a]In August 2006, the International Astronomical Union adopted a definition of a planet that separates Pluto from the other eight planets. Pluto is now defined as a "dwarf planet" (like the asteroid Ceres).

Physical Data Often Used

Average Earth–Moon distance	3.84×10^8 m
Average Earth–Sun distance	1.496×10^{11} m
Average radius of the Earth	6.37×10^6 m
Density of air (20°C and 1 atm)	1.20 kg/m^3
Density of air (0°C and 1 atm)	1.29 kg/m^3
Density of water (20°C and 1 atm)	1.00×10^3 kg/m^3
Free-fall acceleration	9.80 m/s^2
Mass of the Earth	5.97×10^{24} kg
Mass of the Moon	7.35×10^{22} kg
Mass of the Sun	1.99×10^{30} kg
Standard atmospheric pressure	1.013×10^5 Pa

Note: These values are the ones used in the text.

Some Prefixes for Powers of Ten

Power	Prefix	Abbreviation	Power	Prefix	Abbreviation
10^{-24}	yocto	y	10^1	deka	da
10^{-21}	zepto	z	10^2	hecto	h
10^{-18}	atto	a	10^3	kilo	k
10^{-15}	femto	f	10^6	mega	M
10^{-12}	pico	p	10^9	giga	G
10^{-9}	nano	n	10^{12}	tera	T
10^{-6}	micro	μ	10^{15}	peta	P
10^{-3}	milli	m	10^{18}	exa	E
10^{-2}	centi	c	10^{21}	zetta	Z
10^{-1}	deci	d	10^{24}	yotta	Y

KFIFTH EDITION

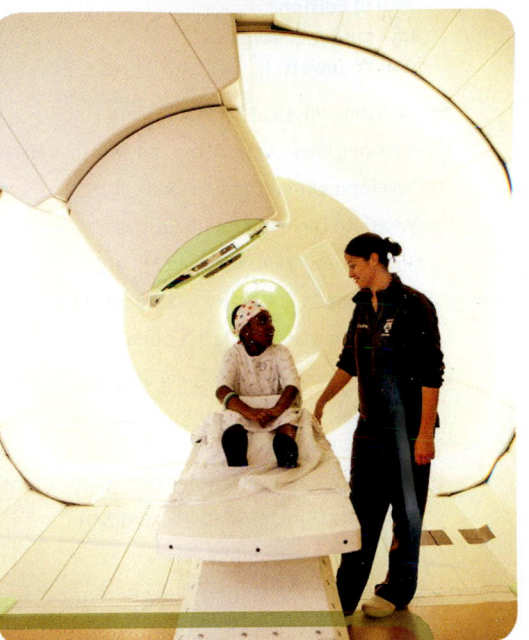

Principles of Physics

A CALCULUS-BASED TEXT

HYBRID EDITION

Raymond A. Serway
Emeritus, James Madison University

John W. Jewett, Jr.
Emeritus, California State Polytechnic University, Pomona

BROOKS/COLE
CENGAGE Learning®

Australia • Brazil • Japan • Korea • Mexico • Singapore • Spain • United Kingdom • United States

BROOKS/COLE
CENGAGE Learning

Principles of Physics, **Fifth Edition**
Hybrid Edition
Raymond A. Serway
John W. Jewett, Jr.

Publisher, Physical Sciences: Mary Finch

Publisher, Physics and Astronomy: Charles Hartford

Development Editor: Ed Dodd

Associate Development Editor: Brandi Kirksey

Editorial Assistant: Brendan Killion

Senior Media Editor: Rebecca Berardy Schwartz

Marketing Manager: Jack Cooney

Marketing Coordinator: Julie Stefani

Marketing Communications Manager:
    Darlene Macanan

Senior Content Project Manager: Cathy Brooks

Senior Art Director: Cate Rickard Barr

Print Buyer: Diane Gibbons

Manufacturing Planner: Doug Bertke

Rights Acquisition Specialist:
    Shalice Shah-Caldwell

Production Service: MPS Limited, a Macmillan
    Company

Compositor: MPS Limited, a Macmillan Company

Text Designer: Brian Salisbury

Cover Designer: Brian Salisbury

Cover Images:

• Child prepares for proton therapy at the Roberts
Proton Therapy Center in Philadelphia, PA:
Copyright © 2011 Ed Cunicelli

• A full moon rises through an iridescent lenticular
cloud at Mt. Rainier: © Caren Brinkema/
Science Faction/Getty Images

• Supernova remnant: NASA

• Infinity Bridge on the River Tees: © Paul Downing/
Flickr Select/Getty Images

For product information and technology assistance, contact us at
Cengage Learning Customer & Sales Support, 1-800-354-9706

For permission to use material from this text or product,
submit all requests online at www.cengage.com/permissions.
Further permissions questions can be emailed to
permissionrequest@cengage.com

Library of Congress Control Number: 2011937185

ISBN-13: 978-1-305-58687-1

ISBN-10: 1-305-58687-5

Brooks/Cole
20 Channel Center Street
Boston, MA 02210
USA

Cengage Learning is a leading provider of customized learning solutions with
office locations around the globe, including Singapore, the United Kingdom,
Australia, Mexico, Brazil and Japan. Locate your local office at
international.cengage.com/region

Cengage Learning products are represented in Canada by Nelson Education, Ltd.

For your course and learning solutions, visit **www.cengage.com**.

Purchase any of our products at your local college store or at our preferred
online store **www.cengagebrain.com**.

Instructors: Please visit **login.cengage.com** and log in to access
instructor-specific resources.

We dedicate this book to our wives Elizabeth and
Lisa and all our children and grandchildren for
their loving understanding when we spent time on
writing instead of being with them.

Printed in the United States of America
3 4 5 6 17 16 15 14

Contents

About the Authors

Raymond A. Serway received his doctorate at Illinois Institute of Technology and is Professor Emeritus at James Madison University. In 2011, he was awarded with an honorary doctorate degree from his alma mater, Utica College. He received the 1990 Madison Scholar Award at James Madison University, where he taught for 17 years. Dr. Serway began his teaching career at Clarkson University, where he conducted research and taught from 1967 to 1980. He was the recipient of the Distinguished Teaching Award at Clarkson University in 1977 and the Alumni Achievement Award from Utica College in 1985. As Guest Scientist at the IBM Research Laboratory in Zurich, Switzerland, he worked with K. Alex Müller, 1987 Nobel Prize recipient. Dr. Serway also was a visiting scientist at Argonne National Laboratory, where he collaborated with his mentor and friend, the late Dr. Sam Marshall. Dr. Serway is the coauthor of *College Physics,* ninth edition; *Physics for Scientists and Engineers,* eighth edition; *Essentials of College Physics; Modern Physics,* third edition; and the high school textbook *Physics,* published by Holt McDougal. In addition, Dr. Serway has published more than 40 research papers in the field of condensed matter physics and has given more than 60 presentations at professional meetings. Dr. Serway and his wife Elizabeth enjoy traveling, playing golf, fishing, gardening, singing in the church choir, and especially spending quality time with their four children, nine grandchildren, and a recent great grandson.

John W. Jewett, Jr. earned his undergraduate degree in physics at Drexel University and his doctorate at Ohio State University, specializing in optical and magnetic properties of condensed matter. Dr. Jewett began his academic career at Richard Stockton College of New Jersey, where he taught from 1974 to 1984. He is currently Emeritus Professor of Physics at California State Polytechnic University, Pomona. Throughout his teaching career, Dr. Jewett has been active in promoting effective physics education. In addition to receiving four National Science Foundation grants, he helped found and direct the Southern California Area Modern Physics Institute (SCAMPI) and Science IMPACT (Institute for Modern Pedagogy and Creative Teaching). Dr. Jewett's honors include the Stockton Merit Award at Richard Stockton College in 1980, selection as Outstanding Professor at California State Polytechnic University for 1991–1992, and the Excellence in Undergraduate Physics Teaching Award from the American Association of Physics Teachers (AAPT) in 1998. In 2010, he received an Alumni Lifetime Achievement Award from Drexel University in recognition of his contributions in physics education. He has given more than 100 presentations both domestically and abroad, including multiple presentations at national meetings of the AAPT. Dr. Jewett is the author of *The World of Physics: Mysteries, Magic, and Myth,* which provides many connections between physics and everyday experiences. In addition to his work as the coauthor for *Principles of Physics* he is also the coauthor on *Physics for Scientists and Engineers,* eighth edition, as well as *Global Issues,* a four-volume set of instruction manuals in integrated science for high school. Dr. Jewett enjoys playing keyboard with his all-physicist band, traveling, underwater photography, learning foreign languages, and collecting antique quack medical devices that can be used as demonstration apparatus in physics lectures. Most importantly, he relishes spending time with his wife Lisa and their children and grandchildren.

Principles of Physics is designed for a one-year introductory calculus-based physics course for engineering and science students and for premed students taking a rigorous physics course. This fifth edition contains many new pedagogical features—most notably, an integrated Web-based learning system and a structured problem-solving strategy that uses a modeling approach. Based on comments from users of the fourth edition and reviewers' suggestions, a major effort was made to improve organization, clarity of presentation, precision of language, and accuracy throughout.

This textbook was initially conceived because of well-known problems in teaching the introductory calculus-based physics course. The course content (and hence the size of textbooks) continues to grow, while the number of contact hours with students has either dropped or remained unchanged. Furthermore, traditional one-year courses cover little if any physics beyond the 19th century.

In preparing this textbook, we were motivated by the spreading interest in reforming the teaching and learning of physics through physics education research (PER). One effort in this direction was the Introductory University Physics Project (IUPP), sponsored by the American Association of Physics Teachers and the American Institute of Physics. The primary goals and guidelines of this project are to:

- Reduce course content following the "less may be more" theme;
- Incorporate contemporary physics naturally into the course;
- Organize the course in the context of one or more "story lines";
- Treat all students equitably.

Recognizing a need for a textbook that could meet these guidelines several years ago, we studied the various proposed IUPP models and the many reports from IUPP committees. Eventually, one of us (RAS) became actively involved in the review and planning of one specific model, initially developed at the U.S. Air Force Academy, entitled "A Particles Approach to Introductory Physics." An extended visit at the Academy was spent working with Colonel James Head and Lt. Col. Rolf Enger, the primary authors of the Particles model, and other members of that department. This most useful collaboration was the starting point of this project.

The other author (JWJ) became involved with the IUPP model called "Physics in Context," developed by John Rigden (American Institute of Physics), David Griffiths (Oregon State University), and Lawrence Coleman (University of Arkansas at Little Rock). This involvement led to National Science Foundation (NSF) grant support for the development of new contextual approaches and eventually to the contextual overlay that is used in this book and described in detail later in the Preface.

The combined IUPP approach in this book has the following features:

- It is an evolutionary approach (rather than a revolutionary approach), which should meet the current demands of the physics community.
- It deletes many topics in classical physics (such as alternating current circuits and optical instruments) and places less emphasis on rigid object motion, optics, and thermodynamics.
- Some topics in contemporary physics, such as fundamental forces, special relativity, energy quantization, and the Bohr model of the hydrogen atom, are introduced early in the textbook.
- A deliberate attempt is made to show the unity of physics and the global nature of physics principles.
- As a motivational tool, the textbook connects applications of physics principles to interesting biomedical situations, social issues, natural phenomena, and technological advances.

Other efforts to incorporate the results of physics education research have led to several of the features in this textbook described below. These include Quick Quizzes, Objective Questions,

Pitfall Preventions, **What If?** features in worked examples, the use of energy bar charts, the modeling approach to problem solving, and the global energy approach introduced in Chapter 7.

Objectives

This introductory physics textbook has two main objectives: to provide the student with a clear and logical presentation of the basic concepts and principles of physics, and to strengthen an understanding of the concepts and principles through a broad range of interesting applications to the real world. To meet these objectives, we have emphasized sound physical arguments and problem-solving methodology. At the same time, we have attempted to motivate the student through practical examples that demonstrate the role of physics in other disciplines, including engineering, chemistry, and medicine.

Changes in the Fifth Edition

A number of changes and improvements have been made in the fifth edition of this text. Many of these are in response to recent findings in physics education research and to comments and suggestions provided by the reviewers of the manuscript and instructors using the first four editions. The following represent the major changes in the fifth edition:

New Contexts. The context overlay approach is described below under "Organization." The fifth edition introduces two new Contexts: for Chapter 15, "Heart Attacks," and for Chapters 22–23, "Magnetism in Medicine." Both of these new Contexts are aimed at applying the principles of physics to the biomedical field.

In the "Heart Attacks" Context, we study the flow of fluids through a pipe, as an analogy to the flow of blood through blood vessels in the human body. Various details of the blood flow are related to the dangers of cardiovascular disease. In addition, we discuss new developments in the study of blood flow and heart attacks using nanoparticles and computer imaging.

The "Magnetism in Medicine" Context explores the application of the principles of electromagnetism to diagnostic and therapeutic procedures in medicine. We begin by looking at historical uses of magnetism, including several "quack" medical devices. More modern applications include remote magnetic navigation in cardiac catheter ablation procedures for atrial fibrillation, transcranial magnetic stimulation for the treatment of depression, and magnetic resonance imaging as a diagnostic tool.

Worked Examples. All in-text worked examples have been recast and are now presented in a two-column format to better reinforce physical concepts. The left column shows textual information that describes the steps for solving the problem. The right column shows the mathematical manipulations and results of taking these steps. This layout facilitates matching the concept with its mathematical execution and helps students organize their work. The examples closely follow the General Problem-Solving Strategy introduced in Chapter 1 to reinforce effective problem-solving habits. In almost all cases, examples are solved symbolically until the end, when numerical values are substituted into the final symbolic result. This procedure allows students to analyze the symbolic result to see how the result depends on the parameters in the problem, or to take limits to test the final result for correctness. Most worked examples in the text may be assigned for homework in Enhanced WebAssign. A sample of a worked example can be found on the next page.

Line-by-Line Revision of the Questions and Problems Set. For the Fifth Edition, the authors reviewed each question and problem and incorporated revisions designed to improve both readability and assignability. To make problems clearer to both students and instructors, this extensive process involved editing problems for clarity, editing for length, adding figures where appropriate, and introducing better problem architecture by breaking up problems into clearly defined parts.

Data from Enhanced WebAssign Used to Improve Questions and Problems. As part of the full-scale analysis and revision of the questions and problems sets, the authors utilized extensive user data gathered by WebAssign, from both instructors who assigned and students who worked on problems from previous editions of *Principles of Physics*. These data helped tremendously, indicating when the phrasing in problems could be clearer, thus providing

WebAssign Most worked examples are also available to be assigned as interactive examples in the Enhanced WebAssign homework management system.

Each solution has been written to closely follow the General Problem-Solving Strategy as outlined on pages 25–26 in Chapter 1, so as to reinforce good problem-solving habits.

Each step of the solution is detailed in a two-column format. The left column provides an explanation for each mathematical step in the right column, to better reinforce the physical concepts.

Example 6.6 | A Block Pulled on a Frictionless Surface

A 6.0-kg block initially at rest is pulled to the right along a frictionless, horizontal surface by a constant horizontal force of 12 N. Find the block's speed after it has moved 3.0 m.

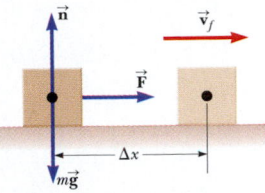

Figure 6.14 (Example 6.6) A block pulled to the right on a frictionless surface by a constant horizontal force.

SOLUTION

Conceptualize Figure 6.14 illustrates this situation. Imagine pulling a toy car across a table with a horizontal rubber band attached to the front of the car. The force is maintained constant by ensuring that the stretched rubber band always has the same length.

Categorize We could apply the equations of kinematics to determine the answer, but let us practice the energy approach. The block is the system, and three external forces act on the system. The normal force balances the gravitational force on the block, and neither of these vertically acting forces does work on the block because their points of application are horizontally displaced.

Analyze The net external force acting on the block is the horizontal 12-N force.

Use the work–kinetic energy theorem for the block, noting that its initial kinetic energy is zero:

$$W_{ext} = K_f - K_i = \tfrac{1}{2}mv_f^2 - 0 = \tfrac{1}{2}mv_f^2$$

Solve for v_f and use Equation 6.1 for the work done on the block by $\vec{F}$:

$$v_f = \sqrt{\frac{2W_{ext}}{m}} = \sqrt{\frac{2F\Delta x}{m}}$$

Substitute numerical values:

$$v_f = \sqrt{\frac{2(12\ N)(3.0\ m)}{6.0\ kg}} = 3.5\ m/s$$

Finalize It would be useful for you to solve this problem again by modeling the block as a particle under a net force to find its acceleration and then as a particle under constant acceleration to find its final velocity.

What If? Suppose the magnitude of the force in this example is doubled to $F' = 2F$. The 6.0-kg block accelerates to 3.5 m/s due to this applied force while moving through a displacement $\Delta x'$. How does the displacement $\Delta x'$ compare with the original displacement Δx?

Answer If we pull harder, the block should accelerate to a given speed in a shorter distance, so we expect that $\Delta x' < \Delta x$. In both cases, the block experiences the same change in kinetic energy ΔK. Mathematically, from the work–kinetic energy theorem, we find that

$$W_{ext} = F'\Delta x' = \Delta K = F\Delta x$$

$$\Delta x' = \frac{F}{F'}\Delta x = \frac{F}{2F}\Delta x = \tfrac{1}{2}\Delta x$$

and the distance is shorter as suggested by our conceptual argument.

What If? statements appear in about 1/3 of the worked examples and offer a variation on the situation posed in the text of the example. For instance, this feature might explore the effects of changing the conditions of the situation, determine what happens when a quantity is taken to a particular limiting value, or question whether additional information can be determined about the problem situation. This feature encourages students to think about the results of the example and assists in conceptual understanding of the principles.

The final result is symbolic; numerical values are substituted into the final result.

guidance on how to revise problems so that they are more easily understandable for students and more easily assignable by instructors in Enhanced WebAssign. Finally, the data were used to ensure that the problems most often assigned were retained for this new edition.

Revised Questions Organization. We reorganized the end-of-chapter questions for this new edition. The previous edition's Questions section is now divided into two sections: *Objective Questions* and *Conceptual Questions*.

Objective Questions are multiple-choice, true/false, ranking, or other multiple guess-type questions. Some require calculations designed to facilitate students' familiarity with the equations, the variables used, the concepts the variables represent, and the relationships between the concepts. Others are more conceptual in nature and are designed to encourage conceptual thinking. Objective Questions are also written with the personal response system user in mind, and most of the questions could easily be used in these systems.

Conceptual Questions are more traditional short-answer and essay-type questions that require students to think conceptually about a physical situation.

New Types of Problems. We have introduced four new problem types for this edition:

Quantitative/Conceptual problems contain parts that ask students to think both quantitatively and conceptually. An example of a Quantitative/Conceptual problem appears here:

55. A horizontal spring attached to a wall has a force constant of $k = 850$ N/m. A block of mass $m = 1.00$ kg is attached to the spring and rests on a frictionless, horizontal surface as in Figure P7.55. (a) The block is pulled to a position $x_i = 6.00$ cm from equilibrium and released. Find the elastic potential energy stored in the spring when the block is 6.00 cm from equilibrium and when the block passes through equilibrium. (b) Find the speed of the block as it passes through the equilibrium point. (c) What is the speed of the block when it is at a position $x_i/2 = 3.00$ cm? (d) Why isn't the answer to part (c) half the answer to part (b)?

Parts (a)–(c) of the problem ask for quantitative calculations.

Part (d) asks a conceptual question about the situation.

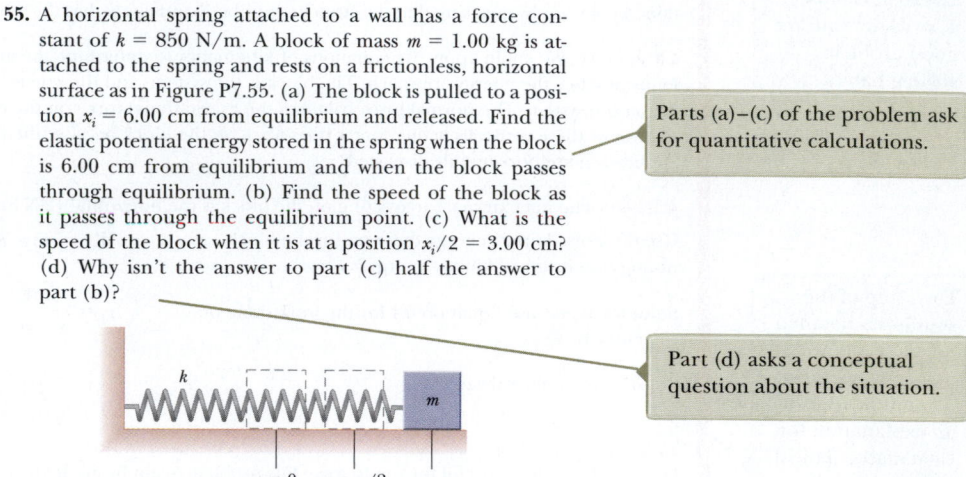

Figure P7.55

Symbolic problems ask students to solve a problem using only symbolic manipulation. A majority of survey respondents asked specifically for an increase in the number of symbolic problems found in the text because it better reflects the way instructors want their students to think when solving physics problems. An example of a Symbolic problem appears here:

57. **Review.** A uniform board of length L is sliding along a smooth, frictionless, horizontal plane as shown in Figure P7.57a (page 226). The board then slides across the boundary with a rough horizontal surface. The coefficient of kinetic friction between the board and the second surface is μ_k. (a) Find the acceleration of the board at the moment its front end has traveled a distance x beyond the boundary. (b) The board stops at the moment its back end reaches the boundary as shown in Figure P7.57b. Find the initial speed v of the board.

No numerical values appear in the problem statement.

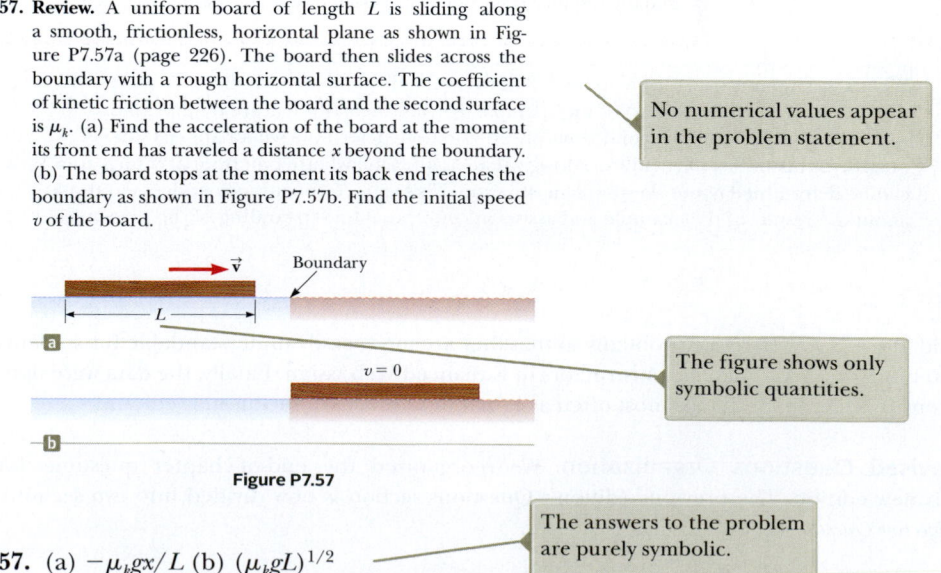

The figure shows only symbolic quantities.

Figure P7.57

The answers to the problem are purely symbolic.

57. (a) $-\mu_k gx/L$ (b) $(\mu_k gL)^{1/2}$

Guided Problems help students break problems into steps. A physics problem typically asks for one physical quantity in a given context. Often, however, several concepts must be used and a number of calculations are required to obtain that final answer. Many students are not accustomed to this level of complexity and often don't know where to start. A Guided Problem breaks a standard problem into smaller steps, enabling students to grasp all the concepts and strategies required to arrive at a correct solution. Unlike standard physics problems, guidance is often built into the problem statement. Guided Problems are reminiscent of how a student might interact with a professor in an office visit. These problems (there is one in every chapter of the text) help train students to break down complex problems into a series of simpler problems, an essential problem-solving skill. An example of a Guided Problem appears here:

28. A uniform beam resting on two pivots has a length $L = 6.00$ m and mass $M = 90.0$ kg. The pivot under the left end exerts a normal force n_1 on the beam, and the second pivot located a distance $\ell = 4.00$ m from the left end exerts a normal force n_2. A woman of mass $m = 55.0$ kg steps onto the left end of the beam and begins walking to the right as in Figure P10.28. The goal is to find the woman's position when the beam begins to tip. (a) What is the appropriate analysis model for the beam before it begins to tip? (b) Sketch a force diagram for the beam, labeling the gravitational and normal forces acting on the beam and placing the woman a distance x to the right of the first pivot, which is the origin. (c) Where is the woman when the normal force n_1 is the greatest? (d) What is n_1 when the beam is about to tip? (e) Use Equation 10.27 to find the value of n_2 when the beam is about to tip. (f) Using the result of part (d) and Equation 10.28, with torques computed around the second pivot, find the woman's position x when the beam is about to tip. (g) Check the answer to part (e) by computing torques around the first pivot point.

> The goal of the problem is identified.

> Analysis begins by identifying the appropriate analysis model.

> Students are provided with suggestions for steps to solve the problem.

> The calculation associated with the goal is requested.

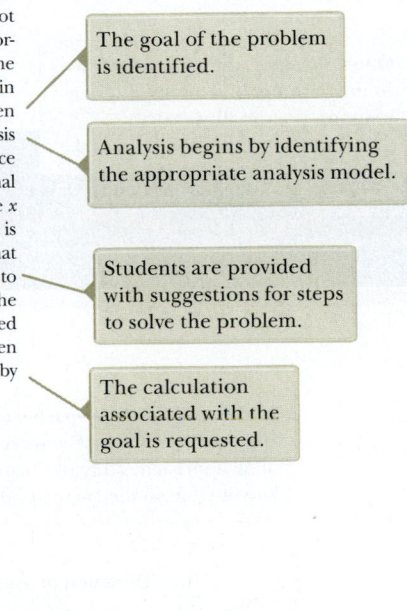
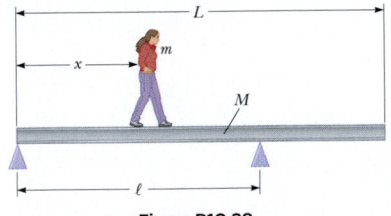

Figure P10.28

Impossibility Problems. Physics education research has focused heavily on the problem-solving skills of students. Although most problems in this text are structured in the form of providing data and asking for a result of computation, two problems in each chapter, on average, are structured as impossibility problems. They begin with the phrase *Why is the following situation impossible?* That is followed by the description of a situation. The striking aspect of these problems is that no question is asked of the students, other than that in the initial italics. The student must determine what questions need to be asked and what calculations need to be performed. Based on the results of these calculations, the student must determine why the situation described is not possible. This determination may require information from personal experience, common sense, Internet or print research, measurement, mathematical skills, knowledge of human norms, or scientific thinking.

These problems can be assigned to build critical thinking skills in students. They are also fun, having the aspect of physics "mysteries" to be solved by students individually or in groups. An example of an impossibility problem appears here:

> The initial phrase in italics signals an impossibility problem.

> A situation is described.

51. *Why is the following situation impossible?* Albert Pujols hits a home run so that the baseball just clears the top row of bleachers, 24.0 m high, located 130 m from home plate. The ball is hit at 41.7 m/s at an angle of 35.0° to the horizontal, and air resistance is negligible.

> No question is asked. The student must determine what needs to be calculated and why the situation is impossible.

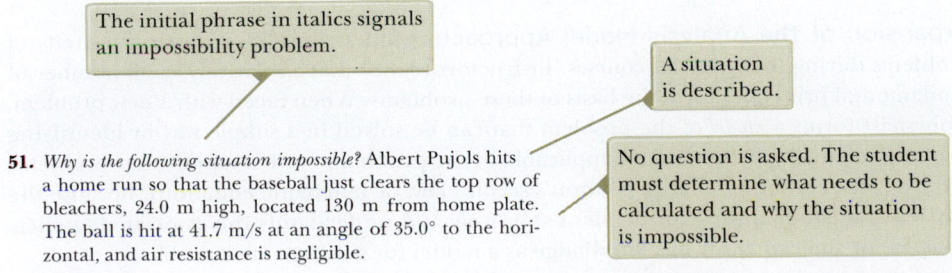

Increased Number of Paired Problems. Based on the positive feedback we received in a survey of the market, we have increased the number of paired problems in this edition. These problems are otherwise identical, one asking for a numerical solution and one asking for a symbolic derivation. There are now three pairs of these problems in most chapters.

Thorough Revision of Artwork. Every piece of artwork in the Fifth Edition was revised in a new and modern style that helps express the physics principles at work in a clear and precise fashion. Every piece of art was also revised to make certain that the physical situations presented correspond exactly to the text discussion at hand.

Also added for this edition is a new feature: "focus pointers" that either point out important aspects of a figure or guide students through a process illustrated by the artwork or photo. This format helps those students who are more visual learners. Examples of figures with focus pointers appear below:

Figure 10.28 Two points on a rolling object take different paths through space.

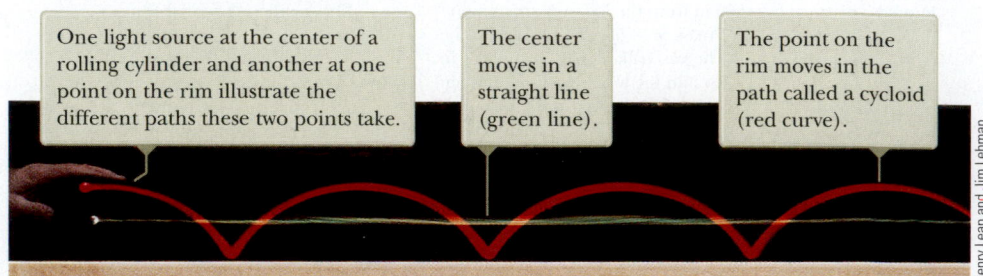

One light source at the center of a rolling cylinder and another at one point on the rim illustrate the different paths these two points take.

The center moves in a straight line (green line).

The point on the rim moves in the path called a cycloid (red curve).

Henry Leap and Jim Lehman

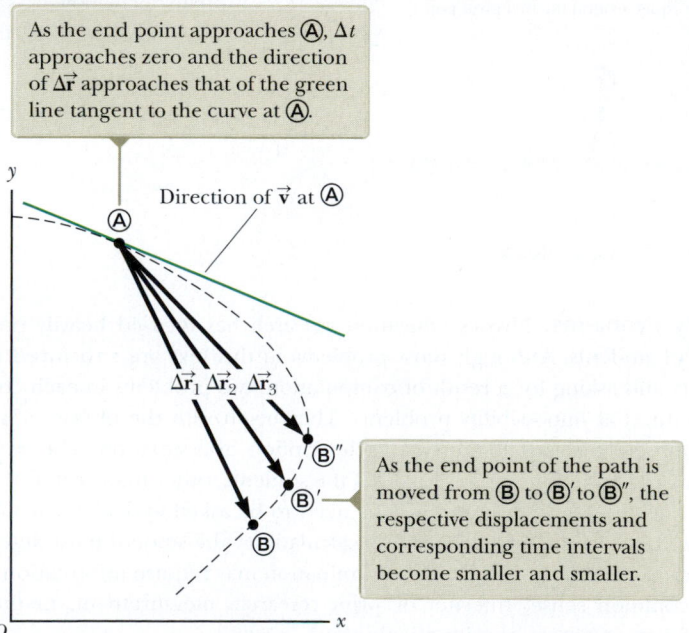

As the end point approaches Ⓐ, Δt approaches zero and the direction of $\Delta \vec{r}$ approaches that of the green line tangent to the curve at Ⓐ.

Direction of $\vec{v}$ at Ⓐ

$\Delta \vec{r}_1$ $\Delta \vec{r}_2$ $\Delta \vec{r}_3$

As the end point of the path is moved from Ⓑ to Ⓑ′ to Ⓑ″, the respective displacements and corresponding time intervals become smaller and smaller.

Figure 3.2 As a particle moves between two points, its average velocity is in the direction of the displacement vector $\Delta \vec{r}$. By definition, the instantaneous velocity at Ⓐ is directed along the line tangent to the curve at Ⓐ.

Expansion of the Analysis Model Approach. Students are faced with hundreds of problems during their physics courses. Instructors realize that a relatively small number of fundamental principles form the basis of these problems. When faced with a new problem, a physicist forms a *model* of the problem that can be solved in a simple way by identifying the fundamental principle that is applicable in the problem. For example, many problems involve conservation of energy, Newton's second law, or kinematic equations. Because the physicist has studied these principles extensively and understands the associated applications, he or she can apply this knowledge as a model for solving a new problem.

Although it would be ideal for students to follow this same process, most students have difficulty becoming familiar with the entire palette of fundamental principles that are available. It is easier for students to identify a *situation* rather than a fundamental principle. The *Analysis Model* approach we focus on in this revision lays out a standard set of situations that appear in most physics problems. These situations are based on an entity in one of four simplification models: particle, system, rigid object, and wave.

Once the simplification model is identified, the student thinks about what the entity is doing or how it interacts with its environment, which leads the student to identify a particular analysis model for the problem. For example, if an object is falling, the object is modeled as a particle. What it is doing is undergoing a constant acceleration due to gravity. The student has learned that this situation is described by the analysis model of a particle under constant acceleration. Furthermore, this model has a small number of equations associated with it for use in solving problems, the kinematic equations in Chapter 2. Therefore, an understanding of the situation has led to an analysis model, which then identifies a very small number of equations to solve the problem, rather than the myriad equations that students see in the chapter. In this way, the use of analysis models leads the student to the fundamental principle the physicist would identify. As the student gains more experience, he or she will lean less on the analysis model approach and begin to identify fundamental principles directly, more like the physicist does. This approach is further reinforced in the end-of-chapter summary under the heading *Analysis Models for Problem Solving*.

Content Changes. The content and organization of the textbook are essentially the same as in the fourth edition. Several sections in various chapters have been streamlined, deleted, or combined with other sections to allow for a more balanced presentation. Chapters 6 and 7 have been completely reorganized to prepare students for a unified approach to energy that is used throughout the text. Updates have been added to reflect the current status of several areas of research and application of physics, including information on discoveries of new Kuiper belt objects (Chapter 11), comparisons of competing theories of pitch perception in humans (Chapter 14), progress in using grating light valves for optical applications (Chapter 27), new experiments to search for the cosmic background radiation (Chapter 28), developments in the search for evidence of a quark–gluon plasma (Chapter 31), and the status of the Large Hadron Collider (Chapter 31).

Organization

We have incorporated a "context overlay" scheme into the textbook, in response to the "Physics in Context" approach in the IUPP. This feature adds interesting applications of the material to real issues. We have developed this feature to be flexible; it is an "overlay" in the sense that the instructor who does not wish to follow the contextual approach can simply ignore the additional contextual features without sacrificing complete coverage of the existing material. We believe, though, that the benefits students will gain from this approach will be many.

The context overlay organization divides the text into nine sections, or "Contexts," after Chapter 1, as follows:

Context Number	Context	Physics Topics	Chapters
1	Alternative-Fuel Vehicles	Classical mechanics	2–7
2	Mission to Mars	Classical mechanics	8–11
3	Earthquakes	Vibrations and waves	12–14
4	Heart Attacks	Fluids	15
5	Global Warming	Thermodynamics	16–18
6	Lightning	Electricity	19–21
7	Magnetism in Medicine	Magnetism	22–23
8	Lasers	Optics	24–27
9	The Cosmic Connection	Modern physics	28–31

Each Context begins with an introductory section that provides some historical background or makes a connection between the topic of the Context and associated social issues. The introductory section ends with a "central question" that motivates study within the Context. The final section of each chapter is a "Context Connection," which discusses how the specific material in the chapter relates to the Context and to the central question. The final chapter in each Context is followed by a "Context Conclusion." Each Conclusion applies a combination of the principles learned in the various chapters of the Context to respond fully to the central question. Each chapter, as well as the Context Conclusions, includes problems related to the context material.

Text Features

Most instructors believe that the textbook selected for a course should be the student's primary guide for understanding and learning the subject matter. Furthermore, the textbook should be easily accessible and should be styled and written to facilitate instruction and learning. With these points in mind, we have included many pedagogical features, listed below, that are intended to enhance its usefulness to both students and instructors.

Problem Solving and Conceptual Understanding

General Problem-Solving Strategy. A general strategy outlined at the end of Chapter 1 (pages 25–26) provides students with a structured process for solving problems. In all remaining chapters, the strategy is employed explicitly in every example so that students learn how it is applied. Students are encouraged to follow this strategy when working problems.

In most chapters, more specific strategies and suggestions are included for solving the types of problems featured in Enhanced WebAssign. This feature helps students identify the essential steps in solving problems and increases their skills as problem solvers.

Thinking Physics. We have included many Thinking Physics examples throughout each chapter. These questions relate the physics concepts to common experiences or extend the concepts beyond what is discussed in the textual material. Immediately following each of these questions is a "Reasoning" section that responds to the question. Ideally, the student will use these features to better understand physical concepts before being presented with quantitative examples and working homework problems.

Active Figures. Many diagrams from the text have been animated to become Active Figures (identified in the figure legend), part of the Enhanced WebAssign online homework system. By viewing animations of phenomena and processes that cannot be fully represented on a static page, students greatly increase their conceptual understanding. In addition to viewing animations of the figures, students can see the outcome of changing variables, conduct suggested explorations of the principles involved in the figure, and take and receive feedback on quizzes related to the figure.

Quick Quizzes. Students are provided an opportunity to test their understanding of the physical concepts presented through Quick Quizzes. The questions require students to make decisions on the basis of sound reasoning, and some of the questions have been written to help students overcome common misconceptions. Quick Quizzes have been cast in an objective format, including multiple choice, true–false, and ranking. Answers to all Quick Quiz questions are found at the end of the text. Many instructors choose to use such questions in a "peer instruction" teaching style or with the use of personal response system "clickers," but they can be used in standard quiz format as well. An example of a Quick Quiz follows below.

QUICK QUIZ 6.5 A dart is inserted into a spring-loaded dart gun by pushing the spring in by a distance x. For the next loading, the spring is compressed a distance $2x$. How much faster does the second dart leave the gun compared with the first? (**a**) four times as fast (**b**) two times as fast (**c**) the same (**d**) half as fast (**e**) one-fourth as fast

Pitfall Preventions. Over 150 Pitfall Preventions (such as the one to the right) are provided to help students avoid common mistakes and misunderstandings. These features, which are placed in the margins of the text, address both common student misconceptions and situations in which students often follow unproductive paths.

Summaries. Each chapter contains a summary that reviews the important concepts and equations discussed in that chapter. New for the Fifth Edition is the Analysis Models for Problem Solving section of the Summary, which highlights the relevant analysis models presented in a given chapter.

Questions. As mentioned previously, the previous edition's Questions section is now divided into two sections: *Objective Questions* and *Conceptual Questions*. The instructor may select items to assign as homework or use in the classroom, possibly with "peer instruction" methods and possibly with personal response systems. More than seven hundred Objective and Conceptual Questions are included in this edition. Answers for selected questions are included in the *Student Solutions Manual/Study Guide,* and answers for all questions are found in the *Instructor's Solutions Manual.*

Problems. An extensive set of problems is available in Enhanced WebAssign; in all, over 2 200 problems. Full solutions for approximately 20% of the problems are included in the *Student Solutions Manual/Study Guide,* and solutions for all problems are found in the *Instructor's Solutions Manual.*

In addition to the new problem types mentioned previously, there are several other kinds of problems featured in each chapter's problems set in Enhanced WebAssign:

- **Biomedical problems.** We added a number of problems related to biomedical situations in this edition to highlight the relevance of physics principles to those students taking this course who are majoring in one of the life sciences.
- **Paired Problems.** As an aid for students learning to solve problems symbolically, paired numerical and symbolic problems are included in all chapters of the text.
- **Review problems.** Many chapters include review problems requiring the student to combine concepts covered in the chapter with those discussed in previous chapters. These problems (marked **Review**) reflect the cohesive nature of the principles in the text and verify that physics is not a scattered set of ideas. When facing a real-world issue such as global warming or nuclear weapons, it may be necessary to call on ideas in physics from several parts of a textbook such as this one.
- **"Fermi problems."** One or more problems in most chapters ask the student to reason in order-of-magnitude terms.
- **Design problems.** Several chapters contain problems that ask the student to determine design parameters for a practical device so that it can function as required.
- **Calculus-based problems.** Most chapters contain at least one problem applying ideas and methods from differential calculus and one problem using integral calculus.

The instructor's Web site, **www.cengage.com/physics/serway,** provides lists of all the various problem types, including problems most often assigned in Enhanced WebAssign, symbolic problems, quantitative/conceptual problems, Master It tutorials, Watch It solution videos, impossibility problems, paired problems, problems using calculus, problems encouraging or requiring computer use, problems with **What If?** parts, problems referred to in the chapter text, problems based on experimental data, order-of-magnitude problems, problems about biological applications, design problems, review problems, problems reflecting historical reasoning, and ranking questions.

Alternative Representations. We emphasize alternative representations of information, including mental, pictorial, graphical, tabular, and mathematical representations. Many problems are easier to solve if the information is presented in alternative ways, to reach the many different methods students use to learn.

Math Appendix. The math appendix (Appendix B), a valuable tool for students, shows the math tools in a physics context. This resource is ideal for students who need a quick review on topics such as algebra, trigonometry, and calculus.

Helpful Features

Style. To facilitate rapid comprehension, we have written the book in a clear, logical, and engaging style. We have chosen a writing style that is somewhat informal and relaxed so that

---

Pitfall Prevention | 13.2
Two Kinds of Speed/Velocity
Do not confuse v, the speed of the wave as it propagates along the string, with v_y, the transverse velocity of a point on the string. The speed v is constant for a uniform medium, whereas v_y varies sinusoidally.

students will find the text appealing and enjoyable to read. New terms are carefully defined, and we have avoided the use of jargon.

Important Definitions and Equations. Most important definitions are set in **boldface** or are set off from the paragraph in centered text for added emphasis and ease of review. Similarly, important equations are highlighted with a background screen to facilitate location.

Marginal Notes. Comments and notes appearing in the margin with a ▶ icon can be used to locate important statements, equations, and concepts in the text.

Pedagogical Use of Color. Readers should consult the **pedagogical color chart** (inside the front cover) for a listing of the color-coded symbols used in the text diagrams. This system is followed consistently throughout the text.

Mathematical Level. We have introduced calculus gradually, keeping in mind that students often take introductory courses in calculus and physics concurrently. Most steps are shown when basic equations are developed, and reference is often made to mathematical appendices near the end of the textbook. Although vectors are discussed in detail in Chapter 1, vector products are introduced later in the text, where they are needed in physical applications. The dot product is introduced in Chapter 6, which addresses energy of a system; the cross product is introduced in Chapter 10, which deals with angular momentum.

Significant Figures. In both worked examples and problems, significant figures have been handled with care. Most numerical examples are worked to either two or three significant figures, depending on the precision of the data provided. Problems regularly state data and answers to three-digit precision. When carrying out estimation calculations, we shall typically work with a single significant figure. (More discussion of significant figures can be found in Chapter 1, pages 10–12.)

Units. The international system of units (SI) is used throughout the text. The U.S. customary system of units is used only to a limited extent in the chapters on mechanics and thermodynamics.

Appendices and Endpapers. Several appendices are provided near the end of the textbook. Most of the appendix material represents a review of mathematical concepts and techniques used in the text, including scientific notation, algebra, geometry, trigonometry, differential calculus, and integral calculus. Reference to these appendices is made throughout the text. Most mathematical review sections in the appendices include worked examples and exercises with answers. In addition to the mathematical reviews, the appendices contain tables of physical data, conversion factors, and the SI units of physical quantities as well as a periodic table of the elements. Other useful information—fundamental constants and physical data, planetary data, a list of standard prefixes, mathematical symbols, the Greek alphabet, and standard abbreviations of units of measure—appears on the endpapers.

TextChoice Custom Options for *Principles of Physics*

Cengage Learning's digital library, TextChoice, enables you to build your custom version of Serway and Jewett's *Principles of Physics* from scratch. You may pick and choose the content you want to include in your text and even add your own original materials creating a unique, all-in-one learning solution. This all happens from the convenience of your desktop. Visit **www.textchoice.com** to start building your book today.

Cengage Learning offers the fastest and easiest way to create unique customized learning materials delivered the way you want. For more information about custom publishing options, visit **www.cengage.com/custom** or contact your local Cengage Learning representative.

Course Solutions That Fit Your Teaching Goals and Your Students' Learning Needs

Recent advances in educational technology have made homework management systems and audience response systems powerful and affordable tools to enhance the way you teach your course. Whether you offer a more traditional text-based course, are interested in using or are currently using an online homework management system such as Enhanced WebAssign, or are ready to turn your lecture into an interactive learning environment with JoinIn, you can be confident that the text's proven content provides the foundation for each and every component of our technology and ancillary package.

Homework Management Systems

Enhanced WebAssign for *Principles of Physics,* Fifth Edition.

Exclusively from Cengage Learning, Enhanced WebAssign offers an extensive online program for physics to encourage the practice that's so critical for concept mastery. The meticulously crafted pedagogy and exercises in our proven texts become even more effective in Enhanced WebAssign. Enhanced WebAssign includes the Cengage YouBook, a highly customizable, interactive eBook. WebAssign includes:

- All of the quantitative problems from the fifth edition
- Selected problems enhanced with targeted feedback. An example of targeted feedback appears below:

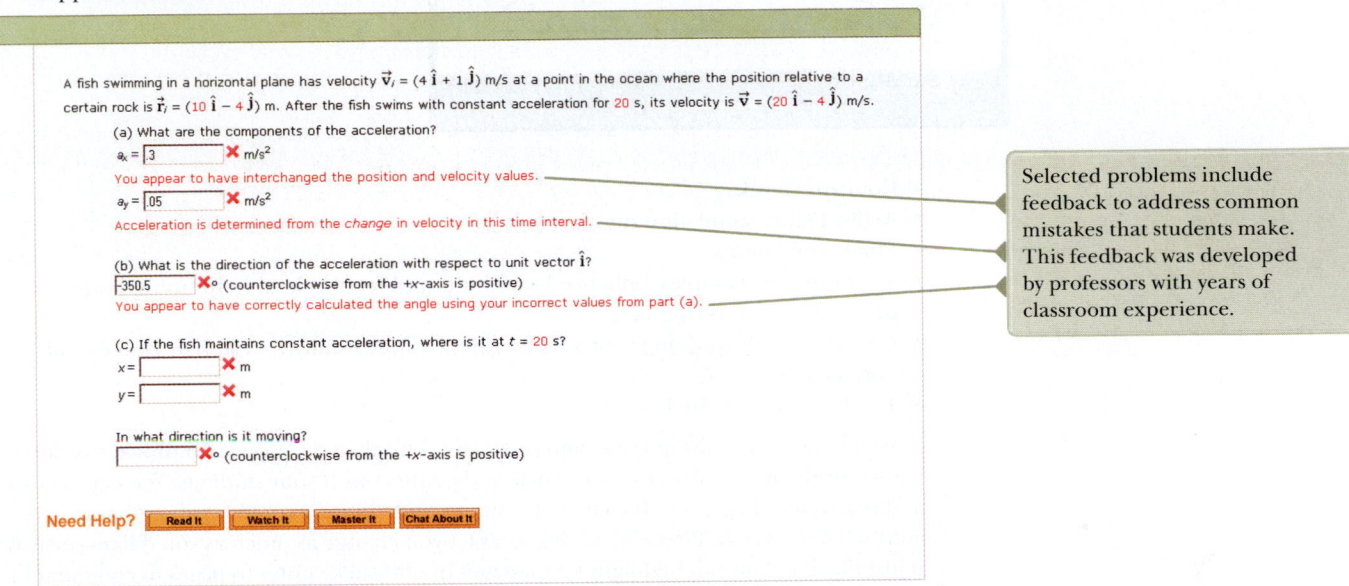

Selected problems include feedback to address common mistakes that students make. This feedback was developed by professors with years of classroom experience.

- Master It tutorials to help students work through the problem one step at a time. An example of a Master It tutorial appears below:

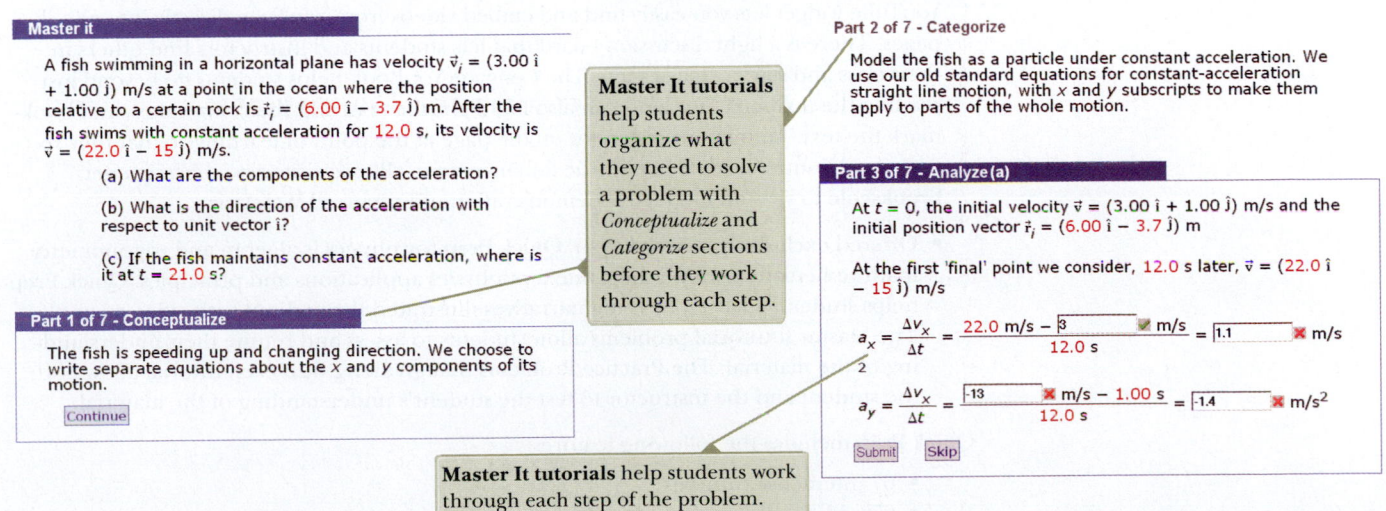

Master It tutorials help students organize what they need to solve a problem with *Conceptualize* and *Categorize* sections before they work through each step.

Master It tutorials help students work through each step of the problem.

- Watch It solution videos that explain fundamental problem-solving strategies, to help students step through the problem. In addition, instructors can choose to include video hints of problem-solving strategies. A screen shot from a Watch It solution video appears below:

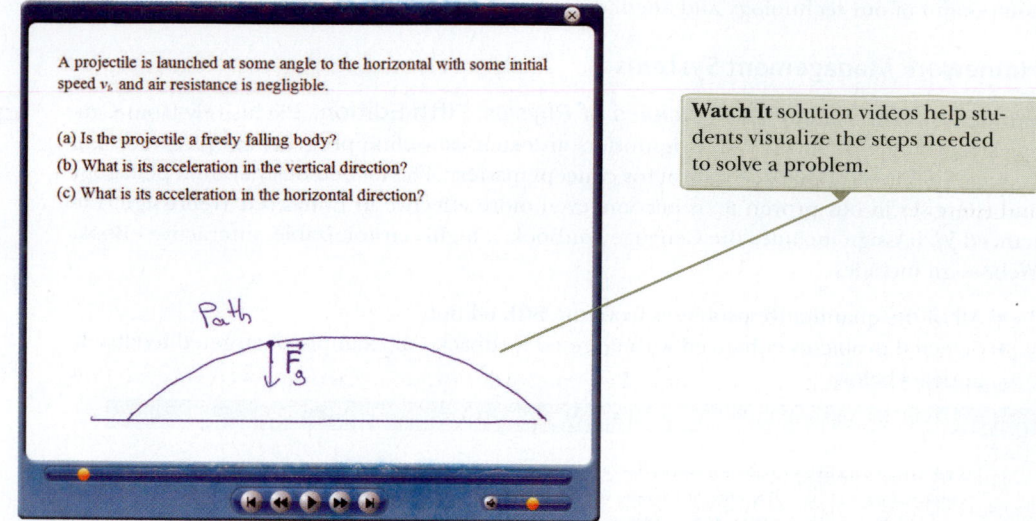

A projectile is launched at some angle to the horizontal with some initial speed v_i, and air resistance is negligible.

(a) Is the projectile a freely falling body?

(b) What is its acceleration in the vertical direction?

(c) What is its acceleration in the horizontal direction?

Watch It solution videos help students visualize the steps needed to solve a problem.

- Concept Checks
- Active Figure simulation tutorials
- PhET simulations
- Most worked examples, enhanced with hints and feedback, to help strengthen students' problem-solving skills
- Every Quick Quiz, giving your students ample opportunity to test their conceptual understanding
- The Cengage YouBook

WebAssign has a customizable and interactive eBook, the **Cengage YouBook**, that lets you tailor the textbook to fit your course and connect with your students. You can remove and rearrange chapters in the table of contents and tailor assigned readings that match your syllabus exactly. Powerful editing tools let you change as much as you'd like—or leave it just like it is. You can highlight key passages or add sticky notes to pages to comment on a concept in the reading, and then share any of these individual notes and highlights with your students, or keep them personal. You can also edit narrative content in the textbook by adding a text box or striking out text. With a handy link tool, you can drop in an icon at any point in the eBook that lets you link to your own lecture notes, audio summaries, video lectures, or other files on a personal website or anywhere on the web. A simple YouTube widget lets you easily find and embed videos from YouTube directly into eBook pages. There is a light discussion board that lets students and instructors find others in their class and start a chat session. The Cengage YouBook helps students go beyond just reading the textbook. Students can also highlight the text, add their own notes, and bookmark the text. Animations play right on the page at the point of learning so that they're not speed bumps to reading but true enhancements. Please visit **www.webassign.net/ brookscole** to view an interactive demonstration of Enhanced WebAssign.

- Offered exclusively in WebAssign, **Quick Prep** for physics is algebra and trigonometry math remediation within the context of physics applications and principles. Quick Prep helps students succeed by using narratives illustrated throughout with video examples. The Master It tutorial problems allow students to assess and retune their understanding of the material. The Practice Problems that go along with each tutorial allow both the student and the instructor to test the student's understanding of the material.

Quick Prep includes the following features:

- 67 interactive tutorials
- 67 additional practice problems
- Thorough overview of each topic that includes video examples
- Taken before the semester begins or during the first few weeks of the course
- Can also be assigned alongside each chapter for "just in time" remediation

Topics include: units, scientific notation, and significant figures; the motion of objects along a line; functions; approximation and graphing; probability and error; vectors, displacement, and velocity; spheres; force and vector projections.

CengageBrain.com

On **CengageBrain.com** students will be able to save up to 60% on their course materials through our full spectrum of options. Students will have the option to rent their textbooks, purchase print textbooks, e-textbooks, or individual e-chapters and audio books all for substantial savings over average retail prices. **CengageBrain.com** also includes access to Cengage Learning's broad range of homework and study tools and features a selection of free content.

Lecture Presentation Resources

PowerLecture with ExamView® and JoinIn for *Principles of Physics,* **Fifth Edition.** Bringing physics principles and concepts to life in your lectures has never been easier! The full-featured, two-volume **PowerLecture** Instructor's Resource DVD-ROM (Volume 1: Chapters 1–15; Volume 2: Chapters 16–31) provides everything you need for *Principles of Physics,* fifth edition. Key content includes the *Instructor's Solutions Manual* solutions, art and images from the text, pre-made chapter-specific PowerPoint lectures, ExamView test generator software with pre-loaded test questions, JoinIn response-system "clickers," Active Figures animations, and a physics movie library.

JoinIn. *Assessing to Learn in the Classroom* questions developed at the University of Massachusetts Amherst. This collection of 250 advanced conceptual questions has been tested in the classroom for more than ten years and takes peer learning to a new level. JoinIn helps you turn your lectures into an interactive learning environment that promotes conceptual understanding. Available exclusively for higher education from our partnership with Turning Technologies, JoinIn™ is the easiest way to turn your lecture hall into a personal, fully interactive experience for your students!

Assessment and Course Preparation Resources

A number of resources listed below will assist with your assessment and preparation processes.

Instructor's Solutions Manual by Vahe Peroomian (University of California at Los Angeles). Thoroughly revised for this edition, the *Instructor's Solutions Manual* contains complete worked solutions to all problems as well as answers to the even-numbered problems and all the questions. The solutions to problems new to the fifth edition are marked for easy identification. Volume 1 contains Chapters 1 through 15, and Volume 2 contains Chapters 16 through 31. Electronic files of the *Instructor's Solutions Manual* are available on the *PowerLecture*™ DVD-ROM.

Test Bank by Susan English (Durham Technical College). The test bank is available on the two-volume *PowerLecture*™ DVD-ROM via the ExamView® test software. This two-volume test bank contains approximately 2 000 multiple-choice questions. Instructors may print and duplicate pages for distribution to students. Volume 1 contains Chapters 1 through 15, and Volume 2 contains Chapters 16 through 31. WebCT and Blackboard versions of the test bank are available on the instructor's companion site at **www.cengage/physics/serway.**

Instructor's Companion Web Site. Consult the instructor's site by pointing your browser to **www.cengage.com/physics/serway** for a problem correlation guide, PowerPoint lectures, and JoinIn audience response content. Instructors adopting the fifth edition of *Principles of Physics* may download these materials after securing the appropriate password from their local sales representative.

Supporting Materials for the Instructor

Supporting instructor materials are available to qualified adopters. Please consult your local Cengage Learning, Brooks/Cole representative for details. Visit **www.cengage.com/physics/serway** to:

- request a desk copy
- locate your local representative
- download electronic files of select support materials

CENGAGE**brain**.com

Student Resources

Visit the *Principles of Physics* Web site at **www.cengagebrain.com/shop/ISBN/9781133104261** to see samples of select student supplements. Go to **CengageBrain.com** to purchase and access this product at Cengage Learning's preferred online store.

Student Solutions Manual/Study Guide by John R. Gordon, Vahe Peroomian, Raymond A. Serway, and John W. Jewett, Jr. This two-volume manual features detailed solutions to 20% of the problems from this edition. The manual also features a list of important equations, concepts, and notes from key sections of the text in addition to answers to selected questions. Volume 1 contains Chapters 1 through 15, and Volume 2 contains Chapters 16 through 31.

Physics Laboratory Manual, Third Edition by David Loyd (Angelo State University) supplements the learning of basic physical principles while introducing laboratory procedures and equipment. Each chapter includes a prelaboratory assignment, objectives, an equipment list, the theory behind the experiment, experimental procedures, graphing exercises, and questions. A laboratory report form is included with each experiment so that the student can record data, calculations, and experimental results. Students are encouraged to apply statistical analysis to their data. A complete *Instructor's Manual* is also available to facilitate use of this lab manual.

Physics Laboratory Experiments, Seventh Edition by Jerry D. Wilson (Lander College) and Cecilia A. Hernández (American River College). This market-leading manual for the first-year physics laboratory course offers a wide range of class-tested experiments designed specifically for use in small to midsize lab programs. A series of integrated experiments emphasizes the use of computerized instrumentation and includes a set of "computer-assisted experiments" to allow students and instructors to gain experience with modern equipment. This option also enables instructors to determine the appropriate balance between traditional and computer-based experiments for their courses. By analyzing data through two different methods, students gain a greater understanding of the concepts behind the experiments. The seventh edition is updated with the latest information and techniques involving state-of-the-art equipment and a new Guided Learning feature addresses the growing interest in guided-inquiry pedagogy. Fourteen additional experiments are also available through custom printing.

Teaching Options

Although some topics found in traditional textbooks have been omitted from this textbook, instructors may find that the current text still contains more material than can be covered in a two-semester sequence. For this reason, we would like to offer the following suggestions. If you wish to place more emphasis on contemporary topics in physics, you should consider omitting parts or all of Chapters 15, 16, 17, 18, 24, 25, and 26. On the other hand, if you wish to follow a more traditional approach that places more emphasis on classical physics, you could omit Chapters 9, 11, 28, 29, 30, and 31. Either approach can be used without any loss in continuity. Other teaching options would fall somewhere between these two extremes by choosing to omit some or all of the following sections, which can be considered optional:

3.6	Relative Velocity and Relative Acceleration
6.10	Energy Diagrams and Equilibrium of a System
9.9	General Relativity
10.12	Rolling Motion of Rigid Objects
12.6	Damped Oscillations
12.7	Forced Oscillations
14.6	Nonsinusoidal Wave Patterns
14.7	The Ear and Theories of Pitch Perception
15.8	Other Applications of Fluid Dynamics
16.6	Distribution of Molecular Speeds
17.7	Molar Specific Heats of Ideal Gases
17.8	Adiabatic Processes for an Ideal Gas
17.9	Molar Specific Heats and the Equipartition of Energy
20.10	Capacitors with Dielectrics
22.11	Magnetism in Matter
26.5	The Eye
27.9	Diffraction of X-Rays by Crystals
28.13	Tunneling Through a Potential Energy Barrier

Acknowledgments

Prior to our work on this revision, we conducted two separate surveys of professors to gauge their textbook needs in the introductory calculus-based physics market. We were overwhelmed not only by the number of professors who wanted to take part in the survey but also by their insightful comments. Their feedback and suggestions helped shape the revision of this edition; we thank them.

We also thank the following people for their suggestions and assistance during the preparation of earlier editions of this textbook:

Edward Adelson, *Ohio State University;* Anthony Aguirre, *University of California at Santa Cruz;* Yildirim M. Aktas, *University of North Carolina–Charlotte;* Alfonso M. Albano, *Bryn Mawr College;* Royal Albridge, *Vanderbilt University;* Subash Antani, *Edgewood College;* Michael Bass, *University of Central Florida;* Harry Bingham, *University of California, Berkeley;* Billy E. Bonner, *Rice University;* Anthony Buffa, *California Polytechnic State University, San Luis Obispo;* Richard Cardenas, *St. Mary's University;* James Carolan, *University of British Columbia;* Kapila Clara Castoldi, *Oakland University;* Ralph V. Chamberlin, *Arizona State University;* Christopher R. Church, *Miami University (Ohio);* Gary G. DeLeo, *Lehigh University;* Michael Dennin, *University of California, Irvine;* Alan J. DeWeerd, *Creighton University;* Madi Dogariu, *University of Central Florida;* Gordon Emslie, *University of Alabama at Huntsville;* Donald Erbsloe, *United States Air Force Academy;* William Fairbank, *Colorado State University;* Marco Fatuzzo, *University of Arizona;* Philip Fraundorf, *University of Missouri–St. Louis;* Patrick Gleeson, *Delaware State University;* Christopher M. Gould, *University of Southern California;* James D. Gruber, *Harrisburg Area Community College;* John B. Gruber, *San Jose State University;* Todd Hann, *United States Military Academy;* Gail Hanson, *Indiana University;* Gerald Hart, *Moorhead State University;* Dieter H. Hartmann, *Clemson University;* Richard W. Henry, *Bucknell University;* Athula Herat, *Northern Kentucky University;* Laurent Hodges, *Iowa State University;* Michael J. Hones, *Villanova University;* Huan Z. Huang, *University of California at Los Angeles;* Joey Huston, *Michigan State University;* George Igo, *University of California at Los Angeles;* Herb Jaeger, *Miami University;* David Judd, *Broward Community College;* Thomas H. Keil, *Worcester Polytechnic Institute;* V. Gordon Lind, *Utah State University;* Edwin Lo; Michael J. Longo, *University of Michigan;*

Rafael Lopez-Mobilia, *University of Texas at San Antonio;* Roger M. Mabe, *United States Naval Academy;* David Markowitz, *University of Connecticut;* Thomas P. Marvin, *Southern Oregon University;* Bruce Mason, *University of Oklahoma at Norman;* Martin S. Mason, *College of the Desert;* Wesley N. Mathews, Jr., *Georgetown University;* Ian S. McLean, *University of California at Los Angeles;* John W. McClory, *United States Military Academy;* L. C. McIntyre, Jr., *University of Arizona;* Alan S. Meltzer, *Rensselaer Polytechnic Institute;* Ken Mendelson, *Marquette University;* Roy Middleton, *University of Pennsylvania;* Allen Miller, *Syracuse University;* Clement J. Moses, *Utica College of Syracuse University;* John W. Norbury, *University of Wisconsin–Milwaukee;* Anthony Novaco, *Lafayette College;* Romulo Ochoa, *The College of New Jersey;* Melvyn Oremland, *Pace University;* Desmond Penny, *Southern Utah University;* Steven J. Pollock, *University of Colorado–Boulder;* Prabha Ramakrishnan, *North Carolina State University;* Rex D. Ramsier, *The University of Akron;* Ralf Rapp, *Texas A&M University;* Rogers Redding, *University of North Texas;* Charles R. Rhyner, *University of Wisconsin–Green Bay;* Perry Rice, *Miami University;* Dennis Rioux, *University of Wisconsin–Oshkosh;* Richard Rolleigh, *Hendrix College;* Janet E. Seger, *Creighton University;* Gregory D. Severn, *University of San Diego;* Satinder S. Sidhu, *Washington College;* Antony Simpson, *Dalhousie University;* Harold Slusher, *University of Texas at El Paso;* J. Clinton Sprott, *University of Wisconsin at Madison;* Shirvel Stanislaus, *Valparaiso University;* Randall Tagg, *University of Colorado at Denver;* Cecil Thompson, *University of Texas at Arlington;* Harry W. K. Tom, *University of California at Riverside;* Chris Vuille, *Embry–Riddle Aeronautical University;* Fiona Waterhouse, *University of California at Berkeley;* Robert Watkins, *University of Virginia;* James Whitmore, *Pennsylvania State University*

Principles of Physics, fifth edition, was carefully checked for accuracy by Grant Hart (Brigham Young University), James E. Rutledge (University of California at Irvine), and Som Tyagi (Drexel University).

We are indebted to the developers of the IUPP models "A Particles Approach to Introductory Physics" and "Physics in Context," upon which much of the pedagogical approach in this textbook is based.

Vahe Peroomian wrote the initial draft of the new context on Heart Attacks, and we are very grateful for his efforts. He provided further assistance by reviewing early drafts of questions and problems sets.

We are grateful to John R. Gordon and Vahe Peroomian for writing the *Student Solutions Manual and Study Guide,* and to Vahe Peroomian for preparing an excellent *Instructor's Solutions Manual.* During the development of this text, the authors benefited from many useful discussions with colleagues and other physics instructors, including Robert Bauman, William Beston, Don Chodrow, Jerry Faughn, John R. Gordon, Kevin Giovanetti, Dick Jacobs, Harvey Leff, John Mallinckrodt, Clem Moses, Dorn Peterson, Joseph Rudmin, and Gerald Taylor.

Special thanks and recognition go to the professional staff at the Brooks/Cole Publishing Company—in particular, Charles Hartford, Ed Dodd, Brandi Kirksey, Rebecca Berardy-Schwartz, Jack Cooney, Cathy Brooks, Cate Barr, and Brendan Killion—for their fine work during the development and production of this textbook. We recognize the skilled production service provided by Jill Traut and the staff at Macmillan Solutions and the dedicated photo research efforts of Josh Garvin of the Bill Smith Group.

Finally, we are deeply indebted to our wives and children for their love, support, and long-term sacrifices.

RAYMOND A. SERWAY
St. Petersburg, Florida

JOHN W. JEWETT, JR.
Anaheim, California

To the Student

It is appropriate to offer some words of advice that should be of benefit to you, the student. Before doing so, we assume you have read the Preface, which describes the various features of the text and support materials that will help you through the course.

How to Study

Instructors are often asked, "How should I study physics and prepare for examinations?" There is no simple answer to this question, but we can offer some suggestions based on our own experiences in learning and teaching over the years.

First and foremost, maintain a positive attitude toward the subject matter, keeping in mind that physics is the most fundamental of all natural sciences. Other science courses that follow will use the same physical principles, so it is important that you understand and are able to apply the various concepts and theories discussed in the text.

Concepts and Principles

It is essential that you understand the basic concepts and principles before attempting to solve assigned problems. You can best accomplish this goal by carefully reading the textbook before you attend your lecture on the covered material. When reading the text, you should jot down those points that are not clear to you. Also be sure to make a diligent attempt at answering the questions in the Quick Quizzes as you come to them in your reading. We have worked hard to prepare questions that help you judge for yourself how well you understand the material. Study the **What If?** features that appear in many of the worked examples carefully. They will help you extend your understanding beyond the simple act of arriving at a numerical result. The Pitfall Preventions will also help guide you away from common misunderstandings about physics. During class, take careful notes and ask questions about those ideas that are unclear to you. Keep in mind that few people are able to absorb the full meaning of scientific material after only one reading; several readings of the text and your notes may be necessary. Your lectures and laboratory work supplement the textbook and should clarify some of the more difficult material. You should minimize your memorization of material. Successful memorization of passages from the text, equations, and derivations does not necessarily indicate that you understand the material. Your understanding of the material will be enhanced through a combination of efficient study habits, discussions with other students and with instructors, and your ability to solve the problems presented in Enhanced WebAssign. Ask questions whenever you believe that clarification of a concept is necessary.

Study Schedule

It is important that you set up a regular study schedule, preferably a daily one. Make sure that you read the syllabus for the course and adhere to the schedule set by your instructor. The lectures will make much more sense if you read the corresponding text material *before* attending them. As a general rule, you should devote about two hours of study time for each hour you are in class. If you are having trouble with the course, seek the advice of the instructor or other students who have taken the course. You may find it necessary to seek further instruction from experienced students. Very often, instructors offer review sessions in addition to regular class periods. Avoid the practice of delaying study until a day or two before an exam. More often than not, this approach has disastrous results. Rather than undertake an all-night study session before a test, briefly review the basic concepts and equations, and then get a good night's rest. If you believe that you need additional help in understanding the concepts, in preparing for exams, or in problem solving, we suggest that you acquire a copy of the *Student Solutions Manual/Study Guide* that accompanies this textbook.

Visit the *Principles of Physics* Web site at **www.cengagebrain.com/shop/ISBN/9781133104261** to see samples of select student supplements. You can purchase any Cengage Learning product at your local college store or at our preferred online store **CengageBrain.com.**

Use the Features

You should make full use of the various features of the text discussed in the Preface. For example, marginal notes are useful for locating and describing important equations and concepts, and **boldface** indicates important definitions. Many useful tables are contained in the appendices, but most are incorporated in the text where they are most often referenced. Appendix B is a convenient review of mathematical tools used in the text.

Answers to Quick Quizzes and Context Conclusion problems are given at the end of the textbook, and solutions to selected questions and problems are provided in the *Student Solutions Manual/Study Guide*. The table of contents provides an overview of the entire text, and the index enables you to locate specific material quickly. Footnotes are sometimes used to supplement the text or to cite other references on the subject discussed.

After reading a chapter, you should be able to define any new quantities introduced in that chapter and discuss the principles and assumptions that were used to arrive at certain key relations. The chapter summaries and the review sections of the *Student Solutions Manual/Study Guide* should help you in this regard. In some cases, you may find it necessary to refer to the textbook's index to locate certain topics. You should be able to associate with each physical quantity the correct symbol used to represent that quantity and the unit in which the quantity is specified. Furthermore, you should be able to express each important equation in concise and accurate prose.

Problem Solving

R. P. Feynman, Nobel laureate in physics, once said, "You do not know anything until you have practiced." In keeping with this statement, we strongly advise you to develop the skills necessary to solve a wide range of problems. Your ability to solve problems will be one of the main tests of your knowledge of physics; therefore, you should try to solve as many problems as possible. It is essential that you understand basic concepts and principles before attempting to solve problems. It is good practice to try to find alternate solutions to the same problem. For example, you can solve problems in mechanics using Newton's laws, but very often an alternative method that draws on energy considerations is more direct. You should not deceive yourself into thinking that you understand a problem merely because you have seen it solved in class. You must be able to solve the problem and similar problems on your own.

The approach to solving problems should be carefully planned. A systematic plan is especially important when a problem involves several concepts. First, read the problem several times until you are confident you understand what is being asked. Look for any key words that will help you interpret the problem and perhaps allow you to make certain assumptions. Your ability to interpret a question properly is an integral part of problem solving. Second, you should acquire the habit of writing down the information given in a problem and those quantities that need to be found; for example, you might construct a table listing both the quantities given and the quantities to be found. This procedure is sometimes used in the worked examples of the textbook. Finally, after you have decided on the method you believe is appropriate for a given problem, proceed with your solution. The General Problem-Solving Strategy will guide you through complex problems. If you follow the steps of this procedure (*Conceptualize, Categorize, Analyze, Finalize*), you will find it easier to come up with a solution and gain more from your efforts. This strategy, located at the end of Chapter 1 (pages 25–26), is used in all worked examples in the remaining chapters so that you can learn how to apply it. Specific problem-solving strategies for certain types of situations are included in the text and appear with a special heading. These specific strategies follow the outline of the General Problem-Solving Strategy.

Often, students fail to recognize the limitations of certain equations or physical laws in a particular situation. It is very important that you understand and remember the assumptions that underlie a particular theory or formalism. For example, certain equations in kinematics apply only to a particle moving with constant acceleration. These equations are not valid for describing motion whose acceleration is not constant, such as the motion of an object connected to a spring or the motion of an object through a fluid. Study the Analysis Models for Problem Solving in the chapter summaries carefully so that you know how each model can be applied to a specific situation. The analysis models provide you with a logical structure for solving problems and help you develop your thinking skills to become more like those of a physicist. Use the analysis model approach to save you hours of looking for the correct equation and to make you a faster and more efficient problem solver.

Experiments

Physics is a science based on experimental observations. Therefore, we recommend that you try to supplement the text by performing various types of "hands-on" experiments either at home or in the laboratory. These experiments can be used to test ideas and models discussed in class or in the textbook. For example, the common Slinky toy is excellent for studying traveling waves, a ball swinging on the end of a long string can be used to investigate pendulum motion, various masses attached to the end of a vertical spring or rubber band can be used to determine its elastic nature, an old pair of polarized sunglasses and some discarded lenses and a magnifying glass are the components of various experiments in optics, and an approximate measure of the free-fall acceleration can be determined simply by measuring with a stopwatch the time interval required for a ball to drop from a known height. The list of such experiments is endless. When physical models are not available, be imaginative and try to develop models of your own.

New Media

If available, we strongly encourage you to use the **Enhanced WebAssign** product that is available with this textbook. It is far easier to understand physics if you see it in action, and the materials available in Enhanced WebAssign will enable you to become a part of that action.

It is our sincere hope that you will find physics an exciting and enjoyable experience and that you will benefit from this experience, regardless of your chosen profession. Welcome to the exciting world of physics!

The scientist does not study nature because it is useful; he studies it because he delights in it, and he delights in it because it is beautiful. If nature were not beautiful, it would not be worth knowing, and if nature were not worth knowing, life would not be worth living.

—**HENRI POINCARÉ**

Life Science Applications and Problems

An Invitation to Physics

Stonehenge, in southern England, was built thousands of years ago. Various theories have been proposed about its function, including a burial ground, a healing site, and a place for ancestor worship. One of the more intriguing theories suggests that Stonehenge was an observatory, allowing for predictions of celestial events such as eclipses, solstices, and equinoxes.

Physics, the most fundamental physical science, is concerned with the basic principles of the universe. It is the foundation on which engineering, technology, and the other sciences—astronomy, biology, chemistry, and geology—are based. The beauty of physics lies in the simplicity of its fundamental theories and in the manner in which just a small number of basic concepts, equations, and assumptions can alter and expand our view of the world around us.

Classical physics, developed prior to 1900, includes the theories, concepts, laws, and experiments in classical mechanics, thermodynamics, electromagnetism, and optics. For example, Galileo Galilei (1564–1642) made significant contributions to classical mechanics through his work on the laws of motion with constant acceleration. In the same era, Johannes Kepler (1571–1630) used astronomical observations to develop empirical laws for the motions of planetary bodies.

The most important contributions to classical mechanics, however, were provided by Isaac Newton (1642–1727), who developed classical mechanics as a systematic theory and was one of the originators of calculus as a mathematical tool. Although major developments in classical physics continued in the 18th century, thermodynamics and electromagnetism were not developed until the latter part of the 19th century, principally because the apparatus for controlled experiments was either too crude or unavailable until then. Although many electric and magnetic phenomena had been studied earlier, the work of James Clerk Maxwell (1831–1879) provided a unified theory of electromagnetism. In this text, we shall treat the various disciplines of classical physics in separate sections; we will see, however, that the disciplines of mechanics and electromagnetism are basic to all the branches of physics.

A major revolution in physics, usually referred to as *modern physics,* began near the end of the 19th century. Modern physics developed mainly because many physical phenomena could not be explained by classical physics. The two most important developments in this modern era were the theories of relativity and quantum mechanics. Albert Einstein's theory of relativity

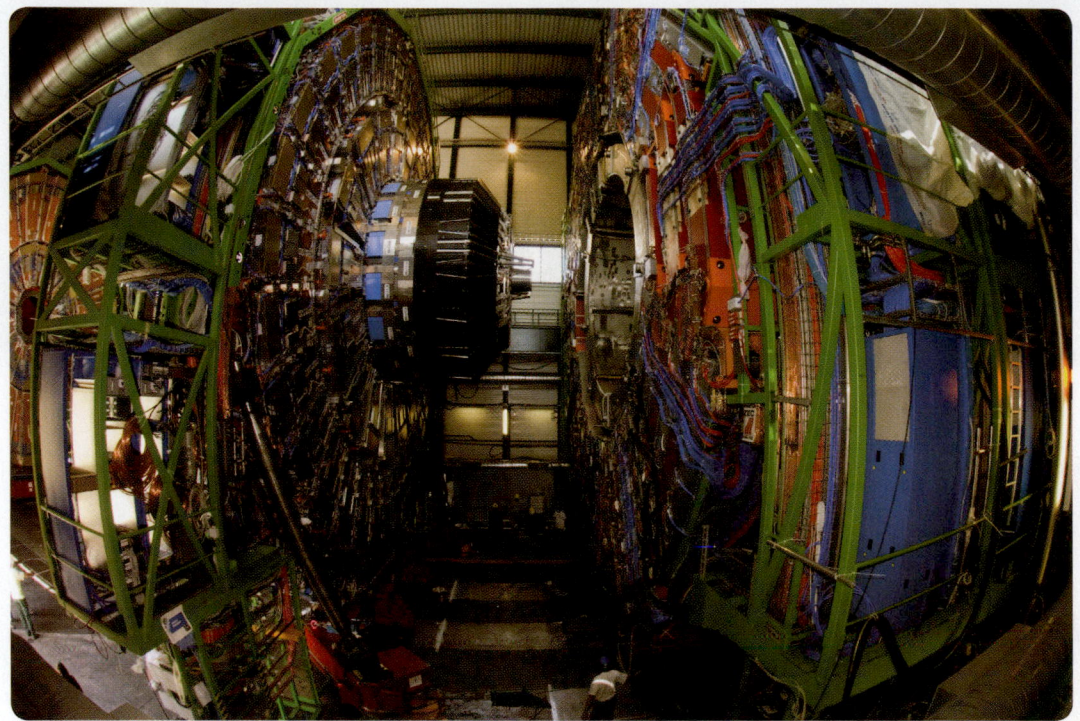

© 2011 CERN

The Compact Muon Solenoid (CMS) detector, part of the Large Hadron Collider operated by CERN (Conseil Européen pour la Recherche Nucléaire). The system is designed to detect and measure particles created in collisions of high-energy protons. Despite the word *compact* in the name, the detector is 15 meters in diameter. For a sense of scale, notice the worker in the blue helmet at the bottom of the photo as well as other workers in yellow helmets on the far side of the detector.

completely revolutionized the traditional concepts of space, time, and energy. This theory correctly describes the motion of objects moving at speeds comparable to the speed of light. The theory of relativity also shows that the speed of light is the upper limit of the speed of an object and that mass and energy are related. Quantum mechanics was formulated by a number of distinguished scientists to provide descriptions of physical phenomena at the atomic level.

Scientists continually work at improving our understanding of fundamental laws, and new discoveries are made every day. In many research areas, a great deal of overlap exists among physics, chemistry, and biology. Evidence for this overlap is seen in the names of some subspecialties in science: biophysics, biochemistry, chemical physics, biotechnology, and so on. Numerous technological advances in recent times are the result of the efforts of many scientists, engineers, and technicians. Some of the most notable developments in the latter half of the 20th century were (1) space missions to the Moon and other planets, (2) microcircuitry and high-speed computers, (3) sophisticated imaging techniques used in scientific research and medicine, and (4) several remarkable accomplishments in genetic engineering. The early years of the 21st century have seen additional developments. Materials such as carbon nanotubes are now experiencing a variety of new applications. The 2010 Nobel Prize in Physics was

awarded for experiments performed on graphene, a two-dimensional material formed from carbon atoms. Potential applications include incorporation into a variety of electrical components and biodevices such as those used in DNA sequencing. The impact of such developments and discoveries on society has indeed been great, and future discoveries and developments will very likely be exciting, challenging, and of great benefit to humanity.

To investigate the impact of physics on developments in our society, we will use a *contextual* approach to the study of the content in this textbook. The book is divided into nine *Contexts,* which relate the physics to social issues, natural phenomena, or medical/technological applications, as outlined here:

Chapters	Context
2–7	Alternative-Fuel Vehicles
8–11	Mission to Mars
12–14	Earthquakes
15	Heart Attacks
16–18	Global Warming
19–21	Lightning
22–23	Magnetism in Medicine
24–27	Lasers
28–31	The Cosmic Connection

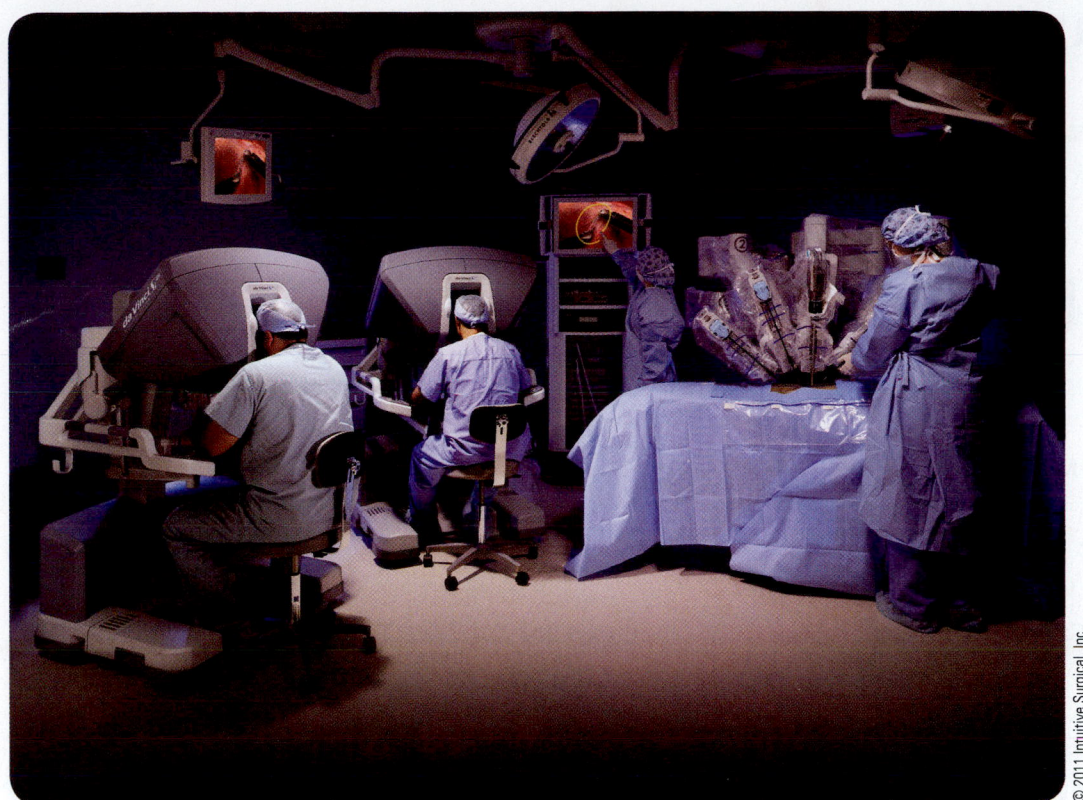

© 2011 Intuitive Surgical, Inc.

Physics is being used extensively today in the biomedical field. Shown here is the da Vinci Surgical System, a robotic device used to perform procedures such as prostatectomies, hysterectomies, mitral valve repairs, and coronary artery anastomosis. The surgeon sits at the console on the left and views a stereoscopic image of the surgery site. The movements of his hands are translated by a computer into movements of the robotic arms seen above the operating table at the right.

The Contexts provide a story line for each section of the text, which will help provide relevance and motivation for studying the material.

Each Context begins with a discussion of the topic, culminating in a *central question*, which forms the focus for the study of the physics in the Context. The final section of each chapter is a Context Connection, in which the material in the chapter is explored with the central question in mind. At the end of each Context, a Context Conclusion brings together all the principles necessary to respond as fully as possible to the central question.

In Chapter 1, we investigate some of the mathematical fundamentals and problem-solving strategies that we will use in our study of physics. The first Context, *Alternative-Fuel Vehicles*, is introduced just before Chapter 2; in this Context, the principles of classical mechanics are applied to the problem of designing, developing, producing, and marketing a vehicle that will help to reduce dependence on foreign oil and emit fewer harmful by-products into the atmosphere than current gasoline engines.

Introduction and Vectors

Chapter Outline

Raymond A. Serway

A signpost in Saint Petersburg, Florida, shows the distance and direction to several cities. Quantities that are defined by both a magnitude and a direction are called *vector quantities*.

The goal of physics is to provide a quantitative understanding of certain basic phenomena that occur in our Universe. Physics is a science based on experimental observations and mathematical analyses. The main objectives behind such experiments and analyses are to develop theories that explain the phenomenon being studied and to relate those theories to other established theories. Fortunately, it is possible to explain the behavior of various physical systems using relatively few fundamental laws. Analytical procedures require the expression of those laws in the language of mathematics, the tool that provides a bridge between theory and experiment. In this chapter, we shall discuss a few mathematical concepts and techniques that will be used throughout the text. In addition, we will outline an effective problem-solving strategy that should be adopted and used in your problem-solving activities throughout the text.

ENHANCED
WebAssign Interactive content from this and other chapters may be assigned online in Enhanced WebAssign.

1.1 | Standards of Length, Mass, and Time

To describe natural phenomena, we must make measurements associated with physical quantities, such as the length of an object. The laws of physics can be expressed as mathematical relationships among physical quantities that will be introduced and discussed throughout the book. In mechanics, the three fundamental quantities are length, mass, and time. All other quantities in mechanics can be expressed in terms of these three.

If we measure a certain quantity and wish to describe it to someone, a unit for the quantity must be specified and defined. For example, it would be meaningless for a visitor from another planet to talk to us about a length of 8.0 "glitches" if we did not know the meaning of the unit glitch. On the other hand, if someone familiar with our system of measurement reports that a wall is 2.0 meters high and our unit of length is defined to be 1.0 meter, we then know that the height of the wall is twice our fundamental unit of length. An international committee has agreed on a system of definitions and standards to describe fundamental physical quantities. It is called the **SI system** (Système International) of units. Its units of length, mass, and time are the meter, kilogram, and second, respectively.

Length

In A.D. 1120, King Henry I of England decreed that the standard of length in his country would be the yard and that the yard would be precisely equal to the distance from the tip of his nose to the end of his outstretched arm. Similarly, the original standard for the foot adopted by the French was the length of the royal foot of King Louis XIV. This standard prevailed until 1799, when the legal standard of length in France became the **meter,** defined as one ten-millionth of the distance from the equator to the North Pole.

Many other systems have been developed in addition to those just discussed, but the advantages of the French system have caused it to prevail in most countries and in scientific circles everywhere. Until 1960, the length of the meter was defined as the distance between two lines on a specific bar of platinum–iridium alloy stored under controlled conditions. This standard was abandoned for several reasons, a principal one being that the limited accuracy with which the separation between the lines can be determined does not meet the current requirements of science and technology. The definition of the meter was modified to be equal to 1 650 763.73 wavelengths of orange–red light emitted from a krypton-86 lamp. In October 1983, the meter was redefined to be **the distance traveled by light in a vacuum during a time interval of 1/299 792 458 second.** This value arises from the establishment of the speed of light in a vacuum as exactly 299 792 458 meters per second. We will use the standard scientific notation for numbers with more than three digits in which groups of three digits are separated by spaces rather than commas. Therefore, 1 650 763.73 and 299 792 458 in this paragraph are the same as the more popular American cultural notations of 1,650,763.73 and 299,792,458. Similarly, $\pi = 3.14159265$ is written as 3.141 592 65.

▶ Definition of the meter

Mass

Mass represents a measure of the resistance of an object to changes in its motion. The SI unit of mass, the **kilogram,** is defined as **the mass of a specific platinum–iridium alloy cylinder kept at the International Bureau of Weights and Measures at Sèvres, France.** At this point, we should add a word of caution. Many beginning students of physics tend to confuse the physical quantities called *weight* and *mass*. For the present we shall not discuss the distinction between them; they will be clearly defined in later chapters. For now you should note that they are distinctly different quantities.

▶ Definition of the kilogram

Time

Before 1967, the standard of time was defined in terms of the average length of a *mean solar day.* (A solar day is the time interval between successive appearances of the Sun at the highest point it reaches in the sky each day.) The basic unit of time, the **second,** was defined to be $(1/60)(1/60)(1/24) = 1/86\ 400$ of a mean solar day. In 1967, the second was redefined to take advantage of the great precision obtainable with a device known as an atomic clock (Fig. 1.1), which uses the characteristic frequency of the

Figure 1.1 A cesium fountain atomic clock. The clock will neither gain nor lose a second in 20 million years.

◄TABLE 1.1 | **Approximate Values of Some Measured Lengths**

	Length (m)
Distance from the Earth to the most remote quasar	1.4×10^{26}
Distance from the Earth to the most remote normal galaxies	9×10^{25}
Distance from the Earth to the nearest large galaxy (M 31, the Andromeda galaxy)	2×10^{22}
Distance from the Sun to the nearest star (Proxima Centauri)	4×10^{16}
One light-year	9.46×10^{15}
Mean orbit radius of the Earth	1.50×10^{11}
Mean distance from the Earth to the Moon	3.84×10^{8}
Distance from the equator to the North Pole	1.00×10^{7}
Mean radius of the Earth	6.37×10^{6}
Typical altitude (above the surface) of a satellite orbiting the Earth	2×10^{5}
Length of a football field	9.1×10^{1}
Length of this textbook	2.8×10^{-1}
Length of a housefly	5×10^{-3}
Size of smallest visible dust particles	$\sim 10^{-4}$
Size of cells of most living organisms	$\sim 10^{-5}$
Diameter of a hydrogen atom	$\sim 10^{-10}$
Diameter of a uranium nucleus	$\sim 10^{-14}$
Diameter of a proton	$\sim 10^{-15}$

◄TABLE 1.2 | **Masses of Various Objects (Approximate Values)**

	Mass (kg)
Visible Universe	$\sim 10^{52}$
Milky Way galaxy	$\sim 10^{42}$
Sun	1.99×10^{30}
Earth	5.98×10^{24}
Moon	7.36×10^{22}
Shark	$\sim 10^{3}$
Human	$\sim 10^{2}$
Frog	$\sim 10^{-1}$
Mosquito	$\sim 10^{-5}$
Bacterium	$\sim 10^{-15}$
Hydrogen atom	1.67×10^{-27}
Electron	9.11×10^{-31}

► Definition of the second

cesium-133 atom as the "reference clock." The second is now defined as **9 192 631 770 times the period of oscillation of radiation from the cesium atom.** It is possible today to purchase clocks and watches that receive radio signals from an atomic clock in Colorado, which the clock or watch uses to continuously reset itself to the correct time.

Approximate Values for Length, Mass, and Time

Approximate values of various lengths, masses, and time intervals are presented in Tables 1.1, 1.2, and 1.3, respectively. Note the wide range of values for these quantities.[1] You should study the tables and begin to generate an intuition for what is meant by a mass of 100 kilograms, for example, or by a time interval of 3.2×10^{7} seconds.

Systems of units commonly used in science, commerce, manufacturing, and everyday life are (1) the *SI system,* in which the units of length, mass, and time are the meter (m), kilogram (kg), and second (s), respectively; and (2) the *U.S. customary system,* in which the units of length, mass, and time are the foot (ft), slug, and second, respectively. Throughout most of this text we shall use SI units because they are almost universally accepted in science and industry. We will make limited use of U.S. customary units in the study of classical mechanics.

Some of the most frequently used prefixes for the powers of ten and their abbreviations are listed in Table 1.4. For example, 10^{-3} m is equivalent to 1 millimeter (mm), and 10^{3} m is 1 kilometer (km). Likewise, 1 kg is 10^{3} grams (g), and 1 megavolt (MV) is 10^{6} volts (V).

The variables length, time, and mass are examples of *fundamental quantities.* A much larger list of variables contains *derived quantities,* or quantities that can be expressed as a mathematical combination of fundamental quantities. Common examples are *area,* which is a product of two lengths, and *speed,* which is a ratio of a length to a time interval.

Another example of a derived quantity is **density.** The density ρ (Greek letter rho; a table of the letters in the Greek alphabet is provided at the back of the book) of any substance is defined as its *mass per unit volume:*

► Definition of density

$$\rho \equiv \frac{m}{V}$$

1.1◄

Pitfall Prevention | 1.1
Reasonable Values
Generating intuition about typical values of quantities when solving problems is important because you must think about your end result and determine if it seems reasonable. For example, if you are calculating the mass of a housefly and arrive at a value of 100 kg, this answer is *unreasonable* and there is an error somewhere.

---

[1] If you are unfamiliar with the use of powers of ten (scientific notation), you should review Appendix B.1.

TABLE 1.3 | Approximate Values of Some Time Intervals

	Time Interval (s)
Age of the Universe	4×10^{17}
Age of the Earth	1.3×10^{17}
Time interval since the fall of the Roman empire	5×10^{12}
Average age of a college student	6.3×10^{8}
One year	3.2×10^{7}
One day (time interval for one revolution of the Earth about its axis)	8.6×10^{4}
One class period	3.0×10^{3}
Time interval between normal heartbeats	8×10^{-1}
Period of audible sound waves	$\sim 10^{-3}$
Period of typical radio waves	$\sim 10^{-6}$
Period of vibration of an atom in a solid	$\sim 10^{-13}$
Period of visible light waves	$\sim 10^{-15}$
Duration of a nuclear collision	$\sim 10^{-22}$
Time interval for light to cross a proton	$\sim 10^{-24}$

TABLE 1.4 | Some Prefixes for Powers of Ten

Power	Prefix	Abbreviation
10^{-24}	yocto	y
10^{-21}	zepto	z
10^{-18}	atto	a
10^{-15}	femto	f
10^{-12}	pico	p
10^{-9}	nano	n
10^{-6}	micro	μ
10^{-3}	milli	m
10^{-2}	centi	c
10^{-1}	deci	d
10^{3}	kilo	k
10^{6}	mega	M
10^{9}	giga	G
10^{12}	tera	T
10^{15}	peta	P
10^{18}	exa	E
10^{21}	zetta	Z
10^{24}	yotta	Y

which is a ratio of mass to a product of three lengths. For example, aluminum has a density of $2.70 \times 10^{3} \text{ kg/m}^{3}$, and lead has a density of $11.3 \times 10^{3} \text{ kg/m}^{3}$. An extreme difference in density can be imagined by thinking about holding a 10-centimeter (cm) cube of Styrofoam in one hand and a 10-cm cube of lead in the other.

1.2 | Dimensional Analysis

In physics, the word *dimension* denotes the physical nature of a quantity. The distance between two points, for example, can be measured in feet, meters, or furlongs, which are all different ways of expressing the dimension of length.

The symbols used in this book to specify the dimensions[2] of length, mass, and time are L, M, and T, respectively. We shall often use square brackets [ ] to denote the dimensions of a physical quantity. For example, in this notation the dimensions of speed v are written $[v] = \text{L/T}$, and the dimensions of area A are $[A] = \text{L}^{2}$. The dimensions of area, volume, speed, and acceleration are listed in Table 1.5, along with their units in the two common systems. The dimensions of other quantities, such as force and energy, will be described as they are introduced in the text.

In many situations, you may be faced with having to derive or check a specific equation. Although you may have forgotten the details of the derivation, a useful and powerful procedure called **dimensional analysis** can be used as a consistency check, to assist in the derivation, or to check your final expression. Dimensional analysis makes use of the fact that dimensions can be treated as algebraic quantities. For example, quantities can be added or subtracted only if they have the same dimensions. Furthermore, the terms on both sides of an equation must have the same dimensions. By following these simple rules, you can use dimensional analysis to help determine

Pitfall Prevention | 1.2
Symbols for Quantities
Some quantities have a small number of symbols that represent them. For example, the symbol for time is almost always t. Other quantities might have various symbols depending on the usage. Length may be described with symbols such as x, y, and z (for position); r (for radius); a, b, and c (for the legs of a right triangle); ℓ (for the length of an object); d (for a distance); h (for a height); and so forth.

TABLE 1.5 | Dimensions and Units of Four Derived Quantities

Quantity	Area (A)	Volume (V)	Speed (v)	Acceleration (a)
Dimensions	L^{2}	L^{3}	L/T	L/T^{2}
SI units	m^{2}	m^{3}	m/s	m/s^{2}
U.S. customary units	ft^{2}	ft^{3}	ft/s	ft/s^{2}

[2] The *dimensions* of a variable will be symbolized by a capitalized, nonitalic letter, such as, in the case of length, L. The *symbol* for the variable itself will be italicized, such as L for the length of an object or t for time.

whether an expression has the correct form because the relationship can be correct only if the dimensions on the two sides of the equation are the same.

To illustrate this procedure, suppose you wish to derive an expression for the position x of a car at a time t if the car starts from rest at $t = 0$ and moves with constant acceleration a. In Chapter 2, we shall find that the correct expression for this special case is $x = \frac{1}{2}at^2$. Let us check the validity of this expression from a dimensional analysis approach.

The quantity x on the left side has the dimension of length. For the equation to be dimensionally correct, the quantity on the right side must also have the dimension of length. We can perform a dimensional check by substituting the basic dimensions for acceleration, L/T^2 (Table 1.5), and time, T, into the equation $x = \frac{1}{2}at^2$. That is, the dimensional form of the equation $x = \frac{1}{2}at^2$ can be written as

$$[x] = \frac{\text{L}}{\text{T}^2}\,\text{T}^2 = \text{L}$$

The dimensions of time cancel as shown, leaving the dimension of length, which is the correct dimension for the position x. Notice that the number $\frac{1}{2}$ in the equation has no units, so it does not enter into the dimensional analysis.

> ◀ **QUICK QUIZ 1.1** **True or False:** Dimensional analysis can give you the numerical value of constants of proportionality that may appear in an algebraic expression.

◀ *Example* **1.1** | **Analysis of an Equation**

Show that the expression $v = at$, where v represents speed, a acceleration, and t an instant of time, is dimensionally correct.

SOLUTION

Identify the dimensions of v from Table 1.5:

$$[v] = \frac{\text{L}}{\text{T}}$$

Identify the dimensions of a from Table 1.5 and multiply by the dimensions of t:

$$[at] = \frac{\text{L}}{\text{T}^2}\,\text{T} = \frac{\text{L}}{\text{T}}$$

Therefore, $v = at$ is dimensionally correct because we have the same dimensions on both sides. (If the expression were given as $v = at^2$, it would be dimensionally *incorrect*. Try it and see!)

◀ **1.3** | **Conversion of Units**

Sometimes it is necessary to convert units from one system to another or to convert within a system, for example, from kilometers to meters. Equalities between SI and U.S. customary units of length are as follows:

1 mile (mi) = 1 609 m = 1.609 km	1 ft = 0.304 8 m = 30.48 cm
1 m = 39.37 in. = 3.281 ft	1 inch (in.) = 0.025 4 m = 2.54 cm

A more complete list of equalities can be found in Appendix A.

Units can be treated as algebraic quantities that can cancel each other. To perform a conversion, a quantity can be multiplied by a **conversion factor,** which is a fraction equal to 1, with numerator and denominator having different units, to provide the desired units in the final result. For example, suppose we wish to convert 15.0 in. to centimeters. Because 1 in. = 2.54 cm, we multiply by a conversion factor that is the appropriate ratio of these equal quantities and find that

$$15.0 \text{ in.} = (15.0 \text{ in.}) \left(\frac{2.54 \text{ cm}}{1 \text{ in.}}\right) = 38.1 \text{ cm}$$

Pitfall Prevention | 1.3
Always Include Units
When performing calculations, make it a habit to include the units with every quantity and carry the units through the entire calculation. Avoid the temptation to drop the units during the calculation steps and then apply the expected unit to the number that results for an answer. By including the units in every step, you can detect errors if the units for the answer are incorrect.

where the ratio in parentheses is equal to 1. Notice that we express 1 as 2.54 cm/1 in. (rather than 1 in./2.54 cm) so that the inch cancels with the unit in the original quantity. The remaining unit is the centimeter, which is our desired result.

QUICK QUIZ 1.2 The distance between two cities is 100 mi. What is the number of kilometers between the two cities? (**a**) smaller than 100 (**b**) larger than 100 (**c**) equal to 100

Example **1.2** | **Is He Speeding?**

On an interstate highway in a rural region of Wyoming, a car is traveling at a speed of 38.0 m/s. Is the driver exceeding the speed limit of 75.0 mi/h?

SOLUTION

Convert meters in the speed to miles:
$$(38.0 \text{ m/s})\left(\frac{1 \text{ mi}}{1\ 609 \text{ m}}\right) = 2.36 \times 10^{-2} \text{ mi/s}$$

Convert seconds to hours:
$$(2.36 \times 10^{-2} \text{ mi/s})\left(\frac{60 \text{ s}}{1 \text{ min}}\right)\left(\frac{60 \text{ min}}{1 \text{ h}}\right) = 85.0 \text{ mi/h}$$

The driver is indeed exceeding the speed limit and should slow down.

What If? What if the driver were from outside the United States and is familiar with speeds measured in kilometers per hour? What is the speed of the car in km/h?

Answer We can convert our final answer to the appropriate units:

$$(85.0 \text{ mi/h})\left(\frac{1.609 \text{ km}}{1 \text{ mi}}\right) = 137 \text{ km/h}$$

Figure 1.2 shows an automobile speedometer displaying speeds in both mi/h and km/h. Can you check the conversion we just performed using this photograph?

Figure 1.2 (Example 1.2) The speedometer of a vehicle that shows speeds in both miles per hour and kilometers per hour.

© Cengage Learning/Ed Dodd

1.4 | Order-of-Magnitude Calculations

Suppose someone asks you the number of bits of data on a typical musical compact disc. In response, it is not generally expected that you would provide the exact number but rather an estimate, which may be expressed in scientific notation. The estimate may be made even more approximate by expressing it as an order of *magnitude,* which is a power of ten determined as follows:

1. Express the number in scientific notation, with the multiplier of the power of ten between 1 and 10 and a unit.
2. If the multiplier is less than 3.162 (the square root of ten), the order of magnitude of the number is the power of ten in the scientific notation. If the multiplier is greater than 3.162, the order of magnitude is one larger than the power of ten in the scientific notation.

We use the symbol ~ for "is on the order of." Use the procedure above to verify the orders of magnitude for the following lengths:

$$0.008\ 6 \text{ m} \sim 10^{-2}\text{ m} \qquad 0.002\ 1 \text{ m} \sim 10^{-3}\text{ m} \qquad 720 \text{ m} \sim 10^{3}\text{ m}$$

Usually, when an order-of-magnitude estimate is made, the results are reliable to within about a factor of ten. If a quantity increases in value by three orders of magnitude, its value increases by a factor of about $10^{3} = 1\ 000$.

Example 1.3 | The Number of Atoms in a Solid

Estimate the number of atoms in 1 cm^3 of a solid.

SOLUTION

From Table 1.1 we note that the diameter d of an atom is about 10^{-10} m. Let us assume that the atoms in the solid are spheres of this diameter. Then the volume of each sphere is about 10^{-30} m^3 (more precisely, volume $= 4\pi r^3/3 = \pi d^3/6$, where $r = d/2$). Therefore, because 1 cm$^3 = 10^{-6}$ m^3, the number of atoms in the solid is on the order of $10^{-6}/10^{-30} = 10^{24}$ atoms.

A more precise calculation would require additional knowledge that we could find in tables. Our estimate, however, agrees with the more precise calculation to within a factor of 10.

Example 1.4 | Breaths in a Lifetime

Estimate the number of breaths taken during an average human lifetime.

SOLUTION

We start by guessing that the typical human lifetime is about 70 years. Think about the average number of breaths that a person takes in 1 min. This number varies depending on whether the person is exercising, sleeping, angry, serene, and so forth. To the nearest order of magnitude, we shall choose 10 breaths per minute as our estimate. (This estimate is certainly closer to the true average value than an estimate of 1 breath per minute or 100 breaths per minute.)

Find the approximate number of minutes in a year:
$$1 \text{ yr} \left(\frac{400 \text{ days}}{1 \text{ yr}} \right) \left(\frac{25 \text{ h}}{1 \text{ day}} \right) \left(\frac{60 \text{ min}}{1 \text{ h}} \right) = 6 \times 10^5 \text{ min}$$

Find the approximate number of minutes in a 70-year lifetime:
$$\text{number of minutes} = (70 \text{ yr})(6 \times 10^5 \text{ min/yr})$$
$$= 4 \times 10^7 \text{ min}$$

Find the approximate number of breaths in a lifetime:
$$\text{number of breaths} = (10 \text{ breaths/min})(4 \times 10^7 \text{ min})$$
$$= \boxed{4 \times 10^8 \text{ breaths}}$$

Therefore, a person takes on the order of 10^9 breaths in a lifetime. Notice how much simpler it is in the first calculation above to multiply 400×25 than it is to work with the more accurate 365×24.

What If? What if the average lifetime were estimated as 80 years instead of 70? Would that change our final estimate?

Answer We could claim that $(80 \text{ yr})(6 \times 10^5 \text{ min/yr}) = 5 \times 10^7$ min, so our final estimate should be 5×10^8 breaths. This answer is still on the order of 10^9 breaths, so an order-of-magnitude estimate would be unchanged.

1.5 | Significant Figures

When certain quantities are measured, the measured values are known only to within the limits of the experimental uncertainty. The value of this uncertainty can depend on various factors, such as the quality of the apparatus, the skill of the experimenter, and the number of measurements performed. The number of **significant figures** in a measurement can be used to express something about the uncertainty. The number of significant figures is related to the number of numerical digits used to express the measurement, as we discuss below.

As an example of significant figures, suppose we are asked to measure the radius of a compact disc using a meterstick as a measuring instrument. Let us assume the accuracy to which we can measure the radius of the disc is ± 0.1 cm. Because of the uncertainty of ± 0.1 cm, if the radius is measured to be 6.0 cm, we can claim only that its radius lies somewhere between 5.9 cm and 6.1 cm. In this case, we say that the measured value of 6.0 cm has two significant figures. Note that *the significant figures include the first estimated digit.* Therefore, we could write the radius as (6.0 ± 0.1) cm.

Zeros may or may not be significant figures. Those used to position the decimal point in such numbers as 0.03 and 0.007 5 are not significant. Therefore, there are one and two significant figures, respectively, in these two values. When the zeros come after other digits, however, there is the possibility of misinterpretation. For example, suppose the mass of an object is given as 1 500 g. This value is ambiguous because we do not know whether the last two zeros are being used to locate the decimal point or whether they represent significant figures in the measurement. To remove this ambiguity, it is common to use scientific notation to indicate the number of significant figures. In this case, we would express the mass as 1.5×10^3 g if there are two significant figures in the measured value, 1.50×10^3 g if there are three significant figures, and 1.500×10^3 g if there are four. The same rule holds for numbers less than 1, so 2.3×10^{-4} has two significant figures (and therefore could be written 0.000 23) and 2.30×10^{-4} has three significant figures (also written as 0.000 230).

In problem solving, we often combine quantities mathematically through multiplication, division, addition, subtraction, and so forth. When doing so, you must make sure that the result has the appropriate number of significant figures. A good rule of thumb to use in determining the number of significant figures that can be claimed in a multiplication or a division is as follows:

When multiplying several quantities, the number of significant figures in the final answer is the same as the number of significant figures in the quantity having the smallest number of significant figures. The same rule applies to division.

Let's apply this rule to find the area of the compact disc whose radius we measured above. Using the equation for the area of a circle,

$$A = \pi r^2 = \pi (6.0 \text{ cm})^2 = 1.1 \times 10^2 \text{ cm}^2$$

If you perform this calculation on your calculator, you will likely see 113.097 335 5. It should be clear that you don't want to keep all of these digits, but you might be tempted to report the result as 113 cm². This result is not justified because it has three significant figures, whereas the radius only has two. Therefore, we must report the result with only two significant figures as shown above.

For addition and subtraction, you must consider the number of decimal places when you are determining how many significant figures to report:

When numbers are added or subtracted, the number of decimal places in the result should equal the smallest number of decimal places of any term in the sum or difference.

As an example of this rule, consider the sum

$$23.2 + 5.174 = 28.4$$

Notice that we do not report the answer as 28.374 because the lowest number of decimal places is one, for 23.2. Therefore, our answer must have only one decimal place.

The rules for addition and subtraction can often result in answers that have a different number of significant figures than the quantities with which you start. For example, consider these operations that satisfy the rule:

$$1.000 1 + 0.000 3 = 1.000 4$$

$$1.002 - 0.998 = 0.004$$

In the first example, the result has five significant figures even though one of the terms, 0.000 3, has only one significant figure. Similarly, in the second calculation, the result has only one significant figure even though the numbers being subtracted have four and three, respectively.

Pitfall Prevention | 1.4
Read Carefully
Notice that the rule for addition and subtraction is different from that for multiplication and division. For addition and subtraction, the important consideration is the number of *decimal places,* not the number of *significant figures.*

▶ Significant figure guidelines used in this book

In this book, most of the numerical examples and end-of-chapter problems will yield answers having three significant figures. When carrying out estimation calculations, we shall typically work with a single significant figure.

If the number of significant figures in the result of a calculation must be reduced, there is a general rule for rounding numbers: the last digit retained is increased by 1 if the last digit dropped is greater than 5. (For example, 1.346 becomes 1.35.) If the last digit dropped is less than 5, the last digit retained remains as it is. (For example, 1.343 becomes 1.34.) If the last digit dropped is equal to 5, the remaining digit should be rounded to the nearest even number. (This rule helps avoid accumulation of errors in long arithmetic processes.)

A technique for avoiding error accumulation is to delay the rounding of numbers in a long calculation until you have the final result. Wait until you are ready to copy the final answer from your calculator before rounding to the correct number of significant figures. In this book, we display numerical values rounded off to two or three significant figures. This occasionally makes some mathematical manipulations look odd or incorrect. For instance, looking ahead to Example 1.8 on page 21, you will see the operation -17.7 km $+ 34.6$ km $= 17.0$ km. This looks like an incorrect subtraction, but that is only because we have rounded the numbers 17.7 km and 34.6 km for display. If all digits in these two intermediate numbers are retained and the rounding is only performed on the final number, the correct three-digit result of 17.0 km is obtained.

Example 1.5 | Installing a Carpet

A carpet is to be installed in a rectangular room whose length is measured to be 12.71 m and whose width is measured to be 3.46 m. Find the area of the room.

SOLUTION

If you multiply 12.71 m by 3.46 m on your calculator, you will see an answer of 43.976 6 m². How many of these numbers should you claim? Our rule of thumb for multiplication tells us that you can claim only the number of significant figures in your answer as are present in the measured quantity having the lowest number of significant figures. In this example, the lowest number of significant figures is three in 3.46 m, so we should express our final answer as 44.0 m².

◀1.6 | Coordinate Systems

Many aspects of physics deal in some way or another with locations in space. For example, the mathematical description of the motion of an object requires a method for specifying the object's position. Therefore, we first discuss how to describe the position of a point in space by means of coordinates in a graphical representation. A point on a line can be located with one coordinate, a point in a plane is located with two coordinates, and three coordinates are required to locate a point in space.

A coordinate system used to specify locations in space consists of

- A fixed reference point O, called the origin
- A set of specified axes or directions with an appropriate scale and labels on the axes
- Instructions that tell us how to label a point in space relative to the origin and axes

One convenient coordinate system that we will use frequently is the *Cartesian coordinate system*, sometimes called the *rectangular coordinate system*. Such a system in two dimensions is illustrated in Figure 1.3. An arbitrary point in this system is labeled with the coordinates (x, y). Positive x is taken to the right of the origin, and positive y is upward from the origin. Negative x is to the left of the origin, and negative y is downward from the origin. For example, the point P, which has coordinates (5, 3), may be reached by going first 5 m to the right of the origin and then 3 m above the origin

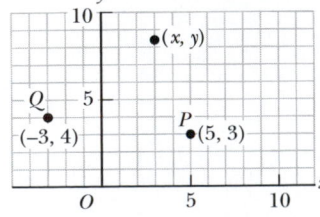

Figure 1.3 Designation of points in a Cartesian coordinate system. Each square in the xy plane is 1 m on a side. Every point is labeled with coordinates (x, y).

(*or* by going 3 m above the origin and then 5 m to the right). Similarly, the point Q has coordinates $(-3, 4)$, which correspond to going 3 m to the left of the origin and 4 m above the origin.

Sometimes it is more convenient to represent a point in a plane by its *plane polar coordinates* (r, θ), as in Active Figure 1.4a. In this coordinate system, r is the length of the line from the origin to the point, and θ is the angle between that line and a fixed axis, usually the positive x axis, with θ measured counterclockwise. From the right triangle in Active Figure 1.4b, we find that $\sin \theta = y/r$ and $\cos \theta = x/r$. (A review of trigonometric functions is given in Appendix B.4.) Therefore, starting with plane polar coordinates, one can obtain the Cartesian coordinates through the equations

$$x = r \cos \theta \qquad \text{1.2} \blacktriangleleft$$

$$y = r \sin \theta \qquad \text{1.3} \blacktriangleleft$$

Furthermore, if we know the Cartesian coordinates, the definitions of trigonometry tell us that

$$\tan \theta = \frac{y}{x} \qquad \text{1.4} \blacktriangleleft$$

and

$$r = \sqrt{x^2 + y^2} \qquad \text{1.5} \blacktriangleleft$$

You should note that these expressions relating the coordinates (x, y) to the coordinates (r, θ) apply only when θ is defined as in Active Figure 1.4a, where positive θ is an angle measured *counterclockwise* from the positive x axis. Other choices are made in navigation and astronomy. If the reference axis for the polar angle θ is chosen to be other than the positive x axis or if the sense of increasing θ is chosen differently, the corresponding expressions relating the two sets of coordinates will change.

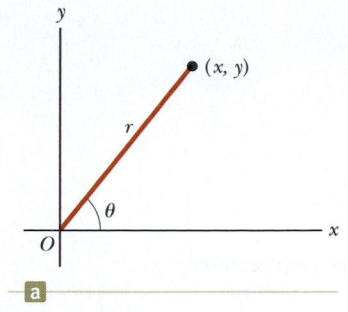

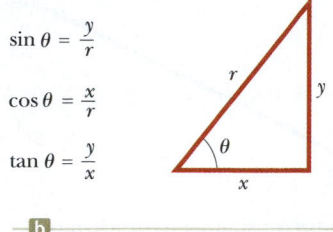

Active Figure 1.4 (a) The plane polar coordinates of a point are represented by the distance r and the angle θ, where θ is measured in a counterclockwise direction from the positive x axis. (b) The right triangle used to relate (x, y) to (r, θ).

1.7 | Vectors and Scalars

Each of the physical quantities that we shall encounter in this text can be placed in one of two categories, either a scalar or a vector. A **scalar** is a quantity that is completely specified by a positive or negative number with appropriate units. On the other hand, a **vector** is a physical quantity that must be specified by both magnitude and direction.

The number of grapes in a bunch (Fig. 1.5a) is an example of a scalar quantity. If you are told that there are 38 grapes in the bunch, this statement completely specifies

Figure 1.5 (a) The number of grapes in this bunch is one example of a scalar quantity. Can you think of other examples? (b) This helpful person pointing in the correct direction tells us to travel five blocks north to reach the courthouse. A vector is a physical quantity that is specified by both magnitude and direction.

the information; no specification of direction is required. Other examples of scalars are temperature, volume, mass, and time intervals. The rules of ordinary arithmetic are used to manipulate scalar quantities; they can be freely added and subtracted (assuming that they have the same units!), multiplied and divided.

Force is an example of a vector quantity. To describe the force on an object completely, we must specify both the direction of the applied force and the magnitude of the force.

▶ Displacement

Another simple example of a vector quantity is the **displacement** of a particle, defined as its *change in position*. The person in Figure 1.5b is pointing out the direction of your desired displacement vector if you would like to reach a destination such as the courthouse. She will also tell you the magnitude of the displacement along with the direction, for example, "5 blocks north."

Suppose a particle moves from some point Ⓐ to a point Ⓑ along a straight path, as in Figure 1.6. This displacement can be represented by drawing an arrow from Ⓐ to Ⓑ, where the arrowhead represents the direction of the displacement and the length of the arrow represents the magnitude of the displacement. If the particle travels along some other path from Ⓐ to Ⓑ, such as the broken line in Figure 1.6, its displacement is still the vector from Ⓐ to Ⓑ. The vector displacement along any indirect path from Ⓐ to Ⓑ is defined as being equivalent to the displacement represented by the direct path from Ⓐ to Ⓑ. The magnitude of the displacement is the shortest distance between the end points. Therefore, **the displacement of a particle is completely known if its initial and final coordinates are known.** The path need not be specified. In other words, the **displacement is independent of the path** if the end points of the path are fixed.

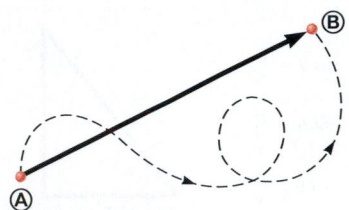

Figure 1.6 After a particle moves from Ⓐ to Ⓑ along an arbitrary path represented by the broken line, its displacement is a vector quantity shown by the arrow drawn from Ⓐ to Ⓑ.

▶ Distance

Note that the **distance** traveled by a particle is distinctly different from its displacement. The distance traveled (a scalar quantity) is the length of the path, which in general can be much greater than the magnitude of the displacement. In Figure 1.6, the length of the curved broken path is much larger than the magnitude of the solid black displacement vector.

If the particle moves along the x axis from position x_i to position x_f, as in Figure 1.7, its displacement is given by $x_f - x_i$. (The indices i and f refer to the initial and final values.) We use the Greek letter delta (Δ) to denote the *change* in a quantity. Therefore, we define the change in the position of the particle (the displacement) as

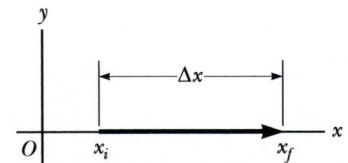

Figure 1.7 A particle moving along the x axis from x_i to x_f undergoes a displacement $\Delta x = x_f - x_i$.

$$\Delta x \equiv x_f - x_i \qquad \text{1.6} \blacktriangleleft$$

From this definition we see that Δx is positive if x_f is greater than x_i and negative if x_f is less than x_i. For example, if a particle changes its position from $x_i = -5$ m to $x_f = 3$ m, its displacement is $\Delta x = +8$ m.

Many physical quantities in addition to displacement are vectors. They include velocity, acceleration, force, and momentum, all of which will be defined in later chapters. In this text, we will use boldface letters with an arrow over the letter, such as $\vec{A}$, to represent vectors. Another common notation for vectors with which you should be familiar is a simple boldface character: **A**.

The magnitude of the vector $\vec{A}$ is written with an italic letter A or, alternatively, $|\vec{A}|$. The magnitude of a vector is always positive and carries the units of the quantity that the vector represents, such as meters for displacement or meters per second for velocity. Vectors combine according to special rules, which will be discussed in Sections 1.8 and 1.9.

◀ **QUICK QUIZ 1.3** Which of the following are vector quantities and which are scalar quantities?   **(a)** your age   **(b)** acceleration   **(c)** velocity   **(d)** speed   **(e)** mass

THINKING PHYSICS 1.1

Consider your commute to work or school in the morning. Which is larger, the distance you travel or the magnitude of the displacement vector?

Reasoning Unless you have a very unusual commute, the distance traveled *must* be larger than the magnitude of the displacement vector. The distance includes the results of all the twists and turns you make in following the roads from home to work or school. On the other hand, the magnitude of the displacement vector is the length of a straight line from your home to work or school. This length is often described informally as "the distance as the crow flies." The only way that the distance could be the same as the magnitude of the displacement vector is if your commute is a perfect straight line, which is highly unlikely! The distance could *never* be less than the magnitude of the displacement vector because the shortest distance between two points is a straight line. ◄

1.8 | Some Properties of Vectors

Equality of Two Vectors

Two vectors $\vec{A}$ and $\vec{B}$ are defined to be equal if they have the same units, the same magnitude, and the same direction. That is, $\vec{A} = \vec{B}$ only if $A = B$ *and* $\vec{A}$ and $\vec{B}$ point in the same direction. For example, all the vectors in Figure 1.8 are equal even though they have different starting points. This property allows us to translate a vector parallel to itself in a diagram without affecting the vector.

Figure 1.8 These four representations of vectors are equal because all four vectors have the same magnitude and point in the same direction.

Addition

The rules for vector sums are conveniently described using a graphical method. To add vector $\vec{B}$ to vector $\vec{A}$, first draw a diagram of vector $\vec{A}$ on graph paper, with its magnitude represented by a convenient scale, and then draw vector $\vec{B}$ to the same scale with its tail starting from the tip of $\vec{A}$, as in Active Figure 1.9a. The *resultant vector* $\vec{R} = \vec{A} + \vec{B}$ is the vector drawn from the tail of $\vec{A}$ to the tip of $\vec{B}$. The technique for adding two vectors is often called the "head-to-tail method."

When vectors are added, the sum is independent of the order of the addition. This independence can be seen for two vectors from the geometric construction in Active Figure 1.9b and is known as the **commutative law of addition:**

$$\vec{A} + \vec{B} = \vec{B} + \vec{A} \qquad \qquad 1.7 ◄$$

Pitfall Prevention | 1.6
Vector Addition Versus Scalar Addition
Keep in mind that $\vec{A} + \vec{B} = \vec{C}$ is very different from $A + B = C$. The first equation is a vector sum, which must be handled carefully, such as with the graphical method described in Active Figure 1.9. The second equation is a simple algebraic addition of numbers that is handled with the normal rules of arithmetic.

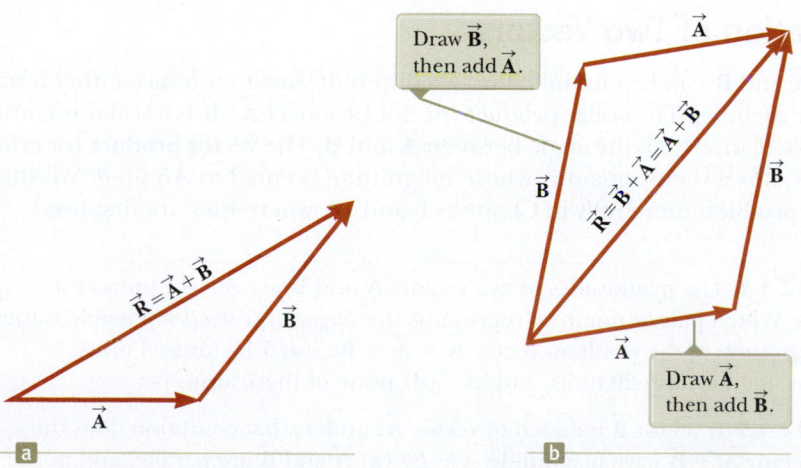

Active Figure 1.9 (a) When vector $\vec{B}$ is added to vector $\vec{A}$, the resultant $\vec{R}$ is the vector that runs from the tail of $\vec{A}$ to the tip of $\vec{B}$. (b) This construction shows that $\vec{A} + \vec{B} = \vec{B} + \vec{A}$; vector addition is commutative.

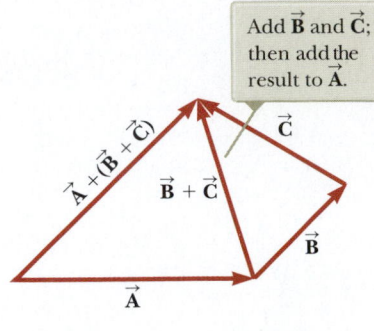

Add $\vec{\mathbf{B}}$ and $\vec{\mathbf{C}}$; then add the result to $\vec{\mathbf{A}}$.

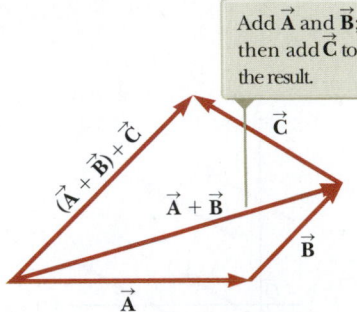

Add $\vec{\mathbf{A}}$ and $\vec{\mathbf{B}}$; then add $\vec{\mathbf{C}}$ to the result.

Figure 1.10 Geometric constructions for verifying the associative law of addition.

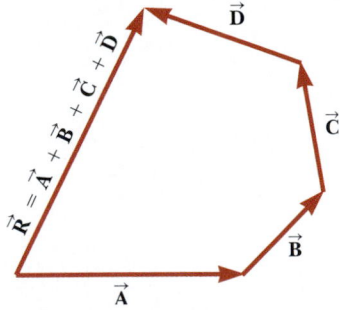

Figure 1.11 Geometric construction for summing four vectors. The resultant vector $\vec{\mathbf{R}}$ closes the polygon and points from the tail of the first vector to the tip of the final vector.

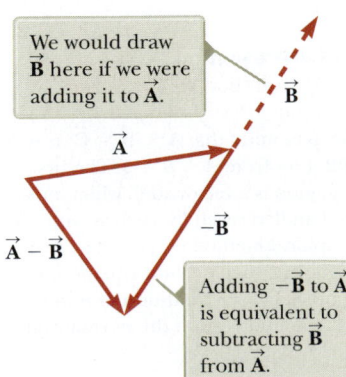

We would draw $\vec{\mathbf{B}}$ here if we were adding it to $\vec{\mathbf{A}}$.

Adding $-\vec{\mathbf{B}}$ to $\vec{\mathbf{A}}$ is equivalent to subtracting $\vec{\mathbf{B}}$ from $\vec{\mathbf{A}}$.

Figure 1.12 Subtracting vector $\vec{\mathbf{B}}$ from vector $\vec{\mathbf{A}}$. The vector $-\vec{\mathbf{B}}$ is equal in magnitude to vector $\vec{\mathbf{B}}$ and points in the opposite direction.

If three or more vectors are added, their sum is independent of the way in which they are grouped. A geometric demonstration of this property for three vectors is given in Figure 1.10. This property is called the **associative law of addition:**

$$\vec{\mathbf{A}} + (\vec{\mathbf{B}} + \vec{\mathbf{C}}) = (\vec{\mathbf{A}} + \vec{\mathbf{B}}) + \vec{\mathbf{C}} \qquad \text{1.8} \blacktriangleleft$$

Geometric constructions can also be used to add more than three vectors, as shown in Figure 1.11 for the case of four vectors. The resultant vector $\vec{\mathbf{R}} = \vec{\mathbf{A}} + \vec{\mathbf{B}} + \vec{\mathbf{C}} + \vec{\mathbf{D}}$ is the vector that closes the polygon formed by the vectors being added. In other words, $\vec{\mathbf{R}}$ is the vector drawn from the tail of the first vector to the tip of the last vector. Again, the order of the summation is unimportant.

In summary, **a vector quantity has both magnitude and direction and also obeys the laws of vector addition** as described in Active Figure 1.9 and Figures 1.10 and 1.11. When two or more vectors are added together, they must all have the same units and they must all be the same type of quantity. It would be meaningless to add a velocity vector (for example, 60 km/h to the east) to a displacement vector (for example, 200 km to the north) because these vectors represent different physical quantities. The same rule also applies to scalars. For example, it would be meaningless to add time intervals to temperatures.

Negative of a Vector

The negative of the vector $\vec{\mathbf{A}}$ is defined as the vector that, when added to $\vec{\mathbf{A}}$, gives zero for the vector sum. That is, $\vec{\mathbf{A}} + (-\vec{\mathbf{A}}) = 0$. The vectors $\vec{\mathbf{A}}$ and $-\vec{\mathbf{A}}$ have the same magnitude but point in opposite directions.

Subtraction of Vectors

The operation of vector subtraction makes use of the definition of the negative of a vector. We define the operation $\vec{\mathbf{A}} - \vec{\mathbf{B}}$ as vector $-\vec{\mathbf{B}}$ added to vector $\vec{\mathbf{A}}$:

$$\vec{\mathbf{A}} - \vec{\mathbf{B}} = \vec{\mathbf{A}} + (-\vec{\mathbf{B}}) \qquad \text{1.9} \blacktriangleleft$$

The geometric construction for subtracting two vectors is illustrated in Figure 1.12.

Multiplication of a Vector by a Scalar

If a vector $\vec{\mathbf{A}}$ is multiplied by a positive scalar quantity s, the product $s\vec{\mathbf{A}}$ is a vector that has the same direction as $\vec{\mathbf{A}}$ and magnitude sA. If s is a negative scalar quantity, the vector $s\vec{\mathbf{A}}$ is directed opposite to $\vec{\mathbf{A}}$. For example, the vector $5\vec{\mathbf{A}}$ is five times longer than $\vec{\mathbf{A}}$ and has the same direction as $\vec{\mathbf{A}}$; the vector $-\frac{1}{3}\vec{\mathbf{A}}$ has one-third the magnitude of $\vec{\mathbf{A}}$ and points in the direction opposite $\vec{\mathbf{A}}$.

Multiplication of Two Vectors

Two vectors $\vec{\mathbf{A}}$ and $\vec{\mathbf{B}}$ can be multiplied in two different ways to produce either a scalar or a vector quantity. The **scalar product** (or dot product) $\vec{\mathbf{A}} \cdot \vec{\mathbf{B}}$ is a scalar quantity equal to $AB\cos\theta$, where θ is the angle between $\vec{\mathbf{A}}$ and $\vec{\mathbf{B}}$. The **vector product** (or cross product) $\vec{\mathbf{A}} \times \vec{\mathbf{B}}$ is a vector quantity whose magnitude is equal to $AB\sin\theta$. We shall discuss these products more fully in Chapters 6 and 10, where they are first used.

QUICK QUIZ 1.4 The magnitudes of two vectors $\vec{\mathbf{A}}$ and $\vec{\mathbf{B}}$ are $A = 12$ units and $B = 8$ units. Which pair of numbers represents the *largest* and *smallest* possible values for the magnitude of the resultant vector $\vec{\mathbf{R}} = \vec{\mathbf{A}} + \vec{\mathbf{B}}$? (**a**) 14.4 units, 4 units (**b**) 12 units, 8 units (**c**) 20 units, 4 units (**d**) none of these answers

QUICK QUIZ 1.5 If vector $\vec{\mathbf{B}}$ is added to vector $\vec{\mathbf{A}}$, under what condition does the resultant vector $\vec{\mathbf{A}} + \vec{\mathbf{B}}$ have magnitude $A + B$? (**a**) $\vec{\mathbf{A}}$ and $\vec{\mathbf{B}}$ are parallel and in the same direction. (**b**) $\vec{\mathbf{A}}$ and $\vec{\mathbf{B}}$ are parallel and in opposite directions. (**c**) $\vec{\mathbf{A}}$ and $\vec{\mathbf{B}}$ are perpendicular.

⟨**1.9** | Components of a Vector and Unit Vectors

The graphical method of adding vectors is not recommended whenever high accuracy is required or in three-dimensional problems. In this section, we describe a method of adding vectors that makes use of the projections of vectors along coordinate axes. These projections are called the **components** of the vector or its **rectangular components.** Any vector can be completely described by its components.

Consider a vector $\vec{A}$ lying in the xy plane and making an arbitrary angle θ with the positive x axis as shown in Figure 1.13a. This vector can be expressed as the sum of two other *component vectors* $\vec{A}_x$, which is parallel to the x axis, and $\vec{A}_y$, which is parallel to the y axis. From Figure 1.13b, we see that the three vectors form a right triangle and that $\vec{A} = \vec{A}_x + \vec{A}_y$. We shall often refer to the "components of a vector $\vec{A}$," written A_x and A_y (without the boldface notation). The component A_x represents the projection of $\vec{A}$ along the x axis, and the component A_y represents the projection of $\vec{A}$ along the y axis. These components can be positive or negative. The component A_x is positive if the component vector $\vec{A}_x$ points in the positive x direction and is negative if $\vec{A}_x$ points in the negative x direction. A similar statement is made for the component A_y.

From Figure 1.13b and the definition of the sine and cosine of an angle, we see that $\cos \theta = A_x/A$ and $\sin \theta = A_y/A$. Hence, the components of $\vec{A}$ are given by

$$A_x = A \cos \theta \quad \text{and} \quad A_y = A \sin \theta \quad \text{1.10} ◀$$

The magnitudes of these components are the lengths of the two sides of a right triangle with a hypotenuse of length A. Therefore, the magnitude and direction of $\vec{A}$ are related to its components through the expressions

$$A = \sqrt{A_x^2 + A_y^2} \quad \text{1.11} ◀ \qquad ▶ \text{Magnitude of } \vec{A}$$

$$\tan \theta = \frac{A_y}{A_x} \quad \text{1.12} ◀ \qquad ▶ \text{Direction of } \vec{A}$$

To solve for θ, we can write $\theta = \tan^{-1}(A_y/A_x)$, which is read "$\theta$ equals the angle whose tangent is the ratio A_y/A_x." *Note that the signs of the components A_x and A_y depend on the angle θ.* For example, if $\theta = 120°$, A_x is negative and A_y is positive. If $\theta = 225°$, both A_x and A_y are negative. Figure 1.14 summarizes the signs of the components when $\vec{A}$ lies in the various quadrants.

If you choose reference axes or an angle other than those shown in Figure 1.13, the components of the vector must be modified accordingly. In many applications, it is more convenient to express the components of a vector in a coordinate system having axes that are not horizontal and vertical but are still perpendicular to each other.

Pitfall Prevention | 1.7
***x* and *y* Components**
Equation 1.10 associates the cosine of the angle with the x component and the sine of the angle with the y component. This association is true *only* because we measured the angle θ with respect to the x axis, so do not memorize these equations. If θ is measured with respect to the y axis (as in some problems), these equations will be incorrect. Think about which side of the triangle containing the components is adjacent to the angle and which side is opposite and then assign the cosine and sine accordingly.

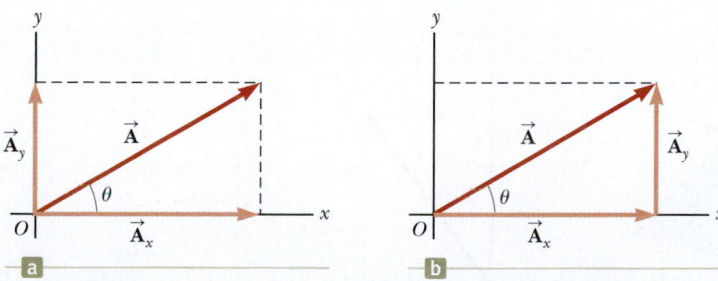

Figure 1.13 (a) A vector $\vec{A}$ lying in the xy plane can be represented by its component vectors $\vec{A}_x$ and $\vec{A}_y$. (b) The y component vector $\vec{A}_y$ can be moved to the right so that it adds to $\vec{A}_x$. The vector sum of the component vectors is $\vec{A}$. These three vectors form a right triangle.

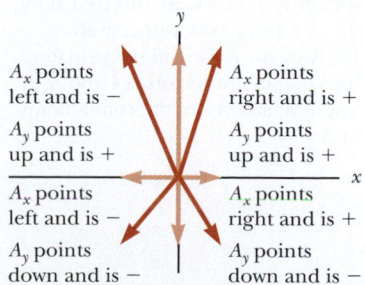

A_x points left and is $-$
A_y points up and is $+$

A_x points right and is $+$
A_y points up and is $+$

A_x points left and is $-$
A_y points down and is $-$

A_x points right and is $+$
A_y points down and is $-$

Figure 1.14 The signs of the components of a vector $\vec{A}$ depend on the quadrant in which the vector is located.

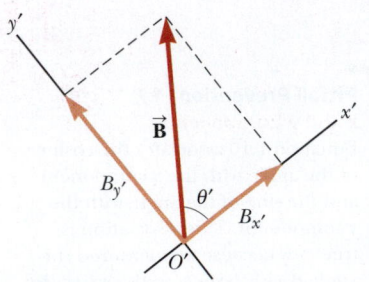

Figure 1.15 The component vectors of vector $\vec{\mathbf{B}}$ in a coordinate system that is tilted.

Suppose a vector $\vec{\mathbf{B}}$ makes an angle θ' with the x' axis defined in Figure 1.15. The components of $\vec{\mathbf{B}}$ along these axes are given by $B_{x'} = B\cos\theta'$ and $B_{y'} = B\sin\theta'$, as in Equation 1.10. The magnitude and direction of $\vec{\mathbf{B}}$ are obtained from expressions equivalent to Equations 1.11 and 1.12. Therefore, we can express the components of a vector in *any* coordinate system that is convenient for a particular situation.

> **QUICK QUIZ 1.6** Choose the correct response to make the sentence true: A component of a vector is (**a**) always, (**b**) never, or (**c**) sometimes larger than the magnitude of the vector.

Unit Vectors

Vector quantities often are expressed in terms of unit vectors. A **unit vector** is a dimensionless vector having a magnitude of exactly 1. Unit vectors are used to specify a given direction and have no other physical significance. We shall use the symbols $\hat{\mathbf{i}}$, $\hat{\mathbf{j}}$, and $\hat{\mathbf{k}}$ to represent unit vectors pointing in the x, y, and z directions, respectively. The "hat" over the letters is a common notation for a unit vector; for example, $\hat{\mathbf{i}}$ is called "i-hat." The unit vectors $\hat{\mathbf{i}}$, $\hat{\mathbf{j}}$, and $\hat{\mathbf{k}}$ form a set of mutually perpendicular vectors as shown in Active Figure 1.16a, where the magnitude of each unit vector equals 1; that is, $|\hat{\mathbf{i}}| = |\hat{\mathbf{j}}| = |\hat{\mathbf{k}}| = 1$.

Consider a vector $\vec{\mathbf{A}}$ lying in the xy plane, as in Active Figure 1.16b. The product of the component A_x and the unit vector $\hat{\mathbf{i}}$ is the component vector $\vec{\mathbf{A}}_x = A_x\hat{\mathbf{i}}$, which lies on the x axis and has magnitude A_x. Likewise, $A_y\hat{\mathbf{j}}$ is a component vector of magnitude A_y lying on the y axis. Therefore, the unit-vector notation for the vector $\vec{\mathbf{A}}$ is

$$\vec{\mathbf{A}} = A_x\hat{\mathbf{i}} + A_y\hat{\mathbf{j}} \qquad \textbf{1.13} \blacktriangleleft$$

Now suppose we wish to add vector $\vec{\mathbf{B}}$ to vector $\vec{\mathbf{A}}$, where $\vec{\mathbf{B}}$ has components B_x and B_y. The procedure for performing this sum is simply to add the x and y components separately. The resultant vector $\vec{\mathbf{R}} = \vec{\mathbf{A}} + \vec{\mathbf{B}}$ is therefore

$$\vec{\mathbf{R}} = (A_x + B_x)\hat{\mathbf{i}} + (A_y + B_y)\hat{\mathbf{j}} \qquad \textbf{1.14} \blacktriangleleft$$

From this equation, the components of the resultant vector are given by

$$R_x = A_x + B_x$$
$$R_y = A_y + B_y \qquad \textbf{1.15} \blacktriangleleft$$

Therefore, we see that in the component method of adding vectors, we add all the x components to find the x component of the resultant vector and use the same

Active Figure 1.16 (a) The unit vectors $\hat{\mathbf{i}}$, $\hat{\mathbf{j}}$, and $\hat{\mathbf{k}}$ are directed along the x, y, and z axes, respectively. (b) A vector $\vec{\mathbf{A}}$ lying in the xy plane has component vectors $A_x\hat{\mathbf{i}}$ and $A_y\hat{\mathbf{j}}$, where A_x and A_y are the components of $\vec{\mathbf{A}}$.

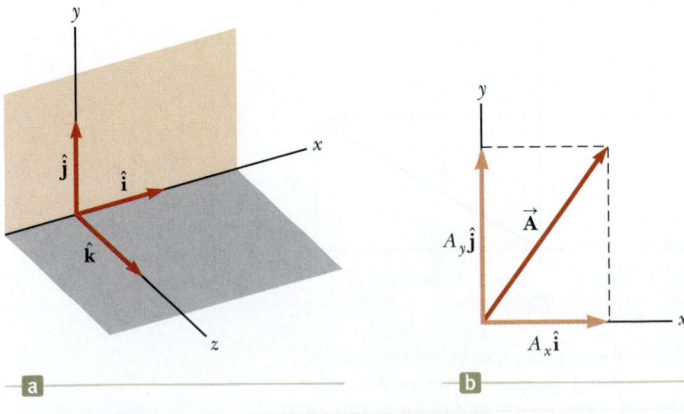

process for the y components. The procedure just described for adding two vectors $\vec{A}$ and $\vec{B}$ using the component method can be checked using a diagram like Figure 1.17.

The magnitude of $\vec{R}$ and the angle it makes with the x axis can be obtained from its components using the relationships

$$R = \sqrt{R_x^{\,2} + R_y^{\,2}} = \sqrt{(A_x + B_x)^2 + (A_y + B_y)^2} \qquad \textbf{1.16}\blacktriangleleft$$

$$\tan \theta = \frac{R_y}{R_x} = \frac{A_y + B_y}{A_x + B_x} \qquad \textbf{1.17}\blacktriangleleft$$

The extension of these methods to three-dimensional vectors is straightforward. If $\vec{A}$ and $\vec{B}$ both have x, y, and z components, we express them in the form

$$\vec{A} = A_x\hat{\mathbf{i}} + A_y\hat{\mathbf{j}} + A_z\hat{\mathbf{k}}$$

$$\vec{B} = B_x\hat{\mathbf{i}} + B_y\hat{\mathbf{j}} + B_z\hat{\mathbf{k}}$$

The sum of $\vec{A}$ and $\vec{B}$ is

$$\vec{R} = \vec{A} + \vec{B} = (A_x + B_x)\hat{\mathbf{i}} + (A_y + B_y)\hat{\mathbf{j}} + (A_z + B_z)\hat{\mathbf{k}} \qquad \textbf{1.18}\blacktriangleleft$$

If a vector $\vec{R}$ has x, y, and z components, the magnitude of the vector is

$$R = \sqrt{R_x^{\,2} + R_y^{\,2} + R_z^{\,2}}$$

The angle θ_x that $\vec{R}$ makes with the x axis is given by

$$\cos \theta_x = \frac{R_x}{R}$$

with similar expressions for the angles with respect to the y and z axes.

The extension of our method to adding more than two vectors is also straightforward. For example, $\vec{A} + \vec{B} + \vec{C} = (A_x + B_x + C_x)\hat{\mathbf{i}} + (A_y + B_y + C_y)\hat{\mathbf{j}} + (A_z + B_z + C_z)\hat{\mathbf{k}}$. Adding displacement vectors is relatively easy to visualize. We can also add other types of vectors, such as velocity, force, and electric field vectors, which we will do in later chapters.

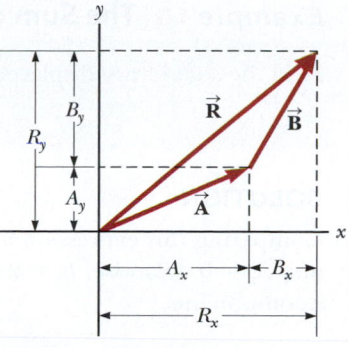

Figure 1.17 A geometric construction showing the relation between the components of the resultant $\vec{R}$ of two vectors and the individual components.

> **Pitfall Prevention | 1.8**
> **Tangents on Calculators**
> Equation 1.17 involves the calculation of an angle by means of a tangent function. Generally, the inverse tangent function on calculators provides an angle between $-90°$ and $+90°$. As a consequence, if the vector you are studying lies in the second or third quadrant, the angle measured from the positive x axis will be the angle your calculator returns plus 180°.

QUICK QUIZ 1.7 If at least one component of a vector is a positive number, the vector cannot **(a)** have any component that is negative, **(b)** be zero, **(c)** have three dimensions.

QUICK QUIZ 1.8 If $\vec{A} + \vec{B} = 0$, the corresponding components of the two vectors $\vec{A}$ and $\vec{B}$ must be **(a)** equal, **(b)** positive, **(c)** negative, **(d)** of opposite sign.

> ### THINKING PHYSICS 1.2
>
> You may have asked someone directions to a destination in a city and been told something like, "Walk 3 blocks east and then 5 blocks south." If so, are you experienced with vector components?
>
> **Reasoning** Yes, you are! Although you may not have thought of vector component language when you heard these directions, that is exactly what the directions represent. The perpendicular streets of the city reflect an xy coordinate system; we can assign the x axis to the east–west streets, and the y axis to the north–south streets. Therefore, the comment of the person giving you directions can be translated as, "Undergo a displacement vector that has an x component of $+3$ blocks and a y component of -5 blocks." You would arrive at the same destination by undergoing the y component first, followed by the x component, demonstrating the commutative law of addition. ◀

Example **1.6** | **The Sum of Two Vectors**

Find the sum of two displacement vectors $\vec{A}$ and $\vec{B}$ lying in the xy plane and given by

$$\vec{A} = (2.0\hat{i} + 2.0\hat{j})\text{ m} \quad \text{and} \quad \vec{B} = (2.0\hat{i} - 4.0\hat{j})\text{ m}$$

SOLUTION

Comparing this expression for $\vec{A}$ with the general expression $\vec{A} = A_x\hat{i} + A_y\hat{j} + A_z\hat{k}$, we see that $A_x = 2.0$ m, $A_y = 2.0$ m, and $A_z = 0$. Likewise, $B_x = 2.0$ m, $B_y = -4.0$ m, and $B_z = 0$. We can use a two-dimensional approach because there are no z components.

Use Equation 1.14 to obtain the resultant vector $\vec{R}$: $\vec{R} = \vec{A} + \vec{B} = (2.0 + 2.0)\hat{i}\text{ m} + (2.0 - 4.0)\hat{j}\text{ m}$

Evaluate the components of $\vec{R}$: $R_x = 4.0\text{ m} \qquad R_y = -2.0\text{ m}$

Use Equation 1.16 to find the magnitude of $\vec{R}$: $R = \sqrt{R_x^2 + R_y^2} = \sqrt{(4.0\text{ m})^2 + (-2.0\text{ m})^2} = \sqrt{20}\text{ m} = \boxed{4.5\text{ m}}$

Find the direction of $\vec{R}$ from Equation 1.17: $\tan\theta = \dfrac{R_y}{R_x} = \dfrac{-2.0\text{ m}}{4.0\text{ m}} = -0.50$

Your calculator likely gives the answer $-27°$ for $\theta = \tan^{-1}(-0.50)$. This answer is correct if we interpret it to mean $27°$ clockwise from the x axis. Our standard form has been to quote the angles measured counterclockwise from the $+x$ axis, and that angle for this vector is $\theta = \boxed{333.°}$

Example **1.7** | **The Resultant Displacement**

A particle undergoes three consecutive displacements: $\Delta\vec{r}_1 = (15\hat{i} + 30\hat{j} + 12\hat{k})$ cm, $\Delta\vec{r}_2 = (23\hat{i} - 14\hat{j} - 5.0\hat{k})$ cm, and $\Delta\vec{r}_3 = (-13\hat{i} + 15\hat{j})$ cm. Find unit-vector notation for the resultant displacement and its magnitude.

SOLUTION

Although x is sufficient to locate a point in one dimension, we need a vector $\vec{r}$ to locate a point in two or three dimensions. The notation $\Delta\vec{r}$ is a generalization of the one-dimensional displacement Δx. Three-dimensional displacements are more difficult to conceptualize than those in two dimensions because the latter can be drawn on paper.

For this problem, let us imagine that you start with your pencil at the origin of a piece of graph paper on which you have drawn x and y axes. Move your pencil 15 cm to the right along the x axis, then 30 cm upward along the y axis, and then 12 cm *perpendicularly toward you away* from the graph paper. This procedure provides the displacement described by $\Delta\vec{r}_1$. From this point, move your pencil 23 cm to the right parallel to the x axis, then 14 cm parallel to the graph paper in the $-y$ direction, and then 5.0 cm perpendicularly away from you toward the graph paper. You are now at the displacement from the origin described by $\Delta\vec{r}_1 + \Delta\vec{r}_2$. From this point, move your pencil 13 cm to the left in the $-x$ direction, and (finally!) 15 cm parallel to the graph paper along the y axis. Your final position is at a displacement $\Delta\vec{r}_1 + \Delta\vec{r}_2 + \Delta\vec{r}_3$ from the origin.

To find the resultant displacement, add the three vectors:

$$\begin{aligned}\Delta\vec{r} &= \Delta\vec{r}_1 + \Delta\vec{r}_2 + \Delta\vec{r}_3 \\ &= (15 + 23 - 13)\hat{i}\text{ cm} + (30 - 14 + 15)\hat{j}\text{ cm} \\ &\quad + (12 - 5.0 + 0)\hat{k}\text{ cm} \\ &= \boxed{(25\hat{i} + 31\hat{j} + 7.0\hat{k})\text{ cm}}\end{aligned}$$

Find the magnitude of the resultant vector:

$$\begin{aligned}R &= \sqrt{R_x^2 + R_y^2 + R_z^2} \\ &= \sqrt{(25\text{ cm})^2 + (31\text{ cm})^2 + (7.0\text{ cm})^2} = \boxed{40\text{ cm}}\end{aligned}$$

Example **1.8** | Taking a Hike

A hiker begins a trip by first walking 25.0 km southeast from her car. She stops and sets up her tent for the night. On the second day, she walks 40.0 km in a direction 60.0° north of east, at which point she discovers a forest ranger's tower.

(A) Determine the components of the hiker's displacement for each day.

SOLUTION

If we denote the displacement vectors on the first and second days by $\vec{A}$ and $\vec{B}$, respectively, and use the car as the origin of coordinates, we obtain the vectors shown in Figure 1.18. Drawing the resultant $\vec{R}$, we see that this problem is one we've solved before: an addition of two vectors.

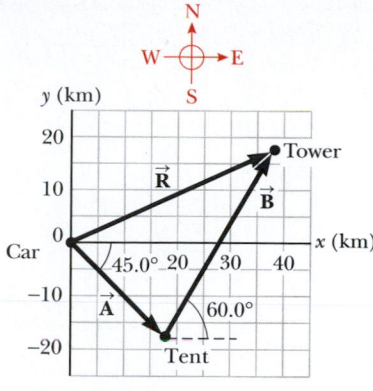

Figure 1.18 (Example 1.8) The total displacement of the hiker is the vector $\vec{R} = \vec{A} + \vec{B}$.

Displacement $\vec{A}$ has a magnitude of 25.0 km and is directed 45.0° below the positive *x* axis.

Find the components of $\vec{A}$ using Equation 1.10:

$$A_x = A \cos(-45.0°) = (25.0 \text{ km})(0.707) = \boxed{17.7 \text{ km}}$$

$$A_y = A \sin(-45.0°) = (25.0 \text{ km})(-0.707) = \boxed{-17.7 \text{ km}}$$

The negative value of A_y indicates that the hiker walks in the negative *y* direction on the first day. The signs of A_x and A_y also are evident from Figure 1.18.

Find the components of $\vec{B}$ using Equation 1.10:

$$B_x = B \cos 60.0° = (40.0 \text{ km})(0.500) = \boxed{20.0 \text{ km}}$$

$$B_y = B \sin 60.0° = (40.0 \text{ km})(0.866) = \boxed{34.6 \text{ km}}$$

(B) Determine the components of the hiker's resultant displacement $\vec{R}$ for the trip. Find an expression for $\vec{R}$ in terms of unit vectors.

SOLUTION

Use Equation 1.15 to find the components of the resultant displacement $\vec{R} = \vec{A} + \vec{B}$:

$$R_x = A_x + B_x = 17.7 \text{ km} + 20.0 \text{ km} = \boxed{37.7 \text{ km}}$$

$$R_y = A_y + B_y = -17.7 \text{ km} + 34.6 \text{ km} = \boxed{17.0 \text{ km}}$$

Write the total displacement in unit-vector form:

$$\vec{R} = \boxed{(37.7\hat{i} + 17.0\hat{j}) \text{ km}}$$

Looking at the graphical representation in Figure 1.18, we estimate the position of the tower to be about (38 km, 17 km), which is consistent with the components of $\vec{R}$ in our result for the final position of the hiker. Also, both components of $\vec{R}$ are positive, putting the final position in the first quadrant of the coordinate system, which is also consistent with Figure 1.18.

What If? After reaching the tower, the hiker wishes to return to her car along a single straight line. What are the components of the vector representing this hike? What should the direction of the hike be?

Answer The desired vector $\vec{R}_{car}$ is the negative of vector $\vec{R}$:

$$\vec{R}_{car} = -\vec{R} = (-37.7\hat{i} - 17.0\hat{j}) \text{ km}$$

The direction is found by calculating the angle that the vector makes with the *x* axis:

$$\tan \theta = \frac{R_{car,y}}{R_{car,x}} = \frac{-17.0 \text{ km}}{-37.7 \text{ km}} = 0.450$$

which gives an angle of $\theta = 204.2°$, or 24.2° south of west.

❮1.10 | Modeling, Alternative Representations, and Problem-Solving Strategy

Most courses in general physics require the student to learn the skills of problem solving, and examinations usually include problems that test such skills. This section describes some useful ideas that will enable you to enhance your understanding of physical concepts, increase your accuracy in solving problems, eliminate initial panic or lack of direction in approaching a problem, and organize your work.

One of the primary problem-solving methods in physics is to form an appropriate **model** of the problem. **A model is a simplified substitute for the real problem that allows us to solve the problem in a relatively simple way.** As long as the predictions of the model agree to our satisfaction with the actual behavior of the real system, the model is valid. If the predictions do not agree, the model must be refined or replaced with another model. The power of modeling is in its ability to reduce a wide variety of very complex problems to a limited number of classes of problems that can be approached in similar ways.

In science, a model is very different from, for example, an architect's scale model of a proposed building, which appears as a smaller version of what it represents. A scientific model is a theoretical construct and may have no visual similarity to the physical problem. A simple application of modeling is presented in Example 1.9, and we shall encounter many more examples of models as the text progresses.

Models are needed because the actual operation of the Universe is extremely complicated. Suppose, for example, we are asked to solve a problem about the Earth's motion around the Sun. The Earth is very complicated, with many processes occurring simultaneously. These processes include weather, seismic activity, and ocean movements as well as the multitude of processes involving human activity. Trying to maintain knowledge and understanding of all these processes is an impossible task.

The modeling approach recognizes that none of these processes affects the motion of the Earth around the Sun to a measurable degree. Therefore, these details are all ignored. In addition, as we shall find in Chapter 11, the size of the Earth does not affect the gravitational force between the Earth and the Sun; only the masses of the Earth and Sun and the distance between them determine this force. In a simplified model, the Earth is imagined to be a particle, an object with mass but zero size. This replacement of an extended object by a particle is called the **particle model,** which is used extensively in physics. By analyzing the motion of a particle with the mass of the Earth in orbit around the Sun, we find that the predictions of the particle's motion are in excellent agreement with the actual motion of the Earth.

The two primary conditions for using the particle model are as follows:

- The size of the actual object is of no consequence in the analysis of its motion.
- Any internal processes occurring in the object are of no consequence in the analysis of its motion.

Both of these conditions are in action in modeling the Earth as a particle. Its radius is not a factor in determining its motion, and internal processes such as thunderstorms, earthquakes, and manufacturing processes can be ignored.

Four categories of models used in this book will help us understand and solve physics problems. The first category is the **geometric model.** In this model, we form a geometric construction that represents the real situation. We then set aside the real problem and perform an analysis of the geometric construction. Consider a popular problem in elementary trigonometry, as in the following example.

Example 1.9 | Finding the Height of a Tree

You wish to find the height of a tree but cannot measure it directly. You stand 50.0 m from the tree and determine that a line of sight from the ground to the top of the tree makes an angle of 25.0° with the ground. How tall is the tree?

SOLUTION

Figure 1.19 shows the tree and a right triangle corresponding to the information in the problem superimposed over it. (We assume that the tree is exactly perpendicular to a perfectly flat ground.) In the triangle, we know the length of the horizontal leg and the angle between the hypotenuse and the horizontal leg. We can find the height of the tree by calculating the length of the vertical leg. We do so with the tangent function:

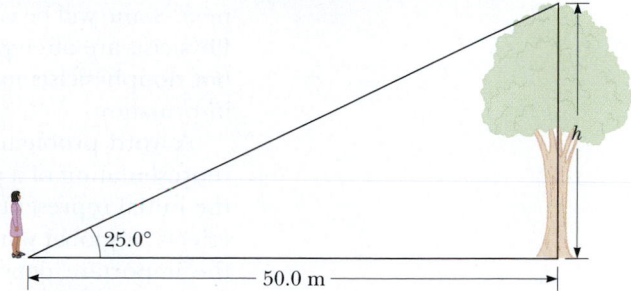

Figure 1.19 (Example 1.9) The height of a tree can be found by measuring the distance from the tree and the angle of sight to the top above the ground. This problem is a simple example of geometrically *modeling* the actual problem.

$$\tan \theta = \frac{\text{opposite side}}{\text{adjacent side}} = \frac{h}{50.0 \text{ m}}$$

$$h = (50.0 \text{ m}) \tan \theta = (50.0 \text{ m}) \tan 25.0° = \boxed{23.3 \text{ m}}$$

You may have solved a problem very similar to Example 1.9 but never thought about the notion of modeling. From the modeling approach, however, once we draw the triangle in Figure 1.19, the triangle is a geometric model of the real problem; it is a *substitute*. Until we reach the end of the problem, we do not imagine the problem to be about a *tree* but to be about a *triangle*. We use trigonometry to find the vertical leg of the triangle, leading to a value of 23.3 m. Because this leg *represents* the height of the tree, we can now return to the original problem and claim that the height of the tree is 23.3 m.

Other examples of geometric models include modeling the Earth as a perfect sphere, a pizza as a perfect disk, a meter stick as a long rod with no thickness, and an electric wire as a long, straight cylinder.

The particle model is an example of the second category of models, which we will call **simplification models.** In a simplification model, details that are not significant in determining the outcome of the problem are ignored. When we study rotation in Chapter 10, objects will be modeled as *rigid objects*. All the molecules in a rigid object maintain their exact positions with respect to one another. We adopt this simplification model because a spinning rock is much easier to analyze than a spinning block of gelatin, which is *not* a rigid object. Other simplification models will assume that quantities such as friction forces are negligible, remain constant, or are proportional to some power of the object's speed.

The third category is that of **analysis models,** which are general types of problems that we have solved before. An important technique in problem solving is to cast a new problem into a form similar to one we have already solved and which can be used as a model. As we shall see, there are about two dozen analysis models that can be used to solve most of the problems you will encounter. We will see our first analysis models in Chapter 2, where we will discuss them in more detail.

The fourth category of models is **structural models.** These models are generally used to understand the behavior of a system that is far different in scale from our macroscopic world—either much smaller or much larger—so that we cannot interact with it directly. As an example, the notion of a hydrogen atom as an electron in a circular orbit around a proton is a structural model of the atom. We will discuss this model and structural models in general in Chapter 11.

Intimately related to the notion of modeling is that of forming **alternative representations** of the problem. **A representation is a method of viewing or presenting the information related to the problem.** Scientists must be able to communicate complex ideas to individuals without scientific backgrounds. The best representation to use in conveying the information successfully will vary from one individual to the next. Some will be convinced by a well-drawn graph, and others will require a picture. Physicists are often persuaded to agree with a point of view by examining an equation, but nonphysicists may not be convinced by this mathematical representation of the information.

A word problem, such as those at the ends of the chapters in this book, is one representation of a problem. In the "real world" that you will enter after graduation, the initial representation of a problem may be just an existing situation, such as the effects of global warming or a patient in danger of dying. You may have to identify the important data and information, and then cast the situation yourself into an equivalent word problem!

Considering alternative representations can help you think about the information in the problem in several different ways to help you understand and solve it. Several types of representations can be of assistance in this endeavor:

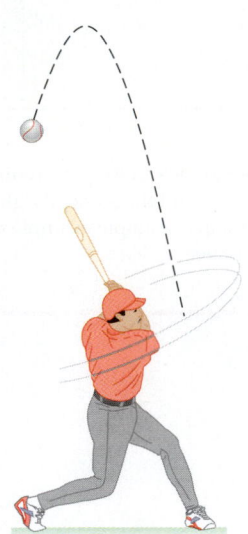

Figure 1.20 A pictorial representation of a pop foul being hit by a baseball player.

- **Mental representation.** From the description of the problem, imagine a scene that describes what is happening in the word problem, then let time progress so that you understand the situation and can predict what changes will occur in the situation. This step is critical in approaching *every* problem.
- **Pictorial representation.** Drawing a picture of the situation described in the word problem can be of great assistance in understanding the problem. In Example 1.9, the pictorial representation in Figure 1.19 allows us to identify the triangle as a geometric model of the problem. In architecture, a blueprint is a pictorial representation of a proposed building.

  Generally, a pictorial representation describes *what you would see* if you were observing the situation in the problem. For example, Figure 1.20 shows a pictorial representation of a baseball player hitting a short pop foul. Any coordinate axes included in your pictorial representation will be in two dimensions: x and y axes.

- **Simplified pictorial representation.** It is often useful to redraw the pictorial representation without complicating details by applying a simplification model. This process is similar to the discussion of the particle model described earlier. In a pictorial representation of the Earth in orbit around the Sun, you might draw the Earth and the Sun as spheres, with possibly some attempt to draw continents to identify which sphere is the Earth. In the simplified pictorial representation, the Earth and the Sun would be drawn simply as dots, representing particles. Figure 1.21 shows a simplified pictorial representation corresponding to the pictorial representation of the baseball trajectory in Figure 1.20. The notations v_x and v_y refer to the components of the velocity vector for the baseball. We shall use such simplified pictorial representations throughout the book.

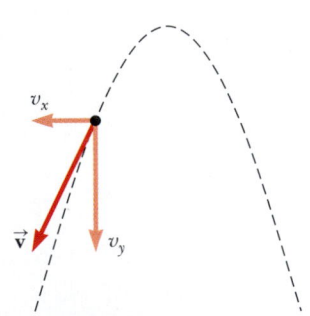

Figure 1.21 A simplified pictorial representation for the situation shown in Figure 1.20.

- **Graphical representation.** In some problems, drawing a graph that describes the situation can be very helpful. In mechanics, for example, position–time graphs can be of great assistance. Similarly, in thermodynamics, pressure–volume graphs are essential to understanding. Figure 1.22 shows a graphical representation of the position as a function of time of a block on the end of a vertical spring as it oscillates up and down. Such a graph is helpful for understanding simple harmonic motion, which we study in Chapter 12.

  A graphical representation is different from a pictorial representation, which is also a two-dimensional display of information but whose axes, if any, represent *length* coordinates. In a graphical representation, the axes may represent *any* two related variables. For example, a graphical representation may have axes for temperature and time. Therefore, in comparison to a pictorial representation, a

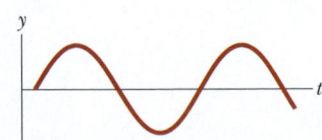

Figure 1.22 A graphical representation of the position as a function of time of a block hanging from a spring and oscillating.

graphical representation is generally *not* something you would see when observing the situation in the problem with your eyes.

- **Tabular representation.** It is sometimes helpful to organize the information in tabular form to help make it clearer. For example, some students find that making tables of known quantities and unknown quantities is helpful. The periodic table of the elements is an extremely useful tabular representation of information in chemistry and physics.
- **Mathematical representation.** The ultimate goal in solving a problem is often the mathematical representation. You want to move from the information contained in the word problem, through various representations of the problem that allow you to understand what is happening, to one or more equations that represent the situation in the problem and that can be solved mathematically for the desired result.

Besides what you might expect to learn about physics concepts, a very valuable skill you should acquire from your physics course is the ability to solve complicated problems. The way physicists approach complex situations and break them into manageable pieces is extremely useful. The following is a general problem-solving strategy to guide you through the steps. To help you remember the steps of the strategy, they are *Conceptualize, Categorize, Analyze,* **and** *Finalize.*

 GENERAL PROBLEM-SOLVING STRATEGY

Conceptualize

- The first things to do when approaching a problem are to *think about* and *understand* the situation. Study carefully any representations of the information (for example, diagrams, graphs, tables, or photographs) that accompany the problem. Imagine a movie, running in your mind, of what happens in the problem.
- If a pictorial representation is not provided, you should almost always make a quick drawing of the situation. Indicate any known values, perhaps in a table or directly on your sketch.
- Now focus on what algebraic or numerical information is given in the problem. Carefully read the problem statement, looking for key phrases such as "starts from rest" ($v_i = 0$) or "stops" ($v_f = 0$).
- Now focus on the expected result of solving the problem. Exactly what is the question asking? Will the final result be numerical or algebraic? Do you know what units to expect?
- Don't forget to incorporate information from your own experiences and common sense. What should a reasonable answer look like? For example, you wouldn't expect to calculate the speed of an automobile to be 5×10^6 m/s.

Categorize

- Once you have a good idea of what the problem is about, you need to *simplify* the problem. Remove the details that are not important to the solution. For example, model a moving object as a particle. If appropriate, ignore air resistance or friction between a sliding object and a surface.
- Once the problem is simplified, it is important to *categorize* the problem. Is it a simple *substitution problem* such that numbers can be substituted into an equation? If so, the problem is likely to be finished when this substitution is done. If not, you face what we call an *analysis problem:* the situation must be analyzed more deeply to reach a solution.
- If it is an analysis problem, it needs to be categorized further. Have you seen this type of problem before? Does it fall into the growing list of types of problems that you have solved previously? If so, identify any analysis

model(s) appropriate for the problem to prepare for the Analyze step below. Being able to classify a problem with an analysis model can make it much easier to lay out a plan to solve it. For example, if your simplification shows that the problem can be treated as a particle under constant acceleration and you have already solved such a problem (such as the examples we shall see in Section 2.6), the solution to the present problem follows a similar pattern.

Analyze

- Now you must analyze the problem and strive for a mathematical solution. Because you have already categorized the problem and identified an analysis model, it should not be too difficult to select relevant equations that apply to the type of situation in the problem. For example, if the problem involves a particle under constant acceleration (which we will study in Section 2.6), Equations 2.10 to 2.14 are relevant.
- Use algebra (and calculus, if necessary) to solve symbolically for the unknown variable in terms of what is given. Substitute in the appropriate numbers, calculate the result, and round it to the proper number of significant figures.

Finalize

- Examine your numerical answer. Does it have the correct units? Does it meet your expectations from your conceptualization of the problem? What about the algebraic form of the result? Does it make sense? Examine the variables in the problem to see whether the answer would change in a physically meaningful way if the variables were drastically increased or decreased or even became zero. Looking at limiting cases to see whether they yield expected values is a very useful way to make sure that you are obtaining reasonable results.
- Think about how this problem compared with others you have solved. How was it similar? In what critical ways did it differ? Why was this problem assigned? Can you figure out what you have learned by doing it? If it is a new category of problem, be sure you understand it so that you can use it as a model for solving similar problems in the future.

When solving complex problems, you may need to identify a series of subproblems and apply the problem-solving strategy to each. For simple problems, you probably don't need this strategy. When you are trying to solve a problem and you don't know what to do next, however, remember the steps in the strategy and use them as a guide.

In the rest of this book, we will label the *Conceptualize, Categorize, Analyze,* and *Finalize* steps explicitly in the worked examples. Many chapters in this book include a section labeled Problem-Solving Strategy that should help you through the rough spots. These sections are organized according to the General Problem-Solving Strategy outlined above and are tailored to the specific types of problems addressed in that chapter.

To clarify how this Strategy works, we repeat Example 1.8 on the next page with the particular steps of the Strategy identified.

When you **Conceptualize** a problem, try to understand the situation that is presented in the problem statement. Study carefully any representations of the information (for example, diagrams, graphs, tables, or photographs) that accompany the problem. Imagine a movie, running in your mind, of what happens in the problem.

Simplify the problem. Remove the details that are not important to the solution. Then **Categorize** the problem. Is it a simple substitution problem such that numbers can be substituted into an equation? If not, you face an analysis problem. In this case, identify the appropriate analysis model. (Analysis models will be introduced in Chapter 2.)

Now **Analyze** the problem. Select relevant equations from the analysis model. Solve symbolically for the unknown variable in terms of what is given. Substitute in the appropriate numbers, calculate the result, and round it to the proper number of significant figures.

Example **1.8** | Taking a Hike

A hiker begins a trip by first walking 25.0 km southeast from her car. She stops and sets up her tent for the night. On the second day, she walks 40.0 km in a direction 60.0° north of east, at which point she discovers a forest ranger's tower.

(A) Determine the components of the hiker's displacement for each day.

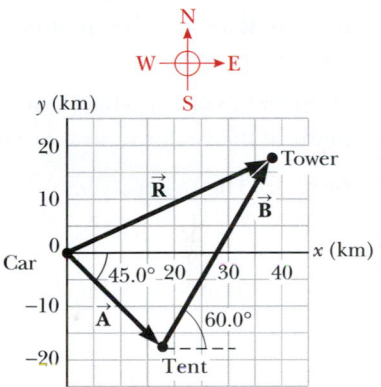

Figure 1.18 (Example 1.8) The total displacement of the hiker is the vector $\vec{R} = \vec{A} + \vec{B}$.

SOLUTION

Conceptualize We conceptualize the problem by drawing a sketch as in Figure 1.18. If we denote the displacement vectors on the first and second days by $\vec{A}$ and $\vec{B}$, respectively, and use the car as the origin of coordinates, we obtain the vectors shown in Figure 1.18.

Categorize Drawing the resultant $\vec{R}$, we can now categorize this problem as one we've solved before: an addition of two vectors. You should now have a hint of the power of categorization in that many new problems are very similar to problems we have already solved if we are careful to conceptualize them. Once we have drawn the displacement vectors and categorized the problem, this problem is no longer about a hiker, a walk, a car, a tent, or a tower. It is a problem about vector addition, one that we have already solved.

Analyze Displacement $\vec{A}$ has a magnitude of 25.0 km and is directed 45.0° below the positive x axis.

Find the components of $\vec{A}$ using Equation 1.10:

$$A_x = A\cos(-45.0°) = (25.0 \text{ km})(0.707) = \boxed{17.7 \text{ km}}$$

$$A_y = A\sin(-45.0°) = (25.0 \text{ km})(-0.707) = \boxed{-17.7 \text{ km}}$$

The negative value of A_y indicates that the hiker walks in the negative y direction on the first day. The signs of A_x and A_y also are evident from Figure 1.18.

Find the components of $\vec{B}$ using Equation 1.10:

$$B_x = B\cos 60.0° = (40.0 \text{ km})(0.500) = \boxed{20.0 \text{ km}}$$

$$B_y = B\sin 60.0° = (40.0 \text{ km})(0.866) = \boxed{34.6 \text{ km}}$$

(B) Determine the components of the hiker's resultant displacement $\vec{R}$ for the trip. Find an expression for $\vec{R}$ in terms of unit vectors.

SOLUTION

Use Equation 1.15 to find the components of the resultant displacement $\vec{R} = \vec{A} + \vec{B}$:

$$R_x = A_x + B_x = 17.7 \text{ km} + 20.0 \text{ km} = \boxed{37.7 \text{ km}}$$

$$R_y = A_y + B_y = -17.7 \text{ km} + 34.6 \text{ km} = \boxed{17.0 \text{ km}}$$

Write the total displacement in unit-vector form:

$$\vec{R} = \boxed{(37.7\hat{\mathbf{i}} + 17.0\hat{\mathbf{j}}) \text{ km}}$$

Finalize the problem. Examine the numerical answer. Does it have the correct units? Does it meet your expectations from your conceptualization of the problem? Does the answer make sense? What about the algebraic form of the result? Examine the variables in the problem to see whether the answer would change in a physically meaningful way if the variables were drastically increased or decreased or even became zero.

What If? questions will appear in many examples in the text, and offer a variation on the situation just explored. This feature encourages you to think about the results of the example and assists in conceptual understanding of the principles.

1.8 *cont.*

Finalize Looking at the graphical representation in Figure 1.18, we estimate the position of the tower to be about (38 km, 17 km), which is consistent with the components of $\vec{R}$ in our result for the final position of the hiker. Also, both components of $\vec{R}$ are positive, putting the final position in the first quadrant of the coordinate system, which is also consistent with Figure 1.18.

What If? After reaching the tower, the hiker wishes to return to her car along a single straight line. What are the components of the vector representing this hike? What should the direction of the hike be?

Answer The desired vector $\vec{R}_{car}$ is the negative of vector $\vec{R}$:

$$\vec{R}_{car} = -\vec{R} = (-37.7\hat{i} - 17.0\hat{j}) \text{ km}$$

The direction is found by calculating the angle that the vector makes with the x axis:

$$\tan \theta = \frac{R_{car,y}}{R_{car,x}} = \frac{-17.0 \text{ km}}{-37.7 \text{ km}} = 0.450$$

which gives an angle of $\theta = 204.2°$, or 24.2° south of west.

SUMMARY

Mechanical quantities can be expressed in terms of three fundamental quantities—**length, mass,** and **time**—which in the SI system have the units **meters** (m), **kilograms** (kg), and **seconds** (s), respectively. It is often useful to use the method of **dimensional analysis** to check equations and to assist in deriving expressions.

The **density** of a substance is defined as its mass per unit volume:

$$\rho \equiv \frac{m}{V} \qquad \qquad \text{1.1} \blacktriangleleft$$

Vectors are quantities that have both magnitude and direction and obey the vector law of addition. **Scalars** are quantities that add algebraically.

Two vectors $\vec{A}$ and $\vec{B}$ can be added using the triangle method. In this method (see Active Fig. 1.9), the vector $\vec{R} = \vec{A} + \vec{B}$ runs from the tail of $\vec{A}$ to the tip of $\vec{B}$.

The x component A_x of the vector $\vec{A}$ is equal to its projection along the x axis of a coordinate system, where $A_x = A \cos \theta$ and where θ is the angle $\vec{A}$ makes with the x axis. Likewise, the y component A_y of $\vec{A}$ is its projection along the y axis, where $A_y = A \sin \theta$.

If a vector $\vec{A}$ has an x component equal to A_x and a y component equal to A_y, the vector can be expressed in unit-vector form as $\vec{A} = (A_x\hat{i} + A_y\hat{j})$. In this notation, $\hat{i}$ is a unit vector in the positive x direction and $\hat{j}$ is a unit vector in the positive y direction. Because $\hat{i}$ and $\hat{j}$ are unit vectors, $|\hat{i}| = |\hat{j}| = 1$. In three dimensions, a vector can be expressed as $\vec{A} = (A_x\hat{i} + A_y\hat{j} + A_z\hat{k})$, where $\hat{k}$ is a unit vector in the z direction.

The resultant of two or more vectors can be found by resolving all vectors into their x, y, and z components and adding their components:

$$\vec{R} = \vec{A} + \vec{B} = (A_x + B_x)\hat{i} + (A_y + B_y)\hat{j} + (A_z + B_z)\hat{k} \qquad \text{1.18} \blacktriangleleft$$

Problem-solving skills and physical understanding can be improved by **modeling** the problem and by constructing **alternative representations** of the problem. Models helpful in solving problems include **geometric, simplification,** and **analysis models.** Scientists use **structural models** to understand systems larger or smaller in scale than those with which we normally have direct experience. Helpful representations include the **mental, pictorial, simplified pictorial, graphical, tabular,** and **mathematical representations.**

Complicated problems are best approached in an organized manner. Recall and apply the *Conceptualize, Categorize, Analyze,* and *Finalize* steps of the **General Problem-Solving Strategy** when you need them.

OBJECTIVE QUESTIONS

1. Answer each question yes or no. Must two quantities have the same dimensions (a) if you are adding them? (b) If you are multiplying them? (c) If you are subtracting them? (d) If you are dividing them? (e) If you are equating them?

2. The price of gasoline at a particular station is 1.5 euros per liter. An American student can use 33 euros to buy gasoline. Knowing that 4 quarts make a gallon and that 1 liter is close to 1 quart, she quickly reasons that she can buy how many gallons of gasoline? (a) less than 1 gallon (b) about 5 gallons (c) about 8 gallons (d) more than 10 gallons

3. Rank the following five quantities in order from the largest to the smallest. If two of the quantities are equal, give them equal rank in your list. (a) 0.032 kg (b) 15 g (c) 2.7×10^5 mg (d) 4.1×10^{-8} Gg (e) 2.7×10^8 μg

4. What is the y component of the vector $(3\hat{\mathbf{i}} - 8\hat{\mathbf{k}})$ m/s? (a) 3 m/s (b) −8 m/s (c) 0 (d) 8 m/s (e) none of those answers

5. Which of the following is the best estimate for the mass of all the people living on the Earth? (a) 2×10^8 kg (b) 1×10^9 kg (c) 2×10^{10} kg (d) 3×10^{11} kg (e) 4×10^{12} kg

6. What is the sum of the measured values 21.4 s + 15 s + 17.17 s + 4.00 3 s? (a) 57.573 s (b) 57.57 s (c) 57.6 s (d) 58 s (e) 60 s

7. One student uses a meterstick to measure the thickness of a textbook and obtains 4.3 cm ± 0.1 cm. Other students measure the thickness with vernier calipers and obtain four different measurements: (a) 4.32 cm ± 0.01 cm, (b) 4.31 cm ± 0.01 cm, (c) 4.24 cm ± 0.01 cm, and (d) 4.43 cm ± 0.01 cm. Which of these four measurements, if any, agree with that obtained by the first student?

8. Newton's second law of motion (Chapter 4) says that the mass of an object times its acceleration is equal to the net force on the object. Which of the following gives the correct units for force? (a) kg · m/s² (b) kg · m²/s² (c) kg/m · s² (d) kg · m²/s (e) none of those answers

9. What is the x component of the vector shown in Figure OQ1.9? (a) 3 cm (b) 6 cm (c) −4 cm (d) −6 cm (e) none of those answers

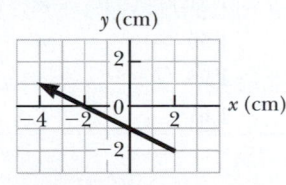

Figure OQ1.9 Objective Questions 9 and 10.

10. What is the y component of the vector shown in Figure OQ1.9? (a) 3 cm (b) 6 cm (c) −4 cm (d) −6 cm (e) none of those answers

11. Yes or no: Is each of the following quantities a vector? (a) force (b) temperature (c) the volume of water in a can (d) the ratings of a TV show (e) the height of a building (f) the velocity of a sports car (g) the age of the Universe

12. A vector lying in the xy plane has components of opposite sign. The vector must lie in which quadrant? (a) the first quadrant (b) the second quadrant (c) the third quadrant (d) the fourth quadrant (e) either the second or the fourth quadrant

13. Figure OQ1.13 shows two vectors $\vec{\mathbf{D}}_1$ and $\vec{\mathbf{D}}_2$. Which of the possibilities (a) through (d) is the vector $\vec{\mathbf{D}}_2 - 2\vec{\mathbf{D}}_1$, or (e) is it none of them?

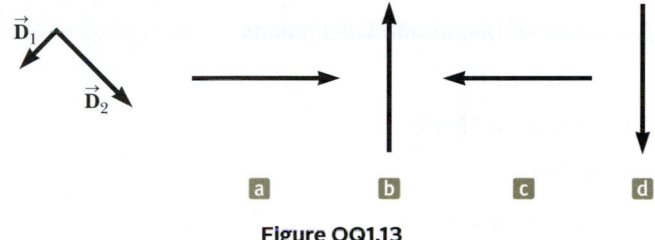

Figure OQ1.13

14. A vector points from the origin into the second quadrant of the xy plane. What can you conclude about its components? (a) Both components are positive. (b) The x component is positive, and the y component is negative. (c) The x component is negative, and the y component is positive. (d) Both components are negative. (e) More than one answer is possible.

15. What is the magnitude of the vector $(10\hat{\mathbf{i}} - 10\hat{\mathbf{k}})$ m/s? (a) 0 (b) 10 m/s (c) −10 m/s (d) 10 (e) 14.1 m/s

16. Vector $\vec{\mathbf{A}}$ lies in the xy plane. Both of its components will be negative if it points from the origin into which quadrant? (a) the first quadrant (b) the second quadrant (c) the third quadrant (d) the fourth quadrant (e) the second or fourth quadrants

CONCEPTUAL QUESTIONS

1. What natural phenomena could serve as alternative time standards?

2. Express the following quantities using the prefixes given in Table 1.4. (a) 3×10^{-4} m (b) 5×10^{-5} s (c) 72×10^2 g

3. Suppose the three fundamental standards of the metric system were length, *density*, and time rather than length, *mass*, and time. The standard of density in this system is to be defined as that of water. What considerations about water would you need to address to make sure that the standard of density is as accurate as possible?

4. If the component of vector $\vec{\mathbf{A}}$ along the direction of vector $\vec{\mathbf{B}}$ is zero, what can you conclude about the two vectors?

5. A book is moved once around the perimeter of a tabletop with the dimensions 1.0 m by 2.0 m. The book ends up at its

initial position. (a) What is its displacement? (b) What is the distance traveled?

6. Can the magnitude of a vector have a negative value? Explain.

7. On a certain calculator, the inverse tangent function returns a value between −90° and +90°. In what cases will this value correctly state the direction of a vector in the *xy* plane, by giving its angle measured counterclockwise from the positive *x* axis? In what cases will it be incorrect?

8. Is it possible to add a vector quantity to a scalar quantity? Explain.

PROBLEMS AVAILABLE IN

1.1 Standards of Length, Mass, and Time
Problems 1–4

1.2 Dimensional Analysis
Problems 5–7

1.3 Conversion of Units
Problems 8–16

1.4 Order-of-Magnitude Calculations
Problems 17–20

1.5 Significant Figures
Problems 21–29

1.6 Coordinate Systems
Problems 30–33

1.7 Vectors and Scalars
1.8 Some Properties of Vectors
Problems 34–37

1.9 Components of a Vector and Unit Vectors
Problems 38–52

1.10 Modeling, Alternative Representations, and Problem-Solving Strategy
Problems 53–55

Additional Problems
Problems 56–71

List of Enhanced Problems

Problem Number	Targeted Feedback in Enhanced WebAssign	Master It in Enhanced WebAssign	Watch It in Enhanced WebAssign
1.1	✓		✓
1.6	✓		✓
1.9	✓	✓	
1.10	✓		✓
1.13		✓	
1.15	✓		✓
1.22	✓		✓
1.23	✓		✓
1.31	✓		✓
1.33	✓		✓
1.34		✓	
1.38	✓		✓
1.41	✓		✓
1.43		✓	
1.45		✓	
1.46	✓		✓
1.47		✓	
1.50	✓		✓
1.51	✓	✓	
1.55	✓		✓
1.58		✓	
1.67	✓		✓

Solutions to the following Problems are available in the *Student Solutions Manual/Study Guide*:

1.5, 1.9, 1.13, 1.17, 1.19, 1.23, 1.31, 1.34, 1.37, 1.41, 1.43, 1.47, 1.51, 1.56, and 1.58

Alternative-Fuel Vehicles

The idea of self-propelled vehicles has been part of the human imagination for centuries. Leonardo da Vinci drew plans for a vehicle powered by a wound spring in 1478. This vehicle was never built although models have been constructed from his plans and appear in museums. Isaac Newton developed a vehicle in 1680 that operated by ejecting steam out the back, similar to a rocket engine. This invention did not develop into a useful device. Despite these and other attempts, self-propelled vehicles did not succeed; that is, they did not begin to replace the horse as a primary means of transportation until the 19th century.

The history of *successful* self-propelled vehicles begins in 1769 with the invention of a military tractor by Nicolas Joseph Cugnot in France. This vehicle, as well as Cugnot's follow-up vehicles, was powered by a steam engine. During the remainder of the 18th century and for most of the 19th century, additional steam-driven vehicles were developed in France, Great Britain, and the United States.

After the invention of the electric battery by Italian Alessandro Volta at the beginning of the 19th century and its further development over three decades came the invention of early electric vehicles in the 1830s. The development in 1859 of the storage battery, which could be recharged, provided significant impetus to the development of electric vehicles. By the early 20th century,

electric cars with a range of about 20 miles and a top speed of 15 miles per hour had been developed.

An internal combustion engine was designed but never built by Dutch physicist Christiaan Huygens in 1680. The invention of modern gasoline-powered internal combustion vehicles is generally credited to Gottlieb Daimler in 1885 and Karl Benz in 1886. Several earlier vehicles, dating back to 1807, however, used internal combustion engines operating on various fuels, including coal gas and primitive gasoline.

At the beginning of the 20th century, steam-powered, gasoline-powered, and electric cars shared the roadways in the United States. Electric cars did not possess the vibration, smell, and noise of gasoline-powered cars and did not suffer from the long start-up time intervals, up to 45 minutes, of steam-powered cars on cold mornings. Electric cars were especially preferred by women, who did not enjoy the difficult task of cranking a gasoline-powered car to start the engine. The limited range of electric cars was not a significant problem because the only roads that existed were in highly populated areas and cars were primarily used for short trips in town.

The end of electric cars in the early 20th century began with the following developments:

- 1901: A major discovery of crude oil in Texas reduced prices of gasoline to widely affordable levels.
- 1912: The electric starter for gasoline engines was invented, removing the physical task of cranking the engine.
- During the 1910s: Henry Ford successfully introduced mass production of internal combustion vehicles, resulting in a drop in the price of these vehicles to significantly less than that of an electric car.
- By the early 1920s: Roadways in the United States were of much better quality than in the previous decades, and now connected cities with each other, requiring vehicles with a longer range than when roadways existed only within city limits.

Because of these factors, the roadways were ruled by gasoline-powered cars almost exclusively by the 1920s. Gasoline, however, is a finite and short-lived commodity. We are approaching the end of our ability to use gasoline

Figure 1 A model of a spring-drive car designed by Leonardo da Vinci.

Figure 2 This magazine advertisement for an electric car is typical of this popular type of car in the early 20th century.

in transportation; some experts predict that diminishing supplies of crude oil will push the cost of gasoline to prohibitively high levels within two more decades. Furthermore, gasoline and diesel fuel result in serious

Figure 3 A bus running on natural gas operates in Port Huron, Michigan. Several cities such as Port Huron have established natural gas refueling centers so that a large percentage of their fleet can be operated with this fuel that is less expensive than diesel and emits fewer particulates into the atmosphere.

Figure 4 Modern electric cars can take advantage of an infrastructure set up in some localities to provide charging stations in parking lots.

tailpipe emissions that are harmful to the environment. As we look for a replacement for gasoline, we also want to pursue fuels that will be kinder to the atmosphere. Such fuels will help reduce the climate change effects of global warming, which we will study in Context 5.

What do the steam engine, the electric motor, and the internal combustion engine have in common? That is, what do they each extract from a source, be it a type of fuel or an electric battery? The answer to this question is *energy*. Regardless of the type of automobile, some source of energy must be provided. Energy is one of the physical concepts that we will investigate in this Context. A fuel such as gasoline contains energy due to its chemical composition and its ability to undergo a combustion process. The battery in an electric car also contains energy, again related to chemical composition, but in this case it is associated with an ability to produce an electric current.

One difficult social aspect of developing a new energy source for automobiles is that there must be a synchronized development of the new automobile along with the infrastructure for delivering the new source of energy. This aspect requires close cooperation between automotive corporations and energy manufacturers and suppliers. For example, electric cars cannot be used to travel long distances unless an infrastructure of charging stations develops in parallel with the development of electric cars.

As we draw near to the time when we run out of gasoline, our central question in this first Context is an important one for our future development:

> **What source besides gasoline can be used to provide energy for an automobile while reducing environmentally damaging emissions?**

Motion in One Dimension

iStockphoto.com/technottechnot·

One of the physical quantities we will study in this chapter is the velocity of an object moving in a straight line. Downhill skiers can reach velocities with a magnitude greater than 100 km/h.

To begin our study of motion, it is important to be able to *describe* motion using the concepts of space and time without regard to the causes of the motion. This portion of mechanics is called *kinematics* (from the same root as the word *cinema*). In this chapter, we shall consider motion along a straight line, that is, one-dimensional motion. Chapter 3 extends our discussion to two-dimensional motion.

From everyday experience we recognize that motion represents continuous change in the position of an object. For example, if you are driving from your home to a destination, your position on the Earth's surface is changing.

The movement of an object through space (translation) may be accompanied by the rotation or vibration of the object. Such motions can be quite complex. It is often possible to simplify matters, however, by temporarily ignoring rotation and internal motions of the moving object. The result is the simplification model that we call the particle model, discussed in Chapter 1. In many situations, an object can be treated as a particle if the only motion being considered is translation through space. We will use the particle model extensively throughout this book..

❮ **2.1** | Average Velocity

We begin our study of kinematics with the notion of average velocity. You may be familiar with a similar notion, average speed, from experiences with driving. If you drive your car 100 miles according to your odometer and it takes 2.0 hours to do so, your average speed is $(100 \text{ mi})/(2.0 \text{ h}) = 50 \text{ mi/h}$. For a particle moving through a distance d in a time interval Δt, the **average speed** v_{avg} is mathematically defined as

▶ Definition of average speed

$$v_{\text{avg}} \equiv \frac{d}{\Delta t} \qquad \qquad \text{2.1} \blacktriangleleft$$

Speed is not a vector, so there is no direction associated with average speed.

Average velocity may be a little less familiar to you due to its vector nature. Let us start by imagining the motion of a particle, which, through the particle model, can represent the motion of many types of objects. We shall restrict our study at this point to one-dimensional motion along the x axis.

The motion of a particle is completely specified if the position of the particle in space is known at all times. Consider a car moving back and forth along the x axis and imagine that we take data on the position of the car every 10 s. Active Figure 2.1a is a *pictorial* representation of this one-dimensional motion that shows the positions of the car at 10-s intervals. The six data points we have recorded are represented by the letters Ⓐ through Ⓕ. Table 2.1 is a *tabular* representation of the motion. It lists the data as entries for position at each time. The black dots in Active Figure 2.1b show a *graphical* representation of the motion. Such a plot is often called a **position– time graph.** The curved line in Active Figure 2.1b cannot be unambiguously drawn

Active Figure 2.1 A car moves back and forth along a straight line. Because we are interested only in the car's translational motion, we can model it as a particle. Several representations of the information about the motion of the car can be used. Table 2.1 is a tabular representation of the information. (a) A pictorial representation of the motion of the car. (b) A graphical representation, known as a position–time graph, of the car's motion in part (a). The average velocity $v_{x,\text{avg}}$ in the interval $t = 0$ to $t = 10$ s is obtained from the slope of the straight line connecting points Ⓐ and Ⓑ. (c) A velocity–time graph of the motion of the car in part (a).

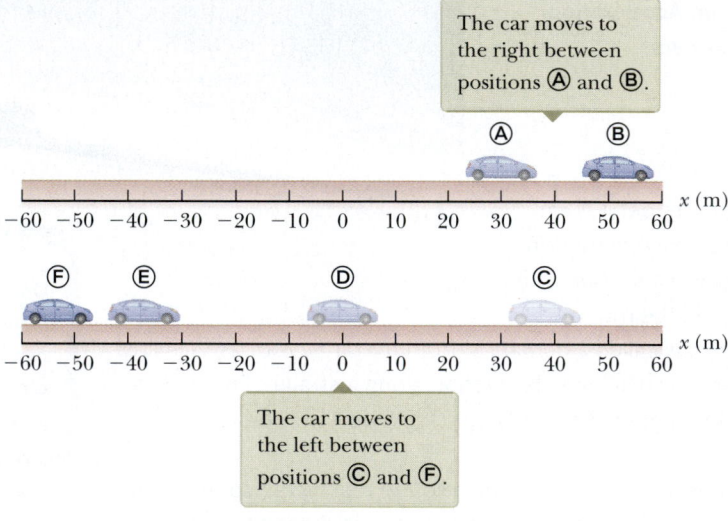

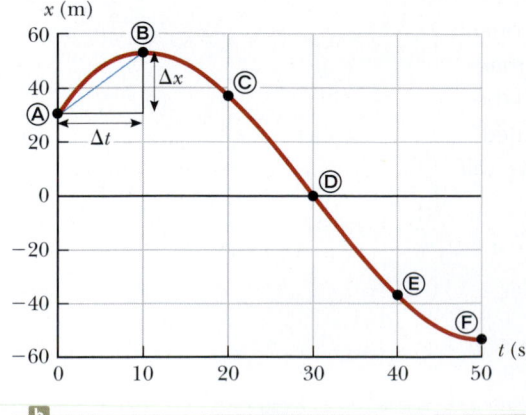

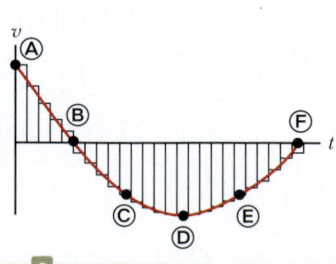

❮ **TABLE 2.1** |
Positions of the Car at Various Times

Position	t (s)	x (m)
Ⓐ	0	30
Ⓑ	10	52
Ⓒ	20	38
Ⓓ	30	0
Ⓔ	40	−37
Ⓕ	50	−53

through our six data points because we have no information about what happened between these points. The curved line is, however, a *possible* graphical representation of the position of the car at all instants of time during the 50 s.

If a particle is moving during a time interval $\Delta t = t_f - t_i$, the displacement of the particle is described as $\Delta \vec{\mathbf{x}} = \vec{\mathbf{x}}_f - \vec{\mathbf{x}}_i = (x_f - x_i)\hat{\mathbf{i}}$. (Recall from Chapter 1 that displacement is defined as the change in the position of the particle, which is equal to its final position value minus its initial position value.) Because we are considering only one-dimensional motion in this chapter, we shall drop the vector notation at this point and pick it up again in Chapter 3. The direction of a vector in this chapter will be indicated by means of a positive or negative sign.

The **average velocity** $v_{x,\text{avg}}$ of the particle is defined as the ratio of its displacement Δx to the time interval Δt during which the displacement takes place:

$$v_{x,\text{avg}} \equiv \frac{\Delta x}{\Delta t} = \frac{x_f - x_i}{t_f - t_i} \qquad \textbf{2.2} \blacktriangleleft$$

▶ Definition of average velocity

where the subscript x indicates motion along the x axis. From this definition we see that average velocity has the dimensions of length divided by time: meters per second in SI units and feet per second in U.S. customary units. The average velocity is *independent* of the path taken between the initial and final points. This independence is a major difference from the average speed discussed at the beginning of this section. The average velocity is independent of path because it is proportional to the displacement Δx, which depends only on the initial and final coordinates of the particle. Average speed (a scalar) is found by dividing the *distance* traveled by the time interval, whereas average velocity (a vector) is the *displacement* divided by the time interval. Therefore, average velocity gives us no details of the motion; rather, it only gives us the result of the motion. Finally, note that the average velocity in one dimension can be positive or negative, depending on the sign of the displacement. (The time interval Δt is always positive.) If the x coordinate of the particle increases during the time interval (i.e., if $x_f > x_i$), Δx is positive and $v_{x,\text{avg}}$ is positive, which corresponds to an average velocity in the positive x direction. On the other hand, if the coordinate decreases over time ($x_f < x_i$), Δx is negative; hence, $v_{x,\text{avg}}$ is negative, which corresponds to an average velocity in the negative x direction.

QUICK QUIZ 2.1 Under which of the following conditions is the magnitude of the average velocity of a particle moving in one dimension smaller than the average speed over some time interval? (**a**) A particle moves in the $+x$ direction without reversing. (**b**) A particle moves in the $-x$ direction without reversing. (**c**) A particle moves in the $+x$ direction and then reverses the direction of its motion. (**d**) There are no conditions for which it is true.

The average velocity can also be interpreted geometrically, as seen in the graphical representation in Active Figure 2.1b. A straight line can be drawn between any two points on the curve. Active Figure 2.1b shows such a line drawn between points Ⓐ and Ⓑ. Using a geometric model, this line forms the hypotenuse of a right triangle of height Δx and base Δt. The slope of the hypotenuse is the ratio $\Delta x / \Delta t$. Therefore, we see that the average velocity of the particle during the time interval t_i to t_f is equal to the slope of the straight line joining the initial and final points on the position–time graph. For example, the average velocity of the car between points Ⓐ and Ⓑ is $v_{x,\text{avg}} = (52 \text{ m} - 30 \text{ m})/(10 \text{ s} - 0) = 2.2 \text{ m/s}$.

We can also identify a geometric interpretation for the total displacement during the time interval. Active Figure 2.1c shows the velocity–time graphical representation of the motion in Active Figures 2.1a and 2.1b. The total time interval for the motion has been divided into small increments of duration Δt_n. During each of these increments, if we model the velocity as constant during the short increment, the displacement of the particle is given by $\Delta x_n = v_n \Delta t_n$.

Geometrically, the product on the right side of this expression represents the area of a thin rectangle associated with each time increment in Active Figure 2.1c;

Pitfall Prevention | 2.1
Average Speed and Average Velocity
The magnitude of the average velocity is *not* the average speed. Consider a particle moving from the origin to $x = 10$ m and then back to the origin in a time interval of 4.0 s. The magnitude of the average velocity is zero because the particle ends the time interval at the same position at which it started; the displacement is zero. The average speed, however, is the total distance divided by the time interval: 20 m/4.0 s = 5.0 m/s.

Pitfall Prevention | 2.2
Slopes of Graphs
In any graph of physical data, the slope represents the ratio of the change in the quantity represented on the vertical axis to the change in the quantity represented on the horizontal axis. Remember that a slope has units (unless both axes have the same units). The units of slope in Active Figures 2.1b and 2.2 (page 38) are meters per second, the units of velocity.

the height of the rectangle (measured from the time axis) is v_n, and the width is Δt_n. The total displacement of the particle will be the sum of the displacements during each of the increments:

$$\Delta x \approx \sum_n \Delta x_n = \sum_n v_n \Delta t_n$$

This sum is an approximation because we have modeled the velocity as constant in each increment, which is not the case. The term on the right represents the total area of all the thin rectangles. Now let us take the limit of this expression as the time increments shrink to zero, in which case the approximation becomes exact:

$$\Delta x = \lim_{\Delta t_n \to 0} \sum_n \Delta x_n = \lim_{\Delta t_n \to 0} \sum_n v_n \Delta t_n$$

In this limit, the sum of the areas of all the very thin rectangles becomes equal to the total area under the curve. Therefore, the displacement of a particle during the time interval t_i to t_f is equal to the area under the curve between the initial and final points on the velocity–time graph. We will make use of this geometric interpretation in Section 2.6.

Example 2.1 | Calculating the Average Velocity and Speed

Find the displacement, average velocity, and average speed of the car in Active Figure 2.1a between positions Ⓐ and Ⓕ.

SOLUTION

Conceptualize Consult the pictorial representation in Active Figure 2.1 to form a mental image of the car and its motion. Active Figure 2.1b shows a graphical representation of the motion in the form of a position–time graph for the particle.

Categorize We model the car as a particle. We will be substituting numerical values into definitions that we have seen, so this problem will be categorized as a substitution problem.

Analyze From the position–time graph given in Active Figure 2.1b, notice that $x_Ⓐ = 30$ m at $t_Ⓐ = 0$ s and that $x_Ⓕ = -53$ m at $t_Ⓕ = 50$ s.

Use Equation 1.6 to find the displacement of the car:

$$\Delta x = x_Ⓕ - x_Ⓐ = -53 \text{ m} - 30 \text{ m} = \boxed{-83 \text{ m}}$$

Use Equation 2.2 to find the car's average velocity:

$$v_{x,\text{avg}} = \frac{x_Ⓕ - x_Ⓐ}{t_Ⓕ - t_Ⓐ}$$

$$= \frac{-53 \text{ m} - 30 \text{ m}}{50 \text{ s} - 0 \text{ s}} = \frac{-83 \text{ m}}{50 \text{ s}} = \boxed{-1.7 \text{ m/s}}$$

We cannot unambiguously find the average speed of the car from the data in Table 2.1, because we do not have information about the positions of the car between the data points. If we adopt the assumption that the details of the car's position are described by the curve in Active Figure 2.1b, the distance traveled is 22 m (from Ⓐ to Ⓑ) plus 105 m (from Ⓑ to Ⓕ), for a total of 127 m.

Use Equation 2.1 to find the car's average speed:

$$v_{\text{avg}} = \frac{127 \text{ m}}{50 \text{ s}} = \boxed{2.5 \text{ m/s}}$$

Finalize The first result means that the car ends up 83 m in the negative direction (to the left, in this case) from where it started. This number has the correct units and is of the same order of magnitude as the supplied data. A quick look at Active Figure 2.1a indicates that it is the correct answer. The fact that the car ends up to the left of its initial position also makes it reasonable that the average velocity is negative.

Notice that the average speed is positive, as it must be. Suppose the red-brown curve in Active Figure 2.1b were different so that between 0 s and 10 s it went from Ⓐ up to 100 m and then came back down to Ⓑ. The average speed of the car would change because the distance is different, but the average velocity would not change.

Substitution problems generally do not have an extensive Analyze section other than the substitution of numbers into a given equation. Similarly, the Finalize step consists primarily of checking the units and making sure that the answer is reasonable. Therefore, for substitution problems after this one, we will not label Analyze or Finalize steps. We include those labels in this first example of a substitution problem just to demonstrate the process.

Example 2.2 | Motion of a Jogger

A jogger runs in a straight line, with an average velocity of magnitude 5.00 m/s for 4.00 min and then with an average velocity of magnitude 4.00 m/s for 3.00 min.

(A) What is the magnitude of the final displacement from her initial position?

SOLUTION

Conceptualize From your experience, imagine a runner running on a track. Notice that the runner runs more slowly on the average during the second time interval, as if he or she is tiring.

Categorize That this problem involves a jogger is not important; we model the jogger as a particle.

Analyze From the data for the two separate portions of the motion, find the displacement for each portion, using Equation 2.2:

$$v_{x,avg} = \frac{\Delta x}{\Delta t} \quad \rightarrow \quad \Delta x = v_{x,avg} \Delta t$$

$$\Delta x_{portion\ 1} = (5.00\ \text{m/s})(4.00\ \text{min})\left(\frac{60\ \text{s}}{1\ \text{min}}\right)$$

$$= 1.20 \times 10^3\ \text{m}$$

$$\Delta x_{portion\ 2} = (4.00\ \text{m/s})(3.00\ \text{min})\left(\frac{60\ \text{s}}{1\ \text{min}}\right)$$

$$= 7.20 \times 10^2\ \text{m}$$

We add these two displacements to find the total displacement of 1.92×10^3 m.

(B) What is the magnitude of her average velocity during this entire time interval of 7.00 min?

SOLUTION

Find the average velocity for the entire time interval using Equation 2.2:

$$v_{x,avg} = \frac{\Delta x}{\Delta t} = \frac{1.92 \times 10^3\ \text{m}}{7.00\ \text{min}}\left(\frac{1\ \text{min}}{60\ \text{s}}\right) = 4.57\ \text{m/s}$$

Finalize Notice that the average velocity is between the two velocities given in the problem, as expected, but is *not* the arithmetic mean of these two velocities.

2.2 | Instantaneous Velocity

Suppose you drive your car through a displacement of magnitude 40 miles and it takes exactly 1 hour to do so, from 1:00:00 P.M. to 2:00:00 P.M. Then the magnitude of your average velocity is 40 mi/h for the 1-h interval. How fast, though, were you going at the particular *instant* of time 1:20:00 P.M.? It is likely that your velocity varied during the trip, owing to hills, traffic lights, slow drivers ahead of you, and the like, so that there was not a single velocity maintained during the entire hour of travel. The velocity of a particle at any instant of time is called the *instantaneous velocity*.

Consider again the motion of the car shown in Active Figure 2.1a. Active Figure 2.2a (page 38) is the graphical representation again, with two blue lines representing average velocities over very different time intervals. One blue line represents the average velocity we calculated earlier over the interval from Ⓐ to Ⓑ. The second blue line represents the average velocity over the much longer interval Ⓐ to Ⓕ. How well does either of these represent the instantaneous velocity at point Ⓐ? In Active Figure 2.1a, the car begins to move to the right, which we identify as a positive velocity. The average velocity from Ⓐ to Ⓕ is *negative* (because the slope of the line from Ⓐ to Ⓕ

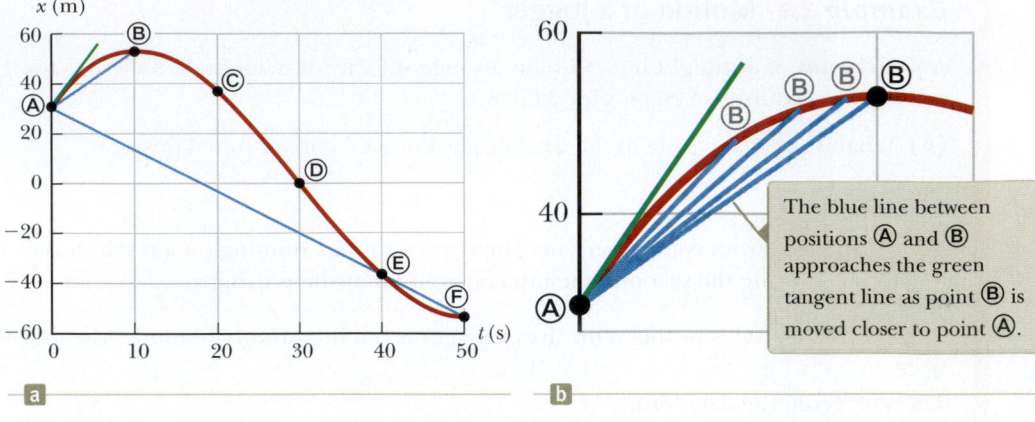

a

b

The blue line between positions Ⓐ and Ⓑ approaches the green tangent line as point Ⓑ is moved closer to point Ⓐ.

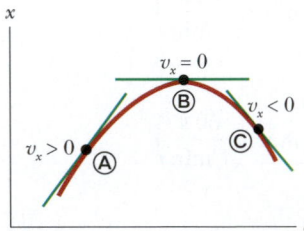

Figure 2.3 In the position–time graph shown, the velocity is positive at Ⓐ, where the slope of the tangent line is positive; the velocity is zero at Ⓑ, where the slope of the tangent line is zero; and the velocity is negative at Ⓒ, where the slope of the tangent line is negative.

▶ Definition of instantaneous velocity

Pitfall Prevention | 2.3
Instantaneous Speed and Instantaneous Velocity
In Pitfall Prevention 2.1, we argued that the magnitude of the average velocity is not the average speed. The magnitude of the instantaneous velocity, however, *is* the instantaneous speed. In an infinitesimal time interval, the magnitude of the displacement is equal to the distance traveled by the particle.

is negative), so this velocity clearly is not an accurate representation of the instantaneous velocity at Ⓐ. The average velocity from interval Ⓐ to Ⓑ is *positive*, so this velocity at least has the right sign.

In Active Figure 2.2b, we show the result of drawing the lines representing the average velocity of the car as point Ⓑ is brought closer and closer to point Ⓐ. As that occurs, the slope of the blue line approaches that of the green line, which is the line drawn tangent to the curve at point Ⓐ. As Ⓑ approaches Ⓐ, the time interval that includes point Ⓐ becomes infinitesimally small. Therefore, the average velocity over this interval as the interval shrinks to zero can be interpreted as the instantaneous velocity at point Ⓐ. Furthermore, the slope of the line tangent to the curve at Ⓐ is the instantaneous velocity at the time $t_{Ⓐ}$. In other words, the **instantaneous velocity** v_x equals the limiting value of the ratio $\Delta x/\Delta t$ as Δt approaches zero:[1]

$$v_x \equiv \lim_{\Delta t \to 0} \frac{\Delta x}{\Delta t}$$

In calculus notation, this limit is called the *derivative* of x with respect to t, written dx/dt:

$$v_x \equiv \lim_{\Delta t \to 0} \frac{\Delta x}{\Delta t} = \frac{dx}{dt} \qquad \text{2.3} \blacktriangleleft$$

The instantaneous velocity can be positive, negative, or zero. When the slope of the position–time graph is positive, such as at point Ⓐ in Figure 2.3, v_x is positive. At point Ⓒ, v_x is negative because the slope is negative. Finally, the instantaneous velocity is zero at the peak Ⓑ (the turning point), where the slope is zero. From here on, we shall usually use the word *velocity* to designate instantaneous velocity. ✦

The **instantaneous speed** of a particle is defined as the magnitude of the instantaneous velocity vector. Hence, by definition, *speed* can never be negative.

▌ **QUICK QUIZ 2.2** Are members of the highway patrol more interested in **(a)** your average speed or **(b)** your instantaneous speed as you drive?

If you are familiar with calculus, you should recognize that specific rules exist for taking the derivatives of functions. These rules, which are listed in Appendix B.6, enable us to evaluate derivatives quickly.

Suppose x is proportional to some power of t, such as

$$x = At^n$$

where A and n are constants. (This equation is a very common functional form.) The derivative of x with respect to t is

$$\frac{dx}{dt} = nAt^{n-1}$$

For example, if $x = 5t^3$, we see that $dx/dt = 3(5)t^{3-1} = 15t^2$.

---

[1]Note that the displacement Δx also approaches zero as Δt approaches zero. As Δx and Δt become smaller and smaller, however, the ratio $\Delta x/\Delta t$ approaches a value equal to the *true* slope of the line tangent to the x versus t curve.

THINKING PHYSICS 2.1

Consider the following motions of an object in one dimension. (**a**) A ball is thrown directly upward, rises to its highest point, and falls back into the thrower's hand. (**b**) A race car starts from rest and speeds up to 100 m/s along a straight line. (**c**) A spacecraft on the way to another star drifts through empty space at constant velocity. Are there any instants of time in the motion of these objects at which the instantaneous velocity at the instant and the average velocity over the entire interval are the same? If so, identify the point(s).

Reasoning (**a**) The average velocity over the entire interval for the thrown ball is zero; the ball returns to the starting point at the end of the time interval. There is one point—at the top of the motion—at which the instantaneous velocity is zero. (**b**) The average velocity for the motion of the race car cannot be evaluated unambiguously with the information given, but its magnitude must be some value between 0 and 100 m/s. Because the magnitude of the instantaneous velocity of the car will have every value between 0 and 100 m/s at some time during the interval, there must be some instant at which the instantaneous velocity is equal to the average velocity over the entire interval. (**c**) Because the instantaneous velocity of the spacecraft is constant, its instantaneous velocity at *any* time and its average velocity over *any* time interval are the same. ◀

Example 2.3 | The Limiting Process

The position of a particle moving along the x axis varies in time according to the expression[2] $x = 3t^2$, where x is in meters and t is in seconds. Find the velocity in terms of t at any time.

SOLUTION

Conceptualize The position–time graphical representation for this motion is shown in Figure 2.4. Before beginning the calculation, imagine the motion of the particle on the x axis. Does the particle ever reverse direction?

Categorize The entity in motion is already presented as a particle, so no simplification model is needed.

Analyze We can compute the velocity at any time t by using the definition of the instantaneous velocity.

If the initial coordinate of the particle at time t is $x_i = 3t^2$, find the coordinate at a later time $t + \Delta t$:

$$x_f = 3(t + \Delta t)^2 = 3[t^2 + 2t\,\Delta t + (\Delta t)^2]$$
$$= 3t^2 + 6t\,\Delta t + 3(\Delta t)^2$$

Find the displacement in the time interval Δt:

$$\Delta x = x_f - x_i = (3t^2 + 6t\,\Delta t + 3(\Delta t)^2) - (3t^2)$$
$$= 6t\,\Delta t + 3(\Delta t)^2$$

Find the average velocity in this time interval:

$$v_{x,avg} = \frac{\Delta x}{\Delta t} = \frac{6t\,\Delta t + 3(\Delta t)^2}{\Delta t} = 6t + 3\,\Delta t$$

To find the instantaneous velocity, take the limit of this expression as Δt approaches zero:

$$v_x = \lim_{\Delta t \to 0} \frac{\Delta x}{\Delta t} = 6t + 3(0) = \boxed{6t}$$

continued

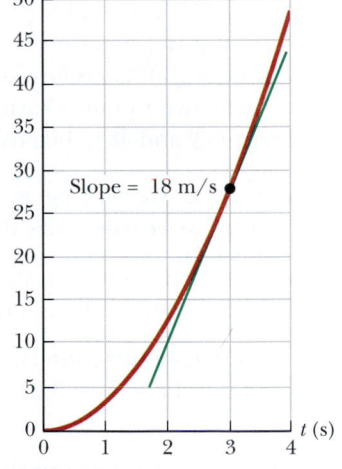

Figure 2.4 (Example 2.3) Position–time graph for a particle having an x coordinate that varies in time according to $x = 3t^2$. Note that the instantaneous velocity at $t = 3.0$ s is obtained from the slope of the green line tangent to the curve at this point.

Slope = 18 m/s

---

[2] Simply to make it easier to read, we write the equation as $x = 3t^2$ rather than as $x = (3.00 \text{ m/s}^2)\,t^{2.00}$. When an equation summarizes measurements, consider its coefficients to have as many significant digits as other data quoted in a problem. Also consider its coefficients to have the units required for dimensional consistency. When we start our clocks at $t = 0$, we usually do not mean to limit precision to a single digit. Consider any zero value in this book to have as many significant figures as you need.

2.3 cont.

Finalize Notice that this expression gives us the velocity at *any* general time *t*. It tells us that v_x is increasing linearly in time. It is then a straightforward matter to find the velocity at some specific time from the expression $v_x = 6t$ by substituting the value of the time. For example, at $t = 3.0$ s, the velocity is $v_x = 6(3) = 18$ m/s. Again, this answer can be checked from the slope at $t = 3.0$ s (the green line in Fig. 2.4).

We can also find v_x by taking the first derivative of x with respect to time, as in Equation 2.3. In this example, $x = 3t^2$, and we see that $v_x = dx/dt = 6t$, in agreement with our result from taking the limit explicitly.

Example 2.4 | Average and Instantaneous Velocity

A particle moves along the *x* axis. Its position varies with time according to the expression $x = -4t + 2t^2$, where *x* is in meters and *t* is in seconds. The position–time graph for this motion is shown in Figure 2.5a. Because the position of the particle is given by a mathematical function, the motion of the particle is completely known, unlike that of the car in Active Figure 2.1. Notice that the particle moves in the negative *x* direction for the first second of motion, is momentarily at rest at the moment $t = 1$ s, and moves in the positive *x* direction at times $t > 1$ s.

(A) Determine the displacement of the particle in the time intervals $t = 0$ to $t = 1$ s and $t = 1$ s to $t = 3$ s.

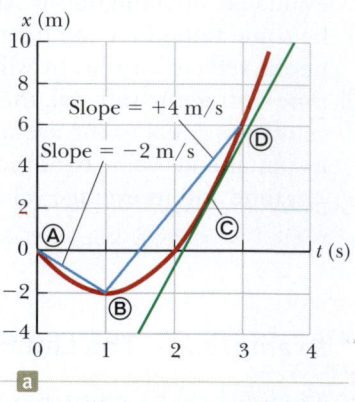

SOLUTION

Conceptualize From the graph in Figure 2.5a, form a mental representation of the particle's motion. Keep in mind that the particle does not move in a curved path in space such as that shown by the red-brown curve in the graphical representation. The particle moves only along the *x* axis in one dimension as shown in Figure 2.5b. At $t = 0$, is it moving to the right or to the left?

During the first time interval, the slope is negative and hence the average velocity is negative. Therefore, we know that the displacement between Ⓐ and Ⓑ must be a negative number having units of meters. Similarly, we expect the displacement between Ⓑ and Ⓓ to be positive.

Categorize We will be evaluating results from definitions given in the first two chapters, so we categorize this example as a substitution problem.

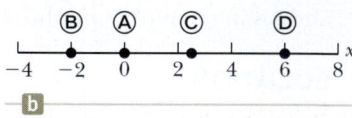

Figure 2.5 (Example 2.4) (a) Position–time graph for a particle having an *x* coordinate that varies in time according to the expression $x = -4t + 2t^2$. (b) The particle moves in one dimension along the *x* axis.

In the first time interval, set $t_i = t_Ⓐ = 0$ and $t_f = t_Ⓑ = 1$ s and use Equation 1.6 to find the displacement:

$$\Delta x_{Ⓐ \to Ⓑ} = x_f - x_i = x_Ⓑ - x_Ⓐ$$
$$= [-4(1) + 2(1)^2] - [-4(0) + 2(0)^2] = \boxed{-2 \text{ m}}$$

For the second time interval ($t = 1$ s to $t = 3$ s), set $t_i = t_Ⓑ = 1$ s and $t_f = t_Ⓓ = 3$ s:

$$\Delta x_{Ⓑ \to Ⓓ} = x_f - x_i = x_Ⓓ - x_Ⓑ$$
$$= [-4(3) + 2(3)^2] - [-4(1) + 2(1)^2] = \boxed{+8 \text{ m}}$$

These displacements can also be read directly from the position–time graph.

(B) Calculate the average velocity during these two time intervals.

SOLUTION

In the first time interval, use Equation 2.2 with $\Delta t = t_f - t_i = t_Ⓑ - t_Ⓐ = 1$ s:

$$v_{x,\text{avg}(Ⓐ \to Ⓑ)} = \frac{\Delta x_{Ⓐ \to Ⓑ}}{\Delta t} = \frac{-2 \text{ m}}{1 \text{ s}} = \boxed{-2 \text{ m/s}}$$

In the second time interval, $\Delta t = 2$ s:

$$v_{x,\text{avg}(Ⓑ \to Ⓓ)} = \frac{\Delta x_{Ⓑ \to Ⓓ}}{\Delta t} = \frac{8 \text{ m}}{2 \text{ s}} = \boxed{+4 \text{ m/s}}$$

These values are the same as the slopes of the blue lines joining these points in Figure 2.5a.

2.4 *cont.*

(C) Find the instantaneous velocity of the particle at $t = 2.5$ s.

SOLUTION

Measure the slope of the green line at $t = 2.5$ s (point **©**) in Figure 2.5a:

$$v_x = \frac{10 \text{ m} - (-4 \text{ m})}{3.8 \text{ s} - 1.5 \text{ s}} = +6 \text{ m/s}$$

Notice that this instantaneous velocity is on the same order of magnitude as our previous results, that is, a few meters per second. Is that what you would have expected? Do you see any symmetry in the motion? For example, are there points at which the speed is the same? Is the velocity the same at these points?

2.3 | Analysis Model: Particle Under Constant Velocity

As mentioned in Section 1.10, the third category of models used in this book is that of *analysis models*. Such models help us analyze the situation in a physics problem and guide us toward the solution. **An analysis model is a problem we have solved before.** It is a description of either (1) the behavior of some physical entity or (2) the interaction between that entity and the environment. When you encounter a new problem, you should identify the fundamental details of the problem and attempt to recognize which, if any, of the types of problems you have already solved might be used as a model for the new problem.

This method is somewhat similar to the common practice in the legal profession of finding "legal precedents." If a previously resolved case can be found that is very similar legally to the present one, it is offered as a model and an argument is made in court to link them logically. The finding in the previous case can then be used to sway the finding in the present case. We will do something similar in physics. For a given problem, we search for a "physics precedent," a model with which we are already familiar and that can be applied to the present problem.

We shall generate analysis models based on four fundamental simplification models. The first simplification model is the particle model discussed in Chapter 1. We will look at a particle under various behaviors and environmental interactions. Further analysis models are introduced in later chapters based on simplification models of a *system*, a *rigid object*, and a *wave*. Once we have introduced these analysis models, we shall see that they appear over and over again later in the book in different situations.

When solving a problem, you should avoid browsing through the chapter looking for an equation that contains the unknown variable that is requested in the problem. In many cases, the equation you find may have nothing to do with the problem you are attempting to solve. It is *much* better to take this first step: **Identify the analysis model that is appropriate for the problem.** Think carefully about what is going on in the problem and match it to a situation you have seen before. What simplification model is appropriate for the entity involved in the problem? Is it a particle, a system, a rigid object, or a wave? Second, what is the entity doing or how is it interacting with its environment? For example, the analysis model in the title of this section indicates that we modeled the entity of interest as a particle. Furthermore, we determined that the particle is moving with constant velocity.

Once the analysis model is identified, there are a small number of equations from which to choose that are appropriate for that model. Therefore, **the model tells you which equation(s) to use for the mathematical representation.** In this section, we will learn what mathematical equations are associated with the particle under constant

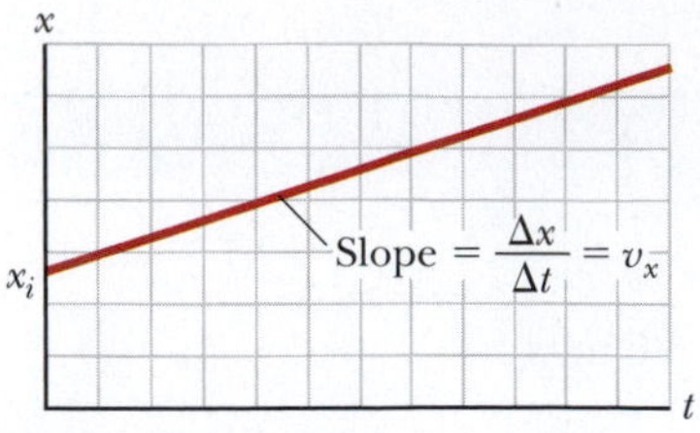

Figure 2.6 Position–time graph for a particle under constant velocity. The value of the constant velocity is the slope of the line.

velocity analysis model. In the future, when you identify the appropriate model in a problem as the particle under constant velocity, you will immediately know which equations to use to solve the problem.

Let us use Equation 2.2 to build our first analysis model. We imagine a particle moving with a constant velocity. The analysis model of a **particle under constant velocity** can be applied in *any* situation in which an entity that can be modeled as a particle is moving with constant velocity. This situation occurs frequently, so it is an important model.

If the velocity of a particle is constant, its instantaneous velocity at any instant during a time interval is the same as the average velocity over the interval, $v_x = v_{x,\text{avg}}$. Therefore, we start with Equation 2.2 to generate an equation to be used in the mathematical representation of this situation:

$$v_x = v_{x,\text{avg}} = \frac{\Delta x}{\Delta t} \qquad\qquad \textbf{2.4} \blacktriangleleft$$

Remembering that $\Delta x = x_f - x_i$, we see that $v_x = (x_f - x_i)/\Delta t$, or

$$x_f = x_i + v_x \Delta t$$

This equation tells us that the position of the particle is given by the sum of its original position x_i plus the displacement $v_x \Delta t$ that occurs during the time interval Δt. In practice, we usually choose the time at the beginning of the interval to be $t_i = 0$ and the time at the end of the interval to be $t_f = t$, so our equation becomes

▶ Position as a function of time for the particle under constant velocity model

$$\boxed{x_f = x_i + v_x t} \qquad \text{(for constant } v_x) \qquad\qquad \textbf{2.5} \blacktriangleleft$$

Equations 2.4 and 2.5 are the primary equations used in the model of a particle under constant velocity. They can be applied to particles or objects that can be modeled as particles. In the future, once you have identified a problem as requiring the particle under constant velocity model, either of these equations can be used to solve the problem.

Figure 2.6 is a graphical representation of the particle under constant velocity. On the position–time graph, the slope of the line representing the motion is constant and equal to the velocity. It is consistent with the mathematical representation, Equation 2.5, which is the equation of a straight line. The slope of the straight line is v_x and the y intercept is x_i in both representations.

Example 2.5 | Modeling a Runner as a Particle BIO

A kinesiologist is studying the biomechanics of the human body. (*Kinesiology* is the study of the movement of the human body. Notice the connection to the word *kinematics*.) She determines the velocity of an experimental subject while he runs along a straight line at a constant rate. The kinesiologist starts the stopwatch at the moment the runner passes a given point and stops it after the runner has passed another point 20 m away. The time interval indicated on the stopwatch is 4.0 s.

(A) What is the runner's velocity?

SOLUTION

Conceptualize You have probably watched track and field events at some point in your life, so it should be easy to conceptualize this situation.

Categorize We model the moving runner as a particle because the size of the runner and the movement of arms and legs are unnecessary details. Because the problem states that the subject runs at a constant rate, we can model him as a particle under constant velocity.

....

Analyze Having identified the model, we can use Equation 2.4 to find the constant velocity of the runner:

$$v_x = \frac{\Delta x}{\Delta t} = \frac{x_f - x_i}{\Delta t} = \frac{20 \text{ m} - 0}{4.0 \text{ s}} = \boxed{5.0 \text{ m/s}}$$

2.5 *cont.*

(B) If the runner continues his motion after the stopwatch is stopped, what is his position after 10 s has passed?

SOLUTION

Use Equation 2.5 and the velocity found in part (A) to find the position of the particle at time $t = 10$ s:

$$x_f = x_i + v_x t = 0 + (5.0 \text{ m/s})(10 \text{ s}) = \boxed{50 \text{ m}}$$

Finalize Is the result for part (A) a reasonable speed for a human? How does it compare to world-record speeds in 100-m and 200-m sprints? Notice that the value in part (B) is more than twice that of the 20-m position at which the stopwatch was stopped. Is this value consistent with the time of 10 s being more than twice the time of 4.0 s?

The mathematical manipulations for the particle under constant velocity stem from Equation 2.4 and its descendent, Equation 2.5. These equations can be used to solve for any variable in the equations that happens to be unknown if the other variables are known. For example, in part (B) of Example 2.5, we find the position when the velocity and the time are known. Similarly, if we know the velocity and the final position, we could use Equation 2.5 to find the time at which the runner is at this position. We shall present more examples of a particle under constant velocity in Chapter 3.

A particle under constant velocity moves with a constant speed along a straight line. Now consider a particle moving with a constant speed along a curved path. It can be represented with the **particle under constant speed model.** The primary equation for this model is Equation 2.1, with the average speed v_{avg} replaced by the constant speed v:

$$v \equiv \frac{d}{\Delta t} \qquad \qquad \textbf{2.6} \blacktriangleleft$$

As an example, imagine a particle moving at a constant speed in a circular path. If the speed is 5.00 m/s and the radius of the path is 10.0 m, we can calculate the time interval required to complete one trip around the circle:

$$v = \frac{d}{\Delta t} \rightarrow \Delta t = \frac{d}{v} = \frac{2\pi r}{v} = \frac{2\pi (10.0 \text{ m})}{5.00 \text{ m/s}} = 12.6 \text{ s}$$

❮ 2.4 | Acceleration

When the velocity of a particle changes with time, the particle is said to be *accelerating*. For example, the speed of a car increases when you "step on the gas," the car slows down when you apply the brakes, and it changes direction when you turn the wheel; these changes are all accelerations. We will need a precise definition of acceleration for our studies of motion.

Suppose a particle moving along the *x* axis has a velocity v_{xi} at time t_i and a velocity v_{xf} at time t_f. The **average acceleration** $a_{x,\text{avg}}$ of the particle in the time interval $\Delta t = t_f - t_i$ is defined as the ratio $\Delta v_x / \Delta t$, where $\Delta v_x = v_{xf} - v_{xi}$ is the *change* in velocity of the particle in this time interval:

$$a_{x,\text{avg}} \equiv \frac{v_{xf} - v_{xi}}{t_f - t_i} = \frac{\Delta v_x}{\Delta t} \qquad \qquad \textbf{2.7} \blacktriangleleft \qquad \blacktriangleright \text{ Definition of average acceleration}$$

Therefore, acceleration is a measure of how rapidly the velocity is changing. Acceleration is a vector quantity having dimensions of length divided by (time)², or L/T^2. Some of the common units of acceleration are meters per second per second (m/s²) and feet per second per second (ft/s²). For example, an acceleration of 2 m/s² means that the velocity changes by 2 m/s during each second of time that passes.

In some situations, the value of the average acceleration may be different for different time intervals. It is therefore useful to define the **instantaneous acceleration** as the limit of the average acceleration as Δt approaches zero, analogous to the definition of instantaneous velocity discussed in Section 2.2:

▶ Definition of instantaneous acceleration

$$a_x \equiv \lim_{\Delta t \to 0} \frac{\Delta v_x}{\Delta t} = \frac{dv_x}{dt}$$ 2.8 ◀

That is, the instantaneous acceleration equals the derivative of the velocity with respect to time, which by definition is the slope of the velocity–time graph. Note that if a_x is positive, the acceleration is in the positive x direction, whereas negative a_x implies acceleration in the negative x direction. A negative acceleration does not necessarily mean that the particle is *moving* in the negative x direction, a point we shall address in more detail shortly. From now on, we use the term *acceleration* to mean instantaneous acceleration.

Because $v_x = dx/dt$, the acceleration can also be written

$$a_x = \frac{dv_x}{dt} = \frac{d}{dt}\left(\frac{dx}{dt}\right) = \frac{d^2x}{dt^2}$$ 2.9 ◀

This equation shows that the acceleration equals the *second derivative* of the position with respect to time.

Figure 2.7 shows how the acceleration–time curve in a graphical representation can be derived from the velocity–time curve. In these diagrams, the acceleration of a particle at any time is simply the slope of the velocity–time graph at that time. Positive values of the acceleration correspond to those points (between 0 and $t_{Ⓑ}$) where the velocity in the positive x direction is increasing in magnitude (the particle is speeding up). The acceleration reaches a maximum at time $t_{Ⓐ}$, when the slope of the velocity–time graph is a maximum. The acceleration then goes to zero at time $t_{Ⓑ}$, when the velocity is a maximum (i.e., when the velocity is momentarily not changing and the slope of the v versus t graph is zero). Finally, the acceleration is negative when the velocity in the positive x direction is decreasing in magnitude (between $t_{Ⓑ}$ and $t_{Ⓒ}$).

Pitfall Prevention | 2.4

Negative Acceleration

Keep in mind that *negative acceleration does not necessarily mean that an object is slowing down.* If the acceleration is negative and the velocity is negative, the object is speeding up!

Pitfall Prevention | 2.5

Deceleration

The word *deceleration* has a common popular connotation as *slowing down.* When combined with the misconception in Pitfall Prevention 2.4 that negative acceleration means slowing down, the situation can be further confused by the use of the word *deceleration.* We will not use this word in this text.

QUICK QUIZ 2.3 Using Active Figure 2.8, match each v_x–t graph on the top with the a_x–t graph on the bottom that best describes the motion.

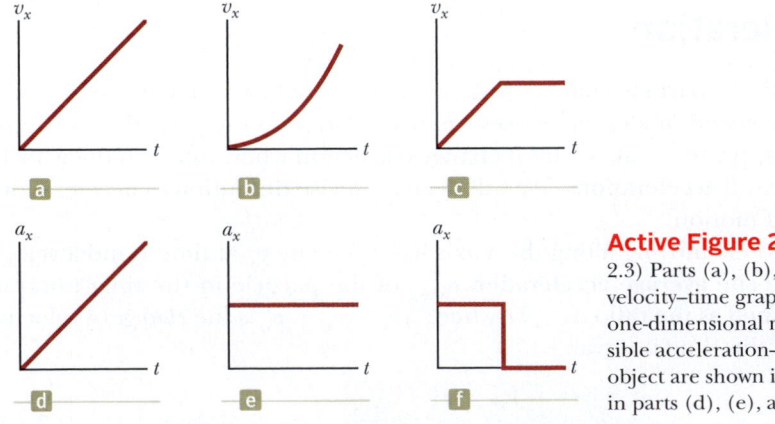

Active Figure 2.8 (Quick Quiz 2.3) Parts (a), (b), and (c) are velocity–time graphs of objects in one-dimensional motion. The possible acceleration–time graphs of each object are shown in scrambled order in parts (d), (e), and (f).

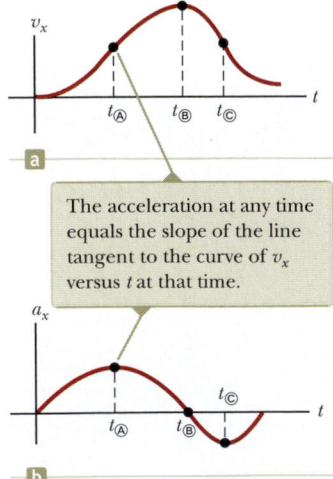

Figure 2.7 (a) The velocity–time graph for a particle moving along the x axis. (b) The instantaneous acceleration can be obtained from the velocity–time graph.

The acceleration at any time equals the slope of the line tangent to the curve of v_x versus t at that time.

As an example of the computation of acceleration, consider the pictorial representation of a car's motion in Figure 2.9. In this case, the velocity of the car has changed from an initial value of 30 m/s to a final value of 15 m/s in a time interval of 2.0 s. The average acceleration during this time interval is

$$a_{x,\text{avg}} = \frac{15 \text{ m/s} - 30 \text{ m/s}}{2.0 \text{ s}} = -7.5 \text{ m/s}^2$$

The negative sign in this example indicates that the acceleration vector is in the negative x direction (to the left in Figure 2.9). For the case of motion in a straight line, the direction of the velocity of an object and the direction of its acceleration are related as follows. When the object's velocity and acceleration are in the same direction, the object is speeding up in that direction. On the other hand, when the object's velocity and acceleration are in opposite directions, the speed of the object decreases in time.

To help with this discussion of the signs of velocity and acceleration, let us take a peek ahead to Chapter 4, where we shall relate the acceleration of an object to the *force* on the object. We will save the details until that later discussion, but for now, let us borrow the notion that **the force on an object is proportional to the acceleration of the object:**

$$\vec{F} \propto \vec{a}$$

This proportionality indicates that acceleration is caused by force. What's more, as indicated by the vector notation in the proportionality, force and acceleration are in the same direction. Therefore, let us think about the signs of velocity and acceleration by forming a mental representation in which a force is applied to the object to cause the acceleration. Again consider the case in which the velocity and acceleration are in the same direction. This situation is equivalent to an object moving in a given direction and experiencing a force that pulls on it in the same direction. It is clear in this case that the object speeds up! If the velocity and acceleration are in opposite directions, the object moves one way and a force pulls in the opposite direction. In this case, the object slows down! It is very useful to equate the direction of the acceleration in these situations to the direction of a force because it is easier from our everyday experience to think about what effect a force will have on an object than to think only in terms of the direction of the acceleration.

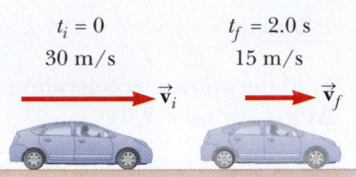

Figure 2.9 The velocity of the car decreases from 30 m/s to 15 m/s in a time interval of 2.0 s.

QUICK QUIZ 2.4 If a car is traveling eastward and slowing down, what is the direction of the force on the car that causes it to slow down? **(a)** eastward **(b)** westward **(c)** neither of these directions

---

Example 2.6 | Average and Instantaneous Acceleration

The velocity of a particle moving along the x axis varies according to the expression $v_x = 40 - 5t^2$, where v_x is in meters per second and t is in seconds.

(A) Find the average acceleration in the time interval $t = 0$ to $t = 2.0$ s.

SOLUTION

Conceptualize Think about what the particle is doing from the mathematical representation. Is it moving at $t = 0$? In which direction? Does it speed up or slow down? Figure 2.10 is a v_x–t graph that was created from the velocity versus time expression given in the problem statement. Because the slope of the entire v_x–t curve is negative, we expect the acceleration to be negative.

Categorize While this problem does not involve an analysis model, it does involve taking a limit of a function, so it is a bit more sophisticated than a pure substitution problem.

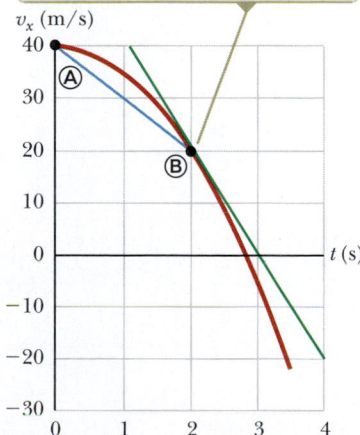

The acceleration at Ⓑ is equal to the slope of the green tangent line at $t = 2$ s, which is -20 m/s².

Figure 2.10 (Example 2.6) The velocity–time graph for a particle moving along the x axis according to the expression $v_x = 40 - 5t^2$.

Analyze Find the velocities at $t_i = t_Ⓐ = 0$ and $t_f = t_Ⓑ = 2.0$ s by substituting these values of t into the expression for the velocity:

$v_{xⓐ} = 40 - 5t_Ⓐ^2 = 40 - 5(0)^2 = +40$ m/s
$v_{xⓑ} = 40 - 5t_Ⓑ^2 = 40 - 5(2.0)^2 = +20$ m/s

continued

2.6 *cont.*

Find the average acceleration in the specified time interval $\Delta t = t_{\circledB} - t_{\circledA} = 2.0$ s:

$$a_{x,\text{avg}} = \frac{v_{xf} - v_{xi}}{t_f - t_i} = \frac{v_{x\circledB} - v_{x\circledA}}{t_{\circledB} - t_{\circledA}} = \frac{20 \text{ m/s} - 40 \text{ m/s}}{2.0 \text{ s} - 0 \text{ s}}$$

$$= -10 \text{ m/s}^2$$

Finalize The negative sign is consistent with our expectations: the average acceleration, represented by the slope of the blue line joining the initial and final points on the velocity–time graph, is negative.

(B) Determine the acceleration at $t = 2.0$ s.

SOLUTION

Analyze

Knowing that the initial velocity at any time t is $v_{xi} = 40 - 5t^2$, find the velocity at any later time $t + \Delta t$:

$$v_{xf} = 40 - 5(t + \Delta t)^2 = 40 - 5t^2 - 10t\,\Delta t - 5(\Delta t)^2$$

Find the change in velocity over the time interval Δt:

$$\Delta v_x = v_{xf} - v_{xi} = -10t\,\Delta t - 5(\Delta t)^2$$

To find the acceleration at any time t, divide this expression by Δt and take the limit of the result as Δt approaches zero:

$$a_x = \lim_{\Delta t \to 0} \frac{\Delta v_x}{\Delta t} = \lim_{\Delta t \to 0} (-10t - 5\Delta t) = -10t$$

Substitute $t = 2.0$ s:

$$a_x = (-10)(2.0) \text{ m/s}^2 = -20 \text{ m/s}^2$$

Finalize Because the velocity of the particle is positive and the acceleration is negative at this instant, the particle is slowing down.

Notice that the answers to parts (A) and (B) are different. The average acceleration in part (A) is the slope of the blue line in Figure 2.10 connecting points Ⓐ and Ⓑ. The instantaneous acceleration in part (B) is the slope of the green line tangent to the curve at point Ⓑ. Notice also that the acceleration is *not* constant in this example. Situations involving constant acceleration are treated in Section 2.6.

⫷ 2.5 | Motion Diagrams

The concepts of velocity and acceleration are often confused with each other, but in fact they are quite different quantities. It is instructive to make use of the specialized pictorial representation called a **motion diagram** to describe the velocity and acceleration vectors while an object is in motion.

A *stroboscopic photograph* of a moving object shows several images of the object taken as the strobe light flashes at a constant rate. Figure 2.1a is a motion diagram for the car studied in Section 2.1. Active Figure 2.11 represents three sets of strobe photographs of cars moving along a straight roadway in a single direction, from left to right. The time intervals between flashes of the stroboscope are equal in each part of the diagram. To distinguish between the two vector quantities, we use red arrows for velocity vectors and purple arrows for acceleration vectors in Active Figure 2.11. The vectors are sketched at several instants during the motion of the object. Let us describe the motion of the car in each diagram.

In Active Figure 2.11a, the images of the car are equally spaced, and the car moves through the same displacement in each time interval. Therefore, the car moves with *constant positive velocity* and has *zero acceleration*. We could model the car as a particle and describe it using the particle under constant velocity analysis model.

In Active Figure 2.11b, the images of the car become farther apart as time progresses. In this case, the velocity vector increases in time because the car's displacement between adjacent positions increases as time progresses. Therefore, the car is moving with a *positive velocity* and a *positive acceleration*. The velocity and acceleration are in the same direction. In terms of our earlier force discussion, imagine a force pulling on the car in the same direction it is moving: it speeds up.

In Active Figure 2.11c, we interpret the car as slowing down as it moves to the right because its displacement between adjacent positions decreases as time progresses. In

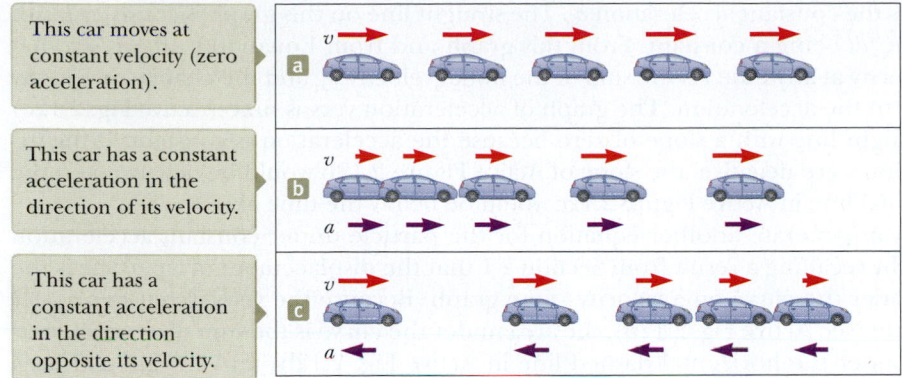

Active Figure 2.11 Motion diagrams of a car moving along a straight roadway in a single direction. The velocity at each instant is indicated by a red arrow, and the constant acceleration is indicated by a purple arrow.

this case, the car moves initially to the right with a *positive velocity* and a *negative acceleration*. The velocity vector decreases in time and eventually reaches zero. (This type of motion is exhibited by a car that skids to a stop after its brakes are applied.) From this diagram we see that the acceleration and velocity vectors are *not* in the same direction. The velocity and acceleration are in opposite directions. In terms of our earlier force discussion, imagine a force pulling on the car opposite to the direction it is moving: it slows down.

The purple acceleration vectors in Active Figures 2.11b and 2.11c are all the same length. Therefore, these diagrams represent a motion with constant acceleration. This important type of motion is discussed in the next section.

QUICK QUIZ 2.5 Which of the following statements is true? **(a)** If a car is traveling eastward, its acceleration must be eastward. **(b)** If a car is slowing down, its acceleration must be negative. **(c)** A particle with constant acceleration can never stop and stay stopped.

2.6 | Analysis Model: Particle Under Constant Acceleration

If the acceleration of a particle varies in time, the motion may be complex and difficult to analyze. A very common and simple type of one-dimensional motion occurs when the acceleration is constant, such as for the motion of the cars in Active Figures 2.11b and 2.11c. In this case, the average acceleration over any time interval equals the instantaneous acceleration at any instant of time within the interval. Consequently, the velocity increases or decreases at the same rate throughout the motion. The **particle under constant acceleration** is a common analysis model that we can apply to appropriate problems. It is often used to model situations such as falling objects and braking cars.

If we replace $a_{x,\text{avg}}$ with the constant a_x in Equation 2.7, we find that

$$a_x = \frac{v_{xf} - v_{xi}}{t_f - t_i}$$

For convenience, let $t_i = 0$ and t_f be any arbitrary time t. With this notation, we can solve for v_{xf}:

$$v_{xf} = v_{xi} + a_x t \qquad \text{(for constant } a_x) \qquad \textbf{2.10} \blacktriangleleft$$

This expression enables us to predict the velocity at *any* time t if the initial velocity and constant acceleration are known. It is the first of four equations that can be used to solve problems using the particle under constant acceleration model. A graphical representation of position versus time for this motion is shown in Active Figure 2.12a. The velocity–time graph shown in Active Figure 2.12b is a straight line, the slope of

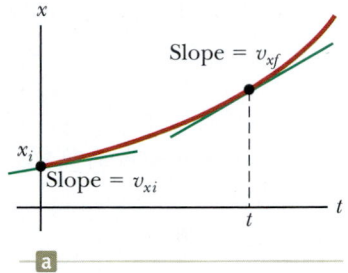

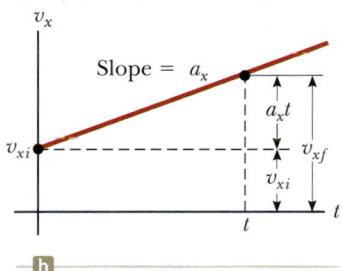

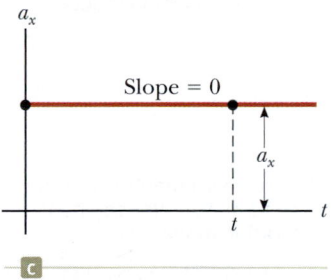

Active Figure 2.12 Graphical representations of a particle moving along the x axis with constant acceleration a_x. (a) The position–time graph, (b) the velocity–time graph, and (c) the acceleration–time graph.

▶ Velocity as a function of time for the particle under constant acceleration model

which is the constant acceleration a_x. The straight line on this graph is consistent with $a_x = dv_x/dt$ being a constant. From this graph and from Equation 2.10, we see that the velocity at any time t is the sum of the initial velocity v_{xi} and the change in velocity a_xt due to the acceleration. The graph of acceleration versus time (Active Fig. 2.12c) is a straight line with a slope of zero because the acceleration is constant. If the acceleration were negative, the slope of Active Figure 2.12b would be negative and the horizontal line in Active Figure 2.12c would be below the time axis.

We can generate another equation for the particle under constant acceleration model by recalling a result from Section 2.1 that the displacement of a particle is the area under the curve on a velocity–time graph. Because the velocity varies linearly with time (see Active Fig. 2.12b), the area under the curve is the sum of a rectangular area (under the horizontal dashed line in Active Fig. 2.12b) and a triangular area (from the horizontal dashed line upward to the curve). Therefore,

$$\Delta x = v_{xi}\Delta t + \tfrac{1}{2}(v_{xf} - v_{xi})\Delta t$$

which can be simplified as follows:

$$\Delta x = (v_{xi} + \tfrac{1}{2}v_{xf} - \tfrac{1}{2}v_{xi})\Delta t = \tfrac{1}{2}(v_{xi} + v_{xf})\Delta t$$

In general, from Equation 2.2, the displacement for a time interval is

$$\Delta x = v_{x,\,\text{avg}}\Delta t$$

Comparing these last two equations, we find that the average velocity in any time interval is the arithmetic mean of the initial velocity v_{xi} and the final velocity v_{xf}:

▶ Average velocity for the particle under constant acceleration model

$$v_{x,\,\text{avg}} = \tfrac{1}{2}(v_{xi} + v_{xf}) \qquad \text{(for constant } a_x) \qquad \text{2.11} ◀$$

Remember that this expression is valid only when the acceleration is constant, that is, when the velocity varies linearly with time.

We now use Equations 2.2 and 2.11 to obtain the position as a function of time. Again we choose $t_i = 0$, at which time the initial position is x_i, which gives

$$\Delta x = v_{x,\,\text{avg}}\Delta t = \tfrac{1}{2}(v_{xi} + v_{xf})t$$

▶ Position as a function of velocity and time for the particle under constant acceleration model

$$x_f = x_i + \tfrac{1}{2}(v_{xi} + v_{xf})t \qquad \text{(for constant } a_x) \qquad \text{2.12} ◀$$

We can obtain another useful expression for the position by substituting Equation 2.10 for v_{xf} in Equation 2.12:

$$x_f = x_i + \tfrac{1}{2}[v_{xi} + (v_{xi} + a_xt)]t$$

▶ Position as a function of time for the particle under constant acceleration model

$$x_f = x_i + v_{xi}t + \tfrac{1}{2}a_xt^2 \qquad \text{(for constant } a_x) \qquad \text{2.13} ◀$$

Note that the position at any time t is the sum of the initial position x_i, the displacement $v_{xi}t$ that would result if the velocity remained constant at the initial velocity, and the displacement $\tfrac{1}{2}a_xt^2$ because the particle is accelerating. Consider again the position–time graph for motion under constant acceleration shown in Active Figure 2.12a. The curve representing Equation 2.13 is a parabola, as shown by the t^2 dependence in the equation. The slope of the tangent to this curve at $t = 0$ equals the initial velocity v_{xi}, and the slope of the tangent line at any time t equals the velocity at that time.

Finally, we can obtain an expression that does not contain the time by substituting the value of t from Equation 2.10 into Equation 2.12, which gives

$$x_f = x_i + \tfrac{1}{2}(v_{xi} + v_{xf})\left(\frac{v_{xf} - v_{xi}}{a_x}\right) = x_i + \frac{v_{xf}^2 - v_{xi}^2}{2a_x}$$

▶ Velocity as a function of position for the particle under constant acceleration model

$$v_{xf}^2 = v_{xi}^2 + 2a_x(x_f - x_i) \qquad \text{(for constant } a_x) \qquad \text{2.14} ◀$$

TABLE 2.2 | Kinematic Equations for Motion of a Particle Under Constant Acceleration

Equation Number	Equation	Information Given by Equation
2.10	$v_{xf} = v_{xi} + a_x t$	Velocity as a function of time
2.12	$x_f = x_i + \frac{1}{2}(v_{xf} + v_{xi})t$	Position as a function of velocity and time
2.13	$x_f = x_i + v_{xi}t + \frac{1}{2}a_x t^2$	Position as a function of time
2.14	$v_{xf}^2 = v_{xi}^2 + 2a_x(x_f - x_i)$	Velocity as a function of position

Note: Motion is along the *x* axis. At *t* = 0, the position of the particle is x_i and its velocity is v_{xi}.

This expression is *not* an independent equation because it arises from combining Equations 2.10 and 2.12. It is useful, however, for those problems in which a value for the time is not involved.

If motion occurs in which the constant value of the acceleration is *zero*, Equations 2.10 and 2.13 become

$$\left. \begin{array}{l} v_{xf} = v_{xi} \\ x_f = x_i + v_{xi}t \end{array} \right\} \quad \text{when } a_x = 0$$

That is, when the acceleration is zero, the velocity remains constant and the position changes linearly with time. In this case, the particle under constant *acceleration* model reduces to the particle under constant *velocity* model.

Equations 2.10, 2.12, 2.13, and 2.14 are four **kinematic equations** that may be used to solve any problem in one-dimensional motion of a particle (or an object that can be modeled as a particle) under constant acceleration. If your analysis of a problem indicates that the particle under constant acceleration is the appropriate analysis model, select from these four equations to solve the problem. Keep in mind that these relationships were derived from the definitions of velocity and acceleration together with some simple algebraic manipulations and the requirement that the acceleration be constant. It is often convenient to choose the initial position of the particle as the origin of the motion so that x_i = 0 at *t* = 0. We will see cases, however, in which we must choose the value of x_i to be something other than zero.

The four kinematic equations for the particle under constant acceleration are listed in Table 2.2 for convenience. The choice of which kinematic equation or equations you should use in a given situation depends on what is known beforehand. Sometimes it is necessary to use two of these equations to solve for two unknowns, such as the position and velocity at some instant. You should recognize that the quantities that vary during the motion are velocity v_{xf}, position x_f, and time *t*. The other quantities—x_i, v_{xi}, and a_x—are *parameters* of the motion and remain constant.

 PROBLEM-SOLVING STRATEGY: Particle Under Constant Acceleration

The following procedure is recommended for solving problems that involve an object undergoing a constant acceleration. As mentioned in Chapter 1, individual strategies such as this one will follow the outline of the General Problem-Solving Strategy from Chapter 1, with specific hints regarding the application of the general strategy to the material in the individual chapters.

1. **Conceptualize** Think about what is going on physically in the problem. Establish the mental representation.

2. **Categorize** Simplify the problem as much as possible. Confirm that the problem involves either a particle or an object that can be modeled as a particle and that it is moving with a constant acceleration. Construct an appropriate

pictorial representation, such as a motion diagram, or a graphical representation. Make sure all the units in the problem are consistent. That is, if positions are measured in meters, be sure that velocities have units of m/s and accelerations have units of m/s². Choose a coordinate system to be used throughout the problem.

3. **Analyze** Set up the mathematical representation. Choose an instant to call the "initial" time $t = 0$ and another to call the "final" time t. Let your choice be guided by what you know about the particle and what you want to know about it. The initial instant need not be when the particle starts to move, and the final instant will only rarely be when the particle stops moving. Identify all the quantities given in the problem and a separate list of those to be determined. A tabular representation of these quantities may be helpful to you. Select from the list of kinematic equations the one or ones that will enable you to determine the unknowns. Solve the equations.

4. **Finalize** Once you have determined your result, check to see if your answers are consistent with the mental and pictorial representations and that your results are realistic.

Example 2.7 | Carrier Landing

A jet lands on an aircraft carrier at a speed of 140 mi/h ($\approx$ 63 m/s).

(A) What is its acceleration (assumed constant) if it stops in 2.0 s due to an arresting cable that snags the jet and brings it to a stop?

SOLUTION

Conceptualize You might have seen movies or television shows in which a jet lands on an aircraft carrier and is brought to rest surprisingly fast by an arresting cable. A careful reading of the problem reveals that in addition to being given the initial speed of 63 m/s, we also know that the final speed is zero. We define our x axis as the direction of motion of the jet. Notice that we have no information about the change in position of the jet while it is slowing down.

Categorize Because the acceleration of the jet is assumed constant, we model it as a particle under constant acceleration.

Analyze
Equation 2.10 is the only equation in Table 2.2 that does not involve position, so we use it to find the acceleration of the jet, modeled as a particle:

$$a_x = \frac{v_{xf} - v_{xi}}{t} \approx \frac{0 - 63 \text{ m/s}}{2.0 \text{ s}}$$

$$= \boxed{-32 \text{ m/s}^2}$$

(B) If the jet touches down at position $x_i = 0$, what is its final position?

SOLUTION

Use Equation 2.12 to solve for the final position: $x_f = x_i + \frac{1}{2}(v_{xi} + v_{xf})t = 0 + \frac{1}{2}(63 \text{ m/s} + 0)(2.0 \text{ s}) = \boxed{63 \text{ m}}$

Finalize Given the size of aircraft carriers, a length of 63 m seems reasonable for stopping the jet. The idea of using arresting cables to slow down landing aircraft and enable them to land safely on ships originated at about the time of World War I. The cables are still a vital part of the operation of modern aircraft carriers.

What If? Suppose the jet lands on the deck of the aircraft carrier with a speed faster than 63 m/s but has the same acceleration due to the cable as that calculated in part (A). How will that change the answer to part (B)?

Answer If the jet is traveling faster at the beginning, it will stop farther away from its starting point, so the answer to part (B) should be larger. Mathematically, we see in Equation 2.12 that if v_{xi} is larger, then x_f will be larger.

Example **2.8** | Watch Out for the Speed Limit!

A car traveling at a constant speed of 45.0 m/s passes a trooper on a motorcycle hidden behind a billboard. One second after the speeding car passes the billboard, the trooper sets out from the billboard to catch the car, accelerating at a constant rate of 3.00 m/s². How long does it take her to overtake the car?

SOLUTION

Conceptualize A pictorial representation (Fig. 2.13) helps clarify the sequence of events.

Categorize The car is modeled as a particle under constant velocity, and the trooper is modeled as a particle under constant acceleration.

Analyze First, we write expressions for the position of each vehicle as a function of time. It is convenient to choose the position of the billboard as the origin and to set $t_{Ⓑ} = 0$ as the time the trooper begins moving. At that instant, the car has already traveled a distance of 45.0 m from the billboard because it has traveled at a constant speed of $v_x = 45.0$ m/s for 1 s. Therefore, the initial position of the speeding car is $x_{Ⓑ} = 45.0$ m.

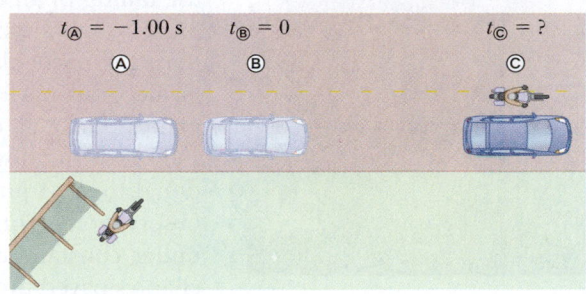

$t_{Ⓐ} = -1.00$ s $t_{Ⓑ} = 0$ $t_{Ⓒ} = ?$

Figure 2.13 (Example 2.8) A speeding car passes a hidden trooper.

Using the particle under constant velocity model, apply Equation 2.5 to give the car's position at any time t:

$$x_{car} = x_{Ⓑ} + v_{x\,car}t$$

A quick check shows that at $t = 0$, this expression gives the car's correct initial position when the trooper begins to move: $x_{car} = x_{Ⓑ} = 45.0$ m.

The trooper starts from rest at $t_{Ⓑ} = 0$ and accelerates at $a_x = 3.00$ m/s² away from the origin. Use Equation 2.13 to give her position at any time t:

$$x_f = x_i + v_{xi}t + \tfrac{1}{2}a_xt^2$$
$$x_{trooper} = 0 + (0)t + \tfrac{1}{2}a_xt^2 = \tfrac{1}{2}a_xt^2$$

Set the positions of the car and trooper equal to represent the trooper overtaking the car at position Ⓒ:

$$x_{trooper} = x_{car}$$
$$\tfrac{1}{2}a_xt^2 = x_{Ⓑ} + v_{x\,car}t$$

Rearrange to give a quadratic equation:

$$\tfrac{1}{2}a_xt^2 - v_{x\,car}t - x_{Ⓑ} = 0$$

Solve the quadratic equation for the time at which the trooper catches the car (for help in solving quadratic equations, see Appendix B.2.):

$$t = \frac{v_{x\,car} \pm \sqrt{v_{x\,car}^2 + 2a_xx_{Ⓑ}}}{a_x}$$

$$(1)\ t = \frac{v_{x\,car}}{a_x} \pm \sqrt{\frac{v_{x\,car}^2}{a_x^2} + \frac{2x_{Ⓑ}}{a_x}}$$

Evaluate the solution, choosing the positive root because that is the only choice consistent with a time $t > 0$:

$$t = \frac{45.0\ \text{m/s}}{3.00\ \text{m/s}^2} + \sqrt{\frac{(45.0\ \text{m/s})^2}{(3.00\ \text{m/s}^2)^2} + \frac{2(45.0\ \text{m})}{3.00\ \text{m/s}^2}} = \boxed{31.0\ \text{s}}$$

Finalize Why didn't we choose $t = 0$ as the time at which the car passes the trooper? If we did so, we would not be able to use the particle under constant acceleration model for the trooper. Her acceleration would be zero for the first second and then 3.00 m/s² for the remaining time. By defining the time $t = 0$ as when the trooper begins moving, we can use the particle under constant acceleration model for her movement for all positive times.

What If? What if the trooper had a more powerful motorcycle with a larger acceleration? How would that change the time at which the trooper catches the car?

Answer If the motorcycle has a larger acceleration, the trooper should catch up to the car sooner, so the answer for the time should be less than 31 s. Because all terms on the right side of Equation (1) have the acceleration a_x in the denominator, we see symbolically that increasing the acceleration will decrease the time at which the trooper catches the car.

▌2.7 | Freely Falling Objects

Galileo Galilei
**Italian physicist and astronomer
(1564–1642)**
Galileo formulated the laws that govern
the motion of objects in free fall and
made many other significant discoveries
in physics and astronomy. Galileo publicly
defended Nicolaus Copernicus's assertion
that the Sun is at the center of the Universe
(the heliocentric system). He published
*Dialogue Concerning Two New World
Systems* to support the Copernican model,
a view that the Catholic Church declared to
be heretical.

Pitfall Prevention | 2.6
_g_ and g
Be sure not to confuse the italic
symbol *g* for free-fall acceleration
with the nonitalic symbol g used as
the abbreviation for the unit gram.

It is well known that all objects, when dropped, fall toward the Earth with nearly constant acceleration. Legend has it that Galileo Galilei first discovered this fact by observing that two different weights dropped simultaneously from the Leaning Tower of Pisa hit the ground at approximately the same time. (Air resistance plays a role in the falling of an object, but for now we shall model falling objects as if they are falling through a vacuum; this is a simplification model.) Although there is some doubt that this particular experiment was actually carried out, it is well established that Galileo did perform many systematic experiments on objects moving on inclined planes. Through careful measurements of distances and time intervals, he was able to show that the displacement from an origin of an object starting from rest is proportional to the square of the time interval during which the object is in motion. This observation is consistent with one of the kinematic equations we derived for a particle under constant acceleration (Eq. 2.13, with $v_{xi} = 0$). Galileo's achievements in mechanics paved the way for Newton in his development of the laws of motion.

If a coin and a crumpled-up piece of paper are dropped simultaneously from the same height, there will be a small time difference between their arrivals at the floor. If this same experiment could be conducted in a good vacuum, however, where air friction is truly negligible, the paper and coin would fall with the same acceleration, regardless of the shape or weight of the paper, even if the paper were still flat. In the idealized case, where air resistance is ignored, such motion is referred to as *free-fall*. This point is illustrated very convincingly in Figure 2.14, which is a photograph of an apple and a feather falling in a vacuum. On August 2, 1971, such an experiment was conducted on the Moon by astronaut David Scott. He simultaneously released a geologist's hammer and a falcon's feather, and in unison they fell to the lunar surface. This demonstration surely would have pleased Galileo!

We shall denote the magnitude of the free-fall acceleration with the symbol *g*, representing a vector acceleration $\vec{\mathbf{g}}$. At the surface of the Earth, *g* is approximately 9.80 m/s², or 980 cm/s², or 32 ft/s². Unless stated otherwise, we shall use the value 9.80 m/s² when doing calculations. Furthermore, we shall assume that the vector $\vec{\mathbf{g}}$ is directed downward toward the center of the Earth.

When we use the expression *freely falling object*, we do not necessarily mean an object dropped from rest. A freely falling object is an object moving freely under the influence of gravity alone, regardless of its initial motion. Therefore, objects thrown upward or downward and those released from rest are all freely falling objects once they are

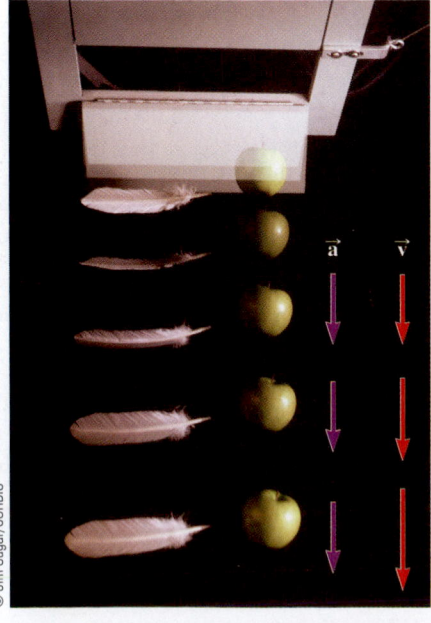

Figure 2.14 An apple and a feather, released from rest in a vacuum chamber, fall at the same rate, regardless of their masses. Ignoring air resistance, all objects fall to the Earth with the same acceleration of magnitude 9.80 m/s², as indicated by the purple arrows in this multiflash photograph. The velocity of the two objects increases linearly with time, as indicated by the series of red arrows.

released! Because the value of g is constant as long as we are close to the surface of the Earth, we can model a freely falling object as a particle under constant acceleration.

In previous examples in this chapter, the particles were undergoing constant acceleration, as stated in the problem. Therefore, it may have been difficult to understand the need for modeling. We can now begin to see the need for modeling; we are *modeling* a real falling object with an analysis model. Notice that we are (1) ignoring air resistance and (2) assuming that the free-fall acceleration is constant. Therefore, the model of a particle under constant acceleration is a *replacement* for the real problem, which could be more complicated. If air resistance and any variation in g are small, however, the model should make predictions that agree closely with the real situation.

The equations developed in Section 2.6 for the particle under constant acceleration model can be applied to the falling object. The only necessary modification that we need to make in these equations for freely falling objects is to note that the motion is in the vertical direction, so we will use y instead of x, and that the acceleration is downward and of magnitude 9.80 m/s². Therefore, for a freely falling object we commonly take $a_y = -g = -9.80$ m/s², where the negative sign indicates that the acceleration of the object is downward. The choice of negative for the downward direction is arbitrary, but common.

QUICK QUIZ 2.6 A ball is thrown upward. While the ball is in free-fall, does its acceleration (**a**) increase, (**b**) decrease, (**c**) increase and then decrease, (**d**) decrease and then increase, or (**e**) remain constant?

Pitfall Prevention | 2.7
Acceleration at the Top of the Motion
A common misconception is that the acceleration of a projectile at the top of its trajectory is zero. Although the velocity at the top of the motion of an object thrown upward momentarily goes to zero, *the acceleration is still that due to gravity* at this point. If the velocity and acceleration were both zero, the projectile would stay at the top.

Pitfall Prevention | 2.8
The sign of g
Keep in mind that g is a *positive number*. It is tempting to substitute -9.80 m/s² for g, but resist the temptation. That the gravitational acceleration is downward is indicated explicitly by stating the acceleration as $a_y = -g$.

THINKING PHYSICS 2.2

A skydiver steps out of a stationary helicopter. A few seconds later, another skydiver steps out, so that both skydivers fall along the same vertical line. Ignore air resistance, so that both skydivers fall with the same acceleration, and model the skydivers as particles under constant acceleration. Does the vertical separation distance between them stay the same? Does the difference in their speeds stay the same?

Reasoning At any given instant of time, the speeds of the skydivers are definitely different, because one had a head start over the other. In any time interval, however, each skydiver increases his or her speed by the same amount, because they have the same acceleration. Therefore, the difference in speeds remains the same. The first skydiver will always be moving with a higher speed than the second. In a given time interval, then, the first skydiver will have a larger displacement than the second. Therefore, the separation distance between them increases. ◄

Example 2.9 | Not a Bad Throw for a Rookie!

A stone thrown from the top of a building is given an initial velocity of 20.0 m/s straight upward. The stone is launched 50.0 m above the ground, and the stone just misses the edge of the roof on its way down as shown in Figure 2.15 (page 54).

(A) Using $t_\text{Ⓐ} = 0$ as the time the stone leaves the thrower's hand at position Ⓐ, determine the time at which the stone reaches its maximum height.

SOLUTION

Conceptualize You most likely have experience with dropping objects or throwing them upward and watching them fall, so this problem should describe a familiar experience. To simulate this situation, toss a small object upward and notice the time interval required for it to fall to the floor. Now imagine throwing that object upward from the roof of a building.

Recognize that the initial velocity is positive because the stone is launched upward. The velocity will change sign after the stone reaches its highest point, but the acceleration of the stone will *always* be downward.

Categorize Because the stone is in free fall, it is modeled as a particle under constant acceleration due to gravity.

continued

2.9 *cont.*

Analyze Choose an initial point just after the stone leaves the person's hand and a final point at the top of its flight.

Use Equation 2.10 to calculate the time at which the stone reaches its maximum height:

$$v_{yf} = v_{yi} + a_y t \quad \rightarrow \quad t = \frac{v_{yf} - v_{yi}}{a_y}$$

Substitute numerical values:

$$t = t_{\text{Ⓑ}} = \frac{0 - 20.0 \text{ m/s}}{-9.80 \text{ m/s}^2} = \boxed{2.04 \text{ s}}$$

(B) Find the maximum height of the stone.

SOLUTION

As in part (A), choose the initial and final points at the beginning and the end of the upward flight.

Set $y_{\text{Ⓐ}} = 0$ and substitute the time from part (A) into Equation 2.13 to find the maximum height:

$$y_{\max} = y_{\text{Ⓑ}} = y_{\text{Ⓐ}} + v_{x\text{Ⓐ}}t + \tfrac{1}{2}a_y t^2$$
$$y_{\text{Ⓑ}} = 0 + (20.0 \text{ m/s})(2.04 \text{ s}) +$$
$$\tfrac{1}{2}(-9.80 \text{ m/s}^2)(2.04 \text{ s})^2 = \boxed{20.4 \text{ m}}$$

(C) Determine the velocity of the stone when it returns to the height from which it was thrown.

SOLUTION

Choose the initial point where the stone is launched and the final point when it passes this position coming down.

Substitute known values into Equation 2.14:

$$v_{y\text{Ⓒ}}^2 = v_{y\text{Ⓐ}}^2 + 2a_y(y_{\text{Ⓒ}} - y_{\text{Ⓐ}})$$
$$v_{y\text{Ⓒ}}^2 = (20.0 \text{ m/s})^2 + 2(-9.80 \text{ m/s}^2)(0 - 0) = 400 \text{ m}^2/\text{s}^2$$
$$v_{y\text{Ⓒ}} = \boxed{-20.0 \text{ m/s}}$$

When taking the square root, we could choose either a positive or a negative root. We choose the negative root because we know that the stone is moving downward at point Ⓒ. The velocity of the stone when it arrives back at its original height is equal in magnitude to its initial velocity but is opposite in direction.

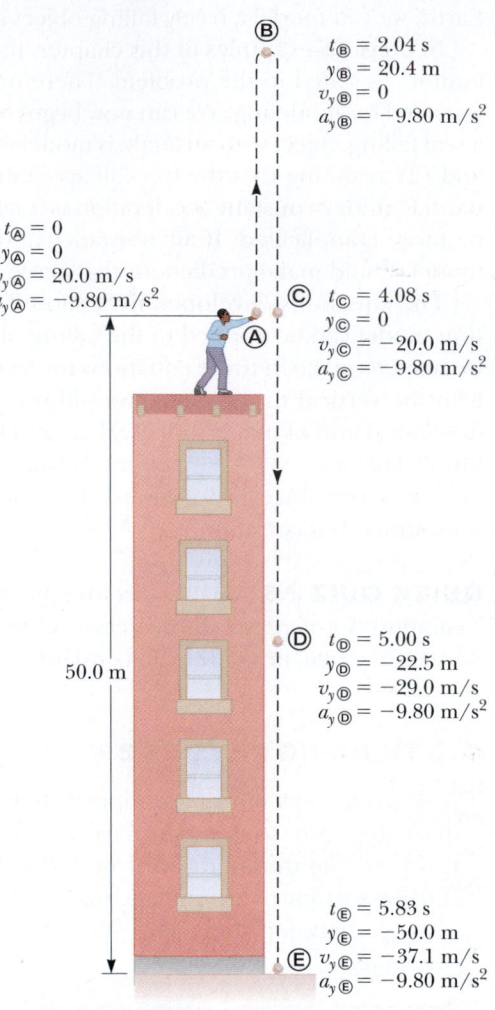

Figure 2.15 (Example 2.9) Position and velocity versus time for a freely falling stone thrown initially upward with a velocity $v_{yi} = 20.0$ m/s. Many of the quantities in the labels for points in the motion of the stone are calculated in the example. Can you verify the other values that are not?

(D) Find the velocity and position of the stone at $t = 5.00$ s.

SOLUTION

Choose the initial point just after the throw and the final point 5.00 s later.

Calculate the velocity at Ⓓ from Equation 2.10:

$$v_{y\text{Ⓓ}} = v_{y\text{Ⓐ}} + a_y t = 20.0 \text{ m/s} + (-9.80 \text{ m/s}^2)(5.00 \text{ s}) = \boxed{-29.0 \text{ m/s}}$$

Use Equation 2.13 to find the position of the stone at $t_{\text{Ⓓ}} = 5.00$ s:

$$y_{\text{Ⓓ}} = y_{\text{Ⓐ}} + v_{y\text{Ⓐ}}t + \tfrac{1}{2}a_y t^2$$
$$= 0 + (20.0 \text{ m/s})(5.00 \text{ s}) + \tfrac{1}{2}(-9.80 \text{ m/s}^2)(5.00 \text{ s})^2$$
$$= \boxed{-22.5 \text{ m}}$$

Finalize The choice of the time defined as $t = 0$ is arbitrary and up to you to select as the problem solver. As an example of this arbitrariness, choose $t = 0$ as the time at which the stone is at the highest point in its motion. Then solve parts (C) and (D) again using this new initial instant and notice that your answers are the same as those above.

What If? What if the throw were from 30.0 m above the ground instead of 50.0 m? Which answers in parts (A) to (D) would change?

Answer None of the answers would change. All the motion takes place in the air during the first 5.00 s. (Notice that even for a throw from 30.0 m, the stone is above the ground at $t = 5.00$ s.) Therefore, the height of the throw is not an issue. Mathematically, if we look back over our calculations, we see that we never entered the height of the throw into any equation.

◄ **2.8** | **Context Connection: Acceleration Required by Consumers**

We now have our first opportunity to address a Context in a closing section, as we will do in each remaining chapter. Our present Context is *Alternative-Fuel Vehicles*, and our central question is, *What source besides gasoline can be used to provide energy for an automobile while reducing environmentally damaging emissions?*

Consumers have been driving gasoline-powered vehicles for decades and have become used to a certain amount of acceleration. In addition, roadway features such as the lengths of freeway on-ramps have been designed with the expectation of a minimum acceleration required for a vehicle to merge with existing traffic. These experiences raise the question as to what kind of acceleration today's consumer would expect for an alternative-fuel vehicle that might replace a gasoline-powered vehicle. In turn, developers of alternative-fuel vehicles should strive for such an acceleration so as to satisfy consumer expectations and hope to generate a demand for the new vehicle.

If we consider published time intervals for accelerations from 0 to 60 mi/h for a number of automobile models, we find the data shown in the third column of Table 2.3 (page 56). The average acceleration of each vehicle is calculated from these data using Equation 2.7. It is clear from the upper part of this table (*Very expensive vehicles*) that acceleration upward of 20 mi/h · s is very expensive. The highest acceleration is 23.1 mi/h · s for the Bugatti Veyron 16.4 Super Sport and costs over two million dollars. A slightly lower acceleration can be had with the Shelby SuperCars Ultimate Aero for a bargain price of $654,000. The *Performance vehicles* between $44,000 and $102,000 show an average acceleration of 14.1 mi/h · s compared to 19.1 mi/h · s for the *Very expensive vehicles*. For the less affluent driver, the accelerations in the third section of the table (*Traditional vehicles*) have an average value of 6.9 mi/h · s. This number is typical of consumer-oriented gasoline-powered vehicles and provides an approximate standard for the acceleration desired in an alternative-fuel vehicle.

In the lower part of Table 2.3, we see data for five alternative vehicles. The average acceleration of these five cars is 6.2 mi/h · s, about 90% of the average value for the traditional vehicles. This acceleration is sufficiently large that it satisfies consumer demand for a car with "get-up-and-go." Figure 2.16 shows a graph of the cost of the vehicles in Table 2.3 versus acceleration. The graph clearly shows the skyrocketing cost of accelerations larger than 20 mi/h · s.

The Honda CR-Z, Honda Insight, and Toyota Prius are *hybrid vehicles*, which we will discuss further in the Context Conclusion. These vehicles combine a gasoline engine and an electric motor, both directly driving the wheels. The accelerations for these vehicles are among the lowest in the table. The disadvantage of the low acceleration is

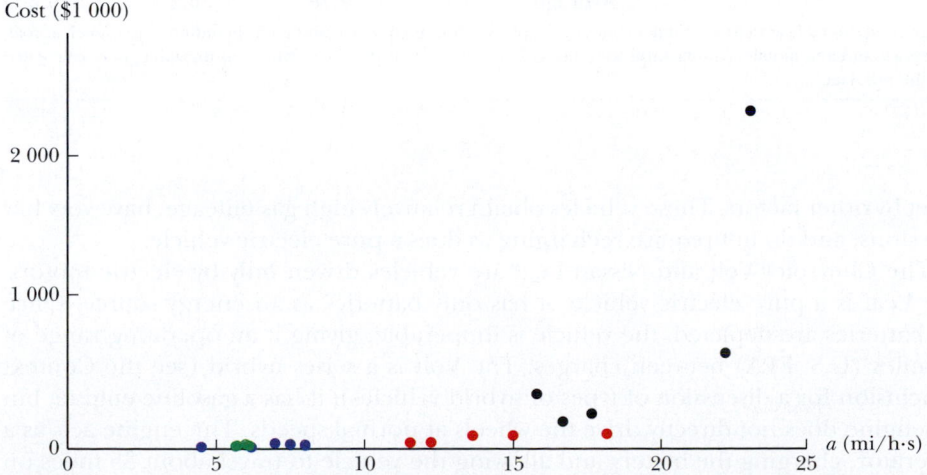

Figure 2.16 The cost to obtain a certain acceleration for alternative vehicles (in green), traditional vehicles (in blue), performance vehicles (in red), and very expensive vehicles (in black).

TABLE 2.3 | Accelerations of Various Vehicles, 0–60 mph

Automobile	Model Year	Time Interval, 0 to 60 mi/h (s)	Average Acceleration (mi/h · s)	Price
Very expensive vehicles:				
Bugatti Veyron 16.4 Super Sport	2011	2.60	23.1	$2,300,000
Lamborghini LP 570-4 Superleggera	2011	3.40	17.6	$240,000
Lexus LFA	2011	3.80	15.8	$375,000
Mercedes-Benz SLS AMG	2011	3.60	16.7	$186,000
Shelby SuperCars Ultimate Aero	2009	2.70	22.2	$654,000
Average		**3.22**	**19.1**	**$751,000**
Performance vehicles:				
Chevrolet Corvette ZR1	2010	3.30	18.2	$102,000
Dodge Viper SRT10	2010	4.00	15.0	$91,000
Jaguar XJL Supercharged	2011	4.40	13.6	$90,500
Acura TL SH-AWD	2009	5.20	11.5	$44,000
Dodge Challenger SRT8	2010	4.90	12.2	$45,000
Average		**4.36**	**14.1**	**$74,500**
Traditional vehicles:				
Buick Regal CXL Turbo	2011	7.50	8.0	$30,000
Chevrolet Tahoe 1500 LS (SUV)	2011	8.60	7.0	$40,000
Ford Fiesta SES	2010	9.70	6.2	$14,000
Hummer H3 (SUV)	2010	8.00	7.5	$34,000
Hyundai Sonata SE	2010	7.50	8.0	$25,000
Smart ForTwo	2010	13.30	4.5	$16,000
Average		**9.10**	**6.9**	**$26,500**
Alternative vehicles:				
Chevrolet Volt (hybrid)	2011	8.00	7.5	$41,000
Nissan Leaf (electric)	2011	10.00	6.0	$34,000
Honda CR-Z (hybrid)	2011	10.50	5.7	$25,000
Honda Insight (hybrid)	2010	10.60	5.7	$21,000
Toyota Prius (hybrid)	2010	9.80	6.1	$24,000
Average		**9.78**	**6.2**	**$29,000**

Note: Data given in this table as well as in similar tables in Chapters 3 through 6 were gathered from online sources such as road test reports and automobile manufacturer websites. Other data, such as the accelerations in this table, were calculated from the raw data.

offset by other factors. These vehicles obtain relatively high gas mileage, have very low emissions, and do not require recharging as does a pure electric vehicle.

The Chevrolet Volt and Nissan Leaf are vehicles driven only by electric motors. The Leaf is a pure electric vehicle: it has only batteries as an energy source. Once the batteries are depleted, the vehicle is inoperable, giving it an operating range of 73 miles (U.S. EPA) between charges. The Volt is a series hybrid (see the Context Conclusion for a discussion of types of hybrid vehicles): it has a gasoline engine, but the engine does not directly drive the wheels at normal speeds. The engine acts as a generator, charging the battery and allowing the vehicle to travel about 35 miles on electricity alone but over 350 miles between charges.

In comparison to the vehicles in Table 2.3, consider the acceleration of an even higher-level "performance vehicle," a typical drag racer, as shown in Figure 2.17.

Figure 2.17 In drag racing, acceleration is a highly desired quantity. In a distance of 1/4 mile, speeds of over 320 mi/h are reached, with the entire distance being covered in under 5 s.

Typical data show that such a vehicle covers a distance of 0.25 mi in 5.0 s, starting from rest. We can find the acceleration from Equation 2.13:

$$x_f = x_i + v_i t + \tfrac{1}{2} a_x t^2 = 0 + 0(t) + \tfrac{1}{2}(a_x)(t)^2 \rightarrow a_x = \frac{2x_f}{t^2}$$

$$a_x = \frac{2(0.25 \text{ mi})}{(5.0 \text{ s})^2} = 0.020 \text{ mi/s}^2 \left(\frac{3\,600 \text{ s}}{1 \text{ h}} \right) = 72 \text{ mi/h} \cdot \text{s}$$

This value is much larger than any accelerations in the table, as would be expected. We can show that the acceleration due to gravity has the following value in units of mi/h · s:

$$g = 9.80 \text{ m/s}^2 = 21.9 \text{ mi/h} \cdot \text{s}$$

Therefore, the drag racer is moving horizontally with 3.3 times as much acceleration as it would move vertically if you pushed it off a cliff! (Of course, the horizontal acceleration can only be maintained for a very short time interval.)

As we investigate two-dimensional motion in the next chapter, we shall consider a different type of acceleration for vehicles, that associated with the vehicle turning in a sharp circle at high speed.

▶ SUMMARY

The **average speed** of a particle during some time interval is equal to the ratio of the distance d traveled by the particle and the time interval Δt:

$$v_{\text{avg}} \equiv \frac{d}{\Delta t} \qquad \text{2.1} \blacktriangleleft$$

The **average velocity** of a particle moving in one dimension during some time interval is equal to the ratio of the displacement Δx and the time interval Δt:

$$v_{x,\text{avg}} \equiv \frac{\Delta x}{\Delta t} \qquad \text{2.2} \blacktriangleleft$$

The **instantaneous velocity** of a particle is defined as the limit of the ratio $\Delta x / \Delta t$ as Δt approaches zero:

$$v_x \equiv \lim_{\Delta t \to 0} \frac{\Delta x}{\Delta t} = \frac{dx}{dt} \qquad \text{2.3} \blacktriangleleft$$

The **instantaneous speed** of a particle is defined as the magnitude of the instantaneous velocity vector.

The **average acceleration** of a particle moving in one dimension during some time interval is defined as the ratio of the change in its velocity Δv_x and the time interval Δt:

$$a_{x,\text{avg}} \equiv \frac{\Delta v_x}{\Delta t} \qquad \text{2.7} \blacktriangleleft$$

The **instantaneous acceleration** is equal to the limit of the ratio $\Delta v_x / \Delta t$ as $\Delta t \to 0$. By definition, this limit equals the derivative of v_x with respect to t, or the time rate of change of the velocity:

$$a_x \equiv \lim_{\Delta t \to 0} \frac{\Delta v_x}{\Delta t} = \frac{dv_x}{dt} \qquad \text{2.8} \blacktriangleleft$$

The slope of the tangent to the x versus t curve at any instant gives the instantaneous velocity of the particle.

The slope of the tangent to the v versus t curve gives the instantaneous acceleration of the particle.

An object falling freely experiences an acceleration directed toward the center of the Earth. If air friction is ignored and if the altitude of the motion is small compared with the Earth's radius, one can assume that the magnitude of the free-fall acceleration g is constant over the range of motion, where g is equal to 9.80 m/s², or 32 ft/s². Assuming y to be positive upward, the acceleration is given by $-g$, and the equations of kinematics for an object in free-fall are the same as those already given, with the substitutions $x \to y$ and $a_y \to -g$.

George Lepp/Stone/Getty Images

Analysis Models for Problem-Solving

Particle Under Constant Velocity. If a particle moves in a straight line with a constant speed v_x, its constant velocity is given by

$$v_x = \frac{\Delta x}{\Delta t}$$

2.4◀

and its position is given by

$$x_f = x_i + v_x t$$

2.5◀

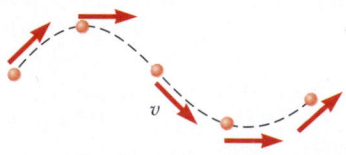

Particle Under Constant Speed. If a particle moves a distance d along a curved or straight path with a constant speed, its constant speed is given by

$$v = \frac{d}{\Delta t}$$

2.6◀

Particle Under Constant Acceleration. If a particle moves in a straight line with a constant acceleration a_x, its motion is described by the kinematic equations:

$$v_{xf} = v_{xi} + a_x t$$

2.10◀

$$v_{x,\text{avg}} = \frac{v_{xi} + v_{xf}}{2}$$

2.11◀

$$x_f = x_i + \tfrac{1}{2}(v_{xi} + v_{xf})t$$

2.12◀

$$x_f = x_i + v_{xi}t + \tfrac{1}{2}a_x t^2$$

2.13◀

$$v_{xf}^2 = v_{xi}^2 + 2a_x(x_f - x_i)$$

2.14◀

OBJECTIVE QUESTIONS

1. One drop of oil falls straight down onto the road from the engine of a moving car every 5 s. Figure OQ2.1 shows the pattern of the drops left behind on the pavement. What is the average speed of the car over this section of its motion? (a) 20 m/s (b) 24 m/s (c) 30 m/s (d) 100 m/s (e) 120 m/s

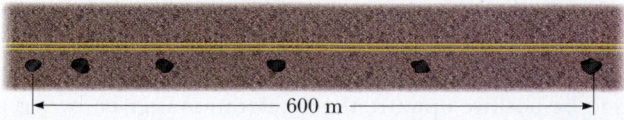

Figure OQ2.1

2. When applying the equations of kinematics for an object moving in one dimension, which of the following statements *must* be true? (a) The velocity of the object must remain constant. (b) The acceleration of the object must remain constant. (c) The velocity of the object must increase with time. (d) The position of the object must increase with time. (e) The velocity of the object must always be in the same direction as its acceleration.

3. A juggler throws a bowling pin straight up in the air. After the pin leaves his hand and while it is in the air, which statement is true? (a) The velocity of the pin is always in the same direction as its acceleration. (b) The velocity of the pin is never in the same direction as its acceleration. (c) The acceleration of the pin is zero. (d) The velocity of the pin is opposite its acceleration on the way up. (e) The velocity of the pin is in the same direction as its acceleration on the way up.

4. An arrow is shot straight up in the air at an initial speed of 15.0 m/s. After how much time is the arrow moving downward at a speed of 8.00 m/s? (a) 0.714 s (b) 1.24 s (c) 1.87 s (d) 2.35 s (e) 3.22 s

5. When the pilot reverses the propeller in a boat moving north, the boat moves with an acceleration directed south. Assume the acceleration of the boat remains constant in magnitude and direction. What happens to the boat? (a) It eventually stops and remains stopped. (b) It eventually stops and then speeds up in the forward direction. (c) It eventually stops and then speeds up in the reverse direction. (d) It never stops but loses speed more and more slowly forever. (e) It never stops but continues to speed up in the forward direction.

6. A pebble is dropped from rest from the top of a tall cliff and falls 4.9 m after 1.0 s has elapsed. How much farther does it drop in the next 2.0 s? (a) 9.8 m (b) 19.6 m (c) 39 m (d) 44 m (e) none of the above

7. A student at the top of a building of height h throws one ball upward with a speed of v_i and then throws a second ball downward with the same initial speed v_i. Just before it reaches the ground, is the final speed of the ball thrown upward (a) larger, (b) smaller, or (c) the same in magnitude, compared with the final speed of the ball thrown downward?

8. A rock is thrown downward from the top of a 40.0-m-tall tower with an initial speed of 12 m/s. Assuming negligible air resistance, what is the speed of the rock just before hitting the ground? (a) 28 m/s (b) 30 m/s (c) 56 m/s (d) 784 m/s (e) More information is needed.

9. As an object moves along the x axis, many measurements are made of its position, enough to generate a smooth, accurate graph of x versus t. Which of the following quantities for the object *cannot* be obtained from this graph *alone*? (a) the velocity at any instant (b) the acceleration at any instant (c) the displacement during some time interval (d) the average velocity during some time interval (e) the speed at any instant

10. You drop a ball from a window located on an upper floor of a building. It strikes the ground with speed v. You now repeat the drop, but your friend down on the ground throws another ball upward at the same speed v, releasing her ball at the same moment that you drop yours from the window. At some location, the balls pass each other. Is this location (a) *at* the halfway point between window and ground, (b) *above* this point, or (c) *below* this point?

11. A skateboarder starts from rest and moves down a hill with constant acceleration in a straight line, traveling for 6 s. In a second trial, he starts from rest and moves along the same straight line with the same acceleration for only 2 s. How does his displacement from his starting point in this second trial compare with that from the first trial? (a) one-third as large (b) three times larger (c) one-ninth as large (d) nine times larger (e) $1/\sqrt{3}$ times as large

12. A ball is thrown straight up in the air. For which situation are both the instantaneous velocity and the acceleration zero? (a) on the way up (b) at the top of its flight path (c) on the way down (d) halfway up and halfway down (e) none of the above

13. A hard rubber ball, not affected by air resistance in its motion, is tossed upward from shoulder height, falls to the sidewalk, rebounds to a smaller maximum height, and is caught on its way down again. This motion is represented in Figure OQ2.13, where the successive positions of the ball Ⓐ through Ⓔ are not equally spaced in time. At point Ⓓ the center of the ball is at its lowest point in the motion. The motion of the ball is along a straight, vertical line, but the diagram shows successive positions offset to the right to avoid overlapping. Choose the positive y direction to be upward. (a) Rank the situations Ⓐ through Ⓔ according to the speed of the ball $|v_y|$ at each point, with the largest speed first. (b) Rank the same situations according to the acceleration a_y of the ball at each point. (In both rankings, remember that zero is greater than a negative value. If two values are equal, show that they are equal in your ranking.)

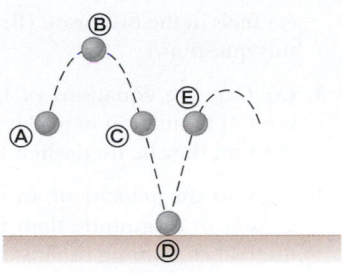

Figure OQ2.13

14. Each of the strobe photographs (a), (b), and (c) in Figure OQ2.14 was taken of a single disk moving toward the right, which we take as the positive direction. Within each photograph, the time interval between images is constant. (i) Which photograph shows motion with zero acceleration? (ii) Which photograph shows motion with positive acceleration? (iii) Which photograph shows motion with negative acceleration?

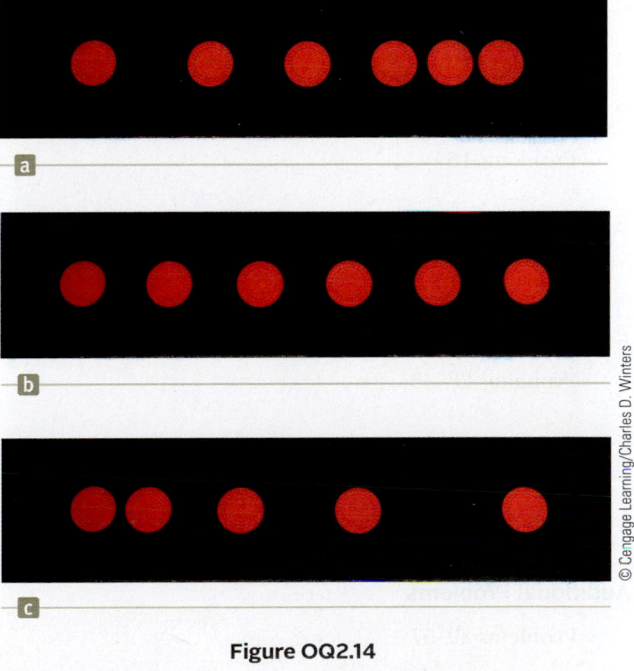

Figure OQ2.14

© Cengage Learning/Charles D. Winters

CONCEPTUAL QUESTIONS

☐ denotes answer available in *Student Solutions Manual/Study Guide*

1. If a car is traveling eastward, can its acceleration be westward? Explain.

2. Try the following experiment away from traffic where you can do it safely. With the car you are driving moving slowly on a straight, level road, shift the transmission into neutral and let the car coast. At the moment the car comes to a complete stop, step hard on the brake and notice what you feel. Now repeat the same experiment on a fairly gentle, uphill slope. Explain the difference in what a person riding in the car feels in the two cases. (Brian Popp suggested the idea for this question.)

3. (a) Can the equations of kinematics (Eqs. 2.10–2.14) be used in a situation in which the acceleration varies in time? (b) Can they be used when the acceleration is zero?

4. (a) Can the velocity of an object at an instant of time be greater in magnitude than the average velocity over a time interval containing the instant? (b) Can it be less?

5. If the average velocity of an object is zero in some time interval, what can you say about the displacement of the object for that interval?

6. If the velocity of a particle is zero, can the particle's acceleration be zero? Explain.

7. If the velocity of a particle is nonzero, can the particle's acceleration be zero? Explain.

8. You throw a ball vertically upward so that it leaves the ground with velocity +5.00 m/s. (a) What is its velocity when it reaches its maximum altitude? (b) What is its acceleration at this point? (c) What is the velocity with which it returns to ground level? (d) What is its acceleration at this point?

9. Two cars are moving in the same direction in parallel lanes along a highway. At some instant, the velocity of car A exceeds the velocity of car B. Does that mean that the acceleration of car A is greater than that of car B? Explain.

> PROBLEMS AVAILABLE IN ENHANCED WebAssign

2.1 Average Velocity

Problems 1–4

2.2 Instantaneous Velocity

Problems 5–8

2.3 Analysis Model: Particle Under Constant Velocity

Problem 9

2.4 Acceleration

Problems 10–15

2.5 Motion Diagrams

Problems 16

2.6 Analysis Model: Particle Under Constant Acceleration

Problems 17–27

2.7 Freely Falling Objects

Problems 28–35

2.8 Context Connection: Acceleration Required by Consumers

Problems 36–38

Additional Problems

Problems 39–57

Solutions to the following Problems are available in the *Student Solutions Manual/Study Guide*:

2.1, 2.4, 2.5, 2.11, 2.17, 2.20, 2.30, 2.31, 2.33, 2.41, and 2.53

List of Enhanced Problems

Problem Number	Targeted Feedback in Enhanced WebAssign	Master It in Enhanced WebAssign	Watch It in Enhanced WebAssign
2.1	✓		✓
2.2	✓		✓
2.4	✓	✓	
2.7	✓		✓
2.10	✓		✓
2.11	✓	✓	
2.13	✓		✓
2.14	✓		✓
2.17		✓	
2.19	✓		✓
2.20	✓	✓	
2.23	✓		✓
2.26	✓		✓
2.30	✓		✓
2.33		✓	
2.35	✓		✓
2.41		✓	
2.53		✓	

Motion in Two Dimensions

Chapter Outline

Photo courtesy of Laservision

I n this chapter, we shall study the kinematics of an object that can be modeled as a particle moving in a plane. This motion is two dimensional. Some common examples of motion in a plane are the motions of satellites in orbit around the Earth, projectiles such as a thrown baseball, and the motion of electrons in uniform electric fields. We shall also study a particle in uniform circular motion and discuss various aspects of particles moving in curved paths.

The Musical Fountain of Eternal Life at the Swaminarayan Akshardham, a Hindu temple complex in New Delhi, India, presents a twelve-minute water, sound, and light show each evening. In this chapter, we will learn why the water arcs in the fountain have the shapes of parabolas.

❰3.1 | The Position, Velocity, and Acceleration Vectors

In Chapter 2, we found that the motion of a particle moving along a straight line such as the x axis is completely specified if its position is known as a function of time. Now let us extend this idea to motion in the xy plane. We will find equations for position and velocity that are the same as those in Chapter 2 except for their vector nature.

We begin by describing the position of a particle with a **position vector $\vec{r}$**, drawn from the origin of a coordinate system to the location of the particle in the xy plane, as in Figure 3.1 (page 62). At time t_i, the particle is at the point Ⓐ, and at some later time t_f, the particle is at Ⓑ, where the subscripts i and f refer to initial and final values. As the particle moves from Ⓐ to Ⓑ in the time interval $\Delta t = t_f - t_i$, the position vector changes from $\vec{r}_i$ to $\vec{r}_f$. As we learned in Chapter 2, the displacement of a particle is the difference between its final position and its initial position:

$$\Delta\vec{\mathbf{r}} \equiv \vec{\mathbf{r}}_f - \vec{\mathbf{r}}_i \qquad \textbf{3.1}\blacktriangleleft$$

The direction of $\Delta\vec{\mathbf{r}}$ is indicated in Figure 3.1.

The **average velocity** $\vec{\mathbf{v}}_{\text{avg}}$ of the particle during the time interval Δt is defined as the ratio of the displacement to the time interval:

▶ Definition of average velocity

$$\vec{\mathbf{v}}_{\text{avg}} \equiv \frac{\Delta\vec{\mathbf{r}}}{\Delta t} \qquad \textbf{3.2}\blacktriangleleft$$

Because displacement is a vector quantity and the time interval is a scalar quantity, we conclude that the average velocity is a *vector* quantity in the same direction as $\Delta\vec{\mathbf{r}}$. Compare Equation 3.2 with its one-dimensional counterpart, Equation 2.2. The average velocity between points Ⓐ and Ⓑ is *independent of the path* between the two points. That is true because the average velocity is proportional to the displacement, which in turn depends only on the initial and final position vectors and not on the path taken between those two points. As with one-dimensional motion, if a particle starts its motion at some point and returns to this point via any path, its average velocity is zero for this trip because its displacement is zero.

Consider again the motion of a particle between two points in the *xy* plane, as shown in Figure 3.2. As the time intervals over which we observe the motion become smaller and smaller, the direction of the displacement approaches that of the line tangent to the path at the point Ⓐ.

The **instantaneous velocity** $\vec{\mathbf{v}}$ is defined as the limit of the average velocity $\Delta\vec{\mathbf{r}}/\Delta t$ as Δt approaches zero:

$$\vec{\mathbf{v}} \equiv \lim_{\Delta t \to 0} \frac{\Delta\vec{\mathbf{r}}}{\Delta t} = \frac{d\vec{\mathbf{r}}}{dt} \qquad \textbf{3.3}\blacktriangleleft$$

That is, the instantaneous velocity equals the derivative of the position vector with respect to time. The direction of the instantaneous velocity vector at any point in a particle's path is along a line that is tangent to the path at that point and in the direction of motion. The magnitude of the instantaneous velocity is called the *speed*.

As a particle moves from point Ⓐ to point Ⓑ along some path as in Figure 3.3, its instantaneous velocity changes from $\vec{\mathbf{v}}_i$ at time t_i to $\vec{\mathbf{v}}_f$ at time t_f. The **average**

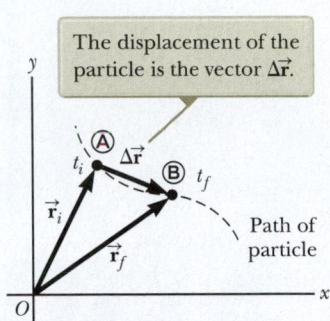

Figure 3.1 A particle moving in the *xy* plane is located with the position vector $\vec{\mathbf{r}}$ drawn from the origin to the particle. The displacement of the particle as it moves from Ⓐ to Ⓑ in the time interval $\Delta t = t_f - t_i$ is equal to the vector $\Delta\vec{\mathbf{r}} \equiv \vec{\mathbf{r}}_f - \vec{\mathbf{r}}_i$.

The displacement of the particle is the vector $\Delta\vec{\mathbf{r}}$.

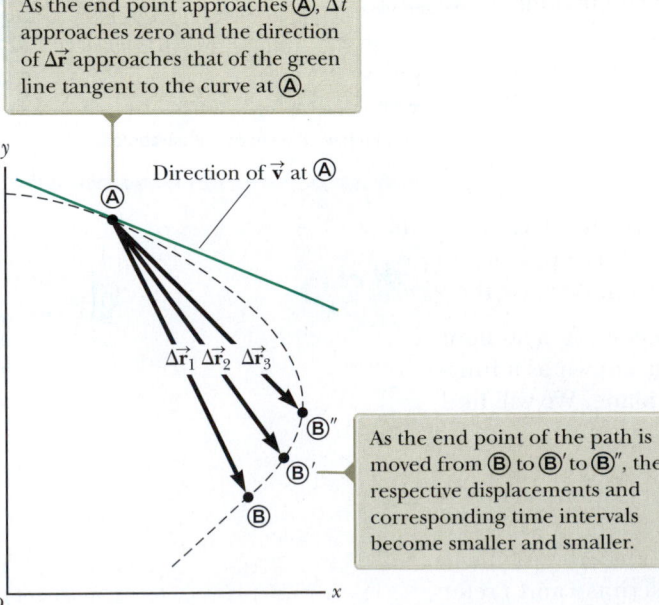

As the end point approaches Ⓐ, Δt approaches zero and the direction of $\Delta\vec{\mathbf{r}}$ approaches that of the green line tangent to the curve at Ⓐ.

As the end point of the path is moved from Ⓑ to Ⓑ' to Ⓑ", the respective displacements and corresponding time intervals become smaller and smaller.

Figure 3.2 As a particle moves between two points, its average velocity is in the direction of the displacement vector $\Delta\vec{\mathbf{r}}$. By definition, the instantaneous velocity at Ⓐ is directed along the line tangent to the curve at Ⓐ.

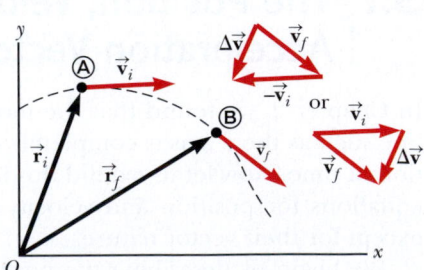

Figure 3.3 A particle moves from position Ⓐ to position Ⓑ. Its velocity vector changes from $\vec{\mathbf{v}}_i$ at time t_i to $\vec{\mathbf{v}}_f$ at time t_f. The vector addition diagrams at the upper right show two ways of determining the vector $\Delta\vec{\mathbf{v}}$ from the initial and final velocities.

acceleration $\vec{a}_{avg}$ of a particle over a time interval is defined as the ratio of the change in the instantaneous velocity $\Delta \vec{v}$ to the time interval Δt:

$$\vec{a}_{avg} \equiv \frac{\vec{v}_f - \vec{v}_i}{t_f - t_i} = \frac{\Delta \vec{v}}{\Delta t} \qquad 3.4 \blacktriangleleft$$

▶ Definition of average acceleration

Because the average acceleration is the ratio of a vector quantity $\Delta \vec{v}$ and a scalar quantity Δt, we conclude that $\vec{a}_{avg}$ is a vector quantity in the same direction as $\Delta \vec{v}$. Compare Equation 3.4 with the corresponding one-dimensional version, Equation 2.7. As indicated in Figure 3.3, the direction of $\Delta \vec{v}$ is found by adding the vector $-\vec{v}_i$ (the negative of $\vec{v}_i$) to the vector $\vec{v}_f$ because by definition $\Delta \vec{v} = \vec{v}_f - \vec{v}_i$.

The **instantaneous acceleration** $\vec{a}$ is defined as the limiting value of the ratio $\Delta \vec{v}/\Delta t$ as Δt approaches zero:

$$\vec{a} \equiv \lim_{\Delta t \to 0} \frac{\Delta \vec{v}}{\Delta t} = \frac{d \vec{v}}{dt} \qquad 3.5 \blacktriangleleft$$

▶ Definition of instantaneous acceleration

That is, the instantaneous acceleration equals the derivative of the velocity vector with respect to time. Compare Equation 3.5 with Equation 2.8.

It is important to recognize that various changes can occur that represent a particle undergoing an acceleration. First, the magnitude of the velocity vector (the speed) may change with time as in straight-line (one-dimensional) motion. Second, the direction of the velocity vector may change with time as its magnitude remains constant. Finally, both the magnitude and the direction of the velocity vector may change.

QUICK QUIZ 3.1 Consider the following controls in an automobile in motion: gas pedal, brake, steering wheel. What are the controls in this list that cause an acceleration of the car? (a) all three controls (b) the gas pedal and the brake (c) only the brake (d) only the gas pedal (e) only the steering wheel

Pitfall Prevention | 3.1
Vector Addition
Although the vector addition discussed in Chapter 1 involves *displacement* vectors, vector addition can be applied to *any* type of vector quantity. Figure 3.3, for example, shows the addition of *velocity* vectors using the graphical approach.

3.2 | Two-Dimensional Motion with Constant Acceleration

Let us consider two-dimensional motion during which the magnitude and direction of the acceleration remain unchanged. In this situation, we shall investigate motion as a two-dimensional version of the analysis in Section 2.6.

Before embarking on this investigation, we need to emphasize an important point regarding two-dimensional motion. Imagine an air hockey puck moving in a straight line along a perfectly level, friction-free surface of an air hockey table. Figure 3.4a shows a motion diagram from an overhead point of view of this puck. Recall that in Section 2.4 we related the acceleration of an object to a force on the object. Because there are no forces on the puck in the horizontal plane, it moves with constant velocity in the x direction. Now suppose you blow a puff of air on the puck as it passes your position, with the force from your puff of air *exactly* in the y direction. Because the force

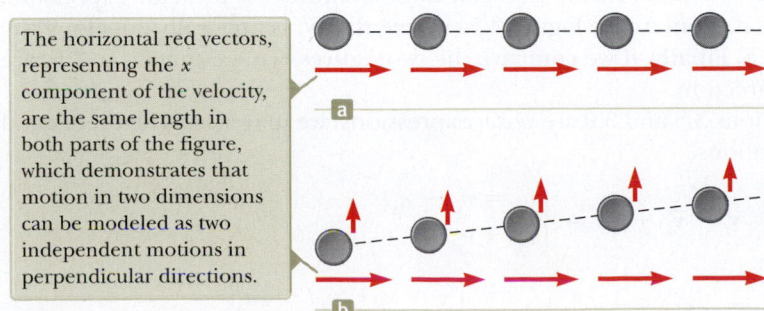

The horizontal red vectors, representing the x component of the velocity, are the same length in both parts of the figure, which demonstrates that motion in two dimensions can be modeled as two independent motions in perpendicular directions.

Figure 3.4 (a) A puck moves across a horizontal air hockey table at constant velocity in the x direction. (b) After a puff of air in the y direction is applied to the puck, the puck has gained a y component of velocity, but the x component is unaffected by the force in the perpendicular direction.

from this puff of air has no component in the x direction, it causes no acceleration in the x direction. It only causes a momentary acceleration in the y direction, causing the puck to have a constant y component of velocity once the force from the puff of air is removed. After your puff of air on the puck, its velocity component in the x direction is unchanged as shown in Figure 3.4b. The generalization of this simple experiment is that **motion in two dimensions can be modeled as two *independent* motions in each of the two perpendicular directions associated with the x and y axes. That is, any influence in the y direction does not affect the motion in the x direction and vice versa.**

The motion of a particle can be determined if its position vector $\vec{r}$ is known at all times. The position vector for a particle moving in the xy plane can be written

$$\vec{r} = x\hat{\mathbf{i}} + y\hat{\mathbf{j}} \qquad \text{3.6} \blacktriangleleft$$

where x, y, and $\vec{r}$ change with time as the particle moves. If the position vector is known, the velocity of the particle can be obtained from Equations 3.3 and 3.6:

$$\vec{v} = \frac{d\vec{r}}{dt} = \frac{dx}{dt}\hat{\mathbf{i}} + \frac{dy}{dt}\hat{\mathbf{j}} = v_x\hat{\mathbf{i}} + v_y\hat{\mathbf{j}} \qquad \text{3.7} \blacktriangleleft$$

Because we are assuming that $\vec{a}$ is constant in this discussion, its components a_x and a_y are also constants. Therefore, we can apply the equations of kinematics to the x and y components of the velocity vector separately. Substituting $v_x = v_{xf} = v_{xi} + a_xt$ and $v_y = v_{yf} = v_{yi} + a_yt$ into Equation 3.7 gives

$$\vec{v}_f = (v_{xi} + a_xt)\hat{\mathbf{i}} + (v_{yi} + a_yt)\hat{\mathbf{j}}$$

$$= (v_{xi}\hat{\mathbf{i}} + v_{yi}\hat{\mathbf{j}}) + (a_x\hat{\mathbf{i}} + a_y\hat{\mathbf{j}})t$$

▶ Velocity vector as a function of time for a particle under constant acceleration

$$\vec{v}_f = \vec{v}_i + \vec{a}t \qquad \text{3.8} \blacktriangleleft$$

This result states that the velocity $\vec{v}_f$ of a particle at some time t equals the vector sum of its initial velocity $\vec{v}_i$ and the additional velocity $\vec{a}t$ acquired at time t as a result of its constant acceleration. This result is the same as Equation 2.10, except for its vector nature.

Similarly, from Equation 2.13 we know that the x and y coordinates of a particle moving with constant acceleration are

$$x_f = x_i + v_{xi}t + \tfrac{1}{2}a_xt^2 \quad \text{and} \quad y_f = y_i + v_{yi}t + \tfrac{1}{2}a_yt^2$$

Substituting these expressions into Equation 3.6 gives

$$\vec{r}_f = (x_i + v_{xi}t + \tfrac{1}{2}a_xt^2)\hat{\mathbf{i}} + (y_i + v_{yi}t + \tfrac{1}{2}a_yt^2)\hat{\mathbf{j}}$$

$$= (x_i\hat{\mathbf{i}} + y_i\hat{\mathbf{j}}) + (v_{xi}\hat{\mathbf{i}} + v_{yi}\hat{\mathbf{j}})t + \tfrac{1}{2}(a_x\hat{\mathbf{i}} + a_y\hat{\mathbf{j}})t^2$$

▶ Position vector as a function of time for a particle under constant acceleration

$$\vec{r}_f = \vec{r}_i + \vec{v}_it + \tfrac{1}{2}\vec{a}t^2 \qquad \text{3.9} \blacktriangleleft$$

This equation implies that the final position vector $\vec{r}_f$ is the vector sum of the initial position vector $\vec{r}_i$ plus a displacement $\vec{v}_it$, arising from the initial velocity of the particle, and a displacement $\tfrac{1}{2}\vec{a}t^2$, resulting from the uniform acceleration of the particle. It is the same as Equation 2.13 except for its vector nature.

Graphical representations of Equations 3.8 and 3.9 are shown in Active Figures 3.5a and 3.5b. Note from Active Figure 3.5b that $\vec{r}_f$ is generally not along the direction of $\vec{r}_i$, $\vec{v}_i$, or $\vec{a}$ because the relationship between these quantities is a vector expression. For the same reason, from Active Figure 3.5a we see that $\vec{v}_f$ is generally not along the direction of $\vec{v}_i$ or $\vec{a}$. Finally, if we compare the two figures, we see that $\vec{v}_f$ and $\vec{r}_f$ are not in the same direction.

Because Equations 3.8 and 3.9 are *vector* expressions, we may also write their x and y component equations:

$$\vec{v}_f = \vec{v}_i + \vec{a}t \rightarrow \begin{cases} v_{xf} = v_{xi} + a_xt \\ v_{yf} = v_{yi} + a_yt \end{cases}$$

$$\vec{r}_f = \vec{r}_i + \vec{v}_it + \tfrac{1}{2}\vec{a}t^2 \rightarrow \begin{cases} x_f = x_i + v_{xi}t + \tfrac{1}{2}a_xt^2 \\ y_f = y_i + v_{yi}t + \tfrac{1}{2}a_yt^2 \end{cases}$$

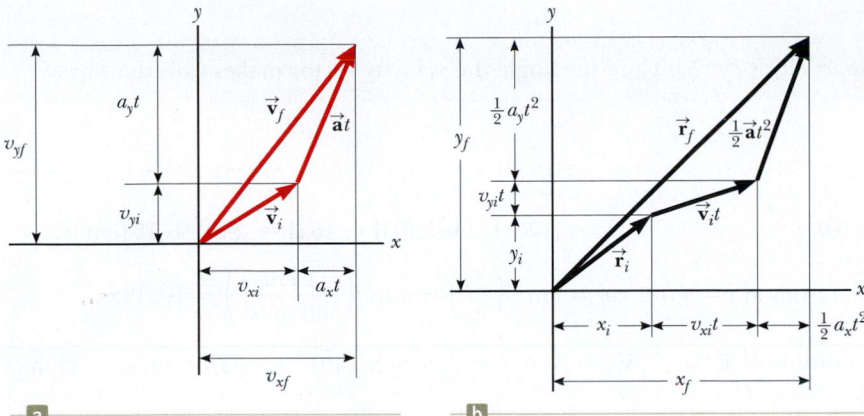

Active Figure 3.5 Vector representations and components of (a) the velocity and (b) the position of a particle moving with a constant acceleration $\vec{a}$.

These components are illustrated in Active Figure 3.5. Consistent with our discussion related to Figure 3.4, two-dimensional motion having constant acceleration is equivalent to two *independent* motions in the x and y directions having constant accelerations a_x and a_y. Therefore, there is no new model for a particle under two-dimensional constant acceleration; the appropriate model is just the one-dimensional particle under constant acceleration applied twice, in the x and y directions separately!

Example **3.1** | **Motion in a Plane**

A particle moves in the xy plane, starting from the origin at $t = 0$ with an initial velocity having an x component of 20 m/s and a y component of -15 m/s. The particle experiences an acceleration in the x direction, given by $a_x = 4.0$ m/s^2.

(A) Determine the total velocity vector at any time.

SOLUTION

Conceptualize The components of the initial velocity tell us that the particle starts by moving toward the right and downward. The x component of velocity starts at 20 m/s and increases by 4.0 m/s every second. The y component of velocity never changes from its initial value of -15 m/s. We sketch a motion diagram of the situation in Figure 3.6. Because the particle is accelerating in the $+x$ direction, its velocity component in this direction increases and the path curves as shown in the diagram. Notice that the spacing between successive images increases as time goes on because the speed is increasing. The placement of the acceleration and velocity vectors in Figure 3.6 helps us further conceptualize the situation.

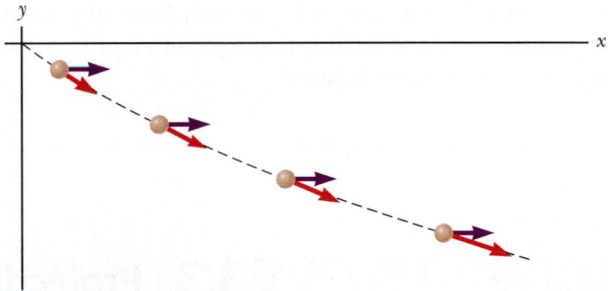

Figure 3.6 (Example 3.1) Motion diagram for the particle.

Categorize Because the initial velocity has components in both the x and y directions, we categorize this problem as one involving a particle moving in two dimensions. Because the particle only has an x component of acceleration, we model it as a particle under constant acceleration in the x direction and a particle under constant velocity in the y direction.

⋯⋯

Analyze To begin the mathematical analysis, we set $v_{xi} = 20$ m/s, $v_{yi} = -15$ m/s, $a_x = 4.0$ m/s^2, and $a_y = 0$.

Use Equation 3.8 for the velocity vector:

$$\vec{v}_f = \vec{v}_i + \vec{a}t = (v_{xi} + a_x t)\hat{i} + (v_{yi} + a_y t)\hat{j}$$

Substitute numerical values with the velocity in meters per second and the time in seconds:

$$\vec{v}_f = [20 + (4.0)t]\hat{i} + [-15 + (0)t]\hat{j}$$

$$(1)\quad \vec{v}_f = \boxed{[(20 + 4.0t)\hat{i} - 15\hat{j}]}$$

⋯⋯

Finalize Notice that the x component of velocity increases in time while the y component remains constant; this result is consistent with our prediction.

continued

3.1 cont.

(B) Calculate the velocity and speed of the particle at $t = 5.0$ s and the angle the velocity vector makes with the x axis.

SOLUTION

Analyze

Evaluate the result from Equation (1) at $t = 5.0$ s:

$$\vec{v}_f = [(20 + 4.0(5.0))\hat{i} - 15\hat{j}] = (40\hat{i} - 15\hat{j})\,\text{m/s}$$

Determine the angle θ that $\vec{v}_f$ makes with the x axis at $t = 5.0$ s:

$$\theta = \tan^{-1}\left(\frac{v_{yf}}{v_{xf}}\right) = \tan^{-1}\left(\frac{-15\,\text{m/s}}{40\,\text{m/s}}\right) = -21°$$

Evaluate the speed of the particle as the magnitude of $\vec{v}_f$:

$$v_f = |\vec{v}_f| = \sqrt{v_{xf}^2 + v_{yf}^2} = \sqrt{(40)^2 + (-15)^2}\,\text{m/s} = 43\,\text{m/s}$$

Finalize The negative sign for the angle θ indicates that the velocity vector is directed at an angle of 21° below the positive x axis. Notice that if we calculate v_i from the x and y components of $\vec{v}_i$, we find that $v_f > v_i$. Is that consistent with our prediction?

(C) Determine the x and y coordinates of the particle at any time t and its position vector at this time.

SOLUTION

Analyze

Use the components of Equation 3.9 with $x_i = y_i = 0$ at $t = 0$ with x and y in meters and t in seconds:

$$x_f = v_{xi}t + \tfrac{1}{2}a_x t^2 = 20t + 2.0t^2$$

$$y_f = v_{yi}t = -15t$$

Express the position vector of the particle at any time t:

$$\vec{r}_f = x_f\hat{i} + y_f\hat{j} = (20t + 2.0t^2)\,\hat{i} - 15t\,\hat{j}$$

Finalize Let us now consider a limiting case for very large values of t.

What If? What if we wait a very long time and then observe the motion of the particle? How would we describe the motion of the particle for large values of the time?

Answer Looking at Figure 3.6, we see the path of the particle curving toward the x axis. There is no reason to assume this tendency will change, which suggests that the path will become more and more parallel to the x axis as time grows large. Mathematically, Equation (1) shows that the y component of the velocity remains constant while the x component grows linearly with t. Therefore, when t is very large, the x component of the velocity will be much larger than the y component, suggesting that the velocity vector becomes more and more parallel to the x axis. Both x_f and y_f continue to grow with time, although x_f grows much faster.

❮3.3 | Projectile Motion

Anyone who has observed a baseball in motion (or, for that matter, any object thrown into the air) has observed projectile motion. The ball moves in a curved path when thrown at some angle with respect to the Earth's surface. **Projectile motion** of an object is surprisingly simple to analyze if the following two assumptions are made when building a model for these types of problems: (1) the free-fall acceleration g is constant over the range of motion and is directed downward,[1] and (2) the effect of air resistance is negligible.[2] With these assumptions, the path of a projectile, called

---

[1]In effect, this approximation is equivalent to assuming that the Earth is flat within the range of motion considered and that the maximum height of the object is small compared to the radius of the Earth.

[2]This approximation is often *not* justified, especially at high velocities. In addition, the spin of a projectile, such as a baseball, can give rise to some very interesting effects associated with aerodynamic forces (for example, a curve ball thrown by a pitcher).

The *y* component of velocity is zero at the peak of the path.

$v_y = 0$ $\vec{v}_C$

$\vec{g}$

The *x* component of velocity remains constant because there is no acceleration in the *x* direction.

$\vec{v}_B$

v_y

θ

v_{xi}

$\vec{v}_i$

v_{xi}

v_{yi}

θ_i

$\vec{v}_D$ v_{xi}

θ

v_y

$\vec{v}_D$

v_{xi}

θ_i

v_y

$\vec{v}_E$

Active Figure 3.7 The parabolic path of a projectile that leaves the origin (point Ⓐ) with a velocity $\vec{v}_i$. The velocity vector $\vec{v}$ changes with time in both magnitude and direction. This change is the result of acceleration $\vec{a} = \vec{g}$ in the negative *y* direction.

its *trajectory*, is *always* a parabola. **We shall use a simplification model based on these assumptions throughout this chapter.**

If we choose our reference frame such that the *y* direction is vertical and positive upward, $a_y = -g$ (as in one-dimensional free-fall) and $a_x = 0$ (because the only possible horizontal acceleration is due to air resistance, and it is ignored). Furthermore, let us assume that at $t = 0$, the projectile leaves the origin (point Ⓐ, $x_i = y_i = 0$) with speed v_i, as in Active Figure 3.7. If the vector $\vec{v}_i$ makes an angle θ_i with the horizontal, we can identify a right triangle in the diagram as a geometric model, and from the definitions of the cosine and sine functions we have

$$\cos \theta_i = \frac{v_{xi}}{v_i} \quad \text{and} \quad \sin \theta_i = \frac{v_{yi}}{v_i}$$

Therefore, the initial *x* and *y* components of velocity are

$$v_{xi} = v_i \cos \theta_i \quad \text{and} \quad v_{yi} = v_i \sin \theta_i$$

Substituting these expressions into Equations 3.8 and 3.9 with $a_x = 0$ and $a_y = -g$ gives the velocity components and position coordinates for the projectile at any time *t*:

$$v_{xf} = v_{xi} = v_i \cos \theta_i = \text{constant} \qquad \textbf{3.10} \blacktriangleleft$$

$$v_{yf} = v_{yi} - gt = v_i \sin \theta_i - gt \qquad \textbf{3.11} \blacktriangleleft$$

$$x_f = x_i + v_{xi}t = (v_i \cos \theta_i)t \qquad \textbf{3.12} \blacktriangleleft$$

$$y_f = y_i + v_{yi}t - \tfrac{1}{2}gt^2 = (v_i \sin \theta_i)t - \tfrac{1}{2}gt^2 \qquad \textbf{3.13} \blacktriangleleft$$

From Equation 3.10 we see that v_{xf} remains constant in time and is equal to v_{xi}; there is no horizontal component of acceleration. Therefore, we model the horizontal motion as that of a particle under constant velocity. For the *y* motion, note that the equations for v_{yf} and y_f are similar to Equations 2.10 and 2.13 for freely falling objects. Therefore, we can apply the model of a particle under constant acceleration to the *y* component. In fact, *all* the equations of kinematics developed in Chapter 2 are applicable to projectile motion.

If we solve for *t* in Equation 3.12 and substitute this expression for *t* into Equation 3.13, we find that

$$y_f = (\tan \theta_i)x_f - \left(\frac{g}{2v_i^2 \cos^2 \theta_i}\right)x_f^2 \qquad \textbf{3.14} \blacktriangleleft$$

which is valid for angles in the range $0 < \theta_i < \pi/2$. This expression is of the form $y = ax - bx^2$, which is the equation of a parabola that passes through the origin. Therefore, we have proven that the trajectory of a projectile can be geometrically modeled as a parabola. The trajectory is *completely* specified if v_i and θ_i are known.

Pitfall Prevention | 3.2

Acceleration at the Highest Point
As discussed in Pitfall Prevention 2.7, many people claim that the acceleration of a projectile at the topmost point of its trajectory is zero. This mistake arises from confusion between zero vertical velocity and zero acceleration. If the projectile were to experience zero acceleration at the highest point, its velocity at that point would not change; rather, the projectile would move horizontally at constant speed from then on! That does not happen, however, because the acceleration is *not* zero anywhere along the trajectory.

A welder cuts holes through a heavy metal construction beam with a hot torch. The sparks generated in the process follow parabolic paths.

Lester Lefkowitz/Taxi/Getty Images

The vector expression for the position of the projectile as a function of time follows directly from Equation 3.9, with $\vec{\mathbf{a}} = \vec{\mathbf{g}}$:

$$\vec{\mathbf{r}}_f = \vec{\mathbf{r}}_i + \vec{\mathbf{v}}_i t + \tfrac{1}{2}\vec{\mathbf{g}}t^2$$

This equation gives the same information as the combination of Equations 3.12 and 3.13 and is plotted in Figure 3.8. Note that this expression for $\vec{\mathbf{r}}_f$ is consistent with Equation 3.13 because the expression for $\vec{\mathbf{r}}_f$ is a vector equation and $\vec{\mathbf{a}} = \vec{\mathbf{g}} = -g\hat{\mathbf{j}}$ when the upward direction is taken to be positive.

The position of a particle can be considered the sum of its original position $\vec{\mathbf{r}}_i$, the term $\vec{\mathbf{v}}_i t$, which would be the displacement if no acceleration were present, and the term $\tfrac{1}{2}\vec{\mathbf{g}}t^2$, which arises from the acceleration caused by gravity. In other words, if no gravitational acceleration occurred, the particle would continue to move along a straight path in the direction of $\vec{\mathbf{v}}_i$.

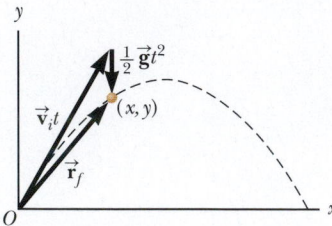

Figure 3.8 The position vector $\vec{\mathbf{r}}_f$ of a projectile whose initial velocity at the origin is $\vec{\mathbf{v}}_i$. The vector $\vec{\mathbf{v}}_i t$ would be the position vector of the projectile if gravity were absent and the vector $\tfrac{1}{2}\vec{\mathbf{g}}t^2$ is the particle's vertical displacement due to its downward gravitational acceleration.

QUICK QUIZ 3.2 (i) As a projectile thrown upward moves in its parabolic path (such as in Fig. 3.8), at what point along its path are the velocity and acceleration vectors for the projectile perpendicular to each other? (a) nowhere (b) the highest point (c) the launch point (ii) From the same choices, at what point are the velocity and acceleration vectors for the projectile parallel to each other?

Horizontal Range and Maximum Height of a Projectile

Let us assume that a projectile is launched over flat ground from the origin at $t = 0$ with a positive v_y component, as in Figure 3.9. This a common situation in sports, where baseballs, footballs, and golf balls often land at the same level from which they were launched.

There are two special points in this motion that are interesting to analyze: the peak point Ⓐ, which has Cartesian coordinates $(R/2, h)$, and the landing point Ⓑ, having coordinates $(R, 0)$. The distance R is called the *horizontal range* of the projectile, and h is its *maximum height*. Because of the symmetry of the trajectory, the projectile is at the maximum height h when its x position is half the range R. Let us find h and R in terms of v_i, θ_i, and g.

We can determine h by noting that at the peak $v_{y\text{Ⓐ}} = 0$. Therefore, Equation 3.11 can be used to determine the time $t_\text{Ⓐ}$ at which the projectile reaches the peak:

$$t_\text{Ⓐ} = \frac{v_i \sin \theta_i}{g}$$

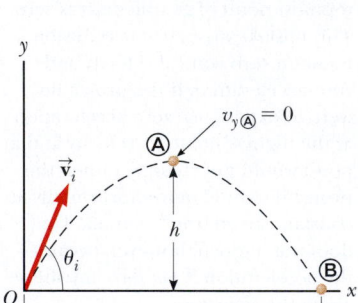

Figure 3.9 A projectile launched from the origin at $t = 0$ with an initial velocity $\vec{\mathbf{v}}_i$. The maximum height of the projectile is h, and its horizontal range is R. At Ⓐ, the peak of the trajectory, the projectile has coordinates $(R/2, h)$.

Substituting this expression for $t_\text{Ⓐ}$ into Equation 3.13 and replacing y_f with h gives h in terms of v_i and θ_i:

$$h = (v_i \sin \theta_i)\frac{v_i \sin \theta_i}{g} - \tfrac{1}{2}g\left(\frac{v_i \sin \theta_i}{g}\right)^2$$

$$h = \frac{v_i^2 \sin^2 \theta_i}{2g} \qquad\qquad \textbf{3.15} \blacktriangleleft$$

Notice from the mathematical representation how you could increase the maximum height h: You could launch the projectile with a larger initial velocity, at a higher angle, or at a location with lower free-fall acceleration, such as on the Moon. Is that consistent with your mental representation of this situation?

The range R is the horizontal distance traveled in twice the time interval required to reach the peak. Equivalently, we are seeking the position of the projectile at a time $2t_\text{Ⓐ}$. Using Equation 3.12 and noting that $x_f = R$ at $t = 2t_\text{Ⓐ}$, we find that

$$R = (v_i \cos \theta_i)2t_\text{Ⓐ} = (v_i \cos \theta_i)\frac{2v_i \sin \theta_i}{g} = \frac{2v_i^2 \sin \theta_i \cos \theta_i}{g}$$

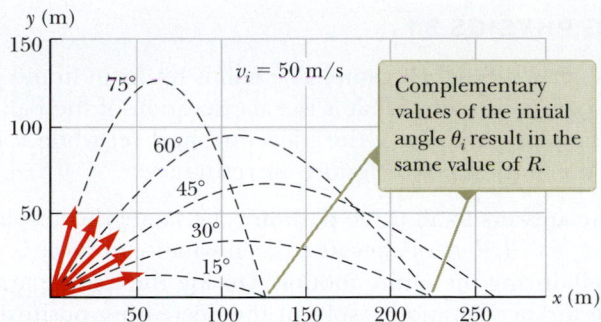

Active Figure 3.10 A projectile launched from the origin with an initial speed of 50 m/s at various angles of projection.

Because $\sin 2\theta = 2 \sin \theta \cos \theta$, R can be written in the more compact form

$$R = \frac{v_i^2 \sin 2\theta_i}{g}$$ 3.16 ◄

Notice from the mathematical expression how you could increase the range R: You could launch the projectile with a larger initial velocity or at a location with lower free-fall acceleration, such as on the Moon. Is that consistent with your mental representation of this situation?

The range also depends on the angle of the initial velocity vector. The maximum possible value of R from Equation 3.16 is given by $R_{max} = v_i^2/g$. This result follows from the maximum value of $\sin 2\theta_i$ being unity, which occurs when $2\theta_i = 90°$. Therefore, R is a maximum when $\theta_i = 45°$.

Active Figure 3.10 illustrates various trajectories for a projectile of a given initial speed. As you can see, the range is a maximum for $\theta_i = 45°$. In addition, for any θ_i other than 45°, a point with coordinates $(R, 0)$ can be reached by using either one of two complementary values of θ_i, such as 75° and 15°. Of course, the maximum height and the time of flight will be different for these two values of θ_i.

Pitfall Prevention | 3.3
The Height and Range Equations
Keep in mind that Equations 3.15 and 3.16 are useful for calculating h and R only for a symmetric path, as shown in Figure 3.9. If the path is not symmetric, *do not use these equations.* The general expressions given by Equations 3.10 through 3.13 are the *more important* results because they give the coordinates and velocity components of the projectile at any time t for any trajectory.

▸ **QUICK QUIZ 3.3** Rank the launch angles for the five paths in Active Figure 3.10 with respect to time of flight from the shortest time of flight to the longest.

▸ **PROBLEM-SOLVING STRATEGY: Projectile Motion**

We suggest that you use the following approach when solving projectile motion problems:

1. **Conceptualize** Think about what is going on physically in the problem. Establish the mental representation by imagining the projectile moving along its trajectory.

2. **Categorize** Confirm that the problem involves a particle in free-fall and that air resistance is neglected. Select a coordinate system with x in the horizontal direction and y in the vertical direction.

3. **Analyze** If the initial velocity vector is given, resolve it into x and y components. Treat the horizontal motion and the vertical motion independently. Analyze the horizontal motion of the projectile with the particle under constant velocity model. Analyze the vertical motion of the projectile with the particle under constant acceleration model.

4. **Finalize** Once you have determined your result, check to see if your answers are consistent with the mental and pictorial representations and that your results are realistic.

> **THINKING PHYSICS 3.1**
>
> A home run is hit in a baseball game. The ball is hit from home plate into the stands along a parabolic path. What is the acceleration of the ball **(a)** while it is rising, **(b)** at the highest point of the trajectory, and **(c)** while it is descending after reaching the highest point? Ignore air resistance.
>
> **Reasoning** The answers to all three parts are the same: the acceleration is that due to gravity, $a_y = -9.80 \text{ m/s}^2$, because the gravitational force is pulling downward on the ball during the entire motion. During the rising part of the trajectory, the downward acceleration results in the decreasing positive values of the vertical component of the ball's velocity. During the falling part of the trajectory, the downward acceleration results in the increasing negative values of the vertical component of the velocity. ◄

Example 3.2 | That's Quite an Arm!

A stone is thrown from the top of a building upward at an angle of 30.0° to the horizontal with an initial speed of 20.0 m/s as shown in Figure 3.11. The height from which the stone is thrown is 45.0 m above the ground.

(A) How long does it take the stone to reach the ground?

SOLUTION

Conceptualize Study Figure 3.11, in which we have indicated the trajectory and various parameters of the motion of the stone.

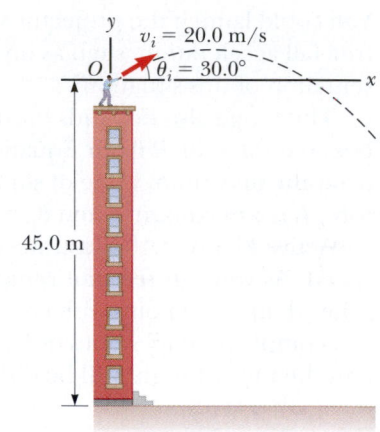

Figure 3.11
(Example 3.2) A stone is thrown from the top of a building.

Categorize We categorize this problem as a projectile motion problem. The stone is modeled as a particle under constant acceleration in the y direction and a particle under constant velocity in the x direction.

Analyze We have the information $x_i = y_i = 0$, $y_f = -45.0$ m, $a_y = -g$, and $v_i = 20.0$ m/s (the numerical value of y_f is negative because we have chosen the point of the throw as the origin).

Find the initial x and y components of the stone's velocity:

$$v_{xi} = v_i \cos\theta_i = (20.0 \text{ m/s}) \cos 30.0° = 17.3 \text{ m/s}$$
$$v_{yi} = v_i \sin\theta_i = (20.0 \text{ m/s}) \sin 30.0° = 10.0 \text{ m/s}$$

Express the vertical position of the stone from the vertical component of Equation 3.9:

$$y_f = y_i + v_{yi}t + \tfrac{1}{2}a_y t^2$$

Substitute numerical values:

$$-45.0 \text{ m} = 0 + (10.0 \text{ m/s})t + \tfrac{1}{2}(-9.80 \text{ m/s}^2)t^2$$

Solve the quadratic equation for t:

$$t = \boxed{4.22 \text{ s}}$$

(B) What is the speed of the stone just before it strikes the ground?

SOLUTION

Analyze Use the y component of Equation 3.8 to obtain the y component of the velocity of the stone just before it strikes the ground:

$$v_{yf} = v_{yi} + a_y t$$

Substitute numerical values, using $t = 4.22$ s:

$$v_{yf} = 10.0 \text{ m/s} + (-9.80 \text{ m/s}^2)(4.22 \text{ s}) = -31.3 \text{ m/s}$$

Use this component with the horizontal component $v_{xf} = v_{xi} = 17.3$ m/s to find the speed of the stone at $t = 4.22$ s:

$$v_f = \sqrt{v_{xf}^2 + v_{yf}^2} = \sqrt{(17.3 \text{ m/s})^2 + (-31.3 \text{ m/s})^2} = \boxed{35.8 \text{ m/s}}$$

3.2 cont.

Finalize Is it reasonable that the y component of the final velocity is negative? Is it reasonable that the final speed is larger than the initial speed of 20.0 m/s?

What If? What if a horizontal wind is blowing in the same direction as the stone is thrown and it causes the stone to have a horizontal acceleration component $a_x = 0.500$ m/s²? Which part of this example, (A) or (B), will have a different answer?

Answer Recall that the motions in the x and y directions are independent. Therefore, the horizontal wind cannot affect the vertical motion. The vertical motion determines the time of the projectile in the air, so the answer to part (A) does not change. The wind causes the horizontal velocity component to increase with time, so the final speed will be larger in part (B). Taking $a_x = 0.500$ m/s², we find $v_{xf} = 19.4$ m/s and $v_f = 36.9$ m/s.

Example 3.3 | The End of the Ski Jump

A ski jumper leaves the ski track moving in the horizontal direction with a speed of 25.0 m/s as shown in Figure 3.12. The landing incline below her falls off with a slope of 35.0°. Where does she land on the incline?

SOLUTION

Conceptualize We can conceptualize this problem based on memories of observing winter Olympic ski competitions. We estimate the skier to be air-borne for perhaps 4 s and to travel a distance of about 100 m horizontally. We should expect the value of d, the distance traveled along the incline, to be of the same order of magnitude.

Categorize We categorize the problem as one of a particle in projectile motion.

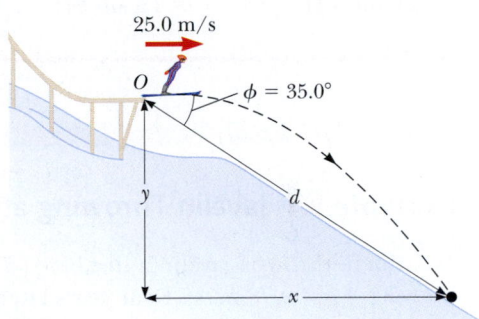

Figure 3.12 (Example 3.3) A ski jumper leaves the track moving in a horizontal direction.

Analyze It is convenient to select the beginning of the jump as the origin. The initial velocity components are $v_{xi} = 25.0$ m/s and $v_{yi} = 0$. From the right triangle in Figure 3.12, we see that the jumper's x and y coordinates at the landing point are given by $x_f = d \cos \phi$ and $y_f = -d \sin \phi$.

Express the coordinates of the jumper as a function of time:

(1) $x_f = v_{xi} t$

(2) $y_f = v_{yi} t + \frac{1}{2} a_y t^2 = -\frac{1}{2} g t^2$

Substitute the values of x_f and y_f at the landing point:

(3) $d \cos \phi = v_{xi} t$

(4) $-d \sin \phi = -\frac{1}{2} g t^2$

Solve Equation (3) for t and substitute the result into Equation (4):

$-d \sin \phi = -\frac{1}{2} g \left(\dfrac{d \cos \phi}{v_{xi}} \right)^2$

Solve for d:

$d = \dfrac{2 v_{xi}^2 \sin \phi}{g \cos^2 \phi} = \dfrac{2(25.0 \text{ m/s})^2 \sin 35.0°}{(9.80 \text{ m/s}^2) \cos^2 35.0°} = 109 \text{ m}$

Evaluate the x and y coordinates of the point at which the skier lands:

$x_f = d \cos \phi = (109 \text{ m}) \cos 35.0° = \boxed{89.3 \text{ m}}$

$y_f = -d \sin \phi = -(109 \text{ m}) \sin 35.0° = \boxed{-62.5 \text{ m}}$

Finalize Let us compare these results with our expectations. We expected the horizontal distance to be on the order of 100 m, and our result of 89.3 m is indeed on this order of magnitude. It might be useful to calculate the time interval that the jumper is in the air and compare it with our estimate of about 4 s.

What If? Suppose everything in this example is the same except the ski jump is curved so that the jumper is projected upward at an angle from the end of the track. Is this design better in terms of maximizing the length of the jump?

Answer If the initial velocity has an upward component, the skier will be in the air longer and should therefore travel farther. Tilting the initial velocity vector upward, however, will reduce the horizontal component of the initial velocity.

continued

3.3 cont.

Therefore, angling the end of the ski track upward at a *large* angle may actually *reduce* the distance. Consider the extreme case: the skier is projected at 90° to the horizontal and simply goes up and comes back down at the end of the ski track! This argument suggests that there must be an optimal angle between 0° and 90° that represents a balance between making the flight time longer and the horizontal velocity component smaller.

Let us find this optimal angle mathematically. We modify Equations (1) through (4) in the following way, assuming the skier is projected at an angle θ with respect to the horizontal over a landing incline sloped with an arbitrary angle ϕ:

(1) and (3) $\rightarrow$ $x_f = (v_i \cos \theta) t = d \cos \phi$

(2) and (4) $\rightarrow$ $y_f = (v_i \sin \theta) t - \frac{1}{2} g t^2 = -d \sin \phi$

By eliminating the time t between these equations and using differentiation to maximize d in terms of θ, we arrive at the following equation for the angle θ that gives the maximum value of d:

$$\theta = 45° - \frac{\phi}{2}$$

For the slope angle in Figure 3.12, $\phi = 35.0°$; this equation results in an optimal launch angle of $\theta = 27.5°$. For a slope angle of $\phi = 0°$, which represents a horizontal plane, this equation gives an optimal launch angle of $\theta = 45°$, as we would expect (see Active Figure 3.10).

Example 3.4 | Javelin Throwing at the Olympics

An athlete throws a javelin a distance of 80.0 m at the Olympics held at the equator, where $g = 9.78$ m/s². Four years later the Olympics are held at the North Pole, where $g = 9.83$ m/s². Assuming that the thrower provides the javelin with exactly the same initial velocity as she did at the equator, how far does the javelin travel at the North Pole?

A javelin can be thrown over a very long distance by a world class athlete.

SOLUTION

Conceptualize In traveling between these two locations, we most likely would not feel any difference in the weight of an object. The increased gravity at the North Pole, however, will cause the javelin to return to the ground sooner and shorten its range compared to the throw at the equator.

Categorize In the absence of any information about how the javelin is affected by moving through the air, we adopt the free-fall model for the javelin. Track and field events are normally held on flat fields. Therefore, we surmise that the javelin returns to the same vertical position from which it was thrown and therefore that the trajectory is symmetric. These assumptions allow us to use Equations 3.15 and 3.16 to analyze the motion. The difference in range is due to the difference in the free-fall acceleration at the two locations.

Analyze To solve this problem, we will set up a ratio based on the range of the projectile being mathematically related to the acceleration due to gravity. This technique of solving by ratios is very powerful and should be studied and understood so that it can be applied in future problem solving.

Use Equation 3.16 to express the range of the particle at each of the two locations:

$$R_{\text{North Pole}} = \frac{v_i^2 \sin 2\theta_i}{g_{\text{North Pole}}}$$

$$R_{\text{equator}} = \frac{v_i^2 \sin 2\theta_i}{g_{\text{equator}}}$$

Divide the first equation by the second to establish a relationship between the ratio of the ranges and the ratio of the free-fall accelerations. Note that the problem states that the same initial velocity is provided to the javelin at both locations, so v_i and θ_i are the same in the numerator and denominator of the ratio:

$$\frac{R_{\text{North Pole}}}{R_{\text{equator}}} = \frac{\left(\dfrac{v_i^2 \sin 2\theta_i}{g_{\text{North Pole}}}\right)}{\left(\dfrac{v_i^2 \sin 2\theta_i}{g_{\text{equator}}}\right)} = \frac{g_{\text{equator}}}{g_{\text{North Pole}}}$$

Solve this equation for the range at the North Pole and substitute the numerical values:

$$R_{\text{North Pole}} = \frac{g_{\text{equator}}}{g_{\text{North Pole}}} R_{\text{equator}} = \frac{9.78 \text{ m/s}^2}{9.83 \text{ m/s}^2} (80.0 \text{ m})$$

$$= \boxed{79.6 \text{ m}}$$

Finalize Notice one of the advantages of this powerful technique of setting up ratios; we do not need to know the magnitude (v_i) nor the direction (θ_i) of the initial velocity. As long as they are the same at both locations, they cancel in the ratio.

3.4 | Analysis Model: Particle in Uniform Circular Motion

Figure 3.13a shows a car moving in a circular path; we describe this motion by calling it **circular motion**. If the car is moving on this path with *constant speed v,* we call it **uniform circular motion**. Because it occurs so often, this type of motion is recognized as an analysis model called the **particle in uniform circular motion**. We discuss this model in this section.

It is often surprising to students to find that even though an object moves at a constant speed in a circular path, *it still has an acceleration.* To see why, consider the defining equation for average acceleration, $\vec{a}_{\text{avg}} = \Delta\vec{v}/\Delta t$ (Eq. 3.4). The acceleration depends on *the change in the velocity vector.* Because velocity is a vector quantity, an acceleration can be produced in two ways, as mentioned in Section 3.1: by a change in the *magnitude* of the velocity or by a change in the *direction* of the velocity. The latter situation is occurring for an object moving with constant speed in a circular path. The constant-magnitude velocity vector is always tangent to the path of the object and perpendicular to the radius of the circular path. Therefore, the velocity vector is constantly *changing.* We now show that the acceleration vector in uniform circular motion is always perpendicular to the path and always points toward the center of the circle.

Let us first argue conceptually that the acceleration must be perpendicular to the path followed by the particle. If not, there would be a component of the acceleration parallel to the path and therefore parallel to the velocity vector. Such an acceleration component would lead to a change in the speed of the object, which we model as a particle, along the path. This change, however, is inconsistent with our setup of the problem in which the particle moves with constant speed along the path. Therefore, for *uniform* circular motion, the acceleration vector can only have a component perpendicular to the path, which is toward the center of the circle.

Let us now find the magnitude of the acceleration of the particle. Consider the pictorial representation of the position and velocity vectors for the car modeled as a particle in Figure 3.13b. In addition, the figure shows the vector representing the change in position, $\Delta\vec{r}$. The particle follows a circular path, part of which is shown by the dashed

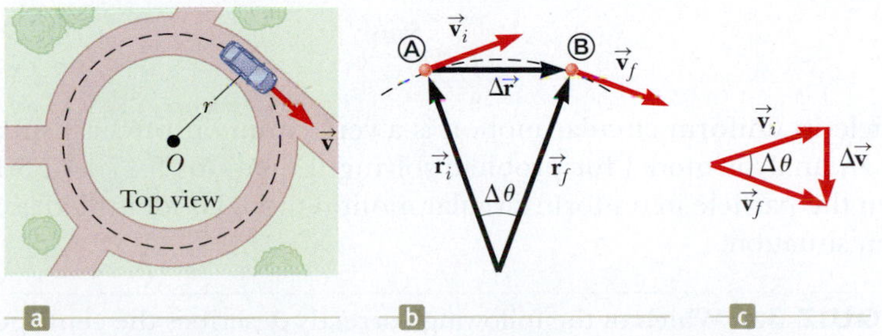

Pitfall Prevention | 3.4
Acceleration of a Particle in Uniform Circular Motion
Remember that acceleration in physics is defined as a change in the *velocity*, not a change in the *speed* (contrary to the everyday interpretation). In circular motion, the velocity vector is changing in direction, so there is indeed an acceleration.

Figure 3.13 (a) A car moving along a circular path at constant speed is in uniform circular motion. (b) As the particle moves from Ⓐ to Ⓑ, its velocity vector changes from $\vec{v}_i$ to $\vec{v}_f$. (c) The construction for determining the direction of the change in velocity $\Delta\vec{v}$, which is toward the center of the circle for small $\Delta\theta$.

curve. The particle is at Ⓐ at time t_i, and its velocity at that time is $\vec{\mathbf{v}}_i$; it is at Ⓑ at some later time t_f, and its velocity at that time is $\vec{\mathbf{v}}_f$. Let us also assume that $\vec{\mathbf{v}}_i$ and $\vec{\mathbf{v}}_f$ differ only in direction; their magnitudes are the same (i.e., $v_i = v_f = v$, because it is *uniform circular motion*). To calculate the acceleration of the particle, let us begin with the defining equation for average acceleration (Eq. 3.4):

$$\vec{\mathbf{a}}_{\text{avg}} = \frac{\vec{\mathbf{v}}_f - \vec{\mathbf{v}}_i}{t_f - t_i} = \frac{\Delta \vec{\mathbf{v}}}{\Delta t}$$

In Figure 3.13c, the velocity vectors in Figure 3.13b have been redrawn tail to tail. The vector $\Delta \vec{\mathbf{v}}$ connects the tips of the vectors, representing the vector addition, $\vec{\mathbf{v}}_f = \vec{\mathbf{v}}_i + \Delta \vec{\mathbf{v}}$. In Figures 3.13b and 3.13c, we can identify triangles that can serve as geometric models to help us analyze the motion. The angle $\Delta \theta$ between the two position vectors in Figure 3.13b is the same as the angle between the velocity vectors in Figure 3.13c because the velocity vector $\vec{\mathbf{v}}$ is always perpendicular to the position vector $\vec{\mathbf{r}}$. Therefore, the two triangles are *similar*. (Two triangles are similar if the angle between any two sides is the same for both triangles and if the ratio of the lengths of these sides is the same.) This similarity enables us to write a relationship between the lengths of the sides for the two triangles:

$$\frac{|\Delta \vec{\mathbf{v}}|}{v} = \frac{|\Delta \vec{\mathbf{r}}|}{r}$$

where $v = v_i = v_f$ and $r = r_i = r_f$. This equation can be solved for $|\Delta \vec{\mathbf{v}}|$ and the expression so obtained can be substituted into $\vec{\mathbf{a}}_{\text{avg}} = \Delta \vec{\mathbf{v}}/\Delta t$ (Eq. 3.4) to give the magnitude of the average acceleration over the time interval for the particle to move from Ⓐ to Ⓑ:

$$|\vec{\mathbf{a}}_{\text{avg}}| = \frac{v |\Delta \vec{\mathbf{r}}|}{r \, \Delta t}$$

Now imagine that we bring points Ⓐ and Ⓑ in Figure 3.13b very close together. As Ⓐ and Ⓑ approach each other, Δt approaches zero and the ratio $|\Delta \vec{\mathbf{r}}|/\Delta t$ approaches the speed v. In addition, the average acceleration becomes the instantaneous acceleration at point Ⓐ. Hence, in the limit $\Delta t \to 0$, the magnitude of the acceleration is

▶ Magnitude of centripetal acceleration

$$a_c = \frac{v^2}{r} \qquad \text{3.17} \blacktriangleleft$$

An acceleration of this nature is called a **centripetal acceleration** (*centripetal* means *center-seeking*). The subscript on the acceleration symbol reminds us that the acceleration is centripetal.

In many situations, it is convenient to describe the motion of a particle moving with constant speed in a circle of radius r in terms of the **period** T, which is defined as the time interval required for one complete revolution. In the time interval T, the particle moves a distance of $2\pi r$, which is equal to the circumference of the particle's circular path. Therefore, because its speed is equal to the circumference of the circular path divided by the period, or $v = 2\pi r/T$, it follows that

▶ Period of a particle in uniform circular motion

$$T = \frac{2\pi r}{v} \qquad \text{3.18} \blacktriangleleft$$

The particle in uniform circular motion is a very common physical situation and is useful as an analysis model for problem solving. Equations 3.17 and 3.18 are to be used when the particle in uniform circular motion model is identified as appropriate for a given situation.

Pitfall Prevention | 3.5
Centripetal Acceleration Is Not Constant
The magnitude of the centripetal acceleration vector is constant for uniform circular motion, but *the centripetal acceleration vector is not constant.* It always points toward the center of the circle, so it continuously changes direction as the particle moves.

▌ **QUICK QUIZ 3.4** Which of the following correctly describes the centripetal acceleration vector for a particle moving in a circular path? **(a)** constant and always perpendicular to the velocity vector for the particle **(b)** constant and always parallel to the velocity vector for the particle **(c)** of constant magnitude and always perpendicular to the velocity vector for the particle **(d)** of constant magnitude and always parallel to the velocity vector for the particle

> **THINKING PHYSICS 3.2**

An airplane travels from Los Angeles to Sydney, Australia. After cruising altitude is reached, the instruments on the plane indicate that the ground speed holds rock-steady at 700 km/h and that the heading of the airplane does not change. Is the velocity of the airplane constant during the flight?

Reasoning The velocity is not constant because of the curvature of the Earth. Even though the speed does not change and the heading is always toward Sydney (is that actually true?), the airplane travels around a significant portion of the Earth's circumference. Therefore, the direction of the velocity vector does indeed change. We could extend this situation by imagining that the airplane passes over Sydney and continues (assuming it has enough fuel!) around the Earth until it arrives at Los Angeles again. It is impossible for an airplane to have a constant velocity (relative to the Universe, not to the Earth's surface) and return to its starting point. ◀

Example **3.5** | **The Centripetal Acceleration of the Earth**

What is the centripetal acceleration of the Earth as it moves in its orbit around the Sun?

SOLUTION

Conceptualize Think about a mental image of the Earth in a circular orbit around the Sun. We will model the Earth as a particle and approximate the Earth's orbit as circular (it's actually slightly elliptical, as we discuss in Chapter 11).

Categorize The Conceptualize step allows us to categorize this problem as one of a particle in uniform circular motion.

Analyze We do not know the orbital speed of the Earth to substitute into Equation 3.17. With the help of Equation 3.18, however, we can recast Equation 3.17 in terms of the period of the Earth's orbit, which we know is one year, and the radius of the Earth's orbit around the Sun, which is 1.496×10^{11} m.

Combine Equations 3.17 and 3.18:

$$a_c = \frac{v^2}{r} = \frac{\left(\dfrac{2\pi r}{T}\right)^2}{r} = \frac{4\pi^2 r}{T^2}$$

Substitute numerical values:

$$a_c = \frac{4\pi^2(1.496 \times 10^{11} \text{ m})}{(1 \text{ yr})^2}\left(\frac{1 \text{ yr}}{3.156 \times 10^7 \text{ s}}\right)^2 = 5.93 \times 10^{-3} \text{ m/s}^2$$

Finalize This acceleration is much smaller than the free-fall acceleration on the surface of the Earth. An important technique we learned here is replacing the speed v in Equation 3.17 in terms of the period T of the motion. In many problems, it is more likely that T is known rather than v.

3.5 | Tangential and Radial Acceleration

Let us consider a more general motion than that presented in Section 3.4. Consider a particle moving to the right along a curved path where the velocity changes both in direction *and* in magnitude, as described in Active Figure 3.14 (page 76). In this situation, the velocity vector is always tangent to the path; the acceleration vector $\vec{a}$, however, is at some angle to the path. At each instant, the particle can be modeled as if it were moving on a circular path. The radius of the circular path is the radius of curvature of the path at that instant. In the next instant, the particle is moving as if on a different circular path, with a different center and a different radius than the previous one. At each of three points Ⓐ, Ⓑ, and Ⓒ in Active Figure 3.14, we see the dashed circles that form geometric models of circular paths for the actual path at each point.

Active Figure 3.14 The motion of a particle along an arbitrary curved path lying in the xy plane. If the velocity vector $\vec{v}$ (always tangent to the path) changes in direction and magnitude, the acceleration vector $\vec{a}$ has a tangential component a_t and a radial component a_r.

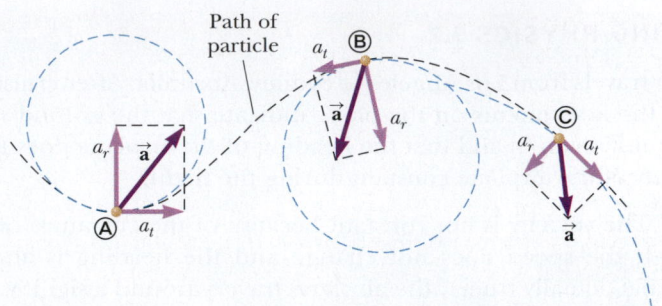

As the particle moves along the curved path in Active Figure 3.14, the direction of the total acceleration vector $\vec{a}$ changes from point to point. This vector can be resolved into two components based on an origin at the center of the model circle: a radial component a_r along the radius of the model circle and a tangential component a_t perpendicular to this radius. The *total* acceleration vector $\vec{a}$ can be written as the vector sum of the component vectors:

$$\vec{a} = \vec{a}_r + \vec{a}_t \qquad \qquad \textbf{3.19} \blacktriangleleft$$

The tangential acceleration arises from the change in the speed of the particle and is given by

▶ Tangential acceleration

$$a_t = \frac{d|\vec{v}|}{dt} \qquad \qquad \textbf{3.20} \blacktriangleleft$$

The radial acceleration is a result of the change in direction of the velocity vector and is given by

▶ Radial acceleration

$$a_r = -a_c = -\frac{v^2}{r}$$

where r is the radius of curvature of the path at the point in question, which is the radius of the model circle. We recognize the magnitude of the radial component of the acceleration as the centripetal acceleration discussed in Section 3.4. The negative sign indicates that the direction of the centripetal acceleration is toward the center of the model circle, opposite the direction of the radial unit vector $\hat{\mathbf{r}}$, which always points away from the center of the circle.

Because $\vec{a}_r$ and $\vec{a}_t$ are perpendicular component vectors of $\vec{a}$, it follows that $a = \sqrt{a_r^2 + a_t^2}$. At a given speed, a_r is large when the radius of curvature is small (as at points Ⓐ and Ⓑ in Active Fig. 3.14) and small when r is large (such as at point Ⓒ). The direction of $\vec{a}_t$ is either in the same direction as $\vec{v}$ (if v is increasing) or opposite $\vec{v}$ (if v is decreasing, as at point Ⓑ).

In the case of uniform circular motion, where v is constant, $a_t = 0$ and the acceleration is always radial, as described in Section 3.4. In other words, uniform circular motion is a special case of motion along a curved path. Furthermore, if the direction of $\vec{v}$ doesn't change, no radial acceleration occurs and the motion is one dimensional ($a_r = 0$, but a_t may not be zero).

QUICK QUIZ 3.5 A particle moves along a path, and its speed increases with time. **(i)** In which of the following cases are its acceleration and velocity vectors parallel? (a) when the path is circular. (b) when the path is straight. (c) when the path is a parabola. (d) never. **(ii)** From the same choices, in which case are its acceleration and velocity vectors perpendicular everywhere along the path?

◢ 3.6 | Relative Velocity and Relative Acceleration

In this section, we describe how observations made by different observers in different frames of reference are related to one another. A frame of reference can be described by a Cartesian coordinate system for which an observer is at rest with respect to the origin.

Let us conceptualize a sample situation in which there will be different observations for different observers. Consider the two observers A and B along the number line in Figure 3.15a. Observer A is located at the origin of a one-dimensional x_A axis, while observer B is at the position $x_A = -5$. We denote the position variable as x_A because observer A is at the origin of this axis. Both observers measure the position of point P, which is located at $x_A = +5$. Suppose observer B decides that he is located at the origin of an x_B axis as in Figure 3.15b. Notice that the two observers disagree on the value of the position of point P. Observer A claims point P is located at a position with a value of +5, whereas observer B claims it is located at a position with a value of +10. Both observers are correct, even though they make different measurements. Their measurements differ because they are making the measurement from different frames of reference.

Imagine now that observer B in Figure 3.15b is moving to the right along the x_B axis. Now the two measurements are even more different. Observer A claims point P remains at rest at a position with a value of +5, whereas observer B claims the position of P continuously changes with time, even passing him and moving behind him! Again, both observers are correct, with the difference in their measurements arising from their different frames of reference.

We explore this phenomenon further by considering two observers watching a man walking on a moving beltway at an airport in Figure 3.16. The woman standing on the moving beltway sees the man moving at a normal walking speed. The woman observing from the stationary floor sees the man moving with a higher speed because the beltway speed combines with his walking speed. Both observers look at the same man and arrive at different values for his speed. Both are correct; the difference in their measurements results from the relative velocity of their frames of reference.

In a more general situation, consider a particle located at point P in Figure 3.17 (page 78). Imagine that the motion of this particle is being described by two observers, observer A in a reference frame S_A fixed relative to the Earth and a second observer B in a reference frame S_B moving to the right relative to S_A (and therefore relative to the Earth) with a constant velocity $\vec{\mathbf{v}}_{BA}$. In this discussion of relative velocity, we use a double-subscript notation; the first subscript represents what is being observed, and the second represents who is doing the observing. Therefore, the notation $\vec{\mathbf{v}}_{BA}$ means the velocity of observer B (and the attached frame S_B) as measured by observer A. With this notation, observer B measures A to be moving to the left with a velocity $\vec{\mathbf{v}}_{AB} = -\vec{\mathbf{v}}_{BA}$. For purposes of this discussion, let us place each observer at her or his respective origin.

We define the time $t = 0$ as the instant at which the origins of the two reference frames coincide in space. Therefore, at time t, the origins of the reference frames will be separated by a distance $v_{BA}t$. We label the position P of the particle relative to observer A with the position vector $\vec{\mathbf{r}}_{PA}$ and that relative to observer B with the position vector $\vec{\mathbf{r}}_{PB}$, both at time t. From Figure 3.17, we see that the vectors $\vec{\mathbf{r}}_{PA}$ and $\vec{\mathbf{r}}_{PB}$ are related to each other through the expression

$$\vec{\mathbf{r}}_{PA} = \vec{\mathbf{r}}_{PB} + \vec{\mathbf{v}}_{BA}t \qquad \textbf{3.21} \blacktriangleleft$$

By differentiating Equation 3.21 with respect to time, noting that $\vec{\mathbf{v}}_{BA}$ is constant, we obtain

$$\frac{d\vec{\mathbf{r}}_{PA}}{dt} = \frac{d\vec{\mathbf{r}}_{PB}}{dt} + \vec{\mathbf{v}}_{BA}$$

$$\vec{\mathbf{u}}_{PA} = \vec{\mathbf{u}}_{PB} + \vec{\mathbf{v}}_{BA} \qquad \textbf{3.22} \blacktriangleleft \qquad \blacktriangleright \text{ Galilean velocity transformation}$$

where $\vec{\mathbf{u}}_{PA}$ is the velocity of the particle at P measured by observer A and $\vec{\mathbf{u}}_{PB}$ is its velocity measured by B. (We use the symbol $\vec{\mathbf{u}}$ for particle velocity rather than $\vec{\mathbf{v}}$, which we have already used for the relative velocity of two reference frames.) Equations 3.21 and 3.22 are known as **Galilean transformation equations.** They relate the position and velocity of a particle as measured by observers in relative motion. Notice the pattern of the subscripts in Equation 3.22. When relative velocities are

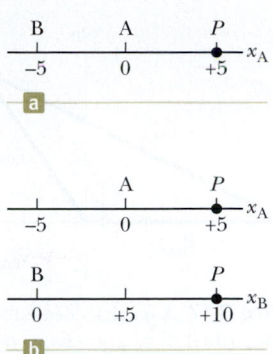

Figure 3.15 Different observers make different measurements. (a) Observer A is located at the origin, and Observer B is at a position of −5. Both observers measure the position of a particle at P. (b) If both observers see themselves at the origin of their own coordinate system, they disagree on the value of the position of the particle at P.

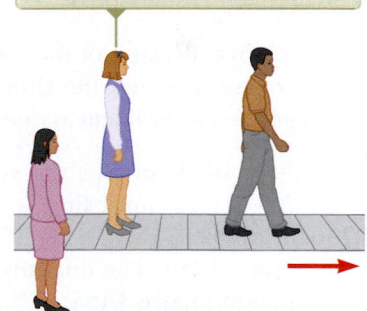

The woman standing on the beltway sees the man moving with a slower speed than does the other woman observing the man from the stationary floor.

Figure 3.16 Two observers measure the speed of a man walking on a moving beltway.

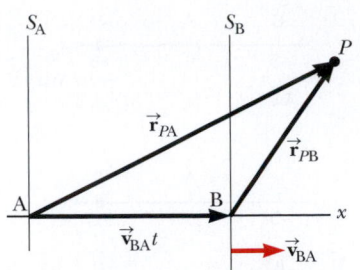

Figure 3.17 A particle located at P is described by two observers, one in the fixed frame of reference S_A and the other in the frame S_B, which moves to the right with a constant velocity $\vec{\mathbf{v}}_{BA}$. The vector $\vec{\mathbf{r}}_{PA}$ is the particle's position vector relative to S_A, and $\vec{\mathbf{r}}_{PB}$ is its position vector relative to S_B.

added, the inner subscripts (B) are the same and the outer ones (P, A) match the subscripts on the velocity on the left of the equation.

Although observers in two frames measure different velocities for the particle, they measure the *same acceleration* when $\vec{\mathbf{v}}_{BA}$ is constant. We can verify that by taking the time derivative of Equation 3.22:

$$\frac{d\vec{\mathbf{u}}_{PA}}{dt} = \frac{d\vec{\mathbf{u}}_{PB}}{dt} + \frac{d\vec{\mathbf{v}}_{BA}}{dt}$$

Because $\vec{\mathbf{v}}_{BA}$ is constant, $d\vec{\mathbf{v}}_{BA}/dt = 0$. Therefore, we conclude that $\vec{\mathbf{a}}_{PA} = \vec{\mathbf{a}}_{PB}$ because $\vec{\mathbf{a}}_{PA} = d\vec{\mathbf{u}}_{PA}/dt$ and $\vec{\mathbf{a}}_{PB} = d\vec{\mathbf{u}}_{PB}/dt$. That is, the acceleration of the particle measured by an observer in one frame of reference is the same as that measured by any other observer moving with constant velocity relative to the first frame.

Example 3.6 | A Boat Crossing a River

A boat crossing a wide river moves with a speed of 10.0 km/h relative to the water. The water in the river has a uniform speed of 5.00 km/h due east relative to the Earth.

(A) If the boat heads due north, determine the velocity of the boat relative to an observer standing on either bank.

SOLUTION

Conceptualize Imagine moving in a boat across a river while the current pushes you down the river. You will not be able to move directly across the river, but will end up downstream as suggested in Figure 3.18a.

Categorize Because of the combined velocities of you relative to the river and the river relative to the Earth, we can categorize this problem as one involving relative velocities.

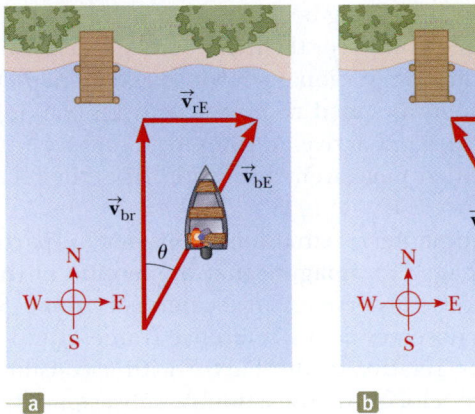

Figure 3.18 (Example 3.6) (a) A boat aims directly across a river and ends up downstream. (b) To move directly across the river, the boat must aim upstream.

Analyze We know $\vec{\mathbf{v}}_{br}$, the velocity of the *boat* relative to the *river*, and $\vec{\mathbf{v}}_{rE}$, the velocity of the *river* relative to the *Earth*. What we must find is $\vec{\mathbf{v}}_{bE}$, the velocity of the *boat* relative to the *Earth*. The relationship between these three quantities is $\vec{\mathbf{v}}_{bE} = \vec{\mathbf{v}}_{br} + \vec{\mathbf{v}}_{rE}$. The terms in the equation must be manipulated as vector quantities; the vectors are shown in Figure 3.18a. The quantity $\vec{\mathbf{v}}_{br}$ is due north; $\vec{\mathbf{v}}_{rE}$ is due east; and the vector sum of the two, $\vec{\mathbf{v}}_{bE}$, is at an angle θ as defined in Figure 3.18a.

Find the speed v_{bE} of the boat relative to the Earth using the Pythagorean theorem:

$$v_{bE} = \sqrt{v_{br}^2 + v_{rE}^2} = \sqrt{(10.0 \text{ km/h})^2 + (5.00 \text{ km/h})^2}$$
$$= 11.2 \text{ km/h}$$

Find the direction of $\vec{\mathbf{v}}_{bE}$:

$$\theta = \tan^{-1}\left(\frac{v_{rE}}{v_{br}}\right) = \tan^{-1}\left(\frac{5.00}{10.0}\right) = 26.6°$$

Finalize The boat is moving at a speed of 11.2 km/h in the direction 26.6° east of north relative to the Earth. Notice that the speed of 11.2 km/h is faster than your boat speed of 10.0 km/h. The current velocity adds to yours to give you a higher speed. Notice in Figure 3.18a that your resultant velocity is at an angle to the direction straight across the river, so you will end up downstream, as we predicted.

(B) If the boat travels with the same speed of 10.0 km/h relative to the river and is to travel due north as shown in Figure 3.18b, what should its heading be?

SOLUTION

Conceptualize/Categorize This question is an extension of part (A), so we have already conceptualized and categorized the problem. In this case, however, we must aim the boat upstream so as to go straight across the river.

3.6 *cont.*

Analyze The analysis now involves the new triangle shown in Figure 3.18b. As in part (A), we know $\vec{v}_{rE}$ and the magnitude of the vector $\vec{v}_{br}$, and we want $\vec{v}_{bE}$ to be directed across the river. Notice the difference between the triangle in Figure 3.18a and the one in Figure 3.18b: the hypotenuse in Figure 3.18b is no longer $\vec{v}_{bE}$.

Use the Pythagorean theorem to find v_{bE}:

$$v_{bE} = \sqrt{v_{br}^2 - v_{rE}^2} = \sqrt{(10.0 \text{ km/h})^2 - (5.00 \text{ km/h})^2} = 8.66 \text{ km/h}$$

Find the direction in which the boat is heading:

$$\theta = \tan^{-1}\left(\frac{v_{rE}}{v_{bE}}\right) = \tan^{-1}\left(\frac{5.00}{8.66}\right) = \boxed{30.0°}$$

Finalize The boat must head upstream so as to travel directly northward across the river. For the given situation, the boat must steer a course 30.0° west of north. For faster currents, the boat must be aimed upstream at larger angles.

What If? Imagine that the two boats in parts (A) and (B) are racing across the river. Which boat arrives at the opposite bank first?

Answer In part (A), the velocity of 10 km/h is aimed directly across the river. In part (B), the velocity that is directed across the river has a magnitude of only 8.66 km/h. Therefore, the boat in part (A) has a larger velocity component directly across the river and arrives first.

3.7 | Context Connection: Lateral Acceleration of Automobiles

An automobile does not travel in a straight line. It follows a two-dimensional path on a flat Earth surface and a three-dimensional path if there are hills and valleys. Let us restrict our thinking at this point to an automobile traveling in two dimensions on a flat roadway. During a turn, the automobile can be modeled as following an arc of a circular path at each point in its motion. Consequently, the automobile will have a centripetal acceleration.

A desired characteristic of automobiles is that they can negotiate a curve without rolling over. This characteristic depends on the centripetal acceleration. Imagine standing a book upright on a strip of sandpaper. If the sandpaper is moved slowly across the surface of a table with a very small acceleration, the book will stay upright. If the sandpaper is moved with a large acceleration, however, the book will fall over. That is what we would like to avoid in a car.

Imagine that instead of accelerating a book in one dimension we are centripetally accelerating a car in a circular path. The effect is the same. If there is too much centripetal acceleration, the car will "fall over" and will go into a sideways roll. The maximum possible centripetal acceleration that a car can exhibit without rolling over in a turn is called *lateral acceleration*. Two contributions to the lateral acceleration of a car are the height of the center of mass of the car above the ground and the side-to-side distance between the wheels. (We will study center of mass in Chapter 8.) The book in our demonstration has a relatively large ratio of the height of the center of mass to the width of the book upon which it is sitting, so it falls over relatively easily at low accelerations. An automobile has a much lower ratio of the height of the center of mass to the distance between the wheels. Therefore, it can withstand higher accelerations.

Consider the documented lateral acceleration of the vehicles from Table 2.3 listed in Table 3.1 (page 80). These values are given as multiples of g, the acceleration due to gravity. Notice that most of the very expensive vehicles and performance vehicles have a lateral acceleration close to that due to gravity and that the lateral acceleration of the Bugatti is 40% larger than that due to gravity. The Bugatti is a very stable vehicle!

In contrast, the lateral acceleration of nonperformance cars is lower because they generally are not designed to travel around turns at such a high speed as the performance cars. For example, the Buick Regal has a lateral acceleration of 0.85g. The two sport utility vehicles in the table have lateral accelerations lower than this value and such vehicles can have values as low as 0.62g. As a result, they are highly prone to rollovers in emergency maneuvers.

TABLE 3.1 | Lateral Acceleration of Automobiles

Automobile	Lateral Acceleration (g)	Automobile	Lateral Acceleration (g)
Very expensive vehicles:		*Traditional vehicles:*	
Bugatti Veyron 16.4 Super Sport	1.40	Buick Regal CXL Turbo	0.85
Lamborghini LP 570-4 Superleggera	0.98	Chevrolet Tahoe 1500 LS (SUV)	0.70
Lexus LFA	1.04	Ford Fiesta SES	0.84
Mercedes-Benz SLS AMG	0.96	Hummer H3 (SUV)	0.66
Shelby SuperCars Ultimate Aero	1.05	Hyundai Sonata SE	0.85
Average	**1.09**	Smart ForTwo	0.72
		Average	**0.77**
Performance vehicles:		*Alternative vehicles:*	
Chevrolet Corvette ZR1	1.07	Chevrolet Volt (hybrid)	0.83
Dodge Viper SRT10	1.06	Nissan Leaf (electric)	0.79
Jaguar XJL Supercharged	0.88	Honda CR-Z (hybrid)	0.83
Acura TL SH-AWD	0.91	Honda Insight (hybrid)	0.74
Dodge Challenger SRT8	0.88	Toyota Prius (hybrid)	0.76
Average	**0.96**	**Average**	**0.79**

SUMMARY

If a particle moves with *constant* acceleration $\vec{a}$ and has velocity $\vec{v}_i$ and position $\vec{r}_i$ at $t = 0$, its velocity and position vectors at some later time t are

$$\vec{v}_f = \vec{v}_i + \vec{a}t \qquad \textbf{3.8} \blacktriangleleft$$

$$\vec{r}_f = \vec{r}_i + \vec{v}_i t + \tfrac{1}{2}\vec{a}t^2 \qquad \textbf{3.9} \blacktriangleleft$$

For two-dimensional motion in the xy plane under constant acceleration, these vector expressions are equivalent to two component expressions, one for the motion along x and one for the motion along y.

Projectile motion is a special case of two-dimensional motion under constant acceleration, where $a_x = 0$ and $a_y = -g$. In this case, the horizontal components of Equations 3.8 and 3.9 reduce to those of a particle under constant velocity:

$$v_{xf} = v_{xi} = \text{constant} \qquad \textbf{3.10} \blacktriangleleft$$

$$x_f = x_i + v_{xi}t \qquad \textbf{3.12} \blacktriangleleft$$

The vertical components of Equations 3.8 and 3.9 are those of a particle under constant acceleration:

$$v_{yf} = v_{yi} - gt \qquad \textbf{3.11} \blacktriangleleft$$

$$y_f = y_i + v_{yi}t - \tfrac{1}{2}gt^2 \qquad \textbf{3.13} \blacktriangleleft$$

where $v_{xi} = v_i \cos\theta_i$, $v_{yi} = v_i \sin\theta_i$, v_i is the initial speed of the projectile, and θ_i is the angle $\vec{v}_i$ makes with the positive x axis.

If a particle moves along a curved path in such a way that the magnitude and direction of $\vec{v}$ change in time, the particle has an acceleration vector that can be described by two components: (1) a radial component a_r arising from the change in direction of $\vec{v}$ and (2) a tangential component a_t arising from the change in magnitude of $\vec{v}$. The radial acceleration is called **centripetal acceleration,** and its direction is always toward the center of the circular path.

If an observer B is moving with velocity $\vec{v}_{BA}$ with respect to observer A, their measurements of the velocity of a particle located at point P are related according to

$$\vec{u}_{PA} = \vec{u}_{PB} + \vec{v}_{BA} \qquad \textbf{3.22} \blacktriangleleft$$

Equation 3.22 is the **Galilean transformation equation** for velocities and indicates that different observers will measure different velocities for the same particle.

Analysis Model for Problem Solving

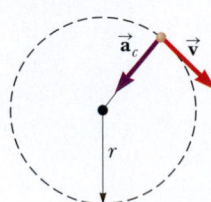

Particle in Uniform Circular Motion If a particle moves in a circular path of radius r with a constant speed v, the magnitude of its centripetal acceleration is given by

$$a_c = \frac{v^2}{r} \qquad \textbf{3.17} \blacktriangleleft$$

and the **period** of the particle's motion is given by

$$T = \frac{2\pi r}{v} \qquad \textbf{3.18} \blacktriangleleft$$

▶ OBJECTIVE QUESTIONS |

1. In which of the following situations is the moving object appropriately modeled as a projectile? Choose all correct answers. (a) A shoe is tossed in an arbitrary direction. (b) A jet airplane crosses the sky with its engines thrusting the plane forward. (c) A rocket leaves the launch pad. (d) A rocket moves through the sky, at much less than the speed of sound, after its fuel has been used up. (e) A diver throws a stone under water.

2. A rubber stopper on the end of a string is swung steadily in a horizontal circle. In one trial, it moves at speed v in a circle of radius r. In a second trial, it moves at a higher speed $3v$ in a circle of radius $3r$. In this second trial, is its acceleration (a) the same as in the first trial, (b) three times larger, (c) one-third as large, (d) nine times larger, or (e) one-ninth as large?

3. Figure OQ3.3 shows a bird's-eye view of a car going around a highway curve. As the car moves from point 1 to point 2, its speed doubles. Which of the vectors (a) through (e) shows the direction of the car's average acceleration between these two points?

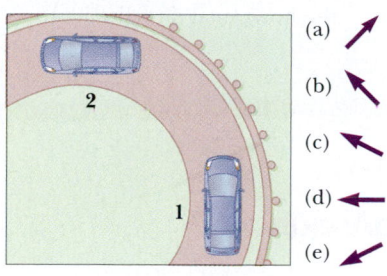

Figure OQ3.3

4. Entering his dorm room, a student tosses his book bag to the right and upward at an angle of 45° with the horizontal (Fig. OQ3.4). Air resistance does not affect the bag. The bag moves through point Ⓐ immediately after it leaves the student's hand, through point Ⓑ at the top of its flight, and through point Ⓒ immediately before it lands on the top bunk bed. **(i)** Rank the following horizontal and vertical velocity components from the largest to the smallest. (a) $v_{Ⓐx}$ (b) $v_{Ⓐy}$ (c) $v_{Ⓑx}$ (d) $v_{Ⓑy}$ (e) $v_{Ⓒy}$. Note that zero is larger than a negative number. If two quantities are equal, show them as equal in your list. If any quantity is equal to zero, show that fact in your list. **(ii)** Similarly, rank the following acceleration components. (a) $a_{Ⓐx}$ (b) $a_{Ⓐy}$ (c) $a_{Ⓑx}$ (d) $a_{Ⓑy}$ (e) $a_{Ⓒy}$.

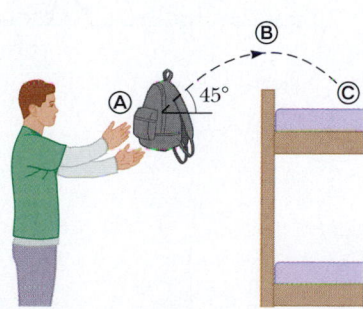

Figure OQ3.4

5. Does a car moving around a circular track with constant speed have (a) zero acceleration, (b) an acceleration in the direction of its velocity, (c) an acceleration directed away from the center of its path, (d) an acceleration directed toward the center of its path, or (e) an acceleration with a direction that cannot be determined from the given information?

6. An astronaut hits a golf ball on the Moon. Which of the following quantities, if any, remain constant as a ball travels through the vacuum there? (a) speed (b) acceleration (c) horizontal component of velocity (d) vertical component of velocity (e) velocity

7. A projectile is launched on the Earth with a certain initial velocity and moves without air resistance. Another projectile is launched with the same initial velocity on the Moon, where the acceleration due to gravity is one-sixth as large. How does the range of the projectile on the Moon compare with that of the projectile on the Earth? (a) It is one-sixth as large. (b) It is the same. (c) It is $\sqrt{6}$ times larger. (d) It is 6 times larger. (e) It is 36 times larger.

8. A baseball is thrown from the outfield toward the catcher. When the ball reaches its highest point, which statement is true? (a) Its velocity and its acceleration are both zero. (b) Its velocity is not zero, but its acceleration is zero. (c) Its velocity is perpendicular to its acceleration. (d) Its acceleration depends on the angle at which the ball was thrown. (e) None of statements (a) through (d) is true.

9. A student throws a heavy red ball horizontally from a balcony of a tall building with an initial speed v_i. At the same time, a second student drops a lighter blue ball from the balcony. Neglecting air resistance, which statement is true? (a) The blue ball reaches the ground first. (b) The balls reach the ground at the same instant. (c) The red ball reaches the ground first. (d) Both balls hit the ground with the same speed. (e) None of statements (a) through (d) is true.

10. A sailor drops a wrench from the top of a sailboat's vertical mast while the boat is moving rapidly and steadily straight forward. Where will the wrench hit the deck? (a) ahead of the base of the mast (b) at the base of the mast (c) behind the base of the mast (d) on the windward side of the base of the mast (e) None of the choices (a) through (d) is true.

11. A set of keys on the end of a string is swung steadily in a horizontal circle. In one trial, it moves at speed v in a circle of radius r. In a second trial, it moves at a higher speed $4v$ in a circle of radius $4r$. In the second trial, how does the period of its motion compare with its period in the first trial? (a) It is the same as in the first trial. (b) It is 4 times larger. (c) It is one-fourth as large. (d) It is 16 times larger. (e) It is one-sixteenth as large.

12. A certain light truck can go around a curve having a radius of 150 m with a maximum speed of 32.0 m/s. To have the same acceleration, at what maximum speed can it go around a curve having a radius of 75.0 m? (a) 64 m/s (b) 45 m/s (c) 32 m/s (d) 23 m/s (e) 16 m/s

CONCEPTUAL QUESTIONS

denotes answer available in *Student Solutions Manual/Study Guide*

1. Explain whether or not the following particles have an acceleration: (a) a particle moving in a straight line with constant speed and (b) a particle moving around a curve with constant speed.

2. Describe how a driver can steer a car traveling at constant speed so that (a) the acceleration is zero or (b) the magnitude of the acceleration remains constant.

3. If you know the position vectors of a particle at two points along its path and also know the time interval during which it moved from one point to the other, can you determine the particle's instantaneous velocity? Its average velocity? Explain.

4. Construct motion diagrams showing the velocity and acceleration of a projectile at several points along its path, assuming (a) the projectile is launched horizontally and (b) the projectile is launched at an angle θ with the horizontal.

5. A spacecraft drifts through space at a constant velocity. Suddenly, a gas leak in the side of the spacecraft gives it a constant acceleration in a direction perpendicular to the initial velocity. The orientation of the spacecraft does not change, so the acceleration remains perpendicular to the original direction of the velocity. What is the shape of the path followed by the spacecraft in this situation?

6. An ice skater is executing a figure eight, consisting of two identically shaped, tangent circular paths. Throughout the first loop she increases her speed uniformly, and during the second loop she moves at a constant speed. Draw a motion diagram showing her velocity and acceleration vectors at several points along the path of motion.

7. A projectile is launched at some angle to the horizontal with some initial speed v_i, and air resistance is negligible. (a) Is the projectile a freely falling body? (b) What is its acceleration in the vertical direction? (c) What is its acceleration in the horizontal direction?

PROBLEMS AVAILABLE IN ENHANCED WebAssign

3.1 The Position, Velocity, and Acceleration Vectors

Problems 1–2

3.2 Two-Dimensional Motion with Constant Acceleration

Problems 3–6

3.3 Projectile Motion

Problems 7–22

3.4 Analysis Model: Particle in Uniform Circular Motion

Problems 23–28

3.5 Tangential and Radial Acceleration

Problems 29–32

3.6 Relative Velocity and Relative Acceleration

Problems 33–40

3.7 Context Connection: Lateral Acceleration of Automobiles

Problem 41

Additional Problems

Problems 42–62

Solutions to the following Problems are available in the *Student Solutions Manual/Study Guide*:

3.1, 3.5, 3.11, 3.13, 3.23, 3.29, 3.31, 3.35, 3.39, 3.51, 3.52, 3.53, and 3.55

List of Enhanced Problems

Problem Number	Targeted Feedback in Enhanced WebAssign	Master It in Enhanced WebAssign	Watch It in Enhanced WebAssign
3.3	✓		✓
3.5	✓	✓	
3.11	✓	✓	
3.13	✓	✓	
3.14	✓		✓
3.17	✓		✓
3.29		✓	
3.31	✓		✓
3.35		✓	
3.39		✓	
3.55		✓	

The Laws of Motion

AL PARKER PHOTOGRAPHY/Shutterstock.com

In the preceding two chapters on kinematics, we *described* the motion of particles based on the definitions of position, velocity, and acceleration. Aside from our discussion of gravity for objects in free-fall, we did not address what factors might *influence* an object to move as it does. We would like to be able to answer general questions related to the influences on motion, such as "What mechanism causes changes in motion?" and "Why do some objects accelerate at higher rates than others?" In this first chapter on *dynamics*, we shall discuss the causes of the change in motion of particles using the concepts of force and mass. We will discuss the three fundamental laws of motion, which are based on experimental observations and were formulated about three centuries ago by Sir Isaac Newton.

By intuitively applying Newton's laws of motion, these two big horn sheep compete for dominance. They each exert forces against the Earth through muscular exertions of their legs, aided by the friction forces that keep them from slipping. The reaction forces of the Earth act back on the sheep and cause them to surge forward and butt heads. The goal is to force the opposing sheep out of equilibrium.

4.1 | The Concept of Force

As a result of everyday experiences, everyone has a basic understanding of the concept of force. When you push or pull an object, you exert a force on it. You exert a force when you throw or kick a ball. In these examples, the word *force* is associated with the result of muscular activity and with some change in the state of motion of an object. Forces do not always cause an object to move, however. For example, as you sit reading this book, the gravitational force acts on your body and yet you remain stationary. You can push on a heavy block of stone and yet fail to move it.

Figure 4.1 Some examples of forces applied to various objects. In each case, a force is exerted on the particle or object within the boxed area. The environment external to the boxed area provides this force.

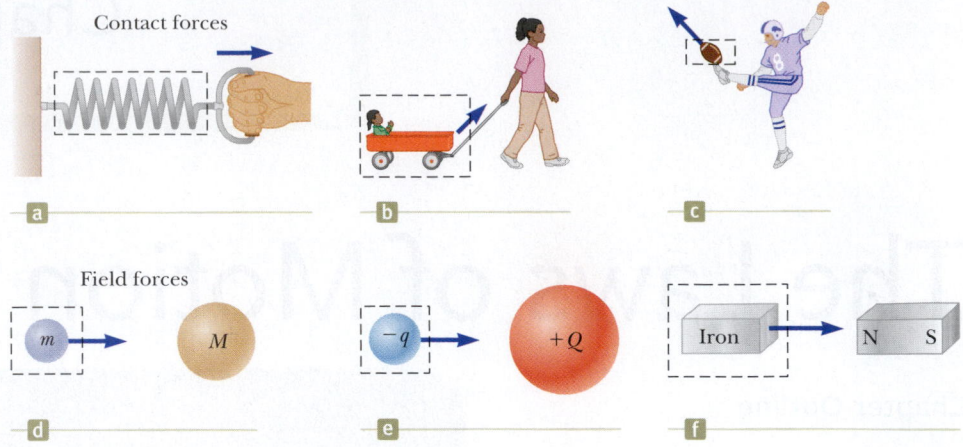

Isaac Newton
English Physicist and Mathematician (1642–1727)
Isaac Newton was one of the most brilliant scientists in history. Before the age of 30, he formulated the basic concepts and laws of mechanics, discovered the law of universal gravitation, and invented the mathematical methods of calculus. As a consequence of his theories, Newton was able to explain the motions of the planets, the ebb and flow of the tides, and many special features of the motions of the Moon and the Earth. He also interpreted many fundamental observations concerning the nature of light. His contributions to physical theories dominated scientific thought for two centuries and remain important today.

This chapter is concerned with the relation between the force on an object and the change in motion of that object. If you pull on a spring, as in Figure 4.1a, the spring stretches. If the spring is calibrated, the distance it stretches can be used to measure the strength of the force. If a child pulls on a wagon, as in Figure 4.1b, the wagon moves. When a football is kicked, as in Figure 4.1c, it is both deformed and set in motion. These examples all show the results of a class of forces called *contact forces*. That is, these forces represent the result of physical contact between two objects.

There exist other forces that do not involve physical contact between two objects. These forces, known as *field forces*, can act through empty space. The gravitational force between two objects that causes the free-fall acceleration described in Chapters 2 and 3 is an example of this type of force and is illustrated in Figure 4.1d. This gravitational force keeps objects bound to the Earth and gives rise to what we commonly call the *weight* of an object. The planets of our solar system are bound to the Sun under the action of gravitational forces. Another common example of a field force is the electric force that one electric charge exerts on another electric charge, as in Figure 4.1e. These charges might be an electron and a proton forming a hydrogen atom. A third example of a field force is the force that a bar magnet exerts on a piece of iron, as shown in Figure 4.1f.

The distinction between contact forces and field forces is not as sharp as you may have been led to believe by the preceding discussion. At the atomic level, all the forces classified as contact forces turn out to be caused by electric (field) forces similar in nature to the attractive electric force illustrated in Figure 4.1e. Nevertheless, in understanding macroscopic phenomena, it is convenient to use both classifications of forces.

We can use the linear deformation of a spring to measure force, as in the case of a common spring scale. Suppose a force is applied vertically to a spring that has a fixed upper end, as in Figure 4.2a. The spring can be calibrated by defining the unit force $\vec{F}_1$ as the force that produces an elongation of 1.00 cm. If a force $\vec{F}_2$, applied as in Figure 4.2b, produces an elongation of 2.00 cm, the magnitude of $\vec{F}_2$ is 2.00 units. If the two forces $\vec{F}_1$ and $\vec{F}_2$ are applied simultaneously, as in Figure 4.2c, the elongation of the spring is 3.00 cm because the forces are applied in the same direction and their magnitudes add. If the two forces $\vec{F}_1$ and $\vec{F}_2$ are applied in perpendicular directions, as in Figure 4.2d, the elongation is $\sqrt{(1.00)^2 + (2.00)^2}$ cm $= \sqrt{5.00}$ cm $= 2.24$ cm. The single force $\vec{F}$ that would produce this same elongation is the vector sum of $\vec{F}_1$ and $\vec{F}_2$, as described in Figure 4.2d. That is, $|\vec{F}| = \sqrt{F_1^2 + F_2^2} = 2.24$ units, and its direction is $\theta = \tan^{-1}(-0.500) = -26.6°$. Because forces have been experimentally verified to behave as vectors, you *must* use the rules of vector addition to obtain the total force on an object.

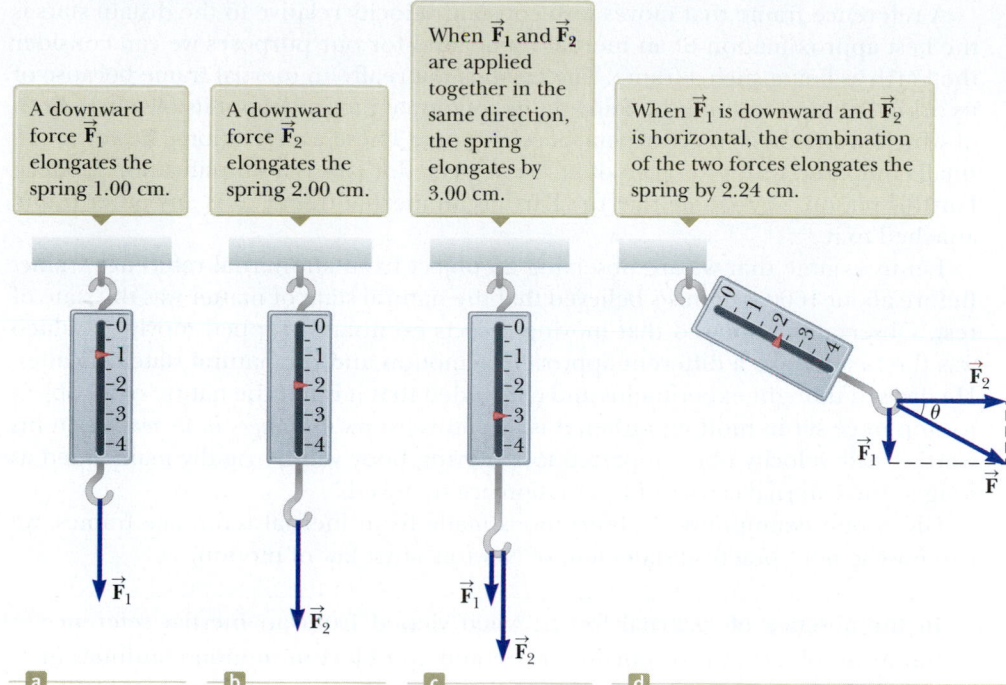

A downward force $\vec{F}_1$ elongates the spring 1.00 cm.

A downward force $\vec{F}_2$ elongates the spring 2.00 cm.

When $\vec{F}_1$ and $\vec{F}_2$ are applied together in the same direction, the spring elongates by 3.00 cm.

When $\vec{F}_1$ is downward and $\vec{F}_2$ is horizontal, the combination of the two forces elongates the spring by 2.24 cm.

Figure 4.2 The vector nature of a force is tested with a spring scale.

4.2 | Newton's First Law

We begin our study of forces by imagining that you place a puck on a perfectly level air hockey table (Fig. 4.3). You expect that the puck will remain stationary when it is placed gently at rest on the table. Now imagine putting your air hockey table on a train moving with constant velocity. If the puck is placed on the table, the puck again remains where it is placed. If the train were to accelerate, however, the puck would start moving along the table, just as a set of papers on your dashboard slides onto the front seat of your car when you step on the gas.

As we saw in Section 3.6, a moving object can be observed from any number of reference frames. **Newton's first law of motion**, sometimes called the *law of inertia*, defines a special set of reference frames called *inertial frames*. This law can be stated as follows:

> If an object does not interact with other objects, it is possible to identify a reference frame in which the object has zero acceleration.

Such a reference frame is called an **inertial frame of reference**. When the puck is on the air hockey table located on the ground, you are observing it from an inertial reference frame; there are no horizontal interactions of the puck with any other objects, and you observe it to have zero acceleration in the horizontal direction. When you are on the train moving at constant velocity, you are also observing the puck from an inertial reference frame. Any reference frame that moves with constant velocity relative to an inertial frame is itself an inertial frame. When the train accelerates, however, you are observing the puck from a **noninertial reference frame** because you and the train are accelerating relative to the inertial reference frame of the surface of the Earth. Although the puck appears to be accelerating according to your observations, we can identify a reference frame in which the puck has zero acceleration. For example, an observer standing outside the train on the ground sees the puck sliding relative to the table but always moving with the same velocity with respect to the ground as the train had before it started to accelerate (because there is almost no friction to "tie" the puck and the train together). Therefore, Newton's first law is still satisfied even though your observations say otherwise.

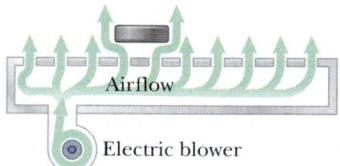

Airflow

Electric blower

Figure 4.3 On an air hockey table, air blown through holes in the surface allows the puck to move almost without friction. If the table is not accelerating, a puck placed on the table will remain at rest with respect to the table if there are no horizontal forces acting on it.

▶ Newton's first law

▶ Inertial frame of reference

A reference frame that moves with constant velocity relative to the distant stars is the best approximation of an inertial frame, and for our purposes we can consider the Earth as being such a frame. The Earth is not really an inertial frame because of its orbital motion around the Sun and its rotational motion about its own axis, both of which are related to centripetal accelerations. These accelerations, however, are small compared with g and can often be neglected. (This is a simplification model.) For this reason, we assume that the Earth is an inertial frame, as is any other frame attached to it.

Let us assume that we are observing an object from an inertial reference frame. Before about 1600, scientists believed that the natural state of matter was the state of rest. Observations showed that moving objects eventually stopped moving. Galileo was the first to take a different approach to motion and the natural state of matter. He devised thought experiments and concluded that it is not the nature of an object to stop once set in motion; rather, it is its nature to *resist changes in its motion*. In his words, "Any velocity once imparted to a moving body will be rigidly maintained as long as the external causes of retardation are removed."

Given our assumption of observations made from inertial reference frames, we can pose a more practical statement of Newton's first law of motion:

▶ Another statement of Newton's first law

In the absence of external forces, when viewed from an inertial reference frame, an object at rest remains at rest and an object in motion continues in motion with a constant velocity (that is, with a constant speed in a straight line).

In simpler terms, we can say that **when no force acts on an object, the acceleration of the object is zero.** If nothing acts to change the object's motion, its velocity does not change. From the first law, we conclude that any *isolated object* (one that does not interact with its environment) is either at rest or moving with constant velocity. The tendency of an object to resist any attempt to change its velocity is called **inertia.**

Consider a spacecraft traveling in space, far removed from any planets or other matter. The spacecraft requires some propulsion system to change its velocity. If the propulsion system is turned off when the spacecraft reaches a velocity $\vec{v}$, however, the spacecraft "coasts" in space with that velocity and the astronauts enjoy a "free ride" (i.e., no propulsion system is required to keep them moving at the velocity $\vec{v}$).

Finally, recall our discussion in Chapter 2 about the proportionality between force and acceleration:

$$\vec{F} \propto \vec{a}$$

Newton's first law tells us that the velocity of an object remains constant if no force acts on an object; the object maintains its state of motion. The preceding proportionality tells us that if a force *does* act, a change does occur in the motion, measured by the acceleration. This notion will form the basis of Newton's second law, and we shall provide more details on this concept shortly.

> **Pitfall Prevention | 4.1**
> **Newton's First Law**
> Newton's first law does *not* say what happens for an object with *zero net force*, that is, multiple forces that cancel; it says what happens *in the absence of external forces*. This subtle but important difference allows us to define force as that which causes a change in the motion. The description of an object under the effect of forces that balance is covered by Newton's second law.

> **QUICK QUIZ 4.1** Which of the following statements is most correct? (**a**) It is possible for an object to have motion in the absence of forces on the object. (**b**) It is possible to have forces on an object in the absence of motion of the object. (**c**) Neither statement (**a**) nor statement (**b**) is correct. (**d**) Both statements (**a**) and (**b**) are correct.

4.3 | Mass

Imagine playing catch with either a table-tennis ball or a bowling ball. Which ball is more likely to keep moving when you try to catch it? Which ball has the greater tendency to remain motionless when you try to throw it? The bowling ball is more resistant to changes in its velocity than the table-tennis ball. How can we quantify this concept?

▶ Definition of mass

Mass is that property of an object that specifies how much resistance an object exhibits to changes in its velocity, and as we learned in Section 1.1, the SI unit of mass is the kilogram. The greater the mass of an object, the less that object accelerates under the action of a given applied force.

To describe mass quantitatively, we begin by experimentally comparing the accelerations a given force produces on different objects. Suppose a force acting on an object of mass m_1 produces a change in motion of the object that we can quantify with the object's acceleration $\vec{a}_1$ and the *same force* acting on an object of mass m_2 produces an acceleration $\vec{a}_2$. The ratio of the two masses is defined as the *inverse* ratio of the magnitudes of the accelerations produced by the force:

$$\frac{m_1}{m_2} \equiv \frac{a_2}{a_1}$$

4.1 ◀

For example, if a given force acting on a 3-kg object produces an acceleration of 4 m/s^2, the same force applied to a 6-kg object produces an acceleration of 2 m/s^2. If one object has a known mass, the mass of the other object can be obtained from acceleration measurements.

Mass is an inherent property of an object and is independent of the object's surroundings and of the method used to measure it. Also, mass is a scalar quantity and therefore obeys the rules of ordinary arithmetic. That is, several masses can be combined in simple numerical fashion. For example, if you combine a 3-kg mass with a 5-kg mass, their total mass is 8 kg. We can verify this result experimentally by comparing the acceleration that a known force gives to several objects separately with the acceleration that the same force gives to the same objects combined as one unit.

Mass should not be confused with weight. Mass and weight are two different quantities. As we shall see later in this chapter, the weight of an object is equal to the magnitude of the gravitational force exerted on the object and varies with location. For example, a person who weighs 180 lb on the Earth weighs only about 30 lb on the Moon. On the other hand, the mass of an object is the same everywhere. An object having a mass of 2 kg on Earth also has a mass of 2 kg on the Moon.

▶ Mass and weight are different quantities

4.4 | Newton's Second Law

Newton's first law explains what happens to an object when no force acts on it: It either remains at rest or moves in a straight line with constant speed. This law allows us to define an inertial frame of reference. It also allows us to identify force as that which changes motion. Newton's second law answers the question of what happens to an object when it has one or more forces acting on it, based on our discussion of mass in the preceding section.

Imagine you are pushing a block of ice across a frictionless horizontal surface. When you exert some horizontal force $\vec{F}$, the block moves with some acceleration $\vec{a}$. Experiments show that if you apply a force twice as large to the same object, the acceleration doubles. If you increase the applied force to $3\vec{F}$, the original acceleration is tripled, and so on. From such observations, we conclude that the acceleration of an object is directly proportional to the net force acting on it. We alluded to this proportionality in our discussion of acceleration in Chapter 2. We also know from the preceding section that the magnitude of the acceleration of an object is inversely proportional to its mass: $|\vec{a}| \propto 1/m$.

These experimental observations are summarized in **Newton's second law:**

When viewed from an inertial reference frame, the acceleration of an object is directly proportional to the net force acting on it and inversely proportional to its mass.

▶ Newton's second law

We write this law as

$$\vec{a} \propto \frac{\sum \vec{F}}{m}$$

where $\Sigma\vec{F}$ is the **net force,** which is the vector sum of *all* forces acting on the object of mass m. If the object consists of a system of individual elements, the net force is the vector sum of all forces *external* to the system. Any *internal* forces—that is, forces between elements of the system—are not included because they do not affect the motion of the entire system. The net force is sometimes called the *resultant force*, the *sum of the forces*, the *total force*, or the *unbalanced force*.

Newton's second law in mathematical form is a statement of this relationship that makes the preceding proportionality an equality:[1]

▶ Mathematical representation of Newton's second law

$$\Sigma\vec{F} = m\vec{a} \qquad \qquad \textbf{4.2}◀$$

Note that Equation 4.2 is a *vector* expression and hence is equivalent to the following three component equations:

▶ Newton's second law in component form

$$\Sigma F_x = ma_x \quad \Sigma F_y = ma_y \quad \Sigma F_z = ma_z \qquad \textbf{4.3}◀$$

Newton's second law introduces us to a new analysis model, the particle under a net force. If a particle, or an object that can be modeled as a particle, is under the influence of a net force, Equation 4.2, the mathematical statement of Newton's second law, can be used to describe its motion. The acceleration is constant if the net force is constant. Therefore, the particle under a constant net force will have its motion described as a particle under constant acceleration. Of course, not all forces are constant, and when they are not, the particle cannot be modeled as one under constant acceleration. We shall investigate situations in this chapter and the next involving both constant and varying forces.

> **Pitfall Prevention | 4.3**
> **$m\vec{a}$ Is Not a Force**
> Equation 4.2 does *not* say that the product $m\vec{a}$ is a force. All forces on an object are added vectorially to generate the net force on the left side of the equation. This net force is then equated to the product of the mass of the object and the acceleration that results from the net force. Do *not* include an "$m\vec{a}$ force" in your analysis of the forces on an object.

QUICK QUIZ 4.2 An object experiences no acceleration. Which of the following *cannot* be true for the object? (**a**) A single force acts on the object. (**b**) No forces act on the object. (**c**) Forces act on the object, but the forces cancel.

QUICK QUIZ 4.3 You push an object, initially at rest, across a frictionless floor with a constant force for a time interval Δt, resulting in a final speed of v for the object. You then repeat the experiment, but with a force that is twice as large. What time interval is now required to reach the same final speed v?   (**a**) $4\,\Delta t$ (**b**) $2\,\Delta t$ (**c**) Δt (**d**) $\Delta t/2$ (**e**) $\Delta t/4$

Unit of Force

The SI unit of force is the **newton,** which is defined as the force that, when acting on a 1-kg mass, produces an acceleration of 1 m/s^2.

From this definition and Newton's second law, we see that the newton can be expressed in terms of the fundamental units of mass, length, and time:

▶ Definition of the newton

$$1\,\text{N} \equiv 1\,\text{kg} \cdot \text{m/s}^2 \qquad \qquad \textbf{4.4}◀$$

The units of mass, acceleration, and force are summarized in Table 4.1. Most of the calculations we shall make in our study of mechanics will be in SI units. Equalities between units in the SI and U.S. customary systems are given in Appendix A.

TABLE 4.1 | Units of Mass, Acceleration, and Force

System of Units	Mass (M)	Acceleration (L/T^2)	Force (ML/T^2)
SI	kg	m/s^2	N = kg · m/s^2
U.S. customary	slug	ft/s^2	lb = slug · ft/s^2

---

[1]Equation 4.2 is valid only when the speed of the object is much less than the speed of light. We will treat the relativistic situation in Chapter 9.

In a train, the cars are connected by *couplers*. The couplers between the cars exert forces on the cars as the train is pulled by the locomotive in the front. Imagine that the train is speeding up in the forward direction. As you imagine moving from the locomotive to the last car, does the force exerted by the couplers *increase, decrease,* or *stay the same?* What if the engineer applies the brakes? How does the force vary from locomotive to last car in this case? (Assume that the only brakes applied are those on the engine.)

Reasoning The force *decreases* from the front of the train to the back. The coupler between the locomotive and the first car must apply enough force to accelerate all the remaining cars. As we move back along the train, each coupler is accelerating less mass behind it. The last coupler only has to accelerate the last car, so it exerts the smallest force. If the brakes are applied, the force decreases from front to back of the train also. The first coupler, at the back of the locomotive, must apply a large force to slow down all the remaining cars. The final coupler must only apply a force large enough to slow down the mass of the last car. ◄

Example 4.1 | An Accelerating Hockey Puck

A hockey puck having a mass of 0.30 kg slides on the frictionless, horizontal surface of an ice rink. Two hockey sticks strike the puck simultaneously, exerting the forces on the puck shown in Figure 4.4. The force $\vec{F}_1$ has a magnitude of 5.0 N, and the force $\vec{F}_2$ has a magnitude of 8.0 N. Determine both the magnitude and the direction of the puck's acceleration.

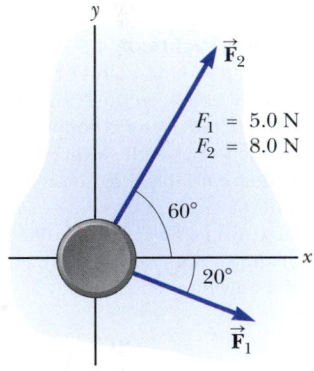

SOLUTION

Conceptualize Study Figure 4.4. Using your expertise in vector addition from Chapter 1, predict the approximate direction of the net force vector on the puck. The acceleration of the puck will be in the same direction.

Categorize Because we can determine a net force and we want an acceleration, this problem is categorized as one that may be solved using Newton's second law.

Figure 4.4 (Example 4.1) A hockey puck moving on a frictionless surface is subject to two forces $\vec{F}_1$ and $\vec{F}_2$.

Analyze Find the component of the net force acting on the puck in the *x* direction:

$$\sum F_x = F_{1x} + F_{2x} = F_1 \cos(-20°) + F_2 \cos 60°$$
$$= (5.0\ \text{N})(0.940) + (8.0\ \text{N})(0.500) = 8.7\ \text{N}$$

Find the component of the net force acting on the puck in the *y* direction:

$$\sum F_y = F_{1y} + F_{2y} = F_1 \sin(-20°) + F_2 \sin 60°$$
$$= (5.0\ \text{N})(-0.342) + (8.0\ \text{N})(0.866) = 5.2\ \text{N}$$

Use Newton's second law in component form (Eq. 4.3) to find the *x* and *y* components of the puck's acceleration:

$$a_x = \frac{\sum F_x}{m} = \frac{8.7\ \text{N}}{0.30\ \text{kg}} = 29\ \text{m/s}^2$$

$$a_y = \frac{\sum F_y}{m} = \frac{5.2\ \text{N}}{0.30\ \text{kg}} = 17\ \text{m/s}^2$$

Find the magnitude of the acceleration:

$$a = \sqrt{(29\ \text{m/s}^2)^2 + (17\ \text{m/s}^2)^2} = \boxed{34\ \text{m/s}^2}$$

Find the direction of the acceleration relative to the positive *x* axis:

$$\theta = \tan^{-1}\left(\frac{a_y}{a_x}\right) = \tan^{-1}\left(\frac{17}{29}\right) = \boxed{31°}$$

Finalize The vectors in Figure 4.4 can be added graphically to check the reasonableness of our answer. Because the acceleration vector is along the direction of the resultant force, a drawing showing the resultant force vector helps us check the validity of the answer. (Try it!)

continued

4.1 *cont.*

What If? Suppose three hockey sticks strike the puck simultaneously, with two of them exerting the forces shown in Figure 4.4. The result of the three forces is that the hockey puck shows *no* acceleration. What must be the components of the third force?

Answer If there is zero acceleration, the net force acting on the puck must be zero. Therefore, the three forces must cancel. We have found the components of the combination of the first two forces. The components of the third force must be of equal magnitude and opposite sign so that all the components add to zero. Therefore, $F_{3x} = -8.7$ N and $F_{3y} = -5.2$ N.

Pitfall Prevention | 4.4
"Weight of an Object"
We are familiar with the everyday phrase, the "weight of an object." Weight, however, is not an inherent property of an object; rather, it is a measure of the gravitational force between the object and the Earth (or other planet). Therefore, weight is a property of a *system* of items: the object and the Earth.

Pitfall Prevention | 4.5
Kilogram Is Not a Unit of Weight
You may have seen the "conversion" 1 kg = 2.2 lb. Despite popular statements of weights expressed in kilograms, the kilogram is not a unit of *weight*, it is a unit of *mass*. The conversion statement is not an equality; it is an *equivalence* that is valid only on the Earth's surface.

4.5 | The Gravitational Force and Weight

We are well aware that all objects are attracted to the Earth. The force exerted by the Earth on an object is the **gravitational force** $\vec{\mathbf{F}}_g$. This force is directed toward the center of the Earth.[2] The magnitude of the gravitational force is called the **weight** F_g of the object.

We have seen in Chapters 2 and 3 that a freely falling object experiences an acceleration $\vec{\mathbf{g}}$ directed toward the center of the Earth. A freely falling object has only one force on it, the gravitational force, so the net force on the object in this situation is equal to the gravitational force:

$$\sum \vec{\mathbf{F}} = \vec{\mathbf{F}}_g$$

Because the acceleration of a freely falling object is equal to the free-fall acceleration $\vec{\mathbf{g}}$, it follows that

$$\sum \vec{\mathbf{F}} = m\vec{\mathbf{a}} \quad \rightarrow \quad \vec{\mathbf{F}}_g = m\vec{\mathbf{g}}$$

or, in magnitude,

$$F_g = mg \qquad \qquad \text{4.5} \blacktriangleleft$$

Because it depends on g, weight varies with location, as we mentioned in Section 4.3. Objects weigh less at higher altitudes than at sea level because g decreases with increasing distance from the center of the Earth. Hence, weight, unlike mass, is not an inherent property of an object. It is a property of the *system* of the object and the Earth. For example, if an object has a mass of 70 kg, its weight in a location where $g = 9.80$ m/s^2 is $mg = 686$ N. At the top of a mountain where $g = 9.76$ m/s^2, the object's weight would be 683 N. Therefore, if you want to lose weight without going on a diet, climb a mountain or weigh yourself at 30 000 ft during an airplane flight.

Because $F_g = mg$, we can compare the masses of two objects by measuring their weights with a spring scale. At a given location (so that g is fixed) the ratio of the weights of two objects equals the ratio of their masses.

Equation 4.5 quantifies the gravitational force on the object, but notice that this equation does not require the object to be moving. Even for a stationary object, or an object on which several forces act, Equation 4.5 can be used to calculate the magnitude of the gravitational force. This observation results in a subtle shift in the interpretation of m in the equation. The mass m in Equation 4.5 is playing the role of determining the strength of the gravitational attraction between the object and the Earth. This role is completely different from that previously described for mass, that of measuring the resistance to changes in motion in response to an external force. In that role, mass is also called **inertial mass.** We call m in Equation 4.5 the **gravitational mass.** Despite this quantity being different from inertial mass, it is one of the experimental conclusions in Newtonian dynamics that gravitational mass and inertial mass have the same value at the present level of experimental refinement.

The life-support unit strapped to the back of astronaut Harrison Schmitt weighed 300 lb on the Earth and had a mass of 136 kg. During his training, a 50-lb mock-up with a mass of 23 kg was used. Although this strategy effectively simulated the reduced weight the unit would have on the Moon, it did not correctly mimic the unchanging mass. It was more difficult to accelerate the 136-kg unit (perhaps by jumping or twisting suddenly) on the Moon than it was to accelerate the 23-kg unit on the Earth.

Eugene Cernan/NASA

---

[2] This statement represents a simplification model in that it ignores that the mass distribution of the Earth is not perfectly spherical.

4.6 | Newton's Third Law

Newton's third law conveys the notion that forces are always interactions between two objects:

If two objects interact, the force $\vec{\mathbf{F}}_{12}$ exerted by object 1 on object 2 is equal in magnitude but opposite in direction to the force $\vec{\mathbf{F}}_{21}$ exerted by object 2 on object 1:

$$\vec{\mathbf{F}}_{12} = -\vec{\mathbf{F}}_{21}$$ **4.6** ◄   ► Newton's third law

When it is important to designate forces as interactions between two objects, we will use this subscript notation, where $\vec{\mathbf{F}}_{ab}$ means "the force exerted *by* a *on* b." The third law, illustrated in Figure 4.5a, is equivalent to stating that **forces always occur in pairs** or that a **single isolated force cannot exist.** The force that object 1 exerts on object 2 may be called the *action force,* and the force of object 2 on object 1 may be called the *reaction force.* In reality, either force can be labeled the action or reaction force. The action force is equal in magnitude to the reaction force and opposite in direction. In all cases, the action and reaction forces act on different objects and must be of the same type. For example, the force acting on a freely falling projectile is the gravitational force exerted by the Earth on the projectile $\vec{\mathbf{F}}_g = \vec{\mathbf{F}}_{Ep}$ (E = Earth, p = projectile), and the magnitude of this force is *mg*. The reaction to this force is the gravitational force exerted by the projectile on the Earth $\vec{\mathbf{F}}_{pE} = -\vec{\mathbf{F}}_{Ep}$. The reaction force $\vec{\mathbf{F}}_{pE}$ must accelerate the Earth toward the projectile just as the action force $\vec{\mathbf{F}}_{Ep}$ accelerates the projectile toward the Earth. Because the Earth has such a large mass, however, its acceleration as a result of this reaction force is negligibly small.

Another example of Newton's third law in action is shown in Figure 4.5b. The force $\vec{\mathbf{F}}_{hn}$ exerted by the hammer on the nail (the action) is equal in magnitude and opposite the force $\vec{\mathbf{F}}_{nh}$ exerted by the nail on the hammer (the reaction). This latter force stops the forward motion of the hammer when it strikes the nail.

Pitfall Prevention | 4.6
Newton's Third Law
Newton's third law is such an important and often misunderstood notion that it is repeated here in a Pitfall Prevention. In Newton's third law, action and reaction forces act on *different* objects. Two forces acting on the same object, even if they are equal in magnitude and opposite in direction, *cannot* be an action–reaction pair.

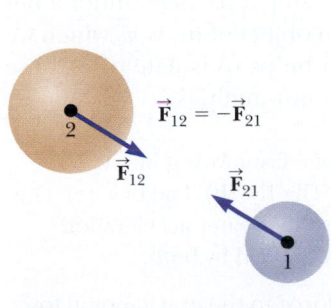

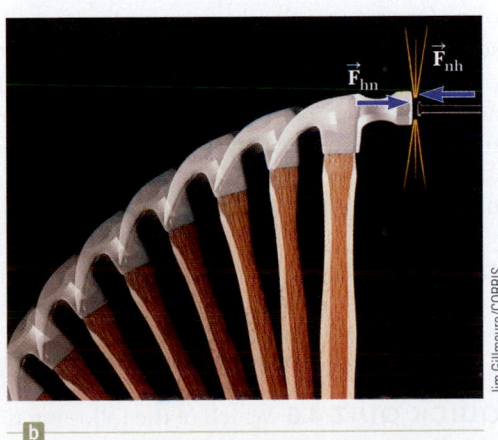

Jim Gillmoure/CORBIS

Figure 4.5 Newton's third law. (a) The force $\vec{\mathbf{F}}_{12}$ exerted by object 1 on object 2 is equal in magnitude and opposite in direction to the force $\vec{\mathbf{F}}_{21}$ exerted by object 2 on object 1. (b) The force $\vec{\mathbf{F}}_{hn}$ exerted by the hammer on the nail is equal in magnitude and opposite in direction to the force $\vec{\mathbf{F}}_{nh}$ exerted by the nail on the hammer.

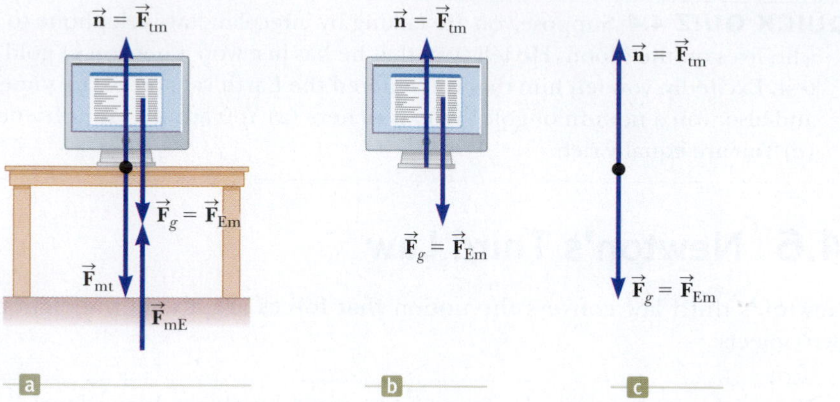

Figure 4.6 (a) When a computer monitor is at rest on a table, the forces acting on the monitor are the normal force $\vec{\mathbf{n}}$ and the gravitational force $\vec{\mathbf{F}}_g$. The reaction to $\vec{\mathbf{n}}$ is the force $\vec{\mathbf{F}}_{mt}$ exerted by the monitor on the table. The reaction to $\vec{\mathbf{F}}_g$ is the force $\vec{\mathbf{F}}_{mE}$ exerted by the monitor on the Earth. (b) A diagram showing the forces on the monitor. (c) A free-body diagram shows the monitor as a black dot with the forces acting on it.

▶ Normal force

Pitfall Prevention | 4.7
n Does Not Always Equal _mg_
In the situation shown in Figure 4.6 and in many others, we find that $n = mg$ (the normal force has the same magnitude as the gravitational force). This result, however, is not generally true. If an object is on an incline, if there are applied forces with vertical components, or if there is a vertical acceleration of the system, then $n \neq mg$. *Always* apply Newton's second law to find the relationship between n and mg.

Pitfall Prevention | 4.8
Free-Body Diagrams
The *most important* step in solving a problem using Newton's laws is to draw a proper simplified pictorial representation, the free-body diagram. Be sure to draw only those forces that act on the object you are isolating. Be sure to draw *all* forces acting on the object, including any field forces, such as the gravitational force.

The Earth exerts a gravitational force $\vec{\mathbf{F}}_g$ on any object. If the object is a computer monitor at rest on a table, as in the pictorial representation in Figure 4.6a, the reaction force to $\vec{\mathbf{F}}_g = \vec{\mathbf{F}}_{Em}$ is the force exerted by the monitor on the Earth $\vec{\mathbf{F}}_{mE} = -\vec{\mathbf{F}}_{Em}$. The monitor does not accelerate because it is held up by the table. The table exerts on the monitor an upward force $\vec{\mathbf{n}} = \vec{\mathbf{F}}_{tm}$, called the **normal force**.[3] This force prevents the monitor from falling through the table; it can have any value needed, up to the point at which the table breaks. From Newton's second law we see that, because the monitor has zero acceleration, it follows that $\Sigma\vec{\mathbf{F}} = \vec{\mathbf{n}} + \vec{\mathbf{F}}_g = 0$, or $n = mg$. The normal force balances the gravitational force on the monitor, so the net force on the monitor is zero. The reaction to $\mathbf{n}$ is the force exerted by the monitor downward on the table, $\vec{\mathbf{F}}_{mt} = -\vec{\mathbf{F}}_{tm}$.

Note that the forces acting on the monitor are $\vec{\mathbf{F}}_g$ and $\vec{\mathbf{n}}$, as shown in Figure 4.6b. The two reaction forces $\vec{\mathbf{F}}_{mE}$ and $\vec{\mathbf{F}}_{mt}$ are exerted by the monitor on the Earth and the table, respectively. Remember that the two forces in an action–reaction pair always act on two different objects.

Figure 4.6 illustrates an extremely important difference between a pictorial representation and a simplified pictorial representation for solving problems involving forces. Figure 4.6a shows many of the forces in the situation: those acting on the monitor, one acting on the table, and one acting on the Earth. Figure 4.6b, by contrast, shows only the forces on *one object*, the monitor, and is called a **force diagram,** or a *diagram showing the forces on the object.* The important simplified pictorial representation in Figure 4.6c is called a **free-body diagram.** In a free-body diagram, the particle model is used by representing the object as a dot and showing the forces that act on the object as being applied to the dot. When analyzing a particle under a net force, we are interested in the net force on one object, an object of mass m, which we will model as a particle. Therefore, a free-body diagram helps us isolate only those forces on the object and eliminate the other forces from our analysis.

QUICK QUIZ 4.5 (i) If a fly collides with the windshield of a fast-moving bus, which experiences an impact force with a larger magnitude? (a) The fly. (b) The bus. (c) The same force is experienced by both. (ii) Which experiences the greater acceleration? (a) The fly. (b) The bus. (c) The same acceleration is experienced by both.

QUICK QUIZ 4.6 Which of the following is the reaction force to the gravitational force acting on your body as you sit in your desk chair? (a) the normal force from the chair (b) the force you apply downward on the seat of the chair (c) neither of these forces

[3]The word *normal* is used because the direction of $\vec{\mathbf{n}}$ is always *perpendicular* to the surface.

Figure 4.7 (Thinking Physics 4.2) (a) A horse pulls a sled through the snow. (b) The forces on the sled. (c) The forces on the horse.

> **THINKING PHYSICS 4.2**
>
> A horse pulls on a sled with a horizontal force, causing the sled to accelerate as in Figure 4.7a. Newton's third law says that the sled exerts a force of equal magnitude and opposite direction on the horse. In view of this situation, how can the sled accelerate? Don't these forces cancel?
>
> **Reasoning** When applying Newton's third law, it is important to remember that the forces involved act on different objects. Notice that the force exerted by the horse acts *on the sled*, whereas the force exerted by the sled acts *on the horse*. Because these forces act on different objects, they cannot cancel.
>
> The horizontal forces exerted on the *sled* alone are the forward force $\vec{\mathbf{F}}_{hs}$ exerted by the horse and the backward force of friction $\vec{\mathbf{f}}_{sled}$ between sled and surface (Fig. 4.7b). When $\vec{\mathbf{F}}_{hs}$ exceeds $\vec{\mathbf{f}}_{sled}$, the sled accelerates to the right.
>
> The horizontal forces exerted on the *horse* alone are the forward friction force $\vec{\mathbf{f}}_{horse}$ from the ground and the backward force $\vec{\mathbf{F}}_{sh}$ exerted by the sled (Fig. 4.7c). The resultant of these two forces causes the horse to accelerate. When $\vec{\mathbf{f}}_{horse}$ exceeds $\vec{\mathbf{F}}_{sh}$, the horse accelerates to the right. ◄

4.7 | Analysis Models Using Newton's Second Law

In this section, we discuss two analysis models for solving problems in which objects are either in equilibrium ($\vec{\mathbf{a}} = 0$) or are accelerating under the action of constant external forces. We shall assume that the objects behave as particles so that we need not worry about rotational motion or other complications. In this section, we also apply some additional simplification models. We ignore the effects of friction for those problems involving motion, which is equivalent to stating that the surfaces are *frictionless*. We usually ignore the masses of any ropes or strings involved. In this approximation, the magnitude of the force exerted at any point along a string is the same. In problem statements, the terms *light* and *of negligible mass* are used to indicate that a mass is to be ignored when you work the problem. These two terms are synonymous in this context.

Analysis Model: Particle in Equilibrium

Objects that are either at rest or moving with constant velocity are treated with the **particle in equilibrium** model. From Newton's second law with $\vec{\mathbf{a}} = 0$, this condition of equilibrium can be expressed as

$$\sum \vec{\mathbf{F}} = 0 \qquad \qquad 4.7 ◄$$

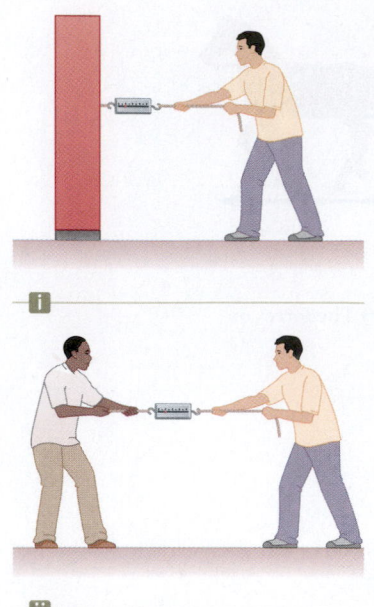

Figure 4.8 (Quick Quiz 4.7) (i) An individual pulls with a force of magnitude F on a spring scale attached to a wall. (ii) Two individuals pull with forces of magnitude F in opposite directions on a spring scale attached between two ropes.

This statement signifies that the vector sum of all the forces (the net force) acting on an object in equilibrium is zero.[4] If a particle is subject to forces but exhibits an acceleration of zero, we use Equation 4.7 to analyze the situation, as we shall see in some of the following examples.

Usually, the problems we encounter in our study of equilibrium are easier to solve if we work with Equation 4.7 in terms of the components of the external forces acting on an object. In other words, in a two-dimensional problem, the sum of all the external forces in the x and y directions must separately equal zero; that is,

$$\sum F_x = 0 \qquad \sum F_y = 0 \qquad \text{4.8} \blacktriangleleft$$

The extension of Equations 4.8 to a three-dimensional situation can be made by adding a third component equation, $\sum F_z = 0$.

In a given situation, we may have balanced forces on an object in one direction but unbalanced forces in the other. Therefore, for a given problem, we may need to model the object as a particle in equilibrium for one component and a particle under a net force for the other.

> **QUICK QUIZ 4.7** Consider the two situations shown in Figure 4.8, in which no acceleration occurs. In both cases, all individuals pull with a force of magnitude F on a rope attached to a spring scale. Is the reading on the spring scale in part (i) of the figure (a) greater than, (b) less than, or (c) equal to the reading in part (ii)?

Analysis Model: Particle Under a Net Force

If an object experiences an acceleration, its motion can be analyzed with the **particle under a net force** model. The appropriate equation for this model is Newton's second law, Equation 4.2:

$$\sum \vec{F} = m\vec{a} \qquad \text{4.2} \blacktriangleleft$$

Consider a crate being pulled to the right on a frictionless, horizontal floor as in Figure 4.9a. Of course, the floor directly under the boy must have friction; otherwise, his feet would simply slip when he tries to pull on the crate! Suppose you wish to find the acceleration of the crate and the force the floor exerts on it. The forces acting on the crate are illustrated in the free-body diagram in Figure 4.9b. Notice that the horizontal force $\vec{T}$ being applied to the crate acts through the rope. The magnitude of $\vec{T}$ is equal to the tension in the rope. In addition to the force $\vec{T}$, the free-body diagram for the crate includes the gravitational force $\vec{F}_g$ and the normal force $\vec{n}$ exerted by the floor on the crate.

We can now apply Newton's second law in component form to the crate. The only force acting in the x direction is $\vec{T}$. Applying $\sum F_x = ma_x$ to the horizontal motion gives

$$\sum F_x = T = ma_x \quad \text{or} \quad a_x = \frac{T}{m}$$

No acceleration occurs in the y direction because the crate moves only horizontally. Therefore, we use the particle in equilibrium model in the y direction. Applying the y component of Equation 4.7 yields

$$\sum F_y = n + (-F_g) = 0 \quad \text{or} \quad n = F_g$$

That is, the normal force has the same magnitude as the gravitational force but acts in the opposite direction.

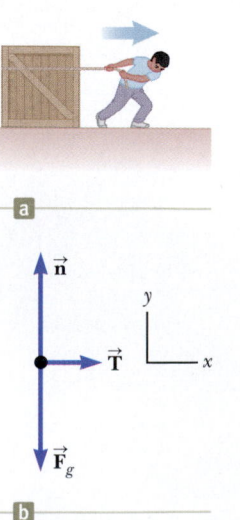

Figure 4.9 (a) A crate being pulled to the right on a frictionless floor. (b) The free-body diagram representing the external forces acting on the crate.

---

[4]This statement is only one condition of equilibrium for an object. An object moving through space is said to be in translational motion. If the object is spinning, it is said to be in rotational motion. A second condition of equilibrium is a statement of rotational equilibrium. This condition will be discussed in Chapter 10 when we discuss spinning objects. Equation 4.7 is sufficient for analyzing particle-like objects in translational motion, which are those of interest to us at this point.

If $\vec{T}$ is a constant force, the acceleration $a_x = T/m$ also is constant. Hence, the crate is also modeled as a particle under constant acceleration in the x direction, and the equations of kinematics from Chapter 2 can be used to obtain the crate's position x and velocity v_x as functions of time.

> ## PROBLEM-SOLVING STRATEGY: Applying Newton's Laws
>
> The following procedure is recommended when dealing with problems involving Newton's laws.
>
> 1. **Conceptualize** Draw a simple, neat diagram of the system to help establish the mental representation. Establish convenient coordinate axes for each object in the system.
>
> 2. **Categorize** If an acceleration component for an object is zero, it is modeled as a particle in equilibrium in this direction and $\Sigma F = 0$. If not, the object is modeled as a particle under a net force in this direction and $\Sigma F = ma$.
>
> 3. **Analyze** Isolate the object whose motion is being analyzed. Draw a free-body diagram for this object. For systems containing more than one object, draw *separate* free-body diagrams for each object. *Do not* include in the free-body diagram forces exerted by the object on its surroundings.
>
>   Find the components of the forces along the coordinate axes. Apply Newton's second law, $\Sigma \vec{F} = m\vec{a}$, in component form. Check your dimensions to make sure all terms have units of force.
>
>   Solve the component equations for the unknowns. Remember that to obtain a complete solution, you generally must have as many independent equations as you have unknowns.
>
> 4. **Finalize** Make sure your results are consistent with the free-body diagram. Also check the predictions of your solutions for extreme values of the variables. By doing so, you can often detect errors in your results.

Example 4.2 | A Traffic Light at Rest

A traffic light weighing 122 N hangs from a cable tied to two other cables fastened to a support as in Figure 4.10a. The upper cables make angles of 37.0° and 53.0° with the horizontal. These upper cables are not as strong as the vertical cable and will break if the tension in them exceeds 100 N. Does the traffic light remain hanging in this situation, or will one of the cables break?

SOLUTION

Conceptualize Inspect the drawing in Figure 4.10a. Let us assume the cables do not break and nothing is moving.

Categorize If nothing is moving, no part of the system is accelerating. We can now model the light as a particle in equilibrium on which the net force is zero. Similarly, the net force on the knot (Fig. 4.10c) is zero.

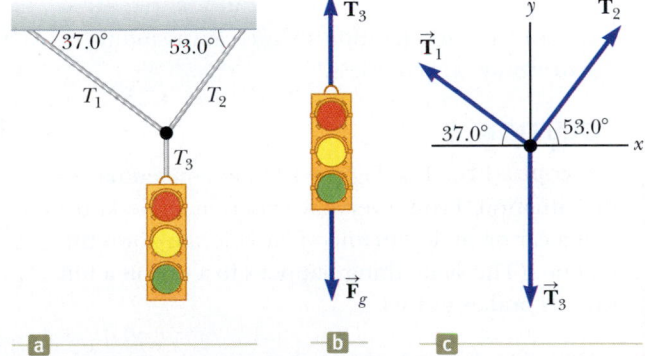

Figure 4.10 (Example 4.2) (a) A traffic light suspended by cables. (b) The forces acting on the traffic light. (c) The free-body diagram for the knot where the three cables are joined.

Analyze We construct a diagram of the forces acting on the traffic light, shown in Figure 4.10b, and a free-body diagram for the knot that holds the three cables together, shown in Figure 4.10c. This knot is a convenient object to choose because all the forces of interest act along lines passing through the knot.

Apply Equation 4.8 for the traffic light in the y direction:
$$\Sigma F_y = 0 \quad \rightarrow \quad T_3 - F_g = 0$$

$$T_3 = F_g = 122 \text{ N}$$

continued

4.2 *cont.*

Choose the coordinate axes as shown in Figure 4.10c and resolve the forces acting on the knot into their components:

Force	x Component	y Component
$\vec{\mathbf{T}}_1$	$-T_1 \cos 37.0°$	$T_1 \sin 37.0°$
$\vec{\mathbf{T}}_2$	$T_2 \cos 53.0°$	$T_2 \sin 53.0°$
$\vec{\mathbf{T}}_3$	0	-122 N

Apply the particle in equilibrium model to the knot:

(1) $\sum F_x = -T_1 \cos 37.0° + T_2 \cos 53.0° = 0$

(2) $\sum F_y = T_1 \sin 37.0° + T_2 \sin 53.0° + (-122 \text{ N}) = 0$

Equation (1) shows that the horizontal components of $\vec{\mathbf{T}}_1$ and $\vec{\mathbf{T}}_2$ must be equal in magnitude, and Equation (2) shows that the sum of the vertical components of $\vec{\mathbf{T}}_1$ and $\vec{\mathbf{T}}_2$ must balance the downward force $\vec{\mathbf{T}}_3$, which is equal in magnitude to the weight of the light.

Solve Equation (1) for T_2 in terms of T_1:

$$T_2 = T_1 \left(\frac{\cos 37.0°}{\cos 53.0°} \right) = 1.33 T_1$$

Substitute this value for T_2 into Equation (2):

$$T_1 \sin 37.0° + (1.33 T_1)(\sin 53.0°) - 122 \text{ N} = 0$$

$$T_1 = 73.4 \text{ N}$$

$$T_2 = 1.33 T_1 = 97.4 \text{ N}$$

Both values are less than 100 N (just barely for T_2), so the cables will not break.

Finalize Imagine changing some of the variables in the problem. What variables could be changed and what would their values have to be so that the cable would break? Suppose the two angles in Figure 4.10a are equal. What would be the relationship between T_1 and T_2?

Example 4.3 | The Runaway Car

A car of mass m is on an icy driveway inclined at an angle θ as in Figure 4.11a.

(A) Find the acceleration of the car, assuming the driveway is frictionless.

SOLUTION

Conceptualize Use Figure 4.11a to conceptualize the situation. From everyday experience, we know that a car on an icy incline will accelerate down the incline. (The same thing happens to a car on a hill with its brakes not set.)

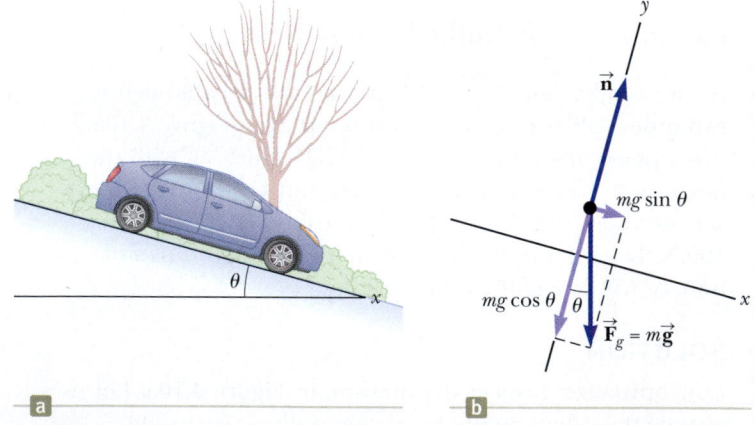

Categorize We categorize the car as a particle under a net force because it accelerates. Furthermore, this example belongs to a very common category of problems in which an object moves under the influence of gravity on an inclined plane.

Figure 4.11 (Example 4.3) (a) A car on a frictionless incline. (b) The free-body diagram for the car. The black dot represents the position of the center of mass of the car. We will learn about center of mass in Chapter 8.

Analyze Figure 4.11b shows the free-body diagram for the car. The only forces acting on the car are the normal force $\vec{\mathbf{n}}$ exerted by the inclined plane, which acts perpendicular to the plane, and the gravitational force $\vec{\mathbf{F}}_g = m\vec{\mathbf{g}}$, which acts vertically downward. For problems involving inclined planes, it is convenient to choose the coordinate axes with x along the incline and y perpendicular to it as in Figure 4.11b. With these axes, we represent the gravitational force by a component of magnitude $mg \sin \theta$ along the positive x axis and one of magnitude $mg \cos \theta$ along the negative y axis. Our choice of axes results in the car being modeled as a particle under a net force in the x direction and a particle in equilibrium in the y direction.

4.3 *cont.*

Apply these models to the car:

(1) $\sum F_x = mg \sin \theta = ma_x$

(2) $\sum F_y = n - mg \cos \theta = 0$

Solve Equation (1) for a_x:

(3) $a_x = \boxed{g \sin \theta}$

Finalize Note that the acceleration component a_x is independent of the mass of the car! It depends only on the angle of inclination and on g.

From Equation (2), we conclude that the component of $\vec{F}_g$ perpendicular to the incline is balanced by the normal force; that is, $n = mg \cos \theta$. This situation is a case in which the normal force is *not* equal in magnitude to the weight of the object (as discussed in Pitfall Prevention 4.7 on page 92).

It is possible, although inconvenient, to solve the problem with "standard" horizontal and vertical axes. You may want to try it, just for practice.

(B) Suppose the car is released from rest at the top of the incline and the distance from the car's front bumper to the bottom of the incline is d. How long does it take the front bumper to reach the bottom of the hill, and what is the car's speed as it arrives there?

SOLUTION

Conceptualize Imagine that the car is sliding down the hill and you use a stopwatch to measure the entire time interval until it reaches the bottom.

Categorize This part of the problem belongs to kinematics rather than to dynamics, and Equation (3) shows that the acceleration a_x is constant. Therefore, you should categorize the car in this part of the problem as a particle under constant acceleration.

Analyze Defining the initial position of the front bumper as $x_i = 0$ and its final position as $x_f = d$, and recognizing that $v_{xi} = 0$, apply Equation 2.13, $x_f = x_i + v_{xi}t + \frac{1}{2}a_x t^2$:

$d = \frac{1}{2}a_x t^2$

Solve for t:

(4) $t = \sqrt{\dfrac{2d}{a_x}} = \boxed{\sqrt{\dfrac{2d}{g \sin \theta}}}$

Use Equation 2.14, with $v_{xi} = 0$, to find the final velocity of the car:

$v_{xf}^2 = 2a_x d$

(5) $v_{xf} = \sqrt{2a_x d} = \boxed{\sqrt{2gd \sin \theta}}$

Finalize We see from Equations (4) and (5) that the time t at which the car reaches the bottom and its final speed v_{xf} are independent of the car's mass, as was its acceleration. Notice that we have combined techniques from Chapter 2 with new techniques from this chapter in this example. As we learn more techniques in later chapters, this process of combining analysis models and information from several parts of the book will occur more often. In these cases, use the General Problem-Solving Strategy to help you identify what analysis models you will need.

What If? What previously solved problem does this situation become if $\theta = 90°$?

Answer Imagine θ going to 90° in Figure 4.11. The inclined plane becomes vertical, and the car is an object in free fall! Equation (3) becomes

$$a_x = g \sin \theta = g \sin 90° = g$$

which is indeed the free-fall acceleration. (We find $a_x = g$ rather than $a_x = -g$ because we have chosen positive x to be downward in Fig. 4.11.) Notice also that the condition $n = mg \cos \theta$ gives us $n = mg \cos 90° = 0$. That is consistent with the car falling downward *next to* the vertical plane, in which case there is no contact force between the car and the plane.

Example 4.4 | **The Atwood Machine**

When two objects of unequal mass are hung vertically over a frictionless pulley of negligible mass as in Active Figure 4.12a, the arrangement is called an *Atwood machine*. The device is sometimes used in the laboratory to determine the value of g. Determine the magnitude of the acceleration of the two objects and the tension in the lightweight string.

SOLUTION

Conceptualize Imagine the situation pictured in Active Figure 4.12a in action: as one object moves upward, the other object moves downward. Because the objects are connected by an inextensible string, their accelerations must be of equal magnitude.

Categorize The objects in the Atwood machine are subject to the gravitational force as well as to the forces exerted by the strings connected to them. Therefore, we can categorize this problem as one involving two particles under a net force.

Active Figure 4.12 (Example 4.4) The Atwood machine. (a) Two objects connected by a massless inextensible string over a frictionless pulley. (b) The free-body diagrams for the two objects.

Analyze The free-body diagrams for the two objects are shown in Active Figure 4.12b. Two forces act on each object: the upward force $\vec{\mathbf{T}}$ exerted by the string and the downward gravitational force. In problems such as this one in which the pulley is modeled as massless and frictionless, the tension in the string on both sides of the pulley is the same. If the pulley has mass or is subject to friction, the tensions on either side are not the same and the situation requires techniques we will learn in Chapter 10.

We must be very careful with signs in problems such as this one. In Active Figure 4.12a, notice that if object 1 accelerates upward, object 2 accelerates downward. Therefore, for consistency with signs, if we define the upward direction as positive for object 1, we must define the downward direction as positive for object 2. With this sign convention, both objects accelerate in the same direction as defined by the choice of sign. Furthermore, according to this sign convention, the y component of the net force exerted on object 1 is $T - m_1 g$, and the y component of the net force exerted on object 2 is $m_2 g - T$.

Apply Newton's second law to object 1:

$$(1) \quad \sum F_y = T - m_1 g = m_1 a_y$$

Apply Newton's second law to object 2:

$$(2) \quad \sum F_y = m_2 g - T = m_2 a_y$$

Add Equation (2) to Equation (1), noticing that T cancels:

$$- m_1 g + m_2 g = m_1 a_y + m_2 a_y$$

Solve for the acceleration:

$$(3) \quad a_y = \left(\frac{m_2 - m_1}{m_1 + m_2} \right) g$$

Substitute Equation (3) into Equation (1) to find T:

$$(4) \quad T = m_1 (g + a_y) = \left(\frac{2 m_1 m_2}{m_1 + m_2} \right) g$$

Finalize The acceleration given by Equation (3) can be interpreted as the ratio of the magnitude of the unbalanced force on the system $(m_2 - m_1)g$ to the total mass of the system $(m_1 + m_2)$, as expected from Newton's second law. Notice that the sign of the acceleration depends on the relative masses of the two objects.

What If? Describe the motion of the system if the objects have equal masses, that is, $m_1 = m_2$.

Answer If we have the same mass on both sides, the system is balanced and should not accelerate. Mathematically, we see that if $m_1 = m_2$, Equation (3) gives us $a_y = 0$.

What If? What if one of the masses is much larger than the other: $m_1 \gg m_2$?

Answer In the case in which one mass is infinitely larger than the other, we can ignore the effect of the smaller mass. Therefore, the larger mass should simply fall as if the smaller mass were not there. We see that if $m_1 \gg m_2$, Equation (3) gives us $a_y = -g$.

Example 4.5 | One Block Pushes Another

Two blocks of masses m_1 and m_2, with $m_1 > m_2$, are placed in contact with each other on a frictionless, horizontal surface as in Active Figure 4.13a. A constant horizontal force $\vec{F}$ is applied to m_1 as shown.

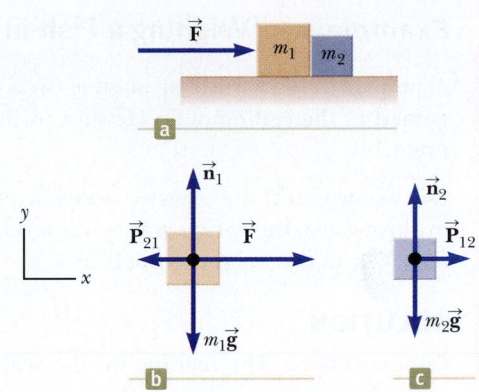

(A) Find the magnitude of the acceleration of the system.

SOLUTION

Conceptualize Conceptualize the situation by using Active Figure 4.13a and realize that both blocks must experience the *same* acceleration because they are in contact with each other and remain in contact throughout the motion.

Categorize We categorize this problem as one involving a particle under a net force because a force is applied to a system of blocks and we are looking for the acceleration of the system.

Active Figure 4.13 (Example 4.5) (a) A force is applied to a block of mass m_1, which pushes on a second block of mass m_2. (b) The forces acting on m_1. (c) The forces acting on m_2.

Analyze First model the combination of two blocks as a single particle under a net force. Apply Newton's second law to the combination in the x direction to find the acceleration:

$$\sum F_x = F = (m_1 + m_2)a_x$$

$$(1)\quad a_x = \frac{F}{m_1 + m_2}$$

Finalize The acceleration given by Equation (1) is the same as that of a single object of mass $m_1 + m_2$ and subject to the same force.

(B) Determine the magnitude of the contact force between the two blocks.

SOLUTION

Conceptualize The contact force is internal to the system of two blocks. Therefore, we cannot find this force by modeling the whole system (the two blocks) as a single particle.

Categorize Now consider each of the two blocks individually by categorizing each as a particle under a net force.

Analyze We construct a diagram of forces acting on the object for each block as shown in Active Figures 4.13b and 4.13c, where the contact force is denoted by $\vec{P}$. From Active Figure 4.13c, we see that the only horizontal force acting on m_2 is the contact force $\vec{P}_{12}$ (the force exerted by m_1 on m_2), which is directed to the right.

Apply Newton's second law to m_2:

$$(2)\quad \sum F_x = P_{12} = m_2 a_x$$

Substitute the value of the acceleration a_x given by Equation (1) into Equation (2):

$$(3)\quad P_{12} = m_2 a_x = \left(\frac{m_2}{m_1 + m_2}\right)F$$

Finalize This result shows that the contact force P_{12} is *less* than the applied force F. The force required to accelerate block 2 alone must be less than the force required to produce the same acceleration for the two-block system.

To finalize further, let us check this expression for P_{12} by considering the forces acting on m_1, shown in Active Figure 4.13b. The horizontal forces acting on m_1 are the applied force $\vec{F}$ to the right and the contact force $\vec{P}_{21}$ to the left (the force exerted by m_2 on m_1). From Newton's third law, $\vec{P}_{21}$ is the reaction force to $\vec{P}_{12}$, so $P_{21} = P_{12}$.

Apply Newton's second law to m_1:

$$(4)\quad \sum F_x = F - P_{21} = F - P_{12} = m_1 a_x$$

Solve for P_{12} and substitute the value of a_x from Equation (1):

$$P_{12} = F - m_1 a_x = F - m_1\left(\frac{F}{m_1 + m_2}\right) = \left(\frac{m_2}{m_1 + m_2}\right)F$$

This result agrees with Equation (3), as it must.

What If? Imagine that the force $\vec{F}$ in Active Figure 4.13 is applied toward the left on the right-hand block of mass m_2. Is the magnitude of the force $\vec{P}_{12}$ the same as it was when the force was applied toward the right on m_1?

Answer When the force is applied toward the left on m_2, the contact force must accelerate m_1. In the original situation, the contact force accelerates m_2. Because $m_1 > m_2$, more force is required, so the magnitude of $\vec{P}_{12}$ is greater than in the original situation.

Example 4.6 | Weighing a Fish in an Elevator

A person weighs a fish of mass m on a spring scale attached to the ceiling of an elevator as illustrated in Figure 4.14.

(A) Show that if the elevator accelerates either upward or downward, the spring scale gives a reading that is different from the weight of the fish.

SOLUTION

Conceptualize The reading on the scale is related to the extension of the spring in the scale, which is related to the force on the end of the spring as in Figure 4.2. Imagine that the fish is hanging on a string attached to the end of the spring. In this case, the magnitude of the force exerted on the spring is equal to the tension T in the string. Therefore, we are looking for T. The force $\vec{T}$ pulls down on the string and pulls up on the fish.

Categorize We can categorize this problem by identifying the fish as a particle under a net force.

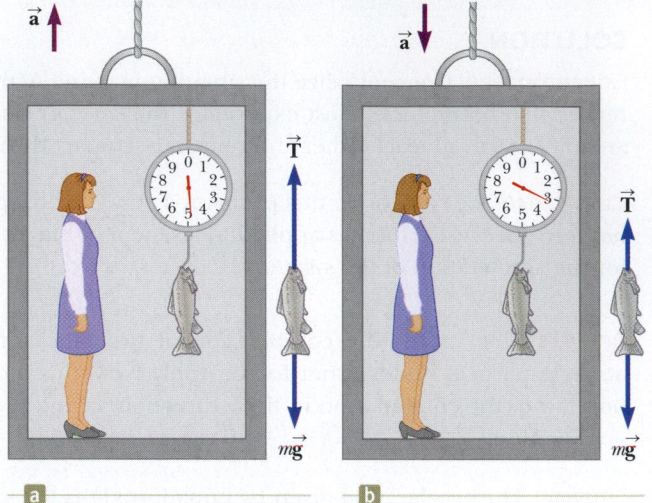

When the elevator accelerates upward, the spring scale reads a value greater than the weight of the fish.

When the elevator accelerates downward, the spring scale reads a value less than the weight of the fish.

Figure 4.14 (Example 4.6) A fish is weighed on a spring scale in an accelerating elevator car.

Analyze Inspect the diagrams of the forces acting on the fish in Figure 4.14 and notice that the external forces acting on the fish are the downward gravitational force $\vec{F}_g = m\vec{g}$ and the force $\vec{T}$ exerted by the string. If the elevator is either at rest or moving at constant velocity, the fish is a particle in equilibrium, so $\Sigma F_y = T - F_g = 0$ or $T = F_g = mg$. (Remember that the scalar mg is the weight of the fish.)

Now suppose the elevator is moving with an acceleration $\vec{a}$ relative to an observer standing outside the elevator in an inertial frame. The fish is now a particle under a net force.

Apply Newton's second law to the fish:

$$\sum F_y = T - mg = ma_y$$

Solve for T:

$$(1)\ T = ma_y + mg = mg\left(\frac{a_y}{g} + 1\right) = F_g\left(\frac{a_y}{g} + 1\right)$$

where we have chosen upward as the positive y direction. We conclude from Equation (1) that the scale reading T is greater than the fish's weight mg if $\vec{a}$ is upward, so a_y is positive (Fig. 4.14a), and that the reading is less than mg if $\vec{a}$ is downward, so a_y is negative (Fig. 4.14b).

(B) Evaluate the scale readings for a 40.0-N fish if the elevator moves with an acceleration $a_y = \pm 2.00$ m/s^2.

SOLUTION

Evaluate the scale reading from Equation (1) if $\vec{a}$ is upward:

$$T = (40.0\ \text{N})\left(\frac{2.00\ \text{m/s}^2}{9.80\ \text{m/s}^2} + 1\right) = 48.2\ \text{N}$$

Evaluate the scale reading from Equation (1) if $\vec{a}$ is downward:

$$T = (40.0\ \text{N})\left(\frac{-2.00\ \text{m/s}^2}{9.80\ \text{m/s}^2} + 1\right) = 31.8\ \text{N}$$

Finalize Take this advice: if you buy a fish in an elevator, make sure the fish is weighed while the elevator is either at rest or accelerating downward! Furthermore, notice that from the information given here, one cannot determine the direction of motion of the elevator.

What If? Suppose the elevator cable breaks and the elevator and its contents are in free fall. What happens to the reading on the scale?

Answer If the elevator falls freely, its acceleration is $a_y = -g$. We see from Equation (1) that the scale reading T is zero in this case; that is, the fish *appears* to be weightless.

❰ 4.8 | Context Connection: Forces on Automobiles

In the Context Connections of Chapters 2 and 3, we focused on two types of acceleration exhibited by a number of vehicles. In this chapter, we learned how the acceleration of an object is related to the force on the object. Let us apply this understanding to an investigation of the forces that are applied to automobiles when they are exhibiting their maximum acceleration in speeding up from rest to 60 mi/h.

The force that accelerates an automobile is the friction force from the ground. (We will study friction forces in detail in Chapter 5.) The engine applies a force to the wheels, attempting to rotate them so that the bottoms of the tires apply forces backward on the road surface. By Newton's third law, the road surface applies forces in the forward direction on the tires, causing the car to move forward. If we ignore air resistance, this force can be modeled as the net force on the automobile in the horizontal direction.

In Chapter 2, we investigated the 0 to 60 mi/h acceleration of a number of vehicles. Table 4.2 repeats this acceleration information and also shows the weight of the vehicle in pounds and the mass in kilograms. With both the acceleration and the mass, we can find the force driving the car forward, as shown in the last column of Table 4.2.

❰ TABLE 4.2 | Driving Forces on Various Vehicles

Automobile	Model Year	Average Acceleration (mi/h·s)	Weight (lb)	Mass (kg)	Force ($\times 10^3$ N)
Very expensive vehicles:					
Bugatti Veyron 16.4 Super Sport	2011	23.1	4160	1887	19.5
Lamborghini LP 570-4 Superleggera	2011	17.6	2954	1340	10.5
Lexus LFA	2011	15.8	3580	1624	11.5
Mercedes-Benz SLS AMG	2011	16.7	3795	1721	12.8
Shelby SuperCars Ultimate Aero	2009	22.2	2750	1247	12.4
Average		**19.1**	**3448**	**1564**	**13.3**
Performance vehicles:					
Chevrolet Corvette ZR1	2010	18.2	3333	1512	12.3
Dodge Viper SRT10	2010	15.0	3460	1569	10.5
Jaguar XJL Supercharged	2011	13.6	4323	1961	11.9
Acura TL SH-AWD	2009	11.5	3860	1751	9.0
Dodge Challenger SRT8	2010	12.2	4140	1878	10.2
Average		**14.1**	**3823**	**1734**	**10.8**
Traditional vehicles:					
Buick Regal CXL Turbo	2011	8.0	3671	1665	6.0
Chevrolet Tahoe 1500 LS (SUV)	2011	7.0	5636	2556	8.0
Ford Fiesta SES	2010	6.2	2330	1057	2.9
Hummer H3 (SUV)	2010	7.5	4695	2130	7.1
Hyundai Sonata SE	2010	8.0	3340	1515	5.4
Smart ForTwo	2010	4.5	1825	828	1.7
Average		**6.9**	**3583**	**1625**	**5.2**
Alternative vehicles:					
Chevrolet Volt (hybrid)	2011	7.5	3500	1588	5.3
Nissan Leaf (electric)	2011	6.0	3500	1588	4.3
Honda CR-Z (hybrid)	2011	5.7	2637	1196	3.0
Honda Insight (hybrid)	2010	5.7	2723	1235	3.2
Toyota Prius (hybrid)	2010	6.1	3042	1380	3.8
Average		**6.2**	**3080**	**1397**	**3.9**

We can see some interesting results in Table 4.2. The forces in the very expensive and performance vehicle sections are all large compared with the forces in the other parts of the table. Furthermore, the average masses of very expensive and performance vehicles are less than 10% larger than those in the traditional vehicle portion of the table. Therefore, the large forces of the very expensive and performance vehicles translate into the very large accelerations exhibited by these vehicles. One standout in the very expensive vehicles is the Bugatti Veyron 16.4 Super Sport. It is the most massive car in the group, but the huge force generated on the tires results in its having the highest acceleration in the group. The second-highest acceleration in the group is the Shelby SuperCars Ultimate Aero. This vehicle has only 66% of the mass of the Bugatti, representing much less resistance to being accelerated. The force on the Shelby, however, is only 64% of that on the Bugatti, resulting in a lower acceleration of the Shelby, despite its smaller mass.

As expected, the forces exerted on the traditional vehicles are smaller than those of the very expensive and performance vehicles, corresponding to the smaller accelerations of this group. Notice, however, that the forces for the two SUVs are large. Because these two vehicles have accelerations that are somewhat similar to those of the other vehicles in this portion of the table, we can identify these large forces as being required to accelerate the larger mass of the SUVs.

Also as expected, the forces driving the alternative vehicles have the lowest average in the table. This finding is consistent with the accelerations of these vehicles being lower than those elsewhere in the table.

Another interesting entry in the table is the Smart ForTwo in the traditional vehicles. Its force is by far the smallest in the table, but its mass is also the smallest in the table. As a result, its acceleration is 4.5 mi/h · s, which, although not impressive, is enough to satisfy some consumers who are looking for other advantages offered by the Smart car, such as increased fuel efficiency.

❯ SUMMARY |

Newton's first law states that if an object does not interact with other objects, it is possible to identify a reference frame in which the object has zero acceleration. Therefore, if we observe an object from such a frame and no force is exerted on the object, an object at rest remains at rest and an object in uniform motion in a straight line maintains that motion.

Newton's first law defines an **inertial frame of reference,** which is a frame in which Newton's first law is valid.

Newton's second law states that the acceleration of an object is directly proportional to the net force acting on the object and inversely proportional to the object's mass.

Newton's third law states that if two objects interact, the force exerted by object 1 on object 2 is equal in magnitude but opposite in direction to the force exerted by object 2 on object 1. Therefore, an isolated force cannot exist in nature.

The **weight** of an object is equal to the product of its mass (a scalar quantity) and the magnitude of the free-fall acceleration, or

$$F_g = mg \qquad\qquad \textbf{4.5} \blacktriangleleft$$

❯ Analysis Models for Problem Solving |

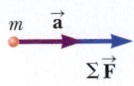

Particle Under a Net Force If a particle of mass m experiences a nonzero net force, its acceleration is related to the net force by Newton's second law:

$$\sum \vec{\mathbf{F}} = m\vec{\mathbf{a}} \qquad\qquad \textbf{4.2} \blacktriangleleft$$

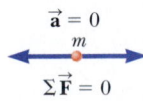

Particle in Equilibrium If a particle maintains a constant velocity (so that $\vec{\mathbf{a}} = 0$), which could include a velocity of zero, the forces on the particle balance and Newton's second law reduces to

$$\sum \vec{\mathbf{F}} = 0 \qquad\qquad \textbf{4.8} \blacktriangleleft$$

OBJECTIVE QUESTIONS

1. The third graders are on one side of a schoolyard, and the fourth graders are on the other. They are throwing snowballs at each other. Between them, snowballs of various masses are moving with different velocities as shown in Figure OQ4.1. Rank the snowballs (a) through (e) according to the magnitude of the total force exerted on each one. Ignore air resistance. If two snowballs rank together, make that fact clear.

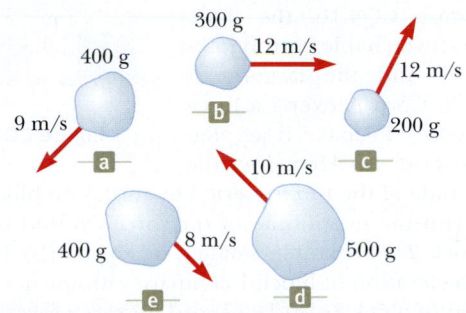

Figure OQ4.1

2. In Figure OQ4.2, a locomotive has broken through the wall of a train station. During the collision, what can be said about the force exerted by the locomotive on the wall? (a) The force exerted by the locomotive on the wall was larger than the force the wall could exert on the locomotive. (b) The force exerted by the locomotive on the wall was the same in magnitude as the force exerted by the wall on the locomotive. (c) The force exerted by the

Figure OQ4.2

locomotive on the wall was less than the force exerted by the wall on the locomotive. (d) The wall cannot be said to "exert" a force; after all, it broke.

3. An experiment is performed on a puck on a level air hockey table, where friction is negligible. A constant horizontal force is applied to the puck, and the puck's acceleration is measured. Now the same puck is transported far into outer space, where both friction and gravity are negligible. The same constant force is applied to the puck (through a spring scale that stretches the same amount), and the puck's acceleration (relative to the distant stars) is measured. What is the puck's acceleration in outer space? (a) It is somewhat greater than its acceleration on the Earth. (b) It is the same as its acceleration on the Earth. (c) It is less than its acceleration on the Earth. (d) It is infinite because neither friction nor gravity constrains it. (e) It is very large because acceleration is inversely proportional to weight and the puck's weight is very small but not zero.

4. A truck loaded with sand accelerates along a highway. The driving force on the truck remains constant. What happens to the acceleration of the truck if its trailer leaks sand at a constant rate through a hole in its bottom? (a) It decreases at a steady rate. (b) It increases at a steady rate. (c) It increases and then decreases. (d) It decreases and then increases. (e) It remains constant.

5. If an object is in equilibrium, which of the following statements is *not* true? (a) The speed of the object remains constant. (b) The acceleration of the object is zero. (c) The net force acting on the object is zero. (d) The object must be at rest. (e) There are at least two forces acting on the object.

6. Two objects are connected by a string that passes over a frictionless pulley as in Active Figure 4.12a, where $m_1 < m_2$ and a_1 and a_2 are the magnitudes of the respective accelerations. Which mathematical statement is true regarding the magnitude of the acceleration a_2 of the mass m_2? (a) $a_2 < g$ (b) $a_2 > g$ (c) $a_2 = g$ (d) $a_2 < a_1$ (e) $a_2 > a_1$

CONCEPTUAL QUESTIONS

1. A passenger sitting in the rear of a bus claims that she was injured when the driver slammed on the brakes, causing a suitcase to come flying toward her from the front of the bus. If you were the judge in this case, what disposition would you make? Why?

2. If a car is traveling due westward with a constant speed of 20 m/s, what is the resultant force acting on it?

3. A person holds a ball in her hand. (a) Identify all the external forces acting on the ball and the Newton's third-law reaction force to each one. (b) If the ball is dropped, what

force is exerted on it while it is falling? Identify the reaction force in this case. (Ignore air resistance.)

4. In the motion picture *It Happened One Night* (Columbia Pictures, 1934), Clark Gable is standing inside a stationary bus in front of Claudette Colbert, who is seated. The bus suddenly starts moving forward and Clark falls into Claudette's lap. Why did this happen?

5. If you hold a horizontal metal bar several centimeters above the ground and move it through grass, each leaf of grass bends out of the way. If you increase the speed

of the bar, each leaf of grass will bend more quickly. How then does a rotary power lawn mower manage to cut grass? How can it exert enough force on a leaf of grass to shear it off?

6. A spherical rubber balloon inflated with air is held stationary, with its opening, on the west side, pinched shut. (a) Describe the forces exerted by the air inside and outside the balloon on sections of the rubber. (b) After the balloon is released, it takes off toward the east, gaining speed rapidly. Explain this motion in terms of the forces now acting on the rubber. (c) Account for the motion of a skyrocket taking off from its launch pad.

7. A rubber ball is dropped onto the floor. What force causes the ball to bounce?

8. The mayor of a city reprimands some city employees because they will not remove the obvious sags from the cables that support the city traffic lights. What explanation can the employees give? How do you think the case will be settled in mediation?

9. Balancing carefully, three boys inch out onto a horizontal tree branch above a pond, each planning to dive in separately. The third boy in line notices that the branch is barely strong enough to support them. He decides to jump straight up and land back on the branch to break it, spilling all three into the pond. When he starts to carry out his plan, at what precise moment does the branch break? Explain. *Suggestion:* Pretend to be the third boy and imitate what he does in slow motion. If you are still unsure, stand on a bathroom scale and repeat the suggestion.

10. A child tosses a ball straight up. She says that the ball is moving away from her hand because the ball feels an upward "force of the throw" as well as the gravitational force. (a) Can the "force of the throw" exceed the gravitational force? How would the ball move if it did? (b) Can the "force of the throw" be equal in magnitude to the gravitational force? Explain. (c) What strength can accurately be attributed to the "force of the throw"? Explain. (d) Why does the ball move away from the child's hand?

11. A weightlifter stands on a bathroom scale. He pumps a barbell up and down. What happens to the reading on the scale as he does so? **What If?** What if he is strong enough to actually *throw* the barbell upward? How does the reading on the scale vary now?

12. When you push on a box with a 200-N force instead of a 50-N force, you can feel that you are making a greater effort. When a table exerts a 200-N normal force instead of one of smaller magnitude, is the table really doing anything differently?

13. Identify action–reaction pairs in the following situations: (a) a man takes a step (b) a snowball hits a girl in the back (c) a baseball player catches a ball (d) a gust of wind strikes a window

14. Give reasons for the answers to each of the following questions: (a) Can a normal force be horizontal? (b) Can a normal force be directed vertically downward? (c) Consider a tennis ball in contact with a stationary floor and with nothing else. Can the normal force be different in magnitude from the gravitational force exerted on the

ball? (d) Can the force exerted by the floor on the ball be different in magnitude from the force the ball exerts on the floor?

15. An athlete grips a light rope that passes over a low-friction pulley attached to the ceiling of a gym. A sack of sand precisely equal in weight to the athlete is tied to the other end of the rope. Both the sand and the athlete are initially at rest. The athlete climbs the rope, sometimes speeding up and slowing down as he does so. What happens to the sack of sand? Explain.

16. In Figure CQ4.16, the light, taut, unstretchable cord B joins block 1 and the larger-mass block 2. Cord A exerts a force on block 1 to make it accelerate forward. (a) How does the magnitude of the force exerted by cord A on block 1 compare with the magnitude of the force exerted by cord B on block 2? Is it larger, smaller, or equal? (b) How does the acceleration of block 1 compare with the acceleration (if any) of block 2? (c) Does cord B exert a force on block 1? If so, is it forward or backward? Is it larger, smaller, or equal in magnitude to the force exerted by cord B on block 2?

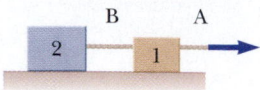

Figure CQ4.16

17. Twenty people participate in a tug-of-war. The two teams of ten people are so evenly matched that neither team wins. After the game they notice that a car is stuck in the mud. They attach the tug-of-war rope to the bumper of the car, and all the people pull on the rope. The heavy car has just moved a couple of decimeters when the rope breaks. Why did the rope break in this situation when it did not break when the same twenty people pulled on it in a tug-of-war?

18. Can an object exert a force on itself? Argue for your answer.

19. As shown in Figure CQ4.19, student A, a 55-kg girl, sits on one chair with metal runners, at rest on a classroom floor. Student B, an 80-kg boy, sits on an identical chair. Both students keep their feet off the floor. A rope runs from student A's hands around a light pulley and then over her shoulder to the hands of a teacher standing on the floor behind her. The low-friction axle of the pulley is attached to a second rope held by student B. All ropes run parallel to the chair runners. (a) If student A pulls on her end of the rope, will her chair or will B's chair slide on the floor? Explain why. (b) If instead the teacher pulls on his rope end, which chair slides? Why this one? (c) If student B pulls on his rope, which chair slides? Why? (d) Now the teacher ties his end of the rope to student A's chair. Student A pulls on the end of the rope in her hands. Which chair slides and why?

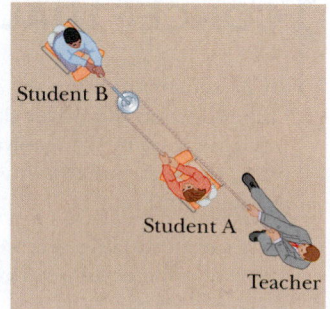

Figure CQ4.19

▶ PROBLEMS AVAILABLE IN ⬯ENHANCED Web**Assign**

4.3 Mass

Problems 1–2

4.4 Newton's Second Law

Problems 3–8

4.5 The Gravitational Force and Weight

Problems 9–14

4.6 Newton's Third Law

Problems 15–17

4.7 Analysis Models Using Newton's Second Law

Problems 18–40

4.8 Context Connection: Forces on Automobiles

Problems 41–42

Additional Problems

Problems 43–59

Solutions to the following Problems are available in the
Student Solutions Manual/Study Guide:

4.3, 4.7, 4.11, 4.24, 4.25, 4.29, 4.31, 4.37, 4.43, 4.45, 4.51, and 4.57

List of Enhanced Problems

Problem Number	Targeted Feedback in Enhanced WebAssign	Master It in Enhanced WebAssign	Watch It in Enhanced WebAssign
4.1	✓		✓
4.3	✓	✓	
4.5	✓		✓
4.6	✓		✓
4.7	✓	✓	
4.8	✓		✓
4.11		✓	
4.18	✓		✓
4.22	✓		✓
4.23	✓		✓
4.25		✓	
4.28	✓		✓
4.37		✓	
4.57		✓	

Chapter 5

More Applications of Newton's Laws

Chapter Outline

Chris Graythen/Getty Images

In Chapter 4, we introduced Newton's laws of motion and applied them to situations in which we ignored friction. In this chapter, we shall expand our investigation to objects moving in the presence of friction, which will allow us to model situations more realistically. Such objects include those sliding on rough surfaces and those moving through viscous media such as liquids and air. We also apply Newton's laws to the dynamics of circular motion so that we can understand more about objects moving in circular paths under the influence of various types of forces.

Kyle Busch, driver of the #18 Snickers Toyota, leads Jeff Gordon, driver of the #24 Dupont Chevrolet, during the NASCAR Sprint Cup Series Kobalt Tools 500 at the Atlanta Motor Speedway on March 9, 2008, in Hampton, Georgia. The cars travel on a banked roadway to help them undergo circular motion on the turns.

5.1 | Forces of Friction

When an object moves either on a surface or through a viscous medium such as air or water, there is resistance to the motion because the object interacts with its surroundings. We call such resistance a **force of friction.** Forces of friction are very important in our everyday lives. They allow us to walk or run and are necessary for the motion of wheeled vehicles.

Imagine you are working in your garden and have filled a trash can with yard clippings. You then try to drag the trash can across the surface of your concrete patio as in Active Figure 5.1a (page 108). The patio surface is *real*, not an idealized, frictionless surface in a simplification model. If we apply an

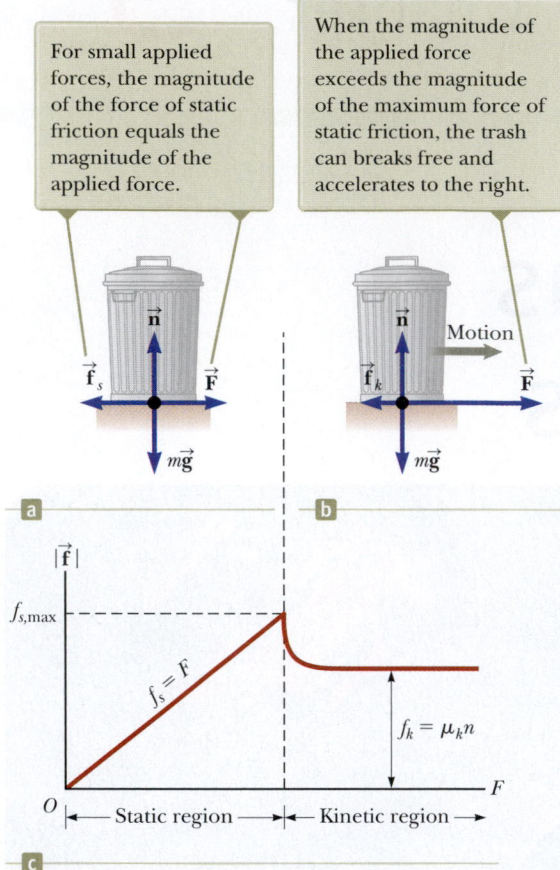

For small applied forces, the magnitude of the force of static friction equals the magnitude of the applied force.

When the magnitude of the applied force exceeds the magnitude of the maximum force of static friction, the trash can breaks free and accelerates to the right.

Active Figure 5.1 (a) and (b) When pulling on a trash can, the direction of the force of friction $\vec{\mathbf{f}}$ between the can and a rough surface is opposite the direction of the applied force $\vec{\mathbf{F}}$. (c) A graph of friction force versus applied force. Notice that $f_{s,\text{max}} > f_k$.

external horizontal force $\vec{\mathbf{F}}$ to the trash can, acting to the right, the trash can remains stationary if $\vec{\mathbf{F}}$ is small. The force that counteracts $\vec{\mathbf{F}}$ and keeps the trash can from moving is applied at the base of the can by the surface and acts to the left. It is called the **force of static friction** $\vec{\mathbf{f}}_s$. As long as the trash can is not moving, it is modeled as a particle in equilibrium and $f_s = F$. Therefore, if $\vec{\mathbf{F}}$ is increased in magnitude, the magnitude of $\vec{\mathbf{f}}_s$ also increases. Likewise, if $\vec{\mathbf{F}}$ decreases, $\vec{\mathbf{f}}_s$ also decreases.

Experiments show that the friction force arises from the nature of the two surfaces; because of their roughness, contact is made only at a few locations where peaks of the material touch. At these locations, the friction force arises in part because one peak physically blocks the motion of a peak from the opposing surface and in part from chemical bonding ("spot welds") of opposing peaks as they come into contact. Although the details of friction are quite complex at the atomic level, this force ultimately involves an electrical interaction between atoms or molecules.

If we increase the magnitude of $\vec{\mathbf{F}}$, as in Active Figure 5.1b, the trash can eventually slips. When the trash can is on the verge of slipping, f_s is a maximum as shown in Active Figure 5.1c. If F exceeds $f_{s,\text{max}}$, the trash can moves and accelerates to the right. While the trash can is in motion, the friction force is less than $f_{s,\text{max}}$ (Active Fig. 5.1c). We call the friction force for an object in motion the **force of kinetic friction** $\vec{\mathbf{f}}_k$. The net force $F - f_k$ in the x direction produces an acceleration to the right, according to Newton's second law. If we reduce the magnitude of $\vec{\mathbf{F}}$ so that $F = f_k$, the acceleration is zero and the trash can moves to the right with constant speed. If the applied force is removed, the friction force acting to the left provides an acceleration of the trash can in the $-x$ direction and eventually brings it to rest.

Experimentally, one finds that, to a good approximation, both $f_{s,\text{max}}$ and f_k for an object on a surface are proportional to the normal force exerted by the surface on the object; therefore, we adopt a simplification model in which this approximation is assumed to be exact. The assumptions in this simplification model can be summarized as follows:

- The magnitude of the force of static friction between any two surfaces in contact can have the values

▶ Force of static friction

$$f_s \leq \mu_s n \qquad\qquad \textbf{5.1}◀$$

where the dimensionless constant μ_s is called the **coefficient of static friction** and n is the magnitude of the normal force. The equality in Equation 5.1 holds when the surfaces are on the verge of slipping, that is, when $f_s = f_{s,\text{max}} \equiv \mu_s n$. This situation is called *impending motion*. The inequality holds when the component of the applied force parallel to the surfaces is less than this value.

- The magnitude of the force of kinetic friction acting between two surfaces is

▶ Force of kinetic friction

$$f_k = \mu_k n \qquad\qquad \textbf{5.2}◀$$

where μ_k is the **coefficient of kinetic friction.** In our simplification model, this coefficient is independent of the relative speed of the surfaces.

- The values of μ_k and μ_s depend on the nature of the surfaces, but μ_k is generally less than μ_s. Table 5.1 lists some measured values.

- The direction of the friction force on an object is opposite to the actual motion (kinetic friction) or the impending motion (static friction) of the object relative to the surface with which it is in contact.

The approximate nature of Equations 5.1 and 5.2 is easily demonstrated by trying to arrange for an object to slide down an incline at constant speed. Especially at low speeds, the motion is likely to be characterized by alternate stick and slip episodes.

Pitfall Prevention | 5.1
The Equal Sign Is Used in Limited Situations
In Equation 5.1, the equal sign is used *only* when the surfaces are just about to break free and begin sliding. Do not fall into the common trap of using $f_s = \mu_s n$ in *any* static situation.

TABLE 5.1 | Coefficients of Friction

	μ_s	μ_k
Rubber on concrete	1.0	0.8
Steel on steel	0.74	0.57
Aluminium on steel	0.61	0.47
Glass on glass	0.94	0.4
Copper on steel	0.53	0.36
Wood on wood	0.25–0.5	0.2
Waxed wood on wet snow	0.14	0.1
Waxed wood on dry snow	—	0.04
Metal on metal (lubricated)	0.15	0.06
Teflon on Teflon	0.04	0.04
Ice on ice	0.1	0.03
Synovial joints in humans	0.01	0.003

Note: All values are approximate. In some cases, the coefficient of friction can exceed 1.0.

The simplification model described in the bulleted list above has been developed so that we can solve problems involving friction in a relatively straightforward way.

Now that we have identified the characteristics of the friction force, we can include the friction force in the net force on an object in the model of a particle under a net force.

QUICK QUIZ 5.1 You press your physics textbook flat against a vertical wall with your hand, which applies a normal force perpendicular to the book. What is the direction of the friction force on the book due to the wall? (**a**) downward (**b**) upward (**c**) out from the wall (**d**) into the wall

QUICK QUIZ 5.2 A crate is located in the center of a flatbed truck. The truck accelerates to the east and the crate moves with it, not sliding at all. What is the direction of the friction force exerted by the truck on the crate? (**a**) It is to the west. (**b**) It is to the east. (**c**) No friction force exists because the crate is not sliding.

QUICK QUIZ 5.3 You are playing with your daughter in the snow. She sits on a sled and asks you to slide her across a flat, horizontal field. You have a choice of (**a**) pushing her from behind, by applying a force downward on her shoulders at 30° below the horizontal (Fig. 5.2a) or (**b**) attaching a rope to the front of the sled and pulling with a force at 30° above the horizontal (Fig 5.2b). Which would be easier for you and why?

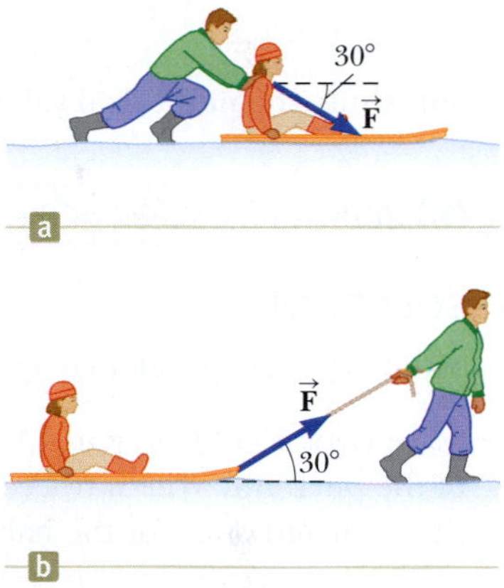

Figure 5.2 (Quick Quiz 5.3) A father tries to slide his daughter on a sled over snow by (a) pushing downward on her shoulders or (b) pulling upward on a rope attached to the sled. Which is easier?

Example 5.1 | The Sliding Hockey Puck

A hockey puck on a frozen pond is given an initial speed of 20.0 m/s.

(A) If the puck always remains on the ice and slides 115 m before coming to rest, determine the coefficient of kinetic friction between the puck and ice.

SOLUTION

Conceptualize Imagine that the puck in Figure 5.3 slides to the right and eventually comes to rest due to the force of kinetic friction.

Categorize The forces acting on the puck are identified in Figure 5.3, but the text of the problem provides kinematic variables. Therefore, we categorize the problem in two ways. First, it involves a particle under a net force: kinetic friction causes the puck to accelerate. Furthermore, because we model the force of kinetic friction as independent of speed, the acceleration of the puck is constant. So, we can also categorize this problem as one involving a particle under constant acceleration.

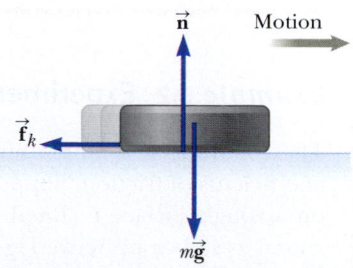

Figure 5.3 (Example 5.1) After the puck is given an initial velocity to the right, the only external forces acting on it are the gravitational force $m\vec{g}$, the normal force $\vec{n}$, and the force of kinetic friction $\vec{f}_k$.

continued

5.1 *cont.*

....

Analyze First, let's find the acceleration algebraically in terms of the coefficient of kinetic friction, using Newton's second law. Once we know the acceleration of the puck and the distance it travels, the equations of kinematics can be used to find the numerical value of the coefficient of kinetic friction. The diagram in Figure 5.3 shows the forces on the puck.

Apply the particle under a net force model in the *x* direction to the puck:

(1) $\sum F_x = -f_k = ma_x$

Apply the particle in equilibrium model in the *y* direction to the puck:

(2) $\sum F_y = n - mg = 0$

Substitute $n = mg$ from Equation (2) and $f_k = \mu_k n$ into Equation (1):

$-\mu_k n = -\mu_k mg = ma_x$
$a_x = -\mu_k g$

The negative sign means the acceleration is to the left in Figure 5.3. Because the velocity of the puck is to the right, the puck is slowing down. The acceleration is independent of the mass of the puck and is constant because we assume μ_k remains constant.

Apply the particle under constant acceleration model to the puck, using Equation 2.14, $v_{xf}^2 = v_{xi}^2 + 2a_x(x_f - x_i)$, with $x_i = 0$ and $v_f = 0$:

$0 = v_{xi}^2 + 2a_x x_f = v_{xi}^2 - 2\mu_k g x_f$

Solve for the coefficient of kinetic friction:

(3) $\mu_k = \dfrac{v_{xi}^2}{2gx_f}$

Substitute the numerical values:

$\mu_k = \dfrac{(20.0 \text{ m/s})^2}{2(9.80 \text{ m/s}^2)(115 \text{ m})} = \boxed{0.177}$

(B) If the initial speed of the puck is halved, what would be the sliding distance?

SOLUTION

This part of the problem is a comparison problem and can be solved by a ratio technique such as that used in Example 3.4.

Solve Equation (3) in part (A) for the final position x_f of the puck and write it twice, once for the original situation and once for the halved initial velocity:

$x_{f1} = \dfrac{v_{1xi}^2}{2\mu_k g}$

$x_{f2} = \dfrac{v_{2xi}^2}{2\mu_k g} = \dfrac{\left(\frac{1}{2}v_{1xi}\right)^2}{2\mu_k g} = \dfrac{1}{4}\dfrac{v_{1xi}^2}{2\mu_k g}$

Divide the first equation by the second:

$\dfrac{x_{f1}}{x_{f2}} = 4 \quad \rightarrow \quad x_{f2} = \dfrac{1}{4}x_{f1}$

....

Finalize Notice in part (A) that μ_k is dimensionless, as it should be, and that it has a low value, consistent with an object sliding on ice. We learn in part (B) that halving the initial velocity of the puck reduces the sliding distance by 75%! Applying this idea to a sliding vehicle, we see that driving at low speeds on slippery roads is an important safety consideration.

Example **5.2** | **Experimental Determination of μ_s and μ_k**

The following is a simple method of measuring coefficients of friction. Suppose a block is placed on a rough surface inclined relative to the horizontal as shown in Active Figure 5.4. The incline angle is increased until the block starts to move. Show that you can obtain μ_s by measuring the critical angle θ_c at which this slipping just occurs.

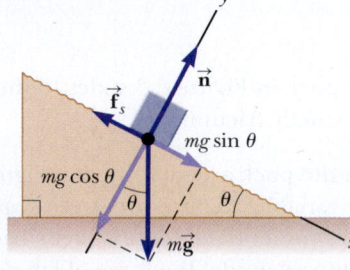

Active Figure 5.4 (Example 5.2) The external forces exerted on a block lying on a rough incline are the gravitational force $m\vec{g}$, the normal force $\vec{n}$, and the force of friction $\vec{f}_s$. For convenience, the gravitational force is resolved into a component $mg \sin \theta$ along the incline and a component $mg \cos \theta$ perpendicular to the incline.

SOLUTION

Conceptualize Consider Active Figure 5.4 and imagine that the block tends to slide down the incline due to the gravitational force. To simulate the situation, place a coin on this book's cover and tilt the book until the coin begins to slide.

5.2 *cont.*

Notice how this example differs from Example 4.3. When there is no friction on an incline, *any* angle of the incline will cause a stationary object to begin moving. When there is friction, however, there is no movement of the object for angles less than the critical angle.

Categorize   The block is subject to various forces. Because we are raising the plane to the angle at which the block is just ready to begin to move but is not moving, we categorize the block as a particle in equilibrium.

Analyze   The diagram in Active Figure 5.4 shows the forces on the block: the gravitational force $m\vec{\mathbf{g}}$, the normal force $\vec{\mathbf{n}}$, and the force of static friction $\vec{\mathbf{f}}_s$. We choose x to be parallel to the plane and y perpendicular to it.

Apply Equation 4.7 to the block in both the x and y directions:

(1) $\sum F_x = mg \sin \theta - f_s = 0$

(2) $\sum F_y = n - mg \cos \theta = 0$

Substitute $mg = n/\cos \theta$ from Equation (2) into Equation (1):

(3) $f_s = mg \sin \theta = \left(\dfrac{n}{\cos \theta}\right) \sin \theta = n \tan \theta$

When the incline angle is increased until the block is on the verge of slipping, the force of static friction has reached its maximum value $\mu_s n$. The angle θ in this situation is the critical angle θ_c. Make these substitutions in Equation (3):

$\mu_s n = n \tan \theta_c$

$\mu_s = \tan \theta_c$

For example, if the block just slips at $\theta_c = 20.0°$, we find that $\mu_s = \tan 20.0° = 0.364$.

Finalize   Once the block starts to move at $\theta \geq \theta_c$, it accelerates down the incline and the force of friction is $f_k = \mu_k n$. If θ is reduced to a value less than θ_c, however, it may be possible to find an angle θ'_c such that the block moves down the incline with constant speed as a particle in equilibrium again ($a_x = 0$). In this case, use Equations (1) and (2) with f_s replaced by f_k to find μ_k: $\mu_k = \tan \theta'_c$, where $\theta'_c < \theta_c$.

Example 5.3 | Acceleration of Two Connected Objects When Friction Is Present

A block of mass m_2 on a rough, horizontal surface is connected to a ball of mass m_1 by a lightweight cord over a lightweight, frictionless pulley as shown in Figure 5.5a. A force of magnitude F at an angle θ with the horizontal is applied to the block as shown, and the block slides to the right. The coefficient of kinetic friction between the block and surface is μ_k. Determine the magnitude of the acceleration of the two objects.

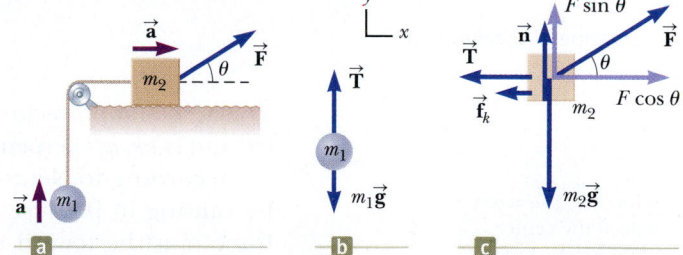

Figure 5.5 (Example 5.3) (a) The external force $\vec{\mathbf{F}}$ applied as shown can cause the block to accelerate to the right. (b, c) Diagrams showing the forces on the two objects, assuming the block accelerates to the right and the ball accelerates upward.

SOLUTION

Conceptualize Imagine what happens as $\vec{\mathbf{F}}$ is applied to the block. Assuming $\vec{\mathbf{F}}$ is not large enough to lift the block, the block slides to the right and the ball rises.

Categorize   We can identify forces and we want an acceleration, so we categorize this problem as one involving two particles under a net force, the ball and the block.

Analyze   First draw force diagrams for the two objects as shown in Figures 5.5b and 5.5c. Notice that the string exerts a force of magnitude T on both objects. The applied force $\vec{\mathbf{F}}$ has x and y components $F \cos \theta$ and $F \sin \theta$, respectively. Because the two objects are connected, we can equate the magnitudes of the x component of the acceleration of the block and the y component of the acceleration of the ball and call them both a. Let us assume the motion of the block is to the right.

continued

5.3 *cont.*

Apply the particle under a net force model to the block in the horizontal direction:

(1) $\sum F_x = F\cos\theta - f_k - T = m_2 a_x = m_2 a$

Because the block moves only horizontally, apply the particle in equilibrium model to the block in the vertical direction:

(2) $\sum F_y = n + F\sin\theta - m_2 g = 0$

Apply the particle under a net force model to the ball in the vertical direction:

(3) $\sum F_y = T - m_1 g = m_1 a_y = m_1 a$

Solve Equation (2) for n:

$n = m_2 g - F\sin\theta$

Substitute n into $f_k = \mu_k n$ from Equation 5.2:

(4) $f_k = \mu_k(m_2 g - F\sin\theta)$

Substitute Equation (4) and the value of T from Equation (3) into Equation (1):

$F\cos\theta - \mu_k(m_2 g - F\sin\theta) - m_1(a + g) = m_2 a$

Solve for a:

(5) $a = \dfrac{F(\cos\theta + \mu_k \sin\theta) - (m_1 + \mu_k m_2)g}{m_1 + m_2}$

Finalize The acceleration of the block can be either to the right or to the left depending on the sign of the numerator in Equation (5). If the velocity is to the left, we must reverse the sign of f_k in Equation (1) because the force of kinetic friction must oppose the motion of the block relative to the surface. In this case, the value of a is the same as in Equation (5), with the two plus signs in the numerator changed to minus signs.

5.2 | Extending the Particle in Uniform Circular Motion Model

Solving problems involving friction is just one of many applications of Newton's second law. Let us now consider another common situation, associated with a particle in uniform circular motion. In Chapter 3, we found that a particle moving in a circular path of radius r with uniform speed v experiences a centripetal acceleration of magnitude

▶ Centripetal acceleration

$$a_c = \frac{v^2}{r}$$

The acceleration vector with this magnitude is directed toward the center of the circle and is *always* perpendicular to $\vec{v}$.

According to Newton's second law, if an acceleration occurs, a net force must be causing it. Because the acceleration is toward the center of the circle, the net force must be toward the center of the circle. Therefore, when a particle travels in a circular path, a force must be acting *inward* on the particle that causes the circular motion. We investigate the forces causing this type of acceleration in this section.

Consider a puck of mass m that is tied to a string of length r and is moving at constant speed in a horizontal, circular path as illustrated in Figure 5.6. Its weight is supported by a frictionless table, and the string is anchored to a peg at the center of the circular path of the puck. Why does the puck move in a circle? The natural tendency of the puck is to move in a straight-line path, according to Newton's first law; the string, however, prevents this motion along a straight line by exerting a radial force $\vec{F}_r$ on the puck to make it follow a circular path. This force, whose magnitude is the tension in the string, is directed along the length of the string toward the center of the circle as shown in Figure 5.6.

In this discussion, the tension in the string causes the circular motion of the puck. Other forces also cause objects to move in circular paths. For example, friction forces

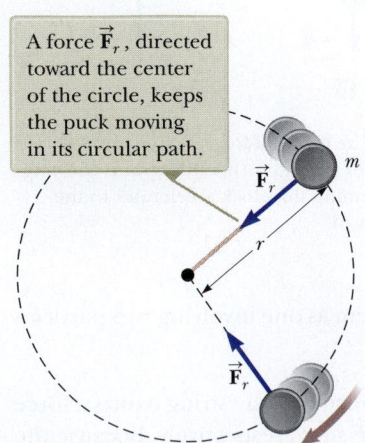

A force $\vec{F}_r$, directed toward the center of the circle, keeps the puck moving in its circular path.

Figure 5.6 An overhead view of a puck moving in a circular path in a horizontal plane.

cause automobiles to travel around curved roadways and the gravitational force causes a planet to orbit the Sun.

Regardless of the nature of the force acting on the particle in circular motion, we can apply Newton's second law to the particle along the radial direction:

$$\sum F = ma_c = m\frac{v^2}{r} \qquad \textbf{5.3} \blacktriangleleft$$

In general, an object can move in a circular path under the influence of various types of forces, or a *combination* of forces, as we shall see in some of the examples that follow.

If the force acting on an object vanishes, the object no longer moves in its circular path; instead, it moves along a straight-line path tangent to the circle. This idea is illustrated in Active Figure 5.7 for the case of the puck moving in a circular path at the end of a string in a horizontal plane. If the string breaks at some instant, the puck moves along the straight-line path tangent to the circle at the position of the puck at this instant.

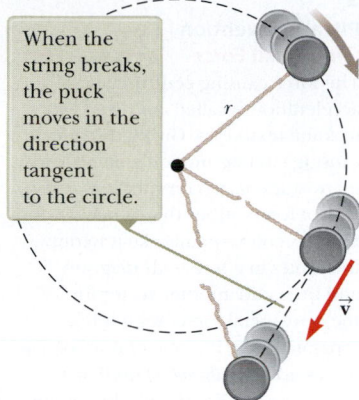

When the string breaks, the puck moves in the direction tangent to the circle.

Active Figure 5.7 The string holding the puck in its circular path breaks.

 QUICK QUIZ 5.4 You are riding on a Ferris wheel (Fig. 5.8) that is rotating with constant speed. The car in which you are riding always maintains its correct upward orientation: it does not invert. (**i**) What is the direction of the normal force on you from the seat when you are at the top of the wheel? (**a**) upward (**b**) downward (**c**) impossible to determine. (**ii**) From the same choices, what is the direction of the net force on you when you are at the top of the wheel?

Pitfall Prevention | 5.3
Direction of Travel When the String Is Cut
Study Active Figure 5.7 carefully. Many students have a misconception that the puck moves *radially* away from the center of the circle when the string is cut. The velocity of the puck is *tangent* to the circle. By Newton's first law, the puck simply continues to move in the direction that it is moving just as the force from the string disappears.

Pitfall Prevention | 5.4
Centrifugal Force
The commonly heard phrase "centrifugal force" is described as a force pulling *outward* on an object moving in a circular path. If you are experiencing a "centrifugal force" on a rotating carnival ride, what is the other object with which you are interacting? You cannot identify another object because centrifugal force is a fictitious force.

> ## THINKING PHYSICS 5.1
>
> The Copernican theory of the solar system is a structural model in which the planets are assumed to travel around the Sun in circular orbits. Historically, this theory was a break from the Ptolemaic theory, a structural model in which the Earth was at the center. When the Copernican theory was proposed, a natural question arose: What keeps the Earth and other planets moving in their paths around the Sun? An interesting response to this question comes from Richard Feynman: "In those days, one of the theories proposed was that the planets went around because behind them there were invisible angels, beating their wings and driving the planets forward. . . . It turns out that in order to keep the planets going around, the invisible angels must fly in a different direction."[1] What did Feynman mean by this statement?
>
> **Reasoning** The question asked by those at the time of Copernicus indicates that they did not have a proper understanding of inertia as described by Newton's first law. At that time in history, before Galileo and Newton, the interpretation was that *motion* was caused by force. This interpretation is different from our current understanding that *changes in motion* are caused by force. Therefore, it was natural for Copernicus's contemporaries to ask what force propelled a planet in its orbit. According to our current understanding, it is equally natural for us to realize that no force tangent to the orbit is necessary, that the motion simply continues owing to inertia.
>
> Therefore, in Feynman's imagery, the angels do not have to push the planet *from behind.* The angels must push *inward,* to provide the centripetal acceleration associated with the orbital motion of the planet. Of course, the angels are not real from a scientific point of view, but are a metaphor for the *gravitational force.* ◀

Figure 5.8 (Quick Quiz 5.4) A Ferris wheel.

Thomas Barrat /Shutterstock.com

---

[1]R. P. Feynman, R. B. Leighton, and M. Sands, *The Feynman Lectures on Physics,* Vol. 1 (Reading, MA: Addison-Wesley, 1963), p. 7-2.

Pitfall Prevention | 5.5
Centripetal Force

The force causing centripetal acceleration is called *centripetal force* in some textbooks. Giving the force causing circular motion a name leads many students to consider it as a new *kind* of force rather than a new *role* for force. A common mistake is to draw the forces in a free-body diagram and then add another vector for the centripetal force. Yet it is not a separate force; it is one of our familiar forces *acting in the role of causing a circular motion*. For the motion of the Earth around the Sun, for example, the "centripetal force" is *gravity*. For a rock whirled on the end of a string, the "centripetal force" is the *tension* in the string. After this discussion, we shall no longer use the phrase *centripetal force*.

Robin Smith/GettyImages

The cars of a corkscrew roller coaster must travel in tight loops. The normal force from the track contributes to the centripetal acceleration. The gravitational force, because it remains constant in direction, is sometimes in the same direction as the normal force, but is sometimes in the opposite direction.

Example 5.4 | How Fast Can It Spin?

A puck of mass 0.500 kg is attached to the end of a cord 1.50 m long. The puck moves in a horizontal circle as shown in Figure 5.6. If the cord can withstand a maximum tension of 50.0 N, what is the maximum speed at which the puck can move before the cord breaks? Assume the string remains horizontal during the motion.

SOLUTION

Conceptualize It makes sense that the stronger the cord, the faster the puck can move before the cord breaks. Also, we expect a more massive puck to break the cord at a lower speed. (Imagine whirling a bowling ball on the cord!)

Categorize Because the puck moves in a circular path, we model it as a particle in uniform circular motion.

Analyze Incorporate the tension and the centripetal acceleration into Newton's second law as described by Equation 5.3:

$$T = m\frac{v^2}{r}$$

Solve for v:

$$(1) \quad v = \sqrt{\frac{Tr}{m}}$$

Find the maximum speed the puck can have, which corresponds to the maximum tension the string can withstand:

$$v_{max} = \sqrt{\frac{T_{max}r}{m}} = \sqrt{\frac{(50.0\ \text{N})(1.50\ \text{m})}{0.500\ \text{kg}}} = \boxed{12.2\ \text{m/s}}$$

Finalize Equation (1) shows that v increases with T and decreases with larger m, as we expected from our conceptualization of the problem.

What If? Suppose the puck moves in a circle of larger radius at the same speed v. Is the cord more likely or less likely to break?

Answer The larger radius means that the change in the direction of the velocity vector will be smaller in a given time interval. Therefore, the acceleration is smaller and the required tension in the string is smaller. As a result, the string is less likely to break when the puck travels in a circle of larger radius.

Example 5.5 | The Conical Pendulum

A small ball of mass m is suspended from a string of length L. The ball revolves with constant speed v in a horizontal circle of radius r as shown in Figure 5.9. (Because the string sweeps out the surface of a cone, the system is known as a *conical pendulum*.) Find an expression for v.

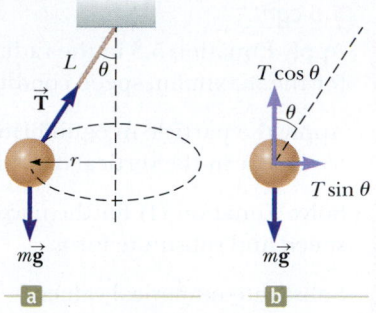

Figure 5.9 (Example 5.5) (a) A conical pendulum. The path of the ball is a horizontal circle. (b) The forces acting on the ball.

SOLUTION

Conceptualize Imagine the motion of the ball in Figure 5.9a and convince yourself that the string sweeps out a cone and that the ball moves in a horizontal circle.

Categorize The ball in Figure 5.9 does not accelerate vertically. Therefore, we model it as a particle in equilibrium in the vertical direction. It experiences a centripetal acceleration in the horizontal direction, so it is modeled as a particle in uniform circular motion in this direction.

Analyze Let θ represent the angle between the string and the vertical. In the diagram of forces acting on the ball in Figure 5.9b, the force $\vec{T}$ exerted by the string on the ball is resolved into a vertical component $T \cos \theta$ and a horizontal component $T \sin \theta$ acting toward the center of the circular path.

Apply the particle in equilibrium model in the vertical direction:

$$\sum F_y = T \cos \theta - mg = 0$$

$$(1) \quad T \cos \theta = mg$$

Use Equation 5.3 from the particle in uniform circular motion model in the horizontal direction:

$$(2) \quad \sum F_x = T \sin \theta = ma_c = \frac{mv^2}{r}$$

Divide Equation (2) by Equation (1) and use $\sin \theta / \cos \theta = \tan \theta$:

$$\tan \theta = \frac{v^2}{rg}$$

Solve for v:

$$v = \sqrt{rg \tan \theta}$$

Incorporate $r = L \sin \theta$ from the geometry in Figure 5.9a:

$$v = \sqrt{Lg \sin \theta \tan \theta}$$

Finalize Notice that the speed is independent of the mass of the ball. Consider what happens when θ goes to 90° so that the string is horizontal. Because the tangent of 90° is infinite, the speed v is infinite, which tells us the string cannot possibly be horizontal. If it were, there would be no vertical component of the force $\vec{T}$ to balance the gravitational force on the ball. That is why we mentioned in regard to Figure 5.6 that the puck's weight in the figure is supported by a frictionless table.

Example 5.6 | What Is the Maximum Speed of the Car?

A 1 500-kg car moving on a flat, horizontal road negotiates a curve as shown in Figure 5.10a (page 116). If the radius of the curve is 35.0 m and the coefficient of static friction between the tires and dry pavement is 0.523, find the maximum speed the car can have and still make the turn successfully.

SOLUTION

Conceptualize Imagine that the curved roadway is part of a large circle so that the car is moving in a circular path.

Categorize Based on the conceptualize step of the problem, we model the car as a particle in uniform circular motion in the horizontal direction. The car is not accelerating vertically, so it is modeled as a particle in equilibrium in the vertical direction.

Analyze Figure 5.10b shows the forces on the car. The force that enables the car to remain in its circular path is the force of static friction. (It is *static* because no slipping occurs at the point of contact between road and tires. If this force of static friction were zero—for example, if the car were on an icy road—the car would continue in a straight line and slide off the curved road.) The maximum speed v_{max} the car can have around the curve is the speed at which it is on the verge of skidding outward. At this point, the friction force has its maximum value $f_{s,\text{max}} = \mu_s n$.

continued

5.6 *cont.*

Apply Equation 5.3 in the radial direction for the maximum speed condition:

$$(1) \quad f_{s,max} = \mu_s n = m\frac{v_{max}^2}{r}$$

Apply the particle in equilibrium model to the car in the vertical direction:

$$\sum F_y = 0 \quad \rightarrow \quad n - mg = 0 \quad \rightarrow \quad n = mg$$

Solve Equation (1) for the maximum speed and substitute for n:

$$(2) \quad v_{max} = \sqrt{\frac{\mu_s nr}{m}} = \sqrt{\frac{\mu_s mgr}{m}} = \sqrt{\mu_s gr}$$

Substitute numerical values:

$$v_{max} = \sqrt{(0.523)(9.80 \text{ m/s}^2)(35.0 \text{ m})} = \boxed{13.4 \text{ m/s}}$$

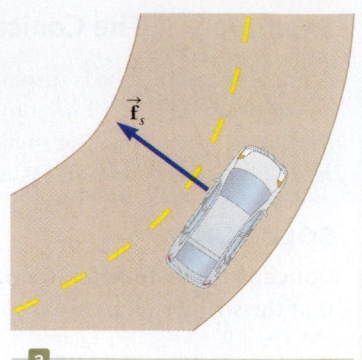

Finalize This speed is equivalent to 30.0 mi/h. Therefore, if the speed limit on this roadway is higher than 30 mi/h, this roadway could benefit greatly from some banking, as in the next example! Notice that the maximum speed does not depend on the mass of the car, which is why curved highways do not need multiple speed limits to cover the various masses of vehicles using the road.

What If? Suppose a car travels this curve on a wet day and begins to skid on the curve when its speed reaches only 8.00 m/s. What can we say about the coefficient of static friction in this case?

Answer The coefficient of static friction between the tires and a wet road should be smaller than that between the tires and a dry road. This expectation is consistent with experience with driving because a skid is more likely on a wet road than a dry road.

To check our suspicion, we can solve Equation (2) for the coefficient of static friction:

$$\mu_s = \frac{v_{max}^2}{gr}$$

Substituting the numerical values gives

$$\mu_s = \frac{v_{max}^2}{gr} = \frac{(8.00 \text{ m/s})^2}{(9.80 \text{ m/s}^2)(35.0 \text{ m})} = 0.187$$

which is indeed smaller than the coefficient of 0.523 for the dry road.

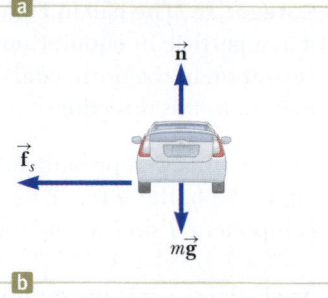

Figure 5.10 (Example 5.6) (a) The force of static friction directed toward the center of the curve keeps the car moving in a circular path. (b) The forces acting on the car.

Example 5.7 | The Banked Roadway

A civil engineer wishes to redesign the curved roadway in Example 5.6 in such a way that a car will not have to rely on friction to round the curve without skidding. In other words, a car moving at the designated speed can negotiate the curve even when the road is covered with ice. Such a road is usually *banked*, which means that the roadway is tilted toward the inside of the curve as seen in the opening photograph for this chapter. Suppose the designated speed for the ramp is to be 13.4 m/s (30.0 mi/h) and the radius of the curve is 35.0 m. At what angle should the curve be banked?

SOLUTION

Conceptualize The difference between this example and Example 5.6 is that the car is no longer moving on a flat roadway. Figure 5.11 shows the banked roadway, with the center of the circular path of the car far to the left of the figure. Notice that the horizontal component of the normal force participates in causing the car's centripetal acceleration.

Categorize As in Example 5.6, the car is modeled as a particle in equilibrium in the vertical direction and a particle in uniform circular motion in the horizontal direction.

Analyze On a level (unbanked) road, the force that causes the centripetal acceleration is the force of static friction between car and road as we saw in the preceding example. If the road is banked at an angle θ as in Figure 5.11, however, the normal force $\vec{\mathbf{n}}$ has a

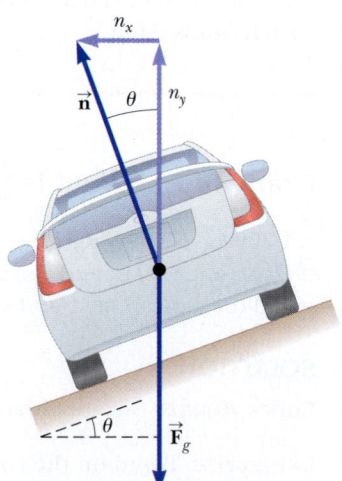

Figure 5.11 (Example 5.7) A car moves into the page and is rounding a curve on a road banked at an angle θ to the horizontal. When friction is neglected, the force that causes the centripetal acceleration and keeps the car moving in its circular path is the horizontal component of the normal force.

5.7 *cont.*

horizontal component toward the center of the curve. Because the ramp is to be designed so that the force of static friction is zero, only the component $n_x = n \sin \theta$ causes the centripetal acceleration.

Write Newton's second law for the car in the radial direction, which is the x direction:

$$(1) \quad \sum F_r = n \sin \theta = \frac{mv^2}{r}$$

Apply the particle in equilibrium model to the car in the vertical direction:

$$\sum F_y = n \cos \theta - mg = 0$$
$$(2) \quad n \cos \theta = mg$$

Divide Equation (1) by Equation (2):

$$(3) \quad \tan \theta = \frac{v^2}{rg}$$

Solve for the angle θ:

$$\theta = \tan^{-1} \left[\frac{(13.4 \text{ m/s})^2}{(35.0 \text{ m})(9.80 \text{ m/s}^2)} \right] = \boxed{27.6°}$$

Finalize Equation (3) shows that the banking angle is independent of the mass of the vehicle negotiating the curve. If a car rounds the curve at a speed less than 13.4 m/s, friction is needed to keep it from sliding down the bank (to the left in Fig. 5.11). A driver attempting to negotiate the curve at a speed greater than 13.4 m/s has to depend on friction to keep from sliding up the bank (to the right in Fig. 5.11).

What If? Imagine that this same roadway were built on Mars in the future to connect different colony centers. Could it be traveled at the same speed?

Answer The reduced gravitational force on Mars would mean that the car is not pressed as tightly to the roadway. The reduced normal force results in a smaller component of the normal force toward the center of the circle. This smaller component would not be sufficient to provide the centripetal acceleration associated with the original speed. The centripetal acceleration must be reduced, which can be done by reducing the speed v.

Mathematically, notice that Equation (3) shows that the speed v is proportional to the square root of g for a roadway of fixed radius r banked at a fixed angle θ. Therefore, if g is smaller, as it is on Mars, the speed v with which the roadway can be safely traveled is also smaller.

Example 5.8 | Riding the Ferris Wheel

A child of mass m rides on a Ferris wheel as shown in Figure 5.12a. The child moves in a vertical circle of radius 10.0 m at a constant speed of 3.00 m/s.

(A) Determine the force exerted by the seat on the child at the bottom of the ride. Express your answer in terms of the weight of the child, mg.

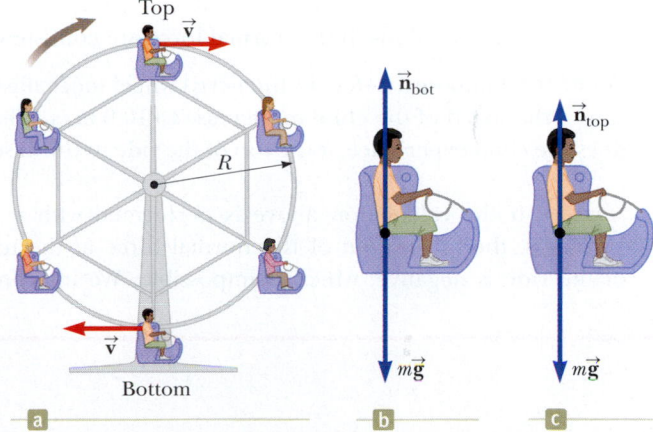

Figure 5.12 (Example 5.8) (a) A child rides on a Ferris wheel. (b) The forces acting on the child at the bottom of the path. (c) The forces acting on the child at the top of the path.

SOLUTION

Conceptualize Look carefully at Figure 5.12a. Based on experiences you may have had on a Ferris wheel or driving over small hills on a roadway, you would expect to feel lighter at the top of the path. Similarly, you would expect to feel heavier at the bottom of the path. At both the bottom of the path and the top, the normal and gravitational forces on the child act in *opposite* directions. The vector sum of these two forces gives a force of constant magnitude that keeps the child moving in a circular path at a constant speed. To yield net force vectors with the same magnitude, the normal force at the bottom must be greater than that at the top.

Categorize Because the speed of the child is constant, we can categorize this problem as one involving a particle (the child) in uniform circular motion, complicated by the gravitational force acting at all times on the child.

continued

5.8 *cont.*

Analyze We draw a diagram of forces acting on the child at the bottom of the ride as shown in Figure 5.12b. The only forces acting on him are the downward gravitational force $\vec{\mathbf{F}}_g = m\vec{\mathbf{g}}$ and the upward force $\vec{\mathbf{n}}_{bot}$ exerted by the seat. The net upward force on the child that provides his centripetal acceleration has a magnitude $n_{bot} - mg$.

Apply Newton's second law to the child in the radial direction when he is at the bottom of the ride:

$$\sum F = n_{bot} - mg = m\frac{v^2}{r}$$

Solve for the force exerted by the seat on the child:

$$n_{bot} = mg + m\frac{v^2}{r} = mg\left(1 + \frac{v^2}{rg}\right)$$

Substitute the values given for the speed and radius:

$$n_{bot} = mg\left[1 + \frac{(3.00 \text{ m/s})^2}{(10.0 \text{ m})(9.80 \text{ m/s}^2)}\right]$$

$$= 1.09\,mg$$

Hence, the magnitude of the force $\vec{\mathbf{n}}_{bot}$ exerted by the seat on the child is *greater* than the weight of the child by a factor of 1.09. So, the child experiences an apparent weight that is greater than his true weight by a factor of 1.09.

(B) Determine the force exerted by the seat on the child at the top of the ride.

SOLUTION

Analyze The diagram of forces acting on the child at the top of the ride is shown in Figure 5.12c. The net downward force that provides the centripetal acceleration has a magnitude $mg - n_{top}$.

Apply Newton's second law to the child at this position:

$$\sum F = mg - n_{top} = m\frac{v^2}{r}$$

Solve for the force exerted by the seat on the child:

$$n_{top} = mg - m\frac{v^2}{r} = mg\left(1 - \frac{v^2}{rg}\right)$$

Substitute numerical values:

$$n_{top} = mg\left[1 - \frac{(3.00 \text{ m/s})^2}{(10.0 \text{ m})(9.80 \text{ m/s}^2)}\right]$$

$$= 0.908\,mg$$

In this case, the magnitude of the force exerted by the seat on the child is *less* than his true weight by a factor of 0.908, and the child feels lighter.

Finalize The variations in the normal force are consistent with our prediction in the Conceptualize step of the problem.

What If? Suppose a defect in the Ferris wheel mechanism causes the speed of the child to increase to 10.0 m/s. What does the child experience at the top of the ride in this case?

Answer If the calculation above is performed with $v = 10.0$ m/s, the magnitude of the normal force at the top of the ride is negative, which is impossible. We interpret it to mean that the required centripetal acceleration of the child is larger than that due to gravity. As a result, the child will lose contact with the seat and will only stay in his circular path if there is a safety bar that provides a downward force on him to keep him in his seat. At the bottom of the ride, the normal force is $2.02\,mg$, which would be uncomfortable.

❬ **5.3** | Nonuniform Circular Motion

In Chapter 3, we found that if a particle moves with varying speed in a circular path, there is, in addition to the radial component of acceleration, a tangential component of magnitude dv/dt. Therefore, the net force acting on the particle must also have a radial and a tangential component as shown in Active Figure 5.13. That is, because the total acceleration is $\vec{\mathbf{a}} = \vec{\mathbf{a}}_r + \vec{\mathbf{a}}_t$, the total force exerted on the particle is $\sum\vec{\mathbf{F}} = \sum\vec{\mathbf{F}}_r + \sum\vec{\mathbf{F}}_t$. (We express the radial and tangential forces as net forces with the summation notation because each force could consist of multiple forces that combine.) The component vector $\Sigma\vec{\mathbf{F}}_r$ is directed toward the center of the circle and is responsible for the centripetal acceleration. The component vector $\Sigma\vec{\mathbf{F}}_t$ tangent to the circle is responsible for the tangential acceleration, which causes the speed of the particle to change with time.

▶ **QUICK QUIZ 5.5** Which of the following is *impossible* for a car moving in a circular path? Assume that the car is never at rest. (**a**) The car has tangential acceleration but no centripetal acceleration. (**b**) The car has centripetal acceleration but no tangential acceleration. (**c**) The car has both centripetal acceleration and tangential acceleration.

▶ **QUICK QUIZ 5.6** A bead slides freely along a curved wire lying on a horizontal surface at constant speed as shown by Figure 5.14. (**a**) Draw the vectors representing the force exerted by the wire on the bead at points Ⓐ, Ⓑ, and Ⓒ. (**b**) Suppose the bead in Figure 5.14 speeds up with constant tangential acceleration as it moves toward the right. Draw the vectors representing the force on the bead at points Ⓐ, Ⓑ, and Ⓒ.

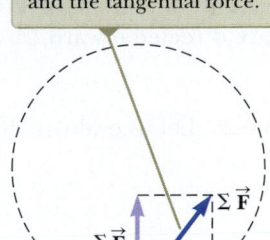

The net force exerted on the particle is the vector sum of the radial force and the tangential force.

Active Figure 5.13 When the net force acting on a particle moving in a circular path has a tangential component vector $\Sigma\vec{\mathbf{F}}_t$, its speed changes.

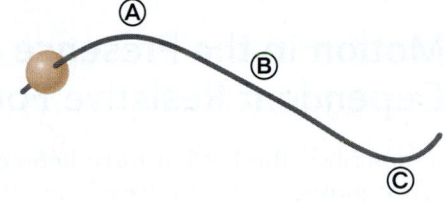

Figure 5.14 (Quick Quiz 5.6) A bead slides along a curved wire.

Example 5.9 | **Keep Your Eye on the Ball**

A small sphere of mass m is attached to the end of a cord of length R and set into motion in a *vertical* circle about a fixed point O as illustrated in Figure 5.15. Determine the tangential acceleration of the sphere and the tension in the cord at any instant when the speed of the sphere is v and the cord makes an angle θ with the vertical.

SOLUTION

Conceptualize Compare the motion of the sphere in Figure 5.15 with that of the child in Figure 5.12a associated with Example 5.8. Both objects travel in a circular path. Unlike the child in Example 5.8, however, the speed of the sphere is *not* uniform in this example because, at most points along the path, a tangential component of acceleration arises from the gravitational force exerted on the sphere.

Categorize We model the sphere as a particle under a net force and moving in a circular path, but it is not a particle in *uniform* circular motion. We need to use the techniques discussed in this section on nonuniform circular motion.

· ·

Analyze From the force diagram in Figure 5.15, we see that the only forces acting on the sphere are the gravitational force $\vec{\mathbf{F}}_g = m\vec{\mathbf{g}}$ exerted by the Earth and the force $\vec{\mathbf{T}}$ exerted by the cord. We resolve $\vec{\mathbf{F}}_g$ into a tangential component $mg \sin\theta$ and a radial component $mg \cos\theta$.

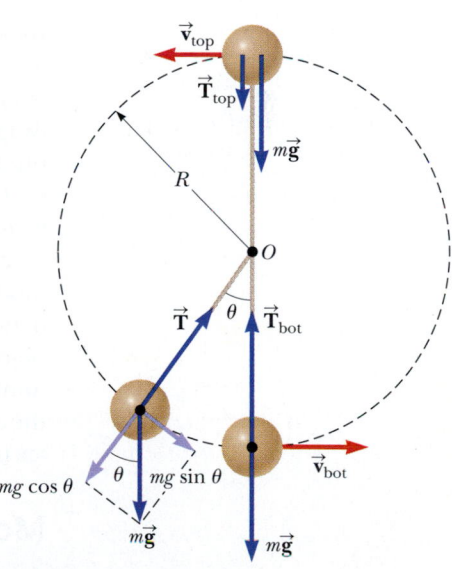

Figure 5.15 (Example 5.9) The forces acting on a sphere of mass m connected to a cord of length R and rotating in a vertical circle centered at O. Forces acting on the sphere are shown when the sphere is at the top and bottom of the circle and at an arbitrary location.

Apply Newton's second law to the sphere in the tangential direction:

$$\Sigma F_t = mg \sin\theta = ma_t$$
$$a_t = \boxed{g \sin\theta}$$

continued

5.9 cont.

Apply Newton's second law to the forces acting on the sphere in the radial direction, noting that both $\vec{T}$ and $\vec{a}_r$ are directed toward O:

$$\sum F_r = T - mg \cos\theta = \frac{mv^2}{R}$$

$$T = mg\left(\frac{v^2}{Rg} + \cos\theta\right)$$

Finalize Let us evaluate this result at the top and bottom of the circular path (Fig. 5.15):

$$T_{\text{top}} = mg\left(\frac{v_{\text{top}}^2}{Rg} - 1\right) \qquad T_{\text{bot}} = mg\left(\frac{v_{\text{bot}}^2}{Rg} + 1\right)$$

These results have similar mathematical forms as those for the normal forces n_{top} and n_{bot} on the child in Example 5.8, which is consistent with the normal force on the child playing a similar physical role in Example 5.8 as the tension in the string plays in this example. Keep in mind, however, that the normal force $\vec{n}$ on the child in Example 5.8 is always upward, whereas the force $\vec{T}$ in this example changes direction because it must always point inward along the string. Also note that v in the expressions above varies for different positions of the sphere, as indicated by the subscripts, whereas v in Example 5.8 is constant.

▌5.4 | Motion in the Presence of Velocity-Dependent Resistive Forces

Earlier, we described the friction force between a moving object and the surface along which it moves. So far, we have ignored any interaction between the object and the *medium* through which it moves. Let us now consider the effect of a medium such as a liquid or gas. The medium exerts a **resistive force** $\vec{R}$ on the object moving through it. You feel this force if you ride in a car at high speed with your hand out the window; the force you feel pushing your hand backward is the resistive force of the air rushing past the car. The magnitude of this force depends on the relative speed between the object and the medium, and the direction of $\vec{R}$ on the object is always opposite the direction of the object's motion relative to the medium. Some examples are the air resistance associated with moving vehicles (sometimes called air drag), the force of the wind on the sails of a sailboat, and the viscous forces that act on objects sinking through a liquid.

Generally, the magnitude of the resistive force increases with increasing speed. The resistive force can have a complicated speed dependence. In the following discussions, we consider two simplification models that allow us to analyze these situations. The first model assumes that the resistive force is proportional to the velocity, which is approximately the case for objects that fall through a liquid with low speed and for very small objects, such as dust particles, that move through air. The second model treats situations for which we assume that the magnitude of the resistive force is proportional to the square of the speed of the object. Large objects, such as a skydiver moving through air in free-fall, experience such a force.

Model 1: Resistive Force Proportional to Object Velocity

At low speeds, the resistive force acting on an object that is moving through a viscous medium is effectively modeled as being proportional to the object's velocity. The mathematical representation of the resistive force can be expressed as

$$\vec{R} = -b\vec{v} \qquad \qquad \textbf{5.4} \blacktriangleleft$$

where $\vec{v}$ is the velocity of the object relative to the medium and b is a constant that depends on the properties of the medium and on the shape and dimensions of the object. The negative sign represents that the resistive force is opposite the velocity of the object relative to the medium.

Consider a sphere of mass m released from rest in a liquid, as in Active Figure 5.16a. We assume that the only forces acting on the sphere are the resistive force $\vec{R}$

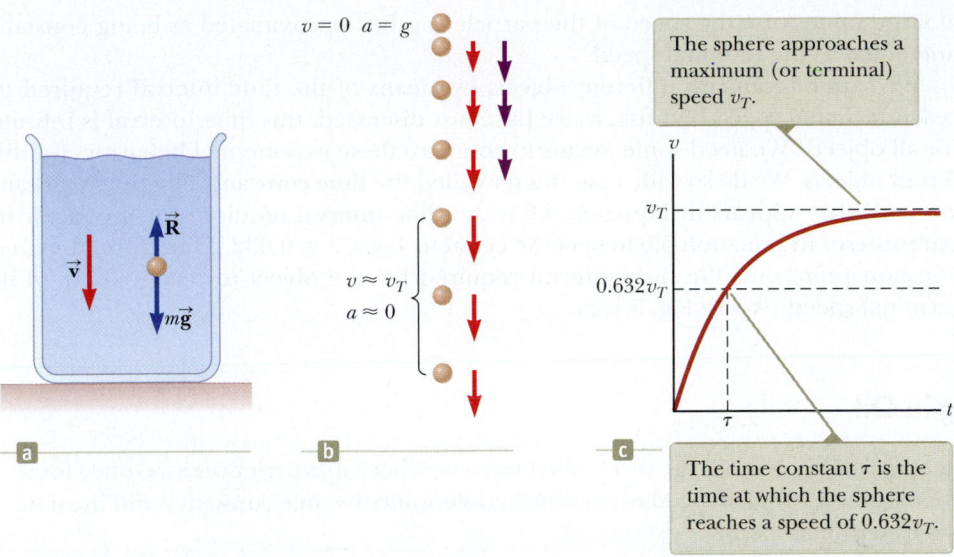

The sphere approaches a maximum (or terminal) speed v_T.

The time constant τ is the time at which the sphere reaches a speed of $0.632v_T$.

and the weight $m\vec{\mathbf{g}}$, and we describe its motion using Newton's second law.[2] Considering the vertical motion and choosing the downward direction to be positive, we have

$$\sum F_y = ma_y \quad \rightarrow \quad mg - bv = m\frac{dv}{dt}$$

Dividing this equation by the mass m gives

$$\frac{dv}{dt} = g - \frac{b}{m}v \qquad \qquad \textbf{5.5} \blacktriangleleft$$

Equation 5.5 is called a *differential equation*; it includes both the speed v and the derivative of the speed. The methods of solving such an equation may not be familiar to you as yet. Note, however, that if we define $t = 0$ when $v = 0$, the resistive force is zero at this time and the acceleration dv/dt is simply g. As t increases, the speed increases, the resistive force increases, and the acceleration decreases. Therefore, this situation is one in which neither the velocity nor the acceleration of the particle is constant.

The acceleration becomes zero when the increasing resistive force eventually balances the weight. At this point, the object reaches its **terminal speed** v_T and from then on it continues to move with zero acceleration. The motion diagram in Active Figure 5.16b shows the sphere accelerating over the early part of its motion and then reaching terminal speed later on. After the object reaches terminal speed, its motion is that of a particle under constant velocity. The terminal speed can be obtained from Equation 5.5 by setting $a = dv/dt = 0$, which gives

$$mg - bv_T = 0 \quad \rightarrow \quad v_T = \frac{mg}{b}$$

The expression for v that satisfies Equation 5.5 with $v = 0$ at $t = 0$ is

$$v = \frac{mg}{b}(1 - e^{-bt/m}) = v_T(1 - e^{-t/\tau}) \qquad \qquad \textbf{5.6} \blacktriangleleft$$

where $v_T = mg/b$, $\tau = m/b$, and $e = 2.718\,28$ is the base of the natural logarithm. This expression for v can be verified by substituting it back into Equation 5.5. (Try it!) This function is plotted in Active Figure 5.16c.

The mathematical representation of the motion (Eq. 5.6) indicates that the terminal speed is never reached because the exponential function is never exactly equal to zero. For all practical purposes, however, when the exponential function is very small

---

[2] A *buoyant* force also acts on any object surrounded by a fluid. This force is constant and equal to the weight of the displaced fluid, as will be discussed in Chapter 15. The effect of this force can be modeled by changing the apparent weight of the sphere by a constant factor, so we can ignore it here.

at large values of t, the speed of the particle can be approximated as being constant and equal to the terminal speed.

We cannot compare different objects by means of the time interval required to reach terminal speed because, as we have just discussed, this time interval is infinite for all objects! We need some means to compare these exponential behaviors for different objects. We do so with a parameter called the **time constant.** The time constant $\tau = m/b$ that appears in Equation 5.6 is the time interval required for the factor in parentheses in Equation 5.6 to become equal to $1 - e^{-1} = 0.632$. Therefore, the time constant represents the time interval required for the object to reach 63.2% of its terminal speed (Active Fig. 5.16c).

Example 5.10 | Sphere Falling in Oil

A small sphere of mass 2.00 g is released from rest in a large vessel filled with oil, where it experiences a resistive force proportional to its speed. The sphere reaches a terminal speed of 5.00 cm/s. Determine the time constant τ and the time at which the sphere reaches 90.0% of its terminal speed.

SOLUTION

Conceptualize With the help of Active Figure 5.16, imagine dropping the sphere into the oil and watching it sink to the bottom of the vessel. If you have some thick shampoo in a clear container, drop a marble in it and observe the motion of the marble.

Categorize We model the sphere as a particle under a net force, with one of the forces being a resistive force that depends on the speed of the sphere.

. . . .

Analyze From $v_T = mg/b$, evaluate the coefficient b:

$$b = \frac{mg}{v_T} = \frac{(2.00 \text{ g})(980 \text{ cm/s}^2)}{5.00 \text{ cm/s}} = 392 \text{ g/s}$$

Evaluate the time constant τ:

$$\tau = \frac{m}{b} = \frac{2.00 \text{ g}}{392 \text{ g/s}} = \boxed{5.10 \times 10^{-3} \text{ s}}$$

Find the time t at which the sphere reaches a speed of $0.900v_T$ by setting $v = 0.900v_T$ in Equation 5.6 and solving for t:

$$0.900v_T = v_T(1 - e^{-t/\tau})$$

$$1 - e^{-t/\tau} = 0.900$$

$$e^{-t/\tau} = 0.100$$

$$-\frac{t}{\tau} = \ln(0.100) = -2.30$$

$$t = 2.30\tau = 2.30(5.10 \times 10^{-3} \text{ s}) = 11.7 \times 10^{-3} \text{ s}$$

$$= \boxed{11.7 \text{ ms}}$$

. . . .

Finalize The sphere reaches 90.0% of its terminal speed in a very short time interval. You should have also seen this behavior if you performed the activity with the marble and the shampoo. Because of the short time interval required to reach terminal velocity, you may not have noticed the time interval at all. The marble may have appeared to immediately begin moving through the shampoo at a constant velocity.

Model 2: Resistive Force Proportional to Object Speed Squared

For large objects moving at high speeds through air, such as airplanes, skydivers, and baseballs, the magnitude of the resistive force is modeled as being proportional to the square of the speed:

$$R = \tfrac{1}{2}D\rho Av^2 \qquad \blacktriangleleft \text{5.7}$$

where ρ is the density of air, A is the cross-sectional area of the moving object measured in a plane perpendicular to its velocity, and D is a dimensionless empirical quantity called the *drag coefficient*. The drag coefficient has a value of about 0.5 for spherical objects moving through air but can be as high as 2 for irregularly shaped objects.

Consider an airplane in flight that experiences such a resistive force. Equation 5.7 shows that the force is proportional to the density of air and hence decreases with decreasing air density. Because air density decreases with increasing altitude, the resistive force on a jet airplane flying at a given speed will decrease with increasing altitude. Therefore, airplanes tend to fly at very high altitudes to take advantage of this reduced resistive force, which allows them to fly faster for a given engine thrust. Of course, this higher speed *increases* the resistive force, in proportion to the square of the speed, so a balance is struck between fuel economy and higher speed.

Now let us analyze the motion of a falling object subject to an upward air resistive force whose magnitude is given by Equation 5.7. Suppose an object of mass m is released from rest, as in Figure 5.17, from the position $y = 0$. The object experiences two external forces: the downward gravitational force $m\vec{g}$ and the upward resistive force $\vec{R}$. Hence, using Newton's second law,

$$\sum F = ma \quad \rightarrow \quad mg - \tfrac{1}{2}D\rho Av^2 = ma \qquad \text{5.8} \blacktriangleleft$$

Solving for a, we find that the object has a downward acceleration of magnitude

$$a = g - \left(\frac{D\rho A}{2m}\right)v^2 \qquad \text{5.9} \blacktriangleleft$$

Because $a = dv/dt$, Equation 5.9 is another differential equation that provides us with the speed as a function of time.

Again, we can calculate the terminal speed v_T because when the gravitational force is balanced by the resistive force, the net force is zero and therefore the acceleration is zero. Setting $a = 0$ in Equation 5.9 gives

$$g - \left(\frac{D\rho A}{2m}\right)v_T^2 = 0$$

$$v_T = \sqrt{\frac{2mg}{D\rho A}} \qquad \text{5.10} \blacktriangleleft$$

Table 5.2 lists the terminal speeds for several objects falling through air, all computed on the assumption that the drag coefficient is 0.5.

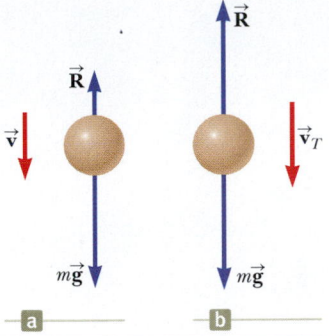

Figure 5.17 (a) An object falling through air experiences a resistive drag force $\vec{R}$ and a gravitational force $\vec{F}_g = m\vec{g}$. (b) The object reaches terminal speed when the net force acting on it is zero, that is, when $\vec{R} = -\vec{F}_g$, or $R = mg$. Before that occurs, the acceleration varies with speed according to Equation 5.9.

QUICK QUIZ 5.7 Consider a sky surfer falling through air, as in Figure 5.18, before reaching his terminal speed. As the speed of the sky surfer increases, the magnitude of his acceleration (a) remains constant, (b) decreases until it reaches a constant nonzero value, or (c) decreases until it reaches zero.

Figure 5.18 (Quick Quiz 5.7) A sky surfer takes advantage of the upward force of the air on his board.

TABLE 5.2 | Terminal Speeds for Various Objects Falling Through Air

Object	Mass (kg)	Cross-sectional Area (m²)	v_T (m/s)[a]
Skydiver	75	0.70	60
Baseball (radius 3.7 cm)	0.145	4.2×10^{-3}	33
Golf ball (radius 2.1 cm)	0.046	1.4×10^{-3}	32
Hailstone (radius 0.50 cm)	4.8×10^{-4}	7.9×10^{-5}	14
Raindrop (radius 0.20 cm)	3.4×10^{-5}	1.3×10^{-5}	9.0

[a]The drag coefficient D is assumed to be 0.5 in each case.

⟨5.5 | The Fundamental Forces of Nature

We have described a variety of forces experienced in our everyday activities, such as the gravitational force acting on all objects at or near the Earth's surface and the force of friction as one surface slides over another. Newton's second law tells us how to relate the forces to the object's or particle's acceleration.

In addition to these familiar macroscopic forces in nature, forces also act in the atomic and subatomic world. For example, atomic forces within the atom are responsible for holding its constituents together and nuclear forces act on different parts of the nucleus to keep its parts from separating.

Until recently, physicists believed that there were four fundamental forces in nature: the gravitational force, the electromagnetic force, the strong force, and the weak force. We shall discuss these forces individually and then consider the current view of fundamental forces.

The Gravitational Force

The **gravitational force** is the mutual force of attraction between any two objects in the Universe. It is interesting and rather curious that although the gravitational force can be very strong between macroscopic objects, it is inherently the weakest of all the fundamental forces. For example, the gravitational force between the electron and proton in the hydrogen atom has a magnitude on the order of 10^{-46} N, whereas the electromagnetic force between these same two particles is on the order of 10^{-7} N.

In addition to his contributions to the understanding of motion, Newton studied gravity extensively. **Newton's law of universal gravitation** states that every particle in the Universe attracts every other particle with a force that is directly proportional to the product of the masses of the particles and inversely proportional to the square of the distance between them. If the particles have masses m_1 and m_2 and are separated by a distance r, as in Figure 5.19, the magnitude of the gravitational force is

▶ Newton's law of universal gravitation

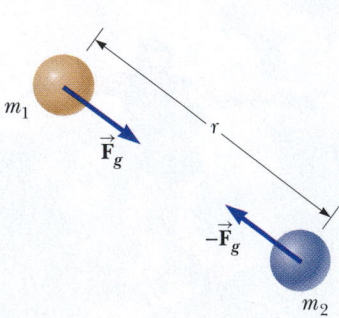

$$F_g = G\frac{m_1 m_2}{r^2} \qquad \text{5.11}◀$$

where $G = 6.674 \times 10^{-11}$ N·m²/kg² is the **universal gravitational constant.** More detail on the gravitational force will be provided in Chapter 11.

Figure 5.19 Two particles with masses m_1 and m_2 attract each other with a force of magnitude Gm_1m_2/r^2.

The Electromagnetic Force

The **electromagnetic force** is the force that binds atoms and molecules in compounds to form ordinary matter. It is much stronger than the gravitational force. The force that causes a rubbed comb to attract bits of paper and the force that a magnet exerts on an iron nail are electromagnetic forces. Essentially all forces at work in our macroscopic world, apart from the gravitational force, are manifestations of the electromagnetic force. For example, friction forces, contact forces, tension forces, and forces in elongated springs are consequences of electromagnetic forces between charged particles in proximity.

The electromagnetic force involves two types of particles: those with positive charge and those with negative charge. (More information on these two types of charge is provided in Chapter 19.) Unlike the gravitational force, which is always an attractive interaction, the electromagnetic force can be either attractive or repulsive, depending on the charges on the particles.

▶ Coulomb's law

Coulomb's law expresses the magnitude of the *electrostatic force*[3] F_e between two charged particles separated by a distance r:

$$F_e = k_e\frac{q_1 q_2}{r^2} \qquad \text{5.12}◀$$

---

[3]The electrostatic force is the electromagnetic force between two electric charges that are at rest. If the charges are moving, magnetic forces are also present; these forces will be studied in Chapter 22.

where q_1 and q_2 are the charges on the two particles, measured in units called *coulombs* (C), and k_e (= 8.99 × 10^9 N·m²/C²) is the **Coulomb constant.** Note that the electrostatic force has the same mathematical form as Newton's law of universal gravitation (see Eq. 5.11), with charge playing the mathematical role of mass and the Coulomb constant being used in place of the universal gravitational constant. The electrostatic force is attractive if the two charges have opposite signs and is repulsive if the two charges have the same sign, as indicated in Figure 5.20.

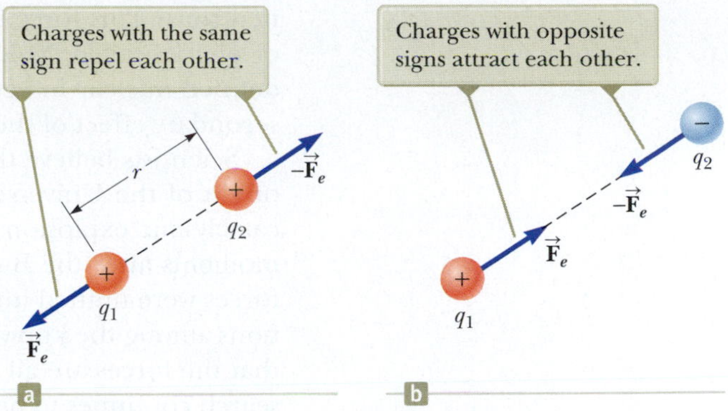

The smallest amount of isolated charge found in nature (so far) is the charge on an electron or proton. This fundamental unit of charge is given the symbol e and has the magnitude $e = 1.60 × 10^{-19}$ C. An electron has charge $-e$, whereas a proton has charge $+e$. Theories developed in the latter half of the 20th century propose that protons and neutrons are made up of smaller particles called **quarks,** which have charges of either $\frac{2}{3}e$ or $-\frac{1}{3}e$ (discussed further in Chapter 31). Although experimental evidence has been found for such particles inside nuclear matter, free quarks have never been detected.

Figure 5.20 Two point charges separated by a distance r exert an electrostatic force on each other given by Coulomb's law.

The Strong Force

An atom, as we currently model it, consists of an extremely dense positively charged nucleus surrounded by a cloud of negatively charged electrons, with the electrons attracted to the nucleus by the electric force. All nuclei except those of hydrogen are combinations of positively charged protons and neutral neutrons (collectively called nucleons), yet why does the repulsive electrostatic force between the protons not cause nuclei to break apart? Clearly, there must be an attractive force that counteracts the strong electrostatic repulsive force and is responsible for the stability of nuclei. This force that binds the nucleons to form a nucleus is called the **nuclear force.** It is one manifestation of the **strong force,** which is the force between quarks, which we will discuss in Chapter 31. Unlike the gravitational and electromagnetic forces, which depend on distance in an inverse-square fashion, the nuclear force is extremely short range; its strength decreases very rapidly outside the nucleus and is negligible for separations greater than approximately 10^{-14} m.

The Weak Force

The **weak force** is a short-range force that tends to produce instability in certain nuclei. It was first observed in naturally occurring radioactive substances and was later found to play a key role in most radioactive decay reactions. The weak force is about 10^{34} times stronger than the gravitational force and about 10^3 times weaker than the electromagnetic force.

The Current View of Fundamental Forces

For years, physicists have searched for a simplification scheme that would reduce the number of fundamental forces needed to describe physical phenomena. In 1967, physicists predicted that the electromagnetic force and the weak force, originally thought to be independent of each other and both fundamental, are in fact manifestations of one force, now called the **electroweak** force. This prediction was confirmed experimentally in 1984. We shall discuss it more fully in Chapter 31.

We also now know that protons and neutrons are not fundamental particles; current models of protons and neutrons theorize that they are composed of simpler particles called quarks, as mentioned previously. The quark model has led to a modification of our understanding of the nuclear force. Scientists now define the strong force as the force that binds the quarks to one another in a nucleon (proton or neutron). This force is also referred to as a **color force,** in reference to a property of quarks called "color," which we shall investigate in Chapter 31. The previously defined nuclear force, the force that acts between nucleons, is now interpreted as a secondary effect of the strong force between the quarks.

Scientists believe that the fundamental forces of nature are closely related to the origin of the Universe. The Big Bang theory states that the Universe began with a cataclysmic explosion about 14 billion years ago. According to this theory, the first moments after the Big Bang saw such extremes of energy that all the fundamental forces were unified into one force. Physicists are continuing their search for connections among the known fundamental forces, connections that could eventually prove that the forces are all merely different forms of a single superforce. This fascinating search continues to be at the forefront of physics.

⦗5.6 | Context Connection: Drag Coefficients of Automobiles

In the Context Connection of Chapter 4, we ignored air resistance and assumed that the driving force on the tires was the only force on the vehicle in the horizontal direction. Given our understanding of velocity-dependent forces from Section 5.4, we should understand now that air resistance could be a significant factor in the design of an automobile.

Table 5.3 shows the drag coefficients for the vehicles that we have investigated in previous chapters. Notice that the coefficients for the very expensive, performance, and traditional vehicles vary from 0.27 to 0.43, with the average coefficient in the three portions of the table almost the same. A look at the alternative vehicles shows that this parameter is the lowest on the average for all the vehicles, with the Chevrolet Volt and Toyota Prius having the lowest values in the entire table. The discontinued GM EV1, an electric car produced between 1996 and 1999, had a remarkable coefficient of just 0.19.

Designers of alternative-fuel vehicles try to squeeze every last mile of travel out of the energy that is stored in the vehicle in the form of fuel or an electric battery. A significant method of doing so is to reduce the force of air resistance so that the net force driving the car forward is as large as possible.

⦗TABLE 5.3 | Drag Coefficients of Various Vehicles

Automobile	Drag Coefficient	Automobile	Drag Coefficient
Very expensive vehicles:		*Traditional vehicles:*	
Bugatti Veyron 16.4 Super Sport	0.36	Buick Regal CXL Turbo	0.27
Lamborghini LP 570-4 Superleggera	0.31	Chevrolet Tahoe 1500 LS (SUV)	0.42
Lexus LFA	0.31	Ford Fiesta SES	0.33
Mercedes-Benz SLS AMG	0.36	Hummer H3 (SUV)	0.43
Shelby SuperCars Ultimate Aero	0.36	Hyundai Sonata SE	0.32
Average	**0.34**	Smart ForTwo	0.34
		Average	**0.35**
Performance vehicles:		*Alternative vehicles:*	
Chevrolet Corvette ZR1	0.28	Chevrolet Volt (hybrid)	0.26
Dodge Viper SRT10	0.40	Nissan Leaf (electric)	0.29
Jaguar XJL Supercharged	0.29	Honda CR-Z (hybrid)	0.30
Acura TL SH-AWD	0.29	Honda Insight (hybrid)	0.28
Dodge Challenger SRT8	0.35	Toyota Prius (hybrid)	0.25
Average	**0.32**	**Average**	**0.28**

Figure 5.21 (a) The Chevrolet Corvette ZR1 has a streamlined shape that contributes to its low drag coefficient of 0.28. (b) The Hummer H3 is not streamlined like the Corvette and consequently has a much higher drag coefficient of 0.43.

A number of techniques can be used to reduce the drag coefficient. Two factors that help are a small frontal area and smooth curves from the front of the vehicle to the back. For example, the Chevrolet Corvette ZR1 shown in Figure 5.21a exhibits a streamlined shape that contributes to its low drag coefficient. As a comparison, consider a large, boxy vehicle, such as the Hummer H3 in Figure 5.21b. The drag coefficient for this vehicle is 0.43. (This is an improvement over the previous model, the H2, which had a coefficient of 0.57.) Another factor includes elimination or minimization of as many irregularities in the surfaces as possible, including door handles that project from the body, windshield wipers, wheel wells, and rough surfaces on headlamps and grills. An important consideration is the underside of the carriage. As air rushes beneath the car, there are many irregular surfaces associated with brakes, drive trains, suspension components, and so on. The drag coefficient can be made lower by assuring that the overall surface of the car's undercarriage is as smooth as possible.

SUMMARY

Forces of friction are complicated, but we design a simplification model for friction that allows us to analyze motion that includes the effects of friction. The **maximum force of static friction** $f_{s,\max}$ between two surfaces is proportional to the normal force between the surfaces. This maximum force occurs when the surfaces are on the verge of slipping. In general, $f_s \leq \mu_s n$, where μ_s is the **coefficient of static friction** and n is the magnitude of the normal force. When an object slides over a rough surface, the **force of kinetic friction** $\vec{f}_k$ is opposite the direction of the velocity of the object relative to the surface and its magnitude is proportional to the magnitude of the normal force on the object. The magnitude is given by $f_k = \mu_k n$, where μ_k is the **coefficient of kinetic friction**. Usually, $\mu_k < \mu_s$.

An object moving through a liquid or gas experiences a **resistive force** that is velocity dependent. This resistive force, which is opposite the velocity of the object relative to the medium, generally increases with speed. The force depends on the object's shape and on the properties of the medium through which the object is moving. In the limiting case for a falling object, when the resistive force balances the weight ($a = 0$), the object reaches its **terminal speed.**

The fundamental forces existing in nature can be expressed as the following four: the gravitational force, the electromagnetic force, the strong force, and the weak force.

▶Analysis Model for Problem Solving

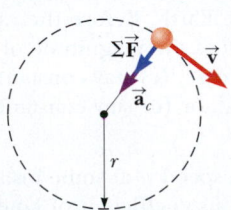

Particle in Uniform Circular Motion (Extension) With our new knowledge of forces, we can extend the model of a particle in uniform circular motion, first introduced in Chapter 3. Newton's second law applied to a particle moving in uniform circular motion states that the net force causing the particle to undergo a centripetal acceleration (Eq. 3.17) is related to the acceleration according to

$$\sum F = ma_c = m\frac{v^2}{r}$$

5.3 ◀

OBJECTIVE QUESTIONS

1. The driver of a speeding empty truck slams on the brakes and skids to a stop through a distance d. On a second trial, the truck carries a load that doubles its mass. What will now be the truck's "skidding distance"? (a) $4d$ (b) $2d$ (c) $\sqrt{2}d$ (d) d (e) $d/2$

2. The manager of a department store is pushing horizontally with a force of magnitude 200 N on a box of shirts. The box is sliding across the horizontal floor with a forward acceleration. Nothing else touches the box. What must be true about the magnitude of the force of kinetic friction acting on the box (choose one)? (a) It is greater than 200 N. (b) It is less than 200 N. (c) It is equal to 200 N. (d) None of those statements is necessarily true.

3. An object of mass m moves with acceleration $\vec{a}$ down a rough incline. Which of the following forces should appear in a free-body diagram of the object? Choose all correct answers. (a) the gravitational force exerted by the planet (b) $m\vec{a}$ in the direction of motion (c) the normal force exerted by the incline (d) the friction force exerted by the incline (e) the force exerted by the object on the incline

4. An office door is given a sharp push and swings open against a pneumatic device that slows the door down and then reverses its motion. At the moment the door is open the widest, (a) does the doorknob have a centripetal acceleration? (b) Does it have a tangential acceleration?

5. A crate remains stationary after it has been placed on a ramp inclined at an angle with the horizontal. Which of the following statements is or are correct about the magnitude of the friction force that acts on the crate? Choose all that are true. (a) It is larger than the weight of the crate. (b) It is equal to $\mu_s n$. (c) It is greater than the component of the gravitational force acting down the ramp. (d) It is equal to the component of the gravitational force acting down the ramp. (e) It is less than the component of the gravitational force acting down the ramp.

6. A pendulum consists of a small object called a bob hanging from a light cord of fixed length, with the top end of the cord fixed, as represented in Figure OQ5.6. The bob moves without friction, swinging equally high on both sides. It moves from its turning point A through point B and reaches its maximum speed at point C. (a) Of these points, is there a point where the bob has nonzero radial acceleration and zero tangential acceleration? If so, which point? What is the direction of its total acceleration at this point? (b) Of these points, is there a point where the bob has nonzero tangential acceleration and zero radial acceleration? If so, which point? What is the direction of its total acceleration at this point? (c) Is there a point where the bob has no acceleration? If so, which point? (d) Is there a point where the bob has both nonzero tangential and radial acceleration? If so, which point? What is the direction of its total acceleration at this point?

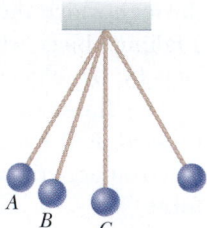

Figure OQ5.6

7. A door in a hospital has a pneumatic closer that pulls the door shut such that the doorknob moves with constant speed over most of its path. In this part of its motion, (a) does the doorknob experience a centripetal acceleration? (b) Does it experience a tangential acceleration?

8. The driver of a speeding truck slams on the brakes and skids to a stop through a distance d. On another trial, the initial speed of the truck is half as large. What now will be the truck's skidding distance? (a) $2d$ (b) $\sqrt{2}d$ (c) d (d) $d/2$ (e) $d/4$

9. A child is practicing for a BMX race. His speed remains constant as he goes counterclockwise around a level track with two straight sections and two nearly semicircular sections as shown in the aerial view of Figure OQ5.9. (a) Rank the magnitudes of his acceleration at the points A, B, C, D, and E from largest to smallest. If his acceleration is the same size at two points, display that fact in your ranking. If his acceleration is zero, display that fact. (b) What are the directions of his velocity at points A, B, and C? For each point, choose one: north, south, east, west, or nonexistent. (c) What are the directions of his acceleration at points A, B, and C?

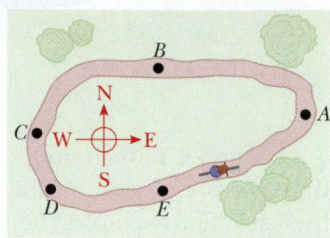

Figure OQ5.9

10. A large crate of mass m is placed on the flatbed of a truck but not tied down. As the truck accelerates forward with acceleration a, the crate remains at rest relative to the truck. What force causes the crate to accelerate? (a) the normal force (b) the gravitational force (c) the friction force (d) the ma force exerted by the crate (e) No force is required.

11. Before takeoff on an airplane, an inquisitive student on the plane dangles an iPod by its earphone wire. It hangs straight down as the plane is at rest waiting to take off. The plane then gains speed rapidly as it moves down the runway. (i) Relative to the student's hand, does the iPod (a) shift toward the front of the plane, (b) continue to hang straight down, or (c) shift toward the back of the plane? (ii) The speed of the plane increases at a constant rate over a time interval of several seconds. During this interval, does the angle the earphone wire makes with the vertical (a) increase, (b) stay constant, or (c) decrease?

12. Consider a skydiver who has stepped from a helicopter and is falling through air. Before she reaches terminal speed and long before she opens her parachute, does her speed (a) increase, (b) decrease, or (c) stay constant?

13. As a raindrop falls through the atmosphere, its speed initially changes as it falls toward the Earth. Before the raindrop reaches its terminal speed, does the magnitude of its acceleration (a) increase, (b) decrease, (c) stay constant at zero, (d) stay constant at 9.80 m/s^2, or (e) stay constant at some other value?

14. An object of mass m is sliding with speed v_i at some instant across a level tabletop, with which its coefficient of kinetic friction is μ. It then moves through a distance d and comes to rest. Which of the following equations for the speed v_i is reasonable? (a) $v_i = \sqrt{-2\mu mgd}$ (b) $v_i = \sqrt{2\mu mgd}$ (c) $v_i = \sqrt{-2\mu gd}$ (d) $v_i = \sqrt{2\mu gd}$ (e) $v_i = \sqrt{2\mu d}$

CONCEPTUAL QUESTIONS

1. A car is moving forward slowly and is speeding up. A student claims that "the car exerts a force on itself" or that "the car's engine exerts a force on the car." (a) Argue that this idea cannot be accurate and that friction exerted by the road is the propulsive force on the car. Make your evidence and reasoning as persuasive as possible. (b) Is it static or kinetic friction? *Suggestions:* Consider a road covered with light gravel. Consider a sharp print of the tire tread on an asphalt road, obtained by coating the tread with dust.

2. Your hands are wet, and the restroom towel dispenser is empty. What do you do to get drops of water off your hands? How does the motion of the drops exemplify one of Newton's laws? Which one?

3. Describe two examples in which the force of friction exerted on an object is in the direction of motion of the object.

4. Describe the path of a moving body in the event that (a) its acceleration is constant in magnitude at all times and perpendicular to the velocity, and (b) its acceleration is constant in magnitude at all times and parallel to the velocity.

5. An object executes circular motion with constant speed whenever a net force of constant magnitude acts perpendicular to the velocity. What happens to the speed if the force is not perpendicular to the velocity?

6. Consider a small raindrop and a large raindrop falling through the atmosphere. (a) Compare their terminal speeds. (b) What are their accelerations when they reach terminal speed?

7. Suppose you are driving a classic car. Why should you avoid slamming on your brakes when you want to stop in the shortest possible distance? (Many modern cars have anti-lock brakes that avoid this problem.)

8. A falling skydiver reaches terminal speed with her parachute closed. After the parachute is opened, what parameters change to decrease this terminal speed?

9. What forces cause (a) an automobile, (b) a propeller-driven airplane, and (c) a rowboat to move?

10. A pail of water can be whirled in a vertical path such that no water is spilled. Why does the water stay in the pail, even when the pail is above your head?

11. It has been suggested that rotating cylinders about 20 km in length and 8 km in diameter be placed in space and used as colonies. The purpose of the rotation is to simulate gravity for the inhabitants. Explain this concept for producing an effective imitation of gravity.

12. If someone told you that astronauts are weightless in orbit because they are beyond the pull of gravity, would you accept the statement? Explain.

13. **BIO** Why does a pilot tend to black out when pulling out of a steep dive?

PROBLEMS AVAILABLE IN

5.1 Forces of Friction

Problems 1–15

5.2 Extending the Particle in Uniform Circular Motion Model

Problems 16–21

5.3 Nonuniform Circular Motion

Problems 22–28

5.4 Motion in the Presence of Velocity-Dependent Resistive Forces

Problems 29–34

5.5 The Fundamental Forces of Nature

Problems 35–38

5.6 Context Connection: Drag Coefficients of Automobiles

Problems 39–40

Additional Problems

Problems 41–64

Solutions to the following Problems are available in the *Student Solutions Manual/Study Guide*:

List of Enhanced Problems

Problem Number	Targeted Feedback in Enhanced WebAssign	Master It in Enhanced WebAssign	Watch It in Enhanced WebAssign
5.3	✓		✓
5.7	✓		✓
5.9	✓	✓	
5.10	✓		✓
5.13	✓	✓	
5.17		✓	
5.19	✓		✓
5.21	✓		✓
5.26	✓		✓
5.27	✓	✓	
5.29		✓	
5.31	✓		✓
5.47	✓		✓
5.51	✓		✓
5.57		✓	
5.63		✓	

Chapter 6

Energy of a System

Chapter Outline

Christopher Furlong/Getty Images

On a wind farm at the mouth of the River Mersey in Liverpool, England, the moving air does work on the blades of the windmills, causing the blades and the rotor of an electrical generator to rotate. Energy is transferred out of the system of the windmill by means of electricity.

The definitions of quantities such as position, velocity, acceleration, and force and associated principles such as Newton's second law have allowed us to solve a variety of problems. Some problems that could theoretically be solved with Newton's laws, however, are very difficult in practice, but they can be made much simpler with a different approach. Here and in the following chapters, we will investigate this new approach, which will include definitions of quantities that may not be familiar to you. Other quantities may sound familiar, but they may have more specific meanings in physics than in everyday life. We begin this discussion by exploring the notion of *energy*.

The concept of energy is one of the most important topics in science and engineering. In everyday life, we think of energy in terms of fuel for transportation and heating, electricity for lights and appliances, and foods for consumption. These ideas, however, do not truly define energy. They merely tell us that fuels are needed to do a job and that those fuels provide us with something we call energy.

Energy is present in the Universe in various forms. *Every* physical process that occurs in the Universe involves energy and energy transfers or transformations. Unfortunately, despite its extreme importance, energy cannot be easily defined. The variables in previous chapters were relatively concrete; we have everyday experience with velocities and forces, for example. Although we have *experiences* with energy, such as running out of gasoline or losing our electrical service following a violent storm, the *notion* of energy is more abstract.

The concept of energy can be applied to mechanical systems without resorting to Newton's laws. Furthermore, the energy approach allows us to understand thermal and electrical phenomena in later chapters of the book.

Our analysis models presented in earlier chapters were based on the motion of a *particle* or an object that could be modeled as a particle. We begin our new approach by focusing our attention on a *system* and analysis models based on the model of a system. These analysis models will be formally introduced in Chapter 7. In this chapter, we introduce systems and three ways to store energy in a system.

6.1 | Systems and Environments

In the system model, we focus our attention on a small portion of the Universe—the **system**—and ignore details of the rest of the Universe outside of the system. A critical skill in applying the system model to problems is *identifying the system.*

A valid system

- may be a single object or particle
- may be a collection of objects or particles
- may be a region of space (such as the interior of an automobile engine combustion cylinder)
- may vary with time in size and shape (such as a rubber ball, which deforms upon striking a wall)

Identifying the need for a system approach to solving a problem (as opposed to a particle approach) is part of the Categorize step in the General Problem-Solving Strategy outlined in Chapter 1. Identifying the particular system is a second part of this step.

No matter what the particular system is in a given problem, we identify a **system boundary**, an imaginary surface (not necessarily coinciding with a physical surface) that divides the Universe into the system and the **environment** surrounding the system.

As an example, imagine a force applied to an object in empty space. We can define the object as the system and its surface as the system boundary. The force applied to it is an influence on the system from the environment that acts across the system boundary. We will see how to analyze this situation from a system approach in a subsequent section of this chapter.

Another example was seen in Example 5.3, where the system can be defined as the combination of the ball, the block, and the cord. The influence from the environment includes the gravitational forces on the ball and the block, the normal and friction forces on the block, the force exerted by the pulley on the cord, and the applied force of magnitude F. The forces exerted by the cord on the ball and the block are internal to the system and therefore are not included as an influence from the environment.

There are a number of mechanisms by which a system can be influenced by its environment. The first one we shall investigate is *work*.

Pitfall Prevention | 6.1
Identify the System
The most important *first* step to take in solving a problem using the energy approach is to identify the appropriate system of interest.

a   b   c

© Cengage Learning/Charles D. Winters

Figure 6.1 An eraser being pushed along a chalkboard tray by a force acting at different angles with respect to the horizontal direction.

6.2 | Work Done by a Constant Force

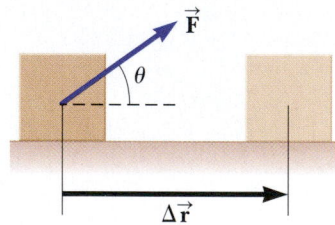

Figure 6.2 An object undergoes a displacement $\Delta\vec{r}$ under the action of a constant force $\vec{F}$.

▶ Work done by a constant force

Almost all the terms we have used thus far—velocity, acceleration, force, and so on—convey a similar meaning in physics as they do in everyday life. Now, however, we encounter a term whose meaning in physics is distinctly different from its everyday meaning: work.

To understand what work as an influence on a system means to the physicist, consider the situation illustrated in Figure 6.1. A force $\vec{F}$ is applied to a chalkboard eraser, which we identify as the system, and the eraser slides along the tray. If we want to know how effective the force is in moving the eraser, we must consider not only the magnitude of the force but also its direction. Notice that the finger in Figure 6.1 applies forces in three different directions on the eraser. Assuming the magnitude of the applied force is the same in all three photographs, the push applied in Figure 6.1b does more to move the eraser than the push in Figure 6.1a. On the other hand, Figure 6.1c shows a situation in which the applied force does not move the eraser at all, regardless of how hard it is pushed (unless, of course, we apply a force so great that we break the chalkboard tray!). These results suggest that when analyzing forces to determine the influence they have on the system, we must consider the vector nature of forces. We must also consider the magnitude of the force. Moving a force of magnitude 2 N through a displacement represents a greater influence on the system than moving a force of magnitude 1 N through the same displacement. The magnitude of the displacement is also important. Moving the eraser 3 m along the tray represents a greater influence than moving it 2 cm if the same force is used in both cases.

Let us examine the situation in Figure 6.2, where the object (the system) undergoes a displacement along a straight line while acted on by a constant force of magnitude F that makes an angle θ with the direction of the displacement.

The **work** W done on a system by an agent exerting a constant force on the system is the product of the magnitude F of the force, the magnitude Δr of the displacement of the point of application of the force, and $\cos\theta$, where θ is the angle between the force and displacement vectors:

$$W \equiv F\,\Delta r\cos\theta \qquad\qquad 6.1$$

Notice in Equation 6.1 that work is a scalar, even though it is defined in terms of two vectors, a force $\vec{F}$ and a displacement $\Delta\vec{r}$. In Section 6.3, we explore how to combine two vectors to generate a scalar quantity.

Notice also that the displacement in Equation 6.1 is that of *the point of application of the force.* If the force is applied to a particle or a rigid object that can be modeled as a particle, this displacement is the same as that of the particle. For a deformable system, however, these displacements are not the same. For example, imagine pressing in on the sides of a balloon with both hands. The center of the balloon moves through zero displacement. The points of application of the forces from your hands on the sides of the balloon, however, do indeed move through a displacement as the balloon is compressed, and that is the displacement to be used in Equation 6.1. We will see other examples of deformable systems, such as springs and samples of gas contained in a vessel.

As an example of the distinction between the definition of work and our everyday understanding of the word, consider holding a heavy chair at arm's length for 3 min. At the end of this time interval, your tired arms may lead you to think you have done a considerable amount of work on the chair. According to our definition, however, you have done no work on it whatsoever. You exert a force to support the chair, but you do not move it. A force does no work on an object if the force does not move through a displacement. If $\Delta r = 0$, Equation 6.1 gives $W = 0$, which is the situation depicted in Figure 6.1c.

Equation 6.1 also shows that the work done by a force on a moving object is zero when the force applied is perpendicular to the displacement of its point of application. That is, if $\theta = 90°$, then $W = 0$ because $\cos 90° = 0$. For example, in Figure 6.3, the work done by the normal force on the object and the work done by the gravitational force on the object are both zero because both forces are perpendicular to the displacement and have zero components along an axis in the direction of $\Delta\vec{r}$.

The sign of the work also depends on the direction of $\vec{F}$ relative to $\Delta\vec{r}$. The work done by the applied force on a system is positive when the projection of $\vec{F}$ onto $\Delta\vec{r}$ is in the same direction as the displacement. For example, when an object is lifted, the work done by the applied force on the object is positive because the direction of that force is upward, in the same direction as the displacement of its point of application. When the projection of $\vec{F}$ onto $\Delta\vec{r}$ is in the direction opposite the displacement, W is negative. For example, as an object is lifted, the work done by the gravitational force on the object is negative. The factor $\cos \theta$ in the definition of W (Eq. 6.1) automatically takes care of the sign.

If an applied force $\vec{F}$ is in the same direction as the displacement $\Delta\vec{r}$, then $\theta = 0$ and $\cos 0 = 1$. In this case, Equation 6.1 gives

$$W = F\Delta r$$

The units of work are those of force multiplied by those of length. Therefore, the SI unit of work is the **newton · meter** ($N \cdot m = kg \cdot m^2/s^2$). This combination of units is used so frequently that it has been given a name of its own, the **joule** (J).

An important consideration for a system approach to problems is that **work is an energy transfer.** If W is the work done on a system and W is positive, energy is transferred *to* the system; if W is negative, energy is transferred *from* the system. Therefore, if a system interacts with its environment, this interaction can be described as a transfer of energy across the system boundary. The result is a change in the energy stored in the system. We will learn about the first type of energy storage in Section 6.5, after we investigate more aspects of work.

QUICK QUIZ 6.1 The gravitational force exerted by the Sun on the Earth holds the Earth in an orbit around the Sun. Let us assume that the orbit is perfectly circular. The work done by this gravitational force during a short time interval in which the Earth moves through a displacement in its orbital path is (a) zero (b) positive (c) negative (d) impossible to determine

QUICK QUIZ 6.2 Figure 6.4 shows four situations in which a force is applied to an object. In all four cases, the force has the same magnitude, and the displacement of the object is to the right and of the same magnitude. Rank the situations in order of the work done by the force on the object, from most positive to most negative.

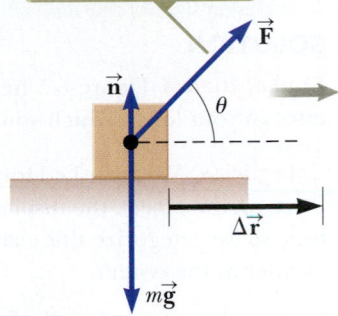

$\vec{F}$ is the only force that does work on the block in this situation.

Figure 6.3 An object is displaced on a frictionless, horizontal surface. The normal force $\vec{n}$ and the gravitational force $m\vec{g}$ do no work on the object.

Pitfall Prevention | 6.3
Cause of the Displacement
We can calculate the work done by a force on an object, but that force is *not* necessarily the cause of the object's displacement. For example, if you lift an object, (negative) work is done on the object by the gravitational force, although gravity is not the cause of the object moving upward!

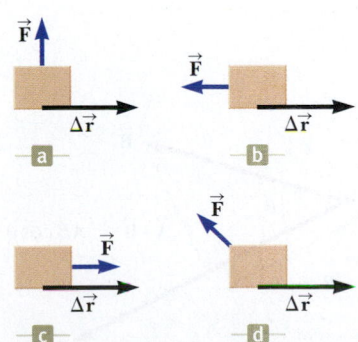

Figure 6.4 (Quick Quiz 6.2) A block is pulled by a force in four different directions. In each case, the displacement of the block is to the right and of the same magnitude.

Example 6.1 | Mr. Clean

A man cleaning a floor pulls a vacuum cleaner with a force of magnitude $F = 50.0$ N at an angle of 30.0° with the horizontal (Fig. 6.5). Calculate the work done by the force on the vacuum cleaner as the vacuum cleaner is displaced 3.00 m to the right.

Figure 6.5 (Example 6.1) A vacuum cleaner being pulled at an angle of 30.0° from the horizontal.

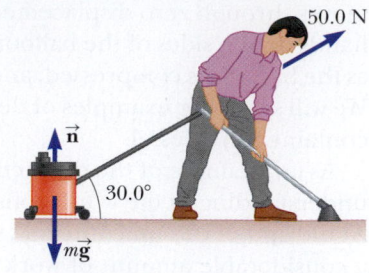

SOLUTION

Conceptualize Figure 6.5 helps conceptualize the situation. Think about an experience in your life in which you pulled an object across the floor with a rope or cord.

Categorize We are asked for the work done on an object by a force and are given the force on the object, the displacement of the object, and the angle between the two vectors, so we categorize this example as a substitution problem. We identify the vacuum cleaner as the system.

Use the definition of work (Eq. 6.1):

$$W = F\Delta r\cos\theta = (50.0\text{ N})(3.00\text{ m})(\cos 30.0°)$$
$$= 130\text{ J}$$

Notice in this situation that the normal force $\vec{n}$ and the gravitational $\vec{F}_g = m\vec{g}$ do no work on the vacuum cleaner because these forces are perpendicular to the displacements of their points of application. Furthermore, there was no mention of whether there was friction between the vacuum cleaner and the floor. The presence or absence of friction is not important when calculating the work done by the applied force. In addition, this work does not depend on whether the vacuum moved at constant velocity or if it accelerated.

Pitfall Prevention | 6.4
Work Is a Scalar
Although Equation 6.3 defines the work in terms of two vectors, *work is a scalar;* there is no direction associated with it. *All* types of energy and energy transfer are scalars. This fact is a major advantage of the energy approach because we often don't need vector calculations!

6.3 | The Scalar Product of Two Vectors

Because of the way the force and displacement vectors are combined in Equation 6.1, it is helpful to use a convenient mathematical tool called the **scalar product** of two vectors. We write this scalar product of vectors $\vec{A}$ and $\vec{B}$ as $\vec{A}\cdot\vec{B}$. (Because of the dot symbol, the scalar product is often called the **dot product.**)

The scalar product of any two vectors $\vec{A}$ and $\vec{B}$ is defined as a scalar quantity equal to the product of the magnitudes of the two vectors and the cosine of the angle θ between them:

▶ Scalar product of any two vectors $\vec{A}$ and $\vec{B}$

$$\vec{A}\cdot\vec{B} = AB\cos\theta \qquad 6.2$$

As is the case with any multiplication, $\vec{A}$ and $\vec{B}$ need not have the same units.

By comparing this definition with Equation 6.1, we can express Equation 6.1 as a scalar product:

$$W = F\Delta r\cos\theta = \vec{F}\cdot\Delta\vec{r} \qquad 6.3$$

In other words, $\vec{F}\cdot\Delta\vec{r}$ is a shorthand notation for $F\Delta r\cos\theta$.

Before continuing with our discussion of work, let us investigate some properties of the dot product. Figure 6.6 shows two vectors $\vec{A}$ and $\vec{B}$ and the angle θ between them used in the definition of the dot product. In Figure 6.6, $B\cos\theta$ is the projection of $\vec{B}$ onto $\vec{A}$. Therefore, Equation 6.2 means that $\vec{A}\cdot\vec{B}$ is the product of the magnitude of $\vec{A}$ and the projection of $\vec{B}$ onto $\vec{A}$.[1]

From the right-hand side of Equation 6.2, we also see that the scalar product is **commutative.**[2] That is,

$$\vec{A}\cdot\vec{B} = \vec{B}\cdot\vec{A}$$

Figure 6.6 The scalar product $\vec{A}\cdot\vec{B}$ equals the magnitude of $\vec{A}$ multiplied by $B\cos\theta$, which is the projection of $\vec{B}$ onto $\vec{A}$.

[1] This statement is equivalent to stating that $\vec{A}\cdot\vec{B}$ equals the product of the magnitude of $\vec{B}$ and the projection of $\vec{A}$ onto $\vec{B}$.

[2] In Chapter 10, you will see another way of combining vectors that proves useful in physics and is not commutative.

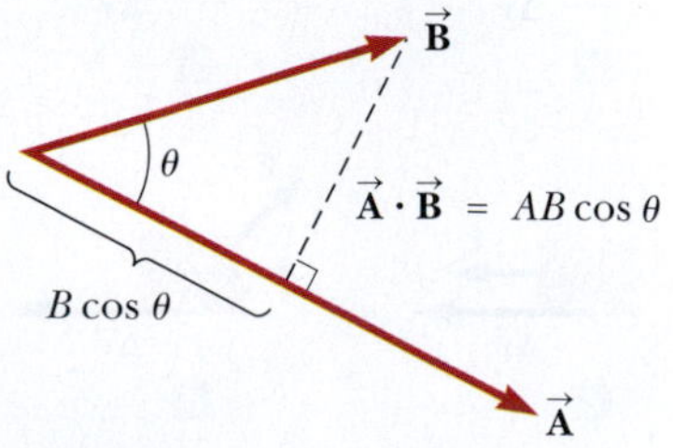

Finally, the scalar product obeys the **distributive law of multiplication,** so

$$\vec{A} \cdot (\vec{B} + \vec{C}) = \vec{A} \cdot \vec{B} + \vec{A} \cdot \vec{C}$$

The scalar product is simple to evaluate from Equation 6.2 when $\vec{A}$ is either perpendicular or parallel to $\vec{B}$. If $\vec{A}$ is perpendicular to $\vec{B}$ ($\theta = 90°$), then $\vec{A} \cdot \vec{B} = 0$. (The equality $\vec{A} \cdot \vec{B} = 0$ also holds in the more trivial case in which either $\vec{A}$ or $\vec{B}$ is zero.) If vector $\vec{A}$ is parallel to vector $\vec{B}$ and the two point in the same direction ($\theta = 0$), then $\vec{A} \cdot \vec{B} = AB$. If vector $\vec{A}$ is parallel to vector $\vec{B}$ but the two point in opposite directions ($\theta = 180°$), then $\vec{A} \cdot \vec{B} = -AB$. The scalar product is negative when $90° < \theta \leq 180°$.

The unit vectors $\hat{\mathbf{i}}$, $\hat{\mathbf{j}}$, and $\hat{\mathbf{k}}$, which were defined in Chapter 1, lie in the positive x, y, and z directions, respectively, of a right-handed coordinate system. Therefore, it follows from the definition of $\vec{A} \cdot \vec{B}$ that the scalar products of these unit vectors are

$$\hat{\mathbf{i}} \cdot \hat{\mathbf{i}} = \hat{\mathbf{j}} \cdot \hat{\mathbf{j}} = \hat{\mathbf{k}} \cdot \hat{\mathbf{k}} = 1 \qquad \qquad \textbf{6.4} \blacktriangleleft$$

$$\hat{\mathbf{i}} \cdot \hat{\mathbf{j}} = \hat{\mathbf{i}} \cdot \hat{\mathbf{k}} = \hat{\mathbf{j}} \cdot \hat{\mathbf{k}} = 0 \qquad \qquad \textbf{6.5} \blacktriangleleft$$

▶ Scalar products of unit vectors

According to Section 1.9, two vectors $\vec{A}$ and $\vec{B}$ can be expressed in unit-vector form as

$$\vec{A} = A_x\hat{\mathbf{i}} + A_y\hat{\mathbf{j}} + A_z\hat{\mathbf{k}}$$

$$\vec{B} = B_x\hat{\mathbf{i}} + B_y\hat{\mathbf{j}} + B_z\hat{\mathbf{k}}$$

Using these expressions for the vectors and the information given in Equations 6.4 and 6.5 shows that the scalar product of $\vec{A}$ and $\vec{B}$ reduces to

$$\vec{A} \cdot \vec{B} = A_xB_x + A_yB_y + A_zB_z \qquad \qquad \textbf{6.6} \blacktriangleleft$$

(Details of the derivation are left for you in Problem 6.8 in Enhanced WebAssign.) In the special case in which $\vec{A} = \vec{B}$, we see that

$$\vec{A} \cdot \vec{A} = A_x^2 + A_y^2 + A_z^2 = A^2$$

QUICK QUIZ 6.3 Which of the following statements is true about the relationship between the dot product of two vectors and the product of the magnitudes of the vectors? (a) $\vec{A} \cdot \vec{B}$ is larger than AB. (b) $\vec{A} \cdot \vec{B}$ is smaller than AB. (c) $\vec{A} \cdot \vec{B}$ could be larger or smaller than AB, depending on the angle between the vectors. (d) $\vec{A} \cdot \vec{B}$ could be equal to AB.

Example 6.2 | The Scalar Product

The vectors $\vec{A}$ and $\vec{B}$ are given by $\vec{A} = 2\hat{\mathbf{i}} + 3\hat{\mathbf{j}}$ and $\vec{B} = -\hat{\mathbf{i}} + 2\hat{\mathbf{j}}$.

(A) Determine the scalar product $\vec{A} \cdot \vec{B}$.

SOLUTION

Conceptualize There is no physical system to imagine here. Rather, it is purely a mathematical exercise involving two vectors.

Categorize Because we have a definition for the scalar product, we categorize this example as a substitution problem.

Substitute the specific vector expressions for $\vec{A}$ and $\vec{B}$:

$$\vec{A} \cdot \vec{B} = (2\hat{\mathbf{i}} + 3\hat{\mathbf{j}}) \cdot (-\hat{\mathbf{i}} + 2\hat{\mathbf{j}})$$

$$= -2\hat{\mathbf{i}} \cdot \hat{\mathbf{i}} + 2\hat{\mathbf{i}} \cdot 2\hat{\mathbf{j}} - 3\hat{\mathbf{j}} \cdot \hat{\mathbf{i}} + 3\hat{\mathbf{j}} \cdot 2\hat{\mathbf{j}}$$

$$= -2(1) + 4(0) - 3(0) + 6(1) = -2 + 6 = \boxed{4}$$

The same result is obtained when we use Equation 6.6 directly, where $A_x = 2$, $A_y = 3$, $B_x = -1$, and $B_y = 2$.

continued

6.2 *cont.*

(B) Find the angle θ between $\vec{\mathbf{A}}$ and $\vec{\mathbf{B}}$.

SOLUTION

Evaluate the magnitudes of $\vec{\mathbf{A}}$ and $\vec{\mathbf{B}}$ using the Pythagorean theorem:

$$A = \sqrt{A_x^2 + A_y^2} = \sqrt{(2)^2 + (3)^2} = \sqrt{13}$$
$$B = \sqrt{B_x^2 + B_y^2} = \sqrt{(-1)^2 + (2)^2} = \sqrt{5}$$

Use Equation 6.2 and the result from part (A) to find the angle:

$$\cos \theta = \frac{\vec{\mathbf{A}} \cdot \vec{\mathbf{B}}}{AB} = \frac{4}{\sqrt{13}\sqrt{5}} = \frac{4}{\sqrt{65}}$$

$$\theta = \cos^{-1} \frac{4}{\sqrt{65}} = \boxed{60.3°}$$

Example **6.3** | **Work Done by a Constant Force**

A particle moving in the xy plane undergoes a displacement given by $\Delta \vec{\mathbf{r}} = (2.0\hat{\mathbf{i}} + 3.0\hat{\mathbf{j}})$ m as a constant force $\vec{\mathbf{F}} = (5.0\hat{\mathbf{i}} + 2.0\hat{\mathbf{j}})$ N acts on the particle. Calculate the work done by $\vec{\mathbf{F}}$ on the particle.

SOLUTION

Conceptualize Although this example is a little more physical than the previous one in that it identifies a force and a displacement, it is similar in terms of its mathematical structure.

Categorize Because we are given force and displacement vectors and asked to find the work done by this force on the particle, we categorize this example as a substitution problem.

Substitute the expressions for $\vec{\mathbf{F}}$ and $\Delta \vec{\mathbf{r}}$ into Equation 6.3 and use Equations 6.4 and 6.5:

$$W = \vec{\mathbf{F}} \cdot \Delta \vec{\mathbf{r}} = [(5.0\hat{\mathbf{i}} + 2.0\hat{\mathbf{j}}) \text{ N}] \cdot [(2.0\hat{\mathbf{i}} + 3.0\hat{\mathbf{j}}) \text{ m}]$$
$$= (5.0\hat{\mathbf{i}} \cdot 2.0\hat{\mathbf{i}} + 5.0\hat{\mathbf{i}} \cdot 3.0\hat{\mathbf{j}} + 2.0\hat{\mathbf{j}} \cdot 2.0\hat{\mathbf{i}} + 2.0\hat{\mathbf{j}} \cdot 3.0\hat{\mathbf{j}}) \text{ N} \cdot \text{m}$$
$$= [10 + 0 + 0 + 6] \text{ N} \cdot \text{m} = \boxed{16 \text{ J}}$$

6.4 | Work Done by a Varying Force

Consider a particle being displaced along the x axis under the action of a force that varies with position. The particle is displaced in the direction of increasing x from $x = x_i$ to $x = x_f$. In such a situation, we cannot use $W = F \Delta r \cos \theta$ to calculate the work done by the force because this relationship applies only when $\vec{\mathbf{F}}$ is constant in magnitude and direction. If, however, we imagine that the particle undergoes a very small displacement Δx, shown in Figure 6.7a, the x component F_x of the force is approximately constant over this small interval; for this small displacement, we can approximate the work done on the particle by the force as

$$W \approx F_x \Delta x$$

which is the area of the shaded rectangle in Figure 6.7a. If we imagine the F_x versus x curve divided into a large number of such intervals, the total work done for the displacement from x_i to x_f is approximately equal to the sum of a large number of such terms:

$$W \approx \sum_{x_i}^{x_f} F_x \Delta x$$

If the size of the small displacements is allowed to approach zero, the number of terms in the sum increases without limit but the value of the sum approaches a definite value equal to the area bounded by the F_x curve and the x axis:

$$\lim_{\Delta x \to 0} \sum_{x_i}^{x_f} F_x \, \Delta x = \int_{x_i}^{x_f} F_x \, dx$$

Therefore, we can express the work done by F_x on the particle as it moves from x_i to x_f as

$$W = \int_{x_i}^{x_f} F_x \, dx \qquad \textbf{6.7} \blacktriangleleft$$

This equation reduces to Equation 6.1 when the component $F_x = F \cos \theta$ remains constant.

If more than one force acts on a system *and the system can be modeled as a particle,* the total work done on the system is just the work done by the net force. If we express the net force in the x direction as ΣF_x, the total work, or *net work,* done as the particle moves from x_i to x_f is

$$\sum W = W_{\text{ext}} = \int_{x_i}^{x_f} \left(\sum F_x \right) dx \quad \text{(particle)}$$

For the general case of a net force $\Sigma \vec{F}$ whose magnitude and direction may vary, we use the scalar product,

$$\sum W = W_{\text{ext}} = \int \left(\sum \vec{F} \right) \cdot d\vec{r} \quad \text{(particle)} \qquad \textbf{6.8} \blacktriangleleft$$

where the integral is calculated over the path that the particle takes through space. The subscript "ext" on work reminds us that the net work is done by an *external* agent on the system. We will use this notation in this chapter as a reminder and to differentiate this work from an *internal* work to be described shortly.

If the system cannot be modeled as a particle (for example, if the system is deformable), we cannot use Equation 6.8 because different forces on the system may move through different displacements. In this case, we must evaluate the work done by each force separately and then add the works algebraically to find the net work done on the system:

$$\sum W = W_{\text{ext}} = \sum_{\text{forces}} \left(\int \vec{F} \cdot d\vec{r} \right) \quad \text{(deformable system)}$$

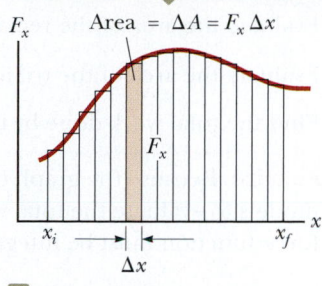

The total work done for the displacement from x_i to x_f is approximately equal to the sum of the areas of all the rectangles.

a

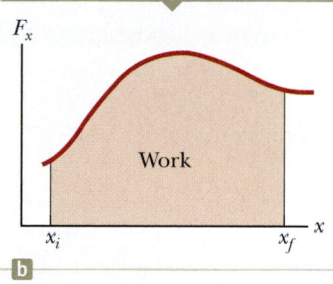

The work done by the component F_x of the varying force as the particle moves from x_i to x_f is *exactly* equal to the area under the curve.

b

Figure 6.7 (a) The work done on a particle by the force component F_x for the small displacement Δx is $F_x \Delta x$, which equals the area of the shaded rectangle. (b) The width Δx of each rectangle is shrunk to zero.

Example 6.4 | Calculating Total Work Done from a Graph

A force acting on a particle varies with x as shown in Figure 6.8. Calculate the work done by the force on the particle as it moves from $x = 0$ to $x = 6.0$ m.

SOLUTION

Conceptualize Imagine a particle subject to the force in Figure 6.8. The force remains constant as the particle moves through the first 4.0 m and then decreases linearly to zero at 6.0 m. In terms of earlier discussions of motion, the particle could be modeled as a particle under constant acceleration for the first 4.0 m because the force is constant. Between 4.0 m and 6.0 m, however, the motion does not fit into one of our earlier analysis models because the acceleration of the particle is changing. If the particle starts from rest, its speed increases throughout the motion and the particle is always moving in the positive x direction. These details about its speed and direction are not necessary for the calculation of the work done, however.

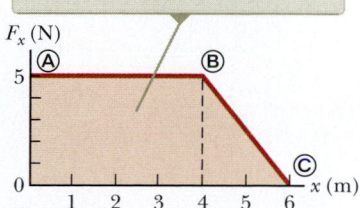

The net work done by this force is the area under the curve.

Figure 6.8 (Example 6.4) The force acting on a particle is constant for the first 4.0 m of motion and then decreases linearly with x from $x_{\text{B}} = 4.0$ m to $x_{\text{C}} = 6.0$ m.

Categorize Because the force varies during the entire motion of the particle, we must use the techniques for work done by varying forces. In this case, the graphical representation in Figure 6.8 can be used to evaluate the work done.

continued

6.4 *cont.*

....

Analyze The work done by the force is equal to the area under the curve from $x_\text{Ⓐ} = 0$ to $x_\text{Ⓒ} = 6.0$ m. This area is equal to the area of the rectangular section from Ⓐ to Ⓑ plus the area of the triangular section from Ⓑ to Ⓒ.

Evaluate the area of the rectangle: $\qquad\qquad\qquad W_\text{Ⓐ to Ⓑ} = (5.0\text{ N})(4.0\text{ m}) = 20\text{ J}$

Evaluate the area of the triangle: $\qquad\qquad\qquad W_\text{Ⓑ to Ⓒ} = \frac{1}{2}(5.0\text{ N})(2.0\text{ m}) = 5.0\text{ J}$

Find the total work done by the force on the particle: $\qquad W_\text{Ⓐ to Ⓒ} = W_\text{Ⓐ to Ⓑ} + W_\text{Ⓑ to Ⓒ} = 20\text{ J} + 5.0\text{ J} = \boxed{25\text{ J}}$

....

Finalize Because the graph of the force consists of straight lines, we can use rules for finding the areas of simple geometric models to evaluate the total work done in this example. If a force does not vary linearly, such rules cannot be used and the force function must be integrated as in Equation 6.7 or 6.8.

Work Done by a Spring

A model of a common physical system on which the force varies with position is shown in Active Figure 6.9. The system is a block on a frictionless, horizontal surface and connected to a spring. For many springs, if the spring is either stretched or compressed a small distance from its unstretched (equilibrium) configuration, it exerts on the block a force that can be mathematically modeled as

▶ Spring force

$$F_s = -kx \qquad\qquad\qquad \text{6.9} \blacktriangleleft$$

where x is the position of the block relative to its equilibrium ($x = 0$) position and k is a positive constant called the **force constant** or the **spring constant** of the spring. In other words, the force required to stretch or compress a spring is proportional to the amount of stretch or compression x. This force law for springs is known as **Hooke's law.** The value of k is a measure of the *stiffness* of the spring. Stiff

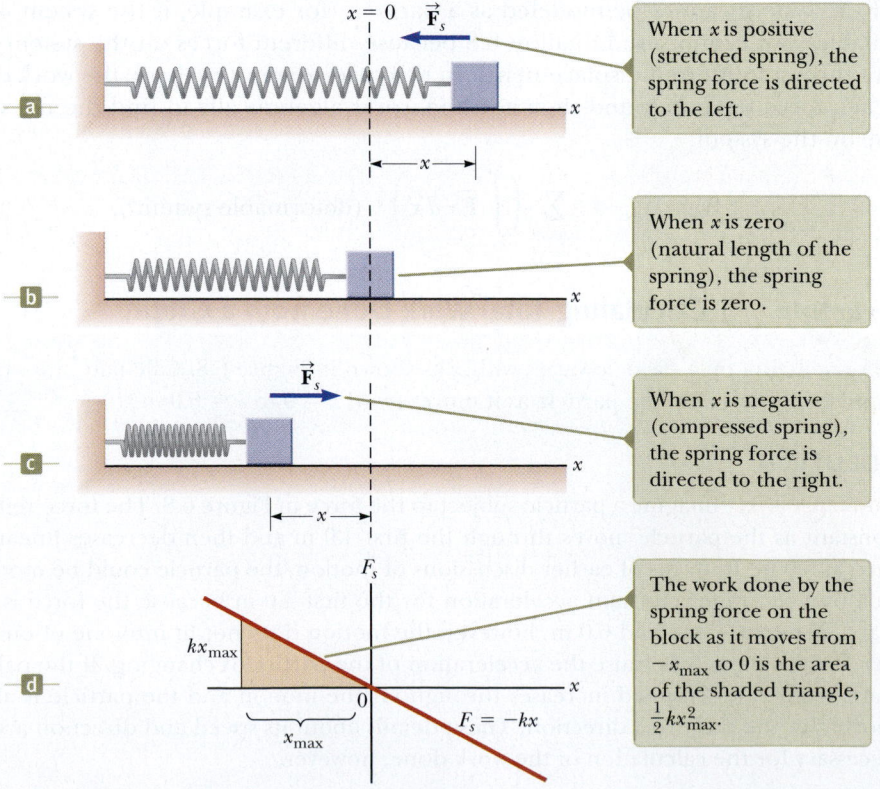

Active Figure 6.9 The force exerted by a spring on a block varies with the block's position x relative to the equilibrium position $x = 0$. (a) x is positive. (b) x is zero. (c) x is negative. (d) Graph of F_s versus x for the block–spring system.

springs have large k values, and soft springs have small k values. As can be seen from Equation 6.9, the units of k are N/m.

The vector form of Equation 6.9 is

$$\vec{F}_s = F_s\hat{i} = -kx\hat{i} \qquad \textbf{6.10}◀$$

where we have chosen the x axis to lie along the direction the spring extends or compresses.

The negative sign in Equations 6.9 and 6.10 signifies that the force exerted by the spring is always directed *opposite* the displacement from equilibrium. When $x > 0$ as in Active Figure 6.9a so that the block is to the right of the equilibrium position, the spring force is directed to the left, in the negative x direction. When $x < 0$ as in Active Figure 6.9c, the block is to the left of equilibrium and the spring force is directed to the right, in the positive x direction. When $x = 0$ as in Active Figure 6.9b, the spring is unstretched and $F_s = 0$. Because the spring force always acts toward the equilibrium position ($x = 0$), it is sometimes called a *restoring force*.

If the spring is compressed until the block is at the point $-x_{max}$ and is then released, the block moves from $-x_{max}$ through zero to $+x_{max}$. It then reverses direction, returns to $-x_{max}$, and continues oscillating back and forth. We will study these oscillations in more detail in Chapter 12. For now, let's investigate the work done by the spring on the block over small portions of one oscillation.

Suppose the block is pushed to the left to a position $-x_{max}$ and is then released. We identify the block as our system and calculate the work W_s done by the spring force on the block as the block moves from $x_i = -x_{max}$ to $x_f = 0$. Applying Equation 6.8 and assuming the block may be modeled as a particle, we obtain

$$W_s = \int \vec{F}_s \cdot d\vec{r} = \int_{x_i}^{x_f} (-kx\hat{i}) \cdot (dx\hat{i}) = \int_{-x_{max}}^{0} (-kx)\, dx = \tfrac{1}{2}kx_{max}^2 \qquad \textbf{6.11}◀$$

where we have used the integral $\int x^n\, dx = x^{n+1}/(n+1)$ with $n = 1$. The work done by the spring force is positive because the force is in the same direction as its displacement (both are to the right). Because the block arrives at $x = 0$ with some speed, it will continue moving until it reaches a position $+x_{max}$. The work done by the spring force on the block as it moves from $x_i = 0$ to $x_f = x_{max}$ is $W_s = -\tfrac{1}{2}kx_{max}^2$. The work is negative because for this part of the motion the spring force is to the left and its displacement is to the right. Therefore, the *net* work done by the spring force on the block as it moves from $x_i = -x_{max}$ to $x_f = x_{max}$ is zero.

Active Figure 6.9d is a plot of F_s versus x. The work calculated in Equation 6.11 is the area of the shaded triangle, corresponding to the displacement from $-x_{max}$ to 0. Because the triangle has base x_{max} and height kx_{max}, its area is $\tfrac{1}{2}kx_{max}^2$, agreeing with the work done by the spring as given by Equation 6.11.

If the block undergoes an arbitrary displacement from $x = x_i$ to $x = x_f$, the work done by the spring force on the block is

$$W_s = \int_{x_i}^{x_f} (-kx)\, dx = \tfrac{1}{2}kx_i^2 - \tfrac{1}{2}kx_f^2 \qquad \textbf{6.12}◀$$

From Equation 6.12, we see that the work done by the spring force is zero for any motion that ends where it began ($x_i = x_f$). We shall make use of this important result in Chapter 7 when we describe the motion of this system in greater detail.

Equations 6.11 and 6.12 describe the work done by the spring on the block. Now let us consider the work done on the block by an *external agent* as the agent applies a force on the block and the block moves *very slowly* from $x_i = -x_{max}$ to $x_f = 0$ as in Figure 6.10. We can calculate this work by noting that at any value of the position, the *applied force* $\vec{F}_{app}$ is equal in magnitude and opposite in direction to the spring force $\vec{F}_s$, so $\vec{F}_{app} = F_{app}\hat{i} = -\vec{F}_s = -(-kx\hat{i}) = kx\hat{i}$. Therefore, the work done by this applied force (the external agent) on the system of the block is

$$W_{ext} = \int \vec{F}_{app} \cdot d\vec{r} = \int_{x_i}^{x_f} (kx\hat{i}) \cdot (dx\hat{i}) = \int_{-x_{max}}^{0} kx\, dx = -\tfrac{1}{2}kx_{max}^2$$

If the process of moving the block is carried out very slowly, then $\vec{F}_{app}$ is equal in magnitude and opposite in direction to $\vec{F}_s$ at all times.

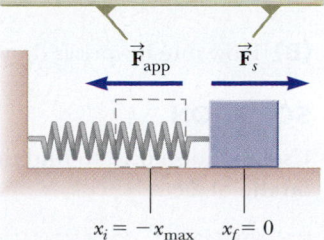

$$x_i = -x_{max} \qquad x_f = 0$$

Figure 6.10 A block moves from $x_i = -x_{max}$ to $x_f = 0$ on a frictionless surface as a force $\vec{F}_{app}$ is applied to the block.

This work is equal to the negative of the work done by the spring force for this displacement (Eq. 6.11). The work is negative because the external agent must push inward on the spring to prevent it from expanding, and this direction is opposite the direction of the displacement of the point of application of the force as the block moves from $-x_{max}$ to 0.

For an arbitrary displacement of the block, the work done on the system by the external agent is

$$W_{ext} = \int_{x_i}^{x_f} kx \, dx = \tfrac{1}{2}kx_f^2 - \tfrac{1}{2}kx_i^2 \qquad \text{6.13} \blacktriangleleft$$

Notice that this equation is the negative of Equation 6.12.

> **QUICK QUIZ 6.4** A dart is inserted into a spring-loaded dart gun by pushing the spring in by a distance x. For the next loading, the spring is compressed a distance $2x$. How much work is required to load the second dart compared with that required to load the first? **(a)** four times as much **(b)** two times as much **(c)** the same **(d)** half as much **(e)** one-fourth as much

Example 6.5 | Measuring *k* for a Spring

A common technique used to measure the force constant of a spring is demonstrated by the setup in Figure 6.11. The spring is hung vertically (Fig. 6.11a), and an object of mass m is attached to its lower end. Under the action of the "load" mg, the spring stretches a distance d from its equilibrium position (Fig. 6.11b).

(A) If a spring is stretched 2.0 cm by a suspended object having a mass of 0.55 kg, what is the force constant of the spring?

SOLUTION

Conceptualize Figure 6.11b shows what happens to the spring when the object is attached to it. Simulate this situation by hanging an object on a rubber band.

Categorize The object in Figure 6.11b is not accelerating, so it is modeled as a particle in equilibrium.

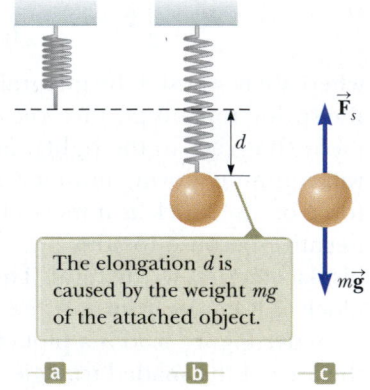

The elongation d is caused by the weight mg of the attached object.

Figure 6.11 (Example 6.5) Determining the force constant k of a spring.

Analyze Because the object is in equilibrium, the net force on it is zero and the upward spring force balances the downward gravitational force $m\vec{g}$ (Fig. 6.11c).

Apply the particle in equilibrium model to the object:

$$\vec{F}_s + m\vec{g} = 0 \rightarrow F_s - mg = 0 \rightarrow F_s = mg$$

Apply Hooke's law to give $F_s = kd$ and solve for k:

$$k = \frac{mg}{d} = \frac{(0.55 \text{ kg})(9.80 \text{ m/s}^2)}{2.0 \times 10^{-2} \text{ m}} = \boxed{2.7 \times 10^2 \text{ N/m}}$$

(B) How much work is done by the spring on the object as it stretches through this distance?

SOLUTION

Use Equation 6.12 to find the work done by the spring on the object:

$$W_s = 0 - \tfrac{1}{2}kd^2 = -\tfrac{1}{2}(2.7 \times 10^2 \text{ N/m})(2.0 \times 10^{-2} \text{ m})^2$$
$$= \boxed{-5.4 \times 10^{-2} \text{ J}}$$

Finalize As the object moves through the 2.0-cm distance, the gravitational force also does work on it. This work is positive because the gravitational force is downward and so is the displacement of the point of application of this force. Based on Equation 6.12 and the discussion afterward, would we expect the work done by the gravitational force to be $+5.4 \times 10^{-2}$ J? Let's find out.

6.5 *cont.*

Evaluate the work done by the gravitational force on the object:

$$W = \vec{\mathbf{F}} \cdot \Delta \vec{\mathbf{r}} = (mg)(d) \cos 0 = mgd$$

$$= (0.55 \text{ kg})(9.80 \text{ m/s}^2)(2.0 \times 10^{-2} \text{ m}) = 1.1 \times 10^{-1} \text{ J}$$

If you expected the work done by gravity simply to be that done by the spring with a positive sign, you may be surprised by this result! To understand why that is not the case, we need to explore further, as we do in the next section.

6.5 | Kinetic Energy and the Work–Kinetic Energy Theorem

We have investigated work and identified it as a mechanism for transferring energy into a system. We have stated that work is an influence on a system from the environment, but we have not yet discussed the *result* of this influence on the system. One possible result of doing work on a system is that the system changes its speed. In this section, we investigate this situation and introduce our first type of energy that a system can possess, called *kinetic energy.*

Consider a system consisting of a single object. Figure 6.12 shows a block of mass *m* moving through a displacement directed to the right under the action of a net force $\Sigma\vec{\mathbf{F}}$, also directed to the right. We know from Newton's second law that the block moves with an acceleration $\vec{\mathbf{a}}$. If the block (and therefore the force) moves through a displacement $\Delta\vec{\mathbf{r}} = \Delta x\hat{\mathbf{i}} = (x_f - x_i)\hat{\mathbf{i}}$, the net work done on the block by the external net force $\Sigma\vec{\mathbf{F}}$ is

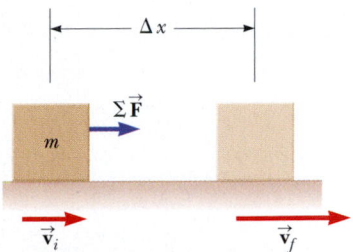

Figure 6.12 An object undergoing a displacement $\Delta\vec{\mathbf{r}} = \Delta x\hat{\mathbf{i}}$ and a change in velocity under the action of a constant net force $\Sigma\vec{\mathbf{F}}$.

$$W_{\text{ext}} = \int_{x_i}^{x_f} \Sigma F \, dx \qquad \textbf{6.14} \blacktriangleleft$$

Using Newton's second law, we substitute for the magnitude of the net force $\Sigma F = ma$ and then perform the following chain-rule manipulations on the integrand:

$$W_{\text{ext}} = \int_{x_i}^{x_f} ma \, dx = \int_{x_i}^{x_f} m\frac{dv}{dt} \, dx = \int_{x_i}^{x_f} m\frac{dv}{dx}\frac{dx}{dt} \, dx = \int_{v_i}^{v_f} mv \, dv$$

$$W_{\text{ext}} = \tfrac{1}{2}mv_f^2 - \tfrac{1}{2}mv_i^2 \qquad \textbf{6.15} \blacktriangleleft$$

where v_i is the speed of the block at $x = x_i$ and v_f is its speed at x_f.

Equation 6.15 was generated for the specific situation of one-dimensional motion, but it is a general result. It tells us that the work done by the net force on a particle of mass *m* is equal to the difference between the initial and final values of a quantity $\frac{1}{2}mv^2$. This quantity is so important that it has been given a special name, **kinetic energy:**

$$K \equiv \tfrac{1}{2}mv^2 \qquad \textbf{6.16} \blacktriangleleft \qquad \blacktriangleright \text{ Kinetic energy}$$

Kinetic energy represents the energy associated with the motion of the particle. Kinetic energy is a scalar quantity and has the same units as work. For example, a 2.0-kg object moving with a speed of 4.0 m/s has a kinetic energy of 16 J. Table 6.1 (page 142) lists the kinetic energies for various objects.

Equation 6.15 states that the work done on a particle by a net force $\Sigma\vec{\mathbf{F}}$ acting on it equals the change in kinetic energy of the particle. It is often convenient to write Equation 6.15 in the form

$$W_{\text{ext}} = K_f - K_i = \Delta K \qquad \textbf{6.17} \blacktriangleleft$$

Another way to write it is $K_f = K_i + W_{\text{ext}}$, which tells us that the final kinetic energy of an object is equal to its initial kinetic energy plus the change in energy due to the net work done on it.

TABLE 6.1 | Kinetic Energies for Various Objects

Object	Mass (kg)	Speed (m/s)	Kinetic Energy (J)
Earth orbiting the Sun	5.97×10^{24}	2.98×10^4	2.65×10^{33}
Moon orbiting the Earth	7.35×10^{22}	1.02×10^3	3.82×10^{28}
Rocket moving at escape speed[a]	500	1.12×10^4	3.14×10^{10}
Automobile at 65 mi/h	2 000	29	8.4×10^5
Running athlete	70	10	3 500
Stone dropped from 10 m	1.0	14	98
Golf ball at terminal speed	0.046	44	45
Raindrop at terminal speed	3.5×10^{-5}	9.0	1.4×10^{-3}
Oxygen molecule in air	5.3×10^{-26}	500	6.6×10^{-21}

[a]Escape speed is the minimum speed an object must reach near the Earth's surface to move infinitely far away from the Earth.

We have generated Equation 6.17 by imagining doing work on a particle. We could also do work on a deformable system, in which parts of the system move with respect to one another. In this case, we also find that Equation 6.17 is valid as long as the net work is found by adding up the works done by each force and adding, as discussed earlier with regard to Equation 6.8.

Equation 6.17 is an important result known as the **work–kinetic energy theorem:**

▶ Work–kinetic energy theorem

> When work is done on a system and the only change in the system is in its speed, the net work done on the system equals the change in kinetic energy of the system.

The work–kinetic energy theorem indicates that the speed of a system *increases* if the net work done on it is *positive* because the final kinetic energy is greater than the initial kinetic energy. The speed *decreases* if the net work is *negative* because the final kinetic energy is less than the initial kinetic energy.

Because we have so far only investigated translational motion through space, we arrived at the work–kinetic energy theorem by analyzing situations involving translational motion. Another type of motion is *rotational motion,* in which an object spins about an axis. We will study this type of motion in Chapter 10. The work–kinetic energy theorem is also valid for systems that undergo a change in the rotational speed due to work done on the system. The windmill in the photograph at the beginning of this chapter is an example of work causing rotational motion.

The work–kinetic energy theorem will clarify a result seen earlier in this chapter that may have seemed odd. In Section 6.4, we arrived at a result of zero net work done when we let a spring push a block from $x_i = -x_{max}$ to $x_f = x_{max}$. Notice that because the speed of the block is continually changing, it may seem complicated to analyze this process. The quantity ΔK in the work–kinetic energy theorem, however, only refers to the initial and final points for the speeds; it does not depend on details of the path followed between these points. Therefore, because the speed is zero at both the initial and final points of the motion, the net work done on the block is zero. We will often see this concept of path independence in similar approaches to problems.

Let us also return to the mystery in the Finalize step at the end of Example 6.5. Why was the work done by gravity not just the value of the work done by the spring with a positive sign? Notice that the work done by gravity is larger than the magnitude of the work done by the spring. Therefore, the total work done by all forces on the object is positive. Imagine now how to create the situation in which the *only* forces on the object are the spring force and the gravitational force. You must support the object at the highest point and then remove your hand and let the object fall. If you do so, you know that when the object reaches a position 2.0 cm below your hand, it will be *moving,* which is consistent with Equation 6.17. Positive net work is

Pitfall Prevention | 6.5
Conditions for the Work–Kinetic Energy Theorem
The work–kinetic energy theorem is important but limited in its application; it is not a general principle. In many situations, other changes in the system occur besides its speed, and there are other interactions with the environment besides work. A more general principle involving energy is *conservation of energy* in Section 7.1.

Pitfall Prevention | 6.6
The Work–Kinetic Energy Theorem: Speed, Not Velocity
The work–kinetic energy theorem relates work to a change in the *speed* of a system, not a change in its velocity. For example, if an object is in uniform circular motion, its speed is constant. Even though its velocity is changing, no work is done on the object by the force causing the circular motion.

done on the object, and the result is that it has a kinetic energy as it passes through the 2.0-cm point.

The only way to prevent the object from having a kinetic energy after moving through 2.0 cm is to slowly lower it with your hand. Then, however, there is a third force doing work on the object, the normal force from your hand. If this work is calculated and added to that done by the spring force and the gravitational force, the net work done on the object is zero, which is consistent because the object is not moving at the 2.0-cm point.

Earlier, we indicated that work can be considered as a mechanism for transferring energy into a system. Equation 6.17 is a mathematical statement of this concept. When work W_{ext} is done on a system, the result is a transfer of energy across the boundary of the system. The result on the system, in the case of Equation 6.17, is a change ΔK in kinetic energy. In the next section, we investigate another type of energy that can be stored in a system as a result of doing work on the system.

QUICK QUIZ 6.5 A dart is inserted into a spring-loaded dart gun by pushing the spring in by a distance x. For the next loading, the spring is compressed a distance $2x$. How much faster does the second dart leave the gun compared with the first? (**a**) four times as fast (**b**) two times as fast (**c**) the same (**d**) half as fast (**e**) one-fourth as fast

---

> **THINKING PHYSICS 6.1**

A man wishes to load a refrigerator onto a truck using a ramp at angle θ as shown in Figure 6.13. He claims that less work would be required to load the truck if the length L of the ramp were increased. Is his claim valid?

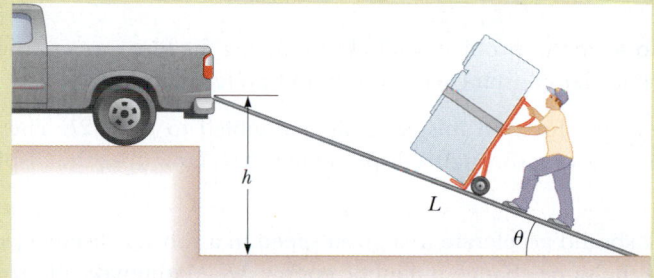

Figure 6.13 (Thinking Physics 6.1) A refrigerator attached to a frictionless, wheeled hand truck is moved up a ramp at constant speed.

Reasoning No. Suppose the refrigerator is wheeled on a hand truck up the ramp at constant speed. In this case, for the system of the refrigerator and the hand truck, $\Delta K = 0$. The normal force exerted by the ramp on the system is directed at 90° to the displacement of its point of application and so does no work on the system. Because $\Delta K = 0$, the work–kinetic energy theorem gives

$$W_{ext} = W_{by\ man} + W_{by\ gravity} = 0$$

The work done by the gravitational force equals the product of the weight mg of the system, the distance L through which the refrigerator is displaced, and $\cos(\theta + 90°)$. Therefore,

$$W_{by\ man} = -W_{by\ gravity} = -(mg)(L)[\cos(\theta + 90°)]$$

$$= mgL\sin\theta = mgh$$

where $h = L\sin\theta$ is the height of the ramp. Therefore, the man must do the same amount of work mgh on the system *regardless* of the length of the ramp. The work depends only on the height of the ramp. Although less force is required with a longer ramp, the point of application of that force moves through a greater displacement. ◄

Example 6.6 | A Block Pulled on a Frictionless Surface

A 6.0-kg block initially at rest is pulled to the right along a frictionless, horizontal surface by a constant horizontal force of 12 N. Find the block's speed after it has moved 3.0 m.

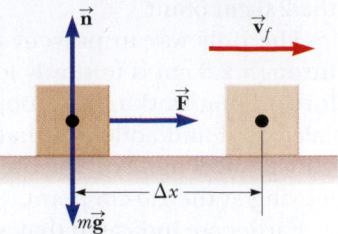

Figure 6.14 (Example 6.6) A block pulled to the right on a frictionless surface by a constant horizontal force.

SOLUTION

Conceptualize Figure 6.14 illustrates this situation. Imagine pulling a toy car across a table with a horizontal rubber band attached to the front of the car. The force is maintained constant by ensuring that the stretched rubber band always has the same length.

Categorize We could apply the equations of kinematics to determine the answer, but let us practice the energy approach. The block is the system, and three external forces act on the system. The normal force balances the gravitational force on the block, and neither of these vertically acting forces does work on the block because their points of application are horizontally displaced.

Analyze The net external force acting on the block is the horizontal 12-N force.

Use the work–kinetic energy theorem for the block, noting that its initial kinetic energy is zero:

$$W_{ext} = K_f - K_i = \tfrac{1}{2}mv_f^2 - 0 = \tfrac{1}{2}mv_f^2$$

Solve for v_f and use Equation 6.1 for the work done on the block by $\vec{F}$:

$$v_f = \sqrt{\frac{2W_{ext}}{m}} = \sqrt{\frac{2F\Delta x}{m}}$$

Substitute numerical values:

$$v_f = \sqrt{\frac{2(12 \text{ N})(3.0 \text{ m})}{6.0 \text{ kg}}} = \boxed{3.5 \text{ m/s}}$$

Finalize It would be useful for you to solve this problem again by modeling the block as a particle under a net force to find its acceleration and then as a particle under constant acceleration to find its final velocity.

What If? Suppose the magnitude of the force in this example is doubled to $F' = 2F$. The 6.0-kg block accelerates to 3.5 m/s due to this applied force while moving through a displacement $\Delta x'$. How does the displacement $\Delta x'$ compare with the original displacement Δx?

Answer If we pull harder, the block should accelerate to a given speed in a shorter distance, so we expect that $\Delta x' < \Delta x$. In both cases, the block experiences the same change in kinetic energy ΔK. Mathematically, from the work–kinetic energy theorem, we find that

$$W_{ext} = F'\Delta x' = \Delta K = F\Delta x$$

$$\Delta x' = \frac{F}{F'}\Delta x = \frac{F}{2F}\Delta x = \tfrac{1}{2}\Delta x$$

and the distance is shorter as suggested by our conceptual argument.

6.6 | Potential Energy of a System

So far in this chapter, we have defined a system in general, but have focused our attention primarily on single particles or objects under the influence of external forces. Let us now consider systems of two or more particles or objects interacting via a force that is *internal* to the system. The kinetic energy of such a system is the algebraic sum of the kinetic energies of all members of the system. There may be systems, however, in which one object is so massive that it can be modeled as stationary and its kinetic energy can be neglected. For example, if we consider a ball–Earth system as the ball falls to the Earth, the kinetic energy of the system can be considered as just the kinetic energy of the ball. The Earth moves so slowly in this process that we can ignore its kinetic energy. On the other hand, the kinetic energy of a system of two electrons must include the kinetic energies of both particles.

Imagine a system consisting of a book and the Earth, interacting via the gravitational force. We do some work on the system by lifting the book slowly from rest through a vertical displacement $\Delta\vec{r} = (y_f - y_i)\hat{j}$ as in Active Figure 6.15. According to our discussion of work as an energy transfer, this work done on the system must appear as an increase in energy of the system. The book is at rest before we perform the work and is at rest after we perform the work. Therefore, there is no change in the kinetic energy of the system.

Because the energy change of the system is not in the form of kinetic energy, it must appear as some other form of energy storage. After lifting the book, we could release it and let it fall back to the position y_i. Notice that the book (and therefore, the system) now has kinetic energy and that its source is in the work that was done in lifting the book. While the book was at the highest point, the system had the *potential* to possess kinetic energy, but it did not do so until the book was allowed to fall. Therefore, we call the energy storage mechanism before the book is released **potential energy.** We will find that the potential energy of a system can only be associated with specific types of forces acting between members of a system. The amount of potential energy in the system is determined by the *configuration* of the system. Moving members of the system to different positions or rotating them may change the configuration of the system and therefore its potential energy.

Let us now derive an expression for the potential energy associated with an object at a given location above the surface of the Earth. Consider an external agent lifting an object of mass m from an initial height y_i above the ground to a final height y_f as in Active Figure 6.15. We assume the lifting is done slowly, with no acceleration, so the applied force from the agent is equal in magnitude to the gravitational force on the object: the object is modeled as a particle in equilibrium moving at constant velocity. The work done by the external agent on the system (object and the Earth) as the object undergoes this upward displacement is given by the product of the upward applied force $\vec{F}_{app}$ and the upward displacement of this force, $\Delta\vec{r} = \Delta y\hat{j}$:

$$W_{ext} = (\vec{F}_{app}) \cdot \Delta\vec{r} = (mg\hat{j}) \cdot [(y_f - y_i)\hat{j}] = mgy_f - mgy_i \qquad \textbf{6.18} \blacktriangleleft$$

This result is the net work done on the system because the applied force is the only force on the system from the environment. (Remember that the gravitational force is *internal* to the system.) Notice the similarity between Equation 6.18 and Equation 6.15. In each equation, the work done on a system equals a difference between the final and initial values of a quantity. In Equation 6.15, the work represents a transfer of energy into the system and the increase in energy of the system is kinetic in form. In Equation 6.18, the work represents a transfer of energy into the system and the system energy appears in a different form, which we have called potential energy.

Therefore, we can identify the quantity mgy as the **gravitational potential energy** U_g:

$$U_g \equiv mgy \qquad \textbf{6.19} \blacktriangleleft \qquad \blacktriangleright \text{ Gravitational potential energy}$$

The units of gravitational potential energy are joules, the same as the units of work and kinetic energy. Potential energy, like work and kinetic energy, is a scalar quantity. Notice that Equation 6.19 is valid only for objects near the surface of the Earth, where g is approximately constant.[3]

Using our definition of gravitational potential energy, Equation 6.18 can now be rewritten as

$$W_{ext} = \Delta U_g \qquad \textbf{6.20} \blacktriangleleft$$

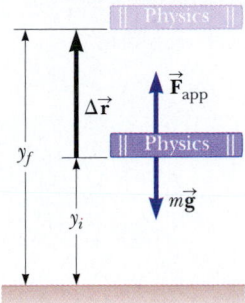

The work done by the agent on the book–Earth system is $mgy_f - mgy_i$.

Active Figure 6.15 An external agent lifts a book slowly from a height y_i to a height y_f.

Pitfall Prevention | 6.7
Potential Energy
The phrase *potential energy* does not refer to something that has the potential to become energy. Potential energy *is* energy.

Pitfall Prevention | 6.8
Potential Energy Belongs to a System
Potential energy is always associated with a *system* of two or more interacting objects. When a small object moves near the surface of the Earth under the influence of gravity, we may sometimes refer to the potential energy "associated with the object" rather than the more proper "associated with the system" because the Earth does not move significantly. We will not, however, refer to the potential energy "of the object" because this wording ignores the role of the Earth.

---

[3]The assumption that g is constant is valid as long as the vertical displacement of the object is small compared with the Earth's radius.

which mathematically describes that the net external work done on the system in this situation appears as a change in the gravitational potential energy of the system.

Gravitational potential energy depends only on the vertical height of the object above the surface of the Earth. The same amount of work must be done on an object–Earth system whether the object is lifted vertically from the Earth or is pushed starting from the same point up a frictionless incline, ending up at the same height. We verified this statement for a specific situation of rolling a refrigerator up a ramp in Thinking Physics 6.1. This statement can be shown to be true in general by calculating the work done on an object by an agent moving the object through a displacement having both vertical and horizontal components:

$$W_{\text{ext}} = (\vec{\mathbf{F}}_{\text{app}}) \cdot \Delta \vec{\mathbf{r}} = (mg\hat{\mathbf{j}}) \cdot [(x_f - x_i)\hat{\mathbf{i}} + (y_f - y_i)\hat{\mathbf{j}}] = mgy_f - mgy_i$$

where there is no term involving x in the final result because $\hat{\mathbf{j}} \cdot \hat{\mathbf{i}} = 0$.

In solving problems, you must choose a reference configuration for which the gravitational potential energy of the system is set equal to some reference value, which is normally zero. The choice of reference configuration is completely arbitrary because the important quantity is the *difference* in potential energy, and this difference is independent of the choice of reference configuration.

It is often convenient to choose as the reference configuration for zero gravitational potential energy the configuration in which an object is at the surface of the Earth, but this choice is not essential. Often, the statement of the problem suggests a convenient configuration to use.

> **QUICK QUIZ 6.6** Choose the correct answer. The gravitational potential energy of a system (**a**) is always positive (**b**) is always negative (**c**) can be negative or positive

---

Example 6.7 | The Proud Athlete and the Sore Toe

A trophy being shown off by a careless athlete slips from the athlete's hands and drops on his toe. Choosing floor level as the $y = 0$ point of your coordinate system, estimate the change in gravitational potential energy of the trophy–Earth system as the trophy falls. Repeat the calculation, using the top of the athlete's head as the origin of coordinates.

SOLUTION

Conceptualize The trophy changes its vertical position with respect to the surface of the Earth. Associated with this change in position is a change in the gravitational potential energy of the trophy–Earth system.

Categorize We evaluate a change in gravitational potential energy defined in this section, so we categorize this example as a substitution problem. Because there are no numbers, it is also an estimation problem.

The problem statement tells us that the reference configuration of the trophy–Earth system corresponding to zero potential energy is when the bottom of the trophy is at the floor. To find the change in potential energy for the system, we need to estimate a few values. Let's say the trophy has a mass of approximately 2 kg, and the top of a person's toe is about 0.03 m above the floor. Also, let's assume the trophy falls from a height of 0.5 m.

Calculate the gravitational potential energy of the trophy–Earth system just before the trophy is released:
$$U_i = mgy_i = (2 \text{ kg})(9.80 \text{ m/s}^2)(0.5 \text{ m}) = 9.80 \text{ J}$$

Calculate the gravitational potential energy of the trophy–Earth system when the trophy reaches the athlete's toe:
$$U_f = mgy_f = (2 \text{ kg})(9.80 \text{ m/s}^2)(0.03 \text{ m}) = 0.588 \text{ J}$$

Evaluate the change in gravitational potential energy of the trophy–Earth system:
$$\Delta U_g = 0.588 \text{ J} - 9.80 \text{ J} = -9.21 \text{ J}$$

We should probably keep only one digit because of the roughness of our estimates; therefore, we estimate that the change in gravitational potential energy is -9 J. The system had about 10 J of gravitational potential energy before the trophy began its fall and approximately 1 J of potential energy as the trophy reaches the top of the toe.

6.7 *cont.*

The second case presented indicates that the reference configuration of the system for zero potential energy is chosen to be when the trophy is at the athlete's head (even though the trophy is never at this position in its motion). We estimate this position to be 1.50 m above the floor).

Calculate the gravitational potential energy of the trophy–Earth system just before the trophy is released from its position 1 m below the athlete's head:

$$U_i = mgy_i = (2 \text{ kg})(9.80 \text{ m/s}^2)(-1 \text{ m}) = -19.6 \text{ J}$$

Calculate the gravitational potential energy of the trophy–Earth system when the trophy reaches the athlete's toe located 1.47 m below the athlete's head:

$$U_f = mgy_f = (2 \text{ kg})(9.80 \text{ m/s}^2)(-1.47 \text{ m}) = -28.8 \text{ J}$$

Evaluate the change in gravitational potential energy of the trophy–Earth system:

$$\Delta U_g = -28.8 \text{ J} - (-19.6 \text{ J}) = -9.2 \text{ J} \approx \boxed{-9 \text{ J}}$$

This value is the same as before, as it must be.

Elastic Potential Energy

Because members of a system can interact with one another by means of different types of forces, it is possible that there are different types of potential energy in a system. We are familiar with gravitational potential energy of a system in which members interact via the gravitational force. Let's explore a second type of potential energy that a system can possess.

Consider a system consisting of a block and a spring as shown in Active Figure 6.16 (page 148). In Section 6.4, we identified *only* the block as the system. Now we include both the block and the spring in the system and recognize that the spring force is the interaction between the two members of the system. The force that the spring exerts on the block is given by $F_s = -kx$ (Eq. 6.9). The work done by an external applied force F_{app} on a system consisting of a block connected to the spring is given by Equation 6.13:

$$W_{\text{ext}} = \tfrac{1}{2}kx_f^2 - \tfrac{1}{2}kx_i^2 \qquad \textbf{6.21}\blacktriangleleft$$

In this situation, the initial and final x coordinates of the block are measured from its equilibrium position, $x = 0$. Again (as in the gravitational case) we see that the work done on the system is equal to the difference between the initial and final values of an expression related to the system's configuration. The **elastic potential energy** function associated with the block–spring system is defined by

$$U_s \equiv \tfrac{1}{2}kx^2 \qquad \textbf{6.22}\blacktriangleleft \qquad \blacktriangleright \text{ Elastic potential energy}$$

The elastic potential energy of the system can be thought of as the energy stored in the deformed spring (one that is either compressed or stretched from its equilibrium position). The elastic potential energy stored in a spring is zero whenever the spring is undeformed ($x = 0$). Energy is stored in the spring only when the spring is either stretched or compressed. Because the elastic potential energy is proportional to x^2, we see that U_s is always positive in a deformed spring. Everyday examples of the storage of elastic potential energy can be found in old-style clocks or watches that operate from a wound-up spring and small wind-up toys for children.

Consider Active Figure 6.16, which shows a spring on a frictionless, horizontal surface. When a block is pushed against the spring by an external agent, the elastic potential energy and the total energy of the system increase as indicated in Figure 6.16b. When the spring is compressed a distance x_{max} (Active Fig. 6.16c), the elastic potential energy stored in the spring is $\tfrac{1}{2}kx_{\text{max}}^2$. When the block is released from rest, the spring exerts a force on the block and pushes the block to the right.

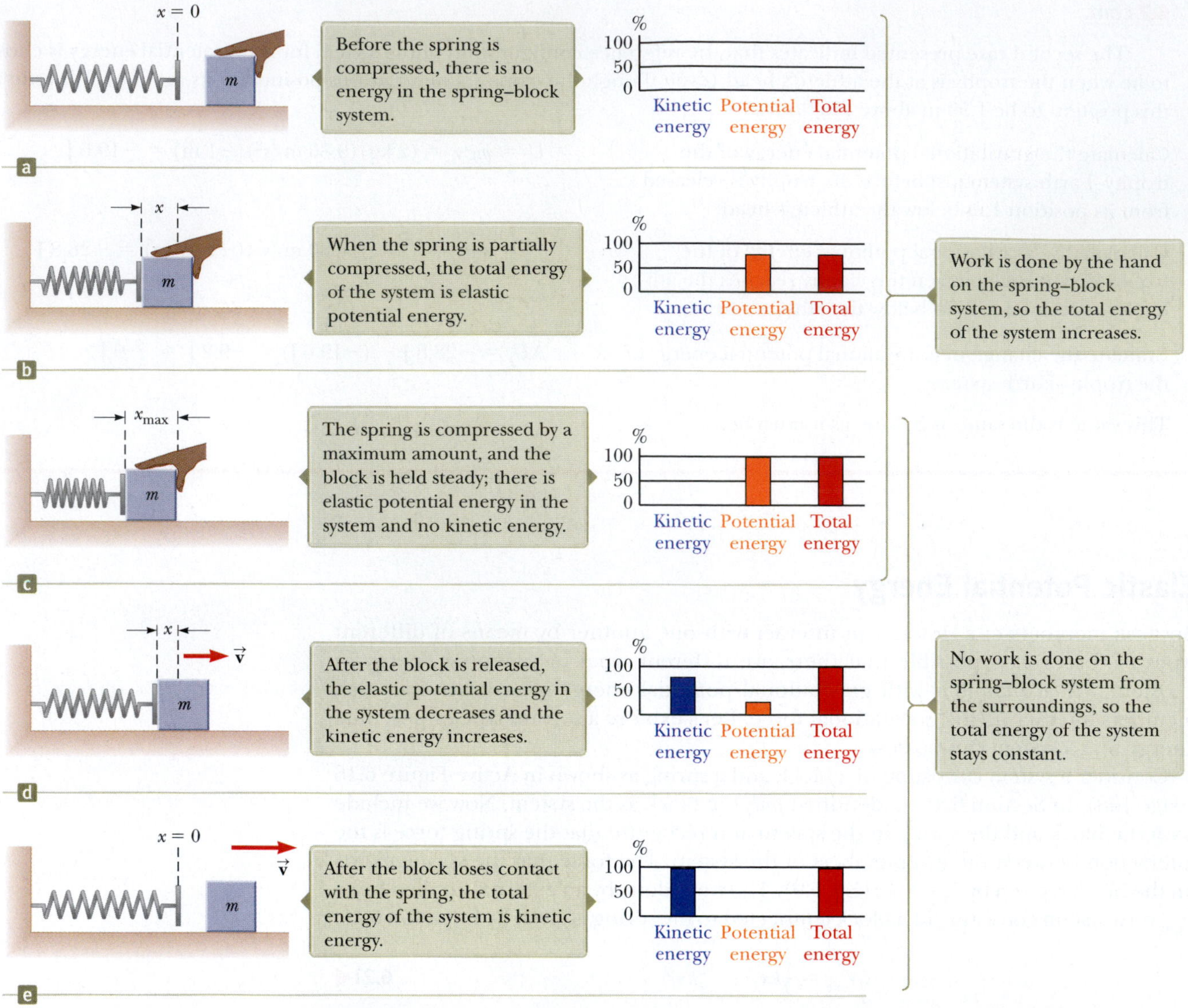

$x = 0$

a Before the spring is compressed, there is no energy in the spring–block system.

Kinetic energy	Potential energy	Total energy

x

b When the spring is partially compressed, the total energy of the system is elastic potential energy.

Kinetic energy	Potential energy	Total energy

Work is done by the hand on the spring–block system, so the total energy of the system increases.

x_{max}

c The spring is compressed by a maximum amount, and the block is held steady; there is elastic potential energy in the system and no kinetic energy.

Kinetic energy	Potential energy	Total energy

x

$\vec{v}$

d After the block is released, the elastic potential energy in the system decreases and the kinetic energy increases.

Kinetic energy	Potential energy	Total energy

No work is done on the spring–block system from the surroundings, so the total energy of the system stays constant.

$x = 0$

$\vec{v}$

e After the block loses contact with the spring, the total energy of the system is kinetic energy.

Kinetic energy	Potential energy	Total energy

Active Figure 6.16 A spring on a frictionless, horizontal surface is compressed a distance x_{max} when a block of mass m is pushed against it. The block is then released and the spring pushes it to the right, where the block eventually loses contact with the spring. Parts (a) through (e) show various instants in the process. Energy bar charts on the right of each part of the figure help keep track of the energy in the system.

The elastic potential energy of the system decreases, whereas the kinetic energy increases and the total energy remains fixed (Fig. 6.16d). When the spring returns to its original length, the stored elastic potential energy is completely transformed into kinetic energy of the block (Active Fig. 6.16e).

Energy Bar Charts

Active Figure 6.16 shows an important graphical representation of information related to energy of systems called an **energy bar chart.** The vertical axis represents the amount of energy of a given type in the system. The horizontal axis shows the types of energy in the system. The bar chart in Active Figure 6.16a shows that the system contains zero energy because the spring is relaxed and the block is not moving. Between Active Figure 6.16a and Active Figure 6.16c, the hand does work on the system, compressing the spring and storing elastic potential energy in the system. In Active Figure 6.16d, the block has been released and is moving to the right while still in contact with the spring. The height of the bar for the elastic potential energy of the system decreases, the kinetic energy bar increases, and the total energy bar remains fixed. In Active Figure 6.16e, the spring has returned to its relaxed length and the system now contains only kinetic energy associated with the moving block.

Energy bar charts can be a very useful representation for keeping track of the various types of energy in a system. For practice, try making energy bar charts for the book–Earth system in Active Figure 6.15 when the book is dropped from the higher position. Figure 6.17 associated with Quick Quiz 6.7 shows another system for which drawing an energy bar chart would be a good exercise. We will show energy bar charts in some figures in this chapter. Some Active Figures will not show a bar chart in the text but will include one in the animation in Enhanced WebAssign.

Figure 6.17 (Quick Quiz 6.7) A ball connected to a massless spring suspended vertically. What forms of potential energy are associated with the system when the ball is displaced downward?

> **QUICK QUIZ 6.7** A ball is connected to a light spring suspended vertically as shown in Figure 6.17. When pulled downward from its equilibrium position and released, the ball oscillates up and down. **(i)** In the system of *the ball, the spring, and the Earth,* what forms of energy are there during the motion? (a) kinetic and elastic potential (b) kinetic and gravitational potential (c) kinetic, elastic potential, and gravitational potential (d) elastic potential and gravitational potential **(ii)** In the system of *the ball and the spring,* what forms of energy are there during the motion? Choose from the same possibilities (a) through (d).

6.7 | Conservative and Nonconservative Forces

We now introduce a third type of energy that a system can possess. Imagine that the book in Active Figure 6.18a has been accelerated by your hand and is now sliding to the right on the surface of a heavy table and slowing down due to the friction force. Suppose the *surface* is the system. Then the friction force from the sliding book does work on the surface. The force on the surface is to the right and the displacement of the point of application of the force is to the right because the book has moved to the right. The work done on the surface is positive, but the surface is not moving after the book has stopped. Positive work has been done on the surface, yet there is no increase in the surface's kinetic energy or the potential energy of any system.

From your everyday experience with sliding over surfaces with friction, you can probably guess that the surface will be *warmer* after the book slides over it. (Rub your hands together briskly to find out!) The work that was done on the surface has gone into warming the surface rather than increasing its speed or changing the configuration of a system. We call the energy associated with the temperature of a system its **internal energy,** symbolized E_{int}. (We will define internal energy more generally in Chapter 17.) In this case, the work done on the surface does indeed represent energy transferred into the system, but it appears in the system as internal energy rather than kinetic or potential energy.

Consider the book and the surface in Active Figure 6.18a together as a system. Initially, the system has kinetic energy because the book is moving. While the book is sliding, the internal energy of the system increases: the book and the surface are warmer than before. When the book stops, the kinetic energy has been completely transformed to internal energy. We can consider the work done by friction within the system—that is, between the book and the surface—as a *transformation mechanism* for energy. This work transforms the kinetic energy of the system into internal energy. Similarly, when a book falls straight down with no air resistance, the work done by the gravitational force within the book–Earth system transforms gravitational potential energy of the system to kinetic energy.

Active Figures 6.18b through 6.18d show energy bar charts for the situation in Active Figure 6.18a. In Active Figure 6.18b, the bar chart shows that the system

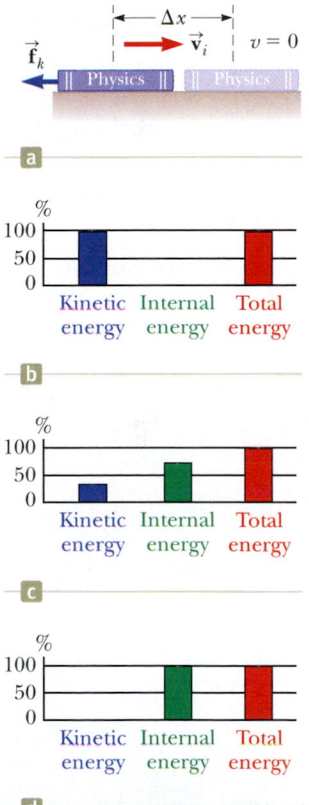

Active Figure 6.18 (a) A book sliding to the right on a horizontal surface slows down in the presence of a force of kinetic friction acting to the left. (b) An energy bar chart showing the energy in the system of the book and the surface at the initial instant of time. The energy of the system is all kinetic energy. (c) While the book is sliding, the kinetic energy of the system decreases as it is transformed to internal energy. (d) After the book has stopped, the energy of the system is all internal energy.

contains kinetic energy at the instant the book is released by your hand. We define the reference amount of internal energy in the system as zero at this instant. Active Figure 6.18c shows the kinetic energy transforming to internal energy as the book slows down due to the friction force. In Active Figure 6.18d, after the book has stopped sliding, the kinetic energy is zero and the system now contains only internal energy. Notice that the total energy bar in red has not changed during the process. The amount of internal energy in the system after the book has stopped is equal to the amount of kinetic energy in the system at the initial instant. This equality is described by an important principle called *conservation of energy*. We will explore this principle in Chapter 7.

Now consider in more detail an object moving downward near the surface of the Earth. The work done by the gravitational force on the object does not depend on whether it falls vertically or slides down a sloping incline with friction. All that matters is the change in the object's elevation. The energy transformation to internal energy due to friction on that incline, however, depends very much on the distance the object slides. The longer the incline, the more potential energy is transformed to internal energy. In other words, the path makes no difference when we consider the work done by the gravitational force, but it does make a difference when we consider the energy transformation due to friction forces. We can use this varying dependence on path to classify forces as either conservative or nonconservative. Of the two forces just mentioned, the gravitational force is conservative and the friction force is nonconservative.

Conservative Forces

▶ Properties of conservative forces

Conservative forces have these two equivalent properties:

1. The work done by a conservative force on a particle moving between any two points is independent of the path taken by the particle.
2. The work done by a conservative force on a particle moving through any closed path is zero. (A closed path is one for which the beginning point and the endpoint are identical.)

The gravitational force is one example of a conservative force; the force that an ideal spring exerts on any object attached to the spring is another. The work done by the gravitational force on an object moving between any two points near the Earth's surface is $W_g = -mg\hat{\mathbf{j}} \cdot [(y_f - y_i)\hat{\mathbf{j}}] = mgy_i - mgy_f$. From this equation, notice that W_g depends only on the initial and final y coordinates of the object and hence is independent of the path. Furthermore, W_g is zero when the object moves over any closed path (where $y_i = y_f$).

For the case of the object–spring system, the work W_s done by the spring force is given by $W_s = \frac{1}{2}kx_i^2 - \frac{1}{2}kx_f^2$ (Eq. 6.12). We see that the spring force is conservative because W_s depends only on the initial and final x coordinates of the object and is zero for any closed path.

Pitfall Prevention | 6.9
Similar Equation Warning
Compare Equation 6.23 with Equation 6.20. These equations are similar except for the negative sign, which is a common source of confusion. Equation 6.20 tells us that positive work done *by an outside agent* on a system causes an increase in the potential energy of the system (with no change in the kinetic or internal energy). Equation 6.23 states that work done *on a component of a system by a conservative force internal to the system* causes a decrease in the potential energy of the system.

We can associate a potential energy for a system with a force acting between members of the system, but we can do so only if the force is conservative. In general, the work W_{int} done by a conservative force on an object that is a member of a system as the system changes from one configuration to another is equal to the initial value of the potential energy of the system minus the final value:

$$W_{\text{int}} = U_i - U_f = -\Delta U \qquad \qquad \textbf{6.23} \blacktriangleleft$$

We use the subscript "int" in Equation 6.23 to remind us that the work we are discussing is done by one member of the system on another member and is therefore *internal* to the system. It is different from the work W_{ext} done *on* the system as a whole by an external agent. As an example, compare Equation 6.23 with the specific equation for the work done by the spring force (Eq. 6.12) as the extension of the spring changes.

Nonconservative Forces

A force is **nonconservative** if it does not satisfy properties 1 and 2 for conservative forces. We define the sum of the kinetic and potential energies of a system as the **mechanical energy** of the system:

$$E_{mech} \equiv K + U \qquad \textbf{6.24} \blacktriangleleft$$

where K includes the kinetic energy of all moving members of the system and U includes all types of potential energy in the system. For a book falling under the action of the gravitational force, the mechanical energy of the book–Earth system remains fixed; gravitational potential energy transforms to kinetic energy, and the total mechanical energy of the system remains constant. Nonconservative forces acting within a system, however, cause a *change* in the mechanical energy of the system. For example, for a book sent sliding on a horizontal surface that is not frictionless, the mechanical energy of the book–surface system is transformed to internal energy as we discussed earlier. Only part of the book's kinetic energy is transformed to internal energy in the book. The rest appears as internal energy in the surface. (When you trip and slide across a gymnasium floor, not only does the skin on your knees warm up, so does the floor!) Because the force of kinetic friction transforms the mechanical energy of a system into internal energy, it is a nonconservative force.

As an example of the path dependence of the work for a nonconservative force, consider Figure 6.19. Suppose you displace a book between two points on a table. If the book is displaced in a straight line along the blue path between points Ⓐ and Ⓑ in Figure 6.19, you do a certain amount of work against the kinetic friction force to keep the book moving at a constant speed. Now, imagine that you push the book along the brown semicircular path in Figure 6.19. You perform more work against friction along this curved path than along the straight path because the curved path is longer. The work done on the book depends on the path, so the friction force *cannot* be conservative.

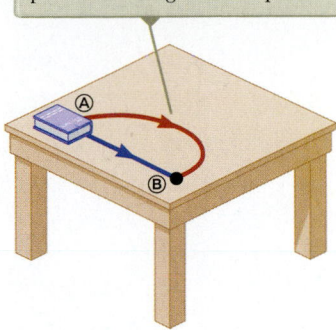

The work done in moving the book is greater along the brown path than along the blue path.

Figure 6.19 The work done against the force of kinetic friction depends on the path taken as the book is moved from Ⓐ to Ⓑ.

⫷6.8 | Relationship Between Conservative Forces and Potential Energy

In the preceding section, we found that the work done on a member of a system by a conservative force between the members of the system does not depend on the path taken by the moving member. The work depends only on the initial and final coordinates. For such a system, we can define a **potential energy function** U such that the work done within the system by the conservative force equals the negative of the change in the potential energy of the system. Let us imagine a system of particles in which a conservative force $\vec{F}$ acts between the particles. Imagine also that the configuration of the system changes due to the motion of one particle along the x axis. The work done by the force $\vec{F}$ as the particle moves along the x axis is[4]

$$W_{int} = \int_{x_i}^{x_f} F_x \, dx = -\Delta U \qquad \textbf{6.25} \blacktriangleleft$$

where F_x is the component of $\vec{F}$ in the direction of the displacement. That is, the work done by a conservative force acting between members of a system equals the negative of the change in the potential energy of the system associated with that force when the system's configuration changes. We can also express Equation 6.25 as

$$\Delta U = U_f - U_i = -\int_{x_i}^{x_f} F_x \, dx \qquad \textbf{6.26} \blacktriangleleft$$

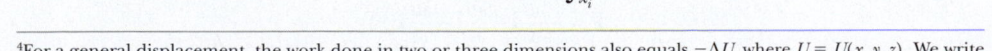

[4]For a general displacement, the work done in two or three dimensions also equals $-\Delta U$, where $U = U(x, y, z)$. We write this equation formally as $W_{int} = \int_i^f \vec{F} \cdot d\vec{r} = U_i - U_f$.

Therefore, ΔU is negative when F_x and dx are in the same direction, as when an object is lowered in a gravitational field or when a spring pushes an object toward equilibrium.

It is often convenient to establish some particular location x_i of one member of a system as representing a reference configuration and measure all potential energy differences with respect to it. We can then define the potential energy function as

$$U_f(x) = -\int_{x_i}^{x_f} F_x \, dx + U_i \qquad \textbf{6.27} \blacktriangleleft$$

The value of U_i is often taken to be zero for the reference configuration. It does not matter what value we assign to U_i because any nonzero value merely shifts $U_f(x)$ by a constant amount and only the *change* in potential energy is physically meaningful.

If the point of application of the force undergoes an infinitesimal displacement dx, we can express the infinitesimal change in the potential energy of the system dU as

$$dU = -F_x \, dx$$

Therefore, the conservative force is related to the potential energy function through the relationship[5]

▶ Relation of force between members of a system to the potential energy of the system

$$F_x = -\frac{dU}{dx} \qquad \textbf{6.28} \blacktriangleleft$$

That is, the x component of a conservative force acting on a member within a system equals the negative derivative of the potential energy of the system with respect to x.

We can easily check Equation 6.28 for the two examples already discussed. In the case of the deformed spring, $U_s = \frac{1}{2}kx^2$; therefore,

$$F_s = -\frac{dU_s}{dx} = -\frac{d}{dx}\left(\frac{1}{2}kx^2\right) = -kx$$

which corresponds to the restoring force in the spring (Hooke's law). Because the gravitational potential energy function is $U_g = mgy$, it follows from Equation 6.28 that $F_g = -mg$ when we differentiate U_g with respect to y instead of x.

We now see that U is an important function because a conservative force can be derived from it. Furthermore, Equation 6.28 should clarify that adding a constant to the potential energy is unimportant because the derivative of a constant is zero.

QUICK QUIZ 6.8 What does the slope of a graph of $U(x)$ versus x represent? (**a**) the magnitude of the force on the object (**b**) the negative of the magnitude of the force on the object (**c**) the x component of the force on the object (**d**) the negative of the x component of the force on the object

6.9 | Potential Energy for Gravitational and Electric Forces

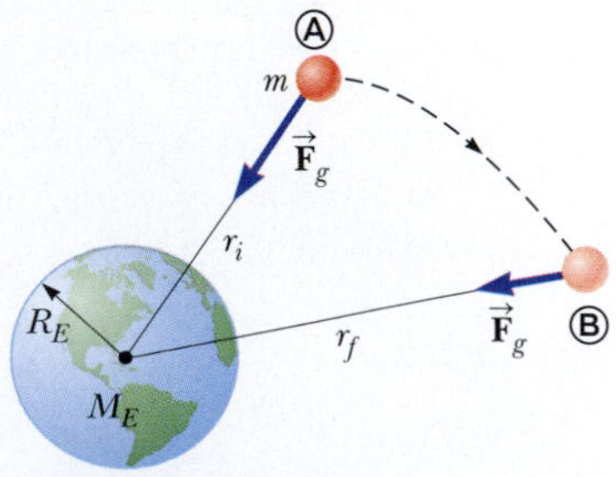

Earlier in this chapter we introduced the concept of gravitational potential energy, that is, the energy associated with a system of objects interacting via the gravitational force. We emphasized that the gravitational potential energy function, Equation 6.19, is valid only when the object of mass m is near the Earth's surface. We would like to find a more general expression for the gravitational potential energy that is valid for

Figure 6.20 As a particle of mass m moves from Ⓐ to Ⓑ above the Earth's surface, the potential energy of the particle–Earth system, given by Equation 6.31, changes because of the change in the particle–Earth separation distance r from r_i to r_f.

[5]In three dimensions, the expression is

$$\vec{\mathbf{F}} = -\frac{\partial U}{\partial x}\hat{\mathbf{i}} - \frac{\partial U}{\partial y}\hat{\mathbf{j}} - \frac{\partial U}{\partial z}\hat{\mathbf{k}}$$

where $(\partial U/\partial x)$ and so forth are partial derivatives. In the language of vector calculus, $\vec{\mathbf{F}}$ equals the negative of the *gradient* of the scalar quantity $U(x, y, z)$.

all separation distances. Because the value of g varies with height, it follows that the general dependence of the potential energy function of the system on separation distance is more complicated than our simple expression, Equation 6.19.

Consider a particle of mass m moving between two points Ⓐ and Ⓑ above the Earth's surface as in Figure 6.20. The gravitational force on the particle due to the Earth, first introduced in Section 5.5, can be written in vector form as

$$\vec{F}_g = -G\frac{M_E m}{r^2}\hat{r} \qquad \textbf{6.29}\blacktriangleleft$$

where $\hat{r}$ is a unit vector directed from the Earth toward the particle and the negative sign indicates that the force is downward toward the Earth. This expression shows that the gravitational force depends on the radial coordinate r. Furthermore, the gravitational force is conservative. Equation 6.27 gives

$$U_f = -\int_{r_i}^{r_f} F(r)\,dr + U_i = GM_E m \int_{r_i}^{r_f} \frac{dr}{r^2} + U_i = GM_E m \left(-\frac{1}{r}\right)\Big|_{r_i}^{r_f} + U_i$$

or

$$U_f = -GM_E m \left(\frac{1}{r_f} - \frac{1}{r_i}\right) + U_i \qquad \textbf{6.30}\blacktriangleleft$$

As always, the choice of a reference configuration for the potential energy is completely arbitrary. It is customary to define the reference configuration as that for which the force is zero. Letting $U_i \to 0$ as $r_i \to \infty$, we obtain the important result

$$U_g = -G\frac{M_E m}{r} \qquad \textbf{6.31}\blacktriangleleft$$

for separation distances $r > R_E$, the radius of the Earth. Because of our choice of reference configuration for zero potential energy, the function U_g is always negative (Fig. 6.21).

Although Equation 6.31 was derived for the particle–Earth system, it can be applied to *any* two particles. For *any pair* of particles of masses m_1 and m_2 separated by a distance r, the gravitational force of attraction is given by Equation 5.11 and the gravitational potential energy of the system of two particles is

$$U_g = -G\frac{m_1 m_2}{r} \qquad \textbf{6.32}\blacktriangleleft$$

This expression also applies to larger objects *if their mass distributions are spherically symmetric*, as first shown by Newton. In this case, r is measured between the centers of the spherical objects.

Equation 6.32 shows that the gravitational potential energy for any pair of particles varies as $1/r$ (whereas the force between them varies as $1/r^2$). Furthermore, the potential energy is *negative* because the force is attractive and we have chosen the potential energy to be zero when the particle separation is infinity. Because the force between the particles is attractive, we know that an external agent must do positive work to increase the separation between the two particles. The work done by the external agent produces an increase in the potential energy as the two particles are separated. That is, U_g becomes less negative as r increases.

We can extend this concept to three or more particles. In this case, the total potential energy of the system is the sum over all *pairs* of particles. Each pair contributes a term of the form given by Equation 6.32. For example, if the system contains three particles, as in Figure 6.22, we find that

$$U_{\text{total}} = U_{12} + U_{13} + U_{23} = -G\left(\frac{m_1 m_2}{r_{12}} + \frac{m_1 m_3}{r_{13}} + \frac{m_2 m_3}{r_{23}}\right) \qquad \textbf{6.33}\blacktriangleleft$$

The absolute value of U_{total} represents the work needed to separate all three particles by an infinite distance.

Pitfall Prevention | 6.10
What is r?
In Section 5.5, we discussed the gravitational force between two *particles*. In Equation 6.29, we present the gravitational force between a particle and an extended object, the Earth. We could also express the gravitational force between two extended objects, such as the Earth and the Sun. In these kinds of situations, remember that r is measured *between the centers* of the objects. Be sure *not* to measure r from the surface of the Earth.

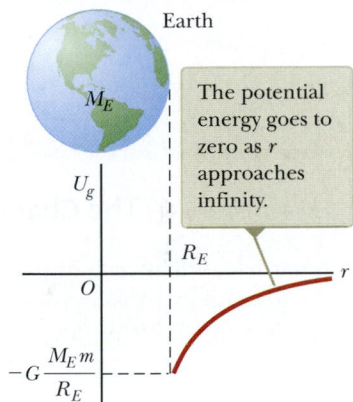

Figure 6.21 Graph of the gravitational potential energy U_g versus r for a particle above the Earth's surface.

Pitfall Prevention | 6.11
Gravitational Potential Energy
Be careful! Equation 6.32 looks similar to Equation 5.11 for the gravitational force, but there are two major differences. The gravitational force is a vector, whereas the gravitational potential energy is a scalar. The gravitational force varies as the *inverse square* of the separation distance, whereas the gravitational potential energy varies as the simple *inverse* of the separation distance.

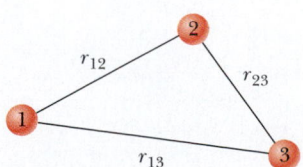

Figure 6.22 Three interacting particles.

> **THINKING PHYSICS 6.2**
>
> Why is the Sun hot?
>
> **Reasoning** The Sun was formed when a cloud of gas and dust coalesced, because of gravitational attraction, into a massive astronomical object. Let us define this cloud as our system and model the gas and dust as particles. Initially, the particles of the system were widely scattered, representing a large amount of gravitational potential energy. As the particles moved together to form the Sun, the gravitational potential energy of the system decreased. This potential energy was transformed to kinetic energy as the particles fell toward the center. As the speeds of the particles increased, many collisions occurred between particles, randomizing their motion and transforming the kinetic energy to internal energy, which represented an increase in temperature. As the particles came together, the temperature rose to a point at which nuclear reactions occurred. These reactions release huge amounts of energy that maintain the high temperature of the Sun. This process has occurred for every star in the Universe. ◄

Example 6.8 | The Change in Potential Energy

A particle of mass m is displaced through a small vertical distance Δy near the Earth's surface. Show that in this situation the general expression for the change in gravitational potential energy given by Equation 6.30 reduces to the familiar relationship $\Delta U = mg\,\Delta y$.

SOLUTION

Conceptualize Compare the two different situations for which we have developed expressions for gravitational potential energy: (1) a planet and an object that are far apart for which the energy expression is Equation 6.30 and (2) a small object at the surface of a planet for which the energy expression is Equation 6.19. We wish to show that these two expressions are equivalent.

Categorize This example is a substitution problem.

Combine the fractions in Equation 6.30:

$$(1) \quad \Delta U = -GM_E m \left(\frac{1}{r_f} - \frac{1}{r_i} \right) = GM_E m \left(\frac{r_f - r_i}{r_i r_f} \right)$$

Evaluate $r_f - r_i$ and $r_i r_f$ if both the initial and final positions of the particle are close to the Earth's surface:

$$r_f - r_i = \Delta y \qquad r_i r_f \approx R_E^2$$

Substitute these expressions into Equation (1):

$$\Delta U \approx \frac{GM_E m}{R_E^2}\,\Delta y$$

Use Equation 6.29 to express $GM_E m / R_E^2$ as the magnitude of the gravitational force F_g on an object of mass m at the Earth's surface:

$$\Delta U \approx F_g \Delta y$$

Use Equation 4.5 to express the gravitational force in terms of the acceleration due to gravity:

$$\Delta U \approx mg\,\Delta y$$

In Chapter 5, we discussed the electrostatic force between two point particles, which is given by Coulomb's law,

$$F_e = k_e \frac{q_1 q_2}{r^2} \qquad \text{6.34} ◄$$

Because this expression looks so similar to Newton's law of universal gravitation, we would expect that the generation of a potential energy function for this force would proceed in a similar way. That is indeed the case, and this procedure results in the **electric potential energy** function,

$$U_e = k_e \frac{q_1 q_2}{r}$$

6.35 ◀

As with the gravitational potential energy, the electric potential energy is defined as zero when the charges are infinitely far apart. Comparing this expression with that for the gravitational potential energy, we see the obvious differences in the constants and the use of charges instead of masses, but there is one more difference. The gravitational expression has a negative sign, but the electrical expression doesn't. For systems of objects that experience an attractive force, the potential energy decreases as the objects are brought closer together. Because we have defined zero potential energy at infinite separation, all real separations are finite and the energy must decrease from a value of zero. Therefore, all potential energies for systems of objects that attract must be negative. In the gravitational case, attraction is the only possibility. The constant, the masses, and the separation distance are all positive, so the negative sign must be included explicitly, as it is in Equation 6.32.

The electric force can be either attractive or repulsive. Attraction occurs between charges of opposite sign. Therefore, for the two charges in Equation 6.35, one is positive and one is negative if the force is attractive. The product of the charges provides the negative sign for the potential energy mathematically, and we do not need an explicit negative sign in the potential energy expression. In the case of charges with the same sign, either a product of two negative charges or two positive charges will be positive, leading to a positive potential energy. This conclusion is reasonable because to cause repelling particles to move together from infinite separation requires work to be done on the system, so the potential energy increases.

6.10 | Energy Diagrams and Equilibrium of a System

The motion of a system can often be understood qualitatively through a graph of its potential energy versus the position of a member of the system. Consider the potential energy function for a block–spring system, given by $U_s = \frac{1}{2}kx^2$. This function is plotted versus x in Active Figure 6.23a, where x is the position of the block. The force F_s exerted by the spring on the block is related to U_s through Equation 6.28:

$$F_s = -\frac{dU_s}{dx} = -kx$$

As we saw in Quick Quiz 6.8, the x component of the force is equal to the negative of the slope of the U-versus-x curve. When the block is placed at rest at the equilibrium position of the spring ($x = 0$), where $F_s = 0$, it will remain there unless some external force F_{ext} acts on it. If this external force stretches the spring from equilibrium, x is positive and the slope dU/dx is positive; therefore, the force F_s exerted by the spring is negative and the block accelerates back toward $x = 0$ when released. If the external force compresses the spring, x is negative and the slope is negative; therefore, F_s is positive and again the mass accelerates toward $x = 0$ upon release.

From this analysis, we conclude that the $x = 0$ position for a block–spring system is one of **stable equilibrium.** That is, any movement away from this position results in a force directed back toward $x = 0$. In general, configurations of a system in stable equilibrium correspond to those for which $U(x)$ for the system is a minimum.

If the block in Active Figure 6.23 is moved to an initial position x_{max} and then released from rest, its total energy initially is the potential energy $\frac{1}{2}kx_{max}^2$ stored in the spring. As the block starts to move, the system acquires kinetic energy and loses potential energy. The block oscillates (moves back and forth) between the two

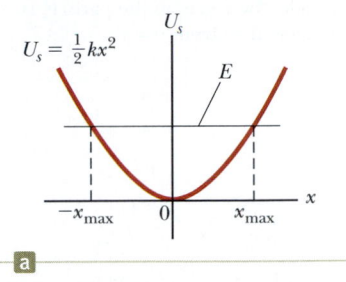

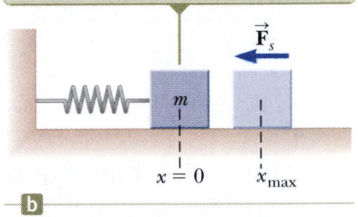

The restoring force exerted by the spring always acts toward $x = 0$, the position of stable equilibrium.

Active Figure 6.23 (a) Potential energy as a function of x for the frictionless block–spring system shown in (b). For a given energy E of the system, the block oscillates between the turning points, which have the coordinates $x = \pm x_{max}$.

Pitfall Prevention | 6.12
Energy Diagrams
A common mistake is to think that potential energy on the graph in an energy diagram represents the height of some object. For example, that is not the case in Active Figure 6.23, where the block is only moving horizontally.

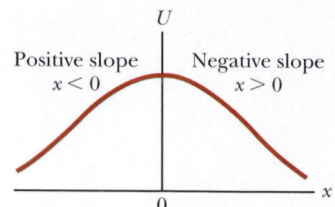

Figure 6.24 A plot of U versus x for a particle that has a position of unstable equilibrium located at $x = 0$. For any finite displacement of the particle, the force on the particle is directed away from $x = 0$.

points $x = -x_{max}$ and $x = +x_{max}$, called the *turning points*. In fact, because no energy is transformed to internal energy due to friction, the block oscillates between $-x_{max}$ and $+x_{max}$ forever. (We will discuss these oscillations further in Chapter 12.)

Another simple mechanical system with a configuration of stable equilibrium is a ball rolling about in the bottom of a bowl. Anytime the ball is displaced from its lowest position, it tends to return to that position when released.

Now consider a particle moving along the x axis under the influence of a conservative force F_x, where the U-versus-x curve is as shown in Figure 6.24. Once again, $F_x = 0$ at $x = 0$, and so the particle is in equilibrium at this point. This position, however, is one of **unstable equilibrium** for the following reason. Suppose the particle is displaced to the right ($x > 0$). Because the slope is negative for $x > 0$, $F_x = -dU/dx$ is positive and the particle accelerates away from $x = 0$. If instead the particle is at $x = 0$ and is displaced to the left ($x < 0$), the force is negative because the slope is positive for $x < 0$ and the particle again accelerates away from the equilibrium position. The position $x = 0$ in this situation is one of unstable equilibrium because for any displacement from this point, the force pushes the particle farther away from equilibrium and toward a position of lower potential energy. A pencil balanced on its point is in a position of unstable equilibrium. If the pencil is displaced slightly from its absolutely vertical position and is then released, it will surely fall over. In general, configurations of a system in unstable equilibrium correspond to those for which $U(x)$ for the system is a maximum.

Finally, a configuration called **neutral equilibrium** arises when U is constant over some region. Small displacements of an object from a position in this region produce neither restoring nor disrupting forces. A ball lying on a flat, horizontal surface is an example of an object in neutral equilibrium.

Example **6.9** | **Force and Energy on an Atomic Scale**

The potential energy associated with the force between two neutral atoms in a molecule can be modeled by the Lennard–Jones potential energy function:

$$U(x) = 4\epsilon\left[\left(\frac{\sigma}{x}\right)^{12} - \left(\frac{\sigma}{x}\right)^{6}\right]$$

where x is the separation of the atoms. The function $U(x)$ contains two parameters σ and ϵ that are determined from experiments. Sample values for the interaction between two atoms in a molecule are $\sigma = 0.263$ nm and $\epsilon = 1.51 \times 10^{-22}$ J. Using a spreadsheet or similar tool, graph this function and find the most likely distance between the two atoms.

SOLUTION

Conceptualize We identify the two atoms in the molecule as a system. Based on our understanding that stable molecules exist, we expect to find stable equilibrium when the two atoms are separated by some equilibrium distance.

Categorize Because a potential energy function exists, we categorize the force between the atoms as conservative. For a conservative force, Equation 6.28 describes the relationship between the force and the potential energy function.

Analyze Stable equilibrium exists for a separation distance at which the potential energy of the system of two atoms (the molecule) is a minimum.

Take the derivative of the function $U(x)$:

$$\frac{dU(x)}{dx} = 4\epsilon\frac{d}{dx}\left[\left(\frac{\sigma}{x}\right)^{12} - \left(\frac{\sigma}{x}\right)^{6}\right] = 4\epsilon\left[\frac{-12\sigma^{12}}{x^{13}} + \frac{6\sigma^{6}}{x^{7}}\right]$$

Minimize the function $U(x)$ by setting its derivative equal to zero:

$$4\epsilon\left[\frac{-12\sigma^{12}}{x_{eq}^{13}} + \frac{6\sigma^{6}}{x_{eq}^{7}}\right] = 0 \rightarrow x_{eq} = (2)^{1/6}\sigma$$

Evaluate x_{eq}, the equilibrium separation of the two atoms in the molecule:

$$x_{eq} = (2)^{1/6}(0.263 \text{ nm}) = \boxed{2.95 \times 10^{-10} \text{ m}}$$

6.9 *cont.*

We graph the Lennard–Jones function on both sides of this critical value to create our energy diagram as shown in Figure 6.25.

....

Finalize Notice that $U(x)$ is extremely large when the atoms are very close together, is a minimum when the atoms are at their critical separation, and then increases again as the atoms move apart. When $U(x)$ is a minimum, the atoms are in stable equilibrium, indicating that the most likely separation between them occurs at this point.

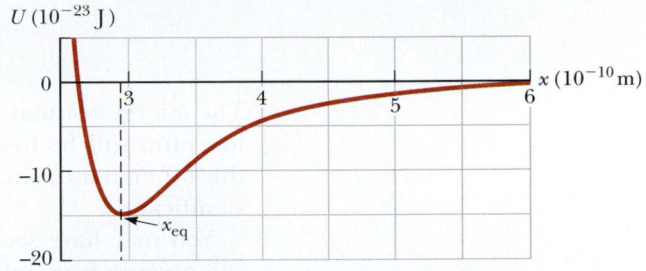

Figure 6.25 (Example 6.9) Potential energy curve associated with a molecule. The distance x is the separation between the two atoms making up the molecule.

6.11 | Context Connection: Potential Energy in Fuels

Fuel represents a storage mechanism for potential energy to be used to make a vehicle move. The standard fuel for automobiles for several decades has been *gasoline*. Gasoline is refined from crude oil that is present in the Earth. This oil represents the decay products of plant life that existed on the Earth, primarily from 100 to 600 million years ago. The source of energy in crude oil is hydrocarbons produced from molecules in the ancient plants.

The primary chemical reactions occurring in an internal combustion engine involve the oxidation of carbon and hydrogen:

$$C + O_2 \rightarrow CO_2$$

$$4H + O_2 \rightarrow 2H_2O$$

Both reactions release energy that is used to operate the automobile.

Notice the final products in these reactions. One is water, which is not harmful to the environment. Carbon dioxide, however, contributes to the greenhouse effect, which leads to global warming, which we will study in Context 5. The incomplete combustion of carbon and oxygen can form CO, carbon monoxide, which is a poisonous gas. Because air contains other elements besides oxygen, other harmful emission products, such as oxides of nitrogen, exist.

The amount of potential energy stored in a fuel and available from the fuel is typically called the *heat of combustion*, even though this term is a misuse of the word *heat*. For automotive gasoline, this value is about 44 MJ/kg. Because the efficiency of the engine is not 100%, only part of this energy eventually finds its way into kinetic energy of the car. We will study efficiencies of engines in Context 5.

Another common fuel is *diesel fuel*. The heat of combustion for diesel fuel is 42.5 MJ/kg, slightly lower than that for gasoline. Diesel engines, however, operate at a higher efficiency than gasoline engines, so they can extract a larger percentage of the available energy.

A number of additional fuels have been developed to operate internal combustion engines with minimal modifications. They are described briefly below.

Ethanol

Ethanol is the most widely used alternative fuel and is used for commercial fleet vehicles and increasingly for private vehicles. It is an alcohol made from such crops as corn, wheat, and barley. Because these crops can be grown, ethanol is renewable. The use of ethanol reduces carbon monoxide and carbon dioxide emissions compared with the use of normal gasoline.

Ethanol is mixed with gasoline to form the following mixtures:

E10: 10% ethanol, 90% gasoline

E85: 85% ethanol, 15% gasoline

The energy content of E85 is about 70% of that for gasoline, so the miles per gallon ratio will be lower than that for a vehicle powered by straight gasoline. On the other hand, the renewable nature of ethanol counteracts this disadvantage significantly.

You may have seen some automobiles labeled as "FLEXFUEL." These vehicles can operate from ethanol fuel over a continuous range from pure gasoline to E85. Whatever the mixture, the fuel is stored in a single fuel tank and sensors in the fuel system determine the amount of ethanol, automatically adjusting the fuel injection and spark timing accordingly.

Biodiesel

Biodiesel fuel is formed by a chemical reaction between alcohol and oils from field crops as well as vegetable oil, fat, and grease from commercial sources. Pacific Biodiesel in Hawaii makes biodiesel from used restaurant cooking oil, providing a usable fuel as well as diverting this used oil from landfills.

Biodiesel is available in the following forms:

B20: 20% biodiesel, 80% gasoline

B100: 100% biodiesel

B100 is nontoxic and biodegradable. The use of biodiesel reduces environmentally harmful tailpipe emissions significantly. Furthermore, tests have shown that the emission of cancer-causing particulate matter is reduced by 94% with the use of pure biodiesel.

The energy content of B100 is about 90% of that for conventional diesel. As with ethanol, the renewable nature of biodiesel counteracts this disadvantage significantly.

Natural Gas

Natural gas is a fossil fuel, originating from gas wells or as a by-product of the refining process for crude oil. It is primarily methane (CH_4), with smaller amounts of nitrogen, ethane, propane, and other gases. It burns cleanly and generates much lower amounts of harmful tailpipe emissions than gasoline. Natural gas vehicles are used in many fleets of buses, delivery trucks, and refuse haulers.

Although ethanol and biodiesel mixtures can be used in conventional engines with minimal modifications, a natural gas engine is much more heavily modified. In addition, the gas must be carried on board the vehicle in one of two ways that require higher-level technology than a simple fuel tank. One possibility is to liquefy the gas, requiring a well-insulated storage container to keep the gas at $-190°C$. The other possibility is to compress the gas to about 200 times atmospheric pressure and carry it in the vehicle in a high-pressure storage tank.

The energy content of natural gas is 48 MJ/kg, a bit higher than that for gasoline. Note that natural gas, like gasoline, is *not* a renewable source.

Propane

Propane is available commercially as liquefied petroleum gas, which is actually a mixture of propane, propylene, butane, and butylenes. It is a by-product of natural gas processing and refining of crude oil. Propane is the most widely accessible alternative fuel, with fueling facilities in all states of the United States.

Tailpipe emissions for propane-fueled vehicles are significantly lower than those for gasoline-powered vehicles. Tests show that carbon monoxide is reduced by 30% to 90%.

As with natural gas, high-pressure tanks are necessary to carry the fuel. In addition, propane is a nonrenewable resource. The energy content of propane is 46 MJ/kg, slightly higher than that of gasoline.

Electric Vehicles

In the Context introduction before Chapter 2, we discussed the electric cars that were on the roadways in the early part of the 20th century. As mentioned, these electric cars virtually disappeared around the 1920s due to several factors. One was that oil was plentiful during the 20th century and there was little incentive to operate vehicles on anything other than gasoline or diesel.

In the early 1970s, difficulties arose with regard to the availability of oil from the Middle East, leading to shortages at gas stations. At this time, interest arose anew in electric-powered vehicles. An early attempt to market a new electric vehicle was the Electrovette, an electric version of the Chevrolet Chevette.

Although the oil crisis eased somewhat, political instabilities in the Middle East created uncertainty in the availability of oil and interest in electric cars continued, albeit on a small scale. In the late 1980s, General Motors developed a prototype called the Impact, an electric car that could accelerate from 0 to 60 in 8 s and had a drag coefficient of 0.19, much lower than that of traditional cars. The Impact was the hit of the 1990 Los Angeles Auto Show. In the 1990s, the Impact became commercially available as the EV1. General Motors canceled the EV1 program in 2001 and recalled the vehicles. While a few of the EV1 vehicles were retained for museums, the vast majority of the vehicles were destroyed by being crushed.

Two major disadvantages of electric cars are the limited range, 70 to 100 mi, on a single charging of the batteries, and the several hours of time required to recharge the batteries. Despite these difficulties, new electric vehicles are available for the public, including the Nissan Leaf, discussed in Section 2.8, and the Tesla Roadster, a high-priced electric sports car that can accelerate from rest to 60 mi/h in 3.7 s. In addition, the Chevrolet Volt, also discussed in Section 2.8, is operated as an electric car over short trips. It solves the limited-range and charging-time problems for longer trips by incorporating a gasoline engine to charge the battery once the original charge has been depleted.

▶ SUMMARY

A **system** is most often a single particle, a collection of particles, or a region of space, and may vary in size and shape. A **system boundary** separates the system from the **environment.**

The **work** W done on a system by an agent exerting a constant force $\vec{F}$ on the system is the product of the magnitude Δr of the displacement of the point of application of the force and the component $F \cos \theta$ of the force along the direction of the displacement $\Delta \vec{r}$:

$$W \equiv F \Delta r \cos \theta \qquad \text{6.1} \blacktriangleleft$$

The **scalar product** (dot product) of two vectors $\vec{A}$ and $\vec{B}$ is defined by the relationship

$$\vec{A} \cdot \vec{B} \equiv AB \cos \theta \qquad \text{6.2} \blacktriangleleft$$

where the result is a scalar quantity and θ is the angle between the two vectors. The scalar product obeys the commutative and distributive laws.

If a varying force does work on a particle as the particle moves along the x axis from x_i to x_f, the work done by the force on the particle is given by

$$W = \int_{x_i}^{x_f} F_x \, dx \qquad \text{6.7} \blacktriangleleft$$

where F_x is the component of force in the x direction.

The **kinetic energy** of a particle of mass m moving with a speed v is

$$K \equiv \tfrac{1}{2} m v^2 \qquad \text{6.16} \blacktriangleleft$$

The **work–kinetic energy theorem** states that if work is done on a system by external forces and the only change in the system is in its speed,

$$W_{\text{ext}} = K_f - K_i = \Delta K = \tfrac{1}{2} m v_f^2 - \tfrac{1}{2} m v_i^2 \qquad \text{6.15, 6.17} \blacktriangleleft$$

If a particle of mass m is at a distance y above the Earth's surface, the **gravitational potential energy** of the particle–Earth system is

$$U_g \equiv mgy \qquad \text{6.19} \blacktriangleleft$$

The **elastic potential energy** stored in a spring of force constant k is

$$U_s \equiv \tfrac{1}{2}kx^2 \qquad \text{6.22} \blacktriangleleft$$

The **total mechanical energy of a system** is defined as the sum of the kinetic energy and the potential energy:

$$E_{\text{mech}} \equiv K + U \qquad \text{6.24} \blacktriangleleft$$

A force is **conservative** if the work it does on a particle that is a member of the system as the particle moves between two points is independent of the path the particle takes between the two points. Furthermore, a force is conservative if the work it does on a particle is zero when the particle moves

through an arbitrary closed path and returns to its initial position. A force that does not meet these criteria is said to be **nonconservative.**

A **potential energy function** U can be associated only with a conservative force. If a conservative force $\vec{F}$ acts between members of a system while one member moves along the x axis from x_i to x_f, the change in the potential energy of the system equals the negative of the work done by that force:

$$U_f - U_i = -\int_{x_i}^{x_f} F_x \, dx \qquad \text{6.26} \blacktriangleleft$$

Systems can be in three types of equilibrium configurations when the net force on a member of the system is zero. Configurations of **stable equilibrium** correspond to those for which $U(x)$ is a minimum. Configurations of **unstable equilibrium** correspond to those for which $U(x)$ is a maximum. **Neutral equilibrium** arises when U is constant as a member of the system moves over some region.

▶ OBJECTIVE QUESTIONS

☐ denotes answer available in *Student Solutions Manual/Study Guide*

1. Alex and John are loading identical cabinets onto a truck. Alex lifts his cabinet straight up from the ground to the bed of the truck, whereas John slides his cabinet up a rough ramp to the truck. Which statement is correct about the work done on the cabinet–Earth system? (a) Alex and John do the same amount of work. (b) Alex does more work than John. (c) John does more work than Alex. (d) None of those statements is necessarily true because the force of friction is unknown. (e) None of those statements is necessarily true because the angle of the incline is unknown.

2. Is the work required to be done by an external force on an object on a frictionless, horizontal surface to accelerate it from a speed v to a speed $2v$ (a) equal to the work required to accelerate the object from $v = 0$ to v, (b) twice the work required to accelerate the object from $v = 0$ to v, (c) three times the work required to accelerate the object from $v = 0$ to v, (d) four times the work required to accelerate the object from 0 to v, or (e) not known without knowledge of the acceleration?

3. A worker pushes a wheelbarrow with a horizontal force of 50 N on level ground over a distance of 5.0 m. If a friction force of 43 N acts on the wheelbarrow in a direction opposite that of the worker, what work is done on the wheelbarrow by the worker? (a) 250 J (b) 215 J (c) 35 J (d) 10 J (e) None of those answers is correct.

4. Mark and David are loading identical cement blocks onto David's pickup truck. Mark lifts his block straight up from the ground to the truck, whereas David slides his block up a

ramp containing frictionless rollers. Which statement is true about the work done on the block–Earth system? (a) Mark does more work than David. (b) Mark and David do the same amount of work. (c) David does more work than Mark. (d) None of those statements is necessarily true because the angle of the incline is unknown. (e) None of those statements is necessarily true because the mass of one block is not given.

5. Bullet 2 has twice the mass of bullet 1. Both are fired so that they have the same speed. If the kinetic energy of bullet 1 is K, is the kinetic energy of bullet 2 (a) $0.25K$, (b) $0.5K$, (c) $0.71K$, (d) K, or (e) $2K$?

6. As a simple pendulum swings back and forth, the forces acting on the suspended object are (a) the gravitational force, (b) the tension in the supporting cord, and (c) air resistance. (**i**) Which of these forces, if any, does no work on the pendulum at any time? (**ii**) Which of these forces does negative work on the pendulum at all times during its motion?

7. A block of mass m is dropped from the fourth floor of an office building and hits the sidewalk below at speed v. From what floor should the mass be dropped to double that impact speed? (a) the sixth floor (b) the eighth floor (c) the tenth floor (d) the twelfth floor (e) the sixteenth floor

8. If the net work done by external forces on a particle is zero, which of the following statements about the particle must be true? (a) Its velocity is zero. (b) Its velocity is decreased.

(c) Its velocity is unchanged. (d) Its speed is unchanged. (e) More information is needed.

9. Let $\hat{\mathbf{N}}$ represent the direction horizontally north, $\widehat{\mathbf{NE}}$ represent northeast (halfway between north and east), and so on. Each direction specification can be thought of as a unit vector. Rank from the largest to the smallest the following dot products. Note that zero is larger than a negative number. If two quantities are equal, display that fact in your ranking. (a) $\hat{\mathbf{N}} \cdot \hat{\mathbf{N}}$ (b) $\hat{\mathbf{N}} \cdot \widehat{\mathbf{NE}}$ (c) $\hat{\mathbf{N}} \cdot \hat{\mathbf{S}}$ (d) $\hat{\mathbf{N}} \cdot \hat{\mathbf{E}}$ (e) $\widehat{\mathbf{SE}} \cdot \hat{\mathbf{S}}$

10. Figure OQ6.10 shows a light extended spring exerting a force F_s to the left on a block. (i) Does the block exert a force on the spring? Choose every correct answer. (a) No, it doesn't. (b) Yes, it does, to the left. (c) Yes, it does, to the right. (d) Yes, it does, and its magnitude is larger than F_s. (e) Yes, it does, and its magnitude is equal to F_s. (ii) Does the spring exert a force on the wall? Choose your answers from the same list (a) through (e).

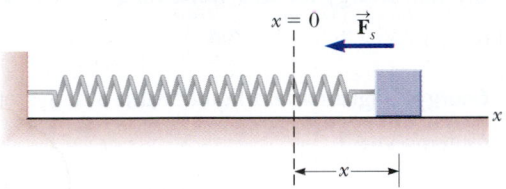

Figure OQ6.10

11. If the speed of a particle is doubled, what happens to its kinetic energy? (a) It becomes four times larger. (b) It becomes two times larger. (c) It becomes $\sqrt{2}$ times larger. (d) It is unchanged. (e) It becomes half as large.

12. A cart is set rolling across a level table, at the same speed on every trial. If it runs into a patch of sand, the cart exerts on the sand an average horizontal force of 6 N and travels a distance of 6 cm through the sand as it comes to

a stop. If instead the cart runs into a patch of gravel on which the cart exerts an average horizontal force of 9 N, how far into the gravel will the cart roll before stopping? (a) 9 cm (b) 6 cm (c) 4 cm (d) 3 cm (e) none of those answers

13. A cart is set rolling across a level table, at the same speed on every trial. If it runs into a patch of sand, the cart exerts on the sand an average horizontal force of 6 N and travels a distance of 6 cm through the sand as it comes to a stop. If instead the cart runs into a patch of flour, it rolls an average of 18 cm before stopping. What is the average magnitude of the horizontal force the cart exerts on the flour? (a) 2 N (b) 3 N (c) 6 N (d) 18 N (e) none of those answers

14. A certain spring that obeys Hooke's law is stretched by an external agent. The work done in stretching the spring by 10 cm is 4 J. How much additional work is required to stretch the spring an additional 10 cm? (a) 2 J (b) 4 J (c) 8 J (d) 12 J (e) 16 J

15. (i) Rank the gravitational accelerations you would measure for the following falling objects: (a) a 2-kg object 5 cm above the floor, (b) a 2-kg object 120 cm above the floor, (c) a 3-kg object 120 cm above the floor, and (d) a 3-kg object 80 cm above the floor. List the one with the largest magnitude of acceleration first. If any are equal, show their equality in your list. (ii) Rank the gravitational forces on the same four objects, listing the one with the largest magnitude first. (iii) Rank the gravitational potential energies (of the object–Earth system) for the same four objects, largest first, taking $y = 0$ at the floor.

16. An ice cube has been given a push and slides without friction on a level table. Which is correct? (a) It is in stable equilibrium. (b) It is in unstable equilibrium. (c) It is in neutral equilibrium. (d) It is not in equilibrium.

▶ CONCEPTUAL QUESTIONS |

☐ denotes answer available in *Student Solutions Manual/Study Guide*

1. Discuss whether any work is being done by each of the following agents and, if so, whether the work is positive or negative. (a) a chicken scratching the ground (b) a person studying (c) a crane lifting a bucket of concrete (d) the gravitational force on the bucket in part (c) (e) the leg muscles of a person in the act of sitting down

2. Discuss the work done by a pitcher throwing a baseball. What is the approximate distance through which the force acts as the ball is thrown?

3. A certain uniform spring has spring constant k. Now the spring is cut in half. What is the relationship between k and the spring constant k' of each resulting smaller spring? Explain your reasoning.

4. (a) For what values of the angle θ between two vectors is their scalar product positive? (b) For what values of θ is their scalar product negative?

5. Can kinetic energy be negative? Explain.

6. Cite two examples in which a force is exerted on an object without doing any work on the object.

7. Does the kinetic energy of an object depend on the frame of reference in which its motion is measured? Provide an example to prove this point.

8. If only one external force acts on a particle, does it necessarily change the particle's (a) kinetic energy? (b) Its velocity?

9. Can a normal force do work? If not, why not? If so, give an example.

10. You are reshelving books in a library. You lift a book from the floor to the top shelf. The kinetic energy of the book on the floor was zero and the kinetic energy of the book on the top shelf is zero, so no change occurs in the kinetic energy, yet you did some work in lifting the book. Is the work–kinetic energy theorem violated? Explain.

11. A student has the idea that the total work done on an object is equal to its final kinetic energy. Is this idea true

always, sometimes, or never? If it is sometimes true, under what circumstances? If it is always or never true, explain why.

12. Object 1 pushes on object 2 as the objects move together, like a bulldozer pushing a stone. Assume object 1 does 15.0 J of work on object 2. Does object 2 do work on object 1? Explain your answer. If possible, determine how much work and explain your reasoning.

▶ PROBLEMS AVAILABLE IN

List of Enhanced Problems

Problem Number	Targeted Feedback in Enhanced WebAssign	Master It in Enhanced WebAssign	Watch It in Enhanced WebAssign
6.2	✓		✓
6.3	✓	✓	
6.5		✓	
6.9		✓	
6.11	✓		✓
6.14	✓		✓
6.15	✓		✓
6.17	✓	✓	
6.25	✓		✓
6.31	✓		✓
6.32	✓		✓
6.33	✓		✓
6.35		✓	
6.40	✓		✓
6.42	✓	✓	
6.43	✓	✓	
6.47		✓	

Conservation of Energy

Chapter Outline

Jade Lee/Getty Images

Three youngsters enjoy the transformation of potential energy to kinetic energy on a waterslide. We can analyze processes such as these with the techniques developed in this chapter.

In Chapter 6, we introduced three methods for storing energy in a system: kinetic energy, associated with movement of members of the system; potential energy, determined by the configuration of the system; and internal energy, which is related to the temperature of the system.

We now consider analyzing physical situations using the energy approach for two types of systems: *nonisolated* and *isolated* systems. For nonisolated systems, we shall investigate ways that energy can cross the boundary of the system, resulting in a change in the system's total energy. This analysis leads to a critically important principle called *conservation of energy*. The conservation of energy principle extends well beyond physics and can be applied to biological organisms, technological systems, and engineering situations.

In isolated systems, energy does not cross the boundary of the system. For these systems, the total energy of the system is constant. If no nonconservative forces act within the system, we can use *conservation of mechanical energy* to solve a variety of problems.

Situations involving the transformation of mechanical energy to internal energy due to nonconservative forces require special handling. We investigate the procedures for these types of problems.

Finally, we recognize that energy can cross the boundary of a system at different rates. We describe the rate of energy transfer with the quantity *power*.

❰7.1❱ | Analysis Model: Nonisolated System (Energy)

As we have seen, an object, modeled as a particle, can be acted on by various forces, resulting in a change in its kinetic energy. If we choose the object as the system, this very simple situation is the first example of a *nonisolated system,* for which energy crosses the boundary of the system during some time interval due to an interaction with the environment. This scenario is common in physics problems. If a system does not interact with its environment, it is an isolated system, which we will study in Section 7.2.

The work–kinetic energy theorem from Chapter 6 is our first example of an energy equation appropriate for a nonisolated system. In the case of that theorem, the interaction of the system with its environment is the work done by the external force, and the quantity in the system that changes is the kinetic energy.

So far, we have seen only one way to transfer energy into a system: work. We mention below a few other ways to transfer energy into or out of a system. The details of these processes will be studied in other sections of the book. We illustrate mechanisms to transfer energy in Figure 7.1 and summarize them as follows.

Work, as we have learned in Chapter 6, is a method of transferring energy to a system by applying a force to the system such that the point of application of the force undergoes a displacement (Fig. 7.1a).

Mechanical waves (Chapters 13–14) are a means of transferring energy by allowing a disturbance to propagate through air or another medium. It is the method by which energy (which you detect as sound) leaves the system of your clock radio through the loudspeaker and enters your ears to stimulate the hearing process (Fig. 7.1b). Other examples of mechanical waves are seismic waves and ocean waves.

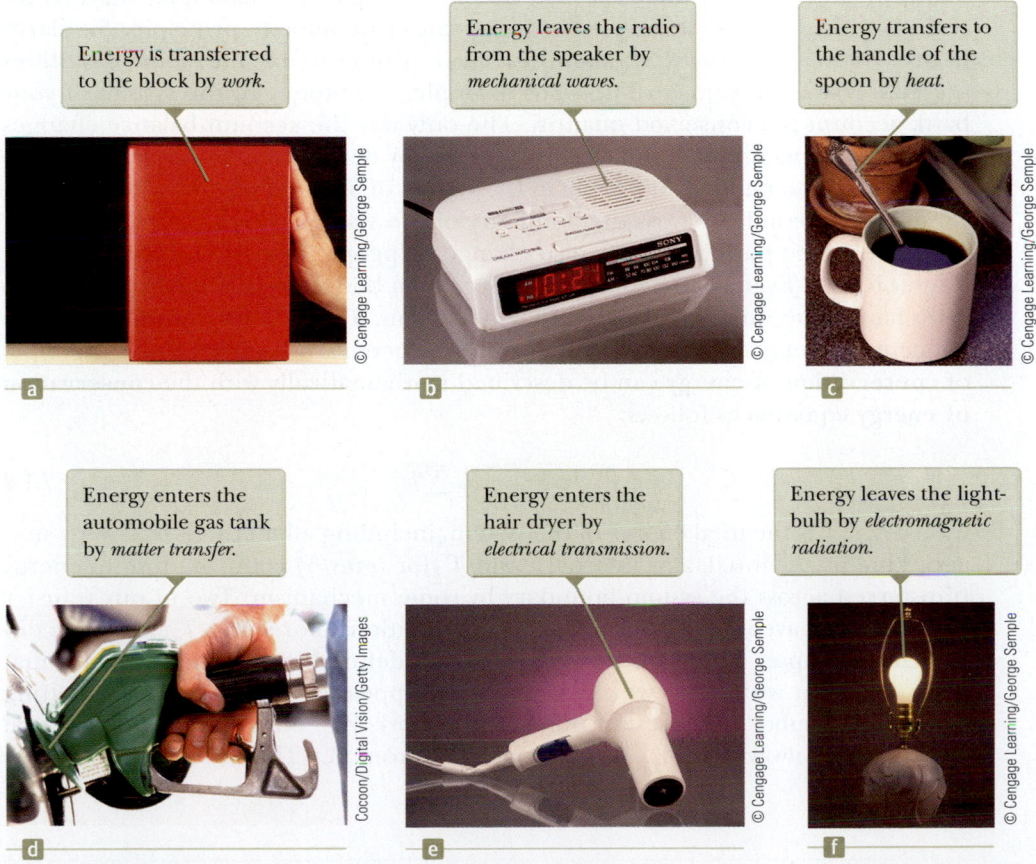

Figure 7.1 Energy transfer mechanisms. In each case, the system into which or from which energy is transferred is indicated.

Heat (Chapter 17) is a mechanism of energy transfer that is driven by a temperature difference between a system and its environment. For example, imagine dividing a metal spoon into two parts: the handle, which we identify as the system, and the portion submerged in a cup of coffee, which is part of the environment (Fig. 7.1c). The handle of the spoon becomes hot because fast-moving electrons and atoms in the submerged portion bump into slower ones in the nearby part of the handle. These particles move faster because of the collisions and bump into the next group of slow particles. Therefore, the internal energy of the spoon handle rises from energy transfer due to this collision process.

Matter transfer (Chapter 17) involves situations in which matter physically crosses the boundary of a system, carrying energy with it. Examples include filling your automobile tank with gasoline (Fig. 7.1d) and carrying energy to the rooms of your home by circulating warm air from the furnace, a process called *convection*.

Electrical transmission (Chapter 21) involves energy transfer into or out of a system by means of electric currents. It is how energy transfers into your hair dryer (Fig. 7.1e), stereo system, or any other electrical device.

Electromagnetic radiation (Chapter 24) refers to electromagnetic waves such as light (Fig. 7.1f), microwaves, and radio waves crossing the boundary of a system. Examples of this method of transfer include cooking a baked potato in your microwave oven and light energy traveling from the Sun to the Earth through space.[1]

A central feature of the energy approach is the notion that we can neither create nor destroy energy, that energy is always *conserved*. This feature has been tested in countless experiments, and no experiment has ever shown this statement to be incorrect. Therefore, **if the total amount of energy in a system changes, it can *only* be because energy has crossed the boundary of the system by a transfer mechanism such as one of the methods listed above.**

Energy is one of several quantities in physics that are conserved. We will see other conserved quantities in subsequent chapters. There are many physical quantities that do not obey a conservation principle. For example, there is no conservation of force principle or conservation of velocity principle. Similarly, in areas other than physical quantities, such as in everyday life, some quantities are conserved and some are not. For example, the money in the system of your bank account is a conserved quantity. The only way the account balance changes is if money crosses the boundary of the system by deposits or withdrawals. On the other hand, the number of people in the system of a country is not conserved. Although people indeed cross the boundary of the system, which changes the total population, the population can also change by people dying and by giving birth to new babies. Even if no people cross the system boundary, the births and deaths will change the number of people in the system. There is no equivalent in the concept of energy to dying or giving birth. The general statement of the principle of **conservation of energy** can be described mathematically with the **conservation of energy equation** as follows:

► Conservation of energy

$$\Delta E_{\text{system}} = \sum T \qquad \text{7.1} ◄$$

where E_{system} is the total energy of the system, including all methods of energy storage (kinetic, potential, and internal), and T (for *transfer*) is the amount of energy transferred across the system boundary by some mechanism. Two of our transfer mechanisms have well-established symbolic notations. For work, $T_{\text{work}} = W$ as discussed in Chapter 6, and for heat, $T_{\text{heat}} = Q$ as defined in Chapter 17. (Now that we are familiar with work, we can simplify the appearance of equations by letting the simple symbol W represent the external work W_{ext} on a system. For internal work, we will always use W_{int} to differentiate it from W.) The other four members

---

[1]Electromagnetic radiation and work done by field forces are the only energy transfer mechanisms that do not require molecules of the environment to be available at the system boundary. Therefore, systems surrounded by a vacuum (such as planets) can only exchange energy with the environment by means of these two possibilities.

of our list do not have established symbols, so we will call them T_{MW} (mechanical waves), T_{MT} (matter transfer), T_{ET} (electrical transmission), and T_{ER} (electromagnetic radiation).

The full expansion of Equation 7.1 is

$$\Delta K + \Delta U + \Delta E_{int} = W + Q + T_{MW} + T_{MT} + T_{ET} + T_{ER} \qquad \textbf{7.2} \blacktriangleleft$$

which is the primary mathematical representation of the energy version of the analysis model of the **nonisolated system.** (We will see other versions of the nonisolated system model, involving linear momentum and angular momentum, in later chapters.) In most cases, Equation 7.2 reduces to a much simpler one because some of the terms are zero. If, for a given system, all terms on the right side of the conservation of energy equation are zero, the system is an *isolated system,* which we study in the next section.

The conservation of energy equation is no more complicated in theory than the process of balancing your checking account statement. If your account is the system, the change in the account balance for a given month is the sum of all the transfers: deposits, withdrawals, fees, interest, and checks written. You may find it useful to think of energy as the *currency of nature!*

Suppose a force is applied to a nonisolated system and the point of application of the force moves through a displacement. Then suppose the only effect on the system is to change its speed. In this case, the only transfer mechanism is work (so that the right side of Eq. 7.2 reduces to just W) and the only kind of energy in the system that changes is the kinetic energy (so that ΔE_{system} reduces to just ΔK). Equation 7.2 then becomes

$$\Delta K = W$$

which is the work–kinetic energy theorem. This theorem is a special case of the more general principle of conservation of energy. We shall see several more special cases in future chapters.

> **QUICK QUIZ 7.1** By what transfer mechanisms does energy enter and leave **(a)** your television set? **(b)** Your gasoline-powered lawn mower? **(c)** Your hand-cranked pencil sharpener?

> **QUICK QUIZ 7.2** Consider a block sliding over a horizontal surface with friction. Ignore any sound the sliding might make. **(i)** If the system is the *block*, this system is (a) isolated (b) nonisolated (c) impossible to determine **(ii)** If the system is the *surface*, describe the system from the same set of choices. **(iii)** If the system is the *block and the surface*, describe the system from the same set of choices.

7.2 | Analysis Model: Isolated System (Energy)

In this section, we study another very common scenario in physics problems: a system is chosen such that no energy crosses the system boundary by any method. We begin by considering a gravitational situation. Think about the book–Earth system in Active Figure 6.15 in the preceding chapter. After we have lifted the book, there is gravitational potential energy stored in the system, which can be calculated from the work done by the external agent on the system, using $W = \Delta U_g$.

Let us now shift our focus to the work done *on the book alone* by the gravitational force (Fig. 7.2) as the book falls back to its original height. As the book falls from y_i to y_f the work done by the gravitational force on the book is

$$W_{on\ book} = (m\vec{\mathbf{g}}) \cdot \Delta \vec{\mathbf{r}} = (-mg\hat{\mathbf{j}}) \cdot [(y_f - y_i)\hat{\mathbf{j}}] = mgy_i - mgy_f \qquad \textbf{7.3} \blacktriangleleft$$

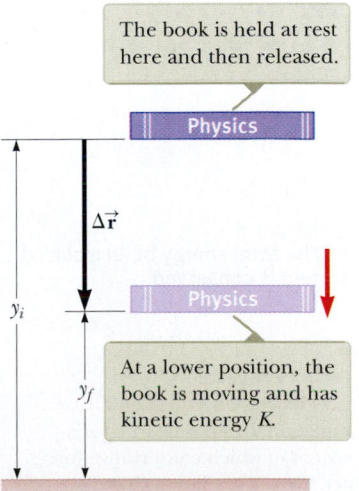

The book is held at rest here and then released.

$\Delta\vec{\mathbf{r}}$

At a lower position, the book is moving and has kinetic energy K.

Figure 7.2 A book is released from rest and falls due to work done by the gravitational force on the book.

From the work–kinetic energy theorem of Chapter 6, the work done on the book is equal to the change in the kinetic energy of the book:

$$W_{\text{on book}} = \Delta K_{\text{book}}$$

We can equate these two expressions for the work done on the book:

$$\Delta K_{\text{book}} = mgy_i - mgy_f \qquad \textbf{7.4} \blacktriangleleft$$

Let us now relate each side of this equation to the *system* of the book and the Earth. For the right-hand side of Equation 7.4,

$$mgy_i - mgy_f = -(mgy_f - mgy_i) = -\Delta U_g$$

where $U_g = mgy$ is the gravitational potential energy of the system. For the left-hand side of Equation 7.4, because the book is the only part of the system that is moving, we see that $\Delta K_{\text{book}} = \Delta K$, where K is the kinetic energy of the system. Therefore, with each side of Equation 7.4 replaced with its system equivalent, the equation becomes

$$\Delta K = -\Delta U_g \qquad \textbf{7.5} \blacktriangleleft$$

This equation can be manipulated to provide a very important general result for solving problems. First, we move the change in potential energy to the left side of the equation:

$$\Delta K + \Delta U_g = 0$$

The left side represents a sum of changes of the energy stored in the system. The right-hand side is zero because there are no transfers of energy across the boundary of the system; the book–Earth system is *isolated* from the environment. We developed this equation for a gravitational system, but it can be shown to be valid for a system with any type of potential energy. Therefore, for an isolated system,

$$\Delta K + \Delta U = 0 \qquad \textbf{7.6} \blacktriangleleft$$

We defined in Chapter 6 the sum of the kinetic and potential energies of a system as its mechanical energy:

▶ Mechanical energy of a system

$$E_{\text{mech}} \equiv K + U \qquad \textbf{7.7} \blacktriangleleft$$

where U represents the total of *all* types of potential energy. Because the system under consideration is isolated, Equations 7.6 and 7.7 tell us that the mechanical energy of the system is conserved:

▶ The mechanical energy of an isolated system with no nonconservative forces acting is conserved.

$$\Delta E_{\text{mech}} = 0 \qquad \textbf{7.8} \blacktriangleleft$$

Equation 7.8 is a statement of **conservation of mechanical energy** for an isolated system with no nonconservative forces acting. The mechanical energy in such a system is conserved: the sum of the kinetic and potential energies remains constant.

If there are nonconservative forces acting within the system, mechanical energy is transformed to internal energy as discussed in Section 6.7. If nonconservative forces act in an isolated system, the total energy of the system is conserved although the mechanical energy is not. In that case, we can express the conservation of energy of the system as

▶ The total energy of an isolated system is conserved.

$$\Delta E_{\text{system}} = 0 \qquad \textbf{7.9} \blacktriangleleft$$

where E_{system} includes all kinetic, potential, and internal energies. This equation is the most general statement of the energy version of the **isolated system** model. It is equivalent to Equation 7.2 with all terms on the right-hand side equal to zero.

Let us now write the changes in energy in Equation 7.6 explicitly:

$$(K_f - K_i) + (U_f - U_i) = 0$$

$$K_f + U_f = K_i + U_i \qquad \textbf{7.10} \blacktriangleleft$$

Pitfall Prevention | 7.2
Conditions on Equation 7.10
Equation 7.10 is only true for a system in which conservative forces act. We will see how to handle nonconservative forces in Sections 7.4 and 7.5.

For the gravitational situation of the falling book, Equation 7.10 can be written as

$$\tfrac{1}{2}mv_f^2 + mgy_f = \tfrac{1}{2}mv_i^2 + mgy_i$$

As the book falls to the Earth, the book–Earth system loses potential energy and gains kinetic energy such that the total of the two types of energy always remains constant.

QUICK QUIZ 7.3 A rock of mass m is dropped to the ground from a height h. A second rock, with mass $2m$, is dropped from the same height. When the second rock strikes the ground, what is its kinetic energy? (a) twice that of the first rock (b) four times that of the first rock (c) the same as that of the first rock (d) half as much as that of the first rock (e) impossible to determine

QUICK QUIZ 7.4 Three identical balls are thrown from the top of a building, all with the same initial speed. As shown in Active Figure 7.3, the first is thrown horizontally, the second at some angle above the horizontal, and the third at some angle below the horizontal. Neglecting air resistance, rank the speeds of the balls at the instant each hits the ground.

Active Figure 7.3 (Quick Quiz 7.4) Three identical balls are thrown with the same initial speed from the top of a building.

> **PROBLEM-SOLVING STRATEGY: Isolated Systems with No Nonconservative Forces: Conservation of Mechanical Energy**

Many problems in physics can be solved using the principle of conservation of energy for an isolated system. The following procedure should be used when you apply this principle:

1. **Conceptualize.** Study the physical situation carefully and form a mental representation of what is happening. As you become more proficient working energy problems, you will begin to be comfortable imagining the types of energy that are changing in the system.

2. **Categorize.** Define your system, which may consist of more than one object and may or may not include springs or other possibilities for storing potential energy. Determine if any energy transfers occur across the boundary of your system. If so, use the nonisolated system model, $\Delta E_{system} = \Sigma T$, from Section 7.1. If not, use the isolated system model, $\Delta E_{system} = 0$.

   Determine whether any nonconservative forces are present within the system. If so, use the techniques of Sections 7.4 and 7.5. If not, use the principle of conservation of mechanical energy as outlined below.

3. **Analyze.** Choose configurations to represent the initial and final conditions of the system. For each object that changes elevation, select a reference position for the object that defines the zero configuration of gravitational potential energy for the system. For an object on a spring, the zero configuration for elastic potential energy is when the object is at its equilibrium position. If there is more than one conservative force, write an expression for the potential energy associated with each force.

   Write the total initial mechanical energy E_i of the system for some configuration as the sum of the kinetic and potential energies associated with the configuration. Then write a similar expression for the total mechanical energy E_f of the system for the final configuration that is of interest. Because mechanical energy is *conserved*, equate the two total energies and solve for the quantity that is unknown.

4. **Finalize.** Make sure your results are consistent with your mental representation. Also make sure the values of your results are reasonable and consistent with connections to everyday experience.

Example 7.1 | Ball in Free Fall

A ball of mass m is dropped from a height h above the ground as shown in Active Figure 7.4.

(A) Neglecting air resistance, determine the speed of the ball when it is at a height y above the ground.

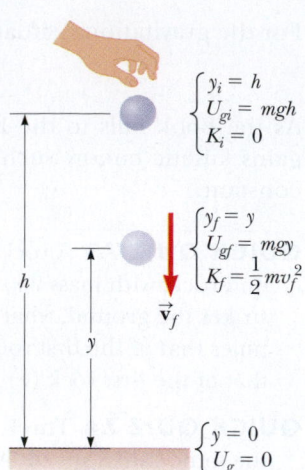

$$\begin{cases} y_i = h \\ U_{gi} = mgh \\ K_i = 0 \end{cases}$$

$$\begin{cases} y_f = y \\ U_{gf} = mgy \\ K_f = \frac{1}{2}mv_f^2 \end{cases}$$

$$\begin{cases} y = 0 \\ U_g = 0 \end{cases}$$

Active Figure 7.4 (Example 7.1) A ball is dropped from a height h above the ground. Initially, the total energy of the ball–Earth system is gravitational potential energy, equal to mgh relative to the ground. At the position y, the total energy is the sum of the kinetic and potential energies.

SOLUTION

Conceptualize Active Figure 7.4 and our everyday experience with falling objects allow us to conceptualize the situation. Although we can readily solve this problem with the techniques of Chapter 2, let us practice an energy approach.

Categorize We identify the system as the ball and the Earth. Because there is neither air resistance nor any other interaction between the system and the environment, the system is isolated and we use the isolated system model. The only force between members of the system is the gravitational force, which is conservative.

Analyze Because the system is isolated and there are no nonconservative forces acting within the system, we apply the principle of conservation of mechanical energy to the ball–Earth system. At the instant the ball is released, its kinetic energy is $K_i = 0$ and the gravitational potential energy of the system is $U_{gi} = mgh$. When the ball is at a position y above the ground, its kinetic energy is $K_f = \frac{1}{2}mv_f^2$ and the potential energy relative to the ground is $U_{gf} = mgy$.

Apply Equation 7.10:

$$K_f + U_{gf} = K_i + U_{gi}$$

$$\tfrac{1}{2}mv_f^2 + mgy = 0 + mgh$$

Solve for v_f:

$$v_f^2 = 2g(h - y) \quad \rightarrow \quad v_f = \boxed{\sqrt{2g(h - y)}}$$

The speed is always positive. If you had been asked to find the ball's velocity, you would use the negative value of the square root as the y component to indicate the downward motion.

(B) Determine the speed of the ball at y if at the instant of release it already has an initial upward speed v_i at the initial altitude h.

SOLUTION

Analyze In this case, the initial energy includes kinetic energy equal to $\frac{1}{2}mv_i^2$.

Apply Equation 7.10:

$$\tfrac{1}{2}mv_f^2 + mgy = \tfrac{1}{2}mv_i^2 + mgh$$

Solve for v_f:

$$v_f^2 = v_i^2 + 2g(h - y) \quad \rightarrow \quad v_f = \boxed{\sqrt{v_i^2 + 2g(h - y)}}$$

Finalize This result for the final speed is consistent with the expression $v_{yf}^2 = v_{yi}^2 - 2g(y_f - y_i)$ from the particle under constant acceleration model for a falling object, where $y_i = h$. Furthermore, this result is valid even if the initial velocity is at an angle to the horizontal (Quick Quiz 7.4) for two reasons: (1) the kinetic energy, a scalar, depends only on the magnitude of the velocity; and (2) the change in the gravitational potential energy of the system depends only on the change in position of the ball in the vertical direction.

What If? What if the initial velocity $\vec{v}_i$ in part (B) were downward? How would that affect the speed of the ball at position y?

Answer You might claim that throwing the ball downward would result in it having a higher speed at y than if you threw it upward. Conservation of mechanical energy, however, depends on kinetic and potential energies, which are scalars. Therefore, the direction of the initial velocity vector has no bearing on the final speed.

Example 7.2 | A Grand Entrance

You are designing an apparatus to support an actor of mass 65 kg who is to "fly" down to the stage during the performance of a play. You attach the actor's harness to a 130-kg sandbag by means of a lightweight steel cable running smoothly over two frictionless pulleys as in Figure 7.5a. You need 3.0 m of cable between the harness and the nearest pulley so that the pulley can be hidden behind a curtain. For the apparatus to work successfully, the sandbag must never lift above the floor as the actor swings from above the stage to the floor. Let us call the initial angle that the actor's cable makes with the vertical θ. What is the maximum value θ can have before the sandbag lifts off the floor?

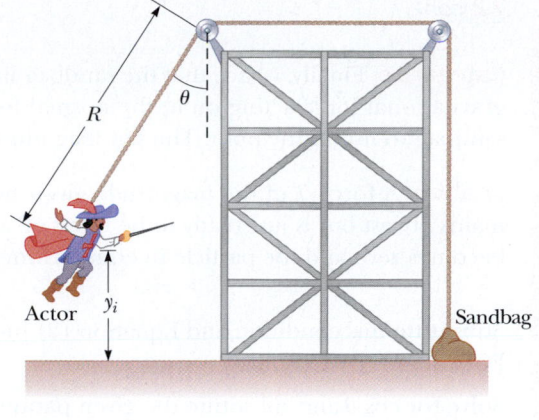

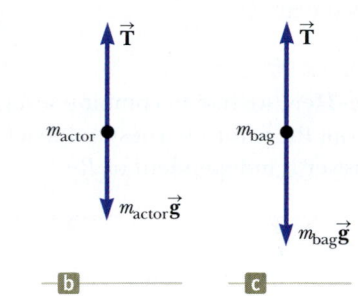

Figure 7.5 (Example 7.2) (a) An actor uses some clever staging to make his entrance. (b) The free-body diagram for the actor at the bottom of the circular path. (c) The free-body diagram for the sandbag if the normal force from the floor goes to zero.

SOLUTION

Conceptualize We must use several concepts to solve this problem. Imagine what happens as the actor approaches the bottom of the swing. At the bottom, the cable is vertical and must support his weight as well as provide centripetal acceleration of his body in the upward direction. At this point in his swing, the tension in the cable is the highest and the sandbag is most likely to lift off the floor.

Categorize Looking first at the swinging of the actor from the initial point to the lowest point, we model the actor and the Earth as an isolated system. We ignore air resistance, so there are no nonconservative forces acting. You might initially be tempted to model the system as nonisolated because of the interaction of the system with the cable, which is in the environment. The force applied to the actor by the cable, however, is always perpendicular to each element of the displacement of the actor and hence does no work. Therefore, in terms of energy transfers across the boundary, the system is isolated.

Analyze We first find the actor's speed as he arrives at the floor as a function of the initial angle θ and the radius R of the circular path through which he swings.

From the isolated system model, apply conservation of mechanical energy to the actor–Earth system:

$$K_f + U_f = K_i + U_i$$

Let y_i be the initial height of the actor above the floor and v_f be his speed at the instant before he lands. (Notice that $K_i = 0$ because the actor starts from rest and that $U_f = 0$ because we define the configuration of the actor at the floor as having a gravitational potential energy of zero.)

$$(1) \quad \tfrac{1}{2}m_{actor}\,v_f^{\,2} + 0 = 0 + m_{actor}\,gy_i$$

From the geometry in Figure 7.5a, notice that $y_f = 0$, so $y_i = R - R\cos\theta = R(1 - \cos\theta)$. Use this relationship in Equation (1) and solve for $v_f^{\,2}$:

$$(2) \quad v_f^{\,2} = 2gR(1 - \cos\theta)$$

Categorize Next, focus on the instant the actor is at the lowest point. Because the tension in the cable is transferred as a force applied to the sandbag, we model the actor at this instant as a particle under a net force. Because the actor moves along a circular arc, he experiences at the bottom of the swing a centripetal acceleration of $v_f^{\,2}/r$ directed upward.

Analyze Apply Newton's second law from the particle under a net force model to the actor at the bottom of his path, using the free-body diagram in Figure 7.5b as a guide:

$$\sum F_y = T - m_{actor}g = m_{actor}\frac{v_f^{\,2}}{R}$$

$$(3) \quad T = m_{actor}g + m_{actor}\frac{v_f^{\,2}}{R}$$

continued

7.2 *cont.*

Categorize Finally, notice that the sandbag lifts off the floor when the upward force exerted on it by the cable exceeds the gravitational force acting on it; the normal force from the floor is zero when that happens. We do *not*, however, want the sandbag to lift off the floor. The sandbag must remain at rest, so we model it as a particle in equilibrium.

Analyze A force *T* of the magnitude given by Equation (3) is transmitted by the cable to the sandbag. If the sandbag re-mains at rest but is just ready to be lifted off the floor if any more force were applied by the cable, the normal force on it becomes zero and the particle in equilibrium model tells us that $T = m_{\text{bag}}g$ as in Figure 7.5c.

Substitute this condition and Equation (2) into Equation (3):

$$m_{\text{bag}}g = m_{\text{actor}}g + m_{\text{actor}}\frac{2gR(1 - \cos\theta)}{R}$$

Solve for $\cos\theta$ and substitute the given parameters:

$$\cos\theta = \frac{3m_{\text{actor}} - m_{\text{bag}}}{2m_{\text{actor}}} = \frac{3(65\text{ kg}) - 130\text{ kg}}{2(65\text{ kg})} = 0.50$$

$$\theta = \boxed{60°}$$

Finalize Here we had to combine several analysis models from different areas of our study. Notice that the length *R* of the cable from the actor's harness to the leftmost pulley did not appear in the final algebraic equation for $\cos\theta$. Therefore, the final answer is independent of *R*.

Example 7.3 | The Spring-Loaded Popgun

The launching mechanism of a popgun consists of a trigger-released spring (Active Fig. 7.6a). The spring is compressed to a position $y_{Ⓐ}$, and the trigger is fired. The projectile of mass *m* rises to a position $y_{Ⓒ}$ above the position at which it leaves the spring, indicated in Active Figure 7.6b as position $y_{Ⓑ} = 0$. Consider a firing of the gun for which $m = 35.0$ g, $y_{Ⓐ} = -0.120$ m, and $y_{Ⓒ} = 20.0$ m.

(A) Neglecting all resistive forces, determine the spring constant.

SOLUTION

Conceptualize Imagine the process illustrated in parts (a) and (b) of Active Figure 7.6. The projectile starts from rest, speeds up as the spring pushes upward on it, leaves the spring, and then slows down as the gravitational force pulls down-ward on it.

Categorize We identify the system as the projectile, the spring, and the Earth. We ignore both air resistance on the projec-tile and friction in the gun, so we model the system as isolated with no nonconservative forces acting.

Analyze Because the projectile starts from rest, its initial kinetic energy is zero. We choose the zero configuration for the gravitational potential energy of the system to be when the projectile leaves the spring. For this configuration, the elastic potential energy is also zero.

  After the gun is fired, the projectile rises to a maximum height $y_{Ⓒ}$. The final kinetic energy of the projectile is zero.

From the isolated system model, write a conser-vation of mechanical energy equation for the system between points Ⓐ and Ⓒ:

$$K_{Ⓒ} + U_{gⒸ} + U_{sⓒ} = K_{Ⓐ} + U_{gⒶ} + U_{sⒶ}$$

Substitute for each energy:

$$0 + mgy_{Ⓒ} + 0 = 0 + mgy_{Ⓐ} + \tfrac{1}{2}kx^2$$

Solve for *k*:

$$k = \frac{2mg(y_{Ⓒ} - y_{Ⓐ})}{x^2}$$

Substitute numerical values:

$$k = \frac{2(0.035\,0\text{ kg})(9.80\text{ m/s}^2)[20.0\text{ m} - (-0.120\text{ m})]}{(0.120\text{ m})^2} = \boxed{958\text{ N/m}}$$

7.3 *cont.*

Active Figure 7.6 (Example 7.3) A spring-loaded popgun (a) before firing and (b) when the spring extends to its relaxed length. (c) An energy bar chart for the popgun–projectile–Earth system before the popgun is loaded. The energy in the system is zero. (d) The popgun is loaded by means of an external agent doing work on the system to push the spring downward. Therefore the system is nonisolated during this process. After the popgun is loaded, elastic potential energy is stored in the spring and the gravitational potential energy of the system is lower because the projectile is below point Ⓑ. (e) As the projectile passes through point Ⓑ, all of the energy of the isolated system is kinetic. (f) When the projectile reaches point Ⓒ, all of the energy of the isolated system is gravitational potential.

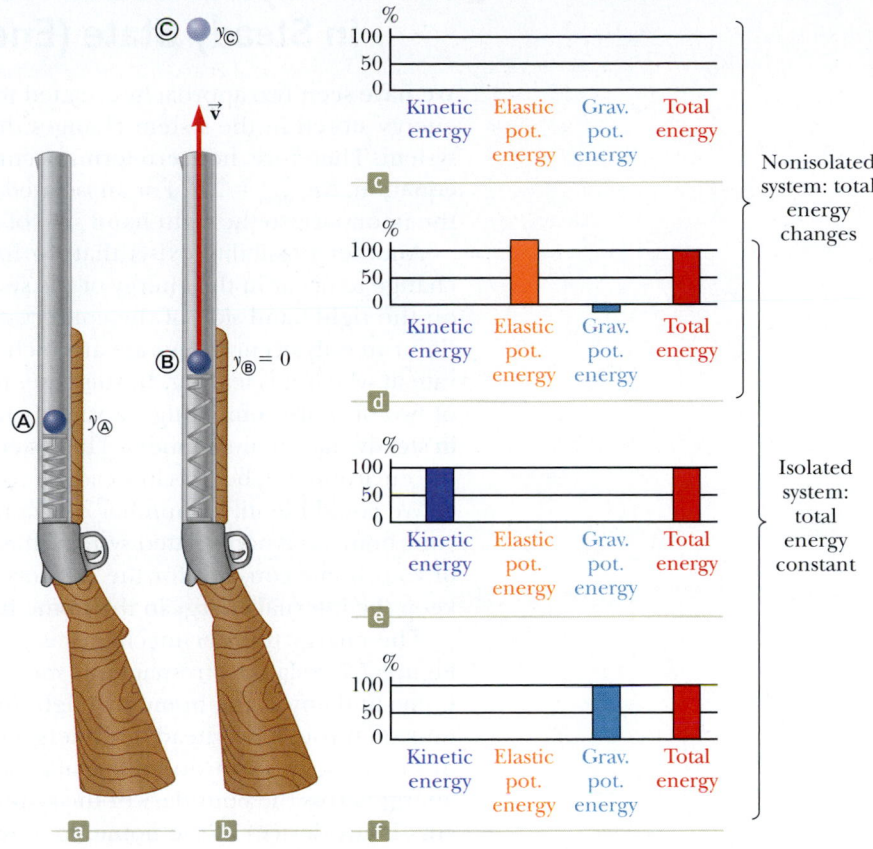

(B) Find the speed of the projectile as it moves through the equilibrium position Ⓑ of the spring as shown in Active Figure 7.6b.

SOLUTION

Analyze The energy of the system as the projectile moves through the equilibrium position of the spring includes only the kinetic energy of the projectile $\frac{1}{2}mv_Ⓑ^2$. Both types of potential energy are equal to zero for this configuration of the system.

Write a conservation of mechanical energy equation for the system between points Ⓐ and Ⓑ:

$$K_Ⓑ + U_{gⒷ} + U_{sⒷ} = K_Ⓐ + U_{gⒶ} + U_{sⒶ}$$

Substitute for each energy:

$$\tfrac{1}{2}mv_Ⓑ^2 + 0 + 0 = 0 + mgy_Ⓐ + \tfrac{1}{2}kx^2$$

Solve for $v_Ⓑ$:

$$v_Ⓑ = \sqrt{\frac{kx^2}{m} + 2gy_Ⓐ}$$

Substitute numerical values:

$$v_Ⓑ = \sqrt{\frac{(958 \text{ N/m})(0.120 \text{ m})^2}{(0.035\,0 \text{ kg})} + 2(9.80 \text{ m/s}^2)(-0.120 \text{ m})} = \boxed{19.8 \text{ m/s}}$$

Finalize This example is the first one we have seen in which we must include two different types of potential energy. Notice in part (A) that we never needed to consider anything about the speed of the ball between points Ⓐ and Ⓒ, which is part of the power of the energy approach: changes in kinetic and potential energy depend only on the initial and final values, not on what happens between the configurations corresponding to these values.

7.3 | Analysis Model: Nonisolated System in Steady State (Energy)

We have seen two approaches related to systems so far. In a nonisolated system, the energy stored in the system changes due to transfers across the boundaries of the system. Therefore, nonzero terms occur on both sides of the conservation of energy equation, $\Delta E_{system} = \Sigma T$. For an isolated system, no energy transfer takes place across the boundary, so the right-hand side of the equation is zero; that is, $\Delta E_{system} = 0$.

Another possibility exists that we have not yet addressed. It is possible for no change to occur in the energy of the system even though nonzero terms are present on the right-hand side of the conservation of energy equation, $0 = \Sigma T$. This situation can only occur if the rate at which energy is entering the system is equal to the rate at which it is leaving. In this case, the system is in steady state under the effects of two or more competing transfers, which we describe with the **nonisolated system in steady state** analysis model. The system is nonisolated because it is interacting with the environment, but it is in steady state because the system energy remains constant.

We could identify a number of examples of this type of situation. First, consider your home as a nonisolated system. Ideally, you would like to keep the temperature of your home constant for the comfort of the occupants. Therefore, your goal is to keep the internal energy in the home fixed.

The energy transfer mechanisms for the home are numerous, as we can see in Figure 7.7. Solar electromagnetic radiation is absorbed by the roof and walls of the home and enters the home through the windows. Energy enters by electrical transmission through overhead or underground wires to operate electrical devices. Leaks in the walls, windows, and doors allow warm or cold air to enter and leave, carrying energy across the boundary of the system by matter transfer. Matter transfer also occurs if any devices in the home operate from natural gas because energy is carried in with the gas. Energy transfer by heat occurs through the walls, windows, floor, and roof as a result of temperature differences between the inside and outside of the home. Therefore, we have a variety of transfers, but the energy in the home remains constant in the idealized case. In reality, the home is a system in *quasi-steady state* because some small temperature variations actually occur over a 24-h period, but we can imagine an idealized situation that conforms to the nonisolated system in steady-state model.

As a second example, consider the Earth and its atmosphere as a system. Because this system is located in the vacuum of space, the only possible types of energy

Figure 7.7 Energy enters and leaves a home by several mechanisms. The home can be modeled as a nonisolated system in steady state.

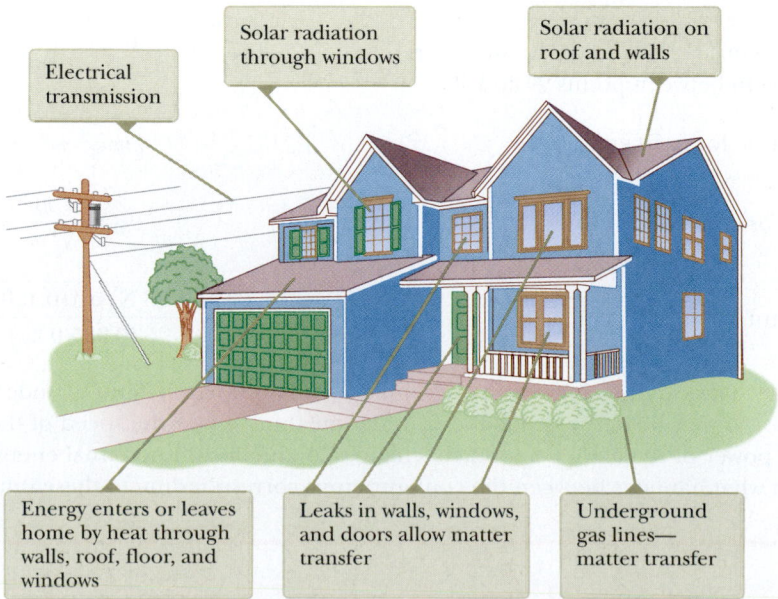

Electrical transmission

Solar radiation through windows

Solar radiation on roof and walls

Energy enters or leaves home by heat through walls, roof, floor, and windows

Leaks in walls, windows, and doors allow matter transfer

Underground gas lines—matter transfer

transfers are those that involve no contact between the system and external molecules in the environment. As mentioned in the footnote on page 166, only two types of transfer do not depend on contact with molecules: work done by field forces and electromagnetic radiation. The Earth–atmosphere system exchanges energy with the rest of the Universe only by means of electromagnetic radiation (ignoring work done by field forces and ignoring some small matter transfer as a result of cosmic ray particles and meteoroids entering the system and spacecraft leaving the system!). The primary input radiation is that from the Sun, and the output radiation is primarily infrared radiation emitted from the atmosphere and the ground. Ideally, these transfers are balanced so that the Earth maintains a constant temperature. In reality, however, the transfers are not *exactly* balanced, so the Earth is in quasi-steady state; measurements of the temperature show that it does appear to be changing. The change in temperature is very gradual and currently appears to be in the positive direction. This change is the essence of the social issue of global warming. (See Context 5, beginning on page 439.)

If we consider a time interval of several days, the human body can be modeled as another nonisolated system in steady state. If the body is at rest at the beginning and end of the time interval, there is no change in kinetic energy. Assuming that no major weight gain or loss occurs during this time interval, the amount of potential energy stored in the body as food in the stomach and fat remains constant on the average. If no fevers are experienced during this time interval, the internal energy of the body remains constant. Therefore, the change in the energy of the system is zero. Energy transfer methods during this time interval include work (you apply forces on objects that move), heat (your body is warmer than the surrounding air), matter transfer (breathing, eating), mechanical waves (you speak and hear), and electromagnetic radiation (you see, as well as absorb and emit radiation from your skin). Table 7.1 shows the amount of energy leaving the body by all methods during one hour of various activities.

TABLE 7.1 |
Energy Output for One Hour of Various Activities

Activity	Energy Output in One Hour (MJ)
Sleeping	0.27
Sitting at rest	0.42
Standing relaxed	0.44
Getting dressed	0.49
Typing	0.59
Walking on level ground (2.6 mi/h)	0.84
Painting a house	1.00
Bicycling on level ground (5.5 mi/h)	1.27
Shoveling snow	2.01
Swimming	2.09
Jogging (5.3 mi/h)	2.39
Rowing (20 strokes/min)	3.47
Walking up stairs	4.60

Adapted from L. Sherwood, *Fundamentals of Human Physiology*, 4th ed. (Belmont, CA: Brooks/Cole, 2012), p. 480

7.4 | Situations Involving Kinetic Friction

Consider again the book in Active Figure 6.18a sliding to the right on the surface of a heavy table and slowing down due to the friction force. Work is done by the friction force because there is a force and a displacement. Keep in mind, however, that our equations for work involve the displacement *of the point of application of the force*. A simple model of the friction force between the book and the surface is shown in Figure 7.8a. We have represented the entire friction force between the book and surface as being due to two identical teeth that have been spot-welded together.[2] One tooth projects upward from the surface, the other downward from the book, and they are welded at the points where they touch. The friction force acts at the junction of the two teeth. Imagine that the book slides a small distance d to the right as in Figure 7.8b. Because the teeth are modeled as identical, the junction of the teeth moves to the right by a distance $d/2$. Therefore, the displacement of the point of application of the friction force is $d/2$, but the displacement of the book is d!

In reality, the friction force is spread out over the entire contact area of an object sliding on a surface, so the force is not localized at a point. In addition, because the magnitudes of the friction forces at various points are constantly changing as individual spot welds occur, the surface and the book deform locally, and so on, the displacement of the point of application of the friction force is not at all the same as the displacement of the book. In fact, the displacement of the point of application of the friction force is not calculable and so neither is the work done by the friction force.

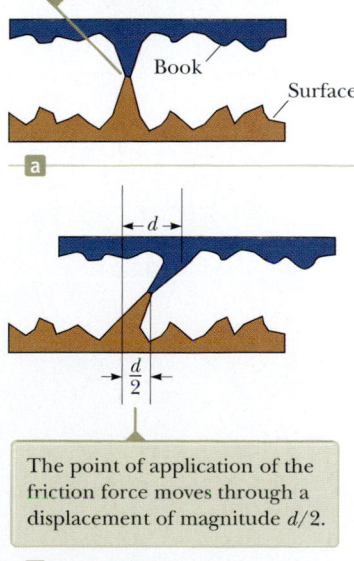

The entire friction force is modeled to be applied at the interface between two identical teeth projecting from the book and the surface.

Book

Surface

a

The point of application of the friction force moves through a displacement of magnitude $d/2$.

b

Figure 7.8 (a) A simplified model of friction between a book and a surface. (b) The book is moved to the right by a distance d.

---

[2]Figure 7.8 and its discussion are inspired by a classic article on friction: B. A. Sherwood and W. H. Bernard, "Work and heat transfer in the presence of sliding friction," *American Journal of Physics*, **52**:1001, 1984.

The work–kinetic energy theorem is valid for a particle or an object that can be modeled as a particle. When a friction force acts, however, we cannot calculate the work done by friction. For such situations, Newton's second law is still valid for the system even though the work–kinetic energy theorem is not. The case of a non-deformable object like our book sliding on the surface[3] with friction can be handled in a relatively straightforward way.

Starting from a situation in which forces, including friction, are applied to the book, we can follow a similar procedure to that done in developing Equation 6.17. Let us start by writing Equation 6.8 for all forces other than friction:

$$\sum W_{\text{other forces}} = \int \left(\sum \vec{\mathbf{F}}_{\text{other forces}} \right) \cdot d\vec{\mathbf{r}} \qquad \text{7.11} \blacktriangleleft$$

The $d\vec{\mathbf{r}}$ in this equation is the displacement of the object because for forces other than friction, under the assumption that these forces do not deform the object, this displacement is the same as the displacement of the point of application of the forces. To each side of Equation 7.11 let us add the integral of the scalar product of the force of kinetic friction and $d\vec{\mathbf{r}}$. In doing so, we are not defining this quantity as work! We are simply saying that it is a quantity that can be calculated mathematically and will turn out to be useful to us in what follows.

$$\sum W_{\text{other forces}} + \int \vec{\mathbf{f}}_k \cdot d\vec{\mathbf{r}} = \int \left(\sum \vec{\mathbf{F}}_{\text{other forces}} \right) \cdot d\vec{\mathbf{r}} + \int \vec{\mathbf{f}}_k \cdot d\vec{\mathbf{r}}$$

$$= \int \left(\sum \vec{\mathbf{F}}_{\text{other forces}} + \vec{\mathbf{f}}_k \right) \cdot d\vec{\mathbf{r}}$$

The integrand on the right side of this equation is the net force $\sum \vec{\mathbf{F}}$, so

$$\sum W_{\text{other forces}} + \int \vec{\mathbf{f}}_k \cdot d\vec{\mathbf{r}} = \int \sum \vec{\mathbf{F}} \cdot d\vec{\mathbf{r}}$$

Incorporating Newton's second law $\sum \vec{\mathbf{F}} = m\vec{\mathbf{a}}$ gives

$$\sum W_{\text{other forces}} + \int \vec{\mathbf{f}}_k \cdot d\vec{\mathbf{r}} = \int m\vec{\mathbf{a}} \cdot d\vec{\mathbf{r}} = \int m\frac{d\vec{\mathbf{v}}}{dt} \cdot d\vec{\mathbf{r}} = \int_{t_i}^{t_f} m\frac{d\vec{\mathbf{v}}}{dt} \cdot \vec{\mathbf{v}}\, dt \qquad \text{7.12} \blacktriangleleft$$

where we have used Equation 3.3 to rewrite $d\vec{\mathbf{r}}$ as $\vec{\mathbf{v}}\, dt$. The scalar product obeys the product rule for differentiation (see Eq. B.30 in Appendix B.6), so the derivative of the scalar product of $\vec{\mathbf{v}}$ with itself can be written

$$\frac{d}{dt}(\vec{\mathbf{v}} \cdot \vec{\mathbf{v}}) = \frac{d\vec{\mathbf{v}}}{dt} \cdot \vec{\mathbf{v}} + \vec{\mathbf{v}} \cdot \frac{d\vec{\mathbf{v}}}{dt} = 2\frac{d\vec{\mathbf{v}}}{dt} \cdot \vec{\mathbf{v}}$$

where we have used the commutative property of the scalar product to justify the final expression in this equation. Consequently,

$$\frac{d\vec{\mathbf{v}}}{dt} \cdot \vec{\mathbf{v}} = \frac{1}{2}\frac{d}{dt}(\vec{\mathbf{v}} \cdot \vec{\mathbf{v}}) = \frac{1}{2}\frac{dv^2}{dt}$$

Substituting this result into Equation 7.12 gives

$$\sum W_{\text{other forces}} + \int \vec{\mathbf{f}}_k \cdot d\vec{\mathbf{r}} = \int_{t_i}^{t_f} m\left(\frac{1}{2}\frac{dv^2}{dt}\right) dt = \frac{1}{2}m\int_{v_i}^{v_f} d(v^2) = \frac{1}{2}mv_f^2 - \frac{1}{2}mv_i^2 = \Delta K$$

---

[3]The overall shape of the book remains the same, which is why we say it is nondeformable. On a microscopic level, however, there is deformation of the book's face as it slides over the surface.

Looking at the left side of this equation, notice that in the inertial frame of the surface, $\vec{f}_k$ and $d\vec{r}$ will be in opposite directions for every increment $d\vec{r}$ of the path followed by the object. Therefore, $\vec{f}_k \cdot d\vec{r} = -f_k\,dr$. The previous expression now becomes

$$\sum W_{\text{other forces}} - \int f_k\,dr = \Delta K$$

In our model for friction, the magnitude of the kinetic friction force is constant, so f_k can be brought out of the integral. The remaining integral $\int dr$ is simply the sum of increments of length along the path, which is the total path length d. Therefore,

$$\sum W_{\text{other forces}} - f_k d = \Delta K \qquad \textbf{7.13} \blacktriangleleft$$

or

$$K_f = K_i - f_k d + \sum W_{\text{other forces}} \qquad \textbf{7.14} \blacktriangleleft$$

Equation 7.13 can be used when a friction force acts on an object. The change in kinetic energy is equal to the work done by all forces other than friction minus a term $f_k d$ associated with the friction force.

Considering the sliding book situation again, let's identify the larger system of the book *and* the surface as the book slows down under the influence of a friction force alone. There is no work done across the boundary of this system because the system does not interact with the environment. There are no other types of energy transfer occurring across the boundary of the system, assuming we ignore the inevitable sound the sliding book makes! In this case, Equation 7.2 becomes

$$\Delta E_{\text{system}} = \Delta K + \Delta E_{\text{int}} = 0$$

The change in kinetic energy of this book–surface system is the same as the change in kinetic energy of the book alone because the book is the only part of the system that is moving. Therefore, incorporating Equation 7.13 gives

$$-f_k d + \Delta E_{\text{int}} = 0$$

$$\Delta E_{\text{int}} = f_k d \qquad \textbf{7.15} \blacktriangleleft$$

▶ Change in internal energy due to a constant friction force within the system

The increase in internal energy of the system is therefore equal to the product of the friction force and the path length through which the block moves. In summary, a friction force transforms kinetic energy in a system to internal energy, and the increase in internal energy of the system is equal to its decrease in kinetic energy. Equation 7.13, with the help of Equation 7.15, can be written as

$$\sum W_{\text{other forces}} = W = \Delta K + \Delta E_{\text{int}}$$

which is a reduced form of Equation 7.2 and represents the nonisolated system model for a system within which a nonconservative force acts.

QUICK QUIZ 7.5 You are traveling along a freeway at 65 mi/h. Your car has kinetic energy. You suddenly skid to a stop because of congestion in traffic. Where is the kinetic energy your car once had? **(a)** It is all in internal energy in the road. **(b)** It is all in internal energy in the tires. **(c)** Some of it has transformed to internal energy and some of it transferred away by mechanical waves. **(d)** It is all transferred away from your car by various mechanisms.

> **THINKING PHYSICS 7.1**
>
> A car traveling at an initial speed v slides a distance d to a halt after its brakes lock. If the car's initial speed is instead $2v$ at the moment the brakes lock, estimate the distance it slides.
>
> **Reasoning** Let us assume the force of kinetic friction between the car and the road surface is constant and the same for both speeds. According to Equation 7.14, the friction force multiplied by the distance d is equal to the initial kinetic energy of the car (because $K_f = 0$ and there is no work done by other forces). If the speed is doubled, as it is in this example, the kinetic energy is quadrupled. For a given friction force, the distance traveled is four times as great when the initial speed is doubled, and so the estimated distance the car slides is $4d$. This result agrees with that in part (b) of Example 5.1, but it is determined by using energy techniques rather than force techniques. ◀

Example 7.4 | A Block Pulled on a Rough Surface

A 6.0-kg block initially at rest is pulled to the right along a horizontal surface by a constant horizontal force of 12 N.

(A) Find the speed of the block after it has moved 3.0 m if the surfaces in contact have a coefficient of kinetic friction of 0.15.

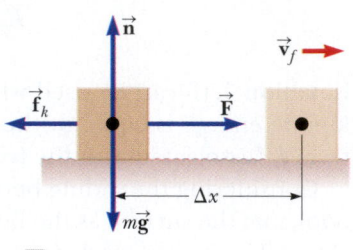

SOLUTION

Conceptualize This example is Example 6.6 (page 144), modified so that the surface is no longer frictionless. The rough surface applies a friction force on the block opposite to the applied force. As a result, we expect the speed to be lower than that found in Example 6.6.

Categorize The block is pulled by a force and the surface is rough, so we model the block–surface system as nonisolated with a nonconservative force acting.

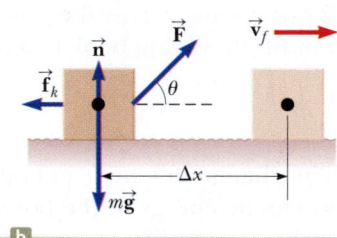

Active Figure 7.9 (Example 7.4)
(a) A block pulled to the right on a rough surface by a constant horizontal force. (b) The applied force is at an angle θ to the horizontal.

Analyze Active Figure 7.9a illustrates this situation. Neither the normal force nor the gravitational force does work on the system because their points of application are displaced horizontally.

Find the work done on the system by the applied force just as in Example 6.6:

$$\sum W_{\text{other forces}} = W_F = F\Delta x$$

Apply the particle in equilibrium model to the block in the vertical direction:

$$\sum F_y = 0 \;\rightarrow\; n - mg = 0 \;\rightarrow\; n = mg$$

Find the magnitude of the friction force:

$$f_k = \mu_k n = \mu_k mg = (0.15)(6.0 \text{ kg})(9.80 \text{ m/s}^2) = 8.82 \text{ N}$$

Find the final speed of the block from Equation 7.14:

$$\tfrac{1}{2}mv_f^2 = \tfrac{1}{2}mv_i^2 - f_k d + W_F$$

$$v_f = \sqrt{v_i^2 + \frac{2}{m}(-f_k d + F\Delta x)}$$

Substitute numerical values:

$$v_f = \sqrt{0 + \frac{2}{6.0 \text{ kg}}[-(8.82 \text{ N})(3.0 \text{ m}) + (12 \text{ N})(3.0 \text{ m})]} = \boxed{1.8 \text{ m/s}}$$

Finalize As expected, this value is less than the 3.5 m/s found in the case of the block sliding on a frictionless surface (see Example 6.6). The difference in kinetic energies between the block in Example 6.6 and the block in this example is equal to the increase in internal energy of the block–surface system in this example.

(B) Suppose the force $\vec{F}$ is applied at an angle θ as shown in Active Figure 7.9b. At what angle should the force be applied to achieve the largest possible speed after the block has moved 3.0 m to the right?

7.4 *cont.*

SOLUTION

Conceptualize You might guess that $\theta = 0$ would give the largest speed because the force would have the largest component possible in the direction parallel to the surface. Think about $\vec{F}$ applied at an arbitrary nonzero angle, however. Although the horizontal component of the force would be reduced, the vertical component of the force would reduce the normal force, in turn reducing the force of friction, which suggests that the speed could be maximized by pulling at an angle other than $\theta = 0$.

Categorize As in part (A), we model the block–surface system as nonisolated with a nonconservative force acting.

. . . .

Analyze Find the work done by the applied force, noting that $\Delta x = d$ because the path followed by the block is a straight line:

$$\sum W_{\text{other forces}} = W_F = F\Delta x \cos\theta = Fd\cos\theta$$

Apply the particle in equilibrium model to the block in the vertical direction:

$$\sum F_y = n + F\sin\theta - mg = 0$$

Solve for n:

$$n = mg - F\sin\theta$$

Use Equation 7.14 to find the final kinetic energy for this situation:

$$K_f = K_i - f_k d + W_F$$
$$= 0 - \mu_k nd + Fd\cos\theta = -\mu_k(mg - F\sin\theta)d + Fd\cos\theta$$

Maximizing the speed is equivalent to maximizing the final kinetic energy. Consequently, differentiate K_f with respect to θ and set the result equal to zero:

$$\frac{dK_f}{d\theta} = -\mu_k(0 - F\cos\theta)d - Fd\sin\theta = 0$$

$$\mu_k\cos\theta - \sin\theta = 0$$

$$\tan\theta = \mu_k$$

Evaluate θ for $\mu_k = 0.15$:

$$\theta = \tan^{-1}(\mu_k) = \tan^{-1}(0.15) = \boxed{8.5°}$$

. . . .

Finalize Notice that the angle at which the speed of the block is a maximum is indeed not $\theta = 0$. When the angle exceeds 8.5°, the horizontal component of the applied force is too small to be compensated by the reduced friction force and the speed of the block begins to decrease from its maximum value.

Example 7.5 | A Block–Spring System

A block of mass 1.6 kg is attached to a horizontal spring that has a force constant of 1 000 N/m as shown in Figure 7.10. The spring is compressed 2.0 cm and is then released from rest.

(A) Calculate the speed of the block as it passes through the equilibrium position $x = 0$ if the surface is frictionless.

SOLUTION

Conceptualize This situation has been discussed before, and it is easy to visualize the block being pushed to the right by the spring and moving with some speed at $x = 0$.

Categorize We identify the system as the block and model the block as a nonisolated system.

. . . .

Analyze In this situation, the block starts with $v_i = 0$ at $x_i = -2.0$ cm, and we want to find v_f at $x_f = 0$.

Use Equation 6.11 to find the work done by the spring on the system with $x_{\text{max}} = x_i$:

$$\sum W_{\text{other forces}} = W_s = \tfrac{1}{2}kx_{\text{max}}^2$$

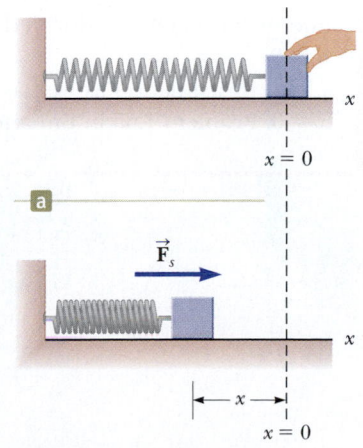

Figure 7.10 (Example 7.5) (a) A block attached to a spring is pushed inward from an initial position $x = 0$ by an external agent. (b) At position x, the block is released from rest and the spring pushes it to the right.

continued

7.5 *cont.*

Work is done on the block, and its speed changes. The conservation of energy equation, Equation 7.2, reduces to the work–kinetic energy theorem. Use that theorem to find the speed at $x = 0$:

$$W_s = \tfrac{1}{2}mv_f^2 - \tfrac{1}{2}mv_i^2$$

$$v_f = \sqrt{v_i^2 + \frac{2}{m}W_s} = \sqrt{v_i^2 + \frac{2}{m}\left(\tfrac{1}{2}kx_{max}^2\right)}$$

Substitute numerical values:

$$v_f = \sqrt{0 + \frac{2}{1.6\text{ kg}}\left[\tfrac{1}{2}(1\,000\text{ N/m})(0.020\text{ m})^2\right]} = \boxed{0.50\text{ m/s}}$$

Finalize Although this problem could have been solved in Chapter 6, it is presented here to provide contrast with the following part (B), which requires the techniques of this chapter.

(B) Calculate the speed of the block as it passes through the equilibrium position if a constant friction force of 4.0 N retards its motion from the moment it is released.

SOLUTION

Conceptualize The correct answer must be less than that found in part (A) because the friction force retards the motion.

Categorize We identify the system as the block and the surface. The system is nonisolated because of the work done by the spring, and there is a nonconservative force acting: the friction between the block and the surface.

Analyze Write Equation 7.14:

$$(1)\ K_f = K_i - f_k d + W_s$$

Substitute numerical values:

$$K_f = 0 - (4.0\text{ N})(0.020\text{ m}) + \tfrac{1}{2}(1\,000\text{ N/m})(0.020\text{ m})^2 = 0.12\text{ J}$$

Write the definition of kinetic energy:

$$K_f = \tfrac{1}{2}mv_f^2$$

Solve for v_f and substitute numerical values:

$$v_f = \sqrt{\frac{2K_f}{m}} = \sqrt{\frac{2(0.12\text{ J})}{1.6\text{ kg}}} = \boxed{0.39\text{ m/s}}$$

Finalize As expected, this value is less than the 0.50 m/s found in part (A).

What If? What if the friction force were increased to 10.0 N? What is the block's speed at $x = 0$?

Answer In this case, the value of $f_k d$ as the block moves to $x = 0$ is

$$f_k d = (10.0\text{ N})(0.020\text{ m}) = 0.20\text{ J}$$

which is equal in magnitude to the kinetic energy at $x = 0$ for the frictionless case. (Verify it!) Therefore, all the kinetic energy has been transformed to internal energy by friction when the block arrives at $x = 0$, and its speed at this point is $v = 0$.

In this situation as well as that in part (B), the speed of the block reaches a maximum at some position other than $x = 0$. Problem 7.65 in Enhanced WebAssign asks you to locate these positions.

❮7.5❯ Changes in Mechanical Energy for Nonconservative Forces

Consider the book sliding across the surface in the preceding section. As the book moves through a distance d, the only force that does work on it is the force of kinetic friction. This force causes a change $-f_k d$ in the kinetic energy of the book as described by Equation 7.13.

Now, however, suppose the book is part of a system that also exhibits a change in potential energy. In this case, $-f_k d$ is the amount by which the mechanical energy of the system changes because of the force of kinetic friction. For example, if the book moves on an incline that is not frictionless, there is a change in both the kinetic energy and the gravitational potential energy of the book–Earth system. Consequently,

$$\Delta E_{mech} = \Delta K + \Delta U_g = -f_k d$$

In general, if a friction force acts within an isolated system,

$$\Delta E_{mech} = \Delta K + \Delta U = -f_k d \qquad \textbf{7.16} \blacktriangleleft$$

▶ Change in mechanical energy of a system due to friction within the system

where ΔU is the change in all forms of potential energy. Notice that Equation 7.16 reduces to Equation 7.10 if the friction force is zero.

If the system in which nonconservative forces act is nonisolated and the external influence on the system is by means of work, the generalization of Equation 7.13 is

$$\Delta E_{mech} = -f_k d + \sum W_{other\ forces} \qquad \textbf{7.17} \blacktriangleleft$$

Equation 7.17, with the help of Equations 7.7 and 7.15, can be written as

$$\sum W_{other\ forces} = W = \Delta K + \Delta U + \Delta E_{int}$$

This reduced form of Equation 7.2 represents the nonisolated system model for a system that possesses potential energy and within which a nonconservative force acts. In practice during problem solving, you do not need to use equations like Equation 7.15 or Equation 7.17. You can simply use Equation 7.2 and keep only those terms in the equation that correspond to the physical situation. See Example 7.8 for a sample of this approach.

> ### PROBLEM-SOLVING STRATEGY: Systems with Nonconservative Forces
>
> The following procedure should be used when you face a problem involving a system in which nonconservative forces act:
>
> 1. **Conceptualize.** Study the physical situation carefully and form a mental representation of what is happening.
>
> 2. **Categorize.** Define your system, which may consist of more than one object. The system could include springs or other possibilities for storage of potential energy. Determine whether any nonconservative forces are present. If not, use the principle of conservation of mechanical energy as outlined in Section 7.2. If so, use the procedure discussed below.
>
>   Determine if any work is done across the boundary of your system by forces other than friction. If so, use Equation 7.17 to analyze the problem. If not, use Equation 7.16.
>
> 3. **Analyze.** Choose configurations to represent the initial and final conditions of the system. For each object that changes elevation, select a reference position for the object that defines the zero configuration of gravitational potential energy for the system. For an object on a spring, the zero configuration for elastic potential energy is when the object is at its equilibrium position. If there is more than one conservative force, write an expression for the potential energy associated with each force.
>
>   Use either Equation 7.16 or Equation 7.17 to establish a mathematical representation of the problem. Solve for the unknown.
>
> 4. **Finalize.** Make sure your results are consistent with your mental representation. Also make sure the values of your results are reasonable and consistent with connections to everyday experience.

Example 7.6 | Crate Sliding Down a Ramp

A 3.00-kg crate slides down a ramp. The ramp is 1.00 m in length and inclined at an angle of 30.0° as shown in Figure 7.11. The crate starts from rest at the top, experiences a constant friction force of magnitude 5.00 N, and continues to move a short distance on the horizontal floor after it leaves the ramp.

(A) Use energy methods to determine the speed of the crate at the bottom of the ramp.

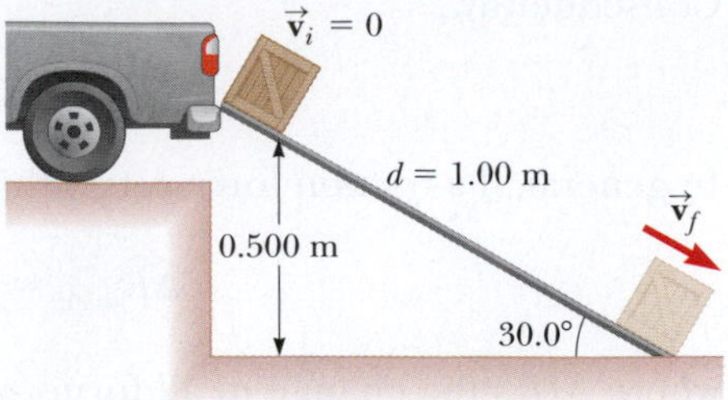

Figure 7.11 (Example 7.6) A crate slides down a ramp under the influence of gravity. The potential energy of the system decreases, whereas the kinetic energy increases.

SOLUTION

Conceptualize Imagine the crate sliding down the ramp in Figure 7.11. The larger the friction force, the more slowly the crate will slide.

Categorize We identify the crate, the surface, and the Earth as the system. The system is categorized as isolated with a nonconservative force acting.

Analyze Because $v_i = 0$, the initial kinetic energy of the system when the crate is at the top of the ramp is zero. If the y coordinate is measured from the bottom of the ramp (the final position of the crate, for which we choose the gravitational potential energy of the system to be zero) with the upward direction being positive, then $y_i = 0.500$ m.

Write the expression for the total mechanical energy of the system when the crate is at the top:

$$E_i = K_i + U_i = 0 + U_i = mgy_i$$

Write an expression for the final mechanical energy:

$$E_f = K_f + U_f = \tfrac{1}{2}mv_f^2 + 0 = \tfrac{1}{2}mv_f^2$$

Apply Equation 7.16:

$$\Delta E_{\text{mech}} = E_f - E_i = \tfrac{1}{2}mv_f^2 - mgy_i = -f_k d$$

Solve for v_f:

$$(1)\ v_f = \sqrt{\frac{2}{m}(mgy_i - f_k d)}$$

Substitute numerical values:

$$v_f = \sqrt{\frac{2}{3.00\ \text{kg}}[(3.00\ \text{kg})(9.80\ \text{m/s}^2)(0.500\ \text{m}) - (5.00\ \text{N})(1.00\ \text{m})]} = \boxed{2.54\ \text{m/s}}$$

(B) How far does the crate slide on the horizontal floor if it continues to experience a friction force of magnitude 5.00 N?

SOLUTION

Analyze This part of the problem is handled in exactly the same way as part (A), but in this case we can consider the mechanical energy of the system to consist only of kinetic energy because the potential energy of the system remains fixed.

Write an expression for the mechanical energy of the system when the crate leaves the bottom of the ramp:

$$E_i = K_i = \tfrac{1}{2}mv_i^2$$

Apply Equation 7.16 with $E_f = 0$:

$$E_f - E_i = 0 - \tfrac{1}{2}mv_i^2 = -f_k d \quad \rightarrow \quad \tfrac{1}{2}mv_i^2 = f_k d$$

Solve for the distance d and substitute numerical values:

$$d = \frac{mv_i^2}{2f_k} = \frac{(3.00\ \text{kg})(2.54\ \text{m/s})^2}{2(5.00\ \text{N})} = \boxed{1.94\ \text{m}}$$

Finalize For comparison, you may want to calculate the speed of the crate at the bottom of the ramp in the case in which the ramp is frictionless. Also notice that the increase in internal energy of the system as the crate slides down the ramp is $f_k d = (5.00\ \text{N})(1.00\ \text{m}) = 5.00\ \text{J}$. This energy is shared between the crate and the surface, each of which is a bit warmer than before.

Also notice that the distance d the object slides on the horizontal surface is infinite if the surface is frictionless. Is that consistent with your conceptualization of the situation?

What If? A cautious worker decides that the speed of the crate when it arrives at the bottom of the ramp may be so large that its contents may be damaged. Therefore, he replaces the ramp with a longer one such that the new ramp makes an angle of 25.0° with the ground. Does this new ramp reduce the speed of the crate as it reaches the ground?

7.6 cont.

Answer Because the ramp is longer, the friction force acts over a longer distance and transforms more of the mechanical energy into internal energy. The result is a reduction in the kinetic energy of the crate, and we expect a lower speed as it reaches the ground.

Find the length d of the new ramp:

$$\sin 25.0° = \frac{0.500 \text{ m}}{d} \quad \rightarrow \quad d = \frac{0.500 \text{ m}}{\sin 25.0°} = 1.18 \text{ m}$$

Find v_f from Equation (1) in part (A):

$$v_f = \sqrt{\frac{2}{3.00 \text{ kg}}[(3.00 \text{ kg})(9.80 \text{ m/s}^2)(0.500 \text{ m}) - (5.00 \text{ N})(1.18 \text{ m})]} = 2.42 \text{ m/s}$$

The final speed is indeed lower than in the higher-angle case.

Example 7.7 | Block–Spring Collision

A block having a mass of 0.80 kg is given an initial velocity $v_{Ⓐ} = 1.2$ m/s to the right and collides with a spring whose mass is negligible and whose force constant is $k = 50$ N/m as shown in Figure 7.12.

(A) Assuming the surface to be frictionless, calculate the maximum compression of the spring after the collision.

SOLUTION

Conceptualize The various parts of Figure 7.12 help us imagine what the block will do in this situation. All motion takes place in a horizontal plane, so we do not need to consider changes in gravitational potential energy.

Categorize We identify the system to be the block and the spring. The block–spring system is isolated with no nonconservative forces acting.

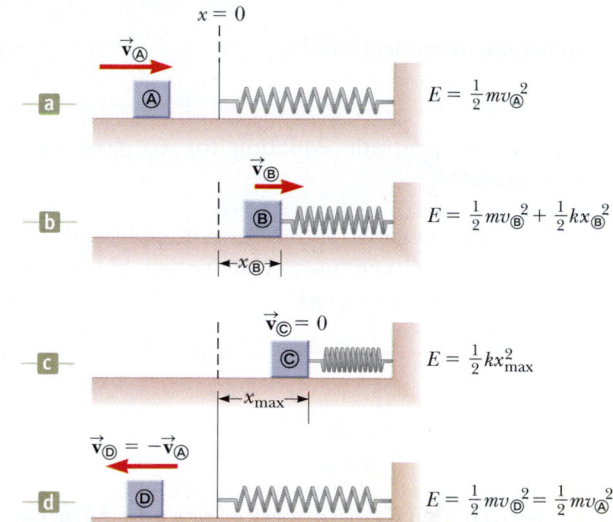

Figure 7.12 (Example 7.7) A block sliding on a frictionless, horizontal surface collides with a light spring. (a) Initially, the mechanical energy is all kinetic energy. (b) The mechanical energy is the sum of the kinetic energy of the block and the elastic potential energy in the spring. (c) The energy is entirely potential energy. (d) The energy is transformed back to the kinetic energy of the block. The total energy of the system remains constant throughout the motion.

Analyze Before the collision, when the block is at Ⓐ, it has kinetic energy and the spring is uncompressed, so the elastic potential energy stored in the system is zero. Therefore, the total mechanical energy of the system before the collision is just $\frac{1}{2}mv_{Ⓐ}^2$. After the collision, when the block is at Ⓒ, the spring is fully compressed; now the block is at rest and so has zero kinetic energy. The elastic potential energy stored in the system, however, has its maximum value $\frac{1}{2}kx^2 = \frac{1}{2}kx_{max}^2$, where the origin of coordinates $x = 0$ is chosen to be the equilibrium position of the spring and x_{max} is the maximum compression of the spring, which in this case happens to be $x_{Ⓒ}$. The total mechanical energy of the system is conserved because no nonconservative forces act on objects within the isolated system.

Write a conservation of mechanical energy equation:

$$K_{Ⓒ} + U_{sⒸ} = K_{Ⓐ} + U_{sⒶ}$$

$$0 + \frac{1}{2}kx_{max}^2 = \frac{1}{2}mv_{Ⓐ}^2 + 0$$

Solve for x_{max} and evaluate:

$$x_{max} = \sqrt{\frac{m}{k}}v_{Ⓐ} = \sqrt{\frac{0.80 \text{ kg}}{50 \text{ N/m}}}(1.2 \text{ m/s}) = \boxed{0.15 \text{ m}}$$

(B) Suppose a constant force of kinetic friction acts between the block and the surface, with $\mu_k = 0.50$. If the speed of the block at the moment it collides with the spring is $v_{Ⓐ} = 1.2$ m/s, what is the maximum compression $x_{Ⓒ}$ in the spring?

continued

7.7 *cont.*

SOLUTION

Conceptualize Because of the friction force, we expect the compression of the spring to be smaller than in part (A) because some of the block's kinetic energy is transformed to internal energy in the block and the surface.

Categorize We identify the system as the block, the surface, and the spring. This system is isolated but now involves a nonconservative force.

· ·

Analyze In this case, the mechanical energy $E_{mech} = K + U_s$ of the system is *not* conserved because a friction force acts on the block. From the particle in equilibrium model in the vertical direction, we see that $n = mg$.

Evaluate the magnitude of the friction force: $\qquad f_k = \mu_k n = \mu_k mg$

Write the change in the mechanical energy of the system $\qquad \Delta E_{mech} = -f_k x_{©}$
due to friction as the block is displaced from $x = 0$ to $x_{©}$:

Substitute the initial and final energies: $\qquad \Delta E_{mech} = E_f - E_i = (0 + \frac{1}{2}kx_{©}^2) - (\frac{1}{2}mv_{Ⓐ}^2 + 0) = -f_k x_{©}$

$$\frac{1}{2}kx_{©}^2 - \frac{1}{2}mv_{Ⓐ}^2 = -\mu_k mg x_{©}$$

Substitute numerical values: $\qquad \frac{1}{2}(50)x_{©}^2 - \frac{1}{2}(0.80)(1.2)^2 = -(0.50)(0.80)(9.80)x_{©}$

$$25x_{©}^2 + 3.9x_{©} - 0.58 = 0$$

Solving the quadratic equation for $x_{©}$ gives $x_{©} = 0.093$ m and $x_{©} = -0.25$ m. The physically meaningful root is $x_{©} = \boxed{0.093 \text{ m.}}$

· ·

Finalize The negative root does not apply to this situation because the block must be to the right of the origin (positive value of x) when it comes to rest. Notice that the value of 0.093 m is less than the distance obtained in the frictionless case of part (A) as we expected.

Example 7.8 | Connected Blocks in Motion

Two blocks are connected by a light string that passes over a frictionless pulley as shown in Figure 7.13. The block of mass m_1 lies on a horizontal surface and is connected to a spring of force constant k. The system is released from rest when the spring is unstretched. If the hanging block of mass m_2 falls a distance h before coming to rest, calculate the coefficient of kinetic friction between the block of mass m_1 and the surface.

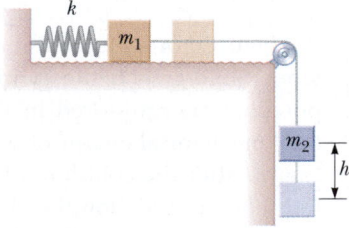

SOLUTION

Conceptualize The key word *rest* appears twice in the problem statement. This word suggests that the configurations of the system associated with rest are good candidates for the initial and final configurations because the kinetic energy of the system is zero for these configurations.

Categorize In this situation, the system consists of the two blocks, the spring, the surface, and the Earth. The system is isolated with a nonconservative force acting. We also model the sliding block as a particle in equilibrium in the vertical direction, leading to $n = m_1 g$.

Figure 7.13 (Example 7.8) As the hanging block moves from its highest elevation to its lowest, the system loses gravitational potential energy but gains elastic potential energy in the spring. Some mechanical energy is transformed to internal energy because of friction between the sliding block and the surface.

· ·

Analyze We need to consider two forms of potential energy for the system, gravitational and elastic: $\Delta U_g = U_{gf} - U_{gi}$ is the change in the system's gravitational potential energy, and $\Delta U_s = U_{sf} - U_{si}$ is the change in the system's elastic potential energy. The change in the gravitational potential energy of the system is associated with only the falling block because the vertical coordinate of the horizontally sliding block does not change. The initial and final kinetic energies of the system are zero, so $\Delta K = 0$.

7.8 *cont.*

For this example, let us start from Equation 7.2 to show how this approach would work in practice. Because the system is isolated, the entire right side of Equation 7.2 is zero. Based on the physical situation described in the problem, we see that there could be changes of kinetic energy, potential energy, and internal energy in the system. Write the corresponding reduction of Equation 7.2:

$$\Delta K + \Delta U + \Delta E_{int} = 0$$

Incorporate into this equation that $\Delta K = 0$ and that there are two types of potential energy:

$$(1)\ \Delta U_g + \Delta U_s + \Delta E_{int} = 0$$

Use Equation 7.15 to find the change in internal energy in the system due to friction between the horizontally sliding block and the surface, noticing that as the hanging block falls a distance h, the horizontally moving block moves the same distance h to the right:

$$(2)\ \Delta E_{int} = f_k h = (\mu_k n) h = \mu_k m_1 gh$$

Evaluate the change in gravitational potential energy of the system, choosing the configuration with the hanging block at the lowest position to represent zero potential energy:

$$(3)\ \Delta U_g = U_{gf} - U_{gi} = 0 - m_2 gh$$

Evaluate the change in the elastic potential energy of the system:

$$(4)\ \Delta U_s = U_{sf} - U_{si} = \tfrac{1}{2}kh^2 - 0$$

Substitute Equations (2), (3), and (4) into Equation (1):

$$-m_2 gh + \tfrac{1}{2}kh^2 + \mu_k m_1 gh = 0$$

Solve for μ_k:

$$\mu_k = \frac{m_2 g - \tfrac{1}{2}kh}{m_1 g}$$

Finalize This setup represents a method of measuring the coefficient of kinetic friction between an object and some surface. Notice that we do not need to remember which energy equation goes with which type of problem with this approach. You can always begin with Equation 7.2 and then tailor it to the physical situation. This process may include deleting terms, such as the kinetic energy term and all terms on the right-hand side in this example. It can also include expanding terms, such as rewriting ΔU due to two types of potential energy in this example.

THINKING PHYSICS 7.2

The energy bar charts in Figure 7.14 show three instants in the motion of the system in Figure 7.13 and described in Example 7.8. For each bar chart, identify the configuration of the system that corresponds to the chart.

Reasoning In Figure 7.14a, there is no kinetic energy in the system. Therefore, nothing in the system is moving. The bar chart shows that the system contains only gravitational potential energy and no internal energy yet, which corresponds to the configuration with the darker blocks in Figure 7.13 and represents the instant just after the system is released.

In Figure 7.14b, the system contains four types of energy. The height of the gravitational potential energy bar is at 50%, which tells us that the hanging block has moved halfway between its position corresponding to Figure 7.14a and the position defined as $y = 0$. Therefore, in this configuration, the hanging block is between the dark and light images of the hanging block in Figure 7.13. The system has gained kinetic energy because the blocks are

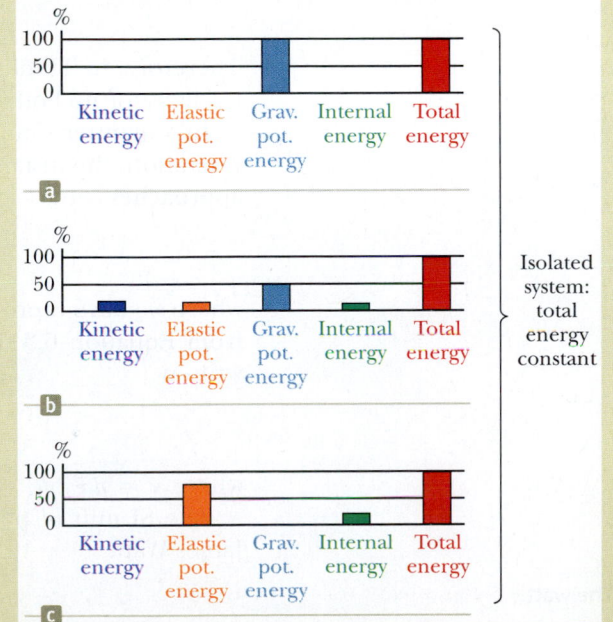

Figure 7.14 (Thinking Physics 7.2) Three energy bar charts are shown for the system in Figure 7.13.

moving, elastic potential energy because the spring is stretching, and internal energy because of friction between the block of mass m_1 and the surface.

In Figure 7.14c, the height of the gravitational potential energy bar is zero, telling us that the hanging block is at $y = 0$. In addition, the height of the kinetic energy bar is zero, indicating that the blocks have stopped moving momentarily. Therefore, the configuration of the system is that shown by the light images of the blocks in Figure 7.13. The height of the elastic potential energy bar is high because the spring is stretched its maximum amount. The height of the internal energy bar is higher than in Figure 7.14b because the block of mass m_1 has continued to slide over the surface. ◄

◤7.6 | Power

Consider Thinking Physics 6.1 again, which involved rolling a refrigerator up a ramp into a truck. Suppose the man is not convinced the work is the same regardless of the ramp's length and sets up a long ramp with a gentle rise. Although he does the same amount of work as someone using a shorter ramp, he takes longer to do the work because he has to move the refrigerator over a greater distance. Although the work done on both ramps is the same, there is *something* different about the tasks: the *time interval* during which the work is done.

The time rate of energy transfer is called the **instantaneous power** P and is defined as

▶ Definition of power

$$P \equiv \frac{dE}{dt}$$ 7.18◄

We will focus on work as the energy transfer method in this discussion, but keep in mind that the notion of power is valid for *any* means of energy transfer discussed in Section 7.1. If an external force is applied to an object (which we model as a particle) and if the work done by this force on the object in the time interval Δt is W, the **average power** during this interval is

$$P_{avg} = \frac{W}{\Delta t}$$

Therefore, in Thinking Physics 6.1, although the same work is done in rolling the refrigerator up both ramps, less power is required for the longer ramp.

In a manner similar to the way we approached the definition of velocity and acceleration, the instantaneous power is the limiting value of the average power as Δt approaches zero:

$$P = \lim_{\Delta t \to 0} \frac{W}{\Delta t} = \frac{dW}{dt}$$

where we have represented the infinitesimal value of the work done by dW. We find from Equation 6.3 that $dW = \vec{F} \cdot d\vec{r}$. Therefore, the instantaneous power can be written

$$P = \frac{dW}{dt} = \vec{F} \cdot \frac{d\vec{r}}{dt} = \vec{F} \cdot \vec{v}$$ 7.19◄

where $\vec{v} = d\vec{r}/dt$.

The SI unit of power is joules per second (J/s), also called the **watt** (W) after James Watt:

▶ The watt

$$1\ \text{W} = 1\ \text{J/s} = 1\ \text{kg} \cdot \text{m}^2/\text{s}^3$$

A unit of power in the U.S. customary system is the **horsepower** (hp):

$$1\ \text{hp} = 746\ \text{W}$$

A unit of energy (or work) can now be defined in terms of the unit of power. One **kilowatt-hour** (kWh) is the energy transferred in 1 h at the constant rate of 1 kW = 1 000 J/s. The amount of energy represented by 1 kWh is

$$1 \text{ kWh} = (10^3 \text{ W})(3\ 600 \text{ s}) = 3.60 \times 10^6 \text{ J}$$

A kilowatt-hour is a unit of energy, not power. When you pay your electric bill, you are buying energy, and the amount of energy transferred by electrical transmission into a home during the period represented by the electric bill is usually expressed in kilowatt-hours. For example, your bill may state that you used 900 kWh of energy during a month and that you are being charged at the rate of 10¢ per kilowatt-hour. Your obligation is then $90 for this amount of energy. As another example, suppose an electric bulb is rated at 100 W. In 1.00 h of operation, it would have energy transferred to it by electrical transmission in the amount of (0.100 kW)(1.00 h) = 0.100 kWh = 3.60×10^5 J.

> **Pitfall Prevention | 7.3**
> **W, W, and Watts**
> Do not confuse the symbol W for the watt with the italic symbol *W* for work. Also, remember that the watt already represents a rate of energy transfer, so "watts per second" does not make sense. The watt is *the same as* a joule per second.

Example 7.9 | Power Delivered by an Elevator Motor

An elevator car (Fig. 7.15a) has a mass of 1 600 kg and is carrying passengers having a combined mass of 200 kg. A constant friction force of 4 000 N retards its motion.

(A) How much power must a motor deliver to lift the elevator car and its passengers at a constant speed of 3.00 m/s?

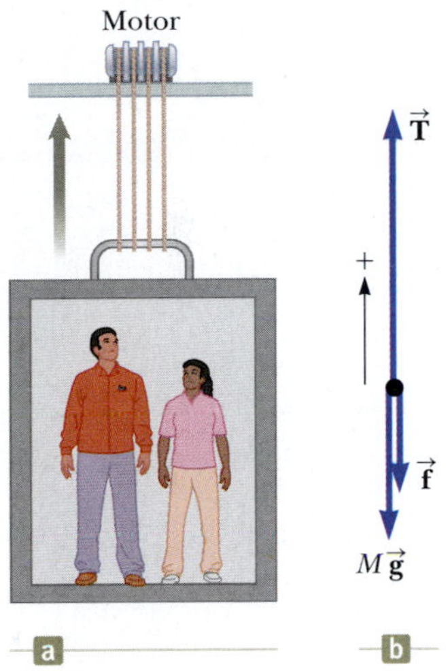

SOLUTION

Conceptualize The motor must supply the force of magnitude *T* that pulls the elevator car upward.

Categorize The friction force increases the power necessary to lift the elevator. The problem states that the speed of the elevator is constant, which tells us that $a = 0$. We model the elevator as a particle in equilibrium.

....

Analyze The free-body diagram in Figure 7.15b specifies the upward direction as positive. The *total* mass *M* of the system (car plus passengers) is equal to 1 800 kg.

Using the particle in equilibrium model, apply Newton's second law to the car:

$$\sum F_y = T - f - Mg = 0$$

Solve for *T*:

$$T = f + Mg$$

Use Equation 7.19 and that $\vec{\mathbf{T}}$ is in the same direction as $\vec{\mathbf{v}}$ to find the power:

$$P = \vec{\mathbf{T}} \cdot \vec{\mathbf{v}} = Tv = (f + Mg)v$$

Substitute numerical values:

$$P = [(4\ 000 \text{ N}) + (1\ 800 \text{ kg})(9.80 \text{ m/s}^2)](3.00 \text{ m/s}) = \boxed{6.49 \times 10^4 \text{ W}}$$

Figure 7.15 (Example 7.9) (a) The motor exerts an upward force $\vec{\mathbf{T}}$ on the elevator car. The magnitude of this force is the total tension *T* in the cables connecting the car and motor. The downward forces acting on the car are a friction force $\vec{\mathbf{f}}$ and the gravitational force $\vec{\mathbf{F}}_g = M\vec{\mathbf{g}}$. (b) The free-body diagram for the elevator car.

(B) What power must the motor deliver at the instant the speed of the elevator is *v* if the motor is designed to provide the elevator car with an upward acceleration of 1.00 m/s²?

SOLUTION

Conceptualize In this case, the motor must supply the force of magnitude *T* that pulls the elevator car upward with an increasing speed. We expect that more power will be required to do that than in part (A) because the motor must now perform the additional task of accelerating the car.

Categorize In this case, we model the elevator car as a particle under a net force because it is accelerating.

....

Analyze Using the particle under a net force model, apply Newton's second law to the car:

$$\sum F_y = T - f - Mg = Ma$$

Solve for *T*:

$$T = M(a + g) + f$$

7.9 *cont.*

Use Equation 7.19 to obtain the required power: $P = Tv = [M(a + g) + f]v$

Substitute numerical values: $P = [(1\ 800\ \text{kg})(1.00\ \text{m/s}^2 + 9.80\ \text{m/s}^2) + 4\ 000\ \text{N}]v$

$$= (2.34 \times 10^4)v$$

where v is the instantaneous speed of the car in meters per second and P is in watts.

Finalize To compare with part (A), let $v = 3.00$ m/s, giving a power of

$$P = (2.34 \times 10^4\ \text{N})(3.00\ \text{m/s}) = 7.02 \times 10^4\ \text{W}$$

which is larger than the power found in part (A), as expected.

◄7.7 | Context Connection: Horsepower Ratings of Automobiles

As discussed in Section 4.8, an automobile moves because of Newton's third law. The engine attempts to rotate the wheels in such a direction as to push the Earth toward the back of the car because of the friction force between the wheels and the roadway. By Newton's third law, the Earth pushes in the opposite direction on the wheels, which is toward the front of the car. Because the Earth is much more massive than the car, the Earth remains stationary while the car moves forward.

This principle is the same one humans use for walking. By pushing your leg backward while your foot is on the ground, you apply a friction force backward on the surface of the Earth. By Newton's third law, the surface applies a forward friction force on you, which causes your body to move forward.

The strength of the friction force $\vec{f}$ exerted on a car by the roadway is related to the rate at which energy is transferred to the wheels to set them into rotation, which is the power of the engine:

$$P_{\text{avg}} = \frac{\Delta E}{\Delta t} = \frac{f\Delta x}{\Delta t} = fv \quad \rightarrow \quad P \leftrightarrow f$$

where the symbol $\leftrightarrow$ implies a relationship between the variables that is not necessarily an exact proportionality. In turn, the magnitude of the driving force is related to the acceleration of the car owing to Newton's second law:

$$f = ma \quad \rightarrow \quad f \propto a$$

Consequently, there should be a close relationship between the power rating of a vehicle and the possible acceleration of the vehicle:

$$P \leftrightarrow a$$

Let us see if this relationship exists for actual data. For automobiles, a common unit for power is the *horsepower* (hp), defined in Section 7.6. Table 7.2 shows the gasoline-powered automobiles we have studied in the preceding chapters. The third column provides the published horsepower rating of each vehicle. In Figure 7.16, we graph the acceleration values against the horsepower ratings for the vehicles. From this graph, we see a clear correlation between the acceleration and the horsepower as proposed above: as the horsepower rating goes up, the maximum possible acceleration increases. The two black data points at the far right of the graph lie below a line that could be drawn through the other data points. These two points represent the Bugatti Veyron 16.4 Super Sport and the Shelby SuperCars Ultimate Aero. These

TABLE 7.2 | Horsepower Ratings and Accelerations of Various Vehicles

Automobile	Average Acceleration (mi/h·s)	Horsepower Rating (hp)	Ratio HP to Acceleration (hp/mi/h·s)
Very expensive vehicles:			
Bugatti Veyron 16.4 Super Sport	23.1	1200	52
Lamborghini LP 570-4 Superleggera	17.6	570	32
Lexus LFA	15.8	560	35
Mercedes-Benz SLS AMG	16.7	563	34
Shelby SuperCars Ultimate Aero	22.2	1287	58
Average	**19.1**	**836**	**42.3**
Performance vehicles:			
Chevrolet Corvette ZR1	18.2	638	35
Dodge Viper SRT10	15.0	600	40
Jaguar XJL Supercharged	13.6	470	35
Acura TL SH-AWD	11.5	305	27
Dodge Challenger SRT8	12.2	425	35
Average	**14.1**	**488**	**34.2**
Traditional vehicles:			
Buick Regal CXL Turbo	8.0	220	28
Chevrolet Tahoe 1500 LS (SUV)	7.0	326	47
Ford Fiesta SES	6.2	120	19
Hummer H3 (SUV)	7.5	300	40
Hyundai Sonata SE	8.0	200	25
Smart ForTwo	4.5	70	16
Average	**6.9**	**206**	**29.0**
Alternative vehicles:			
Chevrolet Volt (hybrid)	7.5	74	10
Nissan Leaf (electric)	6.0	110	18
Honda CR-Z (hybrid)	5.7	122	21
Honda Insight (hybrid)	5.7	98	17
Toyota Prius (hybrid)	6.1	134	22
Average	**6.2**	**108**	**17.8**

are very high-powered vehicles, each sporting a horsepower rating of 1200 hp or more. The graph shows that this large increase in horsepower over the other vehicles results in a relatively modest increase in acceleration. This behavior is similar to that in Figure 2.16, in which a huge increase in cost is required for a relatively small increase in acceleration. Perhaps there is an upper limit of acceleration beyond which horsepower and money cannot take us.

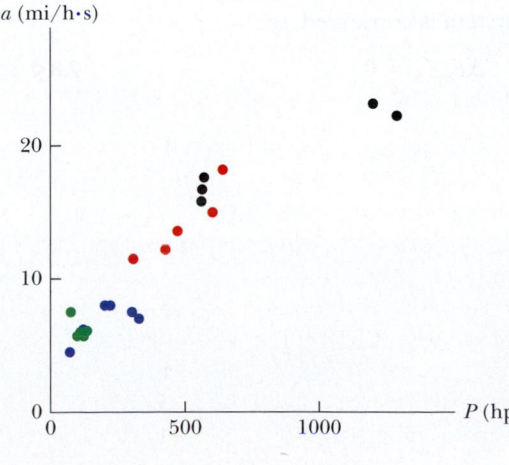

Figure 7.16 The acceleration as a function of horsepower rating for alternative vehicles (in green), traditional vehicles (in blue), performance vehicles (in red), and very expensive vehicles (in black).

SUMMARY

A **nonisolated system** is one for which energy crosses the boundary of the system. An **isolated system** is one for which no energy crosses the boundary of the system.

For a nonisolated system, we can equate the change in the total energy stored in the system to the sum of all the transfers of energy across the system boundary, which is a statement of **conservation of energy.** For an isolated system, the total energy is constant.

If a system is isolated and if no nonconservative forces are acting on objects inside the system, the total mechanical energy of the system is constant:

$$K_f + U_f = K_i + U_i \qquad \textbf{7.10}\blacktriangleleft$$

If nonconservative forces (such as friction) act between objects inside a system, mechanical energy is not conserved. In these situations, the difference between the total final mechanical energy and the total initial mechanical energy of the system equals the energy transformed to internal energy by the nonconservative forces.

If a friction force acts within an isolated system, the mechanical energy of the system is reduced and the appropriate equation to be applied is

$$\Delta E_{mech} = \Delta K + \Delta U = -f_k d \qquad \textbf{7.16}\blacktriangleleft$$

If a friction force acts within a nonisolated system, the appropriate equation to be applied is

$$\Delta E_{mech} = -f_k d + \sum W_{other\ forces} \qquad \textbf{7.17}\blacktriangleleft$$

The **instantaneous power** P is defined as the time rate of energy transfer:

$$P \equiv \frac{dE}{dt} \qquad \textbf{7.18}\blacktriangleleft$$

Analysis Models for Problem Solving

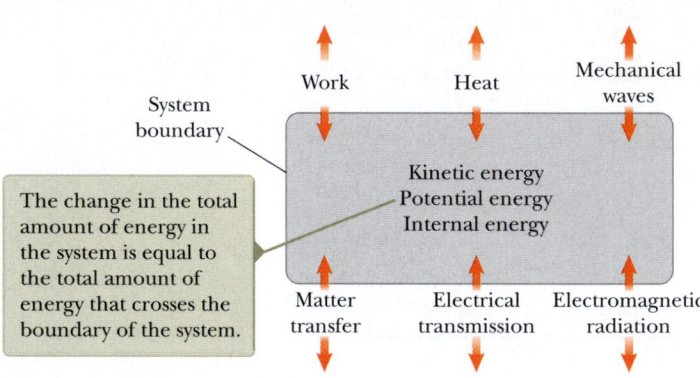

The change in the total amount of energy in the system is equal to the total amount of energy that crosses the boundary of the system.

Nonisolated System (Energy). The most general statement describing the behavior of a nonisolated system is the **conservation of energy equation:**

$$\Delta E_{system} = \sum T \qquad \textbf{7.1}\blacktriangleleft$$

Including the types of energy storage and energy transfer that we have discussed gives

$$\Delta K + \Delta U + \Delta E_{int} = \\ W + Q + T_{MW} + T_{MT} + T_{ET} + T_{ER} \qquad \textbf{7.2}\blacktriangleleft$$

For a specific problem, this equation is generally reduced to a smaller number of terms by eliminating the terms that are not appropriate to the situation.

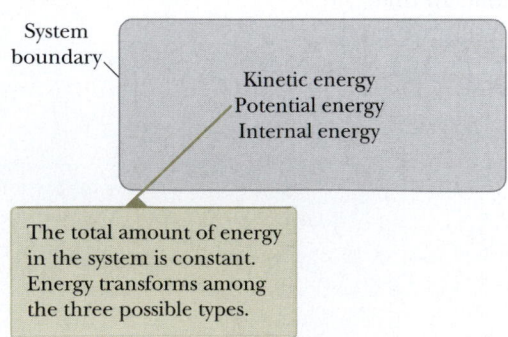

The total amount of energy in the system is constant. Energy transforms among the three possible types.

Isolated System (Energy). The total energy of an isolated system is conserved, so

$$\Delta E_{system} = 0 \qquad \textbf{7.9}\blacktriangleleft$$

If no nonconservative forces act within the isolated system, the mechanical energy of the system is conserved, so

$$\Delta E_{mech} = 0 \qquad \textbf{7.8}\blacktriangleleft$$

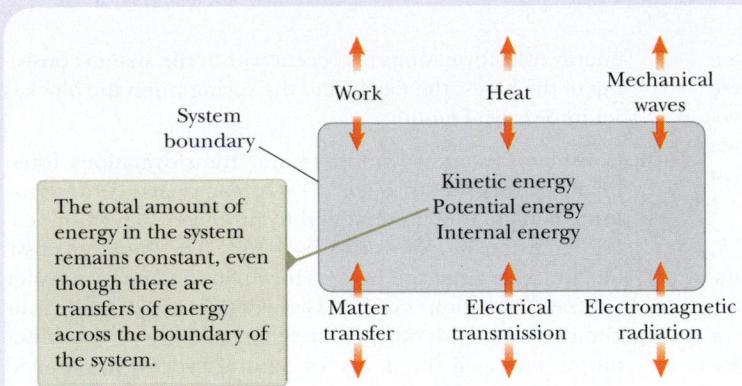

Nonisolated System in Steady State (Energy). If there are energy inputs and outputs across the boundary of the system, but they are in balance, then the change in energy of the system is zero:

$$0 = \sum T$$

OBJECTIVE QUESTIONS

☐ denotes answer available in *Student Solutions Manual/Study Guide*

1. You hold a slingshot at arm's length, pull the light elastic band back to your chin, and release it to launch a pebble horizontally with speed 200 cm/s. With the same procedure, you fire a bean with speed 600 cm/s. What is the ratio of the mass of the bean to the mass of the pebble? (a) $\frac{1}{9}$ (b) $\frac{1}{3}$ (c) 1 (d) 3 (e) 9

2. An athlete jumping vertically on a trampoline leaves the surface with a velocity of 8.5 m/s upward. What maximum height does she reach? (a) 13 m (b) 2.3 m (c) 3.7 m (d) 0.27 m (e) The answer can't be determined because the mass of the athlete isn't given.

3. A pile driver drives posts into the ground by repeatedly dropping a heavy object on them. Assume the object is dropped from the same height each time. By what factor does the energy of the pile driver–Earth system change when the mass of the object being dropped is doubled? (a) $\frac{1}{2}$ (b) 1; the energy is the same (c) 2 (d) 4

4. Two children stand on a platform at the top of a curving slide next to a backyard swimming pool. At the same moment the smaller child hops off to jump straight down into the pool, the bigger child releases herself at the top of the frictionless slide. (**i**) Upon reaching the water, the kinetic energy of the smaller child compared with that of the larger child is (a) greater (b) less (c) equal. (**ii**) Upon reaching the water, the speed of the smaller child compared with that of the larger child is (a) greater (b) less (c) equal. (**iii**) During their motions from the platform to the water, the average acceleration of the smaller child compared with that of the larger child is (a) greater (b) less (c) equal.

5. Answer yes or no to each of the following questions. (a) Can an object–Earth system have kinetic energy and not gravitational potential energy? (b) Can it have gravitational potential energy and not kinetic energy? (c) Can it have both types of energy at the same moment? (d) Can it have neither?

6. A ball of clay falls freely to the hard floor. It does not bounce noticeably, and it very quickly comes to rest. What, then, has happened to the energy the ball had while it was falling? (a) It has been used up in producing the downward motion. (b) It has been transformed back into potential energy. (c) It has been transferred into the ball by heat. (d) It is in the ball and floor (and walls) as energy of invisible molecular motion. (e) Most of it went into sound.

7. What average power is generated by a 70.0-kg mountain climber who climbs a summit of height 325 m in 95.0 min? (a) 39.1 W (b) 54.6 W (c) 25.5 W (d) 67.0 W (e) 88.4 W

8. In a laboratory model of cars skidding to a stop, data are measured for four trials using two blocks. The blocks have identical masses but different coefficients of kinetic friction with a table: $\mu_k = 0.2$ and 0.8. Each block is launched with speed $v_i = 1$ m/s and slides across the level table as the block comes to rest. This process represents the first two trials. For the next two trials, the procedure is repeated but the blocks are launched with speed $v_i = 2$ m/s. Rank the four trials (a) through (d) according to the stopping distance from largest to smallest. If the stopping distance is the same in two cases, give them equal rank. (a) $v_i = 1$ m/s, $\mu_k = 0.2$ (b) $v_i = 1$ m/s, $\mu_k = 0.8$ (c) $v_i = 2$ m/s, $\mu_k = 0.2$ (d) $v_i = 2$ m/s, $\mu_k = 0.8$

9. At the bottom of an air track tilted at angle θ, a glider of mass m is given a push to make it coast a distance d up the slope as it slows down and stops. Then the glider comes back down the track to its starting point. Now the experiment is repeated with the same original speed but with a second identical glider set on top of the first. The airflow from the track is strong enough to support the stacked pair of gliders so that the combination moves over the track with negligible friction. Static friction holds the second glider stationary relative to the first glider throughout the motion. The coefficient of static friction between the two gliders is μ_s. What is the change in mechanical energy of the two-glider–Earth system in the up- and down-slope motion after the pair of gliders is released? Choose one. (a) $-2\mu_s mg$ (b) $-2mgd \cos \theta$ (c) $-2\mu_s mgd \cos \theta$ (d) 0 (e) $+2\mu_s mgd \cos \theta$

CONCEPTUAL QUESTIONS

1. One person drops a ball from the top of a building while another person at the bottom observes its motion. Will these two people agree (a) on the value of the gravitational potential energy of the ball–Earth system? (b) On the change in potential energy? (c) On the kinetic energy of the ball at some point in its motion?

2. In Chapter 6 the work–kinetic energy theorem, $W_{ext} = \Delta K$, was introduced. This equation states that work done on a system appears as a change in kinetic energy. It is a special-case equation, valid if there are no changes in any other type of energy such as potential or internal. Give two or three examples in which work is done on a system but the change in energy of the system is not a change in kinetic energy.

3. Does everything have energy? Give the reasoning for your answer.

4. Can a force of static friction do work? If not, why not? If so, give an example.

5. A bowling ball is suspended from the ceiling of a lecture hall by a strong cord. The ball is drawn away from its equilibrium position and released from rest at the tip of the demonstrator's nose as shown in Figure CQ7.5. The demonstrator remains stationary. (a) Explain why the ball does not strike her on its return swing. (b) Would this demonstrator be safe if the ball were given a push from its starting position at her nose?

6. You ride a bicycle. In what sense is your bicycle solar powered?

Figure CQ7.5

7. A block is connected to a spring that is suspended from the ceiling. Assuming air resistance is ignored, describe the energy transformations that occur within the system consisting of the block, the Earth, and the spring when the block is set into vertical motion.

8. Consider the energy transfers and transformations listed below in parts (a) through (e). For each part, (i) describe human-made devices designed to produce each of the energy transfers or transformations and, (ii) whenever possible, describe a natural process in which the energy transfer or transformation occurs. Give details to defend your choices, such as identifying the system and identifying other output energy if the device or natural process has limited efficiency. (a) Chemical potential energy transforms into internal energy. (b) Energy transferred by electrical transmission becomes gravitational potential energy. (c) Elastic potential energy transfers out of a system by heat. (d) Energy transferred by mechanical waves does work on a system. (e) Energy carried by electromagnetic waves becomes kinetic energy in a system.

9. In the general conservation of energy equation, state which terms predominate in describing each of the following devices and processes. For a process going on continuously, you may consider what happens in a 10-s time interval. State which terms in the equation represent original and final forms of energy, which would be inputs, and which outputs. (a) a slingshot firing a pebble (b) a fire burning (c) a portable radio operating (d) a car braking to a stop (e) the surface of the Sun shining visibly (f) a person jumping up onto a chair

10. A car salesperson claims that a 300-hp engine is a necessary option in a compact car, in place of the conventional 130-hp engine. Suppose you intend to drive the car within speed limits (≤ 65 mi/h) on flat terrain. How would you counter this sales pitch?

PROBLEMS AVAILABLE IN

7.1 Analysis Model: Nonisolated System (Energy)

Problems 1–2

7.2 Analysis Model: Isolated System (Energy)

Problems 3–11

7.4 Situations Involving Kinetic Friction

Problems 12–17

7.5 Changes in Mechanical Energy for Nonconservative Forces

Problems 18–27

7.6 Power

Problems 28–41

7.7 Context Connection: Horsepower Ratings of Automobiles

Problems 42–43

Additional Problems

Problems 44–83

Solutions to the following Problems are available in the *Student Solutions Manual/Study Guide*:

7.3, 7.5, 7.7, 7.12, 7.13, 7.16, 7.21, 7.22, 7.23, 7.29, 7.37, 7.46, 7.51, 7.71, 7.73, 7.76, and 7.79

List of Enhanced Problems

Problem Number	Targeted Feedback in Enhanced WebAssign	Master It in Enhanced WebAssign	Watch It in Enhanced WebAssign
7.3	✓	✓	
7.5	✓		✓
7.6	✓		✓
7.7	✓	✓	
7.10	✓		✓
7.12	✓	✓	
7.15	✓		✓
7.21	✓	✓	
7.22	✓		✓
7.23		✓	
7.25	✓		✓
7.29	✓		✓
7.51		✓	
7.68	✓		✓
7.71	✓	✓	
7.73		✓	

Present and Future Possibilities

Now that we have explored some fundamental principles of classical mechanics, let us return to our central question for the *Alternative-Fuel Vehicles* Context:

> **What source besides gasoline can be used to provide energy for an automobile while reducing environmentally damaging emissions?**

Available Now—The Hybrid Electric Vehicle

As discussed in Section 6.11, a few purely electric vehicles are currently available, but they suffer from difficulties such as limited range and long charging times. Also available and more widely used by consumers are a growing number of **hybrid electric vehicles.** In these automobiles, a gasoline engine and an electric motor are combined to increase the fuel economy of the vehicle and reduce its emissions. Currently available models include the Toyota Prius and Honda Insight, which are originally designed hybrid vehicles, as well as other existing traditional gasoline-powered models that have been modified with a hybrid drive system.

Two major categories of hybrid vehicles are the **series hybrid** and the **parallel hybrid.** In a series hybrid, such as the Chevrolet Volt (Fig. 1) operating at low speeds, the gasoline engine does not provide propulsion energy to the transmission directly. The engine turns a generator, which in turn either charges the batteries or powers the electric motor. Only the electric motor is connected directly to the transmission to propel the car.

In a parallel hybrid, both the engine and the motor are connected to the transmission, so either one can provide propulsion energy for the car. The Honda Insight is a parallel hybrid. Both the engine and the motor provide power to the transmission, and the engine is running at all times while the car is moving. The goal of the development of this hybrid is maximum mileage, which is achieved through a number of design features. Because the engine is small, the Insight has lower emissions than a traditional gasoline-powered vehicle. Because the engine is running at all vehicle speeds, however, its emissions are not as low as those of the Toyota Prius.

Figure 2 shows the third-generation Toyota Prius, which is a series/parallel combination. At higher speeds, power to the wheels comes from both the gasoline engine and the electric motor. The vehicle has some aspects of a series hybrid, however, in that the electric motor alone accelerates the vehicle from rest until it is moving at a speed of about 15 mi/h (24 km/h). During this acceleration period, the engine is not running, so gasoline is not used and there is no emission. As a result, the average tailpipe

Figure 1 The Chevrolet Volt.

BERNARD TRONCALE/Birmingham News/Landov

emissions are lower than those of the Insight. The Chevrolet Volt has the possibility of the lowest tailpipe emissions because, for repeated cycles of short trips alternating with recharging, the gasoline engine may not operate at all.

When a hybrid vehicle brakes, the motor acts as a generator and returns some of the kinetic energy of the vehicle back to the battery as electric potential energy. In a normal vehicle, this kinetic energy is not recoverable because it is transformed to internal energy in the brakes and roadway.

Gas mileage for hybrid vehicles is in the range of 40 to 55 mi/gal and emissions are far below those of a standard gasoline engine. A hybrid vehicle does not need to be charged like a purely electric vehicle. The battery that drives the electric motor is charged while the gasoline engine is running. Conse-

Figure 2 The third-generation Toyota Prius.

quently, even though the hybrid vehicle has an electric motor like a pure electric vehicle, it can simply be filled at a gas station like a normal vehicle.

Hybrid electric vehicles are not strictly alternative-fuel vehicles because they use the same fuel as normal vehicles, gasoline. They do, however, represent an important step toward more efficient cars with lower emissions, and the increased mileage helps conserve crude oil.

In the Future—The Fuel Cell Vehicle

In an internal combustion engine, the chemical potential energy in the fuel is transformed to internal energy during an explosion initiated by a spark plug. The resulting expanding gases do work on pistons, directing energy to the wheels of the vehicle. In current development is the **fuel cell**, in which the conversion of the energy in the fuel to internal energy is not required. The fuel (hydrogen) is oxidized, and energy leaves the fuel cell by electrical transmission. The energy is used by an electric motor to drive the vehicle.

The advantages of this type of vehicle are many. There is no internal combustion engine to generate harmful emissions, so the vehicle is emission-free. Other than the energy used to power the vehicle, the only by-products are internal energy and water. The fuel is hydrogen, which is the most abundant element in the universe. The efficiency of a fuel cell is much higher than that of an internal combustion engine, so more of the potential energy in the fuel can be extracted.

That is all good news. The bad news is that fuel cell vehicles are only in the early prototype stage. Honda has a production model fuel-cell vehicle, the Honda FCX Clarity (Fig. 3). The Clarity is only available in the United States in southern California, where there are a few hydrogen filling stations, and only about 20 vehicles were operating there as of 2010. It will be many years before fuel cell vehicles are widely available to consumers. During these years, fuel cells must be perfected to operate in weather extremes, manufacturing infrastructure must be set up to supply the hydrogen, and a fueling infrastructure must be established to allow transfer of hydrogen into individual vehicles.

Problems

1. When a conventional car brakes to a stop, all (100%) its kinetic energy is converted into internal energy. None of this energy is available to get the car moving again. Consider a hybrid electric car of mass 1 300 kg moving at 22.0 m/s. (a) Calculate its kinetic energy. (b) The car uses its regenerative braking system to come to a stop at a red light. Assume that the motor-generator converts 70.0% of the car's kinetic energy into energy delivered to the battery by electrical transmission. The other 30.0% becomes internal energy. Compute the amount of energy charging up the battery. (c) Assume that the battery can give back 85.0% of the

Figure 3 The fuel filler inlet for hydrogen in the Honda FCX Clarity.

energy chemically stored in it. Compute the amount of this energy. The other 15.0% becomes internal energy. (d) When the light turns green, the car's motor-generator runs as a motor to convert 68.0% of the energy from the battery into kinetic energy of the car. Compute the amount of this energy and (e) the speed at which the car will be set moving with no other energy input. (f) Compute the overall efficiency of the braking-and-starting process. (g) Compute the net amount of internal energy produced.

2. In both a conventional car and a hybrid electric car, the gasoline engine is the original source of all the energy the car uses to push through the air and against rolling resistance of the road. In city traffic, a conventional gasoline engine must run at a wide variety of rotation rates and fuel inputs. That is, it must run at a wide variety of tachometer and throttle settings. It is almost never running at its maximum-efficiency point. In a hybrid electric car, on the other hand, the gasoline engine can run at maximum efficiency whenever it is on. A simple model can reveal the distinction numerically. Assume that the two cars both do 66.0 MJ of "useful" work in making the same trip to the drugstore. Let the conventional car run at 7.00% efficiency as it puts out useful energy 33.0 MJ and let it run at 30.0% efficiency as it puts out 33.0 MJ. Let the hybrid car run at 30.0% efficiency all the time. Compute (a) the required energy input for each car and (b) the overall efficiency of each.

Mission to Mars

In this Context, we shall investigate the physics necessary to send a spacecraft from Earth to Mars. If the two planets were sitting still in space, millions of kilometers apart, it would be a difficult enough proposition, but keep in mind that we are launching the spacecraft from a moving object, the Earth, and are aiming at a moving target, Mars. Furthermore, the spacecraft's motion is influenced by gravitational forces from the Earth, the Sun, and Mars as well as from any other massive objects in the vicinity. Despite these apparent difficulties, we can use the principles of physics to plan a successful mission.

In the 1970s, the Viking Project landed spacecraft on Mars to analyze the soil for signs of life. These tests were inconclusive. The United States returned to Mars in the 1990s with the Mars Global Surveyor, designed to perform careful mapping of the Martian surface, and Mars Pathfinder, which landed on Mars and deployed a roving robot to analyze rocks and soil. Not all trips have been successful. In 1999, Mars Polar Lander was launched to land near the polar ice cap and search for water. As it entered the Martian atmosphere, it sent its last data and was never heard from again. Mars Climate Orbiter was also lost in 1999 due to communication errors between the builder of the spacecraft and the mission control team.

In late 2003 and early 2004, arrivals of spacecraft at Mars were expected by three space agencies, the National Aeronautics and Space Administration (NASA) in the United States, the European Space Agency (ESA) in Europe, and the Japanese Aerospace Exploration Agency (JAXA) in Japan. The Japanese mission ended in failure when a stuck valve and electrical circuit problems affected a critical midcourse correction, resulting in the inability of the spacecraft, named *Nozomi*, to achieve an orbit around Mars. It passed about 1 000 km above the Martian surface on December 14, 2003, and then left the planet to continue its orbit around the Sun.

The European effort resulted in a successful injection of their *Mars Express* spacecraft into an orbit around Mars. A lander, named *Beagle 2*, descended to the surface. Unfortunately, no signals from the lander have

Figure 1 The Mars rover *Spirit* is tested in a clean room at the Jet Propulsion Laboratory in Pasadena, California.

NASA/JPL

Courtesy of NASA/JPL/Cornell

Figure 2 An image from a camera on the Mars rover *Opportunity* shows a rock called the "Berry Bowl." The "berries" are sphere-like grains containing hematite, which scientists used to confirm the earlier presence of water on the surface. The circular area on the rock is the result of using the rover's rock abrasion tool to remove a layer of dust. In this way, a clean surface of the rock was available for spectral analysis by the rover's spectrometers.

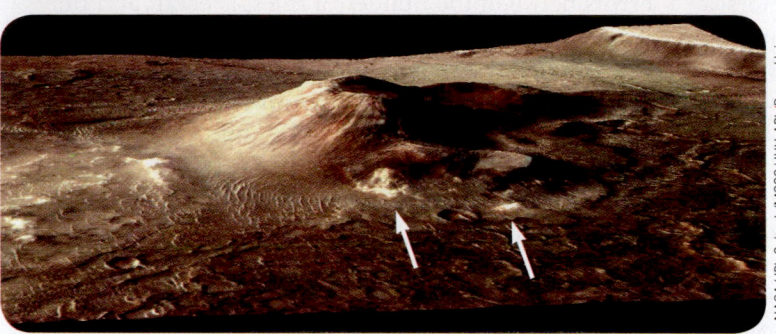

NASA/JPL-Caltech/MSSS/JHU-APL/Brown Univ.

Figure 3 This volcanic cone on Mars has hydrothermal mineral deposits on the southern flanks and nearby terrains. Two of the largest deposits are marked by arrows, and the entire field of light-toned material on the left of the cone is hydrothermal deposits.

been detected and it is presumed lost. The *Mars Express* orbiter continues to send data and is equipped to perform scientific analyses from orbit.

The NASA effort was the most successful of the three missions, with the *Spirit* rover landing successfully on the surface of Mars on January 4, 2004. Its twin, *Opportunity,* also landed successfully, on January 24, 2004, on the opposite side of the planet from *Spirit.* Amazingly, *Opportunity* landed inside a crater, providing scientists with a wonderful opportunity to study the geology of an impact crater. Aside from a computer glitch that was successfully repaired, both rovers performed excellently and sent back very high-quality photographs of the Martian surface as well as large amounts of data including verification of water that once existed on the surface.

More recent observations in 2010 by NASA's Mars Reconnaissance Orbiter revealed a volcanic cone containing hydrothermal mineral deposits on the flanks of the cone. Researchers have identified one of the minerals as hydrated silica, and the new results suggest that in some regions, Mars may have supported microbial life. Excellent photographs near the North Pole of Mars were obtained using a camera mounted on the Orbiter as part of the High Resolution Imaging Science Experiment, HiRISE. The images show only small patches of ice at the surface whose structure is typical of icy permafrost that expands and contracts with changing seasons.

Many individuals dream of one day establishing colonies on Mars. This dream is far in the future; we are still learning

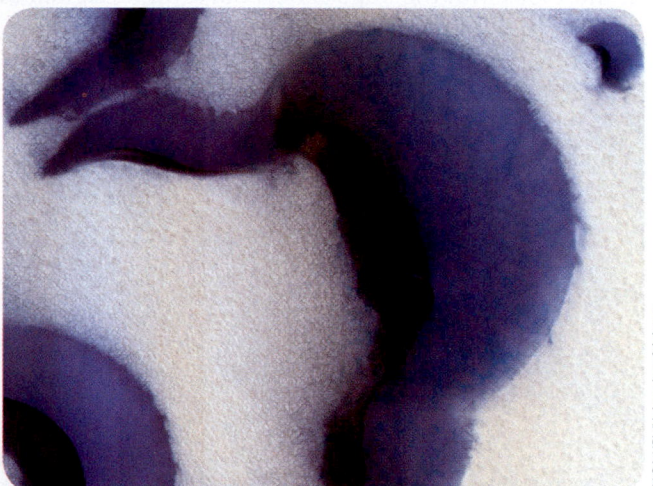

NASA/JPL/University of Arizona

Figure 4 This HiRISE image shows some patches of surface ice near the North Pole of Mars.

much about Mars today and have yet taken only a handful of trips to the planet. Travel to Mars is still not an everyday occurrence, although we learn more from each mission. In this Context, we address the central question,

> **How can we undertake a successful transfer of a spacecraft from Earth to Mars?**

Momentum and Collisions

Chapter Outline

AP photos/Keystone/Regina Kuehne

The concept of momentum allows the analysis of car collisions even without detailed knowledge of the forces involved. Such analysis can determine the relative velocity of the cars before the collision, and in addition aid engineers in designing safer vehicles. (The English translation of the text on the side of the trailer in the background is: "Pit stop for your vehicle.")

Consider what happens when two cars collide as in the opening photograph for this chapter. Both cars change their motion from having a very large velocity to being at rest because of the collision. Because each car experiences a large change in velocity over a very short time interval, the average force on it is very large. By Newton's third law, each of the cars experiences a force of the same magnitude. By Newton's second law, the results of those forces on the motion of the car depends on the mass of the car.

One main objective of this chapter is to enable you to understand and analyze such events. As a first step, we shall introduce the concept of *momentum*, a term used to describe objects in motion. The concept of momentum leads us to a new conservation law and new analysis models incorporating momentum approaches for isolated and nonisolated systems. This conservation law is especially useful for treating problems that involve collisions between objects.

8.1 | Linear Momentum

In the preceding two chapters, we studied situations that are difficult to analyze with Newton's laws. We were able to solve problems involving these situations

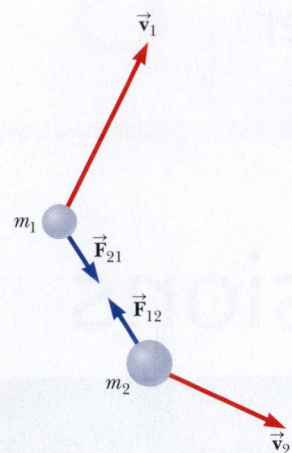

Figure 8.1 Two particles interact with each other. According to Newton's third law, we must have $\vec{\mathbf{F}}_{12} = -\vec{\mathbf{F}}_{21}$.

by applying a conservation principle, conservation of energy. Let us consider another situation and see if we can solve it with the models we have developed so far:

A 60-kg archer stands at rest on frictionless ice and fires a 0.030-kg arrow horizontally at 85 m/s. With what velocity does the archer move across the ice after firing the arrow?

From Newton's third law, we know that the force that the bow exerts on the arrow is matched by a force in the opposite direction on the bow (and the archer). This force causes the archer to slide backward on the ice with the speed requested in the problem. We cannot determine this speed using motion models such as the particle under constant acceleration because we don't have any information about the acceleration of the archer. We cannot use force models such as the particle under a net force because we don't know anything about forces in this situation. Energy models are of no help because we know nothing about the work done in pulling the bowstring back or the elastic potential energy in the system related to the taut bowstring.

Despite our inability to solve the archer problem using models learned so far, this problem is very simple to solve if we introduce a new quantity that describes motion, *linear momentum*. To generate this new quantity, consider an isolated system of two particles (Fig. 8.1) with masses m_1 and m_2 moving with velocities $\vec{\mathbf{v}}_1$ and $\vec{\mathbf{v}}_2$ at an instant of time. Because the system is isolated, the only force on one particle is that from the other particle, and we can categorize this situation as one in which Newton's laws can be applied. If a force from particle 1 (for example, a gravitational force) acts on particle 2, there must be a second force—equal in magnitude but opposite in direction—that particle 2 exerts on particle 1. That is, the forces form a Newton's third law action–reaction pair so that $\vec{\mathbf{F}}_{12} = -\vec{\mathbf{F}}_{21}$. We can express this condition as a statement about the *system* of two particles as follows:

$$\vec{\mathbf{F}}_{21} + \vec{\mathbf{F}}_{12} = 0$$

Let us further analyze this situation by incorporating Newton's second law. At the instant shown in Figure 8.1, the interacting particles have accelerations corresponding to the forces on them. Therefore, replacing the force on each particle with $m\vec{\mathbf{a}}$ for the particle gives

$$m_1\vec{\mathbf{a}}_1 + m_2\vec{\mathbf{a}}_2 = 0$$

Now we replace the acceleration with its definition from Equation 3.5:

$$m_1\frac{d\vec{\mathbf{v}}_1}{dt} + m_2\frac{d\vec{\mathbf{v}}_2}{dt} = 0$$

If the masses m_1 and m_2 are constant, we can bring them inside the derivative operation, which gives

$$\frac{d(m_1\vec{\mathbf{v}}_1)}{dt} + \frac{d(m_2\vec{\mathbf{v}}_2)}{dt} = 0$$

$$\frac{d}{dt}(m_1\vec{\mathbf{v}}_1 + m_2\vec{\mathbf{v}}_2) = 0 \qquad \textbf{8.1} \blacktriangleleft$$

Notice that the derivative of the sum $m_1\vec{\mathbf{v}}_1 + m_2\vec{\mathbf{v}}_2$ with respect to time is zero. Consequently, this sum must be constant. We learn from this discussion that the quantity $m\vec{\mathbf{v}}$ for a particle is important in that the sum of the values of this quantity for the particles in an isolated system is conserved. We call this quantity *linear momentum*:

The **linear momentum** $\vec{\mathbf{p}}$ of a particle or an object that can be modeled as a particle of mass m moving with a velocity $\vec{\mathbf{v}}$ is defined to be the product of the mass and velocity:[1]

▶ Definition of linear momentum of a particle

$$\vec{\mathbf{p}} \equiv m\vec{\mathbf{v}} \qquad \textbf{8.2} \blacktriangleleft$$

---

[1]This expression is nonrelativistic and is valid only when $v \ll c$, where c is the speed of light. In the next chapter, we discuss momentum for high-speed particles.

Linear momentum is a vector quantity because it equals the product of a scalar m and a vector $\vec{\mathbf{v}}$. Its direction is along $\vec{\mathbf{v}}$, it has dimensions ML/T, and its SI unit is kg·m/s.

If a particle is moving in an arbitrary direction in three-dimensional space, $\vec{\mathbf{p}}$ has three components and Equation 8.2 is equivalent to the component equations

$$p_x = mv_x \quad p_y = mv_y \quad p_z = mv_z \qquad \text{8.3}\blacktriangleleft$$

As you can see from its definition, the concept of momentum provides a quantitative distinction between objects of different masses moving at the same velocity. For example, the momentum of a truck moving at 2 m/s is much greater in magnitude than that of a Ping-Pong ball moving at the same speed. Newton called the product $m\vec{\mathbf{v}}$ the *quantity of motion*, perhaps a more graphic description than *momentum*, which comes from the Latin word for movement.

QUICK QUIZ 8.1 Two objects have equal kinetic energies. How do the magnitudes of their momenta compare? (a) $p_1 < p_2$ (b) $p_1 = p_2$ (c) $p_1 > p_2$ (d) not enough information to tell

QUICK QUIZ 8.2 Your physical education teacher throws a baseball to you at a certain speed and you catch it. The teacher is next going to throw you a medicine ball whose mass is ten times the mass of the baseball. You are given the following choices: You can have the medicine ball thrown with (a) the same speed as the baseball, (b) the same momentum, or (c) the same kinetic energy. Rank these choices from easiest to hardest to catch.

Let us use the particle model for an object in motion. By using Newton's second law of motion, we can relate the linear momentum of a particle to the net force acting on the particle. In Chapter 4, we learned that Newton's second law can be written as $\Sigma\vec{\mathbf{F}} = m\vec{\mathbf{a}}$. This form applies only when the mass of the particle remains constant, however. In situations where the mass is changing with time, one must use an alternative statement of Newton's second law: **The time rate of change of momentum of a particle is equal to the net force acting on the particle**, or

$$\Sigma\vec{\mathbf{F}} = \frac{d\vec{\mathbf{p}}}{dt} \qquad \text{8.4}\blacktriangleleft$$

▶ Newton's second law for a particle

If the mass of the particle is constant, the preceding equation reduces to our previous expression for Newton's second law:

$$\Sigma\vec{\mathbf{F}} = \frac{d\vec{\mathbf{p}}}{dt} = \frac{d(m\vec{\mathbf{v}})}{dt} = m\frac{d\vec{\mathbf{v}}}{dt} = m\vec{\mathbf{a}}$$

It is difficult to imagine a particle whose mass is changing, but if we consider objects, a number of examples emerge. These examples include a rocket that is ejecting its fuel as it operates, a snowball rolling down a hill and picking up additional snow, and a watertight pickup truck whose bed is collecting water as it moves in the rain.

From Equation 8.4 we see that if the net force on an object is zero, the time derivative of the momentum is zero and therefore the momentum of the object must be constant. This conclusion should sound familiar because it is the model of a particle in equilibrium, expressed in terms of momentum. Of course, if the particle is *isolated* (that is, if it does not interact with its environment), no forces act on it and $\vec{\mathbf{p}}$ remains unchanged, which is Newton's first law.

8.2 | Analysis Model: Isolated System (Momentum)

Using the definition of momentum, Equation 8.1 can be written as

$$\frac{d}{dt}(\vec{\mathbf{p}}_1 + \vec{\mathbf{p}}_2) = 0$$

Because the time derivative of the total system momentum $\vec{\mathbf{p}}_{\text{tot}} = \vec{\mathbf{p}}_1 + \vec{\mathbf{p}}_2$ is *zero*, we conclude that the *total* momentum $\vec{\mathbf{p}}_{\text{tot}}$ must remain constant:

$$\vec{\mathbf{p}}_{\text{tot}} = \text{constant} \qquad \text{8.5}\blacktriangleleft$$

▶ Conservation of momentum for an isolated system

or, equivalently,

$$\vec{\mathbf{p}}_{1i} + \vec{\mathbf{p}}_{2i} = \vec{\mathbf{p}}_{1f} + \vec{\mathbf{p}}_{2f} \qquad \textbf{8.6} \blacktriangleleft$$

where $\vec{\mathbf{p}}_{1i}$ and $\vec{\mathbf{p}}_{2i}$ are initial values and $\vec{\mathbf{p}}_{1f}$ and $\vec{\mathbf{p}}_{2f}$ are final values of the momentum during a period over which the particles interact. Equation 8.6 in component form states that the momentum components of the isolated system in the x, y, and z directions are all *independently constant;* that is,

$$\sum_{\text{system}} p_{ix} = \sum_{\text{system}} p_{fx} \qquad \sum_{\text{system}} p_{iy} = \sum_{\text{system}} p_{fy} \qquad \sum_{\text{system}} p_{iz} = \sum_{\text{system}} p_{fz} \qquad \textbf{8.7} \blacktriangleleft$$

Equation 8.6 is the mathematical statement of a new analysis model, the **isolated system (momentum).** It can be extended to any number of particles in an isolated system as we show in Section 8.7. We studied the energy version of the isolated system model in Chapter 7 and now we have a momentum version. In general, Equation 8.6 can be stated in words as follows:

Whenever two or more particles in an isolated system interact, the total momentum of the system remains constant.

> **Pitfall Prevention | 8.1**
> **Momentum of an Isolated *System* Is Conserved**
> Although the momentum of an isolated system is conserved, the momentum of one particle within an isolated system is not necessarily conserved, because other particles in the system may be interacting with it. Avoid applying conservation of momentum to a single particle.

Notice that we have made no statement concerning the nature of the forces acting between members of the system. The only requirement is that the forces must be *internal* to the system. Therefore, momentum is conserved for an isolated system *regardless* of the nature of the internal forces, *even if the forces are nonconservative.*

Example 8.1 | Can We Really Ignore the Kinetic Energy of the Earth?

In Section 6.6, we claimed that we can ignore the kinetic energy of the Earth when considering the energy of a system consisting of the Earth and a dropped ball. Verify this claim.

SOLUTION

Conceptualize Imagine dropping a ball at the surface of the Earth. From your point of view, the ball falls while the Earth remains stationary. By Newton's third law, however, the Earth experiences an upward force and therefore an upward acceleration while the ball falls. In the calculation below, we will show that this motion is extremely small and can be ignored.

Categorize We identify the system as the ball and the Earth. We assume there are no forces on the system from outer space, so the system is isolated. Let's use the momentum version of the isolated system model.

Analyze We begin by setting up a ratio of the kinetic energy of the Earth to that of the ball. We identify v_E and v_b as the speeds of the Earth and the ball, respectively, after the ball has fallen through some distance.

Use the definition of kinetic energy to set up this ratio:

$$(1) \quad \frac{K_E}{K_b} = \frac{\frac{1}{2}m_E v_E^2}{\frac{1}{2}m_b v_b^2} = \left(\frac{m_E}{m_b}\right)\left(\frac{v_E}{v_b}\right)^2$$

Apply the isolated system (momentum) model: the initial momentum of the system is zero, so set the final momentum equal to zero:

$$p_i = p_f \quad \rightarrow \quad 0 = m_b v_b + m_E v_E$$

Solve the equation for the ratio of speeds:

$$\frac{v_E}{v_b} = -\frac{m_b}{m_E}$$

Substitute this expression for v_E/v_b in Equation (1):

$$\frac{K_E}{K_b} = \left(\frac{m_E}{m_b}\right)\left(-\frac{m_b}{m_E}\right)^2 = \frac{m_b}{m_E}$$

Substitute order-of-magnitude numbers for the masses:

$$\frac{K_E}{K_b} = \frac{m_b}{m_E} \sim \frac{1\text{ kg}}{10^{25}\text{ kg}} \sim 10^{-25}$$

Finalize The kinetic energy of the Earth is a very small fraction of the kinetic energy of the ball, so we are justified in ignoring it in the kinetic energy of the system.

Example 8.2 | The Archer

Let us consider the situation proposed at the beginning of Section 8.1. A 60-kg archer stands at rest on frictionless ice and fires a 0.030-kg arrow horizontally at 85 m/s (Fig. 8.2). With what velocity does the archer move across the ice after firing the arrow?

Figure 8.2 (Example 8.2) An archer fires an arrow horizontally to the right. Because he is standing on frictionless ice, he will begin to slide to the left across the ice.

SOLUTION

Conceptualize You may have conceptualized this problem already when it was introduced at the beginning of Section 8.1. Imagine the arrow being fired one way and the archer recoiling in the opposite direction.

Categorize As discussed in Section 8.1, we cannot solve this problem with models based on motion, force, or energy. Nonetheless, we *can* solve this problem very easily with an approach involving momentum.

Let us take the system to consist of the archer (including the bow) and the arrow. The system is not isolated because the gravitational force and the normal force from the ice act on the system. These forces, however, are vertical and perpendicular to the motion of the system. Therefore, there are no external forces in the horizontal direction, and we can apply the isolated system (momentum) model in terms of momentum components in this direction.

Analyze The total horizontal momentum of the system before the arrow is fired is zero because nothing in the system is moving. Therefore, the total horizontal momentum of the system after the arrow is fired must also be zero. We choose the direction of firing of the arrow as the positive x direction. Identifying the archer as particle 1 and the arrow as particle 2, we have $m_1 = 60$ kg, $m_2 = 0.030$ kg, and $\vec{\mathbf{v}}_{2f} = 85\hat{\mathbf{i}}$ m/s.

Using the isolated system (momentum) model, set the final momentum of the system equal to the initial value of zero:

$$m_1\vec{\mathbf{v}}_{1f} + m_2\vec{\mathbf{v}}_{2f} = 0$$

Solve this equation for $\vec{\mathbf{v}}_{1f}$ and substitute numerical values:

$$\vec{\mathbf{v}}_{1f} = -\frac{m_2}{m_1}\vec{\mathbf{v}}_{2f} = -\left(\frac{0.030 \text{ kg}}{60 \text{ kg}}\right)(85\hat{\mathbf{i}} \text{ m/s}) = \boxed{-0.042\hat{\mathbf{i}} \text{ m/s}}$$

Finalize The negative sign for $\vec{\mathbf{v}}_{1f}$ indicates that the archer is moving to the left in Figure 8.2 after the arrow is fired, in the direction opposite the direction of motion of the arrow, in accordance with Newton's third law. Because the archer is much more massive than the arrow, his acceleration and consequent velocity are much smaller than the acceleration and velocity of the arrow. Notice that this problem sounds very simple, but we could not solve it with models based on motion, force, or energy. Our new momentum model, however, shows us that it not only *sounds* simple, it *is* simple!

What If? What if the arrow were fired in a direction that makes an angle θ with the horizontal? How will that change the recoil velocity of the archer?

Answer The recoil velocity should decrease in magnitude because only a component of the velocity of the arrow is in the x direction. Conservation of momentum in the x direction gives

$$m_1 v_{1f} + m_2 v_{2f} \cos\theta = 0$$

leading to

$$v_{1f} = -\frac{m_2}{m_1} v_{2f}\cos\theta$$

For $\theta = 0$, $\cos\theta = 1$ and the final velocity of the archer reduces to the value when the arrow is fired horizontally. For nonzero values of θ, the cosine function is less than 1 and the recoil velocity is less than the value calculated for $\theta = 0$. If $\theta = 90°$, then $\cos\theta = 0$ and $v_{1f} = 0$, so there is no recoil velocity. In this case, the archer is simply pushed downward harder against the ice as the arrow is fired.

Example 8.3 | Decay of the Kaon at Rest

One type of nuclear particle, called the *neutral kaon* (K^0), decays into a pair of other particles called *pions* (π^+ and π^-), which are oppositely charged but equal in mass, as in Figure 8.3. Assuming that the kaon is initially at rest, prove that the two pions must have momenta that are equal in magnitude and opposite in direction.

SOLUTION

Conceptualize Study Figure 8.3 carefully and imagine the kaon at rest decaying into two moving particles. Compare Figure 8.3 with Figure 8.2 and correlate the arrow and the archer with the individual pions.

Categorize Because the kaon does not interact with its surroundings, we model it as an isolated system. The system after the decay is the two pions.

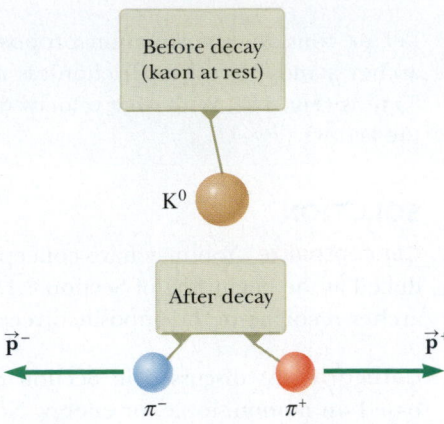

Figure 8.3 (Example 8.3) A kaon at rest decays into a pair of oppositely charged pions. The pions move apart with momenta of equal magnitudes but opposite directions.

Analyze Write an expression for decay of the kaon, represented in Figure 8.3:

$$K^0 \rightarrow \pi^+ + \pi^-$$

Let $\vec{\mathbf{p}}^+$ be the momentum of the positive pion and $\vec{\mathbf{p}}^-$ be the momentum of the negative pion after the decay, and find an expression for the final momentum $\vec{\mathbf{p}}_f$ of the isolated system of two pions:

$$\vec{\mathbf{p}}_f = \vec{\mathbf{p}}^+ + \vec{\mathbf{p}}^-$$

Because the kaon is at rest before the decay, we know that the initial system momentum $\vec{\mathbf{p}}_i = 0$. Furthermore, because the momentum of the isolated system is conserved, $\vec{\mathbf{p}}_i = \vec{\mathbf{p}}_f = 0$.

Incorporate this result in the previous equation:

$$0 = \vec{\mathbf{p}}^+ + \vec{\mathbf{p}}^- \rightarrow \vec{\mathbf{p}}^+ = -\vec{\mathbf{p}}^-$$

Finalize Therefore, we see that the two momentum vectors of the pions are equal in magnitude and opposite in direction.

8.3 | Analysis Model: Nonisolated System (Momentum)

As described by Equation 8.4, the momentum of a particle changes if a net force acts on the particle. Let us assume that a net force $\Sigma\vec{\mathbf{F}}$ acts on a particle and that this force may vary with time. According to Equation 8.4,

$$d\vec{\mathbf{p}} = \Sigma\vec{\mathbf{F}}\, dt \qquad \text{8.8} \blacktriangleleft$$

We can integrate this expression to find the change in the momentum of a particle during the time interval $\Delta t = t_f - t_i$. Integrating Equation 8.8 gives

$$\Delta\vec{\mathbf{p}} = \vec{\mathbf{p}}_f - \vec{\mathbf{p}}_i = \int_{t_i}^{t_f} \Sigma\vec{\mathbf{F}}\, dt \qquad \text{8.9} \blacktriangleleft$$

The integral of a force over the time interval during which it acts is called the **impulse** of the force. The impulse of the net force $\Sigma\vec{\mathbf{F}}$ is a vector defined by

▶ Impulse of a net force

$$\vec{\mathbf{I}} \equiv \int_{t_i}^{t_f} \Sigma\vec{\mathbf{F}}\, dt \qquad \text{8.10} \blacktriangleleft$$

From its definition, we see that impulse $\vec{\mathbf{I}}$ is a vector quantity having a magnitude equal to the area under the force–time curve as described in Figure 8.4a. It is assumed the force varies in time in the general manner shown in the figure and is nonzero in the time interval $\Delta t = t_f - t_i$. The direction of the impulse vector is the same as the direction of the change in momentum. Impulse has the dimensions of momentum, that is, ML/T. Impulse is *not* a property of a particle; rather, it is a measure of the degree to which an external force changes the particle's momentum.

Combining Equations 8.9 and 8.10 gives us an important statement known as the **impulse–momentum theorem:**

The change in the momentum of a particle is equal to the impulse of the net force acting on the particle:

$$\Delta \vec{p} = \vec{I} \qquad \text{8.11} \blacktriangleleft$$

◀ Impulse–momentum theorem for a particle

This statement is equivalent to Newton's second law. When we say that an impulse is given to a particle, we mean that momentum is transferred from an external agent to that particle. Equation 8.11 is identical in form to the conservation of energy equation, Equation 7.1, and its full expansion, Equation 7.2. Equation 8.11 is the most general statement of the principle of **conservation of momentum** and is called the **conservation of momentum equation.** In the case of a momentum approach, isolated systems tend to appear in problems more often than nonisolated systems, so, in practice, the conservation of momentum equation is often identified as the special case shown in Equation 8.6.

The left side of Equation 8.11 represents the change in the momentum of the system, which in this case is a single particle. The right side is a measure of how much momentum crosses the boundary of the system due to the net force being applied to the system. Equation 8.11 is the mathematical statement of a new analysis model, the **nonisolated system (momentum)** model. Although this equation is similar in form to Equation 7.1, there are several differences in its application to problems. First, Equation 8.11 is a vector equation, whereas Equation 7.1 is a scalar equation. Therefore, directions are important for Equation 8.11. Second, there is only one type of linear momentum and therefore only one way to store momentum in a system. In contrast, as we see from Equation 7.2, there are three ways to store energy in a system: kinetic, potential, and internal. Third, there is only one way to transfer momentum into a system: by the application of a force on the system over a time interval. Equation 7.2 shows six ways we have identified as transferring energy into a system. Therefore, there is no expansion of Equation 8.11 analogous to Equation 7.2.

Because the net force on a particle can generally vary in time as in Figure 8.4a, it is convenient to define a time-averaged net force $(\Sigma \vec{F})_{avg}$ given by

$$\left(\Sigma \vec{F}\right)_{avg} \equiv \frac{1}{\Delta t} \int_{t_i}^{t_f} \Sigma \vec{F} \, dt \qquad \text{8.12} \blacktriangleleft$$

where $\Delta t = t_f - t_i$. Therefore, we can express Equation 8.10 as

$$\vec{I} = \left(\Sigma \vec{F}\right)_{avg} \Delta t \qquad \text{8.13} \blacktriangleleft$$

The magnitude of this average net force, described in Figure 8.4b, can be thought of as the magnitude of the constant net force that would give the same impulse to the particle in the time interval Δt as the actual time-varying net force gives over this same interval.

In principle, if $\Sigma \vec{F}$ is known as a function of time, the impulse can be calculated from Equation 8.10. The calculation becomes especially simple if the net force acting on the particle is constant. In this case, $(\Sigma \vec{F})_{avg}$ over a time interval is the same as the constant $\Sigma \vec{F}$ at any instant within the interval, and Equation 8.13 becomes

$$\vec{I} = \Sigma \vec{F} \Delta t \qquad \text{8.14} \blacktriangleleft$$

In many physical situations, we shall use what is called the **impulse approximation,** in which we assume that one of the forces exerted on a particle acts for a short time but is much greater than any other force present. This simplification model

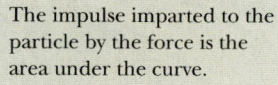

The impulse imparted to the particle by the force is the area under the curve.

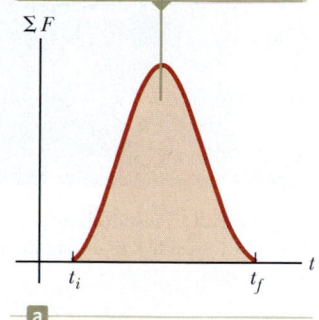

a

The time-averaged net force gives the same impulse to a particle as does the time-varying force in (a).

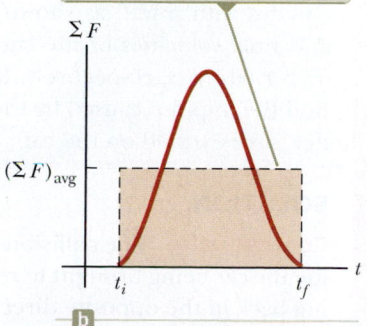

b

Figure 8.4 (a) A net force acting on a particle may vary in time. (b) The value of the constant force $(\Sigma F)_{avg}$ (horizontal dashed line) is chosen so that the area $(\Sigma F)_{avg} \Delta t$ of the rectangle is the same as the area under the curve in (a).

allows us to ignore the effects of other forces because these effects are small for the short time interval during which the large force acts. This approximation is especially useful in treating collisions in which the duration of the collision is very short. When this approximation is made, we refer to the force that is greater as an *impulsive force*. For example, when a baseball is struck with a bat, the duration of the collision is about 0.01 s and the average force the bat exerts on the ball during this time interval is typically several thousand newtons. This average force is much greater than the gravitational force, so we ignore any change in velocity related to the gravitational force during the collision. It is important to remember that $\vec{\mathbf{p}}_i$ and $\vec{\mathbf{p}}_f$ represent the momenta *immediately* before and after the collision, respectively. Therefore in the impulse approximation, very little motion of the particle takes place during the collision.

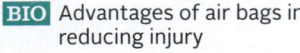

BIO Advantages of air bags in reducing injury

The concept of impulse helps us understand the value of air bags in stopping a passenger in an automobile accident (Fig. 8.5). The passenger experiences the same change in momentum and therefore the same impulse in a collision whether the car has air bags or not. The air bag allows the passenger to experience that change in momentum over a longer time interval, however, reducing the peak force on the passenger and increasing the chances of escaping without injury. Without the air bag, the passenger's head could move forward and be brought to rest in a short time interval by the steering wheel or the dashboard. In this case, the passenger undergoes the same change in momentum, but the short time interval results in a very large force that could cause severe head injury. Such injuries often result in spinal cord nerve damage where the nerves enter the base of the brain.

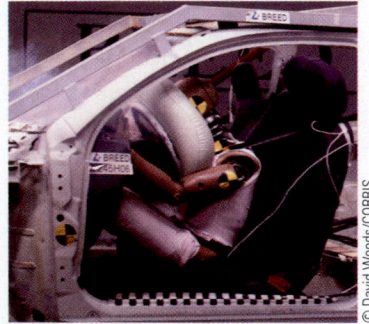

Figure 8.5 A test dummy is brought to rest by an air bag in an automobile.

QUICK QUIZ 8.3 Two objects are at rest on a frictionless surface. Object 1 has a greater mass than object 2. (**i**) When a constant force is applied to object 1, it accelerates through a distance d in a straight line. The force is removed from object 1 and is applied to object 2. At the moment when object 2 has accelerated through the same distance d, which statements are true? (**a**) $p_1 < p_2$ (**b**) $p_1 = p_2$ (**c**) $p_1 > p_2$ (**d**) $K_1 < K_2$ (**e**) $K_1 = K_2$ (**f**) $K_1 > K_2$ (**ii**) When a constant force is applied to object 1, it accelerates for a time interval Δt. The force is removed from object 1 and is applied to object 2. From the same list of choices, which statements are true after object 2 has accelerated for the same time interval Δt?

Example 8.4 | How Good Are the Bumpers?

In a particular crash test, a car of mass 1 500 kg collides with a wall as shown in Figure 8.6. The initial and final velocities of the car are $\vec{\mathbf{v}}_i = -15.0\hat{\mathbf{i}}$ m/s and $\vec{\mathbf{v}}_f = 2.60\hat{\mathbf{i}}$ m/s, respectively. If the collision lasts 0.150 s, find the impulse caused by the collision and the average net force exerted on the car.

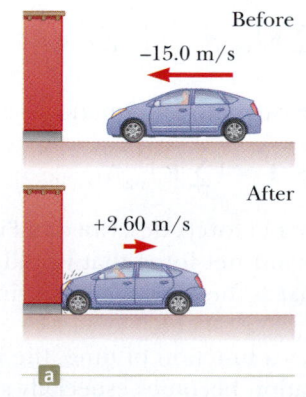

Before

−15.0 m/s

After

+2.60 m/s

a

b

Figure 8.6 (Example 8.4) (a) This car's momentum changes as a result of its collision with the wall. (b) In a crash test, much of the car's initial kinetic energy is transformed into energy associated with the damage to the car.

SOLUTION

Conceptualize The collision time is short, so we can imagine the car being brought to rest very rapidly and then moving back in the opposite direction with a reduced speed.

Categorize Let us assume the net force exerted on the car by the wall and friction from the ground is large compared with other forces on the car (such as air resistance). Furthermore, the gravitational force and the normal force exerted by the road on the car are perpendicular to the motion and therefore do not affect the horizontal momentum. Therefore, we categorize the problem as one in which we can apply the impulse approximation in the horizontal direction. We also see that the car's momentum changes due to an impulse from the environment. Therefore, we can apply the nonisolated system (momentum) model.

8.4 *cont.*

Analyze

Evaluate the initial and final momenta of the car:

$$\vec{p}_i = m\vec{v}_i = (1\,500 \text{ kg})(-15.0\hat{i} \text{ m/s}) = -2.25 \times 10^4\hat{i} \text{ kg} \cdot \text{m/s}$$

$$\vec{p}_f = m\vec{v}_f = (1\,500 \text{ kg})(2.60\hat{i} \text{ m/s}) = 0.39 \times 10^4\hat{i} \text{ kg} \cdot \text{m/s}$$

Use Equation 8.11 to find the impulse on the car:

$$\vec{I} = \Delta\vec{p} = \vec{p}_f - \vec{p}_i = 0.39 \times 10^4\hat{i} \text{ kg} \cdot \text{m/s} - (-2.25 \times 10^4\hat{i} \text{ kg} \cdot \text{m/s})$$

$$= 2.64 \times 10^4\hat{i} \text{ kg} \cdot \text{m/s}$$

Use Equation 8.13 to evaluate the average net force exerted on the car:

$$\left(\sum\vec{F}\right)_{avg} = \frac{\vec{I}}{\Delta t} = \frac{2.64 \times 10^4\hat{i} \text{ kg} \cdot \text{m/s}}{0.150 \text{ s}} = 1.76 \times 10^5\hat{i} \text{ N}$$

Finalize The net force found above is a combination of the normal force on the car from the wall and any friction force between the tires and the ground as the front of the car crumples. If the brakes are not operating while the crash occurs and the crumpling metal does not interfere with the free rotation of the tires, this friction force could be relatively small due to the freely rotating wheels. Notice that the signs of the velocities in this example indicate the reversal of directions. What would the mathematics be describing if both the initial and final velocities had the same sign?

What If? What if the car did not rebound from the wall? Suppose the final velocity of the car is zero and the time interval of the collision remains at 0.150 s. Would that represent a larger or a smaller net force on the car?

Answer In the original situation in which the car rebounds, the net force on the car does two things during the time interval: (1) it stops the car, and (2) it causes the car to move away from the wall at 2.60 m/s after the collision. If the car does not rebound, the net force is only doing the first of these steps—stopping the car—which requires a *smaller* force.

Mathematically, in the case of the car that does not rebound, the impulse is

$$\vec{I} = \Delta\vec{p} = \vec{p}_f - \vec{p}_i = 0 - (-2.25 \times 10^4\hat{i} \text{ kg} \cdot \text{m/s}) = 2.25 \times 10^4\hat{i} \text{ kg} \cdot \text{m/s}$$

The average net force exerted on the car is

$$\left(\sum\vec{F}\right)_{avg} = \frac{\vec{I}}{\Delta t} = \frac{2.25 \times 10^4\hat{i} \text{ kg} \cdot \text{m/s}}{0.150 \text{ s}} = 1.50 \times 10^5\hat{i} \text{ N}$$

which is indeed smaller than the previously calculated value, as was argued conceptually.

8.4 | Collisions in One Dimension

In this section, we use the law of conservation of momentum to describe what happens when two objects collide. The term **collision** represents an event during which two particles come close to each other and interact by means of forces. The forces due to the collision are assumed to be much larger than any external forces present, so we use the simplification model we call the impulse approximation. The general goal in collision problems is to relate the final conditions of the system to the initial conditions.

A collision may be the result of physical contact between two objects, as described in Figure 8.7a. This observation is common when two macroscopic objects collide, such as two billiard balls or a baseball and a bat.

The notion of what we mean by *collision* must be generalized because "contact" on a microscopic scale is ill defined. To understand the distinction between macroscopic and microscopic collisions, consider the collision of a proton with an alpha particle (the nucleus of the helium atom), illustrated in Figure 8.7b. Because the two particles are positively charged, they repel each other. A collision has occurred, but the colliding particles were never in "contact."

When two particles of masses m_1 and m_2 collide, the collision forces may vary in time in a complicated way, as seen in Figure 8.4. As a result, an analysis of the situation with Newton's second law could be very complicated. We find, however, that

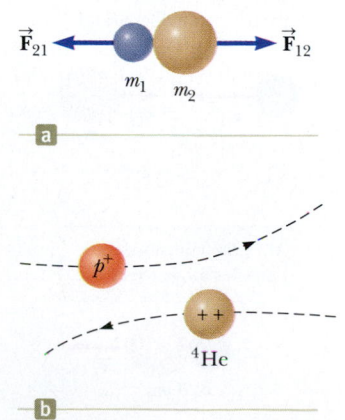

Figure 8.7 (a) A collision between two objects as the result of direct contact. (b) A "collision" between two charged particles that do not make contact.

the momentum concept is similar to the energy concept in Chapters 6 and 7 in that it provides us with a much easier method to solve problems involving isolated systems.

According to Equation 8.5, the momentum of an isolated system is conserved during some interaction event, such as a collision. The kinetic energy of the system, however, is generally *not* conserved in a collision. We define an **inelastic collision** as one in which the kinetic energy of the system is not conserved (even though momentum is conserved). The collision of a rubber ball with a hard surface is inelastic because some of the kinetic energy of the ball is transformed to internal energy when the ball is deformed while in contact with the surface.

BIO Glaucoma testing

A practical example of an inelastic collision is used to detect glaucoma, a disease in which the pressure inside the eye builds up and leads to blindness by damaging the cells of the retina. In this application, medical professionals use a device called a *tonometer* to measure the pressure inside the eye. This device releases a puff of air against the outer surface of the eye and measures the speed of the air after reflection from the eye. At normal pressure, the eye is slightly spongy and the pulse is reflected at low speed. As the pressure inside the eye increases, the outer surface becomes more rigid and the speed of the reflected pulse increases. Therefore, the speed of the reflected puff of air is used to measure the internal pressure of the eye.

Pitfall Prevention | 8.2
Inelastic Collisions
Generally, inelastic collisions are hard to analyze without additional information. Lack of this information appears in the mathematical representation as having more unknowns than equations.

When two objects collide and stick together after a collision, the maximum possible fraction of the initial kinetic energy is transformed or transferred away (by sound, for example); this collision is called a **perfectly inelastic collision.** For example, if two vehicles collide and become entangled, they move with some common velocity after the perfectly inelastic collision. If a meteorite collides with the Earth, it becomes buried in the ground and the collision is perfectly inelastic.

An **elastic collision** is defined as one in which the kinetic energy of the system is conserved (as well as momentum). Real collisions in the macroscopic world, such as those between billiard balls, are only approximately elastic because some transformation of kinetic energy takes place and some energy leaves the system by mechanical waves, sound. Imagine a billiard game with truly elastic collisions. The opening break would be completely silent! Truly elastic collisions do occur between atomic and subatomic particles. Elastic and perfectly inelastic collisions are *limiting* cases; a large number of collisions fall in the range between them.

In the remainder of this section, we treat collisions in one dimension and consider the two extreme cases: perfectly inelastic collisions and elastic collisions. The important distinction between these two types of collisions is that the momentum of the system is conserved in all cases, but the kinetic energy is conserved only in elastic collisions. When analyzing one-dimensional collisions, we can drop the vector notation and use positive and negative signs for velocities to denote directions, as we did in Chapter 2.

Before the collision, the particles move separately.

After the collision, the particles move together.

Active Figure 8.8 Schematic representation of a perfectly inelastic head-on collision between two particles.

Perfectly Inelastic Collisions

Consider two objects of masses m_1 and m_2 moving with initial velocities v_{1i} and v_{2i} along a straight line as in Active Figure 8.8. If the two objects collide head-on, stick together, and move with some common velocity v_f after the collision, the collision is perfectly inelastic. Because the total momentum of the two-object isolated system before the collision equals the total momentum of the combined-object system after the collision, we have

$$m_1 v_{1i} + m_2 v_{2i} = (m_1 + m_2) v_f \qquad \text{8.15} \blacktriangleleft$$

$$v_f = \frac{m_1 v_{1i} + m_2 v_{2i}}{m_1 + m_2} \qquad \text{8.16} \blacktriangleleft$$

Therefore, if we know the initial velocities of the two objects, we can use this single equation to determine the final common velocity.

Elastic Collisions

Now consider two objects that undergo an elastic head-on collision (Active Fig. 8.9) in one dimension. In this collision, both momentum and kinetic energy are conserved; therefore, we can write[2]

$$m_1 v_{1i} + m_2 v_{2i} = m_1 v_{1f} + m_2 v_{2f} \qquad \textbf{8.17}\blacktriangleleft$$

$$\tfrac{1}{2} m_1 v_{1i}^2 + \tfrac{1}{2} m_2 v_{2i}^2 = \tfrac{1}{2} m_1 v_{1f}^2 + \tfrac{1}{2} m_2 v_{2f}^2 \qquad \textbf{8.18}\blacktriangleleft$$

In a typical problem involving elastic collisions, two unknown quantities occur (such as v_{1f} and v_{2f}), and Equations 8.17 and 8.18 can be solved simultaneously to find them. An alternative approach, employing a little mathematical manipulation of Equation 8.18, often simplifies this process. Let us cancel the factor of $\tfrac{1}{2}$ in Equation 8.18 and rewrite the equation as

$$m_1(v_{1i}^2 - v_{1f}^2) = m_2(v_{2f}^2 - v_{2i}^2)$$

Here we have moved the terms containing m_1 to one side of the equation and those containing m_2 to the other. Next, let us factor both sides:

$$m_1(v_{1i} - v_{1f})(v_{1i} + v_{1f}) = m_2(v_{2f} - v_{2i})(v_{2f} + v_{2i}) \qquad \textbf{8.19}\blacktriangleleft$$

We now separate the terms containing m_1 and m_2 in the equation for conservation of momentum (Eq. 8.17) to obtain

$$m_1(v_{1i} - v_{1f}) = m_2(v_{2f} - v_{2i}) \qquad \textbf{8.20}\blacktriangleleft$$

To obtain our final result, we divide Equation 8.19 by Equation 8.20 and obtain

$$v_{1i} + v_{1f} = v_{2f} + v_{2i}$$

or, gathering initial and final values on opposite sides of the equation,

$$v_{1i} - v_{2i} = -(v_{1f} - v_{2f}) \qquad \textbf{8.21}\blacktriangleleft$$

This equation, in combination with the condition for conservation of momentum, Equation 8.17, can be used to solve problems dealing with one-dimensional elastic collisions between two objects. According to Equation 8.21, the relative speed[3] $v_{1i} - v_{2i}$ of the two objects before the collision equals the negative of their relative speed after the collision, $-(v_{1f} - v_{2f})$.

Suppose the masses and the initial velocities of both objects are known. Equations 8.17 and 8.21 can be solved for the final velocities in terms of the initial values because we have two equations and two unknowns:

$$v_{1f} = \left(\frac{m_1 - m_2}{m_1 + m_2}\right) v_{1i} + \left(\frac{2 m_2}{m_1 + m_2}\right) v_{2i} \qquad \textbf{8.22}\blacktriangleleft$$

$$v_{2f} = \left(\frac{2 m_1}{m_1 + m_2}\right) v_{1i} + \left(\frac{m_2 - m_1}{m_1 + m_2}\right) v_{2i} \qquad \textbf{8.23}\blacktriangleleft$$

It is important to remember that the appropriate signs for the numerical values of velocities v_{1i} and v_{2i} must be included in Equations 8.22 and 8.23. For example, if m_2 is moving to the left initially, as in Active Figure 8.9a, v_{2i} is negative.

Before the collision, the particles move separately.

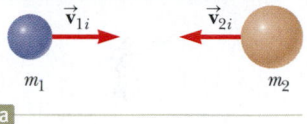

m_1 m_2

a

After the collision, the particles continue to move separately with new velocities.

b

Active Figure 8.9 Schematic representation of an elastic head-on collision between two particles.

---

[2]Notice that the kinetic energy of the system is the sum of the kinetic energies of the two particles. In our energy conservation examples in Chapter 7 involving a falling object and the Earth, we ignored the kinetic energy of the Earth because it is so small. Therefore, the kinetic energy of the *system* is just the kinetic energy of the falling *object*. That is a special case in which the mass of one of the objects (the Earth) is so immense that ignoring its kinetic energy introduces no measurable error. For problems such as those described here, however, and for the particle decay problems we will see in Chapters 30 and 31, we need to include the kinetic energies of *all* particles in the system.

[3]See Section 3.6 for a review of relative speed.

Let us consider some special cases. If $m_1 = m_2$, Equations 8.22 and 8.23 show us that $v_{1f} = v_{2i}$ and $v_{2f} = v_{1i}$. That is, the objects exchange speeds if they have equal masses. That is what one observes in head-on billiard ball collisions, assuming there is no spin on the ball: The initially moving ball stops and the initially stationary ball moves away with approximately the same speed.

If m_2 is initially at rest, $v_{2i} = 0$ and Equations 8.22 and 8.23 become

▶ Elastic collision in one dimension: particle 2 initially at rest

$$v_{1f} = \left(\frac{m_1 - m_2}{m_1 + m_2} \right) v_{1i} \qquad \text{8.24} \blacktriangleleft$$

$$v_{2f} = \left(\frac{2m_1}{m_1 + m_2} \right) v_{1i} \qquad \text{8.25} \blacktriangleleft$$

If m_1 is very large compared with m_2, we see from Equations 8.24 and 8.25 that $v_{1f} \approx v_{1i}$ and $v_{2f} \approx 2v_{1i}$. That is, when a very heavy object collides head-on with a very light one initially at rest, the heavy object continues its motion unaltered after the collision but the light object rebounds with a speed equal to about twice the initial speed of the heavy object. An example of such a collision is that of a moving heavy atom, such as uranium, with a light atom, such as hydrogen.

If m_2 is much larger than m_1 and if m_2 is initially at rest, we find from Equations 8.24 and 8.25 that $v_{1f} \approx -v_{1i}$ and $v_{2f} \approx 0$. That is, when a very light object collides head-on with a very heavy object initially at rest, the velocity of the light object is reversed and the heavy object remains approximately at rest. For example, imagine what happens when a table-tennis ball hits a stationary bowling ball.

▶ **QUICK QUIZ 8.4** A table-tennis ball is thrown at a stationary bowling ball. The table-tennis ball makes a one-dimensional elastic collision and bounces back along the same line. Compared with the bowling ball after the collision, does the table-tennis ball have (a) a larger magnitude of momentum and more kinetic energy, (b) a smaller magnitude of momentum and more kinetic energy, (c) a larger magnitude of momentum and less kinetic energy, (d) a smaller magnitude of momentum and less kinetic energy, or (e) the same magnitude of momentum and the same kinetic energy

▶ **PROBLEM-SOLVING STRATEGY: One-Dimensional Collisions**

You should use the following approach when solving collision problems in one dimension:

1. **Conceptualize.** Imagine the collision occurring in your mind. Draw simple diagrams of the particles before and after the collision and include appropriate velocity vectors. At first, you may have to guess at the directions of the final velocity vectors.

2. **Categorize.** Is the system of particles isolated? If so, categorize the collision as elastic, inelastic, or perfectly inelastic.

3. **Analyze.** Set up the appropriate mathematical representation for the problem. If the collision is perfectly inelastic, use Equation 8.15. If the collision is elastic, use Equations 8.17 and 8.21. If the collision is inelastic, use Equation 8.17. To find the final velocities in this case, you will need some additional information.

4. **Finalize.** Once you have determined your result, check to see if your answers are consistent with the mental and pictorial representations and that your results are realistic.

Example 8.5 | **Kinetic Energy in a Perfectly Inelastic Collision**

We claimed that the maximum amount of kinetic energy was transformed to other forms in a perfectly inelastic collision. Prove this statement mathematically for a one-dimensional two-particle collision.

SOLUTION

Conceptualize We will assume that the maximum kinetic energy is transformed and prove that the collision must be perfectly inelastic.

Categorize We categorize the system of two particles as an isolated system. We also categorize the collision as one-dimensional.

Analyze Find an expression for the ratio of the final kinetic energy after the collision to the initial kinetic energy:

$$f = \frac{K_f}{K_i} = \frac{\frac{1}{2}m_1 v_{1f}^2 + \frac{1}{2}m_2 v_{2f}^2}{\frac{1}{2}m_1 v_{1i}^2 + \frac{1}{2}m_2 v_{2i}^2} = \frac{m_1 v_{1f}^2 + m_2 v_{2f}^2}{m_1 v_{1i}^2 + m_2 v_{2i}^2}$$

The *maximum* amount of energy transformed to other forms corresponds to the *minimum* value of f. For fixed initial conditions, imagine that the final velocities v_{1f} and v_{2f} are variables. Minimize the fraction f by taking the derivative of f with respect to v_{1f} and setting the result equal to zero:

$$\frac{df}{dv_{1f}} = \frac{d}{dv_{1f}}\left(\frac{m_1 v_{1f}^2 + m_2 v_{2f}^2}{m_1 v_{1i}^2 + m_2 v_{2i}^2}\right)$$

$$= \frac{2m_1 v_{1f} + 2m_2 v_{2f}\dfrac{dv_{2f}}{dv_{1f}}}{m_1 v_{1i}^2 + m_2 v_{2i}^2} = 0$$

$$\rightarrow \quad (1) \quad m_1 v_{1f} + m_2 v_{2f}\frac{dv_{2f}}{dv_{1f}} = 0$$

From the conservation of momentum condition, we can evaluate the derivative in (1). Differentiate Equation 8.17 with respect to v_{1f}:

$$\frac{d}{dv_{1f}}(m_1 v_{1i} + m_2 v_{2i}) = \frac{d}{dv_{1f}}(m_1 v_{1f} + m_2 v_{2f})$$

$$\rightarrow \quad 0 = m_1 + m_2 \frac{dv_{2f}}{dv_{1f}} \quad \rightarrow \quad \frac{dv_{2f}}{dv_{1f}} = -\frac{m_1}{m_2}$$

Substitute this expression for the derivative into (1):

$$m_1 v_{1f} - m_2 v_{2f}\frac{m_1}{m_2} = 0 \quad \rightarrow \quad v_{1f} = v_{2f}$$

Finalize If the particles come out of the collision with the same velocities, they are joined together and it is a perfectly inelastic collision, which is what we set out to prove.

Example 8.6 | **Carry Collision Insurance!**

An 1 800-kg car stopped at a traffic light is struck from the rear by a 900-kg car. The two cars become entangled, moving along the same path as that of the originally moving car. If the smaller car were moving at 20.0 m/s before the collision, what is the velocity of the entangled cars after the collision?

SOLUTION

Conceptualize This kind of collision is easily visualized, and one can predict that after the collision both cars will be moving in the same direction as that of the initially moving car. Because the initially moving car has only half the mass of the stationary car, we expect the final velocity of the cars to be relatively small.

Categorize We identify the system of two cars as isolated in terms of momentum in the horizontal direction and apply the impulse approximation during the short time interval of the collision. The phrase "become entangled" tells us to categorize the collision as perfectly inelastic.

Analyze The magnitude of the total momentum of the system before the collision is equal to that of the smaller car because the larger car is initially at rest.

continued

8.6 *cont.*

Set the initial momentum of the system equal to the final momentum of the system:

$$p_i = p_f \quad \rightarrow \quad m_1 v_i = (m_1 + m_2) v_f$$

Solve for v_f and substitute numerical values:

$$v_f = \frac{m_1 v_i}{m_1 + m_2} = \frac{(900 \text{ kg})(20.0 \text{ m/s})}{900 \text{ kg} + 1\,800 \text{ kg}} = 6.67 \text{ m/s}$$

Finalize Because the final velocity is positive, the direction of the final velocity of the combination is the same as the velocity of the initially moving car as predicted. The speed of the combination is also much lower than the initial speed of the moving car.

What If? Suppose we reverse the masses of the cars. What if a stationary 900-kg car is struck by a moving 1 800-kg car? Is the final speed the same as before?

Answer Intuitively, we can guess that the final speed of the combination is higher than 6.67 m/s if the initially moving car is the more massive car. Mathematically, that should be the case because the system has a larger momentum if the initially moving car is the more massive one. Solving for the new final velocity, we find

$$v_f = \frac{m_1 v_i}{m_1 + m_2} = \frac{(1\,800 \text{ kg})(20.0 \text{ m/s})}{1\,800 \text{ kg} + 900 \text{ kg}} = 13.3 \text{ m/s}$$

which is two times greater than the previous final velocity.

Example 8.7 | Slowing Down Neutrons by Collisions

In a nuclear reactor, neutrons are produced when $^{235}_{92}U$ atoms split in a process called *fission*. These neutrons are moving at about 10^7 m/s and must be slowed down to about 10^3 m/s before they take part in another fission event. They are slowed down by being passed through a solid or liquid material called a *moderator*. The slowing-down process involves elastic collisions. Let us show that a neutron can lose most of its kinetic energy if it collides elastically with a moderator containing light nuclei, such as deuterium (in "heavy water," D_2O).

SOLUTION

Conceptualize Imagine a single neutron passing through the moderator material and repeatedly colliding with nuclei. The kinetic energy of the neutron will decrease in each collision and the neutron will eventually slow down to the desired 10^3 m/s.

Categorize We identify the neutron and a particular moderator nucleus as an isolated system and use the momentum version of the isolated system model. Let us assume that the moderator nucleus of mass m_m is at rest initially and that the neutron of mass m_n and initial speed v_{ni} collides head-on with it. Because the momentum and kinetic energy of this system are conserved in an elastic collision, Equations 8.24 and 8.25 can be applied to a one-dimensional collision of these two particles.

Analyze

Find an expression for the initial kinetic energy of the neutron:

$$K_{ni} = \tfrac{1}{2} m_n v_{ni}^2$$

Using Equation 8.24, find an expression for the final kinetic energy of the neutron:

$$K_{nf} = \tfrac{1}{2} m_n v_{nf}^2 = \tfrac{1}{2} m_n \left(\frac{m_n - m_m}{m_n + m_m}\right)^2 v_{ni}^2$$

Now find an expression for the fraction of the total kinetic energy possessed by the neutron after the collision:

$$(1) \quad f_n = \frac{K_{nf}}{K_{ni}} = \frac{\tfrac{1}{2} m_n \left(\dfrac{m_n - m_m}{m_n + m_m}\right)^2 v_{ni}^2}{\tfrac{1}{2} m_n v_{ni}^2} = \left(\frac{m_n - m_m}{m_n + m_m}\right)^2$$

Find an expression for the kinetic energy of the moderator nucleus after the collision using Equation 8.25:

$$(2) \quad K_{mf} = \tfrac{1}{2} m_m v_{mf}^2 = \frac{2 m_n^2 m_m}{(m_n + m_m)^2} v_{ni}^2$$

Use Equation (2) to find an expression for the fraction of the total kinetic energy transferred to the moderator nucleus:

$$(3) \quad f_{\text{trans}} = \frac{K_{mf}}{K_{ni}} = \frac{\dfrac{2 m_n^2 m_m}{(m_n + m_m)^2} v_{ni}^2}{\tfrac{1}{2} m_n v_{ni}^2} = \frac{4 m_n m_m}{(m_n + m_m)^2}$$

8.7 *cont.*

Finalize If $m_m \approx m_n$, we see that $f_{\text{trans}} \approx 1 = 100\%$. Because the system's kinetic energy is conserved, Equation (3) can also be obtained from Equation (1) with the condition that $f_n + f_m = 1$, so that $f_m = 1 - f_n$.

For collisions of the neutrons with deuterium nuclei in D_2O ($m_m = 2m_n$), $f_n = 1/9$ and $f_{\text{trans}} = 8/9$. That is, 89% of the neutron's kinetic energy is transferred to the deuterium nucleus. In practice, the moderator efficiency is reduced because head-on collisions are very unlikely to occur.

Example 8.8 | A Two-Body Collision with a Spring

A block of mass $m_1 = 1.60$ kg initially moving to the right with a speed of 4.00 m/s on a frictionless, horizontal track collides with a light spring attached to a second block of mass $m_2 = 2.10$ kg initially moving to the left with a speed of 2.50 m/s as shown in Figure 8.10a. The spring constant is 600 N/m.

(A) Find the velocities of the two blocks after the collision.

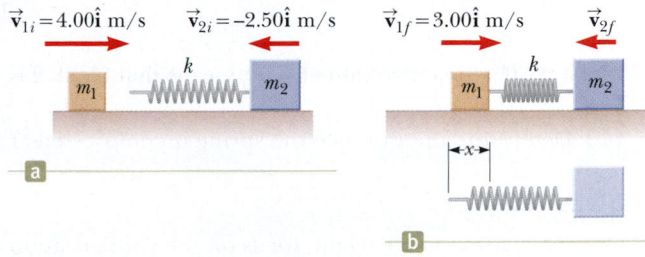

Figure 8.10 (Example 8.8) A moving block approaches a second moving block that is attached to a spring.

SOLUTION

Conceptualize With the help of Figure 8.10a, run an animation of the collision in your mind. Figure 8.10b shows an instant during the collision when the spring is compressed. Eventually, block 1 and the spring will again separate, so the system will look like Figure 8.10a again but with different velocity vectors for the two blocks.

Categorize Because the spring force is conservative, kinetic energy in the system of two blocks and the spring is not transformed to internal energy during the compression of the spring. Ignoring any sound made when the block hits the spring, we can categorize the collision as being elastic and the system as being isolated for both energy and momentum.

Analyze Because momentum of the system is conserved, apply Equation 8.17:

$$(1) \quad m_1 v_{1i} + m_2 v_{2i} = m_1 v_{1f} + m_2 v_{2f}$$

Because the collision is elastic, apply Equation 8.21:

$$(2) \quad v_{1i} - v_{2i} = -(v_{1f} - v_{2f})$$

Multiply Equation (2) by m_1:

$$(3) \quad m_1 v_{1i} - m_1 v_{2i} = -m_1 v_{1f} + m_1 v_{2f}$$

Add Equations (1) and (3):

$$2m_1 v_{1i} + (m_2 - m_1) v_{2i} = (m_1 + m_2) v_{2f}$$

Solve for v_{2f}:

$$v_{2f} = \frac{2m_1 v_{1i} + (m_2 - m_1) v_{2i}}{m_1 + m_2}$$

Substitute numerical values:

$$v_{2f} = \frac{2(1.60 \text{ kg})(4.00 \text{ m/s}) + (2.10 \text{ kg} - 1.60 \text{ kg})(-2.50 \text{ m/s})}{1.60 \text{ kg} + 2.10 \text{ kg}} = 3.12 \text{ m/s}$$

Solve Equation (2) for v_{1f} and substitute numerical values:

$$v_{1f} = v_{2f} - v_{1i} + v_{2i} = 3.12 \text{ m/s} - 4.00 \text{ m/s} + (-2.50 \text{ m/s}) = -3.38 \text{ m/s}$$

(B) Determine the velocity of block 2 during the collision, at the instant block 1 is moving to the right with a velocity of +3.00 m/s as in Figure 8.10b.

SOLUTION

Conceptualize Focus your attention now on Figure 8.10b, which represents the final configuration of the system for the time interval of interest.

continued

8.8 *cont.*

Categorize Because the momentum and mechanical energy of the system of two blocks and the spring are conserved *throughout* the collision, the collision can be categorized as elastic for *any* final instant of time. Let us now choose the final instant to be when block 1 is moving with a velocity of +3.00 m/s.

Analyze Apply Equation 8.17:

$$m_1 v_{1i} + m_2 v_{2i} = m_1 v_{1f} + m_2 v_{2f}$$

Solve for v_{2f}:

$$v_{2f} = \frac{m_1 v_{1i} + m_2 v_{2i} - m_1 v_{1f}}{m_2}$$

Substitute numerical values:

$$v_{2f} = \frac{(1.60 \text{ kg})(4.00 \text{ m/s}) + (2.10 \text{ kg})(-2.50 \text{ m/s}) - (1.60 \text{ kg})(3.00 \text{ m/s})}{2.10 \text{ kg}}$$

$$= -1.74 \text{ m/s}$$

Finalize The negative value for v_{2f} means that block 2 is still moving to the left at the instant we are considering.

(C) Determine the distance the spring is compressed at that instant.

SOLUTION

Conceptualize Once again, focus on the configuration of the system shown in Figure 8.10b.

Categorize For the system of the spring and two blocks, no friction or other nonconservative forces act within the system. Therefore, we categorize the system as isolated in terms of energy with no nonconservative forces acting. The system also remains isolated in terms of momentum.

Analyze We choose the initial configuration of the system to be that existing immediately before block 1 strikes the spring and the final configuration to be that when block 1 is moving to the right at 3.00 m/s.

Write a conservation of mechanical energy equation for the system:

$$K_i + U_i = K_f + U_f$$

Evaluate the energies, recognizing that two objects in the system have kinetic energy and that the potential energy is elastic:

$$\tfrac{1}{2}m_1 v_{1i}^2 + \tfrac{1}{2}m_2 v_{2i}^2 + 0 = \tfrac{1}{2}m_1 v_{1f}^2 + \tfrac{1}{2}m_2 v_{2f}^2 + \tfrac{1}{2}kx^2$$

Substitute the known values and the result of part (B):

$$\tfrac{1}{2}(1.60 \text{ kg})(4.00 \text{ m/s})^2 + \tfrac{1}{2}(2.10 \text{ kg})(2.50 \text{ m/s})^2 + 0$$
$$= \tfrac{1}{2}(1.60 \text{ kg})(3.00 \text{ m/s})^2 + \tfrac{1}{2}(2.10 \text{ kg})(1.74 \text{ m/s})^2 + \tfrac{1}{2}(600 \text{ N/m})x^2$$

Solve for x:

$$x = 0.173 \text{ m}$$

Finalize This answer is not the maximum compression of the spring because the two blocks are still moving toward each other at the instant shown in Figure 8.10b. Can you determine the maximum compression of the spring?

8.5 | Collisions in Two Dimensions

In Section 8.1, we showed that the total momentum of a system is conserved when the system is isolated (i.e., when no external forces act on the system). For a general collision of two objects in three-dimensional space, the principle of conservation of momentum implies that the total momentum in each direction is conserved. An important subset of collisions takes place in a plane. The game of billiards is a familiar example involving multiple collisions of objects moving on a two-dimensional surface. Let us restrict our attention to a single two-dimensional collision between two objects that takes place in a plane. For such collisions, we obtain two component equations for the conservation of momentum:

$$m_1 v_{1ix} + m_2 v_{2ix} = m_1 v_{1fx} + m_2 v_{2fx}$$
$$m_1 v_{1iy} + m_2 v_{2iy} = m_1 v_{1fy} + m_2 v_{2fy}$$

where we use three subscripts in this general equation to represent, respectively, (1) the identification of the object, (2) initial and final values, and (3) the velocity component in the x or y direction.

Consider a two-dimensional problem in which an object of mass m_1 collides with an object of mass m_2 that is initially at rest as in Active Figure 8.11. After the collision, m_1 moves at an angle θ with respect to the horizontal and m_2 moves at an angle ϕ with respect to the horizontal. This collision is called a *glancing* collision. Applying the law of conservation of momentum in component form and noting that the initial y component of the momentum of the system is zero, we have

x component: $\quad m_1 v_{1i} + 0 = m_1 v_{1f} \cos\theta + m_2 v_{2f} \cos\phi$ **8.26** ◀

y component: $\quad 0 + 0 = m_1 v_{1f} \sin\theta - m_2 v_{2f} \sin\phi$ **8.27** ◀

If the collision is elastic, we can write a third equation for conservation of kinetic energy in the form

$$\tfrac{1}{2} m_1 v_{1i}^2 = \tfrac{1}{2} m_1 v_{1f}^2 + \tfrac{1}{2} m_2 v_{2f}^2 \qquad \textbf{8.28} \blacktriangleleft$$

If we know the initial velocity v_{1i} and the masses, we are left with four unknowns (v_{1f}, v_{2f}, θ, and ϕ). Because we have only three equations, one of the four remaining quantities must be given to determine the motion after the collision from conservation principles alone.

If the collision is inelastic, kinetic energy is *not* conserved and Equation 8.28 does *not* apply.

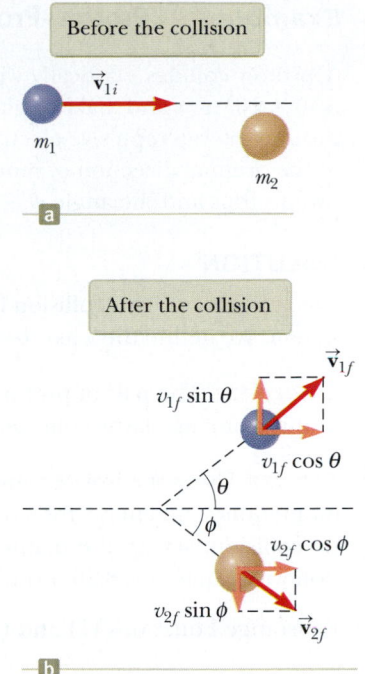

Active Figure 8.11 A glancing collision between two particles.

> **PROBLEM-SOLVING STRATEGY: Two-Dimensional Collisions**

The following procedure is recommended when dealing with problems involving collisions between two particles in two dimensions.

1. **Conceptualize.** Imagine the collisions occurring and predict the approximate directions in which the particles will move after the collision. Set up a coordinate system and define your velocities in terms of that system. It is convenient to have the x axis coincide with one of the initial velocities. Sketch the coordinate system, draw and label all velocity vectors, and include all the given information.

2. **Categorize.** Is the system of particles truly isolated? If so, categorize the collision as elastic, inelastic, or perfectly inelastic.

3. **Analyze.** Write expressions for the x and y components of the momentum of each object before and after the collision. Remember to include the appropriate signs for the components of the velocity vectors and pay careful attention to signs throughout the calculation.

    Write expressions for the *total* momentum in the x direction *before* and *after* the collision and equate the two. Repeat this procedure for the total momentum in the y direction.

    Proceed to solve the momentum equations for the unknown quantities. If the collision is inelastic, kinetic energy is *not* conserved and additional information is probably required. If the collision is perfectly inelastic, the final velocities of the two objects are equal.

    If the collision is elastic, kinetic energy is conserved and you can equate the total kinetic energy of the system before the collision to the total kinetic energy after the collision, providing an additional relationship between the velocity magnitudes.

4. **Finalize.** Once you have determined your result, check to see if your answers are consistent with the mental and pictorial representations and that your results are realistic.

Example 8.9 | Proton–Proton Collision

A proton collides elastically with another proton that is initially at rest. The incoming proton has an initial speed of 3.50×10^5 m/s and makes a glancing collision with the second proton as in Active Figure 8.11. (At close separations, the protons exert a repulsive electrostatic force on each other.) After the collision, one proton moves off at an angle of 37.0° to the original direction of motion and the second deflects at an angle of ϕ to the same axis. Find the final speeds of the two protons and the angle ϕ.

SOLUTION

Conceptualize This collision is like that shown in Active Figure 8.11, which will help you conceptualize the behavior of the system. We define the x axis to be along the direction of the velocity vector of the initially moving proton.

Categorize The pair of protons form an isolated system. Both momentum and kinetic energy of the system are conserved in this glancing elastic collision.

Analyze Using the isolated system model for both momentum and energy for a two-dimensional elastic collision, set up the mathematical representation with Equations 8.26 through 8.28:

$$(1) \quad v_{1f} \cos \theta + v_{2f} \cos \phi = v_{1i}$$

$$(2) \quad v_{1f} \sin \theta - v_{2f} \sin \phi = 0$$

$$(3) \quad v_{1f}^2 + v_{2f}^2 = v_{1i}^2$$

Rearrange Equations (1) and (2):

$$v_{2f} \cos \phi = v_{1i} - v_{1f} \cos \theta$$

$$v_{2f} \sin \phi = v_{1f} \sin \theta$$

Square these two equations and add them:

$$v_{2f}^2 \cos^2 \phi + v_{2f}^2 \sin^2 \phi = v_{1i}^2 - 2v_{1i}v_{1f} \cos \theta + v_{1f}^2 \cos^2 \theta + v_{1f}^2 \sin^2 \theta$$

Incorporate that the sum of the squares of sine and cosine for *any* angle is equal to 1:

$$(4) \quad v_{2f}^2 = v_{1i}^2 - 2v_{1i}v_{1f} \cos \theta + v_{1f}^2$$

Substitute Equation (4) into Equation (3):

$$v_{1f}^2 + (v_{1i}^2 - 2v_{1i}v_{1f} \cos \theta + v_{1f}^2) = v_{1i}^2$$

$$(5) \quad v_{1f}^2 - v_{1i}v_{1f} \cos \theta = 0$$

One possible solution of Equation (5) is $v_{1f} = 0$, which corresponds to a head-on, one-dimensional collision in which the first proton stops and the second continues with the same speed in the same direction. That is not the solution we want.

Divide both sides of Equation (5) by v_{1f} and solve for the remaining factor of v_{1f}:

$$v_{1f} = v_{1i} \cos \theta = (3.50 \times 10^5 \text{ m/s}) \cos 37.0° = \boxed{2.80 \times 10^5 \text{ m/s}}$$

Use Equation (3) to find v_{2f}:

$$v_{2f} = \sqrt{v_{1i}^2 - v_{1f}^2} = \sqrt{(3.50 \times 10^5 \text{ m/s})^2 - (2.80 \times 10^5 \text{ m/s})^2}$$

$$= \boxed{2.11 \times 10^5 \text{ m/s}}$$

Use Equation (2) to find ϕ:

$$(2) \quad \phi = \sin^{-1}\left(\frac{v_{1f} \sin \theta}{v_{2f}}\right) = \sin^{-1}\left[\frac{(2.80 \times 10^5 \text{ m/s}) \sin 37.0°}{(2.11 \times 10^5 \text{ m/s})}\right]$$

$$= \boxed{53.0°}$$

Finalize It is interesting that $\theta + \phi = 90°$. This result is *not* accidental. Whenever two objects of equal mass collide elastically in a glancing collision and one of them is initially at rest, their final velocities are perpendicular to each other.

Example 8.10 | Collision at an Intersection

A 1 500-kg car traveling east with a speed of 25.0 m/s collides at an intersection with a 2 500-kg truck traveling north at a speed of 20.0 m/s as shown in Figure 8.12. Find the direction and magnitude of the velocity of the wreckage after the collision, assuming the vehicles stick together after the collision.

SOLUTION

Conceptualize Figure 8.12 should help you conceptualize the situation before and after the collision. Let us choose east to be along the positive x direction and north to be along the positive y direction.

8.10 cont.

Categorize Because we consider moments immediately before and immediately after the collision as defining our time interval, we ignore the small effect that friction would have on the wheels of the vehicles and model the system of two vehicles as isolated in terms of momentum. We also ignore the vehicles' sizes and model them as particles. The collision is perfectly inelastic because the car and the truck stick together after the collision.

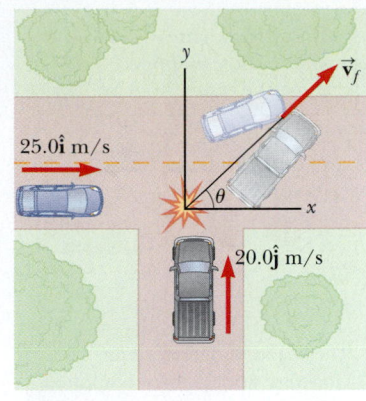

Figure 8.12 (Example 8.10) An eastbound car colliding with a northbound truck.

Analyze Before the collision, the only object having momentum in the x direction is the car. Therefore, the magnitude of the total initial momentum of the system (car plus truck) in the x direction is that of only the car. Similarly, the total initial momentum of the system in the y direction is that of the truck. After the collision, let us assume the wreckage moves at an angle θ with respect to the x axis with speed v_f.

Equate the initial and final momenta of the system in the x direction:

$$\sum p_{xi} = \sum p_{xf} \quad \rightarrow \quad (1) \ m_1 v_{1i} = (m_1 + m_2) v_f \cos \theta$$

Equate the initial and final momenta of the system in the y direction:

$$\sum p_{yi} = \sum p_{yf} \quad \rightarrow \quad (2) \ m_2 v_{2i} = (m_1 + m_2) v_f \sin \theta$$

Divide Equation (2) by Equation (1):

$$\frac{m_2 v_{2i}}{m_1 v_{1i}} = \frac{\sin \theta}{\cos \theta} = \tan \theta$$

Solve for θ and substitute numerical values:

$$\theta = \tan^{-1}\left(\frac{m_2 v_{2i}}{m_1 v_{1i}}\right) = \tan^{-1}\left[\frac{(2\,500 \text{ kg})(20.0 \text{ m/s})}{(1\,500 \text{ kg})(25.0 \text{ m/s})}\right] = \boxed{53.1°}$$

Use Equation (2) to find the value of v_f and substitute numerical values:

$$v_f = \frac{m_2 v_{2i}}{(m_1 + m_2)\sin \theta} = \frac{(2\,500 \text{ kg})(20.0 \text{ m/s})}{(1\,500 \text{ kg} + 2\,500 \text{ kg})\sin 53.1°} = \boxed{15.6 \text{ m/s}}$$

Finalize Notice that the angle θ is qualitatively in agreement with Figure 8.12. Also notice that the final speed of the combination is less than the initial speeds of the two cars. This result is consistent with the kinetic energy of the system being reduced in an inelastic collision. It might help if you draw the momentum vectors of each vehicle before the collision and the two vehicles together after the collision.

8.6 | The Center of Mass

In this section, we describe the overall motion of a system of particles in terms of a very special point called the **center of mass** of the system. This notion gives us confidence in the particle model because we will see that the center of mass accelerates as if all the system's mass were concentrated at that point and all external forces act there.

Consider a system consisting of a pair of particles connected by a light, rigid rod (Active Fig. 8.13, page 218). The center of mass as indicated in the figure is located on the rod and is closer to the larger mass in the figure; we will see why soon. If a single force is applied at some point on the rod that is above the center of mass, the system rotates clockwise (Active Fig. 8.13a) as it translates through space. If the force is applied at a point on the rod below the center of mass, the system rotates counterclockwise (Active Fig. 8.13b). If the force is applied exactly at the center of mass, the system moves in the direction of $\vec{\mathbf{F}}$ without rotating (Active Fig. 8.13c) as if the system is behaving as a particle. Therefore, in theory, the center of mass can be located with this experiment.

If we were to analyze the motion in Active Figure 8.13c, we would find that the system moves as if all its mass were concentrated at the center of mass. Furthermore, if the external net force on the system is $\sum \vec{\mathbf{F}}$ and the total mass of the system is M, the

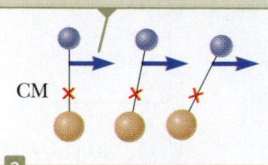

The system rotates clockwise when a force is applied above the center of mass.

CM

a

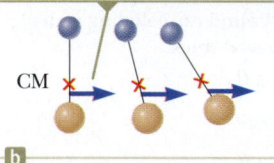

The system rotates counter-clockwise when a force is applied below the center of mass.

CM

b

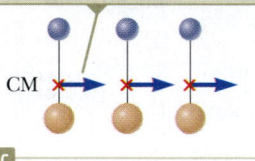

The system moves in the direction of the force without rotating when a force is applied at the center of mass.

CM

c

Active Figure 8.13 A force is applied to a system of two particles of unequal mass connected by a light, rigid rod.

center of mass moves with an acceleration given by $\vec{a} = \Sigma\vec{F}/M$. That is, the system moves as if the resultant external force were applied to a single particle of mass M located at the center of mass, which justifies our particle model for extended objects. We have ignored all rotational effects for extended objects so far, implicitly assuming that forces were provided at just the right position so as to cause no rotation. We will study rotational motion in Chapter 10, where we will apply forces that do not pass through the center of mass.

The position of the center of mass of a system can be described as being the *average position* of the system's mass. For example, the center of mass of the pair of particles described in Active Figure 8.14 is located on the x axis, somewhere between the particles. The x coordinate of the center of mass in this case is

$$x_{CM} = \frac{m_1 x_1 + m_2 x_2}{m_1 + m_2}$$ **8.29**◀

For example, if $x_1 = 0$, $x_2 = d$, and $m_2 = 2m_1$, we find that $x_{CM} = \frac{2}{3}d$. That is, the center of mass lies closer to the more massive particle. If the two masses are equal, the center of mass lies midway between the particles.

We can extend the concept of center of mass to a system of many particles in three dimensions. The x coordinate of the center of mass of n particles is defined to be

$$x_{CM} \equiv \frac{m_1 x_1 + m_2 x_2 + m_3 x_3 + \cdots + m_n x_n}{m_1 + m_2 + m_3 + \cdots + m_n} = \frac{\sum_i m_i x_i}{\sum_i m_i} = \frac{\sum_i m_i x_i}{M}$$ **8.30**◀

where x_i is the x coordinate of the ith particle and M is the *total mass* of the system. The y and z coordinates of the center of mass are similarly defined by the equations

$$y_{CM} \equiv \frac{\sum_i m_i y_i}{M} \quad \text{and} \quad z_{CM} \equiv \frac{\sum_i m_i z_i}{M}$$ **8.31**◀

The center of mass can also be located by its position vector, $\vec{r}_{CM}$. The rectangular coordinates of this vector are x_{CM}, y_{CM}, and z_{CM}, defined in Equations 8.30 and 8.31. Therefore,

$$\vec{r}_{CM} = x_{CM}\hat{i} + y_{CM}\hat{j} + z_{CM}\hat{k} = \frac{\sum_i m_i x_i \hat{i} + \sum_i m_i y_i \hat{j} + \sum_i m_i z_i \hat{k}}{M}$$

$$\vec{r}_{CM} = \frac{\sum_i m_i \vec{r}_i}{M}$$ **8.32**◀

where $\vec{r}_i$ is the position vector of the ith particle, defined by

$$\vec{r}_i \equiv x_i\hat{i} + y_i\hat{j} + z_i\hat{k}$$

Equation 8.32 is useful for finding the center of mass of a relatively small number of discrete particles. What about an extended object, which has a *continuous* distribution of mass? Although locating the center of mass for an extended object is somewhat more cumbersome than locating the center of mass of a system of particles, this location is based on the same fundamental ideas. We can model the extended object as a system containing a large number of elements (Fig. 8.15). Each element is modeled as a particle of mass Δm_i, with coordinates x_i, y_i, z_i. The particle separation is very small, so this model is a good representation of the continuous mass distribution of the object. The x coordinate of the center of mass of the particles representing the object, and therefore of the approximate center of mass of the object, is

$$x_{CM} \approx \frac{\sum_i x_i \Delta m_i}{M}$$

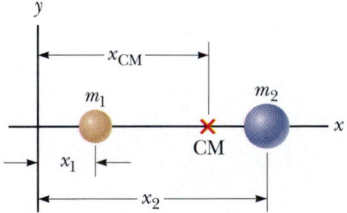

Active Figure 8.14 The center of mass of two particles having unequal mass is located on the x axis at x_{CM}, a point between the particles, closer to the one having the larger mass.

with similar expressions for y_{CM} and z_{CM}. If we let the number of elements approach infinity (and, as a consequence, the size and mass of each element approach zero), the model becomes indistinguishable from the continuous mass distribution and x_{CM} is given precisely. In this limit, we replace the sum by an integral and Δm_i by the differential element dm:

$$x_{CM} = \lim_{\Delta m_i \to 0} \frac{\sum_i x_i \Delta m_i}{M} = \frac{1}{M} \int x \, dm \qquad \text{8.33} \blacktriangleleft$$

where the integration is over the length of the object in the x direction. Likewise, for y_{CM} and z_{CM} we obtain

$$y_{CM} = \frac{1}{M} \int y \, dm \quad \text{and} \quad z_{CM} = \frac{1}{M} \int z \, dm \qquad \text{8.34} \blacktriangleleft$$

We can express the vector position of the center of mass of an extended object as

$$\vec{\mathbf{r}}_{CM} = \frac{1}{M} \int \vec{\mathbf{r}} \, dm \qquad \text{8.35} \blacktriangleleft$$

which is equivalent to the three expressions in Equations 8.33 and 8.34.

The center of mass of a homogeneous, symmetric object must lie on an axis of symmetry. For example, the center of mass of a homogeneous rod must lie midway between the ends of the rod. The center of mass of a homogeneous sphere or a homogeneous cube must lie at the geometric center of the object.

The center of mass of a system is often confused with the **center of gravity** of a system. Each portion of a system is acted on by the gravitational force. The net effect of all these forces is equivalent to the effect of a single force $M\vec{\mathbf{g}}$ acting at a special point called the center of gravity. The center of gravity is the average position of the gravitational forces on all parts of the object. If $\vec{\mathbf{g}}$ is uniform over the system, the center of gravity coincides with the center of mass. If the gravitational field over the system is not uniform, the center of gravity and the center of mass are different. In most cases, for objects or systems of reasonable size, the two points can be considered to be coincident.

One can experimentally determine the center of gravity of an irregularly shaped object, such as a wrench, by suspending the wrench from two different points (Fig. 8.16). An object of this size has virtually no variation in the gravitational field over its dimensions, so this method also locates the center of mass. The wrench is first hung from point A, and a vertical line AB is drawn (which can be established with a plumb bob) when the wrench is in equilibrium. The wrench is then hung from point C, and a second vertical line CD is drawn. The center of mass coincides with the intersection of these two lines. In fact, if the wrench is hung freely from any point, the vertical line through that point will pass through the center of mass.

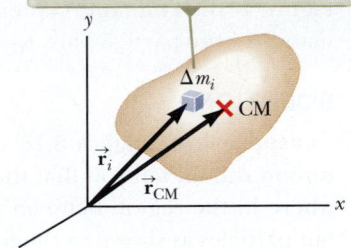

An extended object can be considered to be a distribution of small elements of mass Δm_i.

Figure 8.15 The center of mass of the object is located at the vector position $\vec{\mathbf{r}}_{CM}$, which has coordinates x_{CM}, y_{CM}, and z_{CM}.

▶ Center of mass of a continuous mass distribution

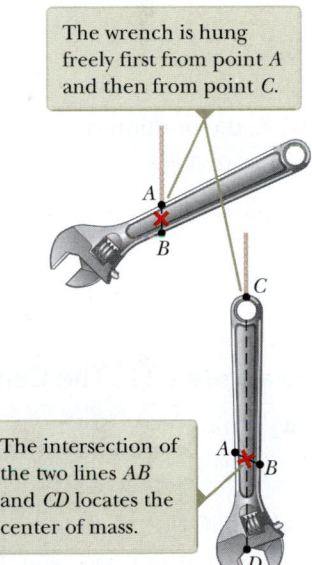

The wrench is hung freely first from point A and then from point C.

The intersection of the two lines AB and CD locates the center of mass.

Figure 8.16 An experimental technique for determining the center of mass of a wrench.

QUICK QUIZ 8.5 A baseball bat of uniform denisty is cut at the location of its center of mass as shown in Figure 8.17. Which piece has the smaller mass? (a) the piece on the right (b) the piece on the left (c) both pieces have the same mass (d) impossible to determine

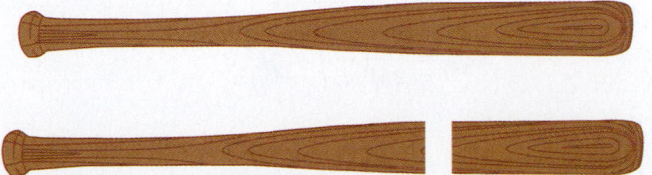

Figure 8.17 (Quick Quiz 8.5) A baseball bat cut at the location of its center of mass.

Example 8.11 | The Center of Mass of Three Particles

A system consists of three particles located as shown in Figure 8.18. Find the center of mass of the system. The masses of the particles are $m_1 = m_2 = 1.0$ kg and $m_3 = 2.0$ kg.

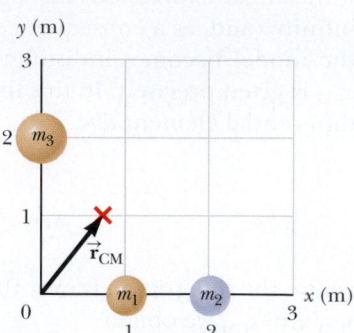

Figure 8.18
(Example 8.11) Two particles are located on the x axis, and a single particle is located on the y axis as shown. The vector $\vec{r}_{CM}$ indicates the location of the system's center of mass.

SOLUTION

Conceptualize Figure 8.18 shows the three masses. Your intuition should tell you that the center of mass is located somewhere in the region between the blue particle and the pair of tan particles as shown in the figure.

Categorize We categorize this example as a substitution problem because we will be using the equations for the center of mass developed in this section.

Use the defining equations for the coordinates of the center of mass and notice that $z_{CM} = 0$:

$$x_{CM} = \frac{1}{M}\sum_i m_i x_i = \frac{m_1 x_1 + m_2 x_2 + m_3 x_3}{m_1 + m_2 + m_3}$$

$$= \frac{(1.0 \text{ kg})(1.0 \text{ m}) + (1.0 \text{ kg})(2.0 \text{ m}) + (2.0 \text{ kg})(0)}{1.0 \text{ kg} + 1.0 \text{ kg} + 2.0 \text{ kg}} = \frac{3.0 \text{ kg} \cdot \text{m}}{4.0 \text{ kg}} = 0.75 \text{ m}$$

$$y_{CM} = \frac{1}{M}\sum_i m_i y_i = \frac{m_1 y_1 + m_2 y_2 + m_3 y_3}{m_1 + m_2 + m_3}$$

$$= \frac{(1.0 \text{ kg})(0) + (1.0 \text{ kg})(0) + (2.0 \text{ kg})(2.0 \text{ m})}{4.0 \text{ kg}} = \frac{4.0 \text{ kg} \cdot \text{m}}{4.0 \text{ kg}} = 1.0 \text{ m}$$

Write the position vector of the center of mass:

$$\vec{r}_{CM} \equiv x_{CM}\hat{i} + y_{CM}\hat{j} = (0.75\hat{i} + 1.0\hat{j}) \text{ m}$$

Example 8.12 | The Center of Mass of a Rod

(A) Show that the center of mass of a rod of mass M and length L lies midway between its ends, assuming the rod has a uniform mass per unit length.

SOLUTION

Conceptualize The rod is shown aligned along the x axis in Figure 8.19, so $y_{CM} = z_{CM} = 0$.

Categorize We categorize this example as an analysis problem because we need to divide the rod into small mass elements to perform the integration in Equation 8.33.

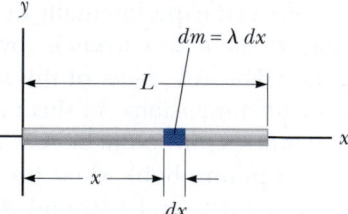

Figure 8.19 (Example 8.12) The geometry used to find the center of mass of a uniform rod.

Analyze The mass per unit length (this quantity is called the *linear mass density*) can be written as $\lambda = M/L$ for the uniform rod. If the rod is divided into elements of length dx, the mass of each element is $dm = \lambda \, dx$.

Use Equation 8.33 to find an expression for x_{CM}:

$$x_{CM} = \frac{1}{M}\int x \, dm = \frac{1}{M}\int_0^L x\lambda \, dx = \frac{\lambda}{M}\frac{x^2}{2}\Big|_0^L = \frac{\lambda L^2}{2M}$$

Substitute $\lambda = M/L$:

$$x_{CM} = \frac{L^2}{2M}\left(\frac{M}{L}\right) = \tfrac{1}{2}L$$

One can also use symmetry arguments to obtain the same result.

(B) Suppose a rod is *nonuniform* such that its mass per unit length varies linearly with x according to the expression $\lambda = \alpha x$, where α is a constant. Find the x coordinate of the center of mass as a fraction of L.

8.12 *cont.*

SOLUTION

Conceptualize Because the mass per unit length is not constant in this case but is proportional to *x*, elements of the rod to the right are more massive than elements near the left end of the rod.

Categorize This problem is categorized similarly to part (A), with the added twist that the linear mass density is not constant.

Analyze In this case, we replace *dm* in Equation 8.33 by $\lambda\,dx$, where $\lambda = \alpha x$.

Use Equation 8.33 to find an expression for x_{CM}:

$$x_{CM} = \frac{1}{M}\int x\,dm = \frac{1}{M}\int_0^L x\lambda\,dx = \frac{1}{M}\int_0^L x\alpha x\,dx$$

$$= \frac{\alpha}{M}\int_0^L x^2\,dx = \frac{\alpha L^3}{3M}$$

Find the total mass of the rod:

$$M = \int dm = \int_0^L \lambda\,dx = \int_0^L \alpha x\,dx = \frac{\alpha L^2}{2}$$

Substitute *M* into the expression for x_{CM}:

$$x_{CM} = \frac{\alpha L^3}{3\alpha L^2/2} = \boxed{\tfrac{2}{3}L}$$

Finalize Notice that the center of mass in part (B) is farther to the right than that in part (A). That result is reasonable because the elements of the rod become more massive as one moves to the right along the rod in part (B).

8.7 | Motion of a System of Particles

We can begin to understand the physical significance and utility of the center of mass concept by taking the time derivative of the position vector $\vec{r}_{CM}$ of the center of mass, given by Equation 8.32. Assuming that *M* remains constant—that is, no particles enter or leave the system—we find the following expression for the **velocity of the center of mass** of the system:

$$\vec{v}_{CM} = \frac{d\vec{r}_{CM}}{dt} = \frac{1}{M}\sum_i m_i \frac{d\vec{r}_i}{dt} = \frac{1}{M}\sum_i m_i\vec{v}_i \qquad \textbf{8.36} \blacktriangleleft$$

▶ Velocity of the center of mass for a system of particles

where $\vec{v}_i$ is the velocity of the *i*th particle. Rearranging Equation 8.36 gives

$$M\vec{v}_{CM} = \sum_i m_i\vec{v}_i = \sum_i \vec{p}_i = \vec{p}_{tot} \qquad \textbf{8.37} \blacktriangleleft$$

This result tells us that the total momentum of the system equals its total mass multiplied by the velocity of its center of mass. In other words, the total momentum of the system is equal to the momentum of a single particle of mass *M* moving with a velocity $\vec{v}_{CM}$; this is the particle model.

If we now differentiate Equation 8.36 with respect to time, we find the **acceleration of the center of mass** of the system:

$$\vec{a}_{CM} = \frac{d\vec{v}_{CM}}{dt} = \frac{1}{M}\sum_i m_i \frac{d\vec{v}_i}{dt} = \frac{1}{M}\sum_i m_i\vec{a}_i \qquad \textbf{8.38} \blacktriangleleft$$

▶ Acceleration of the center of mass for a system of particles

Rearranging this expression and using Newton's second law, we have

$$M\vec{a}_{CM} = \sum_i m_i\vec{a}_i = \sum_i \vec{F}_i \qquad \textbf{8.39} \blacktriangleleft$$

where $\vec{F}_i$ is the force on particle *i*.

The forces on any particle in the system may include both external and internal forces. By Newton's third law, however, the force exerted by particle 1 on particle 2, for example, is equal in magnitude and opposite the force exerted by particle 2 on particle 1. When we sum over all internal forces in Equation 8.39, they cancel in pairs. Therefore, the net force on the system is due *only* to external forces and we can write Equation 8.39 in the form

◀ Newton's second law for a system of particles

$$\sum \vec{\mathbf{F}}_{\text{ext}} = M \vec{\mathbf{a}}_{\text{CM}} = \frac{d\vec{\mathbf{P}}_{\text{tot}}}{dt} \qquad 8.40 \blacktriangleleft$$

That is, the external net force on the system of particles equals the total mass of the system multiplied by the acceleration of the center of mass, or the time rate of change of the momentum of the system. Comparing Equation 8.40 with Newton's second law for a single particle, we see that the particle model we have used in several chapters can be described in terms of the center of mass:

> The center of mass of a system of particles having combined mass M moves like an equivalent particle of mass M would move under the influence of the net external force on the system.

Let us integrate Equation 8.40 over a finite time interval:

$$\int \sum \vec{\mathbf{F}}_{\text{ext}} \, dt = \int M \vec{\mathbf{a}}_{\text{CM}} \, dt = \int M \frac{d\vec{\mathbf{v}}_{\text{CM}}}{dt} \, dt = M \int d\vec{\mathbf{v}}_{\text{CM}} = M \Delta \vec{\mathbf{v}}_{\text{CM}}$$

Notice that this equation can be written as

◀ Impulse–momentum theorem for a system of particles

$$\Delta \vec{\mathbf{P}}_{\text{tot}} = \vec{\mathbf{I}} \qquad 8.41 \blacktriangleleft$$

where $\vec{\mathbf{I}}$ is the impulse imparted to the system by external forces and $\vec{\mathbf{P}}_{\text{tot}}$ is the momentum of the system. Equation 8.41 is the generalization of the impulse–momentum theorem for a particle (Eq. 8.11) to a system of many particles. It is also the mathematical representation of the nonisolated system (momentum) model for a system of many particles.

In the absence of external forces, the center of mass moves with uniform velocity as in the case of the translating and rotating wrench in Figure 8.20. If the net force acts along a line through the center of mass of an extended object such as the wrench, the object is accelerated without rotation. If the net force does not act through the center of mass, the object will undergo rotation in addition to translation. The linear acceleration of the center of mass is the same in either case, as given by Equation 8.40.

Finally, we see that if the external net force is zero, from Equation 8.40 it follows that

$$\frac{d\vec{\mathbf{P}}_{\text{tot}}}{dt} = M \vec{\mathbf{a}}_{\text{CM}} = 0$$

so that

$$\vec{\mathbf{P}}_{\text{tot}} = M \vec{\mathbf{v}}_{\text{CM}} = \text{constant} \quad (\text{when } \sum \vec{\mathbf{F}}_{\text{ext}} = 0) \qquad 8.42 \blacktriangleleft$$

That is, the total linear momentum of a system of particles is constant if no external forces act on the system. It follows that, for an *isolated* system of particles, the total momentum is conserved. The law of conservation of momentum that was derived in Section 8.1 for a two-particle system is thus generalized to a many-particle system.

The center of mass of the wrench (marked with a white dot) moves in a straight line as the wrench rotates about this point.

Note the decreasing distance between the white dots.

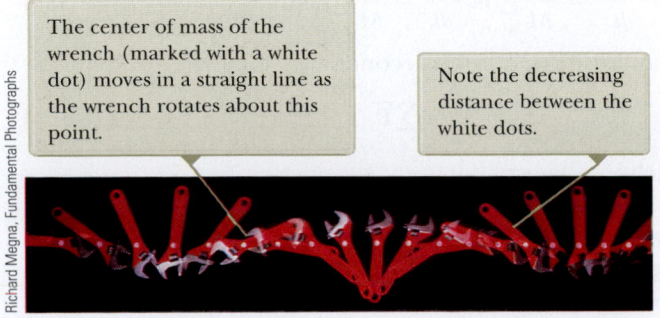

Figure 8.20 Strobe photograph showing an overhead view of a wrench moving on a horizontal surface. The wrench moves from left to right in the photograph and is slowing down due to friction between the wrench and the supporting surface.

Richard Megna, Fundamental Photographs

Figure 8.21 (Thinking Physics 8.1) A boy takes a step in a boat. What happens to the boat?

THINKING PHYSICS 8.1

A boy stands at one end of a boat that is stationary relative to the dock (Fig. 8.21). He then walks to the opposite end of the boat, away from the dock. Does the boat move?

Reasoning Yes, the boat moves toward the dock. Ignoring friction between the boat and water, no horizontal force acts on the system consisting of the boy and boat. The center of mass of the system therefore remains fixed relative to the dock (or any stationary point). As the boy moves away from the dock, the boat must move toward the dock such that the center of mass of the system remains fixed in position. ◀

QUICK QUIZ 8.6 A cruise ship is moving at constant speed through the water. The vacationers on the ship are eager to arrive at their next destination. They decide to try to speed up the cruise ship by gathering at the bow (the front) and running together toward the stern (the back) of the ship. (**i**) While they are running toward the stern, is the speed of the ship (a) higher than it was before, (b) unchanged, (c) lower than it was before, or (d) impossible to determine? (**ii**) The vacationers stop running when they reach the stern of the ship. After they have all stopped running, is the speed of the ship (a) higher than it was before they started running, (b) unchanged from what it was before they started running, (c) lower than it was before they started running, or (d) impossible to determine?

Example 8.13 | The Exploding Rocket

A rocket is fired vertically upward. At the instant it reaches an altitude of 1 000 m and a speed of $v_i = 300$ m/s, it explodes into three fragments having equal mass. One fragment moves upward with a speed of $v_1 = 450$ m/s following the explosion. The second fragment has a speed of $v_2 = 240$ m/s and is moving east right after the explosion. What is the velocity of the third fragment immediately after the explosion?

SOLUTION

Conceptualize Picture the explosion in your mind, with one piece going upward and a second piece moving horizontally toward the east. Do you have an intuitive feeling about the direction in which the third piece moves?

Categorize This example is a two-dimensional problem because we have two fragments moving in perpendicular directions after the explosion as well as a third fragment moving in an unknown direction in the plane defined by the velocity vectors of the other two fragments. We assume the time interval of the explosion is very short, so we use the impulse approximation in which we ignore the gravitational force and air resistance. Because the forces of the explosion are internal to the system (the rocket), the system is modeled as isolated in terms of momentum. Therefore, the total momentum $\vec{\mathbf{p}}_i$ of the rocket immediately before the explosion must equal the total momentum $\vec{\mathbf{p}}_f$ of the fragments immediately after the explosion.

Analyze Because the three fragments have equal mass, the mass of each fragment is $M/3$, where M is the total mass of the rocket. We will let $\vec{\mathbf{v}}_3$ represent the unknown velocity of the third fragment.

Using the isolated system (momentum) model, equate the initial and final momenta of the system and express the momenta in terms of masses and velocities:

$$\vec{\mathbf{p}}_i = \vec{\mathbf{p}}_f \quad \rightarrow \quad M\vec{\mathbf{v}}_i = \frac{M}{3}\vec{\mathbf{v}}_1 + \frac{M}{3}\vec{\mathbf{v}}_2 + \frac{M}{3}\vec{\mathbf{v}}_3$$

continued

8.13 *cont.*

Solve for $\vec{\mathbf{v}}_3$:

$$\vec{\mathbf{v}}_3 = 3\vec{\mathbf{v}}_i - \vec{\mathbf{v}}_1 - \vec{\mathbf{v}}_2$$

Substitute the numerical values:

$$\vec{\mathbf{v}}_3 = 3(300\hat{\mathbf{j}} \text{ m/s}) - (450\hat{\mathbf{j}} \text{ m/s}) - (240\hat{\mathbf{i}} \text{ m/s}) = \boxed{(-240\hat{\mathbf{i}} + 450\hat{\mathbf{j}}) \text{ m/s}}$$

Finalize Notice that this event is the reverse of a perfectly inelastic collision. There is one object before the collision and three objects afterward. Imagine running a movie of the event backward: the three objects would come together and become a single object. In a perfectly inelastic collision, the kinetic energy of the system decreases. If you were to calculate the kinetic energy before and after the event in this example, you would find that the kinetic energy of the system increases. (Try it!) This increase in kinetic energy comes from the potential energy stored in whatever fuel exploded to cause the breakup of the rocket.

8.8 | Context Connection: Rocket Propulsion

On our trip to Mars, we will need to control our spacecraft by firing the rocket engines. When ordinary vehicles, such as the automobiles in Context 1, are propelled, the driving force for the motion is the friction force exerted by the road on the car. A rocket moving in space, however, has no road to "push" against. The source of the propulsion of a rocket must therefore be different. The operation of a rocket depends on the law of conservation of momentum as applied to a system, where the system is the rocket plus its ejected fuel.

The propulsion of a rocket can be understood by first considering the archer on ice in Example 8.2. As an arrow is fired from the bow, the arrow receives momentum $m\vec{\mathbf{v}}$ in one direction and the archer receives a momentum of equal magnitude in the opposite direction. As additional arrows are fired, the archer moves faster, so a large velocity of the archer can be established by firing many arrows.

In a similar manner, as a rocket moves in free space (a vacuum), its momentum changes when some of its mass is released in the form of ejected gases. Because the ejected gases acquire some momentum, the rocket receives a compensating momentum in the opposite direction. The rocket therefore is accelerated as a result of the "push," or thrust, from the exhaust gases. Note that the rocket represents the *inverse* of an inelastic collision; that is, momentum is conserved, but the kinetic energy of the system is *increased* (at the expense of energy stored in the fuel of the rocket).

Suppose at some time t the magnitude of the momentum of the rocket plus the fuel is $(M + \Delta m)v$ (Fig. 8.22a). During a short time interval Δt, the rocket ejects fuel of mass Δm and the rocket's speed therefore increases to $v + \Delta v$ (Fig. 8.22b). If the fuel is ejected with velocity $\vec{\mathbf{v}}_e$ *relative to the rocket*, the speed of the fuel relative to a stationary frame of reference is $v - v_e$ according to our discussion of relative velocity in Section 3.6. Therefore, if we equate the total initial momentum of the system with the total final momentum, we have

$$(M + \Delta m)v = M(v + \Delta v) + \Delta m(v - v_e)$$

Simplifying this expression gives

$$M\Delta v = \Delta m(v_e)$$

If we now take the limit as Δt goes to zero, $\Delta v \rightarrow dv$ and $\Delta m \rightarrow dm$. Furthermore, the increase dm in the exhaust mass corresponds to an equal decrease in the rocket mass, so $dm = -dM$. Note that the negative sign is introduced into the equation because dM represents a decrease in mass. Using this fact, we have

$$M\,dv = -v_e\,dM \qquad\qquad \textbf{8.43} \blacktriangleleft$$

Integrating this equation and taking the initial mass of the rocket plus fuel to be M_i and the final mass of the rocket plus its remaining fuel to be M_f, we have

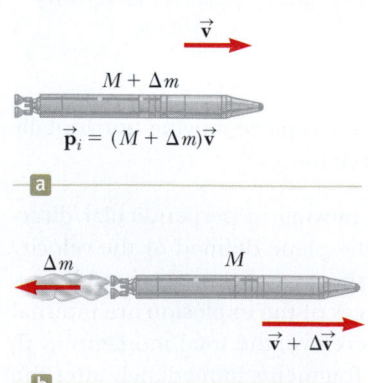

Figure 8.22 Rocket propulsion. (a) The initial mass of the rocket and all its fuel is $M + \Delta m$ at a time t, and its speed is v. (b) At a time $t + \Delta t$, the rocket's mass has been reduced to M, and an amount of fuel Δm has been ejected. The rocket's speed increases by an amount Δv.

$$\int_{v_i}^{v_f} dv = -v_e \int_{M_i}^{M_f} \frac{dM}{M}$$

$$v_f - v_i = v_e \ln\left(\frac{M_i}{M_f}\right) \qquad \text{8.44} \blacktriangleleft$$

▶ Velocity change in rocket propulsion

which is the basic expression for rocket propulsion. It tells us that the increase in speed is proportional to the exhaust speed v_e. The exhaust speed should therefore be very high.

The **thrust** on the rocket is the force exerted on the rocket by the ejected exhaust gases. We can obtain an expression for the thrust from Equation 8.43:

$$\text{Thrust} = Ma = M\frac{dv}{dt} = \left| v_e \frac{dM}{dt} \right| \qquad \text{8.45} \blacktriangleleft$$

▶ Rocket thrust

Here we see that the thrust increases as the exhaust speed increases and as the rate of change of mass (burn rate) increases.

We can now determine the amount of fuel needed to set us on our journey to Mars. The fuel requirements are well within the capabilities of current technology, as evidenced by the several missions to Mars that have already been accomplished. What if we wanted to visit another *star*, however, rather than another *planet*? This question raises many new technological challenges, including the requirement to consider the effects of relativity, which we investigate in the next chapter.

THINKING PHYSICS 8.2

When Robert Goddard proposed the possibility of rocket-propelled vehicles, the *New York Times* agreed that such vehicles would be useful and successful within the Earth's atmosphere ("Topics of the Times," *New York Times*, January 13, 1920, p. 12). The *Times,* however, balked at the idea of using such a rocket in the vacuum of space, noting that "its flight would be neither accelerated nor maintained by the explosion of the charges it then might have left. To claim that it would be is to deny a fundamental law of dynamics, and only Dr. Einstein and his chosen dozen, so few and fit, are licensed to do that. . . . That Professor Goddard, with his 'chair' in Clark College and the countenancing of the Smithsonian Institution, does not know the relation of action to reaction, and of the need to have something better than a vacuum against which to react—to say that would be absurd. Of course, he only seems to lack the knowledge ladled out daily in high schools." What did the writer of this passage overlook?

Reasoning The writer of this passage was making a common mistake in believing that a rocket works by expelling gases that push on something, propelling the rocket forward. With this belief, it is impossible to see how a rocket fired in empty space would work.

Gases do not need to push on anything; it is the act itself of expelling the gases that pushes the rocket forward. This point can be argued from Newton's third law: The rocket pushes the gases backward, resulting in the gases pushing the rocket forward. It can also be argued from conservation of momentum: As the gases gain momentum in one direction, the rocket must gain momentum in the opposite direction to conserve the original momentum of the rocket–gas system.

The *New York Times* did publish a retraction 49 years later ("A Correction," *New York Times,* July 17, 1969, p. 43) while the *Apollo 11* astronauts were on their way to the Moon. It appeared on a page with two other articles entitled "Fundamentals of Space Travel" and "Spacecraft, Like Squid, Maneuver by 'Squirts'" and contained the following passages: "an editorial feature of the *New York Times* dismissed the notion that a rocket could function in a vacuum and commented on the ideas of Robert H. Goddard. . . . Further investigation and experimentation have confirmed the findings of Isaac Newton in the 17th century, and it is now definitely established that a rocket can function in a vacuum as well as in an atmosphere. The *Times* regrets the error." ◀

Example 8.14 | A Rocket in Space

A rocket moving in space, far from all other objects, has a speed of 3.0×10^3 m/s relative to the Earth. Its engines are turned on, and fuel is ejected in a direction opposite the rocket's motion at a speed of 5.0×10^3 m/s relative to the rocket.

(A) What is the speed of the rocket relative to the Earth once the rocket's mass is reduced to half its mass before ignition?

SOLUTION

Conceptualize Figure 8.22 shows the situation in this problem. From the discussion in this section and scenes from science fiction movies, we can easily imagine the rocket accelerating to a higher speed as the engine operates.

Categorize This problem is a substitution problem in which we use given values in the equations derived in this section.

Solve Equation 8.44 for the final velocity and substitute the known values:

$$v_f = v_i + v_e \ln\left(\frac{M_i}{M_f}\right)$$

$$= 3.0 \times 10^3 \text{ m/s} + (5.0 \times 10^3 \text{ m/s})\ln\left(\frac{M_i}{0.50M_i}\right)$$

$$= 6.5 \times 10^3 \text{ m/s}$$

(B) What is the thrust on the rocket if it burns fuel at the rate of 50 kg/s?

SOLUTION

Use Equation 8.45 and the result from part (A), noting that $dM/dt = 50$ kg/s:

$$\text{Thrust} = \left|v_e \frac{dM}{dt}\right| = (5.0 \times 10^3 \text{ m/s})(50 \text{ kg/s}) = 2.5 \times 10^5 \text{ N}$$

▶ SUMMARY

The linear momentum of any object of mass m moving with a velocity $\vec{v}$ is

$$\vec{p} \equiv m\vec{v} \qquad \text{8.2} \blacktriangleleft$$

The **impulse** imparted to a particle by a net force $\Sigma\vec{F}$ is equal to the time integral of the force:

$$\vec{I} = \int_{t_i}^{t_f} \Sigma\vec{F}\, dt \qquad \text{8.10} \blacktriangleleft$$

When two objects collide, the total momentum of the isolated system before the collision always equals the total momentum after the collision, regardless of the nature of the collision. An **inelastic collision** is one in which kinetic energy is not conserved. A **perfectly inelastic collision** is one in which the colliding objects stick together after the collision. An **elastic collision** is one in which both momentum and kinetic energy are conserved.

In a two- or three-dimensional collision, the components of momentum in each of the directions are conserved independently.

The vector position of the center of mass of a system of particles is defined as

$$\vec{r}_{CM} = \frac{\sum_i m_i \vec{r}_i}{M} \qquad \text{8.32} \blacktriangleleft$$

where M is the total mass of the system and $\vec{r}_i$ is the position vector of the ith particle.

The **velocity of the center of mass for a system of particles** is

$$\vec{v}_{CM} = \frac{1}{M}\sum_i m_i \vec{v}_i \qquad \text{8.36} \blacktriangleleft$$

The total momentum of a system of particles equals the total mass multiplied by the velocity of the center of mass; that is, $\vec{p}_{tot} = M\vec{v}_{CM}$.

Newton's second law applied to a system of particles is

$$\sum\vec{F}_{ext} = M\vec{a}_{CM} = \frac{d\vec{p}_{tot}}{dt} \qquad \text{8.40} \blacktriangleleft$$

where $\vec{a}_{CM}$ is the acceleration of the center of mass and the sum is over all external forces. The center of mass therefore moves like an imaginary particle of mass M under the influence of the resultant external force on the system.

Analysis Models for Problem Solving

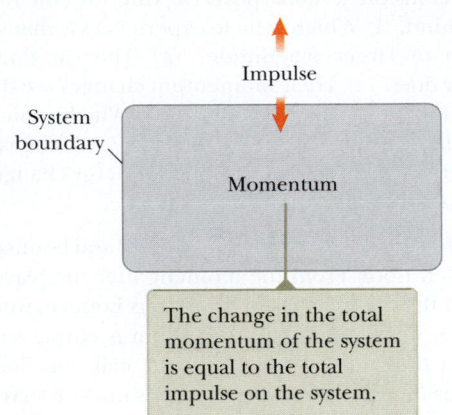

Nonisolated System (Momentum). If a system interacts with its environment in the sense that there is an external force on the system, the behavior of the system is described by the **impulse–momentum theorem:**

$$\Delta \vec{\mathbf{p}}_{tot} = \vec{\mathbf{I}} \qquad \text{8.11} \blacktriangleleft$$

The change in the total momentum of the system is equal to the total impulse on the system.

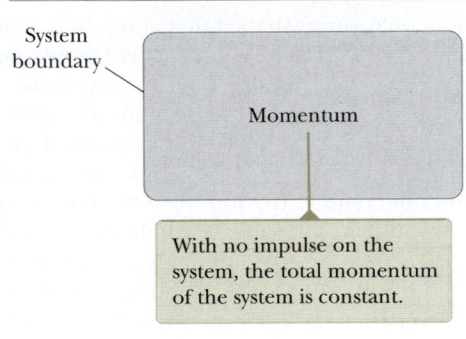

Isolated System (Momentum). The principle of **conservation of linear momentum** indicates that the total momentum of an isolated system (no external forces) is conserved regardless of the nature of the forces between the members of the system:

$$\vec{\mathbf{p}}_{tot} = M\vec{\mathbf{v}}_{CM} = \text{constant} \quad (\text{when } \sum \vec{\mathbf{F}}_{ext} = 0) \qquad \text{8.42} \blacktriangleleft$$

In the case of a two-particle system, this principle can be expressed as

$$\vec{\mathbf{p}}_{1i} + \vec{\mathbf{p}}_{2i} = \vec{\mathbf{p}}_{1f} + \vec{\mathbf{p}}_{2f} \qquad \text{8.6} \blacktriangleleft$$

The system may be isolated in terms of momentum but nonisolated in terms of energy, as in the case of inelastic collisions.

With no impulse on the system, the total momentum of the system is constant.

OBJECTIVE QUESTIONS

☐ denotes answer available in *Student Solutions Manual/Study Guide*

1. A 3-kg object moving to the right on a frictionless, horizontal surface with a speed of 2 m/s collides head-on and sticks to a 2-kg object that is initially moving to the left with a speed of 4 m/s. After the collision, which statement is true? (a) The kinetic energy of the system is 20 J. (b) The momentum of the system is 14 kg · m/s. (c) The kinetic energy of the system is greater than 5 J but less than 20 J. (d) The momentum of the system is −2 kg · m/s. (e) The momentum of the system is less than the momentum of the system before the collision.

2. A head-on, elastic collision occurs between two billiard balls of equal mass. If a red ball is traveling to the right with speed v and a blue ball is traveling to the left with speed $3v$ before the collision, what statement is true concerning their velocities subsequent to the collision? Neglect any effects of spin. (a) The red ball travels to the left with speed v, while the blue ball travels to the right with speed $3v$. (b) The red ball travels to the left with speed v, while the blue ball continues to move to the left with a speed $2v$. (c) The red ball travels to the left with speed $3v$, while the blue ball travels to the right with speed v. (d) Their final velocities cannot be determined because mo-

mentum is not conserved in the collision. (e) The velocities cannot be determined without knowing the mass of each ball.

3. A car of mass m traveling at speed v crashes into the rear of a truck of mass $2m$ that is at rest and in neutral at an intersection. If the collision is perfectly inelastic, what is the speed of the combined car and truck after the collision? (a) v (b) $v/2$ (c) $v/3$ (d) $2v$ (e) None of those answers is correct.

4. A 57.0-g tennis ball is traveling straight at a player at 21.0 m/s. The player volleys the ball straight back at 25.0 m/s. If the ball remains in contact with the racket for 0.060 s, what average force acts on the ball? (a) 22.6 N (b) 32.5 N (c) 43.7 N (d) 72.1 N (e) 102 N

5. A 5-kg cart moving to the right with a speed of 6 m/s collides with a concrete wall and rebounds with a speed of 2 m/s. What is the change in momentum of the cart? (a) 0 (b) 40 kg · m/s (c) −40 kg · m/s (d) −30 kg · m/s (e) −10 kg · m/s

6. A 2-kg object moving to the right with a speed of 4 m/s makes a head-on, elastic collision with a 1-kg object that is initially at rest. The velocity of the 1-kg object after the

collision is (a) greater than 4 m/s, (b) less than 4 m/s, (c) equal to 4 m/s, (d) zero, or (e) impossible to say based on the information provided.

7. The momentum of an object is increased by a factor of 4 in magnitude. By what factor is its kinetic energy changed? (a) 16 (b) 8 (c) 4 (d) 2 (e) 1

8. The kinetic energy of an object is increased by a factor of 4. By what factor is the magnitude of its momentum changed? (a) 16 (b) 8 (c) 4 (d) 2 (e) 1

9. A 10.0-g bullet is fired into a 200-g block of wood at rest on a horizontal surface. After impact, the block slides 8.00 m before coming to rest. If the coefficient of friction between the block and the surface is 0.400, what is the speed of the bullet before impact? (a) 106 m/s (b) 166 m/s (c) 226 m/s (d) 286 m/s (e) none of those answers is correct.

10. If two particles have equal kinetic energies, are their momenta equal? (a) yes, always (b) no, never (c) yes, as long as their masses are equal (d) yes, if both their masses and directions of motion are the same (e) yes, as long as they move along parallel lines

11. If two particles have equal momenta, are their kinetic energies equal? (a) yes, always (b) no, never (c) no, except when their speeds are the same (d) yes, as long as they move along parallel lines

12. Two particles of different mass start from rest. The same net force acts on both of them as they move over equal distances. How do their final kinetic energies compare? (a) The particle of larger mass has more kinetic energy. (b) The particle of smaller mass has more kinetic energy. (c) The particles have equal kinetic energies. (d) Either particle might have more kinetic energy.

13. Two particles of different mass start from rest. The same net force acts on both of them as they move over equal distances. How do the magnitudes of their final momenta compare? (a) The particle of larger mass has more momentum. (b) The particle of smaller mass has more momentum. (c) The particles have equal momenta. (d) Either particle might have more momentum.

14. A ball is suspended by a string that is tied to a fixed point above a wooden block standing on end. The ball is pulled back as shown in Figure OQ8.14 and released. In trial A, the ball rebounds elastically from the block. In trial B, two-sided tape causes the ball to stick to the block. In which case is the ball more likely to knock the

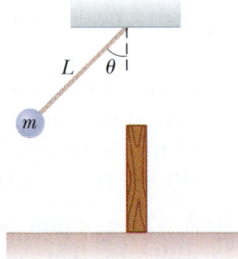

Figure OQ8.14

block over? (a) It is more likely in trial A. (b) It is more likely in trial B. (c) It makes no difference. (d) It could be either case, depending on other factors.

15. A massive tractor is rolling down a country road. In a perfectly inelastic collision, a small sports car runs into the machine from behind. (i) Which vehicle experiences a change in momentum of larger magnitude? (a) The car does. (b) The tractor does. (c) Their momentum changes are the same size. (d) It could be either vehicle. (ii) Which vehicle experiences a larger change in kinetic energy? (a) The car does. (b) The tractor does. (c) Their kinetic energy changes are the same size. (d) It could be either vehicle.

16. A basketball is tossed up into the air, falls freely, and bounces from the wooden floor. From the moment after the player releases it until the ball reaches the top of its bounce, what is the smallest system for which momentum is conserved? (a) the ball (b) the ball plus player (c) the ball plus floor (d) the ball plus the Earth (e) momentum is not conserved for any system.

17. You are standing on a saucer-shaped sled at rest in the middle of a frictionless ice rink. Your lab partner throws you a heavy Frisbee. You take different actions in successive experimental trials. Rank the following situations according to your final speed from largest to smallest. If your final speed is the same in two cases, give them equal rank. (a) You catch the Frisbee and hold onto it. (b) You catch the Frisbee and throw it back to your partner. (c) You bobble the catch, just touching the Frisbee so that it continues in its original direction more slowly. (d) You catch the Frisbee and throw it so that it moves vertically upward above your head. (e) You catch the Frisbee and set it down so that it remains at rest on the ice.

18. A boxcar at a rail yard is set into motion at the top of a hump. The car rolls down quietly and without friction onto a straight, level track where it couples with a flatcar of smaller mass, originally at rest, so that the two cars then roll together without friction. Consider the two cars as a system from the moment of release of the boxcar until both are rolling together. Answer the following questions yes or no. (a) Is mechanical energy of the system conserved? (b) Is momentum of the system conserved? Next, consider only the process of the boxcar gaining speed as it rolls down the hump. For the boxcar and the Earth as a system, (c) is mechanical energy conserved? (d) Is momentum conserved? Finally, consider the two cars as a system as the boxcar is slowing down in the coupling process. (e) Is mechanical energy of this system conserved? (f) Is momentum of this system conserved?

CONCEPTUAL QUESTIONS

1. Does a larger net force exerted on an object always produce a larger change in the momentum of the object compared with a smaller net force? Explain.

2. While in motion, a pitched baseball carries kinetic energy and momentum. (a) Can we say that it carries a force that it can exert on any object it strikes? (b) Can the baseball

deliver more kinetic energy to the bat and batter than the ball carries initially? (c) Can the baseball deliver to the bat and batter more momentum than the ball carries initially? Explain each of your answers.

3. A bomb, initially at rest, explodes into several pieces. (a) Is linear momentum of the system (the bomb before the

explosion, the pieces after the explosion) conserved? Explain. (b) Is kinetic energy of the system conserved? Explain.

4. Does a larger net force always produce a larger change in kinetic energy than a smaller net force? Explain.

5. You are standing perfectly still and then take a step forward. Before the step, your momentum was zero, but afterward you have some momentum. Is the principle of conservation of momentum violated in this case? Explain your answer.

6. A juggler juggles three balls in a continuous cycle. Any one ball is in contact with one of his hands for one fifth of the time. (a) Describe the motion of the center of mass of the three balls. (b) What average force does the juggler exert on one ball while he is touching it?

7. Two students hold a large bed sheet vertically between them. A third student, who happens to be the star pitcher on the school baseball team, throws a raw egg at the center of the sheet. Explain why the egg does not break when it hits the sheet, regardless of its initial speed.

8. A sharpshooter fires a rifle while standing with the butt of the gun against her shoulder. If the forward momentum of a bullet is the same as the backward momentum of the gun, why isn't it as dangerous to be hit by the gun as by the bullet?

9. An airbag in an automobile inflates when a collision occurs, which protects the passenger from serious injury (see the photo on page 206). Why does the airbag soften the blow? Discuss the physics involved in this dramatic photograph.

10. On the subject of the following positions, state your own view and argue to support it. (a) The best theory of motion is that force causes acceleration. (b) The true measure of a force's effectiveness is the work it does, and the best theory of motion is that work done on an object changes its energy. (c) The true measure of a force's effect is impulse, and the best theory of motion is that impulse imparted to an object changes its momentum.

11. (a) Does the center of mass of a rocket in free space accelerate? Explain. (b) Can the speed of a rocket exceed the exhaust speed of the fuel? Explain.

12. In golf, novice players are often advised to be sure to "follow through" with their swing. Why does this advice make the ball travel a longer distance? If a shot is taken near the green, very little follow-through is required. Why?

13. An open box slides across a frictionless, icy surface of a frozen lake. What happens to the speed of the box as water from a rain shower falls vertically downward into the box? Explain.

▶ PROBLEMS AVAILABLE IN

8.1 Linear Momentum
8.2 Analysis Model: Isolated System (Momentum)

Problems 1–7

8.3 Analysis Model: Nonisolated System (Momentum)

Problems 8–14

8.4 Collisions in One Dimension

Problems 15–24

8.5 Collisions in Two Dimensions

Problems 25–33

8.6 The Center of Mass

Problems 34–38

8.7 Motion of a System of Particles

Problems 38–42

8.8 Context Connection: Rocket Propulsion

Problems 43–47

Additional Problems

Problems 48–65

Solutions to the following Problems are available in the *Student Solutions Manual/Study Guide*:

8.5, 8.9, 8.11, 8.17, 8.21, 8.25, 8.29, 8.31, 8.34, 8.39, 8.41, 8.45, and 8.57

List of Enhanced Problems

Problem Number	Targeted Feedback in Enhanced WebAssign	Master It in Enhanced WebAssign	Watch It in Enhanced WebAssign
8.5		✓	
8.7	✓		✓
8.9	✓	✓	
8.10	✓		✓
8.11	✓		✓
8.17		✓	
8.19	✓		✓
8.21		✓	
8.25	✓		✓
8.27	✓		✓
8.29	✓	✓	
8.30	✓		✓
8.31		✓	
8.35	✓		✓
8.39	✓		
8.41		✓	
8.55	✓		✓
8.57		✓	

Relativity

Chapter Outline

Emily Serway

Standing on the shoulders of a giant. David Serway, son of one of the authors, watches over his children, Nathan and Kaitlyn, as they frolic in the arms of Albert Einstein at the Einstein memorial in Washington, D.C. It is well known that Einstein, the principal architect of relativity, was very fond of children.

Our everyday experiences and observations are associated with objects that move at speeds much less than that of light in a vacuum, $c = 3.00 \times 10^8$ m/s. Analysis models and definitions of quantities based on Newtonian mechanics and early concepts of space and time were formulated to describe the motion of such objects. This formalism is very successful in describing a wide range of phenomena that occur at low speeds, as we have seen in previous chapters. It fails, however, when applied to objects whose speeds approach that of light. Experimentally, the predictions of Newtonian theory can be tested by accelerating electrons or other particles to very high speeds. For example, it is possible to accelerate an electron to a speed of 0.99c. According to the Newtonian definition of kinetic energy, if the energy transferred to such an electron were increased by a factor of 4, the electron speed should double to 1.98c. Relativistic calculations, however, show that the speed of the electron—as well as the speeds of all other objects in the Universe—remains less than the speed of light. Because it places no upper limit on speed, Newtonian mechanics is contrary to modern theoretical predictions and experimental results, and the Newtonian models that we have developed are limited to objects moving much slower than the speed of light. Because Newtonian mechanics does not correctly predict the results of experiments carried out on objects moving at high speeds, we need a new formalism that is valid for these objects.

In 1905, at the age of only 26, Albert Einstein published his *special theory of relativity*, which is the subject of most of this chapter. Regarding the theory, Einstein wrote:

> The relativity theory arose from necessity, from serious and deep contradictions in the old theory from which there seemed no escape. The strength of the new theory lies in the consistency and simplicity with which it solves all these difficulties, using only a few very convincing assumptions.[1]

Although Einstein made many important contributions to science, special relativity alone represents one of the greatest intellectual achievements of the 20th century. With special relativity, experimental observations can be correctly predicted for objects over the range of all possible speeds, from rest to speeds approaching the speed of light. This chapter gives an introduction to special relativity, with emphasis on some of its consequences.

9.1 | The Principle of Galilean Relativity

We will begin by considering the notion of relativity at low speeds. This discussion was actually begun in Section 3.6 when we discussed relative velocity. At that time, we discussed the importance of the observer and the significance of his or her motion with respect to what is being observed. In a similar way here, we will generate equations that allow us to express one observer's measurements in terms of the other's. This process will lead to some rather unexpected and startling results about our understanding of space and time.

As we have mentioned previously, it is necessary to establish a frame of reference when describing a physical event. You should recall from Chapter 4 that an inertial frame is one in which an object is measured to have no acceleration if no forces act on it. Furthermore, any frame moving with constant velocity with respect to an inertial frame must also be an inertial frame. The laws predicting the results of an experiment performed in a vehicle moving with uniform velocity will be identical for the driver of the vehicle and a hitchhiker on the side of the road. The formal statement of this result is called the **principle of Galilean relativity:**

The laws of mechanics must be the same in all inertial frames of reference.

▶ Principle of Galilean relativity

The following observation illustrates the equivalence of the laws of mechanics in different inertial frames. Consider a pickup truck moving with a constant velocity as in Figure 9.1a. If a passenger in the truck throws a ball straight up in the air, the

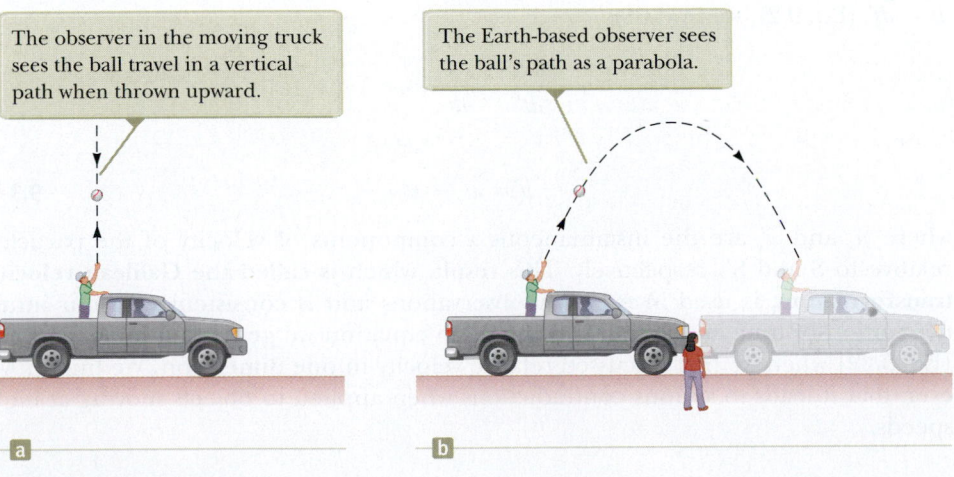

The observer in the moving truck sees the ball travel in a vertical path when thrown upward.

The Earth-based observer sees the ball's path as a parabola.

a

b

Figure 9.1 Two observers watch the path of a thrown ball and obtain different results.

---

[1]A. Einstein and L. Infeld, *The Evolution of Physics* (New York, Simon and Schuster, 1966), p. 192.

passenger observes that the ball moves in a vertical path (ignoring air resistance). The motion of the ball appears to be precisely the same as if the ball were thrown by a person at rest on the Earth and observed by that person. The kinematic equations of Chapter 2 describe the results correctly whether the truck is at rest or in uniform motion. Now consider the ball thrown in the truck as viewed by an observer at rest on the Earth. This observer sees the path of the ball as a parabola as in Figure 9.1b. Furthermore, according to this observer, the ball has a horizontal component of velocity equal to the speed of the truck. Although the two observers measure different velocities and see different paths of the ball, they see the same forces on the ball and agree on the validity of Newton's laws as well as classical principles such as conservation of energy and conservation of momentum. Their measurements differ, but the measurements they make satisfy the same laws. All differences between the two views stem from the relative motion of one frame with respect to the other.

Suppose some physical phenomenon, which we call an **event**, occurs. The event's location in space and time of occurrence can be specified by an observer with the coordinates (x, y, z, t). We would like to be able to transform these coordinates from one inertial frame to another moving with uniform relative velocity, which will allow us to express one observer's measurements in terms of the other's.

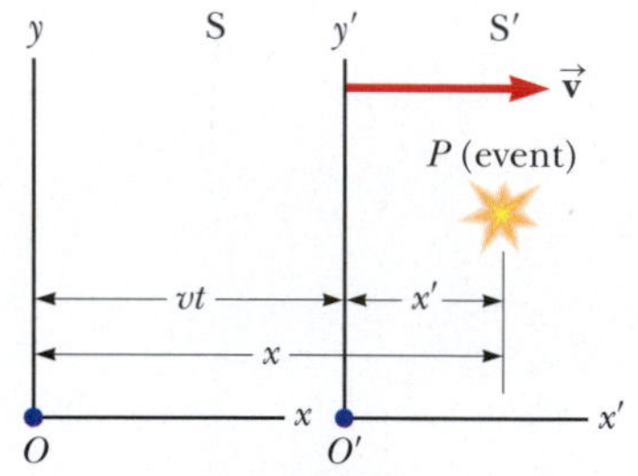

Consider two inertial frames S and S' (Fig. 9.2). The frame S' moves with a constant velocity $\vec{v}$ along the common x and x' axes, where $\vec{v}$ is measured relative to S. We assume that the origins of S and S' coincide at $t = 0$. Therefore, at time t, the origin of frame S' is to the right of the origin of S by a distance vt. An event occurs at point P and time t. An observer in S describes the event with space–time coordinates (x, y, z, t), and an observer in S' describes the same event with coordinates (x', y', z', t'). As we can see from Figure 9.2, a simple geometric argument shows that the space coordinates are related by the equations

Figure 9.2 An event occurs at a point P and time t. The event is seen by two observers O and O' in inertial frames S and S', where S' moves with a velocity $\vec{v}$ relative to S.

$$x' = x - vt \qquad y' = y \qquad z' = z \qquad \text{9.1} \blacktriangleleft$$

Time is assumed to be the same in both inertial frames. That is, within the framework of classical mechanics, all clocks run at the same rate, regardless of their velocity, so that the time at which an event occurs for an observer in S is the same as the time for the same event in S':

$$t' = t \qquad \text{9.2} \blacktriangleleft$$

Equations 9.1 and 9.2 constitute what is known as the **Galilean transformation of coordinates.**

Now suppose a particle moves through a displacement dx in a time interval dt as measured by an observer in S. It follows from the first of Equations 9.1 that the corresponding displacement dx' measured by an observer in S' is $dx' = dx - v\,dt$. Because $dt = dt'$ (Eq. 9.2), we find that

$$\frac{dx'}{dt'} = \frac{dx}{dt} - v$$

or

$$u_x' = u_x - v \qquad \text{9.3} \blacktriangleleft$$

where u_x and u_x' are the instantaneous x components of velocity of the particle[2] relative to S and S', respectively. This result, which is called the **Galilean velocity transformation,** is used in everyday observations and is consistent with our intuitive notion of time and space. It is the same equation we generated in Section 3.6 (Eq. 3.22) when we first discussed relative velocity in one dimension. We find, however, that it leads to serious contradictions when applied to objects moving at high speeds.

Pitfall Prevention | 9.1
The Relationship Between the S and S' Frames
Many of the mathematical representations in this chapter are true only for the specified relationship between the S and S' frames. The x and x' axes coincide, except their origins are different. The y and y' axes (and the z and z' axes) are parallel, but they only coincide at one instant due to the time-varying displacement of the origin of S' with respect to that of S. We choose the time $t = 0$ to be the instant at which the origins of the two coordinate systems coincide. If the S' frame is moving in the positive x direction relative to S, then v is positive; otherwise, it is negative.

---

[2]We have used v for the speed of the S' frame relative to the S frame. To avoid confusion, we will use u for the speed of an object or particle.

9.2 | The Michelson–Morley Experiment

Many experiments similar to throwing the ball in the pickup truck, described in the preceding section, show us that the laws of classical mechanics are the same in all inertial frames of reference. When similar inquiries are made into the laws of other branches of physics, however, the results are contradictory. In particular, the laws of electricity and magnetism are found to depend on the frame of reference used. It might be argued that these laws are wrong, but that is difficult to accept because the laws are in total agreement with known experimental results. The Michelson–Morley experiment was one of many attempts to investigate this dilemma.

The experiment stemmed from a misconception early physicists had concerning the manner in which light propagates. The properties of mechanical waves, such as water and sound waves, were well known, and all these waves require a *medium* to support the propagation of the disturbance, as we shall discuss in Chapter 13. For sound from your stereo system, the medium is the air, and for ocean waves, the medium is the water surface. In the 19th century, physicists subscribed to a model for light in which electromagnetic waves also require a medium through which to propagate. They proposed that such a medium exists, filling all space, and they named it the **luminiferous ether.** The ether would define an **absolute frame of reference** in which the speed of light is c.

The most famous experiment designed to show the presence of the ether was performed in 1887 by A. A. Michelson (1852–1931) and E. W. Morley (1838–1923). The objective was to determine the speed of the Earth through space with respect to the ether, and the experimental tool used was a device called the *interferometer*, shown schematically in Active Figure 9.3.

Light from the source at the left encounters a beam splitter M_0, which is a partially silvered mirror. Part of the light passes through toward mirror M_2, and the other part is reflected upward toward mirror M_1. Both mirrors are the same distance from the beam splitter. After reflecting from these mirrors, the light returns to the beam splitter, and part of each light beam propagates toward the observer at the bottom.

Suppose one arm of the interferometer (Arm 2, in Active Fig. 9.3) is aligned along the direction of the velocity $\vec{v}$ of the Earth through space and therefore through the ether. The "ether wind" blowing in the direction opposite the Earth's motion should cause the speed of light, as measured in the Earth's frame of reference, to be $c - v$ as the light approaches mirror M_2 in Active Figure 9.3 and $c + v$ after reflection.

The other arm (Arm 1) is perpendicular to the ether wind. For light to travel in this direction, the vector $\vec{c}$ must be aimed "upstream" so that the vector addition of $\vec{c}$ and $\vec{v}$ gives the speed of the light perpendicular to the ether wind as $\sqrt{c^2 - v^2}$. This situation is similar to Example 3.6, in which a boat crosses a river with a current. The boat is a model for the light beam in the Michelson–Morley experiment, and the river current is a model for the ether wind.

Because they travel in perpendicular directions with different speeds, light beams leaving the beam splitter simultaneously will arrive back at the beam splitter at different times. The interferometer is designed to detect this time difference. Measurements failed, however, to show any time difference! The Michelson–Morley experiment was repeated by other researchers under varying conditions and at different locations, but the results were always the same: *No time difference of the magnitude required was ever observed.*[3]

The negative result of the Michelson–Morley experiment not only contradicted the ether hypothesis, but it also meant that it was impossible to measure the absolute speed of the Earth with respect to the ether frame. From a theoretical

---

[3]From an Earth observer's point of view, changes in the Earth's speed and direction of motion in the course of a year are viewed as ether wind shifts. Even if the speed of the Earth with respect to the ether were zero at some time, six months later the Earth is moving in the opposite direction, the speed of the Earth with respect to the ether would be nonzero, and a clear time difference should be detected. None has ever been observed, however.

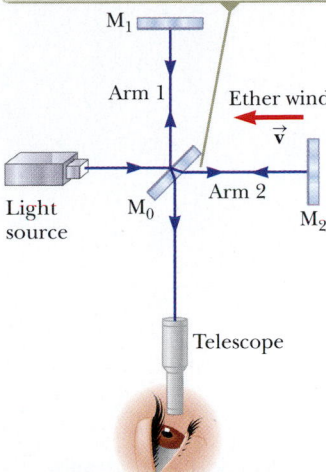

According to the ether wind theory, the speed of light should be $c - v$ as the beam approaches mirror M_2 and $c + v$ after reflection.

Active Figure 9.3 In the Michelson interferometer, the ether theory claims that the time interval for a light beam to travel from the beam splitter to mirror M_1 and back will be different from that for a light beam to travel from the beam splitter to mirror M_2 and back. The interferometer is sufficiently sensitive to detect this difference.

© AstroLab/National Park of the Mount-Mégantic

Albert A. Michelson
(1852–1931)
Michelson was born in Prussia in a town that later became part of Poland. He moved to the United States as a small child and spent much of his adult life making accurate measurements of the speed of light. In 1907, he was the first American to be awarded the Nobel Prize in Physics, which he received for his work in optics. His most famous experiment, conducted with Edward Morley in 1887, indicated that it was impossible to measure the absolute velocity of the Earth with respect to the ether.

viewpoint, it was impossible to find the absolute frame. As we shall see in the next section, however, Einstein offered a postulate that places a different interpretation on the negative result. In later years, when more was known about the nature of light, the idea of an ether that permeates all space was abandoned. Light is now understood to be an electromagnetic wave that requires no medium for its propagation. As a result, an ether through which light travels is an unnecessary construct.

Modern versions of the Michelson–Morley experiment have placed an upper limit of about 5 cm/s = 0.05 m/s on ether wind velocity. We can show that the speed of the Earth in its orbit around the Sun is 2.97×10^4 m/s, six orders of magnitude larger than the upper limit of ether wind velocity! These results have shown quite conclusively that the motion of the Earth has no effect on the measured speed of light.

9.3 | Einstein's Principle of Relativity

In the preceding section, we noted the failure of experiments to measure the relative speed between the ether and the Earth. Einstein proposed a theory that boldly removed these difficulties and at the same time completely altered our notion of space and time.[4] He based his relativity theory on two postulates:

1. **The principle of relativity:** All the laws of physics are the same in all inertial reference frames.
2. **The constancy of the speed of light:** The speed of light in vacuum has the same value in all inertial frames, regardless of the velocity of the observer or the velocity of the source emitting the light.

These postulates form the basis of **special relativity,** which is the relativity theory applied to observers moving with constant velocity. The first postulate asserts that *all* the laws of physics—those dealing with mechanics, electricity and magnetism, optics, thermodynamics, and so on—are the same in all reference frames moving with constant velocity relative to each other. This postulate is a sweeping generalization of the principle of Galilean relativity, which only refers to the laws of mechanics. From an experimental point of view, Einstein's principle of relativity means that any kind of experiment performed in a laboratory at rest must agree with the same laws of physics as when performed in a laboratory moving at constant velocity relative to the first one. Hence, no preferred inertial reference frame exists and it is impossible to detect absolute motion.

Note that postulate 2, the principle of the constancy of the speed of light, is required by postulate 1: If the speed of light were not the same in all inertial frames, it would be possible to experimentally distinguish between inertial frames and a preferred, absolute frame in which the speed of light is c, in contradiction to postulate 1. Postulate 2 also eliminates the problem of measuring the speed of the ether by denying the existence of the ether and boldly asserting that light always moves with speed c relative to all inertial observers.

9.4 | Consequences of Special Relativity

If we accept the postulates of special relativity, we must conclude that relative motion is unimportant when measuring the speed of light, which is the lesson of the Michelson–Morley experiment. At the same time, we must alter our commonsense notion of space and time and be prepared for some very unexpected consequences, as we shall see now.

© Mary Evans Picture Library/Alamy

Albert Einstein
German-American Physicist (1879–1955)
Einstein, one of the greatest physicists of all time, was born in Ulm, Germany. In 1905, at age 26, he published four scientific papers that revolutionized physics. Two of these papers were concerned with what is now considered his most important contribution: the special theory of relativity.

In 1916, Einstein published his work on the general theory of relativity. The most dramatic prediction of this theory is the degree to which light is deflected by a gravitational field. Measurements made by astronomers on bright stars in the vicinity of the eclipsed Sun in 1919 confirmed Einstein's prediction, and Einstein became a world celebrity as a result. Einstein was deeply disturbed by the development of quantum mechanics in the 1920s despite his own role as a scientific revolutionary. In particular, he could never accept the probabilistic view of events in nature that is a central feature of quantum theory. The last few decades of his life were devoted to an unsuccessful search for a unified theory that would combine gravitation and electromagnetism.

---

[4]A. Einstein, "On the Electrodynamics of Moving Bodies," *Ann. Physik* 17:891, 1905. For an English translation of this article and other publications by Einstein, see the book by H. Lorentz, A. Einstein, H. Minkowski, and H. Weyl, *The Principle of Relativity* (New York: Dover, 1958).

Simultaneity and the Relativity of Time

A basic premise of Newtonian mechanics is that a universal time scale exists that is the same for all observers. In fact, Newton wrote, "Absolute, true, and mathematical time, of itself, and from its own nature, flows equably without relation to anything external." Therefore, Newton and his followers simply took simultaneity for granted. In his development of special relativity, Einstein abandoned the notion that two events that appear simultaneous to one observer appear simultaneous to all observers. According to Einstein, a time measurement depends on the reference frame in which the measurement is made.

Einstein devised the following thought experiment to illustrate this point. A boxcar moves with uniform velocity and two lightning bolts strike its ends, as in Figure 9.4a, leaving marks on the boxcar and on the ground. The marks on the boxcar are labeled A' and B', and those on the ground are labeled A and B. An observer at O' moving with the boxcar is midway between A' and B', and a ground observer at O is midway between A and B. The events recorded by the observers are the arrivals of light signals from the lightning bolts.

The two light signals reach observer O at the same time as indicated in Figure 9.4b. As a result, O concludes that the events at A and B occurred simultaneously. Now consider the same events as viewed by the observer on the boxcar at O'. From our frame of reference, at rest with respect to the tracks in Figure 9.4, we see the lightning strikes occur as A' passes A, O' passes O, and B' passes B. By the time the light has reached observer O, observer O' has moved as indicated in Figure 9.4b. Therefore, the light signal from B' has already swept past O' because it had less distance to travel, but the light from A' has not yet reached O'. According to Einstein, observer O' must find that light travels at the same speed as that measured by observer O. Observer O' therefore concludes that the lightning struck the front of the boxcar before it struck the back. This thought experiment clearly demonstrates that the two events, which appear to be simultaneous to observer O, do not appear to be simultaneous to observer O'. In general, two events separated in space and observed to be simultaneous in one reference frame are not observed to be simultaneous in a second frame moving relative to the first. That is, simultaneity is not an absolute concept but one that depends on the state of motion of the observer.

Einstein's thought experiment demonstrates that two observers can disagree on the simultaneity of two events. This disagreement, however, depends on the transit time of light to the observers and therefore does *not* demonstrate the deeper meaning of relativity. In relativistic analyses of high-speed situations, relativity shows that simultaneity is relative *even when the transit time is subtracted out*. In fact, all the relativistic effects that we will discuss from here on will assume that we are ignoring differences caused by the transit time of light to the observers.

> **Pitfall Prevention | 9.2**
> **Who's Right?**
> At this point, you might wonder which observer in Figure 9.4 is correct concerning the two events. *Both are correct* because the principle of relativity states that *no inertial frame of reference is preferred.* Although the two observers reach different conclusions, both are correct in their own reference frame because the concept of simultaneity is not absolute. In fact, the central point of relativity is that any uniformly moving frame of reference can be used to describe events and do physics.

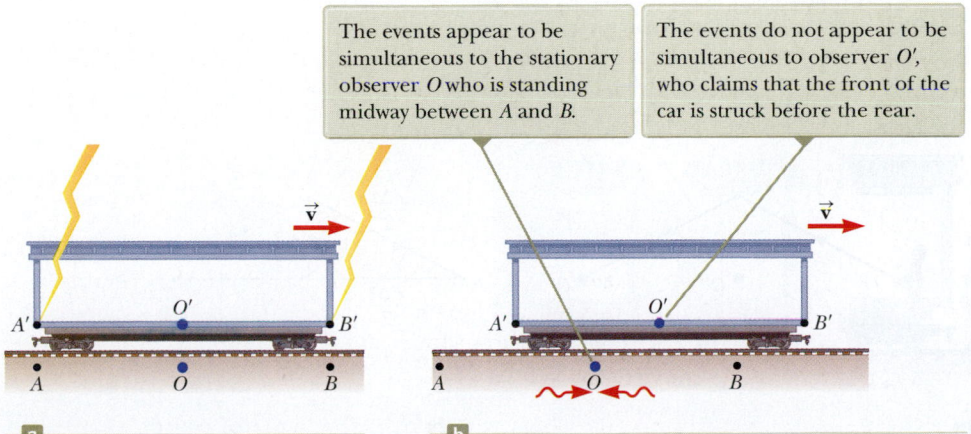

> The events appear to be simultaneous to the stationary observer O who is standing midway between A and B.

> The events do not appear to be simultaneous to observer O', who claims that the front of the car is struck before the rear.

Figure 9.4 (a) Two lightning bolts strike the ends of a moving boxcar. (b) Note that the leftward-traveling light signal from B' has already passed observer O', but the rightward-traveling light signal from A' has not yet reached O'.

Time Dilation

According to the preceding paragraph, observers in different inertial frames measure different time intervals between a pair of events, independent of the transit time of the light. This situation can be illustrated by considering a vehicle moving to the right with a speed v as in the pictorial representation in Active Figure 9.5a. A mirror is fixed to the ceiling of the vehicle, and observer O', at rest in a frame attached to the vehicle, holds a flashlight a distance d below the mirror. At some instant, the flashlight is turned on momentarily and emits a pulse of light (event 1) directed toward the mirror. At some later time after reflecting from the mirror, the pulse arrives back at the flashlight (event 2). Observer O' carries a clock that she uses to measure the time interval Δt_p between these two events. (The subscript p stands for "proper," as will be discussed shortly.) Because the light pulse has a constant speed c, the time interval required for the pulse to travel from O' to the mirror and back to O' (a distance of $2d$) can be found by modeling the light pulse as a particle under constant speed as discussed in Chapter 2:

$$\Delta t_p = \frac{2d}{c} \qquad \textbf{9.4} \blacktriangleleft$$

This time interval Δt_p is measured by O', for whom the two events occur at the same spatial position.

Now consider the same pair of events as viewed by observer O at rest with respect to a second frame attached to the ground as in Active Figure 9.5b. According to this observer, the mirror and flashlight are moving to the right with a speed v. The geometry appears to be entirely different as viewed by this observer. By the time the light from the flashlight reaches the mirror, the mirror has moved horizontally a distance $v\Delta t/2$, where Δt is the time interval required for the light to travel from the flashlight to the mirror and back to the flashlight as measured by observer O. In other words, the second observer concludes that because of the motion of the vehicle, if the light is to hit the mirror, it must leave the flashlight at an angle with respect to the vertical direction. Comparing Active Figures 9.5a and 9.5b, we see that the light must travel farther to arrive back at the mirror when observed in the second frame than in the first frame.

According to the second postulate of special relativity, both observers must measure c for the speed of light. Because the light travels farther in the second frame but at the same speed, it follows that the time interval Δt measured by the observer in the second frame is longer than the time interval Δt_p measured by the observer in

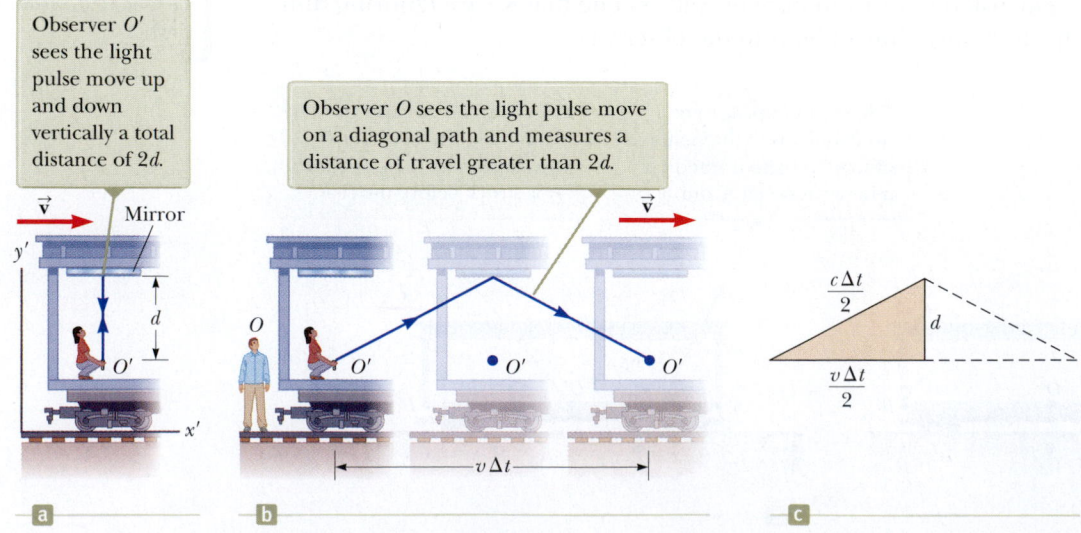

Active Figure 9.5 (a) A mirror is fixed to a moving vehicle, and a light pulse is sent out by observer O' at rest in the vehicle. (b) Relative to a stationary observer O standing alongside the vehicle, the mirror and O' move with a speed v. (c) The right triangle for calculating the relationship between Δt and Δt_p.

the first frame. To obtain a relationship between these two time intervals, it is convenient to use the right triangle geometric model shown in Active Figure 9.5c. The Pythagorean theorem applied to the triangle gives

$$\left(\frac{c\,\Delta t}{2}\right)^2 = \left(\frac{v\,\Delta t}{2}\right)^2 + d^2$$

Solving for Δt gives

$$\Delta t = \frac{2d}{\sqrt{c^2 - v^2}} = \frac{2d}{c\sqrt{1 - \dfrac{v^2}{c^2}}} \qquad \textbf{9.5} \blacktriangleleft$$

Because $\Delta t_p = 2d/c$, we can express Equation 9.5 as

$$\Delta t = \frac{\Delta t_p}{\sqrt{1 - \dfrac{v^2}{c^2}}} = \gamma \Delta t_p \qquad \textbf{9.6} \blacktriangleleft$$

where $\gamma = (1 - v^2/c^2)^{-1/2}$. This result says that the time interval Δt measured by O is longer than the time interval Δt_p measured by O' because γ is always greater than one. That is, $\Delta t > \Delta t_p$. This effect is known as **time dilation.**

We can see that time dilation is not observed in our everyday lives by considering the factor γ. This factor deviates significantly from a value of 1 only for very high speeds, as shown in Table 9.1. For example, for a speed of $0.1c$, the value of γ is 1.005. Therefore, a time dilation of only 0.5% occurs at one-tenth the speed of light. Speeds we encounter on an everyday basis are far slower than that, so we do not see time dilation in normal situations.

The time interval Δt_p in Equation 9.6 is called the **proper time interval.** In general, the proper time interval is defined as **the time interval between two events as measured by an observer for whom the events occur at the same point in space.** In our case, observer O' measures the proper time interval. For us to be able to use Equation 9.6, the events must occur at the same spatial position in *some* inertial frame. Therefore, for instance, this equation cannot be used to relate the measurements made by the two observers in the lightning example described at the beginning of this section because the lightning strikes occur at different positions for both observers.

If a clock is moving with respect to you, the time interval between ticks of the moving clock is measured to be longer than the time interval between ticks of an identical clock in your reference frame. Therefore, it is often said that a moving clock is measured to run more slowly than a clock in your reference frame by a factor γ. That is true for mechanical clocks as well as for the light clock just described. We can generalize this result by stating that all physical processes, including chemical and biological ones, slow down relative to a stationary clock when those processes occur in a frame moving with respect to the clock. For example, the heartbeat of an astronaut moving through space would keep time with a clock inside the spaceship. Both the astronaut's clock and heartbeat would be measured to be slowed down according to an observer on the Earth comparing time intervals with his own clock at rest with respect to him (although the astronaut would have no sensation of life slowing down in the spaceship).

Time dilation is a verifiable phenomenon; let us look at one situation in which the effects of time dilation can be observed and that served as an important historical confirmation of the predictions of relativity. Muons are unstable elementary particles that have a charge equal to that of the electron and a mass 207 times that of the electron. They decay into electrons and neutrinos, which we will study in Chapters 30 and 31. Muons can be produced as a result of collisions of cosmic radiation with atoms high in the atmosphere. Slow-moving muons in the laboratory have a lifetime measured to be the proper time interval $\Delta t_p = 2.2\ \mu s$. If we assume that the speed of atmospheric muons is close to the speed of

TABLE 9.1 |
Approximate Values for γ at Various Speeds

v/c	γ
0	1
0.001 0	1.000 000 5
0.010	1.000 05
0.10	1.005
0.20	1.021
0.30	1.048
0.40	1.091
0.50	1.155
0.60	1.250
0.70	1.400
0.80	1.667
0.90	2.294
0.92	2.552
0.94	2.931
0.96	3.571
0.98	5.025
0.99	7.089
0.995	10.01
0.999	22.37

Figure 9.6 Travel of muons according to an Earth-based observer.

Without relativistic considerations, according to an observer on the Earth, muons created in the atmosphere and traveling downward with a speed close to c travel only about 6.6×10^2 m before decaying with an average lifetime of 2.2 μs. Therefore, very few muons would reach the surface of the Earth.

With relativistic considerations, the muon's lifetime is dilated according to an observer on the Earth. Hence, according to this observer, the muon can travel about 4.8×10^3 m before decaying. The result is many of them arriving at the surface.

Muon is created

$\approx 6.6 \times 10^2$ m

Muon decays

Muon is created

$\approx 4.8 \times 10^3$ m

Muon decays

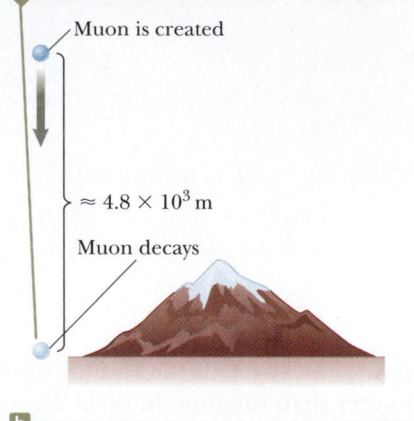

a

b

light, we find that these particles can travel a distance during their lifetime of approximately $(3.0 \times 10^8 \text{ m/s})(2.2 \times 10^{-6} \text{ s}) \approx 6.6 \times 10^2$ m before they decay (Fig. 9.6a). Hence, they are unlikely to reach the surface of the Earth from high in the atmosphere where they are produced; nonetheless, experiments show that a large number of muons *do* reach the surface. The phenomenon of time dilation explains this effect. As measured by an observer on the Earth, the muons have a dilated lifetime equal to $\gamma \Delta t_p$. For example, for $v = 0.99c$, $\gamma \approx 7.1$ and $\gamma \Delta t_p \approx 16$ μs. Hence, the average distance traveled by the muons in this time interval as measured by an observer on the Earth is approximately $(3.0 \times 10^8 \text{ m/s})$ $(16 \times 10^{-6} \text{ s}) \approx 4.8 \times 10^3$ m, as shown in Figure 9.6b.

The results of an experiment reported by J. C. Hafele and R. E. Keating provided direct evidence of time dilation.[5] The experiment involved the use of very stable atomic clocks. Time intervals measured with four such clocks in jet flight were compared with time intervals measured by reference clocks located at the U.S. Naval Observatory. Their results were in good agreement with the predictions of special relativity and can be explained in terms of the relative motion between the Earth's rotation and the jet aircraft. In their paper, Hafele and Keating report the following: "Relative to the atomic time scale of the U.S. Naval Observatory, the flying clocks lost 59 ± 10 ns during the eastward trip and gained 273 ± 7 ns during the westward trip."

In a more recent experiment, Chou, Hume, Rosenband, and Wineland[6] demonstrated time dilation with speeds as low as 10 m/s. Their experimental design included laser cooling of trapped ions, which we will discuss in Chapter 24.

> **QUICK QUIZ 9.1** Suppose the observer O' on the train in Active Figure 9.5 aims her flashlight at the far wall of the boxcar and turns it on and off, sending a pulse of light toward the far wall. Both O' and O measure the time interval between when the pulse leaves the flashlight and when it hits the far wall. Which observer measures the proper time interval between these two events? (a) O' (b) O (c) both observers (d) neither observer

[5] J. C. Hafele and R. E. Keating, "Around the World Atomic Clocks: Relativistic Time Gains Observed," *Science,* July 14, 1972, p. 168.

[6] C. Chou, D. Hume, T. Rosenband, and D. Wineland, "Optical Clocks and Relativity," *Science,* September 24, 2010, p. 1630

> **QUICK QUIZ 9.2** A crew on a spacecraft watches a movie that is two hours long. The spacecraft is moving at high speed through space. Does an Earth-based observer watching the movie screen on the spacecraft through a powerful telescope measure the duration of the movie to be (**a**) longer than, (**b**) shorter than, or (**c**) equal to two hours?

The Twin Paradox

An intriguing consequence of time dilation is the so-called twin paradox (Fig. 9.7). Consider an experiment involving a set of twins named Speedo and Goslo. At age 20, Speedo, the more adventuresome of the two, sets out on an epic journey to Planet X, located 20 light-years (ly) from the Earth. (Note that 1 ly is the distance light travels through free space in 1 year. It is equal to 9.46×10^{15} m.) Furthermore, his spaceship is capable of reaching a speed of $0.95c$ relative to the inertial frame of his twin brother back home. After reaching Planet X, Speedo becomes homesick and immediately returns to the Earth at the same speed $0.95c$. Upon his return, Speedo is shocked to discover that Goslo has aged 42 yr and is now 62 yr old. Speedo, on the other hand, has aged only 13 yr.

At this point, it is fair to raise the following question: Which twin is the traveler and which is really younger as a result of this experiment? From Goslo's frame of reference, he was at rest while his brother traveled at a high speed away from him and then came back. According to Speedo, however, he himself remained stationary while Goslo and the Earth raced away from him and then headed back. There is an apparent contradiction due to the apparent symmetry of the observations. Which twin has developed signs of excess aging?

The situation in this problem is actually not symmetrical. To resolve this apparent paradox, recall that the special theory of relativity describes observations made in inertial frames of reference moving relative to each other. Speedo, the space traveler, must experience a series of accelerations during his journey because he must fire his rocket engines to slow down and start moving back toward the Earth. As a result, his speed is not always uniform, and consequently he is not always in a single inertial frame. Therefore, there is no paradox because only Goslo, who is always in a single inertial frame, can make correct predictions based on special relativity. During each passing year noted by Goslo, slightly less than 4 months elapses for Speedo.

Only Goslo, who is in a single inertial frame, can apply the simple time dilation equation to Speedo's trip. Therefore, Goslo finds that instead of aging 42 yr, Speedo ages only $(1 - v^2/c^2)^{1/2}(42 \text{ yr}) = 13 \text{ yr}$. According to both twins, Speedo spends 6.5 yr traveling to Planet X and 6.5 yr returning, for a total travel time of 13 yr.

As Speedo (on the left) leaves his brother on Earth, both twins are the same age.

When Speedo returns from his journey, Goslo (on the right) is much older than Speedo.

a

b

Figure 9.7 The twin paradox. Speedo takes a journey to a star 20 light-years away and returns to the Earth.

QUICK QUIZ 9.3 Suppose astronauts are paid according to the amount of time they spend traveling in space. After a long voyage traveling at a speed approaching c, would a crew rather be paid according to (**a**) an Earth-based clock, (**b**) their spacecraft's clock, or (**c**) either clock?

THINKING PHYSICS 9.1

Suppose a student explains time dilation with the following argument: If I start running away from a clock at 12:00 at a speed very close to the speed of light, I would not see the time change, because the light from the clock representing 12:01 would never reach me. What is the flaw in this argument?

Reasoning The implication in this argument is that the velocity of light relative to the runner is approximately *zero* because "the light . . . would never reach me." In this Galilean point of view, the relative velocity is a simple subtraction of running velocity from the light velocity. From the point of view of special relativity, one of the fundamental postulates is that the speed of light is the same for all observers, *including one running away from the light source at the speed of light*. Therefore, the light from 12:01 will move toward the runner at the speed of light, as measured by all observers, including the runner. ◄

Example 9.1 | What Is the Period of the Pendulum?

The period of a pendulum is measured to be 3.00 s in the reference frame of the pendulum. What is the period when measured by an observer moving at a speed of 0.960c relative to the pendulum?

SOLUTION

Conceptualize Let's change frames of reference. Instead of the observer moving at 0.960c, we can take the equivalent point of view that the observer is at rest and the pendulum is moving at 0.960c past the stationary observer. Hence, the pendulum is an example of a clock moving at high speed with respect to an observer.

Categorize Based on the Conceptualize step, we can categorize this problem as one involving time dilation.

Analyze The proper time interval, measured in the rest frame of the pendulum, is $\Delta t_p = 3.00$ s.

Use Equation 9.6 to find the dilated time interval:

$$\Delta t = \gamma \Delta t_p = \frac{1}{\sqrt{1 - \dfrac{(0.960c)^2}{c^2}}} \Delta t_p = \frac{1}{\sqrt{1 - 0.921\,6}} \Delta t_p$$

$$= 3.57(3.00 \text{ s}) = \boxed{10.7 \text{ s}}$$

Finalize This result shows that a moving pendulum is indeed measured to take longer to complete a period than a pendulum at rest does. The period increases by a factor of $\gamma = 3.57$.

What If? What if the speed of the observer increases by 4.00%? Does the dilated time interval increase by 4.00%?

Answer Based on the highly nonlinear behavior of γ as a function of v exhibited in Table 9.1, we would guess that the increase in Δt would be different from 4.00%.

Find the new speed if it increases by 4.00%:

$$v_{\text{new}} = (1.040\,0)(0.960c) = 0.998\,4c$$

Perform the time dilation calculation again:

$$\Delta t = \gamma \Delta t_p = \frac{1}{\sqrt{1 - \dfrac{(0.998\,4c)^2}{c^2}}} \Delta t_p = \frac{1}{\sqrt{1 - 0.996\,8}} \Delta t_p$$

$$= 17.68(3.00 \text{ s}) = 53.1 \text{ s}$$

Therefore, the 4.00% increase in speed results in almost a 400% increase in the dilated time!

Length Contraction

The measured distance between two points also depends on the frame of reference. The **proper length** of an object is defined as **the distance in space between the end points of the object measured by someone who is at rest relative to the object.** An observer in a reference frame that is moving with respect to the object will measure a length along the direction of the velocity that is always less than the proper length. This effect is known as **length contraction.** Although we have introduced this effect through the mental representation of an object, the object is not necessary. The distance between *any* two points in space is measured by an observer to be contracted along the direction of the velocity of the observer relative to the points.

Consider a spacecraft traveling with a speed v from one star to another. We will consider the time interval between two events: (1) the leaving of the spacecraft from the first star and (2) the arrival of the spacecraft at the second star. There are two observers: one on the Earth and the other in the spacecraft. The observer at rest on the Earth (and also at rest with respect to the two stars) measures the distance between the stars to be L_p, the proper length. Using the particle under constant velocity model, according to this observer, the time interval required for the spacecraft to complete the voyage is $\Delta t = L_p/v$. What does an observer in the moving spacecraft measure for the distance between the stars? This observer measures the proper time interval because the passage of each of the two stars by his spacecraft occurs at the same position in his reference frame, at his spacecraft. Therefore, because of time dilation, the time interval required to travel between the stars as measured by the space traveler will be smaller than that for the Earth-bound observer, who is in motion with respect to the space traveler. Using the time dilation expression, the proper time interval between events is $\Delta t_p = \Delta t/\gamma$. The space traveler claims to be at rest and sees the destination star moving toward the spacecraft with speed v. Because the space traveler reaches the star in the time interval $\Delta t_p < \Delta t$, he concludes that the distance L between the stars is shorter than L_p. This distance measured by the space traveler is

$$L = v\,\Delta t_p = v\frac{\Delta t}{\gamma}$$

Because $L_p = v\Delta t$, we see that

$$L = \frac{L_p}{\gamma} = L_p\sqrt{1 - \frac{v^2}{c^2}} \qquad \textbf{9.7} \blacktriangleleft$$

Because $(1 - v^2/c^2)^{1/2}$ is less than 1, the space traveler measures a length that is shorter than the proper length. Therefore, an observer in motion with respect to two points in space measures the length L between the points (along the direction of motion) to be shorter than the length L_p measured by an observer at rest with respect to the points (the proper length).

Note that length contraction takes place only along the direction of motion. For example, suppose a meterstick moves past an Earth observer with speed v as in Active Figure 9.8. The length of the meterstick as measured by an observer in a frame attached to the stick is the proper length L_p as in Active Figure 9.8a. The length L of the stick measured by the Earth observer is shorter than L_p by the factor $(1 - v^2/c^2)^{1/2}$, but the width is the same. Furthermore, length contraction is a symmetric effect. If the stick is at rest on the Earth, an observer in the moving frame would also measure its length to be shorter by the same factor $(1 - v^2/c^2)^{1/2}$.

It is important to emphasize that the proper length and proper time interval are defined differently. The proper length is measured by an observer at rest with respect to the end points of the length. The proper time interval between two events is measured by someone for whom the events occur at the same position. Often, the proper time interval and the proper length are not measured by the same observer. As an example, let us return to the decaying muons moving at speeds close to the speed of light. An observer in the muon's reference frame would measure the proper lifetime, and an Earth-based observer would measure the proper length

A meterstick measured by an observer in a frame attached to the stick has its proper length L_p.

A meterstick measured by an observer in a frame in which the stick has a velocity relative to the frame is measured to be shorter than its proper length.

Active Figure 9.8 The length of a meterstick is measured by two observers.

(the distance from creation to decay in Fig. 9.6). In the muon's reference frame, no time dilation occurs, but the distance of travel is observed to be shorter when measured in this frame. Likewise, in the Earth observer's reference frame, a time dilation does occur, but the distance of travel is measured to be the proper length. Therefore, when calculations on the muon are performed in both frames, the outcome of the experiment in one frame is the same as the outcome in the other frame: More muons reach the surface than would be predicted without relativistic calculations.

> **QUICK QUIZ 9.4** You are packing for a trip to another star. During the journey, you will be traveling at $0.99c$. You are trying to decide whether you should buy smaller sizes of your clothing because you will be thinner on your trip due to length contraction. You also plan to save money by reserving a smaller cabin to sleep in because you will be shorter when you lie down. Should you (**a**) buy smaller sizes of clothing, (**b**) reserve a smaller cabin, (**c**) do neither of these things, or (**d**) do both of these things?

Example 9.2 | A Voyage to Sirius

An astronaut takes a trip to Sirius, which is located a distance of 8 light-years from the Earth. The astronaut measures the time of the one-way journey to be 6 years. If the spaceship moves at a constant speed of $0.8c$, how can the 8-ly distance be reconciled with the 6-year trip time measured by the astronaut?

SOLUTION

Conceptualize An observer on the Earth measures light to require 8 years to travel between Sirius and the Earth. The astronaut measures a time interval for his travel of only 6 years. Is the astronaut traveling faster than light?

Categorize Because the astronaut is measuring a length of space between the Earth and Sirius that is in motion with respect to her, we categorize this example as a length contraction problem. We also model the astronaut as a particle moving with constant velocity.

Analyze The distance of 8 ly represents the proper length from the Earth to Sirius measured by an observer on the Earth seeing both objects nearly at rest.

Calculate the contracted length measured by the astronaut using Equation 9.7:

$$L = \frac{8 \text{ ly}}{\gamma} = (8 \text{ ly})\sqrt{1 - \frac{v^2}{c^2}} = (8 \text{ ly})\sqrt{1 - \frac{(0.8c)^2}{c^2}} = 5 \text{ ly}$$

Use the particle under constant velocity model to find the travel time measured on the astronaut's clock:

$$\Delta t = \frac{L}{v} = \frac{5 \text{ ly}}{0.8c} = \frac{5 \text{ ly}}{0.8(1 \text{ ly/yr})} = 6 \text{ yr}$$

Finalize Notice that we have used the value for the speed of light as $c = 1$ ly/yr. The trip takes a time interval shorter than 8 years for the astronaut because, to her, the distance between the Earth and Sirius is measured to be shorter.

What If? What if this trip is observed with a very powerful telescope by a technician in Mission Control on the Earth? At what time will this technician *see* that the astronaut has arrived at Sirius?

Answer The time interval the technician measures for the astronaut to arrive is

$$\Delta t = \frac{L_p}{v} = \frac{8 \text{ ly}}{0.8c} = 10 \text{ yr}$$

For the technician to *see* the arrival, the light from the scene of the arrival must travel back to the Earth and enter the telescope. This travel requires a time interval of

$$\Delta t = \frac{L_p}{v} = \frac{8 \text{ ly}}{c} = 8 \text{ yr}$$

Therefore, the technician sees the arrival after 10 yr + 8 yr = 18 yr. If the astronaut immediately turns around and comes back home, she arrives, according to the technician, 20 years after leaving, only 2 years *after the technician saw her arrive!* In addition, the astronaut would have aged by only 12 years.

Example 9.3 | Speedy Plunge

An observer on Earth sees a spaceship at an altitude of 4 350 km moving downward toward the Earth with a speed of $0.970c$.

(A) What is the distance from the spaceship to the Earth as measured by the spaceship's captain?

SOLUTION

Conceptualize Imagine you are the captain, at rest in a reference frame attached to the spaceship: the Earth is rushing toward you at $0.970c$; hence, the distance between the spaceship and the Earth is contracted.

Cagegorize We have an observer (the captain) and a moving length in space (the Earth–spaceship distance), so we categorize this example as a length contraction problem. The proper length is 4 350 km as measured by the Earth observer.

Analyze Use Equation 9.7 to find the contracted length, which represents the altitude of the spaceship above the surface of the Earth as measured by the captain:

$$L = L_p\sqrt{1 - v^2/c^2} = (4\ 350\ \text{km})\sqrt{1 - (0.970c)^2/c^2}$$

$$= 1.06 \times 10^3\ \text{km}$$

(B) After firing his engines for a time interval to slow down, the captain measures the ship's altitude as 267 km, whereas the observer on Earth measures it to be 625 km. What is the speed of the spaceship at this instant?

SOLUTION

Analyze Write the length-contraction equation (Eq.9.7):

$$L = L_p\sqrt{1 - v^2/c^2}$$

Square both sides of this equation and solve for v:

$$L^2 = L_p^2(1 - v^2/c^2) \quad \rightarrow \quad 1 - v^2/c^2 = \left(\frac{L}{L_p}\right)^2$$

$$v = c\sqrt{1 - (L/L_p)^2} = c\sqrt{1 - (267\ \text{km}/625\ \text{km})^2}$$

$$v = 0.904c$$

Finalize The answers are consistent with our expectations. The length in part (A) is shorter than the proper length, as we would expect from the length contraction phenomenon. In part (B), the calculated speed is indeed lower than the original speed, consistent with the fact that the captain fired the rocket engines to slow down.

9.5 | The Lorentz Transformation Equations

Suppose an event that occurs at some location and time is reported by two observers: one at rest in a frame S and another in a frame S′ that is moving to the right with speed v as in Figure 9.9. The observer in S reports the event with space–time coordinates (x, y, z, t), and the observer in S′ reports the same event using the coordinates (x', y', z', t'). If two events occur at P and Q in Figure 9.9, Equation 9.1 predicts that $\Delta x = \Delta x'$; that is, the distance between the two points in space at which the events occur does not depend on the motion of the observer. Because this notion is contradictory to that of length contraction, the Galilean transformation is not valid when v approaches the speed of light. In this section, we state the correct transformation equations that apply for all speeds in the range $0 \leq v < c$.

The equations that relate these measurements and enable us to transform coordinates from S to S′ are the **Lorentz transformation equations:**

$$x' = \gamma(x - vt) \qquad y' = y \qquad z' = z \qquad t' = \gamma\left(t - \frac{v}{c^2}x\right) \qquad \textbf{9.8} \blacktriangleleft$$

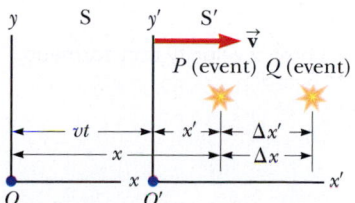

Figure 9.9 Events occur at points P and Q and are observed by an observer at rest in the S frame and another in the S′ frame, which is moving to the right with a speed v.

► Lorentz transformation for
S → S′

These transformation equations were developed by Hendrik A. Lorentz (1853–1928) in 1890 in connection with electromagnetism. Einstein, however, recognized their physical significance and took the bold step of interpreting them within the framework of special relativity.

We see that the value for t' assigned to an event by observer O' depends both on the time t and on the coordinate x as measured by observer O. Therefore, in relativity, space and time are not separate concepts but rather are closely interwoven with each other in what we call **space–time.** This case is unlike that of the Galilean transformation in which $t = t'$.

If we wish to transform coordinates in the S′ frame to coordinates in the S frame, we simply replace v by $-v$ and interchange the primed and unprimed coordinates in Equation 9.8:

▶ Inverse Lorentz transformation
for S′ → S

$$x = \gamma(x' + vt') \qquad y = y' \qquad z = z' \qquad t = \gamma\left(t' + \frac{v}{c^2}x'\right) \qquad \textbf{9.9} \blacktriangleleft$$

When $v \ll c$, the Lorentz transformation reduces to the Galilean transformation. To check, note that if $v \ll c$, $v^2/c^2 \ll 1$, so γ approaches 1 and Equation 9.8 reduces in this limit to Equations 9.1 and 9.2:

$$x' = x - vt \qquad y' = y \qquad z' = z \qquad t' = t$$

Lorentz Velocity Transformation

Let us now derive the **Lorentz velocity transformation,** which is the relativistic counterpart of the Galilean velocity transformation, Equation 9.3. Once again S′ is a frame of reference that moves at a speed v relative to another frame S along the common x and x' axes. Suppose an object is measured in S′ to have an instantaneous velocity component u'_x given by

$$u'_x = \frac{dx'}{dt'} \qquad \textbf{9.10} \blacktriangleleft$$

Using Equations 9.8, we have

$$dx' = \gamma(dx - v\,dt) \qquad \text{and} \qquad dt' = \gamma\left(dt - \frac{v}{c^2}dx\right)$$

Substituting these values into Equation 9.10 gives

$$u'_x = \frac{dx'}{dt'} = \frac{dx - v\,dt}{dt - \frac{v}{c^2}dx} = \frac{\frac{dx}{dt} - v}{1 - \frac{v}{c^2}\frac{dx}{dt}}$$

Note, though, that dx/dt is the velocity component u_x of the object measured in S, so this expression becomes

▶ Lorentz velocity transformation
for S → S′

$$u'_x = \frac{u_x - v}{1 - \frac{u_x v}{c^2}} \qquad \textbf{9.11} \blacktriangleleft$$

Similarly, if the object has velocity components along y and z, the components in S′ are

$$u'_y = \frac{u_y}{\gamma\left(1 - \frac{u_x v}{c^2}\right)} \qquad \text{and} \qquad u'_z = \frac{u_z}{\gamma\left(1 - \frac{u_x v}{c^2}\right)} \qquad \textbf{9.12} \blacktriangleleft$$

When u_x or v is much smaller than c (the nonrelativistic case), the denominator of Equation 9.11 approaches unity and so $u'_x \approx u_x - v$. This result corresponds to the Galilean velocity transformations. In the other extreme, when $u_x = c$, Equation 9.11 becomes

$$u'_x = \frac{c - v}{1 - \frac{cv}{c^2}} = \frac{c\left(1 - \frac{v}{c}\right)}{1 - \frac{v}{c}} = c$$

From this result, we see that an object whose speed approaches c relative to an observer in S also has a speed approaching c relative to an observer in S', independent of the relative motion of S and S'. Note that this conclusion is consistent with Einstein's second postulate, namely, that the speed of light must be c relative to all inertial frames of reference.

To obtain u_x in terms of u'_x, we replace v by $-v$ in Equation 9.11 and interchange the roles of primed and unprimed variables:

$$u_x = \frac{u'_x + v}{1 + \dfrac{u'_x v}{c^2}}$$ **9.13** ◀   ▶ Inverse Lorentz velocity transformation for S' → S

QUICK QUIZ 9.5 You are driving on a freeway at a relativistic speed. Straight ahead of you, a technician standing on the ground turns on a searchlight and a beam of light moves exactly vertically upward, as seen by the technician. As you observe the beam of light, you measure the magnitude of the vertical component of its velocity as (a) equal to c, (b) greater than c, or (c) less than c. If the technician aims the searchlight directly at you instead of upward, you measure the magnitude of the horizontal component of its velocity as (d) equal to c, (e) greater than c, or (f) less than c.

Example 9.4 | Relative Velocity of Two Spacecraft

Two spacecraft A and B are moving in opposite directions as shown in Figure 9.10. An observer on the Earth measures the speed of spacecraft A to be $0.750c$ and the speed of spacecraft B to be $0.850c$. Find the velocity of spacecraft B as observed by the crew on spacecraft A.

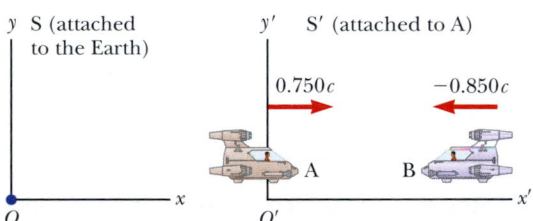

Figure 9.10 (Example 9.4) Two spacecraft A and B move in opposite directions. The speed of spacecraft B relative to spacecraft A is *less* than c and is obtained from the relativistic velocity transformation equation.

SOLUTION

Conceptualize There are two observers, one (*O*) on the Earth and one (*O'*) on spacecraft A. The event is the motion of spacecraft B.

Categorize Because the problem asks to find an observed velocity, we categorize this example as one requiring the Lorentz velocity transformation.

Analyze The Earth-based observer at rest in the S frame makes two measurements, one of each spacecraft. We want to find the velocity of spacecraft B as measured by the crew on spacecraft A. Therefore, $u_x = -0.850c$. The velocity of spacecraft A is also the velocity of the observer at rest in spacecraft A (the S' frame) relative to the observer at rest on the Earth. Therefore, $v = 0.750c$.

Obtain the velocity u'_x of spacecraft B relative to spacecraft A using Equation 9.11:

$$u'_x = \frac{u_x - v}{1 - \dfrac{u_x v}{c^2}} = \frac{-0.850c - 0.750c}{1 - \dfrac{(-0.850c)(0.750c)}{c^2}} = -0.977c$$

Finalize The negative sign indicates that spacecraft B is moving in the negative x direction as observed by the crew on spacecraft A. Is that consistent with your expectation from Figure 9.10? Notice that the speed is less than c. That is, an object whose speed is less than c in one frame of reference must have a speed less than c in any other frame. (Had you used the Galilean velocity transformation equation in this example, you would have found that $u'_x = u_x - v = -0.850c - 0.750c = -1.60c$, which is impossible. The Galilean transformation equation does not work in relativistic situations.)

What If? What if the two spacecraft pass each other? What is their relative speed now?

Answer The calculation using Equation 9.11 involves only the velocities of the two spacecraft and does not depend on their locations. After they pass each other, they have the same velocities, so the velocity of spacecraft B as observed by the crew on spacecraft A is the same, $-0.977c$. The only difference after they pass is that spacecraft B is receding from spacecraft A, whereas it was approaching spacecraft A before it passed.

Example 9.5 | Relativistic Leaders of the Pack

Two motorcycle pack leaders named David and Emily are racing at relativistic speeds along perpendicular paths as shown in Figure 9.11. How fast does Emily recede as seen by David over his right shoulder?

SOLUTION

Conceptualize The two observers are David and the police officer in Figure 9.11. The event is the motion of Emily. Figure 9.11 represents the situation as seen by the police officer at rest in frame S. Frame S′ moves along with David.

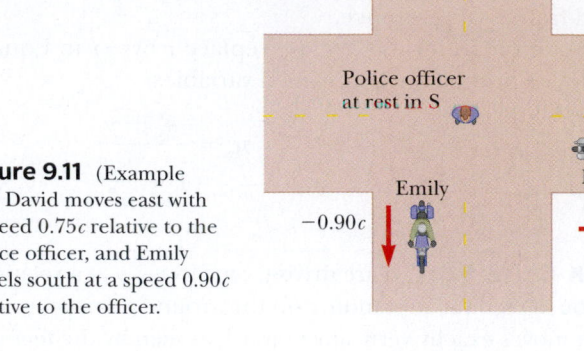

Figure 9.11 (Example 9.5) David moves east with a speed $0.75c$ relative to the police officer, and Emily travels south at a speed $0.90c$ relative to the officer.

Categorize Because the problem asks to find an observed velocity, we categorize this problem as one requiring the Lorentz velocity transformation. The motion takes place in two dimensions.

Analyze Identify the velocity components for David and Emily according to the police officer:

David: $v_x = v = 0.75c$ $v_y = 0$

Emily: $u_x = 0$ $u_y = -0.90c$

Using Equations 9.11 and 9.12, calculate u'_x and u'_y for Emily as measured by David:

$$u'_x = \frac{u_x - v}{1 - \frac{u_x v}{c^2}} = \frac{0 - 0.75c}{1 - \frac{(0)(0.75c)}{c^2}} = -0.75c$$

$$u'_y = \frac{u_y}{\gamma\left(1 - \frac{u_x v}{c^2}\right)} = \frac{\sqrt{1 - \frac{(0.75c)^2}{c^2}}\,(-0.90c)}{1 - \frac{(0)(0.75c)}{c^2}} = -0.60c$$

Using the Pythagorean theorem, find the speed of Emily as measured by David:

$$u' = \sqrt{(u'_x)^2 + (u'_y)^2} = \sqrt{(-0.75c)^2 + (-0.60c)^2} = \boxed{0.96c}$$

Finalize This speed is less than c, as required by the special theory of relativity.

▌9.6 | Relativistic Momentum and the Relativistic Form of Newton's Laws

We have seen that to describe the motion of particles within the framework of special relativity properly, the Galilean transformation must be replaced by the Lorentz transformation. Because the laws of physics must remain unchanged under the Lorentz transformation, we must generalize Newton's laws and the definitions of momentum and energy to conform to the Lorentz transformation and the principle of relativity. These generalized definitions should reduce to the classical (nonrelativistic) definitions for $v \ll c$ or $u \ll c$. (As we have done previously, we will use v for the speed of one reference frame relative to another and u for the speed of a particle.)

First, recall one of our isolated system models: the total momentum of an isolated system of particles is conserved. Suppose a collision between two particles is described in a reference frame S in which the momentum of the system is measured to be conserved. If the velocities in a second reference frame S′ are calculated using the Lorentz transformation and the Newtonian definition of momentum, $\vec{\mathbf{p}} = m\vec{\mathbf{u}}$, is used, it is found that the momentum of the system is *not* measured to be conserved in the second reference frame. This finding violates one of Einstein's postulates: The laws of physics are the same in all inertial frames. Therefore, assuming the Lorentz transformation to be correct, we must modify the definition of momentum.

The relativistic equation for the momentum of a particle of mass m that maintains the principle of conservation of momentum is

$$\vec{\mathbf{p}} \equiv \frac{m\vec{\mathbf{u}}}{\sqrt{1 - \dfrac{u^2}{c^2}}}$$

9.14 ◄

► Definition of relativistic momentum

where $\vec{\mathbf{u}}$ is the velocity of the particle. When u is much less than c, the denominator of Equation 9.14 approaches unity, so $\vec{\mathbf{p}}$ approaches $m\vec{\mathbf{u}}$. Therefore, the relativistic equation for $\vec{\mathbf{p}}$ reduces to the classical expression when u is small compared with c. Equation 9.14 is often written in simpler form as

$$\vec{\mathbf{p}} = \gamma m\vec{\mathbf{u}}$$

9.15 ◄

using our previously defined expression[7] for γ.

The relativistic force $\vec{\mathbf{F}}$ on a particle whose momentum is $\vec{\mathbf{p}}$ is defined as

$$\vec{\mathbf{F}} \equiv \frac{d\vec{\mathbf{p}}}{dt}$$

9.16 ◄

where $\vec{\mathbf{p}}$ is given by Equation 9.14. This expression preserves both classical mechanics in the limit of low velocities and conservation of momentum for an isolated system ($\Sigma \vec{\mathbf{F}}_{ext} = 0$) both relativistically and classically.

We leave it to Problem 9.56 in Enhanced WebAssign to show that the acceleration $\vec{\mathbf{a}}$ of a particle decreases under the action of a constant force, in which case $a \propto (1 - u^2/c^2)^{3/2}$. From this proportionality, note that as the particle's speed approaches c, the acceleration caused by any finite force approaches zero. It is therefore impossible to accelerate a particle from rest to a speed $u \geq c$.

Hence, c is an upper limit for the speed of any particle. In fact, it is possible to show that no *matter, energy,* or *information* can travel through space faster than c. Note that the relative speeds of the two spacecraft in Example 9.4 and the two motorcyclists in Example 9.5 were both less than c. If we had attempted to solve these examples with Galilean transformations, we would have obtained relative speeds larger than c in both cases.

Example 9.6 | Linear Momentum of an Electron

An electron, which has a mass of 9.11×10^{-31} kg, moves with a speed of $0.750c$. Find the magnitude of its relativistic momentum and compare this value with the momentum calculated from the classical expression.

SOLUTION

Conceptualize Imagine an electron moving with high speed. The electron carries momentum, but the magnitude of its momentum is not given by $p = mu$ because the speed is relativistic.

Categorize We categorize this example as a substitution problem involving a relativistic equation.

Use Equation 9.14 with $u = 0.750c$ to find the magnitude of the momentum:

$$p = \frac{m_e u}{\sqrt{1 - \dfrac{u^2}{c^2}}}$$

$$p = \frac{(9.11 \times 10^{-31}\ \text{kg})(0.750)(3.00 \times 10^8\ \text{m/s})}{\sqrt{1 - \dfrac{(0.750c)^2}{c^2}}}$$

$$= 3.10 \times 10^{-22}\ \text{kg} \cdot \text{m/s}$$

The classical expression (used incorrectly here) gives $p_{classical} = m_e u = 2.05 \times 10^{-22}$ kg · m/s. Hence, the correct relativistic result is 50% greater than the classical result!

---

[7] We defined γ previously in terms of the speed v of one frame relative to another frame. The same symbol is also used for $(1 - u^2/c^2)^{-1/2}$, where u is the speed of a particle.

◀9.7 | Relativistic Energy

We have seen that the definition of momentum requires generalization to make it compatible with the principle of relativity. We find that the definition of kinetic energy must also be modified.

To derive the relativistic form of the work–kinetic energy theorem, let us start with the definition of the work done by a force of magnitude F on a particle initially at rest. Recall from Chapter 6 that the work–kinetic energy theorem states that in the appropriate simple situation, the work done by a net force acting on a particle equals the change in kinetic energy of the particle. Because the initial kinetic energy is zero, we conclude that the work W done in accelerating a particle from rest is equivalent to the relativistic kinetic energy K of the particle:

$$W = \Delta K = K - 0 = K = \int_{x_1}^{x_2} F\, dx = \int_{x_1}^{x_2} \frac{dp}{dt}\, dx \qquad \textbf{9.17}◀$$

where we are considering the special case of force and displacement vectors along the x axis for simplicity. To perform this integration and find the relativistic kinetic energy as a function of u, we first evaluate dp/dt, using Equation 9.14:

$$\frac{dp}{dt} = \frac{d}{dt}\frac{mu}{\sqrt{1 - \dfrac{u^2}{c^2}}} = \frac{m(du/dt)}{\left(1 - \dfrac{u^2}{c^2}\right)^{3/2}}$$

Substituting this expression for dp/dt and $dx = u\, dt$ into Equation 9.17 gives

$$K = \int_0^t \frac{m(du/dt)\, u\, dt}{\left(1 - \dfrac{u^2}{c^2}\right)^{3/2}} = m \int_0^u \frac{u}{\left(1 - \dfrac{u^2}{c^2}\right)^{3/2}}\, du$$

Evaluating the integral, we find that

▶ Relativistic kinetic energy

$$K = \frac{mc^2}{\sqrt{1 - \dfrac{u^2}{c^2}}} - mc^2 = \gamma mc^2 - mc^2 = (\gamma - 1)mc^2 \qquad \textbf{9.18}◀$$

> The relativistic calculation, using Equation 9.18, shows correctly that u is always less than c.

> The nonrelativistic calculation, using $K = \frac{1}{2}mu^2$, predicts a parabolic curve and the speed u grows without limit.

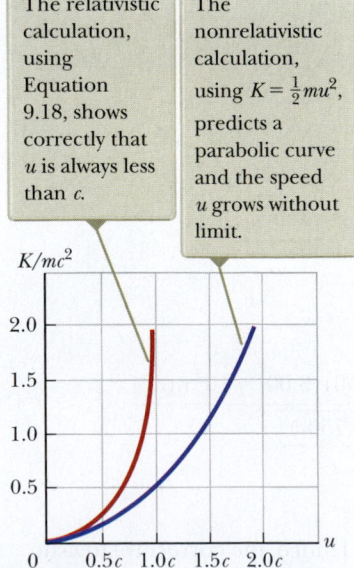

Figure 9.12 A graph comparing relativistic and nonrelativistic kinetic energy of a moving particle. The energies are plotted as a function of speed u.

At low speeds, where $u/c \ll 1$, Equation 9.18 should reduce to the classical expression $K = \frac{1}{2}mu^2$. We can show this reduction by using the binomial expansion $(1 - x^2)^{-1/2} \approx 1 + \frac{1}{2}x^2 + \cdots$ for $x \ll 1$, where the higher-order powers of x are ignored in the expansion because they are so small. In our case, $x = u/c$, so

$$\gamma = \frac{1}{\sqrt{1 - \dfrac{u^2}{c^2}}} = \left(1 - \frac{u^2}{c^2}\right)^{-1/2} \approx 1 + \frac{1}{2}\frac{u^2}{c^2} + \cdots$$

Substituting into Equation 9.18 gives

$$K \approx \left(1 + \frac{1}{2}\frac{u^2}{c^2} + \cdots\right)mc^2 - mc^2 = \frac{1}{2}mu^2$$

which agrees with the classical result. Figure 9.12 shows a comparison of the speed–kinetic energy relationships for a particle using the nonrelativistic expression for K (the blue curve) and the relativistic expression for K (the brown curve). The curves are in good agreement at low speeds, but deviate at higher speeds. The nonrelativistic expression indicates a violation of special relativity because it suggests that sufficient energy can be added to the particle to accelerate it to a speed larger than c. In the relativistic case, the particle speed never exceeds c, regardless of the kinetic energy, which is consistent with experimental results. When an object's speed is less than one-tenth the speed of light, the classical kinetic energy equation differs by less than 1% from the relativistic equation (which is experimentally verified at all speeds). Therefore, for practical calculations it is valid to use the classical equation when the object's speed is less than $0.1c$.

The constant term mc^2 in Equation 9.18, which is independent of the speed, is called the **rest energy** E_R of the particle:

$$E_R = mc^2 \qquad \text{9.19} \blacktriangleleft$$

▶ Rest energy

The term γmc^2 in Equation 9.18 depends on the particle speed and is the sum of the kinetic and rest energies. We define γmc^2 to be the **total energy** E; that is, total energy = kinetic energy + rest energy:

$$E = \gamma mc^2 = K + mc^2 = K + E_R \qquad \text{9.20} \blacktriangleleft$$

or, when γ is replaced by its equivalent,

$$E = \frac{mc^2}{\sqrt{1 - \dfrac{u^2}{c^2}}} \qquad \text{9.21} \blacktriangleleft$$

▶ Total energy of a relativistic particle

The relation $E_R = mc^2$ shows that **mass is a manifestation of energy.** It also shows that a small mass corresponds to an enormous amount of energy. This concept is fundamental to much of the field of nuclear physics.

In many situations, the momentum or energy of a particle is measured rather than its speed. It is therefore useful to have an expression relating the total energy E to the relativistic momentum p, which is accomplished by using the expressions $E = \gamma mc^2$ and $p = \gamma mu$. By squaring these equations and subtracting, we can eliminate u (see Problem 9.36 in Enhanced WebAssign). The result, after some algebra, is

$$E^2 = p^2c^2 + (mc^2)^2 \qquad \text{9.22} \blacktriangleleft$$

▶ Energy–momentum relationship for a relativistic particle

When the particle is at rest, $p = 0$, and so $E = E_R = mc^2$. That is, the total energy equals the rest energy.

For the case of particles that have zero mass, such as photons (massless, charge-less particles of light to be discussed further in Chapter 28), we set $m = 0$ in Equation 9.22 and see that

$$E = pc \qquad \text{9.23} \blacktriangleleft$$

This equation is an exact expression relating energy and momentum for photons, which always travel at the speed of light.

When dealing with subatomic particles, it is convenient to express their energy in a unit called an *electron volt* (eV). The equality between electron volts and our standard energy unit is

$$1 \text{ eV} = 1.602 \times 10^{-19} \text{ J}$$

For example, the mass of an electron is 9.11×10^{-31} kg. Hence, the rest energy of the electron is

$$E_R = m_e c^2 = (9.11 \times 10^{-31} \text{ kg})(3.00 \times 10^8 \text{ m/s})^2 = 8.20 \times 10^{-14} \text{ J}$$

Converting to eV, we have

$$E_R = m_e c^2 = (8.20 \times 10^{-14} \text{ J}) \left(\frac{1 \text{ eV}}{1.602 \times 10^{-19} \text{ J}} \right) = 0.511 \text{ MeV}$$

> **QUICK QUIZ 9.6** The following *pairs* of energies—particle 1: E, $2E$; particle 2: E, $3E$; particle 3: $2E$, $4E$—represent the rest energy and total energy of three different particles. Rank the particles from greatest to least according to their (**a**) mass, (**b**) kinetic energy, and (**c**) speed.

Example 9.7 | The Energy of a Speedy Proton

(A) Find the rest energy of a proton in units of electron volts.

SOLUTION

Conceptualize Even if the proton is not moving, it has energy associated with its mass. If it moves, the proton possesses more energy, with the total energy being the sum of its rest energy and its kinetic energy.

Categorize The phrase "rest energy" suggests we must take a relativistic rather than a classical approach to this problem.

Analyze Use Equation 9.19 to find the rest energy:

$$E_R = m_p c^2 = (1.673 \times 10^{-27} \text{ kg})(2.998 \times 10^8 \text{ m/s})^2$$

$$= (1.504 \times 10^{-10} \text{ J})\left(\frac{1.00 \text{ eV}}{1.602 \times 10^{-19} \text{ J}}\right) = \boxed{938 \text{ MeV}}$$

(B) If the total energy of a proton is three times its rest energy, what is the speed of the proton?

SOLUTION

Use Equation 9.21 to relate the total energy of the proton to the rest energy:

$$E = 3m_p c^2 = \frac{m_p c^2}{\sqrt{1 - \dfrac{u^2}{c^2}}} \quad \rightarrow \quad 3 = \frac{1}{\sqrt{1 - \dfrac{u^2}{c^2}}}$$

Solve for u:

$$1 - \frac{u^2}{c^2} = \frac{1}{9} \quad \rightarrow \quad \frac{u^2}{c^2} = \frac{8}{9}$$

$$u = \frac{\sqrt{8}}{3} c = 0.943c = \boxed{2.83 \times 10^8 \text{ m/s}}$$

(C) Determine the kinetic energy of the proton in units of electron volts.

SOLUTION

Use Equation 9.20 to find the kinetic energy of the proton:

$$K = E - m_p c^2 = 3m_p c^2 - m_p c^2 = 2m_p c^2$$

$$= 2(938 \text{ MeV}) = \boxed{1.88 \times 10^3 \text{ MeV}}$$

(D) What is the proton's momentum?

SOLUTION

Use Equation 9.22 to calculate the momentum:

$$E^2 = p^2 c^2 + (m_p c^2)^2 = (3m_p c^2)^2$$

$$p^2 c^2 = 9(m_p c^2)^2 - (m_p c^2)^2 = 8(m_p c^2)^2$$

$$p = \sqrt{8} \frac{m_p c^2}{c} = \sqrt{8} \frac{938 \text{ MeV}}{c} = \boxed{2.65 \times 10^3 \text{ MeV}/c}$$

Finalize The unit of momentum in part (D) is written MeV/c, which is a common unit in particle physics. For comparison, you might want to solve this example using classical equations.

9.8 | Mass and Energy

Equation 9.20, $E = \gamma mc^2$, which represents the total energy of a particle, suggests that even when a particle is at rest ($\gamma = 1$) it still possesses enormous energy through its mass. The clearest experimental proof of the equivalence of mass and energy occurs in nuclear and elementary particle interactions in which the conversion of mass into kinetic energy takes place. Hence, we cannot use the principle of conservation of energy in relativistic situations exactly as it is outlined in Chapter 7. We must include rest energy as another form of energy storage.

This concept is important in atomic and nuclear processes, in which the change in mass during the process is on the order of the initial mass. For example, in a conventional nuclear reactor, the uranium nucleus undergoes *fission*, a reaction that results in several lighter fragments having considerable kinetic energy. In the case of a ^{235}U atom, which is used as fuel in nuclear power plants, the fragments are two lighter nuclei and a few neutrons. The total mass of the fragments is less than that of the ^{235}U by an amount Δm. The corresponding energy Δmc^2 associated with this mass difference is exactly equal to the total kinetic energy of the fragments. The kinetic energy is transferred by collisions with water molecules as the fragments move through water, raising the internal energy of the water. This internal energy is used to produce steam for the generation of electric power.

Next, consider a basic *fusion* reaction in which two deuterium atoms combine to form one helium atom. The decrease in mass that results from the creation of one helium atom from two deuterium atoms is $\Delta m = 4.25 \times 10^{-29}$ kg. Hence, the corresponding energy that results from one fusion reaction is calculated to be $\Delta mc^2 = 3.83 \times 10^{-12}$ J $= 23.9$ MeV. To appreciate the magnitude of this result, consider that if 1 g of deuterium is converted to helium, the energy released is on the order of 10^{12} J! At the year 2012 cost of energy transferred from power plants in the United States by electrical transmission, this energy would be worth about $32 000. We will see more details of these nuclear processes in Chapter 30.

Example 9.8 | Mass Change in a Radioactive Decay

The ^{216}Po nucleus is unstable and exhibits radioactivity (Chapter 30). It decays to ^{212}Pb by emitting an alpha particle, which is a helium nucleus, ^{4}He. The relevant masses are $m_i = m(^{216}\text{Po}) = 216.001\ 915$ u and $m_f = m(^{212}\text{Pb}) + m(^4\text{He}) = 211.991\ 898$ u $+ 4.002\ 603$ u. The unit u is an *atomic mass unit*, where 1 u $= 1.660 \times 10^{-27}$ kg.

(A) Find the mass change of the system in this decay.

SOLUTION

Conceptualize The initial system is the ^{216}Po nucleus. Imagine the mass of the system decreasing during the decay and transforming to kinetic energy of the alpha particle and the ^{212}Pb nucleus after the decay.

Categorize We use concepts discussed in this section, so we categorize this example as a substitution problem.

Calculate the change in mass using the mass values given in the problem statement:

$$\Delta m = 216.001\ 915\ \text{u} - (211.991\ 898\ \text{u} + 4.002\ 603\ \text{u})$$

$$= 0.007\ 414\ \text{u} = 1.23 \times 10^{-29}\ \text{kg}$$

(B) Find the energy this mass change represents.

SOLUTION

Use Equation 9.19 to find the energy associated with this mass change:

$$E = \Delta mc^2 = (1.23 \times 10^{-29}\ \text{kg})(3.00 \times 10^8\ \text{m/s})^2$$

$$= 1.11 \times 10^{-12}\ \text{J} = 6.92\ \text{MeV}$$

9.9 | General Relativity

Up to this point, we have sidestepped a curious puzzle. Mass has two seemingly different properties: It determines a force of mutual gravitational attraction between two objects (Newton's law of universal gravitation), and it also represents the resistance of a single object to being accelerated (Newton's second law), regardless of the type of force producing the acceleration. How can one quantity have two such different properties? An answer to this question, which puzzled

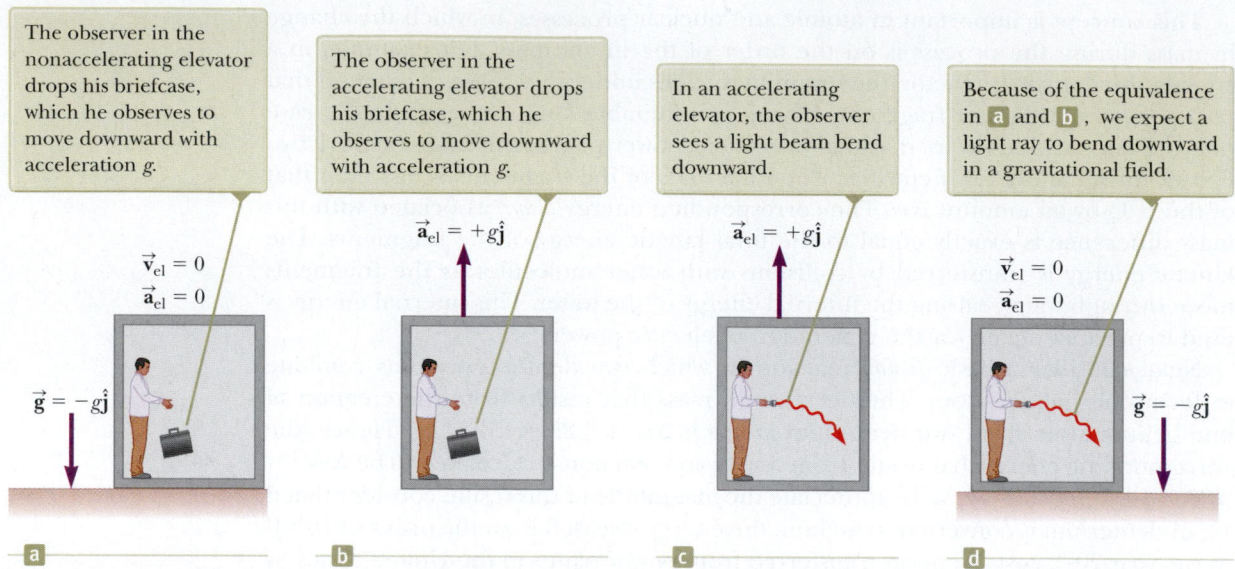

The observer in the nonaccelerating elevator drops his briefcase, which he observes to move downward with acceleration g.

The observer in the accelerating elevator drops his briefcase, which he observes to move downward with acceleration g.

In an accelerating elevator, the observer sees a light beam bend downward.

Because of the equivalence in **a** and **b** , we expect a light ray to bend downward in a gravitational field.

Figure 9.13 (a) The observer is at rest in an elevator in a uniform gravitational field $\vec{\mathbf{g}} = -g\hat{\mathbf{j}}$, directed downward. (b) The observer is in a region where gravity is negligible, but the elevator moves upward with an acceleration $\vec{\mathbf{a}}_{el} = +g\hat{\mathbf{j}}$. According to Einstein, the frames of reference in parts (a) and (b) are equivalent in every way. No local experiment can distinguish any difference between the two frames. (c) An observer watches a beam of light in an accelerating elevator. (d) Einstein's prediction of the behavior of a beam of light in a gravitational field.

Newton and many other physicists over the years, was provided when Einstein published his theory of gravitation, known as *general relativity*, in 1916. Because it is a mathematically complex theory, we offer merely a hint of its elegance and insight.

In Einstein's view, the dual behavior of mass was evidence for a very intimate and basic connection between the two behaviors. He pointed out that no mechanical experiment (e.g., dropping an object) could distinguish between the two situations illustrated in Figures 9.13a and 9.13b. In Figure 9.13a, a person is standing in an elevator on the surface of a planet and feels pressed into the floor due to the gravitational force. If he releases his briefcase, he observes it moving toward the floor with acceleration $\vec{\mathbf{g}} = -g\hat{\mathbf{j}}$. In Figure 9.13b, the person is in an elevator in empty space accelerating upward with $\vec{\mathbf{a}}_{el} = +g\hat{\mathbf{j}}$. The person feels pressed into the floor with the same force as in Figure 9.13a. If he releases his briefcase, he observes it moving toward the floor with acceleration g, just as in the previous situation. In each case, an object released by the observer undergoes a downward acceleration of magnitude g relative to the floor. In Figure 9.13a, the person is at rest in an inertial frame in a gravitational field due to the planet. (A gravitational field exists around any object with mass, such as a planet. We will define the gravitational field formally in Chapter 11.) In Figure 9.13b, the person is in a noninertial frame accelerating in gravity-free space. Einstein's claim is that these two situations are completely equivalent.

Einstein carried this idea further and proposed that *no* experiment, mechanical or otherwise, could distinguish between the two cases. This extension to include all phenomena (not just mechanical ones) has interesting consequences. For example, suppose a light pulse is sent horizontally across the elevator as in Figure 9.13c, in which the elevator is accelerating upward in empty space. From the point of view of an observer in an inertial frame outside the elevator, the light travels in a straight line while the floor of the elevator accelerates upward. According to the observer on the elevator, however, the trajectory of the light pulse bends downward as the floor of the elevator (and the observer) accelerates upward. Therefore, based on the equality of parts (a) and (b) of the figure for all phenomena, Einstein proposed that a beam of light should also be bent downward by a gravitational field, as in Figure 9.13d.

The two postulates of Einstein's **general theory of relativity** are as follows:

- All the laws of nature have the same form for observers in any frame of reference, whether accelerated or not.
- In the vicinity of any given point, a gravitational field is equivalent to an accelerated frame of reference in the absence of gravitational effects. (This postulate is known as the **principle of equivalence.**)

One interesting effect predicted by general relativity is that the passage of time is altered by gravity. A clock in the presence of gravity runs more slowly than one for which gravity is negligible. Consequently, the frequencies of radiation emitted by atoms in the presence of a strong gravitational field are shifted to lower values compared with the same emissions in a weak field. This gravitational shift has been detected in light emitted by atoms in massive stars. It has also been verified on the Earth by comparing the frequencies of radiation emitted from laser-cooled ions separated vertically by less than 1 m.

The second postulate suggests that a gravitational field may be "transformed away" at any point if we choose an appropriate accelerated frame of reference, a freely falling one. Einstein developed an ingenious method of describing the acceleration necessary to make the gravitational field "disappear." He specified a certain quantity, the *curvature of space–time*, that describes the gravitational effect of a mass. In fact, the curvature of space–time completely replaces Newton's gravitational theory. According to Einstein, there is no such thing as a gravitational force. Rather, the presence of a mass causes a curvature of space–time in the vicinity of the mass, and this curvature dictates the space–time path that all freely moving objects must follow.

One important test of general relativity is the prediction that a light ray passing near the Sun should be deflected by some angle. This prediction was confirmed by astronomers as the bending of starlight during a total solar eclipse shortly following World War I (Fig. 9.14).

As an example of the effects of curved space–time, imagine two travelers moving on parallel paths a few meters apart on the surface of the Earth and maintaining an exact northward heading along two longitude lines. As they observe each other near the equator, they will claim that their paths are exactly parallel. As they approach the North Pole, however, they will notice that they are moving closer together and that they will actually meet at the North Pole. Therefore, they will claim that they moved along parallel paths, but moved toward each other, *as if there were an attractive force between them.* They will make this conclusion based on their everyday experience of moving on flat surfaces. From our mental representation, however, we realize that they are walking on a curved surface, and the geometry of the curved surface, rather than an attractive force, causes them to converge. In a similar way, general relativity replaces the notion of forces with the movement of objects through curved space–time.

If a concentration of mass in space becomes very great, as is believed to occur when a large star exhausts its nuclear fuel and collapses to a very small volume, a **black hole** may form. Here the curvature of space–time is so extreme that, within a certain distance from the center of the black hole, all matter and light become trapped. We will say more about black holes in Chapter 11.

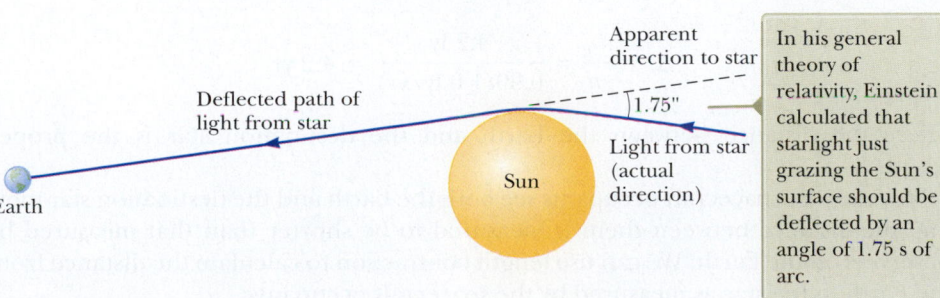

Figure 9.14 Deflection of starlight passing near the Sun. Because of this effect, the Sun or other remote objects can act as a *gravitational lens.*

> **THINKING PHYSICS 9.2**
>
> Atomic clocks are extremely accurate; in fact, an error of 1 s in 3 million years is typical. This error can be described as about 1 part in 10^{14}. On the other hand, the atomic clock in Boulder, Colorado, near Denver, is often 15 ns faster than the one in Washington, D.C., after only one day. This error is one of about 1 part in 6×10^{12}, which is about 17 times larger than the previously expressed error. If atomic clocks are so accurate, why does a clock in Boulder not remain in synchronization with one in Washington, D.C.? (*Hint:* Denver is known as the Mile High City.)
>
> **Reasoning** According to the general theory of relativity, the passage of time depends on gravity. Time is measured to run more slowly in strong gravitational fields. Washington, D.C., is at an elevation very close to sea level, but Boulder is about a mile higher in altitude. This difference results in a weaker gravitational field at Boulder than at Washington, D.C. As a result, time is measured to run more rapidly in Boulder than in Washington, D.C. ◀

9.10 | Context Connection: From Mars to the Stars

In this chapter, we have discussed the strange effects of traveling at high speeds. Do we need to consider these effects in our planned mission to Mars?

To answer this question, let us consider a typical spacecraft speed necessary to travel from the Earth to Mars. This speed is on the order of 10^4 m/s. Let us evaluate γ for this speed:

$$\gamma = \frac{1}{\sqrt{1 - \dfrac{u^2}{c^2}}} = \frac{1}{\sqrt{1 - \dfrac{(10^4 \text{ m/s})^2}{(3.00 \times 10^8 \text{ m/s})^2}}} = 1.000\,000\,000\,6$$

where we have completely ignored the rules of significant figures so that we could find the first nonzero digit to the right of the decimal place!

It is clear from this result that relativistic considerations are not critically important for our trip to Mars. Yet what about deeper travels into space? Suppose we wish to travel to another star. This distance is several orders of magnitude larger. The nearest star is about 4.2 ly from the Earth. In comparison, Mars is 4.0×10^{-5} ly at its farthest from the Earth. Therefore, we are talking about a distance to the nearest star that is five orders of magnitude larger than the distance to Mars. Very long time intervals will be needed to reach even the nearest star. At the escape speed from the Sun, for example, assuming that this speed is maintained during the entire trip, the time interval is 30 000 years for travel to the nearest star. This time interval is clearly prohibitive, especially if we would like the people who leave the Earth to be the same people who arrive at the star!

We can use the principles of relativity to reduce this time interval significantly by traveling at very high speeds. Suppose our spacecraft travels at a constant speed of 0.99c. The time interval as measured by an observer on the Earth then is

$$\Delta t = \frac{L_p}{u} = \frac{4.2 \text{ ly}}{0.99(1.0 \text{ ly/yr})} = 4.2 \text{ yr}$$

where the distance between the Earth and the destination star is the proper length L_p.

Because the spacecraft occupants see both the Earth and the destination star moving, the distance between them is measured to be shorter than that measured by observers on the Earth. We can use length contraction to calculate the distance from the Earth to the star as measured by the spacecraft occupants:

$$L = \frac{L_p}{\gamma} = L_p\sqrt{1 - \frac{u^2}{c^2}} = (4.2\ \text{ly})\sqrt{1 - \frac{(0.99c)^2}{c^2}} = 0.59\ \text{ly}$$

The time interval required to reach the star is now

$$\Delta t = \frac{L}{u} = \frac{0.59\ \text{ly}}{0.99(1.0\ \text{ly/yr})} = 0.60\ \text{yr}$$

which is clearly a reduction from the low-speed trip!

There are three major technical problems with this scenario, however. The first is the technological challenge of designing and building a spacecraft and rocket engine assembly that can attain a speed of $0.99c$. Second is the design of a safety system that will provide early warnings about running into asteroids, meteoroids, or other bits of matter while traveling at almost light speed through space. Even a small piece of rock could be disastrous if struck at $0.99c$. The third problem is related to the twin paradox discussed earlier in this chapter. During the trip to the star, 4.2 yr will pass on the Earth. If the travelers return to the Earth, another 4.2 yr will pass. Therefore, the travelers will have aged by only $2(0.6\ \text{yr}) = 1.2\ \text{yr}$, but 8.4 yr will have passed on the Earth. For stars farther away than the nearest star, these effects could result in the personnel assisting with the liftoff from the Earth no longer being alive when the travelers return. We see that travel to the stars will be an enormous challenge!

There is also a biological consideration associated with the prospect of traveling to a star at $0.99c$. Attaining this speed will require a large acceleration in order that the time interval to reach $0.99c$ is short compared to the time interval traveling at this speed. The human body has certain limits, however, to being accelerated. As found in Problem 2.41 in Enhanced WebAssign, an Air Force officer experienced accelerations of magnitude $20g$ for very short time intervals as his rocket sled was brought to rest. In other experiments, he survived brief accelerations up to $46g$. In our proposed space travel, the inhabitants of the spacecraft will have to experience *sustained* large accelerations.

BIO Human limits on acceleration

If the spacecraft inhabitants have the tops of their heads pointing in the direction of the acceleration, blood will be driven downward toward their feet, just as if they were in an increased gravitational field. This can cause loss of consciousness at accelerations as low as $5g$. If the spacecraft inhabitants are traveling feetfirst, the limits are even lower. Accelerations in the range of $2g$ to $3g$ can drive blood to the head and cause capillaries in the eyes to burst.

Sustained accelerations can cause serious symptoms, including death, if the acceleration is higher than about $10g$. Pilots have used specially designed suits and have learned muscle techniques that allow tolerance of accelerations of up to $9g$. Further technological advances will need to be made, however, in order for space travelers to be protected from injury during accelerations up to $0.99c$.

SUMMARY

The two basic postulates of **special relativity** are:

- All the laws of physics are the same in all inertial reference frames.
- The speed of light in vacuum has the same value, in all inertial frames, regardless of the velocity of the observer or the velocity of the source emitting the light.

Three consequences of special relativity are:

- Events that are simultaneous for one observer may not be simultaneous for another observer who is in motion relative to the first.
- Clocks in motion relative to an observer are measured to be slowed down by a factor γ. This phenomenon is known as **time dilation**.

- Lengths of objects in motion are measured to be shorter in the direction of motion. This phenomenon is known as **length contraction**.

To satisfy the postulates of special relativity, the Galilean transformations must be replaced by the **Lorentz transformation equations**:

$$x' = \gamma(x - vt)$$
$$y' = y$$
$$z' = z$$

9.8◀

$$t' = \gamma\left(t - \frac{v}{c^2}x\right)$$

where $\gamma = (1 - v^2/c^2)^{-1/2}$.

The relativistic form of the **Lorentz velocity transformation** is

$$u'_x = \frac{u_x - v}{1 - \frac{u_x v}{c^2}} \qquad \textbf{9.11} \blacktriangleleft$$

where u_x is the speed of an object as measured in the S frame and u'_x is its speed measured in the S' frame.

The relativistic expression for the momentum of a particle moving with a velocity $\vec{u}$ is

$$\vec{p} \equiv \frac{m\vec{u}}{\sqrt{1 - \frac{u^2}{c^2}}} = \gamma m\vec{u} \qquad \textbf{9.14, 9.15} \blacktriangleleft$$

The relativistic expression for the kinetic energy of a particle is

$$K = \gamma mc^2 - mc^2 = (\gamma - 1)mc^2 \qquad \textbf{9.18} \blacktriangleleft$$

where $E_R = mc^2$ is the **rest energy** of the particle.

The **total energy** E of a particle is given by the expression

$$E = \frac{mc^2}{\sqrt{1 - \frac{u^2}{c^2}}} \qquad \textbf{9.21} \blacktriangleleft$$

The total energy of a particle is the sum of its rest energy and its kinetic energy: $E = E_R + K$.

The relativistic momentum of a particle is related to its total energy through the equation

$$E^2 = p^2c^2 + (mc^2)^2 \qquad \textbf{9.22} \blacktriangleleft$$

The **general theory of relativity** claims that no experiment can distinguish between a gravitational field and an accelerating reference frame. It correctly predicts that the path of light is affected by a gravitational field.

▶ OBJECTIVE QUESTIONS

☐ denotes answer available in *Student Solutions Manual/Study Guide*

1. An astronaut is traveling in a spacecraft in outer space in a straight line at a constant speed of $0.500c$. Which of the following effects would she experience? (a) She would feel heavier. (b) She would find it harder to breathe. (c) Her heart rate would change. (d) Some of the dimensions of her spacecraft would be shorter. (e) None of those answers is correct.

2. A distant astronomical object (a quasar) is moving away from us at half the speed of light. What is the speed of the light we receive from this quasar? (a) greater than c (b) c (c) between $c/2$ and c (d) $c/2$ (e) between 0 and $c/2$

3. As a car heads down a highway traveling at a speed v away from a ground observer, which of the following statements are true about the measured speed of the light beam from the car's headlights? More than one statement may be correct. (a) The ground observer measures the light speed to be $c + v$. (b) The driver measures the light speed to be c. (c) The ground observer measures the light speed to be c. (d) The driver measures the light speed to be $c - v$. (e) The ground observer measures the light speed to be $c - v$.

4. A spacecraft zooms past the Earth with a constant velocity. An observer on the Earth measures that an undamaged clock on the spacecraft is ticking at one-third the rate of an identical clock on the Earth. What does an observer on the spacecraft measure about the Earth-based clock's ticking rate? (a) It runs more than three times faster than his own clock. (b) It runs three times faster than his own. (c) It runs at the same rate as his own. (d) It runs at one-third the rate of his own. (e) It runs at less than one-third the rate of his own.

5. Which of the following statements are fundamental postulates of the special theory of relativity? More than one statement may be correct. (a) Light moves through a substance called the ether. (b) The speed of light depends on the inertial reference frame in which it is measured. (c) The laws of physics depend on the inertial reference frame in which they are used. (d) The laws of physics are the same in all inertial reference frames. (e) The speed of light is independent of the inertial reference frame in which it is measured.

6. A spacecraft built in the shape of a sphere moves past an observer on the Earth with a speed of $0.500c$. What shape does the observer measure for the spacecraft as it goes by? (a) a sphere (b) a cigar shape, elongated along the direction of motion (c) a round pillow shape, flattened along the direction of motion (d) a conical shape, pointing in the direction of motion

7. (i) Does the speed of an electron have an upper limit? (a) yes, the speed of light c (b) yes, with another value (c) no (ii) Does the magnitude of an electron's momentum have an upper limit? (a) yes, $m_e c$ (b) yes, with another value (c) no (iii) Does the electron's kinetic energy have an upper limit? (a) yes, $m_e c^2$ (b) yes, $\frac{1}{2}m_e c^2$ (c) yes, with another value (d) no

8. The following three particles all have the same total energy E: (a) a photon, (b) a proton, and (c) an electron. Rank the magnitudes of the particles' momenta from greatest to smallest.

9. Two identical clocks are set side by side and synchronized. One remains on the Earth. The other is put into orbit around the Earth moving rapidly toward the east. (i) As measured by an observer on the Earth, does the orbiting clock (a) run faster than the Earth-based clock, (b) run at the same rate, or (c) run slower? (ii) The orbiting clock is returned to its original location and brought to rest relative to the Earth-based clock. Thereafter, what happens? (a) Its reading lags farther and farther behind the Earth-based clock. (b) It lags behind the Earth-based clock by a constant amount. (c) It is synchronous with the Earth-based clock. (d) It is ahead of the Earth-based clock by a constant amount. (e) It gets farther and farther ahead of the Earth-based clock.

10. You measure the volume of a cube at rest to be V_0. You then measure the volume of the same cube as it passes you in a direction parallel to one side of the cube. The speed of the cube is $0.980c$, so $\gamma \approx 5$. Is the volume you measure close to (a) $V_0/25$, (b) $V_0/5$, (c) V_0, (d) $5V_0$, or (e) $25V_0$?

CONCEPTUAL QUESTIONS

1. A train is approaching you at very high speed as you stand next to the tracks. Just as an observer on the train passes you, you both begin to play the same recorded version of a Beethoven symphony on identical MP3 players. (a) According to you, whose MP3 player finishes the symphony first? (b) **What If?** According to the observer on the train, whose MP3 player finishes the symphony first? (c) Whose MP3 player actually finishes the symphony first?

2. Explain why, when defining the length of a rod, it is necessary to specify that the positions of the ends of the rod are to be measured simultaneously.

3. The speed of light in water is 230 Mm/s. Suppose an electron is moving through water at 250 Mm/s. Does that violate the principle of relativity? Explain.

4. A particle is moving at a speed less than $c/2$. If the speed of the particle is doubled, what happens to its momentum?

5. Two identical clocks are in the same house, one upstairs in a bedroom and the other downstairs in the kitchen. Which clock runs slower? Explain.

6. List three ways our day-to-day lives would change if the speed of light were only 50 m/s.

7. It is said that Einstein, in his teenage years, asked the question, "What would I see in a mirror if I carried it in my hands and ran at a speed near that of light?" How would you answer this question?

8. (a) "Newtonian mechanics correctly describes objects moving at ordinary speeds, and relativistic mechanics correctly describes objects moving very fast." (b) "Relativistic mechanics must make a smooth transition as it reduces to Newtonian mechanics in a case in which the speed of an object becomes small compared with the speed of light." Argue for or against statements (a) and (b).

9. Give a physical argument that shows it is impossible to accelerate an object of mass m to the speed of light, even with a continuous force acting on it.

10. (i) An object is placed at a position $p > f$ from a concave mirror as shown in Figure CQ9.10a, where f is the focal length of the mirror. In such a situation, an image is formed at a distance q from the mirror, as we discuss in Chapter 26. The distances are related by the mirror equation:

$$\frac{1}{p} + \frac{1}{q} = \frac{1}{f}$$

In a finite time interval, the object is moved to the right to a position at the focal point F of the mirror. Show that the image of the object moves at a speed greater than the speed of light. (ii) A laser pointer is suspended in a horizontal plane and set into rapid rotation as shown in Figure CQ9.10b. Show that the spot of light it produces on a distant screen can move across the screen at a speed greater than the speed of light. (If you carry out this experiment, make sure the direct laser light cannot enter a person's eyes.) (iii) Argue that the experiments in parts (i) and (ii) do not invalidate the principle that no material, no energy, and no information can move faster than light moves in a vacuum.

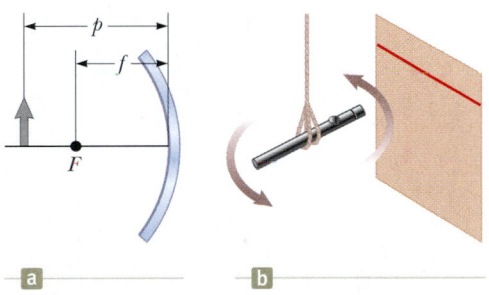

Figure CQ9.10

11. With regard to reference frames, how does general relativity differ from special relativity?

PROBLEMS AVAILABLE IN ENHANCED WebAssign

9.1 The Principle of Galilean Relativity

Problems 1–2

9.2 The Michelson–Morley Experiment
9.3 Einstein's Principle of Relativity
9.4 Consequences of Special Relativity

Problems 3–15

9.5 The Lorentz Transformation Equations

Problems 16–21

9.6 Relativistic Momentum and the Relativistic Form of Newton's Laws

Problems 22–27

9.7 Relativistic Energy

Problems 28–39

9.8 Mass and Energy

Problems 40–42

9.9 General Relativity

Problem 43

9.10 Context Connection: From Mars to the Stars

Problems 44–45

Additional Problems

Problems 46–65

Solutions to the following Problems are available in the Student Solutions Manual/Study Guide:

9.1, 9.9, 9.11, 9.21, 9.27, 9.35, 9.36, 9.37, 9.39, 9.41, 9.51, 9.56, and 9.59

List of Enhanced Problems

Problem Number	Targeted Feedback in Enhanced WebAssign	Master It in Enhanced WebAssign	Watch It in Enhanced WebAssign
9.3	✓		✓
9.4	✓		
9.5	✓		✓
9.11	✓	✓	
9.15	✓		
9.17	✓		✓
9.21	✓	✓	
9.23	✓		✓
9.27	✓	✓	
9.37		✓	
9.39		✓	
9.40	✓		✓
9.42	✓		✓
9.47		✓	
9.51		✓	
9.59		✓	

Rotational Motion

Chapter Outline

Courtesy Tourism Malaysia

The Malaysian pastime of *gasing* involves the spinning of tops that can have masses up to 5 kg. Professional spinners can spin their tops so that they might rotate for 1 to 2 h before stopping. We will study the rotational motion of objects such as these tops in this chapter.

When an extended object, such as a wheel, rotates about its axis, the motion cannot be analyzed by modeling the object as a particle because at any given time different parts of the object are moving with different speeds and in different directions. We can, however, analyze the motion by considering an extended object to be composed of a *collection* of moving particles.

In dealing with a rotating object, analysis is greatly simplified by assuming that the object is rigid. A **rigid object** is one that is nondeformable; that is, it is an object in which the relative locations of all particles of which the object is composed remain constant. All real objects are deformable to some extent; our rigid-object model, however, is useful in many situations in which deformation is negligible.

K 10.1 | Angular Position, Speed, and Acceleration

▶ The radian

We began our study of translational motion in Chapter 2 by defining the terms *position*, *velocity*, and *acceleration*. For example, we locate a particle in one-dimensional space with the position variable x. In this chapter, we will insert the word *translational* before our previously studied kinematic variables to distinguish them from the analogous *rotational* variables that we will develop.

Figure 10.1 illustrates an overhead view of a rotating compact disc. The disc rotates about a fixed axis that is perpendicular to the page and passes through the center of the disc at O. A particle at P is at a fixed distance r from the origin and rotates about it in a circle of radius r. (In fact, *every* particle on the disc undergoes circular motion about O.) It is convenient to represent the position of P with its polar coordinates (r, θ), where r is the distance from the origin to P and θ is measured *counterclockwise* from some reference line shown in Figure 10.1. In this representation, the only coordinate for the particle that changes in time is the angle θ, while r remains constant. As the particle moves along the circular path from the reference line, which is at $\theta = 0$, it moves through an arc of length s as in Figure 10.1b. The arc length s is related to the angle θ through the relationship

$$s = r\theta \qquad\qquad \textbf{10.1a} \blacktriangleleft$$

$$\theta = \frac{s}{r} \qquad\qquad \textbf{10.1b} \blacktriangleleft$$

Because θ is the ratio of an arc length and the radius of the circle, it is a pure number. Usually, we give θ the artificial unit **radian** (rad), where one radian is the angle subtended by an arc length equal to the radius of the arc. Because the circumference of a circle is $2\pi r$, it follows from Equation 10.1b that $360°$ corresponds to an angle of $(2\pi r/r)$ rad $= 2\pi$ rad. (Also note that 2π rad corresponds to one complete revolution.) Hence, 1 rad $= 360°/2\pi \approx 57.3°$. To convert an angle in degrees to an angle in radians, we use π rad $= 180°$, so

$$\theta(\text{rad}) = \frac{\pi}{180°}\theta(\text{deg})$$

For example, $60°$ equals $\pi/3$ rad, and $45°$ equals $\pi/4$ rad.

We will focus much of our attention in this chapter on rigid objects. Approximating a real object as a rigid object is a simplification model, the **rigid object model.** We will base several analysis models on this simplification model, as we did for the particle model.

Because the disc in Figure 10.1 is a rigid object, as the particle at P moves along the circular path from the reference line every other particle on the object rotates through the same angle θ. Therefore, we can associate the angle θ with the entire rigid object as well as with an individual particle, which allows us to define the angular position of a rigid object in its rotational motion. We choose a radial line on the object, such as a line connecting O and a chosen particle on the object. The **angular position** of the rigid object is the angle θ between this radial line on the object and the fixed reference line in space, which is often chosen as the x axis. This process is similar to the way we identify the position of an object in translational motion as the distance x between the object and the reference position, which is the origin, $x = 0$. Therefore, the angle θ plays the same role in rotational motion that the position x does in translational motion.

As a particle on a rigid object travels from position Ⓐ to position Ⓑ in a time interval Δt as in Figure 10.2, the reference line fixed to the object sweeps out an angle $\Delta\theta = \theta_f - \theta_i$. This quantity $\Delta\theta$ is defined as the **angular displacement** of the rigid object:

$$\Delta\theta \equiv \theta_f - \theta_i$$

To define angular position for the disc, a fixed reference line is chosen. A particle at P is located at a distance r from the rotation axis through O.

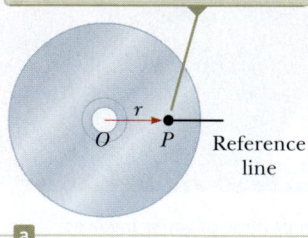

a

As the disc rotates, a particle at P moves through an arc length s on a circular path of radius r. The angular position of P is θ.

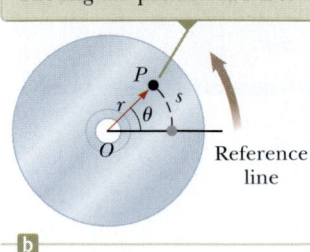

b

Figure 10.1 A compact disc rotating about a fixed axis through O perpendicular to the plane of the figure.

The rate at which this angular displacement occurs can vary. If the rigid object spins rapidly, this displacement can occur in a short time interval. If it rotates slowly, this displacement occurs in a longer time interval. These different rotation rates can be quantified by defining the **average angular speed** ω_{avg} (Greek letter omega) as the ratio of the angular displacement of a rigid object to the time interval Δt during which the displacement occurs:

$$\omega_{avg} \equiv \frac{\theta_f - \theta_i}{t_f - t_i} = \frac{\Delta\theta}{\Delta t} \qquad \textbf{10.2} \blacktriangleleft \qquad \blacktriangleright \text{ Average angular speed}$$

In analogy to instantaneous translational speed, the **instantaneous angular speed** ω is defined as the limit of the average angular speed as Δt approaches zero:

$$\omega \equiv \lim_{\Delta t \to 0} \frac{\Delta\theta}{\Delta t} = \frac{d\theta}{dt} \qquad \textbf{10.3} \blacktriangleleft \qquad \blacktriangleright \text{ Instantaneous angular speed}$$

Angular speed has units of rad/s, which can be written as s^{-1} because radians are not dimensional. We take ω to be positive when θ is increasing (counterclockwise motion in Fig. 10.2) and negative when θ is decreasing (clockwise motion in Fig. 10.2).

If the instantaneous angular speed of an object changes from ω_i to ω_f in the time interval Δt, the object has an angular acceleration. The **average angular acceleration** α_{avg} (Greek letter alpha) of a rotating rigid object is defined as the ratio of the change in the angular speed to the time interval Δt during which the change in angular speed occurs:

$$\alpha_{avg} \equiv \frac{\omega_f - \omega_i}{t_f - t_i} = \frac{\Delta\omega}{\Delta t} \qquad \textbf{10.4} \blacktriangleleft \qquad \blacktriangleright \text{ Average angular acceleration}$$

In analogy to instantaneous translational acceleration, the **instantaneous angular acceleration** is defined as the limit of the average angular acceleration as Δt approaches zero:

$$\alpha \equiv \lim_{\Delta t \to 0} \frac{\Delta\omega}{\Delta t} = \frac{d\omega}{dt} \qquad \textbf{10.5} \blacktriangleleft \qquad \blacktriangleright \text{ Instantaneous angular acceleration}$$

Angular acceleration has units of radians per second squared (rad/s^2), or simply s^{-2}. Notice that α is positive when a rigid object rotating counterclockwise is speeding up or when a rigid object rotating clockwise is slowing down during some time interval.

When a rigid object is rotating about a *fixed* axis, every particle on the object rotates about that axis through the same angle in a given time interval and has the same angular speed and the same angular acceleration. That is, the quantities θ, ω, and α characterize the rotational motion of the entire rigid object as well as individual particles in the object.

The angular position (θ), angular speed (ω), and angular acceleration (α) of a rigid object are analogous to translational position (x), translational speed (v), and translational acceleration (a), respectively, for the corresponding one-dimensional motion of a particle discussed in Chapter 2. The variables θ, ω, and α differ dimensionally from the variables x, v, and a only by a factor having the unit of length. (See Section 10.3.)

We have not associated any direction with the angular speed and angular acceleration.[1] Strictly speaking, ω and α are the magnitudes of the angular velocity and angular acceleration vectors $\vec{\omega}$ and $\vec{\alpha}$ and they should always be positive. Because we are considering rotation about a fixed axis, we can indicate the directions of these vectors by assigning a positive or negative sign to ω and α, as discussed for ω after Equation 10.3. For rotation about a fixed axis, the only direction in space that uniquely specifies the rotational motion is the direction along the axis of rotation. Therefore, the direction of $\vec{\omega}$ is along this axis. If a particle rotates in the xy plane

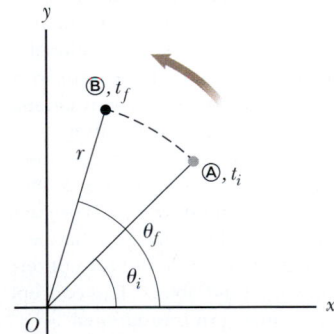

Figure 10.2 A particle on a rotating rigid object moves from Ⓐ to Ⓑ along the arc of a circle. In the time interval $\Delta t = t_f - t_i$, the radial line of length r sweeps out an angle $\Delta\theta = \theta_f - \theta_i$.

---

[1]Although we do not verify it here, the instantaneous angular velocity and instantaneous angular acceleration are vector quantities, but the corresponding average values are not because angular displacements do not add as vector quantities for finite rotations.

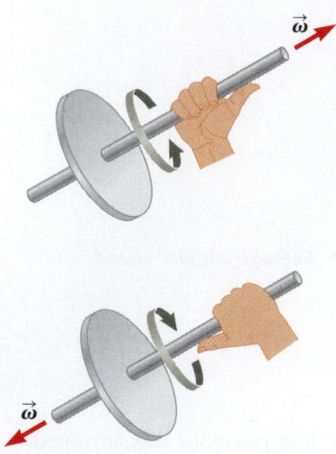

Figure 10.3 The right-hand rule for determining the direction of the angular velocity vector.

as in Figure 10.2, the direction of $\vec{\omega}$ is *out of* the plane of the diagram when the rotation is counterclockwise and *into* the plane of the diagram when the rotation is clockwise. To illustrate this convention, it is convenient to use the **right-hand rule** demonstrated in Figure 10.3. When the four fingers of the right hand are wrapped in the direction of the rotation, the extended right thumb points in the direction of $\vec{\omega}$. The direction of $\vec{\alpha}$ follows from its definition $d\vec{\omega}/dt$. The direction of $\vec{\alpha}$ is the same as $\vec{\omega}$ if the angular speed is increasing in time and is antiparallel to $\vec{\omega}$ if the angular speed is decreasing in time.

> **QUICK QUIZ 10.1** A rigid object is rotating in a counterclockwise sense around a fixed axis. Each of the following pairs of quantities represents an initial angular position and a final angular position of the rigid object. (i) Which of the sets can *only* occur if the rigid object rotates through more than 180°? (a) 3 rad, 6 rad (b) −1 rad, 1 rad (c) 1 rad, 5 rad (ii) Suppose the change in angular position for each of these pairs of values occurs in 1 s. Which choice represents the lowest average angular speed?

10.2 | Analysis Model: Rigid Object Under Constant Angular Acceleration

Imagine that a rigid object rotates about a fixed axis and that it has a constant angular acceleration. In this case, we generate a new analysis model for rotational motion called the **rigid object under constant angular acceleration.** This model is the rotational analog to the particle under constant acceleration model. We develop kinematic relationships for this model in this section. Writing Equation 10.5 in the form $d\omega = \alpha\, dt$ and integrating from $t_i = 0$ to $t_f = t$ gives

$$\omega_f = \omega_i + \alpha t \quad \text{(for constant } \alpha\text{)} \qquad \textbf{10.6} \blacktriangleleft$$

where ω_i is the angular speed of the rigid object at time $t = 0$. Equation 10.6 allows us to find the angular speed ω_f of the object at any later time t. Substituting Equation 10.6 into Equation 10.3 and integrating once more, we obtain

$$\theta_f = \theta_i + \omega_i t + \tfrac{1}{2}\alpha t^2 \quad \text{(for constant } \alpha\text{)} \qquad \textbf{10.7} \blacktriangleleft$$

where θ_i is the angular position of the rigid object at time $t = 0$. Equation 10.7 allows us to find the angular position θ_f of the object at any later time t. Eliminating t from Equations 10.6 and 10.7 gives

$$\omega_f^2 = \omega_i^2 + 2\alpha(\theta_f - \theta_i) \quad \text{(for constant } \alpha\text{)} \qquad \textbf{10.8} \blacktriangleleft$$

This equation allows us to find the angular speed ω_f of the rigid object for any value of its angular position θ_f. If we eliminate α between Equations 10.6 and 10.7, we obtain

$$\theta_f = \theta_i + \tfrac{1}{2}(\omega_i + \omega_f)t \quad \text{(for constant } \alpha\text{)} \qquad \textbf{10.9} \blacktriangleleft$$

Notice that these kinematic expressions for the rigid object under constant angular acceleration are of the same mathematical form as those for a particle under constant acceleration (Chapter 2). They can be generated from the equations for translational motion by making the substitutions $x \rightarrow \theta$, $v \rightarrow \omega$, and $a \rightarrow \alpha$. Table 10.1 compares the kinematic equations for rotational and translational motion.

Pitfall Prevention | 10.3
Just Like Translation?
Equations 10.6 to 10.9 and Table 10.1 might suggest that rotational kinematics is just like translational kinematics. That is almost true, with two key differences. (1) In rotational kinematics, you must specify a rotation axis (per Pitfall Prevention 10.1). (2) In rotational motion, the object keeps returning to its original orientation; therefore, you may be asked for the number of revolutions made by a rigid object. This concept has no analog in translational motion.

TABLE 10.1 | Kinematic Equations for Rotational and Translational Motion

Rigid Body Under Constant Angular Acceleration	Particle Under Constant Acceleration
$\omega_f = \omega_i + \alpha t$	$v_f = v_i + at$
$\theta_f = \theta_i + \omega_i t + \tfrac{1}{2}\alpha t^2$	$x_f = x_i + v_i t + \tfrac{1}{2}at^2$
$\omega_f^2 = \omega_i^2 + 2\alpha(\theta_f - \theta_i)$	$v_f^2 = v_i^2 + 2a(x_f - x_i)$
$\theta_f = \theta_i + \tfrac{1}{2}(\omega_i + \omega_f)t$	$x_f = x_i + \tfrac{1}{2}(v_i + v_f)t$

consider again the pairs of angular positions for the rigid object in Quick Quiz 10.1. If the object starts from rest at the initial angular position, moves counterclockwise with constant angular acceleration, and arrives at the final angular position with the same angular speed in all three cases, for which choice is the angular acceleration the highest?

Example 10.1 | Rotating Wheel

A wheel rotates with a constant angular acceleration of 3.50 rad/s².

(A) If the angular speed of the wheel is 2.00 rad/s at $t = 0$, through what angular displacement does the wheel rotate in 2.00 s?

SOLUTION

Conceptualize Look again at Figure 10.1. Imagine that the compact disc rotates with its angular speed increasing at a constant rate. You start your stopwatch when the disc is rotating at 2.00 rad/s. This mental image is a model for the motion of the wheel in this example.

Categorize The phrase "with a constant angular acceleration" tells us to use the rigid object under constant angular acceleration model.

Analyze

Arrange Equation 10.7 so that it expresses the angular displacement of the object:

$$\Delta\theta = \theta_f - \theta_i = \omega_i t + \tfrac{1}{2}\alpha t^2$$

Substitute the known values to find the angular displacement at $t = 2.00$ s:

$$\Delta\theta = (2.00 \text{ rad/s})(2.00 \text{ s}) + \tfrac{1}{2}(3.50 \text{ rad/s}^2)(2.00 \text{ s})^2$$
$$= \boxed{11.0 \text{ rad}} = (11.0 \text{ rad})(180°/\pi \text{ rad}) = \boxed{630°}$$

(B) Through how many revolutions has the wheel turned during this time interval?

SOLUTION

Multiply the angular displacement found in part (A) by a conversion factor to find the number of revolutions:

$$\Delta\theta = 630°\left(\frac{1 \text{ rev}}{360°}\right) = \boxed{1.75 \text{ rev}}$$

(C) What is the angular speed of the wheel at $t = 2.00$ s?

SOLUTION

Use Equation 10.6 to find the angular speed at $t = 2.00$ s:

$$\omega_f = \omega_i + \alpha t = 2.00 \text{ rad/s} + (3.50 \text{ rad/s}^2)(2.00 \text{ s})$$
$$= \boxed{9.00 \text{ rad/s}}$$

Finalize We could also obtain this result using Equation 10.8 and the results of part (A). (Try it!)

What If? Suppose a particle moves along a straight line with a constant acceleration of 3.50 m/s². If the velocity of the particle is 2.00 m/s at $t = 0$, through what displacement does the particle move in 2.00 s? What is the velocity of the particle at $t = 2.00$ s?

Answer Notice that these questions are translational analogs to parts (A) and (C) of the original problem. The mathematical solution follows exactly the same form. For the displacement,

$$\Delta x = x_f - x_i = v_i t + \tfrac{1}{2}at^2$$
$$= (2.00 \text{ m/s})(2.00 \text{ s}) + \tfrac{1}{2}(3.50 \text{ m/s}^2)(2.00 \text{ s})^2 = \boxed{11.0 \text{ m}}$$

and for the velocity,

$$v_f = v_i + at = 2.00 \text{ m/s} + (3.50 \text{ m/s}^2)(2.00 \text{ s}) = \boxed{9.00 \text{ m/s}}$$

There is no translational analog to part (B) because translational motion under constant acceleration is not repetitive.

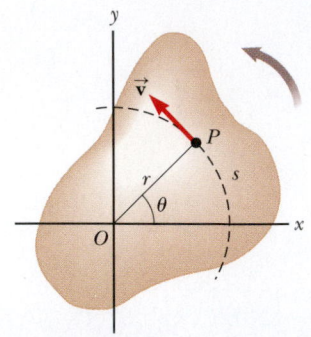

Active Figure 10.4 As a rigid object rotates about the fixed axis through O, the point P has a tangential velocity $\vec{v}$ that is always tangent to the circular path of radius r.

10.3 | Relations Between Rotational and Translational Quantities

In this section, we derive some useful relations between the angular speed and angular acceleration of a rotating rigid object and the translational speed and translational acceleration of a point in the object. To do so, we must keep in mind that when a rigid object rotates about a fixed axis, as in Active Figure 10.4, *every* particle of the object moves in a circle whose center is on the axis of rotation.

Because the point P in Active Figure 10.4 moves in a circle of radius r, its translational velocity vector $\vec{v}$ is always tangent to the path and hence is called **tangential velocity.** The magnitude of the tangential velocity of the particle is, by definition, the **tangential speed,** $v = ds/dt$, where s is the distance traveled by the particle along the circular path. Recalling that $s = r\theta$ (Eq. 10.1a) and noting that r is a constant, we obtain

$$v = \frac{ds}{dt} = r\frac{d\theta}{dt}$$

$$v = r\omega \qquad \text{10.10} \blacktriangleleft$$

That is, the tangential speed of a point on a rotating rigid object equals the perpendicular distance of that point from the axis of rotation multiplied by the angular speed. Therefore, although every point on the rigid object has the same *angular* speed, not every point has the same *tangential* speed because r is not the same for all points on the object. Equation 10.10 shows that the tangential speed of a point on the rotating object increases as one moves outward from the center of rotation, as we would intuitively expect. For example, the outer end of a swinging golf club moves much faster than the handle.

We can relate the angular acceleration of the rotating rigid object to the tangential acceleration of the point P by taking the time derivative of v:

$$a_t = \frac{dv}{dt} = r\frac{d\omega}{dt}$$

$$a_t = r\alpha \qquad \text{10.11} \blacktriangleleft$$

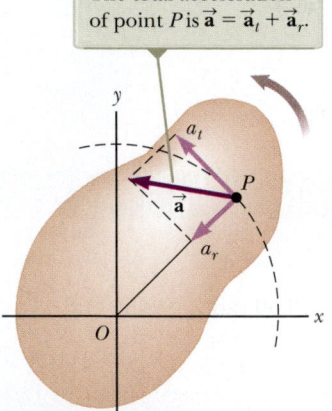

The total acceleration of point P is $\vec{a} = \vec{a}_t + \vec{a}_r$.

Figure 10.5 As a rigid object rotates about a fixed axis through O, the point P experiences a tangential component a_t and a radial component a_r of translational acceleration.

That is, the tangential component of the translational acceleration of a point on a rotating object equals the point's perpendicular distance from the axis of rotation multiplied by the angular acceleration.

In Chapter 3, we found that a particle rotating in a circular path undergoes a centripetal, or radial, acceleration of magnitude v^2/r directed toward the center of rotation (Fig. 10.5). Because $v = r\omega$, we can express the centripetal acceleration of that point in terms of the angular speed as

$$a_c = \frac{v^2}{r} = r\omega^2 \qquad \text{10.12} \blacktriangleleft$$

The total acceleration vector at the point is $\vec{a} = \vec{a}_t + \vec{a}_r$, where the magnitude of $\vec{a}_r$ is the centripetal acceleration a_c. Because $\vec{a}$ is a vector having a radial and a tangential component, the magnitude of $\vec{a}$ at the point P on the rotating rigid object is

$$a = \sqrt{a_t^2 + a_r^2} = \sqrt{r^2\alpha^2 + r^2\omega^4} = r\sqrt{\alpha^2 + \omega^4} \qquad \text{10.13} \blacktriangleleft$$

QUICK QUIZ 10.3 Ethan and Joseph are riding on a merry-go-round. Ethan rides on a horse at the outer rim of the circular platform, twice as far from the center of the circular platform as Joseph, who rides on an inner horse. **(i)** When the merry-go-round is rotating at a constant angular speed, what is Ethan's angular speed? **(a)** twice Joseph's **(b)** the same as Joseph's **(c)** half of Joseph's **(d)** impossible to determine **(ii)** When the merry-go-round is rotating at a constant angular speed, describe Ethan's tangential speed from the same list of choices.

> **THINKING PHYSICS 10.1**

A phonograph record (LP, for *long-playing*) rotates at a constant *angular* speed. A compact disc (CD) rotates so that the surface sweeps past the laser at a constant *tangential* speed. Consider two circular grooves of information on an LP, one near the outer edge and one near the inner edge. Suppose the outer groove "contains" 1.8 s of music. Does the inner groove also contain 1.8 s of music? And for the CD, do the inner and outer "grooves" contain the same time interval of music?

Reasoning On the LP the inner and outer grooves must both rotate once in the same time interval. Therefore, each groove, regardless of where it is on the record, contains the same time interval of information. Of course, on the inner grooves, this same information must be compressed into a smaller circumference. On a CD, the constant tangential speed requires that no such compression occur; the digital pits representing the information are spaced uniformly everywhere on the surface. Therefore, there is more information in an outer "groove," because of its larger circumference and, as a result, a longer time interval of music than in the inner "groove." ◄

> **THINKING PHYSICS 10.2**

The launch area for the European Space Agency is not in Europe, but rather in South America. Why?

Reasoning Placing a satellite in Earth orbit requires providing a large tangential speed to the satellite, which is the task of the rocket propulsion system. Anything that reduces the requirements on the propulsion system is a welcome contribution. The surface of the Earth is already traveling toward the east at a high speed due to the rotation of the Earth. Therefore, if rockets are launched toward the east, the rotation of the Earth provides some initial tangential speed, reducing somewhat the requirements on the propulsion system. If rockets were launched from Europe, which is at a relatively large latitude, the contribution of the Earth's rotation is relatively small because the distance between Europe and the rotation axis of the Earth is relatively small. The ideal place for launching is at the equator, which is as far as one can be from the rotation axis of the Earth and still be on the surface of the Earth. This location results in the largest possible tangential speed due to the Earth's rotation. The European Space Agency exploits this advantage by launching from French Guiana, which is only a few degrees north of the equator.

A second advantage of this location is that launching toward the east takes the spacecraft over water. In the event of an accident or a failure, the wreckage will fall into the ocean rather than into populated areas as it would if launched to the east from Europe. Similarly, the United States launches spacecraft from Florida rather than California, despite the more favorable weather conditions in California. ◄

❮ 10.4 | Rotational Kinetic Energy

Let us consider an object as a system of particles and assume it rotates about a fixed z axis with an angular speed ω. Figure 10.6 shows the rotating object and identifies one particle on the object located at a distance r_i from the rotation axis. If the mass of the ith particle is m_i and its tangential speed is v_i, its kinetic energy is

$$K_i = \tfrac{1}{2} m_i v_i^2$$

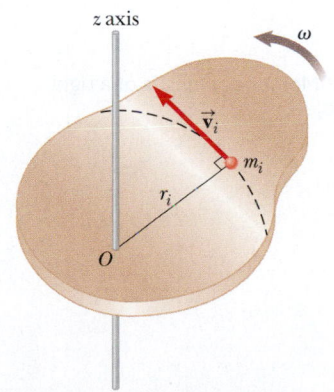

Figure 10.6 A rigid object rotating about the z axis with angular speed ω. The kinetic energy of the particle of mass m_i is $\frac{1}{2}m_i v_i^2$. The total kinetic energy of the rigid object is called its rotational kinetic energy.

To proceed further, recall that although every particle in the rigid object has the same angular speed ω, the individual tangential speeds depend on the distance r_i from the axis of rotation according to Equation 10.10. The *total* kinetic energy of the rotating rigid object is the sum of the kinetic energies of the individual particles:

$$K_R = \sum_i K_i = \sum_i \tfrac{1}{2} m_i v_i^2 = \tfrac{1}{2} \sum_i m_i r_i^2 \omega^2$$

We can write this expression in the form

$$K_R = \tfrac{1}{2} \left(\sum_i m_i r_i^2 \right) \omega^2 \qquad \textbf{10.14} \blacktriangleleft$$

where we have factored ω^2 from the sum because it is common to every particle. We simplify this expression by defining the quantity in parentheses as the **moment of inertia** I of the rigid object:

▶ Moment of inertia

$$I \equiv \sum_i m_i r_i^2 \qquad \textbf{10.15} \blacktriangleleft$$

From the definition of moment of inertia,[2] we see that it has dimensions of ML^2 ($\text{kg} \cdot \text{m}^2$ in SI units). With this notation, Equation 10.14 becomes

▶ Rotational kinetic energy

$$K_R = \tfrac{1}{2} I \omega^2 \qquad \textbf{10.16} \blacktriangleleft$$

The moment of inertia is a measure of an object's *resistance to change in its angular speed*. Therefore, it plays a role in rotational motion identical to the role mass plays in translational motion. Notice that moment of inertia depends not only on the mass of the rigid object but also on *how the mass is distributed around the rotation axis*.

Although we shall commonly refer to the quantity $\tfrac{1}{2} I \omega^2$ in Equation 10.16 as the **rotational kinetic energy**, it is not a new form of energy. It is ordinary kinetic energy because it was derived from a sum over individual kinetic energies of the particles contained in the rigid object. It is a new role for kinetic energy for us, however, because we have only considered kinetic energy associated with translation through space so far. On the storage side of the conservation of energy equation (see Eq. 7.2), we should now consider that the kinetic energy term should be the sum of the changes in both translational and rotational kinetic energy. Therefore, in energy versions of system models, we should keep in mind the possibility of rotational kinetic energy.

The moment of inertia of a system of discrete particles can be calculated in a straightforward way with Equation 10.15. We can evaluate the moment of inertia of a continuous rigid object by imagining the object to be divided into many small elements, each of which has mass Δm_i. We use the definition $I = \sum_i r_i^2 \Delta m_i$ and take the limit of this sum as $\Delta m_i \to 0$. In this limit, the sum becomes an integral over the volume of the object:

▶ Moment of inertia of a rigid object

$$I = \lim_{\Delta m_i \to 0} \sum_i r_i^2 \Delta m_i = \int r^2 \, dm \qquad \textbf{10.17} \blacktriangleleft$$

It is usually easier to calculate moments of inertia in terms of the volume of the elements rather than their mass, and we can easily make this change by using Equation 1.1, $\rho = m/V$, where ρ is the density of the object and V is its volume. We can express the mass of an element by writing Equation 1.1 in differential form, $dm = \rho \, dV$. Substituting this result into Equation 10.17 gives

$$I = \int \rho r^2 \, dV \qquad \textbf{10.18} \blacktriangleleft$$

If the object is homogenous, ρ is uniform over the volume of the object and the integral can be evaluated for a known geometry. If ρ is not uniform over the volume of the object, its variation with position must be known in order to perform the integration.

Pitfall Prevention | 10.4
No Single Moment of Inertia
We have pointed out that moment of inertia is analogous to mass, but there is one major difference. Mass is an inherent property of an object and has a single value. The moment of inertia of an object depends on your choice of rotation axis; therefore, an object has no single value of the moment of inertia. An object does have a *minimum* value of the moment of inertia, which is that calculated around an axis passing through the center of mass of the object.

---

[2] Civil engineers use moment of inertia to characterize the elastic properties (rigidity) of such structures as loaded beams. Hence, it is often useful even in a nonrotational context.

TABLE 10.2 |
Moments of Inertia of Homogeneous Rigid Objects with Different Geometries

Hoop or thin cylindrical shell
$I_{CM} = MR^2$

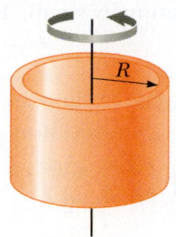

Hollow cylinder
$I_{CM} = \frac{1}{2}M(R_1^2 + R_2^2)$

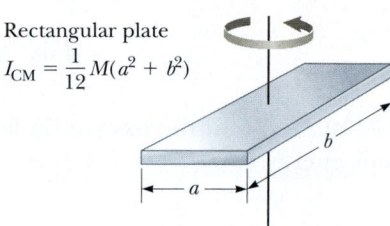

Solid cylinder or disk
$I_{CM} = \frac{1}{2}MR^2$

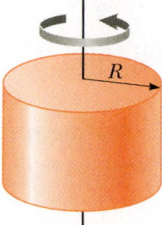

Rectangular plate
$I_{CM} = \frac{1}{12}M(a^2 + b^2)$

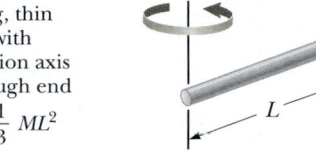

Long, thin rod with rotation axis through center
$I_{CM} = \frac{1}{12}ML^2$

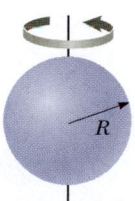

Long, thin rod with rotation axis through end
$I = \frac{1}{3}ML^2$

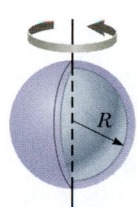

Solid sphere
$I_{CM} = \frac{2}{5}MR^2$

Thin spherical shell
$I_{CM} = \frac{2}{3}MR^2$

For symmetric objects, the moment of inertia can be expressed in terms of the total mass of the object and one or more dimensions of the object. Table 10.2 shows the moments of inertia of various common symmetric objects.

QUICK QUIZ 10.4 A section of hollow pipe and a solid cylinder have the same radius, mass, and length. They both rotate about their long central axes with the same angular speed. Which object has the higher rotational kinetic energy? **(a)** The hollow pipe does. **(b)** The solid cylinder does. **(c)** They have the same rotational kinetic energy. **(d)** It is impossible to determine.

Example 10.2 | The Oxygen Molecule

Consider the diatomic oxygen molecule O_2, which is rotating in the xy plane about the z axis passing through its center, perpendicular to its length. The mass of each oxygen atom is 2.66×10^{-26} kg, and at room temperature, the average separation between the two oxygen atoms is $d = 1.21 \times 10^{-10}$ m.

(A) Calculate the moment of inertia of the molecule about the z axis.

continued

10.2 *cont.*

SOLUTION

Conceptualize Imagine the thin rod rotating about its center on the left side of Table 10.2. Now imagine placing two identical small spheres on each end of the rod and letting the mass of the rod become infinitesimally small. The result of this imaginary process is a macroscopic mental model for the oxygen molecule.

Categorize We model the molecule as a rigid object, consisting of two particles (the two oxygen atoms), in rotation. We will evaluate results from definitions developed in this section, so we categorize this example as a substitution problem.

Note that the distance of each particle from the z axis is $d/2$. Find the moment of inertia about the z axis:

$$I = \sum_i m_i r_i^2 = m\left(\frac{d}{2}\right)^2 + m\left(\frac{d}{2}\right)^2 = \frac{md^2}{2}$$

$$= \frac{(2.66 \times 10^{-26}\ \text{kg})(1.21 \times 10^{-10}\ \text{m})^2}{2}$$

$$= 1.95 \times 10^{-46}\ \text{kg} \cdot \text{m}^2$$

(B) A typical angular speed of a molecule is 4.60×10^{12} rad/s. If the oxygen molecule is rotating with this angular speed about the z axis, what is its rotational kinetic energy?

SOLUTION

Use Equation 10.16 to find the rotational kinetic energy:

$$K_R = \tfrac{1}{2}I\omega^2$$

$$= \tfrac{1}{2}(1.95 \times 10^{-46}\ \text{kg} \cdot \text{m}^2)(4.60 \times 10^{12}\ \text{rad/s})^2$$

$$= 2.06 \times 10^{-21}\ \text{J}$$

Example 10.3 | An Unusual Baton

Four tiny spheres are fastened to the ends of two rods of negligible mass lying in the xy plane to form an unusual baton (Fig. 10.7). We shall assume the radii of the spheres are small compared with the dimensions of the rods.

(A) If the system rotates about the y axis (Fig. 10.7a) with an angular speed ω, find the moment of inertia and the rotational kinetic energy of the system about this axis.

SOLUTION

Conceptualize Figure 10.7 is a pictorial representation that helps conceptualize the system of spheres and how it spins.

Categorize This example is a substitution problem because it is a straightforward application of the definitions discussed in this section.

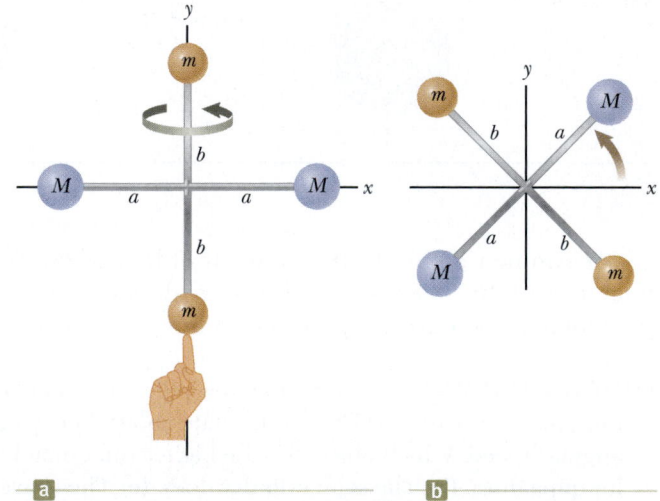

Figure 10.7 (Example 10.3) Four spheres form an unusual baton. (a) The baton is rotated about the y axis. (b) The baton is rotated about the z axis.

Apply Equation 10.15 to the system:

$$I_y = \sum_i m_i r_i^2 = Ma^2 + Ma^2 = 2Ma^2$$

Evaluate the rotational kinetic energy using Equation 10.16:

$$K_R = \tfrac{1}{2}I_y\omega^2 = \tfrac{1}{2}(2Ma^2)\omega^2 = Ma^2\omega^2$$

That the two spheres of mass m do not enter into this result makes sense because they have no motion about the axis of rotation; hence, they have no rotational kinetic energy. By similar logic, we expect the moment of inertia about the x axis to be $I_x = 2mb^2$ with a rotational kinetic energy about that axis of $K_R = mb^2\omega^2$.

10.3 *cont.*

(B) Suppose the system rotates in the *xy* plane about an axis (the *z* axis) through the center of the baton (Fig. 10.7b). Calculate the moment of inertia and rotational kinetic energy about this axis.

SOLUTION

Apply Equation 10.15 for this new rotation axis:

$$I_z = \sum_i m_i r_i^2 = Ma^2 + Ma^2 + mb^2 + mb^2 = \boxed{2Ma^2 + 2mb^2}$$

Evaluate the rotational kinetic energy using Equation 10.16:

$$K_R = \tfrac{1}{2} I_z \omega^2 = \tfrac{1}{2}(2Ma^2 + 2mb^2)\omega^2 = \boxed{(Ma^2 + mb^2)\omega^2}$$

Comparing the results for parts (A) and (B), we conclude that the moment of inertia and therefore the rotational kinetic energy associated with a given angular speed depend on the axis of rotation. In part (B), we expect the result to include all four spheres and distances because all four spheres are rotating in the *xy* plane. Based on the work–kinetic energy theorem, the smaller rotational kinetic energy in part (A) than in part (B) indicates it would require less work to set the system into rotation about the *y* axis than about the *z* axis.

What If? What if the mass *M* is much larger than *m*? How do the answers to parts (A) and (B) compare?

Answer If $M \gg m$, then *m* can be neglected and the moment of inertia and the rotational kinetic energy in part (B) become

$$I_z = 2Ma^2 \quad \text{and} \quad K_R = Ma^2\omega^2$$

which are the same as the answers in part (A). If the masses *m* of the two tan spheres in Figure 10.7 are negligible, these spheres can be removed from the figure and rotations about the *y* and *z* axes are equivalent.

Example 10.4 | Uniform Rigid Rod

Calculate the moment of inertia of a uniform rigid rod of length *L* and mass *M* (Fig. 10.8) about an axis perpendicular to the rod (the *y′* axis) and passing through its center of mass.

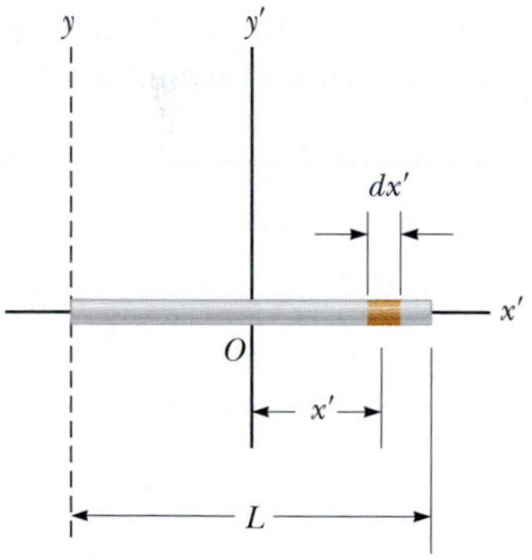

SOLUTION

Conceptualize Imagine twirling the rod in Figure 10.8 with your fingers around its midpoint. If you have a meterstick handy, use it to simulate the spinning of a thin rod and feel the resistance it offers to being spun.

Figure 10.8 (Example 10.4) A uniform rigid rod of length *L*. The moment of inertia about the *y′* axis is less than that about the *y* axis.

Categorize This example is a substitution problem, using the definition of moment of inertia in Equation 10.17. As with any calculus problem, the solution involves reducing the integrand to a single variable.

The shaded length element *dx′* in Figure 10.8 has a mass *dm* equal to the mass per unit length λ multiplied by *dx′*.

Express *dm* in terms of *dx′*:

$$dm = \lambda \, dx' = \frac{M}{L} \, dx'$$

Substitute this expression into Equation 10.17, with $r^2 = (x')^2$:

$$I_y = \int r^2 \, dm = \int_{-L/2}^{L/2} (x')^2 \frac{M}{L} \, dx' = \frac{M}{L} \int_{-L/2}^{L/2} (x')^2 \, dx'$$

$$= \frac{M}{L}\left[\frac{(x')^3}{3} \right]_{-L/2}^{L/2} = \boxed{\tfrac{1}{12} ML^2}$$

Check this result in Table 10.2. For practice, calculate the moment of inertia about an axis *y* passing through the end of the rod in Figure 10.8.

Example 10.5 | Uniform Solid Cylinder

A uniform solid cylinder has a radius R, mass M, and length L. Calculate its moment of inertia about its central axis (the z axis in Fig. 10.9).

SOLUTION

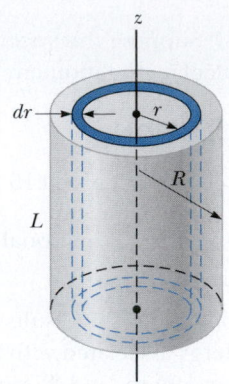

Figure 10.9 (Example 10.5) Calculating I about the z axis for a uniform solid cylinder.

Conceptualize To simulate this situation, imagine twirling a can of frozen juice around its central axis. Don't twirl a can of vegetable soup; it is not a rigid object! The liquid is able to move relative to the metal can.

Categorize This example is a substitution problem, using the definition of moment of inertia. As with Example 10.4, we must reduce the integrand to a single variable.

It is convenient to divide the cylinder into many cylindrical shells, each having radius r, thickness dr, and length L as shown in Figure 10.9. The density of the cylinder is ρ. The volume dV of each shell is its cross-sectional area multiplied by its length: $dV = L\, dA = L(2\pi r)\, dr$.

Express dm in terms of dr:

$$dm = \rho\, dV = \rho L(2\pi r)\, dr$$

Substitute this expression into Equation 10.17:

$$I_z = \int r^2\, dm = \int r^2 [\rho L(2\pi r)\, dr] = 2\pi\rho L \int_0^R r^3\, dr = \tfrac{1}{2}\pi\rho LR^4$$

Use the total volume $\pi R^2 L$ of the cylinder to express its density:

$$\rho = \frac{M}{V} = \frac{M}{\pi R^2 L}$$

Substitute this value into the expression for I_z:

$$I_z = \tfrac{1}{2}\pi\left(\frac{M}{\pi R^2 L}\right)LR^4 = \tfrac{1}{2}MR^2$$

Check this result in Table 10.2.

What If? What if the length of the cylinder in Figure 10.9 is increased to $2L$, while the mass M and radius R are held fixed? How does that change the moment of inertia of the cylinder?

Answer Notice that the result for the moment of inertia of a cylinder does not depend on L, the length of the cylinder. It applies equally well to a long cylinder and a flat disk having the same mass M and radius R. Therefore, the moment of inertia of the cylinder would not be affected by changing its length.

10.5 | Torque and the Vector Product

When a net force is exerted on a rigid object pivoted about some axis and the line of action[3] of the force does not pass through the pivot, the object tends to rotate about that axis. For example, when you push on a door, the door rotates about an axis through the hinges. The tendency of a force to rotate an object about some axis is measured by a vector quantity called **torque**. Torque is the cause of changes in rotational motion and is analogous to force, which causes changes in translational motion. Consider the wrench pivoted about the axis through O in Figure 10.10. The applied force $\vec{\mathbf{F}}$ generally can act at an angle ϕ with respect to the position vector $\vec{\mathbf{r}}$ locating the point of application of the force. We define the torque τ resulting from the force $\vec{\mathbf{F}}$ with the expression[4]

$$\tau \equiv rF\sin\phi \qquad \qquad \textbf{10.19}\blacktriangleleft$$

Torque has units of newton · meters (N · m) in the SI system.[5]

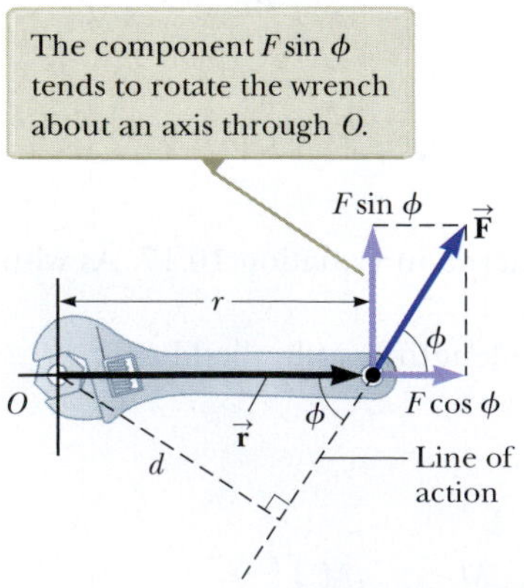

The component $F\sin\phi$ tends to rotate the wrench about an axis through O.

Figure 10.10 A force $\vec{\mathbf{F}}$ is applied to a wrench in an effort to loosen a bolt. The force has a greater rotating tendency about an axis through O as F increases and as the moment arm d increases.

---

[3]The line of action of a force is an imaginary line colinear with the force vector and extending to infinity in both directions.

[4]In general, torque is a vector. For rotation about a fixed axis, however, we will use italic, nonbold notation and specify the direction with a positive or a negative sign as we did for angular speed and acceleration in Section 10.1. We will briefly treat the vector nature of torque in a short while.

[5]In Chapter 6, we saw the product of newtons and meters when we defined work and called this product a *joule*. We do not use this term here because the joule is only to be used when discussing energy. For torque, the unit is simply the newton · meter, or N · m.

It is very important to recognize that torque is defined only when a reference axis is specified, from which the distance r is determined. We can interpret Equation 10.19 in two different ways. Looking at the force components in Figure 10.10, we see that the component $F\cos\phi$ parallel to $\vec{r}$ will not cause a rotation of the wrench around the pivot point because its line of action passes right through the pivot point. Similarly, you cannot open a door by pushing on the hinges! Therefore, only the perpendicular component $F\sin\phi$ causes a rotation of the wrench about the pivot. In this case, we can write Equation 10.19 as

$$\tau = r(F\sin\phi)$$

so that the torque is the product of the distance to the point of application of the force and the perpendicular component of the force. In some problems, this method is the easiest way to interpret the calculation of the torque.

The second way to interpret Equation 10.19 is to associate the sine function with the distance r so that we can write

$$\tau = F(r\sin\phi) = Fd$$

The quantity $d = r\sin\phi$, called the **moment arm** (or *lever arm*) of the force $\vec{F}$, represents the perpendicular distance from the rotation axis to the line of action of $\vec{F}$. In some problems, this approach to the calculation of the torque is easier than that of resolving the force into components.

If two or more forces are acting on a rigid object, as in Active Figure 10.11, each has a tendency to produce a rotation about the pivot at O. For example, if the object is initially at rest, $\vec{F}_2$ tends to rotate the object clockwise and $\vec{F}_1$ tends to rotate the object counterclockwise. We shall use the convention that the sign of the torque resulting from a force is positive if its turning tendency is counterclockwise around the rotation axis and negative if its turning tendency is clockwise. For example, in Active Figure 10.11, the torque resulting from $\vec{F}_1$, which has a moment arm of d_1, is *positive* and equal to $+F_1d_1$; the torque from $\vec{F}_2$ is *negative* and equal to $-F_2d_2$. Hence, the *net* torque acting on the rigid object about an axis through O is

$$\tau_{net} = \tau_1 + \tau_2 = F_1d_1 - F_2d_2$$

From the definition of torque, we see that the rotating tendency increases as F increases and as d increases. For example, we cause more rotation of a door if (a) we push harder or (b) we push at the doorknob rather than at a point close to the hinges. Torque should *not* be confused with force. Torque *depends* on force, but it also depends on *where the force is applied.*

So far, we have not discussed the vector nature of torque aside from assigning a positive or negative value to τ. Consider a force $\vec{F}$ acting on a particle located at the vector position $\vec{r}$ (Active Fig. 10.12). The *magnitude* of the torque due to this force relative to an axis through the origin is $|rF\sin\phi|$, where ϕ is the angle between $\vec{r}$ and $\vec{F}$. The axis about which $\vec{F}$ would tend to produce rotation is perpendicular to the plane formed by $\vec{r}$ and $\vec{F}$. If the force lies in the xy plane, as in Active Figure 10.12, the torque is represented by a vector parallel to the z axis. The force in Active Figure 10.12 creates a torque that tends to rotate the particle counterclockwise when we are looking down the z axis. We define the direction of torque such that the vector $\vec{\tau}$ is in the positive z direction (i.e., coming toward your eyes). If we reverse the direction of $\vec{F}$ in Active Figure 10.12, $\vec{\tau}$ is in the negative z direction. With this choice, the torque vector can be defined to be equal to the **vector product,** or **cross product,** of $\vec{r}$ and $\vec{F}$:

$$\vec{\tau} \equiv \vec{r} \times \vec{F} \qquad \text{10.20} \blacktriangleleft$$

We now give a formal definition of the vector product, first introduced in Section 1.8. Given any two vectors $\vec{A}$ and $\vec{B}$, the vector product $\vec{A} \times \vec{B}$ is defined as a

Pitfall Prevention | 10.5
Torque Depends on Your Choice of Axis
Like moment of inertia, there is no unique value of the torque on an object. Its value depends on your choice of rotation axis.

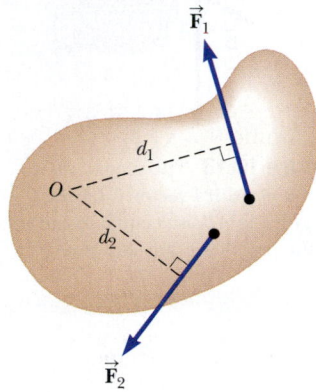

Active Figure 10.11 The force $\vec{F}_1$ tends to rotate the object counterclockwise about an axis through O, and $\vec{F}_2$ tends to rotate the object clockwise.

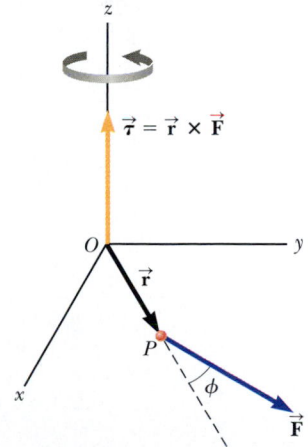

Active Figure 10.12 The torque vector $\vec{\tau}$ lies in a direction perpendicular to the plane formed by the position vector $\vec{r}$ and the applied force vector $\vec{F}$.

▶ Definition of torque using the cross product

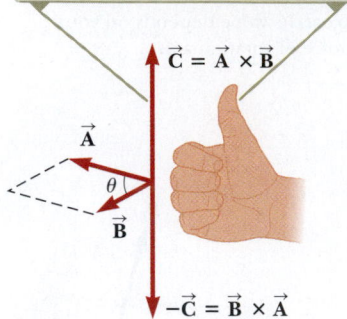

The direction of $\vec{C}$ is perpendicular to the plane formed by $\vec{A}$ and $\vec{B}$, and its direction is determined by the right-hand rule.

$\vec{C} = \vec{A} \times \vec{B}$

$\vec{A}$

θ

$\vec{B}$

$-\vec{C} = \vec{B} \times \vec{A}$

Figure 10.13 The vector product $\vec{A} \times \vec{B}$ is a third vector $\vec{C}$ having a magnitude $AB \sin \theta$ equal to the area of the parallelogram shown.

third vector $\vec{C}$, the *magnitude* of which is $AB \sin \theta$, where θ is the angle between $\vec{A}$ and $\vec{B}$:

$$\vec{C} = \vec{A} \times \vec{B} \qquad \qquad \text{10.21} \blacktriangleleft$$

$$C = |\vec{C}| \equiv AB \sin \theta \qquad \qquad \text{10.22} \blacktriangleleft$$

Note that the quantity $AB \sin \theta$ is equal to the area of the parallelogram formed by $\vec{A}$ and $\vec{B}$, as shown in Figure 10.13. The *direction* of $\vec{A} \times \vec{B}$ is perpendicular to the plane formed by $\vec{A}$ and $\vec{B}$ and is determined by the right-hand rule illustrated in Figure 10.13. The four fingers of the right hand are pointed along $\vec{A}$ and then "wrapped" into $\vec{B}$ through the angle θ. The direction of the upright thumb is the direction of $\vec{A} \times \vec{B}$. Because of the notation, $\vec{A} \times \vec{B}$ is often read "$\vec{A}$ cross $\vec{B}$," hence the term *cross product*.

Some properties of the vector product follow from its definition:

- Unlike the case of the scalar product, the vector product is not commutative; in fact,

$$\vec{A} \times \vec{B} = -\vec{B} \times \vec{A} \qquad \qquad \text{10.23} \blacktriangleleft$$

Therefore, if you change the order of the vector product, you must change the sign. You can easily verify this relation with the right-hand rule (see Fig. 10.13).
- If $\vec{A}$ is parallel to $\vec{B}$ ($\theta = 0°$ or $180°$), then $\vec{A} \times \vec{B} = 0$; therefore, it follows that $\vec{A} \times \vec{A} = 0$.
- If $\vec{A}$ is perpendicular to $\vec{B}$, then $|\vec{A} \times \vec{B}| = AB$.
- The vector product obeys the distributive law:

$$\vec{A} \times (\vec{B} + \vec{C}) = \vec{A} \times \vec{B} + \vec{A} \times \vec{C} \qquad \qquad \text{10.24} \blacktriangleleft$$

- The derivative of the vector product with respect to some variable such as t is

$$\frac{d}{dt}(\vec{A} \times \vec{B}) = \frac{d\vec{A}}{dt} \times \vec{B} + \vec{A} \times \frac{d\vec{B}}{dt} \qquad \qquad \text{10.25} \blacktriangleleft$$

where it is important to preserve the multiplicative order of the terms on the right side in view of Equation 10.23.

It is left to Problem 10.26 in Enhanced WebAssign to show, from Equations 10.21 and 10.22 and the definition of unit vectors, that the vector products of the unit vectors $\hat{i}$, $\hat{j}$, and $\hat{k}$ obey the following expressions:

$$\hat{i} \times \hat{i} = \hat{j} \times \hat{j} = \hat{k} \times \hat{k} = 0$$

$$\hat{i} \times \hat{j} = -\hat{j} \times \hat{i} = \hat{k} \qquad \qquad \text{10.26} \blacktriangleleft$$

$$\hat{j} \times \hat{k} = -\hat{k} \times \hat{j} = \hat{i}$$

$$\hat{k} \times \hat{i} = -\hat{i} \times \hat{k} = \hat{j}$$

Signs are interchangeable. For example, $\hat{i} \times (-\hat{j}) = -\hat{i} \times \hat{j} = -\hat{k}$.

▸ **QUICK QUIZ 10.5** **(i)** If you are trying to loosen a stubborn screw from a piece of wood with a screwdriver and fail, should you find a screwdriver for which the handle is (a) longer or (b) fatter? **(ii)** If you are trying to loosen a stubborn bolt from a piece of metal with a wrench and fail, should you find a wrench for which the handle is (a) longer or (b) fatter?

Example **10.6** | **The Net Torque on a Cylinder**

A one-piece cylinder is shaped as shown in Figure 10.14, with a core section protruding from the larger drum. The cylinder is free to rotate about the central z axis shown in the drawing. A rope wrapped around the drum, which has radius R_1, exerts a force $\vec{\mathbf{T}}_1$ to the right on the cylinder. A rope wrapped around the core, which has radius R_2, exerts a force $\vec{\mathbf{T}}_2$ downward on the cylinder.

(A) What is the net torque acting on the cylinder about the rotation axis (which is the z axis in Fig. 10.14)?

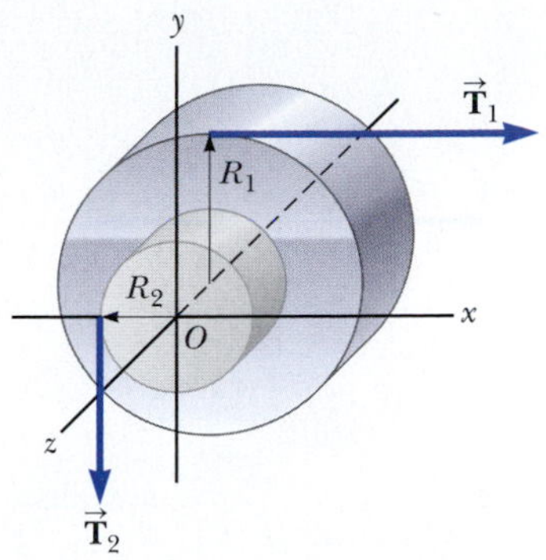

SOLUTION

Conceptualize Imagine that the cylinder in Figure 10.14 is a shaft in a machine. The force $\vec{\mathbf{T}}_1$ could be applied by a drive belt wrapped around the drum. The force $\vec{\mathbf{T}}_2$ could be applied by a friction brake at the surface of the core.

Figure 10.14 (Example 10.6) A solid cylinder pivoted about the z axis through O. The moment arm of $\vec{\mathbf{T}}_1$ is R_1, and the moment arm of $\vec{\mathbf{T}}_2$ is R_2.

Categorize This example is a substitution problem in which we evaluate the net torque using Equation 10.19.

The torque due to $\vec{\mathbf{T}}_1$ about the rotation axis is $-R_1 T_1$. (The sign is negative because the torque tends to produce clockwise rotation.) The torque due to $\vec{\mathbf{T}}_2$ is $+R_2 T_2$. (The sign is positive because the torque tends to produce counterclockwise rotation of the cylinder.)

Evaluate the net torque about the rotation axis:

$$\sum \tau = \tau_1 + \tau_2 = \boxed{R_2 T_2 - R_1 T_1}$$

As a quick check, notice that if the two forces are of equal magnitude, the net torque is negative because $R_1 > R_2$. Starting from rest with both forces of equal magnitude acting on it, the cylinder would rotate clockwise because $\vec{\mathbf{T}}_1$ would be more effective at turning it than would $\vec{\mathbf{T}}_2$.

(B) Suppose $T_1 = 5.0$ N, $R_1 = 1.0$ m, $T_2 = 15$ N, and $R_2 = 0.50$ m. What is the net torque about the rotation axis, and which way does the cylinder rotate starting from rest?

SOLUTION

Substitute the given values: $\sum \tau = (0.50 \text{ m})(15 \text{ N}) - (1.0 \text{ m})(5.0 \text{ N}) = \boxed{2.5 \text{ N} \cdot \text{m}}$

Because this net torque is positive, the cylinder begins to rotate in the counterclockwise direction.

Example **10.7** | **The Vector Product**

Two vectors lying in the xy plane are given by the equations $\vec{\mathbf{A}} = 2\hat{\mathbf{i}} + 3\hat{\mathbf{j}}$ and $\vec{\mathbf{B}} = -\hat{\mathbf{i}} + 2\hat{\mathbf{j}}$. Find $\vec{\mathbf{A}} \times \vec{\mathbf{B}}$ and verify that $\vec{\mathbf{A}} \times \vec{\mathbf{B}} = -\vec{\mathbf{B}} \times \vec{\mathbf{A}}$.

SOLUTION

Conceptualize Given the unit-vector notations of the vectors, think about the directions the vectors point in space. Imagine the parallelogram shown in Figure 10.13 for these vectors.

Categorize Because we use the definition of the cross product discussed in this section, we categorize this example as a substitution problem.

Write the cross product of the two vectors: $\vec{\mathbf{A}} \times \vec{\mathbf{B}} = (2\hat{\mathbf{i}} + 3\hat{\mathbf{j}}) \times (-\hat{\mathbf{i}} + 2\hat{\mathbf{j}})$

Perform the multiplication: $\vec{\mathbf{A}} \times \vec{\mathbf{B}} = 2\hat{\mathbf{i}} \times (-\hat{\mathbf{i}}) + 2\hat{\mathbf{i}} \times 2\hat{\mathbf{j}} + 3\hat{\mathbf{j}} \times (-\hat{\mathbf{i}}) + 3\hat{\mathbf{j}} \times 2\hat{\mathbf{j}}$

Use Equation 10.26 to evaluate the various terms: $\vec{\mathbf{A}} \times \vec{\mathbf{B}} = 0 + 4\hat{\mathbf{k}} + 3\hat{\mathbf{k}} + 0 = \boxed{7\hat{\mathbf{k}}}$

To verify that $\vec{\mathbf{A}} \times \vec{\mathbf{B}} = -\vec{\mathbf{B}} \times \vec{\mathbf{A}}$, evaluate $\vec{\mathbf{B}} \times \vec{\mathbf{A}}$: $\vec{\mathbf{B}} \times \vec{\mathbf{A}} = (-\hat{\mathbf{i}} + 2\hat{\mathbf{j}}) \times (2\hat{\mathbf{i}} + 3\hat{\mathbf{j}})$

Perform the multiplication: $\vec{\mathbf{B}} \times \vec{\mathbf{A}} = (-\hat{\mathbf{i}}) \times 2\hat{\mathbf{i}} + (-\hat{\mathbf{i}}) \times 3\hat{\mathbf{j}} + 2\hat{\mathbf{j}} \times 2\hat{\mathbf{i}} + 2\hat{\mathbf{j}} \times 3\hat{\mathbf{j}}$

Use Equation 10.26 to evaluate the various terms: $\vec{\mathbf{B}} \times \vec{\mathbf{A}} = 0 - 3\hat{\mathbf{k}} - 4\hat{\mathbf{k}} + 0 = \boxed{-7\hat{\mathbf{k}}}$

Therefore, $\vec{\mathbf{A}} \times \vec{\mathbf{B}} = -\vec{\mathbf{B}} \times \vec{\mathbf{A}}$.

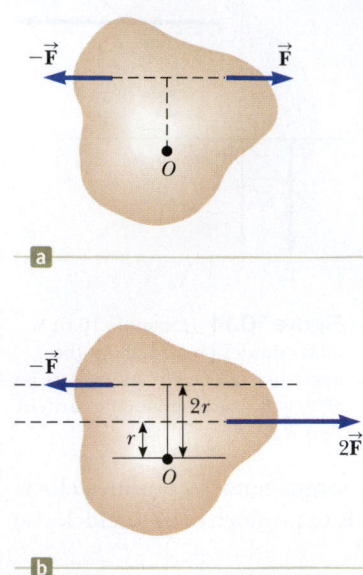

Figure 10.15 (a) The two forces acting on the object are equal in magnitude and opposite in direction. Because they also act along the same line of action, the net torque is zero and the object is in equilibrium. (b) Another situation in which two forces act on an object to produce zero net torque about O (but *not* zero net force).

⟨10.6 | Analysis Model: Rigid Object in Equilibrium

We have defined a rigid object and have discussed torque as the cause of changes in rotational motion of a rigid object. We can now establish analysis models for a rigid object subject to torques that are analogous to those for a particle subject to forces. We begin by imagining a rigid object with balanced torques, which will give us an analysis model that we call the **rigid object in equilibrium.**

Consider two forces of equal magnitude and opposite direction applied to an object as shown in Figure 10.15a. The force directed to the right tends to rotate the object clockwise about an axis perpendicular to the diagram through O, whereas the force directed to the left tends to rotate it counterclockwise about that axis. Because the forces are of equal magnitude and act at the same perpendicular distance from O, their torques are equal in magnitude. Therefore, the net torque on the rigid object is zero. The situation shown in Figure 10.15b is another case in which the net torque about O is zero (although the net *force* on the object is not zero), and we can devise many more cases.

With no net torque, no change occurs in rotational motion and the rotational motion of the rigid object remains in its original state. This state is an equilibrium situation, analogous to translational equilibrium, discussed in Chapter 4.

We now have two conditions for complete equilibrium of an object, which can be stated as follows:

- The net external force must equal zero:

$$\sum \vec{F}_{ext} = 0 \qquad \textbf{10.27} ◀$$

- The net external torque must be zero about *any* axis:

$$\sum \vec{\tau}_{ext} = 0 \qquad \textbf{10.28} ◀$$

The first condition is a statement of translational equilibrium. The second condition is a statement of rotational equilibrium. In the special case of **static equilibrium**, the object is at rest, so it has no translational or angular speed (i.e., $v_{CM} = 0$ and $\omega = 0$).

The two vector expressions given by Equations 10.27 and 10.28 are equivalent, in general, to six scalar equations: three from the first condition of equilibrium and three from the second (corresponding to x, y, and z components). Hence, in a complex system involving several forces acting in various directions, you would be faced with solving a set of equations with many unknowns. Here, we restrict our discussion to situations in which all the forces on an object lie in the xy plane. (Forces whose vector representations are in the same plane are said to be *coplanar.*) With this restriction, we need to deal with only three scalar equations. Two of them come from balancing the forces on the object in the x and y directions. The third comes from the torque equation, namely, that the net torque on the object about an axis through *any* point in the xy plane must be zero. Hence, the two conditions of equilibrium provide the equations

$$\sum F_x = 0 \qquad \sum F_y = 0 \qquad \sum \tau_z = 0 \qquad \textbf{10.29} ◀$$

where the axis of the torque equation is arbitrary. Use these equations when you determine that the rigid object in equilibrium model is appropriate and the forces on the rigid object are in the xy plane.

In working static equilibrium problems, it is important to recognize all external forces acting on the object. Failure to do so will result in an incorrect analysis. The following procedure is recommended when analyzing an object in equilibrium under the action of several external forces:

> **PROBLEM-SOLVING STRATEGY:** Rigid Object in Equilibrium

When analyzing a rigid object in equilibrium under the action of several external forces, use the following procedure.

1. **Conceptualize.** Think about the object that is in equilibrium and identify all the forces on it. Imagine what effect each force would have on the rotation of the object if it were the only force acting.

2. **Categorize.** Confirm that the object under consideration is indeed a rigid object in equilibrium. The object must have zero translational acceleration and zero angular acceleration.

3. **Analyze.** Draw a diagram and label all external forces acting on the object. Try to guess the correct direction for any forces that are not specified. When using the particle under a net force model, the object on which forces act can be represented in a free-body diagram with a dot because it does not matter where on the object the forces are applied. When using the rigid object in equilibrium model, however, we cannot use a dot to represent the object because the location where forces act is important in the calculation. Therefore, in a diagram showing the forces on an object, we must show the actual object or a simplified version of it.

   Resolve all forces into rectangular components, choosing a convenient coordinate system. Then apply the first condition for equilibrium, Equation 10.27. Remember to keep track of the signs of the various force components.

   Choose a convenient axis for calculating the net torque on the rigid object. Remember that the choice of the axis for the torque equation is arbitrary; therefore, choose an axis that simplifies your calculation as much as possible. Usually, the most convenient axis for calculating torques is one through a point at which several forces act, so their torques around this axis are zero. If you don't know a force or don't need to know a force, it is often beneficial to choose an axis through the point at which this force acts. Apply the second condition for equilibrium, Equation 10.28.

   Solve the simultaneous equations for the unknowns in terms of the known quantities.

4. **Finalize.** Make sure your results are consistent with your diagram. If you selected a direction that leads to a negative sign in your solution for a force, do not be alarmed; it merely means that the direction of the force is the opposite of what you guessed. Add up the vertical and horizontal forces on the object and confirm that each set of components adds to zero. Add up the torques on the object and confirm that the sum equals zero.

Example 10.8 | Standing on a Horizontal Beam

A uniform horizontal beam with a length of $\ell = 8.00$ m and a weight of $W_b = 200$ N is attached to a wall by a pin connection. Its far end is supported by a cable that makes an angle of $\phi = 53.0°$ with the beam (Fig. 10.16a, page 276). A person of weight $W_p = 600$ N stands a distance $d = 2.00$ m from the wall. Find the tension in the cable as well as the magnitude and direction of the force exerted by the wall on the beam.

SOLUTION

Conceptualize Imagine that the person in Figure 10.16a moves outward on the beam. It seems reasonable that the farther he moves outward, the larger the torque he applies about the pivot and the larger the tension in the cable must be to balance this torque.

Categorize Because the system is at rest, we categorize the beam as a rigid object in equilibrium.

continued

10.8 *cont.*

Analyze We identify all the external forces acting on the beam: the 200-N gravitational force, the force $\vec{T}$ exerted by the cable, the force $\vec{R}$ exerted by the wall at the pivot, and the 600-N force that the person exerts on the beam. These forces are all indicated in the force diagram for the beam shown in Figure 10.16b. When we assign directions for forces, it is sometimes helpful to imagine what would happen if a force were suddenly removed. For example, if the wall were to vanish suddenly, the left end of the beam would move to the left as it begins to fall. This scenario tells us that the wall is not only holding the beam up but is also pressing outward against it. Therefore, we draw the vector $\vec{R}$ in the direction shown in Figure 10.16b. Figure 10.16c shows the horizontal and vertical components of $\vec{T}$ and $\vec{R}$.

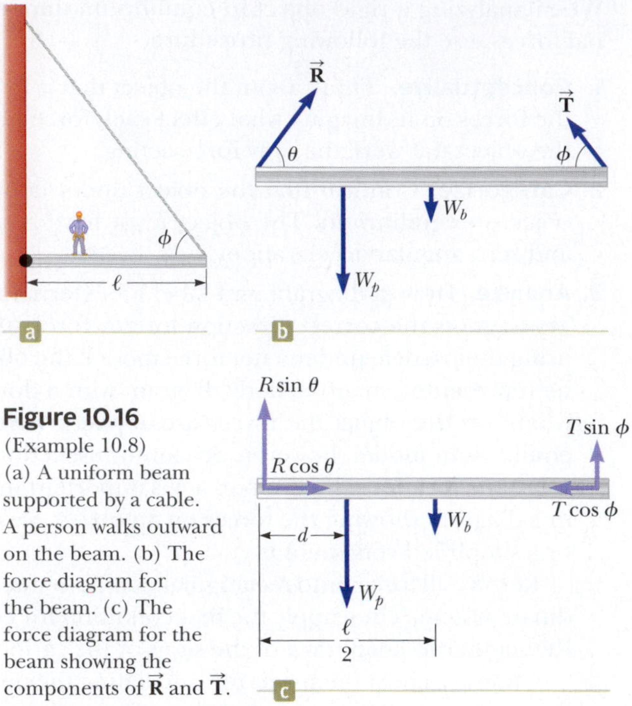

Substitute expressions for the forces on the beam into Equation 10.27:

$$(1) \quad \sum F_x = R \cos \theta - T \cos \phi = 0$$

$$(2) \quad \sum F_y = R \sin \theta + T \sin \phi - W_p - W_b = 0$$

where we have chosen rightward and upward as our positive directions. Because R, T, and θ are all unknown, we cannot obtain a solution from these expressions alone. (To solve for the unknowns, the number of simultaneous equations must generally equal the number of unknowns.)

Now let's invoke the condition for rotational equilibrium. A convenient axis to choose for our torque equation is the one that passes through the pin connection. The feature that makes this axis so convenient is that the force $\vec{R}$ and the horizontal component of $\vec{T}$ both have a moment arm of zero; hence, these forces produce no torque about this axis.

Figure 10.16
(Example 10.8)
(a) A uniform beam supported by a cable. A person walks outward on the beam. (b) The force diagram for the beam. (c) The force diagram for the beam showing the components of $\vec{R}$ and $\vec{T}$.

Substitute expressions for the torques on the beam into Equation 10.28:

$$\sum \tau_z = (T \sin \phi)(\ell) - W_p d - W_b \left(\frac{\ell}{2}\right) = 0$$

This equation contains only T as an unknown because of our choice of rotation axis. Solve for T and substitute numerical values:

$$T = \frac{W_p d + W_b (\ell/2)}{\ell \sin \phi} = \frac{(600 \text{ N})(2.00 \text{ m}) + (200 \text{ N})(4.00 \text{ m})}{(8.00 \text{ m}) \sin 53.0°} = \boxed{313 \text{ N}}$$

Rearrange Equations (1) and (2) and then divide:

$$\frac{R \sin \theta}{R \cos \theta} = \tan \theta = \frac{W_p + W_b - T \sin \phi}{T \cos \phi}$$

Solve for θ and substitute numerical values:

$$\theta = \tan^{-1}\left(\frac{W_p + W_b - T \sin \phi}{T \cos \phi}\right)$$

$$= \tan^{-1}\left[\frac{600 \text{ N} + 200 \text{ N} - (313 \text{ N}) \sin 53.0°}{(313 \text{ N}) \cos 53.0°}\right] = \boxed{71.1°}$$

Solve Equation (1) for R and substitute numerical values:

$$R = \frac{T \cos \phi}{\cos \theta} = \frac{(313 \text{ N}) \cos 53.0°}{\cos 71.1°} = \boxed{581 \text{ N}}$$

Finalize The positive value for the angle θ indicates that our estimate of the direction of $\vec{R}$ was accurate.

Had we selected some other axis for the torque equation, the solution might differ in the details but the answers would be the same. For example, had we chosen an axis through the center of gravity of the beam, the torque equation would involve both T and R. This equation, coupled with Equations (1) and (2), however, could still be solved for the unknowns. Try it!

Example 10.9 | The Leaning Ladder

A uniform ladder of length ℓ rests against a smooth, vertical wall (Fig. 10.17a). The mass of the ladder is m, and the coefficient of static friction between the ladder and the ground is $\mu_s = 0.40$. Find the minimum angle θ_{min} at which the ladder does not slip.

SOLUTION

Conceptualize Think about any ladders you have climbed. Do you want a large friction force between the bottom of the ladder and the surface or a small one? If the friction force is zero, will the ladder stay up? Simulate a ladder with a ruler leaning against a vertical surface. Does the ruler slip at some angles and stay up at others?

Categorize We do not wish the ladder to slip, so we model it as a rigid object in equilibrium.

Analyze A diagram showing all the external forces acting on the ladder is illustrated in Figure 10.17b. The force exerted by the ground on the ladder is the vector sum of a normal force $\vec{n}$ and the force of static friction $\vec{f}_s$. The force $\vec{P}$ exerted by the wall on the ladder is horizontal because the wall is frictionless.

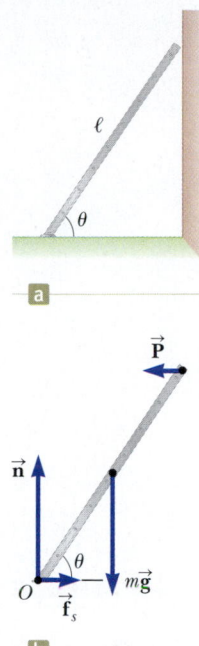

Figure 10.17 (Example 10.9) (a) A uniform ladder at rest, leaning against a smooth wall. The ground is rough. (b) The forces on the ladder.

Apply the first condition for equilibrium to the ladder:	(1) $\sum F_x = f_s - P = 0$
	(2) $\sum F_y = n - mg = 0$
Solve Equation (1) for P:	(3) $P = f_s$
Solve Equation (2) for n:	(4) $n = mg$
When the ladder is on the verge of slipping, the force of static friction must have its maximum value, which is given by $f_{s,max} = \mu_s n$. Combine this equation with Equations (3) and (4):	(5) $P = f_{s,max} = \mu_s n = \mu_s mg$
Apply the second condition for equilibrium to the ladder, taking torques about an axis through O:	$\sum \tau_O = P\ell \sin \theta_{min} - mg\dfrac{\ell}{2}\cos \theta_{min} = 0$
Solve for $\tan \theta_{min}$ and substitute for P from Equation (5):	$\dfrac{\sin \theta_{min}}{\cos \theta_{min}} = \tan \theta_{min} = \dfrac{mg}{2P} = \dfrac{mg}{2\mu_s mg} = \dfrac{1}{2\mu_s}$
Solve for the angle θ_{min}:	$\theta_{min} = \tan^{-1}\left(\dfrac{1}{2\mu_s}\right) = \tan^{-1}\left[\dfrac{1}{2(0.40)}\right] = 51°$

Finalize Notice that the angle depends only on the coefficient of friction, not on the mass or length of the ladder.

10.7 | Analysis Model: Rigid Object Under a Net Torque

In the preceding section, we investigated the equilibrium situation in which the net torque on a rigid object is zero. What if the net torque on a rigid object is not zero? In analogy with Newton's second law for translational motion, we should expect the angular speed of the rigid object to change. The net torque will cause angular acceleration of the rigid object. We describe this situation with a new analysis model, the **rigid object under a net torque,** and investigate this model in this section.

Let us imagine a rotating rigid object again as a collection of particles. The rigid object will be subject to a number of forces applied at various locations on the rigid object at which individual particles will be located. Therefore, we can imagine that the forces on the rigid object are exerted on individual particles of the rigid object. We will calculate the net torque on the object due to the torques resulting from these forces around the rotation axis of the rotating object. Any applied force can be represented by its radial component and its tangential component. The radial component of an applied force provides no torque because its line of action goes through the rotation axis. Therefore, only the tangential component of an applied force contributes to the torque.

On any given particle, described by index variable i, within the rigid object, we can use Newton's second law to describe the tangential acceleration of the particle:

$$F_{ti} = m_i a_{ti}$$

where the t subscript refers to tangential components. Let us multiply both sides of this expression by r_i, the distance of the particle from the rotation axis:

$$r_i F_{ti} = r_i m_i a_{ti}$$

Using Equation 10.11 and recognizing the definition of torque ($\tau = rF \sin \phi = rF_t$ in this case), we can rewrite this expression as

$$\tau_i = m_i r_i^2 \alpha_i$$

Now, let us add up the torques on all particles of the rigid object:

$$\sum_i \tau_i = \sum_i m_i r_i^2 \alpha_i$$

The left side is the net torque on all particles of the rigid object. The net torque associated with *internal* forces is zero, however. To understand why, recall that Newton's third law tells us that the internal forces occur in equal and opposite pairs that lie along the line of separation of each pair of particles. The torque due to each action–reaction force pair is therefore zero. On summation of all torques, we see that the *net internal torque vanishes.* The term on the left, then, reduces to the net *external* torque.

On the right, we adopt the rigid object model by demanding that all particles have the same angular acceleration α. Therefore, this equation becomes

$$\sum \tau_{ext} = \left(\sum_i m_i r_i^2 \right) \alpha$$

where the torque and angular acceleration no longer have subscripts because they refer to quantities associated with the rigid object as a whole rather than to individual particles. We recognize the quantity in parentheses as the moment of inertia I of the rigid object. Therefore,

▶ Rotational analog to Newton's second law

$$\sum \tau_{ext} = I\alpha \qquad \qquad \textbf{10.30} ◀$$

That is, the net torque acting on the rigid object is proportional to its angular acceleration, and the proportionality constant is the moment of inertia. It is important to note that $\sum \tau_{ext} = I\alpha$ is the rotational analog of Newton's second law of motion for a system of particles (Eq. 8.40), $\sum F_{ext} = Ma_{CM}$.

◀ **QUICK QUIZ 10.6** You turn off your electric drill and find that the time interval for the rotating bit to come to rest due to frictional torque in the drill is Δt. You replace the bit with a larger one that results in a doubling of the moment of inertia of the drill's entire rotating mechanism. When this larger bit is rotated at the same angular speed as the first, and the drill is turned off, the frictional torque remains the same as that for the previous situation. What is the time interval for this second bit to come to rest?   **(a)** $4\Delta t$ **(b)** $2\Delta t$ **(c)** Δt **(d)** $0.5\Delta t$ **(e)** $0.25\Delta t$ **(f)** impossible to determine

Example **10.10** | **Angular Acceleration of a Wheel**

A wheel of radius R, mass M, and moment of inertia I is mounted on a frictionless, horizontal axle as in Figure 10.18. A light cord wrapped around the wheel supports an object of mass m. When the wheel is released, the object accelerates downward, the cord unwraps off the wheel, and the wheel rotates with an angular acceleration. Find expressions for the angular acceleration of the wheel, the translational acceleration of the object, and the tension in the cord.

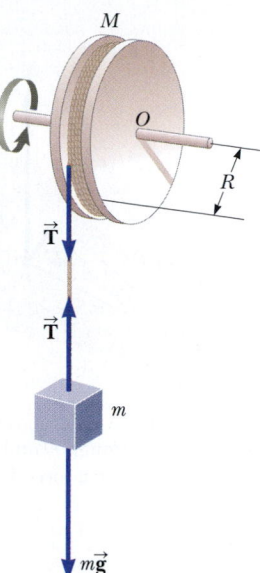

Figure 10.18 (Example 10.10) An object hangs from a cord wrapped around a wheel.

SOLUTION

Conceptualize Imagine that the object is a bucket in an old-fashioned wishing well. It is tied to a cord that passes around a cylinder equipped with a crank for raising the bucket. After the bucket has been raised, the system is released and the bucket accelerates downward while the cord unwinds off the cylinder.

Categorize We apply two analysis models here. The object is modeled as a particle under a net force. The wheel is modeled as a rigid object under a net torque.

Analyze The magnitude of the torque acting on the wheel about its axis of rotation is $\tau = TR$, where T is the force exerted by the cord on the rim of the wheel. (The gravitational force exerted by the Earth on the wheel and the normal force exerted by the axle on the wheel both pass through the axis of rotation and therefore produce no torque.)

Write Equation 10.30:

$$\sum \tau_{\text{ext}} = I\alpha$$

Solve for α and substitute the net torque:

$$(1)\ \alpha = \frac{\sum \tau_{\text{ext}}}{I} = \frac{TR}{I}$$

Apply Newton's second law to the motion of the object, taking the downward direction to be positive:

$$\sum F_y = mg - T = ma$$

Solve for the acceleration a:

$$(2)\ a = \frac{mg - T}{m}$$

Equations (1) and (2) have three unknowns: α, a, and T. Because the object and wheel are connected by a cord that does not slip, the translational acceleration of the suspended object is equal to the tangential acceleration of a point on the wheel's rim. Therefore, the angular acceleration α of the wheel and the translational acceleration of the object are related by $a = R\alpha$.

Use this fact together with Equations (1) and (2):

$$(3)\ a = R\alpha = \frac{TR^2}{I} = \frac{mg - T}{m}$$

Solve for the tension T:

$$(4)\ T = \frac{mg}{1 + (mR^2/I)}$$

Substitute Equation (4) into Equation (2) and solve for a:

$$(5)\ a = \frac{g}{1 + (I/mR^2)}$$

Use $a = R\alpha$ and Equation (5) to solve for α:

$$\alpha = \frac{a}{R} = \frac{g}{R + (I/mR)}$$

Finalize We finalize this problem by imagining the behavior of the system in some extreme limits.

What If? What if the wheel were to become very massive so that I becomes very large? What happens to the acceleration a of the object and the tension T?

Answer If the wheel becomes infinitely massive, we can imagine that the object of mass m will simply hang from the cord without causing the wheel to rotate.

We can show that mathematically by taking the limit $I \rightarrow \infty$. Equation (5) then becomes

$$a = \frac{g}{1 + (I/mR^2)} \rightarrow 0$$

which agrees with our conceptual conclusion that the object will hang at rest. Also, Equation (4) becomes

$$T = \frac{mg}{1 + (mR^2/I)} \rightarrow mg$$

which is consistent because the object simply hangs at rest in equilibrium between the gravitational force and the tension in the string.

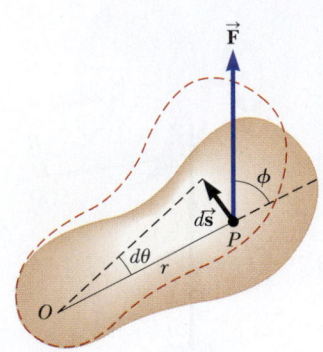

Figure 10.19 A rigid object rotates about an axis through O under the action of an external force $\vec{F}$ applied at P.

10.8 | Energy Considerations in Rotational Motion

In translational motion, we found energy concepts, and in particular the reduction of the conservation of energy equation called the work–kinetic energy theorem, to be extremely useful in describing the motion of a system. Energy concepts can be equally useful in simplifying the analysis of rotational motion. From the conservation of energy equation, we expect that for rotation of an object about a fixed axis, the work done by external forces on the object will equal the change in the rotational kinetic energy as long as energy is not stored by any other means. To show that this case is in fact true, we begin by finding an expression for the work done by a torque.

Consider a rigid object pivoted at the point O in Figure 10.19. Suppose a single external force $\vec{F}$ is applied at the point P and $d\vec{s}$ is the displacement of the point of application of the force. The small amount of work dW done on the object by $\vec{F}$ as the point of application rotates through an infinitesimal distance $ds = r\,d\theta$ in a time interval dt is

$$dW = \vec{F} \cdot d\vec{s} = (F\sin\phi)\,r\,d\theta$$

where $F\sin\phi$ is the tangential component of $\vec{F}$, or the component of the force along the displacement. Note from Figure 10.19 that the radial component of $\vec{F}$ does no work because it is perpendicular to the displacement of the point of application of the force.

Because the magnitude of the torque due to $\vec{F}$ about the origin is defined as $rF\sin\phi$, we can write the work done on the object for the infinitesimal rotation in the form

$$dW = \tau\,d\theta \qquad\qquad \textbf{10.31}\blacktriangleleft$$

Notice that this expression is the product of torque and angular displacement, making it analogous to the work done on the object in translational motion, which is the product of force and translational displacement.

Now, we will combine this result with the rotational form of Newton's second law, $\tau = I\alpha$. Using the chain rule from calculus, we can express the torque as

$$\tau = I\alpha = I\frac{d\omega}{dt} = I\frac{d\omega}{d\theta}\frac{d\theta}{dt} = I\frac{d\omega}{d\theta}\omega$$

Rearranging this expression and noting that $\tau\,d\theta = dW$ from Equation 10.31, we have

$$\tau\,d\theta = dW = I\omega\,d\omega$$

Integrating this expression, we find the total work done by the torque:

$$W = \int_{\theta_i}^{\theta_f} \tau\,d\theta = \int_{\omega_i}^{\omega_f} I\omega\,d\omega$$

▶ Work–kinetic energy theorem for pure rotation

$$W = \tfrac{1}{2}I\omega_f^2 - \tfrac{1}{2}I\omega_i^2 = \Delta K_R \qquad\qquad \textbf{10.32}\blacktriangleleft$$

Notice that this equation has exactly the same mathematical form as the work–kinetic energy theorem for translation. Equation 10.32 is a form of the nonisolated system (energy) model discussed in Chapter 7. Work is done on the system of the rigid object, which represents a transfer of energy across the boundary of the system that appears as an increase in the object's rotational kinetic energy.

In general, we can combine this theorem with the translational form of the work–kinetic energy theorem from Chapter 6. Therefore, the net work done by external forces on an object is the change in its *total* kinetic energy, which is the sum of the translational and rotational kinetic energies. For example, when a pitcher throws a baseball, the work done by the pitcher's hands appears as kinetic energy associated with the ball moving through space as well as rotational kinetic energy associated with the spinning of the ball.

In addition to the work–kinetic energy theorem, other energy principles can also be applied to rotational situations. For example, if a system involving rotating objects is isolated and no nonconservative forces act within the system, the isolated system model and the principle of conservation of mechanical energy can be used to analyze the system as in Example 10.11 below.

We finish this discussion of energy concepts for rotation by investigating the *rate* at which work is being done by $\vec{F}$ on an object rotating about a fixed axis. This rate is obtained by dividing the left and right sides of Equation 10.31 by dt:

$$\frac{dW}{dt} = \tau \frac{d\theta}{dt} \qquad \textbf{10.33} \blacktriangleleft$$

The quantity dW/dt is, by definition, the instantaneous power P delivered by the force. Furthermore, because $d\theta/dt = \omega$, Equation 10.33 reduces to

$$P = \tau\omega \qquad \textbf{10.34} \blacktriangleleft \qquad \blacktriangleright \text{ Power delivered to a rotating object}$$

This expression is analogous to $P = Fv$ in the case of translational motion.

Example 10.11 | Rotating Rod

A uniform rod of length L and mass M is free to rotate on a frictionless pin passing through one end (Fig. 10.20). The rod is released from rest in the horizontal position.

(A) What is its angular speed when the rod reaches its lowest position?

SOLUTION

Conceptualize Consider Figure 10.20 and imagine the rod rotating downward through a quarter turn about the pivot at the left end.

Categorize The angular acceleration of the rod is not constant. Therefore, the kinematic equations for rotation (Section 10.2) cannot be used to solve this example. We categorize the system of the rod and the Earth as an isolated system in terms of energy with no nonconservative forces acting and use the principle of conservation of mechanical energy.

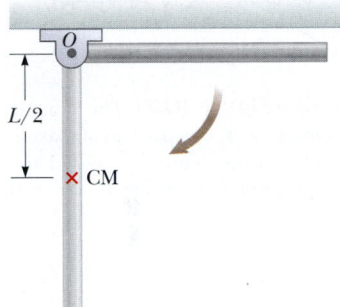

Figure 10.20 (Example 10.11) A uniform rigid rod pivoted at O rotates in a vertical plane under the action of the gravitational force.

Analyze We choose the configuration in which the rod is hanging straight down as the reference configuration for gravitational potential energy and assign a value of zero for this configuration. When the rod is in the horizontal position, it has no rotational kinetic energy. The potential energy of the system in this configuration relative to the reference configuration is $MgL/2$ because the center of mass of the rod is at a height $L/2$ higher than its position in the reference configuration. When the rod reaches its lowest position, the energy of the system is entirely rotational energy $\frac{1}{2}I\omega^2$, where I is the moment of inertia of the rod about an axis passing through the pivot.

Using the isolated system (energy) model, write a conservation of mechanical energy equation for the system:

$$K_f + U_f = K_i + U_i$$

Substitute for each of the energies:

$$\tfrac{1}{2}I\omega^2 + 0 = 0 + \tfrac{1}{2}MgL$$

Solve for ω and use $I = \frac{1}{3}ML^2$ (see Table 10.2) for the rod:

$$\omega = \sqrt{\frac{MgL}{I}} = \sqrt{\frac{MgL}{\frac{1}{3}ML^2}} = \boxed{\sqrt{\frac{3g}{L}}}$$

(B) Determine the tangential speed of the center of mass and the tangential speed of the lowest point on the rod when it is in the vertical position.

SOLUTION

Use Equation 10.10 and the result from part (A):

$$v_{CM} = r\omega = \frac{L}{2}\omega = \boxed{\tfrac{1}{2}\sqrt{3gL}}$$

Because r for the lowest point on the rod is twice what it is for the center of mass, the lowest point has a tangential speed twice that of the center of mass:

$$v = 2v_{CM} = \boxed{\sqrt{3gL}}$$

Finalize Applying an energy approach allows us to find the angular speed of the rod at the lowest point. Convince yourself that you could find the angular speed of the rod at any angular position by knowing the location of the center of mass at this position.

🢐10.9 | Analysis Model: Nonisolated System (Angular Momentum)

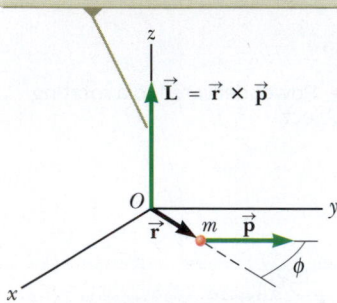

The angular momentum $\vec{\mathbf{L}}$ depends on the origin about which it is measured and is a vector perpendicular to both $\vec{\mathbf{r}}$ and $\vec{\mathbf{p}}$.

$$\vec{\mathbf{L}} = \vec{\mathbf{r}} \times \vec{\mathbf{p}}$$

Active Figure 10.21 The angular momentum $\vec{\mathbf{L}}$ of a particle of mass m and linear momentum $\vec{\mathbf{p}}$ located at the position $\vec{\mathbf{r}}$ is given by $\vec{\mathbf{L}} = \vec{\mathbf{r}} \times \vec{\mathbf{p}}$.

Pitfall Prevention | 10.6
Is Rotation Necessary for Angular Momentum?
We can define angular momentum even if the particle is not moving in a circular path. Even a particle moving in a straight line has angular momentum about any axis displaced from the path of the particle.

Imagine an object rotating in space with no motion of its center of mass. Each particle in the object is moving in a circular path, so momentum is associated with the motion of each particle. Although the object has no linear momentum (its center of mass is not moving through space), particles in the object are in motion, so a "quantity of motion" is associated with its rotation. We will investigate the **angular momentum** that the object has in this section.

Consider a particle of mass m, situated at the vector position $\vec{\mathbf{r}}$ and moving with a momentum $\vec{\mathbf{p}}$, as shown in Active Figure 10.21. For now, we don't consider it as a particle on a rigid object; it is any particle moving with momentum $\vec{\mathbf{p}}$. We will apply the result to a rotating rigid object shortly. The **instantaneous angular momentum** $\vec{\mathbf{L}}$ of the particle relative to the origin O is defined by the vector product of its instantaneous position vector $\vec{\mathbf{r}}$ and the instantaneous linear momentum $\vec{\mathbf{p}}$:

$$\vec{\mathbf{L}} \equiv \vec{\mathbf{r}} \times \vec{\mathbf{p}} \qquad \text{10.35} \blacktriangleleft$$

The SI units of angular momentum are $kg \cdot m^2/s$. Note that both the magnitude and the direction of $\vec{\mathbf{L}}$ depend on the choice of origin. The direction of $\vec{\mathbf{L}}$ is perpendicular to the plane formed by $\vec{\mathbf{r}}$ and $\vec{\mathbf{p}}$, and the sense of $\vec{\mathbf{L}}$ is governed by the right-hand rule. For example, in Active Figure 10.21, $\vec{\mathbf{r}}$ and $\vec{\mathbf{p}}$ are assumed to be in the xy plane and $\vec{\mathbf{L}}$ points in the z direction. Because $\vec{\mathbf{p}} = m\vec{\mathbf{v}}$, the magnitude of $\vec{\mathbf{L}}$ is

$$L = mvr \sin \phi \qquad \text{10.36} \blacktriangleleft$$

where ϕ is the angle between $\vec{\mathbf{r}}$ and $\vec{\mathbf{p}}$. It follows that $\vec{\mathbf{L}}$ is zero when $\vec{\mathbf{r}}$ is parallel to $\vec{\mathbf{p}}$ ($\phi = 0°$ or $180°$). In other words, when the particle moves along a line that passes through the origin, it has zero angular momentum with respect to the origin, which is equivalent to stating that the momentum vector is not tangent to *any* circle drawn about the origin. On the other hand, if $\vec{\mathbf{r}}$ is perpendicular to $\vec{\mathbf{p}}$ ($\phi = 90°$), L is a maximum and equal to mvr. In fact, at that instant the particle moves exactly as though it were on the rim of a wheel of radius r rotating at angular speed $\omega = v/r$ about an axis through the origin in a plane defined by $\vec{\mathbf{r}}$ and $\vec{\mathbf{p}}$. A particle has nonzero angular momentum about some point if the position vector of the particle measured from that point rotates about the point as the particle moves.

For translational motion, we found that the net force on a particle equals the time rate of change of the particle's linear momentum (Eq. 8.4). We shall now show that Newton's second law implies an analogous situation for rotation: that the net torque acting on a particle equals the time rate of change of the particle's angular momentum. Let us start by writing the torque on the particle in the form

$$\vec{\boldsymbol{\tau}} = \vec{\mathbf{r}} \times \vec{\mathbf{F}} = \vec{\mathbf{r}} \times \frac{d\vec{\mathbf{p}}}{dt} \qquad \text{10.37} \blacktriangleleft$$

where we have used $\vec{\mathbf{F}} = d\vec{\mathbf{p}}/dt$ (Eq. 8.4). Now let us differentiate Equation 10.35 with respect to time, using the product rule for differentiation (Eq. 10.25):

$$\frac{d\vec{\mathbf{L}}}{dt} = \frac{d}{dt}(\vec{\mathbf{r}} \times \vec{\mathbf{p}}) = \frac{d\vec{\mathbf{r}}}{dt} \times \vec{\mathbf{p}} + \vec{\mathbf{r}} \times \frac{d\vec{\mathbf{p}}}{dt}$$

It is important to adhere to the order of factors in the vector product because the vector product is not commutative, as we saw in Section 10.5.

The first term on the right in the preceding equation is zero because $\vec{\mathbf{v}} = d\vec{\mathbf{r}}/dt$ is parallel to $\vec{\mathbf{p}}$. Therefore,

$$\frac{d\vec{\mathbf{L}}}{dt} = \vec{\mathbf{r}} \times \frac{d\vec{\mathbf{p}}}{dt} \qquad \text{10.38} \blacktriangleleft$$

Comparing Equations 10.37 and 10.38, we see that

$$\vec{\boldsymbol{\tau}} = \frac{d\vec{\mathbf{L}}}{dt} \qquad \text{10.39} \blacktriangleleft$$

▶ Torque on a particle equals time rate of change of angular momentum of the particle

This result is the rotational analog of Newton's second law, $\vec{\mathbf{F}} = d\vec{\mathbf{p}}/dt$. Equation 10.39 says that the torque acting on a particle is equal to the time rate of change of the particle's angular momentum. Note that Equation 10.39 is valid only if the axes used to define $\vec{\tau}$ and $\vec{\mathbf{L}}$ are the *same*. Equation 10.39 is also valid when several forces are acting on the particle, in which case $\vec{\tau}$ is the *net* torque on the particle. Of course, the same origin must be used in calculating all torques as well as the angular momentum.

Now, let us apply these ideas to a system of particles. The total angular momentum $\vec{\mathbf{L}}$ of the system of particles about some point is defined as the vector sum of the angular momenta of the individual particles:

$$\vec{\mathbf{L}} = \vec{\mathbf{L}}_1 + \vec{\mathbf{L}}_2 + \cdots + \vec{\mathbf{L}}_n = \sum_i \vec{\mathbf{L}}_i$$

where the vector sum is over all the n particles in the system.

Because the individual angular momenta of the particles may change in time, the total angular momentum may also vary in time. In fact, the time rate of change of the total angular momentum of the system equals the vector sum of *all* torques, including those associated with internal forces between particles and those associated with external forces.

As we found in our discussion of the rigid object under a net torque, however, the sum of the internal torques is zero. Therefore, we conclude that the total angular momentum can vary with time *only* if there is a net *external* torque on the system, so that we have

$$\sum \vec{\tau}_{\text{ext}} = \sum_i \frac{d\vec{\mathbf{L}}_i}{dt} = \frac{d}{dt} \sum_i \vec{\mathbf{L}}_i$$

$$\sum \vec{\tau}_{\text{ext}} = \frac{d\vec{\mathbf{L}}_{\text{tot}}}{dt}$$

10.40 ◄   ► Net external torque on a system equals time rate of change of angular momentum of the system

That is, the time rate of change of the total angular momentum of the system about some origin in an inertial frame equals the net external torque acting on the system about that origin. Note that Equation 10.40 is the rotational analog of $\sum \vec{\mathbf{F}}_{\text{ext}} = d\vec{\mathbf{p}}_{\text{tot}}/dt$ (Eq. 8.40) for a system of particles.

This result is valid for a system of particles that change their positions with respect to one another, that is, a nonrigid object. In this discussion of angular momentum of a system of particles, notice that we never imposed the rigid-object condition.

Equation 10.40 is the primary equation in the **angular momentum version of the nonisolated system model.** The system's angular momentum changes in response to an interaction with the environment, described by means of the net torque on the system.

One final result can be obtained for angular momentum, which will serve as an analog to the definition of linear momentum. Let us imagine a rigid object rotating about an axis. Each particle of mass m_i in the rigid object moves in a circular path of radius r_i with a tangential speed v_i. Therefore, the total angular momentum of the rigid object is

$$L = \sum_i m_i v_i r_i$$

Let us now replace the tangential speed with the product of the radial distance and the angular speed (Eq. 10.10):

$$L = \sum_i m_i v_i r_i = \sum_i m_i (r_i \omega) r_i = \left(\sum_i m_i r_i^2 \right) \omega$$

We recognize the combination in the parentheses as the moment of inertia, so we can write the angular momentum of the rigid object as

$$L = I\omega$$

10.41 ◄   ► Angular momentum of an object with moment of inertia I

▌**TABLE 10.3** | A Comparison of Equations for Rotational and Translational Motion: Dynamic Equations

	Rotational Motion About a Fixed Axis	Translational Motion
Kinetic energy	$K_R = \frac{1}{2}I\omega^2$	$K = \frac{1}{2}mv^2$
Equilibrium	$\sum \vec{\boldsymbol{\tau}}_{\text{ext}} = 0$	$\sum \vec{\mathbf{F}}_{\text{ext}} = 0$
Newton's second law	$\sum \tau_{\text{ext}} = I\alpha$	$\sum \vec{\mathbf{F}}_{\text{ext}} = m\vec{\mathbf{a}}$
Nonisolated system	$\vec{\boldsymbol{\tau}}_{\text{ext}} = \dfrac{d\vec{\mathbf{L}}_{\text{tot}}}{dt}$	$\vec{\mathbf{F}}_{\text{ext}} = \dfrac{d\vec{\mathbf{p}}_{\text{tot}}}{dt}$
Momentum	$L = I\omega$	$\vec{\mathbf{p}} = m\vec{\mathbf{v}}$
Isolated system	$\vec{\mathbf{L}}_i = \vec{\mathbf{L}}_f$	$\vec{\mathbf{p}}_i = \vec{\mathbf{p}}_f$
Power	$P = \tau\omega$	$P = Fv$

Note: Equations in translation motion expressed in terms of vectors have rotational analogs in terms of vectors. Because the full vector treatment of rotation is beyond the scope of this book, however, some rotational equations are given in nonvector form.

which is the rotational analog to $p = mv$. Table 10.3 is a continuation of Table 10.1, with additional translational and rotational analogs that we have developed in the past few sections and one that we will develop in the next section.

▌**QUICK QUIZ 10.7** A solid sphere and a hollow sphere have the same mass and radius. They are rotating with the same angular speed. Which is the one with the higher angular momentum? (a) the solid sphere (b) the hollow sphere (c) both have the same angular momentum (d) impossible to determine

Example **10.12** | A System of Objects

A sphere of mass m_1 and a block of mass m_2 are connected by a light cord that passes over a pulley as shown in Figure 10.22. The radius of the pulley is R, and the mass of the thin rim is M. The spokes of the pulley have negligible mass. The block slides on a frictionless, horizontal surface. Find an expression for the linear acceleration of the two objects, using the concepts of angular momentum and torque.

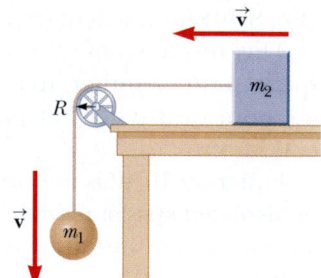

Figure 10.22 (Example 10.12) When the system is released, the sphere moves downward and the block moves to the left.

SOLUTION

Conceptualize When the system is released, the block slides to the left, the sphere drops downward, and the pulley rotates counterclockwise. This situation is similar to problems we have solved earlier except that now we want to use an angular momentum approach.

Categorize We identify the block, pulley, and sphere as a nonisolated system, subject to the external torque due to the gravitational force on the sphere. We shall calculate the angular momentum about an axis that coincides with the axle of the pulley. The angular momentum of the system includes that of two objects moving translationally (the sphere and the block) and one object undergoing pure rotation (the pulley).

Analyze At any instant of time, the sphere and the block have a common speed v, so the angular momentum of the sphere is $m_1 vR$ and that of the block is $m_2 vR$. At the same instant, all points on the rim of the pulley also move with speed v, so the angular momentum of the pulley is MvR.

Now let's address the total external torque acting on the system about the pulley axle. Because it has a moment arm of zero, the force exerted by the axle on the pulley does not contribute to the torque. Furthermore, the normal force acting on the block is balanced by the gravitational force $m_2\vec{\mathbf{g}}$, so these forces do not contribute to the torque. The gravitational force $m_1\vec{\mathbf{g}}$ acting on the sphere produces a torque about the axle equal in magnitude to $m_1 gR$, where R is the moment arm of the force about the axle. This result is the total external torque about the pulley axle; that is, $\sum \tau_{\text{ext}} = m_1 gR$.

10.12 *cont.*

Write an expression for the total angular momentum of the system:

$$(1) \quad L_{tot} = m_1vR + m_2vR + MvR = (m_1 + m_2 + M)vR$$

Substitute this expression and the total external torque into Equation 10.40:

$$\sum \tau_{ext} = \frac{dL_{tot}}{dt}$$

$$m_1gR = \frac{d}{dt}[(m_1 + m_2 + M)vR]$$

$$(2) \quad m_1gR = (m_1 + m_2 + M)R\frac{dv}{dt}$$

Recognizing that $dv/dt = a$, solve Equation (2) for a:

$$(3) \quad a = \frac{m_1g}{m_1 + m_2 + M}$$

Finalize When we evaluated the net torque about the axle, we did not include the forces that the cord exerts on the objects because these forces are internal to the system under consideration. Instead, we analyzed the system as a whole. Only *external* torques contribute to the change in the system's angular momentum.

10.10 | Analysis Model: Isolated System (Angular Momentum)

In Chapter 8, we found that the total linear momentum of a system of particles remains constant if the system is isolated, that is, if the net external force acting on the system is zero. We have an analogous conservation law in rotational motion:

The total angular momentum of a system is constant in both magnitude and direction if the net external torque acting on the system is zero, that is, if the system is isolated.

▶ Conservation of angular momentum

This statement is often called[6] the principle of **conservation of angular momentum** and is the basis of the **angular momentum version of the isolated system model.** This principle follows directly from Equation 10.40, which indicates that if

$$\sum \vec{\tau}_{ext} = \frac{d\vec{L}_{tot}}{dt} = 0 \qquad\qquad \textbf{10.42} \blacktriangleleft$$

then

$$\vec{L}_{tot} = \text{constant} \quad \text{or} \quad \vec{L}_i = \vec{L}_f \qquad\qquad \textbf{10.43} \blacktriangleleft$$

For an isolated system consisting of a number of particles, we write this conservation law as $\vec{L}_{tot} = \sum \vec{L}_n = \text{constant}$, where the index n denotes the nth particle in the system.

If an isolated rotating system is deformable so that its mass undergoes redistribution in some way, the system's moment of inertia changes. Because the magnitude of the angular momentum of the system is $L = I\omega$ (Eq. 10.41), conservation of angular momentum requires that the product of I and ω must remain constant. Therefore, a change in I for an isolated system requires a change in ω. In this case, we can express the principle of conservation of angular momentum as

$$I_i\omega_i = I_f\omega_f = \text{constant} \qquad\qquad \textbf{10.44} \blacktriangleleft$$

This expression is valid both for rotation about a fixed axis and for rotation about an axis through the center of mass of a moving system as long as that axis remains fixed in direction. We require only that the net external torque be zero.

---

[6]The most general conservation of angular momentum equation is Equation 10.40, which describes how the system interacts with its environment.

When his arms and legs are close to his body, the skater's moment of inertia is small and his angular speed is large.

To slow down for the finish of his spin, the skater moves his arms and legs outward, increasing his moment of inertia.

Figure 10.23 Angular momentum is conserved as Russian gold medalist Evgeni Plushenko performs during the Turin 2006 Winter Olympic Games.

Many examples demonstrate conservation of angular momentum for a deformable system. You may have observed a figure skater spinning in the finale of a program (Fig. 10.23). The angular speed of the skater is large when his hands and feet are close to the trunk of his body. (Notice the skater's hair!) Ignoring friction between skater and ice, there are no external torques on the skater. The moment of inertia of his body increases as his hands and feet are moved away from his body at the finish of the spin. According to the principle of conservation of angular momentum, his angular speed must decrease. In a similar way, when divers or acrobats wish to make several somersaults, they pull their hands and feet close to their bodies to rotate at a higher rate. In these cases, the external force due to gravity acts through the center of mass and hence exerts no torque about an axis through this point. Therefore, the angular momentum about the center of mass must be conserved; that is, $I_i\omega_i = I_f\omega_f$. For example, when divers wish to double their angular speed, they must reduce their moment of inertia to half its initial value.

In Equation 10.43, we have a third version of the isolated system model. We can now state that the energy, linear momentum, and angular momentum of an isolated system are all constant:

$$E_i = E_f \qquad \text{(if there are no energy transfers across the system boundary)}$$

$$\vec{\mathbf{p}}_i = \vec{\mathbf{p}}_f \qquad \text{(if the net external force on the system is zero)}$$

$$\vec{\mathbf{L}}_i = \vec{\mathbf{L}}_f \qquad \text{(if the net external torque on the system is zero)}$$

A system may be isolated in terms of one of these quantities but not in terms of another. If a system is nonisolated in terms of momentum or angular momentum, it will often be nonisolated also in terms of energy because the system has a net force or torque on it and the net force or torque will do work on the system. We can, however, identify systems that are nonisolated in terms of energy but isolated in terms of momentum. For example, imagine pushing inward on a balloon (the system) between your hands. Work is done in compressing the balloon, so the system is nonisolated in terms of energy, but there is zero net force on the system, so the system is isolated in terms of momentum. A similar statement could be made about twisting the ends of a long, springy piece of metal with both hands. Work is done on the metal (the system), so energy is stored in the nonisolated system as elastic potential energy, but the net torque on the system is zero. Therefore, the system is isolated in terms of angular momentum. Other examples are collisions of macroscopic objects, which represent isolated systems in terms of momentum but nonisolated systems in terms of energy because of the output of energy from the system by mechanical waves (sound).

An interesting astrophysical example of conservation of angular momentum occurs when, at the end of its lifetime, a massive star uses up all its fuel and collapses under the influence of gravitational forces, causing a gigantic outburst of energy called a supernova explosion. The best-studied example of a remnant of a supernova explosion is the Crab Nebula, a chaotic, expanding mass of gas (Fig. 10.24). In a supernova, part of the star's mass is released into space, where it eventually condenses into new stars and planets. Most of what is left behind typically collapses into a **neutron star,** an extremely dense sphere of matter with a diameter of about 10 km in comparison with the 10^6-km diameter of the original star and containing a large fraction of the star's original mass. As the moment of inertia of the system decreases during the collapse, the star's rotational speed increases, similar to the change in speed of the skater in Figure 10.23. About 2 000 rapidly rotating neutron stars have been identified since the first discovery of such astronomical bodies in 1967, with periods of rotation ranging from a millisecond to several seconds. The neutron star—an object with a mass greater than the Sun, rotating about its axis many times each second—is a most dramatic system!

We can also detect the effects of conservation of momentum on the rotation of the Earth when an earthquake occurs. An earthquake causes the mass distribution of the Earth to change, and the result is a change in the moment of inertia of the Earth. Just as with the spinning skater, this change will cause the angular speed of the Earth

Figure 10.24 The Crab Nebula, in the constellation Taurus. This nebula is the remnant of a supernova explosion, which was seen on Earth in the year 1054. It is located some 6 300 light-years away and is approximately 6 light-years in diameter, still expanding outward.

to change. An earthquake of magnitude 8.8 in Chile in February 2010 caused the Earth's period to decrease by 1.3 μs. Similarly, the magnitude-9.0 earthquake off the coast of Japan in March 2011 caused a further decrease by 1.8 μs.

> **QUICK QUIZ 10.8** A competitive diver leaves the diving board and falls toward the water with her body straight and rotating slowly. She pulls her arms and legs into a tight tuck position. What happens to her rotational kinetic energy? (**a**) It increases. (**b**) It decreases. (**c**) It stays the same. (**d**) It is impossible to determine.

Example 10.13 | A Revolving Puck on a Horizontal, Frictionless Surface

A puck of mass m on a horizontal, frictionless table is connected to a string that passes through a small hole in the table. The puck is set into circular motion of radius R, at which time its speed is v_i (Fig. 10.25).

(A) If the string is pulled from the bottom so that the radius of the circular path is decreased to r, find an expression for the final speed v_f of the puck.

SOLUTION

Conceptualize Imagine the puck in Figure 10.25 moving in its circular path. Now imagine pulling downward on the string so that the puck moves into a circular path with a smaller radius. Do you expect it to move faster or slower? What happens to a spinning skater when he brings his arms in close to his body?

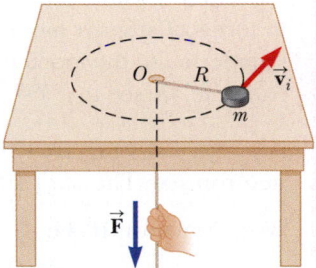

Figure 10.25 (Example 10.13) When the string is pulled downward, the speed of the puck changes.

Categorize We identify the puck as the system. Is the system isolated or nonisolated? The gravitational force acting on the puck is balanced by the upward normal force, so these forces cancel, resulting in zero net torque from these forces. The force $\vec{\mathbf{F}}$ of the string on the puck acts toward the center of rotation, and the vector position $\vec{\mathbf{r}}$ is directed away from O. Therefore, we see that the torque about the center of rotation due to this force is $\vec{\boldsymbol{\tau}} = \vec{\mathbf{r}} \times \vec{\mathbf{F}} = 0$. Although three forces act on the puck, the net torque exerted on it is zero. Therefore, the puck is an isolated system in terms of angular momentum!

....

Analyze

From the isolated system model, set the initial angular momentum equal to the final angular momentum:

$$L = mv_i R = mv_f r$$

continued

10.13 *cont.*

Solve for the final speed:
$$v_f = \frac{v_i R}{r}$$

From this result we see that as r decreases, the speed v increases.

(B) Show that the kinetic energy of the puck is not conserved in this process.

SOLUTION

Find an expression for the ratio of the final kinetic energy to the initial kinetic energy:
$$\frac{K_f}{K_i} = \frac{\frac{1}{2}mv_f^2}{\frac{1}{2}mv_i^2} = \frac{1}{v_i^2}\left(\frac{v_i R}{r}\right)^2 = \frac{R^2}{r^2}$$

Because this ratio is not equal to 1, kinetic energy is not conserved.

Finalize Because $R > r$, the kinetic energy of the puck has increased. This increase corresponds to energy entering the system of the puck by means of the work done by the person pulling the string. Even though the system is isolated in terms of angular momentum, it is nonisolated in terms of energy!

Example 10.14 | Formation of a Neutron Star

A star rotates with a period of 30 days about an axis through its center. The period is the time interval required for a point on the star's equator to make one complete revolution around the axis of rotation. After the star undergoes a supernova explosion, the stellar core, which had a radius of 1.0×10^4 km, collapses into a neutron star of radius 3.0 km. Determine the period of rotation of the neutron star.

SOLUTION

Conceptualize The change in the neutron star's motion is similar to that of the skater described earlier, but in the reverse direction. As the mass of the star moves closer to the rotation axis, we expect the star to spin faster.

Categorize Let us assume that during the collapse of the stellar core, (1) no external torque acts on it, (2) it remains spherical with the same relative mass distribution, and (3) its mass remains constant. We categorize the star as an isolated system in terms of angular momentum. We do not know the mass distribution of the star, but we have assumed the distribution is symmetric, so the moment of inertia can be expressed as kMR^2, where k is some numerical constant. (From Table 10.2, for example, we see that $k = \frac{2}{5}$ for a solid sphere and $k = \frac{2}{3}$ for a spherical shell.)

Analyze Let's use the symbol T for the period, with T_i being the initial period of the star and T_f being the period of the neutron star. The star's angular speed is given by $\omega = 2\pi/T$.

Write Equation 10.44:
$$I_i\omega_i = I_f\omega_f$$

Use $\omega = 2\pi/T$ to rewrite this equation in terms of the initial and final periods:
$$I_i\left(\frac{2\pi}{T_i}\right) = I_f\left(\frac{2\pi}{T_f}\right)$$

Substitute the moments of inertia in the preceding equation:
$$kMR_i^2\left(\frac{2\pi}{T_i}\right) = kMR_f^2\left(\frac{2\pi}{T_f}\right)$$

Solve for the final period of the star:
$$T_f = \left(\frac{R_f}{R_i}\right)^2 T_i$$

Substitute numerical values:
$$T_f = \left(\frac{3.0 \text{ km}}{1.0 \times 10^4 \text{ km}}\right)^2 (30 \text{ days}) = 2.7 \times 10^{-6} \text{ days} = \boxed{0.23 \text{ s}}$$

Finalize The neutron star does indeed rotate faster after it collapses, as predicted. It moves very fast, in fact, rotating about four times each second.

10.11 | Precessional Motion of Gyroscopes

An unusual and fascinating type of motion you have probably observed is that of a top spinning about its axis of symmetry as shown in Figure 10.26a. If the top spins rapidly, the symmetry axis rotates about the z axis, sweeping out a cone (see Fig. 10.26b). The motion of the symmetry axis about the vertical—known as **precessional motion**—is usually slow relative to the spinning motion of the top.

It is quite natural to wonder why the top does not fall over. Because the center of mass is not directly above the pivot point O, a net torque is acting on the top about an axis passing through O, a torque resulting from the gravitational force $M\vec{g}$. The top would certainly fall over if it were not spinning. Because it is spinning, however, it has an angular momentum $\vec{L}$ directed along its symmetry axis. We shall show that this symmetry axis moves about the z axis (precessional motion occurs) because the torque produces a change in the *direction* of the symmetry axis. This illustration is an excellent example of the importance of the vector nature of angular momentum.

The essential features of precessional motion can be illustrated by considering the simple gyroscope shown in Figure 10.27a. The two forces acting on the gyroscope are shown in Figure 10.27b: the downward gravitational force $M\vec{g}$ and the normal force $\vec{n}$ acting upward at the pivot point O. The normal force produces no torque about an axis passing through the pivot because its moment arm through that point is zero. The gravitational force, however, produces a torque $\vec{\tau} = \vec{r} \times M\vec{g}$ about an axis passing through O, where the direction of $\vec{\tau}$ is perpendicular to the plane formed by $\vec{r}$ and $M\vec{g}$. By necessity, the vector $\vec{\tau}$ lies in a horizontal xy plane perpendicular to the angular momentum vector. The net torque and angular momentum of the gyroscope are related through Equation 10.40:

$$\sum \vec{\tau}_{\text{ext}} = \frac{d\vec{L}}{dt}$$

This expression shows that in the infinitesimal time interval dt, the nonzero torque produces a change in angular momentum $d\vec{L}$, a change that is in the same direction as $\vec{\tau}$. Therefore, like the torque vector, $d\vec{L}$ must also be perpendicular to $\vec{L}$. Figure 10.27c illustrates the resulting precessional motion of the symmetry axis of the gyroscope. In a time interval dt, the change in angular momentum is $d\vec{L} = \vec{L}_f - \vec{L}_i = \vec{\tau} \, dt$. Because $d\vec{L}$ is perpendicular to $\vec{L}$, the magnitude of $\vec{L}$ does not change ($|\vec{L}_i| = |\vec{L}_f|$). Rather, what is changing is the *direction* of $\vec{L}$. Because the change in

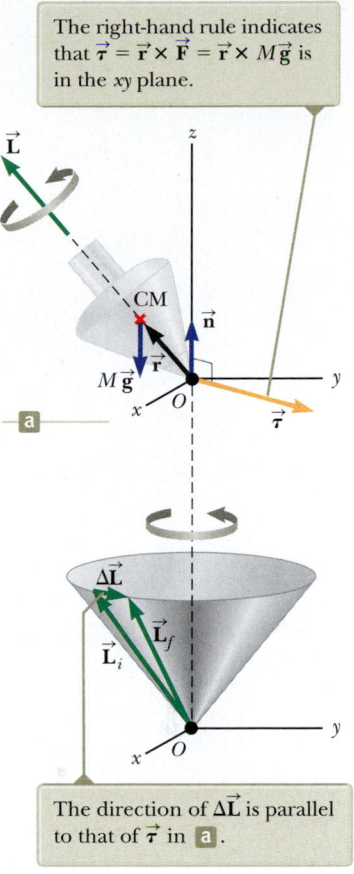

The right-hand rule indicates that $\vec{\tau} = \vec{r} \times \vec{F} = \vec{r} \times M\vec{g}$ is in the xy plane.

The direction of $\Delta\vec{L}$ is parallel to that of $\vec{\tau}$ in **a**.

Figure 10.26 Precessional motion of a top spinning about its symmetry axis. (a) The only external forces acting on the top are the normal force $\vec{n}$ and the gravitational force $M\vec{g}$. The direction of the angular momentum $\vec{L}$ is along the axis of symmetry. (b) Because $\vec{L}_f = \Delta\vec{L} + \vec{L}_i$, the top precesses about the z axis.

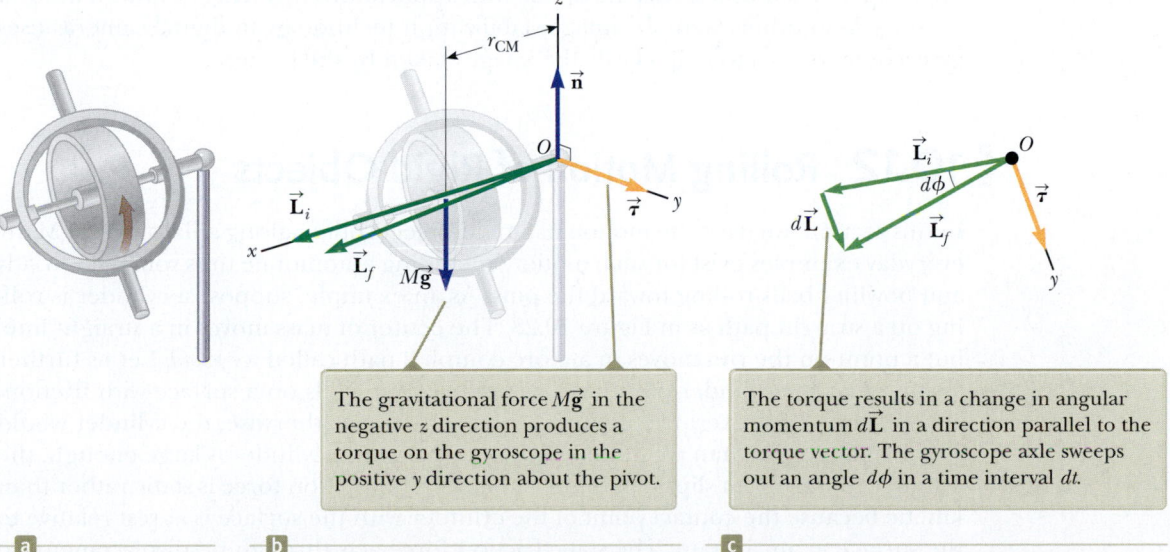

The gravitational force $M\vec{g}$ in the negative z direction produces a torque on the gyroscope in the positive y direction about the pivot.

The torque results in a change in angular momentum $d\vec{L}$ in a direction parallel to the torque vector. The gyroscope axle sweeps out an angle $d\phi$ in a time interval dt.

Figure 10.27 (a) A spinning gyroscope is placed on a pivot at the right end. (b) Diagram for the spinning gyroscope showing forces, torque, and angular momentum. (c) Overhead view (looking down the z axis) of the gyroscope's initial and final angular momentum vectors for an infinitesimal time interval dt.

angular momentum $d\vec{\mathbf{L}}$ is in the direction of $\vec{\boldsymbol{\tau}}$, which lies in the xy plane, the gyroscope undergoes precessional motion.

To simplify the description of the system, we assume the total angular momentum of the precessing wheel is the sum of the angular momentum $I\vec{\boldsymbol{\omega}}$ due to the spinning and the angular momentum due to the motion of the center of mass about the pivot. In our treatment, we shall neglect the contribution from the center-of-mass motion and take the total angular momentum to be simply $I\vec{\boldsymbol{\omega}}$. In practice, this approximation is good if $\vec{\boldsymbol{\omega}}$ is made very large.

The vector diagram in Figure 10.27c shows that in the time interval dt, the angular momentum vector rotates through an angle $d\phi$, which is also the angle through which the gyroscope axle rotates. From the vector triangle formed by the vectors $\vec{\mathbf{L}}_i$, $\vec{\mathbf{L}}_f$, and $d\vec{\mathbf{L}}$, we see that

$$d\phi = \frac{dL}{L} = \frac{\sum \tau_{\text{ext}} \, dt}{L} = \frac{(Mgr_{\text{CM}}) \, dt}{L}$$

Dividing through by dt and using the relationship $L = I\omega$, we find that the rate at which the axle rotates about the vertical axis is

$$\omega_p = \frac{d\phi}{dt} = \frac{Mgr_{\text{CM}}}{I\omega} \qquad \qquad \textbf{10.45} \blacktriangleleft$$

The angular speed ω_p is called the **precessional frequency.** This result is valid only when $\omega_p \ll \omega$. Otherwise, a much more complicated motion is involved. As you can see from Equation 10.45, the condition $\omega_p \ll \omega$ is met when ω is large, that is, when the wheel spins rapidly. Furthermore, notice that the precessional frequency decreases as ω increases, that is, as the wheel spins faster about its axis of symmetry.

With careful manufacturing tolerances, precession due to gravitational torque can be made very small and gyroscopes can be used for guidance systems in vehicles, whereby a change in the direction of the velocity of a vehicle is detected as a change between the direction of the angular momentum of the gyroscope and a reference direction attached to the vehicle. With proper electronic feedback, the deviation from the desired direction of motion can be removed, bringing the angular momentum back in line with the reference direction. Precession rates for highly specialized military gyroscopes are as low as 0.02° per day.

Gyroscopes are becoming increasingly involved in everyday life applications. Anyone who has ridden a Segway electric vehicle has been kept upright by a system of five gyroscopes in the control system of the device. The Apple iPhone 4 includes a gyroscopic sensor that assists the device with applications involving advanced motion sensing. As another example, image stabilization technology in digital cameras uses gyroscopic sensors to help clarify the images taken by the camera.

❰10.12 | Rolling Motion of Rigid Objects

In this section, we treat the motion of a rigid object rolling along a flat surface. Many everyday examples exist for such motion, including automobile tires rolling on roads and bowling balls rolling toward the pins. As an example, suppose a cylinder is rolling on a straight path as in Figure 10.28. The center of mass moves in a straight line, but a point on the rim moves in a more complex path called a *cycloid*. Let us further assume that the cylinder of radius R is uniform and rolls on a surface with friction. The surfaces must exert friction forces on each other; otherwise, the cylinder would simply slide rather than roll. If the friction force on the cylinder is large enough, the cylinder rolls without slipping. In this situation, the friction force is static rather than kinetic because the contact point of the cylinder with the surface is at rest relative to the surface at any instant. The static friction force acts through no displacement, so it does no work on the cylinder and causes no decrease in mechanical energy of the cylinder. In real rolling objects, deformations of the surfaces result in some rolling resistance. If both surfaces are hard, however, they will deform very little, and rolling

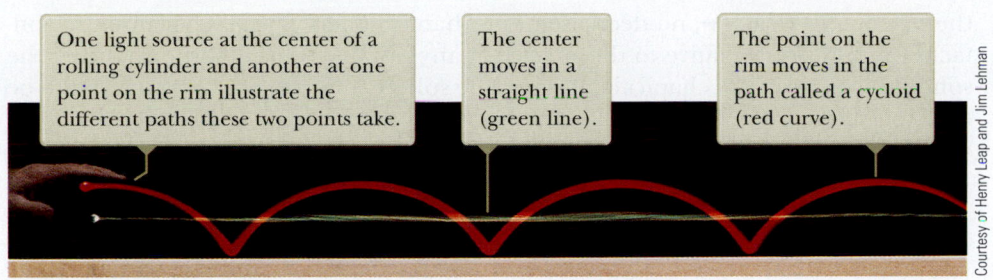

One light source at the center of a rolling cylinder and another at one point on the rim illustrate the different paths these two points take.

The center moves in a straight line (green line).

The point on the rim moves in the path called a cycloid (red curve).

Figure 10.28 Two points on a rolling object take different paths through space.

Courtesy of Henry Leap and Jim Lehman

resistance can be negligibly small. Therefore, we can model the rolling motion as maintaining constant mechanical energy. The wheel was a great invention!

As the cylinder rotates through an angle θ, its center of mass moves a distance of $s = r\theta$. Therefore, the speed and acceleration of the center of mass for pure rolling motion are

$$v_{CM} = \frac{ds}{dt} = R\frac{d\theta}{dt} = R\omega \qquad \text{10.46} \blacktriangleleft$$

▶ Relations between translational and rotational variables for a rolling object

$$a_{CM} = \frac{dv_{CM}}{dt} = R\frac{d\omega}{dt} = R\alpha \qquad \text{10.47} \blacktriangleleft$$

The translational velocities of various points on the rolling cylinder are illustrated in Figure 10.29. Note that the translational velocity of any point is in a direction perpendicular to the line from that point to the contact point. At any instant, the point P is at rest relative to the surface because sliding does not occur.

We can express the **total kinetic energy** of a rolling object of mass M and moment of inertia I as the combination of the rotational kinetic energy around the center of mass plus the translational kinetic energy of the center of mass:

$$K = \tfrac{1}{2}I_{CM}\omega^2 + \tfrac{1}{2}Mv_{CM}^2 \qquad \text{10.48} \blacktriangleleft$$

▶ Total kinetic energy of a rolling object

A useful theorem called the **parallel axis theorem** enables us to express this energy in terms of the moment of inertia I_p through any axis parallel to the axis through the center of mass of an object. This theorem states that

$$I_p = I_{CM} + MD^2 \qquad \text{10.49} \blacktriangleleft$$

where D is the distance from the center-of-mass axis to the parallel axis and M is the total mass of the object. Let us use this theorem to express the moment of inertia around an axis passing through the contact point P between the rolling object and the surface. The distance from this point to the center of mass of the symmetric object is its radius, so

$$I_P = I_{CM} + MR^2$$

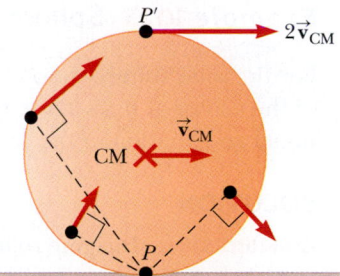

If we write the translational speed of the center of mass of the object in Equation 10.48 in terms of the angular speed, we have

$$K = \tfrac{1}{2}I_{CM}\omega^2 + \tfrac{1}{2}MR^2\omega^2 = \tfrac{1}{2}(I_{CM} + MR^2)\omega^2 = \tfrac{1}{2}I_P\omega^2 \qquad \text{10.50} \blacktriangleleft$$

Figure 10.29 All points on a rolling object move in a direction perpendicular to a line through the instantaneous point of contact P. The center of the object moves with a velocity $\vec{v}_{CM}$, whereas the point P' moves with a velocity $2\vec{v}_{CM}$.

Therefore, the kinetic energy of the rolling object can be considered as equivalent to a purely rotational kinetic energy of the object rotating around its contact point.

We can use the energy version of the isolated system model to treat a class of problems concerning the rolling motion of a rigid object down a rough incline. In these types of problems, gravitational potential energy of the object–Earth system decreases as the rotational and translational kinetic energies of the object increase. For example, consider a sphere rolling without slipping after being released from rest at the top of an incline and descending through a vertical height h. Note that accelerated rolling motion is possible only if a friction force is present between the sphere and the incline to produce a net torque about the center of mass. Despite the presence of friction, no decrease in mechanical energy occurs because the contact point is at rest relative to the surface at any instant. (On the other hand, if the sphere were to slip, mechanical energy of the sphere–incline–Earth system would be transformed to internal energy due to the nonconservative force of kinetic friction.)

Using $v_{CM} = R\omega$ for pure rolling motion, we can express Equation 10.48 as

$$K = \tfrac{1}{2}I_{CM}\left(\frac{v_{CM}}{R}\right)^2 + \tfrac{1}{2}Mv_{CM}{}^2$$

$$K = \tfrac{1}{2}\left(\frac{I_{CM}}{R^2} + M\right)v_{CM}{}^2 \qquad\qquad \textbf{10.51}\blacktriangleleft$$

For the system of the sphere and the Earth, we define the zero configuration of gravitational potential energy to be when the sphere is at the bottom of the incline. Therefore, conservation of mechanical energy gives us

$$K_f + U_f = K_i + U_i$$

$$\tfrac{1}{2}\left(\frac{I_{CM}}{R^2} + M\right)v_{CM}{}^2 + 0 = 0 + Mgh$$

$$v_{CM} = \left(\frac{2gh}{1 + I_{CM}/MR^2}\right)^{1/2} \qquad\qquad \textbf{10.52}\blacktriangleleft$$

> **QUICK QUIZ 10.9** Two items A and B are placed at the top of an incline and released from rest. For *each* of the three pairs of items in (i), (ii), and (iii), which item arrives at the bottom of the incline first? (**i**) a ball A rolling without slipping and a box B sliding on a frictionless portion of the incline (**ii**) a sphere A that has twice the mass and twice the radius of a sphere B, where both roll without slipping (**iii**) a sphere A that has the same mass and radius as a sphere B, but sphere A is solid while sphere B is hollow and both roll without slipping. Choose from the following list for each of the three pairs of items. (**a**) item A (**b**) item B (**c**) items A and B arrive at the same time (**d**) impossible to determine

---

◀ *Example* **10.15** | **Sphere Rolling Down an Incline**

For the solid sphere shown in Active Figure 10.30, calculate the translational speed of the center of mass at the bottom of the incline and the magnitude of the translational acceleration of the center of mass.

SOLUTION

Conceptualize Imagine rolling the sphere down the incline. Compare it in your mind to a book sliding down a frictionless incline. You probably have experience with objects rolling down inclines and may be tempted to think that the sphere would move down the incline faster than the book. You do *not*, however, have experience with objects sliding down *frictionless* inclines! So, which object will reach the bottom first?

Categorize We model the sphere and the Earth as an isolated system in terms of energy with no nonconservative forces acting. This model is the one that led to Equation 10.52, so we can use that result.

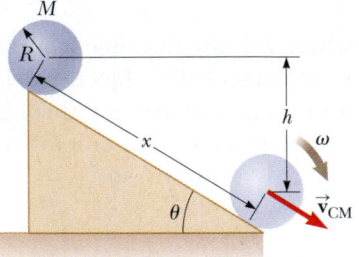

Figure 10.30 (Example 10.15) A sphere rolling down an incline. Mechanical energy of the sphere–Earth system is conserved if no slipping occurs.

Analyze Evaluate the speed of the center of mass of the sphere from Equation 10.52:

$$(1) \quad v_{CM} = \left[\frac{2gh}{1 + \left(\tfrac{2}{5}MR^2/MR^2\right)}\right]^{1/2} = \left(\tfrac{10}{7}gh\right)^{1/2}$$

This result is less than $\sqrt{2gh}$, which is the speed an object would have if it simply slid down the incline without rotating. (Eliminate the rotation by setting $I_{CM} = 0$ in Eq. 10.52.)

To calculate the translational acceleration of the center of mass, notice that the vertical displacement of the sphere is related to the distance x it moves along the incline through the relationship $h = x \sin \theta$.

10.15 *cont.*

Use this relationship to rewrite Equation (1):

$$v_{CM}^2 = \tfrac{10}{7}gx\sin\theta$$

Write Equation 2.14 for an object starting from rest and moving through a distance x under constant acceleration:

$$v_{CM}^2 = 2a_{CM}x$$

Equate the preceding two expressions to find a_{CM}:

$$a_{CM} = \tfrac{5}{7}g\sin\theta$$

Finalize Both the speed and the acceleration of the center of mass are *independent* of the mass and the radius of the sphere. That is, all homogeneous solid spheres experience the same speed and acceleration on a given incline. Try to verify this statement experimentally with balls of different sizes, such as a marble and a croquet ball.

If we were to repeat the acceleration calculation for a hollow sphere, a solid cylinder, or a hoop, we would obtain similar results in which only the factor in front of $g\sin\theta$ would differ. The constant factors that appear in the expressions for v_{CM} and a_{CM} depend only on the moment of inertia about the center of mass for the specific object. In all cases, the acceleration of the center of mass is *less* than $g\sin\theta$, the value the acceleration would have if the incline were frictionless and no rolling occurred.

10.13 | Context Connection: Turning the Spacecraft

In the Context Connection of Chapter 8, we discussed how to make a spacecraft move in empty space by firing its rocket engines. Let us now consider how to make the spacecraft turn in empty space.

One way to change the orientation of a spacecraft is to have small rocket engines that fire perpendicularly out the side of the spacecraft, providing a torque around its center of mass. This torque causes an angular acceleration around the center of mass of the spacecraft and therefore an angular speed. This rotation can be stopped to give the spacecraft the desired final orientation by firing the sideward-mounted rocket engines in the opposite direction. This option is desirable, and many spacecraft have such sideward-mounted rocket engines. An undesirable feature of this technique is that it consumes nonrenewable fuel on the spacecraft, both to initiate and to stop the rotation.

The torque exerted on the spacecraft in this situation is not an external torque, so this is not an example of the rigid object under a net torque model. The torque on the spacecraft arises from internal forces between components of the system. The spacecraft exerts forces on the exhaust gases to expel them from the spacecraft, and, by Newton's third law, the gases exert a force back on the spacecraft. Therefore, this is an application of the isolated system model for angular momentum. The gases are given an angular momentum in one direction and the spacecraft turns in the other direction. It is a rotational analog to the archer discussed in Example 8.2 or the rocket propulsion discussed in Section 8.8.

Let us consider another possibility related to the angular momentum version of the isolated system model that does not involve expelling gases. Suppose the spacecraft carries a gyroscope that is not rotating, as in Figure 10.31a (page 294). In this case, the angular momentum of the spacecraft about its center of mass is zero. Suppose the gyroscope is set into rotation. Now, it would appear that the spacecraft system has a nonzero angular momentum because of the rotation of the gyroscope. Yet there is no external torque on the system, so the angular momentum of the isolated system must remain zero according to the principle of conservation of angular momentum. This principle can be satisfied by realizing that the spacecraft will turn in the direction opposite to that of the gyroscope so that the angular momentum vectors of the gyroscope and the spacecraft cancel, resulting in no angular momentum of the system. The result of rotating the gyroscope, as in Figure 10.31b, is that the spacecraft turns! By including three gyroscopes with mutually perpendicular axles, any desired rotation in space can be achieved. Once the desired orientation is achieved, the rotation of the gyroscope is halted.

This effect occurred in an undesirable situation with the *Voyager 2* spacecraft during its flight. The spacecraft carried a tape recorder whose reels rotated at high speeds.

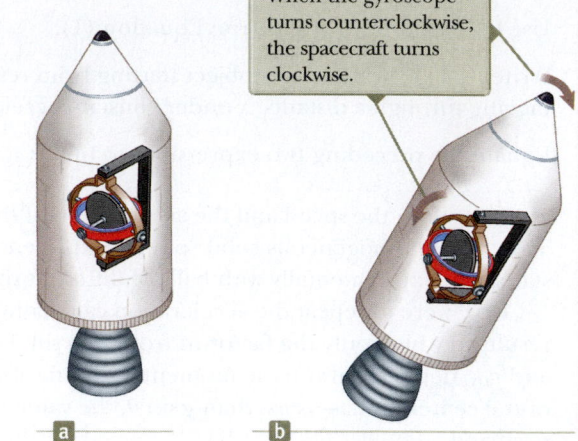

When the gyroscope turns counterclockwise, the spacecraft turns clockwise.

Figure 10.31 (a) A spacecraft carries a gyroscope that is not spinning. (b) When the gyroscope is set into rotation, the spacecraft turns the other way so that the angular momentum of the system is conserved.

Each time the tape recorder was turned on, the reels acted as gyroscopes and the spacecraft started an undesirable rotation in the opposite direction. This rotation had to be counteracted by Mission Control by using the sideward-firing jets to stop the rotation!

▶ SUMMARY

The **instantaneous angular speed** of a particle rotating in a circle or of a rigid object rotating about a fixed axis is

$$\omega \equiv \frac{d\theta}{dt} \qquad \text{10.3} \blacktriangleleft$$

where ω is in rad/s or s^{-1}.

The **instantaneous angular acceleration** of a particle rotating in a circle or of a rigid object rotating about a fixed axis is

$$\alpha \equiv \frac{d\omega}{dt} \qquad \text{10.5} \blacktriangleleft$$

and has units of rad/s^2 or s^{-2}.

When a rigid object rotates about a fixed axis, every part of the object has the same angular speed and the same angular acceleration. Different parts of the object, in general, have different translational speeds and different translational accelerations, however.

When a particle rotates about a fixed axis, the angular position, the angular speed, and the angular acceleration are related to the tangential position, the tangential speed, and the tangential acceleration through the relationships

$$s = r\theta \qquad \text{10.1a} \blacktriangleleft$$

$$v = r\omega \qquad \text{10.10} \blacktriangleleft$$

$$a_t = r\alpha \qquad \text{10.11} \blacktriangleleft$$

The **moment of inertia** of a system of particles is

$$I = \sum_i m_i r_i^2 \qquad \text{10.15} \blacktriangleleft$$

If a rigid object rotates about a fixed axis with angular speed ω, its **rotational kinetic energy** can be written

$$K_R = \tfrac{1}{2} I \omega^2 \qquad \text{10.16} \blacktriangleleft$$

where I is the moment of inertia about the axis of rotation.

The moment of inertia of a continuous object of density ρ is

$$I = \int \rho r^2 \, dV \qquad \text{10.18} \blacktriangleleft$$

The **torque** $\vec{\tau}$ due to a force $\vec{F}$ about an origin in an inertial frame is defined to be

$$\vec{\tau} \equiv \vec{r} \times \vec{F} \qquad \text{10.20} \blacktriangleleft$$

where $\vec{r}$ is the position vector of the point of application of the force.

Given two vectors $\vec{A}$ and $\vec{B}$, their **vector product** or **cross product** $\vec{A} \times \vec{B}$ is a vector $\vec{C}$ having the magnitude

$$C \equiv AB \sin \theta \qquad \text{10.22} \blacktriangleleft$$

where θ is the angle between $\vec{A}$ and $\vec{B}$. The direction of $\vec{C}$ is perpendicular to the plane formed by $\vec{A}$ and $\vec{B}$, and is determined by the right-hand rule.

The **angular momentum** $\vec{L}$ of a particle with linear momentum $\vec{p} = m\vec{v}$ is

$$\vec{L} \equiv \vec{r} \times \vec{p} \qquad \text{10.35} \blacktriangleleft$$

where $\vec{r}$ is the vector position of the particle relative to the origin. If ϕ is the angle between $\vec{r}$ and $\vec{p}$, the magnitude of $\vec{L}$ is

$$L = mvr \sin \phi \qquad \text{10.36} \blacktriangleleft$$

The **total kinetic energy** of a rigid object, such as a cylinder, that is rolling on a rough surface without slipping equals the rotational kinetic energy $\tfrac{1}{2} I_{CM} \omega^2$ about the object's center of mass plus the translational kinetic energy $\tfrac{1}{2} M v_{CM}^2$ of the center of mass:

$$K = \tfrac{1}{2} I_{CM} \omega^2 + \tfrac{1}{2} M v_{CM}^2 \qquad \text{10.48} \blacktriangleleft$$

In this expression, v_{CM} is the speed of the center of mass and $v_{CM} = R\omega$ for pure rolling motion.

Analysis Models for Problem Solving

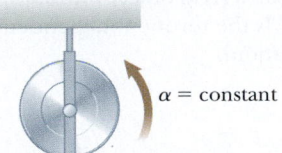

α = constant

Rigid Object Under Constant Angular Acceleration. If a rigid object rotates about a fixed axis under constant angular acceleration, one can apply equations of kinematics that are analogous to those for translational motion of a particle under constant acceleration:

$$\omega_f = \omega_i + \alpha t \qquad \textbf{10.6} \blacktriangleleft$$

$$\theta_f = \theta_i + \omega_i t + \tfrac{1}{2}\alpha t^2 \qquad \textbf{10.7} \blacktriangleleft$$

$$\omega_f{}^2 = \omega_i{}^2 + 2\alpha(\theta_f - \theta_i) \qquad \textbf{10.8} \blacktriangleleft$$

$$\theta_f = \theta_i + \tfrac{1}{2}(\omega_i + \omega_f)t \qquad \textbf{10.9} \blacktriangleleft$$

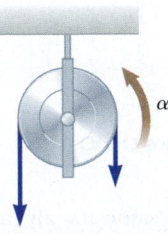

α

Rigid Object Under a Net Torque. If a rigid object free to rotate about a fixed axis has a net external torque acting on it, the object undergoes an angular acceleration α, where

$$\sum \tau_{\text{ext}} = I\alpha \qquad \textbf{10.30} \blacktriangleleft$$

This equation is the rotational analog to Newton's second law in the particle under a net force model.

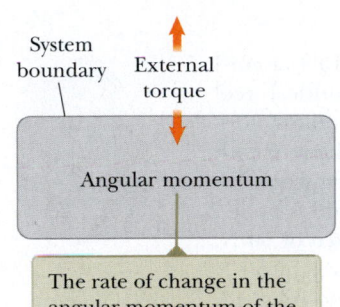

System boundary   External torque

Angular momentum

The rate of change in the angular momentum of the nonisolated system is equal to the net external torque on the system.

Nonisolated System (Angular Momentum). If a system interacts with its environment in the sense that there is an external torque on the system, the net external torque acting on the system is equal to the time rate of change of its angular momentum:

$$\sum \vec{\tau}_{\text{ext}} = \frac{d\vec{L}_{\text{tot}}}{dt} \qquad \textbf{10.40} \blacktriangleleft$$

System boundary

Angular momentum

The angular momentum of the isolated system is constant.

Isolated System (Angular Momentum). If a system experiences no external torque from the environment, the total angular momentum of the system is conserved:

$$\vec{L}_i = \vec{L}_f \qquad \textbf{10.43} \blacktriangleleft$$

Applying this law of conservation of angular momentum to a system whose moment of inertia changes gives

$$I_i\omega_i = I_f\omega_f = \text{constant} \qquad \textbf{10.44} \blacktriangleleft$$

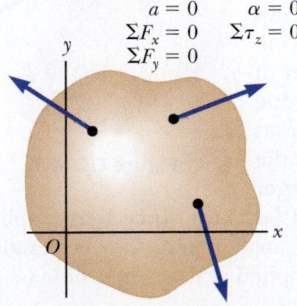

$a = 0 \qquad \alpha = 0$
$\Sigma F_x = 0 \qquad \Sigma \tau_z = 0$
$\Sigma F_y = 0$

y

O — x

Rigid Object in Equilibrium A rigid object in equilibrium exhibits no translational or angular acceleration. The net external force acting on it is zero, and the net external torque on it is zero about any axis:

$$\sum \vec{F}_{\text{ext}} = 0 \qquad \textbf{10.27} \blacktriangleleft$$

$$\sum \vec{\tau}_{\text{ext}} = 0 \qquad \textbf{10.28} \blacktriangleleft$$

The first condition is the condition for translational equilibrium, and the second is the condition for rotational equilibrium.

OBJECTIVE QUESTIONS

1. A cyclist rides a bicycle with a wheel radius of 0.500 m across campus. A piece of plastic on the front rim makes a clicking sound every time it passes through the fork. If the cyclist counts 320 clicks between her apartment and the cafeteria, how far has she traveled? (a) 0.50 km (b) 0.80 km (c) 1.0 km (d) 1.5 km (e) 1.8 km

2. A grindstone increases in angular speed from 4.00 rad/s to 12.00 rad/s in 4.00 s. Through what angle does it turn during that time interval if the angular acceleration is constant? (a) 8.00 rad (b) 12.0 rad (c) 16.0 rad (d) 32.0 rad (e) 64.0 rad

3. A wheel is rotating about a fixed axis with constant angular acceleration 3 rad/s². At different moments, its angular speed is −2 rad/s, 0, and +2 rad/s. For a point on the rim of the wheel, consider at these moments the magnitude of the tangential component of acceleration and the magnitude of the radial component of acceleration. Rank the following five items from largest to smallest: (a) $|a_t|$ when $\omega = -2$ rad/s, (b) $|a_r|$ when $\omega = -2$ rad/s, (c) $|a_r|$ when $\omega = 0$, (d) $|a_t|$ when $\omega = 2$ rad/s, and (e) $|a_r|$ when $\omega = 2$ rad/s. If two items are equal, show them as equal in your ranking. If a quantity is equal to zero, show that fact in your ranking.

4. Vector $\vec{A}$ is in the negative y direction, and vector $\vec{B}$ is in the negative x direction. (i) What is the direction of $\vec{A} \times \vec{B}$? (a) no direction because it is a scalar (b) x (c) $-y$ (d) z (e) $-z$ (ii) What is the direction of $\vec{B} \times \vec{A}$? Choose from the same possibilities (a) through (e).

5. Assume a single 300-N force is exerted on a bicycle frame as shown in Figure OQ10.5. Consider the torque produced by this force about axes perpendicular to the plane of the paper and through each of the points A through E, where E is the center of mass of the frame. Rank the torques τ_A, τ_B, τ_C, τ_D, and τ_E from largest to smallest, noting that zero is greater than a negative quantity. If two torques are equal, note their equality in your ranking.

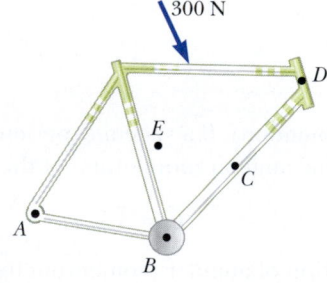

Figure OQ10.5

6. Consider an object on a rotating disk a distance r from its center, held in place on the disk by static friction. Which of the following statements is *not* true concerning this object? (a) If the angular speed is constant, the object must have constant tangential speed. (b) If the angular speed is constant, the object is not accelerated. (c) The object has a tangential acceleration only if the disk has an angular acceleration. (d) If the disk has an angular acceleration, the object has both a centripetal acceleration and a tangential acceleration. (e) The object always has a centripetal acceleration except when the angular speed is zero.

7. Answer yes or no to the following questions. (a) Is it possible to calculate the torque acting on a rigid object without specifying an axis of rotation? (b) Is the torque independent of the location of the axis of rotation?

8. Figure OQ10.8 shows a system of four particles joined by light, rigid rods. Assume $a = b$ and M is larger than m. About which of the coordinate axes does the system have (i) the smallest and (ii) the largest moment of inertia? (a) the x axis (b) the y axis (c) the z axis (d) The moment of inertia has the same small value for two axes. (e) The moment of inertia is the same for all three axes.

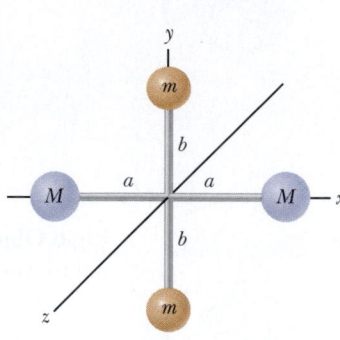

Figure OQ10.8

9. As shown in Figure OQ10.9, a cord is wrapped onto a cylindrical reel mounted on a fixed, frictionless, horizontal axle. When does the reel have a greater magnitude of angular acceleration? (a) When the cord is pulled down with a constant force of 50 N. (b) When an object of weight 50 N is hung from the cord and released. (c) The angular accelerations in parts (a) and (b) are equal. (d) It is impossible to determine.

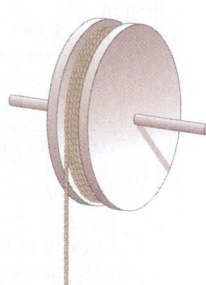

Figure OQ10.9

10. Two forces are acting on an object. Which of the following statements is correct? (a) The object is in equilibrium if the forces are equal in magnitude and opposite in direction. (b) The object is in equilibrium if the net torque on the object is zero. (c) The object is in equilibrium if the forces act at the same point on the object. (d) The object is in equilibrium if the net force and the net torque on the object are both zero. (e) The object cannot be in equilibrium because more than one force acts on it.

11. Consider the object in Figure OQ10.11. A single force is exerted on the object. The line of action of the force does not pass through the object's center of mass. The acceleration of the object's center of mass due to this force (a) is the same as if the force were applied at the center of mass, (b) is larger than the acceleration would be if the force were applied at the center of mass, (c) is smaller than the acceleration would be if the force were applied at the center of mass, or

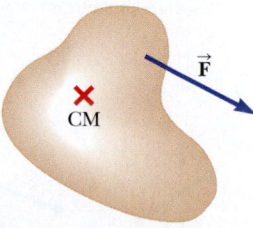

Figure OQ10.11

(d) is zero because the force causes only angular acceleration about the center of mass.

12. A constant net torque is exerted on an object. Which of the following quantities for the object cannot be constant? Choose all that apply. (a) angular position (b) angular velocity (c) angular acceleration (d) moment of inertia (e) kinetic energy

13. Let us name three perpendicular directions as right, up, and toward you as you might name them when you are facing a television screen that lies in a vertical plane. Unit vectors for these directions are $\hat{\mathbf{r}}$, $\hat{\mathbf{u}}$, and $\hat{\mathbf{t}}$, respectively. Consider the quantity $(-3\hat{\mathbf{u}} \times 2\hat{\mathbf{t}})$. (**i**) Is the magnitude of this vector (a) 6, (b) 3, (c) 2, or (d) 0? (**ii**) Is the direction of this vector (a) down, (b) toward you, (c) up, (d) away from you, or (e) left?

14. A rod 7.0 m long is pivoted at a point 2.0 m from the left end. A downward force of 50 N acts at the left end, and

a downward force of 200 N acts at the right end. At what distance to the right of the pivot can a third force of 300 N acting upward be placed to produce rotational equilibrium? *Note:* Neglect the weight of the rod. (a) 1.0 m (b) 2.0 m (c) 3.0 m (d) 4.0 m (e) 3.5 m

15. An ice skater starts a spin with her arms stretched out to the sides. She balances on the tip of one skate to turn without friction. She then pulls her arms in so that her moment of inertia decreases by a factor of 2. In the process of her doing so, what happens to her kinetic energy? (a) It increases by a factor of 4. (b) It increases by a factor of 2. (c) It remains constant. (d) It decreases by a factor of 2. (e) It decreases by a factor of 4.

16. A 20.0-kg horizontal plank 4.00 m long rests on two supports, one at the left end and a second 1.00 m from the right end. What is the magnitude of the force exerted on the plank by the support near the right end? (a) 32.0 N (b) 45.2 N (c) 112 N (d) 131 N (e) 98.2 N

CONCEPTUAL QUESTIONS |

☐ denotes answer available in *Student Solutions Manual/Study Guide*

1. (a) What is the angular speed of the second hand of a clock? (b) What is the direction of $\vec{\omega}$ as you view a clock hanging on a vertical wall? (c) What is the magnitude of the angular acceleration vector $\vec{\alpha}$ of the second hand?

2. Explain why changing the axis of rotation of an object changes its moment of inertia.

3. Why does a long pole help a tightrope walker stay balanced?

4. Which of the entries in Table 10.2 applies to finding the moment of inertia (a) of a long, straight sewer pipe rotating about its axis of symmetry? (b) Of an embroidery hoop rotating about an axis through its center and perpendicular to its plane? (c) Of a uniform door turning on its hinges? (d) Of a coin turning about an axis through its center and perpendicular to its faces?

5. (a) Give an example in which the net force acting on an object is zero and yet the net torque is nonzero. (b) Give an example in which the net torque acting on an object is zero and yet the net force is nonzero.

6. Suppose just two external forces act on a stationary, rigid object and the two forces are equal in magnitude and opposite in direction. Under what condition does the object start to rotate?

7. If you see an object rotating, is there necessarily a net torque acting on it?

8. A scientist arriving at a hotel asks a bellhop to carry a heavy suitcase. When the bellhop rounds a corner, the suitcase suddenly swings away from him for some unknown reason.

The alarmed bellhop drops the suitcase and runs away. What might be in the suitcase?

9. Three objects of uniform density—a solid sphere, a solid cylinder, and a hollow cylinder—are placed at the top of an incline (Fig. CQ10.9). They are all released from rest at the same elevation and roll without slipping. (a) Which object reaches the bottom first? (b) Which reaches it last? *Note:* The result is independent of the masses and the radii of the objects. (Try this activity at home!)

Figure CQ10.9

10. A person balances a meterstick in a horizontal position on the extended index fingers of her right and left hands. She slowly brings the two fingers together. The stick remains balanced, and the two fingers always meet at the 50-cm mark regardless of their original positions. (Try it!) Explain why that occurs.

11. If the torque acting on a particle about an axis through a certain origin is zero, what can you say about its angular momentum about that axis?

12. A cat usually lands on its feet regardless of the position from which it is dropped. A slow-motion film of a cat falling shows that the upper half of its body twists in one direction while the lower half twists in the opposite direction. (See Fig. CQ10.12.) Why does this type of rotation occur?

Figure CQ10.12

13. Stars originate as large bodies of slowly rotating gas. Because of gravity, these clumps of gas slowly decrease in size. What happens to the angular speed of a star as it shrinks? Explain.

14. A girl has a large, docile dog she wishes to weigh on a small bathroom scale. She reasons that she can determine her dog's weight with the following method. First she puts the dog's two front feet on the scale and records the scale reading. Then she places only the dog's two back feet on the scale and records the reading. She thinks that the sum of the readings will be the dog's weight. Is she correct? Explain your answer.

15. If global warming continues over the next one hundred years, it is likely that some polar ice will melt and the water will be distributed closer to the equator. (a) How would that change the moment of inertia of the Earth? (b) Would the duration of the day (one revolution) increase or decrease?

16. Can an object be in equilibrium if it is in motion? Explain.

17. A ladder stands on the ground, leaning against a wall. Would you feel safer climbing up the ladder if you were told that the ground is frictionless but the wall is rough or if you were told that the wall is frictionless but the ground is rough? Explain your answer.

> PROBLEMS AVAILABLE IN

Solutions to the following Problems are available in the *Student Solutions Manual/Study Guide*:

List of Enhanced Problems

Problem Number	Targeted Feedback in Enhanced WebAssign	Master It in Enhanced WebAssign	Watch It in Enhanced WebAssign
10.1	✓		✓
10.7	✓	✓	
10.9	✓		✓
10.10		✓	
10.11	✓	✓	
10.13	✓		✓
10.16	✓		✓
10.18	✓		✓
10.22	✓		✓
10.23	✓	✓	
10.25	✓		✓
10.29	✓		✓
10.31	✓		✓
10.33	✓	✓	
10.38			✓
10.41	✓		✓
10.46	✓		✓
10.47		✓	
10.50	✓		✓
10.51	✓	✓	
10.56	✓		✓
10.57		✓	
10.59	✓	✓	
10.67	✓	✓	
10.72	✓		✓
10.73	✓	✓	

Chapter 11

Gravity, Planetary Orbits, and the Hydrogen Atom

Chapter Outline

I n Chapter 1, we introduced the notion of modeling and defined four categories of models: geometric, simplification, analysis, and structural. In this chapter, we apply our analysis models to two very common *structural models:* a structural model for a large system—the Solar System—and a structural model for a small system—the hydrogen atom.

We return to Newton's law of universal gravitation—one of the fundamental force laws in nature discussed in Chapter 5—and show how it, together with our analysis models, enables us to understand the motions of planets, moons, and artificial Earth satellites.

We conclude this chapter with a discussion of Niels Bohr's model of the hydrogen atom, which represents an interesting mixture of classical and nonclassical physics. Despite the hybrid nature of the model, some of its predictions agree with experimental measurements made on hydrogen atoms. This discussion will be our first major venture into the area of *quantum physics,* which we will continue in Chapter 28.

Hubble image of the Whirlpool Galaxy, M51, taken in 2005. The arms of this spiral galaxy compress hydrogen gas and create new clusters of stars. Some astronomers believe that the arms are prominent due to a close encounter with the small, yellow galaxy, NGC 5195, at the tip of one of its arms.

NASA, Hubble Heritage Team, (STScI/AURA), ESA, S. Beckwith (STScI). Additional Processing: Robert Gendler

11.1 | Newton's Law of Universal Gravitation Revisited

Before 1687, a large amount of data had been collected on the motions of the Moon and the planets, but a clear understanding of the forces involved with the motions was not yet attainable. In that year, Isaac Newton provided the key that unlocked the secrets of the heavens. He knew, from the first law of motion, that a net force had to be acting on the Moon. If not, the Moon would move in a straight-line path rather than in its almost circular orbit. Newton reasoned that this force between the Moon and the Earth was an attractive force. He realized that the forces involved in the Earth–Moon attraction and in the Sun–planet attraction were not something special to those systems, but rather were particular cases of a general and universal attraction between objects.

As you should recall from Chapter 5, every particle in the Universe attracts every other particle with a force that is directly proportional to the product of their masses and inversely proportional to the square of the distance between them. If two particles have masses m_1 and m_2 and are separated by a distance r, the magnitude of the gravitational force between them is

$$F_g = G \frac{m_1 m_2}{r^2}$$ **11.1** ◀

where G is the **universal gravitational constant** whose value in SI units is

$$G = 6.674 \times 10^{-11} \ \text{N} \cdot \text{m}^2/\text{kg}^2$$ **11.2** ◀

The force law given by Equation 11.1 is often referred to as an **inverse-square law** because the magnitude of the force varies as the inverse square of the separation of the particles. We can express this attractive force in vector form by defining a unit vector $\hat{\mathbf{r}}_{12}$ directed from m_1 toward m_2 as shown in Active Figure 11.1. The force exerted by m_1 on m_2 is

$$\vec{\mathbf{F}}_{12} = -G \frac{m_1 m_2}{r^2} \hat{\mathbf{r}}_{12}$$ **11.3** ◀

where the negative sign indicates that particle 2 is attracted toward particle 1. Likewise, by Newton's third law, the force exerted by m_2 on m_1, designated $\vec{\mathbf{F}}_{21}$, is equal in magnitude to $\vec{\mathbf{F}}_{12}$ and in the opposite direction. That is, these forces form an action–reaction pair, and $\vec{\mathbf{F}}_{21} = -\vec{\mathbf{F}}_{12}$.

As Newton demonstrated, the gravitational force exerted by a finite-sized, spherically symmetric mass distribution on a particle outside the distribution is the same as if the entire mass of the distribution were concentrated at its center. For example, the force on a particle of mass m at the Earth's surface has the magnitude

$$F_g = G \frac{M_E m}{R_E^2}$$

where M_E is the Earth's mass and R_E is the Earth's radius. This force is directed toward the center of the Earth.

Measurement of the Gravitational Constant

The universal gravitational constant G was first evaluated in the late 19th century, based on the results of an important experiment by Sir Henry Cavendish in 1798. The law of universal gravitation was not expressed by Newton in the form of Equation 11.1, and Newton did not mention a constant such as G. In fact, even by the time of Cavendish, a unit of force had not yet been included in the existing system of units. Cavendish's goal was to measure the density of the Earth. His results were then used by other scientists 100 years later to generate a value for G.

The apparatus he used consists of two small spheres, each of mass m, fixed to the ends of a light horizontal rod suspended by a thin wire as in Figure 11.2. Two

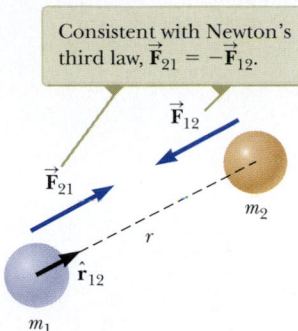

Consistent with Newton's third law, $\vec{\mathbf{F}}_{21} = -\vec{\mathbf{F}}_{12}$.

Active Figure 11.1 The gravitational force between two particles is attractive. The unit vector $\hat{\mathbf{r}}_{12}$ is directed from particle 1 toward particle 2.

Pitfall Prevention | 11.1
Be Clear on g and G
The symbol g represents the magnitude of the free-fall acceleration near a planet. At the surface of the Earth, g has an average value of 9.80 m/s². On the other hand, G is a universal constant that has the same value everywhere in the Universe.

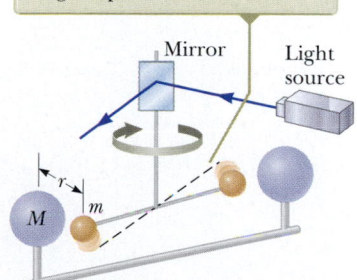

The dashed line represents the original position of the rod.

Mirror   Light source

Figure 11.2 Schematic diagram of the Cavendish apparatus. As the small spheres of mass m are attracted to the large spheres of mass M, the rod rotates through a small angle. A light beam reflected from a mirror on the rotating apparatus measures the angle of rotation. (In reality, the length of wire above the mirror is much larger than that below it.)

large spheres, each of mass M, are then placed near the smaller spheres. The attractive force between the smaller and larger spheres causes the rod to rotate and twist the wire. If the system is oriented as shown in Figure 11.2, the rod rotates clockwise when viewed from the top. The angle through which it rotates is measured by the deflection of a light beam that is reflected from a mirror attached to the wire. The experiment is carefully repeated with different masses at various separations.

It is interesting that G is the least well known of the fundamental constants, with a percentage uncertainty thousands of times larger than those for other constants such as the mass m_e of the electron and the fundamental electric charge e. Several recent measurements of G vary significantly from the previous values and from one another! The search for a more precise value of G continues to be an area of active research. A 2006 experiment measured weight changes in a stationary object as a second object was brought near, resulting in a value of G of $6.674\ 3 \times 10^{-11}\ \mathrm{m^3/kg \cdot s^2}$, with an uncertainty of $\pm 0.001\ 5\%$. A 2007 experiment measured G using a gravity gradiometer based on atom interferometry. The result of this experiment was $6.693 \times 10^{-11}\ \mathrm{m^3/kg \cdot s^2}$, with an uncertainty of $\pm 0.3\%$. The 2006 result is barely within the relatively large uncertainty range of the 2007 result!

> **QUICK QUIZ 11.1** A planet has two moons of equal mass. Moon 1 is in a circular orbit of radius r. Moon 2 is in a circular orbit of radius $2r$. What is the magnitude of the gravitational force exerted by the planet on Moon 2? (**a**) four times as large as that on Moon 1 (**b**) twice as large as that on Moon 1 (**c**) equal to that on Moon 1 (**d**) half as large as that on Moon 1 (**e**) one-fourth as large as that on Moon 1

The Gravitational Field

When Newton first published his theory of gravitation, his contemporaries found it difficult to accept the concept of a force that one object could exert on another without anything happening in the space between them. They asked how it was possible for two objects with mass to interact even though they were not in contact with each other. Although Newton himself could not answer this question, his theory was considered a success because it satisfactorily explained the motions of the planets.

▶ Gravitational field

An alternative mental representation of the gravitational force is to think of the gravitational interaction as a two-step process involving a *field*, as discussed in Section 4.1. First, one object (a *source mass*) creates a **gravitational field** $\vec{g}$ throughout the space around it. Then, a second object (a *test mass*) of mass m residing in this field experiences a force $\vec{F}_g = m\vec{g}$. In other words, we model the *field* as exerting a force on the test mass rather than the source mass exerting the force directly. The gravitational field is defined by

$$\vec{g} \equiv \frac{\vec{F}_g}{m} \qquad \qquad \textbf{11.4} \blacktriangleleft$$

That is, the gravitational field at a point in space equals the gravitational force that a test mass m experiences at that point divided by the mass. Consequently, if $\vec{g}$ is known at some point in space, a particle of mass m experiences a gravitational force $\vec{F}_g = m\vec{g}$ when placed at that point. We will also see the model of a particle in a field for electricity and magnetism in later chapters, where it plays a much larger role than it does for gravity.

As an example, consider an object of mass m near the Earth's surface. The gravitational force on the object is directed toward the center of the Earth and has a magnitude mg. Therefore, we see that the gravitational field experienced by the object at some point has a magnitude equal to the free-fall acceleration at that point. Because the gravitational force on the object has a magnitude GM_Em/r^2 (where M_E is the mass of the Earth), the field $\vec{\mathbf{g}}$ at a distance r from the center of the Earth is given by

$$\vec{\mathbf{g}} = \frac{\vec{\mathbf{F}}_g}{m} = -\frac{GM_E}{r^2}\hat{\mathbf{r}}$$ **11.5**◀

where $\hat{\mathbf{r}}$ is a unit vector pointing radially outward from the Earth and the negative sign indicates that the field vector points toward the center of the Earth as shown in Figure 11.3a. Note that the field vectors at different points surrounding the spherical mass vary in both direction and magnitude. In a small region near the Earth's surface, $\vec{\mathbf{g}}$ is approximately constant and the downward field is uniform as indicated in Figure 11.3b. Equation 11.5 is valid at all points outside the Earth's surface, assuming that the Earth is spherical and that rotation can be neglected. At the Earth's surface, where $r = R_E$, $\vec{\mathbf{g}}$ has a magnitude of 9.80 m/s².

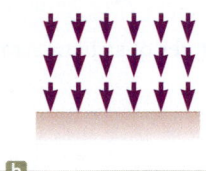

The field vectors point in the direction of the acceleration a particle would experience if it were placed in the field. The magnitude of the field vector at any location is the magnitude of the free-fall acceleration at that location.

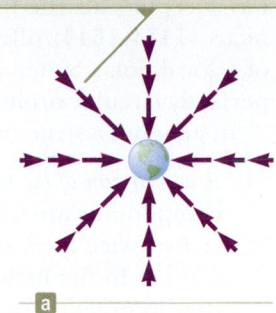

Figure 11.3 (a) The gravitational field vectors in the vicinity of a uniform spherical mass vary in both direction and magnitude. (b) The gravitational field vectors in a small region near the Earth's surface are uniform; that is, they all have the same direction and magnitude.

Example 11.1 | The Density of the Earth

Using the known radius of the Earth and that $g = 9.80$ m/s² at the Earth's surface, find the average density of the Earth.

SOLUTION

Conceptualize Assume the Earth is a perfect sphere. The density of material in the Earth varies, but let's adopt a simplified model in which we assume the density to be uniform throughout the Earth. The resulting density is the average density of the Earth.

Categorize This example is a relatively simple substitution problem.

Using the magnitudes in Equation 11.5 at the surface of the Earth, solve for the mass of the Earth:

$$M_E = \frac{gR_E^2}{G}$$

Substitute this mass into the definition of density (Eq. 1.1):

$$\rho_E = \frac{M_E}{V_E} = \frac{gR_E^2/G}{\frac{4}{3}\pi R_E^3} = \frac{3}{4}\frac{g}{\pi GR_E}$$

$$= \frac{3}{4}\frac{9.80 \text{ m/s}^2}{\pi(6.67 \times 10^{-11} \text{ N} \cdot \text{m}^2/\text{kg}^2)(6.37 \times 10^6 \text{ m})} = \boxed{5.51 \times 10^3 \text{ kg/m}^3}$$

What If? What if you were told that a typical density of granite at the Earth's surface is 2.75×10^3 kg/m³? What would you conclude about the density of the material in the Earth's interior?

Answer Because this value is about half the density we calculated as an average for the entire Earth, we would conclude that the inner core of the Earth has a density much higher than the average value. It is most amazing that the Cavendish experiment—which can be used to determine G and can be done today on a tabletop—combined with simple free-fall measurements of g provides information about the core of the Earth!

❰11.2 | Structural Models

In Chapter 1, we mentioned that we would discuss four categories of models. The fourth category is **structural models.** In these models, we propose theoretical structures in an attempt to understand the behavior of a system with which we cannot interact directly because it is far different in scale—either much smaller or much larger—from our macroscopic world.

One of the earliest structural models to be explored was that of the place of the Earth in the Universe. The movements of the planets, stars, and other celestial bodies have been observed by people for thousands of years. Early in history, scientists regarded the Earth as the center of the Universe because it appeared that objects in the sky moved around the Earth. This organization of the Earth and other objects is a structural model for the Universe called the *geocentric model*. It was elaborated and formalized by the Greek astronomer Claudius Ptolemy in the 2nd century A.D. and was accepted for the next 1400 years. In 1543, Polish astronomer Nicolaus Copernicus (1473–1543) offered a different structural model in which the Earth is part of a local Solar System, suggesting that the Earth and the other planets revolve in perfectly circular orbits about the Sun (the *heliocentric model*).

In general, a structural model contains the following features:

▶ Features of structural models

1. *A description of the physical components of the system:* In the heliocentric model, the components are the planets and the Sun.
2. *A description of where the components are located relative to one another and how they interact:* In the heliocentric model, the planets are in orbit around the Sun and they interact via the gravitational force.
3. *A description of the time evolution of the system:* The heliocentric model assumes a steady-state Solar System, with planets revolving in orbits around the Sun with fixed periods.
4. *A description of the agreement between predictions of the model and actual observations and, possibly, predictions of new effects that have not yet been observed:* The heliocentric model predicts Earth-based observations of Mars that are in agreement with historical and present measurements. The geocentric model was also able to find agreement between predictions and observations, but only at the expense of a very complicated structural model in which the planets moved in circles built on other circles. The heliocentric model, along with Newton's law of universal gravitation, predicted that a spacecraft could be sent from the Earth to Mars long before it was actually first done in the 1970s.

In Sections 11.3 and 11.4, we explore some of the details of the heliocentric model of the Solar System and supplement the description above for this structural model. In Section 11.5, we investigate a structural model of the hydrogen atom. We will use the components of structural models listed above many times throughout the book.

◖11.3 | Kepler's Laws

Danish astronomer Tycho Brahe (1546–1601) made accurate astronomical measurements over a period of 20 years and provided the basis for the currently accepted structural model of the Solar System. These precise observations, made on the planets and 777 stars, were carried out with nothing more elaborate than a large sextant and compass; the telescope had not yet been invented.

German astronomer Johannes Kepler, who was Brahe's assistant, acquired Brahe's astronomical data and spent about 16 years trying to deduce a mathematical model for the motions of the planets. After many laborious calculations, he found that Brahe's precise data on the revolution of Mars about the Sun provided the answer. Kepler's analysis first showed that the concept of circular orbits about the Sun in the heliocentric model had to be abandoned. He discovered that the orbit of Mars could be accurately described by a curve called an *ellipse*. He then generalized this analysis to include the motions of all planets. The complete analysis is summarized in three statements, known as **Kepler's laws of planetary motion,** each of which is discussed in the following sections.

Newton demonstrated that these laws are consequences of the gravitational force that exists between any two masses. Newton's law of universal gravitation, together with his laws of motion, provides the basis for a full mathematical representation of the motion of planets and satellites.

Johannes Kepler
German astronomer (1571–1630)
Kepler is best known for developing the laws of planetary motion based on the careful observations of Tycho Brahe.

iStockphoto.com/GeorgiosArt

Kepler's First Law

Kepler's first law indicates that the circular orbit of an object around a gravitational force center is a very special case and that elliptical orbits are the general situation:[1]

All planets move in elliptical orbits with the Sun at one focus.

▶ Kepler's first law

Active Figure 11.4 shows the geometry of an ellipse, which serves as our geometric model for the elliptical orbit of a planet.[2] An ellipse is mathematically defined by choosing two points, F_1 and F_2, each of which is a called a **focus,** and then drawing a curve through points for which the sum of the distances r_1 and r_2 from F_1 and F_2 is a constant. The longest distance through the center between points on the ellipse (and passing through each focus) is called the **major axis,** and this distance is $2a$. In Active Figure 11.4, the major axis is drawn along the x direction. The distance a is called the **semimajor axis.** Similarly, the shortest distance through the center between points on the ellipse is called the **minor axis** of length $2b$, where the distance b is the **semiminor axis.** Either focus of the ellipse is located at a distance c from the center of the ellipse, where $a^2 = b^2 + c^2$. In the elliptical orbit of a planet around the Sun, the Sun is at one focus of the ellipse. There is nothing at the other focus.

The **eccentricity** of an ellipse is defined as $e \equiv c/a$ and it describes the general shape of the ellipse. For a circle, $c = 0$ and the eccentricity is therefore zero. The smaller b is compared with a, the shorter the ellipse is along the y direction compared with its extent in the x direction in Active Figure 11.4. As b decreases, c increases and the eccentricity e increases. Therefore, higher values of eccentricity correspond to longer and thinner ellipses. The range of values of the eccentricity for an ellipse is $0 < e < 1$.

Eccentricities for planetary orbits vary widely in the Solar System. The eccentricity of the Earth's orbit is 0.017, which makes it nearly circular. On the other hand, the eccentricity of Mercury's orbit is 0.21, the highest of all the eight planets. Figure 11.5a shows an ellipse with the eccentricity of that of Mercury's orbit. Notice that even this highest-eccentricity orbit is difficult to distinguish from a circle, which is one reason Kepler's first law is an admirable accomplishment. The eccentricity of the orbit of Comet Halley is 0.97, describing an orbit whose major axis is much longer than its minor axis as shown in Figure 11.5b. As a result, Comet Halley spends much of its 76-year period far from the Sun and invisible from the Earth. It is only visible to the naked eye during a small part of its orbit when it is near the Sun.

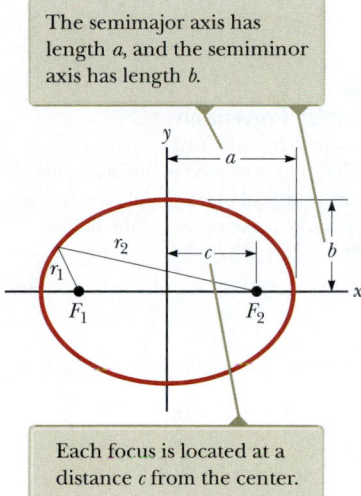

The semimajor axis has length a, and the semiminor axis has length b.

Each focus is located at a distance c from the center.

Active Figure 11.4 Plot of an ellipse.

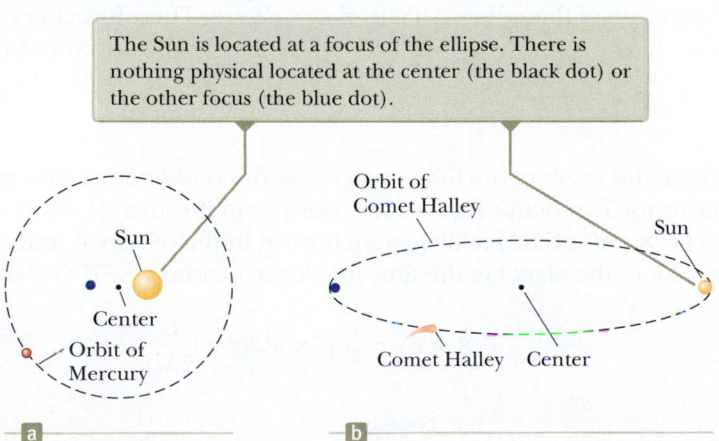

The Sun is located at a focus of the ellipse. There is nothing physical located at the center (the black dot) or the other focus (the blue dot).

Sun

Center

Orbit of Mercury

Orbit of Comet Halley

Sun

Comet Halley   Center

a

b

Figure 11.5 (a) The shape of the orbit of Mercury, which has the highest eccentricity ($e = 0.21$) among the eight planets in the solar system. (b) The shape of the orbit of Comet Halley. The shape of the orbit is correct; the comet and the Sun are shown larger than in reality for clarity.

---

[1]We choose a simplification model in which a body of mass m is in orbit around a body of mass M, with $M \gg m$. In this way, we model the body of mass M to be stationary. In reality, that is not true; both M and m move around the center of mass of the system of two objects. That is how we indirectly detect planets around other stars; we see the "wobbling" motion of the star as the planet and the star rotate about the center of mass.

[2]Actual orbits show perturbations due to moons in orbit around the planet and passages of the planet near other planets. We will ignore these perturbations and adopt a simplification model in which the planet follows a perfectly elliptical orbit.

Now imagine a planet in an elliptical orbit such as that shown in Active Figure 11.4 with the Sun at focus F_2. When the planet is at the far left in the diagram, the distance between the planet and the Sun is $a + c$. At this point, called the *aphelion*, the planet is at its maximum distance from the Sun. (For an object in orbit around the Earth, this point is called the *apogee*.) Conversely, when the planet is at the right end of the ellipse, the distance between the planet and the Sun is $a - c$. At this point, called the *perihelion* (for an Earth orbit, the *perigee*), the planet is at its minimum distance from the Sun.

Kepler's first law is a direct result of the inverse-square nature of the gravitational force. We have discussed circular and elliptical orbits, which are the allowed shapes of orbits for objects that are *bound* to the gravitational force center. These objects include planets, asteroids, and comets that move repeatedly around the Sun, as well as moons orbiting a planet. *Unbound* objects might also occur, such as a meteoroid from deep space that might pass by the Sun once and then never return. The gravitational force between the Sun and these objects also varies as the inverse square of the separation distance, and the allowed paths for these objects include parabolas ($e = 1$) and hyperbolas ($e > 1$).

> **Pitfall Prevention | 11.2**
> **Where Is The Sun?**
> The Sun is located at one focus of the elliptical orbit of a planet. It is *not* located at the center of the ellipse.

Kepler's Second Law

Let us now look at Kepler's second law:

▶ Kepler's second law

> The radius vector drawn from the Sun to any planet sweeps out equal areas in equal time intervals.

This law can be shown to be a consequence of angular momentum conservation for an isolated system as follows. Consider a planet of mass M_p moving about the Sun in an elliptical orbit (Active Fig. 11.6a). Let us consider the planet as a system. We model the Sun to be so much more massive than the planet that the Sun does not move. The gravitational force acting on the planet is a central force, always directed along the radius vector toward the Sun. The torque on the planet due to this central force is zero because $\vec{\mathbf{F}}_g$ is parallel to $\vec{\mathbf{r}}$. That is,

$$\vec{\boldsymbol{\tau}}_{\text{ext}} \equiv \vec{\mathbf{r}} \times \vec{\mathbf{F}}_g = \vec{\mathbf{r}} \times F_g(r)\hat{\mathbf{r}} = 0$$

Recall that the external net torque on a system equals the time rate of change of angular momentum of the system; that is, $\vec{\boldsymbol{\tau}}_{\text{ext}} = d\vec{\mathbf{L}}/dt$. Therefore, because $\vec{\boldsymbol{\tau}}_{\text{ext}} = 0$ for the planet, the angular momentum $\vec{\mathbf{L}}$ of the planet is a constant of the motion:

$$\vec{\mathbf{L}} = \vec{\mathbf{r}} \times \vec{\mathbf{p}} = M_p \vec{\mathbf{r}} \times \vec{\mathbf{v}} = \text{constant}$$

We can relate this result to the following geometric consideration. In a time interval dt, the radius vector $\vec{\mathbf{r}}$ in Active Figure 11.6b sweeps out the area dA, which equals one-half the area $|\vec{\mathbf{r}} \times d\vec{\mathbf{r}}|$ of the parallelogram formed by the vectors $\vec{\mathbf{r}}$ and $d\vec{\mathbf{r}}$. Because the displacement of the planet in the time interval dt is given by $d\vec{\mathbf{r}} = \vec{\mathbf{v}} dt$, we have

$$dA = \tfrac{1}{2}|\vec{\mathbf{r}} \times d\vec{\mathbf{r}}| = \tfrac{1}{2}|\vec{\mathbf{r}} \times \vec{\mathbf{v}} dt| = \frac{L}{2M_p} dt$$

$$\frac{dA}{dt} = \frac{L}{2M_p} = \text{constant} \qquad \textbf{11.6} \blacktriangleleft$$

where L and M_p are both constants. Therefore, we conclude that the radius vector from the Sun to any planet sweeps out equal areas in equal times.

This conclusion is a consequence of the gravitational force being a central force, which in turn implies that angular momentum of the planet is constant. Therefore, the law applies to *any* situation that involves a central force, whether inverse-square or not.

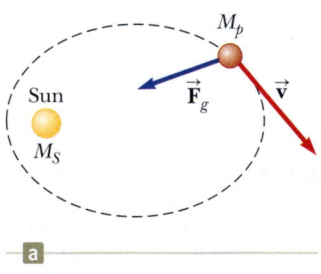

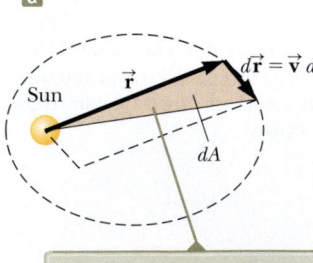

The area swept out by $\vec{\mathbf{r}}$ in a time interval dt is half the area of the parallelogram.

Active Figure 11.6 (a) The gravitational force acting on a planet acts toward the Sun, along the radius vector. (b) During a time interval dt, the vectors form a parallelogram.

> **THINKING PHYSICS 11.1**

The Earth is closer to the Sun when it is winter in the Northern Hemisphere than when it is summer. July and January both have 31 days. In which month, if either, does the Earth move through a longer distance in its orbit?

Reasoning The Earth is in a slightly elliptical orbit around the Sun. Because of angular momentum conservation, the Earth moves more rapidly when it is close to the Sun and more slowly when it is farther away. Therefore, because it is closer to the Sun in January, it is moving faster and will cover more distance in its orbit than it will in July. ◄

Kepler's Third Law

Kepler's third law reads as follows:

The square of the orbital period of any planet is proportional to the cube of the semimajor axis of the elliptical orbit.

Kepler's third law can be predicted from the inverse-square law for circular orbits. Consider a planet of mass M_p that is assumed to be moving about the Sun (mass M_S) in a circular orbit as in Active Figure 11.7. Because the gravitational force provides the centripetal acceleration of the planet as it moves in a circle, we model the planet as a particle in uniform circular motion and incorporate Newton's law of universal gravitation:

$$F_g = M_p a \quad \rightarrow \quad \frac{GM_S M_p}{r^2} = \frac{M_p v^2}{r}$$

The orbital speed of the planet is $2\pi r/T$, where T is the period; therefore, the preceding expression becomes

$$\frac{GM_S}{r^2} = \frac{(2\pi r/T)^2}{r}$$

$$T^2 = \left(\frac{4\pi^2}{GM_S}\right) r^3 = K_S r^3$$

where K_S is a constant given by

$$K_S = \frac{4\pi^2}{GM_S} = 2.97 \times 10^{-19} \ \mathrm{s^2/m^3}$$

This equation is also valid for elliptical orbits if we replace r with the length a of the semimajor axis (see Active Fig. 11.4):

$$T^2 = \left(\frac{4\pi^2}{GM_S}\right) a^3 = K_S a^3$$

11.7 ◄ ► Kepler's third law

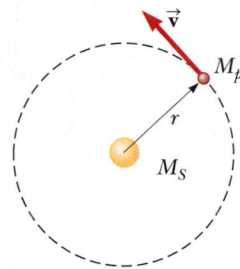

Active Figure 11.7 A planet of mass M_p moving in a circular orbit about the Sun. Kepler's third law relates the period of the orbit to the radius. The orbits of all planets except Mercury are nearly circular.

Equation 11.7 is Kepler's third law. Because the semimajor axis of a circular orbit is its radius, Equation 11.7 is valid for both circular and elliptical orbits. Note that the constant of proportionality K_S is independent of the mass of the planet. Equation 11.7 is therefore valid for *any* planet. If we were to consider the orbit of a satellite about the Earth, such as the Moon, the constant would have a different value, with the Sun's mass replaced by the Earth's mass; that is, $K_E = 4\pi^2/GM_E$.

Table 11.1 (page 308) is a collection of useful planetary data. The last column verifies that the ratio T^2/r^3 is constant. The small variations in the values in this column are because of uncertainties in the data measured for the periods and semimajor axes of the planets.

Recent astronomical work has revealed the existence of a large number of solar system objects beyond the orbit of Neptune. In general, these objects lie in the *Kuiper belt*, a region that extends from about 30 AU (the orbital radius of Neptune)

◄ TABLE 11.1 | Useful Planetary Data

Body	Mass (kg)	Mean Radius (m)	Period of Revolution (s)	Mean Distance from the Sun (m)	$\frac{T^2}{r^3}$ (s^2/m^3)
Mercury	3.30×10^{23}	2.44×10^6	7.60×10^6	5.79×10^{10}	2.98×10^{-19}
Venus	4.87×10^{24}	6.05×10^6	1.94×10^7	1.08×10^{11}	2.99×10^{-19}
Earth	5.97×10^{24}	6.37×10^6	3.156×10^7	1.496×10^{11}	2.97×10^{-19}
Mars	6.42×10^{23}	3.39×10^6	5.94×10^7	2.28×10^{11}	2.98×10^{-19}
Jupiter	1.90×10^{27}	6.99×10^7	3.74×10^8	7.78×10^{11}	2.97×10^{-19}
Saturn	5.68×10^{26}	5.82×10^7	9.29×10^8	1.43×10^{12}	2.95×10^{-19}
Uranus	8.68×10^{25}	2.54×10^7	2.65×10^9	2.87×10^{12}	2.97×10^{-19}
Neptune	1.02×10^{26}	2.46×10^7	5.18×10^9	4.50×10^{12}	2.94×10^{-19}
Pluto[a]	1.25×10^{22}	1.20×10^6	7.82×10^9	5.91×10^{12}	2.96×10^{-19}
Moon	7.35×10^{22}	1.74×10^6	—	—	—
Sun	1.989×10^{30}	6.96×10^8	—	—	—

[a]In August, 2006, the International Astronomical Union adopted a definition of a planet that separates Pluto the other the eight planets. Pluto is now defined as a "dwarf planet" like the asteroid Ceres.

to 50 AU. (An AU is an *astronomical unit,* equal to the radius of the Earth's orbit.) Current estimates identify at least 70 000 objects in this region with diameters larger than 100 km. The first Kuiper belt object (KBO) is Pluto, discovered in 1930 and formerly classified as a planet. Starting in 1992, many more have been detected. Several have diameters in the 1 000-km range, such as Varuna (discovered in 2000), Ixion (2001), Quaoar (2002), Sedna (2003), Haumea (2004), Orcus (2004), and Makemake (2005). One KBO, Eris, discovered in 2005, is believed to be significantly larger than Pluto. Other KBOs do not yet have names, but they are currently indicated by their year of discovery and a code, such as 2009 YE7 and 2010 EK139.

A subset of about 1 400 KBOs are called "Plutinos" because, like Pluto, they exhibit a resonance phenomenon, orbiting the Sun two times in the same time interval as Neptune revolves three times. The contemporary application of Kepler's laws suggest the excitement of this active area of current research.

◄ QUICK QUIZ 11.2 An asteroid is in a highly eccentric elliptical orbit around the Sun. The period of the asteroid's orbit is 90 days. Which of the following statements is true about the possibility of a collision between this asteroid and the Earth? **(a)** There is no possible danger of a collision **(b)** There is a possibility of a collision. **(c)** There is not enough information to determine whether there is a danger of a collision.

THINKING PHYSICS 11.2

The novel *Icebound,* by Dean Koontz (Bantam Books, 2000), is a story of a group of scientists trapped on a floating iceberg near the North Pole. One of the devices the scientists have with them is a transmitter with which they can fix their position with "the aid of a geosynchronous polar satellite." Can a satellite in a *polar* orbit be *geosynchronous?*

Reasoning A geosynchronous satellite is one that stays over one location on the Earth's surface at all times. Therefore, an antenna on the surface that receives signals from the satellite, such as a television dish, can stay pointed in a fixed direction toward the sky. The satellite must be in an orbit with the correct radius such that its orbital period is the same as that of the Earth's rotation.

This orbit results in the satellite appearing to have no east–west motion relative to the observer at the chosen location. Another requirement is that a geosynchronous satellite *must be in orbit over the equator.* Otherwise it would appear to undergo a north–south oscillation during one orbit. Therefore, it would be impossible to have a geosynchronous satellite in a *polar* orbit. Even if such a satellite were at the proper distance from the Earth, it would be moving rapidly in the north–south direction, resulting in the necessity of accurate tracking equipment. What's more, it would be below the horizon for long periods of time, making it useless for determining one's position. ◄

Example 11.2 | A Geosynchronous Satellite

Consider a satellite of mass m moving in a circular orbit around the Earth at a constant speed v and at an altitude h above the Earth's surface as illustrated in Figure 11.8.

(A) Determine the speed of satellite in terms of G, h, R_E (the radius of the Earth), and M_E (the mass of the Earth).

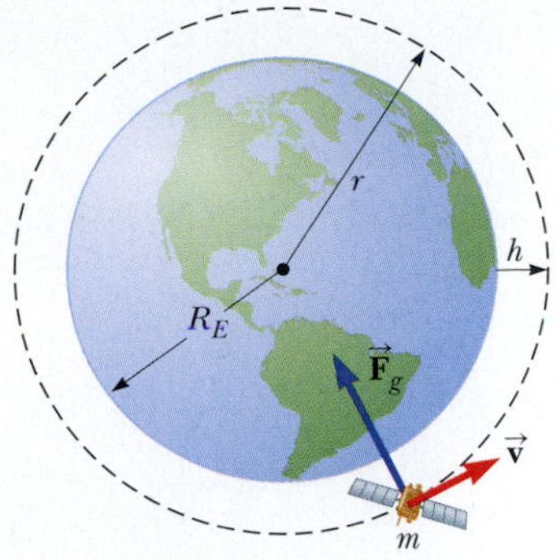

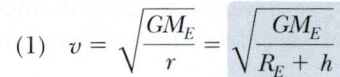

SOLUTION

Conceptualize Imagine the satellite moving around the Earth in a circular orbit under the influence of the gravitational force. This motion is similar to that of the International Space Station, the Hubble Space Telescope, and other objects in orbit around the Earth.

Figure 11.8 (Example 11.2) A satellite of mass m moving around the Earth in a circular orbit of radius r with constant speed v. The only force acting on the satellite is the gravitational force $\vec{F}_g$. (Not drawn to scale.)

Categorize The satellite must have a centripetal acceleration. Therefore, we categorize the satellite as a particle under a net force and a particle in uniform circular motion.

Analyze The only external force acting on the satellite is the gravitational force, which acts toward the center of the Earth and keeps the satellite in its circular orbit.

Apply the particle under a net force and particle in uniform circular motion models to the satellite:

$$F_g = ma \rightarrow G\frac{M_E m}{r^2} = m\left(\frac{v^2}{r}\right)$$

Solve for v, noting that the distance r from the center of the Earth to the satellite is $r = R_E + h$:

$$(1) \quad v = \sqrt{\frac{GM_E}{r}} = \sqrt{\frac{GM_E}{R_E + h}}$$

(B) If the satellite is to be *geosynchronous* (that is, appearing to remain over a fixed position on the Earth), how fast is it moving through space?

SOLUTION

To appear to remain over a fixed position on the Earth, the period of the satellite must be 24 h = 86 400 s and the satellite must be in orbit directly over the equator.

Solve Kepler's third law (Equation 11.7, with $a = r$ and $M_S \rightarrow M_E$) for r:

$$r = \left(\frac{GM_E T^2}{4\pi^2}\right)^{1/3}$$

Substitute numerical values:

$$r = \left[\frac{(6.67 \times 10^{-11}\ \text{N} \cdot \text{m}^2/\text{kg}^2)(5.97 \times 10^{24}\ \text{kg})(86\ 400\ \text{s})^2}{4\pi^2}\right]^{1/3}$$

$$= 4.22 \times 10^7\ \text{m}$$

Use Equation (1) to find the speed of the satellite:

$$v = \sqrt{\frac{(6.67 \times 10^{-11}\ \text{N} \cdot \text{m}^2/\text{kg}^2)(5.97 \times 10^{24}\ \text{kg})}{4.22 \times 10^7\ \text{m}}}$$

$$= 3.07 \times 10^3\ \text{m/s}$$

Finalize The value of r calculated here translates to a height of the satellite above the surface of the Earth of almost 36 000 km. Therefore, geosynchronous satellites have the advantage of allowing an earthbound antenna to be aimed in a fixed direction, but there is a disadvantage in that the signals between the Earth and the satellite must travel a long distance. It is difficult to use geosynchronous satellites for optical observation of the Earth's surface because of their high altitude.

What If? What if the satellite motion in part (A) were taking place at height h above the surface of another planet more massive than the Earth but of the same radius? Would the satellite be moving at a higher speed or a lower speed than it does around the Earth?

Answer If the planet exerts a larger gravitational force on the satellite due to its larger mass, the satellite must move with a higher speed to avoid moving toward the surface. This conclusion is consistent with the predictions of Equation (1), which shows that because the speed v is proportional to the square root of the mass of the planet, the speed increases as the mass of the planet increases.

❮11.4❯ Energy Considerations in Planetary and Satellite Motion

So far we have approached orbital mechanics from the point of view of forces and angular momentum. Let us now investigate the motion of planets in orbit from the *energy* point of view.

Consider an object of mass m moving with a speed v in the vicinity of a massive object of mass $M \gg m$. This two-object system might be a planet moving around the Sun, a satellite orbiting the Earth, or a comet making a one-time flyby past the Sun. We will model the two objects of mass m and M as an isolated system. If we assume that M is at rest in an inertial reference frame (because $M \gg m$), the total energy E of the two-object system is the sum of the kinetic energy of the object of mass m and the gravitational potential energy of the system:

$$E = K + U_g$$

Recall from Section 6.9 that the gravitational potential energy U_g associated with *any pair* of particles of masses m_1 and m_2 separated by a distance r is given by

$$U_g = -\frac{Gm_1 m_2}{r}$$

where we have defined $U_g \rightarrow 0$ as $r \rightarrow \infty$; therefore, in our case, the total energy of the system of m and M is

$$E = \tfrac{1}{2}mv^2 - \frac{GMm}{r} \qquad \text{11.8} \blacktriangleleft$$

Equation 11.8 shows that E may be positive, negative, or zero, depending on the value of v at a particular separation distance r. If we consider the energy diagram method of Section 6.10, we can show the potential and total energies of the system as a function of r as in Figure 11.9. A planet moving around the Sun and a satellite in orbit around the Earth are *bound* systems, such as those we discussed in Section 11.3; the Earth will always stay near the Sun and the satellite near the Earth. In Figure 11.9, these systems are represented by a total energy E that is negative. The point at which the total energy line intersects the potential energy curve is a turning point, the maximum separation distance r_{max} between the two bound objects.

A one-time meteoroid flyby represents an unbound system. The meteoroid interacts with the Sun but is not bound to it. Therefore, the meteoroid can in theory move infinitely far away from the Sun as represented in Figure 11.9 by a total energy line in the positive region of the graph. This line never intersects the potential energy curve, so all values of r are possible.

For a bound system, such as the Earth and Sun, E is necessarily less than zero because we have chosen the convention that $U_g \rightarrow 0$ as $r \rightarrow \infty$. We can easily establish that $E < 0$ for the system consisting of an object of mass m moving in a circular orbit

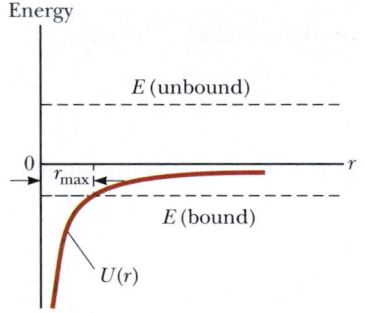

Figure 11.9 The lower total energy line represents a bound system. The separation distance r between the two gravitationally bound objects never exceeds r_{max}. The upper total energy line represents an unbound system of two objects interacting gravitationally. The separation distance r between the two objects can have any value.

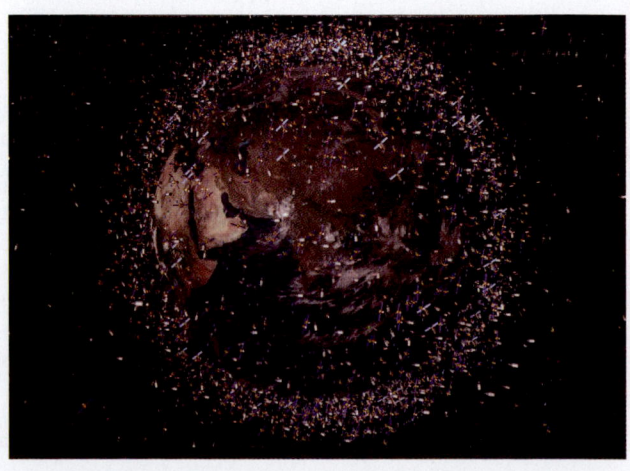

Many artificial satellites have been placed in orbit about the Earth. This diagram shows many satellites in low Earth orbits. This region of space is becoming very crowded: in 2009, a U.S. commercial Iridium satellite collided with an inactive Russian Kosmos satellite, destroying both. (The debris field shown in the image is an artist's impression based on actual data. However, the debris objects are shown at an exaggerated size to make them visible at the scale shown.)

ESA

about an object of mass $M \gg m$ (Fig. 11.8). Applying Newton's second law to the object of mass m gives

$$\sum F = ma \quad \rightarrow \quad \frac{GMm}{r^2} = \frac{mv^2}{r}$$

Multiplying both sides by r and dividing by 2 gives

$$\frac{1}{2}mv^2 = \frac{GMm}{2r} \qquad \textbf{11.9} \blacktriangleleft$$

Substituting this result into Equation 11.8, we obtain

$$E = \frac{GMm}{2r} - \frac{GMm}{r}$$

$$E = -\frac{GMm}{2r} \quad \text{(circular orbits)} \qquad \textbf{11.10} \blacktriangleleft$$

This result clearly shows that the total energy must be negative in the case of circular orbits. Furthermore, Equation 11.9 shows that the kinetic energy of an object in a circular orbit is equal to one-half the magnitude of the potential energy of the system (when the potential energy is chosen to be zero at infinite separation).

The total energy is also negative in the case of elliptical orbits. The expression for E for elliptical orbits is the same as Equation 11.10, with r replaced by the semimajor axis a:

$$E = -\frac{GMm}{2a} \quad \text{(elliptical orbits)} \qquad \textbf{11.11} \blacktriangleleft$$

▶ Total energy of a planet–star system

Combining this statement of energy conservation with our earlier discussion of conservation of angular momentum, we see that both the total energy and the total angular momentum of a gravitationally bound, two-object system are constants of the motion.

QUICK QUIZ 11.3 A comet moves in an elliptical orbit around the Sun. Which point in its orbit (perihelion or aphelion) represents the highest value of (**a**) the speed of the comet, (**b**) the potential energy of the comet–Sun system, (**c**) the kinetic energy of the comet, and (**d**) the total energy of the comet–Sun system?

Example 11.3 | Changing the Orbit of a Satellite

A space transportation vehicle releases a 470-kg communications satellite while in an orbit 280 km above the surface of the Earth. A rocket engine on the satellite boosts it into a geosynchronous orbit. How much energy does the engine have to provide?

SOLUTION

Conceptualize Notice that the height of 280 km is much lower than that for a geosynchronous satellite, 36 000 km, as mentioned in Example 11.2. Therefore, energy must be expended to raise the satellite to this much higher position.

Categorize This example is a substitution problem.

Find the initial radius of the satellite's orbit when it is still in the vehicle's cargo bay:

$$r_i = R_E + 280 \text{ km} = 6.65 \times 10^6 \text{ m}$$

Use Equation 11.10 to find the difference in energies for the satellite–Earth system with the satellite at the initial and final radii:

$$\Delta E = E_f - E_i = -\frac{GM_E m}{2r_f} - \left(-\frac{GM_E m}{2r_i}\right) = -\frac{GM_E m}{2}\left(\frac{1}{r_f} - \frac{1}{r_i}\right)$$

continued

11.3 *cont.*

Substitute numerical values, using $r_f = 4.22 \times 10^7$ m from Example 11.2:

$$\Delta E = -\frac{(6.67 \times 10^{-11} \text{ N} \cdot \text{m}^2/\text{kg}^2)(5.97 \times 10^{24} \text{ kg})(470 \text{ kg})}{2}$$

$$\times \left(\frac{1}{4.22 \times 10^7 \text{ m}} - \frac{1}{6.65 \times 10^6 \text{ m}}\right)$$

$$= \boxed{1.19 \times 10^{10} \text{ J}}$$

which is the energy equivalent of 89 gal of gasoline. NASA engineers must account for the changing mass of the spacecraft as it ejects burned fuel, something we have not done here. Would you expect the calculation that includes the effect of this changing mass to yield a greater or a lesser amount of energy required from the engine?

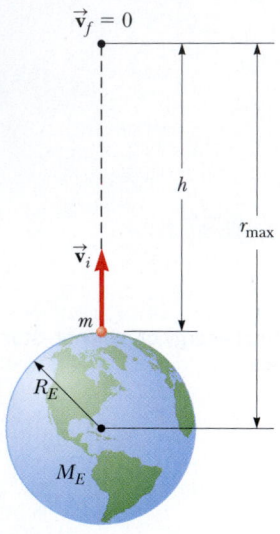

Figure 11.10 An object of mass m projected upward from the Earth's surface with an initial speed v_i reaches a maximum altitude $h = r_{max} - R_E$.

Escape Speed

Suppose an object of mass m is projected vertically upward from the Earth's surface with an initial speed v_i as in Figure 11.10. We can use energy considerations to find the minimum value of the initial speed such that the object will continue to move infinitely far away from the Earth. Equation 11.8 gives the total energy of the object–Earth system at any point when the speed of the object and its distance from the center of the Earth are known. At the surface of the Earth, $r_i = R_E$. When the object reaches its maximum altitude, $v_f = 0$ and $r_f = r_{max}$. Because the total energy of the isolated object–Earth system is conserved, substitution of these conditions into Equation 11.8 gives

$$\tfrac{1}{2}mv_i^2 - \frac{GM_E m}{R_E} = -\frac{GM_E m}{r_{max}}$$

Solving for v_i^2 gives

$$v_i^2 = 2GM_E\left(\frac{1}{R_E} - \frac{1}{r_{max}}\right) \qquad \textbf{11.12} \blacktriangleleft$$

For a given maximum altitude $h = r_{max} - R_E$, we can use this equation to find the required initial speed.

We are now in a position to calculate **escape speed,** which is the minimum speed the object must have at the Earth's surface to continue to move away forever. Traveling at this minimum speed, the object continues to move away from the Earth as its speed asymptotically approaches zero. Letting $r_{max} \rightarrow \infty$ in Equation 11.12 and setting $v_i = v_{esc}$, we have

$$v_{esc} = \sqrt{\frac{2GM_E}{R_E}} \qquad \textbf{11.13} \blacktriangleleft$$

This expression for v_{esc} is independent of the mass of the object. For example, a spacecraft has the same escape speed as a molecule. Furthermore, the result is independent of the direction of the velocity and ignores air resistance.

Note also that Equations 11.12 and 11.13 can be applied to objects projected from *any* planet. That is, in general, the escape speed from any planet of mass M and radius R is

$$v_{esc} = \sqrt{\frac{2GM}{R}} \qquad \textbf{11.14} \blacktriangleleft$$

A list of escape speeds for the planets, the Moon, and the Sun is given in Table 11.2. Note that the values vary from 2.3 km/s for the Moon to about 618 km/s for the Sun. These results, together with some ideas from the kinetic theory of gases (Chapter 16), explain why our atmosphere does not contain significant amounts of hydrogen, which is the most abundant element in the Universe. As we shall see later, gas

TABLE 11.2 |

Escape Speeds from the Surfaces of the Planets, the Moon, and the Sun

Planet	v_{esc} (km/s)
Mercury	4.3
Venus	10.3
Earth	11.2
Mars	5.0
Jupiter	60
Saturn	36
Uranus	22
Neptune	24
Moon	2.4
Sun	618

molecules have an average kinetic energy that depends on the temperature of the gas. Lighter molecules in an atmosphere have translational speeds that are closer to the escape speed than more massive molecules, so they have a higher probability of escaping from the planet and the lighter molecules diffuse into space. This mechanism explains why the Earth does not retain hydrogen molecules and helium atoms in its atmosphere but does retain much heavier molecules, such as oxygen and nitrogen. On the other hand, Jupiter has a very large escape speed (60 km/s), which enables it to retain hydrogen, the primary constituent of its atmosphere.

> **Pitfall Prevention | 11.3**
> **You Can't Really Escape**
> Although Equation 11.13 provides the "escape speed" from the Earth, complete escape from the Earth's gravitational influence is impossible because the gravitational force is of infinite range. No matter how far away you are, you will always feel some gravitational force due to the Earth.

Example 11.4 | Escape Speed of a Rocket

Calculate the escape speed from the Earth for a 5 000-kg spacecraft and determine the kinetic energy it must have at the Earth's surface to move infinitely far away from the Earth.

SOLUTION

Conceptualize Imagine projecting the spacecraft from the Earth's surface so that it moves farther and farther away, traveling more and more slowly, with its speed approaching zero. Its speed will never reach zero, however, so the object will never turn around and come back.

Categorize This example is a substitution problem.

Use Equation 11.13 to find the escape speed:

$$v_{esc} = \sqrt{\frac{2GM_E}{R_E}} = \sqrt{\frac{2(6.67 \times 10^{-11}\ \text{N} \cdot \text{m}^2/\text{kg}^2)(5.97 \times 10^{24}\ \text{kg})}{6.37 \times 10^6\ \text{m}}}$$

$$= 1.12 \times 10^4\ \text{m/s}$$

Evaluate the kinetic energy of the spacecraft from Equation 6.16:

$$K = \tfrac{1}{2}mv_{esc}^2 = \tfrac{1}{2}(5.00 \times 10^3\ \text{kg})(1.12 \times 10^4\ \text{m/s})^2$$

$$= 3.13 \times 10^{11}\ \text{J}$$

The calculated escape speed corresponds to about 25 000 mi/h. The kinetic energy of the spacecraft is equivalent to the energy released by the combustion of about 2 300 gal of gasoline.

Black Holes

In Chapter 10, we briefly described a rare event called a supernova, the catastrophic explosion of a very massive star. The material that remains in the central core of such an object continues to collapse, and the core's ultimate fate depends on its mass. If the core has a mass less than 1.4 times the mass of our Sun, it gradually cools down and ends its life as a white dwarf star. If, however, the core's mass is greater than that, it may collapse further due to gravitational forces. What remains is a neutron star, discussed in Chapter 10, in which the mass of a star is compressed to a radius of about 10 km. (On the Earth, a teaspoon of this material would weigh about 5 billion tons!)

An even more unusual star death may occur when the core has a mass greater than about three solar masses. The collapse may continue until the star becomes a very small object in space, commonly referred to as a **black hole.** In effect, black holes are the remains of stars that have collapsed under their own gravitational force. If an object such as a spacecraft comes close to a black hole, it experiences an extremely strong gravitational force and is trapped forever.

The escape speed from any spherical body depends on the mass and radius of the body. The escape speed for a black hole is very high because of the concentration of the star's mass into a sphere of very small radius. If the escape speed exceeds the speed of light c, radiation from the body (e.g. visible light) cannot escape and the body appears to be black, hence the origin of the term *black hole*. The critical radius R_S at which the escape speed is c is called the **Schwarzschild radius** (Fig. 11.11). The imaginary surface of a sphere of this radius surrounding the black hole is called the

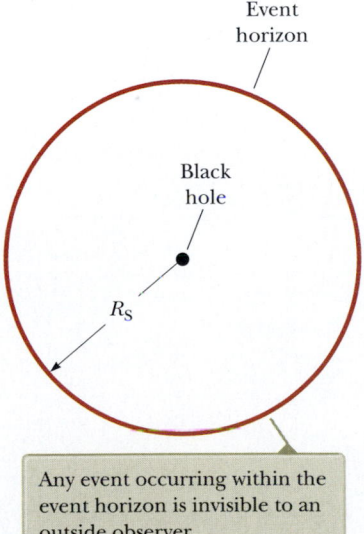

Event horizon

Black hole

R_S

> Any event occurring within the event horizon is invisible to an outside observer.

Figure 11.11 A black hole. The distance R_S equals the Schwarzschild radius.

Figure 11.12 Hubble Space Telescope image of the galaxy M107. This galaxy contains about 800 billion stars and is 28 million light years from the Earth. Scientists believe that a supermassive black hole exists at the center of the galaxy.

event horizon, which is the limit of how close you can approach the black hole and hope to be able to escape.

Although light from a black hole cannot escape, light from events taking place near the black hole should be visible. For example, it is possible for a binary star system to consist of one normal star and one black hole. Material surrounding the ordinary star can be pulled into the black hole, forming an **accretion disk** around the black hole. Friction among particles in the accretion disk results in transformation of mechanical energy into internal energy. As a result, the orbital height of the material above the event horizon decreases and the temperature rises. This high-temperature material emits a large amount of radiation, extending well into the x-ray region of the electromagnetic spectrum. These x-rays are characteristic of a black hole. Several possible candidates for black holes have been identified by observation of these x-rays.

Figure 11.12 shows a Hubble Space Telescope photograph of M107, known as the Sombrero galaxy. Scientists have shown that the speed of revolution of the stars could not be maintained unless a mass one billion times the mass of the Sun is present at its center. This is strong evidence for a supermassive black hole at the center of the galaxy.

Black holes are of considerable interest to those searching for **gravity waves,** which are ripples in space–time caused by changes in a gravitational system. These ripples can be caused by a star collapsing into a black hole, a binary star consisting of a black hole and a visible companion, and supermassive black holes at a galaxy center. A gravity wave detector, the Laser Interferometer Gravitational Wave Observatory (LIGO), is currently being built and tested in the United States, and hopes are high for detecting gravitational waves with this instrument.

11.5 | Atomic Spectra and the Bohr Theory of Hydrogen

In the preceding sections, we described a structural model for a large-scale system, the Solar System. Let us now do the same for a very small-scale system, the hydrogen atom. We shall find that a Solar System model of the atom, with a few extra features, provides explanations for some of the experimental observations made on the hydrogen atom.

As you may have already learned in a chemistry course, the hydrogen atom is the simplest known atomic system and an especially important one to understand. Much of what is learned about the hydrogen atom (which consists of one proton and one electron) can be extended to single-electron ions such as He^+ and Li^{2+}. Furthermore, a thorough understanding of the physics underlying the hydrogen atom can then be used to describe more complex atoms and the periodic table of the elements.

In this section, we will investigate the changes in the structural model of the hydrogen atom during the second decade of the 20th century. Early in that decade, the structural model had the following components, following the format outlined in Section 11.2:

1. *A description of the physical components of the system:* In the hydrogen atom model, the physical components are the electron and a positive charge distribution.

2. *A description of where the components are located relative to one another and how they interact:* The hydrogen atom model at the time was the Rutherford model, to be discussed in Chapter 29. In this model, the positive charge is concentrated in a small region of space called the *nucleus*. The electron is in orbit around the nucleus. The particle-like nature of the positive charge and the word *proton* were not yet understood at the time, so we will avoid referring to the nucleus as a proton in this discussion. The interaction between the electron and the nucleus is the electric force.

3. *A description of the time evolution of the system:* In the hydrogen atom model of the early 20th century, the time evolution was unclear and not understood.

4. *A description of the agreement between predictions of the model and actual observations and, possibly, predictions of new effects that have not yet been observed:* Rutherford's model was unable to explain the spectral lines exhibited by hydrogen that had been observed experimentally and are discussed below. Furthermore, as we discuss in Chapter 29, the model predicted an unstable atom, clearly at odds with reality.

Atomic systems can be investigated by observing *electromagnetic waves* emitted from the atom. Our eyes are sensitive to visible light, one type of electromagnetic wave. A wave, which is a disturbance that propagates through space, will be one of our four simplification models around which we will identify analysis models, as we have done for a particle, a system, and a rigid object. Think of ocean waves as an example; they represent disturbances of the ocean surface and move across the surface toward the shore. A common form of periodic wave is the sinusoidal wave, whose shape is depicted in Figure 11.13. If this graph represents an electromagnetic wave, the vertical axis represents the magnitude of the electric field. (We will study electric fields in Chapter 19.) The horizontal axis is position in the direction of travel of the wave. The distance between two consecutive crests of the wave is called the **wavelength** λ. As the wave travels to the right with a speed v, any point on the wave travels a distance of one wavelength in a time interval of one period T (the time interval for one cycle), so the wave speed is given by $v = \lambda / T$. The inverse of the period, $1/T$, is called the **frequency** f of the wave; it represents the number of cycles per second. Therefore, the speed of the wave is often written as $v = \lambda f$. In this section, because we shall deal with electromagnetic waves—which travel at the speed of light c—the appropriate relation is

$$c = \lambda f \qquad \qquad \textbf{11.15} \blacktriangleleft$$

> Any point on the wave moves a distance of one wavelength λ in a time interval equal to the period T of the wave.

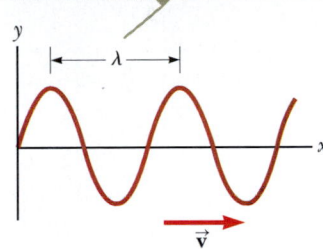

Figure 11.13 A sinusoidal wave traveling to the right with wave speed v.

▶ Relation between wavelength, frequency, and wave speed

Suppose an evacuated glass tube is filled with hydrogen (or some other gas). If a voltage applied between metal electrodes in the tube is great enough to produce an electric current in the gas, the tube emits light with colors that are characteristic of the gas. (That is how a neon sign works.) When the emitted light is analyzed with a device called a spectroscope, in which the light passes through a narrow slit, a series of discrete **spectral lines** is observed, each line corresponding to a different wavelength, or color, of light. Such a series of spectral lines is commonly referred to as an **emission spectrum.** The wavelengths contained in a given spectrum are characteristic of the element emitting the light. Figure 11.14 (page 316) is a semigraphical representation of the spectra of various elements. It is semigraphical because the horizontal axis is linear in wavelength, but the vertical axis has no significance. Because no two elements emit the same line spectrum, this phenomenon represents a marvelous and reliable technique for identifying elements in a substance.

In addition to emitting light at specific wavelengths, an element can also absorb light at specific wavelengths. The spectral lines corresponding to this process form what is known as an **absorption spectrum.** An absorption spectrum can be obtained by passing a continuous radiation spectrum (one containing all wavelengths) through a vapor of the element being analyzed. The absorption spectrum consists of a series of dark lines superimposed on the otherwise continuous spectrum (Fig. 11.14b).

Figure 11.14 Visible spectra. (a) Line spectra produced by emission in the visible range for the elements hydrogen, mercury, and neon. (b) The absorption spectrum for hydrogen. The dark absorption lines occur at the same wavelengths as the emission lines for hydrogen shown in (a).

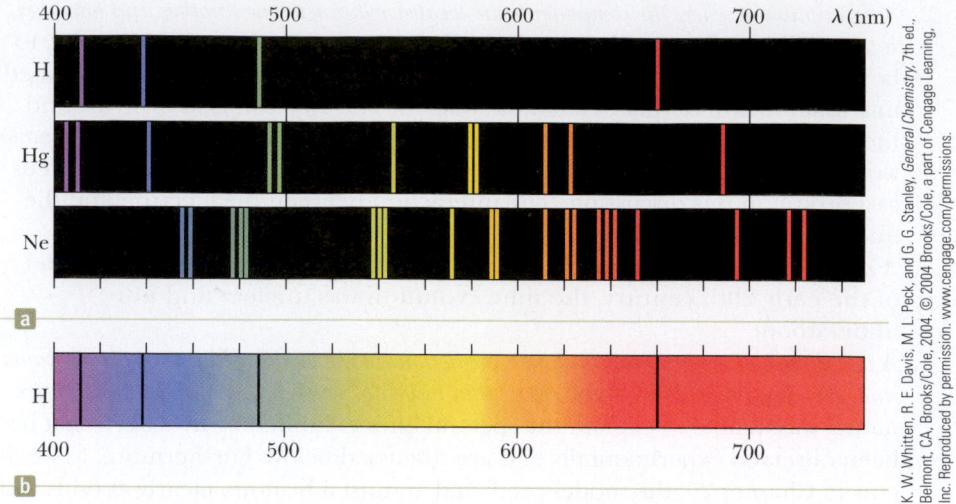

K. W. Whitten, R. E. Davis, M. L. Peck, and G. G. Stanley, *General Chemistry*, 7th ed., Belmont, CA, Brooks/Cole, 2004. © 2004 Brooks/Cole, a part of Cengage Learning, Inc. Reproduced by permission. www.cengage.com/permissions.

The emission spectrum of hydrogen shown in Figure 11.15 includes four visible lines that occur at wavelengths of 656.3 nm, 486.1 nm, 434.1 nm, and 410.2 nm. In 1885, Johann Balmer (1825–1898) found that the wavelengths of these and other invisible lines can be described by the following simple empirical equation:

$$\lambda = 364.56\,\frac{n^2}{n^2 - 4} \qquad n = 3, 4, 5, \ldots$$

in which n is an integer starting at 3 and the wavelengths given by this expression are in nanometers. These spectral lines are called the **Balmer series.** The first line in the Balmer series, at 656.3 nm, corresponds to $n = 3$, the line at 486.1 nm corresponds to $n = 4$, and so on. At the time this equation was formulated, it had no valid theoretical basis; it simply predicted the wavelengths correctly. Therefore, this equation is not based on a model but is simply a trial-and-error equation that happens to work. A few years later, Johannes Rydberg (1854–1919) recast the equation in the following form:

▶ Rydberg equation

$$\frac{1}{\lambda} = R_{\text{H}}\left(\frac{1}{2^2} - \frac{1}{n^2}\right) \qquad n = 3, 4, 5, \ldots \qquad \textbf{11.16} ◀$$

where n may have integral values of 3, 4, 5, . . . and R_{H} is a constant, now called the **Rydberg constant,** with a value of $R_{\text{H}} = 1.097\ 373\ 2 \times 10^7\ \text{m}^{-1}$. Equation 11.16 is no more based on a model than is Balmer's equation. In this form, however, we can compare it with the predictions of a structural model of the hydrogen atom that is described below.

At the beginning of the 20th century, scientists were perplexed by the failure of classical physics to explain the characteristics of atomic spectra. Why did atoms of a given element emit only certain wavelengths of radiation so that the emission spectrum displayed discrete lines? Furthermore, why did the atoms absorb many of the same wavelengths that they emitted? In 1913, Niels Bohr provided an explanation of atomic spectra that includes some features of the currently accepted theory. Using the simplest atom, hydrogen, Bohr described a structural model for the atom called the **Bohr theory of the hydrogen atom**. His model of the hydrogen atom contains some classical features that can be related to our analysis models as well as some revolutionary postulates that could not be justified within the framework of classical physics. The components of Bohr's structural model as it applies to the hydrogen atom are as follows:

1. *A description of the physical components of the system:* In the hydrogen atom model, the physical components are the electron and a positive charge distribution, just as in Rutherford's model.

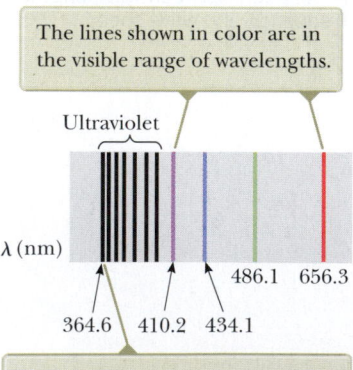

The lines shown in color are in the visible range of wavelengths.

Figure 11.15 The Balmer series of spectral lines for atomic hydrogen, with several lines marked with the wavelength in nanometers. (The horizontal wavelength axis is not to scale.)

2. *A description of where the components are located relative to one another and how they interact:* The electron moves in a circular orbit about the nucleus under the influence of the electric force of attraction as in Figure 11.16. This notion is again consistent with the Rutherford model.

3. *A description of the time evolution of the system:* Here is where Bohr's model deviates from Rutherford's. We discuss three major parts of the theory:

   (a) Bohr's model claims that only certain electron orbits are stable, and they are the only orbits in which we find the electron. In these orbits, the hydrogen atom does not emit energy in the form of radiation. Hence, the total energy of the atom remains constant, and classical mechanics can be used to describe the electron's motion. This restriction to certain orbits is a new idea that is not consistent with classical physics. As we shall see in Chapter 24, an accelerating electron should emit energy by electromagnetic radiation. Therefore, according to the conservation of energy equation, the emission of radiation from the atom should result in a decrease in the energy of the atom. Bohr's postulate boldly claims that this radiation simply does not happen.

   (b) The size of the stable electron orbits is determined by a condition imposed on the electron's orbital angular momentum. The allowed orbits are those for which the electron's orbital angular momentum about the nucleus is an integral multiple of $\hbar \equiv h/2\pi$:

   $$m_e vr = n\hbar \qquad n = 1, 2, 3, \ldots \qquad \text{11.17} \blacktriangleleft$$

   where h is **Planck's constant** ($h = 6.63 \times 10^{-34}$ J $\cdot$ s; we will see Planck's constant extensively in our studies of modern physics). This new idea cannot be related to any of the models we have developed so far.

   It can be related, however, to a model that will be developed in later chapters, and we shall return to this idea at that time to see how it is predicted by the model. This concept is our first introduction to a notion from **quantum mechanics,** which describes the behavior of microscopic particles. The orbital radii are *quantized.*

   (c) Radiation is emitted by the hydrogen atom when the atom makes a transition from a more energetic initial state to a lower state. The transition cannot be visualized or treated classically. In particular, the frequency f of the radiation emitted in the transition is related to the change in the atom's energy. The frequency of the emitted radiation is found from

   $$E_i - E_f = hf \qquad \text{11.18} \blacktriangleleft$$

   where E_i is the energy of the initial state, E_f is the energy of the final state, and $E_i > E_f$. The notion of energy being emitted only when a transition occurs is nonclassical. Given this notion, however, Equation 11.18 is simply the conservation of energy equation $\Delta E = \Sigma T \rightarrow E_f - E_i = -hf$. On the left is the change in energy of the system—the atom—and on the right is the energy transferred out of the system by electromagnetic radiation.

4. *A description of the agreement between predictions of the model and actual observations and, possibly, predictions of new effects that have not yet been observed:* In the discussion below, we see how the structural model makes predictions and agrees with some experimental results.

The electric potential energy of the system shown in Figure 11.16 is found from Equation 6.35, $U_e = -k_e e^2/r$, where k_e is the electric constant, e is the magnitude of the charge on the electron, and r is the electron–nucleus separation. Therefore,

Niels Bohr
(1885–1962)
Bohr, a Danish physicist, was an active participant in the early development of quantum mechanics and provided much of its philosophical framework. During the 1920s and 1930s, Bohr headed the Institute for Advanced Studies in Copenhagen. The institute was a magnet for many of the world's best physicists and provided a forum for the exchange of ideas. Bohr was awarded the 1922 Nobel Prize in Physics for his investigation of the structure of atoms and of the radiation emanating from them.

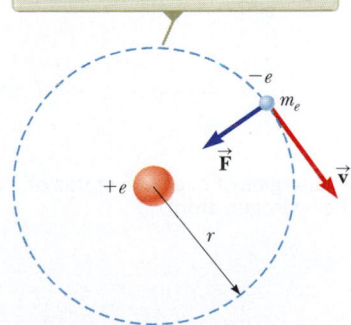

The orbiting electron is allowed to be only in specific orbits of discrete radii.

Figure 11.16 Diagram representing Bohr's model of the hydrogen atom.

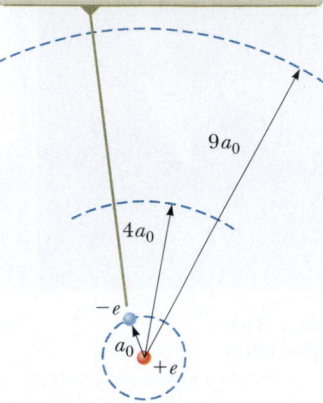

The electron is shown in the lowest-energy orbit, but it could be in any of the allowed orbits.

$9a_0$

$4a_0$

$-e$

a_0

$+e$

Active Figure 11.17 The first three circular orbits predicted by the Bohr model for hydrogen.

▶ Total energy of the hydrogen atom

▶ Radii of Bohr orbits in hydrogen

▶ The Bohr radius

▶ Energies of quantum states of the hydrogen atom

the total energy of the atom, which contains both kinetic and potential energy terms, is

$$E = K + U_e = \tfrac{1}{2}m_e v^2 - k_e \frac{e^2}{r}$$ **11.19◀**

According to component 3(a) of the structural model, the energy of the system remains constant; the system is isolated because the structural model does not allow for electromagnetic radiation for a given orbit.

Applying Newton's second law to this system, we see that the magnitude of the attractive electric force on the electron, $k_e e^2/r^2$ (Eq. 5.12), is equal to the product of its mass and its centripetal acceleration ($a_c = v^2/r$):

$$\frac{k_e e^2}{r^2} = \frac{m_e v^2}{r}$$

From this expression, the kinetic energy of the electron is found to be

$$K = \tfrac{1}{2}m_e v^2 = \frac{k_e e^2}{2r}$$ **11.20◀**

Substituting this value of K into Equation 11.19 gives the following expression for the total energy E of the hydrogen atom:

$$E = -\frac{k_e e^2}{2r}$$ **11.21◀**

Note that the total energy is negative,[3] indicating a bound electron–proton system. Therefore, energy in the amount of $k_e e^2/2r$ must be added to the atom just to separate the electron and proton by an infinite distance and make the total energy zero.[4] An expression for r, the radius of the allowed orbits, can be obtained by eliminating v by substitution between Equation 11.17 from component 3(b) of the structural model and Equation 11.20:

$$r_n = \frac{n^2 \hbar^2}{m_e k_e e^2} \qquad n = 1, 2, 3 \ldots$$ **11.22◀**

This result determines the discrete radii of the electron orbits. The integer n is called a **quantum number** and specifies the particular allowed **quantum state** of the atomic system.

The orbit for which $n = 1$ has the smallest radius; it is called the **Bohr radius** a_0 and has the value

$$a_0 = \frac{\hbar^2}{m_e k_e e^2} = 0.052\,9 \text{ nm}$$ **11.23◀**

The first three Bohr orbits are shown to scale in Active Figure 11.17.

The quantization of the orbit radii immediately leads to quantization of the energy of the atom, which can be seen by substituting $r_n = n^2 a_0$ into Equation 11.21. The allowed energies of the atom are

$$E_n = -\frac{k_e e^2}{2a_0}\left(\frac{1}{n^2}\right) \qquad n = 1, 2, 3, \ldots$$ **11.24◀**

Insertion of numerical values into Equation 11.24 gives

$$E_n = -\frac{13.606 \text{ eV}}{n^2} \qquad n = 1, 2, 3, \ldots$$ **11.25◀**

---

[3]Compare this expression with Equation 11.10 for a gravitational system.

[4]This process is called *ionizing* the atom. In theory, ionization requires separating the electron and proton by an infinite distance. In reality, however, the electron and proton are in an environment with huge numbers of other particles. Therefore, ionization means separating the electron and proton by a distance large enough so that the interaction of these particles with other entities in their environment is larger than the remaining interaction between them.

(Recall from Section 9.7 that $1 \text{ eV} = 1.60 \times 10^{-19} \text{ J}$.) The lowest quantum state, corresponding to $n = 1$, is called the **ground state** and has an energy of $E_1 = -13.606 \text{ eV}$. The next state, the **first excited state**, has $n = 2$ and an energy of $E_2 = E_1/2^2 = -3.401 \text{ eV}$. Active Figure 11.18 is an **energy level diagram** showing the energies of these discrete energy states and the corresponding quantum numbers. This diagram is another semigraphical representation. The vertical axis is linear in energy, but the horizontal axis has no significance. The horizontal lines correspond to the allowed energies. The atomic system cannot have any energies other than those represented by the lines. The vertical lines with arrowheads represent transitions between states, during which energy is emitted.

The upper limit of the quantized levels, corresponding to $n \rightarrow \infty$ (or $r \rightarrow \infty$) and $E \rightarrow 0$, represents the state for which the electron is removed from the atom.[5] Above this energy is a continuum of available states for the ionized atom. The minimum energy required to ionize the atom is called the **ionization energy.** As can be seen from Active Figure 11.18, the ionization energy for hydrogen, predicted by Bohr's structural model, is 13.6 eV. This finding constituted a major achievement for the Bohr model because the ionization energy for hydrogen had already been measured to be 13.6 eV!

Active Figure 11.18 also shows various transitions of the atom from one state to a lower state, as referred to in component 3(c) of the structural model. As the energy of the atom decreases in a transition, the difference in energy between the states is carried away by electromagnetic radiation as described by Equation 11.18. Those transitions ending on $n = 2$ are shown in color, corresponding to the color of the light they represent. The transitions ending on $n = 2$ form the Balmer series of spectral lines, the wavelengths of which are correctly predicted by the Rydberg equation (see Eq. 11.16). Active Figure 11.18 also shows other spectral series (the Lyman series and the Paschen series) that were found after Balmer's discovery.

Equation 11.24, together with Equation 11.18, can be used to calculate the frequency of the radiation that is emitted when the atom makes a transition[6] from a high-energy state to a low-energy state:

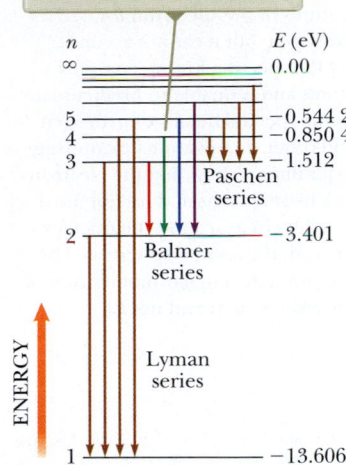

The colored arrows for the Balmer series indicate that this series results in the emission of visible light.

Active Figure 11.18 An energy-level diagram for hydrogen. The discrete allowed energies are plotted on the vertical axis. Nothing is plotted on the horizontal axis, but the horizontal extent of the diagram is made large enough to show allowed transitions. Quantum numbers are given on the left and energies (in electron volts) on the right. Vertical arrows represent the four lowest energy transitions in each of the spectral series shown.

$$f = \frac{E_i - E_f}{h} = \frac{k_e e^2}{2 a_0 h}\left(\frac{1}{n_f^2} - \frac{1}{n_i^2}\right)$$ **11.26** ◄   ► Frequency of radiation emitted from hydrogen

Because the quantity expressed in the Rydberg equation is wavelength, it is convenient to convert frequency to wavelength, using $c = f\lambda$, to obtain

$$\frac{1}{\lambda} = \frac{f}{c} = \frac{k_e e^2}{2 a_0 h c}\left(\frac{1}{n_f^2} - \frac{1}{n_i^2}\right)$$ **11.27** ◄   ► Emission wavelengths of hydrogen

Notice that the *theoretical* expression, Equation 11.27, is identical to the *empirical* Rydberg equation (Equation 11.16), provided that the combination of constants $k_e e^2/2 a_0 h c$ is equal to the experimentally determined Rydberg constant and that $n_f = 2$. After Bohr demonstrated the agreement of the constants in these two equations to a precision of about 1%, it was soon recognized as the crowning achievement of his structural model of the atom.

One question remains: What is the significance of $n_f = 2$? Its importance is simply because those transitions ending on $n_f = 2$ result in radiation that happens to lie in the visible; therefore, they were easily observed! As seen in Active Figure 11.18, other series of lines end on other final states. These lines lie in regions of the spectrum not

---

[5] The phrase "the electron is removed from the atom" is very commonly used, but, of course, we realize that we mean that the electron and nucleus are separated *from each other.*

[6] The phrase "the electron makes a transition" is also commonly used, but we will use "the atom makes a transition" to emphasize that the energy belongs to the system of the atom, not just to the electron. This wording is similar to our discussion in Chapter 6 of gravitational potential energy belonging to the system of an object and the Earth, not to the object alone.

visible to the eye, the infrared and ultraviolet. The generalized Rydberg equation for any initial and final states is

$$\frac{1}{\lambda} = R_{\text{H}}\left(\frac{1}{n_f^2} - \frac{1}{n_i^2}\right) \qquad \textbf{11.28} \blacktriangleleft$$

In this equation, different series correspond to different values of n_f and different lines within a series correspond to varying values of n_i.

Bohr immediately extended his structural model for hydrogen to other elements in which all but one electron had been removed. Ionized elements such as He^+, Li^{2+}, and Be^{3+} were suspected to exist in hot stellar atmospheres, where frequent atomic collisions occur with enough energy to completely remove one or more atomic electrons. Bohr showed that many mysterious lines observed in the Sun and several stars could not be due to hydrogen, but were correctly predicted by his theory if attributed to singly ionized helium.

> **QUICK QUIZ 11.4** A hydrogen atom makes a transition from the $n = 3$ level to the $n = 2$ level. It then makes a transition from the $n = 2$ level to the $n = 1$ level. Which transition results in emission of the longest-wavelength photon? (**a**) the first transition (**b**) the second transition (**c**) neither transition because the wavelengths are the same for both

---

Example 11.5 | Electronic Transitions in Hydrogen

The electron in a hydrogen atom makes a transition from the $n = 2$ energy level to the ground level ($n = 1$). Find the wavelength and frequency of the emitted photon.

SOLUTION

Conceptualize Imagine the electron in a circular orbit about the nucleus as in the Bohr model in Figure 11.16. When the electron makes a transition to a lower stationary state, it emits a photon with a given frequency.

Categorize We evaluate the results using equations developed in this section, so we categorize this example as a substitution problem.

Use Equation 11.28 to obtain λ, with $n_i = 2$ and $n_f = 1$:

$$\frac{1}{\lambda} = R_{\text{H}}\left(\frac{1}{1^2} - \frac{1}{2^2}\right) = \frac{3R_{\text{H}}}{4}$$

$$\lambda = \frac{4}{3R_{\text{H}}} = \frac{4}{3(1.097 \times 10^7 \text{ m}^{-1})} = 1.22 \times 10^{-7} \text{ m} = \boxed{122 \text{ nm}}$$

Use Equation 11.15 to find the frequency of the photon:

$$f = \frac{c}{\lambda} = \frac{3.00 \times 10^8 \text{ m/s}}{1.22 \times 10^{-7} \text{ m}} = \boxed{2.47 \times 10^{15} \text{ Hz}}$$

❮ 11.6 | Context Connection: Changing from a Circular to an Elliptical Orbit

In part (A) of Example 11.2, we discussed a spacecraft in a circular orbit around the Earth. From our studies of Kepler's laws in this chapter, we are also aware that an elliptical orbit is possible for our spacecraft. Let us investigate how the motion of our spacecraft can be changed from a circular to an elliptical orbit, which will set us up for the conclusion to our *Mission to Mars* Context.

Let us identify the system as the spacecraft and the Earth, *but not the portion of the fuel in the spacecraft that we use to change the orbit*. In a given orbit, the total energy of the spacecraft–Earth system is given by Equation 11.10,

$$E = -\frac{GMm}{2r}$$

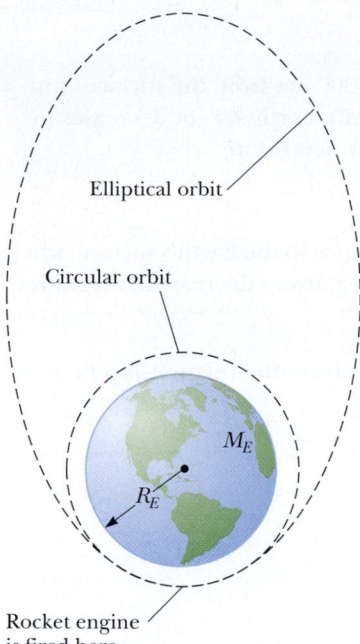

Figure 11.19 A spacecraft, originally in a circular orbit about the Earth, fires its engines and enters an elliptical orbit about the Earth.

This energy includes the kinetic energy of the spacecraft and the potential energy associated with the gravitational force between the spacecraft and the Earth. If the rocket engines are fired, the exhausted fuel can be seen as doing work on the spacecraft–Earth system because the thrust force moves through a displacement. As a result, the total energy of the spacecraft–Earth system increases.

The spacecraft has a new, higher energy but is constrained to be in an orbit that includes the original starting point. It cannot be in a higher-energy circular orbit having a larger radius because this orbit would not contain the starting point. The only possibility is that the orbit is elliptical. Figure 11.19 shows the change from the original circular orbit to the new elliptical orbit for our spacecraft.

Equation 11.11 gives the energy of the spacecraft–Earth system for an elliptical orbit. Therefore, if we know the new energy of the orbit, we can find the semimajor axis of the elliptical orbit. Conversely, if we know the semimajor axis of an elliptical orbit we would like to achieve, we can calculate how much additional energy is required from the rocket engines. This information can then be converted to a required burn time for the rockets.

Larger amounts of energy increase supplied by the rocket engines will move the spacecraft into elliptical orbits with larger semimajor axes. What happens if the burn time of the engines is so long that the total energy of the spacecraft–Earth system becomes positive? A positive energy refers to an *unbound* system. Therefore, in this case, the spacecraft will *escape* from the Earth, going into a hyperbolic path that would not bring it back to the Earth.

This process is the essence of what must be done to transfer to Mars. Our rocket engines must be fired to leave the circular parking orbit and escape the Earth. At this point, our thinking must shift to a spacecraft–Sun system rather than a spacecraft–Earth system. From this point of view, the spacecraft in orbit around the Earth can also be considered to be in a circular orbit around the Sun, moving along with the Earth, as shown in Figure 11.20. The orbit is not a perfect circle because there are perturbations corresponding to its extra motion around the Earth, but these perturbations are small compared with the radius of the orbit around the Sun. When our engines are fired to escape from the Earth, our orbit around the Sun changes from a circular orbit (ignoring the perturbations) to an elliptical one with the Sun at one focus. We shall choose the semimajor axis of our elliptical orbit so that it intersects the orbit of Mars! In the Context 2 Conclusion, we shall look at more details of this process.

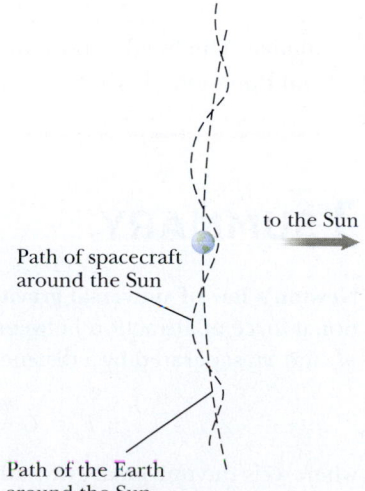

to the Sun

Path of spacecraft around the Sun

Path of the Earth around the Sun

Figure 11.20 A spacecraft in orbit about the Earth can be modeled as one in a circular orbit about the Sun, with its orbit about the Earth appearing as small perturbations from the circular orbit.

Example 11.6 | How High Do We Go?

Imagine that you are in a spacecraft in circular orbit around the Earth, at a height of $h = 300$ km from the surface. You fire your rocket engines, and as a result the magnitude of the total energy of the spacecraft–Earth system decreases by 10.0%. What is the greatest height of your spacecraft above the surface of the Earth in your new orbit?

SOLUTION

Conceptualize Study Figure 11.19, which represents this situation. We wish to find the distance to the Earth's surface, when the spacecraft is at the topmost point in the figure. Notice that because the total energy is negative, a decrease in magnitude represents an increase in energy.

Categorize We don't need an analysis model for this problem. We can evaluate the result from the relationship between orbit energy and semimajor axis as exhibited by Equation 11.11.

....

Analyze

Set up a ratio of the energies of the two orbits, using Equations 11.10 and 11.11 for circular and elliptical orbits:

$$\frac{E_{\text{elliptical}}}{E_{\text{circular}}} = \frac{\left(-\dfrac{GMm}{2a}\right)}{\left(-\dfrac{GMm}{2r}\right)} = \frac{r}{a} = f$$

The ratio f is equal to 0.900 because of the 10.0% decrease in magnitude of the total energy. Find a in terms of r:

$$a = \frac{r}{f}$$

Substitute for the orbital radius in terms of the Earth's radius and the initial height of the spacecraft above the surface:

$$a = \frac{1}{f}(R_E + h)$$

The maximum distance from the center of the Earth will occur when the spacecraft is at apogee and is given by $r_{\text{max}} = 2a - r$. Find the value of this maximum distance:

$$r_{\text{max}} = 2a - r = \frac{2}{f}(R_E + h) - (R_E + h) = \left(\frac{2}{f} - 1\right)(R_E + h)$$

Subtract the radius of the Earth from r_{max} to find the maximum height above the Earth's surface:

$$(1) \quad h_{\text{max}} = \left(\frac{2}{f} - 1\right)(R_E + h) - R_E$$

Substitute numerical values:

$$h_{\text{max}} = \left(\frac{2}{0.900} - 1\right)(6.37 \times 10^3 \text{ km} + 300 \text{ km}) - 6.37 \times 10^3 \text{ km}$$

$$= \boxed{1.78 \times 10^3 \text{ km}}$$

....

Finalize The height above the Earth's surface has been increased by a factor of almost 6 for this fuel expenditure. Notice from Equation (1) that h_{max} increases as f decreases (representing more fuel expenditure), but not in a simple way.

▶ SUMMARY

Newton's law of universal gravitation states that the gravitational force of attraction between any two particles of masses m_1 and m_2 separated by a distance r has the magnitude

$$F_g = G\frac{m_1 m_2}{r^2} \qquad \textbf{11.1}\blacktriangleleft$$

where G is the **universal gravitational constant** whose value is 6.674×10^{-11} N · m²/kg².

Rather than considering the gravitational force as a direct interaction between two objects, we can imagine that one object sets up a **gravitational field** in space:

$$\vec{\mathbf{g}} \equiv \frac{\vec{\mathbf{F}}_g}{m} \qquad \textbf{11.4}\blacktriangleleft$$

A second object in this field experiences a force $\vec{\mathbf{F}}_g = m\vec{\mathbf{g}}$ when placed in this field.

Kepler's laws of planetary motion state the following:

1. Each planet in the Solar System moves in an elliptical orbit with the Sun at one focus.
2. The radius vector drawn from the Sun to any planet sweeps out equal areas in equal time intervals.
3. The square of the orbital period of any planet is proportional to the cube of the semimajor axis of the elliptical orbit.

Kepler's first law is a consequence of the inverse-square nature of the law of universal gravitation. The **semimajor axis**

of an ellipse is a, where $2a$ is the longest dimension of the ellipse. The **semiminor axis** of the ellipse is b, where $2b$ is the shortest dimension of the ellipse. The **eccentricity** of the ellipse is $e = c/a$, where c is the distance between the center and a focus and $a^2 = b^2 + c^2$.

Kepler's second law is a consequence of the gravitational force being a central force. For a central force, the angular momentum of the planet is conserved.

Kepler's third law is a consequence of the inverse-square nature of the universal law of gravitation. Newton's second law, together with the force law given by Equation 11.1, verifies that the period T and semimajor axis a of the orbit of a planet about the Sun are related by

$$T^2 = \left(\frac{4\pi^2}{GM_S}\right)a^3 \qquad \textbf{11.7}\blacktriangleleft$$

where M_S is the mass of the Sun.

If an isolated system consists of a particle of mass m moving with a speed v in the vicinity of a massive body of mass M, the total energy of the system is constant and is

$$E = \tfrac{1}{2}mv^2 - \frac{GMm}{r} \qquad \textbf{11.8}\blacktriangleleft$$

If m moves in an elliptical orbit of major axis $2a$ about M, where $M \gg m$, the total energy of the system is

$$E = -\frac{GMm}{2a} \qquad \textbf{11.11}\blacktriangleleft$$

The total energy is negative for any bound system, that is, one in which the orbit is closed, such as a circular or an elliptical orbit.

The Bohr model of the atom successfully describes the spectra of atomic hydrogen and hydrogen-like ions. One basic assumption of this structural model is that the electron can exist only in discrete orbits such that the angular momentum $m_e vr$ is an integral multiple of $\hbar \equiv h/2\pi$. Assuming circular orbits and a simple electrical attraction between the electron and proton, the energies of the quantum states for hydrogen are calculated to be

$$E_n = -\frac{k_e e^2}{2a_0}\left(\frac{1}{n^2}\right) \qquad n = 1, 2, 3, \ldots \qquad \textbf{11.24}\blacktriangleleft$$

where k_e is the Coulomb constant, e is the fundamental electric charge, n is a positive integer called a **quantum number,** and $a_0 = 0.052\,9$ nm is the **Bohr radius.**

If the hydrogen atom makes a transition from a state whose quantum number is n_i to one whose quantum number is n_f, where $n_f < n_i$, the frequency of the radiation emitted by the atom is

$$f = \frac{k_e e^2}{2a_0 h}\left(\frac{1}{n_f^2} - \frac{1}{n_i^2}\right) \qquad \textbf{11.26}\blacktriangleleft$$

Using $E_i - E_f = hf = hc/\lambda$, one can calculate the wavelengths of the radiation for various transitions. The calculated wavelengths are in excellent agreement with those in observed atomic spectra.

▶ OBJECTIVE QUESTIONS |

☐ denotes answer available in *Student Solutions Manual/Study Guide*

1. Imagine that nitrogen and other atmospheric gases were more soluble in water so that the atmosphere of the Earth is entirely absorbed by the oceans. Atmospheric pressure would then be zero, and outer space would start at the planet's surface. Would the Earth then have a gravitational field? (a) Yes, and at the surface it would be larger in magnitude than 9.8 N/kg. (b) Yes, and it would be essentially the same as the current value. (c) Yes, and it would be somewhat less than 9.8 N/kg. (d) Yes, and it would be much less than 9.8 N/kg. (e) No, it would not.

2. The gravitational force exerted on an astronaut on the Earth's surface is 650 N directed downward. When she is in the space station in orbit around the Earth, is the gravitational force on her (a) larger, (b) exactly the same, (c) smaller, (d) nearly but not exactly zero, or (e) exactly zero?

3. Rank the magnitudes of the following gravitational forces from largest to smallest. If two forces are equal, show their equality in your list. (a) the force exerted by a 2-kg object on a 3-kg object 1 m away (b) the force exerted by a 2-kg object on a 9-kg object 1 m away (c) the force exerted by a 2-kg object on a 9-kg object 2 m away (d) the force exerted by a 9-kg object on a 2-kg object 2 m away (e) the force exerted by a 4-kg object on another 4-kg object 2 m away

4. An object of mass m is located on the surface of a spherical planet of mass M and radius R. The escape speed from the planet does not depend on which of the following? (a) M (b) m (c) the density of the planet (d) R (e) the acceleration due to gravity on that planet

5. A system consists of five particles. How many terms appear in the expression for the total gravitational potential energy of the system? (a) 4 (b) 5 (c) 10 (d) 20 (e) 25

6. Suppose the gravitational acceleration at the surface of a certain moon A of Jupiter is 2 m/s². Moon B has twice the mass and twice the radius of moon A. What is the gravitational acceleration at its surface? Neglect the gravitational acceleration due to Jupiter. (a) 8 m/s² (b) 4 m/s² (c) 2 m/s² (d) 1 m/s² (e) 0.5 m/s²

7. A satellite originally moves in a circular orbit of radius R around the Earth. Suppose it is moved into a circular orbit of radius $4R$. **(i)** What does the force exerted on the satellite then become? (a) eight times larger (b) four times larger (c) one-half as large (d) one-eighth as large (e) one-sixteenth as large **(ii)** What happens to the satellite's speed? Choose from the same possibilities (a) through (e). **(iii)** What happens to its period? Choose from the same possibilities (a) through (e).

8. (a) Can a hydrogen atom in the ground state absorb a photon of energy less than 13.6 eV? (b) Can this atom absorb a photon of energy greater than 13.6 eV?

9. A satellite moves in a circular orbit at a constant speed around the Earth. Which of the following statements is true? (a) No force acts on the satellite. (b) The satellite moves at constant speed and hence doesn't accelerate. (c) The satellite has an acceleration directed away from the Earth. (d) The satellite

has an acceleration directed toward the Earth. (e) Work is done on the satellite by the gravitational force.

10. Rank the following quantities of energy from largest to the smallest. State if any are equal. (a) the absolute value of the average potential energy of the Sun–Earth system (b) the average kinetic energy of the Earth in its orbital motion relative to the Sun (c) the absolute value of the total energy of the Sun–Earth system

11. Halley's comet has a period of approximately 76 years, and it moves in an elliptical orbit in which its distance from the Sun at closest approach is a small fraction of its maximum distance. Estimate the comet's maximum distance from the Sun in astronomical units (AU, the distance from the Earth to the Sun). (a) 6 AU (b) 12 AU (c) 20 AU (d) 28 AU (e) 35 AU

12. The vernal equinox and the autumnal equinox are associated with two points 180° apart in the Earth's orbit. That is, the Earth is on precisely opposite sides of the Sun when it passes through these two points. From the vernal equinox, 185.4 days elapse before the autumnal equinox. Only 179.8 days elapse from the autumnal equinox until the next vernal equinox. Why is the interval from the March (vernal) to the September (autumnal) equinox (which contains the summer solstice) longer than the interval from the September to the March equinox rather than being equal to that interval? Choose one of the following reasons. (a) They are really the same, but the Earth spins faster during the "summer" interval, so the days are shorter. (b) Over the "summer" interval, the Earth moves slower because it is farther from the Sun. (c) Over the March-to-September interval, the Earth moves slower because it is closer to the Sun. (d) The Earth has less kinetic energy when it is warmer. (e) The Earth has less orbital angular momentum when it is warmer.

13. (i) Rank the following transitions for a hydrogen atom from the transition with the greatest gain in energy to that with the greatest loss, showing any cases of equality. (a) $n_i = 2$; $n_f = 5$; (b) $n_i = 5$; $n_f = 3$; (c) $n_i = 7$; $n_f = 4$; (d) $n_i = 4$; $n_f = 7$ (ii) Rank the same transitions as in part (i) according to the wavelength of the photon absorbed or emitted by an otherwise isolated atom from greatest wavelength to smallest.

14. Let $-E$ represent the energy of a hydrogen atom. (i) What is the kinetic energy of the electron? (a) $2E$ (b) E (c) 0 (d) $-E$ (e) $-2E$ (ii) What is the potential energy of the atom? Choose from the same possibilities (a) through (e).

☐ denotes answer available in *Student Solutions Manual/Study Guide*

CONCEPTUAL QUESTIONS

1. A satellite in low-Earth orbit is not truly traveling through a vacuum. Rather, it moves through very thin air. Does the resulting air friction cause the satellite to slow down?

2. (a) Explain why the force exerted on a particle by a uniform sphere must be directed toward the center of the sphere. (b) Would this statement be true if the mass distribution of the sphere were not spherically symmetric? Explain.

3. Why don't we put a geosynchronous weather satellite in orbit around the 45th parallel? Wouldn't such a satellite be more useful in the United States than one in orbit around the equator?

4. Explain why it takes more fuel for a spacecraft to travel from the Earth to the Moon than for the return trip. Estimate the difference.

5. (a) If a hole could be dug to the center of the Earth, would the force on an object of mass m still obey Equation 11.1 there? (b) What do you think the force on m would be at the center of the Earth?

6. You are given the mass and radius of planet X. How would you calculate the free-fall acceleration on this planet's surface?

7. (a) At what position in its elliptical orbit is the speed of a planet a maximum? (b) At what position is the speed a minimum?

8. Each *Voyager* spacecraft was accelerated toward escape speed from the Sun by the gravitational force exerted by Jupiter on the spacecraft. (a) Is the gravitational force a conservative or a nonconservative force? (b) Does the interaction of the spacecraft with Jupiter meet the definition of an elastic collision? (c) How could the spacecraft be moving faster after the collision?

9. In his 1798 experiment, Cavendish was said to have "weighed the Earth." Explain this statement.

PROBLEMS AVAILABLE IN ENHANCED WebAssign

11.1 Newton's Law of Universal Gravitation Revisited

Problems 1–13

11.3 Kepler's Laws

Problems 14–21

11.4 Energy Considerations in Planetary and Satellite Motion

Problems 22–36

11.5 Atomic Spectra and the Bohr Theory of Hydrogen

Problems 37–43

11.6 Context Connection: Changing from a Circular to an Elliptical Orbit

Problems 44–45

Additional Problems

Problems 46–65

Solutions to the following Problems are available in the *Student Solutions Manual/Study Guide*:

11.5, 11.7, 11.12, 11.13, 11.17, 11.19, 11.23, 11.24, 11.41, 11.52, 11.55, 11.57, and 11.61

List of Enhanced Problems

Problem Number	Targeted Feedback in Enhanced WebAssign	Master It in Enhanced WebAssign	Watch It in Enhanced WebAssign
11.3	✓		✓
11.5		✓	
11.7	✓		✓
11.10	✓		✓
11.11	✓		✓
11.13	✓	✓	
11.19	✓	✓	
11.21	✓		✓
11.26		✓	
11.29	✓		✓
11.41	✓		✓
11.57		✓	

A Successful Mission Plan

Now that we have explored the physics of classical mechanics, let us return to our central question for the *Mission to Mars* Context:

> **How can we undertake a successful transfer of a spacecraft from the Earth to Mars?**

We make use of the physical principles that we now understand and apply them to our journey from the Earth to Mars.

Let us start with a more modest proposition. Suppose a spacecraft is in a circular orbit around the Earth and you are a passenger on the spacecraft. If you toss a wrench in the direction of travel, tangent to the circular path, what orbital path will the wrench follow?

Let us adopt a simplification model in which the spacecraft is much more massive than the wrench. Conservation of momentum for the isolated system of the wrench and the spacecraft tells us that the spacecraft must slow down slightly once the wrench is thrown. Because of the mass difference between the wrench and spacecraft, however, we can ignore the small change in the spacecraft's speed. The wrench now enters a new orbit, from its perigee position, and the wrench–Earth system has more energy than it had when the wrench was in the circular orbit. Because the orbital energy is related to the major axis, the wrench is injected into an elliptical orbit as discussed in the Context Connection of Chapter 11 and as shown in Figure 1. Therefore, the path of the wrench is changed from a circular orbit to an elliptical orbit by providing the wrench–Earth system with extra energy. The energy is provided by the force you apply to the wrench tangent to the circular orbit because you have done work on the system. The elliptical orbit will take the wrench farther from the Earth than the circular orbit. If there were another spacecraft in a higher circular orbit than your spacecraft, you could throw the wrench so that it transfers from one spacecraft to another as shown in Figure 2. For that to occur, the elliptical orbit of the wrench must intersect with the higher spacecraft orbit. Furthermore, the wrench and the second spacecraft must arrive at the same point at the same time.

This scenario is the essence of our planned mission from the Earth to Mars. Rather than transferring a wrench between two spacecraft in orbit around the Earth, we will transfer a spacecraft between two planets in orbit around the Sun. Kinetic energy is added to the wrench–Earth system by throwing the wrench. Kinetic energy is added to the spacecraft–Sun system by firing the engines.

What if you were to throw the wrench harder and harder in the previous example? The wrench would be placed in a larger and larger elliptical orbit around the Earth. As you increased the launch velocity, you could inject the wrench into a *hyperbolic* escape orbit, relative to the Earth, and into an *elliptical* orbit around the *Sun*. This approach is the one we will take for the trip from the Earth to Mars; we will break free from a circular *parking* orbit around the Earth and

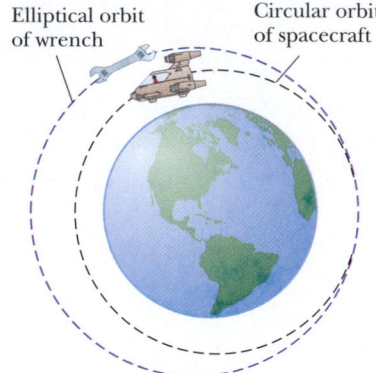

Elliptical orbit of wrench

Circular orbit of spacecraft

Figure 1 A wrench thrown tangent to the circular orbit of a spacecraft enters an elliptical orbit.

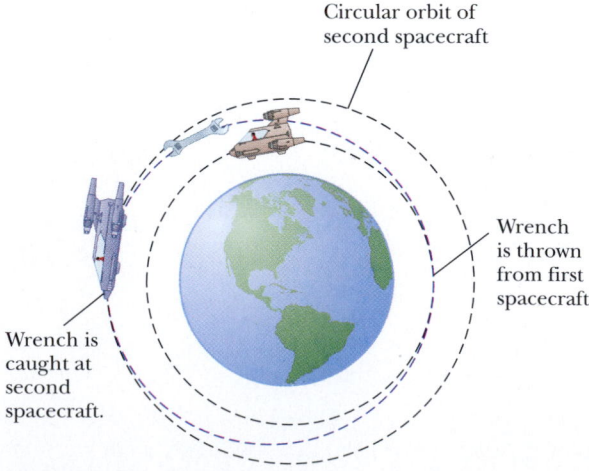

Circular orbit of second spacecraft

Wrench is thrown from first spacecraft

Wrench is caught at second spacecraft.

Figure 2 If a second spacecraft were in a higher circular orbit, the wrench could be carefully thrown so as to be transferred from one spacecraft to the other.

move into an elliptical *transfer* orbit around the Sun. The spacecraft will then continue on its journey to Mars, where it will enter a new parking orbit.

Now let us focus our attention on the transfer orbit part of the journey. One simple transfer orbit is called a **Hohmann transfer,** the type of transfer imparted to the wrench shown in Figure 2. The Hohmann transfer involves the least energy expenditure and therefore requires the smallest amount of fuel. As might be expected for a lowest-energy transfer, the transfer time for a Hohmann transfer is longer than for other types of orbits. We shall investigate the Hohmann transfer because of its simplicity and its general usefulness in planetary transfers.

The rocket engine on the spacecraft is fired from the parking orbit such that the spacecraft enters an elliptical orbit around the Sun at its perihelion and encounters the planet at the spacecraft's aphelion. Therefore, the spacecraft makes exactly one-half of a revolution about its elliptical path during the transfer as shown in Figure 3.

This process is energy efficient because fuel is expended only at the beginning and the end. The movement between parking orbits around the Earth and Mars is free; the spacecraft simply follows Kepler's laws while in an elliptical orbit around the Sun.

Let us perform a simple numerical calculation to see how to apply the mechanical laws to this process. We assume that the spacecraft is in a parking orbit above the Earth's surface. Notice also that the spacecraft is in orbit around the Sun, with a perturbation in its orbit caused by the Earth. Therefore, if we calculate the tangential speed of the Earth about the Sun, we can let this speed represent the average speed of the spacecraft around the Sun. This speed is calculated from Newton's second law for a particle in uniform circular motion:

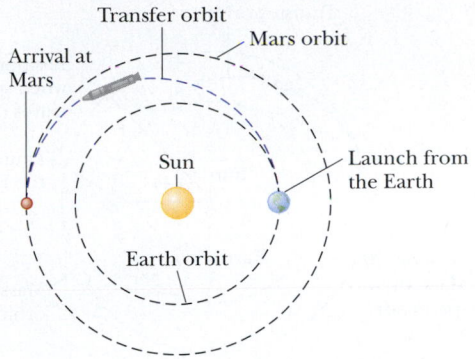

Figure 3 The Hohmann transfer orbit from the Earth to Mars. It is similar to transferring the wrench from one spacecraft to another in Figure 2, but here we are transferring a spacecraft from one planet to another.

$$F = ma \quad \rightarrow \quad G\frac{M_{Sun}\,m_{Earth}}{r^2} = m_{Earth}\frac{v^2}{r}$$

$$\rightarrow \quad v = \sqrt{\frac{GM_{Sun}}{r}} = \sqrt{\frac{(6.67 \times 10^{-11}\,N \cdot m^2/kg^2)(1.99 \times 10^{30}\,kg)}{1.50 \times 10^{11}\,m}}$$

$$= 2.97 \times 10^4\,m/s$$

This result is the original speed of the spacecraft, to which we add a change Δv to inject the spacecraft into the transfer orbit.

The major axis of the elliptical transfer orbit is found by adding together the orbit radii of the Earth and Mars (see Fig. 3):

$$\text{Major axis} = 2a = r_{Earth} + r_{Mars}$$

$$= 1.50 \times 10^{11}\,m + 2.28 \times 10^{11}\,m = 3.78 \times 10^{11}\,m$$

Therefore, the semimajor axis is half this value:

$$a = 1.89 \times 10^{11}\,m$$

From this value, Kepler's third law is used to find the travel time, which is one-half of the period of the orbit:

$$\Delta t_{travel} = \tfrac{1}{2}T = \tfrac{1}{2}\sqrt{\frac{4\pi^2}{GM_{Sun}}a^3}$$

$$= \tfrac{1}{2}\sqrt{\frac{4\pi^2}{(6.67 \times 10^{-11}\,N \cdot m^2/kg^2)(1.99 \times 10^{30}\,kg)}(1.89 \times 10^{11}\,m)^3}$$

$$= 2.24 \times 10^7\,s = 0.710\,yr = 259\,d$$

Therefore, the journey to Mars will require 259 Earth days. We can also determine where in their orbits Mars and the Earth must be so that the planet will be there when the spacecraft arrives. Mars has an orbital period of 687 Earth days. During the transfer time, the angular position *change* of Mars is

$$\Delta\theta_{Mars} = \frac{259\,d}{687\,d}(2\pi) = 2.37\,rad = 136°$$

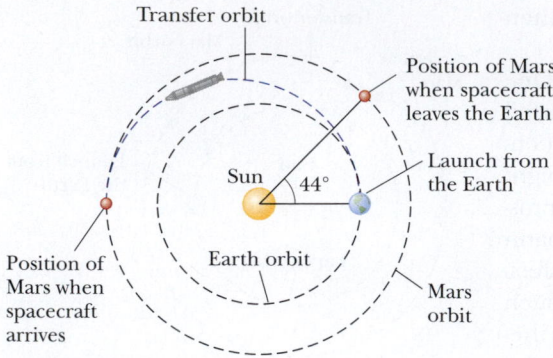

Figure 4 The spacecraft must be launched when Mars is 44° ahead of the Earth in its orbit.

BIO Effects of space travel on human health

Therefore, for the spacecraft and Mars to arrive at the same point at the same time, the spacecraft must be launched when Mars is $180° - 136° = 44°$ ahead of the Earth in its orbit. This geometry is shown in Figure 4.

With relatively simple mathematics, that is as far as we can go in describing the details of a trip to Mars. We have found the desired path, the time for the trip, and the position of Mars at launch time. Another important issue for the spacecraft captain would be that of the amount of fuel required for the trip. This question is related to the speed changes necessary to put us into a transfer orbit. These types of calculations involve energy considerations and are explored in Problem 3.

Our experiences so far with space travel have indicated a number of biological issues that will need to be addressed on the journey to Mars. In the absence of a gravitational field, the middle ear can no longer perceive a downward direction, and muscles are no longer required to maintain posture. Results include motion sickness and illusions about being right side up or upside down. The absence of gravity also results in bodily fluids distributing throughout the body, leading to symptoms similar to a cold. A serious problem in space travel is the atrophy of muscles and the loss of bone tissue due to the very different requirements of moving the body in a gravity-free environment. Bone loss occurs within 10 days of being in space because excessive amounts of calcium and phosphorus are released by the body, which could cause urinary stones and bone fracture. These issues can be addressed by spinning a spacecraft with a circular cross section around its axis so that the space travelers experience a centripetal acceleration that is equivalent to a gravitational field (see Section 9.9 on general relativity).

A difficulty that cannot be addressed by spinning the spacecraft is that of radiation. Being outside the atmosphere and magnetosphere of Earth exposes the space travelers to cosmic rays and other types of radiation. This exposure can lead to several detrimental health conditions, including cancer, cataracts, and suppression of the immune system. It is not clear at present whether protective shielding or pharmaceuticals would be sufficient to avoid these effects.

Although many considerations for a successful mission to Mars have not been addressed, we have successfully designed a transfer orbit from the Earth to Mars that is consistent with the laws of mechanics. We consequently declare success for our endeavor and bring this *Mission to Mars* Context to a close.

Questions

1. Some science fiction stories describe a twin planet to the Earth. It is exactly 180° ahead of us in the same orbit as the Earth, so we will never see it because it is on the other side of the Sun. Assuming you are in a spacecraft in orbit around the Earth, describe conceptually how you could visit this planet by altering your orbit.

2. You are in an orbiting spacecraft. Another spacecraft is in precisely the same orbit but is 1 km ahead of you, moving in the same direction around the circle. Through an oversight, your food supplies have been exhausted, but there is more than enough food in the other spacecraft. The commander of the other spacecraft is going to throw, from her spacecraft to yours, a picnic basket full of sandwiches. Give a qualitative description of how she should throw it.

Problems

1. Consider a Hohmann transfer from the Earth to Venus. (a) How long will this transfer take? (b) Should Venus be ahead of or behind the Earth in its orbit when the spacecraft leaves the Earth on its way to the rendezvous? How many degrees should Venus be ahead of or behind the Earth?

2. You are on a space station in a circular orbit 500 km above the surface of the Earth. Your passenger and guest is a large, strong, intelligent extraterrestrial. You try to teach her to play golf. Walking on the space station surface with magnetic shoes, you demonstrate a drive. The alien tees up a golf ball and hits it with incredible power, sending it off with speed Δv, relative to the space station, in a direction parallel to the instantaneous velocity vector of the space station. You notice that after you then complete precisely 2.00 orbits of the Earth, the golf ball also returns to the same location, so you reach up and catch the ball as it is passing the space station. With what speed Δv was the golf ball hit?

3. Investigate what the engine has to do to make a spacecraft follow the Hohmann transfer orbit from the Earth to Mars described in the text. Short-duration burns of our rocket engine are required to change the speed of our spacecraft whenever we alter our orbit. There are no brakes in space, so fuel is required both to increase and to decrease the speed of the spacecraft. First, ignore the gravitational attraction between the spacecraft and the planets. (a) Calculate the speed change required for switching the craft from a circular orbit around the Sun at the Earth's distance to the transfer orbit to Mars. (b) Calculate the speed change required for switching from the transfer orbit to a circular orbit around the Sun at the distance of Mars. Now consider the effects of the two planets' gravity. (c) Calculate the speed change required to carry the craft from the Earth's surface to its own independent orbit around the Sun. You may suppose the craft is launched from the Earth's equator toward the east. (d) Model the craft as falling to the surface of Mars from solar orbit. Calculate the magnitude of the speed change required to make a soft landing on Mars at the end of the fall. Mars rotates on its axis with a period of 24.6 h.

Context | 3

Earthquakes

Earthquakes result in massive movement of the ground, as evidenced by the accompanying photograph of severe damage caused by a magnitude 7.0 earthquake in Port-au-Prince, Haiti, in 2010. One of the most devastating events ever recorded was the magnitude 9.0 earthquake that took place on March 11, 2011, on the east coast of Japan. The earthquake triggered a devastating and widespread tsunami that killed thousands and caused major damage to buildings and several nuclear power plants.

Whereas earthquakes in Japan are relatively common, another earthquake in 2011 was far rarer. A magnitude 5.8 earthquake struck in August 2011 in the Appalachian Mountain region of Virginia in the United States. East coast earthquakes in the United States are not common. Shaking from the earthquake was felt as far north as Quebec in Canada and as far south as Atlanta, Georgia. Only minor damage was reported in towns surrounding the epicenter, although the White House and Capitol Building in Washington, D.C. were evacuated as a precaution. The National Cathedral, the Washington Monument, and the Smithsonian Castle all reported damage to structural components of the buildings.

Anyone who has experienced a serious earthquake can attest to the violent shaking it produces. In this

Figure 2 A secondary effect of some earthquakes that occur in the ocean is a tsunami. The tsunami caused by the Japanese earthquake of March 2011 caused extensive damage to the east coast of the country. This photo shows houses that have been swept off their foundations by the water as well as fires from ruptured gas lines.

AP Photo/Yasushi Kanno, The Yomiuri Shimbun

Context, we shall focus on earthquakes as an application of our study of the physics of vibrations and waves.

The cause of an earthquake is a release of energy within the Earth at a point called the *focus*, or *hypocenter*, of the earthquake. The point on the Earth's surface radially above the focus is called the *epicenter*. As the energy from the focus reaches the surface, it spreads out along the surface of the Earth.

Earthquakes generally originate along a *fault*, which is a fracture or discontinuity in the rock beneath the Earth's surface. When there is sudden relative movement between the material on either side of a fault, an earthquake occurs. U.S. Geological Survey studies have shown a direct correlation between the magnitude of an earthquake and the size of nearby faults. Furthermore, such studies indicate that large-magnitude earthquakes can last as long as 2 minutes.

We might expect that the risk of damage in an earthquake decreases as one moves farther from the epicenter, and over long distances that assumption is correct. For example, structures in Kansas are not affected by earthquakes in California. In regions close to the earthquake, however, the notion of decrease in risk with

Figure 1 A day after the magnitude 7.0 earthquake in Port-au-Prince, Haiti, on January 13, 2010, a young woman climbs over the rubble of a collapsed store.

ANTHONY BELIZAIRE/AFP/Getty Images

distance is not consistent. Consider, for example, the following comparisons describing local and distant effects resulting from two different earthquakes.

With regard to the magnitude-7.9 Michoacán earthquake, September 19, 1985:[1]

An earthquake rattled the coast of Mexico in the state of Michoacán, about 400 kilometers west of Mexico City. Near the coast, the shaking of the ground was mild and caused little damage. As the seismic waves raced inland, the ground shook even less, and by the time the waves were 100 kilometers from Mexico City, the shaking had nearly subsided. Nevertheless, the seismic waves induced severe shaking in the city, and some areas continued to shake for several minutes after the seismic waves had passed. Some 300 buildings collapsed and more than 20,000 people died.

A magnitude-6.3 earthquake ocurred on February 22, 2011, 10 km southeast of Christchurch, New Zealand. Flight crews from the New York Air National Guard were at Christchurch International Airport, 12 km northwest of the city, when the quake struck, and reported that they were safe and unharmed, and that the airport had water and electricity.

Consider, however, a contrasting situation much farther away, 200 km from Christchurch:[2]

The magnitude 6.3 earthquake . . . was strong enough to shake 30 million tonnes of ice loose from Tasman Glacier at Aoraki Mt Cook National Park. Passengers of two explorer boats were hit with waves of up to 3.5 metres as the ice crashed into Terminal Lake under the Tasman Glacier at the mountain.

Figure 3 Severe damage occurred in localized regions of Mexico City in 1985, even though the epicenter of the Michoacán earthquake was hundreds of kilometers away.

It is clear from these comparisons that the notion of a simple decrease in risk with distance is misleading. We will use these comparisons as motivation in our study of the physics of vibrations and waves so that we can better analyze the risk of damage to structures in an earthquake. Our study here will also be important when we investigate electromagnetic waves in Chapters 24 through 27. In this Context, we shall address the central question:

> **How can we choose locations and build structures to minimize the risk of damage in an earthquake?**

---

[1] *American Scientist*, November–December 1992, p. 566.

[2] *New Zealand Herald*, 22 February 2011

Oscillatory Motion

Chapter Outline

© Ranjit Doroszkeiwicz/Alamy

To reduce swaying in tall buildings because of the wind, tuned dampers are placed near the top of the building. These mechanisms include an object of large mass that oscillates under computer control at the same frequency as the building, reducing the swaying. The 730-ton suspended sphere in the photograph above is part of the tuned damper system of the Taipei Financial Center, at one time the world's tallest building.

Y ou are most likely familiar with several examples of *periodic* motion, such as the oscillations of an object on a spring, the motion of a pendulum, and the vibrations of a stringed musical instrument. Numerous other systems exhibit periodic behavior. For example, the molecules in a solid oscillate about their equilibrium positions; electromagnetic waves, such as light waves, radar, and radio waves, are characterized by oscillating electric and magnetic field vectors; in alternating-current circuits, such as in your household electrical service, voltage and current vary periodically with time. In this chapter, we will investigate mechanical systems that exhibit periodic motion.

We have experienced a number of situations in which the net force on a particle is constant. In these situations, the acceleration of the particle is also constant and we can describe the motion of the particle using the particle under constant acceleration model and the kinematic equations of Chapter 2. If a force acting on a particle varies in time, the acceleration of the particle also changes with time and so the kinematic equations cannot be used.

A special kind of periodic motion occurs when the force that acts on a particle is always directed toward an equilibrium position and is proportional to the position of the particle relative to the equilibrium position. We shall study this special type of varying force in this chapter. When this type of force acts on a particle, the particle exhibits *simple harmonic motion,* which will serve as an analysis model for a large class of oscillation problems.

12.1 | Motion of an Object Attached to a Spring

As a model for simple harmonic motion, consider a block of mass m attached to the end of a spring, with the block free to move on a frictionless, horizontal surface (Active Fig. 12.1). When the spring is neither stretched nor compressed, the block is at rest at the position called the **equilibrium position** of the system, which we identify as $x = 0$ (Active Fig. 12.1b). We know from experience that such a system oscillates back and forth if disturbed from its equilibrium position.

We can understand the oscillating motion of the block in Active Figure 12.1 qualitatively by first recalling that when the block is displaced to a position x, the spring exerts on the block a force that is proportional to the position and given by **Hooke's law** (see Section 6.4):

$$F_s = -kx \qquad\qquad \textbf{12.1} \blacktriangleleft \qquad \blacktriangleright \text{Hooke's law}$$

We call F_s a **restoring force** because it is always directed toward the equilibrium position and therefore *opposite* the displacement of the block from equilibrium. That is, when the block is displaced to the right of $x = 0$ in Active Figure 12.1a, the position is positive and the restoring force is directed to the left. When the block is displaced to the left of $x = 0$ as in Figure 12.1c, the position is negative and the restoring force is directed to the right.

When the block is displaced from the equilibrium point and released, it is a particle under a net force and consequently undergoes an acceleration. Applying Newton's second law to the motion of the block, with Equation 12.1 providing the net force in the x direction, we obtain

$$-kx = ma_x$$

$$a_x = -\frac{k}{m}x \qquad\qquad \textbf{12.2} \blacktriangleleft$$

That is, the acceleration of the block is proportional to its position, and the direction of the acceleration is opposite the direction of the displacement of the block from equilibrium. Systems that behave in this way are said to exhibit **simple harmonic motion.** An object moves with simple harmonic motion whenever its acceleration is proportional to its position and is oppositely directed to the displacement from equilibrium.

If the block in Active Figure 12.1 is displaced to a position $x = A$ and released from rest, its *initial* acceleration is $-kA/m$. When the block passes through the equilibrium position $x = 0$, its acceleration is zero. At this instant, its speed is a maximum because the acceleration changes sign. The block then continues to travel to the left of equilibrium with a positive acceleration and finally reaches $x = -A$, at which time its acceleration is $+kA/m$ and its speed is again zero as discussed in Sections 6.4 and 6.6. The block completes a full cycle of its motion by returning to the original position, again passing through $x = 0$ with maximum speed. Therefore, the block oscillates between the turning points $x = \pm A$. In the absence of friction, this idealized motion will continue forever because the force exerted by the spring is conservative. Real systems are generally subject to friction, so they do not oscillate forever. We shall explore the details of the situation with friction in Section 12.6.

> **Pitfall Prevention | 12.1**
> **The Orientation of the Spring**
> Active Figure 12.1 shows a *horizontal* spring, with an attached block sliding on a frictionless surface. Another possibility is a block hanging from a *vertical* spring. All the results we discuss for the horizontal spring are the same for the vertical spring with one exception: when the block is placed on the vertical spring, its weight causes the spring to extend. If the resting position of the block is defined as $x = 0$, the results of this chapter also apply to this vertical system.

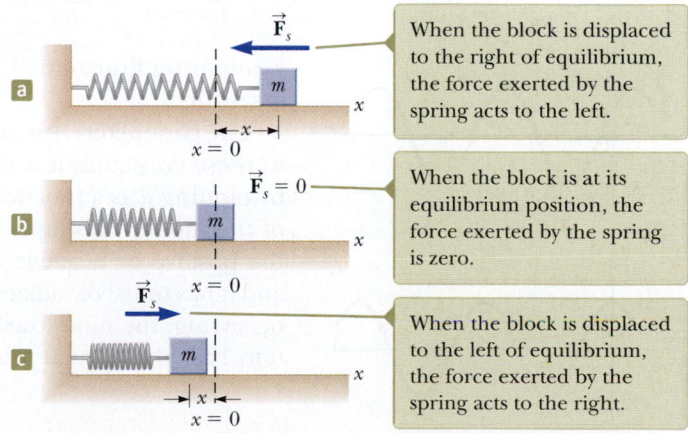

When the block is displaced to the right of equilibrium, the force exerted by the spring acts to the left.

When the block is at its equilibrium position, the force exerted by the spring is zero.

When the block is displaced to the left of equilibrium, the force exerted by the spring acts to the right.

Active Figure 12.1 A block attached to a spring moving on a frictionless surface.

QUICK QUIZ 12.1 A block on the end of a spring is pulled to position $x = A$ and released from rest. In one full cycle of its motion, through what total distance does it travel? **(a)** $A/2$ **(b)** A **(c)** $2A$ **(d)** $4A$

❮12.2 | Analysis Model: Particle in Simple Harmonic Motion

The motion described in the preceding section occurs so often that we identify the **particle in simple harmonic motion** model to represent such situations. To develop a mathematical representation for this model, we will generally choose x as the axis along which the oscillation occurs; hence, we will drop the subscript-x notation in this discussion. Recall that, by definition, $a = dv/dt = d^2x/dt^2$, so we can express Equation 12.2 as

Pitfall Prevention | 12.2
A Nonconstant Acceleration
The acceleration of a particle in simple harmonic motion is not constant. Equation 12.3 shows that its acceleration varies with position x. Therefore, we *cannot* apply the kinematic equations of Chapter 2 in this situation.

$$\frac{d^2x}{dt^2} = -\frac{k}{m}x \qquad \text{12.3} \blacktriangleleft$$

If we denote the ratio k/m with the symbol ω^2 (we choose ω^2 rather than ω so as to make the solution we develop below simpler in form), then

$$\omega^2 = \frac{k}{m} \qquad \text{12.4} \blacktriangleleft$$

and Equation 12.3 can be written in the form

$$\frac{d^2x}{dt^2} = -\omega^2 x \qquad \text{12.5} \blacktriangleleft$$

Pitfall Prevention | 12.3
Where's the Triangle?
Equation 12.6 includes a trigonometric function, a *mathematical function* that can be used whether it refers to a triangle or not. In this case, the cosine function happens to have the correct behavior for representing the position of a particle in simple harmonic motion.

Let's now find a mathematical solution to Equation 12.5, that is, a function $x(t)$ that satisfies this second-order differential equation and is a mathematical representation of the position of the particle as a function of time. We seek a function whose second derivative is the same as the original function with a negative sign and multiplied by ω^2. The trigonometric functions sine and cosine exhibit this behavior, so we can build a solution around one or both of them. The following cosine function is a solution to the differential equation:

$$x(t) = A\cos(\omega t + \phi) \qquad \text{12.6} \blacktriangleleft$$

▶ Position versus time for a particle in simple harmonic motion

where A, ω, and ϕ are constants. To show explicitly that this solution satisfies Equation 12.5, notice that

$$\frac{dx}{dt} = A\frac{d}{dt}\cos(\omega t + \phi) = -\omega A\sin(\omega t + \phi) \qquad \text{12.7} \blacktriangleleft$$

$$\frac{d^2x}{dt^2} = -\omega A\frac{d}{dt}\sin(\omega t + \phi) = -\omega^2 A\cos(\omega t + \phi) \qquad \text{12.8} \blacktriangleleft$$

Comparing Equations 12.6 and 12.8, we see that $d^2x/dt^2 = -\omega^2 x$ and Equation 12.5 is satisfied.

The parameters A, ω, and ϕ are constants of the motion. To give physical significance to these constants, it is convenient to form a graphical representation of the motion by plotting x as a function of t as in Active Figure 12.2a. First, A, called the **amplitude** of the motion, is simply the maximum value of the position of the particle in either the positive or negative x direction. The constant ω is called the **angular frequency,** and it has units[1] of radians per second. It is a measure of how rapidly the oscillations are occurring; the more oscillations per unit time, the higher the value of ω. From Equation 12.4, the angular frequency is

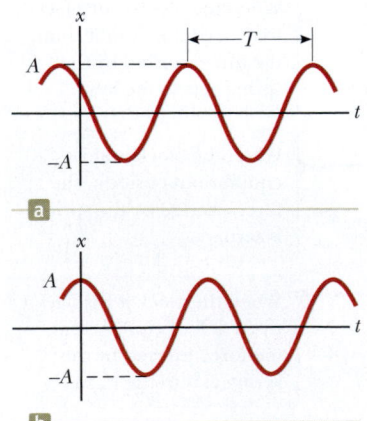

$$\omega = \sqrt{\frac{k}{m}} \qquad \text{12.9} \blacktriangleleft$$

Active Figure 12.2 (a) An x–t graph for a particle undergoing simple harmonic motion. The amplitude of the motion is A, and the period (defined in Eq. 12.10) is T. (b) The x–t graph for the special case in which $x = A$ at $t = 0$ and hence $\phi = 0$.

[1]We have seen many examples in earlier chapters in which we evaluate a trigonometric function of an angle. The argument of a trigonometric function, such as sine or cosine, *must* be a pure number. The radian is a pure number because it is a ratio of lengths. Angles in degrees are pure numbers because the degree is an artificial "unit"; it is not related to measurements of lengths. The argument of the trigonometric function in Equation 12.6 must be a pure number. Therefore, ω *must* be expressed in radians per second (and not, for example, in revolutions per second) if t is expressed in seconds. Furthermore, other types of functions such as logarithms and exponential functions require arguments that are pure numbers.

The constant angle ϕ is called the **phase constant** (or initial phase angle) and, along with the amplitude A, is determined uniquely by the position and velocity of the particle at $t = 0$. If the particle is at its maximum position $x = A$ at $t = 0$, the phase constant is $\phi = 0$ and the graphical representation of the motion is as shown in Active Figure 12.2b. The quantity $(\omega t + \phi)$ is called the **phase** of the motion. Notice that the function $x(t)$ is periodic and its value is the same each time ωt increases by 2π radians.

Equations 12.1, 12.5, and 12.6 form the basis of the mathematical representation of the particle in simple harmonic motion model. If you are analyzing a situation and find that the force on an object modeled as a particle is of the mathematical form of Equation 12.1, you know the motion is that of a simple harmonic oscillator and the position of the particle is described by Equation 12.6. If you analyze a system and find that it is described by a differential equation of the form of Equation 12.5, the motion is that of a simple harmonic oscillator. If you analyze a situation and find that the position of a particle is described by Equation 12.6, you know the particle undergoes simple harmonic motion.

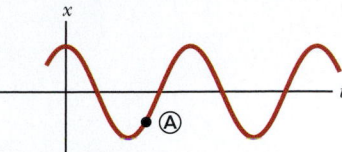

Figure 12.3 (Quick Quiz 12.2) An x–t graph for a particle undergoing simple harmonic motion. At a particular time, the particle's position is indicated by Ⓐ in the graph.

QUICK QUIZ 12.2 Consider a graphical representation (Fig. 12.3) of simple harmonic motion as described mathematically in Equation 12.6. When the particle is at point Ⓐ on the graph, what can you say about its position and velocity? (a) The position and velocity are both positive. (b) The position and velocity are both negative. (c) The position is positive, and the velocity is zero. (d) The position is negative, and the velocity is zero. (e) The position is positive, and the velocity is negative. (f) The position is negative, and the velocity is positive.

QUICK QUIZ 12.3 Figure 12.4 shows two curves representing particles undergoing simple harmonic motion. The correct description of these two motions is that the simple harmonic motion of particle B is (a) of larger angular frequency and larger amplitude than that of particle A, (b) of larger angular frequency and smaller amplitude than that of particle A, (c) of smaller angular frequency and larger amplitude than that of particle A, or (d) of smaller angular frequency and smaller amplitude than that of particle A.

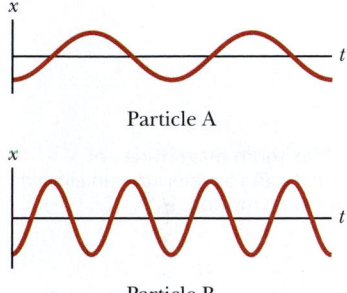

Particle A

Particle B

Figure 12.4 (Quick Quiz 12.3) Two x–t graphs for particles undergoing simple harmonic motion. The amplitudes and frequencies are different for the two particles.

Let us investigate further the mathematical description of simple harmonic motion. The **period** T of the motion is the time interval required for the particle to go through one full cycle of its motion (Active Fig. 12.2a). That is, the values of x and v for the particle at time t equal the values of x and v at time $t + T$. Because the phase increases by 2π radians in a time interval of T,

$$[\omega(t + T) + \phi] - (\omega t + \phi) = 2\pi$$

Simplifying this expression gives $\omega T = 2\pi$, or

$$T = \frac{2\pi}{\omega} \qquad \text{12.10} \blacktriangleleft$$

The inverse of the period is called the **frequency** f of the motion. Whereas the period is the time interval per oscillation, the frequency represents the number of oscillations the particle undergoes per unit time interval:

$$f = \frac{1}{T} = \frac{\omega}{2\pi} \qquad \text{12.11} \blacktriangleleft$$

The units of f are cycles per second, or **hertz** (Hz). Rearranging Equation 12.11 gives

$$\omega = 2\pi f = \frac{2\pi}{T} \qquad \text{12.12} \blacktriangleleft$$

Equations 12.9 through 12.11 can be used to express the period and frequency of the motion for the particle in simple harmonic motion in terms of the characteristics m and k of the system as

$$T = \frac{2\pi}{\omega} = 2\pi\sqrt{\frac{m}{k}} \qquad \text{12.13} \blacktriangleleft \qquad \blacktriangleright \text{Period}$$

Pitfall Prevention | 12.4
Two Kinds of Frequency
We identify two kinds of frequency for a simple harmonic oscillator: f, called simply the *frequency*, is measured in hertz, and ω, the *angular frequency*, is measured in radians per second. Be sure you are clear about which frequency is being discussed or requested in a given problem. Equations 12.11 and 12.12 show the relationship between the two frequencies.

▶ Frequency

$$f = \frac{1}{T} = \frac{1}{2\pi}\sqrt{\frac{k}{m}} \qquad \text{12.14} \blacktriangleleft$$

That is, the period and frequency depend *only* on the mass of the particle and the force constant of the spring and *not* on the parameters of the motion, such as A or ϕ. As we might expect, the frequency is larger for a stiffer spring (larger value of k) and decreases with increasing mass of the particle.

We can obtain the velocity and acceleration[2] of a particle undergoing simple harmonic motion from Equations 12.7 and 12.8:

▶ Velocity of a particle in simple harmonic motion

$$v = \frac{dx}{dt} = -\omega A \sin(\omega t + \phi) \qquad \text{12.15} \blacktriangleleft$$

▶ Acceleration of a particle in simple harmonic motion

$$a = \frac{d^2x}{dt^2} = -\omega^2 A \cos(\omega t + \phi) \qquad \text{12.16} \blacktriangleleft$$

From Equation 12.15, we see that because the sine and cosine functions oscillate between ± 1, the extreme values of the velocity v are $\pm \omega A$. Likewise, Equation 12.16 shows that the extreme values of the acceleration a are $\pm \omega^2 A$. Therefore, the *maximum* values of the magnitudes of the velocity and acceleration are

▶ Maximum magnitudes of velocity and acceleration in simple harmonic motion

$$v_{max} = \omega A = \sqrt{\frac{k}{m}}\,A \qquad \text{12.17} \blacktriangleleft$$

$$a_{max} = \omega^2 A = \frac{k}{m}\,A \qquad \text{12.18} \blacktriangleleft$$

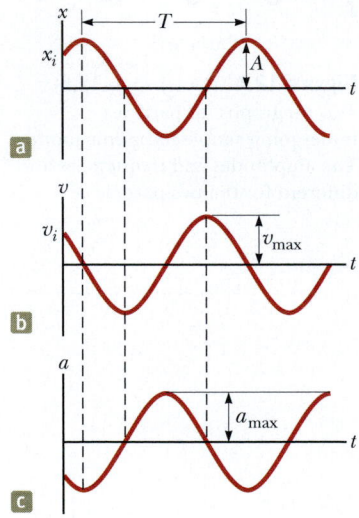

Figure 12.5a plots position versus time for an arbitrary value of the phase constant. The associated velocity–time and acceleration–time curves are illustrated in Figures 12.5b and 12.5c, respectively. They show that the phase of the velocity differs from the phase of the position by $\pi/2$ rad, or 90°. That is, when x is a maximum or a minimum, the velocity is zero. Likewise, when x is zero, the speed is a maximum. Furthermore, notice that the phase of the acceleration differs from the phase of the position by π radians, or 180°. For example, when x is a maximum, a has a maximum magnitude in the opposite direction.

QUICK QUIZ 12.4 An object of mass m is hung from a spring and set into oscillation. The period of the oscillation is measured and recorded as T. The object of mass m is removed and replaced with an object of mass $2m$. When this object is set into oscillation, what is the period of the motion? (a) $2T$ (b) $\sqrt{2}\,T$ (c) T (d) $T/\sqrt{2}$ (e) $T/2$

Figure 12.5 Graphical representation of simple harmonic motion. (a) Position versus time. (b) Velocity versus time. (c) Acceleration versus time. Notice that at any specified time the velocity is 90° out of phase with the position and the acceleration is 180° out of phase with the position.

Equation 12.6 describes simple harmonic motion of a particle in general. Let's now see how to evaluate the constants of the motion. The angular frequency ω is evaluated using Equation 12.9. The constants A and ϕ are evaluated from the initial conditions, that is, the state of the oscillator at $t = 0$.

Suppose a block is set into motion by pulling it from equilibrium by a distance A and releasing it from rest at $t = 0$ as in Active Figure 12.6. We must then require our solutions for $x(t)$ and $v(t)$ (Eqs. 12.6 and 12.15) to obey the initial conditions that $x(0) = A$ and $v(0) = 0$:

$$x(0) = A \cos\phi = A$$

$$v(0) = -\omega A \sin\phi = 0$$

These conditions are met if $\phi = 0$, giving $x = A \cos\omega t$ as our solution. To check this solution, notice that it satisfies the condition that $x(0) = A$ because $\cos 0 = 1$.

The position, velocity, and acceleration of the block versus time are plotted in Figure 12.7a for this special case. The acceleration reaches extreme values of $\mp\omega^2 A$ when the position has extreme values of $\pm A$. Furthermore, the velocity has extreme

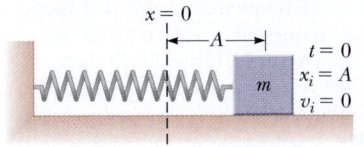

Active Figure 12.6 A block–spring system that begins its motion from rest with the block at $x = A$ at $t = 0$.

---

[2]Because the motion of a simple harmonic oscillator takes place in one dimension, we denote velocity as v and acceleration as a, with the direction indicated by a positive or negative sign as in Chapter 2.

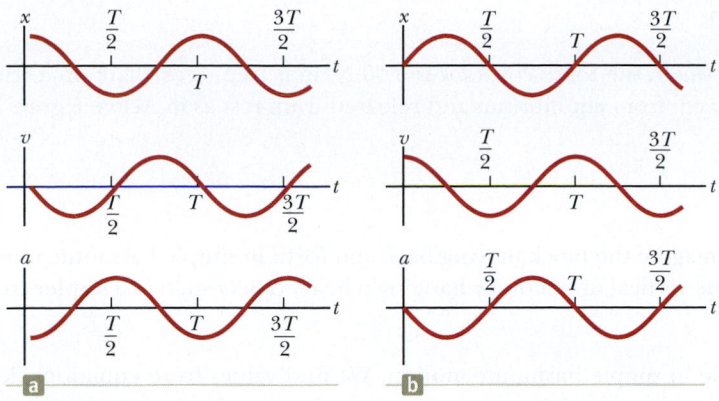

Figure 12.7 (a) Position, velocity, and acceleration versus time for the block in Active Figure 12.6 under the initial conditions that at $t = 0$, $x(0) = A$, and $v(0) = 0$. (b) Position, velocity, and acceleration versus time for the block in Active Figure 12.8 under the initial conditions that at $t = 0$, $x(0) = 0$, and $v(0) = v_i$.

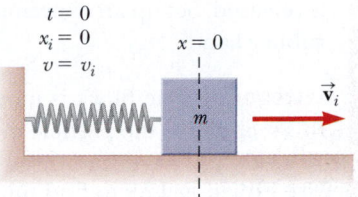

Active Figure 12.8 The block–spring system is undergoing oscillation, and $t = 0$ is defined at an instant when the block passes through the equilibrium position $x = 0$ and is moving to the right with speed v_i.

values of $\pm\omega A$, which both occur at $x = 0$. Hence, the quantitative solution agrees with our qualitative description of this system.

Let's consider another possibility. Suppose the system is oscillating and we define $t = 0$ as the instant the block passes through the unstretched position of the spring while moving to the right (Active Fig. 12.8). In this case, our solutions for $x(t)$ and $v(t)$ must obey the initial conditions that $x(0) = 0$ and $v(0) = v_i$:

$$x(0) = A \cos \phi = 0$$

$$v(0) = -\omega A \sin \phi = v_i$$

The first of these conditions tells us that $\phi = \pm\pi/2$. With these choices for ϕ, the second condition tells us that $A = \mp v_i/\omega$. Because the initial velocity is positive and the amplitude must be positive, we must have $\phi = -\pi/2$. Hence, the solution is

$$x = \frac{v_i}{\omega} \cos\left(\omega t - \frac{\pi}{2}\right)$$

The graphs of position, velocity, and acceleration versus time for this choice of $t = 0$ are shown in Figure 12.7b. Notice that these curves are the same as those in Figure 12.7a, but shifted to the right by one-fourth of a cycle. This shift is described mathematically by the phase constant $\phi = -\pi/2$, which is one-fourth of a full cycle of 2π.

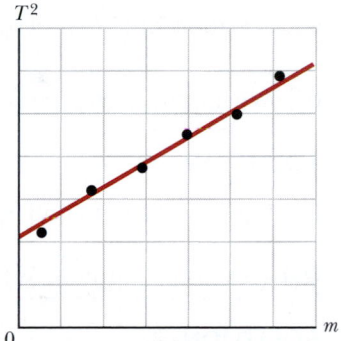

Figure 12.9 (Thinking Physics 12.1) A graph of experimental data: the square of the period versus mass of a block in a block–spring system.

> ### THINKING PHYSICS 12.1

We know that the period of oscillation of an object attached to a spring is proportional to the square root of the mass of the object (Eq. 12.13). Therefore, if we perform an experiment in which we place objects with a range of masses on the end of a spring and measure the period of oscillation of each object–spring system, a graph of the square of the period versus the mass will result in a straight line as suggested in Figure 12.9. We find, however, that the line does not go through the origin. Why not?

Reasoning The line does not go through the origin because the spring itself has mass. Therefore, the resistance to changes in motion of the system is a combination of the mass of the object on the end of the spring and the mass of the oscillating spring coils. The entire mass of the spring is not oscillating in the same way, however. The coil of the spring attached to the object is oscillating over the same amplitude as the object, but the coil at the fixed end of the spring is not oscillating at all. For a cylindrical spring, energy arguments can be used to show that the effective additional mass representing the oscillations of the spring is one-third of the mass of the spring. The square of the period is proportional to the total oscillating mass, but the graph in Figure 12.9 shows the square of the period versus only the mass of the object on the spring. A graph of period squared versus total mass (mass of the object on the spring plus the effective oscillating mass of the spring) would pass through the origin. ◀

Example 12.1 | A Block–Spring System

A 200-g block connected to a light spring for which the force constant is 5.00 N/m is free to oscillate on a frictionless, horizontal surface. The block is displaced 5.00 cm from equilibrium and released from rest as in Active Figure 12.6.

(A) Find the period of its motion.

SOLUTION

Conceptualize Study Active Figure 12.6 and imagine the block moving back and forth in simple harmonic motion once it is released. Set up an experimental model in the vertical direction by hanging a heavy object such as a stapler from a strong rubber band.

Categorize The block is modeled as a particle in simple harmonic motion. We find values from equations developed in this section for the particle in simple harmonic motion model, so we categorize this example as a substitution problem.

Use Equation 12.9 to find the angular frequency of the block–spring system:

$$\omega = \sqrt{\frac{k}{m}} = \sqrt{\frac{5.00 \text{ N/m}}{200 \times 10^{-3} \text{ kg}}} = 5.00 \text{ rad/s}$$

Use Equation 12.13 to find the period of the system:

$$T = \frac{2\pi}{\omega} = \frac{2\pi}{5.00 \text{ rad/s}} = \boxed{1.26 \text{ s}}$$

(B) Determine the maximum speed of the block.

SOLUTION

Use Equation 12.17 to find v_{max}:

$$v_{max} = \omega A = (5.00 \text{ rad/s})(5.00 \times 10^{-2} \text{ m}) = \boxed{0.250 \text{ m/s}}$$

(C) What is the maximum acceleration of the block?

SOLUTION

Use Equation 12.18 to find a_{max}:

$$a_{max} = \omega^2 A = (5.00 \text{ rad/s})^2(5.00 \times 10^{-2} \text{ m}) = \boxed{1.25 \text{ m/s}^2}$$

(D) Express the position, velocity, and acceleration as functions of time in SI units.

SOLUTION

Find the phase constant from the initial condition that $x = A$ at $t = 0$:

$$x(0) = A \cos \phi = A \quad \rightarrow \quad \phi = 0$$

Use Equation 12.6 to write an expression for $x(t)$:

$$x = A \cos (\omega t + \phi) = \boxed{0.050\ 0 \cos 5.00t}$$

Use Equation 12.15 to write an expression for $v(t)$:

$$v = -\omega A \sin (\omega t + \phi) = \boxed{-0.250 \sin 5.00t}$$

Use Equation 12.16 to write an expression for $a(t)$:

$$a = -\omega^2 A \cos (\omega t + \phi) = \boxed{-1.25 \cos 5.00t}$$

Example 12.2 | Watch Out for Potholes!

A car with a mass of 1 300 kg is constructed so that its frame is supported by four springs. Each spring has a force constant of 20 000 N/m. Two people riding in the car have a combined mass of 160 kg. Find the frequency of vibration of the car after it is driven over a pothole in the road and the car oscillates vertically.

SOLUTION

Conceptualize Think about your experiences with automobiles. When you sit in a car, it moves downward a small distance because your weight is compressing the springs further. If you push down on the front bumper and release it, the front of the car oscillates a few times.

Categorize We imagine the car as being supported by a single spring and model the car as a particle in simple harmonic motion.

12.2 *cont.*

Analyze First, let's determine the effective spring constant of the four springs combined. For a given extension x of the springs, the combined force on the car is the sum of the forces from the individual springs.

Find an expression for the total force on the car:

$$F_{\text{total}} = \sum(-kx) = -\left(\sum k\right)x$$

In this expression, x has been factored from the sum because it is the same for all four springs. The effective spring constant for the combined springs is the sum of the individual spring constants.

Evaluate the effective spring constant:

$$k_{\text{eff}} = \sum k = 4 \times 20\,000 \text{ N/m} = 80\,000 \text{ N/m}$$

Use Equation 12.14 to find the frequency of vibration:

$$f = \frac{1}{2\pi}\sqrt{\frac{k_{\text{eff}}}{m}} = \frac{1}{2\pi}\sqrt{\frac{80\,000 \text{ N/m}}{1\,460 \text{ kg}}} = \boxed{1.18 \text{ Hz}}$$

Finalize The mass we used here is that of the car plus the people because that is the total mass that is oscillating. Also notice that we have explored only up-and-down motion of the car. If an oscillation is established in which the car rocks back and forth such that the front end goes up when the back end goes down, the frequency will be different.

What If? Suppose the car stops on the side of the road and the two people exit the car. One of them pushes downward on the car and releases it so that it oscillates vertically. Is the frequency of the oscillation the same as the value we just calculated?

Answer The suspension system of the car is the same, but the mass that is oscillating is smaller: it no longer includes the mass of the two people. Therefore, the frequency should be higher. Let's calculate the new frequency, taking the mass to be 1 300 kg:

$$f = \frac{1}{2\pi}\sqrt{\frac{k_{\text{eff}}}{m}} = \frac{1}{2\pi}\sqrt{\frac{80\,000 \text{ N/m}}{1\,300 \text{ kg}}} = 1.25 \text{ Hz}$$

As predicted, the new frequency is a bit higher.

12.3 | Energy of the Simple Harmonic Oscillator

Let us examine the mechanical energy of a system in which a particle undergoes simple harmonic motion, such as the block–spring system illustrated in Active Figure 12.1. Because the surface is frictionless, the system is isolated and we expect the total mechanical energy of the system to be constant. We assume a massless spring, so the kinetic energy of the system corresponds only to that of the block. We can use Equation 12.15 to express the kinetic energy of the block as

$$K = \tfrac{1}{2}mv^2 = \tfrac{1}{2}m\omega^2 A^2 \sin^2(\omega t + \phi) \qquad \textbf{12.19} \blacktriangleleft$$

▶ Kinetic energy of a simple harmonic oscillator

The elastic potential energy stored in the spring for any elongation x is given by $\tfrac{1}{2}kx^2$ (see Eq. 6.22). Using Equation 12.6 gives

$$U = \tfrac{1}{2}kx^2 = \tfrac{1}{2}kA^2 \cos^2(\omega t + \phi) \qquad \textbf{12.20} \blacktriangleleft$$

▶ Potential energy of a simple harmonic oscillator

We see that K and U are *always* positive quantities or zero. Because $\omega^2 = k/m$, we can express the total mechanical energy of the simple harmonic oscillator as

$$E = K + U = \tfrac{1}{2}kA^2[\sin^2(\omega t + \phi) + \cos^2(\omega t + \phi)]$$

From the identity $\sin^2\theta + \cos^2\theta = 1$, we see that the quantity in square brackets is unity. Therefore, this equation reduces to

$$E = \tfrac{1}{2}kA^2 \qquad \textbf{12.21} \blacktriangleleft$$

▶ Total energy of a simple harmonic oscillator

That is, the total mechanical energy of a simple harmonic oscillator is a constant of the motion and is proportional to the square of the amplitude. The total mechanical energy is equal to the maximum potential energy stored in the spring when $x = \pm A$

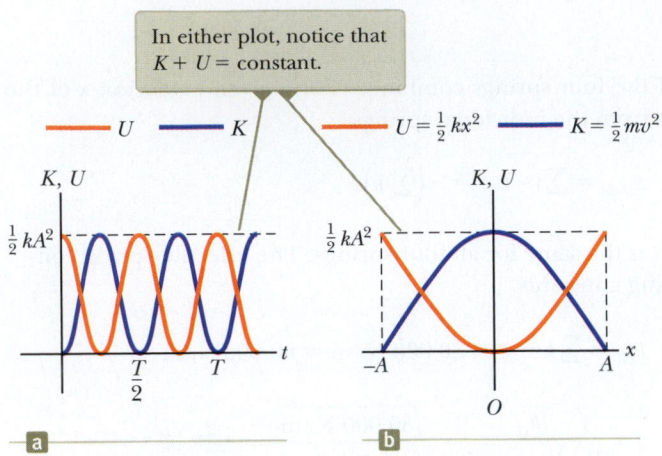

In either plot, notice that $K + U =$ constant.

—— U —— K —— $U = \frac{1}{2}kx^2$ —— $K = \frac{1}{2}mv^2$

Active Figure 12.10 (a) Kinetic energy and potential energy versus time for a simple harmonic oscillator with $\phi = 0$. (b) Kinetic energy and potential energy versus position for a simple harmonic oscillator.

because $v = 0$ at these points and there is no kinetic energy. At the equilibrium position, where $U = 0$ because $x = 0$, the total energy, all in the form of kinetic energy, is again $\frac{1}{2}kA^2$.

Plots of the kinetic and potential energies versus time appear in Active Figure 12.10a, where we have taken $\phi = 0$. At all times, the sum of the kinetic and potential energies is a constant equal to $\frac{1}{2}kA^2$, the total energy of the system.

The variations of K and U with the position x of the block are plotted in Active Figure 12.10b. Energy is continuously being transformed between potential energy stored in the spring and kinetic energy of the block.

Active Figure 12.11 illustrates the position, velocity, acceleration, kinetic energy, and potential energy of the block–spring system for one full period of the motion. Most of the ideas discussed so far are incorporated in this important figure. Study it carefully.

Finally, we can obtain the velocity of the block at an arbitrary position by expressing the total energy of the system at some arbitrary position x as

$$E = K + U = \tfrac{1}{2}mv^2 + \tfrac{1}{2}kx^2 = \tfrac{1}{2}kA^2$$

and then solving for v:

▶ Velocity as a function of position for a simple harmonic oscillator

$$v = \pm\sqrt{\frac{k}{m}(A^2 - x^2)} = \pm\omega\sqrt{A^2 - x^2} \qquad \textbf{12.22} \blacktriangleleft$$

When you check Equation 12.22 to see whether it agrees with known cases, you find that it verifies that the speed is a maximum at $x = 0$ and is zero at the turning points $x = \pm A$.

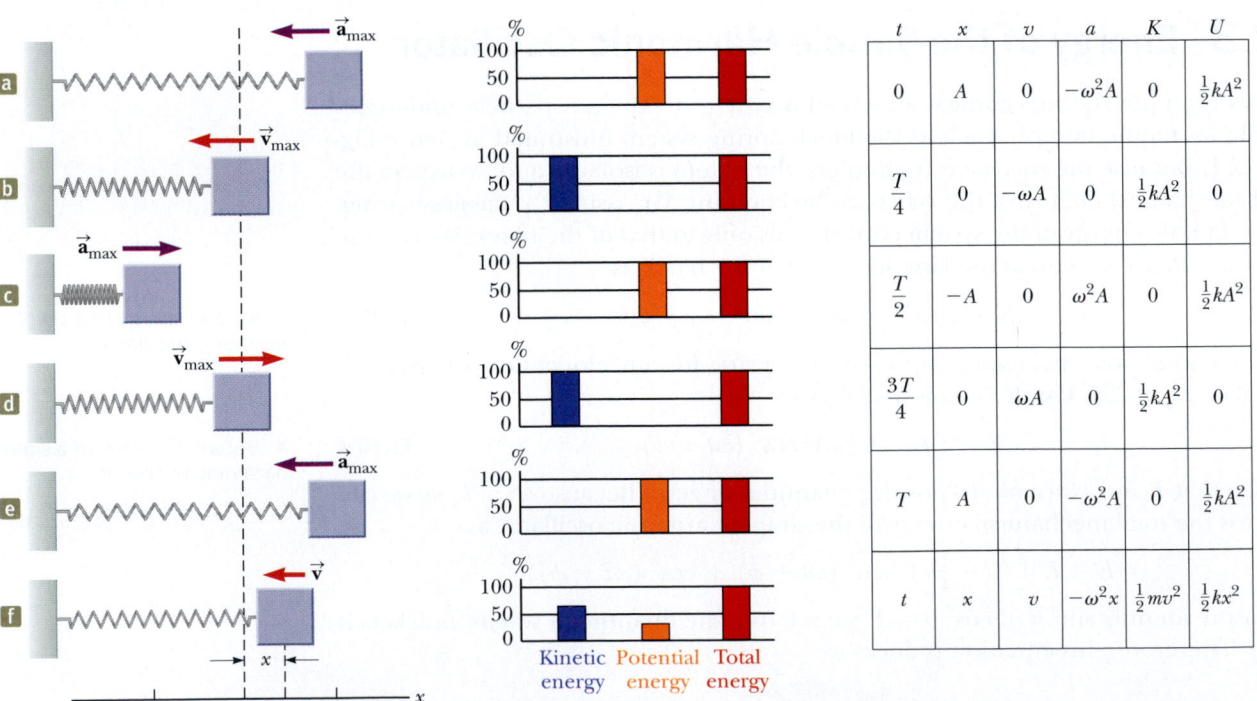

			t	x	v	a	K	U
			0	A	0	$-\omega^2 A$	0	$\frac{1}{2}kA^2$
			$\dfrac{T}{4}$	0	$-\omega A$	0	$\frac{1}{2}kA^2$	0
			$\dfrac{T}{2}$	$-A$	0	$\omega^2 A$	0	$\frac{1}{2}kA^2$
			$\dfrac{3T}{4}$	0	ωA	0	$\frac{1}{2}kA^2$	0
			T	A	0	$-\omega^2 A$	0	$\frac{1}{2}kA^2$
			t	x	v	$-\omega^2 x$	$\frac{1}{2}mv^2$	$\frac{1}{2}kx^2$

Kinetic energy   Potential energy   Total energy

Active Figure 12.11 (a) through (e) Several instants in the simple harmonic motion for a block–spring system. Energy bar graphs show the distribution of the energy of the system at each instant. The parameters in the table at the right refer to the block–spring system, assuming at $t = 0$, $x = A$; hence, $x = A\cos\omega t$. For these five special instants, one of the types of energy is zero. (f) An arbitrary point in the motion of the oscillator. The system possesses both kinetic energy and potential energy at this instant as shown in the bar graph.

> **THINKING PHYSICS 12.2**

An object oscillating on the end of a horizontal spring slides back and forth over a frictionless surface. During one oscillation, you set an identical object at the maximum displacement point, with instant-acting glue on its surface. Just as the oscillating object reaches its largest displacement and is momentarily at rest, it adheres to the new object by means of the glue and the two objects continue the oscillation together. Does the period of the oscillation change? Does the amplitude of oscillation change? Does the energy of the oscillation change?

Reasoning The period of oscillation changes because the period depends on the mass that is oscillating (Eq. 12.13). The amplitude does not change. Because the new object was added under the special condition that

the original object was at rest, the combined objects are at rest at this point also, defining the amplitude as the same as in the original oscillation. The energy does not change either. At the maximum displacement point, the energy is all potential energy stored in the spring, which depends only on the force constant and the amplitude, not on the mass of the object. The object of increased mass will pass through the equilibrium point with lower speed than in the original oscillation but with the same kinetic energy. Another approach is to think about how energy could be transferred into the oscillating system. No work was done on the system (nor did any other form of energy transfer occur), so the energy in the system cannot change. ◄

You may wonder why we are spending so much time studying simple harmonic oscillators. We do so because they are good models of a wide variety of physical phenomena. For example, recall the Lennard–Jones potential discussed in Example 6.9. This complicated function describes the forces holding atoms together. Figure 12.12a shows that for small displacements from the equilibrium position, the potential energy curve for this function approximates a parabola, which represents the potential energy function for a simple harmonic oscillator. Therefore, we can model the complex atomic binding forces as being due to tiny springs as depicted in Figure 12.12b.

The ideas presented in this chapter apply not only to block–spring systems and atoms, but also to a wide range of situations that include bungee jumping, playing a musical instrument, and viewing the light emitted by a laser. You will see more examples of simple harmonic oscillators as you work through this book.

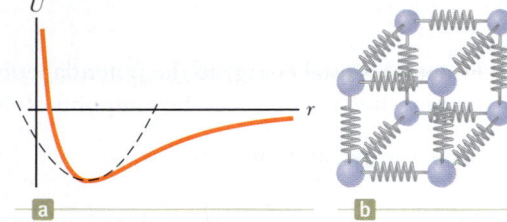

Figure 12.12 (a) If the atoms in a molecule do not move too far from their equilibrium positions, a graph of potential energy versus separation distance between atoms is similar to the graph of potential energy versus position for a simple harmonic oscillator (dashed black curve). (b) The forces between atoms in a solid can be modeled by imagining springs between neighboring atoms.

Example 12.3 | Oscillations on a Horizontal Surface

A 0.500-kg cart connected to a light spring for which the force constant is 20.0 N/m oscillates on a frictionless, horizontal air track.

(A) Calculate the maximum speed of the cart if the amplitude of the motion is 3.00 cm.

SOLUTION

Conceptualize The system oscillates in exactly the same way as the block in Active Figure 12.11, so use that figure in your mental image of the motion.

Categorize The cart is modeled as a particle in simple harmonic motion.

Analyze Use Equation 12.21 to express the total energy of the oscillator system and equate it to the kinetic energy of the system when the cart is at $x = 0$:

$$E = \tfrac{1}{2}kA^2 = \tfrac{1}{2}mv_{\text{max}}^2$$

Solve for the maximum speed and substitute numerical values:

$$v_{\text{max}} = \sqrt{\frac{k}{m}}\, A = \sqrt{\frac{20.0 \text{ N/m}}{0.500 \text{ kg}}}(0.030\,0 \text{ m}) = \boxed{0.190 \text{ m/s}}$$

(B) What is the velocity of the cart when the position is 2.00 cm?

SOLUTION

Use Equation 12.22 to evaluate the velocity:

$$v = \pm\sqrt{\frac{k}{m}(A^2 - x^2)}$$

$$= \pm\sqrt{\frac{20.0 \text{ N/m}}{0.500 \text{ kg}}[(0.030\,0 \text{ m})^2 - (0.020\,0 \text{ m})^2]} = \boxed{\pm 0.141 \text{ m/s}}$$

The positive and negative signs indicate that the cart could be moving to either the right or the left at this instant.

continued

12.3 *cont.*

(C) Compute the kinetic and potential energies of the system when the position of the cart is 2.00 cm.

SOLUTION

Use the result of part (B) to evaluate the kinetic energy at $x = 0.020\ 0$ m:

$$K = \tfrac{1}{2}mv^2 = \tfrac{1}{2}(0.500\ \text{kg})(0.141\ \text{m/s})^2 = \boxed{5.00 \times 10^{-3}\ \text{J}}$$

Evaluate the elastic potential energy at $x = 0.020\ 0$ m:

$$U = \tfrac{1}{2}kx^2 = \tfrac{1}{2}(20.0\ \text{N/m})(0.0200\ \text{m})^2 = \boxed{4.00 \times 10^{-3}\ \text{J}}$$

Finalize The sum of the kinetic and potential energies in part (C) is equal to the total energy, which can be found from Equation 12.21. That must be true for *any* position of the cart.

What If? The cart in this example could have been set into motion by releasing the cart from rest at $x = 3.00$ cm. What if the cart were released from the same position, but with an initial velocity of $v = -0.100$ m/s? What are the new amplitude and maximum speed of the cart?

Answer We can respond to this question by applying an energy approach

First calculate the total energy of the system at $t = 0$:

$$E = \tfrac{1}{2}mv^2 + \tfrac{1}{2}kx^2$$
$$= \tfrac{1}{2}(0.500\ \text{kg})(-0.100\ \text{m/s})^2 + \tfrac{1}{2}(20.0\ \text{N/m})(0.030\ 0\ \text{m})^2$$
$$= 1.15 \times 10^{-2}\ \text{J}$$

Equate this total energy to the potential energy of the system when the cart is at the endpoint of the motion:

$$E = \tfrac{1}{2}kA^2$$

Solve for the amplitude A:

$$A = \sqrt{\frac{2E}{k}} = \sqrt{\frac{2(1.15 \times 10^{-2}\ \text{J})}{20.0\ \text{N/m}}} = 0.033\ 9\ \text{m}$$

Equate the total energy to the kinetic energy of the system when the cart is at the equilibrium position:

$$E = \tfrac{1}{2}mv_{\text{max}}^2$$

Solve for the maximum speed:

$$v_{\text{max}} = \sqrt{\frac{2E}{m}} = \sqrt{\frac{2(1.15 \times 10^{-2}\ \text{J})}{0.500\ \text{kg}}} = 0.214\ \text{m/s}$$

The amplitude and maximum velocity are larger than the previous values because the cart was given an initial velocity at $t = 0$.

▸12.4 | The Simple Pendulum

The **simple pendulum** is another mechanical system that exhibits periodic motion. It consists of a particle-like bob of mass m suspended by a light string of length L that is fixed at the upper end as in Active Figure 12.13. For a real object, as long as the size of the object is small relative to the length of the string, the pendulum can be modeled as a simple pendulum, so we adopt the particle model. When the bob is pulled to the side and released, it oscillates about the lowest point, which is the equilibrium position. The motion occurs in a vertical plane and is driven by the gravitational force.

The forces acting on the bob are the force $\vec{\mathbf{T}}$ acting along the string and the gravitational force $m\vec{\mathbf{g}}$. The component vector of the gravitational force tangent to the curved path of the bob and of magnitude $mg \sin \theta$ always acts toward $\theta = 0$, opposite the displacement of the bob from the lowest position. The gravitational force is therefore a restoring force, and we can use Newton's second law to write the equation of motion in the tangential direction as

$$F_t = ma_t \quad \rightarrow \quad -mg \sin \theta = m\frac{d^2s}{dt^2}$$

where s is the position measured along the circular arc in Active Figure 12.13 and the negative sign indicates that F_t acts toward the equilibrium position. Because $s = L\theta$ (Eq. 10.1a) and L is constant, this equation reduces to

$$\frac{d^2\theta}{dt^2} = -\frac{g}{L}\sin \theta$$

Considering θ as the position, let us compare this equation to Equation 12.5, which is of a similar, but not identical, mathematical form. The right side is proportional to $\sin \theta$ rather than to θ; hence, we conclude that the motion is *not* simple

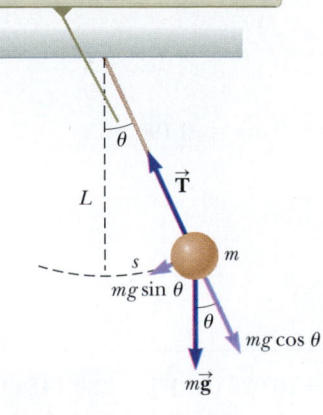

When θ is small, a simple pendulum's motion can be modeled as simple harmonic motion about the equilibrium position $\theta = 0$.

Active Figure 12.13 A simple pendulum.

harmonic motion because the equation describing the motion is not of the form of Equation 12.5. If we assume that θ is *small* (less than about 10° or 0.2 rad), however, we can use a simplification model called the **small angle approximation**, in which $\sin \theta \approx \theta$, where θ is measured in radians. Table 12.1 shows angles, in degrees and radians, and the sines of these angles. As long as θ is less than about 10°, the angle in radians and its sine are the same, at least to within an accuracy of less than 1.0%.

Therefore, for small angles, the equation of motion becomes

$$\frac{d^2\theta}{dt^2} = -\frac{g}{L}\theta$$

12.23◄

Now we have an expression with exactly the same mathematical form as Equation 12.5, with $\omega^2 = g/L$, and so we conclude that the motion is approximately simple harmonic motion for small amplitudes. Modeling the solution after Equation 12.6, θ can therefore be written as $\theta = \theta_{max} \cos(\omega t + \phi)$, where θ_{max} is the *maximum angular position* and the angular frequency ω is

$$\omega = \sqrt{\frac{g}{L}}$$

12.24◄ ► Angular frequency for a simple pendulum

The period of the motion is

$$T = \frac{2\pi}{\omega} = 2\pi \sqrt{\frac{L}{g}}$$

12.25◄ ► Period for a simple pendulum

We see that the period and frequency of a simple pendulum oscillating at small angles depend only on the length of the string and the acceleration due to gravity. Because the period is *independent* of the mass, we conclude that *all* simple pendula that are of equal length and are at the same location (so that g is constant) oscillate with the same period. Experiments show that this conclusion is correct.

Notice the importance of our modeling technique in this discussion. Equation 12.23 is a mathematical representation of the simple pendulum. This representation has exactly the same *mathematical* form as Equation 12.5 for the block on a spring, despite the fact that there are clear *physical* differences between the two systems. Despite the physical differences, because the mathematical representations are the same, we can immediately write the solution of the angular position θ for the pendulum and identify its angular frequency ω as in Equation 12.24. This is a very powerful technique, made possible by the fact that we are forming a mathematical model of the physical system.

TABLE 12.1 | Angles and Sines of Angles

Angle in Degrees	Angle in Radians	Sine of Angle	Percent Difference
0°	0.000 0	0.000 0	0.0%
1°	0.017 5	0.017 5	0.0%
2°	0.034 9	0.034 9	0.0%
3°	0.052 4	0.052 3	0.0%
5°	0.087 3	0.087 2	0.1%
10°	0.174 5	0.173 6	0.5%
15°	0.261 8	0.258 8	1.2%
20°	0.349 1	0.342 0	2.1%
30°	0.523 6	0.500 0	4.7%

Pitfall Prevention | 12.4
Not True Simple Harmonic Motion
The pendulum *does not* exhibit true simple harmonic motion for *any* angle. If the angle is less than about 10°, the motion is close to and can be *modeled* as simple harmonic.

QUICK QUIZ 12.5 A grandfather clock depends on the period of a pendulum to keep correct time. **(i)** Suppose a grandfather clock is calibrated correctly and then a mischievous child slides the bob of the pendulum downward on the oscillating rod. Does the grandfather clock run **(a)** slow, **(b)** fast, or **(c)** correctly? **(ii)** Suppose the grandfather clock is calibrated correctly at sea level and is then taken to the top of a very tall mountain. Does the grandfather clock run **(a)** slow, **(b)** fast, or **(c)** correctly?

▶ **THINKING PHYSICS 12.3**

You set up two oscillating systems: a simple pendulum and a block hanging from a vertical spring. You carefully adjust the length of the pendulum so that both oscillators have the same period. You now take the two oscillators to the Moon. Will they still have the same period as each other? What happens if you observe the two oscillators in an orbiting spacecraft? (Assume that the spring is one with open space between the coils when it is unstretched, so the spring can be both stretched and compressed.)

Reasoning The block hanging from the spring will have the same period on the Moon that it had on the Earth because the period depends on the mass of the block and the force constant of the spring, neither of which

have changed. The pendulum's period on the Moon will be different from its period on the Earth because the period of the pendulum depends on the value of g. Because g is smaller on the Moon than on the Earth, the pendulum will oscillate with a longer period.

In the orbiting spacecraft, the block–spring system will oscillate with the same period as on the Earth when it is set into motion because the period does not depend on gravity. The pendulum will not oscillate at all; if you pull it to the side from a direction you define as "vertical" and release it, it stays there. Because the spacecraft is in free-fall while in orbit around the Earth, the effective gravity is zero and there is no restoring force on the pendulum. ◄

Example 12.4 | A Connection Between Length and Time

Christian Huygens (1629–1695), the greatest clockmaker in history, suggested that an international unit of length could be defined as the length of a simple pendulum having a period of exactly 1 s. How much shorter would our length unit be if his suggestion had been followed?

SOLUTION

Conceptualize Imagine a pendulum that swings back and forth in exactly 1 second. Based on your experience in observing swinging objects, can you make an estimate of the required length? Hang a small object from a string and simulate the 1-s pendulum.

Categorize This example involves a simple pendulum, so we categorize it as an application of the concepts introduced in this section.

Analyze Solve Equation 12.25 for the length and substitute the known values:

$$L = \frac{T^2 g}{4\pi^2} = \frac{(1.00 \text{ s})^2 (9.80 \text{ m/s}^2)}{4\pi^2} = \boxed{0.248 \text{ m}}$$

Finalize The meter's length would be slightly less than one-fourth of its current length. Also, the number of significant digits depends only on how precisely we know g because the time has been defined to be exactly 1 s.

❰12.5 | The Physical Pendulum

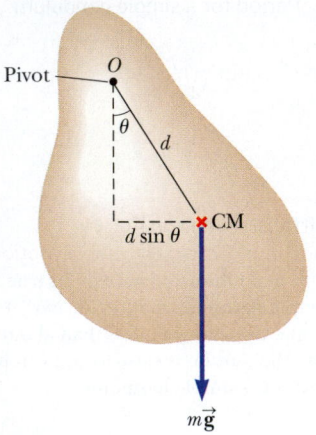

Figure 12.14 A physical pendulum pivoted at *O*.

Suppose you balance a wire coat hanger so that the hook is supported by your extended index finger. When you give the hanger a small angular displacement with your other hand and then release it, it oscillates. If a hanging object oscillates about a fixed axis that does not pass through its center of mass and the object cannot be approximated as a point mass, we cannot treat the system as a simple pendulum. In this case, the system is called a **physical pendulum.**

Consider a rigid object pivoted at a point *O* that is a distance *d* from the center of mass (Fig. 12.14). The gravitational force provides a torque about an axis through *O*, and the magnitude of that torque is $mgd \sin\theta$, where θ is as shown in Figure 12.14. We model the object as a rigid object under a net torque and use the rotational form of Newton's second law, $\Sigma\tau_{ext} = I\alpha$, where I is the moment of inertia of the object about the axis through *O*. The result is

$$-mgd \sin\theta = I \frac{d^2\theta}{dt^2}$$

The negative sign indicates that the torque about *O* tends to decrease θ. That is, the gravitational force produces a restoring torque. If we again assume θ is small, the approximation $\sin\theta \approx \theta$ is valid and the equation of motion reduces to

$$\frac{d^2\theta}{dt^2} = -\left(\frac{mgd}{I}\right)\theta = -\omega^2\theta \qquad \textbf{12.26} \blacktriangleleft$$

Because this equation is of the same mathematical form as Equation 12.5, its solution is that of the simple harmonic oscillator. That is, the solution of Equation 12.26 is given by $\theta = \theta_{max} \cos(\omega t + \phi)$, where θ_{max} is the maximum angular position and

$$\omega = \sqrt{\frac{mgd}{I}}$$

The period is

▶ Period of a physical pendulum

$$T = \frac{2\pi}{\omega} = 2\pi\sqrt{\frac{I}{mgd}} \qquad \textbf{12.27} \blacktriangleleft$$

This result can be used to measure the moment of inertia of a flat, rigid object. If the location of the center of mass—and hence the value of *d*—is known, the moment of inertia can be obtained by measuring the period. Finally, notice that Equation 12.27 reduces to the period of a simple pendulum (Eq. 12.25) when $I = md^2$, that is, when all the mass is concentrated at the center of mass.

Notice again the importance of modeling here, as discussed for the simple pendulum. Because the mathematical representation in Equation 12.26 is identical in form to that in Equation 12.5, we were able to immediately write the solution for the physical pendulum.

QUICK QUIZ 12.6 Two students, Alex and Brian, are in a museum watching the swinging of a pendulum with a large bob. Alex says, "I'm going to sneak past the fence and stick some chewing gum on the top of the pendulum bob to change its period of oscillation." Brian says, "That won't change the period. The period of a pendulum is independent of mass." Which student is correct?   (**a**) Alex   (**b**) Brian

---

Example 12.5 | A Swinging Rod

A uniform rod of mass M and length L is pivoted about one end and oscillates in a vertical plane (Fig. 12.15). Find the period of oscillation if the amplitude of the motion is small.

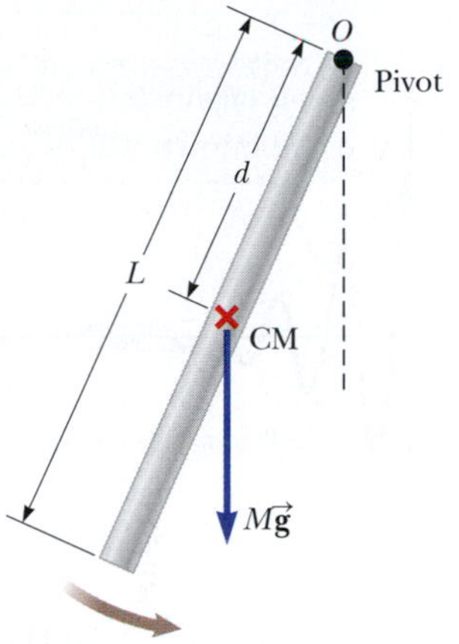

SOLUTION

Conceptualize Imagine a rod swinging back and forth when pivoted at one end. Try it with a meterstick or a scrap piece of wood.

Categorize Because the rod is not a point particle, we categorize it as a physical pendulum.

Figure 12.15 (Example 12.5) A rigid rod oscillating about a pivot through one end is a physical pendulum with $d = L/2$.

Analyze In Chapter 10, we found that the moment of inertia of a uniform rod about an axis through one end is $\frac{1}{3}ML^2$. The distance d from the pivot to the center of mass of the rod is $L/2$.

Substitute these quantities into Equation 12.27:

$$T = 2\pi\sqrt{\frac{\frac{1}{3}ML^2}{Mg(L/2)}} = 2\pi\sqrt{\frac{2L}{3g}}$$

Finalize In one of the Moon landings, an astronaut walking on the Moon's surface had a belt hanging from his space suit, and the belt oscillated as a physical pendulum. A scientist on the Earth observed this motion on television and used it to estimate the free-fall acceleration on the Moon. How did the scientist make this calculation?

12.6 | Damped Oscillations

The oscillatory motions we have considered so far have been for ideal systems, that is, systems that oscillate indefinitely under the action of only one force, a linear restoring force. In many real systems, nonconservative forces such as friction or air resistance retard the motion. Consequently, the mechanical energy of the system diminishes in time, and the motion is described as a **damped oscillation.**

Consider an object moving through a medium such as a liquid or a gas. One common type of retarding force on the object, which we discussed in Chapter 5, is proportional to the velocity of the object and acts in the direction opposite that of the object's velocity relative to the medium. This type of force is often observed when an object is oscillating slowly in air, for instance. Because the resistive force can be expressed as $\vec{R} = -b\vec{v}$, where b is a constant related to the strength of the resistive force, and the restoring force exerted on the system is $-kx$, Newton's second law gives us

$$\sum F_x = -kx - bv = ma_x$$

$$-kx - b\frac{dx}{dt} = m\frac{d^2x}{dt^2} \qquad \text{12.28} \blacktriangleleft$$

The solution of this differential equation requires mathematics that may not yet be familiar to you, so it will simply be stated without proof. When the parameters of the system are such that $b < \sqrt{4mk}$ so that the resistive force is small, the solution to Equation 12.28 is

$$x = (Ae^{-(b/2m)t})\cos(\omega t + \phi) \qquad \text{12.29} \blacktriangleleft$$

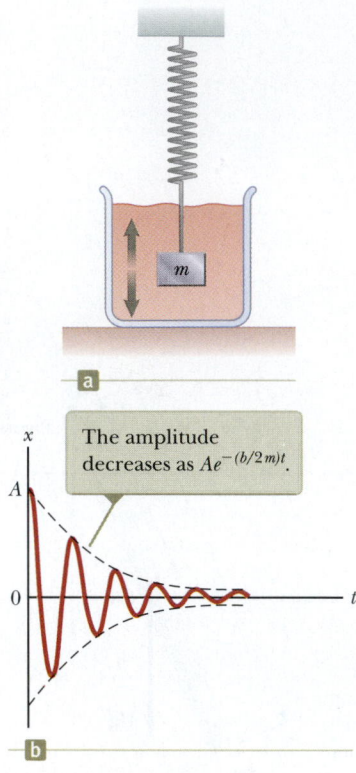

Active Figure 12.16 (a) One example of a damped oscillator is an object attached to a spring and submersed in a viscous liquid. (b) Graph of position versus time for a damped oscillator.

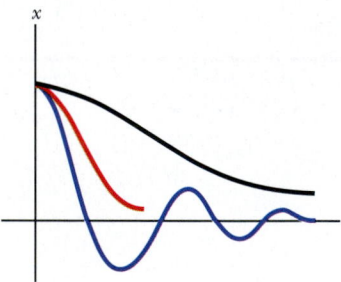

Figure 12.17 Graphs of position versus time for an underdamped oscillator (blue curve), a critically damped oscillator (red curve), and an overdamped oscillator (black curve).

where the angular frequency of the oscillation is

$$\omega = \sqrt{\frac{k}{m} - \left(\frac{b}{2m}\right)^2}$$ **12.30**◀

This result can be verified by substituting Equation 12.29 into Equation 12.28. It is convenient to express the angular frequency of a damped oscillator in the form

$$\omega = \sqrt{\omega_0^2 - \left(\frac{b}{2m}\right)^2}$$

where $\omega_0 = \sqrt{k/m}$ represents the angular frequency in the absence of a retarding force (the undamped oscillator) and is called the **natural frequency**[3] of the system.

In Active Figure 12.16a, we see one example of a damped system. The object suspended from the spring experiences both a force from the spring and a resistive force from the surrounding liquid. Active Figure 12.16b shows the position as a function of time for an object oscillating in the presence of a retarding force. When the retarding force is small, the oscillatory character of the motion is preserved but the amplitude decreases exponentially in time, with the result that the motion ultimately becomes undetectable. Any system that behaves in this way is known as a **damped oscillator.** The dashed black lines in Active Figure 12.16b, which define the *envelope* of the oscillatory curve, represent the exponential factor in Equation 12.29. This envelope shows that the amplitude decays exponentially with time. For motion with a given spring constant and object mass, the oscillations dampen more rapidly for larger values of the retarding force.

When the magnitude of the retarding force is small such that $b/2m < \omega_0$, the system is said to be **underdamped.** The resulting motion is represented by the blue curve in Figure 12.17. As the value of b increases, the amplitude of the oscillations decreases more and more rapidly. When b reaches a critical value b_c such that $b_c/2m = \omega_0$, the system does not oscillate and is said to be **critically damped.** In this case, the system, once released from rest at some nonequilibrium position, approaches but does not pass through the equilibrium position. The graph of position versus time for this case is the red curve in Figure 12.17.

If the medium is so viscous that the retarding force is large compared with the restoring force—that is, if $b/2m > \omega_0$—the system is **overdamped.** Again, the displaced system, when free to move, does not oscillate but rather simply returns to its equilibrium position. As the damping increases, the time interval required for the system to approach equilibrium also increases as indicated by the black curve in Figure 12.17. For critically damped and overdamped systems, there is no angular frequency ω and the solution in Equation 12.29 is not valid.

▌12.7 | Forced Oscillations

We have seen that the mechanical energy of a damped oscillator decreases in time as a result of the resistive force. It is possible to compensate for this energy decrease by applying an external force that does positive work on the system. Such an oscillator then undergoes **forced oscillations.** At any instant, energy can be transferred into the system by an applied force that acts in the direction of motion of the oscillator. For example, a child on a swing can be kept in motion by appropriately timed "pushes." The amplitude of motion remains constant if the energy input per cycle of motion exactly equals the decrease in mechanical energy in each cycle that results from resistive forces.

A common example of a forced oscillator is a damped oscillator driven by an external force $F(t)$ that varies periodically, such as $F(t) = F_0 \sin \omega t$, where ω is the angular frequency of the driving force and F_0 is a constant. In general, the frequency ω of the driving force is different from the natural frequency ω_0 of the oscillator. Newton's second law in this situation gives

$$\sum F_x = ma_x \quad \rightarrow \quad F_0 \sin \omega t - b\frac{dx}{dt} - kx = m\frac{d^2x}{dt^2}$$ **12.31**◀

---

[3]In practice, both ω_0 and $f_0 = \omega_0/2\pi$ are described as the natural frequency. The context of the discussion will help you determine which frequency is being discussed.

Again, the solution of this equation is rather lengthy and will not be presented. After the driving force on an initially stationary object begins to act, the amplitude of the oscillation will increase. After a sufficiently long time interval, when the energy input per cycle from the driving force equals the amount of mechanical energy transformed to internal energy for each cycle, a steady-state condition is reached in which the oscillations proceed with constant amplitude. In this case, Equation 12.31 has the solution

$$x = A \cos(\omega t + \phi) \qquad \textbf{12.32} \blacktriangleleft$$

where

$$A = \frac{F_0/m}{\sqrt{\left(\omega^2 - \omega_0^2\right)^2 + \left(\dfrac{b\omega}{m}\right)^2}} \qquad \textbf{12.33} \blacktriangleleft$$

and where $\omega_0 = \sqrt{k/m}$ is the natural frequency of the undamped oscillator ($b = 0$).

Equation 12.33 shows that the amplitude of the forced oscillator is constant for a given driving force because it is being driven in steady state by an external force. For small damping the amplitude becomes large when the frequency of the driving force is near the natural frequency of oscillation, or when $\omega \approx \omega_0$ as can be seen in Equation 12.33. The dramatic increase in amplitude near the natural frequency is called **resonance**, and the natural frequency ω_0 is called the **resonance frequency** of the system.

Figure 12.18 is a graph of amplitude as a function of frequency for the forced oscillator, with varying resistive forces. Note that the amplitude increases with decreasing damping ($b \rightarrow 0$) and that the resonance curve flattens as the damping increases. In the absence of a damping force ($b = 0$), we see from Equation 12.33 that the steady-state amplitude approaches infinity as $\omega \rightarrow \omega_0$. In other words, if there are no resistive forces in the system and we continue to drive an oscillator with a sinusoidal force at the resonance frequency, the amplitude of motion will build up without limit. This situation does not occur in practice because some damping is always present in real oscillators.

Resonance appears in many areas of physics. For example, certain electric circuits have resonance frequencies. This fact is exploited in radio tuners, which allow you to select the station you wish to hear. Vibrating strings and columns of air also have resonance frequencies, which allow them to be used for musical instruments, which we shall discuss in Chapter 14.

◀ **12.8** | Context Connection: Resonance in Structures

In the preceding section, we investigated the phenomenon of resonance in which an oscillating system exhibits its maximum response to a periodic driving force when the frequency of the driving force matches the oscillator's natural frequency. We now apply this understanding to the interaction between the shaking of the ground during an earthquake and structures attached to the ground. The structure is the oscillator. It has a set of natural frequencies, determined by its stiffness, its mass, and the details of its construction. The periodic driving force is supplied by the shaking of the ground.

A disastrous result can occur if a natural frequency of the building matches a frequency contained in the ground shaking. In this case, the resonance vibrations of the building can build to a very large amplitude, large enough to damage or destroy the building. This result can be avoided in two ways. The first involves designing the structure so that natural frequencies of the building lie outside the range of earthquake frequencies. (A typical range of earthquake frequencies is 0–15 Hz.) Such a building can be designed by varying its size or mass structure. The second method involves incorporating sufficient damping in the building. This method may not change the resonance frequency significantly, but it will lower the response to the natural frequency as in Figure 12.18. It will also flatten the resonance curve, so the building will respond to a wide range of frequencies but with relatively small amplitude at any given frequency.

We now describe two examples involving resonance excitations in bridge structures. One example of such a structural resonance occurred in 1940, when the Tacoma Narrows Bridge in Washington State was destroyed by resonant vibrations (Fig. 12.19). The winds were not particularly strong on that occasion, but the bridge

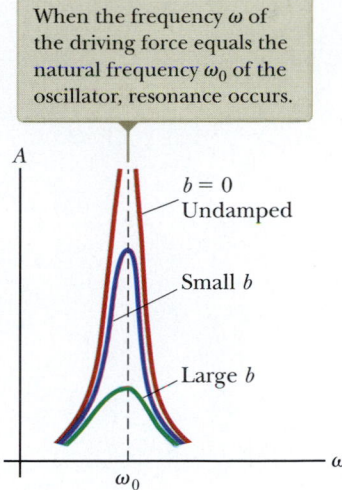

When the frequency ω of the driving force equals the natural frequency ω_0 of the oscillator, resonance occurs.

$b = 0$
Undamped

Small b

Large b

Figure 12.18 Graph of amplitude versus frequency for a damped oscillator when a periodic driving force is present. Notice that the shape of the resonance curve depends on the size of the damping coefficient b.

Figure 12.19 (a) In 1940, turbulent winds set up torsional vibrations in the Tacoma Narrows Bridge, causing it to oscillate at a frequency near one of the natural frequencies of the bridge structure. (b) Once established, this resonance condition led to the bridge's collapse. (Mathematicians and physicists are currently challenging this interpretation.)

still collapsed because vortices (turbulences) generated by the wind blowing through the bridge occurred at a frequency that matched a natural frequency of the bridge. The flapping of this wind across the roadway (similar to the flapping of a flag in a strong breeze) provided the periodic driving force that brought the bridge down into the river.

As a second example, consider soldiers marching across a bridge. They are commanded to break step while on the bridge because of resonance. If the marching frequency of the soldiers matches that of the bridge, the bridge could be set into resonance oscillation. If the amplitude becomes large enough, the bridge could actually collapse. Just such a situation occurred on April 14, 1831, when the Broughton suspension bridge in England collapsed while troops marched over it. Investigations after the accident showed that the bridge was near failure, and the resonance vibration induced by the marching soldiers caused it to fail sooner than it otherwise might have.

Resonance gives us our first clue to responding to the central question for this Context. Suppose a building is far from the epicenter of an earthquake so that the ground shaking is small. If the shaking frequency matches a natural frequency of the building, a very effective energy coupling occurs between the ground and the building. Therefore, even for relatively small shaking, the ground, by resonance, can feed energy into the building efficiently enough to cause the failure of the structure. The structure must be carefully designed so as to reduce the resonance response.

❯ SUMMARY

An object attached to the end of a spring moves with a motion called **simple harmonic motion,** and the system is called a **simple harmonic oscillator.**

The time for one complete oscillation of the object is called the **period** T of the motion. The inverse of the period is the **frequency** f of the motion, which equals the number of oscillations per second:

$$f = \frac{1}{T} = \frac{\omega}{2\pi} \qquad \text{12.11} ◀$$

The velocity and acceleration of an object in simple harmonic oscillation are

$$v = \frac{dx}{dt} = -\omega A \sin(\omega t + \phi) \qquad \text{12.15} ◀$$

$$a = \frac{d^2x}{dt^2} = -\omega^2 A \cos(\omega t + \phi) \qquad \text{12.16} ◀$$

Therefore, the maximum speed of the object is ωA and its maximum acceleration is of magnitude $\omega^2 A$. The speed is zero when the object is at its turning points, $x = \pm A$, and the speed is a maximum at the equilibrium position, $x = 0$. The magnitude of the acceleration is a maximum at the turning points and is zero at the equilibrium position.

The kinetic energy and potential energy of a simple harmonic oscillator vary with time and are given by

$$K = \tfrac{1}{2}mv^2 = \tfrac{1}{2}m\omega^2 A^2 \sin^2(\omega t + \phi) \qquad \text{12.19} ◀$$

$$U = \tfrac{1}{2}kx^2 = \tfrac{1}{2}kA^2 \cos^2(\omega t + \phi) \qquad \text{12.20} ◀$$

The **total energy** of a simple harmonic oscillator is a constant of the motion and is

$$E = \tfrac{1}{2}kA^2 \qquad \text{12.21} ◀$$

The potential energy of a simple harmonic oscillator is a maximum when the particle is at its turning points (maximum displacement from equilibrium) and is zero at the equilibrium position. The kinetic energy is zero at the turning points and is a maximum at the equilibrium position.

A **simple pendulum** of length L exhibits motion that can be modeled as simple harmonic for small angular displacements from the vertical, with a period of

$$T = 2\pi\sqrt{\frac{L}{g}} \qquad \text{12.25} ◀$$

The period of a simple pendulum is independent of the mass of the suspended object.

A **physical pendulum** exhibits motion that can be modeled as simple harmonic for small angular displacements from equilibrium about a pivot that does not go through the center of mass. The period of this motion is

$$T = 2\pi\sqrt{\frac{I}{mgd}} \qquad \text{12.27} ◀$$

where I is the moment of inertia about an axis through the pivot and d is the distance from the pivot to the center of mass.

Damped oscillations occur in a system in which a resistive force opposes the motion of the oscillating object. If such a system is set in motion and then left to itself, its mechanical energy decreases in time because of the presence of the nonconservative resistive force. It is possible to compensate for this transformation of energy by driving the system with an external periodic force. The oscillator in this case is undergoing **forced oscillations.** When the frequency of the driving force matches the natural frequency of the *undamped* oscillator, energy is efficiently transferred to the oscillator and its steady-state amplitude is a maximum. This situation is called **resonance.**

Analysis Model for Problem Solving

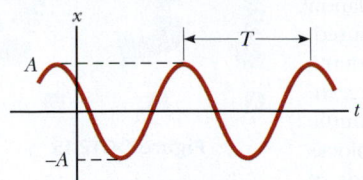

Particle in Simple Harmonic Motion If a particle is subject to a force of the form of Hooke's law $F = -kx$, the particle exhibits **simple harmonic motion.** Its position is described by

$$x(t) = A \cos (\omega t + \phi)$$ 12.6◀

where A is the **amplitude** of the motion, ω is the **angular frequency,** and ϕ is the **phase constant.** The value of ϕ depends on the initial position and initial velocity of the oscillator.

The period of the oscillation is related to the parameters of a block–spring system according to

$$T = \frac{2\pi}{\omega} = 2\pi\sqrt{\frac{m}{k}}$$ 12.13◀

OBJECTIVE QUESTIONS

□ denotes answer available in *Student Solutions Manual/Study Guide*

1. Which of the following statements is *not* true regarding a mass–spring system that moves with simple harmonic motion in the absence of friction? (a) The total energy of the system remains constant. (b) The energy of the system is continually transformed between kinetic and potential energy. (c) The total energy of the system is proportional to the square of the amplitude. (d) The potential energy stored in the system is greatest when the mass passes through the equilibrium position. (e) The velocity of the oscillating mass has its maximum value when the mass passes through the equilibrium position.

2. The position of an object moving with simple harmonic motion is given by $x = 4 \cos (6\pi t)$, where x is in meters and t is in seconds. What is the period of the oscillating system? (a) 4 s (b) $\frac{1}{6}$ s (c) $\frac{1}{3}$ s (d) 6π s (e) impossible to determine from the information given

3. A block–spring system vibrating on a frictionless, horizontal surface with an amplitude of 6.0 cm has an energy of 12 J. If the block is replaced by one whose mass is twice the mass of the original block and the amplitude of the motion is again 6.0 cm, what is the energy of the system? (a) 12 J (b) 24 J (c) 6 J (d) 48 J (e) none of those answers

4. A runaway railroad car, with mass 3.0×10^5 kg, coasts across a level track at 2.0 m/s when it collides elastically with a spring-loaded bumper at the end of the track. If the spring constant of the bumper is 2.0×10^6 N/m, what is the maximum compression of the spring during the collision? (a) 0.77 m (b) 0.58 m (c) 0.34 m (d) 1.07 m (e) 1.24 m

5. An object of mass 0.40 kg, hanging from a spring with a spring constant of 8.0 N/m, is set into an up-and-down simple harmonic motion. What is the magnitude of the acceleration of the object when it is at its maximum displacement of 0.10 m? (a) zero (b) 0.45 m/s^2 (c) 1.0 m/s^2 (d) 2.0 m/s^2 (e) 2.4 m/s^2

6. If an object of mass m attached to a light spring is replaced by one of mass $9m$, the frequency of the vibrating system changes by what factor? (a) $\frac{1}{9}$ (b) $\frac{1}{3}$ (c) 3.0 (d) 9.0 (e) 6.0

7. If a simple pendulum oscillates with small amplitude and its length is doubled, what happens to the frequency of its motion? (a) It doubles. (b) It becomes $\sqrt{2}$ times as large. (c) It becomes half as large. (d) It becomes $1/\sqrt{2}$ times as large. (e) It remains the same.

8. An object–spring system moving with simple harmonic motion has an amplitude A. When the kinetic energy of the object equals twice the potential energy stored in the spring, what is the position x of the object? (a) A (b) $\frac{1}{3}A$ (c) $A/\sqrt{3}$ (d) 0 (e) none of those answers

9. A mass–spring system moves with simple harmonic motion along the x axis between turning points at $x_1 = 20$ cm and $x_2 = 60$ cm. For parts (i) through (iii), choose from the same five possibilities. (i) At which position does the particle have the greatest magnitude of momentum? (a) 20 cm (b) 30 cm (c) 40 cm (d) some other position (e) The greatest value occurs at multiple points. (ii) At which position does the particle have greatest kinetic energy? (iii) At which position does the particle–spring system have the greatest total energy?

10. A particle on a spring moves in simple harmonic motion along the x axis between turning points at $x_1 = 100$ cm and $x_2 = 140$ cm. (i) At which of the following positions does the particle have maximum speed? (a) 100 cm (b) 110 cm (c) 120 cm (d) at none of those positions (ii) At which position does it have maximum acceleration? Choose from the same possibilities as in part (i). (iii) At which position is the greatest net force exerted on the particle? Choose from the same possibilities as in part (i).

11. A block with mass $m = 0.1$ kg oscillates with amplitude $A = 0.1$ m at the end of a spring with force constant $k = 10$ N/m on a frictionless, horizontal surface. Rank the periods of the following situations from greatest to smallest. If any periods are equal, show their equality in your ranking. (a) The system is as described above. (b) The system is as described in situation (a) except the amplitude is 0.2 m. (c) The situation is as described in situation (a) except the mass is 0.2 kg. (d) The situation is as described in situation (a) except the spring has force constant 20 N/m. (e) A small resistive force makes the motion underdamped.

12. For a simple harmonic oscillator, answer yes or no to the following questions. (a) Can the quantities position and velocity have the same sign? (b) Can velocity and acceleration have the same sign? (c) Can position and acceleration have the same sign?

13. A simple pendulum has a period of 2.5 s. (i) What is its period if its length is made four times larger? (a) 1.25 s (b) 1.77 s (c) 2.5 s (d) 3.54 s (e) 5 s (ii) What is its period if the length is held constant at its initial value and the mass of the suspended bob is made four times larger? Choose from the same possibilities.

14. You attach a block to the bottom end of a spring hanging vertically. You slowly let the block move down and find that it hangs at rest with the spring stretched by 15.0 cm. Next, you lift the block back up to the initial position and release it from rest with the spring unstretched. What maximum distance does it move down? (a) 7.5 cm (b) 15.0 cm (c) 30.0 cm (d) 60.0 cm (e) The distance cannot be determined without knowing the mass and spring constant.

15. The top end of a spring is held fixed. A block is hung on the bottom end as in Figure OQ12.15a, and the frequency f of the oscillation of the system is measured. The block, a second identical block, and the spring are carried up in a space shuttle to Earth orbit. The two blocks are attached to the ends of the spring. The spring is compressed without making adjacent coils touch (Fig. OQ12.15b), and the system is released to oscillate while floating within the shuttle cabin (Fig. OQ12.15c). What is the frequency of oscillation for this system in terms of f? (a) $f/2$ (b) $f/\sqrt{2}$ (c) f (d) $\sqrt{2}f$ (e) $2f$

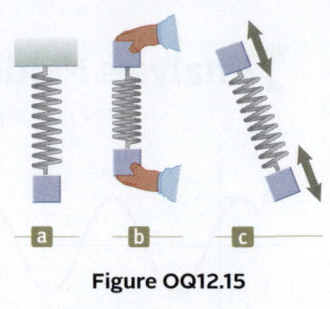

Figure OQ12.15

CONCEPTUAL QUESTIONS

☐ denotes answer available in *Student Solutions Manual/Study Guide*

1. (a) If the coordinate of a particle varies as $x = -A \cos \omega t$, what is the phase constant in Equation 12.6? (b) At what position is the particle at $t = 0$?

2. The equations listed in Table 2.2 give position as a function of time, velocity as a function of time, and velocity as a function of position for an object moving in a straight line with constant acceleration. The quantity v_{xi} appears in every equation. (a) Do any of these equations apply to an object moving in a straight line with simple harmonic motion? (b) Using a similar format, make a table of equations describing simple harmonic motion. Include equations giving acceleration as a function of time and acceleration as a function of position. State the equations in such a form that they apply equally to a block–spring system, to a pendulum, and to other vibrating systems. (c) What quantity appears in every equation?

3. Is a bouncing ball an example of simple harmonic motion? Is the daily movement of a student from home to school and back simple harmonic motion? Why or why not?

4. Figure CQ12.4 shows graphs of the potential energy of four different systems versus the position of a particle in each system. Each particle is set into motion with a push at an arbitrarily chosen location. Describe its subsequent motion in each case (a), (b), (c), and (d).

6. Is it possible to have damped oscillations when a system is at resonance? Explain.

7. The mechanical energy of an undamped block–spring system is constant as kinetic energy transforms to elastic potential energy and vice versa. For comparison, explain what happens to the energy of a damped oscillator in terms of the mechanical, potential, and kinetic energies.

8. A student thinks that any real vibration must be damped. Is the student correct? If so, give convincing reasoning. If not, give an example of a real vibration that keeps constant amplitude forever if the system is isolated.

9. Will damped oscillations occur for any values of b and k? Explain.

10. If a pendulum clock keeps perfect time at the base of a mountain, will it also keep perfect time when it is moved to the top of the mountain? Explain.

11. A pendulum bob is made from a sphere filled with water. What would happen to the frequency of vibration of this pendulum if there were a hole in the sphere that allowed the water to leak out slowly?

12. You are looking at a small, leafy tree. You do not notice any breeze, and most of the leaves on the tree are motionless. One leaf, however, is fluttering back and forth wildly. After a while, that leaf stops moving and you notice a different leaf moving much more than all the others. Explain what could cause the large motion of one particular leaf.

13. Consider the simplified single-piston engine in Figure CQ12.13. Assuming the wheel rotates with constant angular speed, explain why the piston rod oscillates in simple harmonic motion.

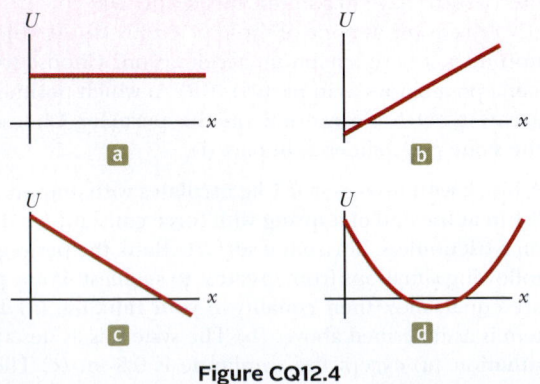

Figure CQ12.4

5. A simple pendulum can be modeled as exhibiting simple harmonic motion when θ is small. Is the motion periodic when θ is large?

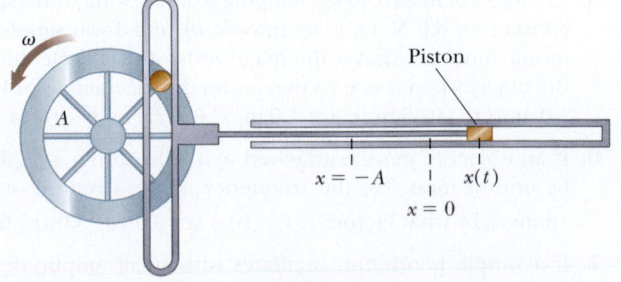

Figure CQ12.13

PROBLEMS AVAILABLE IN  WebAssign

12.1 Motion of an Object Attached to a Spring

Problems 1–2

12.2 Analysis Model: Particle in Simple Harmonic Motion

Problems 3–15

12.3 Energy of the Simple Harmonic Oscillator

Problems 16–25

12.4 The Simple Pendulum
12.5 The Physical Pendulum

Problems 26–35

12.6 Damped Oscillations

Problems 36–38

12.7 Forced Oscillations

Problems 39–43

12.8 Context Connection: Resonance in Structures

Problems 44–45

Additional Problems

Problems 46–69

**Solutions to the following Problems are available in the
Student Solutions Manual/Study Guide:**

12.3, 12.5, 12.9, 12.13, 12.27, 12.32, 12.35, 12.43, 12.47,
12.55, 12.59, 12.62, 12.64, and 12.65

List of Enhanced Problems

Problem Number	Targeted Feedback in Enhanced WebAssign	Master It in Enhanced WebAssign	Watch It in Enhanced WebAssign
12.3	✓	✓	
12.5	✓		✓
12.10	✓		✓
12.12	✓		✓
12.13	✓	✓	
12.14	✓		✓
12.19		✓	
12.21	✓		✓
12.27	✓	✓	
12.33	✓		✓
12.37	✓		✓
12.43		✓	
12.55	✓		✓
12.59		✓	

Chapter | **13**

Mechanical Waves

Chapter Outline

Stefano Cellai/AGE fotostock

Three musicians play the alpenhorn in Valais, Switzerland. In this chapter, we explore the behavior of sound waves such as those coming from these large musical instruments.

Most of us experienced waves as children when we dropped pebbles into a pond. The disturbance created by a pebble manifests itself as ripples that move outward from the point at which the pebble lands in the water. If you were to carefully examine the motion of a leaf floating near the point where the pebble enters the water, you would see that the leaf moves up and down and back and forth about its original position but does not undergo any net displacement away from or toward the source of the disturbance. The *disturbance* in the water moves over a long distance, but a given small *element of the water* oscillates only over a very small distance. This behavior is the essence of wave motion.

The world is full of other kinds of waves, including sound waves, waves on strings, seismic waves, radio waves, and x-rays. Most waves can be placed in one of two categories. **Mechanical waves** are waves that disturb and propagate through a medium; the ripple in the water because of the pebble and a sound wave, for which air is the medium, are examples of mechanical waves. The opening photograph shows an example of one possible source of sound waves in air: blowing on very large pipes of different dimensions. **Electromagnetic waves** are a special class of waves that do not require a medium to propagate, as discussed with regard to the absence of the ether in Section 9.2; light waves and radio waves are two familiar examples. In this chapter, we shall confine our attention to the study of mechanical waves, deferring our study of electromagnetic waves to Chapter 24.

13.1 | Propagation of a Disturbance

In the introduction, we alluded to the essence of wave motion: the transfer of a *disturbance* through space without the accompanying transfer of *matter*. The propagation of the disturbance also represents a transfer of energy; therefore, we can view waves as a means of energy transfer. In the list of energy transfer mechanisms in Section 7.1, we see two entries that depend on waves: mechanical waves and electromagnetic radiation. These entries are to be contrasted with another entry—matter transfer—in which the energy transfer is accompanied by a movement of matter through space.

All waves carry energy, but the amount of energy transmitted through a medium and the mechanism responsible for the energy transport differ from case to case. For instance, the power of ocean waves during a storm is much greater than that of sound waves generated by a musical instrument.

All mechanical waves require (1) some source of disturbance, (2) a medium that can be disturbed, and (3) some physical mechanism through which elements of the medium can influence one another. This final requirement assures that a disturbance to one element will cause a disturbance to the next so that the disturbance will indeed propagate through the medium.

One way to demonstrate wave motion is to flip the free end of a long rope that is under tension and has its opposite end fixed as in Figure 13.1. In this manner, a single **pulse** is formed and travels (to the right in Fig. 13.1) with a definite speed. The rope is the medium through which the pulse travels. Figure 13.1 represents consecutive "snapshots" of the traveling pulse. The shape of the pulse changes very little as it travels along the rope.

As the pulse travels, each rope element that is disturbed moves in a direction perpendicular to the direction of propagation. Figure 13.2 illustrates this point for a particular element, labeled *P*. Note that there is no motion of any part of the rope that is in the direction of the wave. A disturbance such as this one in which the elements of the disturbed medium move perpendicularly to the direction of propagation is called a **transverse wave.**

In another class of mechanical waves, called **longitudinal waves,** the elements of the medium undergo displacements *parallel* to the direction of propagation. Sound waves in air, for instance, are longitudinal. Their disturbance corresponds to a series of high- and low-pressure regions that may travel through air or through any material medium with a certain speed. A longitudinal pulse can be easily produced in a stretched spring as in Figure 13.3. A group of coils at the free end is pushed forward and pulled back. This action produces a pulse in the form of a compressed region of coils that travels along the spring.

So far, we have provided pictorial representations of a traveling pulse and hope you have begun to develop a mental representation of such a pulse. Let us now develop a mathematical representation for the propagation of this pulse. Consider a pulse traveling to the right with constant speed v on a long, stretched string as in Figure 13.4. The pulse moves along the x axis (the axis of the string), and the transverse (up-and-down) displacement of the elements of the string is described by means of the position y.

Figure 13.4a represents the shape and position of the pulse at time $t = 0$. At this time, the shape of the pulse, whatever it may be, can be represented by some mathematical function that we will write as $y(x, 0) = f(x)$. This function describes the vertical position y of the element of the string located at each value of x at time

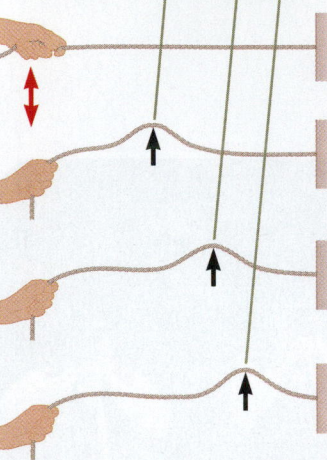

As the pulse moves along the string, new elements of the string are displaced from their equilibrium positions.

Figure 13.1 A hand moves the end of a stretched string up and down once (red arrow), causing a pulse to travel along the string.

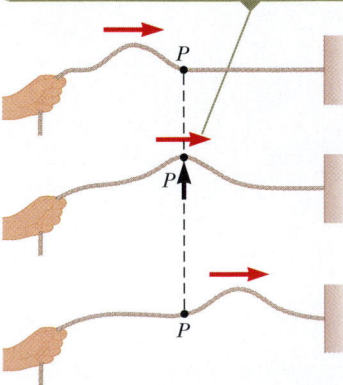

The direction of the displacement of any element at a point *P* on the string is perpendicular to the direction of propagation (red arrow).

Figure 13.2 The displacement of a particular string element for a transverse pulse traveling on a stretched string.

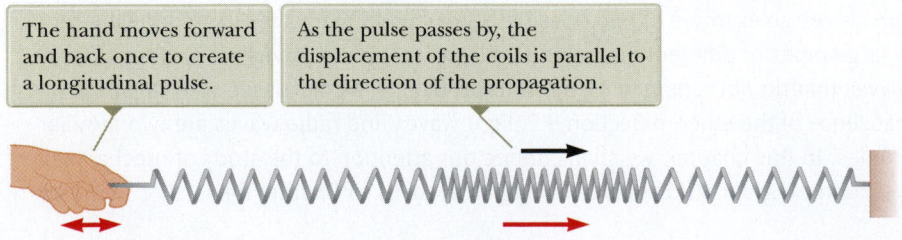

The hand moves forward and back once to create a longitudinal pulse.

As the pulse passes by, the displacement of the coils is parallel to the direction of the propagation.

Figure 13.3 A longitudinal pulse along a stretched spring.

$t = 0$. Because the speed of the pulse is v, the pulse has traveled to the right a distance vt at time t (Fig. 13.4b). We adopt a simplification model in which the shape of the pulse does not change with time.[1] Therefore, at time t, the shape of the pulse is the same as it was at time $t = 0$, as in Figure 13.4a. Consequently, an element of the string at x at this time has the same y position as an element located at $x - vt$ had at time $t = 0$:

$$y(x, t) = y(x - vt, 0)$$

In general, then, we can represent the position y for all values of x and t, measured in a stationary frame with the origin at O, as

$$y(x, t) = f(x - vt) \qquad \text{(pulse traveling to the right)} \qquad \textbf{13.1} \blacktriangleleft$$

If the pulse travels to the left, the position of an element of the string is described by

$$y(x, t) = f(x + vt) \qquad \text{(pulse traveling to the left)} \qquad \textbf{13.2} \blacktriangleleft$$

The function y, sometimes called the **wave function,** depends on the two variables x and t. For this reason, it is often written $y(x, t)$, which is read "y as a function of x and t."

It is important to understand the meaning of y. Consider a point P on the string, identified by a particular value of its x coordinate as in Figure 13.4. As the pulse passes through P, the y coordinate of this point increases, reaches a maximum, and then decreases to zero. The wave function $y(x, t)$ represents the y position of any element of string located at position x at any time t. Furthermore, if t is fixed (e.g., in the case of taking a snapshot of the pulse), the wave function y as a function of x, sometimes called the **waveform,** defines a curve representing the actual geometric shape of the pulse at that time.

▸ **QUICK QUIZ 13.1** (i) In a long line of people waiting to buy tickets, the first person leaves and a pulse of motion occurs as people step forward to fill the gap. As each person steps forward, the gap moves through the line. Is the propagation of this gap (a) transverse or (b) longitudinal? (ii) Consider "the wave" at a baseball game when people stand up and shout as the wave arrives at their location and the resultant pulse moves around the stadium. Is this wave (a) transverse or (b) longitudinal?

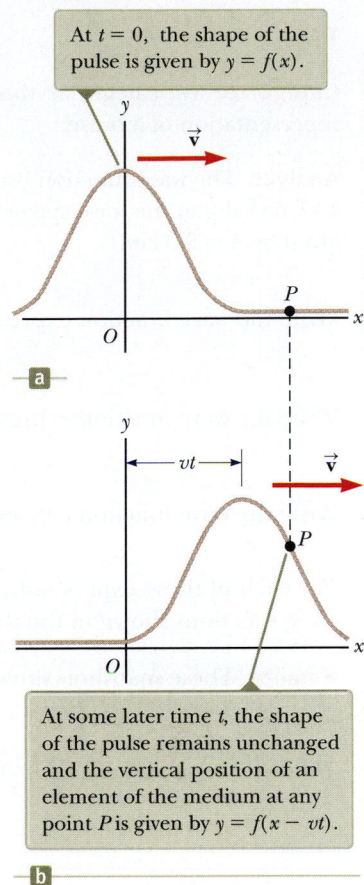

At $t = 0$, the shape of the pulse is given by $y = f(x)$.

At some later time t, the shape of the pulse remains unchanged and the vertical position of an element of the medium at any point P is given by $y = f(x - vt)$.

Figure 13.4 A one-dimensional pulse traveling to the right with a speed v.

Example 13.1 | A Pulse Moving to the Right

A pulse moving to the right along the x axis is represented by the wave function

$$y(x, t) = \frac{2}{(x - 3.0t)^2 + 1}$$

where x and y are measured in centimeters and t is measured in seconds. Find expressions for the wave function at $t = 0$, $t = 1.0$ s, and $t = 2.0$ s.

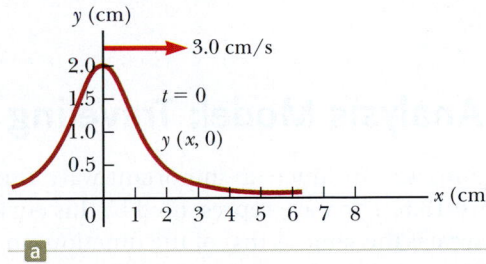

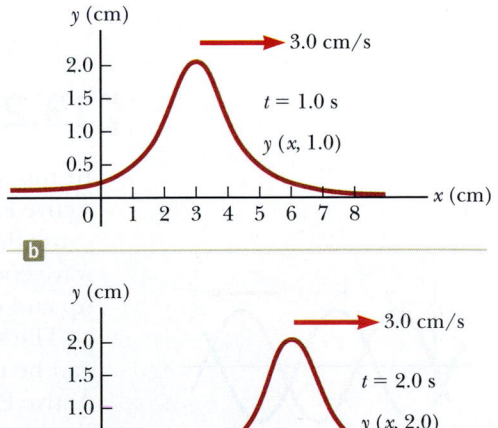

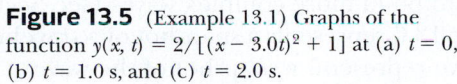

Figure 13.5 (Example 13.1) Graphs of the function $y(x, t) = 2/[(x - 3.0t)^2 + 1]$ at (a) $t = 0$, (b) $t = 1.0$ s, and (c) $t = 2.0$ s.

SOLUTION

Conceptualize Figure 13.5a shows the pulse represented by this wave function at $t = 0$. Imagine this pulse moving to the right and maintaining its shape as suggested by Figures 13.5b and 13.5c.

continued

---

[1]In reality, the pulse changes its shape and gradually spreads out during the motion. This effect, called *dispersion*, is common to many mechanical waves, but we adopt a simplification model that ignores this effect.

13.1 *cont.*

Categorize We categorize this example as a relatively simple analysis problem in which we interpret the mathematical representation of a pulse.

....

Analyze The wave function is of the form $y = f(x - vt)$. Inspection of the expression for $y(x, t)$ and comparison to Equation 13.1 reveal that the wave speed is $v = 3.0$ cm/s. Furthermore, by letting $x - 3.0t = 0$, we find that the maximum value of y is given by $A = 2.0$ cm.

Write the wave function expression at $t = 0$:

$$y(x, 0) = \frac{2}{x^2 + 1}$$

Write the wave function expression at $t = 1.0$ s:

$$y(x, 1.0) = \frac{2}{(x - 3.0)^2 + 1}$$

Write the wave function expression at $t = 2.0$ s:

$$y(x, 2.0) = \frac{2}{(x - 6.0)^2 + 1}$$

For each of these expressions, we can substitute various values of x and plot the wave function. This procedure yields the wave functions shown in the three parts of Figure 13.5.

....

Finalize These snapshots show that the pulse moves to the right without changing its shape and that it has a constant speed of 3.0 cm/s.

What If? What if the wave function were

$$y(x, t) = \frac{4}{(x + 3.0t)^2 + 1}$$

How would that change the situation?

Answer One new feature in this expression is the plus sign in the denominator rather than the minus sign. The new expression represents a pulse with a similar shape as that in Figure 13.5, but moving to the left as time progresses. Another new feature here is the numerator of 4 rather than 2. Therefore, the new expression represents a pulse with twice the height of that in Figure 13.5.

13.2 | Analysis Model: Traveling Wave

In this section, we introduce an important wave function whose shape is shown in Active Figure 13.6. The wave represented by this curve is called a **sinusoidal wave** because the curve is the same as that of the function $\sin \theta$ plotted against θ. A sinusoidal wave could be established on the rope in Figure 13.1 by shaking the end of the rope up and down in simple harmonic motion.

The sinusoidal wave is the simplest example of a periodic continuous wave and can be used to build more complex waves (see Section 14.6). The brown curve in Active Figure 13.6 represents a snapshot of a traveling sinusoidal wave at $t = 0$, and the blue curve represents a snapshot of the wave at some later time t. Imagine two types of motion that can occur. First, the entire waveform in Active Figure 13.6 moves to the right so that the brown curve moves toward the right and eventually reaches the position of the blue curve. This movement is the motion of the *wave*. If we focus on one element of the medium, such as the element at $x = 0$, we see that each element moves up and down along the y axis in simple harmonic motion. This movement is the motion of the *elements of the medium*. It is important to differentiate between the motion of the wave and the motion of the elements of the medium.

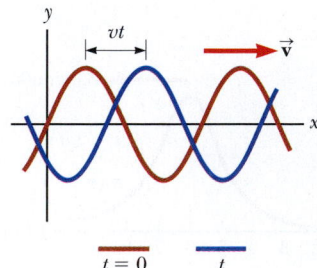

Active Figure 13.6 A one-dimensional sinusoidal wave traveling to the right with a speed v. The brown curve represents a snapshot of the wave at $t = 0$, and the blue curve represents a snapshot at some later time t.

In the early chapters of this book, we developed several analysis models based on three simplification models: the particle, the system, and the rigid object. With our introduction to waves, we can develop a new simplification model, the **wave**, that will allow us to explore more analysis models for solving problems. An ideal particle has zero size. We can build physical objects with nonzero size as combinations of particles. Therefore, the particle can be considered a basic building block. An ideal wave has a single frequency and is infinitely long; that is, the wave exists throughout the Universe. (A wave of finite length must necessarily have a mixture of frequencies.) When this concept is explored in Section 14.6, we will find that ideal waves can be combined to build complex waves, just as we combined particles.

In what follows, we will develop the principal features and mathematical representations of the analysis model of a **traveling wave.** This model is used in situations in which a wave moves through space without interacting with other waves or particles.

Active Figure 13.7a shows a snapshot of a wave moving through a medium. Active Figure 13.7b shows a graph of the position of one element of the medium as a function of time. A point in Active Figure 13.7a at which the displacement of the element from its normal position is highest is called the **crest** of the wave. The lowest point is called the **trough.** The distance from one crest to the next is called the **wavelength** λ (Greek letter lambda). More generally, the wavelength is the minimum distance between any two identical points on adjacent waves as shown in Active Figure 13.7a.

If you count the number of seconds between the arrivals of two adjacent crests at a given point in space, you measure the **period** T of the waves. In general, the period is the time interval required for two identical points of adjacent waves to pass by a point as shown in Active Figure 13.7b. The period of the wave is the same as the period of the simple harmonic oscillation of one element of the medium.

The same information is more often given by the inverse of the period, which is called the **frequency** f. In general, the frequency of a periodic wave is the number of crests (or troughs, or any other point on the wave) that pass a given point in a unit time interval. The frequency of a sinusoidal wave is related to the period by the expression

$$f = \frac{1}{T}$$ 13.3 ◀

The frequency of the wave is the same as the frequency of the simple harmonic oscillation of one element of the medium. The most common unit for frequency, as we learned in Chapter 12, is s^{-1}, or **hertz** (Hz). The corresponding unit for T is seconds.

The maximum position of an element of the medium relative to its equilibrium position is called the **amplitude** A of the wave as indicated in Active Figure 13.7.

Waves travel with a specific speed, and this speed depends on the properties of the medium being disturbed. For instance, sound waves travel through room-temperature air with a speed of about 343 m/s (781 mi/h), whereas they travel through most solids with a speed greater than 343 m/s.

Consider the sinusoidal wave in Active Figure 13.7a, which shows the position of the wave at $t = 0$. Because the wave is sinusoidal, we expect the wave function at this instant to be expressed as $y(x, 0) = A \sin ax$, where A is the amplitude and a is a constant to be determined. At $x = 0$, we see that $y(0, 0) = A \sin a(0) = 0$, consistent with Active Figure 13.7a. The next value of x for which y is zero is $x = \lambda/2$. Therefore,

$$y\left(\frac{\lambda}{2}, 0\right) = A \sin\left(a\frac{\lambda}{2}\right) = 0$$

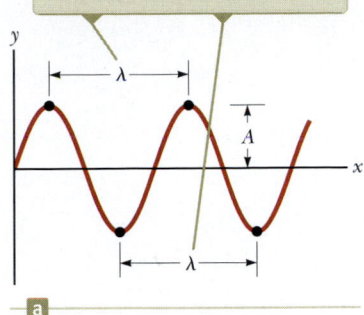

The wavelength λ of a wave is the distance between adjacent crests or adjacent troughs.

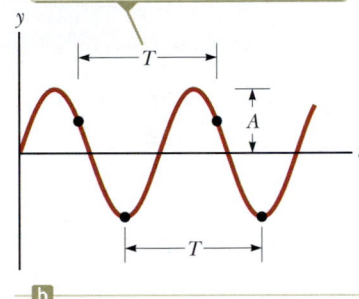

The period T of a wave is the time interval required for the element to complete one cycle of its oscillation and for the wave to travel one wavelength.

Active Figure 13.7 (a) A snapshot of a sinusoidal wave. (b) The position of one element of the medium as a function of time.

Pitfall Prevention | 13.1
What's the Difference Between Active Figures 13.7a and 13.7b?
Notice the visual similarity between Active Figures 13.7a and 13.7b. The shapes are the same, but (a) is a graph of vertical position versus horizontal position, whereas (b) is vertical position versus time. Active Figure 13.7a is a pictorial representation of the wave *for a series of elements of the medium;* it is what you would see at an instant of time. Active Figure 13.7b is a graphical representation of the position of *one element of the medium* as a function of time. That both figures have the identical shape represents Equation 13.1: a wave is the *same* function of both x and t.

For this equation to be true, we must have $a\lambda/2 = \pi$, or $a = 2\pi/\lambda$. Therefore, the function describing the positions of the elements of the medium through which the sinusoidal wave is traveling can be written

$$y(x, 0) = A \sin\left(\frac{2\pi}{\lambda} x\right) \qquad \text{13.4} \blacktriangleleft$$

where the constant A represents the wave amplitude and the constant λ is the wavelength. Notice that the vertical position of an element of the medium is the same whenever x is increased by an integral multiple of λ. Based on our discussion of Equation 13.1, if the wave moves to the right with a speed v, the wave function at some later time t is

$$y(x, t) = A \sin\left[\frac{2\pi}{\lambda}(x - vt)\right] \qquad \text{13.5} \blacktriangleleft$$

If the wave were traveling to the left, the quantity $x - vt$ would be replaced by $x + vt$ as we learned when we developed Equations 13.1 and 13.2.

By definition, the wave travels through a displacement Δx equal to one wavelength λ in a time interval Δt of one period T. Therefore, the wave speed, wavelength, and period are related by the expression

$$v = \frac{\Delta x}{\Delta t} = \frac{\lambda}{T} \qquad \text{13.6} \blacktriangleleft$$

Substituting this expression for v into Equation 13.5 gives

$$y = A \sin\left[2\pi\left(\frac{x}{\lambda} - \frac{t}{T}\right)\right] \qquad \text{13.7} \blacktriangleleft$$

This form of the wave function shows the *periodic* nature of y. Note that we will often use y rather than $y(x, t)$ as a shorthand notation. At any given time t, y has the *same* value at the positions x, $x + \lambda$, $x + 2\lambda$, and so on. Furthermore, at any given position x, the value of y is the same at times t, $t + T$, $t + 2T$, and so on.

We can express the wave function in a convenient form by defining two other quantities, the **angular wave number** k (usually called simply the **wave number**) and the **angular frequency** ω:

▶ Angular wave number

$$k \equiv \frac{2\pi}{\lambda} \qquad \text{13.8} \blacktriangleleft$$

▶ Angular frequency

$$\omega \equiv \frac{2\pi}{T} = 2\pi f \qquad \text{13.9} \blacktriangleleft$$

Using these definitions, Equation 13.7 can be written in the more compact form

▶ Wave function for a sinusoidal wave

$$y = A \sin(kx - \omega t) \qquad \text{13.10} \blacktriangleleft$$

Using Equations 13.3, 13.8, and 13.9, the wave speed v originally given in Equation 13.6 can be expressed in the following alternative forms:

$$v = \frac{\omega}{k} \qquad \text{13.11} \blacktriangleleft$$

▶ Speed of a sinusoidal wave

$$v = \lambda f \qquad \text{13.12} \blacktriangleleft$$

The wave function given by Equation 13.10 assumes the vertical position y of an element of the medium is zero at $x = 0$ and $t = 0$. That need not be the case. If it is not, we generally express the wave function in the form

▶ General expression for a sinusoidal wave

$$y = A \sin(kx - \omega t + \phi) \qquad \text{13.13} \blacktriangleleft$$

where ϕ is the **phase constant,** just as we learned in our study of periodic motion in Chapter 12. This constant can be determined from the initial conditions. The primary equations in the mathematical representation of the traveling wave analysis model are Equations 13.3, 13.10, and 13.12.

QUICK QUIZ 13.2 A sinusoidal wave of frequency f is traveling along a stretched string. The string is brought to rest, and a second traveling wave of frequency $2f$ is established on the string. **(i)** What is the wave speed of the second wave? (a) twice that of the first wave (b) half that of the first wave (c) the same as that of the first wave (d) impossible to determine **(ii)** From the same choices, describe the wavelength of the second wave. **(iii)** From the same choices, describe the amplitude of the second wave.

Example 13.2 | A Traveling Sinusoidal Wave

A sinusoidal wave traveling in the positive x direction has an amplitude of 15.0 cm, a wavelength of 40.0 cm, and a frequency of 8.00 Hz. The vertical position of an element of the medium at $t = 0$ and $x = 0$ is also 15.0 cm as shown in Figure 13.8.

(A) Find the wave number k, period T, angular frequency ω, and speed v of the wave.

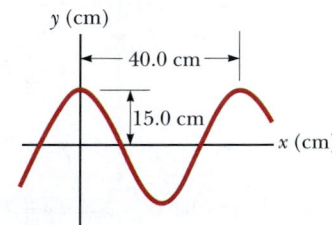

Figure 13.8 (Example 13.2) A sinusoidal wave of wavelength $\lambda = 40.0$ cm and amplitude $A = 15.0$ cm.

SOLUTION

Conceptualize Figure 13.8 shows the wave at $t = 0$. Imagine this wave moving to the right and maintaining its shape.

Categorize We will evaluate parameters of the wave using equations generated in the preceding discussion, so we categorize this example as a substitution problem.

Evaluate the wave number from Equation 13.8:

$$k = \frac{2\pi}{\lambda} = \frac{2\pi \text{ rad}}{40.0 \text{ cm}} = \boxed{15.7 \text{ rad/m}}$$

Evaluate the period of the wave from Equation 13.3:

$$T = \frac{1}{f} = \frac{1}{8.00 \text{ s}^{-1}} = \boxed{0.125 \text{ s}}$$

Evaluate the angular frequency of the wave from Equation 13.9:

$$\omega = 2\pi f = 2\pi (8.00 \text{ s}^{-1}) = \boxed{50.3 \text{ rad/s}}$$

Evaluate the wave speed from Equation 13.12:

$$v = \lambda f = (40.0 \text{ cm})(8.00 \text{ s}^{-1}) = \boxed{3.20 \text{ m/s}}$$

(B) Determine the phase constant ϕ and write a general expression for the wave function.

SOLUTION

Substitute $A = 15.0$ cm, $y = 15.0$ cm, $x = 0$, and $t = 0$ into Equation 13.13:

$$15.0 = (15.0) \sin \phi \quad \rightarrow \quad \sin \phi = 1 \quad \rightarrow \quad \phi = \frac{\pi}{2} \text{ rad}$$

Write the wave function:

$$y = A \sin \left(kx - \omega t + \frac{\pi}{2} \right) = A \cos (kx - \omega t)$$

Substitute the values for A, k, and ω in SI units into this expression:

$$y = \boxed{0.150 \cos (15.7x - 50.3t)}$$

The Linear Wave Equation

In Figure 13.1, we demonstrated how to create a pulse by jerking a taut string up and down once. To create a series of such pulses—a wave—let's replace the hand with an oscillating blade vibrating in simple harmonic motion. Active Figure 13.9 (page 360) represents snapshots of the wave created in this way at intervals of $T/4$. Because the end of the blade oscillates in simple harmonic motion, each element of the string, such as that at P, also oscillates vertically with simple harmonic motion. Therefore, every element of the string can be treated as a simple

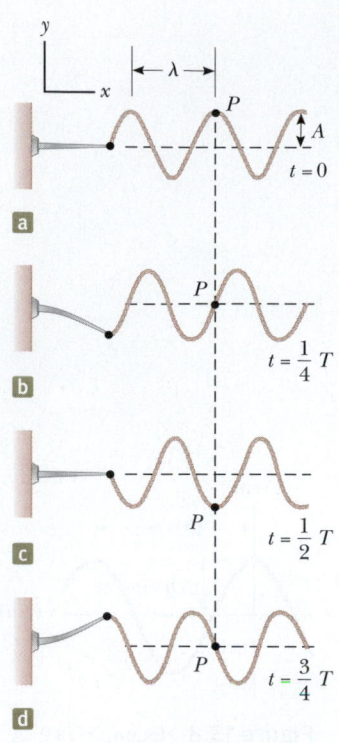

Active Figure 13.9 One method for producing a sinusoidal wave on a string. The left end of the string is connected to a blade that is set into oscillation. Every element of the string, such as that at point P, oscillates with simple harmonic motion in the vertical direction.

Pitfall Prevention | 13.2
Two Kinds of Speed/Velocity
Do not confuse v, the speed of the wave as it propagates along the string, with v_y, the transverse velocity of a point on the string. The speed v is constant for a uniform medium, whereas v_y varies sinusoidally.

---

harmonic oscillator vibrating with a frequency equal to the frequency of oscillation of the blade.[2] Notice that while each element oscillates in the y direction, the wave travels in the x direction with a speed v. Of course, that is the definition of a transverse wave.

If we define $t = 0$ as the time for which the configuration of the string is as shown in Active Figure 13.9a, the wave function can be written as in Equation 13.10:

$$y = A \sin(kx - \omega t)$$

We can use this expression to describe the motion of any element of the string. An element at point P (or any other element of the string) moves only vertically, and so its x coordinate remains constant. Therefore, the **transverse speed** v_y (not to be confused with the wave speed v) and the **transverse acceleration** a_y of elements of the string are

$$v_y = \frac{dy}{dt}\bigg]_{x=\text{constant}} = \frac{\partial y}{\partial t} = -\omega A \cos(kx - \omega t) \qquad \textbf{13.14}\blacktriangleleft$$

$$a_y = \frac{dv_y}{dt}\bigg]_{x=\text{constant}} = \frac{\partial v_y}{\partial t} = \frac{\partial^2 y}{\partial t^2} = -\omega^2 A \sin(kx - \omega t) \qquad \textbf{13.15}\blacktriangleleft$$

These expressions incorporate partial derivatives because y depends on both x and t. In the operation $\partial y/\partial t$, for example, we take a derivative with respect to t while holding x constant. The maximum magnitudes of the transverse speed and transverse acceleration are simply the absolute values of the coefficients of the cosine and sine functions:

$$v_{y,\text{max}} = \omega A \qquad \textbf{13.16}\blacktriangleleft$$

$$a_{y,\text{max}} = \omega^2 A \qquad \textbf{13.17}\blacktriangleleft$$

The transverse speed and transverse acceleration of elements of the string do not reach their maximum values simultaneously. The transverse speed reaches its maximum value (ωA) when $y = 0$, whereas the magnitude of the transverse acceleration reaches its maximum value ($\omega^2 A$) when $y = \pm A$. Finally, Equations 13.16 and 13.17 are identical in mathematical form to the corresponding equations for simple harmonic motion, Equations 12.17 and 12.18.

QUICK QUIZ 13.3 The amplitude of a wave is doubled, with no other changes made to the wave. As a result of this doubling, which of the following statements is correct? (a) The speed of the wave changes. (b) The frequency of the wave changes. (c) The maximum transverse speed of an element of the medium changes. (d) Statements (a) through (c) are all true. (e) None of statements (a) through (c) is true.

Now let's take derivatives of our wave function (Eq. 13.10) with respect to position at a fixed time, similar to the process by which we took derivatives with respect to time in Equations 13.14 and 13.15:

$$\frac{dy}{dx}\bigg]_{t=\text{constant}} = \frac{\partial y}{\partial x} = kA \cos(kx - \omega t) \qquad \textbf{13.18}\blacktriangleleft$$

$$\frac{d^2 y}{dx^2}\bigg]_{t=\text{constant}} = \frac{\partial^2 y}{\partial x^2} = -k^2 A \sin(kx - \omega t) \qquad \textbf{13.19}\blacktriangleleft$$

Comparing Equations 13.15 and 13.19, we see that

$$A \sin(kx - \omega t) = -\frac{1}{k^2}\frac{\partial^2 y}{\partial x^2} = -\frac{1}{\omega^2}\frac{\partial^2 y}{\partial t^2} \quad \longrightarrow \quad \frac{\partial^2 y}{\partial x^2} = \frac{k^2}{\omega^2}\frac{\partial^2 y}{\partial t^2}$$

---

[2]In this arrangement, we are assuming that a string element always oscillates in a vertical line. The tension in the string would vary if an element were allowed to move sideways. Such motion would make the analysis very complex.

Using Equation 13.11, we can rewrite this expression as

$$\frac{\partial^2 y}{\partial x^2} = \frac{1}{v^2}\frac{\partial^2 y}{\partial t^2}$$

13.20 ◀ ▶ Linear wave equation

which is known as the **linear wave equation.** If we analyze a situation and find this kind of relationship between derivatives of a function describing the situation, wave motion is occurring. Equation 13.20 is a differential equation representation of the traveling wave model. The solutions to the equation describe **linear mechanical waves.** We have developed the linear wave equation from a sinusoidal mechanical wave traveling through a medium, but it is much more general. The linear wave equation successfully describes waves on strings, sound waves, and also electromagnetic waves.[3] What's more, although the sinusoidal wave that we have studied is a solution to Equation 13.20, the general solution to the equation is *any* function of the form $y(x, t) = f(x \pm vt)$ as discussed in Section 13.1.

Nonlinear waves are more difficult to analyze, but they are an important area of current research, especially in optics. An example of a nonlinear mechanical wave is one for which the amplitude is not small compared with the wavelength.

Example **13.3** | A Solution to the Linear Wave Equation

Verify that the wave function presented in Example 13.1 is a solution to the linear wave equation.

SOLUTION

Conceptualize Look back at Figure 13.5 for a pictorial representation of the pulse. Imagine the pulse moving to the right as suggested by the three parts of the figure.

Categorize This is not an example of the traveling wave model because the moving entity is a single pulse, with no discernible wavelength or frequency. The linear wave equation, however, applies to both waves and pulses.

Analyze Write an expression for the wave function:

$$y(x, t) = \frac{2}{(x - 3.0t)^2 + 1}$$

Take partial derivatives of this function with respect to x and to t:

$$(1)\quad \frac{\partial^2 y}{\partial x^2} = \frac{12(x - 3.0t)^2 - 4.0}{[(x - 3.0t)^2 + 1]^3}$$

$$(2)\quad \frac{\partial^2 y}{\partial t^2} = \frac{108(x - 3.0t)^2 - 36}{[(x - 3.0t)^2 + 1]^3} = 9.0\frac{[12(x - 3.0t)^2 - 4.0]}{[(x - 3.0t)^2 + 1]^3}$$

Use Equations (1) and (2) to find a relationship between the left sides of these expressions:

$$\frac{\partial^2 y}{\partial x^2} = \frac{1}{9.0}\frac{\partial^2 y}{\partial t^2}$$

Finalize Comparing this result with Equation 13.20, we see that the wave function is a solution to the linear wave equation if the speed at which the pulse moves is 3.0 cm/s. We have already determined in Example 13.1 that this speed is indeed the speed of the pulse, so we have proven what we set out to do.

13.3 | The Speed of Transverse Waves on Strings

An aspect of the behavior of linear mechanical waves is that the wave speed depends only on the properties of the medium through which the wave travels. Waves for which the amplitude A is small relative to the wavelength λ can be represented as linear waves. In this section, we determine the speed of a transverse wave traveling on a stretched string.

---

[3]In the case of electromagnetic waves, y is interpreted to represent an electric field, which we will study in Chapter 24.

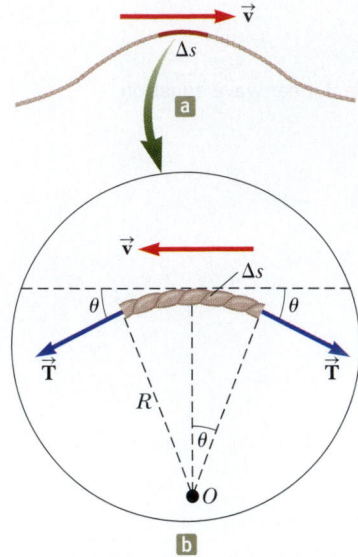

Figure 13.10 (a) In the reference frame of the Earth, a pulse moves to the right on a string with speed v. (b) In a frame of reference moving to the right with the pulse, the small element of length Δs moves to the left with speed v.

Let us use a mechanical analysis to derive the expression for the speed of a pulse traveling on a stretched string under tension T. Consider a pulse moving to the right with a uniform speed v, measured relative to a stationary (with respect to the Earth) inertial reference frame as shown in Figure 13.10a. Recall from Chapter 9 that Newton's laws are valid in any inertial reference frame. Therefore, let us view this pulse from a different inertial reference frame, one that moves along with the pulse at the same speed so that the pulse appears to be at rest in the frame as in Figure 13.10b. In this reference frame, the pulse remains fixed and each element of the string moves to the left through the pulse shape.

A short element of the string, of length Δs, forms an approximate arc of a circle of radius R as shown in the magnified view in Figure 13.10b. In our moving frame of reference, the element of the string moves to the left with speed v. As it travels through the arc, we can model the element as a particle in uniform circular motion. This element has a centripetal acceleration of v^2/R, which is supplied by components of the force $\vec{T}$ whose magnitude is the tension in the string. The force $\vec{T}$ acts on each side of the element, tangent to the arc, as in Figure 13.10b. The horizontal components of $\vec{T}$ cancel, and each vertical component $T\sin\theta$ acts downward. Hence, the magnitude of the total radial force on the element is $2T\sin\theta$. Because the element is small, θ is small and we can use the small-angle approximation $\sin\theta \approx \theta$. Therefore, the magnitude of the total radial force is

$$F_r = 2T\sin\theta \approx 2T\theta$$

The element has mass $m = \mu\,\Delta s$, where μ is the mass per unit length of the string. Because the element forms part of a circle and subtends an angle of 2θ at the center, $\Delta s = R(2\theta)$, and

$$m = \mu\,\Delta s = 2\mu R\theta$$

Applying Newton's second law to this element in the radial direction gives

$$F_r = \frac{mv^2}{R} \quad\rightarrow\quad 2T\theta = \frac{2\mu R\theta v^2}{R} \quad\rightarrow\quad T = \mu v^2$$

Solving for v gives

▶ Speed of a wave on a stretched string

$$v = \sqrt{\frac{T}{\mu}}$$

13.21 ◀

Notice that this derivation is based on the assumption that the pulse height is small relative to the length of the pulse. Using this assumption, we were able to use the approximation $\sin\theta \approx \theta$. Furthermore, the model assumes that the tension T is not affected by the presence of the pulse, so T is the same at all points on the pulse. Finally, this proof does *not* assume any particular shape for the pulse. We therefore conclude that a pulse of *any shape* will travel on the string with speed $v = \sqrt{T/\mu}$, without any change in pulse shape.

Pitfall Prevention | 13.3
Multiple T's
Do not confuse the T in Equation 13.21 for the tension with the symbol T used in this chapter for the period of a wave. The context of the equation should help you identify which quantity is meant. There simply aren't enough letters in the alphabet to assign a unique letter to each variable!

QUICK QUIZ 13.4 Suppose you create a pulse by moving the free end of a taut string up and down once with your hand beginning at $t = 0$. The string is attached at its other end to a distant wall. The pulse reaches the wall at time t. Which of the following actions, taken by itself, decreases the time interval required for the pulse to reach the wall? More than one choice may be correct. (**a**) moving your hand more quickly, but still only up and down once by the same amount (**b**) moving your hand more slowly, but still only up and down once by the same amount (**c**) moving your hand a greater distance up and down in the same amount of time (**d**) moving your hand a lesser distance up and down in the same amount of time (**e**) using a heavier string of the same length and under the same tension (**f**) using a lighter string of the same length and under the same tension (**g**) using a string of the same linear mass density but under decreased tension (**h**) using a string of the same linear mass density but under increased tension

THINKING PHYSICS 13.1

A secret agent is trapped in a building on top of an elevator car at a lower floor. He attempts to signal a fellow agent on the roof by tapping a message in Morse code on the elevator cable so that transverse pulses move upward on the cable. As the pulses move up the cable toward the accomplice, does the speed with which they move stay the same, increase, or decrease? If the pulses are sent 1 s apart, are they received 1 s apart by the agent on the roof?

Reasoning The elevator cable can be modeled as a vertical string. The speed of waves on the cable is a function of the tension in the cable. As the waves move higher on the cable, they encounter increased tension because each higher point on the cable must support the weight of all the cable below it (and the elevator). Therefore, the speed of the pulses increases as they move higher on the cable. The frequency of the pulses will not be affected because each pulse takes the same time interval to reach the top. They will still arrive at the top of the cable at intervals of 1 s. ◀

Example 13.4 | The Speed of a Pulse on a Cord

A uniform string has a mass of 0.300 kg and a length of 6.00 m. The string passes over a pulley and supports a 2.00-kg object (Fig. 13.11). Find the speed of a pulse traveling along this string.

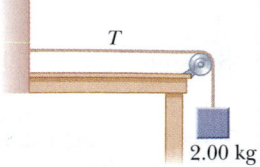

SOLUTION

Conceptualize In Figure 13.11, the hanging block establishes a tension in the horizontal string. This tension determines the speed with which waves move on the string.

Figure 13.11 (Example 13.4) The tension T in the cord is maintained by the suspended object. The speed of any wave traveling along the cord is given by $v = \sqrt{T/\mu}$.

Categorize To find the tension in the string, we model the hanging block as a particle in equilibrium. Then we use the tension to evaluate the wave speed on the string using Equation 13.21.

Analyze Apply the particle in equilibrium model to the block:

$$\sum F_y = T - m_{block}g = 0$$

Solve for the tension in the string:

$$T = m_{block}g$$

Use Equation 13.21 to find the wave speed, using $\mu = m_{string}/\ell$ for the linear mass density of the string:

$$v = \sqrt{\frac{T}{\mu}} = \sqrt{\frac{m_{block}g\,\ell}{m_{string}}}$$

Evaluate the wave speed:

$$v = \sqrt{\frac{(2.00\ \text{kg})(9.80\ \text{m/s}^2)(6.00\ \text{m})}{0.300\ \text{kg}}} = 19.8\ \text{m/s}$$

Finalize The calculation of the tension neglects the small mass of the string. Strictly speaking, the string can never be exactly straight; therefore, the tension is not uniform.

Example 13.5 | Rescuing the Hiker

An 80.0-kg hiker is trapped on a mountain ledge following a storm. A helicopter rescues the hiker by hovering above him and lowering a cable to him. The mass of the cable is 8.00 kg, and its length is 15.0 m. A sling of mass 70.0 kg is attached to the end of the cable. The hiker attaches himself to the sling, and the helicopter then accelerates upward. Terrified by hanging from the cable in midair, the hiker tries to signal the pilot by sending transverse pulses up the cable. A pulse takes 0.250 s to travel the length of the cable. What is the acceleration of the helicopter? Assume the tension in the cable is uniform.

SOLUTION

Conceptualize Imagine the effect of the acceleration of the helicopter on the cable. The greater the upward acceleration, the larger the tension in the cable. In turn, the larger the tension, the higher the speed of pulses on the cable.

continued

13.5 *cont.*

Categorize This problem is a combination of one involving the speed of pulses on a string and one in which the hiker and sling are modeled as a particle under a net force.

Analyze Use the time interval for the pulse to travel from the hiker to the helicopter to find the speed of the pulses on the cable:

$$v = \frac{\Delta x}{\Delta t} = \frac{15.0 \text{ m}}{0.250 \text{ s}} = 60.0 \text{ m/s}$$

Solve Equation 13.21 for the tension in the cable:

$$v = \sqrt{\frac{T}{\mu}} \rightarrow T = \mu v^2$$

Model the hiker and sling as a particle under a net force, noting that the acceleration of this particle of mass m is the same as the acceleration of the helicopter:

$$\sum F = ma \rightarrow T - mg = ma$$

Solve for the acceleration:

$$a = \frac{T}{m} - g = \frac{\mu v^2}{m} - g = \frac{m_{cable} v^2}{\ell_{cable} m} - g$$

Substitute numerical values:

$$a = \frac{(8.00 \text{ kg})(60.0 \text{ m/s})^2}{(15.0 \text{ m})(150.0 \text{ kg})} - 9.80 \text{ m/s}^2 = 3.00 \text{ m/s}^2$$

Finalize A real cable has stiffness in addition to tension. Stiffness tends to return a wire to its original straight-line shape even when it is not under tension. For example, a piano wire straightens if released from a curved shape; package-wrapping string does not.

Stiffness represents a restoring force in addition to tension and increases the wave speed. Consequently, for a real cable, the speed of 60.0 m/s that we determined is most likely associated with a smaller acceleration of the helicopter.

❰13.4 | Reflection and Transmission

The traveling wave model describes waves traveling through a uniform medium without interacting with anything along the way. We now consider how a traveling wave is affected when it encounters a change in the medium. For example, consider a pulse traveling on a string that is rigidly attached to a support at one end as in Active Figure 13.12. When the pulse reaches the support, a severe change in the medium occurs: the string ends. As a result, the pulse undergoes **reflection;** that is, the pulse moves back along the string in the opposite direction.

Notice that the reflected pulse is *inverted*. This inversion can be explained as follows. When the pulse reaches the fixed end of the string, the string produces an upward force on the support. By Newton's third law, the support must exert an equal-magnitude and oppositely directed (downward) reaction force on the string. This downward force causes the pulse to invert upon reflection.

Now consider another case. This time, the pulse arrives at the end of a string that is free to move vertically as in Active Figure 13.13. The tension at the free end is maintained because the string is tied to a ring of negligible mass that is free to slide vertically on a smooth post without friction. Again, the pulse is reflected, but this time it is not inverted. When it reaches the post, the pulse exerts a force on the free end of the string, causing the ring to accelerate upward. The ring rises as high as the incoming pulse, and then the downward component of the tension force pulls the ring back down. This

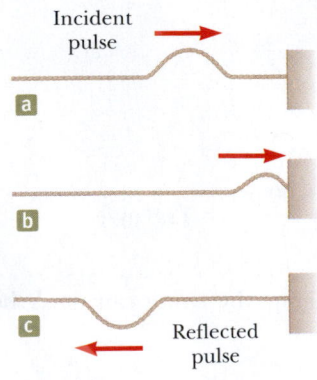

Incident pulse

a

b

c

Reflected pulse

Active Figure 13.12 The reflection of a traveling pulse at the fixed end of a stretched string. The reflected pulse is inverted, but its shape is otherwise unchanged.

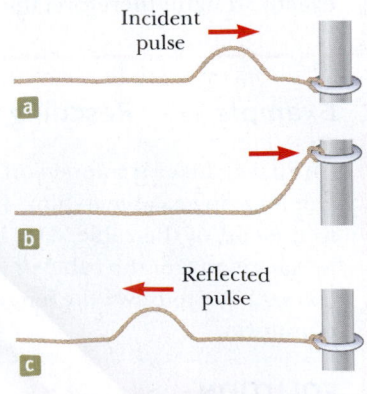

Incident pulse

a

b

Reflected pulse

c

Active Figure 13.13 The reflection of a traveling pulse at the free end of a stretched string. The reflected pulse is not inverted.

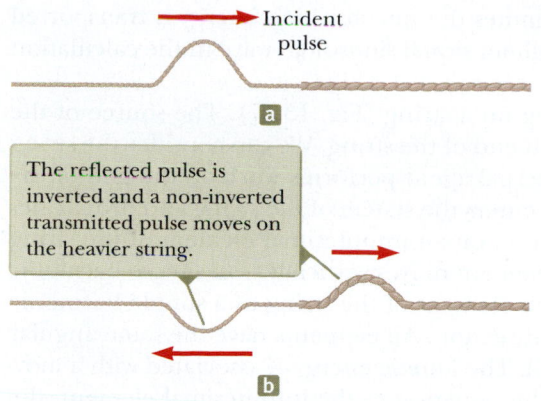

Incident pulse

a

The reflected pulse is inverted and a non-inverted transmitted pulse moves on the heavier string.

b

Active Figure 13.14 (a) A pulse traveling to the right on a light string approaches the junction with a heavier string. (b) The situation after the pulse reaches the junction.

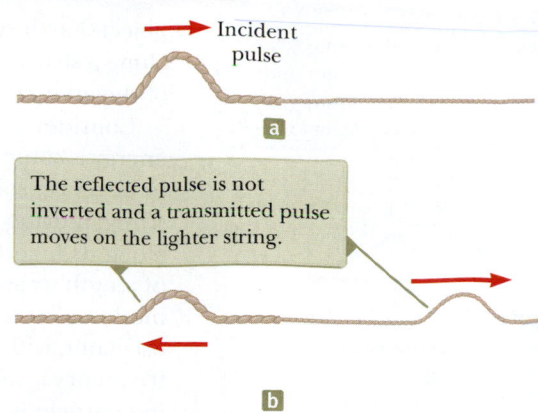

Incident pulse

a

The reflected pulse is not inverted and a transmitted pulse moves on the lighter string.

b

Active Figure 13.15 (a) A pulse traveling to the right on a heavy string approaches the junction with a lighter string. (b) The situation after the pulse reaches the junction.

movement of the ring produces a reflected pulse that is not inverted and that has the same amplitude as the incoming pulse.

Finally, consider a situation in which the boundary is intermediate between these two extremes. In this case, part of the energy in the incident pulse is reflected and part undergoes **transmission;** that is, some of the energy passes through the boundary. For instance, suppose a light string is attached to a heavier string as in Active Figure 13.14. When a pulse traveling on the light string reaches the boundary between the two strings, part of the pulse is reflected and inverted and part is transmitted to the heavier string. The reflected pulse is inverted for the same reasons described earlier in the case of the string rigidly attached to a support.

The reflected pulse has a smaller amplitude than the incident pulse. In Section 13.5, we show that the energy carried by a wave is related to its amplitude. According to the principle of the conservation of energy, when the pulse breaks up into a reflected pulse and a transmitted pulse at the boundary, the sum of the energies of these two pulses must equal the energy of the incident pulse. Because the reflected pulse contains only part of the energy of the incident pulse, its amplitude must be smaller.

When a pulse traveling on a heavy string strikes the boundary between the heavy string and a lighter one as in Active Figure 13.15, again part is reflected and part is transmitted. In this case, the reflected pulse is not inverted.

In either case, the relative heights of the reflected and transmitted pulses depend on the relative densities of the two strings. If the strings are identical, there is no discontinuity at the boundary and no reflection takes place.

According to Equation 13.21, the speed of a wave on a string decreases as the mass per unit length of the string increases. In other words, a wave travels more slowly on a heavy string than on a light string if both are under the same tension. The following general rules apply to reflected waves: When a wave or pulse travels from medium A to medium B and $v_A > v_B$ (that is, when B is denser than A), it is inverted upon reflection. When a wave or pulse travels from medium A to medium B and $v_A < v_B$ (that is, when A is denser than B), it is not inverted upon reflection.

13.5 | Rate of Energy Transfer by Sinusoidal Waves on Strings

Waves transport energy through a medium as they propagate. For example, suppose an object is hanging on a stretched string and a pulse is sent down the string as in Figure 13.16a. When the pulse meets the suspended object, the object is momentarily displaced upward as in Figure 13.16b. In the process, energy is transferred to the object and appears as an increase in the gravitational potential energy of the

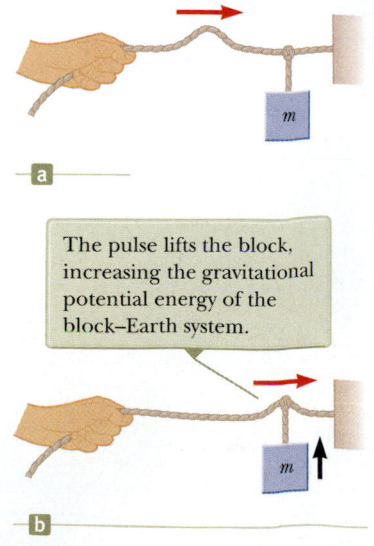

a

The pulse lifts the block, increasing the gravitational potential energy of the block–Earth system.

b

Figure 13.16 (a) A pulse travels to the right on a stretched string, carrying energy with it. (b) The energy of the pulse arrives at the hanging block.

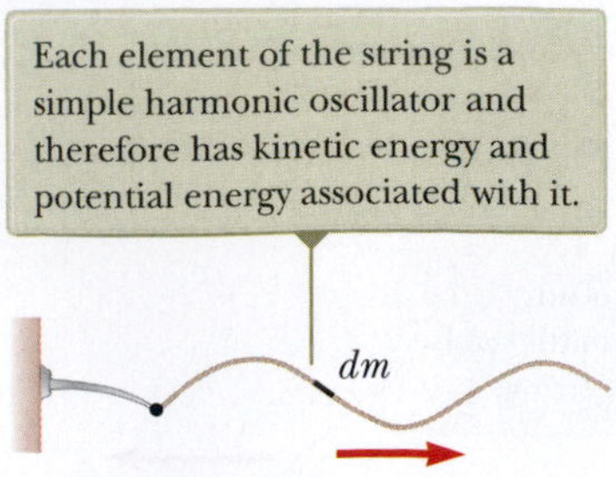

Each element of the string is a simple harmonic oscillator and therefore has kinetic energy and potential energy associated with it.

Figure 13.17 A sinusoidal wave traveling along the x axis on a stretched string.

object–Earth system. This section examines the rate at which energy is transported along a string. We shall assume a one-dimensional sinusoidal wave in the calculation of the energy transferred.

Consider a sinusoidal wave traveling on a string (Fig. 13.17). The source of the energy is some external agent at the left end of the string. We can consider the string to be a nonisolated system. As the external agent performs work on the end of the string, moving it up and down, energy enters the system of the string and propagates along its length. Let's focus our attention on an infinitesimal element of the string of length dx and mass dm. Each such element moves vertically with simple harmonic motion. Therefore, we can model each element of the string as a simple harmonic oscillator, with the oscillation in the y direction. All elements have the same angular frequency ω and the same amplitude A. The kinetic energy K associated with a moving particle is $K = \frac{1}{2}mv^2$. If we apply this equation to the infinitesimal element, the kinetic energy dK associated with the up and down motion of this element is

$$dK = \tfrac{1}{2}(dm)v_y^2$$

where v_y is the transverse speed of the element. If μ is the mass per unit length of the string, the mass dm of the element of length dx is equal to $\mu \, dx$. Hence, we can express the kinetic energy of an element of the string as

$$dK = \tfrac{1}{2}(\mu \, dx)v_y^2 \qquad \textbf{13.22} \blacktriangleleft$$

Substituting for the general transverse speed of an element of the medium using Equation 13.14 gives

$$dK = \tfrac{1}{2}\mu[-\omega A \cos(kx - \omega t)]^2 \, dx = \tfrac{1}{2}\mu\omega^2 A^2 \cos^2(kx - \omega t)\,dx$$

If we take a snapshot of the wave at time $t = 0$, the kinetic energy of a given element is

$$dK = \tfrac{1}{2}\mu\omega^2 A^2 \cos^2 kx \, dx$$

Integrating this expression over all the string elements in a wavelength of the wave gives the total kinetic energy K_λ in one wavelength:

$$K_\lambda = \int dK = \int_0^\lambda \tfrac{1}{2}\mu\omega^2 A^2 \cos^2 kx \, dx = \tfrac{1}{2}\mu\omega^2 A^2 \int_0^\lambda \cos^2 kx \, dx$$

$$= \tfrac{1}{2}\mu\omega^2 A^2 \left[\tfrac{1}{2}x + \frac{1}{4k}\sin 2kx\right]_0^\lambda = \tfrac{1}{2}\mu\omega^2 A^2 \left[\tfrac{1}{2}\lambda\right] = \tfrac{1}{4}\mu\omega^2 A^2\lambda$$

In addition to kinetic energy, there is potential energy associated with each element of the string due to its displacement from the equilibrium position and the restoring forces from neighboring elements. A similar analysis to that above for the total potential energy U_λ in one wavelength gives exactly the same result:

$$U_\lambda = \tfrac{1}{4}\mu\omega^2 A^2\lambda$$

The total energy in one wavelength of the wave is the sum of the potential and kinetic energies:

$$E_\lambda = U_\lambda + K_\lambda = \tfrac{1}{2}\mu\omega^2 A^2\lambda \qquad \textbf{13.23} \blacktriangleleft$$

As the wave moves along the string, this amount of energy passes by a given point on the string during a time interval of one period of the oscillation. Therefore, the power P, or rate of energy transfer T_{MW} associated with the mechanical wave, is

$$P = \frac{T_{MW}}{\Delta t} = \frac{E_\lambda}{T} = \frac{\tfrac{1}{2}\mu\omega^2 A^2\lambda}{T} = \tfrac{1}{2}\mu\omega^2 A^2 \left(\frac{\lambda}{T}\right)$$

▶ Power of a wave

$$P = \tfrac{1}{2}\mu\omega^2 A^2 v \qquad \textbf{13.24} \blacktriangleleft$$

Equation 13.24 shows that the rate of energy transfer by a sinusoidal wave on a string is proportional to (a) the square of the frequency, (b) the square of the amplitude, and (c) the wave speed. In fact, the rate of energy transfer in *any* sinusoidal wave is proportional to the square of the angular frequency and to the square of the amplitude.

QUICK QUIZ 13.5 Which of the following, taken by itself, would be most effective in increasing the rate at which energy is transferred by a wave traveling along a string? (**a**) reducing the linear mass density of the string by one-half (**b**) doubling the wavelength of the wave (**c**) doubling the tension in the string (**d**) doubling the amplitude of the wave

---

Example 13.6 | Power Supplied to a Vibrating String

A taut string for which $\mu = 5.00 \times 10^{-2}$ kg/m is under a tension of 80.0 N. How much power must be supplied to the string to generate sinusoidal waves at a frequency of 60.0 Hz and an amplitude of 6.00 cm?

SOLUTION

Conceptualize Consider Active Figure 13.9 again and notice that the vibrating blade supplies energy to the string at a certain rate. This energy then propagates to the right along the string.

Categorize We evaluate quantities from equations developed in the chapter, so we categorize this example as a substitution problem.

Use Equation 13.24 to evaluate the power:

$$P = \tfrac{1}{2}\mu\omega^2 A^2 v$$

Use Equations 13.9 and 13.21 to substitute for ω and v:

$$P = \tfrac{1}{2}\mu(2\pi f)^2 A^2 \left(\sqrt{\frac{T}{\mu}}\right) = 2\pi^2 f^2 A^2 \sqrt{\mu T}$$

Substitute numerical values:

$$P = 2\pi^2 (60.0 \text{ Hz})^2 (0.060\ 0 \text{ m})^2 \sqrt{(0.050\ 0 \text{ kg/m})(80.0 \text{ N})} = \boxed{512 \text{ W}}$$

What If? What if the string is to transfer energy at a rate of 1 000 W? What must be the required amplitude if all other parameters remain the same?

Answer Let us set up a ratio of the new and old power, reflecting only a change in the amplitude:

$$\frac{P_{\text{new}}}{P_{\text{old}}} = \frac{\tfrac{1}{2}\mu\omega^2 A_{\text{new}}^2 v}{\tfrac{1}{2}\mu\omega^2 A_{\text{old}}^2 v} = \frac{A_{\text{new}}^2}{A_{\text{old}}^2}$$

Solving for the new amplitude gives

$$A_{\text{new}} = A_{\text{old}}\sqrt{\frac{P_{\text{new}}}{P_{\text{old}}}} = (6.00 \text{ cm})\sqrt{\frac{1\ 000 \text{ W}}{512 \text{ W}}} = 8.39 \text{ cm}$$

13.6 | Sound Waves

Let us turn our attention from transverse waves to longitudinal waves. As stated in Section 13.1, for longitudinal waves the elements of the medium undergo displacements parallel to the direction of wave motion. Sound waves in air are the most important examples of longitudinal waves. Sound waves can travel through any material medium, however, and their speed depends on the properties of that medium. Table 13.1 (page 368) provides examples of the speed of sound in different media.

In Section 13.1, we began our investigation of waves by imagining the creation of a single pulse that traveled down a string (Figure 13.1) or a spring (Figure 13.3).

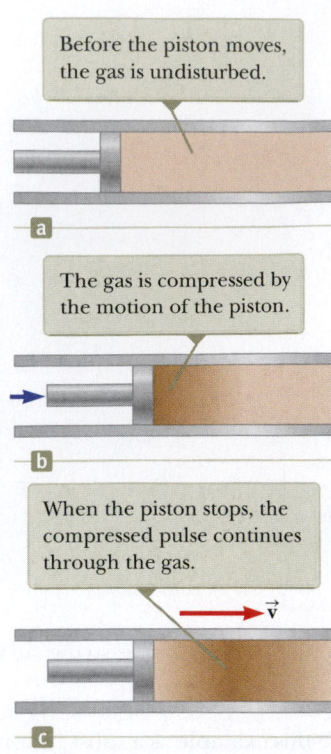

Before the piston moves, the gas is undisturbed.

The gas is compressed by the motion of the piston.

When the piston stops, the compressed pulse continues through the gas.

Figure 13.18 Motion of a longitudinal pulse through a compressible gas. The compression (darker region) is produced by the moving piston.

TABLE 13.1 | **Speed of Sound in Various Media**

Medium	v (m/s)	Medium	v (m/s)	Medium	v (m/s)
Gases		**Liquids at 25°C**		**Solids**[a]	
Hydrogen (0°C)	1 286	Glycerol	1 904	Pyrex glass	5 640
Helium (0°C)	972	Sea water	1 533	Iron	5 950
Air (20°C)	343	Water	1 493	Aluminum	6 420
Air (0°C)	331	Mercury	1 450	Brass	4 700
Oxygen (0°C)	317	Kerosene	1 324	Copper	5 010
		Methyl alcohol	1 143	Gold	3 240
		Carbon tetrachloride	926	Lucite	2 680
				Lead	1 960
				Rubber	1 600

[a]Values given are for propagation of longitudinal waves in bulk media. Speeds for longitudinal waves in thin rods are smaller, and speeds of transverse waves in bulk are smaller yet.

Let's do something similar for sound. We describe pictorially the motion of a one-dimensional longitudinal sound pulse moving through a long tube containing a compressible gas as shown in Figure 13.18. A piston at the left end can be quickly moved to the right to compress the gas and create the pulse. Before the piston is moved, the gas is undisturbed and of uniform density as represented by the uniformly shaded region in Figure 13.18a. When the piston is pushed to the right (Fig. 13.18b), the gas just in front of it is compressed (as represented by the more heavily shaded region); the pressure and density in this region are now higher than they were before the piston moved. When the piston comes to rest (Fig. 13.18c), the compressed region of the gas continues to move to the right, corresponding to a longitudinal pulse traveling through the tube with speed v.

One can produce a one-dimensional *periodic* sound wave in the tube of gas in Figure 13.18 by causing the piston to move in simple harmonic motion. The results are shown in Active Figure 13.19. The darker parts of the colored areas in this figure represent regions in which the gas is compressed and the density and pressure are above their equilibrium values. A compressed region is formed whenever the piston is pushed into the tube. This compressed region, called a **compression,** moves through the tube, continuously compressing the region just in front of itself. When the piston is pulled back, the gas in front of it expands and the pressure and density in this region fall below their equilibrium values (represented by the lighter parts of the colored areas in Active Fig. 13.19). These low-pressure regions, called **rarefactions,** also propagate along the tube, following the compressions. Both regions move at the speed of sound in the medium.

As the piston oscillates back and forth in a sinusoidal fashion, regions of compression and rarefaction are continuously set up. The distance between two successive compressions (or two successive rarefactions) equals the wavelength λ. As these regions travel along the tube, any small element of the medium moves with simple harmonic motion parallel to the direction of the wave (in other words, longitudinally). If $s(x, t)$ is the position of a small element measured relative to its equilibrium position,[4] we can express this position function as

$$s(x, t) = s_{\max} \cos(kx - \omega t) \qquad \text{13.25}$$

where $s_{\max}$ is the maximum position relative to equilibrium, often called the **displacement amplitude.** Equation 13.25 represents the **displacement wave,** where k is the wave number and ω is the angular frequency of the piston. The variation ΔP in

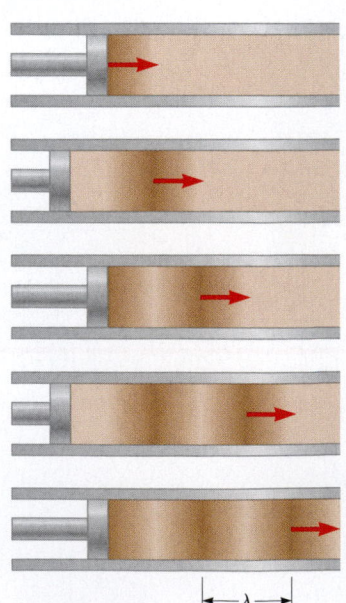

Active Figure 13.19 A longitudinal wave propagating through a gas-filled tube. The source of the wave is an oscillating piston at the left.

[4]We use $s(x, t)$ here instead of $y(x, t)$ because the displacement of elements in the medium is not perpendicular to the x direction.

the pressure[5] of the gas measured from its equilibrium value is also sinusoidal; it is given by

$$\Delta P = \Delta P_{max} \sin(kx - \omega t)$$ **13.26**◀

The **pressure amplitude** ΔP_{max} is the maximum change in pressure from the equilibrium value, and Equation 13.26 represents the **pressure wave.** The pressure amplitude is proportional to the displacement amplitude s_{max}:

$$\Delta P_{max} = \rho v \omega s_{max}$$ **13.27**◀

where ρ is the density of the medium, v is the wave speed, and ωs_{max} is the maximum longitudinal speed of an element of the medium. It is these pressure variations in a sound wave that result in an oscillating force on the eardrum, leading to the sensation of hearing.

This discussion shows that a sound wave may be described equally well in terms of either pressure or displacement. A comparison of Equations 13.25 and 13.26 shows that the pressure wave is 90° out of phase with the displacement wave. Graphs of these functions are shown in Figure 13.20. Note that the change in pressure from equilibrium is a maximum when the displacement is zero, whereas the displacement is a maximum when the pressure change is zero.

Note that Figure 13.20 presents two graphical representations of the longitudinal wave: one for position of the elements of the medium and the other for pressure variation. They are *not* pictorial representations for longitudinal waves, however. For transverse waves, the element displacement is perpendicular to the direction of propagation and the pictorial and graphical representations look the same because the perpendicularity of the oscillations and propagation is matched by the perpendicularity of x and y axes. For longitudinal waves, the oscillations and propagation exhibit no perpendicularity, so those pictorial representations look like Active Figure 13.19.

The speed of sound depends on the temperature of the medium. For sound traveling through air, the relationship between wave speed and air temperature is

$$v = 331\sqrt{1 + \frac{T_C}{273}}$$ **13.28**◀

where v is in meters/second, 331 m/s is the speed of sound in air at 0°C, and T_C is the air temperature in degrees Celsius. Using this equation, one finds that at 20°C, the speed of sound in air is approximately 343 m/s.

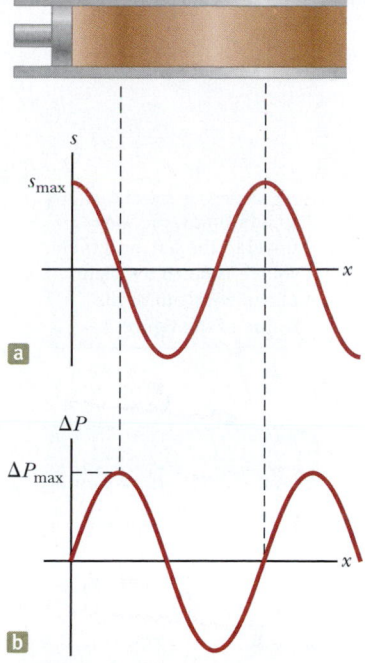

Figure 13.20 (a) Displacement amplitude versus position and (b) pressure amplitude versus position for a sinusoidal longitudinal wave. The displacement wave is 90° out of phase with the pressure wave.

▶ THINKING PHYSICS 13.2

Why does thunder produce an extended "rolling" sound when its source, a lightning strike, occurs in a fraction of a second? How does lightning produce thunder in the first place?

Reasoning Let us assume that we are at ground level and ignore ground reflections. When cloud-to-ground lightning strikes, a channel of ionized air carries a very large electric current from the cloud to the ground. (We will study electric current in Chapter 21.) The result is a very rapid temperature increase of this channel of air as it carries the current. The temperature increase causes a sudden expansion of the air. This expansion is so sudden and so intense that a tremendous disturbance is produced in the air: thunder. The thunder rolls because the lightning channel is a long, extended source; the entire length of the channel produces the sound at essentially the same instant of time. Sound produced at the end of the channel nearest you reaches you first, but sounds from progressively farther portions of the channel reach you shortly thereafter. If the lightning channel were a perfectly straight line, the resulting sound might be a steady roar, but the zigzagged shape of the path results in the rolling variation in loudness. ◀

---

[5]We will formally introduce pressure in Chapter 15. In the case of longitudinal waves in a gas, each compressed area is a region of higher-than-average pressure and density, and each stretched region is a region of lower-than-average pressure and density.

[13.7] The Doppler Effect

When someone honks the horn of a vehicle as it travels along a highway, the frequency of the sound you hear is higher as the vehicle approaches you than it is as the vehicle moves away from you. This change is one example of the **Doppler effect,** named after Christian Johann Doppler (1803–1853), an Austrian physicist.

The Doppler effect for sound is experienced whenever there is relative motion between the source of sound and the observer. Motion of the source or observer toward the other results in the observer's hearing a frequency that is higher than the true frequency of the source. Motion of the source or observer away from the other results in the observer hearing a frequency that is lower than the true frequency of the source.

Although we shall restrict our attention to the Doppler effect for sound waves, it is associated with waves of all types. The Doppler effect for electromagnetic waves is used in police radar systems to measure the speeds of motor vehicles. Likewise, astronomers use the effect to determine the relative motions of stars, galaxies, and other celestial objects. In 1842, Doppler first reported the frequency shift in connection with light emitted by two stars revolving about each other in double-star systems. In the early 20th century, the Doppler effect for light from galaxies was used to argue for the expansion of the Universe, which led to the Big Bang theory, discussed in Chapter 31.

To see what causes this apparent frequency change, imagine you are in a boat lying at anchor on a gentle sea where the waves have a period of $T = 2.0$ s. Therefore, every 2.0 s a crest hits your boat. Figure 13.21a shows this situation with the water waves moving toward the left. If you start a stopwatch at $t = 0$ just as one crest hits, the stopwatch reads 2.0 s when the next crest hits, 4.0 s when the third crest hits, and so on. From these observations you conclude that the wave frequency is $f = 1/T = 0.50$ Hz. Now suppose you start your motor and head directly into the oncoming waves as shown in Figure 13.21b. Again you set your stopwatch to $t = 0$ as a crest hits the bow of your boat. This time, however, because you are moving toward the next wave crest as it moves toward you, it hits you less than 2.0 s after the first hit. In other words, the period you observe is shorter than the 2.0-s period you observed when you were stationary. Because $f = 1/T$, you observe a higher wave frequency than when you were at rest.

If you turn around and move in the same direction as the waves (Fig. 13.21c), you observe the opposite effect. You set your watch to $t = 0$ as a crest hits the stern of the boat. Because you are now moving away from the next crest, more than 2.0 s has elapsed on your watch by the time that crest catches you. Therefore, you observe a lower frequency than when you were at rest.

These effects occur because the *relative* speed between your boat and the crest of a wave depends on the direction of travel and on the speed of your boat. When you are moving toward the right in Figure 13.21b, this relative speed is higher than that of the wave speed, which leads to the observation of an increased frequency. When you turn around and move to the left, the relative speed is lower, as is the observed frequency of the water waves.

Let's now examine an analogous situation with sound waves in which we replace the water waves with sound waves, the water becomes the air, and the person on the boat becomes an observer listening to the sound. In this case, an observer O is moving with a speed of v_O and a sound source S is stationary. For simplicity, we assume that the air is also stationary and that the observer moves directly toward the source.

The circles in Active Figure 13.22 represent curves connecting the crests of sound waves moving away from the source. Therefore, the radial distance between adjacent circles is one wavelength. We shall take the frequency of the source to be f, the wavelength to be λ, and the speed of sound to be v. A stationary observer would detect a frequency f, where $f = v/\lambda$ (i.e., when the source and observer are both at rest, the observed frequency must equal the true frequency of the source). If the observer

In all frames, the waves travel to the left, and their source is far to the right of the boat, out of the frame of the figure.

$\vec{v}_{waves}$

a

$\vec{v}_{boat}$

$\vec{v}_{waves}$

b

$\vec{v}_{boat}$

$\vec{v}_{waves}$

c

Figure 13.21 (a) Waves moving toward a stationary boat. (b) The boat moving toward the wave source. (c) The boat moving away from the wave source.

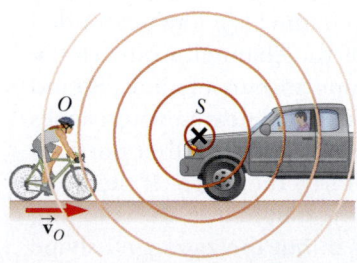

O S

$\vec{v}_O$

Active Figure 13.22 An observer O (the cyclist) moving with a speed v_O toward a stationary point source S, the horn of a parked truck. The observer hears a frequency f' that is greater than the source frequency.

moves toward the source with the speed v_O, however, the relative speed of sound experienced by the observer is higher than the speed of sound in air. Using our relative speed discussion of Section 3.6, if the sound is coming toward the observer at v and the observer is moving toward the sound at v_O, the relative speed of sound as measured by the observer is

$$v' = v + v_O$$

The frequency of sound heard by the observer is based on this apparent speed of sound:

$$f' = \frac{v'}{\lambda} = \frac{v + v_O}{\lambda} = \left(\frac{v + v_O}{v}\right)f \qquad \text{(observer moving toward source)} \qquad 13.29 \blacktriangleleft$$

Now consider the situation in which the source moves with a speed of v_S relative to the medium and the observer is at rest. Active Figure 13.23a shows this situation. If the source moves directly toward observer A in Active Figure 13.23a, the crests detected by the observer along a line between the source and observer are closer to one another than they would be if the source were at rest. (Active Figure 13.23b shows this effect for waves moving on the surface of water.) As a result, the wavelength λ' measured by observer A is shorter than the wavelength λ of the source. During each vibration, which lasts for a time interval T (the period), the source moves a distance $v_S T = v_S/f$ and the wavelength is *shortened* by this amount. Therefore, the observed wavelength is $\lambda' = \lambda - v_S/f$. Because $\lambda = v/f$, the frequency f' heard by observer A is

$$f' = \frac{v}{\lambda'} = \left(\frac{v}{v - v_S}\right)f \qquad \text{(source moving toward observer)} \qquad 13.30 \blacktriangleleft$$

That is, the frequency is *increased* when the source moves toward the observer. In a similar manner, if the source moves away from observer B at rest, the sign of v_S is reversed in Equation 13.30 and the frequency is lower.

In Equation 13.30, notice that the denominator approaches zero when the speed of the source approaches the speed of sound, resulting in the frequency f' approaching infinity. Such a situation results in waves that cannot escape from the source in the direction of motion of the source. This concentration of energy in front of the source results in a *shock wave*. Such a disturbance is noted when a jet aircraft flying at a speed equal to or greater than the speed of sound produces a *sonic boom*.

Finally, if both the source and the observer are in motion, the following general equation for the observed frequency is found:

$$f' = \left(\frac{v + v_O}{v - v_S}\right)f \qquad 13.31 \blacktriangleleft$$

In this expression, the signs for the values substituted for v_O and v_S depend on the direction of the velocity. A positive value is used for motion of the observer or the source *toward* the other, and a negative sign is used for motion of one *away from* the other.

When working with any Doppler effect problem, remember the following rule concerning signs: The word *toward* is associated with an *increase* in the observed frequency, and the words *away from* are associated with a *decrease* in the observed frequency.

Doppler Sonography BIO

The Doppler effect is used in medicine to study many different systems. For example, a technique called *Doppler sonography* is a non-invasive diagnostic procedure that can measure the speed of blood in arteries and detect turbulence in blood flow. The

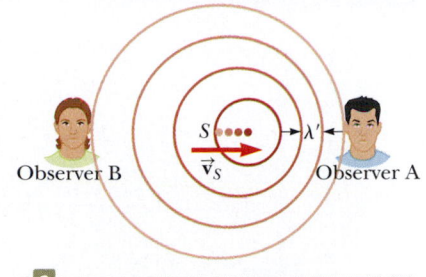

a

A point source is moving to the right with speed v_S.

b

Courtesy of the Educational Development Center, Newton, MA

Active Figure 13.23 (a) A source S moving with a speed v_S toward a stationary observer A and away from a stationary observer B. Observer A hears an increased frequency, and observer B hears a decreased frequency. (b) The Doppler effect in water, observed in a ripple tank. The vibrating source is moving to the right. Letters shown in the photo refer to Quick Quiz 13.6.

▶ General Doppler-shift expression

Pitfall Prevention | 13.4
Doppler Effect Does Not Depend on Distance
A common misconception about the Doppler effect is that it depends on the distance between the source and the observer. Although the *intensity* of a sound will vary as the distance changes, the apparent *frequency* will not; the frequency depends only on the relative speed of source and observer.

sound waves reflect from moving blood cells, undergoing a frequency shift based on the speed of the cells. The instrumentation detects the reflected sound waves and converts the frequency information to a speed of flow of the blood. By viewing the image of the heart, physicians can monitor carotid artery diseases and detect problems with heart valves. Typical diagnostic sonographic devices operate at frequencies ranging from 1 MHz to 18 MHz. Sonography is an effective method for imaging soft body tissues such as muscles, tendons, breasts, and neonatal brain at higher frequencies (7 MHz to 18 MHz). Lower frequencies (1 MHz to 6 MHz) must be used to image deeper structures in the body such as the liver and the kidney. However, the lower frequencies used in imaging these deeper structures result in lower-resolution images.

> **QUICK QUIZ 13.6** Consider detectors of water waves at three locations A, B, and C in Active Figure 13.23b. Which of the following statements is true? (**a**) The wave speed is highest at location A. (**b**) The wave speed is highest at location C. (**c**) The detected wavelength is largest at location B. (**d**) The detected wavelength is largest at location C. (**e**) The detected frequency is highest at location C. (**f**) The detected frequency is highest at location A.

> **QUICK QUIZ 13.7** You stand on a platform at a train station and listen to a train approaching the station at a constant velocity. While the train approaches, but before it arrives, what do you hear? (**a**) the intensity and the frequency of the sound both increasing (**b**) the intensity and the frequency of the sound both decreasing (**c**) the intensity increasing and the frequency decreasing (**d**) the intensity decreasing and the frequency increasing (**e**) the intensity increasing and the frequency remaining the same (**f**) the intensity decreasing and the frequency remaining the same

Example 13.7 | Doppler Submarines

A submarine (sub A) travels through water at a speed of 8.00 m/s, emitting a sonar wave at a frequency of 1 400 Hz. The speed of sound in the water is 1 533 m/s. A second submarine (sub B) is located such that both submarines are traveling directly toward each other. The second submarine is moving at 9.00 m/s.

(A) What frequency is detected by an observer riding on sub B as the subs approach each other?

SOLUTION

Conceptualize Even though the problem involves subs moving in water, there is a Doppler effect just like there is when you are in a moving car and listening to a sound moving through the air from another car.

Categorize Because both subs are moving, we categorize this problem as one involving the Doppler effect for both a moving source and a moving observer.

Analyze Use Equation 13.31 to find the Doppler-shifted frequency heard by the observer in sub B, being careful with the signs assigned to the source and observer speeds:

$$f' = \left(\frac{v + v_O}{v - v_S}\right)f$$

$$f' = \left[\frac{1\ 533\ \text{m/s} + (+9.00\ \text{m/s})}{1\ 533\ \text{m/s} - (+8.00\ \text{m/s})}\right](1\ 400\ \text{Hz}) = \boxed{1\ 416\ \text{Hz}}$$

(B) The subs barely miss each other and pass. What frequency is detected by an observer riding on sub B as the subs recede from each other?

SOLUTION

Use Equation 13.31 to find the Doppler-shifted frequency heard by the observer in sub B, again being careful with the signs assigned to the source and observer speeds:

$$f' = \left(\frac{v + v_O}{v - v_S}\right)f$$

$$f' = \left[\frac{1\ 533\ \text{m/s} + (-9.00\ \text{m/s})}{1\ 533\ \text{m/s} - (-8.00\ \text{m/s})}\right](1\ 400\ \text{Hz}) = \boxed{1\ 385\ \text{Hz}}$$

13.7 *cont.*

Notice that the frequency drops from 1 416 Hz to 1 385 Hz as the subs pass. This effect is similar to the drop in frequency you hear when a car passes by you while blowing its horn.

(C) While the subs are approaching each other, some of the sound from sub A reflects from sub B and returns to sub A. If this sound were to be detected by an observer on sub A, what is its frequency?

SOLUTION

The sound of apparent frequency 1 416 Hz found in part (A) is reflected from a moving source (sub B) and then detected by a moving observer (sub A). Find the frequency detected by sub A:

$$f'' = \left(\frac{v + v_O}{v - v_S}\right)f'$$

$$= \left[\frac{1\ 533\ \text{m/s} + (+8.00\ \text{m/s})}{1\ 533\ \text{m/s} - (+9.00\ \text{m/s})}\right](1\ 416\ \text{Hz}) = 1\ 432\ \text{Hz}$$

Finalize This technique is used by police officers to measure the speed of a moving car. Microwaves are emitted from the police car and reflected by the moving car. By detecting the Doppler-shifted frequency of the reflected microwaves, the police officer can determine the speed of the moving car.

13.8 | Context Connection: Seismic Waves

As mentioned in the Context introduction, the release of energy in an earthquake occurs at the **focus** or **hypocenter** of the earthquake. The **epicenter** is the point on the Earth's surface radially above the hypocenter. The released energy will propagate away from the focus of the earthquake by means of **seismic waves.** Seismic waves are like the sound waves that we have studied in the later sections of this chapter in that they are mechanical disturbances moving through a medium.

In discussing mechanical waves in this chapter, we identified two types: transverse and longitudinal. In the case of mechanical waves moving through air, we have only a longitudinal possibility. For mechanical waves moving through a solid, however, both possibilities are available because of the strong interatomic forces between elements of the solid. Therefore, in the case of seismic waves, energy propagates away from the focus both by longitudinal and transverse waves.

In the language used in earthquake studies, these two types of waves are named according to the order of their arrival at a seismograph. The longitudinal wave travels at a higher speed than the transverse wave. As a result, the longitudinal wave arrives at a seismograph first and is therefore called the **P wave,** where P stands for *primary*. The slower moving transverse wave arrives next, so it is called the **S wave,** or *secondary* wave.

Let us see why longitudinal waves travel faster than transverse waves. The speed of *all* mechanical waves follows an expression of the general form

$$v = \sqrt{\frac{\text{elastic property}}{\text{inertial property}}}$$

13.32 ◀

For a wave traveling on a string, the speed is given by Equation 13.21:

$$v = \sqrt{\frac{T}{\mu}}$$

where the elastic property is the tension in the string. It is the tension in the string that returns a displaced element of the string to equilibrium. The appropriate inertial property is the linear mass density of the string.

For a transverse wave moving in a bulk solid, the elastic property is the *shear modulus S* of the material.[6] The shear modulus is a parameter that measures the deformation of a solid to a shear force, a force in the sideways direction. For example, lay your textbook down on a table and place your hand flat on the cover. Now, move

[6]For details on various elastic moduli for materials see R. A. Serway and J. W. Jewett Jr., *Physics for Scientists and Engineers*, 8th ed. (Belmont, CA, Brooks-Cole: 2010), Section 12.4.

your hand in a direction away from the book spine. The book will deform so that its cross section changes from a rectangle to a parallelogram. The amount by which the book deforms under a given force from your hand is related to the shear modulus of the book. The speed of a transverse wave (an S wave) in a bulk solid is

$$v_S = \sqrt{\frac{S}{\rho}}$$ 13.33◀

where ρ is the density and S is the shear modulus of the material.

For a longitudinal wave moving in a gas or liquid, the elastic property in Equation 13.32 is the *bulk modulus B* of the material. The bulk modulus is a parameter that measures the change in volume of a sample of material due to a force compressing it that is uniform over a surface area. The speed of sound in a gas is given by

$$v = \sqrt{\frac{B}{\rho}}$$ 13.34◀

where B is the bulk modulus of the gas and ρ is the gas density.

Now we consider longitudinal waves moving through a bulk solid. As a wave passes through a sample of the material, the material is compressed, so the wave speed should depend on the bulk modulus. As the material is compressed along the direction of travel of the wave, however, it is also distorted in the perpendicular direction. (Imagine a partially inflated balloon that is pressed downward against a table. It spreads out in the direction parallel to the table.) The result is a shear distortion of the sample of material. Therefore, the wave speed should depend on both the bulk modulus and the shear modulus! Careful analysis shows that this wave speed is

$$v_P = \sqrt{\frac{B + \frac{4}{3}S}{\rho}}$$ 13.35◀

Notice that this equation for the speed of a P wave gives a value that is larger than that for the S wave in Equation 13.33.

The wave speed for a seismic wave depends on the medium through which it travels. Typical values are 8 km/s for a P wave and 5 km/s for an S wave. Figure 13.24 shows typical seismograph traces of a distant earthquake at two seismograph stations, with the S wave clearly arriving after the P wave.

Figure 13.24 An earthquake occurs at time $t = 0$. Two seismograph traces record the arrival of seismic waves from the earthquake. The bottom trace is for a seismograph located a few hundred miles from the epicenter. The top trace shows the waves arriving at a seismograph thousands of miles from the epicenter. The time interval between arrival of the P and S waves can be used to determine the distance from the epicenter to the seismograph station.

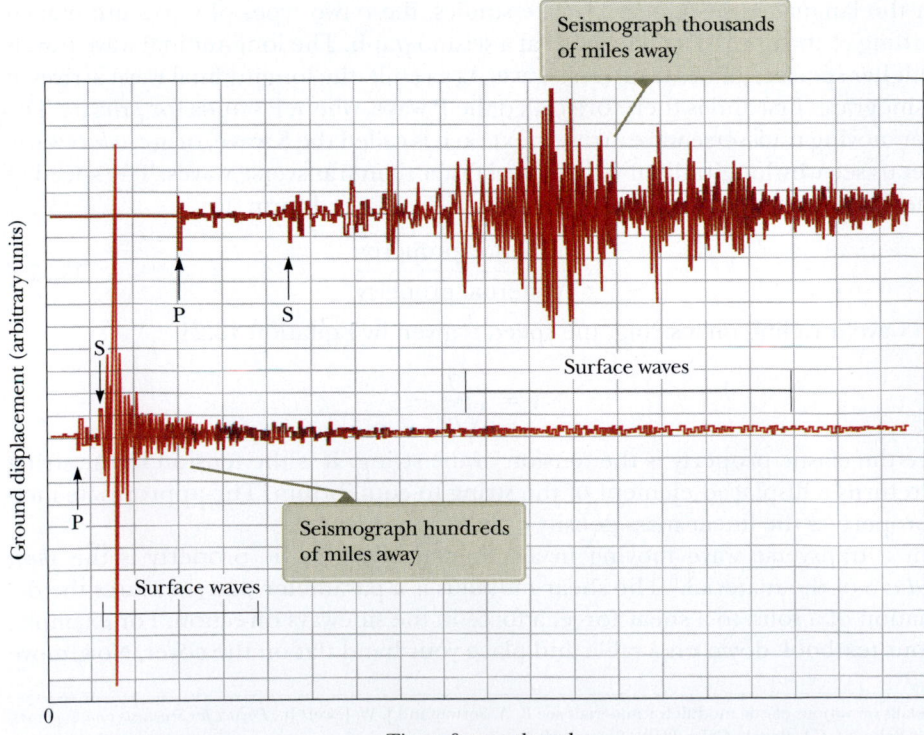

The P and S waves move through the body of the Earth. Once these waves reach the surface, the energy can propagate by additional types of waves along the surface. In a *Rayleigh wave,* the motion of the elements of the medium at the surface is a combination of longitudinal and transverse displacements so that the net motion of a point on the surface is circular or elliptical. This motion is similar to the path followed by elements of water on the ocean surface as a wave passes by, as in Active Figure 13.25. The *Love wave* is a transverse surface wave in which the transverse oscillations are parallel to the surface. Therefore, no vertical displacement of the surface occurs in a Love wave.

It is possible to use the P and S waves traveling through the body of the Earth to gain information about the structure of the Earth's interior. Measurements of a given earthquake by seismographs at various locations on the surface indicate that the Earth has an interior region that allows the passage of P waves but not S waves. This fact can be understood if this particular region is modeled as having liquid characteristics. Similar to a gas, a liquid cannot sustain a transverse force. Therefore, the transverse S waves cannot pass through this region. This information leads us to a structural model in which the Earth has a **liquid core** between radii of approximately 1.2×10^3 km and 3.5×10^3 km.

Other measurements of seismic waves allow additional interpretations of layers within the interior of the Earth, including a **solid core** at the center, a rocky region called the **mantle,** and a relatively thin outer layer called the **crust.** Figure 13.26 shows this structure. Using x-rays or ultrasound in medicine to provide information about the interior of the human body is somewhat similar to using seismic waves to provide information about the interior of the Earth.

As P and S waves propagate in the interior of the Earth, they will encounter variations in the medium. At each boundary at which the properties of the medium change, reflection and transmission occur. When the seismic wave arrives at the surface of the Earth, a small amount of the energy is transmitted into the air as low-frequency sound waves. Some of the energy spreads out along the surface in the form of Rayleigh and Love waves. The remaining wave energy is reflected back into the interior. As a result, seismic waves can travel over long distances within the Earth

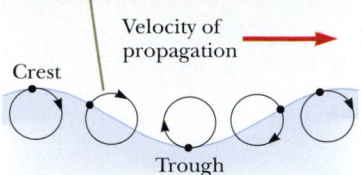

The elements at the surface move in nearly circular paths. Each element is displaced both horizontally and vertically from its equilibrium position.

Active Figure 13.25 The motion of water elements on the surface of deep water in which a wave is propagating is a combination of transverse and longitudinal displacements.

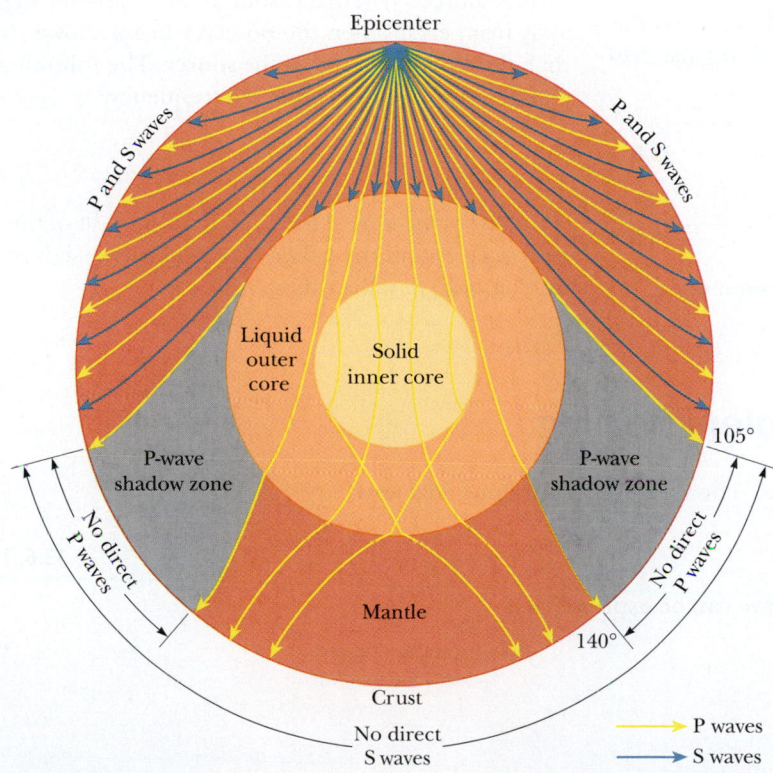

Figure 13.26 Cross section of the Earth showing paths of waves produced by an earthquake. Only P waves (yellow) can propagate in the liquid core. The S waves (blue) do not enter the liquid core. When the P waves transmit from one region to another, such as from the mantle to the liquid core, they experience *refraction,* a change in the direction of propagation. We will study refraction for light in Chapter 25. Because of the refraction for seismic waves, there is a "shadow" zone between 105° and 140° from the epicenter in which no waves following a direct path (i.e., a path with no reflections) arrive.

and can be detected at seismographs at many locations around the globe. In addition, because a relatively large fraction of the wave energy continues to be reflected at each encounter with the surface, the wave can propagate for a long time. Data are available showing seismograph activity for several hours after an earthquake, a result of the repeated reflections of seismic waves from the surface.

Another example of the reflection of seismic waves is available in the technology of oil exploration. A "thumper truck" applies large impulsive forces to the ground, resulting in low-energy seismic waves propagating into the Earth. Specialized microphones are used to detect the waves reflected from various boundaries between layers under the surface. By using computers to map out the underground structure corresponding to these layers, it is possible to detect layers likely to contain oil.

⟩ SUMMARY

A **transverse wave** is a wave in which the elements of the medium move in a direction perpendicular to the direction of the wave velocity. An example is a wave moving along a stretched string.

Longitudinal waves are waves in which the elements of the medium move back and forth parallel to the direction of the wave velocity. Sound waves in air are longitudinal.

Any one-dimensional wave traveling with a speed of v in the positive x direction can be represented by a **wave function** of the form $y = f(x - vt)$. Likewise, the wave function for a wave traveling in the negative x direction has the form $y = f(x + vt)$.

The wave function for a one-dimensional sinusoidal wave traveling to the right can be expressed as

$$y = A \sin\left[\frac{2\pi}{\lambda}(x - vt)\right] \qquad 13.5 ◀$$

where A is the **amplitude**, λ is the **wavelength**, and v is the **wave speed.** The **angular wave number** k and **angular frequency** ω are defined as follows:

$$k \equiv \frac{2\pi}{\lambda} \qquad 13.8 ◀$$

$$\omega \equiv \frac{2\pi}{T} = 2\pi f \qquad 13.9 ◀$$

where T is the **period** of the wave and f is its **frequency.**

The speed of a transverse wave traveling on a stretched string of mass per unit length μ and tension T is

$$v = \sqrt{\frac{T}{\mu}} \qquad 13.21 ◀$$

When a pulse traveling on a string meets a fixed end, the pulse is reflected and inverted. If the pulse reaches a free end, it is reflected but not inverted.

The **power** transmitted by a sinusoidal wave on a stretched string is

$$P = \tfrac{1}{2}\mu\omega^2 A^2 v \qquad 13.24 ◀$$

The change in frequency of a sound wave heard by an observer whenever there is relative motion between a wave source and the observer is called the **Doppler effect.** When the source and observer are moving toward each other, the observer hears a higher frequency than the true frequency of the source. When the source and observer are moving away from each other, the observer hears a lower frequency than the true frequency of the source. The following general equation provides the observed frequency:

$$f' = \left(\frac{v + v_O}{v - v_S}\right)f \qquad 13.31 ◀$$

A positive value is used for v_O or v_S for motion of the observer or source *toward* the other, and a negative sign is used for motion *away from* the other.

⟩ Analysis Model for Problem Solving

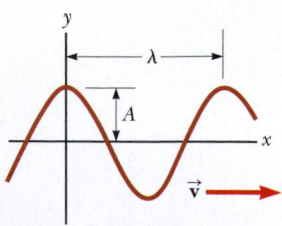

Traveling Wave. The wave speed of a sinusoidal wave is

$$v = \frac{\lambda}{T} = \lambda f \qquad 13.6, 13.12 ◀$$

A sinusoidal wave can be expressed as

$$y = A \sin(kx - \omega t) \qquad 13.10 ◀$$

OBJECTIVE QUESTIONS

1. Which of the following statements is not necessarily true regarding mechanical waves? (a) They are formed by some source of disturbance. (b) They are sinusoidal in nature. (c) They carry energy. (d) They require a medium through which to propagate. (e) The wave speed depends on the properties of the medium in which they travel.

2. The distance between two successive peaks of a sinusoidal wave traveling along a string is 2 m. If the frequency of this wave is 4 Hz, what is the speed of the wave? (a) 4 m/s (b) 1 m/s (c) 8 m/s (d) 2 m/s (e) impossible to answer from the information given

3. Rank the waves represented by the following functions from the largest to the smallest according to (i) their amplitudes, (ii) their wavelengths, (iii) their frequencies, (iv) their periods, and (v) their speeds. If the values of a quantity are equal for two waves, show them as having equal rank. For all functions, x and y are in meters and t is in seconds. (a) $y = 4 \sin (3x - 15t)$ (b) $y = 6 \cos (3x + 15t - 2)$ (c) $y = 8 \sin (2x + 15t)$ (d) $y = 8 \cos (4x + 20t)$ (e) $y = 7 \sin (6x - 24t)$

4. If you stretch a rubber hose and pluck it, you can observe a pulse traveling up and down the hose. (i) What happens to the speed of the pulse if you stretch the hose more tightly? (a) It increases. (b) It decreases. (c) It is constant. (d) It changes unpredictably. (ii) What happens to the speed if you fill the hose with water? Choose from the same possibilities.

5. When all the strings on a guitar (Fig. OQ13.5) are stretched to the same tension, will the speed of a wave along the most massive bass string be (a) faster, (b) slower, or (c) the same as the speed of a wave on the lighter strings? Alternatively, (d) is the speed on the bass string not necessarily any of these answers?

Figure OQ13.5

Aaron Graubart/Getty Images

6. By what factor would you have to multiply the tension in a stretched string so as to double the wave speed? Assume the string does not stretch. (a) a factor of 8 (b) a factor of 4 (c) a factor of 2 (d) a factor of 0.5 (e) You could not change the speed by a predictable factor by changing the tension.

7. A sound wave can be characterized as (a) a transverse wave, (b) a longitudinal wave, (c) a transverse wave or a longitudinal wave, depending on the nature of its source, (d) one that carries no energy, or (e) a wave that does not require a medium to be transmitted from one place to the other.

8. Two sirens A and B are sounding so that the frequency from A is twice the frequency from B. Compared with the speed of sound from A, is the speed of sound from B (a) twice as fast, (b) half as fast, (c) four times as fast, (d) one-fourth as fast, or (e) the same?

9. Table 13.1 shows the speed of sound is typically an order of magnitude larger in solids than in gases. To what can this higher value be most directly attributed? (a) the difference in density between solids and gases (b) the difference in compressibility between solids and gases (c) the limited size of a solid object compared to a free gas (d) the impossibility of holding a gas under significant tension

10. A source vibrating at constant frequency generates a sinusoidal wave on a string under constant tension. If the power delivered to the string is doubled, by what factor does the amplitude change? (a) a factor of 4 (b) a factor of 2 (c) a factor of $\sqrt{2}$ (d) a factor of 0.707 (e) cannot be predicted

11. A source of sound vibrates with constant frequency. Rank the frequency of sound observed in the following cases from highest to the lowest. If two frequencies are equal, show their equality in your ranking. All the motions mentioned have the same speed, 25 m/s. (a) The source and observer are stationary. (b) The source is moving toward a stationary observer. (c) The source is moving away from a stationary observer. (d) The observer is moving toward a stationary source. (e) The observer is moving away from a stationary source.

12. (a) Can a wave on a string move with a wave speed that is greater than the maximum transverse speed $v_{y,\text{max}}$ of an element of the string? (b) Can the wave speed be much greater than the maximum element speed? (c) Can the wave speed be equal to the maximum element speed? (d) Can the wave speed be less than $v_{y,\text{max}}$?

13. If one end of a heavy rope is attached to one end of a lightweight rope, a wave can move from the heavy rope into the lighter one. (i) What happens to the speed of the wave? (a) It increases. (b) It decreases. (c) It is constant. (d) It changes unpredictably. (ii) What happens to the frequency? Choose from the same possibilities. (iii) What happens to the wavelength? Choose from the same possibilities.

14. If a 1.00-kHz sound source moves at a speed of 50.0 m/s toward a listener who moves at a speed of 30.0 m/s in a direction away from the source, what is the apparent frequency heard by the listener? (a) 796 Hz (b) 949 Hz (c) 1 000 Hz (d) 1 068 Hz (e) 1 273 Hz

15. As you travel down the highway in your car, an ambulance approaches you from the rear at a high speed (Fig. OQ13.15) sounding its siren at a frequency of 500 Hz. Which statement is correct? (a) You hear a frequency less than 500 Hz. (b) You hear a frequency equal to 500 Hz. (c) You hear a frequency greater than

Figure OQ13.15

© Anthony Redpath/Corbis

500 Hz. (d) You hear a frequency greater than 500 Hz, whereas the ambulance driver hears a frequency lower than 500 Hz. (e) You hear a frequency less than 500 Hz, whereas the ambulance driver hears a frequency of 500 Hz.

16. Assume a change at the source of sound reduces the wavelength of a sound wave in air by a factor of 2. **(i)** What happens to its frequency? (a) It increases by a factor of 4. (b) It increases by a factor of 2. (c) It is unchanged. (d) It decreases by a factor of 2. (e) It changes by an unpredictable factor. **(ii)** What happens to its speed? Choose from the same possibilities as in part (i).

17. Suppose an observer and a source of sound are both at rest relative to the ground and a strong wind is blowing away from the source toward the observer. **(i)** What effect does the wind have on the observed frequency? (a) It causes an increase. (b) It causes a decrease. (c) It causes no change. **(ii)** What effect does the wind have on the observed wavelength? Choose from the same possibilities as in part (i). **(iii)** What effect does the wind have on the observed speed of the wave? Choose from the same possibilities as in part (i).

CONCEPTUAL QUESTIONS

1. *The Tunguska event.* On June 30, 1908, a meteor burned up and exploded in the atmosphere above the Tunguska River valley in Siberia. It knocked down trees over thousands of square kilometers and started a forest fire, but produced no crater and apparently caused no human casualties. A witness sitting on his doorstep outside the zone of falling trees recalled events in the following sequence. He saw a moving light in the sky, brighter than the Sun and descending at a low angle to the horizon. He felt his face become warm. He felt the ground shake. An invisible agent picked him up and immediately dropped him about a meter from where he had been seated. He heard a very loud protracted rumbling. Suggest an explanation for these observations and for the order in which they happened.

2. (a) How would you create a longitudinal wave in a stretched spring? (b) Would it be possible to create a transverse wave in a spring?

3. Why is a pulse on a string considered to be transverse?

4. Older auto-focus cameras sent out a pulse of sound and measured the time interval required for the pulse to reach an object, reflect off of it, and return to be detected. Can air temperature affect the camera's focus? New cameras use a more reliable infrared system.

5. When a pulse travels on a taut string, does it always invert upon reflection? Explain.

6. Does the vertical speed of an element of a horizontal, taut string, through which a wave is traveling, depend on the wave speed? Explain.

7. Explain how the distance to a lightning bolt (Fig. CQ13.7) can be determined by counting the seconds between the flash and the sound of thunder.

Figure CQ13.7

8. You are driving toward a cliff and honk your horn. Is there a Doppler shift of the sound when you hear the echo? If so, is it like a moving source or a moving observer? What if the reflection occurs not from a cliff, but from the forward edge of a huge alien spacecraft moving toward you as you drive?

9. If you steadily shake one end of a taut rope three times each second, what would be the period of the sinusoidal wave set up in the rope?

10. (a) If a long rope is hung from a ceiling and waves are sent up the rope from its lower end, why does the speed of the waves change as they ascend? (b) Does the speed of the ascending waves increase or decrease? Explain.

11. The radar systems used by police to detect speeders are sensitive to the Doppler shift of a pulse of microwaves. Discuss how this sensitivity can be used to measure the speed of a car.

12. How can an object move with respect to an observer so that the sound from it is not shifted in frequency?

13. In an earthquake, both S (transverse) and P (longitudinal) waves propagate from the focus of the earthquake. The focus is in the ground radially below the epicenter on the surface (Fig. CQ13.13). Assume the waves move in straight lines through uniform material. The S waves travel through the Earth more slowly than the P waves (at about 5 km/s versus 8 km/s). By detecting the time of arrival of the waves at a seismograph, (a) how can one determine the distance to the focus of the earthquake? (b) How many detection stations are necessary to locate the focus unambiguously?

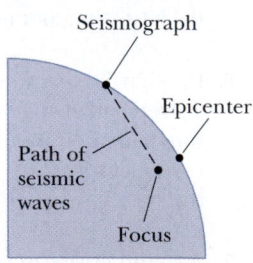

Figure CQ13.13

 **PROBLEMS AVAILABLE IN**

13.1 Propagation of a Disturbance

Problems 1–2

13.2 Analysis Model: Traveling Wave

Problems 3–14

13.3 The Speed of Transverse Waves on Strings

Problems 15–20

13.4 Reflection and Transmission

Problem 21

13.5 Rate of Energy Transfer by Sinusoidal Waves on Strings

Problems 22–26

13.6 Sound Waves

Problems 27–38

13.7 The Doppler Effect

Problems 39–46

13.8 Context Connection: Seismic Waves

Problems 47–48

Additional Problems

Problems 49–71

Solutions to the following Problems are available in the *Student Solutions Manual/Study Guide*:

13.1, 13.3, 13.8, 13.11, 13.15, 13.19, 13.25, 13.28, 13.31, 13.34, 13.41, 13.45, 13.60, and 13.63

List of Enhanced Problems

Problem Number	Targeted Feedback in Enhanced WebAssign	Master It in Enhanced WebAssign	Watch It in Enhanced WebAssign
13.3		✓	
13.5	✓		✓
13.8	✓	✓	
13.10	✓		✓
13.11	✓	✓	
13.13	✓		✓
13.15		✓	
13.17	✓		✓
13.19		✓	
13.20	✓		✓
13.23	✓		✓
13.24	✓		✓
13.25		✓	
13.26		✓	
13.28	✓		✓
13.34	✓	✓	
13.35	✓		✓
13.37	✓		✓
13.38	✓		✓
13.41	✓	✓	
13.47	✓		✓
13.57	✓		✓
13.63		✓	

Chapter 14

Superposition and Standing Waves

Chapter Outline

Blues master B. B. King takes advantage of standing waves on strings. He changes to higher notes on the guitar by pushing the strings against the frets on the fingerboard, shortening the lengths of the portions of the strings that vibrate.

AP Photo/Danny Moloshok

I n Chapter 13, we introduced the wave model. We have seen that waves are very different from particles. An ideal particle is of zero size, but an ideal wave is of infinite length. Another important difference between waves and particles is that we can explore the possibility of two or more waves combining at one point in the same medium. We can combine particles to form extended objects, but the particles must be at different locations. In contrast, two waves can both be present at a given location, and the ramifications of this possibility are explored in this chapter.

One ramification of the combination of waves is that only certain allowed frequencies can exist in systems with boundary conditions; that is, the frequencies are *quantized*. In Chapter 11, we learned about quantized energies of the hydrogen atom. Quantization is at the heart of quantum mechanics, a subject that is introduced formally in Chapter 28. We shall see that waves under boundary conditions explain many of the quantum phenomena. For our present purposes in this chapter, quantization enables us to understand the behavior of the wide array of musical instruments that are based on strings and air columns.

14.1 | Analysis Model: Waves in Interference

Many interesting wave phenomena in nature cannot be described by a single traveling wave. Instead, one must analyze these phenomena in terms of a combination of traveling waves. As noted in the introduction, waves have a remarkable difference from particles in that waves can be combined at the *same* location in space. To analyze such wave combinations, we make use of the **superposition principle:**

▶ Superposition principle

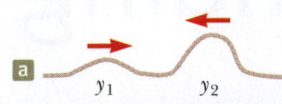

When the pulses overlap, the wave function is the sum of the individual wave functions.

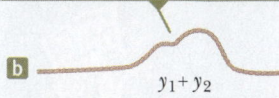

When the crests of the two pulses align, the amplitude is the sum of the individual amplitudes.

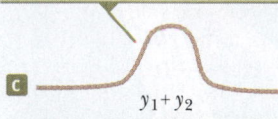

When the pulses no longer overlap, they have not been permanently affected by the interference.

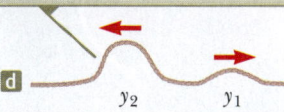

Active Figure 14.1 Constructive interference. Two positive pulses travel on a stretched string in opposite directions and overlap.

▶ Constructive interference

▶ Destructive interference

> If two or more traveling waves are moving through a medium, the resultant value of the wave function at any point is the algebraic sum of the values of the wave functions of the individual waves.

Waves that obey this principle are called *linear waves.* In the case of mechanical waves, linear waves are generally characterized by having amplitudes much smaller than their wavelengths. Waves that violate the superposition principle are called *nonlinear waves* and are often characterized by large amplitudes. In this book, we deal only with linear waves.

One consequence of the superposition principle is that two traveling waves can pass through each other without being destroyed or even altered. For instance, when two pebbles are thrown into a pond and hit the surface at different locations, the expanding circular surface waves from the two locations simply pass through each other with no permanent effect. The resulting complex pattern can be viewed as two independent sets of expanding circles.

Active Figure 14.1 is a pictorial representation of the superposition of two pulses. The wave function for the pulse moving to the right is y_1, and the wave function for the pulse moving to the left is y_2. The pulses have the same speed but different shapes, and the displacement of the elements of the medium is in the positive y direction for both pulses. When the waves overlap (Active Fig. 14.1b), the wave function for the resulting complex wave is given by $y_1 + y_2$. When the crests of the pulses coincide (Active Fig. 14.1c), the resulting wave given by $y_1 + y_2$ has a larger amplitude than that of the individual pulses. The two pulses finally separate and continue moving in their original directions (Active Fig. 14.1d). Notice that the pulse shapes remain unchanged after the interaction, as if the two pulses had never met!

The combination of separate waves in the same region of space to produce a resultant wave is called **interference.** For the two pulses shown in Active Figure 14.1, the displacement of the elements of the medium is in the positive y direction for both pulses, and the resultant pulse (created when the individual pulses overlap) exhibits an amplitude greater than that of either individual pulse. Because the displacements caused by the two pulses are in the same direction, we refer to their superposition as **constructive interference.**

Now consider two pulses traveling in opposite directions on a taut string where one pulse is inverted relative to the other as illustrated in Active Figure 14.2. When these pulses begin to overlap, the resultant pulse is given by $y_1 + y_2$, but the values of the function y_2 are negative. Again, the two pulses pass through each other; because the displacements caused by the two pulses are in opposite directions, however, we refer to their superposition as **destructive interference.**

The superposition principle is the centerpiece of the analysis model called **waves in interference.** In many situations, both in acoustics and optics, waves combine according to this principle and exhibit interesting phenomena with practical applications.

▶ **QUICK QUIZ 14.1** Two symmetric pulses move in opposite directions on a string and are identical in shape except that one has positive displacements of the elements of the string and the other has negative displacements. At the moment the two pulses completely overlap on the string, what happens? (**a**) The energy associated with the pulses has disappeared. (**b**) The string is not moving. (**c**) The string forms a straight line. (**d**) The pulses have vanished and will not reappear.

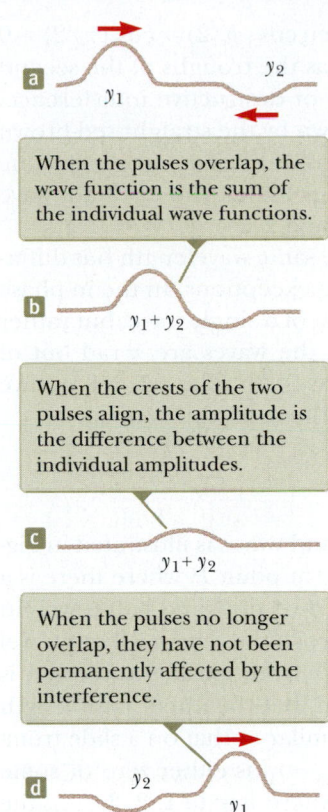

When the pulses overlap, the wave function is the sum of the individual wave functions.

When the crests of the two pulses align, the amplitude is the difference between the individual amplitudes.

When the pulses no longer overlap, they have not been permanently affected by the interference.

Active Figure 14.2 Destructive interference. Two pulses, one positive and one negative, travel on a stretched string in opposite directions and overlap.

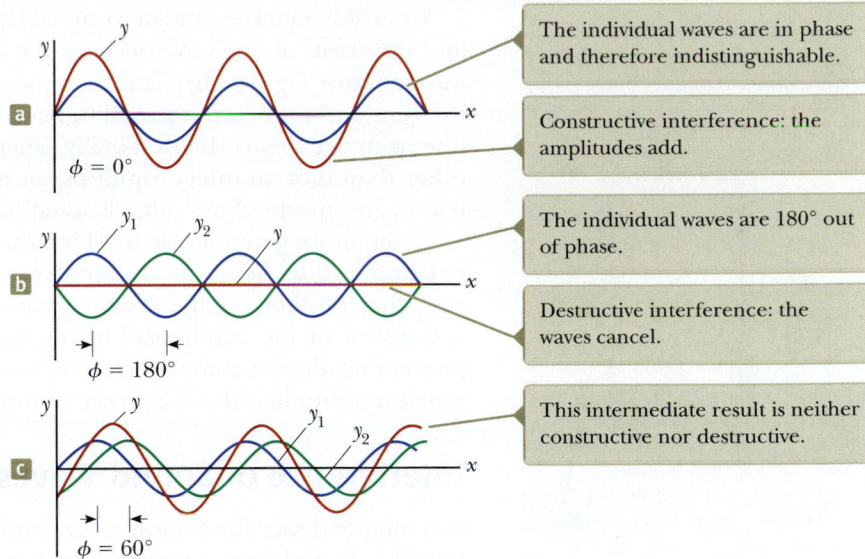

The individual waves are in phase and therefore indistinguishable.

Constructive interference: the amplitudes add.

The individual waves are 180° out of phase.

Destructive interference: the waves cancel.

This intermediate result is neither constructive nor destructive.

Active Figure 14.3 The superposition of two identical waves y_1 and y_2 (blue and green, respectively) to yield a resultant wave (red-brown).

Superposition of Sinusoidal Waves

Let us now apply the principle of superposition to two sinusoidal waves traveling in the same direction in a linear medium. If the two waves are traveling to the right and have the same frequency, wavelength, and amplitude but differ in phase, we can express their individual wave functions as

$$y_1 = A \sin (kx - \omega t) \quad y_2 = A \sin (kx - \omega t + \phi)$$

where, as usual, $k = 2\pi/\lambda$, $\omega = 2\pi f$, and ϕ is the phase constant as discussed in Section 13.2. Hence, the resultant wave function y is

$$y = y_1 + y_2 = A [\sin (kx - \omega t) + \sin (kx - \omega t + \phi)]$$

To simplify this expression, we use the trigonometric identity

$$\sin a + \sin b = 2 \cos \left(\frac{a - b}{2}\right) \sin \left(\frac{a + b}{2}\right)$$

Letting $a = kx - \omega t$ and $b = kx - \omega t + \phi$, we find that the resultant wave function y reduces to

$$y = 2A \cos \left(\frac{\phi}{2}\right) \sin \left(kx - \omega t + \frac{\phi}{2}\right) \qquad \textbf{14.1} \blacktriangleleft$$

▶ Resultant of two traveling sinusoidal waves

> **Pitfall Prevention | 14.1**
> **Do Waves Actually *Interfere*?**
> In popular usage, the word *interfere* implies that an agent affects a situation in some way so as to preclude something from happening. For example, in American football, *pass interference* means that a defending player has affected the receiver so that the receiver is unable to catch the ball. This usage is very different from its use in physics, where waves pass through each other and interfere, but do not affect each other in any way. In physics, interference is similar to the notion of *combination* as described in this chapter.

This result has several important features. The resultant wave function y also is sinusoidal and has the same frequency and wavelength as the individual waves because the sine function incorporates the same values of k and ω that appear in the original wave functions. The amplitude of the resultant wave is $2A \cos (\phi/2)$, and its phase is $\phi/2$. If the phase constant ϕ equals 0, then $\cos (\phi/2) = \cos 0 = 1$ and the amplitude of the resultant wave is $2A$, twice the amplitude of either individual wave. In this case, the crests of the two waves are at the same locations in space and the waves are said to be everywhere *in phase* and therefore interfere constructively. The individual waves y_1 and y_2 combine to form the red-brown curve y of amplitude $2A$ shown in Active Figure 14.3a. Because the individual waves are in phase, they are indistinguishable in Active Figure 14.3a, where they appear as a single blue curve. In general, constructive interference occurs when $\cos (\phi/2) = \pm 1$. That is true, for example, when $\phi = 0$, 2π, 4π, ... rad, that is, when ϕ is an *even* multiple of π.

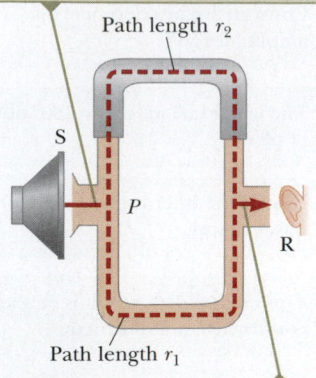

A sound wave from the speaker (S) propagates into the tube and splits into two parts at point P.

Path length r_2

S

P

R

Path length r_1

The two waves, which combine at the opposite side, are detected at the receiver (R).

Figure 14.4 An acoustical system for demonstrating interference of sound waves. The upper path length r_2 can be varied by sliding the upper section.

When ϕ is equal to π rad or to any *odd* multiple of π, then $\cos(\phi/2) = \cos(\pi/2) = 0$ and the crests of one wave occur at the same positions as the troughs of the second wave (Active Fig. 14.3b). Therefore, as a consequence of destructive interference, the resultant wave has *zero* amplitude everywhere as shown by the straight red-brown line in Active Figure 14.3b. Finally, when the phase constant has an arbitrary value other than 0 or an integer multiple of π rad (Active Fig. 14.3c), the resultant wave has an amplitude whose value is somewhere between 0 and $2A$.

In the more general case in which the waves have the same wavelength but different amplitudes, the results are similar with the following exceptions. In the in-phase case, the amplitude of the resultant wave is not twice that of a single wave, but rather is the sum of the amplitudes of the two waves. When the waves are π rad out of phase, they do not completely cancel as in Active Figure 14.3b. The result is a wave whose amplitude is the difference in the amplitudes of the individual waves.

Interference of Sound Waves

One simple device for demonstrating interference of sound waves is illustrated in Figure 14.4. Sound from a loudspeaker S is sent into a tube at point P, where there is a T-shaped junction. Half the sound energy travels in one direction, and half travels in the opposite direction. Therefore, the sound waves that reach the receiver R can travel along either of the two paths. The distance along any path from speaker to receiver is called the **path length** r. The lower path length r_1 is fixed, but the upper path length r_2 can be varied by sliding the U-shaped tube, which is similar to that on a slide trombone. When the difference in the path lengths $\Delta r = |r_2 - r_1|$ is either zero or some integer multiple of the wavelength λ (that is, $\Delta r = n\lambda$, where $n = 0, 1, 2, 3, \ldots$), the two waves reaching the receiver at any instant are in phase and interfere constructively as shown in Active Figure 14.3a. For this case, a maximum in the sound intensity is detected at the receiver. If the path length r_2 is adjusted such that the path difference $\Delta r = \lambda/2, 3\lambda/2, \ldots, n\lambda/2$ (for n odd), the two waves are exactly π rad, or 180°, out of phase at the receiver and hence cancel each other. In this case of destructive interference, no sound is detected at the receiver. This simple experiment demonstrates that a phase difference may arise between two waves generated by the same source when they travel along paths of unequal lengths. This important phenomenon will be indispensable in our investigation of the interference of light waves in Chapter 27.

> ### THINKING PHYSICS 14.1

If stereo speakers are connected to the amplifier "out of phase," one speaker is moving outward when the other is moving inward. The result is a weakness in the bass notes, which can be corrected by reversing the wires on one of the speaker connections. Why are only the bass notes affected in this case and not the treble notes? For help in answering this question, note that the range of wavelengths of sound from a standard piano is from 0.082 m for the highest C to 13 m for the lowest A.

Reasoning Imagine that you are sitting in front of the speakers, midway between them. Then, the sound from each speaker travels the same distance to you, so there is no phase difference in the sound due to a path difference. Because the speakers are connected out of phase, the sound waves are half a wavelength out of phase on leaving the speaker and, consequently, on arriving at your ear. As a result, the sound for all frequencies cancels in the simplification model of a zero-size head located exactly on the midpoint between the speakers.

If the ideal head were moved off the centerline, an additional phase difference is introduced by the path length difference for the sound from the two speakers. In the case of low-frequency, long-wavelength bass notes, the path length differences are a small fraction of a wavelength, so significant cancellation still occurs. For the high-frequency, short-wavelength treble notes, a small movement of the ideal head results in a much larger fraction of a wavelength in path length difference or even multiple wavelengths. Therefore, the treble notes could be in phase with this head movement. If we now add that the head is not of zero size and that it has two ears, we can see that complete cancellation is not possible and, with even small movements of the head, one or both ears will be at or near maxima for the treble notes. The size of the head is much smaller than bass wavelengths, however, so the bass notes are significantly weakened over much of the region in front of the speakers. ◄

Example 14.1 | Two Speakers Driven by the Same Source

Two identical loudspeakers placed 3.00 m apart are driven by the same oscillator (Fig. 14.5). A listener is originally at point O, located 8.00 m from the center of the line connecting the two speakers. The listener then moves to point P, which is a perpendicular distance 0.350 m from O, and she experiences the *first minimum* in sound intensity. What is the frequency of the oscillator?

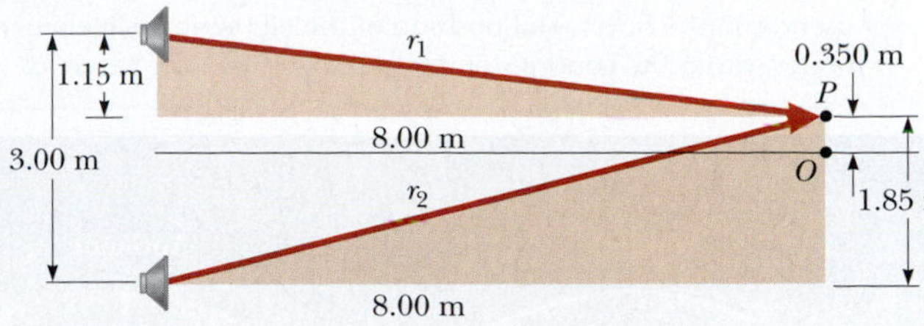

Figure 14.5 (Example 14.1) Two identical loudspeakers emit sound waves to a listener at P.

SOLUTION

Conceptualize In Figure 14.4, a sound wave enters a tube and is then *acoustically* split into two different paths before re-combining at the other end. In this example, a signal representing the sound is *electrically* split and sent to two different loudspeakers. After leaving the speakers, the sound waves recombine at the position of the listener. Despite the difference in how the splitting occurs, the path difference discussion related to Figure 14.4 can be applied here.

Categorize Because the sound waves from two separate sources combine, we apply the waves in interference analysis model.

Analyze Figure 14.5 shows the physical arrangement of the speakers, along with two shaded right triangles that can be drawn on the basis of the lengths described in the problem. The first minimum occurs when the two waves reaching the listener at point P are 180° out of phase, in other words, when their path difference Δr equals $\lambda/2$.

From the shaded triangles, find the path lengths from the speakers to the listener:

$$r_1 = \sqrt{(8.00 \text{ m})^2 + (1.15 \text{ m})^2} = 8.08 \text{ m}$$

$$r_2 = \sqrt{(8.00 \text{ m})^2 + (1.85 \text{ m})^2} = 8.21 \text{ m}$$

Hence, the path difference is $r_2 - r_1 = 0.13$ m. Because this path difference must equal $\lambda/2$ for the first minimum, $\lambda = 0.26$ m.

To obtain the oscillator frequency, use Equation 13.12, $v = \lambda f$, where v is the speed of sound in air, 343 m/s:

$$f = \frac{v}{\lambda} = \frac{343 \text{ m/s}}{0.26 \text{ m}} = 1.3 \text{ kHz}$$

Finalize This example enables us to understand why the speaker wires in a stereo system should be connected properly. When connected the wrong way—that is, when the positive (or red) wire is connected to the negative (or black) terminal on one of the speakers and the other is correctly wired—the speakers are said to be "out of phase," with one speaker moving outward while the other moves inward, as discussed in Thinking Physics 14.1. As a consequence, the sound wave coming from one speaker destructively interferes with the wave coming from the other at point O in Figure 14.5. A rarefaction region due to one speaker is superposed on a compression region from the other speaker. Although the two sounds probably do not completely cancel each other (because the left and right stereo signals are usually not identical), a substantial loss of sound quality occurs at point O.

What If? What if the speakers were connected out of phase? What happens at point P in Figure 14.5?

Answer In this situation, the path difference of $\lambda/2$ combines with a phase difference of $\lambda/2$ due to the incorrect wiring to give a full phase difference of λ. As a result, the waves are in phase and there is a *maximum* intensity at point P.

14.2 | Standing Waves

The sound waves from the pair of loudspeakers in Example 14.1 leave the speakers in the forward direction, and we considered interference at a point in front of the speakers. Suppose we turn the speakers so that they face each other and then have them emit sound of the same frequency and amplitude. In this situation, two identical waves travel in opposite directions in the same medium as in Figure 14.6. These waves combine in accordance with the waves in interference model.

We can analyze such a situation by considering wave functions for two transverse sinusoidal waves having the same amplitude, frequency, and wavelength but traveling in opposite directions in the same medium:

$$y_1 = A \sin(kx - \omega t) \qquad \text{and} \qquad y_2 = A \sin(kx + \omega t)$$

Figure 14.6 Two speakers emit sound waves toward each other. Between the speakers, identical waves traveling in opposite directions combine to form standing waves.

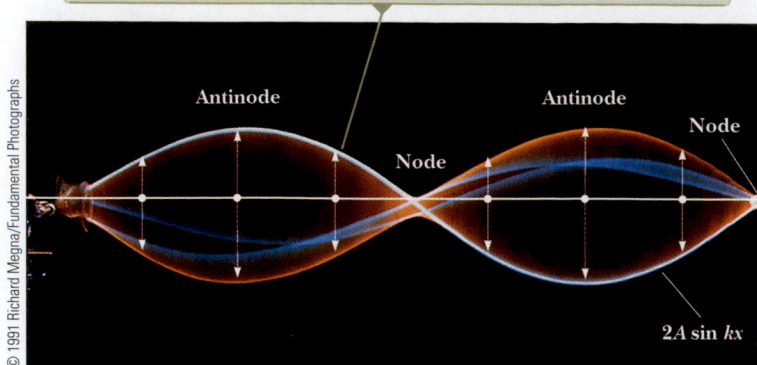

The amplitude of the vertical oscillation of any element of the string depends on the horizontal position of the element. Each element vibrates within the confines of the envelope function 2A sin kx.

Antinode   Antinode

Node

Node

2A sin kx

© 1991 Richard Megna/Fundamental Photographs

Figure 14.7 Multiflash photograph of a standing wave on a string. The vertical displacement from equilibrium of an individual element of the string is proportional to cos ωt. That is, each element vibrates at an angular frequency ω.

▶ Positions of nodes

where y_1 represents a wave traveling in the $+x$ direction and y_2 represents a wave traveling in the $-x$ direction. According to the principle of superposition, adding these two functions gives the resultant wave function y:

$$y = y_1 + y_2 = A \sin(kx - \omega t) + A \sin(kx + \omega t)$$

Using the trigonometric identity $\sin(a \pm b) = \sin a \cos b \pm \cos a \sin b$, this expression reduces to

$$y = (2A \sin kx) \cos \omega t \qquad \textbf{14.2} \blacktriangleleft$$

Notice that this function does not look mathematically like a traveling wave because there is no function of $kx - \omega t$. Equation 14.2 represents the wave function of a **standing wave** such as that shown in Figure 14.7. A standing wave is an oscillation pattern that results from two waves traveling in opposite directions. Mathematically, this equation looks more like simple harmonic motion than wave motion for traveling waves. Every element of the medium vibrates in simple harmonic motion with the same angular frequency ω (according to the factor $\cos \omega t$). The amplitude of motion of a given element (the factor $2A \sin kx$), however, depends on its position along the medium, described by the variable x. From this result, we see that the simple harmonic motion of every element has an angular frequency of ω and a position-dependent amplitude of $2A \sin kx$.

Because the amplitude of the simple harmonic motion of an element at any value of x is equal to $2A \sin kx$, we see that the *maximum* amplitude of the simple harmonic motion has the value $2A$. This maximum amplitude is described as the amplitude of the standing wave. It occurs when the coordinate x for an element satisfies the condition $\sin kx = 1$, or when

$$kx = \frac{\pi}{2}, \frac{3\pi}{2}, \frac{5\pi}{2}, \dots$$

Because $k = 2\pi/\lambda$, the positions of maximum amplitude, called **antinodes,** are

$$x = \frac{\lambda}{4}, \frac{3\lambda}{4}, \frac{5\lambda}{4}, \dots = \frac{n\lambda}{4} \qquad n = 1, 3, 5, \dots \qquad \textbf{14.3} \blacktriangleleft$$

Note that adjacent antinodes are separated by a distance $\lambda/2$.

Similarly, the simple harmonic motion has a *minimum* amplitude of zero when x satisfies the condition $\sin kx = 0$, or when $kx = \pi, 2\pi, 3\pi, \dots$, giving

$$x = 0, \frac{\lambda}{2}, \lambda, \frac{3\lambda}{2}, \dots = \frac{n\lambda}{2} \qquad n = 0, 1, 2, 3, \dots \qquad \textbf{14.4} \blacktriangleleft$$

These points of zero amplitude, called **nodes,** are also spaced by $\lambda/2$. The distance between a node and an adjacent antinode is $\lambda/4$. The standing wave patterns produced at various times by two waves traveling in opposite directions are represented graphically in Active Figure 14.8. The upper part of each figure represents the individual traveling waves and the lower part represents the standing wave patterns. The nodes of the standing wave are labeled N and the antinodes are labeled A. At $t = 0$ (Active Fig. 14.8a), the two waves are in phase, giving a wave pattern with amplitude $2A$. One-quarter of a period later, at $t = T/4$ (Active Fig. 14.8b), the individual waves have moved one-quarter of a wavelength (one to the right and the other to the left). At this time, the waves are 180° out of phase. The individual displacements of the elements of the medium from their equilibrium positions are of equal magnitude and opposite direction for all values of x; hence, the resultant wave has zero displacement everywhere. At $t = T/2$ (Active Fig. 14.8c), the individual waves are again in phase, producing a wave pattern that is inverted relative to the $t = 0$ pattern. In the

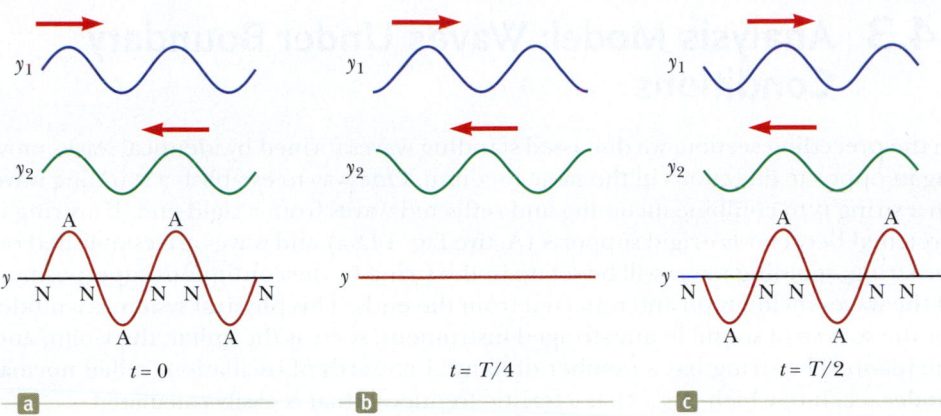

Active Figure 14.8 Standing wave patterns produced at various times by two waves of equal amplitude traveling in opposite directions. For the resultant wave y, the nodes (N) are points of zero displacement and the antinodes (A) are points of maximum displacement.

standing wave, the elements of the medium alternate in time between the extremes shown in Active Figures 14.8a and 14.8c.

QUICK QUIZ 14.2 Consider Active Figure 14.8 as representing a standing wave on a string. Define the velocity of elements of the string as positive if they are moving upward in the figure. (**i**) At the moment the string has the shape shown by the red-brown curve in Active Figure 14.8a, what is the instantaneous velocity of elements along the string? (**a**) zero for all elements (**b**) positive for all elements (**c**) negative for all elements (**d**) varies with the position of the element (**ii**) From the same set of choices, at the moment the string has the shape shown by the red-brown curve in Active Figure 14.8b, what is the instantaneous velocity of elements along the string?

Example 14.2 | Formation of a Standing Wave

Two waves traveling in opposite directions produce a standing wave. The individual wave functions are

$$y_1 = 4.0 \sin (3.0x - 2.0t) \qquad\qquad y_2 = 4.0 \sin (3.0x + 2.0t)$$

where x and y are measured in centimeters and t is in seconds.

(A) Find the amplitude of the simple harmonic motion of the element of the medium located at $x = 2.3$ cm.

SOLUTION

Conceptualize The waves described by the given equations are identical except for their directions of travel, so they indeed combine to form a standing wave as discussed in this section. We can represent the waves graphically by the blue and green curves in Active Figure 14.8.

Categorize We will substitute values into equations developed in this section, so we categorize this example as a substitution problem.

From the equations for the waves, we see that $A = 4.0$ cm, $k = 3.0$ rad/cm, and $\omega = 2.0$ rad/s. Use Equation 14.2 to write an expression for the standing wave:

$$y = (2A \sin kx) \cos \omega t = 8.0 \sin 3.0x \cos 2.0t$$

Find the amplitude of the simple harmonic motion of the element at the position $x = 2.3$ cm by evaluating the coefficient of the cosine function at this position:

$$y_{max} = (8.0 \text{ cm}) \sin 3.0x \big|_{x = 2.3}$$
$$= (8.0 \text{ cm}) \sin (6.9 \text{ rad}) = \boxed{4.6 \text{ cm}}$$

(B) Find the positions of the nodes and antinodes if one end of the string is at $x = 0$.

SOLUTION

Find the wavelength of the traveling waves:

$$k = \frac{2\pi}{\lambda} = 3.0 \text{ rad/cm} \quad \rightarrow \quad \lambda = \frac{2\pi}{3.0} \text{ cm}$$

Use Equation 14.4 to find the locations of the nodes:

$$x = n\frac{\lambda}{2} = n\left(\frac{\pi}{3.0}\right) \text{ cm} \quad n = 0, 1, 2, 3, \ldots$$

Use Equation 14.3 to find the locations of the antinodes:

$$x = n\frac{\lambda}{4} = n\left(\frac{\pi}{6.0}\right) \text{ cm} \quad n = 1, 3, 5, 7, \ldots$$

14.3 | Analysis Model: Waves Under Boundary Conditions

In the preceding section, we discussed standing waves formed by identical waves moving in opposite directions in the same medium. One way to establish a standing wave on a string is to combine incoming and reflected waves from a rigid end. If a string is stretched between *two* rigid supports (Active Fig. 14.9a) and waves are established on the string, standing waves will be set up in the string by the continuous superposition of the waves incident on and reflected from the ends. This physical system is a model for the source of sound in any stringed instrument, such as the guitar, the violin, and the piano. The string has a number of natural patterns of oscillation, called **normal modes,** each of which has a characteristic frequency that is easily calculated.

This discussion is our first introduction to an important analysis model, the **wave under boundary conditions.** When boundary conditions are applied to a wave, we find very interesting behavior that has no analog in the physics of particles. The most prominent aspect of this behavior is **quantization.** We shall find that only certain waves—those that satisfy the boundary conditions—are allowed. The notion of quantization was introduced in Chapter 11 when we discussed the Bohr model of the atom. In that model, angular momentum was quantized. As we shall see in Chapter 29, this quantization is just an application of the wave under boundary conditions model.

In the standing wave pattern on a stretched string, the ends of the string must be nodes because these points are fixed, establishing the boundary condition on the waves. The rest of the pattern can be built from this boundary condition along with the requirement that nodes and antinodes are equally spaced and separated by one-fourth of a wavelength. The simplest pattern that satisfies these conditions has the required nodes at the ends of the string and an antinode at the center point (Active Fig. 14.9b). For this normal mode, the length of the string equals $\lambda/2$ (the distance between adjacent nodes):

$$L = \frac{\lambda_1}{2} \quad \text{or} \quad \lambda_1 = 2L$$

The next normal mode, of wavelength λ_2 (Active Fig. 14.9c), occurs when the length of the string equals one wavelength, that is, when $\lambda_2 = L$. In this mode, the two halves of the string are moving in opposite directions at a given instant, and we sometimes say that two *loops* occur. The third normal mode (Active Fig. 14.9d) corresponds to the case when the length equals $3\lambda/2$; therefore, $\lambda_3 = 2L/3$. In general, the wavelengths of the various normal modes can be conveniently expressed as

▶ Wavelengths of nomal modes

$$\lambda_n = \frac{2L}{n} \qquad n = 1, 2, 3, \ldots \qquad \textbf{14.5} \blacktriangleleft$$

Active Figure 14.9 (a) A string of length L fixed at both ends. (b)–(d) The normal modes of vibration of the string in Active Figure 14.9a form a harmonic series. The string vibrates between the extremes shown.

Second harmonic

N   A   N   A   N
f_2
$n = 2$ $L = \lambda_2$

Fundamental, or first harmonic

A
N   N
f_1
$n = 1$ $L = \frac{1}{2}\lambda_1$

Third harmonic

N A N A N A N
f_3
$n = 3$ $L = \frac{3}{2}\lambda_3$

where the index n refers to the nth mode of oscillation. The natural frequencies associated with these modes are obtained from the relationship $f = v/\lambda$, where the wave speed v is determined by the tension T and linear mass density μ of the string and therefore is the same for all frequencies. Using Equation 14.5, we find that the frequencies of the normal modes are

$$f_n = \frac{v}{\lambda_n} = \frac{n}{2L} v \qquad n = 1, 2, 3, \ldots \qquad \textbf{14.6} \blacktriangleleft$$

▶ Frequencies of normal modes as functions of wave speed and length of string

Because $v = \sqrt{T/\mu}$ (Equation 13.21), we can express the natural frequencies of a stretched string as

$$f_n = \frac{n}{2L} \sqrt{\frac{T}{\mu}} \qquad n = 1, 2, 3, \ldots \qquad \textbf{14.7} \blacktriangleleft$$

▶ Frequencies of normal modes as functions of string tension and linear mass density

Equation 14.7 demonstrates the quantization that we mentioned as a feature of the wave under boundary conditions model. The frequencies are quantized because only certain frequencies of waves satisfy the boundary conditions and can exist on the string. The lowest frequency, corresponding to $n = 1$, is called the **fundamental frequency** f_1 and is

$$f_1 = \frac{1}{2L} \sqrt{\frac{T}{\mu}} \qquad \textbf{14.8} \blacktriangleleft$$

▶ Fundamental frequency of a taut string

The frequencies of the remaining normal modes are integer multiples of the fundamental frequency. Frequencies of normal modes that exhibit such an integer-multiple relationship form a **harmonic series,** and the normal modes are called **harmonics.** The fundamental frequency f_1 is the frequency of the first harmonic, the frequency $f_2 = 2f_1$ is the frequency of the second harmonic, and the frequency $f_n = nf_1$ is the frequency of the nth harmonic. Other oscillating systems, such as a drumhead, exhibit normal modes, but the frequencies are not related as integer multiples of a fundamental. Therefore, we do not use the term *harmonic* in association with those types of systems.

Let us examine further how the various harmonics are created in a string. To excite only a single harmonic, the string must be distorted into a shape that corresponds to that of the desired harmonic. After being released, the string vibrates at the frequency of that harmonic. This maneuver is difficult to perform, however, and is not how a string of a musical instrument is excited. If the string is distorted such that its shape is not that of just one harmonic, the resulting vibration includes a combination of various harmonics. Such a distortion occurs in musical instruments when the string is plucked (as in a guitar), bowed (as in a cello), or struck (as in a piano). When the string is distorted into a nonsinusoidal shape, only waves that satisfy the boundary conditions can persist on the string. These waves are the harmonics.

The frequency of a string that defines the musical note that it plays is that of the fundamental. The string's frequency can be varied by changing either the string's tension or its length. For example, the tension in guitar and violin strings is varied by a screw adjustment mechanism or by tuning pegs located on the neck of the instrument. As the tension is increased, the frequency of the normal modes increases in accordance with Equation 14.7. Once the instrument is "tuned," players vary the frequency by moving their fingers along the neck, thereby changing the length of the oscillating portion of the string. As the length is shortened, the frequency increases because, as Equation 14.7 specifies, the normal-mode frequencies are inversely proportional to string length.

Imagine that we have several strings of the same length under the same tension but varying linear mass density μ. The strings will have different wave speeds and therefore different fundamental frequencies. The linear mass density can be changed either by varying the diameter of the string or by wrapping extra mass around the string. Both of these possibilities can be seen on the guitar, on which the higher-frequency strings vary in diameter and the lower-frequency strings have additional wire wrapped around them.

> **QUICK QUIZ 14.3** When a standing wave is set up on a string fixed at both ends, which of the following statements is true? (**a**) The number of nodes is equal to the number of antinodes. (**b**) The wavelength is equal to the length of the string divided by an integer. (**c**) The frequency is equal to the number of nodes times the fundamental frequency. (**d**) The shape of the string at any instant shows a symmetry about the midpoint of the string.

Example 14.3 | Give Me a C Note!

The middle C string on a piano has a fundamental frequency of 262 Hz, and the string for the first A above middle C has a fundamental frequency of 440 Hz.

(A) Calculate the frequencies of the next two harmonics of the C string.

SOLUTION

Conceptualize Remember that the harmonics of a vibrating string have frequencies that are related by integer multiples of the fundamental.

Categorize This first part of the example is a simple substitution problem.

Knowing that the fundamental frequency is $f_1 = 262$ Hz, find the frequencies of the next harmonics by multiplying by integers:

$$f_2 = 2f_1 = \boxed{524 \text{ Hz}}$$
$$f_3 = 3f_1 = \boxed{786 \text{ Hz}}$$

(B) If the A and C strings have the same linear mass density μ and length L, determine the ratio of tensions in the two strings.

SOLUTION

Categorize This part of the example is more of an analysis problem than is part (A).

Analyze Use Equation 14.8 to write expressions for the fundamental frequencies of the two strings:

$$f_{1A} = \frac{1}{2L}\sqrt{\frac{T_A}{\mu}} \quad \text{and} \quad f_{1C} = \frac{1}{2L}\sqrt{\frac{T_C}{\mu}}$$

Divide the first equation by the second and solve for the ratio of tensions:

$$\frac{f_{1A}}{f_{1C}} = \sqrt{\frac{T_A}{T_C}} \quad \rightarrow \quad \frac{T_A}{T_C} = \left(\frac{f_{1A}}{f_{1C}}\right)^2 = \left(\frac{440}{262}\right)^2 = \boxed{2.82}$$

Finalize If the frequencies of piano strings were determined solely by tension, this result suggests that the ratio of tensions from the lowest string to the highest string on the piano would be enormous. Such large tensions would make it difficult to design a frame to support the strings. In reality, the frequencies of piano strings vary due to additional parameters, including the mass per unit length and the length of the string. The What If? below explores a variation in length.

What If? If you look inside a real piano, you'll see that the assumption made in part (B) is only partially true. The strings are not likely to have the same length. The string densities for the given notes might be equal, but suppose the length of the A string is only 64% of the length of the C string. What is the ratio of their tensions?

Answer Using Equation 14.8 again, we set up the ratio of frequencies:

$$\frac{f_{1A}}{f_{1C}} = \frac{L_C}{L_A}\sqrt{\frac{T_A}{T_C}} \quad \rightarrow \quad \frac{T_A}{T_C} = \left(\frac{L_A}{L_C}\right)^2\left(\frac{f_{1A}}{f_{1C}}\right)^2$$

$$\frac{T_A}{T_C} = (0.64)^2\left(\frac{440}{262}\right)^2 = 1.16$$

Notice that this result represents only a 16% increase in tension, compared with the 182% increase in part (B).

14.4 | Standing Waves in Air Columns

We have discussed musical instruments that use strings, which include guitars, violins, and pianos. What about instruments classified as brasses or woodwinds? These instruments produce music using a column of air. The waves under boundary conditions model can be applied to sound waves in a column of air such as that inside an organ pipe or a clarinet. Standing waves are the result of interference between longitudinal sound waves traveling in opposite directions.

Whether a node or an antinode occurs at the end of an air column depends on whether that end is open or closed. The closed end of an air column is a **displacement node,** just as the fixed end of a vibrating string is a displacement node. Furthermore, because the pressure wave is 90° out of phase with the displacement wave (Section 13.6), the closed end of an air column corresponds to a **pressure antinode** (i.e., a point of maximum pressure variation). On the other hand, the open end of an air column is approximately a **displacement antinode** and a **pressure node.**

You may wonder how a sound wave can reflect from an open end because there may not appear to be a change in the medium at this point. It is indeed true that the medium through which the sound wave moves is air both inside and outside the pipe. Sound is a pressure wave, however, and a compression region of the sound wave is constrained by the sides of the pipe as long as the region is inside the pipe. As the compression region exits at the open end of the pipe, the constraint of the pipe is removed and the compressed air is free to expand into the atmosphere. Therefore, there is a change in the *character* of the medium between the inside of the pipe and the outside even though there is no change in the *material* of the medium. This change in character is sufficient to allow some reflection.[1]

We can determine the modes of vibration of an air column by applying the appropriate boundary condition at the end of the column, along with the requirement that nodes and antinodes be separated by one fourth of a wavelength. We shall find that the frequency for sound waves in air columns is quantized, similar to the results found for waves on strings under boundary conditions.

The first three modes of vibration of a pipe that is open at both ends are shown in Figure 14.10a. Note that the ends are displacement antinodes (approximately). In

Figure 14.10 Graphical representations of the motion of elements of air in standing longitudinal waves in (a) a column open at both ends and (b) a column closed at one end.

In a pipe open at both ends, the ends are displacement antinodes and the harmonic series contains all integer multiples of the fundamental.

In a pipe closed at one end, the open end is a displacement antinode and the closed end is a node. The harmonic series contains only odd integer multiples of the fundamental.

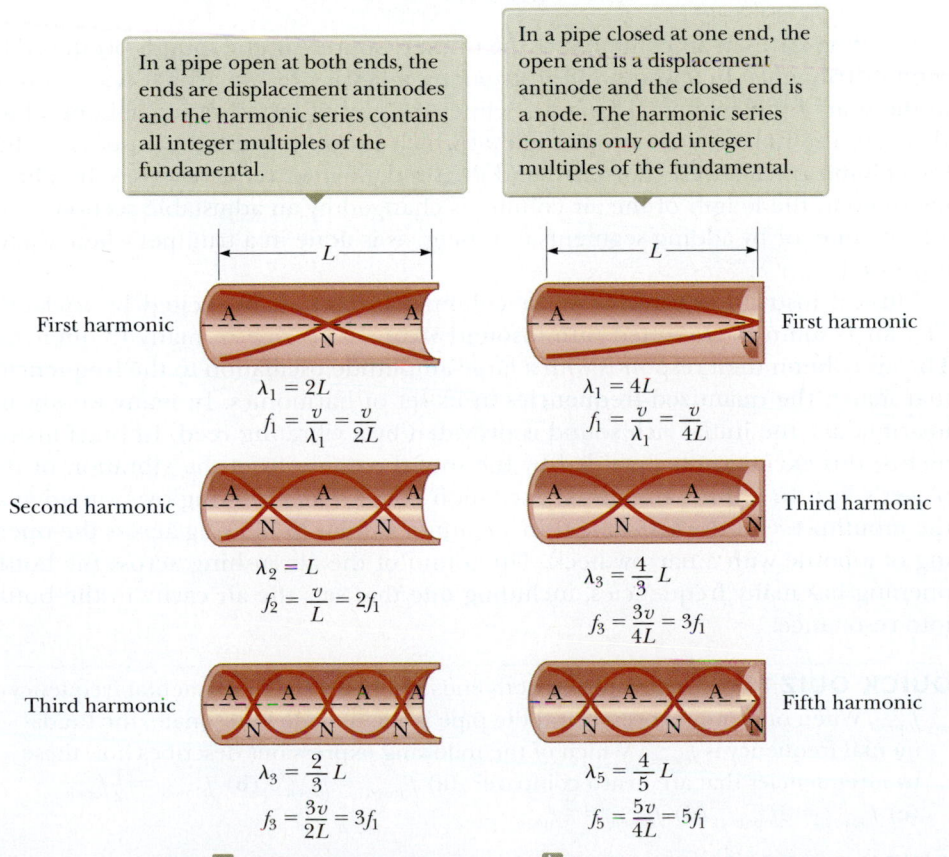

First harmonic
$$\lambda_1 = 2L$$
$$f_1 = \frac{v}{\lambda_1} = \frac{v}{2L}$$

Second harmonic
$$\lambda_2 = L$$
$$f_2 = \frac{v}{L} = 2f_1$$

Third harmonic
$$\lambda_3 = \frac{2}{3}L$$
$$f_3 = \frac{3v}{2L} = 3f_1$$

First harmonic
$$\lambda_1 = 4L$$
$$f_1 = \frac{v}{\lambda_1} = \frac{v}{4L}$$

Third harmonic
$$\lambda_3 = \frac{4}{3}L$$
$$f_3 = \frac{3v}{4L} = 3f_1$$

Fifth harmonic
$$\lambda_5 = \frac{4}{5}L$$
$$f_5 = \frac{5v}{4L} = 5f_1$$

a   b

Pitfall Prevention | 14.3
Sound Waves In Air Are Not Transverse
Note that the standing longitudinal waves are drawn as transverse waves in Figure 14.10. It is difficult to draw longitudinal displacements because they are in the same direction as the propagation. Therefore, it is best to interpret the curves in Figure 14.10 as a graphical representation of the waves (our diagrams of string waves are pictorial representations), with the vertical axis representing horizontal position of the elements of the medium.

[1]Strictly speaking, the open end of an air column is not exactly a displacement antinode. A compression reaching an open does not reflect until it passes beyond the end. For a tube of circular cross section, an end correction equal to approximately 0.6R, where R is the tube's radius, must be added to the length of the air column. Hence, the effective length of the air column is longer than the true length L. We ignore this end correction in this discussion.

the fundamental mode, the standing wave extends between two adjacent antinodes, which is a distance of half a wavelength. Therefore, the wavelength is twice the length of the pipe, and the frequency of the fundamental f_1 is $v/2L$. As Figure 14.10a shows, the frequencies of the higher harmonics are $2f_1$, $3f_1$, ... Therefore,

> in a pipe open at both ends, the natural frequencies of vibration form a harmonic series that includes all integer multiples of the fundamental frequency.

Because all harmonics are present, we can express the natural frequencies of vibration as

▶ Natural frequencies of a pipe open at both ends

$$f_n = n \frac{v}{2L} \qquad n = 1, 2, 3, \dots$$ **14.9**◀

where v is the speed of sound in air.

If a pipe is closed at one end and open at the other, the closed end is a displacement node and the open end is a displacement antinode (Fig. 14.10b). In this case, the wavelength for the fundamental mode is four times the length of the column. Hence, the fundamental frequency f_1 is equal to $v/4L$, and the frequencies of the higher harmonics are equal to $3f_1$, $5f_1$, ... That is,

> in a pipe that is closed at one end, the natural frequencies of oscillation form a harmonic series that includes only odd integer multiples of the fundamental frequency.

We express this result mathematically as

▶ Natural frequencies of a pipe closed at one end and open at the other

$$f_n = n \frac{v}{4L} \qquad n = 1, 3, 5, \dots$$ **14.10**◀

Standing waves in air columns are the primary sources of the sounds produced by wind instruments. In a woodwind instrument, a key is pressed, which opens a hole in the side of the column. This hole defines the end of the vibrating column of air (because the hole acts as an open end at which pressure can be released), so that the column is effectively shortened and the fundamental frequency rises. In a brass instrument, the length of the air column is changed by an adjustable section, as in a trombone, or by adding segments of tubing, as is done in a trumpet when a valve is pressed.

Musical instruments based on air columns are generally excited by *resonance*. The air column is presented with a sound wave that is rich in many frequencies. The air column then responds with a large-amplitude oscillation to the frequencies that match the quantized frequencies in its set of harmonics. In many woodwind instruments, the initial rich sound is provided by a vibrating reed. In brass instruments, this excitation is provided by the sound coming from the vibration of the player's lips. In a flute, the initial excitation comes from blowing over an edge at the mouthpiece of the instrument in a manner similar to blowing across the opening of a bottle with a narrow neck. The sound of the air rushing across the bottle opening has many frequencies, including one that sets the air cavity in the bottle into resonance.

> **QUICK QUIZ 14.4** A pipe open at both ends resonates at a fundamental frequency f_{open}. When one end is covered and the pipe is again made to resonate, the fundamental frequency is f_{closed}. Which of the following expressions describes how these two frequencies that are heard compare? (**a**) $f_{closed} = f_{open}$ (**b**) $f_{closed} = \frac{1}{2}f_{open}$ (**c**) $f_{closed} = 2f_{open}$ (**d**) $f_{closed} = \frac{3}{2}f_{open}$

> **QUICK QUIZ 14.5** Balboa Park in San Diego has an outdoor organ. When the air temperature increases, the fundamental frequency of one of the organ pipes (**a**) stays the same, (**b**) goes down, (**c**) goes up, or (**d**) is impossible to determine.

THINKING PHYSICS 14.2

A bugle has no valves, keys, slides, or finger holes. How can it play a song?

Reasoning Songs for the bugle are limited to harmonics of the fundamental frequency because the bugle has no control over frequencies by means of valves, keys, slides, or finger holes. The player obtains different notes by changing the tension in the lips as the bugle is played to excite different harmonics. The normal playing range of a bugle is among the third, fourth, fifth, and sixth harmonics of the fundamental. As examples, "Reveille" is played with just the three notes D (294 Hz), G (392 Hz), and B (490 Hz), and "Taps" is played with these same three notes and the D one octave above the lower D (588 Hz). Note that the frequencies of these four notes are, respectively, three, four, five, and six times the fundamental of 98 Hz. ◀

THINKING PHYSICS 14.3

If an orchestra doesn't warm up before a performance, the strings go flat and the wind instruments go sharp during the performance. Why?

Reasoning Without warming up, all the instruments will be at room temperature at the beginning of the concert. As the wind instruments are played, they fill with warm air from the player's exhalation. The increase in temperature of the air in the instrument causes an increase in the speed of sound, which raises the fundamental frequencies of the air columns. As a result, the wind instruments go sharp. The strings on the stringed instruments also increase in temperature due to the friction of rubbing with the bow. This increase in temperature results in thermal expansion, which causes a decrease in the tension in the strings. (We will study thermal expansion in Chapter 16.) With a decrease in tension, the wave speed on the strings drops and the fundamental frequencies decrease. Therefore, the stringed instruments go flat. ◀

Example 14.4 | Wind in a Culvert

A section of drainage culvert 1.23 m in length makes a howling noise when the wind blows across its open ends.

(A) Determine the frequencies of the first three harmonics of the culvert if it is cylindrical in shape and open at both ends. Take $v = 343$ m/s as the speed of sound in air.

SOLUTION

Conceptualize The sound of the wind blowing across the end of the pipe contains many frequencies, and the culvert responds to the sound by vibrating at the natural frequencies of the air column.

Categorize This example is a relatively simple substitution problem.

Find the frequency of the first harmonic of the culvert, modeling it as an air column open at both ends:

$$f_1 = \frac{v}{2L} = \frac{343 \text{ m/s}}{2(1.23 \text{ m})} = \boxed{139 \text{ Hz}}$$

Find the next harmonics by multiplying by integers:

$$f_2 = 2f_1 = \boxed{279 \text{ Hz}}$$

$$f_3 = 3f_1 = \boxed{418 \text{ Hz}}$$

(B) What are the three lowest natural frequencies of the culvert if it is blocked at one end?

SOLUTION

Find the frequency of the first harmonic of the culvert, modeling it as an air column closed at one end:

$$f_1 = \frac{v}{4L} = \frac{343 \text{ m/s}}{4(1.23 \text{ m})} = \boxed{69.7 \text{ Hz}}$$

Find the next two harmonics by multiplying by odd integers:

$$f_3 = 3f_1 = \boxed{209 \text{ Hz}}$$

$$f_5 = 5f_1 = \boxed{349 \text{ Hz}}$$

Example 14.5 | Measuring the Frequency of a Tuning Fork

A simple apparatus for demonstrating resonance in an air column is depicted in Figure 14.11. A vertical pipe open at both ends is partially submerged in water, and a tuning fork vibrating at an unknown frequency is placed near the top of the pipe. The length L of the air column can be adjusted by moving the pipe vertically. The sound waves generated by the fork are reinforced when L corresponds to one of the resonance frequencies of the pipe. For a certain pipe, the smallest value of L for which a peak occurs in the sound intensity is 9.00 cm.

(A) What is the frequency of the tuning fork?

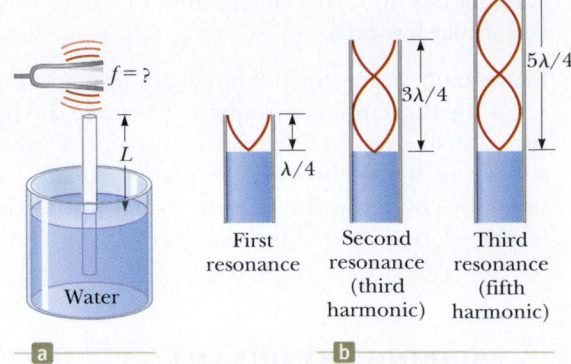

Figure 14.11 (Example 14.5) (a) Apparatus for demonstrating the resonance of sound waves in a pipe closed at one end. The length L of the air column is varied by moving the pipe vertically while it is partially submerged in water. (b) The first three normal modes of the system shown in (a).

SOLUTION

Conceptualize Consider how this problem differs from the preceding example. In the culvert, the length was fixed and the air column was presented with a mixture of very many frequencies. The pipe in this example is presented with one single frequency from the tuning fork, and the length of the pipe is varied until resonance is achieved.

Categorize This example is a simple substitution problem. Although the pipe is open at its lower end to allow the water to enter, the water's surface acts like a barrier. Therefore, this setup can be modeled as an air column closed at one end.

Use Equation 14.10 to find the fundamental frequency for $L = 0.090\ 0$ m:

$$f_1 = \frac{v}{4L} = \frac{343 \text{ m/s}}{4(0.090\ 0 \text{ m})} = \boxed{953 \text{ Hz}}$$

Because the tuning fork causes the air column to resonate at this frequency, this frequency must also be that of the tuning fork.

(B) What are the values of L for the next two resonance conditions?

SOLUTION

Use Equation 13.12 to find the wavelength of the sound wave from the tuning fork:

$$\lambda = \frac{v}{f} = \frac{343 \text{ m/s}}{953 \text{ Hz}} = 0.360 \text{ m}$$

Notice from Figure 14.11b that the length of the air column for the second resonance is $3\lambda/4$:

$$L = 3\lambda/4 = \boxed{0.270 \text{ m}}$$

Notice from Figure 14.11b that the length of the air column for the third resonance is $5\lambda/4$:

$$L = 5\lambda/4 = \boxed{0.450 \text{ m}}$$

◄ 14.5 | Beats: Interference in Time

The interference phenomena that we have discussed so far involve the superposition of two or more waves with the same frequency. Because the resultant displacement of an element in the medium in this case depends on the position of the element, we can refer to the phenomenon as *spatial interference*. Standing waves in strings and air columns are common examples of spatial interference.

We now consider another type of interference effect, one that results from the superposition of two waves with slightly *different* frequencies. In this case, when the two waves of amplitudes A_1 and A_2 are observed at a given point, they are alternately in and out of phase. We refer to this phenomenon as *interference in time* or *temporal interference*. When the waves are in phase, the combined amplitude is $A_1 + A_2$. When they are out of phase, the combined amplitude is $|A_1 - A_2|$. The combination therefore varies between small and large amplitudes, resulting in a phenomenon called **beating.**

Although beats occur for all types of waves, they are particularly noticeable for sound waves. For example, if two tuning forks of slightly different frequencies are struck, you hear a sound of pulsating intensity.

The number of beats you hear per second, the *beat frequency*, equals the difference in frequency between the two sources. The maximum beat frequency that the human ear can detect is about 20 beats/s. When the beat frequency exceeds this value, it blends with the sounds producing the beats.

One can use beats to tune a stringed instrument, such as a piano, by beating a note against a reference tone of known frequency. The frequency of the string can then be adjusted to equal the frequency of the reference by changing the string's tension until the beats disappear; the two frequencies are then the same.

Let us look at the mathematical representation of beats. Consider two waves with equal amplitudes traveling through a medium with slightly different frequencies f_1 and f_2. We can represent the position of an element of the medium associated with each wave at a fixed point, which we choose as $x = 0$, as

$$y_1 = A \cos 2\pi f_1 t \quad \text{and} \quad y_2 = A \cos 2\pi f_2 t$$

Using the superposition principle, we find that the resultant position at that point is given by

$$y = y_1 + y_2 = A(\cos 2\pi f_1 t + \cos 2\pi f_2 t)$$

It is convenient to write this expression in a form that uses the trigonometric identity

$$\cos a + \cos b = 2 \cos \left(\frac{a - b}{2} \right) \cos \left(\frac{a + b}{2} \right)$$

Letting $a = 2\pi f_1 t$ and $b = 2\pi f_2 t$, we find that

$$y = \left[2A \cos 2\pi \left(\frac{f_1 - f_2}{2} \right) t \right] \cos 2\pi \left(\frac{f_1 + f_2}{2} \right) t \qquad \textbf{14.11} \blacktriangleleft$$

Graphs demonstrating the individual waves as well as the resultant wave are shown in Active Figure 14.12. From the factors in Equation 14.11, we see that the resultant wave has an effective frequency equal to the average frequency $(f_1 + f_2)/2$ and an amplitude of

$$A_{x=0} = 2A \cos 2\pi \left(\frac{f_1 - f_2}{2} \right) t \qquad \textbf{14.12} \blacktriangleleft$$

That is, the *amplitude varies in time* with a frequency of $(f_1 - f_2)/2$. When f_1 is close to f_2, this amplitude variation is slow compared with the frequency of the individual waves, as illustrated by the envelope (broken line) of the resultant wave in Active Figure 14.12b.

Note that a maximum in amplitude will be detected whenever

$$\cos 2\pi \left(\frac{f_1 - f_2}{2} \right) t = \pm 1$$

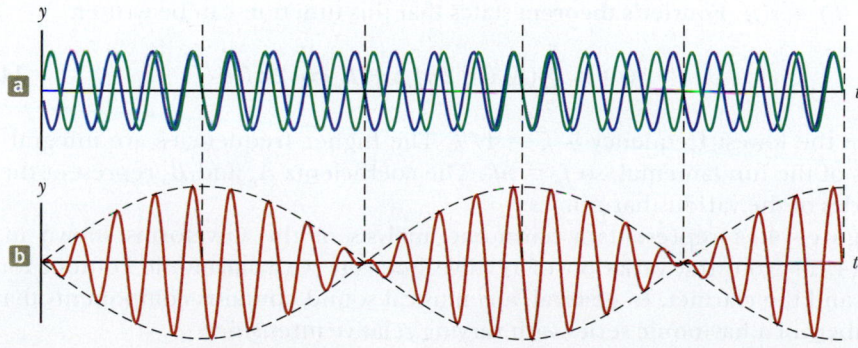

Active Figure 14.12 Beats are formed by the combination of two waves of slightly different frequencies. (a) The blue and green curves represent the individual waves. (b) The combined wave has an amplitude (dashed line) that oscillates in time.

That is, the amplitude maximizes twice in each cycle of the function on the left in the preceding expression. Therefore, the number of beats per second, or the beat frequency f_b, is twice the frequency of this function:

▶ Beat frequency

$$f_b = |f_1 - f_2| \qquad \textbf{14.13} ◀$$

For instance, if two tuning forks vibrate individually at frequencies of 438 Hz and 442 Hz, respectively, the resultant sound wave of the combination has a frequency of $(f_1 + f_2)/2 = 440$ Hz (the musical note A) and a beat frequency of $|f_1 - f_2| = 4$ Hz. That is, the listener hears the 440-Hz sound wave go through an intensity maximum four times every second.

> **QUICK QUIZ 14.6** You are tuning a guitar by comparing the sound of the string with that of a standard tuning fork. You notice a beat frequency of 5 Hz when both sounds are present. You tighten the guitar string and the beat frequency rises to 8 Hz. To tune the string exactly to the tuning fork, what should you do? (**a**) Continue to tighten the string. (**b**) Loosen the string. (**c**) It is impossible to determine.

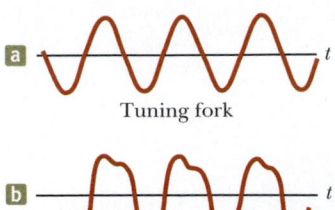

Tuning fork

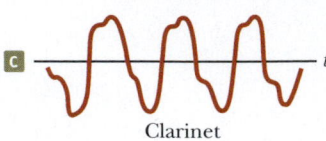

Flute

Clarinet

Figure 14.13 Waveforms of sound produced by (a) a tuning fork, (b) a flute, and (c) a clarinet, each at approximately the same frequency.

14.6 | Nonsinusoidal Wave Patterns

The sound wave patterns produced by most instruments are not sinusoidal. Some characteristic waveforms produced by a tuning fork, a flute, and a clarinet are shown in Figure 14.13. Although each instrument has its own characteristic pattern, Figure 14.13 shows that all three waveforms are periodic. A struck tuning fork produces primarily one harmonic (the fundamental), whereas the flute and clarinet produce many frequencies, which include the fundamental and various harmonics. The nonsinusoidal waveforms produced by a violin or clarinet, and the corresponding richness of musical tones, are the result of the superposition of various harmonics.

This phenomenon is in contrast to a percussive musical instrument, such as the drum, in which the combination of frequencies does not form a harmonic series. When frequencies that are integer multiples of a fundamental frequency are combined, the result is a *musical* sound. A listener can assign a pitch to the sound based on the fundamental frequency. Pitch is a psychological reaction to a sound that allows the listener to place the sound on a scale of low to high (bass to treble). Combinations of frequencies that are not integer multiples of a fundamental result in a *noise* rather than a musical sound. It is much harder for a listener to assign a pitch to a noise than to a musical sound.

Analysis of nonsinusoidal waveforms appears at first sight to be a formidable task. If the waveform is periodic, however, it can be represented with arbitrary precision by the combination of a sufficiently large number of sinusoidal waves that form a harmonic series. In fact, one can represent any periodic function or any function over a finite interval as a series of sine and cosine terms by using a mathematical technique based on *Fourier's theorem*. The corresponding sum of terms that represents the periodic waveform is called a **Fourier series.**

Let $y(t)$ be any function that is periodic in time, with a period of T, so that $y(t + T) = y(t)$. **Fourier's theorem** states that this function can be written

▶ Fourier's theorem

$$y(t) = \sum_n (A_n \sin 2\pi f_n t + B_n \cos 2\pi f_n t) \qquad \textbf{14.14} ◀$$

where the lowest frequency is $f_1 = 1/T$. The higher frequencies are integral multiples of the fundamental, so $f_n = nf_1$. The coefficients A_n and B_n represent the amplitudes of the various harmonics.

Figure 14.14 represents a harmonic analysis of the waveforms shown in Figure 14.13. Note the variation of relative intensity with harmonic content for the flute and the clarinet. In general, any musical sound contains components that are members of a harmonic series with varying relative intensities.

Pitfall Prevention | 14.4
Pitch Versus Frequency
A very common mistake made in speech when talking about sound is to use the term *pitch* when one means *frequency*. Frequency is the physical measurement of the number of oscillations per second, as we have defined. Pitch is a psychological reaction of humans to sound that enables a human to place the sound on a scale from high to low or from treble to bass. Therefore, frequency is the stimulus and pitch is the response. Although pitch is related mostly (but not completely) to frequency, they are not the same. A phrase such as "the pitch of the sound" is incorrect because pitch is not a physical property of the sound.

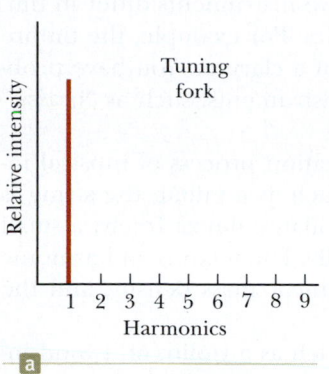
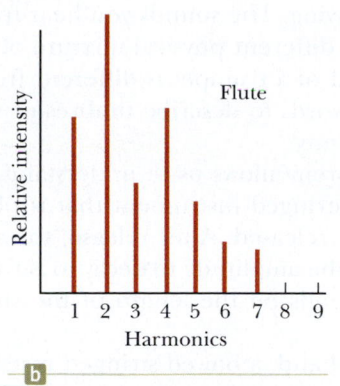
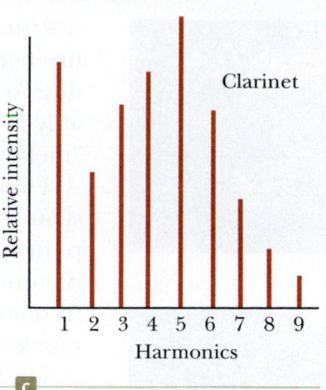

Figure 14.14 Harmonics of the waveforms shown in Figure 14.13. Note the variations in intensity of the various harmonics. Parts (a), (b), and (c) correspond to those in Figure 14.13.

We have discussed the *analysis* of a wave pattern using Fourier's theorem. The analysis involves determining the coefficients of the trigonometric functions in Equation 14.14 from a knowledge of the wave pattern. We can also perform the reverse process, *Fourier synthesis*. In this process, the various harmonics are added together to form a resultant wave pattern. As an example of Fourier synthesis, consider the building of a square wave as shown in Active Figure 14.15. The symmetry of the square wave results in only odd multiples of the fundamental combining in the synthesis. In Active Figure 14.15a, the blue curve shows the combination of f and $3f$. In Active Figure 14.15b, we have added $5f$ to the combination and obtained the green curve. Notice how the general shape of the square wave is approximated, even though the upper and lower portions are not as flat as they should be.

Active Figure 14.15c shows the result of adding odd frequencies up to $9f$, the red-brown curve. This approximation to the square wave (black curve) is better than in parts (a) and (b). To approximate the square wave as closely as possible, we would need to add all odd multiples of the fundamental frequency up to infinite frequency.

The physical mixture of harmonics can be described as the **spectrum** of the sound, with the spectrum displayed in a graphical representation such as Figure 14.14. The psychological reaction to changes in the spectrum of a sound is the detection of a change in the **timbre** or the **quality** of the sound. If a clarinet and a trumpet are both playing the same note, you will assign the same pitch to the two notes. Yet if only one of the instruments then plays the note, you will likely be able to tell which

Active Figure 14.15 Fourier synthesis of a square wave represented by the sum of odd multiples of the first harmonic, which has frequency f.

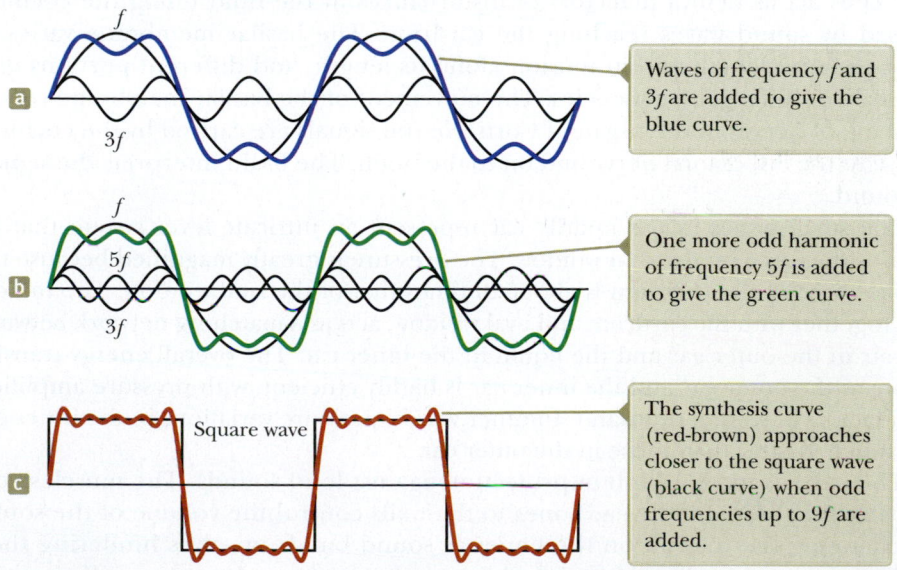

Waves of frequency f and $3f$ are added to give the blue curve.

One more odd harmonic of frequency $5f$ is added to give the green curve.

The synthesis curve (red-brown) approaches closer to the square wave (black curve) when odd frequencies up to $9f$ are added.

Each musical instrument has its own characteristic sound and mixture of harmonics. Instruments shown are (a) the violin, (b) the saxophone, and (c) the trumpet.

instrument is playing. The sounds you hear from the two instruments differ in timbre because of a different physical mixture of harmonics. For example, the timbre due to the sound of a trumpet is different from that of a clarinet. You have probably developed words to describe timbres of various instruments, such as "brassy," "mellow," and "tinny."

Fourier's theorem allows us to understand the excitation process of musical instruments. In a stringed instrument that is plucked, such as a guitar, the string is pulled aside and released. After release, the string oscillates almost freely; a small damping causes the amplitude to decay to zero eventually. The mixture of harmonic frequencies depends on the length of the string, its linear mass density, and the plucking point.

On the other hand, a bowed stringed instrument, such as a violin, or a wind instrument is a forced oscillator. In the case of the violin, the alternate sticking and slipping of the bow on the string provides the periodic driving force. In the case of a wind instrument, the vibration of a reed (in a woodwind), of the lips of the player (in a brass), or the blowing of air across an edge (as in a flute) provides the periodic driving force. According to Fourier's theorem, these periodic driving forces contain a mixture of harmonic frequencies. The violin string or the air column in a wind instrument is therefore driven with a wide variety of frequencies. The frequency actually played is determined by *resonance*, which we studied in Chapter 12. The maximum response of the instrument will be to those frequencies that match or are very close to the harmonic frequencies of the instrument. The spectrum of the instrument therefore depends heavily on the strengths of the various harmonics in the initial periodic driving force.

14.7 | The Ear and Theories of Pitch Perception BIO

The human ear (Fig. 14.16) is divided into three regions: the outer ear, the middle ear, and the inner ear. The *outer ear* consists of the ear canal (which is open to the atmosphere), terminating at the eardrum (tympanum). Sound waves travel down the ear canal to the eardrum, which vibrates in response to the alternating high and low pressures of the waves. Behind the eardrum are three small bones of the *middle ear*, called the hammer, the anvil, and the stirrup because of their shapes. These bones transmit the vibration to the *inner ear*, which contains the cochlea, a snail-shaped tube about 2 cm long. The cochlea makes contact with the stirrup at the oval window and is divided along its length by the basilar membrane into an upper compartment and a lower compartment. Resting on the basilar membrane is the *organ of Corti*, which consists of about 15 000 auditory hair cells (cilia). The hair cells act as neural detectors of disturbances in the fluid filling the cochlea, caused by sound waves reaching the eardrum. The basilar membrane varies in mass per unit length and in tension along its length, and different portions of it resonate at different frequencies. The movement of the basilar membrane results in firing of nerves in the organ of Corti. Neural signals are carried by the cochlear nerve to the 8th cranial nerve and on to the brain. The brain interprets the signals as sound.

The small bones in the middle ear represent an intricate lever system that increases the force on the oval window. The pressure is greatly magnified because the surface area of the eardrum is about 20 times that of the oval window. The middle ear, together with the eardrum and oval window, acts as a matching network between the air in the outer ear and the liquid in the inner ear. The overall energy transfer between the outer ear and the inner ear is highly efficient, with pressure amplification factors of several thousand. In other words, pressure variations in the inner ear are much greater than those in the outer ear.

The ear has its own built-in protection against loud sounds. The muscles connecting the three middle-ear bones to the walls control the volume of the sound by changing the tension on the bones as sound builds up, thus hindering their ability to transmit vibrations. In addition, the eardrum becomes stiffer as the

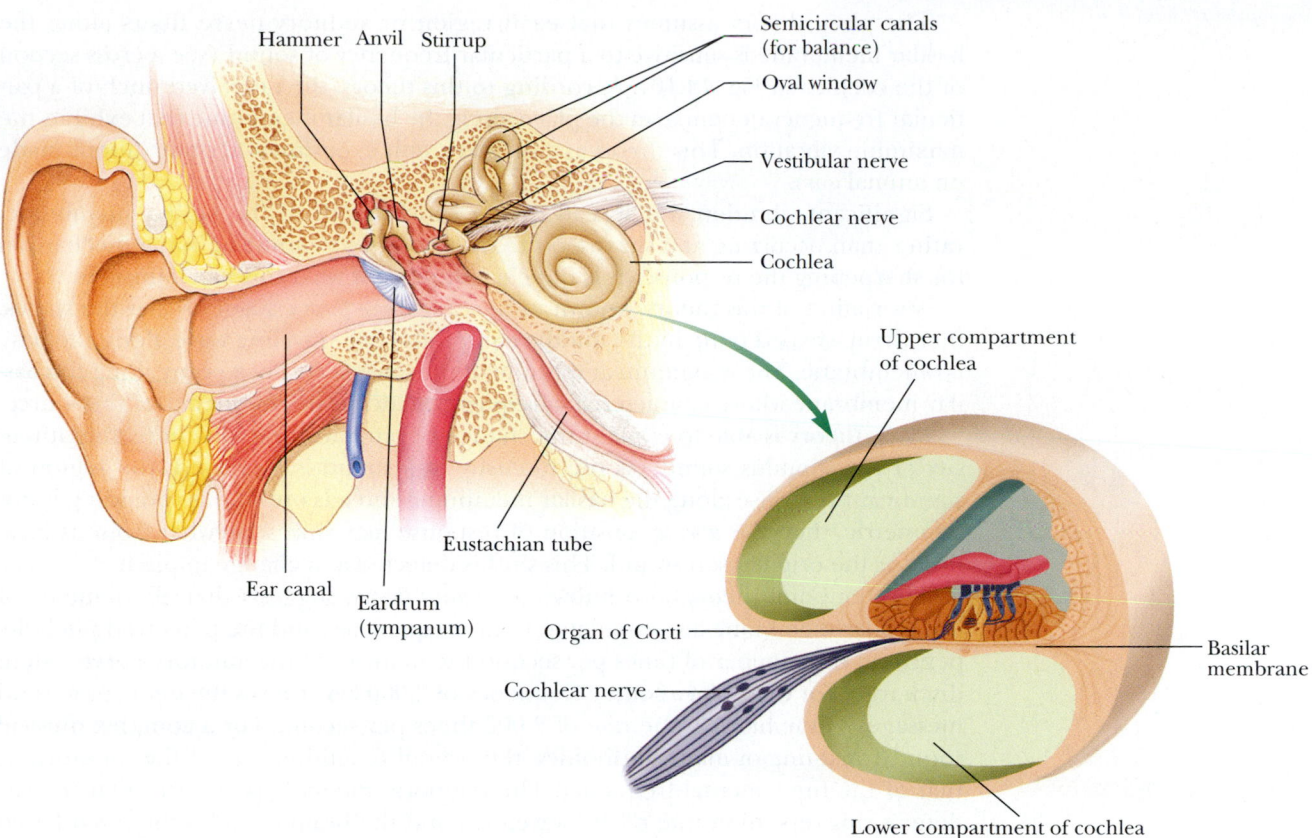

Figure 14.16 The structure of the human ear. The three tiny bones (hammer, anvil, and stirrup) that connect the eardrum to the window of the cochlea act as a double-lever system to decrease the amplitude of vibration and hence increase the pressure on the fluid in the cochlea. The right side of this figure shows a magnified cross section of the cochlea. (Magnified view from Sherwood, *Fundamentals of Human Physiology*, 4th ed., Brooks/Cole.)

sound intensity increases. These two events make the ear less sensitive to loud incoming sounds. There is a time delay between the onset of a loud sound and the ear's protective reaction, however, so a very sudden loud sound can still damage the ear.

Theories of Pitch Perception

As noted in Pitfall Prevention 14.4, *pitch* is a human psychological response to the stimulus of a sound wave. Several theories in the field of *psychoacoustics* have been proposed regarding the manner in which pitch is perceived by the human ear. We shall briefly describe the two most common and accepted theories of hearing, *place theory* and *temporal theory*.

The biggest influence on pitch perception is the frequency of the sound. The pitch can change, however, even though the frequency remains constant. For example, a large fraction of individuals perceive a change in the pitch associated with a sound when its intensity is increased rapidly while its frequency remains fixed. Any successful theory of pitch perception must address this effect. Another interesting experimental result is the *missing fundamental*. Suppose a sound consisting of a mixture of harmonics is presented to a listener. The overall pitch of the sound will be that associated with the fundamental frequency, as noted for vibrating strings in Section 14.3. Now imagine that the fundamental frequency of the sound is filtered out. Experiments show that the pitch of the sound remains the same, even though the frequency associated with that pitch is no longer present. This effect must also be predicted by a theory of pitch perception.

The place theory assumes that each region of auditory nerve fibers along the basilar membrane is sensitive to a particular frequency of sound (see a cross section of the cochlea in Fig. 14.16). According to this theory, the perceived pitch of a particular frequency depends on the place along the basilar membrane that exhibits the maximum vibration. This theory is consistent with experimental observations made on animal ears.

Significant vibration will spread over a short length of the basilar membrane, rather than occurring at a sharp point. Therefore, there must be some mechanism for sharpening the response. It is not clear at present what this mechanism is.

A variation of this theory, known as *traveling wave theory*, suggests that the stapes, the stirrup-shaped bone in the middle ear, produces a traveling wave along the basilar membrane. The maximum amplitude of the wave occurs at a point along the basilar membrane whose characteristic frequency matches the frequency of the source.

Place theory is able to explain the change of pitch with rapidly increasing intensity of the stimulus sound. As the intensity of a sound is increased, the region of significant response along the basilar membrane spreads out. If this spreading is not symmetric, then the *average* position of response may shift somewhat from its location for the original soft sound. This shift is detected as a change in pitch.

The temporal theory, also known as *timing theory*, suggests that all elements of the basilar membrane are stimulated by all frequencies, and the perceived pitch depends on the number of times per second the neurons in the auditory nerve system discharge. For example, a source frequency of 2.000 kHz causes the neurons to send messages to the brain at the rate of 2 000 times per second. For a complex musical sound consisting of many harmonics, the overall repetition rate of the waveform is that of the fundamental frequency. The temporal theory hypothesizes that the ear detects this repetition rate of the waveform and the brain decodes the pitch based on that information.

Temporal theory explains the phenomenon of the missing fundamental. Even after the fundamental frequency is filtered out of a complex sound, *the repetition rate of the combination of harmonics is still that of the fundamental*. For example, if the first harmonic of the clarinet in Figure 14.14c is filtered out, the clarinet waveform in Figure 14.13c will change shape, but it will still repeat at the same frequency, that of the fundamental. Based on this concept, the ear sends a signal to the brain related to the repetition rate of the sound and a pitch associated with the fundamental is assigned even though the fundamental is not present.

BIO Cochlear implants

One of the most amazing medical advances in recent decades is the cochlear implant, allowing some deaf individuals to hear. Deafness can occur when the hair-like sensors (cilia) in the cochlea break off over a lifetime or sometimes because of prolonged exposure to loud sounds. Because the cilia don't grow back, the ear loses sensitivity to certain frequencies of sound. The cochlear implant stimulates the nerves in the ear electronically to restore hearing loss that is due to damaged or absent cilia.

Research using modern cochlear implants suggests that the perception of pitch may depend on both location of response along the basilar membrane *and* the rate at which the neurons fire, a combination of both theories. Place theory may be dominant for frequencies above the maximum firing rate of the neurons. Research continues in this interesting area that combines physics, biology, and psychology.

▌14.8 | Context Connection: Building on Antinodes

As an example of the application of standing waves to earthquakes, we consider the effects of standing waves in *sedimentary basins*. Many of the world's major cities are built on sedimentary basins, which are topographic depressions that over geologic time have filled with sediment. These areas provide large expanses of flat land, often surrounded by attractive mountains, as in the Los Angeles basin. Flat land for building and attractive scenery attracted early settlers and led to today's cities.

Destruction from an earthquake can increase dramatically if the natural frequencies of buildings or other structures coincide with the resonant frequencies of the underlying basin. These resonant frequencies are associated with three-dimensional standing waves, formed from seismic waves reflecting from the boundaries of the basin.

To understand these standing waves, let us assume a simple model of a basin shaped like a half-ellipsoid, similar to an egg sliced in half along its long diameter. Four possible normal modes associated with ground motion in such a basin are shown in the pictorial representation in Figure 14.17. The long axis of the ellipsoid is designated x and the short axis is y. In Figure 14.17a, the entire surface of the ground moves up and down (that is, in and out of the page) except at a nodal curve running around the edge of the basin.

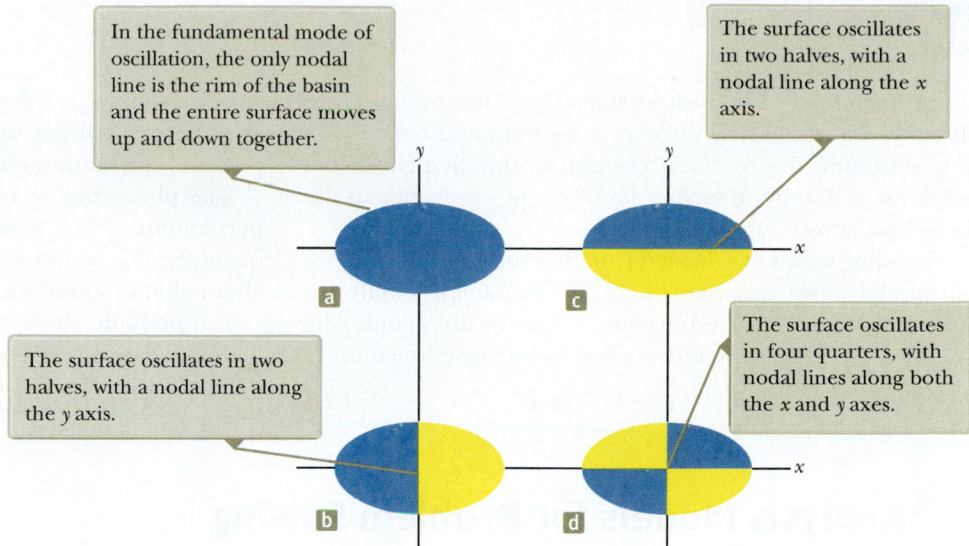

In the fundamental mode of oscillation, the only nodal line is the rim of the basin and the entire surface moves up and down together.

The surface oscillates in two halves, with a nodal line along the x axis.

The surface oscillates in two halves, with a nodal line along the y axis.

The surface oscillates in four quarters, with nodal lines along both the x and y axes.

Figure 14.17 Overhead views of standing waves in a basin shaped like a half-ellipsoid. In each case, if the blue element is above the plane of the page at an instant of time, the yellow element is below the plane of the page.

In Figures 14.17b and 14.17c, half the ground surface lies above and half lies below the equilibrium position, and each half oscillates up and down on either side of a nodal line. The nodal line is along the y axis in Figure 14.17b and along the x axis in Figure 14.17c. In Figure 14.17d, nodal lines occur along both the x and y axes and the surface oscillates in four segments, with two above the equilibrium position at any time and the other two below.

The standing wave patterns in a basin arise from seismic waves traveling horizontally between the boundaries of the basin. For structures built on sedimentary basins, the degree of seismic risk will depend on the standing wave modes excited by the interference of seismic waves trapped in the basin. It is clear that structures built on regions of maximum ground motion (i.e., the antinodes) will suffer maximum shaking, whereas structures residing near nodes will experience relatively mild ground motion. These considerations appear to have played an important role in the selective destruction that occurred in Mexico City in the Michoacán earthquake in 1985 and in the 1989 Loma Prieta earthquake, which caused the collapse of a section of the Nimitz Freeway in Oakland, California.

A similar effect occurs in bounded bodies of water, such as harbors and bays. A standing wave pattern established in such a body of water is called a **seiche**. This wave pattern can result in variations in the water level that exhibit a period of several minutes, superposed on the longer-period tidal variations. Seiches can be caused by earthquakes, tsunamis, winds, or weather disturbances. You can create a seiche in your bathtub by sliding back and forth at just the right frequency such that the water sloshes back and forth at such a large amplitude that much of it spills out onto the floor.

During the Northridge earthquake of 1994, swimming pools throughout southern California overflowed as a result of seiches set up by the shaking of the ground. Seismic events can also cause seiches very far away from the epicenter. The magnitude 8.8 Chile earthquake of February 27, 2010, caused a measureable seiche in Lake Pontchartrain, Louisiana, of height 0.15 m. A more dramatic example is the magnitude 9.0 earthquake in Japan on March 11, 2011. It caused a seiche measured at 1.8 m in Sognefjorden, the largest fjord in Norway!

We have now considered the role of standing waves in the damage caused by an earthquake. In the Context Conclusion, we will gather together the principles of vibrations and waves that we have learned to respond more fully to the central question of this Context.

SUMMARY

The **principle of superposition** states that if two or more traveling waves are moving through a medium and combine at a given point, the resultant position of the element of the medium at that point is the sum of the positions due to the individual waves.

Standing waves are formed from the superposition of two sinusoidal waves that have the same frequency, amplitude, and wavelength but are traveling in *opposite* directions. The resultant standing wave is described by the wave function

$$y = (2A \sin kx) \cos \omega t \qquad \textbf{14.2} \blacktriangleleft$$

The maximum amplitude points (called **antinodes**) are separated by a distance $\lambda/2$. Halfway between antinodes are points of zero amplitude (called **nodes**).

The phenomenon of **beats** occurs as a result of the superposition of two traveling waves of slightly different frequencies. For sound waves at a given point, one hears an alternation in sound intensity with time.

Any periodic waveform can be represented by the combination of sinusoidal waves that form a harmonic series. The process is based on **Fourier's theorem.**

Analysis Models for Problem Solving

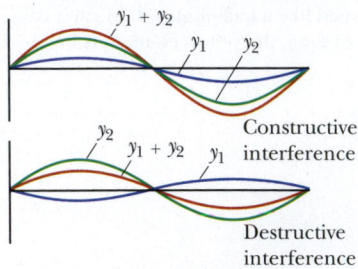

Waves in Interference. When two traveling waves having equal frequencies superimpose, the resultant wave has an amplitude that depends on the phase angle ϕ between the two waves. **Constructive interference** occurs when the two waves are in phase, corresponding to $\phi = 0, 2\pi, 4\pi, \ldots$ rad. **Destructive interference** occurs when the two waves are 180° out of phase, corresponding to $\phi = \pi, 3\pi, 5\pi, \ldots$ rad.

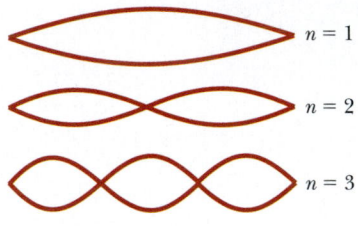

Waves Under Boundary Conditions. When a wave is subject to boundary conditions, only certain natural frequencies are allowed; we say that the frequencies are quantized.

For waves on a string fixed at both ends, the natural frequencies are

$$f_n = \frac{n}{2L}\sqrt{\frac{T}{\mu}} \quad n = 1, 2, 3, \ldots \qquad \textbf{14.7} \blacktriangleleft$$

where T is the tension in the string and μ is its linear mass density.

For sound waves with speed v in an air column of length L open at both ends, the natural frequencies are

$$f_n = n\frac{v}{2L} \quad n = 1, 2, 3, \ldots \qquad \textbf{14.9} \blacktriangleleft$$

If an air column is open at one end and closed at the other, only odd harmonics are present and the natural frequencies are

$$f_n = n\frac{v}{4L} \quad n = 1, 3, 5, \ldots \qquad \textbf{14.10} \blacktriangleleft$$

OBJECTIVE QUESTIONS

☐ denotes answer available in *Student Solutions Manual/Study Guide*

1. A flute has a length of 58.0 cm. If the speed of sound in air is 343 m/s, what is the fundamental frequency of the flute, assuming it is a tube closed at one end and open at the other? (a) 148 Hz (b) 296 Hz (c) 444 Hz (d) 591 Hz (e) none of those answers

2. A string of length L, mass per unit length μ, and tension T is vibrating at its fundamental frequency. **(i)** If the length of the string is doubled, with all other factors held constant, what is the effect on the fundamental frequency? (a) It becomes two times larger. (b) It becomes $\sqrt{2}$ times larger. (c) It is unchanged. (d) It becomes $1/\sqrt{2}$ times as large. (e) It becomes one-half as large. **(ii)** If the mass per unit length is doubled, with all other factors held constant, what is the effect on the fundamental frequency? Choose from the same possibilities as in part (i). **(iii)** If the tension is doubled, with all other factors held constant, what is the effect

on the fundamental frequency? Choose from the same possibilities as in part (i).

3. In Figure OQ14.3, a sound wave of wavelength 0.8 m divides into two equal parts that recombine to interfere constructively, with the original difference between their path lengths being $|r_2 - r_1| = 0.8$ m. Rank the following situations according to the intensity of sound at the receiver from the highest to the lowest. Assume the tube walls absorb no sound energy. Give equal ranks to situations in which the intensity is equal. (a) From its original position, the sliding section is moved out by 0.1 m. (b) Next it slides out an additional 0.1 m. (c) It slides out still another 0.1 m. (d) It slides out 0.1 m more.

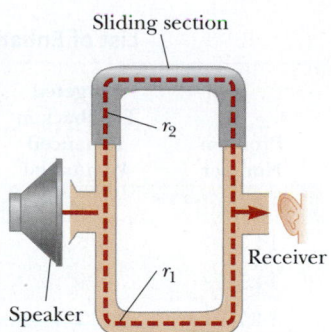

Figure OQ14.3

4. A series of pulses, each of amplitude 0.1 m, is sent down a string that is attached to a post at one end. The pulses are reflected at the post and travel back along the string without loss of amplitude. **(i)** What is the net displacement at a point on the string where two pulses are crossing? Assume the string is rigidly attached to the post. (a) 0.4 m (b) 0.3 m (c) 0.2 m (d) 0.1 m (e) 0 **(ii)** Next assume the end at which reflection occurs is free to slide up and down. Now what is the net displacement at a point on the string where two pulses are crossing? Choose your answer from the same possibilities as in part (i).

5. Suppose all six equal-length strings of an acoustic guitar are played without fingering, that is, without being pressed down at any frets. What quantities are the same for all six strings? Choose all correct answers. (a) the fundamental frequency (b) the fundamental wavelength of the string wave (c) the fundamental wavelength of the sound emitted (d) the speed of the string wave (e) the speed of the sound emitted

6. When two tuning forks are sounded at the same time, a beat frequency of 5 Hz occurs. If one of the tuning forks has a frequency of 245 Hz, what is the frequency of the other tuning fork? (a) 240 Hz (b) 242.5 Hz (c) 247.5 Hz (d) 250 Hz (e) More than one answer could be correct.

7. As oppositely moving pulses of the same shape (one upward, one downward) on a string pass through each other, at one particular instant the string shows no displacement from the equilibrium position at any point. What has happened to the energy carried by the pulses at this instant of time? (a) It was used up in producing the previous motion. (b) It is all potential energy. (c) It is all internal energy. (d) It is all kinetic energy. (e) The positive energy of one pulse adds to zero with the negative energy of the other pulse.

8. Assume two identical sinusoidal waves are moving through the same medium in the same direction. Under what condition will the amplitude of the resultant wave be greater than either of the two original waves? (a) in all cases (b) only if the waves have no difference in phase (c) only if the phase difference is less than $90°$ (d) only if the phase difference is less than $120°$ (e) only if the phase difference is less than $180°$

9. An archer shoots an arrow horizontally from the center of the string of a bow held vertically. After the arrow leaves it, the string of the bow will vibrate as a superposition of what standing-wave harmonics? (a) It vibrates only in harmonic number 1, the fundamental. (b) It vibrates only in the second harmonic. (c) It vibrates only in the odd-numbered harmonics 1, 3, 5, 7, ... (d) It vibrates only in the even-numbered harmonics 2, 4, 6, 8, ... (e) It vibrates in all harmonics.

10. A tuning fork is known to vibrate with frequency 262 Hz. When it is sounded along with a mandolin string, four beats are heard every second. Next, a bit of tape is put onto each tine of the tuning fork, and the tuning fork now produces five beats per second with the same mandolin string. What is the frequency of the string? (a) 257 Hz (b) 258 Hz (c) 262 Hz (d) 266 Hz (e) 267 Hz

11. A standing wave having three nodes is set up in a string fixed at both ends. If the frequency of the wave is doubled, how many antinodes will there be? (a) 2 (b) 3 (c) 4 (d) 5 (e) 6

☐ denotes answer available in *Student Solutions Manual/Study Guide*

CONCEPTUAL QUESTIONS

1. A crude model of the human throat is that of a pipe open at both ends with a vibrating source to introduce the sound into the pipe at one end. Assuming the vibrating source produces a range of frequencies, discuss the effect of changing the pipe's length.

2. When two waves interfere constructively or destructively, is there any gain or loss in energy in the system of the waves? Explain.

3. Explain how a musical instrument such as a piano may be tuned by using the phenomenon of beats.

4. Does the phenomenon of wave interference apply only to sinusoidal waves?

5. What limits the amplitude of motion of a real vibrating system that is driven at one of its resonant frequencies?

6. An airplane mechanic notices that the sound from a twin-engine aircraft rapidly varies in loudness when both engines are running. What could be causing this variation from loud to soft?

7. A soft-drink bottle resonates as air is blown across its top. What happens to the resonance frequency as the level of fluid in the bottle decreases?

8. A tuning fork by itself produces a faint sound. Explain how each of the following methods can be used to obtain a louder sound from it. Explain also any effect on the time interval for which the fork vibrates audibly. (a) holding the edge of a sheet of paper against one vibrating tine (b) pressing the handle of the tuning fork against a chalkboard or a tabletop (c) holding the tuning fork above a column of air of properly chosen length as in Example 14.5 (d) holding the tuning fork close to an open slot cut in a sheet of foam plastic or cardboard (with the slot similar in size and shape to one tine of the fork and the motion of the tines perpendicular to the sheet)

PROBLEMS AVAILABLE IN

14.1 Analysis Model: Waves in Interference

Problems 1–12

14.2 Standing Waves

Problems 13–17

14.3 Analysis Model: Waves Under Boundary Conditions

Problems 18–28

14.4 Standing Waves in Air Columns

Problems 29–43

14.5 Beats: Interference in Time

Problems 44–46

14.6 Nonsinusoidal Wave Patterns

Problems 47–48

14.7 The Ear and Theories of Pitch Perception

Problems 49–50

14.8 Context Connection: Building on Antinodes

Problems 51–52

Additional Problems

Problems 53–69

Solutions to the following Problems are available in the *Student Solutions Manual/Study Guide*:

14.3, 14.7, 14.9, 14.13, 14.17, 14.19, 14.27, 14.29, 14.34, 14.43, 14.46, 14.57, 14.59, and 14.62

List of Enhanced Problems

Problem Number	Targeted Feedback in Enhanced WebAssign	Master It in Enhanced WebAssign	Watch It in Enhanced WebAssign
14.1	✓		✓
14.3		✓	
14.5	✓		✓
14.7	✓		
14.9		✓	
14.10	✓		✓
14.13	✓	✓	
14.15	✓		✓
14.17	✓	✓	
14.18	✓		✓
14.27	✓		✓
14.32	✓		✓
14.35		✓	
14.40	✓		
14.44	✓		✓
14.45		✓	
14.46		✓	
14.59	✓		

Minimizing the Risk

We have explored the physics of vibrations and waves. Let us now return to our central question for this *Earthquakes* Context:

> **How can we choose locations and build structures to minimize the risk of damage in an earthquake?**

To answer this question, we shall use the physical principles that we now understand more clearly and apply them to our choices of locations and structural design.

In our discussion of simple harmonic oscillation, we learned about resonance. Resonance is one of the most important considerations in designing buildings with regard to earthquake safety. Designers of structures in earthquake-prone areas need to pay careful attention to resonance vibrations from shaking of the ground. The design features to be considered include ensuring that the resonance frequencies of the building do not match typical earthquake frequencies. In addition, the structural details should include sufficient damping to ensure that the amplitude of resonance vibration does not destroy the structure.

Resonance is a prime consideration for the design of a structure; what about, as suggested by our central question, the *location* of the structure? In Chapter 13, we discussed the role of the medium in the propagation of a wave. For seismic waves

© Lloyd Cluff/CORBIS

Figure 1 Portions of the double-decked Nimitz Freeway in Oakland, California, collapsed during the Loma Prieta earthquake of 1989.

moving across the surface of the Earth, the soil on the surface is the medium. Because soil varies from one location to another, the speed of seismic waves will vary at different locations. A particularly dangerous situation exists for structures built on loose soil or mudfill. In these types of media, the interparticle forces are much weaker than in a more solid foundation such as granite bedrock. As a result, the wave speed is less in loose soil than in bedrock.

Consider Equation 13.24, which provides an expression for the rate of energy transfer by waves. This equation was derived for waves on strings, but the proportionality to the square of the amplitude and the speed is general. Because of conservation of energy, the rate of energy transfer for a wave must remain constant regardless of the medium. Therefore, according to Equation 13.24, if the wave speed decreases, as it does for seismic waves moving from rock into loose soil, the amplitude must increase. As a result, the shaking of structures built on loose soil is of larger magnitude than for those built on solid bedrock.

This factor contributed to the collapse of the Nimitz Freeway during the Loma Prieta earthquake, near San Francisco, in 1989. Figure 1 shows the results of the earthquake on the freeway. The portion of the freeway that collapsed was built on mudfill, but the surviving portion was built on bedrock. The amplitude of oscillation in the portion built on mudfill was more than five times as large as the amplitude of other portions.

Another danger for structures on loose soil is the possibility of **liquefaction** of the soil. When soil is shaken, the elements of soil can move with respect to one another and the soil tends to act like a liquid rather than a solid. It is possible for the structure to sink into the soil during an earthquake. If the liquefaction is not uniform over the foundation of the structure, the structure can deviate from its vertical orientation, as seen in the case of the Japanese police station in Figure 2. In some cases, buildings can tip over completely, as happened to some apartment buildings during a Japanese earthquake in 1964. As a result, even if the earthquake vibrations are not sufficient to damage the structure, it will be unusable in its leaning orientation.

As discussed in Section 14.8, constructing buildings or other structures where standing seismic waves can be established is dangerous. Such construction was a factor in the Michoacán earthquake of 1985. The shape of the bedrock under Mexico City resulted in standing waves, with severe damage to buildings located at antinodes.

In summary, to minimize risk of damage in an earthquake, architects and engineers must design structures to prevent destructive resonances, avoid building on loose soil, and pay attention to the underground rock formations so as to be aware of possible standing wave patterns. Other precautions can also be taken. For example, buildings can be constructed with **seismic isolation** from the ground. This method involves mounting the structure on **isolation dampers,** heavy-duty bearings that dampen the oscillations of the building, resulting in reduced amplitude of vibration. Figure 3 shows the results of the 2011 earthquake in Christchurch, New Zealand, on a building that did not take advantage of isolation dampers. Many older buildings have been retrofitted with dampers, including several in California (Los Angeles City Hall, San Francisco City Hall, Oakland City Hall) as well as in other parts of the world, such as the New Zealand Parliament Buildings. Additional measures include tuned dampers, such as that in the opening photograph of Chapter 12, shear trusses, external bracing, and other techniques.

Koki Nagahama/Getty Images

Figure 2 A police station leans to one side due to liquefaction of the underlying soil during the Japanese earthquake of March 2011.

Hannah Johnston/Getty Images

Figure 3 Damage to a garage building in Christchurch, New Zealand, after the magnitude 6.3 earthquake on February 22, 2011. The garage did not have isolation dampers installed to isolate it from the ground.

We have not addressed many other considerations for earthquake safety in structures, but we have been able to apply many of our concepts from oscillations and waves so as to understand some aspects of logical choices in locating and designing structures.

Problems

1. For seismic waves spreading out from a point (the epicenter) on the surface of the Earth, the intensity of the waves decreases with distance according to an inverse proportionality to distance. That is, the wave intensity is proportional to $1/r$, where r is the distance from the epicenter to the observation point. This rule applies if the medium is uniform. The intensity of the wave is proportional to the rate of energy transfer for the wave. Furthermore, we have shown that the energy of vibration of an oscillator is proportional to the square of the amplitude of the vibration. Assume that a particular earthquake causes ground shaking with an amplitude of 5.0 cm at a distance of 10 km from the epicenter. If the medium is uniform, what is the amplitude of the ground shaking at a point 20 km from the epicenter?

2. As mentioned in the text, the amplitude of oscillation during the Loma Prieta earthquake of 1989 was five times greater in areas of mudfill than in areas of bedrock. From this information, find the factor by which the seismic wave speed changed as the waves moved from the bedrock to the mudfill. Ignore any reflection of wave energy and any change in density between the two media.

3. Figure 4 is a graphical representation of the travel time for P and S waves from the epicenter of an earthquake to a seismograph as a function of the distance of travel. The following table shows the measured times of day for arrival of P waves from a particular earthquake at three seismograph locations. In the last column, fill in the times of day for the arrival of the S waves at the three seismograph locations.

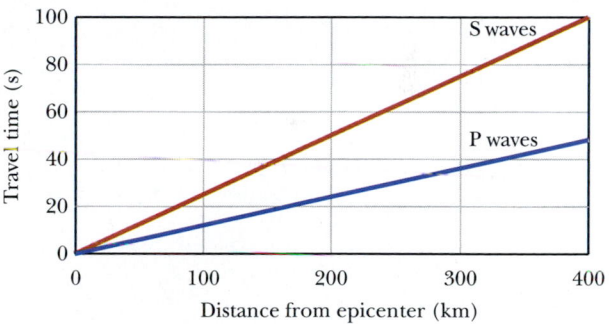

Figure 4 A graph of travel time versus distance from the epicenter for P and S waves.

Seismograph Station	Distance from Epicenter (km)	P Wave Arrival Time	S Wave Arrival Time
#1	200	15:46:06	
#2	160	15:46:01	
#3	105	15:45:54	

Heart Attacks

During an average lifetime, the human heart beats over three billion times without rest, pumping over one million barrels of blood (there are 42 gallons, or 159 liters, in a barrel). This rhythm of life, however, is sometimes interrupted by a heart attack, or a *myocardial infarction* (as it is known medically), one of the leading causes of death in the world. A heart attack occurs when there is an interruption of blood flow to the heart, often resulting in permanent damage to this vital organ. The term *cardiovascular disease* (CVD) refers to diseases affecting the heart and the blood vessels. Figure 1 shows the prevalence of deaths attributed to cardiovascular disease and total deaths per year per 100 000 men of age 35 to 74 in several developed countries. The percentage

of all deaths due to CVD ranges from a low of 19.7% in France to 48% for the Russian Federation. Cardiovascular disease accounts for 31% of all deaths in the United States each year for men of age 35 to 74. The corresponding rate for women in the same age bracket is 25%.

The human *cardiovascular system,* or *circulatory system,* has been the subject of scientific interest for millennia. The Ebers Papyrus from the 16th century BC proposed a connection between the heart and the arteries. In the second century, Galen, a prominent Greek physician famous for attempting cataract surgeries, identified the roles of the blood carried by arteries and veins. Ibn Al-Nafis, a 13th-century Arab physician, correctly

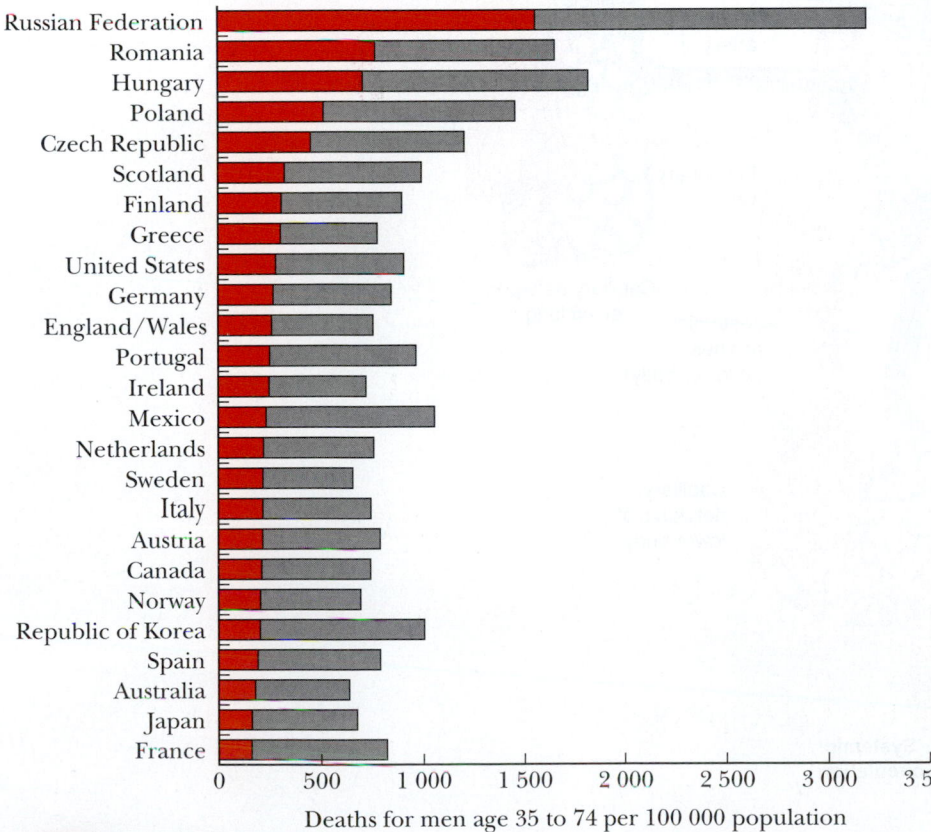

Deaths for men age 35 to 74 per 100 000 population

■ CVD deaths
■ Deaths from all causes

Figure 1 Annual deaths per 100 000 men from cardiovascular disease (red) and from all causes (gray) in selected countries. The largest percentage due to cardiovascular disease compared to all deaths occurs in the Russian Federation. (Graph derived from Table 2–3, p. e42, International Death Rates (Revised 2008): Death Rates (Per 100 000 Population) for Total Cardiovascular Disease, Coronary Heart Disease, Stroke, and Total Deaths in Selected Countries (Most Recent Year Available) by WRITING GROUP MEMBERS et al. for American Heart Association Statistics Committee and Stroke Statistics Subcommittee, "Heart Disease and Stroke Statistics—2009 Update: A Report From the American Heart Association Statistics Committee and Stroke Statistics Subcommittee" Circulation 119 (3): e21–e181.)

described the pulmonary circulation system, the portion of the cardiovascular system that delivers blood from the heart to the lungs and back. Building on the work of his predecessors, William Harvey is credited with the discovery and nearly complete description of the circulatory system in a 1628 publication, as well as with the realization that the heart was responsible for pumping blood throughout the body. Harvey correctly explained the roles of pulmonary circulation in oxygenating the blood and disposing of the carbon dioxide produced by cell metabolism, and of systemic circulation (see Fig. 2) in carrying oxygenated blood to vital organs.

While pumping oxygenated blood through its chambers, the heart itself relies on a network of vessels and capillaries surrounding its outer surface for its own oxygen supply, and is the third largest consumer of oxygen in the human body (roughly 12% of the total oxygen intake), after the liver (20%) and the brain (18%). Figure 3 shows the heart's surface and the network of blood vessels that provide oxygen to the heart.

We can identify several examples of systems that depend on the flow of fluids for proper operation. For example, if a water pipe in a home ruptures, the water supply to sinks, showers, and washing machines is

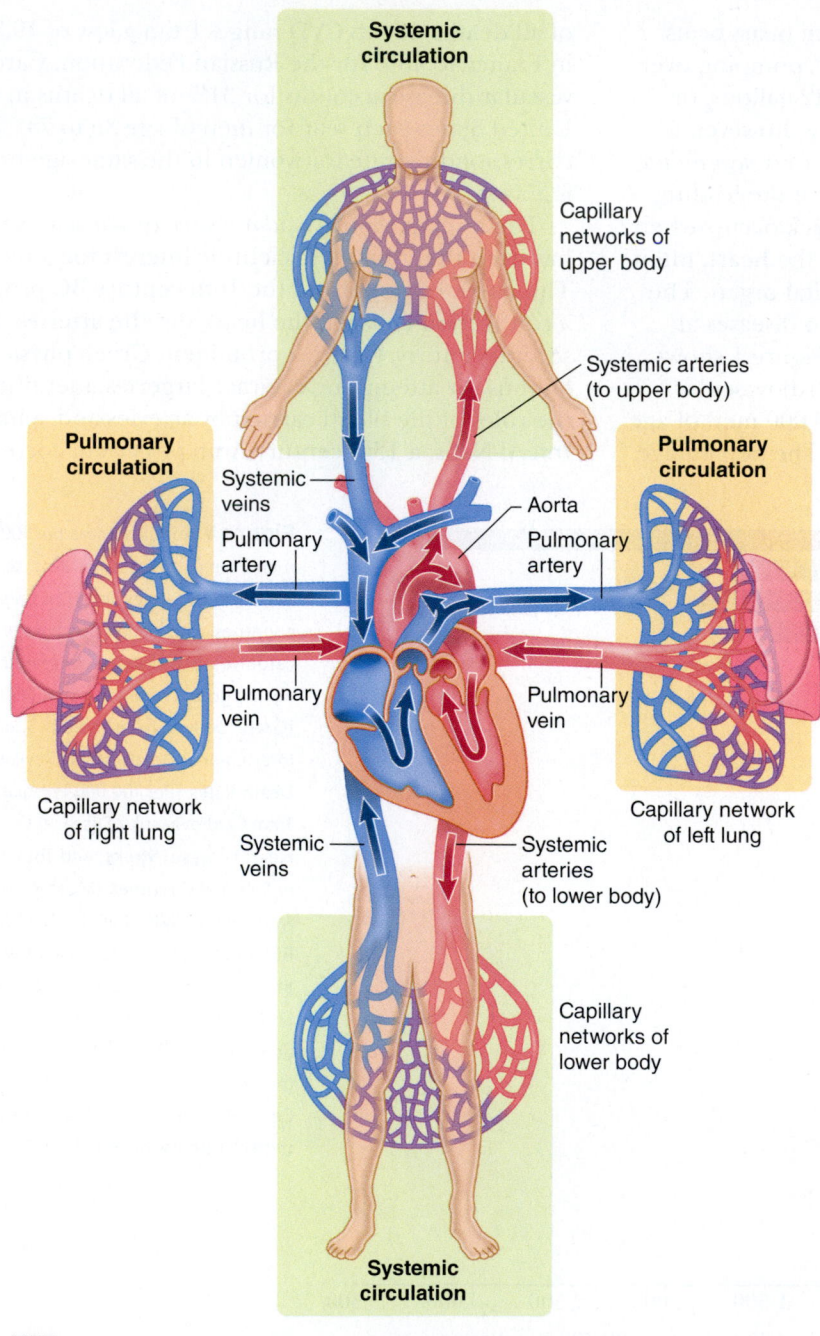

Figure 2 The human circulatory system consists of two separate loops. The pulmonary circulation system exchanges blood between the heart and the lungs. The systemic circulation system exchanges blood between the heart and the other organs of the body. (From Sherwood, *Fundamentals of Human Physiology*, 4th ed., 2012, Brooks/Cole, Figure 9.1, p. 230)

Systemic circulation

Capillary networks of upper body

Systemic arteries (to upper body)

Pulmonary circulation

Systemic veins

Aorta

Pulmonary artery

Pulmonary artery

Pulmonary circulation

Pulmonary vein

Pulmonary vein

Capillary network of right lung

Systemic veins

Systemic arteries (to lower body)

Capillary network of left lung

Capillary networks of lower body

Systemic circulation

KEY

= O$_2$-rich blood   = O$_2$-poor blood

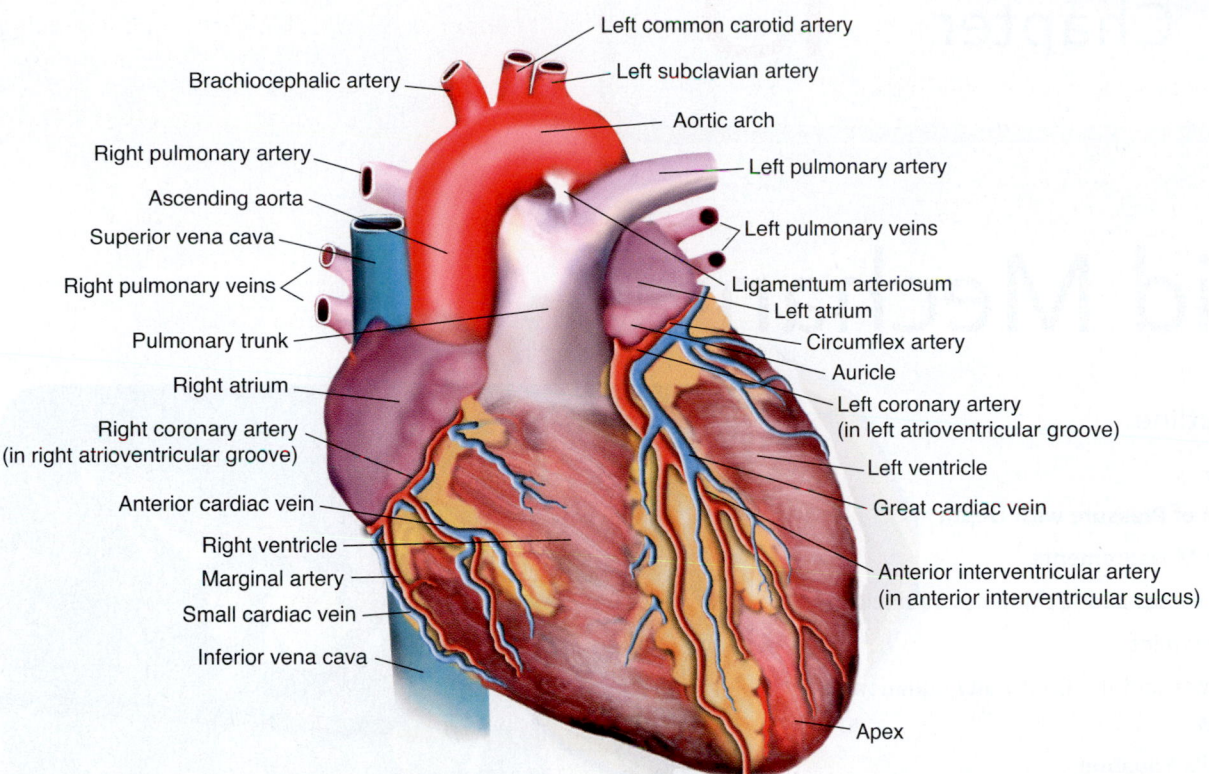

Figure 3 The human heart. In this diagram, we see sections of the major blood vessels carrying blood to the rest of the body as well as the system of vessels supplying blood to the heart itself. [from Des Jardins, *Cardiopulmonary Anatomy and Physiology: Essentials for Respiratory Care*, 5th ed., 2008, Delmar, p. 189 (Fig. 5-2, part (A)]

affected. If a hydraulic line of an automobile's brake system ruptures, the brakes may fail to operate. In a similar way, a defect in the blood vessels that affects the flow of blood to the heart can cause a number of dangerous medical conditions, including heart attacks.

Although heart attacks are sudden occurrences, in most cases they are the consequence of years of plaque buildup in arteries. During a heart attack, plaque in the arterial network of the heart ruptures, resulting in the formation of a blood clot that can cause the disruption or stoppage of blood flow to a portion of the heart, depriving it of oxygen. If oxygen deprivation lasts too long, the cells in the affected portion of the heart die, resulting in permanent heart damage.

Even if a patient survives a heart attack, it is a life-changing event. A number of lifestyle changes are necessary to reduce the risk of a subsequent heart attack, including incorporation of exercise in one's daily schedule, alterations in the diet, cessation of smoking, and a variety of medications. In addition, careful blood pressure management is necessary—again suggesting the importance of the flow of fluid within the circulatory system.

Having seen this introduction to the circulatory system and some of the impacts of heart disease and heart attacks on human lives, we will explore the physics of fluids in this Context. We will apply the principles that we learn to this question:

> **How can the principles of physics be applied in medicine to help prevent heart attacks?**

Chapter 15

Fluid Mechanics

Chapter Outline

Ed Robinson/Pacific Stock/Photolibrary

Fish congregate around a reef in Hawaii searching for food. How do fish such as the yellow butterflyfish in the front control their movements up and down in the water? We'll find out in this chapter.

Matter is normally classified as being in one of three states: solid, liquid, or gas. Everyday experience tells us that a solid has a definite volume and shape. A brick maintains its familiar shape and size over a long time. We also know that a liquid has a definite volume but no definite shape. For example, a cup of liquid water has a fixed volume but assumes the shape of its container. Finally, an unconfined gas has neither definite volume nor definite shape. For example, if there is a leak in the natural gas supply in your home, the escaping gas continues to expand into the surrounding atmosphere. These definitions help us picture the states of matter, but they are somewhat artificial. For example, asphalt, glass, and plastics are normally considered solids, but over a long time interval they tend to flow like liquids. Likewise, most substances can be a solid, liquid, or gas (or combinations of these states), depending on the temperature and pressure. In general, the time interval required for a particular substance to change its shape in response to an external force determines whether we treat the substance as a solid, liquid, or gas.

A **fluid** is a collection of molecules that are randomly arranged and held together by weak cohesive forces between molecules and forces exerted by the walls of a container. Both liquids and gases are fluids. In our treatment of the mechanics of fluids, we shall see that no new physical principles are needed to explain such effects as the buoyant force on a submerged object and a curve ball in baseball. In this chapter, we shall apply a number of familiar analysis models to the physics of fluids.

15.1 | Pressure

Our first task in understanding the physics of fluids is to define a new quantity to describe fluids. Imagine applying a force to the surface of an object, with the force having components both parallel to and perpendicular to the surface.

If the object is a solid at rest on a table, the force component perpendicular to the surface may cause the object to flatten, depending on how hard the object is. Assuming that the object does not slide on the table, the component of the force parallel to the surface of the object will cause the object to distort. As an example, suppose you place your physics book flat on a table and apply a force with your hand parallel to the front cover and perpendicular to the spine. The book will distort, with the bottom pages staying fixed at their original location and the top pages shifting horizontally by some distance. The cross section of the book changes from a rectangle to a parallelogram. This kind of force parallel to the surface is called a *shearing force*.

We shall adopt a simplification model in which the fluids we study will be nonviscous; that is, no friction exists between adjacent layers of the fluid. Nonviscous fluids and static fluids do not sustain shearing forces. If you imagine placing your hand on a water surface and pushing parallel to the surface, your hand simply slides over the water; you cannot distort the water as you did the book. This phenomenon occurs because the interatomic forces in a fluid are not strong enough to lock atoms in place with respect to one another. The fluid cannot be modeled as a rigid object as in Chapter 10. If we try to apply a shearing force, the molecules of the fluid simply slide past one another.

Therefore, the only type of force that can exist in a fluid is one that is perpendicular to a surface. For example, the forces exerted by the fluid on the object in Figure 15.1 are everywhere perpendicular to the surfaces of the object.

The force that a fluid exerts on a surface originates in the collisions of molecules of the fluid with the surface. Each collision results in the reversal of the component of the velocity vector of the molecule perpendicular to the surface. By the impulse–momentum theorem and Newton's third law, each collision results in a force on the surface. A huge number of these impulsive forces occur every second, resulting in a constant macroscopic force on the surface. This force is spread out over the area of the surface and is related to a new quantity called *pressure*.

The pressure at a specific point in a fluid can be measured with the device pictured in Figure 15.2. The device consists of an evacuated cylinder enclosing a light piston connected to a spring. As the device is submerged in a fluid, the fluid presses in on the top of the piston and compresses the spring until the inward force of the fluid is balanced by the outward force of the spring. The force exerted on the piston by the fluid can be measured if the spring is calibrated in advance.

If F is the magnitude of the force exerted by the fluid on the piston and A is the surface area of the piston, the **pressure** P of the fluid at the level to which the device has been submerged is defined as the ratio of force to area:

$$P \equiv \frac{F}{A}$$

15.1 ◄   ▶ Definition of pressure

Although we have defined pressure in terms of our device in Figure 15.2, the definition is general. Because pressure is force per unit area, it has units of newtons per square meter in the SI system. Another name for the SI unit of pressure is the **pascal** (Pa):

$$1 \text{ Pa} \equiv 1 \text{ N/m}^2$$

15.2 ◄   ▶ The pascal

Notice that pressure and force are different quantities. We can have a very large pressure from a relatively small force by making the area over which the force is applied small. Such is the case with hypodermic needles. The area of the tip of the needle is very small, so a small force pushing on the needle is sufficient to cause a pressure large enough to puncture the skin. We can also create a small pressure from a large force by enlarging the area over which the force acts. Such is the principle behind the design of snowshoes. If a person were to walk on deep snow with regular shoes, it is possible for his or her feet to break through the snow and sink. Snowshoes, however, allow the force on the snow due to the weight of the person to spread out over a larger area, reducing the pressure enough so that the snow surface is not broken (Fig. 15.3, page 414).

The atmosphere exerts a pressure on the surface of the Earth and all objects at the surface. This pressure is responsible for the action of suction cups, drinking

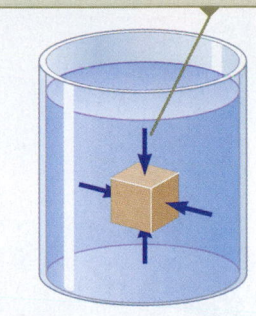

At any point on the surface of the object, the force exerted by the fluid is perpendicular to the surface of the object.

Figure 15.1 The forces exerted by a fluid on the surfaces of a submerged object. (The forces on the front and back sides of the object are not shown.)

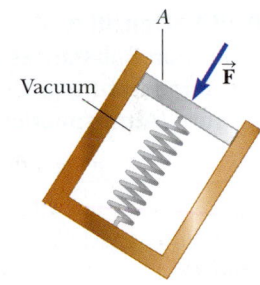

Figure 15.2 A simple device for measuring the pressure exerted by a fluid.

Pitfall Prevention | 15.1
Force and Pressure
Equation 15.1 makes a clear distinction between force and pressure. Another important distinction is that *force is a vector* and *pressure is a scalar*. There is no direction associated with pressure, but the direction of the force associated with the pressure is perpendicular to the surface on which the pressure acts.

Figure 15.3 Snowshoes keep you from sinking into soft snow because they spread the downward force you exert on the snow over a large area, reducing the pressure on the snow surface.

TABLE 15.1 | Densities of Some Common Substances at Standard Temperature (0°C) and Pressure (Atmospheric)

Substance	ρ (kg/m^3)
Air	1.29
Air (at 20°C and atmospheric pressure)	1.20
Aluminum	2.70×10^3
Benzene	0.879×10^3
Brass	8.4×10^3
Copper	8.92×10^3
Ethyl alcohol	0.806×10^3
Fresh water	1.00×10^3
Glycerin	1.26×10^3
Gold	19.3×10^3
Helium gas	1.79×10^{-1}
Hydrogen gas	8.99×10^{-2}
Ice	0.917×10^3
Iron	7.86×10^3
Lead	11.3×10^3
Mercury	13.6×10^3
Nitrogen gas	1.25
Oak	0.710×10^3
Osmium	22.6×10^3
Oxygen gas	1.43
Pine	0.373×10^3
Platinum	21.4×10^3
Seawater	1.03×10^3
Silver	10.5×10^3
Tin	7.30×10^3
Uranium	19.1×10^3

straws, vacuum cleaners, and many other devices. In our calculations and end-of-chapter problems, we usually take atmospheric pressure to be

$$P_0 = 1.00 \text{ atm} \approx 1.013 \times 10^5 \text{ Pa} \qquad \textbf{15.3}$$

Pressures higher than atmospheric are used in *hyperbaric medicine,* or *hyperbaric oxygen therapy* (HBOT). This type of therapy was initially developed for the treatment of disorders associated with diving accidents, such as decompression sickness and air embolisms. Today it is used for a wider range of medical situations.

To receive hyperbaric oxygen therapy, the patient reclines in a special chamber. Modern chambers are transparent, allowing the patient to see the therapist outside. Patients may read a book, listen to music, watch a movie, or simply rest during the procedure. The pressure in the chamber is slowly increased, up to as much as three times atmospheric pressure. The patient experiences the increased pressure for a time interval determined by the therapist, and then the pressure is decreased, the entire session requiring one to two hours.

Many cancer patients undergo radiation treatments. Radiation applied to the pelvic region can cause *radiation cystitis,* resulting in bladder infections, sometimes occurring years after the radiation therapy. Since 1985, hyperbaric oxygen therapy has been used to treat this condition. The therapy stimulates *angiogenesis,* the growth of new blood vessels. This growth reverses the vascular changes induced by the radiation, thereby healing the radiation-induced bladder injury.

Another area in which HBOT is used is for problem wounds, such as those associated with diabetes or amputations. The increased pressure assists with tissue oxygenation in the wound and stimulates angiogenesis in the damaged tissue. It has also been shown that the increase in pressure helps to kill various types of bacteria in the wound area.

QUICK QUIZ 15.1 Suppose you are standing directly behind someone who steps back and accidentally stomps on your foot with the heel of one shoe. Would you be better off if that person were **(a)** a large male professional basketball player wearing sneakers or **(b)** a petite woman wearing spike-heeled shoes?

THINKING PHYSICS 15.1

Suction cups can be used to hold objects onto surfaces. Why don't astronauts use suction cups to hold onto the outside surface of an orbiting spacecraft?

Reasoning A suction cup works because air is pushed out from under the cup when it is pressed against a surface. When the cup is released, it tends to spring back a bit, causing the trapped air under the cup to expand. This expansion causes a reduced pressure inside the cup. Therefore, the difference between the atmospheric pressure on the outside of the cup and the reduced pressure inside provides a net force pushing the cup against the surface. For astronauts in orbit around the Earth, almost no air exists outside the surface of the spacecraft. Therefore, if a suction cup were to be pressed against the outside surface of the spacecraft, the pressure differential needed to press the cup to the surface is not present. ◄

15.2 | Variation of Pressure with Depth

The study of fluid mechanics involves the density of a substance, defined in Equation 1.1 as the mass per unit volume for the substance. Table 15.1 lists the densities of various substances. These values vary slightly with temperature because the volume of a substance is temperature-dependent (as we shall see in Chapter 16). Note that under standard conditions (0°C and atmospheric pressure) the densities of gases are on the order of 1/1 000 the densities of solids and liquids. This difference implies that the average molecular spacing in a gas under these conditions is about ten times greater in each dimension than in a solid or liquid.

As divers know well, the pressure in the sea or a lake increases as they dive to greater depths. Likewise, atmospheric pressure decreases with increasing altitude. For this reason, aircraft flying at high altitudes must have pressurized cabins to provide sufficient oxygen for the passengers.

We now show mathematically how the pressure in a liquid increases with depth. Consider a liquid of density ρ at rest as in Figure 15.4. Let us select a sample of the liquid contained within an imaginary cylinder of cross-sectional area A extending from depth d to depth $d + h$. This sample of liquid is in equilibrium and at rest. Therefore, according to the particle in equilibrium model, the net force on the sample must be equal to zero. We will investigate the forces on the sample related to the pressure on it.

The liquid external to our sample exerts forces at all points on the sample's surface, perpendicular to it. On the sides of the sample of liquid in Figure 15.4, forces due to the pressure act horizontally and cancel in pairs on opposite sides of the sample for a net horizontal force of zero. The pressure exerted by the liquid on the sample's bottom face is P and the pressure on the top face is P_0. Therefore, from Equation 15.1, the magnitude of the upward force exerted by the liquid on the bottom of the sample is PA, and the magnitude of the downward force exerted by the liquid on the top is P_0A. In addition, a gravitational force is exerted on the sample. Because the sample is in equilibrium, the net force in the vertical direction must be zero:

$$\sum F_y = 0 \quad \rightarrow \quad PA - P_0A - Mg = 0$$

Because the mass of liquid in the sample is $M = \rho V = \rho Ah$, the gravitational force on the liquid in the sample is $Mg = \rho gAh$. Therefore,

$$PA = P_0A + \rho g Ah$$

or

$$P = P_0 + \rho gh \qquad \text{15.4} \blacktriangleleft$$

If the top surface of our sample is at $d = 0$ so that it is open to the atmosphere, P_0 is atmospheric pressure. Equation 15.4 indicates that the pressure in a liquid depends only on the depth h within the liquid. The pressure is therefore the same at all points having the same depth, independent of the shape of the container.

In view of Equation 15.4, any increase in pressure at the surface must be transmitted to every point in the liquid. This behavior was first recognized by French scientist Blaise Pascal (1623–1662) and is called **Pascal's law:**

A change in the pressure applied to an enclosed fluid is transmitted undiminished to every point of the fluid and to the walls of the container.

▶ Pascal's law

You use Pascal's law when you squeeze the sides of your toothpaste tube. The increase in pressure on the sides of the tube increases the pressure everywhere, which pushes a stream of toothpaste out of the opening.

An important application of Pascal's law is the hydraulic press illustrated by Figure 15.5 (page 416). A force $\vec{F}_1$ is applied to a small piston of area A_1. The pressure is transmitted through a liquid to a larger piston of area A_2, and force $\vec{F}_2$ is exerted by the liquid on this piston. Because the pressure is the same at both pistons, we see that $P = F_1/A_1 = F_2/A_2$. The force magnitude F_2 is therefore larger than F_1 by the multiplying factor A_2/A_1. Hydraulic brakes, car lifts, hydraulic jacks, and forklifts all make use of this principle.

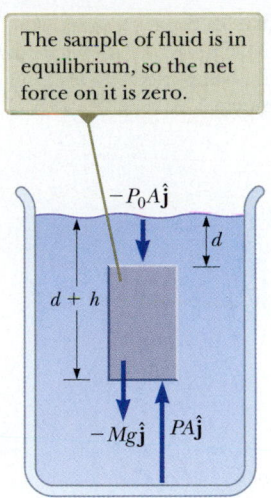

The sample of fluid is in equilibrium, so the net force on it is zero.

$-P_0A\hat{j}$

d

$d + h$

$-Mg\hat{j}$ $PA\hat{j}$

Figure 15.4 A sample of fluid in a larger volume of fluid is singled out.

▶ Variation of pressure with depth in a liquid

QUICK QUIZ 15.2 The pressure at the bottom of a filled glass of water ($\rho = 1\,000$ kg/m^3) is P. The water is poured out, and the glass is filled with ethyl alcohol ($\rho = 806$ kg/m^3). What is the pressure at the bottom of the glass? **(a)** smaller than P **(b)** equal to P **(c)** larger than P **(d)** indeterminate

Figure 15.5 (a) Diagram of a hydraulic press. (b) A vehicle under repair is supported by a hydraulic lift in a garage.

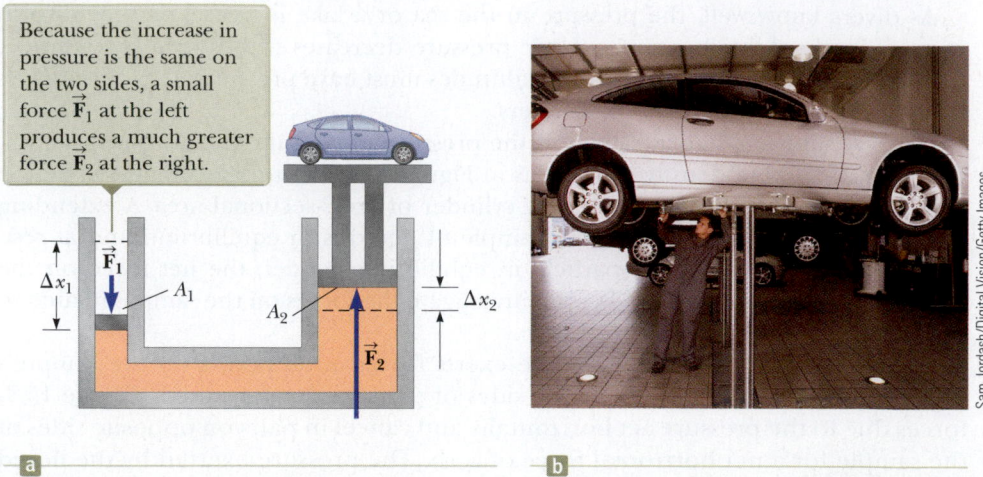

Because the increase in pressure is the same on the two sides, a small force $\vec{F}_1$ at the left produces a much greater force $\vec{F}_2$ at the right.

THINKING PHYSICS 15.2

BIO Measuring blood pressure

Blood pressure is normally measured with the cuff of the sphygmomanometer around the arm. Suppose the blood pressure were measured with the cuff around the calf of the leg of a standing person. Would the reading of the blood pressure be the same here as it is for the arm?

Reasoning The blood pressure measured at the calf would be higher than that measured at the arm. If we imagine the vascular system of the body to be a vessel containing a liquid (blood), the pressure in the liquid will increase with depth. The blood at the calf is deeper in the liquid than that at the arm and is at a higher pressure.

Blood pressures are normally taken at the arm because it is at approximately the same height as the heart. If blood pressures at the calf were used as a standard, adjustments would need to be made for the height of the person and the blood pressure would be different if the person were lying down. ◀

Example 15.1 | The Car Lift

In a car lift used in a service station (Fig. 15.5), compressed air exerts a force on a small piston that has a circular cross section and a radius of 5.00 cm. This pressure is transmitted by a liquid to a piston that has a radius of 15.0 cm.

(A) What force must the compressed air exert to lift a car weighing 13 300 N?

SOLUTION

Conceptualize Review the material just discussed about Pascal's law to understand the operation of a car lift.

Categorize This example is a substitution problem.

Solve $F_1/A_1 = F_2/A_2$ for F_1:

$$F_1 = \left(\frac{A_1}{A_2}\right)F_2 = \frac{\pi(5.00 \times 10^{-2}\text{ m})^2}{\pi(15.0 \times 10^{-2}\text{ m})^2}(1.33 \times 10^4\text{ N})$$

$$= \boxed{1.48 \times 10^3\text{ N}}$$

(B) What air pressure produces this force?

SOLUTION

Use Equation 15.1 to find the air pressure that produces this force:

$$P = \frac{F_1}{A_1} = \frac{1.48 \times 10^3\text{ N}}{\pi(5.00 \times 10^{-2}\text{ m})^2}$$

$$= \boxed{1.88 \times 10^5\text{ Pa}}$$

This pressure is approximately twice atmospheric pressure.

(C) Consider the lift as a nonisolated system and show that the input energy transfer is equal in magnitude to the output energy transfer.

15.1 *cont.*

SOLUTION

The energy input and output are by means of work done by the forces as the pistons move. To determine the work done, we must find the magnitude of the displacement through which each force acts. Because the liquid is modeled to be incompressible, the volume of the cylinder through which the input piston moves must equal that through which the output piston moves. The lengths of these cylinders are the magnitudes Δx_1 and Δx_2 of the displacements of the forces (see Fig. 15.5a).

Set the volumes through which the pistons move equal:

$$V_1 = V_2 \rightarrow A_1\Delta x_1 = A_2\Delta x_2$$

$$\frac{A_1}{A_2} = \frac{\Delta x_2}{\Delta x_1}$$

Evaluate the ratio of the input work to the output work:

$$\frac{W_1}{W_2} = \frac{F_1\,\Delta x_1}{F_2\,\Delta x_2} = \left(\frac{F_1}{F_2}\right)\left(\frac{\Delta x_1}{\Delta x_2}\right) = \left(\frac{A_1}{A_2}\right)\left(\frac{A_2}{A_1}\right) = 1$$

This result verifies that the work input and output are the same, as they must be to conserve energy.

Example 15.2 | The Force on a Dam

Water is filled to a height H behind a dam of width w (Fig. 15.6). Determine the resultant force exerted by the water on the dam.

SOLUTION

Conceptualize Because pressure varies with depth, we cannot calculate the force simply by multiplying the area by the pressure. As the pressure in the water increases with depth, the force on the adjacent portion of the dam also increases.

Categorize Because of the variation of pressure with depth, we must use integration to solve this example, so we categorize it as an analysis problem.

Analyze Let's imagine a vertical y axis, with $y = 0$ at the bottom of the dam. We divide the face of the dam into narrow horizontal strips at a distance y above the bottom, such as the red strip in Figure 15.6. The pressure on each such strip is due only to the water; atmospheric pressure acts on both sides of the dam.

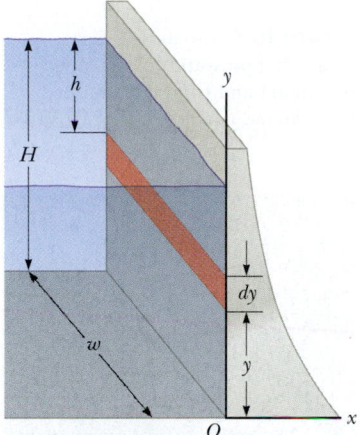

Figure 15.6 (Example 15.2) Water exerts a force on a dam.

Use Equation 15.4 to calculate the pressure due to the water at the depth h:

$$P = \rho gh = \rho g(H - y)$$

Use Equation 15.1 to find the force exerted on the shaded strip of area $dA = w\,dy$:

$$dF = P\,dA = \rho g(H - y)\,w\,dy$$

Integrate to find the total force on the dam:

$$F = \int P\,dA = \int_0^H \rho g(H - y)\,w\,dy = \tfrac{1}{2}\rho g w H^2$$

Finalize Notice that the thickness of the dam shown in Figure 15.6 increases with depth. This design accounts for the greater force the water exerts on the dam at greater depths.

What If? What if you were asked to find this force without using calculus? How could you determine its value?

Answer We know from Equation 15.4 that pressure varies linearly with depth. Therefore, the average pressure due to the water over the face of the dam is the average of the pressure at the top and the pressure at the bottom:

$$P_{\text{avg}} = \frac{P_{\text{top}} + P_{\text{bottom}}}{2} = \frac{0 + \rho gH}{2} = \tfrac{1}{2}\rho gH$$

The total force on the dam is equal to the product of the average pressure and the area of the face of the dam:

$$F = P_{\text{avg}}A = (\tfrac{1}{2}\rho gH)(Hw) = \tfrac{1}{2}\rho g w H^2$$

which is the same result we obtained using calculus.

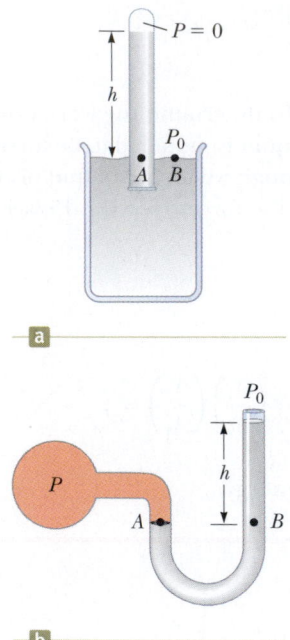

Figure 15.7 Two devices for measuring pressure: (a) a mercury barometer and (b) an open-tube manometer.

Archimedes
Greek Mathematician, Physicist, and Engineer (c. 287–212 BC)
Archimedes was perhaps the greatest scientist of antiquity. He was the first to compute accurately the ratio of a circle's circumference to its diameter, and he also showed how to calculate the volume and surface area of spheres, cylinders, and other geometric shapes. He is well known for discovering the nature of the buoyant force and was also a gifted inventor. One of his practical inventions, still in use today, is Archimedes's screw, an inclined, rotating, coiled tube used originally to lift water from the holds of ships. He also invented the catapult and devised systems of levers, pulleys, and weights for raising heavy loads. Such inventions were successfully used to defend his native city, Syracuse, during a two-year siege by Romans.

15.3 | Pressure Measurements

During the weather report on a television news program, the *barometric pressure* is often provided. Barometric pressure is the current pressure of the atmosphere, which varies over a small range from the standard value provided in Equation 15.3. How is this pressure measured?

One instrument used to measure atmospheric pressure is the common barometer, invented by Evangelista Torricelli (1608–1647). A long tube closed at one end is filled with mercury and then inverted into a dish of mercury (Fig. 15.7a). The closed end of the tube is nearly a vacuum, so the pressure at the top of the mercury column can be taken as zero. In Figure 15.7a, the pressure at point A due to the column of mercury must equal the pressure at point B due to the atmosphere. If that were not the case, a net force would move mercury from one point to the other until equilibrium was established. It therefore follows that $P_0 = \rho_{Hg}gh$, where ρ_{Hg} is the density of the mercury and h is the height of the mercury column. As atmospheric pressure varies, the height of the mercury column varies, so the height can be calibrated to measure atmospheric pressure. Let us determine the height of a mercury column for one atmosphere of pressure, $P_0 = 1$ atm $= 1.013 \times 10^5$ Pa:

$$P_0 = \rho_{Hg}gh \quad \rightarrow \quad h = \frac{P_0}{\rho_{Hg}g} = \frac{1.013 \times 10^5 \text{ Pa}}{(13.6 \times 10^3 \text{ kg/m}^3)(9.80 \text{ m/s}^2)} = 0.760 \text{ m}$$

Based on a calculation such as this one, one atmosphere of pressure is defined as the pressure equivalent of a column of mercury that is exactly 0.760 0 m in height at 0°C.

The open-tube manometer illustrated in Figure 15.7b is a device for measuring the pressure of a gas contained in a vessel. One end of a U-shaped tube containing a liquid is open to the atmosphere, and the other end is connected to a system of unknown pressure P. The pressures at points A and B must be the same (otherwise, the curved portion of the liquid would experience a net force and would accelerate), and the pressure at A is the unknown pressure of the gas. Therefore, equating the unknown pressure P to the pressure at point B, we see that $P = P_0 + \rho gh$. The difference in pressure $P - P_0$ is equal to ρgh. Pressure P is called the **absolute pressure,** and the difference $P - P_0$ is called the **gauge pressure.** For example, the pressure you measure in your bicycle tire is gauge pressure.

15.4 | Buoyant Forces and Archimedes's Principle

Have you ever tried to push a beach ball down under water (Fig. 15.8a)? It is extremely difficult to do because of the large upward force exerted by the water on the ball. The upward force exerted by a fluid on any immersed object is called a **buoyant force.** We can determine the magnitude of a buoyant force by applying some logic. Imagine a beach ball–sized parcel of water beneath the water surface as in Figure 15.8b. Because this parcel is in equilibrium, there must be an upward force that balances the downward gravitational force on the parcel. This upward force is

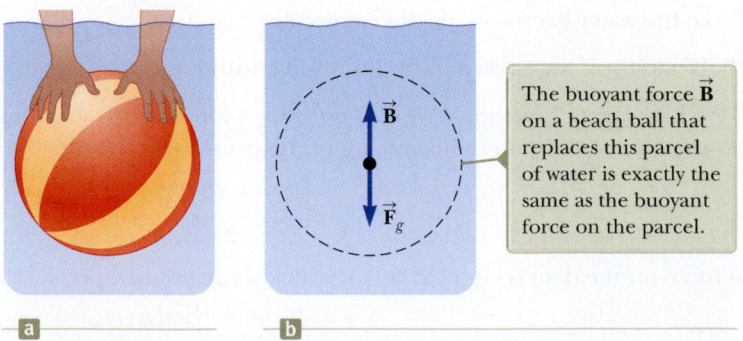

Figure 15.8 (a) A swimmer pushes a beach ball under water. (b) The forces on a beach ball–sized parcel of water.

the buoyant force, and its magnitude is equal to the weight of the water in the parcel. The buoyant force is the resultant force on the parcel due to all forces applied by the fluid surrounding the parcel.

Now imagine replacing the beach ball–sized parcel of water with a beach ball of the same size. The net force applied by the fluid surrounding the beach ball is the same, regardless of whether it is applied to a beach ball or to a parcel of water. Consequently, **the magnitude of the buoyant force on an object always equals the weight of the fluid displaced by the object.** This statement is known as **Archimedes's principle.**

With the beach ball under water, the buoyant force, equal to the weight of a beach ball–sized parcel of water, is much larger than the weight of the beach ball. Therefore, there is a large net upward force, which explains why it is so hard to hold the beach ball under the water. Note that Archimedes's principle does not refer to the makeup of the object experiencing the buoyant force. The object's composition is not a factor in the buoyant force because the buoyant force is exerted by the surrounding fluid.

To better understand the origin of the buoyant force, consider a cube of solid material immersed in a liquid as in Figure 15.9. According to Equation 15.4, the pressure P_{bot} at the bottom of the cube is greater than the pressure P_{top} at the top by an amount $\rho_{fluid}gh$, where h is the height of the cube and ρ_{fluid} is the density of the fluid. The pressure at the bottom of the cube causes an *upward* force equal to $P_{bot}A$, where A is the area of the bottom face. The pressure at the top of the cube causes a *downward* force equal to $P_{top}A$. The resultant of these two forces is the buoyant force $\vec{B}$ with magnitude

$$B = (P_{bot} - P_{top})A = (\rho_{fluid}gh)A$$

$$B = \rho_{fluid}gV_{disp}$$ 15.5 ◀

where $V_{disp} = Ah$ is the volume of the fluid displaced by the cube. Because the product $\rho_{fluid}V_{disp}$ is equal to the mass of fluid displaced by the object,

$$B = Mg$$

where Mg is the weight of the fluid displaced by the cube. This result is consistent with our initial statement about Archimedes's principle above, based on the discussion of the beach ball.

Before proceeding with a few examples, it is instructive to compare two common cases: the buoyant force acting on a totally submerged object and that acting on a floating object.

Case I: A Totally Submerged Object

When an object is totally submerged in a fluid of density ρ_{fluid}, the volume V_{disp} of the displaced fluid is equal to the volume V_{obj} of the object; so, from Equation 15.5, the magnitude of the upward buoyant force is $B = \rho_{fluid}gV_{obj}$. If the object has a mass M and density ρ_{obj}, its weight is equal to $F_g = Mg = \rho_{obj}gV_{obj}$, and the net force on the object is $B - F_g = (\rho_{fluid} - \rho_{obj})gV_{obj}$. Hence, if the density of the object is less than the density of the fluid, the downward gravitational force is less than the buoyant force and the unsupported object accelerates upward (Active Fig. 15.10a). If the density of the object is greater than the density of the fluid, the upward buoyant force is less than the downward gravitational force and the unsupported object sinks (Active Fig. 15.10b). If the density of the submerged object equals the density of the fluid, the net force on the object is zero and the object remains in equilibrium. Therefore, the direction of motion of an object submerged in a fluid is determined *only* by the densities of the object and the fluid.

The same behavior is exhibited by an object immersed in a gas, such as the air in the atmosphere.[1] If the object is less dense than air, like a helium-filled balloon, the object floats upward. If it is denser, like a rock, it falls downward.

---

[1]The general behavior is the same, but the buoyant force varies with height in the atmosphere due to the variation in density of the air.

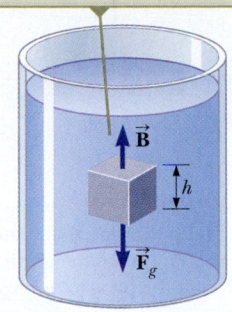

Figure 15.9 The external forces acting on an immersed cube are the gravitational force $\vec{F}_g$ and the buoyant force $\vec{B}$.

> The buoyant force on the cube is the resultant of the forces exerted on its top and bottom faces by the liquid.

Pitfall Prevention | 15.2
Buoyant Force Is Exerted by the Fluid
Remember that **the buoyant force is exerted by the fluid.** It is not determined by properties of the object except for the amount of fluid displaced by the object. Therefore, if several objects of different densities but the same volume are immersed in a fluid, they will all experience the same buoyant force. Whether they sink or float is determined by the relationship between the buoyant force and the gravitational force.

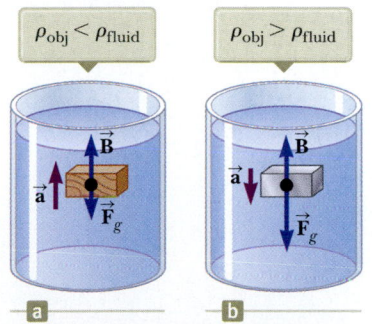

Active Figure 15.10 (a) A totally submerged object that is less dense than the fluid in which it is submerged experiences a net upward force and rises to the surface after it is released. (b) A totally submerged object that is denser than the fluid experiences a net downward force and sinks.

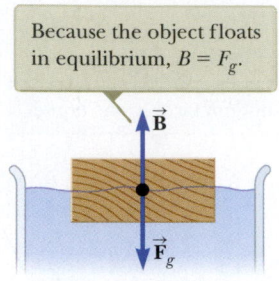

Because the object floats in equilibrium, $B = F_g$.

Active Figure 15.11 An object floating on the surface of a liquid experiences two forces, the gravitational force $\vec{\mathbf{F}}_g$ and the buoyant force $\vec{\mathbf{B}}$.

These hot-air balloons float on air because they are filled with air at high temperature. The buoyant force on a balloon due to the surrounding air is equal to the weight of the balloon, resulting in a net force of zero.

Case II: A Floating Object

Now consider an object of volume V_{obj} and density $\rho_{\text{obj}} < \rho_{\text{fluid}}$ in static equilibrium floating on the surface of a fluid, that is, an object that is only *partially* submerged (Active Fig. 15.11). In this case, the upward buoyant force is balanced by the downward gravitational force acting on the object. If V_{disp} is the volume of the fluid displaced by the object (this volume is the same as the volume of that part of the object beneath the surface of the fluid), the buoyant force has a magnitude $B = \rho_{\text{fluid}} g V_{\text{disp}}$. Because the weight of the object is $F_g = Mg = \rho_{\text{obj}} g V_{\text{obj}}$ and because $F_g = B$, we see that $\rho_{\text{fluid}} g V_{\text{disp}} = \rho_{\text{obj}} g V_{\text{obj}}$, or

$$\frac{V_{\text{disp}}}{V_{\text{obj}}} = \frac{\rho_{\text{obj}}}{\rho_{\text{fluid}}} \qquad \text{15.6} \blacktriangleleft$$

Therefore, the fraction of the volume of the object under the liquid surface is equal to the ratio of the object density to the liquid density.

Let us consider examples of both cases. Under normal conditions, the average density of a fish in the opening photograph of this chapter is slightly greater than the density of water. That being the case, a fish would sink if it did not have some mechanism to counteract the net downward force. The fish does so by internally regulating the size of its swim bladder, a gas-filled cavity within the fish's body. Increasing its size increases the amount of water displaced, which increases the buoyant force. In this manner, fish are able to swim to various depths. Because the fish is totally submerged in the water, this example illustrates Case I.

As an example of Case II, imagine a large cargo ship. When the ship is at rest, the upward buoyant force from the water balances the weight so that the ship is in equilibrium. Only part of the volume of the ship is under water. If the ship is loaded with heavy cargo, it sinks deeper into the water. The increased weight of the ship due to the cargo is balanced by the extra buoyant force related to the extra volume of the ship that is now beneath the water surface.

▸ **QUICK QUIZ 15.3** An apple is held completely submerged just below the surface of a container of water. The apple is then moved to a deeper point in the water. Compared with the force needed to hold the apple just below the surface, what is the force needed to hold it at a deeper point? **(a)** larger **(b)** the same **(c)** smaller **(d)** impossible to determine

▸ **QUICK QUIZ 15.4** You are shipwrecked and floating in the middle of the ocean on a raft. Your cargo on the raft includes a treasure chest full of gold that you found before your ship sank and the raft is just barely afloat. To keep you floating as high as possible in the water, should you **(a)** leave the treasure chest on top of the raft, **(b)** secure the treasure chest to the underside of the raft, or **(c)** hang the treasure chest in the water with a rope attached to the raft? (Assume that throwing the treasure chest overboard is not an option you wish to consider!)

▸ **THINKING PHYSICS 15.3**

A florist delivery person is delivering a flower basket to a home. The basket includes an attached helium-filled balloon, which suddenly comes loose from the basket and begins to accelerate upward toward the sky. Startled by the release of the balloon, the delivery person drops the flower basket. As the basket falls, the basket–Earth system experiences an increase in kinetic energy and a decrease in gravitational potential energy, consistent with conservation of mechanical energy. The balloon–Earth system, however, experiences an increase in *both* gravitational potential energy and kinetic energy. Is that consistent with the principle of conservation of mechanical energy? If not, from where is the extra energy coming?

Reasoning In the case of the system of the flower basket and the Earth, a good approximation to the motion of the basket can be made by ignoring the effects of the air. Therefore, the basket–Earth system can be analyzed with the isolated system model and mechanical energy is conserved. For the balloon–Earth system, we cannot ignore the effects of the air because it is the buoyant force of the air that causes the balloon to rise. Therefore, the balloon–Earth system is analyzed with the nonisolated system model. The buoyant force of the air does work across the boundary of the system, and that work results in an increase in both the kinetic and gravitational potential energies of the system. ◀

Example 15.3 | Eureka!

Archimedes supposedly was asked to determine whether a crown made for the king consisted of pure gold. According to legend, he solved this problem by weighing the crown first in air and then in water as shown in Figure 15.12. Suppose the scale read 7.84 N when the crown was in air and 6.84 N when it was in water. What should Archimedes have told the king?

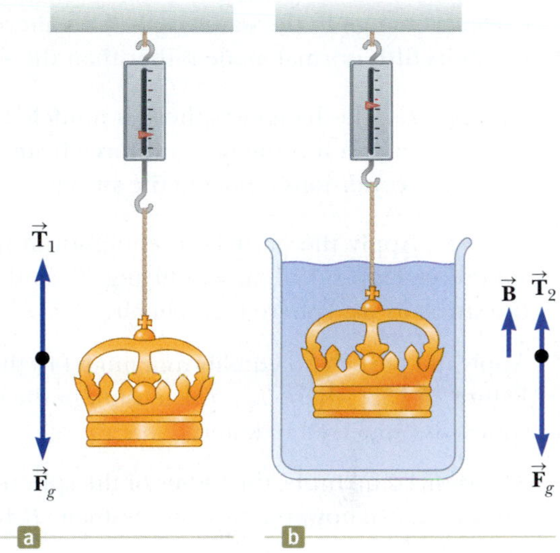

SOLUTION

Conceptualize Figure 15.12 helps us imagine what is happening in this example. Because of the buoyant force, the scale reading is smaller in Figure 15.12b than in Figure 15.12a.

Categorize This problem is an example of Case 1 discussed earlier because the crown is completely submerged. The scale reading is a measure of one of the forces on the crown, and the crown is stationary. Therefore, we can categorize the crown as a particle in equilibrium.

Figure 15.12 (Example 15.3) (a) When the crown is suspended in air, the scale reads its true weight because $T_1 = F_g$ (the buoyancy of air is negligible). (b) When the crown is immersed in water, the buoyant force $\vec{B}$ changes the scale reading to a lower value $T_2 = F_g - B$.

Analyze When the crown is suspended in air, the scale reads the true weight $T_1 = F_g$ (neglecting the small buoyant force due to the surrounding air). When the crown is immersed in water, the buoyant force $\vec{B}$ reduces the scale reading to an *apparent* weight of $T_2 = F_g - B$.

Apply the particle in equilibrium model to the crown in water:

$$\sum F = B + T_2 - F_g = 0$$

Solve for B and substitute the known values:

$$B = F_g - T_2 = 7.84 \text{ N} - 6.84 \text{ N} = 1.00 \text{ N}$$

Because this buoyant force is equal in magnitude to the weight of the displaced water, $B = \rho_w g V_{\text{disp}}$, where V_{disp} is the volume of the displaced water and ρ_w is its density. Also, the volume of the crown V_c is equal to the volume of the displaced water because the crown is completely submerged, so $B = \rho_w g V_c$.

Find the density of the crown from Equation 1.1:

$$\rho_c = \frac{m_c}{V_c} = \frac{m_c g}{V_c g} = \frac{m_c g}{(B/\rho_w)} = \frac{m_c g \rho_w}{B}$$

Substitute numerical values:

$$\rho_c = \frac{(7.84 \text{ N})(1\,000 \text{ kg/m}^3)}{1.00 \text{ N}} = 7.84 \times 10^3 \text{ kg/m}^3$$

Finalize From Table 15.1, we see that the density of gold is $19.3 \times 10^3 \text{ kg/m}^3$. Therefore, Archimedes should have reported that the king had been cheated. Either the crown was hollow, or it was not made of pure gold.

What If? Suppose the crown has the same weight but is indeed pure gold and not hollow. What would the scale reading be when the crown is immersed in water?

Answer Find the buoyant force on the crown:

$$B = \rho_w g V_w = \rho_w g V_c = \rho_w g\left(\frac{m_c}{\rho_c}\right) = \rho_w\left(\frac{m_c g}{\rho_c}\right)$$

Substitute numerical values:

$$B = (1.00 \times 10^3 \text{ kg/m}^3)\frac{7.84 \text{ N}}{19.3 \times 10^3 \text{ kg/m}^3} = 0.406 \text{ N}$$

Find the tension in the string hanging from the scale:

$$T_2 = F_g - B = 7.84 \text{ N} - 0.406 \text{ N} = 7.43 \text{ N}$$

Example 15.4 | Changing String Vibration with Water

One end of a horizontal string is attached to a vibrating blade, and the other end passes over a pulley as in Figure 15.13a. A sphere of mass 2.00 kg hangs on the end of the string. The string is vibrating in its second harmonic. A container of water is raised under the sphere so that the sphere is completely submerged. In this configuration, the string vibrates in its fifth harmonic as shown in Figure 15.13b. What is the radius of the sphere?

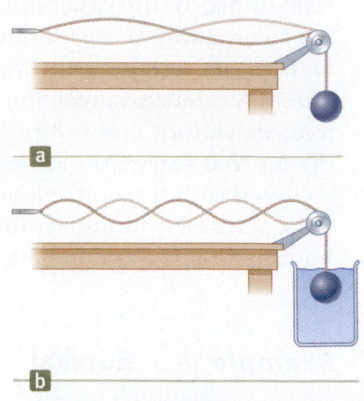

SOLUTION

Conceptualize Imagine what happens when the sphere is immersed in the water. The buoyant force acts upward on the sphere, reducing the tension in the string. The change in tension causes a change in the speed of waves on the string, which in turn causes a change in the wavelength. This altered wavelength results in the string vibrating in its fifth normal mode rather than the second.

Figure 15.13 (Example 15.4) (a) When the sphere hangs in air, the string vibrates in its second harmonic. (b) When the sphere is immersed in water, the string vibrates in its fifth harmonic.

Categorize The hanging sphere is modeled as a particle in equilibrium. One of the forces acting on it is the buoyant force from the water. We also apply the waves under boundary conditions model to the string.

Analyze Apply the particle in equilibrium model to the sphere in Figure 15.13a, identifying T_1 as the tension in the string as the sphere hangs in air:

$$\sum F = T_1 - mg = 0$$

$$T_1 = mg$$

Apply the particle in equilibrium model to the sphere in Figure 15.13b, where T_2 is the tension in the string as the sphere is immersed in water:

$$T_2 + B - mg = 0$$

$$(1) \quad B = mg - T_2$$

The desired quantity, the radius of the sphere, will appear in the expression for the buoyant force B. Before proceeding in this direction, however, we must evaluate T_2 from the information about the standing wave.

Write the equation for the frequency of a standing wave on a string (Eq. 14.7) twice, once before the sphere is immersed and once after. Notice that the frequency f is the same in both cases because it is determined by the vibrating blade. In addition, the linear mass density μ and the length L of the vibrating portion of the string are the same in both cases. Divide the equations:

$$f = \frac{n_1}{2L}\sqrt{\frac{T_1}{\mu}}$$

$$f = \frac{n_2}{2L}\sqrt{\frac{T_2}{\mu}} \quad \rightarrow \quad 1 = \frac{n_1}{n_2}\sqrt{\frac{T_1}{T_2}}$$

Solve for T_2:

$$T_2 = \left(\frac{n_1}{n_2}\right)^2 T_1 = \left(\frac{n_1}{n_2}\right)^2 mg$$

Substitute this result into Equation (1):

$$(2) \quad B = mg - \left(\frac{n_1}{n_2}\right)^2 mg = mg\left[1 - \left(\frac{n_1}{n_2}\right)^2\right]$$

Using Equation 15.5, express the buoyant force in terms of the radius of the sphere:

$$B = \rho_{water}gV_{sphere} = \rho_{water}g\left(\tfrac{4}{3}\pi r^3\right)$$

Solve for the radius of the sphere and substitute from Equation (2):

$$r = \left(\frac{3B}{4\pi\rho_{water}g}\right)^{1/3} = \left\{\frac{3m}{4\pi\rho_{water}}\left[1 - \left(\frac{n_1}{n_2}\right)^2\right]\right\}^{1/3}$$

Substitute numerical values:

$$r = \left\{\frac{3(2.00\text{ kg})}{4\pi(1\,000\text{ kg/m}^3)}\left[1 - \left(\frac{2}{5}\right)^2\right]\right\}^{1/3}$$

$$= 0.073\,7\text{ m} = \boxed{7.37\text{ cm}}$$

Finalize Notice that only certain radii of the sphere will result in the string vibrating in a normal mode; the speed of waves on the string must be changed to a value such that the length of the string is an integer multiple of half wavelengths. This limitation is a feature of the *quantization* that was introduced in Chapters 11 and 14: the sphere radii that cause the string to vibrate in a normal mode are *quantized*.

15.5 | Fluid Dynamics

Thus far, our study of fluids has been restricted to fluids at rest, or **fluid statics.** We now turn our attention to **fluid dynamics,** the study of fluids in motion. Instead of trying to study the motion of each particle of the fluid as a function of time, we describe the properties of the fluid as a whole.

Flow Characteristics

When fluid is in motion, its flow is of one of two main types. The flow is said to be **steady,** or **laminar,** if each particle of the fluid follows a smooth path so that the paths of different particles never cross each other as in Figure 15.14. Therefore, in steady flow, the velocity of the fluid at any point remains constant in time.

Above a certain critical speed, fluid flow becomes **turbulent.** Turbulent flow is an irregular flow characterized by small, whirlpool-like regions as in Figure 15.15. As an example, the flow of water in a river becomes turbulent in regions where rocks and other obstructions are encountered, often forming "white-water" rapids.

The term **viscosity** is commonly used in fluid flow to characterize the degree of internal friction in the fluid. This internal friction, or viscous force, is associated with the resistance of two adjacent layers of the fluid against moving relative to each other. Because viscosity represents a nonconservative force, part of a fluid's kinetic energy is converted to internal energy when layers of fluid slide past one another. This conversion is similar to the mechanism by which an object sliding on a rough horizontal surface experiences a transformation of kinetic energy to internal energy.

Because the motion of a real fluid is very complex and not yet fully understood, we adopt a simplification model. As we shall see, many features of real fluids in motion can be understood by considering the behavior of an ideal fluid. In our simplification model, we make the following four assumptions:

1. *Nonviscous fluid.* In a nonviscous fluid, internal friction is ignored. An object moving through the fluid experiences no viscous force.
2. *Incompressible fluid.* The density of the fluid is assumed to remain constant regardless of the pressure in the fluid.
3. *Steady flow.* In steady flow, we assume that the velocity of the fluid at each point remains constant in time.
4. *Irrotational flow.* Fluid flow is irrotational if the fluid has no angular momentum about any point. If a small paddle wheel placed anywhere in the fluid does not rotate about the wheel's center of mass, the flow is irrotational. (If the wheel were to rotate, as it would if turbulence were present, the flow would be rotational.)

The first two assumptions in our simplification model are properties of our ideal fluid. The last two are descriptions of the way that the fluid flows.

Figure 15.14 An illustration of steady flow around an automobile in a test wind tunnel. The streamlines in the airflow are made visible by smoke particles.

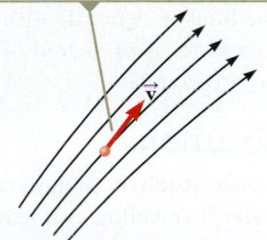

Figure 15.15 Hot gases from a cigarette made visible by smoke particles. The smoke first moves in laminar flow at the bottom and then in turbulent flow above.

15.6 | Streamlines and the Continuity Equation for Fluids

If you are watering your garden and your garden hose is too short, you might do one of two things to help you reach the garden with the water (before you look for a longer hose!). You might attach a nozzle to the end of the hose, or, in the absence of a nozzle, you might place your thumb over the end of the hose, allowing the water to come out of a narrower opening. Why does either of these techniques cause the water to come out faster so that it can be projected over a longer range? We shall see the answer to this question in this section.

The path taken by a fluid particle under steady flow is called a **streamline.** The velocity of the particle is always tangent to the streamline as shown in Figure 15.16. A set of streamlines like the ones shown in Figure 15.16 form a *tube of flow.* Fluid

At each point along its path, the particle's velocity is tangent to the streamline.

Figure 15.16 A particle in laminar flow follows a streamline.

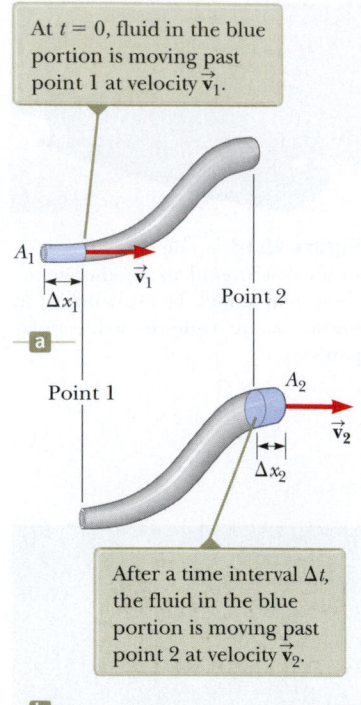

At $t = 0$, fluid in the blue portion is moving past point 1 at velocity $\vec{\mathbf{v}}_1$.

a

After a time interval Δt, the fluid in the blue portion is moving past point 2 at velocity $\vec{\mathbf{v}}_2$.

b

Figure 15.17 A fluid moving with steady flow through a pipe of varying cross-sectional area. (a) At $t = 0$, the small blue-colored portion of the fluid at the left is moving through area A_1. (b) After a time interval Δt, the blue-colored portion is that fluid that has moved through area A_2.

© Cengage Learning/George Semple

Figure 15.18 The speed of water spraying from the end of a garden hose increases as the size of the opening is decreased with the thumb.

particles cannot flow into or out of the sides of this tube; if they could, the streamlines would cross one another.

Consider ideal fluid flow through a pipe of nonuniform size as illustrated in Figure 15.17. Let's focus our attention on a segment of fluid in the pipe. Figure 15.17a shows the segment at time $t = 0$ consisting of the gray portion between point 1 and point 2 and the short blue portion to the left of point 1. At this time, the fluid in the short blue portion is flowing through a cross section of area A_1 at speed v_1. During the time interval Δt, the small length Δx_1 of fluid in the blue portion moves past point 1. During the same time, fluid moves past point 2 at the other end of the pipe. Figure 15.17b shows the situation at the end of the time interval Δt. The blue portion at the right end represents the fluid that has moved past point 2 through an area A_2 at a speed v_2.

The mass of fluid contained in the blue portion in Figure 15.17a is given by $m_1 = \rho A_1 \, \Delta x_1 = \rho A_1 v_1 \, \Delta t$, where ρ is the (unchanging) density of the ideal fluid. Similarly, the fluid in the blue portion in Figure 15.17b has a mass $m_2 = \rho A_2 \, \Delta x_2 = \rho A_2 v_2 \, \Delta t$. Because the fluid is incompressible and the flow is steady, however, the mass of fluid that passes point 1 in a time interval Δt must equal the mass that passes point 2 in the same time interval. That is, $m_1 = m_2$ or $\rho A_1 v_1 \, \Delta t = \rho A_2 v_2 \, \Delta t$, which means that

$$A_1 v_1 = A_2 v_2 = \text{constant} \qquad 15.7 \blacktriangleleft$$

This expression is called the **continuity equation for fluids.** It states that the product of the area and the fluid speed at all points along a pipe is constant for an incompressible fluid. Equation 15.7 shows that the speed is high where the tube is constricted (small A) and low where the tube is wide (large A). The product Av, which has the dimensions of volume per unit time, is called either the *volume flux* or the *flow rate*. The condition $Av = $ constant is equivalent to the statement that the volume of fluid that enters one end of a tube in a given time interval equals the volume leaving the other end of the tube in the same time interval if no leaks are present.

You demonstrate the continuity equation each time you water your garden with your thumb over the end of a garden hose as in Figure 15.18. By partially blocking the opening with your thumb, you reduce the cross-sectional area through which the water passes. As a result, the speed of the water increases as it exits the hose, and the water can be sprayed over a long distance.

▶ **QUICK QUIZ 15.5** You tape two different soda straws together end to end to make a longer straw with no leaks. The two straws have radii of 3 mm and 5 mm. You drink a soda through your combination straw. In which straw is the speed of the liquid higher? **(a)** It is higher in whichever one is nearest your mouth. **(b)** It is higher in the one of radius 3 mm. **(c)** It is higher in the one of radius 5 mm. **(d)** Neither, because the speed is the same in both straws.

▶ *Example* **15.5** | **Watering a Garden**

A gardener uses a water hose 2.50 cm in diameter to fill a 30.0-L bucket. The gardener notes that it takes 1.00 min to fill the bucket. A nozzle with an opening of cross-sectional area 0.500 cm² is then attached to the hose. The nozzle is held so that water is projected horizontally from a point 1.00 m above the ground. Over what horizontal distance can the water be projected?

SOLUTION

Conceptualize Imagine any past experience you have with projecting water from a horizontal hose or a pipe. The faster the water is traveling as it leaves the hose, the farther it will land on the ground from the end of the hose.

Categorize Once the water leaves the hose, it is in free fall. Therefore, we categorize a given element of the water as a projectile. The element is modeled as a particle under constant acceleration (due to gravity) in the vertical direction and

15.5 *cont.*

a particle under constant velocity in the horizontal direction. The horizontal distance over which the element is projected depends on the speed with which it is projected. This example involves a change in area for the pipe, so we also categorize it as one in which we use the continuity equation for fluids.

Analyze We first find the speed of the water in the hose from the bucket-filling information.

Find the cross-sectional area of the hose:

$$A = \pi r^2 = \pi \frac{d^2}{4} = \pi \left[\frac{(2.50 \text{ cm})^2}{4} \right] = 4.91 \text{ cm}^2$$

Evaluate the volume flow rate:

$$A v_1 = 30.0 \text{ L/min} = \frac{30.0 \times 10^3 \text{ cm}^3}{60.0 \text{ s}} = 500 \text{ cm}^3/\text{s}$$

Solve for the speed of the water in the hose:

$$v_1 = \frac{500 \text{ cm}^3/\text{s}}{A} = \frac{500 \text{ cm}^3/\text{s}}{4.91 \text{ cm}^2} = 102 \text{ cm/s} = 1.02 \text{ m/s}$$

We have labeled this speed v_1 because we identify point 1 within the hose. We identify point 2 in the air just outside the nozzle. We must find the speed $v_2 = v_{xi}$ with which the water exits the nozzle. The subscript i anticipates that it will be the *initial* velocity component of the water projected from the hose, and the subscript x indicates that the initial velocity vector of the projected water is horizontal.

Solve the continuity equation for fluids for v_2:

$$v_2 = v_{xi} = \frac{A_1}{A_2} v_1$$

Substitute numerical values:

$$v_{xi} = \frac{4.91 \text{ cm}^2}{0.500 \text{ cm}^2} (1.02 \text{ m/s}) = 10.0 \text{ m/s}$$

We now shift our thinking away from fluids and to projectile motion. In the vertical direction, an element of the water starts from rest and falls through a vertical distance of 1.00 m.

Write Equation 2.13 for the vertical position of an element of water, modeled as a particle under constant acceleration:

$$y_f = y_i + v_{yi} t - \tfrac{1}{2} g t^2$$

Substitute numerical values:

$$-1.00 \text{ m} = 0 + 0 - \tfrac{1}{2} (9.80 \text{ m/s}^2) t^2$$

Solve for the time at which the element of water lands on the ground:

$$t = \sqrt{\frac{2(1.00 \text{ m})}{9.80 \text{ m/s}^2}} = 0.452 \text{ s}$$

Use Equation 2.5 to find the horizontal position of the element at this time, modeled as a particle under constant velocity:

$$x_f = x_i + v_{xi} t = 0 + (10.0 \text{ m/s})(0.452 \text{ s}) = \boxed{4.52 \text{ m}}$$

Finalize The time interval for the element of water to fall to the ground is unchanged if the projection speed is changed because the projection is horizontal. Increasing the projection speed results in the water hitting the ground farther from the end of the hose, but requires the same time interval to strike the ground.

15.7 | Bernoulli's Equation

You have probably experienced driving on a highway and having a large truck pass you at high speed. In this situation, you may have had the frightening feeling that your car was being pulled in toward the truck as it passed. We will investigate the origin of this effect in this section.

As a fluid moves through a region where its speed or elevation above the Earth's surface changes, the pressure in the fluid varies with these changes. The relationship between fluid speed, pressure, and elevation was first derived in 1738 by Swiss physicist Daniel Bernoulli. Consider the flow of a segment of an ideal fluid through a nonuniform pipe in a time interval Δt as illustrated in Figure 15.19.

istockphoto.com/ZU_09

Daniel Bernoulli
Swiss physicist (1700–1782)
Bernoulli made important discoveries in fluid dynamics. Bernoulli's most famous work, *Hydrodynamica*, was published in 1738; it is both a theoretical and a practical study of equilibrium, pressure, and speed in fluids. He showed that as the speed of a fluid increases, its pressure decreases. Referred to as "Bernoulli's principle," Bernoulli's work is used to produce a partial vacuum in chemical laboratories by connecting a vessel to a tube through which water is running rapidly.

This figure is very similar to Figure 15.17, which we used to develop the continuity equation. We have added two features: the forces on the outer ends of the blue portions of fluid and the heights of these portions above the reference position $y = 0$.

The force exerted by the fluid to the left of the blue portion in Figure 15.19a has a magnitude P_1A_1. The work done by this force on the segment in a time interval Δt is $W_1 = F_1 \Delta x_1 = P_1A_1 \Delta x_1 = P_1V$, where V is the volume of the blue portion of fluid passing point 1 in Figure 15.19a. In a similar manner, the work done by the fluid to the right of the segment in the same time interval Δt is $W_2 = -P_2A_2 \Delta x_2 = -P_2V$, where V is the volume of the blue portion of fluid passing point 2 in Figure 15.19b. (The volumes of the blue portions of fluid in Figures 15.19a and 15.19b are equal because the fluid is incompressible.) This work is negative because the force on the segment of fluid is to the left and the displacement of the point of application of the force is to the right. Therefore, the net work done on the segment by these forces in the time interval Δt is

$$W = (P_1 - P_2)V$$

Part of this work goes into changing the kinetic energy of the segment of fluid, and part goes into changing the gravitational potential energy of the segment–Earth system. Because we are assuming streamline flow, the kinetic energy K_{gray} of the gray portion of the segment is the same in both parts of Figure 15.19. Therefore, the change in the kinetic energy of the segment of fluid is

$$\Delta K = \left(\tfrac{1}{2}mv_2^2 + K_{\text{gray}}\right) - \left(\tfrac{1}{2}mv_1^2 + K_{\text{gray}}\right) = \tfrac{1}{2}mv_2^2 - \tfrac{1}{2}mv_1^2$$

where m is the mass of the blue portions of fluid in both parts of Figure 15.19. (Because the volumes of both portions are the same, they also have the same mass.)

Considering the gravitational potential energy of the segment–Earth system, once again there is no change during the time interval for the gravitational potential energy U_{gray} associated with the gray portion of the fluid. Consequently, the change in gravitational potential energy of the system is

$$\Delta U = (mgy_2 + U_{\text{gray}}) - (mgy_1 + U_{\text{gray}}) = mgy_2 - mgy_1$$

From Equation 7.2, the total work done on the segment–Earth system by the fluid outside the segment is equal to the change in mechanical energy of the system: $W = \Delta K + \Delta U$. Substituting for each of these terms gives

$$(P_1 - P_2)V = \tfrac{1}{2}mv_2^2 - \tfrac{1}{2}mv_1^2 + mgy_2 - mgy_1 \qquad \textbf{15.8} \blacktriangleleft$$

If we divide each term by the portion volume V and recall that $\rho = m/V$, this expression reduces to

$$P_1 - P_2 = \tfrac{1}{2}\rho v_2^2 - \tfrac{1}{2}\rho v_1^2 + \rho gy_2 - \rho gy_1$$

Rearranging terms, we obtain

$$\boxed{P_1 + \tfrac{1}{2}\rho v_1^2 + \rho gy_1 = P_2 + \tfrac{1}{2}\rho v_2^2 + \rho gy_2} \qquad \textbf{15.9} \blacktriangleleft$$

which is **Bernoulli's equation** applied to an ideal fluid. It is often expressed as

$$P + \tfrac{1}{2}\rho v^2 + \rho gy = \text{constant} \qquad \textbf{15.10} \blacktriangleleft$$

Bernoulli's equation says that the sum of the pressure P, the kinetic energy per unit volume $\tfrac{1}{2}\rho v^2$, and gravitational potential energy per unit volume ρgy has the same value at all points along a streamline.

When the fluid is at rest, $v_1 = v_2 = 0$ and Equation 15.9 becomes

$$P_1 - P_2 = \rho g(y_2 - y_1) = \rho gh$$

which agrees with Equation 15.4.

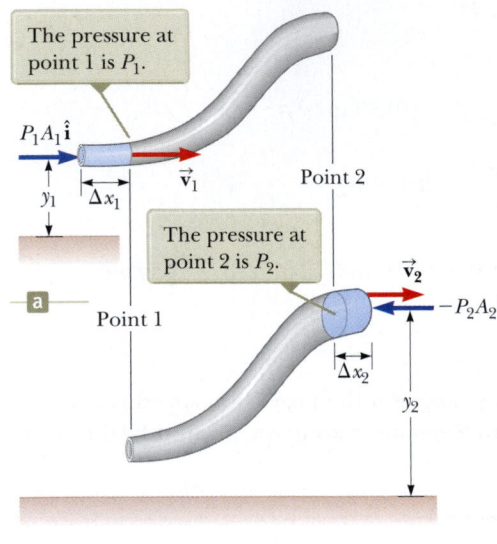

The pressure at point 1 is P_1.

$P_1A_1\hat{\mathbf{i}}$

$\vec{\mathbf{v}}_1$

y_1 Δx_1

Point 2

The pressure at point 2 is P_2.

$\vec{\mathbf{v}}_2$

$-P_2A_2\hat{\mathbf{i}}$

a   Point 1

Δx_2

y_2

b

Figure 15.19 A fluid in laminar flow through a pipe. (a) A segment of the fluid at time $t = 0$. A small portion of the blue-colored fluid is at height y_1 above a reference position. (b) After a time interval Δt, the entire segment has moved to the right. The blue-colored portion of the fluid is at height y_2.

Although Equation 15.10 was derived for an incompressible fluid, the general behavior of pressure with speed is true even for gases: as the speed increases, the pressure decreases. This *Bernoulli effect* explains the experience with the truck on the highway at the opening of this section. As air passes between your car and the truck, it must pass through a relatively narrow channel. According to the continuity equation, the speed of the air is higher. According to the Bernoulli effect, this higher-speed air exerts less pressure on your car than the slower-moving air on the other side of your car. Therefore, there is a net force pushing you toward the truck.

QUICK QUIZ 15.6 You observe two helium balloons floating next to each other at the ends of strings secured to a table. The facing surfaces of the balloons are separated by 1–2 cm. You blow through the small space between the balloons. What happens to the balloons? **(a)** They move toward each other. **(b)** They move away from each other. **(c)** They are unaffected.

Example 15.6 | Sinking the Cruise Ship

A scuba diver is hunting for fish with a spear gun. He accidentally fires the gun so that a spear punctures the side of a cruise ship. The hole is located at a depth of 10.0 m below the water surface. With what speed does the water enter the cruise ship through the hole?

SOLUTION

Conceptualize Imagine the water streaming in through the hole. The deeper the hole is below the surface, the higher the pressure, and therefore, the higher the speed of the incoming water.

Categorize We will evaluate the result directly from Bernoulli's equation, so this is a substitution problem.

We identify point 2 as the water surface outside the ship, which we will assign as $y = 0$. At this point, the water is static, so $v_2 = 0$. We identify point 1 as a point just inside the hole in the interior of the ship because that is the point at which we wish to evaluate the speed of the water. This point is at a depth $y = -h = -10.0$ m below the water surface. We use Bernoulli's equation to compare these two points. At both points, the water is open to atmospheric pressure, so $P_1 = P_2 = P_0$.

Based on this argument, solve Bernoulli's equation for the speed v_1 of the water entering the ship and evaluate:

$$P_0 + \tfrac{1}{2}\rho(0)^2 + \rho g(0) = P_0 + \tfrac{1}{2}\rho v_1^2 + \rho g(-h) \rightarrow$$
$$v_1 = \sqrt{2gh} = \sqrt{2(9.80 \text{ m/s}^2)(10.0 \text{ m})} = \boxed{14 \text{ m/s}}$$

Example 15.7 | Torricelli's Law

An enclosed tank containing a liquid of density ρ has a hole in its side at a distance y_1 from the tank's bottom (Fig. 15.20). The hole is open to the atmosphere, and its diameter is much smaller than the diameter of the tank. The air above the liquid is maintained at a pressure P. Determine the speed of the liquid as it leaves the hole when the liquid's level is a distance h above the hole.

SOLUTION

Conceptualize Imagine that the tank is a fire extinguisher. When the hole is opened, liquid leaves the hole with a certain speed. If the pressure P at the top of the liquid is increased, the liquid leaves with a higher speed. If the pressure P falls too low, the liquid leaves with a low speed and the extinguisher must be replaced.

Categorize Looking at Figure 15.20, we know the pressure at two points and the velocity at one of those points. We wish to find the velocity at the second point. Therefore, we can categorize this example as one in which we can apply Bernoulli's equation.

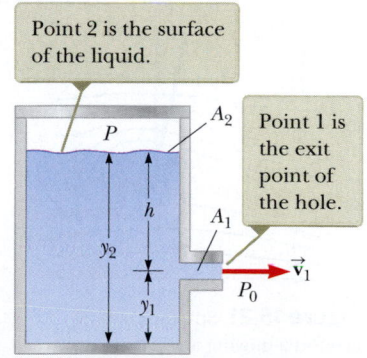

Figure 15.20 (Example 15.7) A liquid leaves a hole in a tank at speed v_1.

Analyze Because $A_2 \gg A_1$, the liquid is approximately at rest at the top of the tank, where the pressure is P. At the hole, P_1 is equal to atmospheric pressure P_0.

continued

15.7 cont.

Apply Bernoulli's equation between points 1 and 2 and solve for v_1, noting that $y_2 - y_1 = h$:

$$P_0 + \tfrac{1}{2}\rho v_1^2 + \rho g y_1 = P + \rho g y_2 \;\rightarrow\; v_1 = \sqrt{\dfrac{2(P - P_0)}{\rho} + 2gh}$$

Finalize When P is much greater than P_0 (so that the term $2gh$ can be neglected), the exit speed of the water is mainly a function of P. If the tank is open to the atmosphere, then $P = P_0$ and $v_1 = \sqrt{2gh}$, as in Example 15.6. In other words, for an open tank, the speed of the liquid leaving a hole a distance h below the surface is equal to that acquired by an object falling freely through a vertical distance h. This phenomenon is known as **Torricelli's law.**

What If? What if the position of the hole in Figure 15.20 could be adjusted vertically? If the tank is open to the atmosphere and sitting on a table, what position of the hole would cause the water to land on the table at the farthest distance from the tank?

Answer Model a parcel of water exiting the hole as a projectile. Find the time at which the parcel strikes the table from a hole at an arbitrary position y_1:

$$y_f = y_i + v_{yi}t - \tfrac{1}{2}gt^2 \;\rightarrow\; 0 = y_1 + 0 - \tfrac{1}{2}gt^2$$

$$t = \sqrt{\dfrac{2y_1}{g}}$$

Find the horizontal position of the parcel at the time it strikes the table:

$$x_f = x_i + v_{xi}t = 0 + \sqrt{2g(y_2 - y_1)}\,\sqrt{\dfrac{2y_1}{g}} = 2\sqrt{(y_2 y_1 - y_1^2)}$$

Maximize the horizontal position by taking the derivative of x_f with respect to y_1 (because y_1, the height of the hole, is the variable that can be adjusted) and setting it equal to zero. Solve for y_1:

$$\dfrac{dx_f}{dy_1} = \tfrac{1}{2}(2)(y_2 y_1 - y_1^2)^{-1/2}(y_2 - 2y_1) = 0 \;\rightarrow\; y_1 = \tfrac{1}{2}y_2$$

Therefore, to maximize the horizontal distance, the hole should be halfway between the bottom of the tank and the upper surface of the water. Below this location, the water is projected at a higher speed but falls for a short time interval, reducing the horizontal range. Above this point, the water is in the air for a longer time interval but is projected with a smaller horizontal speed.

⟨ 15.8 | Other Applications of Fluid Dynamics

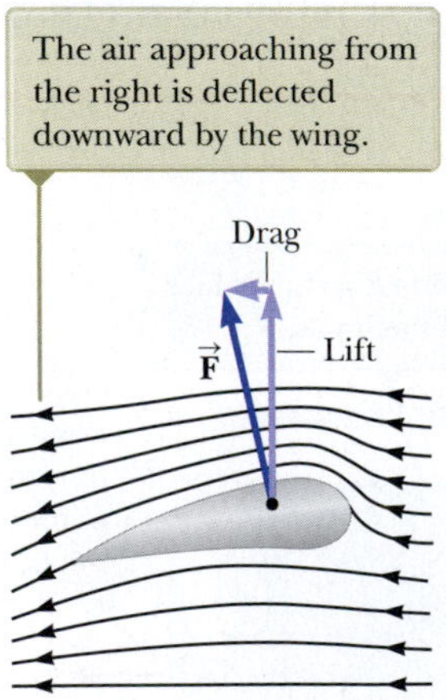

Consider the streamlines that flow around an airplane wing as shown in Figure 15.21. Let us assume that the airstream approaches the wing horizontally from the right. The tilt of the wing causes the airstream to be deflected downward. Because the airstream is deflected by the wing, the wing must exert a force on the airstream. According to Newton's third law, the airstream must exert an equal and opposite force $\vec{F}$ on the wing. This force has a vertical component called the **lift** (or aerodynamic lift) and a horizontal component called **drag.** The lift depends on several factors, such as the speed of the airplane, the area of the wing, its curvature, and the angle between the wing and the horizontal. As this angle increases, turbulent flow can set in above the wing to reduce the lift.

In general, an object experiences lift by any effect that causes the fluid to change its direction as it flows past the object. Some factors that influence lift are the shape of the object, its orientation with respect to the fluid flow, spinning motion (for example, a curve ball thrown in a baseball game due to the spinning of the baseball), and the texture of the object's surface.

A number of devices operate in a manner similar to the *atomizer* in Figure 15.22. A stream of air passing over an open tube reduces the pressure above the tube. This reduction in pressure causes the liquid to rise into the air stream. The liquid is then dispersed into a fine spray of droplets. This type of system is used in perfume bottles and paint sprayers.

The air approaching from the right is deflected downward by the wing.

Figure 15.21 Streamline flow around a moving airplane wing. By Newton's third law, the air deflected by the wing results in an upward force on the wing from the air: lift. Because of air resistance, there is also a force opposite the velocity of the wing: drag.

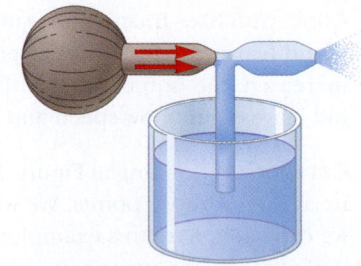

Figure 15.22 A stream of air passing over a tube dipped into a liquid causes the liquid to rise in the tube.

15.9 | Context Connection: Turbulent Flow of Blood

Fluids play a dominant role in the transport of nutrients and other materials in the human body. The circulatory system transports nutrients to cells and removes waste products, the respiratory system provides the oxygen needed for cells to consume nutrients and removes the carbon dioxide produced in these reactions, and the gastrointestinal system takes in food and removes waste from the body. Each of these systems represents a complex fluid dynamic system with unique properties.

With each heartbeat, the blood in our bodies is moved along arteries, veins, and the vast network of capillaries that comprise the blood circulation system. Blood flow in straight, healthy portions of arteries would be simple to analyze if the flow could be modeled as being *laminar*, as discussed in Section 15.5. However, such a simplification model is incorrect for at least two reasons. First, the flow of blood is unsteady because the beating heart causes a time-varying pressure differential in the arteries. Second, turbulent eddies are created as the flowing blood interacts with the walls of the arteries and smaller vessels.

Each liter of blood contains 4×10^{12} to 6×10^{12} red blood cells having diameters ranging from 6 mm to 8 mm. These cells are sufficiently large so that each side of the pancake-shaped cells experiences a different force from the surrounding fluid. The resulting torque on the cells causes them to spin, creating turbulence in the fluid. As the velocity of the blood increases, the gradients in the flow velocity between the walls and the center of the vessel become larger. These larger gradients cause the red blood cells to spin faster, stirring the blood and causing it to become even more turbulent.

The flowing blood also interacts chemically with the cells lining the walls of blood vessels. The interior surfaces of the heart and blood vessels are covered with a one-cell-thick layer of *endothelial cells*, which reduce friction between the cells and the vessel walls. These cells also have a significant role in the extraction of minerals from the blood, the passage of white blood cells into and out of the bloodstream, and in the formation of blood clots. In laminar flow, endothelial cells are football-shaped and align themselves along the direction of blood flow, creating a protective layer over the vessel walls. Blood vessels continually expand and contract because of environmental factors such as changes in temperature of the surroundings. Inherited blood vessel defects, disturbances in the nerves controlling vessel contraction, injuries, and drugs can cause blood vessels to contract. According to the continuity equation for fluids (Eq. 15.7), this will result in an increased flow velocity of the blood. When the blood flows faster and becomes more turbulent, endothelial cells respond by becoming rounded in shape, and dividing much faster than normal. The division of endothelial cells creates gaps in the coating of the blood vessel, allowing blood platelets and cholesterol-carrying lipoproteins to attach themselves to the vessel wall, and plaque begins to form in the blood vessel. The smooth walls of the blood vessel become rough, further disrupting the blood flow, which in turn affects endothelial cells downstream from the rough patches, creating even more plaque. As plaque builds up in the blood vessel, the flow channel further constricts, increasing the flow velocity and affecting large numbers of endothelial cells (Fig. 15.23).

The gradual buildup of plaque, called *atherosclerosis,* can become catastrophic when the plaque becomes unstable and ruptures. In this case, the blood is exposed to the collagen in the tissue cap of the plaque buildup. This exposure causes the blood to clot at the point of rupture, forming what is called a *thrombus*. The thrombus may continue to grow until it completely blocks the blood vessel. On the other hand, the thrombus may break loose from the site of the plaque and flow with the blood until it blocks a

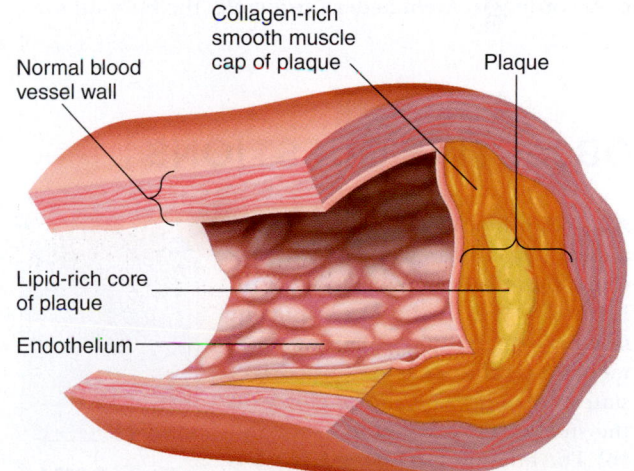

Figure 15.23 The left side of this cutaway section of a coronary vessel is normal. At the right side, atherosclerosis has led to the development of a plaque, bulging into the interior of the vessel and affecting the flow of blood. (from Sherwood, *Fundamentals of Human Physiology,* 4th ed., 2012, Brooks/Cole, top half of Fig. 9–24, p. 253, © 2012 Brooks/Cole, a part of Cengage Learning, Inc. Reproduced by permission. www.cengage.com/permissions.)

Labels in figure: Normal blood vessel wall; Collagen-rich smooth muscle cap of plaque; Plaque; Lipid-rich core of plaque; Endothelium

smaller vessel somewhere downstream. Blocked blood vessels in the arms, legs, or the pelvis can result in numbness, pain, and the onset of infections such as *gangrene*. A severe blockage in a coronary blood vessel, however, can lead to a heart attack, and a blockage in a carotid artery can lead to stroke. The severity of damage to cardiac muscle during a heart attack depends on the location of the blockage. The left coronary artery (see Fig. 3 in the Context Introduction) supplies blood to 85% of the cardiac tissue, so that a blockage in this vessel at a point high on the heart near the pulmonary veins could cause great damage.

In this Context Connection, we investigated the role of turbulent flow and used the continuity equation for fluids to clarify the process of cardiovascular plaque buildup. In the Context Conclusion, we will explore an application of Bernoulli's principle to the diagnosis and prevention of cardiovascular disease and heart attacks.

SUMMARY

The **pressure** P in a fluid is the force per unit area that the fluid exerts on a surface:

$$P \equiv \frac{F}{A} \qquad \text{15.1} \blacktriangleleft$$

In the SI system, pressure has units of newtons per square meter, and $1 \text{ N/m}^2 = 1$ pascal (Pa).

The pressure in a liquid varies with depth h according to the expression

$$P = P_0 + \rho g h \qquad \text{15.4} \blacktriangleleft$$

where P_0 is the pressure at the surface of the liquid and ρ is the density of the liquid, assumed uniform.

Pascal's law states that when a change in pressure is applied to a fluid, the change in pressure is transmitted undiminished to every point in the fluid and to every point on the walls of the container.

When an object is partially or fully submerged in a fluid, the fluid exerts an upward force on the object called the **buoyant force.** According to **Archimedes's principle,** the buoyant force

is equal to the weight of the fluid displaced by the object:

$$B = \rho_{\text{fluid}} g V_{\text{disp}} \qquad \text{15.5} \blacktriangleleft$$

Various aspects of fluid dynamics can be understood by adopting a simplification model in which the fluid is nonviscous and incompressible and the fluid motion is a steady flow with no turbulence.

Using this model, two important results regarding fluid flow through a pipe of nonuniform size can be obtained:

1. The flow rate through the pipe is a constant, which is equivalent to stating that the product of the cross-sectional area A and the speed v at any point is a constant. This behavior is described by the **continuity equation for fluids:**

$$A_1 v_1 = A_2 v_2 = \text{constant} \qquad \text{15.7} \blacktriangleleft$$

2. The sum of the pressure, kinetic energy per unit volume, and gravitational potential energy per unit volume has the same value at all points along a streamline. This behavior is described by **Bernoulli's equation:**

$$P_1 + \tfrac{1}{2}\rho v_1{}^2 + \rho g y_1 = P_2 + \tfrac{1}{2}\rho v_2{}^2 + \rho g y_2 \qquad \text{15.9} \blacktriangleleft$$

OBJECTIVE QUESTIONS

☐ denotes answer available in *Student Solutions Manual/Study Guide*

1. A wooden block floats in water, and a steel object is attached to the bottom of the block by a string as in Figure OQ15.1. If the block remains floating, which of the following statements are valid? (Choose all correct statements.) (a) The buoyant force on the steel object is equal to its weight. (b) The buoyant force on the block is equal to its weight. (c) The tension in the string is equal to the weight of the steel object. (d) The tension in the string is less than the weight of the steel object. (e) The buoyant force on the block is equal to the volume of water it displaces.

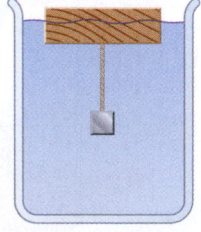

Figure OQ15.1

2. A beach ball filled with air is pushed about 1 m below the surface of a swimming pool and released from rest. Which of the following statements are valid, assuming the size of the ball remains the same? (Choose all correct statements.) (a) As the ball rises in the pool, the buoyant force on it increases. (b) When the ball is released, the buoyant force exceeds the gravitational force, and the ball accelerates upward. (c) The buoyant force on the ball decreases as the ball approaches the surface of the pool. (d) The buoyant force on the ball equals its weight and remains constant as the ball rises. (e) The buoyant force on the ball while it is submerged is approximately equal to the weight of a volume of water that could fill the ball.

3. Figure OQ15.3 shows aerial views from directly above two dams. Both dams are equally wide (the vertical dimension in the diagram) and equally high (into the page in the diagram). The dam on the left holds back a very large lake, and the dam on the right holds back a narrow river. Which dam has to be built more strongly? (a) the dam on the left (b) the dam on the right (c) both the same (d) cannot be predicted

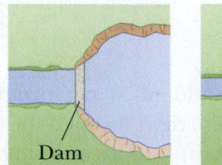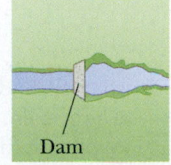

Figure OQ15.3

4. A boat develops a leak and, after its passengers are rescued, eventually sinks to the bottom of a lake. When the boat is at the bottom, what is the force of the lake bottom on the boat? (a) greater than the weight of the boat (b) equal to the weight of the boat (c) less than the weight of the boat (d) equal to the weight of the displaced water (e) equal to the buoyant force on the boat

5. A solid iron sphere and a solid lead sphere of the same size are each suspended by strings and are submerged in a tank of water. (Note that the density of lead is greater than that of iron.) Which of the following statements are valid? (Choose all correct statements.) (a) The buoyant force on each is the same. (b) The buoyant force on the lead sphere is greater than the buoyant force on the iron sphere because lead has the greater density. (c) The tension in the string supporting the lead sphere is greater than the tension in the string supporting the iron sphere. (d) The buoyant force on the iron sphere is greater than the buoyant force on the lead sphere because lead displaces more water. (e) None of those statements is true.

6. Rank the buoyant forces exerted on the following five objects of equal volume from the largest to the smallest. Assume the objects have been dropped into a swimming pool and allowed to come to mechanical equilibrium. If any buoyant forces are equal, state that in your ranking. (a) a block of solid oak (b) an aluminum block (c) a beach ball made of thin plastic and inflated with air (d) an iron block (e) a thin-walled, sealed bottle of water.

7. A person in a boat floating in a small pond throws an anchor overboard. What happens to the level of the pond? (a) It rises. (b) It falls. (c) It remains the same.

8. Three vessels of different shapes are filled to the same level with water as in Figure OQ15.8. The area of the base is the same for all three vessels. Which of the following statements are valid? (Choose all correct statements.) (a) The pressure at the top surface of vessel A is greatest because it has the largest surface area. (b) The pressure at the bottom of vessel A is greatest because it contains the most water. (c) The pressure at the bottom of each vessel is the same. (d) The force on the bottom of each vessel is not the same. (e) At a given depth below the surface of each vessel, the pressure on the side of vessel A is greatest because of its slope.

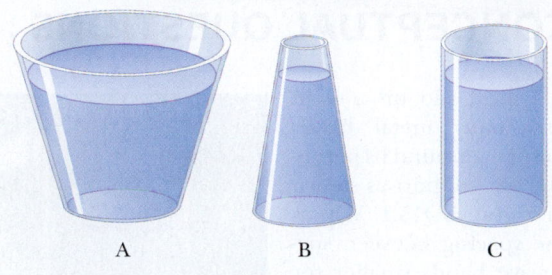

A   B   C

Figure OQ15.8

9. An ideal fluid flows through a horizontal pipe whose diameter varies along its length. Measurements would indicate that the sum of the kinetic energy per unit volume and pressure at different sections of the pipe would (a) decrease as the pipe diameter increases, (b) increase as the pipe diameter increases, (c) increase as the pipe diameter decreases, (d) decrease as the pipe diameter decreases, or (e) remain the same as the pipe diameter changes.

10. A beach ball is made of thin plastic. It has been inflated with air, but the plastic is not stretched. By swimming with fins on, you manage to take the ball from the surface of a pool to the bottom. Once the ball is completely submerged, what happens to the buoyant force exerted on the beach ball as you take it deeper? (a) It increases. (b) It remains constant. (c) It decreases. (d) It is impossible to determine.

11. One of the predicted problems due to global warming is that ice in the polar ice caps will melt and raise sea levels everywhere in the world. Is that more of a worry for ice (a) at the north pole, where most of the ice floats on water; (b) at the south pole, where most of the ice sits on land; (c) both at the north and south pole equally; or (d) at neither pole?

12. A small piece of steel is tied to a block of wood. When the wood is placed in a tub of water with the steel on top, half of the block is submerged. Now the block is inverted so that the steel is under water. (i) Does the amount of the block submerged (a) increase, (b) decrease, or (c) remain the same? (ii) What happens to the water level in the tub when the block is inverted? (a) It rises. (b) It falls. (c) It remains the same.

13. A piece of unpainted porous wood barely floats in an open container partly filled with water. The container is then sealed and pressurized above atmospheric pressure. What happens to the wood? (a) It rises in the water. (b) It sinks lower in the water. (c) It remains at the same level.

14. A glass of water contains floating ice cubes. When the ice melts, does the water level in the glass (a) go up, (b) go down, or (c) remain the same?

15. A water supply maintains a constant rate of flow for water in a hose. You want to change the opening of the nozzle so that water leaving the nozzle will reach a height that is four times the current maximum height the water reaches with the nozzle vertical. To do so, should you (a) decrease the area of the opening by a factor of 16, (b) decrease the area by a factor of 8, (c) decrease the area by a factor of 4, (d) decrease the area by a factor of 2, or (e) give up because it cannot be done?

CONCEPTUAL QUESTIONS

1. A typical silo on a farm has many metal bands wrapped around its perimeter for support as shown in Figure CQ15.1. Why is the spacing between successive bands smaller for the lower portions of the silo on the left, and why are double bands used at lower portions of the silo on the right?

Figure CQ15.1

2. **BIO** Because atmospheric pressure is about 10^5 N/m^2 and the area of a person's chest is about 0.13 m^2, the force of the atmosphere on one's chest is around 13 000 N. In view of this enormous force, why don't our bodies collapse?

3. Two thin-walled drinking glasses having equal base areas but different shapes, with very different cross-sectional areas above the base, are filled to the same level with water. According to the expression $P = P_0 + \rho g h$, the pressure is the same at the bottom of both glasses. In view of this equality, why does one weigh more than the other?

4. Why do airplane pilots prefer to take off with the airplane facing into the wind?

5. Prairie dogs ventilate their burrows by building a mound around one entrance, which is open to a stream of air when wind blows from any direction. A second entrance at ground level is open to almost stagnant air. How does this construction create an airflow through the burrow?

6. A fish rests on the bottom of a bucket of water while the bucket is being weighed on a scale. When the fish begins to swim around, does the scale reading change? Explain your answer.

7. When an object is immersed in a liquid at rest, why is the net force on the object in the horizontal direction equal to zero?

8. In Figure CQ15.8, an airstream moves from right to left through a tube that is constricted at the middle. Three table-tennis balls are levitated in equilibrium above the vertical columns through which the air escapes. (a) Why is the ball at the right higher than the one in the middle? (b) Why is the ball at the left lower than the ball at the right even though the horizontal tube has the same dimensions at these two points?

Figure CQ15.8

9. You are a passenger on a spacecraft. For your survival and comfort, the interior contains air just like that at the surface of the Earth. The craft is coasting through a very empty region of space. That is, a nearly perfect vacuum exists just outside the wall. Suddenly, a meteoroid pokes a hole, about the size of a large coin, right through the wall next to your seat. (a) What happens? (b) Is there anything you can or should do about it?

10. Does a ship float higher in the water of an inland lake or in the ocean? Why?

11. A water tower is a common sight in many communities. Figure CQ15.11 shows a collection of colorful water towers in Kuwait City, Kuwait. Notice that the large weight of the water results in the center of mass of the system being high above the ground. Why is it desirable for a water tower to have this highly unstable shape rather than being shaped as a tall cylinder?

Figure CQ15.11

12. If the airstream from a hair dryer is directed over a table-tennis ball, the ball can be levitated. Explain.

13. (a) Is the buoyant force a conservative force? (b) Is a potential energy associated with the buoyant force? (c) Explain your answers to parts (a) and (b).

14. An empty metal soap dish barely floats in water. A bar of Ivory soap floats in water. When the soap is stuck in the soap dish, the combination sinks. Explain why.

15. If you release a ball while inside a freely falling elevator, the ball remains in front of you rather than falling to the floor because the ball, the elevator, and you all experience the same downward gravitational acceleration. What happens if you repeat this experiment with a helium-filled balloon?

16. How would you determine the density of an irregularly shaped rock?

17. The water supply for a city is often provided from reservoirs built on high ground. Water flows from the reservoir, through pipes, and into your home when you turn the tap on your faucet. Why does water flow more rapidly out of a faucet on the first floor of a building than in an apartment on a higher floor?

18. Place two cans of soft drinks, one regular and one diet, in a container of water. You will find that the diet drink floats while the regular one sinks. Use Archimedes's principle to devise an explanation.

19. When ski jumpers are airborne (Fig. CQ15.19), they bend their bodies forward and keep their hands at their sides. Why?

Figure CQ15.19

PROBLEMS AVAILABLE IN

15.1 Pressure

Problems 1–4

15.2 Variation of Pressure with Depth

Problems 5–16

15.3 Pressure Measurements

Problems 17–21

15.4 Buoyant Forces and Archimedes's Principle

Problems 22–36

15.5 Fluid Dynamics
15.6 Streamlines and the Continuity Equation for Fluids
15.7 Bernoulli's Equation

Problems 37–48

15.8 Other Applications of Fluid Dynamics

Problems 49–51

15.9 Context Connection: Turbulent Flow of Blood

Problem 52

Additional Problems

Problems 53–70

Solutions to the following Problems are available in the _Student Solutions Manual/Study Guide_:

15.2, 15.3, 15.5, 15.11, 15.18, 15.23, 15.27, 15.29, 15.31, 15.39, 15.43, 15.54, 15.60, and 15.67

List of Enhanced Problems

Problem Number	Targeted Feedback in Enhanced WebAssign	Master It in Enhanced WebAssign	Watch It in Enhanced WebAssign
15.2	✓		✓
15.3		✓	
15.5		✓	
15.6	✓		✓
15.7			✓
15.8	✓		✓
15.9	✓		✓
15.11		✓	
15.17	✓		✓
15.27	✓	✓	
15.29		✓	
15.31		✓	
15.36	✓		✓
15.43	✓	✓	
Context Conclusion 4.4		✓	

Detecting Atherosclerosis and Preventing Heart Attacks

We have explored the physics of fluids and can return to our central question for this *Heart Attacks* Context:

> How can the principles of physics be applied in medicine to help prevent heart attacks?

We shall apply our understanding of fluid dynamics to explore research into the causes of cardiovascular disease and heart attacks.

Traditionally, the prevention and treatment of cardiovascular disease and heart attacks has focused on a regimen of exercise, a healthy diet aimed at reducing cholesterol and preventing high blood pressure, cessation of smoking, reducing stress, and medications to reduce the patient's cholesterol and blood pressure, and to prevent blood clots from forming. In severe cases, *angioplasty,* a procedure to widen constricted arteries, or the placement of a *stent,* a mesh tube that acts as a scaffold to keep the arteries open, may also be used. This procedure is shown in Figure 1, which illustrates the use of a balloon to open a constricted artery and the placement of a stent.

Although the causes of atherosclerosis (Section 15.9) are still not known, fluid dynamics has helped illuminate many of the factors that can result in the hardening of arteries and plaque buildup that characterize this condition. As discussed in Section 15.9, recent research has focused on the response of endothelial cells lining arterial walls to turbulent flows and the buildup of plaque in constricted blood vessels.

Let us further examine the physics of an arterial constriction, or *stenosis.* Blood flow in the healthy blood vessel is laminar, and research has shown that plaque does not build up under these circumstances. The situation is very different in the constricted artery; in cases with severe narrowing, the arterial cross-section can be reduced by as much as 75%, to one-fourth of its original area. From the continuity equation for fluids (Eq. 15.7), with $A_2 = \frac{1}{4}A_1$, we obtain

$$A_1 v_1 = (\tfrac{1}{4}A_1)\, v_2 \quad \rightarrow \quad v_2 = 4v_1 \qquad \text{1}\blacktriangleleft$$

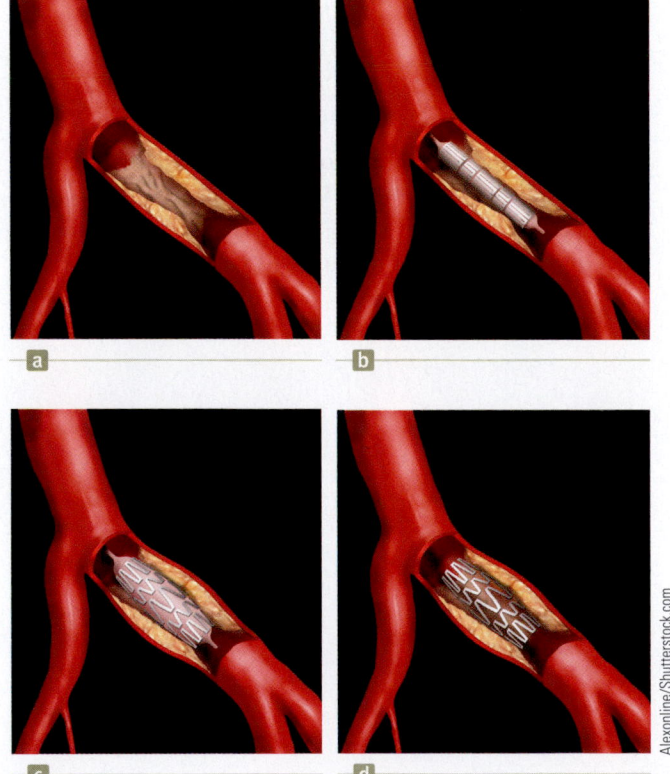

Alexonline/Shutterstock.com

Figure 1 The placement of a stent to improve the flow of blood in a vessel. (a) An artery has a plaque buildup that restricts the flow of blood. (b) A balloon catheter with a deflated balloon and collapsed stent is inserted into the artery and guided to the site of the plaque. (c) The balloon is inflated, expanding the stent and the walls of the vessel. (d) The balloon is deflated and the catheter is removed. The stent keeps the artery open.

We see that the blood flows four times faster in the constricted region of the artery. Similar to the situation shown in Figure 15.15, as a fluid speeds up, its flow becomes turbulent, so that the blood flow immediately downstream of the stenosis exhibits whirlpool-like regions. As discussed in Section 15.9, this turbulence can lead to further problems.

If we assume streamline flow for the moment, we can relate the pressure in the constricted portion 2 of the blood vessel to that in an open portion 1 by using Bernoulli's equation for a horizontal segment of the artery (Eq. 15.9):

$$P_1 + \tfrac{1}{2}\rho v_1^2 = P_2 + \tfrac{1}{2}\rho v_2^2$$

Solving for the pressure difference between these two locations,

$$\Delta P = P_2 - P_1 = \tfrac{1}{2}\rho v_1^2 - \tfrac{1}{2}\rho v_2^2 = \tfrac{1}{2}\rho(v_1^2 - v_2^2)$$

Substituting for v_2 from Equation (1),

$$\Delta P = \tfrac{1}{2}\rho\left[v_1^2 - (4v_1)^2\right] = -\tfrac{15}{2}\rho v_1^2 \qquad \textbf{2} \blacktriangleleft$$

The average *systolic* blood pressure (the pressure during the contraction of the heart) is 120 mm of mercury, or 15.7 kPa (note that this is a gauge pressure, not the absolute pressure in the blood vessel). Also, on average, blood ($\rho = 1.05 \times 10^3 \,\text{kg/m}^3$) flows at a speed of $v_1 = 0.40 \,\text{m/s}$. Evaluating the pressure difference in Equation (2) numerically,

$$\Delta P = -\tfrac{15}{2}(1.05 \times 10^3 \,\text{kg/m}^3)(0.40 \,\text{m/s})^2 = -1.3 \times 10^3 \,\text{Pa}$$

Comparing this result to the initial pressure, we find

$$\frac{\Delta P}{P_0} = \frac{-1.3 \times 10^3 \,\text{Pa}}{15.7 \times 10^3 \,\text{Pa}} = -8.0\%$$

This 8% drop in pressure can be enough that the pressure difference between the tissue outside the vessel and that in the constriction can cause the vessel to collapse, causing a momentary interruption in blood flow. At this point, the speed of the blood goes to zero, its pressure rises again, and the vessel reopens. As the blood rushes through the constricted artery, the internal pressure drops and again the artery closes. Such a phenomenon is called *vascular flutter*. It can be heard by a physician with a stethoscope and is an indication of advanced atherosclerotic disease. Furthermore, the continued opening and closing of the artery can contribute even more to turbulence of the blood and its effects. Therefore, vascular flutter should be taken very seriously and recognized as a strong indication that medical care is needed to avoid a heart attack.

The relationship between turbulent flow occurring downstream of arterial constrictions and the build-up of plaque has been demonstrated by a number of recent studies that combine medical research, physics, and engineering to find the physical and biochemical causes of atherosclerosis and cardiovascular disease. In these studies, platelet-rich animal blood tagged with radioactive indium-111 was circulated through flow tubes with various constriction geometries. The location and amount of platelet deposits were then recorded using a device for measuring the gamma radiation emitted by the radioactive platelets. (We will study gamma radiation in Chapters 24 and 30.) This allowed researchers to determine the locations of maximum platelet deposition and their relationship to the blood flow in the artery.[1]

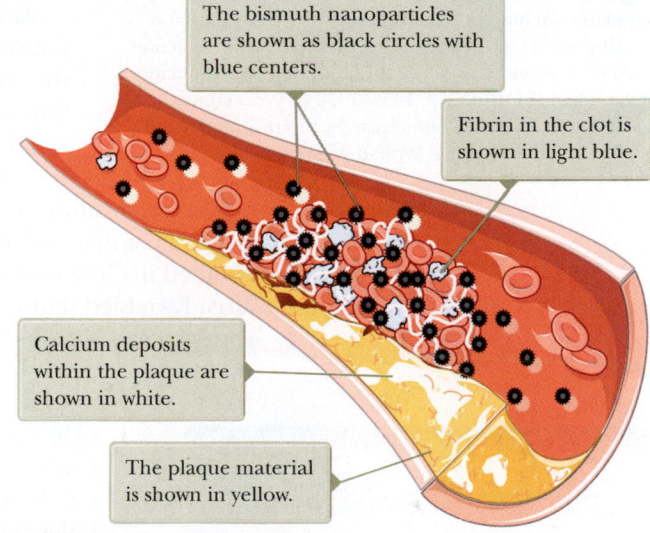

The bismuth nanoparticles are shown as black circles with blue centers.

Fibrin in the clot is shown in light blue.

Calcium deposits within the plaque are shown in white.

The plaque material is shown in yellow.

a

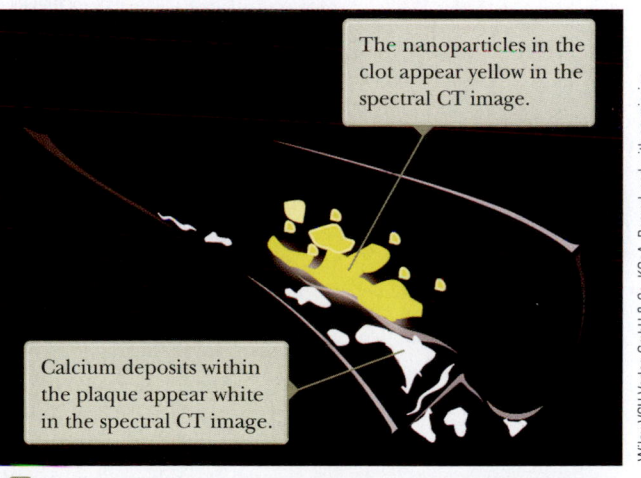

The nanoparticles in the clot appear yellow in the spectral CT image.

Calcium deposits within the plaque appear white in the spectral CT image.

b

Wiley-VCH Verlag GmbH & Co. KGaA. Reproduced with permission

Figure 2 (a) A blood vessel with ruptured atherosclerotic plaque and calcium deposits is developing a blood clot. A traditional CT image would not differentiate between the blood clot and the calcium in the plaque, making it unclear whether the image shows a clot that should be treated. To help make this differentiation, bismuth nanoparticles are targeted to a protein in the blood clot called fibrin. (b) A spectral CT image shows the nanoparticles targeted to fibrin in yellow, differentiating it from calcium, still shown in white, in the plaque.

[1]Schoephoerster, R. T., et al., "Effects of local geometry and fluid dynamics on regional platelet deposition on artificial surfaces," *Arterioscler. Thromb. Vasc. Biol.*, 1993, 13, 1806–1813.

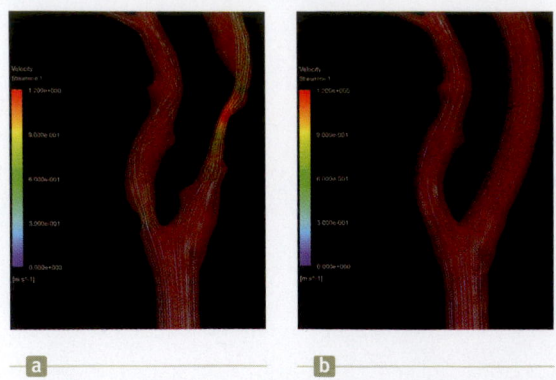

Figure 3 A computational fluid dynamic (CFD) simulation of blood flow through constricted (left) and healthy (right) arteries. (From Ding, S, Tu, J and Cheung, C 2007, "Geometric model generation for CFD simulation of blood and air flows," p. 1335–1338 in Proceedings of the 1st International Conference on Bioinformatics and Biomedical Engineering, Wuhan, China, 6–8 July 2007.)

Similar studies have used fluid dynamics to examine the deposition of plaque on artificial heart valves and helped design more hydrodynamic valves that are also less prone to plaque build-up.

More recently, researchers have begun using radioactive nanoparticles that attach themselves to plaque particles forming in the bloodstream, allowing for the earlier detection of plaque formation than with current techniques (Fig. 2). Using magnetic resonance imaging (MRI), researchers can also follow the flow of these nanoparticles to obtain unprecedented details of the flow of blood in the complicated geometry of the human cardiovascular system. (We will discuss MRIs in more detail in Context 7.) By combining nanoparticles with adult stem cells, arterial plaque in pig hearts has actually been burned away when the nanoparticles are illuminated with laser light.[2]

At the same time, advances in computing speed and memory capability have allowed computational fluid dynamic models to simulate the complex fluid interactions taking place in blood vessels (Fig. 3). One advantage of computer models is that very complicated geometries can be examined, with the ultimate goal of simulating the entire human cardiovascular system from the heart to the tiniest of capillaries.

The application of fluid dynamics to the human body, the use of magnetic resonance imaging, and the use of lasers in exploring the cardiovascular system have resulted in close cooperation between physics and medicine. This collaboration has already yielded significant results, and is certain to drive important advances in medicine in years to come.

Problems

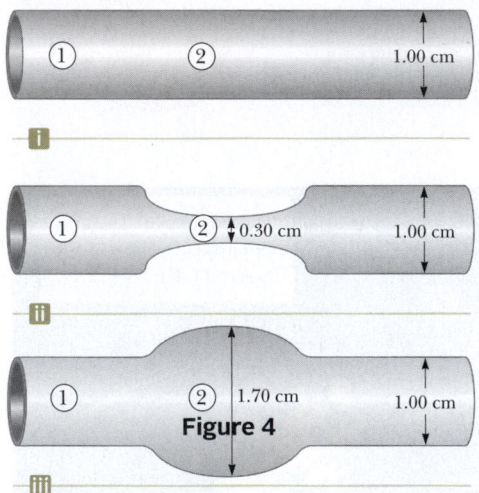

Figure 4

1. **BIO** Three arterial geometries are shown in Figure 4, including a healthy blood vessel (Figure 4, top), a constricted blood vessel (Figure 4, middle), and a blood vessel with an aneurysm (a balloon-like bulge, Figure 4, bottom). The speed of the blood at point 1 in all three vessels is the same. (a) In which of the three blood vessels would the blood be flowing at a higher speed at point 2? (b) What would be the ratio of the speeds of the blood at point 2 in blood vessels (ii) and (iii)?

2. **BIO** There are many situations in physics that can be described by an equation of the form,

(1)   (driving influence) = (resistance)(result of influence)

In many cases, the driving influence can be expressed as the difference between two values of a variable at different locations in space. For example, when we study thermodynamics we will find that the rate P of energy transfer by heat through a material of cross sectional area A and length L is related to the temperature difference ΔT between the two ends of the material as follows:

$$(2)\quad \Delta T = \left(\frac{L}{kA}\right)P$$

where k is the *thermal conductivity* of the material (see Section 17.10). The temperature difference ΔT is the driving influence. The result is the rate of energy transfer P. The quantity L/kA represents resistance to the transfer of energy by heat. The ratio L/k is called the *R-value* (representing thermal resistance) for materials used to provide thermal insulation in homes and buildings.

When we study electricity, we will find that the *potential difference* ΔV between two ends of a material is a driving influence that results in a *current I* in the material as given by

$$(3)\quad \Delta V = \left(\frac{\ell}{\sigma A}\right)I$$

[2]American Heart Association. "Nanoparticles plus adult stem cells demolish plaque, study finds." *Science Daily* 21 July 2010.

The current exists in a piece of material of length ℓ and cross-sectional area A, with electrical conductivity σ (see Section 21.2). The combination in parentheses represents the resistance to the influence, which in this case is called *electrical resistance*.

Let's now think about the flow of blood in an artery. What drives the flow? What resists the flow? The flow of blood is driven by a pressure difference ΔP along an artery, just like the flow of water is driven by a pressure difference along a pipe. The following equation describes the flow of any liquid in a pipe:

$$(4) \quad \Delta P = \left(8\pi\mu\frac{L}{A} \right)v$$

where v, the speed of the fluid, is the result of the pressure difference and the resistance is related to the *viscosity* μ of the fluid. Viscosity is a measure of the internal resistance of a liquid to flow and has units of Pa · s. Honey, for example, is a more viscous fluid than water. Peanut butter is a tremendously viscous fluid. Notice the striking resemblance of Equation (4) to Equations (2) and (3).

A pulmonary artery carries deoxygenated blood from the heart to a lung. (See Figure 2 in the *Heart Attacks* Context Introduction.) Suppose a pulmonary artery has a length of 9.00 cm and a radius of 3.00 mm. A pressure difference of 400 Pa exists along the length of the artery and between its ends. (a) Find the speed of the flow of blood through this artery if the blood has a viscosity of 3.00×10^{-3} Pa · s. (b) Blood is not a simple liquid. It contains red blood cells as well as other cells. The percentage of blood volume occupied by red blood cells is called *hematocrit*. Some illnesses, for example, *polycythemia*, are characterized by increased hematocrit levels. The increased percentage of red blood cells can increase the viscosity of the blood. Suppose a patient with an increased hematocrit level has blood with a viscosity 1.80 times that of the blood in (a). What difference in pressure is required across the length of the pulmonary artery to provide the same blood speed?

3. **BIO** **Q|C** The human brain and spinal cord are immersed in the cerebrospinal fluid. The fluid is normally continuous between the cranial and spinal cavities and exerts a pressure of 100 to 200 mm of H_2O above the prevailing atmospheric pressure. In medical work, pressures are often measured in units of millimeters of H_2O because body fluids, including the cerebrospinal fluid, typically have the same density as water. The pressure of the cerebrospinal fluid can be measured by means of a *spinal tap* as illustrated in Figure 5. A hollow tube is inserted into the spinal column, and the height to which the fluid rises is observed. If the fluid rises to a height of 160 mm, we write its gauge pressure as 160 mm H_2O. (a) Express this pressure in pascals, in atmospheres, and in millimeters of mercury. (b) Some conditions that block or inhibit the flow of cerebrospinal fluid can be investigated by means of *Queckenstedt's test*. In this procedure, the veins in the patient's neck are compressed to make the blood pressure rise in the brain, which in turn should be transmitted to the cerebrospinal fluid. Explain how the level of fluid in the spinal tap can be used as a diagnostic tool for the condition of the patient's spine.

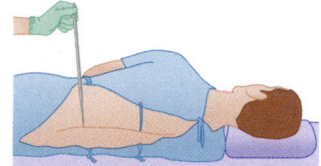

Figure 5

4. **BIO** **M** A hypodermic syringe contains a medicine with the density of water (Figure 6). The barrel of the syringe has a cross-sectional area $A = 2.50 \times 10^{-5}$ m², and the needle has a cross-sectional area $a = 1.00 \times 10^{-8}$ m². In the absence of a force on the plunger, the pressure everywhere is 1.00 atm. A force $\vec{F}$ of magnitude 2.00 N acts on the plunger, making medicine squirt horizontally from the needle. Determine the speed of the medicine as it leaves the needle's tip.

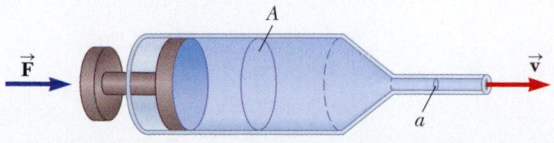

Figure 6

Global Warming

Numerous scientific studies have detailed the effects of an increase in temperature of the Earth, including melting of ice from the polar ice caps and changes in climate and the corresponding effects on vegetation. Data taken over the past few decades show a measurable global temperature increase. Life on this planet depends on a delicate balance that keeps the global temperature in a narrow range necessary for our survival. How is this temperature determined? What factors need to be in balance to keep the temperature constant? If we can devise an adequate structural model that predicts the correct surface temperature of the Earth, we can use the model to predict changes in the temperature as we vary the parameters.

You most likely have an intuitive sense for the temperature of an object, and as long as the object is small (and the object is not undergoing combustion or some other rapid process) no significant temperature variation occurs between different points on the object. What about a huge object like the Earth, though? It is clear that no single temperature describes the entire planet; we know that it is summer in Australia when it is winter in Canada. The polar ice caps clearly have different temperatures from the tropical regions. Variations also occur in temperature within a single large body of water such as an ocean. Temperature varies greatly with altitude in a relatively local region, such as in and near Palm Springs, California, as shown in Figure 1. Therefore, when we speak of the temperature of the Earth, we will refer to an *average* surface temperature, taking into account all the variations across the surface. It is this average temperature that we would like to calculate by building a structural model of the atmosphere and comparing its prediction with the measured surface temperature.

A primary factor in determining the surface temperature of the Earth is the existence of our atmosphere. The atmosphere is a relatively thin (compared with the radius of the Earth) layer of gas above the surface that provides us with life-supporting oxygen. In addition to providing this important element for life, the atmosphere plays a major role in the energy balance that determines the average temperature. As we proceed with

Figure 1 Temperature variations with altitude can exist in a local region on the Earth. Here in Palm Springs, California, palm trees grow in the city while snow is present at the top of the local mountains.

this Context, we shall focus on the physics of gases and apply the principles we learn to the atmosphere.

One important component of the global warming problem is the concentration of carbon dioxide in the atmosphere. Carbon dioxide plays an important role in absorbing energy and raising the temperature of the atmosphere. As seen in Figure 2, the

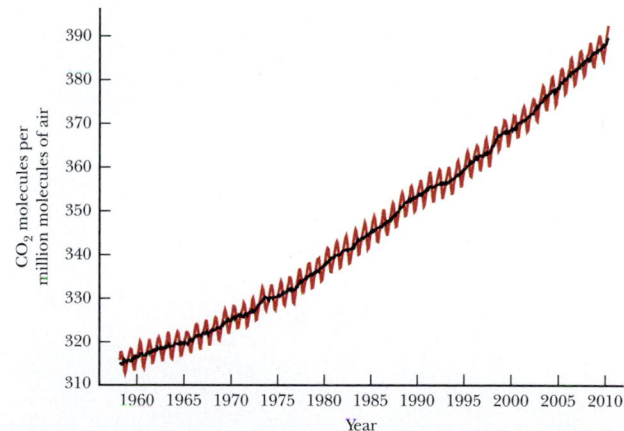

Figure 2 The concentration of atmospheric carbon dioxide in parts per million (ppm) of dry air as a function of time. These data were recorded at the Mauna Loa Observatory in Hawaii. The yearly variations (*red-brown curve*) coincide with growing seasons because vegetation absorbs carbon dioxide from the air. The steady increase in the average concentration (*black curve*) is of concern to scientists.

Courtesy of NASA

Figure 3 In this Context, we explore the energy balance of the Earth, which is a large, nonisolated system that interacts with its environment solely by means of electromagnetic radiation.

The Intergovernmental Panel on Climate Change (IPCC) is a scientific body that assesses the available information related to global warming and associated effects related to climate change. It was originally established in 1988 by two United Nations organizations, the World Meteorological Organization and the United Nations Environment Programme. The IPCC has published four assessment reports on climate change, the most recent in 2007, and a fifth report is scheduled to be released in 2014. The 2007 report concludes that there is a probability of greater than 90% that the increased global temperature measured by scientists is due to the placement of greenhouse gases such as carbon dioxide in the atmosphere by humans. The report also predicts a global temperature increase between 1 and 6°C in the 21st century, sea level rises from 18 to 59 cm, and very high probabilities of weather extremes, including heat waves, droughts, cyclones, and heavy rainfall.

In addition to its scientific aspects, global warming is a social issue with many facets. These facets encompass international politics and economics, because global warming is a worldwide problem. Changing our policies requires real costs to solve the problem. Global warming also has technological aspects, and new methods of manufacturing, transportation, and energy supply must be designed to slow down or reverse the increase in temperature. We shall restrict our attention to the physical aspects of global warming as we address this central question:

> **Can we build a structural model of the atmosphere that predicts the average temperature at the Earth's surface?**

amount of carbon dioxide in the atmosphere has been steadily increasing since the middle of the 20th century. This graph shows hard data that indicate that the atmosphere is undergoing a distinct change, although not all scientists agree on the interpretation of what that change means in terms of global temperatures.

Chapter 16

Temperature and the Kinetic Theory of Gases

Chapter Outline

A bubble in one of the many mud pots in Yellowstone National Park is caught just at the moment of popping. A mud pot is a pool of bubbling hot mud that demonstrates the existence of high temperatures below the Earth's surface.

© Adambooth | Dreamstime.com

Our study thus far has focused mainly on Newtonian mechanics, which explains a wide range of phenomena such as the motion of baseballs, rockets, and planets. We have applied these principles to isolated and nonisolated systems, oscillating systems, the propagation of mechanical waves through a medium, and the properties of fluids at rest and in motion. In Chapter 6, we introduced the notions of temperature and internal energy. We now extend the study of such notions as we focus on **thermodynamics**, which is concerned with the concepts of energy transfers between a system and its environment and the resulting variations in temperature or changes of state. As we shall see, thermodynamics explains the bulk properties of matter and the correlation between these properties and the mechanics of atoms and molecules.

Have you ever wondered how a refrigerator cools or what types of transformations occur in an automobile engine or why a bicycle pump becomes warm as you inflate a tire? The laws of thermodynamics enable us to answer such questions. In general, thermodynamics deals with the physical and chemical transformations of matter in various states: solid, liquid, and gas.

This chapter concludes with a study of ideal gases, which we shall approach on two levels. The first examines ideal gases on the macroscopic scale. Here we shall be concerned with the relationships among such quantities as pressure, volume, and temperature of the gas. On the second level, we shall examine gases on a microscopic (molecular) scale, using a structural model that treats the gas as a collection of particles. The latter approach will help us understand how behavior on the atomic level affects such macroscopic properties as pressure and temperature.

◤ 16.1 | Temperature and the Zeroth Law of Thermodynamics

We often associate the concept of temperature with how hot or cold an object feels to the touch. Our sense of touch provides us with a qualitative indication of temperature. Our senses are unreliable and often misleading, however. For example, if you stand with one bare foot on a tile floor and the other on an adjacent carpeted floor, the tile floor feels colder to your foot than the carpet even though the two are at the same temperature. That is because the properties of the tile are such that the transfer of energy (by heat) to the tile floor from your foot is more rapid than to the carpet. Your skin is sensitive to the *rate* of energy transfer—power—not the temperature of the object. Of course, the larger the difference in temperature between that of the object and that of your hand, the faster the energy transfer, so temperature and your sense of touch are related in *some* way. Recent research suggests that part of the temperature sensation in the skin is related to the TRPV3 protein present in sensory neurons that terminate in the skin. What we need is a reliable and reproducible method for establishing the relative "hotness" or "coldness" of objects that is related solely to the temperature of the object. Scientists have developed a variety of thermometers for making such quantitative measurements.

BIO Sense of warm and cold

We are all familiar with experiences in which two objects at different initial temperatures eventually reach some intermediate temperature when placed in contact with each other. For example, if you combine hot water and cold water in a bathtub from separate faucets, the combined water reaches an equilibrium temperature between the temperatures of the hot water and the cold water. Likewise, if an ice cube is placed in a cup of hot coffee, the ice eventually melts and the temperature of the coffee decreases.

We shall use these familiar examples to develop the scientific notion of temperature. Imagine two objects placed in an insulated container so that they form an isolated system. If the objects are at different temperatures, energy can be exchanged between them by, for example, heat or electromagnetic radiation. Objects that can exchange energy with each other in this way are said to be in **thermal contact.** Eventually, the temperatures of the two objects will become the same, one becoming warmer and the other cooler, as in our preceding examples. **Thermal equilibrium** is the situation in which two objects in thermal contact cease to have any net exchange of energy by heat or electromagnetic radiation.

Using these ideas, we can develop a formal definition of temperature. Consider two objects A and B that are not in thermal contact and a third object C that will be our **thermometer,** a device calibrated to measure the temperature of an object. We wish to determine whether A and B would be in thermal equilibrium if they were placed in thermal contact. The thermometer is first placed in thermal contact with A and its reading is recorded, as shown in Figure 16.1a. The thermometer is then placed in thermal contact with B and its reading is recorded (Fig. 16.1b). If the two readings are the same, A and B are in thermal equilibrium with each other. If they are placed in thermal contact with each other, as in Figure 16.1c, there is no net transfer of energy between them.

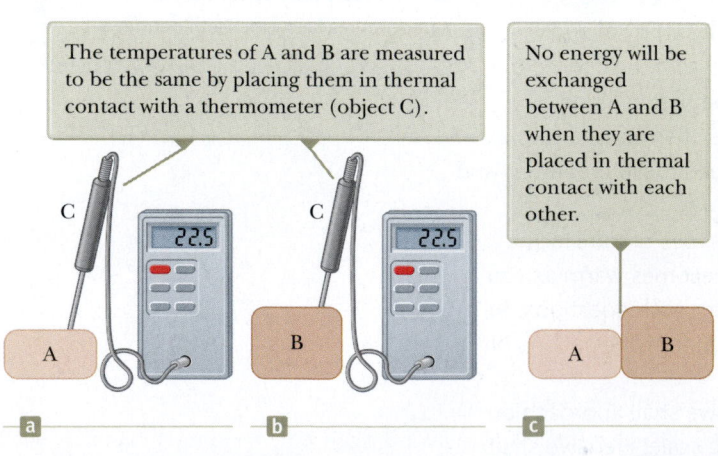

The temperatures of A and B are measured to be the same by placing them in thermal contact with a thermometer (object C).

No energy will be exchanged between A and B when they are placed in thermal contact with each other.

Figure 16.1 The zeroth law of thermodynamics.

We can summarize these results in a statement known as the **zeroth law of thermodynamics:**

If objects A and B are separately in thermal equilibrium with a third object C, then A and B are in thermal equilibrium with each other.

▶ Zeroth law of thermodynamics

This statement, elementary as it may seem, is very important because it can be used to define the notion of temperature and is easily proved experimentally. We can think of temperature as the property that determines whether an object is in thermal equilibrium with other objects: Two objects in thermal equilibrium with each other are at the same temperature.

16.2 | Thermometers and Temperature Scales

In our discussion of the zeroth law, we mentioned a thermometer. Thermometers are devices used to measure the temperature of an object or a system with which the thermometer is in thermal equilibrium. All thermometers make use of some physical property that exhibits a change with temperature that can be calibrated to make the temperature measurable. Some of the physical properties used are (1) the volume of a liquid, (2) the length of a solid, (3) the pressure of a gas held at constant volume, (4) the volume of a gas held at constant pressure, (5) the electric resistance of a conductor, and (6) the color of a hot object.

A common thermometer in everyday use consists of a liquid—usually mercury or alcohol—that expands into a glass capillary tube when its temperature rises (Fig. 16.2). In this case, the physical property that changes is the volume of a liquid. Because the cross-sectional area of the capillary tube is uniform, the change in volume of the liquid varies linearly with its length along the tube. We can then define a temperature to be related to the length of the liquid column.

The thermometer can be calibrated by placing it in thermal contact with some environments that remain at constant temperature and marking the end of the liquid column on the thermometer. One such environment is a mixture of water and ice in thermal equilibrium with each other at atmospheric pressure. Once we have marked the ends of the liquid column for our chosen environments on our thermometer, we need to define a scale of numbers associated with various temperatures. One such scale is the **Celsius temperature scale.** On the Celsius scale, the temperature of the ice–water mixture is defined as zero degrees Celsius, written 0°C; this temperature is called the **ice point** or **freezing point** of water. Another commonly used environment is a mixture of water and steam in thermal equilibrium with each other at atmospheric pressure. On the Celsius scale, this temperature is defined as 100°C, the **steam point** or **boiling point** of water. Once the ends of the liquid column in the thermometer have been marked at these two points, the distance between the marks is divided into 100 equal segments, each denoting a change in temperature of one degree Celsius.

Thermometers calibrated in this way present problems when extremely accurate readings are needed. For instance, an alcohol thermometer calibrated at the ice and steam points of water might agree with a mercury thermometer only at the calibration points. Because mercury and alcohol have different thermal expansion properties (the expansion may not be perfectly linear with temperature), when one indicates a given temperature, the other may indicate a slightly different value. The discrepancies between different types of thermometers are especially large when the temperatures to be measured are far from the calibration points.

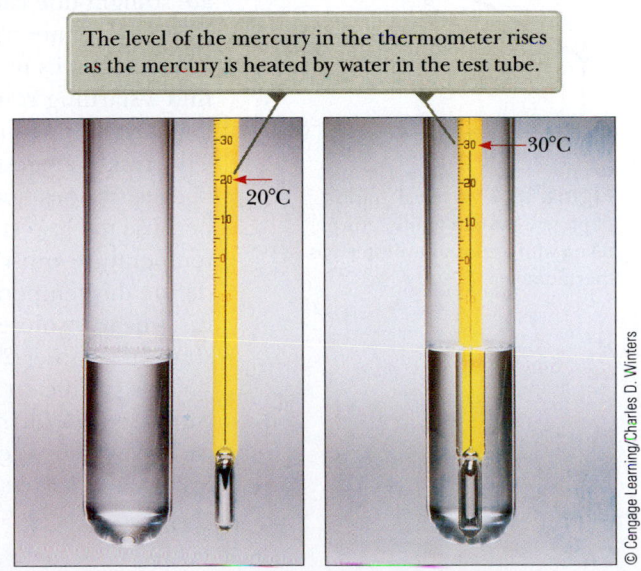

The level of the mercury in the thermometer rises as the mercury is heated by water in the test tube.

© Cengage Learning/Charles D. Winters

Figure 16.2 A mercury thermometer before and after increasing its temperature.

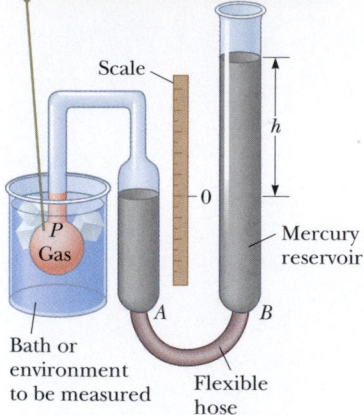

The volume of gas in the flask is kept constant by raising or lowering reservoir B to keep the mercury level in column A constant.

Figure 16.3 A constant-volume gas thermometer measures the pressure of the gas contained in the flask immersed in the bath.

The two dots represent known reference temperatures (the ice and steam points of water).

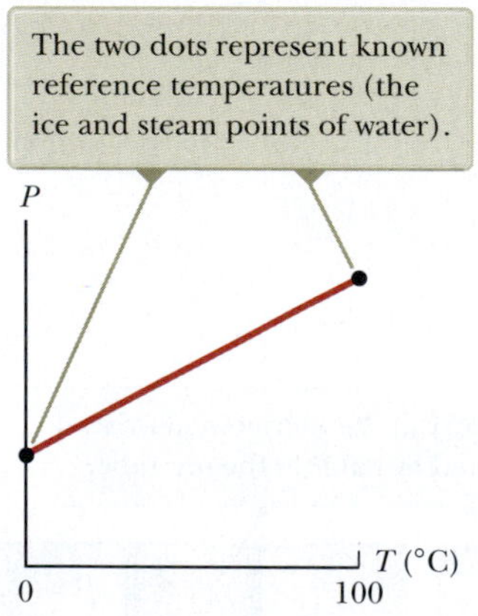

Figure 16.4 A typical graph of pressure versus temperature taken with a constant-volume gas thermometer.

The Constant-Volume Gas Thermometer and the Kelvin Scale

Although practical devices such as the mercury thermometer can measure temperature, they do not define it in a fundamental way. Only one thermometer, the **gas thermometer,** offers a way to define temperature and relate it to internal energy directly. In a gas thermometer, the temperature readings are nearly independent of the substance used in the thermometer. One type of gas thermometer is the constant-volume example shown in Figure 16.3. The behavior observed in this device is the pressure variation with temperature of a fixed volume of gas.

When the constant-volume gas thermometer was developed, it was calibrated using the ice and steam points of water as follows. (A different calibration procedure, to be discussed shortly, is now used.) The gas flask is inserted into an ice bath, and mercury reservoir B is raised or lowered until the volume of the confined gas is at some value, indicated by the zero point on the scale. The height h, the difference between the levels in the reservoir and column A, indicates the pressure in the flask at 0°C, according to Equation 15.4. The flask is inserted into water at the steam point, and reservoir B is readjusted until the height in column A is again brought to zero on the scale, ensuring that the gas volume is the same as it had been in the ice bath (hence the designation "constant-volume"). A measure of the new value for h gives a value for the pressure at 100°C. These pressure and temperature values are then plotted on a graph, as in Figure 16.4. Based on experimental observations that the pressure of a gas varies linearly with its temperature, which is discussed in more detail in Section 16.4, we draw a straight line through our two points. The line connecting the two points serves as a calibration curve for measuring unknown temperatures. If we want to measure the temperature of a substance, we place the gas flask in thermal contact with the substance and adjust the column of mercury until the level in column A again returns to zero. The height of the mercury column tells us the pressure of the gas, and we can then find the temperature of the substance from the calibration curve.

Now suppose temperatures are measured with various gas thermometers containing different gases. Experiments show that the thermometer readings are nearly independent of the type of gas used as long as the gas pressure is low and the temperature is well above the point at which the gas liquefies.

We can also perform the temperature measurements with the gas in the flask at different starting pressures at 0°C. As long as the pressure is low, we will generate straight-line calibration curves for each different starting pressure, as shown for three experimental trials (solid lines) in Figure 16.5.

If the curves in Figure 16.5 are extended back toward negative temperatures, we find a startling result. In every case, regardless of the type of gas or the value of the low starting pressure, *the pressure extrapolates to zero when the temperature is* −273.15°C! This result suggests that this particular temperature is universal in its importance because it does not depend on the substance used in the thermometer. In addition, because the lowest possible pressure is $P = 0$, which would be a perfect vacuum, this temperature must represent a lower bound for physical processes. Therefore, we define this temperature as **absolute zero.** Some interesting effects occur at temperatures near absolute zero, such as the phenomenon of *superconductivity,* which we shall study in Chapter 21.

This significant temperature is used as the basis for the **Kelvin temperature scale,** which sets −273.15°C as its zero point (0 K). The size of a degree on the Kelvin scale is chosen to be identical to the size of a degree on the Celsius scale. Therefore, the following relationship enables conversion between these temperatures:

$$T_C = T - 273.15 \qquad \textbf{16.1}\blacktriangleleft$$

where T_C is the Celsius temperature and T is the Kelvin temperature (sometimes called the **absolute temperature**). The primary difference between these two temperature scales is a shift in the zero of the scale. The zero of the Celsius scale is arbitrary; it depends on a property associated with only one substance, water. The zero on the Kelvin scale is not arbitrary because it is characteristic of a behavior associated with all substances. Consequently, when an equation contains T as a variable, the absolute temperature must be used. Similarly, a ratio of temperatures is only meaningful if the temperatures are expressed on the Kelvin scale.

Equation 16.1 shows that the Celsius temperature T_C is shifted from the absolute temperature T by 273.15°. Because the size of a degree is the same on the two scales, a temperature difference of 5°C is equal to a temperature difference of 5 K. The two scales differ only in the choice of the zero point. Therefore, the ice point (273.15 K) corresponds to 0.00°C, and the steam point (373.15 K) is equivalent to 100.00°C.

Early gas thermometers made use of ice and steam points according to the procedure just described. These points are experimentally difficult to duplicate, however. For this reason, a new procedure based on two new points was adopted in 1954 by the International Committee on Weights and Measures. The first point is absolute zero. The second point is the **triple point of water**, which corresponds to the single temperature and pressure at which water, water vapor, and ice can coexist in equilibrium. This point is a convenient and reproducible reference temperature for the Kelvin scale. It occurs at a temperature of 0.01°C and a very low pressure of 4.58 mm of mercury. The temperature at the triple point of water on the Kelvin scale has a value of 273.16 K. Therefore, the SI unit of temperature, the **kelvin**, is defined as **1/273.16 of the temperature of the triple point of water.**

Figure 16.6 shows the Kelvin temperatures for various physical processes and conditions. As the figure reveals, absolute zero has never been achieved, although laboratory experiments have created conditions that are very close to absolute zero.

What would happen to a gas if its temperature could reach 0 K? As Figure 16.5 indicates (if we ignore the liquefaction and solidification of the substance), the pressure it would exert on the container's walls would be zero. In Section 16.5, we shall show that the pressure of a gas is proportional to the kinetic energy of the molecules of that gas. Therefore, according to classical physics, the kinetic energy of the gas would go to zero and there would be no motion at all of the individual components of the gas; hence, the molecules would settle out on the bottom of the container. Quantum theory, to be discussed in Chapter 28, modifies this statement to indicate that there would be some residual energy, called the *zero-point energy*, at this low temperature.

The Fahrenheit Scale

The most common temperature scale in everyday use in the United States is the **Fahrenheit scale.** This scale sets the temperature of the ice point at 32°F and the temperature of the steam point at 212°F. The relationship between the Celsius and Fahrenheit temperature scales is

$$T_F = \tfrac{9}{5} T_C + 32°F \qquad \blacktriangleleft \text{16.2}$$

Equation 16.2 can easily be used to find a relationship between changes in temperature on the Celsius and Fahrenheit scales. It is left as a problem in Enhanced WebAssign for you to show that if the Celsius temperature changes by ΔT_C, the Fahrenheit temperature changes by an amount ΔT_F given by

$$\Delta T_F = \tfrac{9}{5} \Delta T_C \qquad \blacktriangleleft \text{16.3}$$

QUICK QUIZ 16.1 Consider the following pairs of materials. Which pair represents two materials, one of which is twice as hot as the other? **(a)** boiling water at 100°C, a glass of water at 50°C **(b)** boiling water at 100°C, frozen methane at −50°C **(c)** an ice cube at −20°C, flames from a circus fire-eater at 233°C **(d)** none of those pairs

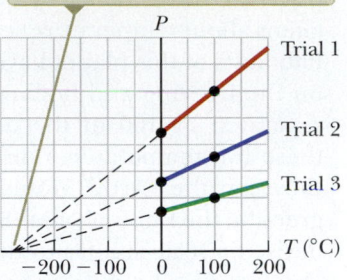

For all three trials, the pressure extrapolates to zero at the temperature −273.15°C.

Figure 16.5 Pressure versus temperature for experimental trials in which gases have different pressures in a constant volume gas thermometer.

Pitfall Prevention | 16.1
A Matter of Degree
Note that notations for temperatures in the Kelvin scale do not use the degree sign. The unit for a Kelvin temperature is simply "kelvins" and not "degrees Kelvin."

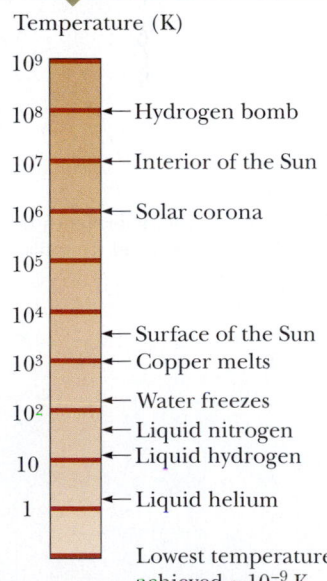

Note that the scale is logarithmic.

Temperature (K)

10^9	
10^8	← Hydrogen bomb
10^7	← Interior of the Sun
10^6	← Solar corona
10^5	
10^4	
10^3	← Surface of the Sun / ← Copper melts
10^2	← Water freezes / ← Liquid nitrogen
10	← Liquid hydrogen
1	← Liquid helium

Lowest temperature achieved ∼ 10^{-9} K

Figure 16.6 Absolute temperatures at which various selected physical processes occur.

> ### THINKING PHYSICS 16.1
>
> A group of future astronauts lands on an inhabited planet. The astronauts strike up a conversation with the aliens about temperature scales. It turns out that the inhabitants of this planet have a temperature scale based on the freezing and boiling points of water, which are separated by 100 of the inhabitants' degrees. Would these two temperatures on this planet be the same as those on the Earth? Would the size of the aliens' degrees be the same as ours? Suppose the aliens have also devised a scale similar to the Kelvin scale. Would their absolute zero be the same as ours?
>
> **Reasoning** The values of 0°C and 100°C for the freezing and boiling points of water are defined at atmospheric pressure. On another planet, it is unlikely that atmospheric pressure would be exactly the same as that on the Earth. Therefore, water would freeze and boil at different temperatures on the alien planet. The aliens may call these temperatures 0° and 100°, but they would not be the same temperatures as our 0°C and 100°C. If the aliens did assign values of 0° and 100° for these temperatures, their degrees would not be the same size as our Celsius degrees (unless their atmospheric pressure were the same as ours). For an alien version of the Kelvin scale, the absolute zero would be the same as ours because it is based on a natural, universal definition rather than being associated with a particular substance or a given atmospheric pressure. ◀

Example 16.1 | Converting Temperatures

On a day when the temperature reaches 50°F, what is the temperature in degrees Celsius and in kelvins?

SOLUTION

Conceptualize In the United States, a temperature of 50°F is well understood. In many other parts of the world, however, this temperature might be meaningless because people are familiar with the Celsius temperature scale.

Categorize This example is a simple substitution problem.

Substitute the given temperature into Equation 16.2:

$$T_C = \tfrac{5}{9}(T_F - 32) = \tfrac{5}{9}(50 - 32) = \boxed{10°C}$$

Use Equation 16.1 to find the Kelvin temperature:

$$T = T_C + 273.15 = 10°C + 273.15 = \boxed{283 \text{ K}}$$

A convenient set of weather-related temperature equivalents to keep in mind is that 0°C is (literally) freezing at 32°F, 10°C is cool at 50°F, 20°C is room temperature, 30°C is warm at 86°F, and 40°C is a hot day at 104°F.

16.3 | Thermal Expansion of Solids and Liquids

Our discussion of the liquid thermometer makes use of one of the best known changes that occur in most substances, that as the temperature of a substance increases, its volume increases. This phenomenon, known as **thermal expansion,** plays an important role in numerous applications. For example, thermal expansion joints (Fig. 16.7) must

Figure 16.7 Thermal-expansion joints in (a) bridges and (b) walls.

Without these joints to separate sections of roadway on bridges, the surface would buckle due to thermal expansion on very hot days or crack due to contraction on very cold days.

The long, vertical joint is filled with a soft material that allows the wall to expand and contract as the temperature of the bricks changes.

© Cengage Learning/George Semple

© Cengage Learning/George Semple

be included in buildings, concrete highways, railroad tracks, and bridges to compensate for changes in dimensions with temperature variations.

The overall thermal expansion of an object is a consequence of the change in the average separation between its constituent atoms or molecules. To understand this concept, consider how the atoms in a solid substance behave. These atoms are located at fixed equilibrium positions; if an atom is pulled away from its position, a restoring force pulls it back. We can build a structural model in which we imagine that the atoms are particles at their equilibrium positions connected by springs to their neighboring atoms (Fig. 16.8). If an atom is pulled away from its equilibrium position, the distortion of the springs provides a restoring force. If the atom is released, it oscillates, and we can apply the particle in simple harmonic motion model to it. A number of macroscopic properties of the substance can be understood with this type of structural model on the atomic level.

In Chapter 6, we introduced the notion of internal energy and pointed out that it is related to the temperature of a system. For a solid, the internal energy is associated with the kinetic and potential energy of the vibrations of the atoms around their equilibrium positions. At ordinary temperatures, the atoms vibrate with an amplitude of about 10^{-11} m, and the average spacing between the atoms is about 10^{-10} m. As the temperature of the solid increases, the average separation between atoms increases. The increase in average separation with increasing temperature (and subsequent thermal expansion) is the result of a breakdown in the model of simple harmonic motion. Active Figure 6.23a in Chapter 6 shows the potential energy curve for an ideal simple harmonic oscillator. The potential energy curve for atoms in a solid is similar but not exactly the same as that one; it is slightly asymmetric around the equilibrium position. It is this asymmetry that leads to thermal expansion.

If the thermal expansion of an object is sufficiently small compared with the object's initial dimensions, the change in any dimension is, to a good approximation, dependent on the first power of the temperature change. For most situations, we can adopt a simplification model in which this dependence is true. Suppose an object has an initial length L_i along some direction at some temperature. The length increases by ΔL for a change in temperature ΔT. Experiments show that when ΔT is small enough, ΔL is proportional to ΔT and to L_i:

$$\Delta L = \alpha L_i \Delta T \qquad \textbf{16.4} \blacktriangleleft$$

or

$$L_f - L_i = \alpha L_i (T_f - T_i) \qquad \textbf{16.5} \blacktriangleleft$$

where L_f is the final length, T_f is the final temperature, and the proportionality constant α is called the **average coefficient of linear expansion** for a given material and has units of inverse degrees Celsius, or $(°C)^{-1}$.

It may be helpful to think of thermal expansion as a magnification or a photographic enlargement. For example, as a metal washer is heated (Active Fig. 16.9), all dimensions, including the radius of the hole, increase according to Equation 16.4. Because the linear dimensions of an object change with temperature, it follows that volume and surface area also change with temperature. Consider a cube having an initial edge length L_i and therefore an initial volume $V_i = L_i^3$. As the temperature is increased, the length of each side increases to

$$L_f = L_i + \alpha L_i \Delta T$$

The new volume, $V_f = L_f^3$, is

$$L_f^3 = (L_i + \alpha L_i \Delta T)^3 = L_i^3 + 3\alpha L_i^3 \Delta T + 3\alpha^2 L_i^3 (\Delta T)^2 + \alpha^3 L_i^3 (\Delta T)^3$$

The last two terms in this expression contain the quantity $\alpha \Delta T$ raised to the second and third powers. Because $\alpha \Delta T$ is a pure number much less than 1, raising it

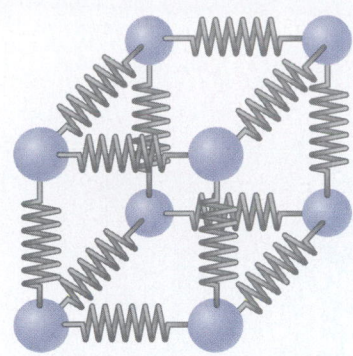

Figure 16.8 A structural model of the atomic configuration in a solid. The atoms (spheres) are imagined to be connected to one another by springs that reflect the elastic nature of the interatomic forces.

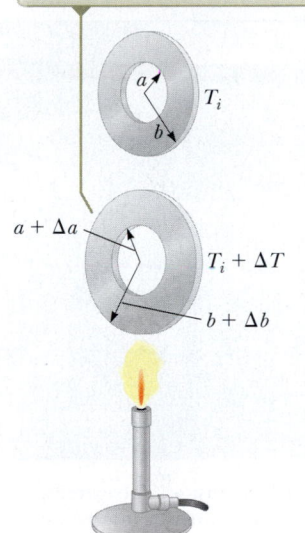

As the washer is heated, all dimensions increase, including the radius of the hole.

Active Figure 16.9 Thermal expansion of a homogeneous metal washer. (The expansion is exaggerated in this figure.)

Pitfall Prevention | 16.2
Do Holes Become Larger or Smaller?
When an object's temperature is raised, every linear dimension increases in size. Included are any holes in the material, which expand in the same way as if the hole were filled with the material, as shown in Active Figure 16.9. Keep in mind the notion of thermal expansion as being similar to a photographic enlargement.

> **TABLE 16.1 | Average Expansion Coefficients for Some Materials Near Room Temperature**

Material	Average Coefficient of Linear Expansion $(\alpha)(°C^{-1})$	Material	Average Coefficient of Volume Expansion $(\beta)(°C^{-1})$
Aluminum	24×10^{-6}	Acetone	1.5×10^{-4}
Brass and bronze	19×10^{-6}	Alcohol, ethyl	1.12×10^{-4}
Concrete	12×10^{-6}	Benzene	1.24×10^{-4}
Copper	17×10^{-6}	Gasoline	9.6×10^{-4}
Glass (ordinary)	9×10^{-6}	Glycerin	4.85×10^{-4}
Glass (Pyrex)	3.2×10^{-6}	Mercury	1.82×10^{-4}
Invar (Ni–Fe alloy)	0.9×10^{-6}	Turpentine	9.0×10^{-4}
Lead	29×10^{-6}	Air[a] at 0°C	3.67×10^{-3}
Steel	11×10^{-6}	Helium[a]	3.665×10^{-3}

[a]Gases do not have a specific value for the volume expansion coefficient because the amount of expansion depends on the type of process through which the gas is taken. The values given here assume the gas undergoes an expansion at constant pressure.

to a power makes it even smaller. Therefore, we can ignore these terms to obtain a simpler expression:

$$V_f = L_f{}^3 = L_i{}^3 + 3\alpha L_i{}^3 \Delta T = V_i + 3\alpha V_i \Delta T$$

or

$$\Delta V = V_f - V_i = \beta V_i \Delta T \qquad \textbf{16.6} \blacktriangleleft$$

where $\beta = 3\alpha$. The quantity β is called the **average coefficient of volume expansion.** We considered a cubic shape in deriving this equation, but Equation 16.6 describes a sample of any shape as long as the average coefficient of linear expansion is the same in all directions.

By a similar procedure, we can show that the *increase in area* of an object accompanying an increase in temperature is

$$\Delta A = \gamma A_i \Delta T \qquad \textbf{16.7} \blacktriangleleft$$

where γ, the **average coefficient of area expansion,** is given by $\gamma = 2\alpha$.

Table 16.1 lists the average coefficient of linear expansion for various materials. Note that for these materials α is positive, indicating an increase in length with increasing temperature, but that is not always the case. For example, some substances, such as calcite ($CaCO_3$), expand along one dimension (positive α) and contract along another (negative α) with increasing temperature.

AP Photo

Thermal expansion. The extremely high temperature of a July day in Asbury Park, New Jersey, caused these railroad tracks to buckle.

> **QUICK QUIZ 16.2** Two spheres are made of the same metal and have the same radius, but one is hollow and the other is solid. The spheres are taken through the same temperature increase. Which sphere expands more? (**a**) The solid sphere expands more. (**b**) The hollow sphere expands more. (**c**) They expand by the same amount. (**d**) There is not enough information to say.

> **THINKING PHYSICS 16.2**
>
> As a homeowner is painting a ceiling, a drop of paint falls from the brush onto an operating incandescent lightbulb. The bulb breaks. Why?
>
> **Reasoning** The glass envelope of an incandescent lightbulb receives energy on the inside surface by electromagnetic radiation from the very hot filament. In addition, because the bulb contains gas, the glass envelope receives energy by matter transfer related to the movement of the hot gas near the filament to the colder glass. Therefore, the glass can become very hot. If a drop of relatively cold

paint falls onto the glass, that portion of the glass envelope suddenly becomes colder than the other portions, and the contraction of this region can cause thermal stresses that might break the glass. ◀

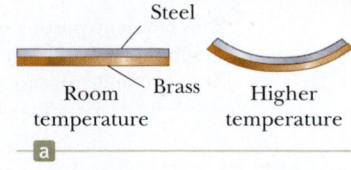

As Table 16.1 indicates, each substance has its own characteristic coefficients of expansion. For example, when the temperatures of a brass rod and a steel rod of equal length are raised by the same amount from some common initial value, the brass rod expands more than the steel rod because brass has a larger coefficient of expansion than steel. A simple device called a bimetallic strip that demonstrates this principle is found in practical devices such as thermostats in home furnace systems. The strip is made by securely bonding two different metals together along their surfaces. As the temperature of the strip increases, the two metals expand by different amounts and the strip bends as in Figure 16.10.

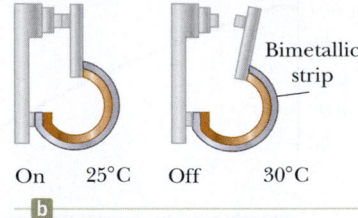

Figure 16.10 (a) A bimetallic strip bends as the temperature changes because the two metals have different expansion coefficients. (b) A bimetallic strip used in a thermostat to make or break electrical contact.

Example 16.2 | The Thermal Electrical Short

A poorly designed electronic device has two bolts attached to different parts of the device that almost touch each other in its interior as in Figure 16.11. The steel and brass bolts are at different electric potentials, and if they touch, a short circuit will develop, damaging the device. (We will study electric potential in Chapter 20.) The initial gap between the ends of the bolts is 5.0 μm at 27°C. At what temperature will the bolts touch? Assume the distance between the walls of the device is not affected by the temperature change.

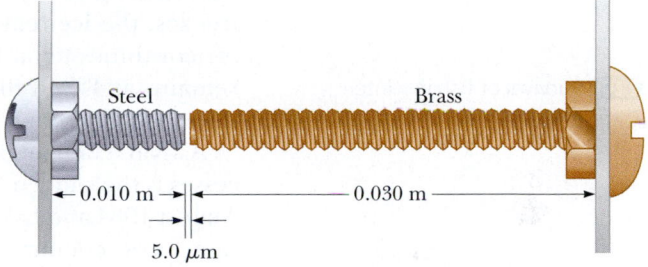

Figure 16.11 (Example 16.2) Two bolts attached to different parts of an electrical device are almost touching when the temperature is 27°C. As the temperature increases, the ends of the bolts move toward each other.

SOLUTION

Conceptualize Imagine the ends of both bolts expanding into the gap between them as the temperature rises.

Categorize We categorize this example as a thermal expansion problem in which the *sum* of the changes in length of the two bolts must equal the length of the initial gap between the ends.

Analyze Set the sum of the length changes equal to the width of the gap:

$$\Delta L_{br} + \Delta L_{st} = \alpha_{br} L_{i,br} \Delta T + \alpha_{st} L_{i,st} \Delta T = 5.0 \times 10^{-6} \text{ m}$$

Solve for ΔT:

$$\Delta T = \frac{5.0 \times 10^{-6} \text{ m}}{\alpha_{br} L_{i,br} + \alpha_{st} L_{i,st}}$$

$$= \frac{5.0 \times 10^{-6} \text{ m}}{[19 \times 10^{-6} \text{ (°C)}^{-1}](0.030 \text{ m}) + [11 \times 10^{-6} \text{ (°C)}^{-1}](0.010 \text{ m})} = 7.4°C$$

Find the temperature at which the bolts touch:

$$T = 27°C + 7.4°C = \boxed{34°C}$$

Finalize This temperature is possible if the air conditioning in the building housing the device fails for a long period on a very hot summer day.

The Unusual Behavior of Water

Liquids generally increase in volume with increasing temperature and have volume expansion coefficients on the order ten times greater than those of solids. Water is an exception to this rule over a small temperature range, as we can see from its

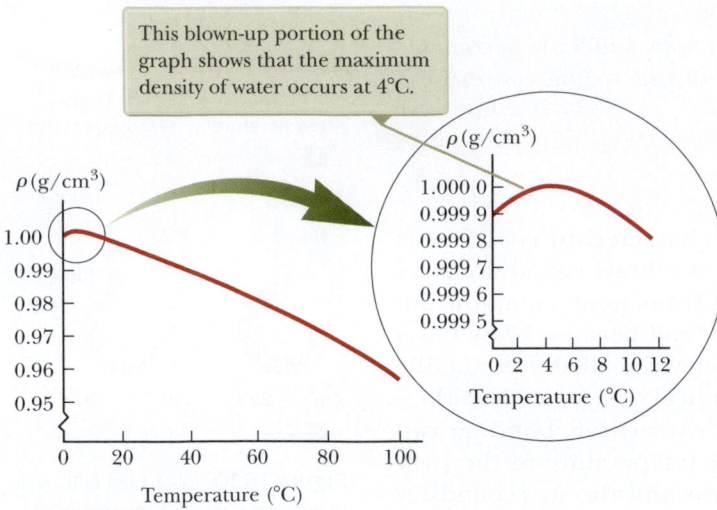

This blown-up portion of the graph shows that the maximum density of water occurs at 4°C.

Figure 16.12 The variation of density with temperature for water at atmospheric pressure.

BIO Survival of fish in winter

density versus temperature curve in Figure 16.12. As the temperature increases from 0°C to 4°C, water contracts and therefore its density increases. Above 4°C, water exhibits the expected expansion with increasing temperature. Therefore, the density of water reaches a maximum value of 1 000 kg/m³ at 4°C.

We can use this unusual thermal expansion behavior of water to explain why a pond freezes at the surface. When the atmospheric temperature drops from 7°C to 6°C, for example, the water at the surface of the pond also cools and consequently decreases in volume. Hence, the surface water is denser than the water below it, which has not cooled and has not decreased in volume. As a result, the surface water sinks and warmer water from below moves to the surface to be cooled in a process called upwelling. When the atmospheric temperature is between 4°C and 0°C, however, the surface water expands as it cools, becoming less dense than the water below it. The sinking process stops, and eventually the surface water freezes. As the water freezes, the ice remains on the surface because ice is less dense than water. The ice continues to build up on the surface, while water near the bottom of the pool remains at 4°C. If that did not happen, fish and other forms of marine life would not survive through the winter.

A vivid example of the dangers of the absence of the upwelling and mixing processes is the sudden and deadly release of carbon dioxide gas from Lake Monoun in August 1984 and Lake Nyos in August 1986 (Fig. 16.13). Both lakes are located in the rain forest country of Cameroon in Africa. More than 1 700 natives of Cameroon died in these events.

In a lake located in a temperate zone such as the United States, significant temperature variations occur during the day and during the entire year. For example, imagine the Sun going down in the evening. As the temperature of the surface water drops because of the absence of sunlight, the sinking process tends to mix the upper and lower layers of water.

This mixing process does not normally occur in Lake Monoun and Lake Nyos because of two characteristics that contributed significantly to the disasters. First, the lakes are very deep, so mixing the various layers of water over such a large vertical distance is difficult. This factor also results in such very large pressure at the bottom

Figure 16.13 (a) Lake Nyos, in Cameroon, after an explosive outpouring of carbon dioxide. (b) The carbon dioxide caused many deaths, both of humans and animals, such as the cattle shown here.

of the lake that a large amount of carbon dioxide from local rocks and deep springs dissolves into the water. Second, both lakes are located in an equatorial rain forest region where the temperature variation is much smaller than in temperate zones, which results in little driving force to mix the layers of water in the lakes. Water near the bottom of the lake stays there for a long time and collects a large amount of dissolved carbon dioxide. In the absence of a mixing process, this carbon dioxide cannot be brought to the surface and released safely. It simply continues to increase in concentration.

The situation described is explosive. If the carbon dioxide–laden water is brought to the surface where the pressure is much lower, the gas expands and comes out of the solution rapidly. Once the carbon dioxide comes out of the solution, bubbles of carbon dioxide rise through the water and cause more mixing of layers.

Suppose the temperature of the surface water were to decrease; this water would become denser and sink, possibly triggering the release of carbon dioxide and the beginning of the explosive situation just described. The monsoon season in Cameroon occurs in August. Monsoon clouds block the sunlight, resulting in lower surface water temperatures, which may be the reason the disasters occurred in August. Climate data for Cameroon show lower than normal temperatures and higher than normal rainfall in the mid-1980s. The resulting decrease in surface temperature could explain why these events occurred in 1984 and 1986. The exact reasons for the sudden release of carbon dioxide are unknown and remain an area of active research.

Finally, once the carbon dioxide was released from the lakes, it stayed near the ground because carbon dioxide is denser than air. Therefore, a layer of carbon dioxide gas spread out over the land around the lake, representing a deadly suffocating gas for all humans and animals in its path.

BIO Suffocation by explosive release of carbon dioxide

16.4 | Macroscopic Description of an Ideal Gas

The properties of gases are very important in a number of thermal processes. Our everyday weather is a perfect example of the types of processes that depend on the behavior of gases.

If we introduce a gas into a container, it expands to fill the container uniformly. Therefore, the gas does not have a fixed volume or pressure. Its volume is that of the container, and its pressure depends on the size of the container. In this section, we shall be concerned with the properties of a gas with pressure P and temperature T, confined to a container of volume V. It is useful to know how these quantities are related. In general, the equation that interrelates these quantities, called the **equation of state,** can be complicated. If the gas is maintained at a very low pressure (or low density), however, the equation of state is found experimentally to be relatively simple. Such a low-density gas is commonly referred to as an **ideal gas.** Most gases at room temperature and atmospheric pressure behave approximately as ideal gases. We shall adopt a simplification model, called the **ideal gas model,** for these types of studies. In this model, an ideal gas is a collection of atoms or molecules that (1) move randomly, (2) exert no long-range forces on one another, and (3) are so small that they occupy a negligible fraction of the volume of their container.

It is convenient to express the amount of gas in a given volume in terms of the number of moles. One **mole** of any substance is that mass of the substance that contains **Avogadro's number,** $N_A = 6.022 \times 10^{23}$, of molecules. The number of moles n of a substance in a sample is related to its mass m through the expression

$$n = \frac{m}{M}$$

16.8 ◀

where M is the **molar mass** of the substance, usually expressed in grams per mole. For example, the molar mass of molecular oxygen O_2 is 32.0 g/mol. The mass of one mole of oxygen is therefore 32.0 g. We can calculate the mass m_0 of one molecule by dividing the molar mass by the number of molecules, which is Avogadro's number. Therefore, for oxygen,

$$m_0 = \frac{M}{N_A} = \frac{32.0 \times 10^{-3} \text{ kg/mol}}{6.02 \times 10^{23} \text{ molecule/mol}} = 5.32 \times 10^{-26} \text{ kg/molecule}$$

Now suppose an ideal gas is confined to a cylindrical container whose volume can be varied by means of a movable piston, as in Active Figure 16.14. We shall assume that the cylinder does not leak, so the number of moles of gas remains constant. For such a system, experiments provide the following information:

- When the gas is kept at a constant temperature, its pressure is inversely proportional to the volume. (This behavior is described historically as Boyle's law.)
- When the pressure of the gas is kept constant, the volume is directly proportional to the temperature. (This behavior is described historically as Charles's law.)
- When the volume of the gas is kept constant, the pressure is directly proportional to the temperature. (This behavior is described historically as Gay-Lussac's law.)

These observations can be summarized by the following equation of state, known as the **ideal gas law:**

▶ Ideal gas law

$$PV = nRT \qquad\qquad \textbf{16.9} \blacktriangleleft$$

In this expression, R is a constant for a specific gas that can be determined from experiments and T is the absolute temperature in kelvins. Experiments on several gases show that as the pressure approaches zero, the quantity PV/nT approaches the same value of R for *all* gases. For this reason, R is called the **universal gas constant.** In SI units, where pressure is expressed in pascals and volume in cubic meters, R has the value

▶ The universal gas constant

$$R = 8.314 \text{ J/mol} \cdot \text{K} \qquad\qquad \textbf{16.10} \blacktriangleleft$$

If the pressure is expressed in atmospheres and the volume in liters ($1 \text{ L} = 10^3 \text{ cm}^3 = 10^{-3} \text{ m}^3$), R has the value

$$R = 0.082\ 1 \text{ L} \cdot \text{atm/mol} \cdot \text{K}$$

Using this value of R and Equation 16.9, one finds that the volume occupied by 1 mol of any gas at atmospheric pressure and 0°C (273 K) is 22.4 L.

The ideal gas law is often expressed in terms of the total number of molecules N rather than the number of moles n. Because the total number of molecules equals the product of the number of moles and Avogadro's number N_A, we can write Equation 16.9 as

$$PV = nRT = \frac{N}{N_A} RT$$

$$PV = Nk_B T \qquad\qquad \textbf{16.11} \blacktriangleleft$$

where k_B is called **Boltzmann's constant** and has the value

$$k_B = \frac{R}{N_A} = 1.38 \times 10^{-23} \text{ J/K} \qquad\qquad \textbf{16.12} \blacktriangleleft$$

Active Figure 16.14 An ideal gas confined to a cylinder whose volume can be varied with a movable piston.

Pitfall Prevention | 16.3
So Many k's
In a variety of situations in physics, the letter k is used. We have seen two uses previously, the force constant for a spring (Chapter 12) and the wave number for a mechanical wave (Chapter 13). We also saw k_e, the Coulomb constant, in Chapter 5. Boltzmann's constant is another k, and we will see k used for thermal conductivity in Chapter 17. To make some sense of this confusing state of affairs, we will use a subscript for Boltzmann's constant to help us recognize it. In this book, we will see Boltzmann's constant as k_B, but you may see Boltzmann's constant in other resources as simply k.

QUICK QUIZ 16.3 A common material for cushioning objects in packages is made by trapping bubbles of air between sheets of plastic. Is this material more effective at keeping the contents of the package from moving around inside the package on (a) a hot day, (b) a cold day, or (c) either hot or cold days?

QUICK QUIZ 16.4 On a winter day, you turn on your furnace and the temperature of the air inside your home increases. Assume that your home has the normal amount of leakage between inside air and outside air. Is the number of moles of air in your room at the higher temperature (a) larger than before, (b) smaller than before, or (c) the same as before?

Example 16.3 | Heating a Spray Can

A spray can containing a propellant gas at twice atmospheric pressure (202 kPa) and having a volume of 125.00 cm³ is at 22°C. It is then tossed into an open fire. (*Warning:* Do not do this experiment; it is very dangerous.) When the temperature of the gas in the can reaches 195°C, what is the pressure inside the can? Assume any change in the volume of the can is negligible.

SOLUTION

Conceptualize Intuitively, you should expect that the pressure of the gas in the container increases because of the increasing temperature.

Categorize We model the gas in the can as ideal and use the ideal gas law to calculate the new pressure.

Analyze Rearrange Equation 16.9:

$$(1) \quad \frac{PV}{T} = nR$$

No air escapes during the compression, so n, and therefore nR, remains constant. Hence, set the initial value of the left side of Equation (1) equal to the final value:

$$(2) \quad \frac{P_i V_i}{T_i} = \frac{P_f V_f}{T_f}$$

Because the initial and final volumes of the gas are assumed to be equal, cancel the volumes:

$$(3) \quad \frac{P_i}{T_i} = \frac{P_f}{T_f}$$

Solve for P_f:

$$P_f = \left(\frac{T_f}{T_i}\right) P_i = \left(\frac{468 \text{ K}}{295 \text{ K}}\right)(202 \text{ kPa}) = \boxed{320 \text{ kPa}}$$

Finalize The higher the temperature, the higher the pressure exerted by the trapped gas as expected. If the pressure increases sufficiently, the can may explode. Because of this possibility, you should never dispose of spray cans in a fire.

What If? Suppose we include a volume change due to thermal expansion of the steel can as the temperature increases. Does that alter our answer for the final pressure significantly?

Answer Because the thermal expansion coefficient of steel is very small, we do not expect much of an effect on our final answer.

Find the change in the volume of the can using Equation 16.6 and the value for α for steel from Table 16.1:

$$\Delta V = \beta V_i \Delta T = 3\alpha V_i \Delta T$$
$$= 3[11 \times 10^{-6} \, (°C)^{-1}](125.00 \text{ cm}^3)(173°C) = 0.71 \text{ cm}^3$$

Start from Equation (2) again and find an equation for the final pressure:

$$P_f = \left(\frac{T_f}{T_i}\right)\left(\frac{V_i}{V_f}\right) P_i$$

This result differs from Equation (3) only in the factor V_i/V_f. Evaluate this factor:

$$\frac{V_i}{V_f} = \frac{125.00 \text{ cm}^3}{(125.00 \text{ cm}^3 + 0.71 \text{ cm}^3)} = 0.994 = 99.4\%$$

Therefore, the final pressure will differ by only 0.6% from the value calculated without considering the thermal expansion of the can. Taking 99.4% of the previous final pressure, the final pressure including thermal expansion is 318 kPa.

16.5 | The Kinetic Theory of Gases

In the preceding section, we discussed the macroscopic properties of an ideal gas using such quantities as pressure, volume, number of moles, and temperature. From a *macroscopic* point of view, the mathematical representation of the ideal gas model is the ideal gas law. In this section, we consider the *microscopic* point of view of the ideal gas model. We shall show that the macroscopic properties can be understood on the basis of what is happening on the atomic scale.

Using the ideal gas model, we shall build a structural model of a gas enclosed in a container. The mathematical structure and the predictions made by this model

© INTERFOTO/Alamy

Ludwig Boltzmann
Austrian physicist (1844–1906)
Boltzmann made many important contributions to the development of the kinetic theory of gases, electromagnetism, and thermodynamics. His pioneering work in the field of kinetic theory led to the branch of physics known as statistical mechanics.

constitute what is known as the **kinetic theory of gases.** With this theory, we shall interpret the pressure and temperature of an ideal gas in terms of microscopic variables. Our structural model will include the following components:

1. *A description of the physical components of the system*: The gas consists of a number of identical molecules within a cubic container of side length d. The number of molecules in the gas is large, and the average separation between them is large compared with their dimensions. Therefore, the molecules occupy a negligible volume in the container. This assumption is consistent with the ideal gas model, in which we imagine the molecules to be point-like.

2. *A description of where the components are located relative to one another and how they interact*: The molecules are distributed uniformly throughout the container and behave as follows:
   (a) The molecules obey Newton's laws of motion, but as a whole their motion is isotropic: any molecule can move in any direction with any speed.
   (b) The molecules interact only by short-range forces during elastic collisions. This assumption is consistent with the ideal gas model, in which the molecules exert no long-range forces on one another.
   (c) The molecules make elastic collisions with the walls.

3. *A description of the time evolution of the system*: The system has reached a steady-state situation so that macroscopic descriptions of the gas (volume, temperature, pressure, etc.) remain fixed. The velocities of individual molecules are constantly changing.

4. *A description of the agreement between predictions of the model and actual observations and, possibly, predictions of new effects that have not yet been observed*: Our structural model should make some specific predictions relating macroscopic measurements to microscopic behavior. In particular, we would like to predict how pressure and temperature are related to the microscopic parameters associated with the molecules.

Although we often picture an ideal gas as consisting of single atoms, molecular gases exhibit equally good approximations to ideal gas behavior at low pressures. Effects associated with molecular structure have no influence on the motions considered here. Therefore, we can apply the results of the following development to molecular gases as well as to monatomic gases.

Molecular Interpretation of the Pressure of an Ideal Gas

For our first application of kinetic theory, let us derive an expression for the pressure of N molecules of an ideal gas in a container of volume V in terms of microscopic quantities. As outlined in our structural model, the container is a cube with edges of length d (Fig. 16.15). We shall focus our attention on one of these molecules of mass m_0 and assumed to be moving so that its component of velocity in the x direction is v_{xi} as in Active Figure 16.16. (The subscript i here refers to the ith molecule, not to an initial value. We will combine the effects of all of the molecules shortly.) As the molecule collides elastically with any wall, as proposed in structural model component 2(c), its velocity component perpendicular to the wall is reversed because the mass of the wall is far greater than the mass of the molecule. Because the momentum component p_{xi} of the molecule is $m_0 v_{xi}$ before the collision and $-m_0 v_{xi}$ after the collision, the change in momentum of the molecule in the x direction is

$$\Delta p_{xi} = -m_0 v_{xi} - (m_0 v_{xi}) = -2m_0 v_{xi}$$

Applying the impulse–momentum theorem (Eq. 8.11) to the molecule gives

$$\overline{F}_{i,\text{on molecule}} \, \Delta t_{\text{collision}} = \Delta p_{xi} = -2m_0 v_{xi}$$

where $\overline{F}_{i,\text{on molecule}}$ is the average force component,[1] perpendicular to the wall, for the force that the wall exerts on the molecule during the collision and $\Delta t_{\text{collision}}$ is the

One molecule of the gas moves with velocity $\vec{v}$ on its way toward a collision with the wall.

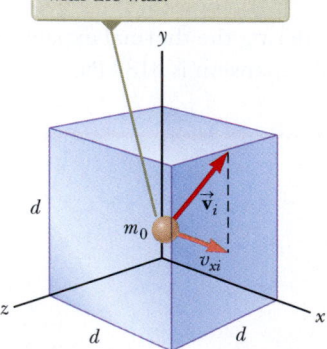

Figure 16.15 A cubical box with sides of length d containing an ideal gas.

---

[1]For this discussion, we will use a bar over a variable to represent the average value of the variable, such as $\overline{F}$ for the average force, rather than the subscript "avg" that we have used before. This notation saves confusion because we will already have a number of subscripts on variables.

duration of the collision. For the molecule to make another collision with the same wall after this first collision, it must travel a distance of $2d$ in the x direction (across the container and back). The time interval between two collisions with the same wall is therefore

$$\Delta t = \frac{2d}{v_{xi}}$$

The force that causes the change in momentum of the molecule in the collision with the wall occurs only during the collision. We can, however, average the force over the time interval for the molecule to move across the cube and back. Sometime during this time interval the collision occurs, so the change in momentum for this time interval is the same as that for the short duration of the collision. Therefore, we can rewrite the impulse–momentum theorem as

$$\bar{F_i}\,\Delta t = -2m_0 v_{xi}$$

where $\bar{F_i}$ is interpreted as the average force component on the molecule over the time for the molecule to move across the cube and back. Because exactly one collision occurs for each such time interval, it is also the long-term average force component on the molecule, over long time intervals containing any number of multiples of Δt.

The substitution of Δt into the impulse–momentum equation enables us to express the long-term average force component of the wall on the molecule:

$$\bar{F_i} = \frac{-2m_0 v_{xi}}{\Delta t} = \frac{-2m_0 v_{xi}^2}{2d} = \frac{-m_0 v_{xi}^2}{d}$$

Now, by Newton's third law, the force component of the molecule on the wall is equal in magnitude and opposite in direction:

$$\bar{F}_{i,\text{on wall}} = -\bar{F_i} = -\left(\frac{-m_0 v_{xi}^2}{d}\right) = \frac{m_0 v_{xi}^2}{d}$$

The magnitude of the total average force $\bar{F}$ exerted on the wall by the gas is found by adding the average force components exerted by the individual molecules. We add terms such as those shown in the preceding equation for all molecules:

$$\bar{F} = \sum_{i=1}^{N} \frac{m_0 v_{xi}^2}{d} = \frac{m_0}{d}\sum_{i=1}^{N} v_{xi}^2$$

where we have factored out the length of the box and the mass m_0 because structural model component 1 tells us that all the molecules are the same. We now impose the condition that the number of molecules is large. For a small number of molecules, the actual force on the wall would vary with time. It would be nonzero during the short interval of a collision of a molecule with the wall and zero when no molecule happens to be hitting the wall. For a very large number of molecules, however, such as Avogadro's number, these variations in force are smoothed out, so the average force is the same over *any* time interval. Therefore, the *constant* force F on the wall due to the molecular collisions is the same as the average force $\bar{F}$ and is of magnitude

$$F = \frac{m_0}{d}\sum_{i=1}^{N} v_{xi}^2$$

To proceed further, let us consider how we express the average value of the square of the x component of the velocity for the N molecules. The traditional average of a value is the sum of the values over the number of values:

$$\overline{v_x^2} = \frac{\displaystyle\sum_{i=1}^{N} v_{xi}^2}{N}$$

The numerator of this expression is contained in the right-hand side of the previous equation. Therefore, by combining the two expressions the total force on the wall can be written

$$F = \frac{m_0}{d} N \overline{v_x^2}$$

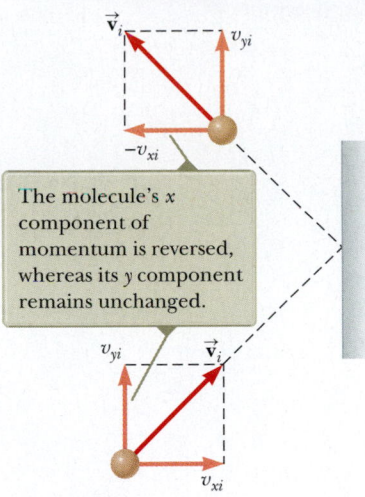

Active Figure 16.16 A molecule makes an elastic collision with the wall of the container. In this construction, we assume that the molecule moves in the xy plane.

The molecule's x component of momentum is reversed, whereas its y component remains unchanged.

Now let us focus again on one molecule with velocity components v_{xi}, v_{yi}, and v_{zi}. The Pythagorean theorem relates the square of the speed of the molecule to the squares of the velocity components:

$$v_i^2 = v_{xi}^2 + v_{yi}^2 + v_{zi}^2$$

If we take an average of both sides of this equation (sum over all particles and divide by N), the average value of v^2 for all the molecules in the container is related to the average values of v_x^2, v_y^2, and v_z^2 according to the expression

$$\overline{v^2} = \overline{v_x^2} + \overline{v_y^2} + \overline{v_z^2}$$

Now we use structural model component 2(a), that the motion is completely isotropic, which implies that no direction is preferred. On the average, the x, y, and z directions are equivalent, so

$$\overline{v_x^2} = \overline{v_y^2} = \overline{v_z^2}$$

which allows us to write

$$\overline{v^2} = 3\overline{v_x^2}$$

Therefore, the total force on the wall is

$$F = \frac{m_0}{d} N \left(\tfrac{1}{3}\overline{v^2}\right) = \frac{N}{3}\left(\frac{m_0 \overline{v^2}}{d}\right)$$

From this expression, we can make our prediction of the pressure exerted on the wall by dividing this force by the area of the wall:

$$P = \frac{F}{A} = \frac{F}{d^2} = \tfrac{1}{3}\frac{N}{d^3}\left(m_0 \overline{v^2}\right) = \tfrac{1}{3}\left(\frac{N}{V}\right)\left(m_0 \overline{v^2}\right)$$

▶ Pressure of an ideal gas

$$P = \tfrac{2}{3}\left(\frac{N}{V}\right)\left(\tfrac{1}{2}m_0 \overline{v^2}\right) \qquad \qquad \textbf{16.13}◀$$

This prediction proposes that the pressure is proportional to (1) the number of molecules per unit volume and (2) the average translational kinetic energy of the molecules, $\tfrac{1}{2}m_0\overline{v^2}$. With this structural model of an ideal gas, we have arrived at an important result that relates the macroscopic quantity of pressure to a microscopic quantity, the average value of the molecular translational kinetic energy. Therefore, we have a key link between the atomic world and the large-scale world.

Let's see how this prediction of the structural model compares to reality. Equation 16.13 verifies some features of pressure that are probably familiar to you. One way to increase the pressure inside a container is to increase the number of molecules per unit volume in the container (N/V). You do so when you add air to a tire.

The pressure in the tire can also be increased by increasing the average translational kinetic energy of the molecules in the tire. As we shall see shortly, that can be accomplished by increasing the temperature of the gas inside the tire. Hence, the pressure inside a tire increases as the tire warms up during long trips. The continuous flexing of the tires as they move along the road surface results in work done as parts of the tire distort and in an increase in internal energy of the rubber. The increased temperature of the rubber results in transfer of energy into the air by heat, increasing the average translational kinetic energy of the molecules, which in turn produces an increase in pressure.

Molecular Interpretation of the Temperature of an Ideal Gas

We have related the pressure to the average kinetic energy of molecules; let us now relate temperature to a microscopic description of the gas. We can obtain some insight into the meaning of temperature by first writing Equation 16.13 in the form

$$PV = \tfrac{2}{3}N \left(\tfrac{1}{2}m_0\overline{v^2}\right)$$

Let us now compare this equation with the equation of state for an ideal gas:

$$PV = Nk_B T$$

The left-hand sides of these two equations are identical. Equating the right-hand sides of these expressions gives us a prediction from the structural model regarding temperature:

$$T = \frac{2}{3k_B}\left(\tfrac{1}{2}m_0\overline{v^2}\right)$$ **16.14** ◄

▶ Temperature is proportional to average kinetic energy

This equation tells us that the temperature of a gas is a direct measure of average translational molecular kinetic energy. Therefore, as the temperature of a gas increases, the molecules move with higher average kinetic energy.

By rearranging Equation 16.14, we can relate the average translational molecular kinetic energy to the temperature:

$$\tfrac{1}{2}m_0\overline{v^2} = \tfrac{3}{2}k_B T$$ **16.15** ◄

▶ Average kinetic energy per molecule

That is, the average translational kinetic energy per molecule is $\frac{3}{2}k_B T$. Because $\overline{v_x^2} = \frac{1}{3}\overline{v^2}$, it follows that

$$\tfrac{1}{2}m_0\overline{v_x^2} = \tfrac{1}{2}k_B T$$ **16.16** ◄

In a similar manner, for the y and z motions we find that

$$\tfrac{1}{2}m_0\overline{v_y^2} = \tfrac{1}{2}k_B T \quad \text{and} \quad \tfrac{1}{2}m_0\overline{v_z^2} = \tfrac{1}{2}k_B T$$

Therefore, each translational degree of freedom contributes an equal amount of energy to the gas, namely $\frac{1}{2}k_B T$ per molecule. (In general, the phrase *degrees of freedom* refers to the number of independent means by which a molecule can possess energy.) A generalization of this result, known as the **theorem of equipartition of energy,** states that

each degree of freedom contributes $\frac{1}{2}k_B T$ to the energy of a system, where possible degrees of freedom are those associated with translation, rotation, and vibration of molecules.

▶ Theorem of equipartition of energy

The total translational kinetic energy of N molecules of gas is simply N times the average translational kinetic energy per molecule, which is given by Equation 16.15:

$$E_{total} = N\left(\tfrac{1}{2}m_0\overline{v^2}\right) = \tfrac{3}{2}Nk_B T = \tfrac{3}{2}nRT$$ **16.17** ◄

▶ Total kinetic energy of N molecules

where we have used $k_B = R/N_A$ for Boltzmann's constant and $n = N/N_A$ for the number of moles of gas. From this result we see that the total translational kinetic energy of a system of molecules is proportional to the absolute temperature of the system and depends *only* on temperature.

For a monatomic gas, translational kinetic energy is the only type of energy the particles of the gas can have. Therefore, Equation 16.17 gives **the internal energy for a monatomic gas:**

$$E_{int} = \tfrac{3}{2}nRT \quad \text{(monatomic gas)}$$ **16.18** ◄

This equation mathematically justifies our claim that internal energy is related to the temperature of a system, which we introduced in Chapter 6. For diatomic and polyatomic molecules, additional possibilities for energy storage are available in the vibration and rotation of the molecule, but a proportionality between E_{int} and T remains.

The square root of $\overline{v^2}$ is called the **root-mean-square** (rms) **speed** of the molecules. From Equation 16.15 we find for the rms speed that

$$v_{rms} = \sqrt{\overline{v^2}} = \sqrt{\frac{3k_B T}{m_0}} = \sqrt{\frac{3RT}{M}}$$ **16.19** ◄

▶ Root-mean-square speed

where M is the molar mass in kilograms per mole. This expression shows that, at a given temperature, lighter molecules move faster, on the average, than heavier

TABLE 16.2 |
Some rms Speeds

Gas	Molar Mass (g/mol)	v_{rms} at 20°C (m/s)
H_2	2.02	1 902
He	4.00	1 352
H_2O	18.0	637
Ne	20.2	602
N_2 or CO	28.0	511
NO	30.0	494
O_2	32.0	478
CO_2	44.0	408
SO_2	64.1	338

molecules. For example, hydrogen, with a molar mass of 2.0×10^{-3} kg/mol, moves four times as fast as oxygen, whose molar mass is 32×10^{-3} kg/mol. If we calculate the rms speed for hydrogen at room temperature ($\approx$ 300 K), we find that

$$v_{rms} = \sqrt{\frac{3RT}{M}} = \sqrt{\frac{3(8.31 \text{ J/mol·K})(300 \text{ K})}{2.0 \times 10^{-3} \text{ kg/mol}}} = 1.9 \times 10^3 \text{ m/s}$$

This value is about 17% of the escape speed for the Earth, which we calculated in Chapter 11. Because this value is an average speed, a large number of molecules having speeds much higher than the average can escape from the Earth's atmosphere. Therefore, the Earth's atmosphere does not at present contain hydrogen because it has all bled off into space.

Table 16.2 lists the rms speeds for various molecules at 20°C.

QUICK QUIZ 16.5 Two containers hold an ideal gas at the same temperature and pressure. Both containers hold the same type of gas, but container B has twice the volume of container A. (i) What is the average translational kinetic energy per molecule in container B? (a) twice that of container A (b) the same as that of container A (c) half that of container A (d) impossible to determine (ii) From the same choices, describe the internal energy of the gas in container B.

Example 16.4 | A Tank of Helium

A tank used for filling helium balloons has a volume of 0.300 m³ and contains 2.00 mol of helium gas at 20.0°C. Assume the helium behaves like an ideal gas.

(A) What is the total translational kinetic energy of the gas molecules?

SOLUTION

Conceptualize Imagine a microscopic model of a gas in which you can watch the molecules move about the container more rapidly as the temperature increases. Because the gas is monatomic, the total translational kinetic energy of the molecules is the internal energy of the gas.

Categorize We evaluate parameters with equations developed in the preceding discussion, so this example is a substitution problem.

Use Equation 16.18 with n = 2.00 mol and T = 293 K:

$$E_{int} = \tfrac{3}{2}nRT = \tfrac{3}{2}(2.00 \text{ mol})(8.31 \text{ J/mol · K})(293 \text{ K})$$
$$= 7.30 \times 10^3 \text{ J}$$

(B) What is the average kinetic energy per molecule?

SOLUTION

Use Equation 16.15:

$$\tfrac{1}{2}m_0\overline{v^2} = \tfrac{3}{2}k_B T = \tfrac{3}{2}(1.38 \times 10^{-23} \text{ J/K})(293 \text{ K})$$
$$= 6.07 \times 10^{-21} \text{ J}$$

What If? What if the temperature is raised from 20.0°C to 40.0°C? Because 40.0 is twice as large as 20.0, is the total translational energy of the molecules of the gas twice as large at the higher temperature?

Answer The expression for the total translational energy depends on the temperature, and the value for the temperature must be expressed in kelvins, not in degrees Celsius. Therefore, the ratio of 40.0 to 20.0 is *not* the appropriate ratio. Converting the Celsius temperatures to kelvins, 20.0°C is 293 K and 40.0°C is 313 K. Therefore, the total translational energy increases by a factor of only 313 K/293 K = 1.07.

16.6 | Distribution of Molecular Speeds

In the preceding section, we derived an expression for the average speed of a gas molecule but made no mention of the actual distribution of molecular speeds among all possible values. In the 1860s, James Clerk Maxwell (1831–1879) developed a structural model that predicts this distribution of molecular speeds. His work and developments by other scientists shortly thereafter were highly controversial because experiments at that time could not directly detect molecules. About 60 years later, however, experiments confirmed Maxwell's predictions.

Consider a container of gas whose molecules have some distribution of speeds. Suppose we want to determine how many gas molecules have a speed in the range from, for example, 400 to 410 m/s. Intuitively, we expect that the speed distribution depends on temperature. Furthermore, we expect that the distribution peaks in the vicinity of v_{rms}. That is, few molecules are expected to have speeds much less than or much greater than v_{rms} because these extreme speeds will result only from an unlikely chain of collisions.

The observed speed distribution of gas molecules in thermal equilibrium is shown in Active Figure 16.17. The quantity N_v, called the **Maxwell–Boltzmann distribution function,** is defined as follows. If N is the total number of molecules, the number of molecules with speeds between v and $v + dv$ is $dN = N_v\, dv$. This number is also equal to the area of the shaded rectangle in Active Figure 16.17. Furthermore, the fraction of molecules with speeds between v and $v + dv$ is $N_v\, dv / N$. This fraction is also equal to the *probability* that a molecule has a speed in the range v to $v + dv$.

The fundamental expression that describes the distribution of speeds of N gas molecules is

$$N_v = 4\pi N \left(\frac{m_0}{2\pi k_B T}\right)^{3/2} v^2 e^{-m_0 v^2 / 2 k_B T} \qquad \textbf{16.20} \blacktriangleleft$$

▶ Maxwell–Boltzmann distribution function

where m_0 is the mass of a gas molecule, k_B is Boltzmann's constant, and T is the absolute temperature.[2]

As indicated in Active Figure 16.17, the average speed v_{avg} is somewhat lower than the rms speed. The most probable speed v_{mp} is the speed at which the distribution curve reaches a peak. Using Equation 16.20, we find that

$$v_{rms} = \sqrt{\overline{v^2}} = \sqrt{\frac{3 k_B T}{m_0}} = 1.73 \sqrt{\frac{k_B T}{m_0}} \qquad \textbf{16.21} \blacktriangleleft$$

$$v_{avg} = \sqrt{\frac{8 k_B T}{\pi m_0}} = 1.60 \sqrt{\frac{k_B T}{m_0}} \qquad \textbf{16.22} \blacktriangleleft$$

$$v_{mp} = \sqrt{\frac{2 k_B T}{m_0}} = 1.41 \sqrt{\frac{k_B T}{m_0}} \qquad \textbf{16.23} \blacktriangleleft$$

From these equations we see that $v_{rms} > v_{avg} > v_{mp}$.

Active Figure 16.18 (page 460) represents speed distribution curves for nitrogen molecules. The curves were obtained using Equation 16.20 to evaluate the distribution function at various speeds and at two temperatures. Note that the peak in the curve shifts to the right as T increases, indicating that the average speed increases with increasing temperature, as expected. In addition, the overall width of the curve increases with temperature. The shape of the curves is asymmetrical because the lowest speed possible is zero, whereas the upper classical limit of the speed is infinity.

The number of molecules having speeds ranging from v to $v + dv$ equals the area of the shaded rectangle, $N_v\, dv$.

Active Figure 16.17 The speed distribution of gas molecules at some temperature. The function N_v approaches zero as v approaches infinity.

---

[2]For the derivation of this expression, see a text on thermodynamics such as that by R. P. Bauman, *Modern Thermodynamics and Statistical Mechanics* (New York: Macmillan, 1992).

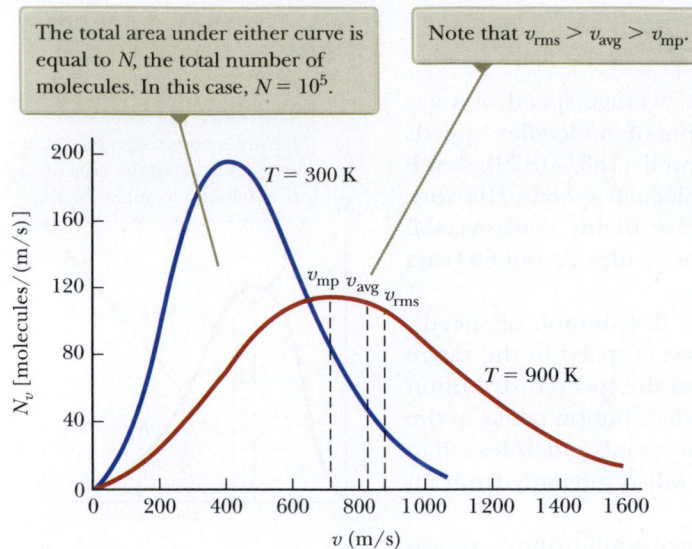

The total area under either curve is equal to N, the total number of molecules. In this case, $N = 10^5$.

Note that $v_{rms} > v_{avg} > v_{mp}$.

Active Figure 16.18 The speed distribution function for 10^5 nitrogen molecules at 300 K and 900 K.

The speed distribution curve for molecules in a liquid is similar to those shown in Active Figure 16.18. The phenomenon of evaporation of a liquid can be understood from this distribution in speeds because some molecules in the liquid are more energetic than others. Some of the faster-moving molecules in the liquid penetrate the surface and leave the liquid even at temperatures well below the boiling point. The molecules that escape the liquid by evaporation are those that have sufficient energy to overcome the attractive forces of the molecules in the liquid phase. Consequently, the molecules left behind in the liquid phase have a lower average kinetic energy, causing the temperature of the liquid to decrease. Hence, evaporation is a cooling process. For example, an alcohol-soaked cloth is often placed on a feverish head to cool and comfort the patient. Alcohol has a high rate of evaporation due to its high vapor pressure and low boiling point compared to water.

QUICK QUIZ 16.6 Consider the qualitative shapes of the two curves in Active Figure 16.18, without regard for the numerical values or labels in the graph. Suppose you have two containers of gas *at the same temperature*. Container A has 10^5 nitrogen molecules and container B has 10^5 hydrogen molecules. What is the correct qualitative matching between the containers and the two curves in Active Figure 16.18? **(a)** Container A corresponds to the blue curve and container B to the brown curve. **(b)** Container B corresponds to the blue curve and container. A to the brown curve. **(c)** Both containers will correspond to the same curve.

Example 16.5 | Molecular Speeds in a Hydrogen Gas

A 0.500-mol sample of hydrogen gas is at 300 K.

(A) Find the average speed, the rms speed, and the most probable speed of the hydrogen molecules.

SOLUTION

Conceptualize Imagine a huge number of particles in a real gas, all moving in random directions with different speeds.

Categorize We cannot calculate the average by adding the speeds and dividing by the number of particles because the individual speeds of the particles are not known. We are dealing with a very large number of particles, however, so we can use the Maxwell–Boltzmann speed distribution function.

Analyze Use Equation 16.22 to find the average speed:

$$v_{avg} = 1.60\sqrt{\frac{k_B T}{m_0}} = 1.60\sqrt{\frac{(1.38 \times 10^{-23} \text{ J/K})(300 \text{ K})}{2(1.67 \times 10^{-27} \text{ kg})}}$$

$$= 1.78 \times 10^3 \text{ m/s}$$

Use Equation 16.21 to find the rms speed:

$$v_{rms} = 1.73\sqrt{\frac{k_B T}{m_0}} = 1.73\sqrt{\frac{(1.38 \times 10^{-23} \text{ J/K})(300 \text{ K})}{2(1.67 \times 10^{-27} \text{ kg})}}$$

$$= 1.93 \times 10^3 \text{ m/s}$$

Use Equation 16.23 to find the most probable speed:

$$v_{mp} = 1.41\sqrt{\frac{k_B T}{m_0}} = 1.41\sqrt{\frac{(1.38 \times 10^{-23} \text{ J/K})(300 \text{ K})}{2(1.67 \times 10^{-27} \text{ kg})}}$$

$$= 1.57 \times 10^3 \text{ m/s}$$

16.5 *cont.*

(B) Find the number of molecules with speeds between 400 m/s and 401 m/s.

SOLUTION

Use Equation 16.20 to evaluate the number of molecules in a narrow speed range between v and $v + dv$.

$$(1) \quad N_v \, dv = 4\pi N \left(\frac{m_0}{2\pi k_{\mathrm{B}} T}\right)^{3/2} v^2 e^{-m_0 v^2/2k_{\mathrm{B}} T} \, dv$$

Evaluate the constant in front of v^2:

$$4\pi N \left(\frac{m_0}{2\pi k_{\mathrm{B}} T}\right)^{3/2} = 4\pi n N_{\mathrm{A}} \left(\frac{m_0}{2\pi k_{\mathrm{B}} T}\right)^{3/2}$$

$$= 4\pi (0.500 \text{ mol}) (6.02 \times 10^{23} \text{ mol}^{-1}) \left[\frac{2(1.67 \times 10^{-27} \text{ kg})}{2\pi (1.38 \times 10^{-23} \text{ J/K}) (300 \text{ K})}\right]^{3/2}$$

$$= 1.74 \times 10^{14} \text{ s}^3/\text{m}^3$$

Evaluate the exponent of e that appears in Equation (1):

$$-\frac{m_0 v^2}{2k_{\mathrm{B}} T} = -\frac{2(1.67 \times 10^{-27} \text{ kg}) (400 \text{ m/s})^2}{2(1.38 \times 10^{-23} \text{ J/K}) (300 \text{ K})} = -0.064 \, 5$$

Evaluate $N_v \, dv$ using Equation (1):

$$N_v \, dv = (1.74 \times 10^{14} \text{ s}^3/\text{m}^3) (400 \text{ m/s})^2 e^{-0.064 \, 5} (1 \text{ m/s})$$

$$= 2.61 \times 10^{19} \text{ molecules}$$

Finalize In this evaluation, we could calculate the result without integration because $dv = 1$ m/s is much smaller than $v = 400$ m/s. Had we sought the number of particles between, say, 400 m/s and 500 m/s, we would need to integrate Equation (1) between these speed limits.

⟨16.7 | Context Connection: The Atmospheric Lapse Rate

We have discussed the temperature of a gas with the assumption that all parts of the gas are at the same temperature. For small volumes of gas, this assumption is relatively good. What about a *huge* volume of gas, however, such as the atmosphere? It is clear that the assumption of a uniform temperature throughout the gas is not valid in this case. When it is a hot summer day in Los Angeles, it is a cold winter day in Melbourne; different parts of the atmosphere are clearly at different temperatures.

We can address this question, as we discussed in the opening section of this Context, by considering the global average of the air temperature at the surface of the Earth. Yet variations also occur in temperature at different *heights* in the atmosphere. It is this variation of temperature with height that we explore here.

Figure 16.19 shows graphical representations of average air temperature in January at various heights in six American states. These data are taken at locations on the surface of the Earth, but at varying elevations, such as at sea level and on mountains. For all six states we see a scattering of the data points (related to factors other than elevation), but a clear indication that the temperature decreases as we move to higher elevations. Of course, one look at snow-capped mountains tells us that is the case.

We can argue conceptually why the temperature decreases with height. Imagine a parcel of air moving upward

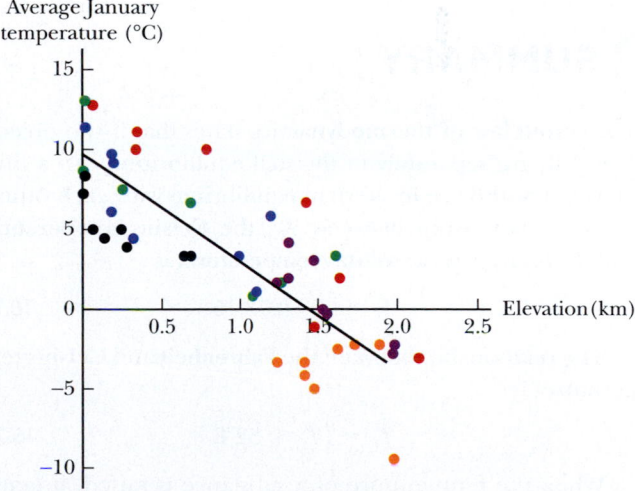

Figure 16.19 Variation of average January temperature with elevation for eight locations in each of six American states: Arizona (red), California (green), Colorado (orange), New Mexico (purple), North Carolina (black), and Texas (blue). The best-fit line shown in black has a slope of −6.2°C/km. Data from www.noaa.gov (U.S. Department of Commerce/National Oceanic and Atmospheric Administration, Physical Sciences Division).

along the slope of a mountain. As this parcel rises into higher elevations, the pressure on it from the surrounding air decreases. The pressure difference between the interior and the exterior of the parcel causes the parcel to expand. In doing so, the parcel is pushing the surrounding air outward, doing work on it. Because the system (the parcel of air) is doing work on the environment, the work done *on* the system is negative and the internal energy in the parcel decreases. The decreased internal energy is manifested as a decrease in temperature.

If this process is reversed so that the parcel moves toward lower elevations, work is done on the parcel, which increases its internal energy so that it becomes warmer. This situation occurs during Santa Ana wind conditions in the Los Angeles basin in which air is pushed from the mountains down into the low elevations of the basin, resulting in hot, dry winds. Similar conditions go by other names in other regions, such as the *chinook* from the Rocky Mountains and the *foehn* from the Swiss Alps.

Imagine drawing best-fit straight lines through each set of colored data points in Figure 16.19. You would find that the slopes of all six lines would be similar. This similarity suggests that the decrease in temperature with height—called the **atmospheric lapse rate**—is similar at various locations across the surface of the Earth, so we might define an average lapse rate for the entire surface.

That is indeed the case, and we find that the average global lapse rate is about $-6.5°C/km$. The data in Figure 16.19 is limited to only a few locations within the United States and to elevations reachable from the ground, but the average lapse rate of this data of $-6.2°C/km$ is close to the global average.

The linear decrease with temperature only occurs in the lower part of the atmosphere called the **troposphere,** the part of the atmosphere in which weather occurs and airplanes fly. Above the troposphere is the **stratosphere,** with an imaginary boundary called the **tropopause** separating the two layers. In the stratosphere, temperature tends to be relatively constant with height.

The decrease in temperature with height in the troposphere is one component of a structural model of the atmosphere that will allow us to predict the surface temperature of the Earth. If we can find the temperature of the stratosphere and the height of the tropopause, we can extrapolate to the surface, using the lapse rate to find the temperature at the surface. The lapse rate and the height of the tropopause can be measured. To find the temperature of the stratosphere, we need to know more about energy exchanges in the Earth's atmosphere, which we will investigate in the next chapter.

❭ SUMMARY |

The **zeroth law of thermodynamics** states that if two objects, A and B, are separately in thermal equilibrium with a third object, A and B are in thermal equilibrium with each other.

The relationship between T_C, the **Celsius temperature,** and T, the **Kelvin (absolute) temperature,** is

$$T_C = T - 273.15 \qquad \textbf{16.1}◀$$

The relationship between the **Fahrenheit** and Celsius temperatures is

$$T_F = \tfrac{9}{5}T_C + 32°F \qquad \textbf{16.2}◀$$

When the temperature of a substance is raised, it generally expands. If an object has an initial length of L_i at some temperature and undergoes a change in temperature ΔT, its length changes by the amount ΔL, which is proportional to the object's initial length and the temperature change:

$$\Delta L = \alpha L_i \Delta T \qquad \textbf{16.4}◀$$

The parameter α is called the **average coefficient of linear expansion.**

The change in volume of most substances is proportional to the initial volume V_i and the temperature change ΔT:

$$\Delta V = \beta V_i \Delta T \qquad \textbf{16.6}◀$$

where β is the **average coefficient of volume expansion** and is equal to 3α.

The change in area of a substance is given by

$$\Delta A = \gamma A_i \Delta T \qquad \textbf{16.7}◀$$

where γ is the **average coefficient of area expansion** and is equal to 2α.

The **ideal gas model** refers to a collection of gas molecules that move randomly and are of negligible size. An ideal gas obeys the equation

$$PV = nRT \qquad \textbf{16.9}◀$$

where P is the pressure of the gas, V is its volume, n is the number of moles of gas, R is the universal gas constant ($8.314\,\text{J/mol}\cdot\text{K}$), and T is the absolute temperature in kelvins. A real gas at very low pressures behaves approximately as an ideal gas.

The pressure of N molecules of an ideal gas contained in a volume V is given by

$$P = \frac{2}{3}\left(\frac{N}{V}\right)\left(\frac{1}{2}m_0\overline{v^2}\right) \qquad \textbf{16.13}\blacktriangleleft$$

where $\frac{1}{2}m_0\overline{v^2}$ is the **average translational kinetic energy per molecule.**

The average kinetic energy of the molecules of a gas is directly proportional to the absolute temperature of the gas:

$$\frac{1}{2}m_0\overline{v^2} = \frac{3}{2}k_B T \qquad \textbf{16.15}\blacktriangleleft$$

where k_B is **Boltzmann's constant** (1.38×10^{-23} J/K).

For a monatomic gas, the internal energy of the gas is the total translational kinetic energy

$$E_{\text{int}} = \frac{3}{2}nRT \qquad \text{(monatomic gas)} \qquad \textbf{16.18}\blacktriangleleft$$

The **root-mean-square** (rms) **speed** of the molecules of a gas is

$$v_{\text{rms}} = \sqrt{\overline{v^2}} = \sqrt{\frac{3k_B T}{m_0}} = \sqrt{\frac{3RT}{M}} \qquad \textbf{16.19}\blacktriangleleft$$

The **Maxwell–Boltzmann distribution function** describes the distribution of speeds of N gas molecules:

$$N_v = 4\pi N\left(\frac{m_0}{2\pi k_B T}\right)^{3/2} v^2 e^{-m_0 v^2/2k_B T} \qquad \textbf{16.20}\blacktriangleleft$$

where m_0 is the mass of a gas molecule, k_B is Boltzmann's constant, and T is the absolute temperature.

▶ OBJECTIVE QUESTIONS

☐ denotes answer available in *Student Solutions Manual/Study Guide*

1. If the volume of an ideal gas is doubled while its temperature is quadrupled, does the pressure (a) remain the same, (b) decrease by a factor of 2, (c) decrease by a factor of 4, (d) increase by a factor of 2, or (e) increase by a factor of 4?

2. A cylinder with a piston holds 0.50 m³ of oxygen at an absolute pressure of 4.0 atm. The piston is pulled outward, increasing the volume of the gas until the pressure drops to 1.0 atm. If the temperature stays constant, what new volume does the gas occupy? (a) 1.0 m³ (b) 1.5 m³ (c) 2.0 m³ (d) 0.12 m³ (e) 2.5 m³

3. A hole is drilled in a metal plate. When the metal is raised to a higher temperature, what happens to the diameter of the hole? (a) It decreases. (b) It increases. (c) It remains the same. (d) The answer depends on the initial temperature of the metal. (e) None of those answers is correct.

4. When a certain gas under a pressure of 5.00×10^6 Pa at 25.0°C is allowed to expand to 3.00 times its original volume, its final pressure is 1.07×10^6 Pa. What is its final temperature? (a) 450 K (b) 233 K (c) 212 K (d) 191 K (e) 115 K

5. A temperature of 162°F is equivalent to what temperature in kelvins? (a) 373 K (b) 288 K (c) 345 K (d) 201 K (e) 308 K

6. Which of the assumptions below is *not* made in the kinetic theory of gases? (a) The number of molecules is very large. (b) The molecules obey Newton's laws of motion. (c) The forces between molecules are long range. (d) The gas is a pure substance. (e) The average separation between molecules is large compared to their dimensions.

7. What would happen if the glass of a thermometer expanded more on warming than did the liquid in the tube? (a) The thermometer would break. (b) It could be used only for temperatures below room temperature. (c) You would have to hold it with the bulb on top. (d) The scale on the thermometer is reversed so that higher temperature values would be found closer to the bulb. (e) The numbers would not be evenly spaced.

8. An ideal gas is maintained at constant pressure. If the temperature of the gas is increased from 200 K to 600 K, what happens to the rms speed of the molecules? (a) It increases by a factor of 3. (b) It remains the same. (c) It is one-third the original speed. (d) It is $\sqrt{3}$ times the original speed. (e) It increases by a factor of 6.

9. A gas is at 200 K. If we wish to double the rms speed of the molecules of the gas, to what value must we raise its temperature? (a) 283 K (b) 400 K (c) 566 K (d) 800 K (e) 1130 K

10. Rank the following from largest to smallest, noting any cases of equality. (a) the average speed of molecules in a particular sample of ideal gas (b) the most probable speed (c) the root-mean-square speed (d) the average vector velocity of the molecules

11. The coefficient of linear expansion of copper is 17×10^{-6} (°C)$^{-1}$. The Statue of Liberty is 93 m tall on a summer morning when the temperature is 25°C. Assume the copper plates covering the statue are mounted edge to edge without expansion joints and do not buckle or bind on the framework supporting them as the day grows hot. What is the order of magnitude of the statue's increase in height? (a) 0.1 mm (b) 1 mm (c) 1 cm (d) 10 cm (e) 1 m

12. A rubber balloon is filled with 1 L of air at 1 atm and 300 K and is then put into a cryogenic refrigerator at 100 K. The rubber remains flexible as it cools. **(i)** What happens to the volume of the balloon? (a) It decreases to $\frac{1}{3}$ L. (b) It decreases to $1/\sqrt{3}$ L. (c) It is constant. (d) It increases to $\sqrt{3}$ L. (e) It increases to 3 L. **(ii)** What happens to the pressure of the air in the balloon? (a) It decreases to $\frac{1}{3}$ atm. (b) It decreases to $1/\sqrt{3}$ atm. (c) It is constant. (d) It increases to $\sqrt{3}$ atm. (e) It increases to 3 atm.

13. Two cylinders A and B at the same temperature contain the same quantity of the same kind of gas. Cylinder A has three times the volume of cylinder B. What can you conclude about the pressures the gases exert? (a) We can conclude nothing about the pressures. (b) The pressure in A is three times the pressure in B. (c) The pressures must be equal. (d) The pressure in A must be one-third the pressure in B.

14. An ideal gas is contained in a vessel at 300 K. The temperature of the gas is then increased to 900 K. **(i)** By what factor does the average kinetic energy of the molecules change, (a) a factor of 9, (b) a factor of 3, (c) a factor of $\sqrt{3}$, (d) a factor of 1, or (e) a factor of $\frac{1}{3}$? Using the same choices as in part (i), by what factor does each of the following change: **(ii)** the rms molecular speed of the molecules, **(iii)** the average momentum change that one molecule undergoes in a collision with one particular wall, **(iv)** the rate of collisions of molecules with walls, and **(v)** the pressure of the gas?

15. Cylinder A contains oxygen (O_2) gas, and cylinder B contains nitrogen (N_2) gas. If the molecules in the two cylinders have the same rms speeds, which of the following statements is *false*? (a) The two gases have different temperatures. (b) The temperature of cylinder B is less than the temperature of cylinder A. (c) The temperature of cylinder B is greater than the temperature of cylinder A. (d) The average kinetic energy of the nitrogen molecules is less than the average kinetic energy of the oxygen molecules.

16. A cylinder with a piston contains a sample of a thin gas. The kind of gas and the sample size can be changed. The cylinder can be placed in different constant-temperature baths, and the piston can be held in different positions. Rank the following cases according to the pressure of the gas from the highest to the lowest, displaying any cases of equality. (a) A 0.002-mol sample of oxygen is held at 300 K in a 100-cm³ container. (b) A 0.002-mol sample of oxygen is held at 600 K in a 200-cm³ container. (c) A 0.002-mol sample of oxygen is held at 600 K in a 300-cm³ container. (d) A 0.004-mol sample of helium is held at 300 K in a 200-cm³ container. (e) A 0.004-mol sample of helium is held at 250 K in a 200-cm³ container.

17. Markings to indicate length are placed on a steel tape in a room that is at a temperature of 22°C. Measurements are then made with the same tape on a day when the temperature is 27°C. Assume the objects you are measuring have a smaller coefficient of linear expansion than steel. Are the measurements (a) too long, (b) too short, or (c) accurate?

18. A sample of gas with a thermometer immersed in the gas is held over a hot plate. A student is asked to give a step-by-step account of what makes our observation of the temperature of the gas increase. His response includes the following steps. (a) The molecules speed up. (b) Then the molecules collide with one another more often. (c) Internal friction makes the collisions inelastic. (d) Heat is produced in the collisions. (e) The molecules of the gas transfer more energy to the thermometer when they strike it, so we observe that the temperature has gone up. (f) The same process can take place without the use of a hot plate if you quickly push in the piston in an insulated cylinder containing the gas. **(i)** Which of the parts (a) through (f) of this account are correct statements necessary for a clear and complete explanation? **(ii)** Which are correct statements that are not necessary to account for the higher thermometer reading? **(iii)** Which are incorrect statements?

19. Two samples of the same ideal gas have the same pressure and density. Sample B has twice the volume of sample A. What is the rms speed of the molecules in sample B? (a) twice that in sample A (b) equal to that in sample A (c) half that in sample A (d) impossible to determine

CONCEPTUAL QUESTIONS

☐ denotes answer available in *Student Solutions Manual/Study Guide*

1. Is it possible for two objects to be in thermal equilibrium if they are not in contact with each other? Explain.

2. One container is filled with helium gas and another with argon gas. Both containers are at the same temperature. Which molecules have the higher rms speed? Explain.

3. Why does a diatomic gas have a greater energy content per mole than a monatomic gas at the same temperature?

4. A piece of copper is dropped into a beaker of water. (a) If the water's temperature rises, what happens to the temperature of the copper? (b) Under what conditions are the water and copper in thermal equilibrium?

5. In describing his upcoming trip to the Moon, and as portrayed in the movie *Apollo 13* (Universal, 1995), astronaut Jim Lovell said, "I'll be walking in a place where there's a 400-degree difference between sunlight and shadow." Suppose an astronaut standing on the Moon holds a thermometer in his gloved hand. (a) Is the thermometer reading the temperature of the vacuum at the Moon's surface? (b) Does it read any temperature? If so, what object or substance has that temperature?

6. What happens to a helium-filled latex balloon released into the air? Does it expand or contract? Does it stop rising at some height?

7. (a) What does the ideal gas law predict about the volume of a sample of gas at absolute zero? (b) Why is this prediction incorrect?

8. Use a periodic table of the elements (see Appendix C) to determine the number of grams in one mole of (a) hydrogen, which has diatomic molecules; (b) helium; and (c) carbon monoxide.

9. An automobile radiator is filled to the brim with water when the engine is cool. (a) What happens to the water when the engine is running and the water has been raised to a high temperature? (b) What do modern automobiles have in their cooling systems to prevent the loss of coolants?

10. Metal lids on glass jars can often be loosened by running hot water over them. Why does that work?

11. Common thermometers are made of a mercury column in a glass tube. Based on the operation of these thermometers, which has the larger coefficient of linear expansion, glass or mercury? (Don't answer the question by looking in a table.)

12. When the metal ring and metal sphere in Figure CQ16.12 are both at room temperature, the sphere can barely be passed through the ring. (a) After the sphere is warmed in a flame, it cannot be passed through the ring. Explain. (b) **What If?** What if the ring is warmed and the sphere is left at room temperature? Does the sphere pass through the ring?

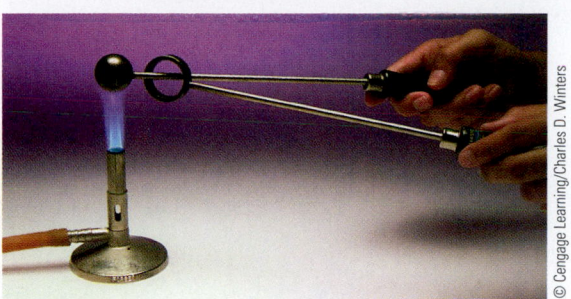

Figure CQ16.12

13. Some picnickers stop at a convenience store to buy some food, including bags of potato chips. They then drive up into the mountains to their picnic site. When they unload the food, they notice that the bags of chips are puffed up like balloons. Why did that happen?

▶ PROBLEMS AVAILABLE IN

16.2 Thermometers and Temperature Scales

Problems 1–6

16.3 Thermal Expansion of Solids and Liquids

Problems 7–19

16.4 Macroscopic Description of an Ideal Gas

Problems 20–36

16.5 The Kinetic Theory of Gases

Problems 37–44

16.6 Distribution of Molecular Speeds

Problems 45–48

16.7 Context Connection: The Atmospheric Lapse Rate

Problems 49–50

Additional Problems

Problems 51–75

Solutions to the following Problems are available in the *Student Solutions Manual/Study Guide*:

16.4, 16.5, 16.15, 16.25, 16.29, 16.36, 16.40, 16.44, 16.45, 16.46, 16.53, 16.56, 16.60, 16.62, and 16.67

List of Enhanced Problems

Problem Number	Targeted Feedback in Enhanced WebAssign	Master It in Enhanced WebAssign	Watch It in Enhanced WebAssign
16.4		✓	
16.8	✓		✓
16.13	✓	✓	
16.14	✓		✓
16.15	✓	✓	
16.21	✓		✓
16.24	✓		✓
16.25		✓	
16.27		✓	
16.29	✓	✓	
16.33	✓		✓
16.35	✓		✓
16.36		✓	
16.37	✓		✓
16.39	✓		✓
16.40		✓	
16.41	✓	✓	
16.43	✓		✓
16.44	✓	✓	
16.45		✓	
16.53		✓	
16.73	✓		✓

Chapter 17

Energy in Thermal Processes: The First Law of Thermodynamics

Chapter Outline

© iStockphoto.com/KingWu

In this photograph of the Mt. Baker area near Bellingham, Washington, we see evidence of water in all three phases. In the lake is liquid water, and solid water in the form of snow appears on the ground. The clouds in the sky consist of liquid water droplets that have condensed from the gaseous water vapor in the air. Changes of a substance from one phase to another are a result of energy transfer.

In Chapters 6 and 7, we introduced the relationship between energy in mechanics and energy in thermodynamics. We discussed the transformation of mechanical energy to internal energy in cases in which a nonconservative force such as friction is acting. In Chapter 16, we discussed additional concepts of the relationship between internal energy and temperature. In this chapter, we extend these discussions into a complete treatment of energy in thermal processes.

Until around 1850, the fields of thermodynamics and mechanics were considered to be two distinct branches of science, and the law of conservation of energy seemed to describe only certain kinds of mechanical systems. Mid-19th-century experiments performed by English physicist James Joule (1818–1889) and others showed that energy may enter or leave a system by heat and by work. Today, as we discussed in

James Prescott Joule
British Physicist (1818–1889)
Joule received some formal education in mathematics, philosophy, and chemistry from John Dalton but was in large part self-educated. Joule's research led to the establishment of the principle of conservation of energy. His study of the quantitative relationship among electrical, mechanical, and chemical effects of heat culminated in his announcement in 1843 of the amount of work required to produce a unit of energy, called the mechanical equivalent of heat.

Pitfall Prevention | 17.1
Heat, Temperature, and Internal Energy Are Different
As you read the newspaper or listen to the radio, be alert for incorrectly used phrases including the word *heat* and think about the proper word to be used in place of it. "As the truck braked to a stop, a large amount of heat was generated by friction" and "The heat of a hot summer day . . ." are two examples.

Chapter 6, internal energy is treated as a form of energy that can be transformed into mechanical energy and vice versa. Once the concept of energy was broadened to include internal energy, the law of conservation of energy emerged as a universal law of nature.

This chapter focuses on developing the concept of heat, extending our concept of work to thermal processes, introducing the first law of thermodynamics, and investigating some important applications.

17.1 | Heat and Internal Energy

A major distinction must be made between internal energy and heat because these terms tend to be used interchangeably in everyday communication. You should read the following descriptions carefully and try to use these terms correctly because they are not interchangeable. They have very different meanings.

We introduced internal energy in Chapter 6, and we formally define it here:

Internal energy E_{int} is the energy associated with the microscopic components of a system—atoms and molecules—when viewed from a reference frame at rest with respect to the system. It includes kinetic and potential energy associated with the random translational, rotational, and vibrational motion of the atoms or molecules that make up the system as well as intermolecular potential energy.

In Chapter 16, we showed that the internal energy of a monatomic ideal gas is associated with the translational motion of its atoms. In this special case, the internal energy is simply the total translational kinetic energy of the atoms; the higher the temperature of the gas, the greater the kinetic energy of the atoms and the greater the internal energy of the gas. For more complicated diatomic and polyatomic gases, internal energy includes other forms of molecular energy, such as rotational kinetic energy and the kinetic and potential energy associated with molecular vibrations.

Heat was introduced in Chapter 7 as one possible method of energy transfer, and we provide a formal definition here:

Heat is a mechanism by which energy is transferred between a system and its environment because of a temperature difference between them. It is also the amount of energy Q transferred by this mechanism.

Figure 17.1 shows a pan of water in contact with a gas flame. Energy enters the water by heat from the hot gases in the flame, and the internal energy of the water increases as a result. It is *incorrect* to say that the water has more heat as time goes by.

As further clarification of the use of the word *heat*, consider the distinction between work and energy. The work done on (or by) a system is a measure of the amount of energy transferred between the system and its surroundings, whereas the mechanical energy of the system (kinetic or potential) is a consequence of its motion and coordinates. Therefore, when a person does work on a system, energy is transferred from the person to the system. It makes no sense to talk about the work *in* a system; one refers only to the work done *on* or *by* a system when some process has occurred in which energy has been transferred to or from the system. Likewise, it makes no sense to use the term *heat* unless energy has been transferred as a result of a temperature difference.

Units of Heat

Early in the development of thermodynamics, before scientists recognized the connection between thermodynamics and mechanics, heat was defined in terms of the temperature changes it produced in an object, and a separate unit of energy, the calorie, was used for heat. The **calorie** (cal) was defined as the amount of energy

Figure 17.1 A pan of boiling water is warmed by a gas flame. Energy enters the water through the bottom of the pan by heat.

transfer necessary to raise the temperature of 1 g of water[1] from 14.5°C to 15.5°C. (The "Calorie," with a capital C, used in describing the energy content of foods, is actually a kilocalorie.) Likewise, the unit of heat in the U.S. customary system, the **British thermal unit** (Btu), was defined as the amount of energy transfer required to raise the temperature of 1 lb of water from 63°F to 64°F.

In 1948, scientists agreed that because heat (like work) is a measure of the transfer of energy, its SI unit should be the joule. The calorie is now defined to be exactly 4.186 J:

$$1 \text{ cal} \equiv 4.186 \text{ J} \qquad \qquad \textbf{17.1} \blacktriangleleft$$

▶ Mechanical equivalent of heat

Note that this definition makes no reference to the heating of water. The calorie is a general energy unit. We could have used it in Chapter 6 for the kinetic energy of an object, for example. It is introduced here for historical reasons, but we shall make little use of it as an energy unit. The definition in Equation 17.1 is known historically as the **mechanical equivalent of heat.**

Example 17.1 | Losing Weight the Hard Way BIO

A student eats a dinner rated at 2 000 Calories. He wishes to do an equivalent amount of work in the gymnasium by lifting a 50.0-kg barbell. How many times must he raise the barbell to expend this much energy? Assume he raises the barbell 2.00 m each time he lifts it and he regains no energy when he lowers the barbell.

SOLUTION

Conceptualize Imagine the student raising the barbell. He is doing work on the system of the barbell and the Earth, so energy is leaving his body. The total amount of work that the student must do is 2 000 Calories.

Categorize We model the system of the barbell and the Earth as a nonisolated system.

Analyze Reduce the conservation of energy equation, Equation 7.2, to the appropriate expression for the system of the barbell and the Earth:

$$(1) \quad \Delta U_{\text{total}} = W_{\text{total}}$$

Express the change in gravitational potential energy of the system after the barbell is raised once:

$$\Delta U = mgh$$

Express the total amount of energy that must be transferred into the system by work for lifting the barbell n times, assuming energy is not regained when the barbell is lowered:

$$(2) \quad \Delta U_{\text{total}} = nmgh$$

Substitute Equation (2) into Equation (1):

$$nmgh = W_{\text{total}}$$

Solve for n:

$$n = \frac{W_{\text{total}}}{mgh}$$

$$= \frac{(2\ 000 \text{ Cal})}{(50.0 \text{ kg})(9.80 \text{ m/s}^2)(2.00 \text{ m})}\left(\frac{1.00 \times 10^3 \text{ cal}}{\text{Calorie}}\right)\left(\frac{4.186 \text{ J}}{1 \text{ cal}}\right)$$

$$= 8.54 \times 10^3 \text{ times}$$

Finalize If the student is in good shape and lifts the barbell once every 5 s, it will take him about 12 h to perform this feat. Clearly, it is much easier for this student to lose weight by dieting.

In reality, the human body is not 100% efficient. Therefore, not all the energy transformed within the body from the dinner transfers out of the body by work done on the barbell. Some of this energy is used to pump blood and perform other functions within the body. Therefore, the 2 000 Calories can be worked off in less time than 12 h when these other energy processes are included.

[1]Originally, the calorie was defined as the heat necessary to raise the temperature of 1 g of water by 1°C at any initial temperature. Careful measurements, however, showed that the energy required depends somewhat on temperature; hence, a more precise definition evolved.

17.2 | Specific Heat

The definition of the calorie indicates the amount of energy necessary to raise the temperature of 1 g of a specific substance—water—by 1°C, which is 4.186 J. To raise the temperature of 1 kg of water by 1°C, we need to transfer 4 186 J of energy to it from the environment. The quantity of energy required to raise the temperature of 1 kg of an arbitrary substance by 1°C varies with the substance. For example, the energy required to raise the temperature of 1 kg of copper by 1°C is 387 J, which is significantly less than that required for water. Every substance requires a unique amount of energy per unit mass to change the temperature of that substance by 1°C.

Suppose a quantity of energy Q is transferred to a mass m of a substance, thereby changing its temperature by ΔT. The **specific heat** c of the substance is defined as

$$c \equiv \frac{Q}{m \, \Delta T} \qquad \text{17.2} \blacktriangleleft$$

The units of specific heat are joules per kilogram-degree Celsius, or $J/kg \cdot {}^\circ C$. Table 17.1 lists specific heats for several substances. From the definition of the calorie, the specific heat of water is 4 186 $J/kg \cdot {}^\circ C$.

From this definition, we can express the energy Q transferred between a system of mass m and its surroundings in terms of the resulting temperature change ΔT as

$$Q = mc \, \Delta T \qquad \text{17.3} \blacktriangleleft$$

For example, the energy required to raise the temperature of 0.500 kg of water by 3.00°C is $Q = (0.500 \text{ kg})(4\,186 \text{ J/kg} \cdot {}^\circ C)(3.00{}^\circ C) = 6.28 \times 10^3$ J. Note that when the temperature increases, ΔT and Q are taken to be *positive*, corresponding to energy flowing *into* the system. When the temperature decreases, ΔT and Q are *negative* and energy flows *out* of the system. These sign conventions are consistent with those in our discussion of the conservation of energy equation, Equation 7.2.

Table 17.1 shows that water has a high specific heat relative to most other common substances (the specific heats of hydrogen and helium are higher). The high specific heat of water is responsible for the moderate temperatures found in regions near large bodies of water. As the temperature of a body of water decreases during the winter, the water transfers energy to the air, which carries the energy landward when prevailing winds are toward the land. For example, the prevailing winds off the western coast of the United States are toward the land, and the energy liberated by the Pacific Ocean as it cools keeps coastal areas much warmer than they would be otherwise. Therefore, the western coastal states generally have warmer winter weather than the eastern coastal states, where the winds do not transfer energy toward land.

That the specific heat of water is higher than that of sand accounts for the pattern of air flow at a beach. During the day, the Sun adds roughly equal amounts of energy to beach and water, but the lower specific heat of sand causes the beach to reach a higher temperature than the water. As a result, the air above the land reaches a higher temperature than the air above the water. The denser cold air pushes the less dense hot air upward (due to Archimedes's principle), which results in a breeze from ocean to land during the day. During the night, the sand cools more quickly than the water, and the direction of the breeze reverses because the hotter air is now over the water. These offshore and onshore breezes are well known to sailors.

> **Pitfall Prevention | 17.2**
> **An Unfortunate Choice of Terminology**
> The name *specific heat* is an unfortunate holdover from the days when thermodynamics and mechanics developed separately. A better name would be *specific energy transfer*, but the existing term is too entrenched to be replaced.

TABLE 17.1 | Specific Heats of Some Substances at 25°C and Atmospheric Pressure

Substance	Specific Heat c	
	$J/kg \cdot {}^\circ C$	$cal/g \cdot {}^\circ C$
Elemental Solids		
Aluminum	900	0.215
Beryllium	1 830	0.436
Cadmium	230	0.055
Copper	387	0.092 4
Germanium	322	0.077
Gold	129	0.030 8
Iron	448	0.107
Lead	128	0.030 5
Silicon	703	0.168
Silver	234	0.056
Other Solids		
Brass	380	0.092
Glass	837	0.200
Ice (−5°C)	2 090	0.50
Marble	860	0.21
Wood	1 700	0.41
Liquids		
Alcohol (ethyl)	2 400	0.58
Mercury	140	0.033
Water (15°C)	4 186	1.00
Gas		
Steam (100°C)	2 010	0.48

QUICK QUIZ 17.1 Imagine you have 1 kg each of iron, glass, and water and that all three samples are at 10°C. **(a)** Rank the samples from highest to lowest temperature after 100 J of energy is added to each sample. **(b)** Rank the samples from greatest to least amount of energy transferred by heat if each sample increases in temperature by 20.0°C.

Calorimetry

One technique for measuring the specific heat of a solid or liquid is to raise the temperature of the substance to some value, place it into a vessel containing water of known mass and temperature, and measure the temperature of the combination after equilibrium is reached. Let us define the system as the substance and the water. If the vessel is assumed to be a good insulator so that energy does not leave the system by heat (nor by any other means), we can use the isolated system model. A vessel having this property is called a **calorimeter,** and the analysis performed by using such a vessel is called **calorimetry.** Figure 17.2 shows the hot sample in the cold water and the resulting energy transfer by heat from the high-temperature part of the system to the low-temperature part.

The principle of conservation of energy for this isolated system requires that the energy leaving by heat from the warmer substance (of unknown specific heat) equals the energy entering the water.[2] Therefore, we can write

$$Q_{\text{cold}} = -Q_{\text{hot}} \qquad \text{17.4} \blacktriangleleft$$

To see how to set up a calorimetry problem, suppose m_x is the mass of a substance whose specific heat we wish to determine, c_x its specific heat, and T_x its initial temperature. Let m_w, c_w, and T_w represent corresponding values for the water. If T is the final equilibrium temperature after the substance and the water are combined, from Equation 17.3 we find that the energy gained by the water is $m_w c_w (T - T_w)$ and that the energy transferred from the substance of unknown specific heat is $m_x c_x (T - T_x)$. Substituting these values into Equation 17.4, we have

$$m_w c_w (T - T_w) = -m_x c_x (T - T_x)$$

This equation can be solved for the unknown specific heat c_x.

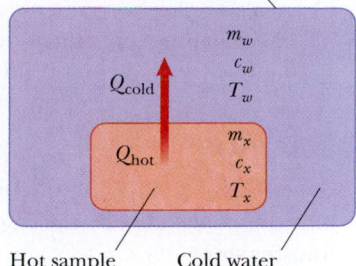

Figure 17.2 In a calorimetry experiment, a hot sample whose specific heat is unknown is placed in cold water in a container that isolates the system from the environment.

> **Pitfall Prevention | 17.3**
> **Remember the Negative Sign**
> It is *critical* to include the negative sign in Equation 17.4. The negative sign in the equation is necessary for consistency with our sign convention for energy transfer. The energy transfer Q_{hot} is negative because energy is leaving the hot substance. The negative sign in the equation ensures that the right-hand side is a positive number, consistent with the left-hand side, which is positive because energy is entering the cold substance.

> **Pitfall Prevention | 17.4**
> **Celsius versus Kelvin**
> In equations in which T appears (e.g., the ideal gas law), the Kelvin temperature *must* be used. In equations involving ΔT, such as calorimetry equations, it is *possible* to use Celsius temperatures because a change in temperature is the same on both scales. It is *safest*, however, to use Kelvin temperatures *consistently* in all equations involving T or ΔT.

> ## THINKING PHYSICS 17.1
>
> The equation $Q = mc \, \Delta T$ indicates the relationship between energy Q transferred to an object of mass m and specific heat c by means of heat and the resulting temperature change ΔT. In reality, the energy transfer on the left-hand side of the equation could be made by any method, not just heat. Give a few examples in which this equation could be used to calculate a temperature change of an object due to an energy transfer process other than heat.
>
> **Reasoning** The following are a few of several possible examples.
>
> During the first few seconds after turning on a toaster, the temperature of the electrical coils rises. The transfer mechanism here is *electrical transmission* of energy through the power cord.
>
> The temperature of a potato in a microwave oven increases due to the absorption of microwaves. In this case, the energy transfer mechanism is by *electromagnetic radiation*, the microwaves.
>
> A carpenter attempts to use a dull drill bit to bore a hole in a piece of wood. The bit fails to make much headway but becomes very warm. The increase in temperature in this case is due to *work* done on the bit by the wood.
>
> In each of these cases, as well as many other possibilities, the Q on the left of the equation of interest is not a measure of heat but, rather, is replaced with the energy transferred or transformed by other means. Even though heat is not involved, the equation can still be used to calculate the temperature change. ◄

---

[2]For precise measurements, the container holding the water should be included in the calculations because it also changes temperature. Doing so would require a knowledge of its mass and composition. If, however, the mass of the water is large compared with that of the container, we can adopt a simplification model in which we ignore the energy gained by the container.

Example 17.2 | Cooling a Hot Ingot

A 0.050 0-kg ingot of metal is heated to 200.0°C and then dropped into a calorimeter containing 0.400 kg of water initially at 20.0°C. The final equilibrium temperature of the mixed system is 22.4°C. Find the specific heat of the metal.

SOLUTION

Conceptualize Imagine the process occurring in the isolated system of Figure 17.2. Energy leaves the hot ingot and goes into the cold water, so the ingot cools off and the water warms up. Once both are at the same temperature, the energy transfer stops.

Categorize We use an equation developed in this section, so we categorize this example as a substitution problem.

Use Equation 17.3 to evaluate each side of Equation 17.4:

$$m_w c_w (T_f - T_w) = -m_x c_x (T_f - T_x)$$

Solve for c_x:

$$c_x = \frac{m_w c_w (T_f - T_w)}{m_x (T_x - T_f)}$$

Substitute numerical values:

$$c_x = \frac{(0.400 \text{ kg})(4\,186 \text{ J/kg} \cdot {}^\circ\text{C})(22.4{}^\circ\text{C} - 20.0{}^\circ\text{C})}{(0.050\,0 \text{ kg})(200.0{}^\circ\text{C} - 22.4{}^\circ\text{C})}$$

$$= 453 \text{ J/kg} \cdot {}^\circ\text{C}$$

The ingot is most likely iron as you can see by comparing this result with the data given in Table 17.1. The temperature of the ingot is initially above the steam point. Therefore, some of the water may vaporize when the ingot is dropped into the water. We assume the system is sealed and this steam cannot escape. Because the final equilibrium temperature is lower than the steam point, any steam that does result recondenses back into water.

What If? Suppose you are performing an experiment in the laboratory that uses this technique to determine the specific heat of a sample and you wish to decrease the overall uncertainty in your final result for c_x. Of the data given in this example, changing which value would be most effective in decreasing the uncertainty?

....

Answer The largest experimental uncertainty is associated with the small difference in temperature of 2.4°C for the water. For example, using the rules for propagation of uncertainty in Appendix Section B.8, an uncertainty of 0.1°C in each of T_f and T_w leads to an 8% uncertainty in their difference. For this temperature difference to be larger experimentally, the most effective change is to *decrease the amount of water*.

⟨17.3 | Latent Heat

As we have seen in the preceding section, a substance can undergo a change in temperature when energy is transferred between it and its surroundings. In some situations, however, the transfer of energy does not result in a change in temperature. That is the case whenever the physical characteristics of the substance change from one form to another; such a change is commonly referred to as a **phase change.** Two common phase changes are from solid to liquid (melting) and from liquid to gas (boiling); another is a change in the crystalline structure of a solid. All such phase changes involve a change in the system's internal energy but no change in its temperature. The increase in internal energy in boiling, for example, is represented by the breaking of bonds between molecules in the liquid state; this bond breaking allows the molecules to move farther apart in the gaseous state, with a corresponding increase in intermolecular potential energy.

As you might expect, different substances respond differently to the addition or removal of energy as they change phase because their internal molecular arrangements vary. Also, the amount of energy transferred during a phase change depends on the amount of substance involved. (It takes less energy to melt an ice cube than it does to thaw a frozen lake.) When discussing two phases of a material, we will use the term *higher-phase material* to mean the material existing at the higher temperature. So, for example, if we discuss water and ice, water is the higher-phase material, whereas steam is the higher-phase material in a discussion of steam and water. Consider a system containing a substance in two phases in equilibrium such as water and ice. The initial amount of the higher-phase material, water, in the system is m_i. Now imagine that energy Q enters the system. As a result, the final amount of water is m_f due to the melting of some of the ice. Therefore, the amount of ice that melted, equal to the amount of *new* water, is $\Delta m = m_f - m_i$. We define the **latent heat** for this phase change as

$$L \equiv \frac{Q}{\Delta m}$$ **17.5** ◄

This parameter is called latent heat (literally, the "hidden" heat) because this added or removed energy does not result in a temperature change. The value of L for a substance depends on the nature of the phase change as well as on the properties of the substance. If the entire amount of the lower-phase material undergoes a phase change, the change in mass Δm of the higher-phase material is equal to the initial mass of the lower-phase material. For example, if an ice cube of mass m on a plate melts completely, the change in mass of the water is $m_f - 0 = m$, which is the mass of new water and is also equal to the initial mass of the ice cube.

From the definition of latent heat, and again choosing heat as our energy transfer mechanism, the energy required to change the phase of a pure substance is

$$Q = L \Delta m$$ **17.6** ◄

where Δm is the change in mass of the higher-phase material.

Latent heat of fusion L_f is the term used when the phase change is from solid to liquid (*to fuse* means "to combine by melting"), and **latent heat of vaporization** L_v is the term used when the phase change is from liquid to gas (the liquid "vaporizes").[3] The latent heats of various substances vary considerably as data in Table 17.2 show. When energy enters a system, causing melting or vaporization, the amount of the higher-phase material increases, so Δm is positive and Q is positive, consistent with our sign convention. When energy is extracted from a system, causing freezing or condensation, the amount of the higher-phase material decreases, so Δm is negative and Q is negative, again consistent with our sign convention. Keep in mind that Δm in Equation 17.6 always refers to the higher-phase material.

► Energy transferred to a substance during a phase change

Pitfall Prevention | 17.5
Signs Are Critical
Sign errors occur very often when students apply calorimetry equations. For phase changes, remember that Δm in Equation 17.6 is always the change in mass of the higher-phase material. In Equation 17.3, be sure your ΔT is *always* the final temperature minus the initial temperature. In addition, you must *always* include the negative sign on the right side of Equation 17.4.

◄ **TABLE 17.2** | Latent Heats of Fusion and Vaporization

Substance	Melting Point (°C)	Latent Heat of Fusion (J/kg)	Boiling Point (°C)	Latent Heat of Vaporization (J/kg)
Helium[a]	−272.2	5.23×10^3	−268.93	2.09×10^4
Oxygen	−218.79	1.38×10^4	−182.97	2.13×10^5
Nitrogen	−209.97	2.55×10^4	−195.81	2.01×10^5
Ethyl alcohol	−114	1.04×10^5	78	8.54×10^5
Water	0.00	3.33×10^5	100.00	2.26×10^6
Sulfur	119	3.81×10^4	444.60	3.26×10^5
Lead	327.3	2.45×10^4	1 750	8.70×10^5
Aluminum	660	3.97×10^5	2 450	1.14×10^7
Silver	960.80	8.82×10^4	2 193	2.33×10^6
Gold	1 063.00	6.44×10^4	2 660	1.58×10^6
Copper	1 083	1.34×10^5	1 187	5.06×10^6

[a]Helium does not solidify at atmospheric pressure. Therefore, its melting point is given under the conditions that the pressure is 2.5 MPa.

[3]When a gas cools, it eventually *condenses*; that is, it returns to the liquid phase. The energy given up per unit mass is called the *latent heat of condensation* and is numerically equal to the latent heat of vaporization. Likewise, when a liquid cools, it eventually solidifies, and the *latent heat of solidification* is numerically equal to the latent heat of fusion.

Figure 17.3 A plot of temperature versus energy added when 1.00 g of ice initially at −30.0°C is converted to steam at 120.0°C.

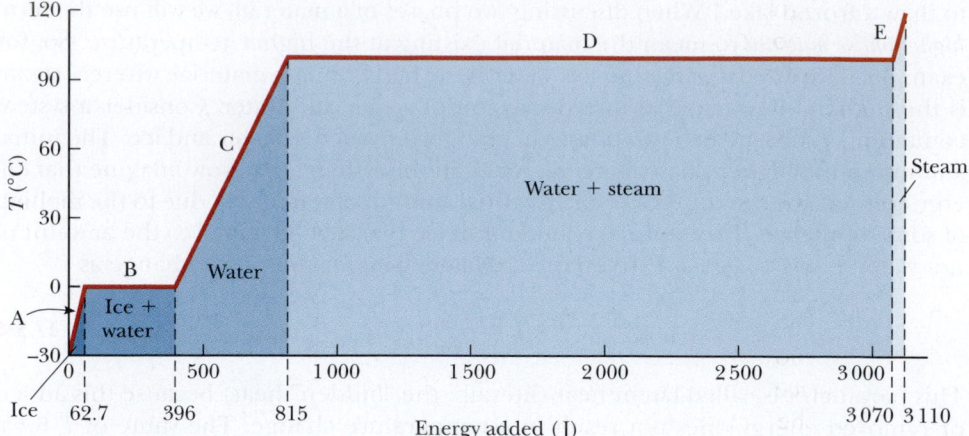

To understand the role of latent heat in phase changes, consider the energy required to convert a 1.00-g cube of ice at −30.0°C to steam at 120.0°C. Figure 17.3 indicates the experimental results obtained when energy is gradually added to the ice. The results are presented as a graph of temperature of the system of the ice cube versus energy added to the system. Let's examine each portion of the red-brown curve, which is divided into parts A through E.

Part A. On this portion of the curve, the temperature of the ice changes from −30.0°C to 0.0°C. Equation 17.3 indicates that the temperature varies linearly with the energy added, so the experimental result is a straight line on the graph. Because the specific heat of ice is 2 090 J/kg · °C, we can calculate the amount of energy added by using Equation 17.3:

$$Q = m_i c_i \Delta T = (1.00 \times 10^{-3} \text{ kg})(2\,090 \text{ J/kg} \cdot °\text{C})(30.0°\text{C}) = 62.7 \text{ J}$$

Part B. When the temperature of the ice reaches 0.0°C, the ice–water mixture remains at this temperature—even though energy is being added—until all the ice melts. The energy required to melt 1.00 g of ice at 0.0°C is, from Equation 17.6,

$$Q = L_f \Delta m_w = L_f m_i = (3.33 \times 10^5 \text{ J/kg})(1.00 \times 10^{-3} \text{ kg}) = 333 \text{ J}$$

At this point, we have moved to the 396 J (= 62.7 J + 333 J) mark on the energy axis in Figure 17.3.

Part C. Between 0.0°C and 100.0°C, nothing surprising happens. No phase change occurs, and so all energy added to the water is used to increase its temperature. The amount of energy necessary to increase the temperature from 0.0°C to 100.0°C is

$$Q = m_w c_w \Delta T = (1.00 \times 10^{-3} \text{ kg})(4.19 \times 10^3 \text{ J/kg} \cdot °\text{C})(100.0°\text{C}) = 419 \text{ J}$$

Part D. At 100.0°C, another phase change occurs as the water changes from water at 100.0°C to steam at 100.0°C. Similar to the ice–water mixture in part B, the water–steam mixture remains at 100.0°C—even though energy is being added—until all the liquid has been converted to steam. The energy required to convert 1.00 g of water to steam at 100.0°C is

$$Q = L_v \Delta m_s = L_v m_w = (2.26 \times 10^6 \text{ J/kg})(1.00 \times 10^{-3} \text{ kg}) = 2.26 \times 10^3 \text{ J}$$

Part E. On this portion of the curve, as in parts A and C, no phase change occurs; therefore, all energy added is used to increase the temperature of the steam. The energy that must be added to raise the temperature of the steam from 100.0°C to 120.0°C is

$$Q = m_s c_s \Delta T = (1.00 \times 10^{-3} \text{ kg})(2.01 \times 10^3 \text{ J/kg} \cdot °\text{C})(20.0°\text{C}) = 40.2 \text{ J}$$

The total amount of energy that must be added to change 1 g of ice at −30.0°C to steam at 120.0°C is the sum of the results from all five parts of the curve, which is 3.11×10^3 J. Conversely, to cool 1 g of steam at 120.0°C to ice at −30.0°C, we must remove 3.11×10^3 J of energy.

Notice in Figure 17.3 the relatively large amount of energy that is transferred into the water to vaporize it to steam. Imagine reversing this process, with a large amount of energy transferred out of steam to condense it into water. That is why a burn to your skin from steam at 100°C is much more damaging than exposure of your skin to water at 100°C. A very large amount of energy enters your skin from the steam, and the steam remains at 100°C for a long time while it condenses. Conversely, when your skin makes contact with water at 100°C, the water immediately begins to drop in temperature as energy transfers from the water to your skin.

If liquid water is held perfectly still in a very clean container, it is possible for the water to drop below 0°C without freezing into ice. This phenomenon, called **supercooling,** arises because the water requires a disturbance of some sort for the molecules to move apart and start forming the large, open ice structure that makes the density of ice lower than that of water as discussed in Section 16.3. If supercooled water is disturbed, it suddenly freezes. The system drops into the lower-energy configuration of bound molecules of the ice structure, and the energy released raises the temperature back to 0°C.

Commercial hand warmers consist of liquid sodium acetate in a sealed plastic pouch. The solution in the pouch is in a stable supercooled state. When a disk in the pouch is clicked by your fingers, the liquid solidifies and the temperature increases, just like the supercooled water just mentioned. In this case, however, the freezing point of the liquid is higher than body temperature, so the pouch feels warm to the touch. To reuse the hand warmer, the pouch must be boiled until the solid liquefies. Then, as it cools, it passes below its freezing point into the supercooled state.

It is also possible to create **superheating.** For example, clean water in a very clean cup placed in a microwave oven can sometimes rise in temperature beyond 100°C without boiling because the formation of a bubble of steam in the water requires scratches in the cup or some type of impurity in the water to serve as a nucleation site. When the cup is removed from the microwave oven, the superheated water can become explosive as bubbles form immediately and the hot water is forced upward out of the cup.

A classic prank related to phase changes is to fashion a spoon made of pure gallium. The melting point of gallium is 29.8°C. Therefore, when the spoon is used to stir hot tea, the submerged portion of the spoon turns into a liquid and drops to the bottom of the cup. Picking up the spoon and beginning to stir must be done quickly, because the melting point of the gallium is lower than normal body temperature, so the spoon will melt in one's hand!

> **QUICK QUIZ 17.2** Suppose the same process of adding energy to the ice cube is performed as discussed above, but instead we graph the internal energy of the system as a function of energy input. What would this graph look like?

> **QUICK QUIZ 17.3** Calculate the slopes for the A, C, and E portions of Figure 17.3. Rank the slopes from least steep to steepest, and explain what this ordering means.

Example 17.3 | Cooling the Steam

What mass of steam initially at 130°C is needed to warm 200 g of water in a 100-g glass container from 20.0°C to 50.0°C?

SOLUTION

Conceptualize Imagine placing water and steam together in a closed insulated container. The system eventually reaches a uniform state of water with a final temperature of 50.0°C.

Categorize Based on our conceptualization of this situation, we categorize this example as one involving calorimetry in which a phase change occurs.

Analyze Write Equation 17.4 to describe the calorimetry process:

$$(1) \quad Q_{cold} = -Q_{hot}$$

continued

17.3 *cont.*

The steam undergoes three processes: first a decrease in temperature to 100°C, then condensation into liquid water, and finally a decrease in temperature of the water to 50.0°C. Find the energy transfer in the first process using the unknown mass m_s of the steam:

$$Q_1 = m_s c_s \, \Delta T_s$$

Find the energy transfer in the second process:

$$Q_2 = L_v \, \Delta m_s = L_v(0 - m_s) = -m_s L_v$$

Find the energy transfer in the third process:

$$Q_3 = m_s c_w \, \Delta T_{\text{hot water}}$$

Add the energy transfers in these three stages:

$$(2) \quad Q_{\text{hot}} = Q_1 + Q_2 + Q_3 = m_s(c_s \, \Delta T_s - L_v + c_w \, \Delta T_{\text{hot water}})$$

The 20.0°C water and the glass undergo only one process, an increase in temperature to 50.0°C. Find the energy transfer in this process:

$$(3) \quad Q_{\text{cold}} = m_w c_w \, \Delta T_{\text{cold water}} + m_g c_g \, \Delta T_{\text{glass}}$$

Substitute Equations (2) and (3) into Equation (1):

$$m_w c_w \, \Delta T_{\text{cold water}} + m_g c_g \, \Delta T_{\text{glass}} = -m_s(c_s \, \Delta T_s - L_v + c_w \, \Delta T_{\text{hot water}})$$

Solve for m_s:

$$m_s = -\frac{m_w c_w \, \Delta T_{\text{cold water}} + m_g c_g \, \Delta T_{\text{glass}}}{c_s \, \Delta T_s - L_v + c_w \, \Delta T_{\text{hot water}}}$$

Substitute numerical values:

$$m_s = -\frac{(0.200 \text{ kg})(4\,186 \text{ J/kg} \cdot {}^\circ\text{C})(50.0^\circ\text{C} - 20.0^\circ\text{C}) + (0.100 \text{ kg})(837 \text{ J/kg} \cdot {}^\circ\text{C})(50.0^\circ\text{C} - 20.0^\circ\text{C})}{(2\,010 \text{ J/kg} \cdot {}^\circ\text{C})(100^\circ\text{C} - 130^\circ\text{C}) - (2.26 \times 10^6 \text{ J/kg}) + (4\,186 \text{ J/kg} \cdot {}^\circ\text{C})(50.0^\circ\text{C} - 100^\circ\text{C})}$$

$$= 1.09 \times 10^{-2} \text{ kg} = \boxed{10.9 \text{ g}}$$

What If? What if the final state of the system is water at 100°C? Would we need more steam or less steam? How would the analysis above change?

Answer More steam would be needed to raise the temperature of the water and glass to 100°C instead of 50.0°C. There would be two major changes in the analysis. First, we would not have a term Q_3 for the steam because the water that condenses from the steam does not cool below 100°C. Second, in Q_{cold}, the temperature change would be 80.0°C instead of 30.0°C. For practice, show that the result is a required mass of steam of 31.8 g.

◀ 17.4 | Work in Thermodynamic Processes

▶ State variables

In the macroscopic approach to thermodynamics, we describe the *state* of a system with such quantities as pressure, volume, temperature, and internal energy. As a result, these quantities belong to a category called **state variables.** For any given condition of the system, we can identify values of the state variables. It is important to note, however, that a macroscopic state of a system can be specified only if the system is in internal thermal equilibrium. In the case of a gas in a container, internal thermal equilibrium requires that every part of the gas be at the same pressure and temperature. If the temperature varies from one part of the gas to another, for example, we cannot specify a single temperature for the entire gas to be used in the ideal gas law.

▶ Transfer variables

A second category of variables in situations involving energy is **transfer variables.** These variables only have a value if a process occurs in which energy is transferred across the boundary of the system. Because a transfer of energy across the boundary represents a change in the system, transfer variables are not associated with a given state of the system, but with a *change* in the state of the system. In the previous sections, we discussed heat as a transfer variable. For a given set of conditions of a system, the heat has no defined value. We can assign a value to the heat only if energy crosses the boundary by heat, resulting in a change in the system. State variables are characteristic of a system in internal thermal equilibrium. Transfer variables are characteristic of a process in which energy is transferred between a system and its environment.

We have seen this notion before, but we have not used the language of state variables and transfer variables. In the conservation of energy equation, $\Delta E_{\text{system}} = \Sigma T$, we can identify the terms on the right-hand side as transfer variables: work, heat, mechanical waves, matter transfer, electromagnetic radiation, and electrical transmission. The left side of the conservation of energy equation represents *changes* in state variables: kinetic energy, potential energy, and internal energy. For a gas, we have additional state variables that are not energies, such as pressure, volume, and temperature.

In this section, we study an important transfer variable for thermodynamic systems, work. Work performed on particles was studied extensively in Chapter 6, and here we investigate the work done on a deformable system, a gas. Consider a gas contained in a cylinder fitted with a frictionless, movable piston of face area A (Fig. 17.4) and in thermal equilibrium. The gas occupies a volume V and exerts a uniform pressure P on the cylinder walls and the piston. Now let us adopt a simplification model in which the gas is compressed in a **quasi-static process,** that is, slowly enough to allow the system to remain in thermal equilibrium at all times. As the piston is pushed inward by an external force $\vec{F}_{ext}$, its point of application on the gas (the bottom face of the piston) moves through a displacement $d\vec{r} = dy\hat{j}$ (Fig. 17.4b). Therefore, the work done on the gas is, according to our definition of work in Chapter 6,

$$dW = \vec{F}_{ext} \cdot d\vec{r} = \vec{F}_{ext} \cdot dy\hat{j}$$

Because the piston is in equilibrium at all times during the process, the external force has the same magnitude as the force exerted on it by the gas but is in the opposite direction:

$$\vec{F}_{ext} = -\vec{F}_{gas} = -PA\hat{j}$$

where we have set the magnitude of the force exerted by the gas equal to PA. The work done by the external force can now be expressed as

$$dW = -PA\hat{j} \cdot dy\hat{j} = -PA\,dy$$

Because $A\,dy$ is the change in volume of the gas dV, we can express the work done *on* the gas as

$$dW = -P\,dV$$

If the gas is compressed, dV is negative and the work done on the gas is positive. If the gas expands, dV is positive and the work done on the gas is negative. If the volume remains constant, the work done on the gas is zero. The total **work** done on the gas as its volume changes from V_i to V_f is given by the integral of dW above:

$$W = -\int_{V_i}^{V_f} P\,dV \qquad \textbf{17.7} \blacktriangleleft$$

▶ Work done on a gas

To evaluate this integral, one must know how the pressure varies with volume during the expansion process.

In general, the pressure is not constant during a process that takes a gas from its initial state to its final state but depends on the volume and temperature. If the pressure and volume are known at each step of the process, the state of the gas at each step can be plotted on a specialized graphical representation—a **PV diagram,** as in Active Figure 17.5—that is very important in thermodynamics. This type of diagram allows us to visualize a process through which a gas is progressing. The curve on such a graphical representation is called the *path* taken between the initial and final states.

Considering the integral in Equation 17.7 and recognizing the significance of the integral as an area under a curve, we can identify an important use for *PV* diagrams:

The work done on a gas in a quasi-static process that takes the gas from an initial state to a final state is the negative of the area under the curve on a *PV* diagram, evaluated between the initial and final states.

For our process of compressing a gas in the cylinder, as Active Figure 17.5 suggests, the work done depends on the particular path taken between the initial and final states. To illustrate this important point, consider several different paths connecting i and f (Active Fig. 17.6, page 478). In the process depicted in Active Figure 17.6a, the volume of the gas is first reduced from V_i to V_f at constant pressure P_i and the pressure of the gas then increases

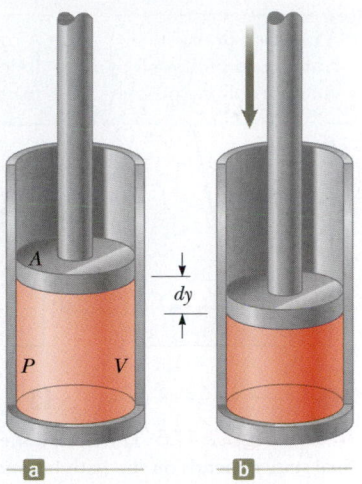

Figure 17.4 Work is done on a gas contained in a cylinder at pressure P as the piston is pushed downward so that the gas is compressed.

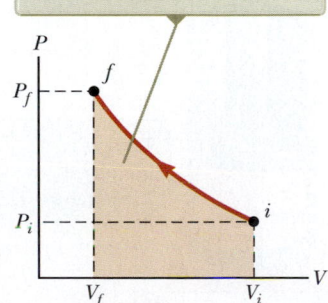

The work done on a gas equals the negative of the area under the *PV* curve. The area is negative here because the volume is decreasing, resulting in positive work.

Active Figure 17.5 A gas is compressed quasi-statically (slowly) from state i to state f. An outside agent must do positive work on the gas to compress it.

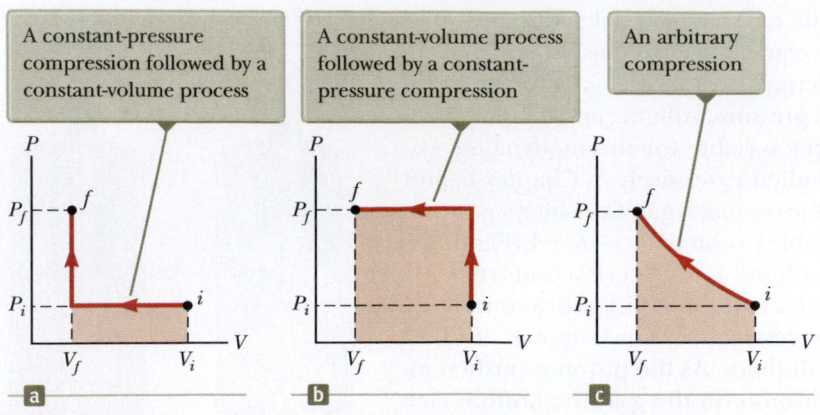

A constant-pressure compression followed by a constant-volume process

A constant-volume process followed by a constant-pressure compression

An arbitrary compression

a   **b**   **c**

Active Figure 17.6 The work done on a gas as it is taken from an initial state to a final state depends on the path between these states.

from P_i to P_f by heating at constant volume V_f. The work done on the gas along this path is $-P_i(V_f - V_i)$. In Active Figure 17.6b, the pressure of the gas is increased from P_i to P_f at constant volume V_i and then the volume of the gas is reduced from V_i to V_f at constant pressure P_f. The work done on the gas along this path is $-P_f(V_f - V_i)$, which is greater in magnitude than that for the process described in Active Figure 17.6a because the piston is displaced through the same distance by a larger force than for the situation in Active Figure 17.6a. Finally, for the process described in Active Figure 17.6c, where both P and V change continuously, the work done on the gas has some value intermediate between the values obtained in the first two processes.

The energy transfer Q into or out of a system by heat also depends on the process. Consider the situations depicted in Figure 17.7. In both processes that are illustrated, the gas has the same initial volume, temperature, and pressure, and is assumed to be ideal. In Figure 17.7a, the gas is thermally insulated from its surroundings except at the bottom of the gas-filled region, where it is in thermal contact with an energy reservoir. An *energy reservoir* is a source of energy that is considered to be so great that a finite transfer of energy to or from the reservoir does not change its temperature. The piston is held at its initial position by an external agent such as a hand. When the force holding the piston is reduced slightly, the piston rises very slowly to its final position shown in Figure 17.7b. Because the piston is moving upward, the gas is doing work on the piston. During this expansion to the final volume V_f, just enough energy is transferred by heat from the reservoir to the gas to maintain a constant temperature T_i.

Now consider the completely thermally insulated system shown in Figure 17.7c. When the membrane is broken, the gas expands rapidly into the vacuum until it occupies a volume V_f and is at a pressure P_f. The final state of the gas is shown in Figure 17.7d. In this case, the gas does no work because it does not apply a force; no force is required to expand into a vacuum. Furthermore, no energy is transferred by heat through the insulating wall.

As we discuss in Section 17.6, experiments show that the temperature of the ideal gas does not change in the process indicated in Figures 17.7c and 17.7d. Therefore,

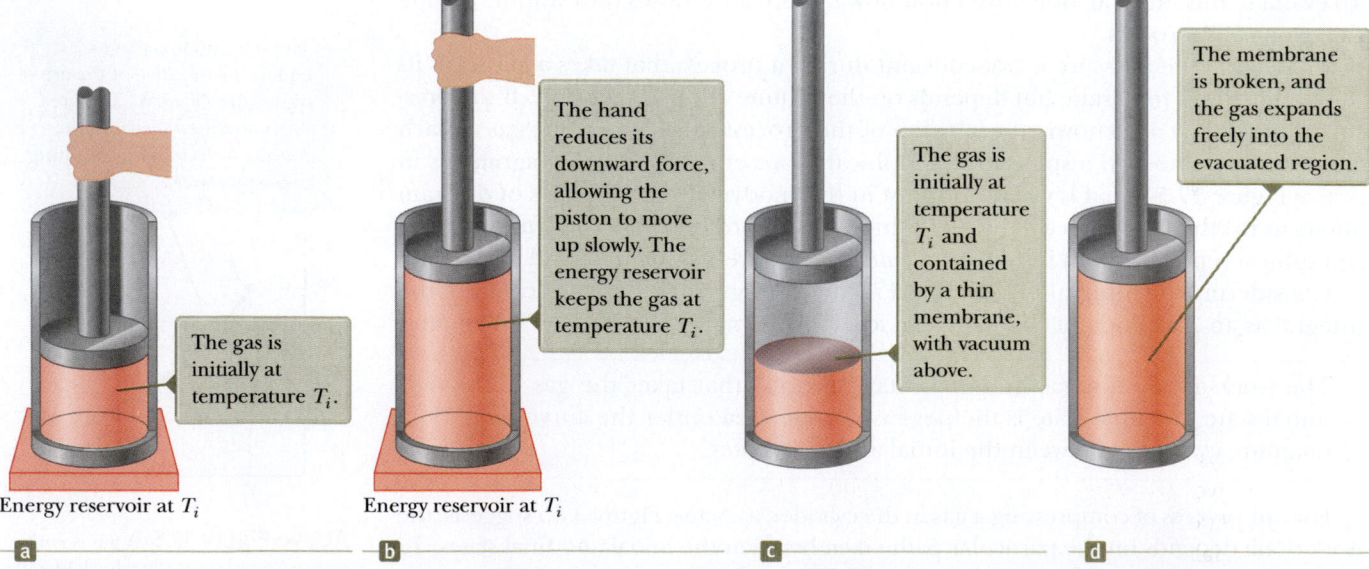

Figure 17.7 Gas in a cylinder. (a) The gas is in contact with an energy reservoir. The walls of the cylinder are perfectly insulating, but the base in contact with the reservoir is conducting. (b) The gas expands slowly to a larger volume. (c) The gas is contained by a membrane in half of a volume, with vacuum in the other half. The entire cylinder is perfectly insulating. (d) The gas expands freely into the larger volume.

the initial and final states of the ideal gas in Figures 17.7a and 17.7b are identical to the initial and final states in Figures 17.7c and 17.7d, but the paths are different. In the first case, the gas does work on the piston and energy is transferred slowly to the gas by heat. In the second case, no energy is transferred by heat and the value of the work done is zero. Therefore, energy transfer by heat, like work done, depends on the initial, final, and intermediate states of the system. In other words, because heat and work depend on the path, neither quantity is determined solely by the end-points of a thermodynamic process.

Example 17.4 | Comparing Processes

An ideal gas is taken through two processes in which $P_f = 1.00 \times 10^5$ Pa, $V_f = 2.00$ m^3, $P_i = 0.200 \times 10^5$ Pa, and $V_i = 10.0$ m^3. For process 1 shown in Active Figure 17.6c, the temperature remains constant. For process 2 shown in Active Figure 17.6a, the pressure remains constant and then the volume remains constant. What is the ratio of the work W_1 done on the gas in the first process to the work W_2 done in the second process?

SOLUTION

Conceptualize In Figure 17.6a (process 2), the displacement occurs at a fixed pressure equal to the initial pressure. In Figure 17.6c (process 1), the displacement occurs at an ever increasing pressure as the piston is moved inward. As a consequence, the force pushing the piston in during process 1 will become larger as the piston moves inward. We therefore expect the work to be larger for process 1 than for process 2.

Categorize We can categorize process 1 as taking place at constant temperature. We categorize process 2 as a combination of processes taking place at constant pressure and then at constant volume. In Section 17.6, we will discuss names for these types of processes.

....

Analyze For process 1, express the pressure as a function of volume using the ideal gas law:

$$P = \frac{nRT}{V}$$

For process 2, no work is done during the portion at constant volume because the piston does not move through a displacement. During the first part of the process, the pressure is constant at $P = P_i$. Use these results and set up the ratio of the work done in the two processes:

$$\frac{W_1}{W_2} = \frac{-\displaystyle\int_{\text{process 1}} P\,dV}{-\displaystyle\int_{\text{process 2}} P\,dV} = \frac{\displaystyle\int_{V_i}^{V_f} \frac{nRT}{V}\,dV}{\displaystyle\int_{V_i}^{V_f} P_i\,dV} = \frac{nRT\displaystyle\int_{V_i}^{V_f} \frac{dV}{V}}{P_i\displaystyle\int_{V_i}^{V_f} dV}$$

$$= \frac{nRT\ln\left(\dfrac{V_f}{V_i}\right)}{P_i(V_f - V_i)} = \frac{P_iV_i\ln\left(\dfrac{V_f}{V_i}\right)}{P_i(V_f - V_i)} = \frac{V_i\ln\left(\dfrac{V_f}{V_i}\right)}{V_f - V_i}$$

Substitute the numerical values for the initial and final volumes:

$$\frac{W_1}{W_2} = \frac{(10.0\text{ m}^3)\ln\left(\dfrac{2.00\text{ m}^3}{10.0\text{ m}^3}\right)}{(2.00\text{ m}^3 - 10.0\text{ m}^3)} = \boxed{2.01}$$

....

Finalize As we expected, the work done in process 1 is larger, by about a factor of 2. How do you think the work done in process 1 would compare to that done in process 3, shown in Figure 17.6b?

17.5 | The First Law of Thermodynamics

In Chapter 7, we discussed the conservation of energy equation, Equation 7.2. Let us consider a special case of this general principle in which the only change in the energy of a system is in its internal energy E_{int} and the only transfer mechanisms are heat Q and work W, which we have discussed in this chapter. This case leads to an equation that can be used to analyze many problems in thermodynamics.

This special case of the conservation of energy equation, called the **first law of thermodynamics,** can be written as

$$\Delta E_{int} = Q + W \qquad \text{17.8} \blacktriangleleft \qquad \blacktriangleright \text{First law of thermodynamics}$$

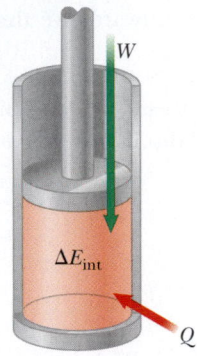

Active Figure 17.8 The first law of thermodynamics equates the change ΔE_{int} in internal energy in a system to the net energy transfer to the system by heat Q and work W. In the situation shown here, the internal energy of the gas increases.

This equation indicates that the change in the internal energy of a system is equal to the sum of the energy transferred across the system boundary by heat and the energy transferred by work.

Active Figure 17.8 shows the energy transfers and change in internal energy for a gas in a cylinder consistent with the first law. Equation 17.8 can be used in a variety of problems in which the only energy considerations are internal energy, heat, and work. We shall consider several examples shortly. Some problems may not fit the conditions of the first law. For example, the internal energy of the coils in your toaster does not increase due to heat or work, but rather due to electrical transmission. Keep in mind that the first law is a special case of the conservation of energy equation, and the latter is the more general equation that covers the widest range of possible situations.

When a system undergoes an infinitesimal change in state, such that a small amount of energy dQ is transferred by heat and a small amount of work dW is done on the system, the internal energy also changes by a small amount dE_{int}. Therefore, for infinitesimal processes we can express the first law as[4]

$$dE_{int} = dQ + dW \qquad\qquad \text{17.9} \blacktriangleleft$$

No practical distinction exists between the results of heat and work on a microscopic scale. Each can produce a change in the internal energy of a system. Although the macroscopic quantities Q and W are *not* properties of a system, they are related to changes of the internal energy of a stationary system through the first law of thermodynamics. Once a process or path is defined, Q and W can be either calculated or measured, and the change in internal energy can be found from the first law.

QUICK QUIZ 17.4 In the last three columns of the following table, fill in the boxes with the correct signs ($-$, $+$, or 0) for Q, W, and ΔE_{int}. For each situation, the system to be considered is identified.

Situation	System	Q	W	ΔE_{int}
(a) Rapidly pumping up a bicycle tire	Air in the pump			
(b) Pan of room-temperature water sitting on a hot stove	Water in the pan			
(c) Air quickly leaking out of a balloon	Air originally in the balloon			

 THINKING PHYSICS 17.2

In the late 1970s, casino gambling was approved in Atlantic City, New Jersey, which can become quite cold in the winter. Energy projections that were performed for the design of the casinos showed that the air-conditioning system would need to operate in the casino even in the middle of a very cold January. Why?

Reasoning If we consider the air in the casino to be the gas to which we apply the first law, imagine a simplification model in which there is no air conditioning and no ventilation so that this gas simply stays in the room. No work is being done on the gas, so we focus on the energy transferred by heat. A casino contains a large number of people, many of whom are active (throwing dice, cheering, etc.) and many of whom are in excited states (celebration, frustration, panic, etc.). As a result, these people have large rates of energy flow by heat from their bodies into the air. This energy results in an increase in internal energy of the air in the casino. With the large number of excited people in a casino (along with the large number of machines and incandescent lights), the temperature of the gas can rise quickly, and to a very high value. To keep the temperature at a comfortable level, energy must be transferred out of the air to compensate for the energy input. Calculations show that energy transfer by heat through the walls even on a 10°F January day is not sufficient to provide the required energy transfer, so the air-conditioning system must be in almost continuous use throughout the year. ◄

---

[4]It should be noted that dQ and dW are not true differential quantities because Q and W are not state variables, although dE_{int} is a true differential. For further details on this point, see R. P. Bauman, *Modern Thermodynamics and Statistical Mechanics* (New York: Macmillan, 1992).

17.6 | Some Applications of the First Law of Thermodynamics

To apply the first law of thermodynamics to specific systems, it is useful to first define some common thermodynamic processes. We shall identify several special processes used as simplification models to approximate real processes. For each of the following processes, we build a mental representation by imagining that the process occurs for the gas in Active Figure 17.8.

During an **adiabatic process,** no energy enters or leaves the system by heat; that is, $Q = 0$. For the piston in Active Figure 17.8, imagine that all surfaces of the piston are perfect insulators so that energy transfer by heat does not exist. (Another way to achieve an adiabatic process is to perform the process very rapidly because energy transfer by heat tends to be relatively slow.) Applying the first law in this case, we see that

$$\Delta E_{int} = W \qquad \textbf{17.10} \blacktriangleleft$$

From this result, we see that when a gas is compressed adiabatically, both W and ΔE_{int} are positive; work is done on the gas, representing a transfer of energy into the system, so the internal energy increases. Conversely, when the gas expands adiabatically, ΔE_{int} is negative.

Adiabatic processes are very important in engineering practice. Common applications include the expansion of hot gases in an internal combustion engine, the liquefaction of gases in a cooling system, and the compression stroke in a diesel engine. We study adiabatic processes in more detail in Section 17.8.

The **free expansion** depicted in Figures 17.7c and 17.7d is a unique adiabatic process in which no work is done on the gas. Because $Q = 0$ and $W = 0$, we see from the first law that $\Delta E_{int} = 0$ for this process. That is, the initial and final internal energies of a gas are equal in a free expansion. As we saw in Chapter 16, the internal energy of an ideal gas depends only on its temperature. Therefore, we expect no change in temperature during an adiabatic free expansion, which is in accord with experiments performed at low pressures. Experiments with real gases at high pressures show a slight increase or decrease in temperature after the expansion because of interactions between molecules.

A process that occurs at constant pressure is called an **isobaric process.** In Active Figure 17.8, as long as the piston is perfectly free to move, the pressure of the gas inside the cylinder is due to atmospheric pressure and the weight of the piston. Hence, the piston can be modeled as a particle in equilibrium. When such a process occurs, the work done on the gas is simply the negative of the constant pressure multiplied by the change in volume, or $-P(V_f - V_i)$. On a PV diagram, an isobaric process appears as a horizontal line, such as the first portion of the process in Active Figure 17.6a or the second portion of the process in Active Figure 17.6b.

A process that takes place at constant volume is called an **isovolumetric process.** In Active Figure 17.8, an isovolumetric process is created by locking the piston in place so that it cannot move. In such a process, the work done is zero because the volume does not change. Hence, the first law applied to an isovolumetric process gives

$$\Delta E_{int} = Q \qquad \textbf{17.11} \blacktriangleleft$$

This equation tells us that if energy is added by heat to a system kept at constant volume, all the energy goes into increasing the internal energy of the system and none enters or leaves the system by work. For example, when an aerosol can is thrown into a fire, energy enters the system (the gas in the can) by heat through the metal walls of the can. Consequently, the temperature and pressure of the gas rise until the can possibly explodes. On a PV diagram, an isovolumetric process appears as a vertical line, such as the second portion of the process in Active Figure 17.6a or the first portion of the process in Active Figure 17.6b.

A process that occurs at constant temperature is called an **isothermal process.** Because the internal energy of an ideal gas is a function of temperature only, in an

Pitfall Prevention | 17.7
The First Law
With our approach to energy in this book, the first law of thermodynamics is a special case of Equation 7.2. Some physicists argue that the first law is the general equation for energy conservation, equivalent to Equation 7.2. In this approach, the first law is applied to a closed system (so that there is no matter transfer), heat is interpreted so as to include electromagnetic radiation, and work is interpreted so as to include electrical transmission ("electrical work") and mechanical waves ("molecular work"). Keep that in mind if you run across the first law in your reading of other physics books.

Pitfall Prevention | 17.8
$Q \neq 0$ in an Isothermal Process
Do not fall into the common trap of thinking there must be no transfer of energy by heat if the temperature does not change as is the case in an isothermal process. Because the cause of temperature change can be either heat or work, the temperature can remain constant even if energy enters the gas by heat, which can only happen if the energy entering the gas by heat leaves by work.

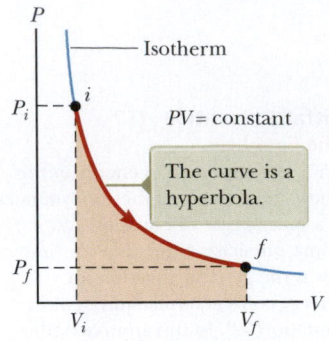

Figure 17.9 The *PV* diagram for an isothermal expansion of an ideal gas from an initial state to a final state.

isothermal process for an ideal gas, $\Delta E_{int} = 0$. Hence, the first law applied to an isothermal process gives

$$Q = -W$$

Whatever energy enters the gas by work leaves the gas by heat in an isothermal process so that the internal energy remains fixed. On a *PV* diagram, an isothermal process appears as a curved line such as that in Figure 17.9. The path of the isothermal process in Figure 17.9 follows along the blue curve, which is an **isotherm,** defined as the curve passing through all points on the *PV* diagram for which the gas has the same temperature. The work done on the ideal gas in an isothermal process was calculated in Example 17.4:

$$W = -nRT \ln\left(\frac{V_f}{V_i}\right) \quad \text{(isothermal process)} \qquad \textbf{17.12} \blacktriangleleft$$

The isothermal process can be analyzed as a model of a nonisolated system in steady state as discussed in Section 7.3. There are transfers of energy across the boundary of the system, but no change occurs in the internal energy of the system. The adiabatic, isobaric, and isovolumetric processes are examples of the nonisolated system model.

Next consider the case in which a nonisolated system is taken through a **cyclic process,** that is, one that originates and ends at the same state. In this case, the change in the internal energy must be zero because internal energy is a state variable and the initial and final states are identical. The energy added by heat to the system must therefore equal the negative of the work done on the system during the cycle. That is, in a cyclic process,

$$\Delta E_{int} = 0 \quad \text{and} \quad Q = -W$$

The net work done per cycle equals the area enclosed by the path representing the process on a *PV* diagram. As we shall see in Chapter 18, cyclic processes are very important in describing the thermodynamics of **heat engines,** thermal devices in which a fraction of the energy added by heat to the system is extracted by mechanical work.

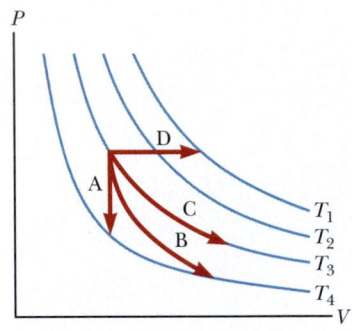

Figure 17.10 (Quick Quiz 17.5) Identify the nature of paths A, B, C, and D.

QUICK QUIZ 17.5 Characterize the paths in Figure 17.10 as isobaric, isovolumetric, isothermal, or adiabatic. For path B, $Q = 0$. The blue curves are isotherms.

Example 17.5 | Cylinder in an Ice-Water Bath

The cylinder in Figure 17.11a has thermally conducting walls and is immersed in an ice-water bath. The gas in the cylinder undergoes three processes: (1) the piston is rapidly pushed downward, compressing the gas in the cylinder; (2) the piston is held at the final position of the previous process while the gas returns to the temperature of the ice-water bath; and (3) the piston is very slowly raised back to the original position.

The work done on the gas during the cycle is 500 J. What mass of ice in the ice-water bath melts during the cycle?

SOLUTION

Conceptualize Imagine grasping the handle of the piston in Figure 17.11a and pushing downward rapidly. You are doing work on the gas, so this action wll cause the temperature of the gas to increase. Then, in process 2, imagine energy flowing from the hot gas to the cooler ice–water mixture by heat as you hold the piston fixed. In process 3, you raise the piston slowly, which would normally cool the gas, but energy flows from the ice–water mixture to the gas to keep its temperature fixed.

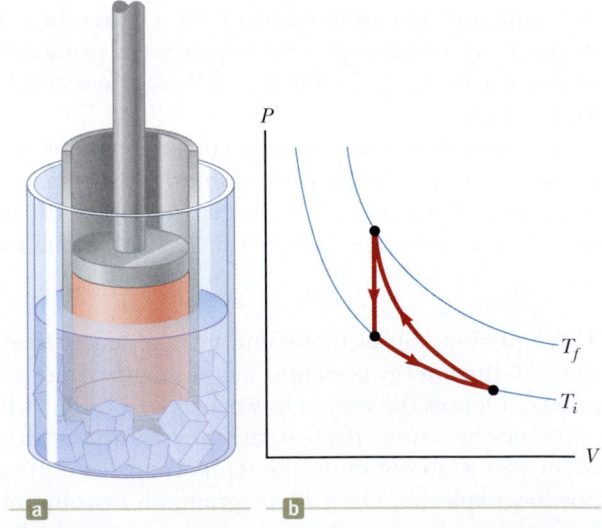

Figure 17.11 (Example 17.5) (a) Cutaway view of a cylinder containing an ideal gas immersed in an ice-water bath. (b) The *PV* diagram for the cycle described.

17.5 *cont.*

Categorize Because process 1 occurs rapidly, it can be modeled as an adiabatic compression. In process 2, the piston is held fixed, so this process is categorized as isovolumetric. In the very slow process 3, the gas and the ice-water bath can be approximated as remaining in thermal equilibrium at all times, so the process is modeled as isothermal. Figure 17.11b shows the *PV* diagram for the entire cycle, a graphical representation that will help us address the problem.

Analyze For the entire cycle, the change in internal energy of the gas is zero. Therefore, from the first law, the energy transfer by heat must equal the negative of the work done on the gas, $Q = -W = -500$ J. This equation indicates that energy leaves the system of the gas by heat during the cycle, entering the ice-water bath (so $Q_{ice} = +500$ J), where it melts some of the ice.

Find the amount of ice that melts using Equation 17.6:

$$Q_{ice} = L_f \Delta m$$

$$\Delta m = \frac{Q_{ice}}{L_f} = \frac{500 \text{ J}}{3.33 \times 10^5 \text{ J/kg}} = 1.5 \times 10^{-3} \text{ kg}$$

$$= \boxed{1.50 \text{ g}}$$

Finalize Based on the interpretation of Δm in Equation 17.6, this is the amount of new water. Of course, that is the same as the amount of ice that melted to form the new water. If we consider the cylinder and the ice-water bath as the system, it is a nonisolated system: the work we do in pushing the piston appears as an increase in the internal energy of the system, represented by the melting of some of the ice. If we consider the gas alone as the system, over a single cycle, it is a nonisolated system in steady state: energy leaves the gas by heat at the same average rate over a cycle as it enters by work. As a result, the internal energy of the gas does not increase over a complete cycle.

Example 17.6 | The Diving Drinking Glass

An empty drinking glass is held upside down just above the surface of water. A scuba diver carefully takes the glass, which remains upside down, to a depth of 10.3 m below the surface so that a sample of air is trapped in the glass. Assume that the temperature of the water remains fixed at 285 K during the descent.

(A) At the depth of 10.3 m, what fraction of the glass's volume is filled with air?

SOLUTION

Conceptualize Imagine the glass held above the water surface just before it enters the water. The pressure of the air in the glass in this situation is atmospheric pressure. As the opening of the glass enters the water, this sample of air is trapped. As the glass moves to a lower position in the water, the pressure of the water will increase. As the water pressure increases, the trapped air is compressed and water enters the open end of the glass.

Categorize We categorize the problem in two ways. First, we will need to use our understanding of the variation of pressure with depth in a fluid from Chapter 15. Second, because the temperature of the water remains fixed, the gas in the glass will also have a fixed temperature, and we categorize the process for the gas as isothermal.

Find the pressure in the water (and of the air in the glass) at the depth of 10.3 m:

$$P = P_{atm} + \rho g h = 1.013 \times 10^5 \text{ Pa} +$$
$$(1\,000 \text{ kg/m}^3)(9.80 \text{ m/s}^2)(10.3 \text{ m})$$
$$= 2.02 \times 10^5 \text{ Pa}$$

Calculate the ratio of volumes of the air in the glass for the final and initial conditions of the isothermal process from the ideal gas law:

$$P_i V_i = P_f V_f \rightarrow \frac{V_f}{V_i} = \frac{P_i}{P_f} = \frac{1.013 \times 10^5 \text{ Pa}}{2.02 \times 10^5 \text{ Pa}} = \boxed{0.500}$$

(B) There are 0.020 0 mol of air trapped in the glass. How much energy crosses the boundary of the system of the air trapped in the glass by heat during the process?

SOLUTION

Analyze Because the process is isothermal, the first law tells us that $\Delta E_{int} = 0$ and the energy flow by heat is equal to the negative of the work done on the gas.

continued

17.6 *cont.*

Use this fact and Equation 17.12 to evaluate the heat:

$$Q = -W = nRT \ln \left(\frac{V_f}{V_i} \right)$$

$$= (0.020\ 0\ \text{mol})(8.314\ \text{J/mol} \cdot \text{K})(285\ \text{K}) \ln (0.500)$$

$$= \boxed{-32.8\ \text{J}}$$

Finalize Notice that because Q is negative, energy comes out of the air by heat. As the air is compressed, the tendency is for its temperature to increase as work is done on it by the surrounding water. As soon as the air temperature rises, the temperature difference between the trapped air and the surrounding water drives a transfer of energy by heat. The transfer of energy out of the air causes its temperature to return to that of the water.

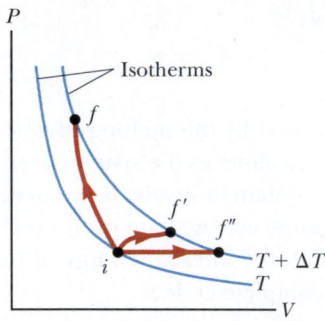

Figure 17.12 An ideal gas is taken from one isotherm at temperature T to another at temperature $T + \Delta T$ along three different paths.

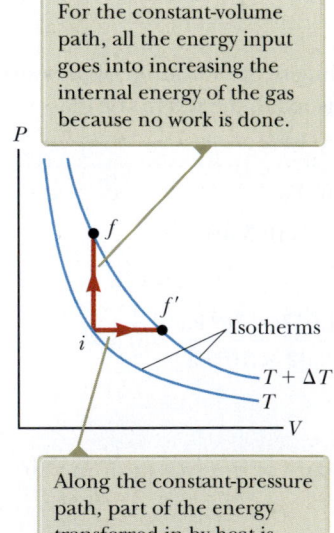

For the constant-volume path, all the energy input goes into increasing the internal energy of the gas because no work is done.

Along the constant-pressure path, part of the energy transferred in by heat is transferred out by work.

Active Figure 17.13 Energy is transferred by heat to an ideal gas in two ways.

17.7 | Molar Specific Heats of Ideal Gases

In Section 17.2, we considered the energy necessary to change the temperature of a mass m of a substance by ΔT. In this section, we focus our attention on ideal gases, and the amount of gas is measured by the number of moles n rather than the mass m. In doing so, some important new connections are found between thermodynamics and mechanics.

The energy transfer by heat required to raise the temperature of n moles of gas from T_i to T_f depends on the path taken between the initial and final states. To understand this concept, consider an ideal gas undergoing several processes such that the change in temperature is $\Delta T = T_f - T_i$ for all processes. The temperature change can be achieved by traveling along a variety of paths from one isotherm to another as in Figure 17.12. Because ΔT is the same for all paths, the change in internal energy ΔE_{int} is the same for all paths. From the first law, $Q = \Delta E_{\text{int}} - W$; we see, however, that the heat Q for each path is different because W (the negative of the area under the curves) is different for each path. Therefore, the heat required to produce a given change in temperature does not have a unique value for a gas.

This difficulty is addressed by defining specific heats for two processes from Section 17.6: isovolumetric and isobaric processes. Modifying Equation 17.3 so that the amount of gas is measured in moles, we define the **molar specific heats** associated with these processes with the following equations:

$$Q = nC_V \Delta T \quad \text{(constant volume)} \qquad \text{17.13} \blacktriangleleft$$

$$Q = nC_P \Delta T \quad \text{(constant pressure)} \qquad \text{17.14} \blacktriangleleft$$

where C_V is the **molar specific heat at constant volume** and C_P is the **molar specific heat at constant pressure.**

In Chapter 16, we found that the temperature of a monatomic gas is a measure of the average translational kinetic energy of the gas molecules. In view of this finding, let us first consider the simplest case of an ideal monatomic gas (i.e., a gas containing one atom per molecule), such as helium, neon, or argon. When energy is added to a monatomic gas in a container of fixed volume (e.g., by heating), all the added energy goes into increasing the translational kinetic energy of the atoms. There is no other way to store the energy in a monatomic gas. The constant-volume process from i to f is described in Active Figure 17.13, where ΔT is the temperature difference between the two isotherms. From Equation 16.18, we see that the total internal energy E_{int} of N molecules (or n mol) of an ideal monatomic gas is

$$E_{\text{int}} = \tfrac{3}{2} nRT \qquad \text{17.15} \blacktriangleleft$$

If energy is transferred by heat to the system at constant volume, the work done on the system is zero. That is, $W = -\int P\,dV = 0$ for a constant-volume process. Hence, from the first law of thermodynamics and Equation 17.15 we find that

$$Q = \Delta E_{\text{int}} = \tfrac{3}{2} nR\,\Delta T \qquad \text{17.16} \blacktriangleleft$$

Substituting the value for Q given by Equation 17.13 into Equation 17.16, we have

$$nC_V \Delta T = \tfrac{3}{2} nR \Delta T$$

$$C_V = \tfrac{3}{2} R = 12.5 \ \text{J/mol} \cdot \text{K} \qquad \text{17.17} \blacktriangleleft$$

This expression predicts a value of $C_V = \tfrac{3}{2} R$ for *all* monatomic gases, regardless of the type of gas. This prediction is based on our structural model of kinetic theory, in which the atoms interact with one another only via short-range forces. The third column of Table 17.3 indicates that this prediction is in excellent agreement with measured values of molar specific heats for monatomic gases. It also indicates that this prediction is not in agreement with values of molar specific heats for diatomic and polyatomic gases. We address these types of gases shortly.

Because no work is done on an ideal gas undergoing an isovolumetric process, the energy transfer by heat is equal to the change in internal energy. Therefore, the change in internal energy can be expressed as

$$\Delta E_{\text{int}} = nC_V \Delta T \qquad \text{17.18} \blacktriangleleft$$

Because internal energy is a state function, the change in internal energy does not depend on the path followed between the initial and final states. Therefore, Equation 17.18 gives the change in internal energy of an ideal gas for *any* process in which the temperature change is ΔT, not just an isovolumetric process. Furthermore, it is true for monatomic, diatomic, and polyatomic gases.

In the case of infinitesimal changes, we can use Equation 17.18 to express the molar specific heat at constant volume as

$$C_V = \frac{1}{n} \frac{dE_{\text{int}}}{dT} \qquad \text{17.19} \blacktriangleleft$$

Now suppose the gas is taken along the constant-pressure path $i \rightarrow f'$ in Active Figure 17.13. Along this path, the temperature again increases by ΔT. The energy transferred to the gas by heat in this process is $Q = nC_P \Delta T$. Because the volume changes in this process, the work done on the gas is $W = -P \Delta V$. Applying the first law to this process gives

$$\Delta E_{\text{int}} = Q + W = nC_P \Delta T - P \Delta V \qquad \text{17.20} \blacktriangleleft$$

The change in internal energy for the process $i \rightarrow f'$ is equal to that for the process $i \rightarrow f$ because E_{int} depends only on temperature for an ideal gas and ΔT is the same for both processes. Because $PV = nRT$, for a constant-pressure process $P \Delta V = nR \Delta T$. Substituting this value for $P \Delta V$ into Equation 17.20 with $\Delta E_{\text{int}} = nC_V \Delta T$ (Eq. 17.18) gives

$$nC_V \Delta T = nC_P \Delta T - nR \Delta T \quad \rightarrow \quad C_P - C_V = R \qquad \text{17.21} \blacktriangleleft$$

▶ Relation between molar specific heats

This expression applies to *any* ideal gas. It shows that the molar specific heat of an ideal gas at constant pressure is greater than the molar specific heat at constant volume by an amount R, the universal gas constant. As shown by the fourth column in Table 17.3, this result is in good agreement with real gases regardless of the number of atoms in the molecule.

Because $C_V = \tfrac{3}{2} R$ for a monatomic ideal gas, Equation 17.21 predicts a value $C_P = \tfrac{5}{2} R = 20.8 \ \text{J/mol} \cdot \text{K}$ for the molar specific heat of a monatomic gas at constant pressure. The second column of Table 17.3 shows the validity of this prediction for monatomic gases.

The ratio of molar specific heats is a dimensionless quantity γ:

$$\gamma = \frac{C_P}{C_V} \qquad \text{17.22} \blacktriangleleft$$

For a monatomic gas, this ratio has the value

$$\gamma = \frac{C_P}{C_V} = \frac{\tfrac{5}{2} R}{\tfrac{3}{2} R} = \frac{5}{3} = 1.67$$

TABLE 17.3 | Molar Specific Heats of Various Gases

Gas	Molar Specific Heat[a] ($\text{J/mol} \cdot \text{K}$)			
	C_P	C_V	$C_P - C_V$	$\gamma = C_P/C_V$
Monatomic Gases				
He	20.8	12.5	8.33	1.67
Ar	20.8	12.5	8.33	1.67
Ne	20.8	12.7	8.12	1.64
Kr	20.8	12.3	8.49	1.69
Diatomic Gases				
H_2	28.8	20.4	8.33	1.41
N_2	29.1	20.8	8.33	1.40
O_2	29.4	21.1	8.33	1.40
CO	29.3	21.0	8.33	1.40
Cl_2	34.7	25.7	8.96	1.35
Polyatomic Gases				
CO_2	37.0	28.5	8.50	1.30
SO_2	40.4	31.4	9.00	1.29
H_2O	35.4	27.0	8.37	1.30
CH_4	35.5	27.1	8.41	1.31

[a]All values except that for water were obtained at 300 K.

The last column in Table 17.3 shows good agreement between this predicted value for γ and experimentally measured values for monatomic gases.

> ◤ **QUICK QUIZ 17.6** (i) How does the internal energy of an ideal gas change as it follows path $i \rightarrow f$ in Active Figure 17.13? (a) E_{int} increases. (b) E_{int} decreases. (c) E_{int} stays the same. (d) There is not enough information to determine how E_{int} changes. (ii) From the same choices, how does the internal energy of a gas change as it follows path $f \rightarrow f'$ along the isotherm labeled $T + \Delta T$ in Active Figure 17.13?

Example 17.7 | Heating a Cylinder of Helium

A cylinder contains 3.00 mol of helium gas at a temperature of 300 K.

(A) If the gas is heated at constant volume, how much energy must be transferred by heat to the gas for its temperature to increase to 500 K?

SOLUTION

Conceptualize Run the process in your mind with the help of the piston–cylinder arrangement in Active Figure 17.8. Imagine that the piston is clamped in position to maintain the constant volume of the gas.

Categorize We evaluate parameters with equations developed in the preceding discussion, so this example is a substitution problem.

Use Equation 17.13 to find the energy transfer: $\qquad Q_1 = nC_V \Delta T$

Substitute the given values: $\qquad Q_1 = (3.00 \text{ mol})(12.5 \text{ J/mol} \cdot \text{K})(500 \text{ K} - 300 \text{ K})$

$$= \boxed{7.50 \times 10^3 \text{ J}}$$

(B) How much energy must be transferred by heat to the gas at constant pressure to raise the temperature to 500 K?

SOLUTION

Use Equation 17.14 to find the energy transfer: $\qquad Q_2 = nC_P \Delta T$

Substitute the given values: $\qquad Q_2 = (3.00 \text{ mol})(20.8 \text{ J/mol} \cdot \text{K})(500 \text{ K} - 300 \text{ K})$

$$= \boxed{12.5 \times 10^3 \text{ J}}$$

This value is larger than Q_1 because of the transfer of energy out of the gas by work to raise the piston in the constant pressure process.

17.8 | Adiabatic Processes for an Ideal Gas

In Section 17.6, we identified several special processes of interest for ideal gases. In three of them, a state variable is held constant: $P =$ constant for an isobaric process, $V =$ constant for an isovolumetric process, and $T =$ constant for an isothermal process. What about the adiabatic process? Is anything constant in this process? As you recall, an adiabatic process is one in which no energy is transferred by heat between a system and its surroundings. In reality, true adiabatic processes on the Earth cannot occur because there is no such thing as a perfect thermal insulator. Some processes, however, are nearly adiabatic. For example, if a gas is compressed (or expanded) very rapidly, very little energy flows out of (or into) the system by heat and so the process is nearly adiabatic.

Suppose an ideal gas undergoes a quasi-static adiabatic expansion. We find that all three variables in the ideal gas law—P, V, and T—change during an adiabatic process. At any time during the process, however, the ideal gas law $PV = nRT$ describes the correct relationship among these variables. Although none of the three variables alone is constant in this process, we find that a *combination* of some of these variables remains constant. This relationship is derived in the following discussion.

Imagine a gas expanding adiabatically in a thermally insulated cylinder so that $Q = 0$. Let us take the infinitesimal change in volume to be dV and the infinitesimal change in temperature to be dT. The work done on the gas is $-P\,dV$. The change in internal energy is given by the differential form of Equation 17.18, $dE_{int} = nC_V\,dT$. Hence, the first law of thermodynamics becomes

$$dE_{int} = dQ + dW \quad \rightarrow \quad nC_V\,dT = 0 - P\,dV \qquad \textbf{17.23} \blacktriangleleft$$

Taking the differential of the equation of state for an ideal gas, $PV = nRT$, gives

$$P\,dV + V\,dP = nR\,dT$$

Eliminating $n\,dT$ from these last two equations, we find that

$$P\,dV + V\,dP = -\frac{R}{C_V}P\,dV$$

From Equation 17.21, we substitute $R = C_P - C_V$ and divide by PV to obtain

$$\frac{dV}{V} + \frac{dP}{P} = -\left(\frac{C_P - C_V}{C_V}\right)\frac{dV}{V} = (1 - \gamma)\frac{dV}{V}$$

$$\frac{dP}{P} + \gamma\frac{dV}{V} = 0$$

Integrating this expression gives

$$\ln P + \gamma \ln V = \text{constant}$$

which we can write as

$$PV^\gamma = \text{constant} \qquad \textbf{17.24} \blacktriangleleft$$

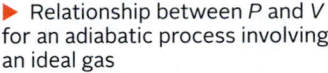

▶ Relationship between P and V for an adiabatic process involving an ideal gas

The PV diagram for an adiabatic expansion is shown in Figure 17.14. Because $\gamma > 1$, the PV curve is steeper than that for an isothermal expansion, in which $PV = \text{constant}$. Equation 17.24 shows that during an adiabatic expansion, ΔE_{int} is negative and so ΔT is also negative. Therefore, the gas cools during an adiabatic expansion. Conversely, the temperature increases if the gas is compressed adiabatically. Equation 17.24 can be expressed in terms of initial and final states as

$$P_iV_i{}^\gamma = P_fV_f{}^\gamma \qquad \textbf{17.25} \blacktriangleleft$$

Using the ideal gas law, Equation 17.24 can also be expressed as

$$TV^{\gamma-1} = \text{constant} \qquad \textbf{17.26} \blacktriangleleft$$

Given the relationship in Equation 17.24, it can be shown that the work done on a gas during an adiabatic process is

$$W = \left(\frac{1}{\gamma - 1}\right)(P_fV_f - P_iV_i) \qquad \textbf{17.27} \blacktriangleleft$$

Problem 17.84 in Enhanced WebAssign invites you to derive this equation.

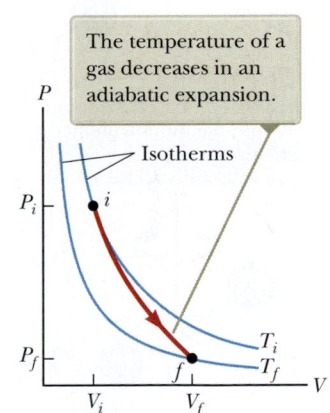

Figure 17.14 The PV diagram for an adiabatic expansion of an ideal gas.

Example 17.8 | A Diesel Engine Cylinder

Air at 20.0°C in the cylinder of a diesel engine is compressed from an initial pressure of 1.00 atm and volume of 800.0 cm³ to a volume of 60.0 cm³. Assume air behaves as an ideal gas with $\gamma = 1.40$ and the compression is adiabatic. Find the final pressure and temperature of the air.

SOLUTION

Conceptualize Imagine what happens if a gas is compressed into a smaller volume. Our discussion above and a reversal of the process in Figure 17.14 tell us that the pressure and temperature both increase.

continued

17.8 *cont.*

Categorize We categorize this example as a problem involving an adiabatic process.

Analyze Use Equation 17.25 to find the final pressure:

$$P_f = P_i\left(\frac{V_i}{V_f}\right)^\gamma = (1.00 \text{ atm})\left(\frac{800.0 \text{ cm}^3}{60.0 \text{ cm}^3}\right)^{1.40}$$

$$= 37.6 \text{ atm}$$

Use the ideal gas law to find the final temperature:

$$\frac{P_i V_i}{T_i} = \frac{P_f V_f}{T_f}$$

$$T_f = \frac{P_f V_f}{P_i V_i}T_i = \frac{(37.6 \text{ atm})(60.0 \text{ cm}^3)}{(1.00 \text{ atm})(800.0 \text{ cm}^3)}(293 \text{ K})$$

$$= 826 \text{ K} = 553°\text{C}$$

Finalize The temperature of the gas increases by a factor of 826 K/293 K = 2.82. The high compression in a diesel engine raises the temperature of the gas enough to cause the combustion of fuel without the use of spark plugs.

17.9 | Molar Specific Heats and the Equipartition of Energy

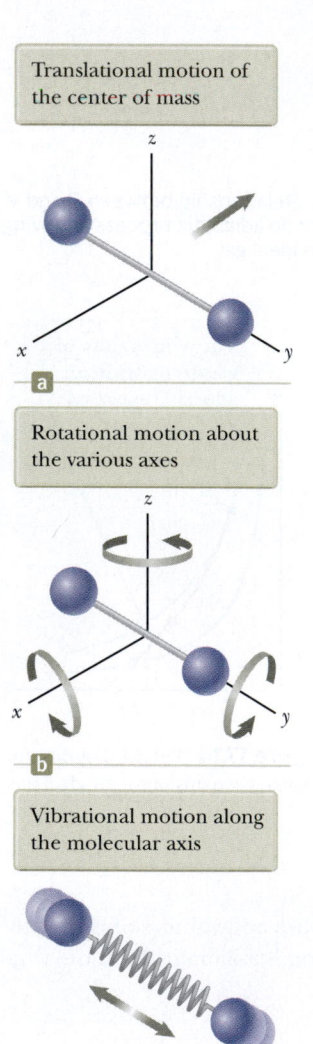

Translational motion of the center of mass

Rotational motion about the various axes

Vibrational motion along the molecular axis

Figure 17.15 Possible motions of a diatomic molecule.

We have found that predictions of molar specific heats based on kinetic theory agree quite well with the behavior of monatomic gases but not with the behavior of complex gases (Table 17.3). To explain the variations in C_V and C_P between monatomic gases and more complex gases, let us explore the origin of specific heat by extending our structural model of kinetic theory in Chapter 16. In Section 16.5, we discussed that the sole contribution to the internal energy of a monatomic gas is the translational kinetic energy of the molecules. We also discussed the theorem of equipartition of energy, which states that, at equilibrium, each degree of freedom contributes, on the average, $\frac{1}{2}k_B T$ of energy per molecule. The monatomic gas has three degrees of freedom, one associated with each of the independent directions of translational motion.

For more complex molecules, other types of motion exist in addition to translation. The internal energy of a diatomic or polyatomic gas includes contributions from the vibrational and rotational motion of the molecules in addition to translation. The rotational and vibrational motions of molecules with structure can be activated by collisions and therefore are "coupled" to the translational motion of the molecules. The branch of physics known as *statistical mechanics* suggests that the average energy for each of these additional degrees of freedom is the same as that for translation, which in turn suggests that the determination of a gas's internal energy is a simple matter of counting the degrees of freedom. We will find that this process works well, although the model must be modified with some notions from quantum physics for us to explain the experimental data completely.

Let us consider a diatomic gas, which we can model as consisting of dumbbell-shaped molecules (Fig. 17.15), and apply concepts that we studied in Chapter 10. In this model, the center of mass of the molecule can translate in the x, y, and z directions (Fig. 17.15a). For this motion, the molecule behaves as a particle, just like an atom in a monatomic gas. In addition, if we consider the molecule as a rigid object, it can rotate about three mutually perpendicular axes (Fig. 17.15b). We can ignore the rotation about the y axis because the moment of inertia and the rotational energy $\frac{1}{2}I\omega^2$ about this axis are negligible compared with those associated with the x and z axes. Therefore, there are five degrees of freedom: three associated with the translational motion and two associated with the rotational motion. Because each degree of freedom contributes, on average, $\frac{1}{2}k_B T$ of energy per molecule, the total internal energy for a diatomic gas consisting of N molecules and considering both translation and rotation is

$$E_{\text{int}} = 3N(\tfrac{1}{2}k_B T) + 2N(\tfrac{1}{2}k_B T) = \tfrac{5}{2}Nk_B T = \tfrac{5}{2}nRT$$

We can use this result and Equation 17.19 to predict the molar specific heat at constant volume:

$$C_V = \frac{1}{n}\frac{dE_{\text{int}}}{dT} = \frac{1}{n}\frac{d}{dT}\left(\tfrac{5}{2}nRT\right) = \tfrac{5}{2}R = 20.8\,\text{J/mol}\cdot\text{K} \qquad \textbf{17.28}\blacktriangleleft$$

From Equations 17.21 and 17.22, we find that the model predicts that

$$C_P = C_V + R = \tfrac{7}{2}R \qquad \textbf{17.29}\blacktriangleleft$$

$$\gamma = \frac{C_P}{C_V} = \frac{\tfrac{7}{2}R}{\tfrac{5}{2}R} = \frac{7}{5} = 1.40 \qquad \textbf{17.30}\blacktriangleleft$$

Let us now incorporate the vibration of the molecule in the model. We use the structural model for the diatomic molecule in which the two atoms are joined by an imaginary spring (Fig. 17.15c) and apply concepts from the particle in simple harmonic motion model in Chapter 12. The vibrational motion has two types of energy associated with the vibrations along the length of the molecule—kinetic energy of the atoms and potential energy in the model spring—which adds two more degrees of freedom for a total of seven for translation, rotation, and vibration. Because each degree of freedom contributes $\tfrac{1}{2}k_B T$ of energy per molecule, the total internal energy for a diatomic gas consisting of N molecules and considering all types of motion is

$$E_{\text{int}} = 3N(\tfrac{1}{2}k_B T) + 2N(\tfrac{1}{2}k_B T) + 2N(\tfrac{1}{2}k_B T) = \tfrac{7}{2}Nk_B T = \tfrac{7}{2}nRT$$

Therefore, the molar specific heat at constant volume is predicted to be

$$C_V = \frac{1}{n}\frac{dE_{\text{int}}}{dT} = \frac{1}{n}\frac{d}{dT}\left(\tfrac{7}{2}nRT\right) = \tfrac{7}{2}R = 29.1\,\text{J/mol}\cdot\text{K} \qquad \textbf{17.31}\blacktriangleleft$$

From Equations 17.21 and 17.22,

$$C_P = C_V + R = \tfrac{9}{2}R \qquad \textbf{17.32}\blacktriangleleft$$

$$\gamma = \frac{C_P}{C_V} = \frac{\tfrac{9}{2}R}{\tfrac{7}{2}R} = \frac{9}{7} = 1.29 \qquad \textbf{17.33}\blacktriangleleft$$

When we compare our predictions with the section of Table 17.3 corresponding to diatomic gases, we find a curious result. For the first four gases—hydrogen, nitrogen, oxygen, and carbon monoxide—the value of C_V is close to that predicted by Equation 17.28, which includes rotation but not vibration. The value for the fifth gas, chlorine, lies between the prediction including rotation and the prediction that includes rotation and vibration. None of the values agrees with Equation 17.31, which is based on the most complete model for motion of the diatomic molecule!

It might seem that our model is a failure for predicting molar specific heats for diatomic gases. We can claim some success for our model, however, if measurements of molar specific heat are made over a wide temperature range rather than at the single temperature that gives us the values in Table 17.3. Figure 17.16 shows the molar specific heat of hydrogen as a function of temperature. The curve has three plateaus and they are at the values of the molar specific heat predicted by Equations 17.17, 17.28, and 17.31! For low temperatures, the diatomic hydrogen gas behaves like a monatomic gas. As the temperature rises to room temperature, its molar specific heat rises to a value for a diatomic gas, consistent with the inclusion of rotation but not vibration. For high temperatures, the molar specific heat is consistent with a model including all types of motion.

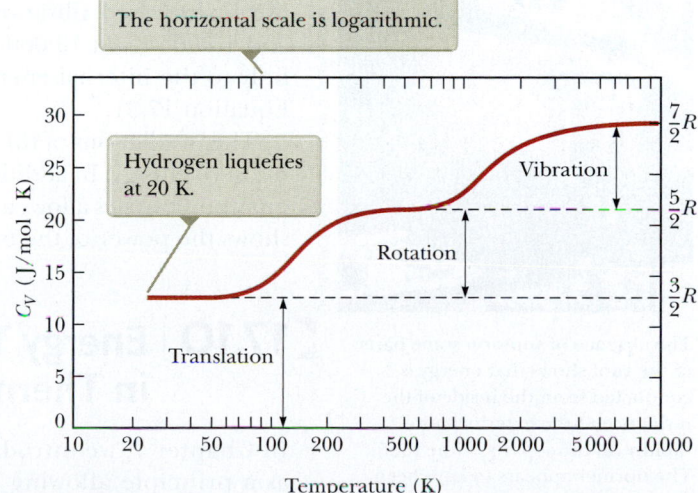

Figure 17.16 The molar specific heat of hydrogen as a function of temperature.

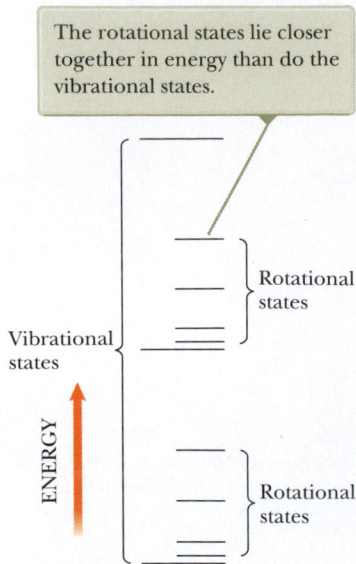

> The rotational states lie closer together in energy than do the vibrational states.

Figure 17.17 An energy level diagram for vibrational and rotational states of a diatomic molecule.

Before addressing the reason for this mysterious behavior, let us make a brief remark about polyatomic gases. For molecules with more than two atoms, the number of degrees of freedom is even larger and the vibrations are more complex than for diatomic molecules. These considerations result in an even higher predicted molar specific heat, which is in qualitative agreement with experiment. For the polyatomic gases shown in Table 17.3, we see that the molar specific heats are higher than those for diatomic gases. The more degrees of freedom available to a molecule, the more "ways" of storing energy are available, resulting in a higher molar specific heat.

A Hint of Energy Quantization

Our model for molar specific heats has been based so far on purely classical notions. It predicts a value of the specific heat for a diatomic gas that, according to Figure 17.16, only agrees with experimental measurements made at high temperatures. To explain why this value is only true at high temperatures and why the plateaus exist in Figure 17.16, we must go beyond classical physics and introduce some quantum physics into the model. In Section 11.5, we discussed energy quantization for the hydrogen atom. Only certain energies are allowed for the system, and an energy level diagram can be drawn to illustrate those allowed energies. For a molecule, quantum physics tells us that the rotational and vibrational energies are quantized. Figure 17.17 shows an energy level diagram for the rotational and vibrational quantum states of a diatomic molecule. Notice that vibrational states are separated by larger energy gaps than rotational states.

At low temperatures, the energy that a molecule gains in collisions with its neighbors is generally not large enough to raise it to the first excited state of either rotation or vibration. All molecules are in the ground state for rotation and vibration. Therefore, at low temperatures, the only contribution to the molecules' average energy is from translation, and the specific heat is that predicted by Equation 17.17.

As the temperature is raised, the average energy of the molecules increases. In some collisions, a molecule may have enough energy transferred to it from another molecule to excite the first rotational state. As the temperature is raised further, more molecules can be excited to this state. The result is that rotation begins to contribute to the internal energy and the molar specific heat rises. At about room temperature in Figure 17.16, the second plateau is reached and rotation contributes fully to the molar specific heat. The molar specific heat is now equal to the value predicted by Equation 17.28.

Vibration contributes nothing at room temperature because the vibrational states are farther apart in energy than the rotational states; the molecules are in the ground vibrational state. The temperature must be even higher to raise the molecules to the first excited vibrational state. That happens in Figure 17.16 between 1 000 K and 10 000 K. At 10 000 K on the right side of the figure, vibration is contributing fully to the internal energy and the molar specific heat has the value predicted by Equation 17.31.

The predictions of this structural model are supportive of the theorem of equipartition of energy. In addition, the inclusion in the model of energy quantization from quantum physics allows a full understanding of Figure 17.16. This excellent example shows the power of the modeling approach.

The absence of snow on some parts of the roof shows that energy is conducted from the inside of the residence to the exterior more rapidly on those parts of the roof. The dormer appears to have been added and insulated. The main roof does not appear to be well insulated.

Dr. Albert A. Bartlett, University of Colorado, Boulder

▌17.10 | Energy Transfer Mechanisms in Thermal Processes

In Chapter 7, we introduced the conservation of energy equation $\Delta E_{\text{system}} = \Sigma T$ as a principle allowing a global approach to energy considerations in physical processes. Earlier in this chapter, we discussed two of the terms on the right-hand side of that equation: work and heat. In this section, we consider more details of heat and two other energy transfer methods that are often related to temperature changes: convection (a form of matter transfer) and electromagnetic radiation.

Conduction

The process of energy transfer by heat can also be called **conduction** or **thermal conduction.** In this process, the transfer mechanism can be viewed on an atomic scale as an exchange of kinetic energy between molecules in which the less energetic molecules gain energy by colliding with the more energetic molecules. For example, if you hold one end of a long metal bar and insert the other end into a flame, the temperature of the metal in your hand soon increases. The energy reaches your hand through conduction. How that happens can be understood by examining what is happening to the atoms in the metal. Initially, before the rod is inserted into the flame, the atoms are vibrating about their equilibrium positions. As the flame provides energy to the rod, those atoms near the flame begin to vibrate with larger and larger amplitudes. These atoms in turn collide with their neighbors and transfer some of their energy in the collisions. Slowly, metal atoms farther and farther from the flame increase their amplitude of vibration until eventually those in the metal near your hand are affected. This increased vibration represents an increase in temperature of the metal (and possibly a burned hand).

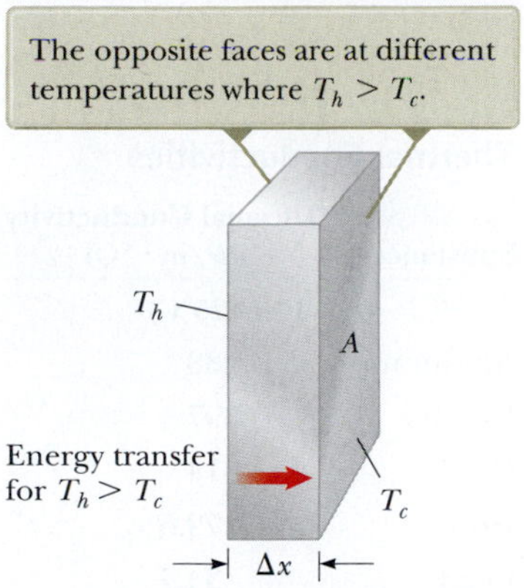

Although the transfer of energy through a material can be partially explained by atomic vibrations, the rate of conduction also depends on the properties of the substance. For example, it is possible to hold a piece of asbestos in a flame indefinitely, which implies that very little energy is being conducted through the asbestos. In general, metals are good thermal conductors because they contain large numbers of electrons that are relatively free to move through the metal and can transport energy from one region to another. Therefore, in a good thermal conductor, such as copper, conduction takes place via the vibration of atoms and via the motion of free electrons. Materials such as asbestos, cork, paper, and fiberglass are poor thermal conductors. Gases also are poor thermal conductors because of the large distance between the molecules.

Conduction occurs only if the temperatures differ between two parts of the conducting medium. This temperature difference drives the flow of energy. Consider a slab of material of thickness Δx and cross-sectional area A with its opposite faces at different temperatures T_c and T_h, where $T_h > T_c$ (Fig. 17.18). The slab allows energy to transfer from the region of high temperature to that of low temperature by thermal conduction. The rate of energy transfer by heat, $P = Q/\Delta t$, is proportional to the cross-sectional area of the slab and the temperature difference and inversely proportional to the thickness of the slab:

$$P = \frac{Q}{\Delta t} \propto A \frac{\Delta T}{\Delta x}$$

Note that P has units of watts when Q is in joules and Δt is in seconds. That is not surprising because P is *power*, the rate of transfer of energy by heat. For a slab of infinitesimal thickness dx and temperature difference dT, we can write the **law of conduction** as

$$P = kA \left| \frac{dT}{dx} \right| \qquad \text{17.34} \blacktriangleleft$$

◀ ▶ Law of conduction

where the proportionality constant k is called the **thermal conductivity** of the material and dT/dx is the **temperature gradient** (the variation of temperature with position). It is the higher thermal conductivity of tile relative to carpet that makes the tile floor feel colder than the carpeted floor in the discussion at the beginning of Chapter 16.

Suppose a substance is in the shape of a long uniform rod of length L as in Figure 17.19 and is insulated so that energy cannot escape by heat from its surface

The opposite faces are at different temperatures where $T_h > T_c$.

Figure 17.18 Energy transfer through a conducting slab with cross-sectional area A and thickness Δx.

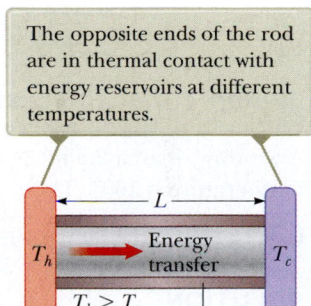

The opposite ends of the rod are in thermal contact with energy reservoirs at different temperatures.

Figure 17.19 Conduction of energy through a uniform, insulated rod of length L.

TABLE 17.4 |
Thermal Conductivities

Substance	Thermal Conductivity (W/m · °C)
Metals (at 25°C)	
Aluminum	238
Copper	397
Gold	314
Iron	79.5
Lead	34.7
Silver	427
Nonmetals (approximate values)	
Asbestos	0.08
Concrete	0.8
Diamond	2 300
Glass	0.8
Ice	2
Rubber	0.2
Water	0.6
Wood	0.08
Gases (at 20°C)	
Air	0.023 4
Helium	0.138
Hydrogen	0.172
Nitrogen	0.023 4
Oxygen	0.023 8

except at the ends, which are in thermal contact with reservoirs having temperatures T_c and T_h. When steady state is reached, the temperature at each point along the rod is constant in time. In this case, the temperature gradient is the same everywhere along the rod and is

$$\left| \frac{dT}{dx} \right| = \frac{T_h - T_c}{L}$$

Therefore, the rate of energy transfer by heat is

$$P = kA \frac{(T_h - T_c)}{L} \qquad \text{17.35} \blacktriangleleft$$

Substances that are good thermal conductors have large thermal conductivity values, whereas good thermal insulators have low thermal conductivity values. Table 17.4 lists thermal conductivities for various substances.

▶ **QUICK QUIZ 17.7** You have two rods of the same length and diameter, but they are formed from different materials. The rods will be used to connect two regions of different temperature such that energy will transfer through the rods by heat. They can be connected in series, as in Figure 17.20a, or in parallel, as in Figure 17.20b. In which case is the rate of energy transfer by heat larger? **(a)** The rate is larger when the rods are in series. **(b)** The rate is larger when the rods are in parallel. **(c)** The rate is the same in both cases.

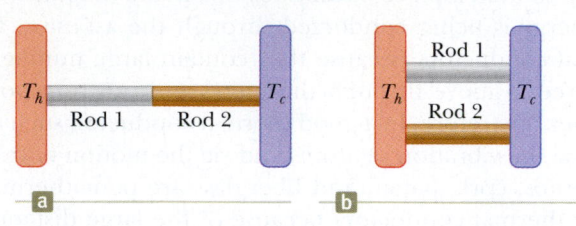

Figure 17.20 (Quick Quiz 17.7) In which case is the rate of energy transfer larger?

▶ **Example 17.9 | The Leaky Window**

A window of area 2.0 m² is glazed with glass of thickness 4.0 mm. The window is in the wall of a house, and the outside temperature is 10°C. The temperature inside the house is 25°C.

(A) How much energy transfers through the window by heat in 1.0 h?

SOLUTION

Conceptualize You have several windows in your home. By placing your hand on the glass of a window on a cold winter day, you may have noticed that the glass is cold compared to the room temperature. The outside surface of the glass is even colder, resulting in a transfer of energy by heat through the glass.

Categorize We categorize the problem as one involving thermal conduction as well as our definition of power from Chapter 7.

....

Analyze Use Equation 17.35 to find the rate of energy transfer by heat:

$$P = kA \frac{(T_h - T_c)}{L}$$

Substitute numerical values, using the value for k for glass from Table 17.4:

$$P = (0.8 \text{ W/m} \cdot °\text{C})(2.0 \text{ m}^2) \frac{(25°\text{C} - 10°\text{C})}{4.0 \times 10^{-3} \text{ m}}$$

$$= 6 \times 10^3 \text{ W}$$

From the definition of power as the rate of energy transfer, find the energy transferred at this rate in 1.0 h:

$$Q = P\Delta t = (6 \times 10^3 \text{ W})(3.6 \times 10^3 \text{ s}) = \boxed{2 \times 10^7 \text{ J}}$$

17.9 cont.

(B) If electrical energy costs 12¢/kWh, how much does the transfer of energy in part (A) cost to replace with electrical heating?

SOLUTION

Cast the answer to part (A) in units of kilowatt-hours:

$$Q = P\,\Delta t = (6 \times 10^3\ \text{W})(1.0\ \text{h}) = 6 \times 10^3\ \text{Wh} = 6\ \text{kWh}$$

Therefore, the cost to replace the energy transferred through the window is $(6\ \text{kWh})(12\text{¢}/\text{kWh}) \approx$ $\boxed{72\text{¢}.}$

Finalize If you imagine paying this much for each hour for each window in your home, your electric bill will be extremely high! For example, for ten such windows, your electric bill would be over $5 000 for one month. It seems like something is wrong here because electric bills are not that high. In reality, a thin layer of air adheres to each of the two surfaces of the window. This air provides additional insulation to that of the glass. As seen in Table 17.4, air is a much poorer thermal conductor than glass, so most of the insulation is performed by the air, not the glass, in a window!

Convection

At one time or another you may have warmed your hands by holding them over an open flame. In this situation, the air directly above the flame is heated and expands. As a result, the density of the air decreases and the air rises. This warmed mass of air transfers energy by heat into your hands as it flows by. The transfer of energy from the flame to your hands is performed by matter transfer because the energy travels with the air. Energy transferred by the movement of a fluid is a process called **convection.** When the movement results from differences in density, as in the example of air around a fire, the process is called **natural convection.** When the fluid is forced to move by a fan or pump, as in some air and water heating systems, the process is called **forced convection.**

The circulating pattern of air flow at a beach (Section 17.2) is an example of convection in nature. The mixing that occurs as water is cooled and eventually freezes at its surface (Section 16.3) is another example.

If it were not for convection currents, it would be very difficult to boil water. As water is heated in a teakettle, the lower layers are warmed first. These regions expand and rise to the top because their density is lower than that of the cooler water. At the same time, the denser cool water falls to the bottom of the kettle so that it can be heated.

The same process occurs near the surface of the Sun. Figure 17.21 shows a close-up view of the solar surface. The granulation that appears is because of *convection cells.* The bright center of a cell is the location at which hot gases rise to the surface, just like the hot water rises to the surface in a pan of boiling water. As the gases cool, they sink back downward along the edges of the cell, forming the darker outline of each cell. The sinking gases appear dark because they are cooler than the gases in the center of the cell. Although the sinking gases emit a tremendous amount of radiation, the filter used to take the photograph in Figure 17.21 makes these areas appear dark relative to the warmer center of the cell.

Convection occurs when a room is heated by a radiator. The radiator warms the air in the lower regions of the room by heat at the interface between the radiator surface and the air. The warm air expands and floats to the ceiling because of its lower density, setting up the continuous air current pattern shown in Figure 17.22.

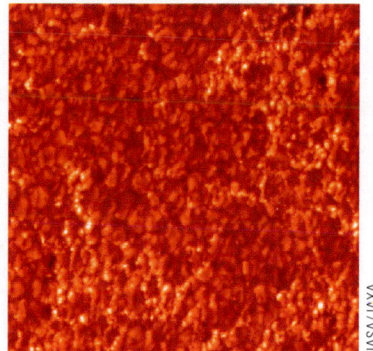

NASA/JAXA

Figure 17.21 The surface of the Sun shows *granulation,* due to the existence of separate convection cells, each carrying energy to the surface by convection.

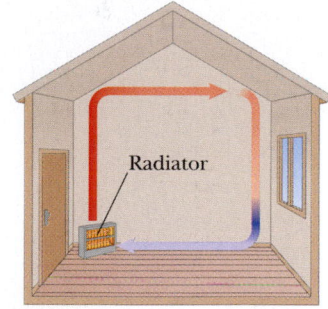

Radiator

Figure 17.22 Convection currents are set up in a room heated by a radiator.

Radiation

Another method of transferring energy that can be related to a temperature change is **electromagnetic radiation.** All objects radiate energy continuously in the form of electromagnetic waves. As we shall find out in Chapter 24, electromagnetic radiation arises from accelerating electric charges. From our discussion of temperature, we know that temperature corresponds to random motion of molecules that are constantly changing direction and therefore are accelerating. Because the molecules contain electric charges, the charges also accelerate. Therefore, any object emits electromagnetic radiation because of the thermal motion of its molecules. This radiation is called **thermal radiation.**

Through electromagnetic radiation, approximately 1 370 J of energy from the Sun strikes each square meter at the top of the Earth's atmosphere every second. Some of this energy is reflected back into space and some is absorbed by the atmosphere, but enough arrives at the surface of the Earth each day to supply all our energy needs on this planet hundreds of times over, if it could be captured and used efficiently. The growth in the number of solar houses in the United States is one example of an attempt to make use of this abundant energy.

The rate at which an object emits energy by thermal radiation from its surface is proportional to the fourth power of its absolute surface temperature. This principle, known as **Stefan's law,** is expressed in equation form as

▶ Stefan's law

$$P = \sigma A e T^4 \qquad \text{17.36} \blacktriangleleft$$

where P is the power radiated by the surface of the object in watts, σ is the **Stefan–Boltzmann constant,** equal to $5.669\,6 \times 10^{-8}\ \text{W/m}^2 \cdot \text{K}^4$, A is the surface area of the object in square meters, e is a constant called the **emissivity,** and T is the surface temperature of the object in kelvins. The value of e can vary between zero and unity, depending on the properties of the surface. The emissivity of a surface is equal to its absorptivity, or the fraction of incoming radiation that the surface absorbs.

At the same time as it radiates, the object also absorbs electromagnetic radiation from the environment. If the latter process did not occur, an object would continuously radiate its energy and its temperature would eventually decrease spontaneously to absolute zero. If an object is at a temperature T and its environment is at a temperature T_0, the net rate of energy change for the object as a result of radiation is

$$P_{\text{net}} = \sigma A e (T^4 - T_0{}^4) \qquad \text{17.37} \blacktriangleleft$$

When an object is in equilibrium with its environment, it radiates and absorbs energy at the same rate and so its temperature remains constant, which is the nonisolated system in steady-state model. When an object is hotter than its surroundings, it radiates more energy than it absorbs and so it cools, which is the nonisolated system model.

BIO Human thermoregulation

The energy transfer methods discussed in this section are important for *thermoregulation* in humans, part of the complex process of *homeostasis*, which refers to the ability of the body to maintain the stability of its inner environment in response to outside influences. Unless the air temperature is very warm, the human body is generally warmer than the air, so energy transfers out of the body through the skin by thermal conduction. Air has a relatively low thermal conductivity, so conduction into the surrounding air is not a very efficient process for cooling the body. Water is a much better thermal conductor, so jumping into a swimming pool that is at the same temperature as the air creates a feeling of "cold water" due to the increased rate of conduction from the skin to the water. Because of its temperature, the body also transfers energy from the skin by electromagnetic radiation. It also receives energy by the same mechanism if exposed to the Sun or other warm environmental source. Energy also leaves the body by convection through the exhalation of warm breath. Convection is involved also in the carrying away of air warmed by conduction from the skin.

BIO The hypothalamus

The part of the brain that regulates body temperature is the *hypothalamus*. This part of the brain also controls other body functions, such as hunger, thirst, and sleep, so it is a very complex region in the brain. The hypothalamus can call on several active mechanisms for regulating the body temperature.

BIO Mechanisms for cooling the body

One important mechanism for maintaining body temperature in warm conditions is the process of *perspiration*. Sweat glands under the skin secrete sweat, which flows onto the surface of the skin. As the sweat evaporates, it cools the skin, similar to the alcohol-soaked cloth mentioned at the end of Section 16.6. During athletic activities, the evaporation of sweat becomes a major factor in cooling the body. Humid weather is uncomfortable because the rate of evaporation into the air is reduced.

Other mechanisms also aid in cooling the body in warm weather. *Arrector pili* muscles under the skin relax so that the hair on the skin lies flat. In this way, the hair does not interfere with air passing close to the skin to carry away warm air and evaporated perspiration. Additional muscles in arterioles relax, causing *vasodilation*, so that blood is redirected into capillaries in the skin. The closeness of the warm blood to the skin surface increases the rate of thermal conduction from the blood, through the skin, and into the cooler surrounding air.

In cold weather, these mechanisms reverse. The hair on the skin stands upright, trapping air on the surface of the skin to act as thermal insulation. The contracted arrector pili muscles are visible on the skin as "goose bumps." *Vasoconstriction* occurs, so that blood is directed away from the skin and closer to the warm center part of the body.

An additional mechanism in very cold weather helps to transform potential energy from previous meals into internal energy in the body. This mechanism involves the degree of tension in skeletal muscles. When necessary, the hypothalamus sends a signal to increase skeletal muscle tone (the constant level of tension in the muscles). The increased metabolic activity in the muscles acts as a source of internal energy within the body because the chemical reactions taking place in the muscle cells are exothermic. If this source of internal energy is not sufficient, the process of *shivering* occurs, where the skeletal muscles undergo rhythmic contractions at a frequency of 10–20 Hertz. The high rate of exothermic chemical reactions in the muscle cells helps to balance the high rate of energy transfer from the skin into the cold air.

BIO Mechanisms for warming the body

> **THINKING PHYSICS 17.3**

If you sit in front of a fire with your eyes closed, you can feel significant warmth in your eyelids. If you now put on a pair of eyeglasses and repeat this activity, your eyelids will not feel nearly so warm. Why?

Reasoning Much of the warmth you feel is because of electromagnetic radiation from the fire. A large fraction of this radiation is in the infrared part of the electromagnetic spectrum. (We will study the electromagnetic spectrum in detail in Chapter 24.) Your eyelids are particularly sensitive to infrared radiation. On the other hand, glass is very opaque to infrared radiation. Therefore, when you put on the glasses, you block much of the radiation from reaching your eyelids and they feel cooler. ◄

> **THINKING PHYSICS 17.4**

If you inspect an incandescent lightbulb that has been operating for a long time, a dark region appears on the inner surface. This region is located on the highest parts of the bulb's glass envelope. What is the origin of this dark region, and why is it located at the high point?

Reasoning The dark region is tungsten that vaporized from the filament of the lightbulb and collected on the inner surface of the glass. Many lightbulbs contain a gas that allows convection to occur within the bulb. The gas near the filament is at a very high temperature, causing it to expand and float upward due to Archimedes's principle. As it floats upward, it carries the vaporized tungsten with it, and the tungsten collects on the surface at the top of the lightbulb. ◄

 # 17.11 | Context Connection: Energy Balance for the Earth

Let us follow up on our discussion of energy transfer by radiation for the Earth from Section 17.10. We will then perform an initial calculation of the temperature of the Earth.

As mentioned previously, energy arrives at the Earth by electromagnetic radiation from the Sun.[5] This energy is absorbed by the surface of the Earth and

---

[5]Some energy arrives at the surface of the Earth from the interior. The source of this energy is radioactive decay (Chapter 30) deep underground. We will ignore this energy because its contribution is much smaller than that due to electromagnetic radiation from the Sun.

is reradiated out into space according to Stefan's law, Equation 17.36. The only type of energy in the system that can change due to radiation is internal energy. Let us assume that any change in temperature of the Earth is so small over a time interval that we can approximate the change in internal energy as zero. This assumption leads to the following reduction of the conservation of energy equation, Equation 7.2:

$$0 = T_{ER}$$

Two energy transfer mechanisms occur by electromagnetic radiation, so we can write this equation as

$$0 = T_{ER} \text{ (in)} + T_{ER} \text{ (out)} \rightarrow T_{ER} \text{ (in)} = -T_{ER} \text{ (out)} \qquad \textbf{17.38} \blacktriangleleft$$

where "in" and "out" refer to energy transfers across the boundary of the system of the Earth. The energy coming into the system is from the Sun, and the energy going out of the system is by thermal radiation emitted from the Earth's surface. Figure 17.23 depicts these energy exchanges. The energy coming in from the Sun comes from only one direction, but the energy radiated out from the Earth's surface leaves in all directions. This distinction will be important in setting up our calculation of the equilibrium temperature.

As mentioned in Section 17.10, the rate of energy transfer per unit area from the Sun is approximately 1 370 W/m² at the top of the atmosphere. The rate of energy transfer per area is called **intensity,** and the intensity of radiation from the Sun at the top of the atmosphere is called the **solar constant** $I_S = 1\ 370$ W/m². A large amount of this energy is in the form of visible radiation, to which the atmosphere is transparent. The radiation emitted from the Earth's surface, however, is not visible. For a radiating object at the temperature of the Earth's surface, the radiation peaks in the infrared, with greatest intensity at a wavelength of about 10 μm. In general, objects with typical household temperatures have wavelength distributions in the infrared, so we do not see them glowing visibly. Only much hotter items emit enough radiation to be seen visibly. An example is a household electric stove burner. When turned off, it emits a small amount of radiation, mostly in the infrared. When turned to its highest setting, its much higher temperature results in significant radiation, with much of it in the visible. As a result, it appears to glow with a reddish color and is described as *red-hot*.

Let us divide Equation 17.38 by the time interval Δt during which the energy transfer occurs, which gives us

$$P_{ER} \text{ (in)} = -P_{ER} \text{ (out)} \qquad \textbf{17.39} \blacktriangleleft$$

We can express the rate of energy transfer into the top of the atmosphere of the Earth in terms of the solar constant I_S:

$$P_{ER} \text{ (in)} = I_S A_c$$

where A_c is the circular cross-sectional area of the Earth. Not all the radiation arriving at the top of the atmosphere reaches the ground. A fraction of it is reflected from clouds and the ground and escapes back into space. For the Earth, this fraction is about 30%, so only 70% of the incident radiation reaches the surface. Using this fact, we modify the input power, assuming that 70.0% reaches the surface:

$$P_{ER} \text{ (in)} = (0.700) I_S A_c$$

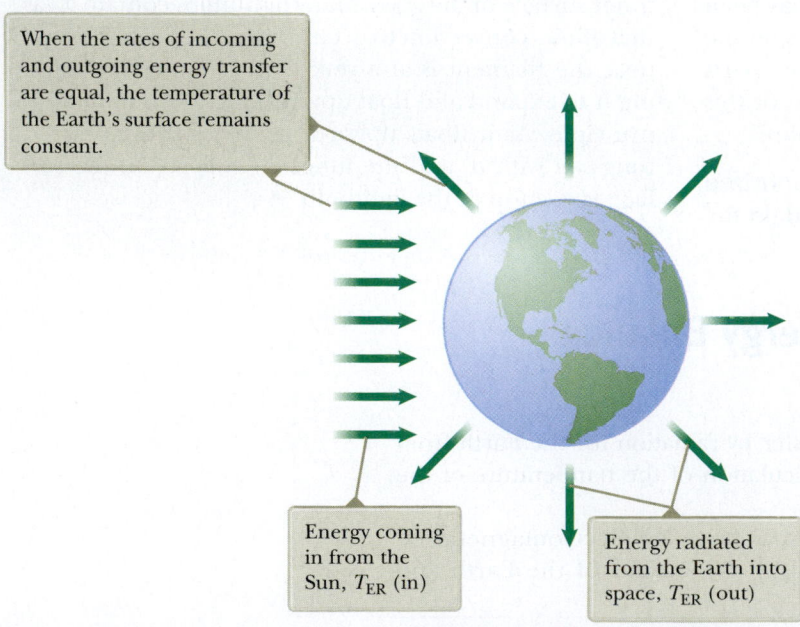

When the rates of incoming and outgoing energy transfer are equal, the temperature of the Earth's surface remains constant.

Energy coming in from the Sun, T_{ER} (in)

Energy radiated from the Earth into space, T_{ER} (out)

Figure 17.23 Energy exchanges by electromagnetic radiation for the Earth. The Sun is far to the left of the diagram and is not visible.

Stefan's law can be used to express the outgoing power, assuming that the Earth is a perfect emitter ($e = 1$):

$$P_{ER} \text{ (out)} = -\sigma A T^4$$

In this expression, A is the surface area of the Earth, T is its surface temperature, and the negative sign indicates that energy is leaving the Earth. Substituting the expressions for the input and output power into Equation 17.39, we have

$$(0.700) I_S A_c = -(-\sigma A T^4)$$

Solving for the temperature of the Earth's surface gives

$$T = \left(\frac{(0.700) I_S A_c}{\sigma A} \right)^{1/4}$$

Substituting the numbers, we find that

$$T = \left(\frac{(0.700)(1\ 370\ \text{W/m}^2)(\pi R_E^2)}{(5.67 \times 10^{-8}\ \text{W/m}^2 \cdot \text{K}^4)(4\pi R_E^2)} \right)^{1/4} = 255\ \text{K} \qquad \blacktriangleleft\ 17.40$$

Measurements show that the average global temperature at the surface of the Earth is 288 K, about 33 K warmer than the temperature from our calculation. This difference indicates that a major factor was left out of our analysis. The major factor is the thermodynamic effects of the atmosphere, which result in additional energy from the Sun being "trapped" in the system of the Earth and raising the temperature. This effect is not included in the simple energy balance calculation we performed. To evaluate it, we must incorporate into our model the principles of thermodynamics of gases for the air in the atmosphere. The details of this incorporation are explored in the Context Conclusion.

❯ SUMMARY

The **internal energy** E_{int} of a system is the total of the kinetic and potential energies of the system associated with its microscopic components. **Heat** is a process by which energy is transferred as a consequence of a temperature difference. It is also the amount of energy Q transferred by this process.

The energy required to change the temperature of a substance by ΔT is

$$Q = mc\,\Delta T \qquad \blacktriangleleft\ 17.3$$

where m is the mass of the substance and c is its **specific heat.**

The energy required to change the phase of a pure substance is

$$Q = L\,\Delta m \qquad \blacktriangleleft\ 17.5$$

where L is **latent heat,** which depends on the substance and the nature of the phase change, and Δm is the change in mass of the higher-phase material.

A **state variable** of a system is a quantity that is defined for a given condition of the system. State variables for a gas include pressure, volume, temperature, and internal energy.

A **quasi-static process** is one that proceeds slowly enough to allow the system to always be in a state of thermal equilibrium.

The **work** done on a gas as its volume changes from some initial value V_i to some final value V_f is

$$W = -\int_{V_i}^{V_f} P\,dV \qquad \blacktriangleleft\ 17.7$$

where P is the pressure, which may vary during the process.

The **first law of thermodynamics** is a special case of the conservation of energy equation, relating the internal energy of a system to energy transfer by heat and work:

$$\Delta E_{int} = Q + W \qquad \blacktriangleleft\ 17.8$$

where Q is the energy transferred across the boundary of the system by heat and W is the work done on the system. Although Q and W both depend on the path taken from the initial state to the final state, internal energy is a state variable, so the quantity ΔE_{int} is independent of the path taken between given initial and final states.

An **adiabatic process** is one in which no energy is transferred by heat between the system and its surroundings ($Q = 0$). In this case, the first law gives $\Delta E_{int} = W$.

An **isobaric process** is one that occurs at constant pressure. The work done on the gas in such a process is $-P(V_f - V_i)$.

An **isovolumetric process** is one that occurs at constant volume. No work is done on the gas in such a process.

An **isothermal process** is one that occurs at constant temperature. The work done on an ideal gas during an isothermal process is

$$W = -nRT \ln\left(\frac{V_f}{V_i} \right) \qquad \blacktriangleleft\ 17.12$$

In a **cyclic process** (one that originates and terminates at the same state), $\Delta E_{int} = 0$, and therefore $Q = -W$.

We define the **molar specific heats** of an ideal gas with the following equations:

$$Q = nC_V \Delta T \quad \text{(constant volume)} \qquad \text{17.13} \blacktriangleleft$$

$$Q = nC_P \Delta T \quad \text{(constant pressure)} \qquad \text{17.14} \blacktriangleleft$$

where C_V is the **molar specific heat at constant volume** and C_P is the **molar specific heat at constant pressure.**

The change in internal energy of an ideal gas for *any* process in which the temperature change is ΔT is

$$\Delta E_{int} = nC_V \Delta T \qquad \text{17.18} \blacktriangleleft$$

The molar specific heat at constant volume is related to internal energy as follows:

$$C_V = \frac{1}{n} \frac{dE_{int}}{dT} \qquad \text{17.19} \blacktriangleleft$$

The molar specific heat at constant volume and molar specific heat at constant pressure for all ideal gases are related as follows:

$$C_P - C_V = R \qquad \text{17.21} \blacktriangleleft$$

For an ideal gas undergoing an adiabatic process, where

$$\gamma = \frac{C_P}{C_V} \qquad \text{17.22} \blacktriangleleft$$

the pressure and volume are related as

$$PV^\gamma = \text{constant} \qquad \text{17.24} \blacktriangleleft$$

The theorem of equipartition of energy can be used to predict the molar specific heat at constant volume for various types of gases. Monatomic gases can only store energy by means of translational motion of the molecules of the gas. Diatomic and polyatomic gases can also store energy by means of rotation and vibration of the molecules. For a given molecule, the rotational and vibrational energies are quantized, so their contribution does not enter into the internal energy until the temperature is raised to a sufficiently high value.

Thermal conduction is the transfer of energy by molecular collisions. It is driven by a temperature difference, and the rate of energy transfer is

$$P = kA \left| \frac{dT}{dx} \right| \qquad \text{17.34} \blacktriangleleft$$

where the constant k is called the **thermal conductivity** of the material and dT/dx is the **temperature gradient** (the variation of temperature with position).

Convection is energy transfer by means of a moving fluid.

All objects emit **electromagnetic radiation** continuously in the form of **thermal radiation,** which depends on temperature according to **Stefan's law:**

$$P = \sigma A e T^4 \qquad \text{17.36} \blacktriangleleft$$

❯ OBJECTIVE QUESTIONS

☐ denotes answer available in *Student Solutions Manual/Study Guide*

1. An amount of energy is added to ice, raising its temperature from $-10°C$ to $-5°C$. A larger amount of energy is added to the same mass of water, raising its temperature from $15°C$ to $20°C$. From these results, what would you conclude? (a) Overcoming the latent heat of fusion of ice requires an input of energy. (b) The latent heat of fusion of ice delivers some energy to the system. (c) The specific heat of ice is less than that of water. (d) The specific heat of ice is greater than that of water. (e) More information is needed to draw any conclusion.

2. A 100-g piece of copper, initially at $95.0°C$, is dropped into 200 g of water contained in a 280-g aluminum can; the water and can are initially at $15.0°C$. What is the final temperature of the system? (Specific heats of copper and aluminum are 0.092 and 0.215 cal/g · °C, respectively.) (a) 16°C (b) 18°C (c) 24°C (d) 26°C (e) none of those answers

3. How long would it take a 1 000-W heater to melt 1.00 kg of ice at $-20.0°C$, assuming all the energy from the heater is absorbed by the ice? (a) 4.18 s (b) 41.8 s (c) 5.55 min (d) 6.25 min (e) 38.4 min

4. The specific heat of substance A is greater than that of substance B. Both A and B are at the same initial temperature

when equal amounts of energy are added to them. Assuming no melting or vaporization occurs, which of the following can be concluded about the final temperature T_A of substance A and the final temperature T_B of substance B? (a) $T_A > T_B$ (b) $T_A < T_B$ (c) $T_A = T_B$ (d) More information is needed.

5. How much energy is required to raise the temperature of 5.00 kg of lead from $20.0°C$ to its melting point of $327°C$? The specific heat of lead is 128 J/kg · °C. (a) 4.04×10^5 J (b) 1.07×10^5 J (c) 8.15×10^4 J (d) 2.13×10^4 J (e) 1.96×10^5 J

6. If a gas undergoes an isobaric process, which of the following statements is true? (a) The temperature of the gas doesn't change. (b) Work is done on or by the gas. (c) No energy is transferred by heat to or from the gas. (d) The volume of the gas remains the same. (e) The pressure of the gas decreases uniformly.

7. Assume you are measuring the specific heat of a sample of originally hot metal by using a calorimeter containing water. Because your calorimeter is not perfectly insulating, energy can transfer by heat between the contents of the calorimeter and the room. To obtain the most accurate result for the specific heat of the metal, you should use water with which initial temperature? (a) slightly lower than room

temperature (b) the same as room temperature (c) slightly higher than room temperature (d) whatever you like because the initial temperature makes no difference

8. Beryllium has roughly one-half the specific heat of water (H_2O). Rank the quantities of energy input required to produce the following changes from the largest to the smallest. In your ranking, note any cases of equality. (a) raising the temperature of 1 kg of H_2O from 20°C to 26°C (b) raising the temperature of 2 kg of H_2O from 20°C to 23°C (c) raising the temperature of 2 kg of H_2O from 1°C to 4°C (d) raising the temperature of 2 kg of beryllium from −1°C to 2°C (e) raising the temperature of 2 kg of H_2O from −1°C to 2°C

9. A poker is a stiff, nonflammable rod used to push burning logs around in a fireplace. For safety and comfort of use, should the poker be made from a material with (a) high specific heat and high thermal conductivity, (b) low specific heat and low thermal conductivity, (c) low specific heat and high thermal conductivity, or (d) high specific heat and low thermal conductivity?

10. A person shakes a sealed insulated bottle containing hot coffee for a few minutes. **(i)** What is the change in the temperature of the coffee? (a) a large decrease (b) a slight decrease (c) no change (d) a slight increase (e) a large increase **(ii)** What is the change in the internal energy of the coffee? Choose from the same possibilities.

11. Star *A* has twice the radius and twice the absolute surface temperature of star *B*. The emissivity of both stars can be assumed to be 1. What is the ratio of the power output of star *A* to that of star *B*? (a) 4 (b) 8 (c) 16 (d) 32 (e) 64

12. If a gas is compressed isothermally, which of the following statements is true? (a) Energy is transferred into the gas by heat. (b) No work is done on the gas. (c) The temperature of the gas increases. (d) The internal energy of the gas remains constant. (e) None of those statements is true.

13. When a gas undergoes an adiabatic expansion, which of the following statements is true? (a) The temperature of the gas does not change. (b) No work is done by the gas. (c) No energy is transferred to the gas by heat. (d) The internal energy of the gas does not change. (e) The pressure increases.

14. Ethyl alcohol has about one-half the specific heat of water. Assume equal amounts of energy are transferred by heat into equal-mass liquid samples of alcohol and water in separate insulated containers. The water rises in temperature by 25°C. How much will the alcohol rise in temperature? (a) It will rise by 12°C. (b) It will rise by 25°C. (c) It will rise by 50°C. (d) It depends on the rate of energy transfer. (e) It will not rise in temperature.

15. An ideal gas is compressed to half its initial volume by means of several possible processes. Which of the following processes results in the most work done on the gas? (a) isothermal (b) adiabatic (c) isobaric (d) The work done is independent of the process.

CONCEPTUAL QUESTIONS

☐ denotes answer available in *Student Solutions Manual/Study Guide*

1. What is wrong with the following statement? "Given any two bodies, the one with the higher temperature contains more heat."

2. In 1801, Humphry Davy rubbed together pieces of ice inside an icehouse. He made sure that nothing in the environment was at a higher temperature than the rubbed pieces. He observed the production of drops of liquid water. Make a table listing this and other experiments or processes to illustrate each of the following situations. (a) A system can absorb energy by heat, increase in internal energy, and increase in temperature. (b) A system can absorb energy by heat and increase in internal energy without an increase in temperature. (c) A system can absorb energy by heat without increasing in temperature or in internal energy. (d) A system can increase in internal energy and in temperature without absorbing energy by heat. (e) A system can increase in internal energy without absorbing energy by heat or increasing in temperature.

3. Pioneers stored fruits and vegetables in underground cellars. In winter, why did the pioneers place an open barrel of water alongside their produce?

4. Why is a person able to remove a piece of dry aluminum foil from a hot oven with bare fingers, whereas a burn results if there is moisture on the foil?

5. Using the first law of thermodynamics, explain why the *total* energy of an isolated system is always constant.

6. Is it possible to convert internal energy to mechanical energy? Explain with examples.

7. It is the morning of a day that will become hot. You just purchased drinks for a picnic and are loading them, with ice, into a chest in the back of your car. (a) You wrap a wool blanket around the chest. Does doing so help to keep the beverages cool, or should you expect the wool blanket to warm them up? Explain your answer. (b) Your younger sister suggests you wrap her up in another wool blanket to keep her cool on the hot day like the ice chest. Explain your response to her.

8. You need to pick up a very hot cooking pot in your kitchen. You have a pair of cotton oven mitts. To pick up the pot most comfortably, should you soak them in cold water or keep them dry?

9. Rub the palm of your hand on a metal surface for about 30 seconds. Place the palm of your other hand on an unrubbed portion of the surface and then on the rubbed portion. The rubbed portion will feel warmer. Now repeat this process on a wood surface. Why does the temperature difference between the rubbed and unrubbed portions of the wood surface seem larger than for the metal surface?

10. When camping in a canyon on a still night, a camper notices that as soon as the sun strikes the surrounding peaks, a breeze begins to stir. What causes the breeze?

11. Suppose you pour hot coffee for your guests, and one of them wants it with cream. He wants the coffee to be as warm as possible several minutes later when he drinks it. To have the warmest coffee, should the person add the cream just after the coffee is poured or just before drinking? Explain.

12. In usually warm climates that experience a hard freeze, fruit growers will spray the fruit trees with water, hoping that a layer of ice will form on the fruit. Why would such a layer be advantageous?

PROBLEMS AVAILABLE IN

List of Enhanced Problems

Problem Number	Targeted Feedback in Enhanced WebAssign	Master It in Enhanced WebAssign	Watch It in Enhanced WebAssign
17.2	✓		✓
17.4	✓	✓	
17.8	✓		✓
17.11	✓	✓	
17.13	✓		✓
17.17	✓	✓	
17.19	✓		✓
17.21	✓	✓	
17.25			✓
17.26	✓		✓
17.28	✓		✓
17.31	✓	✓	
17.35			✓
17.36	✓		
17.37		✓	
17.38	✓		✓
17.39	✓	✓	
17.45	✓		✓
17.46	✓	✓	
17.49	✓		✓
17.51		✓	
17.52	✓		✓
17.57		✓	
17.69		✓	
17.89		✓	

Chapter | 18

Heat Engines, Entropy, and the Second Law of Thermodynamics

Chapter Outline

18.1 Heat Engines and the Second Law
of Thermodynamics

18.2 Reversible and Irreversible Processes

18.3 The Carnot Engine

18.4 Heat Pumps and Refrigerators

18.5 An Alternative Statement of the Second Law

18.6 Entropy

18.7 Entropy and the Second Law of Thermodynamics

18.8 Entropy Changes in Irreversible Processes

18.9 Context Connection: The Atmosphere as a
Heat Engine

SUMMARY

© SSPL/The Image Works

A Stirling engine from the early 19th century. Air is heated in the lower cylinder using an external source. As this happens, the air expands and pushes against a piston, causing it to move. The air is then cooled, allowing the cycle to begin again. This is one example of a heat engine, which we study in this chapter.

The first law of thermodynamics that we studied in Chapter 17 and the more general conservation of energy equation (Eq. 7.2) are statements of the principle of conservation of energy. This principle places no restrictions on the types of energy conversions that can occur. In reality, however, only certain types of energy conversions are observed to take place. Consider the following examples of processes that are consistent with the principle of conservation of energy in either direction but that proceed only in a particular direction in practice.

1. When two objects at different temperatures are placed in thermal contact with each other, energy transfer by heat always occurs from the warmer to the cooler object. We never see energy transfer from the cooler object to the warmer object.

2. A rubber ball dropped to the ground bounces several times and eventually comes to rest, the original gravitational potential energy of the ball–Earth system having been transformed to internal energy in the ball and the

502

ground. A ball lying on the ground, however, never gathers internal energy up from the ground and begins bouncing on its own.

3. If oxygen and nitrogen are maintained in separate halves of a container by a membrane and the membrane is punctured, the oxygen and nitrogen molecules mix together. We never see a mixture of oxygen and nitrogen spontaneously separate into different sides of the container.

These situations all illustrate *irreversible processes;* that is, they occur naturally in only one direction. In this chapter, we investigate a new fundamental principle that allows us to understand why these processes occur in one direction only.[1] The second law of thermodynamics, which is the primary focus of this chapter, establishes which natural processes do and which do not occur.

© Mary Evans Picture Library / Alamy

Lord Kelvin
**British Physicist and Mathematician
(1824–1907)**
Born William Thomson in Belfast, British physicist and mathematician Kelvin was the first to propose the use of an absolute scale of temperature. The Kelvin temperature scale is named in his honor. Kelvin's work in thermodynamics led to the idea that energy cannot pass spontaneously from a colder object to a hotter object.

▌18.1 | Heat Engines and the Second Law of Thermodynamics

One device that is very useful in understanding the second law of thermodynamics is the heat engine. A **heat engine** is a device that takes in energy by heat[2] and, operating in a cycle, expels a fraction of that energy by work. In a typical process for producing electricity in a power plant, for instance, coal or some other fuel is burned and the resulting internal energy is used to convert water to steam. This steam is directed at the blades of a turbine, setting it into rotation. Finally, the mechanical energy associated with this rotation is used to drive an electric generator. In another heat engine, the internal combustion engine in your automobile, energy enters the engine by matter transfer as the fuel is injected into the cylinder and a fraction of this energy is converted to mechanical energy.

In general, a heat engine carries some working substance through cyclic processes[3] during which (1) the working substance absorbs energy by heat from an energy reservoir at a high temperature, (2) work is done by the engine, and (3) energy is expelled by heat to a reservoir at a lower temperature. This output energy is often called wasted energy, exhaust energy, or thermal pollution. As an example, consider the operation of a steam engine in which the working substance is water. The water in the engine is carried through a cycle in which it first evaporates into steam in a boiler and then expands against a piston. After the steam is condensed with cooling water, it is returned to the boiler, and the process is repeated.

It is useful to draw a heat engine schematically as in the pictorial representation in Active Figure 18.1. The engine absorbs a quantity of energy $|Q_h|$ from the hot reservoir. For the mathematical discussion of heat engines, we use absolute value signs to make the values of all energy transfers by heat positive, and the direction of the transfer is indicated with an explicit positive or negative sign. The engine does work W_{eng} (so that *negative* work $W = -W_{eng}$ is done *on* the engine), and then gives up energy $|Q_c|$ to the cold reservoir. Because the working substance goes through a cycle, its initial and final internal energies are equal, so $\Delta E_{int} = 0$. The engine can be modeled as a nonisolated system in steady state. Hence, from the first law,

$$\Delta E_{int} = 0 = Q + W \quad \rightarrow \quad Q_{net} = -W = W_{eng}$$

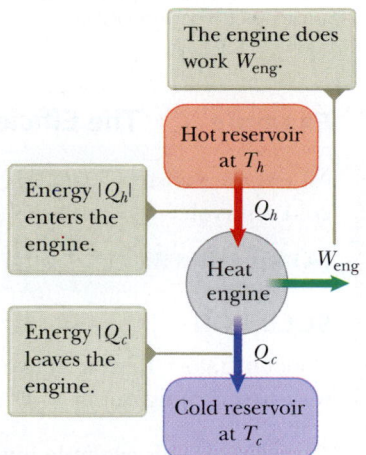

Active Figure 18.1 Schematic representation of a heat engine.

---

[1]As we shall see in this chapter, it is more proper to say that the set of events in the time-reversed sense is highly improbable. From this viewpoint, events in one direction are vastly more probable than those in the opposite direction.

[2]We will use heat as our model for energy transfer into a heat engine. Other methods of energy transfer are also possible in the model of a heat engine, however. For example, as we shall show in Section 18.9, the Earth's atmosphere can be modeled as a heat engine in which the input energy transfer is by means of electromagnetic radiation from the Sun. The output of the atmospheric heat engine causes the wind structure in the atmosphere.

[3]The automobile engine is not strictly a heat engine according to the cyclic process description because the substance (the air–fuel mixture) undergoes only one cycle and is then expelled through the exhaust system.

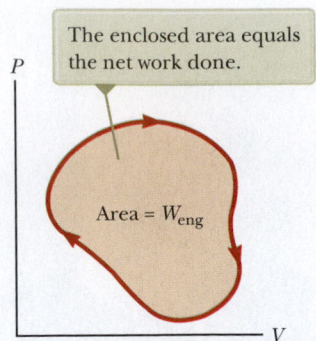

Figure 18.2 The *PV* diagram for an arbitrary cyclic process.

An impossible heat engine

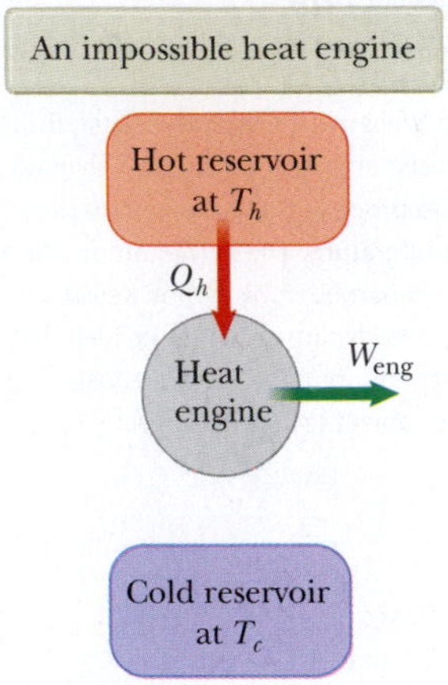

Figure 18.3 Schematic diagram of a heat engine that takes in energy from a hot reservoir and does an equivalent amount of work. It is impossible to construct such a perfect engine.

and we see that the work W_{eng} done by a heat engine equals the net energy absorbed by the engine. As we can see from Active Figure 18.1, $Q_{net} = |Q_h| - |Q_c|$. Therefore,

$$W_{eng} = |Q_h| - |Q_c| \qquad \text{18.1} \blacktriangleleft$$

If the working substance is a gas, the net work done by the engine for a cyclic process is the area enclosed by the curve representing the process on a *PV* diagram. This area is shown for an arbitrary cyclic process in Figure 18.2.

The **thermal efficiency** e of a heat engine is defined as the ratio of the work done by the engine to the energy absorbed at the higher temperature during one cycle:

$$e = \frac{W_{eng}}{|Q_h|} = \frac{|Q_h| - |Q_c|}{|Q_h|} = 1 - \frac{|Q_c|}{|Q_h|} \qquad \text{18.2} \blacktriangleleft$$

We can think of the efficiency as the ratio of what you gain (energy transfer by work) to what you give (energy transfer from the high temperature reservoir). Equation 18.2 shows that a heat engine has 100% efficiency ($e = 1$) only if $Q_c = 0$ (i.e., if no energy is expelled to the cold reservoir). In other words, a heat engine with perfect efficiency would have to expel all the input energy by mechanical work.

The **Kelvin–Planck statement of the second law of thermodynamics** can be stated as follows:

It is impossible to construct a heat engine that, operating in a cycle, produces no effect other than the absorption of energy by heat from a reservoir and the performance of an equal amount of work.

The essence of this form of the second law is that it is theoretically impossible to construct an engine such as that in Figure 18.3 that works with 100% efficiency. All engines must exhaust some energy Q_c to the environment.

▶ **QUICK QUIZ 18.1** The energy input to an engine is 3.00 times greater than the work it performs. **(i)** What is its thermal efficiency? **(a)** 3.00 **(b)** 1.00 **(c)** 0.333 **(d)** impossible to determine **(ii)** What fraction of the energy input is expelled to the cold reservoir? **(a)** 0.333 **(b)** 0.667 **(c)** 1.00 **(d)** impossible to determine

---

Example **18.1** | **The Efficiency of an Engine**

An engine transfers 2.00×10^3 J of energy from a hot reservoir during a cycle and transfers 1.50×10^3 J as exhaust to a cold reservoir.

(A) Find the efficiency of the engine.

SOLUTION

Conceptualize Review Active Figure 18.1; think about energy going into the engine from the hot reservoir and splitting, with part coming out by work and part by heat into the cold reservoir.

Categorize This example involves evaluation of quantities from the equations introduced in this section, so we categorize it as a substitution problem.

Find the efficiency of the engine from Equation 18.2:

$$e = 1 - \frac{|Q_c|}{|Q_h|} = 1 - \frac{1.50 \times 10^3 \text{ J}}{2.00 \times 10^3 \text{ J}} = \boxed{0.250, \text{ or } 25.0\%}$$

(B) How much work does this engine do in one cycle?

SOLUTION

Find the work done by the engine by taking the difference between the input and output energies:

$$W_{eng} = |Q_h| - |Q_c| = 2.00 \times 10^3 \text{ J} - 1.50 \times 10^3 \text{ J}$$
$$= \boxed{5.0 \times 10^2 \text{ J}}$$

18.1 *cont.*

What If? Suppose you were asked for the power output of this engine. Do you have sufficient information to answer this question?

Answer No, you do not have enough information. The power of an engine is the *rate* at which work is done by the engine. You know how much work is done per cycle, but you have no information about the time interval associated with one cycle. If you were told that the engine operates at 2 000 rpm (revolutions per minute), however, you could relate this rate to the period of rotation T of the mechanism of the engine. Assuming there is one thermodynamic cycle per revolution, the power is

$$P = \frac{W_{eng}}{T} = \frac{5.0 \times 10^2 \text{ J}}{\left(\frac{1}{2\,000} \text{ min}\right)}\left(\frac{1 \text{ min}}{60 \text{ s}}\right) = 1.7 \times 10^4 \text{ W}$$

18.2 | Reversible and Irreversible Processes

In the next section, we shall discuss a theoretical heat engine that is the most efficient engine possible. To understand its nature, we must first examine the meaning of reversible and irreversible processes. A **reversible** process is one for which the system can be returned to its initial conditions along the same path and for which every point along the path is an equilibrium state. A process that does not satisfy these requirements is **irreversible.**

Most natural processes are known to be irreversible; the reversible process is an idealization. The three processes described in the introduction to this chapter are irreversible, and we see them proceed in only one direction. The free expansion of a gas discussed in Section 17.6 is irreversible. When the membrane is removed, the gas rushes into the empty half of the vessel and the environment is not changed. No matter how long we watched, we would never see the gas in the full volume spontaneously rush back into only half the volume. The only way we could cause that to happen would be to interact with the gas, perhaps by pushing it inward with a piston, but that method would result in a change in the environment.

If a real process occurs very slowly so that the system is always very nearly in equilibrium, the process can be modeled as reversible. For example, imagine compressing a gas very slowly by dropping some grains of sand onto a frictionless piston as in Figure 18.4. The pressure, volume, and temperature of the gas are well defined during this isothermal compression. Each added grain of sand represents a small change to a new equilibrium state. The process can be reversed by the slow removal of grains of sand from the piston.

18.3 | The Carnot Engine

In 1824, a French engineer named Sadi Carnot described a theoretical engine, now called a **Carnot engine,** that is of great importance from both practical and theoretical viewpoints. He showed that a heat engine operating in an ideal, reversible cycle—called a **Carnot cycle**—between two energy reservoirs is the most efficient engine possible. Such an ideal engine establishes an upper limit on the efficiencies of all real engines. That is, the net work done by a working substance taken through the Carnot cycle is the greatest amount of work possible for a given amount of energy supplied to the substance at the upper temperature.

To describe the Carnot cycle, we shall assume that the working substance in the engine is an ideal gas contained in a cylinder with a movable piston at one end. The cylinder walls and the piston are thermally nonconducting. Four stages of the Carnot cycle are shown in Active Figure 18.5 (page 506); Active Figure 18.6 (page 506) is the *PV* diagram for the cycle, which consists of two adiabatic and two isothermal processes, all reversible:

1. Process $A \rightarrow B$ (Active Fig. 18.5a) is an isothermal expansion at temperature T_h. The gas is placed in thermal contact with an energy reservoir at temperature T_h.

Pitfall Prevention | 18.1
Real Processes Are Irreversible
The reversible process is an idealization. All real processes on Earth are irreversible.

The gas is compressed slowly as individual grains of sand drop onto the piston.

Energy reservoir

Figure 18.4 A method for compressing a gas in a reversible isothermal process.

Pitfall Prevention | 18.2
Don't Shop for a Carnot Engine
The Carnot engine is an idealization; do not expect a Carnot engine to be developed for commercial use. We explore the Carnot engine only for theoretical considerations.

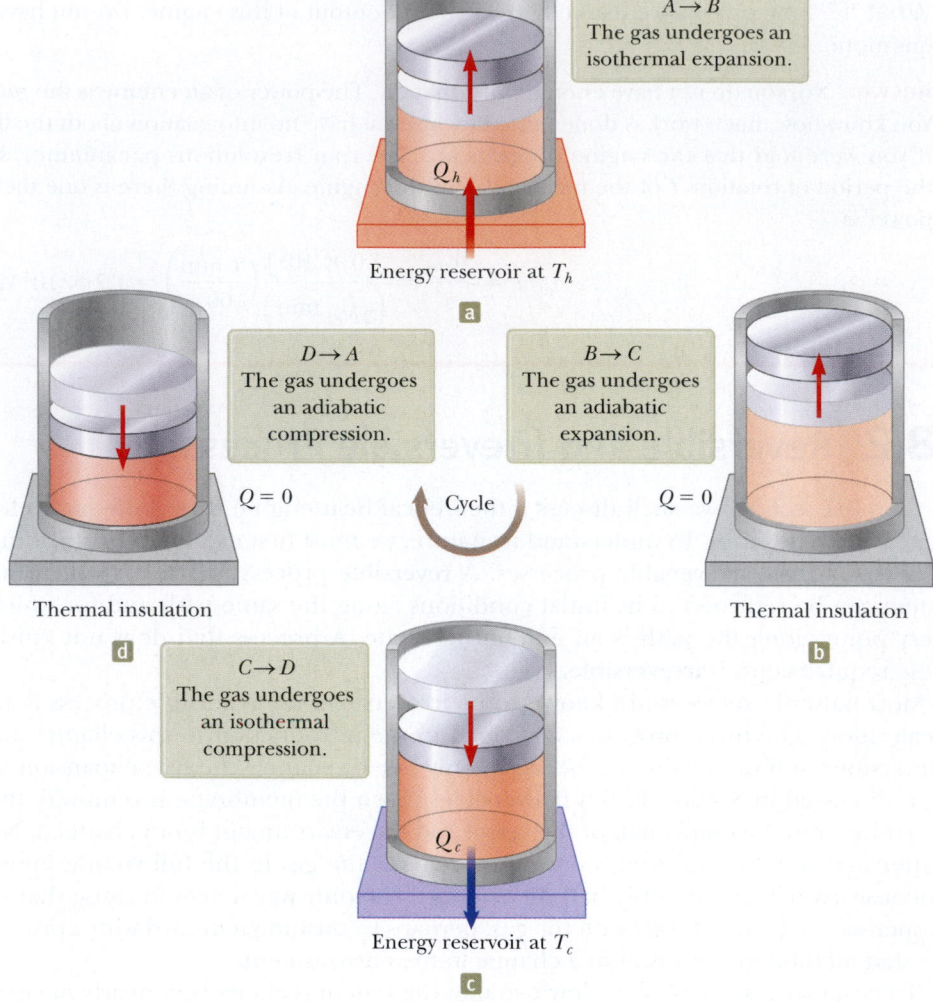

Sadi Carnot
French Engineer (1796–1832)
Carnot was the first to show the quantitative relationship between work and heat. In 1824, he published his only work, *Reflections on the Motive Power of Heat*, which reviewed the industrial, political, and economic importance of the steam engine. In it, he defined work as "weight lifted through a height."

Active Figure 18.5 The Carnot cycle. The letters *A*, *B*, *C*, and *D* refer to the states of the gas shown in Active Figure 18.6. The arrows on the piston indicate the direction of its motion during each process.

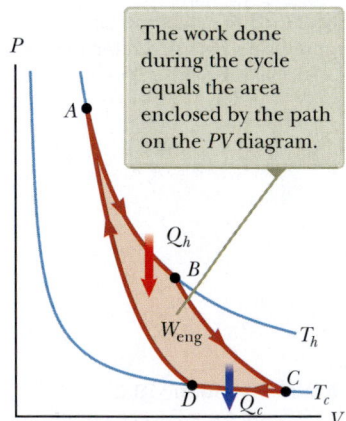

Active Figure 18.6 The *PV* diagram for the Carnot cycle. The net work done W_{eng} equals the net energy transferred into the Carnot engine in one cycle, $|Q_h| - |Q_c|$.

During the expansion, the gas absorbs energy $|Q_h|$ from the reservoir through the base of the cylinder and does work W_{AB} in raising the piston.

2. In process $B \rightarrow C$ (Active Fig. 18.5b), the base of the cylinder is replaced by a thermally nonconducting wall and the gas expands adiabatically; that is, no energy enters or leaves the system by heat. During the expansion, the temperature of the gas decreases from T_h to T_c and the gas does work W_{BC} in raising the piston.

3. In process $C \rightarrow D$ (Active Fig. 18.5c), the gas is placed in thermal contact with an energy reservoir at temperature T_c and is compressed isothermally at temperature T_c. During this time, the gas expels energy $|Q_c|$ to the reservoir and the work done by the piston on the gas is W_{CD}.

4. In the final process $D \rightarrow A$ (Active Fig. 18.5d), the base of the cylinder is replaced by a nonconducting wall and the gas is compressed adiabatically. The temperature of the gas increases to T_h, and the work done by the piston on the gas is W_{DA}.

Carnot showed that for this cycle,

$$\frac{|Q_c|}{|Q_h|} = \frac{T_c}{T_h} \qquad \text{18.3} \blacktriangleleft$$

Therefore, using Equation 18.2, the thermal efficiency of a Carnot engine is

$$e_C = 1 - \frac{T_c}{T_h} \qquad \text{18.4} \blacktriangleleft$$

From this result, we see that all Carnot engines operating between the same two temperatures have the same efficiency.

Equation 18.4 can be applied to any working substance operating in a Carnot cycle between two energy reservoirs. According to this result, the efficiency is zero if $T_c = T_h$, as one would expect. The efficiency increases as T_c is lowered and as T_h is increased. The efficiency can be unity (100%), however, only if $T_c = 0$ K. It is impossible to reach absolute zero,[4] so such reservoirs are not available. Therefore, the maximum efficiency is always less than unity. In most practical cases, the cold reservoir is near room temperature, about 300 K. Therefore, one usually strives to increase the efficiency by raising the temperature of the hot reservoir. All real engines are less efficient than the Carnot engine because they all operate irreversibly so as to complete a cycle in a brief time interval.[5] In addition to this theoretical limitation, real engines are subject to practical difficulties, including friction, that reduce the efficiency further.

QUICK QUIZ 18.2 Three engines operate between reservoirs separated in temperature by 300 K. The reservoir temperatures are as follows: Engine A: $T_h = 1\,000$ K, $T_c = 700$ K; Engine B: $T_h = 800$ K, $T_c = 500$ K; Engine C: $T_h = 600$ K, $T_c = 300$ K. Rank the engines in order of theoretically possible efficiency, from highest to lowest.

Example **18.2** | **The Steam Engine**

A steam engine has a boiler that operates at 500 K. The energy from the burning fuel changes water to steam, and this steam then drives a piston. The cold reservoir's temperature is that of the outside air, approximately 300 K. What is the maximum thermal efficiency of this steam engine?

SOLUTION

Conceptualize In a steam engine, the gas pushing on the piston in Active Figure 18.5 is steam. A real steam engine does not operate in a Carnot cycle, but, to find the maximum possible efficiency, imagine a Carnot steam engine.

Categorize We calculate an efficiency using Equation 18.4, so we categorize this example as a substitution problem.

Substitute the reservoir temperatures into Equation 18.4:
$$e_C = 1 - \frac{T_c}{T_h} = 1 - \frac{300\text{ K}}{500\text{ K}} = \boxed{0.400} \quad \text{or} \quad \boxed{40.0\%}$$

This result is the highest *theoretical* efficiency of the engine. In practice, the efficiency is considerably lower.

What If? Suppose we wished to increase the theoretical efficiency of this engine. This increase can be achieved by raising T_h by ΔT or by decreasing T_c by the same ΔT. Which would be more effective?

Answer A given ΔT would have a larger fractional effect on a smaller temperature, so you would expect a larger change in efficiency if you alter T_c by ΔT. Let's test that numerically. Raising T_h by 50 K, corresponding to $T_h = 550$ K, would give a maximum efficiency of

$$e_C = 1 - \frac{T_c}{T_h} = 1 - \frac{300\text{ K}}{550\text{ K}} = 0.455$$

Decreasing T_c by 50 K, corresponding to $T_c = 250$ K, would give a maximum efficiency of

$$e_C = 1 - \frac{T_c}{T_h} = 1 - \frac{250\text{ K}}{500\text{ K}} = 0.500$$

Although changing T_c is *mathematically* more effective, often changing T_h is *practically* more feasible.

---

[4]The inability to reach absolute zero is known as the *third law of thermodynamics.* It would require an infinite amount of energy to lower the temperature of a substance to absolute zero.

[5]For the processes in the Carnot cycle to be reversible, they must be carried out infinitesimally slowly. Therefore, although the Carnot engine is the most efficient engine possible, it has zero power output because it takes an infinite time interval to complete one cycle! For a real engine, the short time interval for each cycle results in the working substance reaching a high temperature lower than that of the hot reservoir and a low temperature higher than that of the cold reservoir. An engine undergoing a Carnot cycle between this narrower temperature range was analyzed by F. L. Curzon and B. Ahlborn (*Am. J. Phys.,* 43(1):22, 1975), who found that the efficiency at maximum power output depends only on the reservoir temperatures T_c and T_h, and is given by $e_{C-A} = 1 - (T_c/T_h)^{1/2}$. The Curzon–Ahlborn efficiency e_{C-A} provides a closer approximation to the efficiencies of real engines than does the Carnot efficiency.

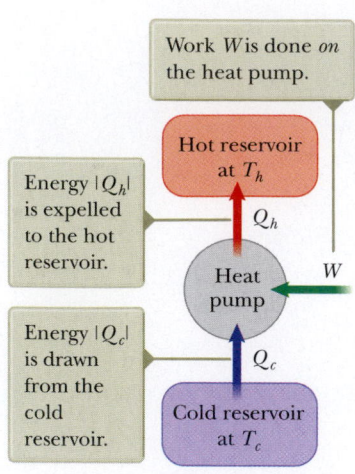

Active Figure 18.7 Schematic representation of a heat pump.

❮18.4 | Heat Pumps and Refrigerators

In a heat engine, the direction of energy transfer is from the hot reservoir to the cold reservoir, which is the natural direction. The role of the heat engine is to process the energy from the hot reservoir so as to do useful work. What if we wanted to transfer energy by heat from the cold reservoir to the hot reservoir? Because this direction is not the natural one, we must transfer some energy into a device to cause it to occur. Devices that perform this task are called **heat pumps** or **refrigerators.**

Active Figure 18.7 is a schematic representation of a heat pump. The cold reservoir temperature is T_c, the hot reservoir temperature is T_h, and the energy absorbed by the heat pump is $|Q_c|$. Energy is transferred into the system, which we model as work[6] W, and the energy transferred out of the pump is $|Q_h|$.

Heat pumps have long been popular for cooling in homes, where they are called *air conditioners,* and are now becoming increasingly popular for heating purposes as well. In the heating mode, a circulating coolant fluid absorbs energy from the outside air (the cold reservoir) and releases energy to the interior of the structure (the hot reservoir). The fluid is usually in the form of a low-pressure vapor when in the coils of the exterior part of the unit, where it absorbs energy from either the air or the ground by heat. This gas is then compressed into a hot, high-pressure vapor and enters the interior part of the unit, where it condenses to a liquid and releases its stored energy. An air conditioner is simply a heat pump installed backward, with "exterior" and "interior" interchanged. The inside of the home is the cold reservoir and the outside air is the hot reservoir.

The effectiveness of a heat pump is described in terms of a number called the **coefficient of performance** COP. In the heating mode, the COP is defined as the ratio of the energy transferred by heat into the hot reservoir to the work required to transfer that energy:

$$\text{COP (heat pump)} \equiv \frac{\text{energy transferred at high temperature}}{\text{work done on heat pump}} \qquad \text{18.5} \blacktriangleleft$$

$$= \frac{|Q_h|}{W}$$

As a practical example, if the outside temperature is −4°C (25°F) or higher, the COP for a typical heat pump is about 4. That is, the energy transferred into the house is about four times greater than the work done by the compressor in the heat pump. As the outside temperature decreases, however, it becomes more difficult for the heat pump to extract sufficient energy from the air, and the COP therefore drops.

A Carnot cycle heat engine operating in reverse constitutes an ideal heat pump, the heat pump with the highest possible COP for the temperatures between which it operates. The maximum coefficient of performance is

$$\text{COP}_C \text{ (heat pump)} = \frac{T_h}{T_h - T_c}$$

Although heat pumps are relatively new products in heating, the refrigerator has been a standard appliance in homes for decades. The refrigerator cools its interior by pumping energy from the food storage compartments into the warmer air outside. During its operation, a refrigerator removes energy $|Q_c|$ from the interior of the refrigerator, and in the process its motor does work W on the coolant fluid. The COP of a refrigerator or of a heat pump used in its cooling cycle is

$$\text{COP (refrigerator)} = \frac{|Q_c|}{W} \qquad \text{18.6} \blacktriangleleft$$

---

[6]The traditional notation is to model the input energy as transferred by work, although most heat pumps operate from electricity, so the more appropriate transfer mechanism *into the device as the system* is *electrical transmission.* If we identify the refrigerant fluid in a heat pump as the system, the energy transfers into the fluid by work done by a piston attached to a compressor operated electrically. In keeping with tradition, we will schematicize the heat pump with input by work regardless of the choice of system.

An efficient refrigerator is one that removes the greatest amount of energy from the cold reservoir with the least amount of work. Therefore, a good refrigerator should have a high coefficient of performance, typically 5 or 6.

The highest possible COP is again that of a refrigerator whose working substance is carried through the Carnot heat engine cycle in reverse:

$$\text{COP}_C \ (\text{refrigerator}) = \frac{T_c}{T_h - T_c}$$

As the difference between the temperatures of the two reservoirs approaches zero, the theoretical coefficient of performance of a Carnot heat pump approaches infinity. In practice, the low temperature of the cooling coils and the high temperature at the compressor limit the COP to values below 10.

> **QUICK QUIZ 18.3** The energy entering an electric heater by electrical transmission can be converted to internal energy with an efficiency of 100%. By what factor does the cost of heating your home change when you replace your electric heating system with an electric heat pump that has a COP of 4.00? Assume that the motor running the heat pump is 100% efficient.   (**a**) 4.00   (**b**) 2.00   (**c**) 0.500   (**d**) 0.250

> **THINKING PHYSICS 18.1**
>
> It is a sweltering summer day and your air-conditioning system is not operating. In your kitchen, you have a working refrigerator and an ice chest full of ice. Which should you open and leave open to cool the room more effectively?
>
> **Reasoning** The high-temperature reservoir for your kitchen refrigerator is the air in the kitchen. If the refrigerator door were left open, energy would be drawn from the air in the kitchen, passed through the refrigeration system and transferred right back into the air. The result would be that the kitchen would become *warmer* because of the addition of the energy coming in by electricity to run the refrigeration system. If the ice chest were opened, energy in the air would enter the ice, raising its temperature and causing it to melt. The transfer of energy from the air would cause its temperature to drop. Therefore, it would be more effective to open the ice chest. ◀

18.5 | An Alternative Statement of the Second Law

Suppose you wish to cool off a hot piece of pizza by placing it on a block of ice. You will certainly be successful because in every similar situation, energy transfer has always taken place from a hot object to a cooler one. Yet nothing in the first law of thermodynamics says that this energy transfer could not proceed in the opposite direction. (Imagine your astonishment if someday you place a piece of hot pizza on ice and the pizza becomes warmer!) It is the second law that determines the directions of such natural phenomena.

An analogy can be made with the impossible sequence of events seen in a movie film running backward, such as a person rising out of a swimming pool and landing back on the diving board, an apple rising from the ground and latching onto the branch of a tree, or a pot of hot water becoming colder as it rests over an open flame. Such events occurring backward in time are impossible because they violate the second law of thermodynamics. Real processes proceed in a preferred direction.

The second law can be stated in several different ways, but all the statements can be shown to be equivalent. Which form you use depends on the application you have in mind. For example, if you were concerned about the energy transfer between pizza and ice, you might choose to concentrate on the **Clausius statement of the second law:**

Energy does not flow spontaneously by heat from a cold object to a hot object.

▶ Second law of thermodynamics; Clausius statement

Figure 18.8 (page 510) shows a heat pump that violates this statement of the second law. Energy is transferring from the cold reservoir to the hot reservoir without an

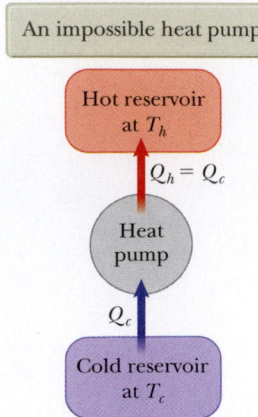

An impossible heat pump

Figure 18.8 Schematic diagram of an impossible heat pump or refrigerator, that is, one that takes in energy from a cold reservoir and expels an equivalent amount of energy to a hot reservoir without the input of energy by work.

Pitfall Prevention | 18.3
Entropy Is Abstract
Entropy is one of the most abstract notions in physics, so follow the discussion in this and the subsequent sections very carefully. Be sure that you do not confuse energy and entropy; even though the names sound similar, they are very different concepts.

input of work. At first glance, this statement of the second law seems to be radically different from that in Section 18.1, but the two are, in fact, equivalent in all respects. Although we shall not prove it here, it can be shown that if either statement of the second law is false, so is the other.

18.6 | Entropy

The zeroth law of thermodynamics involves the concept of temperature and the first law involves the concept of internal energy. Temperature and internal energy are both state variables; that is, they can be used to describe the thermodynamic state of a system. Another state variable, this one related to the second law of thermodynamics, is **entropy** S. In this section, we define entropy on a macroscopic scale as German physicist Rudolf Clausius (1822–1888) first expressed it in 1865.

Equation 18.3, which describes the Carnot engine, can be rewritten as

$$\frac{|Q_c|}{T_c} = \frac{|Q_h|}{T_h}$$

Therefore, the ratio of the energy transfer by heat in a Carnot cycle to the (constant) temperature at which the transfer takes place has the same magnitude for both isothermal processes. To generalize the current discussion beyond heat engines, let us drop the absolute value notation and revive our original sign convention, in which Q_c represents energy leaving the system of the gas and is therefore a negative number. Therefore, we need an explicit negative sign to maintain the equality:

$$-\frac{Q_c}{T_c} = \frac{Q_h}{T_h}$$

We can write this equation as

$$\frac{Q_h}{T_h} + \frac{Q_c}{T_c} = 0 \quad \rightarrow \quad \sum \frac{Q}{T} = 0 \qquad \text{18.7} \blacktriangleleft$$

We have not specified a particular Carnot cycle in generating this equation, so it must be true for all Carnot cycles. Furthermore, by approximating a general reversible cycle with a series of Carnot cycles, we can show that this equation is true for *any* reversible cycle, which suggests that the ratio Q/T may have some special significance. Indeed it does, as we note in the following discussion.

Consider a system undergoing any infinitesimal process between two equilibrium states. If dQ_r is the energy transferred by heat as the system follows a reversible path between the states, the **change in entropy,** regardless of the actual path followed, is equal to this energy transferred by heat along the reversible path divided by the absolute temperature of the system:

▶ Change in entropy for an infinitesimal process

$$dS = \frac{dQ_r}{T} \qquad \text{18.8} \blacktriangleleft$$

The subscript r on the term dQ_r is a reminder that the heat is to be determined along a reversible path, even though the system may actually follow some irreversible path. Therefore, we *must* model a nonreversible process by a reversible process between the same initial and final states to calculate the entropy change. In this case, the model might not be close to the actual process at all, but that is not a concern because entropy is a state variable and the entropy change depends only on the initial and final states. The only requirements are that the model process must be reversible and must connect the given initial and final states.

When energy is absorbed by the system, dQ_r is positive and hence the entropy increases. When energy is expelled by the system, dQ_r is negative and the entropy decreases. Note that Equation 18.8 defines not entropy but rather the *change* in entropy. Hence, the meaningful quantity in a description of a process is the *change* in entropy.

With Equation 18.8, we have a mathematical representation of the change in entropy, but we have developed no mental representation of what entropy means.

In this and the next few sections, we explore various aspects of entropy that will allow us to gain a conceptual understanding of it.

Entropy originally found its place in thermodynamics, but its importance grew tremendously as the field of physics called *statistical mechanics* developed because this method of analysis provided an alternative way of interpreting entropy. In statistical mechanics, a substance's behavior is described in terms of the statistical behavior of its large number of atoms and molecules. Kinetic theory, which we studied in Chapter 16, is an excellent example of the statistical mechanics approach. A main outcome of this treatment is the principle that isolated systems tend toward disorder, and entropy is a measure of that disorder.

To understand this notion, we introduce the distinction between **microstates** and **macrostates** for a system. We can do so by looking at an example far removed from thermodynamics, the throwing of dice at a craps table in a casino. For two dice, a *microstate* is the particular combination of numbers on the upturned faces of the dice; for example, 1–3 and 2–4 are two different microstates (Fig. 18.9). The *macrostate* is the sum of the numbers. Therefore, the macrostates for the two example microstates in Figure 18.9 are 4 and 6. Now, here is the central notion that we will need to understand entropy: The number of microstates associated with a given macrostate is not the same for all macrostates, and the most probable macrostate is that with the largest number of possible microstates. A macrostate of 7 on our pair of dice has six possible microstates: 1–6, 2–5, 3–4, 4–3, 5–2, and 6–1 (Fig. 18.10a). For a macrostate of 2, there is only one possible microstate: 1–1 (Fig. 18.10b). Therefore, a macrostate of 7 has six times as many microstates as a macrostate of 2 and is therefore six times as probable. In fact, the macrostate of 7 is the most probable macrostate for two dice. The game of craps is built around these probabilities of various macrostates.

Consider the low-probability macrostate 2. The *only* way of achieving it is to have a 1 on each die. We say that this macrostate has a high degree of *order;* we *must* have a 1 on each die for this macrostate to exist. Considering the possible microstates for a macrostate of 7, however, we see six possibilities. This macrostate is more *disordered* because several microstates are possible that will result in the same macrostate. Therefore, we conclude that high-probability macrostates are disordered macrostates and low-probability macrostates are ordered macrostates.

As a more physical example, consider the molecules in the air in your room. Let us compare two possible macrostates. Macrostate 1 is the condition in which the oxygen and nitrogen molecules are mixed evenly throughout the room. Macrostate 2 is that in which the oxygen molecules are in the front half of the room and the nitrogen molecules are in the back half. From our everyday experience, it is *extremely unlikely* for macrostate 2 to exist. On the other hand, macrostate 1 is what we would normally expect to see. Let us relate this experience to the microstates, which correspond to the possible positions of molecules of each type. For macrostate 2 to exist, every molecule of oxygen would have to be in one half of the room and every molecule of nitrogen in the other half, which is a highly ordered and unlikely situation. The probability of this occurrence is infinitesimal. For macrostate 1 to exist, both types of molecules are simply distributed evenly around the room, which is a much lower level of order and a highly probable situation. Therefore, the mixed state is much more likely than the separated state, and that is what we normally see.

Let us look now at the notion that isolated systems tend toward disorder. The cause of this tendency toward disorder is easily seen. Let us assume that all microstates for the system are equally probable. When the possible macrostates associated with the microstates are examined, however, far more of them are disordered macrostates with many microstates than ordered macrostates with few microstates. Because each of the microstates is equally probable, it is highly probable that the actual macrostate will be one of the highly disordered macrostates simply because there are more microstates.

In physical systems, we are not talking about microstates of two entities like our pair of dice; we are talking about a number on the order of Avogadro's number of molecules. If you imagine throwing Avogadro's number of dice, the game of craps would be meaningless. You could make an almost perfect prediction of the result when the numbers on the faces are all added up (if the numbers on the face of the

Figure 18.9 Two different microstates for a throw of two dice. These correspond to two macrostates, having values of (a) 4 and (b) 6.

Figure 18.10 Possible two-dice microstates for a macrostate of (a) 7 and (b) 2. The macrostate of 7 is more probable because there are more ways of achieving it; more microstates are associated with a 7 than with a 2.

dice were added once per second, more than 19 thousand trillion years would be needed to tabulate the results for only one throw!) because you are dealing with the statistics of a huge number of dice. We face these kinds of statistics with Avogadro's number of molecules. The macrostate can be well predicted. Even if a system starts off in a very low probability state (e.g., the nitrogen and oxygen molecules separated in a room by a membrane that is then punctured), it quickly develops into a high-probability state (the molecules rapidly mix evenly throughout the room).

We can now present this as a general principle for physical processes: All physical processes tend toward more probable macrostates for the system and its surroundings. The more probable macrostate is always one of higher disorder.

> **QUICK QUIZ 18.4** (a) Suppose you select four cards at random from a standard deck of playing cards and end up with a macrostate of four deuces. How many microstates are associated with this macrostate? (b) Suppose you pick up two cards and end up with a macrostate of two aces. How many microstates are associated with this macrostate?

Now, what does all this talk about dice and states have to do with entropy? To answer this question, we can show that entropy is a measure of the disorder of a state. Then, we shall use these ideas to generate a new statement of the second law of thermodynamics.

As we have seen, entropy can be defined using the macroscopic concepts of heat and temperature. Entropy can also be treated from a microscopic viewpoint through statistical analysis of molecular motions. We can make a connection between entropy and the number of microstates associated with a given macrostate with the following expression:[7]

► Entropy (microscopic definition)

$$S \equiv k_B \ln W \qquad \qquad \textbf{18.9} \blacktriangleleft$$

where W is the number of microstates associated with the macrostate whose entropy is S.

Because the more probable macrostates are the ones with larger numbers of microstates and the larger numbers of microstates are associated with more disorder, Equation 18.9 tells us that entropy is a measure of microscopic disorder.

> **THINKING PHYSICS 18.2**

Suppose you have a bag of 100 marbles of which 50 are red and 50 are green. You are allowed to draw four marbles from the bag according to the following rules. Draw one marble, record its color, and return it to the bag. Shake the bag and then draw another marble. Continue this process until you have drawn and returned four marbles. What are the possible macrostates for this set of events? What is the most likely macrostate? What is the least likely macrostate?

Reasoning Because each marble is returned to the bag before the next one is drawn and the bag is then shaken, the probability of drawing a red marble is always the same as the probability of drawing a green one. All the possible microstates and macrostates are shown in Table 18.1. As this table indicates, there is only one way to draw a macrostate of four red marbles, so there is only one microstate for that macrostate. There are, however, four possible microstates that correspond to the macrostate of one green marble and three red marbles, six microstates that correspond to two green marbles and two

red marbles, four microstates that correspond to three green marbles and one red marble, and one microstate that corresponds to four green marbles. The most likely, and most disordered, macrostate—two red marbles and two green marbles—corresponds to the largest number of microstates. The least likely, most ordered macrostates—four red marbles or four green marbles—correspond to the smallest number of microstates. ◄

TABLE 18.1 | Possible Results of Drawing Four Marbles from a Bag

Macrostate	Possible Microstates	Total Number of Microstates
All R	RRRR	1
1G, 3R	RRRG, RRGR, RGRR, GRRR	4
2G, 2R	RRGG, RGRG, GRRG, RGGR, GRGR, GGRR	6
3G, 1R	GGGR, GGRG, GRGG, RGGG	4
All G	GGGG	1

---

[7]For a derivation of this expression, see Chapter 22 of R. A. Serway and J. W. Jewett Jr., *Physics for Scientists and Engineers*, 8th ed. (Belmont, CA: Brooks-Cole, 2010).

18.7 | Entropy and the Second Law of Thermodynamics

Because entropy is a measure of disorder and physical systems tend toward disordered macrostates, we can state the second law in another way, the **entropy statement of the second law of thermodynamics:**

The entropy of the Universe increases in all natural processes.

▶ Second law of thermodynamics; entropy statement

To calculate the change in entropy for a finite process, we must recognize that T is generally not constant. If dQ_r is the energy transferred reversibly by heat when the system is at temperature T, the change in entropy in an arbitrary reversible process between an initial state and a final state is

$$\Delta S = \int_i^f dS = \int_i^f \frac{dQ_r}{T} \qquad \text{(reversible path)} \qquad \textbf{18.10} \blacktriangleleft$$

▶ Change in entropy for a finite process

The change in entropy of a system depends only on the properties of the initial and final equilibrium states because entropy is a state variable, like internal energy, which is consistent with the relationship of entropy to disorder. For a given macrostate of a system, a given amount of disorder exists, measured by W (Eq. 18.9), the number of microstates corresponding to the macrostate. This number does not depend on the path followed as a system goes from one state to another.

In the case of a reversible, adiabatic process, no energy is transferred by heat between the system and its surroundings, and therefore $\Delta S = 0$. Because no change in entropy occurs, such a process is often referred to as an **isentropic process.**

Consider the changes in entropy that occur in a Carnot heat engine operating between the temperatures T_c and T_h. Equation 18.7 tells us that for a Carnot cycle,

$$\Delta S = 0$$

Now consider a system taken through an arbitrary reversible cycle. Because entropy is a state variable and hence depends only on the properties of a given equilibrium state, we conclude that $\Delta S = 0$ for *any* reversible cycle. In general, we can write this condition in the mathematical form

$$\oint \frac{dQ_r}{T} = 0 \qquad \textbf{18.11} \blacktriangleleft$$

where the symbol $\oint$ indicates that the integration is over a *closed* path.

QUICK QUIZ 18.5 Which of the following is true for the entropy change of a system that undergoes a reversible, adiabatic process? (a) $\Delta S < 0$ (b) $\Delta S = 0$ (c) $\Delta S > 0$

QUICK QUIZ 18.6 An ideal gas is taken from an initial temperature T_i to a higher final temperature T_f along two different reversible paths. Path A is at constant pressure, and path B is at constant volume. What is the relation between the entropy changes of the gas for these paths? (a) $\Delta S_A > \Delta S_B$ (b) $\Delta S_A = \Delta S_B$ (c) $\Delta S_A < \Delta S_B$

One question that often arises is that of the relationship between the second law of thermodynamics and human evolution. The human body is a highly organized system that arose from simple organisms through the evolutionary process. Some individuals claim that the increase in order associated with the evolution of humans on the Earth is inconsistent with the second law.

One point to be made against this claim is that local increases in order are not precluded by the second law, as long as the entire system obeys the second law. We must keep track of *all* energy within a system in order to make a statement about order for the system as a whole. For example, ordered hexagonal snowflakes form spontaneously from water molecules moving randomly in the air. This is a local increase in order, but it does not represent an increase for the

BIO The second law and evolution

Universe. When the water froze to become the snowflake, energy was released from the freezing water into the air. Where did this energy go? The fact that this energy will spread out, as internal energy tends to do, represents an increase in disorder. It is impossible to track this energy exactly, but it and the energy from many other snowflakes will create disorder somewhere that will counteract the order of the snowflake.

What the arguments about evolution also miss is that the Earth is not an isolated system, so, by itself, its entropy doesn't always increase. The Earth is a nonisolated system, so we must consider the Earth *and* its environment. Because a huge amount of energy is continuously arriving at the Earth from the Sun, spontaneous decreases in entropy on the Earth happen often.

Whenever energy enters a system, increases in order are possible. Imagine a set of children's blocks sitting at random locations on the floor. If the blocks are an isolated system, they will never form themselves into an organized stack. But now relax the requirement that the system be isolated. Allow a person to reach in across the boundary of the system, pick up the blocks, and stack them. Energy has entered the system by work done on the blocks by the person, and the system is now more ordered than it was before.

The process of evolution is a version of the snowflake-formation and block-stacking events on a higher and more complicated scale. Because of the huge amount of energy entering the Earth from the Sun, a local increase in order (for example, human evolution) can occur without violating the second law of thermodynamics. The increase in disorder represented by the fusion processes in the Sun and the huge outpouring of energy into space increase the entropy of the Universe at a far higher rate than evolution could possibly decrease it.

The second law predicts that something small and hot (the Big Bang) will become large and cold (the present Universe). When we consider the evolution of ordered life on a tiny planet in one galaxy in an expanding Universe containing billions of galaxies, the second law of thermodynamics is in no danger of being violated.

> **THINKING PHYSICS 18.3**

A box contains five gas molecules, initially spread throughout the box. At some later time, all five are in one half of the box, which is a highly ordered situation. Does this situation violate the second law of thermodynamics? Is the second law valid for this system?

Reasoning Strictly speaking, this situation does violate the second law of thermodynamics. In response to the second question, however, the second law is not valid for small numbers of particles. The second law is based on collections of huge numbers of particles for which disordered states have astronomically higher probabilities than ordered states. Because the macroscopic world is built from these huge numbers of particles, the second law is valid as real processes proceed from order to disorder. In the five-molecule system, the general idea of the second law is valid in that there are more disordered states than ordered ones, but the relatively high probability of the ordered states results in their existence from time to time. ◄

Example 18.3 | Change in Entropy: Melting

A solid that has a latent heat of fusion L_f melts at a temperature T_m. Calculate the change in entropy of this substance when a mass m of the substance melts.

SOLUTION

Conceptualize Imagine placing the substance in a warm environment so that energy enters the substance by heat. The process can be reversed by placing the substance in a cool environment so that energy leaves the substance by heat. The mass m of the substance that melts is equal to Δm, the change in mass of the higher-phase (liquid) substance.

Categorize Because the melting takes place at a fixed temperature, we categorize the process as isothermal.

18.3 *cont.*

Analyze Use Equation 17.6 in Equation 18.10, noting that the temperature remains fixed:

$$\Delta S = \int \frac{dQ_r}{T} = \frac{1}{T_m} \int dQ_r = \frac{Q_r}{T_m} = \frac{L_f \Delta m}{T_m} = \boxed{\frac{L_f m}{T_m}}$$

Finalize Notice that Δm is positive so that ΔS is positive, representing that energy is added to the ice cube.

What If? Suppose you did not have Equation 18.10 available to calculate an entropy change. How could you argue from the statistical description of entropy that the change in entropy should be positive?

Answer When a solid melts, its entropy increases because the molecules are much more disordered in the liquid state than they are in the solid state. The positive value for ΔS also means that the substance in its liquid state does not spontaneously transfer energy from itself to the warm surroundings and freeze because to do so would involve a spontaneous increase in order and a decrease in entropy.

18.8 | Entropy Changes in Irreversible Processes

So far, we have calculated changes in entropy using information about a reversible path connecting the initial and final equilibrium states. We can calculate entropy changes for irreversible processes by devising a reversible process (or a series of reversible processes) between the same two equilibrium states and computing $\int dQ_r/T$ for the reversible process. In irreversible processes, it is critically important to distinguish between Q, the actual energy transfer in the process, and Q_r, the energy that would have been transferred by heat along a reversible path between the same states. Only the second value gives the correct entropy change. For example, as we shall see, if an ideal gas expands adiabatically into a vacuum, $Q = 0$, but $\Delta S \neq 0$ because $Q_r \neq 0$. The reversible path between the same two states is the reversible, isothermal expansion that gives $\Delta S > 0$.

In the statement of the second law of thermodynamics in the previous section, we described the increase in entropy for the entire Universe. We can also investigate the second law for portions of the Universe, that is, for systems. Let us first consider isolated systems. We find that the total entropy of an isolated system that undergoes a change cannot decrease. If the process occurring in the system is irreversible, as most real processes are, the entropy of the system increases. On the other hand, in a reversible adiabatic process, the total entropy of an isolated system remains constant.

When dealing with interacting objects that are not isolated from the environment, we must consider the change of entropy for the system *and* its environment. When two objects interact in an irreversible process, the increase in entropy of one part of the Universe is greater than the decrease in entropy of the other part. Hence, we conclude that the change in entropy of the Universe must be greater than zero for an irreversible process and equal to zero for a reversible adiabatic process. Ultimately, the entropy of the Universe should reach a maximum value. At this point, the Universe will be in a state of uniform temperature and density. All physical, chemical, and biological processes will cease because a state of perfect disorder implies that no energy is available for doing work. This gloomy state of affairs is sometimes referred to as the "heat death" of the Universe.

QUICK QUIZ 18.7 True or False: The entropy change in an adiabatic process must be zero because $Q = 0$.

THINKING PHYSICS 18.4

According to the entropy statement of the second law, the entropy of the Universe increases in irreversible processes. This statement sounds very different from the Kelvin–Planck and Clausius forms of the second law. Can these two statements be made consistent with the entropy interpretation of the second law?

Reasoning These three forms are consistent. In the Kelvin–Planck statement, the energy in the reservoir is disordered internal energy, the random motion of molecules. Performing work results in ordered energy, such as pushing a piston through a displacement. In this case, the motion of all molecules of the piston is in the same direction. If a heat engine absorbed energy by heat and performed an equal amount of work, it would have converted disorder into order, in violation of the entropy statement. In the Clausius statement, we start with an ordered system: higher temperature in the hot object, lower in the cold object. This separation of temperatures is an example of order. Energy transferring spontaneously from the cold object to the hot object, so that the temperatures spread even farther apart, would be an increase in order, in violation of the entropy statement. ◄

Entropy Changes in a Free Expansion

An ideal gas in an insulated container initially occupies a volume of V_i (Fig. 18.11). A membrane separating the gas from an evacuated region is broken so that the gas expands (irreversibly) to a volume V_f. Let us find the change in entropy of the gas and the Universe.

The process is neither reversible nor quasi-static. The work done on the gas is zero, and because the walls are insulating, no energy is transferred by heat during the expansion. That is, $W = 0$ and $Q = 0$. The first law tells us that the change in internal energy ΔE_{int} is zero; therefore, $E_{int,i} = E_{int,f}$. Because the gas is ideal, E_{int} depends on temperature only, so we conclude that $T_i = T_f$.

To apply Equation 18.10, we must find Q_r; that is, we must find an equivalent reversible path that shares the same initial and final states. A simple choice is an isothermal, reversible expansion in which the gas pushes slowly against a piston while energy enters the gas by heat from a reservoir to hold the temperature constant. Because T is constant in this process, Equation 18.10 gives

$$\Delta S = \int \frac{dQ_r}{T} = \frac{1}{T} \int_i^f dQ_r$$

Because we are considering an isothermal process, $\Delta E_{int} = 0$, so the first law of thermodynamics tells us that the energy input by heat is equal to the negative of the work done on the gas, $dQ_r = -dW = P\,dV$. Using this result, we find that

$$\Delta S = \frac{1}{T} \int dQ_r = \frac{1}{T} \int P\,dV = \frac{1}{T} \int \frac{nRT}{V} dV = nR \int_{V_i}^{V_f} \frac{dV}{V}$$

$$\Delta S = nR \ln\left(\frac{V_f}{V_i}\right) \qquad\qquad \textbf{18.12} ◄$$

Because $V_f > V_i$, we conclude that ΔS is positive, and so both the entropy and the disorder of the gas (and the Universe) increase as a result of the irreversible, adiabatic expansion.

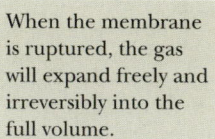

When the membrane is ruptured, the gas will expand freely and irreversibly into the full volume.

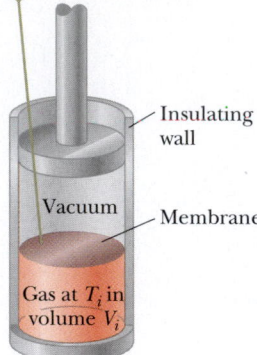

Insulating wall

Vacuum

Membrane

Gas at T_i in volume V_i

Figure 18.11 Adiabatic free expansion of a gas. The container is thermally insulated from its surroundings; therefore, $Q = 0$.

Example **18.4** | **Adiabatic Free Expansion: Revisited**

Let's verify that the macroscopic and microscopic approaches to the calculation of entropy lead to the same conclusion for the adiabatic free expansion of an ideal gas. Suppose an ideal gas expands to four times its initial volume. As we have seen for this process, the initial and final temperatures are the same.

(A) Using a macroscopic approach, calculate the entropy change for the gas.

SOLUTION

Conceptualize Look back at Figure 18.11, which is a diagram of the system before the adiabatic free expansion. Imagine breaking the membrane so that the gas moves into the evacuated area. The expansion is irreversible.

18.4 *cont.*

Categorize We can replace the irreversible process with a reversible isothermal process between the same initial and final states. This approach is macroscopic, so we use a thermodynamic variable, in particular, the volume V.

Analyze Use Equation 18.12 to evaluate the entropy change:

$$\Delta S = nR \ln \left(\frac{V_f}{V_i}\right) = nR \ln \left(\frac{4V_i}{V_i}\right) = nR \ln 4$$

(B) Using statistical considerations, calculate the change in entropy for the gas and show that it agrees with the answer you obtained in part (A).

SOLUTION

Categorize This approach is microscopic, so we use variables related to the individual molecules.

Analyze The number of microstates available to a single molecule in the initial volume V_i is $w_i = V_i/V_m$, where V_m is the microscopic volume occupied by the molecule. Use this number to find the number of available microstates for N molecules:

$$W_i = w_i{}^N = \left(\frac{V_i}{V_m}\right)^N$$

Find the number of available microstates for N molecules in the final volume $V_f = 4V_i$:

$$W_f = \left(\frac{V_f}{V_m}\right)^N = \left(\frac{4V_i}{V_m}\right)^N$$

Use Equation 18.9 to find the entropy change:

$$\Delta S = k_B \ln W_f - k_B \ln W_i = k_B \ln \left(\frac{W_f}{W_i}\right)$$

$$= k_B \ln \left(\frac{4V_i}{V_i}\right)^N = k_B \ln (4^N) = Nk_B \ln 4 = nR \ln 4$$

Finalize The answer is the same as that for part (A), which dealt with macroscopic parameters.

What If? In part (A), we used Equation 18.12, which was based on a reversible isothermal process connecting the initial and final states. Would you arrive at the same result if you chose a different reversible process?

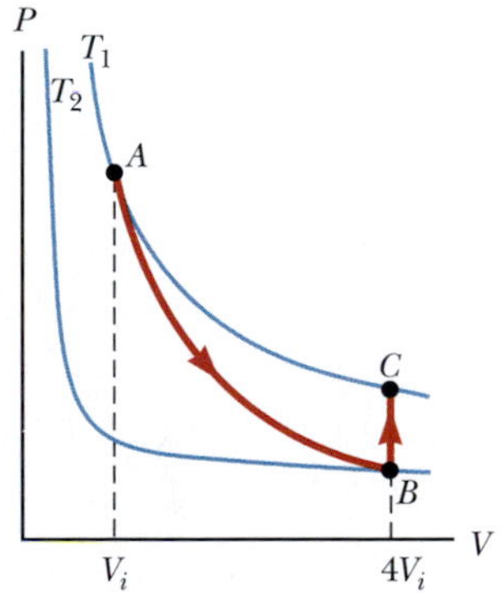

Answer You *must* arrive at the same result because entropy is a state variable. For example, consider the two-step process in Figure 18.12: a reversible adiabatic expansion from V_i to $4V_i$ ($A \to B$) during which the temperature drops from T_1 to T_2 and a reversible isovolumetric process ($B \to C$) that takes the gas back to the initial temperature T_1. During the reversible adiabatic process, $\Delta S = 0$ because $Q_r = 0$.

Figure 18.12
(Example 18.4) A gas expands to four times its initial volume and back to the initial temperature by means of a two-step process.

For the reversible isovolumetric process ($B \to C$), use Equation 18.10:

$$\Delta S = \int_i^f \frac{dQ_r}{T} = \int_{T_2}^{T_1} \frac{nC_V dT}{T} = nC_V \ln \left(\frac{T_1}{T_2}\right)$$

Find the ratio of temperature T_1 to T_2 from Equation 17.26 for the adiabatic process:

$$\frac{T_1}{T_2} = \left(\frac{4V_i}{V_i}\right)^{\gamma-1} = (4)^{\gamma-1}$$

Substitute to find ΔS:

$$\Delta S = nC_V \ln (4)^{\gamma-1} = nC_V(\gamma - 1) \ln 4$$

$$= nC_V \left(\frac{C_P}{C_V} - 1\right) \ln 4 = n(C_P - C_V) \ln 4 = nR \ln 4$$

And you do indeed obtain the exact same result for the entropy change.

18.9 | Context Connection: The Atmosphere as a Heat Engine

In Chapter 17, we predicted a global temperature based on the notion of energy balance between incoming visible radiation from the Sun and outgoing infrared radiation from the Earth. This model leads to a global temperature that is well below the measured temperature. This discrepancy results because atmospheric effects are not included in our model. In this section, we shall introduce some of these effects and show that the atmosphere can be modeled as a heat engine. In the Context Conclusion, we shall use concepts learned in the thermodynamics chapters to build a model that is more successful at predicting the correct temperature of the Earth.

What happens to the energy that enters the atmosphere by radiation from the Sun? Figure 18.13 helps answer this question by showing how the input energy undergoes various processes. If we identify the incoming energy as 100%, we find that 30% of it is reflected back into space, as we mentioned in Chapter 17. This 30% includes 6% back-scattered from air molecules, 20% reflected from clouds, and 4% reflected from the surface of the Earth. The remaining 70% is absorbed by either the air or the surface. Before reaching the surface, 20% of the original radiation is absorbed in the air; 4% by clouds; and 16% by water, dust particles, and ozone in the atmosphere. Of the original radiation striking the top of the atmosphere, the ground absorbs 50%.

The ground emits radiation upward and transfers energy to the atmosphere by several processes. Of the original 100% of the incoming energy, 6% simply passes back through the atmosphere into space (at the right in Fig. 18.13). In addition, 14% of the original incoming energy emitted as radiation from the ground is absorbed by water and carbon dioxide molecules. The air warmed by the surface rises

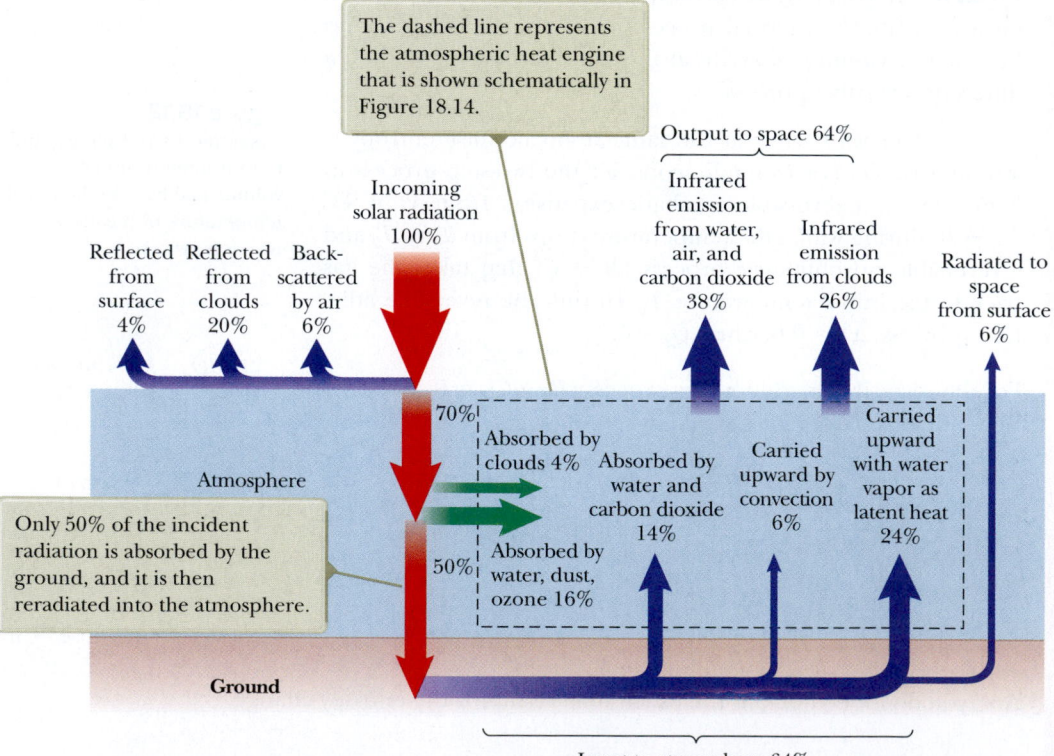

Figure 18.13 Energy input to the atmosphere from the Sun is divided into several components.

upward by convection, carrying 6% of the original energy into the atmosphere. The hydrological cycle results in 24% of the original energy being carried upward as water vapor and released into the atmosphere when the water vapor condenses into liquid water.

These processes result in a total of 64% of the original energy being absorbed in the atmosphere, with another 6% from the surface passing back through into space. Because the atmosphere is in steady state, this 64% is also emitted from the atmosphere into space. The emission is divided into two types. The first is infrared radiation from molecules in the atmosphere, including water vapor, carbon dioxide, and the nitrogen and oxygen molecules of the air, which accounts for emission of 38% of the original energy. The remaining 26% is emitted as infrared radiation from clouds.

Figure 18.13 accounts for all the energy; the amount of energy input equals the amount of energy output, which is the premise used in the Context Connection of Chapter 17. A major difference from our discussion in that chapter, however, is the notion of absorption of the energy by the atmosphere. It is this absorption that creates thermodynamic processes in the atmosphere to raise the surface temperature above the value we determined in Chapter 17. We shall explore more about these processes and the temperature profile of the atmosphere in the Context Conclusion.

To close this chapter, let us discuss one more process that is not included in Figure 18.13. The various processes depicted in that figure result in a small amount of work done on the air, which appears as the kinetic energy of the prevailing winds in the atmosphere.

The amount of the original solar energy that is converted to kinetic energy of prevailing winds is about 0.5%. The process of generating the winds does not change the energy balance shown in Figure 18.13. The kinetic energy of the wind is converted to internal energy as masses of air move past one another. This internal energy produces an increased infrared emission of the atmosphere into space, so the 0.5% is only temporarily in the form of kinetic energy before being emitted as radiation.

We can model the atmosphere as a heat engine, which is indicated in Figure 18.13 by the dotted rectangle. A schematic diagram of this heat engine is shown in Figure 18.14. The warm reservoir is the surface and the atmosphere, and the cold reservoir is empty space. We can calculate the efficiency of the atmospheric engine using Equation 18.2:

$$e = \frac{W_{eng}}{|Q_h|} = \frac{0.5\%}{64\%} = 0.008 = 0.8\%$$

which is a very low efficiency. Keep in mind, however, that a tremendous amount of energy enters the atmosphere from the Sun, so even a very small fraction of it can create a very complex and powerful wind system. Hurricanes represent a vivid example of the energy output of the atmospheric heat engine.

Notice that the output energy in Figure 18.14 is less than that in Figure 18.13 by 0.5%. As noted previously, the 0.5% transferred to the atmosphere by generating winds is eventually transformed to internal energy in the atmosphere by friction and then radiated into space as thermal radiation. We cannot separate the heat engine and the winds in the atmosphere in a diagram because the atmosphere *is* the heat engine and the winds are generated *in* the atmosphere!

We now have all the pieces we need to put together the puzzle of the temperature of the Earth. We shall discuss this subject in the Context Conclusion.

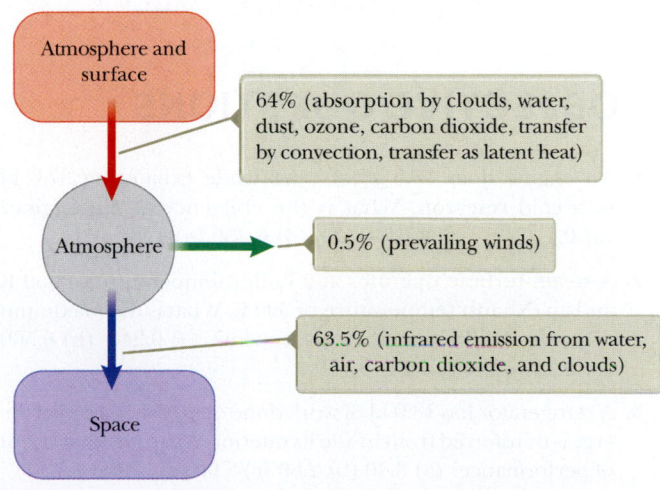

Figure 18.14 A schematic representation of the atmosphere as a heat engine.

SUMMARY

A **heat engine** is a device that takes in energy by heat and, operating in a cycle, expels a fraction of that energy by work. The net work done by a heat engine is

$$W_{eng} = |Q_h| - |Q_c| \qquad \textbf{18.1} \blacktriangleleft$$

where Q_h is the energy absorbed from a hot reservoir and Q_c is the energy expelled to a cold reservoir.

The **thermal efficiency** e of a heat engine is defined as the ratio of the net work done to the energy absorbed per cycle from the higher temperature reservoir:

$$e = \frac{W_{eng}}{|Q_h|} = 1 - \frac{|Q_c|}{|Q_h|} \qquad \textbf{18.2} \blacktriangleleft$$

The **Kelvin–Planck statement of the second law of thermodynamics** can be stated as follows:

- It is impossible to construct a heat engine that, operating in a cycle, produces no effect other than the absorption of energy by heat from a reservoir and the performance of an equal amount of work.

A **reversible** process is one for which the system can be returned to its initial conditions along the same path and for which every point along the path is an equilibrium state. A process that does not satisfy these requirements is **irreversible.**

The thermal efficiency of a heat engine operating in a **Carnot cycle** is given by

$$e_C = 1 - \frac{T_c}{T_h} \qquad \textbf{18.4} \blacktriangleleft$$

where T_c is the absolute temperature of the cold reservoir and T_h is the absolute temperature of the hot reservoir. No real heat engine operating between the temperatures T_c and T_h can be more efficient than an engine operating reversibly in a Carnot cycle between the same two temperatures.

The **Clausius statement of the second law** states that

- Energy will not transfer spontaneously by heat from a cold object to a hot object.

The second law of thermodynamics states that when real (irreversible) processes occur, the degree of disorder in the system plus the surroundings increases. The measure of disorder in a system is called **entropy** S.

The **change in entropy** dS of a system moving through an infinitesimal process between two equilibrium states is

$$dS = \frac{dQ_r}{T} \qquad \textbf{18.8} \blacktriangleleft$$

where dQ_r is the energy transferred by heat in a reversible process between the same states.

From a microscopic viewpoint, the entropy S associated with a macrostate of a system is defined as

$$S \equiv k_B \ln W \qquad \textbf{18.9} \blacktriangleleft$$

where k_B is Boltzmann's constant and W is the number of microstates corresponding to the macrostate whose entropy is S. Therefore, **entropy is a measure of microscopic disorder.** Because of the statistical tendency of systems to proceed toward states of greater probability and greater disorder, all natural processes are irreversible and result in an increase in entropy. Therefore, the **entropy statement of the second law of thermodynamics** is as follows:

- The entropy of the Universe increases in all real processes.

The change in entropy of a system moving between two general equilibrium states is

$$\Delta S = \int_i^f \frac{dQ_r}{T} \qquad \textbf{18.10} \blacktriangleleft$$

The value of ΔS is the same for all paths connecting the initial and final states.

The change in entropy for any reversible, cyclic process is zero, and when such a process occurs, the entropy of the Universe remains constant.

OBJECTIVE QUESTIONS

1. An engine does 15.0 kJ of work while exhausting 37.0 kJ to a cold reservoir. What is the efficiency of the engine? (a) 0.150 (b) 0.288 (c) 0.333 (d) 0.450 (e) 1.20

2. A steam turbine operates at a boiler temperature of 450 K and an exhaust temperature of 300 K. What is the maximum theoretical efficiency of this system? (a) 0.240 (b) 0.500 (c) 0.333 (d) 0.667 (e) 0.150

3. A refrigerator has 18.0 kJ of work done on it while 115 kJ of energy is transferred from inside its interior. What is its coefficient of performance? (a) 3.40 (b) 2.80 (c) 8.90 (d) 6.40 (e) 5.20

4. A compact air-conditioning unit is placed on a table inside a well-insulated apartment and is plugged in and turned on. What happens to the average temperature of the apartment?

(a) It increases. (b) It decreases. (c) It remains constant. (d) It increases until the unit warms up and then decreases. (e) The answer depends on the initial temperature of the apartment.

5. Consider cyclic processes completely characterized by each of the following net energy inputs and outputs. In each case, the energy transfers listed are the *only* ones occurring. Classify each process as (a) possible, (b) impossible according to the first law of thermodynamics, (c) impossible according to the second law of thermodynamics, or (d) impossible according to both the first and second laws. **(i)** Input is 5 J of work, and output is 4 J of work. **(ii)** Input is 5 J of work, and output is 5 J of energy transferred by heat. **(iii)** Input is 5 J of energy transferred by electrical transmission, and output is 6 J of work. **(iv)** Input is 5 J of energy transferred by heat,

and output is 5 J of energy transferred by heat. **(v)** Input is 5 J of energy transferred by heat, and output is 5 J of work. **(vi)** Input is 5 J of energy transferred by heat, and output is 3 J of work plus 2 J of energy transferred by heat.

6. Of the following, which is *not* a statement of the second law of thermodynamics? (a) No heat engine operating in a cycle can absorb energy from a reservoir and use it entirely to do work. (b) No real engine operating between two energy reservoirs can be more efficient than a Carnot engine operating between the same two reservoirs. (c) When a system undergoes a change in state, the change in the internal energy of the system is the sum of the energy transferred to the system by heat and the work done on the system. (d) The entropy of the Universe increases in all natural processes. (e) Energy will not spontaneously transfer by heat from a cold object to a hot object.

7. The arrow *OA* in the *PV* diagram shown in Figure OQ18.7 represents a reversible adiabatic expansion of an ideal gas. The same sample of gas, starting from the same state *O*, now undergoes an adiabatic free expansion to the same final volume. What point on the diagram could represent the final state of the gas? (a) the same point *A* as for the reversible expansion (b) point *B* (c) point *C* (d) any of those choices (e) none of those choices

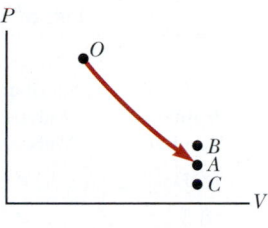

Figure OQ18.7

8. A thermodynamic process occurs in which the entropy of a system changes by −8 J/K. According to the second law of thermodynamics, what can you conclude about the entropy change of the environment? (a) It must be +8 J/K or less. (b) It must be between +8 J/K and 0. (c) It must be equal to +8 J/K. (d) It must be +8 J/K or more. (e) It must be zero.

9. A sample of a monatomic ideal gas is contained in a cylinder with a piston. Its state is represented by the dot in the *PV* diagram shown in Figure OQ18.9. Arrows *A* through *E* represent isobaric, isothermal, adiabatic, and isovolumetric processes that the sample can undergo. In each process except *D*, the volume changes by a factor of 2. All five processes are reversible. Rank the processes according to the change in entropy of the gas from the largest positive value to the largest-magnitude negative value. In your rankings, display any cases of equality.

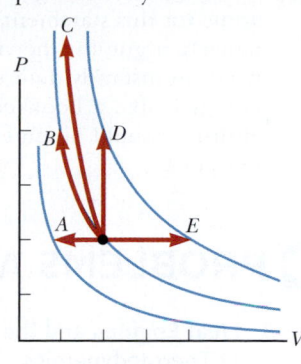

Figure OQ18.9

10. Assume a sample of an ideal gas is at room temperature. What action will *necessarily* make the entropy of the sample increase? (a) Transfer energy into it by heat. (b) Transfer energy into it irreversibly by heat. (c) Do work on it. (d) Increase either its temperature or its volume, without letting the other variable decrease. (e) None of those choices is correct.

11. The second law of thermodynamics implies that the coefficient of performance of a refrigerator must be what? (a) less than 1 (b) less than or equal to 1 (c) greater than or equal to 1 (d) finite (e) greater than 0

CONCEPTUAL QUESTIONS

☐ denotes answer available in *Student Solutions Manual/Study Guide*

1. The device shown in Figure CQ18.1, called a thermoelectric converter, uses a series of semiconductor cells to transform internal energy to electric potential energy, which we will study in Chapter 20. In the photograph on the left, both legs of the device are at the same temperature and no electric potential energy is produced. When one leg is at a higher temperature than the other as shown in the photograph on the right, however, electric potential energy is produced as the device extracts energy from the hot reservoir and drives a small electric motor. (a) Why is the difference in temperature necessary to produce electric potential energy in this demonstration? (b) In what sense does this intriguing experiment demonstrate the second law of thermodynamics?

Figure CQ18.1

2. (a) Give an example of an irreversible process that occurs in nature. (b) Give an example of a process in nature that is nearly reversible.

3. Does the second law of thermodynamics contradict or correct the first law? Argue for your answer.

4. Is it possible to construct a heat engine that creates no thermal pollution? Explain.

5. "The first law of thermodynamics says you can't really win, and the second law says you can't even break even." Explain how this statement applies to a particular device or process; alternatively, argue against the statement.

6. A steam-driven turbine is one major component of an electric power plant. Why is it advantageous to have the temperature of the steam as high as possible?

7. What are some factors that affect the efficiency of automobile engines?

8. (a) If you shake a jar full of jelly beans of different sizes, the larger beans tend to appear near the top and the smaller ones tend to fall to the bottom. Why? (b) Does this process violate the second law of thermodynamics?

9. Discuss the change in entropy of a gas that expands (a) at constant temperature and (b) adiabatically.

10. Suppose your roommate cleans and tidies up your messy room after a big party. Because she is creating more order, does this process represent a violation of the second law of thermodynamics?

11. "Energy is the mistress of the Universe, and entropy is her shadow." Writing for an audience of general readers, argue for this statement with at least two examples. Alternatively, argue for the view that entropy is like an executive who instantly determines what will happen, whereas energy is like a bookkeeper telling us how little we can afford. (Arnold Sommerfeld suggested the idea for this question.)

12. Discuss three different common examples of natural processes that involve an increase in entropy. Be sure to account for all parts of each system under consideration.

13. The energy exhaust from a certain coal-fired electric generating station is carried by "cooling water" into Lake Ontario. The water is warm from the viewpoint of living things in the lake. Some of them congregate around the outlet port and can impede the water flow. (a) Use the theory of heat engines to explain why this action can reduce the electric power output of the station. (b) An engineer says that the electric output is reduced because of "higher back pressure on the turbine blades." Comment on the accuracy of this statement.

PROBLEMS AVAILABLE IN

18.1 Heat Engines and the Second Law of Thermodynamics

Problems 1–5

18.2 Reversible and Irreversible Processes
18.3 The Carnot Engine

Problems 6–16

18.4 Heat Pumps and Refrigerators

Problems 17–26

18.6 Entropy
18.7 Entropy and the Second Law of Thermodynamics

Problems 27–35

18.8 Entropy Changes in Irreversible Processes

Problems 36–41

18.9 Context Connection: The Atmosphere as a Heat Engine

Problems 42–43

Additional Problems

Problems 44–65

Solutions to the following Problems are available in the *Student Solutions Manual/Study Guide*:

18.5, 18.7, 18.11, 18.22, 18.26, 18.29, 18.31, 18.41, 18.45, 18.47, 18.48, 18.58, and 18.62

List of Enhanced Problems

Problem Number	Targeted Feedback in Enhanced WebAssign	Master It in Enhanced WebAssign	Watch It in Enhanced WebAssign
18.1	✓		✓
18.3	✓		✓
18.5	✓	✓	
18.7	✓	✓	
18.9	✓		✓
18.11	✓	✓	
18.14	✓		✓
18.15	✓		✓
18.17	✓		✓
18.26		✓	
18.28	✓		✓
18.36	✓		✓
18.45		✓	

Predicting the Earth's Surface Temperature

Now that we have investigated the principles of thermodynamics, we respond to our central question for the Context on *Global Warming:*

> **Can we build a structural model of the atmosphere that predicts the average temperature at the Earth's surface?**

We discussed some of the factors affecting the temperature—the energy input from the Sun and the energy output by thermal radiation from the surface of the Earth—in Chapter 17. In Chapter 18, we introduced the role of the atmosphere in absorbing radiation by means of various molecules. In the following discussion, we explore how the atmosphere modifies the temperature calculation performed in Chapter 17, which leads to a structural model that predicts a temperature in agreement with observations.

Modeling the Atmosphere

We first ask if the temperature of 255 K that we found in Chapter 17 is valid and, if so, what does it represent? The answer to the first question is *yes*. The energy balance concept is certainly valid, and the Earth, as a system, must emit energy at the same rate as it absorbs energy. The temperature of 255 K is representative of the radiation leaving the atmosphere. A space traveler outside our atmosphere who takes a reading of radiation from the Earth would determine that the temperature representing this radiation is indeed 255 K. This temperature is the one associated with radiation leaving the *top* of the atmosphere, however. It is not the temperature at the surface of the Earth.

As we have mentioned, the atmosphere is almost transparent to the visible radiation from the Sun but not to the infrared radiation emitted by the surface of the Earth. Let us build a model in which we assume that all radiation with wavelengths less than about 5 μm is allowed to pass through the atmosphere. Therefore, almost *all* incoming radiation from the Sun (except for the 30% reflected) reaches the Earth's surface. In addition, let us assume that all radiation with wavelengths above about 5 μm (which is *infrared* radiation, including that emitted by the Earth surface) is absorbed by the atmosphere.

We can identify two layers to the atmosphere in our model (Fig. 1), as discussed in Section 16.7. The lower part of the atmosphere is the *troposphere*. In this layer, the density of the air is relatively high so that the probability of absorption of infrared radiation from the surface by molecules in the air is large. This absorption warms parcels of air near the surface, which then rise upward. As a parcel rises, it expands and its temperature drops. Therefore, the troposphere is the convective region in which

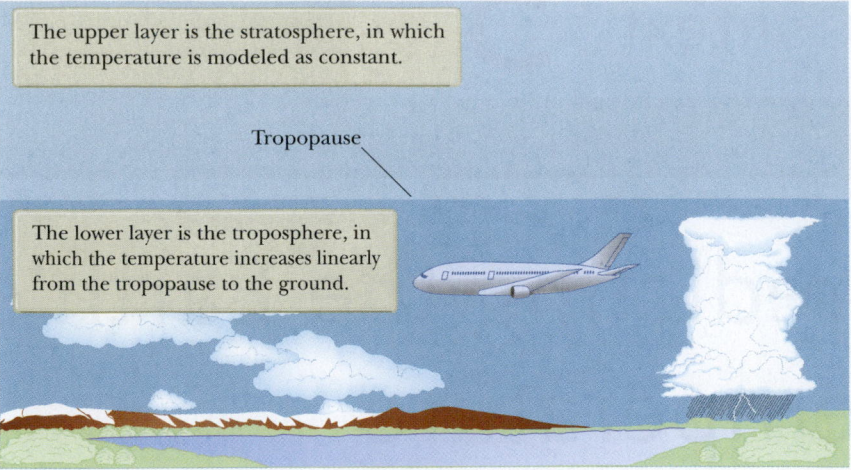

Figure 1 In our structural model of the atmosphere, we consider two layers.

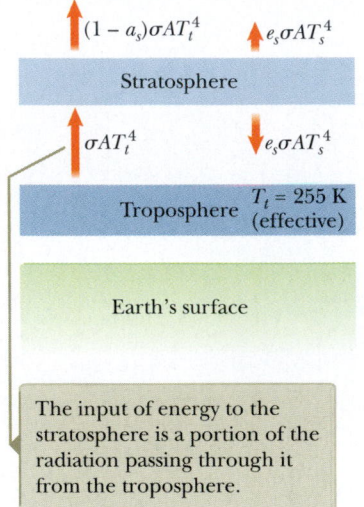

Figure 2 A portion of the stratosphere of area A is modeled as an object with a temperature, emitting thermal radiation from both upper and lower surfaces.

the temperature decreases with height according to the lapse rate, as discussed in Section 16.7. It is also the region of the atmosphere in which our familiar weather occurs. Above the troposphere is the *stratosphere*. In this layer, the density of the air is relatively low so that the probability of absorption of infrared radiation is small. As a result, infrared radiation tends to pass through into space with little absorption. Without this absorption, the temperature in the stratosphere remains approximately constant with height. Between these two layers is the *tropopause,* which is about 11 km from the Earth's surface.[1] In reality, the tropopause is a thin region in which the primary energy transfer mechanism changes continuously from convection to radiation. In our model, we imagine the tropopause to be a sharp boundary.

The first task is to find the temperature, assumed constant, of the stratosphere. We appeal again to Stefan's law and consider the energy transfer into and out of the stratosphere as indicated in Figure 2. Radiation from the troposphere (to which we assign an effective average temperature of $T_t = 255$ K so that it is the temperature associated with the radiation coming through the stratosphere to our imagined outer space observer) passes through the stratosphere, with a fraction a_s absorbed. The stratosphere, at temperature T_s, radiates both upward and downward, according to its emissivity e_s. Therefore, because the stratosphere is in steady state, the power balance equation for the stratosphere is

$$P_{ER} \text{ (in)} = -P_{ER} \text{ (out)}$$

$$a_s \sigma A T_t{}^4 = 2 e_s \sigma A T_s{}^4$$

where the factor of 2 arises from the output radiation of the stratosphere from both top and bottom surfaces. We can solve for the temperature of the stratosphere:

$$T_s = \left(\frac{a_s \sigma T_t{}^4}{2 e_s \sigma} \right)^{1/4} = \left(\frac{a_s}{2 e_s} \right)^{1/4} T_t = \left(\tfrac{1}{2} \right)^{1/4} (255 \text{ K}) = 214 \text{ K}$$

where we have used that the absorptivity and the emissivity of the stratosphere are the same number.

Now we have all the pieces: the temperature of the stratosphere, the height of the tropopause, and the lapse rate. We simply need to extrapolate, using the lapse rate, from the temperature at the tropopause, which is the temperature of the stratosphere, to that at the Earth's surface.

---

[1]The tropopause height of 11 km that we assume here is a simplification model in our structural model. In reality, the tropopause height varies with latitude and with season. At various latitudes and at different times of the year, the tropopause height can vary from less than 8 km to more than 17 km. The height of 11 km is a reasonable average for all latitudes over an entire year.

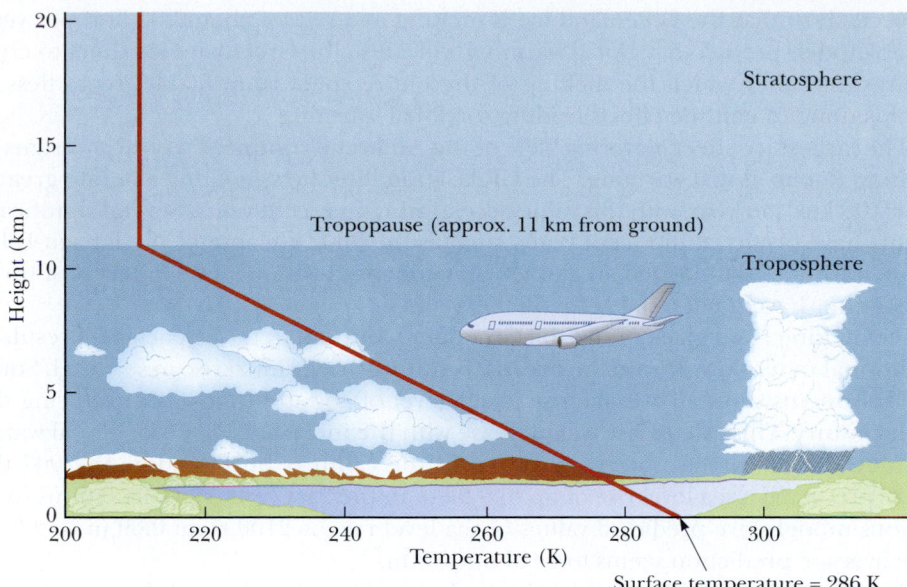

Morton J. Elrod © K. Ross Toole Archives, Mansfield Library, The University of Montana-Missoula

Figure 3 A graphical representation of the temperature variation with altitude in our model atmosphere. The predicted surface temperature agrees with measurements to within 1%.

If the tropopause is 11 km from the surface and the lapse rate is $-6.5°C/km$ (Section 16.7), the net change in temperature from the surface to the tropopause is

$$\Delta T = T_{\text{tropopause}} - T_{\text{surface}} = \left(\frac{\Delta T}{\Delta y}\right)\Delta y = (-6.5°C/km)(11\ km)$$

$$= -72°C = -72\ K$$

Because the tropopause temperature is 214 K, we can now find the surface temperature:

$$\Delta T = T_{\text{tropopause}} - T_{\text{surface}}$$

$$-72\ K = 214\ K - T_{\text{surface}} \rightarrow T_{\text{surface}} = 286\ K$$

which agrees with the measured average temperature of 288 K discussed in Chapter 17 to within less than 1%! Figure 3 shows a graphical representation (height versus temperature) of the temperature in the troposphere.

The absorption of infrared radiation from the Earth's surface is dependent on molecules in the atmosphere. Our industrialized society is changing the atmospheric concentrations of molecules such as water, carbon dioxide, and methane. As a result, we are altering the energy balance and changing the temperature of the Earth's surface. Some data taken since the mid-19th century show a temperature increase of 0.5 to 1.0°C in the last 150 years. As noted in the Context Introduction, the Intergovernmental Panel on Climate Change (IPCC) predicted that a further increase of 1°C–6°C could occur in the 21st century.

Glaring evidence of the effects of rising temperatures is seen in the loss of ice from ice sheets covering Greenland and Antarctica and from glaciers around the globe. Figure 4 shows before and after photographs of the Sperry Glacier in Glacier National Park in Montana. The ice visible in the 1930 photograph has disappeared by the time of the 2008 photograph and the terminus of the glacier has retreated beyond the field of view. Some models predict that all of the glaciers in Glacier National Park will disappear by the year 2030. Measurements of glaciers in other parts of the world show similar loss. Such loss can have catastrophic social effects. For example, a significant portion of the world's population depends on drinking water from Himalayan glaciers. Loss of this drinking water can lead to social upheaval as these populations search to replace the water from other supplies.

Approximately 80% of the surface of Greenland is covered by an ice sheet, second in size only to the ice covering Antarctica. Measurements made from the GRACE (Gravity Recovery and Climate Experiment) satellite, a joint NASA/German Space Agency

Lisa McKeon photo, USGS

Figure 4 The Repeat Photography Project of the United States Geological Survey (USGS) is designed to demonstrate the loss of glacial ice in Glacier National Park (Montana) due to global warming. This pair of photographs of the Sperry Glacier shows an example of such loss. (a) A 1930 photo shows a thick and extensive glacier. (b) The Sperry Glacier has completely disappeared from this field of view by 2008.

project, show that the Greenland ice is melting at a rate of about 200 km^3 per year. Some models predict that global warming will cause the Greenland ice sheet to cross a threshold after which the melting of the entire sheet is inevitable, regardless of what is done to halt the effects leading to global warming.

The largest ice sheet, covering 98% of the Antarctic continent, also shows signs of melting due to global warming. The GRACE satellite shows melting at a rate greater than 100 km^3 per year, with this value accelerating in recent years. Several significant events have occurred in recent years, such as the 2002 collapse of the Larsen B Ice Shelf, a Rhode Island–sized area of ice that collapsed within a time interval of three weeks after having been stable for 12 000 years.

Ice melting from glaciers and the Greenland and Antarctic ice packs will result in additional water flowing into the oceans, resulting in a gradual rise in sea level. Some measurements show average rise in sea level of about 0.18 m–0.20 m during the 20th century. This rate of rise will increase with the increased effects of global warming due to our current society. As noted in the Context Introduction, in 2007 the IPCC predicted a sea-level rise of up to 0.59 m in the 21st century. Calculations from various models give predicted values of sea level rise by 2100 from 0.09 m to 2.0 m. The average prediction seems to be about 0.5 m.

The Maldives is an island nation in the Indian Ocean. It depends heavily on tourism to support its economy. Geographically, the highest natural point on the islands is 2.3 m above sea level. (Areas with major construction have had the ground raised by several meters with fill dirt.) More than 80% of the land area of the Maldives is less than 1.0 m above sea level. As a result, a sea-level rise of 0.5 m would be devastating to the Maldives, with much of its land area underwater and the tourism industry decimated.

The government of Maldives is concerned about its citizens becoming refugees from their homeland as the islands are overrun with water. Plans have been proposed to look for new land in India, Sri Lanka, or Australia for these refugees. In 2009, the government announced a ten-year plan to become the world's first carbon-neutral country by switching to renewable energy sources, such as solar panels and wind turbines. While this will not stop the sea level rise due to carbon emissions from the rest of the world, it may act as a catalyst for other nations to investigate renewable sources more aggressively.

The long-term outlook is possibly bleak. For example, if the Greenland ice sheet were to completely melt over several hundred years, global sea level would rise by about 7 meters, a disastrous result. Models for predicting the effects of global warming, however, are very complicated, and unambiguous predictions are difficult to make. Global warming remains a difficult issue to address, with its combination of influences from science, politics, economics, and sociology.

The model we have described in this Context is successful in predicting the surface temperature. If we extend the model to predict changes in the surface temperature as we add more carbon dioxide to the atmosphere, we find that the predictions are not in agreement with more sophisticated models. The atmosphere is a very complicated entity, and the models used by atmospheric scientists are far more sophisticated than the one we have studied here. For our purposes, however, our successful prediction of the surface temperature is sufficient.

Problems

1. A simple model of absorption in the atmosphere shows that doubling the amount of carbon dioxide in the future will raise the altitude of the tropopause from 11 km to about 13 km. If the stratospheric temperature and the lapse rate remain the same, what is the surface temperature in this case? The result you obtain is much larger than the temperature predicted by sophisticated computer models. This disagreement displays a weakness in our simple model.

2. The stratosphere of Venus has a temperature of about 200 K. The lapse rate in the Venutian troposphere is about $-8.8°C/km$. The measured temperature on the surface of Venus is 732 K. What is the altitude of the Venutian tropopause?

3. **Q|C** Another atmospheric model is based on splitting the atmosphere into N layers of gas. We assume that the atmosphere is transparent to visible light from the Sun but is quite opaque to the infrared light that the planet emits. We choose the depth of each atmospheric layer to be one *radiation thickness*. That is, the probability of absorption of infrared radiation in the layer is just 100%. Because the density of the gas and therefore the probability of absorption vary with altitude, the layers have different geometrical thicknesses. We assume that each layer has uniform temperature T_i, where i runs from 1 for the top layer to N for the layer in contact with the planet surface. Each intermediate layer emits thermal radiation from its top and bottom surfaces and absorbs radiation from the layers above and below it. The lowest layer emits radiation from its bottom surface into the surface of the planet, of temperature T_s, and also absorbs radiation from the planet. The highest layer emits into space from its upper surface but does not have a higher layer from which to absorb infrared radiation. (a) The Earth absorbs 70% of the incident solar radiation, which has an intensity of 1 370 W/m². Show that the temperature T_1 of the top layer is 255 K. (b) For an atmosphere with N layers, show that the surface temperature is $T_s = (N + 1)^{1/4} T_1$. (c) Consider the troposphere and stratosphere of Earth as a two-layer system. What surface temperature does this model predict? (d) Why is this prediction so bad for the Earth? (e) Consider the atmosphere of Venus, from which 77% of incident radiation is reflected. What is the temperature T_1 of the top layer of the Venutian atmosphere? (f) Given that the surface temperature of Venus is 732 K, how many layers are in the Venutian atmosphere? (g) Do you think that the multilayer model will be more successful in describing the atmosphere of Venus than that of the Earth? Why?

Lightning

Lightning occurs all over the world, but more often in some places than others. Florida, for example, experiences lightning storms very often, but lightning is rare in Southern California. We begin this Context by looking at the details of a flash of lightning in a *qualitative* way. As we continue deeper into the Context, we will return to this description and attach more quantitative structure to it.

In general, we shall consider a flash of lightning to be an electric discharge occurring between a charged cloud and the ground or, in other words, an enormous spark. Lightning, however, can occur in *any* situation in which a large electric charge (which we discuss in Chapter 19) can result in electrical breakdown of the air, including snowstorms, sandstorms, and erupting volcanoes. If we consider lightning associated with clouds, we observe cloud-to-ground discharge, cloud-to-cloud discharge, intracloud discharge, and cloud-to-air discharge.

Mark Newman/Photo Researchers, Inc.

Figure 2 During an eruption of the Sakurajima volcano in Japan, lightning is prevalent in the charged atmosphere above the volcano. Although lightning is possible in this as well as many other situations, in this Context we shall study the familiar lightning that occurs in a thunderstorm.

In this Context, we shall consider only the most commonly described discharge, *cloud to ground*. Intracloud discharge actually occurs more often than cloud-to-ground discharge, but it is not the type of lightning that we regularly observe.

Because a flash of lightning occurs in a very short time, the structure of the process is hidden from normal human observation. A *flash* of lightning consists of a number of individual *strokes* of lightning, separated by tens of milliseconds. A typical number of strokes is 3 or 4, although as many as 26 strokes (for a total duration of 2 s) have been measured in a flash.

Although a stroke of lightning may appear as a sudden, single event, several steps are involved in the process. The process begins with an electrical breakdown in the air near the cloud that results in a column of negative charge, called a stepped leader, moving toward the ground at a typical speed of 10^5 m/s. The term *stepped leader* is used because the movement occurs in discrete steps of length about 50 m, with a delay of about 50 μs before the next step. A step occurs whenever the air

Paul and LindaMarie Ambrose/Taxi/Getty Images

Figure 1 Lightning electrically connects a cloud and the ground. In this Context, we shall learn about the details of such a lightning flash and find out how many lightning flashes occur on the Earth in a typical day.

becomes randomly ionized with sufficient free electrons in a short length of air to conduct electricity. The stepped leader is only faintly luminous and is not the bright flash we ordinarily think of as lightning. The radius of the channel of charge carried by the stepped leader is typically several meters.

As the tip of the stepped leader approaches the ground, it can initiate an electrical breakdown in the air near the ground, often at the tip of a pointed object. Negative charges in the ground are repelled by the approaching tip of the column of negative charge in the stepped leader. As a result, the electrical breakdown in the air near the ground results in a column of positive charge beginning to move upward from the ground. (Electrons move downward in this column, which is equivalent to positive charges moving upward.) This process is the beginning of the *return stroke*. At 20 to 100 m above the ground, the return stroke meets the stepped leader, producing an effective short circuit between the cloud and the ground. Electrons pour downward into the ground at high speed, resulting in a very large electric current moving through a channel with a radius measured in centimeters. This high current rapidly raises the temperature of the air, ionizing atoms and providing the bright light flash we associate with lightning. Emission spectra of lightning show many spectral lines from oxygen and nitrogen, the major components of air.

After the return stroke, the conducting channel retains its conductivity for a short time (measured in tens of milliseconds). If more negative charge from the cloud is made available at the top of the conducting channel, this charge can move downward and result in a new stroke. In this case, because the conducting channel is "open," the leader does not move in a stepped fashion but, rather, moves downward smoothly and quickly. For this reason, it is called a *dart leader*. Once again, as the dart leader approaches the ground, a return stroke is initiated and a bright flash of light occurs.

Just after the current has passed through the conducting channel, the air is turned into a plasma at a typical temperature of 30 000 K. As a result, there is a sudden increase of pressure causing a rapid expansion

Figure 3 This photograph shows a lightning stroke as well as the individual components of the stroke. The bright channel represents a lightning stroke in progress, just after a stepped leader and a return stroke have connected and the channel becomes conducting. Several stepped leaders can be seen at the top of the photograph, branching off from the bright channel. They are less luminous than the bright channel because they have not yet connected with return strokes. A return stroke can be seen, just to the left of the bright channel, moving upward from the tree in search of a stepped leader. Another very faint return stroke can be seen leaving the top of the power pole at the left of the photograph.

of the plasma and generating a shock wave in the surrounding gas. This shock wave is the origin of the *thunder* associated with lightning.

Having taken this first qualitative step into the understanding of lightning, let us now seek more details. After investigating the physics of lightning, we shall respond to our central question:

> **How can we determine the number of lightning flashes on the Earth in a typical day?**

Electric Forces and Electric Fields

Chapter Outline

Courtesy of Resonance Research Corporation

Mother and daughter are both enjoying the effects of electrically charging their bodies. Each individual hair on their heads becomes charged and exerts a repulsive force on the other hairs, resulting in the "stand-up" hairdos that you see here.

This chapter is the first of three on *electricity*. You are probably familiar with electrical effects, such as the static cling between articles of clothing removed from the dryer. You may also be familiar with the spark that jumps from your finger to a doorknob after you have walked across a carpet. Much of your daily experience involves working with devices that operate on energy transferred to the device by means of electrical transmission and provided by the electric power company. Even your own body is an electrochemical machine that uses electricity extensively. Nerves carry impulses as electrical signals, and electric forces are involved in the flow of materials across cell membranes.

This chapter begins with a review of some of the basic properties of the electrostatic force that we introduced in Chapter 5 as well as some properties of the electric field associated with stationary charged particles. Our study of electrostatics then continues with the concept of an electric field that is associated with a continuous charge distribution and the effect of this field on other charged particles. Once we understand the electric force exerted on a particle, we can include that force in the particle under a net force model in appropriate situations.

19.1 | Historical Overview

The laws of electricity and magnetism play a central role in the operation of devices such as cell phones, televisions, electric motors, computers, high-energy particle accelerators, and a host of electronic devices used in medicine. More fundamental, however, is that the interatomic and intermolecular forces responsible for the formation of solids and liquids are electric in origin. Furthermore, such forces as the pushes and pulls between objects in contact and the elastic force in a spring arise from electric forces at the atomic level.

Chinese documents suggest that magnetism was recognized as early as about 2000 B.C. The ancient Greeks observed electric and magnetic phenomena possibly as early as 700 B.C. They found that a piece of amber, when rubbed, attracted pieces of straw or feathers. The existence of magnetic forces was known from observations that pieces of a naturally occurring stone called *magnetite* (Fe_3O_4) were attracted to iron. (The word *electric* comes from the Greek word for amber, *elektron*. The word *magnetic* comes from *Magnesia*, a city on the coast of Turkey where magnetite was found.)

In 1600, Englishman William Gilbert discovered that electrification was not limited to amber but was a general phenomenon. Scientists went on to electrify a variety of objects, including people!

It was not until the early part of the 19th century that scientists established that electricity and magnetism are related phenomena. In 1820, Hans Oersted discovered that a compass needle, which is magnetic, is deflected when placed near an electric current. In 1831, Michael Faraday in England and, almost simultaneously, Joseph Henry in the United States showed that when a wire loop is moved near a magnet (or, equivalently, when a magnet is moved near a wire loop) an electric current is observed in the wire. In 1873, James Clerk Maxwell used these observations and other experimental facts as a basis for formulating the laws of electromagnetism as we know them today. Shortly thereafter (around 1888), Heinrich Hertz verified Maxwell's predictions by producing electromagnetic waves in the laboratory. This achievement has been followed by such practical developments as radio, television, cellular telephone systems, Bluetooth™, and Wi-Fi.

Maxwell's contributions to the science of electromagnetism were especially significant because the laws he formulated are basic to *all* forms of electromagnetic phenomena. His work is comparable in importance to Newton's discovery of the laws of motion and the theory of gravitation.

19.2 | Properties of Electric Charges

A number of simple experiments demonstrate the existence of electrostatic forces. For example, after running a comb through your hair, you will find that the comb attracts bits of paper. The attractive electrostatic force is often strong enough to suspend the bits. The same effect occurs with other rubbed materials, such as glass or rubber.

Another simple experiment is to rub an inflated balloon with wool or across your hair (Fig. 19.1). On a dry day, the rubbed balloon will stick to the wall of a room, often for hours. When materials behave this way, they are said to have become electrically charged. You can give your body an electric charge by walking across a wool rug or by sliding across a car seat. You can then feel, and remove, the charge on your body by lightly touching another person or object. Under the right conditions, a visible spark is seen when you touch and a slight tingle is felt by both parties. (Such an experiment works best on a dry day because excessive moisture in the air can provide a pathway for charge to leak off a charged object.)

Figure 19.1 Rubbing a balloon against your hair on a dry day causes the balloon and your hair to become electrically charged.

© Cengage Learning/Charles D. Winters

Experiments also demonstrate that there are two kinds of **electric charge,** given the names **positive** and **negative** by Benjamin Franklin (1706–1790). Figure 19.2 illustrates the interactions of the two kinds of charge. A hard rubber (or plastic) rod that has been rubbed with fur (or an acrylic material) is suspended by a piece of string. When a glass rod that has been rubbed with silk is brought near the rubber rod, the rubber rod is attracted toward the glass rod (Fig. 19.2a). If two charged rubber rods (or two charged glass rods) are brought near each other, as in Figure 19.2b, the force between them is repulsive. This observation demonstrates that the rubber and glass have different kinds of charge. We use the convention suggested by Franklin; the electric charge on the glass rod is called positive and that on the rubber rod is called negative. On the basis of such observations, we conclude that **charges of the same sign repel each other and charges with opposite signs attract each other.**

We know that only two kinds of electric charge exist because any unknown charge that is found experimentally to be attracted to a positive charge is also repelled by a negative charge. No one has ever observed a charged object that is repelled by both a positive and a negative charge or that is attracted to both.

Attractive electric forces are responsible for the behavior of a wide variety of commercial products. For example, the plastic in many contact lenses, *etafilcon*, is made up of molecules that electrically attract the protein molecules in human tears. These protein molecules are absorbed and held by the plastic so that the lens ends up being primarily composed of the wearer's tears. Therefore, the lens does not behave as a foreign object to the wearer's eye and can be worn comfortably. Many cosmetics also take advantage of electric forces by incorporating materials that are electrically attracted to skin or hair, causing the pigments or other chemicals to stay put once they are applied.

Another important characteristic of electric charge is that the net charge in an isolated system is always conserved. This represents the **electric charge version of the isolated system model.** We first introduced isolated system models in Chapter 7 when we discussed conservation of energy; we now see a principle of **conservation of electric charge** for an isolated system. When two initially neutral objects are charged by being rubbed together, charge is not created in the process. The objects become charged because *electrons are transferred* from one object to the other. One object gains some amount of negative charge from the electrons transferred to it while the other loses an equal amount of negative charge and hence is left with a positive charge. For the isolated system of the two objects, no transfer of charge occurs across the boundary of the system. The only change is that charge has been transferred between two members of the system. For example, when a glass rod is rubbed with silk, as in Figure 19.3, the silk obtains a negative charge that is equal in magnitude to the positive charge on the glass rod as negatively charged electrons are transferred from the glass to the silk. Likewise, when rubber is rubbed with fur, electrons are transferred from the fur to the rubber. An *uncharged object* contains an enormous number of electrons (on the order of 10^{23}). For every negative electron, however, a positively charged proton is also present; hence, an uncharged object has no net charge of either sign.

Another property of electric charge is that the total charge on an object is quantized as integral multiples of the elementary charge e. We first saw this charge $e = 1.60 \times 10^{-19}$ C in Chapter 5. The quantization results because the charge on an object must be due to an integral number of excess electrons or a deficiency of an integral number of electrons.

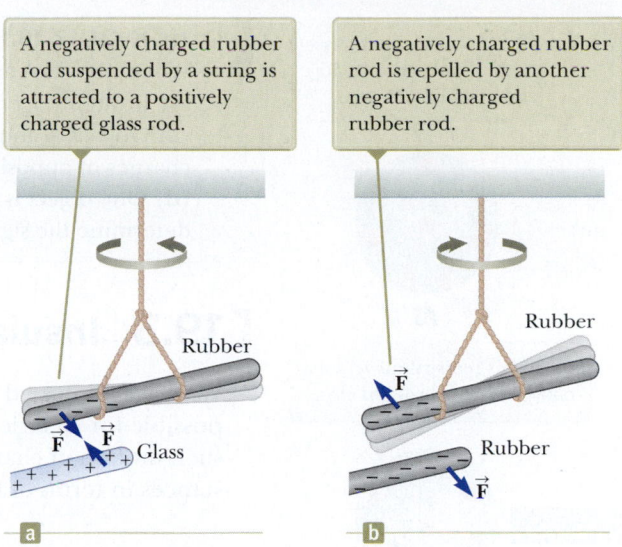

A negatively charged rubber rod suspended by a string is attracted to a positively charged glass rod.

A negatively charged rubber rod is repelled by another negatively charged rubber rod.

Figure 19.2 The electric force between (a) oppositely charged objects and (b) like-charged objects.

BIO Electrical attraction of contact lenses

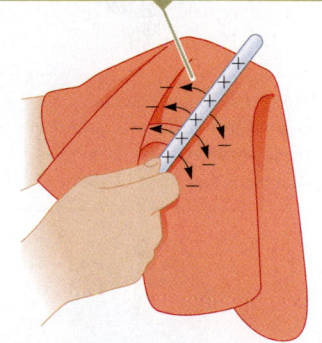

Because of conservation of charge, each electron adds negative charge to the silk and an equal positive charge is left on the glass rod.

Figure 19.3 When a glass rod is rubbed with silk, electrons are transferred from the glass to the silk. Also, because the charges are transferred in discrete bundles, the charges on the two objects are $\pm e$ or $\pm 2e$ or $\pm 3e$, and so on.

The neutral sphere has equal numbers of positive and negative charges.

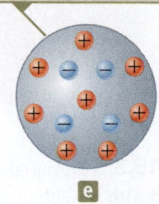

a

Electrons redistribute when a charged rod is brought close.

b

Some electrons leave the grounded sphere through the ground wire.

c

The excess positive charge is nonuniformly distributed.

d

The remaining electrons redistribute uniformly, and there is a net uniform distribution of positive charge on the sphere.

e

Figure 19.4 Charging a metallic object by *induction*. (a) A neutral metallic sphere. (b) A charged rubber rod is placed near the sphere. (c) The sphere is grounded. (d) The ground connection is removed. (e) The rod is removed.

QUICK QUIZ 19.1 Three objects are brought close to one another, two at a time. When objects A and B are brought together, they repel. When objects B and C are brought together, they also repel. Which of the following statements are true? (**a**) Objects A and C possess charges of the same sign. (**b**) Objects A and C possess charges of opposite sign. (**c**) All three objects possess charges of the same sign. (**d**) One object is neutral. (**e**) Additional experiments must be performed to determine the signs of the charges.

19.3 | Insulators and Conductors

We have discussed the transfer of charge from one object to another. It is also possible for electric charges to move from one location to another within an object; such motion of charge is called **electrical conduction.** It is convenient to classify substances in terms of the ability of charges to move within the substance:

Electrical **conductors** are materials in which some of the electrons are free electrons[1] that are not bound to atoms and can move relatively freely through the material; electrical **insulators** are materials in which all electrons are bound to atoms and cannot move freely through the material.

Materials such as glass, rubber, and dry wood are insulators. When such materials are charged by rubbing, only the rubbed area becomes charged; the charge does not tend to move to other regions of the material. In contrast, materials such as copper, aluminum, and silver are good conductors. When such materials are charged in some small region, the charge readily distributes itself over the entire surface of the material. If you hold a copper rod in your hand and rub it with wool or fur, it will not attract a small piece of paper, which might suggest that a metal cannot be charged. If you hold the copper rod by an insulating handle and then rub, however, the rod remains charged and attracts the piece of paper. In the first case, the electric charges produced by rubbing readily move from the copper through your body, which is a conductor, and finally to the Earth. In the second case, the insulating handle prevents the flow of charge to your hand.

Semiconductors are a third class of materials, and their electrical properties are somewhere between those of insulators and those of conductors. Charges can move somewhat freely in a semiconductor, but far fewer charges are moving through a semiconductor than in a conductor. Silicon and germanium are well-known examples of semiconductors that are widely used in the fabrication of a variety of electronic devices. The electrical properties of semiconductors can be changed over many orders of magnitude by adding controlled amounts of certain foreign atoms to the materials.

Charging by Induction

When a conductor is connected to the Earth by means of a conducting wire or pipe, it is said to be **grounded.** For present purposes, the Earth can be modeled as an infinite reservoir for electrons, which means that it can accept or supply an unlimited number of electrons. In this context, the Earth serves a purpose similar to our energy reservoirs introduced in Chapter 17. With that in mind, we can understand how to charge a conductor by a process known as **charging by induction.**

To understand how to charge a conductor by induction, consider a neutral (uncharged) metallic sphere insulated from the ground as shown in Figure 19.4a. There are an equal number of electrons and protons in the sphere if the charge on the sphere is exactly zero. When a negatively charged rubber rod is brought near

---

[1]A metal atom contains one or more outer electrons, which are weakly bound to the nucleus. When many atoms combine to form a metal, the *free electrons* are these outer electrons, which are not bound to any one atom. These electrons move about the metal in a manner similar to that of gas molecules moving in a container.

the sphere, electrons in the region nearest the rod experience a repulsive force and migrate to the opposite side of the sphere. This migration leaves the side of the sphere near the rod with an effective positive charge because of the diminished number of electrons as in Figure 19.4b. (The left side of the sphere in Figure 19.4b is positively charged *as if* positive charges moved into this region, but in a metal it is only electrons that are free to move.) This migration occurs even if the rod never actually touches the sphere. If the same experiment is performed with a conducting wire connected from the sphere to the Earth (Fig. 19.4c), some of the electrons in the conductor are so strongly repelled by the presence of the negative charge in the rod that they move out of the sphere through the wire and into the Earth. The symbol ⏚ at the end of the wire in Figure 19.4c indicates that the wire is connected to **ground,** which means a reservoir such as the Earth. If the wire to ground is then removed (Fig. 19.4d), the conducting sphere contains an excess of *induced* positive charge because it has fewer electrons than it needs to cancel out the positive charge of the protons. When the rubber rod is removed from the vicinity of the sphere (Fig. 19.4e), this induced positive charge remains on the ungrounded sphere. Note that the rubber rod loses none of its negative charge during this process.

Charging an object by induction requires no contact with the object inducing the charge. This behavior is in contrast to charging an object by rubbing, which does require contact between the two objects.

A process similar to the first step in charging by induction in conductors takes place in insulators. In most neutral atoms and molecules, the average position of the positive charge coincides with the average position of the negative charge. In the presence of a charged object, however, these positions may shift slightly because of the attractive and repulsive forces from the charged object, resulting in more positive charge on one side of the molecule than on the other. This effect is known as **polarization.** The polarization of individual molecules produces a layer of charge on the surface of the insulator as shown in Figure 19.5a, in which a charged balloon on the left is placed against a wall on the right. In the figure, the negative charge layer in the wall is closer to the positively charged balloon than the positive charges at the other ends of the molecules. Therefore, the attractive force between the positive and negative charges is larger than the repulsive force between the positive charges. The result is a net attractive force between the charged balloon and the neutral insulator. Your knowledge of induction in insulators should help you explain why a charged rod attracts bits of electrically neutral paper (Fig. 19.5b) or why a balloon that has been rubbed against your hair can stick to a neutral wall.

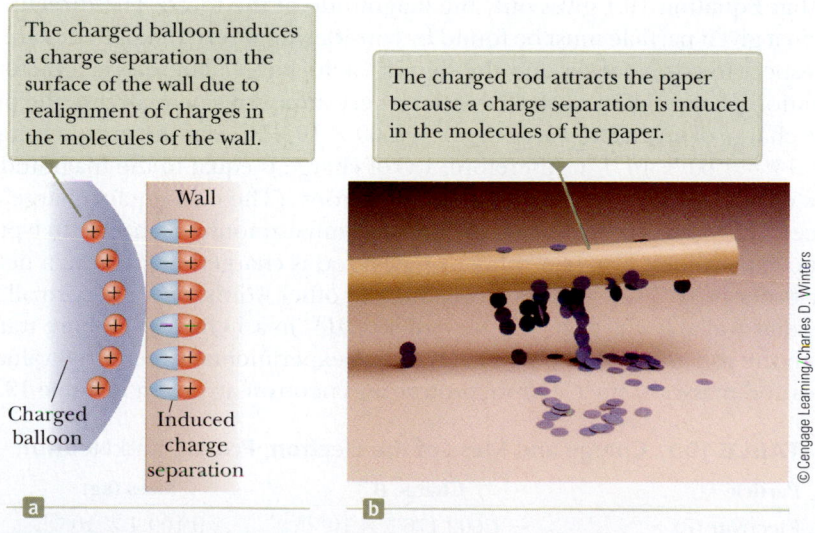

Figure 19.5 (a) A charged balloon is brought near an insulating wall. (b) A charged rod is brought close to bits of paper.

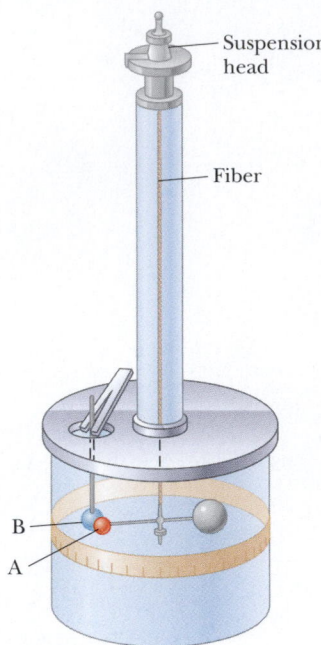

Figure 19.6 Coulomb's torsion balance, which was used to establish the inverse-square law for the electrostatic force between two charges.

© INTERFOTO/Alamy

Charles Coulomb
French Physicist (1736–1806)
Coulomb's major contributions to science were in the areas of electrostatics and magnetism. During his lifetime, he also investigated the strengths of materials and determined the forces that affect objects on beams, thereby contributing to the field of structural mechanics. In the field of ergonomics, his research provided a fundamental understanding of the ways in which people and animals can best do work.

▶ **QUICK QUIZ 19.2** Three objects are brought close to one another, two at a time. When objects A and B are brought together, they attract. When objects B and C are brought together, they repel. Which of the following are necessarily true? (**a**) Objects A and C possess charges of the same sign. (**b**) Objects A and C possess charges of opposite sign. (**c**) All three of the objects possess charges of the same sign. (**d**) One object is neutral. (**e**) Additional experiments must be performed to determine information about the charges on the objects.

19.4 | Coulomb's Law

Electric forces between charged objects were measured quantitatively by Charles Coulomb using the torsion balance, which he invented (Fig. 19.6). Coulomb confirmed that the electric force between two small charged spheres is proportional to the inverse square of their separation distance r, that is, $F_e \propto 1/r^2$. The operating principle of the torsion balance is the same as that of the apparatus used by Sir Henry Cavendish to measure the density of the Earth (Section 11.1), with the electrically neutral spheres replaced by charged ones. The electric force between charged spheres A and B in Figure 19.6 causes the spheres to either attract or repel each other, and the resulting motion causes the suspended fiber to twist. Because the restoring torque of the twisted fiber is proportional to the angle through which it rotates, a measurement of this angle provides a quantitative measure of the electric force of attraction or repulsion. Once the spheres are charged by rubbing, the electric force between them is very large compared with the gravitational attraction, and so the gravitational force can be ignored.

In Chapter 5, we introduced **Coulomb's law,** which describes the magnitude of the electrostatic force between two charged particles with charges q_1 and q_2 and separated by a distance r:

$$F_e = k_e \frac{|q_1||q_2|}{r^2}$$ **19.1** ◀

where k_e ($= 8.987\ 6 \times 10^9\ \text{N} \cdot \text{m}^2/\text{C}^2$) is the **Coulomb constant** and the force is in newtons if the charges are in coulombs and if the separation distance is in meters. The constant k_e is also written as

$$k_e = \frac{1}{4\pi\epsilon_0}$$

where the constant ϵ_0 (Greek letter epsilon), known as the **permittivity of free space,** has the value

$$\epsilon_0 = 8.854\ 2 \times 10^{-12}\ \text{C}^2/\text{N} \cdot \text{m}^2$$

Note that Equation 19.1 gives only the magnitude of the force. The direction of the force on a given particle must be found by considering where the particles are located with respect to one another and the sign of each charge. Therefore, a pictorial representation of a problem in electrostatics is very important in analyzing the problem.

The charge of an electron is $q = -e = -1.60 \times 10^{-19}$ C, and the proton has a charge of $q = +e = 1.60 \times 10^{-19}$ C; therefore, 1 C of charge is equal to the magnitude of the charge of $(1.60 \times 10^{-19})^{-1} = 6.25 \times 10^{18}$ electrons. (The elementary charge e was introduced in Section 5.5.) Note that 1 C is a substantial amount of charge. In typical electrostatic experiments, where a rubber or glass rod is charged by friction, a net charge on the order of 10^{-6} C ($= 1\ \mu$C) is obtained. In other words, only a very small fraction of the total available electrons (on the order of 10^{23} in a 1-cm^3 sample) are transferred between the rod and the rubbing material. The experimentally measured values of the charges and masses of the electron, proton, and neutron are given in Table 19.1.

▶ **TABLE 19.1** | **Charge and Mass of the Electron, Proton, and Neutron**

Particle	Charge (C)	Mass (kg)
Electron (e)	$-1.602\ 176\ 5 \times 10^{-19}$	$9.109\ 4 \times 10^{-31}$
Proton (p)	$+1.602\ 176\ 5 \times 10^{-19}$	$1.672\ 62 \times 10^{-27}$
Neutron (n)	0	$1.674\ 93 \times 10^{-27}$

When dealing with Coulomb's law, remember that force is a *vector* quantity and must be treated accordingly. Furthermore, Coulomb's law applies exactly only to particles.[2] The electrostatic force exerted by q_1 on q_2, written $\vec{\mathbf{F}}_{12}$, can be expressed in vector form as[3]

$$\vec{\mathbf{F}}_{12} = k_e \frac{q_1 q_2}{r^2} \hat{\mathbf{r}}_{12} \qquad \textbf{19.2} \blacktriangleleft$$

where $\hat{\mathbf{r}}_{12}$ is a unit vector directed from q_1 toward q_2 as in Active Figure 19.7a. Equation 19.2 can be used to find the direction of the force in space, although a carefully drawn pictorial representation is needed to clearly identify the direction of $\hat{\mathbf{r}}_{12}$. From Newton's third law, we see that the electric force exerted by q_2 on q_1 is equal in magnitude to the force exerted by q_1 on q_2 and in the opposite direction; that is, $\vec{\mathbf{F}}_{21} = -\vec{\mathbf{F}}_{12}$. From Equation 19.2, we see that if q_1 and q_2 have the same sign, the product $q_1 q_2$ is positive and the force is repulsive as in Active Figure 19.7a. The force on q_2 is in the same direction as $\hat{\mathbf{r}}_{12}$ and is directed away from q_1. If q_1 and q_2 are of opposite sign as in Active Figure 19.7b, the product $q_1 q_2$ is negative and the force is attractive. In this case, the force on q_2 is in the direction opposite to $\hat{\mathbf{r}}_{12}$, directed toward q_1.

When more than two charged particles are present, the force between any pair is given by Equation 19.2. Therefore, the resultant force on any one particle equals the *vector* sum of the individual forces due to all other particles. This **principle of superposition** as applied to electrostatic forces is an experimentally observed fact and simply represents the traditional vector sum of forces introduced in Chapter 4. As an example, if four charged particles are present, the resultant force on particle 1 due to particles 2, 3, and 4 is given by the vector sum

$$\vec{\mathbf{F}}_1 = \vec{\mathbf{F}}_{21} + \vec{\mathbf{F}}_{31} + \vec{\mathbf{F}}_{41}$$

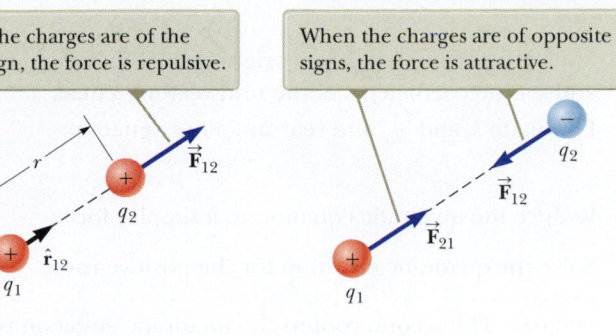

When the charges are of the same sign, the force is repulsive.

When the charges are of opposite signs, the force is attractive.

Active Figure 19.7 Two point charges separated by a distance r exert a force on each other given by Coulomb's law. Note that the force $\vec{\mathbf{F}}_{21}$ exerted by q_2 on q_1 is equal in magnitude and opposite in direction to the force $\vec{\mathbf{F}}_{12}$ exerted by q_1 on q_2.

QUICK QUIZ 19.3 Object A has a charge of $+2\ \mu\text{C}$, and object B has a charge of $+6\ \mu\text{C}$. Which statement is true about the electric forces on the objects?
(a) $\vec{\mathbf{F}}_{AB} = -3\vec{\mathbf{F}}_{BA}$ (b) $\vec{\mathbf{F}}_{AB} = -\vec{\mathbf{F}}_{BA}$ (c) $3\vec{\mathbf{F}}_{AB} = -\vec{\mathbf{F}}_{BA}$ (d) $\vec{\mathbf{F}}_{AB} = 3\vec{\mathbf{F}}_{BA}$
(e) $\vec{\mathbf{F}}_{AB} = \vec{\mathbf{F}}_{BA}$ (f) $3\vec{\mathbf{F}}_{AB} = \vec{\mathbf{F}}_{BA}$

Example 19.1 | **Where Is the Net Force Zero?**

Three point charges lie along the x axis as shown in Figure 19.8. The positive charge $q_1 = 15.0\ \mu\text{C}$ is at $x = 2.00$ m, the positive charge $q_2 = 6.00\ \mu\text{C}$ is at the origin, and the net force acting on q_3 is zero. What is the x coordinate of q_3?

SOLUTION

Figure 19.8 (Example 19.1) Three point charges are placed along the x axis. If the resultant force acting on q_3 is zero, the force $\vec{\mathbf{F}}_{13}$ exerted by q_1 on q_3 must be equal in magnitude and opposite in direction to the force $\vec{\mathbf{F}}_{23}$ exerted by q_2 on q_3.

Conceptualize Because q_3 is near two other charges, it experiences two electric forces. The forces lie along the same line as indicated in Figure 19.8. Because q_3 is negative and q_1 and q_2 are positive, the forces $\vec{\mathbf{F}}_{13}$ and $\vec{\mathbf{F}}_{23}$ are both attractive.

Categorize Because the net force on q_3 is zero, we model the point charge as a particle in equilibrium.

Analyze Write an expression for the net force on charge q_3 when it is in equilibrium:

$$\vec{\mathbf{F}}_3 = \vec{\mathbf{F}}_{23} + \vec{\mathbf{F}}_{13} = -k_e \frac{|q_2||q_3|}{x^2}\hat{\mathbf{i}} + k_e \frac{|q_1||q_3|}{(2.00 - x)^2}\hat{\mathbf{i}} = 0$$

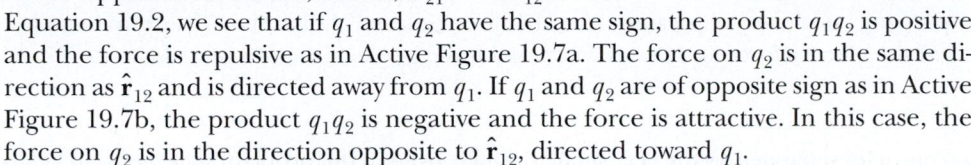

continued

---

[2]Coulomb's law can also be used for larger objects to which the particle model can be applied.

[3]Notice that we use "q_2" as shorthand notation for "the particle with charge q_2." This usage is common when discussing charged particles, similar to the use in mechanics of "m_2" for "the particle with mass m_2." The context of the sentence will tell you whether the symbol represents an amount of charge or a particle with that charge.

19.1 *cont.*

Move the second term to the right side of the equation and set the coefficients of the unit vector $\hat{\mathbf{i}}$ equal:

$$k_e \frac{|q_2||q_3|}{x^2} = k_e \frac{|q_1||q_3|}{(2.00-x)^2}$$

Eliminate k_e and $|q_3|$ and rearrange the equation:

$$(2.00-x)^2|q_2| = x^2|q_1|$$

$$(4.00 - 4.00x + x^2)(6.00 \times 10^{-6}\text{ C}) = x^2(15.0 \times 10^{-6}\text{ C})$$

Reduce the quadratic equation to a simpler form:

$$3.00x^2 + 8.00x - 8.00 = 0$$

Solve the quadratic equation for the positive root:

$$x = 0.775 \text{ m}$$

Finalize The second root to the quadratic equation is $x = -3.44$ m. That is another location where the *magnitudes* of the forces on q_3 are equal, but both forces are in the same direction, so they do not cancel.

Example 19.2 | The Hydrogen Atom

The electron and proton of a hydrogen atom are separated (on the average) by a distance of approximately 5.3×10^{-11} m. Find the magnitudes of the electric force and the gravitational force between the two particles.

SOLUTION

Conceptualize Think about the two particles separated by the very small distance given in the problem statement. In Chapter 5, we mentioned that the gravitational force between an electron and a proton is very small compared to the electric force between them, so we expect this to be the case with the results of this example.

Categorize The electric and gravitational forces will be evaluated from universal force laws, so we categorize this example as a substitution problem.

Use Coulomb's law to find the magnitude of the electric force:

$$F_e = k_e \frac{|e||-e|}{r^2} = (8.99 \times 10^9 \text{ N} \cdot \text{m}^2/\text{C}^2) \frac{(1.60 \times 10^{-19} \text{ C})^2}{(5.3 \times 10^{-11} \text{ m})^2}$$

$$= 8.2 \times 10^{-8} \text{ N}$$

Use Newton's law of universal gravitation and Table 19.1 (for the particle masses) to find the magnitude of the gravitational force:

$$F_g = G \frac{m_e m_p}{r^2}$$

$$= (6.67 \times 10^{-11} \text{ N} \cdot \text{m}^2/\text{kg}^2) \frac{(9.11 \times 10^{-31} \text{ kg})(1.67 \times 10^{-27} \text{ kg})}{(5.3 \times 10^{-11} \text{ m})^2}$$

$$= 3.6 \times 10^{-47} \text{ N}$$

The ratio $F_e/F_g \approx 2 \times 10^{39}$. Therefore, the gravitational force between charged atomic particles is negligible when compared with the electric force. Notice the similar forms of Newton's law of universal gravitation and Coulomb's law of electric forces. Other than the magnitude of the forces between elementary particles, what is a fundamental difference between the two forces?

Example 19.3 | Find the Charge on the Spheres

Two identical small charged spheres, each having a mass of 3.00×10^{-2} kg, hang in equilibrium as shown in Figure 19.9a. The length L of each string is 0.150 m, and the angle θ is 5.00°. Find the magnitude of the charge on each sphere.

SOLUTION

Conceptualize Figure 19.9a helps us conceptualize this example. The two spheres exert repulsive forces on each other. If they are held close to each other and released, they move outward from the center and settle into the configuration in Figure 19.9a after the oscillations have vanished due to air resistance.

Categorize The key phrase "in equilibrium" helps us model each sphere as a particle in equilibrium. This example is similar

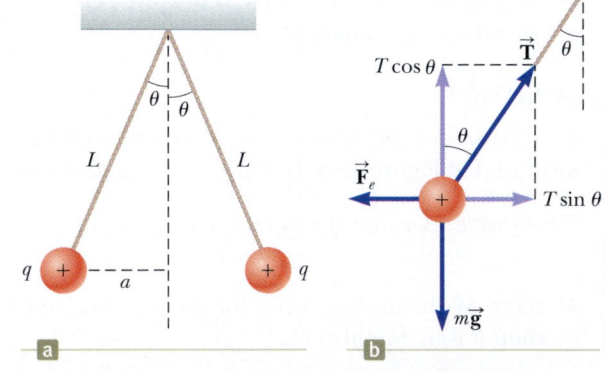

Figure 19.9 (Example 19.3) (a) Two identical spheres, each carrying the same charge q, suspended in equilibrium. (b) Diagram of the forces acting on the sphere on the left part of (a).

19.3 *cont.*

to the particle in equilibrium problems in Chapter 4 with the added feature that one of the forces on a sphere is an electric force.

Analyze The force diagram for the left-hand sphere is shown in Figure 19.9b. The sphere is in equilibrium under the application of the force $\vec{\mathbf{T}}$ from the string, the electric force $\vec{\mathbf{F}}_e$ from the other sphere, and the gravitational force $m\vec{\mathbf{g}}$.

Write Newton's second law for the left-hand sphere in component form:

$$(1) \quad \sum F_x = T \sin\theta - F_e = 0 \quad \rightarrow \quad T \sin\theta = F_e$$

$$(2) \quad \sum F_y = T \cos\theta - mg = 0 \quad \rightarrow \quad T \cos\theta = mg$$

Divide Equation (1) by Equation (2) to find F_e:

$$\tan\theta = \frac{F_e}{mg} \quad \rightarrow \quad F_e = mg\tan\theta$$

Use the geometry of the right triangle in Figure 19.9a to find a relationship between a, L, and θ:

$$\sin\theta = \frac{a}{L} \quad \rightarrow \quad a = L\sin\theta$$

Solve Coulomb's law (Eq. 19.1) for the charge $|q|$ on each sphere:

$$|q| = \sqrt{\frac{F_e r^2}{k_e}} = \sqrt{\frac{F_e(2a)^2}{k_e}} = \sqrt{\frac{mg\tan\theta(2L\sin\theta)^2}{k_e}}$$

Substitute numerical values:

$$|q| = \sqrt{\frac{(3.00\times10^{-2}\,\text{kg})(9.80\,\text{m/s}^2)\tan(5.00°)[2(0.150\,\text{m})\sin(5.00°)]^2}{8.99\times10^9\,\text{N}\cdot\text{m}^2/\text{C}^2}}$$

$$= 4.42\times10^{-8}\,\text{C}$$

Finalize If the sign of the charges were not given in Figure 19.9, we could not determine them. In fact, the sign of the charge is not important. The situation is the same whether both spheres are positively charged or negatively charged.

19.5 | Electric Fields

In Section 4.1, we discussed the differences between contact forces and field forces. Two field forces—the gravitational force in Chapter 11 and the electric force here—have been introduced into our discussions so far. As pointed out earlier, field forces can act through space, producing an effect even when no physical contact occurs between interacting objects. The gravitational field $\vec{\mathbf{g}}$ at a point in space due to a source particle was defined in Section 11.1 to be equal to the gravitational force $\vec{\mathbf{F}}_g$ acting on a test particle of mass m divided by that mass: $\vec{\mathbf{g}} \equiv \vec{\mathbf{F}}_g / m$. The concept of a field was developed by Michael Faraday (1791–1867) in the context of electric forces and is of such practical value that we shall devote much attention to it in the next several chapters. In this approach, an **electric field** is said to exist in the region of space around a charged object, the **source charge.** When another charged object—the **test charge**—enters this electric field, an electric force acts on it. As an example, consider Figure 19.10, which shows a small positive test charge q_0 placed near a second object carrying a much greater positive charge Q. We define the electric field due to the source charge at the location of the test charge to be the electric force on the test charge *per unit charge*, or, to be more specific, the **electric field vector** $\vec{\mathbf{E}}$ at a point in space is defined as the electric force $\vec{\mathbf{F}}_e$ acting on a positive test charge q_0 placed at that point divided by the test charge:[4]

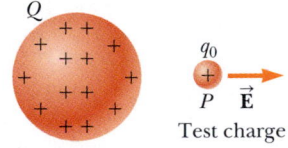

Figure 19.10 A small positive test charge q_0 placed at point P near an object carrying a much larger positive charge Q experiences an electric field $\vec{\mathbf{E}}$ at point P established by the source charge Q. We will *always* assume that the test charge is so small that the field of the source charge is unaffected by its presence.

$$\vec{\mathbf{E}} \equiv \frac{\vec{\mathbf{F}}_e}{q_0} \qquad \qquad 19.3 \blacktriangleleft \qquad \blacktriangleright \text{Definition of electric field}$$

The vector $\vec{\mathbf{E}}$ has the SI units of newtons per coulomb (N/C). The direction of $\vec{\mathbf{E}}$ as shown in Figure 19.10 is the direction of the force a positive test charge experiences when placed in the field. Note that $\vec{\mathbf{E}}$ is the field produced by some charge or charge distribution *separate from* the test charge; it is not the field produced by the

---

[4]When using Equation 19.3, we must assume the test charge q_0 is small enough that it does not disturb the charge distribution responsible for the electric field. If the test charge is great enough, the charge on the metallic sphere is redistributed and the electric field it sets up is different from the field it sets up in the presence of the much smaller test charge.

Active Figure 19.11 (a), (c) When a test charge q_0 is placed near a source charge q, the test charge experiences a force. (b), (d) At a point P near a source charge q, there exists an electric field.

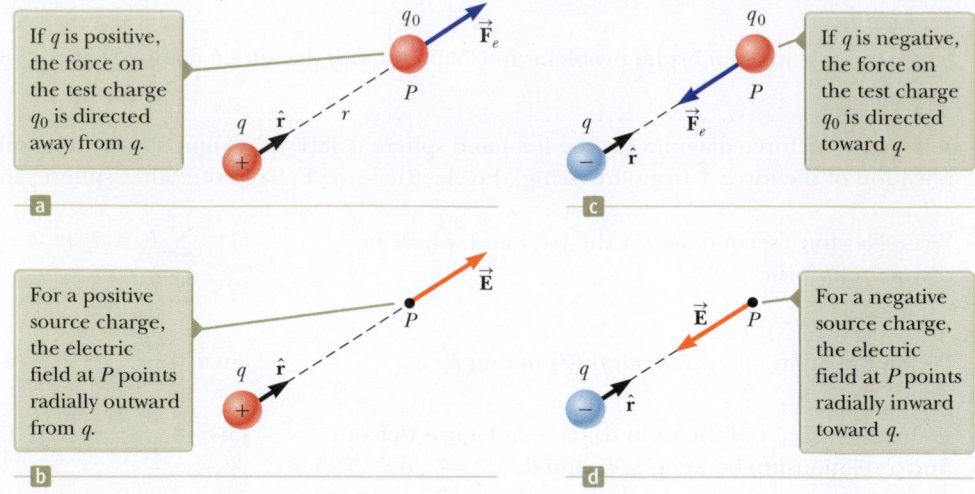

If q is positive, the force on the test charge q_0 is directed away from q.

If q is negative, the force on the test charge q_0 is directed toward q.

For a positive source charge, the electric field at P points radially outward from q.

For a negative source charge, the electric field at P points radially inward toward q.

Pitfall Prevention | 19.1
Particles Only
Equation 19.4 is valid only for a *particle* of charge q, that is, an object of zero size. For a charged object of finite size in an electric field, the field may vary in magnitude and direction over the size of the object, so the corresponding force equation may be more complicated.

test charge itself. Also note that the existence of an electric field is a property of its source; the presence of the test charge is not necessary for the field to exist. The test charge serves as a *detector* of the electric field: an electric field exists at a point if a test charge at that point experiences an electric force.

Once the electric field is known at some point, the force on *any* particle with charge q placed at that point can be calculated from a rearrangement of Equation 19.3:

$$\vec{\mathbf{F}}_e = q\vec{\mathbf{E}} \qquad \text{19.4} \blacktriangleleft$$

Once the electric force on a particle is evaluated, its motion can be determined from the particle under a net force model or the particle in equilibrium model (the electric force may have to be combined with other forces acting on the particle), and the techniques of earlier chapters can be used to find the motion of the particle.

Consider a point charge[5] q located a distance r from a test particle with charge q_0. According to Coulomb's law, the force exerted on the test particle by q is

$$\vec{\mathbf{F}}_e = k_e \frac{qq_0}{r^2} \hat{\mathbf{r}}$$

where $\hat{\mathbf{r}}$ is a unit vector directed from q toward q_0. This force in Active Figure 19.11a is directed away from the source charge q. Because the electric field at P, the position of the test charge, is defined by $\vec{\mathbf{E}} = \vec{\mathbf{F}}_e/q_0$, we find that at P, the electric field created by q is

▶ Electric field due to a point charge

$$\vec{\mathbf{E}} = k_e \frac{q}{r^2} \hat{\mathbf{r}} \qquad \text{19.5} \blacktriangleleft$$

If the source charge q is positive, Active Figure 19.11b shows the situation with the test charge removed; the source charge sets up an electric field at point P, directed away from q. If q is negative as in Active Figure 19.11c, the force on the test charge is toward the source charge, so the electric field at P is directed toward the source charge as in Active Figure 19.11d.

To calculate the electric field at a point P due to a group of point charges, we first calculate the electric field vectors at P individually using Equation 19.5 and then add them vectorially. In other words, the total electric field at a point in space due to a group of charged particles equals the vector sum of the electric fields at that point due to all the particles. This superposition principle applied to fields follows directly from the vector addition property of forces. Therefore, the electric field at point P of a group of source charges can be expressed as

▶ Electric field due to a finite number of point charges

$$\vec{\mathbf{E}} = k_e \sum_i \frac{q_i}{r_i^2} \hat{\mathbf{r}}_i \qquad \text{19.6} \blacktriangleleft$$

[5]We have used the phrase "charged particle" so far. The phrase "point charge" is somewhat misleading because charge is a property of a particle, not a physical entity. It is similar to misleading phrasing in mechanics such as "a mass m is placed . . ." (which we have avoided) rather than "a particle with mass m is placed. . . ." This phrase is so ingrained in physics usage, however, that we will use it and hope that this footnote suffices to clarify its use.

where r_i is the distance from the ith charge q_i to the point P (the location at which the field is to be evaluated) and $\hat{\mathbf{r}}_i$ is a unit vector directed from q_i toward P.

> **QUICK QUIZ 19.4** A test charge of $+3 \, \mu C$ is at a point P where an external electric field is directed to the right and has a magnitude of 4×10^6 N/C. If the test charge is replaced with another charge of $-3 \, \mu C$, what happens to the external electric field at P? (**a**) It is unaffected. (**b**) It reverses direction. (**c**) It changes in a way that cannot be determined.

Example 19.4 | Electric Field of a Dipole

An **electric dipole** consists of a point charge q and a point charge $-q$ separated by a distance of $2a$ as in Figure 19.12. As we shall see in later chapters, neutral atoms and molecules behave as dipoles when placed in an external electric field. Furthermore, many molecules, such as HCl, are permanent dipoles. (HCl can be effectively modeled as an H^+ ion combined with a Cl^- ion.) The effect of such dipoles on the behavior of materials subjected to electric fields is discussed in Chapter 20.

(A) Find the electric field $\vec{\mathbf{E}}$ due to the dipole along the y axis at the point P, which is a distance y from the origin.

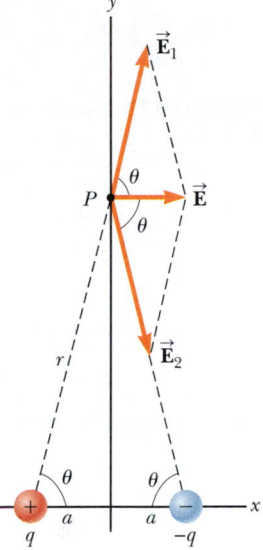

Figure 19.12 (Example 19.4) The total electric field $\vec{\mathbf{E}}$ at P due to two charges of equal magnitude and opposite sign (an electric dipole) equals the vector sum $\vec{\mathbf{E}}_1 + \vec{\mathbf{E}}_2$. The field $\vec{\mathbf{E}}_1$ is due to the positive charge q, and $\vec{\mathbf{E}}_2$ is the field due to the negative charge $-q$.

SOLUTION

Conceptualize In Example 19.1, we added vector forces to find the net force on a particle. Here, we add electric field vectors to find the net electric field at a point in space.

Categorize We have two source charges and wish to find the total electric field, so we categorize this example as one in which we can use the superposition principle represented by Equation 19.6.

· ·

Analyze At P, the fields $\vec{\mathbf{E}}_1$ and $\vec{\mathbf{E}}_2$ due to the two particles are equal in magnitude because P is equidistant from the two charges. The total field at P is $\vec{\mathbf{E}} = \vec{\mathbf{E}}_1 + \vec{\mathbf{E}}_2$.

Find the magnitudes of the fields at P:

$$E_1 = E_2 = k_e \frac{q}{r^2} = k_e \frac{q}{y^2 + a^2}$$

The y components of $\vec{\mathbf{E}}_1$ and $\vec{\mathbf{E}}_2$ are equal in magnitude and opposite in sign, so they cancel. The x components are equal and add because they have the same sign. The total field $\vec{\mathbf{E}}$ is therefore parallel to the x axis.

Find an expression for the magnitude of the electric field at P:

$$(1) \quad E = 2k_e \frac{q}{y^2 + a^2} \cos\theta$$

From the geometry in Figure 19.12 we see that $\cos\theta = a/r = a/(y^2 + a^2)^{1/2}$. Substitute this result into Equation (1):

$$E = 2k_e \frac{q}{(y^2 + a^2)} \frac{a}{(y^2 + a^2)^{1/2}}$$

$$(2) \quad E = k_e \frac{2qa}{(y^2 + a^2)^{3/2}}$$

(B) Find the electric field for points $y \gg a$ far from the dipole.

SOLUTION

Equation (2) gives the value of the electric field on the y axis at all values of y. For points far from the dipole, for which $y \gg a$, neglect a^2 in the denominator and write the expression for E in this case:

$$(3) \quad E \approx k_e \frac{2qa}{y^3}$$

· ·

Finalize Therefore, we see that along the y axis the field of a dipole at a distant point varies as $1/r^3$, whereas the more slowly varying field of a point charge varies as $1/r^2$. (*Note:* In the geometry of this example, $r = y$.) At distant points, the fields of the two charges in the dipole almost cancel each other. The $1/r^3$ variation in E for the dipole is also obtained for a distant point along the x axis (see Problem 19.20 in Enhanced WebAssign) and for a general distant point.

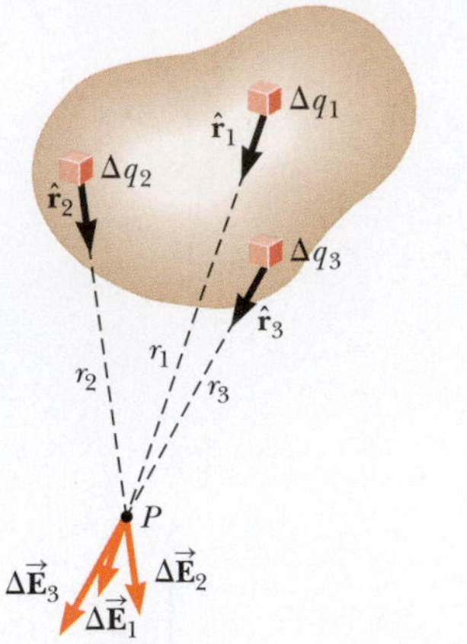

Figure 19.13 The electric field $\vec{E}$ at P due to a continuous charge distribution is the vector sum of the fields $\Delta\vec{E}$ due to all the elements Δq_i of the charge distribution. Three sample elements are shown.

Electric Field Due to Continuous Charge Distributions

In most practical situations (e.g., an object charged by rubbing), the average separation between source charges is small compared with their distances from the point at which the field is to be evaluated. In such cases, the system of source charges can be modeled as *continuous*. That is, we imagine that the system of closely spaced discrete charges is equivalent to a total charge that is continuously distributed through some volume or over some surface.

To evaluate the electric field of a continuous charge distribution, the following procedure is used. First, we divide the charge distribution into small elements, each of which contains a small amount of charge Δq as in Figure 19.13. Next, modeling the element as a point charge, we use Equation 19.5 to calculate the electric field $\Delta\vec{E}$ at a point P due to one of these elements. Finally, we evaluate the total field at P due to the charge distribution by performing a vector sum of the contributions of all the charge elements (i.e., by applying the superposition principle).

The electric field at P in Figure 19.13 due to one element of charge Δq_i is given by

$$\Delta\vec{E}_i = k_e \frac{\Delta q_i}{r_i^2}\hat{\mathbf{r}}_i$$

where the index i refers to the ith element in the distribution, r_i is the distance from the element to point P, and $\hat{\mathbf{r}}_i$ is a unit vector directed from the element toward P. The total electric field $\vec{E}$ at P due to all elements in the charge distribution is approximately

$$\vec{E} \approx k_e \sum_i \frac{\Delta q_i}{r_i^2}\hat{\mathbf{r}}_i$$

Now, we apply the model in which the charge distribution is continuous, and we let the elements of charge become infinitesimally small. With this model, the total field at P in the limit $\Delta q_i \to 0$ becomes

$$\vec{E} = \lim_{\Delta q_i \to 0} k_e \sum_i \frac{\Delta q_i}{r_i^2}\hat{\mathbf{r}}_i = k_e \int \frac{dq}{r^2}\hat{\mathbf{r}} \qquad \text{19.7} \blacktriangleleft$$

where dq is an infinitesimal amount of charge and the integration is over all the charge creating the electric field. The integration is a *vector* operation and must be treated with caution. It can be evaluated in terms of individual components, or perhaps symmetry arguments can be used to reduce it to a scalar integral. We shall illustrate this type of calculation with several examples in which we assume that the charge is *uniformly* distributed on a line or a surface or throughout some volume. When performing such calculations, it is convenient to use the concept of a *charge density* along with the following notations:

- If a total charge Q is uniformly distributed throughout a volume V, the **volume charge density** ρ is defined by

▶ Volume charge density

$$\rho \equiv \frac{Q}{V} \qquad \text{19.8} \blacktriangleleft$$

where ρ has units of coulombs per cubic meter.

- If Q is uniformly distributed on a surface of area A, the **surface charge density** σ is defined by

▶ Surface charge density

$$\sigma \equiv \frac{Q}{A} \qquad \text{19.9} \blacktriangleleft$$

where σ has units of coulombs per square meter.

- If Q is uniformly distributed along a line of length ℓ, the **linear charge density** λ is defined by

▶ Linear charge density

$$\lambda \equiv \frac{Q}{\ell} \qquad \text{19.10} \blacktriangleleft$$

where λ has units of coulombs per meter.

> ## PROBLEM-SOLVING STRATEGY: Calculating the Electric Field

The following procedure is recommended for solving problems that involve the determination of an electric field due to individual charges or a charge distribution.

1. **Conceptualize.** Establish a mental representation of the problem: think carefully about the individual charges or the charge distribution and imagine what type of electric field it would create. Appeal to any symmetry in the arrangement of charges to help you visualize the electric field.

2. **Categorize.** Are you analyzing a group of individual charges or a continuous charge distribution? The answer to this question tells you how to proceed in the Analyze step.

3. **Analyze.**

(a) If you are analyzing a group of individual charges, use the superposition principle: when several point charges are present, the resultant field at a point in space is the *vector sum* of the individual fields due to the individual charges (Eq. 19.6). Be very careful in the manipulation of vector quantities. It may be useful to review the material on vector addition in Chapter 1. Example 19.4 demonstrated this procedure.

(b) If you are analyzing a continuous charge distribution, replace the vector sums for evaluating the total electric field from individual charges by vector integrals. The charge distribution is divided into infinitesimal pieces, and the vector sum is carried out by integrating over the entire charge distribution (Eq. 19.7). Examples 19.5 and 19.6 demonstrate such procedures.

Consider symmetry when dealing with either a distribution of point charges or a continuous charge distribution. Take advantage of any symmetry in the system you observed in the Conceptualize step to simplify your calculations. The cancellation of field components perpendicular to the axis in Example 19.6 is an example of the application of symmetry.

4. **Finalize.** Check to see if your electric field expression is consistent with the mental representation and if it reflects any symmetry that you noted previously. Imagine varying parameters such as the distance of the observation point from the charges or the radius of any circular objects to see if the mathematical result changes in a reasonable way.

Example 19.5 | The Electric Field Due to a Charged Rod

A rod of length ℓ has a uniform positive charge per unit length λ and a total charge Q. Calculate the electric field at a point P that is located along the long axis of the rod and a distance a from one end (Fig. 19.14).

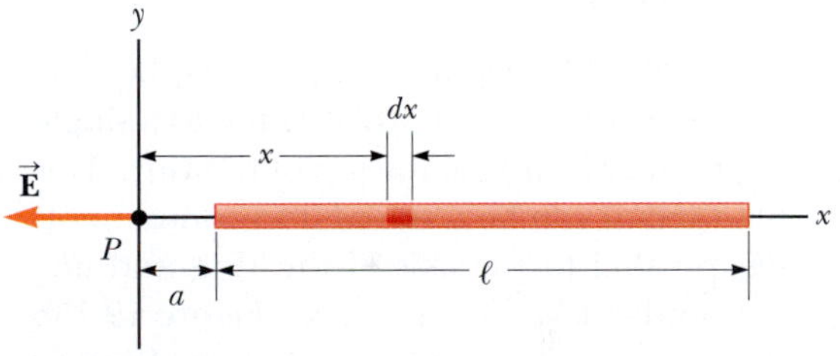

Figure 19.14 (Example 19.5) The electric field at P due to a uniformly charged rod lying along the x axis.

SOLUTION

Conceptualize The field $d\vec{E}$ at P due to each segment of charge on the rod is in the negative x direction because every segment carries a positive charge.

Categorize Because the rod is continuous, we are evaluating the field due to a continuous charge distribution rather than a group of individual charges. Because every segment of the rod produces an electric field in the negative x direction, the sum of their contributions can be handled without the need to add vectors.

....

Analyze Let's assume the rod is lying along the x axis, dx is the length of one small segment, and dq is the charge on that segment. Because the rod has a charge per unit length λ, the charge dq on the small segment is $dq = \lambda\,dx$.

Find the magnitude of the electric field at P due to one segment of the rod having a charge dq:

$$dE = k_e \frac{dq}{x^2} = k_e \frac{\lambda\,dx}{x^2}$$

Find the total field at P using[6] Equation 19.7:

$$E = \int_a^{\ell + a} k_e \lambda \frac{dx}{x^2}$$

continued

---

[6]To carry out integrations such as this one, first express the charge element dq in terms of the other variables in the integral. (In this example, there is one variable, x, so we made the change $dq = \lambda\,dx$.) The integral must be over scalar quantities; therefore, express the electric field in terms of components, if necessary. (In this example, the field has only an x component, so this detail is of no concern.) Then, reduce your expression to an integral over a single variable (or to multiple integrals, each over a single variable). In examples that have spherical or cylindrical symmetry, the single variable is a radial coordinate.

19.5 *cont.*

Noting that k_e and $\lambda = Q/\ell$ are constants and can be removed from the integral, evaluate the integral:

$$E = k_e\lambda \int_a^{\ell+a} \frac{dx}{x^2} = k_e\lambda \left[-\frac{1}{x}\right]_a^{\ell+a}$$

$$(1) \quad E = k_e \frac{Q}{\ell}\left(\frac{1}{a} - \frac{1}{\ell+a}\right) = \boxed{\frac{k_e Q}{a(\ell+a)}}$$

Finalize If $a \to 0$, which corresponds to sliding the bar to the left until its left end is at the origin, then $E \to \infty$. That represents the condition in which the observation point P is at zero distance from the charge at the end of the rod, so the field becomes infinite. We explore large values of a below.

What If? Suppose point P is very far away from the rod. What is the nature of the electric field at such a point?

Answer If P is far from the rod $(a \gg \ell)$, then ℓ in the denominator of Equation (1) can be neglected and $E \approx k_e Q/a^2$. That is exactly the form you would expect for a point charge. Therefore, at large values of a/ℓ, the charge distribution appears to be a point charge of magnitude Q; the point P is so far away from the rod we cannot distinguish that it has a size. The use of the limiting technique $(a/\ell \to \infty)$ is often a good method for checking a mathematical expression.

Example 19.6 | The Electric Field of a Uniform Ring of Charge

A ring of radius a carries a uniformly distributed positive total charge Q. Calculate the electric field due to the ring at a point P lying a distance x from its center along the central axis perpendicular to the plane of the ring (Fig. 19.15a).

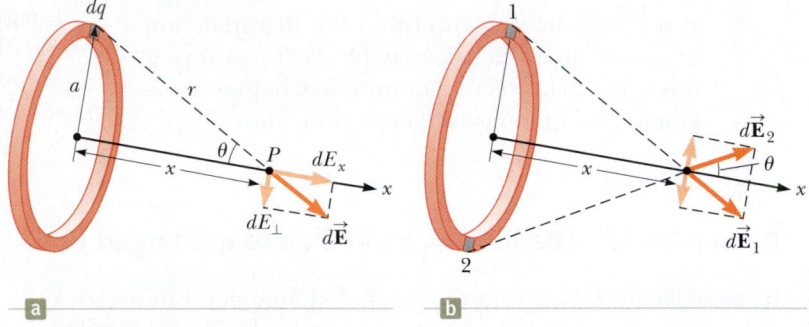

Figure 19.15 (Example 19.6) A uniformly charged ring of radius a. (a) The field at P on the x axis due to an element of charge dq. (b) The total electric field at P is along the x axis. The perpendicular component of the field at P due to segment 1 is canceled by the perpendicular component due to segment 2.

SOLUTION

Conceptualize Figure 19.15a shows the electric field contribution $d\vec{E}$ at P due to a single segment of charge at the top of the ring. This field vector can be resolved into components dE_x parallel to the axis of the ring and $dE_\perp$ perpendicular to the axis. Figure 19.15b shows the electric field contributions from two segments on opposite sides of the ring. Because of the symmetry of the situation, the perpendicular components of the field cancel. That is true for all pairs of segments around the ring, so we can ignore the perpendicular component of the field and focus solely on the parallel components, which simply add.

Categorize Because the ring is continuous, we are evaluating the field due to a continuous charge distribution rather than a group of individual charges.

Analyze Evaluate the parallel component of an electric field contribution from a segment of charge dq on the ring:

$$(1) \quad dE_x = k_e \frac{dq}{r^2}\cos\theta = k_e \frac{dq}{a^2+x^2}\cos\theta$$

From the geometry in Figure 19.15a, evaluate $\cos\theta$:

$$(2) \quad \cos\theta = \frac{x}{r} = \frac{x}{(a^2+x^2)^{1/2}}$$

Substitute Equation (2) into Equation (1):

$$dE_x = k_e \frac{dq}{a^2+x^2}\frac{x}{(a^2+x^2)^{1/2}} = \frac{k_e x}{(a^2+x^2)^{3/2}}dq$$

All segments of the ring make the same contribution to the field at P because they are all equidistant from this point. Integrate to obtain the total field at P:

$$E_x = \int \frac{k_e x}{(a^2+x^2)^{3/2}}dq = \frac{k_e x}{(a^2+x^2)^{3/2}}\int dq$$

$$(3) \quad E = \boxed{\frac{k_e x}{(a^2+x^2)^{3/2}}Q}$$

19.6 *cont.*

Finalize This result shows that the field is zero at $x = 0$. Is that consistent with the symmetry in the problem? Furthermore, notice that Equation (3) reduces to $k_e Q/x^2$ if $x \gg a$, so the ring acts like a point charge for locations far away from the ring.

What If? Suppose a negative charge is placed at the center of the ring in Figure 19.15 and displaced slightly by a distance $x \ll a$ along the x axis. When the charge is released, what type of motion does it exhibit?

Answer In the expression for the field due to a ring of charge, let $x \ll a$, which results in

$$E_x = \frac{k_e Q}{a^3} x$$

Therefore, from Equation 19.4, the force on a charge $-q$ placed near the center of the ring is

$$F_x = -\frac{k_e q Q}{a^3} x$$

Because this force has the form of Hooke's law (Eq. 12.1), the motion of the negative charge is *simple harmonic!*

19.6 | Electric Field Lines

A convenient specialized pictorial representation for visualizing electric field patterns is created by drawing lines showing the direction of the electric field vector at any point. These lines, called **electric field lines,** are related to the electric field in any region of space in the following manner:

- The electric field vector $\vec{E}$ is *tangent* to the electric field line at each point.
- The number of electric field lines per unit area through a surface that is perpendicular to the lines is proportional to the magnitude of the electric field in that region. Therefore, E is large where the field lines are close together and small where they are far apart.

These properties are illustrated in Figure 19.16. The density of lines through surface A is greater than the density of lines through surface B. Therefore, the magnitude of the electric field on surface A is larger than on surface B. Furthermore, the field drawn in Figure 19.16 is nonuniform because the lines at different locations point in different directions.

Some representative electric field lines for a single positive point charge are shown in Figure 19.17a. Note that in this two-dimensional drawing we show only the field lines that lie in the plane of the page. The lines are actually directed radially outward in *all* directions from the charge, somewhat like the needles of a porcupine. Because a positively charged test particle placed in this field would be repelled by the charge q, the lines are directed radially away from q. Similarly, the electric field lines for a single negative point charge are directed toward the charge (Fig. 19.17b). In either case, the lines are radial and extend to infinity. Note that the lines are closer together as they come nearer to the charge, indicating that the magnitude of the field is increasing. The electric field lines end in Figure 19.17a and begin in Figure 19.17b on hypothetical charges we assume to be located infinitely far away.

Is this visualization of the electric field in terms of field lines consistent with Equation 19.5? To answer this question, consider an imaginary spherical surface of radius r, concentric with the charge. From

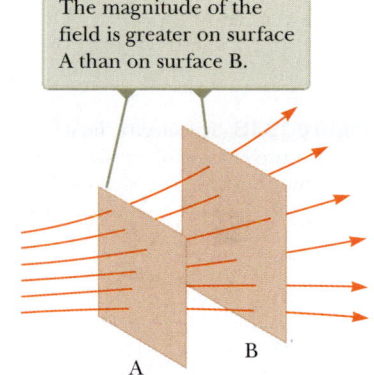

The magnitude of the field is greater on surface A than on surface B.

Figure 19.16 Electric field lines penetrating two surfaces.

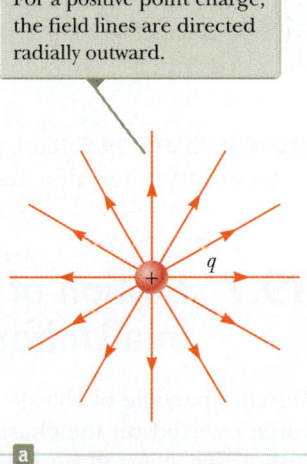

For a positive point charge, the field lines are directed radially outward.

q

a

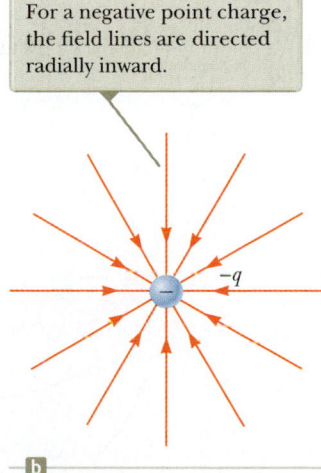

For a negative point charge, the field lines are directed radially inward.

$-q$

b

Figure 19.17 The electric field lines for a point charge. Notice that the figures show only those field lines that lie in the plane of the page.

Pitfall Prevention | 19.2
Electric Field Lines Are Not Paths of Particles
Electric field lines represent the field at various locations. Except in very special cases, they *do not* represent the path of a charged particle released in an electric field.

The number of field lines leaving the positive charge equals the number terminating at the negative charge.

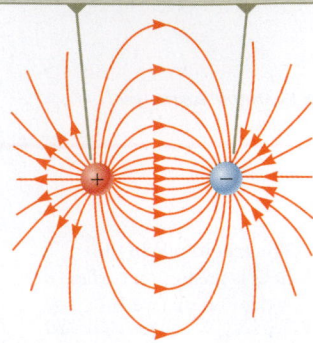

Figure 19.18 The electric field lines for two charges of equal magnitude and opposite sign (an electric dipole).

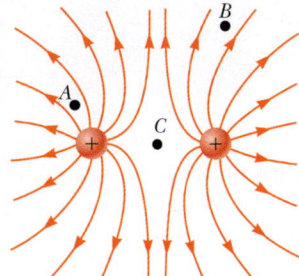

Figure 19.19 The electric field lines for two positive point charges. (The locations A, B, and C are discussed in Quick Quiz 19.5.)

Two field lines leave $+2q$ for every one that terminates on $-q$.

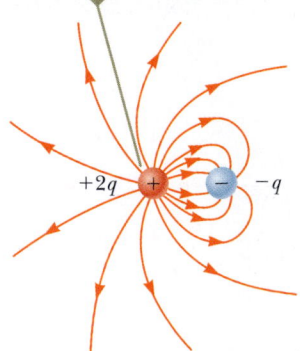

Active Figure 19.20 The electric field lines for a point charge $+2q$ and a second point charge $-q$.

symmetry, we see that the magnitude of the electric field is the same everywhere on the surface of the sphere. The number of lines N emerging from the charge is equal to the number penetrating the spherical surface. Hence, the number of lines per unit area on the sphere is $N/4\pi r^2$ (where the surface area of the sphere is $4\pi r^2$). Because E is proportional to the number of lines per unit area, we see that E varies as $1/r^2$. This result is consistent with that obtained from Equation 19.5; that is, $E = k_e q/r^2$.

The rules for drawing electric field lines for any charge distribution are as follows:

- The lines must begin on a positive charge and terminate on a negative charge. In the case of an excess of one type of charge, some lines will begin or end infinitely far away.
- The number of lines drawn leaving a positive charge or approaching a negative charge is proportional to the magnitude of the charge.
- No two field lines can cross.

Because charge is quantized, the number of lines leaving any positively charged object must be 0, ae, $2ae$, . . ., where a is an arbitrary (but fixed) proportionality constant chosen by the person drawing the lines. Once a is chosen, the number of lines is no longer arbitrary. For example, if object 1 has charge Q_1 and object 2 has charge Q_2, the ratio of the number of lines connected to object 2 to those connected to object 1 is $N_2/N_1 = Q_2/Q_1$.

The electric field lines for two point charges of equal magnitude but opposite signs (the electric dipole) are shown in Figure 19.18. In this case, the number of lines that begin at the positive charge must equal the number that terminate at the negative charge. At points very near the charges, the lines are nearly radial. The high density of lines between the charges indicates a region of strong electric field. The attractive nature of the force between the particles is also suggested by Figure 19.18, with the lines from one particle ending on the other particle.

Figure 19.19 shows the electric field lines in the vicinity of two equal positive point charges. Again, close to either charge the lines are nearly radial. The same number of lines emerges from each particle, because the charges are equal in magnitude, and end on hypothetical charges infinitely far away. At great distances from the particles, the field is approximately equal to that of a single point charge of magnitude $2q$. The repulsive nature of the electric force between particles of like charge is suggested in the figure in that no lines connect the particles and that the lines bend away from the region between the charges.

Finally, we sketch the electric field lines associated with a positive point charge $+2q$ and a negative point charge $-q$ in Active Figure 19.20. In this case, the number of lines leaving $+2q$ is twice the number terminating on $-q$. Hence, only half the lines that leave the positive charge end at the negative charge. The remaining half terminate on hypothetical negative charges infinitely far away. At large distances from the particles (large compared with the particle separation), the electric field lines are equivalent to those of a single point charge $+q$.

▸ **QUICK QUIZ 19.5** Rank the magnitudes of the electric field at points A, B, and C in Figure 19.19 (greatest magnitude first).

▸19.7 | Motion of Charged Particles in a Uniform Electric Field

When a particle of charge q and mass m is placed in an electric field $\vec{\mathbf{E}}$, the electric force exerted on the charge is given by Equation 19.4, $\vec{\mathbf{F}}_e = q\vec{\mathbf{E}}$. If this force is the only force exerted on the particle, it is the net force. If other forces also act on the particle, the electric force is simply added to the other forces vectorially to determine the net force. According to the particle under a net force model from

Chapter 4, the net force causes the particle to accelerate. If the electric force is the only force on the particle, Newton's second law applied to the particle gives

$$\vec{F}_e = q\vec{E} = m\vec{a}$$

The acceleration of the particle is therefore

$$\vec{a} = \frac{q\vec{E}}{m} \qquad\qquad \textbf{19.11} \blacktriangleleft$$

If $\vec{E}$ is uniform (i.e., constant in magnitude and direction), the acceleration is constant and the particle under constant acceleration analysis model can be used to describe the motion of the particle. If the particle has a positive charge, its acceleration is in the direction of the electric field. If the particle has a negative charge, its acceleration is in the direction opposite the electric field.

Example 19.7 | An Accelerating Positive Charge: Two Models

A uniform electric field $\vec{E}$ is directed along the x axis between parallel plates of charge separated by a distance d as shown in Figure 19.21. A positive point charge q of mass m is released from rest at a point Ⓐ next to the positive plate and accelerates to a point Ⓑ next to the negative plate.

(A) Find the speed of the particle at Ⓑ by modeling it as a particle under constant acceleration.

SOLUTION

Conceptualize When the positive charge is placed at Ⓐ, it experiences an electric force toward the right in Figure 19.21 due to the electric field directed toward the right.

Categorize Because the electric field is uniform, a constant electric force acts on the charge. Therefore, as suggested in the problem statement, the point charge can be modeled as a charged particle under constant acceleration.

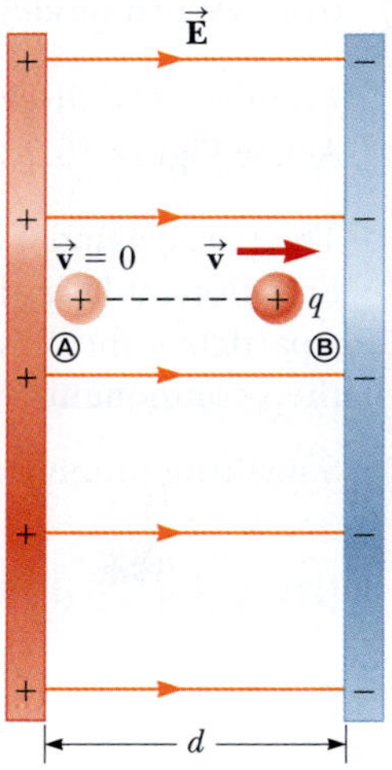

Figure 19.21 (Example 19.7) A positive point charge q in a uniform electric field $\vec{E}$ undergoes constant acceleration in the direction of the field.

Analyze Use Equation 2.14 to express the velocity of the particle as a function of position:

$$v_f^2 = v_i^2 + 2a(x_f - x_i) = 0 + 2a(d - 0) = 2ad$$

Solve for v_f and substitute for the magnitude of the acceleration from Equation 19.11:

$$v_f = \sqrt{2ad} = \sqrt{2\left(\frac{qE}{m}\right)d} = \boxed{\sqrt{\frac{2qEd}{m}}}$$

(B) Find the speed of the particle at Ⓑ by modeling it as a nonisolated system.

SOLUTION

Categorize The problem statement tells us that the charge is a nonisolated system. Energy is transferred to this charge by work done by the electric force exerted on the charge. The initial configuration of the system is when the particle is at Ⓐ, and the final configuration is when it is at Ⓑ.

Analyze Write the appropriate reduction of the conservation of energy equation, Equation 7.2, for the system of the charged particle:

$$W = \Delta K$$

Replace the work and kinetic energies with values appropriate for this situation:

$$F_e \Delta x = K_Ⓑ - K_Ⓐ = \tfrac{1}{2}mv_f^2 - 0 \;\;\rightarrow\;\; v_f = \sqrt{\frac{2F_e \Delta x}{m}}$$

Substitute for the electric force F_e and the displacement Δx:

$$v_f = \sqrt{\frac{2(qE)(d)}{m}} = \boxed{\sqrt{\frac{2qEd}{m}}}$$

Finalize The answer to part (B) is the same as that for part (A), as we expect.

Example 19.8 | An Accelerated Electron

An electron enters the region of a uniform electric field as shown in Active Figure 19.22, with $v_i = 3.00 \times 10^6$ m/s and $E = 200$ N/C. The horizontal length of the plates is $\ell = 0.100$ m.

(A) Find the acceleration of the electron while it is in the electric field.

SOLUTION

Conceptualize This example differs from the preceding one because the velocity of the charged particle is initially perpendicular to the electric field lines. (In Example 19.7, the velocity of the charged particle is always parallel to the electric field lines.) As a result, the electron in this example follows a curved path as shown in Active Figure 19.22.

Categorize Because the electric field is uniform, a constant electric force is exerted on the electron. To find the acceleration of the electron, we can model it as a particle under a net force.

Analyze The direction of the electron's acceleration is downward in Active Figure 19.22, opposite the direction of the electric field lines.

The particle under a net force model was used to develop Equation 19.11 in the case in which the electric force on a particle is the only force. Use this equation to evaluate the y component of the acceleration of the electron:

$$a_y = -\frac{eE}{m_e}$$

Substitute numerical values:

$$a_y = -\frac{(1.60 \times 10^{-19}\text{ C})(200\text{ N/C})}{9.11 \times 10^{-31}\text{ kg}} = -3.51 \times 10^{13}\text{ m/s}^2$$

> The electron undergoes a downward acceleration (opposite $\vec{E}$), and its motion is parabolic while it is between the plates.

Active Figure 19.22 (Example 19.8) An electron is projected horizontally into a uniform electric field produced by two charged plates.

(B) Assuming the electron enters the field at time $t = 0$, find the time at which it leaves the field.

SOLUTION

Categorize Because the electric force acts only in the vertical direction in Active Figure 19.22, the motion of the particle in the horizontal direction can be analyzed by modeling it as a particle under constant velocity.

Analyze Solve Equation 2.5 for the time at which the electron arrives at the right edges of the plates:

$$x_f = x_i + v_x t \rightarrow t = \frac{x_f - x_i}{v_x}$$

Substitute numerical values:

$$t = \frac{\ell - 0}{v_x} = \frac{0.100\text{ m}}{3.00 \times 10^6\text{ m/s}} = 3.33 \times 10^{-8}\text{ s}$$

Finalize We have neglected the gravitational force acting on the electron, which represents a good approximation when dealing with atomic particles. For an electric field of 200 N/C, the ratio of the magnitude of the electric force eE to the magnitude of the gravitational force mg is on the order of 10^{12} for an electron and on the order of 10^9 for a proton.

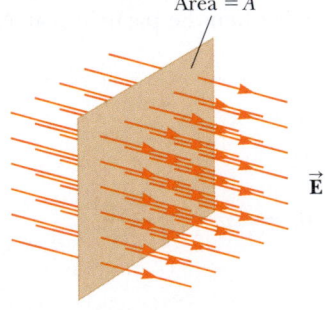

Figure 19.23 Field lines of a uniform electric field penetrating a plane of area A perpendicular to the field. The electric flux Φ_E through this area is equal to EA.

❰19.8 | Electric Flux

Now that we have described the concept of electric field lines qualitatively, let us use a new concept, *electric flux*, to approach electric field lines on a quantitative basis. Electric flux is a quantity proportional to the number of electric field lines penetrating some surface. (We can define only a proportionality because the number of lines we choose to draw is arbitrary.)

First consider an electric field that is uniform in both magnitude and direction as in Figure 19.23. The field lines penetrate a plane rectangular surface of area A, which is perpendicular to the field. Recall that the number of lines per unit area is proportional to the magnitude of the electric field. The number of lines penetrating the surface of area A is therefore proportional to the product EA. The product of the electric field magnitude E and a surface area A perpendicular to the field is called the **electric flux** Φ_E:

$$\Phi_E = EA \qquad \qquad \textbf{19.12} \blacktriangleleft$$

From the SI units of E and A, we see that electric flux has the units $N \cdot m^2/C$.

If the surface under consideration is not perpendicular to the field, the number of lines through it must be less than that given by Equation 19.12. This concept can be understood by considering Figure 19.24, where the normal to the surface of area A is at an angle of θ to the uniform electric field. Note that the number of lines that cross this area is equal to the number that cross the projected area $A_\perp$, which is perpendicular to the field. From Figure 19.24, we see that the two areas are related by $A_\perp = A \cos \theta$. Because the flux through area A equals the flux through $A_\perp$, we conclude that the desired flux is

$$\Phi_E = EA \cos \theta \qquad \qquad \textbf{19.13} \blacktriangleleft$$

From this result, we see that the flux through a surface of fixed area has the maximum value EA when the angle θ between the normal to the surface and the electric field is zero. This situation occurs when the normal is parallel to the field and the surface is perpendicular to the field. The flux is zero when the surface is parallel to the field because the angle θ in Equation 19.13 is then 90°.

In more general situations, the electric field may vary in both magnitude and direction over the surface in question. Unless the field is uniform, our definition of flux given by Equation 19.13 therefore has meaning only over a small element of area. Consider a general surface divided up into a large number of small elements, each of area ΔA. The variation in the electric field over the element can be ignored if the element is small enough. It is convenient to define a vector $\Delta \vec{\mathbf{A}}_i$ whose magnitude represents the area of the ith element and whose direction is defined to be perpendicular to the surface as in Figure 19.25. The electric flux $\Delta \Phi_E$ through this small element is

$$\Delta \Phi_E = E_i \, \Delta A_i \cos \theta_i = \vec{\mathbf{E}}_i \cdot \Delta \vec{\mathbf{A}}_i$$

where we have used the definition of the scalar product of two vectors ($\vec{\mathbf{A}} \cdot \vec{\mathbf{B}} = AB \cos \theta$). By summing the contributions of all elements, we obtain the total flux through the surface. If we let the area of each element approach zero, the number of elements approaches infinity and the sum is replaced by an integral. The general definition of electric flux is therefore

$$\Phi_E \equiv \lim_{\Delta A_i \to 0} \sum \vec{\mathbf{E}}_i \cdot \Delta \vec{\mathbf{A}}_i = \int_{\text{surface}} \vec{\mathbf{E}} \cdot d\vec{\mathbf{A}} \qquad \qquad \textbf{19.14} \blacktriangleleft \qquad \blacktriangleright \text{ Electric flux}$$

Equation 19.14 is a surface integral, which must be evaluated over the surface in question. In general, the value of Φ_E depends both on the field pattern and on the specified surface.

We shall often be interested in evaluating electric flux through a *closed surface*. A closed surface is defined as one that completely divides space into an inside region and an outside region so that movement cannot take place from one region to the other without penetrating the surface. This definition is similar to that of the system boundary in system models, in which the boundary divides space into a region inside the system and the outer region, the environment. The surface of a sphere is an example of a closed surface, whereas a drinking glass is an open surface.

Consider the closed surface in Active Figure 19.26 (page 550). Note that the vectors $\Delta \vec{\mathbf{A}}_i$ point in different directions for the various surface elements. At each point, these vectors are *perpendicular* to the surface and, by convention, always point *outward* from the inside region. At the element labeled ①, $\vec{\mathbf{E}}$ is outward and $\theta_i < 90°$; hence, the flux $\Delta \Phi_E = \vec{\mathbf{E}} \cdot \Delta \vec{\mathbf{A}}_i$ through this element is positive. For element ②, the field lines graze the surface (perpendicular to the vector $\Delta \vec{\mathbf{A}}_i$); therefore, $\theta_i = 90°$ and the

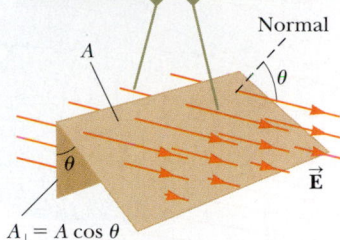

The number of field lines that go through the area $A_\perp$ is the same as the number that go through area A.

$A_\perp = A \cos \theta$

Figure 19.24 Field lines representing a uniform electric field penetrating an area A that is at an angle θ to the field.

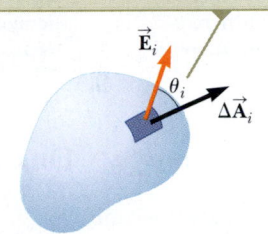

The electric field makes an angle θ_i with the vector $\Delta \vec{\mathbf{A}}_i$, defined as being normal to the surface element.

Figure 19.25 A small element of a surface of area ΔA_i.

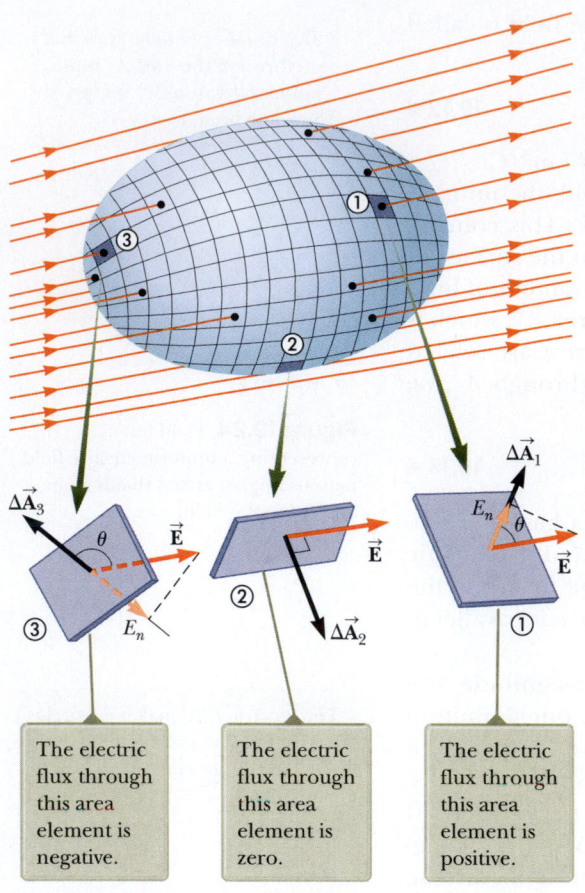

The electric flux through this area element is negative.

The electric flux through this area element is zero.

The electric flux through this area element is positive.

flux is zero. For elements such as ③, where the field lines are crossing the surface from the outside to the inside, $180° > \theta_i > 90°$ and the flux is negative because $\cos \theta_i$ is negative.

The net flux through the surface is proportional to the net number of lines penetrating the surface, where the net number means the number leaving the volume surrounded by the surface minus the number entering the volume. If more lines are leaving the surface than entering, the net flux is positive. If more lines enter than leave the surface, the net flux is negative. Using the symbol $\oint$ to represent an integral over a closed surface, we can write the net flux Φ_E through a closed surface as

$$\Phi_E = \oint \vec{E} \cdot d\vec{A} = \oint E_n \, dA \qquad \textbf{19.15} \blacktriangleleft$$

where E_n represents the component of the electric field normal to the surface.

Evaluating the net flux through a closed surface can be very cumbersome. If the field is perpendicular or parallel to the surface at each point and constant in magnitude, however, the calculation is straightforward. The following example illustrates this point.

Active Figure 19.26 A closed surface in an electric field. The area vectors are, by convention, normal to the surface and point outward.

Example 19.9 | **Flux Through a Cube**

Consider a uniform electric field $\vec{E}$ oriented in the x direction in empty space. A cube of edge length ℓ is placed in the field, oriented as shown in Figure 19.27. Find the net electric flux through the surface of the cube.

SOLUTION

Conceptualize Examine Figure 19.27 carefully. Notice that the electric field lines pass through two faces perpendicularly and are parallel to four other faces of the cube.

Categorize We evaluate the flux from its definition, so we categorize this example as a substitution problem.

The flux through four of the faces (③, ④, and the unnumbered faces) is zero because $\vec{E}$ is parallel to the four faces and therefore perpendicular to $d\vec{A}$ on these faces.

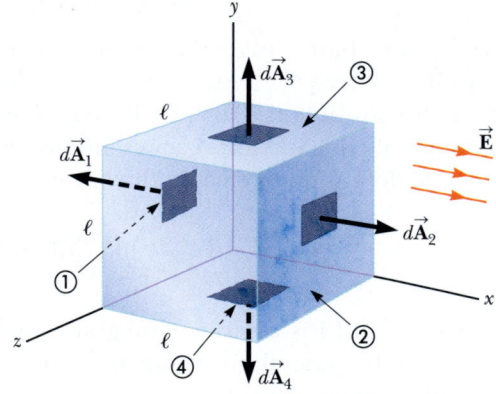

Figure 19.27 (Example 19.9) A closed surface in the shape of a cube in a uniform electric field oriented parallel to the x axis. Side ④ is the bottom of the cube, and side ① is opposite side ②.

Write the integrals for the net flux through faces ① and ②:

$$\Phi_E = \int_1 \vec{E} \cdot d\vec{A} + \int_2 \vec{E} \cdot d\vec{A}$$

For face ①, $\vec{E}$ is constant and directed inward but $d\vec{A}_1$ is directed outward ($\theta = 180°$). Find the flux through this face:

$$\int_1 \vec{E} \cdot d\vec{A} = \int_1 E(\cos 180°) \, dA = -E \int_1 dA = -EA = -E\ell^2$$

For face ②, $\vec{E}$ is constant and outward and in the same direction as $d\vec{A}_2$ ($\theta = 0°$). Find the flux through this face:

$$\int_2 \vec{E} \cdot d\vec{A} = \int_2 E(\cos 0°) \, dA = E \int_2 dA = +EA = E\ell^2$$

Find the net flux by adding the flux over all six faces:

$$\Phi_E = -E\ell^2 + E\ell^2 + 0 + 0 + 0 + 0 = \boxed{0}$$

◄19.9 | Gauss's Law

In this section, we describe a general relation between the net electric flux through a closed surface and the charge *enclosed* by the surface. This relation, known as **Gauss's law,** is of fundamental importance in the study of electrostatic fields.

First, let us consider a positive point charge q located at the center of a spherical surface of radius r as in Figure 19.28. The field lines radiate outward and hence are perpendicular (or normal) to the surface at each point. That is, at each point on the surface, $\vec{E}$ is parallel to the vector $\Delta\vec{A}_i$ representing the local element of area ΔA_i. Therefore, at all points on the surface,

$$\vec{E} \cdot \Delta\vec{A}_i = E_n \Delta A_i = E \Delta A_i$$

and, from Equation 19.15, we find that the net flux through the surface is

$$\Phi_E = \oint E_n \, dA = \oint E \, dA = E \oint dA = EA$$

because E is constant over the surface. From Equation 19.5, we know that the magnitude of the electric field everywhere on the surface of the sphere is $E = k_e q/r^2$. Furthermore, for a spherical surface, $A = 4\pi r^2$ (the surface area of a sphere). Hence, the net flux through the surface is

$$\Phi_E = EA = \left(\frac{k_e q}{r^2}\right)(4\pi r^2) = 4\pi k_e q$$

Recalling that $k_e = 1/4\pi\epsilon_0$, we can write this expression in the form

$$\Phi_E = \frac{q}{\epsilon_0} \qquad \textbf{19.16} ◄$$

This result, which is independent of r, says that the net flux through a spherical surface is proportional to the charge q at the center *inside* the surface. This result mathematically represents that (1) the net flux is proportional to the number of field lines, (2) the number of field lines is proportional to the charge inside the surface, and (3) every field line from the charge must pass through the surface. That the net flux is independent of the radius is a consequence of the inverse-square dependence of the electric field given by Equation 19.5. That is, E varies as $1/r^2$, but the area of the sphere varies as r^2. Their combined effect produces a flux that is independent of r.

Now consider several closed surfaces surrounding a charge q as in Figure 19.29. Surface S_1 is spherical, whereas surfaces S_2 and S_3 are nonspherical. The flux that passes through surface S_1 has the value q/ϵ_0. As we discussed in Section 19.8, the flux is proportional to the number of electric field lines passing through that surface. The construction in Figure 19.29 shows that the number of electric field lines through the spherical surface S_1 is equal to the number of electric field lines through the nonspherical surfaces S_2 and S_3. It is therefore reasonable to conclude that the net flux through any closed surface is independent of the shape of that surface. (One can prove that conclusion using $E \propto 1/r^2$.) In fact,

the net flux through any closed surface surrounding the point charge q is given by q/ϵ_0 and is independent of the position of the charge within the surface.

Now consider a point charge located *outside* a closed surface of arbitrary shape as in Figure 19.30. As you can see from this construction, electric field lines enter the surface and then leave it. Therefore, the number of electric field lines entering the surface equals the number leaving the surface. Consequently, we conclude that the net electric flux through a closed surface that surrounds no net charge is zero. If we apply this result to Example 19.9, we see that the net flux through the cube is zero because there was no charge inside the cube. If there were charge in the cube, the electric field could not be uniform throughout the cube as specified in the example.

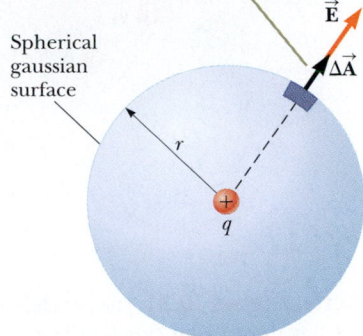

When the charge is at the center of the sphere, the electric field is everywhere normal to the surface and constant in magnitude.

Figure 19.28 A spherical surface of radius r surrounding a point charge q.

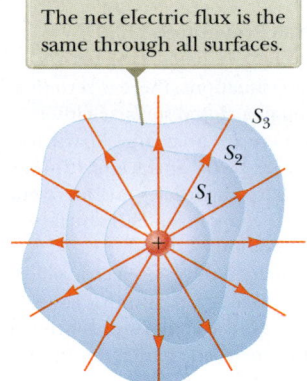

The net electric flux is the same through all surfaces.

Figure 19.29 Closed surfaces of various shapes surrounding a positive charge.

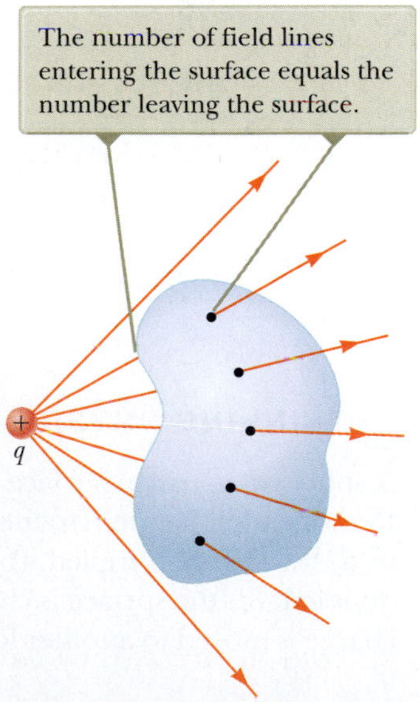

The number of field lines entering the surface equals the number leaving the surface.

Figure 19.30 A point charge located *outside* a closed surface.

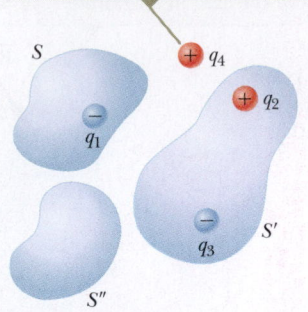

Charge q_4 does not contribute to the flux through any surface because it is outside all surfaces.

Active Figure 19.31 The net electric flux through any closed surface depends only on the charge *inside* that surface. The net flux through surface *S* is q_1/ϵ_0, the net flux through surface *S'* is $(q_2 + q_3)/\epsilon_0$, and the net flux through surface *S''* is zero.

Pitfall Prevention | 19.3
Zero Flux Is Not Zero Field
In two situations, there is zero flux through a closed surface: either (1) there are no charged particles enclosed by the surface or (2) there are charged particles enclosed, but the net charge inside the surface is zero. For either situation, it is *incorrect* to conclude that the electric field on the surface is zero. Gauss's law states that the electric *flux* is proportional to the enclosed charge, not the electric *field*.

Let us extend these arguments to the generalized case of many point charges. We shall again make use of the superposition principle. That is, we can express the net flux through any closed surface as

$$\oint \vec{E} \cdot d\vec{A} = \oint (\vec{E}_1 + \vec{E}_2 + \cdots) \cdot d\vec{A}$$

where $\vec{E}$ is the total electric field at any point on the surface and $\vec{E}_1$, $\vec{E}_2$, . . . are the fields produced by the individual charges at that point. Consider the system of charges shown in Active Figure 19.31. The surface *S* surrounds only one charge, q_1; hence, the net flux through *S* is q_1/ϵ_0. The flux through *S* due to the charges outside it is zero because each electric field line from these charges that enters *S* at one point leaves it at another. The surface *S'* surrounds charges q_2 and q_3; hence, the net flux through *S'* is $(q_2 + q_3)/\epsilon_0$. Finally, the net flux through surface *S''* is zero because no charge exists inside this surface. That is, *all* electric field lines that enter *S''* at one point leave *S''* at another. Notice that charge q_4 does not contribute to the net flux through any of the surfaces because it is outside all the surfaces.

Gauss's law, which is a generalization of the foregoing discussion, states that the net flux through *any* closed surface is

$$\Phi_E = \oint \vec{E} \cdot d\vec{A} = \frac{q_{in}}{\epsilon_0} \qquad \qquad \text{19.17} \blacktriangleleft$$

where q_{in} represents the *net charge inside the surface* and $\vec{E}$ represents the electric field at any point on the surface. In words, Gauss's law states that the net electric flux through any closed surface is equal to the net charge inside the surface divided by ϵ_0. The closed surface used in Gauss's law is called a **gaussian surface.**

Gauss's law is valid for the electric field of any system of charges or continuous distribution of charge. In practice, however, the technique is useful for calculating the electric field only in situations where the degree of symmetry is high. As we shall see in the next section, Gauss's law can be used to evaluate the electric field for charge distributions that have spherical, cylindrical, or plane symmetry. We do so by choosing an appropriate gaussian surface that allows $\vec{E}$ to be removed from the integral in Gauss's law and performing the integration. Note that a gaussian surface is a mathematical surface and need not coincide with any real physical surface.

QUICK QUIZ 19.6 If the net flux through a gaussian surface is zero, the following four statements could be true. Which of the statements must be true? **(a)** There are no charges inside the surface. **(b)** The net charge inside the surface is zero. **(c)** The electric field is zero everywhere on the surface. **(d)** The number of electric field lines entering the surface equals the number leaving the surface.

QUICK QUIZ 19.7 Consider the charge distribution shown in Active Figure 19.31. **(i)** What are the charges contributing to the total electric *flux* through surface *S'*? **(a)** q_1 only **(b)** q_4 only **(c)** q_2 and q_3 **(d)** all four charges **(e)** none of the charges **(ii)** What are the charges contributing to the total electric *field* at a chosen point on the surface *S'*? **(a)** q_1 only **(b)** q_4 only **(c)** q_2 and q_3 **(d)** all four charges **(e)** none of the charges

---

▶ THINKING PHYSICS 19.1

A spherical gaussian surface surrounds a point charge q. Describe what happens to the net flux through the surface if (a) the charge is tripled, (b) the volume of the sphere is doubled, (c) the surface is changed to a cube, and (d) the charge is moved to another location *inside* the surface.

Reasoning (a) If the charge is tripled, the flux through the surface is also tripled because the net flux is proportional to the charge inside the surface. (b) The net flux remains constant when the volume changes because the surface surrounds the same amount of charge, regardless of its volume. (c) The net flux does not change when the shape of the closed surface changes. (d) The net flux through the closed surface remains unchanged as the charge inside the surface is moved to another location as long as the new location remains inside the surface. ◀

19.10 | Application of Gauss's Law to Various Charge Distributions

As mentioned earlier, Gauss's law is useful in determining electric fields when the charge distribution has a high degree of symmetry. The following examples show ways of choosing the gaussian surface over which the surface integral given by Equation 19.17 can be simplified and the electric field determined. The surface should always be chosen to take advantage of the symmetry of the charge distribution so that we can remove E from the integral and solve for it. The crucial step in applying Gauss's law is to determine a useful gaussian surface. Such a surface should be a closed surface for which each portion of the surface satisfies one or more of the following conditions:

1. The value of the electric field can be argued by symmetry to be constant over the portion of the surface.
2. The dot product in Equation 19.17 can be expressed as a simple algebraic product $E\,dA$ because $\vec{\mathbf{E}}$ and $d\vec{\mathbf{A}}$ are parallel.
3. The dot product in Equation 19.17 is zero because $\vec{\mathbf{E}}$ and $d\vec{\mathbf{A}}$ are perpendicular.
4. The electric field is zero over the portion of the surface.

Note that different portions of the gaussian surface can satisfy different conditions as long as every portion satisfies at least one condition. We will see all four of these conditions used in the examples and discussions in the remainder of this chapter. If the charge distribution does not have sufficient symmetry such that a gaussian surface that satisfies these conditions can be found, Gauss's law is not useful for determining the electric field for that charge distribution.

Example 19.10 | A Spherically Symmetric Charge Distribution

An insulating solid sphere of radius a has a uniform volume charge density ρ and carries a total positive charge Q (Fig. 19.32).

(A) Calculate the magnitude of the electric field at a point outside the sphere.

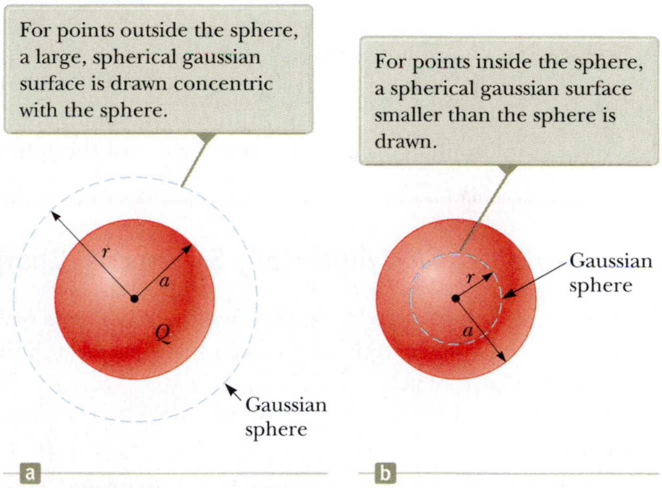

SOLUTION

Conceptualize Notice how this problem differs from our previous discussion of Gauss's law. The electric field due to point charges was discussed in Section 19.9. Now we are considering the electric field due to a distribution of charge. We found the field for various distributions of charge in Section 19.5 by integrating over the distribution. This example demonstrates a difference from our discussions in Section 19.5. In this example, we find the electric field using Gauss's law.

Categorize Because the charge is distributed uniformly throughout the sphere, the charge distribution has spherical symmetry and we can apply Gauss's law to find the electric field.

For points outside the sphere, a large, spherical gaussian surface is drawn concentric with the sphere.

For points inside the sphere, a spherical gaussian surface smaller than the sphere is drawn.

Figure 19.32 (Example 19.10) A uniformly charged insulating sphere of radius a and total charge Q. In diagrams such as this one, the dotted line represents the intersection of the gaussian surface with the plane of the page.

Analyze To reflect the spherical symmetry, let's choose a spherical gaussian surface of radius r, concentric with the sphere, as shown in Figure 19.32a. For this choice, condition (2) is satisfied everywhere on the surface and $\vec{\mathbf{E}} \cdot d\vec{\mathbf{A}} = E\,dA$.

Replace $\vec{\mathbf{E}} \cdot d\vec{\mathbf{A}}$ in Gauss's law with $E\,dA$:

$$\Phi_E = \oint \vec{\mathbf{E}} \cdot d\vec{\mathbf{A}} = \oint E\,dA = \frac{Q}{\epsilon_0}$$

continued

19.10 *cont.*

By symmetry, E is constant everywhere on the surface, which satisfies condition (1), so we can remove E from the integral:

$$\oint E\, dA = E \oint dA = E(4\pi r^2) = \frac{Q}{\epsilon_0}$$

Solve for E:

$$(1) \quad E = \frac{Q}{4\pi\epsilon_0 r^2} = k_e \frac{Q}{r^2} \quad (\text{for } r > a)$$

Finalize This field is identical to that for a point charge. Therefore, **the electric field due to a uniformly charged sphere in the region external to the sphere is** *equivalent* **to that of a point charge located at the center of the sphere.**

(B) Find the magnitude of the electric field at a point inside the sphere.

SOLUTION

Analyze In this case, let's choose a spherical gaussian surface having radius $r < a$, concentric with the insulating sphere (Fig. 19.32b). Let V' be the volume of this smaller sphere. To apply Gauss's law in this situation, recognize that the charge q_{in} within the gaussian surface of volume V' is less than Q.

Calculate q_{in} by using $q_{in} = \rho V'$:

$$q_{in} = \rho V' = \rho\left(\tfrac{4}{3}\pi r^3\right)$$

Notice that conditions (1) and (2) are satisfied everywhere on the gaussian surface in Figure 19.32b. Apply Gauss's law in the region $r < a$:

$$\oint E\, dA = E \oint dA = E(4\pi r^2) = \frac{q_{in}}{\epsilon_0}$$

Solve for E and substitute for q_{in}:

$$E = \frac{q_{in}}{4\pi\epsilon_0 r^2} = \frac{\rho\left(\tfrac{4}{3}\pi r^3\right)}{4\pi\epsilon_0 r^2} = \frac{\rho}{3\epsilon_0} r$$

Substitute $\rho = Q / \tfrac{4}{3}\pi a^3$ and $\epsilon_0 = 1/4\pi k_e$:

$$(2) \quad E = \frac{Q/\tfrac{4}{3}\pi a^3}{3(1/4\pi k_e)} r = k_e \frac{Q}{a^3} r \quad (\text{for } r < a)$$

Finalize This result for E differs from the one obtained in part (A). It shows that $E \to 0$ as $r \to 0$. Therefore, the result eliminates the problem that would exist at $r = 0$ if E varied as $1/r^2$ inside the sphere as it does outside the sphere. That is, if $E \propto 1/r^2$ for $r < a$, the field would be infinite at $r = 0$, which is physically impossible. Notice also that Equations (1) and (2) both give the same value of the field at the surface of the sphere ($r = a$), showing that the field is continuous.

Example 19.11 | A Cylindrically Symmetric Charge Distribution

Find the electric field a distance r from a line of positive charge of infinite length and constant charge per unit length λ (Fig. 19.33a).

SOLUTION

Conceptualize The line of charge is *infinitely* long. Therefore, the field is the same at all points equidistant from the line, regardless of the vertical position of the point in Figure 19.33a.

Categorize Because the charge is distributed uniformly along the line, the charge distribution has cylindrical symmetry and we can apply Gauss's law to find the electric field.

Analyze The symmetry of the charge distribution requires that $\vec{E}$ be perpendicular to the line charge and directed outward as shown in Figure 19.33b. To reflect the symmetry

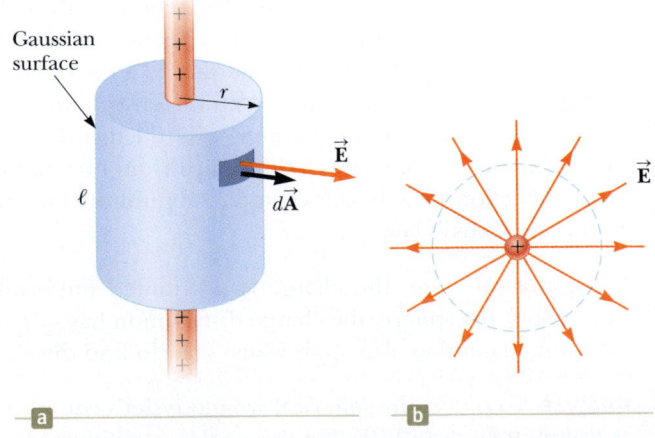

Figure 19.33 (Example 19.11) (a) An infinite line of charge surrounded by a cylindrical gaussian surface concentric with the line. (b) An end view shows that the electric field at the cylindrical surface is constant in magnitude and perpendicular to the surface.

19.11 *cont.*

of the charge distribution, let's choose a cylindrical gaussian surface of radius r and length ℓ that is coaxial with the line charge. For the curved part of this surface, $\vec{E}$ is constant in magnitude and perpendicular to the surface at each point, satisfying conditions (1) and (2). Furthermore, the flux through the ends of the gaussian cylinder is zero because $\vec{E}$ is parallel to these surfaces. That is the first application we have seen of condition (3).

We must take the surface integral in Gauss's law over the entire gaussian surface. Because $\vec{E} \cdot d\vec{A}$ is zero for the flat ends of the cylinder, however, we restrict our attention to only the curved surface of the cylinder.

Apply Gauss's law and conditions (1) and (2) for the curved surface, noting that the total charge inside our gaussian surface is $\lambda\ell$:

$$\Phi_E = \oint \vec{E} \cdot d\vec{A} = E \oint dA = EA = \frac{q_{in}}{\epsilon_0} = \frac{\lambda\ell}{\epsilon_0}$$

Substitute the area $A = 2\pi r\ell$ of the curved surface:

$$E(2\pi r\ell) = \frac{\lambda\ell}{\epsilon_0}$$

Solve for the magnitude of the electric field:

$$E = \frac{\lambda}{2\pi\epsilon_0 r} = 2k_e \frac{\lambda}{r} \qquad \text{19.18} \blacktriangleleft$$

Finalize This result shows that the electric field due to a cylindrically symmetric charge distribution varies as $1/r$, whereas the field external to a spherically symmetric charge distribution varies as $1/r^2$. Equation 19.18 can also be derived by direct integration over the charge distribution. (See Problem 19.18 in Enhanced WebAssign.)

What If? What if the line segment in this example were not infinitely long?

Answer If the line charge in this example were of finite length, the electric field would not be given by Equation 19.18. A finite line charge does not possess sufficient symmetry to make use of Gauss's law because the magnitude of the electric field is no longer constant over the surface of the gaussian cylinder: the field near the ends of the line would be different from that far from the ends. Therefore, condition (1) would not be satisfied in this situation. Furthermore, $\vec{E}$ is not perpendicular to the cylindrical surface at all points: the field vectors near the ends would have a component parallel to the line. Therefore, condition (2) would not be satisfied. For points close to a finite line charge and far from the ends, Equation 19.18 gives a good approximation of the value of the field.

It is left for you to show (see Problem 19.48 in Enhanced WebAssign) that the electric field inside a uniformly charged rod of finite radius and infinite length is proportional to r.

Example **19.12** | **A Plane of Charge**

Find the electric field due to an infinite plane of positive charge with uniform surface charge density σ.

SOLUTION

Conceptualize Notice that the plane of charge is *infinitely* large. Therefore, the electric field should be the same at all points equidistant from the plane.

Categorize Because the charge is distributed uniformly on the plane, the charge distribution is symmetric; hence, we can use Gauss's law to find the electric field.

Analyze By symmetry, $\vec{E}$ must be perpendicular to the plane at all points. The direction of $\vec{E}$ is away from positive charges, indicating that the direction of $\vec{E}$ on one side of the plane must be opposite its direction on the other side as shown in Figure 19.34. A gaussian surface that reflects the symmetry is a small cylinder whose axis is perpendicular to the plane and whose ends each have an area A and are equidistant from the plane. Because $\vec{E}$ is parallel to the curved surface of the cylinder—and therefore perpendicular to $d\vec{A}$ at all points on this surface—condition (3) is satisfied and there is

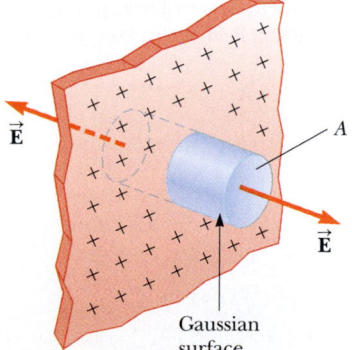

Figure 19.34 (Example 19.12) A cylindrical gaussian surface penetrating an infinite plane of charge. The flux is EA through each end of the gaussian surface and zero through its curved surface.

no contribution to the surface integral from this surface. For the flat ends of the cylinder, conditions (1) and (2) are satisfied. The flux through each end of the cylinder is EA; hence, the total flux through the entire gaussian surface is just that through the ends, $\Phi_E = 2EA$.

continued

19.12 *cont.*

Write Gauss's law for this surface, noting that the enclosed charge is $q_{in} = \sigma A$:

$$\Phi_E = 2EA = \frac{q_{in}}{\epsilon_0} = \frac{\sigma A}{\epsilon_0}$$

Solve for E:

$$E = \frac{\sigma}{2\epsilon_0}$$

19.19 ◀

Finalize Because the distance from each flat end of the cylinder to the plane does not appear in Equation 19.19, we conclude that $E = \sigma/2\epsilon_0$ at *any* distance from the plane. That is, the field is uniform everywhere.

What If? Suppose two infinite planes of charge are parallel to each other, one positively charged and the other negatively charged. Both planes have the same surface charge density. What does the electric field look like in this situation?

Answer The electric fields due to the two planes add in the region between the planes, resulting in a uniform field of magnitude σ/ϵ_0, and cancel elsewhere to give a field of zero. This method is a practical way to achieve uniform electric fields with finite-sized planes placed close to each other.

▌19.11 | Conductors in Electrostatic Equilibrium

A good electrical conductor, such as copper, contains charges (electrons) that are not bound to any atom and are free to move about within the material. When no motion of charge occurs within the conductor (other than thermal motion), the conductor is in **electrostatic equilibrium.** As we shall see, an isolated conductor (one that is insulated from ground) in electrostatic equilibrium has the following properties:

1. The electric field is zero everywhere inside the conductor, whether the conductor is solid or hollow.
2. If the conductor is isolated and carries a charge, the charge resides on its surface.
3. The electric field at a point just outside a charged conductor is perpendicular to the surface of the conductor and has a magnitude σ/ϵ_0, where σ is the surface charge density at that point.
4. On an irregularly shaped conductor, the surface charge density is greatest at locations where the radius of curvature of the surface is smallest.

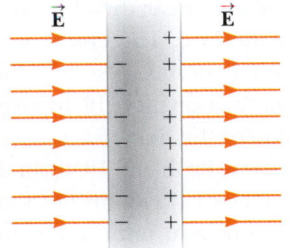

Figure 19.35 A conducting slab in an external electric field $\vec{\mathbf{E}}$. The charges induced on the two surfaces of the slab produce an electric field that opposes the external field, giving a resultant field of zero *inside* the slab.

We will verify the first three properties in the following discussion. The fourth property is presented here so that we have a complete list of properties for conductors in electrostatic equilibrium. The verification of it, however, requires concepts from Chapter 20, so we will postpone its verification until then.

The first property can be understood by considering a conducting slab placed in an external field $\vec{\mathbf{E}}$ (Fig. 19.35). The electric field inside the conductor *must* be zero under the assumption that we have electrostatic equilibrium. If the field were not zero, free charges in the conductor would accelerate under the action of the electric force. This motion of electrons, however, would mean that the conductor is not in electrostatic equilibrium. Therefore, the existence of electrostatic equilibrium is consistent only with a zero field in the conductor.

Let us investigate how this zero field is accomplished. Before the external field is applied, free electrons are uniformly distributed throughout the conductor. When the external field is applied, the free electrons accelerate to the left in Figure 19.35, causing a plane of negative charge to be present on the left surface. The movement of electrons to the left results in a plane of positive charge on the right surface. These planes of charge create an additional electric field inside the conductor that opposes the external field. As the electrons move, the surface charge density increases until the magnitude of the internal field equals that of the external field, giving a net field of zero inside the conductor.

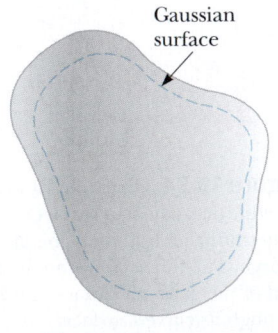

Gaussian surface

Figure 19.36 A conductor of arbitrary shape. The broken line represents a gaussian surface that can be just inside the conductor's surface.

We can use Gauss's law to verify the second property of a conductor in electrostatic equilibrium. Figure 19.36 shows an arbitrarily shaped conductor. A gaussian surface is drawn just inside the conductor and can be as close to the surface as we wish. As we have just shown, the electric field everywhere inside a conductor in electrostatic equilibrium is zero. Therefore, the electric field must be zero at every point on the gaussian surface (condition 4 in Section 19.10). From this result and Gauss's law, we conclude that the net charge inside the gaussian surface is zero. Because there can be no net charge inside the gaussian surface (which is arbitrarily close to the conductor's surface), any net charge on the conductor must reside on its surface. Gauss's law does not tell us how this excess charge is distributed on the surface, only that it must reside on the surface.

Conceptually, we can understand the location of the charges on the surface by imagining placing many charges at the center of the conductor. The mutual repulsion of the charges causes them to move apart. They will move as far as they can, which is to various points on the surface.

To verify the third property, we can also use Gauss's law. We draw a gaussian surface in the shape of a small cylinder having its end faces parallel to the surface (Fig. 19.37). Part of the cylinder is just outside the conductor and part is inside. The field is normal to the surface because the conductor is in electrostatic equilibrium. If $\vec{E}$ had a component parallel to the surface, an electric force would be exerted on the charges parallel to the surface, free charges would move along the surface, and so the conductor would not be in equilibrium. Therefore, we satisfy condition 3 in Section 19.10 for the curved part of the cylinder in that no flux exists through this part of the gaussian surface because $\vec{E}$ is parallel to this part of the surface. No flux exists through the flat face of the cylinder inside the conductor because $\vec{E} = 0$ (condition 4). Hence, the net flux through the gaussian surface is the flux through the flat face outside the conductor where the field is perpendicular to the surface. Using conditions 1 and 2 for this face, the flux is EA, where E is the electric field just outside the conductor and A is the area of the cylinder's face. Applying Gauss's law to this surface gives

$$\Phi_E = \oint E \, dA = EA = \frac{q_{in}}{\epsilon_0} = \frac{\sigma A}{\epsilon_0}$$

where we have used that $q_{in} = \sigma A$. Solving for E gives

$$E = \frac{\sigma}{\epsilon_0} \qquad\qquad \textbf{19.20} \blacktriangleleft$$

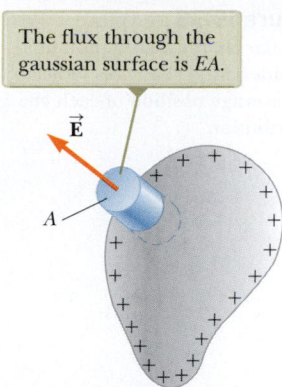

The flux through the gaussian surface is EA.

$\vec{E}$

A

Figure 19.37 A gaussian surface in the shape of a small cylinder is used to calculate the electric field just outside a charged conductor.

> ### THINKING PHYSICS 19.2

Suppose a point charge $+Q$ is in empty space. We surround the charge with a spherical, uncharged conducting shell so that the charge is at the center of the shell. What effect does that have on the field lines from the charge?

Reasoning When the spherical shell is placed around the charge, the free charges in the shell adjust so as to satisfy the rules for a conductor in equilibrium and Gauss's law. A net charge of $-Q$ moves to the interior surface of the conductor, so the electric field within the conductor is zero (a spherical gaussian surface totally within the shell encloses no *net* charge). A net charge of $+Q$ resides on the outer surface, so a gaussian surface outside the sphere encloses a net charge of $+Q$, just as if the shell were not there. Therefore, the only change in the field lines from the initial situation is the absence of field lines over the thickness of the conducting shell. ◄

⟨19.12 | Context Connection: The Atmospheric Electric Field

In this chapter, we discussed the electric field due to various charge distributions. On the surface of the Earth and in the atmosphere, a number of processes create charge distributions, resulting in an electric field in the atmosphere. These processes include cosmic rays entering the atmosphere, radioactive decay at the Earth's surface, and lightning, the focus of our study in this Context.

The result of these processes is an average negative charge distributed over the surface of the Earth of about 5×10^5 C, which is a tremendous amount of charge. (The Earth is neutral overall; the positive charges corresponding to this negative surface charge are spread through the atmosphere, as we shall discuss in

Figure 19.38 A typical tripolar charge distribution in a thundercloud. The dots indicate the average position of each charge distribution.

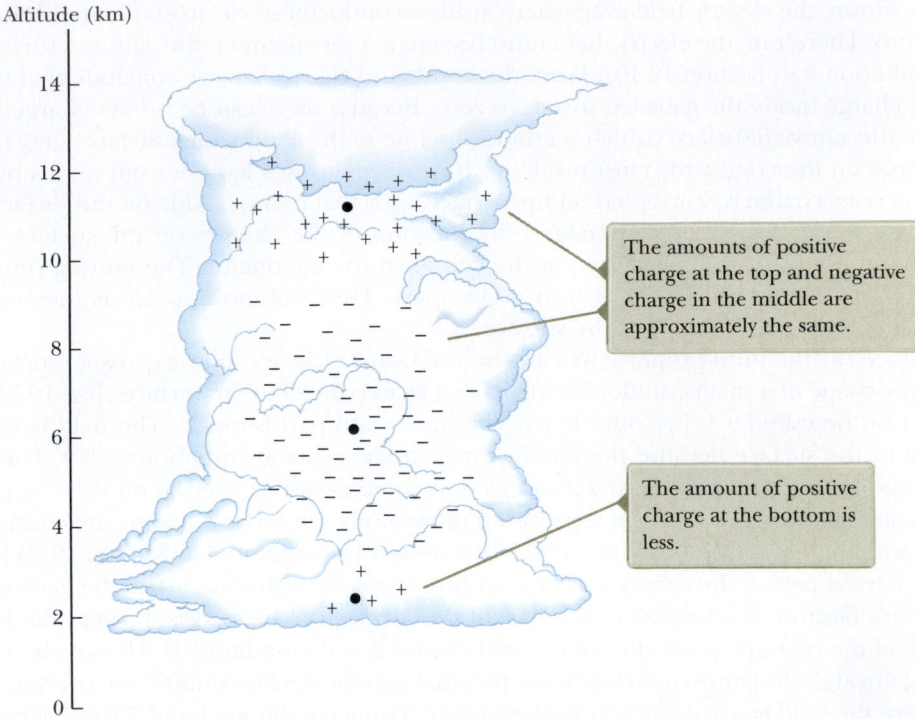

Altitude (km)

The amounts of positive charge at the top and negative charge in the middle are approximately the same.

The amount of positive charge at the bottom is less.

Chapter 20.) We can calculate the average surface charge density over the surface of the Earth:

$$\sigma_{avg} = \frac{Q}{A} = \frac{Q}{4\pi r^2} = \frac{5 \times 10^5 \, C}{4\pi(6.37 \times 10^6 \, m)^2} \sim 10^{-9} \, C/m^2$$

Throughout this Context, we will be adopting a number of simplification models. Consequently, we will consider our calculations to be order-of-magnitude estimates of the actual values, as suggested by the $\sim$ sign above.

The Earth is a good conductor. Therefore, we can use the third property of conductors in Section 19.11 to find the average magnitude of the electric field at the surface of the Earth:

$$E_{avg} = \frac{\sigma_{avg}}{\epsilon_0} = \frac{10^{-9} \, C/m^2}{8.85 \times 10^{-12} \, C^2/N \cdot m^2} \sim 10^2 \, N/C$$

which is a typical value of the **fair-weather electric field** that exists in the absence of a thunderstorm. The direction of the field is downward because the charge on the Earth's surface is negative. During a thunderstorm, the electric field under the thundercloud is significantly higher than the fair-weather electric field, because of the charge distribution in the thundercloud.

Figure 19.38 shows a typical charge distribution in a thundercloud. The charge distribution can be modeled as a *tripole*, although the positive charge at the bottom of the cloud tends to be smaller than the other two charges. The mechanism of charging in thunderclouds is not well understood and continues to be an active area of research.

It is this high concentration of charge in the thundercloud that is responsible for the very strong electric fields that cause lightning discharge between the cloud and the ground. Typical electric fields during a thunderstorm are as high as 25 000 N/C. The distribution of negative charges in the center of the cloud in Figure 19.38 is the source of negative charge that moves downward in a lightning strike.

Transient Luminous Events

Normal lightning is related to atmospheric electric fields in the troposphere between a thundercloud and the ground. Let us consider the effects of electric fields *above* thunderclouds as shown in Figure 19.39. We find a number of visual effects

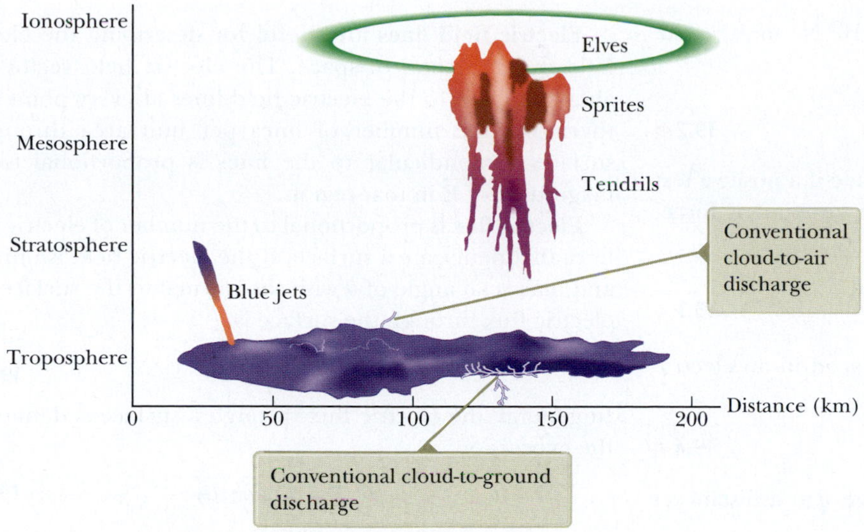

Figure 19.39 A representation of several types of transient luminous events in the atmosphere above thunderclouds.

associated with storms and lightning occurring in this region of the atmosphere. In general, these phenomena are called *transient luminous events*. One type of event is called a *sprite*, which occurs above thunderstorm clouds, with the light from the event originating between 90 and 100 km above the earth's surface. A sprite is triggered by normal tropospheric lightning from the thundercloud below it and appears as a luminous red flash, possibly with vertical tendrils hanging below. These displays last less than a second and are not easily seen with the naked eye. A sprite was first photographed by accident in 1989. Since then, additional photographic evidence has been made available and several astronauts at the International Space Station have reported seeing sprites while they were above a violent storm. Scientists believe that the electric fields at these altitudes are strong enough to ionize air molecules. Red light is released when the electrons recombine with molecular nitrogen ions, in a manner similar to the source of light in a fluorescent lamp.

Other lightning-induced transient luminous events are called *elves*. The light associated with these events lasts for less than 1 ms and has been observed using high speed photometers and CCD cameras. These events, which precede the onset of sprites, appear as expanding halos of light in the ionosphere at altitudes between 75 and 105 km. The expansion of the halos is faster than the speed of light, but there is no violation of the principles of relativity that we studied in Chapter 9 because no particles travel that fast. Current theories relate to an expanding spherical electromagnetic pulse from a lightning strike that interacts with the ionosphere to create the luminous display.

Also seen in Figure 19.39 is an optical event in the stratosphere called a *blue jet*. These displays occur as an upward-propagating ejection from the top of a thundercloud, disappearing at about 40–50 km from the ground. Blue jets are associated with storm clouds but do not appear to be directly triggered by lightning flashes as are the sprites.

Other types of luminous events include *blue starters, trolls, gnomes,* and *pixies.* Research on the origin of transient luminous events is ongoing.

> SUMMARY |

Electric charges have the following important properties:

1. Two kinds of charges exist in nature, **positive** and **negative,** with the property that charges of opposite sign attract each other and charges of the same sign repel each other.
2. The force between charged particles varies as the inverse square of their separation distance.
3. Charge is conserved.
4. Charge is quantized.

Conductors are materials in which charges move relatively freely. **Insulators** are materials in which charges do not move freely.

Coulomb's law states that the electrostatic force between two stationary, charged particles separated by a distance r has the magnitude

$$F_e = k_e \frac{|q_1||q_2|}{r^2} \qquad \textbf{19.1} \blacktriangleleft$$

where the Coulomb constant $k_e = 8.99 \times 10^9$ N $\cdot$ m^2/C^2. The vector form of Coulomb's law is

$$\vec{F}_{12} = k_e \frac{q_1 q_2}{r^2} \hat{r}_{12} \qquad \text{19.2} \blacktriangleleft$$

An **electric field** exists at a point in space if a positive test charge q_0 placed at that point experiences an electric force. The electric field is defined as

$$\vec{E} \equiv \frac{\vec{F}_e}{q_0} \qquad \text{19.3} \blacktriangleleft$$

The force on a particle with charge q placed in an electric field $\vec{E}$ is

$$\vec{F}_e = q\vec{E} \qquad \text{19.4} \blacktriangleleft$$

The electric field due to the point charge q at a distance r from the charge is

$$\vec{E} = k_e \frac{q}{r^2} \hat{r} \qquad \text{19.5} \blacktriangleleft$$

where $\hat{r}$ is a unit vector directed from the charge toward the point in question. The electric field is directed radially outward from a positive charge and is directed toward a negative charge.

The electric field due to a group of charges can be obtained using the superposition principle. That is, the total electric field equals the vector sum of the electric fields of all the charges at some point:

$$\vec{E} = k_e \sum_i \frac{q_i}{r_i^2} \hat{r}_i \qquad \text{19.6} \blacktriangleleft$$

Similarly, the electric field of a continuous charge distribution at some point is

$$\vec{E} = k_e \int \frac{dq}{r^2} \hat{r} \qquad \text{19.7} \blacktriangleleft$$

where dq is the charge on one element of the charge distribution and r is the distance from the element to the point in question.

Electric field lines are useful for describing the electric field in any region of space. The electric field vector $\vec{E}$ is always tangent to the electric field lines at every point. Furthermore, the number of lines per unit area through a surface perpendicular to the lines is proportional to the magnitude of $\vec{E}$ in that region.

Electric flux is proportional to the number of electric field lines that penetrate a surface. If the electric field is uniform and makes an angle of θ with the normal to the surface, the electric flux through the surface is

$$\Phi_E = EA \cos \theta \qquad \text{19.13} \blacktriangleleft$$

In general, the electric flux through a surface is defined by the expression

$$\Phi_E \equiv \int_{\text{surface}} \vec{E} \cdot d\vec{A} \qquad \text{19.14} \blacktriangleleft$$

Gauss's law says that the net electric flux Φ_E through any closed gaussian surface is equal to the *net* charge *inside* the surface divided by ϵ_0:

$$\Phi_E = \oint \vec{E} \cdot d\vec{A} = \frac{q_{\text{in}}}{\epsilon_0} \qquad \text{19.17} \blacktriangleleft$$

Using Gauss's law, one can calculate the electric field due to various symmetric charge distributions.

A conductor in **electrostatic equilibrium** has the following properties:

1. The electric field is zero everywhere inside the conductor, whether the conductor is solid or hollow.
2. If the conductor is isolated and carries a charge, the charge resides on its surface.
3. The electric field at a point just outside a charged conductor is perpendicular to the surface of the conductor and has a magnitude σ/ϵ_0, where σ is the surface charge density at that point.
4. On an irregularly shaped conductor, the surface charge density is greatest at locations where the radius of curvature of the surface is smallest.

OBJECTIVE QUESTIONS

☐ denotes answer available in *Student Solutions Manual/Study Guide*

1. A point charge of -4.00 nC is located at $(0, 1.00)$ m. What is the x component of the electric field due to the point charge at $(4.00, -2.00)$ m? (a) 1.15 N/C (b) -0.864 N/C (c) 1.44 N/C (d) -1.15 N/C (e) 0.864 N/C

2. Charges of 3.00 nC, -2.00 nC, -7.00 nC, and 1.00 nC are contained inside a rectangular box with length 1.00 m, width 2.00 m, and height 2.50 m. Outside the box are charges of 1.00 nC and 4.00 nC. What is the electric flux through the surface of the box? (a) 0 (b) -5.64×10^2 N $\cdot$ m^2/C (c) -1.47×10^3 N $\cdot$ m^2/C (d) 1.47×10^3 N $\cdot$ m^2/C (e) 5.64×10^2 N $\cdot$ m^2/C

3. An object with negative charge is placed in a region of space where the electric field is directed vertically upward. What is the direction of the electric force exerted on this charge?

(a) It is up. (b) It is down. (c) There is no force. (d) The force can be in any direction.

4. A particle with charge q is located inside a cubical gaussian surface. No other charges are nearby. (i) If the particle is at the center of the cube, what is the flux through each one of the faces of the cube? (a) 0 (b) $q/2\epsilon_0$ (c) $q/6\epsilon_0$ (d) $q/8\epsilon_0$ (e) depends on the size of the cube (ii) If the particle can be moved to any point within the cube, what maximum value can the flux through one face approach? Choose from the same possibilities as in part (i).

5. The magnitude of the electric force between two protons is 2.30×10^{-26} N. How far apart are they? (a) 0.100 m (b) 0.022 0 m (c) 3.10 m (d) 0.005 70 m (e) 0.480 m

6. Estimate the magnitude of the electric field due to the proton in a hydrogen atom at a distance of 5.29×10^{-11} m, the expected position of the electron in the atom. (a) 10^{-11} N/C (b) 10^8 N/C (c) 10^{14} N/C (d) 10^6 N/C (e) 10^{12} N/C

7. Rank the electric fluxes through each gaussian surface shown in Figure OQ19.7 from largest to smallest. Display any cases of equality in your ranking.

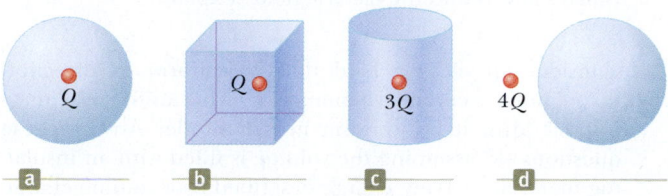

Figure OQ19.7

8. A circular ring of charge with radius b has total charge q uniformly distributed around it. What is the magnitude of the electric field at the center of the ring? (a) 0 (b) $k_e q/b^2$ (c) $k_e q^2/b^2$ (d) $k_e q^2/b$ (e) none of those answers

9. Two solid spheres, both of radius 5 cm, carry identical total charges of 2 μC. Sphere A is a good conductor. Sphere B is an insulator, and its charge is distributed uniformly throughout its volume. (i) How do the magnitudes of the electric fields they separately create at a radial distance of 6 cm compare? (a) $E_A > E_B = 0$ (b) $E_A > E_B > 0$ (c) $E_A = E_B > 0$ (d) $0 < E_A < E_B$ (e) $0 = E_A < E_B$ (ii) How do the magnitudes of the electric fields they separately create at radius 4 cm compare? Choose from the same possibilities as in part (i).

10. An electron with a speed of 3.00×10^6 m/s moves into a uniform electric field of magnitude 1.00×10^3 N/C. The field lines are parallel to the electron's velocity and pointing in the same direction as the velocity. How far does the electron travel before it is brought to rest? (a) 2.56 cm (b) 5.12 cm (c) 11.2 cm (d) 3.34 m (e) 4.24 m

11. A very small ball has a mass of 5.00×10^{-3} kg and a charge of 4.00 μC. What magnitude electric field directed upward will balance the weight of the ball so that the ball is suspended motionless above the ground? (a) 8.21×10^2 N/C (b) 1.22×10^4 N/C (c) 2.00×10^{-2} N/C (d) 5.11×10^6 N/C (e) 3.72×10^3 N/C

12. In which of the following contexts can Gauss's law *not* be readily applied to find the electric field? (a) near a long, uniformly charged wire (b) above a large, uniformly charged plane (c) inside a uniformly charged ball (d) outside a uniformly charged sphere (e) Gauss's law can be readily applied to find the electric field in all these contexts.

13. Two point charges attract each other with an electric force of magnitude F. If the charge on one of the particles is reduced to one-third its original value and the distance between the particles is doubled, what is the resulting magnitude of the electric force between them? (a) $\frac{1}{12}F$ (b) $\frac{1}{3}F$ (c) $\frac{1}{6}F$ (d) $\frac{3}{4}F$ (e) $\frac{3}{2}F$

14. Three charged particles are arranged on corners of a square as shown in Figure OQ19.14, with charge $-Q$ on both the particle at the upper left corner and the particle at the lower right corner and with charge $+2Q$ on the particle at the lower left corner. (i) What is the direction of the electric field at the upper right corner, which is a point in empty space? (a) It is upward and to the right. (b) It is straight to the right. (c) It is straight downward. (d) It is downward and to the left. (e) It is perpendicular to the plane of the picture and outward. (ii) Suppose the $+2Q$ charge at the lower left corner is removed. Then does the magnitude of the field at the upper right corner (a) become larger, (b) become smaller, (c) stay the same, or (d) change unpredictably?

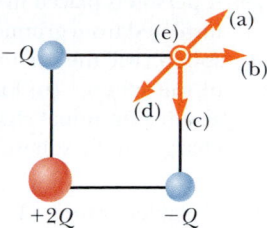

Figure OQ19.14

15. Assume the charged objects in Figure OQ19.15 are fixed. Notice that there is no sight line from the location of q_2 to the location of q_1. If you were at q_1, you would be unable to see q_2 because it is behind q_3. How would you calculate the electric force exerted on the object with charge q_1? (a) Find only the force exerted by q_2 on charge q_1. (b) Find only the force exerted by q_3 on charge q_1. (c) Add the force that q_2 would exert by itself on charge q_1 to the force that q_3 would exert by itself on charge q_1. (d) Add the force that q_3 would exert by itself to a certain fraction of the force that q_2 would exert by itself. (e) There is no definite way to find the force on charge q_1.

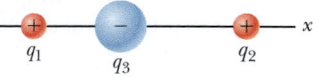

Figure OQ19.15

☐ denotes answer available in *Student Solutions Manual/Study Guide*

CONCEPTUAL QUESTIONS

1. A uniform electric field exists in a region of space containing no charges. What can you conclude about the net electric flux through a gaussian surface placed in this region of space?

2. A glass object receives a positive charge by rubbing it with a silk cloth. In the rubbing process, have protons been added to the object or have electrons been removed from it?

3. If more electric field lines leave a gaussian surface than enter it, what can you conclude about the net charge enclosed by that surface?

4. Why must hospital personnel wear special conducting shoes while working around oxygen in an operating room? What might happen if the personnel wore shoes with rubber soles?

5. Would life be different if the electron were positively charged and the proton was negatively charged? (b) Does the choice of signs have any bearing on physical and chemical interactions? Explain your answers.

6. A student who grew up in a tropical country and is studying in the United States may have no experience with static

electricity sparks and shocks until his or her first American winter. Explain.

7. If a suspended object A is attracted to a charged object B, can we conclude that A is charged? Explain.

8. A cubical surface surrounds a point charge q. Describe what happens to the total flux through the surface if (a) the charge is doubled, (b) the volume of the cube is doubled, (c) the surface is changed to a sphere, (d) the charge is moved to another location inside the surface, and (e) the charge is moved outside the surface.

9. A person is placed in a large, hollow, metallic sphere that is insulated from ground. (a) If a large charge is placed on the sphere, will the person be harmed upon touching the inside of the sphere? (b) Explain what will happen if the person also has an initial charge whose sign is opposite that of the charge on the sphere.

10. Consider point A in Figure CQ19.10 located an arbitrary distance from two positive point charges in otherwise empty space. (a) Is it possible for an electric field to exist at point A in empty space? Explain. (b) Does charge exist at this point? Explain. (c) Does a force exist at this point? Explain.

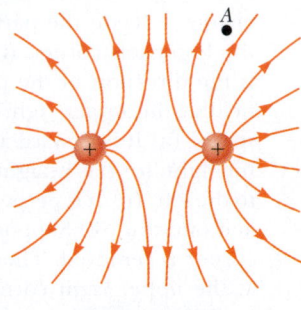

Figure CQ19.10

11. Consider two identical conducting spheres whose surfaces are separated by a small distance. One sphere is given a

large net positive charge, and the other is given a small net positive charge. It is found that the force between the spheres is attractive even though they both have net charges of the same sign. Explain how this attraction is possible.

12. If the total charge inside a closed surface is known but the distribution of the charge is unspecified, can you use Gauss's law to find the electric field? Explain.

13. Consider an electric field that is uniform in direction throughout a certain volume. Can it be uniform in magnitude? Must it be uniform in magnitude? Answer these questions (a) assuming the volume is filled with an insulating material carrying charge described by a volume charge density and (b) assuming the volume is empty space. State reasoning to prove your answers.

14. On the basis of the repulsive nature of the force between like charges and the freedom of motion of charge within a conductor, explain why excess charge on an isolated conductor must reside on its surface.

15. A common demonstration involves charging a rubber balloon, which is an insulator, by rubbing it on your hair and then touching the balloon to a ceiling or wall, which is also an insulator. Because of the electrical attraction between the charged balloon and the neutral wall, the balloon sticks to the wall. Imagine now that we have two infinitely large, flat sheets of insulating material. One is charged, and the other is neutral. If these sheets are brought into contact, does an attractive force exist between them as there was for the balloon and the wall?

▶ PROBLEMS AVAILABLE IN

List of Enhanced Problems

Problem Number	Targeted Feedback in Enhanced WebAssign	Master It in Enhanced WebAssign	Watch It in Enhanced WebAssign
19.2	✓		✓
19.5	✓		✓
19.7	✓		✓
19.9	✓	✓	
19.15	✓	✓	
19.17	✓	✓	
19.23	✓		✓
19.24	✓		✓
19.27		✓	
19.30	✓		✓
19.31	✓		✓
19.33		✓	
19.35		✓	
19.36	✓		✓
19.37	✓	✓	
19.39	✓	✓	
19.40	✓		✓
19.43	✓		✓
19.45	✓		✓
19.46	✓		✓
19.51		✓	
19.52	✓		✓
19.53	✓	✓	
19.54		✓	
19.55		✓	
19.59		✓	

Electric Potential and Capacitance

Chapter Outline

This device is a *variable capacitor*, used to tune radios to a selected station. When one set of metal plates is rotated so as to lie between a fixed set of plates, the *capacitance* of the device changes. Capacitance is a parameter that depends on *electric potential*, the primary topic of this chapter.

The concept of potential energy was introduced in Chapter 6 in connection with such conservative forces as gravity and the elastic force of a spring. By using the principle of conservation of mechanical energy in an isolated system, we are often able to avoid working directly with forces when solving mechanical problems. In this chapter, we shall use the energy concept in our study of electricity. Because the electrostatic force (given by Coulomb's law) is conservative, electrostatic phenomena can conveniently be described in terms of an *electric* potential energy function. This concept enables us to define a quantity called *electric potential*, which is a scalar quantity and which therefore leads to a simpler means of describing some electrostatic phenomena than the electric field method. As we shall see in subsequent chapters, the concept of electric potential is of great practical value in many applications.

This chapter also addresses the properties of capacitors, devices that store charge. The ability of a capacitor to store charge is measured by its *capacitance*. Capacitors are used in common applications such as frequency tuners in radio receivers, filters in power supplies, and energy-storing devices in electronic flash units.

20.1 | Electric Potential and Potential Difference

When a test charge q_0 is placed in an electric field $\vec{E}$ created by some source charge distribution, the electric force acting on the test charge is $q_0\vec{E}$. The force $\vec{F}_e = q_0\vec{E}$ is conservative because the force between charges described by Coulomb's law is conservative. When the test charge is moved in the field at constant velocity by some external agent, the work done by the field on the charge is equal to the negative of the work done by the external agent causing the displacement. This situation is analogous to that of lifting an object with mass in a gravitational field: the work done by the external agent is mgh, and the work done by the gravitational force is $-mgh$.

When analyzing electric and magnetic fields, it is common practice to use the notation $d\vec{s}$ to represent an infinitesimal displacement vector that is oriented tangent to a path through space. This path may be straight or curved, and an integral performed along this path is called either a *path integral* or a *line integral* (the two terms are synonymous).

For an infinitesimal displacement $d\vec{s}$ of a point charge q_0 immersed in an electric field, the work done within the charge–field system by the electric field on the charge is $W_{\text{int}} = \vec{F}_e \cdot d\vec{s} = q_0\vec{E} \cdot d\vec{s}$. As this amount of work is done by the field, the potential energy of the charge–field system is changed by an amount $dU = -W_{\text{int}} = -q_0\vec{E} \cdot d\vec{s}$. For a finite displacement of the charge from point Ⓐ to point Ⓑ, the change in potential energy of the system $\Delta U = U_Ⓑ - U_Ⓐ$ is

$$\Delta U = -q_0 \int_Ⓐ^Ⓑ \vec{E} \cdot d\vec{s} \qquad \textbf{20.1} ◀$$

▶ Change in electric potential energy of a charge–field system

The integration is performed along the path that q_0 follows as it moves from Ⓐ to Ⓑ. Because the force $q_0\vec{E}$ is conservative, this line integral does not depend on the path taken from Ⓐ to Ⓑ.

For a given position of the test charge in the field, the charge–field system has a potential energy U relative to the configuration of the system that is defined as $U = 0$. Dividing the potential energy by the test charge gives a physical quantity that depends only on the source charge distribution and has a value at every point in an electric field. This quantity is called the **electric potential** (or simply the **potential**) V:

$$V = \frac{U}{q_0} \qquad \textbf{20.2} ◀$$

Because potential energy is a scalar quantity, electric potential also is a scalar quantity.

As described by Equation 20.1, if the test charge is moved between two positions Ⓐ and Ⓑ in an electric field, the charge–field system experiences a change in potential energy. The **potential difference** $\Delta V = V_Ⓑ - V_Ⓐ$ between two points Ⓐ and Ⓑ in an electric field is defined as the change in potential energy of the system when a test charge q_0 is moved between the points divided by the test charge:

$$\Delta V \equiv \frac{\Delta U}{q_0} = -\int_Ⓐ^Ⓑ \vec{E} \cdot d\vec{s} \qquad \textbf{20.3} ◀$$

▶ Potential difference between two points

In this definition, the infinitesimal displacement $d\vec{s}$ is interpreted as the displacement between two points in space rather than the displacement of a point charge as in Equation 20.1.

Just as with potential energy, only *differences* in electric potential are meaningful. We often take the value of the electric potential to be zero at some convenient point in an electric field.

Potential difference should not be confused with difference in potential energy. The potential difference between Ⓐ and Ⓑ exists solely because of a source charge and depends on the source charge distribution (consider points Ⓐ and Ⓑ *without* the presence of the test charge). For a potential energy to exist, we must have a *system* of two or more charges. The potential energy belongs to the system and changes only if a charge is moved relative to the rest of the system.

Pitfall Prevention | 20.1
Potential and Potential Energy
The *potential is characteristic of the field only*, independent of a charged test particle that may be placed in the field. *Potential energy is characteristic of the charge–field system* due to an interaction between the field and a charged particle placed in the field.

Pitfall Prevention | 20.2
Voltage
A variety of phrases are used to describe the potential difference between two points, the most common being **voltage**, arising from the unit for potential. A voltage *applied* to a device, such as a television, or *across* a device is the same as the potential difference across the device. Despite popular language, voltage is *not* something that moves *through* a device.

Pitfall Prevention | 20.3
The Electron Volt
The electron volt is a unit of *energy*, NOT of potential. The energy of any system may be expressed in eV, but this unit is most convenient for describing the emission and absorption of visible light from atoms. Energies of nuclear processes are often expressed in MeV.

If an external agent moves a test charge from Ⓐ to Ⓑ without changing the kinetic energy of the test charge, the agent performs work that changes the potential energy of the system: $W = \Delta U$. Imagine an arbitrary charge q located in an electric field. From Equation 20.3, the work done by an external agent in moving a charge q through an electric field at constant velocity is

$$W = q\,\Delta V \qquad \text{20.4} \blacktriangleleft$$

Because electric potential is a measure of potential energy per unit charge, the SI unit of both electric potential and potential difference is joules per coulomb, which is defined as a **volt** (V):

$$1\ \mathrm{V} \equiv 1\ \mathrm{J/C}$$

That is, 1 J of work must be done to move a 1-C charge through a potential difference of 1 V.

Equation 20.3 shows that potential difference also has units of electric field times distance. It follows that the SI unit of electric field (N/C) can also be expressed in volts per meter:

$$1\ \mathrm{N/C} = 1\ \mathrm{V/m}$$

Therefore, we can interpret the electric field as a measure of the rate of change of the electric potential with respect to position.

As discussed in Section 9.7, a unit of energy commonly used in atomic and nuclear physics is the **electron volt** (eV), which is defined as the energy a charge–field system gains or loses when a charge of magnitude e (that is, an electron or a proton) is moved through a potential difference of 1 V. Because $1\ \mathrm{V} = 1\ \mathrm{J/C}$ and the fundamental charge is equal to 1.602×10^{-19} C, the electron volt is related to the joule as follows:

$$1\ \mathrm{eV} = 1.602 \times 10^{-19}\ \mathrm{C \cdot V} = 1.602 \times 10^{-19}\ \mathrm{J} \qquad \text{20.5} \blacktriangleleft$$

For instance, an electron in the beam of a typical dental x-ray machine may have a speed of 1.4×10^8 m/s. This speed corresponds to a kinetic energy 1.1×10^{-14} J (using relativistic calculations as discussed in Chapter 9), which is equivalent to 6.7×10^4 eV. Such an electron has to be accelerated from rest through a potential difference of 67 kV to reach this speed.

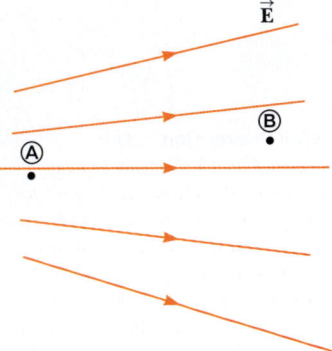

Figure 20.1 (Quick Quiz 20.1) Two points in an electric field.

QUICK QUIZ 20.1 In Figure 20.1, two points Ⓐ and Ⓑ are located within a region in which there is an electric field. **(i)** How would you describe the potential difference $\Delta V = V_\text{Ⓑ} - V_\text{Ⓐ}$? (a) It is positive. (b) It is negative. (c) It is zero. **(ii)** A negative charge is placed at Ⓐ and then moved to Ⓑ. How would you describe the change in potential energy of the charge–field system for this process? Choose from the same possibilities.

20.2 | Potential Difference in a Uniform Electric Field

Equations 20.1 and 20.3 hold in all electric fields, whether uniform or varying, but they can be simplified for a uniform field. First, consider a uniform electric field directed along the negative y axis as shown in Active Figure 20.2a. Let's calculate the potential difference between two points Ⓐ and Ⓑ separated by a distance d, where the displacement $\vec{s}$ points from Ⓐ toward Ⓑ and is parallel to the field lines. Equation 20.3 gives

$$V_\text{Ⓑ} - V_\text{Ⓐ} = \Delta V = -\int_\text{Ⓐ}^\text{Ⓑ} \vec{E} \cdot d\vec{s} = -\int_\text{Ⓐ}^\text{Ⓑ} E\,ds\,(\cos 0°) = -\int_\text{Ⓐ}^\text{Ⓑ} E\,ds$$

Because E is constant, it can be removed from the integral sign, which gives

$$\Delta V = -E \int_\text{Ⓐ}^\text{Ⓑ} ds = -Ed \qquad \text{20.6} \blacktriangleleft$$

▶ Potential difference between two points in a uniform electric field

The negative sign indicates that the electric potential at point Ⓑ is lower than at point Ⓐ; that is, $V_Ⓑ < V_Ⓐ$. Electric field lines *always* point in the direction of decreasing electric potential as shown in Active Figure 20.2a.

Now suppose a test charge q_0 moves from Ⓐ to Ⓑ. We can calculate the change in the potential energy of the charge–field system from Equations 20.3 and 20.6:

$$\Delta U = q_0 \Delta V = -q_0 Ed \qquad \textbf{20.7}◀$$

This result shows that if q_0 is positive, then ΔU is negative. Therefore, in a system consisting of a positive charge and an electric field, the electric potential energy of the system decreases when the charge moves in the direction of the field. Equivalently, an electric field does work on a positive charge when the charge moves in the direction of the electric field. That is analogous to the work done by the gravitational field on a falling object as shown in Active Figure 20.2b. If a positive test charge is released from rest in this electric field, it experiences an electric force $q_0\vec{E}$ in the direction of $\vec{E}$ (downward in Active Fig. 20.2a). Therefore, it accelerates downward, gaining kinetic energy. As the charged particle gains kinetic energy, the potential energy of the charge–field system decreases by an equal amount. This equivalence should not be surprising; it is simply conservation of mechanical energy in an isolated system as introduced in Chapter 7.

The comparison between a system of an electric field with a positive test charge and a gravitational field with a test mass in Active Figure 20.2 is useful for conceptualizing electrical behavior. The electrical situation, however, has one feature that the gravitational situation does not: the test charge can be negative. If q_0 is negative, then ΔU in Equation 20.7 is positive and the situation is reversed. A system consisting of a negative charge and an electric field gains electric potential energy when the charge moves in the direction of the field. If a negative charge is released from rest in an electric field, it accelerates in a direction opposite the direction of the field. For the negative charge to move in the direction of the field, an external agent must apply a force and do positive work on the charge.

Now consider the more general case of a charged particle that moves between Ⓐ and Ⓑ in a uniform electric field such that the vector $\vec{s}$ is *not* parallel to the field lines as shown in Figure 20.3. In this case, Equation 20.3 gives

$$\Delta V = -\int_Ⓐ^Ⓑ \vec{E} \cdot d\vec{s} = -\vec{E} \cdot \int_Ⓐ^Ⓑ d\vec{s} = -\vec{E} \cdot \vec{s} \qquad \textbf{20.8}◀$$

where again $\vec{E}$ was removed from the integral because it is constant. The change in potential energy of the charge–field system is

$$\Delta U = q_0 \Delta V = -q_0 \vec{E} \cdot \vec{s} \qquad \textbf{20.9}◀$$

Finally, we conclude from Equation 20.8 that all points in a plane perpendicular to a uniform electric field are at the same electric potential. We can see that in Figure 20.3, where the potential difference $V_Ⓑ - V_Ⓐ$ is equal to the potential difference $V_Ⓒ - V_Ⓐ$. (Prove this fact to yourself by working out two dot products for $\vec{E} \cdot \vec{s}$: one for $\vec{s}_{Ⓐ→Ⓑ}$, where the angle θ between $\vec{E}$ and $\vec{s}$ is arbitrary as shown in Figure 20.3, and one for $\vec{s}_{Ⓐ→Ⓒ}$, where $\theta = 0$.) Therefore, $V_Ⓑ = V_Ⓒ$. The name **equipotential surface** is given to any surface consisting of a continuous distribution of points having the same electric potential.

When a positive test charge moves from point Ⓐ to point Ⓑ, the electric potential energy of the charge–field system decreases.

When an object with mass moves from point Ⓐ to point Ⓑ, the gravitational potential energy of the object–field system decreases.

Active Figure 20.2 (a) When the electric field $\vec{E}$ is directed downward, point Ⓑ is at a lower electric potential than point Ⓐ. (b) An object of mass m moving downward in a gravitational field $\vec{g}$.

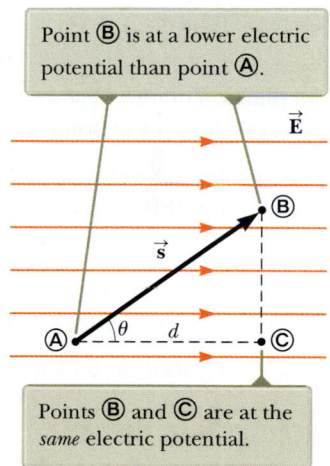

Point Ⓑ is at a lower electric potential than point Ⓐ.

Points Ⓑ and Ⓒ are at the *same* electric potential.

Figure 20.3 A uniform electric field directed along the positive x axis.

▶ Change in potential between two points in a uniform electric field

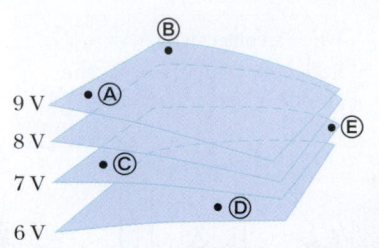

9 V
8 V
7 V
6 V

Figure 20.4 (Quick Quiz 20.2) Four equipotential surfaces.

The equipotential surfaces associated with a uniform electric field consist of a family of parallel planes that are all perpendicular to the field. Equipotential surfaces associated with fields having other symmetries are described in later sections.

> **QUICK QUIZ 20.2** The labeled points in Figure 20.4 are on a series of equipotential surfaces associated with an electric field. Rank (from greatest to least) the work done by the electric field on a positively charged particle that moves from Ⓐ to Ⓑ, from Ⓑ to Ⓒ, from Ⓒ to Ⓓ, and from Ⓓ to Ⓔ.

Example 20.1 | The Electric Field Between Two Parallel Plates of Opposite Charge

A battery has a specified potential difference ΔV between its terminals and establishes that potential difference between conductors attached to the terminals. A 12-V battery is connected between two parallel plates as shown in Figure 20.5. The separation between the plates is $d = 0.30$ cm, and we assume the electric field between the plates to be uniform. (This assumption is reasonable if the plate separation is small relative to the plate dimensions and we do not consider locations near the plate edges.) Find the magnitude of the electric field between the plates.

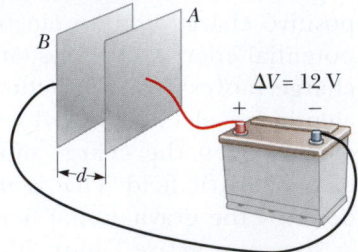

Figure 20.5 (Example 20.1) A 12-V battery connected to two parallel plates. The electric field between the plates has a magnitude given by the potential difference ΔV divided by the plate separation d.

SOLUTION

Conceptualize In Chapter 19 we investigated the uniform electric field between parallel plates. The new feature to this problem is that the electric field is related to the new concept of electric potential.

Categorize The electric field is evaluated from a relationship between field and potential given in this section, so we categorize this example as a substitution problem.

Use Equation 20.6 to evaluate the magnitude of the electric field between the plates:

$$E = \frac{|V_B - V_A|}{d} = \frac{12 \text{ V}}{0.30 \times 10^{-2} \text{ m}} = 4.0 \times 10^3 \text{ V/m}$$

The configuration of plates in Figure 20.5 is called a *parallel-plate capacitor* and is examined in greater detail in Section 20.7.

Example 20.2 | Motion of a Proton in a Uniform Electric Field

A proton is released from rest at point Ⓐ in a uniform electric field that has a magnitude of 8.0×10^4 V/m (Fig. 20.6). The proton undergoes a displacement of magnitude $d = 0.50$ m to point Ⓑ in the direction of $\vec{\mathbf{E}}$. Find the speed of the proton after completing the displacement.

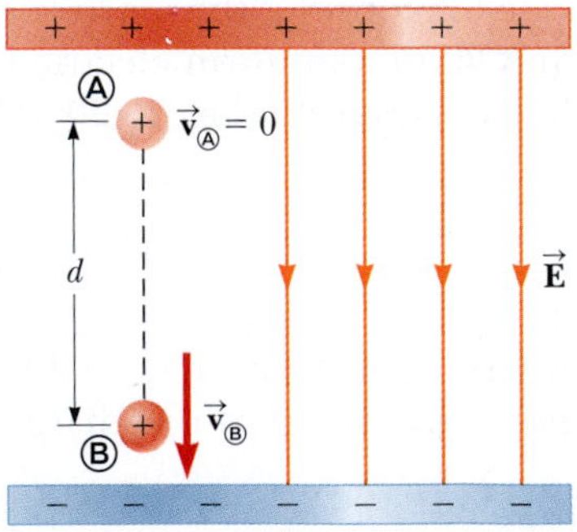

SOLUTION

Conceptualize Visualize the proton in Figure 20.6 moving downward through the potential difference. The situation is analogous to an object falling through a gravitational field.

Figure 20.6 (Example 20.2) A proton accelerates from Ⓐ to Ⓑ in the direction of the electric field.

Categorize The system of the proton and the two plates in Figure 20.6 does not interact with the environment, so we model it as an isolated system.

Analyze Use Equation 20.6 to find the potential difference between points Ⓐ and Ⓑ:

$$\Delta V = -Ed = -(8.0 \times 10^4 \text{ V/m})(0.50 \text{ m}) = -4.0 \times 10^4 \text{ V}$$

Write the appropriate reduction of Equation 7.2, the conservation of energy equation, for the isolated system of the charge and the electric field:

$$\Delta K + \Delta U = 0$$

20.2 cont.

Substitute the changes in energy for both terms:

$$(\tfrac{1}{2}mv^2 - 0) + e\,\Delta V = 0$$

Solve for the final speed of the proton:

$$v = \sqrt{\frac{-2e\,\Delta V}{m}}$$

Substitute numerical values:

$$v = \sqrt{\frac{-2(1.6 \times 10^{-19}\ \text{C})(-4.0 \times 10^4\ \text{V})}{1.67 \times 10^{-27}\ \text{kg}}}$$

$$= \boxed{2.8 \times 10^6\ \text{m/s}}$$

Finalize Because ΔV is negative for the field, ΔU is also negative for the proton–field system. The negative value of ΔU means the potential energy of the system decreases as the proton moves in the direction of the electric field. As the proton accelerates in the direction of the field, it gains kinetic energy while the electric potential energy of the system decreases at the same time.

Figure 20.6 is oriented so that the proton moves downward. The proton's motion is analogous to that of an object falling in a gravitational field. Although the gravitational field is always downward at the surface of the Earth, an electric field can be in any direction, depending on the orientation of the plates creating the field. Therefore, Figure 20.6 could be rotated 90° or 180° and the proton could move horizontally or upward in the electric field!

20.3 | Electric Potential and Potential Energy Due to Point Charges

As discussed in Section 19.6, an isolated positive point charge q produces an electric field directed radially outward from the charge. To find the electric potential at a point located a distance r from the charge, let's begin with the general expression for potential difference,

$$V_{\circledB} - V_{\circledA} = -\int_{\circledA}^{\circledB} \vec{\mathbf{E}} \cdot d\vec{\mathbf{s}}$$

where $\circledA$ and $\circledB$ are the two arbitrary points shown in Figure 20.7. At any point in space, the electric field due to the point charge is $\vec{\mathbf{E}} = (k_e q/r^2)\hat{\mathbf{r}}$ (Eq. 19.5), where $\hat{\mathbf{r}}$ is a unit vector directed radially outward from the charge. The quantity $\vec{\mathbf{E}} \cdot d\vec{\mathbf{s}}$ can be expressed as

$$\vec{\mathbf{E}} \cdot d\vec{\mathbf{s}} = k_e \frac{q}{r^2} \hat{\mathbf{r}} \cdot d\vec{\mathbf{s}}$$

Because the magnitude of $\hat{\mathbf{r}}$ is 1, the dot product $\hat{\mathbf{r}} \cdot d\vec{\mathbf{s}} = ds\cos\theta$, where θ is the angle between $\hat{\mathbf{r}}$ and $d\vec{\mathbf{s}}$. Furthermore, $ds\cos\theta$ is the projection of $d\vec{\mathbf{s}}$ onto $\hat{\mathbf{r}}$; therefore, $ds\cos\theta = dr$. That is, any displacement $d\vec{\mathbf{s}}$ along the path from point $\circledA$ to point $\circledB$ produces a change dr in the magnitude of $\vec{\mathbf{r}}$, the position vector of the point relative to the charge creating the field. Making these substitutions, we find that $\vec{\mathbf{E}} \cdot d\vec{\mathbf{s}} = (k_e q/r^2)\,dr$; hence, the expression for the potential difference becomes

$$V_{\circledB} - V_{\circledA} = -k_e q \int_{r_{\circledA}}^{r_{\circledB}} \frac{dr}{r^2} = k_e \frac{q}{r}\Big|_{r_{\circledA}}^{r_{\circledB}}$$

$$V_{\circledB} - V_{\circledA} = k_e q \left[\frac{1}{r_{\circledB}} - \frac{1}{r_{\circledA}}\right] \qquad \textbf{20.10} \blacktriangleleft$$

Equation 20.10 shows us that the integral of $\vec{\mathbf{E}} \cdot d\vec{\mathbf{s}}$ is *independent* of the path between points $\circledA$ and $\circledB$. Multiplying by a charge q_0 that moves between points $\circledA$ and $\circledB$, we see that the integral of $q_0\vec{\mathbf{E}} \cdot d\vec{\mathbf{s}}$ is also independent of path. This latter integral, which is the work done by the electric force on the charge q_0, shows that the electric force is conservative (see Section 6.7). We define a field that is related to a

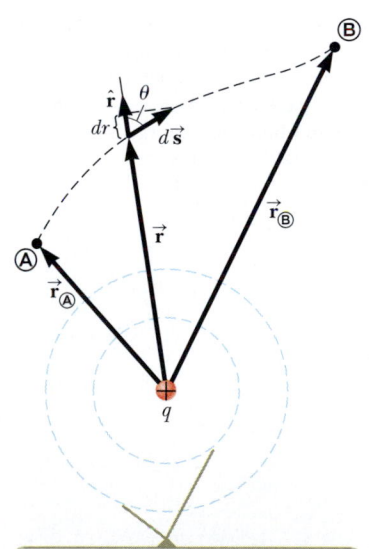

The two dashed circles represent intersections of spherical equipotential surfaces with the page.

Figure 20.7 The potential difference between points $\circledA$ and $\circledB$ due to a point charge q depends *only* on the initial and final radial coordinates $r_{\circledA}$ and $r_{\circledB}$.

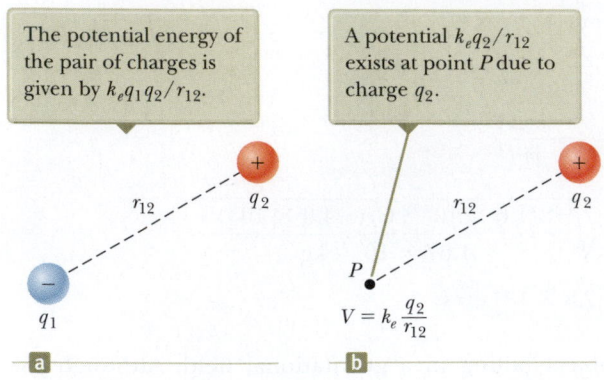

The potential energy of the pair of charges is given by $k_e q_1 q_2 / r_{12}$.

A potential $k_e q_2 / r_{12}$ exists at point P due to charge q_2.

$$V = k_e \frac{q_2}{r_{12}}$$

Active Figure 20.8 (a) Two point charges separated by a distance r_{12}. (b) Charge q_1 is removed.

▶ Electric potential due to several point charges

conservative force as a **conservative field.** Therefore, Equation 20.10 tells us that the electric field of a fixed point charge q is conservative. Furthermore, Equation 20.10 expresses the important result that the potential difference between any two points Ⓐ and Ⓑ in a field created by a point charge depends only on the radial coordinates $r_Ⓐ$ and $r_Ⓑ$. It is customary to choose the reference of electric potential for a point charge to be $V = 0$ at $r_Ⓐ = \infty$. With this reference choice, the electric potential due to a point charge at any distance r from the charge is

$$V = k_e \frac{q}{r} \qquad \text{20.11} ◀$$

We obtain the electric potential resulting from two or more point charges by applying the superposition principle. That is, the total electric potential at some point P due to several point charges is the sum of the potentials due to the individual charges. For a group of point charges, we can write the total electric potential at P as

$$V = k_e \sum_i \frac{q_i}{r_i} \qquad \text{20.12} ◀$$

where the potential is again taken to be zero at infinity and r_i is the distance from the point P to the charge q_i. Notice that the sum in Equation 20.12 is an algebraic sum of scalars rather than a vector sum (which we used to calculate the electric field of a group of charges in Eq. 19.6). Therefore, it is often much easier to evaluate V than $\vec{E}$.

Now consider the potential energy of a system of two charged particles. If V_2 is the electric potential at a point P due to charge q_2, the work an external agent must do to bring a second charge q_1 from infinity to P without acceleration is $q_1 V_2$. This work represents a transfer of energy into the system, and the energy appears in the system as potential energy U when the particles are separated by a distance r_{12} (Active Fig. 20.8a). Therefore, the potential energy of the system can be expressed as[1]

$$U = k_e \frac{q_1 q_2}{r_{12}} \qquad \text{20.13} ◀$$

If the charges are of the same sign, then U is positive. Positive work must be done by an external agent on the system to bring the two charges near each other (because charges of the same sign repel). If the charges are of opposite sign, then U is negative. Negative work is done by an external agent against the attractive force between the charges of opposite sign as they are brought near each other; a force must be applied opposite the displacement to prevent q_1 from accelerating toward q_2.

In Active Figure 20.8b, we have removed the charge q_1. At the position this charge previously occupied, point P, Equations 20.2 and 20.13 can be used to define a potential due to charge q_2 as $V = U/q_1 = k_e q_2 / r_{12}$. This expression is consistent with Equation 20.11.

If the system consists of more than two charged particles, we can obtain the total potential energy of the system by calculating U for every *pair* of charges and summing the terms algebraically. The total electric potential energy of a system of point charges is equal to the work required to bring the charges, one at a time, from an infinite separation to their final positions.

▸ **QUICK QUIZ 20.3** A spherical balloon contains a positively charged object at its center. **(i)** As the balloon is inflated to a greater volume while the charged object remains at the center, does the electric potential at the surface of the balloon (a) increase, (b) decrease, or (c) remain the same? **(ii)** Does the electric flux through the surface of the balloon (a) increase, (b) decrease, or (c) remain the same?

---

[1]The expression for the electric potential energy of a system made up of two point charges, Equation 20.13, is of the *same* form as the equation for the gravitational potential energy of a system made up of two point masses, $-G m_1 m_2 / r$ (see Chapter 11). The similarity is not surprising considering that both expressions are derived from an inverse-square force law.

In Active Figure 20.8a, take q_1 to be a negative source charge and q_2 to be the test charge. (**i**) If q_2 is initially positive and is changed to a charge of the same magnitude but negative, what happens to the potential at the position of q_2 due to q_1? (a) It increases. (b) It decreases. (c) It remains the same. (**ii**) When q_2 is changed from positive to negative, what happens to the potential energy of the two-charge system? Choose from the same possibilities.

Example 20.3 | The Electric Potential Due to Two Point Charges

As shown in Figure 20.9a, a charge $q_1 = 2.00\ \mu C$ is located at the origin and a charge $q_2 = -6.00\ \mu C$ is located at $(0, 3.00)$ m.

(A) Find the total electric potential due to these charges at the point P, whose coordinates are $(4.00, 0)$ m.

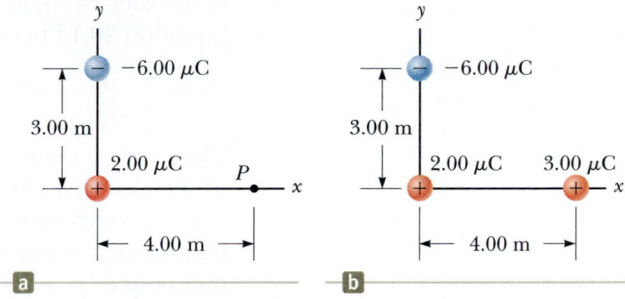

Figure 20.9 (Example 20.3) (a) The electric potential at P due to the two charges q_1 and q_2 is the algebraic sum of the potentials due to the individual charges. (b) A third charge $q_3 = 3.00\ \mu C$ is brought from infinity to point P.

SOLUTION

Conceptualize Recognize first that the 2.00-μC and -6.00-μC charges are source charges and set up an electric field as well as a potential at all points in space, including point P.

Categorize The potential is evaluated using an equation developed in this chapter, so we categorize this example as a substitution problem.

Use Equation 20.12 for the system of two source charges:

$$V_P = k_e\left(\frac{q_1}{r_1} + \frac{q_2}{r_2}\right)$$

Substitute numerical values:

$$V_P = (8.99 \times 10^9\ \text{N} \cdot \text{m}^2/\text{C}^2)\left(\frac{2.00 \times 10^{-6}\ \text{C}}{4.00\ \text{m}} + \frac{-6.00 \times 10^{-6}\ \text{C}}{5.00\ \text{m}}\right)$$

$$= -6.29 \times 10^3\ \text{V}$$

(B) Find the change in potential energy of the system of two charges plus a third charge $q_3 = 3.00\ \mu C$ as the latter charge moves from infinity to point P (Fig. 20.9b).

SOLUTION

Assign $U_i = 0$ for the system to the configuration in which the charge q_3 is at infinity. Use Equation 20.2 to evaluate the potential energy for the configuration in which the charge is at P:

$$U_f = q_3 V_P$$

Substitute numerical values to evaluate ΔU:

$$\Delta U = U_f - U_i = q_3 V_P - 0 = (3.00 \times 10^{-6}\ \text{C})(-6.29 \times 10^3\ \text{V})$$

$$= -1.89 \times 10^{-2}\ \text{J}$$

Therefore, because the potential energy of the system has decreased, an external agent has to do positive work to remove the charge q_3 from point P back to infinity.

What If? You are working through this example with a classmate and she says, "Wait a minute! In part (B), we ignored the potential energy associated with the pair of charges q_1 and q_2!" How would you respond?

Answer Given the statement of the problem, it is not necessary to include this potential energy because part (B) asks for the *change* in potential energy of the system as q_3 is brought in from infinity. Because the configuration of charges q_1 and q_2 does not change in the process, there is no ΔU associated with these charges.

❰20.4 | Obtaining the Value of the Electric Field from the Electric Potential

The electric field $\vec{\mathbf{E}}$ and the electric potential V are related as shown in Equation 20.3, which tells us how to find ΔV if the electric field $\vec{\mathbf{E}}$ is known. We now show how to calculate the value of the electric field if the electric potential is known in a certain region.

From Equation 20.3, we can express the potential difference dV between two points a distance ds apart as

$$dV = -\vec{\mathbf{E}} \cdot d\vec{\mathbf{s}}$$ **20.14** ◀

If the electric field has only one component E_x, then $\vec{\mathbf{E}} \cdot d\vec{\mathbf{s}} = E_x\,dx$. Therefore, Equation 20.14 becomes $dV = -E_x\,dx$, or

$$E_x = -\frac{dV}{dx}$$ **20.15** ◀

That is, the x component of the electric field is equal to the negative of the derivative of the electric potential with respect to x. Similar statements can be made about the y and z components. Equation 20.15 is the mathematical statement of the electric field being a measure of the rate of change with position of the electric potential as mentioned in Section 20.1.

When a test charge undergoes a displacement $d\vec{\mathbf{s}}$ along an equipotential surface, then $dV = 0$ because the potential is constant along an equipotential surface. From Equation 20.14, we see that $dV = -\vec{\mathbf{E}} \cdot d\vec{\mathbf{s}} = 0$; therefore, $\vec{\mathbf{E}}$ must be perpendicular to the displacement along the equipotential surface. This result shows that the equipotential surfaces must always be perpendicular to the electric field lines passing through them.

As mentioned at the end of Section 20.2, the equipotential surfaces associated with a uniform electric field consist of a family of planes perpendicular to the field lines. Figure 20.10a shows some representative equipotential surfaces for this situation.

If the charge distribution creating an electric field has spherical symmetry such that the volume charge density depends only on the radial distance r, the electric field is radial. In this case, $\vec{\mathbf{E}} \cdot d\vec{\mathbf{s}} = E_r\,dr$, and we can express dV as $dV = -E_r\,dr$. Therefore,

$$E_r = -\frac{dV}{dr}$$ **20.16** ◀

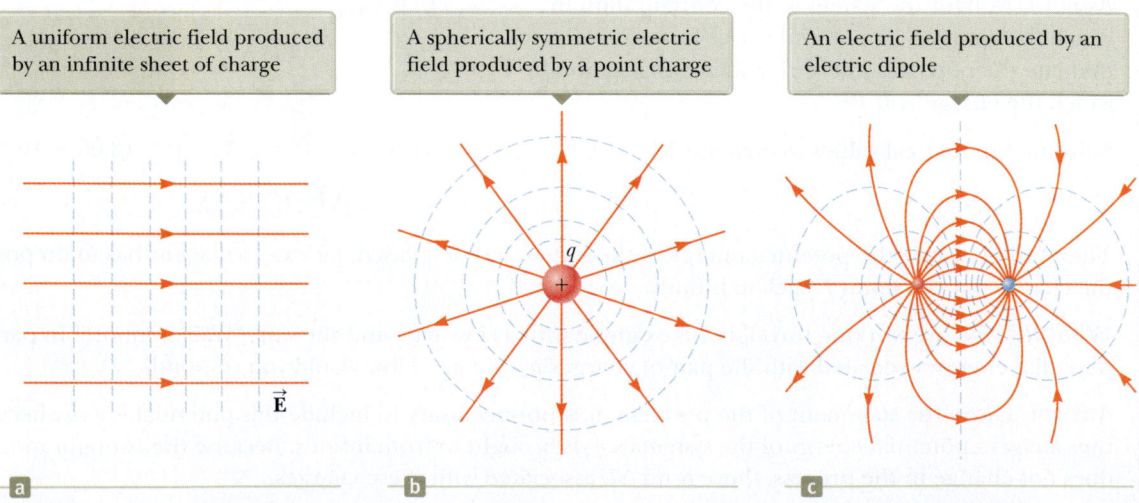

A uniform electric field produced by an infinite sheet of charge

A spherically symmetric electric field produced by a point charge

An electric field produced by an electric dipole

$\vec{\mathbf{E}}$

q
+

a

b

c

Figure 20.10 Equipotential surfaces (the dashed blue lines are intersections of these surfaces with the page) and electric field lines. In all cases, the equipotential surfaces are *perpendicular* to the electric field lines at every point.

For example, the electric potential of a point charge is $V = k_e q/r$. Because V is a function of r only, the potential function has spherical symmetry. Applying Equation 20.16, we find that the magnitude of the electric field due to the point charge is $E_r = k_e q/r^2$, a familiar result. Notice that the potential changes only in the radial direction, not in any direction perpendicular to r. Therefore, V (like E_r) is a function only of r, which is again consistent with the idea that equipotential surfaces are perpendicular to field lines. In this case, the equipotential surfaces are a family of spheres concentric with the spherically symmetric charge distribution (Fig. 20.10b). The equipotential surfaces for an electric dipole are sketched in Figure 20.10c.

In general, the electric potential is a function of all three spatial coordinates. If $V(r)$ is given in terms of the Cartesian coordinates, the electric field components E_x, E_y, and E_z can readily be found from $V(x, y, z)$ as the partial derivatives

$$E_x = -\frac{\partial V}{\partial x} \qquad E_y = -\frac{\partial V}{\partial y} \qquad E_z = -\frac{\partial V}{\partial z} \qquad \textbf{20.17} \blacktriangleleft$$

▶ Finding the electric field from the potential

QUICK QUIZ 20.5 In a certain region of space, the electric potential is zero everywhere along the x axis. (**i**) From this information, you can conclude that the x component of the electric field in this region is (a) zero, (b) in the positive x direction, or (c) in the negative x direction. (**ii**) Suppose the electric potential is $+2$ V everywhere along the x axis. From the same choices, what can you conclude about the x component of the electric field now?

Example 20.4 | The Electric Potential Due to a Dipole

An electric dipole consists of two charges of equal magnitude and opposite sign separated by a distance $2a$ as shown in Figure 20.11. The dipole is along the x axis and is centered at the origin.

(A) Calculate the electric potential at point P on the y axis.

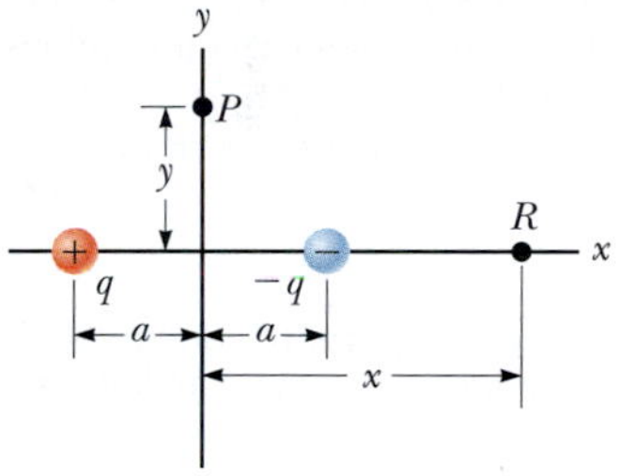

SOLUTION

Conceptualize Compare this situation to that in part (A) of Example 19.4. It is the same situation, but here we are seeking the electric potential rather than the electric field.

Categorize Because the dipole consists of only two source charges, the electric potential can be evaluated by summing the potentials due to the individual charges.

Figure 20.11 (Example 20.4) An electric dipole located on the x axis.

Analyze Use Equation 20.12 to find the electric potential at P due to the two charges:

$$V_P = k_e \sum_i \frac{q_i}{r_i} = k_e \left(\frac{q}{\sqrt{a^2 + y^2}} + \frac{-q}{\sqrt{a^2 + y^2}} \right) = 0$$

(B) Calculate the electric potential at point R on the positive x axis.

SOLUTION

Use Equation 20.12 to find the electric potential at R due to the two charges:

$$V_R = k_e \sum_i \frac{q_i}{r_i} = k_e \left(\frac{-q}{x - a} + \frac{q}{x + a} \right) = -\frac{2k_e qa}{x^2 - a^2}$$

(C) Calculate V and E_x at a point on the x axis far from the dipole.

SOLUTION

For point R far from the dipole such that $x \gg a$, neglect a^2 in the denominator of the answer to part (B) and write V in this limit:

$$V_R = \lim_{x \gg a} \left(-\frac{2k_e qa}{x^2 - a^2} \right) \approx -\frac{2k_e qa}{x^2} \quad (x \gg a)$$

continued

20.4 *cont.*

Use Equation 20.15 and this result to calculate the x component of the electric field at a point on the x axis far from the dipole:

$$E_x = -\frac{dV}{dx} = -\frac{d}{dx}\left(-\frac{2k_e qa}{x^2}\right)$$

$$= 2k_e qa \frac{d}{dx}\left(\frac{1}{x^2}\right) = -\frac{4k_e qa}{x^3} \quad (x \gg a)$$

Finalize The potentials in parts (B) and (C) are negative because points on the positive x axis are closer to the negative charge than to the positive charge. For the same reason, the x component of the electric field is negative.

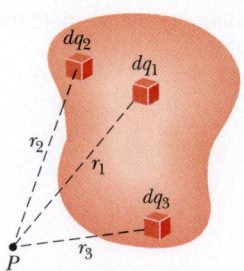

Figure 20.12 The electric potential at point P due to a continuous charge distribution can be calculated by dividing the charge distribution into elements of charge dq and summing the electric potential contributions over all elements. Three sample elements of charge are shown.

20.5 | Electric Potential Due to Continuous Charge Distributions

The electric potential due to a continuous charge distribution can be calculated using two different methods. The first method is as follows. If the charge distribution is known, we consider the potential due to a small charge element dq, treating this element as a point charge (Fig. 20.12). From Equation 20.11, the electric potential dV at some point P due to the charge element dq is

$$dV = k_e \frac{dq}{r} \qquad \text{20.18} \blacktriangleleft$$

where r is the distance from the charge element to point P. To obtain the total potential at point P, we integrate Equation 20.18 to include contributions from all elements of the charge distribution. Because each element is, in general, a different distance from point P and k_e is constant, we can express V as

▶ Electric potential due to a continuous charge distribution

$$V = k_e \int \frac{dq}{r} \qquad \text{20.19} \blacktriangleleft$$

In effect, we have replaced the sum in Equation 20.12 with an integral. In this expression for V, the electric potential is taken to be zero when point P is infinitely far from the charge distribution.

The second method for calculating the electric potential is used if the electric field is already known from other considerations such as Gauss's law. If the charge distribution has sufficient symmetry, we first evaluate $\vec{E}$ using Gauss's law and then substitute the value obtained into Equation 20.3 to determine the potential difference ΔV between any two points. We then choose the electric potential V to be zero at some convenient point.

> ### PROBLEM-SOLVING STRATEGY: Calculating Electric Potential
>
> The following procedure is recommended for solving problems that involve the determination of an electric potential due to a charge distribution.
>
> 1. **Conceptualize** Think carefully about the individual charges or the charge distribution you have in the problem and imagine what type of potential would be created. Appeal to any symmetry in the arrangement of charges to help you visualize the potential.
>
> 2. **Categorize** Are you analyzing a group of individual charges or a continuous charge distribution? The answer to this question will tell you how to proceed in the *Analyze* step.

3. **Analyze** When working problems involving electric potential, remember that it is a *scalar quantity,* so there are no components to consider. Therefore, when using the superposition principle to evaluate the electric potential at a point, simply take the algebraic sum of the potentials due to each charge. You must keep track of signs, however.

As with potential energy in mechanics, only *changes* in electric potential are significant; hence, the point where the potential is set at zero is arbitrary. When dealing with point charges or a finite-sized charge distribution, we usually define $V = 0$ to be at a point infinitely far from the charges. If the charge distribution itself extends to infinity, however, some other nearby point must be selected as the reference point.

(a) *If you are analyzing a group of individual charges:* Use the superposition principle, which states that when several point charges are present, the resultant potential at a point P in space is the *algebraic sum* of the individual potentials at P due to the individual charges (Eq. 20.12). Example 20.4 demonstrated this procedure.

(b) *If you are analyzing a continuous charge distribution:* Replace the sums for evaluating the total potential at some point P from individual charges by integrals (Eq. 20.19). The charge distribution is divided into infinitesimal elements of charge dq located at a distance r from the point P. An element is then treated as a point charge, so the potential at P due to the element is $dV = k_e\, dq/r$. The total potential at P is obtained by integrating over the entire charge distribution. For many problems, it is possible in performing the integration to express dq and r in terms of a single variable. To simplify the integration, give careful consideration to the geometry involved in the problem. Examples 20.5 and 20.6 demonstrate such a procedure.

To obtain the potential from the electric field: Another method used to obtain the potential is to start with the definition of the potential difference given by Equation 20.3. If $\vec{\mathbf{E}}$ is known or can be obtained easily (such as from Gauss's law), the line integral of $\vec{\mathbf{E}} \cdot d\vec{\mathbf{s}}$ can be evaluated.

4. **Finalize** Check to see if your expression for the potential is consistent with the mental representation and reflects any symmetry you noted previously. Imagine varying parameters such as the distance of the observation point from the charges or the radius of any circular objects to see if the mathematical result changes in a reasonable way.

Example 20.5 | Electric Potential Due to a Uniformly Charged Ring

(A) Find an expression for the electric potential at a point P located on the perpendicular central axis of a uniformly charged ring of radius a and total charge Q.

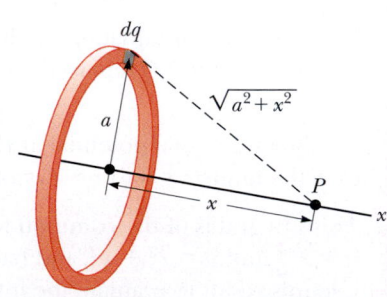

Figure 20.13 (Example 20.5) A uniformly charged ring of radius a lies in a plane perpendicular to the x axis. All elements dq of the ring are the same distance from a point P lying on the x axis.

SOLUTION

Conceptualize Study Figure 20.13, in which the ring is oriented so that its plane is perpendicular to the x axis and its center is at the origin. Notice that the symmetry of the situation means that all the charges on the ring are the same distance from point P.

Categorize Because the ring consists of a continuous distribution of charge rather than a set of discrete charges, we must use the integration technique represented by Equation 20.19 in this example.

Analyze We take point P to be at a distance x from the center of the ring as shown in Figure 20.13.

Use Equation 20.19 to express V in terms of the geometry:

$$V = k_e \int \frac{dq}{r} = k_e \int \frac{dq}{\sqrt{a^2 + x^2}}$$

continued

20.5 *cont.*

Noting that a and x are constants, bring $\sqrt{a^2 + x^2}$ in front of the integral sign and integrate over the ring:

$$(1) \quad V = \frac{k_e}{\sqrt{a^2 + x^2}} \int dq = \frac{k_e Q}{\sqrt{a^2 + x^2}}$$

(B) Find an expression for the magnitude of the electric field at point P.

SOLUTION

From symmetry, notice that along the x axis $\vec{E}$ can have only an x component. Therefore, apply Equation 20.15 to Equation (1):

$$E_x = -\frac{dV}{dx} = -k_e Q \frac{d}{dx}(a^2 + x^2)^{-1/2}$$

$$= -k_e Q(-\tfrac{1}{2})(a^2 + x^2)^{-3/2}(2x)$$

$$E_x = \frac{k_e x}{(a^2 + x^2)^{3/2}} Q$$

Finalize The only variable in the expressions for V and E_x is x. That is not surprising because our calculation is valid only for points along the x axis, where y and z are both zero. This result for the electric field agrees with that obtained by direct integration (see Example 19.6). For practice, use the result of part (B) in Equation 20.3 to verify that the potential is given by the expression in part (A).

Example 20.6 | Electric Potential Due to a Uniformly Charged Disk

A uniformly charged disk has radius R and surface charge density σ.

(A) Find the electric potential at a point P along the perpendicular central axis of the disk.

SOLUTION

Figure 20.14 (Example 20.6) A uniformly charged disk of radius R lies in a plane perpendicular to the x axis. The calculation of the electric potential at any point P on the x axis is simplified by dividing the disk into many rings of radius r and width dr, with area $2\pi r\, dr$.

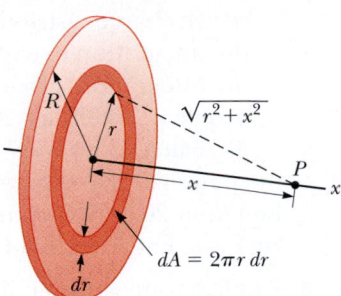

Conceptualize If we consider the disk to be a set of concentric rings, we can use our result from Example 20.5—which gives the potential due to a ring of radius a—and sum the contributions of all rings making up the disk. Because point P is on the central axis of the disk, symmetry again tells us that all points in a given ring are the same distance from P.

Categorize Because the disk is continuous, we evaluate the potential due to a continuous charge distribution rather than a group of individual charges.

Analyze Find the amount of charge dq on a ring of radius r and width dr as shown in Figure 20.14:

$$dq = \sigma\, dA = \sigma(2\pi r\, dr) = 2\pi\sigma r\, dr$$

Use this result in Equation (1) in Example 20.5 (with a replaced by r and Q replaced by dq) to find the potential due to the ring:

$$dV = \frac{k_e\, dq}{\sqrt{r^2 + x^2}} = \frac{k_e 2\pi\sigma r\, dr}{\sqrt{r^2 + x^2}}$$

To obtain the total potential at P, integrate this expression over the limits $r = 0$ to $r = R$, noting that x is a constant:

$$V = \pi k_e \sigma \int_0^R \frac{2r\, dr}{\sqrt{r^2 + x^2}} = \pi k_e \sigma \int_0^R (r^2 + x^2)^{-1/2}\, 2r\, dr$$

This integral is of the common form $\int u^n\, du$, where $n = -\tfrac{1}{2}$ and $u = r^2 + x^2$, and has the value $u^{n+1}/(n+1)$. Use this result to evaluate the integral:

$$(1) \quad V = 2\pi k_e \sigma[(R^2 + x^2)^{1/2} - x]$$

(B) Find the x component of the electric field at a point P along the perpendicular central axis of the disk.

SOLUTION

As in Example 20.5, use Equation 20.15 to find the electric field at any axial point:

$$(2) \quad E_x = -\frac{dV}{dx} = 2\pi k_e \sigma\left[1 - \frac{x}{(R^2 + x^2)^{1/2}}\right]$$

Finalize The calculation of V and $\vec{E}$ for an arbitrary point off the x axis is more difficult to perform because of the absence of symmetry and we do not treat that situation in this book.

20.6 | Electric Potential Due to a Charged Conductor

In Section 19.11, we found that when a solid conductor in equilibrium carries a net charge, the charge resides on the conductor's outer surface. Furthermore, the electric field just outside the conductor is perpendicular to the surface and the field inside is zero.

We now generate another property of a charged conductor, related to electric potential. Consider two points Ⓐ and Ⓑ on the surface of a charged conductor as shown in Figure 20.15. Along a surface path connecting these points, $\vec{E}$ is always perpendicular to the displacement $d\vec{s}$; therefore, $\vec{E} \cdot d\vec{s} = 0$. Using this result and Equation 20.3, we conclude that the potential difference between Ⓐ and Ⓑ is necessarily zero:

$$V_{Ⓑ} - V_{Ⓐ} = -\int_{Ⓐ}^{Ⓑ} \vec{E} \cdot d\vec{s} = 0$$

This result applies to *any* two points on the surface. Therefore, V is constant everywhere on the surface of a charged conductor in equilibrium. That is,

the surface of any charged conductor in electrostatic equilibrium is an equipotential surface: every point on the surface of a charged conductor in equilibrium is at the same electric potential. Furthermore, because the electric field is zero inside the conductor, the electric potential is constant everywhere inside the conductor and equal to its value at the surface.

Because of the constant value of the potential, no work is required to move a test charge from the interior of a charged conductor to its surface.

Consider a solid metal conducting sphere of radius R and total positive charge Q as shown in Figure 20.16a. As determined in part (A) of Example 19.10, the electric field outside the sphere is $k_e Q / r^2$ and points radially outward. Because the field outside a spherically symmetric charge distribution is identical to that of a point charge, we expect the potential to also be that of a point charge, $k_e Q / r$. At the surface of the conducting sphere in Figure 20.16a, the potential must be $k_e Q / R$. Because the entire sphere must be at the same potential, the potential at any point within the sphere must also be $k_e Q / R$. Figure 20.16b is a plot of the electric potential as a function of r, and Figure 20.16c shows how the electric field varies with r.

When a net charge is placed on a spherical conductor, the surface charge density is uniform as indicated in Figure 20.16a. If the conductor is nonspherical as in Figure 20.15, however, the surface charge density is high where the radius of curvature is small (as noted in Section 19.11) and low where the radius of curvature is large. Because the electric field immediately outside the conductor is proportional to the surface charge density, the electric field is large near convex points having small radii of curvature and reaches very high values at sharp points. In Example 20.7, the relationship between electric field and radius of curvature is explored mathematically.

Pitfall Prevention | 20.5
Potential May Not Be Zero
The electric potential inside the conductor is not necessarily zero in Figure 20.15, even though the electric field is zero. Equation 20.14 shows that a zero value of the field results in no *change* in the potential from one point to another inside the conductor. Therefore, the potential everywhere inside the conductor, including the surface, has the same value, which may or may not be zero, depending on where the zero of potential is defined.

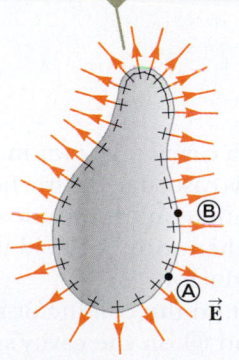

Notice from the spacing of the positive signs that the surface charge density is nonuniform.

Figure 20.15 An arbitrarily shaped conductor carrying a positive charge. When the conductor is in electrostatic equilibrium, all the charge resides at the surface, $\vec{E} = 0$ inside the conductor, and the direction of $\vec{E}$ immediately outside the conductor is perpendicular to the surface. The electric potential is constant inside the conductor and is equal to the potential at the surface.

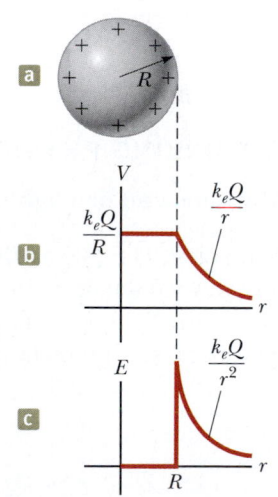

Figure 20.16 (a) The excess charge on a conducting sphere of radius R is uniformly distributed on its surface. (b) Electric potential versus distance r from the center of the charged conducting sphere. (c) Electric field magnitude versus distance r from the center of the charged conducting sphere.

Example 20.7 | Two Connected Charged Spheres

Two spherical conductors of radii r_1 and r_2 are separated by a distance much greater than the radius of either sphere. The spheres are connected by a conducting wire as shown in Figure 20.17. The charges on the spheres in equilibrium are q_1 and q_2, respectively, and they are uniformly charged. Find the ratio of the magnitudes of the electric fields at the surfaces of the spheres.

Figure 20.17
(Example 20.7) Two charged spherical conductors connected by a conducting wire. The spheres are at the *same* electric potential V.

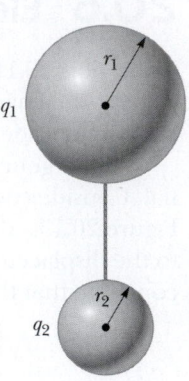

SOLUTION

Conceptualize Imagine the spheres are much farther apart than shown in Figure 20.17. Because they are so far apart, the field of one does not affect the charge distribution on the other. The conducting wire between them ensures that both spheres have the same electric potential.

Categorize Because the spheres are so far apart, we model the charge distribution on them as spherically symmetric, and we can model the field and potential outside the spheres to be that due to point charges.

Analyze Set the electric potentials at the surfaces of the spheres equal to each other:

$$V = k_e \frac{q_1}{r_1} = k_e \frac{q_2}{r_2}$$

Solve for the ratio of charges on the spheres:

$$(1) \quad \frac{q_1}{q_2} = \frac{r_1}{r_2}$$

Write expressions for the magnitudes of the electric fields at the surfaces of the spheres:

$$E_1 = k_e \frac{q_1}{r_1^2} \quad \text{and} \quad E_2 = k_e \frac{q_2}{r_2^2}$$

Evaluate the ratio of these two fields:

$$\frac{E_1}{E_2} = \frac{q_1}{q_2} \frac{r_2^2}{r_1^2}$$

Substitute for the ratio of charges from Equation (1):

$$(2) \quad \frac{E_1}{E_2} = \frac{r_1}{r_2} \frac{r_2^2}{r_1^2} = \frac{r_2}{r_1}$$

Finalize The field is stronger in the vicinity of the smaller sphere even though the electric potentials at the surfaces of both spheres are the same. If $r_2 \to 0$, then $E_2 \to \infty$, verifying the statement above that the electric field is very large at sharp points.

THINKING PHYSICS 20.1

Why is the end of a lightning rod pointed?

Reasoning The role of a lightning rod is to serve as a location at which the lightning strikes so that the charge delivered by the lightning will pass safely to the ground. If the lightning rod is pointed, the electric field due to charges moving between the rod and the ground is very strong near the point because the radius of curvature of the conductor is very small. This large electric field will greatly increase the likelihood that the lightning strike will occur near the tip of the lightning rod rather than elsewhere. ◀

A Cavity Within a Conductor

Suppose a conductor of arbitrary shape contains a cavity as shown in Figure 20.18. Let's assume no charges are inside the cavity. In this case, the electric field inside the cavity must be *zero* regardless of the charge distribution on the outside surface of the conductor as we mentioned in Section 19.11. Furthermore, the field in the cavity is zero even if an electric field exists outside the conductor.

To prove this point, remember that every point on the conductor is at the same electric potential; therefore, any two points Ⓐ and Ⓑ on the cavity's surface must be at the same potential. Now imagine a field $\vec{\mathbf{E}}$ exists in the cavity and evaluate the potential difference $V_{Ⓑ} - V_{Ⓐ}$ defined by Equation 20.3:

$$V_{Ⓑ} - V_{Ⓐ} = -\int_{Ⓐ}^{Ⓑ} \vec{\mathbf{E}} \cdot d\vec{\mathbf{s}}$$

Because $V_{Ⓑ} - V_{Ⓐ} = 0$, the integral of $\vec{\mathbf{E}} \cdot d\vec{\mathbf{s}}$ must be zero for all paths between any two points Ⓐ and Ⓑ on the conductor. The only way that can be true for *all* paths is if $\vec{\mathbf{E}}$ is zero *everywhere* in the cavity. Therefore, a cavity surrounded by conducting walls is a field-free region as long as no charges are inside the cavity.

This result has some interesting applications. For example, it is possible to shield an electronic device or even an entire laboratory from external fields by surrounding it with conducting walls. Shielding is often necessary during highly sensitive electrical measurements. During a thunderstorm, the safest location is inside an automobile. Even if lightning strikes the car, the metal body guarantees that you will not receive a shock inside the car, where $\vec{\mathbf{E}} = 0$.

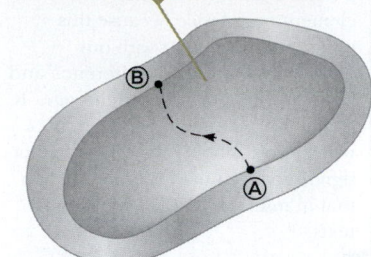

The electric field in the cavity is zero regardless of the charge on the conductor.

Figure 20.18 A conductor in electrostatic equilibrium containing a cavity.

20.7 | Capacitance

As we continue with our discussion of electricity and, in later chapters, magnetism, we shall build *circuits* consisting of *circuit elements.* A circuit generally consists of a number of electrical components (circuit elements) connected together by conducting wires and forming one or more closed loops. These circuits can be considered as systems that exhibit a particular type of behavior. The first circuit element we shall consider is a **capacitor.**

In general, a capacitor consists of two conductors of any shape. Consider two conductors having a potential difference of ΔV between them. Let us assume that the conductors have charges of equal magnitude and opposite sign as in Figure 20.19. This situation can be accomplished by connecting two uncharged conductors to the terminals of a battery. Once that is done and the battery is disconnected, the charges remain on the conductors. We say that the capacitor stores charge.

The potential difference ΔV *across* the capacitor is the magnitude of the potential difference between the two conductors. This potential difference is proportional to the charge Q on the capacitor, which is defined as the magnitude of the charge on *either* of the two conductors. The **capacitance** C of a capacitor is defined as the ratio of the charge on the capacitor to the magnitude of the potential difference across the capacitor:

$$C \equiv \frac{Q}{\Delta V} \qquad \text{20.20} ◀$$

By definition, *capacitance is always a positive quantity.* Because the potential difference is proportional to the charge, the ratio $Q/\Delta V$ is constant for a given capacitor. Equation 20.20 tells us that the capacitance of a system is a measure of the amount of charge that can be stored on the capacitor for a given potential difference.

From Equation 20.20, we see that capacitance has the SI units coulombs per volt, which is called a **farad** (F) in honor of Michael Faraday. The farad is a very large unit of capacitance. In practice, typical devices have capacitances ranging from microfarads to picofarads.

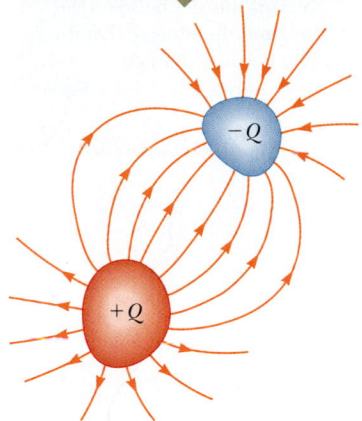

When the capacitor is charged, the conductors carry charges of equal magnitude and opposite sign.

$-Q$

$+Q$

Figure 20.19 A capacitor consists of two conductors electrically isolated from each other and their surroundings.

▶ Definition of capacitance

QUICK QUIZ 20.6 A capacitor stores charge Q at a potential difference ΔV. What happens if the voltage applied to a capacitor by a battery is doubled to $2\,\Delta V$?
(a) The capacitance falls to half its initial value, and the charge remains the same.
(b) The capacitance and the charge both fall to half their initial values. (c) The capacitance and the charge both double. (d) The capacitance remains the same, and the charge doubles.

The capacitance of a device depends on the geometric arrangement of the conductors. To illustrate this point, let us calculate the capacitance of an isolated

Pitfall Prevention | 20.6
Capacitance Is a Capacity
To help you understand the concept of capacitance, think of similar notions that use a similar word. The *capacity* of a milk carton is the volume of milk it can store. The *heat capacity* of an object is the amount of energy an object can store per unit of temperature difference. The *capacitance* of a capacitor is the amount of charge the capacitor can store per unit of potential difference.

spherical conductor of radius R and charge Q. (Based on the shape of the field lines from a single spherical conductor, we can model the second conductor as a concentric spherical shell of infinite radius.) Because the potential of the sphere is simply $k_e Q/R$ (and $V = 0$ for the shell of infinite radius), the capacitance of the sphere is

$$C = \frac{Q}{\Delta V} = \frac{Q}{k_e Q/R} = \frac{R}{k_e} = 4\pi\epsilon_0 R \qquad \textbf{20.21} \blacktriangleleft$$

(Remember from Section 19.4 that the Coulomb constant $k_e = 1/4\pi\epsilon_0$.) Equation 20.21 shows that the capacitance of an isolated charged sphere is proportional to the sphere's radius and is independent of both the charge and the potential difference.

The capacitance of a pair of oppositely charged conductors can be calculated in the following manner. A convenient charge of magnitude Q is assumed, and the potential difference is calculated using the techniques described in Section 20.5. One then uses $C = Q/\Delta V$ to evaluate the capacitance. As you might expect, the calculation is relatively straightforward if the geometry of the capacitor is simple.

Let us illustrate with two familiar geometries: parallel plates and concentric cylinders. In these examples, we shall assume that the charged conductors are separated by a vacuum. (The effect of a material between the conductors will be treated in Section 20.10.)

The Parallel-Plate Capacitor

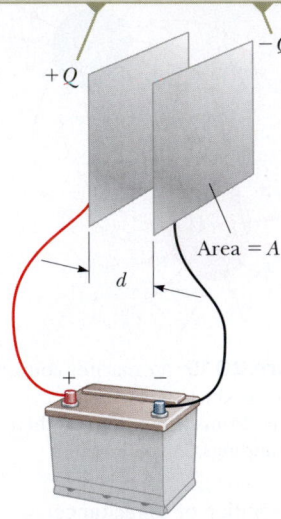

When the capacitor is connected to the terminals of a battery, electrons transfer between the plates and the wires so that the plates become charged.

$+Q$ $-Q$

Area $= A$

d

Figure 20.20 A parallel-plate capacitor consists of two parallel conducting plates, each of area A, separated by a distance d.

A parallel-plate capacitor consists of two parallel plates of equal area A separated by a distance d as in Figure 20.20. If the capacitor is charged, one plate has charge Q and the other, charge $-Q$. The magnitude of the charge per unit area on either plate is $\sigma = Q/A$. If the plates are very close together (compared with their length and width), we adopt a simplification model in which the electric field is uniform between the plates and zero elsewhere, as we discussed in Example 19.12. According to Example 19.12, the magnitude of the electric field between the plates is

$$E = \frac{\sigma}{\epsilon_0} = \frac{Q}{\epsilon_0 A}$$

Because the field is uniform, the potential difference across the capacitor can be found from Equation 20.6. Therefore,

$$\Delta V = Ed = \frac{Qd}{\epsilon_0 A}$$

Substituting this result into Equation 20.20, we find that the capacitance is

$$C = \frac{Q}{\Delta V} = \frac{Q}{Qd/\epsilon_0 A}$$

$$\boxed{C = \frac{\epsilon_0 A}{d}} \qquad \textbf{20.22} \blacktriangleleft$$

That is, the capacitance of a parallel-plate capacitor is proportional to the area of its plates and inversely proportional to the plate separation.

As you can see from the definition of capacitance, $C = Q/\Delta V$, the amount of charge a given capacitor can store for a given potential difference across its plates increases as the capacitance increases. It therefore seems reasonable that a capacitor constructed from plates having large areas should be able to store a large charge.

A careful inspection of the electric field lines for a parallel-plate capacitor reveals that the field is uniform in the central region between the plates, but is nonuniform at the edges of the plates. Figure 20.21 shows a drawing of the electric field pattern of a parallel-plate capacitor, showing the nonuniform field lines at the plates' edges. As long as the separation between the plates is small compared with the dimensions of the plates, the edge effects can be ignored and we can use the simplification model in which the electric field is uniform everywhere between the plates.

Active Figure 20.22 shows a battery connected to a single parallel-plate capacitor with a switch in the circuit. Let us identify the circuit as a system. When the switch is closed, the battery establishes an electric field in the wires and charges flow between the wires and the capacitor. As that occurs, energy is transformed within the system. Before the switch is closed, energy is stored as chemical potential energy in the battery. This type of energy is associated with chemical bonds and is transformed during the chemical reaction that occurs within the battery when it is operating in an electric circuit. When the switch is closed, some of the chemical potential energy in the battery is converted to electric potential energy related to the separation of positive and negative charges on the plates. As a result, we can describe a capacitor as a device that stores *energy* as well as *charge*. We will explore this energy storage in more detail in Section 20.9.

As a biological example of a parallel-plate capacitor, consider the plasma membrane for a neuron. The *plasma membrane* is a lipid bilayer containing a variety of types of molecules. This membrane contains a number of structures, including *ion channels* and *ion pumps*, which control concentrations of various ions on either side of the membrane. These ions include potassium, chlorine, calcium, and sodium. As a result of differences in these concentrations, there is an effective sheet of negative charge on the intracellular side of the membrane and a sheet of positive charge on the extracellular side. This results in a voltage of about 70 to 80 mV across the membrane. The sheets of charge act as parallel plates, so that the membrane can be modeled as a parallel-plate capacitor. The capacitance of the plasma membrane is about 2 μF for each cm^2 of membrane area.

When a neuron is carrying a signal, an event called an *action potential* occurs. Special structures in the cell membrane, called *voltage-gated ion channels*, are normally closed. If the voltage across the membrane capacitor falls in magnitude to a threshold value of about 50 mV, the ion channels open, allowing a flow of sodium ions into the cell. This flow reduces the voltage even further, allowing more sodium ions to enter the cell, thereby reversing the polarity of the voltage across the capacitor in a time interval measured in milliseconds. The voltage-gated ion channels then close and other channels open, allowing movement of ions until the neuron returns to its resting state.

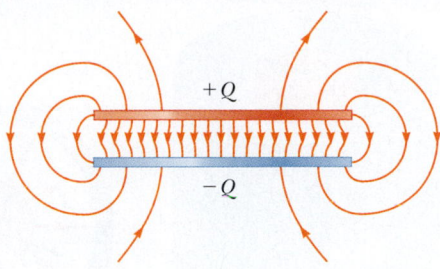

Figure 20.21 The electric field between the plates of a parallel-plate capacitor is uniform near the center but nonuniform near the edges.

BIO Capacitance of cell membranes

BIO Action potential

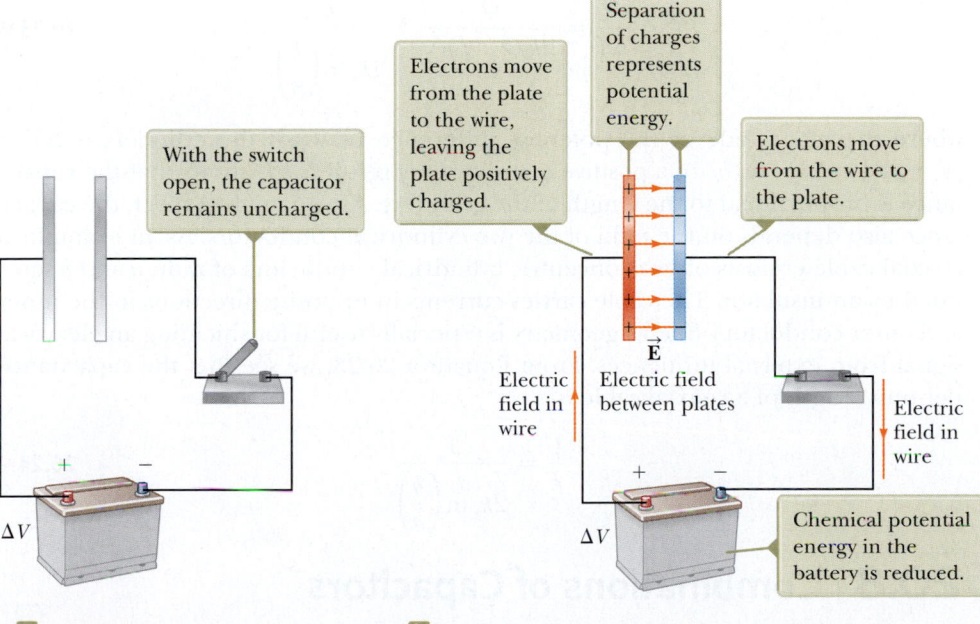

Active Figure 20.22 (a) A circuit consisting of a capacitor, a battery, and a switch. (b) When the switch is closed, the battery establishes an electric field in the wire and the capacitor becomes charged.

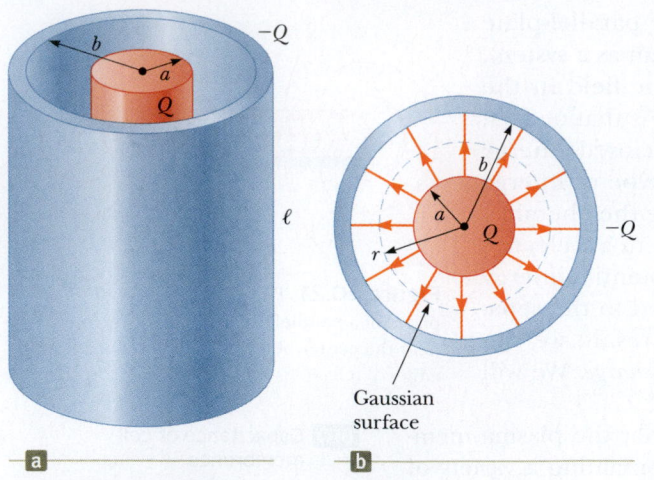

Figure 20.23 (a) A cylindrical capacitor consists of a solid cylindrical conductor of radius a and length ℓ surrounded by a coaxial cylindrical shell of radius b. (b) End view. The dashed line represents the end of the cylindrical gaussian surface of radius r and length ℓ.

This process can disturb neighboring regions of the plasma membrane so that the action potential is propagated along the neuron. In the next chapter, we will see how the capacitance of the plasma membrane combines with another electrical characteristic of the membrane to provide an electrical model for the conduction of a signal along the neuron.

The Cylindrical Capacitor

A cylindrical capacitor consists of a cylindrical conductor of radius a and charge Q coaxial with a larger cylindrical shell of radius b and charge $-Q$ (Fig. 20.23a). Let us find the capacitance of this device if its length is ℓ. If we assume that ℓ is large compared with a and b, we can adopt a simplification model in which we ignore end effects. In this case, the field is perpendicular to the axis of the cylinders and is confined to the region between them (Fig. 20.23b). We first calculate the potential difference between the two cylinders, which is given in general by

$$V_b - V_a = -\int_a^b \vec{\mathbf{E}} \cdot d\vec{\mathbf{s}}$$

where $\vec{\mathbf{E}}$ is the electric field in the region $a < r < b$. In Chapter 19, using Gauss's law, we showed that the electric field of a cylinder with charge per unit length λ has the magnitude $E = 2k_e\lambda/r$. The same result applies here because the outer cylinder does not contribute to the electric field inside it. Using this result and noting that the direction of $\vec{\mathbf{E}}$ is radially away from the inner cylinder in Figure 20.23b, we find that

$$V_b - V_a = -\int_a^b E_r\, dr = -2k_e\lambda \int_a^b \frac{dr}{r} = -2k_e\lambda \ln\!\left(\frac{b}{a}\right)$$

Substituting this result into Equation 20.20 and using that $\lambda = Q/\ell$, we find that

$$C = \frac{Q}{\Delta V} = \frac{Q}{\dfrac{2k_eQ}{\ell}\ln\!\left(\dfrac{b}{a}\right)} = \frac{\ell}{2k_e \ln\!\left(\dfrac{b}{a}\right)} \qquad \textbf{20.23} \blacktriangleleft$$

where the magnitude of the potential difference between the cylinders is $\Delta V = |V_a - V_b| = 2k_e\lambda \ln(b/a)$, a positive quantity. Our result for C shows that the capacitance is proportional to the length of the cylinders. As you might expect, the capacitance also depends on the radii of the two cylindrical conductors. As an example, a coaxial cable consists of two concentric cylindrical conductors of radii a and b separated by an insulator. The cable carries currents in opposite directions in the inner and outer conductors. Such a geometry is especially useful for shielding an electrical signal from external influences. From Equation 20.23, we see that the capacitance per unit length of a coaxial cable is

$$\frac{C}{\ell} = \frac{1}{2k_e \ln\!\left(\dfrac{b}{a}\right)} \qquad \textbf{20.24} \blacktriangleleft$$

◖20.8 | Combinations of Capacitors

Two or more capacitors often are combined in electric circuits. We can calculate the equivalent capacitance of certain combinations using methods described in this section. Throughout this section, we assume the capacitors to be combined are initially uncharged.

In studying electric circuits, we use a simplified pictorial representation called a **circuit diagram.** Such a diagram uses **circuit symbols** to represent various circuit elements. The circuit symbols are connected by straight lines that represent the wires between the circuit elements. The circuit symbols for capacitors, batteries, and switches as well as the color codes used for them in this text are given in Figure 20.24. The symbol for the capacitor reflects the geometry of the most common model for a capacitor, a pair of parallel plates. The positive terminal of the battery is at the higher potential and is represented in the circuit symbol by the longer line.

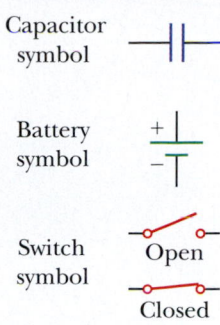

Figure 20.24 Circuit symbols for capacitors, batteries, and switches. Notice that capacitors are in blue, batteries are in green, and switches are in red. The closed switch can carry current, whereas the open one cannot.

Parallel Combination

Two capacitors connected as shown in Active Figure 20.25a are known as a **parallel combination** of capacitors. Active Figure 20.25b shows a circuit diagram for this combination of capacitors. The left plates of the capacitors are connected to the positive terminal of the battery by a conducting wire and are therefore both at the same electric potential as the positive terminal. Likewise, the right plates are connected to the negative terminal and so are both at the same potential as the negative terminal. Therefore, the individual potential differences across capacitors connected in parallel are the same and are equal to the potential difference applied across the combination. That is,

$$\Delta V_1 = \Delta V_2 = \Delta V$$

where ΔV is the battery terminal voltage.

After the battery is attached to the circuit, the capacitors quickly reach their maximum charge. Let's call the maximum charges on the two capacitors Q_1 and Q_2. The *total charge* Q_{tot} stored by the two capacitors is the sum of the charges on the individual capacitors:

$$Q_{tot} = Q_1 + Q_2 \qquad \textbf{20.25} \blacktriangleleft$$

Suppose you wish to replace these two capacitors by one *equivalent capacitor* having a capacitance C_{eq} as in Active Figure 20.25c. The effect this equivalent capacitor has on the circuit must be exactly the same as the effect of the combination of the two individual capacitors. That is, the equivalent capacitor must store charge Q_{tot} when connected to the battery. Active Figure 20.25c shows that the voltage across

Active Figure 20.25 Two capacitors connected in parallel. All three diagrams are equivalent.

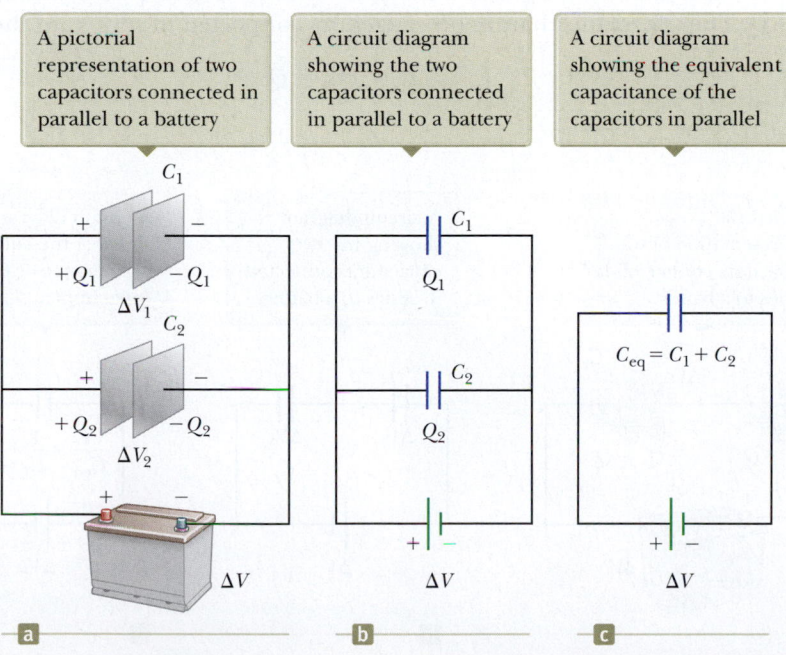

A pictorial representation of two capacitors connected in parallel to a battery

A circuit diagram showing the two capacitors connected in parallel to a battery

A circuit diagram showing the equivalent capacitance of the capacitors in parallel

the equivalent capacitor is ΔV because the equivalent capacitor is connected directly across the battery terminals. Therefore, for the equivalent capacitor,

$$Q_{tot} = C_{eq}\,\Delta V$$

Substituting for the charges in Equation 20.25 gives

$$C_{eq}\,\Delta V = Q_1 + Q_2 = C_1\,\Delta V_1 + C_2\,\Delta V_2$$

$$C_{eq} = C_1 + C_2 \quad \text{(parallel combination)}$$

where we have canceled the voltages because they are all the same. If this treatment is extended to three or more capacitors connected in parallel, the **equivalent capacitance** is found to be

▶ Equivalent capacitance for capacitors in parallel

$$C_{eq} = C_1 + C_2 + C_3 + \cdots \quad \text{(parallel combination)} \qquad \textbf{20.26} ◀$$

Therefore, the equivalent capacitance of a parallel combination of capacitors is (1) the algebraic sum of the individual capacitances and (2) greater than any of the individual capacitances. Statement (2) makes sense because we are essentially combining the areas of all the capacitor plates when they are connected with conducting wire, and capacitance of parallel plates is proportional to area (Eq. 20.22).

Series Combination

Two capacitors connected as shown in Active Figure 20.26a and the equivalent circuit diagram in Active Figure 20.26b are known as a **series combination** of capacitors. The left plate of capacitor 1 and the right plate of capacitor 2 are connected to the terminals of a battery. The other two plates are connected to each other and to nothing else; hence, they form an isolated system that is initially uncharged and must continue to have zero net charge. To analyze this combination, let's first consider the uncharged capacitors and then follow what happens immediately after a battery is connected to the circuit. When the battery is connected, electrons are transferred to the leftmost wire out of the left plate of C_1 and into the right plate of C_2 from the rightmost wire. As this negative charge accumulates on the right plate of C_2, an equivalent amount of negative charge is forced off the left plate of C_2, and this left plate therefore has an excess positive charge. The negative charge leaving the left plate of C_2 causes negative charges to accumulate on the right plate of C_1. As a result, both right plates end up with a charge $-Q$ and both left plates end up with a charge $+Q$. Therefore, the charges on capacitors connected in series are the same:

$$Q_1 = Q_2 = Q$$

Active Figure 20.26 Two capacitors connected in series. All three diagrams are equivalent.

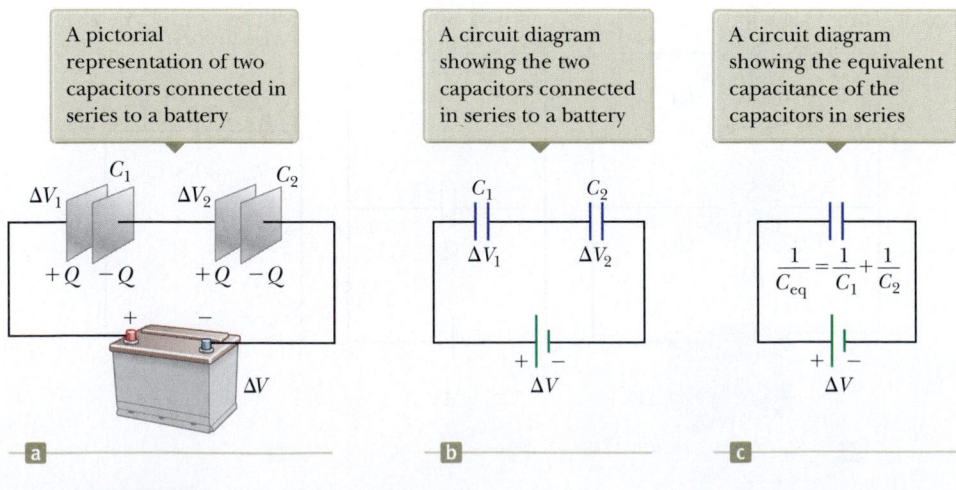

A pictorial representation of two capacitors connected in series to a battery

A circuit diagram showing the two capacitors connected in series to a battery

A circuit diagram showing the equivalent capacitance of the capacitors in series

where Q is the charge that moved between a wire and the connected outside plate of one of the capacitors.

Active Figure 20.26a shows that the total voltage ΔV_{tot} across the combination is split between the two capacitors:

$$\Delta V_{tot} = \Delta V_1 + \Delta V_2 \qquad \textbf{20.27} \blacktriangleleft$$

where ΔV_1 and ΔV_2 are the potential differences across capacitors C_1 and C_2, respectively. In general, the total potential difference across any number of capacitors connected in series is the sum of the potential differences across the individual capacitors.

Suppose the equivalent single capacitor in Active Figure 20.26c has the same effect on the circuit as the series combination when it is connected to the battery. After it is fully charged, the equivalent capacitor must have a charge of $-Q$ on its right plate and a charge of $+Q$ on its left plate. Applying the definition of capacitance to the circuit in Active Figure 20.26c gives

$$\Delta V_{tot} = \frac{Q}{C_{eq}}$$

Substituting for the voltages in Equation 20.27, we have

$$\frac{Q}{C_{eq}} = \Delta V_1 + \Delta V_2 = \frac{Q_1}{C_1} + \frac{Q_2}{C_2}$$

Canceling the charges because they are all the same gives

$$\frac{1}{C_{eq}} = \frac{1}{C_1} + \frac{1}{C_2} \quad \text{(series combination)}$$

When this analysis is applied to three or more capacitors connected in series, the relationship for the **equivalent capacitance** is

$$\frac{1}{C_{eq}} = \frac{1}{C_1} + \frac{1}{C_2} + \frac{1}{C_3} + \cdots \quad \text{(series combination)} \qquad \textbf{20.28} \blacktriangleleft$$

▶ Equivalent capacitance for capacitors in series

This expression shows that (1) the inverse of the equivalent capacitance is the algebraic sum of the inverses of the individual capacitances and (2) the equivalent capacitance of a series combination is always less than any individual capacitance in the combination.

QUICK QUIZ 20.7 Two capacitors are identical. They can be connected in series or in parallel. If you want the *smallest* equivalent capacitance for the combination, how should you connect them? (**a**) in series (**b**) in parallel (**c**) either way because both combinations have the same capacitance

---

Example 20.8 | Equivalent Capacitance

Find the equivalent capacitance between *a* and *b* for the combination of capacitors shown in Figure 20.27a. All capacitances are in microfarads.

SOLUTION

Conceptualize Study Figure 20.27a carefully and make sure you understand how the capacitors are connected.

Categorize Figure 20.27a shows that the circuit contains both series and parallel connections, so we use the rules for series and parallel combinations discussed in this section.

....

Analyze Using Equations 20.26 and 20.28, we reduce the combination step by step as indicated in the figure.

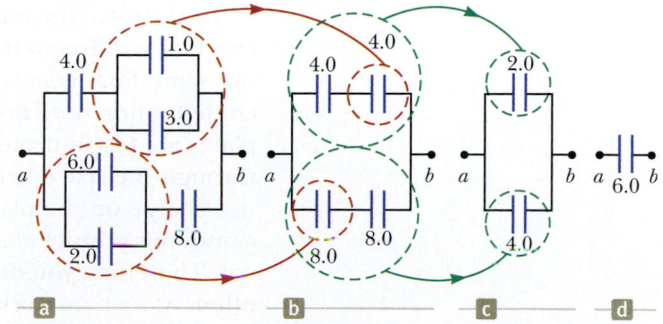

Figure 20.27 (Example 20.8) To find the equivalent capacitance of the capacitors in (a), we reduce the various combinations in steps as indicated in (b), (c), and (d), using the series and parallel rules described in the text. All capacitances are in microfarads.

continued

20.8 *cont.*

The 1.0-μF and 3.0-μF capacitors (upper red-brown circle in Fig. 20.27a) are in parallel. Find the equivalent capacitance from Equation 20.26:

$$C_{eq} = C_1 + C_2 = 4.0\ \mu F$$

The 2.0-μF and 6.0-μF capacitors (lower red-brown circle in Fig. 20.27a) are also in parallel:

$$C_{eq} = C_1 + C_2 = 8.0\ \mu F$$

The circuit now looks like Figure 20.27b. The two 4.0-μF capacitors (upper green circle in Fig. 20.27b) are in series. Find the equivalent capacitance from Equation 20.28:

$$\frac{1}{C_{eq}} = \frac{1}{C_1} + \frac{1}{C_2} = \frac{1}{4.0\ \mu F} + \frac{1}{4.0\ \mu F} = \frac{1}{2.0\ \mu F}$$

$$C_{eq} = 2.0\ \mu F$$

The two 8.0-μF capacitors (lower green circle in Fig. 20.27b) are also in series. Find the equivalent capacitance from Equation 20.28:

$$\frac{1}{C_{eq}} = \frac{1}{C_1} + \frac{1}{C_2} = \frac{1}{8.0\ \mu F} + \frac{1}{8.0\ \mu F} = \frac{1}{4.0\ \mu F}$$

$$C_{eq} = 4.0\ \mu F$$

The circuit now looks like Figure 20.27c. The 2.0-μF and 4.0-μF capacitors are in parallel:

$$C_{eq} = C_1 + C_2 = \boxed{6.0\ \mu F}$$

Finalize This final value is that of the single equivalent capacitor shown in Figure 20.27d. For further practice in treating circuits with combinations of capacitors, imagine a battery is connected between points *a* and *b* in Figure 20.27a so that a potential difference ΔV is established across the combination. Can you find the voltage across and the charge on each capacitor?

20.9 | Energy Stored in a Charged Capacitor

Almost everyone who works with electronic equipment has at some time verified that a capacitor can store energy. If the plates of a charged capacitor are connected by a conductor, such as a wire, charge transfers between the plates and the wire until the two plates are uncharged. The discharge can often be observed as a visible spark. If you accidentally touch the opposite plates of a charged capacitor, your fingers act as pathways by which the capacitor discharges, resulting in an electric shock. The degree of shock depends on the capacitance and the voltage applied to the capacitor. When high voltages are present, such as in the power supply of an electronic instrument, the shock can be fatal.

Consider a parallel-plate capacitor that is initially uncharged so that the initial potential difference across the plates is zero. Now imagine that the capacitor is connected to a battery and develops a charge of Q. The final potential difference across the capacitor is $\Delta V = Q/C$.

To calculate the energy stored in the capacitor, we shall assume a charging process that is different from the actual process described in Section 20.7 but that gives the same final result. This assumption is justified because the energy in the final configuration does not depend on the actual charge-transfer process.[2] Imagine the plates are disconnected from the battery and you transfer the charge mechanically through the space between the plates as follows. You grab a small amount of positive charge on the plate connected to the negative terminal and apply a force that causes this positive charge to move over to the plate connected to the positive terminal. Therefore, you do work on the charge as it is transferred from one plate to the other. At first, no work is required to transfer a small amount of charge dq from one

---

[2]This discussion is similar to that of state variables in thermodynamics. The change in a state variable such as temperature is independent of the path followed between the initial and final states. The potential energy of a capacitor (or any system) is also a state variable, so its change does not depend on the process followed to charge the capacitor.

plate to the other,[3] but once this charge has been transferred, a small potential difference exists between the plates. Therefore, work must be done to move additional charge through this potential difference. As more and more charge is transferred from one plate to the other, the potential difference increases in proportion and more work is required.

The work required to transfer an increment of charge dq from one plate to the other is

$$dW = \Delta V \, dq = \frac{q}{C} dq$$

Therefore, the total work required to charge the capacitor from $q = 0$ to the final charge $q = Q$ is

$$W = \int_0^Q \frac{q}{C} dq = \frac{Q^2}{2C}$$

The capacitor can be modeled as a nonisolated system for this discussion. The work done by the external agent on the system in charging the capacitor appears as potential energy U stored in the capacitor. In reality, of course, this energy is not the result of mechanical work done by an external agent moving charge from one plate to the other, but is due to transformation of chemical energy in the battery. We have used a model of work done by an external agent that gives us a result that is also valid for the actual situation. Using $Q = C\Delta V$, the energy stored in a charged capacitor can be expressed in the following alternative forms:

$$U = \frac{Q^2}{2C} = \tfrac{1}{2} Q \Delta V = \tfrac{1}{2} C (\Delta V)^2 \qquad \textbf{20.29} \blacktriangleleft$$

▶ Energy stored in a charged capacitor

This result applies to *any* capacitor, regardless of its geometry. In practice, the maximum energy (or charge) that can be stored is limited because electric discharge ultimately occurs between the plates of the capacitor at a sufficiently large value of ΔV. For this reason, capacitors are usually labeled with a maximum operating voltage.

For an object on an extended spring, the elastic potential energy can be modeled as being stored *in the spring*. Internal energy of a substance associated with its temperature is located *throughout the substance*. Where is the energy in a capacitor located? The energy stored in a capacitor can be modeled as being stored *in the electric field between the plates of the capacitor*. For a parallel-plate capacitor, the potential difference is related to the electric field through the relationship $\Delta V = Ed$. Furthermore, the capacitance is $C = \epsilon_0 A/d$. Substituting these expressions into Equation 20.29 gives

$$U = \tfrac{1}{2}\left(\frac{\epsilon_0 A}{d}\right)(Ed)^2 = \tfrac{1}{2}(\epsilon_0 Ad)E^2 \qquad \textbf{20.30} \blacktriangleleft$$

Because the volume of a parallel-plate capacitor that is occupied by the electric field is Ad, the energy per unit volume $u_E = U/Ad$, called the **energy density,** is

$$u_E = \tfrac{1}{2}\epsilon_0 E^2 \qquad \textbf{20.31} \blacktriangleleft$$

▶ Energy density in an electric field

Although Equation 20.31 was derived for a parallel-plate capacitor, the expression is generally valid. That is, the energy density in any electric field is proportional to the square of the magnitude of the electric field at a given point.

QUICK QUIZ 20.8 You have three capacitors and a battery. In which of the following combinations of the three capacitors is the maximum possible energy stored when the combination is attached to the battery? **(a)** series **(b)** parallel **(c)** no difference because both combinations store the same amount of energy

---

[3]We shall use lowercase q for the time-varying charge on the capacitor while it is charging to distinguish it from uppercase Q, which is the total charge on the capacitor after it is completely charged.

> **THINKING PHYSICS 20.2**
>
> You charge a capacitor and then remove it from the battery. The capacitor consists of large movable plates, with air between them. You pull the plates farther apart a small distance. What happens to the charge on the capacitor? To the potential difference? To the energy stored in the capacitor? To the capacitance? To the electric field between the plates? Is work done in pulling the plates apart?
>
> **Reasoning** Because the capacitor is removed from the battery, charges on the plates have nowhere to go. Therefore, the charge on the capacitor remains the same as the plates are pulled apart. Because the electric field of large plates is independent of distance for uniform fields, the electric field remains constant. Because the electric field is a measure of the rate of change of potential with distance, the potential difference between the plates increases as the separation distance increases. Because the same charge is stored at a higher potential difference, the capacitance decreases. Because energy stored is proportional to both charge and potential difference, the energy stored in the capacitor increases. This energy must be transferred into the system from somewhere; the plates attract each other, so work is done by you on the system of two plates when you pull them apart. ◀

Example 20.9 | Rewiring Two Charged Capacitors

Two capacitors C_1 and C_2 (where $C_1 > C_2$) are charged to the same initial potential difference ΔV_i. The charged capacitors are removed from the battery, and their plates are connected with opposite polarity as in Figure 20.28a. The switches S_1 and S_2 are then closed as in Figure 20.28b.

(A) Find the final potential difference ΔV_f between a and b after the switches are closed.

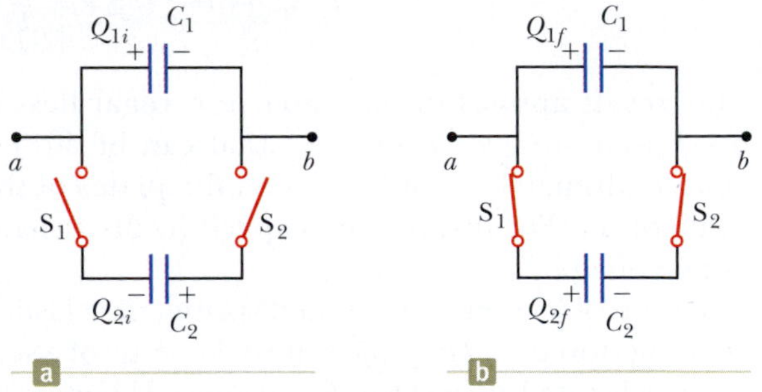

Figure 20.28 (Example 20.9) (a) Two capacitors are charged to the same initial potential difference and connected together with plates of opposite sign to be in contact when the switches are closed. (b) When the switches are closed, the charges redistribute.

SOLUTION

Conceptualize Figure 20.28 helps us understand the initial and final configurations of the system. When the switches are closed, the charge on the system will redistribute between the capacitors until both capacitors have the same potential difference. Because $C_1 > C_2$, more charge exists on C_1 than on C_2, so the final configuration will have positive charge on the left plates as shown in Figure 20.28b.

Categorize In Figure 20.28b, it might appear as if the capacitors are connected in parallel, but there is no battery in this circuit to apply a voltage across the combination. Therefore, we *cannot* categorize this problem as one in which capacitors are connected in parallel. We *can* categorize it as a problem involving an isolated system for electric charge. The left-hand plates of the capacitors form an isolated system because they are not connected to the right-hand plates by conductors.

Analyze Write an expression for the total charge on the left-hand plates of the system before the switches are closed, noting that a negative sign for Q_{2i} is necessary because the charge on the left plate of capacitor C_2 is negative:

$$(1) \quad Q_i = Q_{1i} + Q_{2i} = C_1\,\Delta V_i - C_2\,\Delta V_i = (C_1 - C_2)\Delta V_i$$

After the switches are closed, the charges on the individual capacitors change to new values Q_{1f} and Q_{2f} such that the potential difference is again the same across both capacitors, with a value of ΔV_f. Write an expression for the total charge on the left-hand plates of the system after the switches are closed:

$$(2) \quad Q_f = Q_{1f} + Q_{2f} = C_1\,\Delta V_f + C_2\,\Delta V_f = (C_1 + C_2)\Delta V_f$$

20.9 *cont.*

Because the system is isolated, the initial and final charges on the system must be the same. Use this condition and Equations (1) and (2) to solve for ΔV_f:

$$Q_f = Q_i \;\rightarrow\; (C_1 + C_2)\,\Delta V_f = (C_1 - C_2)\,\Delta V_i$$

$$(3) \quad \Delta V_f = \left(\frac{C_1 - C_2}{C_1 + C_2}\right)\Delta V_i$$

(B) Find the total energy stored in the capacitors before and after the switches are closed and determine the ratio of the final energy to the initial energy.

SOLUTION

Use Equation 20.29 to find an expression for the total energy stored in the capacitors before the switches are closed:

$$(4) \quad U_i = \tfrac{1}{2}C_1(\Delta V_i)^2 + \tfrac{1}{2}C_2(\Delta V_i)^2 = \tfrac{1}{2}(C_1 + C_2)(\Delta V_i)^2$$

Write an expression for the total energy stored in the capacitors after the switches are closed:

$$U_f = \tfrac{1}{2}C_1(\Delta V_f)^2 + \tfrac{1}{2}C_2(\Delta V_f)^2 = \tfrac{1}{2}(C_1 + C_2)(\Delta V_f)^2$$

Use the results of part (A) to rewrite this expression in terms of ΔV_i:

$$(5) \quad U_f = \tfrac{1}{2}(C_1 + C_2)\left[\left(\frac{C_1 - C_2}{C_1 + C_2}\right)\Delta V_i\right]^2 = \frac{1}{2}\frac{(C_1 - C_2)^2(\Delta V_i)^2}{C_1 + C_2}$$

Divide Equation (5) by Equation (4) to obtain the ratio of the energies stored in the system:

$$\frac{U_f}{U_i} = \frac{\tfrac{1}{2}(C_1 - C_2)^2(\Delta V_i)^2/(C_1 + C_2)}{\tfrac{1}{2}(C_1 + C_2)(\Delta V_i)^2}$$

$$(6) \quad \frac{U_f}{U_i} = \left(\frac{C_1 - C_2}{C_1 + C_2}\right)^2$$

Finalize The ratio of energies is *less* than unity, indicating that the final energy is *less* than the initial energy. At first, you might think the law of energy conservation has been violated, but that is not the case. The "missing" energy is transferred out of the system by the mechanism of electromagnetic waves (T_{ER} in Eq. 7.2), as we shall see in Chapter 24. Therefore, this system is isolated for electric charge, but nonisolated for energy.

What If? What if the two capacitors have the same capacitance? What would you expect to happen when the switches are closed?

Answer Because both capacitors have the same initial potential difference applied to them, the charges on the capacitors have the same magnitude. When the capacitors with opposite polarities are connected together, the equal-magnitude charges cancel each other, leaving the capacitors uncharged.

Let's test our results to see if that is the case mathematically. In Equation (1), because the capacitances are equal, the initial charge Q_i on the system of left-hand plates is zero. Equation (3) shows that $\Delta V_f = 0$, which is consistent with uncharged capacitors. Finally, Equation (5) shows that $U_f = 0$, which is also consistent with uncharged capacitors.

20.10 | Capacitors with Dielectrics

A **dielectric** is an insulating material such as rubber, glass, or waxed paper. When a dielectric material is inserted between the plates of a capacitor, the capacitance increases. If the dielectric completely fills the space between the plates, the capacitance increases by the dimensionless factor κ, called the **dielectric constant** of the material.

The following experiment can be performed to illustrate the effect of a dielectric in a capacitor. Consider a parallel-plate capacitor of charge Q_0 and capacitance C_0 in the absence of a dielectric. The potential difference across the capacitor as measured by a voltmeter is $\Delta V_0 = Q_0/C_0$ (Fig. 20.29a, page 590). Notice that the capacitor circuit is *open;* that is, the plates of the capacitor are *not* connected to a battery and charge cannot flow through an ideal voltmeter. Hence, there is *no* path

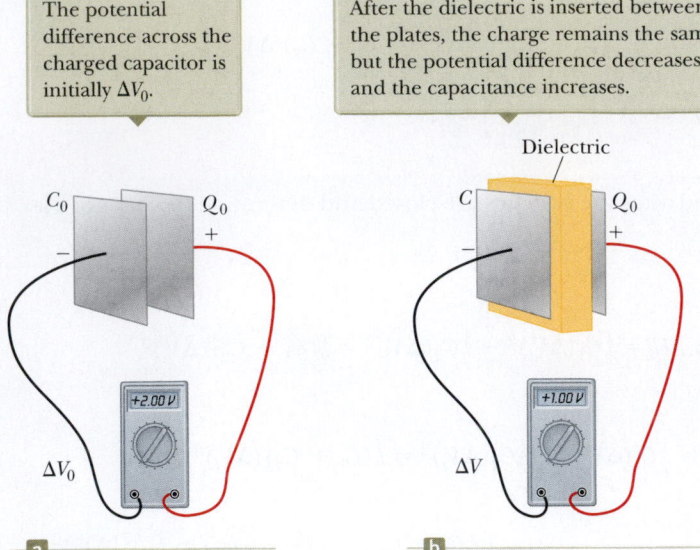

The potential difference across the charged capacitor is initially ΔV_0.

After the dielectric is inserted between the plates, the charge remains the same, but the potential difference decreases and the capacitance increases.

Figure 20.29 A charged capacitor (a) before and (b) after insertion of a dielectric between the plates.

by which charge can flow and alter the charge on the capacitor. If a dielectric is now inserted between the plates as in Figure 20.29b, it is found that the voltmeter reading *decreases* by a factor of κ to the value ΔV, where

$$\Delta V = \frac{\Delta V_0}{\kappa}$$

Because $\Delta V < \Delta V_0$, we see that $\kappa > 1$.

Because the charge Q_0 on the capacitor *does not change*, we conclude that the capacitance must change to the value

$$C = \frac{Q_0}{\Delta V} = \frac{Q_0}{\Delta V_0/\kappa} = \kappa \frac{Q_0}{\Delta V_0}$$

$$C = \kappa C_0 \qquad\qquad \textbf{20.32} \blacktriangleleft$$

where C_0 is the capacitance in the absence of the dielectric. That is, the capacitance *increases* by the factor κ when the dielectric completely fills the region between the plates.[4] For a parallel-plate capacitor, where $C_0 = \epsilon_0 A/d$, we can express the capacitance when the capacitor is filled with a dielectric as

$$C = \kappa \frac{\epsilon_0 A}{d} \qquad\qquad \textbf{20.33} \blacktriangleleft$$

From this result, it would appear that the capacitance could be made very large by decreasing d, the distance between the plates. In practice, however, the lowest value of d is limited by the electric discharge that could occur through the dielectric medium separating the plates. For any given separation d, the maximum voltage that can be applied to a capacitor without causing a discharge depends on the **dielectric strength** (maximum electric field) of the dielectric, which for dry air is equal to 3×10^6 V/m. If the electric field in the medium exceeds the dielectric strength, the insulating properties break down and the medium begins to conduct. Most insulating materials have dielectric strengths and dielectric constants greater than those of air, as Table 20.1 indicates. Therefore, we see that a dielectric provides the following advantages:

- An increase in capacitance
- An increase in maximum operating voltage
- Possible mechanical support between the plates, which allows the plates to be close together without touching, thereby decreasing d and increasing C

We can understand the effects of a dielectric by considering the polarization of molecules that we discussed in Section 19.3. Figure 20.30a shows polarized molecules of a dielectric in random orientations in the absence of an electric field. Figure 20.30b shows the polarization of the molecules when the dielectric is placed between the plates of the charged capacitor and the polarized molecules tend to line up parallel to the field lines. The plates set up an electric field $\vec{\mathbf{E}}_0$ in a direction to the right in Figure 20.30b. In the body of the dielectric, a general homogeneity of charge exists, but look along the edges. There is a layer of negative charge along the left edge of the dielectric and a layer of positive charge along the right edge. These layers of charge can be modeled as additional charged parallel plates, as in Figure 20.30c. Because the polarity is opposite that of the real

---

[4]If another experiment is performed in which the dielectric is introduced while the potential difference is held constant by means of a battery, the charge increases to the value $Q = \kappa Q_0$. The additional charge is transferred from the connecting wires, and the capacitance still increases by the factor κ.

TABLE 20.1 | Approximate Dielectric Constants and Dielectric Strengths of Various Materials at Room Temperature

Material	Dielectric Constant κ	Dielectric Strength[a] (10^6 V/m)
Air (dry)	1.000 59	3
Bakelite	4.9	24
Fused quartz	3.78	8
Mylar	3.2	7
Neoprene rubber	6.7	12
Nylon	3.4	14
Paper	3.7	16
Paraffin-impregnated paper	3.5	11
Polystyrene	2.56	24
Polyvinyl chloride	3.4	40
Porcelain	6	12
Pyrex glass	5.6	14
Silicone oil	2.5	15
Strontium titanate	233	8
Teflon	2.1	60
Vacuum	1.000 00	—
Water	80	—

[a]The dielectric strength equals the maximum electric field that can exist in a dielectric without electrical breakdown. Note that these values depend strongly on the presence of impurities and flaws in the materials.

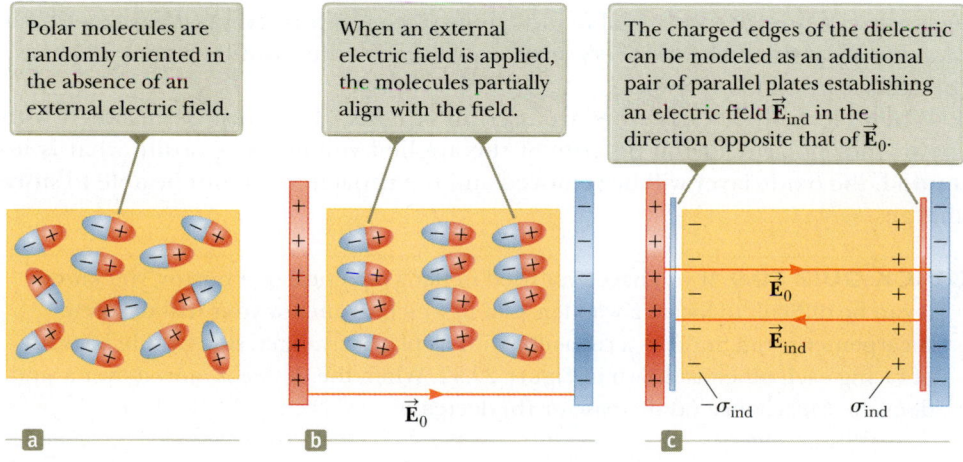

Polar molecules are randomly oriented in the absence of an external electric field.

When an external electric field is applied, the molecules partially align with the field.

The charged edges of the dielectric can be modeled as an additional pair of parallel plates establishing an electric field $\vec{\mathbf{E}}_{ind}$ in the direction opposite that of $\vec{\mathbf{E}}_0$.

$\vec{\mathbf{E}}_0$

$\vec{\mathbf{E}}_0$

$\vec{\mathbf{E}}_{ind}$

$-\sigma_{ind}$ σ_{ind}

a   b   c

Figure 20.30 (a) Polar molecules in a dielectric. (b) An electric field is applied to the dielectric. (c) Details of the electric field inside the dielectric.

plates, these charges set up an induced electric field $\vec{\mathbf{E}}_{ind}$ directed to the left in the diagram that partially cancels the electric field due to the real plates. Therefore, for the charged capacitor removed from a battery, the electric field and hence the voltage between the plates is reduced by the introduction of the dielectric. The charge on the plates is stored at a lower potential difference, so the capacitance increases.

Types of Capacitors

Many capacitors are built into integrated circuit chips, but some electrical devices still use stand-alone capacitors. Commercial capacitors are often made using metal

A collection of capacitors used in a variety of applications.

Chris Vuille

Figure 20.31 Three commercial capacitor designs.

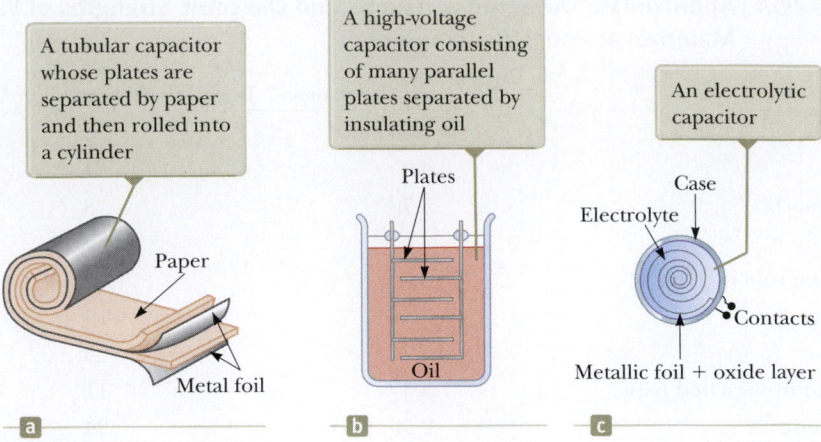

A tubular capacitor whose plates are separated by paper and then rolled into a cylinder

Paper

Metal foil

a

A high-voltage capacitor consisting of many parallel plates separated by insulating oil

Plates

Oil

b

An electrolytic capacitor

Case

Electrolyte

Contacts

Metallic foil + oxide layer

c

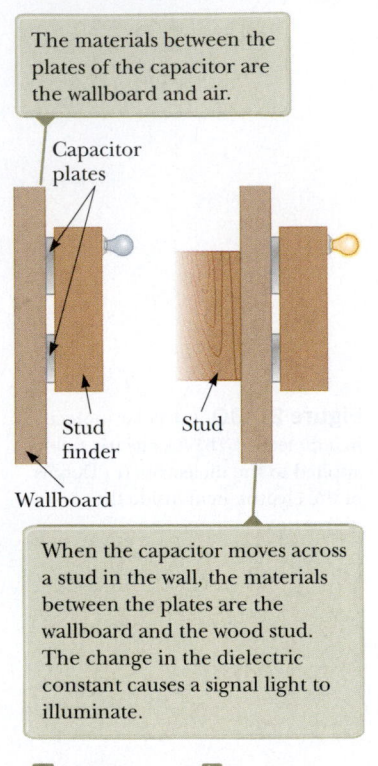

The materials between the plates of the capacitor are the wallboard and air.

Capacitor plates

Stud finder

Stud

Wallboard

When the capacitor moves across a stud in the wall, the materials between the plates are the wallboard and the wood stud. The change in the dielectric constant causes a signal light to illuminate.

a   **b**

Figure 20.32 (Quick Quiz 20.9) An electric stud finder.

foil interlaced with a dielectric such as thin sheets of paraffin-impregnated paper. These alternating layers of metal foil and dielectric are then rolled into the shape of a cylinder to form a small package (Fig. 20.31a). High-voltage capacitors commonly consist of interwoven metal plates immersed in silicone oil (Fig. 20.31b). Small capacitors are often constructed from ceramic materials. Variable capacitors (typically 10–500 pF), such as the one in the chapter-opening photograph, usually consist of two interwoven sets of metal plates, one fixed and the other movable, with air as the dielectric.

An *electrolytic capacitor* is often used to store large amounts of charge at relatively low voltages. This device, shown in Figure 20.31c, consists of a metal foil in contact with an electrolyte, a solution that conducts electricity by virtue of the motion of ions contained in the solution. When a voltage is applied between the foil and the electrolyte, a thin layer of metal oxide (an insulator) is formed on the foil, and this layer serves as the dielectric. Very large capacitance values can be attained because the dielectric layer is very thin.

When electrolytic capacitors are used in circuits, they must be installed with the proper polarity. If the polarity of the applied voltage is opposite what is intended, the oxide layer will be removed and the capacitor will not be able to store charge.

QUICK QUIZ 20.9 If you have ever tried to hang a picture or a mirror, you know it can be difficult to locate a wooden stud in which to anchor your nail or screw. A carpenter's stud finder is a capacitor with its plates arranged side by side instead of facing each other as shown in Figure 20.32. When the device is moved over a stud, does the capacitance **(a)** increase or **(b)** decrease?

Example 20.10 | Energy Stored Before and After

A parallel-plate capacitor is charged with a battery to a charge Q_0. The battery is then removed, and a slab of material that has a dielectric constant κ is inserted between the plates. Identify the system as the capacitor and the dielectric. Find the energy stored in the system before and after the dielectric is inserted.

SOLUTION

Conceptualize Think about what happens when the dielectric is inserted between the plates. Because the battery has been removed, the charge on the capacitor must remain the same. We know from our earlier discussion, however, that the capacitance must change. Therefore, we expect a change in the energy of the system.

Categorize Because we expect the energy of the system to change, we model it as a nonisolated system involving a capacitor and a dielectric.

20.10 *cont.*

Analyze From Equation 20.29, find the energy stored in the absence of the dielectric:

$$U_0 = \frac{Q_0{}^2}{2C_0}$$

Find the energy stored in the capacitor after the dielectric is inserted between the plates:

$$U = \frac{Q_0{}^2}{2C}$$

Use Equation 20.32 to replace the capacitance C:

$$U = \frac{Q_0{}^2}{2\kappa C_0} = \frac{U_0}{\kappa}$$

Finalize Because $\kappa > 1$, the final energy is less than the initial energy. We can account for the decrease in energy of the system by performing an experiment and noting that the dielectric, when inserted, is pulled into the device. To keep the dielectric from accelerating, an external agent must do negative work (W in Eq. 7.2) on the dielectric, which is simply the difference $U - U_0$.

20.11 | Context Connection: The Atmosphere as a Capacitor

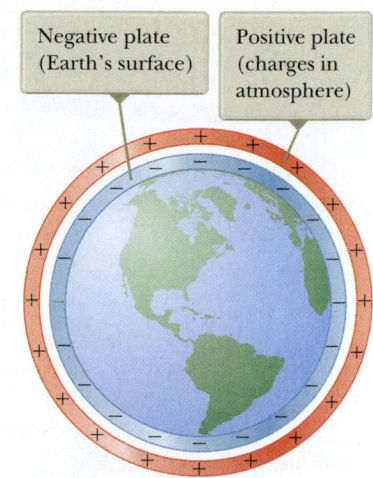

Negative plate (Earth's surface) Positive plate (charges in atmosphere)

Figure 20.33 The atmospheric capacitor.

In the Context Connection of Chapter 19, we mentioned some processes occurring on the surface of the Earth and in the atmosphere that result in charge distributions. These processes result in a negative charge on the Earth's surface and positive charges distributed throughout the air.

This separation of charge can be modeled as a capacitor. The surface of the Earth is one plate and the positive charge in the air is the other plate. The positive charge in the atmosphere is not all located at one height but is spread throughout the atmosphere. Therefore, the single position of the upper plate of the capacitor must be modeled, based on the charge distribution. Models of the atmosphere show that an appropriate effective height of the upper plate is about 5 km from the surface. The model atmospheric capacitor is shown in Figure 20.33.

Considering the charge distribution on the surface of the Earth to be spherically symmetric, we can use Figure 20.16 and its associated discussion in Section 20.6 to argue that the potential at a point above the Earth's surface is

$$V = k_e \frac{Q}{r} = \frac{1}{4\pi\epsilon_0} \frac{Q}{r}$$

where Q is the charge on the surface. The potential difference between the plates of our atmospheric capacitor is

$$\Delta V = \frac{Q}{4\pi\epsilon_0} \left(\frac{1}{r_{surface}} - \frac{1}{r_{upper\ plate}} \right)$$

$$= \frac{Q}{4\pi\epsilon_0} \left(\frac{1}{R_E} - \frac{1}{R_E + h} \right) = \frac{Q}{4\pi\epsilon_0} \left[\frac{h}{R_E(R_E + h)} \right]$$

where R_E is the radius of the Earth and $h = 5$ km. From this expression, we can calculate the capacitance of the atmospheric capacitor:

$$C = \frac{Q}{\Delta V} = \frac{Q}{\dfrac{Q}{4\pi\epsilon_0} \left[\dfrac{h}{R_E(R_E + h)} \right]} = \frac{4\pi\epsilon_0 R_E(R_E + h)}{h}$$

Substituting the numerical values, we have

$$C = \frac{4\pi\epsilon_0 R_E(R_E + h)}{h}$$

$$= \frac{4\pi(8.85 \times 10^{-12} \text{ C}^2/\text{N}\cdot\text{m}^2)(6.4 \times 10^3 \text{ km})(6.4 \times 10^3 \text{ km} + 5 \text{ km})}{5 \text{ km}}\left(\frac{1000 \text{ m}}{1 \text{ km}}\right)$$

$$\approx 0.9 \text{ F}$$

This result is extremely large, compared with the *picofarads* and *microfarads* that are typical values for capacitors in electrical circuits, especially for a capacitor having plates that are 5 km apart! We shall use this model of the atmosphere as a capacitor in our Context Conclusion, in which we calculate the number of lightning strikes on the Earth in one day.

❯ SUMMARY

When a positive test charge q_0 is moved between points Ⓐ and Ⓑ in an electric field $\vec{E}$, the **change in potential energy** of the charge–field system is

$$\Delta U = -q_0 \int_{Ⓐ}^{Ⓑ} \vec{E} \cdot d\vec{s} \qquad \text{20.1} \blacktriangleleft$$

The **potential difference** ΔV between points Ⓐ and Ⓑ in an electric field $\vec{E}$ is defined as the change in potential energy divided by the test charge q_0:

$$\Delta V = \frac{\Delta U}{q_0} = -\int_{Ⓐ}^{Ⓑ} \vec{E} \cdot d\vec{s} \qquad \text{20.3} \blacktriangleleft$$

where **electric potential** V is a scalar and has the units joules per coulomb, defined as 1 **volt** (V).

The potential difference between two points Ⓐ and Ⓑ in a uniform electric field $\vec{E}$ is

$$\Delta V = -E \int_{Ⓐ}^{Ⓑ} ds = -Ed \qquad \text{20.6} \blacktriangleleft$$

where d is the magnitude of the displacement vector between Ⓐ and Ⓑ.

Equipotential surfaces are surfaces on which the electric potential remains constant. Equipotential surfaces are *perpendicular* to electric field lines.

The electric potential due to a point charge q at a distance r from the charge is

$$V = k_e \frac{q}{r} \qquad \text{20.11} \blacktriangleleft$$

The electric potential due to a group of point charges is obtained by summing the potentials due to the individual charges. Because V is a scalar, the sum is a simple algebraic operation.

The **electric potential energy of a pair of point charges** separated by a distance r_{12} is

$$U = k_e \frac{q_1 q_2}{r_{12}} \qquad \text{20.13} \blacktriangleleft$$

which represents the work required to bring the charges from an infinite separation to the separation r_{12}. The potential energy of a distribution of point charges is obtained by summing terms like Equation 20.13 over *all pairs* of particles.

If the electric potential is known as a function of coordinates x, y, and z, the components of the electric field can be obtained by taking the negative derivative of the potential with respect to the coordinates. For example, the x component of an electric field is

$$E_x = -\frac{\partial V}{\partial x} \qquad \text{20.17} \blacktriangleleft$$

The **electric potential due to a continuous charge distribution** is

$$V = k_e \int \frac{dq}{r} \qquad \text{20.19} \blacktriangleleft$$

Every point on the surface of a charged conductor in electrostatic equilibrium is at the same potential. Furthermore, the potential is constant everywhere inside the conductor and is equal to its value at the surface.

A capacitor is a device for storing charge. A charged capacitor consists of two equal and oppositely charged conductors with a potential difference ΔV between them. The **capacitance** C of any capacitor is defined as the ratio of the magnitude of the charge Q on either conductor to the magnitude of the potential difference ΔV:

$$C \equiv \frac{Q}{\Delta V} \qquad \text{20.20} \blacktriangleleft$$

The SI units of capacitance are coulombs per volt, or the **farad** (F), so 1 F = 1 C/V.

If two or more capacitors are connected in parallel, the potential differences across them must be the same. The **equivalent capacitance** of a **parallel combination** of capacitors is

$$C_{eq} = C_1 + C_2 + C_3 + \cdots \qquad \text{20.26} \blacktriangleleft$$

If two or more capacitors are connected in series, the charges on them are the same and the **equivalent capacitance** of the **series combination** is given by

$$\frac{1}{C_{eq}} = \frac{1}{C_1} + \frac{1}{C_2} + \frac{1}{C_3} + \cdots \qquad \textbf{20.28} \blacktriangleleft$$

Energy is required to charge a capacitor because the charging process is equivalent to transferring charges from one conductor at a lower potential to another conductor at a higher potential. The electric potential energy U stored in the capacitor is

$$U = \frac{Q^2}{2C} = \tfrac{1}{2}Q\,\Delta V = \tfrac{1}{2}C(\Delta V)^2 \qquad \textbf{20.29} \blacktriangleleft$$

When a dielectric material is inserted between the plates of a capacitor, the capacitance generally increases by the dimensionless factor κ, called the **dielectric constant**. That is,

$$C = \kappa C_0 \qquad \textbf{20.32} \blacktriangleleft$$

where C_0 is the capacitance in the absence of the dielectric.

❯ OBJECTIVE QUESTIONS |

1. A parallel-plate capacitor is charged and then is disconnected from the battery. By what factor does the stored energy change when the plate separation is then doubled? (a) It becomes four times larger. (b) It becomes two times larger. (c) It stays the same. (d) It becomes one-half as large. (e) It becomes one-fourth as large.

2. The electric potential at $x = 3.00$ m is 120 V, and the electric potential at $x = 5.00$ m is 190 V. What is the x component of the electric field in this region, assuming the field is uniform? (a) 140 N/C (b) -140 N/C (c) 35.0 N/C (d) -35.0 N/C (e) 75.0 N/C

3. A proton is released from rest at the origin in a uniform electric field in the positive x direction with magnitude 850 N/C. What is the change in the electric potential energy of the proton–field system when the proton travels to $x = 2.50$ m? (a) 3.40×10^{-16} J (b) -3.40×10^{-16} J (c) 2.50×10^{-16} J (d) -2.50×10^{-16} J (e) -1.60×10^{-19} J

4. By what factor is the capacitance of a metal sphere multiplied if its volume is tripled? (a) 3 (b) $3^{1/3}$ (c) 1 (d) $3^{-1/3}$ (e) $\tfrac{1}{3}$

5. An electron in an x-ray machine is accelerated through a potential difference of 1.00×10^4 V before it hits the target. What is the kinetic energy of the electron in electron volts? (a) 1.00×10^4 eV (b) 1.60×10^{-15} eV (c) 1.60×10^{-22} eV (d) 6.25×10^{22} eV (e) 1.60×10^{-19} eV

6. Rank the potential energies of the four systems of particles shown in Figure OQ20.6 from largest to smallest. Include equalities if appropriate.

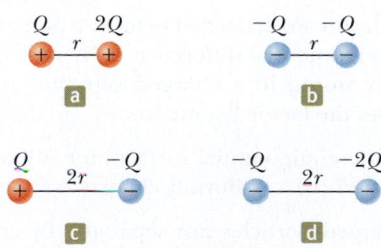

Figure OQ20.6

7. True or False? (a) From the definition of capacitance $C = Q/\Delta V$, it follows that an uncharged capacitor has a capacitance of zero. (b) As described by the definition of capacitance, the potential difference across an uncharged capacitor is zero.

8. In a certain region of space, a uniform electric field is in the x direction. A particle with negative charge is carried from $x = 20.0$ cm to $x = 60.0$ cm. **(i)** Does the electric potential energy of the charge-field system (a) increase, (b) remain constant, (c) decrease, or (d) change unpredictably? **(ii)** Has the particle moved to a position where the electric potential is (a) higher than before, (b) unchanged, (c) lower than before, or (d) unpredictable?

9. Rank the electric potentials at the four points shown in Figure OQ20.9 from largest to smallest.

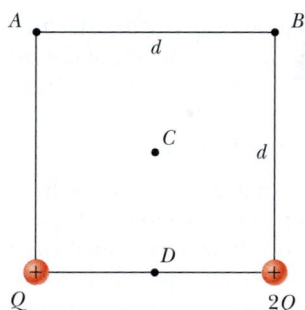

Figure OQ20.9

10. A capacitor with very large capacitance is in series with another capacitor with very small capacitance. What is the equivalent capacitance of the combination? (a) slightly greater than the capacitance of the large capacitor (b) slightly less than the capacitance of the large capacitor (c) slightly greater than the capacitance of the small capacitor (d) slightly less than the capacitance of the small capacitor

11. Consider the equipotential surfaces shown in Figure 20.4. In this region of space, what is the approximate direction of the electric field? (a) It is out of the page. (b) It is into the page (c) It is toward the top of the page. (d) It is toward the bottom of the page. (e) The field is zero.

12. A parallel-plate capacitor is connected to a battery. What happens to the stored energy if the plate separation is doubled while the capacitor remains connected to the battery? (a) It remains the same. (b) It is doubled. (c) It decreases by a factor of 2. (d) It decreases by a factor of 4. (e) It increases by a factor of 4.

13. Rank the electric potential energies of the systems of charges shown in Figure OQ20.13 from largest to smallest. Indicate equalities if appropriate.

Figure OQ20.13

14. Four particles are positioned on the rim of a circle. The charges on the particles are $+0.500\,\mu C$, $+1.50\,\mu C$, $-1.00\,\mu C$, and $-0.500\,\mu C$. If the electric potential at the center of the circle due to the $+0.500\,\mu C$ charge alone is 4.50×10^4 V, what is the total electric potential at the center due to the four charges? (a) 18.0×10^4 V (b) 4.50×10^4 V (c) 0 (d) -4.50×10^4 V (e) 9.00×10^4 V

15. In a certain region of space, the electric field is zero. From this fact, what can you conclude about the electric potential in this region? (a) It is zero. (b) It does not vary with position. (c) It is positive. (d) It is negative. (e) None of those answers is necessarily true.

16. A filament running along the x axis from the origin to $x = 80.0$ cm carries electric charge with uniform density. At the point P with coordinates ($x = 80.0$ cm, $y = 80.0$ cm), this filament creates electric potential 100 V. Now we add another filament along the y axis, running from the origin to $y = 80.0$ cm, carrying the same amount of charge with the same uniform density. At the same point P, is the electric potential created by the pair of filaments (a) greater than 200 V, (b) 200 V, (c) 100 V, (d) between 0 and 200 V, or (e) 0?

17. An electronics technician wishes to construct a parallel plate capacitor using rutile ($\kappa = 100$) as the dielectric. The area of the plates is 1.00 cm². What is the capacitance if the rutile thickness is 1.00 mm? (a) 88.5 pF (b) 177 pF (c) 8.85 μF (d) 100 μF (e) 35.4 μF

18. If three unequal capacitors, initially uncharged, are connected in series across a battery, which of the following statements is true? (a) The equivalent capacitance is greater than any of the individual capacitances. (b) The largest voltage appears across the smallest capacitance. (c) The largest voltage appears across the largest capacitance. (d) The capacitor with the largest capacitance has the greatest charge. (e) The capacitor with the smallest capacitance has the smallest charge.

19. (i) What happens to the magnitude of the charge on each plate of a capacitor if the potential difference between the conductors is doubled? (a) It becomes four times larger. (b) It becomes two times larger. (c) It is unchanged. (d) It becomes one-half as large. (e) It becomes one-fourth as large. (ii) If the potential difference across a capacitor is doubled, what happens to the energy stored? Choose from the same possibilities as in part (i).

20. A parallel-plate capacitor filled with air carries a charge Q. The battery is disconnected, and a slab of material with dielectric constant $\kappa = 2$ is inserted between the plates. Which of the following statements is true? (a) The voltage across the capacitor decreases by a factor of 2. (b) The voltage across the capacitor is doubled. (c) The charge on the plates is doubled. (d) The charge on the plates decreases by a factor of 2. (e) The electric field is doubled.

21. Assume a device is designed to obtain a large potential difference by first charging a bank of capacitors connected in parallel and then activating a switch arrangement that in effect disconnects the capacitors from the charging source and from each other and reconnects them all in a series arrangement. The group of charged capacitors is then discharged in series. What is the maximum potential difference that can be obtained in this manner by using ten $500\text{-}\mu F$ capacitors and an 800-V charging source? (a) 500 V (b) 8.00 kV (c) 400 kV (d) 800 V (e) 0

CONCEPTUAL QUESTIONS

□ denotes answer available in *Student Solutions Manual/Study Guide*

1. Assume you want to increase the maximum operating voltage of a parallel-plate capacitor. Describe how you can do that with a fixed plate separation.

2. Distinguish between electric potential and electric potential energy.

3. Describe the motion of a proton (a) after it is released from rest in a uniform electric field. Describe the changes (if any) in (b) its kinetic energy and (c) the electric potential energy of the proton–field system.

4. Because the charges on the plates of a parallel-plate capacitor are opposite in sign, they attract each other. Hence, it would take positive work to increase the plate separation. What type of energy in the system changes due to the external work done in this process?

5. Explain why the work needed to move a particle with charge Q through a potential difference ΔV is $W = Q\,\Delta V$, whereas the energy stored in a charged capacitor is $U = \frac{1}{2}Q\,\Delta V$. Where does the factor $\frac{1}{2}$ come from?

6. Describe the equipotential surfaces for (a) an infinite line of charge and (b) a uniformly charged sphere.

7. When charged particles are separated by an infinite distance, the electric potential energy of the pair is zero. When the particles are brought close, the electric potential energy of a pair with the same sign is positive, whereas the electric potential energy of a pair with opposite signs is negative. Give a physical explanation of this statement.

8. An air-filled capacitor is charged, then disconnected from the power supply, and finally connected to a voltmeter.

Explain how and why the potential difference changes when a dielectric is inserted between the plates of the capacitor.

9. Why is it dangerous to touch the terminals of a high-voltage capacitor even after the voltage source that charged the capacitor is disconnected from the capacitor? (b) What can be done to make the capacitor safe to handle after the voltage source has been removed?

10. Study Figure 19.4 and the accompanying text discussion of charging by induction. When the grounding wire is touched to the rightmost point on the sphere in Figure 19.4c, electrons are drained away from the sphere to leave the sphere positively charged. Suppose the grounding wire is touched to the leftmost point on the sphere instead. (a) Will electrons still drain away, moving closer to the negatively charged rod as they do so? (b) What kind of charge, if any, remains on the sphere?

11. If you were asked to design a capacitor in which small size and large capacitance were required, what would be the two most important factors in your design?

12. Explain why a dielectric increases the maximum operating voltage of a capacitor even though the physical size of the capacitor doesn't change.

▶ PROBLEMS AVAILABLE IN ᴱᴺᴴᴬᴺᶜᴱᴰ Web**Assign**

20.1 Electric Potential and Potential Difference

Problems 1–3

20.2 Potential Difference in a Uniform Electric Field

Problems 4–6

20.3 Electric Potential and Potential Energy Due to Point Charges

Problems 7–21

20.4 Obtaining the Value of the Electric Field from the Electric Potential

Problems 22–24

20.5 Electric Potential Due to Continuous Charge Distributions

Problems 25–29

20.6 Electric Potential Due to a Charged Conductor

Problems 30–32

20.7 Capacitance

Problems 33–41

20.8 Combinations of Capacitors

Problems 42–52

20.9 Energy Stored in a Charged Capacitor

Problems 53–61

20.10 Capacitors with Dielectrics

Problems 62–66

20.11 Context Connection: The Atmosphere as a Capacitor

Problems 67–68

Additional Problems

Problems 69–87

Solutions to the following Problems are available in the *Student Solutions Manual/Study Guide*:

20.3, 20.5, 20.8, 20.11, 20.20, 20.23, 20.26, 20.32, 20.35, 20.39, 20.45, 20.52, 20.54, 20.64, 20.67, 20.73, 20.81, and 20.82

List of Enhanced Problems

Problem Number	Targeted Feedback in Enhanced WebAssign	Master It in Enhanced WebAssign	Watch It in Enhanced WebAssign
20.1	✓		✓
20.2	✓		✓
20.3	✓	✓	
20.5		✓	
20.9	✓		✓
20.11	✓	✓	
20.13	✓		✓
20.20		✓	
20.23	✓		✓
20.24	✓		✓
20.28	✓		✓
20.29	✓		✓
20.32	✓	✓	
20.33	✓		✓
20.34	✓		✓
20.37	✓	✓	
20.38	✓		
20.39		✓	
20.41	✓		✓
20.42	✓		✓
20.43	✓		✓
20.44	✓		✓
20.45	✓	✓	
20.48	✓		
20.49		✓	
20.52	✓	✓	
20.59	✓		✓
20.63	✓		✓
20.65	✓		✓
20.67		✓	
20.73		✓	
20.81		✓	

Current and Direct Current Circuits

Chapter Outline

Trombax/Shutterstock.com

A technician repairs a connection on a circuit board from a computer. In our lives today, we use many items containing electric circuits, including many with circuit boards much smaller than the board shown in the photograph, including MP3 players, cell phones, and digital cameras. In this chapter, we study simple types of circuits and learn how to analyze them.

Thus far, our discussion of electrical phenomena has focused on charges at rest, or the study of *electrostatics*. We shall now consider situations involving electric charges in motion. The term *electric current*, or simply *current*, is used to describe the flow of charge through some region of space. Most practical applications of electricity involve electric currents. For example, in a flashlight with an incandescent bulb, charges flow in the filament of the lightbulb after the switch is turned on. In most common situations, the flow of charge takes place in a conductor, such as a copper wire. Currents can also exist outside a conductor. For instance, a beam of electrons in a particle accelerator constitutes a current.

In Chapter 20, we introduced the notion of a *circuit*. As we continue our investigations into circuits in this chapter, we introduce the *resistor* as a new circuit element.

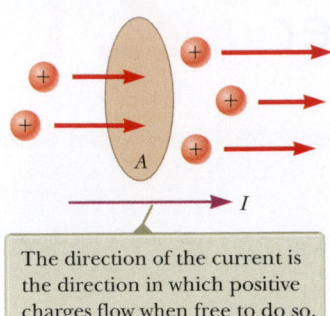

The direction of the current is the direction in which positive charges flow when free to do so.

Figure 21.1 Charges in motion through an area A. The time rate at which charge flows through the area is defined as the current I.

▶ Electric current

Pitfall Prevention | 21.1

"Current Flow" Is Redundant

The phrase *current flow* is commonly used, although it is technically incorrect because current is a flow (of charge). This wording is similar to the phrase *heat transfer*, which is also redundant because heat is a transfer (of energy). We will avoid this phrase and speak of *flow of charge* or *charge flow*.

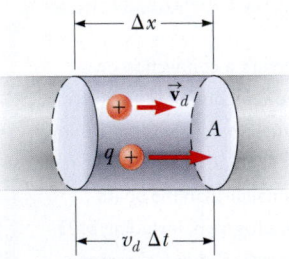

Figure 21.2 A segment of a uniform conductor of cross-sectional area A.

▸ 21.1 | Electric Current

Whenever charge is flowing, an **electric current** is said to exist. To define current mathematically, suppose charged particles are moving perpendicular to a surface of area A as in Figure 21.1. (This area could be the cross-sectional area of a wire, for example.) The current is defined as **the rate at which electric charge flows through this surface.** If ΔQ is the amount of charge that passes through this area in a time interval Δt, the average current I_{avg} over the time interval is the ratio of the charge to the time interval:

$$I_{avg} = \frac{\Delta Q}{\Delta t}$$

21.1 ◀

It is possible for the rate at which charge flows to vary in time. We define the **instantaneous current** I as the limit of the preceding expression as Δt goes to zero:

$$I \equiv \lim_{\Delta t \to 0} \frac{\Delta Q}{\Delta t} = \frac{dQ}{dt}$$

21.2 ◀

The SI unit of current is the **ampere** (A):

$$1 \, A = 1 \, C/s$$

21.3 ◀

That is, 1 A of current is equivalent to 1 C of charge passing through a surface in 1 s.

The particles flowing through a surface as in Figure 21.1 can be charged positively or negatively, or we can have two or more types of particles moving, with charges of both signs in the flow. Conventionally, we define the direction of the current as the direction of flow of positive charge, regardless of the sign of the actual charged particles in motion.[1] In a common conductor such as copper, the current is physically due to the motion of the negatively charged electrons. Therefore, when we speak of current in such a conductor, the direction of the current is opposite the direction of flow of electrons. On the other hand, if one considers a beam of positively charged protons in a particle accelerator, the current is in the direction of motion of the protons. In some cases—gases and electrolytes, for example—the current is the result of the flow of both positive and negative charged particles. It is common to refer to a moving charged particle (whether it is positive or negative) as a mobile **charge carrier**. For example, the charge carriers in a metal are electrons.

We now build a structural model that will allow us to relate the macroscopic current to the motion of the charged particles. Consider identical charged particles moving in a conductor of cross sectional area A (Fig. 21.2). The volume of a segment of the conductor of length Δx (between the two circular cross sections shown in Fig. 21.2) is $A \, \Delta x$. If n represents the number of mobile charge carriers per unit volume (in other words, the charge carrier density), the number of carriers in the segment is $nA \, \Delta x$. Therefore, the total charge ΔQ in this segment is

$$\Delta Q = \text{number of carriers in section} \times \text{charge per carrier} = (nA \, \Delta x) q$$

where q is the charge on each carrier. If the carriers move with an average velocity component v_d in the x direction (along the wire), the displacement they experience in this direction in a time interval Δt is $\Delta x = v_d \Delta t$. The speed v_d of the charge carrier along the wire is an average speed called the **drift speed.** Let us choose Δt to be the time interval required for the charges in the segment to move through a displacement whose magnitude is equal to the length of the segment. This time interval is also that required for all the charges in the segment to pass through the circular area at one end. With this choice, we can write ΔQ in the form

$$\Delta Q = (nAv_d \Delta t) q$$

---

[1]Even though we discuss a direction for current, current is not a vector. As we shall see later in the chapter, currents add algebraically and not vectorially.

If we divide both sides of this equation by Δt, we see that the average current in the conductor is

$$I_{avg} = \frac{\Delta Q}{\Delta t} = nqv_d A \qquad \textbf{21.4} \blacktriangleleft$$

▶ Current in terms of microscopic parameters

Equation 21.4 relates a macroscopically measured average current to the microscopic origin of the current: the density of charge carriers n, the charge per carrier q, and the drift speed v_d.

QUICK QUIZ 21.1 Consider positive and negative charges moving horizontally through the four regions shown in Figure 21.3. Rank the currents in these four regions, from highest to lowest.

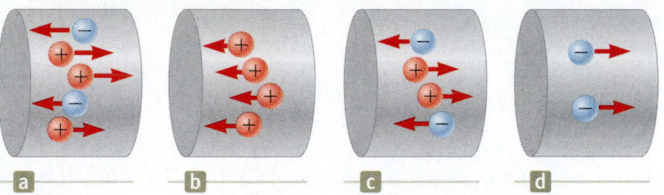

Figure 21.3 (Quick Quiz 21.1) Four groups of charges move through a region.

Let us investigate further the notion of drift speed. We have identified drift speed as an average speed along the wire, but the charge carriers are by no means moving in a straight line with speed v_d. Consider a conductor in which the charge carriers are free electrons. In the absence of a potential difference across the conductor, these electrons undergo random motion similar to that of gas molecules in the structural model of kinetic theory that we studied in Chapter 16. This random motion is related to the temperature of the conductor. The electrons undergo repeated collisions with the metal atoms, and the result is a complicated zigzag motion. When a potential difference is applied across the conductor, an electric field is established in the conductor. The electric field exerts an electric force on the electrons (Eq. 19.4). This force accelerates the electrons and hence produces a current. The motion of the electrons due to the electric force is superimposed on their random motion to provide an average velocity whose magnitude is the drift speed as shown in Active Figure 21.4.

When electrons make collisions with metal atoms during their motion, they transfer energy to the atoms. This energy transfer causes an increase in the vibrational energy of the atoms and a corresponding increase in the temperature of the conductor.[2] This process involves all three types of energy storage in the conservation of energy equation, Equation 7.2. If we consider the system to be the electrons, the metal atoms, and the electric field (which is established by an external source such as a battery), the energy at the instant when the potential difference is applied across the conductor is electric potential energy associated with the field and the electrons. This energy is transformed by work done within the system by the field on the electrons to kinetic energy of electrons. When the electrons strike the metal atoms, some of the kinetic energy is transferred to the atoms, which adds to the internal energy of the system.

The **current density** J in the conductor is defined as the current per unit area. From Equation 21.4, the current density is

$$J \equiv \frac{I}{A} = nqv_d \qquad \textbf{21.5} \blacktriangleleft$$

where J has the SI units amperes per square meter.

Pitfall Prevention | 21.2
Batteries Do Not Supply Electrons
A battery does not supply electrons to the circuit. It establishes the electric field that exerts a force on electrons already in the wires and elements of the circuit.

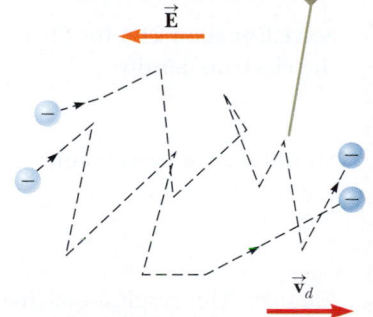

The random motion of the charge carriers is modified by the field, and they have a drift velocity opposite the direction of the electric field.

Active Figure 21.4 A schematic representation of the zigzag motion of negative charge carriers in a conductor. Because of the acceleration of the charge carriers due to the electric force, the paths are actually parabolic. The drift speed, however, is much smaller than the average speed, so the parabolic shape is not visible on this scale.

---

[2]This increase in temperature is sometimes called *Joule heating*, but that term is a misnomer because there is no heat involved. We will not use this wording.

> **THINKING PHYSICS 21.1**
>
> In Chapter 19, we claimed that the electric field inside a conductor is zero. In the preceding discussion, however, we have used the notion of an electric field in a conducting wire that exerts electric forces on electrons, causing them to move with a drift velocity. Is this notion inconsistent with Chapter 19?
>
> **Reasoning** The electric field is zero only in a conductor in *electrostatic equilibrium*, that is, a conductor in which the charges are at rest after having moved to equilibrium positions. In a current-carrying conductor, the charges are not at rest, so the requirement for a zero field is not imposed. The electric field in a conductor in a circuit is due to a distribution of charge over the surface of the conductor that can be quite complicated.[3] ◄

Example 21.1 | Drift Speed in a Copper Wire

The 12-gauge copper wire in a typical residential building has a cross-sectional area of $3.31 \times 10^{-6} \text{ m}^2$. It carries a constant current of 10.0 A. What is the drift speed of the electrons in the wire? Assume each copper atom contributes one free electron to the current. The density of copper is 8.92 g/cm³.

SOLUTION

Conceptualize Imagine electrons following a zigzag motion with a drift velocity parallel to the wire superimposed on the motion as in Active Figure 21.4. As mentioned earlier, the drift speed is small, and this example helps us quantify the speed.

Categorize We evaluate the drift speed using Equation 21.4. Because the current is constant, the average current during any time interval is the same as the constant current: $I_{avg} = I$.

. . . .

Analyze The periodic table of the elements in Appendix C shows that the molar mass of copper is $M = 63.5$ g/mol. Recall that 1 mol of any substance contains Avogadro's number of atoms ($N_A = 6.02 \times 10^{23} \text{ mol}^{-1}$).

Use the molar mass and the density of copper to find the volume of 1 mole of copper:

$$V = \frac{M}{\rho}$$

From the assumption that each copper atom contributes one free electron to the current, find the electron density in copper:

$$n = \frac{N_A}{V} = \frac{N_A \rho}{M}$$

Solve Equation 21.4 for the drift speed and substitute for the electron density:

$$v_d = \frac{I_{avg}}{nqA} = \frac{I}{nqA} = \frac{IM}{qAN_A\rho}$$

Substitute numerical values:

$$v_d = \frac{(10.0 \text{ A})(0.063\ 5 \text{ kg/mol})}{(1.60 \times 10^{-19} \text{ C})(3.31 \times 10^{-6} \text{ m}^2)(6.02 \times 10^{23} \text{ mol}^{-1})(8\ 920 \text{ kg/m}^3)}$$

$$= \boxed{2.23 \times 10^{-4} \text{ m/s}}$$

. . . .

Finalize This result shows that typical drift speeds are very small. For instance, electrons traveling at 2.23×10^{-4} m/s would take about 75 min to travel 1 m! You might therefore wonder why a light turns on almost instantaneously when its switch is thrown. In a conductor, changes in the electric field that drives the free electrons travel through the conductor with a speed close to that of light. So, when you flip on a light switch, electrons already in the filament of an incandescent lightbulb experience electric forces and begin moving after a time interval on the order of nanoseconds.

---

[3]See Chapter 18 in R. Chabay and B. Sherwood, *Matter & Interactions II: Electric and Magnetic Interactions* (Hoboken: Wiley, 2007) for details on this charge distribution.

⟨21.2 | Resistance and Ohm's Law

The drift speed of electrons in a current-carrying wire is related to the electric field in the wire. If the field is increased, the electric force on the electrons is stronger and the drift speed increases. We shall show in Section 21.4 that this relationship is linear and that the drift speed is directly proportional to the electric field. For a uniform field in a conductor of uniform cross-section, the potential difference across the conductor is proportional to the electric field as in Equation 20.6. Therefore, when a potential difference ΔV is applied across the ends of a metallic conductor as in Figure 21.5, the current in the conductor is found to be proportional to the applied voltage; that is, $I \propto \Delta V$. We can write this proportionality as $\Delta V = IR$, where R is called the **resistance** of the conductor. We define this resistance according to the equation we have just written, as the ratio of the voltage across the conductor to the current it carries:

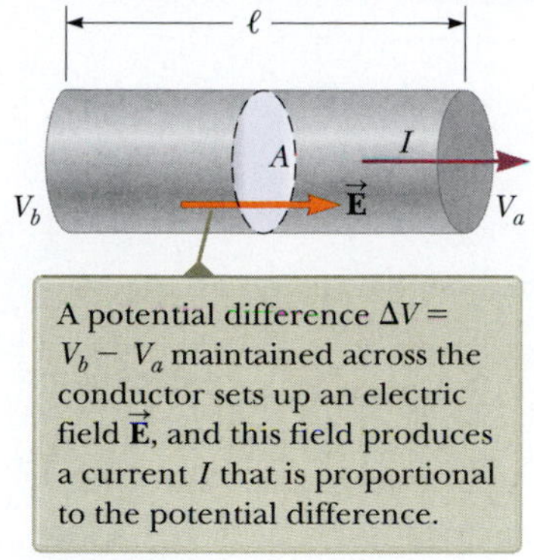

$$R \equiv \frac{\Delta V}{I} \qquad \text{21.6} \blacktriangleleft$$

Resistance has the SI unit volt per ampere, called an **ohm** (Ω). Therefore, if a potential difference of 1 V across a conductor produces a current of 1 A, the resistance of the conductor is 1 Ω. As another example, if an electrical appliance connected to a 120-V source carries a current of 6.0 A, its resistance is 20 Ω.

Resistance is the quantity that determines the current that results due to a voltage in a simple circuit. For a fixed voltage, if the resistance increases, the current decreases. If the resistance decreases, the current increases.

It might be useful for you to build a mental representation for current, voltage, and resistance by comparing these concepts to analogous concepts for the flow of water in a river. As water flows downhill in a river of constant width and depth, the rate of flow of water (analogous to current) depends on the total vertical distance through which the water drops between two points (analogous to voltage) and on the width and depth as well as on the effects of rocks, the riverbank, and other obstructions (analogous to resistance). Likewise, electric current in a uniform conductor depends on the applied voltage, and the resistance of the conductor is caused by collisions of the electrons with atoms in the conductor.

For many materials, including most metals, experiments show that the resistance is constant over a wide range of applied voltages. This behavior is known as **Ohm's law** after Georg Simon Ohm (1787–1854), who was the first to conduct a systematic study of electrical resistance.

Many individuals call Equation 21.6 Ohm's law, but this terminology is incorrect. This equation is simply the definition of resistance, and it provides an important relationship between voltage, current, and resistance. Ohm's law is *not* a fundamental law of nature, but a behavior that is valid only for certain materials and devices, and only over a limited range of conditions. Materials or devices that obey Ohm's law, and hence that have a constant resistance over a wide range of voltages, are said to be **ohmic** (Fig. 21.6a). Materials or devices that do not obey Ohm's law are **nonohmic**. One common semiconducting device that is nonohmic is the *diode*, a circuit element that acts like a one-way valve for current. Its resistance is small for currents in one direction (positive ΔV) and large for currents in the reverse direction (negative ΔV)

A potential difference $\Delta V = V_b - V_a$ maintained across the conductor sets up an electric field $\vec{\mathbf{E}}$, and this field produces a current I that is proportional to the potential difference.

Figure 21.5 A uniform conductor of length ℓ and cross-sectional area A.

▶ Definition of resistance

Pitfall Prevention | 21.3
We've Seen Something Like Equation 21.6 Before
In Chapter 4, we introduced Newton's second law, $\Sigma F = ma$, for a net force on an object of mass m. It can be written as

$$m = \frac{\Sigma F}{a}$$

In Chapter 4, we defined mass as *resistance to a change in motion in response to an external force*. Mass as resistance to changes in motion is analogous to electrical resistance to charge flow, and Equation 21.6 is analogous to the form of Newton's second law above. Each equation states that the resistance (electrical or mechanical) is equal to (1) ΔV, the cause of current, or (2) ΣF, the cause of changes in motion, divided by the result, (1) a charge flow, quantified by current I, or (2) a change in motion, quantified by acceleration a.

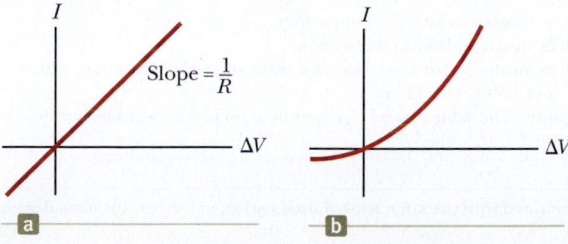

Figure 21.6 (a) The current–potential difference curve for an ohmic material. The curve is linear, and the slope is equal to the inverse of the resistance of the conductor. (b) A nonlinear current–potential difference curve for a semiconducting diode. This device does not obey Ohm's law.

as shown in Figure 21.6b. Most modern electronic devices have nonlinear current–voltage relationships; their operation depends on the particular ways they violate Ohm's law.

> **QUICK QUIZ 21.2** In Figure 21.6b, as the applied voltage increases, does the resistance of the diode **(a)** increase, **(b)** decrease, or **(c)** remain the same?

A **resistor** is a simple circuit element that provides a specified resistance in an electrical circuit. The symbol for a resistor in circuit diagrams is a zigzag red line (—⟋⟍⟋⟍—). We can express Equation 21.6 in the form

$$\Delta V = IR \qquad \textbf{21.7} \blacktriangleleft$$

This equation tells us that the voltage across a resistor is the product of the resistance and the current in the resistor.

The resistance of an ohmic conducting wire such as that shown in Figure 21.5 is found to be proportional to its length ℓ and inversely proportional to its cross-sectional area A. That is,

▶ Resistance of a uniform material of resistivity ρ along a length ℓ

$$R = \rho \frac{\ell}{A} \qquad \textbf{21.8} \blacktriangleleft$$

where the constant of proportionality ρ is called the **resistivity** of the material,[4] which has the unit ohm · meter ($\Omega \cdot$ m). To understand this relationship between resistance and resistivity, note that every ohmic material has a characteristic resistivity, a parameter that depends on the properties of the material and on temperature. On the other hand, as you can see from Equation 21.8, the resistance of a particular conductor depends on its size and shape as well as on the resistivity of the material. Table 21.1 provides a list of resistivities for various materials measured at 20°C.

> **Pitfall Prevention | 21.4**
> **Resistance and Resistivity**
> Resistivity is a property of a *substance*, whereas resistance is a property of an *object*. We have seen similar pairs of variables before. For example, density is a property of a substance, whereas mass is a property of an object. Equation 21.8 relates resistance to resistivity, and we have seen a previous equation (Eq. 1.1) that relates mass to density.

> **TABLE 21.1 | Resistivities and Temperature Coefficients of Resistivity for Various Materials**

Material	Resistivity[a] ($\Omega \cdot$ m)	Temperature Coefficient[b] α [(°C)$^{-1}$]
Silver	1.59×10^{-8}	3.8×10^{-3}
Copper	1.7×10^{-8}	3.9×10^{-3}
Gold	2.44×10^{-8}	3.4×10^{-3}
Aluminum	2.82×10^{-8}	3.9×10^{-3}
Tungsten	5.6×10^{-8}	4.5×10^{-3}
Iron	10×10^{-8}	5.0×10^{-3}
Platinum	11×10^{-8}	3.92×10^{-3}
Lead	22×10^{-8}	3.9×10^{-3}
Nichrome[c]	1.00×10^{-6}	0.4×10^{-3}
Carbon	3.5×10^{-5}	-0.5×10^{-3}
Germanium	0.46	-48×10^{-3}
Silicon[d]	2.3×10^{3}	-75×10^{-3}
Glass	10^{10} to 10^{14}	
Hard rubber	$\sim 10^{13}$	
Sulfur	10^{15}	
Quartz (fused)	75×10^{16}	

[a]All values at 20°C. All elements in this table are assumed to be free of impurities.
[b]The temperature coefficient of resistivity will be discussed later in this section.
[c]A nickel–chromium alloy commonly used in heating elements. The resistivity of Nichrome varies with composition and ranges between 1.00×10^{-6} and 1.50×10^{-6} $\Omega \cdot$ m.
[d]The resistivity of silicon is very sensitive to purity. The value can be changed by several orders of magnitude when it is doped with other atoms.

---

[4]The symbol ρ used for resistivity should not be confused with the same symbol used earlier in the text for mass density and volume charge density.

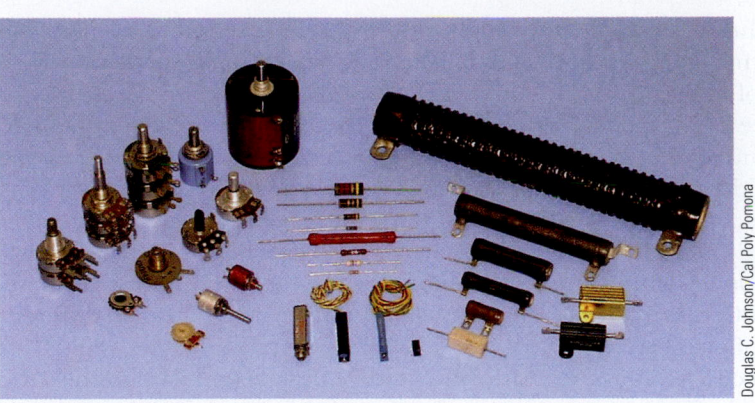

An assortment of resistors used in electric circuits.

The inverse of the resistivity is defined[5] as the **conductivity** σ. Hence, the resistance of an ohmic conductor can be expressed in terms of its conductivity as

$$R = \frac{\ell}{\sigma A} \qquad \text{21.9} \blacktriangleleft$$

where $\sigma = 1/\rho$.

Equation 21.9 shows that the resistance of a conductor is proportional to its length and inversely proportional to its cross-sectional area, similar to the flow of liquid through a pipe. As the length of the pipe is increased and the pressure difference between the ends of the pipe is held constant, the pressure difference between any two points separated by a fixed distance decreases and there is less force pushing the element of fluid between these points through the pipe. Therefore, less fluid flows for a given pressure difference between the ends of the pipe, representing an increased resistance. As its cross-sectional area is increased, the pipe can transport more fluid in a given time interval for a given pressure difference between the ends of the pipe, so its resistance drops.

As another analogy between electrical circuits and our previous studies, let us combine Equations 21.6 and 21.9:

$$R = \frac{\ell}{\sigma A} = \frac{\Delta V}{I} \quad \rightarrow \quad I = \sigma A \frac{\Delta V}{\ell} \quad \rightarrow \quad \frac{q}{\Delta t} = \sigma A \frac{\Delta V}{\ell}$$

where q is the amount of charge transferred in a time interval Δt. Let us compare this equation to Equation 17.35 for conduction of energy through a slab of material of area A, length ℓ, and thermal conductivity k, which we reproduce below:

$$P = kA \frac{(T_h - T_c)}{L} \quad \rightarrow \quad \frac{Q}{\Delta t} = kA \frac{\Delta T}{L}$$

In this equation, Q is the amount of energy transferred by heat in a time interval Δt. Notice the striking similarity between these last two equations.

Another analogy arises in an example that is important in biochemical applications. *Fick's law* describes the rate of transfer of a chemical solute through a solvent by the process of *diffusion*. This transfer occurs because of a difference in concentration of the solute (mass of solute per volume) between the two locations. Fick's law is as follows:

BIO Diffusion in biological systems

$$\frac{n}{\Delta t} = DA \frac{\Delta C}{L}$$

where $n/\Delta t$ is the rate of flow of the solute in moles per second, A is the area through which the solute moves, and L is the length over which the concentration difference is ΔC. The concentration is measured in moles per cubic meter. The parameter D is a diffusion constant (with units of meters squared per second) that describes the

---

[5]Do not confuse the symbol σ for conductivity with the same symbol used earlier for the Stefan–Boltzmann constant and surface charge density.

The colored bands on this resistor are yellow, violet, black, and gold.

Figure 21.7 A close-up view of a circuit board shows the color coding on a resistor.

BIO Electrical activity in the heart

BIO Catheter ablation for atrial fibrillation

rate of diffusion of a solute through the solvent and is similar in nature to electrical or thermal conductivity. Fick's law has important applications in describing the transport of molecules across biological membranes.

All three of the preceding equations have exactly the same mathematical form. Each has a time rate of change on the left, and each has the product of a conductivity, an area, and a ratio of a difference in a variable to a length on the right. This type of equation is a *transport equation* used when we transport energy, charge, or moles of matter. The difference in the variable on the right side of each equation is what drives the transport. A temperature difference drives energy transfer by heat, an electric potential difference drives a transfer of charge, and a concentration difference drives a transfer of matter.

Most electric circuits use resistors to control the current level in the various parts of the circuit. Two common types of resistors are the *composition* resistor containing carbon and the *wire-wound* resistor, which consists of a coil of wire. Resistors are normally color-coded to give their values in ohms, as shown in Figure 21.7 and Table 21.2. As an example, the four colors on the resistor at the bottom of Figure 21.7 are yellow ($= 4$), violet ($= 7$), black ($= 10^0$), and gold ($= 5\%$), and so the resistance value is $47 \times 10^0 \ \Omega = 47 \ \Omega$ with a tolerance value of $5\% = 2 \ \Omega$.

Let's consider the role of electrical resistance in maintaining proper beating of the human heart. The right atrium of the heart contains a specialized set of muscle fibers called the SA (sinoatrial) node that initiates the heartbeat. Electrical impulses that originate in these fibers gradually spread from cell to cell throughout the right and left atrial muscles, causing them to contract. When the impulses reach the atrioventricular (AV) node, the muscles of the atria begin to relax, and the impulses are directed to the ventricular muscles by an assembly of heart muscle cells called the *bundle of His* and the *Purkinje fibers*. After the resulting contraction of the ventricles, the heartbeat is complete and the cycle begins again.

The heart can experience a number of *arrhythmias*, in which the normal heartbeat rhythm is interrupted. Arrhythmias are generally caused by abnormal electrical activity in the heart. The most common cardiac arrhythmia is *atrial fibrillation* (AF). In this condition, the two upper chambers of the heart, the atria, undergo random quivering at a rate that can be greater than 300 per minute, rather than the usual coordinated contractions. In *paroxysmal* AF, the patient goes into episodes of atrial fibrillation that may last from a few minutes to a few days. It some cases, the condition may even become chronic. With episodes longer than a few days, blood can pool in the atria, due to the inefficiency of the quivering action in pumping blood out of the heart. This pooled blood can result in clots, which can then travel to the brain and cause a stroke. Patients with long-lasting episodes of AF are treated with anticoagulants to prevent clots, rate control medications to slow the rate of impulses conducted to the ventricles, and antiarrhythmics to return the heart to its normal rhythm. Defibrillator device paddles are sometimes used to deliver an electric shock to a patient's chest in order to restore normal heart rhythm.

In a large percentage of patients, the source of the chaotic activity is found in the four pulmonary veins leading into the left atrium. Atrial tissue has grown into these veins and can act as electrical triggers competing with the SA node. As a result, the atrial muscles receive electrical signals from a variety of sources rather than the SA node alone, leading to chaotic contractions. Patients whose arrhythmias can not be controlled with medications, as well as those who do not wish to take medications, have an additional option. A procedure known as *cardiac catheter ablation* may be performed by an *electrophysiologist* in an effort to restore normal sinus rhythm. In this procedure, the patient is anesthetized

◀ **TABLE 21.2** | **Color Code for Resistors**

Color	Number	Multiplier	Tolerance
Black	0	1	
Brown	1	10^1	
Red	2	10^2	
Orange	3	10^3	
Yellow	4	10^4	
Green	5	10^5	
Blue	6	10^6	
Violet	7	10^7	
Gray	8	10^8	
White	9	10^9	
Gold		10^{-1}	5%
Silver		10^{-2}	10%
Colorless			20%

and catheters are inserted into a vein in the groin and guided into the right atrium of the heart. The catheter then punctures the septum and enters the left atrium. Figure 21.8 shows an ablation catheter passing into the heart from a vein. The electrophysiologist maps the atrium and then stimulates the heart to determine areas of abnormal electrical activity. Finally, the electrophysiologist *ablates* the tissue around the four pulmonary veins, usually with radiofrequency energy from the tip of one of the catheters. The resulting scar tissue represents a high-resistance path, through which the electrical signals from the AF triggers in the pulmonary veins cannot travel. As a result, the SA node alone again controls the electrical activity of the heart. Because the triggers in the pulmonary veins have been cut off electrically from the rest of the heart, this specific procedure is called a *pulmonary vein isolation*.

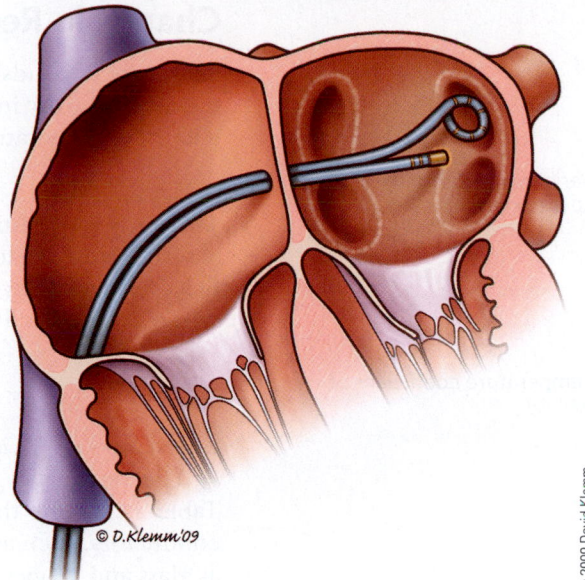

© D.Klemm'09

© 2009 David Klemm

Figure 21.8 During a cardiac catheter ablation procedure, catheters are guided into the left atrium through a vein from the groin. Radiofrequency energy is used to ablate tissue surrounding the pulmonary veins, where abnormal electrical activity is happening.

Example 21.2 | The Resistance of Nichrome Wire

The radius of 22-gauge Nichrome wire is 0.32 mm.

(A) Calculate the resistance per unit length of this wire.

SOLUTION

Conceptualize Table 21.1 shows that Nichrome has a resistivity two orders of magnitude larger than the best conductors in the table. Therefore, we expect it to have some special practical applications that the best conductors may not have.

Categorize We model the wire as a cylinder so that a simple geometric analysis can be applied to find the resistance.

Analyze Use Equation 21.8 and the resistivity of Nichrome from Table 21.1 to find the resistance per unit length:

$$\frac{R}{\ell} = \frac{\rho}{A} = \frac{\rho}{\pi r^2} = \frac{1.0 \times 10^{-6}\,\Omega \cdot \text{m}}{\pi (0.32 \times 10^{-3}\,\text{m})^2} = \boxed{3.1\ \Omega/\text{m}}$$

(B) If a potential difference of 10 V is maintained across a 1.0-m length of the Nichrome wire, what is the current in the wire?

SOLUTION

Analyze Use Equation 21.6 to find the current:

$$I = \frac{\Delta V}{R} = \frac{\Delta V}{(R/\ell)\ell} = \frac{10\ \text{V}}{(3.1\ \Omega/\text{m})(1.0\ \text{m})} = \boxed{3.2\ \text{A}}$$

Finalize Because of its high resistivity and resistance to oxidation, Nichrome is often used for heating elements in toasters, irons, and electric heaters.

What If? What if the wire were composed of copper instead of Nichrome? How would the values of the resistance per unit length and the current change?

Answer Table 21.1 shows us that copper has a resistivity two orders of magnitude smaller than that for Nichrome. Therefore, we expect the answer to part (A) to be smaller and the answer to part (B) to be larger. Calculations show that a copper wire of the same radius would have a resistance per unit length of only 0.053 Ω/m. A 1.0-m length of copper wire of the same radius would carry a current of 190 A with an applied potential difference of 10 V.

Change in Resistivity with Temperature

Resistivity depends on a number of factors, one of which is temperature. For most metals, resistivity increases approximately linearly with increasing temperature over a limited temperature range according to the expression

$$\rho = \rho_0[1 + \alpha(T - T_0)] \qquad \textbf{21.10} \blacktriangleleft$$

▶ Variation of resistivity with temperature

where ρ is the resistivity at some temperature T (in degrees Celsius), ρ_0 is the resistivity at some reference temperature T_0 (usually 20°C), and α is called the **temperature coefficient of resistivity** (not to be confused with the average coefficient of linear expansion α in Chapter 16). From Equation 21.10, we see that α can be expressed as

▶ Temperature coefficient of resistivity

$$\alpha = \frac{1}{\rho_0}\frac{\Delta\rho}{\Delta T} \qquad \textbf{21.11} \blacktriangleleft$$

where $\Delta\rho = \rho - \rho_0$ is the change in resistivity in the temperature interval $\Delta T = T - T_0$.

The resistivities and temperature coefficients of certain materials are listed in Table 21.1. Note the enormous range in resistivities, from very low values for good conductors, such as copper and silver, to very high values for good insulators, such as glass and rubber. An ideal, or "perfect," conductor would have zero resistivity, and an ideal insulator would have infinite resistivity.

Because resistance is proportional to resistivity according to Equation 21.8, the temperature variation of the resistance can be written as

▶ Variation of resistance with temperature

$$R = R_0[1 + \alpha(T - T_0)] \qquad \textbf{21.12} \blacktriangleleft$$

Precise temperature measurements are often made using this property.

> **QUICK QUIZ 21.3** When does an incandescent lightbulb carry more current? **(a)** immediately after it is turned on and the glow of the metal filament is increasing or **(b)** after it has been on for a few milliseconds and the glow is steady?

21.3 | Superconductors

For several metals, resistivity is nearly proportional to temperature as shown in Figure 21.9. In reality, however, there is always a nonlinear region at very low temperatures, and the resistivity usually approaches some finite value near absolute zero (see the magnified inset in Fig. 21.9). This residual resistivity near absolute zero is due primarily to collisions of electrons with impurities and to imperfections in the metal. In contrast, the high temperature resistivity (the linear region) is dominated by collisions of electrons with the vibrating metal atoms. We shall describe this process in more detail in Section 21.4.

There is a class of metals and compounds whose resistance decreases to zero when they are below a certain temperature T_c, known as the **critical temperature.** These materials are known as **superconductors.** The resistance–temperature graph for a superconductor follows that of a normal metal at temperatures above T_c (Fig. 21.10). When the temperature is at or below T_c, the resistivity drops suddenly

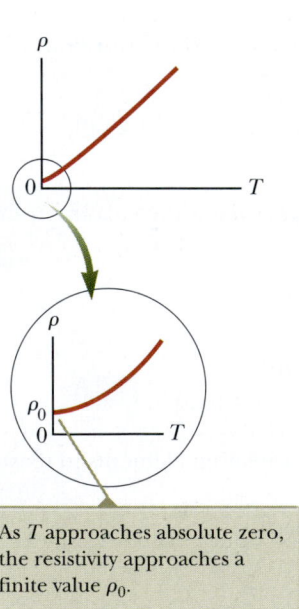

As T approaches absolute zero, the resistivity approaches a finite value ρ_0.

Figure 21.9 Resistivity versus temperature for a normal metal, such as copper. The curve is linear over a wide range of temperatures, and ρ increases with increasing temperature.

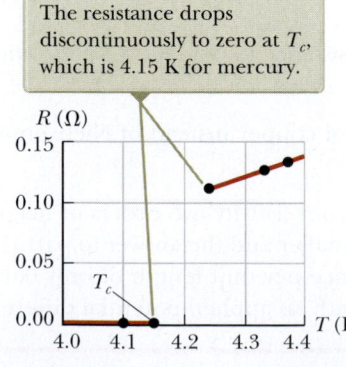

The resistance drops discontinuously to zero at T_c, which is 4.15 K for mercury.

Figure 21.10 Resistance versus temperature for a sample of mercury (Hg). The graph follows that of a normal metal above the critical temperature T_c.

to zero. This phenomenon was discovered in 1911 by Dutch physicist Heike Kamerlingh-Onnes (1853–1926) as he worked with mercury, which is a superconductor below 4.15 K. Measurements have shown that the resistivities of superconductors below their T_c values are less than $4 \times 10^{-25} \ \Omega \cdot m$, or approximately 10^{17} times smaller than the resistivity of copper. In practice, these resistivities are considered to be zero.

Today, thousands of superconductors are known, and as Table 21.3 illustrates, the critical temperatures of recently discovered superconductors are substantially higher than initially thought possible. Two kinds of superconductors are recognized. The more recently identified ones are essentially ceramics with high critical temperatures, whereas superconducting materials such as those observed by Kamerlingh-Onnes are metals. If a room-temperature superconductor is ever identified, its effect on technology could be tremendous.

The value of T_c is sensitive to chemical composition, pressure, and molecular structure. Copper, silver, and gold, which are excellent conductors at room temperature, do not exhibit superconductivity.

One truly remarkable feature of superconductors is that once a current is set up in them, it persists *without any applied potential difference* (because $R = 0$). Steady currents have been observed to persist in superconducting loops for several years with no apparent decay!

An important and useful application of superconductivity is in the development of superconducting magnets, in which the magnitudes of the magnetic field are approximately ten times greater than those produced by the best normal electromagnets. Such superconducting magnets are being considered as a means of storing energy. Superconducting magnets are currently used in medical magnetic resonance imaging, or MRI, units, which produce high-quality images of internal organs without the need for excessive exposure of patients to x-rays or other harmful radiation.

TABLE 21.3 | Critical Temperatures for Various Superconductors

Material	T_c (K)
$HgBa_2Ca_2Cu_3O_8$	134
Tl—Ba—Ca—Cu—O	125
Bi—Sr—Ca—Cu—O	105
$YBa_2Cu_3O_7$	92
Nb_3Ge	23.2
Nb_3Sn	18.05
Nb	9.46
Pb	7.18
Hg	4.15
Sn	3.72
Al	1.19
Zn	0.88

A small permanent magnet levitated above a disk of the superconductor $YBa_2Cu_3O_7$, which is in liquid nitrogen at 77 K.

Courtesy of IBM Research Laboratory

21.4 | A Model for Electrical Conduction

In this section, we describe a classical model of electrical conduction in metals that was first proposed by Paul Drude (1863–1906) in 1900. This structural model leads to Ohm's law and shows that resistivity can be related to the motion of electrons in metals. Although the Drude model described here has limitations, it introduces concepts that are applied in more elaborate treatments.

Using the components of structural models introduced in Section 11.2, we can describe the Drude model as follows.

1. *A description of the physical components of the system:* Consider a conductor as a regular array of ionized atoms plus a collection of free electrons, which are sometimes called *conduction* electrons. The conduction electrons, although bound to their respective atoms when the atoms are not part of a solid, become free when the atoms condense into a solid.

2. *A description of where the components are located relative to one another and how they interact:* The conduction electrons fill the interior of the conductor. In the absence of an electric field, they move in random directions through the conductor. The situation is similar to the motion of gas molecules confined in a vessel. In fact, some scientists refer to conduction electrons in a metal as an *electron gas*. The conduction electrons experience no interaction with the array of ionized atoms except during a collision with one of those atoms.

3. *A description of the time evolution of the system:* When an electric field is applied to the conductor, the conduction electrons drift slowly in a direction opposite that of the electric field (Active Fig. 21.4), with an average drift speed v_d that is much smaller (typically 10^{-4} m/s) than their average speed between collisions

(typically 10^6 m/s). An electron's motion after a collision is independent of its motion before the collision. The kinetic energy acquired by the electrons in the electric field is transferred to the ionized atoms of the conductor when the electrons and atoms collide. The energy transferred to the atoms increases their vibrational energy, which causes the temperature of the conductor to increase.

4. *A description of the agreement between predictions of the model and actual observations and, possibly, predictions of new effects that have not yet been observed:* The test of Drude's model will be this: Can we generate an expression for the resistivity of the conductor that agrees with experimental observations?

We begin to answer the question in (4) above by deriving an expression for the drift velocity. When a free electron of mass m_e and charge q ($= -e$) is subjected to an electric field $\vec{\mathbf{E}}$, it experiences a force $\vec{\mathbf{F}} = q\vec{\mathbf{E}}$. The electron is a particle under a net force, and its acceleration can be found from Newton's second law, $\sum \vec{\mathbf{F}} = m\vec{\mathbf{a}}$:

$$\vec{\mathbf{a}} = \frac{\sum \vec{\mathbf{F}}}{m} = \frac{q\vec{\mathbf{E}}}{m_e} \qquad \text{21.13} \blacktriangleleft$$

Because the electric field is uniform, the electron's acceleration is constant, so the electron can be modeled as a particle under constant acceleration. If $\vec{\mathbf{v}}_i$ is the electron's initial velocity the instant after a collision (which occurs at a time defined as $t = 0$), the velocity of the electron at a very short time t later (immediately before the next collision occurs) is, from Equation 3.8,

$$\vec{\mathbf{v}}_f = \vec{\mathbf{v}}_i + \vec{\mathbf{a}}t = \vec{\mathbf{v}}_i + \frac{q\vec{\mathbf{E}}}{m_e}t \qquad \text{21.14} \blacktriangleleft$$

Let's now take the average value of $\vec{\mathbf{v}}_f$ for all the electrons in the wire over all possible collision times t and all possible values of $\vec{\mathbf{v}}_i$. Assuming the initial velocities are randomly distributed over all possible directions, the average value of $\vec{\mathbf{v}}_i$ is zero. The average value of the second term of Equation 21.14 is $(q\vec{\mathbf{E}}/m_e)\tau$, where τ is the *average time interval between successive collisions*. Because the average value of $\vec{\mathbf{v}}_f$ is equal to the drift velocity,

▶ Drift velocity in terms of
microscopic quantities

$$\vec{\mathbf{v}}_{f,\text{avg}} = \vec{\mathbf{v}}_d = \frac{q\vec{\mathbf{E}}}{m_e}\tau \qquad \text{21.15} \blacktriangleleft$$

Substituting the magnitude of this drift velocity (the drift speed) into Equation 21.4, we have

$$I = nev_d A = ne\left(\frac{eE}{m_e}\tau\right)A = \frac{ne^2E}{m_e}\tau A \qquad \text{21.16} \blacktriangleleft$$

According to Equation 21.6, the current is related to the macroscopic variables of potential difference and resistance:

$$I = \frac{\Delta V}{R}$$

Incorporating Equation 21.8, we can write this expression as

$$I = \frac{\Delta V}{\left(\rho\dfrac{\ell}{A}\right)} = \frac{\Delta V}{\rho\ell}A$$

In the conductor, the electric field is uniform, so we use Equation 20.6, $\Delta V = E\ell$, to substitute for the magnitude of the potential difference across the conductor:

$$I = \frac{E\ell}{\rho\ell}A = \frac{E}{\rho}A \qquad \text{21.17} \blacktriangleleft$$

Setting the two expressions for the current, Equations 21.16 and 21.17, equal, we solve for the resistivity:

$$I = \frac{ne^2E}{m_e}\tau A = \frac{E}{\rho}A \rightarrow \rho = \frac{m_e}{ne^2\tau} \qquad \textbf{21.18} \blacktriangleleft$$

▶ Resistivity in terms of microscopic parameters

According to this structural model, our prediction is that resistivity does not depend on the electric field or, equivalently, on the potential difference, but depends only on fixed parameters associated with the material and the electron. This feature is characteristic of a conductor obeying Ohm's law. The model shows that the resistivity can be calculated from a knowledge of the density of the electrons, their charge and mass, and the average time interval τ between collisions. This time interval is related to the average distance between collisions ℓ_{avg} (the *mean free path*) and the average speed v_{avg} through the expression[6]

$$\tau = \frac{\ell_{avg}}{v_{avg}} \qquad \textbf{21.19} \blacktriangleleft$$

EXAMPLE 21.3 | Electron Collisions in Copper

(A) Using the data from Example 21.1 and the structural model of electron conduction, estimate the average time interval between collisions for electrons in copper at 20°C.

SOLUTION

Conceptualize Imagine the conduction electrons moving in the conductor and making collisions with the array of ionized atoms. Because the speed of the electrons is high, we expect many collisions to occur per unit time interval, so the time interval between collisions should be short.

Categorize We will be using the results of our structural model, so we categorize this problem as a substitution problem.

Solve Equation 21.18 for the average time interval between collisions:

$$(1) \quad \tau = \frac{m_e}{ne^2\rho}$$

In Equation (1), ρ is the *resistivity* of the conductor. From Example 21.1, write the expression for the electron density in a conductor:

$$(2) \quad n = \frac{N_A\rho}{M}$$

In Equation (2), ρ is the *density* of the conductor and M is the molecular mass of the conductor. Substitute numerical values in Equation (2):

$$n = \frac{(6.022 \times 10^{23}\ \text{mol}^{-1})(8\ 920\ \text{kg/m}^3)}{0.063\ 5\ \text{kg/mol}} = 8.46 \times 10^{28}\ \text{m}^{-3}$$

Substitute this result and other numerical values into Equation (1):

$$\tau = \frac{9.109 \times 10^{-31}\ \text{kg}}{(8.46 \times 10^{28}\ \text{m}^{-3})(1.602 \times 10^{-19}\ \text{C})^2(1.7 \times 10^{-8}\ \Omega \cdot \text{m})}$$

$$= 2.5 \times 10^{-14}\ \text{s}$$

Note that this result is a very short time interval so that the electrons make a very large number of collisions per second.

(B) Assuming that the average speed for free electrons in copper is 1.6×10^6 m/s and using the result from part (A), calculate the mean free path for electrons in copper.

SOLUTION

Solve Equation 21.19 for the mean free path and substitute numerical values:

$$\ell_{avg} = v_{avg}\tau = (1.6 \times 10^6\ \text{m/s})(2.5 \times 10^{-14}\ \text{s}) = 4.0 \times 10^{-8}\ \text{m}$$

This result is equivalent to 40 nm (compared with atomic spacings of about 0.2 nm). Therefore, although the time interval between collisions is very short, the electrons travel about 200 atomic distances before colliding with an atom.

---

[6]Recall that the average speed of a group of particles depends on the temperature of the group (Chapter 16) and is not the same as the drift speed v_d.

Although this structural model of conduction is consistent with Ohm's law, it does not correctly predict the values of resistivity or the behavior of the resistivity with temperature. For example, the results of classical calculations for v_{avg} using the ideal gas model for the electrons are about a factor of ten smaller than the actual values, which results in incorrect predictions of values of resistivity from Equation 21.18. Furthermore, according to Equations 21.18 and 21.19, the resistivity is predicted to vary with temperature as does v_{avg}, which according to an ideal-gas model (Chapter 16, Eq. 16.22) is proportional to $\sqrt{T}$. This behavior is in disagreement with the linear dependence of resistivity with temperature for pure metals (Fig. 21.9). Because of these incorrect predictions, we must modify our structural model. We shall call the model that we have developed so far the *classical* model for electrical conduction. To account for the incorrect predictions of the classical model, we will develop it further into a *quantum mechanical* model, which we shall describe briefly.

We discussed two important simplification models in earlier chapters, the particle model and the wave model. Although we discussed these two simplification models separately, quantum physics tells us that this separation is not so clear-cut. As we shall discuss in detail in Chapter 28, particles have wave-like properties. The predictions of some models can only be matched to experimental results if the model includes the wave-like behavior of particles. The structural model for electrical conduction in metals is one of these cases.

Let us imagine that the electrons moving through the metal have wave-like properties. If the array of atoms in a conductor is regularly spaced (that is, periodic), the wave-like character of the electrons makes it possible for them to move freely through the conductor and a collision with an atom is unlikely. For an idealized conductor, no collisions would occur, the mean free path would be infinite, and the resistivity would be zero. Electrons are scattered only if the atomic arrangement is irregular (not periodic), as a result of structural defects or impurities, for example. At low temperatures, the resistivity of metals is dominated by scattering caused by collisions between the electrons and impurities. At high temperatures, the resistivity is dominated by scattering caused by collisions between the electrons and the atoms of the conductor, which are continuously displaced as a result of thermal agitation, destroying the perfect periodicity. The thermal motion of the atoms makes the structure irregular (compared with an atomic array at rest), thereby reducing the electron's mean free path.

Although it is beyond the scope of this text to show this modification in detail, the classical model modified with the wave-like character of the electrons results in predictions of resistivity values that are in agreement with measured values and predicts a linear temperature dependence. When discussing the hydrogen atom in Chapter 11, we had to introduce some quantum notions to understand experimental observations such as atomic spectra. Likewise, we had to introduce quantum notions in Chapter 17 to understand the temperature behavior of molar specific heats of gases. Here we have another case in which quantum physics is necessary for the model to agree with experiment. Although classical physics can explain a tremendous range of phenomena, we continue to see hints that quantum physics must be incorporated into our models. We shall study quantum physics in detail in Chapters 28 through 31.

21.5 | Energy and Power in Electric Circuits

In Section 21.1, we discussed the energy transformations occurring when current exists in a conductor. If a battery is used to establish an electric current in a conductor, there is a continuous transformation of chemical energy in the battery to kinetic energy of the electrons to internal energy in the conductor, resulting in an increase in the temperature of the conductor.

In typical electric circuits, energy is transferred from a source, such as a battery, to some device, such as a lightbulb or a radio receiver by electrical transmission (T_{ET} in Eq. 7.2). Let us determine an expression that will allow us to calculate the rate of this energy transfer. First, consider the simple circuit in Active Figure 21.11,

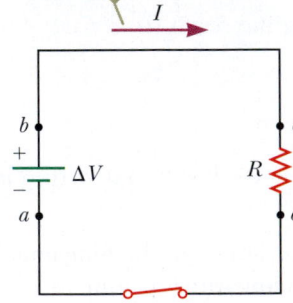

The direction of the effective flow of positive charge is clockwise.

Active Figure 21.11 A circuit consisting of a resistor of resistance R and a battery having a potential difference ΔV across its terminals.

where we imagine that energy is being delivered to a resistor. Because the connecting wires also have resistance, some energy is delivered to the wires and some energy to the resistor. Unless noted otherwise, we will adopt a simplification model in which the resistance of the wires is so small compared with the resistance of the circuit element that we ignore the energy delivered to the wires.

Let us now analyze the energetics of the circuit in which a battery is connected to a resistor of resistance R as in Active Figure 21.11. Imagine following a positive quantity of charge Q around the circuit from point a through the battery and resistor and back to a. Point a is a reference point at which the potential is defined as zero. We identify the entire circuit as our system. As the charge moves from a to b through the battery whose potential difference is ΔV, the electrical potential energy of the system increases by the amount $Q\,\Delta V$, whereas the chemical energy in the battery decreases by the same amount. (Recall from Chapter 20 that $\Delta U = q\,\Delta V$.) As the charge moves from c to d through the resistor, however, the system loses this electrical potential energy during collisions with atoms in the resistor. In this process, the energy is transformed to internal energy corresponding to increased vibrational motion of the atoms in the resistor. Because we have neglected the resistance of the interconnecting wires, no energy transformation occurs for paths bc and da. When the charge returns to point a, the net result is that some of the chemical energy in the battery has been delivered to the resistor and resides in the resistor as internal energy associated with molecular vibration.

The resistor is normally in contact with air, so its increased temperature results in a transfer of energy by heat into the air. In addition, there will be thermal radiation from the resistor, representing another means of escape for the energy. After some time interval has passed, the resistor remains at a constant temperature because the input of energy from the battery is balanced by the output of energy by heat and radiation. Some electrical devices include *heat sinks*[7] connected to parts of the circuit to prevent these parts from reaching dangerously high temperatures. Heat sinks are pieces of metal with many fins. The high thermal conductivity of the metal provides a rapid transfer of energy by heat away from the hot component and the large number of fins provides a large surface area in contact with the air, so energy can transfer by radiation and into the air by heat at a high rate.

Let us consider now the rate at which the system loses electric potential energy as the charge Q passes through the resistor:

$$\frac{dU}{dt} = \frac{d}{dt}(Q\Delta V) = \frac{dQ}{dt}\Delta V = I\Delta V$$

where I is the current in the circuit. Of course, the system regains this potential energy when the charge passes through the battery, at the expense of chemical energy in the battery. The rate at which the system loses potential energy as the charge passes through the resistor is equal to the rate at which the system gains internal energy in the resistor. Therefore, the **power** P, representing the rate at which energy is delivered to the resistor, is

$$P = I\Delta V \qquad\qquad \text{21.20} \blacktriangleleft \qquad \blacktriangleright \text{ Power delivered to a device}$$

We have developed this result by considering a battery delivering energy to a resistor. Equation 21.20, however, can be used to determine the power transferred from a voltage source to *any* device carrying a current I and having a potential difference ΔV between its terminals.

Using Equation 21.20 and that $\Delta V = IR$ for a resistor, we can express the power delivered to the resistor in the alternative forms

$$P = I^2 R = \frac{(\Delta V)^2}{R} \qquad\qquad \text{21.21} \blacktriangleleft$$

---

[7]This terminology is another misuse of the word *heat* that is ingrained in our common language.

Pitfall Prevention | 21.5
Misconceptions about Current
Several common misconceptions are associated with current in a circuit like that in Active Figure 21.11. One is that current comes out of one terminal of the battery and is then "used up" as it passes through the resistor. According to this approach, there is current in only one part of the circuit. The current is actually the same *everywhere* in the circuit. A related misconception has the current coming out of the resistor being smaller than that going in because some of the current is "used up." Yet another misconception has current coming out of both terminals of the battery, in opposite directions, and then "clashing" in the resistor, delivering the energy in this manner. We know that is not the case; charges flow in the same rotational sense at *all* points in the circuit.

Pitfall Prevention | 21.6
Charges Do Not Move All the Way Around a Circuit
Because of the very small magnitude of the drift velocity, it might take *hours* for a single electron to make one complete trip around the circuit. In terms of understanding the energy transfer in a circuit, however, it is useful to *imagine* a charge moving all the way around the circuit.

Pitfall Prevention | 21.7
Energy Is Not "Dissipated"
In some books, you may see Equation 21.21 described as the power "dissipated in" a resistor, suggesting that energy disappears. Instead, we say energy is "delivered to" a resistor. The notion of *dissipation* arises because a warm resistor expels energy by radiation and heat, and energy delivered by the battery leaves the circuit. (It does not disappear!)

The SI unit of power is the watt, introduced in Chapter 7. If you analyze the units in Equations 21.20 and 21.21, you will see that the result of the calculation provides a watt as the unit. The power delivered to a conductor of resistance R is often referred to as an I^2R *loss.*

As we learned in Section 7.6, the unit of energy your electric company uses to calculate energy transfer, the kilowatt-hour, is the amount of energy transferred in 1 h at the constant rate of 1 kW. We learned there that 1 kWh = 3.6×10^6 J.

> **QUICK QUIZ 21.4** For the two incandescent lightbulbs shown in Figure 21.12, rank the currents at points *a* through *f*, from greatest to least.

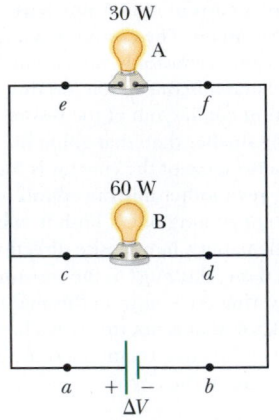

Figure 21.12 (Quick Quiz 21.4 and Thinking Physics 21.2) Two incandescent lightbulbs connected across the same potential difference.

> **THINKING PHYSICS 21.2**
>
> Two incandescent lightbulbs A and B are connected across the same potential difference as in Figure 21.12. The electric input powers to the lightbulbs are shown. Which lightbulb has the higher resistance? Which carries the greater current?
>
> **Reasoning** Because the voltage across each lightbulb is the same and the rate of energy delivered to a resistor is $P = (\Delta V)^2/R$, the lightbulb with the higher resistance exhibits the lower rate of energy transfer. In this case, the resistance of A is larger than that for B. Furthermore, because $P = I\,\Delta V$, we see that the current carried by B is larger than that of A. ◀

> **THINKING PHYSICS 21.3**
>
> When is an incandescent lightbulb more likely to fail, just after it is turned on or after it has been on for a while?
>
> **Reasoning** When the switch is closed, the source voltage is immediately applied across the lightbulb. As the voltage is applied across the cold filament when the lightbulb is first turned on, the resistance of the filament is low. Therefore, the current is high and a relatively large amount of energy is delivered to the bulb per unit time interval. This causes the temperature of the filament to rise rapidly, resulting in thermal stress on the filament that makes it likely to fail at that moment. As the filament warms up in the absence of failure, its resistance rises and the current falls. As a result, the rate of energy delivered to the lightbulb falls. The thermal stress on the filament is reduced so that the failure is less likely to occur after the bulb has been on for a while. ◀

Example 21.4 | Linking Electricity and Thermodynamics

An immersion heater must increase the temperature of 1.50 kg of water from 10.0°C to 50.0°C in 10.0 min while operating at 110 V.

(A) What is the required resistance of the heater?

SOLUTION

Conceptualize An immersion heater is a resistor that is inserted into a container of water. As energy is delivered to the immersion heater, raising its temperature, energy leaves the surface of the resistor by heat, going into the water. When the immersion heater reaches a constant temperature, the rate of energy delivered to the resistance by electrical transmission (T_{ET}) is equal to the rate of energy delivered by heat (Q) to the water.

Categorize This example allows us to link our new understanding of power in electricity with our experience with specific heat in thermodynamics (Chapter 17). The water is a nonisolated system. Its internal energy is rising because of energy transferred into the water by heat from the resistor: $\Delta E_{int} = Q$. In our model, we assume the energy that enters the water from the heater remains in the water.

21.4 cont.

Analyze To simplify the analysis, let's ignore the initial period during which the temperature of the resistor increases and also ignore any variation of resistance with temperature. Therefore, we imagine a constant rate of energy transfer for the entire 10.0 min.

Set the rate of energy delivered to the resistor equal to the rate of energy Q entering the water by heat:

$$P = \frac{(\Delta V)^2}{R} = \frac{Q}{\Delta t}$$

Use Equation 17.3, $Q = mc\,\Delta T$, to relate the energy input by heat to the resulting temperature change of the water and solve for the resistance:

$$\frac{(\Delta V)^2}{R} = \frac{mc\,\Delta T}{\Delta t} \rightarrow R = \frac{(\Delta V)^2\,\Delta t}{mc\,\Delta T}$$

Substitute the values given in the statement of the problem:

$$R = \frac{(110\text{ V})^2(600\text{ s})}{(1.50\text{ kg})(4\ 186\text{ J/kg}\cdot\text{°C})(50.0\text{°C} - 10.0\text{°C})} = \boxed{28.9\ \Omega}$$

(B) Estimate the cost of heating the water.

SOLUTION

Multiply the power by the time interval to find the amount of energy transferred:

$$T_{ET} = P\,\Delta t = \frac{(\Delta V)^2}{R}\,\Delta t = \frac{(110\text{ V})^2}{28.9\ \Omega}(10.0\text{ min})\left(\frac{1\text{ h}}{60.0\text{ min}}\right)$$

$$= 69.8\text{ Wh} = 0.069\ 8\text{ kWh}$$

Find the cost knowing that energy is purchased at an estimated price of 11¢ per kilowatt-hour:

$$\text{Cost} = (0.069\ 8\text{ kWh})(\$0.11/\text{kWh}) = \$0.008 = \boxed{0.8¢}$$

Finalize The cost to heat the water is very low, less than one cent. In reality, the cost is higher because some energy is transferred from the water into the surroundings by heat and electromagnetic radiation while its temperature is increasing. If you have electrical devices in your home with power ratings on them, use this power rating and an approximate time interval of use to estimate the cost for one use of the device.

21.6 | Sources of emf

The entity that maintains the constant voltage in Figure 21.13 is called a **source of emf.**[8] Sources of emf are any devices (such as batteries and generators) that increase the potential energy of a circuit system by maintaining a potential difference between points in the circuit while charges move through the circuit. One can think of a source of emf as a "charge pump." The emf $\mathcal{E}$ of a source describes the work done per unit charge, and hence the SI unit of emf is the volt.

At this point, you may wonder why we need to define a second quantity, emf, with the volt as a unit when we have already defined the potential difference. To see the need for this new quantity, consider the circuit shown in Figure 21.13, consisting of a battery connected to a resistor. We shall assume that the connecting wires have no resistance. We might be tempted to claim that the potential difference across the battery terminals (the terminal voltage) equals the emf of the battery. A real battery, however, always has some **internal resistance** r. As a result, the terminal voltage is not equal to the emf, as we shall show.

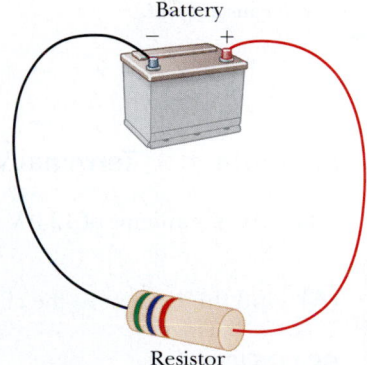

Battery

Figure 21.13 A circuit consisting of a resistor connected to the terminals of a battery.

Resistor

---

[8]The term *emf* was originally an abbreviation for *electromotive force*, but it is not a force, so the long form is discouraged. The name electromotive force was used early in the study of electricity before the understanding of batteries was as sophisticated as it is today.

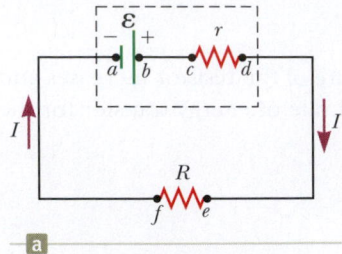

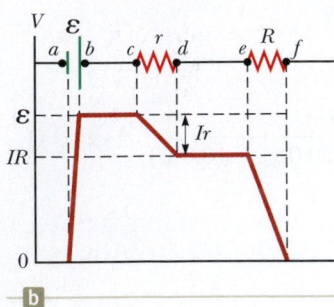

Active Figure 21.14 (a) Circuit diagram of a source of emf $\mathcal{E}$ (in this case, a battery) with internal resistance r, connected to an external resistor of resistance R. (b) Graphical representation showing how the potential changes as the circuit in (a) is traversed clockwise.

Pitfall Prevention | 21.8
What Is Constant in a Battery?
It is a common misconception that a battery is a source of constant current. Equation 21.24 shows that this is not true. The current in the circuit depends on the resistance R connected to the battery. It is also not true that a battery is a source of constant terminal voltage, as shown by Equation 21.22. **A battery is a source of constant emf.**

The circuit shown in Figure 21.13 can be described by the circuit diagram in Active Figure 21.14a. The battery within the dashed rectangle is modeled as an ideal, zero-resistance source of emf $\mathcal{E}$ in series with the internal resistance r. Now imagine moving from a to d in Active Figure 21.14a. As you pass from the negative to the positive terminal within the source of emf the potential increases by $\mathcal{E}$. As you move through the resistance r, however, the potential decreases by an amount Ir, where I is the current in the circuit. Therefore, the terminal voltage $\Delta V = V_d - V_a$ of the battery is[9]

$$\Delta V = \mathcal{E} - Ir \qquad \text{21.22} \blacktriangleleft$$

Note from this expression that $\mathcal{E}$ is equivalent to the **open-circuit voltage,** that is, the terminal voltage when the current is zero. Active Figure 21.14b is a graphical representation of the changes in potential as the circuit is traversed clockwise. By inspecting Active Figure 21.14a, we see that the terminal voltage ΔV must also equal the potential difference across the external resistance R, often called the **load resistance;** that is, $\Delta V = IR$. Combining this expression with Equation 21.22, we see that

$$\mathcal{E} = IR + Ir \qquad \text{21.23} \blacktriangleleft$$

Solving for the current gives

$$I = \frac{\mathcal{E}}{R + r} \qquad \text{21.24} \blacktriangleleft$$

which shows that the current in this simple circuit depends on both the resistance R external to the battery and the internal resistance r. If R is much greater than r, we can adopt a simplification model in which we neglect r in our analysis. In many circuits, we shall adopt this simplification model.

If we multiply Equation 21.23 by the current I, we have

$$I\mathcal{E} = I^2R + I^2r$$

This equation tells us that the total power output $I\mathcal{E}$ of the source of emf is equal to the rate I^2R at which energy is delivered to the load resistance plus the rate I^2r at which energy is delivered to the internal resistance. If $r \ll R$, much more of the energy from the battery is delivered to the load resistance than stays in the battery, although the amount of energy is relatively small because the load resistance is large, resulting in a small current. If $r \gg R$, a significant fraction of the energy from the source of emf stays in the battery package because it is delivered to the internal resistance. For example, if a wire is simply connected between the terminals of a flashlight battery, the battery becomes warm. This warming represents the transfer of energy from the source of emf to the internal resistance, where it appears as internal energy associated with temperature. Problem 21.73 in Enhanced WebAssign explores the conditions under which the largest amount of energy is transferred from the battery to the load resistor.

Example 21.5 | Terminal Voltage of a Battery

A battery has an emf of 12.0 V and an internal resistance of 0.050 0 Ω. Its terminals are connected to a load resistance of 3.00 Ω.

(A) Find the current in the circuit and the terminal voltage of the battery.

SOLUTION

Conceptualize Study Active Figure 21.14a, which shows a circuit consistent with the problem statement. The battery delivers energy to the load resistor.

Categorize This example involves simple calculations from this section, so we categorize it as a substitution problem.

---

[9]The terminal voltage in this case is less than the emf by the amount Ir. In some situations, the terminal voltage may *exceed* the emf by the amount Ir. Such a situation occurs when the direction of the current is *opposite* that of the emf, as when a battery is being charged by another source of emf.

21.5 *cont.*

Use Equation 21.24 to find the current in the circuit:

$$I = \frac{\mathcal{E}}{R + r} = \frac{12.0 \text{ V}}{3.00 \ \Omega + 0.050 \ 0 \ \Omega} = \boxed{3.93 \text{ A}}$$

Use Equation 21.22 to find the terminal voltage:

$$\Delta V = \mathcal{E} - Ir = 12.0 \text{ V} - (3.93 \text{ A})(0.050 \ 0 \ \Omega) = \boxed{11.8 \text{ V}}$$

To check this result, calculate the voltage across the load resistance R:

$$\Delta V = IR = (3.93 \text{ A})(3.00 \ \Omega) = 11.8 \text{ V}$$

(B) Calculate the power delivered to the load resistor, the power delivered to the internal resistance of the battery, and the power delivered by the battery.

SOLUTION

Use Equation 21.21 to find the power delivered to the load resistor:

$$P_R = I^2R = (3.93 \text{ A})^2(3.00 \ \Omega) = \boxed{46.3 \text{ W}}$$

Find the power delivered to the internal resistance:

$$P_r = I^2r = (3.93 \text{ A})^2(0.050 \ 0 \ \Omega) = \boxed{0.772 \text{ W}}$$

Find the power delivered by the battery by adding these quantities:

$$P = P_R + P_r = 46.3 \text{ W} + 0.772 \text{ W} = \boxed{47.1 \text{ W}}$$

What If? As a battery ages, its internal resistance increases. Suppose the internal resistance of this battery rises to $2.00 \ \Omega$ toward the end of its useful life. How does that alter the battery's ability to deliver energy?

Answer Let's connect the same 3.00-Ω load resistor to the battery.

Find the new current in the battery:

$$I = \frac{\mathcal{E}}{R + r} = \frac{12.0 \text{ V}}{3.00 \ \Omega + 2.00 \ \Omega} = 2.40 \text{ A}$$

Find the new terminal voltage:

$$\Delta V = \mathcal{E} - Ir = 12.0 \text{ V} - (2.40 \text{ A})(2.00 \ \Omega) = 7.2 \text{ V}$$

Find the new powers delivered to the load resistor and internal resistance:

$$P_R = I^2R = (2.40 \text{ A})^2(3.00 \ \Omega) = 17.3 \text{ W}$$
$$P_r = I^2r = (2.40 \text{ A})^2(2.00 \ \Omega) = 11.5 \text{ W}$$

The terminal voltage is only 60% of the emf. Notice that 40% of the power from the battery is delivered to the internal resistance when r is $2.00 \ \Omega$. When r is $0.050 \ 0 \ \Omega$ as in part (B), this percentage is only 1.6%. Consequently, even though the emf remains fixed, the increasing internal resistance of the battery significantly reduces the battery's ability to deliver energy to an external load.

21.7 | Resistors in Series and Parallel

When two or more resistors are connected together as are the incandescent lightbulbs in Active Figure 21.15a (page 618), they are said to be in a **series combination.** Active Figure 21.15b is the circuit diagram for the lightbulbs, shown as resistors, and the battery. In a series connection, if an amount of charge Q exits resistor R_1, charge Q must also enter the second resistor R_2. Otherwise, charge would accumulate on the wire between the resistors. Therefore, the same amount of charge passes through both resistors in a given time interval and the currents are the same in both resistors:

$$I = I_1 = I_2$$

where I is the current leaving the battery, I_1 is the current in resistor R_1, and I_2 is the current in resistor R_2.

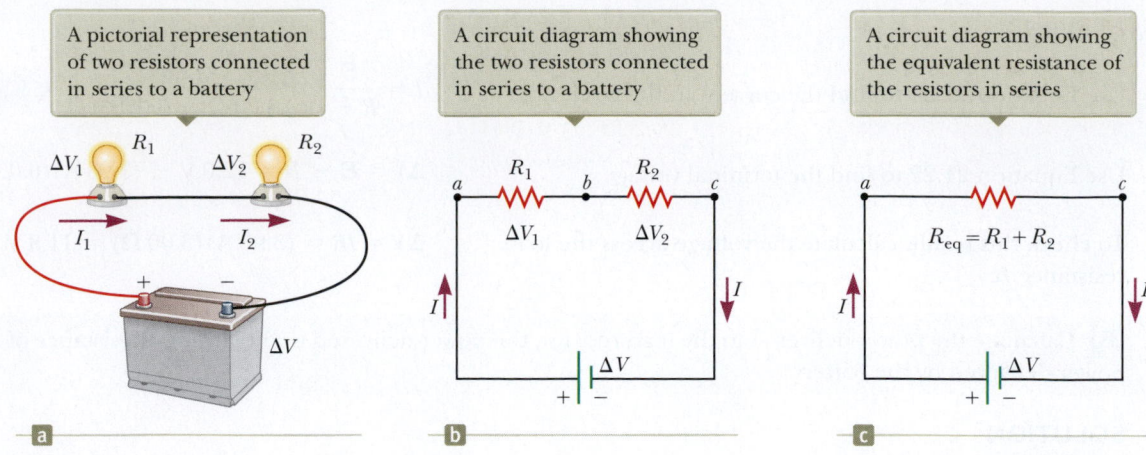

Active Figure 21.15 Two incandescent lightbulbs with resistances R_1 and R_2 connected in series. All three diagrams are equivalent.

The potential difference applied across the series combination of resistors divides between the resistors. In Active Figure 21.15b, because the voltage drop[10] from a to b equals $I_1 R_1$ and the voltage drop from b to c equals $I_2 R_2$, the voltage drop from a to c is

$$\Delta V = \Delta V_1 + \Delta V_2 = I_1 R_1 + I_2 R_2$$

The potential difference across the battery is also applied to the **equivalent resistance** R_{eq} in Active Figure 21.15c:

$$\Delta V = I R_{eq}$$

where the equivalent resistance has the same effect on the circuit as the series combination because it results in the same current I in the battery. Combining these equations for ΔV gives

$$\Delta V = I R_{eq} = I_1 R_1 + I_2 R_2 \quad \rightarrow \quad R_{eq} = R_1 + R_2 \qquad \textbf{21.25} \blacktriangleleft$$

where we have canceled the currents I, I_1, and I_2 because they are all the same. We see that we can replace the two resistors in series with a single equivalent resistance whose value is the *sum* of the individual resistances.

The equivalent resistance of three or more resistors connected in series is

$$R_{eq} = R_1 + R_2 + R_3 + \cdots \qquad \textbf{21.26} \blacktriangleleft$$

This relationship indicates that the equivalent resistance of a series combination of resistors is the numerical sum of the individual resistances and is always greater than any individual resistance.

Looking back at Equation 21.24, we see that the denominator of the right-hand side is the simple algebraic sum of the internal and external resistances. That is consistent with the internal and external resistances being in series in Active Figure 21.14a.

If the filament of one lightbulb in Active Figure 21.15a were to fail, the circuit would no longer be complete (resulting in an open-circuit condition) and the second lightbulb would also go out. This fact is a general feature of a series circuit: if one device in the series creates an open circuit, all devices are inoperative.

▶ The equivalent resistance of a series combination of resistors

Pitfall Prevention | 21.9
Lightbulbs Don't Burn
We will describe the end of the life of an incandescent lightbulb by saying *the filament fails* rather than by saying the lightbulb "burns out." The word *burn* suggests a combustion process, which is not what occurs in a lightbulb. The failure of a lightbulb results from the slow sublimation of tungsten from the very hot filament over the life of the lightbulb. The filament eventually becomes very thin because of this process. The mechanical stress from a sudden temperature increase when the lightbulb is turned on causes the thin filament to break.

---

[10]The term *voltage drop* is synonymous with a decrease in electric potential across a resistor. It is often used by individuals working with electric circuits.

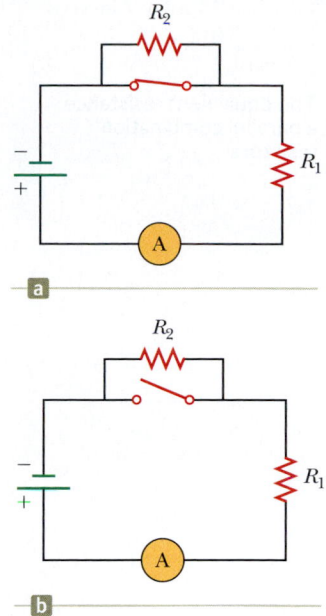

QUICK QUIZ 21.5 With the switch in the circuit of Figure 21.16a closed, there is no current in R_2 because the current has an alternate zero-resistance path through the switch. There is current in R_1, and this current is measured with the ammeter (a device for measuring current) at the bottom of the circuit. If the switch is opened (Figure 21.16b), there is current in R_2. What happens to the reading on the ammeter when the switch is opened? (**a**) The reading goes up. (**b**) The reading goes down. (**c**) The reading does not change.

Now consider two resistors in a **parallel combination** as shown in Active Figure 21.17. Notice that both resistors are connected directly across the terminals of the battery. Therefore, the potential differences across the resistors are the same:

$$\Delta V = \Delta V_1 = \Delta V_2$$

where ΔV is the terminal voltage of the battery.

When charges reach point a in Active Figure 21.17b, they split into two parts, with some going toward R_1 and the rest going toward R_2. A **junction** is any such point in a circuit where a current can split. This split results in less current in each individual resistor than the current leaving the battery. Because electric charge is conserved, the current I that enters point a must equal the total current leaving that point:

$$I = I_1 + I_2 = \frac{\Delta V_1}{R_1} + \frac{\Delta V_2}{R_2}$$

where I_1 is the current in R_1 and I_2 is the current in R_2.

The current in the **equivalent resistance** R_{eq} in Active Figure 21.17c is

$$I = \frac{\Delta V}{R_{eq}}$$

where the equivalent resistance has the same effect on the circuit as the two resistors in parallel; that is, the equivalent resistance draws the same current I from the battery. Combining these equations for I, we see that the equivalent resistance of two resistors in parallel is given by

$$I = \frac{\Delta V}{R_{eq}} = \frac{\Delta V_1}{R_1} + \frac{\Delta V_2}{R_2} \quad \rightarrow \quad \frac{1}{R_{eq}} = \frac{1}{R_1} + \frac{1}{R_2} \qquad \text{21.27} \blacktriangleleft$$

where we have canceled ΔV, ΔV_1, and ΔV_2 because they are all the same.

Figure 21.16 (Quick Quiz 21.5) What happens when the switch is opened?

Pitfall Prevention | 21.10
Local and Global Changes
A local change in one part of a circuit may result in a global change throughout the circuit. For example, if a single resistor is changed in a circuit containing several resistors and batteries, the currents in all resistors and batteries, the terminal voltages of all batteries, and the voltages across all resistors may change as a result.

Pitfall Prevention | 21.11
Current Does Not Take the Path of Least Resistance
You may have heard the phrase "current takes the path of least resistance" (or similar wording) in reference to a parallel combination of current paths such that there are two or more paths for the current to take. Such wording is incorrect. The current takes *all* paths. Those paths with lower resistance have larger currents, but even very high resistance paths carry *some* of the current. In theory, if current has a choice between a zero-resistance path and a finite resistance path, all the current takes the path of zero resistance; a path with zero resistance, however, is an idealization.

A pictorial representation of two resistors connected in parallel to a battery	A circuit diagram showing the two resistors connected in parallel to a battery	A circuit diagram showing the equivalent resistance of the resistors in parallel

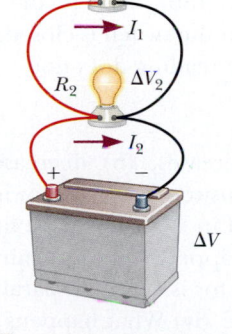

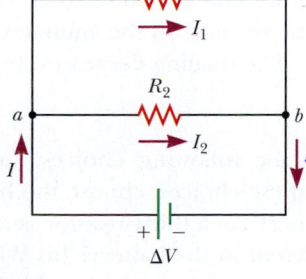

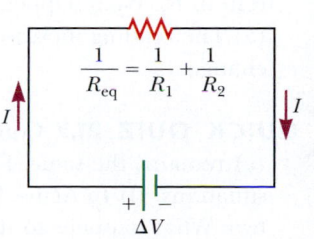

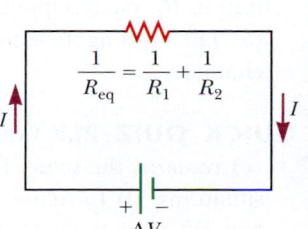

Active Figure 21.17 Two incandescent lightbulbs with resistances R_1 and R_2 connected in parallel. All three diagrams are equivalent.

An extension of this analysis to three or more resistors in parallel gives

◀ The equivalent resistance of a parallel combination of resistors

$$\frac{1}{R_{eq}} = \frac{1}{R_1} + \frac{1}{R_2} + \frac{1}{R_3} + \cdots$$

21.28 ◀

This expression shows that the inverse of the equivalent resistance of two or more resistors in a parallel combination is equal to the sum of the inverses of the individual resistances. Furthermore, the equivalent resistance is always less than the smallest resistance in the group.

A circuit consisting of resistors can often be reduced to a simple circuit containing only one resistor. To do so, examine the initial circuit and replace any resistors in series or any in parallel with equivalent resistances using Equations 21.26 and 21.28. Draw a sketch of the new circuit after these changes have been made. Examine the new circuit and replace any new series or parallel combinations that now exist. Continue this process until a single equivalent resistance is found for the entire circuit. (That may not be possible; if not, see the techniques of Section 21.8.)

If the current in or the potential difference across a resistor in the initial circuit is to be found, start with the final circuit and gradually work your way back through the equivalent circuits. Find currents and voltages across resistors using $\Delta V = IR$ and your understanding of series and parallel combinations.

Household circuits are always wired so that the electrical devices are connected in parallel as in Active Figure 21.17a. In this manner, each device operates independently of the others so that if one is switched off, the others remain on. For example, if one of the lightbulbs in Active Figure 21.17a were removed from its socket, the other would continue to operate. Equally important, each device operates on the same voltage. If devices were connected in series, the voltage applied to the combination would divide among the devices, so the voltage applied to any one device would depend on how many devices were in the combination.

In many household circuits, circuit breakers are used in series with other circuit elements for safety purposes. A circuit breaker is designed to switch off and open the circuit at some maximum current (typically 15 A or 20 A) whose value depends on the nature of the circuit. If a circuit breaker were not used, large currents caused by turning on many devices could result in excessive temperatures in wires and, perhaps, cause a fire. In older home construction, fuses were used in place of circuit breakers. When the current in a circuit exceeds some value, the conductor in a fuse melts and opens the circuit. The disadvantage of fuses is that they are destroyed in the process of opening the circuit, whereas circuit breakers can be reset.

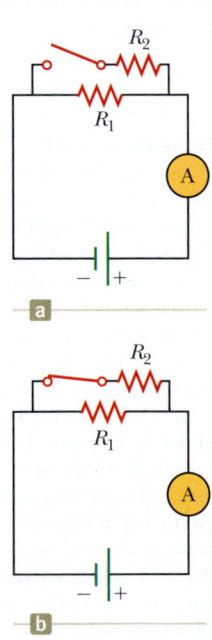

Figure 21.18 (Quick Quiz 21.6) What happens when the switch is closed?

◀ **QUICK QUIZ 21.6** With the switch in the circuit of Figure 21.18a open, there is no current in R_2. There is current in R_1, however, and it is measured with the ammeter at the right side of the circuit. If the switch is closed (Fig. 21.18b), there is current in R_2. What happens to the reading on the ammeter when the switch is closed? (a) The reading increases. (b) The reading decreases. (c) The reading does not change.

◀ **QUICK QUIZ 21.7** Consider the following choices: (a) increases, (b) decreases, (c) remains the same. From these choices, choose the best answer for the following situations. (i) In Active Figure 21.15, a third resistor is added in series with the first two. What happens to the current in the battery? (ii) What happens to the terminal voltage of the battery? (iii) In Active Figure 21.17, a third resistor is added in parallel with the first two. What happens to the current in the battery? (iv) What happens to the terminal voltage of the battery?

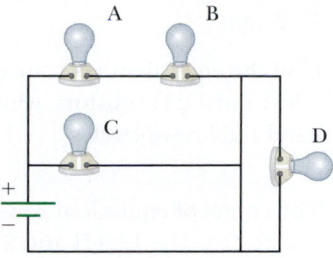

THINKING PHYSICS 21.4

Compare the brightnesses of the four identical lightbulbs in Figure 21.19. What happens if bulb A fails so that it cannot conduct? What if bulb C fails? What if bulb D fails?

Reasoning Bulbs A and B are connected in series across the battery, whereas bulb C is connected by itself. Therefore, the terminal voltage of the battery is split between bulbs A and B. As a result, bulb C will be brighter than bulbs A and B, which should be equally as bright as each other. Bulb D has a wire connected across it. Therefore, there is no potential difference across bulb D and it does not glow at all. If bulb A fails, bulb B goes out but bulb C stays lit. If bulb C fails, there is no effect on the other bulbs. If bulb D fails, the event is undetectable because bulb D was not glowing initially. ◄

Figure 21.19 (Thinking Physics 21.4) What happens to the lightbulbs if one fails?

THINKING PHYSICS 21.5

Figure 21.20 illustrates how a three-way incandescent lightbulb is constructed to provide three levels of light intensity. The socket of the lamp is equipped with a three-way switch for selecting different light intensities. The bulb contains two filaments. Why are the filaments connected in parallel? Explain how the two filaments are used to provide three different light intensities.

Reasoning If the filaments were connected in series and one of them were to fail, there would be no current in the bulb and the bulb would give no illumination, regardless of the switch position. When the filaments are connected in parallel, however, and one of them (say the 75-W filament) fails, the bulb still operates in some switch positions because there is current in the other (100-W) filament. The three light intensities are made possible by selecting one of three values of filament resistance, using a single value of 120 V for the applied voltage. The 75-W filament offers one value of resistance, the 100-W filament offers a second value, and the third resistance is obtained by combining the two filaments in parallel. When switch S_1 is closed and switch S_2 is opened, only the 75-W filament carries current. When switch S_1 is open and switch S_2 is closed, only the 100-W filament carries current. When both switches are closed, both filaments carry current, and a total illumination corresponding to 175 W is obtained. ◄

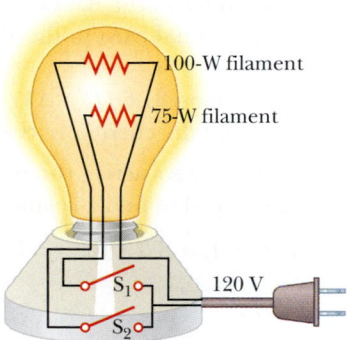

Figure 21.20 (Thinking Physics 21.5) A three-way incandescent lightbulb.

Example **21.6** | **Find the Equivalent Resistance**

Four resistors are connected as shown in Figure 21.21a.

(A) Find the equivalent resistance between points a and c.

SOLUTION

Conceptualize Imagine charges flowing into this combination from the left. All charges must pass through the first two resistors, but the charges split into two different paths when encountering the combination of the 6.0-Ω and the 3.0-Ω resistors.

Categorize Because of the simple nature of the combination of resistors in Figure 21.21, we categorize this example as one for which we can use the rules for series and parallel combinations of resistors.

....

Analyze The combination of resistors can be reduced in steps as shown in Figure 21.21.

Find the equivalent resistance between a and b of the 8.0-Ω and 4.0-Ω resistors, which are in series (left-hand red-brown circles):

$$R_{eq} = 8.0 \ \Omega + 4.0 \ \Omega = 12.0 \ \Omega$$

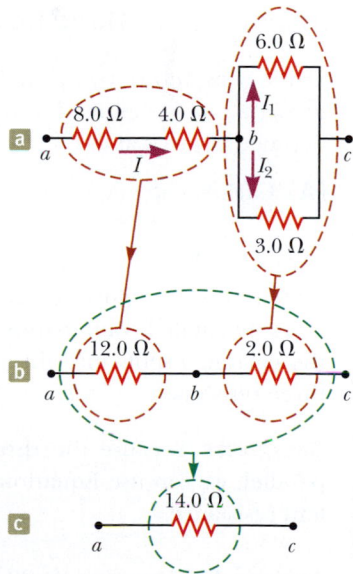

Figure 21.21 (Example 21.6) The original network of resistors is reduced to a single equivalent resistance.

continued

21.6 *cont.*

Find the equivalent resistance between *b* and *c* of the 6.0-Ω and 3.0-Ω resistors, which are in parallel (right-hand red-brown circles):

$$\frac{1}{R_{eq}} = \frac{1}{6.0 \text{ } \Omega} + \frac{1}{3.0 \text{ } \Omega} = \frac{3}{6.0 \text{ } \Omega}$$

$$R_{eq} = \frac{6.0 \text{ } \Omega}{3} = 2.0 \text{ } \Omega$$

The circuit of equivalent resistances now looks like Figure 21.21b. The 12.0-Ω and 2.0-Ω resistors are in series (green circles). Find the equivalent resistance from *a* to *c*:

$$R_{eq} = 12.0 \text{ } \Omega + 2.0 \text{ } \Omega = \boxed{14.0 \text{ } \Omega}$$

This resistance is that of the single equivalent resistor in Figure 21.21c.

(B) What is the current in each resistor if a potential difference of 42 V is maintained between *a* and *c*?

SOLUTION

The currents in the 8.0-Ω and 4.0-Ω resistors are the same because they are in series. In addition, they carry the same current that would exist in the 14.0-Ω equivalent resistor subject to the 42-V potential difference.

Use Equation 21.6 ($R = \Delta V/I$) and the result from part (A) to find the current in the 8.0-Ω and 4.0-Ω resistors:

$$I = \frac{\Delta V_{ac}}{R_{eq}} = \frac{42 \text{ V}}{14.0 \text{ } \Omega} = \boxed{3.0 \text{ A}}$$

Set the voltages across the resistors in parallel in Figure 21.21a equal to find a relationship between the currents:

$$\Delta V_1 = \Delta V_2 \quad \rightarrow \quad (6.0 \text{ } \Omega) I_1 = (3.0 \text{ } \Omega) I_2 \quad \rightarrow \quad I_2 = 2I_1$$

Use $I_1 + I_2 = 3.0$ A to find I_1:

$$I_1 + I_2 = 3.0 \text{ A} \quad \rightarrow \quad I_1 + 2I_1 = 3.0 \text{ A} \quad \rightarrow \quad I_1 = \boxed{1.0 \text{ A}}$$

Find I_2:

$$I_2 = 2I_1 = 2(1.0 \text{ A}) = \boxed{2.0 \text{ A}}$$

Finalize As a final check of our results, note that $\Delta V_{bc} = (6.0 \text{ } \Omega) I_1 = (3.0 \text{ } \Omega) I_2 = 6.0$ V and $\Delta V_{ab} = (12.0 \text{ } \Omega) I = 36$ V; therefore, $\Delta V_{ac} = \Delta V_{ab} + \Delta V_{bc} = 42$ V, as it must.

Example **21.7** | **Three Resistors in Parallel**

Three resistors are connected in parallel as shown in Figure 21.22a. A potential difference of 18.0 V is maintained between points *a* and *b*.

(A) Calculate the equivalent resistance of the circuit.

SOLUTION

Conceptualize Figure 21.22a shows that we are dealing with a simple parallel combination of three resistors. Notice that the current *I* splits into three currents I_1, I_2, and I_3 in the three resistors.

Categorize Because the three resistors are connected in parallel, we can use Equation 21.28 to evaluate the equivalent resistance.

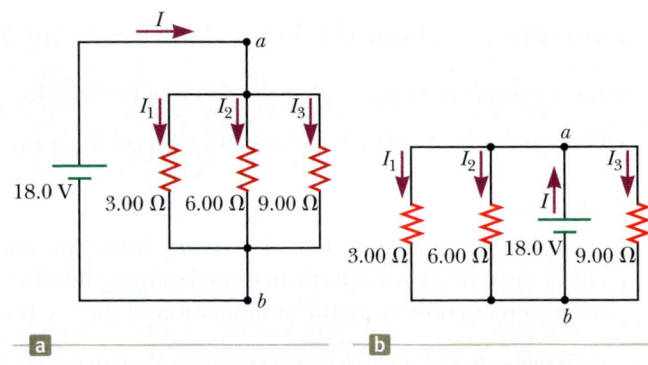

Figure 21.22 (Example 21.7) (a) Three resistors connected in parallel. The voltage across each resistor is 18.0 V. (b) Another circuit with three resistors and a battery. Is it equivalent to the circuit in (a)?

Analyze Use Equation 21.28 to find R_{eq}:

$$\frac{1}{R_{eq}} = \frac{1}{3.00 \text{ } \Omega} + \frac{1}{6.00 \text{ } \Omega} + \frac{1}{9.00 \text{ } \Omega} = \frac{11.0}{18.0 \text{ } \Omega}$$

$$R_{eq} = \frac{18.0 \text{ } \Omega}{11.0} = \boxed{1.64 \text{ } \Omega}$$

21.7 *cont.*

(B) Find the current in each resistor.

SOLUTION

The potential difference across each resistor is 18.0 V. Apply the relationship $\Delta V = IR$ to find the currents:

$$I_1 = \frac{\Delta V}{R_1} = \frac{18.0\text{ V}}{3.00\ \Omega} = \boxed{6.00\text{ A}}$$

$$I_2 = \frac{\Delta V}{R_2} = \frac{18.0\text{ V}}{6.00\ \Omega} = \boxed{3.00\text{ A}}$$

$$I_3 = \frac{\Delta V}{R_3} = \frac{18.0\text{ V}}{9.00\ \Omega} = \boxed{2.00\text{ A}}$$

(C) Calculate the power delivered to each resistor and the total power delivered to the combination of resistors.

SOLUTION

Apply the relationship $P = I^2R$ to each resistor using the currents calculated in part (B):

3.00-Ω: $P_1 = I_1{}^2R_1 = (6.00\text{ A})^2(3.00\ \Omega) = \boxed{108\text{ W}}$

6.00-Ω: $P_2 = I_2{}^2R_2 = (3.00\text{ A})^2(6.00\ \Omega) = \boxed{54\text{ W}}$

9.00-Ω: $P_3 = I_3{}^2R_3 = (2.00\text{ A})^2(9.00\ \Omega) = \boxed{36\text{ W}}$

Finalize Part (C) shows that the smallest resistor receives the most power. Summing the three quantities gives a total power of 198 W. We could have calculated this final result from part (A) by considering the equivalent resistance as follows: $P = (\Delta V)^2/R_{\text{eq}} = (18.0\text{ V})^2/1.64\ \Omega = 198\text{ W}$.

What If? What if the circuit were as shown in Figure 21.22b instead of as in Figure 21.22a? How would that affect the calculation?

Answer There would be no effect on the calculation. The physical placement of the battery is not important. Only the electrical arrangement is important. In Figure 21.22b, the battery still maintains a potential difference of 18.0 V between points *a* and *b,* so the two circuits in the figure are electrically identical.

21.8 | Kirchhoff's Rules

As we saw in the preceding section, combinations of resistors can be simplified and analyzed using the expression $\Delta V = IR$ and the rules for series and parallel combinations of resistors. Very often, however, it is not possible to reduce a circuit to a single loop using these rules. The procedure for analyzing more complex circuits is made possible by using the two following principles, called **Kirchhoff's rules.**

1. **Junction rule.** At any junction, the sum of the currents must equal zero:

$$\sum_{\text{junction}} I = 0 \qquad \text{21.29} \blacktriangleleft$$

2. **Loop rule.** The sum of the potential differences across all elements around any closed circuit loop must be zero:

$$\sum_{\text{loop}} \Delta V = 0 \qquad \text{21.30} \blacktriangleleft$$

Kirchhoff's first rule is a statement of **conservation of electric charge.** All charges that enter a given point in a circuit must leave that point because charge cannot build up at a point. Currents directed into the junction are entered into the junction rule as $+I$, whereas currents directed out of a junction are entered as $-I$. Applying this rule to the junction in Figure 21.23a gives

$$I_1 - I_2 - I_3 = 0$$

Figure 21.23b represents a mechanical analog of this situation, in which water flows through a branched pipe having no leaks. Because water does not build up anywhere

The amount of charge flowing out of the branches on the right must equal the amount flowing into the single branch on the left.

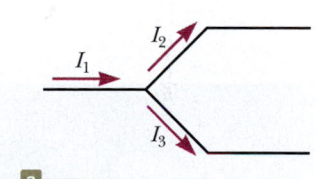

The amount of water flowing out of the branches on the right must equal the amount flowing into the single branch on the left.

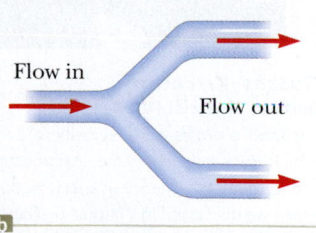

Figure 21.23 (a) Kirchhoff's junction rule. (b) A mechanical analog of the junction rule.

In each diagram, $\Delta V = V_b - V_a$ and the circuit element is traversed from a to b, left to right.

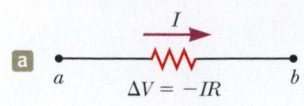

$$\Delta V = -IR$$

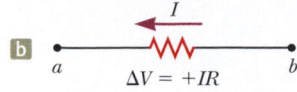

$$\Delta V = +IR$$

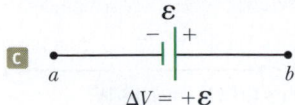

$$\Delta V = +\mathcal{E}$$

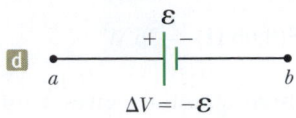

$$\Delta V = -\mathcal{E}$$

Figure 21.24 Rules for determining the potential differences across a resistor and a battery. (The battery is assumed to have no internal resistance.)

in the pipe, the flow rate into the pipe on the left equals the total flow rate out of the two branches on the right.

Kirchhoff's second rule follows from the law of **conservation of energy.** Let's imagine moving a charge around a closed loop of a circuit. When the charge returns to the starting point, the charge–circuit system must have the same total energy as it had before the charge was moved. The sum of the increases in energy as the charge passes through some circuit elements must equal the sum of the decreases in energy as it passes through other elements. The potential energy decreases whenever the charge moves through a potential drop $-IR$ across a resistor or whenever it moves in the reverse direction through a source of emf. The potential energy increases whenever the charge passes through a battery from the negative terminal to the positive terminal.

When applying Kirchhoff's second rule, imagine *traveling* around the loop and consider changes in *electric potential* rather than the changes in *potential energy* described in the preceding paragraph. Imagine traveling through the circuit elements in Figure 21.24 toward the right. The following sign conventions apply when using the second rule:

- Charges move from the high-potential end of a resistor toward the low-potential end, so if a resistor is traversed in the direction of the current, the potential difference ΔV across the resistor is $-IR$ (Fig. 21.24a).
- If a resistor is traversed in the direction *opposite* the current, the potential difference ΔV across the resistor is $+IR$ (Fig. 21.24b).
- If a source of emf (assumed to have zero internal resistance) is traversed in the direction of the emf (from negative to positive), the potential difference ΔV is $+\mathcal{E}$ (Fig. 21.24c).
- If a source of emf (assumed to have zero internal resistance) is traversed in the direction opposite the emf (from positive to negative), the potential difference ΔV is $-\mathcal{E}$ (Fig. 21.24d).

There are limits on the number of times you can usefully apply Kirchhoff's rules in analyzing a circuit. You can use the junction rule as often as you need as long as you include in it a current that has not been used in a preceding junction-rule equation. In general, the number of times you can use the junction rule is one fewer than the number of junction points in the circuit. You can apply the loop rule as often as needed as long as a new circuit element (resistor or battery) or a new current appears in each new equation. In general, to solve a particular circuit problem, the number of independent equations you need to obtain from the two rules equals the number of unknown currents.

Gustav Kirchhoff
German Physicist (1824–1887)
Kirchhoff, a professor at Heidelberg, and Robert Bunsen invented the spectroscope and founded the science of spectroscopy, which we discussed in Chapter 11. They discovered the elements cesium and rubidium and invented astronomical spectroscopy.

> **PROBLEM-SOLVING STRATEGY: Kirchhoff's Rules**

The following procedure is recommended for solving problems that involve circuits that cannot be reduced by the rules for combining resistors in series or parallel.

1. **Conceptualize** Study the circuit diagram and make sure you recognize all elements in the circuit. Identify the polarity of each battery and try to imagine the directions in which the current would exist in the batteries.

2. **Categorize** Determine whether the circuit can be reduced by means of combining series and parallel resistors. If so, use the techniques of Section 21.7. If not, apply Kirchhoff's rules according to the *Analyze* step below.

3. **Analyze** Assign labels to all known quantities and symbols to all unknown quantities. You must assign *directions* to the currents in each part of the circuit. Although the assignment of current directions is arbitrary, you must adhere *rigorously* to the directions you assign when you apply Kirchhoff's rules.

Apply the junction rule (Kirchhoff's first rule) to all junctions in the circuit except one. Now apply the loop rule (Kirchhoff's second rule) to as many loops in the circuit as are needed to obtain, in combination with the equations from the junction rule, as many equations as there are unknowns. To apply this rule, you must choose a direction in which to travel around the loop (either clockwise or counterclockwise) and correctly identify the change in potential as you cross each element. Be careful with signs!

Solve the equations simultaneously for the unknown quantities.

4. Finalize Check your numerical answers for consistency. Do not be alarmed if any of the resulting currents have a negative value. That only means you have guessed the direction of that current incorrectly, but *its magnitude will be correct.*

Example 21.8 | A Multiloop Circuit

Find the currents I_1, I_2, and I_3 in the circuit shown in Figure 21.25.

SOLUTION

Conceptualize Imagine physically rearranging the circuit while keeping it electrically the same. Can you rearrange it so that it consists of simple series or parallel combinations of resistors? You should find that you cannot.

Categorize We cannot simplify the circuit by the rules associated with combining resistances in series and in parallel. (If the 10.0-V battery were removed and replaced by a wire from b to the 6.0-Ω resistor, we could reduce the remaining circuit.) Because the circuit is not a simple series and parallel combination of resistances, this problem is one in which we must use Kirchhoff's rules.

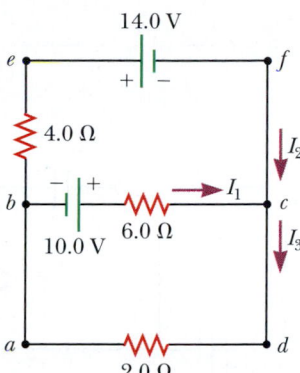

Figure 21.25 (Example 21.8) A circuit containing different branches.

Analyze We arbitrarily choose the directions of the currents as labeled in Figure 21.25.

Apply Kirchhoff's junction rule to junction c:

(1) $\quad I_1 + I_2 - I_3 = 0$

We now have one equation with three unknowns: I_1, I_2, and I_3. There are three loops in the circuit: *abcda*, *befcb*, and *aefda*. We need only two loop equations to determine the unknown currents. (The third equation for loop *aefda* would give no new information.) Let's choose to traverse these loops in the clockwise direction. Apply Kirchhoff's loop rule to loops *abcda* and *befcb*:

abcda: (2) $\quad 10.0\ \text{V} - (6.0\ \Omega)I_1 - (2.0\ \Omega)I_3 = 0$

befcb: $\quad -(4.0\ \Omega)I_2 - 14.0\ \text{V} + (6.0\ \Omega)I_1 - 10.0\ \text{V} = 0$

(3) $\quad -24.0\ \text{V} + (6.0\ \Omega)I_1 - (4.0\ \Omega)I_2 = 0$

Solve Equation (1) for I_3 and substitute into Equation (2):

$10.0\ \text{V} - (6.0\ \Omega)I_1 - (2.0\ \Omega)(I_1 + I_2) = 0$

(4) $\quad 10.0\ \text{V} - (8.0\ \Omega)I_1 - (2.0\ \Omega)I_2 = 0$

Multiply each term in Equation (3) by 4 and each term in Equation (4) by 3:

(5) $\quad -96.0\ \text{V} + (24.0\ \Omega)I_1 - (16.0\ \Omega)I_2 = 0$

(6) $\quad 30.0\ \text{V} - (24.0\ \Omega)I_1 - (6.0\ \Omega)I_2 = 0$

Add Equation (6) to Equation (5) to eliminate I_1 and find I_2:

$-66.0\ \text{V} - (22.0\ \Omega)I_2 = 0$

$I_2 = \boxed{-3.0\ \text{A}}$

Use this value of I_2 in Equation (3) to find I_1:

$-24.0\ \text{V} + (6.0\ \Omega)I_1 - (4.0\ \Omega)(-3.0\ \text{A}) = 0$

$-24.0\ \text{V} + (6.0\ \Omega)I_1 + 12.0\ \text{V} = 0$

$I_1 = \boxed{2.0\ \text{A}}$

Use Equation (1) to find I_3:

$I_3 = I_1 + I_2 = 2.0\ \text{A} - 3.0\ \text{A} = \boxed{-1.0\ \text{A}}$

Finalize Because our values for I_2 and I_3 are negative, the directions of these currents are opposite those indicated in Figure 21.25. The numerical values for the currents are correct. Despite the incorrect direction, we *must* continue to use these negative values in subsequent calculations because our equations were established with our original choice of direction. What would have happened had we left the current directions as labeled in Figure 21.25 but traversed the loops in the opposite direction?

‹21.9 | *RC* Circuits

So far, we have analyzed direct-current circuits in which the current is constant. In DC circuits containing capacitors, the current is always in the same direction but may vary in time. A circuit containing a series combination of a resistor and a capacitor is called an ***RC* circuit.**

Charging a Capacitor

Active Figure 21.26 shows a simple series *RC* circuit. Let's assume the capacitor in this circuit is initially uncharged. There is no current while the switch is open (Active Fig. 21.26a). If the switch is thrown to position *a* at *t* = 0 (Active Fig. 21.26b), however, charge begins to flow, setting up a current in the circuit, and the capacitor begins to charge.[11] Notice that during charging, charges do not jump across the capacitor plates because the gap between the plates represents an open circuit. Instead, charge is transferred between each plate and its connecting wires due to the electric field established in the wires by the battery until the capacitor is fully charged. As the plates are being charged, the potential difference across the capacitor increases. The value of the maximum charge on the plates depends on the voltage of the battery. Once the maximum charge is reached, the current in the circuit is zero because the potential difference across the capacitor matches that supplied by the battery.

To analyze this circuit quantitatively, let's apply Kirchhoff's loop rule to the circuit after the switch is thrown to position *a*. Traversing the loop in Active Figure 21.26b clockwise gives

$$\mathcal{E} - \frac{q}{C} - IR = 0 \qquad \qquad \textbf{21.31}◀$$

where q/C is the potential difference across the capacitor and IR is the potential difference across the resistor. We have used the sign conventions discussed earlier for the signs on $\mathcal{E}$ and IR. The capacitor is traversed in the direction from the positive plate to the negative plate, which represents a decrease in potential. Therefore, we use a negative sign for this potential difference in Equation 21.31. Note that q and I are *instantaneous* values that depend on time (as opposed to steady-state values) as the capacitor is being charged.

We can use Equation 21.31 to find the initial current in the circuit and the maximum charge on the capacitor. At the instant the switch is thrown to position *a* (t = 0), the charge on the capacitor is zero. Equation 21.31 shows that the initial current I_i in the circuit is a maximum and is given by

$$I_i = \frac{\mathcal{E}}{R} \quad \text{(current at } t = 0) \qquad \qquad \textbf{21.32}◀$$

At this time, the potential difference from the battery terminals appears entirely across the resistor. Later, when the capacitor is charged to its maximum value Q, charges cease to flow, the current in the circuit is zero, and the potential difference from the battery terminals appears entirely across the capacitor. Substituting I = 0 into Equation 21.31 gives the maximum charge on the capacitor:

$$Q = C\mathcal{E} \quad \text{(maximum charge)} \qquad \qquad \textbf{21.33}◀$$

To determine analytical expressions for the time dependence of the charge and current, we must solve Equation 21.31, a single equation containing two variables q and I. The current in all parts of the series circuit must be the same. Therefore,

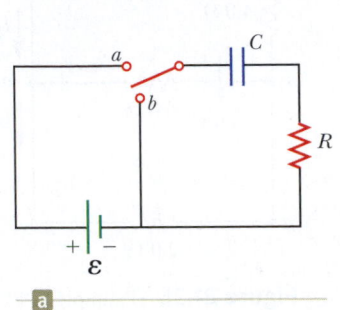

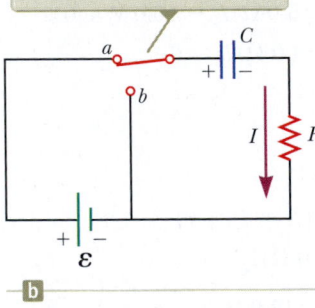
When the switch is thrown to position *a*, the capacitor begins to charge up.

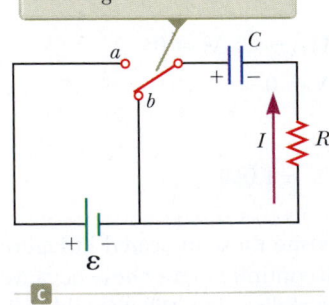
When the switch is thrown to position *b*, the capacitor discharges.

Active Figure 21.26 A capacitor in series with a resistor, switch, and battery.

---

[11]In previous discussions of capacitors, we assumed a steady-state situation, in which no current was present in any branch of the circuit containing a capacitor. Now we are considering the case *before* the steady-state condition is realized; in this situation, charges are moving and a current exists in the wires connected to the capacitor.

the current in the resistance R must be the same as the current between each capacitor plate and the wire connected to it. This current is equal to the time rate of change of the charge on the capacitor plates. Therefore, we substitute $I = dq/dt$ into Equation 21.31 and rearrange the equation:

$$\frac{dq}{dt} = \frac{\mathcal{E}}{R} - \frac{q}{RC}$$

To find an expression for q, we solve this separable differential equation as follows. First combine the terms on the right-hand side:

$$\frac{dq}{dt} = \frac{C\mathcal{E}}{RC} - \frac{q}{RC} = -\frac{q - C\mathcal{E}}{RC}$$

Multiply this equation by dt and divide by $q - C\mathcal{E}$:

$$\frac{dq}{q - C\mathcal{E}} = -\frac{1}{RC} dt$$

Integrate this expression, using $q = 0$ at $t = 0$:

$$\int_0^q \frac{dq}{q - C\mathcal{E}} = -\frac{1}{RC} \int_0^t dt$$

$$\ln\left(\frac{q - C\mathcal{E}}{-C\mathcal{E}}\right) = -\frac{t}{RC}$$

From the definition of the natural logarithm, we can write this expression as

$$q(t) = C\mathcal{E}(1 - e^{-t/RC}) = Q(1 - e^{-t/RC}) \qquad \textbf{21.34} \blacktriangleleft$$

▶ Charge as a function of time for a capacitor being charged

where e is the base of the natural logarithm and we have made the substitution from Equation 21.33.

We can find an expression for the charging current by differentiating Equation 21.34 with respect to time. Using $I = dq/dt$, we find that

$$I(t) = \frac{\mathcal{E}}{R} e^{-t/RC} \qquad \textbf{21.35} \blacktriangleleft$$

▶ Current as a function of time for a capacitor being charged

Plots of capacitor charge and circuit current versus time are shown in Figure 21.27. Notice that the charge is zero at $t = 0$ and approaches the maximum value $C\mathcal{E}$ as $t \to \infty$. The current has its maximum value $I_i = \mathcal{E}/R$ at $t = 0$ and decays exponentially to zero as $t \to \infty$. The quantity RC, which appears in the exponents of Equations 21.34 and 21.35, is called the **time constant** τ of the circuit:

$$\tau = RC \qquad \textbf{21.36} \blacktriangleleft$$

The time constant represents the time interval during which the current decreases to $1/e$ of its initial value; that is, after a time interval τ, the current decreases to $I = e^{-1}I_i = 0.368I_i$. After a time interval 2τ, the current decreases to $I = e^{-2}I_i = 0.135I_i$, and so forth. Likewise, in a time interval τ, the charge increases from zero to $C\mathcal{E}[1 - e^{-1}] = 0.632C\mathcal{E}$.

The energy supplied by the battery during the time interval required to fully charge the capacitor is $Q\mathcal{E} = C\mathcal{E}^2$. After the capacitor is fully charged, the energy stored in the capacitor is $\frac{1}{2}Q\mathcal{E} = \frac{1}{2}C\mathcal{E}^2$, which is only half the energy output of the battery. It is left as a problem (Problem 21.68 in Enhanced WebAssign) to show that the remaining half of the energy supplied by the battery appears as internal energy in the resistor.

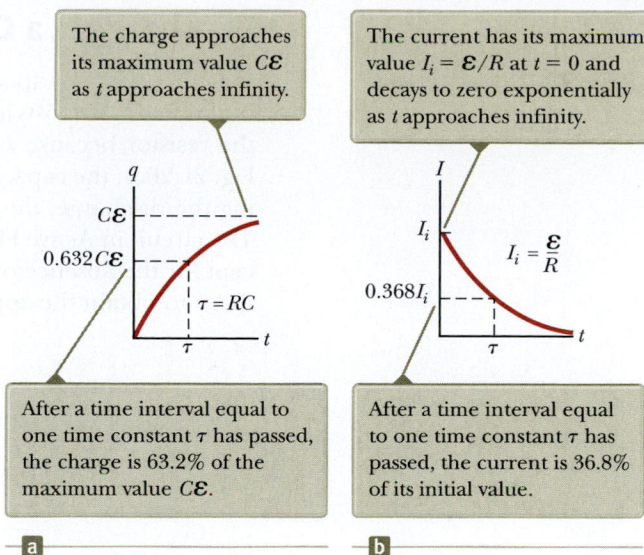

The charge approaches its maximum value $C\mathcal{E}$ as t approaches infinity.

The current has its maximum value $I_i = \mathcal{E}/R$ at $t = 0$ and decays to zero exponentially as t approaches infinity.

After a time interval equal to one time constant τ has passed, the charge is 63.2% of the maximum value $C\mathcal{E}$.

After a time interval equal to one time constant τ has passed, the current is 36.8% of its initial value.

Figure 21.27 (a) Plot of capacitor charge versus time for the circuit shown in Active Figure 21.26b. (b) Plot of current versus time for the circuit shown in Active Figure 21.26b.

Discharging a Capacitor

Imagine that the capacitor in Active Figure 21.26b is completely charged. A potential difference Q/C exists across the capacitor, and there is zero potential difference across the resistor because $I = 0$. If the switch is now thrown to position b at $t = 0$ (Active Fig. 21.26c), the capacitor begins to discharge through the resistor. At some time t during the discharge, the current in the circuit is I and the charge on the capacitor is q. The circuit in Active Figure 21.26c is the same as the circuit in Active Figure 21.26b except for the absence of the battery. Therefore, we eliminate the emf $\mathcal{E}$ from Equation 21.31 to obtain the appropriate loop equation for the circuit in Active Figure 21.26c:

$$-\frac{q}{C} - IR = 0 \qquad \textbf{21.37}\blacktriangleleft$$

When we substitute $I = dq/dt$ into this expression, it becomes

$$-R\frac{dq}{dt} = \frac{q}{C}$$

$$\frac{dq}{q} = -\frac{1}{RC}\,dt$$

Integrating this expression using $q = Q$ at $t = 0$ gives

$$\int_Q^q \frac{dq}{q} = -\frac{1}{RC}\int_0^t dt$$

$$\ln\left(\frac{q}{Q}\right) = -\frac{t}{RC}$$

▶ Charge as a function of time for a discharging capacitor

$$q(t) = Qe^{-t/RC} \qquad \textbf{21.38}\blacktriangleleft$$

Differentiating Equation 21.38 with respect to time gives the instantaneous current as a function of time:

▶ Current as a function of time for a discharging capacitor

$$I(t) = -\frac{Q}{RC}\,e^{-t/RC} \qquad \textbf{21.39}\blacktriangleleft$$

where $Q/RC = I_i$ is the initial current. The negative sign indicates that as the capacitor discharges, the current direction is opposite its direction when the capacitor was being charged. (Compare the current directions in Active Figs. 21.26b and 21.26c.) Both the charge on the capacitor and the current decay exponentially at a rate characterized by the time constant $\tau = RC$.

BIO Cable theory for propagation of an action potential along a nerve

In Section 20.7, we discussed the modeling of a patch of cell membrane as a capacitor. Let us call the capacitance of a given patch of membrane C_m. We also discussed the flow of ions through various ion channels and ion pumps in the membrane. This flow represents a current. The ions cannot move across the membrane unimpeded, so there is a resistance to the current, called the *membrane resistance* R_m. As a result, each small patch of the cell membrane can be modeled as an RC circuit as shown in Figure 21.28.

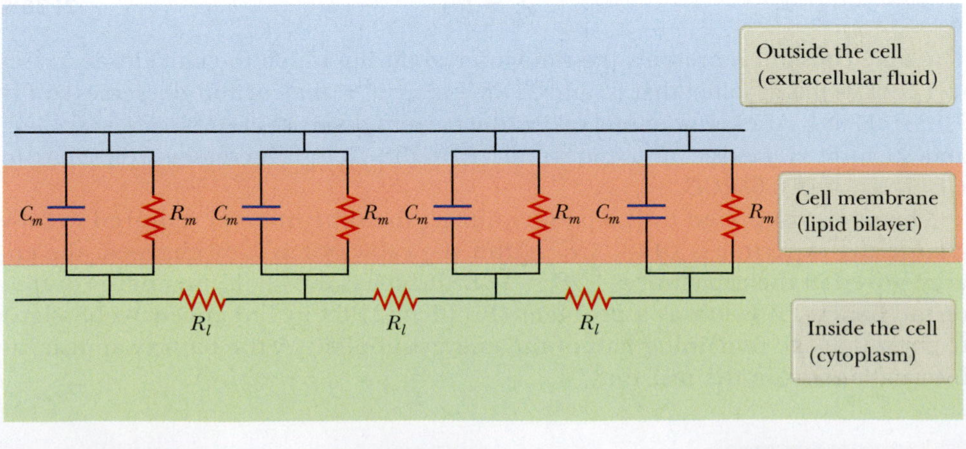

Figure 21.28 Modeling the cell membrane of a neuron using cable theory. Four small patches of the cell membrane are shown, with each patch being modeled electrically as an RC circuit consisting of resistance R_m and capacitance C_m. Adjacent patches are connected electrically by a resistance R_l in the cytoplasm of the cell interior.

A given long structure in a neuron (such as a dendrite or an axon) can be modeled as a series of *RC* circuit modules connected by a longitudinal resistance as shown in Figure 21.28. The *longitudinal resistance R_l* represents resistance to the current along the axis of the neuron through the cytoplasm. This model of a neuron can be analyzed using *cable theory*, first used by Kelvin in the 1850s to analyze the decay of signals in underwater telegraphic cables. In a neuron, we consider the decay of the propagation of an action potential along the neuron.

Using cable theory, we can model the propagation of an action potential along a nerve cell and relate this model to the transfer of information within the human nervous system. The propagation of the action potential is governed by two primary parameters: the time constant and the length constant. The *time constant $\tau = R_m C_m$* for the *RC* circuit associated with each patch of membrane is similar to the time constant discussed above and determines how rapidly the membrane capacitor can charge and discharge. For a given input at a point on the neuron, the membrane voltage along the neuron decays exponentially. The *length constant $\lambda = (R_m/R_l)^{1/2}$* determines a characteristic length along the neuron through which the voltage decays to e^{-1} of its original value. Together, these two parameters describe how efficient the neuron is at transmitting a signal along its length.

The axons of some nerves are wrapped with sections of *myelin*, with each section separated from the next by intervals called the *nodes of Ranvier*. The myelin has the effect of shutting off the transfer of ions across the cell membrane. As a result, the relatively slow patch-to-patch propagation of an action potential as described above does not occur. Instead, the signal is carried primarily within the cell interior, such that an action potential at one node rapidly causes another action potential at the next node. As a result, the signal travels much faster along the neuron, in a process called *saltatory conduction*.

Some diseases cause damage to the myelin sheath around nerve cells, degrading the process of saltatory conduction. As a result, patients suffering from these diseases experience impaired movement, due to the slowness of signals traveling to the muscles. For example, *transverse myelitis* is an autoimmune disease, in which the body attacks the spinal cord, with the resulting inflammation damaging the myelin. In severe cases, patients are left wheelchair-bound and require assistance with daily activities. If the damage to the myelin occurs within the white matter of the brain, the disease is called *multiple sclerosis,* a highly debilitating disease.

BIO The role of myelin in nerve conduction

> **QUICK QUIZ 21.8** Consider the circuit in Figure 21.29 and assume the battery has no internal resistance. (**i**) Just after the switch is closed, what is the current in the battery? (a) 0 (b) $\mathcal{E}/2R$ (c) $2\mathcal{E}/R$ (d) $\mathcal{E}/R$ (e) impossible to determine (**ii**) After a very long time, what is the current in the battery? Choose from the same choices.

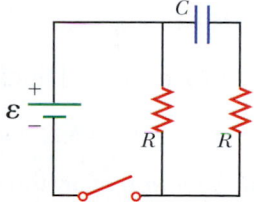

Figure 21.29 (Quick Quiz 21.8) How does the current vary after the switch is closed?

> ### THINKING PHYSICS 21.6
>
> Many roadway construction sites have flashing yellow lights to warn motorists of possible dangers. What causes the lightbulbs to flash?
>
> **Reasoning** A typical circuit for such a flasher is shown in Figure 21.30. The lamp *L* is a gas-filled lamp that acts as an open circuit until a large potential difference causes an electrical discharge in the gas, which gives off a bright light. During this discharge, charges flow through the gas between the electrodes of the lamp. After switch S is closed, the battery charges up the capacitor of capacitance *C*. At the beginning, the current is high and the charge on the capacitor is low, so most of the potential difference appears across the resistance *R*. As the capacitor charges, more potential difference appears across it, reflecting the lower current and therefore lower potential difference across the resistor. Eventually, the potential difference across the capacitor reaches a value at which the lamp will conduct, causing a flash. This discharges the capacitor through the lamp and the process of charging begins again. The period between flashes can be adjusted by changing the time constant of the *RC* circuit. ◄

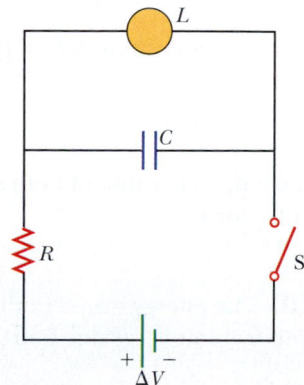

Figure 21.30 (Thinking Physics 21.6) The *RC* circuit in a roadway construction flasher. When the switch is closed, the charge on the capacitor increases until the voltage across the capacitor (and across the flash lamp) is high enough for the lamp to flash, discharging the capacitor.

Example 21.9 | **Charging a Capacitor in an *RC* Circuit**

An uncharged capacitor and a resistor are connected in series to a battery as shown in Active Figure 21.26, where $\mathcal{E} = 12.0$ V, $C = 5.00$ μF, and $R = 8.00 \times 10^5$ Ω. The switch is thrown to position a. Find the time constant of the circuit, the maximum charge on the capacitor, the maximum current in the circuit, and the charge and current as functions of time.

SOLUTION

Conceptualize Study Active Figure 21.26 and imagine throwing the switch to position a as shown in Active Figure 21.26b. Upon doing so, the capacitor begins to charge.

Categorize We evaluate our results using equations developed in this section, so we categorize this example as a substitution problem.

Evaluate the time constant of the circuit from Equation 21.36:

$$\tau = RC = (8.00 \times 10^5 \ \Omega)(5.00 \times 10^{-6} \ \text{F}) = \boxed{4.00 \ \text{s}}$$

Evaluate the maximum charge on the capacitor from Equation 21.33:

$$Q = C\mathcal{E} = (5.00 \ \mu\text{F})(12.0 \ \text{V}) = \boxed{60.0 \ \mu\text{C}}$$

Evaluate the maximum current in the circuit from Equation 21.32:

$$I_i = \frac{\mathcal{E}}{R} = \frac{12.0 \ \text{V}}{8.00 \times 10^5 \ \Omega} = \boxed{15.0 \ \mu\text{A}}$$

Use these values in Equations 21.34 and 21.35 to find the charge and current as functions of time:

$$(1) \quad q(t) = \boxed{60.0(1 - e^{-t/4.00})}$$

$$(2) \quad I(t) = \boxed{15.0e^{-t/4.00}}$$

In Equations (1) and (2), q is in microcoulombs, I is in microamperes, and t is in seconds.

Example 21.10 | **Discharging a Capacitor in an *RC* Circuit**

Consider a capacitor of capacitance C that is being discharged through a resistor of resistance R as shown in Active Figure 21.26c.

(A) After how many time constants is the charge on the capacitor one-fourth its initial value?

SOLUTION

Conceptualize Study Active Figure 21.26 and imagine throwing the switch to position b as shown in Active Figure 21.26c. Upon doing so, the capacitor begins to discharge.

Categorize We categorize the example as one involving a discharging capacitor and use the appropriate equations.

Analyze Substitute $q(t) = Q/4$ into Equation 21.38:

$$\frac{Q}{4} = Qe^{-t/RC}$$

$$\frac{1}{4} = e^{-t/RC}$$

Take the logarithm of both sides of the equation and solve for t:

$$-\ln 4 = -\frac{t}{RC}$$

$$t = RC \ln 4 = 1.39RC = \boxed{1.39\tau}$$

(B) The energy stored in the capacitor decreases with time as the capacitor discharges. After how many time constants is this stored energy one-fourth its initial value?

SOLUTION

Use Equations 20.29 and 21.38 to express the energy stored in the capacitor at any time t:

$$(1) \quad U(t) = \frac{q^2}{2C} = \frac{Q^2}{2C} e^{-2t/RC}$$

Substitute $U(t) = \frac{1}{4}(Q^2/2C)$ into Equation (1):

$$\frac{1}{4}\frac{Q^2}{2C} = \frac{Q^2}{2C} e^{-2t/RC}$$

$$\frac{1}{4} = e^{-2t/RC}$$

21.10 *cont.*

Take the logarithm of both sides of the equation and solve for t:

$$-\ln 4 = -\frac{2t}{RC}$$

$$t = \tfrac{1}{2}RC \ln 4 = 0.693RC = \boxed{0.693\tau}$$

Finalize Notice that because the energy depends on the square of the charge, the energy in the capacitor drops more rapidly than the charge on the capacitor.

21.10 | Context Connection: The Atmosphere as a Conductor

When discussing capacitors with air between the plates in Chapter 20, we adopted the simplification model that air was a perfect insulator. Although that was a good model for typical potential differences encountered in capacitors, we know that it is possible for a current to exist in air. Lightning is a dramatic example of this possibility, but a more mundane example is the common spark that you might receive upon bringing your finger near a doorknob after rubbing your feet across a carpet.

Let us analyze the process that occurs in electrical discharge, which is the same for lightning and the doorknob spark except for the size of the current. Whenever a strong electric field exists in air, it is possible for the air to undergo electrical breakdown in which the effective resistivity of the air drops dramatically and the air becomes a conductor. At any given time, due to cosmic ray collisions and other events, air contains a number of ionized molecules (Fig. 21.31a). For a relatively weak electric field, such as the fair-weather electric field, these ions and freed electrons accelerate slowly due to the electric force. They collide with other molecules with no effect and eventually neutralize as a freed electron ultimately finds an ion and combines with it. In a strong electric field such as that associated with a thunderstorm, however, the freed electrons can accelerate to very high speeds (Fig. 21.31b) before making a collision with a molecule (Fig. 21.31c). If the field is strong enough, the electron may have enough energy to ionize the molecule in this collision (Fig. 21.31d). Now there are two electrons to be accelerated by the field, and each can strike another molecule at high speed (Fig. 21.31e). The result is a very rapid increase in the number of charge carriers available in the air and a corresponding decrease in resistance of the air. Therefore, there can be a large current in the air that tends to neutralize the charges that established the initial potential difference, such as the charges in the cloud and on the ground. When that happens, we have lightning.

Typical currents during lightning strikes can be very high. While the stepped leader is making its way toward the ground, the current is relatively modest, in the range of 200 to 300 A. This current is large compared with typical household currents but small compared with peak currents in lightning discharges. Once the connection is made between the stepped leader and the return stroke, the current rises rapidly to a typical value of 5×10^4 A. Considering that typical potential differences between cloud and ground in a thunderstorm can be measured in hundreds of thousands of volts, the power during a lightning stroke is measured in billions of watts. Much of the energy in the stroke is delivered to the air, resulting in a rapid temperature increase and the resultant flash of light and sound of thunder.

Even in the absence of a thundercloud, there is a flow of charge through the air. The ions in the air make the air a conductor, although not a very good one. Atmospheric measurements indicate a typical potential difference across our atmospheric capacitor (Section 20.11) of about 3×10^5 V. As we shall show in the Context Conclusion, the total resistance of the air between the plates in the atmospheric capacitor is about 300 Ω. Therefore, the average fair-weather current in the air is

$$I = \frac{\Delta V}{R} = \frac{3 \times 10^5 \text{ V}}{300 \ \Omega} \approx 1 \times 10^3 \text{ A}$$

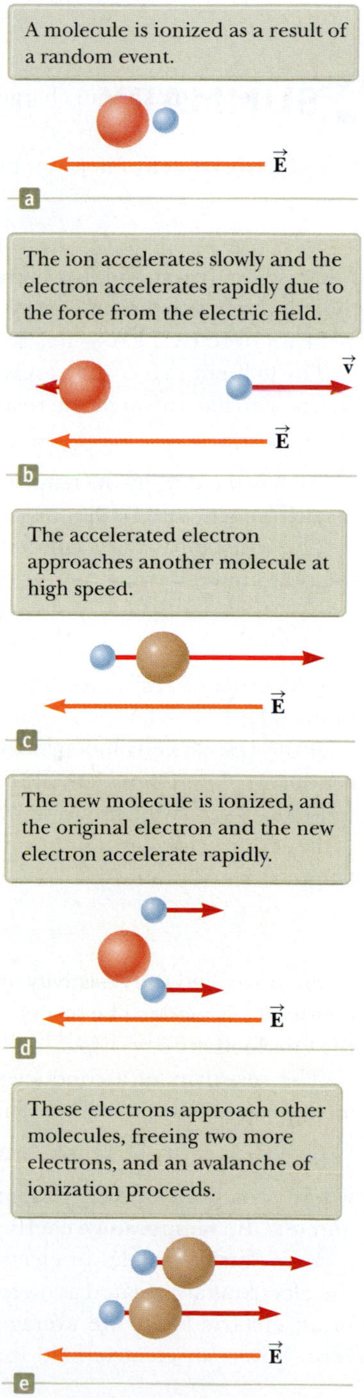

A molecule is ionized as a result of a random event.

The ion accelerates slowly and the electron accelerates rapidly due to the force from the electric field.

The accelerated electron approaches another molecule at high speed.

The new molecule is ionized, and the original electron and the new electron accelerate rapidly.

These electrons approach other molecules, freeing two more electrons, and an avalanche of ionization proceeds.

Figure 21.31 The anatomy of a spark.

A number of simplifying assumptions were made in these calculations, but this result is on the right order of magnitude for the global current. Although the result might seem surprisingly large, remember that this current is spread out over the entire surface area of the Earth. Therefore, the average fair-weather current density is

$$J = \frac{I}{A} = \frac{I}{4\pi R_E^2} = \frac{1 \times 10^3 \,\text{A}}{4\pi(6.4 \times 10^6 \,\text{m})^2} \approx 2 \times 10^{-12} \,\text{A/m}^2$$

In comparison, the current density in a lightning strike is on the order of $10^5 \,\text{A/m}^2$.

The fair-weather current and the lightning current are in opposite directions. The fair-weather current delivers positive charge to the ground, whereas lightning delivers negative charge. These two effects are in balance,[12] which is the principle that we shall use to estimate the average number of lightning strikes on the Earth in the Context Conclusion.

⟩ SUMMARY

The **electric current** I in a conductor is defined as

$$I \equiv \frac{dQ}{dt} \qquad \textbf{21.2}◀$$

where dQ is the charge that passes through a cross-section of the conductor in the time interval dt. The SI unit of current is the ampere (A); $1 \,\text{A} = 1 \,\text{C/s}$.

The current in a conductor is related to the motion of the charge carriers through the relationship

$$I_{\text{avg}} = nqv_d A \qquad \textbf{21.4}◀$$

where n is the density of charge carriers, q is their charge, v_d is the **drift speed**, and A is the cross-sectional area of the conductor.

The **resistance** R of a conductor is defined as the ratio of the potential difference across the conductor to the current:

$$R \equiv \frac{\Delta V}{I} \qquad \textbf{21.6}◀$$

The SI units of resistance are volts per ampere, defined as ohms (Ω); $1 \,\Omega = 1 \,\text{V/A}$.

If the resistance is independent of the applied voltage, the conductor obeys **Ohm's law,** and conductors that have a constant resistance over a wide range of voltages are said to be **ohmic.**

If a conductor has a uniform cross-sectional area A and a length ℓ, its resistance is

$$R = \rho \frac{\ell}{A} \qquad \textbf{21.8}◀$$

where ρ is called the **resistivity** of the material from which the conductor is made. The inverse of the resistivity is defined as the **conductivity** $\sigma = 1/\rho$.

The resistivity of a conductor varies with temperature in an approximately linear fashion; that is,

$$\rho = \rho_0[1 + \alpha(T - T_0)] \qquad \textbf{21.10}◀$$

where ρ_0 is the resistivity at some reference temperature T_0 and α is the **temperature coefficient of resistivity.**

In a classical model of electronic conduction in a metal, the electrons are treated as molecules of a gas. In the absence of an electric field, the average velocity of the electrons is zero. When an electric field is applied, the electrons move (on the average) with a drift velocity $\vec{v}_d$, given by

$$\vec{v}_d = \frac{q\vec{E}}{m_e}\tau \qquad \textbf{21.15}◀$$

where τ is the average time interval between collisions with the atoms of the metal. The resistivity of the material according to this model is

$$\rho = \frac{m_e}{ne^2\tau} \qquad \textbf{21.18}◀$$

where n is the number of free electrons per unit volume.

If a potential difference ΔV is maintained across a circuit element, the **power,** or the rate at which energy is delivered to the circuit element, is

$$P = I\,\Delta V \qquad \textbf{21.20}◀$$

Because the potential difference across a resistor is $\Delta V = IR$, we can express the power delivered to a resistor in the form

$$P = I^2 R = \frac{(\Delta V)^2}{R} \qquad \textbf{21.21}◀$$

The **emf** of a battery is the voltage across its terminals when the current is zero. Because of the voltage drop across the **internal resistance** r of a battery, the **terminal voltage** of the battery is less than the emf when a current exists in the battery.

The **equivalent resistance** of a set of resistors connected in **series** is

$$R_{\text{eq}} = R_1 + R_2 + R_3 + \cdots \qquad \textbf{21.26}◀$$

The **equivalent resistance** of a set of resistors connected in **parallel** is given by

$$\frac{1}{R_{\text{eq}}} = \frac{1}{R_1} + \frac{1}{R_2} + \frac{1}{R_3} + \cdots \qquad \textbf{21.28}◀$$

Circuits involving more than one loop are analyzed using two simple rules called **Kirchhoff's rules:**

- At any junction, the sum of the currents must equal zero:

$$\sum_{\text{junction}} I = 0 \qquad \textbf{21.29}◀$$

---

[12]There are a number of other effects, too, but we will adopt a simplification model in which these are the only two effects. For more information, see E. A. Bering, A. A. Few, and J. R. Benbrook, "The Global Electric Circuit," *Physics Today*, October 1998, pp. 24–30.

• The sum of the potential differences across each element around any closed circuit loop must be zero:

$$\sum_{\text{loop}} \Delta V = 0 \qquad \textbf{21.30} \blacktriangleleft$$

For the junction rule, current in a direction into a junction is $+I$, whereas current with a direction away from a junction is $-I$.

For the loop rule, when a resistor is traversed in the direction of the current, the change in potential ΔV across the resistor is $-IR$. If a resistor is traversed in the direction opposite the current, $\Delta V = +IR$.

If a source of emf is traversed in the direction of the emf (negative to positive), the change in potential is $+\mathcal{E}$. If it is traversed opposite the emf (positive to negative), the change in potential is $-\mathcal{E}$.

If a capacitor is charged with a battery of emf $\mathcal{E}$ through a resistance R, the charge on the capacitor and the current in the circuit vary in time according to the expressions

$$q(t) = Q(1 - e^{-t/RC}) \qquad \textbf{21.34} \blacktriangleleft$$

$$I(t) = \frac{\mathcal{E}}{R} e^{-t/RC} \qquad \textbf{21.35} \blacktriangleleft$$

where $Q = C\mathcal{E}$ is the maximum charge on the capacitor. The product RC is called the **time constant** of the circuit.

If a charged capacitor is discharged through a resistance R, the charge and current decrease exponentially in time according to the expressions

$$q(t) = Qe^{-t/RC} \qquad \textbf{21.38} \blacktriangleleft$$

$$I(t) = -\frac{Q}{RC} e^{-t/RC} \qquad \textbf{21.39} \blacktriangleleft$$

where Q is the initial charge on the capacitor.

❯ OBJECTIVE QUESTIONS |

☐ denotes answer available in *Student Solutions Manual/Study Guide*

1. If the terminals of a battery with zero internal resistance are connected across two identical resistors in series, the total power delivered by the battery is 8.00 W. If the same battery is connected across the same resistors in parallel, what is the total power delivered by the battery? (a) 16.0 W (b) 32.0 W (c) 2.00 W (d) 4.00 W (e) none of those answers

2. Wire B has twice the length and twice the radius of wire A. Both wires are made from the same material. If wire A has a resistance R, what is the resistance of wire B? (a) $4R$ (b) $2R$ (c) R (d) $\frac{1}{2}R$ (e) $\frac{1}{4}R$

3. The current-versus-voltage behavior of a certain electrical device is shown in Figure OQ21.3. When the potential difference across the device is 2 V, what is its resistance? (a) 1 Ω (b) $\frac{3}{4}\Omega$ (c) $\frac{4}{3}\Omega$ (d) undefined (e) none of those answers

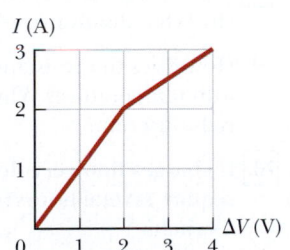

Figure OQ21.3

4. Several resistors are connected in parallel. Which of the following statements are correct? Choose all that are correct. (a) The equivalent resistance is greater than any of the resistances in the group. (b) The equivalent resistance is less than any of the resistances in the group. (c) The equivalent resistance depends on the voltage applied across the group. (d) The equivalent resistance is equal to the sum of the resistances in the group. (e) None of those statements is correct.

5. A potential difference of 1.00 V is maintained across a 10.0-Ω resistor for a period of 20.0 s. What total charge passes by a point in one of the wires connected to the resistor in this time interval? (a) 200 C (b) 20.0 C (c) 2.00 C (d) 0.005 00 C (e) 0.050 0 C

6. Several resistors are connected in series. Which of the following statements is correct? Choose all that are correct.

(a) The equivalent resistance is greater than any of the resistances in the group. (b) The equivalent resistance is less than any of the resistances in the group. (c) The equivalent resistance depends on the voltage applied across the group. (d) The equivalent resistance is equal to the sum of the resistances in the group. (e) None of those statements is correct.

7. A metal wire of resistance R is cut into three equal pieces that are then placed together side by side to form a new cable with a length equal to one-third the original length. What is the resistance of this new cable? (a) $\frac{1}{9}R$ (b) $\frac{1}{3}R$ (c) R (d) $3R$ (e) $9R$

8. The terminals of a battery are connected across two resistors in parallel. The resistances of the resistors are not the same. Which of the following statements is correct? Choose all that are correct. (a) The resistor with the larger resistance carries more current than the other resistor. (b) The resistor with the larger resistance carries less current than the other resistor. (c) The potential difference across each resistor is the same. (d) The potential difference across the larger resistor is greater than the potential difference across the smaller resistor. (e) The potential difference is greater across the resistor closer to the battery.

9. A cylindrical metal wire at room temperature is carrying electric current between its ends. One end is at potential $V_A = 50$ V, and the other end is at potential $V_B = 0$ V. Rank the following actions in terms of the change that each one separately would produce in the current from the greatest increase to the greatest decrease. In your ranking, note any cases of equality. (a) Make $V_A = 150$ V with $V_B = 0$ V. (b) Adjust V_A to triple the power with which the wire converts electrically transmitted energy into internal energy. (c) Double the radius of the wire. (d) Double the length of the wire. (e) Double the Celsius temperature of the wire.

10. Two conducting wires A and B of the same length and radius are connected across the same potential difference.

Conductor A has twice the resistivity of conductor B. What is the ratio of the power delivered to A to the power delivered to B? (a) 2 (b) $\sqrt{2}$ (c) 1 (d) $1/\sqrt{2}$ (e) $\frac{1}{2}$

11. When resistors with different resistances are connected in series, which of the following must be the same for each resistor? Choose all correct answers. (a) potential difference (b) current (c) power delivered (d) charge entering each resistor in a given time interval (e) none of those answers

12. When operating on a 120-V circuit, an electric heater receives 1.30×10^3 W of power, a toaster receives 1.00×10^3 W, and an electric oven receives 1.54×10^3 W. If all three appliances are connected in parallel on a 120-V circuit and turned on, what is the total current drawn from an external source? (a) 24.0 A (b) 32.0 A (c) 40.0 A (d) 48.0 A (e) none of those answers

13. Car batteries are often rated in ampere-hours. Does this information designate the amount of (a) current, (b) power, (c) energy, (d) charge, or (e) potential the battery can supply?

14. The terminals of a battery are connected across two resistors in series. The resistances of the resistors are not the same. Which of the following statements are correct? Choose all that are correct. (a) The resistor with the smaller resistance carries more current than the other resistor. (b) The resistor with the larger resistance carries less current than the other resistor. (c) The current in each resistor is the same. (d) The potential difference across each resistor is the same. (e) The potential difference is greatest across the resistor closest to the positive terminal.

15. In the circuit shown in Figure OQ21.15, each battery is delivering energy to the circuit by electrical transmission. All the resistors have equal resistance. **(i)** Rank the electric potentials at points a, b, c, d, and e from highest to lowest, noting any cases of equality in the ranking. **(ii)** Rank the magnitudes of the currents at the same points from greatest to least, noting any cases of equality.

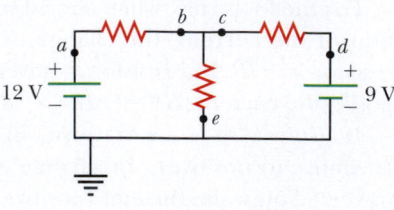

Figure OQ21.15

CONCEPTUAL QUESTIONS

1. Suppose a parachutist lands on a high-voltage wire and grabs the wire as she prepares to be rescued. (a) Will she be electrocuted? (b) If the wire then breaks, should she continue to hold onto the wire as she falls to the ground? Explain.

2. What factors affect the resistance of a conductor?

3. Newspaper articles often contain statements such as "10 000 volts of electricity surged through the victim's body." What is wrong with this statement?

4. Referring to Figure CQ21.4, describe what happens to the lightbulb after the switch is closed. Assume the capacitor has a large capacitance and is initially uncharged. Also assume the light illuminates when connected directly across the battery terminals.

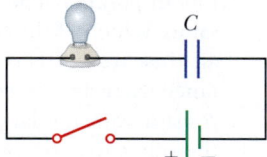

Figure CQ21.4

5. When the potential difference across a certain conductor is doubled, the current is observed to increase by a factor of 3. What can you conclude about the conductor?

6. Use the atomic theory of matter to explain why the resistance of a material should increase as its temperature increases.

7. So that your grandmother can listen to *A Prairie Home Companion*, you take her bedside radio to the hospital where she is staying. You are required to have a maintenance worker test the radio for electrical safety. Finding that it develops 120 V on one of its knobs, he does not let you take it to your grandmother's room. Your grandmother complains that she has had the radio for many years and nobody has ever gotten a shock from it. You end up having to buy a new plastic radio. (a) Why is your grandmother's old radio dangerous in a hospital room? (b) Will the old radio be safe back in her bedroom?

8. (a) What advantage does 120-V operation offer over 240 V? (b) What disadvantages does it have?

9. How does the resistance for copper and for silicon change with temperature? Why are the behaviors of these two materials different?

10. If charges flow very slowly through a metal, why does it not require several hours for a light to come on when you throw a switch?

11. If you were to design an electric heater using Nichrome wire as the heating element, what parameters of the wire could you vary to meet a specific power output such as 1 000 W?

12. Is the direction of current in a battery always from the negative terminal to the positive terminal? Explain.

13. Given three lightbulbs and a battery, sketch as many different electric circuits as you can.

14. A student claims that the second of two lightbulbs in series is less bright than the first because the first lightbulb uses up some of the current. How would you respond to this statement?

15. Why is it possible for a bird to sit on a high-voltage wire without being electrocuted?

▶ PROBLEMS AVAILABLE IN

21.1 Electric Current

Problems 1–7

21.2 Resistance and Ohm's Law

Problems 8–14

21.4 A Model for Electrical Conduction

Problems 15–17

21.5 Energy and Power in Electric Circuits

Problems 18–32

21.6 Sources of emf

Problems 33–35

21.7 Resistors in Series and Parallel

Problems 36–44

21.8 Kirchhoff's Rules

Problems 45–52

21.9 *RC* Circuits

Problems 53–59

21.10 Context Connection: The Atmosphere as a Conductor

Problems 60–61

Additional Problems

Problems 62–77

Solutions to the following Problems are available in the *Student Solutions Manual/Study Guide*:

21.2, 21.5, 21.9, 21.17, 21.21, 21.27, 21.31, 21.33, 21.39, 21.43, 21.47, 21.49, 21.53, 21.59, 21.75, and 21.76

List of Enhanced Problems

Problem Number	Targeted Feedback in Enhanced WebAssign	Master It in Enhanced WebAssign	Watch It in Enhanced WebAssign
21.3	✓		✓
21.5		✓	
21.6	✓		✓
21.7	✓		✓
21.9		✓	
21.10	✓		✓
21.11	✓	✓	
21.17		✓	
21.20	✓		✓
21.21		✓	
21.27		✓	
21.31		✓	
21.32		✓	
21.33		✓	
21.35	✓		✓
21.39	✓	✓	
21.41	✓		✓
21.43	✓		✓
21.45	✓		✓
21.47	✓	✓	
21.51	✓		✓
21.53	✓		✓
21.55	✓		✓
21.57	✓		✓
21.59		✓	
21.67		✓	
21.75		✓	

Determining the Number of Lightning Strikes

Now that we have investigated the principles of electricity, let us respond to our central question for the *Lightning* Context:

> **How can we determine the number of lightning strikes on the Earth in a typical day?**

We must combine several ideas from our knowledge of electricity to perform this calculation. In Chapter 20, the atmosphere was modeled as a capacitor. Such modeling was first done by Lord Kelvin, who modeled the ionosphere as the positive plate several tens of kilometers above the Earth's surface. More sophisticated models have shown the effective height of the positive plate to be the 5 km that we used in our earlier calculation.

The Atmospheric Capacitor Model

The plates of the atmospheric capacitor are separated by a layer of air containing a large number of free ions that can carry current. Air is a good insulator; measurements show that the resistivity of air is about 3×10^{13} $\Omega \cdot$ m. Let us calculate the resistance of the air between our capacitor plates. The shape of the resistor is that of a spherical shell between the plates of the atmospheric capacitor (Fig. 1a). The length of 5 km, however, is very short compared with the radius of 6 400 km. Therefore, we can ignore the spherical shape and approximate the resistor as a 5-km slab of flat material whose area is the surface area of the Earth. Using Equation 21.8,

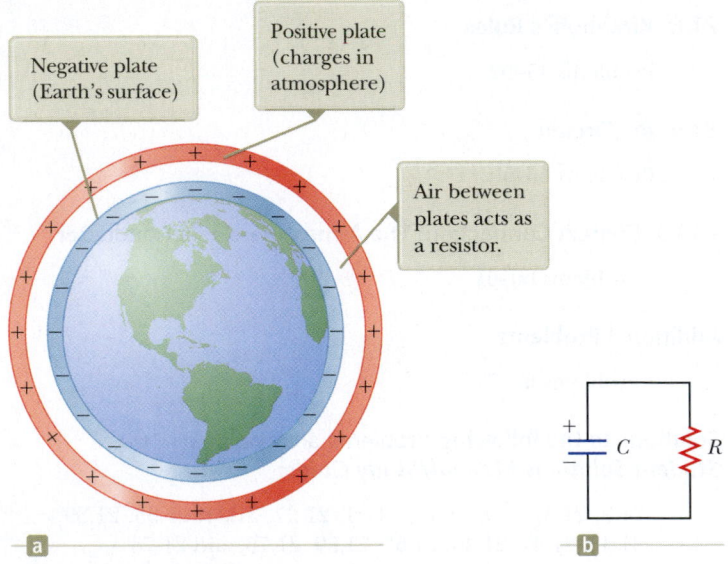

Figure 1 (a) The atmosphere can be modeled as a capacitor, with conductive air between the plates. (b) We can imagine an equivalent *RC* circuit for the atmosphere, with the natural discharge of the capacitor in balance with the charging of the capacitor by lightning.

Positive plate (charges in atmosphere)

Negative plate (Earth's surface)

Air between plates acts as a resistor.

$$R = \rho \frac{\ell}{A} = (3 \times 10^{13} \, \Omega \cdot \text{m}) \frac{5 \times 10^3 \, \text{m}}{4\pi(6.4 \times 10^6 \, \text{m})^2} \approx 3 \times 10^2 \, \Omega$$

The charge on the atmospheric capacitor can pass from the upper plate to the ground by electric current in the air between the plates. Therefore, we can model the atmosphere as an *RC* circuit (Fig. 1b), using the capacitance found in Chapter 20, and the resistance connecting the plates calculated above. The time constant for this *RC* circuit is

$$\tau = RC = (0.9 \, \text{F})(3 \times 10^2 \, \Omega) \approx 3 \times 10^2 \, \text{s} = 5 \, \text{min}$$

Therefore, the charge on the atmospheric capacitor should fall to $e^{-1} = 37\%$ of its original value after only 5 min! After 30 min, less than 0.3% of the charge would remain! Why doesn't that happen? What keeps the atmospheric capacitor charged? The answer is *lightning*. The processes occurring in cloud charging result in lightning strikes that deliver negative charge to the ground to replace that neutralized by the flow of charge through the air. On the average, a net charge on the atmospheric capacitor results from a balance between these two processes.

Now, let's use this balance to numerically answer our central question. We first address the charge on the atmospheric capacitor. In Chapter 19, we mentioned a charge of 5×10^5 C that is spread over the surface of the Earth, which is the charge on the atmospheric capacitor.

A typical lightning strike delivers about 25 C of negative charge to the ground in the process of charging the capacitor. Dividing the charge on the capacitor by the charge per lightning strike tells us the number of lightning strikes required to charge the capacitor:

$$\text{Number of lightning strikes} = \frac{\text{total charge}}{\text{charge per lightning strike}}$$

$$= \frac{5 \times 10^5 \text{ C}}{25 \text{ C per strike}} \approx 2 \times 10^4 \text{ lightning strikes}$$

According to our calculation for the *RC* circuit, the atmospheric capacitor almost completely discharges through the air in about 30 min. Therefore, 2×10^4 lightning strikes must occur every 30 min, or 4×10^4/h, to keep the charging and discharging processes in balance. Multiplying by the number of hours in a day gives us

$$\text{Number of lightning strikes per day} = (4 \times 10^4 \text{ strikes/h})\left(\frac{24 \text{ h}}{1 \text{ d}}\right)$$

$$\approx 1 \times 10^6 \text{ strokes/day}$$

Despite the simplifications that we have adopted in our calculations, this number is on the right order of magnitude for the actual number of lightning strikes on the Earth in a typical day: 1 million!

Problems

1. Consider the atmospheric capacitor described in the text, with the ground as one plate and positive charges in the atmosphere as the other. On one particular day, the capacitance of the atmospheric capacitor is 0.800 F. The effective plate separation distance is 4.00 km, and the resistivity of the air between the plates is $2.00 \times 10^{13} \ \Omega \cdot$ m. If no lightning events occur, the capacitor will discharge through the air. If a charge of 4.00×10^4 C is on the atmospheric capacitor at time $t = 0$, at what later time is the charge reduced (a) to 2.00×10^4 C, (b) to 5.00×10^3 C, and (c) to zero?

2. Consider this alternative line of reasoning to estimate the number of lightning strikes on the Earth in one day. Using the charge on the Earth of 5.00×10^5 C and the atmospheric capacitance of 0.9 F, we find that the potential difference across the capacitor is $\Delta V = Q/C = 5.00 \times 10^5$ C/0.9 F $\approx 6 \times 10^5$ V. The leakage current in the air is $I = \Delta V/R = 6 \times 10^5$ V/300 $\Omega \approx 2$ kA. To keep the capacitor charged, lightning should deliver the same net current in the opposite direction. (a) If each lightning strike delivers 25 C of charge to the ground, what is the average time interval between lightning strikes so that the average current due to lightning is 2 kA? (b) Using this average time interval between lightning strikes, calculate the number of lightning strikes in one day.

3. Consider again the atmospheric capacitor discussed in the text. (a) Assume that atmospheric conditions are such that, for one complete day, the lower 2.50 km of the air between the capacitor plates has resistivity $2.00 \times 10^{13} \ \Omega \cdot$ m and the upper 2.50 km has resistivity $0.500 \times 10^{13} \ \Omega \cdot$ m. How many lightning strikes occur on this day? (b) Assume that atmospheric conditions are such that, for one complete day, resistivity of the air between the plates in the southern hemisphere is $2.00 \times 10^{13} \ \Omega \cdot$ m and the resistivity between the plates in the northern hemisphere is $0.200 \times 10^{13} \ \Omega \cdot$ m. How many lightning strikes occur on this day?

Magnetism in Medicine

Now that we have studied electricity, we turn our attention to the closely related topic of magnetism. Magnetism is prevalent in our everyday life. Magnets are essential for the operation of motors. Magnets in generators provide electricity to homes and businesses. Loudspeaker systems use magnets to convert electrical signals to sound waves. Magnets are also critical in keeping important data securely fixed to refrigerator doors.

Magnetism has also entered the field of medicine with a number of applications that can improve health and save lives. Various medical tests or procedures involve magnets. We will explore some of these important applications in this Context. We begin, however, by exploring some questionable applications of magnetism in medicine between the 18th century and the present day.

You may have heard advertisements or even own a magnetic bracelet, such as those shown in Figure 1. Such a bracelet is just one example of devices that provide purported *magnetic therapy*. Additional such devices include other magnetic jewelry, magnetic straps for various body parts, magnetic shoe inserts, magnetic blankets and mattresses, and magnetic creams. Despite having sales of a billion dollars a year, magnetic therapy has not been shown in any scientific studies to be effective.

Figure 1 Magnetic bracelets are sold to consumers to promote good health and pain relief. Do you think devices such as these bracelets work?

The United States Food and Drug Administration prohibits marketing any magnetic therapy device as having proven medical advantages.

Let us now move back in time and investigate some earlier applications of magnetism in medicine. Some doctors who employed these applications actually believed that their magnetic instruments would help their patients. Other so-called *quack doctors* knew that the instruments would not work, but used them anyway.

Franz Anton Mesmer, from Vienna, was one of the earliest individuals to develop a theory of medicine involving magnetism. In his doctoral thesis (*The Influence of the Planets on the Human Body,* 1767), Mesmer suggested that a universal fluid that he called "animal gravitation" was responsible for all health and illnesses. In 1773, Mesmer began to use magnets to heal diseases. He claimed that he could "cure" some diseases with a combination of stroking the patient with magnets, various forms of wailing, and listening to music from the glass armonica, recently invented by Benjamin Franklin.

By 1776, Mesmer announced that the magnets were not necessary for his treatment—they were only serving as conductors of the universal fluid, which by now had become a magnetic fluid, which he called *animal magnetism.* Mesmer was very careful about the selection of diseases he attempted to cure. For organic diseases, he would refer the patient to a traditional doctor. He used magnets to treat only nervous or hysterical diseases. The startling aspect of Mesmer's practice is that he supposedly restored sight in a blind pianist and relieved many patients from chronic convulsions. Today we realize that his magnetic treatments were not really curing patients. Mesmer was actually *hypnotizing* patients, using stares, stroking, glass armonica music, power of suggestion, etc. In fact, his name is the root of the word *mesmerize.*

Figure 2 (page 640) shows an example of the *Davis and Kidder Magneto-Electric Machine for Nervous Disorders,* which was used from the 1850s into the latter part of the 19th century. It is simply an electromagnetic generator, developed shortly after the magnetic induction discoveries of Michael Faraday. A pair of wire coils was rotated in the vicinity of a permanent magnet. The patient held on to the two metal cylinders, which

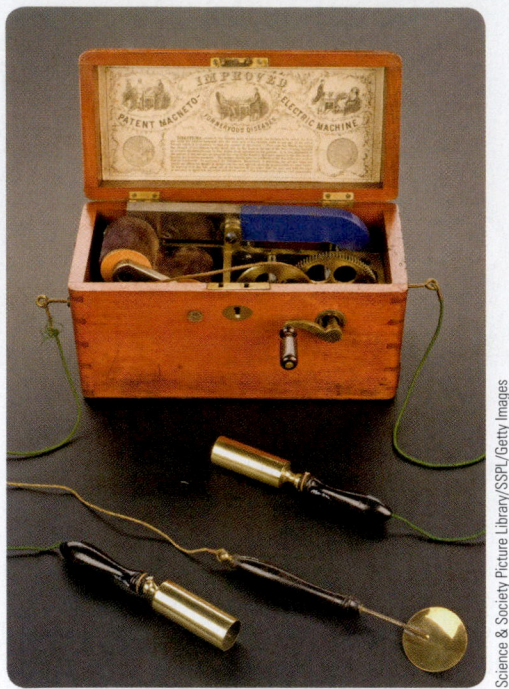

Science & Society Picture Library/SSPL/Getty Images

Figure 2 The Davis & Kidder Magneto-Electric Machine for Nervous Disorders. The patient would hold a brass tube in each hand, while the caretaker turned the crank. The patient received a shock from the voltage generated by the rotating coils in the presence of the magnetic field of the large permanent magnet in the back of the case.

were connected to the generator. The caretaker then turned the crank, providing a jolt of electricity to the patient. Providing jolts of electricity to the patient continued as a supposed treatment until well into the 20th century, and even has its supporters today. The hand-cranked Davis and Kidder device, however, was replaced by plug-in devices such as *violet ray machines* and the various devices of Albert Abrams. (Check out Abrams and his struggle with the American Medical Association on the Internet.)

Another magnetic quack device that arose in the 20th century was originally manufactured under the name of the IONACO, developed by Gaylord Wilshire. A large loop of wire, covered in leather, was plugged into an electrical socket. The goal was to magnetize the blood by wearing the loop around the body. Figure 3 shows a

Dirk Soulis Auctions

Figure 3 The Theronoid magnetic device. The leather-coated loop of wire is worn around the body to "magnetize the blood."

later version of this device, the Theronoid, developed by Philip Ilsey. In its 1933 annual report, the United States Federal Trade Commission (FTC) states, "It was claimed by respondents that the use of said device or appliance . . . was a beneficial therapeutic agent in the aid, relief, prevention, or cure of . . . asthma, arthritis, bladder trouble, bronchitis, catarrh, constipation, diabetes, eczema, heart trouble, hemorrhoids, indigestion, insomnia, lumbago, nervous disorders, neuralgia, neuritis, rheumatism, sciatica, stomach trouble, varicose veins, and high blood pressure." The FTC closes this section of its report by banning advertising of the Theronoid: "the Commission issued an order to . . . cease and desist from representing in any manner whatsoever that the said belt or device or any similar device or appliance . . . has any physiotherapeutic effect upon such subject, or that it is calculated or likely to aid in the prevention, treatment, or cure of any human ailment, sickness, or disease."

In this Context, we will look at scientifically supported uses of magnetism in medicine today as opposed to the unsubstantiated and, in some cases, fraudulent uses discussed here. We will address the central question:

> **How has magnetism entered the field of medicine to diagnose and cure illnesses and save lives?**

Chapter 22

Magnetic Forces and Magnetic Fields

Chapter Outline

CERN

An engineer performs a test on the electronics associated with one of the superconducting magnets in the Large Hadron Collider at the European Laboratory for Particle Physics, operated by the European Organization for Nuclear Research (CERN). The magnets are used to control the motion of charged particles in the accelerator. We will study the effects of magnetic fields on moving charged particles in this chapter.

The list of technological applications of magnetism is very long. For instance, large electromagnets are used to pick up heavy loads in scrap yards. Magnets are used in such devices as meters, motors, and loudspeakers. Magnetic tapes are routinely used in sound and video recording equipment. Magnetic stripes on the backs of credit cards allow our purchase to be completed quickly in a store. Intense magnetic fields generated by superconducting magnets are currently being used as a means of containing plasmas at temperatures on the order of 10^8 K used in controlled nuclear fusion research.

As we investigate magnetism in this chapter, we shall find that the subject cannot be divorced from electricity. For example, magnetic fields affect moving electric charges, and moving charges produce magnetic fields. This close association between electricity and magnetism will justify their union into *electromagnetism* that we explore in this chapter and the next.

⟨22.1 | Historical Overview

Many historians of science believe that the compass, which uses a magnetic needle, was used in China as early as the 13th century B.C., its invention being of Arab or Indian origin. The phenomenon of magnetism was known to the Greeks as early as about 800 B.C. They discovered that certain stones, made of a material now called *magnetite* (Fe_3O_4), attracted pieces of iron.

In 1269, Pierre de Maricourt (c. 1220–?) mapped out the directions taken by a magnetized needle when it was placed at various points on the surface of a spherical natural magnet. He found that the directions formed lines that encircled the sphere and passed through two points diametrically opposite each other, which he called the **poles** of the magnet. Subsequent experiments have shown that every magnet, regardless of its shape, has two poles, called **north** (N) and **south** (S), that exhibit forces on each other in a manner analogous to electric charges. That is, similar poles (N–N or S–S) repel each other and dissimilar poles (N–S) attract each other. The poles received their names because of the behavior of a magnet in the presence of the Earth's magnetic field. If a bar magnet is suspended from its midpoint by a piece of string so that it can swing freely in a horizontal plane, it rotates until its "north" pole points to the north geographic pole of the Earth (which is a south magnetic pole) and its "south" pole points to the Earth's south geographic pole. (The same idea is used to construct a simple compass.)

In 1600, William Gilbert (1544–1603) extended these experiments to a variety of materials. Using the fact that a compass needle orients in preferred directions, Gilbert suggested that magnets are attracted to land masses. In 1750, John Michell (1724–1793) used a torsion balance to show that magnetic poles exert attractive or repulsive forces on each other and that these forces vary as the inverse square of their separation. Although the force between two magnetic poles is similar to the force between two electric charges, an important difference exists. Electric charges can be isolated (witness the electron and proton), whereas magnetic poles cannot be isolated. That is, magnetic poles are always found in pairs. No matter how many times a permanent magnet is cut, each piece always has a north pole and a south pole. (Some theories speculate that magnetic monopoles—isolated north or south poles—may exist in nature, and attempts to detect them currently make up an active experimental field of investigation. None of these attempts has yet proven successful, however.)

The relationship between magnetism and electricity was discovered in 1819 when, while preparing for a lecture demonstration, Danish scientist Hans Christian Oersted found that an electric current in a wire deflected a nearby compass needle. Shortly thereafter, André-Marie Ampère (1775–1836) deduced quantitative laws of magnetic force between current-carrying conductors. He also suggested that electric current loops of molecular size are responsible for *all* magnetic phenomena.

In the 1820s, Faraday and, independently, Joseph Henry (1797–1878) identified further connections between electricity and magnetism. They showed that an electric current could be produced in a circuit either by moving a magnet near the circuit or by changing the current in a nearby circuit. Their observations demonstrated that a changing magnetic field produces an electric field. Years later, theoretical work by James Clerk Maxwell showed that the reverse is also true: a changing electric field gives rise to a magnetic field.

In this chapter, we shall investigate the effects of constant magnetic fields on charges and currents, and study the sources of magnetic fields. In the next chapter, we shall explore the effects of magnetic fields that vary in time.

North Wind Picture Archives

Hans Christian Oersted
Danish Physicist and Chemist (1777–1851)
Oersted is best known for observing that a compass needle deflects when placed near a wire carrying a current. This important discovery was the first evidence of the connection between electric and magnetic phenomena. Oersted was also the first to prepare pure aluminum.

❰22.2 | The Magnetic Field

In earlier chapters, we described the interaction between charged objects in terms of electric fields. Recall that an electric field surrounds any stationary electric charge. The region of space surrounding a *moving* charge includes a **magnetic field** in addition to the electric field. A magnetic field also surrounds any material with permanent magnetism. We find that the magnetic field is a vector field, as is the electric field.

To describe any type of vector field, we must define its magnitude and its direction. The direction of the magnetic field vector $\vec{B}$ at any location is the direction in which the north pole of a compass needle points at that location. Active Figure 22.1 shows how the magnetic field of a bar magnet can be traced with the aid of a compass, defining a **magnetic field line,** similar in many ways to the electric field lines we studied in Chapter 19. Several magnetic field lines of a bar magnet traced out in this manner are shown in the two-dimensional pictorial representation in Active Figure 22.1. Magnetic field patterns can be displayed by small iron filings placed in the vicinity of a magnet, as in Figure 22.2.

We can quantify the magnetic field $\vec{B}$ by using our model of a particle in a field, like the model discussed for gravity in Chapter 11 and for electricity in Chapter 19. The existence of a magnetic field at some point in space can be determined by measuring the **magnetic force** $\vec{F}_B$ exerted on an appropriate test particle placed at that point. This process is the same one we followed in defining the electric field in Chapter 19. Our test particle will be an electrically charged particle such as a proton. If we perform such an experiment, we find the following results that are similar to those for experiments on electric forces:

- The magnetic force is proportional to the charge q of the particle.
- The magnetic force on a negative charge is directed opposite to the force on a positive charge moving in the same direction.
- The magnetic force is proportional to the magnitude of the magnetic field vector $\vec{B}$.

We also find the following results, which are *totally different* from those for experiments on electric forces:

- The magnetic force is proportional to the speed v of the particle.
- If the velocity vector makes an angle θ with the magnetic field, the magnitude of the magnetic force is proportional to $\sin\theta$.
- When a charged particle moves *parallel* to the magnetic field vector, the magnetic force on the charge is zero.
- When a charged particle moves in a direction *not* parallel to the magnetic field vector, the magnetic force acts in a direction perpendicular to both $\vec{v}$ and $\vec{B}$; that is, the magnetic force is perpendicular to the plane formed by $\vec{v}$ and $\vec{B}$.

These results show that the magnetic force on a particle is more complicated than the electric force. The magnetic force is distinctive because it depends on the

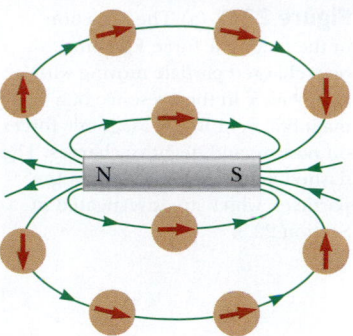

Active Figure 22.1 Compass needles can be used to trace the magnetic field lines in the region outside a bar magnet.

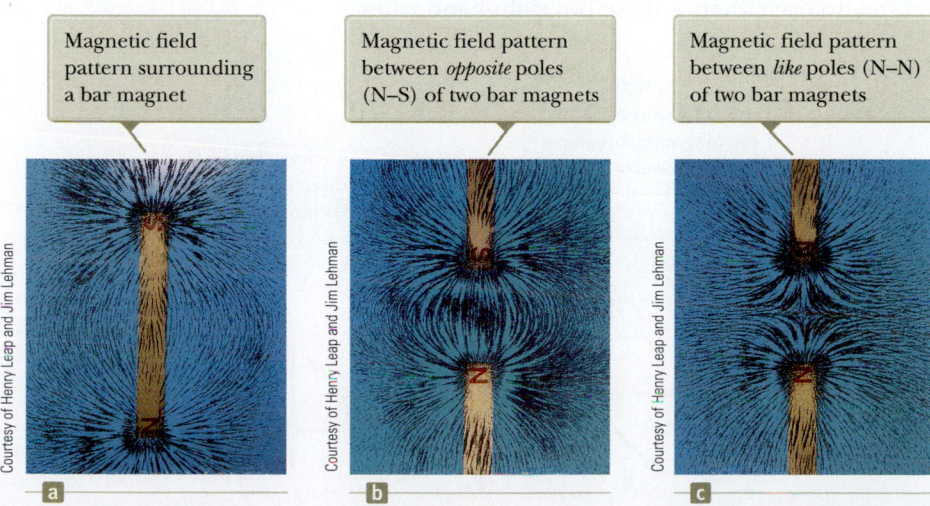

Magnetic field pattern surrounding a bar magnet

Magnetic field pattern between *opposite* poles (N–S) of two bar magnets

Magnetic field pattern between *like* poles (N–N) of two bar magnets

a

b

c

Courtesy of Henry Leap and Jim Lehman

Figure 22.2 Magnetic field patterns can be displayed with iron filings sprinkled on paper near magnets.

Figure 22.3 (a) The direction of the magnetic force $\vec{F}_B$ acting on a charged particle moving with a velocity $\vec{v}$ in the presence of a magnetic field $\vec{B}$. (b) Magnetic forces on positive and negative charges. The dashed lines show the paths of the particles, which are investigated in Section 22.3.

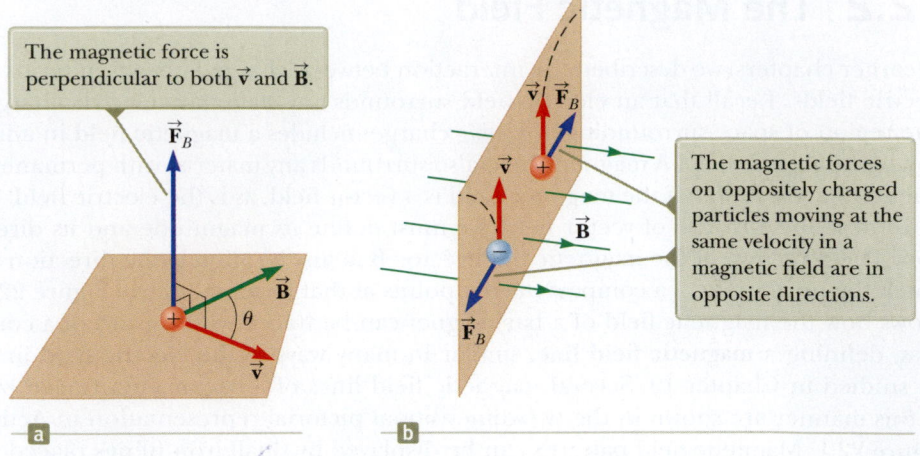

velocity of the particle and because its direction is perpendicular to both $\vec{v}$ and $\vec{B}$. Figure 22.3 shows the details of the direction of the magnetic force on a charged particle. Despite this complicated behavior, these observations can be summarized in a compact way by writing the magnetic force in the form

▶ Vector expression for the magnetic force on a charged particle moving in a magnetic field

$$\vec{F}_B = q\vec{v} \times \vec{B} \qquad \textbf{22.1} ◀$$

where the direction of the magnetic force is that of $\vec{v} \times \vec{B}$, which, by definition of the cross product, is perpendicular to both $\vec{v}$ and $\vec{B}$. Equation 22.1 is analogous to Equation 19.4, $\vec{F}_e = q\vec{E}$, but is clearly more complicated. We can regard Equation 22.1 as an operational definition of the magnetic field at a point in space. The SI unit of magnetic field is the **tesla** (T), where

$$1\,T = 1\,N \cdot s/C \cdot m$$

Figure 22.4 reviews two right-hand rules for determining the direction of the cross product $\vec{v} \times \vec{B}$ and determining the direction of $\vec{F}_B$. The rule in Figure 22.4a depends on our right-hand rule for the cross product in Figure 10.13. You point the four fingers of your right hand along the direction of $\vec{v}$ with the palm facing $\vec{B}$ and curl them toward $\vec{B}$. The extended thumb, which is at a right angle to the fingers, points in the direction of $\vec{v} \times \vec{B}$. Because $\vec{F}_B = q\vec{v} \times \vec{B}$, $\vec{F}_B$ is in the direction of your thumb if q is positive and opposite the direction of your thumb if q is negative.

A second rule is shown in Figure 22.4b. Here the thumb points in the direction of $\vec{v}$ and the extended fingers in the direction of $\vec{B}$. Now, the force $\vec{F}_B$ on a positive charge extends outward from your palm. The advantage of this rule is that the force on the charge is in the direction that you would push on something with your hand, outward from your palm. The force on a negative charge is in the opposite direction. Feel free to use either of these two right-hand rules.

Figure 22.4 Two right-hand rules for determining the direction of the magnetic force $\vec{F}_B = q\vec{v} \times \vec{B}$ acting on a particle with positive charge q moving with a velocity $\vec{v}$ in a magnetic field $\vec{B}$. (a) In this rule, the magnetic force is in the direction in which your thumb points. (b) In this rule, the magnetic force is in the direction of your palm, as if you are pushing the particle with your hand.

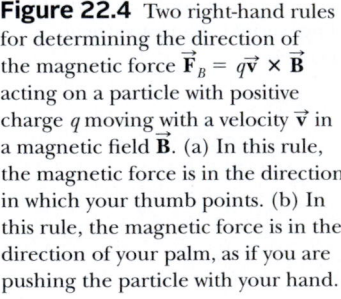

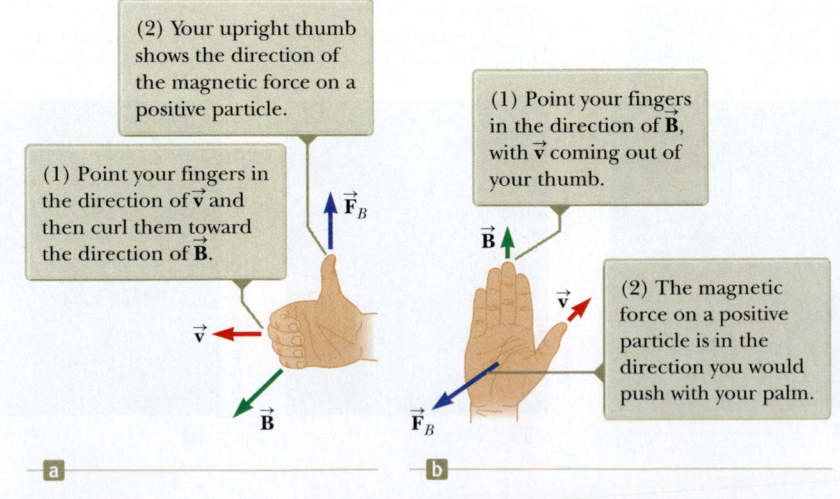

The magnitude of the magnetic force is

$$F_B = |q|vB \sin \theta \qquad \textbf{22.2} \blacktriangleleft$$

▶ Magnitude of the magnetic force on a charged particle moving in a magnetic field

where θ is the angle between $\vec{v}$ and $\vec{B}$. From this expression, we see that F_B is zero when $\vec{v}$ is either parallel or antiparallel to $\vec{B}$ ($\theta = 0$ or $180°$). Furthermore, the force has its maximum value $F_B = |q|vB$ when $\vec{v}$ is perpendicular to $\vec{B}$ ($\theta = 90°$).

Let's summarize the important differences between electric and magnetic forces on charged particles:

- The electric force vector is along the direction of the electric field, whereas the magnetic force vector is perpendicular to the magnetic field.
- The electric force acts on a charged particle regardless of whether the particle is moving, whereas the magnetic force acts on a charged particle only when the particle is in motion.
- The electric force does work in displacing a charged particle, whereas the magnetic force associated with a steady magnetic field does no work when a particle is displaced.

This last statement is true because when a charge moves in a constant magnetic field, the magnetic force is always *perpendicular* to the displacement of its point of application. That is, for a small displacement $d\vec{s}$ of a particle, the work done by the magnetic force on the particle is $dW = \vec{F}_B \cdot d\vec{s} = (\vec{F}_B \cdot \vec{v}) dt = 0$ because the magnetic force is a vector perpendicular to $\vec{v}$. From this property and the work–kinetic energy theorem, we conclude that the kinetic energy of a charged particle *cannot* be altered by a constant magnetic field alone. In other words, when a charge moves with a velocity of $\vec{v}$, an applied magnetic field can alter the direction of the velocity vector, but it cannot change the speed of the particle.

In Figures 22.3 and 22.4, we used green arrows to represent magnetic field vectors, which will be the convention in this book. In Active Figure 22.1, we represented the magnetic field of a bar magnet with green field lines. Studying magnetic fields presents a complication that we avoided in electric fields. In our study of electric fields, we drew all electric field vectors in the plane of the page or used perspective to represent them directed at an angle to the page. The cross product in Equation 22.1 requires us to think in three dimensions for problems in magnetism. Therefore, in addition to drawing vectors pointing left or right and up or down, we will need a method of drawing vectors into or out of the page. These methods of representing the vectors are illustrated in Figure 22.5. A vector coming out of the page is represented by a dot, which we can think of as the tip of the arrowhead representing the vector coming through the paper toward us (Fig. 22.5a). A vector going into the page is represented by a cross, which we can think of as the tail feathers of an arrow going into the page (Fig. 22.5b). This depiction can be used for any type of vector we will encounter: magnetic field, velocity, force, and so on.

> **QUICK QUIZ 22.1** An electron moves in the plane of this paper toward the top of the page. A magnetic field is also in the plane of the page and directed toward the right. What is the direction of the magnetic force on the electron? (**a**) toward the top of the page (**b**) toward the bottom of the page (**c**) toward the left edge of the page (**d**) toward the right edge of the page (**e**) upward out of the page (**f**) downward into the page

Magnetic field lines coming out of the paper are indicated by dots, representing the tips of arrows coming outward.

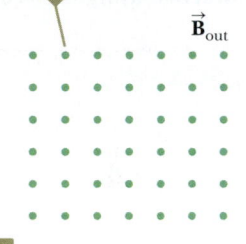

Magnetic field lines going into the paper are indicated by crosses, representing the feathers of arrows going inward.

Figure 22.5 Representations of magnetic field lines perpendicular to the page.

> ## THINKING PHYSICS 22.1
>
> On a business trip to Australia, you take along your U.S.-made compass that you used in your Boy Scout days. Does this compass work correctly in Australia?
>
> **Reasoning** Using the compass in Australia presents no problem. The north pole of the magnet in the compass will be attracted to the south magnetic pole near the north geographic pole, just as it was in the United States. The only difference in the magnetic field lines is that they have an upward component in Australia, whereas they have a downward component in the United States. When you hold the compass in a horizontal plane, it cannot detect the vertical component of the field, however; it only displays the direction of the horizontal component of the magnetic field. ◀

Example 22.1 | **An Electron Moving in a Magnetic Field**

An electron in an old-style television picture tube moves toward the front of the tube with a speed of 8.0×10^6 m/s along the x axis (Fig. 22.6). Surrounding the neck of the tube are coils of wire that create a magnetic field of magnitude 0.025 T, directed at an angle of 60° to the x axis and lying in the xy plane. Calculate the magnetic force on the electron.

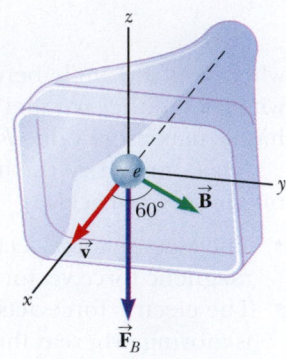

SOLUTION

Conceptualize Recall that the magnetic force on a charged particle is perpendicular to the plane formed by the velocity and magnetic field vectors. Use one of the right-hand rules in Figure 22.4 to convince yourself that the direction of the force on the electron is downward in Figure 22.6.

Categorize We evaluate the magnetic force using an equation developed in this section, so we categorize this example as a substitution problem.

Figure 22.6 (Example 22.1) The magnetic force $\vec{F}_B$ acting on the electron is in the negative z direction when $\vec{v}$ and $\vec{B}$ lie in the xy plane.

Use Equation 22.2 to find the magnitude of the magnetic force:

$$F_B = |q|vB \sin\theta$$
$$= (1.6 \times 10^{-19}\ \text{C})(8.0 \times 10^6\ \text{m/s})(0.025\ \text{T})(\sin 60°)$$
$$= 2.8 \times 10^{-14}\ \text{N}$$

For practice using the vector product, evaluate this force in vector notation using Equation 22.1.

22.3 | Motion of a Charged Particle in a Uniform Magnetic Field

In Section 22.2, we found that the magnetic force acting on a charged particle moving in a magnetic field is perpendicular to the particle's velocity and consequently the work done by the magnetic force on the particle is zero. Now consider the special case of a positively charged particle moving in a uniform magnetic field with the initial velocity vector of the particle perpendicular to the field. Let's assume the direction of the magnetic field is into the page as in Active Figure 22.7. As the particle changes the direction of its velocity in response to the magnetic force, the magnetic force remains perpendicular to the velocity. As we found in Section 5.2, if the force is always perpendicular to the velocity, the path of the particle is a circle! Active Figure 22.7 shows the particle moving in a circle in a plane perpendicular to the magnetic field. Although magnetism and magnetic forces may be new and unfamiliar to you now, we see a magnetic effect that results in something with which we are familiar: the particle in uniform circular motion!

The particle moves in a circle because the magnetic force $\vec{F}_B$ is perpendicular to $\vec{v}$ and $\vec{B}$ and has a constant magnitude qvB. As Active Figure 22.7 illustrates, the rotation is counterclockwise for a positive charge in a magnetic field directed into the page. If q were negative, the rotation would be clockwise. We use the particle under a net force model to write Newton's second law for the particle:

$$\sum F = F_B = ma$$

Because the particle moves in a circle, we also model it as a particle in uniform circular motion and we replace the acceleration with centripetal acceleration:

$$F_B = qvB = \frac{mv^2}{r}$$

This expression leads to the following equation for the radius of the circular path:

$$r = \frac{mv}{qB} \qquad \textbf{22.3} \blacktriangleleft$$

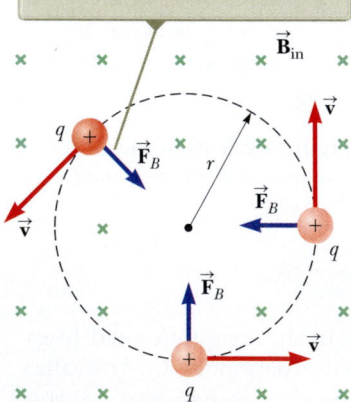

The magnetic force $\vec{F}_B$ acting on the charge is always directed toward the center of the circle.

Active Figure 22.7 When the velocity of a charged particle is perpendicular to a uniform magnetic field, the particle moves in a circular path in a plane perpendicular to $\vec{B}$.

That is, the radius of the path is proportional to the linear momentum mv of the particle and inversely proportional to the magnitude of the charge on the particle and to the magnitude of the magnetic field. The angular speed of the particle is (from Eq. 10.10)

$$\omega = \frac{v}{r} = \frac{qB}{m}$$ 22.4 ◀

The period of the motion (the time interval required for the particle to complete one revolution) is equal to the circumference of the circular path divided by the speed of the particle:

$$T = \frac{2\pi r}{v} = \frac{2\pi}{\omega} = \frac{2\pi m}{qB}$$ 22.5 ◀

These results show that the angular speed of the particle and the period of the circular motion do not depend on the translational speed of the particle or the radius of the orbit for a given particle in a given uniform magnetic field. The angular speed ω is often referred to as the **cyclotron frequency** because charged particles circulate at this angular speed in one type of accelerator called a *cyclotron*, discussed in Section 22.4.

If a charged particle moves in a uniform magnetic field with its velocity at some arbitrary angle to $\vec{B}$, its path is a helix. For example, if the field is in the x direction as in Active Figure 22.8, there is no component of force on the particle in the x direction. As a result, $a_x = 0$, and so the x component of velocity of the particle remains constant. The magnetic force $q\vec{v} \times \vec{B}$ causes the components v_y and v_z to change in time, however, and the resulting motion of the particle is a helix having its axis parallel to the magnetic field. The projection of the path onto the yz plane (viewed along the x axis) is a circle. (The projections of the path onto the xy and xz planes are sinusoids!) Equations 22.3 to 22.5 still apply provided that v is replaced by $v_\perp = \sqrt{v_y^2 + v_z^2}$. In the yz plane, the charged particle is modeled as a particle in uniform circular motion as well as a particle under a net force. In the x direction, the charged particle is modeled as a particle under constant velocity.

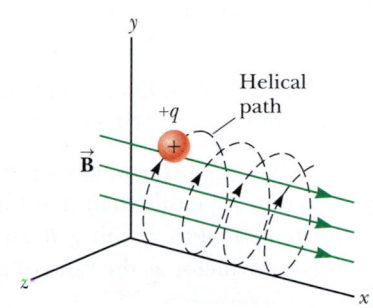

Active Figure 22.8 A charged particle having a velocity vector with a component parallel to a uniform magnetic field moves in a helical path.

▶ **QUICK QUIZ 22.2** A charged particle is moving perpendicular to a magnetic field in a circle with a radius r. (**i**) An identical particle enters the field, with $\vec{v}$ perpendicular to $\vec{B}$, but with a higher speed than the first particle. Compared with the radius of the circle for the first particle, is the radius of the circular path for the second particle (a) smaller, (b) larger, or (c) equal in size? (**ii**) The magnitude of the magnetic field is increased. From the same choices, compare the radius of the new circular path of the first particle with the radius of its initial path.

▶ **THINKING PHYSICS 22.2**

Suppose a uniform magnetic field exists in a finite region of space as in Figure 22.9. Can you inject a charged particle into this region and have it stay trapped in the region by the magnetic force?

Reasoning Consider separately the components of the particle velocity parallel and perpendicular to the field lines in the region. For the component parallel to the field lines, no force is exerted on the particle and it continues to move with the parallel component until it leaves the region of the magnetic field. Now consider the component perpendicular to the field lines. This component results in a magnetic force that is perpendicular to both the field lines and the velocity component. As discussed earlier, if the force acting on a charged particle is always perpendicular to its velocity, the particle moves in a circular path. Therefore, the particle follows half of a circular arc and exits the field on the other side of the circle, as shown in Figure 22.9. Therefore, a particle injected into a uniform magnetic field cannot stay trapped in the field region. ◀

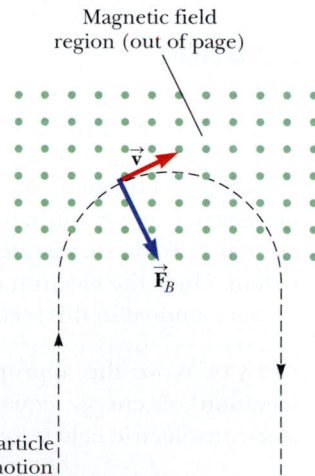

Figure 22.9 (Thinking Physics 22.2) A positively charged particle enters a region of magnetic field directed out of the page.

Example 22.2 | A Proton Moving Perpendicular to a Uniform Magnetic Field

A proton is moving in a circular orbit of radius 14 cm in a uniform 0.35-T magnetic field perpendicular to the velocity of the proton. Find the speed of the proton.

SOLUTION

Conceptualize From our discussion in this section, we know the proton follows a circular path when moving perpendicular to a uniform magnetic field.

Categorize We evaluate the speed of the proton using an equation developed in this section, so we categorize this example as a substitution problem.

Solve Equation 22.3 for the speed of the particle:

$$v = \frac{qBr}{m_p}$$

Substitute numerical values:

$$v = \frac{(1.60 \times 10^{-19}\ \text{C})(0.35\ \text{T})(0.14\ \text{m})}{1.67 \times 10^{-27}\ \text{kg}}$$

$$= 4.7 \times 10^6\ \text{m/s}$$

What If? What if an electron, rather than a proton, moves in a direction perpendicular to the same magnetic field with this same speed? Will the radius of its orbit be different?

Answer An electron has a much smaller mass than a proton, so the magnetic force should be able to change its velocity much more easily than that for the proton. Therefore, we expect the radius to be smaller. Equation 22.3 shows that r is proportional to m with q, B, and v the same for the electron as for the proton. Consequently, the radius will be smaller by the same factor as the ratio of masses m_e/m_p.

Example 22.3 | Bending an Electron Beam

In an experiment designed to measure the magnitude of a uniform magnetic field, electrons are accelerated from rest through a potential difference of 350 V and then enter a uniform magnetic field that is perpendicular to the velocity vector of the electrons. The electrons travel along a curved path because of the magnetic force exerted on them, and the radius of the path is measured to be 7.5 cm. (Such a curved beam of electrons is shown in Fig. 22.10.)

(A) What is the magnitude of the magnetic field?

Figure 22.10
(Example 22.3)
The bending
of an electron
beam in a
magnetic field.

Henry Leap and Jim Lehman

SOLUTION

Conceptualize This example involves electrons accelerating from rest due to an electric force and then moving in a circular path due to a magnetic force. With the help of Active Figure 22.7 and Figure 22.10, visualize the circular motion of the electrons.

Categorize Equation 22.3 shows that we need the speed v of the electron to find the magnetic field magnitude, and v is not given. Consequently, we must find the speed of the electron based on the potential difference through which it is accelerated. To do so, we categorize the first part of the problem by modeling an electron and the electric field as an isolated system. Once the electron enters the magnetic field, we categorize the second part of the problem as one similar to those we have studied in this section.

Analyze Write the appropriate reduction of the conservation of energy equation, Equation 7.2, for the electron–electric field system:

$$\Delta K + \Delta U = 0$$

Substitute the appropriate initial and final energies:

$$(\tfrac{1}{2}m_e v^2 - 0) + (q\,\Delta V) = 0$$

Solve for the speed of the electron:

$$v = \sqrt{\frac{-2q\,\Delta V}{m_e}}$$

22.3 *cont.*

Substitute numerical values:

$$v = \sqrt{\frac{-2(-1.60 \times 10^{-19}\,\text{C})(350\,\text{V})}{9.11 \times 10^{-31}\,\text{kg}}} = 1.11 \times 10^7\,\text{m/s}$$

Now imagine the electron entering the magnetic field with this speed. Solve Equation 22.3 for the magnitude of the magnetic field:

$$B = \frac{m_e v}{er}$$

Substitute numerical values:

$$B = \frac{(9.11 \times 10^{-31}\,\text{kg})(1.11 \times 10^7\,\text{m/s})}{(1.60 \times 10^{-19}\,\text{C})(0.075\,\text{m})} = 8.4 \times 10^{-4}\,\text{T}$$

(B) What is the angular speed of the electrons?

SOLUTION

Use Equation 10.10:

$$\omega = \frac{v}{r} = \frac{1.11 \times 10^7\,\text{m/s}}{0.075\,\text{m}} = 1.5 \times 10^8\,\text{rad/s}$$

Finalize The angular speed can be represented as $\omega = (1.5 \times 10^8\,\text{rad/s})(1\,\text{rev}/2\pi\,\text{rad}) = 2.4 \times 10^7\,\text{rev/s}$. The electrons travel around the circle 24 million times per second! This answer is consistent with the very high speed found in part (A).

What If? What if a sudden voltage surge causes the accelerating voltage to increase to 400 V? How does that affect the angular speed of the electrons, assuming the magnetic field remains constant?

Answer The increase in accelerating voltage ΔV causes the electrons to enter the magnetic field with a higher speed v. This higher speed causes them to travel in a circle with a larger radius r. The angular speed is the ratio of v to r. Both v and r increase by the same factor, so the effects cancel and the angular speed remains the same. Equation 22.4 is an expression for the cyclotron frequency, which is the same as the angular speed of the electrons. The cyclotron frequency depends only on the charge q, the magnetic field B, and the mass m_e, none of which have changed. Therefore, the voltage surge has no effect on the angular speed. (In reality, however, the voltage surge may also increase the magnetic field if the magnetic field is powered by the same source as the accelerating voltage. In that case, the angular speed increases according to Eq. 22.4.)

22.4 | Applications Involving Charged Particles Moving in a Magnetic Field

A charge moving with velocity $\vec{v}$ in the presence of an electric field $\vec{E}$ and a magnetic field $\vec{B}$ experiences both an electric force $q\vec{E}$ and a magnetic force $q\vec{v} \times \vec{B}$. The total force, called the **Lorentz force**, acting on the charge is therefore the vector sum,

$$\vec{F} = q\vec{E} + q\vec{v} \times \vec{B} \qquad \text{22.6} \blacktriangleleft$$

In this section, we look at three applications involving particles experiencing the Lorentz force.

Velocity Selector

In many experiments involving moving charged particles, it is important that all particles move with essentially the same velocity, which can be achieved by applying a combination of an electric field and a magnetic field oriented as shown in Active Figure 22.11. A uniform electric field is directed to the right (in the plane of the page in Active Fig. 22.11), and a uniform magnetic field is applied in the direction perpendicular to the electric field (into the page in Active Fig. 22.11). If q is positive and the velocity $\vec{v}$ is upward, the magnetic force $q\vec{v} \times \vec{B}$ is to the left and the electric force $q\vec{E}$ is to the right. When the magnitudes of the two fields are chosen so that $qE = qvB$, the charged particle is modeled as a particle in equilibrium and moves in a straight vertical line through the region of the fields. From the expression $qE = qvB$, we find that

$$v = \frac{E}{B} \qquad \text{22.7} \blacktriangleleft$$

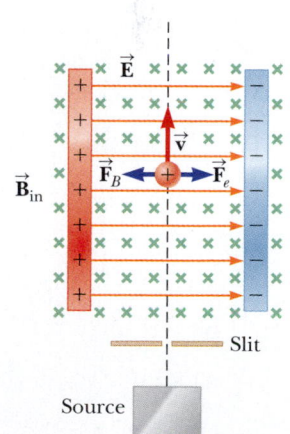

Active Figure 22.11 A velocity selector. When a positively charged particle is moving with velocity $\vec{v}$ in the presence of a magnetic field directed into the page and an electric field directed to the right, it experiences an electric force $q\vec{E}$ to the right and a magnetic force $q\vec{v} \times \vec{B}$ to the left.

Only those particles having this speed pass undeflected through the mutually perpendicular electric and magnetic fields. The magnetic force exerted on particles moving at speeds greater than that is stronger than the electric force, and the particles are deflected to the left. Those moving at slower speeds are deflected to the right.

The Mass Spectrometer

A **mass spectrometer** separates ions according to their mass-to-charge ratio. In one version of this device, known as the *Bainbridge mass spectrometer,* a beam of ions first passes through a velocity selector and then enters a second uniform magnetic field $\vec{\mathbf{B}}_0$ that has the same direction as the magnetic field in the selector (Active Fig. 22.12). Upon entering the second magnetic field, the ions move in a semicircle of radius r before striking a detector array at P. If the ions are positively charged, the beam deflects to the left as Active Figure 22.12 shows. If the ions are negatively charged, the beam deflects to the right. From Equation 22.3, we can express the ratio m/q as

$$\frac{m}{q} = \frac{rB_0}{v}$$

Using Equation 22.7 gives

$$\frac{m}{q} = \frac{rB_0 B}{E}$$

22.8 ◀

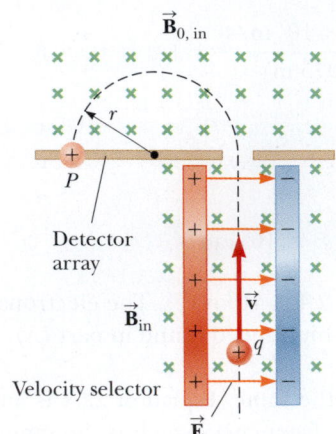

Active Figure 22.12 A mass spectrometer. Positively charged particles are sent first through a velocity selector and then into a region where the magnetic field $\vec{\mathbf{B}}_0$ causes the particles to move in a semicircular path and strike a detector array at P.

Therefore, we can determine m/q by measuring the radius of curvature and knowing the field magnitudes B, B_0, and E. In practice, one usually measures the masses of various isotopes of a given ion, with the ions all carrying the same charge q. In this way, the mass ratios can be determined even if q is unknown.

A variation of this technique was used by J. J. Thomson (1856–1940) in 1897 to measure the ratio e/m_e for electrons. Figure 22.13a shows the basic apparatus he used. Electrons are accelerated from the cathode and pass through two slits. They then drift into a region of perpendicular electric and magnetic fields. The magnitudes of the two fields are first adjusted to produce an undeflected beam. When the magnetic field is turned off, the electric field produces a measurable beam deflection that is recorded on the fluorescent screen. From the size of the deflection and the measured values of E and B, the charge-to-mass ratio can be determined. The results of this crucial experiment represent the discovery of the electron as a fundamental particle of nature.

Electrons are accelerated from the cathode, pass through two slits, and are deflected by both an electric field (formed by the charged deflection plates) and a magnetic field (directed perpendicular to the electric field). The beam of electrons then strikes a fluorescent screen.

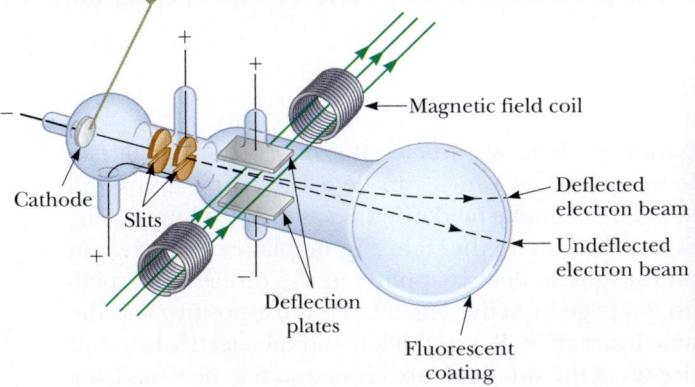

Figure 22.13 (a) Thomson's apparatus for measuring e/m_e. (b) J. J. Thomson (*left*) in the Cavendish Laboratory, University of Cambridge. The man on the right, Frank Baldwin Jewett, is a distant relative of John W. Jewett Jr., coauthor of this text.

The Cyclotron

A **cyclotron** can accelerate charged particles to very high speeds. Both electric and magnetic forces play a key role in its operation. The energetic particles produced are used to bombard atomic nuclei and thereby produce nuclear reactions of interest to researchers. A number of hospitals use cyclotron facilities to produce radioactive substances for diagnosis and treatment as well as beams of high-energy particles for treating cancer. At the time of this printing, there are 37 proton therapy centers around the world. These centers use either cyclotrons or another particle accelerator, called a synchrotron, to accelerate protons to high speeds to be used in external beam radiotherapy for cancer. (See the cover photo.) Conditions that have been treated with proton therapy include prostate cancer, retinoblastoma (a cancer of the eye), head and neck cancers, ocular melanoma, and acoustic neuroma.

BIO Use of cyclotrons in medicine

A schematic drawing of a cyclotron is shown in Figure 22.14a. The charges move inside two hollow metal semicircular containers, D_1 and D_2, referred to as *dees* because they are shaped like the letter D. A high-frequency alternating potential difference is applied to the dees, and a uniform magnetic field is directed perpendicular to them. A positive ion released at P near the center of the magnet moves in a semicircular path in one dee (indicated by the dashed black line in the drawing) and arrives back at the gap in a time interval $T/2$, where T is the time interval needed to make one complete trip around the two dees, given by Equation 22.5. The frequency of the applied potential difference is chosen so that the polarity of the dees is reversed during the time interval in which the ion travels around one dee. If the applied potential difference is adjusted such that D_2 is at a lower electric potential than D_1 by an amount ΔV, the ion accelerates across the gap to D_2 and its kinetic energy increases by an amount $q\Delta V$. It then moves around D_2 in a semicircular path of larger radius (because its speed has increased). After a time interval $T/2$, it again arrives at the gap between the dees. By this time, the polarity across the dees has reversed again and the ion is given another "kick" across the gap. The motion continues so that for each half-circle trip, the ion gains additional kinetic energy equal to $q\Delta V$. When the radius of its path is nearly that of the dees, the energetic ion leaves the system through the exit slit. It is important to note that the operation of the cyclotron is based on T being independent of the speed of the ion and the radius of its circular path (Eq. 22.5).

Pitfall Prevention | 22.1
The Cyclotron Is Not State-of-the-Art Technology
The cyclotron is important historically because it was the first particle accelerator to produce particles with very high speeds. Cyclotrons are still in use in medical applications, but most accelerators currently in research use are not cyclotrons. Research accelerators work on a different principle and are generally called *synchrotrons*.

We can obtain an expression for the kinetic energy of the ion when it exits from the cyclotron in terms of the radius R of the dees. From Equation 22.3 we know that $v = qBR/m$. Hence, the kinetic energy is

$$K = \tfrac{1}{2}mv^2 = \frac{q^2B^2R^2}{2m}$$

22.9 ◀

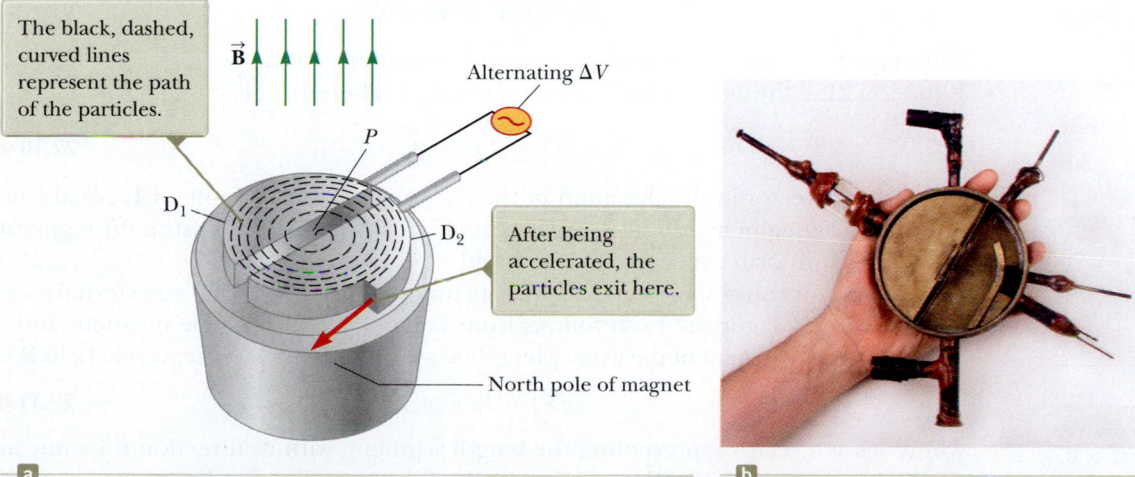

Figure 22.14 (a) A cyclotron consists of an ion source at P, two dees, D_1 and D_2, across which an alternating potential difference is applied, and a uniform magnetic field. (b) The first cyclotron, invented by E. O. Lawrence and M. S. Livingston in 1934.

When the energy of the ions in a cyclotron exceeds about 20 MeV, relativistic effects come into play. For this reason, the moving ions do not remain in phase with the applied potential difference. Some accelerators solve this problem by modifying the frequency of the applied potential difference so that it remains in phase with the moving ions.

⟨ 22.5 | Magnetic Force on a Current-Carrying Conductor

Because a magnetic force is exerted on a single charged particle when it moves through an external magnetic field, it should not surprise you to find that a current-carrying wire also experiences a magnetic force when placed in an external magnetic field because the current represents a collection of many charged particles in motion. Hence, the resultant magnetic force on the wire is due to the sum of the individual magnetic forces on the charged particles. The force on the particles is transmitted to the "bulk" of the wire through collisions with the atoms making up the wire.

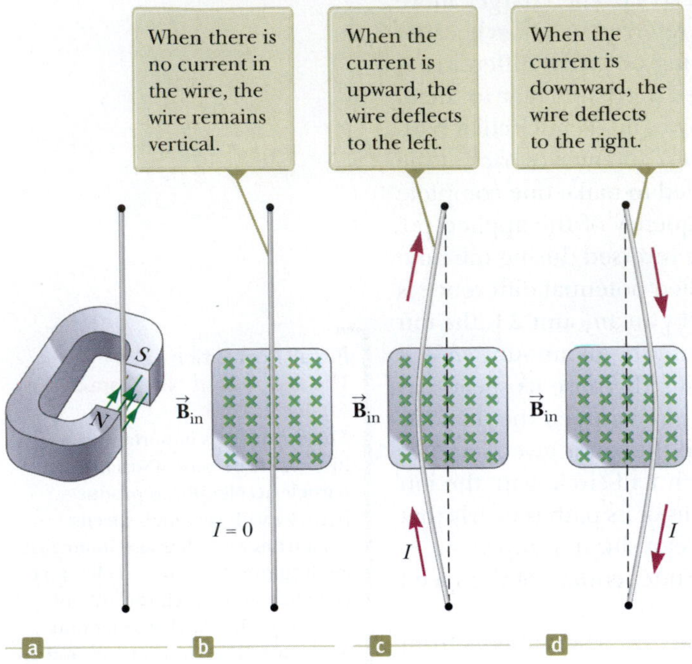

The magnetic force on a current-carrying conductor can be demonstrated by hanging a wire between the poles of a magnet as in Figure 22.15, where the magnetic field is directed into the page. The wire deflects to the left or right when a current is passed through it.

Let us quantify this discussion by considering a straight segment of wire of length L and cross-sectional area A, carrying a current I in a uniform external magnetic field $\vec{\mathbf{B}}$ as in Figure 22.16. As a simplification model, we shall ignore the high-speed zigzag motion of the charges in the wire (which is valid because the net velocity associated with this motion is zero) and assume that the charges simply move with the drift velocity $\vec{\mathbf{v}}_d$. The magnetic force on a charge q moving with drift velocity $\vec{\mathbf{v}}_d$ is $q\vec{\mathbf{v}}_d \times \vec{\mathbf{B}}$. To find the total magnetic force on the wire segment, we multiply the magnetic force on one charge by the number of charges in the segment. Because the volume of the segment is AL, the number of charges in the segment is nAL, where n is the number of charges per unit volume. Hence, the total magnetic force on the wire of length L is

$$\vec{\mathbf{F}}_B = (q\vec{\mathbf{v}}_d \times \vec{\mathbf{B}})\, nAL$$

This equation can be written in a more convenient form by noting that, from Equation 21.4, the current in the wire is $I = nqv_dA$. Therefore, $\vec{\mathbf{F}}_B$ can be expressed as

$$\vec{\mathbf{F}}_B = I\vec{\mathbf{L}} \times \vec{\mathbf{B}} \qquad \text{22.10} \blacktriangleleft$$

where $\vec{\mathbf{L}}$ is a vector in the direction of the current I; the magnitude of $\vec{\mathbf{L}}$ equals the length of the segment. Note that this expression applies only to a straight segment of wire in a uniform external magnetic field.

Now consider an arbitrarily shaped wire of uniform cross-section in an external magnetic field as in Figure 22.17. It follows from Equation 22.10 that the magnetic force on a very small segment of the wire of length ds in the presence of an external field $\vec{\mathbf{B}}$ is

$$d\vec{\mathbf{F}}_B = I\, d\vec{\mathbf{s}} \times \vec{\mathbf{B}} \qquad \text{22.11} \blacktriangleleft$$

where $d\vec{\mathbf{s}}$ is a vector representing the length segment, with its direction the same as that of the current, and $d\vec{\mathbf{F}}_B$ is directed out of the page for the directions assumed in Figure 22.17. We can consider Equation 22.11 as an alternative definition of $\vec{\mathbf{B}}$ to Equation 22.1. That is, the field $\vec{\mathbf{B}}$ can be defined in terms of a measurable force on

When there is no current in the wire, the wire remains vertical.

When the current is upward, the wire deflects to the left.

When the current is downward, the wire deflects to the right.

$\vec{\mathbf{B}}_{in}$ $\vec{\mathbf{B}}_{in}$ $\vec{\mathbf{B}}_{in}$

$I = 0$ I I

| a | b | c | d |

Figure 22.15 (a) A wire suspended vertically between the poles of a magnet. (b) through (d) The setup shown in (a) as seen looking at the south pole of the magnet so that the magnetic field (green crosses) is directed into the page.

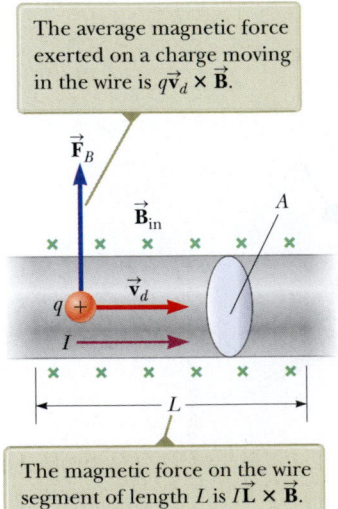

The average magnetic force exerted on a charge moving in the wire is $q\vec{\mathbf{v}}_d \times \vec{\mathbf{B}}$.

$\vec{\mathbf{F}}_B$

$\vec{\mathbf{B}}_{in}$ A

$q\ +$ $\vec{\mathbf{v}}_d$

I

L

The magnetic force on the wire segment of length L is $I\vec{\mathbf{L}} \times \vec{\mathbf{B}}$.

Figure 22.16 A segment of a current-carrying wire in a magnetic field $\vec{\mathbf{B}}$.

a current element, where the force is a maximum when $\vec{\mathbf{B}}$ is perpendicular to the element and zero when $\vec{\mathbf{B}}$ is parallel to the element.

To obtain the total magnetic force $\vec{\mathbf{F}}_B$ on a length of the wire between arbitrary points a and b, we integrate Equation 22.11 over the length of the wire between these points:

$$\vec{\mathbf{F}}_B = I \int_a^b d\vec{\mathbf{s}} \times \vec{\mathbf{B}} \qquad \text{22.12} \blacktriangleleft$$

When this integration is carried out, the magnitude of the magnetic field and the direction of the field relative to the vector $d\vec{\mathbf{s}}$ may vary from point to point.

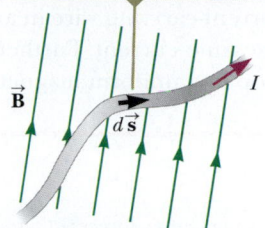

The magnetic force on any segment $d\vec{\mathbf{s}}$ is $I\,d\vec{\mathbf{s}} \times \vec{\mathbf{B}}$ and is directed out of the page.

Figure 22.17 A wire segment of arbitrary shape carrying a current I in a magnetic field $\vec{\mathbf{B}}$ experiences a magnetic force.

> **QUICK QUIZ 22.3** A wire carries current in the plane of this paper toward the top of the page. The wire experiences a magnetic force toward the right edge of the page. Is the direction of the magnetic field causing this force **(a)** in the plane of the page and toward the left edge, **(b)** in the plane of the page and toward the bottom edge, **(c)** upward out of the page, or **(d)** downward into the page?

> ▶ **THINKING PHYSICS 22.3**
>
> In a lightning stroke, negative charge rapidly moves from a cloud to the ground. In what direction is a lightning stroke deflected by the Earth's magnetic field?
>
> **Reasoning** The downward flow of negative charge in a lightning stroke is equivalent to an upward-moving current. Therefore, the vector $d\vec{\mathbf{s}}$ is upward, and the magnetic field vector has a northward component. According to the cross product of the length element and magnetic field vectors (Eq. 22.11), the lightning stroke would be deflected to the *west*. ◀

Example 22.4 | **Force on a Semicircular Conductor**

A wire bent into a semicircle of radius R forms a closed circuit and carries a current I. The wire lies in the xy plane, and a uniform magnetic field is directed along the positive y axis as in Figure 22.18. Find the magnitude and direction of the magnetic force acting on the straight portion of the wire and on the curved portion.

SOLUTION

Conceptualize Using the right-hand rule for cross products, we see that the force $\vec{\mathbf{F}}_1$ on the straight portion of the wire is out of the page and the force $\vec{\mathbf{F}}_2$ on the curved portion is into the page. Is $\vec{\mathbf{F}}_2$ larger in magnitude than $\vec{\mathbf{F}}_1$ because the length of the curved portion is longer than that of the straight portion?

Categorize Because we are dealing with a current-carrying wire in a magnetic field rather than a single charged particle, we must use Equation 22.12 to find the total force on each portion of the wire.

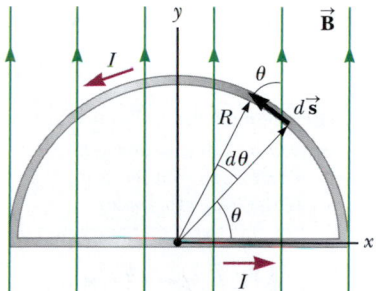

Figure 22.18 (Example 22.4) The magnetic force on the straight portion of the loop is directed out of the page, and the magnetic force on the curved portion is directed into the page.

....

Analyze Notice that $d\vec{\mathbf{s}}$ is perpendicular to $\vec{\mathbf{B}}$ everywhere on the straight portion of the wire. Use Equation 22.12 to find the force on this portion:

$$\vec{\mathbf{F}}_1 = I \int_a^b d\vec{\mathbf{s}} \times \vec{\mathbf{B}} = I \int_{-R}^R B\,dx\,\hat{\mathbf{k}} = \boxed{2IRB\,\hat{\mathbf{k}}}$$

To find the magnetic force on the curved part, first write an expression for the magnetic force $d\vec{\mathbf{F}}_2$ on the element $d\vec{\mathbf{s}}$ in Figure 22.18:

$$(1) \quad d\vec{\mathbf{F}}_2 = I\,d\vec{\mathbf{s}} \times \vec{\mathbf{B}} = -IB \sin\theta\,ds\,\hat{\mathbf{k}}$$

From the geometry in Figure 22.18, write an expression for ds:

$$(2) \quad ds = R\,d\theta$$

Substitute Equation (2) into Equation (1) and integrate over the angle θ from 0 to π:

$$\vec{\mathbf{F}}_2 = -\int_0^\pi IRB \sin\theta\,d\theta\,\hat{\mathbf{k}} = -IRB \int_0^\pi \sin\theta\,d\theta\,\hat{\mathbf{k}} = -IRB[-\cos\theta]_0^\pi\,\hat{\mathbf{k}}$$

$$= IRB(\cos\pi - \cos 0)\hat{\mathbf{k}} = IRB(-1 - 1)\hat{\mathbf{k}} = \boxed{-2IRB\,\hat{\mathbf{k}}}$$

continued

22.4 *cont.*

Finalize Two very important general statements follow from this example. First, the force on the curved portion is the same in magnitude as the force on the straight wire between the same two points. In general, the magnetic force on a curved current-carrying wire in a uniform magnetic field is equal to that on a straight wire connecting the endpoints and carrying the same current. Furthermore, $\vec{F}_1 + \vec{F}_2 = 0$ is also a general result: the net magnetic force acting on any closed current loop in a uniform magnetic field is zero.

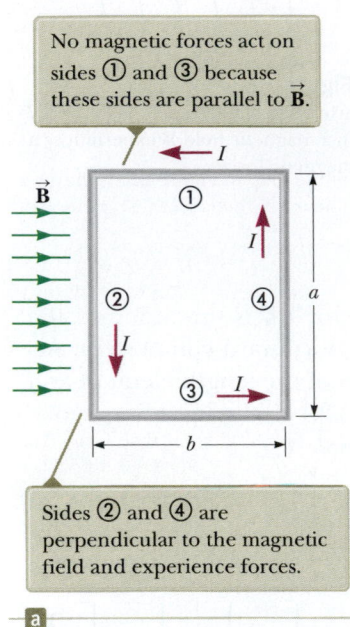

No magnetic forces act on sides ① and ③ because these sides are parallel to $\vec{B}$.

Sides ② and ④ are perpendicular to the magnetic field and experience forces.

a

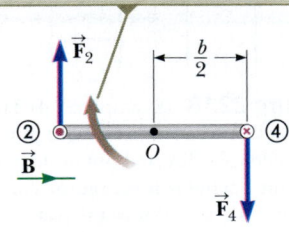

The magnetic forces $\vec{F}_2$ and $\vec{F}_4$ exerted on sides ② and ④ create a torque that tends to rotate the loop clockwise.

b

Figure 22.19 (a) Overhead view of a rectangular current loop in a uniform magnetic field. (b) Edge view of the loop looking from side ③. The purple dot in the left circle represents current in wire ② coming toward you; the purple × in the right circle represents current in wire ④ moving away from you.

22.6 | Torque on a Current Loop in a Uniform Magnetic Field

In the preceding section, we showed how a magnetic force is exerted on a current-carrying conductor when the conductor is placed in an external magnetic field. Starting at this point, we shall show that a *torque* is exerted on a current loop placed in a magnetic field. The results of this analysis are of great practical value in the design of motors and generators.

Consider a rectangular loop carrying a current I in the presence of a uniform external magnetic field *in the plane of the loop* as in Figure 22.19a. The magnetic forces on sides ① and ③, of length b, are zero because these wires are parallel to the field; hence, $d\vec{s} \times \vec{B} = 0$ for these sides. Nonzero magnetic forces act on sides ② and ④, however, because these sides are oriented perpendicular to the field. The magnitude of these forces is

$$F_2 = F_4 = IaB$$

We see that the net force on the loop is zero. The direction of $\vec{F}_2$, the magnetic force on side ②, is out of the paper, and that of $\vec{F}_4$, the magnetic force on side ④, is into the paper. If we look across the plane of the loop from side ③ as in Figure 22.19b, we see the forces on ② and ④ directed as shown. If we assume that the loop is pivoted so that it can rotate about an axis perpendicular to the page and passing through point O, we see that these two magnetic forces produce a net torque about this axis that rotates the loop clockwise. The magnitude of the torque, which we will call τ_{max}, is

$$\tau_{max} = F_2 \frac{b}{2} + F_4 \frac{b}{2} = (IaB)\frac{b}{2} + (IaB)\frac{b}{2} = IabB$$

where the moment arm about this axis is $b/2$ for each force. Because the area of the loop is $A = ab$, the magnitude of the torque can be expressed as

$$\tau_{max} = IAB \qquad\qquad \textbf{22.13} \blacktriangleleft$$

Remember that this torque occurs only when the field $\vec{B}$ is parallel to the plane of the loop. The sense of the rotation is clockwise when the loop is viewed as in Figure 22.19b. If the current were reversed, the magnetic forces would reverse their directions and the rotational tendency would be counterclockwise.

Now suppose the loop is rotated so that a line perpendicular to the plane of the loop makes an angle θ with the uniform magnetic field as in Active Figure 22.20. Notice that $\vec{B}$ is still perpendicular to sides ② and ④. In this case, the magnetic forces on sides ① and ③ cancel each other and produce no torque because they have the same line of action. The magnetic forces $\vec{F}_2$ and $\vec{F}_4$ acting on sides ② and ④, however, both produce a torque about an axis through the center of the loop. Referring to Active Figure 22.20, we note that the moment arm of $\vec{F}_2$ about this axis is $(b/2) \sin \theta$. Likewise, the moment arm of $\vec{F}_4$ is also $(b/2) \sin \theta$. Because $F_2 = F_4 = IaB$, the net torque τ has the magnitude

$$\tau = F_2 \frac{b}{2} \sin \theta + F_4 \frac{b}{2} \sin \theta$$

$$= (IaB)\left(\frac{b}{2} \sin \theta\right) + (IaB)\left(\frac{b}{2} \sin \theta\right) = IabB \sin \theta$$

$$= IAB \sin \theta$$

where $A = ab$ is the area of the loop. This result shows that the torque has its maximum value IAB (Eq. 22.13) when the field is parallel to the plane of the loop ($\theta = 90°$) and is zero when the field is perpendicular to the plane of the loop ($\theta = 0$). As we see in Active Figure 22.20, the loop tends to rotate in the direction of decreasing values of θ (i.e., so that the normal to the plane of the loop rotates toward the direction of the magnetic field). A convenient vector expression for the torque is

$$\vec{\tau} = I\vec{A} \times \vec{B} \qquad \text{22.14} \blacktriangleleft$$

where $\vec{A}$, a vector perpendicular to the plane of the loop (Active Fig. 22.20), has a magnitude equal to the area of the loop. The sense of $\vec{A}$ is determined by the right-hand rule illustrated in Figure 22.21. When the four fingers of the right hand are curled in the direction of the current in the loop, the thumb points in the direction of $\vec{A}$. The product $I\vec{A}$ is defined to be the **magnetic dipole moment** $\vec{\mu}$ (often simply called the "magnetic moment") of the loop:

$$\vec{\mu} = I\vec{A} \qquad \text{22.15} \blacktriangleleft$$

The SI unit of magnetic dipole moment is the ampere · meter² ($A \cdot m^2$). Using this definition, the torque can be expressed as

$$\vec{\tau} = \vec{\mu} \times \vec{B} \qquad \text{22.16} \blacktriangleleft$$

Although the torque was obtained for a particular orientation of $\vec{B}$ with respect to the loop, Equation 22.16 is valid for any orientation. Furthermore, although the torque expression was derived for a rectangular loop, the result is valid for a loop of any shape. Once the torque is determined, the loop can be modeled as a rigid object under a net torque, which we studied in Chapter 10.

If a coil consists of multiple loops such that there are N turns of wire, each carrying the same current and each having the same area, the total magnetic moment of the coil is the product of the number of turns and the magnetic moment for one turn, $\vec{\mu} = NI\vec{A}$. Therefore, the torque on an N-turn coil is N times greater than that on a one-turn loop.

A common electric motor consists of a coil of wire mounted so that it can rotate in the field of a permanent magnet. The torque on the current-carrying coil is used to rotate a shaft that drives a mechanical device such as the power windows in your car, your household fan, or your electric hedge trimmer.

Imagine the loop in Active Figure 22.20 released from rest. The magnetic moment vector (parallel to $\vec{A}$) will begin to rotate clockwise to line up with the magnetic field vector $\vec{B}$. Once $\vec{\mu}$ is aligned with $\vec{B}$, which is the equilibrium configuration, the angular momentum of the loop will carry it past this configuration and it will slow down due to the restoring torque. The result will be an oscillation around the equilibrium configuration. Let's ask a couple of questions about this situation. From where did the energy come that is associated with the oscillation of the loop-field system? It came from the work done by an external agent in initially rotating $\vec{\mu}$ away from the equilibrium position. Now, in what form is the energy in the system before the loop is released? It is in the form of *potential energy*, just like when a block on a spring is moved away from the equilibrium configuration. The potential energy of a system of a magnetic dipole in a magnetic field depends on the orientation of the dipole in the magnetic field and is given by

$$U = -\vec{\mu} \cdot \vec{B} \qquad \text{22.17} \blacktriangleleft$$

This expression shows that the system has its lowest energy $U_{\min} = -\mu B$ when $\vec{\mu}$ points in the same direction as $\vec{B}$. The system has its highest energy $U_{\max} = +\mu B$ when $\vec{\mu}$ points in the direction opposite $\vec{B}$.

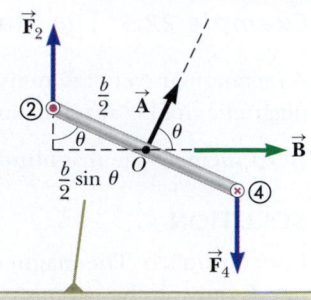

When the normal to the loop makes an angle θ with the magnetic field, the moment arm for the torque is $(b/2) \sin \theta$.

Active Figure 22.20 An end view of the loop in Figure 22.19 with the normal to the loop at an angle θ with respect to the magnetic field.

▶ Torque on a magnetic moment in a magnetic field

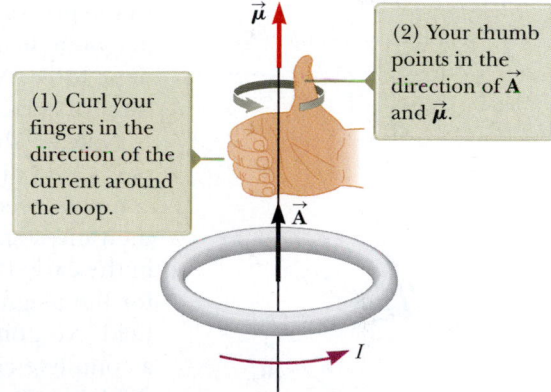

(1) Curl your fingers in the direction of the current around the loop.

(2) Your thumb points in the direction of $\vec{A}$ and $\vec{\mu}$.

Figure 22.21 Right-hand rule for determining the direction of the vector $\vec{A}$. The direction of the magnetic moment $\vec{\mu}$ is the same as the direction of $\vec{A}$.

▶ Potential energy of a system of a magnetic moment in a magnetic field

Example 22.5 | The Magnetic Dipole Moment of a Coil

A rectangular coil of dimensions 5.40 cm × 8.50 cm consists of 25 turns of wire and carries a current of 15.0 mA. A 0.350-T magnetic field is applied parallel to the plane of the coil.

(A) Calculate the magnitude of the magnetic dipole moment of the coil.

SOLUTION

Conceptualize The magnetic moment of the coil is independent of any magnetic field in which the loop resides, so it depends only on the geometry of the loop and the current it carries.

Categorize We evaluate quantities based on equations developed in this section, so we categorize this example as a substitution problem.

Use Equation 22.15 to calculate the magnetic moment associated with a coil consisting of N turns:

$$\mu_{coil} = NIA = (25)(15.0 \times 10^{-3}\,\text{A})(0.054\,0\,\text{m})(0.085\,0\,\text{m})$$
$$= 1.72 \times 10^{-3}\,\text{A}\cdot\text{m}^2$$

(B) What is the magnitude of the torque acting on the loop?

SOLUTION

Use Equation 22.16, noting that $\vec{\mathbf{B}}$ is perpendicular to $\vec{\boldsymbol{\mu}}_{coil}$:

$$\tau = \mu_{coil}B = (1.72 \times 10^{-3}\,\text{A}\cdot\text{m}^2)(0.350\,\text{T})$$
$$= 6.02 \times 10^{-4}\,\text{N}\cdot\text{m}$$

22.7 | The Biot–Savart Law

The direction of the field is out of the page at P.

The direction of the field is into the page at P'.

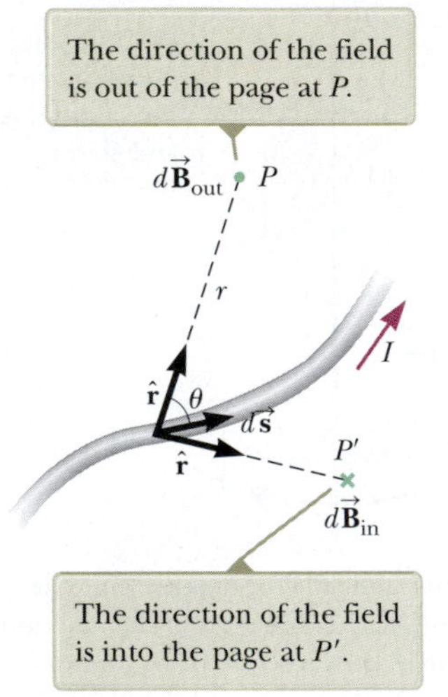

Figure 22.22 The magnetic field $d\vec{\mathbf{B}}$ at a point P due to a current I through a length element $d\vec{\mathbf{s}}$ is given by the Biot–Savart law.

Pitfall Prevention | 22.2
The Biot–Savart Law
The magnetic field described by the Biot–Savart law is the field *due to* a given current-carrying conductor. Do not confuse this field with any *external* field that may be applied to the conductor from some other source.

In the previous sections, we investigated the result of placing an object in an existing magnetic field. When a moving charge is placed in the field, it experiences a magnetic force. A current-carrying wire placed in the field also experiences a magnetic force; a current loop in the field experiences a torque.

Now we shift our thinking and investigate the *source* of the magnetic field. Oersted's 1819 discovery (Section 22.1) that an electric current in a wire deflects a nearby compass needle indicates that a current acts as a source of a magnetic field. From their investigations on the force between a current-carrying conductor and a magnet in the early 19th century, Jean-Baptiste Biot and Félix Savart arrived at an expression for the magnetic field at a point in space in terms of the current that produces the field. No point currents exist comparable to point charges (because we must have a complete circuit for a current to exist). Hence, we must investigate the magnetic field due to an infinitesimally small element of current that is part of a larger current distribution. Suppose the current distribution is a wire carrying a steady current I as in Figure 22.22. Experimental results show that the magnetic field $d\vec{\mathbf{B}}$ at point P created by an element of infinitesimal length ds of the wire has the following properties:

- The vector $d\vec{\mathbf{B}}$ is perpendicular both to $d\vec{\mathbf{s}}$ (which is in the direction of the current) and to the unit vector $\hat{\mathbf{r}}$ directed from the element toward P.
- The magnitude of $d\vec{\mathbf{B}}$ is inversely proportional to r^2, where r is the distance from the element to P.
- The magnitude of $d\vec{\mathbf{B}}$ is proportional to the current I and to the length ds of the element.
- The magnitude of $d\vec{\mathbf{B}}$ is proportional to $\sin\theta$, where θ is the angle between $d\vec{\mathbf{s}}$ and $\hat{\mathbf{r}}$.

The **Biot–Savart law** describes these results and can be summarized in the following compact form:

$$d\vec{\mathbf{B}} = k_m \frac{I\,d\vec{\mathbf{s}} \times \hat{\mathbf{r}}}{r^2} \qquad \text{22.18} \blacktriangleleft$$

where k_m is a constant that in SI units is exactly 10^{-7} T·m/A. The constant k_m is usually written $\mu_0/4\pi$, where μ_0 is another constant, called the **permeability of free space:**

$$\frac{\mu_0}{4\pi} = k_m = 10^{-7} \text{ T} \cdot \text{m/A}$$

$$\mu_0 = 4\pi k_m = 4\pi \times 10^{-7} \text{ T} \cdot \text{m/A}$$

22.19◀   ▶ Permeability of free space

Hence, the Biot–Savart law, Equation 22.18, can also be written

$$d\vec{\mathbf{B}} = \frac{\mu_0}{4\pi} \frac{I \, d\vec{\mathbf{s}} \times \hat{\mathbf{r}}}{r^2}$$

22.20◀   ▶ Biot–Savart law

It is important to note that the Biot–Savart law gives the magnetic field at a point only for a small length element of the conductor. We identify the product $I \, d\vec{\mathbf{s}}$ as a **current element**. To find the total magnetic field $\vec{\mathbf{B}}$ at some point due to a conductor of finite size, we must sum contributions from all current elements making up the conductor. That is, we evaluate $\vec{\mathbf{B}}$ by integrating Equation 22.20 over the entire conductor.

There are two similarities between the Biot–Savart law of magnetism and Equation 19.7 for the electric field of a charge distribution, and there are two important differences. The current element $I \, d\vec{\mathbf{s}}$ produces a magnetic field, and the charge element dq produces an electric field. Furthermore, the magnitude of the magnetic field varies as the inverse square of the distance from the current element, as does the electric field due to a charge element. The directions of the two fields are quite different, however. The electric field due to a charge element is radial; in the case of a positive point charge, $\vec{\mathbf{E}}$ is directed away from the charge. The magnetic field due to a current element is perpendicular to both the current element and the radius vector. Hence, if the conductor lies in the plane of the page, as in Figure 22.22, $d\vec{\mathbf{B}}$ points out of the page at the point P and into the page at P'. Another important difference is that an electric field can be a result either of a single charge or a distribution of charges, but a magnetic field can only be a result of a current distribution.

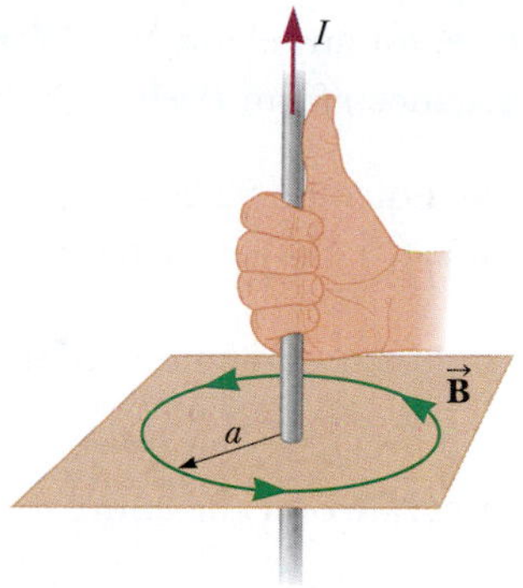

Figure 22.23 shows a convenient right-hand rule for determining the direction of the magnetic field due to a current. Note that the field lines generally encircle the current. In the case of current in a long, straight wire, the field lines form circles that are concentric with the wire and are in a plane perpendicular to the wire. If the wire is grasped in the right hand with the thumb in the direction of the current, the fingers will curl in the direction of $\vec{\mathbf{B}}$.

Although the magnetic field due to an infinitely long, current-carrying wire can be calculated using the Biot–Savart law (Problem 22.70 in Enhanced WebAssign), in Section 22.9 we use a different method to show that the magnitude of this field at a distance r from the wire is

$$B = \frac{\mu_0 I}{2\pi r}$$

22.21◀   ▶ Magnetic field due to a long, straight wire

Figure 22.23 The right-hand rule for determining the direction of the magnetic field surrounding a long, straight wire carrying a current. Note that the magnetic field lines form circles around the wire. The magnitude of the magnetic field at a distance r from the wire is given by Equation 22.21.

QUICK QUIZ 22.4 Consider the magnetic field due to the current in the wire shown in Figure 22.24. Rank the points A, B, and C in terms of magnitude of the magnetic field that is due to the current in just the length element $d\vec{\mathbf{s}}$ shown, from greatest to least.

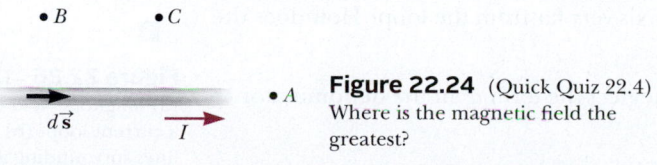

• B • C

• A

$d\vec{\mathbf{s}}$ I

Figure 22.24 (Quick Quiz 22.4) Where is the magnetic field the greatest?

Example 22.6 | Magnetic Field on the Axis of a Circular Current Loop

Consider a circular wire loop of radius a located in the yz plane and carrying a steady current I as in Figure 22.25. Calculate the magnetic field at an axial point P a distance x from the center of the loop.

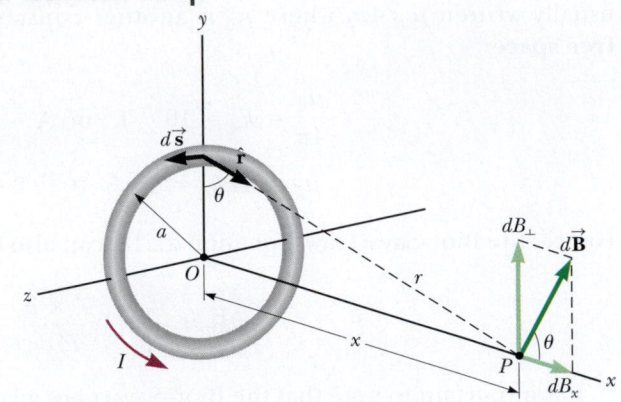

SOLUTION

Conceptualize Figure 22.25 shows the magnetic field contribution $d\vec{B}$ at P due to a single current element at the top of the ring. This field vector can be resolved into components dB_x parallel to the axis of the ring and $dB_\perp$ perpendicular to the axis. Think about the magnetic field contributions from a current element at the bottom of the loop. Because of the symmetry of the situation, the perpendicular components of the field due to elements at the top and bottom of the ring cancel. This cancellation occurs for all pairs of segments around the ring, so we can ignore the perpendicular component of the field and focus solely on the parallel components, which simply add.

Figure 22.25 (Example 22.6) Geometry for calculating the magnetic field at a point P lying on the axis of a current loop. By symmetry, the total field $\vec{B}$ is along this axis.

Categorize We are asked to find the magnetic field due to a simple current distribution, so this example is a typical problem for which the Biot–Savart law is appropriate.

Analyze In this situation, every length element $d\vec{s}$ is perpendicular to the vector $\hat{r}$ at the location of the element. Therefore, for any element, $|d\vec{s} \times \hat{r}| = (ds)(1)\sin 90° = ds$. Furthermore, all length elements around the loop are at the same distance r from P, where $r^2 = a^2 + x^2$.

Use Equation 22.20 to find the magnitude of $d\vec{B}$ due to the current in any length element $d\vec{s}$:

$$dB = \frac{\mu_0 I}{4\pi} \frac{|d\vec{s} \times \hat{r}|}{r^2} = \frac{\mu_0 I}{4\pi} \frac{ds}{(a^2 + x^2)} \qquad \textbf{22.22} \blacktriangleleft$$

Find the x component of the field element:

$$dB_x = \frac{\mu_0 I}{4\pi} \frac{ds}{(a^2 + x^2)} \cos\theta$$

Integrate over the entire loop:

$$B_x = \oint dB_x = \frac{\mu_0 I}{4\pi} \oint \frac{ds \cos\theta}{a^2 + x^2}$$

From the geometry, evaluate $\cos\theta$:

$$\cos\theta = \frac{a}{(a^2 + x^2)^{1/2}}$$

Substitute this expression for $\cos\theta$ into the integral and note that x and a are both constant:

$$B_x = \frac{\mu_0 I}{4\pi} \oint \frac{ds}{a^2 + x^2} \frac{a}{(a^2 + x^2)^{1/2}} = \frac{\mu_0 I}{4\pi} \frac{a}{(a^2 + x^2)^{3/2}} \oint ds$$

Integrate around the loop:

$$B_x = \frac{\mu_0 I}{4\pi} \frac{a}{(a^2 + x^2)^{3/2}}(2\pi a) = \frac{\mu_0 I a^2}{2(a^2 + x^2)^{3/2}} \qquad \textbf{22.23} \blacktriangleleft$$

Finalize To find the magnetic field at the center of the loop, set $x = 0$ in Equation 22.23. At this special point,

$$B = \frac{\mu_0 I}{2a} \text{ (at } x = 0) \qquad \textbf{22.24} \blacktriangleleft$$

The pattern of magnetic field lines for a circular current loop is shown in Figure 22.26a. For clarity, the lines are drawn for only the plane that contains the axis of the loop. The field-line pattern is axially symmetric and looks like the pattern around a bar magnet, which is shown in Figure 22.26b.

What If? What if we consider points on the x axis very far from the loop? How does the magnetic field behave at these distant points?

Answer In this case, in which $x \gg a$, we can neglect the term a^2 in the denominator of Equation 22.23 and obtain

$$B \approx \frac{\mu_0 I a^2}{2x^3} \quad \text{(for } x \gg a) \qquad \textbf{22.25} \blacktriangleleft$$

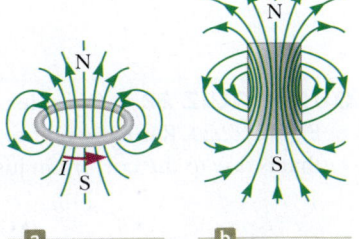

Figure 22.26 (Example 22.6) (a) Magnetic field lines surrounding a current loop. (b) Magnetic field lines surrounding a bar magnet. Notice the similarity between this line pattern and that of a current loop.

22.6 *cont.*

The magnitude of the magnetic moment $\vec{\mu}$ of the loop is defined as the product of current and loop area (see Eq. 22.15): $\mu = I(\pi a^2)$ for our circular loop. We can express Equation 22.25 as

$$B \approx \frac{\mu_0}{2\pi} \frac{\mu}{x^3}$$

22.26 ◀

This result is similar in form to the expression for the electric field due to an electric dipole, $E = k_e(p/y^3)$ (see Example 19.4), where $p = 2aq$ is the electric dipole moment.

22.8 | The Magnetic Force Between Two Parallel Conductors

In Section 22.5, we described the magnetic force that acts on a current-carrying conductor when the conductor is placed in an external magnetic field. Because a current in a conductor sets up its own magnetic field, it is easy to understand that two current-carrying conductors exert magnetic forces on each other. As we shall see, such forces can be used as the basis for defining the ampere and the coulomb.

Consider two long, straight, parallel wires separated by the distance a and carrying currents I_1 and I_2 in the same direction as in Active Figure 22.27. We shall adopt a simplification model in which the radii of the wires are much smaller than a so that the radius plays no role in the calculation. We can determine the force on one wire due to the magnetic field set up by the other wire. Wire 2, which carries current I_2, sets up a magnetic field $\vec{B}_2$ at the position of wire 1. The direction of $\vec{B}_2$ is perpendicular to the wire as shown in Active Figure 22.27. According to Equation 22.10, the magnetic force on a length ℓ of wire 1 is $\vec{F}_1 = I_1 \vec{\ell} \times \vec{B}_2$. Because $\vec{\ell}$ is perpendicular to $\vec{B}_2$, the magnitude of $\vec{F}_1$ is $F_1 = I_1 \ell B_2$. Because the field due to wire 2 is given by Equation 22.21, we see that

$$F_1 = I_1 \ell B_2 = I_1 \ell \left(\frac{\mu_0 I_2}{2\pi a}\right) = \frac{\mu_0 I_1 I_2}{2\pi a} \ell$$

We can rewrite this expression in terms of the force per unit length as

$$\frac{F_1}{\ell} = \frac{\mu_0 I_1 I_2}{2\pi a}$$

The direction of $\vec{F}_1$ is downward, toward wire 2, because $\vec{\ell} \times \vec{B}_2$ is downward. If one considers the field set up at wire 2 due to wire 1, the force $\vec{F}_2$ on wire 2 is found to be equal in magnitude and opposite in direction to $\vec{F}_1$. That is what one would expect because Newton's third law must be obeyed. Therefore, we can drop the force subscript so that the magnetic force per unit length exerted by each long current-carrying wire on the other is

$$\frac{F}{\ell} = \frac{\mu_0 I_1 I_2}{2\pi a}$$

22.27 ◀

▶ Magnetic force per unit length between parallel current-carrying wires

This equation also applies if one of the wires is of finite length. In the discussion above, we used the equation for the magnetic field of an infinite wire carrying current I_2, but did not require that wire 1 be of infinite length.

When the currents are in opposite directions, the magnetic forces are reversed and the wires repel each other. Hence, we find that parallel conductors carrying currents in the same direction attract each other, whereas parallel conductors carrying currents in opposite directions repel each other.

The magnetic force between two parallel wires, each carrying a current, is used to define the **ampere**:

If two long, parallel wires 1 m apart carry the same current and the force per unit length on each wire is 2×10^{-7} N/m, the current is defined to be 1 A.

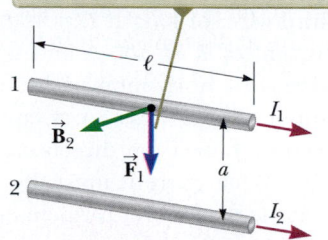

The field $\vec{B}_2$ due to the current in wire 2 exerts a magnetic force of magnitude $F_1 = I_1 \ell B_2$ on wire 1.

Active Figure 22.27 Two parallel wires that each carry a steady current exert a force on each other. The force is attractive if the currents are parallel (as shown) and repulsive if the currents are antiparallel.

The numerical value of 2×10^{-7} N/m is obtained from Equation 22.27, with $I_1 = I_2 = 1$ A and $a = 1$ m.

The SI unit of charge, the **coulomb,** can now be defined in terms of the ampere: If a conductor carries a steady current of 1 A, the quantity of charge that flows through a cross section of the conductor in 1 s is 1 C.

> **QUICK QUIZ 22.5** A loose spiral spring carrying no current is hung from a ceiling. When a switch is thrown so that a current exists in the spring, do the coils **(a)** move closer together, **(b)** move farther apart, or **(c)** not move at all?

Example 22.7 | Suspending a Wire

Two infinitely long, parallel wires are lying on the ground a distance $a = 1.00$ cm apart as shown in Figure 22.28a. A third wire, of length $L = 10.0$ m and mass 400 g, carries a current of $I_1 = 100$ A and is levitated above the first two wires, at a horizontal position midway between them. The infinitely long wires carry equal currents I_2 in the same direction, but in the direction opposite that in the levitated wire. What current must the infinitely long wires carry so that the three wires form an equilateral triangle?

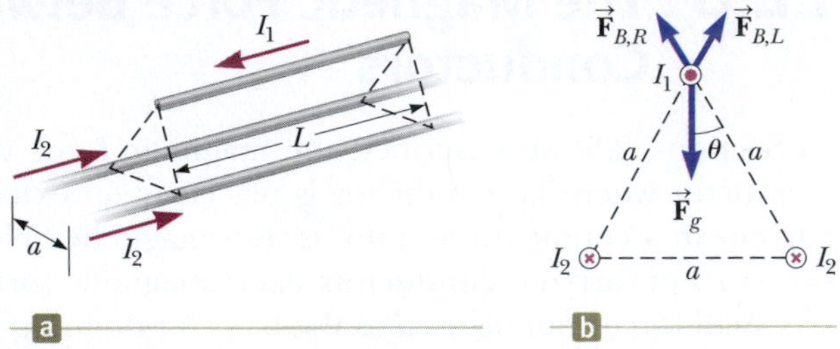

Figure 22.28 (Example 22.7) (a) Two current-carrying wires lie on the ground and suspend a third wire in the air by magnetic forces. (b) End view. In the situation described in the example, the three wires form an equilateral triangle. The two magnetic forces on the levitated wire are $\vec{\mathbf{F}}_{B,L}$, the force due to the left-hand wire on the ground, and $\vec{\mathbf{F}}_{B,R}$, the force due to the right-hand wire. The gravitational force $\vec{\mathbf{F}}_g$ on the levitated wire is also shown.

SOLUTION

Conceptualize Because the current in the short wire is opposite those in the long wires, the short wire is repelled from both of the others. Imagine that the currents in the long wires in Figure 22.28a are increased. The repulsive force becomes stronger, and the levitated wire rises to the point at which the wire is once again levitated in equilibrium at a higher position. Figure 22.28b shows the desired situation with the three wires forming an equilateral triangle.

Categorize Because the levitated wire is subject to forces but does not accelerate, it is modeled as a particle in equilibrium.

Analyze The horizontal components of the magnetic forces on the levitated wire cancel. The vertical components are both positive and add together. Choose the z axis to be upward through the top wire in Figure 22.28b and in the plane of the page.

Find the total magnetic force in the upward direction on the levitated wire:

$$\vec{\mathbf{F}}_B = 2\left(\frac{\mu_0 I_1 I_2}{2\pi a}\ell\right)\cos\theta\,\hat{\mathbf{k}} = \frac{\mu_0 I_1 I_2}{\pi a}\ell\cos\theta\,\hat{\mathbf{k}}$$

Find the gravitational force on the levitated wire:

$$\vec{\mathbf{F}}_g = -mg\hat{\mathbf{k}}$$

Apply the particle in equilibrium model by adding the forces and setting the net force equal to zero:

$$\sum \vec{\mathbf{F}} = \vec{\mathbf{F}}_B + \vec{\mathbf{F}}_g = \frac{\mu_0 I_1 I_2}{\pi a}\ell\cos\theta\,\hat{\mathbf{k}} - mg\hat{\mathbf{k}} = 0$$

Solve for the current in the wires on the ground:

$$I_2 = \frac{mg\pi a}{\mu_0 I_1 \ell \cos\theta}$$

Substitute numerical values:

$$I_2 = \frac{(0.400\ \text{kg})(9.80\ \text{m/s}^2)\pi(0.010\ 0\ \text{m})}{(4\pi \times 10^{-7}\ \text{T}\cdot\text{m/A})(100\ \text{A})(10.0\ \text{m})\cos 30.0°}$$

$$= \boxed{113\ \text{A}}$$

Finalize The currents in all wires are on the order of 10^2 A. Such large currents would require specialized equipment. Therefore, this situation would be difficult to establish in practice. Is the equilibrium of wire 1 stable or unstable?

22.9 | Ampère's Law

A simple experiment first carried out by Oersted in 1820 clearly demonstrates that a current-carrying conductor produces a magnetic field. In this experiment, several compass needles are placed in a horizontal plane near a long vertical wire as

Active Figure 22.29 (a) and (b) Compasses show the effects of the current in a nearby wire. (c) Circular magnetic field lines surrounding a current-carrying conductor, displayed with iron filings.

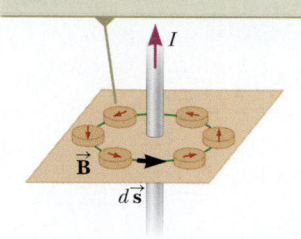

When no current is present in the wire, all compass needles point in the same direction (toward the Earth's north pole).

$I = 0$

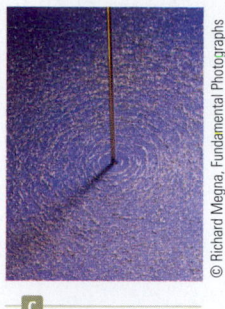

When the wire carries a strong current, the compass needles deflect in a direction tangent to the circle, which is the direction of the magnetic field created by the current.

I

$\vec{\mathbf{B}}$

$d\vec{\mathbf{s}}$

a   b   c

© Richard Megna, Fundamental Photographs

in Active Figure 22.29a. When the wire carries no current, all needles point in the same direction (that of the Earth's magnetic field), as one would expect. When the wire carries a strong, steady current, however, the needles all deflect in a direction tangent to the circle as in Active Figure 22.29b. These observations show that the direction of $\vec{\mathbf{B}}$ is consistent with the right-hand rule described in Section 22.7. When the current is reversed, the needles in Active Figure 22.29b also reverse.

Because the needles point in the direction of $\vec{\mathbf{B}}$, we conclude that the lines of $\vec{\mathbf{B}}$ form circles about the wire, as discussed in Section 22.7. By symmetry, the magnitude of $\vec{\mathbf{B}}$ is the same everywhere on a circular path that is centered on the wire and lies in a plane perpendicular to the wire. By varying the current and distance from the wire, one finds that $\vec{\mathbf{B}}$ is proportional to the current and inversely proportional to the distance from the wire.

In Chapter 19, we investigated Gauss's law, which is a relationship between an electric charge and the electric field it produces. Gauss's law can be used to determine the electric field in highly symmetric situations. We now consider an analogous relationship in magnetism between a current and the magnetic field it produces. This relationship can be used to determine the magnetic field created by a highly symmetric current distribution.

Let us evaluate the product $\vec{\mathbf{B}} \cdot d\vec{\mathbf{s}}$ for a small length element $d\vec{\mathbf{s}}$ on the circular path[1] centered on the wire in Active Figure 22.29b. Along this path, the vectors $d\vec{\mathbf{s}}$ and $\vec{\mathbf{B}}$ are parallel at each point, so $\vec{\mathbf{B}} \cdot d\vec{\mathbf{s}} = B\,ds$. Furthermore, by symmetry, $\vec{\mathbf{B}}$ is constant in magnitude on this circle and is given by Equation 22.21. Therefore, the sum of the products $B\,ds$ over the closed path, which is equivalent to the line integral of $\vec{\mathbf{B}} \cdot d\vec{\mathbf{s}}$, is

$$\oint \vec{\mathbf{B}} \cdot d\vec{\mathbf{s}} = B \oint ds = \frac{\mu_0 I}{2\pi r}(2\pi r) = \mu_0 I \qquad \textbf{22.28} \blacktriangleleft$$

where $\oint ds = 2\pi r$ is the circumference of the circle.

This result was calculated for the special case of a circular path surrounding a wire. It can, however, also be applied in the general case in which a steady current passes through the area surrounded by an arbitrary closed path. The general result is **Ampère's law:**

The line integral of $\vec{\mathbf{B}} \cdot d\vec{\mathbf{s}}$ around any closed path equals $\mu_0 I$, where I is the total steady current passing through any surface bounded by the closed path:

$$\oint \vec{\mathbf{B}} \cdot d\vec{\mathbf{s}} = \mu_0 I \qquad \textbf{22.29} \blacktriangleleft \qquad \blacktriangleright \text{ Ampère's law}$$

---

[1]You may wonder why we would choose to do this evaluation. The origin of Ampère's law is in 19th-century science, in which a "magnetic charge" (the supposed analog to an isolated electric charge) was imagined to be moved around a circular field line. The work done on the charge was related to $\vec{\mathbf{B}} \cdot d\vec{\mathbf{s}}$, just like the work done moving an electric charge in an electric field is related to $\vec{\mathbf{E}} \cdot d\vec{\mathbf{s}}$. Therefore, Ampère's law, a valid and useful principle, arose from an erroneous and abandoned work calculation!

QUICK QUIZ 22.6 Rank the magnitudes of $\oint \vec{\mathbf{B}} \cdot d\vec{\mathbf{s}}$ for the closed paths a through d in Figure 22.30, from greatest to least.

QUICK QUIZ 22.7 Rank the magnitudes of $\oint \vec{\mathbf{B}} \cdot d\vec{\mathbf{s}}$ for the closed paths a through d in Figure 22.31, from greatest to least.

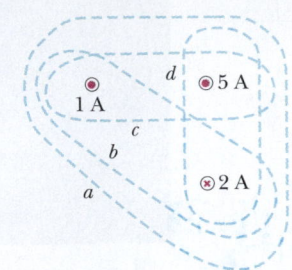

Figure 22.30 (Quick Quiz 22.6)
Four closed paths around three current-carrying wires.

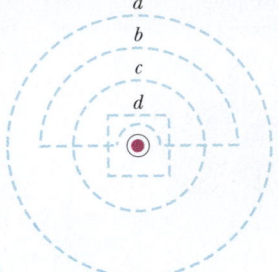

Figure 22.31 (Quick Quiz 22.7)
Four closed paths near a single current-carrying wire.

Andre-Marie Ampère
French Physicist (1775–1836)
Ampère is credited with the discovery of electromagnetism, the relationship between electric currents and magnetic fields. Ampère's genius, particularly in mathematics, became evident by the time he was 12 years old; his personal life, however, was filled with tragedy. His father, a wealthy city official, was guillotined during the French Revolution, and his wife died young, in 1803. Ampère died at the age of 61 of pneumonia. His judgment of his life is clear from the epitaph he chose for his gravestone: *Tandem Felix* (Happy at Last).

Ampère's law is valid only for steady currents. Furthermore, even though Ampère's law is *true* for all current configurations, it is only *useful* for calculating the magnetic fields of configurations with high degrees of symmetry.

In Section 19.10, we provided some conditions to be sought when defining a gaussian surface. Similarly, to apply Equation 22.29 to calculate a magnetic field, we must determine a path of integration (sometimes called an *amperian loop*) such that each portion of the path satisfies one or more of the following conditions:

1. The value of the magnetic field can be argued by symmetry to be constant over the portion of the path.
2. The dot product in Equation 22.29 can be expressed as a simple algebraic product $B\, ds$ because $\vec{\mathbf{B}}$ and $d\vec{\mathbf{s}}$ are parallel.
3. The dot product in Equation 22.29 is zero because $\vec{\mathbf{B}}$ and $d\vec{\mathbf{s}}$ are perpendicular.
4. The magnetic field can be argued to be zero at all points on the portion of the path.

The following examples illustrate some symmetric current configurations for which Ampère's law is useful.

Example 22.8 | The Magnetic Field Created by a Long Current-Carrying Wire

A long, straight wire of radius R carries a steady current I that is uniformly distributed through the cross section of the wire (Fig. 22.32). Calculate the magnetic field a distance r from the center of the wire in the regions $r \geq R$ and $r < R$.

SOLUTION

Conceptualize Study Figure 22.32 to understand the structure of the wire and the current in the wire. The current creates magnetic fields everywhere, both inside and outside the wire.

Categorize Because the wire has a high degree of symmetry, we categorize this example as an Ampère's law problem. For the $r \geq R$ case, we should arrive at the same result as shown in Equation 22.21.

Analyze For the magnetic field exterior to the wire, let us choose for our path of integration circle 1 in Figure 22.32. From symmetry, $\vec{\mathbf{B}}$ must be constant in magnitude and parallel to $d\vec{\mathbf{s}}$ at every point on this circle, satisfying conditions 1 and 2 above.

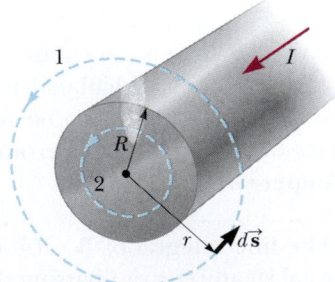

Figure 22.32 (Example 22.8)
A long, straight wire of radius R carrying a steady current I uniformly distributed across the cross section of the wire. The magnetic field at any point can be calculated from Ampère's law using a circular path of radius r, concentric with the wire.

22.8 *cont.*

Note that the total current passing through the plane of the circle is I and apply Ampère's law:

$$\oint \vec{B} \cdot d\vec{s} = B \oint ds = B(2\pi r) = \mu_0 I$$

Solve for B:

$$B = \frac{\mu_0 I}{2\pi r} \quad \text{(for } r \geq R)$$

Now consider the interior of the wire, where $r < R$. Here the current I' passing through the plane of circle 2 is less than the total current I.

Set the ratio of the current I' enclosed by circle 2 to the entire current I equal to the ratio of the area πr^2 enclosed by circle 2 to the cross-sectional area πR^2 of the wire:

$$\frac{I'}{I} = \frac{\pi r^2}{\pi R^2}$$

Solve for I':

$$I' = \frac{r^2}{R^2} I$$

Apply Ampère's law to circle 2:

$$\oint \vec{B} \cdot d\vec{s} = B(2\pi r) = \mu_0 I' = \mu_0 \left(\frac{r^2}{R^2} I\right)$$

Solve for B:

$$B = \left(\frac{\mu_0 I}{2\pi R^2}\right) r \quad \text{(for } r < R) \qquad \text{22.30} \blacktriangleleft$$

Finalize The magnetic field exterior to the wire is identical in form to Equation 22.21. As is often the case in highly symmetric situations, it is much easier to use Ampère's law than the Biot–Savart law to find this result. The magnetic field interior to the wire is similar in form to the expression for the electric field inside a uniformly charged sphere (see Example 19.10). The magnitude of the magnetic field versus r for this configuration is plotted in Figure 22.33. Inside the wire, $B \to 0$ as $r \to 0$. Furthermore, the results for $r > R$ and $r < R$ give the same value of the magnetic field at $r = R$, demonstrating that the magnetic field is continuous at the surface of the wire.

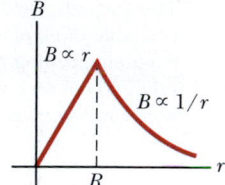

Figure 22.33 (Example 22.8) Magnitude of the magnetic field versus r for the wire shown in Figure 22.32. The field is proportional to r inside the wire and varies as $1/r$ outside the wire.

Example 22.9 | The Magnetic Field Created by a Toroid

A device called a *toroid* (Fig. 22.34) is often used to create an almost uniform magnetic field in some enclosed area. The device consists of a conducting wire wrapped around a ring (a *torus*) made of a nonconducting material. For a toroid having N closely spaced turns of wire, calculate the magnetic field in the region occupied by the torus, a distance r from the center.

SOLUTION

Conceptualize Study Figure 22.34 carefully to understand how the wire is wrapped around the torus. The torus could be a solid material or it could be air, with a stiff wire wrapped into the shape shown in Figure 22.34 to form an empty toroid.

Categorize Because the toroid has a high degree of symmetry, we categorize this example as an Ampère's law problem.

Analyze Consider the circular amperian loop (loop 1) of radius r in the plane of Figure 22.34. By symmetry, the magnitude of the field is constant on this circle and tangent to it, so $\vec{B} \cdot d\vec{s} = B \, ds$. Furthermore, the wire passes through the loop N times, so the total current through the loop is NI.

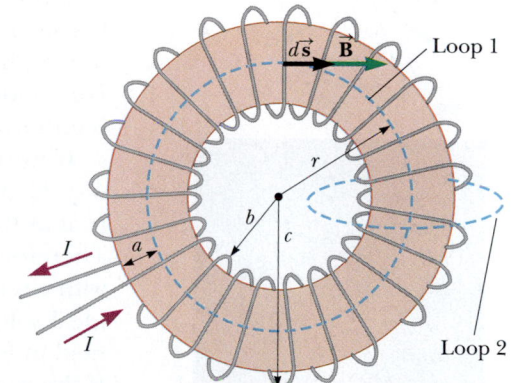

Figure 22.34 (Example 22.9) A toroid consisting of many turns of wire. If the turns are closely spaced, the magnetic field in the interior of the toroid is tangent to the dashed circle (loop 1) and varies as $1/r$. The dimension a is the cross-sectional radius of the torus. The field outside the toroid is very small and can be described by using the amperian loop (loop 2) at the right side, perpendicular to the page.

continued

22.9 *cont.*

Apply Ampère's law to loop 1:

$$\oint \vec{B} \cdot d\vec{s} = B \oint ds = B(2\pi r) = \mu_0 NI$$

Solve for B:

$$B = \frac{\mu_0 NI}{2\pi r} \qquad \qquad \textbf{22.31} \blacktriangleleft$$

Finalize This result shows that B varies as $1/r$ and hence is *nonuniform* in the region occupied by the torus. If, however, r is very large compared with the cross-sectional radius a of the torus, the field is approximately uniform inside the torus.

For an ideal toroid, in which the turns are closely spaced, the external magnetic field is close to zero, but it is not exactly zero. In Figure 22.34, imagine the radius r of amperian loop 1 to be either smaller than b or larger than c. In either case, the loop encloses zero net current, so $\oint \vec{B} \cdot d\vec{s} = 0$. You might think this result proves that $\vec{B} = 0$, but it does not. Consider the amperian loop (loop 2)

on the right side of the toroid in Figure 22.34. The plane of this loop is perpendicular to the page, and the toroid passes through the loop. As charges enter the toroid as indicated by the current directions in Figure 22.34, they work their way counterclockwise around the toroid. Therefore, a current passes through the perpendicular amperian loop! This current is small, but not zero. As a result, the toroid acts as a current loop and produces a weak external field of the form shown in Figure 22.26. The reason $\oint \vec{B} \cdot d\vec{s} = 0$ for amperian loop 1 of radius $r < b$ or $r > c$ is that the field lines are perpendicular to $d\vec{s}$, *not* because $\vec{B} = 0$.

The magnetic field lines resemble those of a bar magnet, meaning that the solenoid effectively has north and south poles.

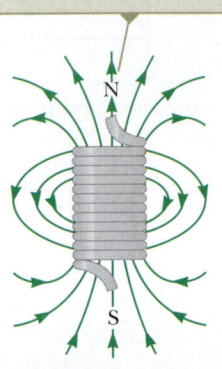

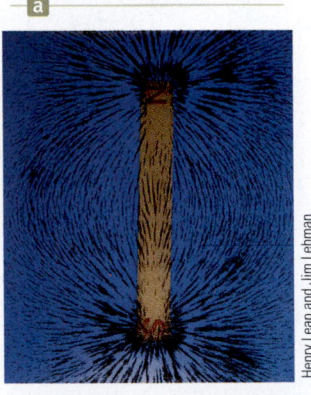

Figure 22.35 (a) Magnetic field lines for a tightly wound solenoid of finite length, carrying a steady current. The field in the interior space is strong and nearly uniform. (b) The magnetic field pattern of a bar magnet, displayed with small iron filings on a sheet of paper.

▌22.10 | The Magnetic Field of a Solenoid

A solenoid is a long wire wound in the form of a helix. If the turns are closely spaced, this configuration can produce a reasonably uniform magnetic field throughout the volume enclosed by the solenoid, except close to its ends. Each of the turns can be modeled as a circular loop, and the net magnetic field is the vector sum of the fields due to all the turns.

If the turns are closely spaced and the solenoid is of finite length, the field lines are as shown in Figure 22.35a. In this case, the field lines diverge from one end and converge at the opposite end. An inspection of this field distribution exterior to the solenoid shows a similarity to the field of a bar magnet (Fig. 22.35b). Hence, one end of the solenoid behaves like the north pole of a magnet and the opposite end behaves like the south pole. As the length of the solenoid increases, the field within it becomes more and more uniform. When the solenoid's turns are closely spaced and its length is large compared with its radius, it approaches the case of an *ideal solenoid*. For an ideal solenoid, the field outside the solenoid is negligible and the field inside is uniform. We will use the ideal solenoid as a simplification model for a real solenoid.

If we consider the amperian loop (loop 1) perpendicular to the page in Figure 22.36, surrounding the ideal solenoid, we see that it it encloses a small current as the charges in the wire move turn by turn along the length of the solenoid. Therefore, there is a nonzero magnetic field outside the solenoid. It is a weak field, with circular field lines, like those due to a line of current as in Figure 22.23. For an ideal solenoid, it is the only field external to the solenoid. We could eliminate this field in Figure 22.36 by adding a second layer of turns of wire outside the first layer. If the first layer of turns is wrapped so that the turns progress from the bottom of Figure 22.36 to the top and the second layer has turns progressing from the top to the bottom, the net current along the axis is zero.

We can use Ampère's law to obtain an expression for the magnetic field inside an ideal solenoid with a single layer of wire. A longitudinal cross-section of part of our ideal solenoid (Fig. 22.36) carries current I. Here, $\vec{B}$ inside the ideal solenoid is uniform and parallel to the axis. Consider a rectangular path (loop 2) of length ℓ and width w as shown in Figure 22.36. We can apply Ampère's law to this path by evaluating the integral of $\vec{B} \cdot d\vec{s}$ over each of the four sides of the rectangle. The contribution along side 3 is zero because the magnetic field lines are perpendicular to the path in this region, which matches condition 3 in Section 22.9. The contributions from sides 2 and 4 are both zero because $\vec{B}$ is perpendicular to $d\vec{s}$ along these paths, both inside and outside the solenoid. Side 1, whose length is ℓ, gives a contribution to the integral because $\vec{\mathbf{B}}$ along this portion of the path is constant in magnitude and parallel to $d\vec{\mathbf{s}}$, which matches conditions 1 and 2. The integral over the closed rectangular path therefore has the value

$$\oint \vec{\mathbf{B}} \cdot d\vec{\mathbf{s}} = \int_{\text{side 1}} \vec{\mathbf{B}} \cdot d\vec{\mathbf{s}} = B \int_{\text{side 1}} ds = B\ell$$

The right side of Ampère's law involves the *total* current that passes through the surface bounded by the path of integration. In our case, the total current through the rectangular path equals the current through each turn of the solenoid multiplied by the number of turns enclosed by the path of integration. If N is the number of turns in the length ℓ, the total current through the rectangle equals NI. Ampère's law applied to this path therefore gives

$$\oint \vec{\mathbf{B}} \cdot d\vec{\mathbf{s}} = B\ell = \mu_0 NI$$

$$B = \mu_0 \frac{N}{\ell} I = \mu_0 nI \qquad \textbf{22.32} \blacktriangleleft$$

where $n = N/\ell$ is the number of turns *per unit length* (not to be confused with N, the number of turns).

We also could obtain this result in a simpler manner by reconsidering the magnetic field of a toroidal coil (Example 22.9). If the radius r of the toroidal coil containing N turns is large compared with its cross-sectional radius a, a short section of the toroidal coil approximates a short section of a solenoid, with $n = N/2\pi r$. In this limit, we see that Equation 22.31 derived for the toroidal coil agrees with Equation 22.32.

Equation 22.32 is valid only for points near the center of a very long solenoid. As you might expect, the field near each end is smaller than the value given by Equation 22.32. At the very end of a long solenoid, the magnitude of the field is about one-half that of the field at the center (see Problem 22.56 in Enhanced WebAssign).

QUICK QUIZ 22.8 Consider a solenoid that is very long compared with its radius. Of the following choices, what is the most effective way to increase the magnetic field in the interior of the solenoid? **(a)** double its length, keeping the number of turns per unit length constant **(b)** reduce its radius by half, keeping the number of turns per unit length constant **(c)** overwrap the entire solenoid with an additional layer of current-carrying wire

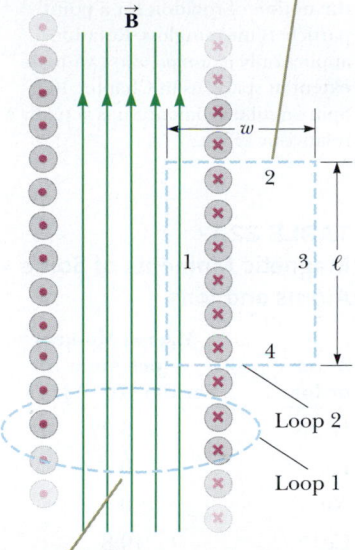

Ampère's law applied to the rectangular dashed path can be used to calculate the magnitude of the interior field.

Ampère's law applied to the circular path whose plane is perpendicular to the page can be used to show that there is a weak field outside the solenoid.

Figure 22.36 Cross-sectional view of an ideal solenoid, where the interior magnetic field is uniform and the exterior field is close to zero.

22.11 | Magnetism in Matter

The magnetic field produced by a current in a loop of wire gives a hint about what causes certain materials to exhibit strong magnetic properties. To understand why some materials are magnetic, it is instructive to begin this discussion with the Bohr structural model of the atom that we discussed in Chapter 11. In this model, electrons are assumed to move in circular orbits about the much more massive nucleus. Figure 22.37 shows the angular momentum associated with the electron. In the Bohr model, each electron, with its charge of magnitude 1.6×10^{-19} C, circles the atom once in about 10^{-16} s. If we divide the electronic charge by this time interval, we find that the orbiting electron is equivalent to a current of 1.6×10^{-3} A. Each orbiting electron is therefore viewed as a tiny current loop with a corresponding magnetic moment. Because the charge of the electron is negative, the magnetic moment is directed opposite to the angular momentum as shown in Figure 22.37.

In most substances, the magnetic moment of one electron in an atom is canceled by that of another electron in the atom, orbiting in the opposite direction. The net result is that the magnetic effect produced by the orbital motion of the electrons is either zero or very small for most materials.

In addition to its orbital angular momentum, an electron has an **intrinsic angular momentum,** called **spin,** which also contributes to its magnetic moment. The spin of

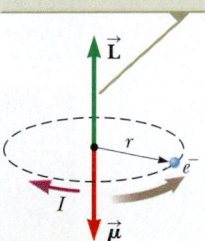

The electron has an angular momentum $\vec{\mathbf{L}}$ in one direction and a magnetic moment $\vec{\boldsymbol{\mu}}$ in the opposite direction.

Figure 22.37 An electron moving in the direction of the gray arrow in a circular orbit of radius r. Because the electron carries a negative charge, the direction of the current due to its motion about the nucleus is opposite the direction of that motion.

TABLE 22.1 |

Magnetic Moments of Some Atoms and Ions

Atom or Ion	Magnet Moment per Atom or Ion (10^{-24} J/T)
H	9.27
He	0
Ne	0
Ce^{3+}	19.8
Yb^{3+}	37.1

an electron is an angular momentum separate from its orbital angular momentum, just as the spin of the Earth is separate from its orbital motion about the Sun. Even if the electron is at rest, it still has an angular momentum associated with spin. We shall investigate spin more deeply in Chapter 29.

In atoms or ions containing multiple electrons, many electrons are paired up with their spins in opposite directions, an arrangement that results in a cancellation of the spin magnetic moments. An atom with an odd number of electrons, however, must have at least one "unpaired" electron and a corresponding spin magnetic moment. The net magnetic moment of the atom leads to various types of magnetic behavior. The magnetic moments of several atoms and ions are listed in Table 22.1.

Ferromagnetic Materials

Iron, cobalt, nickel, gadolinium, and dysprosium are strongly magnetic materials and are said to be **ferromagnetic.** Ferromagnetic substances, used to fabricate permanent magnets, contain atoms with spin magnetic moments that tend to align parallel to each other even in a weak external magnetic field. Once the moments are aligned, the substance remains magnetized after the external field is removed. This permanent alignment is due to strong coupling between neighboring atoms, which can only be understood using quantum physics.

All ferromagnetic materials contain microscopic regions called **domains,** within which all magnetic moments are aligned. The domains range from about 10^{-12} to 10^{-8} m^3 in volume and contain 10^{17} to 10^{21} atoms. The boundaries between domains having different orientations are called **domain walls.** In an unmagnetized sample, the domains are randomly oriented so that the net magnetic moment is zero as in Figure 22.38a. When the sample is placed in an external magnetic field, domains with magnetic moment vectors initially oriented along the external field grow in size at the expense of other domains, which results in a magnetized sample, as in Figures 22.38b and 22.38c. When the external field is removed, the sample may retain most of its magnetism.

The extent to which a ferromagnetic substance retains its magnetism is described by its classification as being magnetically **hard** or **soft.** Soft magnetic materials, such as iron, are easily magnetized but also tend to lose their magnetism easily. When a soft magnetic material is magnetized and the external magnetic field is removed, thermal agitation produces domain motion and the material quickly returns to an unmagnetized state. In contrast, hard magnetic materials, such as cobalt and nickel, are difficult to magnetize but tend to retain their magnetism, and domain alignment persists in them after the external magnetic field is removed. Such hard magnetic materials are referred to as **permanent magnets.** Rare-earth permanent magnets, such as samarium–cobalt, are now regularly used in industry.

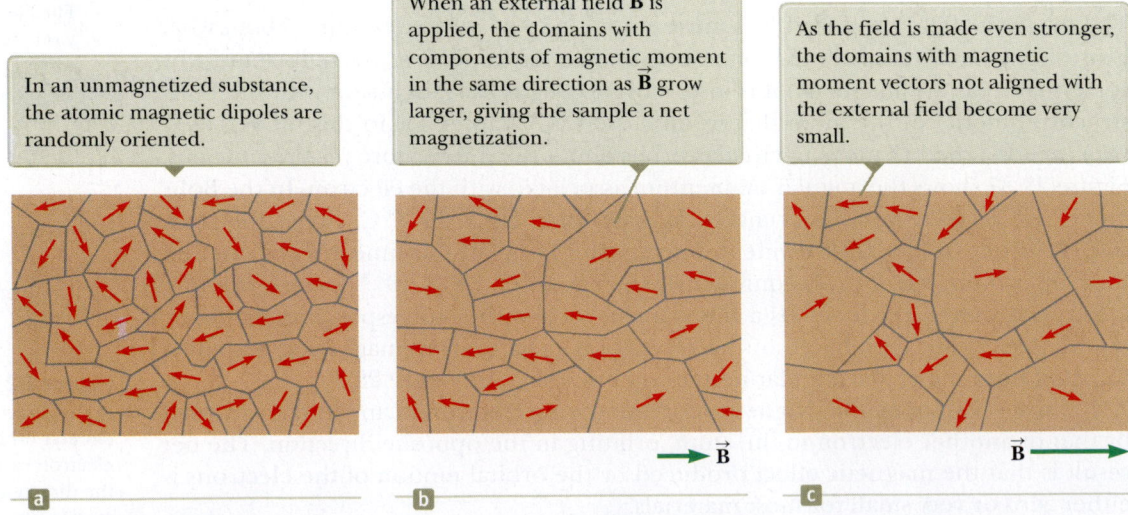

Figure 22.38 Orientation of magnetic dipoles before and after a magnetic field is applied to a ferromagnetic substance.

❮ 22.12 | Context Connection: Remote Magnetic Navigation for Cardiac Catheter Ablation Procedures

In the *Heart Attacks* Context, we studied the role of fluid flow in blood vessels and the dangerous effect of plaque buildup on delivery of blood to the heart. In Section 21.2, we looked at the heart again as we investigated the details of cardiac catheter ablation for a patient who suffers from atrial fibrillation. In this Context Connection, we return to atrial fibrillation in the heart, but consider a newer development in the ablation procedure.

There are a number of risks with a traditional cardiac catheter ablation procedure. One possible outcome is a perforation of the heart wall with one of the catheters. Because the esophagus passes right behind the heart, it is possible to burn through too much tissue during a particular ablation and create an esophageal fistula. Other risks come from exposure to x-rays. In order to observe the positions of the catheters, the electrophysiologist must use x-rays and a fluoroscope to make the heart and the catheters visible. As a result, the patient receives a relatively high dose of radiation during the procedure. In addition, despite the use of lead aprons, the electrophysiologist receives radiation from each ablation procedure during his or her entire career. In addition to the effects of this prolonged radiation exposure, studies have shown that a large percentage of electrophysiologists have been treated for back and neck pain due to the long hours of wearing lead aprons.

One possibility for reducing risks to both patient and doctor is the use of *remote magnetic navigation* in catheter ablation procedures. This procedure uses softer and more flexible catheters than the traditional approach, reducing the risk of perforation and allowing catheters to reach areas of the heart unavailable to the stiffer traditional catheters. The tips of the catheters are guided magnetically with the aid of a computer. The electrophysiologist can sit comfortably at a computer in another room and guide the catheters with a joystick, avoiding exposure to radiation. Figure 22.39 shows a typical computer display that helps the electrophysiologist guide the catheters.

During a catheter ablation procedure using remote magnetic navigation, the patient is located between two strong magnets as shown in Figure 22.40 (page 668). The magnets can be moved over a wide range of positions and orientations relative to the patient. The magnetic field from these magnets is strong, but only about 10% as strong as that used in magnetic resonance imaging (to be discussed in the Context Conclusion). The tip of the catheter includes ferromagnetic material so that its orientation can be precisely controlled by the positions of the external magnets. Once the tip is correctly oriented, it can be advanced mechanically as with the traditional approach.

In addition to the increased safety of the softer catheter and the precise magnetic orientation of its tip, the computer control of the procedure provides additional advantages. For example, locations of ablations can be "memorized" by the computer. The catheter tip can be quickly returned to this exact location for a repeated ablation by calling up the memorized location.

While there are many advantages to remote magnetic navigation, clinical evidence shows one disadvantage. The total procedure time with remote navigation has been measured to be significantly longer than that with the traditional approach.[2] Reasons for this longer time interval include the learning curve for the procedure, "interruption time" because the electrophysiologist is available to other staff in a room separate from the patient, and increased time for the more complicated mapping procedures.

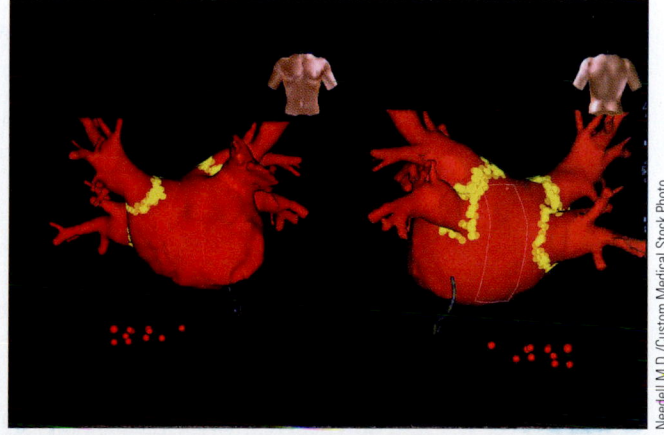

Figure 22.39 In remote magnetic navigation procedures for cardiac catheter ablations, the electrophysiologist views a computer model of the heart such as the front and back images shown here. The yellow dots are lesions around the pulmonary veins made by the ablation process.

Needell M.D./Custom Medical Stock Photo

[2]A. Arya, R. Zaker-Shahrak, P. Sommer, A. Bollmann, U. Wetzel, T. Gaspar, S. Richter, D. Husser, C. Piorkowski, and G. Hindricks, "Catheter Ablation of Atrial Fibrillation Using Remote Magnetic Catheter Navigation: A Case–Control Study," *Europace,* **13,** pp. 45–50 (2011).

Figure 22.40 A cardiac catheterization laboratory using remote magnetic navigation stands ready to receive a patient suffering from atrial fibrillation. The large white objects on either side of the operating table are housings for strong magnets that place the patient in a magnetic field. The electrophysiologist performing a catheter ablation procedure sits at a computer in the room to the left. With guidance from the magnetic field, he or she uses a joystick and other controls to thread the magnetically sensitive tip of a cardiac catheter through blood vessels and into the chambers of the heart.

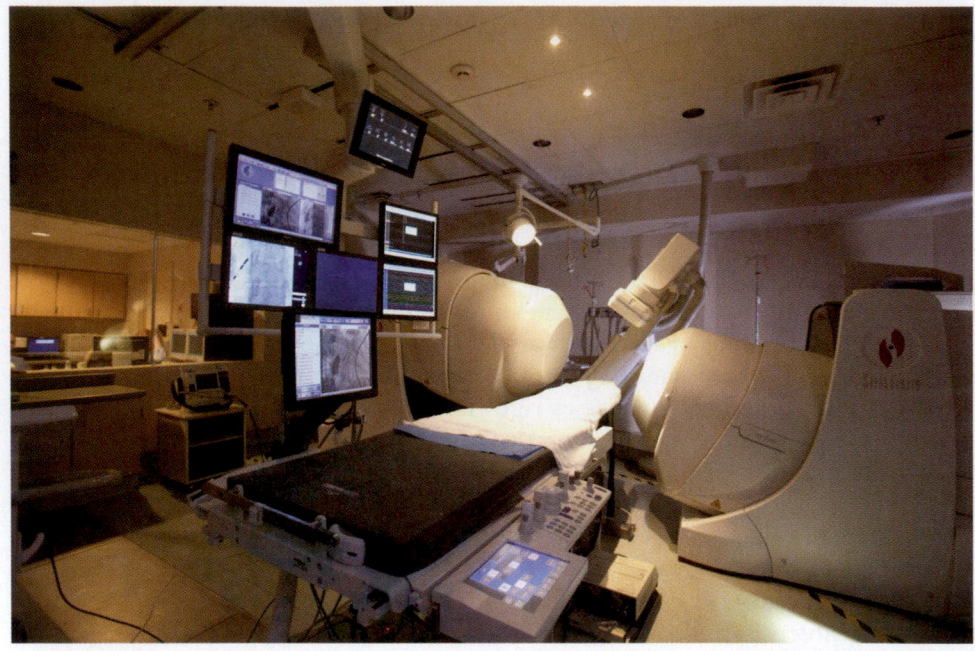

© Courtesy of Stereotaxis Inc.

As more electrophysiologists are trained in remote magnetic navigation and the mapping techniques are streamlined, perhaps the procedure times can be shortened. In that case, the magnetic technique will have a clear advantage over the traditional mechanical approach.

SUMMARY

The **magnetic force** that acts on a charge q moving with velocity $\vec{\mathbf{v}}$ in an external **magnetic field** $\vec{\mathbf{B}}$ is

$$\vec{\mathbf{F}}_B = q\vec{\mathbf{v}} \times \vec{\mathbf{B}} \qquad \textbf{22.1} \blacktriangleleft$$

This force is in a direction perpendicular both to the velocity of the particle and to the magnetic field and given by the right-hand rules shown in Figure 22.4. The magnitude of the magnetic force is

$$F_B = |q|vB\sin\theta \qquad \textbf{22.2} \blacktriangleleft$$

where θ is the angle between $\vec{\mathbf{v}}$ and $\vec{\mathbf{B}}$.

A particle with mass m and charge q moving with velocity $\vec{\mathbf{v}}$ perpendicular to a uniform magnetic field $\vec{\mathbf{B}}$ follows a circular path of radius

$$r = \frac{mv}{qB} \qquad \textbf{22.3} \blacktriangleleft$$

If a straight conductor of length L carries current I, the magnetic force on that conductor when placed in a uniform external magnetic field $\vec{\mathbf{B}}$ is

$$\vec{\mathbf{F}}_B = I\vec{\mathbf{L}} \times \vec{\mathbf{B}} \qquad \textbf{22.10} \blacktriangleleft$$

where $\vec{\mathbf{L}}$ is in the direction of the current and $|\vec{\mathbf{L}}| = L$, the length of the conductor.

If an arbitrarily shaped wire carrying current I is placed in an external magnetic field, the magnetic force on a very small length element $d\vec{\mathbf{s}}$ is

$$d\vec{\mathbf{F}}_B = I\,d\vec{\mathbf{s}} \times \vec{\mathbf{B}} \qquad \textbf{22.11} \blacktriangleleft$$

To determine the total magnetic force on the wire, one must integrate Equation 22.11 over the wire.

The **magnetic dipole moment** $\vec{\boldsymbol{\mu}}$ of a loop carrying current I is

$$\vec{\boldsymbol{\mu}} = I\vec{\mathbf{A}} \qquad \textbf{22.15} \blacktriangleleft$$

where $\vec{\mathbf{A}}$ is perpendicular to the plane of the loop and $|\vec{\mathbf{A}}|$ is equal to the area of the loop. The SI unit of magnetic moment $\vec{\boldsymbol{\mu}}$ is the ampere · meter squared, or A·m².

The torque $\vec{\boldsymbol{\tau}}$ on a current loop when the loop is placed in a uniform external magnetic field $\vec{\mathbf{B}}$ is

$$\vec{\boldsymbol{\tau}} = \vec{\boldsymbol{\mu}} \times \vec{\mathbf{B}} \qquad \textbf{22.16} \blacktriangleleft$$

The potential energy of the system of a magnetic dipole in a magnetic field is

$$U = -\vec{\boldsymbol{\mu}} \cdot \vec{\mathbf{B}} \qquad \textbf{22.17} \blacktriangleleft$$

The **Biot–Savart law** says that the magnetic field $d\vec{\mathbf{B}}$ at a point P due to a wire element $d\vec{\mathbf{s}}$ carrying a steady current I is

$$d\vec{\mathbf{B}} = \frac{\mu_0}{4\pi} \frac{I\,d\vec{\mathbf{s}} \times \hat{\mathbf{r}}}{r^2} \qquad \textbf{22.20} \blacktriangleleft$$

where $\mu_0 = 4\pi \times 10^{-7}\ \text{T}\cdot\text{m/A}$ is the **permeability of free space** and r is the distance from the element to the point P. To find the total field at P due to a current distribution, one must integrate this vector expression over the entire distribution.

The magnitude of the magnetic field at a distance r from a long, straight wire carrying current I is

$$B = \frac{\mu_0 I}{2\pi r} \qquad \text{22.21} \blacktriangleleft$$

The field lines are circles concentric with the wire.

The magnetic force per unit length between two parallel wires (at least one of which is long) separated by a distance a and carrying currents I_1 and I_2 has the magnitude

$$\frac{F}{\ell} = \frac{\mu_0 I_1 I_2}{2\pi a} \qquad \text{22.27} \blacktriangleleft$$

The force is attractive if the currents are in the same direction and repulsive if they are in opposite directions.

Ampère's law says that the line integral of $\vec{\mathbf{B}} \cdot d\vec{\mathbf{s}}$ around any closed path equals $\mu_0 I$, where I is the total steady current passing through any surface bounded by the closed path:

$$\oint \vec{\mathbf{B}} \cdot d\vec{\mathbf{s}} = \mu_0 I \qquad \text{22.29} \blacktriangleleft$$

Using Ampère's law, one finds that the magnitudes of the magnetic fields inside a toroidal coil and solenoid are

$$B = \frac{\mu_0 N I}{2\pi r} \quad \text{(toroid)} \qquad \text{22.31} \blacktriangleleft$$

$$B = \mu_0 \frac{N}{\ell} I = \mu_0 n I \quad \text{(solenoid)} \qquad \text{22.32} \blacktriangleleft$$

where N is the total number of turns and n is the number of turns per unit length.

OBJECTIVE QUESTIONS

☐ denotes answer available in *Student Solutions Manual/Study Guide*

1. A spatially uniform magnetic field cannot exert a magnetic force on a particle in which of the following circumstances? There may be more than one correct statement. (a) The particle is charged. (b) The particle moves perpendicular to the magnetic field. (c) The particle moves parallel to the magnetic field. (d) The magnitude of the magnetic field changes with time. (e) The particle is at rest.

2. What creates a magnetic field? More than one answer may be correct. (a) a stationary object with electric charge (b) a moving object with electric charge (c) a stationary conductor carrying electric current (d) a difference in electric potential (e) a charged capacitor disconnected from a battery and at rest. *Note:* In Chapter 24, we will see that a changing electric field also creates a magnetic field.

3. A charged particle is traveling through a uniform magnetic field. Which of the following statements are true of the magnetic field? There may be more than one correct statement. (a) It exerts a force on the particle parallel to the field. (b) It exerts a force on the particle along the direction of its motion. (c) It increases the kinetic energy of the particle. (d) It exerts a force that is perpendicular to the direction of motion. (e) It does not change the magnitude of the momentum of the particle.

4. A proton moving horizontally enters a region where a uniform magnetic field is directed perpendicular to the proton's velocity as shown in Figure OQ22.4. After the proton enters the field, does it (a) deflect downward, with its speed remaining constant; (b) deflect upward, moving in a semicircular path with constant speed, and exit the field moving to the left; (c) continue to move in the horizontal direction with constant velocity; (d) move in a circular orbit and become trapped by the field; or (e) deflect out of the plane of the paper?

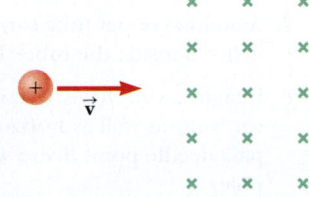

Figure OQ22.4

5. Two long, parallel wires each carry the same current I in the same direction (Fig. OQ22.5). Is the total magnetic field at the point P midway between the wires (a) zero, (b) directed into the page, (c) directed out of the page, (d) directed to the left, or (e) directed to the right?

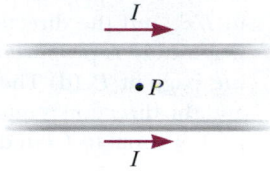

Figure OQ22.5

6. Two long, straight wires cross each other at a right angle, and each carries the same current I (Fig. OQ22.6). Which of the following statements is true regarding the total magnetic field due to the two wires at the various points in the figure? More than one statement may be correct. (a) The field is strongest at points B and D. (b) The field is strongest at points A and C. (c) The field is out of the page at point B and into the page at point D. (d) The field is out of the page at point C and out of the page at point D. (e) The field has the same magnitude at all four points.

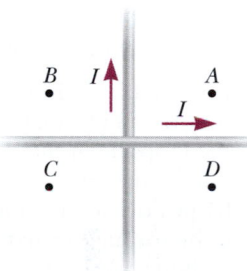

Figure OQ22.6

7. Answer each question yes or no. Assume the motions and currents mentioned are along the x axis and fields are in the y direction. (a) Does an electric field exert a force on a stationary charged object? (b) Does a magnetic field do so? (c) Does an electric field exert a force on a moving charged object? (d) Does a magnetic field do so? (e) Does an electric field exert a force on a straight current-carrying wire? (f) Does a magnetic field do so? (g) Does an electric field exert a force on a beam of moving electrons? (h) Does a magnetic field do so?

8. At a certain instant, a proton is moving in the positive x direction through a magnetic field in the negative z direction. What

is the direction of the magnetic force exerted on the proton? (a) positive z direction (b) negative z direction (c) positive y direction (d) negative y direction (e) The force is zero.

9. Answer each question yes or no. (a) Is it possible for each of three stationary charged particles to exert a force of attraction on the other two? (b) Is it possible for each of three stationary charged particles to repel both of the other particles? (c) Is it possible for each of three current-carrying metal wires to attract the other two wires? (d) Is it possible for each of three current-carrying metal wires to repel the other two wires? André-Marie Ampère's experiments on electromagnetism are models of logical precision and included observation of the phenomena referred to in this question.

10. A long, straight wire carries a current I (Fig. OQ22.10). Which of the following statements is true regarding the magnetic field due to the wire? More than one statement may be correct.

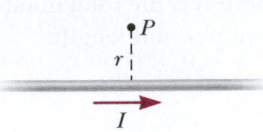

Figure OQ22.10

(a) The magnitude is proportional to I/r, and the direction is out of the page at P. (b) The magnitude is proportional to I/r^2, and the direction is out of the page at P. (c) The magnitude is proportional to I/r, and the direction is into the page at P. (d) The magnitude is proportional to I/r^2, and the direction is into the page at P. (e) The magnitude is proportional to I, but does not depend on r.

11. A thin copper rod 1.00 m long has a mass of 50.0 g. What is the minimum current in the rod that would allow it to levitate above the ground in a magnetic field of magnitude 0.100 T? (a) 1.20 A (b) 2.40 A (c) 4.90 A (d) 9.80 A (e) none of those answers

12. A magnetic field exerts a torque on each of the current carrying single loops of wire shown in Figure OQ22.12. The loops lie in the xy plane, each carrying the same magnitude current, and the uniform magnetic field points in the positive x direction. Rank the loops by the magnitude of the torque exerted on them by the field from largest to smallest

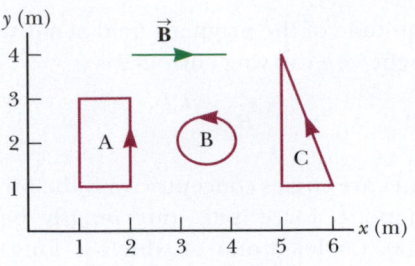

Figure OQ22.12

13. Two long, parallel wires carry currents of 20.0 A and 10.0 A in opposite directions (Fig. OQ22.13). Which of the following statements is true? More than one statement may be correct. (a) In region I, the magnetic field is into the page and is never zero. (b) In region II, the field is into the page and can be zero. (c) In region III, it is possible for the field to be zero. (d) In region I, the magnetic field is out of the page and is never zero. (e) There are no points where the field is zero.

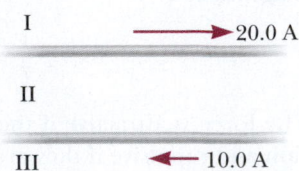

Figure OQ22.13 Objective Questions 13 and 14.

14. Consider the two parallel wires carrying currents in opposite directions in Figure OQ22.13. Due to the magnetic interaction between the wires, does the lower wire experience a magnetic force that is (a) upward, (b) downward, (c) to the left, (d) to the right, or (e) into the paper?

15. A long solenoid with closely spaced turns carries electric current. Does each turn of wire exert (a) an attractive force on the next adjacent turn, (b) a repulsive force on the next adjacent turn, (c) zero force on the next adjacent turn, or (d) either an attractive or a repulsive force on the next turn, depending on the direction of current in the solenoid?

16. Solenoid A has length L and N turns, solenoid B has length $2L$ and N turns, and solenoid C has length $L/2$ and $2N$ turns. If each solenoid carries the same current, rank the magnitudes of the magnetic fields in the centers of the solenoids from largest to smallest.

☐ denotes answer available in *Student Solutions Manual/Study Guide*

CONCEPTUAL QUESTIONS

1. Two charged particles are projected in the same direction into a magnetic field perpendicular to their velocities. If the particles are deflected in opposite directions, what can you say about them?

2. One pole of a magnet attracts a nail. Will the other pole of the magnet attract the nail? Explain. Also explain how a magnet sticks to a refrigerator door.

3. A magnet attracts a piece of iron. The iron can then attract another piece of iron. On the basis of domain alignment, explain what happens in each piece of iron.

4. How can the motion of a moving charged particle be used to distinguish between a magnetic field and an electric field? Give a specific example to justify your argument.

5. Is it possible to orient a current loop in a uniform magnetic field such that the loop does not tend to rotate? Explain.

6. Is Ampère's law valid for all closed paths surrounding a conductor? Why is it not useful for calculating $\vec{B}$ for all such paths?

7. A hollow copper tube carries a current along its length. Why is $B = 0$ inside the tube? Is B nonzero outside the tube?

8. Imagine you have a compass whose needle can rotate vertically as well as horizontally. Which way would the compass needle point if you were at the Earth's north magnetic pole?

9. How can a current loop be used to determine the presence of a magnetic field in a given region of space?

10. Can a constant magnetic field set into motion an electron initially at rest? Explain your answer.

11. Explain why two parallel wires carrying currents in opposite directions repel each other.

12. Figure CQ22.12 shows four permanent magnets, each having a hole through its center. Notice that the blue and yellow magnets are levitated above the red ones. (a) How does this levitation occur? (b) What purpose do the rods serve? (c) What can you say about the poles of the magnets from this observation? (d) If the blue magnet were inverted, what do you suppose would happen?

Cengage Learning Charles D. Winters

Figure CQ22.12

13. Is the magnetic field created by a current loop uniform? Explain.

14. Consider a magnetic field that is uniform in direction throughout a certain volume. (a) Can the field be uniform in magnitude? (b) Must it be uniform in magnitude? Give evidence for your answers.

▶ PROBLEMS AVAILABLE IN ᴱᴺᴴᴬᴺᶜᴱᴰ WebAssign

22.2 The Magnetic Field

Problems 1–8

22.3 Motion of a Charged Particle in a Uniform Magnetic Field

Problems 9–12

22.4 Applications Involving Charged Particles Moving in a Magnetic Field

Problems 13–17

22.5 Magnetic Force on a Current-Carrying Conductor

Problems 18–23

22.6 Torque on a Current Loop in a Uniform Magnetic Field

Problems 24–28

22.7 The Biot–Savart Law

Problems 29–41

22.8 The Magnetic Force Between Two Parallel Conductors

Problems 42–46

22.9 Ampère's Law

Problems 47–54

22.10 The Magnetic Field of a Solenoid

Problems 55–58

22.11 Magnetism in Matter

Problems 59–60

22.12 Context Connection: Remote Magnetic Navigation for Cardiac Catheter Ablation Procedures

Problems 61–62

Additional Problems

Problems 63–81

Solutions to the following Problems are available in the *Student Solutions Manual/Study Guide*:

22.2, 22.7, 22.10, 22.17, 22.18, 22.21, 22.25, 22.30, 22.33, 22.41, 22.43, 22.54, 22.57, 22.59, 22.72, 22.78, and 22.81

List of Enhanced Problems

Problem Number	Targeted Feedback in Enhanced WebAssign	Master It in Enhanced WebAssign	Watch It in Enhanced WebAssign
22.2	✓		✓
22.3	✓		✓
22.4		✓	
22.7		✓	
22.8	✓		✓
22.10		✓	
22.13	✓		✓
22.14		✓	
22.15	✓		✓
22.19	✓		✓
22.21		✓	
22.23	✓		✓
22.25	✓	✓	
22.27	✓		✓
22.29	✓		✓
22.33		✓	
22.38	✓		✓
22.41	✓	✓	
22.43	✓	✓	
22.45	✓		✓
22.47	✓		✓
22.49	✓		✓
22.51	✓		✓
22.54	✓	✓	
22.55	✓		✓
22.57	✓	✓	
22.59		✓	
22.63		✓	
22.77		✓	

Faraday's Law and Inductance

An artist's impression of the Skerries SeaGen Array, a tidal energy generator under development near the island of Anglesey, North Wales. When it is brought on line, possibly in 2015, it will offer 10.5 MW of power from generators turned by tidal streams. The image shows the underwater blades that are driven by the tidal currents. The second blade system has been raised from the water for servicing. We will study generators in this chapter.

Marine Current Turbines TM Ltd.

Our studies in electromagnetism so far have been concerned with the electric fields due to stationary charges and the magnetic fields produced by moving charges. This chapter introduces a new type of electric field, one that is due to a changing magnetic field.

As we learned in Section 19.1, experiments conducted by Michael Faraday in England in the early 1800s and independently by Joseph Henry in the United States showed that an electric current can be induced in a circuit by a changing magnetic field. The results of those experiments led to a very basic and important law of electromagnetism known as *Faraday's law of induction*. Faraday's law explains how generators, as well as other practical devices, work.

Faraday's law is also the basis for a new circuit element, the *inductor*. This new circuit element combines with resistors and capacitors to allow for a variety of useful electric circuits.

23.1 | Faraday's Law of Induction

We begin discussing the concepts in this chapter by considering a simple experiment that builds on material presented in Chapter 22. Imagine that a straight metal conductor resides in a uniform magnetic field directed into the page as in Figure 23.1 (page 674). Within the conductor, there are free electrons. Suppose the conductor is now moved with a velocity $\vec{v}$ toward the right. Equation 22.1

A current is induced in the conductor due to the magnetic force on charged particles in the conductor.

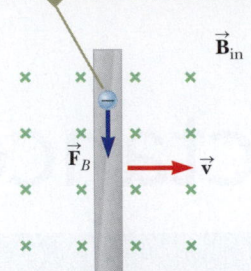

Figure 23.1 A straight electrical conductor moving with a velocity $\vec{v}$ through a uniform magnetic field $\vec{B}$ directed perpendicular to $\vec{v}$.

When a magnet is moved toward a loop of wire connected to a sensitive ammeter, the ammeter shows that a current is induced in the loop.

When the magnet is held stationary, there is no induced current in the loop, even when the magnet is inside the loop.

When the magnet is moved away from the loop, the ammeter shows that the induced current is opposite that shown in part **a**.

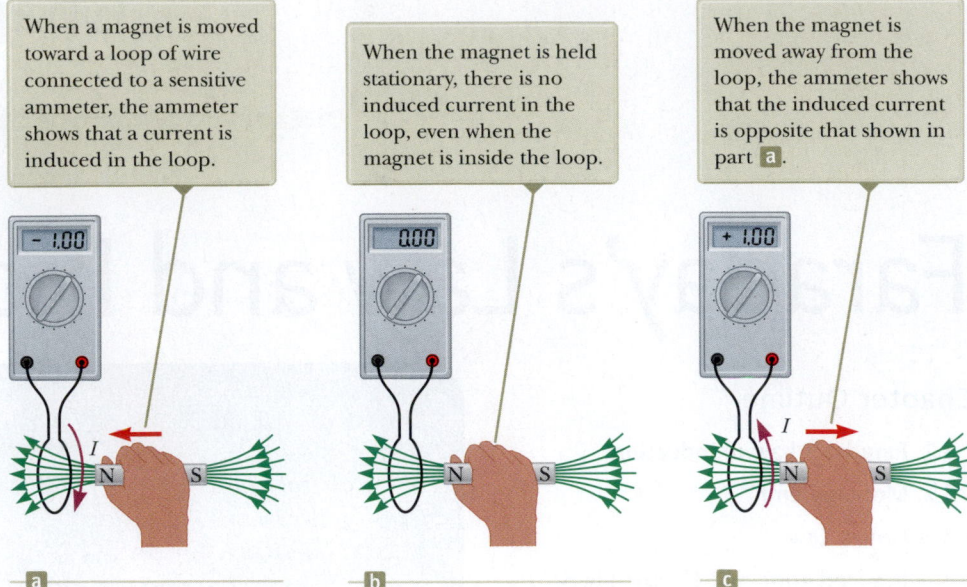

Active Figure 23.2 A simple experiment showing that a current is induced in a loop when a magnet is moved toward or away from the loop.

Michael Faraday
British Physicist and Chemist (1791–1867)
Faraday is often regarded as the greatest experimental scientist of the 1800s. His many contributions to the study of electricity include the invention of the electric motor, the electric generator, and the transformer as well as the discovery of electromagnetic induction and the laws of electrolysis. Greatly influenced by his religious beliefs, he refused to work on the development of poison gas for the British military.

tells us that a magnetic force acts on the electrons in the conductor. Using the right-hand rule, the force on the electrons is downward in Figure 23.1 (remember that the electrons carry a negative charge). Because this direction is along the conductor, the electrons move along the conductor in response to this force. Therefore, a *current* is produced in the conductor as it moves through a magnetic field!

Let us consider another simple experiment that demonstrates that an electric current can be produced by a magnetic field. Consider a loop of wire connected to a sensitive ammeter, a device that measures current, as illustrated in Active Figure 23.2. If a magnet is moved toward the loop, the ammeter display shows the existence of a current as in Active Figure 23.2a. When the magnet is held stationary as in Active Figure 23.2b, the ammeter shows no current. If the magnet is moved away from the loop as in Active Figure 23.2c, the ammeter display shows a current in the opposite direction from that caused by the motion of the magnet toward the ammeter. Finally, if the magnet is held stationary and the loop is moved either toward or away from it, the ammeter shows a current again. From these observations comes the conclusion that an electric current is set up in the loop as long as relative motion occurs between the magnet and the loop.

These results are quite remarkable when we consider that a current exists in a loop of wire even though no batteries are connected to the wire. We call such a current an **induced current,** and it is produced by an **induced emf.**

Another experiment, first conducted by Faraday, is illustrated in Active Figure 23.3. Part of the apparatus consists of a coil of insulated wire connected to a switch and a battery. We shall refer to this coil as the *primary coil* of wire and to the corresponding circuit as the primary circuit. The coil is wrapped around an iron ring to intensify the magnetic field produced by the current through the coil. A second coil of insulated wire at the right is also wrapped around the iron ring and is connected to a sensitive ammeter. We shall refer to this coil as the *secondary coil* and to the corresponding circuit as the secondary circuit. The secondary circuit has no battery, and the secondary coil is not electrically connected to the primary coil. The purpose of this apparatus is to detect any current that might be generated in the secondary circuit by a change in the magnetic field produced by the primary circuit.

Initially, you might guess that no current would ever be detected in the secondary circuit. Something quite surprising happens, however, when the switch in the primary circuit is opened or thrown closed. At the instant the switch is thrown closed, the ammeter display briefly shows a current and then returns to zero. When the switch is opened, the ammeter display shows a current in the opposite direction and then again returns to zero. Finally, the ammeter reads zero when the primary circuit carries a steady current.

As a result of these observations, Faraday concluded that an electric current can be produced by a time-varying magnetic field. A current cannot be produced by a steady magnetic field. In the experiment shown in Active Figure 23.2, the changing magnetic field is a result of the relative motion between the magnet and the loop of wire. As long as the motion persists, the current is maintained. In the experiment shown in Active Figure 23.3, the current produced in the secondary circuit occurs for only an instant after the switch is closed while the magnetic field acting on the secondary coil builds from its zero value to its final value. In effect, the secondary circuit behaves as though a source of emf were connected to it for an instant. It is customary to say that an emf is induced in the secondary circuit by the changing magnetic field produced by the current in the primary circuit.

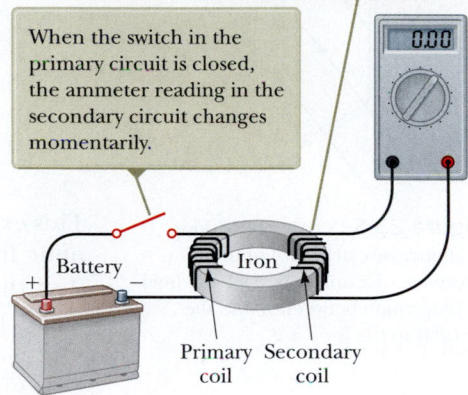

The emf induced in the secondary circuit is caused by the changing magnetic field through the secondary coil.

When the switch in the primary circuit is closed, the ammeter reading in the secondary circuit changes momentarily.

Active Figure 23.3 Faraday's experiment.

To quantify such observations, we define a quantity called **magnetic flux.** The flux associated with a magnetic field is defined in a similar manner to the electric flux (Section 19.8) and is proportional to the number of magnetic field lines passing through an area. Consider an element of area dA on an arbitrarily shaped open surface as in Figure 23.4. If the magnetic field at the location of this element is $\vec{B}$, the magnetic flux through the element is $\vec{B} \cdot d\vec{A}$, where $d\vec{A}$ is a vector perpendicular to the surface whose magnitude equals the area dA. Hence, the total magnetic flux Φ_B through the surface is

$$\Phi_B = \int \vec{B} \cdot d\vec{A} \qquad \textbf{23.1} \blacktriangleleft$$

▶ Magnetic flux

The SI unit of magnetic flux is a tesla·meter squared, which is named the *weber* (Wb); $1 \text{ Wb} = 1 \text{ T} \cdot \text{m}^2$.

The two experiments illustrated in Figures 23.2 and 23.3 have one thing in common. In both cases, an emf is induced in a circuit when the magnetic flux through the surface bounded by the circuit changes with time. A general statement known as **Faraday's law of induction** summarizes such experiments involving induced emfs:

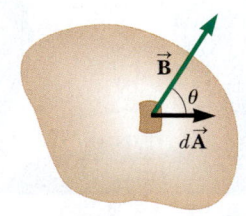

Figure 23.4 The magnetic flux through an area element $d\vec{A}$ is given by $\vec{B} \cdot d\vec{A} = B \, dA \cos\theta$. Note that vector $d\vec{A}$ is perpendicular to the surface.

The emf induced in a circuit is equal to the time rate of change of magnetic flux through the circuit:

$$\varepsilon = -\frac{d\Phi_B}{dt} \qquad \textbf{23.2} \blacktriangleleft$$

▶ Faraday's law

In Equation 23.2, Φ_B is the magnetic flux through the surface bounded by the circuit and is given by Equation 23.1. The negative sign in Equation 23.2 will be discussed in Section 23.3. If the circuit is a coil consisting of N identical and concentric loops and if the field lines pass through all loops, the induced emf is

$$\varepsilon = -N\frac{d\Phi_B}{dt} \qquad \textbf{23.3} \blacktriangleleft$$

The emf is increased by the factor N because all the loops are in series, so the emfs in the individual loops add to give the total emf.

Pitfall Prevention | 23.1
Induced emf Requires a Change
The *existence* of a magnetic flux through an area is not sufficient to create an induced emf. The magnetic flux must *change* to induce an emf.

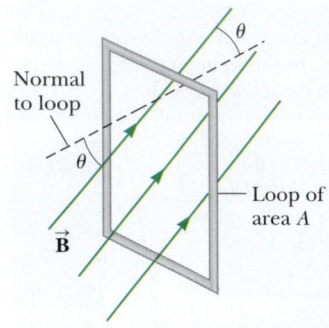

Figure 23.5 A conducting loop that encloses an area A in the presence of a uniform magnetic field $\vec{B}$. The angle between $\vec{B}$ and the normal to the loop is θ.

Suppose a loop enclosing an area A lies in a uniform magnetic field $\vec{B}$ as in Figure 23.5. In this case, the magnetic flux through the loop is

$$\Phi_B = \int \vec{B} \cdot d\vec{A} = \int B \, dA \cos\theta = B \cos\theta \int dA = BA \cos\theta$$

Hence, the induced emf is

$$\varepsilon = -\frac{d}{dt}(BA \cos\theta) \qquad \textbf{23.4} \blacktriangleleft$$

This expression shows that an emf can be induced in a circuit by changing the magnetic flux in several ways: (1) the magnitude of $\vec{B}$ can vary with time, (2) the area A of the circuit can change with time, (3) the angle θ between $\vec{B}$ and the normal to the plane can change with time, and (4) any combination of these changes can occur.

An interesting application of Faraday's law is the production of sound in an electric guitar (Fig. 23.6). The coil in this case, called the *pickup coil,* is placed near the vibrating guitar string, which is made of a metal that can be magnetized. A permanent magnet inside the coil magnetizes the portion of the string nearest the coil. When the string vibrates at some frequency, its magnetized segment produces a changing magnetic flux through the coil. The changing flux induces an emf in the coil that is fed to an amplifier. The output of the amplifier is sent to the loudspeakers, which produce the sound waves we hear.

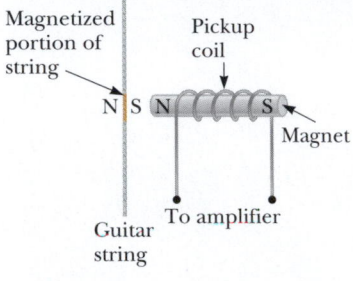

Figure 23.6 (a) In an electric guitar, a vibrating magnetized string induces an emf in a pickup coil. (b) The pickups (the circles beneath the metallic strings) of this electric guitar detect the vibrations of the strings and send this information through an amplifier and into speakers. (A switch on the guitar allows the musician to select which set of six pickups is used.)

▶ **QUICK QUIZ 23.1** A circular loop of wire is held in a uniform magnetic field, with the plane of the loop perpendicular to the field lines. Which of the following will *not* cause a current to be induced in the loop? (**a**) crushing the loop (**b**) rotating the loop about an axis perpendicular to the field lines (**c**) keeping the orientation of the loop fixed and moving it along the field lines (**d**) pulling the loop out of the field

▶ **QUICK QUIZ 23.2** Figure 23.7 shows a graphical representation of the field magnitude versus time for a magnetic field that passes through a fixed loop and that is oriented perpendicular to the plane of the loop. The magnitude of the magnetic field at any time is uniform over the area of the loop. Rank the magnitudes of the emf generated in the loop at the five instants indicated, from largest to smallest.

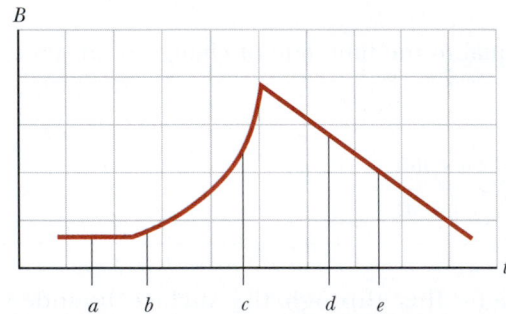

Figure 23.7 (Quick Quiz 23.2) The time behavior of a magnetic field through a loop.

▶ **THINKING PHYSICS 23.1**

The ground fault circuit interrupter (GFCI) is a safety device that protects users of electric power against electric shock when they touch appliances. Its essential parts are shown in Figure 23.8. How does the operation of a GFCI make use of Faraday's law?

Reasoning Wire 1 leads from the wall outlet to the appliance being protected, and wire 2 leads from the appliance back to the wall outlet. An iron ring surrounds the two wires. A sensing coil wrapped around part of the iron ring activates a circuit breaker when changes in magnetic flux occur. Because the currents in the two wires are in opposite directions during normal operation of the appliance, the net magnetic field through the sensing coil due to the currents is zero. A change in magnetic flux through the sensing coil can happen, however, if one of the wires on the appliance loses its insulation and accidentally touches the metal case of the appliance, providing a direct path to ground. When such a short to ground occurs, a net magnetic flux occurs through the sensing coil that alternates in time because household current is alternating. This changing flux produces an induced voltage in the coil, which in turn triggers a circuit breaker, stopping the current before it reaches a level that might be harmful to the person using the appliance. ◄

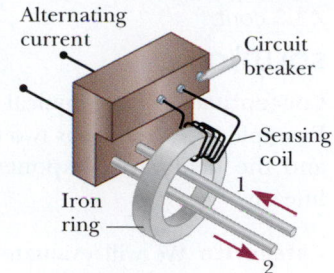

Figure 23.8 (Thinking Physics 23.1) Essential components of a ground fault circuit interrupter.

> ### Example 23.1 | Inducing an emf in a Coil

A coil consists of 200 turns of wire. Each turn is a square of side $d = 18$ cm, and a uniform magnetic field directed perpendicular to the plane of the coil is turned on. If the field changes linearly from 0 to 0.50 T in 0.80 s, what is the magnitude of the induced emf in the coil while the field is changing?

SOLUTION

Conceptualize From the description in the problem, imagine magnetic field lines passing through the coil. Because the magnetic field is changing in magnitude, an emf is induced in the coil.

Categorize We will evaluate the emf using Faraday's law from this section, so we categorize this example as a substitution problem.

Evaluate Equation 23.3 for the situation described here, noting that the magnetic field changes linearly with time:

$$|\varepsilon| = N\frac{\Delta \Phi_B}{\Delta t} = N\frac{\Delta(BA)}{\Delta t} = NA\frac{\Delta B}{\Delta t} = Nd^2\frac{B_f - B_i}{\Delta t}$$

Substitute numerical values:

$$|\varepsilon| = (200)(0.18 \text{ m})^2 \frac{(0.50 \text{ T} - 0)}{0.80 \text{ s}} = \boxed{4.0 \text{ V}}$$

What If? What if you were asked to find the magnitude of the induced current in the coil while the field is changing? Can you answer that question?

Answer If the ends of the coil are not connected to a circuit, the answer to this question is easy: the current is zero! (Charges move within the wire of the coil, but they cannot move into or out of the ends of the coil.) For a steady current to exist, the ends of the coil must be connected to an external circuit. Let's assume the coil is connected to a circuit and the total resistance of the coil and the circuit is 2.0 Ω. Then, the magnitude of the induced current in the coil is

$$I = \frac{|\varepsilon|}{R} = \frac{4.0 \text{ V}}{2.0 \text{ }\Omega} = 2.0 \text{ A}$$

> ### Example 23.2 | An Exponentially Decaying Magnetic Field

A loop of wire enclosing an area A is placed in a region where the magnetic field is perpendicular to the plane of the loop. The magnitude of $\vec{\mathbf{B}}$ varies in time according to the expression $B = B_{max}e^{-at}$, where a is some constant. That is, at $t = 0$, the field is B_{max}, and for $t > 0$, the field decreases exponentially as in Figure 23.9 (page 678). Find the induced emf in the loop as a function of time.

continued

23.2 *cont.*

SOLUTION

Conceptualize The physical situation is similar to that in Example 23.1 except for two things: there is only one loop, and the field varies exponentially with time rather than linearly.

Categorize We will evaluate the emf using Faraday's law from this section, so we categorize this example as a substitution problem.

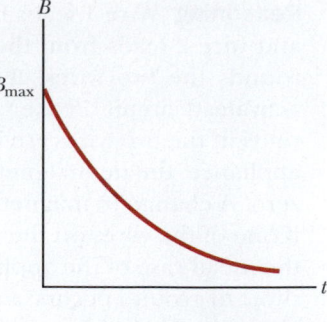

Figure 23.9 (Example 23.2) Exponential decrease in the magnitude of the magnetic field with time. The induced emf and induced current vary with time in the same way.

Evaluate Equation 23.2 for the situation described here:

$$\varepsilon = -\frac{d\Phi_B}{dt} = -\frac{d}{dt}(AB_{max}\,e^{-at}) = -AB_{max}\,\frac{d}{dt}\,e^{-at} = aAB_{max}e^{-at}$$

This expression indicates that the induced emf decays exponentially in time. The maximum emf occurs at $t = 0$, where $\varepsilon_{max} = aAB_{max}$. The plot of ε versus t is similar to the B-versus-t curve shown in Figure 23.9.

❯ 23.2 | Motional emf

Examples 23.1 and 23.2 are cases in which an emf is produced in a circuit when the magnetic field changes with time. In this section, we describe **motional emf,** in which an emf is induced in a conductor moving through a magnetic field. This is the situation described in Figure 23.1 at the beginning of Section 23.1.

Consider a straight conductor of length ℓ moving with constant velocity through a uniform magnetic field directed into the page as in Figure 23.10. For simplicity, we shall assume that the conductor is moving with a velocity that is perpendicular to the field. The electrons in the conductor experience a force along the conductor with magnitude $|\vec{\mathbf{F}}_B| = |q\vec{\mathbf{v}} \times \vec{\mathbf{B}}| = qvB$. According to Newton's second law, the electrons accelerate in response to this force and move along the conductor. Once the electrons move to the lower end of the conductor, they accumulate there, leaving a net positive charge at the upper end. As a result of this charge separation, an electric field $\vec{\mathbf{E}}$ is produced within the conductor. The charge at the ends of the conductor builds up until the magnetic force qvB on an electron in the conductor is balanced by the electric force qE on the electron as shown in Figure 23.10. At this point, charge stops flowing. In this situation, the zero net force on an electron allows us to relate the electric field to the magnetic field:

$$\sum\vec{\mathbf{F}} = \vec{\mathbf{F}}_e - \vec{\mathbf{F}}_B = 0 \quad\rightarrow\quad qE = qvB \quad\rightarrow\quad E = vB$$

Because the electric field produced in the conductor is uniform, it is related to the potential difference across the ends of the conductor according to the relation $\Delta V = E\ell$ (Section 20.2). Therefore,

$$\Delta V = E\ell = B\ell v$$

where the upper end is at a higher potential than the lower end. Therefore, a potential difference is maintained as long as the conductor is moving through the magnetic field. If the motion is reversed, the polarity of ΔV is also reversed.

An interesting situation occurs if we now consider what happens when the moving conductor is part of a closed circuit. Consider a circuit consisting of a conducting bar of length ℓ sliding along two fixed parallel conducting rails as in Active Figure 23.11a. For simplicity, we assume that the moving bar has zero electrical resistance and that the stationary part of the circuit has a resistance R. A uniform and constant magnetic field $\vec{\mathbf{B}}$ is applied perpendicular to the plane of the circuit.

As the bar is pulled to the right with a velocity $\vec{\mathbf{v}}$ under the influence of an applied force $\vec{\mathbf{F}}_{app}$, free charges in the bar experience a magnetic force along the length of

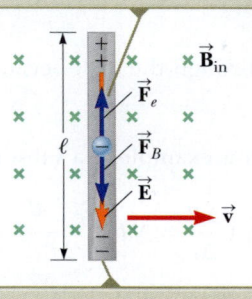

In steady state, the electric and magnetic forces on an electron in the conductor are balanced.

Due to the magnetic force on electrons, the ends of the conductor become oppositely charged, which establishes an electric field in the conductor.

Figure 23.10 A straight electrical conductor of length ℓ moving with a velocity $\vec{\mathbf{v}}$ through a uniform magnetic field $\vec{\mathbf{B}}$ directed perpendicular to $\vec{\mathbf{v}}$.

the bar. Because the moving bar is part of a complete circuit, a continuous current is established in the circuit. In this case, the rate of change of magnetic flux through the loop and the accompanying induced emf across the moving bar are proportional to the change in loop area as the bar moves through the magnetic field.

Because the area of the circuit at any instant is ℓx, the magnetic flux through the circuit is

$$\Phi_B = B\ell x$$

where x is the width of the circuit, a parameter that changes with time. Using Faraday's law, we find that the induced emf is

$$\varepsilon = -\frac{d\Phi_B}{dt} = -\frac{d}{dt}(B\ell x) = -B\ell \frac{dx}{dt}$$

$$\varepsilon = -B\ell v \qquad \qquad \textbf{23.5} \blacktriangleleft$$

Because the resistance of the circuit is R, the magnitude of the induced current is

$$I = \frac{|\varepsilon|}{R} = \frac{B\ell v}{R} \qquad \qquad \textbf{23.6} \blacktriangleleft$$

A counterclockwise current I is induced in the loop. The magnetic force $\vec{F}_B$ on the bar carrying this current opposes the motion.

Active Figure 23.11 (a) A conducting bar sliding with a velocity $\vec{v}$ along two conducting rails under the action of an applied force $\vec{F}_{app}$. (b) The equivalent circuit diagram for the pictorial representation in (a).

The equivalent circuit diagram for this example is shown in Active Figure 23.11b. The moving bar is behaving like a battery in that it is a source of emf as long as the bar continues to move.

Let us examine this situation using energy considerations in the nonisolated system model, with the system being the entire circuit. Because the circuit has no battery, you might wonder about the origin of the induced current and the energy delivered to the resistor. Note that the external force $\vec{F}_{app}$ does work on the conductor, thereby moving charges through a magnetic field, which causes the charges to move along the conductor with some average drift velocity. Hence, a current is established. From the viewpoint of the conservation of energy equation (Eq. 7.2), the total work done on the system by the applied force while the bar moves with constant speed must equal the increase in internal energy in the resistor during this time interval. (This statement assumes that the energy stays in the resistor; in reality, energy leaves the resistor by heat and electromagnetic radiation.)

As the conductor of length ℓ moves through the uniform magnetic field $\vec{B}$, it experiences a magnetic force $\vec{F}_B$ of magnitude $I\ell B$ (Eq. 22.10), where I is the current induced due to its motion. The direction of this force is opposite the motion of the bar, or to the left in Active Figure 23.11a.

If the bar is to move with a *constant* velocity, it is modeled as a particle in equilibrium, so the applied force $\vec{F}_{app}$ must be equal in magnitude and opposite in direction to the magnetic force, or to the right in Active Figure 23.11a. (If the magnetic force acted in the direction of motion, it would cause the bar to accelerate once it was in motion, thereby increasing its speed. This state of affairs would represent a violation of the principle of energy conservation.) Using Equation 23.6 and that $F_{app} = F_B = I\ell B$, we find that the power delivered by the applied force is

$$P = F_{app}v = (I\ell B)v = \frac{B^2\ell^2v^2}{R} = \left(\frac{B\ell v}{R}\right)^2 R = I^2 R \qquad \textbf{23.7} \blacktriangleleft$$

This power is equal to the rate at which energy is delivered to the resistor, as we expect.

QUICK QUIZ 23.3 You wish to move a rectangular loop of wire into a region of uniform magnetic field at a given speed so as to induce an emf in the loop. The plane of the loop must remain perpendicular to the magnetic field lines. In which orientation should you hold the loop while you move it into the region of magnetic field so as to generate the largest emf? (**a**) with the long dimension of the loop parallel to the velocity vector (**b**) with the short dimension of the loop parallel to the velocity vector (**c**) either way because the emf is the same regardless of orientation

QUICK QUIZ 23.4 In Active Figure 23.11, a given applied force of magnitude F_{app} results in a constant speed v and a power input P. Imagine that the force is increased so that the constant speed of the bar is doubled to $2v$. Under these conditions, what are the new force and the new power input? (**a**) $2F$ and $2P$ (**b**) $4F$ and $2P$ (**c**) $2F$ and $4P$ (**d**) $4F$ and $4P$

Example 23.3 | Motional emf Induced in a Rotating Bar

A conducting bar of length ℓ rotates with a constant angular speed ω about a pivot at one end. A uniform magnetic field $\vec{B}$ is directed perpendicular to the plane of rotation as shown in Figure 23.12. Find the motional emf induced between the ends of the bar.

SOLUTION

Conceptualize The rotating bar is different in nature from the sliding bar in Active Figure 23.11. Consider a small segment of the bar, however. It is a short length of conductor moving in a magnetic field and has an emf generated in it like the sliding bar. By thinking of each small segment as a source of emf, we see that all segments are in series and the emfs add.

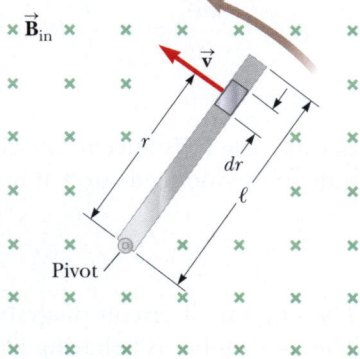

Figure 23.12 (Example 23.3) A conducting bar rotating around a pivot at one end in a uniform magnetic field that is perpendicular to the plane of rotation. A motional emf is induced between the ends of the bar.

Categorize Based on the conceptualization of the problem, we approach this example by modeling each small segment of the bar as a conductor moving in a magnetic field, with the added feature that the short segments of the bar are traveling in circular paths.

....

Analyze Evaluate the magnitude of the emf induced in a segment of the bar of length dr having a velocity $\vec{v}$ from Equation 23.5:

$$d\mathcal{E} = Bv\,dr$$

Find the total emf between the ends of the bar by adding the emfs induced across all segments:

$$\mathcal{E} = \int Bv\,dr$$

The tangential speed v of an element is related to the angular speed ω through the relationship $v = r\omega$ (Eq. 10.10); use that fact and integrate:

$$\mathcal{E} = B \int v\,dr = B\omega \int_0^\ell r\,dr = \tfrac{1}{2} B\omega\ell^2$$

....

Finalize In Equation 23.5 for a sliding bar, we can increase $\mathcal{E}$ by increasing B, ℓ, or v. Increasing any one of these variables by a given factor increases $\mathcal{E}$ by the same factor. Therefore, you would choose whichever of these three variables is most convenient to increase. For the rotating rod, however, there is an advantage to increasing the length of the rod to raise the emf because ℓ is squared. Doubling the length gives four times the emf, whereas doubling the angular speed only doubles the emf.

What If? Suppose, after reading through this example, you come up with a brilliant idea. A Ferris wheel has radial metallic spokes between the hub and the circular rim. These spokes move in the magnetic field of the Earth, so each spoke acts like the bar in Figure 23.12. You plan to use the emf generated by the rotation of the Ferris wheel to power the lightbulbs on the wheel. Will this idea work?

Answer Let's estimate the emf that is generated in this situation. The magnitude of the Earth's magnetic field is about $B = 0.5 \times 10^{-4}$ T. A typical spoke on a Ferris wheel might have a length on the order of 10 m. Suppose the period of rotation is on the order of 10 s.

Determine the angular speed of the spoke:

$$\omega = \frac{2\pi}{T} = \frac{2\pi}{10 \text{ s}} = 0.63 \text{ s}^{-1} \sim 1 \text{ s}^{-1}$$

23.3 *cont.*

Assume the magnetic field lines of the Earth are horizontal at the location of the Ferris wheel and perpendicular to the spokes. Find the emf generated:

$$\varepsilon = \tfrac{1}{2}B\omega\ell^2 = \tfrac{1}{2}(0.5 \times 10^{-4}\,\text{T})(1\,\text{s}^{-1})(10\,\text{m})^2$$
$$= 2.5 \times 10^{-3}\,\text{V} \sim 1\,\text{mV}$$

This value is a tiny emf, far smaller than that required to operate lightbulbs.

An additional difficulty is related to energy. Even assuming you could find lightbulbs that operate using a potential difference on the order of millivolts, a spoke must be part of a circuit to provide a voltage to the lightbulbs. Consequently, the spoke must carry a current. Because this current-carrying spoke is in a magnetic field, a magnetic force is exerted on the spoke in the direction opposite its direction of motion. As a result, the motor of the Ferris wheel must supply more energy to perform work against this magnetic drag force. The motor must ultimately provide the energy that is operating the lightbulbs, and you have not gained anything for free!

Example 23.4 | Magnetic Force Acting on a Sliding Bar

The conducting bar illustrated in Figure 23.13 moves on two frictionless, parallel rails in the presence of a uniform magnetic field directed into the page. The bar has mass m, and its length is ℓ. The bar is given an initial velocity $\vec{v}_i$ to the right and is released at $t = 0$.

(A) Using Newton's laws, find the velocity of the bar as a function of time.

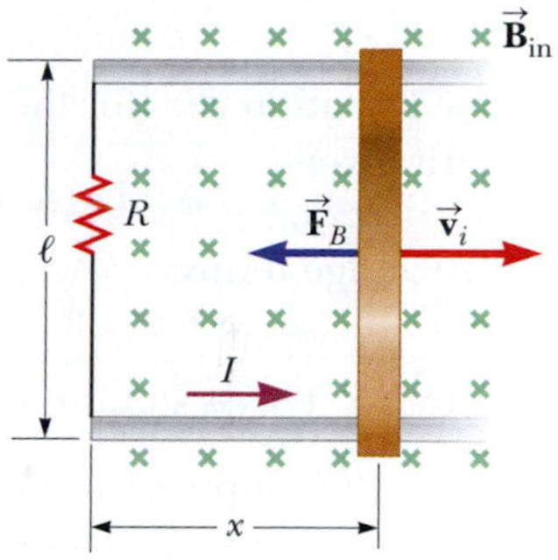

SOLUTION

Conceptualize As the bar slides to the right in Figure 23.13, a counterclockwise current is established in the circuit consisting of the bar, the rails, and the resistor. The upward current in the bar results in a magnetic force to the left on the bar as shown in the figure. Therefore, the bar must slow down, so our mathematical solution should demonstrate that.

Figure 23.13 (Example 23.4) A conducting bar of length ℓ on two fixed conducting rails is given an initial velocity $\vec{v}_i$ to the right.

Categorize The text already categorizes this problem as one that uses Newton's laws. We model the bar as a particle under a net force.

Analyze From Equation 22.10, the magnetic force is $F_B = -I\ell B$, where the negative sign indicates that the force is to the left. The magnetic force is the *only* horizontal force acting on the bar.

Apply Newton's second law to the bar in the horizontal direction:

$$F_x = ma \quad \rightarrow \quad -I\ell B = m\frac{dv}{dt}$$

Substitute $I = B\ell v/R$ from Equation 23.6:

$$m\frac{dv}{dt} = -\frac{B^2\ell^2}{R}v$$

Rearrange the equation so that all occurrences of the variable v are on the left and those of t are on the right:

$$\frac{dv}{v} = -\left(\frac{B^2\ell^2}{mR}\right)dt$$

Integrate this equation using the initial condition that $v = v_i$ at $t = 0$ and noting that $(B^2\ell^2/mR)$ is a constant:

$$\int_{v_i}^{v}\frac{dv}{v} = -\frac{B^2\ell^2}{mR}\int_0^t dt$$

$$\ln\left(\frac{v}{v_i}\right) = -\left(\frac{B^2\ell^2}{mR}\right)t$$

Define the constant $\tau = mR/B^2\ell^2$ and solve for the velocity:

$$(1) \quad v = v_i e^{-t/\tau}$$

continued

23.4 cont.

Finalize This expression for v indicates that the velocity of the bar decreases with time under the action of the magnetic force as expected from our conceptualization of the problem.

(B) Show that the same result is found by using an energy approach.

SOLUTION

Categorize The text of this part of the problem tells us to use an energy approach for the same situation. We model the entire circuit in Figure 23.13 as an isolated system.

Analyze Consider the sliding bar as one system component possessing kinetic energy, which decreases because energy is transferring *out* of the bar by electrical transmission through the rails. The resistor is another system component possessing internal energy, which rises because energy is transferring *into* the resistor. Because energy is not leaving the system, the rate of energy transfer out of the bar equals the rate of energy transfer into the resistor.

Equate the power entering the resistor to that leaving the bar:

$$P_{\text{resistor}} = -P_{\text{bar}}$$

Substitute for the electrical power delivered to the resistor and the time rate of change of kinetic energy for the bar:

$$I^2 R = -\frac{d}{dt}\left(\tfrac{1}{2}mv^2\right)$$

Use Equation 23.6 for the current and carry out the derivative:

$$\frac{B^2\ell^2 v^2}{R} = -mv\frac{dv}{dt}$$

Rearrange terms:

$$\frac{dv}{v} = -\left(\frac{B^2\ell^2}{mR}\right) dt$$

Finalize This result is the same expression to be integrated that we found in part (A).

What If? Suppose you wished to increase the distance through which the bar moves between the time it is initially projected and the time it essentially comes to rest. You can do so by changing one of three variables—v_i, R, or B—by a factor of 2 or $\frac{1}{2}$. Which variable should you change to maximize the distance, and would you double it or halve it?

Answer Increasing v_i would make the bar move farther. Increasing R would decrease the current and therefore the magnetic force, making the bar move farther. Decreasing B would decrease the magnetic force and make the bar move farther. Which method is most effective, though?

Use Equation (1) to find the distance the bar moves by integration:

$$v = \frac{dx}{dt} = v_i e^{-t/\tau}$$

$$x = \int_0^\infty v_i e^{-t/\tau}\, dt = -v_i \tau e^{-t/\tau}\Big|_0^\infty$$

$$= -v_i \tau(0-1) = v_i \tau = v_i\left(\frac{mR}{B^2\ell^2}\right)$$

This expression shows that doubling v_i or R will double the distance. Changing B by a factor of $\frac{1}{2}$, however, causes the distance to be four times as great!

The Alternating-Current Generator

The alternating-current (AC) generator is a device in which energy is transferred in by work and out by electrical transmission. A simplified pictorial representation of an AC generator is shown in Active Figure 23.14a. It consists of a coil of wire rotated in an external magnetic field by some external agent, which represents the work input. In commercial power plants, the energy required to rotate the loop can be derived from a variety of sources. In a hydroelectric plant, for example, falling water directed against the blades of a turbine produces the rotary motion; in a coal-fired plant, the high temperature produced by burning the coal is used to convert water to steam and this steam is directed against turbine blades. As the loop rotates, the magnetic flux through it changes with time, inducing an emf and a current in a circuit connected to the coil.

Suppose the coil has N turns, all of the same area A, and suppose the coil rotates with a constant angular speed ω about an axis perpendicular to the magnetic field. If θ is the angle between the magnetic field and the direction perpendicular to the plane of the coil, the magnetic flux through the loop at any time t is given by

$$\Phi_B = BA \cos \theta = BA \cos \omega t$$

where we have used the relationship between angular position and a constant angular speed, $\theta = \omega t$. (See Eq. 10.7 and set the angular acceleration α equal to zero.) Hence, the induced emf in the coil is

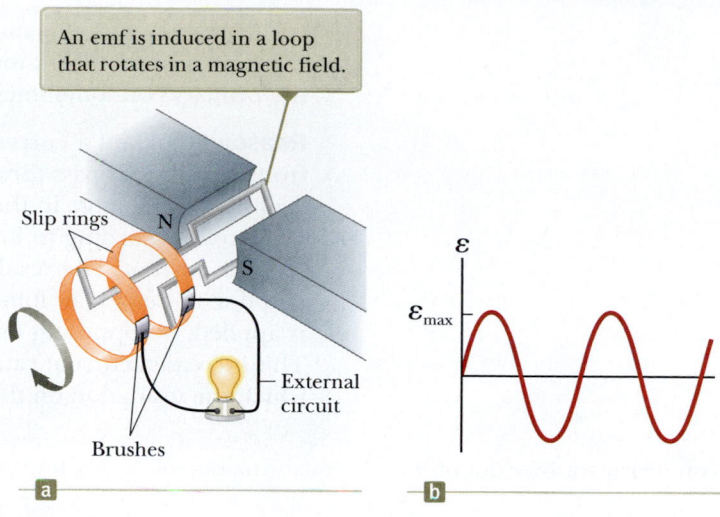

An emf is induced in a loop that rotates in a magnetic field.

Slip rings   N
              S
                          External
                          circuit
Brushes

a

$\mathcal{E}$
$\mathcal{E}_{max}$
                                              t

b

Active Figure 23.14 (a) Schematic diagram of an AC generator. (b) A graphical representation of the alternating emf induced in the loop as a function of time.

$$\mathcal{E} = -N \frac{d\Phi_B}{dt} = -NBA \frac{d}{dt}(\cos \omega t) = NBA\omega \sin \omega t \qquad \textbf{23.8} \blacktriangleleft$$

This result shows that the emf varies sinusoidally with time as shown in Active Figure 23.14b. From Equation 23.8, we see that the maximum emf has the value $\mathcal{E}_{max} = NBA\omega$, which occurs when $\omega t = 90°$ or $270°$. In other words, $\mathcal{E} = \mathcal{E}_{max}$ when the magnetic field is in the plane of the coil, and the time rate of change of flux is a maximum. In this position, the velocity vector for a wire in the loop is perpendicular to the magnetic field vector. Furthermore, the emf is *zero* when $\omega t = 0$ or $180°$—that is, when $\vec{B}$ is perpendicular to the plane of the coil—and the time rate of change of flux is zero. In this orientation, the velocity vector for a wire in the loop is parallel to the magnetic field vector.

The sinusoidally varying emf in Equation 23.8 is the source of *alternating current* delivered to customers of electrical utility companies. It is called **AC voltage** as opposed to the DC voltage from a source such as a battery.

23.3 | Lenz's Law

Let us now address the negative sign in Faraday's law. When a change occurs in the magnetic flux, the direction of the induced emf and induced current can be found from **Lenz's law:**

The induced current in a loop is in the direction that creates a magnetic field that opposes the change in magnetic flux through the area enclosed by the loop. That is, the induced current tends to keep the original magnetic flux through the loop from changing.

Notice that no equation is associated with Lenz's law. The law is in words only and provides a means for determining the direction of the current in a circuit when a magnetic change occurs.

> **THINKING PHYSICS 23.2**
>
> A transformer (Fig. 23.15) consists of a pair of coils wrapped around an iron form. When AC voltage is applied to one coil, the *primary*, the magnetic field lines cutting through the other coil, the *secondary*, induce an emf across a *load* resistor R_L. (This arrangement is used in Faraday's experiment shown in Active Fig. 23.3.) By varying the number of turns of wire on each coil, the AC voltage

An alternating voltage ΔV_1 is applied to the primary coil, and the output voltage ΔV_2 is across the resistor of resistance R_L.

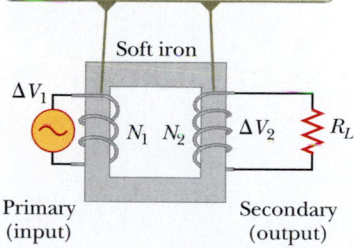

Soft iron

ΔV_1
              N_1 N_2 ΔV_2 R_L

Primary   Secondary
(input)   (output)

Figure 23.15 (Thinking Physics 23.2) An ideal transformer consists of two coils of wire wound on the same iron core.

in the secondary can be made larger or smaller than that in the primary. Clearly, this device cannot work with DC voltage. What's more, if DC voltage is applied, the primary coil sometimes overheats and burns. Why?

Reasoning When a current exists in the primary coil, the magnetic field lines from this current pass through the coil itself. Therefore, any change in the current causes a change in the magnetic field that in turn induces a current in the same coil. According to Lenz's law, this current is in the direction opposite the original current. The result is that when an AC voltage is applied, the opposing emf due to Lenz's law limits the current in the coil to a low value. If DC voltage is applied, no opposing emf occurs and the current can rise to a higher value. This increased current causes the temperature of the coil to rise, to the point at which the insulation on the wire sometimes burns. ◀

Pitfall Prevention | 23.2
Induced Current Opposes the Change
The induced current in a circuit opposes the *change* in the magnetic field, not the field itself. Therefore, in some cases the magnetic field due to the induced current is in the same direction as the changing external magnetic field. Such is the case if the external magnetic field is decreasing in magnitude, for example.

To attain a better understanding of Lenz's law, let us return to the example of a bar moving to the right on two parallel rails in the presence of a uniform magnetic field directed into the page (Fig. 23.16a). As the bar moves to the right, the magnetic flux through the circuit increases with time because the area of the loop increases. Lenz's law says that the induced current must be in such a direction that the magnetic field *it* produces opposes the *change* in the magnetic flux of the external magnetic field. Because the flux is due to an external field *into* the page and is increasing, the induced current, if it is to oppose the change, must produce a magnetic field through the circuit *out* of the page. Hence, the induced current must be counterclockwise when the bar moves to the right to give a counteracting field out of the page in the region inside the loop. (Use the right-hand rule to verify this direction.) If the bar is moving to the left, as in Figure 23.16b, the magnetic flux through the loop decreases with time. Because the magnetic field is into the page, the induced current has to be clockwise to produce a magnetic field into the page inside the loop. In either case, the induced current attempts to maintain the original flux through the circuit.

Let us examine this situation from the viewpoint of energy considerations. Suppose the bar is given a slight push to the right. In the preceding analysis, we found that this motion leads to a counterclockwise current in the loop. What happens if we incorrectly assume that the current is clockwise? For a clockwise

Figure 23.16 (a) Lenz's law can be used to determine the direction of the induced current. (b) When the bar moves to the left, the induced current must be clockwise. Why?

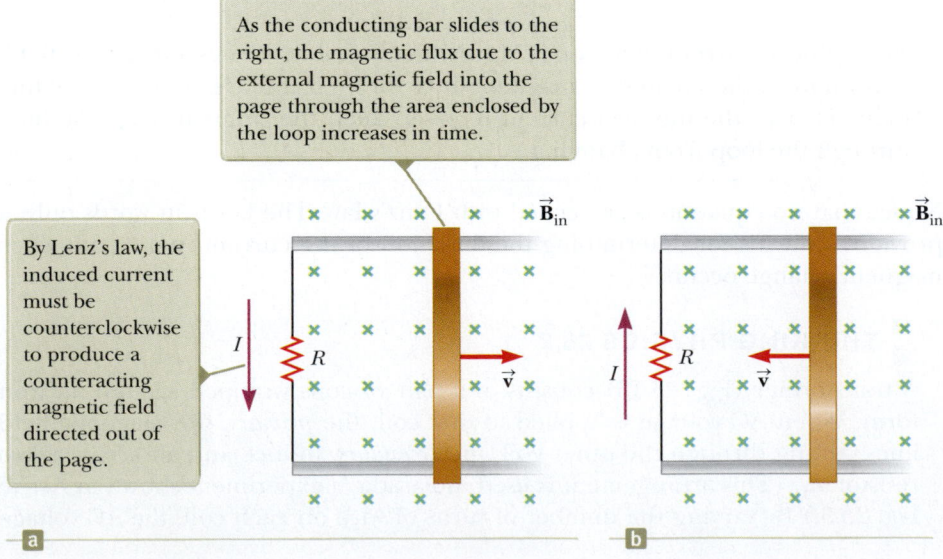

As the conducting bar slides to the right, the magnetic flux due to the external magnetic field into the page through the area enclosed by the loop increases in time.

By Lenz's law, the induced current must be counterclockwise to produce a counteracting magnetic field directed out of the page.

current I, the direction of the magnetic force $I\ell B$ on the sliding bar would be to the right. According to Newton's second law, this force would accelerate the rod and increase its speed, which in turn would cause the area of the loop to increase more rapidly. This increase would increase the induced current, which would increase the force, which would increase the current, and so on. In effect, the system would acquire energy with no additional energy input. This result is clearly inconsistent with all experience and with the conservation of energy equation. Therefore, we are forced to conclude that the current must be counterclockwise.

QUICK QUIZ 23.5 In equal-arm balances from the early 20th century (Fig. 23.17), an aluminum sheet hangs from one of the arms and passes between the poles of a magnet, which causes the oscillations of the balance to decay rapidly. In the absence of such magnetic braking, the oscillation might continue for a long time and the experimenter would have to wait to take a reading. Why do the oscillations decay? (**a**) because the aluminum sheet is attracted to the magnet (**b**) because currents in the aluminum sheet set up a magnetic field that opposes the oscillations (**c**) because aluminum is paramagnetic

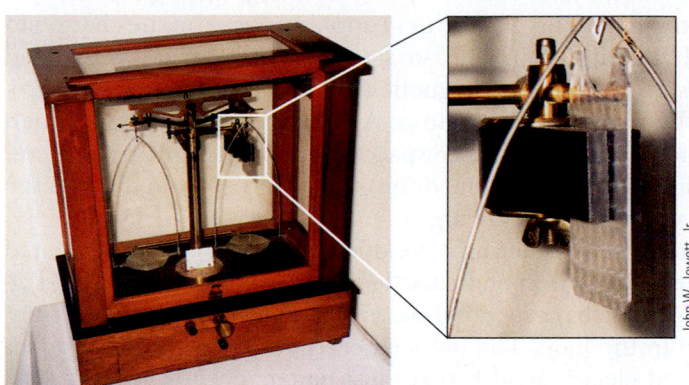

Figure 23.17 (Quick Quiz 23.5) In an old-fashioned equal-arm balance, an aluminum sheet hangs between the poles of a magnet.

John W. Jewett, Jr.

THINKING PHYSICS 23.3

A magnet is placed near a metal loop as shown in Figure 23.18a.

(A) Find the direction of the induced current in the loop when the magnet is pushed toward the loop.

When the magnet is moved toward the stationary conducting loop, a current is induced in the direction shown. The magnetic field lines are due to the bar magnet.

This induced current produces its own magnetic field directed to the left that counteracts the increasing external flux.

When the magnet is moved away from the stationary conducting loop, a current is induced in the direction shown.

This induced current produces a magnetic field directed to the right and so counteracts the decreasing external flux.

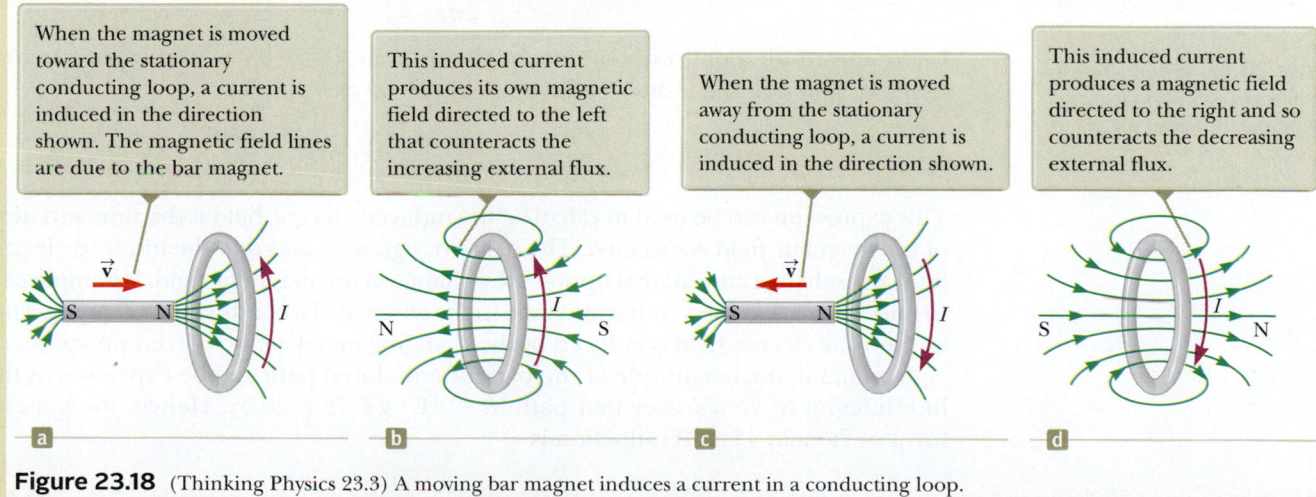

Figure 23.18 (Thinking Physics 23.3) A moving bar magnet induces a current in a conducting loop.

Reasoning As the magnet moves to the right toward the loop, the external magnetic flux through the loop increases with time. To counteract this increase in flux due to a field toward the right, the induced current produces its own magnetic field to the left as illustrated in Figure 23.18b; hence, the induced current is in the direction shown. Knowing that like magnetic poles repel each other, we conclude that the left face of the current loop acts like a north pole and the right face acts like a south pole.

(B) Find the direction of the induced current in the loop when the magnet is pulled away from the loop.

Reasoning If the magnet moves to the left as in Figure 23.18c, its flux through the area enclosed by the loop decreases in time. Now the induced current in the loop is in the direction shown in Figure 23.18d because this current direction produces a magnetic field in the same direction as the external field. In this case, the left face of the loop is a south pole and the right face is a north pole. ◄

23.4 | Induced emfs and Electric Fields

We have seen that a changing magnetic flux induces an emf and a current in a conducting loop. We can also interpret this phenomenon from another point of view. Because the normal flow of charges in a circuit is due to an electric field in the wires set up by a source such as a battery, we can interpret the changing magnetic field as creating an induced electric field. This electric field applies a force on the charges to cause them to move. With this approach, then, we see that an electric field is created in the conductor as a result of changing magnetic flux. In fact, the law of electromagnetic induction can be interpreted as follows: An electric field is always generated by a changing magnetic flux, even in free space where no charges are present. This induced electric field, however, has quite different properties from those of the electrostatic field produced by stationary charges.

Let us illustrate this point by considering a conducting loop of radius r, situated in a uniform magnetic field that is perpendicular to the plane of the loop as in Figure 23.19. If the magnetic field changes with time, Faraday's law tells us that an emf $\varepsilon = -d\Phi_B/dt$ is induced in the loop. The induced current thus produced implies the presence of an induced electric field $\vec{E}$ that must be tangent to the loop so as to provide an electric force on the charges around the loop. The work done by the electric field on the loop in moving a test charge q once around the loop is equal to $W = q\varepsilon$. Because the magnitude of the electric force on the charge is qE, the work done on the charge by the electric field can also be expressed from Equation 6.8 as $W = \int \vec{F} \cdot d\vec{r} = qE(2\pi r)$, where $2\pi r$ is the circumference of the loop. These two expressions for the work must be equal; therefore, we see that

$$q\varepsilon = qE(2\pi r)$$

$$E = \frac{\varepsilon}{2\pi r}$$

Using this result along with Faraday's law and that $\Phi_B = BA = B\pi r^2$ for a circular loop, we find that the induced electric field can be expressed as

$$E = -\frac{1}{2\pi r}\frac{d\Phi_B}{dt} = -\frac{1}{2\pi r}\frac{d}{dt}(B\pi r^2) = -\frac{r}{2}\frac{dB}{dt}$$

This expression can be used to calculate the induced electric field if the time variation of the magnetic field is specified. The negative sign indicates that the induced electric field $\vec{E}$ results in a current that opposes the change in the magnetic field. It is important to understand that this result is also valid in the absence of a conductor or charges. That is, the same electric field is induced by the changing magnetic field in empty space.

In general, the magnitude of the emf for any closed path can be expressed as the line integral of $\vec{E} \cdot d\vec{s}$ over that path: $\varepsilon = \oint \vec{E} \cdot d\vec{s}$ (Eq. 20.3). Hence, the general form of Faraday's law of induction is

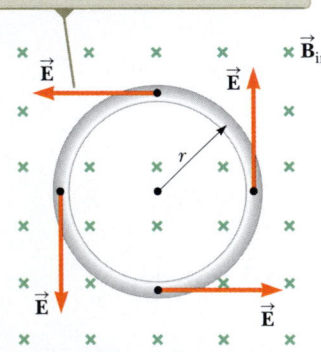

If $\vec{B}$ changes in time, an electric field is induced in a direction tangent to the circumference of the loop.

Figure 23.19 A conducting loop of radius r in a uniform magnetic field perpendicular to the plane of the loop.

► Faraday's law in general form

$$\oint \vec{E} \cdot d\vec{s} = -\frac{d\Phi_B}{dt}$$

23.9 ◄

It is important to recognize that the induced electric field $\vec{E}$ that appears in Equation 23.9 is a nonconservative field that is generated by a changing magnetic field. We call it a nonconservative field because the work done in moving a charge around a closed path (the loop in Fig. 23.19) is not zero. This type of electric field is very different from an electrostatic field.

> **QUICK QUIZ 23.6** In a region of space, a magnetic field is uniform over space but increases at a constant rate. This changing magnetic field induces an electric field that (**a**) increases in time, (**b**) is conservative, (**c**) is in the direction of the magnetic field, or (**d**) has a constant magnitude.

> **THINKING PHYSICS 23.4**
>
> In studying electric fields, we noted that electric field lines begin on positive charges and end on negative charges. Do *all* electric field lines begin and end on charges?
>
> **Reasoning** The statement that electric field lines begin and end on charges is true only for *electrostatic* fields, that is, electric fields due to stationary charges. Electric field lines due to changing magnetic fields form closed loops, with no beginning and no end, and are independent of the presence of charges. ◀

Example 23.5 | Electric Field Induced by a Changing Magnetic Field in a Solenoid

A long solenoid of radius R has n turns of wire per unit length and carries a time-varying current that varies sinusoidally as $I = I_{max} \cos \omega t$, where I_{max} is the maximum current and ω is the angular frequency of the alternating current source (Fig. 23.20).

(A) Determine the magnitude of the induced electric field outside the solenoid at a distance $r > R$ from its long central axis.

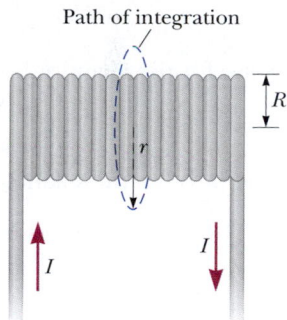

Path of integration

Figure 23.20 (Example 23.5) A long solenoid carrying a time-varying current given by $I = I_{max} \cos \omega t$. An electric field is induced both inside and outside the solenoid.

SOLUTION

Conceptualize Figure 23.20 shows the physical situation. As the current in the coil changes, imagine a changing magnetic field at all points in space as well as an induced electric field.

Categorize Because the current varies in time, the magnetic field is changing, leading to an induced electric field as opposed to the electrostatic electric fields due to stationary electric charges.

....

Analyze First consider an external point and take the path for the line integral to be a circle of radius r centered on the solenoid as illustrated in Figure 23.20.

Evaluate the right side of Equation 23.9, noting that the magnetic field $\vec{B}$ inside the solenoid is perpendicular to the circle bounded by the path of integration:

$$(1) \quad -\frac{d\Phi_B}{dt} = -\frac{d}{dt}(B\pi R^2) = -\pi R^2 \frac{dB}{dt}$$

Evaluate the magnetic field in the solenoid from Equation 22.32:

$$(2) \quad B = \mu_0 nI = \mu_0 n I_{max} \cos \omega t$$

Substitute Equation (2) into Equation (1):

$$(3) \quad -\frac{d\Phi_B}{dt} = -\pi R^2 \mu_0 n I_{max} \frac{d}{dt}(\cos \omega t) = \pi R^2 \mu_0 n I_{max} \omega \sin \omega t$$

Evaluate the left side of Equation 23.9, noting that the magnitude of $\vec{E}$ is constant on the path of integration and $\vec{E}$ is tangent to it:

$$(4) \quad \oint \vec{E} \cdot d\vec{s} = E(2\pi r)$$

23.5 cont.

Substitute Equations (3) and (4) into Equation 23.9: $E(2\pi r) = \pi R^2 \mu_0 n I_{max} \omega \sin \omega t$

Solve for the magnitude of the electric field:

$$E = \frac{\mu_0 n I_{max} \omega R^2}{2r} \sin \omega t \quad \text{(for } r > R)$$

Finalize This result shows that the amplitude of the electric field outside the solenoid falls off as $1/r$ and varies sinusoidally with time. As we will learn in Chapter 24, the time-varying electric field creates an additional contribution to the magnetic field. The magnetic field can be somewhat stronger than we first stated, both inside and outside the solenoid. The correction to the magnetic field is small if the angular frequency ω is small. At high frequencies, however, a new phenomenon can dominate: The electric and magnetic fields, each re-creating the other, constitute an electromagnetic wave radiated by the solenoid as we will study in Chapter 24.

(B) What is the magnitude of the induced electric field inside the solenoid, a distance r from its axis?

SOLUTION

Analyze For an interior point ($r < R$), the magnetic flux through an integration loop is given by $\Phi_B = B\pi r^2$.

Evaluate the right side of Equation 23.9:

$$(5) \quad -\frac{d\Phi_B}{dt} = -\frac{d}{dt}(B\pi r^2) = -\pi r^2 \frac{dB}{dt}$$

Substitute Equation (2) into Equation (5):

$$(6) \quad -\frac{d\Phi_B}{dt} = -\pi r^2 \mu_0 \, n I_{max} \frac{d}{dt}(\cos \omega t) = \pi r^2 \mu_0 n I_{max} \omega \sin \omega t$$

Substitute Equations (4) and (6) into Equation 23.9: $E(2\pi r) = \pi r^2 \mu_0 n I_{max} \omega \sin \omega t$

Solve for the magnitude of the electric field:

$$E = \frac{\mu_0 n I_{max} \omega}{2} r \sin \omega t \quad \text{(for } r < R)$$

Finalize This result shows that the amplitude of the electric field induced inside the solenoid by the changing magnetic flux through the solenoid increases linearly with r and varies sinusoidally with time.

❮23.5 | Inductance

Consider an isolated circuit consisting of a switch, a resistor, and a source of emf as in Figure 23.21. The circuit diagram is represented in perspective so that we can see the orientations of some of the magnetic field lines due to the current in the circuit. When the switch is thrown to its closed position, the current doesn't immediately jump from zero to its maximum value $\mathcal{E}/R$; the law of electromagnetic induction (Faraday's law) describes the actual behavior. As the current increases with time, the magnetic flux through the loop of the circuit itself due to the current also increases with time. This increasing magnetic flux *from* the circuit induces an emf *in* the circuit (sometimes referred to as a *back emf*) that opposes the change in the net magnetic flux through the loop of the circuit. By Lenz's law, the induced electric field in the wires must therefore be opposite the direction of the current, and the opposing emf results in a *gradual* increase in the current. This effect is called *self-induction* because the changing magnetic flux through the circuit arises from the circuit itself. The emf set up in this case is called a **self-induced emf.**

To obtain a quantitative description of self-induction, we recall from Faraday's law that the induced emf is the negative time rate of change of the magnetic flux. The magnetic flux is proportional to the magnetic field, which in turn is proportional to the current in the circuit. Therefore, the self-induced emf is always proportional to the time rate of change of the current. For a closely spaced coil of N turns of fixed

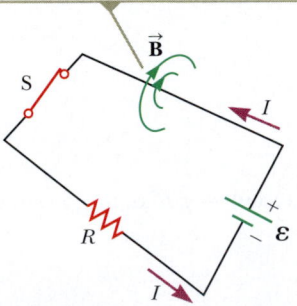

After the switch is closed, the current produces a magnetic flux through the area enclosed by the loop. As the current increases toward its equilibrium value, this magnetic flux changes in time and induces an emf in the loop.

Figure 23.21 Self-induction in a simple circuit

geometry (a toroidal coil or the ideal solenoid), we can express this proportionality as follows:

$$\varepsilon_L = -N\frac{d\Phi_B}{dt} = -L\frac{dI}{dt}$$

23.10 ◀ ▶ Self-induced emf

where L is a proportionality constant, called the **inductance** of the coil, that depends on the geometric features of the coil and other physical characteristics. From this expression, we see that the inductance of a coil containing N turns is

$$L = \frac{N\Phi_B}{I}$$

23.11 ◀

where it is assumed that the same magnetic flux passes through each turn. Later we shall use this equation to calculate the inductance of some special coil geometries.

From Equation 23.10, we can also write the inductance as the ratio

$$L = -\frac{\varepsilon_L}{dI/dt}$$

23.12 ◀

which is usually taken to be the defining equation for the inductance of any coil, regardless of its shape, size, or material characteristics. If we compare Equation 23.10 with Equation 21.6, $R = \Delta V/I$, we see that resistance is a measure of opposition to current, whereas inductance is a measure of opposition to the *change* in current.

The SI unit of inductance is the **henry (H)**, which, from Equation 23.12, is seen to be equal to 1 volt·second per ampere:

$$1\text{ H} = 1\text{ V·s/A}$$

As we shall see, the **inductance of a coil depends on its geometry.** Because inductance calculations can be quite difficult for complicated geometries, the examples we shall explore involve simple situations for which inductances are easily evaluated.

Brady-Handy Collection, Library of Congress Prints and Photographs Division [LC-BH83- 997]

Joseph Henry
American Physicist (1797–1878)
Henry became the first director of the Smithsonian Institution and first president of the Academy of Natural Science. He improved the design of the electromagnet and constructed one of the first motors. He also discovered the phenomenon of self-induction, but he failed to publish his findings. The unit of inductance, the henry, is named in his honor.

Example 23.6 | Inductance of a Solenoid

Consider a uniformly wound solenoid having N turns and length ℓ. Assume ℓ is much longer than the radius of the windings and the core of the solenoid is air.

(A) Find the inductance of the solenoid.

SOLUTION

Conceptualize The magnetic field lines from each turn of the solenoid pass through all the turns, so an induced emf in each coil opposes changes in the current.

Categorize Because the solenoid is long, we can use the results for an ideal solenoid obtained in Chapter 22.

Analyze Find the magnetic flux through each turn of area A in the solenoid, using the expression for the magnetic field from Equation 22.32:

$$\Phi_B = BA = \mu_0 nIA = \mu_0 \frac{N}{\ell}IA$$

Substitute this expression into Equation 23.11:

$$(1) \quad L = \frac{N\Phi_B}{I} = \mu_0 \frac{N^2}{\ell}A$$

(B) Calculate the inductance of the solenoid if it contains 300 turns, its length is 25.0 cm, and its cross-sectional area is 4.00 cm².

SOLUTION

Substitute numerical values into Equation (1):

$$L = (4\pi \times 10^{-7}\text{ T·m/A})\frac{300^2}{25.0 \times 10^{-2}\text{ m}}(4.00 \times 10^{-4}\text{ m}^2)$$

$$= 1.81 \times 10^{-4}\text{ T·m}^2/\text{A} = 0.181\text{ mH}$$

continued

23.6 *cont.*

(C) Calculate the self-induced emf in the solenoid if the current it carries decreases at the rate of 50.0 A/s.

SOLUTION

Substitute $dI/dt = -50.0$ A/s and the answer to part (B) into Equation 23.10:

$$\varepsilon_L = -L\frac{dI}{dt} = -(1.81 \times 10^{-4}\,\text{H})(-50.0\,\text{A/s})$$

$$= 9.05\,\text{mV}$$

Finalize The result for part (A) shows that L depends on geometry and is proportional to the square of the number of turns. Because $N = n\ell$, we can also express the result in the form

$$L = \mu_0 \frac{(n\ell)^2}{\ell}A = \mu_0 n^2 A\ell = \mu_0 n^2 V$$

where $V = A\ell$ is the interior volume of the solenoid.

▌23.6 | *RL* Circuits

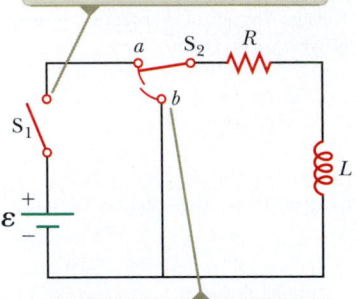

When switch S_1 is thrown closed, the current increases and an emf that opposes the increasing current is induced in the inductor.

When the switch S_2 is thrown to position *b*, the battery is no longer part of the circuit and the current decreases.

Active Figure 23.22 An *RL* circuit. When switch S_2 is in position *a*, the battery is in the circuit.

If a circuit contains a coil such as a solenoid, the inductance of the coil prevents the current in the circuit from increasing or decreasing instantaneously. A circuit element that has a large inductance is called an **inductor** and has the circuit symbol —⟋⟋⟋⟋—. We always assume the inductance of the remainder of a circuit is negligible compared with that of the inductor. Keep in mind, however, that even a circuit without a coil has some inductance that can affect the circuit's behavior.

Because the inductance of an inductor results in a back emf, an inductor in a circuit opposes changes in the current in that circuit. The inductor attempts to keep the current the same as it was before the change occurred. If the battery voltage in the circuit is increased so that the current rises, the inductor opposes this change and the rise is not instantaneous. If the battery voltage is decreased, the inductor causes a slow drop in the current rather than an immediate drop. Therefore, the inductor causes the circuit to be "sluggish" as it reacts to changes in the voltage.

Consider the circuit shown in Active Figure 23.22, which contains a battery of negligible internal resistance. This circuit is an **RL circuit** because the elements connected to the battery are a resistor and an inductor. The curved lines on switch S_2 suggest this switch can never be open; it is always set to either *a* or *b*. (If the switch is connected to neither *a* nor *b*, any current in the circuit suddenly stops.) Suppose S_2 is set to *a* and switch S_1 is open for $t < 0$ and then thrown closed at $t = 0$. The current in the circuit begins to increase, and a back emf (Eq. 23.10) that opposes the increasing current is induced in the inductor.

With this point in mind, let's apply Kirchhoff's loop rule to this circuit, traversing the circuit in the clockwise direction:

$$\varepsilon - IR - L\frac{dI}{dt} = 0 \qquad\qquad \textbf{23.13}◀$$

where IR is the voltage drop across the resistor. (Kirchhoff's rules were developed for circuits with steady currents, but they can also be applied to a circuit in which the current is changing if we imagine them to represent the circuit at one *instant* of time.) Now let's find a solution to this differential equation, which is similar to that for the *RC* circuit (see Section 21.9).

A mathematical solution of Equation 23.13 represents the current in the circuit as a function of time. To find this solution, we change variables for convenience, letting $x = (\varepsilon/R) - I$, so $dx = -dI$. With these substitutions, Equation 23.13 becomes

$$x + \frac{L}{R}\frac{dx}{dt} = 0$$

Rearranging and integrating this last expression gives

$$\int_{x_0}^{x} \frac{dx}{x} = -\frac{R}{L} \int_{0}^{t} dt$$

$$\ln \frac{x}{x_0} = -\frac{R}{L} t$$

where x_0 is the value of x at time $t = 0$. Taking the antilogarithm of this result gives

$$x = x_0 e^{-Rt/L}$$

Because $I = 0$ at $t = 0$, note from the definition of x that $x_0 = \mathcal{E}/R$. Hence, this last expression is equivalent to

$$\frac{\mathcal{E}}{R} - I = \frac{\mathcal{E}}{R} e^{-Rt/L}$$

$$I = \frac{\mathcal{E}}{R} (1 - e^{-Rt/L})$$

This expression shows how the inductor affects the current. The current does not increase instantly to its final equilibrium value when the switch is closed, but instead increases according to an exponential function. If the inductance is removed from the circuit, which corresponds to letting L approach zero, the exponential term becomes zero and there is no time dependence of the current in this case; the current increases instantaneously to its final equilibrium value in the absence of the inductance.

We can also write this expression as

$$I = \frac{\mathcal{E}}{R} (1 - e^{-t/\tau}) \qquad \textbf{23.14} \blacktriangleleft$$

where the constant τ is the **time constant** of the RL circuit:

$$\tau = \frac{L}{R} \qquad \textbf{23.15} \blacktriangleleft$$

Physically, τ is the time interval required for the current in the circuit to reach $(1 - e^{-1}) = 0.632 = 63.2\%$ of its final value $\mathcal{E}/R$. The time constant is a useful parameter for comparing the time responses of various circuits.

Active Figure 23.23 shows a graph of the current versus time in the RL circuit. Notice that the equilibrium value of the current, which occurs as t approaches infinity, is $\mathcal{E}/R$. That can be seen by setting dI/dt equal to zero in Equation 23.13 and solving for the current I. (At equilibrium, the change in the current is zero.) Therefore, the current initially increases very rapidly and then gradually approaches the equilibrium value $\mathcal{E}/R$ as t approaches infinity.

Let's also investigate the time rate of change of the current. Taking the first time derivative of Equation 23.14 gives

$$\frac{dI}{dt} = \frac{\mathcal{E}}{L} e^{-t/\tau} \qquad \textbf{23.16} \blacktriangleleft$$

This result shows that the time rate of change of the current is a maximum (equal to $\mathcal{E}/L$) at $t = 0$ and falls off exponentially to zero as t approaches infinity (Fig. 23.24).

Now consider the RL circuit in Active Figure 23.22 again. Suppose switch S_2 has been set at position a long enough (and switch S_1 remains closed) to allow the current to reach its equilibrium value $\mathcal{E}/R$. In this situation, the circuit is described by the outer loop in Active Figure 23.22. If S_2 is thrown from a to b, the circuit is now described by only the right-hand loop in Active Figure 23.22. Therefore, the battery has been eliminated from the circuit. Setting $\mathcal{E} = 0$ in Equation 23.13 gives

$$IR + L \frac{dI}{dt} = 0 \qquad \textbf{23.17} \blacktriangleleft$$

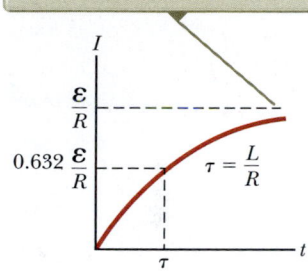

After switch S_1 is thrown closed at $t = 0$, the current increases toward its maximum value $\mathcal{E}/R$.

Active Figure 23.23 Plot of the current versus time for the RL circuit shown in Active Figure 23.22. The time constant τ is the time interval required for I to reach 63.2% of its maximum value.

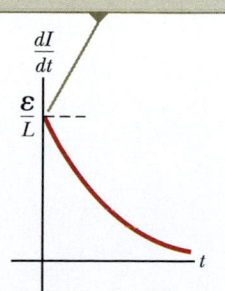

The time rate of change of current is a maximum at $t = 0$, which is the instant at which switch S_1 is thrown closed.

Figure 23.24 Plot of dI/dt versus time for the RL circuit shown in Active Figure 23.22. The rate decreases exponentially with time as I increases toward its maximum value.

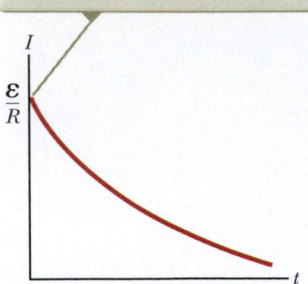

At $t = 0$, the switch is thrown to position b and the current has its maximum value $\mathcal{E}/R$.

Active Figure 23.25 Current versus time for the right-hand loop of the circuit shown in Active Figure 23.22. For $t < 0$, switch S_2 is at position a.

It is left as a problem (Problem 23.38 in Enhanced WebAssign) to show that the solution of this differential equation is

$$I = \frac{\mathcal{E}}{R} e^{-t/\tau} = I_i e^{-t/\tau} \qquad \text{23.18} \blacktriangleleft$$

where $\mathcal{E}$ is the emf of the battery and $I_i = \mathcal{E}/R$ is the initial current at the instant the switch is thrown to b.

If the circuit did not contain an inductor, the current would immediately decrease to zero when the battery is removed. When the inductor is present, it opposes the decrease in the current and causes the current to decrease exponentially. A graph of the current in the circuit versus time (Active Fig. 23.25) shows that the current is continuously decreasing with time.

▸ **QUICK QUIZ 23.7** The circuit in Figure 23.26 includes a power source that provides a sinusoidal voltage. Therefore, the magnetic field in the inductor is constantly changing. The inductor is a simple air-core solenoid. The switch in the circuit is closed and the lightbulb glows steadily. An iron rod is inserted into the interior of the solenoid, which increases the magnitude of the magnetic field in the solenoid. As that happens, the brightness of the lightbulb (**a**) increases, (**b**) decreases, or (**c**) is unaffected.

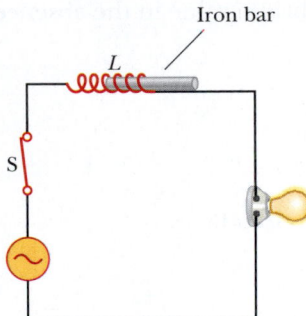

Figure 23.26 (Quick Quiz 23.7) A lightbulb is powered by an AC source with an inductor in the circuit. When the iron bar is inserted into the coil, what happens to the brightness of the lightbulb?

▸ **QUICK QUIZ 23.8** Two circuits like the one shown in Active Figure 23.22 are identical except for the value of L. In circuit A, the inductance of the inductor is L_A, and in circuit B, it is L_B. Switch S_2 has been in position b for both circuits for a long time. At $t = 0$, the switch is thrown to a in both circuits. At $t = 10$ s, the switch is thrown to b in both circuits. The resulting graphical representation of the current as a function of time is shown in Figure 23.27. Assuming that the time constant of each circuit is much less than 10 s, which of the following is true? (a) $L_A > L_B$. (b) $L_A < L_B$. (c) There is not enough information to determine the relative values.

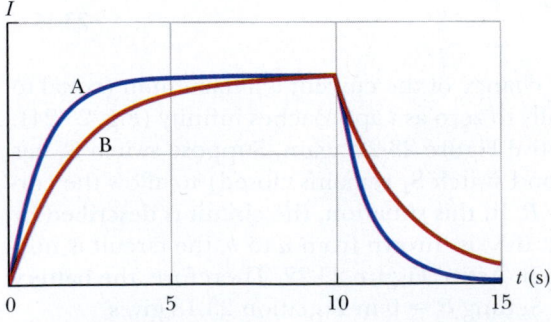

Figure 23.27 (Quick Quiz 23.8) Current–time graphs for two circuits with different inductances.

Example 23.7 | Time Constant of an *RL* Circuit

Consider the circuit in Active Figure 23.22 again. Suppose the circuit elements have the following values: $\mathcal{E} = 12.0$ V, $R = 6.00\ \Omega$, and $L = 30.0$ mH.

(A) Find the time constant of the circuit.

SOLUTION

Conceptualize You should understand the behavior of this circuit from the discussion in this section.

Categorize We evaluate the results using equations developed in this section, so this example is a substitution problem.

Evaluate the time constant from Equation 23.15:

$$\tau = \frac{L}{R} = \frac{30.0 \times 10^{-3}\ \text{H}}{6.00\ \Omega} = 5.00\ \text{ms}$$

(B) Switch S_2 is at position a, and switch S_1 is thrown closed at $t = 0$. Calculate the current in the circuit at $t = 2.00$ ms.

SOLUTION

Evaluate the current at $t = 2.00$ ms from Equation 23.14:

$$I = \frac{\mathcal{E}}{R}(1 - e^{-t/\tau}) = \frac{12.0\ \text{V}}{6.00\ \Omega}(1 - e^{-2.00\ \text{ms}/5.00\ \text{ms}}) = 2.00\ \text{A}\ (1 - e^{-0.400})$$

$$= 0.659\ \text{A}$$

(C) Compare the potential difference across the resistor with that across the inductor.

SOLUTION

At the instant the switch is closed, there is no current and therefore no potential difference across the resistor. At this instant, the battery voltage appears entirely across the inductor in the form of a back emf of 12.0 V as the inductor tries to maintain the zero-current condition. (The top end of the inductor in Active Fig. 23.22 is at a higher electric potential than the bottom end.) As time passes, the emf across the inductor decreases and the current in the resistor (and hence the voltage across it) increases as shown in Figure 23.28. The sum of the two voltages at all times is 12.0 V.

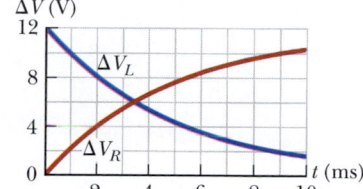

Figure 23.28 (Example 23.7) The time behavior of the voltages across the resistor and inductor in Active Figure 23.22 given the values provided in this example.

❮ 23.7 | Energy Stored in a Magnetic Field

In the preceding section, we found that the induced emf set up by an inductor prevents a battery from establishing an instantaneous current. Part of the energy supplied by the battery goes into internal energy in the resistor, and the remaining energy is stored in the inductor. If we multiply each term in Equation 23.13 by the current I and rearrange the expression, we have

$$I\mathcal{E} = I^2R + LI\frac{dI}{dt} \qquad \text{23.19} \blacktriangleleft$$

This expression tells us that the rate $I\mathcal{E}$ at which energy is supplied by the battery equals the sum of the rate I^2R at which energy is delivered to the resistor and the rate $LI\,(dI/dt)$ at which energy is delivered to the inductor. Therefore, Equation 23.19 is simply an expression of energy conservation for the isolated system of the circuit. (Actually, energy can leave the circuit by thermal conduction into the air and by electromagnetic radiation, so the system need not be completely isolated.) If we let U denote the energy stored in the inductor at any time, the rate dU/dt at which energy is delivered to the inductor can be written as

$$\frac{dU}{dt} = LI\frac{dI}{dt}$$

Pitfall Prevention | 23.3
Capacitors, Resistors, and Inductors Store Energy Differently Different energy-storage mechanisms are at work in capacitors, inductors, and resistors. A charged capacitor stores energy as electrical potential energy. An inductor stores energy as what we could call magnetic potential energy when it carries current. Energy delivered to a resistor is transformed to internal energy.

To find the total energy stored in the inductor at any instant, we can rewrite this expression as $dU = LI\,dI$ and integrate:

$$U = \int_0^U dU = \int_0^I LI\,dI$$

◄ Energy stored in an inductor

$$U = \tfrac{1}{2}LI^2 \qquad \textbf{23.20} ◄$$

where L is constant and so has been removed from the integral. Equation 23.20 represents the energy stored in the magnetic field of the inductor when the current is I.

Equation 23.20 is similar to the equation for the energy stored in the electric field of a capacitor, $U = \tfrac{1}{2}C(\Delta V)^2$ (Eq. 20.29). In either case, we see that energy from a battery is required to establish a field and that energy is stored in the field. In the case of the capacitor, we can conceptually relate the energy stored in the capacitor to the electric potential energy associated with the separated charge on the plates. We have not discussed a magnetic analogy to electric potential energy, so the storage of energy in an inductor is not as easy to conceptualize.

To argue that energy is stored in an inductor, consider the circuit in Figure 23.29a, which is the same circuit as in Active Figure 23.22, with the addition of a switch S_3 across the resistor R. With switch S_2 set to position a and S_3 closed as shown, a current is established in the inductor. Now, as in Figure 23.29b, switch S_2 is thrown to position b. The current persists in this (ideally) resistance-free and battery-free circuit (the right-hand loop in Fig. 23.29b), consisting of only the inductor and a conducting path between its ends. There is no current in the resistor (because the path around it through S_3 is resistance free), so no energy is being delivered to it. The next step is to open switch S_3 as shown in Figure 23.29c, which puts the resistor into the circuit. There is now current in the resistor, and energy is delivered to the resistor. Where is the energy coming from? The only other element in the circuit previous to opening switch S_3 was the inductor. Energy must therefore have been stored in the inductor and is now being delivered to the resistor.

Now let us determine the energy per unit volume, or energy density, stored in a magnetic field. For simplicity, consider a solenoid whose inductance is $L = \mu_0 n^2 A\ell$ (see Example 23.6). The magnetic field of the solenoid is $B = \mu_0 nI$. Substituting the expression for L and $I = B/\mu_0 n$ into Equation 23.20 gives

$$U = \tfrac{1}{2}LI^2 = \tfrac{1}{2}\mu_0 n^2 A\ell \left(\frac{B}{\mu_0 n}\right)^2 = \frac{B^2}{2\mu_0}(A\ell) \qquad \textbf{23.21} ◄$$

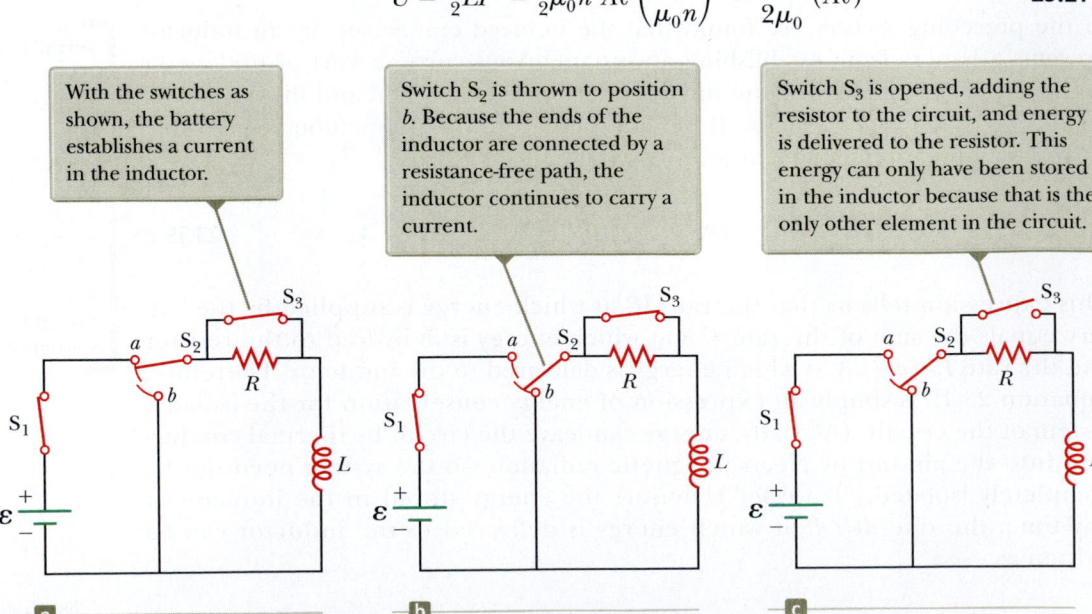

With the switches as shown, the battery establishes a current in the inductor.

Switch S_2 is thrown to position b. Because the ends of the inductor are connected by a resistance-free path, the inductor continues to carry a current.

Switch S_3 is opened, adding the resistor to the circuit, and energy is delivered to the resistor. This energy can only have been stored in the inductor because that is the only other element in the circuit.

Figure 23.29 An RL circuit used for conceptualizing energy storage in an inductor.

Because $A\ell$ is the volume of the solenoid, the energy stored per unit volume in a magnetic field—in other words, the *magnetic energy density*—is

$$u_B = \frac{U}{A\ell} = \frac{B^2}{2\mu_0}$$

23.22 ◀ ▶ Magnetic energy density

Although Equation 23.22 was derived for the special case of a solenoid, it is valid for any region of space in which a magnetic field exists. Note that it is similar to the equation for the energy per unit volume stored in an electric field, given by $\frac{1}{2}\epsilon_0 E^2$ (Eq. 20.31). In both cases, the energy density is proportional to the square of the magnitude of the field.

> **QUICK QUIZ 23.9** You are performing an experiment that requires the highest-possible magnetic energy density in the interior of a very long current-carrying solenoid. Which of the following adjustments increases the energy density? (More than one choice may be correct.) (a) increasing the number of turns per unit length on the solenoid (b) increasing the cross-sectional area of the solenoid (c) increasing only the length of the solenoid while keeping the number of turns per unit length fixed (d) increasing the current in the solenoid

Example 23.8 | What Happens to the Energy in the Inductor?

Consider once again the RL circuit shown in Active Figure 23.22, with switch S_2 at position a and the current having reached its steady-state value. When S_2 is thrown to position b, the current in the right-hand loop decays exponentially with time according to the expression $I = I_i e^{-t/\tau}$, where $I_i = \mathcal{E}/R$ is the initial current in the circuit and $\tau = L/R$ is the time constant. Show that all the energy initially stored in the magnetic field of the inductor appears as internal energy in the resistor as the current decays to zero.

SOLUTION

Conceptualize Before S_2 is thrown to b, energy is being delivered at a constant rate to the resistor from the battery and energy is stored in the magnetic field of the inductor. After $t = 0$, when S_2 is thrown to b, the battery can no longer provide energy and energy is delivered to the resistor only from the inductor.

Categorize We model the right-hand loop of the circuit as an isolated system so that energy is transferred between components of the system but does not leave the system.

Analyze The energy in the magnetic field of the inductor at any time is U. The rate dU/dt at which energy leaves the inductor and is delivered to the resistor is equal to $I^2 R$, where I is the instantaneous current.

Substitute the current given by Equation 23.18 into $dU/dt = I^2 R$:

$$\frac{dU}{dt} = I^2 R = (I_i e^{-Rt/L})^2 R = I_i^2 R e^{-2Rt/L}$$

Solve for dU and integrate this expression over the limits $t = 0$ to $t \to \infty$:

$$U = \int_0^\infty I_i^2 R e^{-2Rt/L}\, dt = I_i^2 R \int_0^\infty e^{-2Rt/L}\, dt$$

The value of the definite integral can be shown to be $L/2R$ (see Problem 23.74 in Enhanced WebAssign). Use this result to evaluate U:

$$U = I_i^2 R\left(\frac{L}{2R}\right) = \tfrac{1}{2} L I_i^2$$

Finalize This result is equal to the initial energy stored in the magnetic field of the inductor, given by Equation 23.20, as we set out to prove.

Example 23.9 | The Coaxial Cable

Coaxial cables are often used to connect electrical devices, such as your video system, and in receiving signals in television cable systems. Model a long coaxial cable as a thin, cylindrical conducting shell of radius b concentric with a solid cylinder of radius a as in Figure 23.30 (page 696). The conductors carry the same current I in opposite directions. Calculate the inductance L of a length ℓ of this cable.

continued

23.9 *cont.*

SOLUTION

Conceptualize Consider Figure 23.30. Although we do not have a visible coil in this geometry, imagine a thin, radial slice of the coaxial cable such as the light gold rectangle in Figure 23.30. If the inner and outer conductors are connected at the ends of the cable (above and below the figure), this slice represents one large conducting loop. The current in the loop sets up a magnetic field between the inner and outer conductors that passes through this loop. If the current changes, the magnetic field changes and the induced emf opposes the original change in the current in the conductors.

Categorize We categorize this situation as one in which we must return to the fundamental definition of inductance, Equation 23.11.

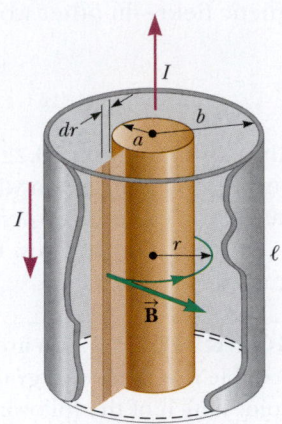

Figure 23.30 (Example 23.9) Section of a long coaxial cable. The inner and outer conductors carry equal currents in opposite directions.

Analyze We must find the magnetic flux through the light gold rectangle in Figure 23.30. Ampère's law (see Section 22.9) tells us that the magnetic field in the region between the conductors is due to the inner conductor alone and that its magnitude is $B = \mu_0 I / 2\pi r$, where r is measured from the common center of the cylinders. A sample circular field line is shown in Figure 23.30, along with a field vector tangent to the field line. The magnetic field is zero outside the outer shell because the net current passing through the area enclosed by a circular path surrounding the cable is zero; hence, from Ampère's law, $\oint \vec{B} \cdot d\vec{s} = 0$.

The magnetic field is perpendicular to the light gold rectangle of length ℓ and width $b - a$, the cross section of interest. Because the magnetic field varies with radial position across this rectangle, we must use calculus to find the total magnetic flux.

Divide the light gold rectangle into strips of width dr such as the darker strip in Figure 23.30. Evaluate the magnetic flux through such a strip:

$$d\Phi_B = B \, dA = B\ell \, dr$$

Substitute for the magnetic field and integrate over the entire light gold rectangle:

$$\Phi_B = \int_a^b \frac{\mu_0 I}{2\pi r} \ell \, dr = \frac{\mu_0 I \ell}{2\pi} \int_a^b \frac{dr}{r} = \frac{\mu_0 I \ell}{2\pi} \ln\left(\frac{b}{a}\right)$$

Use Equation 23.11 to find the inductance of the cable:

$$L = \frac{\Phi_B}{I} = \boxed{\frac{\mu_0 \ell}{2\pi} \ln\left(\frac{b}{a}\right)}$$

Finalize The inductance increases if ℓ increases, if b increases, or if a decreases. This result is consistent with our conceptualization: any of these changes increases the size of the loop represented by our radial slice and through which the magnetic field passes, increasing the inductance.

❮23.8 | Context Connection: The Use of Transcranial Magnetic Stimulation in Depression   BIO

In Sections 20.7 and 21.9, we discussed the electrical characteristics of neurons and the propagation of an action potential along a neuron. In this Context Connection, we discuss a new treatment for depression that is early in its development and is closely related to the propagation of impulses along and between nerves.

Depression is a mental disorder in which patients exhibit decreased self-esteem, low moods, sadness, loss of interest in previously enjoyable activities, and increased possibility of suicidal thoughts and behaviors. It is a complicated disorder whose cause seems to include biological conditions, psychological effects, social interactions, drug and alcohol use, and even genetics. As a result of the large number of possible influences causing depression, the particular treatment plan for a given individual is not clear without extensive counseling of the patient and attempts to use a variety of treatments.

In our discussion, we will focus on the possible biological origins of depression. One hypothesis suggests that depression is related to low levels of *neurotransmitters* (particularly serotonin, norepinephrine, and dopamine) in the synapses between neurons. Antidepressant medications, such as *sertraline*, act to increase the levels of these neurotransmitters.

One of the more controversial treatments for severe depression that has not responded to other treatments is *electroconvulsive therapy* (ECT), in which seizures are induced in a patient under anesthesia. The seizures are induced by placing electrodes on the patient's head and passing a pulsed current between the electrodes. While the effects of this procedure on human brains cannot ethically be studied in detail, results from animal experiments suggest possible new synapse formation from the treatment. Because of the role of levels of neurotransmitters in depression, this growth of synapses may be the reason for the improvement in some depressed patients after undergoing electroconvulsive therapy. ECT was used in the 1940s and 1950s on severely disturbed patients in large mental institutions. Today, its main use is in psychiatric hospitals. There is still continuing controversy about the usage of ECT for patients suffering from mental disorders.

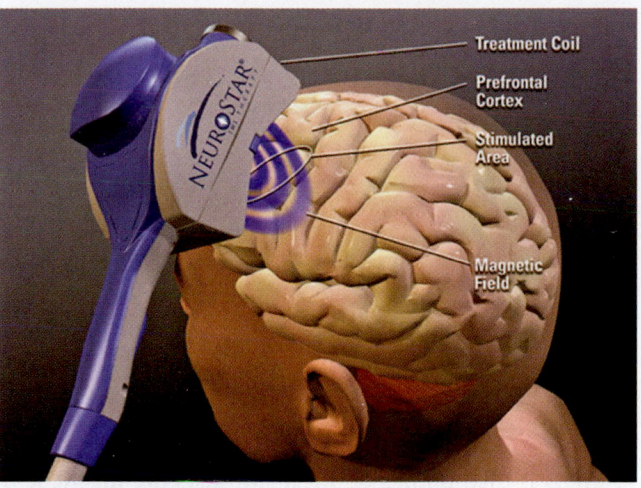

Figure 23.31 The magnetic coil of a Neurostar TMS apparatus is held near the head of a patient.

A newer method of introducing electric current into the brain is **transcranial magnetic stimulation** (TMS). This procedure induces currents in the brain by means of magnetic induction rather than by the application of a large voltage from contact electrodes. A large coil of wire is placed against the scalp of the patient. The coil carries an alternating current, creating an oscillating magnetic field, which induces currents in the nerve cells of the brain. Unlike electroconvulsive therapy, the patient is awake and does not experience seizures. Figure 23.31 shows the coil of a Neurostar TMS machine being applied to a patient's head. While the United States Food and Drug Administration has not approved TMS as a generic procedure at the time of this printing, the agency has cleared the specific Neurostar device that performs TMS.

BIO Transcranial magnetic stimulation

TMS has been used to perform motor cortex mapping, in which connections between the primary motor cortex and various muscles are measured to determine damage from spinal cord injuries, strokes, and motor neuron disease. With this technique, muscular responses in the index finger, forearm, biceps, jaw, and leg can be observed as the magnetic coil is moved to different locations of the cortex.

When the coil is moved over the occipital cortex, some patients report *magnetophosphenes*, which are flashes of light seen even though the eyes are closed. The most common method of inducing phosphenes is mechanical: by rubbing the closed eyes. Phosphenes resulting from a blow to the head are the origin of the phrase "seeing stars" associated with such trauma.

The use of TMS in depression is more recent. Figure 23.32 shows a patient being treated with a TMS apparatus. The patient sits in a chair and the coil is placed against the head. The coil is switched on and off at frequencies up to 10 Hz, inducing currents in the brain. The magnetic field is increased until the patient's fingers begin to twitch, indicating a therapeutic level. Once that level is attained, the treatment portion of the experience continues for about 40 minutes. These treatments are repeated on a daily basis for a period of several weeks.

Studies have reported some effectiveness of the procedure in treating depression, but more studies need to be performed to validate the evidence. One of the factors working against tests of effectiveness is the difficulty in establishing a "fake" TMS experience to be used as a placebo for comparison to the actual treatment. The treatment causes low-level neck pain, headache, and twitching in the scalp, which are difficult to reproduce in a placebo intervention.

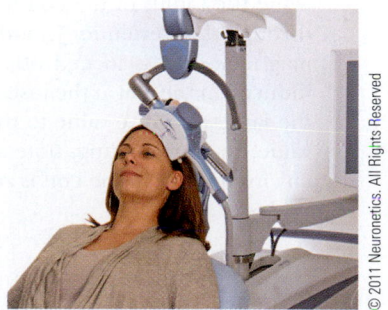

Figure 23.32 A patient is treated with the Neurostar TMS apparatus.

SUMMARY

The **magnetic flux** through a surface associated with a magnetic field $\vec{B}$ is

$$\Phi_B = \int \vec{B} \cdot d\vec{A} \qquad \textbf{23.1} \blacktriangleleft$$

where the integral is over the surface.

Faraday's law of induction states that the emf induced in a circuit is directly proportional to the time rate of change of magnetic flux through the circuit:

$$\varepsilon = -N \frac{d\Phi_B}{dt} \qquad \textbf{23.3} \blacktriangleleft$$

where N is the number of turns and Φ_B is the magnetic flux through each turn.

When a conducting bar of length ℓ moves through a magnetic field $\vec{B}$ with a velocity $\vec{v}$ so that $\vec{v}$ is perpendicular to $\vec{B}$, the emf induced in the bar (called the **motional emf**) is

$$\varepsilon = -B\ell v \qquad \textbf{23.5} \blacktriangleleft$$

Lenz's law states that the induced current and induced emf in a conductor are in such a direction as to oppose the change that produced them.

A general form of Faraday's law of induction is

$$\oint \vec{E} \cdot d\vec{s} = -\frac{d\Phi_B}{dt} \qquad \textbf{23.9} \blacktriangleleft$$

where $\vec{E}$ is a nonconservative electric field produced by the changing magnetic flux.

When the current in a coil changes with time, an emf is induced in the coil according to Faraday's law. The **self-induced emf** is described by the expression

$$\varepsilon_L = -L \frac{dI}{dt} \qquad \textbf{23.10} \blacktriangleleft$$

where L is the *inductance* of the coil. Inductance is a measure of the opposition of a device to a change in current.

The **inductance** of a coil is

$$L = \frac{N\Phi_B}{I} \qquad \textbf{23.11} \blacktriangleleft$$

where Φ_B is the magnetic flux through the coil and N is the total number of turns. Inductance has the SI unit the **henry** (H), where $1\ \text{H} = 1\ \text{V} \cdot \text{s/A}$.

If a resistor and inductor are connected in series to a battery of emf ε as shown in Active Figure 23.22, switch S_2 is set at position a, and switch S_1 is thrown closed at $t = 0$, the current in the circuit varies with time according to the expression

$$I(t) = \frac{\varepsilon}{R}(1 - e^{-t/\tau}) \qquad \textbf{23.14} \blacktriangleleft$$

where $\tau = L/R$ is the **time constant** of the *RL* circuit.

If switch S_2 in Active Figure 23.22 is thrown to position b, the current decays exponentially with time according to the expression

$$I(t) = \frac{\varepsilon}{R} e^{-t/\tau} \qquad \textbf{23.18} \blacktriangleleft$$

where ε/R is the initial current in the circuit.

The energy stored in the magnetic field of an inductor carrying a current I is

$$U = \tfrac{1}{2}LI^2 \qquad \textbf{23.20} \blacktriangleleft$$

The energy per unit volume (or energy density) at a point where the magnetic field is B is

$$u_B = \frac{B^2}{2\mu_0} \qquad \textbf{23.22} \blacktriangleleft$$

OBJECTIVE QUESTIONS

☐ denotes answer available in *Student Solutions Manual/Study Guide*

1. Figure OQ23.1 is a graph of the magnetic flux through a certain coil of wire as a function of time during an interval while the radius of the coil is increased, the coil is rotated through 1.5 revolutions, and the external source of the magnetic field is turned off, in that order. Rank the emf induced in the coil at the instants marked A through E from the largest positive value to the largest-magnitude negative value. In your ranking, note any cases of equality and also any instants when the emf is zero.

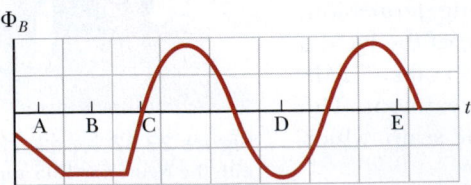

Figure OQ23.1

2. A long, fine wire is wound into a coil with inductance 5 mH. The coil is connected across the terminals of a battery, and the current is measured a few seconds after the connection is made. The wire is unwound and wound again into a different coil with $L = 10$ mH. This second coil is connected across the same battery, and the current is measured in the same way. Compared with the current in the first coil, is the current in the second coil (a) four times as large, (b) twice as large, (c) unchanged, (d) half as large, or (e) one-fourth as large?

3. Two solenoids, A and B, are wound using equal lengths of the same kind of wire. The length of the axis of each solenoid is large compared with its diameter. The axial length of A is twice as large as that of B, and A has twice as many turns as B. What is the ratio of the inductance of solenoid A to that of solenoid B? (a) 4 (b) 2 (c) 1 (d) $\frac{1}{2}$ (e) $\frac{1}{4}$

4. A circular loop of wire with a radius of 4.0 cm is in a uniform magnetic field of magnitude 0.060 T. The plane of the loop is perpendicular to the direction of the magnetic field. In a time interval of 0.50 s, the magnetic field changes to the opposite direction with a magnitude of 0.040 T. What is the magnitude of the average emf induced in the loop? (a) 0.20 V (b) 0.025 V (c) 5.0 mV (d) 1.0 mV (e) 0.20 mV

5. A rectangular conducting loop is placed near a long wire carrying a current I as shown in Figure OQ23.5. If I decreases in time, what can be said of the current induced in the loop? (a) The direction of the current depends on the size of the loop. (b) The current is clockwise. (c) The current is counterclockwise. (d) The current is zero. (e) Nothing can be said about the current in the loop without more information.

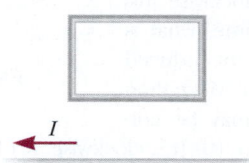

Figure OQ23.5

6. A flat coil of wire is placed in a uniform magnetic field that is in the y direction. **(i)** The magnetic flux through the coil is a maximum if the plane of the coil is where? More than one answer may be correct. (a) in the xy plane (b) in the yz plane (c) in the xz plane (d) in any orientation, because it is a constant **(ii)** For what orientation is the flux zero? Choose from the same possibilities as in part (i).

7. A solenoidal inductor for a printed circuit board is being redesigned. To save weight, the number of turns is reduced by one-half, with the geometric dimensions kept the same. By how much must the current change if the energy stored in the inductor is to remain the same? (a) It must be four times larger. (b) It must be two times larger. (c) It should be left the same. (d) It should be one-half as large. (e) No change in the current can compensate for the reduction in the number of turns.

8. If the current in an inductor is doubled, by what factor is the stored energy multiplied? (a) 4 (b) 2 (c) 1 (d) $\frac{1}{2}$ (e) $\frac{1}{4}$

9. A square, flat loop of wire is pulled at constant velocity through a region of uniform magnetic field directed perpendicular to the plane of the loop as shown in Figure OQ23.9. Which of the following statements are correct? More than one statement may be correct. (a) Current is induced in the loop in the clockwise direction. (b) Current is induced in the loop in the counterclockwise direction. (c) No current is induced in the loop. (d) Charge separation occurs in the loop, with the top edge positive. (e) Charge separation occurs in the loop, with the top edge negative.

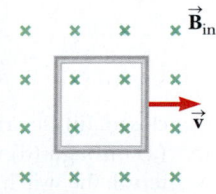

Figure OQ23.9

10. The bar in Figure OQ23.10 moves on rails to the right with a velocity $\vec{v}$, and a uniform, constant magnetic field is directed out of the page. Which of the following statements are correct? More than one statement may be correct. (a) The induced current in the loop is zero. (b) The induced current in the loop is clockwise. (c) The induced current in the loop is counterclockwise. (d) An external force is required to keep the bar moving at constant speed. (e) No force is required to keep the bar moving at constant speed.

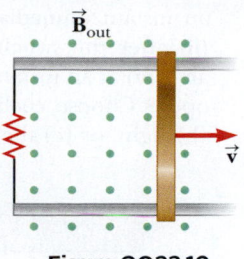

Figure OQ23.10

11. Initially, an inductor with no resistance carries a constant current. Then the current is brought to a new constant value twice as large. *After* this change, when the current is constant at its higher value, what has happened to the emf in the inductor? (a) It is larger than before the change by a factor of 4. (b) It is larger by a factor of 2. (c) It has the same nonzero value. (d) It continues to be zero. (e) It has decreased.

12. In Figure OQ23.12, the switch is left in position a for a long time interval and is then quickly thrown to position b. Rank the magnitudes of the voltages across the four circuit elements a short time thereafter from the largest to the smallest.

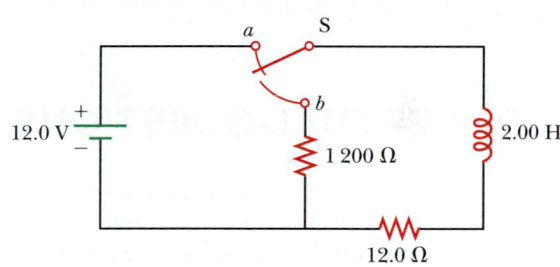

Figure OQ23.12

13. A bar magnet is held in a vertical orientation above a loop of wire that lies in the horizontal plane as shown in Figure OQ23.13. The south end of the magnet is toward the loop. After the magnet is dropped, what is true of the induced current in the loop as viewed from above? (a) It is clockwise as the magnet falls toward the loop. (b) It is counterclockwise as the magnet falls toward the loop. (c) It is clockwise after the magnet has moved through the loop and moves away from it. (d) It is always clockwise. (e) It is first counterclockwise as the magnet approaches the loop and then clockwise after it has passed through the loop.

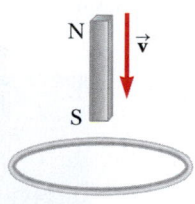

Figure OQ23.13

14. What happens to the amplitude of the induced emf when the rate of rotation of a generator coil is doubled? (a) It becomes four times larger. (b) It becomes two times larger. (c) It is unchanged. (d) It becomes one-half as large. (e) It becomes one-fourth as large.

15. Two coils are placed near each other as shown in Figure OQ23.15. The coil on the left is connected to a battery and a switch, and the coil on the right is connected to a resistor. What is the direction of the current in the resistor **(i)** at

an instant immediately after the switch is thrown closed, **(ii)** after the switch has been closed for several seconds, and **(iii)** at an instant after the switch has then been thrown open? Choose each answer from the possibilities (a) left, (b) right, or (c) the current is zero.

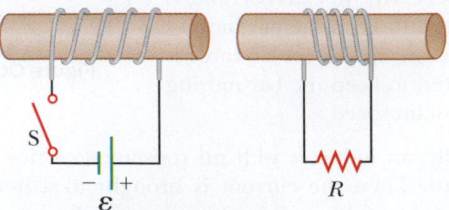

Figure OQ23.15

16. A circuit consists of a conducting movable bar and a light bulb connected to two conducting rails as shown in Figure OQ23.16. An external magnetic field is directed perpendicular to the plane of the circuit. Which of the following actions will make the bulb light up? More than one statement may be correct. (a) The bar is moved to the left. (b) The bar is moved to the right. (c) The magnitude of the magnetic field is increased. (d) The magnitude of

the magnetic field is decreased. (e) The bar is lifted off the rails.

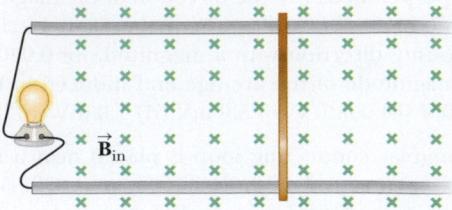

Figure OQ23.16

17. Two rectangular loops of wire lie in the same plane as shown in Figure OQ23.17. If the current I in the outer loop is counterclockwise and increases with time, what is true of the current induced in the inner loop? More than one statement may be correct. (a) It is zero. (b) It is clockwise. (c) It is counterclockwise. (d) Its magnitude depends on the dimensions of the loops. (e) Its direction depends on the dimensions of the loops.

Figure OQ23.17

CONCEPTUAL QUESTIONS

☐ denotes answer available in *Student Solutions Manual/Study Guide*

1. A switch controls the current in a circuit that has a large inductance. The electric arc at the switch (Fig. CQ23.1) can melt and oxidize the contact surfaces, resulting in high resistivity of the contacts and eventual destruction of the switch. Is a spark more likely to be produced at the switch when the switch is being closed, when it is being opened, or does it not matter?

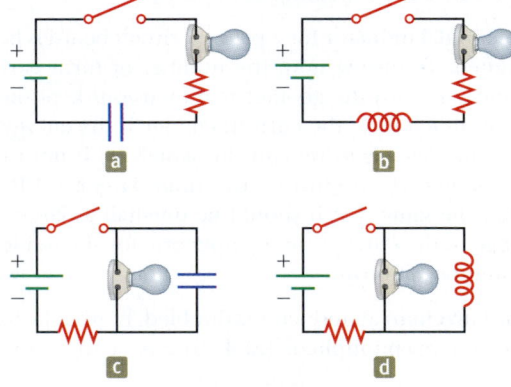

Figure CQ23.2

closed. **(ii)** Describe what the lightbulb does in each of circuits (a) through (d) when, having been closed for a long time interval, the switch is opened.

3. What is the difference between magnetic flux and magnetic field?

4. Discuss the similarities between the energy stored in the electric field of a charged capacitor and the energy stored in the magnetic field of a current-carrying coil.

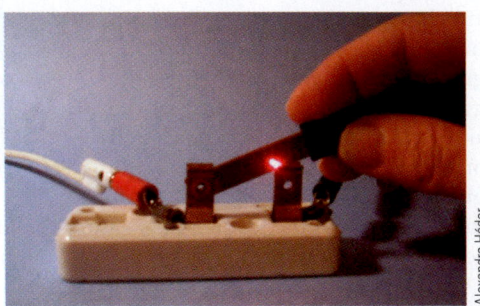

Figure CQ23.1

2. Consider the four circuits shown in Figure CQ23.2, each consisting of a battery, a switch, a lightbulb, a resistor, and either a capacitor or an inductor. Assume the capacitor has a large capacitance and the inductor has a large inductance but no resistance. The lightbulb has high efficiency, glowing whenever it carries electric current. **(i)** Describe what the lightbulb does in each of circuits (a) through (d) after the switch is thrown

5. A spacecraft orbiting the Earth has a coil of wire in it. An astronaut measures a small current in the coil, although there is no battery connected to it and there are no magnets in the spacecraft. What is causing the current?

6. A circular loop of wire is located in a uniform and constant magnetic field. Describe how an emf can be induced in the loop in this situation.

7. A bar magnet is dropped toward a conducting ring lying on the floor. As the magnet falls toward the ring, does it move as a freely falling object? Explain.

8. In a hydroelectric dam, how is energy produced that is then transferred out by electrical transmission? That is, how is the energy of motion of the water converted to energy that is transmitted by AC electricity?

9. A piece of aluminum is dropped vertically downward between the poles of an electromagnet. Does the magnetic field affect the velocity of the aluminum?

10. The current in a circuit containing a coil, a resistor, and a battery has reached a constant value. (a) Does the coil have an inductance? (b) Does the coil affect the value of the current?

11. (a) What parameters affect the inductance of a coil? (b) Does the inductance of a coil depend on the current in the coil?

12. In Section 6.7, we defined conservative and nonconservative forces. In Chapter 19, we stated that an electric charge creates an electric field that produces a conservative force. Argue now that induction creates an electric field that produces a nonconservative force.

13. When the switch in Figure CQ23.13a is closed, a current is set up in the coil and the metal ring springs upward (Fig. CQ23.13b). Explain this behavior.

14. Assume the battery in Figure CQ23.13a is replaced by an AC source and the switch is held closed. If held down, the metal ring on top of the solenoid becomes hot. Why?

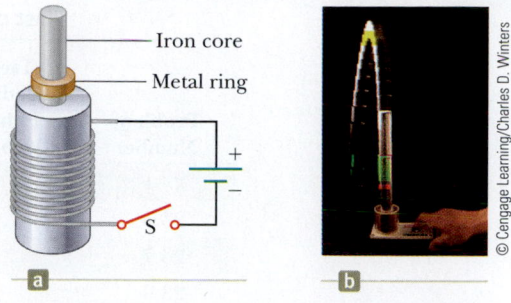
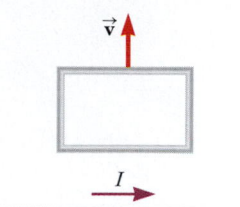

Figure CQ23.13 Conceptual Questions 13 and 14.

15. A loop of wire is moving near a long, straight wire carrying a constant current I as shown in Figure CQ23.15. (a) Determine the direction of the induced current in the loop as it moves away from the wire. (b) What would be the direction of the induced current in the loop if it were moving toward the wire?

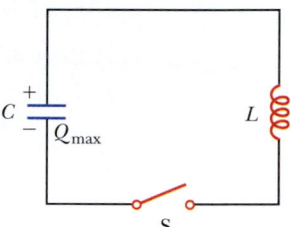

Figure CQ23.15

16. After the switch is closed in the LC circuit shown in Figure CQ23.16, the charge on the capacitor is sometimes zero, but at such instants the current in the circuit is not zero. How is this behavior possible?

Figure CQ23.16

PROBLEMS AVAILABLE IN ENHANCED Web**Assign**

23.1 Faraday's Law of Induction

Problems 1–13

23.2 Motional emf
23.3 Lenz's Law

Problems 14–27

23.4 Induced emfs and Electric Fields

Problems 28–29

23.5 Inductance

Problems 30–36

23.6 *RL* Circuits

Problems 37–47

23.7 Energy Stored in a Magnetic Field

Problems 48–51

23.8 Context Connection: The Use of Transcranial Magnetic Stimulation in Depression

Problems 52–54

Additional Problems

Problems 55–75

Solutions to the following Problems are available in the *Student Solutions Manual/Study Guide*:

23.5, 23.10, 23.15, 23.22, 23.28, 23.33, 23.36, 23.37, 23.47, 23.51, 23.60, 23.61, 23.65, 23.72, and 23.75

List of Enhanced Problems

Problem Number	Targeted Feedback in Enhanced WebAssign	Master It in Enhanced WebAssign	Watch It in Enhanced WebAssign
23.1	✓		✓
23.5		✓	
23.7	✓		✓
23.9			✓
23.10		✓	
23.11	✓		✓
23.13	✓		✓
23.15	✓	✓	
23.21		✓	
23.26	✓		✓
23.28		✓	
23.29	✓		✓
23.33		✓	
23.35	✓		✓
23.36	✓	✓	
23.37		✓	
23.39	✓		
23.45	✓		✓
23.47		✓	
23.49	✓	✓	
23.50	✓		✓
23.61		✓	
23.65		✓	
23.75		✓	

Nuclear Magnetic Resonance and Magnetic Resonance Imaging

In this Context Conclusion, we discuss an application that is now widely used as a noninvasive diagnostic tool in medical practice. This application is known as *MRI*, for *magnetic resonance imaging*.

In Section 22.11, we discussed the spin angular momentum of an electron and the associated magnetic moment of the electron. Spin is a general property of all particles. For example, protons and neutrons in the nucleus of an atom have spin and an associated magnetic dipole moment $\vec{\mu}$. As we saw in Section 22.6, the potential energy of a system consisting of a magnetic dipole moment in an external magnetic field is $U = -\vec{\mu} \cdot \vec{B}$.

When the magnetic moment $\vec{\mu}$ is aligned with the field as closely as quantum physics allows, the potential energy of the dipole–field system has its minimum value E_{min}. When $\vec{\mu}$ is as antiparallel to the field as possible, the potential energy has its maximum value E_{max}. Because the directions of the spin and the magnetic moment for a particle are quantized (see Chapter 29), the energies of the dipole–field system are also quantized. We introduced the concept of quantized states of energy in Chapter 11. In general, there are other allowed energy states between E_{min} and E_{max} corresponding to the quantized directions of the magnetic moment with respect to the field. These states are often called **spin states** because they differ in energy as a result of the direction of the spin.

The number of spin states depends on the spin of the nucleus. The simplest situation is shown in Figure 1, for a nucleus with only two possible spin states having energies E_{min} and E_{max}.

It is possible to observe transitions between these two spin states in a sample using a technique known as **nuclear magnetic resonance** (NMR). A constant magnetic field changes the energy associated with the spin states, splitting them apart in energy as shown in Figure 1. The sample is also exposed to electromagnetic waves in the radio range of the electromagnetic spectrum. When the frequency of the radio waves is adjusted such that the photon energy matches the separation energy between spin states, a resonance condition exists and the photon is absorbed by a nucleus in the ground state, raising the nucleus–magnetic field system to the higher energy spin state. This results in a net absorption of energy by the system, which is detected by the experimental control and measurement system. A diagram of the apparatus used to detect an NMR signal is illustrated in Figure 2 (page 704). The absorbed energy is supplied by the oscillator producing the radio waves. Nuclear magnetic resonance and a related technique called *electron spin resonance* are extremely important methods for studying nuclear and atomic systems and how these systems interact with their surroundings.

A widely used medical diagnostic technique called **MRI,** for **magnetic resonance imaging,** is based on nuclear magnetic resonance. In MRI, the patient is placed

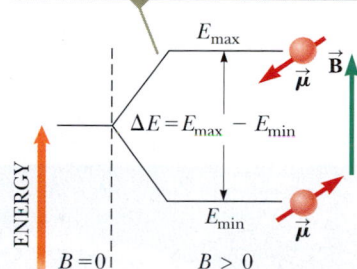

The magnetic field splits a single state of the nucleus into two states.

Figure 1 A nucleus with spin $\frac{1}{2}$ is placed in a magnetic field.

▶ Nuclear magnetic resonance

BIO Magnetic resonance imaging (MRI)

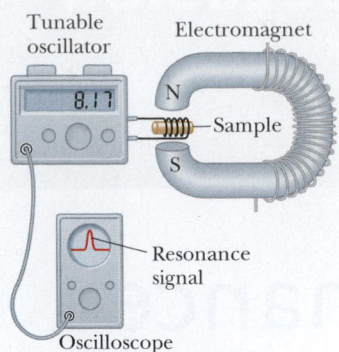

Figure 2 Experimental arrangement for nuclear magnetic resonance. The radio-frequency magnetic field created by the coil surrounding the sample and provided by the variable-frequency oscillator is perpendicular to the constant magnetic field created by the electromagnet. When the nuclei in the sample meet the resonance condition, the nuclei absorb energy from the radio-frequency field of the coil; this absorption changes the characteristics of the circuit in which the coil is included. Most modern NMR spectrometers use superconducting magnets at fixed field strengths and operate at frequencies of approximately 200 MHz.

inside a large solenoid that supplies a spatially varying magnetic field. Because of the variation in the magnetic field across the patient's body, protons in hydrogen atoms in water molecules in different parts of the body have different splittings in energy between spin states, so that the resonance signal can be used to provide information on the positions of the protons. A computer is used to analyze the position information to provide data for constructing a final image. MRI scans showing incredible detail in internal body structure are shown in Figure 3. The main advantage of MRI over other imaging techniques in medical diagnostics is that it does not cause damage to cellular structures as x-rays or gamma rays do. Photons associated with the radio-frequency signals used in MRI have energies of only about 10^{-7} eV. Because molecular bond strengths are much larger (on the order of 1 eV), the radio-frequency radiation cannot cause cellular damage. In comparison, x-rays or gamma rays have energies ranging from 10^4 to 10^6 eV and can cause considerable cellular damage. Therefore, despite some individuals' fears of the word *nuclear* associated with magnetic resonance imaging, the radio-frequency radiation involved is much safer than x-rays or gamma rays!

In this Context, we have seen a number of applications of magnetism to medical procedures. The catheter ablation procedures (Sections 21.2 and 22.12) and MRI scans discussed here have saved many lives through accurate diagnosis and treatment. Transcranial magnetic stimulation (Section 23.8) is a relatively new procedure that may turn out to be extremely helpful in its application to depression. Who knows what other uses of magnetism in medicine may lie in the near future? Keep a watch out in the newspapers and on the Internet and you may see the next application very soon!

Problems

1. The radio frequency at which a nucleus having a magnetic moment of magnitude μ displays resonance absorption between spin states is called the *Larmor frequency* and is given by

$$f = \frac{\Delta E}{h} = \frac{2\mu B}{h}$$

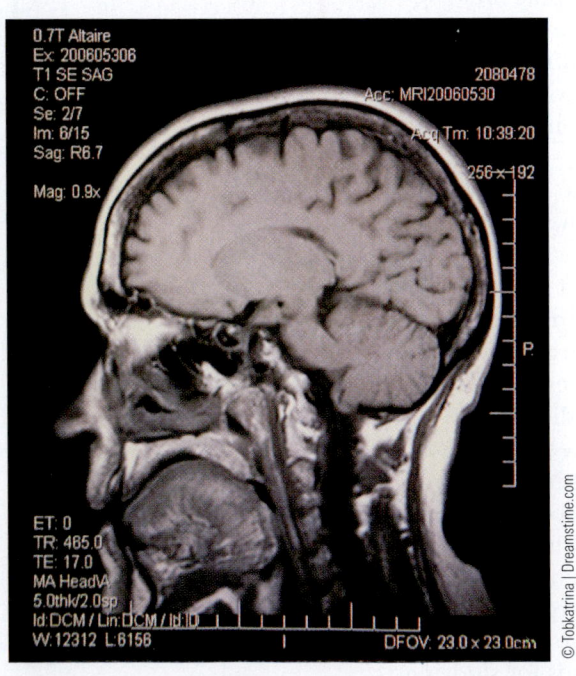

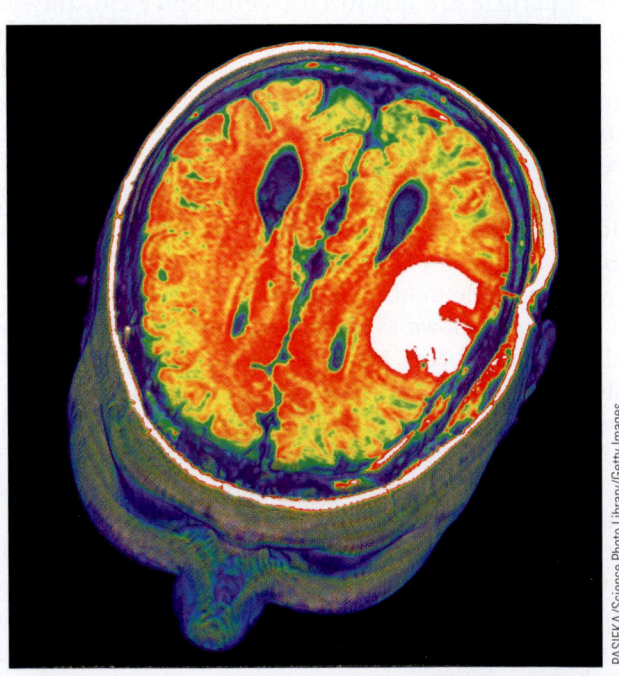

Figure 3 Examples of MRI scans of the human brain. (a) A sagittal section of the human brain, showing highly detailed views of several brain structures. (b) A computer-enhanced view of an axial section through the brain shows a metastatic tumor of the cerebrum, appearing in white.

Calculate the Larmor frequency for (a) free neutrons in a magnetic field of 1.00 T, (b) free protons in a magnetic field of 1.00 T, and (c) free protons in the Earth's magnetic field at a location where the magnitude of the field is 50.0 μT.

2. **BIO** In magnetic resonance imaging (MRI), the patient is placed inside a large solenoid. Suppose an MRI solenoid is 2.40 m long, 0.900 m in diameter, and is wound with a single layer of superconducting niobium-titanium wire of radius 1.00 mm. Each turn of the solenoid is laid right next to the previous turn so that there is no space between the turns. The magnetic field produced by the solenoid is 1.55 T. (a) What current exists in the solenoid to produce this field? (b) What is the magnetic flux through the solenoid? (c) When the machine is turned off, the field reduces linearly to zero in 5.00 s. What emf is induced in the interior of the solenoid while the machine is turning off? (d) What is the total mass of the niobium-titanium wire that is wound to make the solenoid? Assume that the density of the wire is 6.00×10^3 kg/m^3.

The invention of the laser was popularly credited to Arthur L. Schawlow and Charles H. Townes for many years after their publication of a proposal for the laser in a 1958 issue of *Physical Review*. Schawlow and Townes received a patent for the device in 1959. In 1960, the first laser was built and operated by Theodore Maiman. This device used a ruby crystal to create the laser light, which was emitted in pulses from the end of a ruby cylinder. A flash lamp was used to excite the laser action.

In 1977, the first victory in a 30-year-long legal battle was completed in which Gordon Gould, who was a graduate student at Columbia University in the late 1950s, received a patent for inventing the laser in 1957 as well as coining the term. Believing erroneously that he had to have a working prototype before he could file for a patent, he did not file until later in 1959 than had Schawlow and Townes. Gould's legal battles ended in 1987. By this time, Gould's technology was being widely used in industry and medicine. His victory finally resulted in his control of patent rights to perhaps 90% of the lasers used and sold in the United States.

Figure 2 The original ruby laser emitted red light, as did many lasers developed soon afterward. Today, lasers are available in a variety of colors and various regions of the electromagnetic spectrum. In this photograph, a green laser is used to perform scientific research.

Since the development of the first device, laser technology has experienced tremendous growth. Lasers that cover wavelengths in the infrared, visible, and ultraviolet regions are now available. Various types of lasers use solids, liquids, and gases as the active medium. Although the original laser emitted light over a very narrow range around a fixed wavelength, tunable lasers are now available, in which the wavelength can be varied.

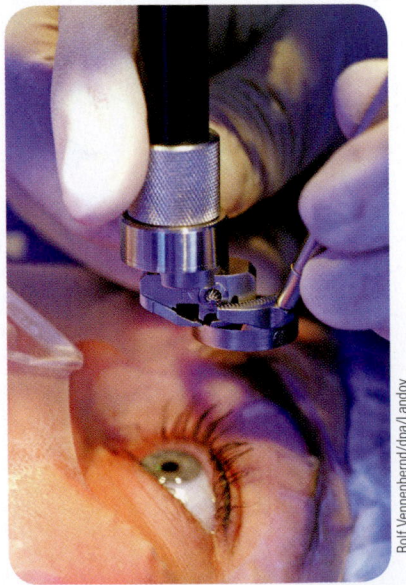

Figure 1 A patient is made ready to receive laser light during eye surgery.

Figure 3 A laser cutting machine cuts through a thick sheet of steel.

The laser is an omnipresent technological tool in our daily life. Applications include LASIK eye surgery (covered in more detail in Chapter 26), surgical "welding" of detached retinas, precision surveying and length measurement, a potential source for inducing nuclear fusion reactions, precision cutting of metals and other materials, and telephone communication along optical fibers.

We also use lasers to read information from compact discs for use in audio entertainment and computer applications. DVD and Blu-ray disc players use lasers to read video information. Lasers are used in retail stores to read price and inventory information from product labels. In the laboratory, lasers can be used to trap atoms and cool them to microkelvins above absolute zero and to move microscopic biological organisms around harmlessly.

These and other applications are possible because of the unique characteristics of laser light. In addition to its being almost monochromatic, laser light is also highly directional and can therefore be sharply focused to produce regions of extreme intensity.

In this Context, we shall investigate the physics of electromagnetic radiation and optics and apply the principles to an understanding of the behavior of laser

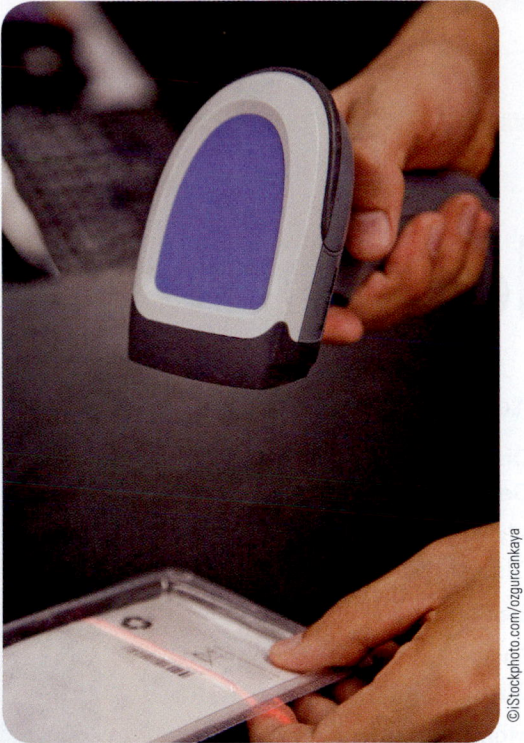

Figure 4 A bar code scanner uses light from a laser to identify products being purchased. The reflections from the bar code on the package are read and entered into the computer to determine the price of the item.

light and its applications. A major focus of our study will be on the technology of optical fibers and how they are used in industry and medicine. We shall study the nature of light as we respond to our central question:

> **What is special about laser light, and how is it used in technological applications?**

Chapter | 24

Electromagnetic Waves

Chapter Outline

NASA/CXC/SAO/F. Seward

Electromagnetic waves cover a broad spectrum of wavelengths, with waves in various wavelength ranges having distinct properties. This photo of the Crab Nebula is made with x-rays, which are like visible light, except for their very short wavelength. In this chapter, we shall study the common features of x-rays, visible light, and other forms of electromagnetic radiation.

Although we are not always aware of their presence, electromagnetic waves permeate our environment. In the form of visible light, they enable us to view the world around us with our eyes. Infrared waves from the surface of the Earth warm our environment, radio-frequency waves carry our favorite radio entertainment, microwaves cook our food and are used in radar communication systems, and the list goes on and on. The waves described in Chapter 13 are mechanical waves, which require a medium through which to propagate. Electromagnetic waves, in contrast, can propagate through a vacuum. Despite this difference between mechanical and electromagnetic waves, much of the behavior in the wave models of Chapters 13 and 14 is similar for electromagnetic waves.

The purpose of this chapter is to explore the properties of electromagnetic waves. The fundamental laws of electricity and magnetism—Maxwell's equations—form the basis of all electromagnetic phenomena. One of these equations predicts that a time-varying electric field produces a magnetic field just as a time-varying magnetic field produces an electric field. From this generalization, Maxwell provided the final important link between electric and magnetic fields. The most dramatic prediction of his equations is the existence of electromagnetic waves that propagate through empty space with the speed of light. This discovery led to many practical applications, such as communication by radio, television, and cellular telephone, and to the realization that light is one form of electromagnetic radiation.

24.1 | Displacement Current and the Generalized Form of Ampère's Law

We have seen that charges in motion, or currents, produce magnetic fields. When a current-carrying conductor has high symmetry, we can calculate the magnetic field using Ampère's law, given by Equation 22.29:

$$\oint \vec{\mathbf{B}} \cdot d\vec{\mathbf{s}} = \mu_0 I$$

where the line integral is over any closed path through which the conduction current passes and the conduction current is defined by $I = dq/dt$.

In this section, we shall use the term *conduction current* to refer to the type of current that we have already discussed, that is, current carried by charged particles in a wire. We use this term to differentiate this current from a different type of current we will introduce shortly. Ampère's law in this form is valid only if the conduction current is continuous in space. Maxwell recognized this limitation and modified Ampère's law to include all possible situations.

This limitation can be understood by considering a capacitor being charged as in Figure 24.1. When conduction current exists in the wires, the charge on the plates changes, but no conduction current exists between the plates. Consider the two surfaces S_1 (a circle, shown in blue) and S_2 (a paraboloid, in tan, passing between the plates) in Figure 24.1 bounded by the same path P. Ampère's law says that the line integral of $\vec{\mathbf{B}} \cdot d\vec{\mathbf{s}}$ around this path must equal $\mu_0 I$, where I is the conduction current through *any* surface bounded by the path P.

When the path P is considered as bounding S_1, the right-hand side of Equation 22.29 is $\mu_0 I$ because the conduction current passes through S_1 while the capacitor is charging. When the path bounds S_2, however, the right-hand side of Equation 22.29 is zero because no conduction current passes through S_2. Therefore, a contradictory situation arises because of the discontinuity of the current! Maxwell solved this problem by postulating an additional term on the right side of Equation 22.29, called the **displacement current** I_d, defined as

$$I_d \equiv \epsilon_0 \frac{d\Phi_E}{dt} \qquad \textbf{24.1} \blacktriangleleft \qquad \blacktriangleright \text{ Displacement current}$$

Recall that Φ_E is the flux of the electric field, defined as $\Phi_E \equiv \oint \vec{\mathbf{E}} \cdot d\vec{\mathbf{A}}$ (Eq. 19.14). (The word *displacement* here does not have the same meaning as in Chapter 2; it is historically entrenched in the language of physics, however, so we continue to use it.)

Equation 24.1 is interpreted as follows. As the capacitor is being charged (or discharged), the changing electric field between the plates may be considered as equivalent to a current between the plates that acts as a continuation of the conduction current in the wire. When the expression for the displacement current given by Equation 24.1 is added to the conduction current on the right side of Ampère's law, the difficulty represented in Figure 24.1 is resolved. No matter what surface bounded by the path P is chosen, either conduction current or displacement current passes through it. With this new notion of displacement current, we can express the general form of Ampère's law (sometimes called the **Ampère–Maxwell law**) as[1]

$$\oint \vec{\mathbf{B}} \cdot d\vec{\mathbf{s}} = \mu_0 (I + I_d) = \mu_0 I + \mu_0 \epsilon_0 \frac{d\Phi_E}{dt} \qquad \textbf{24.2} \blacktriangleleft \qquad \blacktriangleright \text{ Ampère–Maxwell law}$$

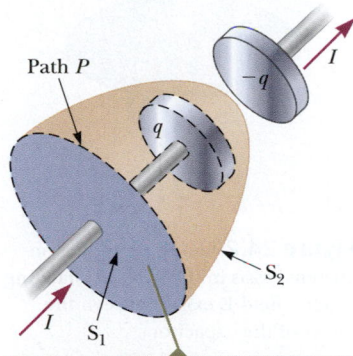

Path P

$-q$

I

q

S_2

I S_1

The conduction current I in the wire passes only through S_1, which leads to a contradiction in Ampère's law that is resolved only if one postulates a displacement current through S_2.

Figure 24.1 Two surfaces S_1 and S_2 near the plate of a capacitor are bounded by the same path P.

---

[1]Strictly speaking, this expression is valid only in a vacuum. If a magnetic material is present, a magnetizing current must also be included on the right side of Equation 24.2 to make Ampère's law fully general.

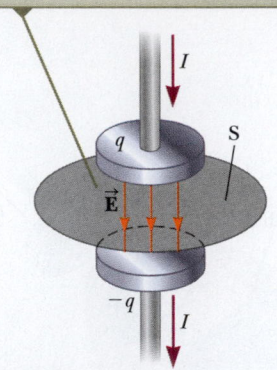

The electric field lines between the plates create an electric flux through surface S.

Figure 24.2 When a conduction current exists in the wires, a changing electric field $\vec{E}$ exists between the plates of the capacitor.

North Wind Picture Archives

James Clerk Maxwell
Scottish Theoretical Physicist (1831–1879)
Maxwell developed the electromagnetic theory of light and the kinetic theory of gases and explained the nature of Saturn's rings and color vision. Maxwell's successful interpretation of the electromagnetic field resulted in the field equations that bear his name. Formidable mathematical ability combined with great insight enabled him to lead the way in the study of electromagnetism and kinetic theory. He died of cancer before he was 50.

We can understand the meaning of this expression by referring to Figure 24.2. The electric flux through surface S is $\Phi_E = \int \vec{E} \cdot d\vec{A} = EA$, where A is the area of the capacitor plates and E is the magnitude of the uniform electric field between the plates. If q is the charge on the plates at any instant, then $E = \sigma/\epsilon_0 = q/(\epsilon_0 A)$ (see Example 19.12). Therefore, the electric flux through S is

$$\Phi_E = EA = \frac{q}{\epsilon_0}$$

Hence, the displacement current through S is

$$I_d = \epsilon_0 \frac{d\Phi_E}{dt} = \frac{dq}{dt} \qquad \textbf{24.3} \blacktriangleleft$$

That is, the displacement current I_d through S is precisely equal to the conduction current I in the wires connected to the capacitor!

By considering surface S, we can identify the displacement current as the source of the magnetic field on the surface boundary. The displacement current has its physical origin in the time-varying electric field. The central point of this formalism is that magnetic fields are produced *both* by conduction currents *and* by time-varying electric fields. This result was a remarkable example of theoretical work by Maxwell, and it contributed to major advances in the understanding of electromagnetism.

QUICK QUIZ 24.1 In an *RC* circuit, the capacitor begins to discharge. **(i)** During the discharge, in the region of space between the plates of the capacitor, is there (a) conduction current but no displacement current, (b) displacement current but no conduction current, (c) both conduction and displacement current, or (d) no current of any type? **(ii)** In the same region of space, is there (a) an electric field but no magnetic field, (b) a magnetic field but no electric field, (c) both electric and magnetic fields, or (d) no fields of any type?

24.2 | Maxwell's Equations and Hertz's Discoveries

We now present four equations that are regarded as the basis of all electrical and magnetic phenomena. These equations, developed by Maxwell, are as fundamental to electromagnetic phenomena as Newton's laws are to mechanical phenomena. In fact, the theory that Maxwell developed was more far-reaching than even he imagined because it turned out to be in agreement with the special theory of relativity, as Einstein showed in 1905.

Maxwell's equations represent the laws of electricity and magnetism that we have already discussed, but they have additional important consequences. For simplicity, we present **Maxwell's equations** as applied to free space, that is, in the absence of any dielectric or magnetic material. The four equations are

▶ Gauss's law

$$\oint \vec{E} \cdot d\vec{A} = \frac{q}{\epsilon_0} \qquad \textbf{24.4} \blacktriangleleft$$

▶ Gauss's law in magnetism

$$\oint \vec{B} \cdot d\vec{A} = 0 \qquad \textbf{24.5} \blacktriangleleft$$

▶ Faraday's law

$$\oint \vec{E} \cdot d\vec{s} = -\frac{d\Phi_B}{dt} \qquad \textbf{24.6} \blacktriangleleft$$

▶ Ampère–Maxwell law

$$\oint \vec{B} \cdot d\vec{s} = \mu_0 I + \epsilon_0 \mu_0 \frac{d\Phi_E}{dt} \qquad \textbf{24.7} \blacktriangleleft$$

Equation 24.4 is Gauss's law: the total electric flux through any closed surface equals the net charge inside that surface divided by ϵ_0. This law relates an electric field to the charge distribution that creates it.

Equation 24.5 is Gauss's law in magnetism, and it states that the net magnetic flux through a closed surface is zero. That is, the number of magnetic field lines that enter a closed volume must equal the number that leave that volume, which implies that magnetic field lines cannot begin or end at any point. If they did, it would mean that isolated magnetic monopoles existed at those points. That isolated magnetic monopoles have not been observed in nature can be taken as a confirmation of Equation 24.5.

Equation 24.6 is Faraday's law of induction, which describes the creation of an electric field by a changing magnetic flux. This law states that the emf, which is the line integral of the electric field around any closed path, equals the rate of change of magnetic flux through any surface bounded by that path. One consequence of Faraday's law is the current induced in a conducting loop placed in a time-varying magnetic field.

Equation 24.7 is the Ampère–Maxwell law, and it describes the creation of a magnetic field by a changing electric field and by electric current: the line integral of the magnetic field around any closed path is the sum of μ_0 multiplied by the net current through that path and $\epsilon_0\mu_0$ multiplied by the rate of change of electric flux through any surface bounded by that path.

Once the electric and magnetic fields are known at some point in space, the force acting on a particle of charge q can be calculated from the expression

$$\vec{F} = q\vec{E} + q\vec{v} \times \vec{B} \qquad \textbf{24.8} \blacktriangleleft$$

▶ Lorentz force law

This relationship is called the **Lorentz force law.** (We saw this relationship earlier as Eq. 22.6.) Maxwell's equations, together with this force law, completely describe all classical electromagnetic interactions in a vacuum.

Notice the symmetry of Maxwell's equations. Equations 24.4 and 24.5 are symmetric, apart from the absence of the term for magnetic monopoles in Equation 24.5. Furthermore, Equations 24.6 and 24.7 are symmetric in that the line integrals of $\vec{E}$ and $\vec{B}$ around a closed path are related to the rate of change of magnetic flux and electric flux, respectively. Maxwell's equations are of fundamental importance not only to electromagnetism, but to all science. Hertz once wrote, "One cannot escape the feeling that these mathematical formulas have an independent existence and an intelligence of their own, that they are wiser than we are, wiser even than their discoverers, that we get more out of them than we put into them."

In the next section, we show that Equations 24.6 and 24.7 can be combined to obtain a wave equation for both the electric field and the magnetic field. In empty space, where $q = 0$ and $I = 0$, the solution to these two equations shows that the speed at which electromagnetic waves travel equals the measured speed of light. This result led Maxwell to predict that light waves are a form of electromagnetic radiation.

Hertz performed experiments that verified Maxwell's prediction. The experimental apparatus Hertz used to generate and detect electromagnetic waves is shown schematically in Figure 24.3. An induction coil is connected to a transmitter made up of two spherical electrodes separated by a narrow gap. The coil provides short voltage surges to the electrodes, making one positive and the other negative. A spark is generated between the spheres when the electric field near either electrode surpasses the dielectric strength for air (3×10^6 V/m; see Table 20.1). Free electrons in a strong electric field are accelerated and gain enough energy to ionize any molecules they strike. This ionization provides more electrons, which can accelerate and cause further ionizations. As the air in the gap is ionized, it becomes a much better conductor and the discharge between the electrodes exhibits an oscillatory behavior at a very high frequency. From an electric-circuit viewpoint, this experimental apparatus is equivalent to an LC circuit in which the inductance is that of the coil and the capacitance is due to the spherical electrodes. By applying Kirchhoff's loop rule to an LC circuit, similar to the way we applied it to an RC circuit in Section 21.9, we can show that the current in an LC circuit oscillates in simple harmonic motion at the frequency

$$\omega = \frac{1}{\sqrt{LC}} \qquad \textbf{24.9} \blacktriangleleft$$

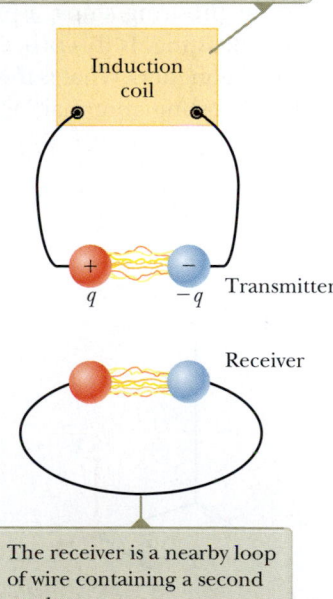

The transmitter consists of two spherical electrodes connected to an induction coil, which provides short voltage surges to the spheres, setting up oscillations in the discharge between the electrodes.

Induction coil

q $-q$ Transmitter

Receiver

The receiver is a nearby loop of wire containing a second spark gap.

Figure 24.3 Schematic diagram of Hertz's apparatus for generating and detecting electromagnetic waves.

© Bettmann/Corbis

Heinrich Rudolf Hertz
German Physicist (1857–1894)
Hertz made his most important discovery of electromagnetic waves in 1887. After finding that the speed of an electromagnetic wave was the same as that of light, Hertz showed that electromagnetic waves, like light waves, could be reflected, refracted, and diffracted. The hertz, equal to one complete vibration or cycle per second, is named after him.

Because L and C are small in Hertz's apparatus, the frequency of oscillation is high, on the order of 100 MHz. Electromagnetic waves are radiated at this frequency as a result of the oscillation (and hence acceleration) of free charges in the transmitter circuit. Hertz was able to detect these waves using a single loop of wire with its own spark gap (the receiver). Such a receiver loop, placed several meters from the transmitter, has its own effective inductance, capacitance, and natural frequency of oscillation. In Hertz's experiment, sparks were induced across the gap of the receiving electrodes when the receiver's frequency was adjusted to match that of the transmitter. In this way, Hertz demonstrated that the oscillating current induced in the receiver was produced by electromagnetic waves radiated by the transmitter. His experiment is analogous to the mechanical phenomenon in which a tuning fork responds to acoustic vibrations from an identical tuning fork that is oscillating.

In addition, Hertz showed in a series of experiments that the radiation generated by his spark-gap device exhibited the wave properties of interference, diffraction, reflection, refraction, and polarization, which are all properties exhibited by light as we shall see in this chapter and Chapters 25–27. Therefore, it became evident that the radio-frequency waves Hertz was generating had properties similar to those of light waves and that they differed only in frequency and wavelength. Perhaps his most convincing experiment was the measurement of the speed of this radiation. Waves of known frequency were reflected from a metal sheet and created a standing-wave interference pattern whose nodal points could be detected. The measured distance between the nodal points enabled determination of the wavelength λ. Using the relationship $v = \lambda f$ (Eq. 13.12), Hertz found that v was close to 3×10^8 m/s, the known speed c of visible light.

> ### THINKING PHYSICS 24.1

In radio transmission, a radio wave serves as a carrier wave and the sound wave is superimposed on the carrier wave. In amplitude modulation (AM radio), the amplitude of the carrier wave varies according to the sound wave. (The word *modulate* means "to change.") In frequency modulation (FM radio), the frequency of the carrier wave varies according to the sound wave. The navy sometimes uses flashing lights to send Morse code to neighboring ships, a process that is similar to radio broadcasting. Is this process AM or FM? What is the carrier frequency? What is the signal frequency? What is the broadcasting antenna? What is the receiving antenna?

Reasoning The flashing of the light according to Morse code is a drastic amplitude modulation because the amplitude is changing between a maximum value and zero. In this sense, it is similar to the on-and-off binary code used in computers and compact discs. The carrier frequency is that of the visible light, on the order of 10^{14} Hz. The signal frequency depends on the skill of the signal operator, but is on the order of a few hertz, as the light is flashed on and off. The broadcasting antenna for this modulated signal is the filament of the lightbulb in the signal source. The receiving antenna is the eye. ◄

24.3 | Electromagnetic Waves

In his unified theory of electromagnetism, Maxwell showed that time-dependent electric and magnetic fields satisfy a linear wave equation. (The linear wave equation for mechanical waves is Equation 13.20.) The most significant outcome of this theory is the prediction of the existence of **electromagnetic waves.**

Maxwell's equations predict that an electromagnetic wave consists of oscillating electric and magnetic fields. The changing fields induce each other, which maintains the propagation of the wave; a changing electric field induces a magnetic field, and a changing magnetic field induces an electric field. The $\vec{\mathbf{E}}$ and $\vec{\mathbf{B}}$ vectors are perpendicular to each other, and to the direction of propagation, as shown in Active Figure 24.4 at one instant of time and one point in space. The direction of propagation is the direction of the vector product $\vec{\mathbf{E}} \times \vec{\mathbf{B}}$, which we shall explore more fully in Section 24.4.

In Active Figure 24.4, we have chosen the direction of propagation of the wave to be the positive x axis. We have also chosen the y axis to be parallel to the electric field

Active Figure 24.4 The fields in an electromagnetic wave traveling at velocity $\vec{\mathbf{c}}$ in the positive x direction at one point on the x axis. These fields depend only on x and t.

vector. Given these choices, it is necessarily true that the magnetic field $\vec{B}$ is in the z direction as in Active Figure 24.4. Waves in which the electric and magnetic fields are restricted to being parallel to certain directions are said to be **linearly polarized waves.** Furthermore, let us assume that at any point in space in Active Figure 24.4, the magnitudes E and B of the fields depend on x and t only, not on the y or z coordinates.

Let us also imagine that the source of the electromagnetic waves is such that a wave radiated from *any* position in the yz plane (not just from the origin as might be suggested by Active Fig. 24.4) propagates in the x direction and that all such waves are emitted in phase. If we define a **ray** as the line along which a wave travels, all rays for these waves are parallel. This whole collection of waves is often called a **plane wave.** A surface connecting points of equal phase on all waves, which we call a **wave front,** is a geometric plane. In comparison, a point source of radiation sends waves out in all directions. A surface connecting points of equal phase for this situation is a sphere, so we call the radiation from a point source a **spherical wave.**

To generate the prediction of electromagnetic waves, we start with Faraday's law, Equation 24.6:

$$\oint \vec{E} \cdot d\vec{s} = -\frac{d\Phi_B}{dt}$$

Let's again assume the electromagnetic wave is traveling in the x direction, with the electric field $\vec{E}$ in the positive y direction and the magnetic field $\vec{B}$ in the positive z direction.

Consider a rectangle of width dx and height ℓ lying in the xy plane as shown in Figure 24.5. To apply Equation 24.6, let's first evaluate the line integral of $\vec{E} \cdot d\vec{s}$ around this rectangle in the counterclockwise direction at an instant of time when the wave is passing through the rectangle. The contributions from the top and bottom of the rectangle are zero because $\vec{E}$ is perpendicular to $d\vec{s}$ for these paths. We can express the electric field on the right side of the rectangle as

$$E(x + dx) \approx E(x) + \frac{dE}{dx}\Big]_{t \text{ constant}} dx = E(x) + \frac{\partial E}{\partial x} dx$$

where $E(x)$ is the field on the left side of the rectangle at this instant.[2] Therefore, the line integral over this rectangle is approximately

$$\oint \vec{E} \cdot d\vec{s} = [E(x + dx)]\ell - [E(x)]\ell \approx \ell\left(\frac{\partial E}{\partial x}\right) dx \qquad \mathbf{24.10} \blacktriangleleft$$

Because the magnetic field is in the z direction, the magnetic flux through the rectangle of area $\ell\, dx$ is approximately $\Phi_B = B\ell\, dx$ (assuming dx is very small compared with the wavelength of the wave). Taking the time derivative of the magnetic flux gives

$$\frac{d\Phi_B}{dt} = \ell\ dx \frac{dB}{dt}\Big]_{x \text{ constant}} = \ell\ dx \frac{\partial B}{\partial t} \qquad \mathbf{24.11} \blacktriangleleft$$

Substituting Equations 24.10 and 24.11 into Equation 24.6 gives

$$\ell\left(\frac{\partial E}{\partial x}\right) dx = -\ell\, dx \frac{\partial B}{\partial t}$$

$$\frac{\partial E}{\partial x} = -\frac{\partial B}{\partial t} \qquad \mathbf{24.12} \blacktriangleleft$$

In a similar manner, we can derive a second equation by starting with Maxwell's fourth equation in empty space (Eq. 24.7). In this case, the line integral of $\vec{B} \cdot d\vec{s}$ is evaluated around a rectangle lying in the xz plane and having width dx and length ℓ as in Figure 24.6 (page 714). Noting that the magnitude of the magnetic field

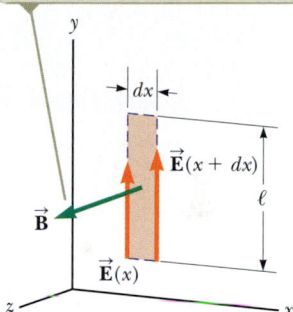
---

[2]Because dE/dx in this equation is expressed as the change in E with x at a given instant t, dE/dx is equivalent to the partial derivative $\partial E/\partial x$. Likewise, dB/dt means the change in B with time at a particular position x; therefore, in Equation 24.11, we can replace dB/dt with $\partial B/\partial t$.

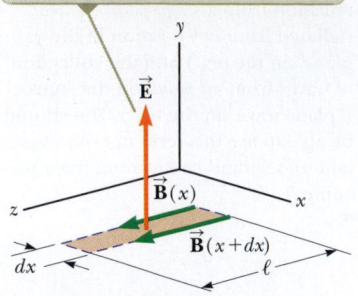

According to Equation 24.15, this spatial variation in $\vec{B}$ gives rise to a time-varying electric field along the y direction.

Figure 24.6 At an instant when a plane wave passes through a rectangular path of width dx lying in the xz plane, the magnetic field in the z direction varies from $\vec{B}(x)$ to $\vec{B}(x + dx)$.

changes from $B(x)$ to $B(x + dx)$ over the width dx and that the direction for taking the line integral is counterclockwise when viewed from above in Figure 24.6, the line integral over this rectangle is found to be approximately

$$\oint \vec{B} \cdot d\vec{s} = [B(x)]\ell - [B(x + dx)]\ell \approx -\ell\left(\frac{\partial B}{\partial x}\right) dx \qquad 24.13 \blacktriangleleft$$

The electric flux through the rectangle is $\Phi_E = E\ell\,dx$, which, when differentiated with respect to time, gives

$$\frac{\partial \Phi_E}{\partial t} = \ell\,dx\,\frac{\partial E}{\partial t} \qquad 24.14 \blacktriangleleft$$

Substituting Equations 24.13 and 24.14 into Equation 24.7 gives

$$-\ell\left(\frac{\partial B}{\partial x}\right) dx = \mu_0\epsilon_0\,\ell\,dx\left(\frac{\partial E}{\partial t}\right)$$

$$\frac{\partial B}{\partial x} = -\mu_0\epsilon_0\frac{\partial E}{\partial t} \qquad 24.15 \blacktriangleleft$$

Taking the derivative of Equation 24.12 with respect to x and combining the result with Equation 24.15 gives

$$\frac{\partial^2 E}{\partial x^2} = -\frac{\partial}{\partial x}\left(\frac{\partial B}{\partial t}\right) = -\frac{\partial}{\partial t}\left(\frac{\partial B}{\partial x}\right) = -\frac{\partial}{\partial t}\left(-\mu_0\epsilon_0\frac{\partial E}{\partial t}\right)$$

$$\frac{\partial^2 E}{\partial x^2} = \mu_0\epsilon_0\frac{\partial^2 E}{\partial t^2} \qquad 24.16 \blacktriangleleft$$

In the same manner, taking the derivative of Equation 24.15 with respect to x and combining it with Equation 24.12 gives

$$\frac{\partial^2 B}{\partial x^2} = \mu_0\epsilon_0\frac{\partial^2 B}{\partial t^2} \qquad 24.17 \blacktriangleleft$$

Equations 24.16 and 24.17 both have the form of the linear wave equation[3] with the wave speed v replaced by c, where

$$\blacktriangleright \text{Speed of electromagnetic waves} \qquad\qquad c = \frac{1}{\sqrt{\mu_0\epsilon_0}} \qquad 24.18 \blacktriangleleft$$

Let's evaluate this speed numerically:

$$c = \frac{1}{\sqrt{(4\pi \times 10^{-7}\,\text{T}\cdot\text{m/A})(8.854\,19 \times 10^{-12}\,\text{C}^2/\text{N}\cdot\text{m}^2)}}$$

$$= 2.997\,92 \times 10^8\,\text{m/s}$$

Because this speed is precisely the same as the speed of light in empty space, we are led to believe (correctly) that light is an electromagnetic wave.

The simplest solution to Equations 24.16 and 24.17 is a sinusoidal wave for which the field magnitudes E and B vary with x and t according to the expressions

$\blacktriangleright$ Sinusoidal electric and magnetic fields

$$E = E_{max} \cos(kx - \omega t) \qquad 24.19 \blacktriangleleft$$
$$B = B_{max} \cos(kx - \omega t) \qquad 24.20 \blacktriangleleft$$

where E_{max} and B_{max} are the maximum values of the fields. The angular wave number is $k = 2\pi/\lambda$, where λ is the wavelength. The angular frequency is $\omega = 2\pi f$, where f is the wave frequency. The ratio ω/k equals the speed of an electromagnetic wave, c:

$$\frac{\omega}{k} = \frac{2\pi f}{2\pi/\lambda} = \lambda f = c$$

[3]The linear wave equation is of the form $(\partial^2 y/\partial x^2) = (1/v^2)(\partial^2 y/\partial t^2)$, where v is the speed of the wave and y is the wave function. The linear wave equation was introduced as Equation 13.20, and we suggest you review Section 13.2.

where we have used Equation 13.12, $v = c = \lambda f$, which relates the speed, frequency, and wavelength of any continuous wave. Therefore, for electromagnetic waves, the wavelength and frequency of these waves are related by

$$\lambda = \frac{c}{f} = \frac{3.00 \times 10^8 \text{ m/s}}{f}$$ **24.21**◀

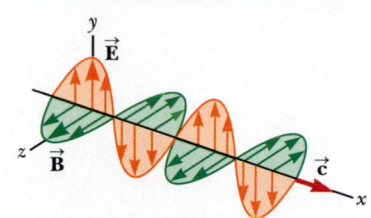

Active Figure 24.7 A sinusoidal electromagnetic wave moves in the positive x direction with a speed c.

Active Figure 24.7 is a pictorial representation, at one instant, of a sinusoidal, linearly polarized electromagnetic wave moving in the positive x direction.

Taking partial derivatives of Equations 24.19 (with respect to x) and 24.20 (with respect to t) gives

$$\frac{\partial E}{\partial x} = -kE_{max} \sin (kx - \omega t)$$

$$\frac{\partial B}{\partial t} = \omega B_{max} \sin (kx - \omega t)$$

Substituting these results into Equation 24.12 shows that, at any instant,

$$kE_{max} = \omega B_{max}$$

$$\frac{E_{max}}{B_{max}} = \frac{\omega}{k} = c$$

Using these results together with Equations 24.19 and 24.20 gives

$$\frac{E_{max}}{B_{max}} = \frac{E}{B} = c$$ **24.22**◀

That is, at every instant, the ratio of the magnitude of the electric field to the magnitude of the magnetic field in an electromagnetic wave equals the speed of light.

Finally, note that electromagnetic waves obey the superposition principle (which we discussed in Section 14.1 with respect to mechanical waves) because the differential equations involving E and B are linear equations. For example, we can add two waves with the same frequency and polarization simply by adding the magnitudes of the two electric fields algebraically.

> **Pitfall Prevention | 24.2**
> **$\vec{E}$ Stronger Than $\vec{B}$?**
> Because the value of c is so large, some students incorrectly interpret Equation 24.22 as meaning that the electric field is much stronger than the magnetic field. Electric and magnetic fields are measured in different units, however, so they cannot be directly compared. In Section 24.4, we find that the electric and magnetic fields contribute equally to the wave's energy.

Doppler Effect for Light

Another feature of electromagnetic waves is that there is a shift in the observed frequency of the waves when there is relative motion between the source of the waves and the observer. This phenomenon, known as the Doppler effect, was introduced in Chapter 13 as it pertains to sound waves. In the case of sound, the motion of the source with respect to the medium of propagation can be distinguished from the motion of the observer with respect to the medium. Light waves must be analyzed differently, however, because they require no medium of propagation, and no method exists for distinguishing the motion of a light source from the motion of the observer.

If a light source and an observer approach each other with a relative speed v, the frequency f' measured by the observer is

$$f' = \sqrt{\frac{c + v}{c - v}} f$$ **24.23**◀

▶ Doppler effect for electromagnetic waves

where f is the frequency of the source measured in its rest frame. This Doppler shift equation, unlike the Doppler shift equation for sound, depends only on the relative speed v of the source and observer and holds for relative speeds as great as c. As you might expect, the equation predicts that $f' > f$ when the source and observer approach each other. We obtain the expression for the case in which the source and observer recede from each other by substituting negative values for v in Equation 24.23.

The most spectacular and dramatic use of the Doppler effect for electromagnetic waves is the measurement of shifts in the frequency of light emitted by a moving astronomical object such as a galaxy. Light emitted by atoms and normally found in the extreme violet region of the spectrum is shifted toward the red end of the spectrum for atoms in other galaxies, indicating that these galaxies are *receding* from us. American astronomer Edwin Hubble (1889–1953) performed extensive measurements of this *red shift* to confirm that most galaxies are moving away from us, indicating that the Universe is expanding.

Example 24.1 | An Electromagnetic Wave

A sinusoidal electromagnetic wave of frequency 40.0 MHz travels in free space in the *x* direction as in Figure 24.8.

(A) Determine the wavelength and period of the wave.

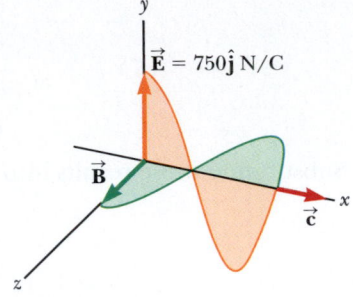

Figure 24.8 (Example 24.1) At some instant, a plane electromagnetic wave moving in the *x* direction has a maximum electric field of 750 N/C in the positive *y* direction.

SOLUTION

Conceptualize Imagine the wave in Figure 24.8 moving to the right along the *x* axis, with the electric and magnetic fields oscillating in phase.

Categorize We determine the results using equations developed in this section, so we categorize this example as a substitution problem.

Use Equation 24.21 to find the wavelength of the wave:
$$\lambda = \frac{c}{f} = \frac{3.00 \times 10^8 \text{ m/s}}{40.0 \times 10^6 \text{ Hz}} = \boxed{7.50 \text{ m}}$$

Find the period *T* of the wave as the inverse of the frequency:
$$T = \frac{1}{f} = \frac{1}{40.0 \times 10^6 \text{ Hz}} = \boxed{2.50 \times 10^{-8} \text{ s}}$$

(B) At some point and at some instant, the electric field has its maximum value of 750 N/C and is directed along the *y* axis. Calculate the magnitude and direction of the magnetic field at this position and time.

SOLUTION

Use Equation 24.22 to find the magnitude of the magnetic field:
$$B_{max} = \frac{E_{max}}{c} = \frac{750 \text{ N/C}}{3.00 \times 10^8 \text{ m/s}} = \boxed{2.50 \times 10^{-6} \text{ T}}$$

Because $\vec{\mathbf{E}}$ and $\vec{\mathbf{B}}$ must be perpendicular to each other and perpendicular to the direction of wave propagation (*x* in this case), we conclude that $\vec{\mathbf{B}}$ is in the *z* direction.

(C) An observer on the *x* axis, far to the right in Figure 24.8, moves to the left along the *x* axis at 0.500*c*. What frequency does this observer measure for the electromagnetic wave?

SOLUTION

Use Equation 24.23 for the Doppler effect to find the observed frequency:
$$f' = \sqrt{\frac{c+v}{c-v}} f = 40.0 \text{ MHz} \sqrt{\frac{c+(+0.500c)}{c-(+0.500c)}}$$
$$= \boxed{69.3 \text{ MHz}}$$

We substituted *v* as a positive number because the observer is moving toward the source.

◄24.4 | Energy Carried by Electromagnetic Waves

In Section 13.5, we found that mechanical waves carry energy. Electromagnetic waves also carry energy, and as they propagate through space they can transfer energy to objects placed in their path. This notion was introduced in Chapter 7 when we discussed the transfer mechanisms in the conservation of energy equation, and it was noted again in Chapter 17 in the discussion of thermal radiation. The rate of flow of energy in an electromagnetic wave is described by a vector $\vec{S}$, called the **Poynting vector,** defined by the expression

$$\vec{S} \equiv \frac{1}{\mu_0} \vec{E} \times \vec{B}$$

24.24 ◄   ► Poynting vector

The magnitude of the Poynting vector represents the rate at which energy flows through a unit surface area perpendicular to the flow and its direction is along the direction of wave propagation (Fig. 24.9). Therefore, the Poynting vector represents *power per unit area*. The SI units of the Poynting vector are $\mathrm{J/s \cdot m^2 = W/m^2}$.

As an example, let us evaluate the magnitude of $\vec{S}$ for a plane electromagnetic wave. We have $|\vec{E} \times \vec{B}| = EB$ because $\vec{E}$ and $\vec{B}$ are perpendicular to each other. In this case,

$$S = \frac{EB}{\mu_0}$$

24.25 ◄

Because $B = E/c$, we can also express the magnitude of the Poynting vector as

$$S = \frac{E^2}{\mu_0 c} = \frac{cB^2}{\mu_0}$$

These equations for S apply at any instant of time.

What is of more interest for a sinusoidal electromagnetic wave (Eqs. 24.19 and 24.20) is the time average of S over one or more cycles, which is the **intensity** I, that is, the average power per unit area. When this average is taken, we obtain an expression involving the time average of $\cos^2(kx - \omega t)$, which equals $\frac{1}{2}$. Therefore, the average value of S (or the intensity of the wave) is

$$I = S_{\text{avg}} = \frac{E_{\max} B_{\max}}{2\mu_0} = \frac{E_{\max}^2}{2\mu_0 c} = \frac{cB_{\max}^2}{2\mu_0}$$

24.26 ◄   ► Wave intensity

Recall that the energy per unit volume u_E, which is the instantaneous energy density associated with an electric field (Section 20.9), is given by Equation 20.31:

$$u_E = \tfrac{1}{2}\epsilon_0 E^2$$

24.27 ◄

and that the instantaneous energy density u_B associated with a magnetic field (Section 23.7) is given by Equation 23.22:

$$u_B = \frac{B^2}{2\mu_0}$$

24.28 ◄

Because E and B vary with time for an electromagnetic wave, the energy densities also vary with time. Using the relationships $B = E/c$ and $c = 1/\sqrt{\epsilon_0 \mu_0}$, Equation 24.28 becomes

$$u_B = \frac{(E/c)^2}{2\mu_0} = \frac{\epsilon_0 \mu_0}{2\mu_0} E^2 = \tfrac{1}{2}\epsilon_0 E^2$$

Comparing this result with the expression for u_E, we see that

$$u_B = u_E$$

That is, for an electromagnetic wave, the instantaneous energy density associated with the magnetic field equals the instantaneous energy density associated with the electric field. Therefore, in a given volume the energy is equally shared by the two fields.

The **total instantaneous energy density** u is equal to the sum of the energy densities associated with the electric and magnetic fields:

$$u = u_E + u_B = \epsilon_0 E^2 = \frac{B^2}{\mu_0}$$

► Total instantaneous energy density of an electromagnetic wave

Pitfall Prevention | 24.3
An Instantaneous Value
The Poynting vector given by Equation 24.24 is time dependent. Its magnitude varies in time, reaching a maximum value at the same instant as the magnitudes of $\vec{E}$ and $\vec{B}$ do.

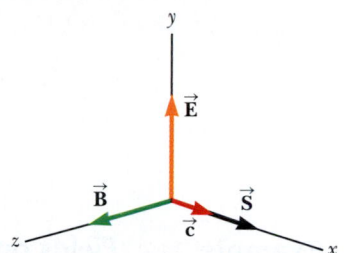

Figure 24.9 The Poynting vector $\vec{S}$ for an electromagnetic wave is along the direction of wave propagation.

When this expression is averaged over one or more cycles of an electromagnetic wave, we again obtain a factor of $\frac{1}{2}$. Therefore, the total average energy per unit volume of an electromagnetic wave is

▶ Average energy density of an electromagnetic wave

$$u_{\text{avg}} = \epsilon_0 (E^2)_{\text{avg}} = \tfrac{1}{2}\epsilon_0 E_{\text{max}}^2 = \frac{B_{\text{max}}^2}{2\mu_0} \qquad \textbf{24.29} \blacktriangleleft$$

Comparing this result with Equation 24.26 for the average value of S, we see that

$$I = S_{\text{avg}} = c u_{\text{avg}} \qquad \textbf{24.30} \blacktriangleleft$$

In other words, the intensity of an electromagnetic wave equals the average energy density multiplied by the speed of light.

> **QUICK QUIZ 24.2** An electromagnetic wave propagates in the negative y direction. The electric field at a point in space is momentarily oriented in the positive x direction. In which direction is the magnetic field at that point momentarily oriented? **(a)** the negative x direction **(b)** the positive y direction **(c)** the positive z direction **(d)** the negative z direction

> **QUICK QUIZ 24.3** Which of the following quantities does not vary in time for plane electromagnetic waves? **(a)** magnitude of the Poynting vector **(b)** energy density u_E **(c)** energy density u_B **(d)** intensity I

Example 24.2 | Fields on the Page

Estimate the maximum magnitudes of the electric and magnetic fields of the light that is incident on this page because of the visible light coming from your desk lamp. Treat the lightbulb as a point source of electromagnetic radiation that is 5% efficient at transforming energy coming in by electrical transmission to energy leaving by visible light.

SOLUTION

Conceptualize The filament in your lightbulb emits electromagnetic radiation. The brighter the light, the larger the magnitudes of the electric and magnetic fields.

Categorize Because the lightbulb is to be treated as a point source, it emits equally in all directions, so the outgoing electromagnetic radiation can be modeled as a spherical wave.

Analyze We mentioned that intensity is equivalent to the average power of radiation per unit area. For a point source emitting uniformly in all directions, the power is distributed evenly over the surface area $4\pi r^2$ of an expanding sphere of increasing radius r, centered on the source. Therefore, $I = P_{\text{avg}}/4\pi r^2$, where P represents power.

Set this expression for I equal to the intensity of an electromagnetic wave given by Equation 24.26:

$$I = \frac{P_{\text{avg}}}{4\pi r^2} = \frac{E_{\text{max}}^2}{2\mu_0 c}$$

Solve for the electric field magnitude:

$$E_{\text{max}} = \sqrt{\frac{\mu_0 c P_{\text{avg}}}{2\pi r^2}}$$

Let's make some assumptions about numbers to enter in this equation. The visible light output of a 60-W lightbulb operating at 5% efficiency is approximately 3.0 W by visible light. (The remaining energy transfers out of the lightbulb by conduction and invisible radiation.) A reasonable distance from the lightbulb to the page might be 0.30 m.

Substitute these values:

$$E_{\text{max}} = \sqrt{\frac{(4\pi \times 10^{-7}\,\text{T} \cdot \text{m/A})(3.00 \times 10^8\,\text{m/s})(3.0\,\text{W})}{2\pi(0.30\,\text{m})^2}}$$

$$= 45\,\text{V/m}$$

Use Equation 24.22 to find the magnetic field magnitude:

$$B_{\text{max}} = \frac{E_{\text{max}}}{c} = \frac{45\,\text{V/m}}{3.00 \times 10^8\,\text{m/s}} = 1.5 \times 10^{-7}\,\text{T}$$

Finalize This value of the magnetic field magnitude is two orders of magnitude smaller than the Earth's magnetic field.

24.5 | Momentum and Radiation Pressure

Electromagnetic waves transport linear momentum as well as energy. Hence, it follows that pressure is exerted on a surface when an electromagnetic wave impinges on it. In what follows, let us assume that the electromagnetic wave strikes a surface at normal incidence and transports a total energy T_{ER} to a surface in a time interval Δt. If the surface absorbs all the incident energy T_{ER} in this time, Maxwell showed that the total momentum $\vec{p}$ delivered to this surface has a magnitude

$$p = \frac{T_{ER}}{c} \quad \text{(complete absorption)} \qquad \textbf{24.31} \blacktriangleleft$$

▶ Momentum delivered to a perfectly absorbing surface

The radiation pressure P exerted on the surface is defined as force per unit area F/A. Let us combine this definition with Newton's second law:

$$P = \frac{F}{A} = \frac{1}{A}\frac{dp}{dt}$$

Pitfall Prevention | 24.4
So Many _p_'s
We have p for momentum and P for pressure, and they are both related to P for power! Be sure to keep all these symbols straight.

If we now replace p, the momentum transported to the surface by radiation, from Equation 24.31, we have

$$P = \frac{1}{A}\frac{dp}{dt} = \frac{1}{A}\frac{d}{dt}\left(\frac{T_{ER}}{c}\right) = \frac{1}{c}\frac{(dT_{ER}/dt)}{A}$$

We recognize $(dT_{ER}/dt)/A$ as the rate at which energy is arriving at the surface per unit area, which is the magnitude of the Poynting vector. Therefore, the radiation pressure P exerted on the perfectly absorbing surface is

$$P = \frac{S}{c} \quad \text{(complete absorption)} \qquad \textbf{24.32} \blacktriangleleft$$

▶ Radiation pressure exerted on a perfectly absorbing surface

An absorbing surface for which all the incident energy is absorbed (none is reflected) is called a **black body.** A more detailed discussion of a black body will be presented in Chapter 28.

As we found in the last section, the intensity of an electromagnetic wave I is equal to the average value of S (Eq. 24.26), so we can express the average radiation pressure as

$$P_{avg} = \frac{S_{avg}}{c} = \frac{I}{c} \quad \text{(complete absorption)} \qquad \textbf{24.33} \blacktriangleleft$$

Furthermore, because S_{avg} represents power per unit area, we find that the average power delivered to a surface of area A is (using "*Power*" to represent power because we also have P for pressure here)

$$(Power)_{avg} = IA \quad \text{(complete absorption)} \qquad \textbf{24.34} \blacktriangleleft$$

If the surface is a perfect reflector, the momentum delivered in a time interval Δt for normal incidence is twice that given by Equation 24.31, or $p = 2T_{ER}/c$. That is, a momentum T_{ER}/c is delivered first by the incident wave and then again by the reflected wave, a situation analogous to a ball colliding elastically with a wall.[4] Finally, the radiation pressure exerted on a perfect reflecting surface for normal incidence of the wave is twice that given by Equation 24.32, or

$$P = \frac{2S}{c} \quad \text{(complete reflection)} \qquad \textbf{24.35} \blacktriangleleft$$

Although radiation pressures are very small (about 5×10^{-6} N/m² for direct sunlight), they have been measured using torsion balances such as the one shown in Figure 24.10. Light is allowed to strike either a mirror or a black disk, both of which are suspended from a fine fiber. Light striking the black disk is completely absorbed, so all its momentum is transferred to the disk. Light striking the mirror (normal

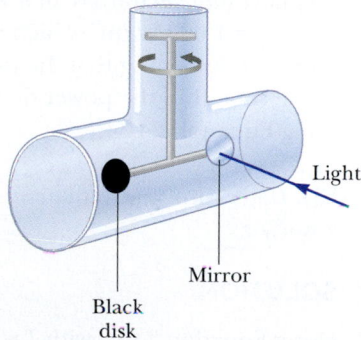

Figure 24.10 An apparatus for measuring the pressure exerted by light. In practice, the system is contained in a high vacuum.

Light
Mirror
Black disk

---

[4]For *oblique* incidence, the momentum transferred is $2T_{ER}\cos\theta/c$ and the pressure is given by $P = 2S\cos^2\theta/c$, where θ is the angle between the normal to the surface and the direction of propagation.

incidence) is totally reflected and hence the momentum transfer is twice as great as that transferred to the disk. The radiation pressure is determined by measuring the angle through which the horizontal connecting rod rotates. The apparatus must be placed in a high vacuum to eliminate the effects of air currents.

 QUICK QUIZ 24.4 In an apparatus such as that in Figure 24.10, suppose the black disk is replaced by one with half the radius. Which of the following are different after the disk is replaced?   (a) radiation pressure on the disk   (b) radiation force on the disk   (c) radiation momentum delivered to the disk in a given time interval

THINKING PHYSICS 24.2

A large amount of dust occurs in the interplanetary space in the Solar System. Although this dust can theoretically have a variety of sizes, from molecular size upward, very little of it is smaller than about 0.2 μm in our Solar System. Why? (*Hint:* The Solar System originally contained dust particles of all sizes.)

Reasoning Dust particles in the Solar System are subject to two forces: the gravitational force toward the Sun and the force from radiation pressure due to sunlight,

which is away from the Sun. The gravitational force is proportional to the cube of the radius of a spherical dust particle because it is proportional to the particle's mass. The radiation force is proportional to the square of the radius because it depends on the circular cross-section of the particle. For large particles, the gravitational force is larger than the force from radiation pressure. For small particles, less than about 0.2 μm, the larger force from radiation pressure sweeps these particles out of the Solar System. ◄

Example 24.3 | Solar Energy

The Sun delivers about 1 000 W/m² of energy to the Earth's surface.

(A) Calculate the total power incident on a roof of dimensions 8.00 m × 20.0 m.

SOLUTION

Conceptualize It should be easy to imagine energy streaming in from the Sun and striking the roof. This transfer of energy is represented by T_{ER} in Equation 7.2.

Categorize We will use equations developed in this section for this problem, so we categorize this as a substitution problem.

The Poynting vector has an average magnitude $I = S_{avg} = 1\,000$ W/m², which represents the power per unit area. Assuming that the radiation is incident normal to the roof, find the power delivered to the whole roof using Equation 24.34:

$$(Power)_{avg} = IA = (1\,000\ \text{W/m}^2)(8.00\ \text{m})(20.0\ \text{m})$$
$$= 1.60 \times 10^5\ \text{W}$$

(B) Determine the radiation pressure and radiation force on the roof, assuming that the roof covering is a perfect absorber.

SOLUTION

Using Equation 24.33 with $I = 1\,000$ W/m², find the average radiation pressure on the roof:

$$P_{avg} = \frac{I}{c} = \frac{1\,000\ \text{W/m}^2}{3.00 \times 10^8\ \text{m/s}} = 3.33 \times 10^{-6}\ \text{N/m}^2$$

Noting that pressure is defined as force per unit area, find the radiation force on the roof:

$$F = P_{avg}A = (3.33 \times 10^{-6}\ \text{N/m}^2)(8.00\ \text{m})(20.0\ \text{m})$$
$$= 5.33 \times 10^{-4}\ \text{N}$$

Example 24.4 | Pressure from a Laser Pointer

When giving presentations, many people use a laser pointer to direct the attention of the audience to information on a screen. If a 3.0-mW pointer creates a spot on a screen that is 2.0 mm in diameter, determine the radiation pressure on a screen that reflects 70% of the light that strikes it. The power 3.0 mW is a time-averaged value.

SOLUTION

Conceptualize Imagine the waves striking the screen and exerting a radiation pressure on it. The pressure should not be very large.

Categorize This problem involves a calculation of radiation pressure using an approach like that leading to Equation 24.32 or Equation 24.35, but it is complicated by the 70% reflection.

Analyze We begin by determining the magnitude of the beam's Poynting vector.

Divide the time-averaged power delivered via the electromagnetic wave by the cross-sectional area of the beam:

$$S_{avg} = \frac{(Power)_{avg}}{A} = \frac{(Power)_{avg}}{\pi r^2} = \frac{3.0 \times 10^{-3} \text{ W}}{\pi \left(\dfrac{2.0 \times 10^{-3} \text{ m}}{2}\right)^2} = 955 \text{ W/m}^2$$

Now let's determine the radiation pressure from the laser beam. Equation 24.35 indicates that a completely reflected beam would apply an average pressure of $P_{avg} = 2S_{avg}/c$. We can model the actual reflection as follows. Imagine that the surface absorbs the beam, resulting in pressure $P_{avg} = S_{avg}/c$. Then the surface emits the beam, resulting in additional pressure $P_{avg} = S_{avg}/c$. If the surface emits only a fraction f of the beam (so that f is the amount of the incident beam reflected), the pressure due to the emitted beam is $P_{avg} = fS_{avg}/c$.

Use this model to find the total pressure on the surface due to absorption and re-emission (reflection):

$$P_{avg} = \frac{S_{avg}}{c} + f\frac{S_{avg}}{c} = (1 + f)\frac{S_{avg}}{c}$$

Evaluate this pressure for a beam that is 70% reflected:

$$P_{avg} = (1 + 0.70)\frac{955 \text{ W/m}^2}{3.0 \times 10^8 \text{ m/s}} = 5.4 \times 10^{-6} \text{ N/m}^2$$

Finalize The pressure has an extremely small value, as expected. (Recall from Section 15.1 that atmospheric pressure is approximately 10^5 N/m^2). Consider the magnitude of the Poynting vector, $S_{avg} = 955$ W/m^2. It is about the same as the intensity of sunlight at the Earth's surface. For this reason, it is not safe to shine the beam of a laser pointer into a person's eyes, which may be more dangerous than looking directly at the Sun.

Space Sailing

When imagining a trip to another planet, we normally think of traditional rocket engines that convert chemical energy in fuel carried on the spacecraft to kinetic energy of the spacecraft. An interesting alternative to this approach is that of **space sailing.** A space-sailing craft includes a very large sail that reflects light. The motion of the spacecraft depends on pressure from light, that is, the force exerted on the sail by the reflection of light from the Sun. Calculations performed (before U.S. government budget cutbacks shelved early space-sailing projects) showed that sailing craft could travel to and from the planets in times similar to those for traditional rockets, but for less cost.

Calculations show that the radiation force from the Sun on a practical sailcraft with large sails could be equal to or slightly larger than the gravitational force on the sailcraft. If these two forces are equal, the sailcraft can be modeled as a particle in equilibrium because the inward gravitational force of the Sun balances the outward force exerted by the light from the Sun. If the sailcraft has an initial velocity in some direction away from the Sun, it would move in a straight line under the action of these two forces, with no necessity for fuel. A traditional spacecraft with its rocket engines turned off, on the other hand, would slow down as a result of the gravitational force on it due to the Sun. Both the force on the sail and the gravitational force from the Sun fall off as the inverse square of the Sun–sailcraft separation. Therefore, in

Wearing sunglasses that do not block ultraviolet (UV) light is worse for your eyes than wearing no sunglasses. The lenses of any sunglasses absorb some visible light, thus causing the wearer's pupils to dilate. If the glasses do not also block UV light, more damage may be done to the eye's lens because of the dilated pupils. If you wear no sunglasses at all, your pupils are contracted, you squint, and much less UV light enters your eyes. High-quality sunglasses block nearly all the eye-damaging UV light.

theory, the straight-line motion of the sailcraft would continue forever with no fuel requirement.

By using just the motion imparted to a sailcraft by the Sun, the craft could reach Alpha Centauri in about 10 000 years. This time interval can be reduced to 30 to 100 years using a *beamed power system*. In this concept, light from the Sun is gathered by a transformation device in orbit around the Earth and is converted to a laser beam or microwave beam aimed at the sailcraft. The force from this intense beam of radiation increases the acceleration of the craft, and the transit time is significantly reduced. Calculations indicate that the sailcraft could achieve design speeds of up to 20% of the speed of light using this technique.

◀ 24.6 | The Spectrum of Electromagnetic Waves

Electromagnetic waves travel through vacuum with speed c, frequency f, and wavelength λ. The various types of electromagnetic waves, all produced by accelerating charges, are shown in Figure 24.11. Note the wide range of frequencies and wavelengths. Let us briefly describe the wave types shown in Figure 24.11.

Radio waves are the result of charges accelerating, for example, through conducting wires in a radio antenna. They are generated by such electronic devices as LC oscillators and are used in radio and television communication systems.

Microwaves (short-wavelength radio waves) have wavelengths ranging between about 1 mm and 30 cm and are also generated by electronic devices. Because of their short wavelengths, they are well suited for radar systems used in aircraft navigation and for studying the atomic and molecular properties of matter. Microwave ovens are a domestic application of these waves.

Infrared waves have wavelengths ranging from about 1 mm to the longest wavelength of visible light, 7×10^{-7} m. These waves, produced by objects at room temperature and by molecules, are readily absorbed by most materials. Infrared radiation has many practical and scientific applications, including physical therapy, infrared photography, and vibrational spectroscopy. Your remote control for your TV or DVD player likely uses an infrared beam to communicate with the video device.

Visible light, the most familiar form of electromagnetic waves, is that part of the spectrum the human eye can detect. Light is produced by hot objects like lightbulb filaments and by the rearrangement of electrons in atoms and molecules. The wavelengths of visible light are classified by color, ranging from violet ($\lambda \approx 4 \times 10^{-7}$ m) to red ($\lambda \approx 7 \times 10^{-7}$ m). The eye's sensitivity is a function of wavelength and is a maximum at a wavelength of about 5.5×10^{-7} m (yellow–green). Table 24.1 provides approximate correspondences between the wavelength of visible light and the color assigned to it by humans. Light is the basis of the science of optics, to be discussed in Chapters 25 through 27.

Ultraviolet light covers wavelengths ranging from about 4×10^{-7} m (400 nm) down to about 6×10^{-10} m (0.6 nm). The Sun is an important source of ultraviolet waves, which are the main cause of suntans and sunburns. Atoms in the stratosphere absorb most of the ultraviolet waves from the Sun (which is fortunate because ultraviolet waves in large quantities have harmful effects on humans). One important constituent of the stratosphere is

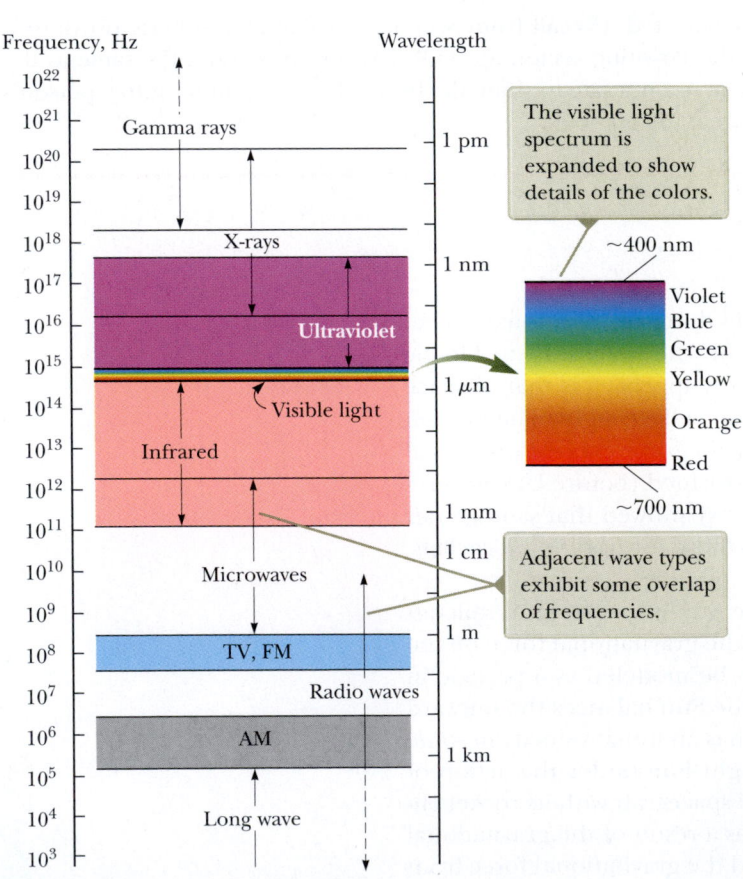

Figure 24.11 The electromagnetic spectrum.

ozone (O_3), which results from reactions of oxygen with ultraviolet radiation. This ozone shield converts lethal high-energy ultraviolet radiation to harmless infrared radiation. A great deal of concern has arisen concerning the depletion of the protective ozone layer by the use of a class of chemicals called chlorofluorocarbons (e.g., Freon) in aerosol spray cans and as refrigerants.

X-rays are electromagnetic waves with wavelengths in the range of about 10^{-8} m (10 nm) down to 10^{-13} m (10^{-4} nm). The most common source of x-rays is the acceleration of high-energy electrons bombarding a metal target. X-rays are used as a diagnostic tool in medicine and as a treatment for certain forms of cancer. Because x-rays damage or destroy living tissues and organisms, care must be taken to avoid unnecessary exposure and overexposure. X-rays are also used in the study of crystal structure; x-ray wavelengths are comparable to the atomic separation distances (≈ 0.1 nm) in solids.

Gamma rays are electromagnetic waves emitted by radioactive nuclei and during certain nuclear reactions. They have wavelengths ranging from about 10^{-10} m to less than 10^{-14} m. Gamma rays are highly penetrating and produce serious damage when absorbed by living tissues. Consequently, those working near such dangerous radiation must be protected with heavily absorbing materials, such as layers of lead.

QUICK QUIZ 24.5 In many kitchens, a microwave oven is used to cook food. The frequency of the microwaves is on the order of 10^{10} Hz. Are the wavelengths of these microwaves on the order of (**a**) kilometers, (**b**) meters, (**c**) centimeters, or (**d**) micrometers?

QUICK QUIZ 24.6 A radio wave of frequency on the order of 10^5 Hz is used to carry a sound wave with a frequency on the order of 10^3 Hz. Is the wavelength of this radio wave on the order of (**a**) kilometers, (**b**) meters, (**c**) centimeters, or (**d**) micrometers?

TABLE 24.1 |

Approximate Correspondence Between Wavelengths of Visible Light and Color

Wavelength Range (nm)	Color Description
400–430	Violet
430–485	Blue
485–560	Green
560–590	Yellow
590–625	Orange
625–700	Red

Note: The wavelength ranges here are approximate. Different people will describe colors differently.

Pitfall Prevention | 24.5
Heat Rays
Infrared rays are often called "heat rays." This terminology is a misnomer. Although infrared radiation is used to raise or maintain temperature, as in the case of keeping food warm with "heat lamps" at a fast-food restaurant, all wavelengths of electromagnetic radiation carry energy that can cause the temperature of a system to increase. As an example, consider using your microwave oven to bake a potato, whose temperature increases because of microwaves.

THINKING PHYSICS 24.3   BIO Center of eyesight sensitivity

The center of sensitivity of our eyes is close to the same frequency as the center of the wavelength distribution of light from the Sun. Is that an amazing coincidence?

Reasoning It is not a coincidence; rather, it is the result of biological evolution. Humans have evolved so as to be most visually sensitive to the wavelengths that are strongest from the Sun. It is an interesting conjecture to imagine aliens from another planet, with a Sun with a different temperature, arriving at Earth. Their eyes would have the center of sensitivity at different wavelengths than ours. How would their vision of the Earth compare with ours? ◄

24.7 | Polarization of Light Waves

As we learned in Section 24.3, the electric and magnetic vectors associated with an electromagnetic wave are perpendicular to each other and also to the direction of wave propagation as shown in Active Figure 24.4. The phenomenon of polarization described in this section is a property that specifies the directions of the electric and magnetic fields associated with an electromagnetic wave.

An ordinary beam of light consists of a large number of waves emitted by the atoms of the light source. Each atom produces a wave with its own orientation of the electric field $\vec{E}$, corresponding to the direction of vibration in the atom. The direction of polarization of the electromagnetic wave is defined to be the direction in which $\vec{E}$ is vibrating. Because all directions of vibration are possible in a group of atoms emitting a beam of light, however, the resultant beam is a superposition of waves produced by the individual atomic sources. The result is an **unpolarized** light

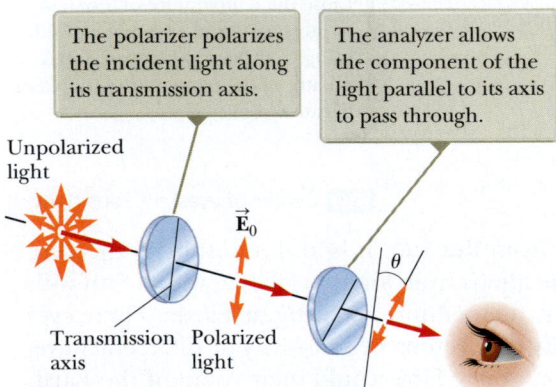

The red dot signifies the velocity vector for the wave coming out of the page.

$\vec{E}$ $\vec{E}$

a   b

Figure 24.12 (a) An unpolarized light beam viewed along the direction of propagation (perpendicular to the page). The time-varying electric field vector can be in any direction in the plane of the page with equal probability. (b) A linearly polarized light beam with the time-varying electric field vector in the vertical direction.

The polarizer polarizes the incident light along its transmission axis.

The analyzer allows the component of the light parallel to its axis to pass through.

Unpolarized light

$\vec{E}_0$

θ

Transmission axis   Polarized light

Active Figure 24.13 Two polarizing sheets whose transmission axes make an angle θ with each other. Only a fraction of the polarized light incident on the analyzer is transmitted through it.

▶ Malus's law

beam, represented schematically in Figure 24.12a. The direction of wave propagation in this figure is perpendicular to the page. The figure suggests that *all* directions of the electric field vector lying in a plane perpendicular to the direction of propagation are equally probable.

A beam of light is said to be **linearly polarized** if the orientation of $\vec{E}$ is the same for all individual waves *at all times* at a particular point as suggested in Figure 24.12b. (Sometimes such a wave is described as **plane polarized.**) The wave described in Active Figure 24.4 is an example of a wave linearly polarized along the *y* axis. As the field propagates in the *x* direction, $\vec{E}$ is always along the *y* axis. The plane formed by $\vec{E}$ and the direction of propagation is called the **plane of polarization** of the wave. In Active Figure 24.4, the plane of polarization is the *xy* plane. It is possible to obtain a linearly polarized wave from an unpolarized wave by removing from the unpolarized wave all components of electric field vectors except those that lie in a single plane.

The most common technique for polarizing light is to send it through a material that passes only components of electric field vectors that are parallel to a characteristic direction of the material called the **polarizing direction.** In 1938, E. H. Land discovered such a material, which he called **Polaroid,** that polarizes light through selective absorption by oriented molecules. This material is fabricated in thin sheets of long-chain hydrocarbons, which are stretched during manufacture so that the molecules align. After a sheet is dipped into a solution containing iodine, the molecules become good electric conductors. The conduction, however, takes place primarily along the hydrocarbon chains because the valence electrons of the molecules can move easily only along the chains (valence electrons are "free" electrons that can readily move through the conductor). As a result, the molecules readily *absorb* light whose electric field vector is parallel to their length and *transmit* light whose electric field vector is perpendicular to their length. It is common to refer to the direction perpendicular to the molecular chains as the **transmission axis.** An ideal polarizer passes the components of electric vectors that are parallel to the transmission axis. Components perpendicular to the transmission axis are absorbed. If light passes through several polarizers, whatever is transmitted has the plane of polarization parallel to the polarizing direction of the last polarizer through which it passed.

Let us now obtain an expression for the intensity of light that passes through a polarizing material. In Active Figure 24.13, an unpolarized light beam is incident on the first polarizing sheet, called the **polarizer,** where the transmission axis is as indicated. The light that passes through this sheet is polarized vertically, and the transmitted electric field vector is $\vec{E}_0$. A second polarizing sheet, called the **analyzer,** intercepts this beam with its transmission axis at an angle of θ to the axis of the polarizer. The component of $\vec{E}_0$ that is perpendicular to the axis of the analyzer is completely absorbed, and the component parallel to that axis is $E_0 \cos \theta$. We know from Equation 24.26 that the transmitted intensity varies as the *square* of the transmitted amplitude, so we conclude that the intensity of the transmitted (polarized) light varies as

$$I = I_{max} \cos^2 \theta \qquad \qquad \textbf{24.36}◀$$

where I_{max} is the intensity of the polarized wave incident on the analyzer. This expression, known as **Malus's law,** applies to any two polarizing materials whose transmission axes are at an angle of θ to each other. From this expression, note that the transmitted intensity is a maximum when the transmission axes are parallel ($\theta = 0$ or 180°) and zero (complete absorption by the analyzer) when the transmission axes are perpendicular to each other. This variation in transmitted intensity through a pair of polarizing sheets is illustrated in Figure 24.14. Because the average value of $\cos^2 \theta$ is $\frac{1}{2}$, the intensity of initially unpolarized light is reduced by a factor of one-half as the light passes through a single ideal polarizer.

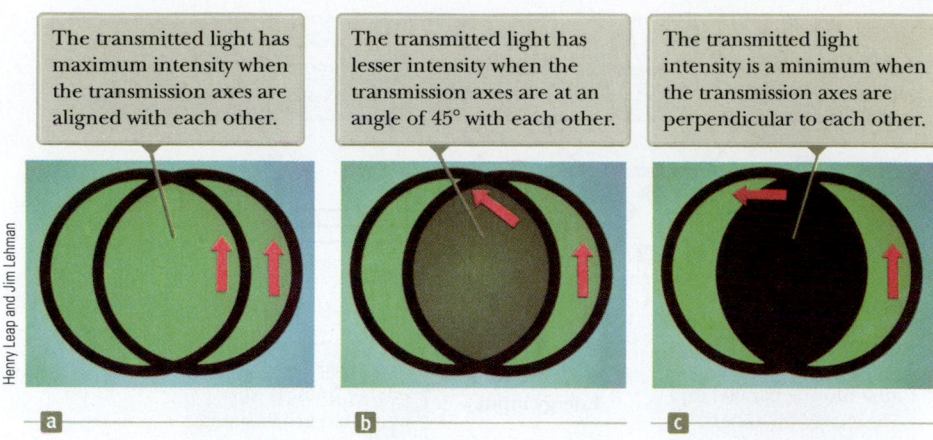

The transmitted light has maximum intensity when the transmission axes are aligned with each other.

The transmitted light has lesser intensity when the transmission axes are at an angle of 45° with each other.

The transmitted light intensity is a minimum when the transmission axes are perpendicular to each other.

Henry Leap and Jim Lehman

Figure 24.14 The intensity of light transmitted through two polarizers depends on the relative orientation of their transmission axes. The red arrows indicate the transmission axes of the polarizers.

QUICK QUIZ 24.7 A polarizer for microwaves can be made as a grid of parallel metal wires approximately 1 cm apart. Is the electric field vector for microwaves transmitted through this polarizer (**a**) parallel or (**b**) perpendicular to the metal wires?

24.8 | Context Connection: The Special Properties of Laser Light

In this chapter and the next three, we shall explore the nature of laser light and a variety of applications of lasers in our technological society. The primary properties of laser light that make it useful in these applications are the following:

- The light is coherent. The individual rays of light in a laser beam maintain a fixed phase relationship with one another, resulting in no destructive interference.
- The light is monochromatic. Laser light has a very small range of wavelengths.
- The light has a small angle of divergence. The beam spreads out very little, even over large distances.

To understand the origin of these properties, let us combine our knowledge of atomic energy levels from Chapter 11 with some special requirements for the atoms that emit laser light.

As we found in Chapter 11, the energies of an atom are quantized. We used a semigraphical representation called an *energy level diagram* in that chapter to help us understand the quantized energies in an atom. The production of laser light depends heavily on the properties of these energy levels in the atoms, the source of the laser light.

The word *laser* is an acronym for **l**ight **a**mplification by **s**timulated **e**mission of **r**adiation. The full name indicates one of the requirements for laser light, that the process of **stimulated emission** must occur to achieve laser action.

Suppose an atom is in the excited state E_2 as in Active Figure 24.15 and a photon with energy $hf = E_2 - E_1$ is incident on it. The incoming photon can stimulate the excited atom to return to the ground state and thereby emit a second photon having the same energy hf and traveling in the same direction. Note that the incident photon is not absorbed, so after the stimulated emission, two identical photons exist: the incident photon and the emitted photon. The emitted photon is in phase with the incident photon. These photons can stimulate other atoms to emit photons in a chain of similar processes. The many photons produced in this fashion are the source of the intense, coherent light in a laser.

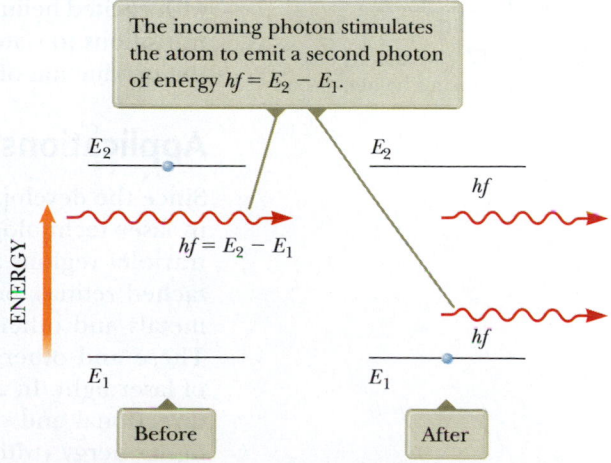

The incoming photon stimulates the atom to emit a second photon of energy $hf = E_2 - E_1$.

E_2 E_2

$hf = E_2 - E_1$ hf

ENERGY

hf

E_1 E_1

Before   After

Active Figure 24.15 Stimulated emission of a photon by an incoming photon of energy hf. Initially, the atom is in the excited state.

Figure 24.16 Schematic diagram of a laser design.

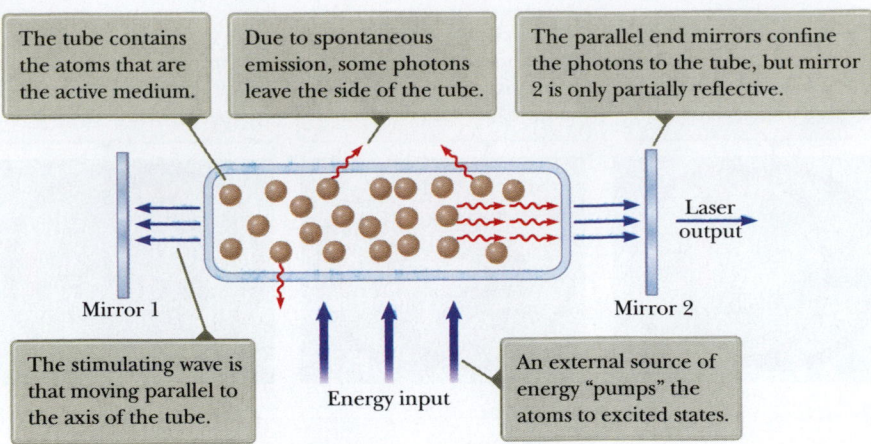

The tube contains the atoms that are the active medium.

Due to spontaneous emission, some photons leave the side of the tube.

The parallel end mirrors confine the photons to the tube, but mirror 2 is only partially reflective.

Mirror 1

Mirror 2

Laser output

The stimulating wave is that moving parallel to the axis of the tube.

Energy input

An external source of energy "pumps" the atoms to excited states.

For the stimulated emission to result in laser light, we must have a buildup of photons in the system. The following three conditions must be satisfied to achieve this buildup:

- The system must be in a state of **population inversion.** More atoms must be in an excited state than in the ground state. Atoms in the ground state can absorb photons, raising them to the excited state. The population inversion assures that we have more emission of photons from excited atoms than absorption by atoms in the ground state.
- The excited state of the system must be a *metastable state,* which means that its lifetime must be long compared with the usually short lifetime of excited states, which is typically 10^{-8} s. In this case, stimulated emission is likely to occur before spontaneous emission. The energy of a metastable state is indicated with an asterisk, E^*.
- The emitted photons must be confined in the system long enough to enable them to stimulate further emission from other excited atoms, which is achieved by using reflecting mirrors at the ends of the system. One end is made totally reflecting, and the other is slightly transparent to allow the laser beam to escape (Fig. 24.16).

One device that exhibits stimulated emission of radiation is the helium–neon gas laser. Figure 24.17 is an energy-level diagram for the neon atom in this system. The mixture of helium and neon is confined to a glass tube that is sealed at the ends by mirrors. A voltage applied across the tube causes electrons to sweep through the tube, colliding with the atoms of the gases and raising them into excited states. Neon atoms are excited to state E_3^* through this process and also as a result of collisions with excited helium atoms. Stimulated emission occurs, causing neon atoms to make transitions to state E_2. Neighboring excited atoms are also stimulated. The result is the production of coherent light at a wavelength of 632.8 nm.

The atom emits 632.8-nm photons through stimulated emission in the transition $E_3^* - E_2$. That is the source of coherent light in the laser.

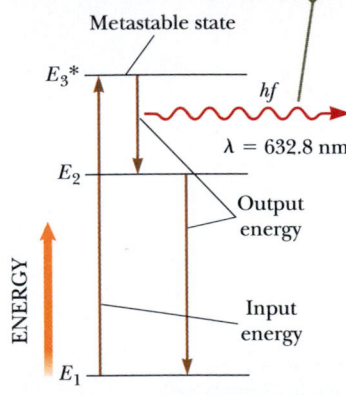

Metastable state

E_3^*

hf

$\lambda = 632.8$ nm

E_2

Output energy

Input energy

ENERGY

E_1

Figure 24.17 Energy-level diagram for a neon atom in a helium–neon laser.

Applications

Since the development of the first laser in 1960, tremendous growth has occurred in laser technology. Lasers that cover wavelengths in the infrared, visible, and ultraviolet regions are now available. Applications include surgical "welding" of detached retinas, precision surveying and length measurement, precision cutting of metals and other materials, and telephone communication along optical fibers. These and other applications are possible because of the unique characteristics of laser light. In addition to being highly monochromatic, laser light is also highly directional and can be sharply focused to produce regions of extremely intense light energy (with energy densities 10^{12} times the density in the flame of a typical cutting torch).

Lasers are used in precision long-range distance measurement (range finding). In recent years, it has become important in astronomy and geophysics to measure as precisely as possible the distances from various points on the surface of the Earth to a point on the Moon's surface. To facilitate these measurements, the *Apollo* astronauts set up a 0.5-m square of reflector prisms on the Moon, which enables laser pulses directed from an Earth-based station to be retroreflected to the same station. Using the known speed of light and the measured round-trip travel time of a laser pulse, the Earth–Moon distance can be determined to a precision of better than 10 cm.

Because various laser wavelengths can be absorbed in specific biological tissues, lasers have a number of medical applications. For example, certain laser procedures have greatly reduced blindness in patients with glaucoma and diabetes. Glaucoma is a widespread eye condition characterized by a high fluid pressure in the eye, a condition that can lead to destruction of the optic nerve. A simple laser operation (iridectomy) can "burn" open a tiny hole in a clogged membrane, relieving the destructive pressure. A serious side effect of diabetes is neovascularization, the proliferation of weak blood vessels, which often leak blood. When neovascularization occurs in the retina, vision deteriorates (diabetic retinopathy) and finally is destroyed. Today, it is possible to direct the green light from an argon ion laser through the clear eye lens and eye fluid, focus on the retina edges, and photocoagulate the leaky vessels. Even people who have only minor vision defects such as nearsightedness are benefiting from the use of lasers to reshape the cornea, changing its focal length and reducing the need for eyeglasses.

BIO Uses of lasers in ophthalmology

Laser surgery is now an everyday occurrence at hospitals and medical clinics around the world. Infrared light at 10 μm from a carbon dioxide laser can cut through muscle tissue, primarily by vaporizing the water contained in cellular material. Laser power of approximately 100 W is required in this technique. The advantage of the "laser knife" over conventional methods is that laser radiation cuts tissue and coagulates blood at the same time, leading to a substantial reduction in blood loss. In addition, the technique virtually eliminates cell migration, an important consideration when tumors are being removed.

BIO Laser surgery

In biological and medical research, it is often important to isolate and collect unusual cells for study and growth. A laser cell separator exploits the tagging of specific cells with fluorescent dyes. All cells are then dropped from a tiny charged nozzle and laser-scanned for the dye tag. If triggered by the correct light-emitting tag, a small voltage applied to parallel plates deflects the falling electrically charged cell into a collection beaker.

BIO Laser cell separator

An exciting area of research and technological applications arose in the 1990s with the development of *laser trapping* of atoms. One scheme, called *optical molasses* and developed by Steven Chu of Stanford University and his colleagues, involves focusing six laser beams onto a small region in which atoms are to be trapped. Each pair of lasers is along one of the *x*, *y*, and *z* axes and emits light in opposite directions (Fig. 24.18). The frequency of the laser light is tuned to be slightly below the absorption frequency of the subject atom. Imagine that an atom has been placed into the trap region and moves along the positive *x* axis toward the laser that is emitting light toward it (the rightmost laser on the *x* axis in Fig. 24.18). Because the atom is moving, the light from the laser appears Doppler-shifted upward in frequency in the reference frame of the atom. Therefore, a match between the Doppler-shifted laser frequency and the absorption frequency of the atom exists and the atom absorbs photons.[5] The momentum carried by these photons results in the atom being pushed back to the center of the trap. By incorporating six lasers, the atoms are pushed back into the trap regardless of which way they move along any axis.

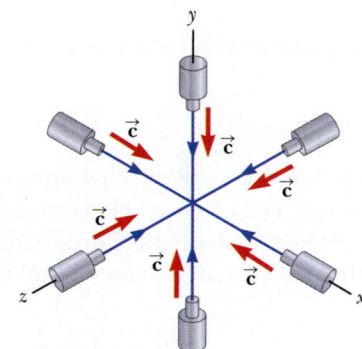

Figure 24.18 An optical trap for atoms is formed at the intersection point of six counterpropagating laser beams along mutually perpendicular axes.

In 1986, Chu developed *optical tweezers,* a device that uses a single tightly focused laser beam to trap and manipulate small particles. In combination with microscopes, optical tweezers have opened up many new possibilities for biologists. Optical tweezers have been used to manipulate live bacteria without damage, move chromosomes within a cell nucleus, and measure the elastic properties of a single DNA molecule.

BIO Optical tweezers

---

[5]The laser light traveling in the same direction as the atom is Doppler-shifted further downward in frequency, so there is no absorption. Therefore, the atom is not pushed out of the trap by the diametrically opposed laser.

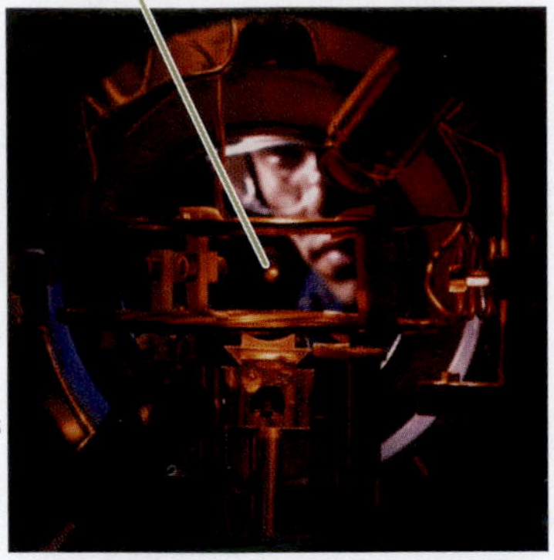

The orange dot is the sample of trapped sodium atoms.

Figure 24.19 A staff member of the National Institute of Standards and Technology views a sample of trapped sodium atoms cooled to a temperature measured in microkelvins.

Chu shared the 1997 Nobel Prize in Physics with two of his colleagues for the development of the techniques of optical trapping.

An extension of laser trapping, *laser cooling*, is possible because the normal high speeds of the atoms are reduced when they are restricted to the region of the trap. As a result, the temperature of the collection of atoms can be reduced to a few microkelvins. The technique of laser cooling allows scientists to study the behavior of atoms at extremely low temperatures (Fig. 24.19).

In the 1920s, Satyendra Nath Bose (1894–1974) was studying photons and investigating collections of identical photons, which can all be in the same quantum state. Einstein followed up on the work of Bose and predicted that a collection of atoms could all be in the same quantum state if the temperature were low enough. The proposed collection of atoms is called a *Bose–Einstein condensate*. In 1995, using laser cooling supplemented with evaporative cooling, the first Bose–Einstein condensate was created in the laboratory by Eric Cornell and Carl Wieman, who won the 2001 Nobel Prize in Physics for their work. Many laboratories are now creating Bose–Einstein condensates and studying their properties and possible applications. One interesting result was reported by a Harvard University group led by Lene Vestergaard Hau in 2001. She and her colleagues announced that they were able to bring a light pulse to a complete stop by using a Bose–Einstein condensate.[6]

More recently, scientists have discovered a new type of Bose–Einstein condensate based on a quasiparticle called the *polariton*.[7] The polariton, which is a superposition of a photon and an electronic excitation in a solid, exists typically for only a few picoseconds in an optical cavity. These condensates are unique because they are extremely light compared to atomic condensates and therefore exhibit quantum effects at higher temperatures.

We have explored general properties of laser light in this chapter. In the Context Connection of Chapter 25, we shall explore the technology of optical fibers, in which lasers are used in a variety of applications.

> SUMMARY

Displacement current I_d is defined as

$$I_d \equiv \epsilon_0 \frac{d\Phi_E}{dt} \qquad \text{24.1} \blacktriangleleft$$

and represents an effective current through a region of space in which an electric field is changing in time.

When used with the Lorentz force law ($\vec{F} = q\vec{E} + q\vec{v} \times \vec{B}$), **Maxwell's equations** describe *all* electromagnetic phenomena:

$$\oint \vec{E} \cdot d\vec{A} = \frac{q}{\epsilon_0} \qquad \text{24.4} \blacktriangleleft$$

$$\oint \vec{B} \cdot d\vec{A} = 0 \qquad \text{24.5} \blacktriangleleft$$

$$\oint \vec{E} \cdot d\vec{s} = -\frac{d\Phi_B}{dt} \qquad \text{24.6} \blacktriangleleft$$

$$\oint \vec{B} \cdot d\vec{s} = \mu_0 I + \epsilon_0 \mu_0 \frac{d\Phi_E}{dt} \qquad \text{24.7} \blacktriangleleft$$

Electromagnetic waves, which are predicted by Maxwell's equations, have the following properties:

- The electric and magnetic fields satisfy the following wave equations, which can be obtained from Maxwell's third and fourth equations:

$$\frac{\partial^2 E}{\partial x^2} = \mu_0 \epsilon_0 \frac{\partial^2 E}{\partial t^2} \qquad \text{24.16} \blacktriangleleft$$

$$\frac{\partial^2 B}{\partial x^2} = \mu_0 \epsilon_0 \frac{\partial^2 B}{\partial t^2} \qquad \text{24.17} \blacktriangleleft$$

- Electromagnetic waves travel through a vacuum with the speed of light $c = 3.00 \times 10^8$ m/s, where

$$c = \frac{1}{\sqrt{\mu_0 \epsilon_0}} \qquad \text{24.18} \blacktriangleleft$$

- The electric and magnetic fields of an electromagnetic wave are perpendicular to each other and perpendicular to the direction of wave propagation; hence, electromagnetic waves are transverse waves. The electric and magnetic

[6]C. Liu, Z. Dutton, C. H. Behroozi, and L. V. Hau, "Observation of Coherent Optical Information Storage in an Atomic Medium Using Halted Light Pulses," *Nature*, 409, 490–493, January 25, 2001.

[7]D. Snoke and P. Littlewood, "Polariton Condensates," *Physics Today*, 42–47, August 2010.

fields of a sinusoidal plane electromagnetic wave propagating in the positive x direction can be written

$$E = E_{max} \cos(kx - \omega t) \qquad \textbf{24.19}\blacktriangleleft$$

$$B = B_{max} \cos(kx - \omega t) \qquad \textbf{24.20}\blacktriangleleft$$

where ω is the angular frequency of the wave and k is the angular wave number. These equations represent special solutions to the wave equations for $\vec{E}$ and $\vec{B}$.

- The instantaneous magnitudes of $\vec{E}$ and $\vec{B}$ in an electromagnetic wave are related by the expression

$$\frac{E}{B} = c \qquad \textbf{24.22}\blacktriangleleft$$

- Electromagnetic waves carry energy. The rate of flow of energy crossing a unit area is described by the **Poynting vector $\vec{S}$**, where

$$\vec{S} \equiv \frac{1}{\mu_0}\vec{E} \times \vec{B} \qquad \textbf{24.24}\blacktriangleleft$$

The average value of the Poynting vector for a plane electromagnetic wave has the magnitude

$$I = S_{avg} = \frac{E_{max}B_{max}}{2\mu_0} = \frac{E_{max}^2}{2\mu_0 c} = \frac{cB_{max}^2}{2\mu_0} \qquad \textbf{24.26}\blacktriangleleft$$

The average power per unit area (intensity) of a sinusoidal plane electromagnetic wave equals the average value of the Poynting vector taken over one or more cycles.

- Electromagnetic waves carry momentum and hence can exert pressure on surfaces. If an electromagnetic wave whose intensity is I is completely absorbed by a surface on which it is normally incident, the radiation pressure on that surface is

$$P = \frac{S}{c} \qquad \text{(complete absorption)} \qquad \textbf{24.32}\blacktriangleleft$$

If the surface totally reflects a normally incident wave, the pressure is doubled.

The **electromagnetic spectrum** includes waves covering a broad range of frequencies and wavelengths.

When polarized light of intensity I_{max} is incident on a polarizing film, the light transmitted through the film has an intensity equal to $I_{max} \cos^2 \theta$, where θ is the angle between the transmission axis of the polarizing film and the electric field vector of the incident light.

▶ OBJECTIVE QUESTIONS |

☐ denotes answer available in *Student Solutions Manual/Study Guide*

1. Which of the following statements are true regarding electromagnetic waves traveling through a vacuum? More than one statement may be correct. (a) All waves have the same wavelength. (b) All waves have the same frequency. (c) All waves travel at 3.00×10^8 m/s. (d) The electric and magnetic fields associated with the waves are perpendicular to each other and to the direction of wave propagation. (e) The speed of the waves depends on their frequency.

2. An electromagnetic wave with a peak magnetic field magnitude of 1.50×10^{-7} T has an associated peak electric field of what magnitude? (a) 0.500×10^{-15} N/C (b) 2.00×10^{-5} N/C (c) 2.20×10^4 N/C (d) 45.0 N/C (e) 22.0 N/C

3. Assume you charge a comb by running it through your hair and then hold the comb next to a bar magnet. Do the electric and magnetic fields produced constitute an electromagnetic wave? (a) Yes they do, necessarily. (b) Yes they do because charged particles are moving inside the bar magnet. (c) They can, but only if the electric field of the comb and the magnetic field of the magnet are perpendicular. (d) They can, but only if both the comb and the magnet are moving. (e) They can, if either the comb or the magnet or both are accelerating.

4. If plane polarized light is sent through two polarizers, the first at 45° to the original plane of polarization and the second at 90° to the original plane of polarization, what fraction of the original polarized intensity passes through the last polarizer? (a) 0 (b) $\frac{1}{4}$ (c) $\frac{1}{2}$ (d) $\frac{1}{8}$ (e) $\frac{1}{10}$

5. A typical microwave oven operates at a frequency of 2.45 GHz. What is the wavelength associated with the electromagnetic waves in the oven? (a) 8.20 m (b) 12.2 cm (c) 1.20×10^8 m (d) 8.20×10^{-9} m (e) none of those answers

6. A student working with a transmitting apparatus like Heinrich Hertz's wishes to adjust the electrodes to generate electromagnetic waves with a frequency half as large as before. **(i)** How large should she make the effective capacitance of the pair of electrodes? (a) four times larger than before (b) two times larger than before (c) one-half as large as before (d) one-fourth as large as before (e) none of those answers **(ii)** After she makes the required adjustment, what will the wavelength of the transmitted wave be? Choose from the same possibilities as in part (i).

7. A small source radiates an electromagnetic wave with a single frequency into vacuum, equally in all directions. **(i)** As the wave moves, does its frequency (a) increase, (b) decrease, or (c) stay constant? Using the same choices, answer the same question about **(ii)** its wavelength, **(iii)** its speed, **(iv)** its intensity, and **(v)** the amplitude of its electric field.

8. A plane electromagnetic wave with a single frequency moves in vacuum in the positive x direction. Its amplitude is uniform over the yz plane. **(i)** As the wave moves, does its frequency (a) increase, (b) decrease, or (c) stay constant? Using the same choices, answer the same question about **(ii)** its wavelength, **(iii)** its speed, **(iv)** its intensity, and **(v)** the amplitude of its magnetic field.

9. **(i)** Rank the following kinds of waves according to their wavelength ranges from those with the largest typical or average wavelength to the smallest, noting any cases of equality: (a) gamma rays (b) microwaves (c) radio waves (d) visible light (e) x-rays **(ii)** Rank the kinds of waves according to their frequencies from highest to lowest. **(iii)** Rank the kinds of waves according to their speeds from fastest to slowest. Choose from the same possibilities as in part (i).

10. Assume the amplitude of the electric field in a plane electromagnetic wave is E_1 and the amplitude of the magnetic field is B_1. The source of the wave is then adjusted so that the amplitude of the electric field doubles to become $2E_1$. **(i)** What happens to the amplitude of the magnetic field in this process? (a) It becomes four times larger. (b) It becomes two times larger. (c) It can stay constant. (d) It becomes one-half as large. (e) It becomes one-fourth as large. **(ii)** What happens to the intensity of the wave? Choose from the same possibilities as in part (i).

11. A spherical interplanetary grain of dust of radius 0.2 mm is at a distance r_1 from the Sun. The gravitational force exerted by the Sun on the grain just balances the force due to radiation pressure from the Sun's light. **(i)** Assume the grain is moved to a distance $2r_1$ from the Sun and released. At this location, what is the net force exerted on the grain? (a) toward the Sun (b) away from the Sun (c) zero (d) impossible to determine without knowing the mass of the grain **(ii)** Now assume the grain is moved back to its original

location at r_1, compressed so that it crystallizes into a sphere with significantly higher density, and then released. In this situation, what is the net force exerted on the grain? Choose from the same possibilities as in part (i).

12. Consider an electromagnetic wave traveling in the positive y direction. The magnetic field associated with the wave at some location at some instant points in the negative x direction as shown in Figure OQ24.12. What is the direction of the electric field at this position and at this instant? (a) the positive x direction (b) the positive y direction (c) the positive z direction (d) the negative z direction (e) the negative y direction

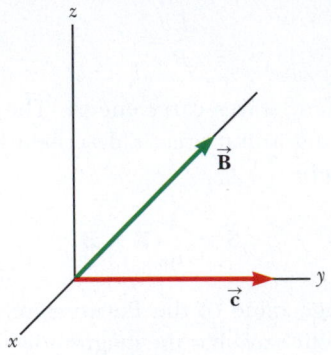

Figure OQ24.12

▶ CONCEPTUAL QUESTIONS

1. Despite the advent of digital television, some viewers still use "rabbit ears" atop their sets (Fig. CQ24.1) instead of purchasing cable television service or satellite dishes. Certain orientations of the receiving antenna on a television set give better reception than others. Furthermore, the best orientation varies from station to station. Explain.

Figure CQ24.1 Conceptual Question 1.

2. List at least three differences between sound waves and light waves.

3. What new concept did Maxwell's generalized form of Ampère's law include?

4. When light (or other electromagnetic radiation) travels across a given region, (a) what is it that oscillates? (b) What is it that is transported?

5. Radio stations often advertise "instant news." If that means you can hear the news the instant the radio announcer speaks it, is the claim true? What approximate time interval is required for a message to travel from Maine to California by radio waves? (Assume the waves can be detected at this range.)

6. For a given incident energy of an electromagnetic wave, why is the radiation pressure on a perfectly reflecting surface twice as great as that on a perfectly absorbing surface?

7. If a high-frequency current exists in a solenoid containing a metallic core, the core becomes warm due to induction. Explain why the material rises in temperature in this situation.

8. An empty plastic or glass dish being removed from a microwave oven can be cool to the touch, even when food on an adjoining dish is hot. How is this phenomenon possible?

9. Describe the physical significance of the Poynting vector.

10. What does a radio wave do to the charges in the receiving antenna to provide a signal for your car radio?

11. Why should an infrared photograph of a person look different from a photograph taken with visible light?

12. Suppose a creature from another planet has eyes that are sensitive to infrared radiation. Describe what the alien would see if it looked around your library. In particular, what would appear bright and what would appear dim?

PROBLEMS AVAILABLE IN

24.1 Displacement Current and the Generalized Form of Ampère's Law

Problems 1–3

24.2 Maxwell's Equations and Hertz's Discoveries

Problems 4–8

24.3 Electromagnetic Waves

Problems 9–22

24.4 Energy Carried by Electromagnetic Waves

Problems 23–31

24.5 Momentum and Radiation Pressure

Problems 32–34

24.6 The Spectrum of Electromagnetic Waves

Problems 35–44

24.7 Polarization of Light Waves

Problems 45–50

24.8 Context Connection: The Special Properties of Laser Light

Problems 51–58

Additional Problems

Problems 59–75

Solutions to the following Problems are available in the *Student Solutions Manual/Study Guide*:

24.3, 24.4, 24.5, 24.13, 24.16, 24.23, 24.29, 24.37, 24.45, 24.57, 24.62, 24.63, 24.68, 24.73, and 24.75

List of Enhanced Problems

Problem Number	Targeted Feedback in Enhanced WebAssign	Master It in Enhanced WebAssign	Watch It in Enhanced WebAssign
24.2	✓		✓
24.3		✓	
24.5		✓	
24.7	✓		✓
24.8	✓		✓
24.12	✓		✓
24.13	✓	✓	
24.16	✓	✓	
24.23		✓	
24.25	✓		✓
24.27	✓		✓
24.29	✓	✓	
24.32	✓		✓
24.33	✓	✓	
24.40	✓		✓
24.45		✓	
24.47	✓		✓
24.51	✓		✓
24.53	✓		✓
24.57	✓	✓	
24.63	✓	✓	

Reflection and Refraction of Light

Chapter Outline

The preceding chapter serves as a bridge between electromagnetism and the area of physics called *optics*. Now that we have established the wave nature of electromagnetic radiation, we shall study the behavior of visible light and apply what we learn to all electromagnetic radiation. Our emphasis in this chapter will be on the behavior of light as it encounters an interface between two media.

So far, we have focused on the wave nature of light and discussed it in terms of our wave simplification model. As we learn more about the behavior of light, however, we shall return to our particle simplification model, especially as we incorporate the notions of quantum physics, beginning in Chapter 28. As we discuss in Section 25.1, a long historical debate took place between proponents of wave and particle models for light.

This photograph of a rainbow shows a distinct secondary rainbow with the colors reversed. The appearance of the rainbow depends on three optical phenomena discussed in this chapter: reflection, refraction, and dispersion.

Patrick J. Endres/Visuals Unlimited

25.1 | The Nature of Light

We encounter light every day, as soon as we open our eyes in the morning. This everyday experience involves a phenomenon that is actually quite complicated. Since the beginning of this book, we have discussed both the particle model and the wave model as simplification models to help us gain understanding of physical phenomena. Both of these models have been applied to the behavior of light. Until the beginning of the 19th century, most scientists thought that light was a stream of particles emitted by a light source. According to this model, the light particles stimulated the sense of sight on entering the eye. The chief architect of this particle model of light was Isaac Newton. The model provided a simple explanation of some known experimental facts concerning the nature of light—namely, the laws of reflection and refraction—to be discussed in this chapter.

Most scientists at the time accepted the particle model of light. During Newton's lifetime, however, another model was proposed—a model that views light as having wave-like properties. In 1678, a Dutch physicist and astronomer, Christian Huygens, showed that a wave model of light can also explain the laws of reflection and refraction. The wave model did not receive immediate acceptance for several reasons. All the waves known at the time (sound, water, and so on) traveled through a medium, but light from the Sun could travel to the Earth through empty space. Even though experimental evidence for the wave nature of light was discovered by Francesco Grimaldi (1618–1663) around 1660, most scientists rejected the wave model for more than a century and adhered to Newton's particle model due, for the most part, to Newton's great reputation as a scientist.

The first clear and convincing demonstration of the wave nature of light was provided in 1801 by Englishman Thomas Young (1773–1829), who showed that under appropriate conditions, light exhibits interference behavior. That is, light waves emitted by a single source and traveling along two different paths can arrive at some point, combine, and cancel each other by destructive interference. Such behavior could not be explained at that time by a particle model, because scientists could not imagine how two or more particles could come together and cancel one another. Additional developments during the 19th century led to the general acceptance of the wave model of light.

A critical development concerning the understanding of light was the work of James Clerk Maxwell, who in 1865 mathematically predicted that light is a form of high-frequency electromagnetic wave. As discussed in Chapter 24, Hertz in 1887 provided experimental confirmation of Maxwell's theory by producing and detecting other electromagnetic waves. Furthermore, Hertz and other investigators showed that these waves exhibited reflection, refraction, and all the other characteristic properties of waves.

Although the electromagnetic wave model seemed to be well established and could explain most known properties of light, some experiments could not be explained by the assumption that light was a wave. The most striking of these was the *photoelectric effect,* discovered by Hertz, in which electrons are ejected from a metal when its surface is exposed to light. We shall explore this experiment in detail in Chapter 28.

In view of these developments, light must be regarded as having a dual nature. In some cases, light acts like a wave, and in others, it acts like a particle. The classical electromagnetic wave model provides an adequate explanation of light propagation and interference, whereas the photoelectric effect and other experiments involving the interaction of light with matter are best explained by assuming that light is a particle. Light is light, to be sure. The question "Is light a wave or a particle?" is inappropriate; in some experiments, we measure its wave properties; in other experiments, we measure its particle properties. This curious dual nature of light may be unsettling at this point, but it will be clarified when we introduce the notion of a *quantum particle.* The photon, a particle of light, is our first example of a quantum particle, which we shall explore more fully in Chapter 28. Until then, we focus our attention on the properties of light that can be satisfactorily explained with the wave model.

25.2 | The Ray Model in Geometric Optics

In the beginning of our study of optics, we shall use a simplification model called the **ray model** or the **ray approximation. A ray** is a straight line drawn along the direction of propagation of a single wave, showing the path of the wave as it travels through space. The ray approximation involves geometric models based on these straight lines. Phenomena explained with the ray approximation do not depend explicitly on the wave nature of light, other than its propagation along a straight line.

A set of light waves can be represented by wave fronts (defined in Section 24.3) as illustrated in the pictorial representation in Figure 25.1 for a plane wave, which was

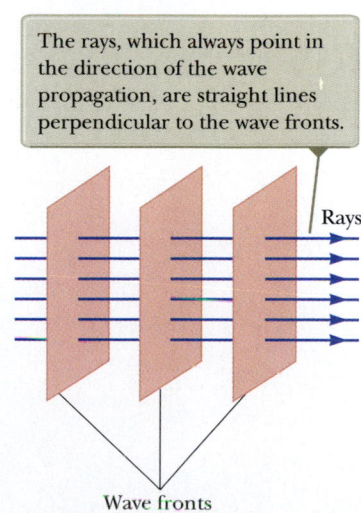

The rays, which always point in the direction of the wave propagation, are straight lines perpendicular to the wave fronts.

Figure 25.1 A plane wave propagating to the right.

Active Figure 25.2 A plane wave of wavelength λ is incident on a barrier in which there is an opening of diameter d.

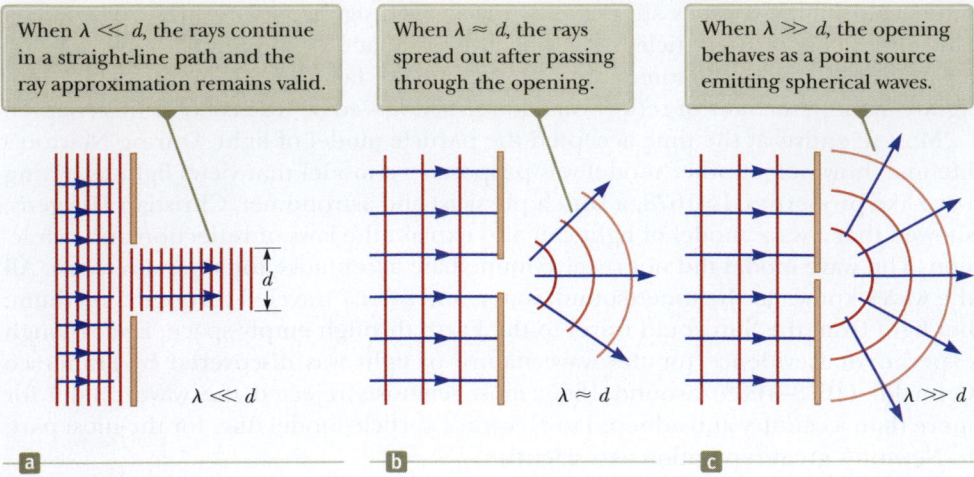

When $\lambda \ll d$, the rays continue in a straight-line path and the ray approximation remains valid.

When $\lambda \approx d$, the rays spread out after passing through the opening.

When $\lambda \gg d$, the opening behaves as a point source emitting spherical waves.

$\lambda \ll d$

$\lambda \approx d$

$\lambda \gg d$

a   b   c

introduced in Section 24.3. The definition of a wave front requires that the rays are perpendicular to the wave front at every location in space.

If a plane wave meets a barrier containing an opening whose size d is large relative to the wavelength λ as in Active Figure 25.2a, the individual waves emerging from the opening continue to move in a straight line (apart from some small edge effects); hence, the ray approximation continues to be valid. If the size of the opening is on the order of the wavelength as in Active Figure 25.2b, the waves (and, consequently, the rays we draw) spread out from the opening in all directions. We say that the incoming plane wave undergoes *diffraction* as it passes through the opening. If the opening is small relative to the wavelength, the diffraction is so strong that the opening can be approximated as a point source of waves (Active Fig. 25.2c). Therefore, diffraction is more pronounced as the ratio d/λ approaches zero.

The ray approximation assumes that $\lambda \ll d$ so that we do not concern ourselves with diffraction effects, which depend on the full wave nature of light. We shall delay studying diffraction until Chapter 27. The ray approximation is used in the current chapter and in Chapter 26. The material in these chapters is often called *geometric optics*. The ray approximation is very good for the study of mirrors, lenses, prisms, and associated optical instruments, such as telescopes, cameras, and eyeglasses.

25.3 | Analysis Model: Wave Under Reflection

In Chapter 13, we introduced a one-dimensional version of the model of a wave under reflection by considering waves on strings. When such a wave meets a discontinuity between strings representing different wave speeds, some of the energy is reflected and some of the energy is transmitted. In that discussion, the waves are constrained to move along the one-dimensional string. In this discussion of optics, we are not subject to that restriction. Light waves can move in three dimensions.

Figure 25.3 shows several rays of light incident on a surface. Unless the surface is perfectly absorbing, some portion of the light is reflected from the surface. (The transmitted portion will be discussed in Section 25.4.) If the surface is very smooth, the reflected rays are parallel as indicated in Figure 25.3a. Reflection of light from such a smooth surface is called **specular reflection.** If the reflecting surface is rough as in Figure 25.3b, it reflects the rays in various directions. Reflection from a rough surface is known as **diffuse reflection.** A surface behaves as a smooth surface as long as the surface variations are small compared with the wavelength of the incident light. For example, light passes through the small holes in a microwave oven door, allowing you to see the interior because the holes are large relative to the wavelengths of visible light. The large-wavelength microwaves, however, reflect from the door as if it were a solid piece of metal.

Figures 25.3c and 25.3d are photographs of specular reflection and diffuse reflection using laser light, made visible by dust in the air, which scatters the light

c, d, Courtesy of Henry Leap and Jim Lehman

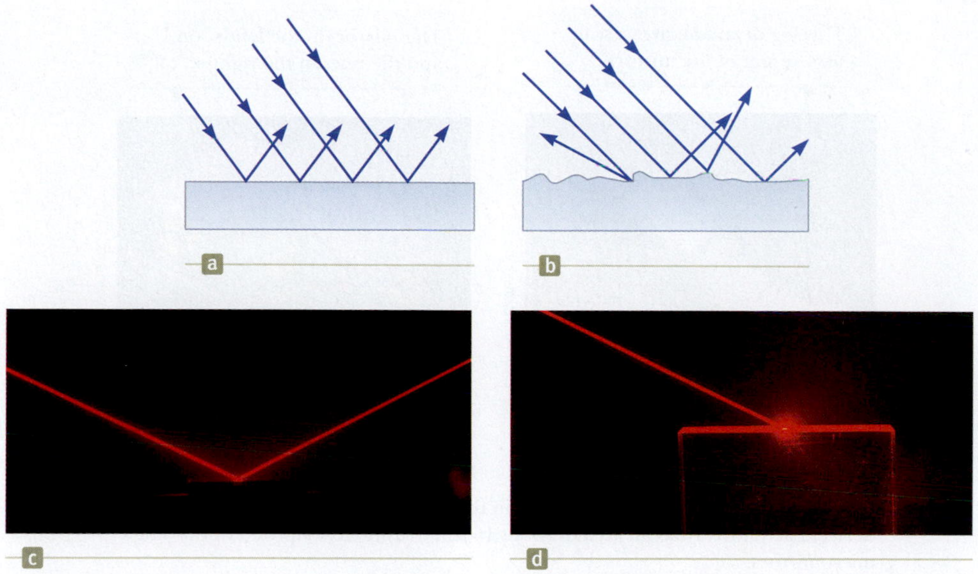

Figure 25.3 Schematic representation of (a) specular reflection, in which the reflected rays are all parallel, and (b) diffuse reflection, in which the reflected rays travel in scattered directions. (c) and (d) Photographs of specular and diffuse reflection using laser light.

toward the camera. The reflected laser beam is clearly visible in Figure 25.3c. In Figure 25.3d, the diffuse reflection has caused the incident beam to be reflected in many directions so that no clear outgoing beam is visible.

Specular reflection is necessary for the formation of clear images from reflecting surfaces, a topic we shall investigate in Chapter 26. Figure 25.4 shows an image resulting from specular reflection from a smooth water surface. If the water surface were rough, diffuse reflection would occur and the reflected image would not be visible.

Both types of reflection can occur from a road surface that you observe when you drive at night. On a dry night, light from oncoming vehicles is scattered off the road in different directions (diffuse reflection) and the road is quite visible. On a rainy night, the small irregularities in the road surface are filled with water. Because the water surface is smooth, the light undergoes specular reflection and the glare from reflected light makes the road less visible.

Let us now develop the mathematical representation for the waves undergoing reflection. Consider a light ray that travels in air and is incident at an angle on a flat, smooth surface as in Active Figure 25.5 (page 736). The incident and reflected rays make angles of θ_1 and θ_1', respectively, with a line drawn normal to the surface at the point where the incident ray strikes the surface. Experiments show that the incident ray, the normal to the surface, and the reflected ray all lie in the same plane and that **the angle of reflection equals the angle of incidence:**

$$\theta_1' = \theta_1 \qquad\qquad \textbf{25.1} \blacktriangleleft$$

Equation 25.1 is called the **law of reflection.** By convention, the angles of incidence and reflection are measured from the normal to the surface rather than from the surface itself. Because reflection of waves from an interface between two media is a common phenomenon, we identify an analysis model for this situation: the **wave under reflection.** Equation 25.1 is the mathematical representation of this model.

In diffuse reflection, the law of reflection is obeyed *with respect to the local normal.* Because of the roughness of the surface, the local normal varies significantly from one location to another. In this book, we shall concern ourselves only with specular reflection and shall use the term *reflection* to mean specular reflection.

As you might guess from Equation 25.1 and the figures we have seen so far, geometric models are used extensively in the study of optics. As we represent physical situations with

topora /Shutterstock.com

Figure 25.4 Waterfront houses in Normandy, France are reflected in the water of Honfleur Harbor. Because the water is so calm, the reflection is specular.

The incident ray, the reflected ray, and the normal all lie in the same plane, and $\theta_1' = \theta_1$.

Normal

Incident ray

Reflected ray

θ_1 θ_1'

Active Figure 25.5 The wave under reflection model.

This leg of an ant gives a scale for the size of the mirrors.

The mirror on the left is "on," and the one on the right is "off."

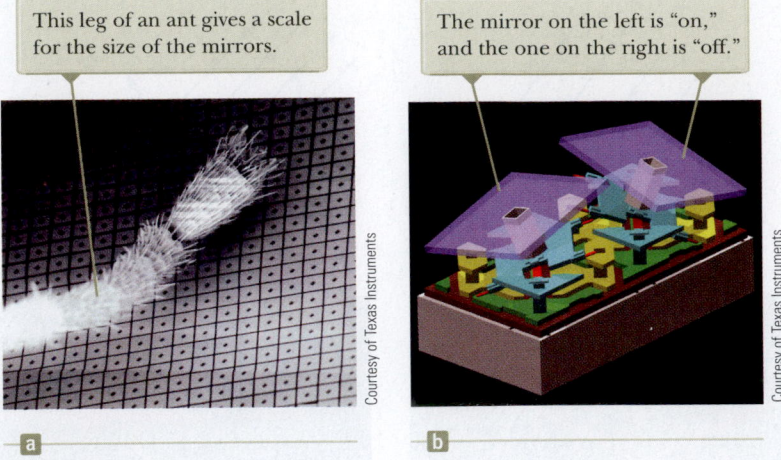

Courtesy of Texas Instruments

a

Courtesy of Texas Instruments

b

Figure 25.6 (a) An array of mirrors on the surface of a digital micromirror device. Each mirror has an area of about 16 μm^2. (b) A close-up view of two single micromirrors.

geometric constructions, the mathematics of triangles and the principles of trigonometry will find many applications.

The path of a light ray is reversible. For example, the ray in Active Figure 25.5 travels from the upper left, reflects from the mirror, and then moves toward a point at the upper right. If the ray originated at the same point at the upper right, it would follow the same path in reverse to reach the same point at the upper left. This reversible property will be useful when we set up geometric constructions for finding the paths of light rays.

A practical application of the law of reflection is the digital projection of movies, television shows, and computer presentations. A digital projector makes use of an optical semiconductor chip called a *digital micromirror device*. This device contains an array of more than one million tiny mirrors (Fig. 25.6a) that can be individually tilted by means of signals to an address electrode underneath the edge of the mirror. Each mirror corresponds to a pixel in the projected image. When the pixel corresponding to a given mirror is to be bright, the mirror is in the "on" position and is oriented so as to reflect light from a source illuminating the array to the screen (Fig. 25.6b). When the pixel for this mirror is to be dark, the mirror is "off" and is tilted so that the light is reflected away from the screen. The brightness of the pixel is determined by the total time interval during which the mirror is in the "on" position during the display of one image.

Digital movie projectors use three micromirror devices, one for each of the primary colors red, blue, and green, so that movies can be displayed with up to 35 trillion colors. Because there is no physical storage mechanism for the movie, a digital movie does not degrade with time as does film. Furthermore, because the movie is entirely in the form of computer software, it can be delivered to theaters by means of satellites, optical discs, or optical fiber networks.

QUICK QUIZ 25.1 In the movies, you sometimes see an actor looking in a mirror and you can see his face in the mirror. It can be said with certainty that during the filming of such a scene, the actor sees in the mirror: **(a)** his face **(b)** your face **(c)** the director's face **(d)** the movie camera **(e)** impossible to determine

---

THINKING PHYSICS 25.1

When looking through a glass window to the outdoors at night, you sometimes see a *double* image of yourself. Why?

Reasoning Reflection occurs whenever light encounters an interface between two optical media.

For the glass in the window, two such interfaces exist. The first is the inner surface of the glass and the second is the outer surface. Each interface results in an image. ◀

Example 25.1 | The Double-Reflected Light Ray

Two mirrors make an angle of 120° with each other as illustrated in Figure 25.7. A ray is incident on mirror M_1 at an angle of 65° to the normal. Find the direction of the ray after it is reflected from mirror M_2.

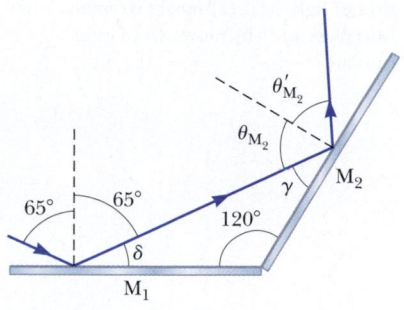

SOLUTION

Conceptualize Figure 25.7 helps conceptualize this situation. The incoming ray reflects from the first mirror, and the reflected ray is directed toward the second mirror. Therefore, there is a second reflection from the second mirror.

Figure 25.7 (Example 25.1) Mirrors M_1 and M_2 make an angle of 120° with each other.

Categorize Because the interactions with both mirrors are simple reflections, we apply the wave under reflection model and some geometry.

Analyze From the law of reflection, the first reflected ray makes an angle of 65° with the normal.

Find the angle the first reflected ray makes with the horizontal:

$$\delta = 90° - 65° = 25°$$

From the triangle made by the first reflected ray and the two mirrors, find the angle the reflected ray makes with M_2:

$$\gamma = 180° - 25° - 120° = 35°$$

Find the angle the first reflected ray makes with the normal to M_2:

$$\theta_{M_2} = 90° - 35° = 55°$$

From the law of reflection, find the angle the second reflected ray makes with the normal to M_2:

$$\theta'_{M_2} = \theta_{M_2} = \boxed{55°}$$

Finalize Notice that this reflection problem, as well as other reflection problems, will involve significant use of principles associated with angles and geometric model triangles. Be sure to consult Appendix B to review some of these principles.

25.4 | Analysis Model: Wave Under Refraction

Referring again to our discussion of string waves in Chapter 13, we discussed that some of the energy of a wave incident on a discontinuity in the string is transmitted through the discontinuity. As a light wave moves through three dimensions, understanding the transmitted light wave involves new principles that we now discuss.

When a ray of light traveling through a transparent medium is obliquely incident on a boundary leading into another transparent medium as in Active Figure 25.8a, part of the ray is reflected but part is transmitted into the second medium. The ray that enters the second medium experiences a change in direction at the boundary

Active Figure 25.8 (a) The wave under refraction model. (b) Light incident on the Lucite block refracts both when it enters the block and when it leaves the block.

All rays and the normal lie in the same plane, and the refracted ray is bent toward the normal because $v_2 < v_1$.

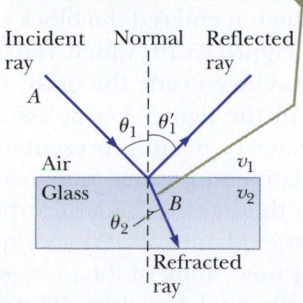

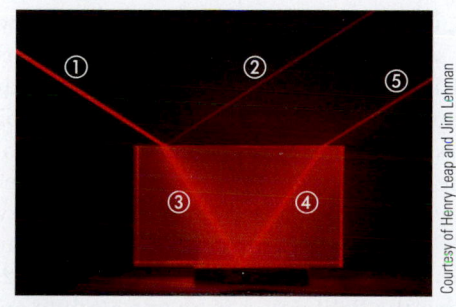

Courtesy of Henry Leap and Jim Lehman

a

b

Active Figure 25.9 The refraction of light as it (a) moves from air into glass and (b) moves from glass into air.

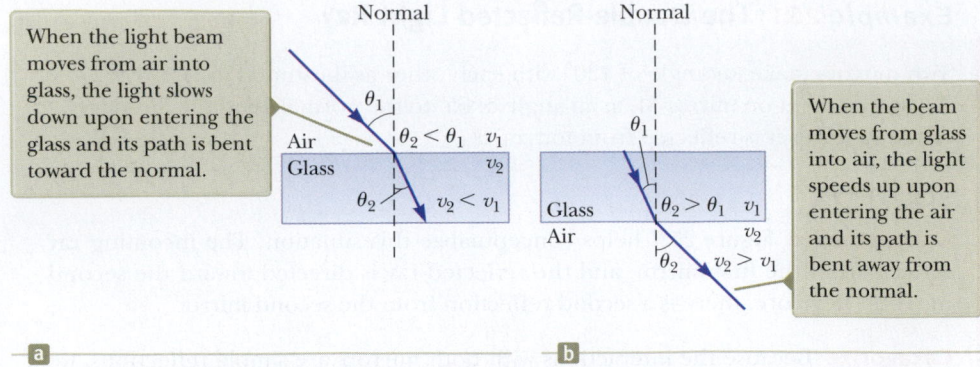

When the light beam moves from air into glass, the light slows down upon entering the glass and its path is bent toward the normal.

When the beam moves from glass into air, the light speeds up upon entering the air and its path is bent away from the normal.

and is said to undergo **refraction.** The incident ray, the reflected ray, and the refracted ray all lie in the same plane. The **angle of refraction** θ_2 in Active Figure 25.8a depends on the properties of the two media and on the angle of incidence through the relationship

$$\frac{\sin \theta_2}{\sin \theta_1} = \frac{v_2}{v_1}$$

25.2 ◄

where v_1 is the speed of light in medium 1 and v_2 is the speed of light in medium 2. Equation 25.2 is a mathematical representation of the wave under refraction model, although we find a more commonly used form in Equation 25.7.

The path of a light ray through a refracting surface is reversible, as was the case for reflection. For example, the ray in Active Figure 25.8a travels from point A to point B. If the ray originated at B, it would follow the same path in reverse to reach point A. In the latter case, however, the reflected ray would be in the glass.

▶ **QUICK QUIZ 25.2** If beam ① is the incoming beam in Active Figure 25.8b, which of the other four red lines are reflected beams and which are refracted beams?

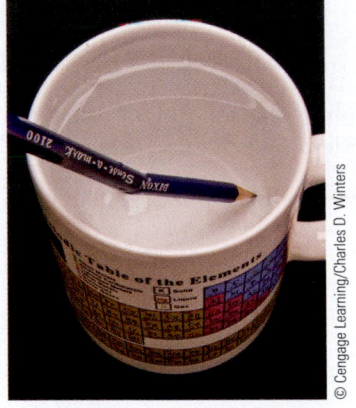

The pencil partially immersed in water appears bent because light from the lower part of the pencil is refracted as it travels across the boundary between water and air.

Equation 25.2 shows that when light moves from a material in which its speed is high to a material in which its speed is lower, the angle of refraction θ_2 is less than the angle of incidence. The refracted ray therefore deviates toward the normal as shown in Active Figure 25.9a. If the ray moves from a material in which it travels slowly to a material in which it travels more rapidly, θ_2 is greater than θ_1, so the ray deviates away from the normal as shown in Active Figure 25.9b.

The behavior of light as it passes from air into another substance and then reemerges into air is often a source of confusion to students. Why is this behavior so different from other occurrences in our daily lives? When light travels in air, its speed is $c = 3.0 \times 10^8$ m/s; on entry into a block of glass, its speed is reduced to approximately 2.0×10^8 m/s. When the light re-emerges into air, its speed increases to its original value 3.0×10^8 m/s. This process is very different from what happens, for example, when a bullet is fired through a block of wood. In that case, the speed of the bullet is reduced as it moves through the wood because some of its original energy is used to tear apart the fibers of the wood. When the bullet enters the air again, it emerges at a speed lower than that with which it entered the block of wood.

To see why light behaves as it does, consider Figure 25.10, which represents a ray of light entering a piece of glass from the left. Once inside the glass, the light may encounter an atom, represented by point A in the figure. Let us assume that light is absorbed by the atom, causing it to oscillate (a detail represented by the double-headed arrows in the drawing). The oscillating atom then radiates (emits) the ray of light toward an atom at point B, where the light is again absorbed. The details of these absorptions and emissions are best explained in terms of quantum physics, a subject we shall study in Chapter 28. For now, think of the process as one in which the light passes from one atom to another through the glass. (The situation is somewhat analogous to a relay race in which a baton is passed between runners on

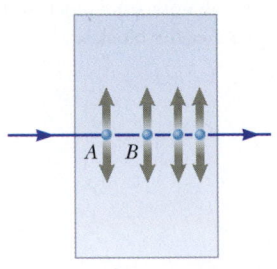

Figure 25.10 Light passing from one atom to another in a medium. The blue spheres are atoms, and the vertical arrows represent their oscillations.

TABLE 25.1 | Indices of Refraction for Various Substances

Substance	Index of Refraction	Substance	Index of Refraction
Solids at 20°C		**Liquids at 20°C**	
Cubic zirconia	2.20	Benzene	1.501
Diamond (C)	2.419	Carbon disulfide	1.628
Fluorite (CaF$_2$)	1.434	Carbon tetrachloride	1.461
Fused quartz (SiO$_2$)	1.458	Corn syrup	2.21
Gallium phosphide	3.50	Ethyl alcohol	1.361
Glass, crown	1.52	Glycerin	1.473
Glass, flint	1.66	Water	1.333
Ice (H$_2$O)	1.309	**Gases at 0°C, 1 atm**	
Polystyrene	1.49	Air	1.000 293
Sodium chloride (NaCl)	1.544	Carbon dioxide	1.000 45

Note: All values are for light having a wavelength of 589 nm in vacuum.

the same team.) Although light travels from one atom to another through the empty space between the atoms with a speed of $c = 3.0 \times 10^8$ m/s, the absorptions and emissions of light by the atoms require time to occur. Therefore, the *average* speed of light through the glass is lower than c. Once the light emerges into the air, the absorptions and emissions cease and the light's average speed returns to its original value.[1] Therefore, whether the light is inside the material or outside, it always travels through vacuum with the same speed.

Light passing from one medium to another is refracted because the average speed of light is different in the two media. In fact, *light travels at its maximum speed in vacuum*. It is convenient to define the **index of refraction** n of a medium to be the ratio

$$n \equiv \frac{\text{speed of light in vacuum}}{\text{speed of light in a medium}} = \frac{c}{v} \qquad \textbf{25.3} \blacktriangleleft$$

▶ Index of refraction

From this definition, we see that the index of refraction is a dimensionless number greater than or equal to unity because v in a medium is less than c. Furthermore, n is equal to unity for vacuum. The indices of refraction for various substances are listed in Table 25.1.

As a wave travels from one medium to another, its frequency does not change. Let us first consider this notion for waves passing from a light string to a heavier string. If the frequencies of the incident and transmitted waves on the two strings at the junction point were different, the strings could not remain tied together because the joined ends of the two pieces of string would not move up and down in unison!

For a light wave passing from one medium to another, the frequency also remains constant. To see why, consider Figure 25.11. Wave fronts pass an observer at point A in medium 1 with a certain frequency and are incident on the boundary between medium 1 and medium 2. The frequency at which the wave fronts pass an observer at point B in medium 2 must equal the frequency at which they arrive at point A. If that were not the case, the wave fronts would either pile up at the boundary or be destroyed or created at the boundary. Because this situation does not occur, the frequency must be a constant as a light ray passes from one medium into another.

Therefore, because the relation $v = \lambda f$ (Eq. 13.12) must be valid in both media and because $f_1 = f_2 = f$, we see that

$$v_1 = \lambda_1 f \qquad \text{and} \qquad v_2 = \lambda_2 f$$

Pitfall Prevention | 25.2
n Is Not an Integer Here
We have seen n used in Chapter 11 to indicate the quantum number of a Bohr orbit and in Chapter 14 to indicate the standing wave mode on a string or in an air column. In those cases, n was an integer. The index of refraction n is *not* an integer.

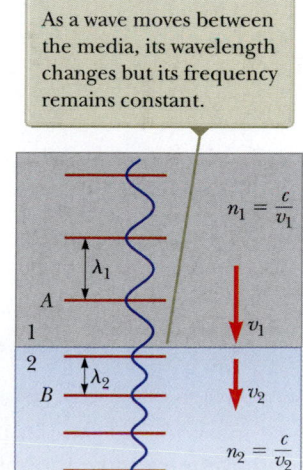

As a wave moves between the media, its wavelength changes but its frequency remains constant.

Figure 25.11 A wave travels from medium 1 to medium 2, in which it moves with lower speed.

---

[1] As an analogy, consider a subway entering a city at a constant speed v and then stopping at several stations in the downtown region of the city. Although the subway may achieve the instantaneous speed v between stations, the *average* speed across the city is less than v. Once the subway leaves the city and makes no stops, it moves again at a constant speed v. The analogy, as is the case with many analogies, is not perfect because the subway requires time to accelerate to the speed v between stations, whereas light achieves speed c immediately as it travels between atoms.

Because $v_1 \neq v_2$, it follows that $\lambda_1 \neq \lambda_2$. A relationship between index of refraction and wavelength can be obtained by dividing these two equations and making use of the definition of the index of refraction given by Equation 25.3:

$$\frac{\lambda_1}{\lambda_2} = \frac{v_1}{v_2} = \frac{c/n_1}{c/n_2} = \frac{n_2}{n_1} \qquad \blacktriangleleft \; 25.4$$

which gives

$$\lambda_1 n_1 = \lambda_2 n_2 \qquad \blacktriangleleft \; 25.5$$

It follows from Equation 25.5 that the index of refraction of any medium can be expressed as the ratio

$$n = \frac{\lambda}{\lambda_n} \qquad \blacktriangleleft \; 25.6$$

where λ is the wavelength of light in vacuum and λ_n is the wavelength in the medium whose index of refraction is n.

We are now in a position to express Equation 25.2 in an alternative form. If we combine Equation 25.3 and Equation 25.2, we find that

$$n_1 \sin \theta_1 = n_2 \sin \theta_2 \qquad \blacktriangleleft \; 25.7$$

▶ Snell's law of refraction

The experimental discovery of this relationship is usually credited to Willebrord Snell (1591–1626), and it is therefore known as **Snell's law of refraction.**[2] Refraction of waves at an interface between two media is a common phenomenon, so we identify an analysis model for this situation: the **wave under refraction.** Equation 25.7 is the mathematical representation of this model for electromagnetic radiation. Other waves, such as seismic waves and sound waves, also exhibit refraction according to this model, and the mathematical representation for these waves is Equation 25.2.

> **Pitfall Prevention | 25.3**
> **An Inverse Relationship**
> The index of refraction is *inversely* proportional to the wave speed. As the wave speed v decreases, the index of refraction n increases. Therefore, the higher the index of refraction of a material, the more it *slows down* light from its speed in vacuum. The more the light slows down, the more θ_2 differs from θ_1 in Equation 25.7.

QUICK QUIZ 25.3 Light passes from a material with index of refraction 1.3 into one with index of refraction 1.2. Compared with the incident ray, what happens to the refracted ray? (**a**) It bends toward the normal. (**b**) It is undeflected. (**c**) It bends away from the normal.

QUICK QUIZ 25.4 As light from the Sun enters the atmosphere, it refracts due to the small difference between the speeds of light in air and in vacuum. The *optical* length of the day is defined as the time interval between the instant when the top of the Sun is just visibly observed above the horizon to the instant at which the top of the Sun just disappears below the horizon. The *geometric* length of the day is defined as the time interval between the instant when a geometric straight line drawn from the observer to the top of the Sun just clears the horizon to the instant at which this line just dips below the horizon. Which is longer, (**a**) the optical length of a day or (**b**) the geometric length of a day?

> ### THINKING PHYSICS 25.2
> **BIO** Underwater vision
>
> Why do face masks make vision clearer under water? A face mask includes a flat piece of glass; the mask does not have lenses like those in eyeglasses.
>
> **Reasoning** The refraction necessary for focused viewing in the eye occurs at the air–cornea interface. The lens of the eye only performs some fine-tuning of this image, allowing for accommodation for objects at various distances. When the eye is opened underwater, the interface is water–cornea rather than air–cornea. Therefore, the light from the scene is not focused on the retina and the scene is blurry. The face mask simply provides a layer of air in front of the eyes so that the air–cornea interface is re-established and the refraction is correct to focus the light on the retina. ◀

---

[2]The same law was deduced from the particle theory of light in 1637 by René Descartes (1596–1650) and hence is known as *Descartes's law* in France.

Example 25.2 | Angle of Refraction for Glass

A light ray of wavelength 589 nm traveling through air is incident on a smooth, flat slab of crown glass at an angle of 30.0° to the normal.

(A) Find the angle of refraction.

SOLUTION

Conceptualize Study Active Figure 25.9a, which illustrates the refraction process occurring in this problem.

Categorize We determine results by using equations developed in this section, so we categorize this example as a substitution problem.

Rearrange Snell's law of refraction to find $\sin \theta_2$:

$$\sin \theta_2 = \frac{n_1}{n_2} \sin \theta_1$$

Solve for θ_2:

$$\theta_2 = \sin^{-1} \left(\frac{n_1}{n_2} \sin \theta_1 \right)$$

Substitute indices of refraction from Table 25.1 and the incident angle:

$$\theta_2 = \sin^{-1} \left(\frac{1.00}{1.52} \sin 30.0° \right) = \boxed{19.2°}$$

(B) Find the speed of this light once it enters the glass.

SOLUTION

Solve Equation 25.3 for the speed of light in the glass:

$$v = \frac{c}{n}$$

Substitute numerical values:

$$v = \frac{3.00 \times 10^8 \text{ m/s}}{1.52} = \boxed{1.97 \times 10^8 \text{ m/s}}$$

(C) What is the wavelength of this light in the glass?

SOLUTION

Use Equation 25.6 to find the wavelength in the glass:

$$\lambda_n = \frac{\lambda}{n} = \frac{589 \text{ nm}}{1.52} = \boxed{388 \text{ nm}}$$

Example 25.3 | Light Passing Through a Slab

A light beam passes from medium 1 to medium 2, with the latter medium being a thick slab of material whose index of refraction is n_2 (Fig. 25.12). Show that the beam emerging into medium 1 from the other side is parallel to the incident beam.

SOLUTION

Conceptualize Follow the path of the light beam as it enters and exits the slab of material in Figure 25.12, where we have assumed that $n_2 > n_1$. The ray bends toward the normal upon entering and away from the normal upon leaving.

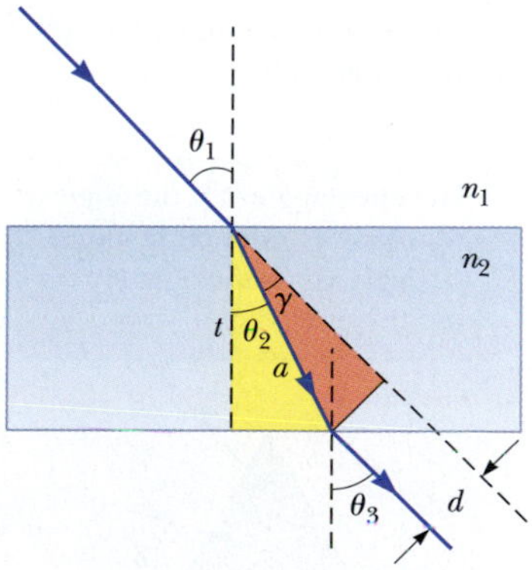

Figure 25.12 (Example 25.3) The dashed line drawn parallel to the ray coming out the bottom of the slab represents the path the light would take were the slab not there.

Categorize We determine results by using equations developed in this section, so we categorize this example as a substitution problem.

Apply Snell's law of refraction to the upper surface:

$$(1) \quad \sin \theta_2 = \frac{n_1}{n_2} \sin \theta_1$$

Apply Snell's law to the lower surface:

$$(2) \quad \sin \theta_3 = \frac{n_2}{n_1} \sin \theta_2$$

continued

25.3 *cont.*

Substitute Equation (1) into Equation (2):
$$\sin \theta_3 = \frac{n_2}{n_1}\left(\frac{n_1}{n_2}\sin \theta_1\right) = \sin \theta_1$$

Therefore, $\theta_3 = \theta_1$ and the slab does not alter the direction of the beam. It does, however, offset the beam parallel to itself by the distance d shown in Figure 25.12.

What If? What if the thickness t of the slab is doubled? Does the offset distance d also double?

Answer Consider the region of the light path within the slab in Figure 25.12. The distance a is the hypotenuse of two right triangles.

Find an expression for a from the yellow triangle:
$$a = \frac{t}{\cos \theta_2}$$

Find an expression for d from the red triangle:
$$d = a \sin \gamma = a \sin (\theta_1 - \theta_2)$$

Combine these equations:
$$d = \frac{t}{\cos \theta_2} \sin (\theta_1 - \theta_2)$$

For a given incident angle θ_1, the refracted angle θ_2 is determined solely by the index of refraction, so the offset distance d is proportional to t. If the thickness doubles, so does the offset distance.

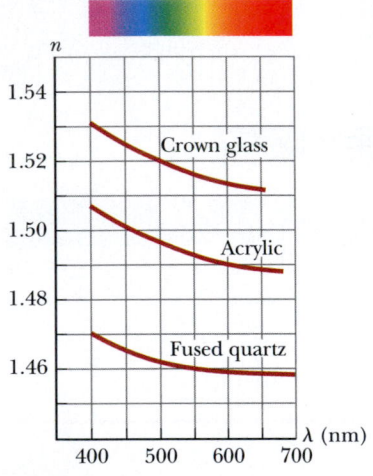

Figure 25.13 Variation of index of refraction with vacuum wavelength for three materials.

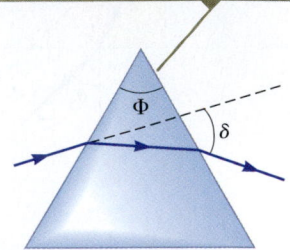

The apex angle Φ is the angle between the sides of the prism through which the light enters and leaves.

Figure 25.14 A prism refracts single-wavelength light and deviates the light through an angle δ.

25.5 | Dispersion and Prisms

In the preceding section, we developed Snell's law, which incorporates the index of refraction of a material. In Table 25.1, we presented index of refraction values for a number of materials. If we make careful measurements, however, we find that the value of the index of refraction in anything but vacuum depends on the wavelength of light. The dependence of the index of refraction on wavelength, which results from the dependence of the wave speed on wavelength, is called **dispersion.** Figure 25.13 is a graphical representation of this variation in index of refraction with wavelength. Because n is a function of wavelength, Snell's law indicates that the angle of refraction when light enters a material depends on the wavelength of the light. As we see from Figure 25.13, the index of refraction for a material generally decreases with increasing wavelength in the visible range. Therefore, violet light ($\lambda \approx 400$ nm) refracts more than red light ($\lambda \approx 650$ nm) when passing from air into a material.

To understand the effects of dispersion on light, consider what happens when a ray of light strikes a prism as in Figure 25.14. The apex angle Φ of the prism is defined as shown in the figure. A ray of light of a single wavelength that is incident on the prism from the left emerges in a direction deviated from its original direction of travel by an angle of deviation δ that depends on the apex angle and the index of refraction of the prism material. Now suppose a beam of white light (a combination of all visible wavelengths) is incident on a prism. Because of dispersion, the different colors refract through different angles of deviation, and the rays that emerge from the second face of the prism spread out in a series of colors known as a **visible spectrum** as shown in Figure 25.15. These colors, in order of decreasing wavelength, are red, orange, yellow, green, blue, and violet.[3] Violet light deviates the most, red light deviates the least, and the remaining colors in the visible spectrum fall between these extremes.

The dispersion of light into a spectrum is demonstrated most vividly in nature through the formation of a rainbow, often seen by an observer positioned between the Sun and a rain shower. To understand how a rainbow is formed, consider Active

---

[3]In Newton's time, the colors we now call teal and blue were called blue and indigo. Your "blue jeans" are dyed with indigo. A mnemonic device for remembering the colors of the spectrum is the acronym ROYGBIV, from the first letters of the colors: red, orange, yellow, green, blue, indigo, violet. Some individuals think of this acronym as the name of a person, Roy G. Biv!

The colors in the refracted beam are separated because dispersion in the prism causes different wavelengths of light to be refracted through different angles.

David Parker/Science Photo Library/Photo Researchers, Inc.

Figure 25.15 White light enters a glass prism at the upper left. A reflected beam of light comes out of the prism just below the incoming beam. The beam moving toward the lower right shows distinct colors. Violet light deviates the most; red light deviates the least.

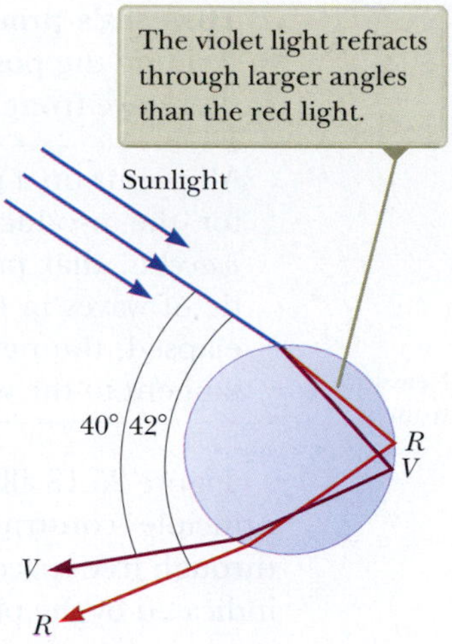

The violet light refracts through larger angles than the red light.

Sunlight

40° 42°

R
V

V

R

Active Figure 25.16 Path of sunlight through a spherical raindrop. Light following this path contributes to the visible rainbow.

Figure 25.16. A ray of light passing overhead strikes a spherical drop of water in the atmosphere and is refracted and reflected as follows. It is first refracted at the front surface of the drop, with the violet light deviating the most and the red light the least. At the back surface of the drop, the light is reflected and returns to the front surface, where it again undergoes refraction as it moves from water into air.

Because light enters the front surface of the raindrop at all locations, there is a range of exit angles for the light leaving the raindrop after reflecting from the back surface. A careful analysis of the spherical shape of the water drop, however, shows that the exit angle of highest light intensity is 42° for the red light and 40° for the violet light. Therefore, the light from the raindrop seen by the observer is brightest for these angles, and the observer sees a rainbow. Figure 25.17 shows the geometry for the observer. The colors of the rainbow are seen in a range of 40° to 42° from the antisolar direction, which is exactly 180° from the Sun. If red light is seen coming from a raindrop high in the sky, the violet light from this drop passes over the observer's head and is not seen. Therefore, the portion of the rainbow in the vicinity of this drop is red. The violet portion of the rainbow seen by an observer is supplied by drops lower in the sky, which send violet light to the observer's eyes and red light below the eyes.

The opening photograph for this chapter shows a *double rainbow.* The secondary rainbow is fainter than the primary rainbow, and its colors are reversed. The secondary rainbow arises from light that makes two reflections from the interior surface before exiting the raindrop. In the laboratory, rainbows have been observed in which the light makes more than 30 reflections before exiting the water drop. Because each reflection involves some loss of light due to refraction out of the water drop, the intensity of these higher-order rainbows is very small.

QUICK QUIZ 25.5 In dispersive materials, the angle of refraction for a light ray depends on the wavelength of the light. True or false: The angle of reflection from the surface of the material depends on the wavelength.

The highest intensity light traveling from higher raindrops toward the eyes of the observer is red, whereas the most intense light from lower drops is violet.

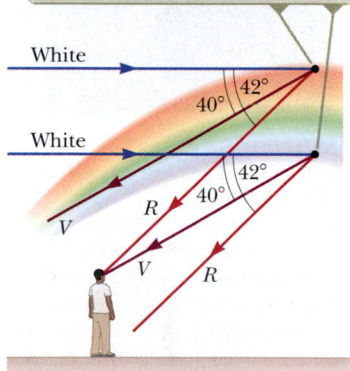

White

40° 42°

White

R   40° 42°

V

V   R

Figure 25.17 The formation of a rainbow seen by an observer standing with the Sun behind his back.

25.6 | Huygens's Principle

In this section, we introduce a geometric construction proposed by Huygens in 1678. Huygens assumed that light consists of waves rather than a stream of particles. He had no knowledge of the electromagnetic character of light. Nevertheless, his geometric model is adequate for understanding many practical aspects of the propagation of light.

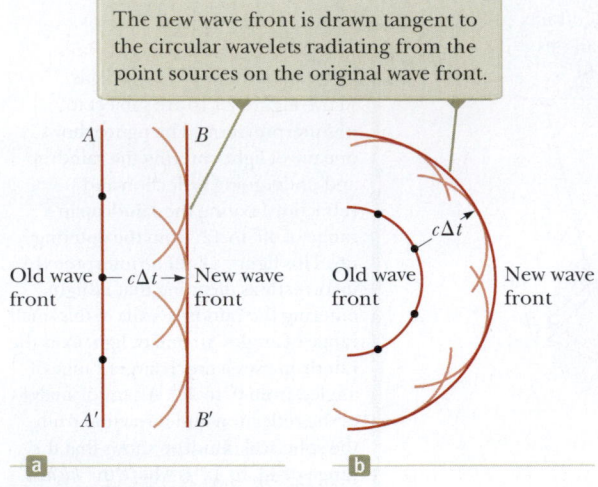

The new wave front is drawn tangent to the circular wavelets radiating from the point sources on the original wave front.

Figure 25.18 Huygens's construction for (a) a plane wave propagating to the right and (b) a spherical wave propagating to the right.

Christian Huygens
Dutch Physicist and Astronomer
(1629–1695)
Huygens is best known for his contributions to the fields of optics and dynamics. To Huygens, light was a type of vibratory motion, spreading out and producing the sensation of light when impinging on the eye. On the basis of this theory, he deduced the laws of reflection and refraction and explained the phenomenon of double refraction.

Huygens's principle is a geometric model that allows us to determine the position of a wave front from a knowledge of an earlier wave front:

> All points on a given wave front are taken as point sources for the production of spherical secondary waves, called *wavelets,* that propagate outward with speeds characteristic of waves in that medium. After some time interval has elapsed, the new position of the wave front is the surface tangent to the wavelets.

Figure 25.18 illustrates two simple examples of a Huygens's principle construction. First, consider a plane wave moving through free space as in Figure 25.18a. At $t = 0$, the wave front is indicated by the plane labeled AA'. Each point on this wave front is a point source for a wavelet. Showing three of these points, we draw arcs of circles, each of radius $c \, \Delta t$, where c is the speed of light in free space and Δt is the time interval during which the wave propagates. The surface drawn tangent to the wavelets is the plane BB', which is parallel to AA'. This plane is the wave front at the end of the time interval Δt. In a similar manner, Figure 25.18b shows Huygens's construction for an outgoing spherical wave.

A convincing demonstration of the existence of Huygens wavelets is obtained with water waves in a shallow tank (called a ripple tank) as in Figure 25.19. Plane waves produced to the left of the slits emerge to the right of the slits as two-dimensional circular waves propagating outward. In the plane wave, each point on the wave front acts as a source of circular waves on the two-dimensional water surface. At a later time, the tangent of the circular wave fronts remains a straight line. As the wave front encounters the barrier, however, waves at all points on the wave front, except those that encounter the openings, are reflected. For very small openings, we can model this situation as if only one source of Huygens wavelets exists at each of the two openings. As a result, the Huygens wavelets from those single sources are seen as the outgoing circular waves in the right portion of Figure 25.19. This is a dramatic example of diffraction that was mentioned in the opening section of this chapter, a phenomenon we shall study in more detail in Chapter 27.

Huygens's Principle Applied to Reflection and Refraction

We now derive the laws of reflection and refraction, using Huygens's principle.

For the law of reflection, refer to Figure 25.20. The line AB represents a plane wave front of the incident light just as ray 1 strikes the surface. At this instant, the wave at A sends out a Huygens wavelet (appearing at a later time as the light brown circular arc passing through D); the reflected light makes an angle γ' with the surface. At the same time, the wave at B emits a Huygens wavelet (the light brown circular arc passing through C) with the incident light making an angle γ with the surface. Figure 25.20 shows these wavelets after a time interval Δt, after which ray 2 strikes the surface. Because both rays 1 and 2 move with the same speed, we must have $AD = BC = c \, \Delta t$.

The remainder of our analysis depends on geometry. Notice that the two triangles ABC and ADC are congruent because they have the same hypotenuse AC and because $AD = BC$. Figure 25.20 shows that

$$\cos \gamma = \frac{BC}{AC} \quad \text{and} \quad \cos \gamma' = \frac{AD}{AC}$$

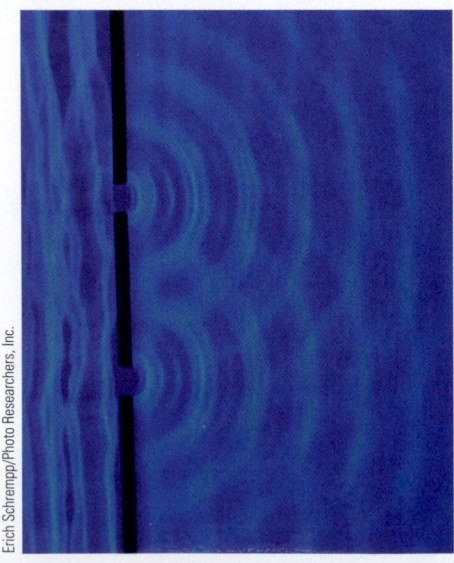

Figure 25.19 Water waves in a ripple tank demonstrate Huygens wavelets. A plane wave is incident on a barrier with two small openings. The openings act as sources of circular wavelets.

where $\gamma = 90° - \theta_1$ and $\gamma' = 90° - \theta_1'$. Because $AD = BC$,

$$\cos \gamma = \cos \gamma'$$

Therefore,

$$\gamma = \gamma'$$

$$90° - \theta_1 = 90° - \theta_1'$$

and

$$\theta_1 = \theta_1'$$

which is the law of reflection.

Now let's use Huygens's principle to derive Snell's law of refraction. We focus our attention on the instant ray 1 strikes the surface and the subsequent time interval until ray 2 strikes the surface as in Figure 25.21. During this time interval, the wave at A sends out a Huygens wavelet (the light brown arc passing through D) and the light refracts into the material, making an angle θ_2 with the normal to the surface. In the same time interval, the wave at B sends out a Huygens wavelet (the light brown arc passing through C) and the light continues to propagate in the same direction. Because these two wavelets travel through different media, the radii of the wavelets are different. The radius of the wavelet from A is $AD = v_2 \Delta t$, where v_2 is the wave speed in the second medium. The radius of the wavelet from B is $BC = v_1 \Delta t$, where v_1 is the wave speed in the original medium.

From triangles ABC and ADC, we find that

$$\sin \theta_1 = \frac{BC}{AC} = \frac{v_1 \Delta t}{AC} \quad \text{and} \quad \sin \theta_2 = \frac{AD}{AC} = \frac{v_2 \Delta t}{AC}$$

Dividing the first equation by the second gives

$$\frac{\sin \theta_1}{\sin \theta_2} = \frac{v_1}{v_2}$$

From Equation 25.3, however, we know that $v_1 = c/n_1$ and $v_2 = c/n_2$. Therefore,

$$\frac{\sin \theta_1}{\sin \theta_2} = \frac{c/n_1}{c/n_2} = \frac{n_2}{n_1}$$

and

$$n_1 \sin \theta_1 = n_2 \sin \theta_2$$

which is Snell's law of refraction.

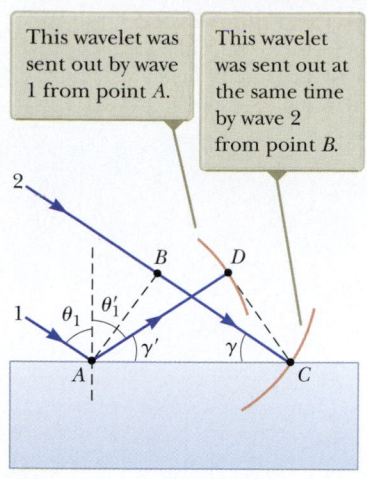

Figure 25.20 Huygens's construction for proving the law of reflection.

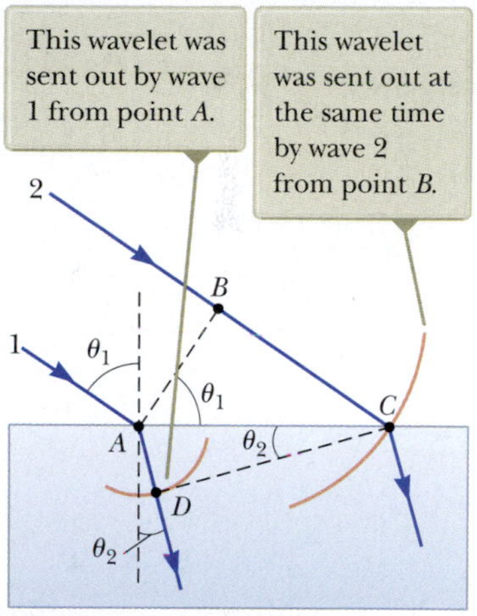

Figure 25.21 Huygens's construction for proving Snell's law of refraction.

25.7 | Total Internal Reflection

An interesting effect called **total internal reflection** can occur when light travels from a medium with a high index of refraction to one with a lower index of refraction. Consider a light ray traveling in medium 1 and meeting the boundary between media 1 and 2, where $n_1 > n_2$ (Active Fig. 25.22a on page 746). Various possible directions of the ray are indicated by rays 1 through 5. The refracted rays are bent away from the normal because $n_1 > n_2$. (Remember that when light refracts at the interface between the two media, it is also partially reflected. These rays are also shown in Active Figure 25.22a.) At some particular angle of incidence θ_c, called the **critical angle,** the refracted light ray moves parallel to the boundary so that $\theta_2 = 90°$ (ray 4 in Active Fig. 25.22a, shown by itself in Active Fig. 25.22b). For angles of incidence greater than θ_c, no ray is refracted and the incident ray is entirely reflected at the boundary, as is ray 5 in Active Figure 25.22a. This ray is reflected at the boundary as though it had struck a perfectly reflecting surface. It obeys the law of reflection; that is, the angle of incidence equals the angle of reflection.

Active Figure 25.22 (a) Rays travel from a medium of index of refraction n_1 into a medium of index of refraction n_2, where $n_1 > n_2$. (b) Ray 4 is singled out.

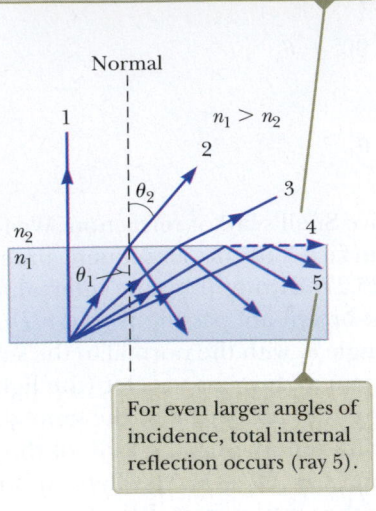

As the angle of incidence θ_1 increases, the angle of refraction θ_2 increases until θ_2 is 90° (ray 4). The dashed line indicates that no energy actually propagates in this direction.

For even larger angles of incidence, total internal reflection occurs (ray 5).

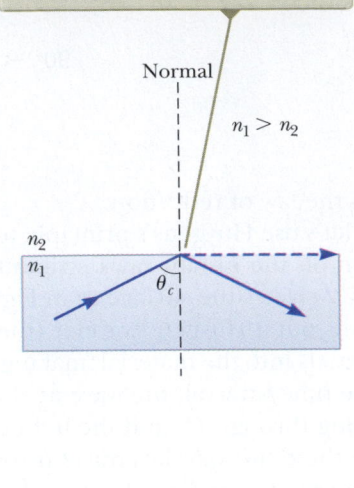

The angle of incidence producing an angle of refraction equal to 90° is the critical angle θ_c. At this angle of incidence, all the energy of the incident light is reflected.

We can use Snell's law to find the critical angle. When $\theta_1 = \theta_c$, $\theta_2 = 90°$, and Snell's law (Eq. 25.7) gives

$$n_1 \sin \theta_c = n_2 \sin 90° = n_2$$

$$\sin \theta_c = \frac{n_2}{n_1} \qquad \text{(for } n_1 > n_2\text{)} \qquad \qquad \textbf{25.8} \blacktriangleleft$$

This equation can be used only when n_1 is greater than n_2. That is, total internal reflection occurs only when light travels from a medium of high index of refraction to a medium of lower index of refraction. That is why the word *internal* is in the name. The light must initially be *inside* a material of higher index of refraction than the medium outside the material. If n_1 were less than n_2, Equation 25.8 would give $\sin \theta_c > 1$, which is meaningless because the sine of an angle can never be greater than unity.

The critical angle for total internal reflection is small when n_1 is considerably larger than n_2. Examples of this situation are diamond ($n = 2.42$ and $\theta_c = 24°$) and crown glass ($n = 1.52$ and $\theta_c = 41°$), where the angles given correspond to light refracting from the material into air. Total internal reflection combined with proper faceting causes diamonds and crystal glass to sparkle when observed in light.

> **QUICK QUIZ 25.6** In Figure 25.23, five light rays enter a glass prism from the left. (i) How many of these rays undergo total internal reflection at the slanted surface of the prism? (a) one (b) two (c) three (d) four (e) five (ii) Suppose the prism in Figure 25.23 can be rotated in the plane of the paper. For *all five* rays to experience total internal reflection from the slanted surface, should the prism be rotated (a) clockwise or (b) counterclockwise?

> **QUICK QUIZ 25.7** A beam of white light is incident on a crown glass–air interface as shown in Active Figure 25.22. The incoming beam is rotated clockwise, so the incident angle θ increases. Because of dispersion in the glass, some colors of light experience total internal reflection (ray 4 in Active Fig. 25.22a) before other colors, so the beam refracting out of the glass is no longer white. What is the last color to refract out of the upper surface? (a) violet (b) green (c) red (d) impossible to determine

Figure 25.23 (Quick Quiz 25.6) Five nonparallel rays of light enter a glass prism from the left.

Example 25.4 | **A View from the Fish's Eye**

Find the critical angle for an air–water boundary. (Assume the index of refraction of water is 1.33.)

SOLUTION

Conceptualize Study Active Figure 25.22 to understand the concept of total internal reflection and the significance of the critical angle.

Categorize We use concepts developed in this section, so we categorize this example as a substitution problem.

Apply Equation 25.8 to the air–water interface:

$$\sin \theta_c = \frac{n_2}{n_1} = \frac{1.00}{1.33} = 0.752$$

$$\theta_c = \boxed{48.8°}$$

What If? What if a fish in a still pond looks upward toward the water's surface at different angles relative to the surface as in Figure 25.24? What does it see?

Answer Because the path of a light ray is reversible, light traveling from medium 2 into medium 1 in Active Figure 25.22a follows the paths shown, but in the *opposite* direction. A fish looking upward toward the water surface as in Figure 25.24 can see out of the water if it looks toward the surface at an angle less than the critical angle. Therefore, when the fish's line of vision makes an angle of $\theta = 40°$ with the normal to the surface, for example, light from above the water reaches the fish's eye. At $\theta = 48.8°$, the critical angle for water, the light has to skim along the water's surface before being refracted to the fish's eye; at this angle, the fish can, in principle, see the entire shore of the pond. At angles greater than the critical angle, the light reaching the fish comes by means of total internal reflection at the surface. Therefore, at $\theta = 60°$, the fish sees a reflection of the bottom of the pond.

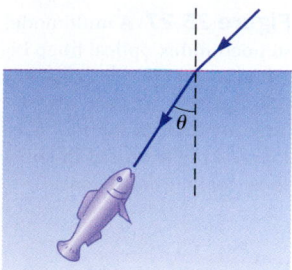

Figure 25.24 (Example 25.4) **What If?** A fish looks upward toward the water surface.

25.8 | Context Connection: Optical Fibers

An interesting application of total internal reflection is the use of glass or transparent plastic rods to "pipe" light from one place to another. In the communication industry, digital pulses of laser light move along these light pipes, carrying information at an extremely high rate. In this Context Connection, we investigate the physics of this technological advance.

As indicated in Figure 25.25, light is confined to traveling within a rod, even around curves, as the result of successive total internal reflections. Such a light pipe is flexible if thin fibers are used rather than thick rods. A flexible light pipe is called an **optical fiber.** If a bundle of parallel fibers is used to construct an optical transmission line, images can be transferred from one point to another. Part of the 2009 Nobel Prize in Physics was awarded to Charles K. Kao (b. 1933) for his discovery of how to transmit light signals over long distances through thin glass fibers. This discovery has led to the development of a sizable industry known as *fiber optics.*

A practical optical fiber consists of a transparent core surrounded by a *cladding,* a material that has a lower index of refraction than the core. The combination may be surrounded by a plastic *jacket* to prevent mechanical damage. Figure 25.26 shows a cutaway view of this construction. Because the index of refraction of the cladding is less than that of the core, light traveling in the core experiences total internal reflection if it arrives at the interface between the core and the cladding at an angle of incidence that exceeds the critical angle. In this case, light "bounces" along the core of the optical fiber, losing very little of its intensity as it travels. Any loss in intensity in an optical fiber is essentially due to reflections from the two ends and absorption by the fiber material.

Optical fiber devices are particularly useful for viewing an object at an inaccessible location. For example, physicians often use such devices to examine internal organs of the body or to perform surgery without making large incisions. Optical

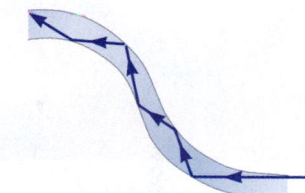

Figure 25.25 Light travels in a curved transparent rod by multiple internal reflections.

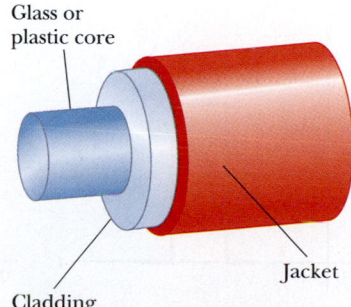

Glass or plastic core

Jacket

Cladding

Figure 25.26 The construction of an optical fiber. Light travels in the core, which is surrounded by a cladding and a protective jacket.

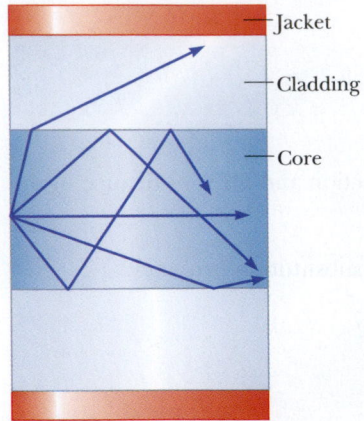

Figure 25.27 A multimode, stepped index optical fiber. Light rays entering over a wide range of angles pass through the core. Those making large angles with the axis take longer to travel the length of the fiber than those making small angles.

Strands of glass optical fibers are used to carry voice, video, and data signals in telecommunication networks. Typical fibers have diameters of 60 μm.

fiber cables are replacing copper wiring and coaxial cables for telecommunications because the fibers can carry a much greater volume of telephone calls or other forms of communication than electrical wires can.

Figure 25.27 shows a cross-sectional view from the side of a simple type of optical fiber known as a *multimode, stepped index fiber*. The term *stepped index* refers to the discontinuity in index of refraction between the core and the cladding, and *multimode* means that light entering the fiber at many angles is transmitted. This type of fiber is acceptable for transmitting signals over a short distance but not long distances because a digital pulse spreads with distance. Let us imagine that we input a perfectly rectangular pulse of laser light to the core of the optical fiber. Active Figure 25.28a shows the idealized time behavior of the laser light intensity for the input pulse. The laser light intensity rises instantaneously to its highest value, stays constant for the duration of the pulse, and then instantaneously drops to zero. The light from the pulse entering along the axis in Figure 25.27 travels the shortest distance and arrives at the other end first. The other light paths represent longer distance of travel because of the angled bounces. As a result, the light from the pulse arrives at the other end over a longer period and the pulse is spread out as in Active Figure 25.28b. If a series of pulses represents zeroes and ones for a binary signal, this spreading could cause the pulses to overlap or might reduce the peak intensity below the detection threshold; either situation would result in obliteration of the information.

One way to improve optical transmission in such a situation is to use a *multimode, graded index fiber* as shown in Figure 25.29. This fiber has a core whose index of refraction is smaller at larger radii from the center. With a graded index core, off-axis rays of light experience continuous refraction and curve gradually away from the edges and back toward the center as shown by the light path in Figure 25.29. Such curving reduces the transit time through the fiber for off-axis rays and also reduces the spreading out of the pulse. The transit time is reduced for two reasons. First, the path length is reduced, and second, much of the time the wave travels in the lower index of refraction region, where the speed of light is higher than at the center.

The spreading effect in Active Figure 25.28 can be further reduced and almost eliminated by designing the fiber with two changes from the multimode, stepped index fiber in Figure 25.27. The core is made very small so that all paths within it are more nearly the same length, and the difference in index of refraction between core and cladding is made relatively small so that off-axis rays enter the cladding and are absorbed. These changes are suggested in Active Figure 25.30. This kind of fiber is called a *single-mode, stepped index fiber*. It can carry information at high bit rates because the pulses are minimally spread out.

In reality, the material of the core is not perfectly transparent. Some absorption and scattering occurs as the light travels down the fiber. Absorption transforms energy being transferred by electromagnetic radiation into increased internal energy in the fiber. Scattering causes light to strike the core–cladding interface at angles less than the critical angle for total internal reflection, resulting in some loss in the cladding or jacket. Even with these problems, optical fibers can transmit about 95% of the input energy over a kilometer. The problems are minimized by using as long a wavelength as possible for which the core material is transparent. The scattering and absorption centers then appear as small as possible to the waves and minimize the probability of interaction. Much of optical fiber communication occurs with light from infrared lasers, having wavelengths of about 1 300 nm.

One common application is the use of optical fibers in telecommunications as mentioned earlier. Optical fibers are also used in "smart buildings." In this application, sensors are located at various points within a building and an optical fiber carries laser light to the sensor, which reflects it back to a control system. If any distortion

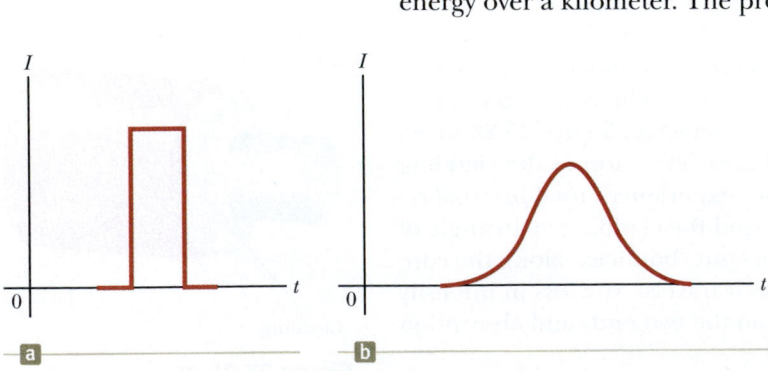

Active Figure 25.28 (a) A rectangular pulse of laser light to be sent into an optical fiber. (b) The output pulse of light, which has been broadened due to light taking different paths through the fiber.

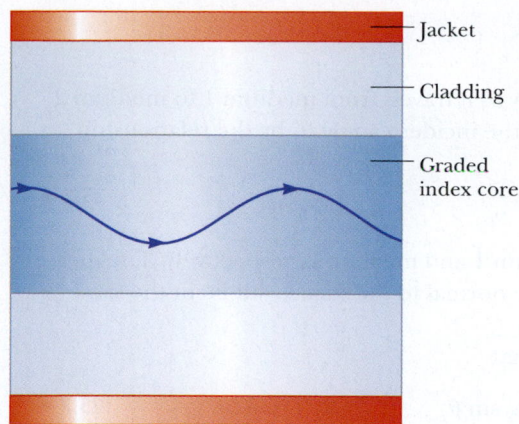

Figure 25.29 A multimode, graded index optical fiber. Because the index of refraction of the core varies radially, off-axis light rays follow curved paths through the core.

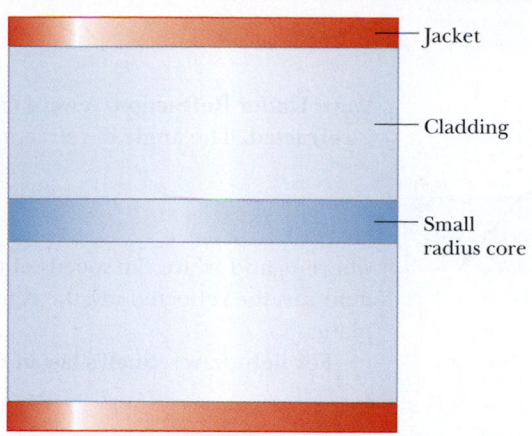

Active Figure 25.30 A single-mode, stepped index optical fiber. The small radius of the core and the small difference between the indices of refraction of the core and cladding reduce the broadening of light pulses.

occurs in the building due to earthquake or other causes, the intensity of the reflected light from the sensor changes and the control system locates the point of distortion by identifying the particular sensor involved.

A single optical fiber can carry a digital signal, as we already described. If it is desired for optical fibers to carry an image of a scene, it is necessary to use a bundle of optical fibers. A popular use of such bundles is in the use of *fiberscopes* in medicine. In the Context Connection of Chapter 26, we shall investigate these devices.

SUMMARY

In geometric optics, we use the **ray approximation** in which we assume that a wave travels through a medium in straight lines in the direction of the rays of that wave. We ignore diffraction effects, which is a good approximation as long as the wavelength is short compared with the size of any openings.

The **index of refraction** n of a material is defined as

$$n \equiv \frac{c}{v} \qquad \text{25.3} \blacktriangleleft$$

where c is the speed of light in a vacuum and v is the speed of light in the material.

In general, n varies with wavelength, which is called **dispersion. Huygens's principle** states that all points on a wave front can be taken as point sources for the production of secondary wavelets. At some later time, the new position of the wave front is the surface tangent to these secondary wavelets.

Total internal reflection can occur when light travels from a medium of high index of refraction to one of lower index of refraction. The **critical angle** of incidence θ_c for which total internal reflection occurs at an interface is

$$\sin \theta_c = \frac{n_2}{n_1} \qquad (\text{for } n_1 > n_2) \qquad \text{25.8} \blacktriangleleft$$

Analysis Models for Problem Solving

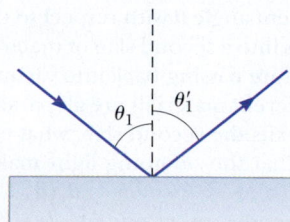

Wave Under Reflection. The **law of reflection** states that for a light ray (or other type of wave) incident on a smooth surface, the angle of reflection θ_1' equals the angle of incidence θ_1:

$$\theta_1' = \theta_1 \qquad \text{25.1} \blacktriangleleft$$

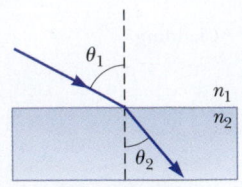

Wave Under Refraction. A wave crossing a boundary as it travels from medium 1 to medium 2 is **refracted.** The angle of refraction θ_2 is related to the incident angle θ_1 by the relationship

$$\frac{\sin \theta_2}{\sin \theta_1} = \frac{v_2}{v_1} \qquad \textbf{25.2} \blacktriangleleft$$

where v_1 and v_2 are the speeds of the wave in medium 1 and medium 2, respectively. The incident ray, the reflected ray, the refracted ray, and the normal to the surface all lie in the same plane.

For light waves, **Snell's law of refraction** states that

$$n_1 \sin \theta_1 = n_2 \sin \theta_2 \qquad \textbf{25.7} \blacktriangleleft$$

where n_1 and n_2 are the indices of refraction in the two media.

OBJECTIVE QUESTIONS

☐ denotes answer available in *Student Solutions Manual/Study Guide*

1. Carbon disulfide ($n = 1.63$) is poured into a container made of crown glass ($n = 1.52$). What is the critical angle for total internal reflection of a light ray in the liquid when it is incident on the liquid-to-glass surface? (a) 89.2° (b) 68.8° (c) 21.2° (d) 1.07° (e) 43.0°

2. In each of the following situations, a wave passes through an opening in an absorbing wall. Rank the situations in order from the one in which the wave is best described by the ray approximation to the one in which the wave coming through the opening spreads out most nearly equally in all directions in the hemisphere beyond the wall. (a) The sound of a low whistle at 1 kHz passes through a doorway 1 m wide. (b) Red light passes through the pupil of your eye. (c) Blue light passes through the pupil of your eye. (d) The wave broadcast by an AM radio station passes through a doorway 1 m wide. (e) An x-ray passes through the space between bones in your elbow joint.

3. What happens to a light wave when it travels from air into glass? (a) Its speed remains the same. (b) Its speed increases. (c) Its wavelength increases. (d) Its wavelength remains the same. (e) Its frequency remains the same.

4. A source emits monochromatic light of wavelength 495 nm in air. When the light passes through a liquid, its wavelength reduces to 434 nm. What is the liquid's index of refraction? (a) 1.26 (b) 1.49 (c) 1.14 (d) 1.33 (e) 2.03

5. The index of refraction for water is about $\frac{4}{3}$. What happens as a beam of light travels from air into water? (a) Its speed increases to $\frac{4}{3}c$, and its frequency decreases. (b) Its speed decreases to $\frac{3}{4}c$, and its wavelength decreases by a factor of $\frac{3}{4}$. (c) Its speed decreases to $\frac{3}{4}c$, and its wavelength increases by a factor of $\frac{4}{3}$. (d) Its speed and frequency remain the same. (e) Its speed decreases to $\frac{3}{4}c$, and its frequency increases.

6. For the following questions, choose from the following possibilities: (a) yes; water (b) no; water (c) yes; air (d) no; air. **(i)** Can light undergo total internal reflection at a smooth interface between air and water? If so, in which medium must it be traveling originally? **(ii)** Can sound undergo total internal reflection at a smooth interface between air and water? If so, in which medium must it be traveling originally?

7. Light traveling in a medium of index of refraction n_1 is incident on another medium having an index of refraction n_2. Under which of the following conditions can total internal reflection occur at the interface of the two media? (a) The indices of refraction have the relation $n_2 > n_1$. (b) The indices of refraction have the relation $n_1 > n_2$. (c) Light travels slower in the second medium than in the first. (d) The angle of incidence is less than the critical angle. (e) The angle of incidence must equal the angle of refraction.

8. Suppose you find experimentally that two colors of light, A and B, originally traveling in the same direction in air, are sent through a glass prism, and A changes direction more than B. Which travels more slowly in the prism, A or B? Alternatively, is there insufficient information to determine which moves more slowly?

9. The core of an optical fiber transmits light with minimal loss if it is surrounded by what? (a) water (b) diamond (c) air (d) glass (e) fused quartz

10. Which color light refracts the most when entering crown glass from air at some incident angle θ with respect to the normal? (a) violet (b) blue (c) green (d) yellow (e) red

11. A light ray travels from vacuum into a slab of material with index of refraction n_1 at incident angle θ with respect to the surface. It subsequently passes into a second slab of material with index of refraction n_2 before passing back into vacuum again. The surfaces of the different materials are all parallel to one another. As the light exits the second slab, what can be said of the final angle ϕ that the outgoing light makes with the normal? (a) $\phi > \theta$ (b) $\phi < \theta$ (c) $\phi = \theta$ (d) The angle depends on the magnitudes of n_1 and n_2. (e) The angle depends on the wavelength of the light.

12. Light can travel from air into water. Some possible paths for the light ray in the water are shown in Figure OQ25.12.

Which path will the light most likely follow? (a) *A* (b) *B* (c) *C* (d) *D* (e) *E*

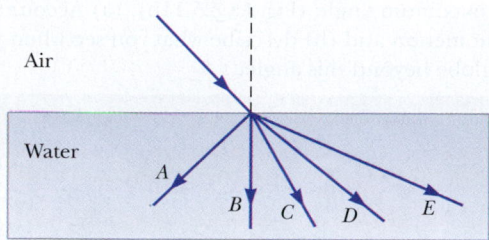

Figure OQ25.12

13. A light wave moves between medium 1 and medium 2. Which of the following are correct statements relating its speed, frequency, and wavelength in the two media, the indices of refraction of the media, and the angles of incidence and refraction? More than one statement may be correct. (a) $v_1/\sin \theta_1 = v_2/\sin \theta_2$ (b) $\csc \theta_1/n_1 = \csc \theta_2/n_2$ (c) $\lambda_1/\sin \theta_1 = \lambda_2/\sin \theta_2$ (d) $f_1/\sin \theta_1 = f_2/\sin \theta_2$ (e) $n_1/\cos \theta_1 = n_2/\cos \theta_2$

14. A light ray containing both blue and red wavelengths is incident at an angle on a slab of glass. Which of the sketches in Figure OQ25.14 represents the most likely outcome? (a) *A* (b) *B* (c) *C* (d) *D* (e) none of them

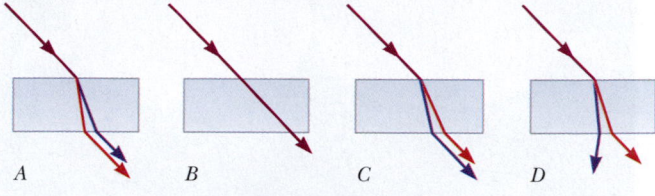

Figure OQ25.14

CONCEPTUAL QUESTIONS

☐ denotes answer available in *Student Solutions Manual/Study Guide*

1. The display windows of some department stores are slanted slightly inward at the bottom. This tilt is to decrease the glare from streetlights and the Sun, which would make it difficult for shoppers to see the display inside. Sketch a light ray reflecting from such a window to show how this design works.

2. The F-117A stealth fighter (Fig. CQ25.2) is specifically designed to be a *non*retroreflector of radar. What aspects of its design help accomplish this purpose?

Courtesy U.S. Air Force

Figure CQ25.2

3. A laser beam passing through a nonhomogeneous sugar solution follows a curved path. Explain.

4. Sound waves have much in common with light waves, including the properties of reflection and refraction. Give an example of each of these phenomena for sound waves.

5. Total internal reflection is applied in the periscope of a submerged submarine to let the user observe events above the water surface. In this device, two prisms are arranged as shown in Figure CQ25.5 so that an incident beam of

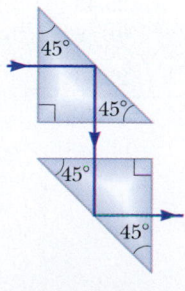

Figure CQ25.5

light follows the path shown. Parallel tilted, silvered mirrors could be used, but glass prisms with no silvered surfaces give higher light throughput. Propose a reason for the higher efficiency.

6. Explain why a diamond sparkles more than a glass crystal of the same shape and size.

7. At one restaurant, a worker uses colored chalk to write the daily specials on a blackboard illuminated with a spotlight. At another restaurant, a worker writes with colored grease pencils on a flat, smooth sheet of transparent acrylic plastic with an index of refraction 1.55. The panel hangs in front of a piece of black felt. Small, bright fluorescent tube lights are installed all along the edges of the sheet, inside an opaque channel. Figure CQ25.7 shows a cutaway view of the sign. (a) Explain why viewers at both restaurants see the letters shining against a black background. (b) Explain why the sign at the second restaurant may use less energy from the electric company than the illuminated blackboard at the first restaurant. (c) What would be a good choice for the index of refraction of the material in the grease pencils?

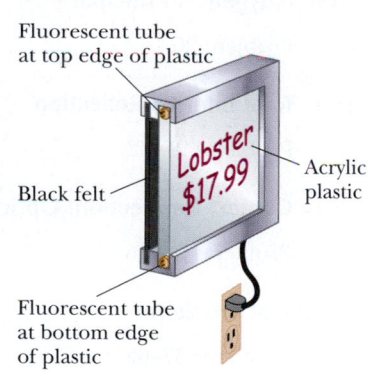

Figure CQ25.7

8. A scientific supply catalog advertises a material having an index of refraction of 0.85. Is that a good product to buy? Why or why not?

9. The level of water in a clear, colorless glass can easily be observed with the naked eye. The level of liquid helium in a clear glass vessel is extremely difficult to see with the naked eye. Explain.

10. Try this simple experiment on your own. Take two opaque cups, place a coin at the bottom of each cup near the edge,

and fill one cup with water. Next, view the cups at some angle from the side so that the coin in water is just visible as shown on the left in Figure CQ25.10. Notice that the coin in air is not visible as shown on the right in Figure CQ25.10. Explain this observation.

Figure CQ25.10

11. Figure CQ25.11a shows a desk ornament globe containing a photograph. The flat photograph is in air, inside a vertical slot located behind a water-filled compartment having the shape of one half of a cylinder. Suppose you are looking at the center of the photograph and then rotate the globe about a vertical axis. You find that the center of the photograph disappears when you rotate the globe beyond a certain maximum angle (Fig. CQ25.11b). (a) Account for this phenomenon and (b) describe what you see when you turn the globe beyond this angle.

a   b

Figure CQ25.11

12. A complete circle of a rainbow can sometimes be seen from an airplane. With a stepladder, a lawn sprinkler, and a sunny day, how can you show the complete circle to children?

PROBLEMS AVAILABLE IN

25.2 The Ray Model in Geometric Optics
25.3 Analysis Model: Wave Under Reflection
25.4 Analysis Model: Wave Under Refraction

   Problems 1–22

25.5 Dispersion and Prisms

   Problems 23–25

25.6 Huygens's Principle

   Problem 26

25.7 Total Internal Reflection

   Problems 27–32

25.8 Context Connection: Optical Fibers

   Problems 33–36

Additional Problems

   Problems 37–62

Solutions to the following Problems are available in the _Student Solutions Manual/Study Guide_:

   25.1, 25.8, 25.16, 25.23, 25.29, 25.31, 25.34, 25.37, 25.41, 25.47, 25.53, and 25.54

List of Enhanced Problems

Problem Number	Targeted Feedback in Enhanced WebAssign	Master It in Enhanced WebAssign	Watch It in Enhanced WebAssign
25.1		✓	
25.3	✓		✓
25.4	✓		✓
25.7	✓		✓
25.8	✓		✓
25.17	✓		✓
25.20	✓		✓
25.22	✓		✓
25.23		✓	
25.27	✓		✓
25.29	✓	✓	
25.31		✓	
25.35	✓		
25.37		✓	
25.41		✓	
25.43	✓		✓
25.45	✓		✓
25.53		✓	

Image Formation by Mirrors and Lenses

Don Hammond Photography Ltd. RF

This chapter is concerned with the images formed when light interacts with flat and curved surfaces. We find that images of an object can be formed by reflection or by refraction and that mirrors and lenses work because of these phenomena.

Images formed by reflection and refraction are used in a variety of everyday devices, such as the rearview mirror in your car, a shaving or makeup mirror, a camera, your eyeglasses, and a magnifying glass. In addition, more scientific devices, such as telescopes and microscopes, take advantage of the image formation principles discussed in this chapter.

We shall make extensive use of geometric models developed from the principles of reflection and refraction. Such constructions allow us to develop mathematical representations for the image locations of various types of mirrors and lenses.

The light rays coming from the leaves in the background of this scene did not form a focused image on the film of the camera that took this photograph. Consequently, the background appears very blurry. Light rays passing though the raindrop, however, have been altered so as to form a focused image of the background leaves on the film. In this chapter, we investigate the formation of images as light rays reflect from mirrors and refract through lenses.

26.1 | Images Formed by Flat Mirrors

We begin by considering the simplest possible mirror, the flat mirror. Consider a point source of light[1] placed at O in Figure 26.1 (page 754), a distance p in front of a flat mirror. The distance p is called the **object distance.** Light rays leave the source and are reflected from the mirror. Upon reflection, the rays continue to diverge (spread apart). The dashed lines in Figure 26.1 are extensions of the diverging rays back to a point of intersection at I.

---

[1]We imagine the object to be a point source of light. It could actually *be* a point source, such as a very small lightbulb, but more often is a single point on some extended object that is illuminated from the exterior by a light source. Therefore, the reflected light leaves the point on the object as if the point were a source of light.

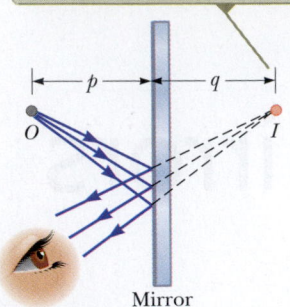

The image point *I* is located behind the mirror a distance *q* from the mirror. The image is virtual.

Figure 26.1 An image formed by reflection from a flat mirror.

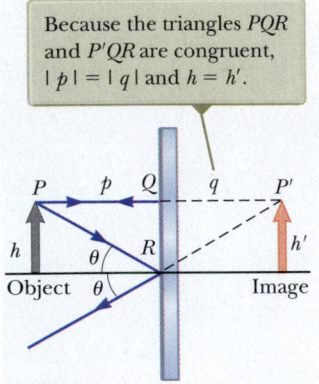

Because the triangles *PQR* and *P'QR* are congruent, | *p* | = | *q* | and *h* = *h'*.

Active Figure 26.2 A geometric construction used to locate the image of an object placed in front of a flat mirror.

▶ Magnification of an image

Pitfall Prevention | 26.1
Magnification Does Not Necessarily Imply Enlargement
For optical elements other than flat mirrors, the magnification defined in Equation 26.1 can result in a number with a magnitude larger *or* smaller than 1. Therefore, despite the cultural usage of the word *magnification* to mean *enlargement*, the image could be smaller than the object. We shall see examples of such a situation in this chapter.

The diverging rays appear to the viewer to come from the point *I* behind the mirror. Point *I* is called the **image** of the object at *O*. Regardless of the system under study, we always locate images by extending diverging rays back to a point at which they intersect.[2] Images are located either at a point from which rays of light *actually* diverge or at a point from which they *appear* to diverge. Because the rays in Figure 26.1 appear to originate at *I*, which is a distance *q* behind the mirror, that is the location of the image. The distance *q* is called the **image distance.**

Images are classified as **real** or **virtual**. A **real image** is formed when light rays pass through and diverge from the image point; a **virtual image** is formed when the light rays do not pass through the image point but only appear to diverge from that point. The image formed by the mirror in Figure 26.1 is virtual. The image of an object seen in a flat mirror is *always* virtual. Real images can be displayed on a screen (as at a movie), but virtual images cannot be displayed on a screen. We shall see an example of a real image in Section 26.2.

Active Figure 26.2 is an example of a specialized pictorial representation, called a **ray diagram,** that is very useful in studies of mirrors and lenses. In a ray diagram, a small number of the myriad rays leaving a point source are drawn, and the location of the image is found by applying the laws of reflection (and refraction, in the case of refracting surfaces and lenses) to these rays. A carefully drawn ray diagram allows us to build a geometric model so that geometry and trigonometry can be used to solve a problem mathematically.

We can use the simple geometry in Active Figure 26.2 to examine the properties of the images of extended objects formed by flat mirrors. Let us locate the image of the tip of the gray arrow. To find out where the image is formed, it is necessary to follow at least two rays of light as they reflect from the mirror. One of those rays starts at *P*, follows the horizontal path *PQ* to the mirror, and reflects back on itself. The second ray follows the oblique path *PR* and reflects at the same angle according to the law of reflection. We can extend the two reflected rays back to the point from which they appear to diverge, point *P'*. A continuation of this process for points other than *P* on the object would result in an image (drawn as a pink arrow) to the right of the mirror. These rays and the extensions of the rays allow us to build a geometric model for the image formation based on triangles *PQR* and *P'QR*. Because these two triangles are identical, *PQ* = *P'Q*, or *p* = |*q*|. (We use the absolute value notation because, as we shall see shortly, a sign convention is associated with the values of *p* and *q*.) Hence, we conclude that the image formed by an object placed in front of a flat mirror is as far behind the mirror as the object is in front of the mirror.

Our geometric model also shows that the object height *h* equals the image height *h'*. We define the **lateral magnification** (or simply the **magnification**) *M* of an image as follows:

$$M \equiv \frac{\text{image height}}{\text{object height}} = \frac{h'}{h}$$

26.1 ◀

which is a general definition of the magnification for any type of image formed by a mirror or lens. Because *h'* = *h* in this case, *M* = 1 for a flat mirror. We also note that the image is *upright* because the image arrow points in the same direction as the object arrow. An upright image is indicated mathematically by a positive value of the magnification. (Later we discuss situations in which *inverted* images, with negative magnifications, are formed.)

Finally, note that a flat mirror produces an image having an *apparent* left–right reversal. This reversal can be seen by standing in front of a mirror and raising your right hand. The image you see raises its left hand. Likewise, your hair appears to be parted on the opposite side, and a mole on your right cheek appears to be on your left cheek.

This reversal is not *actually* a left–right reversal. Imagine, for example, lying on your left side on the floor, with your body parallel to the mirror surface. Now, your head is on the left and your feet are on the right as you face the mirror. If you shake

---

[2]Your eyes and brain interpret diverging light rays as originating at the point from which the rays diverge. Your eye–brain system can detect the rays only *as they enter your eye* and has no access to information about what experiences the rays underwent before reaching your eyes. Therefore, even though the light rays did not *actually originate* at point *I*, they enter the eye *as if they had*, and *I* is the point at which your brain locates the object.

your feet, the image does not shake its head! If you raise your right hand, however, the image raises its left hand. Therefore, it again appears like a left–right reversal, but in an up–down direction!

The apparent left–right reversal is actually a *front–back* reversal caused by the light rays going forward toward the mirror and then reflecting back from it. Figure 26.3 shows a person's right hand and its image in a flat mirror. Notice that no left–right reversal takes place; rather, the thumbs on both the real hand and the image are on the left side. It is the front–back reversal that makes the image of the right hand appear similar to the real left hand at the left side of the photograph.

An interesting experience with front–back reversal is to stand in front of a mirror while holding an overhead transparency in front of you so that you can read the writing on the transparency. You are also able to read the writing on the image of the transparency. You might have had a similar experience if you have a transparent decal with words on it on the rear window of your car. If the decal is placed so that it can be read from outside the car, you can also read it when looking into your rearview mirror from the front seat.

> **QUICK QUIZ 26.1** In the overhead view of Figure 26.4, the image of the stone seen by observer 1 is at *C*. At which of the five points *A*, *B*, *C*, *D*, or *E* does observer 2 see the image?

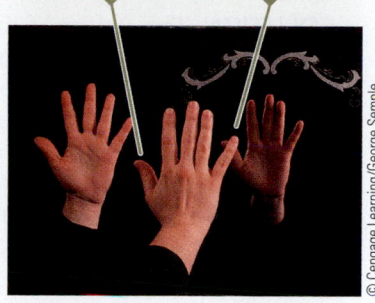

The thumb is on the left side of both real hands and on the left side of the image. That the thumb is not on the right side of the image indicates that there is no left-to-right reversal.

Figure 26.3 The image in the mirror of a person's right hand is reversed front to back, which makes the image in the mirror appear to be a left hand.

© Cengage Learning/George Semple

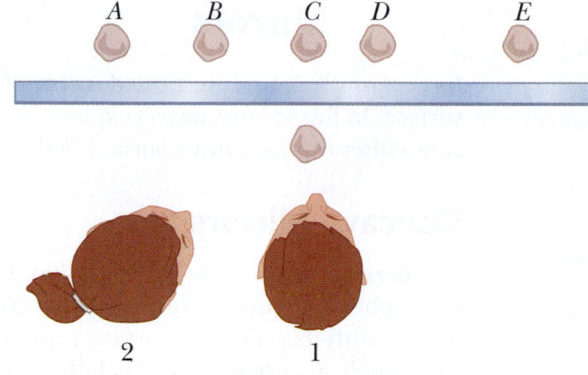

Figure 26.4
(Quick Quiz 26.1)
Where does observer
2 see the image of the
stone?

> **QUICK QUIZ 26.2** You are standing approximately 2 m away from a mirror. The mirror has water spots on its surface. True or False: It is possible for you to see the water spots and your image both in focus at the same time.

> **THINKING PHYSICS 26.1**

Most rearview mirrors in cars have a day setting and a night setting. The night setting greatly diminishes the intensity of the image so that lights from trailing vehicles do not temporarily blind the driver. How does such a mirror work?

SOLUTION Figure 26.5 shows a cross-sectional view of a rearview mirror for each setting. The unit consists of a reflective coating on the back of a wedge of glass. In the day setting (Fig. 26.5a), the light from an object behind the car strikes the glass wedge at point 1. Most of the light enters the wedge, refracting as it crosses the front surface, and reflects from the back surface to return to the front surface, where it is refracted again as it re-enters the air as ray *B* (for *bright*). In addition, a small portion of the light is reflected at the front surface of the glass as indicated by ray *D* (for *dim*).

This dim reflected light is responsible for the image observed when the mirror is in the night setting (Fig. 26.5b).

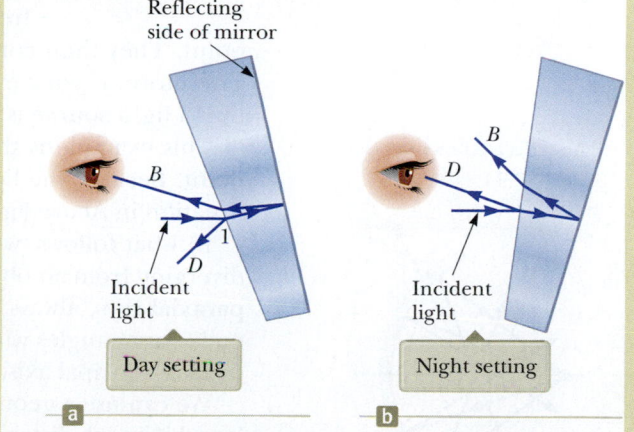

Figure 26.5 (Thinking Physics 26.1) Cross-sectional views of a rearview mirror.

In that case, the wedge is rotated so that the path followed by the bright light (ray *B*) does not lead to the eye. Instead, the dim light reflected from the front surface of the wedge travels to the eye, and the brightness of trailing headlights does not become a hazard. ◄

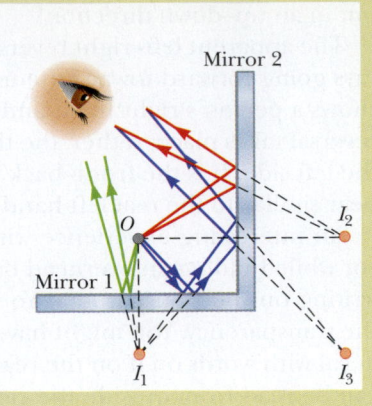
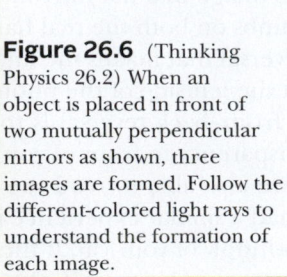

26.2 | Images Formed by Spherical Mirrors

In Section 26.1, we investigated images formed by a flat reflecting surface. In this section, we will explore images formed by curved mirrors, either from a concave surface of the mirror or a convex surface.

Concave Mirrors

A **spherical mirror,** as its name implies, has the shape of a segment of a sphere. Figure 26.7a shows the cross-section of a spherical mirror with its reflecting surface represented by the solid curved line. Such a mirror in which light is reflected from the inner, concave surface is called a **concave mirror.** The mirror's radius of curvature is R, and its center of curvature is at point C. Point V is the center of the spherical segment, and a line drawn through C and V is called the **principal axis** of the mirror.

Now consider a point source of light placed at point O in Figure 26.7b, on the principal axis and outside point C. Two diverging rays that originate at O are shown. After reflecting from the mirror, these rays converge and meet at I, the image point. They then continue to diverge from I as if a source of light existed there. Therefore, if your eyes detect the rays diverging from point I, you would claim that a light source is located at that point.

This example is the second one we have seen of rays diverging from an image point. Because the light rays pass through the image point in this case, unlike the situation in Active Figure 26.2, the image in Figure 26.7b is a real image.

In what follows, we shall adopt a simplification model that assumes that all rays diverging from an object make small angles with the principal axis. Such rays, called **paraxial rays,** always reflect through the image point as in Figure 26.7b. Rays that make large angles with the principal axis as in Figure 26.8 converge at other points on the principal axis, producing a blurred image.

We can use a geometric model based on the ray diagram in Figure 26.9 to calculate the image distance q if we know the object distance p and radius of curvature R. By convention, these distances are measured from point V. Figure 26.9 shows two of the many light rays leaving the tip of the object. One ray passes through the center of curvature C of the mirror, hitting the mirror perpendicular to the mirror surface and reflecting back on itself. The second ray strikes the mirror at the center point V and reflects as shown, obeying the law of reflection. The image of the tip of the

Figure 26.7 (a) A concave mirror of radius R. The center of curvature C is located on the principal axis. (b) A point source of light placed at O in front of a concave spherical mirror of radius R, where O is any point on the principal axis farther than R from the mirror surface, forms a real image at I.

> If the rays diverge from O at small angles, they all reflect through the same image point I.

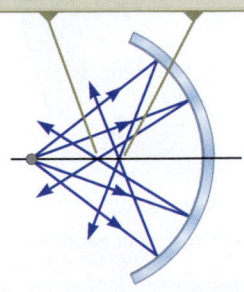

> The reflected rays intersect at different points on the principal axis.

Figure 26.8 A spherical concave mirror produces a blurred image when light rays make large angles with the principal axis.

arrow is at the point at which these two reflected rays intersect. Using these rays, we identify the red and yellow model right triangles in Figure 26.9. From the red triangle, we see that $\tan \theta = h/p$, whereas the yellow triangle gives $\tan \theta = -h'/q$. The negative sign signifies that the image is inverted, so h' is a negative number. Therefore, from Equation 26.1 and these results, we find that the magnification of the image is

$$M = \frac{h'}{h} = \frac{-q \tan \theta}{p \tan \theta} = -\frac{q}{p} \qquad \textbf{26.2}\blacktriangleleft$$

We can identify two additional right triangles in the figure (the green one and the smaller red one), with a common point at C and with angle α. These triangles tell us that

$$\tan \alpha = \frac{h}{p - R} \qquad \text{and} \qquad \tan \alpha = -\frac{h'}{R - q}$$

from which we find that

$$\frac{h'}{h} = -\frac{R - q}{p - R} \qquad \textbf{26.3}\blacktriangleleft$$

If we compare Equations 26.2 and 26.3, we see that

$$\frac{R - q}{p - R} = \frac{q}{p}$$

Algebra reduces this expression to

$$\frac{1}{p} + \frac{1}{q} = \frac{2}{R} \qquad \textbf{26.4}\blacktriangleleft \qquad \blacktriangleright \text{ Mirror equation in terms of the radius of curvature}$$

which is called the **mirror equation.** It is applicable only to the paraxial ray simplification model.

If the object is very far from the mirror—that is, if the object distance p is large compared with R, so that p can be said to approach infinity—$1/p \to 0$, and we see from Equation 26.4 that $q \approx R/2$. In other words, when the object is very far from the mirror, the image point is halfway between the center of curvature and the center of the mirror as in Figure 26.10a. The rays are essentially parallel in this figure because only those few rays traveling parallel to the axis from the distant object encounter the mirror. Rays not parallel to the axis miss the mirror. Figure 26.10b shows an experimental setup of this situation, demonstrating the crossing of the light rays at a single point. The point at which the parallel rays intersect after reflecting from the mirror is called the **focal point** F of the mirror. The focal point is a distance f from the mirror, called the **focal length.** The focal length is a parameter associated with the mirror and is given by

$$f = \frac{R}{2} \qquad \textbf{26.5}\blacktriangleleft$$

The mirror equation can therefore be expressed in terms of the focal length:

$$\boxed{\frac{1}{p} + \frac{1}{q} = \frac{1}{f}} \qquad \textbf{26.6}\blacktriangleleft \qquad \blacktriangleright \text{ Mirror equation in terms of focal length}$$

This equation is the commonly used mirror equation, in terms of the focal length of the mirror rather than its radius of curvature, as in Equation 26.4. We shall see how to use this equation in examples that follow shortly.

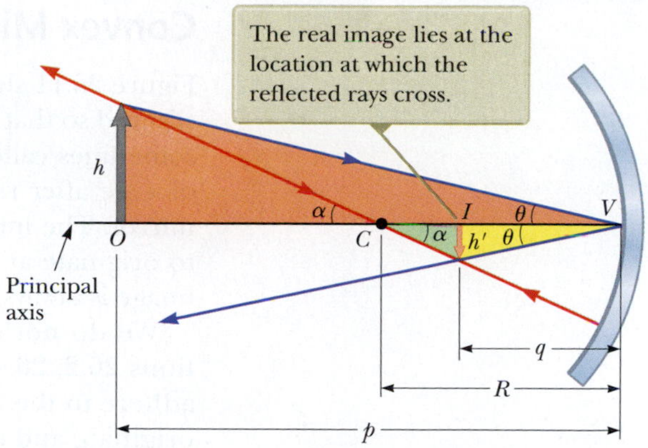

Figure 26.9 The image formed by a spherical concave mirror when the object O lies outside the center of curvature C. This geometric construction is used to derive Equation 26.4.

The real image lies at the location at which the reflected rays cross.

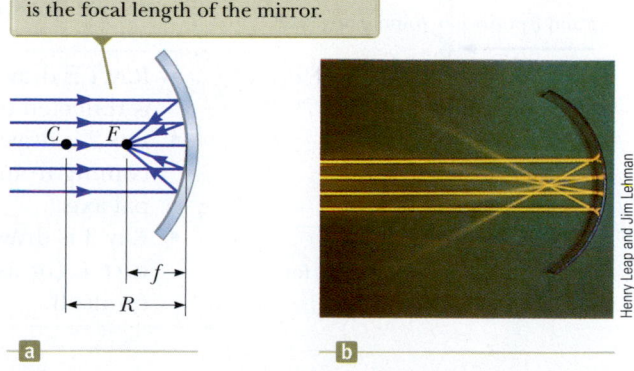

When the object is very far away, the image distance $q \approx R/2 = f$, where f is the focal length of the mirror.

Figure 26.10 (a) Light rays from a distant object ($p \approx \infty$) reflect from a concave mirror through the focal point F. (b) Reflection of parallel rays from a concave mirror.

Henry Leap and Jim Lehman

Convex Mirrors

Figure 26.11 shows the formation of an image by a **convex mirror,** a mirror that is silvered so that light is reflected from the outer, convex surface. Convex mirrors are sometimes called **diverging mirrors** because the rays from any point on an object diverge after reflection as though they were coming from some point behind the mirror. The image in Figure 26.11 is virtual because the reflected rays only appear to originate at the image point as indicated by the dashed lines. Furthermore, the image is always upright and smaller than the object.

We do not derive any equations for convex spherical mirrors because Equations 26.2, 26.4, and 26.6 can be used for either concave or convex mirrors if we adhere to the following procedure. We will refer to the region in which light rays originate and move toward the mirror as the *front side* of the mirror and the other side as the *back side*. For example, in Figures 26.9 and 26.11, the side to the left of the mirrors is the front side and the side to the right of the mirrors is the back side. Figure 26.12 states the sign conventions for object and image distances, and Table 26.1 summarizes the sign conventions for all quantities. One entry in the table, a *virtual object,* is formally introduced in Section 26.4.

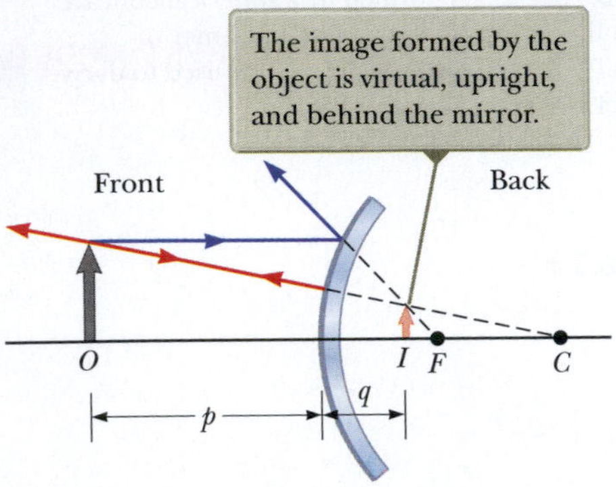

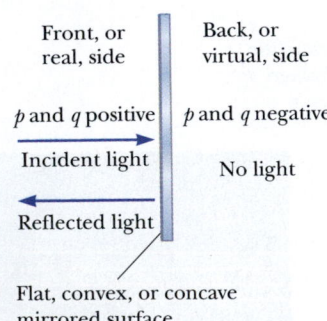

The image formed by the object is virtual, upright, and behind the mirror.

Figure 26.11 Formation of an image by a spherical convex mirror.

Figure 26.12 Signs of p and q for convex and concave mirrors.

Ray Diagrams for Mirrors

We have been using the specialized pictorial representations called ray diagrams to help us locate images for flat and curved mirrors. Let us now formalize the procedure for drawing accurate ray diagrams. To construct such a diagram, we must know the position of the object and the locations of the focal point and center of curvature of the mirror. We will construct three rays in the examples shown in Active Figure 26.13. Only two rays are necessary to locate the image, but we will include a third as a check. In each part of the figure, the right-hand portion shows a photograph of the situation described by the ray diagram in the left-hand portion. All three rays start from the same object point; in these examples, the top of the object arrow is chosen as the starting point. For the concave mirrors in Active Figure 26.13a and 26.13b, the rays are drawn as follows:

- Ray 1 is drawn from the top of the object parallel to the principal axis and is reflected through the focal point F.
- Ray 2 is drawn from the top of the object through the focal point (or as if coming from the focal point if $p < f$) and is reflected parallel to the principal axis.
- Ray 3 is drawn from the top of the object through the center of curvature C (or as if coming from the center C if $p < f$) and is reflected back on itself.

TABLE 26.1 | Sign Conventions for Mirrors

Quantity	Positive when . . .	Negative when . . .
Object location (p)	object is in front of mirror (real object).	object is in back of mirror (virtual object).
Image location (q)	image is in front of mirror (real image).	image is in back of mirror (virtual image).
Image height (h')	image is upright.	image is inverted.
Focal length (f) and radius (R)	mirror is concave.	mirror is convex.
Magnification (M)	image is upright.	image is inverted.

The image point obtained in this fashion must always agree with the value of q calculated from the mirror equation. With concave mirrors, note what happens as the object is moved closer to the mirror from infinity. The real, inverted image in Active Figure 26.13a moves to the left as the object approaches the mirror. When the object is at the center of curvature, the object and image are at the same distance from the mirror and are the same size. When the object is at the focal point, the image is infinitely far to the left. (Check these last three sentences with the mirror equation!)

When the object is located so that the center of curvature lies between the object and a concave mirror surface, the image is real, inverted, and reduced in size.

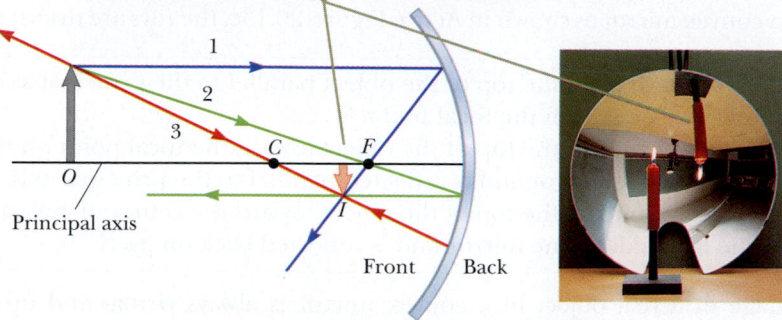

a

A satellite-dish antenna is a concave reflector for television signals from a satellite in orbit around the Earth. Because the satellite is so far away, the signals are carried by microwaves that are parallel when they arrive at the dish. These waves reflect from the dish and are focused on the receiver.

When the object is located between the focal point and a concave mirror surface, the image is virtual, upright, and enlarged.

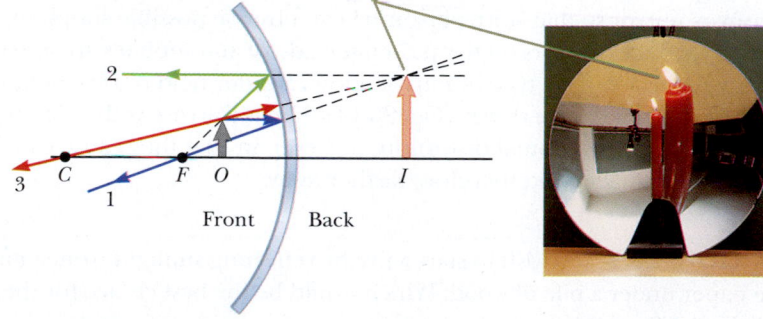

b

When the object is in front of a convex mirror, the image is virtual, upright, and reduced in size.

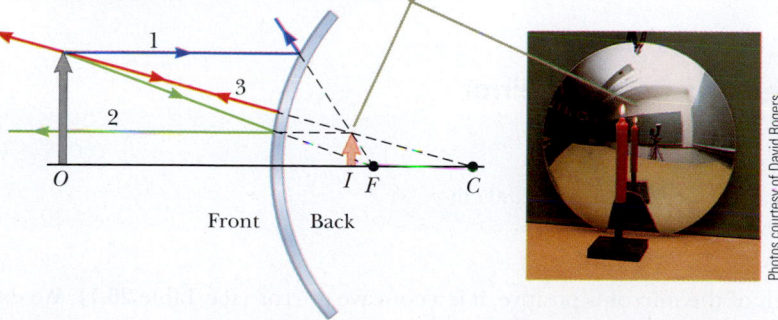

c

Active Figure 26.13 Ray diagrams for spherical mirrors, along with corresponding photographs of the images of candles.

When the object lies between the focal point and the mirror surface as in Active Figure 26.13b, the image is virtual, upright, and located on the back side of the mirror. The image is also larger than the object in this case. This situation illustrates the principle behind a shaving mirror or a makeup mirror. Your face is located closer to the concave mirror than the focal point, so you see an enlarged, upright image of your face, to assist you with shaving or applying makeup. If you have such a mirror, look into it and move your face farther from the mirror. Your head will pass through a point at which the image is indistinct and then the image will reappear with your face upside down as you continue to move farther away. The region where the image is indistinct is where your head passes through the focal point and the image is infinitely far away.

Notice that the image of the camera in Active Figures 26.13a and 26.13b is upside down. Regardless of the position of the candle, the camera remains farther away from the mirror than the focal point, so its image is inverted.

For a convex mirror as shown in Active Figure 26.13c, the rays are drawn as follows:

- Ray 1 is drawn from the top of the object parallel to the principal axis and is reflected away from the focal point *F*.
- Ray 2 is drawn from the top of the object toward the focal point on the back side of the mirror and is reflected parallel to the principal axis.
- Ray 3 is drawn from the top of the object toward the center of curvature *C* on the back side of the mirror and is reflected back on itself.

Figure 26.14 An approaching truck is seen in a convex mirror on the right side of an automobile. Notice that the image of the truck is in focus, but the frame of the mirror is not, which demonstrates that the image is not at the same location as the mirror surface.

© Bo Zaunders/Corbis

The image of a real object in a convex mirror is always virtual and upright. Notice that the images of both the candle and the camera in Active Figure 26.13c are upright. As the object distance increases, the virtual image becomes smaller and approaches the focal point as *p* approaches infinity. You should construct other diagrams to verify how the image position varies with object position.

Convex mirrors are often used as security devices in large stores, where they are hung at a high position on the wall. The large field of view of the store is made smaller by the convex mirror so that store personnel can observe possible shoplifting activity in several aisles at once. Mirrors on the passenger side of automobiles are also often made with a convex surface. This type of mirror allows a wider field of view behind the automobile to be available to the driver (Fig. 26.14) than is the case with a flat mirror. These mirrors introduce a perceptual distortion, however, in that they cause cars behind the viewer to appear smaller and therefore farther away.

QUICK QUIZ 26.3 You wish to start a fire by reflecting sunlight from a mirror onto some paper under a pile of wood. Which would be the best choice for the type of mirror? **(a)** flat **(b)** concave **(c)** convex

QUICK QUIZ 26.4 Consider the image in the mirror in Figure 26.15. Based on the appearance of this image, would you conclude that **(a)** the mirror is concave and the image is real, **(b)** the mirror is concave and the image is virtual, **(c)** the mirror is convex and the image is real, or **(d)** the mirror is convex and the image is virtual?

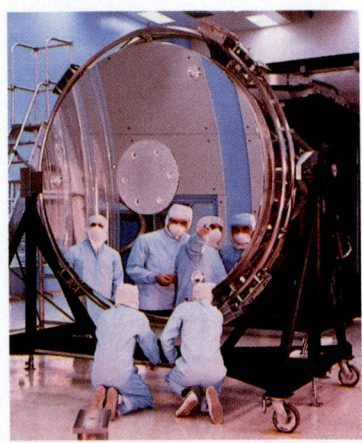

NASA

Figure 26.15 (Quick Quiz 26.4) What type of mirror is shown here?

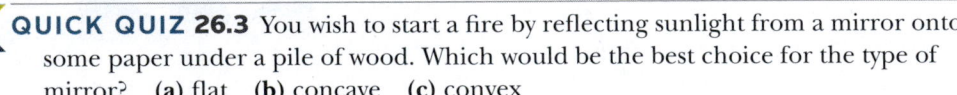

Example 26.1 | The Image Formed by a Concave Mirror

A spherical mirror has a focal length of +10.0 cm.

(A) Locate and describe the image for an object distance of 25.0 cm.

SOLUTION

Conceptualize Because the focal length of the mirror is positive, it is a concave mirror (see Table 26.1). We expect the possibilities of both real and virtual images.

Categorize Because the object distance in this part of the problem is larger than the focal length, we expect the image to be real. This situation is analogous to that in Active Figure 26.13a.

26.1 *cont.*

Analyze Find the image distance by using Equation 26.6:

$$\frac{1}{q} = \frac{1}{f} - \frac{1}{p}$$

$$\frac{1}{q} = \frac{1}{10.0 \text{ cm}} - \frac{1}{25.0 \text{ cm}}$$

$$q = \boxed{16.7 \text{ cm}}$$

Find the magnification of the image from Equation 26.2:

$$M = -\frac{q}{p} = -\frac{16.7 \text{ cm}}{25.0 \text{ cm}} = \boxed{-0.667}$$

Finalize The absolute value of M is less than unity, so the image is smaller than the object, and the negative sign for M tells us that the image is inverted. Because q is positive, the image is located on the front side of the mirror and is real. Look into the bowl of a shiny spoon or stand far away from a shaving mirror to see this image.

(B) Locate and describe the image for an object distance of 10.0 cm.

SOLUTION

Categorize Because the object is at the focal point, we expect the image to be infinitely far away.

Analyze Find the image distance by using Equation 26.6:

$$\frac{1}{q} = \frac{1}{f} - \frac{1}{p}$$

$$\frac{1}{q} = \frac{1}{10.0 \text{ cm}} - \frac{1}{10.0 \text{ cm}}$$

$$q = \boxed{\infty}$$

Finalize This result means that rays originating from an object positioned at the focal point of a mirror are reflected so that the image is formed at an infinite distance from the mirror; that is, the rays travel parallel to one another after reflection. Such is the situation in a flashlight or an automobile headlight, where the bulb filament is placed at the focal point of a reflector, producing a parallel beam of light.

(C) Locate and describe the image for an object distance of 5.00 cm.

SOLUTION

Categorize Because the object distance is smaller than the focal length, we expect the image to be virtual. This situation is analogous to that in Active Figure 26.13b.

Analyze Find the image distance by using Equation 26.6:

$$\frac{1}{q} = \frac{1}{f} - \frac{1}{p}$$

$$\frac{1}{q} = \frac{1}{10.0 \text{ cm}} - \frac{1}{5.00 \text{ cm}}$$

$$q = \boxed{-10.0 \text{ cm}}$$

Find the magnification of the image from Equation 26.2:

$$M = -\frac{q}{p} = -\left(\frac{-10.0 \text{ cm}}{5.00 \text{ cm}}\right) = \boxed{+2.00}$$

Finalize The image is twice as large as the object, and the positive sign for M indicates that the image is upright (see Active Fig. 26.13b). The negative value of the image distance tells us that the image is virtual, as expected. Put your face close to a shaving mirror to see this type of image.

Example 26.2 | **The Image Formed by a Convex Mirror**

An automobile rearview mirror as shown in Figure 26.14 shows an image of a truck located 10.0 m from the mirror. The focal length of the mirror is −0.60 m.

(A) Find the position of the image of the truck.

SOLUTION

Conceptualize This situation is depicted in Active Figure 26.13c.

Categorize Because the mirror is convex, we expect it to form an upright, reduced, virtual image for any object position.

Analyze Find the image distance by using Equation 26.6:

$$\frac{1}{q} = \frac{1}{f} - \frac{1}{p}$$

$$\frac{1}{q} = \frac{1}{-0.60 \text{ m}} - \frac{1}{10.0 \text{ m}}$$

$$q = -0.57 \text{ m}$$

(B) Find the magnification of the image.

SOLUTION

Analyze Use Equation 26.2:

$$M = -\frac{q}{p} = -\left(\frac{-0.57 \text{ m}}{10.0 \text{ m}}\right) = +0.057$$

Finalize The negative value of q in part (A) indicates that the image is virtual, or behind the mirror, as shown in Active Figure 26.13c. The magnification in part (B) indicates that the image is much smaller than the truck and is upright because M is positive. The image is reduced in size, so the truck appears to be farther away than it actually is. Because of the image's small size, these mirrors carry the inscription, "Objects in this mirror are closer than they appear." Look into your rearview mirror or the back side of a shiny spoon to see an image of this type.

26.3 | Images Formed by Refraction

In this section, we describe how images are formed by the refraction of rays at the surface of a transparent material. We shall apply the law of refraction and use the simplification model in which we consider only paraxial rays.

Consider two transparent media with indices of refraction n_1 and n_2, where the boundary between the two media is a spherical surface with radius of curvature R (Fig. 26.16). We shall assume that the object at point O is in the medium with index of refraction n_1. As we shall see, all paraxial rays are refracted at the spherical surface and converge to a single point I, the image point.

Let us proceed by considering the geometric construction in Figure 26.17, which shows a single ray leaving point O and passing through point I. Snell's law applied to this refracted ray gives

$$n_1 \sin \theta_1 = n_2 \sin \theta_2$$

Because the angles θ_1 and θ_2 are small for paraxial rays, we can use the approximation $\sin \theta \approx \theta$ (angles in radians). Therefore, Snell's law becomes

$$n_1 \theta_1 = n_2 \theta_2$$

Now we make use of geometric model triangles and recall that an exterior angle of any triangle equals the sum of the two opposite interior angles. Applying this rule to the triangles OPC and PIC in Figure 26.17 gives

$$\theta_1 = \alpha + \beta$$

$$\beta = \theta_2 + \gamma$$

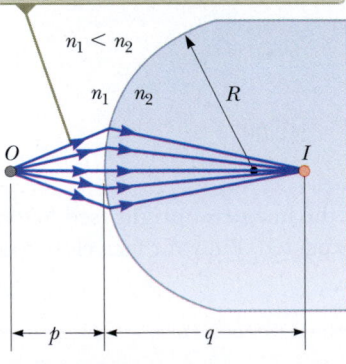

Rays making small angles with the principal axis diverge from a point object at O and are refracted through the image point I.

Figure 26.16 An image formed by refraction at a spherical surface.

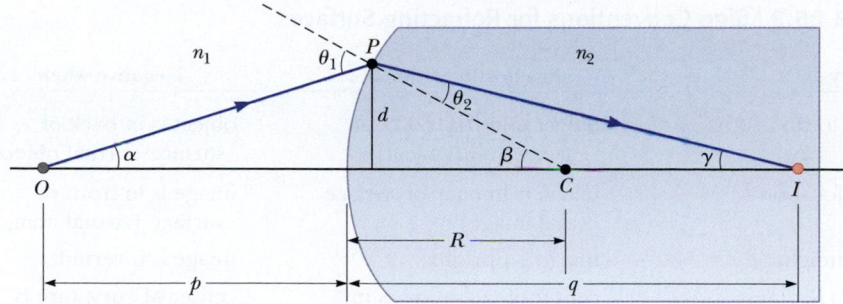

Figure 26.17 Geometry used to derive Equation 26.8, assuming that $n_1 < n_2$. Point C is the center of curvature of the curved refracting surface.

If we combine the last three equations and eliminate θ_1 and θ_2, we find that

$$n_1\alpha + n_2\gamma = (n_2 - n_1)\beta \qquad \text{26.7} \blacktriangleleft$$

In the small angle approximation, $\tan\theta \approx \theta$, and so from Figure 26.17 we can write the approximate relations

$$\tan\alpha \approx \alpha \approx \frac{d}{p} \qquad \tan\beta \approx \beta \approx \frac{d}{R} \qquad \tan\gamma \approx \gamma \approx \frac{d}{q}$$

where d is the distance shown in Figure 26.17. We substitute these equations into Equation 26.7 and divide through by d to give

$$\boxed{\frac{n_1}{p} + \frac{n_2}{q} = \frac{n_2 - n_1}{R}} \qquad \text{26.8} \blacktriangleleft$$

▶ Relation between object and image distance for a refracting surface

Because this expression does not involve any angles, all paraxial rays leaving an object at distance p from the refracting surface will be focused at the same distance q from the surface on the back side.

By setting up a geometric construction with an object and a refracting surface, we can show that the magnification of an image due to a refracting surface is

$$M = -\frac{n_1 q}{n_2 p} \qquad \text{26.9} \blacktriangleleft$$

▶ Magnification of an image formed by a refracting surface

As with mirrors, we must use a sign convention if we are to apply Equations 26.8 and 26.9 to a variety of circumstances. Note that real images are formed on the side of the surface that is *opposite* the side from which the light comes. That is in contrast to mirrors, for which real images are formed on the side where the light originates. Therefore, the sign conventions for spherical refracting surfaces are similar to the conventions for mirrors, recognizing the change in sides of the surface for real and virtual images. For example, in Figure 26.17, p, q, and R are all positive.

The sign conventions for spherical refracting surfaces are summarized in Table 26.2 (page 764). The same conventions will be used for thin lenses discussed in the next section. As with mirrors, we assume that the front of the refracting surface is the side from which the light approaches the surface.

Flat Refracting Surfaces

If the refracting surface is flat, R approaches infinity and Equation 26.8 reduces to

$$\frac{n_1}{p} = -\frac{n_2}{q}$$

or

$$q = -\frac{n_2}{n_1}p \qquad \text{26.10} \blacktriangleleft$$

TABLE 26.2 | Sign Conventions for Refracting Surfaces

Quantity	Positive when . . .	Negative when . . .
Object location (p)	object is in front of surface (real object).	object is in back of surface (virtual object).
Image location (q)	image is in back of surface (real image).	image is in front of surface (virtual image).
Image height (h')	image is upright.	image is inverted.
Radius (R)	center of curvature is in back of surface.	center of curvature is in front of surface.

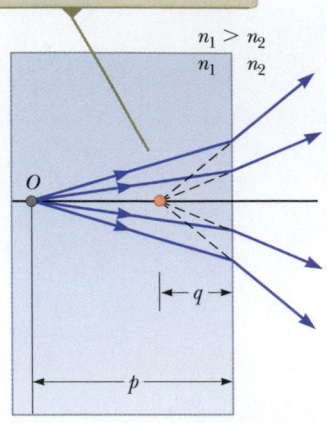

The image is virtual and on the same side of the surface as the object.

$n_1 > n_2$
n_1 n_2

O

q

p

Active Figure 26.18 The image formed by a flat refracting surface. All rays are assumed to be paraxial.

From Equation 26.10, we see that the sign of q is opposite that of p. Therefore, the image formed by a flat refracting surface is on the same side of the surface as the object. This situation is illustrated in Active Figure 26.18 for the case in which n_1 is greater than n_2, where a virtual image is formed between the object and the surface. Note that the refracted ray bends *away* from the normal in this case because $n_1 > n_2$.

The value of q given by Equation 26.10 is always smaller in magnitude than p when $n_1 > n_2$. This fact indicates that the image of an object located within a material with higher index of refraction than that of the material from which it is viewed is always closer to the flat refracting surface than the object. Therefore, transparent bodies of water such as streams and swimming pools always appear shallower than they are because the image of the bottom of the body of water is closer to the surface than the bottom is in reality.

Example 26.3 | Gaze into the Crystal Ball

A set of coins is embedded in a spherical plastic paperweight having a radius of 3.0 cm. The index of refraction of the plastic is $n_1 = 1.50$. One coin is located 2.0 cm from the edge of the sphere (Fig. 26.19). Find the position of the image of the coin.

Figure 26.19
(Example 26.3) Light rays from a coin embedded in a plastic sphere form a virtual image between the surface of the object and the sphere surface. Because the object is inside the sphere, the front of the refracting surface is the *interior* of the sphere.

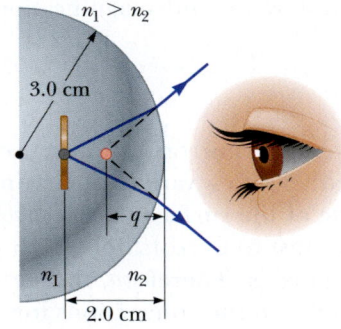

$n_1 > n_2$

3.0 cm

q

n_1 n_2

2.0 cm

SOLUTION

Conceptualize Because $n_1 > n_2$, where $n_2 = 1.00$ is the index of refraction for air, the rays originating from the coin in Figure 26.19 are refracted away from the normal at the surface and diverge outward.

Categorize Because the light rays originate in one material and then pass through a curved surface into another material, this example involves an image formed by refraction.

. . . .

Analyze Apply Equation 26.8, noting from Table 26.2 that R is negative:

$$\frac{n_2}{q} = \frac{n_2 - n_1}{R} - \frac{n_1}{p}$$

$$\frac{1}{q} = \frac{1.00 - 1.50}{-3.0 \text{ cm}} - \frac{1.50}{2.0 \text{ cm}}$$

$$q = \boxed{-1.7 \text{ cm}}$$

. . . .

Finalize The negative sign for q indicates that the image is in front of the surface; in other words, it is in the same medium as the object as shown in Figure 26.19. Therefore, the image must be virtual. (See Table 26.2.) The coin appears to be closer to the paperweight surface than it actually is.

Example 26.4 | The One That Got Away

A small fish is swimming at a depth d below the surface of a pond (Fig. 26.20).

(A) What is the apparent depth of the fish as viewed from directly overhead?

SOLUTION

Conceptualize Because $n_1 > n_2$, where $n_2 = 1.00$ is the index of refraction for air, the rays originating from the fish in Figure 26.20a are refracted away from the normal at the surface and diverge outward.

Categorize Because the refracting surface is flat, R is infinite. Hence, we can use Equation 26.10 to determine the location of the image with $p = d$.

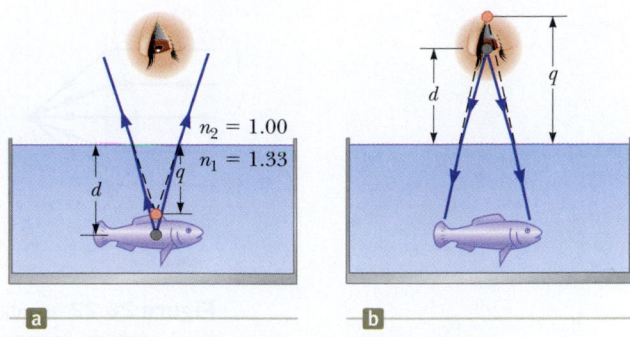

Figure 26.20 (Example 26.4) (a) The apparent depth q of the fish is less than the true depth d. All rays are assumed to be paraxial. (b) Your face appears to the fish to be higher above the surface than it is.

Analyze Use the indices of refraction given in Figure 26.20a in Equation 26.10:

$$q = -\frac{n_2}{n_1}p = -\frac{1.00}{1.33}d = -0.752d$$

Finalize Because q is negative, the image is virtual as indicated by the dashed lines in Figure 26.20a. The apparent depth is approximately three-fourths the actual depth.

(B) If your face is a distance d above the water surface, at what apparent distance above the surface does the fish see your face?

SOLUTION

The light rays from your face are shown in Figure 26.20b.

Conceptualize Because the rays refract toward the normal, your face appears higher above the surface than it actually is.

Categorize Because the refracting surface is flat, R is infinite. Hence, we can use Equation 26.10 to determine the location of the image with $p = d$.

Analyze Use Equation 26.10 to find the image distance:

$$q = -\frac{n_2}{n_1}p = -\frac{1.33}{1.00}d = -1.33d$$

Finalize The negative sign for q indicates that the image is in the medium from which the light originated, which is the air above the water.

❰26.4 | Images Formed by Thin Lenses

A typical **thin lens** consists of a piece of glass or plastic, ground so that its two surfaces are either segments of spheres or planes. Lenses are commonly used in optical instruments such as cameras, telescopes, and microscopes to form images by refraction.

Figure 26.21 shows cross sections of some representative shapes of lenses. These lenses have been placed in two groups. Those in Figure 26.21a are thicker at the center than at the rim, and those in Figure 26.21b are thinner at the center than at the rim. The lenses in the first group are examples of **converging lenses,** and those in the second group are called **diverging lenses.** The reason for these names will become apparent shortly.

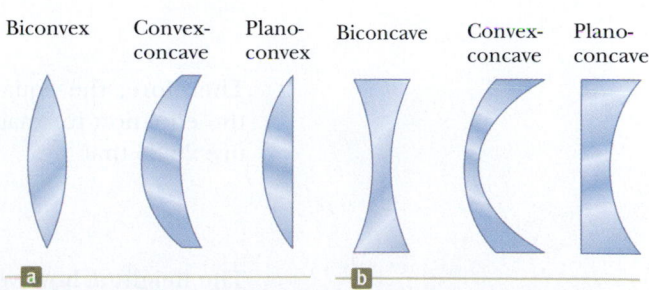

Biconvex   Convex-concave   Plano-convex   Biconcave   Convex-concave   Plano-concave

Figure 26.21 Various lens shapes. (a) Converging lenses have a positive focal length and are thickest at the middle. (b) Diverging lenses have a negative focal length and are thickest at the edges.

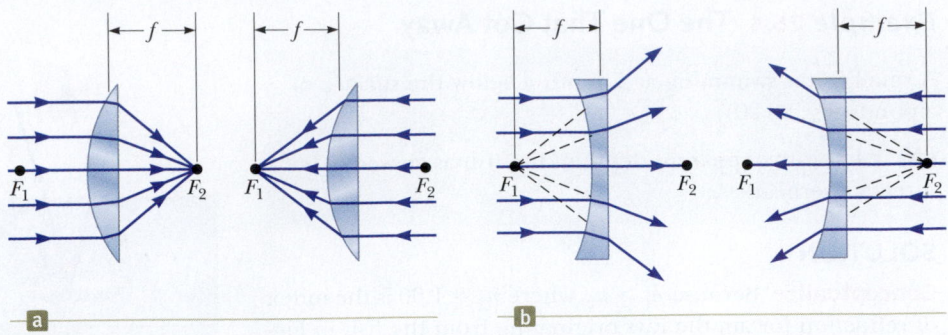

Figure 26.22 Parallel light rays pass through (a) a converging lens and (b) a diverging lens. The focal length is the same for light rays passing through a given lens in either direction. Both focal points F_1 and F_2 are the same distance from the lens.

Pitfall Prevention | 26.4

A Lens Has Two Focal Points but Only One Focal Length

A lens has a focal point on each side, front and back. There is only one focal length, however; each of the two focal points is located the same distance from the lens (Fig. 26.22). As a result, the lens forms an image of an object at the same point if it is turned around. In practice, that might not happen because real lenses are not infinitesimally thin.

As with mirrors, it is convenient to define a point called the **focal point** for a lens. For example, in Figure 26.22a, a group of rays parallel to the principal axis passes through the focal point after being converged by the lens. The distance from the focal point to the lens is again called the **focal length** f. The focal length is the image distance that corresponds to an infinite object distance.

To avoid the complications arising from the thickness of the lens, we adopt a simplification model called the **thin lens approximation,** in which the thickness of the lens is assumed to be negligible. As a result, it makes no difference whether we take the focal length to be the distance from the focal point to the surface of the lens or from the focal point to the center of the lens because the difference in these two lengths is assumed to be negligible. (We will draw lenses in the diagrams with a thickness so that they can be seen.) A thin lens has one focal length and *two* focal points as illustrated in Figure 26.22, corresponding to parallel light rays traveling from the left or right.

Rays parallel to the axis diverge after passing through a lens of the shape shown in Figure 26.22b. In this case, the focal point is defined as the point from which the diverging rays appear to originate, as in Figure 26.22b. Figures 26.22a and 26.22b indicate why the names *converging* and *diverging* are applied to the lenses in Figure 26.21.

Consider now the ray diagram in Figure 26.23 for an object placed a distance p from a converging lens. The red ray from the tip of the object passes through the center of the lens. The blue ray is parallel to the principal axis of the lens (the horizontal axis passing through the center of the lens), and as a result it passes through the focal point F after refraction. The point at which these two rays intersect is the image point at a distance q from the lens.

The tangent of the angle α can be found by using the blue and yellow geometric model triangles in Figure 26.23:

$$\tan \alpha = \frac{h}{p} \quad \text{and} \quad \tan \alpha = -\frac{h'}{q}$$

from which

$$M = \frac{h'}{h} = -\frac{q}{p} \qquad \text{26.11} \blacktriangleleft$$

Therefore, the equation for magnification of an image by a lens is the same as the equation for magnification due to a mirror (Eq. 26.2). We also note from Figure 26.23 that

$$\tan \theta = \frac{d}{f} \quad \text{and} \quad \tan \theta = -\frac{h'}{q - f}$$

The height d, however, is the same as h. Therefore,

$$\frac{h}{f} = -\frac{h'}{q - f} \quad \rightarrow \quad \frac{h'}{h} = -\frac{q - f}{f}$$

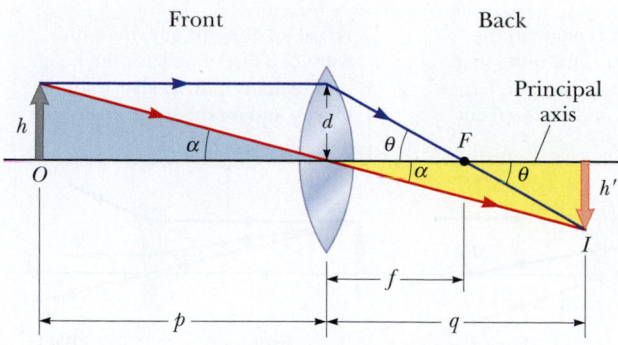

Figure 26.23 A geometric construction for developing the thin lens equation.

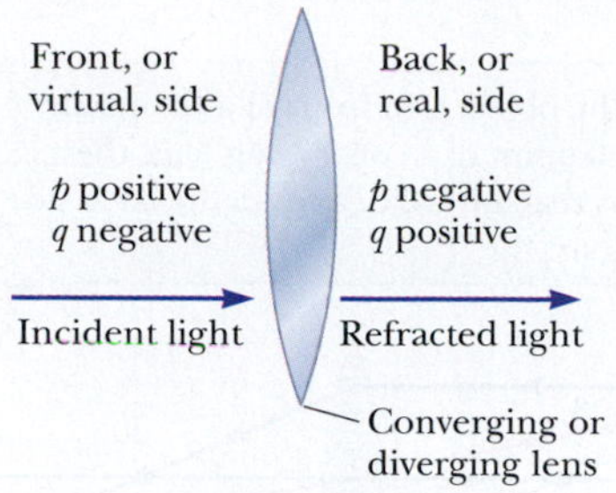

Front ... Back

Principal axis

F

h

O

α

d

θ

α

θ

h'

I

f

p

q

Front, or virtual, side ... Back, or real, side

p positive
q negative

p negative
q positive

Incident light ... Refracted light

Converging or diverging lens

Figure 26.24 A diagram for obtaining the signs of p and q for a thin lens. (This diagram also applies to a refracting surface.)

Using this expression in combination with Equation 26.11 gives us

$$\frac{q}{p} = \frac{q - f}{f}$$

which reduces to

$$\frac{1}{p} + \frac{1}{q} = \frac{1}{f}$$

26.12◄ ▶ Thin lens equation

This equation, called the **thin lens equation** (which is identical to the mirror equation, Eq. 26.6), can be used with either converging or diverging lenses if we adhere to a set of sign conventions. Figure 26.24 is useful for obtaining the signs of p and q. (As with mirrors, we call the side from which the light approaches the *front* of the lens.) The complete sign conventions for lenses are provided in Table 26.3. Note that a converging lens has a positive focal length under this convention and a diverging lens has a negative focal length. Hence, the names *positive* and *negative* are often given to these lenses.

The focal length for a lens in air is related to the curvatures of its surfaces and to the index of refraction n of the lens material by

$$\frac{1}{f} = (n - 1)\left(\frac{1}{R_1} - \frac{1}{R_2}\right)$$

26.13◄ ▶ Lens-makers' equation

where R_1 is the radius of curvature of the front surface and R_2 is the radius of curvature of the back surface. Equation 26.13 enables us to calculate the focal length from the known properties of the lens. It is called the **lens-makers' equation.** Table 26.3 includes the sign conventions for determining the signs of the radii R_1 and R_2.

Ray Diagrams for Thin Lenses

Our specialized pictorial representations called ray diagrams are very convenient for locating the image of a thin lens or system of lenses. They should also help clarify the

◀ TABLE 26.3 | Sign Conventions for Thin Lenses

Quantity	Positive when . . .	Negative when . . .
Object location (p)	object is in front of lens (real object).	object is in back of lens (virtual object).
Image location (q)	image is in back of lens (real image).	image is in front of lens (virtual image).
Image height (h')	image is upright.	image is inverted.
R_1 and R_2	center of curvature is in back of lens.	center of curvature is in front of lens.
Focal length (f)	a converging lens.	a diverging lens.

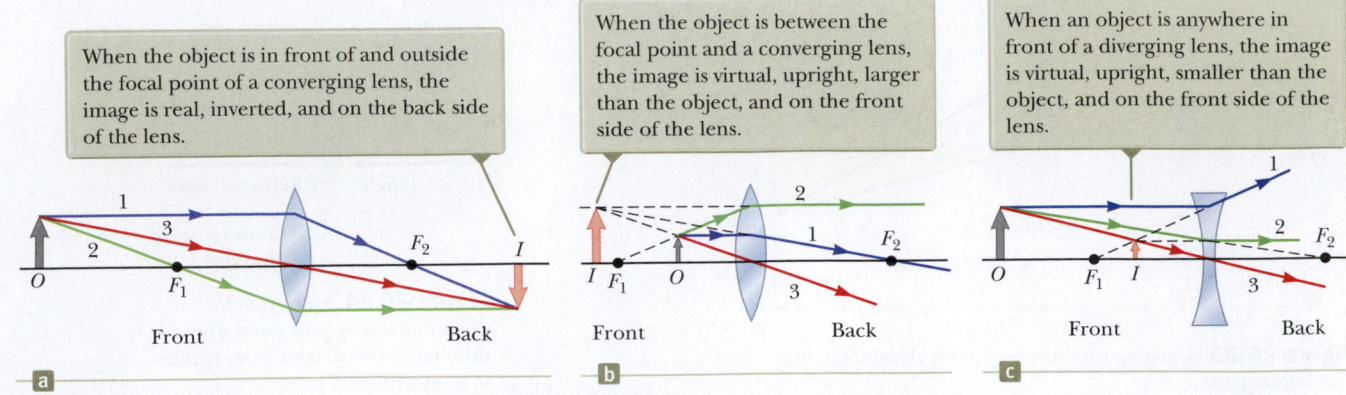

When the object is in front of and outside the focal point of a converging lens, the image is real, inverted, and on the back side of the lens.

When the object is between the focal point and a converging lens, the image is virtual, upright, larger than the object, and on the front side of the lens.

When an object is anywhere in front of a diverging lens, the image is virtual, upright, smaller than the object, and on the front side of the lens.

Active Figure 26.25 Ray diagrams for locating the image formed by a thin lens.

sign conventions we have already discussed. Active Figure 26.25 illustrates this method for three single-lens situations. To locate the image of a converging lens (Active Figs. 26.25a and 26.25b), the following three rays are drawn from the top of the object:

- Ray 1 is drawn parallel to the principal axis. After being refracted by the lens, this ray passes through the focal point on the back side of the lens.
- Ray 2 is drawn through the focal point on the front side of the lens (or as if coming from the focal point if $p < f$) and emerges from the lens parallel to the principal axis.
- Ray 3 is drawn through the center of the lens and continues in a straight line.

To locate the image of a *diverging* lens (Active Fig. 26.25c), the following three rays are drawn from the top of the object:

- Ray 1 is drawn parallel to the principal axis. After being refracted by the lens, this ray emerges directed away from the focal point on the front side of the lens.
- Ray 2 is drawn in the direction toward the focal point on the back side of the lens and emerges from the lens parallel to the principal axis.
- Ray 3 is drawn through the center of the lens and continues in a straight line.

In these ray diagrams, the point of intersection of *any two* of the rays can be used to locate the image. The third ray serves as a check of construction.

For the converging lens in Active Figure 26.25a, where the object is *outside* the front focal point ($p > f$), the image is real and inverted and is located on the back side of the lens. This diagram would be representative of a movie projector, for which the film is the object, the lens is in the projector, and the image is projected on a large screen for the audience to watch. The film is placed in the projector with the scene upside down so that the inverted image is right side up for the audience.

When the object is *inside* the front focal point ($p < f$) as in Active Figure 26.25b, the image is virtual and upright. When used in this way, the lens is acting as a magnifying glass, providing an enlarged upright image for closer study of an object. The object might be a stamp, a fingerprint, or a printed page for someone with failing eyesight.

For the diverging lens of Active Figure 26.25c, the image is virtual and upright for all object locations. A diverging lens is used in a security peephole in a door to give a wide-angle view. Nearsighted individuals use diverging eyeglass lenses or contact lenses. Another use is for a panoramic lens for a camera (although a sophisticated camera "lens" is actually a combination of several lenses). A diverging lens in this application creates a small image of a wide field of view.

QUICK QUIZ 26.5 What is the focal length of a pane of window glass? **(a)** zero **(b)** infinity **(c)** the thickness of the glass **(d)** impossible to determine

QUICK QUIZ 26.6 If you cover the top half of the lens in Active Figure 26.25a with a piece of paper, which of the following happens to the appearance of the image of the object? (a) The bottom half disappears. (b) The top half disappears. (c) The entire image is visible but dimmer. (d) There is no change. (e) The entire image disappears.

THINKING PHYSICS 26.3

BIO Corrective lenses on diving masks

Diving masks often have lenses built into the glass for divers who do not have perfect vision. This kind of mask allows the individual to dive without the necessity for glasses because the lenses in the faceplate perform the necessary refraction to provide clear vision. Normal glasses have lenses that are curved on both the front and rear surfaces. The lenses in a diving mask faceplate often only have curved surfaces on the *inside* of the glass. Why is this design desirable?

Reasoning The main reason for curving only the inner surface of the lenses in the diving mask faceplate is so that the diver can see clearly when looking at objects straight ahead while underwater *and* in the air. Consider light rays approaching the mask along a normal to the plane of the faceplate. If curved surfaces were on both the front and the back of the diving lens on the faceplate, refraction would occur at each surface. The lens could be designed so that these two refractions would give clear vision while the diver is in air. When the diver is underwater, however, the refraction between the water and the glass at the first interface is now different because the index of refraction of water is different from that of air. Therefore, the vision would not be clear underwater.

By making the outer surface of the lens flat, light is not refracted at normal incidence to the faceplate at the outer surface *in either air or water;* all the refraction occurs at the inner glass–air surface. Therefore, the same refractive correction exists in water and in air, and the diver can see clearly in both environments. ◄

Example 26.5 | Images Formed by a Converging Lens

A converging lens has a focal length of 10.0 cm.

(A) An object is placed 30.0 cm from the lens. Construct a ray diagram, find the image distance, and describe the image.

SOLUTION

Conceptualize Because the lens is converging, the focal length is positive (see Table 26.3). We expect the possibilities of both real and virtual images.

Categorize Because the object distance is larger than the focal length, we expect the image to be real. The ray diagram for this situation is shown in Figure 26.26a.

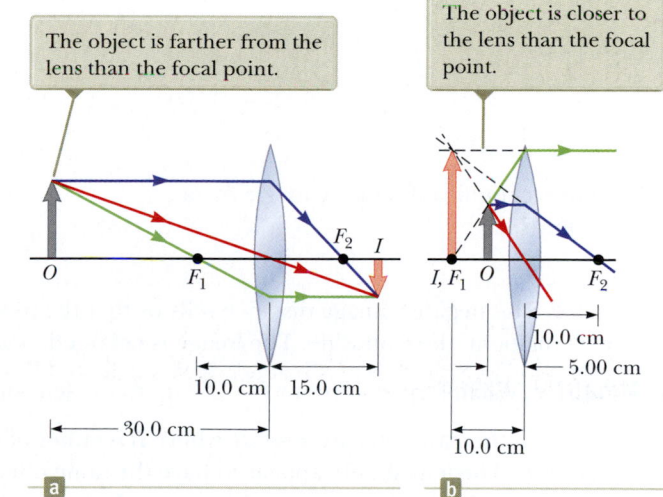

The object is farther from the lens than the focal point.

The object is closer to the lens than the focal point.

Figure 26.26
(Example 26.5) An image is formed by a converging lens.

Analyze Find the image distance by using Equation 26.12:

$$\frac{1}{q} = \frac{1}{f} - \frac{1}{p}$$

$$\frac{1}{q} = \frac{1}{10.0 \text{ cm}} - \frac{1}{30.0 \text{ cm}}$$

$$q = +15.0 \text{ cm}$$

Find the magnification of the image from Equation 26.11:

$$M = -\frac{q}{p} = -\frac{15.0 \text{ cm}}{30.0 \text{ cm}} = -0.500$$

continued

26.5 *cont.*

Finalize The positive sign for the image distance tells us that the image is indeed real and on the back side of the lens. The magnification of the image tells us that the image is reduced in height by one half, and the negative sign for M tells us that the image is inverted.

(B) An object is placed 10.0 cm from the lens. Find the image distance and describe the image.

SOLUTION

Categorize Because the object is at the focal point, we expect the image to be infinitely far away.

Analyze Find the image distance by using Equation 26.12:

$$\frac{1}{q} = \frac{1}{f} - \frac{1}{p}$$

$$\frac{1}{q} = \frac{1}{10.0 \text{ cm}} - \frac{1}{10.0 \text{ cm}}$$

$$q = \infty$$

Finalize This result means that rays originating from an object positioned at the focal point of a lens are refracted so that the image is formed at an infinite distance from the lens; that is, the rays travel parallel to one another after refraction.

(C) An object is placed 5.00 cm from the lens. Construct a ray diagram, find the image distance, and describe the image.

SOLUTION

Categorize Because the object distance is smaller than the focal length, we expect the image to be virtual. The ray diagram for this situation is shown in Figure 26.26b.

Analyze Find the image distance by using Equation 26.12:

$$\frac{1}{q} = \frac{1}{f} - \frac{1}{p}$$

$$\frac{1}{q} = \frac{1}{10.0 \text{ cm}} - \frac{1}{5.00 \text{ cm}}$$

$$q = -10.0 \text{ cm}$$

Find the magnification of the image from Equation 26.11:

$$M = -\frac{q}{p} = -\left(\frac{-10.0 \text{ cm}}{5.00 \text{ cm}}\right) = +2.00$$

Finalize The negative image distance tells us that the image is virtual and formed on the side of the lens from which the light is incident, the front side. The image is enlarged, and the positive sign for M tells us that the image is upright.

What If? What if the object moves right up to the lens surface so that $p \to 0$? Where is the image?

Answer In this case, because $p \ll R$, where R is either of the radii of the surfaces of the lens, the curvature of the lens can be ignored. The lens should appear to have the same effect as a flat piece of material, which suggests that the image is just on the front side of the lens, at $q = 0$. This conclusion can be verified mathematically by rearranging the thin lens equation:

$$\frac{1}{q} = \frac{1}{f} - \frac{1}{p}$$

If we let $p \to 0$, the second term on the right becomes very large compared with the first and we can neglect $1/f$. The equation becomes

$$\frac{1}{q} = -\frac{1}{p} \quad \rightarrow \quad q = -p = 0$$

Therefore, q is on the front side of the lens (because it has the opposite sign as p) and right at the lens surface.

Combinations of Thin Lenses

If two thin lenses are used to form an image, the system can be treated in the following manner. The position of the image of the first lens is calculated as though the second lens were not present. The light then approaches the second lens *as if* it had originally come from the image formed by the first lens. Hence, the image of the first lens is treated as the object of the second lens. The image of the second lens is the final image of the system. If the image of the first lens lies on the back side of the second lens, the image is treated as a *virtual object* for the second lens (i.e., p is negative). The same procedure can be extended to a system of three or more lenses. The overall magnification of a system of thin lenses equals the *product* of the magnifications of the separate lenses.

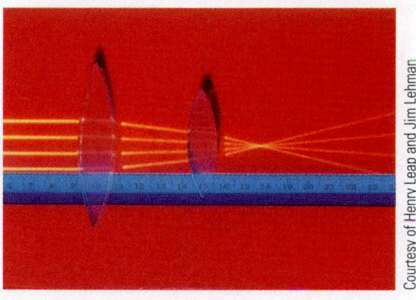

Light from a distant object brought into focus by two converging lenses.

Example 26.6 | Where Is the Final Image?

Two thin converging lenses of focal lengths $f_1 = 10.0$ cm and $f_2 = 20.0$ cm are separated by 20.0 cm as illustrated in Figure 26.27. An object is placed 30.0 cm to the left of lens 1. Find the position and the magnification of the final image.

SOLUTION

Conceptualize Imagine light rays passing through the first lens and forming a real image (because $p > f$) in the absence of a second lens. Figure 26.27 shows these light rays forming the inverted image I_1. Once the light rays converge to the image point, they do not stop. They continue through the image point and interact with the second lens. The rays leaving the image point behave in the same way as the rays

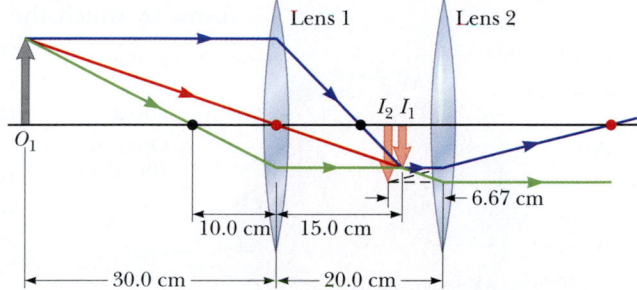

Figure 26.27 (Example 26.6) A combination of two converging lenses. The ray diagram shows the location of the final image (I_2) due to the combination of lenses. The black dots are the focal points of lens 1, and the red dots are the focal points of lens 2.

leaving an object. Therefore, the image of the first lens serves as the object of the second lens.

Categorize We categorize this problem as one in which the thin lens equation is applied in a stepwise fashion to the two lenses.

Analyze Find the location of the image formed by lens 1 from the thin lens equation:

$$\frac{1}{q_1} = \frac{1}{f} - \frac{1}{p_1}$$

$$\frac{1}{q_1} = \frac{1}{10.0 \text{ cm}} - \frac{1}{30.0 \text{ cm}}$$

$$q_1 = +15.0 \text{ cm}$$

Find the magnification of the image from Equation 26.11:

$$M_1 = -\frac{q_1}{p_1} = -\frac{15.0 \text{ cm}}{30.0 \text{ cm}} = -0.500$$

The image formed by this lens acts as the object for the second lens. Therefore, the object distance for the second lens is 20.0 cm − 15.0 cm = 5.00 cm.

Find the location of the image formed by lens 2 from the thin lens equation:

$$\frac{1}{q_2} = \frac{1}{20.0 \text{ cm}} - \frac{1}{5.00 \text{ cm}}$$

$$q_2 = -6.67 \text{ cm}$$

Find the magnification of the image from Equation 26.11:

$$M_2 = -\frac{q_2}{p_2} = -\frac{-6.67 \text{ cm}}{5.00 \text{ cm}} = +1.33$$

The total magnification M of the image due to the two lenses is the product $M_1 M_2$:

$$M = M_1 M_2 = (-0.500)(1.33) = -0.667$$

Finalize The negative sign on the overall magnification indicates that the final image is inverted with respect to the initial object. Because the absolute value of the magnification is less than 1, the final image is smaller than the object. Because q_2 is negative, the final image is on the front, or left, side of lens 2. These conclusions are consistent with the ray diagram in Figure 26.27.

continued

26.6 *cont.*

What If? Suppose you want to create an upright image with this system of two lenses. How must the second lens be moved?

Answer Because the object is farther from the first lens than the focal length of that lens, the first image is inverted. Consequently, the second lens must invert the image once again so that the final image is upright. An inverted image is only formed by a converging lens if the object is outside the focal point. Therefore, the image formed by the first lens must be to the left of the focal point of the second lens in Figure 26.27. To make that happen, you must move the second lens at least as far away from the first lens as the sum $q_1 + f_2 = 15.0 \text{ cm} + 20.0 \text{ cm} = 35.0 \text{ cm}$.

26.5 | The Eye BIO

Like a camera, a normal eye focuses light and produces a sharp image. The mechanisms by which the eye controls the amount of light admitted and adjusts to produce correctly focused images, however, are far more complex, intricate, and effective than those in even the most sophisticated camera. In all respects, the eye is a physiological wonder.

Figure 26.28 shows the basic parts of the human eye. Light entering the eye passes through a transparent structure called the *cornea* (Fig. 26.29), behind which are a clear liquid (the *aqueous humor*), a variable aperture (the *pupil*, which is an opening in the *iris*), and the *crystalline lens*. Most of the refraction occurs at the outer surface of the eye, where the cornea is covered with a film of tears. Relatively little refraction occurs in the crystalline lens because the aqueous humor in contact with the lens has an average index of refraction close to that of the lens. The iris, which is the colored portion of the eye, is a muscular diaphragm that controls pupil size. The iris regulates the amount of light entering the eye by dilating, or opening, the pupil in low-light conditions and contracting, or closing, the pupil in high-light conditions.

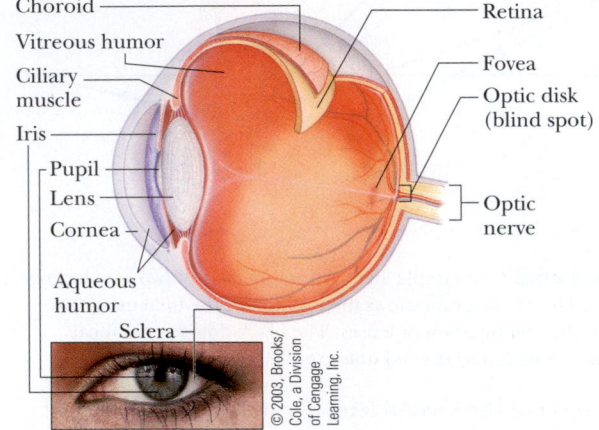

Figure 26.28 Important parts of the eye.

© 2003, Brooks/Cole, a Division of Cengage Learning, Inc.

The cornea–lens system focuses light onto the back surface of the eye, the *retina*, which consists of millions of sensitive receptors called *rods* and *cones*. When stimulated by light, these receptors send impulses via the optic nerve to the brain, where an image is perceived. By this process, a distinct image of an object is observed when the image falls on the retina.

The eye focuses on an object by varying the shape of the pliable crystalline lens through a process called **accommodation.** The lens adjustments take place so swiftly that we are not even aware of the change. Accommodation is limited in that objects very close to the eye produce blurred images. The **near point** is the closest distance for which the lens can accommodate to focus light on the retina. This distance usually increases with age and has an average value of 25 cm. At age 10, the near point of the eye is typically approximately 18 cm. It increases to approximately 25 cm at age 20, to 50 cm at age 40, and to 500 cm or greater at age 60. The **far point** of the eye represents the greatest distance for which the lens of the relaxed eye can focus light on the retina. A person with normal vision can see very distant objects and therefore has a far point that can be approximated as infinity.

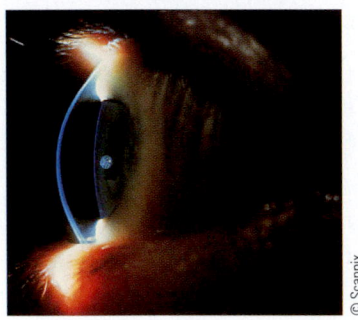

Figure 26.29 Close-up photograph of the cornea of the human eye.

© Scanpix

Only three types of color-sensitive cells are present in the retina. They are called red, green, and blue cones because of the peaks of the color ranges to which they respond (Fig. 26.30). If the red and green cones are stimulated simultaneously (as would be the case if yellow light were shining on them), the brain interprets what is seen as yellow. If all three types of cones are stimulated by the separate colors red, blue, and green, white light is seen. If all three types of cones are stimulated by light that contains *all* colors, such as sunlight, again white light is seen.

Televisions and computer monitors take advantage of this visual illusion by having only red, green, and blue dots on the screen. With specific combinations of brightness in these three primary colors, our eyes can be made to see any color in

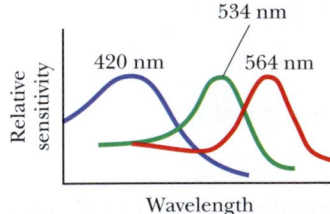

Figure 26.30 Approximate color sensitivity of the three types of cones in the retina.

the rainbow. Therefore, the yellow lemon you see in a television commercial is not actually yellow, it is red and green! The paper on which this page is printed is made of tiny, matted, translucent fibers that scatter light in all directions, and the resultant mixture of colors appears white to the eye. Snow, clouds, and white hair are not actually white. In fact, there is no such thing as a white pigment. The appearance of these things is a consequence of the scattering of light containing all colors, which we interpret as white.

Conditions of the Eye

When the eye suffers a mismatch between the focusing range of the lens–cornea system and the length of the eye, with the result that light rays from a near object reach the retina before they converge to form an image as shown in Figure 26.31a, the condition is known as **farsightedness** (or *hyperopia*). A farsighted person can usually see faraway objects clearly but not nearby objects. Although the near point of a normal eye is approximately 25 cm, the near point of a farsighted person is much farther away. The refracting power in the cornea and lens is insufficient to focus the light from all but distant objects satisfactorily. The condition can be corrected by placing a converging lens in front of the eye as shown in Figure 26.31b. The lens refracts the incoming rays more toward the principal axis before entering the eye, allowing them to converge and focus on the retina.

BIO Hyperopia

A person with **nearsightedness** (or *myopia*), another mismatch condition, can focus on nearby objects but not on faraway objects. The far point of the nearsighted eye is not infinity and may be less than 1 m. The maximum focal length of the nearsighted eye is insufficient to produce a sharp image on the retina, and rays from a distant object converge to a focus in front of the retina. They then continue past that point, diverging before they finally reach the retina and causing blurred vision (Fig. 26.32a). Nearsightedness can be corrected with a diverging lens as shown in Figure 26.32b. The lens refracts the rays away from the principal axis before they enter the eye, allowing them to focus on the retina.

BIO Myopia

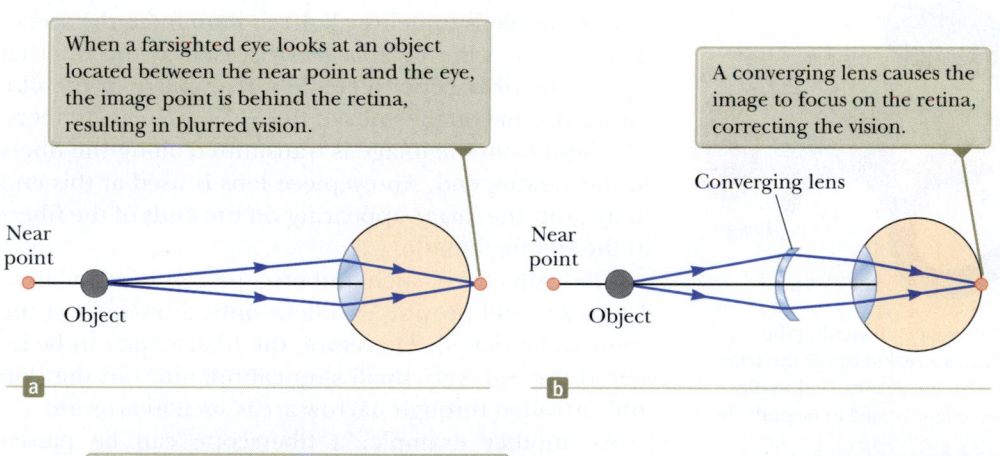

When a farsighted eye looks at an object located between the near point and the eye, the image point is behind the retina, resulting in blurred vision.

A converging lens causes the image to focus on the retina, correcting the vision.

Converging lens

Near point

Object

Near point

Object

a

b

Figure 26.31 (a) An uncorrected farsighted eye. (b) A farsighted eye corrected with a converging lens.

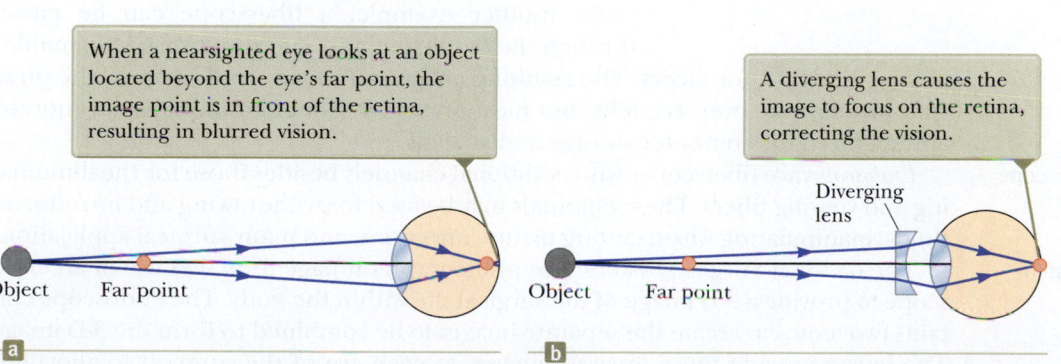

When a nearsighted eye looks at an object located beyond the eye's far point, the image point is in front of the retina, resulting in blurred vision.

A diverging lens causes the image to focus on the retina, correcting the vision.

Diverging lens

Object   Far point

Object   Far point

a

b

Figure 26.32 (a) An uncorrected nearsighted eye. (b) A nearsighted eye corrected with a diverging lens.

BIO Presbyopia

Beginning in middle age, most people lose some of their accommodation ability as their visual muscles weaken and the lens hardens. Unlike farsightedness, which is a mismatch between focusing power and eye length, **presbyopia** (literally, "old-age vision") is due to a reduction in accommodation ability. The cornea and lens do not have sufficient focusing power to bring nearby objects into focus on the retina. The symptoms are the same as those of farsightedness, and the condition can be corrected with converging lenses.

BIO Astigmatism

In eyes having a defect known as **astigmatism,** light from a point source produces a line image on the retina. This condition arises when the cornea, the lens, or both are not perfectly symmetric. Astigmatism can be corrected with lenses that have different curvatures in two mutually perpendicular directions.

Optometrists and ophthalmologists usually prescribe lenses measured in **diopters:** the **power** P of a lens in diopters equals the inverse of the focal length in meters: $P = 1/f$. For example, a converging lens of focal length $+20$ cm has a power of $+5.0$ diopters, and a diverging lens of focal length -40 cm has a power of -2.5 diopters.

> **QUICK QUIZ 26.7** Two campers wish to start a fire during the day. One camper is nearsighted, and one is farsighted. Whose glasses should be used to focus the Sun's rays onto some paper to start the fire? **(a)** either camper **(b)** the nearsighted camper **(c)** the farsighted camper

26.6 | Context Connection: Some Medical Applications BIO

BIO Medical uses of the fiberscope

The first use of optical fibers in medicine appeared with the invention of the *fiberscope* in 1957. Figure 26.33 indicates the construction of a fiberscope, which consists of two bundles of optical fibers. The *illuminating bundle* is an *incoherent* bundle, meaning that no effort is made to match the relative positions of the fibers at the two ends. This matching is not necessary because the sole purpose of this bundle is to deliver light to illuminate the scene. A lens (called the *objective lens*) is used at the internal end of the fiberscope to create a real image of the illuminated scene on the ends of the *viewing bundle* of fibers. The light from the image is transmitted along the fibers to the viewing end. An eyepiece lens is used at this end to magnify the image appearing on the ends of the fibers in the viewing bundle.

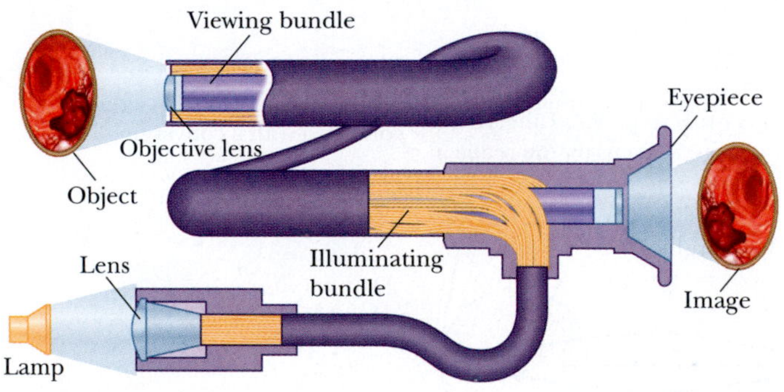

The diameter of such a fiberscope can be as small as 1 mm and still provide excellent optical imaging of the scene to be viewed. Therefore, the fiberscope can be inserted through very small surgical openings in the skin and threaded through narrow areas such as arteries.

As another example, a fiberscope can be passed through the esophagus and into the stomach to enable a physician to look for ulcers. The resulting image can be viewed directly by the physician through the eyepiece lens, but most often it is exhibited on a display monitor and digitized for computer storage and analysis.

Figure 26.33 The construction of a fiberscope for viewing the interior of the body. The objective lens forms a real image of the scene on the end of a bundle of optical fibers. This image is carried to the other end of the bundle, where an eyepiece lens is used to magnify the image for the physician.

BIO Medical uses of the endoscope

Endoscopes are fiberscopes with additional channels besides those for the illuminating and viewing fibers. These channels may be used for withdrawing and introducing fluids, manipulating wires, cutting tissues, injections, and many surgical applications.

BIO The da Vinci Surgical System

The da Vinci Surgical System (see photograph on page 3) makes use of an endoscope to provide a 3-D image of the surgical site within the body. The endoscope contains two lenses to create the separate images to be combined to form the 3-D image. The lenses provide these separate images to each eye of the surgeon to allow him or her to see the image in 3-D. The enhanced image quality provides the surgeon with better visualization of the site, whereas the arms of the robot provide enhanced dexterity and greater precision.

Lasers are used with endoscopes in a variety of medical diagnostic and treatment procedures. As a diagnostic example, the dependence on wavelength of the amount of reflection from a surface allows a fiberscope to be used to make a direct measurement of the blood's oxygen content. Using two laser sources, red light and infrared light are both sent into the blood through optical fibers. Hemoglobin reflects a known fraction of infrared light, regardless of the oxygen carried. Therefore, the measurement of the infrared reflection gives a total hemoglobin count. Red light is reflected much more by hemoglobin carrying oxygen than by hemoglobin that does not. Therefore, the amount of red laser light reflected allows a measurement of the ability of the patient's blood to carry oxygen.

BIO Use of lasers in measuring hemoglobin

Lasers are used to treat medical conditions such as *hydrocephalus,* which occurs in about 0.1% of births. This condition involves an increase in intracranial pressure due to an overproduction of cerebrospinal fluid (CSF), an obstruction of the flow of CSF, or insufficient absorption of CSF. In addition to congenital hydrocephalus, the condition can be acquired later in life due to trauma to the head, brain tumors, or other factors.

BIO Use of lasers in treating hydrocephalus

The older treatment method for obstructive hydrocephalus involved placing a shunt (tube) between ventricular chambers in the brain to allow passage of CSF. A new alternative is *laser-assisted ventriculostomy,* in which a new pathway for CSF is made with an infrared laser beam and an endoscope having a spherical end as shown in Figure 26.34. As the laser beam strikes the spherical end, refraction at the spherical surface causes light waves to spread out in all directions as if the end of the endoscope were a point source of radiation. The result is a rapid decrease in intensity with distance from the sphere, avoiding damage to vital structures in the brain that are close to the area in which a new passageway is to be made. The surface of the spherical end is coated with an infrared radiation-absorbing material, and the absorbed laser energy raises the temperature of the sphere. As the sphere is placed in contact with the location of the desired passageway, the combination of the high temperature and laser radiation leaving the sphere burns a new passageway for the CSF. This treatment requires much less recovery time as well as significantly less postoperative care than that associated with the placement of shunts.

Endoscope covering   Optical fiber bundle   Spherical end

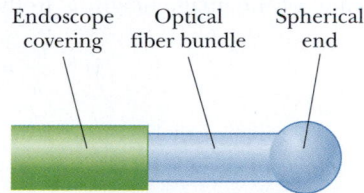

Figure 26.34 An endoscope probe used to open new passageways for cerebrospinal fluid in the treatment of hydrocephalus. Laser light raises the temperature of the sphere and radiates from the sphere to provide energy to tissues for cutting the new passageway.

Lasik (laser-assisted in situ keratomileusis) is a form of refractive eye surgery that uses lasers to correct myopia, hyperopia, and astigmatism. The surgical procedure involves three steps. First, a corneal suction ring is used to immobilize the eye. Once the eye is immobilized, a flap is created on the cornea by using either a metal blade or a laser. The flap is then folded back to reveal the middle section of the cornea called the *stroma.* In the next step of the procedure, the shape of the stroma is modified using a 193-nm wavelength *excimer laser.* The excimer laser vaporizes tissue in a finely controlled manner without damaging adjacent stroma. Finally, once the stroma layer has been reshaped, the flap is repositioned over the treated area and remains in this position by natural adhesion until healing is complete.

BIO Lasik surgery

Tattoos can be removed or modified using a specially designed *Q-switched laser* that penetrates the skin and targets the darker pigments of the tattoo but leaves surrounding tissues unharmed. The Q-switched laser provides very short bursts of energy, measured in nanoseconds, containing a large amount of energy in each burst. Absorption of energy from the laser breaks up the large particles of ink into smaller pieces that can naturally be removed by normal body processes. The short duration of the pulse prevents the energy from spreading into surrounding tissue. It takes several months for the body to eliminate the dissolved tattoo pigments.

BIO Use of lasers in tattoo removal

Patients having an enlarged prostate (benign prostatic hyperplasia) are sometimes treated using laser surgery. In this type of surgery, known by the general name of TURP (transurethral resection of the prostate), the laser removes tissue in the prostate that is blocking the flow of urine. A variety of lasers are used in this procedure, ranging from visible to infrared wavelengths.

BIO Use of lasers in benign prostatic hyperplasia

In Chapter 27, we shall investigate another application of lasers—the technology of *holography*—that has grown tremendously in recent years. In holography, three-dimensional images of objects are recorded on film.

SUMMARY

An **image** of an object is a point from which light either diverges or seems to diverge after interacting with a mirror or lens. If light passes through the image point, the image is a **real image.** If light only appears to diverge from the image point, the image is a **virtual image.**

In the **paraxial ray** simplification model, the object distance p and image distance q for a spherical mirror of radius R are related by the **mirror equation**

$$\frac{1}{p} + \frac{1}{q} = \frac{2}{R} = \frac{1}{f} \qquad \textbf{26.4, 26.6} \blacktriangleleft$$

where $f = R/2$ is the **focal length** of the mirror.

The **magnification** M of a mirror or lens is defined as the ratio of the image height h' to the object height h:

$$M = \frac{h'}{h} = -\frac{q}{p} \qquad \textbf{26.2, 26.11} \blacktriangleleft$$

An image can be formed by refraction from a spherical surface of radius R. The object and image distances for refraction from such a surface are related by

$$\frac{n_1}{p} + \frac{n_2}{q} = \frac{n_2 - n_1}{R} \qquad \textbf{26.8} \blacktriangleleft$$

where the light is incident from the medium of index of refraction n_1 and is refracted in the medium whose index of refraction is n_2.

For a thin lens, and in the paraxial ray approximation, the object and image distances are related by the **thin lens equation:**

$$\frac{1}{p} + \frac{1}{q} = \frac{1}{f} \qquad \textbf{26.12} \blacktriangleleft$$

The **focal length** f of a thin lens in air is related to the curvature of its surfaces and to the index of refraction n of the lens material by

$$\frac{1}{f} = (n - 1)\left(\frac{1}{R_1} - \frac{1}{R_2}\right) \qquad \textbf{26.13} \blacktriangleleft$$

Converging lenses have positive focal lengths, and **diverging lenses** have negative focal lengths.

OBJECTIVE QUESTIONS

☐ denotes answer available in *Student Solutions Manual/Study Guide*

1. **BIO** If a woman's eyes are longer than normal, how is her vision affected and how can her vision be corrected? (a) The woman is farsighted (hyperopia), and her vision can be corrected with a diverging lens. (b) The woman is nearsighted (myopia), and her vision can be corrected with a diverging lens. (c) The woman is farsighted, and her vision can be corrected with a converging lens. (d) The woman is nearsighted, and her vision can be corrected with a converging lens. (e) The woman's vision is not correctible.

2. **(i)** When an image of an object is formed by a plane mirror, which of the following statements is *always* true? More than one statement may be correct. (a) The image is virtual. (b) The image is real. (c) The image is upright. (d) The image is inverted. (e) None of those statements is always true. **(ii)** When the image of an object is formed by a concave mirror, which of the preceding statements are *always* true? **(iii)** When the image of an object is formed by a convex mirror, which of the preceding statements are *always* true?

3. An object is located 50.0 cm from a converging lens having a focal length of 15.0 cm. Which of the following statement is true regarding the image formed by the lens? (a) It is virtual, upright, and larger than the object. (b) It is real, inverted, and smaller than the object. (c) It is virtual, inverted, and smaller than the object. (d) It is real, inverted, and larger than the object. (e) It is real, upright, and larger than the object.

4. **(i)** When an image of an object is formed by a converging lens, which of the following statements is *always* true? More than one statement may be correct. (a) The image is virtual. (b) The image is real. (c) The image is upright. (d) The image is inverted. (e) None of those statements is always true. **(ii)** When the image of an object is formed by a diverging lens, which of the statements is *always* true?

5. **BIO** If a man has eyes that are shorter than normal, how is his vision affected and how can it be corrected? (a) The man is farsighted (hyperopia), and his vision can be corrected with a diverging lens. (b) The man is nearsighted (myopia), and his vision can be corrected with a diverging lens. (c) The man is farsighted, and his vision can be corrected with a converging lens. (d) The man is nearsighted, and his vision can be corrected with a converging lens. (e) The man's vision is not correctible.

6. If Josh's face is 30.0 cm in front of a concave shaving mirror creating an upright image 1.50 times as large as the object, what is the mirror's focal length? (a) 12.0 cm (b) 20.0 cm (c) 70.0 cm (d) 90.0 cm (e) none of those answers

7. A converging lens made of crown glass has a focal length of 15.0 cm when used in air. If the lens is immersed in water, what is its focal length? (a) negative (b) less than 15.0 cm (c) equal to 15.0 cm (d) greater than 15.0 cm (e) none of those answers

8. Two thin lenses of focal lengths $f_1 = 15.0$ and $f_2 = 10.0$ cm, respectively, are separated by 35.0 cm along a common axis. The f_1 lens is located to the left of the f_2 lens. An object is now placed 50.0 cm to the left of the f_1 lens, and a final image due to light passing though both lenses forms. By what factor is the final image different in size from the object? (a) 0.600 (b) 1.20 (c) 2.40 (d) 3.60 (e) none of those answers

9. Lulu looks at her image in a makeup mirror. It is enlarged when she is close to the mirror. As she backs away, the image becomes larger, then impossible to identify when she is 30.0 cm from the mirror, then upside down when she is beyond 30.0 cm, and finally small, clear, and upside down when she is much farther from the mirror. (i) Is the mirror (a) convex, (b) plane, or (c) concave? (ii) Is the magnitude of its focal length (a) 0, (b) 15.0 cm, (c) 30.0 cm, (d) 60.0 cm, or (e) ∞?

10. Model each of the following devices in use as consisting of a single converging lens. Rank the cases according to the ratio of the distance from the object to the lens to the focal length of the lens, from the largest ratio to the smallest. (a) a film-based movie projector showing a movie (b) a magnifying glass being used to examine a postage stamp (c) an astronomical refracting telescope being used to make a sharp image of stars on an electronic detector (d) a searchlight being used to produce a beam of parallel rays from a point source (e) a camera lens being used to photograph a soccer game

11. A converging lens of focal length 8 cm forms a sharp image of an object on a screen. What is the smallest possible distance between the object and the screen? (a) 0 (b) 4 cm (c) 8 cm (d) 16 cm (e) 32 cm

12. An object, represented by a gray arrow, is placed in front of a plane mirror. Which of the diagrams in Figure OQ26.12 correctly describes the image, represented by the pink arrow?

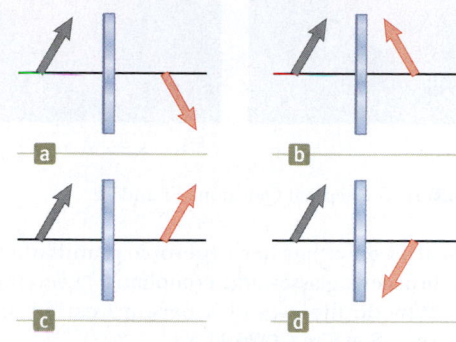

Figure OQ26.12

CONCEPTUAL QUESTIONS

☐ denotes answer available in *Student Solutions Manual/Study Guide*

1. Do the equations $1/p + 1/q = 1/f$ and $M = -q/p$ apply to the image formed by a flat mirror? Explain your answer.

2. **BIO** The optic nerve and the brain invert the image formed on the retina. Why don't we see everything upside down?

3. Consider a spherical concave mirror with the object located to the left of the mirror beyond the focal point. Using ray diagrams, show that the image moves to the left as the object approaches the focal point.

4. In Active Figure 26.25a, assume the gray object arrow is replaced by one that is much taller than the lens. (a) How many rays from the top of the object will strike the lens? (b) How many principal rays can be drawn in a ray diagram?

5. (a) Can a converging lens be made to diverge light if it is placed into a liquid? (b) **What If?** What about a converging mirror?

6. **BIO** Lenses used in eyeglasses, whether converging or diverging, are always designed so that the middle of the lens curves away from the eye like the center lenses of Figures 26.21a and 26.21b. Why?

7. Suppose you want to use a converging lens to project the image of two trees onto a screen. As shown in Figure CQ26.7, one tree is a distance x from the lens and the other is at $2x$. You adjust the screen so that the near tree is in focus. If you now want the far tree to be in focus, do you move the screen toward or away from the lens?

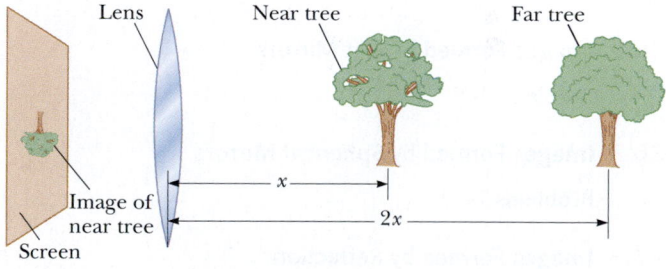

Figure CQ26.7

8. Explain why a fish in a spherical goldfish bowl appears larger than it really is.

9. Why do some emergency vehicles have the symbol ƎƆИAЈU8MA written on the front?

10. Explain this statement: "The focal point of a lens is the location of the image of a point object at infinity." (a) Discuss the notion of infinity in real terms as it applies to object distances. (b) Based on this statement, can you think of a simple method for determining the focal length of a converging lens?

11. **BIO** In Figures CQ26.11a and CQ26.11b, which glasses correct nearsightedness and which correct farsightedness?

Figure CQ26.11 Conceptual Questions 11 and 12.

12. Bethany tries on either her hyperopic grandfather's or her myopic brother's glasses and complains, "Everything looks blurry." Why do the eyes of a person wearing glasses not look blurry? (See Fig. CQ26.11.)

13. **BIO** During LASIK eye surgery (laser-assisted *in situ* keratomileusis), the shape of the cornea is modified by vaporizing some of its material. If the surgery is performed to correct for nearsightedness, how does the cornea need to be reshaped?

14. A solar furnace can be constructed by using a concave mirror to reflect and focus sunlight into a furnace enclosure. What factors in the design of the reflecting mirror would guarantee very high temperatures?

15. Figure CQ26.15 shows a lithograph by M. C. Escher titled *Hand with Reflection Sphere (Self-Portrait in Spherical Mirror)*. Escher said about the work: "The picture shows a spherical mirror, resting on a left hand. But as a print is the reverse of the original drawing on stone, it was my right hand that you see depicted. (Being left-handed, I needed my left hand to make the drawing.) Such a globe reflection collects almost one's whole surroundings in one disk-shaped image. The whole room, four walls, the floor, and the ceiling, everything, albeit distorted, is compressed into that one small circle. Your own head, or more exactly the point between your eyes, is the absolute center. No matter how you turn or twist yourself, you can't get out of that central point. You are immovably the focus, the unshakable core, of your world." Comment on the accuracy of Escher's description.

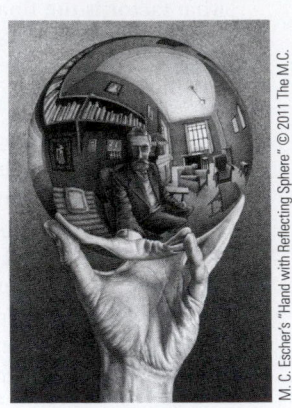

Figure CQ26.15

▶ PROBLEMS AVAILABLE IN ᴱᴺᴴᴬᴺᶜᴱᴰ Web**Assign**

26.1 Images Formed by Flat Mirrors

Problems 1–6

26.2 Images Formed by Spherical Mirrors

Problems 7–22

26.3 Images Formed by Refraction

Problems 23–28

26.4 Images Formed by Thin Lenses

Problems 29–43

26.5 The Eye

Problems 44–50

26.6 Context Connection: Some Medical Applications

Problems 51–52

Additional Problems

Problems 53–73

Solutions to the following Problems are available in the *Student Solutions Manual/Study Guide*:

26.1, 26.7, 26.11, 26.19, 26.24, 26.27, 26.30, 26.33, 26.44, 26.46, 26.49, 26.54, 26.61, 26.67, and 26.69

List of Enhanced Problems

Problem Number	Targeted Feedback in Enhanced WebAssign	Master It in Enhanced WebAssign	Watch It in Enhanced WebAssign
26.1	✓		✓
26.7	✓	✓	
26.11	✓	✓	
26.13	✓		✓
26.17	✓		✓
26.18	✓		✓
26.19		✓	
26.21	✓		✓
26.27		✓	
26.28	✓		✓
26.29	✓		✓
26.30		✓	
26.35	✓		✓
26.41	✓		✓
26.47	✓		✓
26.49	✓		
26.54		✓	
26.61		✓	

Wave Optics

Chapter Outline

Dec Hogan/Shutterstock.com

The colors in many of a hummingbird's feathers are not due to pigment. The iridescence that makes the brilliant colors that often appear on the bird's throat and belly is due to an interference effect caused by structures in the feathers. The colors will vary with the viewing angle.

In Chapters 25 and 26, we used the ray approximation to examine what happens when light reflects from a surface or refracts into a new medium. We used the general term *geometric optics* for these discussions. This chapter is concerned with **wave optics,** a subject that addresses the optical phenomena of interference and diffraction. These phenomena cannot be adequately explained with the ray approximation. We must address the wave nature of light to be able to understand these phenomena.

We introduced the concept of wave interference in Chapter 14 for one-dimensional waves. This phenomenon depends on the principle of superposition, which tells us that when two or more traveling mechanical waves combine at a given point, the resultant displacement of the elements of the medium at that point is the sum of the displacements due to the individual waves.

We shall see the full richness of the waves in interference model in this chapter as we apply it to light. We used one-dimensional waves on strings to introduce interference in Active Figures 14.1 and 14.2. As we discuss the interference of light waves, two major changes from this previous discussion must be noted. First, we shall no longer focus on one-dimensional waves, so we must build geometric models to analyze the situation in two or three dimensions. Second, we shall study electromagnetic waves rather than mechanical waves. Therefore, the principle of superposition needs to be cast in terms of addition of field vectors rather than displacements of the elements of the medium.

27.1 | Conditions for Interference

In our discussion of wave interference for mechanical waves in Chapter 14, we found that two waves can add together constructively or destructively. In constructive interference between waves, the amplitude of the resultant wave is greater than that of either individual wave, whereas in destructive interference, the resultant amplitude is less than that of either individual wave. Electromagnetic waves also undergo interference. Fundamentally, all interference associated with electromagnetic waves arises as a result of combining the electric and magnetic fields that constitute the individual waves.

In Figure 14.4, we described a device that allows interference to be observed for sound waves. Interference effects in visible electromagnetic waves are not easy to observe because of their short wavelengths (from about 4×10^{-7} to 7×10^{-7} m). Two sources producing two waves of identical wavelengths are needed to create interference. To produce a stable interference pattern, however, the individual waves must maintain a constant phase relationship with one another; they must be **coherent.** As an example, the sound waves emitted by two side-by-side loudspeakers driven by a single amplifier can produce interference because the two loudspeakers respond to the amplifier in the same way at the same time.

If two separate light sources are placed side by side, no interference effects are observed because the light waves from one source are emitted independently of the other source; hence, the emissions from the two sources do not maintain a constant phase relationship with each other over the time of observation. An ordinary light source undergoes random changes in time intervals less than a nanosecond. Therefore, the conditions for constructive interference, destructive interference, or some intermediate state are maintained only for such short time intervals. The result is that no interference effects are observed because the eye cannot follow such rapid changes. Such light sources are said to be **incoherent.**

27.2 | Young's Double-Slit Experiment

A common method for producing two coherent light sources is to use a monochromatic source to illuminate a barrier containing two small openings (usually in the shape of slits). The light emerging from the two slits is coherent because a single source produces the original light beam and the two slits serve only to separate the original beam into two parts (which, after all, is what was done to the sound signal from the side-by-side loudspeakers in the preceding section). Any random change in the light emitted by the source occurs in both beams at the same time, and, as a result, interference effects can be observed when the light from the two slits arrives at a viewing screen.

If the light traveled only in its original direction after passing through the slits as shown in Figure 27.1a, the waves would not overlap and no interference pattern would be seen. Instead, as we have discussed in our treatment of Huygens's principle (Section 25.6), the waves spread out from the slits as shown in Figure 27.1b. In other words, the light deviates from a straight-line path and enters the region that would otherwise be shadowed. As noted in Section 25.2, this divergence of light from its initial line of travel is called **diffraction.**

Interference in light waves from two sources was first demonstrated by Thomas Young in 1801. A schematic diagram of the apparatus that Young used is shown in Active Figure 27.2a (page 782). Plane light waves arrive at a barrier that contains two parallel slits S_1 and S_2. These two slits serve as a pair of coherent light sources because waves emerging from them originate from the same wave front and therefore maintain a constant phase relationship. The light from S_1 and S_2 produces on a viewing screen a visible pattern of bright and dark parallel bands

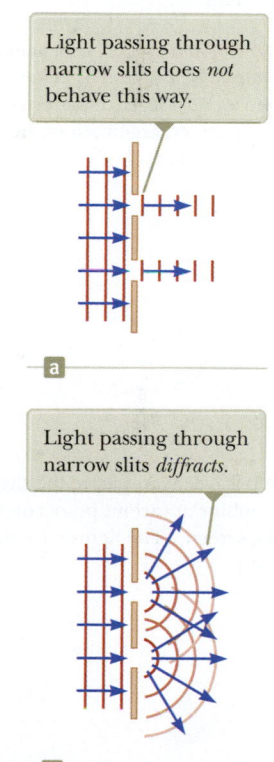

Figure 27.1 (a) If light waves did not spread out after passing through the slits, no interference would occur. (b) The light waves from the two slits overlap as they spread out, filling what we expect to be shadowed regions with light and producing interference fringes on a screen placed to the right of the slits.

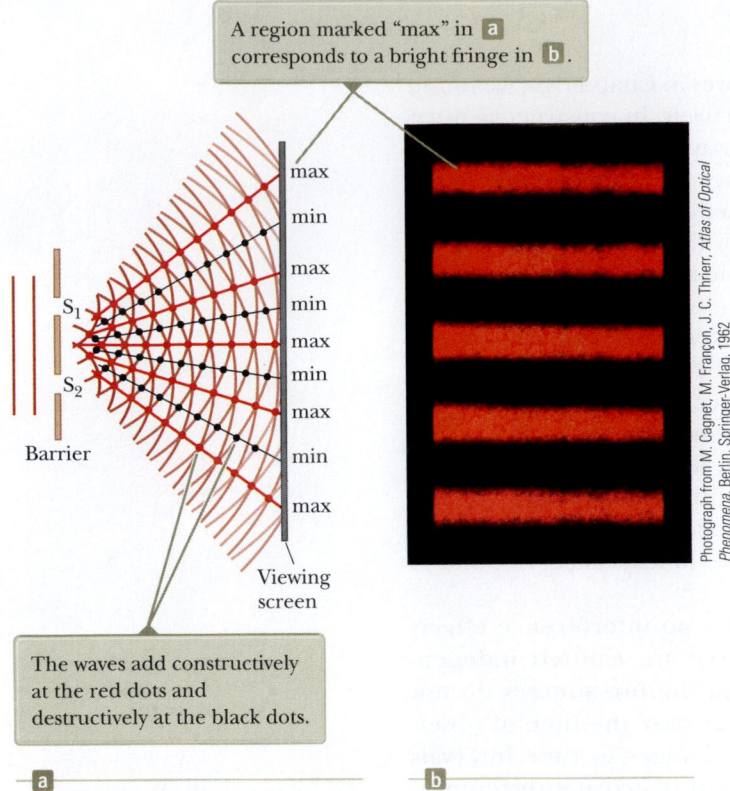

A region marked "max" in **a** corresponds to a bright fringe in **b**.

max
min
max
min
max
min
max
min
max
min
max

S₁

S₂

Barrier

Viewing screen

Photograph from M. Cagnet, M. Françon, J. C. Thrier, *Atlas of Optical Phenomena*, Berlin, Springer-Verlag, 1962

The waves add constructively at the red dots and destructively at the black dots.

a   **b**

Active Figure 27.2 (a) Schematic diagram of Young's double-slit experiment. Slits S₁ and S₂ behave as coherent sources of light waves that produce an interference pattern on the viewing screen (drawing not to scale). (b) An enlargement of the center of a fringe pattern formed on the viewing screen.

called **fringes** (Active Fig. 27.2b). When the light from S_1 and that from S_2 both arrive at a point on the screen such that constructive interference occurs at that location, a bright fringe appears. When the light from the two slits combines destructively at any location on the screen, a dark fringe results.

Figure 27.3 is a schematic diagram that allows us to generate a mathematical representation by modeling the interference as if waves combine at the viewing screen.[1] In Figure 27.3a, two waves leave the two slits in phase and strike the screen at the central point O. Because these waves travel equal distances, they arrive in phase at O. As a result, constructive interference occurs at this location and a bright fringe is observed. In Figure 27.3b, the two light waves again start in phase, but the lower wave has to travel one wavelength farther to reach point P on the screen. Because the lower wave falls behind the upper one by exactly one wavelength, they still arrive in phase at P. Hence, a second bright fringe appears at this location. Now consider point R located between O and P in Figure 27.3c. At this location, the lower wave has fallen half a wavelength behind the upper wave when they arrive at the screen. Hence, the trough from the lower wave overlaps the crest from the upper wave, giving rise to destructive interference at R. For this reason, one observes a dark fringe at this location.

Young's double-slit experiment is the prototype for many interference effects. Interference of waves occurs relatively commonly in technological applications, so this phenomenon represents an important analysis model to understand. In the next section, we develop the mathematical representation for interference of light.

Figure 27.3 Waves leave the slits and combine at various points on the viewing screen. (The figures are not to scale.)

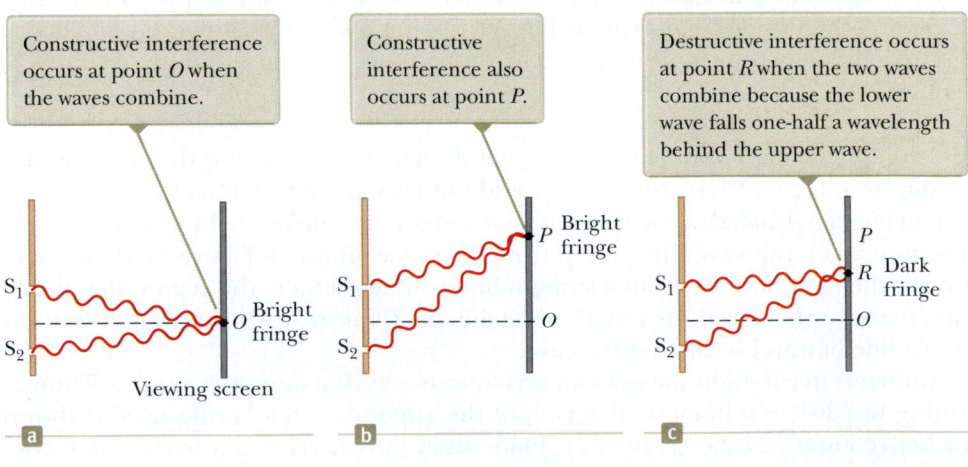

Constructive interference occurs at point O when the waves combine.

S₁

S₂

O Bright fringe

Viewing screen

a

Constructive interference also occurs at point P.

S₁

S₂

P Bright fringe

O

b

Destructive interference occurs at point R when the two waves combine because the lower wave falls one-half a wavelength behind the upper wave.

S₁

S₂

P

R Dark fringe

O

c

[1]The interference occurs everywhere between the slits and the screen, not only at the screen. See Thinking Physics 27.1. The model we have proposed will give us a valid result.

27.3 | Analysis Model: Waves in Interference

We can obtain a quantitative description of Young's experiment with the help of a geometric model constructed from Figure 27.4a. The viewing screen is located a perpendicular distance L from the slits S_1 and S_2, which are separated by a distance d. Consider point P on the screen. Angle θ is measured from a line perpendicular to the screen from the midpoint between the slits and a line from the midpoint to point P. We identify r_1 and r_2 as the distances the waves travel from slit to screen. Let us assume that the source is monochromatic. Under these conditions, the waves emerging from S_1 and S_2 have the same wavelength and amplitude and are in phase. The light intensity on the screen at P is the result of the superposition of the light coming from both slits. Note from the geometric model triangle in yellow in Figure 27.4a that a wave from the lower slit travels farther than a wave from the upper slit by an amount δ (Greek letter delta). This distance is called the **path difference.**

If L is much greater than d, the two paths are very close to being parallel. We shall adopt a simplification model in which the two paths are exactly parallel. In this case, from Figure 27.4b, we see that

$$\delta = r_2 - r_1 = d \sin \theta \qquad \textbf{27.1}\blacktriangleleft \qquad \blacktriangleright \text{ Path difference}$$

In Figure 27.4a, the condition $L \gg d$ is not satisfied because the figure is not to scale; in Figure 27.4b, the rays leave the slits as if the condition is satisfied. As noted earlier, the value of this path difference determines whether the two waves are in phase or out of phase when they arrive at P. If the path difference is either zero or some integral multiple of the wavelength, the two waves are in phase at P and **constructive interference** results. The condition for bright fringes at P is therefore

$$\delta = d \sin \theta_{\text{bright}} = m\lambda \qquad m = 0, \pm 1, \pm 2, \ldots \qquad \textbf{27.2}\blacktriangleleft \qquad \blacktriangleright \text{ Conditions for constructive interference}$$

The number m is an integer called the **order number.** The central bright fringe at $\theta_{\text{bright}} = 0$ is associated with the order number $m = 0$ and is called the **zeroth-order maximum.** The first maximum on either side, for which $m = \pm 1$, is called the **first-order maximum,** and so forth.

Similarly, when the path difference is an odd multiple of $\lambda/2$, the two waves arriving at P are 180° out of phase and give rise to **destructive interference.** Therefore, the condition for dark fringes at P is

$$\delta = d \sin \theta_{\text{dark}} = (m + \tfrac{1}{2})\lambda \qquad m = 0, \pm 1, \pm 2, \ldots \qquad \textbf{27.3}\blacktriangleleft \qquad \blacktriangleright \text{ Conditions for destructive interference}$$

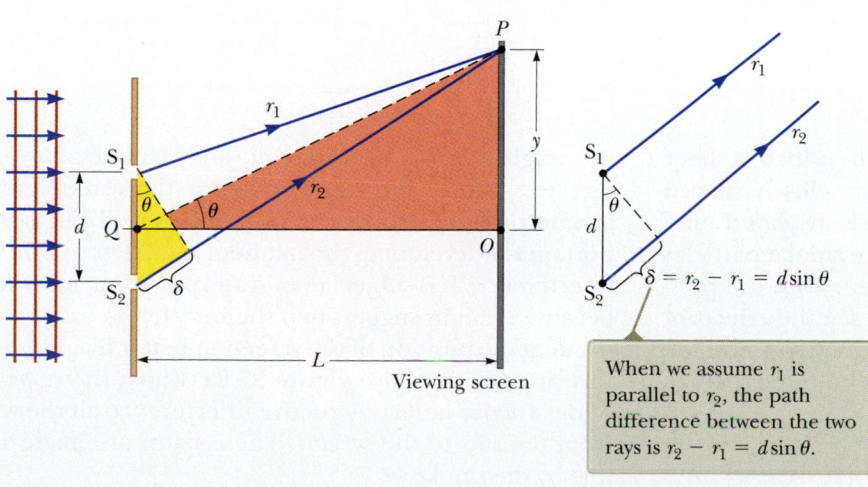

Figure 27.4 (a) Geometric construction for describing Young's double-slit experiment (not to scale). (b) The slits are represented as sources, and the outgoing light rays are assumed to be parallel as they travel to P. To achieve that in practice, it is essential that $L \gg d$.

Viewing screen

When we assume r_1 is parallel to r_2, the path difference between the two rays is $r_2 - r_1 = d \sin \theta$.

These equations provide the *angular* positions of the fringes. It is also useful to obtain expressions for the *linear* positions measured along the screen from O to P. From the geometric model triangle OPQ in Figure 27.4a, we see that

$$\tan \theta = \frac{y}{L} \qquad \qquad \textbf{27.4} \blacktriangleleft$$

Using this result, we can see that the linear positions of bright and dark fringes are given by

$$y_{\text{bright}} = L \tan \theta_{\text{bright}} \qquad \qquad \textbf{27.5} \blacktriangleleft$$

$$y_{\text{dark}} = L \tan \theta_{\text{dark}} \qquad \qquad \textbf{27.6} \blacktriangleleft$$

where θ_{bright} and θ_{dark} are given by Equations 27.2 and 27.3.

When the angles to the fringes are small, the fringes are equally spaced near the center of the pattern. To verify this statement, note that, for small angles, $\tan \theta \approx \sin \theta$ and Equation 27.5 gives the positions of the bright fringes as $y_{\text{bright}} = L \sin \theta_{\text{bright}}$. Incorporating Equation 27.2, we find that

$$y_{\text{bright}} = L\left(\frac{m\lambda}{d}\right) \quad \text{(small angles)} \qquad \qquad \textbf{27.7} \blacktriangleleft$$

and we see that y_{bright} is linear in the order number m, so the fringes are equally spaced. Similarly, for dark fringes,

$$y_{\text{dark}} = L \frac{(m + \frac{1}{2})\lambda}{d} \quad \text{(small angles)} \qquad \qquad \textbf{27.8} \blacktriangleleft$$

As demonstrated in Example 27.1, Young's double-slit experiment provides a method for measuring the wavelength of light. In fact, Young used this technique to do precisely that. In addition, his experiment gave the wave model of light a great deal of credibility. It was inconceivable that particles of light coming through the slits could cancel one another in a way that would explain the dark fringes.

The principles discussed in this section are the basis of the **waves in interference** analysis model. This model was applied to mechanical waves in one dimension in Chapter 14. Here we see the details of applying this model in three dimensions to light.

QUICK QUIZ 27.1 Which of the following causes the fringes in a two-slit interference pattern to move farther apart? (**a**) decreasing the wavelength of the light (**b**) decreasing the screen distance L (**c**) decreasing the slit spacing d (**d**) immersing the entire apparatus in water

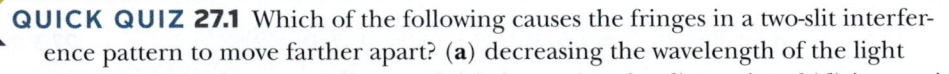

THINKING PHYSICS 27.1

Consider a double-slit experiment in which a laser beam is passed through a pair of very closely spaced slits and a clear interference pattern is displayed on a distant screen. Now suppose you place smoke particles between the double slit and the screen. With the presence of the smoke particles, will you see the effects of the interference in the space between the slits and the screen, or will you only see the effects on the screen?

Reasoning You see the effects in the area filled with smoke. Bright beams of light are directed toward the bright areas on the screen, and dark regions are directed toward the dark areas on the screen. The geometrical construction shown in Figure 27.4a is important for developing the mathematical description of interference. It is subject to misinterpretation, however, because it might suggest that the interference can only occur at the position of the screen. A better diagram for this situation is Active Figure 27.2a, which shows *paths* of destructive and constructive interference all the way from the slits to the screen. These paths are made visible by the smoke. ◄

Example 27.1 | **Measuring the Wavelength of a Light Source**

A viewing screen is separated from a double slit by 4.80 m. The distance between the two slits is 0.030 0 mm. Monochromatic light is directed toward the double slit and forms an interference pattern on the screen. The first dark fringe is 4.50 cm from the center line on the screen.

(A) Determine the wavelength of the light.

SOLUTION

Conceptualize Study Figure 27.4 to be sure you understand the phenomenon of interference of light waves. The distance of 4.50 cm is y in Figure 27.4.

Categorize We determine results using equations developed in this section, so we categorize this example as a substitution problem. Because $L \gg y$, the angles for the fringes are small.

Solve Equation 27.8 for the wavelength and substitute numerical values, taking $m = 0$ for the first dark fringe:

$$\lambda = \frac{y_{\text{dark}} d}{(m + \frac{1}{2})L} = \frac{(4.50 \times 10^{-2}\text{ m})(3.00 \times 10^{-5}\text{ m})}{(0 + \frac{1}{2})(4.80\text{ m})}$$

$$= 5.62 \times 10^{-7}\text{ m} = \boxed{562\text{ nm}}$$

(B) Calculate the distance between adjacent bright fringes.

SOLUTION

Find the distance between adjacent bright fringes from Equation 27.7 and the results of part (A):

$$y_{m+1} - y_m = L\frac{(m+1)\lambda}{d} - L\frac{m\lambda}{d}$$

$$= L\frac{\lambda}{d} = 4.80\text{ m}\left(\frac{5.62 \times 10^{-7}\text{ m}}{3.00 \times 10^{-5}\text{ m}}\right)$$

$$= 9.00 \times 10^{-2}\text{ m} = \boxed{9.00\text{ cm}}$$

For practice, find the wavelength of the sound in Example 14.1 using the procedure in part (A) of this example.

Intensity Distribution of the Double-Slit Interference Pattern

We shall now discuss briefly the distribution of light intensity I (the energy delivered by the light per unit area per unit time) associated with the double-slit interference pattern. Again, suppose the two slits represent coherent sources of sinusoidal waves. In this case, the two waves have the same angular frequency ω and a constant phase difference ϕ. Although the waves have equal phase at the slits, their phase difference ϕ at P depends on the path difference $\delta = r_2 - r_1 = d \sin \theta$. Because a path difference of λ corresponds to a phase difference of 2π rad, we can establish the equality of the ratios:

$$\frac{\delta}{\phi} = \frac{\lambda}{2\pi}$$

$$\phi = \frac{2\pi}{\lambda}\delta = \frac{2\pi}{\lambda} d \sin \theta \qquad \text{27.9}\blacktriangleleft$$

This equation tells us how the phase difference ϕ depends on the angle θ.

Although we shall not prove it here, a careful analysis of the electric fields arriving at the screen from the two very narrow slits shows that the **time-averaged light intensity** at a given angle θ is

$$I = I_{\text{max}} \cos^2\left(\frac{\pi d \sin \theta}{\lambda}\right) \qquad \text{27.10}\blacktriangleleft$$

where I_{max} is the intensity at point O in Figure 27.4a, directly behind the midpoint between the slits. Intensity versus $d \sin \theta$ is plotted in Figure 27.5.

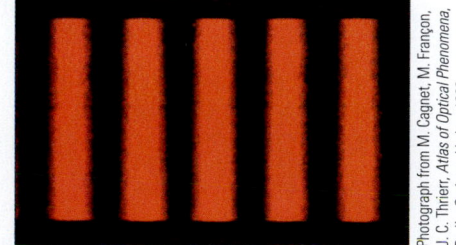

Figure 27.5 Light intensity versus $d \sin \theta$ for the double-slit interference pattern when the screen is far from the two slits ($L \gg d$).

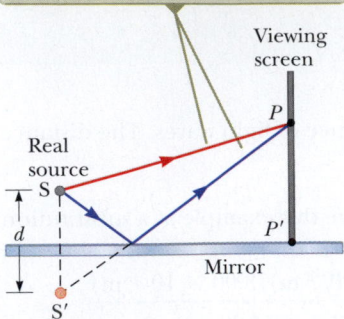

Figure 27.6 Lloyd's mirror. The reflected ray undergoes a phase change of 180°.

27.4 | Change of Phase Due to Reflection

Young's method of producing two coherent light sources involves illuminating a pair of slits with a single source. Another simple arrangement for producing an interference pattern with a single light source is known as *Lloyd's mirror*. A point light source S is placed close to a mirror as illustrated in Figure 27.6. Waves can reach the point *P* either by the direct path from S to *P* or by the indirect path involving reflection from the mirror. The reflected ray strikes the screen as if it originated from a source S′ located below the mirror.

At points far from the source, one would expect an interference pattern due to waves from S and S′, just as is observed for two real coherent sources at these points. An interference pattern is indeed observed. The positions of the dark and bright fringes, however, are *reversed* relative to the pattern of two real coherent sources (Young's experiment) because the coherent sources S and S′ differ in phase by 180°. This 180° phase change is produced on reflection. In general, an electromagnetic wave undergoes a phase change of 180° on reflection from a medium of higher index of refraction than the one in which it is traveling.

It is useful to draw an analogy between reflected light waves and the reflections of a transverse wave on a stretched string when the wave meets a boundary (Section 13.4) as in Figure 27.7. The reflected pulse on a string undergoes a phase change of 180° when it is reflected from a rigid end, and no phase change when it is reflected from a free end, as illustrated in Active Figures 13.12 and 13.13. If the boundary is between two strings, the transmitted wave exhibits no phase change. Similarly, an electromagnetic wave undergoes a 180° phase change when reflected from the boundary of a medium of higher index of refraction than the one in which it is traveling. There is no phase change for the reflected ray when the wave is incident on a boundary leading to a medium of lower index of refraction. In either case, the transmitted wave exhibits no phase change.

27.5 | Interference in Thin Films

Interference effects can be observed in many situations in which one beam of light is split and then recombined. A common occurrence is the appearance of colored bands in a film of oil on water or in a soap bubble illuminated with white light. The colors in these situations result from the interference of waves reflected from the opposite surfaces of the film.

Consider a film of uniform thickness t and index of refraction n as in Figure 27.8. We adopt a simplification model in which the light ray is incident on the film from above and nearly normal to the surface of the film. Two rays are reflected from the film, one from the upper surface and one from the lower surface after the refracted

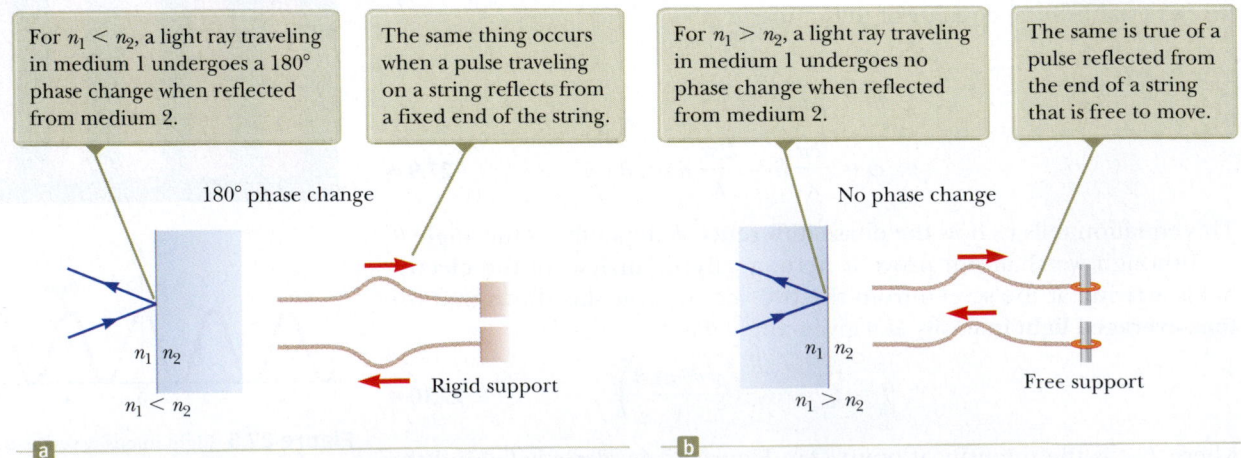

For $n_1 < n_2$, a light ray traveling in medium 1 undergoes a 180° phase change when reflected from medium 2.

The same thing occurs when a pulse traveling on a string reflects from a fixed end of the string.

For $n_1 > n_2$, a light ray traveling in medium 1 undergoes no phase change when reflected from medium 2.

The same is true of a pulse reflected from the end of a string that is free to move.

180° phase change

No phase change

$n_1 \mid n_2$

$n_1 \mid n_2$

$n_1 < n_2$

$n_1 > n_2$

Rigid support

Free support

a

b

Figure 27.7 Comparisons of reflections of light waves and waves on strings.

ray has traveled through the film. Because the film is thin and has parallel sides, the reflected rays are parallel. Hence, rays reflected from the top surface can interfere with rays reflected from the bottom surface. To determine whether the reflected rays interfere constructively or destructively, we first note the following facts:

- An electromagnetic wave traveling from a medium of index of refraction n_1 toward a medium of index of refraction n_2 undergoes a 180° phase change on reflection when $n_2 > n_1$. No phase change occurs in the reflected wave if $n_2 < n_1$.
- The wavelength λ_n of light in a medium whose index of refraction n is

$$\lambda_n = \frac{\lambda}{n} \qquad \text{27.11} \blacktriangleleft$$

where λ is the wavelength of light in free space.

Let us apply these rules to the film of Figure 27.8. According to the first rule, ray 1, which is reflected from the upper surface (A), undergoes a phase change of 180° with respect to the incident wave. Ray 2, which is reflected from the lower surface (B), undergoes no phase change with respect to the incident wave. Therefore, ignoring the path difference for now, outgoing ray 1 is 180° out of phase with respect to ray 2, a phase difference that is equivalent to a path difference of $\lambda_n/2$. We must also consider, however, that ray 2 travels an extra distance approximately equal to $2t$ before the waves recombine. The *total* phase difference arises from a combination of the path difference and the 180° phase change on reflection. For example, if $2t = \lambda_n/2$, rays 1 and 2 will recombine in phase and constructive interference will result. In general, the condition for constructive interference is

$$2t = (m + \tfrac{1}{2})\lambda_n \qquad m = 0, 1, 2, \ldots \qquad \text{27.12} \blacktriangleleft$$

This condition takes into account two factors: (a) the difference in optical path length for the two rays (the term $m\lambda_n$) and (b) the 180° phase change on reflection (the term $\lambda_n/2$). Because $\lambda_n = \lambda/n$, we can write Equation 27.12 in the form

$$2nt = (m + \tfrac{1}{2})\lambda \qquad m = 0, 1, 2, \ldots \qquad \text{27.13} \blacktriangleleft$$

▶ Condition for constructive interference in thin films

If the extra distance $2t$ traveled by ray 2 corresponds to an integer multiple of λ_n, the two waves will combine out of phase and destructive interference results. The general equation for destructive interference is

$$2nt = m\lambda \qquad m = 0, 1, 2, \ldots \qquad \text{27.14} \blacktriangleleft$$

▶ Condition for destructive interference in thin films

The preceding conditions for constructive and destructive interference are valid when the medium above the top surface of the film is the same as the medium below the bottom surface. The surrounding medium may have a refractive index less than or greater than that of the film. In either case, the rays reflected from the two surfaces will be out of phase by 180°. The conditions are also valid if different media are above and below the film and if both have n less than or larger than that of the film.

If the film is placed between two different media, one with $n < n_{film}$ and the other with $n > n_{film}$, the conditions for constructive and destructive interference are reversed. In this case, either a phase change of 180° takes place for both ray 1 reflecting from surface A and ray 2 reflecting from surface B, or no phase change occurs for either ray; hence, the net change in relative phase due to the reflections is zero.

Rays 3 and 4 in Figure 27.8 lead to interference effects in the light transmitted through the thin film. The analysis of these effects is similar to that of the reflected light.

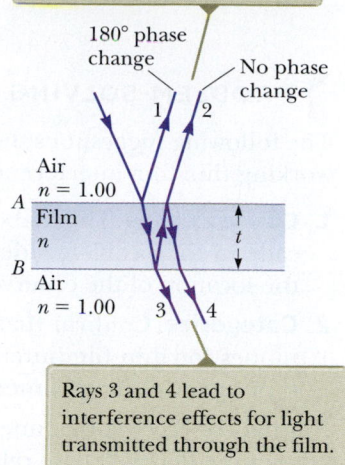

Interference in light reflected from a thin film is due to a combination of rays 1 and 2 reflected from the upper and lower surfaces of the film.

Rays 3 and 4 lead to interference effects for light transmitted through the film.

Figure 27.8 Light paths through a thin film.

QUICK QUIZ 27.2 In a laboratory accident, you spill two liquids onto water, neither of which mixes with the water. They both form thin films on the water surface. When the films become very thin as they spread, you observe that one film becomes bright and the other dark in reflected light. The film that appears dark (**a**) has an index of refraction higher than that of water, (**b**) has an index of refraction lower than that of water, (**c**) has an index of refraction equal to that of water, or (**d**) has an index of refraction lower than that of the bright film.

> **QUICK QUIZ 27.3** One microscope slide is placed on top of another with their left edges in contact and a human hair under the right edge of the upper slide. As a result, a wedge of air exists between the slides. An interference pattern results when monochromatic light is reflected from the wedge. What is at the left edge of the slides? (**a**) a dark fringe   (**b**) a bright fringe   (**c**) impossible to determine

> **PROBLEM-SOLVING STRATEGY: Thin-Film Interference**
>
> The following suggestions should be kept in mind while working thin-film interference problems:
>
> 1. **Conceptualize** Think about what is going on physically in the problem. Identify the light source and the location of the observer.
>
> 2. **Categorize** Confirm that you should use the techniques for thin film interference by identifying the thin film causing the interference.
>
> 3. **Analyze** The type of interference that occurs is determined by the phase relationship between the portion of the wave reflected at the upper surface of the film and the portion reflected at the lower surface. Phase differences between the two portions of the wave have two causes: (a) differences in the distances traveled by the two portions and (b) phase changes occurring on reflection. *Both* causes must be considered when determining which type of interference occurs. If the media above and below the film both have index of refraction larger than that of the film or if both indices are smaller, use Equation 27.13 for constructive interference and Equation 27.14 for destructive interference. If the film is located between two different media, one with $n < n_{film}$ and the other with $n > n_{film}$, reverse these two equations for constructive and destructive interference.
>
> 4. **Finalize** Inspect your final results to see if they make sense physically and are of an appropriate size.

Example 27.2 | Interference in a Soap Film

Calculate the minimum thickness of a soap-bubble film that results in constructive interference in the reflected light if the film is illuminated with light whose wavelength in free space is $\lambda = 600$ nm. The index of refraction of the soap film is 1.33.

SOLUTION

Conceptualize Imagine that the film in Figure 27.8 is soap, with air on both sides.

Categorize We determine the result using an equation from this section, so we categorize this example as a substitution problem.

The minimum film thickness for constructive interference in the reflected light corresponds to $m = 0$ in Equation 27.13. Solve this equation for t and substitute numerical values:

$$t = \frac{(0 + \frac{1}{2})\lambda}{2n} = \frac{\lambda}{4n} = \frac{(600 \text{ nm})}{4(1.33)} = \boxed{113 \text{ nm}}$$

What If? What if the film is twice as thick? Does this situation produce constructive interference?

Answer Using Equation 27.13, we can solve for the thicknesses at which constructive interference occurs:

$$t = (m + \tfrac{1}{2})\frac{\lambda}{2n} = (2m + 1)\frac{\lambda}{4n} \qquad m = 0, 1, 2, \ldots$$

The allowed values of m show that constructive interference occurs for *odd* multiples of the thickness corresponding to $m = 0$, $t = 113$ nm. Therefore, constructive interference does *not* occur for a film that is twice as thick.

Example 27.3 | Nonreflective Coatings for Solar Cells

Solar cells—devices that generate electricity when exposed to sunlight—are often coated with a transparent, thin film of silicon monoxide (SiO, $n = 1.45$) to minimize reflective losses from the surface. Suppose a silicon solar cell ($n = 3.5$) is coated with a thin film of silicon monoxide for this purpose (Fig. 27.9a). Determine the minimum film thickness that produces the least reflection at a wavelength of 550 nm, near the center of the visible spectrum.

27.3 *cont.*

SOLUTION

Conceptualize Figure 27.9a helps us visualize the path of the rays in the SiO film that result in interference in the reflected light.

Categorize Based on the geometry of the SiO layer, we categorize this example as a thin-film interference problem.

Analyze The reflected light is a minimum when rays 1 and 2 in Figure 27.9a meet the condition of destructive interference. In this situation, *both* rays undergo a 180° phase change upon reflection: ray 1 from the upper SiO surface and ray 2 from the lower SiO surface. The net change in phase due to reflection is therefore zero, and the condition for a reflection minimum requires a path difference of $\lambda_n/2$, where λ_n is the wavelength of the light in SiO. Hence, $2nt = \lambda/2$, where λ is the wavelength in air and n is the index of refraction of SiO.

Solve the equation $2nt = \lambda/2$ for t and substitute numerical values:

$$t = \frac{\lambda}{4n} = \frac{550 \text{ nm}}{4(1.45)} = \boxed{94.8 \text{ nm}}$$

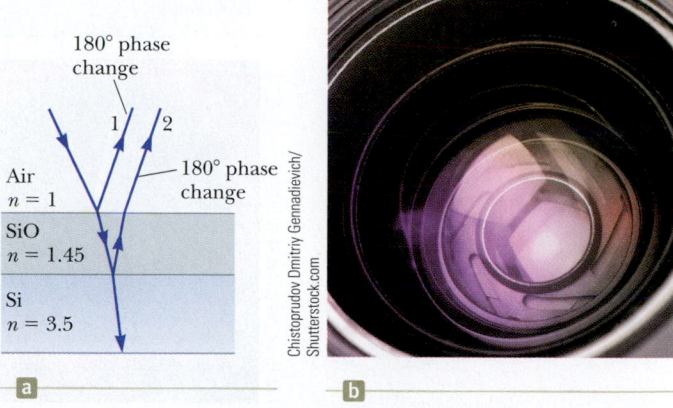

Figure 27.9 (Example 27.3) (a) Reflective losses from a silicon solar cell are minimized by coating the surface of the cell with a thin film of silicon monoxide. (b) The reflected light from a coated camera lens often has a reddish-violet appearance.

Finalize A typical uncoated solar cell has reflective losses as high as 30%, but a coating of SiO can reduce this value to about 10%. This significant decrease in reflective losses increases the cell's efficiency because less reflection means that more sunlight enters the silicon to create charge carriers in the cell. No coating can ever be made perfectly nonreflecting because the required thickness is wavelength-dependent and the incident light covers a wide range of wavelengths.

Glass lenses used in cameras and other optical instruments are usually coated with a transparent thin film to reduce or eliminate unwanted reflection and to enhance the transmission of light through the lenses. The camera lens in Figure 27.9b has several coatings (of different thicknesses) to minimize reflection of light waves having wavelengths near the center of the visible spectrum. As a result, the small amount of light that is reflected by the lens has a greater proportion of the far ends of the spectrum and often appears reddish violet.

27.6 | Diffraction Patterns

In Sections 25.2 and 27.2, we discussed briefly the phenomenon of **diffraction,** and now we shall investigate this phenomenon more fully for light waves. In general, diffraction occurs when waves pass through small openings, around obstacles, or by sharp edges.

We might expect that the light passing through one such small opening would simply result in a broad region of light on a screen due to the spreading of the light as it passes through the opening. We find something more interesting, however. A **diffraction pattern** consisting of light and dark areas is observed, somewhat similar to the interference patterns discussed earlier. For example, when a narrow slit is placed between a distant light source (or a laser beam) and a screen, the light produces a diffraction pattern like that in Figure 27.10. The pattern consists of a broad, intense central band (called the **central maximum**), flanked by a series of narrower, less intense additional bands (called **side maxima**) and a series of dark bands (or **minima**).

Figure 27.11 (page 790) shows the shadow of a penny, which displays bright and dark rings of a diffraction pattern. The bright spot at the center (called the *Arago bright spot* after its discoverer, Dominique Arago) can be explained using the wave theory of light. Waves that diffract from all points on the edge of the penny travel the same distance to the midpoint on the screen. Therefore, the midpoint is a region of constructive interference and a bright spot appears. In contrast, from the viewpoint of geometric optics, the center of the pattern would be completely screened by the penny, and so an approach that does not include the wave nature of light would not predict a central bright spot.

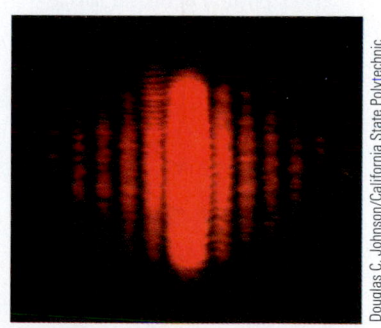

Figure 27.10 The diffraction pattern that appears on a screen when light passes through a narrow vertical slit. The pattern consists of a broad central band and a series of less intense and narrower side bands.

Notice the bright spot at the center.

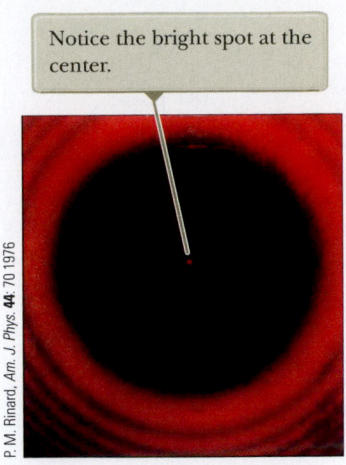

Figure 27.11 Diffraction pattern created by the illumination of a penny, with the penny positioned midway between the screen and light source.

The pattern consists of a central bright fringe flanked by much weaker maxima alternating with dark fringes.

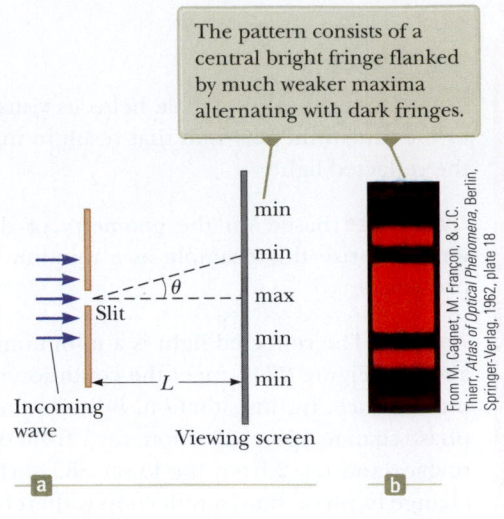

Active Figure 27.12

(a) Geometry for analyzing the Fraunhofer diffraction pattern of a single slit. (Drawing not to scale.) (b) Photograph of a single-slit Fraunhofer diffraction pattern.

Pitfall Prevention | 27.1

Diffraction versus Diffraction Pattern

Diffraction refers to the general behavior of waves spreading out as they pass through a slit. We used diffraction in explaining the existence of an interference pattern. A *diffraction pattern* is actually a misnomer, but it is deeply entrenched in the language of physics. The diffraction pattern seen on a screen when a single slit is illuminated is actually another interference pattern. The interference is between parts of the incident light illuminating different regions of the slit.

Each portion of the slit acts as a point source of light waves.

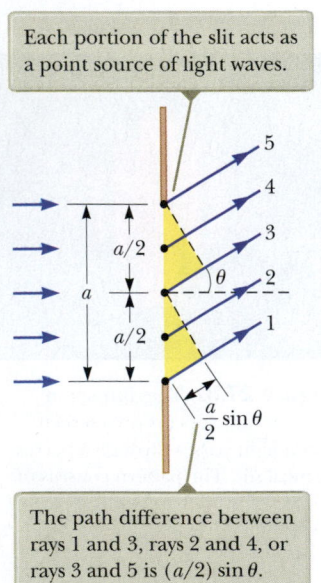

The path difference between rays 1 and 3, rays 2 and 4, or rays 3 and 5 is $(a/2) \sin \theta$.

Figure 27.13 Paths of light rays that encounter a narrow slit of width a and diffract toward a screen in the direction described by angle θ.

Let us consider a common situation, that of light passing through a narrow opening modeled as a slit and projected onto a screen. As a simplification model, we assume that the observing screen is far from the slit so that the rays reaching the screen are approximately parallel. This situation can also be achieved experimentally by using a converging lens to focus the parallel rays on a nearby screen. In this model, the pattern on the screen is called a **Fraunhofer diffraction pattern.**[2]

Active Figure 27.12a shows light entering a single slit from the left and diffracting as it propagates toward a screen. Active Figure 27.12b is a photograph of a single-slit Fraunhofer diffraction pattern. A bright fringe is observed along the axis at $\theta = 0$, with alternating dark and bright fringes on each side of the central bright fringe.

Until now, we assumed that slits act as point sources of light. In this section, we shall determine how their finite widths are the basis for understanding the nature of the Fraunhofer diffraction pattern produced by a single slit. We can deduce some important features of this problem by examining waves coming from various portions of the slit as shown in the geometric model of Figure 27.13. According to Huygens's principle, each portion of the slit acts as a source of waves. Hence, light from one portion of the slit can interfere with light from another portion, and the resultant intensity on the screen depends on the direction θ.

To analyze the diffraction pattern, it is convenient to divide the slit into two halves as in Figure 27.13. All the waves that originate at the slit are in phase. Consider waves 1 and 3, which originate at the bottom and center of the slit, respectively. To reach the same point on the viewing screen, wave 1 travels farther than wave 3 by an amount equal to the path difference $(a/2) \sin \theta$, where a is the width of the slit. Similarly, the path difference between waves 3 and 5 is also $(a/2) \sin \theta$. If the path difference between two waves is exactly one-half of a wavelength (corresponding to a phase difference of 180°), the two waves cancel each other and destructive interference results. That is true, in fact, for any two waves that originate at points separated by half the slit width because the phase difference between two such points is 180°.

---

[2]If the screen were brought close to the slit (and no lens is used), the pattern is a *Fresnel* diffraction pattern. The Fresnel pattern is more difficult to analyze, so we shall restrict our discussion to Fraunhofer diffraction.

Therefore, waves from the upper half of the slit interfere *destructively* with waves from the lower half of the slit when

$$\frac{a}{2} \sin \theta = \pm \frac{\lambda}{2}$$

or when

$$\sin \theta = \pm \frac{\lambda}{a}$$

If we divide the slit into four parts rather than two and use similar reasoning, we find that the screen is also dark when

$$\sin \theta = \pm \frac{2\lambda}{a}$$

Likewise, we can divide the slit into six parts and show that darkness occurs on the screen when

$$\sin \theta = \pm \frac{3\lambda}{a}$$

Therefore, the general condition for destructive interference is

$$\sin \theta_{dark} = m\frac{\lambda}{a} \qquad m = \pm 1, \pm 2, \pm 3, \ldots \qquad \textbf{27.15} \blacktriangleleft$$

▶ Condition for destructive interference in a diffraction pattern

Equation 27.15 gives the values of θ for which the diffraction pattern has zero intensity, that is, a dark fringe is formed. Equation 27.15, however, tells us nothing about the variation in intensity along the screen. The general features of the intensity distribution are shown in Active Figure 27.14: a broad central bright fringe flanked by much weaker, alternating bright fringes. The various dark fringes (points of zero intensity) occur at the values of θ that satisfy Equation 27.15. The position of the points of constructive interference lie approximately halfway between the dark fringes. Note that the central bright fringe is twice as wide as the weaker maxima.

> **QUICK QUIZ 27.4** Suppose the slit width in Active Figure 27.14 is made half as wide. Does the central bright fringe (**a**) become wider, (**b**) remain the same, or (**c**) become narrower?

▶ THINKING PHYSICS 27.2

If a classroom door is open slightly, you can hear sounds coming from the hallway. Yet you cannot see what is happening in the hallway. What accounts for the difference?

Reasoning The space between the slightly open door and the wall is acting as a single slit for waves. Sound waves have wavelengths larger than the slit width, so sound is effectively diffracted by the opening and spread throughout the room. The sound is then reflected from walls, floor, and ceiling, further distributing the sound throughout the room. Light wavelengths are much smaller than the slit width, so virtually no diffraction for the light occurs. You must have a direct line of sight to detect the light waves. ◀

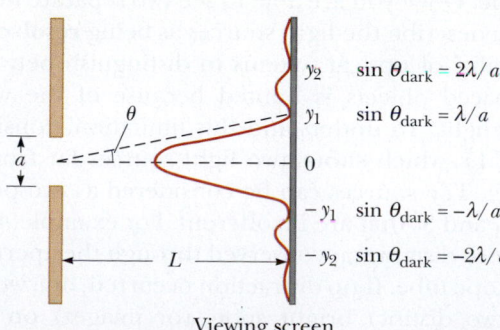

$y_2 \qquad \sin \theta_{dark} = 2\lambda/a$
$y_1 \qquad \sin \theta_{dark} = \lambda/a$
0
$-y_1 \qquad \sin \theta_{dark} = -\lambda/a$
$-y_2 \qquad \sin \theta_{dark} = -2\lambda/a$

Viewing screen

Active Figure 27.14 Light intensity distribution for the Fraunhofer diffraction pattern from a single slit of width a. The positions of two minima on each side of the central maximum are labeled. (Drawing not to scale.)

Example 27.4 | Where Are the Dark Fringes?

Light of wavelength 580 nm is incident on a slit having a width of 0.300 mm. The viewing screen is 2.00 m from the slit. Find the width of the central bright fringe.

SOLUTION

Conceptualize Based on the problem statement, we imagine a single-slit diffraction pattern similar to that in Active Figure 27.14.

Categorize We categorize this example as a straightforward application of our discussion of single-slit diffraction patterns.

Analyze Evaluate Equation 27.15 for the two dark fringes that flank the central bright fringe, which correspond to $m = \pm 1$:

$$\sin \theta_{dark} = \pm \frac{\lambda}{a}$$

Let y represent the vertical position along the viewing screen in Active Figure 27.14, measured from the point on the screen directly behind the slit. Then, $\tan \theta_{dark} = y_1/L$, where the subscript 1 refers to the first dark fringe. Because θ_{dark} is very small, we can use the approximation $\sin \theta_{dark} \approx \tan \theta_{dark}$; therefore, $y_1 = L \sin \theta_{dark}$.

The width of the central bright fringe is twice the absolute value of y_1:

$$2|y_1| = 2|L \sin \theta_{dark}| = 2\left|\pm L\frac{\lambda}{a}\right| = 2L\frac{\lambda}{a} = 2(2.00 \text{ m})\frac{580 \times 10^{-9} \text{ m}}{0.300 \times 10^{-3} \text{ m}}$$

$$= 7.73 \times 10^{-3} \text{ m} = \boxed{7.73 \text{ mm}}$$

Finalize Notice that this value is much greater than the width of the slit. Let's explore below what happens if we change the slit width.

What If? What if the slit width is increased by an order of magnitude to 3.00 mm? What happens to the diffraction pattern?

Answer Based on Equation 27.15, we expect that the angles at which the dark bands appear will decrease as a increases. Therefore, the diffraction pattern narrows.

Repeat the calculation with the larger slit width:

$$2|y_1| = 2L\frac{\lambda}{a} = 2(2.00 \text{ m})\frac{580 \times 10^{-9} \text{ m}}{3.00 \times 10^{-3} \text{ m}} = 7.73 \times 10^{-4} \text{ m} = \boxed{0.773 \text{ mm}}$$

Notice that this result is *smaller* than the width of the slit. In general, for large values of a, the various maxima and minima are so closely spaced that only a large, central bright area resembling the geometric image of the slit is observed. This concept is very important in the performance of optical instruments such as telescopes.

27.7 | Resolution of Single-Slit and Circular Apertures

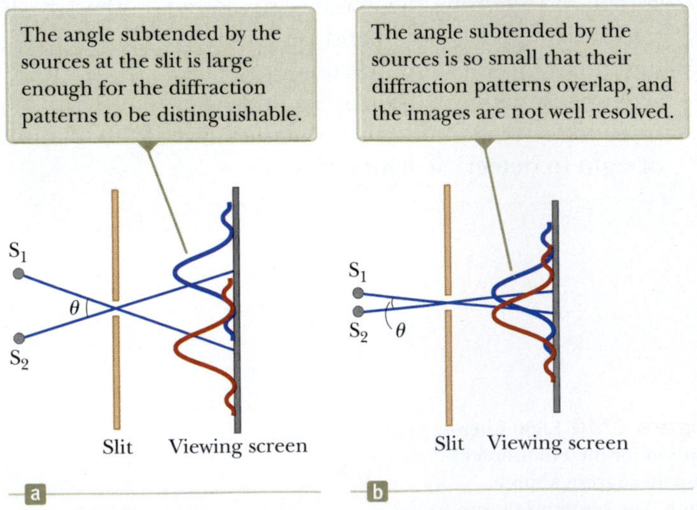

Imagine you are driving in the middle of a dark desert at night, along a road that is perfectly straight and flat for many kilometers. You see another vehicle coming toward you from a distance. When the vehicle is far away, you might be unable to determine whether it is an automobile with two headlights or a motorcycle with one. As it approaches you, at some point you will be able to distinguish the two headlights and determine that it is an automobile. Once you are able to see two separate headlights, you describe the light sources as being **resolved.**

The ability of optical systems to distinguish between closely spaced objects is limited because of the wave nature of light. To understand this limitation, consider Figure 27.15, which shows two light sources far from a narrow slit. The sources can be considered as two point sources S_1 and S_2 that are incoherent. For example, they could be two distant stars observed through the aperture of a telescope tube. If no diffraction occurred, one would observe two distinct bright spots (or images) on the screen at the right in the figure. Because of diffraction,

The angle subtended by the sources at the slit is large enough for the diffraction patterns to be distinguishable.

The angle subtended by the sources is so small that their diffraction patterns overlap, and the images are not well resolved.

Figure 27.15 Two point sources far from a narrow slit each produce a diffraction pattern. (a) The sources are separated by a large angle. (b) The sources are separated by a small angle. (Notice that the angles are greatly exaggerated. The drawing is not to scale.)

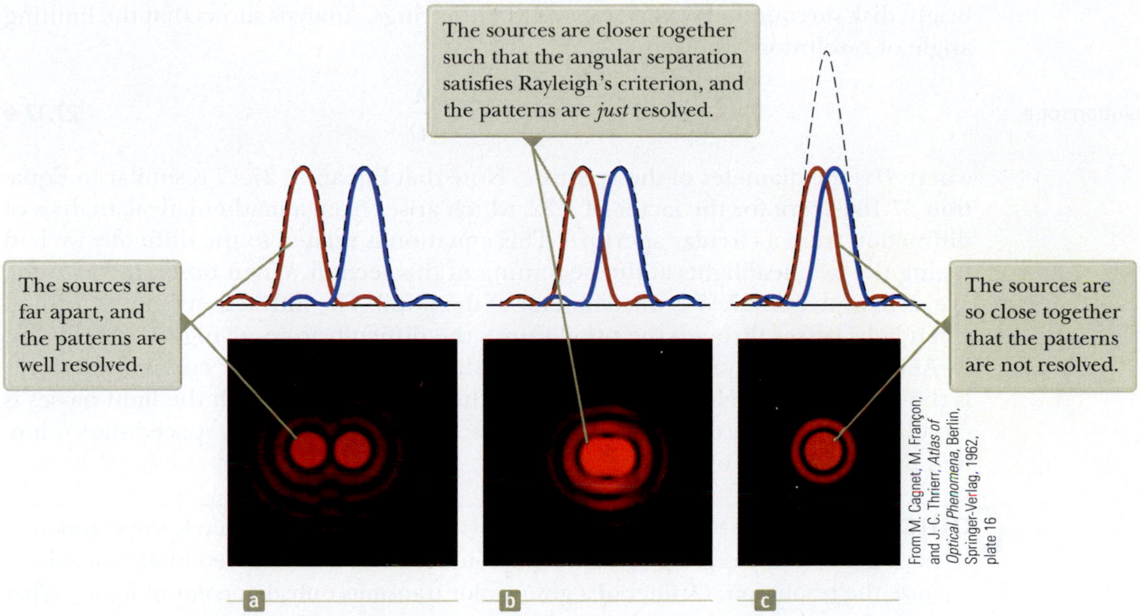

The sources are closer together such that the angular separation satisfies Rayleigh's criterion, and the patterns are *just* resolved.

The sources are far apart, and the patterns are well resolved.

The sources are so close together that the patterns are not resolved.

From M. Cagnet, M. Françon, and J. C. Thierr, *Atlas of Optical Phenomena*, Berlin, Springer-Verlag, 1962, plate 16

Figure 27.16 Individual diffraction patterns of two point sources (solid curves) and the resultant patterns (dashed curves) for various angular separations of the sources as the light passes through a circular aperture. In each case, the dashed curve is the sum of the two solid curves.

however, each source is imaged as a bright central region flanked by weaker bright and dark bands. What is observed on the screen is the sum of two diffraction patterns: one from S_1 and the other from S_2.

If the two sources are far enough apart to ensure that their central maxima do not overlap as in Figure 27.15a, their images can be distinguished and are said to be resolved. If the sources are close together, however, as in Figure 27.15b, the two central maxima may overlap and the sources are not resolved. To decide when two sources are resolved, the following condition is often used:

When the central maximum of one image falls on the first minimum of another image, the images are said to be just resolved. This limiting condition of resolution is known as **Rayleigh's criterion.**

▶ Rayleigh's criterion

Figure 27.16 shows the diffraction patterns from circular apertures for three situations. When the objects are far apart, they are well resolved (Fig. 27.16a). They are just resolved when their angular separation satisfies Rayleigh's criterion (Fig. 27.16b). Finally, the sources are not resolved in Figure 27.16c.

From Rayleigh's criterion, we can determine the minimum angular separation θ_{min} subtended by the sources at a slit such that the sources are just resolved. In Section 27.6, we found that the first minimum in a single-slit diffraction pattern occurs at the angle that satisfies the relationship

$$\sin \theta = \frac{\lambda}{a}$$

where a is the width of the slit. According to Rayleigh's criterion, this expression gives the smallest angular separation for which the two sources are resolved. Because $\lambda \ll a$ in most situations, $\sin \theta$ is small and we can use the approximation $\sin \theta \approx \theta$. Therefore, the limiting angle of resolution for a slit of width a is

$$\theta_{min} = \frac{\lambda}{a}$$

27.16 ◀   ▶ Limiting angle of resolution for a slit

where θ_{min} is expressed in radians. Hence, the angle subtended by the two sources at the slit must be *greater* than λ / a if the sources are to be resolved.

Many optical systems use circular apertures rather than slits. The diffraction pattern of a circular aperture, as seen in Figure 27.16, consists of a central circular bright disk surrounded by progressively fainter rings. Analysis shows that the limiting angle of resolution of the circular aperture is

▶ Limiting angle of resolution for a circular apeture

$$\theta_{min} = 1.22 \frac{\lambda}{D}$$

27.17 ◀

where D is the diameter of the aperture. Note that Equation 27.17 is similar to Equation 27.16 except for the factor of 1.22, which arises from a mathematical analysis of diffraction from a circular aperture. This equation is related to the difficulty we had seeing the two headlights at the beginning of this section. When observing with the eye, D in Equation 27.17 is the diameter of the pupil. The diffraction pattern formed when light passes through the pupil causes the difficulty in resolving the headlights.

Another example of the effect of diffraction on resolution for circular apertures is the astronomical telescope. The end of the tube through which the light passes is circular, so the ability of the telescope to resolve light from closely spaced stars is limited by the diameter of this opening.

▶ **QUICK QUIZ 27.5** Suppose you are observing a binary star with a telescope and are having difficulty resolving the two stars. You decide to use a colored filter to maximize the resolution. (A filter of a given color transmits only that color of light.) What color filter should you choose? (a) blue (b) green (c) yellow (d) red

▶ THINKING PHYSICS 27.3

Cats' eyes have pupils that can be modeled as vertical slits. At night, are cats more successful in resolving headlights on a distant car or vertically separated lights on the mast of a distant boat?

Reasoning The effective slit width in the vertical direction of the cat's eye is larger than that in the horizontal direction. Therefore, the eye has more resolving power for lights separated in the vertical direction and would be more effective at resolving the mast lights on the boat. ◀

Example 27.5 | **Resolution of the Eye** BIO

Light of wavelength 500 nm, near the center of the visible spectrum, enters a human eye. Although pupil diameter varies from person to person, let's estimate a daytime diameter of 2 mm.

(A) Estimate the limiting angle of resolution for this eye, assuming its resolution is limited only by diffraction.

SOLUTION

Conceptualize In Figure 27.16, identify the aperture through which the light travels to be the pupil of the eye. Light passing through this small aperture causes diffraction patterns to occur on the retina.

Categorize We determine the result using equations developed in this section, so we categorize this example as a substitution problem.

Use Equation 27.17, taking $\lambda = 500$ nm and $D = 2$ mm:

$$\theta_{min} = 1.22 \frac{\lambda}{D} = 1.22 \left(\frac{5.00 \times 10^{-7} \text{ m}}{2 \times 10^{-3} \text{ m}} \right)$$

$$= 3 \times 10^{-4} \text{ rad} \approx 1 \text{ min of arc}$$

(B) Determine the minimum separation distance d between two point sources that the eye can distinguish if the point sources are a distance $L = 25$ cm from the observer (Fig. 27.17).

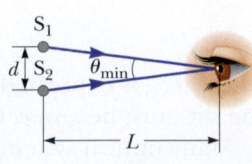

Figure 27.17 (Example 27.5) Two point sources separated by a distance d as observed by the eye.

27.5 cont.

SOLUTION

Noting that θ_{min} is small, find d:

$$\sin \theta_{min} \approx \theta_{min} \approx \frac{d}{L} \quad \rightarrow \quad d = L\theta_{min}$$

Substitute numerical values:

$$d = (25 \text{ cm})(3 \times 10^{-4} \text{ rad}) = \boxed{8 \times 10^{-3} \text{ cm}}$$

This result is approximately equal to the thickness of a human hair.

27.8 | The Diffraction Grating

The **diffraction grating,** a useful device for analyzing light sources, consists of a large number of equally spaced parallel slits. A grating can be made by cutting parallel, equally spaced grooves on a glass or metal plate with a precision ruling machine. In a *transmission grating,* the spaces between lines are transparent to the light and hence act as separate slits. In a *reflection grating,* the spaces between lines are highly reflective. Gratings with many lines very close to one another can have very small slit spacings. For example, a grating ruled with 5 000 lines/cm has a slit spacing of $d = (1/5\,000)$ cm $= 2 \times 10^{-4}$ cm.

Figure 27.18 shows a pictorial representation of a section of a flat diffraction grating. A plane wave is incident from the left, normal to the plane of the grating. The pattern observed on the screen at the right in Figure 27.18 is the result of the combined effects of interference and diffraction. Each slit produces diffraction, and the diffracted beams interfere with one another to produce the final pattern. Each slit acts as a source of waves, and all waves start at the slits in phase. For some arbitrary direction θ measured from the horizontal, however, the waves must travel different path lengths before reaching a particular point on the screen. From Figure 27.18, note that the path difference between waves from any two adjacent slits is equal to $d \sin \theta$. (We assume once again that the distance L to the screen is much larger than d.) If this path difference equals one wavelength or some integral multiple of a wavelength, waves from all slits will be in phase at the screen and a bright line will be observed. When the light is incident normally on the plane of the grating, the condition for *maxima* in the interference pattern at the angle θ is therefore[3]

$$d \sin \theta_{bright} = m\lambda \qquad m = 0, \pm 1, \pm 2, \pm 3, \ldots \qquad \text{27.18} \blacktriangleleft$$

Pitfall Prevention | 27.3
A Diffraction Grating Is an Interference Grating
As with diffraction pattern, diffraction grating is a misnomer, but it is deeply entrenched in the language of physics. The diffraction grating depends on diffraction in the same way as the double slit, spreading the light so that light from different slits can interfere. It would be more correct to call it an interference grating, but diffraction grating is the name in use.

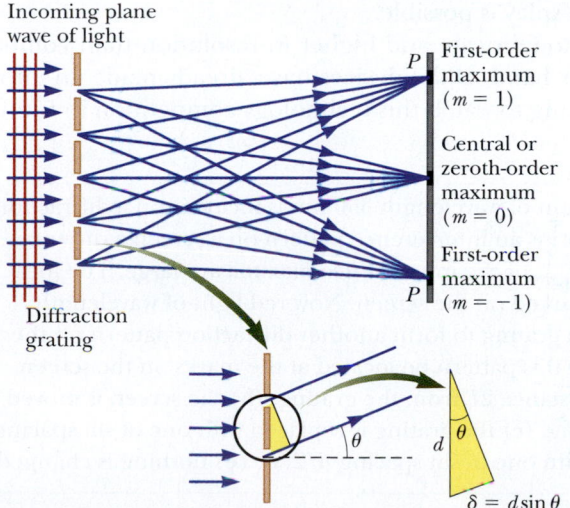

Incoming plane wave of light

First-order maximum ($m = 1$)

Central or zeroth-order maximum ($m = 0$)

First-order maximum ($m = -1$)

Diffraction grating

$\delta = d \sin \theta$

Figure 27.18 Side view of a diffraction grating. The slit separation is d and the path difference between adjacent slits is $d \sin \theta$.

[3]Notice that this equation is identical to Equation 27.2. This equation can be used for a number of slits from two to any number N. The intensity distribution will change with the number of slits, but the locations of the maxima are the same.

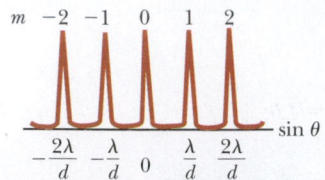

Active Figure 27.19 Intensity versus $\sin\theta$ for a diffraction grating. The zeroth-, first-, and second-order maxima are shown.

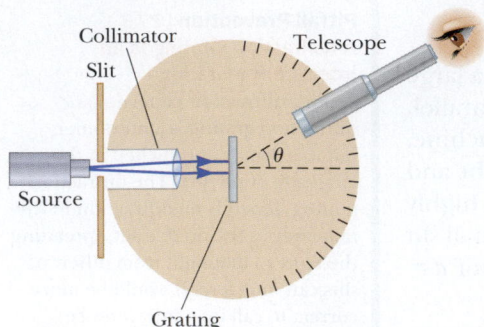

Active Figure 27.20 Diagram of a diffraction grating spectrometer. The collimated beam incident on the grating is spread into its various wavelength components with constructive interference for a particular wavelength occurring at the angles θ_{bright} that satisfy the equation $d\sin\theta_{bright} = m\lambda$, where $m = 0, 1, 2, \ldots$.

Figure 27.21 A small portion of a grating light valve. The alternating reflective ribbons at different levels act as a diffraction grating, offering very high speed control of the direction of light toward a digital display device.

Courtesy Silicon Light Machines

This expression can be used to calculate the wavelength from a knowledge of the grating spacing d and the angle of deviation θ. If the incident radiation contains several wavelengths, the mth-order maximum for each wavelength occurs at an angle determined from Equation 27.18. All wavelengths are mixed together at $\theta = 0$, corresponding to $m = 0$.

The intensity distribution for a diffraction grating is shown in Active Figure 27.19. Note the sharpness of the principal maxima and the broad range of dark areas, which are in contrast to the broad, bright fringes characteristic of the two-slit interference pattern (see Fig. 27.5).

A simple arrangement for measuring the wavelength of light is shown in Active Figure 27.20. This arrangement is called a *diffraction grating spectrometer*. The light to be analyzed passes through a slit,[4] and a parallel beam of light exits from the collimator perpendicular to the grating. The diffracted light leaves the grating and exhibits constructive interference at angles that satisfy Equation 27.18. A telescope is used to view the image of the slit. The wavelength can be determined by measuring the precise angles at which the images of the slit appear for the various orders.

The spectrometer is a useful tool in *atomic spectroscopy*, in which the light from an atom is analyzed to find the wavelength components. These wavelength components can be used to identify the atom as discussed in Section 11.5. We will investigate atomic spectra further in Chapter 29.

Another application of diffraction gratings is in the recently developed *grating light valve* (GLV), which may compete in the near future in video projection with the digital micromirror devices (DMD) discussed in Section 25.3. The grating light valve consists of a silicon microchip fitted with an array of parallel silicon nitride ribbons coated with a thin layer of aluminum (Fig. 27.21). Each ribbon is about 20 μm long and about 5 μm wide and is separated from the silicon substrate by an air gap on the order of 100 nm. With no voltage applied, all ribbons are at the same level. In this situation, the array of ribbons acts as a flat surface, specularly reflecting incident light.

When a voltage is applied between a ribbon and the electrode on the silicon substrate, an electric force pulls the ribbon downward, closer to the substrate. Alternate ribbons can be pulled down, while those in between remain in the higher configuration. As a result, the array of ribbons acts as a diffraction grating, such that the constructive interference for a particular wavelength of light can be directed toward a screen or other optical display system. By using three such devices, one each for red, blue, and green light, full color display is possible.

The GLV tends to be simpler to fabricate and higher in resolution than comparable DMD devices. On the other hand, DMD devices have already made an entry into the market. It will be interesting to watch this technology competition in future years.

QUICK QUIZ 27.6 Ultraviolet light of wavelength 350 nm is incident on a diffraction grating with slit spacing d and forms an interference pattern on a screen a distance L away. The angular positions θ_{bright} of the interference maxima are large. The locations of the bright fringes are marked on the screen. Now red light of wavelength 700 nm is used with a diffraction grating to form another diffraction pattern on the screen. Will the bright fringes of this pattern be located at the marks on the screen if (**a**) the screen is moved to a distance $2L$ from the grating, (**b**) the screen is moved to a distance $L/2$ from the grating, (**c**) the grating is replaced with one of slit spacing $2d$, (**d**) the grating is replaced with one of slit spacing $d/2$, or (**e**) nothing is changed?

---

[4]A long, narrow slit enables us to observe *line* spectra in the light coming from atomic and molecular systems, as discussed in Chapter 11.

> **THINKING PHYSICS 27.4**

White light reflected from the surface of a compact disc has a multicolored appearance as shown in Figure 27.22. Furthermore, the observation depends on the orientation of the disc relative to the eye and the position of the light source. Explain how that works.

Reasoning The surface of a compact disc has a spiral track with a spacing of approximately 1 μm that acts as a reflection grating. The light scattered by these closely spaced tracks interferes constructively in directions that depend on the wavelength and on the direction of the incident light. Any one section of the disc serves as a diffraction grating for white light, sending beams of constructive interference for different colors in different directions. The different colors you see when viewing one section of the disc change as the light source, the disc, or you move to change the angle of incidence or the viewing angle. ◀

Figure 27.22 (Thinking Physics 27.4) A compact disc observed under white light. The colors observed in the reflected light and their intensities depend on the orientation of the disc relative to the eye and relative to the light source.

Example 27.6 | **The Orders of a Diffraction Grating**

Monochromatic light from a helium–neon laser (λ = 632.8 nm) is incident normally on a diffraction grating containing 6 000 grooves per centimeter. Find the angles at which the first- and second-order maxima are observed.

SOLUTION

Conceptualize Study Figure 27.18 and imagine that the light coming from the left originates from the helium–neon laser. Let's evaluate the possible values of the angle θ for constructive interference.

Categorize We determine results using equations developed in this section, so we categorize this example as a substitution problem.

Calculate the slit separation as the inverse of the number of grooves per centimeter:

$$d = \frac{1}{6\,000} \text{ cm} = 1.667 \times 10^{-4} \text{ cm} = 1\,667 \text{ nm}$$

Solve Equation 27.18 for $\sin \theta$ and substitute numerical values for the first-order maximum ($m = 1$) to find θ_1:

$$\sin \theta_1 = \frac{(1)\lambda}{d} = \frac{632.8 \text{ nm}}{1\,667 \text{ nm}} = 0.379\,7$$

$$\theta_1 = \boxed{22.31°}$$

Repeat for the second-order maximum ($m = 2$):

$$\sin \theta_2 = \frac{(2)\lambda}{d} = \frac{2(632.8 \text{ nm})}{1\,667 \text{ nm}} = 0.759\,4$$

$$\theta_2 = \boxed{49.41°}$$

What If? What if you looked for the third-order maximum? Would you find it?

Answer For $m = 3$, we find $\sin \theta_3 = 1.139$. Because $\sin \theta$ cannot exceed unity, this result does not represent a realistic solution. Hence, only zeroth-, first-, and second-order maxima can be observed for this situation.

27.9 | Diffraction of X-Rays by Crystals

In principle, the wavelength of any electromagnetic wave can be determined if a grating of the proper spacing (on the order of λ) is available. **X-rays,** discovered in 1895 by Wilhelm Roentgen (1845–1923), are electromagnetic waves with very short wavelengths (on the order of 10^{-10} m = 0.1 nm). In 1913, Max von Laue (1879–1960) suggested that the regular array of atoms in a crystal, whose spacing is known to be about 10^{-10} m, could act as a three-dimensional diffraction grating for x-rays. Subsequent experiments confirmed his prediction. The observed diffraction patterns are complicated because of the three-dimensional nature of the crystal. Nevertheless, x-ray diffraction is an invaluable technique for elucidating crystalline structures and for understanding the structure of matter.

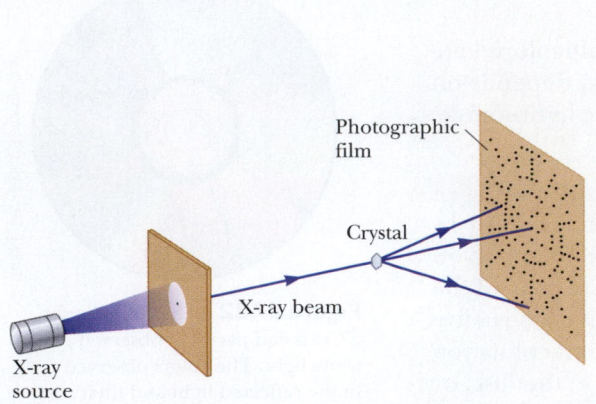

Figure 27.23 Schematic diagram of the technique used to observe the diffraction of x-rays by a crystal. The array of spots formed on the film is called a Laue pattern.

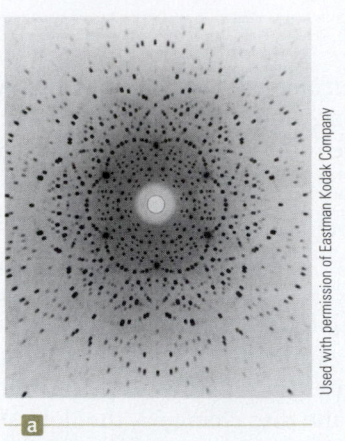

Figure 27.24 (a) A Laue pattern of a single crystal of the mineral beryl (beryllium aluminum silicate). (b) A Laue pattern of the enzyme Rubisco, produced with a wide-band x-ray spectrum. This enzyme is present in plants and takes part in the process of photosynthesis. The Laue pattern is used to determine the crystal structure of Rubisco.

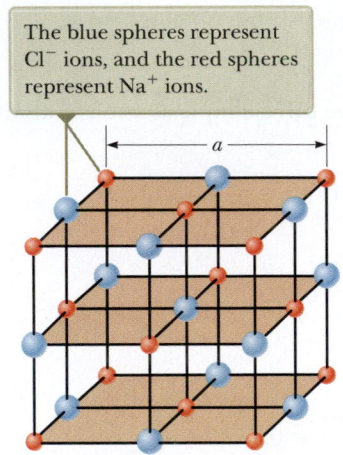

The blue spheres represent Cl⁻ ions, and the red spheres represent Na⁺ ions.

Figure 27.25 Crystalline structure of sodium chloride (NaCl). The length of the cube edge is $a = 0.564$ nm.

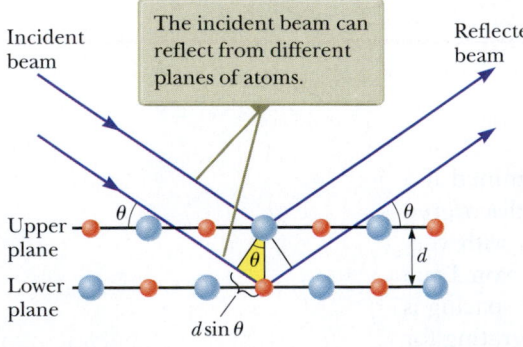

The incident beam can reflect from different planes of atoms.

Incident beam

Reflected beam

Upper plane

Lower plane

$d \sin \theta$

Figure 27.26 A two-dimensional description of the reflection of an x-ray beam from two parallel crystalline planes separated by a distance d. The beam reflected from the lower plane travels farther than the one reflected from the upper plane by a distance equal to $2d \sin \theta$.

Figure 27.23 is one experimental arrangement for observing x-ray diffraction from a crystal. A collimated beam of x-rays with a continuous range of wavelengths is incident on a crystal. The diffracted beams are very intense in certain directions, corresponding to constructive interference from waves reflected from layers of atoms in the crystal. The diffracted beams, which can be detected by a photographic film, form an array of spots known as a *Laue pattern*, as in Figure 27.24a. One can deduce the crystalline structure by analyzing the positions and intensities of the various spots in the pattern. Figure 27.24b shows a Laue pattern from a crystalline enzyme, using a wide range of wavelengths so that a swirling pattern results.

The arrangement of atoms in a crystal of NaCl is shown in Figure 27.25. The red spheres represent Na⁺ ions, and the blue spheres represent Cl⁻ ions. Each unit cell (the geometric shape that repeats through the crystal) contains four Na⁺ and four Cl⁻ ions. The unit cell is a cube whose edge length is a.

The ions in a crystal lie in various planes as shown in Figure 27.26. Suppose an incident x-ray beam makes an angle θ with one of the planes as in Figure 27.26. (Note that the angle θ is traditionally measured from the reflecting surface rather than from the normal, as in the case of the law of reflection in Chapter 25.) The beam can be reflected from both the upper plane and the lower one; the geometric construction in Figure 27.26, however, shows that the beam reflected from the lower surface travels farther than the beam reflected from the upper surface. The path difference between the two beams is $2d \sin \theta$, where d is the distance between the planes. The two beams reinforce each other (constructive interference) when this path difference equals some integral multiple of the wavelength λ. The same is true of reflection from the entire family of parallel planes. Hence, the condition for constructive interference (maxima in the reflected wave) is

$$2d \sin \theta = m\lambda \qquad m = 1, 2, 3, \ldots \qquad \text{27.19} \blacktriangleleft$$

This condition is known as **Bragg's law** after W. Lawrence Bragg (1890–1971), who first derived the relationship. If the wavelength and diffraction angle are measured, Equation 27.19 can be used to calculate the spacing between atomic planes.

27.10 | Context Connection: Holography

One interesting application of the laser is **holography,** the production of three-dimensional images of objects. The physics of holography was developed by Dennis Gabor (1900–1979) in 1948, for which he was awarded the 1971 Nobel Prize in Physics. The requirement of coherent

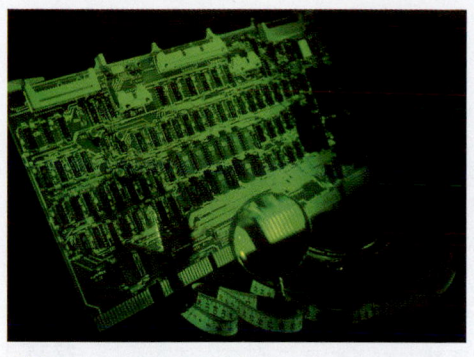

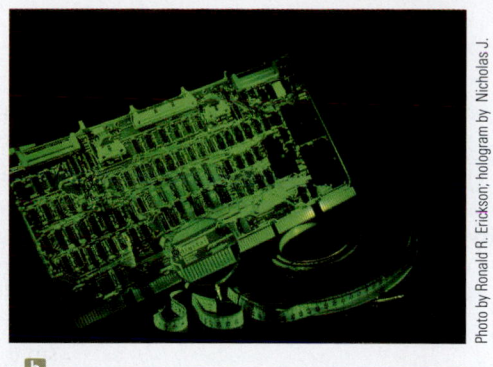

Figure 27.27 In this hologram, a circuit board is shown from two different views. Notice the difference in the appearance of the measuring tape and the view through the magnifying lens in (a) and (b).

Photo by Ronald R. Erickson; hologram by Nicholas J. Phillips

light for holography, however, delayed the realization of holographic images from Gabor's work until the development of lasers in the 1960s. Figure 27.27 shows a hologram and the three-dimensional character of its image.

Figure 27.28 shows how a hologram is made. Light from the laser is split into two parts by a half-silvered mirror at *B*. One part of the beam reflects off the object to be photographed and strikes an ordinary photographic film. The other half of the beam is diverged by lens L_2, reflects from mirrors M_1 and M_2, and finally strikes the film. The two beams overlap to form an extremely complicated interference pattern on the film. Such an interference pattern can be produced only if the phase relationship of the two waves is constant throughout the exposure of the film. This condition is met by illuminating the scene with light coming through a pinhole or with coherent laser radiation. The hologram records not only the intensity of the light scattered from the object (as in a conventional photograph), but also the phase difference between the reference beam and the beam scattered from the object. This phase difference results in an interference pattern that produces an image with full three-dimensional perspective.

In a normal photographic image, a lens is used to focus the image so that each point on the object corresponds to a single point on the film. Notice that no lens is used in Figure 27.28 to focus the light onto the film. Therefore, light from each point on the object reaches *all* points on the film. As a result, each region of the photographic film on which the hologram is recorded contains information about all illuminated points on the object, which leads to a remarkable result: If a small section of the hologram is cut from the film, the complete image can be formed from this small piece!

A hologram is best viewed by allowing coherent light to pass through the developed film as one looks back along the direction from which the beam comes. The interference pattern on the film acts as a diffraction grating. Figure 27.29 shows two rays of light striking the film and passing through. For each ray, the $m = 0$ and $m = \pm 1$ rays in the diffraction pattern are shown emerging from the right side of the film. Notice that the $m = +1$ rays converge to form a real image of the scene, which is not the image that is normally viewed. By extending the light rays corresponding to $m = -1$ back behind the film, we see that there is a virtual image located there, with light coming from it in exactly the same way that light came from the actual object when the film was exposed. This image is the one we see by looking through the holographic film.

Holograms are finding a number of applications in displays and in precision measurements. You may have a hologram on your credit card. This special type of hologram is called a *rainbow hologram*, designed to be viewed in reflected white light.

Holograms represent a means of storing visual information using lasers. In the Context Conclusion, we will investigate means of using lasers to store digital information that can be converted into sound waves or video displays.

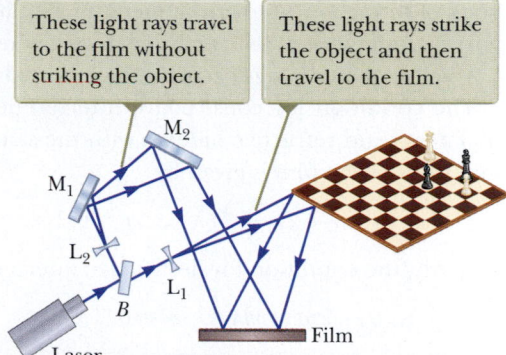

These light rays travel to the film without striking the object.

These light rays strike the object and then travel to the film.

Figure 27.28 Experimental arrangement for producing a hologram.

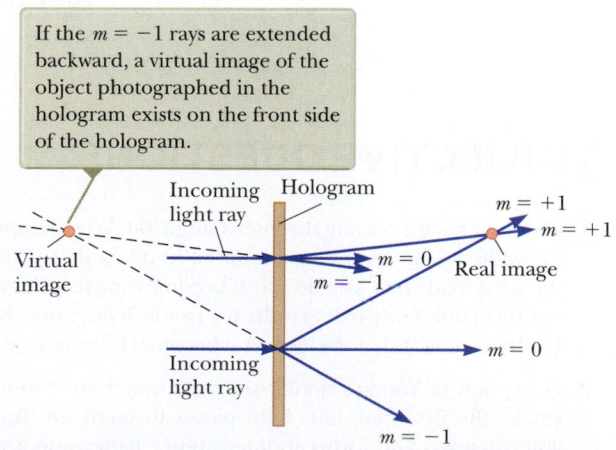

If the $m = -1$ rays are extended backward, a virtual image of the object photographed in the hologram exists on the front side of the hologram.

Figure 27.29 Two light rays strike a hologram at normal incidence. For each ray, outgoing rays corresponding to $m = 0$ and $m = \pm 1$ are shown.

SUMMARY

Interference of light waves is the result of the linear super-position of two or more waves at a given point. A sustained interference pattern is observed if (1) the sources have identical wavelengths and (2) the sources are coherent.

The **time-averaged light intensity** of the double-slit interference pattern is

$$I = I_{max} \cos^2\left(\frac{\pi d \sin\theta}{\lambda}\right) \qquad \textbf{27.8} \blacktriangleleft$$

where I_{max} is the maximum intensity on the screen.

An electromagnetic wave traveling from a medium with an index of refraction n_1 toward a medium with index of refraction n_2 undergoes a 180° phase change on reflection when $n_2 > n_1$. No phase change occurs in the reflected wave if $n_2 < n_1$.

The condition for constructive interference in a film of thickness t and refractive index n with the same medium on both sides of the film is given by

$$2nt = (m + \tfrac{1}{2})\lambda \qquad m = 0, 1, 2, \ldots \qquad \textbf{27.13} \blacktriangleleft$$

Similarly, the condition for destructive interference is

$$2nt = m\lambda \qquad m = 0, 1, 2, \ldots \qquad \textbf{27.14} \blacktriangleleft$$

Diffraction is the spreading of light from a straight-line path when the light passes through an aperture or around obstacles. A **diffraction pattern** can be analyzed as the in-

terference of a large number of coherent Huygens sources spread across the aperture.

The diffraction pattern produced by a single slit of width a on a distant screen consists of a central, bright maximum and alternating bright and dark regions of much lower intensities. The angles θ at which the diffraction pattern has *zero* intensity are given by

$$\sin\theta_{dark} = m\frac{\lambda}{a} \qquad m = \pm1, \pm2, \pm3, \ldots \qquad \textbf{27.15} \blacktriangleleft$$

Rayleigh's criterion, which is a limiting condition of resolution, says that two images formed by an aperture are just distinguishable if the central maximum of the diffraction pattern for one image falls on the first minimum of the other image. The limiting angle of resolution for a slit of width a is given by $\theta_{min} = \lambda/a$, and the limiting angle of resolution for a circular aperture of diameter D is given by $\theta_{min} = 1.22\,\lambda/D$.

A **diffraction grating** consists of a large number of equally spaced, identical slits. The condition for intensity maxima in the interference pattern of a diffraction grating for normal incidence is

$$d\sin\theta_{bright} = m\lambda \qquad m = 0, 1, 2, 3, \ldots \qquad \textbf{27.18} \blacktriangleleft$$

where d is the spacing between adjacent slits and m is the order number of the diffraction maximum.

Analysis Model for Problem Solving

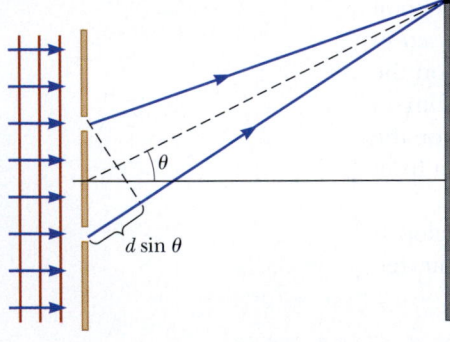

Waves in Interference. Young's double-slit experiment serves as a prototype for interference phenomena involving electromagnetic radiation. In this experiment, two slits separated by a distance d are illuminated by a single-wavelength light source. The condition for bright fringes (**constructive interference**) is

$$d\sin\theta_{bright} = m\lambda \qquad m = 0, \pm1, \pm2, \ldots \qquad \textbf{27.2} \blacktriangleleft$$

The condition for dark fringes (**destructive interference**) is

$$d\sin\theta_{dark} = (m + \tfrac{1}{2})\lambda \qquad m = 0, \pm1, \pm2, \ldots \qquad \textbf{27.3} \blacktriangleleft$$

The number m is called the **order number** of the fringe.

OBJECTIVE QUESTIONS

denotes answer available in *Student Solutions Manual/Study Guide*

1. Consider a wave passing through a single slit. What happens to the width of the central maximum of its diffraction pattern as the slit is made half as wide? (a) It becomes one-fourth as wide. (b) It becomes one-half as wide. (c) Its width does not change. (d) It becomes twice as wide. (e) It becomes four times as wide.

2. Four trials of Young's double-slit experiment are conducted. (a) In the first trial, blue light passes through two fine slits 400 μm apart and forms an interference pattern on a screen 4 m away. (b) In a second trial, red light passes through the same slits and falls on the same screen. (c) A third trial is performed

with red light and the same screen, but with slits 800 μm apart. (d) A final trial is performed with red light, slits 800 μm apart, and a screen 8 m away. **(i)** Rank the trials (a) through (d) from the largest to the smallest value of the angle between the central maximum and the first-order side maximum. In your ranking, note any cases of equality. **(ii)** Rank the same trials according to the distance between the central maximum and the first-order side maximum on the screen.

3. Suppose Young's double-slit experiment is performed in air using red light and then the apparatus is immersed in water.

What happens to the interference pattern on the screen? (a) It disappears. (b) The bright and dark fringes stay in the same locations, but the contrast is reduced. (c) The bright fringes are closer together. (d) The bright fringes are farther apart. (e) No change happens in the interference pattern.

4. Suppose you perform Young's double-slit experiment with the slit separation slightly smaller than the wavelength of the light. As a screen, you use a large half-cylinder with its axis along the midline between the slits. What interference pattern will you see on the interior surface of the cylinder? (a) bright and dark fringes so closely spaced as to be indistinguishable (b) one central bright fringe and two dark fringes only (c) a completely bright screen with no dark fringes (d) one central dark fringe and two bright fringes only (e) a completely dark screen with no bright fringes

5. A plane monochromatic light wave is incident on a double slit as illustrated in Active Figure 27.2. (i) As the viewing screen is moved away from the double slit, what happens to the separation between the interference fringes on the screen? (a) It increases. (b) It decreases. (c) It remains the same. (d) It may increase or decrease, depending on the wavelength of the light. (e) More information is required. (ii) As the slit separation increases, what happens to the separation between the interference fringes on the screen? Select from the same choices.

6. A thin layer of oil ($n = 1.25$) is floating on water ($n = 1.33$). What is the minimum nonzero thickness of the oil in the region that strongly reflects green light ($\lambda = 530$ nm)? (a) 500 nm (b) 313 nm (c) 404 nm (d) 212 nm (e) 285 nm

7. A monochromatic beam of light of wavelength 500 nm illuminates a double slit having a slit separation of 2.00×10^{-5} m. What is the angle of the second-order bright fringe? (a) 0.050 0 rad (b) 0.025 0 rad (c) 0.100 rad (d) 0.250 rad (e) 0.010 0 rad

8. A film of oil on a puddle in a parking lot shows a variety of bright colors in swirled patches. What can you say about the thickness of the oil film? (a) It is much less than the wavelength of visible light. (b) It is on the same order of magnitude as the wavelength of visible light. (c) It is much greater than the wavelength of visible light. (d) It might have any relationship to the wavelength of visible light.

9. What combination of optical phenomena causes the bright colored patterns sometimes seen on wet streets covered with a layer of oil? Choose the best answer. (a) diffraction and polarization (b) interference and diffraction (c) polarization and reflection (d) refraction and diffraction (e) reflection and interference

10. A Fraunhofer diffraction pattern is produced on a screen located 1.00 m from a single slit. If a light source of wavelength 5.00×10^{-7} m is used and the distance from the center of the central bright fringe to the first dark fringe is 5.00×10^{-3} m, what is the slit width? (a) 0.010 0 mm (b) 0.100 mm (c) 0.200 mm (d) 1.00 mm (e) 0.005 00 mm

11. In Active Figure 27.12, assume the slit is in a barrier that is opaque to x-rays as well as to visible light. The photograph in Active Figure 27.12b shows the diffraction pattern produced with visible light. What will happen if the experiment is repeated with x-rays as the incoming wave and with no other changes? (a) The diffraction pattern is similar. (b) There is no noticeable diffraction pattern but rather a projected shadow of high intensity on the screen, having the same width as the slit. (c) The central maximum is much wider, and the minima occur at larger angles than with visible light. (d) No x-rays reach the screen.

12. **BIO** When you receive a chest x-ray at a hospital, the x-rays pass through a set of parallel ribs in your chest. Do your ribs act as a diffraction grating for x-rays? (a) Yes. They produce diffracted beams that can be observed separately. (b) Not to a measurable extent. The ribs are too far apart. (c) Essentially not. The ribs are too close together. (d) Essentially not. The ribs are too few in number. (e) Absolutely not. X-rays cannot diffract.

13. Why is it advantageous to use a large-diameter objective lens in a telescope? (a) It diffracts the light more effectively than smaller-diameter objective lenses. (b) It increases its magnification. (c) It enables you to see more objects in the field of view. (d) It reflects unwanted wavelengths. (e) It increases its resolution.

CONCEPTUAL QUESTIONS

☐ denotes answer available in *Student Solutions Manual/Study Guide*

1. The atoms in a crystal lie in planes separated by a few tenths of a nanometer. Can they produce a diffraction pattern for visible light as they do for x-rays? Explain your answer with reference to Bragg's law.

2. What is the necessary condition on the path length difference between two waves that interfere (a) constructively and (b) destructively?

3. Explain why two flashlights held close together do not produce an interference pattern on a distant screen.

4. A soap film is held vertically in air and is viewed in reflected light as in Figure CQ27.4. Explain why the film appears to be dark at the top.

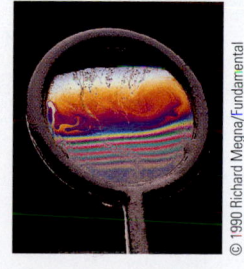

© 1990 Richard Megna/Fundamental Photographs

Figure CQ27.4
Conceptual Question 4.

5. Why is the lens on a good-quality camera coated with a thin film?

6. (a) In Young's double-slit experiment, why do we use monochromatic light? (b) If white light is used, how would the pattern change?

7. Consider a dark fringe in a double-slit interference pattern at which almost no light energy is arriving. Light from both slits is arriving at the location of the dark fringe, but the waves cancel. Where does the energy at the positions of dark fringes go?

8. Why can you hear around corners, but not see around corners?

9. A laser beam is incident at a shallow angle on a horizontal machinist's ruler that has a finely calibrated scale. The engraved rulings on the scale give rise to a diffraction pattern on a vertical screen. Discuss how you can use this technique to obtain a measure of the wavelength of the laser light.

10. Holding your hand at arm's length, you can readily block sunlight from reaching your eyes. Why can you not block sound from reaching your ears this way?

11. If a coin is glued to a glass sheet and this arrangement is held in front of a laser beam, the projected shadow has diffraction rings around its edge and a bright spot in the center. How are these effects possible?

12. A laser produces a beam a few millimeters wide, with uniform intensity across its width. A hair is stretched vertically across the front of the laser to cross the beam. (a) How is the diffraction pattern it produces on a distant screen related to that of a vertical slit equal in width to the hair? (b) How could you determine the width of the hair from measurements of its diffraction pattern?

13. John William Strutt, Lord Rayleigh (1842–1919), invented an improved foghorn. To warn ships of a coastline, a foghorn should radiate sound in a wide horizontal sheet over the ocean's surface. It should not waste energy by broadcasting sound upward or downward. Rayleigh's foghorn trumpet is shown in two possible configurations, horizontal and vertical, in Figure CQ27.13. Which is the correct orientation? Decide whether the long dimension of the rectangular opening should be horizontal or vertical and argue for your decision.

Figure CQ27.13

PROBLEMS AVAILABLE IN

27.1 Conditions for Interference
27.2 Young's Double-Slit Experiment
27.3 Analysis Model: Waves in Interference

Problems 1–19

27.4 Change of Phase Due to Reflection
27.5 Interference in Thin Films

Problems 20–26

27.6 Diffraction Patterns

Problems 27–33

27.7 Resolution of Single-Slit and Circular Apertures

Problems 34–39

27.8 The Diffraction Grating

Problems 40–47

27.9 Diffraction of X-Rays by Crystals

Problems 48–50

27.10 Context Connection: Holography

Problems 51–52

Additional Problems

Problems 53–71

Solutions to the following Problems are available in the *Student Solutions Manual/Study Guide*:

27.1, 27.3, 27.7, 27.13, 27.22, 27.25, 27.29, 27.35, 27.40, 27.42, 27.44, 27.48, 27.58, and 27.65

List of Enhanced Problems

Problem Number	Targeted Feedback in Enhanced WebAssign	Master It in Enhanced WebAssign	Watch It in Enhanced WebAssign
27.1	✓		✓
27.3	✓	✓	
27.4	✓		✓
27.7		✓	
27.8		✓	
27.10	✓		✓
27.13	✓	✓	
27.17	✓		✓
27.22		✓	
27.23	✓		✓
27.25	✓	✓	
27.28	✓		✓
27.29	✓	✓	
27.35	✓	✓	
27.37		✓	
27.39	✓		✓
27.40	✓	✓	
27.41	✓		✓
27.42	✓	✓	
27.47	✓		✓
27.48		✓	

Using Lasers to Record and Read Digital Information

We have now investigated the principles of optics and can respond to our central question for the *Lasers* Context:

> **What is special about laser light and how is it used in technological applications?**

In the Context Connections in Chapters 24 to 27, we discussed several technological applications of lasers. In this Context Conclusion, we will choose one more from the vast number of possibilities, the storage and retrieval of information on compact discs (as well as CD-ROMs and digital video discs, DVDs).

The storage of information in a small volume of space is a goal toward which humans have worked for several decades. In the early days of computing, information was stored on punched cards. This method seems humorous in today's world, especially because the area taken up by laying the cards representing a page of text out on a table was larger than the original page of text.

The magnetic disc recording and storage technique introduced in the 1950s allowed a reduction in space over that taken up by the original data. The beginning of optical storage occurred in the 1970s with the introduction of videodiscs. These plastic discs included encoded pits representing the analog information associated with a video signal. A laser, focused by lenses to a spot about 1 micrometer (μm) in diameter, is used to read the data. When the laser light reflects off the flat area of the disc, the light is reflected back into the system and is detected. When the light encounters a pit, some of it is scattered. The light reflected from the bottom of the pit interferes destructively with that reflected from the surface, and very little of the incident light finds its way back to the detection system.

The next step in the optical recording story involves the digital revolution, exemplified by the introduction of the compact disc, or CD. The reading of the disc is similar to that of the videodisc from the 1970s, but the information is stored in a *digital* format. Musical CDs were rapidly accepted by the public with much more enthusiasm than videodiscs. Shortly after the introduction of CDs, plans were announced to market an optical disc for storage of information for computers, the CD-ROM.

Digital Recording

In digital recording, information is converted to binary code (ones and zeros), similar to the dots and dashes of Morse code. First, the waveform of the sound is *sampled*, typically at the rate of 44 100 times per second. Figure 1

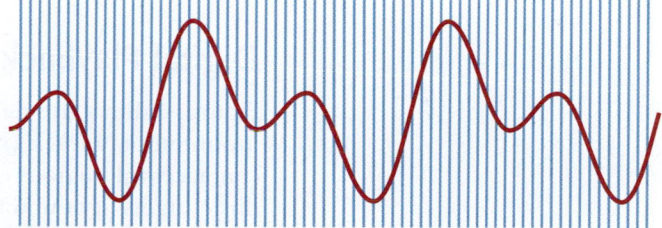

Figure 1 Sound is digitized by sampling the sound waveform at periodic time intervals (between the blue lines). During each interval, a number is recorded for the average voltage during the interval. The sampling rate shown here is much slower than the actual sampling rate of 44 100 per second.

803

TABLE 1 | Sample Binary Numbers

Number in Base 10	Number in Binary	Sum
1	0000000000000001	1
2	0000000000000010	2 + 0
3	0000000000000011	2 + 1
10	0000000000001010	8 + 0 + 2 + 0
37	0000000000100101	32 + 0 + 0 + 4 + 0 + 1
275	0000000100010011	256 + 0 + 0 + 0 + 16 + 0 + 0 + 2 + 1

Figure 2 The surface of a compact disc, showing the pits. Transitions between pits and lands correspond to ones. Regions without transitions correspond to zeros.

illustrates this process. The sampling frequency is much higher than the upper range of hearing, about 20 000 Hz, so all audible frequencies of sound are sampled at this rate. During each sampling, the pressure of the wave is measured and converted to a voltage. Therefore, there are 44 100 numbers associated with each second of the sound being sampled.

These measurements are then converted to *binary numbers,* which are numbers expressed to base 2 rather than base 10. Table 1 shows some sample binary numbers. Generally, voltage measurements are recorded in 16-bit "words," where each bit is a one or a zero. Therefore, the number of different voltage levels that can be assigned codes is $2^{16} = 65\ 536$. The number of bits in 1 second of sound is $16 \times 44\ 100 = 705\ 600$. These strings of ones and zeros, in 16-bit words, are recorded on the surface of a CD.

Figure 2 shows a magnification of the surface of a CD. There are two types of areas that are detected by the laser playback system: *lands* and *pits*. The lands are untouched regions of the disc surface that are highly reflective. The pits are areas that have been burned into the surface by a recording laser. The playback system, described below, converts the pits and lands into binary ones and zeros.

The binary numbers read from the CD are converted back to voltages, and the waveform is reconstructed as shown in Figure 3. Because the sampling rate is so high—44 100 voltage readings each second—the step-wise nature of the reconstructed waveform is not evident in the sound.

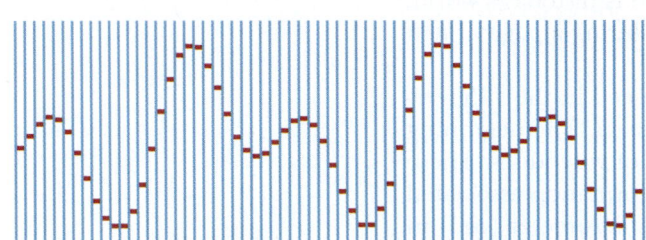

Figure 3 The reconstruction of the sound wave sampled in Figure 1. Notice that the reconstruction is step-wise, rather than the continuous waveform in Figure 1.

The advantage of digital recording is in the high fidelity of the sound. With analog recording, any small imperfection in the record surface or the recording equipment can cause a distortion of the waveform. If all peaks of a maximum in a waveform are clipped off so as to be only 90% as high, for example, there will be a major effect on the spectrum of the sound in an analog recording. With digital recording, however, it takes a major imperfection to turn a one into a zero. If an imperfection causes the magnitude of a one to be 90% of the original value, it still registers as a one and there is no distortion. Another advantage of digital recording is that the information is extracted optically, so there is no mechanical wear on the disc.

Digital Playback

Figure 4 shows the detection system of a CD player. The optical components are mounted on a track (not shown in the figure) that rolls radially so that the system can access all regions of the disc. The laser is located near the bottom of the figure, directing its light upward. The light is collimated by a lens into a parallel beam and passes through a beam splitter. The beam splitter serves no purpose for light on the way up, but it is important for the return light. The laser beam is then focused to a very small spot on the disc by the objective lens.

If the light encounters a pit in the disc, the light is scattered and very little light returns along the original path. If the light encounters a flat region of the disc at which a pit has not been recorded, the light reflects back along its original path. The reflected light moves downward in the diagram, arriving at the beam splitter so that it is partially reflected to the right. Lenses focus the beam, which is then detected by the photocell.

The playback system samples the reflected light 705 600 times per second. When the laser moves from a pit to a land or from a land to a pit, the reflected light changes during the sampling and the bit is recorded as a one. If there is no change during the sampling, the bit is recorded as a zero. The electronic circuitry in the CD player converts the series of zeros and ones back into an audible signal.

The DVD format was introduced in Japan in 1996 and in the United States in 1997. A DVD player uses 650-nm wavelength laser light rather than 780-nm light used in a CD player. Because of this shorter wavelength, DVD discs can store digital information in smaller pits than CD discs, and this is part of the reason the DVD has a larger storage capacity. A more recently developed alternative to the DVD format is the Blu-ray format, which uses even shorter-wavelength 405-nm (blue/violet) light and has a storage capacity of 50 GB for a dual-layer disc. Blu-ray and another high-definition format, HD DVD, were involved in a format war until early 2008, when support for HD DVD was withdrawn.

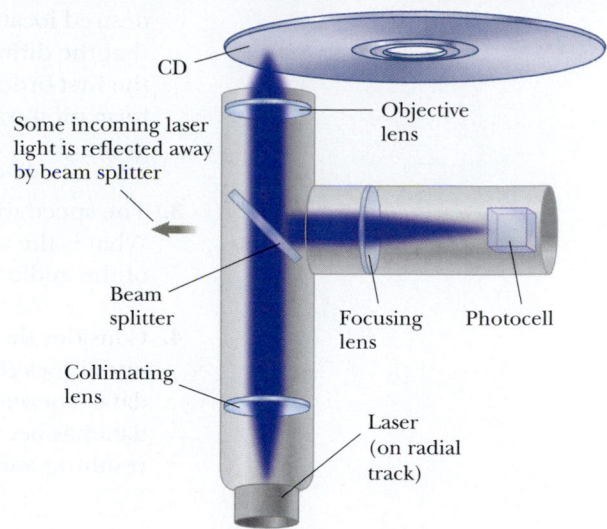

Figure 4 The detection system of a compact disc player. The laser (bottom) sends a beam of light upward. Laser light reflected back from the disc and then reflected to the right by the beam splitter enters a photocell. The digital information entering the photocell as pulses of light is converted to audio information.

Problems

1. Compact disc (CD) and digital video disc (DVD) players use interference to generate a strong signal from a tiny bump. The depth of a pit is chosen to be one-quarter of the wavelength of the laser light used to read the disc. Then light reflected from the pit and light reflected from the adjoining land differ in path length traveled by one-half wavelength, to interfere destructively at the detector. As the disc rotates, the light intensity drops significantly whenever light is reflected from near a pit edge. The space between the leading and trailing edges of a pit determines the time interval between the fluctuations. The series of time intervals is decoded into a series of zeros and ones that carries the stored information. Assume that infrared light with a wavelength of 780 nm in vacuum is used in a CD player. The disc is coated with plastic having a refractive index of 1.50. What should be the depth of each pit? A DVD player uses light of a shorter wavelength, and the pit dimensions are correspondingly smaller, one factor that results in greater storage capacity on a DVD compared with a CD.

2. The laser in a CD player must precisely follow the spiral track, along which the distance between one loop of the spiral and the next is only about 1.25 μm. A feedback mechanism lets the player know if the laser drifts off the track so that the player can steer it back again. Figure 5 shows how a diffraction grating is used to provide information to keep the beam on track. The laser light passes through a diffraction grating just before it reaches the disc. The strong central maximum of the diffraction pattern is used to read the information in the track of pits. The two first-order side maxima are used for steering. The grating is designed so that the first-order maxima fall on the flat surfaces on both sides of the information track. Both side beams are reflected into their own detectors. As long as both beams are reflecting from smooth, nonpitted surfaces they are detected with constant high intensity. If the main beam wanders off the track, however, one of the side beams will begin to strike pits on the information track and the reflected light will diminish. This change is used with an electronic circuit to guide the beam back to the

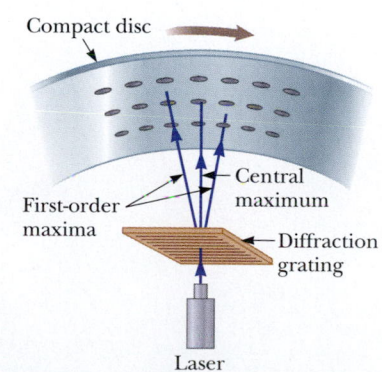

Figure 5 A tracking system in a CD player.

desired location. Assume that the laser light has a wavelength of 780 nm and that the diffraction grating is positioned 6.90 μm from the disc. Assume that the first-order beams are to fall on the disc 0.400 μm on either side of the information track. What should be the number of grooves per millimeter in the grating?

3. The speed with which the surface of a compact disc passes the laser is 1.3 m/s. What is the average length of the audio track on a CD associated with each bit of the audio information?

4. Consider the photograph of the compact disc surface in Figure 2. Audio data undergoes complicated processing to reduce a variety of errors in reading the data. Therefore, an audio "word" is not laid out linearly on the disc. Suppose data has been read from the disc, the error coding has been removed, and the resulting audio word is

$$1\ 0\ 1\ 1\ 1\ 0\ 1\ 1\ 1\ 0\ 1\ 1\ 1\ 0\ 1\ 1$$

What is the decimal number represented by this 16-bit word?

5. Lasers are also used in the recording process for a *magnetooptical disc*. To record a pit, its location on the ferromagnetic layer of the disc must be raised above a minimum temperature called the Curie temperature. Imagine that the surface moves past the laser at a speed on the order of 1 m/s and that the pit is modeled as a cylinder 1 μm deep with a radius of 1 μm. The ferromagnetic material has the following properties: its Curie temperature is 600 K, its specific heat is 300 J/kg · °C, and its density is 2×10^3 kg/m^3. What is the order of magnitude of the intensity of the laser beam necessary to raise the pit above the Curie temperature?

The Cosmic Connection

In this final Context, we investigate the principles included in the area of physics commonly called *modern physics*. Modern physics encompasses the revolution in physics that commenced at the beginning of the 20th century. We began our discussion of modern physics in Chapter 9 with our study of relativity. Other aspects of modern physics have appeared at various locations throughout the book, including atomic spectra and the Bohr model in Chapter 11, black holes in Chapter 11, quantization of molecular rotation and vibration in Chapter 17, black bodies in Chapter 24, and the discussion of the photon in Chapter 24.

In this book, we stress the importance of models in understanding physical phenomena. At the turn of the 20th century, classical physics was well established and provided many principles on which models for phenomena could be built. Many experimental observations, however, could not be brought into agreement with theory using classical models. Attempts to apply the laws of classical physics to atomic systems were consistently unsuccessful in making accurate predictions of the behavior of matter on the atomic scale. Various phenomena such as blackbody radiation, the photoelectric

Figure 2 Supernova 1987A. (*Left*) A region of the Large Magellanic Cloud before the supernova. (*Right*) The supernova appears on February 24, 1987. An understanding of this cosmic explosion is found in the interactions between the microscopic particles within the nucleus.

© Australian Astronomical Observatory, photograph by David Malin from AAT plates.

effect, and the emission of sharp spectral lines by atoms in a gas discharge could not be understood within the framework of classical physics. Between 1900 and 1930, however, new models collectively called *quantum physics* or *quantum mechanics* were highly successful in explaining the behavior of atoms, molecules, and nuclei. Like relativity, quantum physics requires a modification of our ideas concerning the physical world. Quantum mechanics does not, however, directly contradict or invalidate classical mechanics. As with relativity, the equations of quantum physics reduce to classical equations in the appropriate realm, that is, when the quantum equations are used to describe macroscopic systems.

An extensive study of quantum physics is certainly beyond the scope of this book and therefore this Context is simply an introduction to its underlying ideas. One of the true successes of quantum physics is the connection it makes between microscopic phenomena and the structure and evolution of the Universe. Ironically, recent developments in physics that probe smaller and

Figure 1 The Apple iPad, a popular tablet computer. The appearance of information on the display is due to the behavior of microscopic electrons in the circuitry of the microprocessor.

© Kelvintt | Dreamstime.com

NASA, ESA, and M. Livio and the Hubble 20th Anniversary Team (STScI)

Figure 3 This image of a portion of the Carina Nebula, 7 500 light-years from the Earth, was released to celebrate the 20th anniversary of the launching and deployment of the Hubble Space Telescope (April 24, 1990). Significant new star development is causing this display. Several jets of material, such as those at the top of the image, are caused by gravitational attraction of material to the new stars' surfaces. The colors in the image are due to light emitted from individual atoms of oxygen, nitrogen, and hydrogen. The Carina Nebula is a member of our own galaxy. Across the entire sky, it is estimated that the Hubble Space Telescope can detect 100 billion galaxies. It is also estimated that this is a very small fraction of all the galaxies in the visible part of the Universe. To develop a theory of the origin of this tremendously large system, we need to understand quarks, the most fundamental particles in the current theory of particle physics.

smaller scales allow us to advance our understanding of the larger and larger systems that are familiar to us. This connection between the small and the large is the theme of this Context.

Let us consider some examples of macroscopic systems and their connection to the behavior of microscopic particles. Consider your experiences with common electronic devices that are used today to view information on a liquid crystal display: handheld calculators, smartphones, tablet computers, and flat panel televisions. The events you observe—the appearance of numbers, to-do lists, or photographs on an LCD display—are macroscopic, but what controls these macroscopic events? They are controlled by a microprocessor within the electronic device. The operation of the microprocessor depends on the behavior of electrons within the integrated circuit chip. The design and manufacture of the macroscopic electronic device are not possible without an understanding of the behavior of the electrons.

As a second example, a supernova explosion is clearly a macroscopic event; it is a star with a radius on the

order of billions of meters undergoing a violent event. We have been able to advance our understanding of such events by studying the atomic nucleus, which is on the order of 10^{-15} m in size.

If we imagine an even larger system than a star—the entire Universe—we can advance our understanding of its origin by thinking about particles even smaller than the nucleus. Consider the constituents of protons and neutrons, called *quarks*. Models based on quarks provide further understanding of a theory of the origin of the Universe called the *Big Bang*. In this Context, we shall study both quarks and the Big Bang.

It seems that the larger the system we wish to investigate, the smaller are the particles whose behavior we must understand! We shall explore this relationship and study the principles of quantum physics as we respond to our central question:

How can we connect the physics of microscopic particles to the physics of the Universe?

Quantum Physics

Chapter Outline

© Robert Harding Picture Library Ltd./Alamy

A scanning electron microscope photograph shows significant detail of a cheese mite, *Tyrolichus casei*. The mite is so small, with a maximum length of 0.70 mm, that ordinary microscopes do not reveal minute anatomical details. The operation of the electron microscope is based on the wave nature of electrons, a central feature in quantum physics.

I n the earlier chapters of this book, we focused on the physics of particles. The particle model was a simplification model that allowed us to ignore the unnecessary details of an object when studying its behavior. We later combined particles into additional simplification models of systems and rigid objects. In Chapter 13, we introduced the wave as yet another simplification model and found that we could understand the motion of vibrating strings and the intricacies of sound by studying simple waves. In Chapters 24 to 27, we found that the wave model for light helped us understand many phenomena associated with optics.

It is hoped that you now have confidence in your abilities to analyze problems in the very different worlds of particles and waves. Your confidence may have been shaken somewhat by the discussion at the beginning of Chapter 25 in which we indicated that light has both wave-like and particle-like behaviors.

In this chapter, we return to this dual nature of light and study it in more detail. This study leads to a new simplification model, the quantum particle, and a new analysis model, the quantum particle under boundary conditions. A careful analysis of these two models shows that particles and waves are not as unrelated as you might expect.

28.1 | Blackbody Radiation and Planck's Theory

As we discussed in Chapter 17, an object at any temperature emits energy referred to as **thermal radiation.** The characteristics of this radiation depend on the temperature and properties of the surface of the object. If the surface is at room temperature, the wavelengths of the thermal radiation are primarily in the infrared region and hence are not observed by the eye. As the temperature of the surface increases, the object eventually begins to glow red. At sufficiently high temperatures, the object appears to be white, as in the glow of the hot tungsten filament of a lightbulb. A careful study of thermal radiation shows that it consists of a continuous distribution of wavelengths from all portions of the electromagnetic spectrum.

From a classical viewpoint, thermal radiation originates from accelerated charged particles near the surface of the object. The thermally agitated charges can have a distribution of accelerations, which accounts for the continuous spectrum of radiation emitted by the object. By the end of the 19th century, it had become apparent that this classical explanation of thermal radiation was inadequate. The basic problem was in understanding the observed distribution of wavelengths in the radiation emitted by an ideal object called a black body. As mentioned in Chapter 24, a **black body** is an ideal system that absorbs all radiation incident on it. A good approximation of a black body is a small hole leading to the inside of a hollow object as shown in Figure 28.1. The nature of the radiation emitted from the hole depends only on the temperature of the cavity walls.

The wavelength distribution of radiation from cavities was studied extensively in the late 19th century. Experimental data for the distribution of energy in **blackbody radiation** at three temperatures are shown in Active Figure 28.2. The distribution of radiated energy varies with wavelength and temperature. Two regular features of the distribution were noted in these experiments:

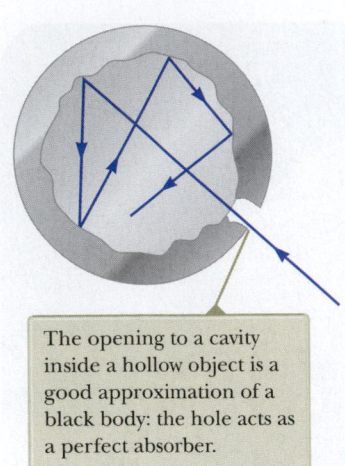

The opening to a cavity inside a hollow object is a good approximation of a black body: the hole acts as a perfect absorber.

Figure 28.1 A physical model of a black body.

1. **The total power of emitted radiation increases with temperature.** We discussed this feature briefly in Chapter 17, where we introduced **Stefan's law,** Equation 17.36, for the power emitted from a surface of area A and temperature T:

$$P = \sigma A e T^4 \qquad \text{28.1} \blacktriangleleft$$

▶ Stefan's law

For a black body, the emissivity is $e = 1$ exactly.

2. **The peak of the wavelength distribution shifts to shorter wavelengths as the temperature increases.** This shift was found experimentally to obey the following relationship, called **Wien's displacement law:**

▶ Wien's displacement law

$$\lambda_{max} T = 2.898 \times 10^{-3} \text{ m} \cdot \text{K} \qquad \text{28.2} \blacktriangleleft$$

The glow emanating from the spaces between these hot charcoal briquettes is, to a close approximation, blackbody radiation. The color of the light depends on the temperature of the briquettes.

The 4 000-K curve has a peak near the visible range. This curve represents an object that would glow with a yellowish-white appearance.

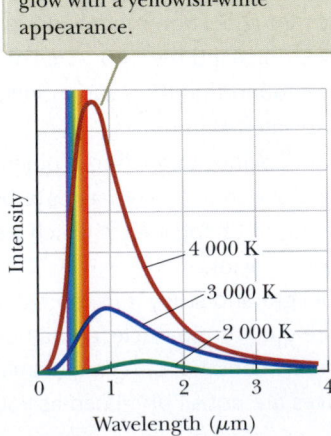

Active Figure 28.2 Intensity of blackbody radiation versus wavelength at three temperatures. The visible range of wavelengths is between 0.4 μm and 0.7 μm. At approximately 6 000 K, the peak is in the center of the visible wavelengths and the object appears white.

where $\lambda_{\max}$ is the wavelength at which the curve in Active Figure 28.2 peaks and T is the absolute temperature of the surface emitting the radiation.

A successful theoretical model for blackbody radiation must predict the shape of the curve in Active Figure 28.2, the dependence of power on temperature expressed in Stefan's law, and the shift of the peak with temperature described by Wien's displacement law. Early attempts to use classical ideas to explain the shapes of the curves in Active Figure 28.2 failed.

Let's consider one of these early attempts. To describe the distribution of energy from a black body, we define $I(\lambda, T) \, d\lambda$ to be the intensity, or power per unit area, emitted in the wavelength interval $d\lambda$. The result of a calculation based on a classical theory of blackbody radiation known as the **Rayleigh–Jeans law** is

$$I(\lambda, T) = \frac{2\pi c k_{\mathrm{B}} T}{\lambda^4}$$ **28.3** ◄

▶ Rayleigh–Jeans law

where k_{B} is Boltzmann's constant. The black body is modeled as the hole leading into a cavity (Fig. 28.1), resulting in many modes of oscillation of the electromagnetic field caused by accelerated charges in the cavity walls and the emission of electromagnetic waves at all wavelengths. In the classical theory used to derive Equation 28.3, the average energy for each wavelength of the standing-wave modes is assumed to be proportional to $k_{\mathrm{B}} T$, based on the theorem of equipartition of energy discussed in Section 16.5.

An experimental plot of the blackbody radiation spectrum, together with the theoretical prediction of the Rayleigh–Jeans law, is shown in Figure 28.3. At long wavelengths, the Rayleigh–Jeans law is in reasonable agreement with experimental data, but at short wavelengths, major disagreement is apparent.

As λ approaches zero, the function $I(\lambda, T)$ given by Equation 28.3 approaches infinity. Hence, according to classical theory, not only should short wavelengths predominate in a blackbody spectrum, but also the energy emitted by any black body should become infinite in the limit of zero wavelength. In contrast to this prediction, the experimental data plotted in Figure 28.3 show that as λ approaches zero, $I(\lambda, T)$ also approaches zero. This mismatch of theory and experiment was so disconcerting that scientists called it the *ultraviolet catastrophe*. (This "catastrophe"—infinite energy—occurs as the wavelength approaches zero; the word *ultraviolet* was applied because ultraviolet wavelengths are short.)

In 1900, Max Planck developed a structural model for blackbody radiation that leads to a theoretical equation for the wavelength distribution that is in complete agreement with experimental results at all wavelengths. His model represents the dawn of **quantum physics.**

Using the components of structural models introduced in Section 11.2, we can describe Planck's model as follows.

1. *A description of the physical components of the system:* Planck identified blackbody radiation as arising from *oscillators,* related to the charged particles within the molecules of the black body. The components of the system are the oscillators and the radiation emitted from them.
2. *A description of where the components are located relative to one another and how they interact:* The oscillators emitting observable blackbody radiation are located at the surface of the black body. The energy of the oscillator is quantized; that is, it can have only certain *discrete* amounts of energy E_n given by

$$E_n = nhf$$ **28.4** ◄

where n is a positive integer called a **quantum number,**[1] f is the oscillator frequency, and h is **Planck's constant,** first introduced in Chapter 11. Because the energy of each oscillator can have only discrete values given by Equation 28.4,

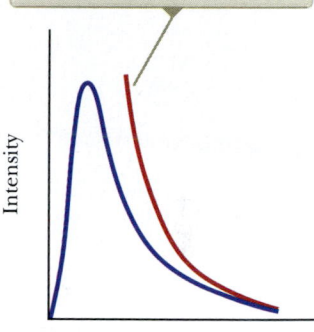

The classical theory (red-brown curve) shows intensity growing without bound for short wavelengths, unlike the experimental data (blue curve).

Intensity

Wavelength

Figure 28.3 Comparison of experimental results and the curve predicted by the Rayleigh–Jeans law for the distribution of blackbody radiation.

© Bettmann/CORBIS

Max Planck
German Physicist (1858–1947)
Planck introduced the concept of a "quantum of action" (Planck's constant, *h*) in an attempt to explain the spectral distribution of blackbody radiation, which laid the foundations for quantum theory. In 1918, he was awarded the Nobel Prize in Physics for this discovery of the quantized nature of energy.

---

[1]We first introduced the notion of a quantum number for microscopic systems in Section 11.5, in which we incorporated it into the Bohr model of the hydrogen atom. We put it in bold again here because it is an important notion for the remaining chapters in this book.

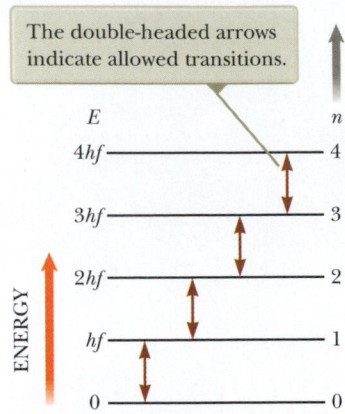

The double-headed arrows indicate allowed transitions.

Figure 28.4 Allowed energy levels for an oscillator with a natural frequency f.

we say that the energy is **quantized.** Each discrete energy value corresponds to a different **quantum state,** represented by the quantum number n. When the oscillator is in the $n = 1$ quantum state, its energy is hf; when it is in the $n = 2$ quantum state, its energy is $2hf$; and so on. An oscillator radiates or absorbs energy only when it changes quantum states. If it remains in one quantum state, no energy is absorbed or emitted.

3. *A description of the time evolution of the system:* As noted at the end of structural model component 2, radiation is emitted or absorbed only when the oscillator makes a transition from one energy state to a different state. The oscillators emit or absorb energy in discrete units, similar to the transitions discussed in the Bohr model in Chapter 11. The entire energy difference between the initial and final states in the transition is emitted as a single quantum of radiation. If the transition is from one state to an adjacent state—say, from the $n = 3$ state to the $n = 2$ state—Equation 28.4 shows that the amount of energy radiated by the oscillator is

$$E = hf \qquad \textbf{28.5} \blacktriangleleft$$

Figure 28.4 shows the quantized energy levels and allowed transitions proposed by Planck.

4. *A description of the agreement between predictions of the model and actual observations and, possibly, predictions of new effects that have not yet been observed:* The test of Planck's model will be this: Does the model predict a wavelength distribution curve that matches the experimental results in Active Figure 28.2 better than the expression in Equation 28.3 does?

These ideas may not sound bold to you because we have seen them in the Bohr model of the hydrogen atom in Chapter 11. It is important to keep in mind, however, that the Bohr model was not introduced until 1913, whereas Planck made his assumptions in 1900. The key point in Planck's theory is the radical assumption of quantized energy states.

Using this approach, Planck generated a theoretical expression for the wavelength distribution that agreed remarkably well with the experimental curves in Active Figure 28.2:

▶ Planck's wavelength distribution function

$$I(\lambda, T) = \frac{2\pi hc^2}{\lambda^5 (e^{hc/\lambda k_\text{B} T} - 1)} \qquad \textbf{28.6} \blacktriangleleft$$

This function includes the parameter h, which Planck adjusted so that his curve matched the experimental data at all wavelengths. The value of this parameter is found to be independent of the material of which the black body is made and independent of the temperature; it is a fundamental constant of nature. The value of h, Planck's constant, which was first introduced in Chapter 11, is

▶ Planck's constant

$$h = 6.626 \times 10^{-34} \, \text{J} \cdot \text{s} \qquad \textbf{28.7} \blacktriangleleft$$

At long wavelengths, Equation 28.6 reduces to the Rayleigh–Jeans expression, Equation 28.3, and at short wavelengths, it predicts an exponential decrease in $I(\lambda, T)$ with decreasing wavelength, in agreement with experimental results.

When Planck presented his theory, most scientists (including Planck!) did not consider the quantum concept realistic. It was believed to be a mathematical trick that happened to predict the correct results. Hence, Planck and others continued to search for what they believed to be a more rational explanation of blackbody radiation. Subsequent developments, however, showed that a theory based on the quantum concept (rather than on classical concepts) was required to explain a number of other phenomena at the atomic level.

We don't see quantum effects on an everyday basis because the energy change in a macroscopic system due to a transition between adjacent states is such a small fraction of the total energy of the system that we could never expect to detect the change. (See Example 28.2 for a numerical example.) Therefore, even though changes in the energy of a macroscopic system are indeed quantized and proceed

Pitfall Prevention | 28.2

n Is Again an Integer

In the preceding chapters on optics, we used the symbol n for the index of refraction, which was not an integer. We are now using n again in the manner in which it was used in Chapter 11 to indicate the quantum number of a Bohr orbit and in Chapter 14 to indicate the standing wave mode on a string or in an air column. In quantum physics, n is often used as an integer quantum number to identify a particular quantum state of a system.

by small quantum jumps, our senses perceive the decrease as continuous. Quantum effects become important and measurable only on the submicroscopic level of atoms and molecules. Furthermore, quantum results must blend smoothly with classical results when the quantum number becomes large. This statement is known as the **correspondence principle.**

You may have had your body temperature measured at the doctor's office by an *ear thermometer,* which can read your temperature in a matter of seconds (Fig. 28.5). This type of thermometer measures the amount of infrared radiation emitted by the eardrum in a fraction of a second. It then converts the amount of radiation into a temperature reading. This thermometer is very sensitive because temperature is raised to the fourth power in Stefan's law. Problem 28.3 in Enhanced WebAssign allows you to explore the sensitivity of this device.

BIO The ear thermometer

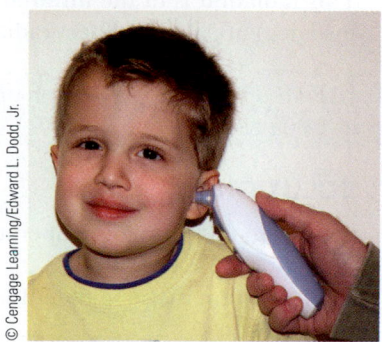

Figure 28.5 An ear thermometer measures a patient's temperature by detecting the intensity of infrared radiation leaving the eardrum.

QUICK QUIZ 28.1 Figure 28.6 shows two stars in the constellation Orion. Betelgeuse appears to glow red, whereas Rigel looks blue in color. Which star has a higher surface temperature?   (**a**) Betelgeuse   (**b**) Rigel   (**c**) both have the same   (**d**) impossible to determine

Figure 28.6 (Quick Quiz 28.1) Which star is hotter, Betelgeuse or Rigel?

THINKING PHYSICS 28.1

You are observing a yellow candle flame, and your laboratory partner claims that the light from the flame is atomic in origin. You disagree, claiming that the candle flame is hot, so the radiation must be thermal in origin. Before this disagreement leads to fisticuffs, how could you determine who is correct?

Reasoning A simple determination could be made by observing the light from the candle flame through a diffraction grating spectrometer, which was discussed in Section 27.8. If the spectrum of the light is continuous, it is thermal in origin. If the spectrum shows discrete lines, it is atomic in origin. The results of the experiment show that the light is primarily thermal in origin and originates in the hot particles of soot in the candle flame. ◄

Example 28.1 | Thermal Radiation from Different Objects   **BIO**

(**A**) Find the peak wavelength of the blackbody radiation emitted by the human body when the skin temperature is 35°C.

SOLUTION

Conceptualize Thermal radiation is emitted from the surface of any object. The peak wavelength is related to the surface temperature through Wien's displacement law (Eq. 28.2).

Categorize We evaluate results using an equation developed in this section, so we categorize this example as a substitution problem.

continued

28.1 *cont.*

Solve Equation 28.2 for λ_{max}:

$$(1) \quad \lambda_{max} = \frac{2.898 \times 10^{-3} \, m \cdot K}{T}$$

Substitute the surface temperature:

$$\lambda_{max} = \frac{2.898 \times 10^{-3} \, m \cdot K}{308 \, K} = \boxed{9.41 \, \mu m}$$

This radiation is in the infrared region of the spectrum and is invisible to the human eye. Some animals (pit vipers, for instance) are able to detect radiation of this wavelength and therefore can locate warm-blooded prey even in the dark.

(B) Find the peak wavelength of the blackbody radiation emitted by the tungsten filament of a lightbulb, which operates at 2 000 K.

SOLUTION

Substitute the filament temperature into Equation (1):

$$\lambda_{max} = \frac{2.898 \times 10^{-3} \, m \cdot K}{2 \, 000 \, K} = \boxed{1.45 \, \mu m}$$

This radiation is also in the infrared, meaning that most of the energy emitted by a lightbulb is not visible to us.

(C) Find the peak wavelength of the blackbody radiation emitted by the Sun, which has a surface temperature of approximately 5 800 K.

SOLUTION

Substitute the surface temperature into Equation (1):

$$\lambda_{max} = \frac{2.898 \times 10^{-3} \, m \cdot K}{5 \, 800 \, K} = \boxed{0.500 \, \mu m}$$

This radiation is near the center of the visible spectrum, near the color of a yellow-green tennis ball. Because it is the most prevalent color in sunlight, our eyes have evolved to be most sensitive to light of approximately this wavelength.

(D) What total power is emitted by your skin, assuming that it emits like a black body?

SOLUTION

We need to make an estimate of the surface area of your skin. Model your body as a rectangular box of height 2 m, width 0.3 m, and depth 0.2 m, and find its total surface area:

$$A = 2(2 \, m)(0.3 \, m) + 2(2 \, m)(0.2 \, m) + 2(0.2 \, m)(0.3 \, m)$$
$$\approx 2 \, m^2$$

Use Stefan's law, Equation 28.1, to find the power of the emitted radiation:

$$P = \sigma A e T^4 \approx (5.7 \times 10^{-8} \, W/m^2 \cdot K^4)(2 \, m^2)(1)(308 \, K)^4$$
$$\approx \boxed{10^3 \, W}$$

(E) Based on your answer to part (D), why don't you glow as brightly as several lightbulbs?

SOLUTION

The answer to part (D) indicates that your skin is radiating energy at approximately the same rate as that which enters ten 100-W lightbulbs by electrical transmission. You are not visibly glowing, however, because most of this radiation is in the infrared range, as we found in part (A), and our eyes are not sensitive to infrared radiation.

Example **28.2** | **The Quantized Oscillator**

A 2.00-kg block is attached to a massless spring that has a force constant of $k = 25.0 \, N/m$. The spring is stretched 0.400 m from its equilibrium position and released from rest.

(A) Find the total energy of the system and the frequency of oscillation according to classical calculations.

SOLUTION

Conceptualize We understand the details of the block's motion from our study of simple harmonic motion in Chapter 12. Review that material if you need to.

28.2 *cont.*

Categorize The phrase "according to classical calculations" tells us to categorize this part of the problem as a classical analysis of the oscillator. We model the block as a particle in simple harmonic motion.

Analyze Based on the way the block is set into motion, its amplitude is 0.400 m.

Evaluate the total energy of the block–spring system using Equation 12.21:

$$E = \tfrac{1}{2}kA^2 = \tfrac{1}{2}(25.0 \text{ N/m})(0.400 \text{ m})^2 = \boxed{2.00 \text{ J}}$$

Evaluate the frequency of oscillation from Equation 12.14:

$$f = \frac{1}{2\pi}\sqrt{\frac{k}{m}} = \frac{1}{2\pi}\sqrt{\frac{25.0 \text{ N/m}}{2.00 \text{ kg}}} = \boxed{0.563 \text{ Hz}}$$

(B) Assuming the energy of the oscillator is quantized, find the quantum number n for the system oscillating with this amplitude.

SOLUTION

Categorize This part of the problem is categorized as a quantum analysis of the oscillator. We model the block–spring system as a Planck oscillator.

Analyze Solve Equation 28.4 for the quantum number n:

$$n = \frac{E_n}{hf}$$

Substitute numerical values:

$$n = \frac{2.00 \text{ J}}{(6.626 \times 10^{-34} \text{ J} \cdot \text{s})(0.563 \text{ Hz})} = \boxed{5.36 \times 10^{33}}$$

Finalize Notice that 5.36×10^{33} is a very large quantum number, which is typical for macroscopic systems. Changes between quantum states for the oscillator are explored next.

What If? Suppose the oscillator makes a transition from the $n = 5.36 \times 10^{33}$ state to the state corresponding to $n = 5.36 \times 10^{33} - 1$. By how much does the energy of the oscillator change in this one-quantum change?

Answer From Equation 28.5 and the result to part (A), the energy carried away due to the transition between states differing in n by 1 is

$$E = hf = (6.626 \times 10^{-34} \text{ J} \cdot \text{s})(0.563 \text{ Hz}) = 3.73 \times 10^{-34} \text{ J}$$

This energy change due to a one-quantum change is fractionally equal to 3.73×10^{-34} J/2.00 J, or on the order of one part in 10^{34}! It is such a small fraction of the total energy of the oscillator that it cannot be detected. Therefore, even though the energy of a macroscopic block–spring system is quantized and does indeed decrease by small quantum jumps, our senses perceive the decrease as continuous. Quantum effects become important and detectable only on the submicroscopic level of atoms and molecules.

28.2 | The Photoelectric Effect

Blackbody radiation was historically the first phenomenon to be explained with a quantum model. In the latter part of the 19th century, at the same time as data were being taken on thermal radiation, experiments showed that light incident on certain metallic surfaces causes electrons to be emitted from the surfaces. As mentioned in Section 25.1, this phenomenon, first discovered by Hertz, is known as the **photoelectric effect.** The emitted electrons are called **photoelectrons.**[2]

Active Figure 28.7 (page 816) is a schematic diagram of a photoelectric effect apparatus. An evacuated glass or quartz tube contains a metal plate E connected to the negative terminal of a battery. Another metal plate C is maintained at a positive potential by the battery. When the tube is kept in the dark, the ammeter reads zero, indicating that there is no current in the circuit. When light of the appropriate

---

[2]Photoelectrons are not different from other electrons. They are given this name solely because of their ejection from the metal by photons in the photoelectric effect.

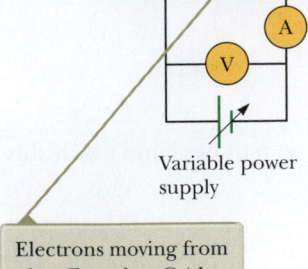

When light strikes plate E (the emitter), photoelectrons are ejected from the plate.

Electrons moving from plate E to plate C (the collector) constitute a current in the circuit.

Active Figure 28.7 A circuit diagram for studying the photoelectric effect.

The current increases with intensity but reaches a saturation level for large values of ΔV.

At voltages equal to or more negative than $-\Delta V_s$, the current is zero.

Active Figure 28.8 Photoelectric current versus applied potential difference for two light intensities.

wavelength shines on plate E, however, a current is detected by the ammeter, indicating a flow of charges across the gap between E and C. This current arises from electrons emitted from the negative plate E (the emitter) and collected at the positive plate C (the collector).

Active Figure 28.8, a graphical representation of the results of a photoelectric experiment, plots the photoelectric current versus the potential difference ΔV between E and C for two light intensities. For large positive values of ΔV, the current reaches a maximum value. In addition, the current increases as the incident light intensity increases, as you might expect. Finally, when ΔV is negative—that is, when the battery polarity is reversed to make E positive and C negative—the current drops because many of the photoelectrons emitted from E are repelled by the negative collecting plate C. Only those electrons ejected from the metal with a kinetic energy greater than $e|\Delta V|$ will reach C, where e is the magnitude of the charge on the electron. When the magnitude of ΔV is equal to ΔV_s, the **stopping potential,** no electrons reach C and the current is zero.

Let us consider the combination of the electric field between the plates and an electron ejected from plate E with the maximum kinetic energy to be an isolated system. Suppose this electron stops just as it reaches plate C. Applying the energy version of the isolated system model, Equation 7.2 becomes:

$$\Delta K + \Delta U = 0 \quad \rightarrow \quad K_f + U_f = K_i + U_i$$

where the initial configuration of the system refers to the instant that the electron leaves the metal with the maximum possible kinetic energy K_{max} and the final configuration is when the electron stops just before touching plate C. If we define the electric potential energy of the system in the initial configuration to be zero, the energy equation above can be written

$$0 + (-e)(-\Delta V_s) = K_{max} + 0$$

$$K_{max} = e\,\Delta V_s \qquad \textbf{28.8} \blacktriangleleft$$

This equation allows us to measure K_{max} experimentally by measuring the voltage at which the current drops to zero.

The following are several features of the photoelectric effect in which the predictions made by a structural model based on a classical approach, using the wave model for light, are compared with the experimental results. Notice the strong contrast between the predictions and the results.

1. Dependence of photoelectron kinetic energy on light intensity

   Classical prediction: Electrons should absorb energy continuously from the electromagnetic waves. As the light intensity incident on a metal is increased, energy should be transferred into the metal at a higher rate and the electrons should be ejected with more kinetic energy.

   Experimental result: The maximum kinetic energy of photoelectrons is *independent* of light intensity as shown in Active Figure 28.8 with both curves falling to zero at the *same* negative voltage. (According to Equation 28.8, the maximum kinetic energy is proportional to the stopping potential.)

2. Time interval between incidence of light and ejection of photoelectrons

   Classical prediction: At low light intensities, a measurable time interval should pass between the instant the light is turned on and the time an electron is ejected from the metal. This time interval is required for the electron to absorb the incident radiation before it acquires enough energy to escape from the metal.

   Experimental result: Electrons are emitted from the surface of the metal almost *instantaneously* (less than 10^{-9} s after the surface is illuminated), even at very low light intensities.

3. Dependence of ejection of electrons on light frequency

   Classical prediction: Electrons should be ejected from the metal at any incident light frequency, as long as the light intensity is high enough, because energy is transferred to the metal regardless of the incident light frequency.

Experimental result: No electrons are emitted if the incident light frequency falls below some **cutoff frequency** f_c, whose value is characteristic of the material being illuminated. No electrons are ejected below this cutoff frequency *regardless* of the light intensity.

4. Dependence of photoelectron kinetic energy on light frequency

Classical prediction: There should be *no* relationship between the frequency of the light and the electron kinetic energy. The kinetic energy should be related to the intensity of the light.

Experimental result: The maximum kinetic energy of the photoelectrons increases with increasing light frequency.

Notice that *all four* predictions of the classical model are incorrect. A successful explanation of the photoelectric effect was given by Einstein in 1905, the same year he published his special theory of relativity. As part of a general paper on electromagnetic radiation, for which he received the Nobel Prize in Physics in 1921, Einstein extended Planck's concept of quantization to electromagnetic waves. He assumed that light (or any other electromagnetic wave) of frequency f can be considered to be a stream of quanta, regardless of the source of the radiation. Today we call these quanta **photons.** Each photon has an energy E given by Equation 28.5, $E = hf$, and moves in a vacuum at the speed of light c, where is $c = 3.00 \times 10^8$ m/s.

QUICK QUIZ 28.2 While standing outdoors one evening, you are exposed to the following four types of electromagnetic radiation: yellow light from a sodium street lamp, radio waves from an AM radio station, radio waves from an FM radio station, and microwaves from an antenna of a communications system. Rank these types of waves in terms of photon energy from highest to lowest.

Let us organize Einstein's model for the photoelectric effect using the components of structural models:

1. *A description of the physical components of the system:* We imagine the system to consist of two physical components: (1) an electron that is to be ejected by an incoming photon and (2) the remainder of the metal.

2. *A description of where the components are located relative to one another and how they interact:* In Einstein's model, a photon of the incident light gives *all* its energy hf to a *single* electron in the metal. Therefore, the absorption of energy by the electrons is not a continuous process as envisioned in the wave model, but rather a discontinuous process in which energy is delivered to the electrons in bundles. The energy transfer is accomplished via a one-photon/one-electron event.

3. *A description of the time evolution of the system:* We can describe the time evolution of the system by applying the nonisolated system model for energy over a time interval that includes the absorption of one photon and the ejection of the corresponding electron. Energy is transferred into the system by electromagnetic radiation, the photon. The system has two types of energy: the potential energy of the metal–electron system and the kinetic energy of the ejected electron. Therefore, we can write the conservation of energy equation (Eq. 7.2) as

$$\Delta K + \Delta U = T_{ER} \qquad \text{28.9} \blacktriangleleft$$

The energy transfer into the system is that of the photon, $T_{ER} = hf$. During the process, the kinetic energy of the electron increases from zero to its final value, which we assume to be the maximum possible value K_{max}. The potential energy of the system increases because the electron is pulled away from the metal to which it is attracted. We define the potential energy of the system when the electron is outside the metal as zero. The potential energy of the system when the electron is in the metal is $U = -\phi$, where ϕ is called the **work function** of the metal. The work function represents the minimum energy with which an electron is bound in the metal and is on the order of a few electron volts. Table 28.1 lists selected values. The increase in potential energy of the system

TABLE 28.1 |

Work Functions of Selected Metals

Metal	ϕ (eV)
Na	2.46
Al	4.08
Cu	4.70
Zn	4.31
Ag	4.73
Pt	6.35
Pb	4.14
Fe	4.50

Note: Values are typical for metals listed. Actual values may vary depending on whether the metal is a single crystal or polycrystalline. Values may also depend on the face from which electrons are ejected from crystalline metals. Furthermore, different experimental procedures may produce differing values.

when the electron is removed from the metal is the work function ϕ. Substituting these energies into Equation 28.9, we have

$$(K_{max} - 0) + [0 - (-\phi)] = hf$$

$$K_{max} + \phi = hf \qquad \textbf{28.10} \blacktriangleleft$$

If the electron makes collisions with other electrons or metal ions as it is being ejected, some of the incoming energy is transferred to the metal and the electron is ejected with less kinetic energy than K_{max}.

4. *A description of the agreement between predictions of the model and actual observations and, possibly, predictions of new effects that have not yet been observed:* The prediction made by Einstein is an equation for the maximum kinetic energy of an ejected electron as a function of frequency of the illuminating radiation. This equation can be found by rearranging Equation 28.10:

▶ Photoelectric effect equation

$$K_{max} = hf - \phi \qquad \textbf{28.11} \blacktriangleleft$$

With Einstein's structural model, one can explain the observed features of the photoelectric effect that cannot be understood using classical concepts:

1. Dependence of photoelectron kinetic energy on light intensity

   Equation 28.11 shows that K_{max} is independent of the light intensity. The maximum kinetic energy of any one electron, which equals $hf - \phi$, depends only on the light frequency and the work function. If the light intensity is doubled, the number of photons arriving per unit time is doubled, which doubles the rate at which photoelectrons are emitted. The maximum kinetic energy of any one photoelectron, however, is unchanged.

2. Time interval between incidence of light and ejection of photoelectrons

   Near-instantaneous emission of electrons is consistent with the photon model of light. The incident energy appears in small packets, and there is a one-to-one interaction between photons and electrons. If the incident light has very low intensity, there are very few photons arriving per unit time interval; each photon, however, can have sufficient energy to eject an electron immediately.

3. Dependence of ejection of electrons on light frequency

   Because the photon must have energy greater than the work function ϕ to eject an electron, the photoelectric effect cannot be observed below a certain cutoff frequency. If the energy of an incoming photon does not satisfy this requirement, an electron cannot be ejected from the surface, even though many photons per unit time are incident on the metal in a very intense light beam.

4. Dependence of photoelectron kinetic energy on light frequency

   A photon of higher frequency carries more energy and therefore ejects a photoelectron with more kinetic energy than does a photon of lower frequency.

   Einstein's theoretical result (Eq. 28.11) predicts a linear relationship between the maximum electron kinetic energy K_{max} and the light frequency f. Experimental observation of such a linear relationship would be a final confirmation of Einstein's theory. Indeed, such a linear relationship is observed as sketched in Active Figure 28.9. The slope of the curves for all metals is Planck's constant h. The absolute value of the intercept on the vertical axis is the work function ϕ, which varies from one metal to another. The intercept on the horizontal axis is the cutoff frequency, which is related to the work function through the relation $f_c = \phi / h$. This cutoff frequency corresponds to a **cutoff wavelength** of

▶ Cutoff wavelength

$$\lambda_c = \frac{c}{f_c} = \frac{c}{\phi / h} = \frac{hc}{\phi} \qquad \textbf{28.12} \blacktriangleleft$$

where c is the speed of light. Light with wavelength *greater* than λ_c incident on a material with a work function of ϕ does not result in the emission of photoelectrons.

The combination hc occurs often when relating the energy of a photon to its wavelength. A common shortcut to use in solving problems is to express this combination in useful units according to the numerical value

$$hc = 1\,240 \text{ eV} \cdot \text{nm}$$

One of the first practical uses of the photoelectric effect was as a detector in a light meter of a camera. Light reflected from the object to be photographed strikes a photoelectric surface in the meter, causing it to emit photoelectrons that then pass through a sensitive ammeter. The magnitude of the current in the ammeter depends on the light intensity.

The phototube, another early application of the photoelectric effect, acts much like a switch in an electric circuit. It produces a current in the circuit when light of sufficiently high frequency falls on a metal plate in the phototube, but produces no current in the dark. Phototubes were used in burglar alarms and in the detection of the soundtrack on motion picture film. Modern semiconductor devices have now replaced older devices based on the photoelectric effect.

The photoelectric effect is used today in the operation of photomultiplier tubes. Figure 28.10 shows the structure of such a device. A photon striking the photocathode ejects an electron by means of the photoelectric effect. This electron is accelerated across the potential difference between the photocathode and the first *dynode*, shown as being at +200 V relative to the photocathode in Figure 28.10. This high-energy electron strikes the dynode and ejects several more electrons. This process is repeated through a series of dynodes at ever higher potentials until an electrical pulse is produced as millions of electrons strike the last dynode. Therefore, the tube is called a *multiplier* because one photon at the input has resulted in millions of electrons at the output.

The photomultiplier tube is used in nuclear detectors to detect the presence of gamma rays emitted from radioactive nuclei, which we will study in Chapter 30. It is also used in astronomy in a technique called *photoelectric photometry*. In this technique, the light collected by a telescope from a single star is allowed to fall on a photomultiplier tube for a time interval. The tube measures the total light energy during the time interval, which can then be converted to a luminosity of the star.

The photomultiplier tube is being replaced in many astronomical observations with a *charge-coupled device* (CCD), which is the same device that is used in a digital camera. In this device, an array of pixels are formed on the silicon surface of an integrated circuit. When the surface is exposed to light from an astronomical scene through a telescope or a terrestrial scene through a digital camera, electrons generated by the photoelectric effect are caught in "traps" beneath the surface. The number of electrons is related to the intensity of the light striking the surface. A signal processor measures the number of electrons associated with each pixel and converts this information into a digital code that a computer can use to reconstruct and display the scene.

The *electron bombardment CCD camera* allows higher sensitivity than a conventional CCD. In this device, electrons ejected from a photocathode by the photoelectric effect are accelerated through a high voltage before striking a CCD array. The higher energy of the electrons results in a very sensitive detector of low-intensity radiation.

The explanation of the photoelectric effect with a quantum model, combined with Planck's quantum model for blackbody radiation, laid a strong foundation for further investigation into quantum physics. In the next section, we present a

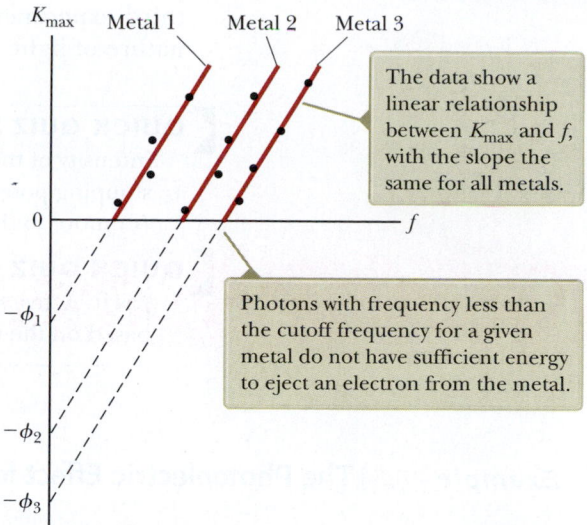

Active Figure 28.9 A plot of results for K_{max} of photoelectrons versus frequency of incident light in a typical photoelectric effect experiment.

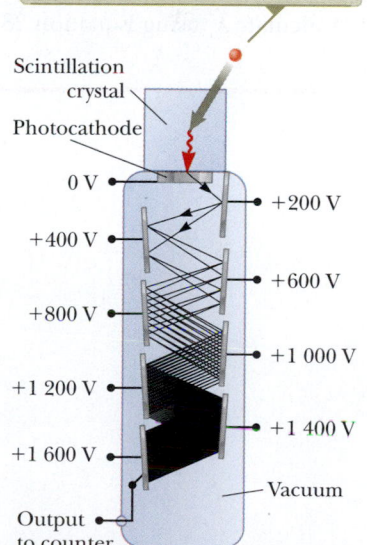

Figure 28.10 The multiplication of electrons in a photomultiplier tube.

third experimental result that provides further strong evidence of the quantum nature of light.

QUICK QUIZ 28.3 Consider one of the curves in Active Figure 28.8. Suppose the intensity of the incident light is held fixed but its frequency is increased. Does the stopping potential in Active Figure 28.8 (**a**) remain fixed, (**b**) move to the right, or (**c**) move to the left?

QUICK QUIZ 28.4 Suppose classical physicists had the idea of plotting K_{max} versus f as in Active Figure 28.9. Draw a graph of what the expected plot would look like, based on the wave model for light.

Example 28.3 | The Photoelectric Effect for Sodium

A sodium surface is illuminated with light having a wavelength of 300 nm. As indicated in Table 28.1, the work function for sodium metal is 2.46 eV.

(A) Find the maximum kinetic energy of the ejected photoelectrons.

SOLUTION

Conceptualize Imagine a photon striking the metal surface and ejecting an electron. The electron with the maximum energy is one near the surface that experiences no interactions with other particles in the metal that would reduce its energy on its way out of the metal.

Categorize We evaluate the results using equations developed in this section, so we categorize this example as a substitution problem.

Find the energy of each photon in the illuminating light beam from Equation 28.5:

$$E = hf = \frac{hc}{\lambda}$$

From Equation 28.11, find the maximum kinetic energy of an electron:

$$K_{max} = \frac{hc}{\lambda} - \phi = \frac{1\ 240\ \text{eV} \cdot \text{nm}}{300\ \text{nm}} - 2.46\ \text{eV} = \boxed{1.67\ \text{eV}}$$

(B) Find the cutoff wavelength λ_c for sodium.

SOLUTION

Calculate λ_c using Equation 28.12:

$$\lambda_c = \frac{hc}{\phi} = \frac{1\ 240\ \text{eV} \cdot \text{nm}}{2.46\ \text{eV}} = \boxed{504\ \text{nm}}$$

28.3 | The Compton Effect

In 1919, Einstein proposed that a photon of energy E carries a momentum equal to $E/c = hf/c$. In 1923, Arthur Holly Compton carried Einstein's idea of photon momentum further with the **Compton effect.**

Prior to 1922, Compton and his coworkers had accumulated evidence that showed that the classical wave theory of light failed to explain the scattering of x-rays from electrons. According to classical theory, incident electromagnetic waves of frequency f_0 should have two effects: (1) the electrons should accelerate in the direction of propagation of the x-ray by radiation pressure (see Section 24.5), and (2) the oscillating electric field should set the electrons into oscillation at the apparent frequency of the radiation as detected by the moving electron. The apparent frequency detected by the electron differs from f_0 due to the Doppler effect (see Section 24.3) because the electron absorbs as a moving particle. The electron then re-radiates as a moving particle, exhibiting another Doppler shift in the frequency of emitted radiation.

Because different electrons move at different speeds, depending on the amount of energy absorbed from the electromagnetic waves, the scattered wave frequency at a given angle should show a distribution of Doppler-shifted values. Contrary to this prediction, Compton's experiment showed that, at a given angle, only *one* frequency of radiation was observed that was different from that of the incident radiation. Compton and his coworkers realized that the scattering of x-ray photons from electrons could be explained by treating photons as point-like particles with energy hf and momentum hf/c and by assuming that the energy and momentum of the isolated system of the photon and the electron are conserved in a two-dimensional collision. By doing so, Compton was adopting a particle model for something that was well known as a wave, as had Einstein in his explanation of the photoelectric effect. Figure 28.11 shows the quantum picture of the exchange of momentum and energy between an individual x-ray photon and an electron. In the classical model, the electron is pushed along the direction of propagation of the incident x-ray by radiation pressure. In the quantum model in Figure 28.11, the electron is scattered through an angle ϕ with respect to this direction as if it were a billiard-ball type collision.

Figure 28.12 is a schematic diagram of the apparatus used by Compton. The x-rays, scattered from a carbon target, were diffracted by a rotating crystal spectrometer, and the intensity was measured with an ionization chamber that generated a current proportional to the intensity. The incident beam consisted of monochromatic x-rays of wavelength $\lambda_0 = 0.071$ nm. The experimental intensity-versus-wavelength plots observed by Compton for four scattering angles (corresponding to θ in Fig. 28.11) are shown in Figure 28.13 (page 822). The graphs for the three nonzero angles show two peaks, one at λ_0 and one at $\lambda' > \lambda_0$. The shifted peak at λ' is caused by the scattering of x-rays from free electrons, which was predicted by Compton to depend on scattering angle as

$$\lambda' - \lambda_0 = \frac{h}{m_e c}(1 - \cos\theta)$$

28.13 ◄   ► Compton shift equation

In this expression, known as the **Compton shift equation**, m_e is the mass of the electron; $h/m_e c$ is called the **Compton wavelength** λ_C for the electron and has the value

$$\lambda_C = \frac{h}{m_e c} = 0.002\,43 \text{ nm}$$

28.14 ◄   ► Compton wavelength

Compton's measurements were in excellent agreement with the predictions of Equation 28.13. They were the first experimental results to convince most physicists of the fundamental validity of the quantum theory!

Arthur Holly Compton
American Physicist (1892–1962)
Compton was born in Wooster, Ohio, and attended Wooster College and Princeton University. He became the director of the laboratory at the University of Chicago, where experimental work concerned with sustained nuclear chain reactions was conducted. This work was of central importance to the construction of the first nuclear weapon. His discovery of the Compton effect led to his sharing of the 1927 Nobel Prize in Physics with Charles Wilson.

© Mary Evans Picture Library/Alamy

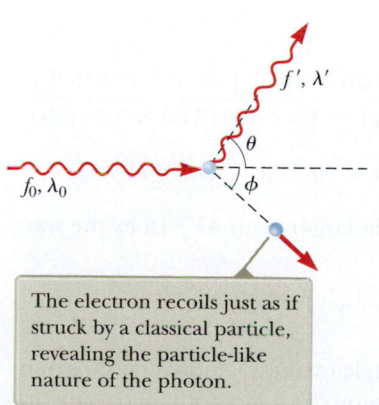

The electron recoils just as if struck by a classical particle, revealing the particle-like nature of the photon.

Figure 28.11 The quantum model for x-ray scattering from an electron.

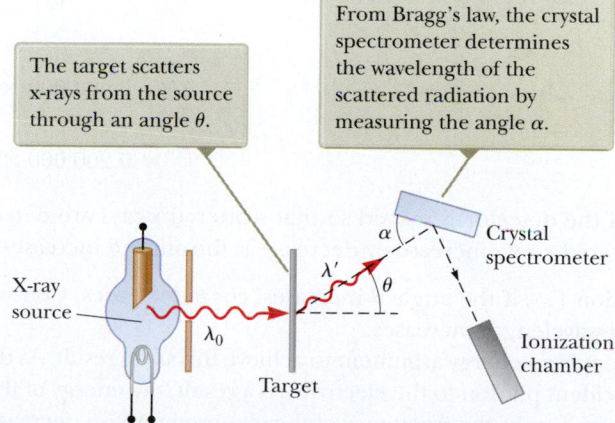

The target scatters x-rays from the source through an angle θ.

From Bragg's law, the crystal spectrometer determines the wavelength of the scattered radiation by measuring the angle α.

Figure 28.12 Schematic diagram of Compton's apparatus.

BIO The Compton effect and x-ray technicians

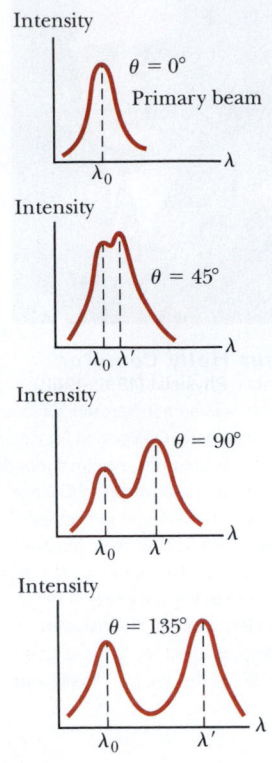

Figure 28.13 Scattered x-ray intensity versus wavelength for Compton scattering at $\theta = 0°$, 45°, 90°, and 135°.

The Compton effect should be kept in mind by x-ray technicians working in hospitals and radiology laboratories. X-rays directed into the patient's body are Compton scattered by electrons in the body in all directions. Equation 28.13 shows that the scattered wavelength is still well within the x-ray region so that these scattered x-rays can damage human tissue. In general, technicians operate the x-ray machine from behind an absorbing wall to avoid exposure to the scattered x-rays. Furthermore, when dental x-rays are taken, a lead apron is placed over the patient to reduce the absorption of scattered x-rays by other parts of the patient's body.

> **THINKING PHYSICS 28.2**
>
> The Compton effect involves a change in wavelength as photons are scattered through different angles. Suppose we illuminate a piece of material with a beam of light and then view the material from different angles relative to the beam of light. Will we see a *color change* corresponding to the change in wavelength of the scattered light?
>
> **Reasoning** Visible light scattered by the material undergoes a change in wavelength, but the change is far too small to detect as a color change. The largest possible change in wavelength, at 180° scattering, is twice the Compton wavelength, about 0.005 nm, which represents a change of less than 0.001% of the wavelength of red light. The Compton effect is only detectable for wavelengths that are very short to begin with, so the Compton wavelength is an appreciable fraction of the incident wavelength. As a result, the usual radiation for observing the Compton effect is in the x-ray range of the electromagnetic spectrum. ◀

◣ *Example* 28.4 | **Compton Scattering at 45°**

X-rays of wavelength $\lambda_0 = 0.200\,000$ nm are scattered from a block of material. The scattered x-rays are observed at an angle of 45.0° to the incident beam. Calculate their wavelength.

SOLUTION

Conceptualize Imagine the process in Figure 28.11, with the photon scattered at 45° to its original direction.

Categorize We evaluate the result using an equation developed in this section, so we categorize this example as a substitution problem.

Solve Equation 28.13 for the wavelength of the scattered x-ray:

$$(1) \quad \lambda' = \lambda_0 + \frac{h(1 - \cos\theta)}{m_e c}$$

Substitute numerical values:

$$\lambda' = 0.200\,000 \times 10^{-9}\ \text{m} + \frac{(6.626 \times 10^{-34}\ \text{J·s})(1 - \cos 45.0°)}{(9.11 \times 10^{-31}\ \text{kg})(3.00 \times 10^8\ \text{m/s})}$$

$$= 0.200\,000 \times 10^{-9}\ \text{m} + 7.10 \times 10^{-13}\ \text{m} = \boxed{0.200\,710\ \text{nm}}$$

What If? What if the detector is moved so that scattered x-rays are detected at an angle larger than 45°? Does the wavelength of the scattered x-rays increase or decrease as the angle θ increases?

Answer In Equation (1), if the angle θ increases, $\cos\theta$ decreases. Consequently, the factor $(1 - \cos\theta)$ increases. Therefore, the scattered wavelength increases.

We could also apply an energy argument to achieve this same result. As the scattering angle increases, more energy is transferred from the incident photon to the electron. As a result, the energy of the scattered photon decreases with increasing scattering angle. Because $E = hf$, the frequency of the scattered photon decreases, and because $\lambda = c/f$, the wavelength increases.

28.4 | Photons and Electromagnetic Waves

The agreement between experimental measurements and theoretical predictions based on quantum models for phenomena such as the photoelectric effect and the Compton effect offers clear evidence that when light and matter interact, the light behaves as if it were composed of particles with energy hf and momentum hf/c. An obvious question at this point is, "How can light be considered a photon when it exhibits wave-like properties?" We describe light in terms of photons having energy and momentum, which are parameters of the particle model. Remember, however, that light and other electromagnetic waves exhibit interference and diffraction effects, which are consistent only with the wave model.

Which model is correct? Is light a wave or a particle? The answer depends on the phenomenon being observed. Some experiments can be explained better, or solely, with the photon model, whereas others are best described, or can only be described, with a wave model. The end result is that we must accept both models and admit that the true nature of light is not describable in terms of any single classical picture. Hence, light has a dual nature in that it exhibits both wave and particle characteristics. You should recognize, however, that the same beam of light that can eject photoelectrons from a metal can also be diffracted by a grating. In other words, the particle model and the wave model of light complement each other.

The success of the particle model of light in explaining the photoelectric effect and the Compton effect raises many other questions. Because a photon is a particle, what is the meaning of its "frequency" and "wavelength," and which determines its energy and momentum? Is light in some sense simultaneously a wave and a particle? Although photons have no rest energy, can some simple expression describe the effective mass of a "moving" photon? If a "moving" photon has mass, do photons experience gravitational attraction? What is the spatial extent of a photon, and how does an electron absorb or scatter one photon? Some of these questions can be answered, but others demand a view of atomic processes that is too pictorial and literal. Furthermore, many of these questions stem from classical analogies such as colliding billiard balls and water waves breaking on a shore. Quantum mechanics gives light a more fluid and flexible nature by treating the particle model and wave model of light as both necessary and complementary. Neither model can be used exclusively to describe all properties of light. A complete understanding of the observed behavior of light can be attained only if the two models are combined in a complementary manner. Before discussing this combination in more detail, we now turn our attention from electromagnetic waves to the behavior of entities that we have called particles.

28.5 | The Wave Properties of Particles

We feel quite comfortable in adopting a particle model for matter because we have studied such concepts as conservation of energy and momentum for particles as well as extended objects. It might therefore be even more difficult than it was for light to accept that *matter* also has a dual nature!

In 1923, in his doctoral dissertation, Louis Victor de Broglie postulated that because photons have wave and particle characteristics, perhaps all forms of matter have wave as well as particle properties. This postulate was a highly revolutionary idea with no experimental confirmation at that time. According to de Broglie, an electron in motion exhibits both wave and particle characteristics. De Broglie explained the source of this assertion in his 1929 Nobel Prize acceptance speech:

> On the one hand the quantum theory of light cannot be considered satisfactory since it defines the energy of a light corpuscle by the equation $E = hf$ containing the frequency f. Now a purely corpuscular theory contains nothing that enables us to define a frequency; for this reason alone, therefore, we are compelled, in the case of light, to introduce the idea of a corpuscle and that of periodicity simultaneously. On the other hand, determination of the stable motion of electrons in the atom introduces integers, and up to this point the

Louis de Broglie
French Physicist (1892–1987)
De Broglie was born in Dieppe, France. At the Sorbonne in Paris, he studied history in preparation for what he hoped would be a career in the diplomatic service. The world of science is lucky he changed his career path to become a theoretical physicist. De Broglie was awarded the Nobel Prize in Physics in 1929 for his prediction of the wave nature of electrons.

SSPL/Getty Images

only phenomena involving integers in physics were those of interference and of normal modes of vibration. This fact suggested to me the idea that electrons too could not be considered simply as corpuscles, but that periodicity must be assigned to them also.

In Chapter 9, we found that the relationship between energy and momentum for a photon is $p = E/c$. We also know from Equation 28.5 that the energy of a photon is $E = hf = hc/\lambda$. Therefore, the momentum of a photon can be expressed as

$$p = \frac{E}{c} = \frac{hf}{c} = \frac{hc}{c\lambda} = \frac{h}{\lambda}$$

From this equation, we see that the photon wavelength can be specified by its momentum: $\lambda = h/p$. De Broglie suggested that material particles of momentum p should also have wave properties and a corresponding wavelength given by the same expression. Because the magnitude of the momentum of a nonrelativistic particle of mass m and speed u is $p = mu$, the **de Broglie wavelength** of that particle is[3]

▶ De Broglie wavelength of a particle

$$\lambda = \frac{h}{p} = \frac{h}{mu} \qquad \textbf{28.15} \blacktriangleleft$$

Furthermore, in analogy with photons, de Broglie postulated that particles obey the Einstein relation $E = hf$, so the frequency of a particle is

▶ Frequency of a particle

$$f = \frac{E}{h} \qquad \textbf{28.16} \blacktriangleleft$$

The dual nature of matter is apparent in these last two equations because each contains both particle concepts (p and E) and wave concepts (λ and f). These relationships are established experimentally for photons. Is there experimental verification for the wave nature of a particle, such as the electron? Let's find out.

The Davisson–Germer Experiment

De Broglie's proposal that any kind of particle exhibits both wave and particle properties was first regarded as pure speculation. If particles such as electrons had wavelike properties, under the correct conditions they should exhibit diffraction effects. In 1927, three years after de Broglie published his work, C. J. Davisson and L. H. Germer of the United States succeeded in observing these diffraction effects and measuring the wavelength of electrons. Their important discovery provided the first experimental confirmation of the wave nature of particles proposed by de Broglie.

Interestingly, the intent of the initial Davisson–Germer experiment was not to confirm the de Broglie hypothesis. In fact, the discovery was made by accident (as is often the case). The experiment involved the scattering of low-energy electrons (≈ 54 eV) projected toward a nickel target in a vacuum. During one experiment, the nickel surface was badly oxidized because of an accidental break in the vacuum system. After the nickel target was heated in a flowing stream of hydrogen to remove the oxide coating, electrons scattered by it exhibited intensity maxima and minima at specific angles. The experimenters finally realized that the nickel had formed large crystal regions on heating and that the regularly spaced planes of atoms in the crystalline regions served as a diffraction grating (Section 27.8) for electrons.

Shortly thereafter, Davisson and Germer performed more extensive diffraction measurements on electrons scattered from single-crystal targets. Their results showed conclusively the wave nature of electrons and confirmed the de Broglie relation $p = h/\lambda$. A year later in 1928, G. P. Thomson of Scotland observed electron diffraction patterns by passing electrons through very thin gold foils. Diffraction patterns have since been observed for helium atoms, hydrogen atoms, and neutrons. Hence, the wave nature of particles has been established in a variety of ways.

Pitfall Prevention | 28.3
What's Waving?
If particles have wave properties, what's waving? You are familiar with waves on strings, which are very concrete. Sound waves are more abstract, but you are likely comfortable with them. Electromagnetic waves are even more abstract, but at least they can be described in terms of physical variables and electric and magnetic fields. In contrast, waves associated with particles are completely abstract and cannot be associated with a physical variable. Later in this chapter, we describe the wave associated with a particle in terms of probability.

---

[3]The de Broglie wavelength for a particle moving at any speed u, including relativistic speeds, is $\lambda = h/\gamma mu$, where $\gamma = (1 - u^2/c^2)^{-1/2}$. Recall that in Chapter 9 we used u for particle speed to distinguish it from v, the speed of a reference frame.

Example 28.5 | Wavelengths for Microscopic and Macroscopic Objects

(A) Calculate the de Broglie wavelength for an electron ($m_e = 9.11 \times 10^{-31}$ kg) moving at 1.00×10^7 m/s.

SOLUTION

Conceptualize Imagine the electron moving through space. From a classical viewpoint, it is a particle under constant velocity. From the quantum viewpoint, the electron has a wavelength associated with it.

Categorize We evaluate the result using an equation developed in this section, so we categorize this example as a substitution problem.

Evaluate the de Broglie wavelength using Equation 28.15:

$$\lambda = \frac{h}{m_e u} = \frac{6.63 \times 10^{-34} \, \text{J} \cdot \text{s}}{(9.11 \times 10^{-31} \, \text{kg})(1.00 \times 10^7 \, \text{m/s})} = 7.27 \times 10^{-11} \, \text{m}$$

The wave nature of this electron could be detected by diffraction techniques such as those in the Davisson–Germer experiment.

(B) A rock of mass 50 g is thrown with a speed of 40 m/s. What is its de Broglie wavelength?

SOLUTION

Evaluate the de Broglie wavelength using Equation 28.15:

$$\lambda = \frac{h}{mu} = \frac{6.63 \times 10^{-34} \, \text{J} \cdot \text{s}}{(50 \times 10^{-3} \, \text{kg})(40 \, \text{m/s})} = 3.3 \times 10^{-34} \, \text{m}$$

This wavelength is much smaller than any aperture through which the rock could possibly pass. Hence, we could not observe diffraction effects, and as a result, the wave properties of large-scale objects cannot be observed.

Example 28.6 | An Accelerated Charge

A particle of charge q and mass m is accelerated from rest through an electric potential difference ΔV. Assuming that the particle moves with a nonrelativistic speed, find its de Broglie wavelength.

SOLUTION

Conceptualize Imagine the motion of the particle. It starts from rest and then accelerates due to the force from the electric field. As the speed of the particle increases, its de Broglie wavelength decreases.

Categorize We identify the system as the particle and the electric field and apply the energy version of the isolated system model. The initial configuration of the system occurs at the instant the particle starts to move and the final configuration is when the particle reaches its final speed after accelerating through the potential difference ΔV. We define the electric potential energy of the system in the initial configuration to be zero.

Analyze

Write the conservation of energy equation (Eq. 7.2) for the isolated system:

$$\Delta K + \Delta U = 0$$

Substitute the initial and final energies, recognizing that a positive charge accelerates in the direction of *decreasing* electric potential:

$$(\tfrac{1}{2}mu^2 - 0) + (-q\Delta V - 0) = 0$$

continued

28.6 *cont.*

Solve for the final speed u:

$$u = \sqrt{\frac{2q\Delta V}{m}}$$

Substitute into Equation 28.15:

$$\lambda = \frac{h}{mu} = \frac{h}{m}\sqrt{\frac{m}{2q\Delta V}} = \frac{h}{\sqrt{2mq\Delta V}}$$

Finalize Notice that increasing the charge of the particle or the potential difference will decrease the wavelength. This result occurs because either of these changes will cause the particle to move with a higher speed. Increasing the mass of the particle would decrease its speed if everything else remains the same, so it might be surprising to see that increasing the mass would also decrease the de Broglie wavelength. Keep in mind, however, that the de Broglie wavelength depends on the *momentum* of the particle. The speed decreases according to the square root of the inverse of the mass in this situation, but the general expression for momentum shows a direct proportionality to the mass. As a result, for this situation, the momentum is proportional to the square root of the mass.

The Electron Microscope BIO

A practical device that relies on the wave characteristics of electrons is the **electron microscope.** A *transmission* electron microscope, used for viewing flat, thin samples, is shown in Figure 28.14. In many respects, it is similar to an optical microscope, but the electron microscope has a much greater resolving power because it can accelerate electrons to very high kinetic energies, giving them very short wavelengths. No microscope can resolve details that are significantly smaller than the wavelength of the waves used to illuminate the object. Typically, the wavelengths of electrons are about 100 times shorter than those of the visible light used in optical microscopes. As a result, an electron microscope with ideal lenses would be able to distinguish details about 100 times smaller than those distinguished by an optical microscope.

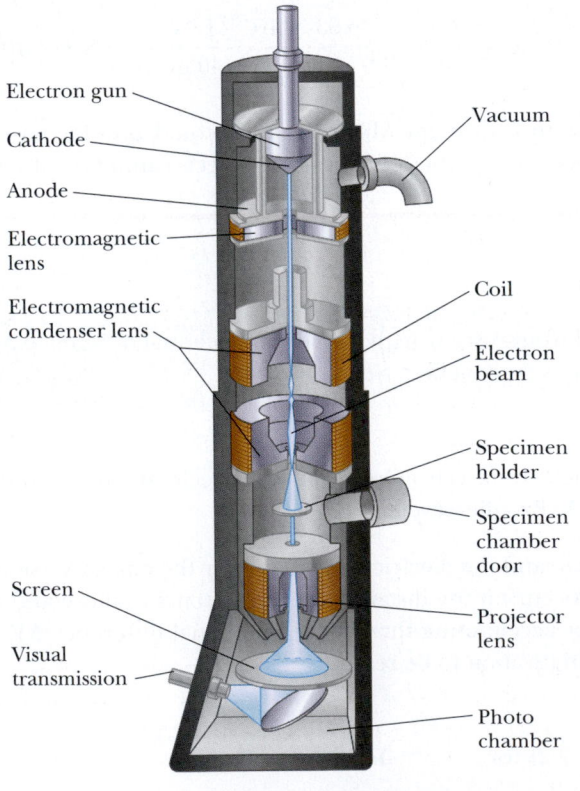

a   b

Figure 28.14 (a) Diagram of a transmission electron microscope for viewing a thinly sectioned sample. The "lenses" that control the electron beam are magnetic deflection coils. (b) An electron microscope in use.

(Electromagnetic radiation of the same wavelength as the electrons in an electron microscope is in the x-ray region of the spectrum.)

The electron beam in an electron microscope is controlled by electrostatic or magnetic deflection, which acts on the electrons to focus the beam and form an image. Rather than examining the image through an eyepiece as in an optical microscope, the viewer looks at an image formed on a monitor or other type of display screen. The photograph at the beginning of this chapter shows the amazing detail available with an electron microscope.

28.6 | A New Model: The Quantum Particle

The discussions presented in previous sections may be quite disturbing because we considered the particle and wave models to be distinct in earlier chapters. The notion that both light and material particles have both particle and wave properties does not fit with this distinction. We have experimental evidence, however, that this dual nature is just what we must accept. This acceptance leads to a new simplification model, the **quantum particle model.** We add the quantum particle to our other simplification models from which we build analysis models: the particle, the system, the rigid object, and the wave. In this model, entities have both particle and wave characteristics, and we must choose one appropriate behavior—particle or wave—to understand a particular phenomenon.

In this section, we shall investigate this model, which might bring you more comfort with this idea. As we shall demonstrate, we can construct from waves an entity that exhibits properties of a particle.

Let us first review the characteristics of ideal particles and waves. In the particle model, an ideal particle has zero size. As mentioned in Section 13.2, in the wave model, an ideal wave has a single frequency and is infinitely long. Therefore, an essential identifying feature of a particle that differentiates it from a wave is that it is *localized* in space. Let us show that we can build a localized entity from infinitely long waves. Imagine drawing one wave along the x axis, with one of its crests located at $x = 0$, as in Figure 28.15a. Now, draw a second wave, of the same amplitude but a different frequency, with one of its crests also at $x = 0$. The result of the superposition of these two waves is a *beat* because the waves are alternately in phase and out of phase. (Beats were discussed in Section 14.5.) Figure 28.15b shows the results of superposing these two waves.

Notice that we have already introduced some localization by doing so. A single wave has the same amplitude everywhere in space; no point in space is any different from any other point. By adding a second wave, however, something is different between the in-phase and the out-of-phase points in space.

Now imagine that more and more waves are added to our original two, each new wave having a new frequency. Each new wave is added so that one of its crests is at $x = 0$. The result at $x = 0$ is that all the waves add constructively. When we consider a large number of waves, the probability of a positive value of a wave function at any point x is equal to the probability of a negative value and destructive interference occurs *everywhere* except near $x = 0$, where we superposed all the crests. The result is shown in Active Figure 28.16. The small region of constructive interference is called a **wave packet.** This wave packet is a localized region of space that is different from all other regions, because the result of the superposition of the waves

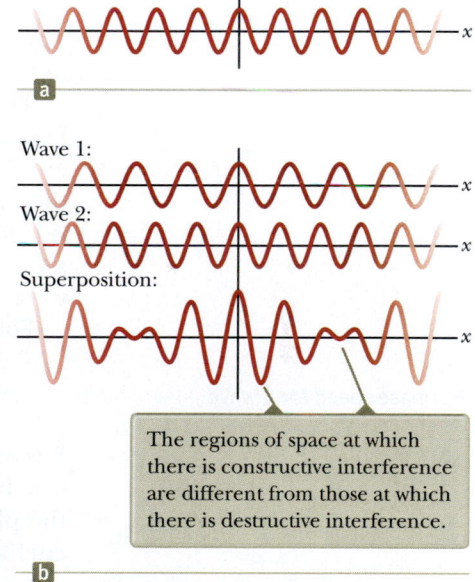

Wave 1:

Wave 2:

Superposition:

The regions of space at which there is constructive interference are different from those at which there is destructive interference.

Figure 28.15 (a) An idealized wave of an exact single frequency is the same throughout space and time. (b) If two ideal waves with slightly different frequencies are combined, beats result (Section 14.5).

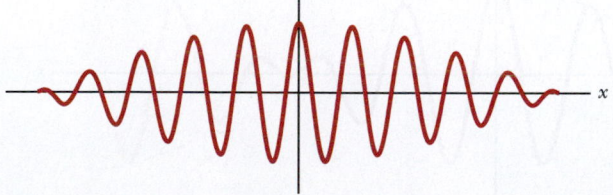

Active Figure 28.16 If a large number of waves are combined, the result is a wave packet, which represents a particle.

everywhere else is zero. We can identify the wave packet as a particle because it has the localized nature of what we have come to recognize as a particle!

The localized nature of this entity is the *only* characteristic of a particle that was generated with this process. We have not addressed how the wave packet might achieve such particle characteristics as mass, electric charge, spin, and so on. Therefore, you may not be completely convinced that we have built a particle. As further evidence that the wave packet can represent the particle, let us show that the wave packet has another characteristic of a particle.

Let us return to our combination of only two waves so as to make the mathematical representation simple. Consider two waves with equal amplitudes but different frequencies f_1 and f_2. We can represent the waves mathematically as

$$y_1 = A \cos(k_1 x - \omega_1 t) \quad \text{and} \quad y_2 = A \cos(k_2 x - \omega_2 t)$$

where, as in Chapter 13, $\omega = 2\pi f$ and $k = 2\pi/\lambda$. Using the superposition principle, we add the waves:

$$y = y_1 + y_2 = A \cos(k_1 x - \omega_1 t) + A \cos(k_2 x - \omega_2 t)$$

It is convenient to write this expression in a form that uses the trigonometric identity

$$\cos a + \cos b = 2 \cos\left(\frac{a-b}{2}\right) \cos\left(\frac{a+b}{2}\right)$$

Letting $a = k_1 x - \omega_1 t$ and $b = k_2 x - \omega_2 t$, we find that

$$y = 2A \cos\left[\frac{(k_1 x - \omega_1 t) - (k_2 x - \omega_2 t)}{2}\right] \cos\left[\frac{(k_1 x - \omega_1 t) + (k_2 x - \omega_2 t)}{2}\right]$$

$$= \left[2A \cos\left(\frac{\Delta k}{2}x - \frac{\Delta \omega}{2}t\right)\right] \cos\left(\frac{k_1 + k_2}{2}x - \frac{\omega_1 + \omega_2}{2}t\right) \qquad \text{28.17} \blacktriangleleft$$

The second cosine factor represents a wave with a wave number and frequency equal to the averages of the values for the individual waves.

The factor in brackets represents the envelope of the wave as shown in Active Figure 28.17. Notice that this factor also has the mathematical form of a wave. This envelope of the combination can travel through space with a different speed than the individual waves. As an extreme example of this possibility, imagine combining two identical waves moving in opposite directions. The two waves move with the same speed, but the envelope has a speed of *zero* because we have built a standing wave, which we studied in Section 14.2.

For an individual wave, the speed is given by Equation 13.11:

▶ Phase speed for a wave

$$v_{\text{phase}} = \frac{\omega}{k}$$

It is called the **phase speed** because it is the rate of advance of a crest on a single wave, which is a point of fixed phase. This equation can be interpreted as the following: the phase speed of a wave is the ratio of the coefficient of the time variable t to the coefficient of the space variable x in the equation for the wave, $y = A \cos(kx - \omega t)$.

Active Figure 28.17 The beat pattern of Figure 28.15b, with an envelope function (black dashed line) superimposed.

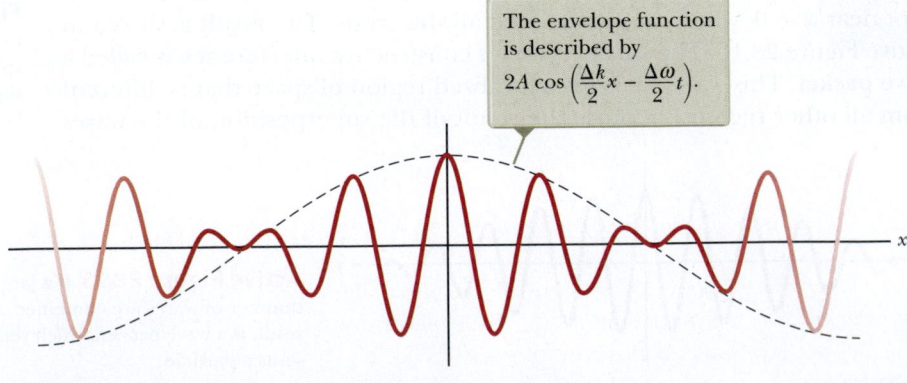

The envelope function is described by $2A \cos\left(\frac{\Delta k}{2}x - \frac{\Delta \omega}{2}t\right)$.

The factor in brackets in Equation 28.17 is of the form of a wave, so it moves with a speed given by this same ratio:

$$v_g = \frac{\text{coefficient of time variable } t}{\text{coefficient of space variable } x} = \frac{(\Delta\omega/2)}{(\Delta k/2)} = \frac{\Delta\omega}{\Delta k}$$

The subscript g on the speed indicates that it is commonly called the **group speed,** or the speed of the wave packet (the *group* of waves) that we have built. We have generated this expression for a simple addition of two waves. For a superposition of a very large number of waves to form a wave packet, this ratio becomes a derivative:

$$v_g = \frac{d\omega}{dk} \qquad \text{28.18} \blacktriangleleft \qquad \blacktriangleright \text{ Group speed for a wave packet}$$

Let us multiply the numerator and the denominator by $\hbar$, where $\hbar = h/2\pi$:

$$v_g = \frac{\hbar \, d\omega}{\hbar \, dk} = \frac{d(\hbar\omega)}{d(\hbar k)} \qquad \text{28.19} \blacktriangleleft$$

We look at the terms in the parentheses in the numerator and denominator in this equation separately. For the numerator

$$\hbar\omega = \frac{h}{2\pi}(2\pi f) = hf = E$$

For the denominator,

$$\hbar k = \frac{h}{2\pi}\left(\frac{2\pi}{\lambda}\right) = \frac{h}{\lambda} = p$$

Therefore, Equation 28.19 can be written as

$$v_g = \frac{d(\hbar\omega)}{d(\hbar k)} = \frac{dE}{dp} \qquad \text{28.20} \blacktriangleleft$$

Because we are exploring the possibility that the envelope of the combined waves represents the particle, consider a free particle moving with a speed u that is small compared with that of light. The energy of the particle is its kinetic energy:

$$E = \tfrac{1}{2}mu^2 = \frac{p^2}{2m}$$

Differentiating this equation with respect to p, where $p = mu$, gives

$$v_g = \frac{dE}{dp} = \frac{d}{dp}\left(\frac{p^2}{2m}\right) = \frac{1}{2m}(2p) = u \qquad \text{28.21} \blacktriangleleft$$

That is, the group speed of the wave packet is identical to the speed of the particle that it is modeled to represent! Therefore, we have further confidence that the wave packet is a reasonable way to model a particle.

QUICK QUIZ 28.7 As an analogy to wave packets, consider an "automobile packet" that occurs near the scene of an accident on a freeway. The phase speed is analogous to the speed of individual automobiles as they move through the backup caused by the accident. The group speed can be identified as the speed of the leading edge of the packet of cars. For the automobile packet, is the group speed (**a**) the same as the phase speed, (**b**) less than the phase speed, or (**c**) greater than the phase speed?

28.7 | The Double-Slit Experiment Revisited

One way to crystallize our ideas about the electron's wave–particle duality is to consider a hypothetical experiment in which electrons are fired at a double slit. Consider a parallel beam of monoenergetic electrons that is incident on a double slit

Figure 28.18 Electron interference. The slit separation d is much greater than the individual slit widths and much less than the distance between the slit and the detector screen.

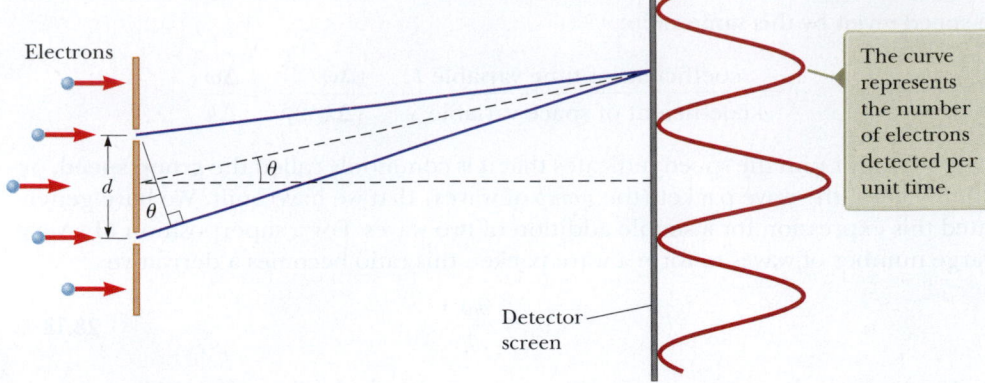

Electrons

The curve represents the number of electrons detected per unit time.

Detector screen

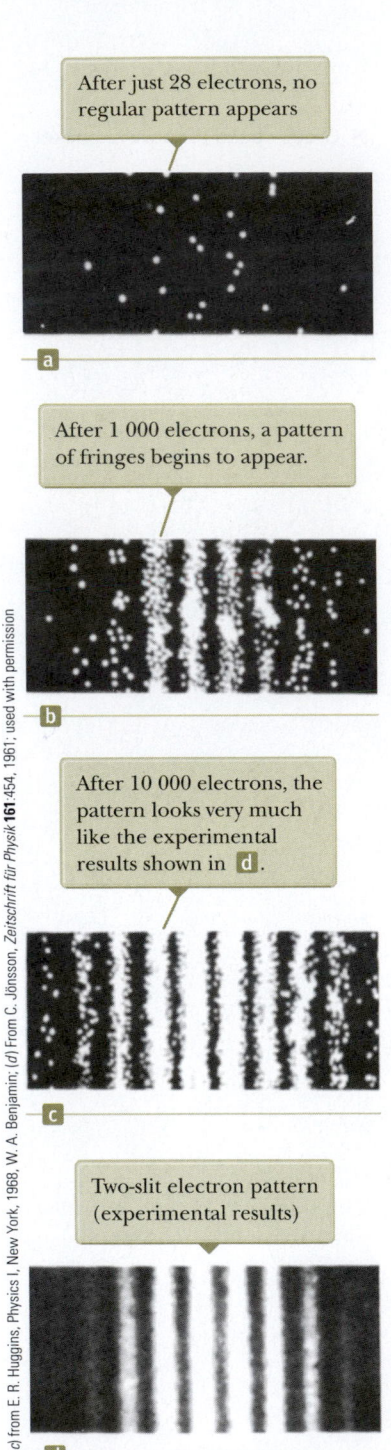

After just 28 electrons, no regular pattern appears

a

After 1 000 electrons, a pattern of fringes begins to appear.

b

After 10 000 electrons, the pattern looks very much like the experimental results shown in **d**.

c

Two-slit electron pattern (experimental results)

d

(a–c) from E. R. Huggins, Physics I, New York, 1968, W. A. Benjamin; (d) From C. Jönsson, Zeitschrift für Physik **161**:454, 1961; used with permission

Active Figure 28.19 (a)–(c) Computer-simulated interference patterns for a beam of electrons incident on a double slit. (d) Photograph of a double-slit interference pattern produced by electrons.

as in Figure 28.18. We shall assume that the slit widths are small compared with the electron wavelength, so we need not worry about diffraction maxima and minima as discussed for light in Section 27.6. An electron detector screen is positioned far from the slits at a distance much greater than the separation distance d of the slits. If the detector screen collects electrons for a long enough time interval, one finds a typical wave interference pattern for the counts per minute, or probability of arrival of electrons. Such an interference pattern would not be expected if the electrons behaved as classical particles. It is clear that electrons are interfering, which is a distinct wave-like behavior.

If we measure the angles θ at which the maximum intensity of electrons arrives at the detector screen in Figure 28.18, we find they are described by exactly the same equation as that for light, $d \sin \theta = m\lambda$ (Eq. 27.2), where m is the order number and λ is the electron wavelength. Therefore, the dual nature of the electron is clearly shown in this experiment: the electrons are detected as particles at a localized spot on the detector screen at some instant of time, but the probability of arrival at that spot is determined by finding the intensity of two interfering waves.

Now imagine that we lower the beam intensity so that one electron at a time arrives at the double slit. It is tempting to assume the electron goes through either slit 1 or slit 2. You might argue that there are no interference effects because there is not a second electron going through the other slit to interfere with the first. This assumption places too much emphasis on the particle model of the electron, however. The interference pattern is still observed if the time interval for the measurement is sufficiently long for many electrons to arrive at the detector screen! This situation is illustrated by the computer-simulated patterns in Active Figure 28.19 where the interference pattern becomes clearer as the number of electrons reaching the detector screen increases. Hence, our assumption that the electron is localized and goes through only one slit when both slits are open must be wrong (a painful conclusion!).

To interpret these results, we are forced to conclude that an electron interacts with both slits *simultaneously*. If you try to determine experimentally which slit the electron goes through, the act of measuring destroys the interference pattern. It is impossible to determine which slit the electron goes through. In effect, we can say only that the electron passes through *both* slits! The same arguments apply to photons.

If we restrict ourselves to a pure particle model, it is an uncomfortable notion that the electron can be present at both slits at once. From the quantum particle model, however, the particle can be considered to be built from waves that exist throughout space. Therefore, the wave components of the electron are present at both slits at the same time, and this model leads to a more comfortable interpretation of this experiment.

28.8 | The Uncertainty Principle

Whenever one measures the position or velocity of a particle at any instant, experimental uncertainties are built into the measurements. According to classical mechanics, there is no fundamental barrier to an ultimate refinement of the apparatus or experimental procedures. In other words, it is possible, in principle, to make such measurements with arbitrarily small uncertainty. Quantum theory predicts, however, that it is fundamentally impossible to make simultaneous measurements of a particle's position and momentum with infinite accuracy.

In 1927, Werner Heisenberg introduced this notion, which is now known as the **Heisenberg uncertainty principle:**

If a measurement of the position of a particle is made with uncertainty Δx and a simultaneous measurement of its momentum is made with uncertainty Δp_x, the product of the two uncertainties can never be smaller than $\hbar/2$:

$$\Delta x \, \Delta p_x \geq \frac{\hbar}{2}$$

28.22 ◀ ▶ Uncertainty principle for momentum and position

Werner Heisenberg
German Theoretical Physicist
(1901–1976)
Heisenberg obtained his Ph.D. in 1923 at the University of Munich. While other physicists tried to develop physical models of quantum phenomena, Heisenberg developed an abstract mathematical model called *matrix mechanics*. The more widely accepted physical models were shown to be equivalent to matrix mechanics. Heisenberg made many other significant contributions to physics, including his famous uncertainty principle for which he received a Nobel Prize in Physics in 1932, the prediction of two forms of molecular hydrogen, and theoretical models of the nucleus.

That is, it is physically impossible to simultaneously measure the exact position and exact momentum of a particle. Heisenberg was careful to point out that the inescapable uncertainties Δx and Δp_x do not arise from imperfections in practical measuring instruments. Furthermore, they do not arise due to any perturbation of the system that we might cause in the measuring process. Rather, the uncertainties arise from the quantum structure of matter.

To understand the uncertainty principle, consider a particle for which we know the wavelength *exactly*. According to the de Broglie relation $\lambda = h/p$, we would know the momentum to infinite accuracy, so $\Delta p_x = 0$.

In reality, as we have mentioned, a single-wavelength wave would exist throughout space. Any region along this wave is the same as any other region (see Fig. 28.15a). If we were to ask, "Where is the particle that this wave represents?" no special location in space along the wave could be identified with the particle because all points along the wave are the same. Therefore, we have *infinite* uncertainty in the position of the particle and we know nothing about where it is. Perfect knowledge of the momentum has cost us all information about the position.

In comparison, now consider a particle with some uncertainty in momentum so that a range of values of momentum are possible. According to the de Broglie relation, the result is a range of wavelengths. Therefore, the particle is not represented by a single wavelength, but a combination of wavelengths within this range. This combination forms a wave packet as we discussed in Section 28.6 and illustrated in Active Figure 28.16. Now, if we are asked to determine the location of the particle, we can only say that it is somewhere in the region defined by the wave packet because a distinct difference exists between this region and the rest of space. Therefore, by losing some information about the momentum of the particle, we have gained information about its position.

If we were to lose all information about the momentum, we would be adding together waves of all possible wavelengths. The result would be a wave packet of zero length. Therefore, if we know nothing about the momentum, we know exactly where the particle is.

The mathematical form of the uncertainty principle argues that the product of the uncertainties in position and momentum will always be larger than some minimum value. This value can be calculated from the types of arguments discussed earlier, which result in the value of $\hbar/2$ in Equation 28.22.

Another form of the uncertainty principle can be generated by reconsidering Active Figure 28.16. Imagine that the horizontal axis is time rather than spatial position x. We can then make the same arguments that we made about knowledge of wavelength and position in the time domain. The corresponding variables would be frequency and time. Because frequency is related to the energy of the particle by $E = hf$, the uncertainty principle in this form is

$$\Delta E \, \Delta t \geq \frac{\hbar}{2}$$

28.23 ◀ ▶ Uncertainty principle for energy and time

Pitfall Prevention | 28.4
The Uncertainty Principle
Some students incorrectly interpret the uncertainty principle as meaning that a measurement interferes with the system. For example, if an electron is observed in a hypothetical experiment using an optical microscope, the photon used to see the electron collides with it and makes it move, giving it an uncertainty in momentum. This scenario does *not* represent the basis of the uncertainty principle. The uncertainty principle is independent of the measurement process and is based on the wave nature of matter.

This form of the uncertainty principle suggests that energy conservation can appear to be violated by an amount ΔE as long as it is only for a short time interval Δt consistent with Equation 28.23. We shall use this notion to estimate the rest energies of particles in Chapter 31.

Example 28.7 | Locating an Electron

The speed of an electron is measured to be 5.00×10^3 m/s to an accuracy of 0.003 00%. Find the minimum uncertainty in determining the position of this electron.

SOLUTION

Conceptualize The fractional value given for the accuracy of the electron's speed can be interpreted as the fractional uncertainty in its momentum. This uncertainty corresponds to a minimum uncertainty in the electron's position through the uncertainty principle.

Categorize We evaluate the result using concepts developed in this section, so we categorize this example as a substitution problem.

Assume the electron is moving along the x axis and find the uncertainty in p_x, letting f represent the accuracy of the measurement of its speed:

$$\Delta p_x = m\,\Delta u_x = mfu_x$$

Solve Equation 28.22 for the uncertainty in the electron's position and substitute numerical values:

$$\Delta x \geq \frac{\hbar}{2\,\Delta p_x} = \frac{\hbar}{2mfu_x} = \frac{1.055 \times 10^{-34}\,\text{J} \cdot \text{s}}{2(9.11 \times 10^{-31}\,\text{kg})(0.000\,030\,0)(5.00 \times 10^3\,\text{m/s})}$$

$$= 3.86 \times 10^{-4}\,\text{m} = \boxed{0.386\,\text{mm}}$$

28.9 | An Interpretation of Quantum Mechanics

We have been introduced to some new and strange ideas so far in this chapter. In an effort to understand the concepts of quantum physics better, let us investigate another bridge between particles and waves. We first think about electromagnetic radiation from the particle point of view. For a particular situation in which electromagnetic radiation exists, the probability per unit volume of finding a photon in a given region of space at an instant of time is proportional to the number of photons per unit volume at that time:

$$\frac{\text{Probability}}{V} \propto \frac{N}{V}$$

The number of photons per unit volume is proportional to the intensity of the radiation:

$$\frac{N}{V} \propto I$$

Now, we form the bridge to the wave model by recalling that the intensity of electromagnetic radiation is proportional to the square of the electric field amplitude for the electromagnetic wave (Section 24.4):

$$I \propto E^2$$

Equating the beginning and the end of this string of proportionalities, we have

$$\frac{\text{Probability}}{V} \propto E^2 \qquad \qquad \text{28.24} \blacktriangleleft$$

Therefore, for electromagnetic radiation, the probability per unit volume of finding a particle associated with this radiation (the photon) is proportional to the square of the amplitude of the wave associated with the particle.

Recognizing the wave–particle duality of both electromagnetic radiation and matter, we should suspect a parallel proportionality for a material particle. That is, the probability per unit volume of finding the particle is proportional to the square of the amplitude of a wave representing the particle. In Section 28.5 we learned that there is a de Broglie wave associated with every particle. The amplitude of the de Broglie wave associated with a particle is not a measurable quantity (because the wave function representing a particle is generally a complex function, as we discuss below). In contrast, the electric field is a measurable quantity for an electromagnetic wave. The matter analog to Equation 28.24 relates the square of the wave's amplitude to the probability per unit volume of finding the particle. As a result, we call the amplitude of the wave associated with the particle the **probability amplitude,** or the **wave function,** and give it the symbol Ψ. In general, the complete wave function Ψ for a system depends on the positions of all the particles in the system and on time; therefore, it can be written $\Psi(\vec{r}_1, \vec{r}_2, \vec{r}_3, \ldots, \vec{r}_j, \ldots, t)$, where $\vec{r}_j$ is the position vector of the jth particle in the system. For many systems of interest, including all those in this text, the wave function Ψ is mathematically separable in space and time and can be written as a product of a space function ψ for one particle of the system and a complex time function:[4]

$$\Psi(\vec{r}_1, \vec{r}_2, \vec{r}_3, \ldots, \vec{r}_j, \ldots, t) = \psi(\vec{r}_j)\, e^{-i\omega t} \qquad \textbf{28.25} \blacktriangleleft$$

▶ Space- and time-dependent wave function Ψ

where $\omega\ (= 2\pi f)$ is the angular frequency of the wave function and $i = \sqrt{-1}$.

For any system in which the potential energy is time-independent and depends only on the positions of particles within the system, the important information about the system is contained within the space part of the wave function. The time part is simply the factor $e^{-i\omega t}$. Therefore, the understanding of ψ is the critical aspect of a given problem.

The wave function ψ is often complex-valued. The quantity $|\psi|^2 = \psi^*\psi$, where ψ^* is the complex conjugate[5] of ψ, is always real and positive, and is proportional to the probability per unit volume of finding a particle at a given point at some instant. The wave function contains within it all the information that can be known about the particle.

This probability interpretation of the wave function was first suggested by Max Born (1882–1970) in 1928. In 1926, Erwin Schrödinger (1887–1961) proposed a wave equation that describes the manner in which the wave function changes in space and time. The *Schrödinger wave equation,* which we shall examine in Section 28.12, represents a key element in the theory of quantum mechanics.

In Section 28.5, we found that the de Broglie equation relates the momentum of a particle to its wavelength through the relation $p = h/\lambda$. If an ideal free particle has a precisely known momentum p_x, its wave function is a sinusoidal wave of wavelength $\lambda = h/p_x$ and the particle has equal probability of being at any point along the x axis. The wave function for such a free particle moving along the x axis can be written as

$$\psi(x) = Ae^{ikx} \qquad \textbf{28.26} \blacktriangleleft$$

▶ Wave function for a free particle

where $k = 2\pi/\lambda$ is the angular wave number and A is a constant amplitude.[6]

Although we cannot measure ψ, we can measure the quantity $|\psi|^2$, the absolute square of ψ, which can be interpreted as follows. If ψ represents a single particle, $|\psi|^2$—called the **probability density**—is the relative probability per unit volume that

Pitfall Prevention | 28.5
The Wave Function Belongs to a System
The common language in quantum mechanics is to associate a wave function with a particle. The wave function, however, is determined by the particle *and* its interaction with its environment, so it more rightfully belongs to a system. In many cases, the particle is the only part of the system that experiences a change, which is why the common language has developed. You will see examples in the future in which it is more proper to think of the system wave function rather than the particle wave function.

---

[4]The standard form of a complex number is $a + ib$. The notation $e^{i\theta}$ is equivalent to the standard form as follows:
$$e^{i\theta} = \cos\theta + i\sin\theta$$
Therefore, the notation $e^{-i\omega t}$ in Equation 28.25 is equivalent to $\cos(-\omega t) + i\sin(-\omega t) = \cos\omega t - i\sin\omega t$.

[5]For a complex number $z = a + ib$, the complex conjugate is found by changing i to $-i$: $z^* = a - ib$. The product of a complex number and its complex conjugate is always real and positive:
$$z^*z = (a - ib)(a + ib) = a^2 - (ib)^2 = a^2 - (i)^2b^2 = a^2 + b^2$$

[6]For the free particle, the full wave function, based on Equation 28.25 is
$$\Psi(x, t) = Ae^{ikx}e^{-i\omega t} = Ae^{i(kx-\omega t)} = A[\cos(kx - \omega t) + i\sin(kx - \omega t)]$$
The real part of this wave function has the same form as the waves that we added together to form wave packets in Section 28.6.

the particle will be found at any given point in the volume. This interpretation can also be stated in the following manner. If dV is a small volume element surrounding some point, the probability of finding the particle in that volume element is $|\psi|^2 \, dV$. In this section, we deal only with one-dimensional systems, where the particle must be located along the x axis, and we therefore replace dV with dx. In this case, the probability $P(x) \, dx$ that the particle will be found in the infinitesimal interval dx around the point x is

$$P(x) \, dx = |\psi|^2 \, dx \qquad \textbf{28.27}\blacktriangleleft$$

Because the particle must be somewhere along the x axis, the sum of the probabilities over all values of x must be 1:

▶ Normalization condition on ψ

$$\int_{-\infty}^{\infty} |\psi|^2 \, dx = 1 \qquad \textbf{28.28}\blacktriangleleft$$

Any wave function satisfying Equation 28.28 is said to be **normalized.** Normalization is simply a statement that the particle exists at some point at all times.

Although it is not possible to specify the position of a particle with complete certainty, it is possible through $|\psi|^2$ to specify the probability of observing it in a small region surrounding a given point. The probability of finding the particle in the arbitrarily sized interval $a \leq x \leq b$ is

$$P_{ab} = \int_a^b |\psi|^2 \, dx \qquad \textbf{28.29}\blacktriangleleft$$

The probability P_{ab} is the area under the curve of $|\psi|^2$ versus x between the points $x = a$ and $x = b$ as in Figure 28.20.

Experimentally, the probability is finite of finding a particle in an interval near some point at some instant. The value of that probability must lie between the limits 0 and 1. For example, if the probability is 0.3, there is a 30% chance of finding the particle in the interval.

The wave function ψ satisfies a wave equation, just as the electric field associated with an electromagnetic wave satisfies a wave equation that follows from Maxwell's equations. As mentioned earlier, the wave equation satisfied by ψ is the Schrödinger equation (Section 28.12), and ψ can be computed from it. Although ψ is not a measurable quantity, all the measurable quantities of a particle, such as its energy and momentum, can be derived from a knowledge of ψ. For example, once the wave function for a particle is known, it is possible to calculate the average position at which you would find the particle after many measurements. This average position is called the **expectation value** of x and is defined by the equation

▶ Expectation value for position x

$$\langle x \rangle \equiv \int_{-\infty}^{\infty} \psi^* x \psi \, dx \qquad \textbf{28.30}\blacktriangleleft$$

where brackets $\langle \; \rangle$ are used to denote expectation values. Furthermore, one can find the expectation value of any function $f(x)$ associated with the particle by using the following equation:

▶ Expectation value for a function $f(x)$

$$\langle f(x) \rangle \equiv \int_{-\infty}^{\infty} \psi^* f(x) \psi \, dx \qquad \textbf{28.31}\blacktriangleleft$$

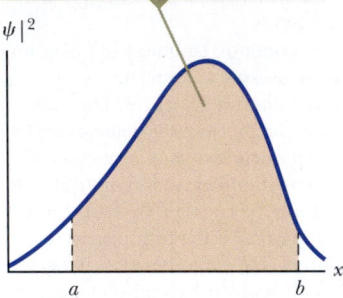

The probability of a particle being in the interval $a \leq x \leq b$ is the area under the probability density curve from a to b.

$|\psi|^2$

Figure 28.20 An arbitrary probability density curve for a particle.

❰ 28.10 | A Particle in a Box

In this section, we shall apply some of the ideas we have developed to a sample problem. Let us choose a simple problem: a particle confined to a one-dimensional region of space, called the *particle in a box* (even though the "box" is one-dimensional!). From a classical viewpoint, if a particle is confined to bouncing back and forth along the *x* axis between two impenetrable walls as in the pictorial representation in Figure 28.21a, its motion is easy to describe. If the speed of the particle is *u*, the magnitude of its momentum *mu* remains constant, as does its kinetic energy. Classical physics places no restrictions on the values of a particle's momentum and energy. The quantum mechanics approach to this problem is quite different and requires that we find the appropriate wave function consistent with the conditions of the situation.[7]

Because the walls are impenetrable, the probability of finding the particle outside the box is zero, so the wave function $\psi(x)$ must be zero for $x < 0$ and for $x > L$, where *L* is the distance between the two walls. A mathematical condition for any wave function is that it must be continuous in space.[8] Therefore, if ψ is zero outside the walls, it must also be zero *at* the walls; that is, $\psi(0) = 0$ and $\psi(L) = 0$. Only those wave functions that satisfy this condition are allowed.

Figure 28.21b shows a graphical representation of the particle in a box problem, which graphs the potential energy of the particle–environment system as a function of the position of the particle. When the particle is inside the box, the potential energy of the system does not depend on the particle's location and we can choose its value to be zero. Outside the box, we have to ensure that the wave function is zero. We can do so by defining the potential energy of the system as infinitely large if the particle were outside the box. Because kinetic energy is necessarily positive, the only way a particle could be outside the box is if the system has an infinite amount of energy.

The wave function for a particle in the box can be expressed as a real sinusoidal function:[9]

$$\psi(x) = A \sin\left(\frac{2\pi x}{\lambda}\right) \qquad \textbf{28.32} \blacktriangleleft$$

This wave function must satisfy the boundary conditions at the walls. The boundary condition at $x = 0$ is satisfied already because the sine function is zero when $x = 0$. For the boundary condition at $x = L$, we have

$$\psi(L) = 0 = A \sin\left(\frac{2\pi L}{\lambda}\right)$$

which can only be true if

$$\frac{2\pi L}{\lambda} = n\pi \quad \rightarrow \quad \lambda = \frac{2L}{n} \qquad \textbf{28.33} \blacktriangleleft$$

where $n = 1, 2, 3, \ldots$. Therefore, only certain wavelengths for the particle are allowed! Each of the allowed wavelengths corresponds to a quantum state for the system, and *n* is the quantum number. Expressing the wave function in terms of the quantum number *n*, we have

$$\psi_n(x) = A \sin\left(\frac{2\pi x}{\lambda}\right) = A \sin\left(\frac{2\pi x}{2L/n}\right) = A \sin\left(\frac{n\pi x}{L}\right) \qquad \textbf{28.34} \blacktriangleleft$$

▶ Wave functions for a particle in a box

Active Figures 28.22a and 28.22b (page 836) are graphical representations of ψ_n versus *x* and $|\psi_n|^2$ versus *x* for $n = 1$, 2, and 3 for the particle in a box. Note that although ψ_n can be positive or negative, $|\psi_n|^2$ is always positive. Because $|\psi_n|^2$ represents a probability density, a negative value for $|\psi_n|^2$ is meaningless.

Further inspection of Active Figure 28.22b shows that $|\psi_n|^2$ is zero at the boundaries, satisfying our boundary condition. In addition, $|\psi_n|^2$ is zero at other points, depending on the value of *n*. For $n = 2$, $|\psi_n|^2 = 0$ at $x = L/2$; for $n = 3$, $|\psi_n|^2 = 0$ at

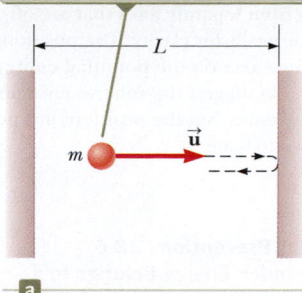

This figure is a *pictorial representation* showing a particle of mass *m* and speed *u* bouncing between two impenetrable walls separated by a distance *L*.

a

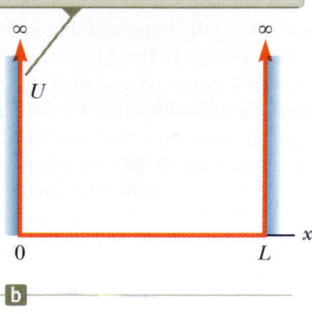

This figure is a *graphical representation* showing the potential energy of the particle–box system. The blue areas are classically forbidden.

b

Figure 28.21 (a) The particle in a box. (b) The potential energy function for the system.

Active Figure 28.22 The first three allowed states for a particle confined to a one-dimensional box. The states are shown superimposed on the potential energy function of Figure 28.21b. The wave functions and probability densities are plotted vertically from separate axes that are offset vertically for clarity. The positions of these axes on the potential energy function suggest the relative energies of the states, but the positions are not shown to scale.

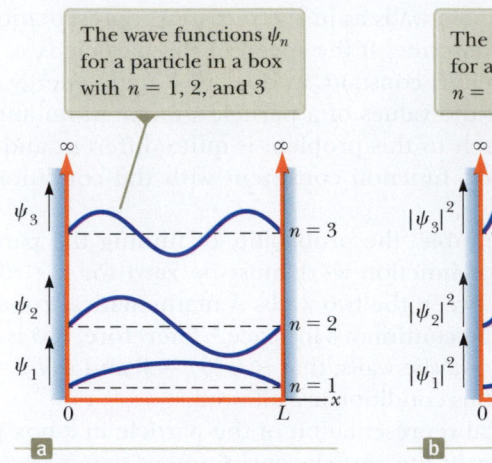

The wave functions ψ_n for a particle in a box with $n = 1, 2,$ and 3

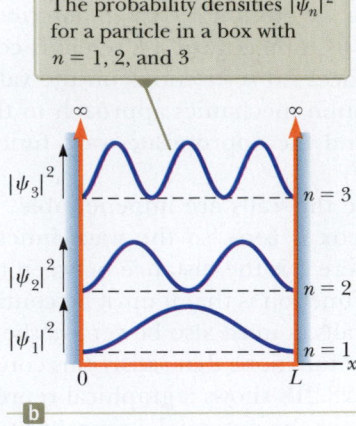

The probability densities $|\psi_n|^2$ for a particle in a box with $n = 1, 2,$ and 3

Pitfall Prevention | 28.6
Reminder: Energy Belongs to a System
We describe Equation 28.35 as representing the energy of the particle; it is commonly used language for the particle in a box problem. In reality, we are analyzing the energy of the *system* of the particle and whatever environment is establishing the impenetrable walls. In the case of a particle in a box, the only nonzero type of energy is kinetic and it belongs to the particle. In general, energies that we calculate using quantum physics are associated with a system of interacting particles, such as the electron and proton in the hydrogen atom studied in Chapter 11.

$x = L/3$ and $x = 2L/3$. The number of zero points increases by one each time the quantum number increases by one.

Because the wavelengths of the particle are restricted by the condition $\lambda = 2L/n$, the magnitude of the momentum of the particle is also restricted to specific values that we can find from the expression for the de Broglie wavelength, Equation 28.15:

$$p = \frac{h}{\lambda} = \frac{h}{2L/n} = \frac{nh}{2L}$$

From this expression, we find that the allowed values of the energy, which is simply the kinetic energy of the particle, are

$$E_n = \tfrac{1}{2}mu^2 = \frac{p^2}{2m} = \frac{(nh/2L)^2}{2m}$$

▶ Quantized energies for a particle in a box

$$E_n = \left(\frac{h^2}{8mL^2}\right)n^2 \qquad n = 1, 2, 3, \ldots \qquad \textbf{28.35} ◀$$

As we see from this expression, the energy of the particle is quantized, similar to our quantization of energy in the hydrogen atom in Chapter 11. The lowest allowed energy corresponds to $n = 1$, for which $E_1 = h^2/8mL^2$. Because $E_n = n^2E_1$, the excited states corresponding to $n = 2, 3, 4, \ldots$ have energies given by $4E_1, 9E_1, 16E_1, \ldots$.

Active Figure 28.23 is an energy level diagram[10] describing the energy values of the allowed states. Note that the state $n = 0$, for which E would be equal to zero, is not allowed. Therefore, according to quantum mechanics, the particle can never be at rest. The least energy it can have, corresponding to $n = 1$, is called the **zero-point energy.** This result is clearly contradictory to the classical viewpoint, in which $E = 0$ is an acceptable state, as are all positive values of E.

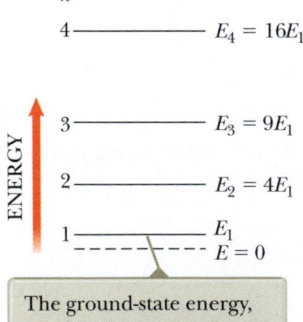

The ground-state energy, which is the lowest allowed energy, is $E_1 = h^2/8mL^2$.

Active Figure 28.23 Energy level diagram for a particle confined to a one-dimensional box of length L.

QUICK QUIZ 28.8 Consider an electron, a proton, and an alpha particle (a helium nucleus), each trapped separately in identical boxes. (**i**) Which particle corresponds to the highest ground-state energy? (a) the electron (b) the proton (c) the alpha particle (d) The ground-state energy is the same in all three cases. (**ii**) Which particle has the longest wavelength when the system is in the ground state? (a) the electron (b) the proton (c) the alpha particle (d) All three particles have the same wavelength.

QUICK QUIZ 28.9 A particle is in a box of length L. Suddenly, the length of the box is increased to $2L$. What happens to the energy levels shown in Active Figure 28.23? (**a**) Nothing; they are unaffected. (**b**) They move farther apart. (**c**) They move closer together.

---

[10]We introduced the energy level diagram as a specialized semigraphical representation in Chapter 11.

Example 28.8 | **Microscopic and Macroscopic Particles in Boxes**

(A) An electron is confined between two impenetrable walls 0.200 nm apart. Determine the energy levels for the states $n = 1, 2,$ and 3.

SOLUTION

Conceptualize In Figure 28.21a, imagine that the particle is an electron and the walls are very close together.

Categorize We evaluate the energy levels using an equation developed in this section, so we categorize this example as a substitution problem.

Use Equation 28.35 for the $n = 1$ state:

$$E_1 = \frac{h^2}{8m_eL^2}(1)^2 = \frac{(6.63 \times 10^{-34}\,\text{J}\cdot\text{s})^2}{8(9.11 \times 10^{-31}\,\text{kg})(2.00 \times 10^{-10}\,\text{m})^2}$$

$$= 1.51 \times 10^{-18}\,\text{J} = 9.42\,\text{eV}$$

Using $E_n = n^2E_1$, find the energies of the $n = 2$ and $n = 3$ states:

$$E_2 = (2)^2E_1 = 4(9.42\,\text{eV}) = 37.7\,\text{eV}$$

$$E_3 = (3)^2E_1 = 9(9.42\,\text{eV}) = 84.8\,\text{eV}$$

(B) Find the speed of the electron in the $n = 1$ state.

SOLUTION

Solve the classical expression for kinetic energy for the particle speed:

$$K = \tfrac{1}{2}m_eu^2 \quad \rightarrow \quad u = \sqrt{\frac{2K}{m_e}}$$

Recognize that the kinetic energy of the particle is equal to the system energy and substitute E_n for K:

$$(1) \quad u = \sqrt{\frac{2E_n}{m_e}}$$

Substitute numerical values from part (A):

$$u = \sqrt{\frac{2(1.51 \times 10^{-18}\,\text{J})}{9.11 \times 10^{-31}\,\text{kg}}} = 1.82 \times 10^6\,\text{m/s}$$

Simply placing the electron in the box results in a *minimum* speed of the electron equal to 0.6% of the speed of light!

(C) A 0.500-kg baseball is confined between two rigid walls of a stadium that can be modeled as a box of length 100 m. Calculate the minimum speed of the baseball.

SOLUTION

Conceptualize In Figure 28.21a, imagine that the particle is a baseball and the walls are those of the stadium.

Categorize This part of the example is a substitution problem in which we apply a quantum approach to a macroscopic object.

Use Equation 28.35 for the $n = 1$ state:

$$E_1 = \frac{h^2}{8mL^2}(1)^2 = \frac{(6.63 \times 10^{-34}\,\text{J}\cdot\text{s})^2}{8(0.500\,\text{kg})(100\,\text{m})^2} = 1.10 \times 10^{-71}\,\text{J}$$

Use Equation (1) to find the speed:

$$u = \sqrt{\frac{2(1.10 \times 10^{-71}\,\text{J})}{0.500\,\text{kg}}} = 6.63 \times 10^{-36}\,\text{m/s}$$

This speed is so small that the object can be considered to be at rest, which is what one would expect for the minimum speed of a macroscopic object.

What If? What if a sharp line drive is hit so that the baseball is moving with a speed of 150 m/s? What is the quantum number of the state in which the baseball now resides?

Answer We expect the quantum number to be very large because the baseball is a macroscopic object.

Evaluate the kinetic energy of the baseball:

$$\tfrac{1}{2}mu^2 = \tfrac{1}{2}(0.500\,\text{kg})(150\,\text{m/s})^2 = 5.62 \times 10^3\,\text{J}$$

From Equation 28.35, calculate the quantum number n:

$$n = \sqrt{\frac{8mL^2E_n}{h^2}} = \sqrt{\frac{8(0.500\,\text{kg})(100\,\text{m})^2(5.62 \times 10^3\,\text{J})}{(6.63 \times 10^{-34}\,\text{J}\cdot\text{s})^2}} = 2.26 \times 10^{37}$$

continued

28.8 *cont.*

This result is a tremendously large quantum number. As the baseball pushes air out of the way, hits the ground, and rolls to a stop, it moves through more than 10^{37} quantum states. These states are so close together in energy that we cannot observe the transitions from one state to the next. Rather, we see what appears to be a smooth variation in the speed of the ball. The quantum nature of the universe is simply not evident in the motion of macroscopic objects.

28.11 | Analysis Model: Quantum Particle Under Boundary Conditions

The discussion of the particle in a box is very similar to the discussion in Chapter 14 of standing waves on strings:

- Because the ends of the string must be nodes, the wave functions for allowed waves must be zero at the boundaries of the string. Because the particle in a box cannot exist outside the box, the allowed wave functions for the particle must be zero at the boundaries.
- The boundary conditions on the string waves lead to quantized wavelengths and frequencies of the waves. The boundary conditions on the wave function for the particle in a box lead to quantized wavelengths and frequencies of the particle.

In quantum mechanics, it is very common for particles to be subject to boundary conditions. We therefore introduce a new analysis model, the **quantum particle under boundary conditions.** In many ways, this model is similar to the waves under boundary conditions model studied in Section 14.3. In fact, the allowed wavelengths for the wave function of a particle in a box (Eq. 28.33) are identical in form to the allowed wavelengths for mechanical waves on a string fixed at both ends (Eq. 14.5).

The quantum particle under boundary conditions model *differs* in some ways from the waves under boundary conditions model:

- In most cases of quantum particles, the wave function is *not* a simple sinusoidal function like the wave function for waves on strings. Furthermore, the wave function for a quantum particle may be a complex function.
- For a quantum particle, frequency is related to energy through $E = hf$, so the quantized frequencies lead to quantized energies.
- There may be no stationary "nodes" associated with the wave function of a quantum particle under boundary conditions. Systems more complicated than the particle in a box have more complicated wave functions, and some boundary conditions may not lead to zeroes of the wave function at fixed points.

In general,

---

an interaction of a quantum particle with its environment represents one or more boundary conditions, and, if the interaction restricts the particle to a finite region of space, results in quantization of the energy of the system.

Boundary conditions on quantum wave functions are related to the coordinates describing the problem. For the particle in a box, the wave function must be zero at two values of x. In the case of a three-dimensional system such as the hydrogen atom we shall discuss in Chapter 29, the problem is best presented in *spherical coordinates*. These coordinates, an extension of the plane polar coordinates introduced in Section 1.6, consist of a radial coordinate r and two angular coordinates. The generation of the wave function and application of the boundary conditions for the hydrogen atom are beyond the scope of this book. We shall, however, examine the behavior of some of the hydrogen-atom wave functions in Chapter 29.

Boundary conditions on wave functions that exist for all values of x require that the wave function approach zero as $x \rightarrow \infty$ (so that the wave function can be normalized) and remain finite as $x \to 0$. One boundary condition on any angular parts of wave functions is that adding 2π radians to the angle must return the wave function to the same value because an addition of 2π results in the same angular position.

28.12 | The Schrödinger Equation

In Section 24.3, we discussed a wave equation for electromagnetic radiation. The waves associated with particles also satisfy a wave equation. We might guess that the wave equation for material particles is different from that associated with photons because material particles have a nonzero rest energy. The appropriate wave equation was developed by Schrödinger in 1926. In applying the quantum particle under boundary conditions model to a quantum system, the approach is to determine a solution to this equation and then apply the appropriate boundary conditions to the solution. The solution yields the allowed wave functions and energy levels of the system under consideration. Proper manipulation of the wave function then enables one to calculate all measurable features of the system.

The Schrödinger equation as it applies to a particle of mass m confined to moving along the x axis and interacting with its environment through a potential energy function $U(x)$ is

$$-\frac{\hbar^2}{2m}\frac{d^2\psi}{dx^2} + U\psi = E\psi$$

28.36 ◄ ▶ Time-independent Schrödinger equation

where E is the total energy of the system (particle and environment). Because this equation is independent of time, it is commonly referred to as the **time-independent Schrödinger equation.** (We shall not discuss the time-dependent Schrödinger equation, whose solution is Ψ, Eq. 28.25, in this text.)

This equation is consistent with the energy version of the isolated system model. The system is the particle and its environment. Problem 54 shows, both for a free particle and a particle in a box, that the first term in the Schrödinger equation reduces to the kinetic energy of the particle multiplied by the wave function. Therefore, Equation 28.36 tells us that the total energy is the sum of the kinetic energy and the potential energy and that the total energy is a constant: $K + U = E =$ constant.

In principle, if the potential energy $U(x)$ for the system is known, one can solve Equation 28.36 and obtain the wave functions and energies for the allowed states of the system. Because U may vary discontinuously with position, it may be necessary to solve the equation separately for various regions. In the process, the wave functions for the different regions must join smoothly at the boundaries and we require that $\psi(x)$ be *continuous*. Furthermore, so that $\psi(x)$ obeys the normalization condition, we require that $\psi(x)$ approach zero as x approaches $\pm\infty$. Finally, $\psi(x)$ must be *single-valued* and $d\psi/dx$ must also be continuous[11] for finite values of $U(x)$.

The task of solving the Schrödinger equation may be very difficult, depending on the form of the potential energy function. As it turns out, the Schrödinger equation has been extremely successful in explaining the behavior of atomic and nuclear systems, whereas classical physics has failed to do so. Furthermore, when quantum mechanics is applied to macroscopic objects, the results agree with classical physics, as required by the correspondence principle.

Erwin Schrödinger
Austrian Theoretical Physicist (1887–1961)
Schrödinger is best known as one of the creators of quantum mechanics. His approach to quantum mechanics was demonstrated to be mathematically equivalent to the more abstract matrix mechanics developed by Heisenberg. Schrödinger also produced important papers in the fields of statistical mechanics, color vision, and general relativity.

© INTERFOTO/Alamy

The Particle in a Box via the Schrödinger Equation

To see how the Schrödinger equation is applied to a problem, let us return to our particle in a one-dimensional box of length L (see Fig. 28.21) and analyze it with the Schrödinger equation. In association with Figure 28.21b, we discussed the potential energy diagram that describes the problem. A potential energy diagram such as this one is a useful representation for understanding and solving problems with the Schrödinger equation.

---

[11]If $d\psi/dx$ were not continuous, we would not be able to evaluate $d^2\psi/dx^2$ in Equation 28.36 at the point of discontinuity.

Because of the shape of the curve in Figure 28.21b, the particle in a box is sometimes said to be in a **square well**,[12] where a **well** is an upward-facing region of the curve in a potential energy diagram. (A downward-facing region is called a *barrier*, which we shall investigate in Section 28.13.)

In the region $0 < x < L$, where $U = 0$, we can express the Schrödinger equation in the form

$$\frac{d^2\psi}{dx^2} = -\frac{2mE}{\hbar^2}\psi = -k^2\psi \qquad \textbf{28.37} \blacktriangleleft$$

where

$$k = \frac{\sqrt{2mE}}{\hbar} \qquad \textbf{28.38} \blacktriangleleft$$

The solution to Equation 28.37 is a function whose second derivative is the negative of the same function multiplied by a constant k^2. We recognize both the sine and cosine functions as satisfying this requirement. Therefore, the most general solution to the equation is a linear combination of both solutions:

$$\psi(x) = A \sin kx + B \cos kx$$

where A and B are constants determined by the boundary conditions.

Our first boundary condition is that $\psi(0) = 0$:

$$\psi(0) = A \sin 0 + B \cos 0 = 0 + B = 0 \quad \rightarrow \quad B = 0$$

Therefore, our solution reduces to

$$\psi(x) = A \sin kx$$

The second boundary condition, $\psi(L) = 0$, when applied to the reduced solution, gives

$$\psi(L) = A \sin kL = 0$$

which is satisfied only if kL is an integral multiple of π, that is, if $kL = n\pi$, where n is an integer. Because $k = \sqrt{2mE}/\hbar$, we have

$$kL = \frac{\sqrt{2mE}}{\hbar}L = n\pi$$

For each integer choice for n, this equation determines a quantized energy E_n. Solving for the allowed energies E_n gives

$$E_n = \left(\frac{h^2}{8mL^2}\right)n^2 \qquad \textbf{28.39} \blacktriangleleft$$

which are identical to the allowed energies in Equation 28.35.

Substituting the values of k in the wave function, the allowed wave functions $\psi_n(x)$ are given by

$$\psi_n(x) = A \sin\left(\frac{n\pi x}{L}\right) \qquad \textbf{28.40} \blacktriangleleft$$

This wave function agrees with Equation 28.34.

Normalizing this relationship shows that $A = \sqrt{(2/L)}$. (See Problem 56 at the end of this chapter.) Therefore, the normalized wave function is

$$\psi_n(x) = \sqrt{\frac{2}{L}} \sin\left(\frac{n\pi x}{L}\right) \qquad \textbf{28.41} \blacktriangleleft$$

The notion of trapping particles in potential wells is used in the burgeoning field of **nanotechnology**, which refers to the design and application of devices having dimensions ranging from 1 to 100 nm. The fabrication of these devices often involves manipulating single atoms or small groups of atoms to form structures such as the quantum corral in Figure 28.24.

One area of nanotechnology of interest to researchers is the **quantum dot.** The quantum dot, a small region that is grown in a silicon crystal, acts as a potential well.

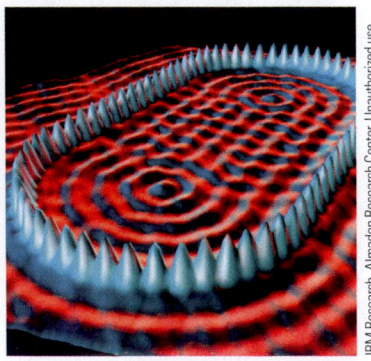

Figure 28.24 This is an image of a quantum corral consisting of a ring of 48 iron atoms located on a copper surface. The diameter of the ring is 143 nm, and the image was obtained using a low-temperature scanning tunneling microscope (STM) as mentioned in Section 28.13. Corrals and other structures are able to confine surface electron waves. The study of such structures will play an important role in determining the future of small electronic devices.

---

[12]It is called a square well even if it has a rectangular shape in a potential energy diagram.

This region can trap electrons into states with quantized energies. The wave functions for a particle in a quantum dot look similar to those in Active Figure 28.22a if L is on the order of nanometers. The storage of binary information using quantum dots is an active field of research. A simple binary scheme would involve associating a one with a quantum dot containing an electron and a zero with an empty dot. Other schemes involve cells of multiple dots such that arrangements of electrons among the dots correspond to ones and zeros. Several research laboratories are studying the properties and potential applications of quantum dots. Information should be forthcoming from these laboratories at a steady rate in the next few years.

Example 28.9 | The Expectation Values for the Particle in a Box

A particle of mass m is confined to a one-dimensional box between $x = 0$ and $x = L$. Find the expectation value of the position x of the particle in the state characterized by quantum number n.

SOLUTION

Conceptualize Active Figure 28.22b shows that the probability for the particle to be at a given location varies with position within the box. Can you predict what the expectation value of x will be from the symmetry of the wave functions?

Categorize The statement of the example categorizes the problem for us: we focus on a quantum particle in a box and on the calculation of its expectation value of x.

Analyze In Equation 28.30, the integration from $-\infty$ to ∞ reduces to the limits 0 to L because $\psi = 0$ everywhere except in the box.

Substitute Equation 28.41 into Equation 28.30 to find the expectation value for x:

$$\langle x \rangle = \int_{-\infty}^{\infty} \psi_n^* x \psi_n \, dx = \int_0^L x \left[\sqrt{\frac{2}{L}} \sin\left(\frac{n\pi x}{L}\right) \right]^2 dx$$

$$= \frac{2}{L} \int_0^L x \sin^2\left(\frac{n\pi x}{L}\right) dx$$

Evaluate the integral by consulting an integral table or by mathematical integration:[13]

$$\langle x \rangle = \frac{2}{L} \left[\frac{x^2}{4} - \frac{x \sin\left(2 \frac{n\pi x}{L}\right)}{4 \frac{n\pi}{L}} - \frac{\cos\left(2 \frac{n\pi x}{L}\right)}{8\left(\frac{n\pi}{L}\right)^2} \right]_0^L$$

$$= \frac{2}{L}\left[\frac{L^2}{4}\right] = \frac{L}{2}$$

Finalize This result shows that the expectation value of x is at the center of the box for all values of n, which you would expect from the symmetry of the square of the wave functions (the probability density) about the center (Active Fig. 28.22b).

The $n = 2$ wave function in Active Figure 28.22b has a value of zero at the midpoint of the box. Can the expectation value of the particle be at a position at which the particle has zero probability of existing? Remember that the expectation value is the *average* position. Therefore, the particle is as likely to be found to the right of the midpoint as to the left, so its average position is at the midpoint even though its probability of being there is zero. As an analogy, consider a group of students for whom the average final examination score is 50%. There is no requirement that some student achieve a score of exactly 50% for the average of all students to be 50%.

28.13 | Tunneling Through a Potential Energy Barrier

Consider the potential energy function shown in Figure 28.25 (page 842), in which the potential energy of the system is zero everywhere except for a region of width L where the potential energy has a constant value of U. This type of potential energy function is called a **square barrier**, and U is called the **barrier height**. A very

---

[13]To integrate this function, first replace $\sin^2(n\pi x/L)$ with $\frac{1}{2}(1 - \cos 2n\pi x/L)$ (refer to Table B.3 in Appendix B), which allows $\langle x \rangle$ to be expressed as two integrals. The second integral can then be evaluated by partial integration (Section B.7 in Appendix B).

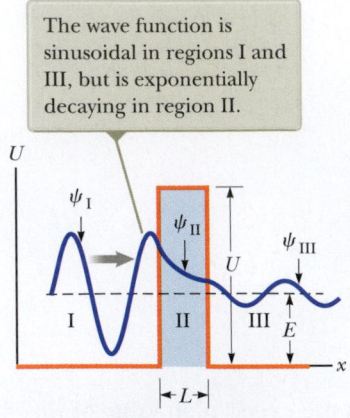

The wave function is sinusoidal in regions I and III, but is exponentially decaying in region II.

Figure 28.25 Wave function ψ for a particle incident from the left on a barrier of height U and width L. The wave function is plotted vertically from an axis positioned at the energy of the particle.

Pitfall Prevention | 28.8
"Height" on an Energy Diagram
The word *height* (as in *barrier height*) refers to an energy in discussions of barriers in potential energy diagrams. For example, we might say the height of the barrier is 10 eV. On the other hand, the barrier *width* refers to the traditional usage of such a word and is an actual physical length measurement between the two locations of the vertical sides of the barrier.

interesting and peculiar phenomenon occurs when a moving particle encounters such a barrier of finite height and width. Consider a particle of energy $E < U$ that is incident on the barrier from the left (see Fig. 28.25). Classically, the particle is reflected by the barrier. If the particle were to exist in region II, its kinetic energy would be negative, which is not allowed classically. Therefore, region II, and in turn region III, are both classically *forbidden* to the particle incident from the left. According to quantum mechanics, however, all regions are accessible to the particle, regardless of its energy. (Although all regions are accessible, the probability of the particle being in a region that is classically forbidden is very low.) According to the uncertainty principle, the particle can be within the barrier as long as the time interval during which it is in the barrier is short and consistent with Equation 28.23. If the barrier is relatively narrow, this short time interval can allow the particle to move across the barrier. Therefore, it is possible for us to understand the passing of the particle through the barrier with the help of the uncertainty principle.

Let us approach this situation using a mathematical representation. The Schrödinger equation has valid solutions in all three regions I, II, and III. The solutions in regions I and III are sinusoidal as in Equation 28.26. In region II, the solution is exponential. Applying the boundary conditions that the wave functions in the three regions must join smoothly at the boundaries, we find that a full solution can be found such as that represented by the curve in Figure 28.25. Therefore, Schrödinger's equation and the boundary conditions are satisfied, which tells us mathematically that such a process can theoretically occur according to the quantum theory.

Because the probability of locating the particle is proportional to $|\psi|^2$, we conclude that the chance of finding the particle beyond the barrier in region III is nonzero. This result is in complete disagreement with classical physics. The movement of the particle to the far side of the barrier is called **tunneling** or **barrier penetration.**

The probability of tunneling can be described with a **transmission coefficient** T and a **reflection coefficient** R. The transmission coefficient represents the probability that the particle penetrates to the other side of the barrier, and the reflection coefficient is the probability that the particle is reflected by the barrier. Because the incident particle is either reflected or transmitted, we require that $T + R = 1$. An approximate expression for the transmission coefficient, obtained when $T \ll 1$ (a very wide barrier or a very high barrier, that is, $U \gg E$), is

$$T \approx e^{-2CL} \qquad \textbf{28.42} \blacktriangleleft$$

where

$$C = \frac{\sqrt{2m(U - E)}}{\hbar} \qquad \textbf{28.43} \blacktriangleleft$$

According to quantum physics, Equation 28.42 tells us that T can be nonzero, which is in contrast to the classical point of view that requires that $T = 0$. That we experimentally observe the phenomenon of tunneling provides further confidence in the principles of quantum physics.

Figure 28.25 shows the wave function of a particle tunneling through a barrier in one dimension. A similar wave function having spherical symmetry describes the barrier penetration of a particle leaving a radioactive nucleus, which we will study in Chapter 30. The wave function exists both inside and outside the nucleus, and its amplitude is constant in time. In this way, the wave function correctly describes the small but constant probability that the nucleus will decay. The moment of decay cannot be predicted. In general, quantum mechanics implies that the future is indeterminate. (This feature is in contrast to classical mechanics, from which the trajectory of an object can be calculated to arbitrarily high precision from precise knowledge of its initial position and velocity and of the forces exerted on it.) We must conclude that the fundamental laws of nature are probabilistic.

A radiation detector can be used to show that a nucleus decays by radiating a particle at a particular moment and in a particular direction. To point out the contrast between this experimental result and the wave function describing it, Schrödinger imagined a box containing a cat, a radioactive sample, a radiation counter, and a vial of poison. When a nucleus in the sample decays, the counter triggers the administration of lethal poison to the cat. Quantum mechanics correctly predicts the probability of finding the cat dead when the box is opened. Before the box is opened, does the animal have a wave function describing it as a fractionally dead cat, with some chance of being alive?

This question is currently under investigation, never with actual cats, but sometimes with interference experiments building upon the experiment described in Section 28.7. Does the act of measurement change the system from a probabilistic to a definite state? When a particle emitted by a radioactive nucleus is detected at one particular location, does the wave function describing the particle drop instantaneously to zero everywhere else in the Universe? (Einstein called such a state change a "spooky action at a distance.") Is there a fundamental difference between a quantum system and a macroscopic system? The answers to these questions are basically unknown.

> **QUICK QUIZ 28.10** Which of the following changes would increase the probability of transmission of a particle through a potential barrier? (You may choose more than one answer.) **(a)** decreasing the width of the barrier **(b)** increasing the width of the barrier **(c)** decreasing the height of the barrier **(d)** increasing the height of the barrier **(e)** decreasing the kinetic energy of the incident particle **(f)** increasing the kinetic energy of the incident particle

---

Example 28.10 | Transmission Coefficient for an Electron

A 30-eV electron is incident on a square barrier of height 40 eV.

(A) What is the probability that the electron tunnels through the barrier if its width is 1.0 nm?

SOLUTION

Conceptualize Because the particle energy is smaller than the height of the potential barrier, we expect the electron to reflect from the barrier with a probability of 100% according to classical physics. Because of the tunneling phenomenon, however, there is a finite probability that the particle can appear on the other side of the barrier.

Categorize We evaluate the probability using an equation developed in this section, so we categorize this example as a substitution problem.

Evaluate the quantity $U - E$ that appears in Equation 28.43:

$$U - E = 40 \text{ eV} - 30 \text{ eV} = 10 \text{ eV} \left(\frac{1.6 \times 10^{-19} \text{ J}}{1 \text{ eV}} \right) = 1.6 \times 10^{-18} \text{ J}$$

Evaluate the quantity $2CL$ using Equation 28.43:

$$(1) \quad 2CL = 2 \frac{\sqrt{2(9.11 \times 10^{-31} \text{ kg})(1.6 \times 10^{-18} \text{ J})}}{1.055 \times 10^{-34} \text{ J} \cdot \text{s}} (1.0 \times 10^{-9} \text{ m}) = 32.4$$

From Equation 28.42, find the probability of tunneling through the barrier:

$$T \approx e^{-2CL} = e^{-32.4} = \boxed{8.5 \times 10^{-15}}$$

(B) What is the probability that the electron tunnels through the barrier if its width is 0.10 nm?

SOLUTION

In this case, the width L in Equation (1) is one-tenth as large, so evaluate the new value of $2CL$:

$$2CL = (0.1)(32.4) = 3.24$$

From Equation 28.42, find the new probability of tunneling through the barrier:

$$T \approx e^{-2CL} = e^{-3.24} = \boxed{0.039}$$

In part (A), the electron has approximately 1 chance in 10^{14} of tunneling through the barrier. In part (B), however, the electron has a much higher probability (3.9%) of penetrating the barrier. Therefore, reducing the width of the barrier by only one order of magnitude increases the probability of tunneling by about 12 orders of magnitude!

Applications of Tunneling

As we have seen, tunneling is a quantum phenomenon, a result of the wave nature of matter. Many applications may be understood only on the basis of tunneling.

- **Alpha decay.** One form of radioactive decay is the emission of alpha particles (the nuclei of helium atoms) by unstable, heavy nuclei (Chapter 30). For an alpha particle to escape from the nucleus, it must penetrate a barrier whose height is several times larger than the energy of the nucleus–alpha particle system. The barrier is due to a combination of the attractive nuclear force (discussed in Chapter 30) and the Coulomb repulsion (discussed in detail in Chapter 19) between the alpha particle and the rest of the nucleus. Occasionally, an alpha particle tunnels through the barrier, which explains the basic mechanism for this type of decay and the large variations in the mean lifetimes of various radioactive nuclei.

- **Nuclear fusion.** The basic reaction that powers the Sun and, indirectly, almost everything else in the solar system is fusion, which we will study in Chapter 30. In one step of the process that occurs at the core of the Sun, protons must approach each other to within such a small distance that they fuse to form a deuterium nucleus. According to classical physics, these protons cannot overcome and penetrate the barrier caused by their mutual electrical repulsion. Quantum-mechanically, however, the protons are able to tunnel through the barrier and fuse together.

- **Scanning tunneling microscope.** The scanning tunneling microscope, or STM, is a remarkable device that uses tunneling to create images of surfaces with resolution comparable to the size of a single atom. A small probe with a very fine tip is made to scan very close to the surface of a specimen. A tunneling current is maintained between the probe and specimen; the current (which is related to the probability of tunneling) is very sensitive to the barrier height (which is related to the separation between the tip and specimen) as seen in Example 28.10. Maintaining a constant tunneling current produces a feedback signal that is used to raise and lower the probe as the surface is scanned. Because the vertical motion of the probe follows the contour of the specimen's surface, an image of the surface is obtained. The image of the quantum corral shown in Figure 28.24 is made with a scanning tunneling microscope.

⫷ 28.14 | Context Connection: The Cosmic Temperature

Now that we have introduced the concepts of quantum physics for microscopic particles and systems, let us see how we can connect these concepts to processes occurring on a cosmic scale. For our first such connection, consider the Universe as a system. It is widely believed that the Universe began with a cataclysmic explosion called the **Big Bang,** first mentioned in Chapter 5. Because of this explosion, all the material in the Universe is moving apart. This expansion causes a Doppler shift in radiation left over from the Big Bang such that the wavelength of the radiation lengthens. In the 1940s, Ralph Alpher, George Gamow, and Robert Hermann developed a structural model of the Universe in which they predicted that the thermal radiation from the Big Bang should still be present and that it should now have a wavelength distribution consistent with a black body with a temperature of a few kelvins.

In 1965, Arno Penzias and Robert Wilson of Bell Telephone Laboratories were measuring radiation from the Milky Way galaxy using a special 20-ft antenna as a radio telescope. They noticed a consistent background "noise" of radiation in the signals from the antenna. Despite their great efforts to test alternative hypotheses for the origin of the noise in terms of interference from the Sun, an unknown source in the Milky Way, structural problems in the antenna, and even the presence of pigeon droppings in the antenna, none of the hypotheses was sufficient to explain the noise.

What Penzias and Wilson were detecting was the thermal radiation from the Big Bang. That it was detected by their system regardless of the direction of the antenna was consistent with the radiation being spread throughout the Universe, as the Big Bang model predicts. A measurement of the intensity of this radiation suggested that the temperature associated with the radiation was about 3 K, consistent with Alpher, Gamow, and Hermann's prediction from the 1940s. Although the measured intensity was consistent with their prediction, the measurement was taken at only a single wavelength. Full agreement with the model of the Universe as a black body would come only if measurements at many wavelengths demonstrated a distribution in wavelengths consistent with Active Figure 28.2.

In the years following Penzias and Wilson's discovery, other researchers made measurements at different wavelengths. In 1989, the COBE (*CO*smic *B*ackground *E*xplorer) satellite was launched by NASA and added critical measurements at wavelengths below 0.1 cm. The results of these measurements led to a Nobel Prize in Physics for the principal investigators in 2006. Several data points from COBE are shown in Figure 28.26. The Wilkinson Microwave Anisotropy Probe, launched in June 2001, exhibits data that allow observation of temperature differences in the cosmos in the microkelvin range. Ongoing observations are also being made from Earth-based facilities, associated with projects such as QUaD, Qubic, and the South Pole Telescope. In addition, the Planck satellite was launched in May 2009 by the European Space Agency. This space-based observatory should measure the cosmic background radiation with higher sensitivity than the Wilkinson probe. The series of measurements taken since 1965 are consistent with thermal radiation associated with a temperature of 2.7 K. The whole story of the cosmic temperature is a remarkable example of science at work: building a model, making a prediction, taking measurements, and testing the measurements against the predictions.

The first chapter of our *Cosmic Connection* Context describes the first example of this connection. By studying the thermal radiation from microscopic vibrating objects, we learn something about the origin of our Universe. In Chapter 29, we shall see more examples of this fascinating connection.

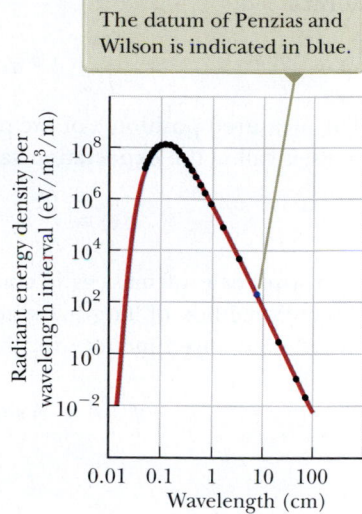

The datum of Penzias and Wilson is indicated in blue.

Figure 28.26 Theoretical blackbody (brown curve) and measured radiation spectra (black points) of the Big Bang. Most of the data were collected from the Cosmic Background Explorer, or COBE, satellite.

SUMMARY

The characteristics of **blackbody radiation** cannot be explained by classical concepts. Planck introduced the first model of **quantum physics** when he argued that the atomic oscillators responsible for this radiation exist only in discrete **quantum states.**

In the **photoelectric effect,** electrons are ejected from a metallic surface when light is incident on that surface. Einstein provided a successful explanation of this effect by extending Planck's quantum theory to the electromagnetic field. In this model, light is viewed as a stream of particles called **photons,** each with energy $E = hf$, where f is the frequency and h is **Planck's constant.** The maximum kinetic energy of the ejected photoelectron is given by

$$K_{max} = hf - \phi \qquad \text{28.11} \blacktriangleleft$$

where ϕ is the **work function** of the metal.

X-rays striking a target are scattered at various angles by electrons in the target. A shift in wavelength is observed for the scattered x-rays, and the phenomenon is known as the **Compton effect.** Classical physics does not correctly explain the experimental results of this effect. If the x-ray is treated as a photon, conservation of energy and momentum applied

to the isolated system of the photon and the electron yields for the Compton shift the expression

$$\lambda' - \lambda_0 = \frac{h}{m_e c}(1 - \cos\theta) \qquad \text{28.13} \blacktriangleleft$$

where m_e is the mass of the electron, c is the speed of light, and θ is the scattering angle for the photon.

Every object of mass m and momentum p has wave-like properties, with a **de Broglie wavelength** given by the relation

$$\lambda = \frac{h}{p} = \frac{h}{mu} \qquad \text{28.15} \blacktriangleleft$$

The wave–particle duality is the basis of the **quantum particle model.** It can be interpreted by imagining a particle to be made up of a combination of a large number of waves. These waves interfere constructively in a small region of space called a **wave packet.**

The **uncertainty principle** states that if a measurement of position is made with uncertainty Δx and a *simultaneous* measurement of momentum is made with uncertainty Δp_x, the product of the two uncertainties can never be less than $\hbar/2$:

$$\Delta x \, \Delta p_x \geq \frac{\hbar}{2} \qquad \text{28.22} \blacktriangleleft$$

The uncertainty principle is a natural outgrowth of the wave packet model.

Particles are represented by a **wave function** $\psi(x, y, z)$. The **probability density** that a particle will be found at a point is $|\psi|^2$. If the particle is confined to moving along the x axis, the probability that it will be located in an interval dx is given by $|\psi|^2\, dx$. Furthermore, the wave function must be **normalized**:

$$\int_{-\infty}^{\infty} |\psi|^2\, dx = 1 \qquad \textbf{28.28} \blacktriangleleft$$

The measured position x of the particle, averaged over many trials, is called the **expectation value** of x and is defined by

$$\langle x \rangle \equiv \int_{-\infty}^{\infty} \psi^* x \psi\, dx \qquad \textbf{28.30} \blacktriangleleft$$

If a particle of mass m is confined to moving in a one-dimensional box of length L whose walls are perfectly rigid, the allowed wave functions for the particle are

$$\psi_n(x) = A \sin\left(\frac{n\pi x}{L}\right) \qquad \textbf{28.34} \blacktriangleleft$$

where n is an integer quantum number starting at 1. The particle has a well-defined wavelength λ whose values are such that the length L of the box is equal to an integral number of half wavelengths, that is, $L = n\lambda/2$. The energies of a particle in a box are quantized and are given by

$$E_n = \left(\frac{h^2}{8mL^2}\right) n^2 \quad n = 1, 2, 3, \ldots \qquad \textbf{28.35} \blacktriangleleft$$

The wave function must satisfy the Schrödinger equation. The **time-independent Schrödinger equation** for a particle confined to moving along the x axis is

$$-\frac{\hbar^2}{2m}\frac{d^2\psi}{dx^2} + U\psi = E\psi \qquad \textbf{28.36} \blacktriangleleft$$

where E is the total energy of the system and U is the potential energy of the system.

When a particle of energy E meets a barrier of height U, where $E < U$, the particle has a finite probability of penetrating the barrier. This process, called **tunneling,** is the basic mechanism that explains the operation of the scanning tunneling microscope and the phenomenon of alpha decay in some radioactive nuclei.

▶ Analysis Model for Problem Solving

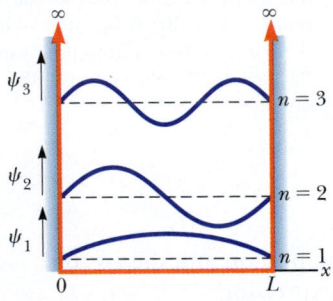

Quantum Particle Under Boundary Conditions. An interaction of a quantum particle with its environment represents one or more boundary conditions. If the interaction restricts the particle to a finite region of space, the energy of the system is quantized. All wave functions must satisfy the following four boundary conditions: (1) $\psi(x)$ must remain finite as x approaches 0, (2) $\psi(x)$ must approach zero as x approaches $\pm\infty$, (3) $\psi(x)$ must be continuous for all values of x, and (4) $d\psi/dx$ must be continuous for all finite values of $U(x)$. If the solution to Equation 28.36 is piecewise, conditions (3) and (4) must be applied at the boundaries between regions of x in which Equation 28.36 has been solved.

▶ OBJECTIVE QUESTIONS

☐ denotes answer available in *Student Solutions Manual/Study Guide*

1. Is each one of the following statements (a) through (e) true or false for an electron? (a) It is a quantum particle, behaving in some experiments like a classical particle and in some experiments like a classical wave. (b) Its rest energy is zero. (c) It carries energy in its motion. (d) It carries momentum in its motion. (e) Its motion is described by a wave function that has a wavelength and satisfies a wave equation.

2. Which of the following phenomena most clearly demonstrates the particle nature of light? (a) diffraction (b) the photoelectric effect (c) polarization (d) interference (e) refraction

3. In a Compton scattering experiment, a photon of energy E is scattered from an electron at rest. After the scattering event occurs, which of the following statements is true? (a) The frequency of the photon is greater than E/h. (b) The energy of the photon is less than E. (c) The wavelength of the photon is less than hc/E. (d) The momentum of the photon increases. (e) None of those statements is true.

4. The probability of finding a certain quantum particle in the section of the x axis between $x = 4$ nm and $x = 7$ nm is 48%. The particle's wave function $\psi(x)$ is constant over this range. What numerical value can be attributed to $\psi(x)$, in units of $\text{nm}^{-1/2}$? (a) 0.48 (b) 0.16 (c) 0.12 (d) 0.69 (e) 0.40

5. A proton, an electron, and a helium nucleus all move at speed v. Rank their de Broglie wavelengths from largest to smallest.

6. Consider (a) an electron (b) a photon, and (c) a proton, all moving in vacuum. Choose all correct answers for each question. **(i)** Which of the three possess rest energy? **(ii)** Which have charge? **(iii)** Which carry energy? **(iv)** Which carry momentum? **(v)** Which move at the speed of light? **(vi)** Which have a wavelength characterizing their motion?

7. In a certain experiment, a filament in an evacuated light-bulb carries a current I_1 and you measure the spectrum of

light emitted by the filament, which behaves as a black body at temperature T_1. The wavelength emitted with highest intensity (symbolized by λ_{max}) has the value λ_1. You then increase the potential difference across the filament by a factor of 8, and the current increases by a factor of 2. **(i)** After this change, what is the new value of the temperature of the filament? (a) $16T_1$ (b) $8T_1$ (c) $4T_1$ (d) $2T_1$ (e) still T_1 **(ii)** What is the new value of the wavelength emitted with highest intensity? (a) $4\lambda_1$ (b) $2\lambda_1$ (c) λ_1 (d) $\frac{1}{2}\lambda_1$ (e) $\frac{1}{4}\lambda_1$

8. What is the de Broglie wavelength of an electron accelerated from rest through a potential difference of 50.0 V? (a) 0.100 nm (b) 0.139 nm (c) 0.174 nm (d) 0.834 nm (e) none of those answers

9. A quantum particle of mass m_1 is in a square well with infinitely high walls and length 3 nm. Rank the situations (a) through (e) according to the particle's energy from highest to lowest, noting any cases of equality. (a) The particle of mass m_1 is in the ground state of the well. (b) The same particle is in the $n = 2$ excited state of the same well. (c) A particle with mass $2m_1$ is in the ground state of the same well. (d) A particle of mass m_1 in the ground state of the same well, and the uncertainty principle has become inoperative; that is, Planck's constant has been reduced to zero. (e) A particle of mass m_1 is in the ground state of a well of length 6 nm.

10. A particle in a rigid box of length L is in the first excited state, for which $n = 2$ (Fig. OQ28.10). Where is the particle most likely to be found? (a) At the center of the box. (b) At either end of the box. (c) All points in the box are equally likely. (d) One-fourth of the way from either end of the box. (e) None of those answers is correct.

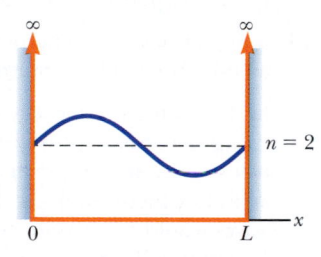

Figure OQ28.10

11. **BIO** Which of the following is most likely to cause sunburn by delivering more energy to individual molecules in skin cells? (a) infrared light (b) visible light (c) ultraviolet light (d) microwaves (e) Choices (a) through (d) are equally likely.

12. An x-ray photon is scattered by an originally stationary electron. Relative to the frequency of the incident photon, is the frequency of the scattered photon (a) lower, (b) higher, or (c) unchanged?

13. A beam of quantum particles with kinetic energy 2.00 eV is reflected from a potential barrier of small width and

original height 3.00 eV. How does the fraction of the particles that are reflected change as the barrier height is reduced to 2.01 eV? (a) It increases. (b) It decreases. (c) It stays constant at zero. (d) It stays constant at 1. (e) It stays constant with some other value.

14. Suppose a tunneling current in an electronic device goes through a potential-energy barrier. The tunneling current is small because the width of the barrier is large and the barrier is high. To increase the current most effectively, what should you do? (a) Reduce the width of the barrier. (b) Reduce the height of the barrier. (c) Either choice (a) or choice (b) is equally effective. (d) Neither choice (a) nor choice (b) increases the current.

15. Figure OQ28.15 represents the wave function for a hypothetical quantum particle in a given region. From the choices a through e, at what value of x is the particle most likely to be found?

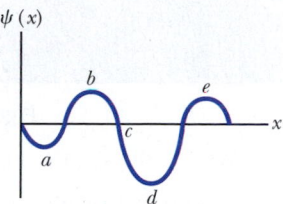

Figure OQ28.15

16. Which of the following statements are true according to the uncertainty principle? More than one statement may be correct. (a) It is impossible to simultaneously determine both the position and the momentum of a particle along the same axis with arbitrary accuracy. (b) It is impossible to simultaneously determine both the energy and momentum of a particle with arbitrary accuracy. (c) It is impossible to determine a particle's energy with arbitrary accuracy in a finite amount of time. (d) It is impossible to measure the position of a particle with arbitrary accuracy in a finite amount of time. (e) It is impossible to simultaneously measure both the energy and position of a particle with arbitrary accuracy.

17. Rank the wavelengths of the following quantum particles from the largest to the smallest. If any have equal wavelengths, display the equality in your ranking. (a) a photon with energy 3 eV (b) an electron with kinetic energy 3 eV (c) a proton with kinetic energy 3 eV (d) a photon with energy 0.3 eV (e) an electron with momentum 3 eV/c

18. Both an electron and a proton are accelerated to the same speed, and the experimental uncertainty in the speed is the same for the two particles. The positions of the two particles are also measured. Is the minimum possible uncertainty in the electron's position (a) less than the minimum possible uncertainty in the proton's position, (b) the same as that for the proton, (c) more than that for the proton, or (d) impossible to tell from the given information?

CONCEPTUAL QUESTIONS

☐ denotes answer available in *Student Solutions Manual/Study Guide*

1. The classical model of blackbody radiation given by the Rayleigh–Jeans law has two major flaws. (a) Identify the flaws and (b) explain how Planck's law deals with them.

2. All objects radiate energy. Why, then, are we not able to see all objects in a dark room?

3. **BIO** *Iridescence* is the phenomenon that gives shining colors to the feathers of peacocks, hummingbirds (see page 780),

resplendent quetzals, and even ducks and grackles. Without pigments, it colors Morpho butterflies (Fig. CQ28.3, page 848), Urania moths, some beetles and flies, rainbow trout, and mother-of-pearl in abalone shells. Iridescent colors change as you turn an object. They are produced by a wide variety of intricate structures in different species. Problem 68 in Chapter 27 describes the structures that produce iridescence in a peacock feather. These structures were all unknown until the invention of the electron microscope. Explain why light microscopes cannot reveal them.

Figure CQ28.3

4. What is the significance of the wave function ψ?

5. Discuss the relationship between ground-state energy and the uncertainty principle.

6. For a quantum particle in a box, the probability density at certain points is zero as seen in Figure CQ28.6. Does this value imply that the particle cannot move across these points? Explain.

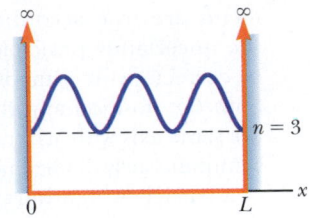

Figure CQ28.6

7. If the photoelectric effect is observed for one metal, can you conclude that the effect will also be observed for another metal under the same conditions? Explain.

8. In the photoelectric effect, explain why the stopping potential depends on the frequency of light but not on the intensity.

9. Why does the existence of a cutoff frequency in the photoelectric effect favor a particle theory for light over a wave theory?

10. In quantum mechanics, it is possible for the energy E of a particle to be less than the potential energy, but classically this condition is not possible. Explain.

11. Consider the wave functions in Figure CQ28.11. Which of them are not physically significant in the interval shown? For those that are not, state why they fail to qualify.

12. Why are the following wave functions not physically possible for all values of x? (a) $\psi(x) = Ae^x$ (b) $\psi(x) = A \tan x$

13. Which has more energy, a photon of ultraviolet radiation or a photon of yellow light? Explain.

14. How does the Compton effect differ from the photoelectric effect?

15. Is an electron a wave or a particle? Support your answer by citing some experimental results.

16. How is the Schrödinger equation useful in describing quantum phenomena?

17. Suppose a photograph were made of a person's face using only a few photons. Would the result be simply a very faint image of the face? Explain your answer.

18. Is light a wave or a particle? Support your answer by citing specific experimental evidence.

19. If matter has a wave nature, why is this wave-like characteristic not observable in our daily experiences?

20. In describing the passage of electrons through a slit and arriving at a screen, physicist Richard Feynman said that "electrons arrive in lumps, like particles, but the probability of arrival of these lumps is determined as the intensity of the waves would be. It is in this sense that the electron behaves sometimes like a particle and sometimes like a wave." Elaborate on this point in your own words. For further discussion, see R. Feynman, *The Character of Physical Law* (Cambridge, MA: MIT Press, 1980), chap. 6.

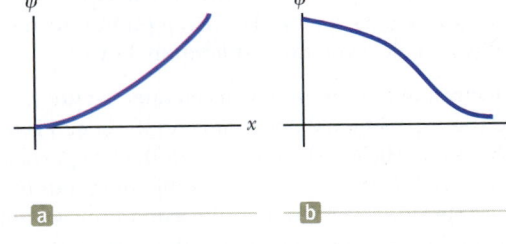
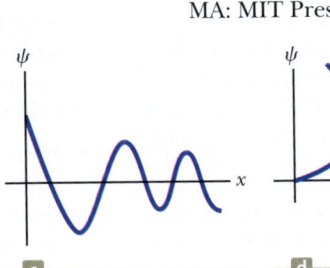
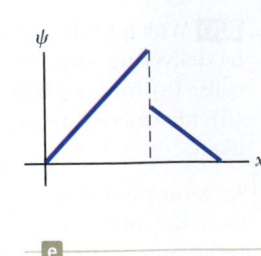

Figure CQ28.11

▶ PROBLEMS AVAILABLE IN WebAssign

28.1 Blackbody Radiation and Planck's Theory

Problems 1–9

28.2 The Photoelectric Effect

Problems 10–15

28.3 The Compton Effect

Problems 16–22

28.4 Photons and Electromagnetic Waves

Problems 23–24

28.5 The Wave Properties of Particles

Problems 25–30

28.6 A New Model: The Quantum Particle

Problems 31–32

28.7 The Double-Slit Experiment Revisited

Problems 33–35

28.8 The Uncertainty Principle

Problems 36–40

28.9 An Interpretation of Quantum Mechanics

Problems 41–42

28.10 A Particle in a Box

Problems 43–50

28.11 Analysis Model: Quantum Particle Under Boundary Conditions

28.12 The Schrödinger Equation

Problems 51–57

28.13 Tunneling Through a Potential Energy Barrier

Problems 58–60

28.14 Context Connection: The Cosmic Temperature

Problems 61–63

Additional Problems

Problems 64–76

Solutions to the following Problems are available in the _Student Solutions Manual/Study Guide_:

28.4, 28.15, 28.21, 28.23, 28.25, 28.28, 28.31, 28.33, 28.37, 28.41, 28.43, 28.47, 28.58, 28.72, and 28.74

List of Enhanced Problems

Problem Number	Targeted Feedback in Enhanced WebAssign	Master It in Enhanced WebAssign	Watch It in Enhanced WebAssign
28.1	✓		✓
28.4		✓	
28.6	✓		✓
28.8		✓	
28.9	✓		✓
28.17	✓		✓
28.23	✓		✓
28.33	✓	✓	
28.34	✓		✓
28.37		✓	
28.41		✓	
28.47		✓	
28.58	✓	✓	
28.59	✓		✓
28.60	✓		✓
28.73	✓		✓

Atomic Physics

Chapter Outline

This fireworks display shows several different colors. The colors are determined by the types of atoms in the material burning in the explosion. Bright white light often comes from oxidizing magnesium or aluminum. Red light often comes from strontium and yellow from sodium. Blue light is more difficult to achieve, but can be obtained by burning a mixture of copper powder, copper chloride, and hexachloroethane. The emission of light from atoms is an important clue that allows us to learn about the structure of the atom.

Deymos/Shutterstock.com

In Chapter 28, we introduced some of the basic concepts and techniques used in quantum physics along with their applications to various simple systems. This chapter describes the application of quantum physics to more sophisticated structural models of atoms than we have seen previously.

We studied the hydrogen atom in Chapter 11 using Bohr's semiclassical approach. In this chapter, we shall analyze the hydrogen atom with a full quantum model. Although the hydrogen atom is the simplest atomic system, it is an especially important system to understand, for several reasons:

- Much of what we learn about the hydrogen atom, with its single electron, can be extended to such single-electron ions as He^+ and Li^{2+}.
- The hydrogen atom is an ideal system for performing precise tests of theory against experiment and for improving our overall understanding of atomic structure.
- The quantum numbers used to characterize the allowed states of hydrogen can be used to qualitatively describe the allowed states of more complex atoms. This characterization enables us to understand the periodic table of the elements, which is one of the greatest triumphs of quantum physics.
- The basic ideas about atomic structure must be well understood before we attempt to deal with the complexities of molecular structures and the electronic structures of solids.

29.1 | Early Structural Models of the Atom

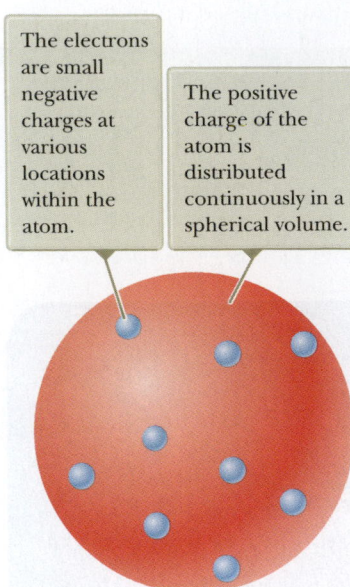

The electrons are small negative charges at various locations within the atom.

The positive charge of the atom is distributed continuously in a spherical volume.

Figure 29.1 Thomson's model of the atom.

The structural model of the atom in Newton's day described the atom as a tiny, hard, indestructible sphere, a particle model that ignored any internal structure of the atom. Although this model was a good basis for the kinetic theory of gases (Chapter 16), new structural models had to be devised when later experiments revealed the electrical nature of atoms. J. J. Thomson suggested a structural model that describes the atom as a continuous volume of positive charge with electrons embedded throughout it (Fig. 29.1).

In 1911, Ernest Rutherford and his students Hans Geiger and Ernst Marsden performed a critical experiment whose results were inconsistent with those predicted by Thomson's model. In this experiment, a beam of positively charged alpha particles was projected into a thin metal foil as in Figure 29.2a. Most of the particles passed through the foil as if it were empty space, which is consistent with the Thomson model. Some of the results of the experiment, however, were astounding. Many alpha particles were deflected from their original direction of travel through large angles. Some particles were even deflected backward, reversing their direction of travel. When Geiger informed Rutherford of these results, Rutherford wrote, "It was quite the most incredible event that has ever happened to me in my life. It was almost as incredible as if you fired a 15-inch [artillery] shell at a piece of tissue paper and it came back and hit you."

Such large deflections were not expected on the basis of Thomson's model. According to this model, a positively charged alpha particle would never come close enough to a sufficiently large concentration of positive charge to cause any large-angle deflections. Furthermore, the electrons in Thomson's model have far too little mass to cause such a large deflection of the massive alpha particles. Rutherford explained his astounding results with a new structural model, as introduced in Section 11.5: he assumed that the positive charge was concentrated in a region that was small relative to the size of the atom. He called this concentration of positive charge the **nucleus** of the atom. Any electrons belonging to the atom were assumed to be outside the nucleus. To explain why these electrons were not pulled into the nucleus by the attractive electric force, Rutherford imagined that the electrons move in orbits about the nucleus in the same manner as the planets orbit the Sun, as in Figure 29.2b.

There are two basic difficulties with Rutherford's planetary structural model. As we saw in Chapter 11, an atom emits discrete characteristic frequencies of electromagnetic radiation and no others; the Rutherford model is unable to explain this phenomenon. A second difficulty is that Rutherford's electrons experience a centripetal acceleration. According to Maxwell's equations in electromagnetism, charges orbiting with frequency f experience centripetal acceleration and therefore should radiate electromagnetic waves of frequency f. Unfortunately, this classical model leads to disaster when applied to the atom. As the electron radiates energy from the electron–proton system, the radius of the orbit of the electron steadily decreases and its frequency of revolution increases. Energy is continuously transferred out of the system by electromagnetic radiation. As a result, the energy of the system decreases, resulting in the decay of the orbit of the electron. This decrease in total

Figure 29.2 (a) Rutherford's technique for observing the scattering of alpha particles from a thin foil target. The source is a naturally occurring radioactive substance, such as radium. (b) Rutherford's planetary model of the atom.

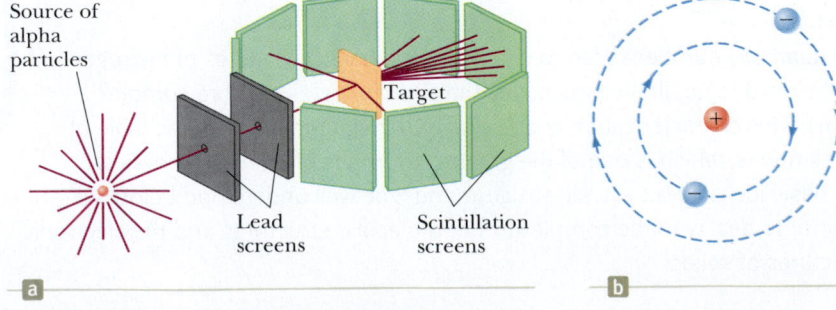

Source of alpha particles

Target

Lead screens

Scintillation screens

a

b

energy leads to an increase in the kinetic energy of the electron,[1] an ever-increasing frequency of emitted radiation, and a rapid collapse of the atom as the electron plunges into the nucleus (Fig. 29.3).

The stage was set for Bohr and his model that we discussed in Chapter 11. To circumvent the unsuccessful predictions of the Rutherford model—electrons falling into the nucleus and a continuous emission spectrum from elements—Bohr postulated that classical radiation theory does not hold for atomic-sized systems. He overcame the problem of an atom that continuously loses energy by applying Planck's ideas of quantized energy levels to orbiting atomic electrons. Therefore, as described in Section 11.5, Bohr postulated that electrons in atoms are generally confined to stable, nonradiating orbits called stationary states. Furthermore, he applied Einstein's concept of the photon to arrive at an expression for the frequency of radiation emitted when the atom makes a transition from one stationary state to another.

While the Bohr model was more successful than Rutherford's model in terms of predicting stable atoms and wavelengths of emitted radiation, it failed to predict more subtle spectral details, as mentioned in Pitfall Prevention 11.4. One of the first indications that the Bohr theory needed modification arose when improved spectroscopic techniques were used to examine the spectral lines of hydrogen. It was found that many of the lines in the Balmer and other series were not single lines at all. Instead, each was a group of closely spaced lines. An additional difficulty arose when it was observed that, in some situations, some single spectral lines were split into three closely spaced lines when the atoms were placed in a strong magnetic field. The Bohr model cannot explain this phenomenon.

Efforts to explain these difficulties with the Bohr model led to improvements in the structural model of the atom. One of the changes introduced was the concept that the electron has an intrinsic angular momentum called *spin,* which we introduced in Chapter 22 in terms of the contribution of spin to the magnetic properties of materials. We shall discuss spin in more detail in this chapter.

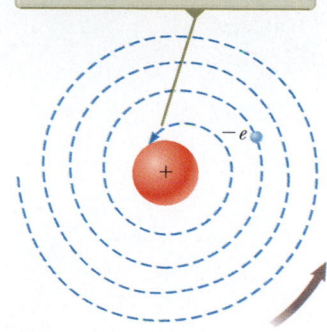

Because the accelerating electron radiates energy, the size of the orbit decreases until the electron falls into the nucleus.

Figure 29.3 The classical model of the nuclear atom predicts that the atom decays.

29.2 | The Hydrogen Atom Revisited

A quantum treatment of the hydrogen atom requires a solution to the Schrödinger equation (Eq. 28.36), with U being the electric potential energy of the electron–proton system. The full mathematical solution of the Schrödinger equation as applied to the hydrogen atom gives a complete and beautiful description of the atom's properties. The mathematical procedures that make up the solution are beyond the scope of this text, however, and so the details shall be omitted. The solutions for some states of hydrogen will be discussed, together with the quantum numbers used to characterize allowed stationary states. We also discuss the physical significance of the quantum numbers.

Let us outline the components of a quantum structural model for the hydrogen atom:

1. *A description of the physical components of the system:* As in the earlier hydrogen atom models, the physical components are the electron and a positive charge, which we model as being concentrated in a very small nucleus. We also model the electron as a localized concentration of charge, but we will find that our results suggest that this assumption must be relaxed.
2. *A description of where the components are located relative to one another and how they interact:* Because of the small size of the atom, we assume that the electron and the nucleus are close together. We also assume that they interact via the electric force. We do *not* assume any sort of orbit for the electron.
3. *A description of the time evolution of the system:* We wish to understand the details of the stable hydrogen atom. We also wish to understand the process that occurs

---

[1] As an orbital system that interacts via an inverse square force law loses energy, the kinetic energy of the orbiting object increases but the potential energy of the system decreases by a larger amount; therefore, the change in the total energy of the system is negative.

in a time interval including an emission or absorption of energy in the form of electromagnetic radiation.

4. *A description of the agreement between predictions of the model and actual observations and, possibly, predictions of new effects that have not yet been observed:* Our model should predict the quantized energies and wavelengths of spectral lines that we have already discussed. In addition, we would like our model to make accurate predictions of subtle details in atomic spectra, the structure of the periodic table, details of x-ray spectra, etc.

Our predictions from the quantum model will be made as follows. We will set up the Schrödinger equation for a system whose physical components are described in 1 and 2 above. We then solve the equation for the *general* wave functions satisfying the equation. Finally, we apply the quantum particle under boundary conditions model by imposing boundary conditions on the general wave functions to determine the *specific* allowed wave functions and energies of the atom.

For the particle in a one-dimensional box in Section 28.10, we found that the imposition of boundary conditions generated a single quantum number. For the three-dimensional system of the hydrogen atom, the application of boundary conditions in each dimension introduces a quantum number, so the model will generate three quantum numbers. We also find the need for a fourth quantum number, representing the spin, that cannot be extracted from the Schrödinger equation.

To set up the Schrödinger equation, we must first specify the potential energy function for the system. For the hydrogen atom, this function is

$$U(r) = -k_e \frac{e^2}{r}$$

$29.1\blacktriangleleft$

where k_e is the Coulomb constant and r is the radial distance between the proton (situated at $r = 0$) and the electron.

The formal procedure for solving the problem of the hydrogen atom is to substitute $U(r)$ into the Schrödinger equation and find appropriate solutions to the equation. We did that for the particle in a box in Section 28.12. The current problem is more complicated, however, because it is three dimensional and because U is not constant. In addition, U depends on the radial coordinate r rather than a Cartesian coordinate x, y, or z. As a result, we must use spherical coordinates. We shall not attempt to carry out these solutions because they are quite complicated. Rather, we shall simply describe their properties and some of their implications with regard to atomic structure.

When the boundary conditions are applied to the solutions of the Schrödinger equation, we find that the energies of the allowed states for the hydrogen atom are

▶ Allowed energies for the hydrogen atom

$$E_n = -\left(\frac{k_e e^2}{2a_0}\right)\frac{1}{n^2} = -\frac{13.606 \text{ eV}}{n^2} \qquad n = 1, 2, 3, \ldots$$

$29.2\blacktriangleleft$

where a_0 is the Bohr radius. This result is in precise agreement with the Bohr model and with observed spectral lines! This agreement is *remarkable* because the Bohr theory and the full quantum theory arrive at the same result from completely different starting points.

Note that the allowed energies in our model depend only on the quantum number n, called the **principal quantum number.** The imposition of boundary conditions also leads to two new quantum numbers that do not appear in the Bohr model. The quantum number ℓ is called the **orbital quantum number,** and m_ℓ is called the **orbital magnetic quantum number.** Whereas n is related to the energy of the atom, the quantum numbers ℓ and m_ℓ are related to the angular momentum of the atom as described in Section 29.4. From the solution to the Schrödinger equation, we find the following allowed values for these three quantum numbers:

- n is an integer that can range from 1 to ∞.

For a particular value of n,

- ℓ is an integer that can range from 0 to $n - 1$.

Pitfall Prevention | 29.1
Energy Depends on *n* Only for Hydrogen
The implication in Equation 29.2 that the energy depends only on the quantum number *n* is true only for the hydrogen atom. For more complicated atoms, we will use the same quantum numbers developed here for hydrogen. The energy levels for these atoms depend primarily on *n*, but they also depend to a lesser degree on other quantum numbers.

TABLE 29.1 | Three Quantum Numbers for the Hydrogen Atom

Quantum Number	Name	Allowed Values	Number of Allowed States
n	Principal quantum number	$1, 2, 3, \ldots$	Any number
ℓ	Orbital quantum number	$0, 1, 2, \ldots, n-1$	n
m_ℓ	Orbital magnetic quantum number	$-\ell, -\ell+1, \ldots,$ $0, \ldots, \ell-1, \ell$	$2\ell + 1$

For a particular value of ℓ,

- m_ℓ is an integer that can range from $-\ell$ to ℓ.

Table 29.1 summarizes the rules for determining the allowed values of ℓ and m_ℓ for a given value of n.

For historical reasons, all states with the same principal quantum number are said to form a **shell**. Shells are identified by the letters K, L, M, . . . , which designate the states for which $n = 1, 2, 3, \ldots$. Likewise, all states with given values of n and ℓ are said to form a **subshell**. Based on early practices in spectroscopy, the letters[2] s, p, d, f, g, h, . . . are used to designate the subshells for which $\ell = 0, 1, 2, 3, 4, 5, \ldots$. For example, the subshell designated by $3p$ has the quantum numbers $n = 3$ and $\ell = 1$; the $2s$ subshell has the quantum numbers $n = 2$ and $\ell = 0$. These notations are summarized in Tables 29.2 and 29.3.

States with quantum numbers that violate the rules given in Table 29.1 cannot exist because they do not satisfy the boundary conditions on the wave function of the system. For instance, a $2d$ state, which would have $n = 2$ and $\ell = 2$, cannot exist; the highest allowed value of ℓ is $n - 1$, or 1 in this case. Therefore, for $n = 2$, $2s$ and $2p$ are allowed states but $2d$, $2f$, . . . are not. For $n = 3$, the allowed subshells are $3s$, $3p$, and $3d$.

TABLE 29.2 |
Atomic Shell Notations

n	Shell Symbol
1	K
2	L
3	M
4	N
5	O
6	P

TABLE 29.3 |
Atomic Subshell Notations

ℓ	Subshell Symbol
0	s
1	p
2	d
3	f
4	g
5	h

QUICK QUIZ 29.1 How many possible subshells are there for the $n = 4$ level of hydrogen? (a) 5 (b) 4 (c) 3 (d) 2 (e) 1

QUICK QUIZ 29.2 When the principal quantum number is $n = 5$, how many different values of (a) ℓ and (b) m_ℓ are possible?

Example 29.1 | The $n = 2$ Level of Hydrogen

For a hydrogen atom, determine the allowed states corresponding to the principal quantum number $n = 2$ and calculate the energies of these states.

SOLUTION

Conceptualize Think about the atom in the $n = 2$ quantum state. There is only one such state in the Bohr theory, but our discussion of the quantum theory allows for more states because of the possible values of ℓ and m_ℓ.

Categorize We evaluate the results using rules discussed in this section, so we categorize this example as a substitution problem.

From Table 29.1, we find that when $n = 2$, ℓ can be 0 or 1. Find the possible values of m_ℓ from Table 29.1:

$$\ell = 0 \quad \rightarrow \quad m_\ell = 0$$
$$\ell = 1 \quad \rightarrow \quad m_\ell = -1, 0, \text{ or } 1$$

Hence, we have one state, designated as the $2s$ state, that is associated with the quantum numbers $n = 2$, $\ell = 0$, and $m_\ell = 0$, and we have three states, designated as $2p$ states, for which the quantum numbers are $n = 2$, $\ell = 1$, and $m_\ell = -1$; $n = 2$, $\ell = 1$, and $m_\ell = 0$; and $n = 2$, $\ell = 1$, and $m_\ell = 1$.

Find the energy for all four of these states with $n = 2$ from Equation 29.2:

$$E_2 = -\frac{13.606 \text{ eV}}{2^2} = \boxed{-3.401 \text{ eV}}$$

---

[2]These seemingly strange letter designations come from descriptions of spectral lines in the early history of spectroscopy: s—sharp; p—principal; d—diffuse; f—fine. After s, p, d, and f, the subsequent letters follow alphabetically from f.

29.3 | The Wave Functions for Hydrogen

The potential energy of the hydrogen atom depends only on the radial distance r between nucleus and electron. We therefore expect that some of the allowed states for this atom can be represented by wave functions that depend only on r, which indeed is the case. (Other wave functions depend on r and on the angular coordinates.) The simplest wave function for the hydrogen atom describes the 1s state and is designated $\psi_{1s}(r)$:

▶ Wave function for hydrogen in its ground state

$$\psi_{1s}(r) = \frac{1}{\sqrt{\pi a_0^3}}\, e^{-r/a_0} \qquad \text{29.3}◀$$

where a_0 is the Bohr radius and the wave function as given is normalized. This wave function satisfies the boundary conditions mentioned in Section 28.11; that is, ψ_{1s} approaches zero as $r \to \infty$ and remains finite as $r \to 0$. Because ψ_{1s} depends only on r, it is spherically symmetric. In fact, all s states have spherical symmetry.

Recall that the probability of finding the electron in any region is equal to an integral of the probability density $|\psi|^2$ over the region, if ψ is normalized. The probability density for the 1s state is

$$|\psi_{1s}|^2 = \left(\frac{1}{\pi a_0^3}\right) e^{-2r/a_0} \qquad \text{29.4}◀$$

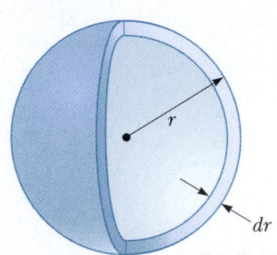

Figure 29.4 A spherical shell of radius r and thickness dr has a volume equal to $4\pi r^2\, dr$.

The probability of finding the electron in a volume element dV is $|\psi|^2\, dV$. It is convenient to define the **radial probability density function** $P(r)$ as the probability per unit radial distance of finding the electron in a spherical shell of radius r and thickness dr. The volume of such a shell equals its surface area $4\pi r^2$ multiplied by the shell thickness dr (Fig. 29.4), so that

$$P(r)\, dr = |\psi|^2\, dV = |\psi|^2 4\pi r^2\, dr \qquad \text{29.5}◀$$

$$P(r) = 4\pi r^2 |\psi|^2 \qquad \text{29.6}◀$$

Substituting Equation 29.4 into Equation 29.6 gives the radial probability density function for the hydrogen atom in its ground state:

▶ Radial probability density for the 1s state of hydrogen

$$P_{1s}(r) = \left(\frac{4r^2}{a_0^3}\right) e^{-2r/a_0} \qquad \text{29.7}◀$$

A graphical representation of the function $P_{1s}(r)$ versus r is presented in Figure 29.5a. The peak of the curve corresponds to the most probable value of r for this particular state. The spherical symmetry of the distribution function is shown in Figure 29.5b.

In Example 29.2, we show that the most probable value of r for the ground state of hydrogen equals the Bohr radius a_0. This is another *remarkable* agreement between the Bohr model and the quantum model. According to quantum mechanics, the atom has no sharply defined boundary. The probability distribution in Figure 29.5a suggests that the charge of the electron is not localized as in a particle model. Rather, the charge is extended throughout a diffuse region of space, commonly referred to as an **electron cloud.** This non-localization of the electron should not be a surprise to us, based on the wave packet concept in the quantum particle model as well as the predictions of the uncertainty principle. The electron cloud model is quite different from the Bohr model, which places the electron at a fixed distance from the nucleus. Figure 29.5b shows the probability density of the electron in a hydrogen atom in the 1s state as a function of position in the xy plane. The darkest portion of the distribution appears at $r = a_0$, corresponding to the most probable value of r for the electron.

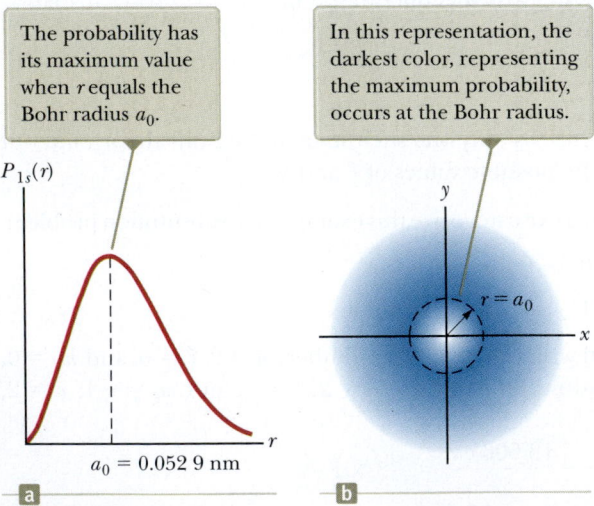

The probability has its maximum value when r equals the Bohr radius a_0.

In this representation, the darkest color, representing the maximum probability, occurs at the Bohr radius.

Figure 29.5 (a) The probability density of finding the electron as a function of distance from the nucleus for the hydrogen atom in the 1s (ground) state. (b) The cross section in the xy plane of the spherical electronic charge distribution for the hydrogen atom in its 1s state.

Let us now address the time evolution of the system, that is, the third component in our structural model of the atom. For an atom in a quantum state that is a solution to the Schrödinger equation, the electron cloud structure remains the same, on the average, over time. Therefore, *the atom does not radiate when it is in one particular quantum state*. This fact removes the problem that plagued the Rutherford model, in which the atom continuously radiates until the electron spirals into the nucleus. Because no change occurs in the charge structure in the electron cloud, the atom does not radiate. Radiation occurs only when a transition is made, so the structure of the electron cloud changes in time.

The next simplest wave function for the hydrogen atom is the one corresponding to the 2s state ($n = 2$, $\ell = 0$). The normalized wave function for this state is

$$\psi_{2s}(r) = \frac{1}{4\sqrt{2\pi}}\left(\frac{1}{a_0}\right)^{3/2}\left[2 - \frac{r}{a_0}\right]e^{-r/2a_0} \qquad \textbf{29.8} \blacktriangleleft$$

Like the ψ_{1s} function, ψ_{2s} depends only on r and is spherically symmetric. The energy corresponding to this state is $E_2 = -(13.6\ \text{eV}/4) = -3.4\ \text{eV}$. This energy level represents the first excited state of hydrogen.

A plot of the radial probability density function for this state in comparison to the 1s state is shown in Active Figure 29.6. The plot for the 2s state has two peaks. In this case, the most probable value corresponds to that value of r that corresponds to the highest value of P_{2s}, which is at $r \approx 5a_0$. An electron in the 2s state would be much farther from the nucleus (on the average) than an electron in the 1s state.

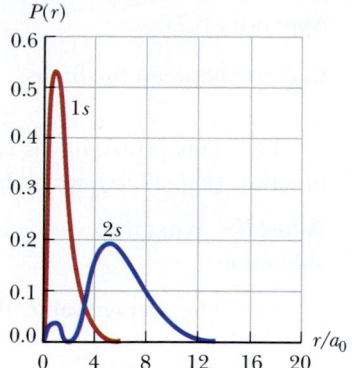

Active Figure 29.6 The radial probability density function versus r/a_0 for the 1s and 2s states of the hydrogen atom.

Example 29.2 | The Ground State of Hydrogen

(A) Calculate the most probable value of r for an electron in the ground state of the hydrogen atom.

SOLUTION

Conceptualize Do not imagine the electron in orbit around the proton as in the Bohr theory of the hydrogen atom. Instead, imagine the charge of the electron spread out in space around the proton in an electron cloud with spherical symmetry.

Categorize Because the statement of the problem asks for the "most probable value of r," we categorize this example as a problem in which the quantum approach is used. (In the Bohr atom, the electron moves in an orbit with an *exact* value of r.)

Analyze The most probable value of r corresponds to the maximum in the plot of $P_{1s}(r)$ versus r. We can evaluate the most probable value of r by setting $dP_{1s}/dr = 0$ and solving for r.

Differentiate Equation 29.7 and set the result equal to zero:

$$\frac{dP_{1s}}{dr} = \frac{d}{dr}\left[\left(\frac{4r^2}{a_0^3}\right)e^{-2r/a_0}\right] = 0$$

$$e^{-2r/a_0}\frac{d}{dr}(r^2) + r^2\frac{d}{dr}(e^{-2r/a_0}) = 0$$

$$2re^{-2r/a_0} + r^2(-2/a_0)e^{-2r/a_0} = 0$$

$$(1)\quad 2r[1 - (r/a_0)]e^{-2r/a_0} = 0$$

Set the bracketed expression equal to zero and solve for r:

$$1 - \frac{r}{a_0} = 0 \qquad \rightarrow \qquad r = a_0$$

Finalize The most probable value of r is the Bohr radius! Equation (1) is also satisfied at $r = 0$ and as $r \rightarrow \infty$. These points are locations of the *minimum* probability, which is equal to zero as seen in Figure 29.5a.

(B) Calculate the probability that the electron in the ground state of hydrogen will be found outside the Bohr radius.

SOLUTION

Analyze The probability is found by integrating the radial probability density function $P_{1s}(r)$ for this state from the Bohr radius a_0 to ∞.

Set up this integral using Equation 29.7:

$$P = \int_{a_0}^{\infty} P_{1s}(r)\ dr = \frac{4}{a_0^3}\int_{a_0}^{\infty} r^2e^{-2r/a_0}\ dr$$

continued

29.2 *cont.*

Put the integral in dimensionless form by changing variables from r to $z = 2r/a_0$, noting that $z = 2$ when $r = a_0$ and that $dr = (a_0/2)\, dz$:

$$P = \frac{4}{a_0^3} \int_2^\infty \left(\frac{za_0}{2}\right)^2 e^{-z}\left(\frac{a_0}{2}\right)\, dz = \frac{1}{2}\int_2^\infty z^2 e^{-z}\, dz$$

Evaluate the integral using partial integration (see Appendix B.7):

$$P = -\tfrac{1}{2}(z^2 + 2z + 2)\,e^{-z}\Big|_2^\infty$$

Evaluate between the limits:

$$P = 0 - [-\tfrac{1}{2}(4 + 4 + 2)\,e^{-2}] = 5e^{-2} = 0.677 \text{ or } 67.7\%$$

Finalize This probability is larger than 50%. The reason for this value is the asymmetry in the radial probability density function (Fig. 29.5a), which has more area to the right of the peak than to the left.

What If? What if you were asked for the *average* value of r for the electron in the ground state rather than the most probable value?

Answer The average value of r is the same as the expectation value for r.

Use Equation 29.7 to evaluate the average value of r:

$$r_{avg} = \langle r \rangle = \int_0^\infty rP(r)\, dr = \int_0^\infty r\left(\frac{4r^2}{a_0^3}\right)e^{-2r/a_0}\, dr$$

$$= \left(\frac{4}{a_0^3}\right)\int_0^\infty r^3 e^{-2r/a_0}\, dr$$

Evaluate the integral with the help of the first integral listed in Table B.6 in Appendix B:

$$r_{avg} = \left(\frac{4}{a_0^3}\right)\left(\frac{3!}{(2/a_0)^4}\right) = \tfrac{3}{2}a_0$$

Again, the average value is larger than the most probable value because of the asymmetry in the wave function as seen in Figure 29.5a.

Example 29.3 | The Quantized Solar System

Consider the Schrödinger equation for a system of two particles interacting via the gravitational force: the Earth and the Sun. What is the quantum number of the system with the Earth in its present orbit?

SOLUTION

Conceptualize Imagine the Earth as the electron and the Sun as the nucleus in a huge atom-like structure.

Categorize Despite the fact that we do not need to use quantum physics to describe the macroscopic motion of objects of planetary size, the statement of the problem suggests that we categorize this problem by using a quantum approach.

Analyze The potential energy function for the system is

$$U(r) = -G\frac{M_E M_S}{r}$$

where M_E is the mass of the Earth and M_S is the mass of the Sun. Comparing this expression with Equation 29.1 for the hydrogen atom, we see that it has the same mathematical form and that the constant $GM_E M_S$ in the expression above plays the role of $k_e e^2$ in Equation 29.1. Therefore, the solutions to the Schrödinger equation for the Earth–Sun system will be the *same* as those of the hydrogen atom with the appropriate change in the constants.

Make the substitution for the constants in Equation 29.2 to find the allowed energies of the quantized states of the Earth–Sun system:

$$E_n = -\left(\frac{k_e e^2}{2a_0}\right)\frac{1}{n^2} \quad \rightarrow \quad E_n = -\left(\frac{GM_E M_S}{2a_0}\right)\frac{1}{n^2}$$

Solve this equation for the quantum number n:

$$(1)\quad n = \sqrt{-\left(\frac{GM_E M_S}{2a_0}\right)\frac{1}{E_n}}$$

From Equation 11.23, make the substitution of constants and find the Bohr radius for the Earth–Sun system:

$$(2)\quad a_0 = \frac{\hbar^2}{m_e k_e e^2} \quad \rightarrow \quad a_0 = \frac{\hbar^2}{M_E(GM_E M_S)} = \frac{\hbar^2}{GM_E^2 M_S}$$

Evaluate the energy of the Earth–Sun system from Equation 11.10, assuming a circular orbit corresponding to quantum number n:

$$(3)\quad E_n = -\frac{GM_E M_S}{2r_n}$$

29.3 *cont.*

Substitute Equations (2) and (3) into Equation (1):

$$n = \sqrt{-\left(\frac{GM_E M_S}{2a_0}\right)\frac{1}{E_n}} = \sqrt{-\left(\frac{GM_E M_S}{2}\right)\left(\frac{GM_E^2 M_S}{\hbar^2}\right)\left(-\frac{2r_n}{GM_E M_S}\right)} = \sqrt{\frac{GM_E^2 M_S r_n}{\hbar^2}}$$

Substitute numerical values:

$$n = \sqrt{\frac{(6.67 \times 10^{-11}\,\text{N}\cdot\text{m}^2/\text{kg}^2)(5.97 \times 10^{24}\,\text{kg})^2(1.99 \times 10^{30}\,\text{kg})(1.50 \times 10^{11}\,\text{m})}{(1.055 \times 10^{-34}\,\text{J}\cdot\text{s})^2}}$$

$$= 2.52 \times 10^{74}$$

Finalize This result is a tremendously large quantum number. Therefore, according to the correspondence principle, classical mechanics describes the Earth's motion as well as quantum mechanics does. The energies of quantum states for adjacent values of n are so close together that we do not see the quantized nature of the energy. For example, if the Earth were to move into the next higher quantum state, calculations show that it would be farther from the Sun by a distance on the order of 10^{-63} m. Even on a nuclear scale of 10^{-15} m, such a small distance is undetectable.

29.4 | Physical Interpretation of the Quantum Numbers

As discussed in Section 29.2, the energy of a particular state in our model depends on the principal quantum number. Now let us see what the other three quantum numbers contribute to the physical nature of our quantum structural model of the atom.

The Orbital Quantum Number ℓ

If a particle moves in a circle of radius r, the magnitude of its angular momentum relative to the center of the circle is $L = mvr$. The direction of $\vec{L}$ is perpendicular to the plane of the circle, and the sense of $\vec{L}$ is given by a right-hand rule.[3] According to classical physics, L can have any value. The Bohr model of hydrogen, however, postulates that the angular momentum is restricted to integer multiples of $\hbar$; that is, $mvr = n\hbar$. This model must be modified because it predicts (incorrectly) that the ground state of hydrogen ($n = 1$) has one unit of angular momentum. Our quantum model shows that the lowest value of the orbital quantum number, which is related to the orbital momentum, is $\ell = 0$, which corresponds to zero angular momentum.

According to the quantum model, an atom in a state whose principal quantum number is n can take on the following *discrete* values for the magnitude of the **orbital angular momentum** vector:[4]

$$L = \sqrt{\ell(\ell + 1)}\,\hbar \qquad \ell = 0, 1, 2, \ldots, n - 1 \qquad \text{29.9}$$ ◀ ▶ Allowed values of L

That L can be zero in this model points out the difficulties inherent in any attempt to describe results based on quantum mechanics in terms of a purely particle-like model. We cannot think in terms of electrons traveling in well-defined orbits of circular shape or any other shape, for that matter. It is more consistent with the probability notions of quantum physics to imagine the electron smeared out in space in an electron cloud, with the "density" of the cloud highest where the probability is highest. In the quantum mechanical interpretation, the electron cloud for the $L = 0$ state is spherically symmetric and has no fundamental axis of rotation.

The Magnetic Orbital Quantum Number m_ℓ

Because angular momentum is a vector, its direction must be specified. Recall from Chapter 22 that a current loop has a corresponding magnetic moment $\vec{\mu} = I\vec{A}$

---

[3]See Sections 10.9 and 10.10 for a review of this material on angular momentum.

[4]Equation 29.9 is a direct result of the mathematical solution of the Schrödinger equation and the application of angular boundary conditions. This development, however, is beyond the scope of this text and will not be presented.

(Eq. 22.15), where I is the current in the loop and $\vec{A}$ is a vector perpendicular to the loop whose magnitude is the area of the loop. Such a moment placed in a magnetic field $\vec{B}$ interacts with the field. Suppose a weak magnetic field applied along the z axis defines a direction in space. According to classical physics, the energy of the loop–field system depends on the direction of the magnetic moment of the loop with respect to the magnetic field as described by Equation 22.17, $U = -\vec{\mu} \cdot \vec{B}$. Any energy between $-\mu B$ and $+\mu B$ is allowed by classical physics.

In the Bohr theory, the circulating electron represents a current loop. In the quantum-mechanical approach to the hydrogen atom, we abandon the circular orbit viewpoint of the Bohr theory, but the atom still possesses an orbital angular momentum. Therefore, there is some sense of rotation of the electron around the nucleus and a magnetic moment is present due to this angular momentum.

As mentioned in Section 29.1, spectral lines from some atoms are observed to split into groups of three closely spaced lines when the atoms are placed in a magnetic field. Suppose the hydrogen atom is located in a magnetic field. According to quantum mechanics, there are *discrete* directions allowed for the magnetic moment vector $\vec{\mu}$ with respect to the magnetic field vector $\vec{B}$. This situation is very different from that in classical physics, in which all directions are allowed.

Because the magnetic moment $\vec{\mu}$ of the atom can be related to the angular momentum vector $\vec{L}$, the discrete directions of $\vec{\mu}$ translate to the direction of $\vec{L}$ being quantized. This quantization means that L_z (the projection of $\vec{L}$ along the z axis) can have only discrete values. The orbital magnetic quantum number m_ℓ specifies the allowed values of the z component of the orbital angular momentum according to the expression

► Allowed values of L_z

$$L_z = m_\ell \hbar$$

29.10◄

The quantization of the possible orientations of $\vec{L}$ with respect to an external magnetic field is often referred to as **space quantization.**

Let us look at the possible orientations of $\vec{L}$ for a given value of ℓ. Recall that m_ℓ can have values ranging from $-\ell$ to ℓ. If $\ell = 0$, then $L = 0$ and there is no vector for which to consider a direction. If $\ell = 1$, then the possible values of m_ℓ are $-1, 0$, and 1, so L_z may be $-\hbar, 0$, or $\hbar$. If $\ell = 2$, m_ℓ can be $-2, -1, 0, 1$, or 2, corresponding to L_z values of $-2\hbar, -\hbar, 0, \hbar$, or $2\hbar$, and so on.

A useful specialized pictorial representation for understanding space quantization is commonly called a **vector model.** A vector model for $\ell = 2$ is shown in Figure 29.7a. Note that $\vec{L}$ can never be aligned parallel or antiparallel to the z axis because L_z must be smaller than the magnitude of the angular momentum $\vec{L}$. The vector $\vec{L}$ can be *perpendicular* to the z axis, which is the case if $m_\ell = 0$. From a three-dimensional viewpoint, $\vec{L}$ can lie on the surfaces of cones that make angles θ with the z axis as shown in Figure 29.7b. From the figure, we see that θ is also quantized and that its values are specified through a relation based on a geometric model triangle with the $\vec{L}$ vector as the hypotenuse and the z component as one leg of the triangle:

$$\cos \theta = \frac{L_z}{L} = \frac{m_\ell}{\sqrt{\ell(\ell + 1)}}$$

29.11◄

Note that m_ℓ is <u>never greater</u> than ℓ, so m_ℓ is always smaller than $\sqrt{\ell(\ell + 1)}$ and therefore θ can never be zero, consistent with our restriction on $\vec{L}$ not being parallel to the z axis.

Because of the uncertainty principle, $\vec{L}$ does not point in a specific direction but rather lies somewhere on a cone as mentioned above. If $\vec{L}$ had a definite direction, all three components L_x, L_y, and

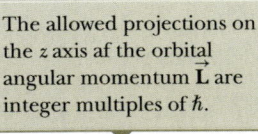

The allowed projections on the z axis of the orbital angular momentum $\vec{L}$ are integer multiples of $\hbar$.

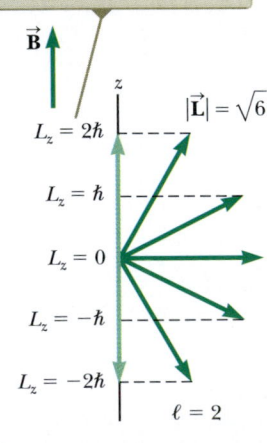

Because the x and y components of the orbital angular momentum vector are not quantized, the vector $\vec{L}$ lies on the surface of a cone.

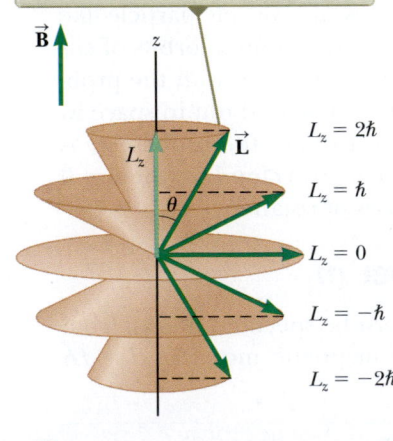

Figure 29.7 A vector model for $\ell = 2$.

L_z would be exactly specified. For the moment, let us assume this case to be true and let us suppose the electron moves in the xy plane, so the uncertainty $\Delta z = 0$. Because the electron moves in the xy plane, $p_z = 0$. Therefore, p_z is *precisely* known, so $\Delta p_z = 0$. The product of these two uncertainties is $\Delta z\, \Delta p_z = 0$, but that is in violation of the uncertainty principle, which requires that $\Delta z\, \Delta p_z \geq \hbar/2$. In reality, only the magnitude of $\vec{L}$ and one component (which is traditionally chosen as L_z) can have definite values at the same time. In other words, quantum mechanics allows us to specify L and L_z but not L_x and L_y. Because the direction of $\vec{L}$ is constantly changing, the average values of L_x and L_y are zero and L_z maintains a fixed value $m_\ell \hbar$.

> **QUICK QUIZ 29.3** Sketch a vector model (shown in Fig. 29.7a for $\ell = 2$) for $\ell = 1$.

Example 29.4 | Space Quantization for Hydrogen

Consider the hydrogen atom in the $\ell = 3$ state. Calculate the magnitude of $\vec{L}$, the allowed values of L_z, and the corresponding angles θ that $\vec{L}$ makes with the z axis.

SOLUTION

Conceptualize Consider Figure 29.7a, which is a vector model for $\ell = 2$. Draw such a vector model for $\ell = 3$ to help with this problem.

Categorize We evaluate results using equations developed in this section, so we categorize this example as a substitution problem.

Calculate the magnitude of the orbital angular momentum using Equation 29.9:

$$L = \sqrt{\ell(\ell + 1)}\hbar = \sqrt{3(3 + 1)}\hbar = \boxed{2\sqrt{3}\,\hbar}$$

Calculate the allowed values of L_z using Equation 29.10 with $m_\ell = -3, -2, -1, 0, 1, 2,$ and 3:

$$L_z = \boxed{-3\hbar,\ -2\hbar,\ -\hbar,\ 0,\ \hbar,\ 2\hbar,\ 3\hbar}$$

Calculate the allowed values of $\cos \theta$ using Equation 29.11:

$$\cos\theta = \frac{\pm 3}{2\sqrt{3}} = \pm 0.866 \quad \cos\theta = \frac{\pm 2}{2\sqrt{3}} = \pm 0.577$$

$$\cos\theta = \frac{\pm 1}{2\sqrt{3}} = \pm 0.289 \quad \cos\theta = \frac{0}{2\sqrt{3}} = 0$$

Find the angles corresponding to these values of $\cos \theta$:

$$\theta = \boxed{30.0°,\ 54.7°,\ 73.2°,\ 90.0°,\ 107°,\ 125°,\ 150°}$$

What If? What if the value of ℓ is an arbitrary integer? For an arbitrary value of ℓ, how many values of m_ℓ are allowed?

Answer For a given value of ℓ, the values of m_ℓ range from $-\ell$ to $+\ell$ in steps of 1. Therefore, there are 2ℓ nonzero values of m_ℓ (specifically, $\pm 1, \pm 2, \ldots, \pm \ell$). In addition, one more value of $m_\ell = 0$ is possible, for a total of $2\ell + 1$ values of m_ℓ. This result is critical in understanding the results of the Stern–Gerlach experiment described below with regard to spin.

The Spin Magnetic Quantum Number m_s

The three quantum numbers n, ℓ, and m_ℓ discussed so far are generated by applying boundary conditions to solutions of the Schrödinger equation, and we can assign a physical interpretation to each of the quantum numbers. This is as far as we can go using the quantum structural model that we developed in Section 29.2. However, we must expand the model by including considerations related to **electron spin.** The results related to spin do *not* come from the Schrödinger equation.

Example 29.1 was presented to give you practice in manipulating quantum numbers, but, as we shall see in this section, there are *eight* electron states for $n = 2$ rather than the four we found. These extra states can be explained by requiring a fourth quantum number for each state, the **spin magnetic quantum number** m_s.

> **Pitfall Prevention | 29.2**
> **The Electron Is Not Spinning**
> Although the concept of a spinning electron is conceptually useful, it should not be taken literally. The spin of the Earth is a mechanical rotation. On the other hand, electron spin is a purely quantum effect that gives the electron an angular momentum as if it were physically spinning.

Evidence of the need for this new quantum number came about because of an unusual feature in the spectra of certain gases such as sodium vapor. Close examination of one of the prominent lines of sodium shows that it is, in fact, two very closely spaced lines called a doublet. The wavelengths of these lines occur in the yellow region at 589.0 nm and 589.6 nm. In 1925, when this doublet was first noticed, atomic models could not explain it. To resolve this dilemma, Samuel Goudsmit and George Uhlenbeck, following a suggestion by the Austrian physicist Wolfgang Pauli, proposed a new quantum number, called the spin quantum number. The origin of this fourth quantum number was shown by Arnold Sommerfeld and Paul Dirac to lie in the relativistic properties of the electron, which requires four quantum numbers to describe it in four-dimensional space–time.

To describe the spin quantum number, it is convenient (but incorrect!) to think of the electron as spinning on its axis as it orbits the nucleus in a planetary model, just as the Earth spins on its axis as it orbits the Sun. The direction in which the spin angular momentum vector can point is quantized; it can have only two directions as shown in Figure 29.8. If the direction of spin is as shown in Figure 29.8a, the electron is said to have "spin up." If the direction of spin is as shown in Figure 29.8b, the electron is said to have "spin down." The spin angular momentum of the charged electron has a magnetic moment associated with it. Therefore, when an atom is in a magnetic field, Equation 22.17 tells us that the energy of the system (the atom and the magnetic field) is slightly different for the two spin directions, and this energy difference accounts for the sodium doublet. The quantum numbers associated with electron spin are $m_s = \frac{1}{2}$ for the spin-up state and $m_s = -\frac{1}{2}$ for the spin-down state. As we shall see shortly, this added quantum number doubles the number of allowed states specified by the quantum numbers n, ℓ, and m_ℓ.

In 1921, Otto Stern (1888–1969) and Walther Gerlach (1889–1979) performed an experiment (Fig. 29.9) that detected the effects of the force on a magnetic moment in a nonuniform magnetic field. The experiment demonstrated that the angular momentum of an atom is quantized. In their experiment, a beam of neutral silver atoms was sent through a nonuniform magnetic field. In such a situation, the atoms experience a force (in the vertical direction in Fig. 29.9) due to their magnetic moments in this field. Classically, we would expect the beam to be spread out into a continuous distribution on the photographic plate in Figure 29.9 because all possible directions of the atomic magnetic moments are allowed. Stern and Gerlach found, however, that the beam split into two *discrete* components. The experiment was repeated using other atoms, and in each case the beam split into two or more discrete components.

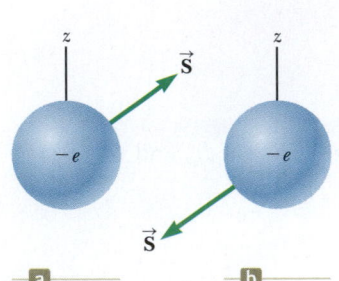

Figure 29.8 The spin of an electron can be either (a) up or (b) down relative to a specified z axis. As in the case of orbital angular momentum, the x and y components of the spin angular momentum vector are not quantized.

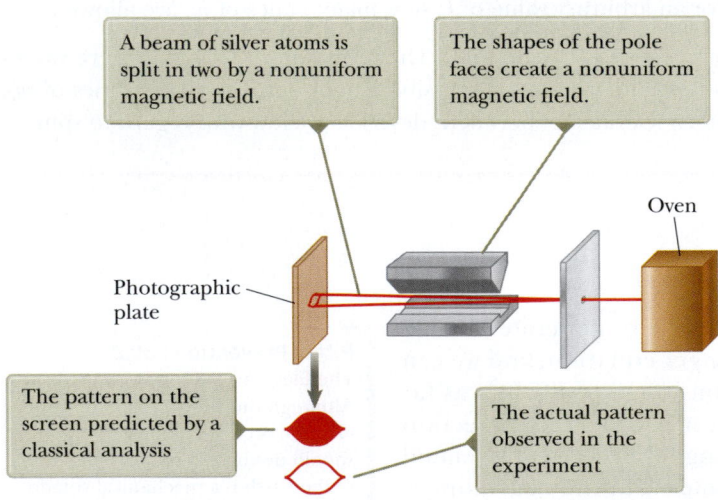

A beam of silver atoms is split in two by a nonuniform magnetic field.

The shapes of the pole faces create a nonuniform magnetic field.

Oven

Photographic plate

The pattern on the screen predicted by a classical analysis

The actual pattern observed in the experiment

Figure 29.9 The technique used by Stern and Gerlach to verify space quantization.

These results are clearly inconsistent with the prediction of a classical model. According to a quantum model, however, the direction of the total angular momentum of the atom, and hence the direction of its magnetic moment $\vec{\mu}$, is quantized. Therefore, the deflected beam has an integral number of discrete components, and the number of components determines the number of possible values of μ_z. Because the Stern–Gerlach experiment showed discrete beams, space quantization was at least qualitatively verified.

For the moment, let us assume that the angular momentum of the atom is due to the orbital angular momentum.[5] Because μ_z is proportional to m_ℓ, the number of possible values of μ_z is $2\ell + 1$. Furthermore, because ℓ is an integer, the number of values of μ_z is always odd. This prediction was not consistent with the observations of Stern and Gerlach, who observed two components, an even number, in the

---

[5]The Stern–Gerlach experiment was performed in 1921, before spin was hypothesized, so orbital angular momentum was the only type of angular momentum in the quantum model at the time.

deflected beam of silver atoms. Therefore, although the Stern–Gerlach experiment demonstrated space quantization, the number of components was not consistent with the quantum model developed at that time.

In 1927, T. E. Phipps and J. B. Taylor repeated the Stern–Gerlach experiment using a beam of hydrogen atoms. This experiment is important because it deals with an atom with a single electron in its ground state, for which the quantum model makes reliable predictions. At room temperature, almost all hydrogen atoms are in the ground state. Recall that $\ell = 0$ for hydrogen in its ground state, and so $m_\ell = 0$. Hence, from the orbital angular momentum approach, one would not expect the beam to be deflected by the field at all because μ_z would be zero. The beam in the Phipps–Taylor experiment, however, was again split into two components. On the basis of this result, one can conclude only one thing: that there is some contribution to the angular momentum of the atom and its magnetic moment other than the orbital angular momentum.

As we learned earlier, Goudsmit and Uhlenbeck had proposed that the electron has an intrinsic angular momentum, spin, apart from its orbital angular momentum. In other words, the total angular momentum of the electron in a particular electronic state contains both an orbital contribution $\vec{\mathbf{L}}$ and a spin contribution $\vec{\mathbf{S}}$. A quantum number s exists for spin that is analogous to ℓ for orbital angular momentum. The value of s for an electron, however, is *always* $s = \frac{1}{2}$, unlike ℓ, which varies for different states of the atom.

Like $\vec{\mathbf{L}}$, the **spin angular momentum** vector $\vec{\mathbf{S}}$ must obey the rules of the quantum model. In analogy with Equation 29.9, the **magnitude of the spin angular momentum** $\vec{\mathbf{S}}$ of an electron is

$$S = \sqrt{s(s+1)}\hbar = \frac{\sqrt{3}}{2}\hbar \qquad \textbf{29.12} \blacktriangleleft$$

This result is the only allowed value for the magnitude of the spin angular momentum vector for an electron, so we usually do not include s in a list of quantum numbers describing states of the atom. Like orbital angular momentum, spin angular momentum is quantized in space as described in Figure 29.10. It can have two orientations, specified by the spin magnetic quantum number m_s, where m_s has two possible values, $\pm\frac{1}{2}$. In analogy with Equation 29.10, the z component of spin angular momentum is

$$S_z = m_s\hbar = \pm\tfrac{1}{2}\hbar \qquad \textbf{29.13} \blacktriangleleft$$

The two values $\pm\hbar/2$ for S_z correspond to the two possible orientations for $\vec{\mathbf{S}}$ shown in Figure 29.10. The quantum number m_s is listed as the fourth quantum number describing a particular state of the atom.

The spin magnetic moment $\vec{\boldsymbol{\mu}}_{\text{spin}}$ of the electron is related to its spin angular momentum $\vec{\mathbf{S}}$ by the expression

$$\vec{\boldsymbol{\mu}}_{\text{spin}} = -\frac{e}{m_e}\vec{\mathbf{S}} \qquad \textbf{29.14} \blacktriangleleft$$

where e is the electronic charge and m_e is the mass of the electron. Because $S_z = \pm\frac{1}{2}\hbar$, the z component of the spin magnetic moment can have the values

$$\vec{\boldsymbol{\mu}}_{\text{spin},z} = \pm\frac{e\hbar}{2m_e} \qquad \textbf{29.15} \blacktriangleleft$$

The quantity $e\hbar/2m_e$ is called the **Bohr magneton** μ_B and has the numerical value 9.274×10^{-24} J/T.

Today physicists explain the outcome of the Stern–Gerlach experiment as follows. The observed moments for both silver and hydrogen are due to spin angular momentum alone and not to orbital angular momentum. (The hydrogen atom in its ground state has $\ell = 0$; for silver, used in the Stern–Gerlach experiment, the net orbital angular momentum for all the electrons is $|\vec{\mathbf{L}}| = 0$.) A single-electron atom such as hydrogen has its electron spin quantized in the magnetic field in such a way that its z component of spin angular momentum is either $\frac{1}{2}\hbar$ or $-\frac{1}{2}\hbar$, corresponding

Wolfgang Pauli and Niels Bohr watch a spinning top. The spin of the electron is analogous to the spin of the top but is different in many ways.

▶ Magnitude of the spin angular momentum of an electron

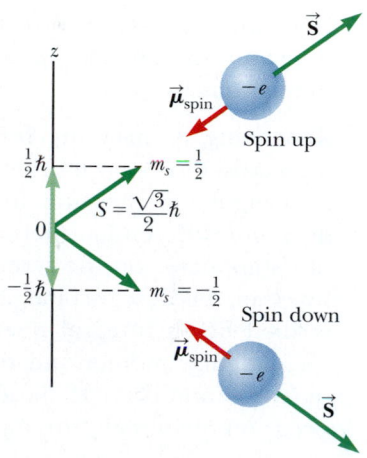

Figure 29.10 Spin angular momentum $\vec{\mathbf{S}}$ exhibits space quantization. This figure shows the two allowed orientations of the spin angular momentum vector $\vec{\mathbf{S}}$ and the spin magnetic moment vector $\vec{\boldsymbol{\mu}}_{\text{spin}}$ for a spin-$\frac{1}{2}$ particle such as the electron.

TABLE 29.4 | Quantum Numbers for the $n = 2$ State of Hydrogen

n	ℓ	m_ℓ	m_s	Subshell	Shell	Number of States in Subshell
2	0	0	$\frac{1}{2}$			
2	0	0	$-\frac{1}{2}$	$2s$	L	2
2	1	1	$\frac{1}{2}$			
2	1	1	$-\frac{1}{2}$			
2	1	0	$\frac{1}{2}$			
2	1	0	$-\frac{1}{2}$	$2p$	L	6
2	1	-1	$\frac{1}{2}$			
2	1	-1	$-\frac{1}{2}$			

to $m_s = \pm\frac{1}{2}$. Electrons with spin $+\frac{1}{2}$ are deflected in one direction by the nonuniform magnetic field, and those with spin $-\frac{1}{2}$ are deflected in the opposite direction.

The Stern–Gerlach experiment provided two important results. First, it verified the concept of space quantization. Second, it showed that spin angular momentum exists even though this property was not recognized until long after the experiments were performed.

As mentioned earlier, there are eight quantum states corresponding to $n = 2$ in the hydrogen atom, not four as found in Example 29.1. Each of the four states in Example 29.1 is actually two states because of the two possible values of m_s. Table 29.4 shows the quantum numbers corresponding to these eight states.

THINKING PHYSICS 29.1

Does the Stern–Gerlach experiment differentiate between orbital angular momentum and spin angular momentum?

Reasoning A magnetic force on the magnetic moment arises from both orbital angular momentum and spin angular momentum. In this sense, the experiment does not differentiate between the two. The number of components on the screen does tell us something, however, because orbital angular momenta are described by an integral quantum number ℓ, whereas spin angular momentum depends on a half-integral quantum number s. If an odd number of components occur on the screen, three possibilities arise: the atom has (1) orbital angular momentum only, (2) an even number of electrons with spin angular momentum, or (3) a combination of orbital angular momentum and an even number of electrons with spin angular momentum. If an even number of components occurs on the screen, at least one unpaired spin angular momentum exists, possibly in combination with orbital angular momentum. The only numbers of components for which we can specify the type of angular momentum are one component (no orbital, no spin) and two components (spin of one electron). Once we see more than two components, multiple possibilities arise because of various combinations of $\vec{\mathbf{L}}$ and $\vec{\mathbf{S}}$. ◄

29.5 | The Exclusion Principle and the Periodic Table

The quantum model for hydrogen generated from the Schrödinger equation, including the notion of electron spin, is based on a system consisting of one electron and one proton. As soon as we consider the next atom, helium, we introduce complicating factors. The two electrons in helium both interact with the nucleus, so we can define a potential energy function for those interactions. They also interact with each other, however. The line of action of the electron–nucleus interaction is along a line between the electron and the nucleus. The line of action of the electron–electron interaction is along the line between the two electrons, which is different from that of the electron–nucleus interaction. Therefore, the Schrödinger equation is extremely difficult to solve. As we consider atoms with more and more electrons, the possibility of an algebraic solution of the Schrödinger equation becomes hopeless.

We find, however, that despite our inability to solve the Schrödinger equation, we can use the same four quantum numbers developed for hydrogen for the electrons in heavier atoms. We are not able to calculate the quantized energy levels easily, but we can gain information about the levels from theoretical models and experimental measurements.

Because a quantum state in any atom is specified by four quantum numbers, n, ℓ, m_ℓ, and m_s, an obvious and important question is, "How many electrons in an atom can have a particular set of quantum numbers?" Pauli provided an answer in 1925 in a powerful statement known as the **exclusion principle**:

> No two electrons in an atom can ever be in the same quantum state; that is, no two electrons in the same atom can have the same set of quantum numbers.

Wolfgang Pauli
Austrian Theoretical Physicist (1900–1958)
An extremely talented Austrian theoretical physicist, Pauli made important contributions in many areas of modern physics. Pauli gained public recognition at the age of 21 with a masterful review article on relativity, which is still considered one of the finest and most comprehensive introductions to the subject. Other major contributions were the discovery of the exclusion principle, the explanation of the connection between particle spin and statistics, and theories of relativistic quantum electrodynamics, the neutrino hypothesis, and the hypothesis of nuclear spin.

It is interesting that if this principle were not valid, every atom would radiate energy by means of photons and end up with all electrons in the lowest energy state. The chemical behavior of the elements would be grossly modified because this behavior depends on the electronic structure of atoms. Nature as we know it would not exist! In reality, we can view the electronic structure of complex atoms as a succession of filled levels increasing in energy, where the outermost electrons are primarily responsible for the chemical properties of the element.

Imagine building an atom by forming the nucleus and then filling in the available quantum states with electrons until the atom is neutral. We shall use the common language here that "electrons go into available states." Keep in mind, however, that the states are those of the *system* of the atom. As a general rule, the order of filling of an atom's subshells with electrons is as follows. Once one subshell is filled, the next electron goes into the vacant subshell that is lowest in energy.

Before we discuss the electronic configurations of some elements, it is convenient to define an **orbital** as the state of an electron characterized by the quantum numbers n, ℓ, and m_ℓ. From the exclusion principle, we see that at most two electrons can be in any orbital. One of these electrons has $m_s = +\frac{1}{2}$ and the other has $m_s = -\frac{1}{2}$. Because each orbital is limited to two electrons, the numbers of electrons that can occupy the shells are also limited.

Table 29.5 shows the allowed quantum states for an atom up to $n = 3$. Each square in the bottom row of the table represents one orbital, with the ↑ arrows representing $m_s = +\frac{1}{2}$ and the ↓ arrows representing $m_s = -\frac{1}{2}$. The $n = 1$ shell can accommodate only two electrons because only one orbital is allowed with $m_\ell = 0$. The $n = 2$ shell has two subshells, with $\ell = 0$ and $\ell = 1$. The $\ell = 0$ subshell is limited to only two electrons because $m_\ell = 0$. The $\ell = 1$ subshell has three allowed orbitals, corresponding to $m_\ell = 1$, 0, and -1. Because each orbital can accommodate two electrons, the $\ell = 1$ subshell can hold six electrons (and the $n = 2$ shell can hold eight). The $n = 3$ shell has three subshells and nine orbitals and can accommodate up to 18 electrons. In general, each shell can accommodate up to $2n^2$ electrons.

The results of the exclusion principle can be illustrated by an examination of the electronic arrangement in a few of the lighter atoms. For example, **hydrogen** has only one electron, which, in its ground state, can be described by either of two sets of quantum numbers: $1, 0, 0, +\frac{1}{2}$ or $1, 0, 0, -\frac{1}{2}$. The electronic configuration of this atom is often designated as $1s^1$. The notation $1s$ refers to a state for which $n = 1$ and $\ell = 0$, and the superscript indicates that one electron is present in the s subshell.

Pitfall Prevention | 29.3
The Exclusion Principle Is More General
The exclusion principle stated here is a limited form of the more general exclusion principle, which states that no two *fermions*, which are *all* particles with half-integral spin $\frac{1}{2}, \frac{3}{2}, \frac{5}{2}, \ldots$ can be in the same quantum state. The present form is satisfactory for our discussions of atomic physics, and we will discuss the general form further in Chapter 31.

TABLE 29.5 | Allowed Quantum States for an Atom Up to $n = 3$

n	1	2			3								
ℓ	0	0	1		0	1			2				
m_ℓ	0	0	1	0	−1	0	1	0	−1	2	1	0	−1	−2
m_s	↑↓	↑↓	↑↓	↑↓	↑↓	↑↓	↑↓	↑↓	↑↓	↑↓	↑↓	↑↓	↑↓	↑↓

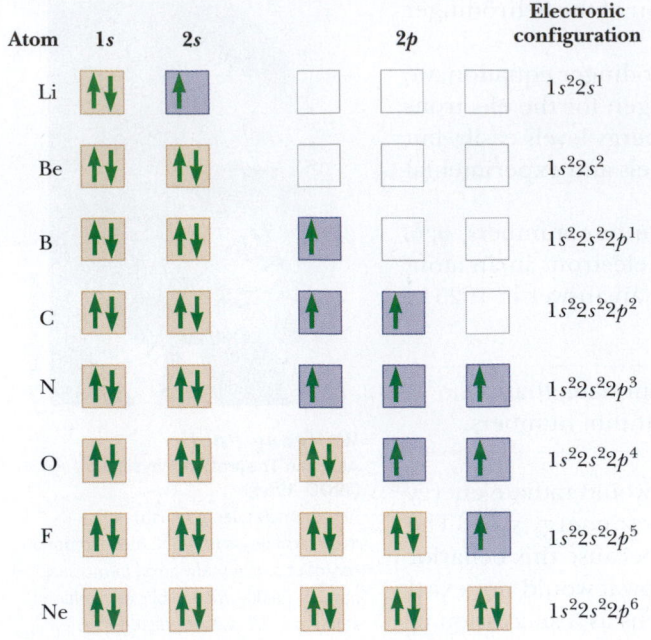

Atom	1s	2s		2p		Electronic configuration
Li	↑↓	↑				$1s^2 2s^1$
Be	↑↓	↑↓				$1s^2 2s^2$
B	↑↓	↑↓	↑			$1s^2 2s^2 2p^1$
C	↑↓	↑↓	↑	↑		$1s^2 2s^2 2p^2$
N	↑↓	↑↓	↑	↑	↑	$1s^2 2s^2 2p^3$
O	↑↓	↑↓	↑↓	↑	↑	$1s^2 2s^2 2p^4$
F	↑↓	↑↓	↑↓	↑↓	↑	$1s^2 2s^2 2p^5$
Ne	↑↓	↑↓	↑↓	↑↓	↑↓	$1s^2 2s^2 2p^6$

Figure 29.11 The filling of electronic states must obey both the exclusion principle and Hund's rules.

Neutral **helium** has two electrons. In the ground state, the quantum numbers for these two electrons are 1, 0, 0, $+\frac{1}{2}$ and 1, 0, 0, $-\frac{1}{2}$. No other combinations of quantum numbers are possible for this level, and we say that the K shell is filled. The electronic configuration of helium is designated as $1s^2$.

The electronic configurations of some successive elements are given in Figure 29.11. Neutral **lithium** has three electrons. In the ground state, two of them are in the $1s$ subshell and the third is in the $2s$ subshell because this subshell is lower in energy than the $2p$ subshell. (In addition to the simple dependence of E on n in Eq. 29.2, there is an additional dependence on ℓ, which will be addressed in Section 29.6.) Hence, the electronic configuration for lithium is $1s^2 2s^1$.

Note that the electronic configuration of **beryllium,** with its four electrons, is $1s^2 2s^2$, and **boron** has a configuration of $1s^2 2s^2 2p^1$. The $2p$ electron in boron may be described by one of six sets of quantum numbers, corresponding to six states of equal energy.

Carbon has six electrons, and a question arises concerning how to assign the two $2p$ electrons. Do they go into the same orbital with paired spins (↑↓), or do they occupy different orbitals with unpaired spins (↑↑ or ↓↓)? Experimental data show that the lowest energy configuration is the latter, where the spins are unpaired. Hence, the two $2p$ electrons in carbon and the three $2p$ electrons in nitrogen have unpaired spins in the ground state (see Fig. 29.11). The general rules that govern such situations throughout the periodic table are called **Hund's rules.** The rule appropriate for elements like carbon is that when an atom has orbitals of equal energy, the order in which they are filled by electrons is such that a maximum number of electrons will have unpaired spins. Some exceptions to this rule occur in elements having subshells close to being filled or half-filled.

An early attempt to find some order among the elements was made by a Russian chemist, Dmitri Mendeleev (1834–1907), in 1871. He developed a tabular representation of the elements, which has become one of the most important, as well as well-recognized, tools of science. He arranged the atoms in a table similar to that shown in Figure 29.12 according to their atomic masses and chemical similarities. The first table Mendeleev proposed contained many blank spaces, and he boldly stated that the gaps were there only because the elements had not yet been discovered. By noting the columns in which these missing elements should be located, he was able to make rough predictions about their chemical properties. Within 20 years of Mendeleev's announcement, the missing elements were indeed discovered. The predictions made possible by this table represent an excellent example of the power of presenting information in an alternative representation.

The elements in the **periodic table** (Fig. 29.12) are arranged so that all those in a vertical column have similar chemical properties. For example, consider the elements in the last column: He (helium), Ne (neon), Ar (argon), Kr (krypton), Xe (xenon), and Rn (radon). The outstanding characteristic of all these elements is that they do not normally take part in chemical reactions; that is, they do not readily join with other atoms to form molecules. They are therefore called *inert gases*.

We can partially understand this behavior by looking at the electronic configurations in Figure 29.12. The element helium is one in which the electronic configuration is $1s^2$; in other words, one shell is filled. Additionally, it is found that the energy associated with this filled shell is considerably lower than the energy of the next available level, the $2s$ level. Next, look at the electronic configuration for neon, $1s^2 2s^2 2p^6$. Again, the outermost shell is filled, and a gap in energy occurs between the $2p$ level and the $3s$ level. Argon has the configuration $1s^2 2s^2 2p^6 3s^2 3p^6$. Here, the $3p$ subshell is filled, and a gap in energy arises between the $3p$ subshell and the $3d$ subshell. We could continue this procedure through all the inert gases; the pattern remains the

Group I	Group II	Transition elements										Group III	Group IV	Group V	Group VI	Group VII	Group 0
H 1 $1s^1$																H 1 $1s^1$	He 2 $1s^2$
Li 3 $2s^1$	Be 4 $2s^2$											B 5 $2p^1$	C 6 $2p^2$	N 7 $2p^3$	O 8 $2p^4$	F 9 $2p^5$	Ne 10 $2p^6$
Na 11 $3s^1$	Mg 12 $3s^2$											Al 13 $3p^1$	Si 14 $3p^2$	P 15 $3p^3$	S 16 $3p^4$	Cl 17 $3p^5$	Ar 18 $3p^6$
K 19 $4s^1$	Ca 20 $4s^2$	Sc 21 $3d^1 4s^2$	Ti 22 $3d^2 4s^2$	V 23 $3d^3 4s^2$	Cr 24 $3d^5 4s^1$	Mn 25 $3d^5 4s^2$	Fe 26 $3d^6 4s^2$	Co 27 $3d^7 4s^2$	Ni 28 $3d^8 4s^2$	Cu 29 $3d^{10} 4s^1$	Zn 30 $3d^{10} 4s^2$	Ga 31 $4p^1$	Ge 32 $4p^2$	As 33 $4p^3$	Se 34 $4p^4$	Br 35 $4p^5$	Kr 36 $4p^6$
Rb 37 $5s^1$	Sr 38 $5s^2$	Y 39 $4d^1 5s^2$	Zr 40 $4d^2 5s^2$	Nb 41 $4d^4 5s^1$	Mo 42 $4d^5 5s^1$	Tc 43 $4d^5 5s^2$	Ru 44 $4d^7 5s^1$	Rh 45 $4d^8 5s^1$	Pd 46 $4d^{10}$	Ag 47 $4d^{10} 5s^1$	Cd 48 $4d^{10} 5s^2$	In 49 $5p^1$	Sn 50 $5p^2$	Sb 51 $5p^3$	Te 52 $5p^4$	I 53 $5p^5$	Xe 54 $5p^6$
Cs 55 $6s^1$	Ba 56 $6s^2$	57–71*	Hf 72 $5d^2 6s^2$	Ta 73 $5d^3 6s^2$	W 74 $5d^4 6s^2$	Re 75 $5d^5 6s^2$	Os 76 $5d^6 6s^2$	Ir 77 $5d^7 6s^2$	Pt 78 $5d^9 6s^1$	Au 79 $5d^{10} 6s^1$	Hg 80 $5d^{10} 6s^2$	Tl 81 $6p^1$	Pb 82 $6p^2$	Bi 83 $6p^3$	Po 84 $6p^4$	At 85 $6p^5$	Rn 86 $6p^6$
Fr 87 $7s^1$	Ra 88 $7s^2$	89–103**	Rf 104 $6d^2 7s^2$	Db 105 $6d^3 7s^2$	Sg 106 $6d^4 7s^2$	Bh 107 $6d^5 7s^2$	Hs 108 $6d^6 7s^2$	Mt 109 $6d^7 7s^2$	Ds 110 $6d^9 7s^1$	Rg 111	Cn 112		114		116		

*Lanthanide series

La 57 $5d^1 6s^2$	Ce 58 $5d^1 4f^1 6s^2$	Pr 59 $4f^3 6s^2$	Nd 60 $4f^4 6s^2$	Pm 61 $4f^5 6s^2$	Sm 62 $4f^6 6s^2$	Eu 63 $4f^7 6s^2$	Gd 64 $5d^1 4f^7 6s^2$	Tb 65 $5d^1 4f^8 6s^2$	Dy 66 $4f^{10} 6s^2$	Ho 67 $4f^{11} 6s^2$	Er 68 $4f^{12} 6s^2$	Tm 69 $4f^{13} 6s^2$	Yb 70 $4f^{14} 6s^2$	Lu 71 $5d^1 4f^{14} 6s^2$

**Actinide series

Ac 89 $6d^1 7s^2$	Th 90 $6d^2 7s^2$	Pa 91 $5f^2 6d^1 7s^2$	U 92 $5f^3 6d^1 7s^2$	Np 93 $5f^4 6d^1 7s^2$	Pu 94 $5f^6 7s^2$	Am 95 $5f^7 7s^2$	Cm 96 $5f^7 6d^1 7s^2$	Bk 97 $5f^8 6d^1 7s^2$	Cf 98 $5f^{10} 7s^2$	Es 99 $5f^{11} 7s^2$	Fm 100 $5f^{12} 7s^2$	Md 101 $5f^{13} 7s^2$	No 102 $5f^{14} 7s^2$	Lr 103 $5f^{14} 6d^1 7s^2$

Figure 29.12 The periodic table of the elements is an organized tabular representation of the elements that shows their periodic chemical behavior. Elements in a given column have similar chemical behavior. This table shows the chemical symbol for the element, the atomic number, and the electronic configuration. A more complete periodic table is available in Appendix C.

same. An inert gas is formed when either a shell or a subshell is filled and a gap in energy occurs before the next possible level is encountered.

If we consider the column to the left of the inert gases in the periodic table, we find a group of elements called the *halogens*: fluorine, chlorine, bromine, iodine, and astatine. At room temperature, fluorine and chlorine are gases, bromine is a liquid, and iodine and astatine are solids. In each of these atoms, the outer subshell is one electron short of being filled. As a result, the halogens are chemically very active, readily accepting an electron from another atom to form a closed shell. The halogens tend to from strong ionic bonds with atoms at the other side of the periodic table. In a halogen lightbulb, bromine or iodine atoms combine with tungsten atoms evaporated from the filament and return them to the filament, resulting in a longer-lasting bulb. In addition, the filament can be operated at a higher temperature than in ordinary lightbulbs, giving a brighter and whiter light.

At the left side of the periodic table, the Group I elements consist of hydrogen and the *alkali metals*, lithium, sodium, potassium, rubidium, cesium, and francium. Each of these atoms contains one electron in a subshell outside of a closed subshell. Therefore, these elements easily form positive ions because the lone electron is bound with a relatively low energy and is easily removed. Therefore, the alkali metal atoms are chemically active and form very strong bonds with halogen atoms. For example, table salt, NaCl, is a combination of an alkali metal and a halogen. Because the outer electron is weakly bound, pure alkali metals tend to be good electrical conductors, although, because of their high chemical activity, pure alkali metals are not generally found in nature.

It is interesting to plot ionization energy versus the atomic number Z as in Figure 29.13 (page 868). Note the pattern of differences in atomic numbers between the peaks in the graph: 8, 8, 18, 18, 32. This pattern follows from the Pauli exclusion

Figure 29.13 Ionization energy of the elements versus atomic number.

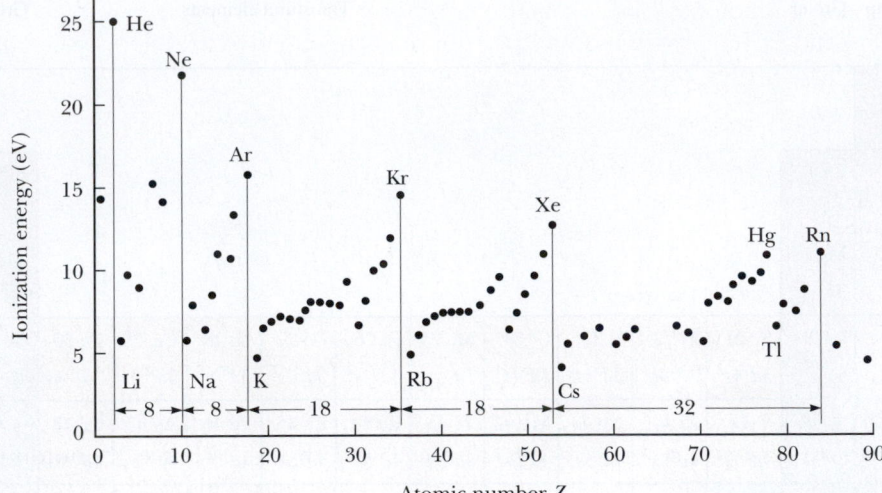

principle and helps explain why the elements repeat their chemical properties in groups. For example, the peaks at $Z = 2$, 10, 18, and 36 correspond to the elements He, Ne, Ar, and Kr, which have filled shells. These elements have similar chemical behavior.

> **QUICK QUIZ 29.4** Rank the energy necessary to remove the outermost electron from the following three elements, smallest to largest: lithium, potassium, cesium.

BIO Treatment of cancers with proton therapy

A variety of treatment plans are available for battling cancerous tumors. Some of these options involving atomic and nuclear phenomena will be discussed in this chapter and the next. One of these treatment procedures is called *proton therapy*. In this procedure, a beam of protons is used to irradiate the cancerous tissue. Protons are one form of *ionizing radiation,* that is, radiation that will ionize atoms of diseased tissue with the goal of destroying the tissue. A major advantage of using protons is that the dose delivered to the tissue, that is, the ionizing energy deposited in the tissue, is a maximum over the last few millimeters of the particle's range. As a result, relatively little ionization occurs along the first part of the protons' path, leaving healthy tissue unharmed. By adjusting the incoming energy of the protons, up to 250 MeV, the depth at which the majority of the energy is delivered can be tuned to coincide with the location of the tumor. Special nozzles at the end of the proton beam shape the beam according to the three-dimensional shape of the tumor, allowing the entire tumor to receive uniform irradiation. As a result, the cancerous tissue is damaged while surrounding healthy tissue experiences much less damage.

Proton therapy has been used for prostate cancer, sarcomas, medically inoperable lung cancer, acoustic neuromas, and a variety of ocular tumors. Proton therapy procedures have been performed since the early 1950s using particle accelerators built for physics research. Beginning in 1990, dedicated hospital-based proton therapy centers were built. At the time of this printing, there are ten such centers in the United States and 37 worldwide.

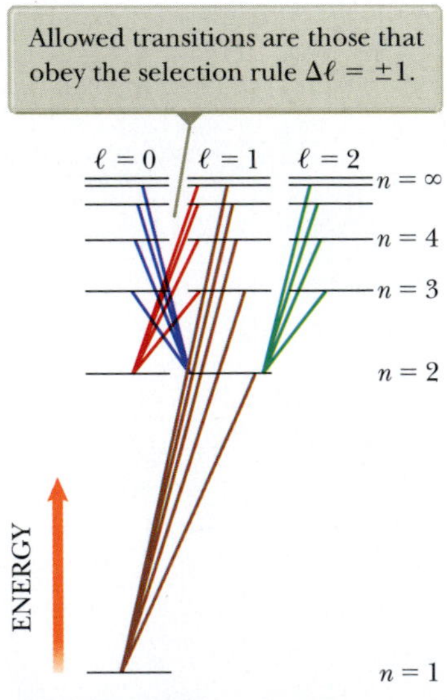

Allowed transitions are those that obey the selection rule $\Delta \ell = \pm 1$.

$\ell = 0$ $\ell = 1$ $\ell = 2$

$n = \infty$
$n = 4$
$n = 3$
$n = 2$

ENERGY

$n = 1$

Figure 29.14 Some allowed electronic transitions for hydrogen, represented by the colored lines.

29.6 | More on Atomic Spectra: Visible and X-Ray

In Chapter 11, we briefly discussed the origin of the spectral lines for hydrogen and hydrogen-like ions. Recall that an atom in an excited state will emit electromagnetic radiation if it makes a transition to a lower energy state.

The energy level diagram for hydrogen is shown in Figure 29.14. This semigraphical representation is different from Active Figure 11.18 in that the individual states corresponding to different values of ℓ within a given value of n are spread out horizontally. Figure 29.14 shows only those states up to $\ell = 2$; the shells from $n = 4$ upward would have more sets of states to the right, which are not shown.

The diagonal lines in Figure 29.14 represent allowed transitions between stationary states. Whenever an atom makes a transition from a higher energy state to a lower one, a photon of light is emitted. The frequency of this photon is $f = \Delta E/h$, where ΔE is the energy difference between the two states and h is Planck's constant. The **selection rules** for the allowed transitions are

$$\Delta \ell = \pm 1 \quad \text{and} \quad \Delta m_\ell = 0 \text{ or } \pm 1 \qquad \text{29.16} \blacktriangleleft$$

▶ Selection rules for allowed atomic transitions

Transitions that do not obey the above selection rules are said to be **forbidden.** (Such transitions can occur, but their probability is negligible relative to the probability of the allowed transitions.) For example, any transition represented by a vertical line in Figure 29.14 is forbidden because the quantum number ℓ does not change.

Because the orbital angular momentum of an atom changes when a photon is emitted or absorbed (i.e., as a result of a transition) and because angular momentum of the isolated system of the atom and the photon must be conserved, we conclude that the photon involved in the process must carry angular momentum. In fact, the photon has an intrinsic angular momentum equivalent to that of a particle with a spin of $s = 1$, compared with the electron with $s = \frac{1}{2}$. Hence, a photon possesses energy, linear momentum, and angular momentum. This example is the first one we have seen of a single particle with *integral* spin.

Equation 29.2 gives the energies of the allowed quantum states for hydrogen. We can also apply the Schrödinger equation to other one-electron systems, such as the He^+ and Li^{++} ions. The primary difference between these ions and the hydrogen atom is the different number of protons Z in the nucleus. The result is a generalization of Equation 29.2 for these other one-electron systems:

$$E_n = -\frac{(13.6 \text{ eV}) Z^2}{n^2} \qquad \text{29.17} \blacktriangleleft$$

For outer electrons in multielectron atoms, the nuclear charge Ze is largely canceled or shielded by the negative charge of the inner-core electrons. Hence, the outer electrons interact with a net charge that is reduced below the actual charge of the nucleus. (According to Gauss's law, the electric field at the position of an outer electron depends on the net charge of the nucleus and the electrons closer to the nucleus.) The expression for the allowed energies for multielectron atoms has the same form as Equation 29.17, with Z replaced by an effective atomic number Z_{eff}. That is,

$$E_n \approx -\frac{(13.6 \text{ eV}) Z_{eff}^2}{n^2} \qquad \text{29.18} \blacktriangleleft$$

where Z_{eff} depends on n and ℓ.

> **THINKING PHYSICS 29.2**

A physics student is watching a meteor shower in the early morning hours. She notices that the streaks of light from the meteoroids entering the very high regions of the atmosphere last for up to 2 or 3 s before fading.

She also notices a lightning storm off in the distance. The streaks of light from the lightning fade away almost immediately after the flash, certainly in much less than 1 s. Both lightning and meteors cause the air to turn into a plasma because of the very high temperatures generated. The light is emitted from both sources when the stripped electrons in the plasma recombine with the ionized molecules. Why would this light last longer for meteors than for lightning?

Reasoning The answer lies in the subtle phrase in the description of the meteoroids "entering the very high regions of the atmosphere." In the very high regions of the atmosphere, the pressure of the air is very low. The *density* of the air is therefore very low, so molecules of the air are relatively far apart. Therefore, after the air is ionized by the passing meteoroid, the probability per unit time interval of freed electrons encountering an ionized molecule with which to recombine is relatively low. As a result, the recombination process for all freed electrons occurs over a relatively long time interval, measured in seconds.

On the other hand, lightning occurs in the lower regions of the atmosphere (the troposphere) where the pressure and density are relatively high. After the ionization by the lightning flash, the freed electrons and ionized molecules are much closer together than in the upper atmosphere. The probability per unit time interval of a recombination is much higher, and the time interval for the recombination of all the electrons and ions to occur is much shorter. ◀

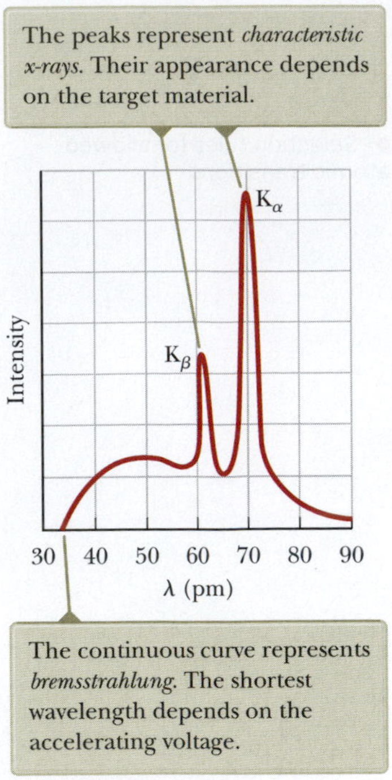

The peaks represent *characteristic x-rays*. Their appearance depends on the target material.

The continuous curve represents *bremsstrahlung*. The shortest wavelength depends on the accelerating voltage.

Figure 29.15 The x-ray spectrum of a metal target. The data shown were obtained when 37-keV electrons bombarded a molybdenum target.

BIO Medical imaging with x-rays

Figure 29.16 Bremsstrahlung is created by this machine and used to treat cancer in a patient.

X-Ray Spectra

X-rays are emitted when high-energy electrons or any other charged particles bombard a metal target. The x-ray spectrum typically consists of a broad continuous band containing a series of sharp lines as shown in Figure 29.15. In Section 24.6, we mentioned that an accelerated electric charge emits electromagnetic radiation. The x-rays in Figure 29.15 are the result of the slowing down of high-energy electrons as they strike the target. It may take several interactions with the atoms of the target before the electron loses all its kinetic energy. The amount of kinetic energy lost in any given interaction can vary from zero up to the entire kinetic energy of the electron. Therefore, the wavelength of radiation from these interactions lies in a continuous range from some minimum value up to infinity. It is this general slowing down of the electrons that provides the continuous curve in Figure 29.15, which shows the cutoff of x-rays below a minimum wavelength value that depends on the kinetic energy of the incoming electrons. X-ray radiation with its origin in the slowing down of electrons is called **bremsstrahlung,** the German word for "braking radiation."

Extremely high-energy bremsstrahlung can be used for the treatment of cancerous tissues in a process known as *external beam radiotherapy*. Figure 29.16 shows a machine that uses a linear accelerator to accelerate electrons up to 18 MeV and smash them into a tungsten target. The result is a beam of photons, up to a maximum energy of 18 MeV, which is actually in the gamma-ray range in Figure 24.11. This radiation is directed at the tumor in the patient.

In the previous section, we discussed treating cancerous tissues with energetic protons. The advantage of that technique is that most of the energy of the protons is delivered within the cancerous tissue, leaving healthy tissue relatively unharmed. The disadvantage is the size and cost of the cyclotron or synchrotron necessary to accelerate the protons up to therapeutic energies. External beam radiotherapy carries a higher probability of damage to healthy tissue, but the equipment necessary to accelerate electrons to therapeutic energies is much smaller and more inexpensive than that for proton therapy.

X-rays have been used for medical imaging since the early 20th century. As x-rays pass through the body, tissues of varying density and composition absorb different amounts of energy. By allowing the x-rays to expose photographic film after passing through the body, the film shows a shadowy image of the internal structure of the body. In an advance called *fluoroscopy*, the photographic film is replaced with a fluorescent screen (first decade of the 20th century) or a detector screen and video monitor (1950s). This procedure allows real-time evaluation of the x-ray image. In the 1970s, a major advance was made with the development of *computed tomography* or *CT scans*. A CT scan is created by a combination of an x-ray source and detector that rotate around the body and a computer to analyze the data. The result is a series of images representing cross-sectional slices through the body. CT scans, including newer versions providing three-dimensional images, are widely used in medical diagnostic procedures. While the use of CT scans has some advantages over the use of MRI (magnetic resonance imaging) scans, such as in the imaging of tumors in the thoracic region, it has a distinct disadvantage in exposing the patient to x-rays, which can cause damage to healthy tissue. MRI scans, on the other hand, expose the patient only to a strong magnetic field and harmless radio waves.

The discrete lines in Figure 29.15, called **characteristic x-rays** and discovered in 1908, have a different origin from that of bremsstrahlung. Their origin remained unexplained until the details of atomic structure were understood. The first step in the production of characteristic x-rays occurs when a bombarding electron collides with a target atom. The incoming electron must have sufficient energy to remove an inner-shell electron from the atom. The vacancy created in the shell is filled when an electron in a higher shell drops down into the shell containing the vacancy. The time interval required for this to happen is very short, less than 10^{-9} s. As usual, this transition is accompanied by the emission of a photon whose energy equals the difference in energy between the two shells. Typically, the energy of such transitions is greater than 1 000 eV, and the emitted x-ray photons have wavelengths in the range of 0.01 to 1 nm.

Let us assume that the incoming electron has dislodged an atomic electron from the innermost shell, the K shell. If the vacancy is filled by an electron dropping from the next higher shell, the L shell, the photon emitted in the process has an energy corresponding to the K_α line on the curve of Figure 29.15. If the vacancy is filled by an electron dropping from the M shell, the line produced is called the K_β line. In this notation, the letter K represents the final shell into which the electron drops and the subscript provides a Greek letter corresponding to the number of the shell above the final shell in which the electron originates. Therefore, K_α indicates that the final shell is the K shell, whereas the initial shell is the first shell above K (because α is the first letter in the Greek alphabet), which is the L shell.

Other characteristic x-ray lines are formed when electrons drop from upper shells to vacancies in shells other than the K shell. For example, L lines are produced when vacancies in the L shell are filled by electrons dropping from higher shells. An L_α line is produced as an electron drops from the M shell to the L shell, and an L_β line is produced by a transition from the N shell to the L shell.

Although multielectron atoms cannot be analyzed exactly using either the Bohr model or the Schrödinger equation, we can apply our knowledge of Gauss's law from Chapter 19 to make some surprisingly accurate estimates of expected x-ray energies and wavelengths. Consider an atom of atomic number Z in which one of the two electrons in the K shell has been ejected. Imagine that we draw a gaussian sphere just inside the most probable radius of the L electrons. The electric field at the position of the L electrons is a combination of that due to the nucleus, the single K electron, the other L electrons, and the outer electrons. The wave functions of the outer electrons are such that they have a very high probability of being farther from the nucleus than the L electrons are. Therefore, they are much more likely to be outside the gaussian surface than inside and, on the average, do not contribute significantly to the electric field at the position of the L electrons. The effective charge inside the gaussian surface is the positive nuclear charge and one negative charge due to the single K electron. If we ignore the interactions between L electrons, a single L electron behaves as if it experiences an electric field due to a charge enclosed by the gaussian surface of $(Z-1)e$. The nuclear charge is in effect shielded by the electron in the K shell such that Z_{eff} in Equation 29.18 is $Z-1$. For higher-level shells, the nuclear charge is shielded by electrons in all the inner shells.

We can now use Equation 29.18 to estimate the energy associated with an electron in the L shell:

$$E_L \approx -(Z-1)^2 \frac{13.6 \text{ eV}}{2^2}$$

After the atom makes the transition, there are two electrons in the K shell. We can approximate the energy associated with one of these electrons as that of a one-electron atom. (In reality, the nuclear charge is reduced somewhat by the negative charge of the other electron, but let's ignore this effect.) Therefore,

$$E_K \approx -Z^2(13.6 \text{ eV}) \qquad \textbf{29.19} \blacktriangleleft$$

As we show in Example 29.5, the energy of the atom with an electron in an M shell can be estimated in a similar fashion. Taking the energy difference between the initial and final levels, the energy and wavelength of the emitted photon can then be calculated.

In 1914, Henry G. J. Moseley (1887–1915) plotted $\sqrt{1/\lambda}$ versus the Z values for a number of elements, where λ is the wavelength of the K_α line of each element. He found that the curve is a straight line as in Figure 29.17. This finding is consistent with rough calculations of the energy levels given by Equation 29.19. From this plot, Moseley was able to determine the Z values of some missing elements, which provided a periodic table in excellent agreement with the known chemical properties of the elements.

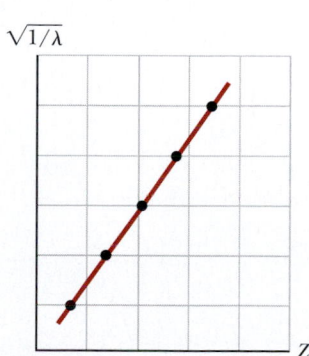

Figure 29.17 A Moseley plot of $\sqrt{1/\lambda}$ versus Z, where λ is the wavelength of the K_α x-ray line of the element with atomic number Z.

> **QUICK QUIZ 29.5** True or False: It is possible for an x-ray spectrum to show the continuous spectrum of x-rays without the presence of the characteristic x-rays.

> **QUICK QUIZ 29.6** In an x-ray tube, as you increase the energy of the electrons striking the metal target, do the wavelengths of the characteristic x-rays (**a**) increase, (**b**) decrease, or (**c**) remain constant?

Example 29.5 | Estimating the Energy of an X-Ray

Estimate the energy of the characteristic x-ray emitted from a tungsten target when an electron drops from an M shell ($n = 3$ state) to a vacancy in the K shell ($n = 1$ state). The atomic number for tungsten is $Z = 74$.

SOLUTION

Conceptualize Imagine an accelerated electron striking a tungsten atom and ejecting an electron from the K shell. Subsequently, an electron in the M shell drops down to fill the vacancy and the energy difference between the states is emitted as an x-ray photon.

Categorize We estimate the results using equations developed in this section, so we categorize this example as a substitution problem.

Use Equation 29.19 and $Z = 74$ for tungsten to estimate the energy associated with the electron in the K shell:

$$E_K \approx -(74)^2(13.6 \text{ eV}) = -7.4 \times 10^4 \text{ eV}$$

Use Equation 29.18 and that nine electrons shield the nuclear charge (eight electrons in the $n = 2$ state and one electron in the $n = 1$ state) to estimate the energy of the M shell:

$$E_M \approx -\frac{(13.6 \text{ eV})(74 - 9)^2}{(3)^2} \approx -6.4 \times 10^3 \text{ eV}$$

Find the energy of the emitted x-ray photon:

$$hf = E_M - E_K \approx -6.4 \times 10^3 \text{ eV} - (-7.4 \times 10^4 \text{ eV})$$

$$\approx 6.8 \times 10^4 \text{ eV} = \boxed{68 \text{ keV}}$$

Consultation of x-ray tables shows that the M–K transition energies in tungsten vary from 66.9 keV to 67.7 keV, where the range of energies is due to slightly different energy values for states of different ℓ. Therefore, our estimate differs from the midpoint of this experimentally measured range by approximately 1%.

29.7 | Context Connection: Atoms in Space

We have spent quite a bit of time on the hydrogen atom in this chapter. Let us now consider hydrogen atoms located in space. Because hydrogen is the most abundant element in the Universe, its role in astronomy and cosmology is very important.

Let us begin by considering pictures of some nebulae you might have seen in an astronomy text, such as Figure 29.18. Time-exposure photographs of these objects show a variety of colors. What causes the colors in these clouds of gas and grains of dust? Let us imagine a cloud of hydrogen atoms in space near a very hot star. The high-energy photons from the star can interact with the hydrogen atoms, either raising them to a high-energy state or ionizing them. As the atoms fall back to the lower states, many atoms emit the Balmer series of wavelengths. Therefore, these atoms provide red, green, blue, and violet colors to the nebula, corresponding to the colors seen in the hydrogen spectrum in Chapter 11.

In practice, nebulae are classified into three groups depending on the transitions occurring in the hydrogen atoms. **Emission nebulae** (Fig. 29.18a) are near a hot star, so hydrogen atoms are excited by light from the star as described above. Therefore, the light from an emission nebula is dominated by discrete emission spectral lines and contains colors. **Reflection nebulae** (Fig. 29.18b) are near a cool star. In these cases, most of the light from the nebula is the starlight reflected from larger grains of material in the nebula rather than emitted by excited atoms. Therefore, the spectrum of the light from the nebula is the same as that from the star: an absorption

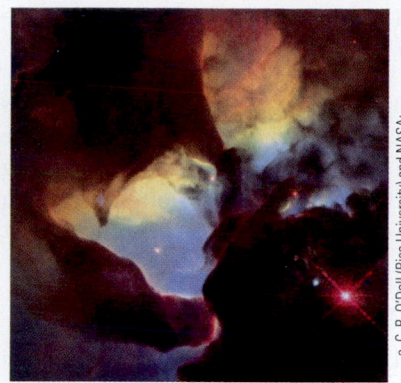

a. C. R. O'Dell (Rice University) and NASA;
b. © Science Photo Library / Alamy;
c. A. Caulet (ST-ECF, ESA) and NASA) 1007 29.16

Figure 29.18 Types of astronomical nebulae. (a) The central part of the Orion Nebula represents an emission nebula, from which colored light is emitted from atoms. (b) The Pleiades. The clouds of light surrounding the stars represent a reflection nebula, from which starlight is reflected by dust particles. (c) The Lagoon Nebula shows the effects of a dark nebula, in which clouds of dust block starlight and appear as a dark silhouette against the light from stars farther away.

spectrum with dark lines corresponding to atoms and ions in the outer regions of the star. The light from these nebulae tends to appear white. Finally, **dark nebulae** (Fig. 29.18c) are not close to a star. Therefore, little radiation is available to excite atoms or reflect from grains of dust. As a result, the material in these nebulae screens out light from stars beyond them, and they appear as black patches against the brightness of the more distant stars.

In addition to hydrogen, some other atoms and ions in space are raised to higher energy states by radiation from stars and proceed to emit various colors. Some of the more prominent colors are violet (373 nm) from the O^+ ion and green (496 nm and 501 nm) from the O^{++} ion. Helium and nitrogen also provide strong colors.

In our discussion of the quantum numbers for the hydrogen atom, we claimed that two states are possible in the 1s shell, corresponding to up or down spin, and that these two states are equivalent in energy in the absence of a magnetic field. If we modify our structural model to include the spin of the proton, however, we find that the two atomic states corresponding to the electron spin are not the same in energy. The state in which the electron and proton spins are parallel is slightly higher in energy than the state in which they are antiparallel. The energy difference is only 5.9×10^{-6} eV. Because these two states differ in energy, it is possible for the atom to make a transition between the states. If the transition is from the parallel state to the antiparallel state, a photon is emitted, with energy equal to the difference in energy between the states. The wavelength of this photon is

$$\lambda = \frac{c}{f} = \frac{hc}{hf} = \frac{hc}{E} = \frac{1\,240 \text{ eV} \cdot \text{nm}}{5.9 \times 10^{-6} \text{ eV}} \left(\frac{10^{-9} \text{ m}}{1 \text{ nm}} \right)$$

$$= 0.21 \text{ m} = 21 \text{ cm}$$

This radiation is called, for obvious reasons, **21-cm radiation.** It is radiation with a wavelength that is identifiable with the hydrogen atom. Therefore, by looking for this radiation in space, we can detect hydrogen atoms. Furthermore, if the wavelength of the observed radiation is not equal to 21 cm, we can infer that it has been Doppler shifted due to relative motion between the Earth and the source. This Doppler shift can then be used to measure the relative speed of the source toward or away from the Earth. This technique has been extensively used to study the hydrogen distribution in the Milky Way galaxy and to detect the presence of spiral arms in our galaxy, similar to the spiral arms in other galaxies.

Our study of atomic physics allows us to understand an important connection between the microscopic world of quantum physics and the macroscopic Universe. Atoms throughout the Universe act as transmitters of information to us about the local conditions. In Chapter 30, which deals with nuclear physics, we shall see how our understanding of microscopic processes helps us understand the local conditions at the center of a star.

SUMMARY

The methods of quantum mechanics can be applied to the hydrogen atom using the appropriate potential energy function $U(r) = -k_e e^2/r$ in the Schrödinger equation. The solution to this equation yields the wave functions for the allowed states and the allowed energies, given by

$$E_n = -\left(\frac{k_e e^2}{2a_0}\right)\frac{1}{n^2} = -\frac{13.606 \text{ eV}}{n^2} \qquad n = 1, 2, 3, \ldots \text{ 29.2} \blacktriangleleft$$

which is precisely the result obtained in the Bohr theory. The allowed energy depends only on the **principal quantum number** n. The allowed wave functions depend on three quantum numbers, n, ℓ, and m_ℓ, where ℓ is the **orbital quantum number** and m_ℓ is the **orbital magnetic quantum number.** The restrictions on the quantum numbers are as follows:

$$n = 1, 2, 3, \ldots$$
$$\ell = 0, 1, 2, \ldots, n - 1$$
$$m_\ell = -\ell, -\ell + 1, \ldots, \ell - 1, \ell$$

All states with the same principal quantum number n form a **shell,** identified by the letters K, L, M, . . . (corresponding to $n = 1, 2, 3, \ldots$). All states with the same values of both n and ℓ form a **subshell,** designated by the letters s, p, d, f, . . . (corresponding to $\ell = 0, 1, 2, 3, \ldots$).

An atom in a state characterized by a specific n can have the following values of **orbital angular momentum** L:

$$L = \sqrt{\ell(\ell + 1)}\hbar \qquad \ell = 0, 1, 2, \ldots, n - 1 \qquad \text{29.9} \blacktriangleleft$$

The allowed values of the projection of the angular momentum vector $\vec{L}$ along the z axis are given by

$$L_z = m_\ell \hbar \qquad \text{29.10} \blacktriangleleft$$

where m_ℓ is restricted to integer values lying between $-\ell$ and ℓ. Only discrete values of L_z are allowed, and they are determined by the restrictions on m_ℓ. This quantization of L_z is referred to as **space quantization.**

To describe a quantum state of the hydrogen atom completely, it is necessary to include a fourth quantum number m_s, called the **spin magnetic quantum number.** This quantum number can have only two values, $\pm\frac{1}{2}$. In effect, this additional quantum number doubles the number of allowed states specified by the quantum numbers n, ℓ, and m_ℓ.

The electron has an intrinsic angular momentum called **spin angular momentum.** That is, the total angular momentum of an atom can have two contributions: one arising from the spin of the electron ($\vec{S}$) and one arising from the orbital motion of the electron ($\vec{L}$).

Electronic spin can be described by a quantum number $s = \frac{1}{2}$. The **magnitude of the spin angular momentum** is

$$S = \frac{\sqrt{3}}{2}\hbar \qquad \text{29.12} \blacktriangleleft$$

and the z component of $\vec{S}$ is

$$S_z = m_s \hbar = \pm\frac{1}{2}\hbar \qquad \text{29.13} \blacktriangleleft$$

The magnetic moment $\vec{\mu}_{\text{spin}}$ associated with the spin angular momentum of an electron is

$$\vec{\mu}_{\text{spin}} = -\frac{e}{m_e}\vec{S} \qquad \text{29.14} \blacktriangleleft$$

The z component of $\vec{\mu}_{\text{spin}}$ can have the values

$$\mu_{\text{spin},z} = \pm\frac{e\hbar}{2m_e} \qquad \text{29.15} \blacktriangleleft$$

The quantity $e\hbar/2m_e$ is called the **Bohr magneton** μ_B and has the numerical value 9.274×10^{-24} J/T.

The **exclusion principle** states that no two electrons in an atom can have the same set of quantum numbers n, ℓ, m_ℓ, and m_s. Using this principle, one can determine the electronic configuration of the elements. This procedure serves as a basis for understanding atomic structure and the chemical properties of the elements.

The allowed electronic transitions between any two states in an atom are governed by the **selection rules**

$$\Delta\ell = \pm 1 \qquad \text{and} \qquad \Delta m_\ell = 0 \text{ or } \pm 1 \qquad \text{29.16} \blacktriangleleft$$

The **x-ray spectrum** of a metal target consists of a set of sharp characteristic lines superimposed on a broad, continuous spectrum. **Bremsstrahlung** is x-radiation with its origin in the slowing down of high-energy electrons as they encounter the target. **Characteristic x-rays** are emitted by atoms when an electron undergoes a transition from an outer shell into an electron vacancy in one of the inner shells.

OBJECTIVE QUESTIONS

☐ denotes answer available in *Student Solutions Manual/Study Guide*

1. Consider the $n = 3$ energy level in a hydrogen atom. How many electrons can be placed in this level? (a) 1 (b) 2 (c) 8 (d) 9 (e) 18

2. When an electron collides with an atom, it can transfer all or some of its energy to the atom. A hydrogen atom is in its ground state. Incident on the atom are several electrons, each having a kinetic energy of 10.5 eV. What is the result? (a) The atom can be excited to a higher allowed state. (b) The atom is ionized. (c) The electrons pass by the atom without interaction.

3. When an atom emits a photon, what happens? (a) One of its electrons leaves the atom. (b) The atom moves to a state of higher energy. (c) The atom moves to a state of lower energy. (d) One of its electrons collides with another particle. (e) None of those events occur.

4. The periodic table is based on which of the following principles? (a) The uncertainty principle. (b) All electrons in an atom must have the same set of quantum numbers. (c) Energy is conserved in all interactions. (d) All electrons in an atom are in orbitals having the same energy.

(e) No two electrons in an atom can have the same set of quantum numbers.

5. If an electron in an atom has the quantum numbers $n = 3$, $\ell = 2$, $m_\ell = 1$, and $m_s = \frac{1}{2}$, what state is it in? (a) $3s$ (b) $3p$ (c) $3d$ (d) $4d$ (e) $3f$

6. What can be concluded about a hydrogen atom with its electron in the d state? (a) The atom is ionized. (b) The orbital quantum number is $\ell = 1$. (c) The principal quantum number is $n = 2$. (d) The atom is in its ground state. (e) The orbital angular momentum of the atom is not zero.

7. Which of the following electronic configurations are *not* allowed for an atom? Choose all correct answers. (a) $2s^2 2p^6$ (b) $3s^2 3p^7$ (c) $3d^7 4s^2$ (d) $3d^{10} 4s^2 4p^6$ (e) $1s^2 2s^2 2d^1$

8. (a) In the hydrogen atom, can the quantum number n increase without limit? (b) Can the frequency of possible discrete lines in the spectrum of hydrogen increase without limit? (c) Can the wavelength of possible discrete lines in the spectrum of hydrogen increase without limit?

9. Consider the quantum numbers (a) n, (b) ℓ, (c) m_ℓ, and (d) m_s. (i) Which of these quantum numbers are fractional as opposed to being integers? (ii) Which can sometimes attain negative values? (iii) Which can be zero?

10. (i) What is the principal quantum number of the initial state of an atom as it emits an M_β line in an x-ray spectrum? (a) 1 (b) 2 (c) 3 (d) 4 (e) 5 (ii) What is the principal quantum number of the final state for this transition? Choose from the same possibilities as in part (i).

▢ denotes answer available in *Student Solutions Manual/Study Guide*

> CONCEPTUAL QUESTIONS |

1. Suppose the electron in the hydrogen atom obeyed classical mechanics rather than quantum mechanics. Why should a gas of such hypothetical atoms emit a continuous spectrum rather than the observed line spectrum?

2. Why are three quantum numbers needed to describe the state of a one-electron atom (ignoring spin)?

3. Compare the Bohr theory and the Schrödinger treatment of the hydrogen atom, specifically commenting on their treatment of total energy and orbital angular momentum of the atom.

4. (a) According to Bohr's model of the hydrogen atom, what is the uncertainty in the radial coordinate of the electron? (b) What is the uncertainty in the radial component of the velocity of the electron? (c) In what way does the model violate the uncertainty principle?

5. Could the Stern–Gerlach experiment be performed with ions rather than neutral atoms? Explain.

6. An energy of about 21 eV is required to excite an electron in a helium atom from the $1s$ state to the $2s$ state. The same transition for the He^+ ion requires approximately twice as much energy. Explain.

7. Why do lithium, potassium, and sodium exhibit similar chemical properties?

8. It is easy to understand how two electrons (one spin up, one spin down) fill the $n = 1$ or K shell for a helium atom. How is it possible that eight more electrons are allowed in the $n = 2$ shell, filling the K and L shells for a neon atom?

9. Why is a *nonuniform* magnetic field used in the Stern–Gerlach experiment?

10. Discuss some consequences of the exclusion principle.

> PROBLEMS AVAILABLE IN ⬭ENHANCED⬭ WebAssign

29.1 Early Structural Models of the Atom

    Problems 1–5

29.2 The Hydrogen Atom Revisited

    Problems 6–11

29.3 The Wave Functions for Hydrogen

    Problems 12–16

29.4 Physical Interpretation of the Quantum Numbers

    Problems 17–26

29.5 The Exclusion Principle and the Periodic Table

    Problems 27–34

29.6 More on Atomic Spectra: Visible and X-Ray

    Problems 35–40

29.7 Context Connection: Atoms in Space

    Problems 41–45

Additional Problems

    Problems 46–66

Solutions to the following Problems are available in the *Student Solutions Manual/Study Guide*:

    29.2, 29.9, 29.10, 29.16, 29.25, 29.32, 29.35, 29.43, 29.50, 29.51, 29.56, 29.57, 29.61, and 29.65

List of Enhanced Problems

Problem Number	Targeted Feedback in Enhanced WebAssign	Master It in Enhanced WebAssign	Watch It in Enhanced WebAssign
29.8	✓		✓
29.18		✓	
29.24	✓		✓
29.35	✓	✓	
29.43	✓	✓	
29.57	✓		
29.59	✓		✓
29.61		✓	

Nuclear Physics

Chapter Outline

© Vienna Report Agency/Sygma/Corbis

Ötzi the Iceman, a Copper Age man, was discovered by German tourists in the Italian Alps in 1991 when a glacier melted enough to expose his remains. Analysis of his corpse has exposed his last meal, illnesses he suffered, and places he lived. Radioactivity was used to determine that he lived in about 3300 BC. Ötzi can be seen today in the Südtiroler Archäologiemuseum (South Tyrol Museum of Archaeology) in Bolzano, Italy.

In 1896, the year that marked the birth of nuclear physics, Antoine-Henri Becquerel (1852–1908) introduced the world of science to radioactivity in uranium compounds by accidentally discovering that uranyl potassium sulfate crystals emit an invisible radiation that can darken a photographic plate when the plate is covered to exclude light. After a series of experiments, he concluded that the radiation emitted by the crystals was of a new type, one that requires no external stimulation and is so penetrating that it can darken protected photographic plates and ionize gases.

A great deal of research followed as scientists attempted to understand the radiation emitted by radioactive nuclei. Pioneering work by Rutherford showed that the radiation was of three types, which he called alpha, beta, and gamma rays. Later experiments showed that alpha rays are helium nuclei, beta rays are electrons or related particles called positrons, and gamma rays are high-energy photons.

As we saw in Section 29.1, the 1911 experiments of Rutherford established that the nucleus of an atom has a very small volume and that most of the atomic mass is contained in the nucleus. Furthermore, such studies demonstrated a new type of force, the nuclear force, first introduced in Section 5.5, that is predominant at distances on the order of 10^{-15} m and essentially zero at distances larger than that.

In this chapter, we discuss the structure of the atomic nucleus. We shall describe the basic properties of nuclei, nuclear forces, nuclear binding energy, the phenomenon of radioactivity, and nuclear reactions.

❮ 30.1 | Some Properties of Nuclei

In the commonly accepted structural model of the nucleus, all nuclei are composed of two types of particles: protons and neutrons. The only exception is the ordinary hydrogen nucleus, which is a single proton with no neutrons. In describing the atomic nucleus, we identify the following integer quantities:

- the **atomic number** Z, which equals the number of protons in the nucleus (sometimes called the *charge number*)
- the **neutron number** N, which equals the number of neutrons in the nucleus
- the **mass number** $A = Z + N$, which equals the number of **nucleons** (neutrons plus protons) in the nucleus

> **Pitfall Prevention | 30.1**
> **Mass Number Is Not Atomic Mass**
> The mass number A should not be confused with the atomic mass. Mass number is an integer specific to an isotope and has no units; it is simply a count of the number of nucleons. Atomic mass has units and is generally not an integer because it is an average of the masses of a given element's naturally occurring isotopes.

A **nuclide** is a specific combination of atomic number and mass number that represents a nucleus. In representing nuclides, is convenient to have a symbolic representation that shows how many protons and neutrons are present. The symbol used is $_{Z}^{A}X$, where X represents the chemical symbol for the element. For example, $_{26}^{56}$Fe (iron) has a mass number of 56 and an atomic number of 26; therefore, it contains 26 protons and 30 neutrons. When no confusion is likely to arise, we omit the subscript Z because the chemical symbol can always be used to determine Z. Therefore, $_{26}^{56}$Fe is the same as ^{56}Fe and can also be expressed as "iron-56."

The nuclei of all atoms of a particular element contain the same number of protons but often contain different numbers of neutrons. Nuclei that are related in this way are called **isotopes.** The isotopes of an element have the same Z value but different N and A values. The natural abundances of isotopes can differ substantially. For example, $_{6}^{11}$C, $_{6}^{12}$C, $_{6}^{13}$C, and $_{6}^{14}$C are four isotopes of carbon. The natural abundance of the $_{6}^{12}$C isotope is about 98.9%, whereas that of the $_{6}^{13}$C isotope is only about 1.1%. ($_{6}^{11}$C and $_{6}^{14}$C exist in trace amounts.) Even the simplest element, hydrogen, has isotopes: $_{1}^{1}$H, the ordinary hydrogen nucleus; $_{1}^{2}$H, deuterium; and $_{1}^{3}$H, tritium. Some isotopes do not occur naturally but can be produced in the laboratory through nuclear reactions.

❮ **QUICK QUIZ 30.1** For each part of this Quick Quiz, choose from the following answers: (a) protons (b) neutrons (c) nucleons. (**i**) The three nuclei ^{12}C, ^{13}N, and ^{14}O have the same number of what type of particle? (**ii**) The three nuclei ^{12}N, ^{13}N, and ^{14}N have the same number of what type of particle? (**iii**) The three nuclei ^{14}C, ^{14}N, and ^{14}O have the same number of what type of particle?

Charge and Mass

The proton carries a single positive charge $+e$ and the electron carries a single negative charge $-e$, where $e = 1.60 \times 10^{-19}$ C. The neutron is electrically neutral, as its name implies. Because the neutron has no charge, it was difficult to detect with early experimental apparatus and techniques. Today we can detect neutrons relatively easily with modern detection devices.

A convenient unit for measuring mass on a nuclear scale is the **atomic mass unit** u. This unit is defined in such a way that the atomic mass of the isotope $_{6}^{12}$C is exactly 12 u, where 1 u $= 1.660\ 539 \times 10^{-27}$ kg. The proton and neutron each have a mass of approximately 1 u, and the electron has a mass that is only a small fraction of an atomic mass unit:

Mass of proton = 1.007 276 u
Mass of neutron = 1.008 665 u
Mass of electron = 0.000 548 6 u

◀ TABLE 30.1 | Masses of Selected Particles in Various Units

Particle	kg	Mass u	MeV/c^2
Proton	$1.672\ 62 \times 10^{-27}$	$1.007\ 276$	938.27
Neutron	$1.674\ 93 \times 10^{-27}$	$1.008\ 665$	939.57
Electron	$9.109\ 38 \times 10^{-31}$	$5.485\ 80 \times 10^{-4}$	$0.510\ 999$
$^{1}_{1}$H atom	$1.673\ 53 \times 10^{-27}$	$1.007\ 825$	938.783
$^{4}_{2}$He nucleus	$6.644\ 66 \times 10^{-27}$	$4.001\ 506$	$3\ 727.38$
$^{12}_{6}$C atom	$1.992\ 65 \times 10^{-27}$	$12.000\ 000$	$11\ 177.9$

Because the rest energy of a particle is given by $E_R = mc^2$ (Section 9.7), it is often convenient to express the atomic mass unit in terms of its rest energy equivalent. For one atomic mass unit, we have

$$E_R = mc^2 = (1.660\ 539 \times 10^{-27}\ \text{kg})(2.997\ 92 \times 10^8\ \text{m/s})^2 = 931.494\ \text{MeV}/c^2$$

where we have used the conversion 1 eV $= 1.602\ 176 \times 10^{-19}$ J. Using this equivalence, nuclear physicists often express mass in terms of the unit MeV/c^2. The masses of several simple particles are given in Table 30.1. The masses and some other properties of selected isotopes are provided in Table A.3 in Appendix A.

The Size of Nuclei

The size and structure of nuclei were first investigated in the scattering experiments of Rutherford, discussed in Section 29.1. Using the principle of conservation of energy, Rutherford found an expression for how close an alpha particle moving directly toward the nucleus can approach the nucleus before being turned around by Coulomb repulsion.

Let us consider the system of the incoming alpha particle ($Z = 2$) and the nucleus (arbitrary Z), and apply the energy version of the isolated system model. Because the nucleus is assumed to be much more massive than the alpha particle, we identify the kinetic energy of the system as the kinetic energy of the alpha particle alone. When the alpha particle and the nucleus are far apart, we can approximate the potential energy of the system as zero. If the collision is head-on, the alpha particle stops momentarily at some point (Active Fig. 30.1) and the energy of the system is entirely potential. Therefore, the initial kinetic energy of the incoming alpha particle is converted completely to electric potential energy of the system when the particle stops:

$$\tfrac{1}{2}mv^2 = k_e \frac{q_1 q_2}{r} = k_e \frac{(2e)(Ze)}{d}$$

where d is the distance of closest approach, Z is the atomic number of the target nucleus, and we have used the nonrelativistic expression for kinetic energy because speeds of alpha particles from radioactive decay are small relative to c. Solving for d, we find that

$$d = \frac{4k_e Ze^2}{mv^2}$$

From this expression, Rutherford found that alpha particles approached to within 3.2×10^{-14} m of a nucleus when the foil was made of gold. Based on this calculation and his analysis of results for collisions that were not head-on, Rutherford argued that the radius of the gold nucleus must be less than this value. For silver atoms, the distance of closest approach was found to be 2×10^{-14} m. From these results, Rutherford reached his conclusion that the positive charge in an atom is concentrated in a small sphere called the nucleus, whose radius is no greater than about 10^{-14} m. Note that this radius is on the order of 10^{-4} of the Bohr radius, corresponding to a nuclear volume which is on the order of 10^{-12} of the volume of a hydrogen atom. The nucleus is an incredibly small part of the atom! Because such small

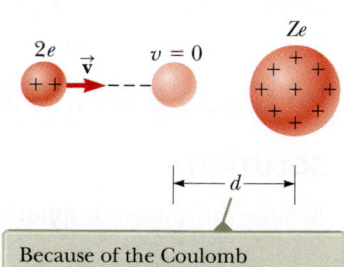

Because of the Coulomb repulsion between the charges of the same sign, the alpha particle approaches to a distance d from the nucleus, called the distance of closest approach.

Active Figure 30.1 An alpha particle on a head-on collision course with a nucleus of charge Ze.

Figure 30.2 A nucleus can be modeled as a cluster of tightly packed spheres, where each sphere is a nucleon.

lengths are common in nuclear physics, a convenient unit of length is the *femtometer* (fm), sometimes called the **fermi**, defined as

$$1 \text{ fm} \equiv 10^{-15} \text{ m}$$

Since the time of Rutherford's scattering experiments, a multitude of other experiments have shown that most nuclei can be geometrically modeled as being approximately spherical with an average radius of

$$r = aA^{1/3} \qquad \qquad \textbf{30.1} \blacktriangleleft$$

where A is the mass number and a is a constant equal to 1.2×10^{-15} m. Because the volume of a sphere is proportional to the cube of the radius, it follows from Equation 30.1 that the volume of a nucleus (assumed to be spherical) is directly proportional to A, the total number of nucleons, which suggests that all nuclei have nearly the same density. Nucleons combine to form a nucleus as though they were tightly packed spheres (Fig. 30.2).

Example **30.1** | **The Volume and Density of a Nucleus**

Consider a nucleus of mass number A.

(A) Find an approximate expression for the mass of the nucleus.

SOLUTION

Conceptualize Imagine the nucleus to be a collection of protons and neutrons as shown in Figure 30.2. The mass number A counts *both* protons and neutrons.

Categorize Let's assume A is large enough that we can imagine the nucleus to be spherical.

. . . .

Analyze The mass of the proton is approximately equal to that of the neutron. Therefore, if the mass of one of these particles is m, the mass of the nucleus is approximately Am.

(B) Find an expression for the volume of this nucleus in terms of A.

SOLUTION

Assume the nucleus is spherical and use Equation 30.1:

$$(1) \quad V_{\text{nucleus}} = \tfrac{4}{3}\pi r^3 = \tfrac{4}{3}\pi a^3 A$$

(C) Find a numerical value for the density of this nucleus.

SOLUTION

Use Equation 1.1 and substitute Equation (1):

$$\rho = \frac{m_{\text{nucleus}}}{V_{\text{nucleus}}} = \frac{Am}{\tfrac{4}{3}\pi a^3 A} = \frac{3m}{4\pi a^3}$$

Substitute numerical values:

$$\rho = \frac{3(1.67 \times 10^{-27} \text{ kg})}{4\pi(1.2 \times 10^{-15} \text{ m})^3} = 2.3 \times 10^{17} \text{ kg/m}^3$$

. . . .

Finalize The nuclear density is approximately 2.3×10^{14} times the density of water ($\rho_{\text{water}} = 1.0 \times 10^3 \text{ kg/m}^3$).

What If? What if the Earth could be compressed until it had this density? How large would it be?

Answer Because this density is so large, we predict that an Earth of this density would be very small.

Use Equation 1.1 and the mass of the Earth to find the volume of the compressed Earth:

$$V = \frac{M_E}{\rho} = \frac{5.97 \times 10^{24} \text{ kg}}{2.3 \times 10^{17} \text{ kg/m}^3} = 2.6 \times 10^7 \text{ m}^3$$

From this volume, find the radius:

$$V = \tfrac{4}{3}\pi r^3 \quad \rightarrow \quad r = \left(\frac{3V}{4\pi}\right)^{1/3} = \left[\frac{3(2.6 \times 10^7 \text{ m}^3)}{4\pi}\right]^{1/3}$$

$$r = 1.8 \times 10^2 \text{ m}$$

An Earth of this radius is indeed a small Earth!

Nuclear Stability

Because the nucleus consists of a closely packed collection of protons and neutrons, you might be surprised that it can exist at all. The very large repulsive electrostatic forces between protons in close proximity should cause the nucleus to fly apart. Nuclei are stable, however, because of the presence of another force, the **nuclear force** (see Section 5.5). This short-range force (it is nonzero only for particle separations less than about 2 fm) is an attractive force that acts between all nucleons.

The nuclear force dominates the Coulomb repulsive force within the nucleus (at short ranges). If that were not the case, stable nuclei would not exist. Moreover, the nuclear force is independent of charge. In other words, the forces associated with the proton–proton, proton–neutron, and neutron–neutron interactions are the same, apart from the additional repulsive Coulomb force for the proton–proton interaction.

Evidence for the limited range of nuclear forces comes from scattering experiments and from studies of nuclear binding energies, which we shall discuss shortly. The short range of the nuclear force is shown in the neutron–proton (n–p) potential energy plot of Figure 30.3a obtained by scattering neutrons from a target containing hydrogen. The depth of the n–p potential energy well is 40 to 50 MeV, and a strong repulsive component prevents the nucleons from approaching much closer than 0.4 fm.

The nuclear force does not affect electrons, enabling energetic electrons to serve as point-like probes of the charge density of nuclei. The charge independence of the nuclear force also means that the main difference between the n–p and p–p interactions is that the p–p potential energy consists of a *superposition* of nuclear and Coulomb interactions as shown in Figure 30.3b. At distances less than 2 fm, the p–p and n–p potential energies are nearly identical, but for distances greater than this, the p–p potential has a positive energy barrier with a maximum at 4 fm.

About 260 stable nuclei exist; hundreds of other nuclei have been observed but are unstable. A useful graphical representation in nuclear physics is a plot of N versus Z for stable nuclei as shown in Figure 30.4. Note that light nuclei are stable if they contain equal numbers of protons and neutrons—that is, if $N = Z$—but heavy nuclei are stable if $N > Z$. This behavior can be partially understood by recognizing that as the number of protons increases, the strength of the Coulomb force increases, which tends to break the nucleus apart. As a result, more neutrons are needed to keep the nucleus stable because neutrons experience only the attractive nuclear force. Eventually, when $Z = 83$, the repulsive forces between protons cannot be compensated by the addition of more neutrons. Elements that contain more than 83 protons do not have stable nuclei.

Interestingly, most stable nuclei have even values of A. In fact, certain values of Z and N correspond to nuclei with unusually high stability. These values of Z and N, called **magic numbers,** are

$$Z \text{ or } N = 2, 8, 20, 28, 50, 82, 126 \qquad \textbf{30.2} \blacktriangleleft$$

For example, the helium nucleus (two protons and two neutrons), which has $Z = 2$ and $N = 2$, is very stable. This stability is reminiscent of the chemical stability of inert gases and suggests quantized nuclear energy levels, which we indeed find to be the case. Some structural models of the nucleus predict a shell structure similar to that for the atom.

Nuclear Spin and Magnetic Moment

In Chapter 29, we discussed that an electron has an intrinsic angular momentum called spin. Protons and neutrons, like electrons,

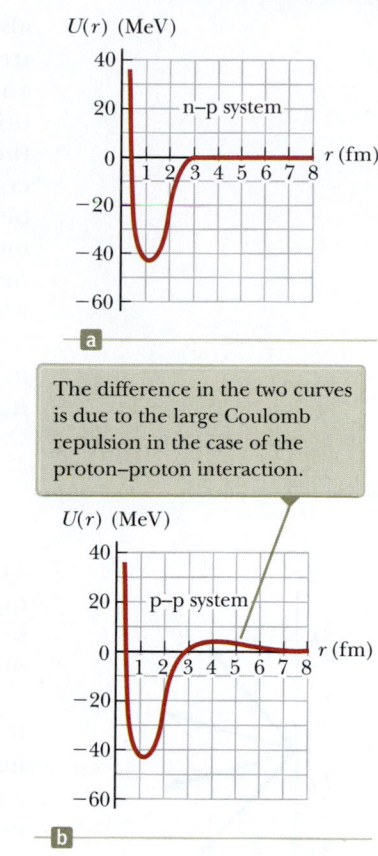

Figure 30.3 (a) Potential energy versus separation distance for the neutron–proton system. (b) Potential energy versus separation distance for the proton–proton system. To display the difference in the curves on this scale, the height of the peak for the proton–proton curve has been exaggerated by a factor of 10.

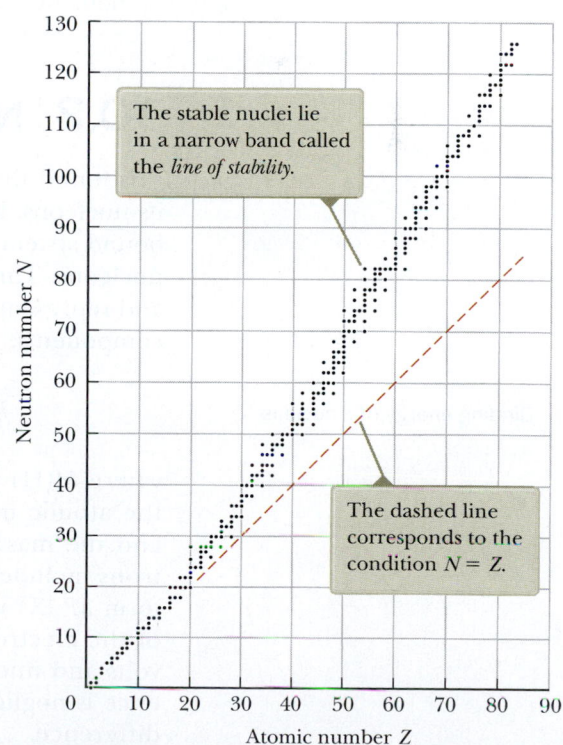

Figure 30.4 Neutron number N versus atomic number Z for the stable nuclei (black dots).

also have an intrinsic angular momentum. Furthermore, a nucleus has a net intrinsic angular momentum that arises from the individual spins of the protons and neutrons. This angular momentum must obey the same quantum rules as orbital angular momentum and spin (Section 29.4). Therefore, the magnitude of the **nuclear angular momentum** is due to the combination of all nucleons and is equal to $\sqrt{I(I + 1)}\hbar$, where I is called the **nuclear spin quantum number** and may be an integer or a half-integer. The maximum component of the nuclear angular momentum projected along any direction is $I\hbar$. Figure 30.5 illustrates the possible orientations of the nuclear spin and its projections along the z axis for the case where $I = \frac{3}{2}$.

The nuclear angular momentum has a nuclear magnetic moment associated with it. The magnetic moment of a nucleus is measured in terms of the **nuclear magneton** μ_n, a unit of magnetic moment defined as

▶ Nuclear magneton

$$\mu_n \equiv \frac{e\hbar}{2m_p} = 5.05 \times 10^{-27} \text{ J/T} \qquad \textbf{30.3}◀$$

This definition is analogous to Equation 29.15 for the z component of the spin magnetic moment for an electron, which is the Bohr magneton μ_B. Note that μ_n is smaller than μ_B by a factor of about 2 000 because of the large difference in masses of the proton and electron.

The magnetic moment of a free proton is $2.792\ 8\mu_n$. Unfortunately, no general theory of nuclear magnetism explains this value. Another surprising point is that a neutron, despite having no electric charge, also has a magnetic moment, which has a value of $-1.913\ 5\mu_n$. The negative sign indicates that the neutron's magnetic moment is opposite its spin angular momentum. Such a magnetic moment for a neutral particle suggests that we need to design a structural model for the neutron that explains such an observation. This structural model, the *quark model*, will be discussed in Chapter 31.

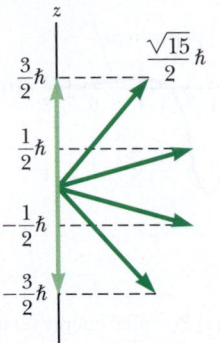

Figure 30.5 A vector model showing possible orientations of the nuclear spin angular momentum vector and its projections along the z axis for the case $I = \frac{3}{2}$.

QUICK QUIZ 30.2 Which do you expect to show very little variation among different isotopes of an element? **(a)** atomic mass **(b)** nuclear spin magnetic moment **(c)** chemical behavior

30.2 | Nuclear Binding Energy

It is found that the mass of a nucleus is always less than the sum of the masses of its nucleons. Because mass is a manifestation of energy, the total rest energy of the bound system (the nucleus) is less than the combined rest energy of the separated nucleons. This difference in energy is called the **binding energy** E_b of the nucleus and represents the energy that must be added to a nucleus to break it apart into its components:

▶ Binding energy of a nucleus

$$E_b(\text{MeV}) = [ZM(\text{H}) + Nm_n - M(_Z^A\text{X})] \times 931.494 \text{ MeV/u} \qquad \textbf{30.4}◀$$

where $M(\text{H})$ is the atomic mass of the neutral hydrogen atom, $M(_Z^A\text{X})$ represents the atomic mass of an atom of the isotope $_Z^A\text{X}$, m_n is the mass of the neutron, and the masses are all in atomic mass units. Note that the mass of the Z electrons included in $M(\text{H})$ cancels with the mass of the Z electrons included in the term $M(_Z^A\text{X})$ within a small difference associated with the atomic binding energy of the electrons. Because atomic binding energies are typically several electron volts and nuclear binding energies are several million electron volts, this difference is negligible, and we adopt a simplification model in which we ignore this difference.

Example 30.2 | The Binding Energy of the Deuteron

Calculate the binding energy of the deuteron (the nucleus of a deuterium atom), which consists of a proton and a neutron, given that the atomic mass of deuterium is 2.014 102 u.

SOLUTION

Conceptualize Imagine how Figure 30.2 would look if there were only two nucleons, as in the nucleus of a deuterium atom. The nucleus would clearly not be spherical. The binding energy expression in Equation 30.4 does not depend on the shape of the nucleus, so it is valid in this situation.

Categorize We will simply be applying Equation 30.4 to find the result, so we categorize this example as a substitution problem.

From Table 30.1, we see that the mass of the hydrogen atom, representing the proton, is $M(H) = 1.007\ 825$ u and that the neutron mass $m_n = 1.008\ 665$ u. From this information, find the binding energy of the deuteron:

$$E_b(\text{MeV}) = [(1)(1.007\ 825\ \text{u}) + (1)(1.008\ 665\ \text{u})$$
$$- 2.014\ 102\ \text{u}] \times 931.494\ \text{MeV/u}$$
$$= \boxed{2.224\ \text{MeV}}$$

This result tells us that separating a deuteron into its constituent proton and neutron requires adding 2.224 MeV of energy to the deuteron. One way of supplying the deuteron with this energy is by bombarding it with energetic particles.

A plot of binding energy per nucleon E_b/A as a function of mass number for various stable nuclei is shown in Figure 30.6. Note that the curve has a maximum in the vicinity of $A = 60$, corresponding to isotopes of iron, cobalt, and nickel. That is, nuclei having mass numbers either greater or less than 60 are not as strongly bound as those near the middle of the periodic table. The higher values of binding energy per nucleon near $A = 60$ imply that energy is released when a heavy nucleus (near the right of the graph) splits, or *fissions*, into two lighter nuclei. Energy is released in fission because the nucleons in each product nucleus are more tightly bound to one another than are the nucleons in the original nucleus. The important process of fission and a second important process of *fusion*, in which energy is released as light nuclei combine, are considered in detail in Section 30.6.

Pitfall Prevention | 30.2
Binding Energy
When separate nucleons are combined to form a nucleus, the energy of the system is reduced. Therefore, the change in energy is negative. The absolute value of this change is called the binding energy. This difference in sign may be confusing. For example, an *increase* in binding energy corresponds to a *decrease* in the energy of the system.

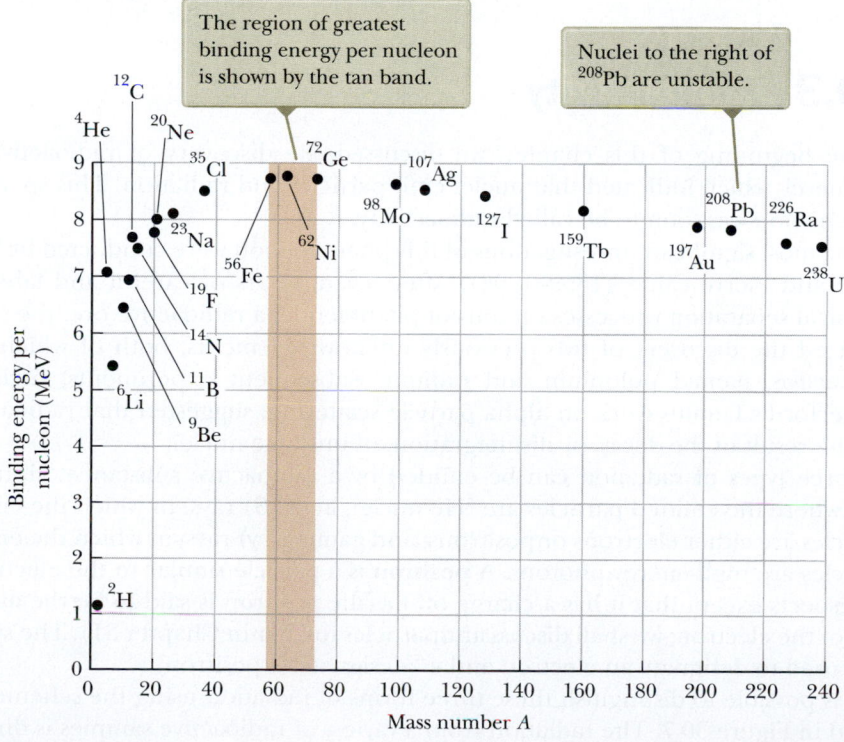

Figure 30.6 Binding energy per nucleon versus mass number for nuclides that lie along the line of stability in Figure 30.4. Some representative nuclei appear as black dots with labels.

The binding energy per nucleon in Figure 30.6 is approximately constant at 8 MeV for $A > 20$. In this case, the nuclear forces between a particular nucleon and all the other nucleons in the nucleus are said to be *saturated*; that is, a particular nucleon interacts with only a limited number of other nucleons because of the short-range character of the nuclear force. These other nucleons can be viewed as being the nearest neighbors in the closely packed structure shown in Figure 30.2.

Figure 30.6 provides insight into fundamental questions about the origin of the chemical elements. In the early life of the Universe, there were only hydrogen and helium. Clouds of cosmic gas coalesced under gravitational forces to form stars. As a star ages, it produces heavier elements from the lighter elements contained within it, beginning by fusing hydrogen atoms to form helium. This process continues as the star becomes older, generating atoms having larger and larger atomic numbers. The nuclide $^{62}_{28}$Ni has the largest binding energy per nucleon of 8.794 5 MeV/nucleon. It takes additional energy to create elements in a star with mass numbers larger than 62 because of their lower binding energies per nucleon. This energy comes from the supernova explosion that occurs at the end of some large stars' lives. Therefore, all the heavy atoms in your body were produced from the explosions of ancient stars. You are literally made of stardust!

▸ THINKING PHYSICS 30.1

Figure 30.6 shows a graph of the average amount of energy necessary to remove a nucleon from the nucleus. Figure 29.13 shows the energy necessary to remove an electron from an atom. Why does Figure 30.6 show an *approximately constant* amount of energy necessary to remove a nucleon (above about $A = 20$), but Figure 29.13 shows *widely varying* amounts of energy necessary to remove an electron from the atom?

Reasoning In the case of Figure 30.6, the approximately constant value of the nuclear binding energy is a result of the short-range nature of the nuclear force. A given nucleon interacts only with its few nearest neighbors, rather than with all the nucleons in the nucleus. Therefore, no matter how many nucleons are present in the nucleus, removing one nucleon involves separating it only from its nearest neighbors. The energy to do so is therefore approximately independent of how many nucleons are present.

On the other hand, the electric force holding the electrons to the nucleus in an atom is a long-range force. An electron in the atom interacts with *all* the protons in the nucleus. When the nuclear charge increases, a stronger attraction occurs between the nucleus and the electrons. As a result, as the nuclear charge increases, more energy is necessary to remove an electron, as demonstrated by the upward tendency of the ionization energy in Figure 29.13 for each period. ◂

▸ 30.3 | Radioactivity

At the beginning of this chapter, we discussed the discovery of radioactivity by Becquerel, which indicated that nuclei emit particles and radiation. This spontaneous emission was soon to be called **radioactivity.**

The most significant investigations of this phenomenon were conducted by Marie Curie and Pierre Curie (1859–1906). After several years of careful and laborious chemical separation processes on tons of pitchblende, a radioactive ore, the Curies reported the discovery of two previously unknown elements, both of which were radioactive, named polonium and radium. Subsequent experiments, including Rutherford's famous work on alpha particle scattering, suggested that radioactivity was the result of the decay, or disintegration, of unstable nuclei.

Three types of radiation can be emitted by a radioactive substance: alpha (α) rays, where the emitted particles are ^{4}He nuclei; beta (β) rays, in which the emitted particles are either electrons or positrons; and gamma (γ) rays, in which the emitted particles are high-energy photons. A **positron** is a particle similar to the electron in all respects except that it has a charge of $+e$ (the positron is said to be the **antiparticle** of the electron; we shall discuss antiparticles further in Chapter 31). The symbol e^- is used to designate an electron and e^+ designates a positron.

It is possible to distinguish these three forms of radiation using the scheme illustrated in Figure 30.7. The radiation from a variety of radioactive samples is directed

Marie Curie
Polish Scientist (1867–1934)
In 1903, Marie Curie shared the Nobel Prize in Physics with her husband, Pierre, and with Becquerel for their studies of radioactive substances. In 1911, she was awarded a Nobel Prize in Chemistry for the discovery of radium and polonium.

into a region with a magnetic field. The radiation is separated into three components by the magnetic field, two bending in opposite directions and the third experiencing no change in direction. From this simple observation, one can conclude that the radiation of the undeflected beam carries no charge (the gamma ray), the component deflected upward corresponds to positively charged particles (alpha particles), and the component deflected downward corresponds to negatively charged particles (e^-). If the beam included positrons (e^+), these particles would be deflected upward with a different radius of curvature from that of the alpha particles.

The three types of radiation have quite different penetrating powers. Alpha particles barely penetrate a sheet of paper, beta particles can penetrate a few millimeters of aluminum, and gamma rays can penetrate several centimeters of lead.

The rate at which a decay process occurs in a radioactive sample is proportional to the number of radioactive nuclei present in the sample (i.e., those nuclei that have not yet decayed). This dependence is similar to the behavior of population growth in that the rate at which babies are born is proportional to the number of people currently alive. If N is the number of radioactive nuclei present at some instant, the rate of change of N is

$$\frac{dN}{dt} = -\lambda N \qquad \textbf{30.5} \blacktriangleleft$$

where λ is called either the **decay constant** or the **disintegration constant** and has a different value for different nuclei. The negative sign indicates that dN/dt is a negative number; that is, N decreases in time.

If we write Equation 30.5 in the form

$$\frac{dN}{N} = -\lambda \ dt$$

we can integrate from an arbitrary initial instant $t = 0$ to a later time t:

$$\int_{N_0}^{N} \frac{dN}{N} = -\lambda \int_{0}^{t} dt$$

$$\ln\left(\frac{N}{N_0}\right) = -\lambda t$$

$$N = N_0 e^{-\lambda t} \qquad \textbf{30.6} \blacktriangleleft$$

The constant N_0 represents the number of undecayed radioactive nuclei at $t = 0$. We have seen exponential behaviors before, for example, with the discharging of a capacitor in Section 21.9. Based on these experiences, we can identify the inverse of the decay constant $1/\lambda$ as the time interval required for the number of undecayed nuclei to fall to $1/e$ of its original value. Therefore, $1/\lambda$ is the **time constant** for this decay, similar to the time constants we investigated for the decay of the current in an RC circuit in Section 21.9 and an RL circuit in Section 23.6.

The **decay rate** R is obtained by differentiating Equation 30.6 with respect to time:

$$R = \left|\frac{dN}{dt}\right| = N_0 \lambda e^{-\lambda t} = R_0 e^{-\lambda t} \qquad \textbf{30.7} \blacktriangleleft$$

where $R = N\lambda$ and $R_0 = N_0\lambda$ is the decay rate at $t = 0$. The decay rate of a sample is often referred to as its **activity**. Note that both N and R decrease exponentially with time. The plot of N versus t in Active Figure 30.8 illustrates the exponential decay law.

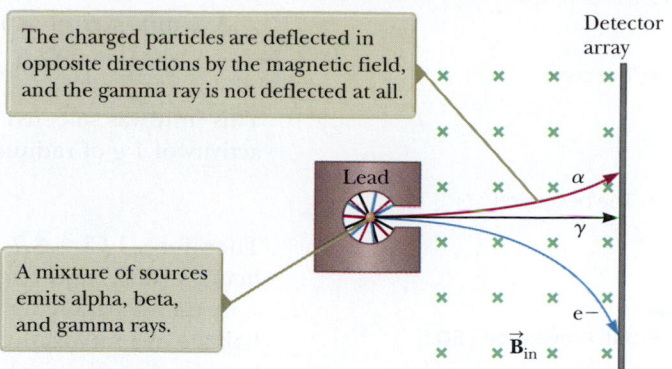

The charged particles are deflected in opposite directions by the magnetic field, and the gamma ray is not deflected at all.

A mixture of sources emits alpha, beta, and gamma rays.

Figure 30.7 The radiation from radioactive sources can be separated into three components by using a magnetic field to deflect the charged particles. The detector array at the right records the events.

Pitfall Prevention | 30.3
Rays or Particles?
Early in the history of nuclear physics, the term radiation was used to describe the emanations from radioactive nuclei. We now know that alpha radiation and beta radiation involve the emission of particles with nonzero rest energy. Even though they are not examples of electromagnetic radiation, the use of the term *radiation* for all three types of emission is deeply entrenched in our language and in the physics community.

Pitfall Prevention | 30.4
Notation Warning
In Section 30.1, we introduced the symbol N as an integer representing the number of neutrons in a nucleus. In this discussion, the symbol N represents the number of undecayed nuclei in a radioactive sample remaining after some time interval. As you read further, be sure to consider the context to determine the appropriate meaning for the symbol N.

▶ Number of undecayed nuclei as a function of time

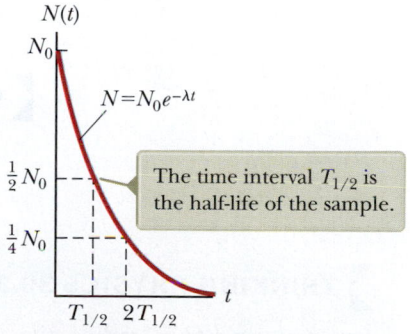

The time interval $T_{1/2}$ is the half-life of the sample.

Active Figure 30.8 Plot of the exponential decay law for radioactive nuclei. The vertical axis represents the number of undecayed radioactive nuclei present at any time t, and the horizontal axis is time.

A common unit of activity for a radioactive sample is the **curie** (Ci), defined as

$$1 \text{ Ci} \equiv 3.7 \times 10^{10} \text{ decays/s}$$

▶ The curie

This unit was selected as the original unit of activity because it is the approximate activity of 1 g of radium. The SI unit of activity is called the **becquerel** (Bq):

$$1 \text{ Bq} \equiv 1 \text{ decay/s}$$

▶ The becquerel

Therefore, $1 \text{ Ci} = 3.7 \times 10^{10}$ Bq. The most commonly used units of activity are millicuries (mCi) and microcuries (μCi).

A useful parameter for characterizing radioactive decay is the **half-life** $T_{1/2}$. The half-life of a radioactive substance is the time interval required for half of a given number of radioactive nuclei to decay. Setting $N = N_0/2$ and $t = T_{1/2}$ in Equation 30.6 gives

$$\frac{N_0}{2} = N_0 e^{-\lambda T_{1/2}}$$

Pitfall Prevention | 30.5
Half-Life
It is *not* true that all the original nuclei have decayed after two half-lives! In one half-life, half of the original nuclei will decay. In the second half-life, half of those remaining will decay, leaving 1/4 of the original number.

Writing this equation in the form $e^{\lambda T_{1/2}} = 2$ and taking the natural logarithm of both sides, we have

$$T_{1/2} = \frac{\ln 2}{\lambda} = \frac{0.693}{\lambda}$$ **30.8** ◀

▶ Half-life

which is a convenient expression relating the half-life to the decay constant. Note that after a time interval of one half-life, $N_0/2$ radioactive nuclei remain (by definition); after two half-lives, half of these have decayed and $N_0/4$ radioactive nuclei remain; after three half-lives, $N_0/8$ remain; and so on. In general, after n half-lives, the number of radioactive nuclei remaining is $N_0/2^n$.

BIO Development of radioimmunoassay

Radioactive isotopes are used in a variety of ways in biomedical fields. One early development in the 1950s was performed by physicist Rosalyn Yalow (1921–2011) and physician Solomon Berson (1918–1972) in the development of the technique of *radioimmunoassay*. They used the technique to make highly sensitive measurements of minute amounts of insulin in human blood. Their procedure involved preparing antibodies against insulin, which they then attached to plastic beads. To this they added free insulin tagged with a known amount of radioactive isotope such as ^{131}I, and the insulin occupied the binding sites on the antibodies. When blood from a patient was added to the mixture, insulin from the blood displaced the radioactively tagged insulin. After washing away the blood, Yalow and Berson were able to measure the activity of the remaining ^{131}I and determine the amount of insulin initially in the blood. Yalow won a share of the Nobel Prize in Physiology or Medicine for this work in 1977. (Berson had already passed away and Nobel Prizes are not awarded posthumously.) Even though other techniques have been developed, radioimmunoassay continues to be used today for measuring small concentrations of a number of antigens.

▌**QUICK QUIZ 30.3** On your birthday, you measure the activity of a sample of ^{210}Bi, which has a half-life of 5.01 days. The activity you measure is 1.000 μCi. What is the activity of this sample on your next birthday?   (a) 1.000 μCi   (b) 0   (c) ~0.2 μCi   (d) ~0.01 μCi   (e) ~10^{-22} μCi

▌**QUICK QUIZ 30.4** Suppose you have a pure radioactive material with a half-life of $T_{1/2}$. You begin with N_0 undecayed nuclei of the material at $t = 0$. At $t = \frac{1}{2}T_{1/2}$, how many of the nuclei *have decayed*?   (a) $\frac{1}{4}N_0$ (b) $\frac{1}{2}N_0$ (c) $\frac{3}{4}N_0$ (d) $0.707N_0$ (e) $0.293N_0$

▌ **THINKING PHYSICS 30.2**

The isotope $^{14}_{6}$C is radioactive and has a half-life of 5 730 years. If you start with a sample of 1 000 carbon-14 nuclei, how many remain (have not decayed) after 17 190 yr?

Reasoning The number of half-lives represented by the time interval of 17 190 years is $(17\,190)/(5\,730) = 3$. Therefore, the number of radioactive nuclei remaining after this time interval is $N_0/2^n = 1\,000/2^3 = 125$.

These numbers represent ideal circumstances. Radioactive decay is a statistical process over a very large number of atoms, and the actual outcome depends on probability. Our original sample in this example contained only 1 000 nuclei, certainly not a very large number when we are dealing with atoms, for which we measure the numbers in macroscopic samples in terms of Avogadro's number. Therefore, if we were actually to count the number remaining after three half-lives for this small sample, it probably would not be exactly 125. ◄

Example 30.3 | The Activity of Carbon

At time $t = 0$, a radioactive sample contains 3.50 μg of pure $^{11}_{6}$C, which has a half-life of 20.4 min.

(A) Determine the number N_0 of nuclei in the sample at $t = 0$.

SOLUTION

Conceptualize The half-life is relatively short, so the number of undecayed nuclei drops rapidly. The molar mass of $^{11}_{6}$C is approximately 11.0 g/mol.

Categorize We evaluate results using equations developed in this section, so we categorize this example as a substitution problem.

Find the number of moles in 3.50 μg of pure $^{11}_{6}$C:

$$n = \frac{3.50 \times 10^{-6}\,\text{g}}{11.0\,\text{g/mol}} = 3.18 \times 10^{-7}\,\text{mol}$$

Find the number of undecayed nuclei in this amount of pure $^{11}_{6}$C:

$$N_0 = (3.18 \times 10^{-7}\,\text{mol})(6.02 \times 10^{23}\,\text{nuclei/mol}) = \boxed{1.92 \times 10^{17}\,\text{nuclei}}$$

(B) What is the activity of the sample initially and after 8.00 h?

SOLUTION

Find the initial activity of the sample using Equation 30.7:

$$R_0 = N_0 \lambda = N_0 \frac{0.693}{T_{1/2}} = (1.92 \times 10^{17}) \frac{0.693}{20.4\,\text{min}} \left(\frac{1\,\text{min}}{60\,\text{s}} \right)$$

$$= (1.92 \times 10^{17})(5.66 \times 10^{-4}\,\text{s}^{-1}) = \boxed{1.09 \times 10^{14}\,\text{Bq}}$$

Use Equation 30.7 to find the activity at $t = 8.00$ h $= 2.88 \times 10^4$ s:

$$R = R_0 e^{-\lambda t} = (1.09 \times 10^{14}\,\text{Bq})\, e^{-(5.66 \times 10^{-4}\,\text{s}^{-1})(2.88 \times 10^4\,\text{s})} = \boxed{8.96 \times 10^6\,\text{Bq}}$$

Example 30.4 | A Radioactive Isotope of Iodine

A sample of the isotope ^{131}I, which has a half-life of 8.04 days, has an activity of 5.0 mCi at the time of shipment. Upon receipt of the sample at a medical laboratory, the activity is 2.1 mCi. How much time has elapsed between the two measurements?

SOLUTION

Conceptualize The sample is continuously decaying as it is in transit. The decrease in the activity is 58% during the time interval between shipment and receipt, so we expect the elapsed time to be greater than the half-life of 8.04 d.

Categorize The stated activity corresponds to many decays per second, so N is large and we can categorize this problem as one in which we can use our statistical analysis of radioactivity.

Analyze Solve Equation 30.7 for the ratio of the final activity to the initial activity:

$$\frac{R}{R_0} = e^{-\lambda t}$$

Take the natural logarithm of both sides:

$$\ln\left(\frac{R}{R_0}\right) = -\lambda t$$

Solve for the time t:

$$(1) \quad t = -\frac{1}{\lambda} \ln\left(\frac{R}{R_0}\right)$$

continued

30.4 cont.

Use Equation 30.8 to substitute for λ:

$$t = -\frac{T_{1/2}}{\ln 2} \ln \left(\frac{R}{R_0} \right)$$

Substitute numerical values:

$$t = -\frac{8.04 \text{ d}}{0.693} \ln \left(\frac{2.1 \text{ mCi}}{5.0 \text{ mCi}} \right) = \boxed{10 \text{ d}}$$

Finalize This result is indeed greater than the half-life, as expected. This example demonstrates the difficulty in shipping radioactive samples with short half-lives. If the shipment is delayed by several days, only a small fraction of the sample might remain upon receipt. This difficulty can be addressed by shipping a combination of isotopes in which the desired isotope is the product of a decay occurring within the sample. It is possible for the desired isotope to be in *equilibrium*, in which case it is created at the same rate as it decays. Therefore, the amount of the desired isotope remains constant during the shipping process and subsequent storage. When needed, the desired isotope can be separated from the rest of the sample; its decay from the initial activity begins at this point rather than upon shipment.

▌30.4 | The Radioactive Decay Processes

When one nucleus changes into another without external influence, the process is called **spontaneous decay.** As we stated in Section 30.3, a radioactive nucleus spontaneously decays by one of three processes: alpha decay, beta decay, or gamma decay. Active Figure 30.9 shows a close-up view of a portion of Figure 30.4 from $Z = 65$ to $Z = 80$. The black circles are the stable nuclei seen in Figure 30.4. In addition, unstable nuclei above and below the line of stability for each value of Z are shown. Above the line of stability, the blue circles show unstable nuclei that are neutron-rich and undergo a beta decay process in which an electron is emitted. Below the black circles are red circles corresponding to proton-rich unstable nuclei that primarily undergo a beta decay process in which a positron is emitted or a competing process called electron capture. Beta decay and electron capture are described in more detail below. Further below the line of stability (with a few exceptions) are tan circles that represent very proton-rich nuclei for which the primary decay mechanism is alpha decay, which we will discuss first.

Alpha Decay

If a nucleus emits an alpha particle ($^4_2$He) in a spontaneous decay, it loses two protons and two neutrons. Therefore, N decreases by 2, Z decreases by 2, and A decreases by 4. The **alpha decay** can be written with a symbolic representation as

$$^A_Z\text{X} \;\rightarrow\; ^{A-4}_{Z-2}\text{Y} + ^4_2\text{He} \qquad \textbf{30.9} \blacktriangleleft$$

where X is called the **parent nucleus** and Y the **daughter nucleus.** As general rules, (1) the sum of the mass numbers must be the same on both sides of the symbolic representation and (2) the sum of the atomic numbers must be the same on both sides. As examples, ^{238}U and ^{226}Ra are both alpha emitters and decay according to the schemes

$$^{238}_{92}\text{U} \;\rightarrow\; ^{234}_{90}\text{Th} + ^4_2\text{He} \qquad \textbf{30.10} \blacktriangleleft$$

$$^{226}_{88}\text{Ra} \;\rightarrow\; ^{222}_{86}\text{Rn} + ^4_2\text{He} \qquad \textbf{30.11} \blacktriangleleft$$

The half-life for ^{238}U decay is 4.47×10^9 years, and the half-life for ^{226}Ra decay is 1.60×10^3 years. In both cases, note that the mass number A of the daughter nucleus is 4 less than that of the parent nucleus. Likewise, the atomic number Z is reduced by 2.

The decay of ^{226}Ra is shown in Active Figure 30.10. In addition to the rules for the mass number and the atomic number, the total energy of the isolated system must be conserved in the decay. If we call M_X the mass of the parent nucleus, M_Y the mass

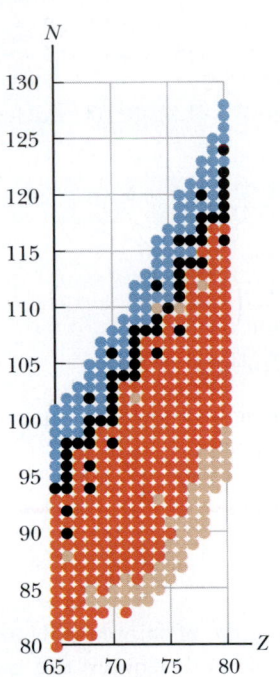

- ● Beta (electron)
- ● Stable
- ● Beta (positron) or electron capture
- ● Alpha

Active Figure 30.9 A close-up view of the line of stability in Figure 30.4 from $Z = 65$ to $Z = 80$. The black dots represent stable nuclei as in Figure 30.4. The other colored dots represent unstable isotopes above and below the line of stability, with the color of the dot indicating the primary means of decay.

of the daughter nucleus, and M_α the mass of the alpha particle, we can define the **disintegration energy** Q:

$$Q \equiv (M_X - M_Y - M_\alpha)c^2 \qquad \textbf{30.12} \blacktriangleleft$$

The value of Q will be in joules if the masses are in kilograms and $c = 3.00 \times 10^8$ m/s. When the nuclear masses are expressed in the more convenient atomic mass unit u, however, the value of Q can be calculated in MeV units using the expression

$$Q = (M_X - M_Y - M_\alpha) \times 931.494 \text{ MeV/u} \qquad \textbf{30.13} \blacktriangleleft$$

The disintegration energy Q represents the decrease in binding energy of the system and appears in the form of kinetic energy of the daughter nucleus and the alpha particle. In this nuclear example of the energy version of the isolated system model, no energy is entering or leaving the system. The energy in the system simply transforms from rest energy to kinetic energy, and Equation 30.13 gives the amount of energy transformed in the process. This quantity is sometimes referred to as the *Q* **value** of the nuclear reaction.

In addition to energy conservation, we can also apply the momentum version of the isolated system model to the decay. Because momentum of the isolated system must be conserved, the lighter alpha particle moves with a much higher speed than the daughter nucleus after the decay occurs. As a result, most of the available kinetic energy is associated with the alpha particle. Generally, light particles carry off most of the energy in nuclear decays.

Equation 30.13 suggests that the alpha particles are emitted with a discrete energy. Such an energy is calculated in Example 30.5. In practice, we find that alpha particles are emitted with a *set* of discrete energies (Active Fig. 30.11), with the *maximum* value calculated as in Example 30.5. This set of energies occurs because the energy of the nucleus is quantized, similar to the quantized energies in an atom. In Equation 30.13, we assume that the daughter nucleus is left in the ground state. If the daughter nucleus is left in an excited state, however, less energy is available for the decay and the alpha particle is emitted with less than the maximum kinetic energy. That the alpha particles have a discrete set of energies is direct evidence for the quantization of energy in the nucleus. This quantization is consistent with the model of a quantum particle under boundary conditions because the nucleons are quantum particles and they are subject to the constraints imposed by their mutual forces.

Finally, it is interesting to note that if one assumes that ^{238}U (or other alpha emitters) decays by emitting protons and neutrons, the mass of the decay products exceeds that of the parent nucleus, corresponding to negative Q values. Because that cannot occur for an isolated system, such spontaneous decays do not occur.

> **QUICK QUIZ 30.5** Which of the following is the correct daughter nucleus associated with the alpha decay of $^{157}_{72}$Hf? (a) $^{153}_{72}$Hf (b) $^{153}_{70}$Yb (c) $^{157}_{70}$Yb

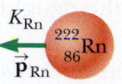

Active Figure 30.10 The alpha decay of radium-226. The radium nucleus is initially at rest. After the decay, the radon nucleus has kinetic energy K_{Rn} and momentum $\vec{p}_{Rn}$, and the alpha particle has kinetic energy K_α and momentum $\vec{p}_\alpha$.

Pitfall Prevention | 30.6
Another Q
We have seen the symbol Q before, but this use is a brand-new meaning for this symbol: the disintegration energy. In this context, it is not heat or charge, for which we have used Q before.

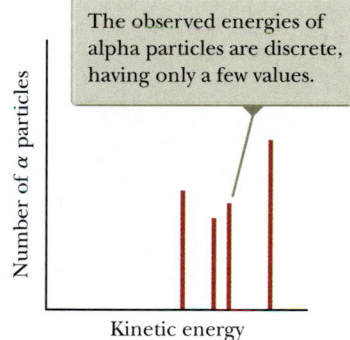

The observed energies of alpha particles are discrete, having only a few values.

Active Figure 30.11 Distribution of alpha-particle energies in a typical alpha decay.

Example 30.5 | The Energy Liberated When Radium Decays

The ^{226}Ra nucleus undergoes alpha decay according to Equation 30.11. Calculate the Q value for this process. From Table A.3 in Appendix A, the masses are 226.025 410 u for ^{226}Ra, 222.017 578 u for ^{222}Rn, and 4.002 603 u for $^4_2$He.

SOLUTION

Conceptualize Study Active Figure 30.10 to understand the process of alpha decay in this nucleus.

Categorize We use an equation developed in this section, so we categorize this example as a substitution problem.

Evaluate Q using Equation 30.13:

$$Q = (M_X - M_Y - M_\alpha) \times 931.494 \text{ MeV/u}$$
$$= (226.025\,410 \text{ u} - 222.017\,578 \text{ u} - 4.002\,603 \text{ u}) \times 931.494 \text{ MeV/u}$$
$$= (0.005\,229 \text{ u}) \times 931.494 \text{ MeV/u} = \boxed{4.87 \text{ MeV}}$$

continued

30.5 *cont.*

What If? Suppose you measured the kinetic energy of the alpha particle from this decay. Would you measure 4.87 MeV?

Answer The value of 4.87 MeV is the disintegration energy for the decay. It includes the kinetic energy of both the alpha particle and the daughter nucleus after the decay. Therefore, the kinetic energy of the alpha particle would be *less* than 4.87 MeV.

Let's determine this kinetic energy mathematically. The parent nucleus is an isolated system that decays into an alpha particle and a daughter nucleus. Therefore, momentum must be conserved for the system.

Set up a conservation of momentum equation, noting that the initial momentum of the system is zero:

$$(1) \quad 0 = M_Y v_Y - M_\alpha v_\alpha$$

Set the disintegration energy equal to the sum of the kinetic energies of the alpha particle and the daughter nucleus (assuming the daughter nucleus is left in the ground state):

$$(2) \quad Q = \tfrac{1}{2} M_\alpha v_\alpha^2 + \tfrac{1}{2} M_Y v_Y^2$$

Solve Equation (1) for v_Y and substitute into Equation (2):

$$Q = \tfrac{1}{2} M_\alpha v_\alpha^2 + \tfrac{1}{2} M_Y \left(\frac{M_\alpha v_\alpha}{M_Y} \right)^2 = \tfrac{1}{2} M_\alpha v_\alpha^2 \left(1 + \frac{M_\alpha}{M_Y} \right)$$

$$Q = K_\alpha \left(\frac{M_Y + M_\alpha}{M_Y} \right)$$

Solve for the kinetic energy of the alpha particle:

$$K_\alpha = Q \left(\frac{M_Y}{M_Y + M_\alpha} \right)$$

Evaluate this kinetic energy for the specific decay of ^{226}Ra that we are exploring in this example:

$$K_\alpha = (4.87 \text{ MeV}) \left(\frac{222}{222 + 4} \right) = 4.78 \text{ MeV}$$

We now turn to a structural model for the mechanism of alpha decay that allows some understanding of the decay process:

1. *A description of the physical components of the system:* Imagine that the alpha particle forms within the parent nucleus so that the parent nucleus is modeled as a system consisting of the alpha particle and the remaining daughter nucleus.

2. *A description of where the components are located relative to one another and how they interact:* The alpha particle and the daughter nucleus interact via the electric force and the nuclear force. Figure 30.12 is a graphical representation of the potential energy of this system as a function of the separation distance r between the alpha particle and the daughter nucleus. The distance R is the range of the nuclear force. The repulsive electric force describes the curve for $r > R$. The attractive nuclear force causes the energy curve to be negative for $r < R$. As we saw in Example 30.5, a typical disintegration energy is a few MeV, which is the approximate kinetic energy of the emitted alpha particle, represented by the lower dashed line in Figure 30.12.

3. *A description of the time evolution of the system:* According to classical physics, the alpha particle is trapped in the potential well of Figure 30.12 forever. We know, however, that alpha decay occurs in nature so that, at some time, the alpha particle will separate from the daughter nucleus. How does it ever escape from the nucleus? The answer to this question was provided by George Gamow (1904–1968) and, independently, Ronald Gurney (1898–1953) and Edward Condon (1902–1974) in 1928, using quantum mechanics. The view of quantum mechanics is that there is always some probability that the particle can *tunnel* through the barrier as we discussed in Section 28.13.

4. *A description of the agreement between predictions of the model and actual observations and, possibly, predictions of new effects that have not yet been observed:* Our model of the potential energy curve, combined with the possibility of tunneling, predicts that the probability of tunneling should increase as the particle energy

increases because of the narrowing of the barrier for higher energies. This increased probability should be reflected as an increased activity and consequently a shorter half-life.

Experimental data show just the relationship predicted in component 4 above: nuclei with higher alpha particle energies have shorter half-lives. If the potential energy curve in Figure 30.12 is modeled as a series of square barriers whose heights vary with particle separation according to the curve, we can generate a theoretical relationship between particle energy and half-life that is in excellent agreement with the experimental results. This particular application of modeling and quantum physics is a very effective demonstration of the power of these approaches.

Beta Decay

When a radioactive nucleus undergoes **beta decay,** the daughter nucleus has the same number of nucleons as the parent nucleus, but the atomic number is changed by 1:

$$_Z^A X \rightarrow \,_{Z+1}^{A}Y + e^- \qquad \text{(incomplete expression)} \qquad \textbf{30.14}\blacktriangleleft$$

$$_Z^A X \rightarrow \,_{Z-1}^{A}Y + e^+ \qquad \text{(incomplete expression)} \qquad \textbf{30.15}\blacktriangleleft$$

Again, note that nucleon number and total charge are both conserved in these decays. As we shall see later, however, these processes are not described completely by these expressions. We shall explain this incomplete description shortly.

The electron or positron involved in these decays is created within the nucleus as an initial step in the decay process. For example, during beta-minus decay, a neutron in the nucleus is transformed into a proton and an electron:

$$n \rightarrow p + e^- \qquad \text{(incomplete expression)}$$

After this process, the electron is ejected from the nucleus. For beta-plus decay, we have a proton transformed into a neutron and a positron:

$$p \rightarrow n + e^+ \qquad \text{(incomplete expression)}$$

After this process, the positron is ejected from the nucleus.

Outside the nucleus, this latter process will not occur because the neutron and electron have more total mass than the proton. This process can occur within the nucleus, however, because we consider the rest energy changes of the entire nuclear system, not just the individual particles. In beta-plus decay, the process $p \rightarrow n + e^+$ does indeed result in a decrease in the mass of the nucleus, so the process does occur spontaneously.

As with alpha decay, the energy of the isolated system of the nucleus and the emitted particle must be conserved in beta decay. Experimentally, one finds that the beta particles are emitted over a continuous range of energies (Active Fig. 30.13), unlike alpha particles, which are emitted with discrete energies (Active Fig. 30.11). The kinetic energy increase of the system must be balanced by the decrease in rest energy of the system; either of these changes is the Q value. Because all decaying nuclei have the same initial mass, however, the Q value must be the same for each decay. Then why do the emitted electrons have a range of kinetic energies? The energy version of the isolated system model seems to make an incorrect prediction! Further experimentation shows that, according to the decay processes given by Equations 30.14 and 30.15, the angular momentum (spin) and linear momentum versions of the isolated system model fail, too, and neither angular momentum nor linear momentum of the system is conserved!

Clearly, the structural model for beta decay must differ from that for alpha decay. After a great deal of experimental and theoretical study, Pauli proposed in 1930 that a third particle must be involved in the decay to account for the "missing" energy and momentum. Enrico Fermi later named this particle the **neutrino** (little neutral one) because it has to be electrically neutral and have little or no rest energy. Although it eluded detection for many years, the neutrino (symbolized by ν, Greek letter nu) was

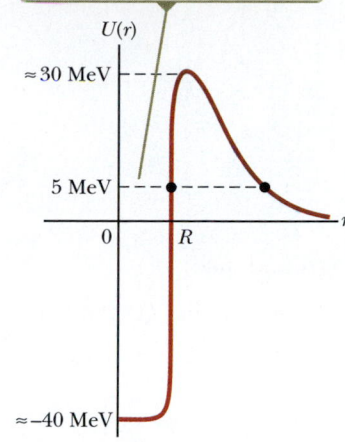

Classically, the 5-MeV energy of the alpha particle is not sufficiently large to overcome the energy barrier, so the particle should not be able to escape from the nucleus.

Figure 30.12 The alpha particle escapes by tunneling through the barrier.

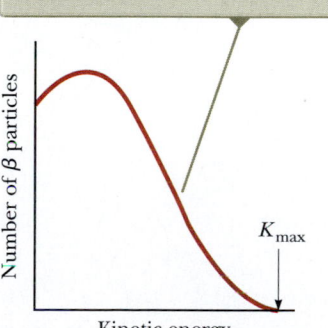

The observed energies of beta particles are continuous, having all values up to a maximum value.

Active Figure 30.13 Distribution of beta particle energies in a typical beta decay. Compare this continuous distribution of energies to the discrete distribution of alpha particle energies in Active Figure 30.11.

finally detected experimentally in 1956 by Frederick Reines (1918–1998) and Clyde Cowan (1919–1974). Reines received the Nobel Prize in Physics in 1995 for this important work. The neutrino has the following properties:

- It has zero electric charge.
- Its mass is much smaller than that of the electron. Recent experiments show that the mass of the neutrino is not 0 and place the upper bound of the neutrino mass at approximately $2 \text{ eV}/c^2$.
- It has a spin of $\frac{1}{2}$, which allows the law of conservation of angular momentum to be satisfied in beta decay.
- It interacts very weakly with matter and is therefore very difficult to detect.

We can now write the beta decay processes (Eqs. 30.14 and 30.15) in their correct form:

$$^A_Z X \rightarrow ^{\;\;A}_{Z+1} Y + e^- + \bar{\nu} \quad \text{(complete expression)} \qquad \mathbf{30.16} \blacktriangleleft$$

$$^A_Z X \rightarrow ^{\;\;A}_{Z-1} Y + e^+ + \nu \quad \text{(complete expression)} \qquad \mathbf{30.17} \blacktriangleleft$$

where $\bar{\nu}$ represents the **antineutrino,** the antiparticle to the neutrino. We shall discuss antiparticles further in Chapter 31. For now, it suffices to say that a neutrino is emitted in positron decay, and an antineutrino is emitted in electron decay. The spin of the neutrino allows angular momentum to be conserved in the decay processes. Despite its small mass, the neutrino does carry momentum, which allows linear momentum to be conserved.

The decays of the neutron and proton within the nucleus are more properly written as

$$\text{n} \rightarrow \text{p} + e^- + \bar{\nu} \quad \text{(complete expression)}$$

$$\text{p} \rightarrow \text{n} + e^+ + \nu \quad \text{(complete expression)}$$

As examples of beta decay, we can write the decay schemes for carbon-14 and nitrogen-12:

$$^{14}_{6}\text{C} \rightarrow ^{14}_{7}\text{N} + e^- + \bar{\nu} \quad \text{(complete expression)} \qquad \mathbf{30.18} \blacktriangleleft$$

$$^{12}_{7}\text{N} \rightarrow ^{12}_{6}\text{C} + e^+ + \nu \quad \text{(complete expression)} \qquad \mathbf{30.19} \blacktriangleleft$$

Active Figure 30.14 shows a pictorial representation of the decays described by Equations 30.18 and 30.19.

In beta-plus decay, the final system consists of the daughter nucleus, the ejected positron and neutrino, and an electron shed from the atom to neutralize the daughter atom. In some cases, this process represents an overall increase in rest energy, so

---

Pitfall Prevention | 30.7
Mass Number of the Electron
An alternative notation for an electron in Equation 30.18 is the symbol $^{\;\;0}_{-1}e$, which does not imply that the electron has zero rest energy. The mass of the electron is so much smaller than that of the lightest nucleon, however, that we approximate it as zero in the context of nuclear decays and reactions.

Active Figure 30.14 (a) The beta decay of carbon-14. (b) The beta decay of nitrogen-12.

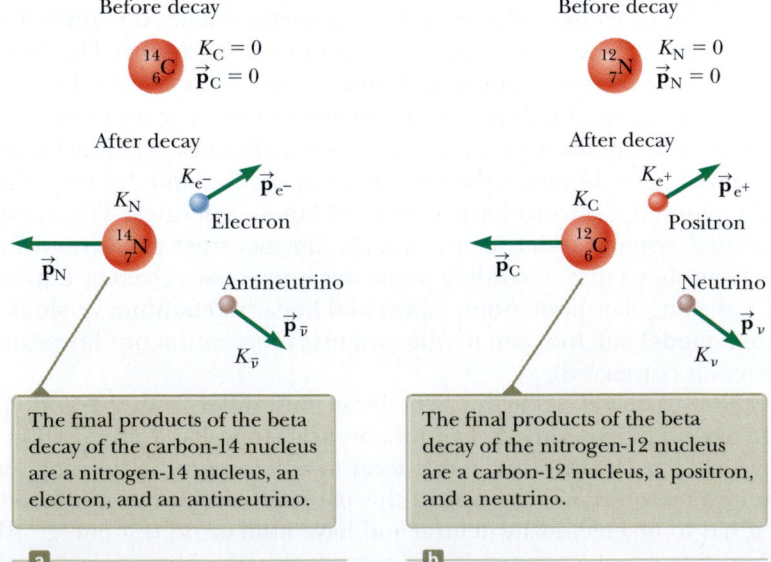

The final products of the beta decay of the carbon-14 nucleus are a nitrogen-14 nucleus, an electron, and an antineutrino.

The final products of the beta decay of the nitrogen-12 nucleus are a carbon-12 nucleus, a positron, and a neutrino.

Figure 30.15 (a) A fragment of the Dead Sea Scrolls, which were discovered in caves located in the West Bank area of the Middle East and visible at the bottom of the photograph (b). The packing material of the scrolls was analyzed by carbon dating to determine their age.

it does not occur. There is an alternative process that allows some proton-rich nuclei to decay and become more stable. This process, called **electron capture,** occurs when a parent nucleus captures one of its own orbital electrons and emits a neutrino. The final product after decay is a nucleus whose charge is $Z - 1$:

$$^A_Z X + e^- \quad \rightarrow \quad ^{A}_{Z-1} Y + \nu \qquad \text{30.20} \blacktriangleleft$$

▶ Electron capture

In most cases, an inner K-shell electron is captured, a process referred to as **K capture.** In this process, the only outgoing particles are the neutrino and x-ray photons, originating in higher-shell electrons falling into the vacancy left by the captured K electron.

> **QUICK QUIZ 30.6** Which of the following is the correct daughter nucleus associated with the beta decay of $^{184}_{72}\text{Hf}$? (a) $^{183}_{72}\text{Hf}$ (b) $^{183}_{73}\text{Ta}$ (c) $^{184}_{73}\text{Ta}$

Carbon Dating BIO

The beta decay of ^{14}C given by Equation 30.18 is commonly used to date organic samples. Cosmic rays (high-energy particles from outer space) in the upper atmosphere cause nuclear reactions that create ^{14}C. The ratio of ^{14}C to ^{12}C in the carbon dioxide molecules of our atmosphere has a constant value of about 1.3×10^{-12}. All living organisms have the same ratio of ^{14}C to ^{12}C because they continuously exchange carbon dioxide with their surroundings. When an organism dies, however, it no longer absorbs ^{14}C from the atmosphere, and so the ratio of ^{14}C to ^{12}C decreases as the result of the beta decay of ^{14}C, which has a half-life of 5 730 yr. It is therefore possible to determine the age of a biological sample by measuring its activity per unit mass due to the decay of ^{14}C. Using carbon dating, samples of wood, charcoal, bone, and shell have been identified as having lived 1 000 to 25 000 years ago.

A particularly interesting example is the dating of the Dead Sea Scrolls, a group of manuscripts discovered by a shepherd in 1947 (Fig. 30.15). Translation showed them to be religious documents, including most of the books of the Old Testament. Because of their historical and religious significance, scholars wanted to know their age. Carbon dating applied to the material in which they were wrapped established their age at approximately 1 950 yr.

Enrico Fermi
Italian Physicist (1901–1954)
Fermi immigrated to the United States to escape the Fascists. Fermi was awarded the Nobel Prize in Physics in 1938 for producing the transuranic elements by neutron irradiation and for his discovery of nuclear reactions brought about by slow neutrons. He made many other outstanding contributions to physics including his theory of beta decay, the free electron theory of metals, and the development of the world's first fission reactor in 1942. Fermi was truly a gifted theoretical and experimental physicist. He was also well known for his ability to present physics in a clear and exciting manner. He wrote, "Whatever Nature has in store for mankind, unpleasant as it may be, men must accept, for ignorance is never better than knowledge."

> ## THINKING PHYSICS 30.3
>
> In 1991, German tourists discovered the well-preserved remains of a man, now called "Ötzi the Iceman," trapped in a glacier in the Italian Alps. (See the photograph at the opening of this chapter.) Radioactive dating with ^{14}C revealed that this person was alive approximately 5 300 years ago. Why did scientists date a sample of Ötzi using ^{14}C rather than ^{11}C, which is a beta emitter having a half-life of 20.4 min?

Reasoning ^{14}C has a long half-life of 5 730 yr, so the fraction of ^{14}C nuclei remaining after one half-life is high enough to measure accurate changes in the sample's activity. The ^{11}C isotope, which has a very short half-life, is not useful because its activity decreases to a vanishingly small value over 5 300 yr, making it impossible to detect.

An isotope used to date a sample must be present in a known amount in the sample when it is formed. As a general rule, the isotope chosen to date a sample should also have a half-life with the same order of magnitude as the age of the sample. If the half-life is much less than the age of the sample, there won't be enough activity left to measure because almost all the original radioactive nuclei will have decayed. If the half-life is much greater than the age of the sample, the reduction in activity that has taken place since the sample died will be too small to measure. ◄

Example 30.6 | **Radioactive Dating**

A piece of charcoal containing 25.0 g of carbon is found in some ruins of an ancient city. The sample shows a ^{14}C activity R of 250 decays/min. How long has the tree from which this charcoal came been dead?

SOLUTION

Conceptualize Because the charcoal was found in ancient ruins, we expect the current activity to be smaller than the initial activity. If we can determine the initial activity, we can find out how long the wood has been dead.

Categorize The text of the question helps us categorize this example as a carbon dating problem.

Analyze Solve Equation 30.7 for t:

$$(1) \quad t = -\frac{1}{\lambda} \ln\left(\frac{R}{R_0}\right)$$

Evaluate the ratio R/R_0 using Equation 30.7, the initial value of the ^{14}C/^{12}C ratio r_0, the number of moles n of carbon, and Avogadro's number N_A:

$$\frac{R}{R_0} = \frac{R}{\lambda N_0(^{14}C)} = \frac{R}{\lambda r_0 N_0(^{12}C)} = \frac{R}{\lambda r_0 n N_A}$$

Replace the number of moles in terms of the molar mass M of carbon and the mass m of the sample and substitute for the decay constant λ:

$$\frac{R}{R_0} = \frac{R}{(\ln 2/T_{1/2}) r_0 (m/M) N_A} = \frac{R M T_{1/2}}{r_0 m N_A \ln 2}$$

Substitute numerical values:

$$\frac{R}{R_0} = \frac{(250 \text{ min}^{-1})(12.0 \text{ g/mol})(5\,730 \text{ yr})}{(1.3 \times 10^{-12})(25.0 \text{ g})(6.022 \times 10^{23} \text{ mol}^{-1}) \ln 2}\left(\frac{3.156 \times 10^7 \text{ s}}{1 \text{ yr}}\right)\left(\frac{1 \text{ min}}{60 \text{ s}}\right)$$

$$= 0.667$$

Substitute this ratio into Equation (1) and substitute for the decay constant λ:

$$t = -\frac{1}{\lambda} \ln\left(\frac{R}{R_0}\right) = -\frac{T_{1/2}}{\ln 2} \ln\left(\frac{R}{R_0}\right)$$

$$= -\frac{5\,730 \text{ yr}}{\ln 2} \ln(0.667) = \boxed{3.4 \times 10^3 \text{ yr}}$$

Finalize Note that the time interval found here is on the same order of magnitude as the half-life, so ^{14}C is a valid isotope to use for this sample, as discussed in Thinking Physics 30.3.

Gamma Decay

Very often, a nucleus that undergoes radioactive decay is left in an excited quantum state. The nucleus can then undergo a second decay, a **gamma decay,** to a lower state, perhaps to the ground state, by emitting a photon:

▶ Gamma decay

$$^A_Z X^* \rightarrow {}^A_Z X + \gamma \qquad \qquad \textbf{30.21}◀$$

where X* indicates a nucleus in an excited state. The typical half-life of an excited nuclear state is 10^{-10} s. Photons emitted in such a de-excitation process are called

TABLE 30.2 | Various Decay Pathways

Alpha decay	$^{A}_{Z}X \rightarrow {}^{A-4}_{Z-2}Y + {}^{4}_{2}He$
Beta decay (e^-)	$^{A}_{Z}X \rightarrow {}^{A}_{Z+1}Y + e^- + \bar{\nu}$
Beta decay (e^+)	$^{A}_{Z}X \rightarrow {}^{A}_{Z-1}Y + e^+ + \nu$
Electron capture	$^{A}_{Z}X + e^- \rightarrow {}^{A}_{Z-1}Y + \nu$
Gamma decay	$^{A}_{Z}X^* \rightarrow {}^{A}_{Z}X + \gamma$

gamma rays. Such photons have very high energy (on the order of 1 MeV or higher) relative to the energy of visible light (on the order of a few electron volts). Recall from Chapter 29 that the energy of photons emitted (or absorbed) by an atom equals the difference in energy between the two atomic quantum states involved in the transition. Similarly, a gamma ray photon has an energy hf that equals the energy difference ΔE between two nuclear quantum states. When a nucleus decays by emitting a gamma ray, it ends up in a lower state, but its atomic mass A and atomic number Z do not change.

A nucleus may reach an excited state as the result of a violent collision with another particle. It is more common, however, for a nucleus to be in an excited state after it has undergone an alpha or beta decay. The following sequence of events represents a typical situation in which gamma decay occurs:

$$^{12}_{5}B \rightarrow {}^{12}_{6}C^* + e^- + \bar{\nu} \qquad \textbf{30.22} \blacktriangleleft$$

$$^{12}_{6}C^* \rightarrow {}^{12}_{6}C + \gamma \qquad \textbf{30.23} \blacktriangleleft$$

Figure 30.16 shows the decay scheme for ^{12}B, which undergoes beta decay with a half-life of 20.4 ms to either of two levels of ^{12}C. It can either (1) decay directly to the ground state of ^{12}C by emitting a 13.4-MeV electron or (2) undergo beta-minus decay to an excited state of $^{12}C^*$, followed by gamma decay to the ground state. The latter process results in the emission of a 9.0-MeV electron and a 4.4-MeV photon. Table 30.2 summarizes the pathways by which radioactive nuclei undergo decay.

In Chapter 29, we discussed the treatment of cancerous tumors with proton therapy and external beam radiotherapy. Yet another alternate treatment for some localized cancers is *brachytherapy*. In this procedure, a radioactive source is placed within or beside the treatment area. This treatment plan results in a high dose of energy to the cancerous tissue in the tumor without the absorption of energy by healthy tissue that is a common side effect of external beam radiotherapy. Common applications of brachytherapy include cancers of the cervix, prostate, breast, and skin.

In treating prostate cancer, for example, up to 150 radioactive seeds are planted within the prostate of a patient with early-stage cancer. A common radioactive source in the seeds is palladium (^{103}Pd), a gamma-ray emitter with a half-life of 17 days. With this source, the radioactivity will have decreased to a very low level over a treatment period of 2 to 3 months. The titanium-encapsulated seeds are left permanently in the body after the treatment. Results from brachytherapy in prostate cancer treatment are comparable to those of radical prostatectomy and external beam radiotherapy.

Another gamma-emitter, ^{99}Tc, is used in a *nuclear bone scan*. In this diagnostic test, the patient is injected with a "tracer" containing about 600 MBq of the technetium isotope, which then spreads throughout the body. In particular, the tracer is absorbed into the bones. A "gamma camera" is then used to detect the gamma decay. Areas appearing dark on the image have absorbed little tracer. This may suggest a lack of blood supply due to cancer in the bone. Bright areas on the image correspond to significant absorption of the tracer, suggesting possible arthritis, fracture, or infection.

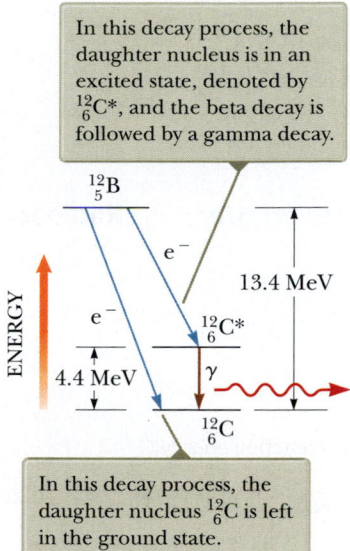

In this decay process, the daughter nucleus is in an excited state, denoted by $^{12}_{6}C^*$, and the beta decay is followed by a gamma decay.

In this decay process, the daughter nucleus $^{12}_{6}C$ is left in the ground state.

Figure 30.16 An energy level diagram showing the initial nuclear state of a ^{12}B nucleus and two possible lower-energy states of the ^{12}C nucleus.

BIO Treatment of cancers with brachytherapy

BIO Nuclear bone scan

30.5 | Nuclear Reactions

In Section 30.4, we discussed the processes by which nuclei can *spontaneously* change to another nucleus by undergoing a radioactive decay process. It is also possible to change the structures and properties of nuclei by bombarding them with energetic particles. Such changes are called **nuclear reactions.** In 1919, Rutherford was the first to observe nuclear reactions, using naturally occurring radioactive sources for the bombarding particles. Since then, thousands of nuclear reactions have been observed following the development of charged-particle accelerators in the 1930s. With today's advanced technology in particle accelerators and particle detectors, it is possible to achieve particle energies of more than 1 000 GeV = 1 TeV. These high-energy particles are used to create new particles whose properties are helping solve the mysteries of the nucleus.

Consider a reaction (Fig. 30.17) in which a target nucleus X is bombarded by an incoming particle a, resulting in a different nucleus Y and an outgoing particle b:

▶ Nuclear reaction

$$a + X \rightarrow Y + b \qquad \textbf{30.24} \blacktriangleleft$$

Sometimes this reaction is written in the equivalent symbolic representation

$$X(a, b)Y$$

In the preceding section, the Q value, or disintegration energy, associated with radioactive decay was defined as the change in the rest energy, which is the amount of the rest energy transformed to kinetic energy during the decay process. In a similar way, we define the **reaction energy** Q associated with a nuclear reaction as the difference between the initial and final rest energies of the system of particles participating in the reaction:

▶ Reaction energy Q

$$Q = (M_a + M_X - M_Y - M_b)\,c^2 \qquad \textbf{30.25} \blacktriangleleft$$

A reaction for which Q is positive is called **exothermic.** After the reaction, the transformed rest energy appears as an increase in kinetic energy of Y and b over that of a and X.

A reaction for which Q is negative is called **endothermic** and represents an increase in rest energy. An endothermic reaction will not occur unless the bombarding particle has a kinetic energy greater than $|Q|$. The minimum kinetic energy of the incoming particle necessary for such a reaction to occur is called the **threshold energy.** The threshold energy is larger than $|Q|$ because we must also conserve linear momentum in the isolated system of the initial and final particles. If an incoming particle has just energy $|Q|$, enough energy is present to increase the rest energy of the system, but none is left over for kinetic energy of the final particles; that is, nothing is moving after the reaction. Therefore, the incoming particle has momentum before the reaction but there is no momentum of the system afterward, which is a violation of the law of conservation of momentum.

If particles a and b in a nuclear reaction are identical so that X and Y are also necessarily identical, the reaction is called a **scattering event.** If the kinetic energy of the system (a and X) before the event is the same as that of the system (b and Y) after the event, it is classified as *elastic scattering*. If the kinetic energies of the system before and after the event are not the same, the reaction is described as *inelastic scattering*. In this case, the difference in energy is accounted for by the target nucleus being raised to an excited state by the event. The final system now consists of b and an excited nucleus Y*, and eventually it will become b, Y, and γ, where γ is the gamma-ray photon that is emitted when the system returns to the ground state. This elastic and inelastic terminology is identical to that used in describing collisions between macroscopic objects (Section 8.4).

In addition to energy and momentum, the total charge and total number of nucleons must be conserved in the system of particles for a nuclear reaction. For example, consider the reaction $^{19}\text{F}(p, \alpha)^{16}\text{O}$, which has a Q value of 8.124 MeV. We can show this reaction more completely as

$$^{1}_{1}\text{H} + ^{19}_{9}\text{F} \rightarrow ^{16}_{8}\text{O} + ^{4}_{2}\text{He}$$

We see that the total number of nucleons before the reaction (1 + 19 = 20) is equal to the total number after the reaction (16 + 4 = 20). Furthermore, the total charge ($Z = 10$) is the same before and after the reaction.

Before the reaction, an incoming particle a moves toward a target nucleus X.

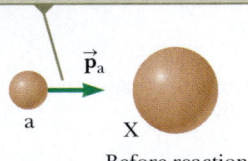

Before reaction

After the reaction, the target nucleus has changed to nucleus Y and an outgoing particle b moves away from the reaction site.

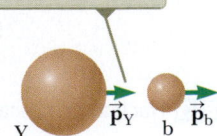

After reaction

Figure 30.17 A nuclear reaction.

K30.6 | Context Connection: The Engine of the Stars

One of the important features of nuclear reactions is that much more energy is released (i.e., converted from rest energy) than in normal chemical reactions such as in the burning of fossil fuels. Let us look back at our binding energy curve (Fig. 30.6) and consider two important nuclear reactions that relate to that curve. If a heavy nucleus at the right of the graph splits into two lighter nuclei, the total binding energy within the system increases, representing energy released from the nuclei. This type of reaction was observed and reported in 1939 by Otto Hahn (1879–1968) and Fritz Strassman (1902–1980). This reaction, known as **fission,** was of great scientific and political interest at the time of World War II because of the development of the first nuclear weapon.

In the fission reaction, a fissionable nucleus (the target nucleus X), which is often ^{235}U, absorbs a slowly moving neutron (the incoming particle a) and the nucleus splits into two smaller nuclei (two nuclei Y_1 and Y_2), releasing energy and more neutrons (several particles b). These neutrons can then go on to be absorbed within other nuclei, causing other fissions. With no means of control, the result is a chain reaction explosion as suggested by Active Figure 30.18. With proper control, the fission process is used in nuclear power generating stations.

While nuclear power stations are successful in generating power worldwide, there are serious safety issues to be considered. A 1986 explosion at the Chernobyl Nuclear Power Plant in what is now Ukraine released significant amounts of radioactive material into the air. Increased radiation readings were measured in most of Europe and much of Russia.

More recently, significant radiation was released from nuclear power plants after the devastating March 2011 earthquake off the coast of Japan. Power plants were automatically shut down after the earthquake, but the subsequent tsunami disabled emergency generators required to keep the cooling pumps operating. These cooling pumps are necessary even after the reactor has been shut down in order to remove internal energy resulting from the decay of fission byproducts. Due to intentional venting to reduce pressures, intentional discharge of cooling water into natural bodies of water, and explosions, the release of radiation placed the Japan nuclear situation in the same category as the Chernobyl accident.

Examining the other end of the binding energy curve, we see that we could also increase the binding energy of the system and release energy by combining two light

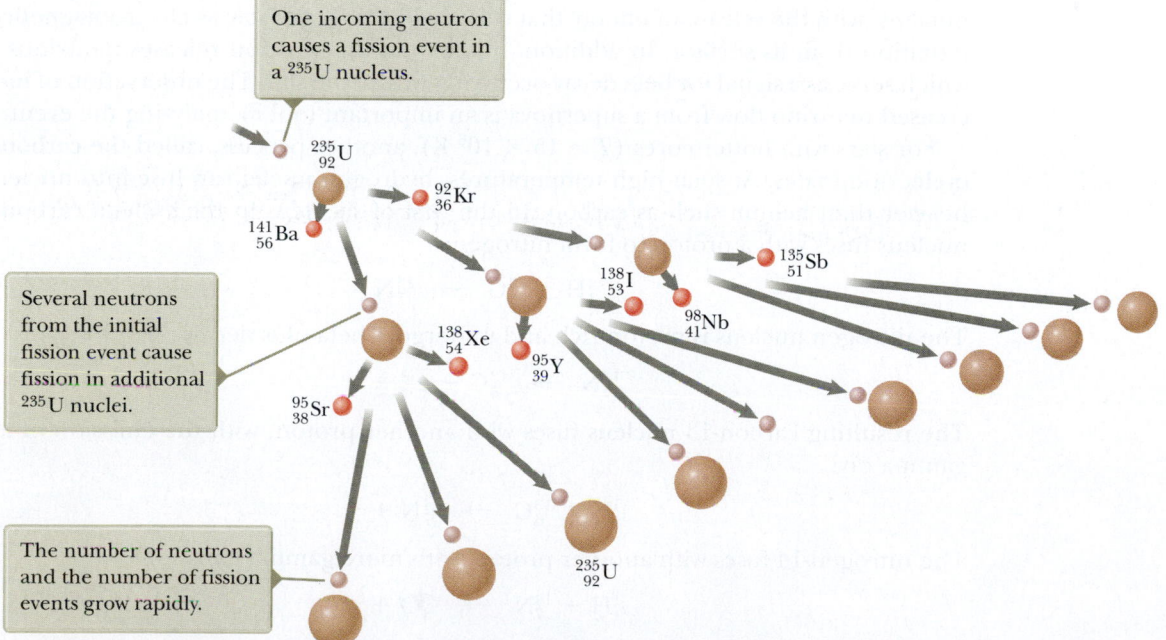

One incoming neutron causes a fission event in a ^{235}U nucleus.

$^{235}_{92}U$

$^{92}_{36}Kr$

$^{141}_{56}Ba$

$^{135}_{51}Sb$

$^{138}_{53}I$

Several neutrons from the initial fission event cause fission in additional ^{235}U nuclei.

$^{138}_{54}Xe$

$^{98}_{41}Nb$

$^{95}_{39}Y$

$^{95}_{38}Sr$

The number of neutrons and the number of fission events grow rapidly.

$^{235}_{92}U$

Active Figure 30.18 A nuclear chain reaction initiated by the capture of a neutron. Uranium nuclei are shown in tan, neutrons in gray, and daughter nuclei in orange.

nuclei. This process of **fusion** is made difficult because the nuclei must overcome a very strong Coulomb repulsion before they become close enough together to fuse. One way to assist the nuclei in overcoming this repulsion is to cause them to move with very high kinetic energy by raising the system of nuclei to a very high temperature. If the density of nuclei is high also, the probability of nuclei colliding is high and fusion can occur. The technological problem of creating very high temperatures and densities is a major challenge in the area of Earth-based controlled fusion research.

At some natural locations (e.g., the cores of stars), the necessary high temperatures and densities exist. Consider a collection of gas and dust somewhere in the Universe to be an isolated system. What happens as this system collapses under its own gravitational attraction? Energy of the system is conserved, and the gravitational potential energy associated with the separated particles decreases while the kinetic energy of the particles increases. As the falling particles collide with the particles that have already fallen into the central region of collapse, their kinetic energy is distributed to the other particles by collisions and randomized; it becomes internal energy, which is related to the temperature of the collection of particles.

If the temperature and density of the system's core rise to the point where fusion can occur, the system becomes a star. The primary constituent of the Universe is hydrogen, so the fusion reaction at the center of a star combines hydrogen nuclei—protons—into helium nuclei. A common reaction process for stars with relatively cool cores ($T < 15 \times 10^6$ K) is the **proton–proton cycle.** In the first step of the process, two protons combine to form deuterium:

$$^1_1\text{H} + ^1_1\text{H} \; \rightarrow \; ^2_1\text{H} + e^+ + \nu$$

Notice the implicit ^2_2He nucleus that is formed but that does not appear in the reaction equation. This nucleus is highly unstable and decays very rapidly by beta-plus decay to the deuterium nucleus, a positron, and a neutrino.

In the next step, the deuterium nucleus undergoes fusion with another proton to form a helium-3 nucleus:

$$^1_1\text{H} + ^2_1\text{H} \; \rightarrow \; ^3_2\text{He} + \gamma$$

Finally, two helium-3 nuclei formed in such reactions can fuse to form helium-4 and two protons:

$$^3_2\text{He} + ^3_2\text{He} \; \rightarrow \; ^4_2\text{He} + ^1_1\text{H} + ^1_1\text{H}$$

The net result of this cycle has been the joining of four protons to form a helium-4 nucleus, with the release of energy that eventually leaves the star as electromagnetic radiation from its surface. In addition, notice that the reaction releases neutrinos, which serve as a signal for beta decay occurring within the star. The observation of increased neutrino flow from a supernova is an important tool in analyzing the event.

For stars with hotter cores ($T > 15 \times 10^6$ K), another process, called the **carbon cycle,** dominates. At such high temperatures, hydrogen nuclei can fuse into nuclei heavier than helium such as carbon. In the first of six steps in the cycle, a carbon nucleus fuses with a proton to form nitrogen:

$$^1_1\text{H} + ^{12}_6\text{C} \; \rightarrow \; ^{13}_7\text{N}$$

The nitrogen nucleus is proton-rich and undergoes beta-plus decay:

$$^{13}_7\text{N} \; \rightarrow \; ^{13}_6\text{C} + e^+ + \nu$$

The resulting carbon-13 nucleus fuses with another proton, with the emission of a gamma ray:

$$^1_1\text{H} + ^{13}_6\text{C} \; \rightarrow \; ^{14}_7\text{N} + \gamma$$

The nitrogen-14 fuses with another proton, with more gamma emission:

$$^1_1\text{H} + ^{14}_7\text{N} \; \rightarrow \; ^{15}_8\text{O} + \gamma$$

The oxygen nucleus undergoes beta-plus decay:

$$^{15}_8\text{O} \; \rightarrow \; ^{15}_7\text{N} + e^+ + \nu$$

Finally, the nitrogen-15 fuses with another proton:

$$^1_1\text{H} + ^{15}_7\text{N} \rightarrow ^{12}_6\text{C} + ^4_2\text{He}$$

Notice that the net effect of this process is to combine four protons into a helium nucleus, just like the proton–proton cycle. The carbon-12 with which we began the process is returned at the end, so it acts only as a catalyst to the process and is not consumed.

Depending on its mass, a star transforms energy in its core at a rate between 10^{23} and 10^{33} W. The energy transformed from the rest energy of the nuclei in the core is transferred outward through the surrounding layers by matter transfer in two forms. First, neutrinos carry energy directly through these layers to space because these particles interact only weakly with matter. Second, energy carried by photons from the core is absorbed by the gases in layers outside the core and slowly works its way to the surface by convection. This energy is eventually radiated from the surface of the star by electromagnetic radiation, mostly in the infrared, visible, and ultraviolet regions of the electromagnetic spectrum. The weight of the layers outside the core keeps the core from exploding. The whole system of a star is stable as long as the supply of hydrogen in the core lasts.

In the previous chapters, we presented examples of the applications of quantum physics and atomic physics to processes in space. In this chapter, we have seen that nuclear processes also have an important role in the cosmos. The formation of stars is a critical process in the development of the Universe. The energy provided by stars is crucial to life on planets such as the Earth. In our next, and final, chapter, we shall discuss the processes that occur on an even smaller scale, the scale of *elementary particles*. We shall find again that looking at a smaller scale allows us to advance our understanding of the largest scale system, the Universe.

SUMMARY

A nuclide can be represented by ^A_ZX, where A is the **mass number,** the total number of nucleons, and Z is the **atomic number,** the total number of protons. The total number of neutrons in a nucleus is the **neutron number** N, where $A = N + Z$. Elements with the same Z but different A and N values are called **isotopes.**

Assuming that a nucleus is spherical, its radius is

$$r = aA^{1/3} \qquad \textbf{30.1} \blacktriangleleft$$

where $a = 1.2$ fm.

Nuclei are stable because of the **nuclear force** between nucleons. This short-range force dominates the Coulomb repulsive force at distances of less than about 2 fm and is independent of charge.

Light nuclei are most stable when the number of protons equals the number of neutrons. Heavy nuclei are most stable when the number of neutrons exceeds the number of protons. In addition, many stable nuclei have Z and N values that are both even. Nuclei with unusually high stability have Z or N values of 2, 8, 20, 28, 50, 82, and 126, called **magic numbers.**

Nuclei have an intrinsic angular momentum (spin) of magnitude $\sqrt{I(I+1)}\,\hbar$, where I is the **nuclear spin quantum number.** The magnetic moment of a nucleus is measured in terms of the **nuclear magneton** μ_n, where

$$\mu_n \equiv \frac{e\hbar}{2m_p} = 5.05 \times 10^{-27}\,\text{J/T} \qquad \textbf{30.3} \blacktriangleleft$$

The difference in mass between the separate nucleons and the nucleus containing these nucleons, when multiplied by c^2, gives the **binding energy** E_b of the nucleus. We can calculate the binding energy of any nucleus ^A_ZX using the expression

$$E_b(\text{MeV}) = [ZM(\text{H}) + Nm_n - M(^A_Z\text{X})] \times 931.494\,\text{MeV/u} \qquad \textbf{30.4} \blacktriangleleft$$

Radioactive processes include alpha decay, beta decay, and gamma decay. An alpha particle is a ^{4}He nucleus, a beta particle is either an electron (e^-) or a positron (e^+), and a gamma particle is a high-energy photon.

If a radioactive material contains N_0 radioactive nuclei at $t = 0$, the number N of nuclei remaining at time t is

$$N = N_0 e^{-\lambda t} \qquad \textbf{30.6} \blacktriangleleft$$

where λ is the **decay constant,** or **disintegration constant. The decay rate,** or **activity,** of a radioactive substance is given by

$$R = \left|\frac{dN}{dt}\right| = N_0\lambda e^{-\lambda t} = R_0 e^{-\lambda t} \qquad \textbf{30.7} \blacktriangleleft$$

where $R_0 = N_0\lambda$ is the activity at $t = 0$. The **half-life** $T_{1/2}$ is defined as the time interval required for half of a given number of radioactive nuclei to decay, where

$$T_{1/2} = \frac{\ln 2}{\lambda} = \frac{0.693}{\lambda} \qquad \textbf{30.8} \blacktriangleleft$$

Alpha decay can occur because according to quantum mechanics some nuclei have barriers that can be penetrated by the alpha particles (the tunneling process). This process is energetically more favorable for those nuclei having large excesses of neutrons. A nucleus can undergo **beta decay** in two ways. It can emit either an electron (e^-) and an antineutrino ($\bar{\nu}$) or a positron (e^+) and a neutrino (ν). In the **electron capture** process, the nucleus of an atom absorbs one of its own electrons (usually from the K shell) and emits a neutrino. In **gamma decay**, a nucleus in an excited state decays to its ground state and emits a gamma ray.

Nuclear reactions can occur when a target nucleus X is bombarded by a particle a, resulting in a nucleus Y and an outgoing particle b:

$$a + X \rightarrow Y + b \quad \text{or} \quad X(a, b)Y \qquad \textbf{30.24} \blacktriangleleft$$

The rest energy transformed to kinetic energy in such a reaction, called the **reaction energy** Q, is

$$Q = (M_a + M_X - M_Y - M_b)c^2 \qquad \textbf{30.25} \blacktriangleleft$$

A reaction for which Q is positive is called **exothermic**. A reaction for which Q is negative is called **endothermic**. The minimum kinetic energy of the incoming particle necessary for such a reaction to occur is called the **threshold energy**.

OBJECTIVE QUESTIONS

☐ denotes answer available in *Student Solutions Manual/Study Guide*

1. When $^{32}_{15}$P decays to $^{32}_{16}$S, which of the following particles is emitted? (a) a proton (b) an alpha particle (c) an electron (d) a gamma ray (e) an antineutrino

2. Two samples of the same radioactive nuclide are prepared. Sample G has twice the initial activity of sample H. **(i)** How does the half-life of G compare with the half-life of H? (a) It is two times larger. (b) It is the same. (c) It is half as large. **(ii)** After each has passed through five half-lives, how do their activities compare? (a) G has more than twice the activity of H. (b) G has twice the activity of H. (c) G and H have the same activity. (d) G has lower activity than H.

3. If a radioactive nuclide ^A_ZX decays by emitting a gamma ray, what happens? (a) The resulting nuclide has a different Z value. (b) The resulting nuclide has the same A and Z values. (c) The resulting nuclide has a different A value. (d) Both A and Z decrease by one. (e) None of those statements is correct.

4. Does a nucleus designated as $^{40}_{18}$X contain (a) 20 neutrons and 20 protons, (b) 22 protons and 18 neutrons, (c) 18 protons and 22 neutrons, (d) 18 protons and 40 neutrons, or (e) 40 protons and 18 neutrons?

5. In the decay $^{234}_{90}$Th $\rightarrow$ ^A_ZRa $+$ $^4_2$He, identify the mass number and the atomic number of the Ra nucleus: (a) $A = 230$, $Z = 92$ (b) $A = 238$, $Z = 88$ (c) $A = 230$, $Z = 88$ (d) $A = 234$, $Z = 88$ (e) $A = 238$, $Z = 86$

6. When $^{144}_{60}$Nd decays to $^{140}_{58}$Ce, identify the particle that is released. (a) a proton (b) an alpha particle (c) an electron (d) a neutron (e) a neutrino

7. When the $^{95}_{36}$Kr nucleus undergoes beta decay by emitting an electron and an antineutrino, does the daughter nucleus (Rb) contain (a) 58 neutrons and 37 protons, (b) 58 protons and 37 neutrons, (c) 54 neutrons and 41 protons, or (d) 55 neutrons and 40 protons?

8. What is the Q value for the reaction ^{9}Be $+ \alpha \rightarrow {}^{12}$C $+$ n? (a) 8.4 MeV (b) 7.3 MeV (c) 6.2 MeV (d) 5.7 MeV (e) 4.2 MeV

9. The half-life of radium-224 is about 3.6 days. What approximate fraction of a sample remains undecayed after two weeks? (a) $\frac{1}{2}$ (b) $\frac{1}{4}$ (c) $\frac{1}{8}$ (d) $\frac{1}{16}$ (e) $\frac{1}{32}$

10. A free neutron has a half-life of 614 s. It undergoes beta decay by emitting an electron. Can a free proton undergo a similar decay? (a) yes, the same decay (b) yes, but by emitting a positron (c) yes, but with a very different half-life (d) no

11. Which of the following quantities represents the reaction energy of a nuclear reaction? (a) (final mass $-$ initial mass)$/c^2$ (b) (initial mass $-$ final mass)$/c^2$ (c) (final mass $-$ initial mass)c^2 (d) (initial mass $-$ final mass)c^2 (e) none of those quantities

12. In the first nuclear weapon test carried out in New Mexico, the energy released was equivalent to approximately 17 kilotons of TNT. Estimate the mass decrease in the nuclear fuel representing the energy converted from rest energy into other forms in this event. *Note:* One ton of TNT has the energy equivalent of 4.2×10^9 J. (a) 1 μg (b) 1 mg (c) 1 g (d) 1 kg (e) 20 kg

CONCEPTUAL QUESTIONS

☐ denotes answer available in *Student Solutions Manual/Study Guide*

1. A student claims that a heavy form of hydrogen decays by alpha emission. How do you respond?

2. In beta decay, the energy of the electron or positron emitted from the nucleus lies somewhere in a relatively large range of possibilities. In alpha decay, however, the alpha-particle energy can only have discrete values. Explain this difference.

3. In Rutherford's experiment, assume an alpha particle is headed directly toward the nucleus of an atom. Why doesn't the alpha particle make physical contact with the nucleus?

4. Explain why nuclei that are well off the line of stability in Figure 30.4 tend to be unstable.

5. Compare and contrast the properties of a photon and a neutrino.

6. Why do nearly all the naturally occurring isotopes lie above the $N = Z$ line in Figure 30.4?

7. Why are very heavy nuclei unstable?

8. "If no more people were to be born, the law of population growth would strongly resemble the radioactive decay law." Discuss this statement.

9. Can carbon-14 dating be used to measure the age of a rock? Explain.

10. In positron decay, a proton in the nucleus becomes a neutron and its positive charge is carried away by the positron. A neutron, though, has a larger rest energy than a proton. How is that possible?

11. Consider two heavy nuclei X and Y having similar mass numbers. If X has the higher binding energy, which nucleus tends to be more unstable? Explain your answer.

12. What fraction of a radioactive sample has decayed after two half-lives have elapsed?

13. Figure CQ30.13 shows a watch from the early 20th century. The numbers and the hands of the watch are painted with a paint that contains a small amount of natural radium $^{226}_{88}$Ra mixed with a phosphorescent material. The decay of the radium causes the phosphorescent material to glow continuously. The radioactive nuclide $^{226}_{88}$Ra has a half-life of approximately 1.60×10^3 years. Being that the solar system is approximately 5 billion years old, why was this isotope still available in the 20th century for use on this watch?

Figure CQ30.13

14. Suppose it could be shown that the cosmic-ray intensity at the Earth's surface was much greater 10 000 years ago. How would this difference affect what we accept as valid carbon-dated values of the age of ancient samples of once-living matter? Explain your answer.

15. (a) How many values of I_z are possible for $I = \frac{5}{2}$? (b) For $I = 3$?

16. Can a nucleus emit alpha particles that have different energies? Explain.

17. If a nucleus such as ^{226}Ra initially at rest undergoes alpha decay, which has more kinetic energy after the decay, the alpha particle or the daughter nucleus? Explain your answer.

PROBLEMS AVAILABLE IN ~~ENHANCED~~ WebAssign

30.1 Some Properties of Nuclei

Problems 1–9

30.2 Nuclear Binding Energy

Problems 10–13

30.3 Radioactivity

Problems 14–22

30.4 The Radioactive Decay Processes

Problems 23–29

30.5 Nuclear Reactions

Problems 30–35

30.6 Context Connection: The Engine of the Stars

Problems 36–41

Additional Problems

Problems 42–70

Solutions to the following Problems are available in the *Student Solutions Manual/Study Guide*:

30.1, 30.7, 30.13, 30.15, 30.18, 30.25, 30.32, 30.35, 30.37, 30.40, 30.51, 30.55, 30.58, and 30.67

List of Enhanced Problems

Problem Number	Targeted Feedback in Enhanced WebAssign	Master It in Enhanced WebAssign	Watch It in Enhanced WebAssign
30.1	✓	✓	
30.15		✓	
30.18	✓	✓	
30.24	✓		✓
30.33	✓		✓
30.35	✓	✓	
30.51	✓		✓
30.55		✓	
30.57	✓		✓
30.67		✓	

Particle Physics

Chapter Outline

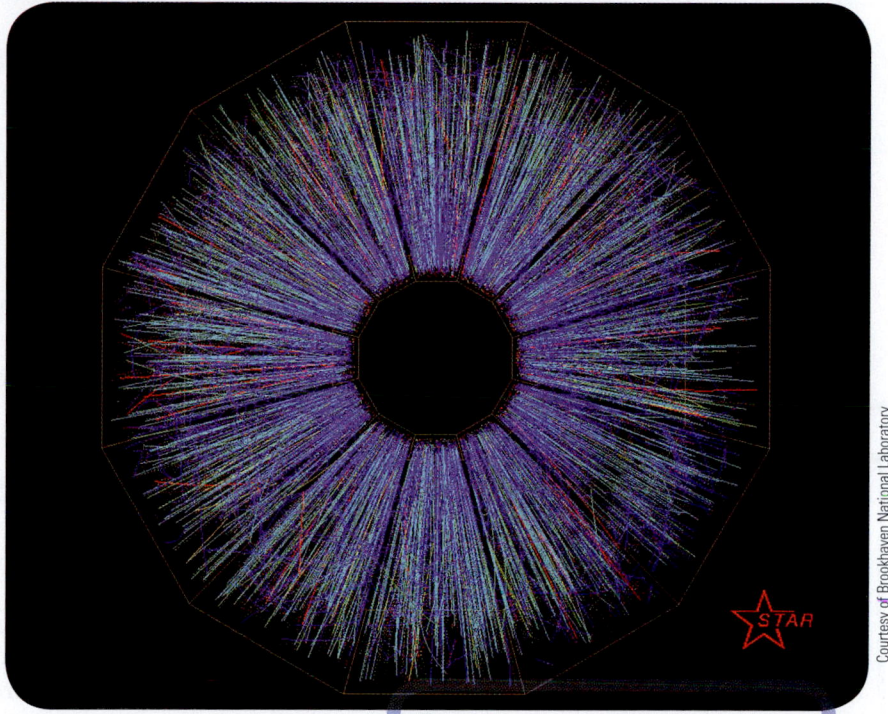

Courtesy of Brookhaven National Laboratory

A shower of particle tracks from a head-on collision of gold nuclei, each moving with energy 100 GeV. This collision occurred at the Relativistic Heavy Ion Collider (RHIC) at Brookhaven National Laboratory and was recorded with the STAR (Solenoidal Tracker at RHIC) detector. The tracks represent many fundamental particles arising from the energy of the collision.

In the early chapters of this book, we discussed the particle model, which treats an object as a particle of zero size with no structure. Some behaviors of objects, such as thermal expansion, can be understood by modeling the object as a collection of particles: atoms. In these models, any internal structure of the atom is ignored. We could not ignore the internal structure of the atom to understand such phenomena as atomic spectra, however. Modeling the hydrogen atom as a system of an electron in orbit about a particle-like nucleus helped in this regard (Section 11.5). In Chapter 30, however, we could not model the nucleus as a particle and ignore its structure to understand behavior such as nuclear stability and radioactive decay. We had to model the nucleus as a collection of smaller particles, nucleons. What about these nuclear constituents, the protons and neutrons? Can we apply the particle model to these entities? As we shall see, even protons and neutrons have structure, which leads to a puzzling question. As we continue to investigate the structure of smaller and smaller "particles," will we ever reach a level at which the building blocks are truly and completely described by the particle model?

In this concluding chapter, we explore this question by examining the properties and classifications of the various known subatomic particles and the fundamental interactions that govern their behavior. We also discuss the current model of elementary particles, in which all matter is believed to be constructed from only two families of particles: quarks and leptons.

The word *atom* is from the Greek *atomos,* which means "indivisible." At one time, atoms were thought to be the indivisible constituents of matter; that is, they were regarded as elementary particles. After 1932, physicists viewed all matter as consisting of only three constituent particles: electrons, protons, and neutrons. (The neutron was observed and identified in 1932.) With the exception of the free neutron (as opposed to a neutron within a nucleus), these particles are very stable. Beginning in 1945, many new particles were discovered in experiments involving high-energy collisions between known particles. These new particles are characteristically very unstable and have very short half-lives, ranging between 10^{-6} s and 10^{-23} s. So far, more than 300 of these unstable, temporary particles have been catalogued.

Since the 1930s, many powerful particle accelerators have been constructed throughout the world, making it possible to observe collisions of highly energetic particles under controlled laboratory conditions so as to reveal the subatomic world in finer detail. Until the 1960s, physicists were bewildered by the large number and variety of subatomic particles being discovered. They wondered if the particles had no systematic relationship connecting them, or whether a pattern was emerging that would provide a better understanding of the elaborate structure in the subnuclear world. Since that time, physicists have advanced our knowledge of the structure of matter tremendously by developing a structural model in which most of the ever-growing number of particles are made of smaller particles called *quarks.* Therefore, protons and neutrons, for example, are not truly elementary but are systems of tightly bound quarks.

◥ 31.1 | The Fundamental Forces in Nature

As we learned in Chapter 5, all natural phenomena can be described by four fundamental forces between particles. In order of decreasing strength, they are the **strong** force, the **electromagnetic** force, the **weak** force, and the **gravitational** force. In current models, the electromagnetic and weak forces are considered to be two manifestations of a single interaction, the **electroweak force,** as discussed in Section 31.11.

The **nuclear force,** as we mentioned in Chapter 30, holds nucleons together. It is very short range and is negligible for separations greater than about 2 fm (about the size of the nucleus). The electromagnetic force, which binds atoms and molecules together to form ordinary matter, has about 10^{-2} times the strength of the nuclear force. It is a long-range force that decreases in strength as the inverse square of the separation between interacting particles. The weak force is a short-range force that accounts for radioactive decay processes such as beta decay, and its strength is only about 10^{-5} times that of the nuclear force. Finally, the gravitational force is a long-range force that has a strength of only about 10^{-39} times that of the nuclear force. Although this familiar interaction is the force that holds the planets, stars, and galaxies together, its effect on elementary particles is negligible.

In modern physics, interactions between particles are often described in terms of a structural model that involves the exchange of **field particles,** or **exchange particles.** Field particles are also called **gauge bosons.**[1] (In general, all particles with integral spin are called *bosons.*) In the case of the familiar electromagnetic interaction, for instance, the field particles are photons. In the language of modern physics, we say that the electromagnetic force is *mediated* by photons and that photons are the quanta of the electromagnetic field. Likewise, the nuclear force is mediated by field particles called **gluons,** the weak force is mediated by particles called the **W and Z bosons,** and the gravitational force is mediated by quanta of the gravitational field called **gravitons.** These forces, their ranges, and their relative strengths are summarized in Table 31.1.

Pitfall Prevention | 31.1
The Nuclear Force and the Strong Force
The nuclear force discussed in Chapter 30 was historically called the strong force. Once the quark theory (Section 31.9) was established, however, the phrase *strong force* was reserved for the force between quarks. We shall follow this convention: the strong force is between quarks or particles built from quarks, and the nuclear force is between nucleons in a nucleus. The nuclear force is a secondary result of the strong force as discussed in Section 31.10. It is sometimes called the residual strong force. Because of this historical development of the names for these forces, other books sometimes refer to the nuclear force as the strong force.

▶ Field particles

---

[1]The word *gauge* comes from *gauge theory,* which is a sophisticated mathematical analysis that is beyond the scope of this book.

TABLE 31.1 | Fundamental Forces

Force	Relative Strength	Range of Force	Mediating Field Particle	Mass of Field Particle (GeV/c^2)
Nuclear/Strong	1	Short (~1 fm)	Gluon	0
Electromagnetic	10^{-2}	∞	Photon	0
Weak	10^{-5}	Short (~10^{-3} fm)	$W^{\pm}$, Z^0 bosons	80.4, 80.4, 91.2
Gravitational	10^{-39}	∞	Graviton	0

31.2 | Positrons and Other Antiparticles

In the 1920s, English theoretical physicist Paul Adrien Maurice Dirac developed a version of quantum mechanics that incorporated special relativity. Dirac's theory explained the origin of electron spin and its magnetic moment. It also presented a major difficulty, however. Dirac's relativistic wave equation required solutions corresponding to negative energy states even for free electrons. If negative energy states existed, however, one would expect an electron in a state of positive energy to make a rapid transition to one of these states, emitting a photon in the process. Dirac avoided this difficulty by postulating a structural model in which all negative energy states are filled. The electrons occupying these negative energy states are collectively called the *Dirac sea*. Electrons in the Dirac sea are not directly observable because the Pauli exclusion principle does not allow them to react to external forces; there are no states available to which an electron can make a transition in response to an external force. Therefore, an electron in such a state acts as an isolated system unless an interaction with the environment is strong enough to excite the electron to a positive energy state. Such an excitation causes one of the negative energy states to be vacant, as in Figure 31.1, leaving a hole in the sea of filled states. (Notice that positive energy states exist only for $E > m_e c^2$, representing the rest energy of the electron. Similarly, negative energy states exist only for $E < -m_e c^2$.) *The hole can react to external forces and is observable.* The hole reacts in a way similar to that of the electron, except that it has a positive charge. It is the **antiparticle** to the electron.

The profound implication of this model is that *every particle has a corresponding antiparticle.* The antiparticle has the same mass as the particle, but the opposite charge. For example, the electron's antiparticle, called a **positron,** has a mass of $0.511\ MeV/c^2$ and a positive charge of $1.60 \times 10^{-19}\ C$.

Carl Anderson (1905–1991) observed and identified the positron in 1932, and in 1936 he was awarded the Nobel Prize in Physics for that achievement. Anderson discovered the positron while examining tracks in a cloud chamber created by electron-like particles of positive charge. (A cloud chamber contains a gas that has been supercooled to just below its usual condensation point. An energetic radioactive particle passing through ionizes the gas and leaves a visible track. These early experiments used cosmic rays—mostly energetic protons passing through interstellar space—to initiate high-energy reactions in the upper atmosphere, which resulted in the production of positrons at ground level.) To discriminate between positive and negative charges, Anderson placed the cloud chamber in a magnetic field, causing moving charged particles to follow curved paths as discussed in Section 22.3. He noted that some of the electron-like tracks deflected in a direction corresponding to a positively charged particle.

Since Anderson's discovery, the positron has been observed in a number of experiments. A common process for producing positrons is **pair production.** In this process, a gamma-ray photon with sufficiently high energy interacts with a nucleus and an electron–positron pair is created. In the Dirac sea model, an electron in a negative energy state is excited to a positive energy state, resulting in a new observable electron and a hole, which is the positron. Because the total rest energy of the electron–positron pair is $2m_e c^2 = 1.022\ MeV$, the photon must have at least this much energy to create an electron–positron pair. Therefore, energy in the form of a

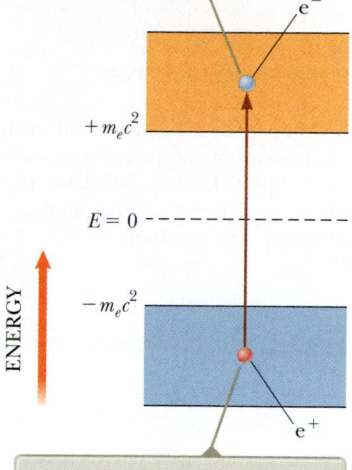

An electron can make a transition out of the Dirac sea only if it is provided with energy equal to or larger than $2m_e c^2$.

An upward transition of an electron leaves a vacancy in the Dirac sea, which can behave as a particle identical to the electron except for its positive charge.

Figure 31.1 Dirac's model for the existence of antielectrons (positrons).

Paul Adrien Maurice Dirac
British Physicist (1902–1984)
Dirac was instrumental in the understanding of antimatter and the unification of quantum mechanics and relativity. He made many contributions to the development of quantum physics and cosmology. In 1933, Dirac won a Nobel Prize in Physics.

Figure 31.2 (a) Bubble-chamber tracks of electron–positron pairs produced by 300-MeV gamma rays striking a lead sheet from the left. (b) The pertinent pair-production events. The positrons deflect upward and the electrons deflect downward in an applied magnetic field.

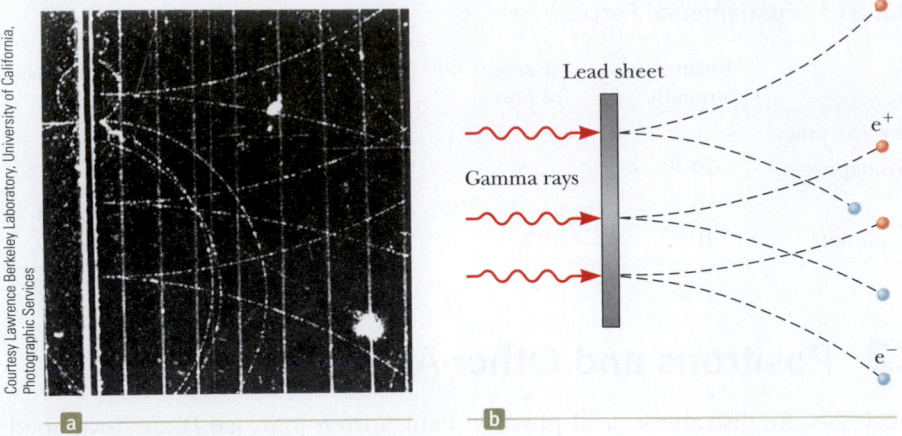

Courtesy Lawrence Berkeley Laboratory, University of California, Photographic Services

Lead sheet

Gamma rays

e^+

e^-

Pitfall Prevention | 31.2

Antiparticles

An antiparticle is not identified solely on the basis of opposite charge; even neutral particles have antiparticles, which are defined in terms of other properties, such as spin.

gamma-ray photon is converted to rest energy in accordance with Einstein's relationship $E_R = mc^2$. We can use the isolated system model to describe this process. The energy of the system of the photon and the nucleus is conserved and transformed to rest energy of the electron and positron, kinetic energy of these particles, and some small amount of kinetic energy associated with the nucleus. Figure 31.2a shows tracks of electron–positron pairs created by 300-MeV gamma rays striking a lead plate.

QUICK QUIZ 31.1 Given the identification of the particles in Figure 31.2b, is the direction of the external magnetic field in Figure 31.2a (**a**) into the page, (**b**) out of the page, (**c**) impossible to determine?

The reverse process can also occur. Under the proper conditions, an electron and positron can annihilate each other to produce two gamma-ray photons (see Thinking Physics 31.1) that have a combined energy of at least 1.022 MeV:

$$e^- + e^+ \rightarrow 2\gamma$$

BIO Positron-emission tomography (PET)

Electron–positron annihilation is used in the medical diagnostic technique called *positron-emission tomography* (PET). The patient is injected with a glucose solution containing a radioactive substance that decays by positron emission (often ^{18}F), and the material is carried by the blood throughout the body. A positron emitted during a decay event in one of the radioactive nuclei in the glucose solution annihilates with an electron in the immediately surrounding tissue, resulting in two gamma-ray photons emitted in opposite directions. A gamma detector surrounding the patient pinpoints the source of the photons and, with the assistance of a computer, displays an image of the sites at which the glucose accumulates. (Glucose is metabolized rapidly in cancerous tumors and accumulates in these sites, providing a strong signal for a PET detector system.) The images from a PET scan can indicate a wide variety of disorders in the brain, including Alzheimer's disease (Fig. 31.3). In addition, because glucose metabolizes more rapidly in active areas of the brain, a PET scan can indicate which areas of the brain are involved when the patient is engaging in such activities as language use, music, or vision.

Prior to 1955, on the basis of the Dirac theory, it was expected that every particle has a corresponding antiparticle, but antiparticles such as the antiproton and antineutron had not been detected experimentally. Because the relativistic Dirac theory had some failures (it predicted the wrong-size magnetic moment for the photon) as well as many successes, it was important to determine whether the antiproton really existed. In 1955, a team led by Emilio Segrè (1905–1989) and Owen Chamberlain (1920–2006) used the Bevatron particle accelerator at the University of California–Berkeley to produce antiprotons and antineutrons. They therefore established

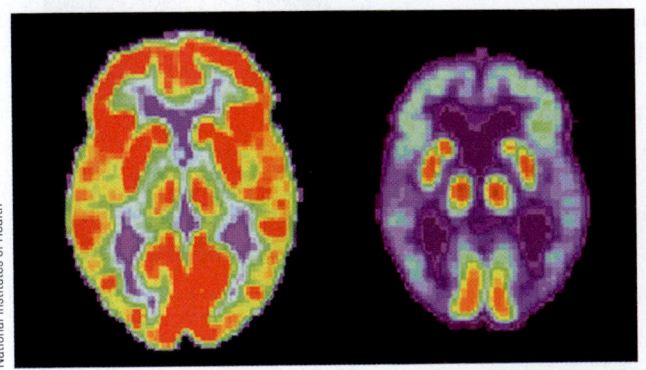

National Institutes of Health

Figure 31.3 PET scans of the brain of a healthy older person (*left*) and that of a patient suffering from Alzheimer's disease (*right*). Lighter regions contain higher concentrations of radioactive glucose, indicating higher metabolism rates and therefore increased brain activity.

with certainty the existence of antiparticles. For this work, Segrè and Chamberlain received the Nobel Prize in Physics in 1959. It is now established that every particle has a corresponding antiparticle with equal mass and spin, and with charge, magnetic moment, and strangeness of equal magnitude but opposite sign. (The property of strangeness is explained in Section 31.6.) The only exception to these rules for particles and antiparticles are the neutral photon, pion, and eta, each of which is its own antiparticle.

An intriguing aspect of the existence of antiparticles is that if we replace every proton, neutron, and electron in an atom with its antiparticle, we can create a stable antiatom; combinations of antiatoms should form antimolecules and eventually antiworlds. As far as we know, everything would behave in the same way in an antiworld as in our world. In principle, it is possible that some distant antimatter galaxies exist, separated from normal-matter galaxies by millions of light-years. Unfortunately, because the photon is its own antiparticle, the light emitted from an antimatter galaxy is no different from that from a normal-matter galaxy, so astronomical observations cannot determine if the galaxy is composed of matter or antimatter. Although no evidence of antimatter galaxies exists at present, it is awe-inspiring to imagine the cosmic spectacle that would result if matter and antimatter galaxies were to collide: a gigantic eruption of jets of annihilation radiation, transforming the entire galactic mass into energetic particles fleeing the collision point.

THINKING PHYSICS 31.1

When an electron and a positron meet at low speed in free space, why are two 0.511-MeV gamma rays produced rather than one gamma ray with an energy of 1.022 MeV?

Reasoning Gamma rays are photons, and photons carry momentum. We apply the momentum version of the isolated system model to the system, which consists initially of the electron and positron. If the system, assumed to be at rest, transformed to only one photon, momentum would not be conserved because the initial momentum of the electron–positron system is zero, whereas the final system consists of a single photon of energy 1.022 MeV and nonzero momentum. On the other hand, the two gamma-ray photons travel in *opposite* directions, so the total momentum of the final system—two photons—is zero, and momentum is conserved. ◄

31.3 | Mesons and the Beginning of Particle Physics

In the mid-1930s, physicists had a fairly simple view of the structure of matter. The building blocks were the proton, the electron, and the neutron. Three other particles were known or had been postulated at the time: the photon, the neutrino, and the positron. These six particles were considered the fundamental constituents of matter. With this marvelously simple picture of the world, however, no one was able to answer an important question. Because many protons in proximity in any nucleus should strongly repel one another due to their positive charges, what is the nature of the force that holds the nucleus together? Scientists recognized that this mysterious force, which we now call the nuclear force, must be much stronger than anything encountered in nature up to that time.

In 1935, Japanese physicist Hideki Yukawa proposed the first theory to successfully explain the nature of the nuclear force, an effort that later earned him the Nobel Prize in Physics. To understand Yukawa's theory, it is useful to first recall that in the modern structural model of electromagnetic interactions, charged particles interact by exchanging photons. Yukawa used this idea to explain the nuclear force by proposing a new particle whose exchange between nucleons in the nucleus produces the nuclear force. Furthermore, he established that the range of the force is inversely proportional to the mass of this particle and predicted that the mass would be about 200 times the mass of the electron. Because the new particle would have a

Hideki Yukawa
Japanese Physicist (1907–1981)
Yukawa was awarded the Nobel Prize in Physics in 1949 for predicting the existence of mesons. This photograph of him at work was taken in 1950 in his office at Columbia University. Yukawa came to Columbia in 1949 after spending the early part of his career in Japan.

© Bettmann/Corbis

mass between that of the electron and that of the proton, it was called a meson (from the Greek *meso*, meaning "middle").

In an effort to substantiate Yukawa's predictions, physicists began an experimental search for the meson by studying cosmic rays entering the Earth's atmosphere. In 1937, Anderson and his collaborators discovered a particle of mass 106 MeV/c^2, about 207 times the mass of the electron. Subsequent experiments showed that the particle interacted very weakly with matter, however, and hence could not be the carrier of the nuclear force. The puzzling situation inspired several theoreticians to propose that two mesons existed with slightly different masses. This idea was confirmed by the discovery in 1947 of the **pi (π) meson**, or simply **pion**, by Cecil Frank Powell (1903–1969) and Giuseppe P. S. Occhialini (1907–1993). The particle discovered by Anderson in 1937, the one thought to be Yukawa's meson, is not really a meson. (We shall discuss the requirements for a particle to be a meson in Section 31.4.) Instead, it takes part in the weak and electromagnetic interactions only and is now called the **muon (μ)**. We first discussed the muon in Section 9.4, with regard to time dilation.

The pion, Yukawa's carrier of the nuclear force, comes in three varieties corresponding to three charge states: π^+, π^-, and π^0. The π^+ and π^- particles have masses of 139.6 MeV/c^2, and the π^0 particle has a mass of 135.0 MeV/c^2. Pions and muons are very unstable particles. For example, the π^-, which has a mean lifetime of 2.6×10^{-8} s, first decays to a muon and an antineutrino. The muon, which has a mean lifetime of 2.2 μs, then decays into an electron, a neutrino, and an antineutrino:

$$\pi^- \rightarrow \mu^- + \bar{\nu}$$

$$\mu^- \rightarrow e^- + \nu + \bar{\nu}$$

31.1 ◀

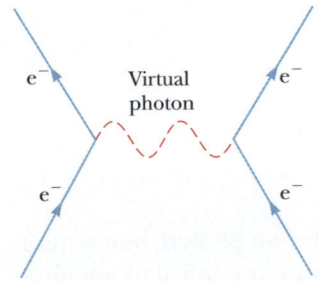

Figure 31.4 Feynman diagram representing a photon mediating the electromagnetic force between two electrons.

Note that for chargeless particles (as well as some charged particles such as the proton), a bar over the symbol indicates an antiparticle.

The interaction between two particles can be represented in a simple qualitative graphical representation called a **Feynman diagram,** developed by American physicist Richard P. Feynman. Figure 31.4 is such a diagram for the electromagnetic interaction between two electrons approaching each other. A Feynman diagram is a qualitative graph of time in the vertical direction versus space in the horizontal direction. It is qualitative in the sense that the actual values of time and space are not important, but the overall appearance of the graph provides a representation of the process. The time evolution of the process can be approximated by starting at the bottom of the diagram and moving your eyes upward.

In the simple case of the electron–electron interaction in Figure 31.4, a photon is the field particle that mediates the electromagnetic force between the electrons. Notice that the entire interaction is represented in such a diagram as if it occurs at a single point in time. Therefore, the paths of the electrons appear to undergo a discontinuous change in direction at the moment of interaction. This representation is correct on a microscopic level over a time interval that includes the exchange of one photon. It is different from the paths produced over the much longer interval during which we watch the interaction from a macroscopic point of view. In this case, the paths would be curved (as in Fig. 31.2) due to the continuous exchange of large numbers of field particles, illustrating another aspect of the qualitative nature of Feynman diagrams.

In the electron–electron interaction, the photon, which transfers energy and momentum from one electron to the other, is called a *virtual photon* because it vanishes during the interaction without having been detected. In Chapter 28, we discussed that a photon has energy $E = hf$, where f is its frequency. Consequently, for a system of two electrons initially at rest, the system has energy $2m_e c^2$ before a virtual photon is released and energy $2m_e c^2 + hf$ after the virtual photon is released (plus any kinetic energy of the electron resulting from the emission of the photon). Is that a violation of the law of conservation of energy for an isolated system? No; this process does *not* violate the law of conservation of energy because the virtual photon has a very short lifetime Δt that makes the uncertainty in the energy $\Delta E \approx \hbar / 2 \, \Delta t$ of the system consisting of two electrons and the photon greater than the photon energy. Therefore,

Richard Feynman
American Physicist (1918–1988)
Inspired by Dirac, Feynman developed quantum electrodynamics, the theory of the interaction of light and matter on a relativistic and quantum basis. In 1965, Feynman won the Nobel Prize in Physics. The prize was shared by Feynman, Julian Schwinger, and Sin Itiro Tomonaga. Early in Feynman's career, he was a leading member of the team developing the first nuclear weapon in the Manhattan Project. Toward the end of his career, he worked on the commission investigating the 1986 *Challenger* tragedy and demonstrated the effects of cold temperatures on the rubber O-rings used in the space shuttle.

within the constraints of the uncertainty principle, the energy of the system is conserved.

Now consider a pion exchange between a proton and a neutron according to Yukawa's model (Fig. 31.5a). The energy ΔE_R needed to create a pion of mass m_π is given by Einstein's equation $\Delta E_R = m_\pi c^2$. As with the photon in Figure 31.4, the very existence of the pion would appear to violate the law of conservation of energy if the particle existed for a time greater than $\Delta t \approx \hbar / 2 \Delta E_R$ (from the uncertainty principle), where Δt is the time interval required for the pion to transfer from one nucleon to the other. Therefore,

$$\Delta t \approx \frac{\hbar}{2 \Delta E_R} = \frac{\hbar}{2 m_\pi c^2}$$

and the rest energy of the pion is

$$m_\pi c^2 = \frac{\hbar}{2 \Delta t} \qquad \text{31.2} \blacktriangleleft$$

Because the pion cannot travel faster than the speed of light, the maximum distance d it can travel in a time interval Δt is $c \Delta t$. Therefore, using Equation 31.2 and $d = c \Delta t$, we find

$$m_\pi c^2 = \frac{\hbar c}{2d} \qquad \text{31.3} \blacktriangleleft$$

From Chapter 30, we know that the range of the nuclear force is on the order of 10^{-15} fm. Using this value for d in Equation 31.3, we estimate the rest energy of the pion to be

$$m_\pi c^2 \approx \frac{(1.055 \times 10^{-34}\,\text{J} \cdot \text{s})(3.00 \times 10^8\,\text{m/s})}{2(1 \times 10^{-15}\,\text{m})}$$

$$= 1.6 \times 10^{-11}\,\text{J} \approx 100\,\text{MeV}$$

which corresponds to a mass of 100 MeV/c^2 (approximately 200 times the mass of the electron). This value is in reasonable agreement with the observed pion mass.

The concept we have just described is quite revolutionary. In effect, it says that a system of two nucleons can change into two nucleons plus a pion as long as it returns to its original state in a very short time interval. (Remember that this model is the older, historical one, which assumes that the pion is the field particle for the nuclear force.) Physicists often say that a nucleon undergoes *fluctuations* as it emits and absorbs pions. As we have seen, these fluctuations are a consequence of a combination of quantum mechanics (through the uncertainty principle) and special relativity (through Einstein's mass–energy relationship $E_R = mc^2$).

This section has dealt with the particles that mediate the nuclear force, pions, and the mediators of the electromagnetic force, photons. Current ideas indicate that the nuclear force is more fundamentally described as an average or residual effect of the force between quarks, as will be explained in Section 31.10. The graviton, which is the mediator of the gravitational force, has yet to be observed. The $W^\pm$ and Z^0 particles that mediate the weak force were discovered in 1983 by Italian physicist Carlo Rubbia (b. 1934) and his associates using a proton–antiproton collider. Rubbia and Simon van der Meer (1925–2011), both at CERN (European Organization for Nuclear Research), shared the 1984 Nobel Prize in Physics for the detection and identification of the $W^\pm$ and Z^0 particles and the development of the proton–antiproton collider. In this accelerator, protons and antiprotons undergo head-on collisions with each other. In some of the collisions, $W^\pm$ and Z^0 particles are produced, which in turn are identified by their decay products. Figure 31.5b shows a Feynman diagram for a weak interaction mediated by a Z^0 boson.

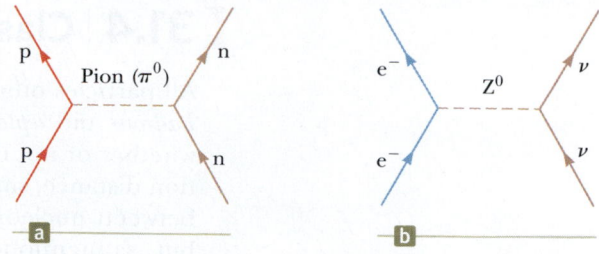

Figure 31.5 (a) Feynman diagram representing a proton and a neutron interacting via the nuclear force with a neutral pion mediating the force. (This model is *not* the most fundamental model for nucleon interaction.) (b) Feynman diagram for an electron and a neutrino interacting via the weak force with a Z^0 boson mediating the force.

31.4 | Classification of Particles

All particles other than field particles can be classified into two broad categories, *hadrons* and *leptons*. The criterion for separating these particles into categories is whether or not they interact via the strong force. This force increases with separation distance, similar to the force exerted by a stretched spring. The nuclear force between nucleons in a nucleus is a particular manifestation of the strong force, but, as mentioned in Pitfall Prevention 31.1, we will use the term *strong force* in general to refer to any interaction between particles made up of more elementary units called quarks. Table 31.2 provides a summary of the properties of some of these particles.

Hadrons

Particles that interact through the strong force are called **hadrons.** The two classes of hadrons—*mesons* and *baryons*—are distinguished by their masses and spins.

TABLE 31.2 | Some Particles and Their Properties

Category	Particle Name	Symbol	Anti-particle	Mass (MeV/c^2)	B	L_e	L_μ	L_τ	S	Lifetime(s)	Spin
Leptons	Electron	e^-	e^+	0.511	0	+1	0	0	0	Stable	$\frac{1}{2}$
	Electron–neutrino	ν_e	$\bar{\nu}_e$	$< 2eV/c^2$	0	+1	0	0	0	Stable	$\frac{1}{2}$
	Muon	μ^-	μ^+	105.7	0	0	+1	0	0	2.20×10^{-6}	$\frac{1}{2}$
	Muon–neutrino	ν_μ	$\bar{\nu}_\mu$	< 0.17	0	0	+1	0	0	Stable	$\frac{1}{2}$
	Tau	τ^-	τ^+	1 784	0	0	0	+1	0	$< 4 \times 10^{-13}$	$\frac{1}{2}$
	Tau–neutrino	ν_τ	$\bar{\nu}_\tau$	< 18	0	0	0	+1	0	Stable	$\frac{1}{2}$
Hadrons **Mesons**	Pion	π^+	π^-	139.6	0	0	0	0	0	2.60×10^{-8}	0
		π^0	Self	135.0	0	0	0	0	0	0.83×10^{-16}	0
	Kaon	K^+	K^-	493.7	0	0	0	0	+1	1.24×10^{-8}	0
		K^0_S	$\bar{K}^0_S$	497.7	0	0	0	0	+1	0.89×10^{-10}	0
		K^0_L	$\bar{K}^0_L$	497.7	0	0	0	0	+1	5.2×10^{-8}	0
	Eta	η	Self	548.8	0	0	0	0	0	$< 10^{-18}$	0
		η'	Self	958	0	0	0	0	0	2.2×10^{-21}	0
Baryons	Proton	p	$\bar{p}$	938.3	+1	0	0	0	0	Stable	$\frac{1}{2}$
	Neutron	n	$\bar{n}$	939.6	+1	0	0	0	0	614	$\frac{1}{2}$
	Lambda	Λ^0	$\bar{\Lambda}^0$	1 115.6	+1	0	0	0	−1	2.6×10^{-10}	$\frac{1}{2}$
	Sigma	Σ^+	$\bar{\Sigma}^-$	1 189.4	+1	0	0	0	−1	0.80×10^{-10}	$\frac{1}{2}$
		Σ^0	$\bar{\Sigma}^0$	1 192.5	+1	0	0	0	−1	6×10^{-20}	$\frac{1}{2}$
		Σ^-	$\bar{\Sigma}^+$	1 197.3	+1	0	0	0	−1	1.5×10^{-10}	$\frac{1}{2}$
	Delta	Δ^{++}	$\bar{\Delta}^{--}$	1 230	+1	0	0	0	0	6×10^{-24}	$\frac{3}{2}$
		Δ^+	$\bar{\Delta}^-$	1 231	+1	0	0	0	0	6×10^{-24}	$\frac{3}{2}$
		Δ^0	$\bar{\Delta}^0$	1 232	+1	0	0	0	0	6×10^{-24}	$\frac{3}{2}$
		Δ^-	$\bar{\Delta}^+$	1 234	+1	0	0	0	0	6×10^{-24}	$\frac{3}{2}$
	Xi	Ξ^0	$\bar{\Xi}^0$	1 315	+1	0	0	0	−2	2.9×10^{-10}	$\frac{1}{2}$
		Ξ^-	Ξ^+	1 321	+1	0	0	0	−2	1.64×10^{-10}	$\frac{1}{2}$
	Omega	Ω^-	Ω^+	1 672	+1	0	0	0	−3	0.82×10^{-10}	$\frac{3}{2}$

Mesons all have zero or integer spin (0 or 1).[2] As indicated in Section 31.3, the origin of the name comes from the expectation that Yukawa's proposed meson mass would lie between the mass of the electron and the mass of the proton. Several meson masses do lie in this range, although there are heavier mesons that have masses larger than that of the proton.

All mesons are known to decay into final products including electrons, positrons, neutrinos, and photons. The pions are the lightest of the known mesons; they have masses of about 140 MeV/c^2 and a spin of 0. Another is the K meson, with a mass of approximately 500 MeV/c^2 and a spin of 0.

Baryons, the second class of hadrons, have masses equal to or greater than the proton mass (*baryon* means "heavy" in Greek), and their spins are always an odd half-integer value ($\frac{1}{2}$ or $\frac{3}{2}$). Protons and neutrons are baryons, as are many other particles. With the exception of the proton, all baryons decay in such a way that the end products include a proton. For example, the baryon called the Ξ hyperon decays to the Λ^0 baryon in about 10^{-10} s. The Λ^0 baryon then decays to a proton and a π^- in approximately 3×10^{-10} s.

Today it is believed that hadrons are not elementary particles, but rather are composed of more elementary units called quarks. We shall discuss quarks in Section 31.9.

Leptons

Leptons (from the Greek *leptos,* meaning "small" or "light") are a group of particles that participate in the electromagnetic (if charged) and weak interactions. All leptons have spins of $\frac{1}{2}$. Unlike hadrons, which have size and structure, leptons appear to be truly elementary particles with no structure.

Quite unlike hadrons, the number of known leptons is small. Currently, scientists believe that only six leptons exist: the electron, the muon, and the tau, e^-, μ^-, τ^-, and a neutrino associated with each, ν_e, ν_μ, ν_τ. The tau lepton, discovered in 1975, has a mass equal to about twice that of the proton. Direct experimental evidence for the neutrino associated with the tau was announced by the Fermi National Accelerator Laboratory (Fermilab) in July 2000. Each of these six leptons has an antiparticle.

Current studies indicate that neutrinos may have a small but nonzero mass. If they do have mass, they cannot travel at the speed of light. Also, so many neutrinos exist that their combined mass may be sufficient to cause all the matter in the Universe to eventually collapse to infinite density and then explode and create a completely new Universe! We shall discuss this concept in more detail in Section 31.12.

31.5 | Conservation Laws

We have seen the importance of conservation laws for isolated systems many times in earlier chapters and have solved problems using conservation of energy, linear momentum, angular momentum, and electric charge. Conservation laws are important in understanding why certain decays and reactions occur but others do not. In general, our familiar conservation laws provide us with a set of rules that all processes must follow.

Certain new conservation laws have been identified through experimentation and are important in the study of elementary particles. The members of the isolated system change identity during a decay or reaction. The initial particles before the decay or reaction are different from the final particles afterward.

---

[2]Therefore, the particle discovered by Anderson in 1937, the muon, is not a meson; the muon has spin $\frac{1}{2}$. It belongs in the *lepton* classification described shortly.

Baryon Number

Experimental results tell us that whenever a baryon is created in a nuclear reaction or decay, an antibaryon is also created. This scheme can be quantified by assigning a baryon number $B = +1$ for all baryons, $B = -1$ for all antibaryons, and $B = 0$ for all other particles. Therefore, the **law of conservation of baryon number** states that

▶ Conservation of baryon number

> whenever a reaction or decay occurs, the sum of the baryon numbers of the system before the process must equal the sum of the baryon numbers after the process.

An equivalent statement is that the net number of baryons remains constant in any process.

If baryon number is absolutely conserved, the proton must be absolutely stable. For example, a decay of the proton to a positron and a neutral pion would satisfy conservation of energy, momentum, and electric charge. Such a decay has never been observed, however. At present, we can say only that the proton has a half-life of at least 10^{33} years (the estimated age of the Universe is only 10^{10} years). Therefore, it is extremely unlikely that one would see a given proton undergo a decay process. If we collect a huge number of protons, however, perhaps we might see *some* proton in the collection undergo a decay, as addressed in Example 31.2.

> **QUICK QUIZ 31.2** Consider the following decays: **(i)** $n \rightarrow \pi^+ + \pi^- + \mu^+ + \mu^-$ and **(ii)** $n \rightarrow p + \pi^-$. From the following choices, which conservation laws are violated by each decay? **(a)** energy **(b)** electric charge **(c)** baryon number **(d)** angular momentum **(e)** no conservation laws

Example 31.1 | Checking Baryon Numbers

Use the law of conservation of baryon number to determine whether each of the following reactions can occur:

(A) $p + n \rightarrow p + p + n + \overline{p}$

SOLUTION

Conceptualize The mass on the right is larger than the mass on the left. Therefore, one might be tempted to claim that the reaction violates energy conservation. The reaction can indeed occur, however, if the initial particles have sufficient kinetic energy to allow for the increase in rest energy of the system.

Categorize We use a conservation law developed in this section, so we categorize this example as a substitution problem.

Evaluate the total baryon number for the left side of the reaction:

$1 + 1 = 2$

Evaluate the total baryon number for the right side of the reaction:

$1 + 1 + 1 + (-1) = 2$

Therefore, baryon number is conserved and the reaction can occur.

(B) $p + n \rightarrow p + p + \overline{p}$

SOLUTION

Evaluate the total baryon number for the left side of the reaction:

$1 + 1 = 2$

Evaluate the total baryon number for the right side of the reaction:

$1 + 1 + (-1) = 1$

Because baryon number is not conserved, the reaction cannot occur.

Example 31.2 | Detecting Proton Decay

Measurements taken at the Super Kamiokande neutrino detection facility (Fig. 31.6) indicate that the half-life of protons is at least 10^{33} yr.

(A) Estimate how long we would have to watch, on average, to see a proton in a glass of water decay.

SOLUTION

Conceptualize Imagine the number of protons in a glass of water. Although this number is huge, the probability of a single proton undergoing decay is small, so we would expect to wait for a long time interval before observing a decay.

Categorize Because a half-life is provided in the problem, we categorize this problem as one in which we can apply our statistical analysis techniques from Section 30.3.

Figure 31.6 (Example 31.2) This detector at the Super Kamiokande neutrino facility in Japan is used to study photons and neutrinos. It holds 50 000 metric tons of highly purified water and 13 000 photomultipliers. The photograph was taken while the detector was being filled. Technicians in a raft clean the photodetectors before they are submerged.

Courtesy of ICRR (Institute for Cosmic Ray Research), University of Tokyo

Analyze Let's estimate that a drinking glass contains a mass $m = 250$ g of water, with a molar mass $M = 18$ g/mol.

Find the number of molecules of water in the glass:

$$N_{molecules} = nN_A = \frac{m}{M} N_A$$

Each water molecule contains one proton in each of its two hydrogen atoms plus eight protons in its oxygen atom, for a total of ten protons. Therefore, there are $N = 10N_{molecules}$ protons in the glass of water.

Find the activity of the protons from Equations 30.5, 30.7, and 30.8:

$$(1) \quad R = \lambda N = \frac{\ln 2}{T_{1/2}}\left(10\, \frac{m}{M} N_A\right) = \frac{\ln 2}{10^{33}\ \text{yr}} (10)\left(\frac{250\ \text{g}}{18\ \text{g/mol}}\right)(6.02 \times 10^{23}\ \text{mol}^{-1})$$

$$= 5.8 \times 10^{-8}\ \text{yr}^{-1}$$

Finalize The decay constant represents the probability that *one* proton decays in one year. The probability that *any* proton in our glass of water decays in the one-year interval is given by Equation (1). Therefore, we must watch our glass of water for $1/R \approx$ 17 million years! That indeed is a long time interval, as expected.

(B) The Super Kamiokande neutrino facility contains 50 000 metric tons of water. Estimate the average time interval between detected proton decays in this much water if the half-life of a proton is 10^{33} yr.

SOLUTION

Analyze The proton decay rate R in a sample of water is proportional to the number N of protons. Set up a ratio of the decay rate in the Super Kamiokande facility to that in a glass of water:

$$\frac{R_{Kamiokande}}{R_{glass}} = \frac{N_{Kamiokande}}{N_{glass}} \quad \rightarrow \quad R_{Kamiokande} = \frac{N_{Kamiokande}}{N_{glass}} R_{glass}$$

The number of protons is proportional to the mass of the sample, so express the decay rate in terms of mass:

$$R_{Kamiokande} = \frac{m_{Kamiokande}}{m_{glass}} R_{glass}$$

Substitute numerical values:

$$R_{Kamiokande} = \left(\frac{50\ 000\ \text{metric tons}}{0.250\ \text{kg}}\right)\left(\frac{1\ 000\ \text{kg}}{1\ \text{metric ton}}\right)(5.8 \times 10^{-8}\ \text{yr}^{-1}) \approx 12\ \text{yr}^{-1}$$

Finalize The average time interval between decays is about one-twelfth of a year, or approximately one month. That is much shorter than the time interval in part (A) due to the tremendous amount of water in the detector facility. Despite this rosy prediction of one proton decay per month, a proton decay has never been observed. This suggests that the half-life of the proton may be larger than 10^{33} years or that proton decay simply does not occur.

Lepton Number

From observations of commonly occurring decays of the electron, muon, and tau, we arrive at three conservation laws involving lepton numbers, one for each variety of lepton. The **law of conservation of electron lepton number** states that

▶ Conservation of electron lepton number

> the sum of the electron lepton numbers of the system before a reaction or decay must equal the sum of the electron lepton numbers after the reaction or decay.

The electron and the electron neutrino are assigned a positive electron lepton number $L_e = +1$, the antileptons e^+ and $\bar{\nu}_e$ are assigned a negative electron lepton number $L_e = -1$; all others have $L_e = 0$. For example, consider the decay of the neutron

$$n \;\rightarrow\; p + e^- + \bar{\nu}_e$$

Before the decay, the electron lepton number is $L_e = 0$; after the decay, it is $0 + 1 + (-1) = 0$. Therefore, the electron lepton number is conserved. It is important to recognize that the baryon number must also be conserved; which can easily be checked by noting that before the decay $B = +1$ and after the decay B is $+1 + 0 + 0 = +1$.

Similarly, when a decay involves muons, the muon lepton number L_μ is conserved. The μ^- and the ν_μ are assigned positive numbers, $L_\mu = +1$, the antimuons μ^+ and $\bar{\nu}_\mu$ are assigned negative numbers, $L_\mu = -1$; all others have $L_\mu = 0$. Finally, the tau lepton number L_τ is conserved, and similar assignments can be made for the tau lepton and its neutrino.

QUICK QUIZ 31.3 Consider the following decay: $\pi^0 \rightarrow \mu^- + e^+ + \nu_\mu$. What conservation laws are violated by this decay? **(a)** energy **(b)** angular momentum **(c)** electric charge **(d)** baryon number **(e)** electron lepton number **(f)** muon lepton number **(g)** tau lepton number **(h)** no conservation laws

QUICK QUIZ 31.4 Suppose a claim is made that the decay of the neutron is given by $n \rightarrow p + e^-$. What conservation laws are violated by this decay? **(a)** energy **(b)** angular momentum **(c)** electric charge **(d)** baryon number **(e)** electron lepton number **(f)** muon lepton number **(g)** tau lepton number **(h)** no conservation laws

---

Example 31.3 | Checking Lepton Numbers

Use the law of conservation of lepton numbers to determine whether each of the following decay schemes (A) and (B) can occur:

(A) $\mu^- \rightarrow e^- + \bar{\nu}_e + \nu_\mu$

SOLUTION

Conceptualize Because this decay involves a muon and an electron, L_μ and L_e must each be conserved separately if the decay is to occur.

Categorize We use a conservation law developed in this section, so we categorize this example as a substitution problem.

Evaluate the lepton numbers before the decay:
$$L_\mu = +1 \qquad L_e = 0$$

Evaluate the total lepton numbers after the decay:
$$L_\mu = 0 + 0 + 1 = +1 \qquad L_e = +1 + (-1) + 0 = 0$$

Therefore, both numbers are conserved and on this basis the decay is possible.

(B) $\pi^+ \rightarrow \mu^+ + \nu_\mu + \nu_e$

31.3 cont.

SOLUTION

Evaluate the lepton numbers before the decay: $L_\mu = 0 \qquad L_e = 0$

Evaluate the total lepton numbers after the decay: $L_\mu = -1 + 1 + 0 = 0 \qquad L_e = 0 + 0 + 1 = 1$

Therefore, the decay is not possible because electron lepton number is not conserved.

31.6 | Strange Particles and Strangeness

Many particles discovered in the 1950s were produced by the nuclear interaction of pions with protons and neutrons in the atmosphere. A group of these particles—the kaon (K), lambda (Λ), and sigma (Σ) particles—exhibited unusual properties in production and decay and hence were called *strange particles*.

One unusual property is that these particles are always produced in pairs. For example, when a pion collides with a proton, two neutral strange particles are produced with high probability:

$$\pi^- + p \rightarrow \Lambda^0 + K^0$$

On the other hand, the reaction $\pi^- + p \rightarrow n^0 + K^0$ in which only one of the final particles is strange never occurs, even though no conservation laws known in the 1950s are violated and the energy of the pion is sufficient to initiate the reaction.

The second peculiar feature of strange particles is that, although they are produced by the strong force at a high rate, they do not decay at a very high rate into particles that interact via the strong force. Instead, they decay very slowly, which is characteristic of the weak interaction. Their half-lives are in the range 10^{-10} s to 10^{-8} s; most other particles that interact via the strong force have very short lifetimes, on the order of 10^{-20} s or less.

Such observations indicate the necessity to make modifications in our model. To explain these unusual properties of strange particles, a new quantum number S, called **strangeness,** was introduced into our model of elementary particles, together with a new conservation law. The strangeness numbers for some particles are given in Table 31.2. The production of strange particles in pairs is handled by assigning $S = +1$ to one of the particles and $S = -1$ to the other. All nonstrange particles are assigned strangeness $S = 0$. The **law of conservation of strangeness** states that

whenever a reaction or decay occurs via the strong force, the sum of the strangeness numbers of the system before the process must equal the sum of the strangeness numbers after the process. In processes that occur via the weak interaction, strangeness may not be conserved.

▶ Conservation of strangeness

The low decay rate of strange particles can be explained by assuming that the nuclear and electromagnetic interactions obey the law of conservation of strangeness, but the weak interaction does not. Because the decay reaction involves the loss of one strange particle, it violates strangeness conservation and hence proceeds slowly via the weak interaction.

Example 31.4 | Is Strangeness Conserved?

(A) Use the law of strangeness conservation to determine whether the reaction $\pi^0 + n \rightarrow K^+ + \Sigma^-$ occurs.

SOLUTION

Conceptualize We recognize that there are strange particles appearing in this reaction, so we see that we will need to investigate conservation of strangeness.

Categorize We use a conservation law developed in this section, so we categorize this example as a substitution problem.

continued

31.4 *cont.*

Evaluate the strangeness for the left side of the reaction using Table 31.2: $S = 0 + 0 = 0$

Evaluate the strangeness for the right side of the reaction: $S = +1 - 1 = 0$

Therefore, strangeness is conserved and the reaction is allowed.

(B) Show that the reaction $\pi^- + p \rightarrow \pi^- + \Sigma^+$ does not conserve strangeness.

SOLUTION

Evaluate the strangeness for the left side of the reaction: $S = 0 + 0 = 0$

Evaluate the strangeness for the right side of the reaction: $S = 0 + (-1) = -1$

Therefore, strangeness is not conserved.

>31.7 | Measuring Particle Lifetimes

The bewildering array of entries in Table 31.2 leaves one yearning for firm ground. In fact, it is natural to wonder about an entry, for example, that shows a particle (Σ^0) that exists for 10^{-20} s and has a mass of 1192.5 MeV/c^2. How is it possible to detect a particle that exists for only 10^{-20} s?

Most particles are unstable and are created in nature only rarely, in cosmic ray showers. In the laboratory, however, large numbers of these particles are created in controlled collisions between high-energy particles and a suitable target. The incident particles must have very high energy, and it takes a considerable time interval for electromagnetic fields to accelerate particles to high energies. Therefore, stable charged particles such as electrons or protons generally make up the incident beam. Similarly, targets must be simple and stable, and the simplest target, hydrogen, serves nicely as both target (the proton) and detector.

Figure 31.7 shows a typical event in which hydrogen in a bubble chamber served as both target source and detector. (A bubble chamber is a device in which the tracks of charged particles are made visible in liquid hydrogen that is maintained near its boiling point.) Many parallel tracks of negative pions are visible entering the photograph from the bottom. As the labels in the inset drawing show, one of the pions has hit a stationary proton in the hydrogen, producing two strange particles, Λ^0 and K^0, according to the reaction

$$\pi^- + p \rightarrow \Lambda^0 + K^0$$

Neither neutral strange particle leaves a track, but their subsequent decays into charged particles can be clearly seen as indicated in Figure 31.7. A magnetic field directed into the plane of the photograph causes the track of each charged particle to curve, and from the measured curvature one can determine the particle's charge and linear momentum. If the mass and momentum of the incident particle are known, we can then usually calculate the product particle mass, kinetic energy, and speed from conservation of momentum and energy. Finally, by combining a product particle's speed with a measurable decay track length, we can calculate the

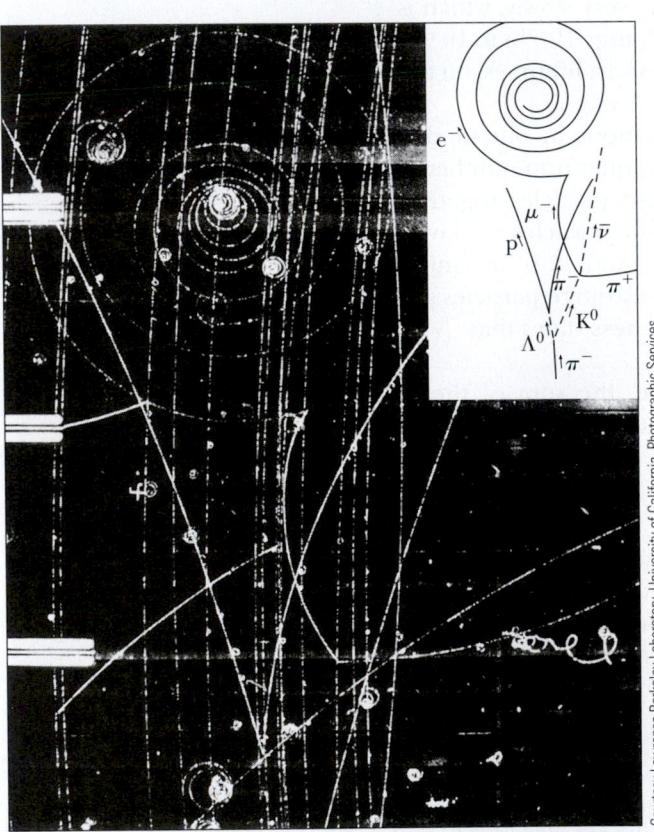

Figure 31.7 This bubble-chamber photograph shows many events, and the inset is a drawing of identified tracks. The strange particles Λ^0 and K^0 are formed at the bottom as the π^- interacts with a proton according to $\pi^- + p \rightarrow \Lambda^0 + K^0$. (Notice that the neutral particles leave no tracks, as indicated by the dashed lines in the insert.) The Λ^0 then decays in the reaction $\Lambda^0 \rightarrow \pi^- + p$ and the K^0 in the reaction $K^0 \rightarrow \pi^0 + \mu^- + \bar{\nu}_\mu$.

Courtesy Lawrence Berkeley Laboratory, University of California, Photographic Services

product particle's lifetime. Figure 31.7 shows that sometimes one can use this lifetime technique even for a neutral particle, which leaves no track. As long as the beginning and end points of the missing track are known as well as the particle speed, one can infer the missing track length and find the lifetime of the neutral particle.

Resonance Particles

With clever experimental technique and much effort, decay track lengths as short as 10^{-6} m can be measured. Therefore, lifetimes as short as 10^{-16} s can be measured for high-energy particles traveling at about the speed of light. We arrive at this result by assuming that a decaying particle travels 1 μm at a speed of $0.99c$ in the reference frame of the laboratory, yielding a lifetime of $\Delta t_{lab} = 1 \times 10^{-6} \, m/0.99c \approx 3.4 \times 10^{-15}$ s. This result is not our final one, however, because we must account for the relativistic effects of time dilation. Because the proper lifetime Δt_p as measured in the decaying particle's reference frame is shorter than the laboratory frame value Δt_{lab} by a factor of $\sqrt{1 - (v^2/c^2)}$ (see Eq. 9.6), we can calculate the proper lifetime:

$$\Delta t_p = \Delta t_{lab} \sqrt{1 - \frac{v^2}{c^2}} = (3.4 \times 10^{-15} \, s) \sqrt{1 - \frac{(0.99c)^2}{c^2}} = 4.8 \times 10^{-16} \, s$$

Unfortunately, even with Einstein's help, the best answer we can obtain with the track length method is several orders of magnitude away from lifetimes of 10^{-20} s. How then can we detect the presence of particles that exist for time intervals like 10^{-20} s? For such short-lived particles, known as **resonance particles,** all we can do is infer their masses, their lifetimes, and, indeed, their very existence from data on their decay products.

31.8 | Finding Patterns in the Particles

A tool scientists use to help understand nature is the detection of patterns in data. One of the best examples of the use of this tool is the development of the periodic table, which provides fundamental understanding of the chemical behavior of the elements. The periodic table explains how more than a hundred elements can be formed from three particles: the electron, proton, and neutron. The number of observed particles and resonances observed by particle physicists is even larger than the number of elements. Is it possible that a small number of entities could exist from which all these particles could be built? Motivated by the success of the periodic table, let us explore the historical search for patterns among the particles.

Many classification schemes have been proposed for grouping particles into families. Consider, for instance, the baryons listed in Table 31.2 that have spins of $\frac{1}{2}$: p, n, $\Lambda^0, \Sigma^+, \Sigma^0, \Sigma^-, \Xi^0$, and Ξ^-. If we plot strangeness versus charge for these baryons using a sloping coordinate system, as in Figure 31.8a, we observe a fascinating pattern. Six of the baryons form a hexagon, and the remaining two are at the hexagon's center.[3]

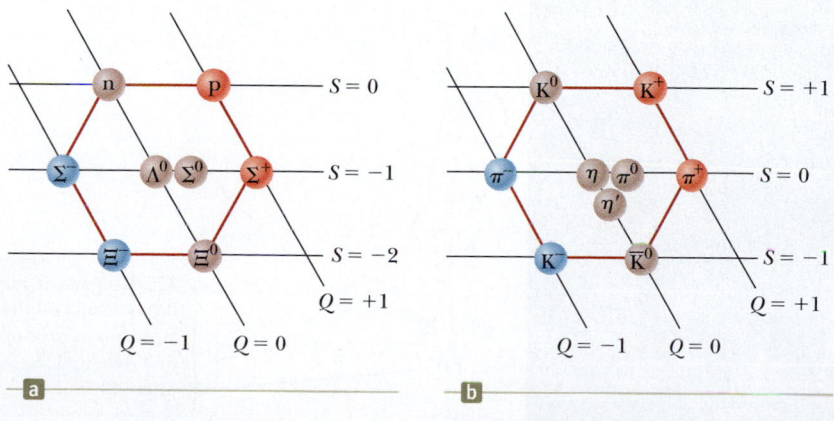

Figure 31.8 (a) The hexagonal eightfold-way pattern for the eight spin-$\frac{1}{2}$ baryons. This strangeness-versus-charge plot uses a sloping axis for charge number Q and a horizontal axis for strangeness S. (b) The eightfold-way pattern for the nine spin-zero mesons.

[3]The reason for the sloping coordinate system is so that a *regular* hexagon is formed, one with equal sides. If a normal orthogonal coordinate system is used, the pattern still appears, but the hexagonal shape does not have equal sides. Try it!

Figure 31.9 The pattern for the higher-mass, spin-$\frac{3}{2}$ baryons known at the time the pattern was proposed.

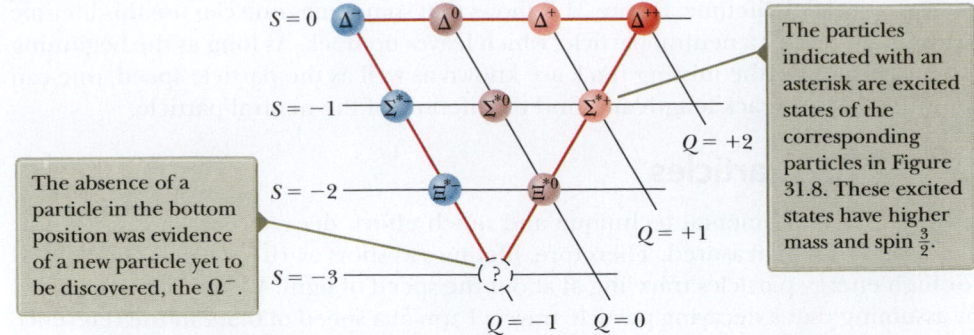

The particles indicated with an asterisk are excited states of the corresponding particles in Figure 31.8. These excited states have higher mass and spin $\frac{3}{2}$.

The absence of a particle in the bottom position was evidence of a new particle yet to be discovered, the Ω^-.

Murray Gell-Mann
American Physicist (b. 1929)
In 1969, Murray Gell-Mann was awarded the Nobel Prize in Physics for his theoretical studies dealing with subatomic particles.

As a second example, consider the following nine spin-zero mesons listed in Table 31.2: π^+, π^0, π^-, K^+, K^0, K^-, η, η', and the antiparticle $\overline{K}^0$. Figure 31.8b is a plot of strangeness versus charge for this family. Again, a hexagonal pattern emerges. In this case, each particle on the perimeter of the hexagon lies opposite its antiparticle, and the remaining three (which form their own antiparticles) are at its center. These and related symmetric patterns were developed independently in 1961 by Murray Gell-Mann and Yuval Ne'eman (1925–2006). Gell-Mann called the patterns the **eightfold way,** after the eightfold path to nirvana in Buddhism.

Groups of baryons and mesons can be displayed in many other symmetric patterns within the framework of the eightfold way. For example, the family of spin-$\frac{3}{2}$ baryons known in 1961 contains nine particles arranged in a pattern like that of the pins in a bowling alley as in Figure 31.9. [The particles Σ^{*+}, Σ^{*0}, Σ^{*-}, Ξ^{*0}, and Ξ^{*-} are excited states of the particles Σ^+, Σ^0, Σ^-, Ξ^0, and Ξ^-. In these higher-energy states, the spins of the three quarks (see Section 31.9) making up the particle are aligned so that the total spin of the particle is $\frac{3}{2}$.] When this pattern was proposed, an empty spot occurred in it (at the bottom position), corresponding to a particle that had never been observed. Gell-Mann predicted that the missing particle, which he called the omega minus (Ω^-), should have spin $\frac{3}{2}$, charge -1, strangeness -3, and rest energy of approximately 1 680 MeV. Shortly thereafter, in 1964, scientists at the Brookhaven National Laboratory found the missing particle through careful analyses of bubble-chamber photographs (Fig. 31.10) and confirmed all its predicted properties.

The prediction of the missing particle from the eightfold way has much in common with the prediction of missing elements in the periodic table. Whenever a vacancy occurs in an organized pattern of information, experimentalists have a guide for their investigations.

Figure 31.10 Discovery of the Ω^- particle. The photograph on the left shows the original bubble-chamber tracks. The drawing on the right isolates the tracks of the important events.

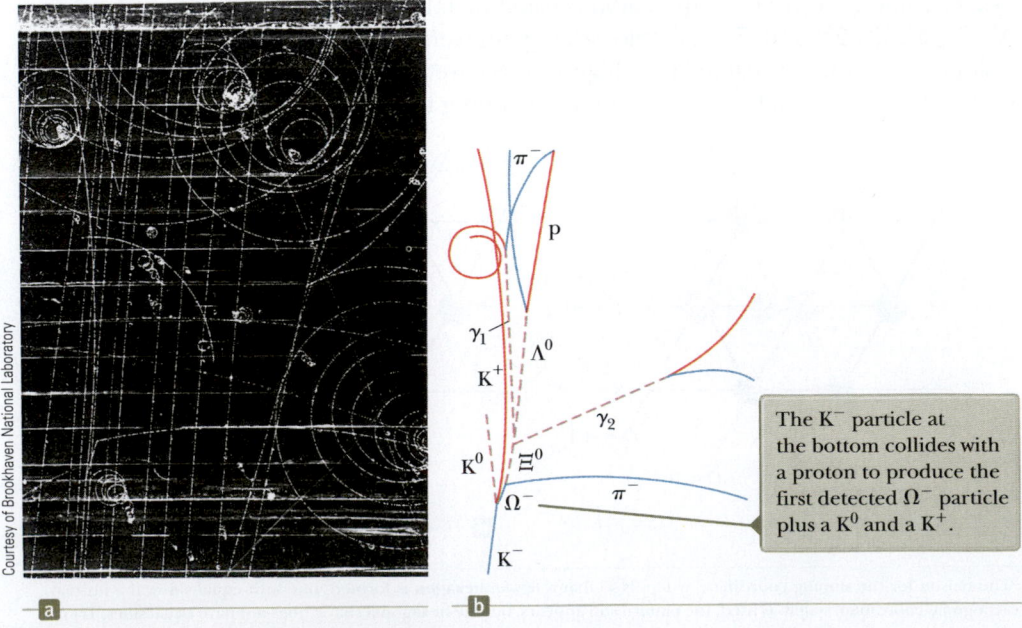

The K^- particle at the bottom collides with a proton to produce the first detected Ω^- particle plus a K^0 and a K^+.

31.9 | Quarks

As we have noted, leptons appear to be truly elementary particles because they occur in a small number of types, have no measurable size or internal structure, and do not seem to break down to smaller units. Hadrons, on the other hand, are complex particles having size and structure. The existence of the eightfold-way patterns suggests that hadrons have a more elemental substructure. Furthermore, we know that hundreds of types of hadrons exist and that many of them decay into other hadrons. These facts strongly suggest that hadrons cannot be truly elementary. In this section, we show that the complexity of hadrons can be explained by a simple substructure.

The Original Quark Model: A Structural Model for Hadrons

In 1963, Gell-Mann and George Zweig (b. 1937) independently proposed that hadrons have a more elemental substructure. According to their structural model, all hadrons are composite systems of two or three fundamental constituents called **quarks** (pronounced to rhyme with *forks*). (Gell-Mann borrowed the word *quark* from the passage "Three quarks for Muster Mark" in James Joyce's *Finnegan's Wake*.) The model proposes that three types of quarks exist, designated by the symbols u, d, and s. They are given the arbitrary names **up, down,** and **strange.** The various types of quarks are called **flavors.** Baryons consist of three quarks, and mesons consist of a quark and an antiquark. Active Figure 31.11 is a pictorial representation of the quark composition of several hadrons.

An unusual property of quarks is that they carry a fractional electronic charge. The u, d, and s quarks have charges of $+\frac{2}{3}e$, $-\frac{1}{3}e$, and $-\frac{1}{3}e$, respectively, where e is the elementary charge 1.6×10^{-19} C. These and other properties of quarks and antiquarks are given in Table 31.3. Notice that quarks have spin $\frac{1}{2}$, which means that all quarks are *fermions,* defined as any particle having half-integral spin. As Table 31.3

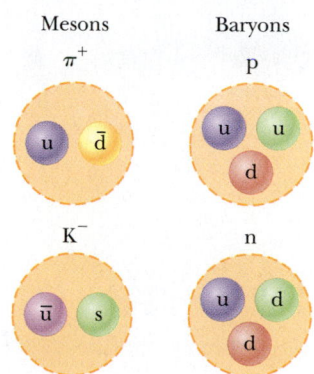

Active Figure 31.11 Quark compositions of two mesons and two baryons.

TABLE 31.3 | Properties of Quarks and Antiquarks

Quarks

Name	Symbol	Spin	Charge	Baryon Number	Strangeness	Charm	Bottomness	Topness
Up	u	$\frac{1}{2}$	$+\frac{2}{3}e$	$\frac{1}{3}$	0	0	0	0
Down	d	$\frac{1}{2}$	$-\frac{1}{3}e$	$\frac{1}{3}$	0	0	0	0
Strange	s	$\frac{1}{2}$	$-\frac{1}{3}e$	$\frac{1}{3}$	-1	0	0	0
Charmed	c	$\frac{1}{2}$	$+\frac{2}{3}e$	$\frac{1}{3}$	0	$+1$	0	0
Bottom	b	$\frac{1}{2}$	$-\frac{1}{3}e$	$\frac{1}{3}$	0	0	$+1$	0
Top	t	$\frac{1}{2}$	$+\frac{2}{3}e$	$\frac{1}{3}$	0	0	0	$+1$

Antiquarks

Name	Symbol	Spin	Charge	Baryon Number	Strangeness	Charm	Bottomness	Topness
Anti-up	$\bar{u}$	$\frac{1}{2}$	$-\frac{2}{3}e$	$-\frac{1}{3}$	0	0	0	0
Anti-down	$\bar{d}$	$\frac{1}{2}$	$+\frac{1}{3}e$	$-\frac{1}{3}$	0	0	0	0
Anti-strange	$\bar{s}$	$\frac{1}{2}$	$+\frac{1}{3}e$	$-\frac{1}{3}$	$+1$	0	0	0
Anti-charmed	$\bar{c}$	$\frac{1}{2}$	$-\frac{2}{3}e$	$-\frac{1}{3}$	0	-1	0	0
Anti-bottom	$\bar{b}$	$\frac{1}{2}$	$+\frac{1}{3}e$	$-\frac{1}{3}$	0	0	-1	0
Anti-top	$\bar{t}$	$\frac{1}{2}$	$-\frac{2}{3}e$	$-\frac{1}{3}$	0	0	0	-1

shows, associated with each quark is an antiquark of opposite charge, baryon number, and strangeness.

The composition of all hadrons known when Gell-Mann and Zweig presented their models can be completely specified by three simple rules:

- A meson consists of one quark and one antiquark, giving it a baryon number of 0, as required.
- A baryon consists of three quarks.
- An antibaryon consists of three antiquarks.

The theory put forth by Gell-Mann and Zweig is referred to as the *original quark model*.

> **QUICK QUIZ 31.5** Using a coordinate system like that in Figure 31.8, draw an eightfold-way diagram for the three quarks in the original quark model.

Charm and Other Developments

Although the original quark model was highly successful in classifying particles into families, some discrepancies were evident between predictions of the model and certain experimental decay rates. It became clear that the structural model needed to be modified to remove these discrepancies. Consequently, several physicists proposed a fourth quark in 1967. They argued that if four leptons exist (as was thought at the time: the electron, the muon, and a neutrino associated with each), four quarks should also exist because of an underlying symmetry in nature. The fourth quark, designated by c, was given a property called **charm**. A **charmed quark** has charge $+\frac{2}{3}e$, but its charm distinguishes it from the other three quarks. This addition introduces a new quantum number C, representing charm. The new quark has charm $C = +1$, its antiquark has charm $C = -1$, and all other quarks have $C = 0$ as indicated in Table 31.3. Charm, like strangeness, is conserved in strong and electromagnetic interactions, but not in weak interactions.

Evidence that the charmed quark exists began to accumulate in 1974 when a new heavy particle called the J/Ψ particle (or simply Ψ) was discovered independently by two groups, one led by Burton Richter (b. 1931) at the Stanford Linear Accelerator (SLAC), and the other led by Samuel Ting (b. 1936) at the Brookhaven National Laboratory. Richter and Ting were awarded the Nobel Prize in Physics in 1976 for this work. The J/Ψ particle does not fit into the three-quark structural model; instead, it has properties of a combination of the proposed charmed quark and its antiquark ($c\bar{c}$). It is much more massive than the other known mesons ($\sim 3\ 100\ \text{MeV}/c^2$), and its lifetime is much longer than the lifetimes of particles that decay via the strong force. Soon, related mesons were discovered, corresponding to such quark combinations as $\bar{c}d$ and cd, which all have large masses and long lifetimes. The existence of these new mesons provided firm evidence for the fourth quark flavor.

In 1975, researchers at Stanford University reported strong evidence for the tau (τ) lepton with a mass of $1\ 784\ \text{MeV}/c^2$. It is the fifth type of lepton to be discovered, which led physicists to propose that more flavors of quarks may exist, based on symmetry arguments similar to those leading to the proposal of the charmed quark. These proposals led to more elaborate quark models and the prediction of two new quarks: **top** (t) and **bottom** (b). To distinguish these quarks from the original four, quantum numbers called *topness* and *bottomness* (with allowed values $+1, 0, -1$) are assigned to all quarks and antiquarks (Table 31.3). In 1977, researchers at the Fermi National Laboratory, under the direction of Leon Lederman (b. 1922), reported the discovery of a very massive new meson Υ whose composition is considered to be $b\bar{b}$, providing evidence for the bottom quark. In March 1995, researchers at Fermilab announced the discovery of the top quark (supposedly the last of the quarks to be found), with a mass of $173\ \text{GeV}/c^2$.

TABLE 31.4 | Quark Composition of Mesons

Antiquarks

		$\bar{b}$		$\bar{c}$		$\bar{s}$		$\bar{d}$		$\bar{u}$	
	b	Υ	$(b\bar{b})$	B_c^-	$(\bar{c}b)$	B_s^0	$(\bar{s}b)$	B_d^0	$(\bar{d}b)$	B^-	$(\bar{u}b)$
	c	B_c^+	$(\bar{b}c)$	J/Ψ	$(\bar{c}c)$	D_s^+	$(\bar{s}c)$	D^+	$(\bar{d}c)$	D^0	$(\bar{u}c)$
Quarks	s	B_s^0	$(\bar{b}s)$	D_s^-	$(\bar{c}s)$	η, η'	$(\bar{s}s)$	$\overline{K}^0$	$(\bar{d}s)$	K^-	$(\bar{u}s)$
	d	B_d^0	$(\bar{b}d)$	D^-	$(\bar{c}d)$	K^0	$(\bar{s}d)$	π^0, η, η'	$(\bar{d}d)$	π^-	$(\bar{u}d)$
	u	B^+	$(\bar{b}u)$	$\overline{D}^0$	$(\bar{c}u)$	K^+	$(\bar{s}u)$	π^+	$(\bar{d}u)$	π^0, η, η'	$(\bar{u}u)$

Note: The top quark does not form mesons because it decays too quickly.

Table 31.4 lists the quark compositions of mesons formed from the up, down, strange, charmed, and bottom quarks. Table 31.5 shows the quark combinations for the baryons listed in Table 31.2. Note that only two flavors of quarks, u and d, are contained in all hadrons encountered in ordinary matter (protons and neutrons).

You are probably wondering if such discoveries will ever end. How many "building blocks" of matter really exist? At present, physicists believe that the fundamental particles in nature are six quarks and six leptons (together with their antiparticles) listed in Table 31.6 and the field particles listed in Table 31.1. Table 31.6 lists the rest energies and charges of the quarks and leptons.

Despite extensive experimental effort, no isolated quark has ever been observed. Physicists now believe that quarks are permanently confined inside hadrons because of the strong force, which prevents them from escaping. Current efforts are under way to form a **quark-gluon plasma,** a state of matter in which the quarks are freed from neutrons and protons. In 2000, scientists at CERN announced evidence for a quark-gluon plasma formed by colliding lead nuclei. In 2005, scientists at the Relativistic Heavy Ion Collider (RHIC) at Brookhaven reported evidence from four experimental studies of a new state of matter that may be a quark-gluon plasma. Neither the CERN nor RHIC results are entirely conclusive and have not been independently verified. Three experimental detectors at the new Large Hadron Collider (LHC) at CERN will look for evidence of the creation of a quark-gluon plasma.

TABLE 31.5 |
Quark Composition of Several Baryons

Particle	Quark Composition
p	uud
n	udd
Λ^0	uds
Σ^+	uus
Σ^0	uds
Σ^-	dds
Δ^{++}	uuu
Δ^+	uud
Δ^0	udd
Δ^-	ddd
Ξ^0	uss
Ξ^-	dss
Ω^-	sss

Note: Some baryons have the same quark composition, such as the p and the Δ^+ and the n and the Δ^0. In these cases, the Δ particles are considered to be excited states of the proton and neutron.

TABLE 31.6 | The Elementary Particles and Their Rest Energies and Charges

Particle	Approximate Rest Energy	Charge
Quarks		
u	2.4 MeV	$+\frac{2}{3}e$
d	4.8 MeV	$-\frac{1}{3}e$
s	104 MeV	$-\frac{1}{3}e$
c	1.27 GeV	$+\frac{2}{3}e$
b	4.2 GeV	$-\frac{1}{3}e$
t	173 GeV	$+\frac{2}{3}e$
Leptons		
e^-	511 keV	$-e$
μ^-	105.7 MeV	$-e$
τ^-	1.78 GeV	$-e$
ν_e	< 2 eV	0
ν_μ	< 0.17 MeV	0
ν_τ	< 18 MeV	0

31.10 | Multicolored Quarks

Shortly after the concept of quarks was proposed, scientists recognized that certain particles had quark compositions that violated the Pauli exclusion principle. As noted in Pitfall Prevention 29.3 in Chapter 29, all fermions obey the exclusion principle. Because all quarks are fermions with spin $\frac{1}{2}$, they are expected to follow the exclusion principle. One example of a particle that appears to violate the exclusion principle is the Ω^- (sss) baryon that contains three s quarks having parallel spins, giving it a total spin of $\frac{3}{2}$. Other examples of baryons that have identical quarks with parallel spins are the Δ^{++} (uuu) and the Δ^- (ddd). To resolve this problem, in 1965 Moo-Young Han (b. 1934) and Yoichiro Nambu (b. 1921) suggested a modification of the structural model of quarks in which quarks possess a new property called **color.** This property is similar in many respects to electric charge except that it occurs in three varieties called **red, green,** and **blue.** The antiquarks have the colors **antired, antigreen,** and **antiblue.** To satisfy the exclusion principle, all three quarks in a baryon must have different colors. Just as a combination of actual colors of light can produce the neutral color white, a combination of three quarks with different colors is also described as white, or colorless. A meson consists of a quark of one color and an antiquark of the corresponding anticolor. The result is that baryons and mesons are always colorless (or white).

Although the concept of color in the quark model was originally conceived to satisfy the exclusion principle, it also provided a better theory for explaining certain experimental results. For example, the modified theory correctly predicts the lifetime of the π^0 meson. The theory of how quarks interact with one another is called **quantum chromodynamics,** or QCD, to parallel quantum electrodynamics (the theory of interaction between electric charges). In QCD, the quark is said to carry a **color charge,** in analogy to electric charge. The strong force between quarks is often called the **color force.**

The color force between quarks is analogous to the electric force between charges; like colors repel and opposite colors attract. Therefore, two green quarks repel each other, but a green quark is attracted to an antigreen quark. The attraction between quarks of opposite color to form a meson ($q\bar{q}$) is indicated in Figure 31.12a. Differently colored quarks also attract one another, but with less strength than opposite colors of quark and antiquark. For example, a cluster of red, blue, and green quarks all attract one another to form a baryon as indicated in Figure 31.12b. Therefore, every baryon contains three quarks of three different colors.

As stated earlier, the strong force between quarks is carried by massless particles that travel at the speed of light called **gluons.** According to QCD, there are eight gluons, all carrying two color charges, a color and an anticolor such as a "blue–antired" gluon. When a quark emits or absorbs a gluon, its color changes. For example, a blue quark that emits a blue–antired gluon becomes a red quark, and a red quark that absorbs this gluon becomes a blue quark.

> **Pitfall Prevention | 31.3**
> **Color Charge Is Not Really Color**
> The description of color for a quark has nothing to do with visual sensation from light. It is simply a convenient name for a property that is analogous to electric charge.

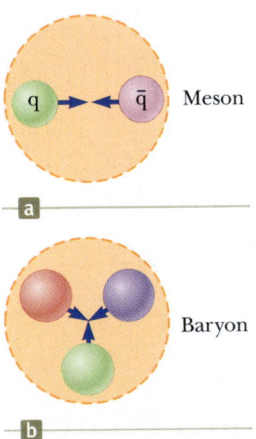

Figure 31.12 (a) A green quark is attracted to an antigreen quark, forming a meson whose quark structure is ($q\bar{q}$). (b) Three quarks of different colors attract one another to form a baryon.

Figure 31.13a shows a Feynman diagram representing the interaction between a neutron and a proton by means of Yukawa's pion, in this case a π^-. In Figure 31.13a, the charged pion carries charge from one nucleon to the other, so the nucleons change identities and the proton becomes a neutron and the neutron becomes a proton. (This process differs from Fig. 31.5a, in which the field particle is a π^0, resulting in no transfer of charge from one nucleon to the other.)

Let us look at the same interaction from the viewpoint of the quark model shown in Figure 31.13b. In this Feynman diagram, the proton and neutron are represented by their quark constituents. Each quark in the neutron and proton is continuously emitting and absorbing gluons. The energy of a gluon can result in the creation of quark–antiquark pairs. This is similar to the creation of electron–positron pairs in pair production, which we investigated in Section 31.2. When the neutron and proton approach to within 1 to 2 fm of each other, these gluons and quarks can be exchanged between the two nucleons, and such exchanges produce the strong force. Figure 31.13b depicts one possibility for the process shown in Figure 31.13a. A down quark in the neutron on the right emits a gluon. The energy of the gluon is then transformed to create a $u\bar{u}$ pair. The u quark stays within the nucleon (which has now changed to a proton), and the recoiling d quark and the $\bar{u}$ antiquark are transmitted to the proton on the left side of the diagram. Here the $\bar{u}$ annihilates a u quark within the proton and the d is captured. Therefore, the net effect is to change a u quark to a d quark, and the proton has changed to a neutron.

As the d quark and $\bar{u}$ antiquark in Figure 31.13b transfer between the nucleons, the d and $\bar{u}$ exchange gluons with each other and can be considered to be bound to each other by means of the strong force. If we look back at Table 31.4, we see that this combination is a π^-, which is Yukawa's field particle! Therefore, the quark model of interactions between nucleons is consistent with the pion-exchange model.

31.11 | The Standard Model

Scientists now believe that there are three classifications of truly elementary particles: leptons, quarks, and field particles. These three particles are further classified as either fermions or bosons. Quarks and leptons have spin $\frac{1}{2}$ and hence are fermions, whereas the field particles have integral spin of 1 or higher and are bosons.

Recall from Section 31.1 that the weak force is believed to be mediated by the W^+, W^-, and Z^0 bosons. These particles are said to have *weak charge* just as quarks have color charge. Therefore, each elementary particle can have mass, electric charge, color charge, and weak charge. Of course, one or more of these could be zero.

In 1979, Sheldon Glashow (b. 1932), Abdus Salam (1926–1996), and Steven Weinberg (b. 1933) won the Nobel Prize in Physics for developing a theory that unified the electromagnetic and weak interactions. This **electroweak theory** postulates that the weak and electromagnetic interactions have the same strength at very high particle energies. The two interactions are viewed as two different manifestations of a single unifying electroweak interaction. The photon and the three massive bosons ($W^\pm$ and Z^0) play a key role in the electroweak theory. The theory makes many concrete predictions, but perhaps the most spectacular is the prediction of the masses of the W and Z particles at about 82 GeV/c^2 and 93 GeV/c^2, respectively. As mentioned earlier, the 1984 Nobel Prize in Physics was awarded to Carlo Rubbia and Simon van der Meer for their work leading to the discovery of these particles at these energies at the CERN Laboratory in Geneva, Switzerland.

The combination of the electroweak theory and QCD for the strong interaction form what is referred to in high-energy physics as the **Standard Model.** Although the details of the Standard Model are complex, its essential ingredients can be summarized with the help of Figure 31.14 on page 924. (The Standard Model does not include the gravitational force at present; we include gravity in Fig. 31.14, however, because physicists hope to eventually incorporate this force into a unified theory.)

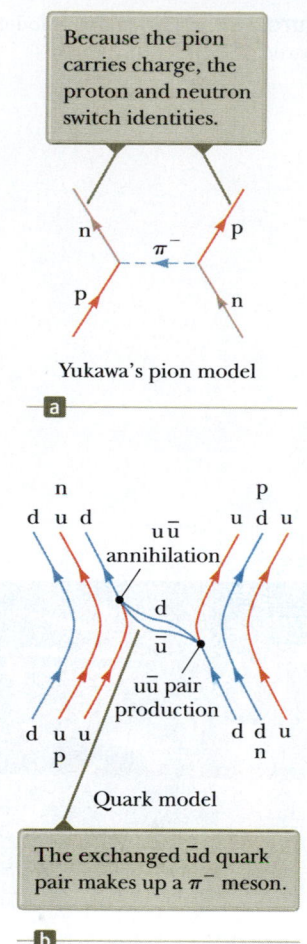

Because the pion carries charge, the proton and neutron switch identities.

Yukawa's pion model

a

Quark model

The exchanged $\bar{u}d$ quark pair makes up a π^- meson.

b

Figure 31.13 (a) A nuclear interaction between a proton and a neutron explained in terms of Yukawa's pion-exchange model. (b) The same interaction, explained in terms of quarks and gluons.

Figure 31.14 The Standard Model of particle physics.

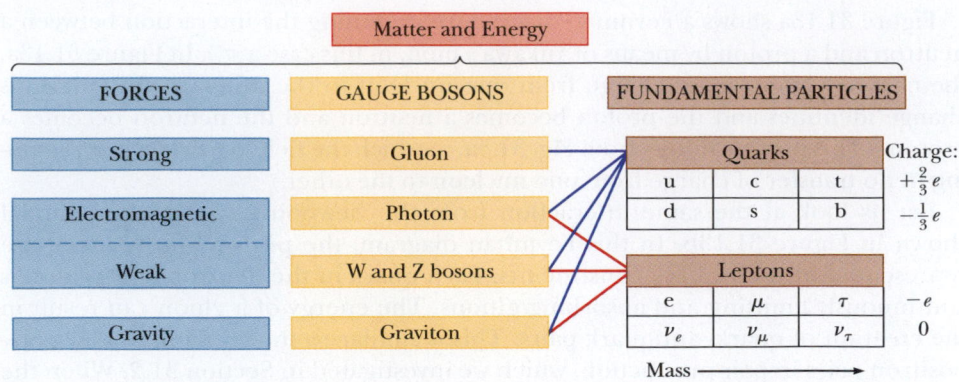

Matter and Energy		

FORCES	GAUGE BOSONS	FUNDAMENTAL PARTICLES

Strong	Gluon	Quarks
Electromagnetic	Photon	
Weak	W and Z bosons	Leptons
Gravity	Graviton	

Quarks — Charge:
| u | c | t | $+\frac{2}{3}e$ |
| d | s | b | $-\frac{1}{3}e$ |

Leptons
| e | μ | τ | $-e$ |
| ν_e | ν_μ | ν_τ | 0 |

Mass →

Figure 31.15 A view from inside the Large Hadron Collider (LHC) tunnel.

CERN

This diagram shows that quarks participate in all the fundamental forces and that leptons participate in all except the strong force.

The Standard Model does not answer all questions. A major question that is still unanswered is why, of the two mediators of the electroweak interaction, the photon has no mass but the W and Z bosons do. Because of this mass difference, the electromagnetic and weak forces are quite distinct at low energies but become similar at very high energies, when the rest energy is negligible relative to the total energy. The behavior as one goes from high to low energies is called *symmetry breaking* because the forces are similar, or symmetric, at high energies but are very different at low energies. The nonzero rest energies of the W and Z bosons raise the question of the origin of particle masses. To resolve this problem, a hypothetical particle called the **Higgs boson,** which provides a mechanism for breaking the electroweak symmetry, has been proposed. The Standard Model, modified to include the Higgs mechanism, provides a logically consistent explanation of the massive nature of the W and Z bosons. Unfortunately, the Higgs boson has not yet been found, but physicists know that its rest energy should be less than 1 TeV. To determine whether the Higgs boson exists, two quarks of at least 1 TeV of energy must collide. Calculations show, however, that this process requires injecting 40 TeV of energy within the volume of a proton.

Scientists are convinced that because the energy available in conventional accelerators using fixed targets is too limited, it is necessary to build colliding-beam accelerators called **colliders.** The concept of colliders is straightforward. Particles with equal masses and kinetic energies, traveling in opposite directions in an accelerator ring, collide head-on to produce the required reaction and the formation of new particles. Because the total momentum of the isolated system of interacting particles is zero, all their kinetic energy is available for the reaction. The Large Electron–Positron (LEP) Collider at CERN, near Geneva, Switzerland, and the Stanford Linear Collider in California were designed to collide both electrons and positrons. The Super Proton Synchrotron at CERN accelerates protons and antiprotons to energies of 300–400 GeV and was used to discover the W and Z bosons in 1983. For many years the world's highest-energy proton accelerator, the Tevatron located at Fermilab in Illinois produces protons at almost 1 000 GeV (1 TeV). CERN completed construction in 2008 on the Large Hadron Collider (LHC), a proton–proton collider that provides a center of mass energy of 14 TeV and allows an exploration of Higgs boson physics. The accelerator is constructed in the same 27-km circumference tunnel that formerly housed the LEP collider (Fig. 31.15).The LHC began operation in late 2009 after repairs were made to magnets damaged in initial runs in 2008. Searches for the Higgs boson as well as the answers to other current mysteries will be carried out during the first few years of its operation.

In addition to increasing energies in modern accelerators, detection techniques have become increasingly sophisticated. Figure 31.16 shows the computer-generated pictorial representation of the tracks of particles after a collision from a modern particle detector.

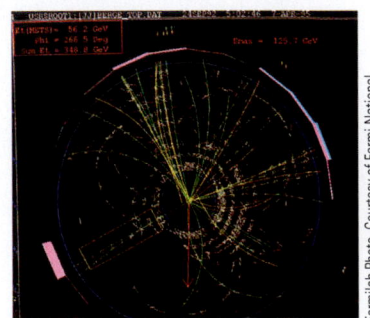

Fermilab Photo, Courtesy of Fermi National Accelerator Laboratory

Figure 31.16 Computers at Fermilab create a pictorial representation such as this one of the paths of particles after a collision.

> **THINKING PHYSICS 31.3**

Consider a car making a head-on collision with an identical car moving in the opposite direction at the same speed. Compare that collision with one of the cars making a collision with the second car at rest. In which collision is the transformation of kinetic energy to other forms larger? How does this example relate to particle accelerators?

Reasoning In the head-on collision with both cars moving, conservation of momentum for the system of two cars requires that the cars come to rest during the collision. Therefore, *all* the original kinetic energy is transformed to other forms. In the collision between a moving car and a stationary car, the cars are still moving with reduced speed after the collision, in the direction of the initially moving car. Therefore, *only part* of the kinetic energy is transformed to other forms.

This example suggests the importance of colliding beams in a particle accelerator as opposed to firing a beam into a stationary target. When particles moving in opposite directions collide, all the kinetic energy is available for transformation into other forms, which in this case is the creation of new particles. When a beam is fired into a stationary target, only part of the energy is available for transformation, so higher mass particles cannot be created. ◄

> 31.12 | Context Connection: Investigating the Smallest System to Understand the Largest

In this section, we shall describe further one of the most fascinating theories in all science—the Big Bang theory of the creation of the Universe, introduced in the Context Connection of Chapter 28—and the experimental evidence that supports it. This theory of cosmology states that the Universe had a beginning and, further, that the beginning was so cataclysmic that it is impossible to look back beyond it. According to this theory, the Universe erupted from a singularity with infinite density about 14 billion years ago. The first few fractions of a second after the Big Bang saw such extremes of energy that all four fundamental forces of physics were believed to be unified and all matter was contained in a quark-gluon plasma.

The evolution of the four fundamental forces from the Big Bang to the present is shown in Figure 31.17 (page 926). During the first 10^{-43} s (the ultrahot epoch, $T \sim 10^{32}$ K), it is presumed that the strong, electroweak, and gravitational forces were joined to form a completely unified force. In the first 10^{-35} s following the Big Bang (the hot epoch, $T \sim 10^{29}$ K), gravity broke free of this unification while the strong and electroweak forces remained unified. During this period, particle energies were so great ($>10^{16}$ GeV) that very massive particles as well as quarks, leptons, and their antiparticles existed. Then, after 10^{-35} s, the Universe rapidly expanded and cooled (the warm epoch, $T \sim 10^{29}$ to 10^{15} K), and the strong and electroweak forces parted company. As the Universe continued to cool, the electroweak force split into the weak force and the electromagnetic force about 10^{-10} s after the Big Bang.

After a few minutes, protons condensed out of the plasma. For half an hour the Universe underwent thermonuclear detonation, exploding like a hydrogen bomb and producing most of the helium nuclei that now exist. The Universe continued to expand and its temperature dropped. Until about 700 000 years after the Big Bang, the Universe was dominated by radiation. Energetic radiation prevented matter from forming single hydrogen atoms because collisions would instantly ionize any atoms that happened to form. Photons experienced continuous Compton scattering from the vast numbers of free electrons, resulting in a Universe that was opaque to radiation. By the time the Universe was about 700 000 years old, it had expanded and cooled to about 3 000 K, and protons could bind to electrons to form neutral

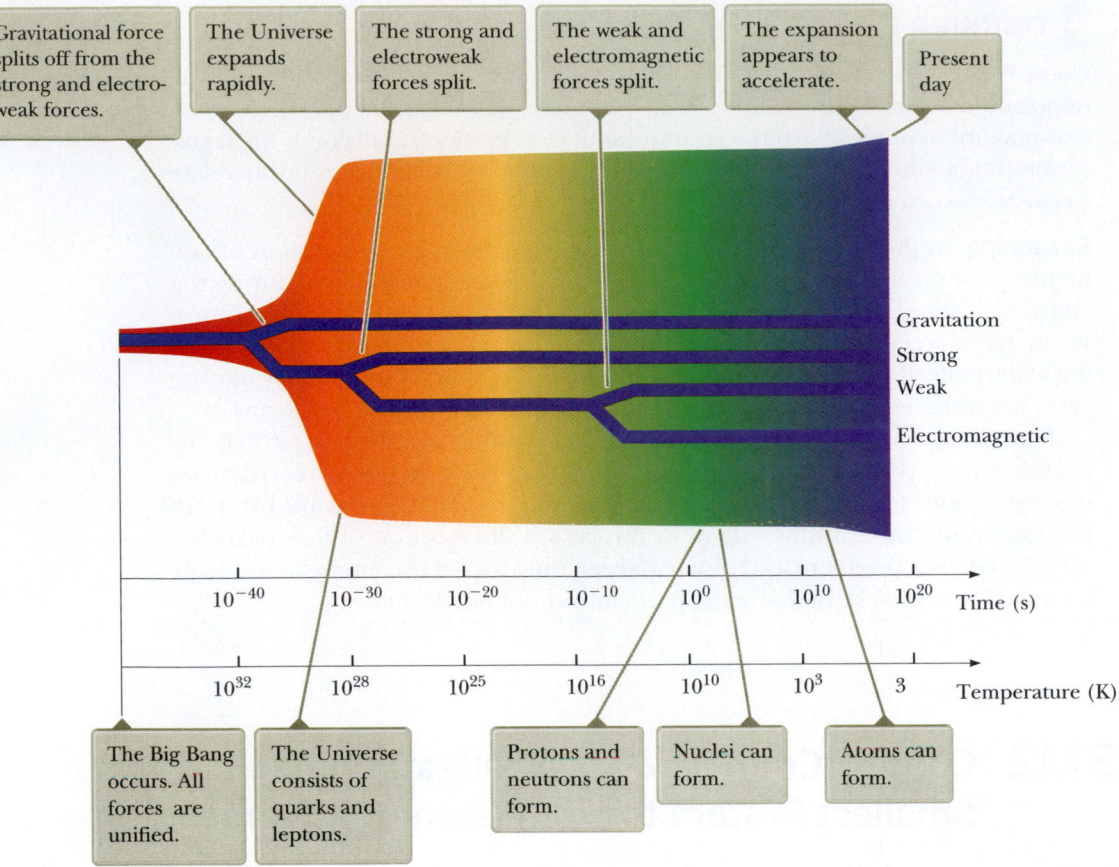

Figure 31.17 A brief history of the Universe from the Big Bang to the present. The four forces became distinguishable during the first nanosecond. Following that, all the quarks combined to form particles that interact via the strong force. The leptons, however, remained separate and to this day exist as individual, observable particles.

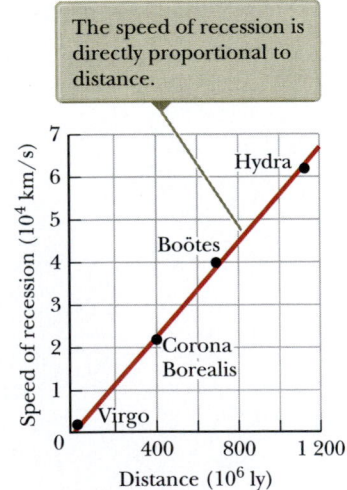

Figure 31.18 Hubble's law. Data points for four galaxies are shown here.

hydrogen atoms. Because of the quantized energies of the atoms, far more wavelengths of radiation were not absorbed by atoms than were, and the Universe suddenly became transparent to photons. Radiation no longer dominated the Universe, and clumps of neutral matter steadily grew, first atoms, followed by molecules, gas clouds, stars, and finally galaxies.

Evidence for the Expanding Universe

In Chapter 28, we discussed the observation of blackbody radiation by Penzias and Wilson that represents the leftover glow from the Big Bang. We discuss here additional relevant astronomical observations. Vesto Melvin Slipher (1875–1969), an American astronomer, reported that most nebulae are receding from the Earth at speeds up to several million miles per hour. Slipher was one of the first to use the methods of Doppler shifts in spectral lines to measure galactic speeds.

In the late 1920s, Edwin P. Hubble (1889–1953) made the bold assertion that the whole Universe is expanding. From 1928 to 1936, he and Milton Humason (1891–1972) toiled at the Mount Wilson Observatory in California to prove this assertion until they reached the limits of that 100-in. telescope. The results of this work and its continuation on a 200-in. telescope in the 1940s showed that the speeds of galaxies increase in direct proportion to their distance R from us (Fig. 31.18). This linear relationship, known as **Hubble's law,** may be written as

▶ Hubble's law

$$v = HR \qquad \textbf{31.4} \blacktriangleleft$$

where H, called the **Hubble parameter,** has the approximate value

$$H \approx 22 \times 10^{-3} \ \text{m/(s} \cdot \text{ly)}$$

Example 31.5 | Recession of a Quasar

A quasar is an object that appears similar to a star and is very distant from the Earth. Its speed can be determined from Doppler-shift measurements in the light it emits. A certain quasar recedes from the Earth at a speed of $0.55c$. How far away is it?

SOLUTION

Conceptualize A common mental representation for the Hubble law is that of raisin bread cooking in an oven. Imagine yourself at the center of the loaf of bread. As the entire loaf of bread expands upon heating, raisins near you move slowly with respect to you. Raisins far away from you on the edge of the loaf move at a higher speed.

Categorize We use a concept developed in this section, so we categorize this example as a substitution problem.

Find the distance through Hubble's law:

$$R = \frac{v}{H} = \frac{(0.55)(3.00 \times 10^8 \text{ m/s})}{22 \times 10^{-3} \text{ m/(s} \cdot \text{ly)}} = \boxed{7.5 \times 10^9 \text{ ly}}$$

What If? Suppose the quasar has moved at this speed ever since the Big Bang. With this assumption, estimate the age of the Universe.

Answer Let's approximate the distance from the Earth to the quasar as the distance the quasar has moved from the singularity since the Big Bang. We can then find the time interval from the particle under constant speed model: $\Delta t = d/v = R/v = 1/H \approx 14$ billion years, which is in approximate agreement with other calculations.

Will the Universe Expand Forever?

In the 1950s and 1960s, Allan R. Sandage (1926–2010) used the 200-in. telescope at the Mount Palomar Observatory in California to measure the speeds of galaxies at distances of up to 6 billion light-years from the Earth. These measurements showed that these very distant galaxies were moving about 10 000 km/s faster than Hubble's law predicted. According to this result, the Universe must have been expanding more rapidly 1 billion years ago, and consequently the expansion is slowing. Today, astronomers and physicists are trying to determine the rate of slowing.

If the average mass density of atoms in the Universe is less than some critical density (about 3 atoms/m³), the galaxies will slow in their outward rush but still escape to infinity. If the average density exceeds the critical value, the expansion will eventually stop and contraction will begin, possibly leading to a new superdense state and another expansion. In this scenario, we have an **oscillating Universe.**

Example 31.6 | The Critical Density of the Universe

(A) Starting from energy conservation, derive an expression for the critical mass density of the Universe ρ_c in terms of the Hubble constant H and the universal gravitational constant G.

SOLUTION

Conceptualize Figure 31.19 shows a large section of the Universe, contained within a sphere of radius R. The total mass in this volume is M. A galaxy of mass $m \ll M$ that has a speed v at a distance R from the center of the sphere escapes to infinity (at which its speed approaches zero) if the sum of its kinetic energy and the gravitational potential energy of the system is zero.

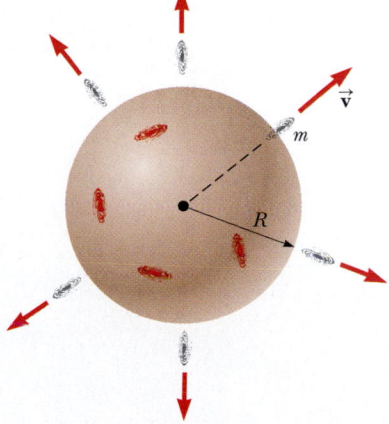

Figure 31.19 (Example 31.6) The galaxy marked with mass m is escaping from a large cluster of galaxies contained within a spherical volume of radius R. Only the mass within R slows the galaxy.

Categorize The Universe may be infinite in spatial extent, but Gauss's law for gravitation (an analog to Gauss's law for electric fields in Chapter 19) implies that only the mass M inside the sphere contributes to the gravitational potential energy of the galaxy–sphere system. Therefore, we categorize this problem as one in which we apply Gauss's law for gravitation. We model the sphere in Figure 31.19 and the escaping galaxy as an isolated system.

continued

31.6 cont.

Analyze Write an expression for the total mechanical energy of the system and set it equal to zero, representing the galaxy moving at the escape speed:

$$E_{\text{total}} = K + U = \tfrac{1}{2}mv^2 - \frac{GmM}{R} = 0$$

Substitute for the mass M contained within the sphere the product of the critical density and the volume of the sphere:

$$\tfrac{1}{2}mv^2 = \frac{Gm(\tfrac{4}{3}\pi R^3 \rho_c)}{R}$$

Solve for the critical density:

$$\rho_c = \frac{3v^2}{8\pi G R^2}$$

From Hubble's law, substitute for the ratio $v/R = H$:

$$(1) \quad \rho_c = \frac{3}{8\pi G}\left(\frac{v}{R}\right)^2 = \boxed{\frac{3H^2}{8\pi G}}$$

(B) Estimate a numerical value for the critical density in grams per cubic centimeter.

SOLUTION

In Equation (1), substitute numerical values for H and G:

$$\rho_c = \frac{3H^2}{8\pi G} = \frac{3[22 \times 10^{-3}\ \text{m}/(\text{s}\cdot\text{ly})]^2}{8\pi(6.67 \times 10^{-11}\ \text{N}\cdot\text{m}^2/\text{kg}^2)} = 8.7 \times 10^5\ \text{kg/m}\cdot(\text{ly})^2$$

Reconcile the units by converting light-years to meters:

$$\rho_c = 8.7 \times 10^5\ \text{kg/m}\cdot(\text{ly})^2\left(\frac{1\ \text{ly}}{9.46 \times 10^{15}\ \text{m}}\right)^2$$

$$= 9.7 \times 10^{-27}\ \text{kg/m}^3 = \boxed{9.7 \times 10^{-30}\ \text{g/cm}^3}$$

Finalize Because the mass of a hydrogen atom is 1.67×10^{-24} g, this value of ρ_c corresponds to 6×10^{-6} hydrogen atoms per cubic centimeter or 6 atoms per cubic meter.

Missing Mass in the Universe?

The luminous matter in galaxies averages out to a Universe density of about 5×10^{-33} g/cm^3. The radiation in the Universe has a mass equivalent of approximately 2% of the visible matter. The total mass of all nonluminous matter (such as interstellar gas and black holes) may be estimated from the speeds of galaxies orbiting one another in a cluster. The higher the galaxy speeds, the more mass in the cluster. Measurements on the Coma cluster of galaxies indicate that the amount of nonluminous matter is 20 to 30 times the amount of luminous matter present in stars and luminous gas clouds. Yet even this large invisible component of dark matter, if extrapolated to the Universe as a whole, leaves the observed mass density a factor of 10 less than ρ_c. The deficit, called *missing mass*, has been the subject of intense theoretical and experimental work. Exotic particles such as axions, photinos, and superstring particles have been suggested as candidates for the missing mass. More mundane proposals argue that the missing mass is present in certain galaxies as neutrinos. In fact, neutrinos are so abundant that a tiny neutrino rest energy on the order of only 20 eV would furnish the missing mass and "close" the Universe. Therefore, current experiments designed to measure the rest energy of the neutrino will affect predictions for the future of the Universe, showing a clear connection between one of the smallest pieces of the Universe and the Universe as a whole!

Mysterious Energy in the Universe?

A surprising twist in the story of the Universe arose in 1998 with the observation of a class of supernovae that have a fixed absolute brightness. By combining the apparent brightness and the redshift of light from these explosions, their distance and speed of recession of the Earth can be determined. These observations led to the conclusion that the expansion of the Universe is not slowing down but rather is accelerating! Observations by other groups also led to the same interpretation.

To explain this acceleration, physicists have proposed *dark energy,* which is energy possessed by the vacuum of space. In the early life of the Universe, gravity dominated over the dark energy. As the Universe expanded and the gravitational force between galaxies became smaller because of the great distances between them, the dark energy became more important. The dark energy results in an effective repulsive force that causes the expansion rate to increase.[4]

Although we have some degree of certainty about the beginning of the Universe, we are uncertain about how the story will end. Will the Universe keep on expanding forever, or will it someday collapse and then expand again, perhaps in an endless series of oscillations? Results and answers to these questions remain inconclusive, and the exciting controversy continues.

⟩ SUMMARY

There are four fundamental forces in nature: **strong, electromagnetic, weak,** and **gravitational.** The strong force is the force between quarks. A residual effect of the strong force is the **nuclear force** between nucleons that keeps the nucleus together. The weak force is responsible for beta decay. The electromagnetic and weak forces are now considered to be manifestations of a single force called the **electroweak force.** Every fundamental interaction is mediated by the exchange of **field particles.** The electromagnetic interaction is mediated by the photon; the weak interaction is mediated by the $W^\pm$ and Z^0 **bosons;** the gravitational interaction is mediated by **gravitons;** the strong interaction is mediated by **gluons.**

An **antiparticle** and a particle have the same mass, but opposite charge, and other properties may have opposite values such as lepton number and baryon number. It is possible to produce particle–antiparticle pairs in nuclear reactions if the available energy is greater than $2mc^2$, where m is the mass of the particle (or antiparticle).

Particles other than field particles are classified as hadrons or leptons. **Hadrons** interact through the strong force. They have size and structure and are not elementary particles. Hadrons are of two types, baryons and mesons. **Mesons** have baryon number zero and have either zero or integral spin. **Baryons,** which generally are the most massive particles, have nonzero baryon number and a spin of $\frac{1}{2}$ or $\frac{3}{2}$. The neutron and proton are examples of baryons.

Leptons have no structure or size and are considered truly elementary. They interact through the weak and electromagnetic forces. The six leptons are the electron e^-, the muon μ^-, the tau τ^-; and their neutrinos ν_e, ν_μ, and ν_τ.

In all reactions and decays, quantities such as energy, linear momentum, angular momentum, electric charge, baryon number, and lepton number are strictly conserved. Certain particles have properties called **strangeness** and **charm.** These unusual properties are conserved only in those reactions and decays that occur via the strong force.

Theories in elementary particle physics have postulated that all hadrons are composed of smaller units known as **quarks.** Quarks have fractional electric charge and come in six "flavors": **up** (u), **down** (d), **strange** (s), **charmed** (c), **top** (t), and **bottom** (b). Each baryon contains three quarks, and each meson contains one quark and one antiquark.

According to the theory of **quantum chromodynamics,** quarks have a property called **color charge,** and the strong force between quarks is referred to as the **color force.**

⟩ OBJECTIVE QUESTIONS

☐ denotes answer available in *Student Solutions Manual/Study Guide*

1. An isolated stationary muon decays into an electron, an electron antineutrino, and a muon neutrino. Is the total kinetic energy of these three particles (a) zero, (b) small, or (c) large compared to their rest energies, or (d) none of those choices are possible?

2. Define the average density of the solar system ρ_{SS} as the total mass of the Sun, planets, satellites, rings, asteroids, icy outliers, and comets, divided by the volume of a sphere around the Sun large enough to contain all these objects. The sphere extends about halfway to the nearest star, with a radius of approximately 2×10^{16} m, about two light-years. How does this average density of the solar system compare with the critical density ρ_c required for the Universe to stop its Hubble's-law expansion? (a) ρ_{SS} is much greater than ρ_c. (b) ρ_{SS} is approximately or precisely equal to ρ_c. (c) ρ_{SS} is much less than ρ_c. (d) It is impossible to determine.

---

[4]For a discussion of dark energy, see S. Perlmutter, "Supernovae, Dark Energy, and the Accelerating Universe," *Physics Today,* 56(4): 53–60, April 2003.

3. What interactions affect protons in an atomic nucleus? More than one answer may be correct. (a) the nuclear interaction (b) the weak interaction (c) the electromagnetic interaction (d) the gravitational interaction

4. Place the following events into the correct sequence from the earliest in the history of the Universe to the latest. (a) Neutral atoms form. (b) Protons and neutrons are no longer annihilated as fast as they form. (c) The Universe is a quark–gluon soup. (d) The Universe is like the core of a normal star today, forming helium by nuclear fusion. (e) The Universe is like the surface of a hot star today, consisting of a plasma of ionized atoms. (f) Polyatomic molecules form. (g) Solid materials form.

5. When an electron and a positron meet at low speed in empty space, they annihilate each other to produce two 0.511-MeV gamma rays. What law would be violated if they produced one gamma ray with an energy of 1.02 MeV? (a) conservation of energy (b) conservation of momentum (c) conservation of charge (d) conservation of baryon number (e) conservation of electron lepton number

6. Which of the following field particles mediates the strong force? (a) photon (b) gluon (c) graviton (d) W^+ and Z bosons (e) none of those field particles

7. The Ω^- particle is a baryon with spin $\frac{3}{2}$. Does the Ω^- particle have (a) three possible spin states in a magnetic field, (b) four possible spin states, (c) three times the charge of a spin $-\frac{1}{2}$ particle, or (d) three times the mass of a spin $-\frac{1}{2}$ particle, or (e) are none of those choices correct?

8. In one experiment, two balls of clay of the same mass travel with the same speed v toward each other. They collide head-on and come to rest. In a second experiment, two clay balls of the same mass are again used. One ball hangs at rest, suspended from the ceiling by a thread. The second ball is fired toward the first at speed v, to collide, stick to the first ball, and continue to move forward. Is the kinetic energy that is transformed into internal energy in the first experiment (a) one-fourth as much as in the second experiment, (b) one-half as much as in the second experiment, (c) the same as in the second experiment, (d) twice as much as in the second experiment, or (e) four times as much as in the second experiment?

CONCEPTUAL QUESTIONS

☐ denotes answer available in *Student Solutions Manual/Study Guide*

1. The Ξ^0 particle decays by the weak interaction according to the decay mode $\Xi^0 \rightarrow \Lambda^0 + \pi^0$. Would you expect this decay to be fast or slow? Explain.

2. Are the laws of conservation of baryon number, lepton number, and strangeness based on fundamental properties of nature (as are the laws of conservation of momentum and energy, for example)? Explain.

3. An antibaryon interacts with a meson. Can a baryon be produced in such an interaction? Explain.

4. Describe the essential features of the Standard Model of particle physics.

5. Name the four fundamental interactions and the field particle that mediates each.

6. What are the differences between hadrons and leptons?

7. Kaons all decay into final states that contain no protons or neutrons. What is the baryon number for kaons?

8. Describe the properties of baryons and mesons and the important differences between them.

9. How many quarks are in each of the following: (a) a baryon, (b) an antibaryon, (c) a meson, (d) an antimeson? (e) How do you explain that baryons have half-integral spins, whereas mesons have spins of 0 or 1?

10. In the theory of quantum chromodynamics, quarks come in three colors. How would you justify the statement that "all baryons and mesons are colorless"?

11. The W and Z bosons were first produced at CERN in 1983 by causing a beam of protons and a beam of antiprotons to meet at high energy. Why was this discovery important?

12. How did Edwin Hubble determine in 1928 that the Universe is expanding?

13. Neutral atoms did not exist until hundreds of thousands of years after the Big Bang. Why?

▶ PROBLEMS AVAILABLE IN ~~ENHANCED~~ Web**Assign**

Solutions to the following Problems are available in the *Student Solutions Manual/Study Guide*:

31.5, 31.7, 31.9, 31.13, 31.17, 31.21, 31.23, 31.29, 31.37, 31.54, 31.61, 31.64, and 31.65

List of Enhanced Problems

Problem Number	Targeted Feedback in Enhanced WebAssign	Master It in Enhanced WebAssign	Watch It in Enhanced WebAssign
31.5		✓	
31.11	✓		✓
31.29	✓	✓	
31.41	✓		✓
31.54		✓	

Context | 9

Problems and Perspectives

We have now investigated the principles of quantum physics and have seen many connections to our central question for the *Cosmic Connection* Context:

> **How can we connect the physics of microscopic particles to the physics of the Universe?**

While particle physicists have been exploring the realm of the very small, cosmologists have been exploring cosmic history back to within the first second after the Big Bang. Observation of events that occur when two particles collide in an accelerator is essential in reconstructing the early moments in cosmic history. The key to understanding the early Universe is first to understand the world of elementary particles. Cosmologists and physicists now find that they have many common goals and are joining hands to attempt to understand the physical world at its most fundamental level.

Problems

We have made great progress in understanding the Universe and its underlying structure, but a multitude of questions remain unanswered. Why does so little antimatter exist in the Universe? Do neutrinos have a small rest energy, and if so, how do they contribute to the "dark matter" of the Universe? Is there "dark energy" in the Universe? Is it possible to unify the strong and electroweak forces in a logical and consistent manner? Can gravity be unified with the other forces? Why do quarks and leptons form three similar but distinct families? Are muons the same as electrons (apart from their difference in mass), or do they have other subtle differences that have not been detected? Why are some particles charged and others neutral? Why do quarks carry a fractional charge? What determines the masses of the fundamental constituents? Can isolated quarks exist? Do leptons and quarks have a substructure?

String Theory: A New Perspective

Let us briefly discuss one current effort at answering some of these questions by proposing a new perspective on particles. As you read this book, you may recall starting off with the particle model and doing quite a bit of physics with it. In the *Earthquakes* Context, we introduced the wave model, and more physics was used to investigate the properties of waves. We used a wave model for light in the *Lasers* Context. Early in this Context, however, we saw the need to return to the particle model for light. Furthermore, we found that material particles had wave-like characteristics. The quantum particle model of Chapter 28 allowed us to build particles out of waves, suggesting that a wave is the fundamental entity. In Chapter 31, however, we discussed the elementary particles as the fundamental entities. It seems as if we cannot make up our mind! In some sense, that is true because the wave–particle duality is still an area of active research. In this final Context Conclusion, we shall discuss a current research effort to build particles out of waves and vibrations.

String theory is an effort to unify the four fundamental forces by modeling all particles as various quantized vibrational modes of a single entity, an incredibly small string. The typical length of such a string is on the order of 10^{-35} m, called the **Planck length.** We have seen quantized modes before with the frequencies of vibrating guitar strings in Chapter 14 and the quantized energy levels of atoms in Chapter 29. In string theory, each quantized mode of vibration of the string corresponds to a different elementary particle in the Standard Model.

One complicating factor in string theory is that it requires space–time to have ten dimensions. Despite the theoretical and conceptual difficulties in dealing with ten dimensions, string theory holds promise in incorporating gravity with the other forces. Four of the ten dimensions are visible to us—three space dimensions and one time dimension—and the other six are *compactified*. In other words, the six dimensions are curled up so tightly that they are not visible in the macroscopic world.

As an analogy, consider a soda straw. We can build a soda straw by cutting a rectangular piece of paper (Fig. 1a), which clearly has two dimensions, and rolling it up into a small tube (Fig. 1b). From far away, the soda straw looks like a one-dimensional straight line. The second dimension has been curled up and is not visible. String theory claims that six space–time dimensions are curled up in an analogous way, with the curling on the size of the Planck length and impossible to see from our viewpoint.

Another complicating factor with string theory is that it is difficult for string theorists to guide experimentalists in how and what to look for in an experiment. The Planck length is so incredibly small that direct experimentation on strings is impossible. Until the theory has been further developed, string theorists are restricted to applying the theory to known results and testing for consistency.

One of the predictions of string theory is called **supersymmetry** (SUSY), which suggests that every elementary particle has a superpartner that has not yet been observed. It is believed that supersymmetry is a broken symmetry (like the broken electroweak symmetry at low energies) and that the masses of the superpartners are above our current capabilities of detection by accelerators. Some theorists claim that the mass of superpartners is the missing mass discussed in the Context Conclusion of Chapter 31. Keeping with the whimsical trend in naming particles and their properties that we saw in Chapter 31, superpartners are given names such as the *squark* (the superpartner to a quark), the *selectron* (electron), and the *gluinos* (gluon).

Other theorists are working on **M-theory,** which is an 11-dimensional theory based on membranes rather than strings. In a way reminiscent of the correspondence principle, M-theory is claimed to reduce to string theory if one compactifies from 11 dimensions to 10.

The questions that we listed at the beginning of this Context Conclusion go on and on. Because of the rapid advances and new discoveries in the field of particle physics, by the time you read this book some of these questions may be resolved and other new questions may emerge.

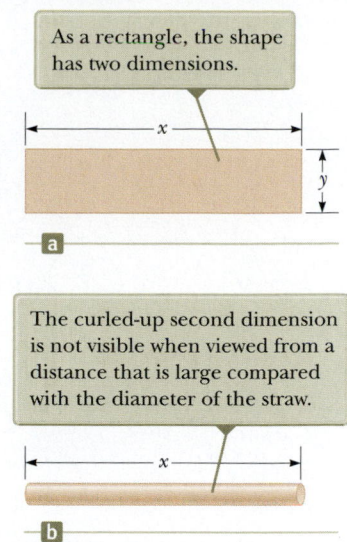

As a rectangle, the shape has two dimensions.

x

y

a

The curled-up second dimension is not visible when viewed from a distance that is large compared with the diameter of the straw.

x

b

Figure 1 (a) A piece of paper is cut into a rectangular shape. (b) The paper is rolled up into a soda straw.

Question

1. **Review question.** A girl and her grandmother grind corn while the woman tells the girl stories about what is most important. A boy keeps crows away from ripening corn while his grandfather sits in the shade and explains to him the Universe and his place in it. What the children do not understand this year they will better understand next year. Now you must take the part of the adults. State the most general, most fundamental, most universal truths that you know. If you need to repeat someone else's ideas, get the best version of those ideas you can and state your source. If there is something you do not understand, make a plan to understand it better within the next year.

Problem

1. Classical general relativity views the structure of space–time as deterministic and well defined down to arbitrarily small distances. On the other hand, quantum general relativity forbids distances smaller than the Planck length given

by $L = (\hbar G/c^3)^{1/2}$. (a) Calculate the value of the Planck length. The quantum limitation suggests that after the Big Bang, when all the presently observable section of the Universe was contained within a point-like singularity, nothing could be observed until that singularity grew larger than the Planck length. Because the size of the singularity grew at the speed of light, we can infer that no observations were possible during the time interval required for light to travel the Planck length. (b) Calculate this time interval, known as the Planck time T, and compare it with the ultrahot epoch mentioned in the text. (c) Does this answer suggest that we may never know what happened between the time $t = 0$ and the time $t = T$?

The Meaning of Success

To earn the respect of intelligent people and to win the affection of children;

To appreciate the beauty in nature and all that surrounds us;

To seek out and nurture the best in others;

To give the gift of yourself to others without the slightest thought of return, for it is in giving that we receive;

To have accomplished a task, whether it be saving a lost soul, healing a sick child, writing a book, or risking your life for a friend;

To have celebrated and laughed with great joy and enthusiasm and sung with exultation;

To have hope even in times of despair, for as long as you have hope, you have life;

To love and be loved;

To be understood and to understand;

To know that even one life has breathed easier because you have lived;

This is the meaning of success.

Ralph Waldo Emerson and modified by Ray Serway

Tables

TABLE A.1 | Conversion Factors

Length

	m	cm	km	in.	ft	mi
1 meter	1	10^2	10^{-3}	39.37	3.281	6.214×10^{-4}
1 centimeter	10^{-2}	1	10^{-5}	0.393 7	3.281×10^{-2}	6.214×10^{-6}
1 kilometer	10^3	10^5	1	3.937×10^4	3.281×10^3	0.621 4
1 inch	2.540×10^{-2}	2.540	2.540×10^{-5}	1	8.333×10^{-2}	1.578×10^{-5}
1 foot	0.304 8	30.48	3.048×10^{-4}	12	1	1.894×10^{-4}
1 mile	1 609	1.609×10^5	1.609	6.336×10^4	5 280	1

Mass

	kg	g	slug	u
1 kilogram	1	10^3	6.852×10^{-2}	6.024×10^{26}
1 gram	10^{-3}	1	6.852×10^{-5}	6.024×10^{23}
1 slug	14.59	1.459×10^4	1	8.789×10^{27}
1 atomic mass unit	1.660×10^{-27}	1.660×10^{-24}	1.137×10^{-28}	1

Note: 1 metric ton = 1 000 kg.

Time

	s	min	h	day	yr
1 second	1	1.667×10^{-2}	2.778×10^{-4}	1.157×10^{-5}	3.169×10^{-8}
1 minute	60	1	1.667×10^{-2}	6.994×10^{-4}	1.901×10^{-6}
1 hour	3 600	60	1	4.167×10^{-2}	1.141×10^{-4}
1 day	8.640×10^4	1 440	24	1	2.738×10^{-5}
1 year	3.156×10^7	5.259×10^5	8.766×10^3	365.2	1

Speed

	m/s	cm/s	ft/s	mi/h
1 meter per second	1	10^2	3.281	2.237
1 centimeter per second	10^{-2}	1	3.281×10^{-2}	2.237×10^{-2}
1 foot per second	0.304 8	30.48	1	0.681 8
1 mile per hour	0.447 0	44.70	1.467	1

Note: 1 mi/min = 60 mi/h = 88 ft/s.

Force

	N	lb
1 newton	1	0.224 8
1 pound	4.448	1

(Continued)

TABLE A.1 | Conversion Factors (*continued*)

Energy, Energy Transfer

	J	ft · lb	eV
1 joule	1	0.737 6	6.242×10^{18}
1 foot-pound	1.356	1	8.464×10^{18}
1 electron volt	1.602×10^{-19}	1.182×10^{-19}	1
1 calorie	4.186	3.087	2.613×10^{19}
1 British thermal unit	1.055×10^{3}	7.779×10^{2}	6.585×10^{21}
1 kilowatt-hour	3.600×10^{6}	2.655×10^{6}	2.247×10^{25}

	cal	Btu	kWh
1 joule	0.238 9	9.481×10^{-4}	2.778×10^{-7}
1 foot-pound	0.323 9	1.285×10^{-3}	3.766×10^{-7}
1 electron volt	3.827×10^{-20}	1.519×10^{-22}	4.450×10^{-26}
1 calorie	1	3.968×10^{-3}	1.163×10^{-6}
1 British thermal unit	2.520×10^{2}	1	2.930×10^{-4}
1 kilowatt-hour	8.601×10^{5}	3.413×10^{2}	1

Pressure

	Pa	atm	
1 pascal	1	9.869×10^{-6}	
1 atmosphere	1.013×10^{5}	1	
1 centimeter mercury[a]	1.333×10^{3}	1.316×10^{-2}	
1 pound per square inch	6.895×10^{3}	6.805×10^{-2}	
1 pound per square foot	47.88	4.725×10^{-4}	

	cm Hg	lb/in.²	lb/ft²
1 pascal	7.501×10^{-4}	1.450×10^{-4}	2.089×10^{-2}
1 atmosphere	76	14.70	2.116×10^{3}
1 centimeter mercury[a]	1	0.194 3	27.85
1 pound per square inch	5.171	1	144
1 pound per square foot	3.591×10^{-2}	6.944×10^{-3}	1

[a]At 0°C and at a location where the free-fall acceleration has its "standard" value, 9.806 65 m/s².

TABLE A.2 | Symbols, Dimensions, and Units of Physical Quantities

Quantity	Common Symbol	Unit[a]	Dimensions[b]	Unit in Terms of Base SI Units
Acceleration	$\vec{a}$	m/s²	L/T^2	m/s²
Amount of substance	n	MOLE		mol
Angle	θ, ϕ	radian (rad)	1	
Angular acceleration	$\vec{\alpha}$	rad/s²	T^{-2}	s^{-2}
Angular frequency	ω	rad/s	T^{-1}	s^{-1}
Angular momentum	$\vec{L}$	kg · m²/s	ML^2/T	kg · m²/s
Angular velocity	$\vec{\omega}$	rad/s	T^{-1}	s^{-1}
Area	A	m²	L^2	m²
Atomic number	Z			
Capacitance	C	farad (F)	Q^2T^2/ML^2	$A^2 \cdot s^4/kg \cdot m^2$
Charge	q, Q, e	coulomb (C)	Q	$A \cdot s$

(Continued)

TABLE A.2 | Symbols, Dimensions, and Units of Physical Quantities (*continued*)

Quantity	Common Symbol	Unit[a]	Dimensions[b]	Unit in Terms of Base SI Units
Charge density				
Line	λ	C/m	Q/L	$A \cdot s/m$
Surface	σ	C/m^2	Q/L^2	$A \cdot s/m^2$
Volume	ρ	C/m^3	Q/L^3	$A \cdot s/m^3$
Conductivity	σ	$1/\Omega \cdot m$	Q^{2}T/ML^3	$A^2 \cdot s^3/kg \cdot m^3$
Current	I	AMPERE	Q/T	A
Current density	J	A/m^2	Q/TL^2	A/m^2
Density	ρ	kg/m^3	M/L^3	kg/m^3
Dielectric constant	κ			
Electric dipole moment	$\vec{p}$	C $\cdot$ m	QL	$A \cdot s \cdot m$
Electric field	$\vec{E}$	V/m	ML/QT^2	$kg \cdot m/A \cdot s^3$
Electric flux	Φ_E	V $\cdot$ m	ML^3/QT^2	$kg \cdot m^3/A \cdot s^3$
Electromotive force	$\mathcal{E}$	volt (V)	ML^2/QT^2	$kg \cdot m^2/A \cdot s^3$
Energy	E, U, K	joule (J)	ML^2/T^2	$kg \cdot m^2/s^2$
Entropy	S	J/K	ML^2/T^{2}K	$kg \cdot m^2/s^2 \cdot K$
Force	$\vec{F}$	newton (N)	ML/T^2	$kg \cdot m/s^2$
Frequency	f	hertz (Hz)	T^{-1}	s^{-1}
Heat	Q	joule (J)	ML^2/T^2	$kg \cdot m^2/s^2$
Inductance	L	henry (H)	ML^2/Q^2	$kg \cdot m^2/A^2 \cdot s^2$
Length	ℓ, L	METER	L	m
Displacement	$\Delta x, \Delta\vec{r}$			
Distance	d, h			
Position	$x, y, z, \vec{r}$			
Magnetic dipole moment	$\vec{\mu}$	N $\cdot$ m/T	QL^2/T	$A \cdot m^2$
Magnetic field	$\vec{B}$	tesla (T) (= Wb/m^2)	M/QT	$kg/A \cdot s^2$
Magnetic flux	Φ_B	weber (Wb)	ML^2/QT	$kg \cdot m^2/A \cdot s^2$
Mass	m, M	KILOGRAM	M	kg
Molar specific heat	C	J/mol $\cdot$ K		$kg \cdot m^2/s^2 \cdot mol \cdot K$
Moment of inertia	I	kg $\cdot$ m^2	ML^2	$kg \cdot m^2$
Momentum	$\vec{p}$	kg $\cdot$ m/s	ML/T	$kg \cdot m/s$
Period	T	s	T	s
Permeability of free space	μ_0	N/A^2 (= H/m)	ML/Q^2	$kg \cdot m/A^2 \cdot s^2$
Permittivity of free space	ϵ_0	C^2/N $\cdot$ m^2 (= F/m)	Q^{2}T^2/ML^3	$A^2 \cdot s^4/kg \cdot m^3$
Potential	V	volt (V) (= J/C)	ML^2/QT^2	$kg \cdot m^2/A \cdot s^3$
Power	P	watt (W) (= J/s)	ML^2/T^3	$kg \cdot m^2/s^3$
Pressure	P	pascal (Pa) (= N/m^2)	M/LT^2	$kg/m \cdot s^2$
Resistance	R	ohm (Ω) (= V/A)	ML^2/Q^{2}T	$kg \cdot m^2/A^2 \cdot s^3$
Specific heat	c	J/kg $\cdot$ K	L^2/T^{2}K	$m^2/s^2 \cdot K$
Speed	v	m/s	L/T	m/s
Temperature	T	KELVIN	K	K
Time	t	SECOND	T	s
Torque	$\vec{\tau}$	N $\cdot$ m	ML^2/T^2	$kg \cdot m^2/s^2$
Velocity	$\vec{v}$	m/s	L/T	m/s
Volume	V	m^3	L^3	m^3
Wavelength	λ	m	L	m
Work	W	joule (J) (= N $\cdot$ m)	ML^2/T^2	$kg \cdot m^2/s^2$

[a]The base SI units are given in uppercase letters.

[b]The symbols M, L, T, K, and Q denote mass, length, time, temperature, and charge, respectively.

TABLE A.3 | Chemical and Nuclear Information for Selected Isotopes

Atomic Number Z	Element	Chemical Symbol	Mass Number A (* means) radioactive)	Mass of Neutral Atom (u)	Percent Abundance	Half-life, if Radioactive $T_{1/2}$
−1	electron	e−	0	0.000 549		
0	neutron	n	1*	1.008 665		614 s
1	hydrogen	$^1H = p$	1	1.007 825	99.988 5	
	[deuterium	$^2H = D$]	2	2.014 102	0.011 5	
	[tritium	$^3H = T$]	3*	3.016 049		12.33 yr
2	helium	He	3	3.016 029	0.000 137	
	[alpha particle	$\alpha = {}^4He$]	4	4.002 603	99.999 863	
			6*	6.018 889		0.81 s
3	lithium	Li	6	6.015 123	7.5	
			7	7.016 005	92.5	
4	beryllium	Be	7*	7.016 930		53.3 d
			8*	8.005 305		10^{-17} s
			9	9.012 182	100	
5	boron	B	10	10.012 937	19.9	
			11	11.009 305	80.1	
6	carbon	C	11*	11.011 434		20.4 min
			12	12.000 000	98.93	
			13	13.003 355	1.07	
			14*	14.003 242		5 730 yr
7	nitrogen	N	13*	13.005 739		9.96 min
			14	14.003 074	99.632	
			15	15.000 109	0.368	
8	oxygen	O	14*	14.008 596		70.6 s
			15*	15.003 066		122 s
			16	15.994 915	99.757	
			17	16.999 132	0.038	
			18	17.999 161	0.205	
9	fluorine	F	18*	18.000 938		109.8 min
			19	18.998 403	100	
10	neon	Ne	20	19.992 440	90.48	
11	sodium	Na	23	22.989 769	100	
12	magnesium	Mg	23*	22.994 124		11.3 s
			24	23.985 042	78.99	
13	aluminum	Al	27	26.981 539	100	
14	silicon	Si	27*	26.986 705		4.2 s
15	phosphorus	P	30*	29.978 314		2.50 min
			31	30.973 762	100	
			32*	31.973 907		14.26 d
16	sulfur	S	32	31.972 071	94.93	
19	potassium	K	39	38.963 707	93.258 1	
			40*	39.963 998	0.011 7	1.28×10^9 yr
20	calcium	Ca	40	39.962 591	96.941	
			42	41.958 618	0.647	
			43	42.958 767	0.135	
25	manganese	Mn	55	54.938 045	100	
26	iron	Fe	56	55.934 938	91.754	
			57	56.935 394	2.119	

(Continued)

TABLE A.3 | Chemical and Nuclear Information for Selected Isotopes (*continued*)

Atomic Number Z	Element	Chemical Symbol	Mass Number A (* means) radioactive)	Mass of Neutral Atom (u)	Percent Abundance	Half-life, if Radioactive $T_{1/2}$
27	cobalt	Co	57*	56.936 291		272 d
			59	58.933 195	100	
			60*	59.933 817		5.27 yr
28	nickel	Ni	58	57.935 343	68.076 9	
			60	59.930 786	26.223 1	
29	copper	Cu	63	62.929 598	69.17	
			64*	63.929 764		12.7 h
			65	64.927 789	30.83	
30	zinc	Zn	64	63.929 142	48.63	
37	rubidium	Rb	87*	86.909 181	27.83	
38	strontium	Sr	87	86.908 877	7.00	
			88	87.905 612	82.58	
			90*	89.907 738		29.1 yr
41	niobium	Nb	93	92.906 378	100	
42	molybdenum	Mo	94	93.905 088	9.25	
44	ruthenium	Ru	98	97.905 287	1.87	
54	xenon	Xe	136*	135.907 219		2.4×10^{21} yr
55	cesium	Cs	137*	136.907 090		30 yr
56	barium	Ba	137	136.905 827	11.232	
58	cerium	Ce	140	139.905 439	88.450	
59	praseodymium	Pr	141	140.907 653	100	
60	neodymium	Nd	144*	143.910 087	23.8	2.3×10^{15} yr
61	promethium	Pm	145*	144.912 749		17.7 yr
79	gold	Au	197	196.966 569	100	
80	mercury	Hg	198	197.966 769	9.97	
			202	201.970 643	29.86	
82	lead	Pb	206	205.974 465	24.1	
			207	206.975 897	22.1	
			208	207.976 652	52.4	
			214*	213.999 805		26.8 min
83	bismuth	Bi	209	208.980 399	100	
84	polonium	Po	210*	209.982 874		138.38 d
			216*	216.001 915		0.145 s
			218*	218.008 973		3.10 min
86	radon	Rn	220*	220.011 394		55.6 s
			222*	222.017 578		3.823 d
88	radium	Ra	226*	226.025 410		1 600 yr
90	thorium	Th	232*	232.038 055	100	1.40×10^{10} yr
			234*	234.043 601		24.1 d
92	uranium	U	234*	234.040 952		2.45×10^{5} yr
			235*	235.043 930	0.720 0	7.04×10^{8} yr
			236*	236.045 568		2.34×10^{7} yr
			238*	238.050 788	99.274 5	4.47×10^{9} yr
93	neptunium	Np	236*	236.046 570		1.15×10^{5} yr
			237*	237.048 173		2.14×10^{6} yr
94	plutonium	Pu	239*	239.052 163		24 120 yr

Source: G. Audi, A. H. Wapstra, and C. Thibault, "The AME2003 Atomic Mass Evaluation," *Nuclear Physics A* **729**: 337–676, 2003.

Appendix B

Mathematics Review

This appendix in mathematics is intended as a brief review of operations and methods. Early in this course, you should be totally familiar with basic algebraic techniques, analytic geometry, and trigonometry. The sections on differential and integral calculus are more detailed and are intended for students who have difficulty applying calculus concepts to physical situations.

B.1 | Scientific Notation

Many quantities used by scientists often have very large or very small values. The speed of light, for example, is about 300 000 000 m/s, and the ink required to make the dot over an i in this textbook has a mass of about 0.000 000 001 kg. Obviously, it is very cumbersome to read, write, and keep track of such numbers. We avoid this problem by using a method incorporating powers of the number 10:

$$10^0 = 1$$
$$10^1 = 10$$
$$10^2 = 10 \times 10 = 100$$
$$10^3 = 10 \times 10 \times 10 = 1\ 000$$
$$10^4 = 10 \times 10 \times 10 \times 10 = 10\ 000$$
$$10^5 = 10 \times 10 \times 10 \times 10 \times 10 = 100\ 000$$

and so on. The number of zeros corresponds to the power to which ten is raised, called the **exponent** of ten. For example, the speed of light, 300 000 000 m/s, can be expressed as 3.00×10^8 m/s.

In this method, some representative numbers smaller than unity are the following:

$$10^{-1} = \frac{1}{10} = 0.1$$

$$10^{-2} = \frac{1}{10 \times 10} = 0.01$$

$$10^{-3} = \frac{1}{10 \times 10 \times 10} = 0.001$$

$$10^{-4} = \frac{1}{10 \times 10 \times 10 \times 10} = 0.000\ 1$$

$$10^{-5} = \frac{1}{10 \times 10 \times 10 \times 10 \times 10} = 0.000\ 01$$

In these cases, the number of places the decimal point is to the left of the digit 1 equals the value of the (negative) exponent. Numbers expressed as some power of ten multiplied by another number between one and ten are said to be in **scientific notation.** For example, the scientific notation for 5 943 000 000 is 5.943×10^9 and that for 0.000 083 2 is 8.32×10^{-5}.

When numbers expressed in scientific notation are being multiplied, the following general rule is very useful:

$$10^n \times 10^m = 10^{n+m}$$

B.1 ◀

where n and m can be *any* numbers (not necessarily integers). For example, $10^2 \times 10^5 = 10^7$. The rule also applies if one of the exponents is negative: $10^3 \times 10^{-8} = 10^{-5}$.

When dividing numbers expressed in scientific notation, note that

$$\frac{10^n}{10^m} = 10^n \times 10^{-m} = 10^{n-m}$$ **B.2◄**

Exercises

With help from the preceding rules, verify the answers to the following equations:

1. $86\ 400 = 8.64 \times 10^4$
2. $9\ 816\ 762.5 = 9.816\ 762\ 5 \times 10^6$
3. $0.000\ 000\ 039\ 8 = 3.98 \times 10^{-8}$
4. $(4.0 \times 10^8)(9.0 \times 10^9) = 3.6 \times 10^{18}$
5. $(3.0 \times 10^7)(6.0 \times 10^{-12}) = 1.8 \times 10^{-4}$
6. $\dfrac{75 \times 10^{-11}}{5.0 \times 10^{-3}} = 1.5 \times 10^{-7}$
7. $\dfrac{(3 \times 10^6)(8 \times 10^{-2})}{(2 \times 10^{17})(6 \times 10^5)} = 2 \times 10^{-18}$

◄ B.2 | Algebra

Some Basic Rules

When algebraic operations are performed, the laws of arithmetic apply. Symbols such as x, y, and z are usually used to represent unspecified quantities, called the **unknowns.**

First, consider the equation

$$8x = 32$$

If we wish to solve for x, we can divide (or multiply) each side of the equation by the same factor without destroying the equality. In this case, if we divide both sides by 8, we have

$$\frac{8x}{8} = \frac{32}{8}$$

$$x = 4$$

Next consider the equation

$$x + 2 = 8$$

In this type of expression, we can add or subtract the same quantity from each side. If we subtract 2 from each side, we have

$$x + 2 - 2 = 8 - 2$$

$$x = 6$$

In general, if $x + a = b$, then $x = b - a$.

Now consider the equation

$$\frac{x}{5} = 9$$

If we multiply each side by 5, we are left with x on the left by itself and 45 on the right:

$$\left(\frac{x}{5}\right)(5) = 9 \times 5$$

$$x = 45$$

In all cases, *whatever operation is performed on the left side of the equality must also be performed on the right side.*

The following rules for multiplying, dividing, adding, and subtracting fractions should be recalled, where a, b, c, and d are four numbers:

	Rule	Example
Multiplying	$\left(\dfrac{a}{b}\right)\left(\dfrac{c}{d}\right) = \dfrac{ac}{bd}$	$\left(\dfrac{2}{3}\right)\left(\dfrac{4}{5}\right) = \dfrac{8}{15}$
Dividing	$\dfrac{(a/b)}{(c/d)} = \dfrac{ad}{bc}$	$\dfrac{2/3}{4/5} = \dfrac{(2)(5)}{(4)(3)} = \dfrac{10}{12}$
Adding	$\dfrac{a}{b} \pm \dfrac{c}{d} = \dfrac{ad \pm bc}{bd}$	$\dfrac{2}{3} - \dfrac{4}{5} = \dfrac{(2)(5) - (4)(3)}{(3)(5)} = -\dfrac{2}{15}$

Exercises

In the following exercises, solve for x.

Answers

1. $a = \dfrac{1}{1 + x}$ $x = \dfrac{1 - a}{a}$

2. $3x - 5 = 13$ $x = 6$

3. $ax - 5 = bx + 2$ $x = \dfrac{7}{a - b}$

4. $\dfrac{5}{2x + 6} = \dfrac{3}{4x + 8}$ $x = -\dfrac{11}{7}$

Powers

When powers of a given quantity x are multiplied, the following rule applies:

$$x^n x^m = x^{n+m} \qquad \text{B.3} \blacktriangleleft$$

For example, $x^2 x^4 = x^{2+4} = x^6$.

When dividing the powers of a given quantity, the rule is

$$\frac{x^n}{x^m} = x^{n-m} \qquad \text{B.4} \blacktriangleleft$$

For example, $x^8 / x^2 = x^{8-2} = x^6$.

A power that is a fraction, such as $\frac{1}{3}$, corresponds to a root as follows:

$$x^{1/n} = \sqrt[n]{x} \qquad \text{B.5} \blacktriangleleft$$

For example, $4^{1/3} = \sqrt[3]{4} = 1.587\,4$. (A scientific calculator is useful for such calculations.)

Finally, any quantity x^n raised to the mth power is

$$(x^n)^m = x^{nm} \qquad \text{B.6} \blacktriangleleft$$

Table B.1 summarizes the rules of exponents.

◀ TABLE B.1 | Rules of Exponents

$$x^0 = 1$$
$$x^1 = x$$
$$x^n x^m = x^{n+m}$$
$$x^n / x^m = x^{n-m}$$
$$x^{1/n} = \sqrt[n]{x}$$
$$(x^n)^m = x^{nm}$$

Exercises

Verify the following equations:

1. $3^2 \times 3^3 = 243$
2. $x^5 x^{-8} = x^{-3}$
3. $x^{10} / x^{-5} = x^{15}$
4. $5^{1/3} = 1.709\,976$ (Use your calculator.)
5. $60^{1/4} = 2.783\,158$ (Use your calculator.)
6. $(x^4)^3 = x^{12}$

Factoring

Some useful formulas for factoring an equation are the following:

$$ax + ay + az = a(x + y + z) \quad \text{common factor}$$

$$a^2 + 2ab + b^2 = (a + b)^2 \quad \text{perfect square}$$

$$a^2 - b^2 = (a + b)(a - b) \quad \text{differences of squares}$$

Quadratic Equations

The general form of a quadratic equation is

$$ax^2 + bx + c = 0 \quad \text{B.7} \blacktriangleleft$$

where x is the unknown quantity and a, b, and c are numerical factors referred to as **coefficients** of the equation. This equation has two roots, given by

$$x = \frac{-b \pm \sqrt{b^2 - 4ac}}{2a} \quad \text{B.8} \blacktriangleleft$$

If $b^2 \geq 4ac$, the roots are real.

> ### Example B.1
>
> The equation $x^2 + 5x + 4 = 0$ has the following roots corresponding to the two signs of the square-root term:
>
> $$x = \frac{-5 \pm \sqrt{5^2 - (4)(1)(4)}}{2(1)} = \frac{-5 \pm \sqrt{9}}{2} = \frac{-5 \pm 3}{2}$$
>
> $$x_+ = \frac{-5 + 3}{2} = -1 \quad x_- = \frac{-5 - 3}{2} = -4$$
>
> where x_+ refers to the root corresponding to the positive sign and x_- refers to the root corresponding to the negative sign.

Exercises

Solve the following quadratic equations:

Answers

1. $x^2 + 2x - 3 = 0$ $x_+ = 1$ $x_- = -3$
2. $2x^2 - 5x + 2 = 0$ $x_+ = 2$ $x_- = \frac{1}{2}$
3. $2x^2 - 4x - 9 = 0$ $x_+ = 1 + \sqrt{22}/2$ $x_- = 1 - \sqrt{22}/2$

Linear Equations

A linear equation has the general form

$$y = mx + b \quad \text{B.9} \blacktriangleleft$$

where m and b are constants. This equation is referred to as linear because the graph of y versus x is a straight line as shown in Figure B.1. The constant b, called the **y-intercept**, represents the value of y at which the straight line intersects the y axis. The constant m is equal to the **slope** of the straight line. If any two points on the straight line are specified by the coordinates (x_1, y_1) and (x_2, y_2) as in Figure B.1, the slope of the straight line can be expressed as

$$\text{Slope} = \frac{y_2 - y_1}{x_2 - x_1} = \frac{\Delta y}{\Delta x} \quad \text{B.10} \blacktriangleleft$$

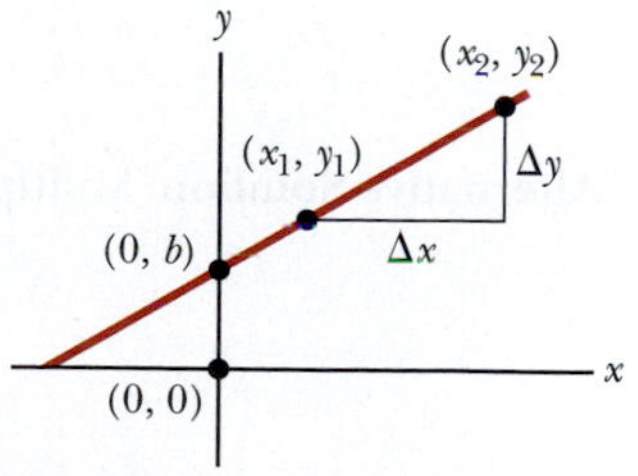

Figure B.1 A straight line graphed on an xy coordinate system. The slope of the line is the ratio of Δy to Δx.

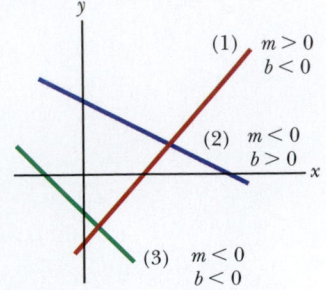

Figure B.2 The brown line has a positive slope and a negative y-intercept. The blue line has a negative slope and a positive y-intercept. The green line has a negative slope and a negative y-intercept.

Note that m and b can have either positive or negative values. If $m > 0$, the straight line has a *positive* slope as in Figure B.1. If $m < 0$, the straight line has a *negative* slope. In Figure B.1, both m and b are positive. Three other possible situations are shown in Figure B.2.

Exercises

1. Draw graphs of the following straight lines: (a) $y = 5x + 3$ (b) $y = -2x + 4$ (c) $y = -3x - 6$
2. Find the slopes of the straight lines described in Exercise 1.

Answers (a) 5 (b) -2 (c) -3

3. Find the slopes of the straight lines that pass through the following sets of points: (a) $(0, -4)$ and $(4, 2)$ (b) $(0, 0)$ and $(2, -5)$ (c) $(-5, 2)$ and $(4, -2)$

Answers (a) $\frac{3}{2}$ (b) $-\frac{5}{2}$ (c) $-\frac{4}{9}$

Solving Simultaneous Linear Equations

Consider the equation $3x + 5y = 15$, which has two unknowns, x and y. Such an equation does not have a unique solution. For example, $(x = 0, y = 3)$, $(x = 5, y = 0)$, and $(x = 2, y = \frac{9}{5})$ are all solutions to this equation.

If a problem has two unknowns, a unique solution is possible only if we have *two* pieces of information. In most common cases, those two pieces of information are equations. In general, if a problem has n unknowns, its solution requires n equations. To solve two simultaneous equations involving two unknowns, x and y, we solve one of the equations for x in terms of y and substitute this expression into the other equation.

In some cases, the two pieces of information may be (1) one equation and (2) a condition on the solutions. For example, suppose we have the equation $m = 3n$ and the condition that m and n must be the smallest positive nonzero integers possible. Then, the single equation does not allow a unique solution, but the addition of the condition gives us that $n = 1$ and $m = 3$.

Example B.2

Solve the two simultaneous equations.

$$(1) \quad 5x + y = -8$$

$$(2) \quad 2x - 2y = 4$$

SOLUTION

From Equation (2), $x = y + 2$. Substitution of this equation into Equation (1) gives

$$5(y + 2) + y = -8$$

$$6y = -18$$

$$y = \boxed{-3}$$

$$x = y + 2 = \boxed{-1}$$

Alternative Solution Multiply each term in Equation (1) by the factor 2 and add the result to Equation (2):

$$10x + 2y = -16$$
$$\underline{2x - 2y = 4}$$
$$12x \qquad\quad = -12$$

$$x = \boxed{-1}$$

$$y = x - 2 = \boxed{-3}$$

Two linear equations containing two unknowns can also be solved by a graphical method. If the straight lines corresponding to the two equations are plotted in a conventional coordinate system, the intersection of the two lines represents the solution. For example, consider the two equations

$$x - y = 2$$

$$x - 2y = -1$$

These equations are plotted in Figure B.3. The intersection of the two lines has the coordinates $x = 5$ and $y = 3$, which represents the solution to the equations. You should check this solution by the analytical technique discussed earlier.

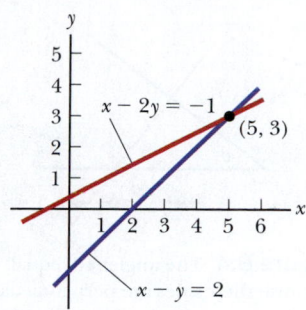

Figure B.3 A graphical solution for two linear equations.

Exercises

Solve the following pairs of simultaneous equations involving two unknowns:

Answers

1. $x + y = 8$ $x = 5, y = 3$
    $x - y = 2$

2. $98 - T = 10a$ $T = 65, a = 3.27$
    $T - 49 = 5a$

3. $6x + 2y = 6$ $x = 2, y = -3$
    $8x - 4y = 28$

Logarithms

Suppose a quantity x is expressed as a power of some quantity a:

$$x = a^y \qquad \text{B.11} \blacktriangleleft$$

The number a is called the **base** number. The **logarithm** of x with respect to the base a is equal to the exponent to which the base must be raised to satisfy the expression $x = a^y$:

$$y = \log_a x \qquad \text{B.12} \blacktriangleleft$$

Conversely, the **antilogarithm** of y is the number x:

$$x = \text{antilog}_a y \qquad \text{B.13} \blacktriangleleft$$

In practice, the two bases most often used are base 10, called the *common* logarithm base, and base $e = 2.718\,282$, called Euler's constant or the *natural* logarithm base. When common logarithms are used,

$$y = \log_{10} x \quad (\text{or } x = 10^y) \qquad \text{B.14} \blacktriangleleft$$

When natural logarithms are used,

$$y = \ln x \quad (\text{or } x = e^y) \qquad \text{B.15} \blacktriangleleft$$

For example, $\log_{10} 52 = 1.716$, so $\text{antilog}_{10} 1.716 = 10^{1.716} = 52$. Likewise, $\ln 52 = 3.951$, so $\text{antiln } 3.951 = e^{3.951} = 52$.

In general, note you can convert between base 10 and base e with the equality

$$\ln x = (2.302\,585) \log_{10} x \qquad \text{B.16} \blacktriangleleft$$

Finally, some useful properties of logarithms are the following:

$$\left.\begin{array}{l} \log(ab) = \log a + \log b \\[4pt] \log(a/b) = \log a - \log b \\[4pt] \log(a^n) = n \log a \end{array}\right\} \quad \text{any base}$$

$$\ln e = 1$$

$$\ln e^a = a$$

$$\ln\left(\frac{1}{a}\right) = -\ln a$$

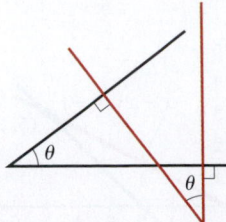

Figure B.4 The angles are equal because their sides are perpendicular.

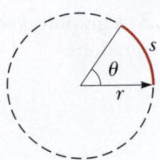

Figure B.5 The angle θ in radians is the ratio of the arc length s to the radius r of the circle.

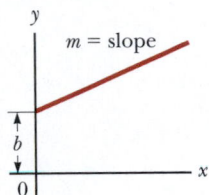

Figure B.6 A straight line with a slope of m and a y-intercept of b.

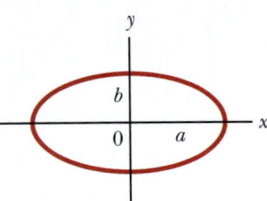

Figure B.7 An ellipse with semimajor axis a and semiminor axis b.

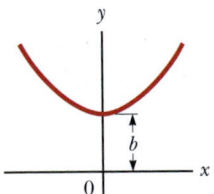

Figure B.8 A parabola with its vertex at $y = b$.

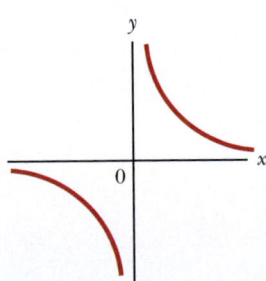

Figure B.9 A hyperbola.

B.3 | Geometry

The **distance** d between two points having coordinates (x_1, y_1) and (x_2, y_2) is

$$d = \sqrt{(x_2 - x_1)^2 + (y_2 - y_1)^2}$$ **B.17**◀

Two angles are equal if their sides are perpendicular, right side to right side and left side to left side. For example, the two angles marked θ in Figure B.4 are the same because of the perpendicularity of the sides of the angles. To distinguish the left and right sides of an angle, imagine standing at the angle's apex and facing into the angle.

Radian measure: The arc length s of a circular arc (Fig. B.5) is proportional to the radius r for a fixed value of θ (in radians):

$$s = r\theta$$

$$\theta = \frac{s}{r}$$ **B.18**◀

Table B.2 gives the **areas** and **volumes** for several geometric shapes used throughout this text.

The equation of a **straight line** (Fig. B.6) is

$$y = mx + b$$ **B.19**◀

where b is the y-intercept and m is the slope of the line.

The equation of a **circle** of radius R centered at the origin is

$$x^2 + y^2 = R^2$$ **B.20**◀

The equation of an **ellipse** having the origin at its center (Fig. B.7) is

$$\frac{x^2}{a^2} + \frac{y^2}{b^2} = 1$$ **B.21**◀

where a is the length of the semimajor axis (the longer one) and b is the length of the semiminor axis (the shorter one).

The equation of a **parabola** the vertex of which is at $y = b$ (Fig. B.8) is

$$y = ax^2 + b$$ **B.22**◀

The equation of a **rectangular hyperbola** (Fig. B.9) is

$$xy = \text{constant}$$ **B.23**◀

◀ **TABLE B.2** | Useful Information for Geometry

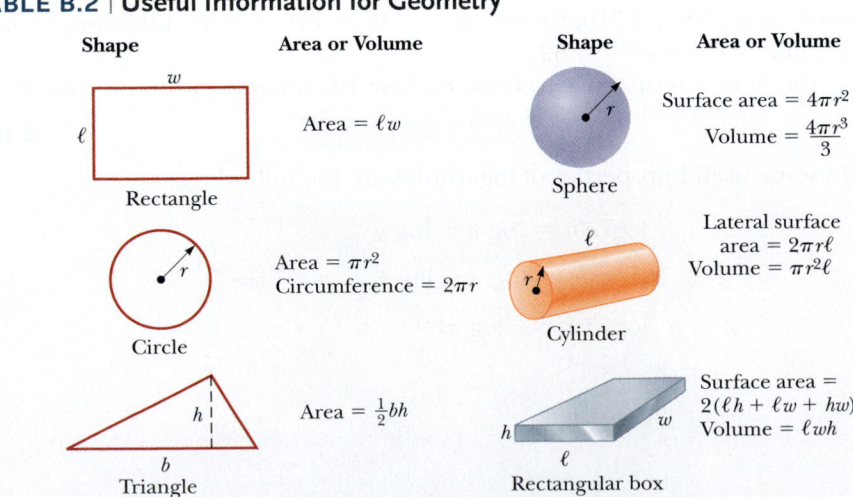

⟨B.4 | Trigonometry

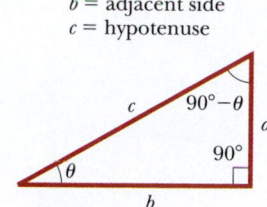

a = opposite side
b = adjacent side
c = hypotenuse

Figure B.10 A right triangle, used to define the basic functions of trigonometry.

That portion of mathematics based on the special properties of the right triangle is called trigonometry. By definition, a right triangle is a triangle containing a 90° angle. Consider the right triangle shown in Figure B.10, where side a is opposite the angle θ, side b is adjacent to the angle θ, and side c is the hypotenuse of the triangle. The three basic trigonometric functions defined by such a triangle are the sine (sin), cosine (cos), and tangent (tan). In terms of the angle θ, these functions are defined as follows:

$$\sin \theta = \frac{\text{side opposite } \theta}{\text{hypotenuse}} = \frac{a}{c} \qquad \textbf{B.24} \blacktriangleleft$$

$$\cos \theta = \frac{\text{side adjacent to } \theta}{\text{hypotenuse}} = \frac{b}{c} \qquad \textbf{B.25} \blacktriangleleft$$

$$\tan \theta = \frac{\text{side opposite } \theta}{\text{side adjacent to } \theta} = \frac{a}{b} \qquad \textbf{B.26} \blacktriangleleft$$

The Pythagorean theorem provides the following relationship among the sides of a right triangle:

$$c^2 = a^2 + b^2 \qquad \textbf{B.27} \blacktriangleleft$$

From the preceding definitions and the Pythagorean theorem, it follows that

$$\sin^2 \theta + \cos^2 \theta = 1$$

$$\tan \theta = \frac{\sin \theta}{\cos \theta}$$

The cosecant, secant, and cotangent functions are defined by

$$\csc \theta = \frac{1}{\sin \theta} \qquad \sec \theta = \frac{1}{\cos \theta} \qquad \cot \theta = \frac{1}{\tan \theta}$$

The following relationships are derived directly from the right triangle shown in Figure B.10:

$$\sin \theta = \cos (90° - \theta)$$

$$\cos \theta = \sin (90° - \theta)$$

$$\cot \theta = \tan (90° - \theta)$$

Some properties of trigonometric functions are the following:

$$\sin (-\theta) = -\sin \theta$$

$$\cos (-\theta) = \cos \theta$$

$$\tan (-\theta) = -\tan \theta$$

The following relationships apply to *any* triangle as shown in Figure B.11:

$$\alpha + \beta + \gamma = 180°$$

Law of cosines
$$\begin{cases} a^2 = b^2 + c^2 - 2bc \cos \alpha \\ b^2 = a^2 + c^2 - 2ac \cos \beta \\ c^2 = a^2 + b^2 - 2ab \cos \gamma \end{cases}$$

Law of sines
$$\frac{a}{\sin \alpha} = \frac{b}{\sin \beta} = \frac{c}{\sin \gamma}$$

Table B.3 (page A-14) lists a number of useful trigonometric identities.

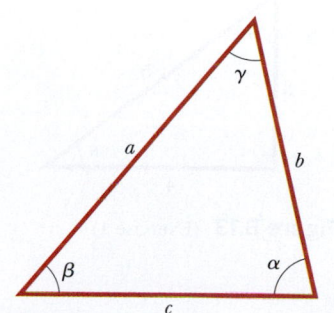

Figure B.11 An arbitrary, nonright triangle.

> **TABLE B.3 | Some Trigonometric Identities**

$$\sin^2 \theta + \cos^2 \theta = 1$$

$$\sec^2 \theta = 1 + \tan^2 \theta$$

$$\sin 2\theta = 2 \sin \theta \cos \theta$$

$$\cos 2\theta = \cos^2 \theta - \sin^2 \theta$$

$$\tan 2\theta = \frac{2 \tan \theta}{1 - \tan^2 \theta}$$

$$\sin (A \pm B) = \sin A \cos B \pm \cos A \sin B$$

$$\cos (A \pm B) = \cos A \cos B \mp \sin A \sin B$$

$$\sin A \pm \sin B = 2 \sin \left[\tfrac{1}{2}(A \pm B)\right] \cos \left[\tfrac{1}{2}(A \mp B)\right]$$

$$\cos A + \cos B = 2 \cos \left[\tfrac{1}{2}(A + B)\right] \cos \left[\tfrac{1}{2}(A - B)\right]$$

$$\cos A - \cos B = 2 \sin \left[\tfrac{1}{2}(A + B)\right] \sin \left[\tfrac{1}{2}(B - A)\right]$$

$$\csc^2 \theta = 1 + \cot^2 \theta$$

$$\sin^2 \frac{\theta}{2} = \tfrac{1}{2}(1 - \cos \theta)$$

$$\cos^2 \frac{\theta}{2} = \tfrac{1}{2}(1 + \cos \theta)$$

$$1 - \cos \theta = 2 \sin^2 \frac{\theta}{2}$$

$$\tan \frac{\theta}{2} = \sqrt{\frac{1 - \cos \theta}{1 + \cos \theta}}$$

Example **B.3**

Consider the right triangle in Figure B.12 in which $a = 2.00$, $b = 5.00$, and c is unknown. From the Pythagorean theorem, we have

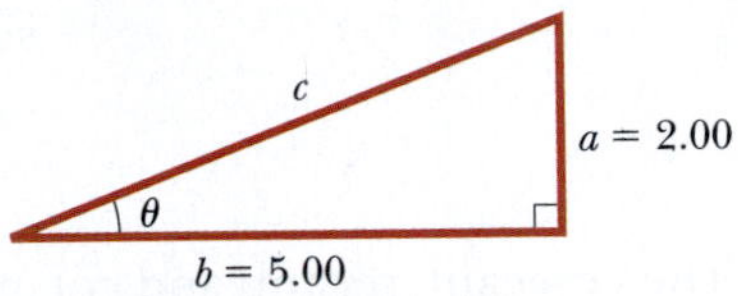

$$c^2 = a^2 + b^2 = 2.00^2 + 5.00^2 = 4.00 + 25.0 = 29.0$$

$$c = \sqrt{29.0} = \boxed{5.39}$$

To find the angle θ, note that

$$\tan \theta = \frac{a}{b} = \frac{2.00}{5.00} = 0.400$$

Using a calculator, we find that

$$\theta = \tan^{-1} (0.400) = \boxed{21.8°}$$

where $\tan^{-1} (0.400)$ is the notation for "angle whose tangent is 0.400," sometimes written as arctan (0.400).

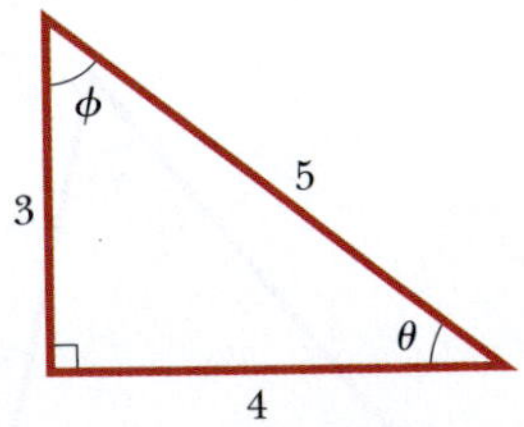

Figure B.12 (Example B.3)

Figure B.13 (Exercise 1)

Exercises

1. In Figure B.13, identify (a) the side opposite θ (b) the side adjacent to ϕ and then find (c) $\cos \theta$, (d) $\sin \phi$, and (e) $\tan \phi$.

   Answers   (a) 3   (b) 3   (c) $\frac{4}{5}$ (d) $\frac{4}{5}$ (e) $\frac{4}{3}$

2. In a certain right triangle, the two sides that are perpendicular to each other are 5.00 m and 7.00 m long. What is the length of the third side?

   Answer   8.60 m

3. A right triangle has a hypotenuse of length 3.0 m, and one of its angles is 30°. (a) What is the length of the side opposite the 30° angle? (b) What is the side adjacent to the 30° angle?

   Answers   (a) 1.5 m   (b) 2.6 m

B.5 | Series Expansions

$$(a + b)^n = a^n + \frac{n}{1!} a^{n-1} b + \frac{n(n-1)}{2!} a^{n-2} b^2 + \cdots$$

$$(1 + x)^n = 1 + nx + \frac{n(n-1)}{2!} x^2 + \cdots$$

$$e^x = 1 + x + \frac{x^2}{2!} + \frac{x^3}{3!} + \cdots$$

$$\ln (1 \pm x) = \pm x - \tfrac{1}{2} x^2 \pm \tfrac{1}{3} x^3 - \cdots$$

$$\sin x = x - \frac{x^3}{3!} + \frac{x^5}{5!} - \cdots$$

$$\cos x = 1 - \frac{x^2}{2!} + \frac{x^4}{4!} - \cdots$$

$$\tan x = x + \frac{x^3}{3} + \frac{2x^5}{15} + \cdots \quad |x| < \frac{\pi}{2}$$

x in radians

For $x \ll 1$, the following approximations can be used:[1]

$$(1 + x)^n \approx 1 + nx \qquad \sin x \approx x$$

$$e^x \approx 1 + x \qquad \cos x \approx 1$$

$$\ln (1 \pm x) \approx \pm x \qquad \tan x \approx x$$

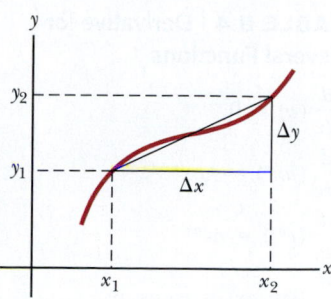

Figure B.14 The lengths Δx and Δy are used to define the derivative of this function at a point.

B.6 | Differential Calculus

In various branches of science, it is sometimes necessary to use the basic tools of calculus, invented by Newton, to describe physical phenomena. The use of calculus is fundamental in the treatment of various problems in Newtonian mechanics, electricity, and magnetism. In this section, we simply state some basic properties and "rules of thumb" that should be a useful review to the student.

First, a **function** must be specified that relates one variable to another (e.g., a coordinate as a function of time). Suppose one of the variables is called y (the dependent variable), and the other x (the independent variable). We might have a function relationship such as

$$y(x) = ax^3 + bx^2 + cx + d$$

If a, b, c, and d are specified constants, y can be calculated for any value of x. We usually deal with continuous functions, that is, those for which y varies "smoothly" with x.

The **derivative** of y with respect to x is defined as the limit as Δx approaches zero of the slopes of chords drawn between two points on the y versus x curve. Mathematically, we write this definition as

$$\frac{dy}{dx} = \lim_{\Delta x \to 0} \frac{\Delta y}{\Delta x} = \lim_{\Delta x \to 0} \frac{y(x + \Delta x) - y(x)}{\Delta x} \qquad \text{B.28} \blacktriangleleft$$

where Δy and Δx are defined as $\Delta x = x_2 - x_1$ and $\Delta y = y_2 - y_1$ (Fig. B.14). Note that dy/dx *does not* mean dy divided by dx, but rather is simply a notation of the limiting process of the derivative as defined by Equation B.28.

A useful expression to remember when $y(x) = ax^n$, where a is a *constant* and n is *any* positive or negative number (integer or fraction), is

$$\frac{dy}{dx} = nax^{n-1} \qquad \text{B.29} \blacktriangleleft$$

---

[1] The approximations for the functions $\sin x$, $\cos x$, and $\tan x$ are for $x \leq 0.1$ rad.

TABLE B.4 | Derivative for Several Functions

$$\frac{d}{dx}(a) = 0$$

$$\frac{d}{dx}(ax^n) = nax^{n-1}$$

$$\frac{d}{dx}(e^{ax}) = ae^{ax}$$

$$\frac{d}{dx}(\sin ax) = a\cos ax$$

$$\frac{d}{dx}(\cos ax) = -a\sin ax$$

$$\frac{d}{dx}(\tan ax) = a\sec^2 ax$$

$$\frac{d}{dx}(\cot ax) = -a\csc^2 ax$$

$$\frac{d}{dx}(\sec x) = \tan x \sec x$$

$$\frac{d}{dx}(\csc x) = -\cot x \csc x$$

$$\frac{d}{dx}(\ln ax) = \frac{1}{x}$$

$$\frac{d}{dx}(\sin^{-1} ax) = \frac{a}{\sqrt{1-a^2x^2}}$$

$$\frac{d}{dx}(\cos^{-1} ax) = \frac{-a}{\sqrt{1-a^2x^2}}$$

$$\frac{d}{dx}(\tan^{-1} ax) = \frac{a}{\sqrt{1+a^2x^2}}$$

Note: The symbols a and n represent constants.

If $y(x)$ is a polynomial or algebraic function of x, we apply Equation B.29 to *each* term in the polynomial and take $d[\text{constant}]/dx = 0$. In Examples B.4 through B.7, we evaluate the derivatives of several functions.

Special Properties of the Derivative

A. Derivative of the product of two functions If a function $f(x)$ is given by the product of two functions—say, $g(x)$ and $h(x)$—the derivative of $f(x)$ is defined as

$$\frac{d}{dx}f(x) = \frac{d}{dx}[g(x)h(x)] = g\frac{dh}{dx} + h\frac{dg}{dx} \qquad \text{B.30} \blacktriangleleft$$

B. Derivative of the sum of two functions If a function $f(x)$ is equal to the sum of two functions, the derivative of the sum is equal to the sum of the derivatives:

$$\frac{d}{dx}f(x) = \frac{d}{dx}[g(x) + h(x)] = \frac{dg}{dx} + \frac{dh}{dx} \qquad \text{B.31} \blacktriangleleft$$

C. Chain rule of differential calculus If $y = f(x)$ and $x = g(z)$, then dy/dz can be written as the product of two derivatives:

$$\frac{dy}{dz} = \frac{dy}{dx}\frac{dx}{dz} \qquad \text{B.32} \blacktriangleleft$$

D. The second derivative The second derivative of y with respect to x is defined as the derivative of the function dy/dx (the derivative of the derivative). It is usually written as

$$\frac{d^2y}{dx^2} = \frac{d}{dx}\left(\frac{dy}{dx}\right) \qquad \text{B.33} \blacktriangleleft$$

Some of the more commonly used derivatives of functions are listed in Table B.4.

Example B.4

Suppose $y(x)$ (that is, y as a function of x) is given by

$$y(x) = ax^3 + bx + c$$

where a and b are constants. It follows that

$$y(x + \Delta x) = a(x + \Delta x)^3 + b(x + \Delta x) + c$$

$$= a(x^3 + 3x^2\,\Delta x + 3x\,\Delta x^2 + \Delta x^3) + b(x + \Delta x) + c$$

so

$$\Delta y = y(x + \Delta x) - y(x) = a(3x^2\,\Delta x + 3x\,\Delta x^2 + \Delta x^3) + b\,\Delta x$$

Substituting this into Equation B.28 gives

$$\frac{dy}{dx} = \lim_{\Delta x \to 0} \frac{\Delta y}{\Delta x} = \lim_{\Delta x \to 0}[3ax^2 + 3ax\,\Delta x + a\,\Delta x^2] + b$$

$$\boxed{\frac{dy}{dx} = 3ax^2 + b}$$

Example **B.5**

Find the derivative of

$$y(x) = 8x^5 + 4x^3 + 2x + 7$$

SOLUTION

Applying Equation B.29 to each term independently and remembering that d/dx (constant) $= 0$, we have

$$\frac{dy}{dx} = 8(5)x^4 + 4(3)x^2 + 2(1)x^0 + 0$$

$$\frac{dy}{dx} = \boxed{40x^4 + 12x^2 + 2}$$

Example **B.6**

Find the derivative of $y(x) = x^3/(x+1)^2$ with respect to x.

SOLUTION

We can rewrite this function as $y(x) = x^3(x+1)^{-2}$ and apply Equation B.30:

$$\frac{dy}{dx} = (x+1)^{-2}\frac{d}{dx}(x^3) + x^3\frac{d}{dx}(x+1)^{-2}$$

$$= (x+1)^{-2}\,3x^2 + x^3(-2)(x+1)^{-3}$$

$$\frac{dy}{dx} = \boxed{\frac{3x^2}{(x+1)^2} - \frac{2x^3}{(x+1)^3}} = \boxed{\frac{x^2(x+3)}{(x+1)^3}}$$

Example **B.7**

A useful formula that follows from Equation B.30 is the derivative of the quotient of two functions. Show that

$$\frac{d}{dx}\left[\frac{g(x)}{h(x)}\right] = \frac{h\dfrac{dg}{dx} - g\dfrac{dh}{dx}}{h^2}$$

SOLUTION

We can write the quotient as gh^{-1} and then apply Equations B.29 and B.30:

$$\frac{d}{dx}\left(\frac{g}{h}\right) = \frac{d}{dx}(gh^{-1}) = g\frac{d}{dx}(h^{-1}) + h^{-1}\frac{d}{dx}(g)$$

$$= -gh^{-2}\frac{dh}{dx} + h^{-1}\frac{dg}{dx}$$

$$= \frac{h\dfrac{dg}{dx} - g\dfrac{dh}{dx}}{h^2}$$

‹B.7 │ Integral Calculus

We think of integration as the inverse of differentiation. As an example, consider the expression

$$f(x) = \frac{dy}{dx} = 3ax^2 + b$$ B.34◀

which was the result of differentiating the function

$$y(x) = ax^3 + bx + c$$

in Example B.4. We can write Equation B.34 as $dy = f(x)\,dx = (3ax^2 + b)\,dx$ and obtain $y(x)$ by "summing" over all values of x. Mathematically, we write this inverse operation as

$$y(x) = \int f(x)\,dx$$

For the function $f(x)$ given by Equation B.34, we have

$$y(x) = \int (3ax^2 + b)\,dx = ax^3 + bx + c$$

where c is a constant of the integration. This type of integral is called an *indefinite integral* because its value depends on the choice of c.

A general **indefinite integral** $I(x)$ is defined as

$$I(x) = \int f(x)\,dx$$ B.35◀

where $f(x)$ is called the *integrand* and $f(x) = dI(x)/dx$.

For a *general continuous* function $f(x)$, the integral can be described as the area under the curve bounded by $f(x)$ and the x axis, between two specified values of x, say, x_1 and x_2, as in Figure B.15.

The area of the blue element in Figure B.15 is approximately $f(x_i)\,\Delta x_i$. If we sum all these area elements between x_1 and x_2 and take the limit of this sum as $\Delta x_i \rightarrow 0$, we obtain the *true* area under the curve bounded by $f(x)$ and the x axis, between the limits x_1 and x_2:

$$\text{Area} = \lim_{\Delta x_i \rightarrow 0} \sum_i f(x_i)\Delta x_i = \int_{x_1}^{x_2} f(x)\,dx$$ B.36◀

Integrals of the type defined by Equation B.36 are called **definite integrals.**

One common integral that arises in practical situations has the form

$$\int x^n\,dx = \frac{x^{n+1}}{n+1} + c \quad (n \neq -1)$$ B.37◀

This result is obvious, being that differentiation of the right-hand side with respect to x gives $f(x) = x^n$ directly. If the limits of the integration are known, this integral becomes a *definite integral* and is written

$$\int_{x_1}^{x_2} x^n\,dx = \frac{x^{n+1}}{n+1}\bigg|_{x_1}^{x_2} = \frac{x_2^{n+1} - x_1^{n+1}}{n+1} \quad (n \neq -1)$$ B.38◀

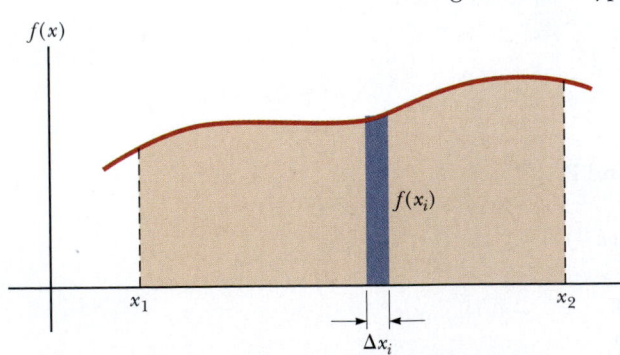

Figure B.15 The definite integral of a function is the area under the curve of the function between the limits x_1 and x_2.

Examples

1. $\displaystyle\int_0^a x^2\,dx = \left.\frac{x^3}{3}\right|_0^a = \frac{a^3}{3}$

2. $\displaystyle\int_0^b x^{3/2}\,dx = \left.\frac{x^{5/2}}{5/2}\right|_0^b = \tfrac{2}{5}\,b^{5/2}$

3. $\displaystyle\int_3^5 x\,dx = \left.\frac{x^2}{2}\right|_3^5 = \frac{5^2 - 3^2}{2} = 8$

Partial Integration

Sometimes it is useful to apply the method of *partial integration* (also called "integrating by parts") to evaluate certain integrals. This method uses the property

$$\int u\,dv = uv - \int v\,du \qquad\qquad \textbf{B.39}\blacktriangleleft$$

where u and v are *carefully* chosen so as to reduce a complex integral to a simpler one. In many cases, several reductions have to be made. Consider the function

$$I(x) = \int x^2\,e^x\,dx$$

which can be evaluated by integrating by parts twice. First, if we choose $u = x^2$, $v = e^x$, we obtain

$$\int x^2\,e^x\,dx = \int x^2\,d(e^x) = x^2\,e^x - 2\int e^x\,x\,dx + c_1$$

Now, in the second term, choose $u = x$, $v = e^x$, which gives

$$\int x^2\,e^x\,dx = x^2\,e^x - 2x\,e^x + 2\int e^x\,dx + c_1$$

or

$$\int x^2\,e^x\,dx = x^2\,e^x - 2xe^x + 2e^x + c_2$$

The Perfect Differential

Another useful method to remember is that of the *perfect differential,* in which we look for a change of variable such that the differential of the function is the differential of the independent variable appearing in the integrand. For example, consider the integral

$$I(x) = \int \cos^2 x \sin x\,dx$$

This integral becomes easy to evaluate if we rewrite the differential as $d(\cos x) = -\sin x\,dx$. The integral then becomes

$$\int \cos^2 x \sin x\,dx = -\int \cos^2 x\,d(\cos x)$$

If we now change variables, letting $y = \cos x$, we obtain

$$\int \cos^2 x \sin x\,dx = -\int y^2\,dy = -\frac{y^3}{3} + c = -\frac{\cos^3 x}{3} + c$$

Table B.5 lists some useful indefinite integrals. Table B.6 gives Gauss's probability integral and other definite integrals. A more complete list can be found in various handbooks, such as *The Handbook of Chemistry and Physics* (Boca Raton, FL: CRC Press, published annually).

TABLE B.5 | Some Indefinite Integrals (An arbitrary constant should be added to each of these integrals.)

$$\int x^n \, dx = \frac{x^{n+1}}{n+1} \text{ (provided } n \neq 1)$$

$$\int \ln ax \, dx = (x \ln ax) - x$$

$$\int \frac{dx}{x} = \int x^{-1} \, dx = \ln x$$

$$\int xe^{ax} \, dx = \frac{e^{ax}}{a^2} (ax - 1)$$

$$\int \frac{dx}{a + bx} = \frac{1}{b} \ln (a + bx)$$

$$\int \frac{dx}{a + be^{cx}} = \frac{x}{a} - \frac{1}{ac} \ln (a + be^{cx})$$

$$\int \frac{x \, dx}{a + bx} = \frac{x}{b} - \frac{a}{b^2} \ln (a + bx)$$

$$\int \sin ax \, dx = -\frac{1}{a} \cos ax$$

$$\int \frac{dx}{x(x + a)} = -\frac{1}{a} \ln \frac{x + a}{x}$$

$$\int \cos ax \, dx = \frac{1}{a} \sin ax$$

$$\int \frac{dx}{(a + bx)^2} = -\frac{1}{b(a + bx)}$$

$$\int \tan ax \, dx = -\frac{1}{a} \ln (\cos ax) = \frac{1}{a} \ln (\sec ax)$$

$$\int \frac{dx}{a^2 + x^2} = \frac{1}{a} \tan^{-1} \frac{x}{a}$$

$$\int \cot ax \, dx = \frac{1}{a} \ln (\sin ax)$$

$$\int \frac{dx}{a^2 - x^2} = \frac{1}{2a} \ln \frac{a + x}{a - x} \left(a^2 - x^2 > 0 \right)$$

$$\int \sec ax \, dx = \frac{1}{a} \ln (\sec ax + \tan ax) = \frac{1}{a} \ln \left[\tan \left(\frac{ax}{2} + \frac{\pi}{4} \right) \right]$$

$$\int \frac{dx}{x^2 - a^2} = \frac{1}{2a} \ln \frac{x - a}{x + a} \left(x^2 - a^2 > 0 \right)$$

$$\int \csc ax \, dx = \frac{1}{a} \ln (\csc ax - \cot ax) = \frac{1}{a} \ln \left(\tan \frac{ax}{2} \right)$$

$$\int \frac{x \, dx}{a^2 \pm x^2} = \pm \frac{1}{2} \ln (a^2 \pm x^2)$$

$$\int \sin^2 ax \, dx = \frac{x}{2} - \frac{\sin 2ax}{4a}$$

$$\int \frac{dx}{\sqrt{a^2 - x^2}} = \sin^{-1} \frac{x}{a} = -\cos^{-1} \frac{x}{a} \left(a^2 - x^2 > 0 \right)$$

$$\int \cos^2 ax \, dx = \frac{x}{2} + \frac{\sin 2ax}{4a}$$

$$\int \frac{dx}{\sqrt{x^2 \pm a^2}} = \ln (x + \sqrt{x^2 \pm a^2})$$

$$\int \frac{dx}{\sin^2 ax} = -\frac{1}{a} \cot ax$$

$$\int \frac{x \, dx}{\sqrt{a^2 - x^2}} = -\sqrt{a^2 - x^2}$$

$$\int \frac{dx}{\cos^2 ax} = \frac{1}{a} \tan ax$$

$$\int \frac{x \, dx}{\sqrt{x^2 \pm a^2}} = \sqrt{x^2 \pm a^2}$$

$$\int \tan^2 ax \, dx = \frac{1}{a} (\tan ax) - x$$

(Continued)

TABLE B.5 | Some Indefinite Integrals (*continued*)

$$\int \sqrt{a^2 - x^2}\ dx = \frac{1}{2}\left(x\sqrt{a^2 - x^2} + a^2 \sin^{-1}\frac{x}{|a|}\right)$$

$$\int \cot^2 ax\ dx = -\frac{1}{a}(\cot ax) - x$$

$$\int x\sqrt{a^2 - x^2}\ dx = -\frac{1}{3}(a^2 - x^2)^{3/2}$$

$$\int \sin^{-1} ax\ dx = x(\sin^{-1} ax) + \frac{\sqrt{1 - a^2x^2}}{a}$$

$$\int \sqrt{x^2 \pm a^2}\ dx = \frac{1}{2}\left[x\sqrt{x^2 \pm a^2} \pm a^2 \ln(x + \sqrt{x^2 \pm a^2})\right]$$

$$\int \cos^{-1} ax\ dx = x(\cos^{-1} ax) - \frac{\sqrt{1 - a^2x^2}}{a}$$

$$\int x(\sqrt{x^2 \pm a^2})\ dx = \frac{1}{3}(x^2 \pm a^2)^{3/2}$$

$$\int \frac{dx}{(x^2 + a^2)^{3/2}} = \frac{x}{a^2\sqrt{x^2 + a^2}}$$

$$\int e^{ax}\ dx = \frac{1}{a}e^{ax}$$

$$\int \frac{x\ dx}{(x^2 + a^2)^{3/2}} = -\frac{1}{\sqrt{x^2 + a^2}}$$

TABLE B.6 | Gauss's Probability Integral and Other Definite Integrals

$$\int_0^\infty x^n e^{-ax}\ dx = \frac{n!}{a^{n+1}}$$

$$I_0 = \int_0^\infty e^{-ax^2}\ dx = \frac{1}{2}\sqrt{\frac{\pi}{a}} \quad \text{(Gauss's probability integral)}$$

$$I_1 = \int_0^\infty xe^{-ax^2}\ dx = \frac{1}{2a}$$

$$I_2 = \int_0^\infty x^2 e^{-ax^2}\ dx = -\frac{dI_0}{da} = \frac{1}{4}\sqrt{\frac{\pi}{a^3}}$$

$$I_3 = \int_0^\infty x^3 e^{-ax^2}\ dx = -\frac{dI_1}{da} = \frac{1}{2a^2}$$

$$I_4 = \int_0^\infty x^4 e^{-ax^2}\ dx = \frac{d^2 I_0}{da^2} = \frac{3}{8}\sqrt{\frac{\pi}{a^5}}$$

$$I_5 = \int_0^\infty x^5 e^{-ax^2}\ dx = \frac{d^2 I_1}{da^2} = \frac{1}{a^3}$$

$$\vdots$$

$$I_{2n} = (-1)^n \frac{d^n}{da^n} I_0$$

$$I_{2n+1} = (-1)^n \frac{d^n}{da^n} I_1$$

⟨B.8 | Propagation of Uncertainty

In laboratory experiments, a common activity is to take measurements that act as raw data. These measurements are of several types—length, time interval, temperature, voltage, and so on—and are taken by a variety of instruments. Regardless of the measurement and the quality of the instrumentation, **there is always uncertainty associated with a physical measurement.** This uncertainty is a combination of that associated with the instrument and that related to the system being measured. An example of the former is the inability to exactly determine the position of a length measurement between the lines on a meterstick. An example of uncertainty related to the system being measured is the variation of temperature within a sample of water so that a single temperature for the sample is difficult to determine.

Uncertainties can be expressed in two ways. **Absolute uncertainty** refers to an uncertainty expressed in the same units as the measurement. Therefore, the length of a computer disk label might be expressed as (5.5 ± 0.1) cm. The uncertainty of ± 0.1 cm by itself is not descriptive enough for some purposes, however. This uncertainty is large if the measurement is 1.0 cm, but it is small if the measurement is 100 m. To give a more descriptive account of the uncertainty, **fractional uncertainty** or **percent uncertainty** is used. In this type of description, the uncertainty is divided by the actual measurement. Therefore, the length of the computer disk label could be expressed as

$$\ell = 5.5 \text{ cm} \pm \frac{0.1 \text{ cm}}{5.5 \text{ cm}} = 5.5 \text{ cm} \pm 0.018 \text{ (fractional uncertainty)}$$

or as

$$\ell = 5.5 \text{ cm} \pm 1.8\% \text{ (percent uncertainty)}$$

When combining measurements in a calculation, the percent uncertainty in the final result is generally larger than the uncertainty in the individual measurements. This is called **propagation of uncertainty** and is one of the challenges of experimental physics.

Some simple rules can provide a reasonable estimate of the uncertainty in a calculated result:

Multiplication and division: When measurements with uncertainties are multiplied or divided, add the *percent uncertainties* to obtain the percent uncertainty in the result.

Example: The Area of a Rectangular Plate

$$A = \ell w = (5.5 \text{ cm} \pm 1.8\%) \times (6.4 \text{ cm} \pm 1.6\%) = 35 \text{ cm}^2 \pm 3.4\%$$

$$= (35 \pm 1) \text{ cm}^2$$

Addition and subtraction: When measurements with uncertainties are added or subtracted, add the *absolute uncertainties* to obtain the absolute uncertainty in the result.

Example: A Change in Temperature

$$\Delta T = T_2 - T_1 = (99.2 \pm 1.5)°\text{C} - (27.6 \pm 1.5)°\text{C} = (71.6 \pm 3.0)°\text{C}$$

$$= 71.6°\text{C} \pm 4.2\%$$

Powers: If a measurement is taken to a power, the percent uncertainty is multiplied by that power to obtain the percent uncertainty in the result.

Example: The Volume of a Sphere

$$V = \tfrac{4}{3}\pi r^3 = \tfrac{4}{3}\pi(6.20 \text{ cm} \pm 2.0\%)^3 = 998 \text{ cm}^3 \pm 6.0\%$$

$$= (998 \pm 60) \text{ cm}^3$$

For complicated calculations, many uncertainties are added together, which can cause the uncertainty in the final result to be undesirably large. Experiments should be designed such that calculations are as simple as possible.

Notice that uncertainties in a calculation always add. As a result, an experiment involving a subtraction should be avoided if possible, especially if the measurements being subtracted are close together. The result of such a calculation is a small difference in the measurements and uncertainties that add together. It is possible that the uncertainty in the result could be larger than the result itself!

Appendix C

Periodic Table of the Elements

	Group I	Group II					Transition elements				

Symbol — **Ca** 20 —**Atomic number**
Atomic mass† —40.078
$4s^2$ —**Electron configuration**

Group I	Group II					Transition elements				
H 1 1.007 9 $1s$										
Li 3 6.941 $2s^1$	**Be** 4 9.0122 $2s^2$									
Na 11 22.990 $3s^1$	**Mg** 12 24.305 $3s^2$									
K 19 39.098 $4s^1$	**Ca** 20 40.078 $4s^2$	**Sc** 21 44.956 $3d^14s^2$	**Ti** 22 47.867 $3d^24s^2$	**V** 23 50.942 $3d^34s^2$	**Cr** 24 51.996 $3d^54s^1$	**Mn** 25 54.938 $3d^54s^2$	**Fe** 26 55.845 $3d^64s^2$	**Co** 27 58.933 $3d^74s^2$
Rb 37 85.468 $5s^1$	**Sr** 38 87.62 $5s^2$	**Y** 39 88.906 $4d^15s^2$	**Zr** 40 91.224 $4d^25s^2$	**Nb** 41 92.906 $4d^45s^1$	**Mo** 42 95.94 $4d^55s^1$	**Tc** 43 (98) $4d^55s^2$	**Ru** 44 101.07 $4d^75s^1$	**Rh** 45 102.91 $4d^85s^1$
Cs 55 132.91 $6s^1$	**Ba** 56 137.33 $6s^2$	57–71*	**Hf** 72 178.49 $5d^26s^2$	**Ta** 73 180.95 $5d^36s^2$	**W** 74 183.84 $5d^46s^2$	**Re** 75 186.21 $5d^56s^2$	**Os** 76 190.23 $5d^66s^2$	**Ir** 77 192.2 $5d^76s^2$
Fr 87 (223) $7s^1$	**Ra** 88 (226) $7s^2$	89–103**	**Rf** 104 (261) $6d^27s^2$	**Db** 105 (262) $6d^37s^2$	**Sg** 106 (266)	**Bh** 107 (264)	**Hs** 108 (277)	**Mt** 109 (268)

*Lanthanide series

La 57 138.91 $5d^16s^2$	**Ce** 58 140.12 $5d^14f^16s^2$	**Pr** 59 140.91 $4f^36s^2$	**Nd** 60 144.24 $4f^46s^2$	**Pm** 61 (145) $4f^56s^2$	**Sm** 62 150.36 $4f^66s^2$
Ac 89 (227) $6d^17s^2$	**Th** 90 232.04 $6d^27s^2$	**Pa** 91 231.04 $5f^26d^17s^2$	**U** 92 238.03 $5f^36d^17s^2$	**Np** 93 (237) $5f^46d^17s^2$	**Pu** 94 (244) $5f^67s^2$

**Actinide series

Note: Atomic mass values given are averaged over isotopes in the percentages in which they exist in nature.
† For an unstable element, mass number of the most stable known isotope is given in parentheses.
†† Elements 114 and 116 have not yet been officially named.
Note: For a description of the atomic data, visit *physics.nist.gov/PhysRefData/Elements/per_text.html*.

Group III	Group IV	Group V	Group VI	Group VII	Group 0

				H 1 1.007 9 $1s^1$	**He** 2 4.002 6 $1s^2$
B 5 10.811 $2p^1$	**C** 6 12.011 $2p^2$	**N** 7 14.007 $2p^3$	**O** 8 15.999 $2p^4$	**F** 9 18.998 $2p^5$	**Ne** 10 20.180 $2p^6$
Al 13 26.982 $3p^1$	**Si** 14 28.086 $3p^2$	**P** 15 30.974 $3p^3$	**S** 16 32.066 $3p^4$	**Cl** 17 35.453 $3p^5$	**Ar** 18 39.948 $3p^6$

Ni 28 58.693 $3d^8 4s^2$	**Cu** 29 63.546 $3d^{10}4s^1$	**Zn** 30 65.41 $3d^{10}4s^2$	**Ga** 31 69.723 $4p^1$	**Ge** 32 72.64 $4p^2$	**As** 33 74.922 $4p^3$	**Se** 34 78.96 $4p^4$	**Br** 35 79.904 $4p^5$	**Kr** 36 83.80 $4p^6$
Pd 46 106.42 $4d^{10}$	**Ag** 47 107.87 $4d^{10}5s^1$	**Cd** 48 112.41 $4d^{10}5s^2$	**In** 49 114.82 $5p^1$	**Sn** 50 118.71 $5p^2$	**Sb** 51 121.76 $5p^3$	**Te** 52 127.60 $5p^4$	**I** 53 126.90 $5p^5$	**Xe** 54 131.29 $5p^6$
Pt 78 195.08 $5d^9 6s^1$	**Au** 79 196.97 $5d^{10}6s^1$	**Hg** 80 200.59 $5d^{10}6s^2$	**Tl** 81 204.38 $6p^1$	**Pb** 82 207.2 $6p^2$	**Bi** 83 208.98 $6p^3$	**Po** 84 (209) $6p^4$	**At** 85 (210) $6p^5$	**Rn** 86 (222) $6p^6$
Ds 110 (271)	**Rg** 111 (272)	**Cn** 112 (285)		114†† (289)		116†† (292)		

Eu 63 151.96 $4f^7 6s^2$	**Gd** 64 157.25 $4f^7 5d^1 6s^2$	**Tb** 65 158.93 $4f^8 5d^1 6s^2$	**Dy** 66 162.50 $4f^{10}6s^2$	**Ho** 67 164.93 $4f^{11}6s^2$	**Er** 68 167.26 $4f^{12}6s^2$	**Tm** 69 168.93 $4f^{13}6s^2$	**Yb** 70 173.04 $4f^{14}6s^2$	**Lu** 71 174.97 $4f^{14}5d^1 6s^2$
Am 95 (243) $5f^7 7s^2$	**Cm** 96 (247) $5f^7 6d^1 7s^2$	**Bk** 97 (247) $5f^8 6d^1 7s^2$	**Cf** 98 (251) $5f^{10}7s^2$	**Es** 99 (252) $5f^{11}7s^2$	**Fm** 100 (257) $5f^{12}7s^2$	**Md** 101 (258) $5f^{13}7s^2$	**No** 102 (259) $5f^{14}7s^2$	**Lr** 103 (262) $5f^{14}6d^1 7s^2$

Appendix D

SI Units

Base Quantity	SI Base Unit	
	Name	Symbol
Length	meter	m
Mass	kilogram	kg
Time	second	s
Electric current	ampere	A
Temperature	kelvin	K
Amount of substance	mole	mol
Luminous intensity	candela	cd

▌TABLE D.2 | Some Derived SI Units

Quantity	Name	Symbol	Expression in Terms of Base Units	Expression in Terms of Other SI Units
Plane angle	radian	rad	m/m	
Frequency	hertz	Hz	s^{-1}	
Force	newton	N	$kg \cdot m/s^2$	J/m
Pressure	pascal	Pa	$kg/m \cdot s^2$	N/m^2
Energy	joule	J	$kg \cdot m^2/s^2$	$N \cdot m$
Power	watt	W	$kg \cdot m^2/s^3$	J/s
Electric charge	coulomb	C	$A \cdot s$	
Electric potential	volt	V	$kg \cdot m^2/A \cdot s^3$	W/A
Capacitance	farad	F	$A^2 \cdot s^4/kg \cdot m^2$	C/V
Electric resistance	ohm	Ω	$kg \cdot m^2/A^2 \cdot s^3$	V/A
Magnetic flux	weber	Wb	$kg \cdot m^2/A \cdot s^2$	$V \cdot s$
Magnetic field	tesla	T	$kg/A \cdot s^2$	
Inductance	henry	H	$kg \cdot m^2/A^2 \cdot s^2$	$T \cdot m^2/A$

Answers to Quick Quizzes and Context Conclusion Problems

CHAPTER 1

Answers to Quick Quizzes

1. False.
2. (b)
3. Scalars: (a), (d), (e). Vectors: (b), (c).
4. (c)
5. (a)
6. (b)
7. (b)
8. (d)

CHAPTER 2

Answers to Quick Quizzes

1. (c)
2. (b)
3. (a)-(e), (b)-(d), (c)-(f)
4. (b)
5. (c)
6. (e)

CHAPTER 3

Answers to Quick Quizzes

1. (a)
2. (i) (b) (ii) (a)
3. 15°, 30°, 45°, 60°, 75°
4. (c)
5. (i) (b) (ii) (d)

CHAPTER 4

Answers to Quick Quizzes

1. (d)
2. (a)
3. (d)
4. (b)
5. (i) (c) (ii) (a)
6. (c)
7. (c)

CHAPTER 5

Answers to Quick Quizzes

1. (b)
2. (b)
3. (b)
4. (i) (a) (ii) (b)
5. (a)
6. (a) Because the speed is constant, the only direction the force can have is that of the centripetal acceleration. The force is larger at Ⓒ than at Ⓐ because the radius at Ⓒ is smaller. There is no force at Ⓑ because the wire is straight. (b) In addition to the forces in the centripetal direction in (a), there are now tangential forces to provide the tangential acceleration. The tangential force is the same at all three points because the tangential acceleration is constant.

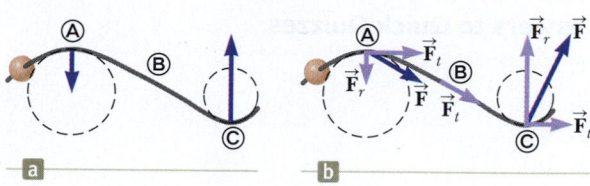

7. (c)

CHAPTER 6

Answers to Quick Quizzes

1. (a)
2. (c), (a), (d), (b)
3. (d)
4. (a)
5. (b)
6. (c)
7. (i) (c) (ii) (a)
8. (d)

CHAPTER 7

Answers to Quick Quizzes

1. (a) For the television set, energy enters by electrical transmission (through the power cord). Energy leaves by heat (from hot surfaces into the air), mechanical waves (sound from the speaker), and electromagnetic radiation (from the screen). (b) For the gasoline-powered lawn mower, energy enters by matter transfer (gasoline). Energy leaves by work (on the blades of grass), mechanical waves (sound), and heat (from hot surfaces into the air). (c) For the hand-cranked pencil sharpener, energy enters by work (from your hand turning the crank). Energy leaves by work (done on the pencil), mechanical waves (sound), and heat due to the temperature increase from friction.
2. (i) (b) (ii) (b) (iii) (a)
3. (a)
4. $v_1 = v_2 = v_3$
5. (c)

Answers to Context 1 Conclusion Problems

1. (a) 315 kJ (b) 220 kJ (c) 187 kJ (d) 127 kJ (e) 14.0 m/s (f) 40.5% (g) 187 kJ
2. (a) Conventional car = 581 MJ; Hybrid car = 220 MJ (b) Conventional car = 11.4%; Hybrid car = 30.0%

CHAPTER 8
Answers to Quick Quizzes
1. (d)
2. (b), (c), (a)
3. (i) (c), (e) (ii) (b), (d)
4. (b)
5. (b)
6. (i) (a) (ii) (b)

CHAPTER 9
Answers to Quick Quizzes
1. (d)
2. (a)
3. (a)
4. (c)
5. (c), (d)
6. (a) $m_3 > m_2 = m_1$ (b) $K_3 = K_2 > K_1$ (c) $u_2 > u_3 = u_1$

CHAPTER 10
Answers to Quick Quizzes
1. (i) (c) (ii) (b)
2. (b)
3. (i) (b) (ii) (a)
4. (a)
5. (i) (b) (ii) (a)
6. (b)
7. (b)
8. (a)
9. (i) (b) (ii) (c) (iii) (a)

CHAPTER 11
Answers to Quick Quizzes
1. (e)
2. (a)
3. (a) Perihelion (b) Aphelion (c) Perihelion (d) All points
4. (a)

Answers to Context 2 Conclusion Problems
1. (1) 146 d (b) Venus 53.9° behind the Earth
2. 1.30×10^3 m/s
3. (a) 2.95 km/s (b) 2.65 km/s (c) 10.7 km/s (d) 4.80 km/s

CHAPTER 12
Answers to Quick Quizzes
1. (d)
2. (f)
3. (a)
4. (b)
5. (i) (a) (ii) (a)
6. (a)

CHAPTER 13
Answers to Quick Quizzes
1. (i) (b) (ii) (a)
2. (i) (c) (ii) (b) (iii) (d)
3. (c)
4. (f) and (h)
5. (d)
6. (e)
7. (e)

CHAPTER 14
Answers to Quick Quizzes
1. (c)
2. (i) (a) (ii) (d)
3. (d)
4. (b)
5. (c)
6. (b)

Answers to Context 3 Conclusion Problems
1. 3.5 cm
2. The speed decreases by a factor of 25
3. Station 1: 15:46:32
   Station 2: 15:46:24
   Station 3: 15:46:09

CHAPTER 15
Answers to Quick Quizzes
1. (a)
2. (a)
3. (b)
4. (b) or (c)
5. (b)
6. (a)

Answers to Context 4 Conclusion Problems
1. (a) The blood in vessel (ii) would have the highest speed at point 2. (b) $v_{ii} = 32v_{iii}$.
2. (a) 1.67 m/s (b) 720 Pa
3. (a) 1.57 kPa, 0.015 5 atm, 11.8 mm (b) Blockage of the fluid within the spinal column or between the skull and the spinal column would prevent the fluid level from rising.
4. 12.6 m/s

CHAPTER 16
Answers to Quick Quizzes
1. (c)
2. (c)
3. (a)
4. (b)
5. (i) (b) (ii) (a)
6. (a)

CHAPTER 17

Answers to Quick Quizzes

1. **(i)** iron, glass, water **(ii)** water, glass, iron
2. The figure shows a graphical representation of the internal energy of the system as a function of energy added. Notice that this graph looks quite different from Figure 17.3 in that it doesn't have the flat portions during the phase changes. Regardless of how the temperature is varying in Figure 17.3, the internal energy of the system simply increases linearly with energy input.

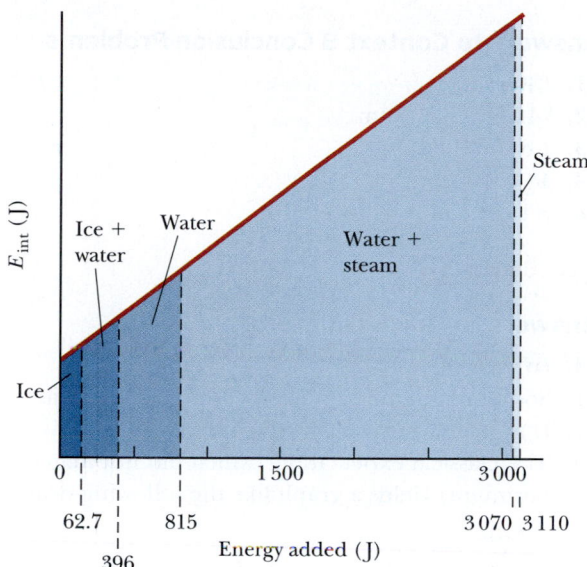

Figure QQ17.2

3. C, A, E. The slope is the ratio of the temperature change to the amount of energy input. Therefore, the slope is proportional to the reciprocal of the specific heat. Liquid water, which has the highest specific heat, has the lowest slope.

4.

Situation	System	Q	W	ΔE_{int}
(a) Rapidly pumping up a bicycle tire	Air in the pump	0	+	+
(b) Pan of room-temperature water sitting on a hot stove	Water in the pan	+	0	+
(c) Air quickly leaking out of a balloon	Air originally in the balloon	0	−	−

5. Path A is isovolumetric, path B is adiabatic, path C is isothermal, and path D is isobaric.
6. **(i)** (a) **(ii)** (c)
7. (b)

CHAPTER 18

Answers to Quick Quizzes

1. **(i)** (c) **(ii)** (b)
2. C, B, A

3. (d)
4. (a) one (b) six
5. (b)
6. (a)
7. false

Answers to Context 5 Conclusion Problems

1. 298 K
2. 60 km
3. (c) 336 K (d) The troposphere and stratosphere are too thick to be accurately modeled as having uniform temperatures. (e) 227 K (f) 107 (g) The multilayer model should be better for Venus than for the Earth. There are many layers, so the temperature of each can reasonably be modeled as uniform.

CHAPTER 19

Answers to Quick Quizzes

1. (a), (c), (e)
2. (e)
3. (b)
4. (a)
5. *A, B, C*
6. (b) and (d)
7. **(i)** (c) **(ii)** (d)

CHAPTER 20

Answers to Quick Quizzes

1. **(i)** (b) **(ii)** (a)
2. Ⓑ to Ⓒ, Ⓒ to Ⓓ, Ⓐ to Ⓑ, Ⓓ to Ⓔ
3. **(i)** (b) **(ii)** (c)
4. **(i)** (c) **(ii)** (a)
5. **(i)** (a) **(ii)** (a)
6. (d)
7. (a)
8. (b)
9. (a)

CHAPTER 21

Answers to Quick Quizzes

1. (a) > (b) = (c) > (d)
2. (b)
3. (a)
4. $I_a = I_b > I_c = I_d > I_e = I_f$
5. (b)
6. (a)
7. **(i)** (b) **(ii)** (a) **(iii)** (a) **(iv)** (b)
8. **(i)** (c) **(ii)** (d)

Answers to Context 6 Conclusion Problems

1. (a) 87.0 s (b) 261 s (c) $t \rightarrow \infty$
2. (a) 0.01 s (b) 7×10^6
3. (a) 3×10^6 (b) 9×10^6

CHAPTER 22

Answers to Quick Quizzes

1. (e)
2. (i) (b) (ii) (a)
3. (c)
4. $B > C > A$
5. (a)
6. $c > a > d > b$
7. $a = c = d > b = 0$
8. (c)

CHAPTER 23

Answers to Quick Quizzes

1. (c)
2. $c, d = e, b, a$
3. (b)
4. (c)
5. (b)
6. (d)
7. (b)
8. (b)
9. (a), (d)

Answers to Context 7 Conclusion Problems

1. (a) 29.2 MHz (b) 42.6 MHz (c) 2.13 kHz
2. (a) 2.47×10^3 A (b) 0.986 T · m^2 (c) 0.197 V (d) 64.0 kg

CHAPTER 24

Answers to Quick Quizzes

1. (i) (b) (ii) (c)
2. (c)
3. (d)
4. (b), (c)
5. (c)
6. (a)
7. (b)

CHAPTER 25

Answers to Quick Quizzes

1. (d)
2. Beams ② and ④ are reflected; beams ③ and ⑤ are refracted.
3. (c)
4. (a)
5. False
6. (i) (b) (ii) (b)
7. (c)

CHAPTER 26

Answers to Quick Quizzes

1. C
2. false
3. (b)
4. (b)

5. (b)
6. (c)
7. (c)

CHAPTER 27

Answers to Quick Quizzes

1. (c)
2. (a)
3. (a)
4. (a)
5. (a)
6. (c)

Answers to Context 8 Conclusion Problems

1. 130 nm
2. 74.2 grooves/mm
3. 1.8 μm/bit
4. 48 059
5. $\sim 10^8$ W/m^2

CHAPTER 28

Answers to Quick Quizzes

1. (b)
2. Sodium light, microwaves, FM radio, AM radio.
3. (c)
4. The classical expectation (which did not match the experiment) yields a graph like the following drawing:

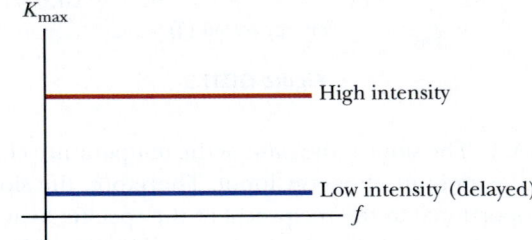

5. (c)
6. (b)
7. (b)
8. (i) (a) (ii) (d)
9. (c)
10. (a), (c), (f)

CHAPTER 29

Answers to Quick Quizzes

1. (b)
2. (a) five (b) nine
3.

4. cesium, potassium, lithium
5. true
6. (c)

CHAPTER 30

Answers to Quick Quizzes

1. **(i)** (b) **(ii)** (a) **(iii)** (c)
2. (c)
3. (e)
4. (e)
5. (b)
6. (c)

CHAPTER 31

Answers to Quick Quizzes

1. (a)
2. **(i)** (c), (d) **(ii)** (a)

3. (b), (e), (f)
4. (b), (e)
5. $S = 0$ $S = -1$

$Q = -\frac{1}{3}$ $Q = +\frac{2}{3}$

Answer to Context 9 Conclusion Problems

1. (a) 1.61×10^{-35} m (b) 5.38×10^{-44} s (c) yes

Index

Page numbers in **bold** indicate a definition; page numbers in *italics* indicate figures; page numbers followed by "n" indicate footnotes; page numbers followed by "*t*" indicate tables.